A TEXTBOOK OF BIOPHYSICS

[For Medical Science, Veterinary Science, Degree Courses in Physiology, Zoology, Botany, Biochemistry, Life Science, other Biological Science Students and Post-Diploma Courses in Medical Electronics and Biomedical Instrumentation and Biomedical Engineering (BTech)]

Dr R N Roy

BSc [Hons], MBBS [Cal], MSc [Gold Medallist], PhD [Med] [Cal]

Formerly, Research Fellow, Medical College and Hospital, Kolkata
Research Fellow, Institute of Child Health, Kolkata
Lecturer, Presidency College, Kolkata
Asst Professor, L T Medical College and Hospital, Mumbai
Faculty Member, Medical Electronics,
Central Calcutta Polytechnic, Kolkata
Examiner and Paper-setter, West Bengal Council of Technical Education

New Central Book Agency (P) Ltd

Co-Published by

Bijay Publications

E-mail : bijaypublications@gmail.com

Exclusively Distributed by

Bijay Pustak Bhandar (P) Ltd

90/6A M. G. Road, Bhawani Dutta Lane, Kolkata-700007

✆ : (033) 4605-7856, 8777898576, 8274916104

E-mail : bijaypustakbhandar@gmail.com

NCBA

REGD OFFICE
8/1 Chintamoni Das Lane, Kolkata 700 009, India
email: ncbakolkata@gmail.com

OVERSEAS
NCBA (UK) Ltd, 149 Park Avenue North, Northampton, NN3 2HY, UK
email: ncbauk@yahoo.com

CORPORATE OFFICE
212 Shahpur Jat, New Delhi 110 049
email: delhincba@gmail.com

BIJAY PUBLICATIONS
90/6A M. G. Road, Bhawani Dutta Lane, Kolkata-7
E-mail : bijaypublications@gmail.com

BIJAY PUSTAK BHANDAR (P) LTD
90/6A M. G. Road, Bhawani Dutta Lane, Kolkata-7
E-mail : bijaypustakbhandar@gmail.com

A TEXTBOOK OF BIOPHYSICS • Dr R N Roy

First Published: February 1996
Reprinted: January 1999, July 2002, January 2004
Revised Edition: January 2005
Reprinted: January 2007
Thoroughly Revised and Updated Second Edition: March 2013
Reprinted: January 2015, July 2017, April 2019, **May 2025**

PUBLISHER • TYPESETTER • PRINTER
New Central Book Agency (P) Ltd
8/1 Chintamoni Das Lane, Kolkata 700 009

CO-PUBLISHED BY
Bijay Publications
90/6A M G Road, Bhawani Dutta Lane, Kolkata 700007

PROOF CHECKING
Dipan Roy, Koushik Samanta

PROJECT TEAM
Swapan Sarkar, Prodip Baidya and Pradip Biswas

ISBN: 978 81 7381 145 6

Price ₹ 985.00

DEDICATED

To

the loving memory of

MY FATHER

for being a great philosopher, guide, aspiration
and
leading into intellectual pursuits

MY WIFE

for consistency of inspiration, enthusiasm, devotion
and
encouragement for flair of writing

MY GRANDDAUGHTER
RANJABATI ROY

for her cooperation and making everything
worthwhile

CONTENTS

Preface to the Second Edition xv

Preface to the First Edition xvi

Chapter 1. Introduction 1-2

Concept of System in Biophysics 1; Divisions of Biophysics 2

Chapter 2. Units, Abbreviations, Prefixes, Physical and Mathematical Constants, Conversion of Units, Expression of Units 3-6

Prefixes 3; Abbreviations 3; Constants 4; Conversion of Units 5; Expression of Units 5

Chapter 3. Thermodynamics 7-25

Thermodynamic approach 7; Scope 7; Limitation 8; Glossary 8; Laws of Thermodynamics and living organisms 10; First Law of Thermodynamics 11; Enthalpy 13; Second Law of Thermodynamics 14; Comparison of 1st and 2nd Law of Thermodynamics 15; Entropy 16; Second Law of Thermodynamics using the concept of Entropy 16; Physiological steady state 17; Efficiency and Free energy in thermodynamic system 18; Concept of Energy in the Biological System in the light of Thermodynamics 19; Living body as a Thermodynamic system 21; Comparison of living and non-living system as a Thermodynamic system 22; Carnot Cycle 23; Energy 25

Chapter 4. True Solutions 26-32

Definition 26; Characteristics 26; Comparison of True Solutions and Colloidal Solutions 27; Solvents for solutions 28; Limitation of Solubility 29; Different types of solutions 29; Methods of expression of Exact concentration of a solution 30; Biological Importance of solution 32

Chapter 5. Biochemical Analytical Methods 33-43

PART I: Biochemical Analysis 33; Photometry 34; Absorption Spectrophotometer 35; Ultraviolent 35; Monochromators 36

PART II: Clinical Electrochemical Methods 39, Introduction 39; Protentiometry 39; Application 39; Voltametry 39; Coulometry 40; Conductimetry 41

Chapter 6. Hydrogen Ion Concentration and pH 44-114

Dissociation (ionization) or protolysis of water 44; Hydrogen Ion Concentration (cH) 46; pH value 48; Experimental Determination of H-ion concentration or (pH) 54; pH Meter 65; Acids, Bases and Salts 69; Reaction of Acids Bases 80; Biological importance of acids, bases and salts 82; Buffers and Buffering system 90; Henderson-Hasselbalch equation 95; Indicators 100; Acid-Base balance 106

Chapter 7. Colloidal System 115-196

Introduction 115; Phases of Colloids 116; Classification 116; Transformation of sol to gel 117; Association Colloids 117; Macromolecules or molecular colloids 117; Properties of colloids 118; Importance of colloids 127; Electrical charge of colloidal particles 130; Emulsion and suspension 133; Gel or gel state 135; Concept of macromolecules 139; Physical Structure of protoplasm 141; Physical State of protoplasm 143; Biomolecules and moleculer organization of cell 145; Micro-dissection 147; Ultramicroscope 148; Birefringence 149; Determination of the size and shape of the calloidal particles 150; Ultracentrifuge 154; Artificial kidney 159; Biological membranes and ultrafiltration 161; Separation of Colloids 170; Electrodialysis 172; Chromatography 185

Chapter 8. Nucleonics in Biology and Medicine **197-310**

PART I: Elementary particles 197; Atomic nucleus 198; Proton 199; Neutron 199; Electron 199; Accommodation of electrons 199

PART II: Radioactivity 200; Energy of radiation 200; Radiation from radioactive materials 201; Detection of nuclear radiations 201; Radioactive decay series 202; Radioelements in the periodic table 203; Rate of radioactive decay and half-life of radioactive element 203; Disintegration schemes 204; Unit of measurement of radioactivity 204; Types of radioactivity 204

PART III: Isotopes 205

PART IV: Biological Application of Radioisotope 207

PART V: Nuclear Radiation Therapy 218

PART VI: Imaging System 219

PART VII: X-rays 220; Metal X-ray tube 223

PART VIII: Image Intensifier tube 227

PART IX: X-ray Machine 228; Protecting housing of X-ray tube 229; Instruments using X-ray energy 232; Factors governing the radiography 232

PART X: Detection and measurement of radioactivity 233; Interventional radiology 234; Protection from X-ray radiation 235

PART XI: Nuclear radiations and interactions with living materials 242

PART XII: Biological effects of radiation 244

PART XIII: Radiation damage in the embryo and Fetus during pregnancy 249

PART XIV: Demerits of different diagnostic and therapeutic methods of nuclear medicine during pregnancy 250

PART XV: Radiation hazards in human being 251

PART XVI: Fluoroscopy 254; Factors for best X-ray image 260

PART XVII: Image contrast 260

PART XVIII: Pattern of X-ray beam and recording 261; Grid 265; Collimators 266

PART XIX: Factors governing exposure 267

PART XX: Xero-radiography 268; Angiogrphy 268

PART XXI: Contrast Media 270

PART XXII: Digital substraction angiography (DSA) 271

PART XXIII: Radiotheraphy 275; Teletherapy 276; Characteristics of various radioactive isotopes for radiotherapy 277

PART XXIV: Tomography 278; Computed Tomography (CT Scan) 282; Collimator 287; Comparison of CT Scan and Conventional Radiography 288; Gantry assemly 289

PART XXV: Magnetic Resonance Imaging (MRI) 294

PART XXVI: Mammography (Soft Tissue Radiography) 307

Chapter 9. Biophysical Principle 311-413

Diffusion 311; Facilitated diffusion 317; Sodium co-transport 319; Active transport 324; Criteria of Carrier mediated transport 326; Osmosis 329; Semipermeable membrane 329; Osmotic pressure 332; Osmoregulation 347; Factors influencing the rate of filtration across the capillary membrane 356; Endocytosis 361; Phagocytosis 361; Receptor-mediated endocytosis 366; Secretion and exocytosis 365; Absorption 368; Reabsorption 375; Adsorption 376; Hydrotropy 384; Viscosity 386; Surface tension 394; Stalagmometer 399; Donnan Membrane equilibria 400; Membrane Hydrolysis 410; Ultrafiltration 412

Chapter 10. Computer in Biology and Medicine 414-420

Computer in Biology and Medicine 414; Units of a Computer 415; Application in Biology and Medicine 416; Computer versus Human Brain 419

Chapter 11. Nuclear Medicine (Diagnostic and Therapeutic Technology) 421-458

Radioimmunoassay method (RIA) 421; Antigen measurement 422; Hormone assay 422; Application of Radioimmunoassay (RIA) 426; Organ scan 427; Liver scan 427; Radionuclide scanning vs. ultrasound 427; Bone scan 428; Brain scan 429; Renal imaging 429; Radioactive Renal Clearance Studies 430; Cardiac imaging 434; Multiple Gated Aquisition Studies 435; Myocardial Scintigraphy 436; Radionuclides for imaging 436; Positron Emission Tomography (PET) 437; Myocardial Imaging 439; Magnetic Resonance Imaging (MRI) 440; Radioiodine (I^{131}) for Diagnosis in thyroid disorder 442; Radioiodine therapy 447; Blood volume determination 448; Determination of red cell volume 450; Laser Beam in Biology, diagnosis and therapy 453; Nuclear Medicine for therapy 458

Chapter 12. Concept of Valency and Chemical Bonding 459-486

Valency or Combining capacity 459; Electronic theory of valency 460; Types of Bonds 460; Chemical Bond Relationship in Biological System 461; Energy concept of Chemical Bonding 461; Strong and Weak bond Relationship in Biological System 462; Biological application 464; Comparison of Micelle and Bimolecular sheet 481; Van der Waal's Forces/Bonds 483

Chapter 13. Light and Associated Phenomena 487-559

PART I: Stereoisomerism 487; Optical isomerism 488; Polarimeter 491; Mutarotation 494; Colorimeter 495; Photometry 497

PART II: Photochemical reactions 499; Photosynthesis 503; Photochemistry of vision 511; Transmitters in vision 517; Bacteriorhodopsin 518

PART III: Biological effects of light 519; Lethal effects on small animals and plants 519; Effects of long continued exposure in large animals and human 520; Phytochrome system 520; Photochemical mechanism 521; Phototropism 521; Photoperiodism in plants 522; Solarization 523; Photodynamic action 524

Part IV: Biophysics of vision 526; Light and its attenuation for vision 526; Light and field of view 527; Light and photopigment 528; Illumination of retina 528; The eye as an optical instrument 528; The constant of the eye 530; The reduced eye 530; Formation of image 531; Critical fusion frequency 531; Changes in the retina on exposure to light 532; Rod electroretinogram 534

Part V: Light and effect on animals 537; Effect of photoperiods on gonads of animals 539; Effect of light on secretion of other anterior pituitary hormones 542

Part VI: Ultraviolet light on living system 543; Photoreactivation 546

Part VII: Light application in therapy 547

Part VIII: Biological light 550

Part IX: Light interaction with biological materials 555

Part X: Light and pigmentary response 557

Chapter 14. Microscopy in Biology and Medicine **560-579**

Phase-contrast Microscope 560; Interference microscope 563; Fluorescence microscope 566; Ultraviolet Microscope 569; Electron Microscope 570; Scanning Electron Microscope 574; Scanning electron-probe microanalysis 576; Low-temperature electron Microscope 578; Centrifuge Microscope 578; Polarising Microscope, Ultra Microscope, Micromanipulator (see Chapter 7); Other microscopes for study 579

Chapter 15. Biophysics of Water **580-585**

Introduction 580; Molecular structure of water 580; Macromolecular structure of water 581; Hydrogen bonds and structure of water 582; Physical properties of water and its Biological applications 582; Summary 585

Chapter 16. Electrochemistry **586-605**

Ionization and Electrolysis Conduction 586; Electrolytes 587; According to the Degree of Ionization 589; Electrolysis 594; Conductance 596; Solubility product 598; Common ion effect 601; Ostwald's Dilution Law 602; Dielectric Constant 603

Chapter 17. Chronobiology **606-617**

Biological rhythm 606; Significance of biorhythm 608; Influence of environmental synchronization (Zeitgeber or Synchronizer) 610; Circadian rhythms 611; Biological clock 613; The Mechanism of Biological clock 615; Master clock 616

Chapter 18. Cybernetics (Control Engineering) **618-620**

Homeostasis or feedback mechanism 618; Examples of homeostatic 618; Principles 618; Requirements for Homeostasis 619; Amplification 619; Positive feedback 619; Oscillation and damping of control system 620; Regulation of Homeostasis and damping of control system 620; Operation of homeostatic mechanism 620

Chapter 19. Biomagnetism and Magneto-Biology **621-625**

Biomagnetism 621; Magnetobiology 621; Generation and nature of biomagnetic fields 621; Ferromagnetic particles in the body 622; The magnetic fields of skin, muscle 622; Neuromagnetic fields 622; Magnetometers (SQUID) 624; Magnetography 624; Magnetocardiography (MCG) 625; Magnetoencephalography (MEG) 626

Chapter 20. Biophysics of the Plasma Membrane 626-660

Salient features of cell membranes 626; Chemical composition of the membrane 627; Molecular structure of cell membrane 633; Modification of cell membranes and its Biophysical importance 636; Ion channel 638; Nuclear Membrane 639; Nuclear membrane vs Cell membrane 639; Function of cell Membrane 640; Physico-chemical properties of the membrane 641; Conformational properties of membranes 644; Permeability of membrane and factors affecting it 645; Membrane receptor 648; Hormone receptor 650; Peptide hormone receptor on the surface of cell membrane 651; Transport of substances across epithelia or through cellular sheet 652; Mitochondrial membrane 653; Factors governing the passage of materials across cell membrane 657

Chapter 21. Bioacoustic 661-735

Sound and its characteristics 661; Functions of ear as a hearing organ 662; Physical basis of Hearing 662; Unit of Intensity of Sound 663; Physical organization of the ear 664; Conduction of sound in the ear 670; Mechanical impedance of the ear structures 672; Transmission of sound waves in the inner ear 673; Electrical responses of cochlea 674; Pitch perception 677; Perception of loudness 680; Mechanism of hearing 681; Beats and dissonance 683; Combination of tones 684; Consonance and Dissonance 685; Sound Analysis 686; Sound Localisation 686; Physical basis of voice (speech) 686; Infrasound 688; Ultrasound 688; General features of the diagnostic ultrasound 691; Biological effect of ultrasonics 694; Importance 695; Clinical ultrasonography 697; Therapeutic 700; Ultrasonic surgery 701; Echocardiography 703; Vascular Ultrasound 703; Ultrasonic transducer 704; Transducer positioning 706; Echocardiography 708; Classification of PZ materials 713; Shock wave 714; Comparison of diagnostic X-ray and diagnostic ultrasound 715; Extracorporeal shock wave (Lithotripsy) 715; Noise 717; Speech Interference Level (SIL) 725; Deafness 729; Pure tone audiometer 729; Hearing Aid 733; Digital hearing Aid 735

Chapter 22. Biophysics of Sense of Balance and Rotation 736-745

Physical organization of vestibular apparatus as mechanical sensors 736; Fluid dynamics of semicircular canals (angular rate sensors) 739; Artificial stimulation of semicircular canal for experiments and clinical diagnosis 740; Utricle and saccule operation (otolith organs as Linear Accelerometer) 742; Sensory neurophysiology of cells 743; Clinical test for function of utricle 743; Function of vestibular apparatus 744; Comparison of semicircular canal with utricle and saccule 745

Chapter 23. Influence of Gravity, G-Force, Acceleration, Deceleration 746-756

Influence of gravity 746; Influence of G-forces 749; Production of artificial G-forces—Human centrifuge 751; Effects of positive G-forces 751; Different cardiovascular phenomenon with mechanism of production 752; Compensatory mechanism 752; Effects of negative G 753; Transverse G 753; Tolerance of G-forces 754; Protection against G-forces 754; Influence of linear acceleration on the body 755; Parachute jump 755; Landing 755; Effects of brief acceleration 756

Chapter 24. Problems and Solutions in Aviation and Space Voyage 757-760

Weightlessness 757; Temperature in Aviation and space 759; Radiation in Aviation and space 759; Atmosphere and climate in the spacecraft 760

Chapter 25. Influence of Barometric Pressure 761-767

High altitude 761; Gradual exposure during mountaineering 763; Chronic mountain sickness 763; High barometric pressure effect on the body 763; Nitrogen at high pressure 764; Oxygen at high-pressure 764; Carbondioxide toxicity 765; Helium and high-pressure Nervous syndrome 765; Decompression 766; Physical factors involved in diving 767

Chapter 26. Vibrations 768-771

Physical basis of vibrations 768; The human body is a mechanical system (oscillatory characteristics) 769; Subjective response 769; Physiological effects 769; Health damage 770; Raynaud's phenomenon 770

Chapter 27. Microwave 772-775

Source and victims to exposure 772; Effect of interaction on living system 772; Penetration and propagation of microwaves within biological target organ 773; Primary interaction of microwave with living system 773; Secondary effect Induced 774

Chapter 28. Haemodynamics 776-849

PART I: Rheology ... 776

Biophysical properties of plasma proteins 776; Viscosity of blood 782; Rheological behaviour of blood (Non-Newtonian fluid) 783; Viscous fluids: Comparison with engineering fluid 786; General concept of viscosity for Newtonian and Non-Newtonian fluid 786; Wall effects 788; Non-Newtonian fluids 788; Plasma skimming 790

PART II: Dynamics of Peripheral Circulation ... 790

Different blood vessels in the vascular system 790; Comparison of Newtonian and Non-Newtonian fluids 791; Windkessel Vessels 792; Flow-pressure curve 793; Poiseuille's Law 794; Difference between Poiseuille's experimental preparation and the circulation 796; Blood flow 798; Flowmeter 800; Turbulent flow of blood, Reynold's number 806; Critical closing pressure 810; Laplace's Law 812; Vascular resistance 816; Pressure in the vascular system 820; Blood pressure 824; Blood pressure Measurement 830; Measurement and Recording of Arterial Blood Pressure 831; Oscillatory Method 840; Indirect Automated and Semiautomated B.P. Instrument 841; Direct Blood Pressure measurement 844; Effect of Hydrostatic Pressure on Venous and Arterial Pressure 845

Chapter 29. Tissue Engineering 850-940

PART I: Cell ... 851

Human Cell 851; Organelles 851; Size, shape, organization 855; Cell membrane 860; Cytoplasm 861; Stem cell 861; Stromal cells 862; Cell differentiation 864; Cancer cell 865; Apoptosis 868; Cell junction 870; Growth factor 871; Growth inhibitor 872; Measurement in cell study (Cytophotometry) 872; Fractionation of cell component 873; Tissue culture 875; Fine needle aspiration Cytology 879; Exfoliative cytopathology 879; Growth and development 880

Part II: Cytokinetics ... 882

Cell adhesion 882; Cell migration 883; Signal reception 884; Cellular communication 885; Signalling of cell 885; Cell to cell recognition 887; Biologic Modulators 888; Cell cycle 890; Chemical mediators of inflammation 892; Interleukins 892; Cytokines 892; Bioreactors 892; Biomaterials 894

Part III: Cell Membrane ... 897

Dynamics of the cell membrane 897; Study of cell membrane proteins 898; Ion channels 899; Membrane Recycling 901; Gap junction 902

Part IV: Tissue Dynamics ... 903

Tissue 903; Mechanism of tissue fluid formation 907; Comparison of different structures of tissue 907; Extracellular matrix (ECM) 908; Organisation of tissue into different functional system or subunits 910; Collagen-rich tissue 912; Tissue dynamics 915; Organisation of tissue at cellular level and diversification 917; Collagen 917; Microenvironment 921; Basal lamina 922; Tissue repair, healing of wounds 923

Part V: Tissue Graft, Transplantation, Cryo-technique ... 924

Transplantation, Grafting 924; Transplantation surgery 925; Rejection of grafts 925; Prosthetic heart valves 927; Cryo-technique 929

Part VI: Bone Marrow and Transplantation ... 934

Bone marrow 934; Bone marrow and transplantation 938; Extracellular matrix of bone marrow 940

Chapter 30. Biomedical Technique **941-960**

Bioelectrical Potential 941; Clinical importance 942; Action Potential 944; Bioelectric signals sensed by biopotential electrodes and their sources 945; Biopotential Electrode 948; Concept of fibre optic 949; Fibre-optic Endoscopy 951; Heart-Lung Machine (Cardio-Pulmonary Bypass (CPB), Heart-Lung Bypass) 955; Centrifugation Technology 958

APPENDIX **961**

BIBLIOGRAPHY **963-967**

INDEX **969-974**

PREFACE TO THE SECOND EDITION

The volume entitled Textbook of Biophysics is intended for the Medical Science and Biological Science students who study Biophysics. Biophysics is not just a branch of Physiology but an intimate part of Physiology. Most of the life-processes are governed by the Biophysical Principles—hence Biophysics may be called 'the physics of the living processes'. Accordingly, an attempt has been made to correlate and explain the Biophysical Principles with each physiological system. Though it is a difficult task for the author to cover the vast course of Biophysics dealing with all their problems, yet an utmost effort has been made encompassing some other subjects like Biochemistry, Histology, Pathology, Bacteriology, and Virology in the present volume.

For the benefit of the students of Biomedical Engineering a new chapter on Tissue Engineering has been incorporated which encompasses cellular biophysics, artificial heart valve, extracorporeal circuit, grafting, biometerials, etc.

For the benefit of the students of Medical Science and Paramedical courses some subject matters have been incorporated in this volume.

This volume has been thoroughly revised and latest and uptodate subjects have been incorporated. This may lead to inevitable errors of omission and commission—the author will appreciate having these brought to attention. Complicated mathematical calculations have been avoided for the benefit of the students and only the utmost essential have been incorporated in a simplified manner.

Biophysics has now developed into a vastly complex science growing at an accelerated pace. This phenomenal development and its relation to the understanding of life-processes is imposing ever-increasing demands upon the students. Considering the problems, an effort has been made to include some chapters, e.g., Nuclear Medicine, Computer in Biology and Medicine, Microscopy in Biology and Medicine, Biomagnetism, and Cybernetics. Some chapters are elaborated such as X-ray—includes fluoroscopy, computed tomography; bioacoustics—includes Piezoelectric crystal, ultrasonography, echocardiography, audiometry, hearing aid; biomagnetism includes magnetic resonance tomography (MRI); and others like fibre optic concept, endoscopy, physiotherapy, heart-lung machine, cryo-technique cardiac catheterisation, and prosthetic heart valve. Though, all these combined, is rendering increasingly difficult the writing of a textbook which encompasses the current facts and theories of the general field of Biophysics.

I am greatly indebted to Late D N Roy and Late T N Banerjee for inspiring me to prepare the volume. I must owe a special debt of gratitude to Mrs S Roy for her sustained encouragement while preparing the volume and taking keen interest for correction of proof, making illustrations, and preparation of index. I am greatly indebted to Miss Ranjabati Roy for going through the manuscript, perusal of references, and preparation of figures. The author appreciates very much the cooperation and help accorded by Mr Amitabha Sen and Mr Prabhat Jas of New Central Book Agency (P) Ltd. Acknowledgements are also due to all production workers of the New Central Book Agency (P) Ltd for bringing out the volume in time.

March 2013
"Geetanjali" Appartement
19/1A/91A Seal Lane
Kolkata 700 015
INDIA

R N Roy

PREFACE TO THE FIRST EDITION

This book is intended for Medical Science and Biological Science students who study Biophysics. Biophysics is not just a branch of Physiology but an intimate part of Physiology. Most of the life-processes are governed by the Biophysical Principles, hence the subject of Biophysics may be called as the Physics of the living processes. Accordingly, an attempt has been made to correlate and explain the Biophysical course of Biophysics dealing with all problems but an utmost effort has been made encompassing some other subjects like Biochemistry, Histology, Pathology, Bacteriology, Virology etc., for examples, in the present volume. This may lead to inevitable error of omission and commission, and the author will appreciate having these brought to attention. Complicated mathematical calculations have been avoided for the benefit of the students and only those which are essential have been incorporated in a simplified manner. Biophysics has now developed into a vastly complex science growing at an accelerated pace. This phenomenal development and its relation to the understanding of life processes is imposing ever-increasing demands upon the students. Considering the problems, an effort has been made to include some chapters—Nuclear Medicine, Computer in Biology and Medicine, Microscopy in Biology and Medicine, Biomagnetism, Cybernetics etc. though it is rendering increasingly difficult the writing of a textbook which encompasses the current facts and theories of the general field of Biophysics.

The volume has been written earlier, but due to delay in publication for unavoidable reasons the author could not incorporate few chapters in this volume. The author expects that the remaining chapters would be published as an appendix to this volume.

I am greatly indebted to Late D N Roy and T N Banerjee for inspiring me to prepare the volume. I am especially grateful to Late S R Maitra, Professor and Head of the Dept. of Physiology, University College of Science and Technology, Calcutta University, and Late A K Mukherjee, Professor and Head of the Dept. of Physiology and Principal, Presidency College, Calcutta, for creating interest in the subject. I must owe a special debt of gratitude to Mrs S Roy for her sustained encouragement while preparing the volume and taking keen interest for correction of proof and preparation of index. My sincere thanks are due to Mr Rana Roy for going through the manuscript, perusal of references and offering valuable suggestions.

I am greatly indebted to many individuals who helped me during the preparation of the volume amongst whom I wish to extend special thanks to Late Amarnath Banerjee, Mr Sibnath Banerjee, Mr Goutam Banerjee, Mrs Jui Roy and Ranjabati Roy for their help with this volume.

I am very grateful to my beloved students Mr Ananda Banerjee, Mr Sabyasachi Banerjee, Mr Soumotosh Chatterjee for their excellent illustration and timely assistance.

The author appreciates very much the cooperation and help accorded by Mr Amitabha Sen and Late Usha Ranjan Bhowmik of New Central Book Agency (P) Ltd.

Acknowledgements are due to all workers of New Central Book Agency (P) Ltd for bringing out the volume in such a short time.

January 1996

R N Roy

'Geetanjali' Appartement
19/1A/91A Seal Lane
Kolkata 700 015
INDIA

1

Introduction

1. Definition, 2. Concept in Biophysics, 3. Division of Biophysics.

1.1 Definition

Biophysics is the branch of science which is concerned about the biological activity under the physical principles.

1.2 Concept of System in Biophysics

The term "system" to the biologists in case of biophysics encompasses the group of materials and activities within the living system.

In other services, the "system" consists of the following elements:

1. **Source of information or detector (or grid input).**
2. **Transmission of information to the centre (or the Grid).**
3. **Control mechanism of expenditure of energy into the central authority to let it know whether required amount of energy is expended or not and to make correction when necessary.**
4. **Storage of information for use when necessary, e.g. modern computer.**

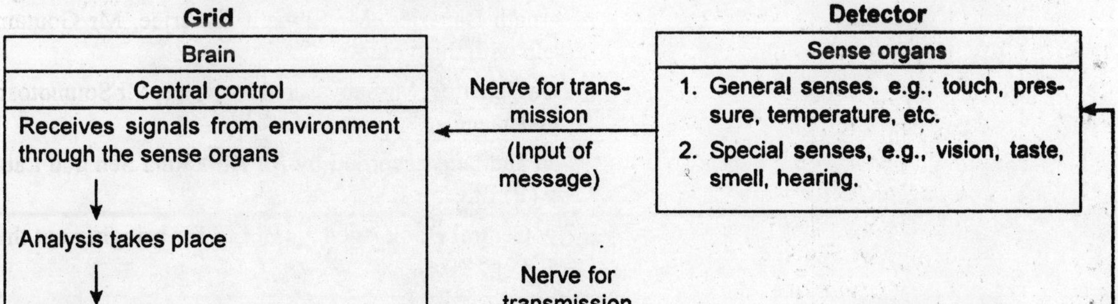

Fig. 1.1: The parts of a living "system"

The biological system, like human beings, is a "self-contained system" because of having all the elements like:

1. Source of information or detectors or Grid input-sensory organs both for general senses, e.g., touch, pressure, temperature, etc. and special senses like vision, taste, smell and hearing, etc.
2. Transmission of information to the brain centre (or the Grid)—through the nerve.
3. Storage of information—Brain stores information, analyzes and absorbs the relevant matter, decides and despatches the signals to the nerve to transmit the message to active parts, e.g., muscles which utilize energy. By the feedback mechanism of energy utilization by muscles and continuous watch by the detectors, e.g., sensory organs, informations are sent to the centre to let it know about the execution of the order, if not, then correct order will be transmitted (Fig. 1.1).

1.3 Divisions of Biophysics

1. **Physical biophysics**: Hydrodynamics, hydrostatic, mechanics, optics, light-effect, sound-effect, radiological and electromagnetic waves, isotopes.
2. **Physicochemical biophysics**: Structure of macromolecules, colloids, gel, thermodynamics, balance and transfer of energy, temperature effect and control, differentiation of membrane, kinetics and mechanism of fluid motion in the circulatory system.
3. **Physiological biophysics**: Electrical nature of the living tissue, i.e., bioelectricity, membrane potential, electrical potential during activity of the tissues, e.g., brain, heart, etc., volume conduction, effect of radiation, space medicine, bioenergetics, system control.
4. **Mathematical biophysics**: Computer and its application in biological services and medicine, cybernetics, biostatistics, growth rates and cycles, concept of system.

2

Units, Abbreviations, Prefixes, Physical and Mathematical Constants, Conversion of Units, Expression of Units

1. Prefixes, 2. Abbreviations, 3. Constant, 4. Conversion of Units, 5. Expression of Units.

2.1 Prefixes and Abbreviations in International System of Units

Prefix	Abbreviation or Symbol	Value or Factor
Tera	T	10^{12}
Giga	G	10^{9}
Mega	M (one million)	10^{6}
Kilo	k (one thousand)	10^{3}
Hecto	h	10^{2}
Deca	da	10^{1}
Deci	d	10^{-1}
Centi	c (one hundredth)	10^{-2}
Milli	m (one thousandth)	10^{-3}
Micro	m (one millionth)	10^{-6}
Nano	n	10^{-9}
Pico	p	10^{-12}
Femto	f	10^{-15}
Atto	a	10^{-18}

2.2 Abbreviations or Symbols used for Units

Unit	Symbol	Value or Factor
angstrom	Å	
angstrom unit	Å. U.	10^{-10}m = 10^{-1}nm
atmosphere	atm	
calorie	cal	
centimetre	cm	
centigram	cg	
coulomb	C	
curie	Ci	
cubic metre	cu. m (or m^3)	
cubic centimetre	cc	
cubic millimetre	cmm	
decimetre	dm	
decilitre	dl	
electron volt	eV	1.602×10^{-19}J
erg	erg	
farad	F	
gram	g	

Contd.

Unit	Symbol	Value or Factor
gramcentimetre	g-cm	
gauss	G	
hour	h	
joule	J	
katal	Kat	
kilogram	kg	
kilogram-metre	kg-m	
kilocalorie	kcal	
kilojoule	kJ	
kelvin	K	
litre	l	10^3ml, 10^{-3}m^3
litre-atmosphere	l.atm	101.328 J
logarithm to the base e	ln	
logarithm to the base 10	log	
metre	m	10^3 mm
micrometre	μm	10^{-3} mm
microlitre	μl	10^{-3} ml
microsecond	μsec (0.001 msec)	10^{-6} sec
microfarad	μF	
micromole	μ mol (0.001 m)	
micron	μ	10^{-3} mm 10^{-6} m
milligram	mg	
millimicron	mμ	10^{-6} mm
millionth micron	μμ	10^{-9} mm
millilitre	ml	
millimetre	mm	
millimetres of mercury pressure	mm Hg	
millimetres of water pressure	mm H_2O	
millimol	mM	M × 10^{-3}
milliosmol	mOsm	
milliequivalent	mEq	
millisecond	ms	10^{-3}sec
minute	min	
mol	mol	
molal concentration	m	1 mol Kg^{-1}
Molar Concentration	M	1 mol dm^{-3}
Molecular weight	MW or Mol. Wt	

Contd.

Unit	Symbol	Value or Factor
nanosecond	nsec	10^{-9} sec
nanometre	nm	10^{-6} mm 10^{-9} m
nanogram	ng	10^{-6} mg
normal concentration	N	1 eq dm^{-3}
ohm	1	VA^{-1}
osmole	Osm	
pH	–log H^+ concentration	
PK^1	–log K^1	
poise	P	1×10^{-1} kgm^{-1} sec^{-1}
picogram	pg	
picosecond	ps	
Revolution	r	
svedberg floatation Unit	Sf	
standard temperature and pressure	STP	
second	sec	
square metre	Sq.m	
watt	W	1Jsec^{-1}
volt	V	1JC^{-1}

2.3 Constants

A. Physical Constants

Physical Constant	Abbre-viation	Value
Acceleration due to gravity	*g*	9.81 m-sec^{-2}
Avogadro's number or constant	N_A or *L*	6.023×10^{33} mol^{-1}
Atomic mass unit (dalton)	amu	1.661×10^{-24} g
Base of natural logar-ithm	*e*	2.7183
Boltzman's constant	*K*	1.381×10^{-23} J deg, $3{,}298\times10^{-24}$ cal deg^{-1}
Curie	Ci	3.70×10^{10} disintegration sec^{-1}
Electron Volt	eV	1.602×10^{-19} J 3.828×10^{-20}cal

Contd.

Physical Constant	Abbreviation	Value
Faraday constant	F	96, 487 C $mole^{-1}$ 23, 062 cal V^{-1} mol^{-1}
Gas constant	R	8.314 J mol^{-1} deg^{-1} 1.987 cal mol^{-1} deg^{-1}
Gravitational constant	G	$6.67\times 10^{-11} m^3 kg^{-1} sec^{-2}$
Ion product of water at 298 K	Kw	1.008×10^{-14}
Plank's constant	h	6.626×10^{-34}J 1.584×10^{-34} cal-sec.
Speed of light in vacuum	c	(a) 2.998×10^{10}cm sec^{-1} (b) 3×10^{8}m Sec^{-1}
Stefan Constant	σ	5.7×10^{-8} $Wm^{-2}K^{-4}$
Speed of sound in air at 0° C		332 m sec^{-1}

B. Mathematical Constants

Mathematical constant	Factor
p	= 3.14159
e	= 2.71828
log e^x	= 2.303 log 10^x
1.0J	= 0.239 cal = 10^7 erg = 1 W-s

2.4 Conversion of Units

Physical quantity	Equivalent
Energy	1 J = 10^7 erg = 0.239 cal = 1 watt sec
Length	1 cm = 10^{-2} m = 10 mm = 10^4 mm = 10^7 nm 1 cm = 10^8Å = 0.3937 inch
Mass	1 g = 10^{-3} kg = 10^3 mg = 10^6 mg 1 g = 3.527×10^{-2} ounce (avoir dupois)
Pressure	1 torr = 1 mm Hg (0°C) = 1.333×10^2 newton/m^2 = 1.333×10^2 pascal = 1.316×10^{-3} atmosphere
Temperature	K = °C + 273.15 °C = 5/9 (°F–32)
Volume	1 cm^3 = $10^{-6}m^3$ = 10^3 mm^3 1 ml = 1 cm^3 = 10^{-3} l = 10^3 ml 1 cm^3 = 6.1×10^{-2} in^3 = 3.53×10^{-5} ft^3

2.5 Expression of Units

1. **Farad: an electrical capacity which holds 1 coulomb when charged to 1 volt.**
2. **Microfarad: one millionth part of 1 farad. (μf)**
3. **Gravitational unit of force: 1 gram-weight = 981.5 dynes.**
4. **Gravitational unit of work: 1 gram-cm = 981.5 ergs = 23.4 micro calories. 1 kg = 2.34 cal (= 0.468 cc O_2 at cal/1).**
5. **Gravitational unit of power: 1 gram-cm per second.**
   1 horse-power = 0.746 kW = 457 kg-m per minute.
6. **1 calories: heat to raise 1 gram of water through 1 °C**
   = 4.2 × 10^7 ergs = 42.7 × 10^3 gram-cms.
   1 micro cal = 42 ergs = 42.7 × 10^{-3} gram cms.
   1 Cal (or 1 kilo cal) = 1000 cal (= 4 B. Th. U. or 0.00004 therms).
7. **Kinetic CGS unit of force: 1 dyne = force which acting on mass of 1 gram, causes it to move with an acceleration of 1 cm per second.**
8. **Kinetic unit of work: 1 erg = 1 dyne-cm**
   = 23.84 × 10^{-9} calories
   = 10.16 × 10^{-4} gram-cm.
9. **Kinetic unit of power: 1 erg per second.**
   1 kilowatt = 10^3 watts = 10^{10} ergs per second.
10. **Per cent by weight: Weight in grams of solute per 100 g of solution.**
11. **Per cent by volume: Weight in grams of solute per 100 ml of solution.**
    Commercial HCl contains 38 per cent HCl by weight having density of 1.19. Then each ml contains 1.19 × 0.38 = 0.45 g of HCl, and 100 ml contain 45 g of HCl, i.e., the solution is 45 per cent HCl by volume.
12. **Molar solution (1 g-molecule per 1) (Molar system of concentration)—**
    Amount of substance in grams equal to the molecular weight of the substance in one litre. It has been established that a gram symbol or formula weight of any substance contains 6.02 × 10^{23} (Avogadro's number) particles.

Molar concentrations are commonly indicated as 1M, 0.5M, 0.1M, etc. according to the number of gram-moles (gram-molecular weights) present per litre of solution. Expression in brackets e.g., [1]1, [0.5], [0.1] are often used in expressing molar concentrations.

13. **Molal solution**: It contains a gram-molecular weight of substance (in gm) in 1,000 gm of solvent. Molal solutions provide a definite ratio of solute to solvent molecules.

14. **A mol (M) or gram-molecular weight**: The amount of substance in grams equal to the molecular weight of the substance = 1 g molecule = 6.02×10^{23} molecules.

15. **Mol fraction**: The solution contains the mol fractions of all constituents of solutes and solvent. The mol fraction of one constituent is equal to the number of mols of the constituent divided by the total number of mols present: Suppose A_1 and A_2 are the mol fractions of the two constituents in the solution and a_1 and a_2 are the numbers of mols of the constituents present.

$$A_1 = \frac{a_1}{a_1 + a_2} \text{ and } A_2 = \frac{a_2}{a_1 + a_2}$$

If 180 gm of fructose are added to 1,000 gm of water to make a solution the mol fraction of fructose is

$$A_1 = \frac{\frac{180}{180}}{\frac{1000}{18} + \frac{180}{180}} = 0.0177$$

The mol fraction of water of the solution is

$$A_2 = \frac{\frac{1000}{18}}{\frac{1000}{18} + \frac{180}{180}} = 0.982$$

16. **Millimol (mM) = M × 10^{-3}**: One millimol is equal to one-thousandth of a mol (M) = 0.001 M = 1 formula weight in milligrams.

Millimols may be calculated by dividing milligrams per litre by the formula weight.

17. **Micromol (μM) = M × 10^{-6}**: One micromol is equal to one-thousandth of a millimol (1/1,000,000 of a mol) = 0.0001 mM = 1 formula weight in microgram.

18. **Millimolar (1 mg molecule per l)**: Molecular weight in mgs of a solute dissolved in 1000 ml water is millimolar.

19. **Osmoles (osm)**: One osmole is equal to the molecular weight of the substance in gm divided by the number of freely moving molecules liberated in the solution. The concentration of osmotically active particles is usually expressed as osmoles.

20. **Milliosmole (M Osm)**: Milliosmole is used in relation with osmotic pressure since osmotic pressure is proportional to the total number of particles in solution. In case of non-electrolyte such as glucose, 1 millimol = 1 milliosmol but in case of electrolytes, one millimol = two milliosmole, e.g., one millimol of NaCl = two milliosmols ($Na^+ + Cl^-$).

21. **Osmolarity**: Osmolarity of a solution having solute concentration in osmoles per litre or dm^3. Plasma usually has an osmolarity of 303.7 mosm dm^{-3} with an osmotic pressure of 7.37 atmospheres at body temperature, whereas for interstitial fluid and intracellular fluid it is 302.2 mosm dm^{-3} and 1.2 mosm dm^{-3}, difference is due to higher protein concentration in plasma than the tissue fluid. It is affected by the solute concentration (e.g. protein in plasma) and temperature.

22. **Osmolality**: Osmolality of a solution having its solute concentration in osmoles per kg of solvent.

23. **Gram equivalent:**

$$\frac{\text{Molecular weight in gm}}{\text{No. of valence of ions present}}$$

24. **Milliequivalent (meq)**: One milliequivalent is one-thousandth of gram equivalent which is the same as millimol when the valence is 1.

3

Thermodynamics

1. Thermodynamic approach, 2. Scope, 3. Limitation, 4. Glossary, 5. Laws of Thermodynamics, 6. First law of Thermodynamics, 7. Enthalpy, 8. Second law of Thermodynamics, 9. Comparison of first and second laws, 10. Entropy, 11. Second law using the concept of entropy, 12. Physiological steady state, 13. Efficiency and free energy in Thermodynamic system, 14. Concept of energy in biological system in the light of Thermodynamics, 15. Living body as a Thermodynamic system, 16. Comparison of living and non-living system as a Thermodynamic system, 17. Carnot cycle.

3.1 Introduction

3.1.1 Thermodynamic approach

The study of a physical or a chemical process may be approached in two different ways:

1. **Kinetic molecular approach**: In this, the process is studied in terms of atoms and molecules present in the system.
2. **Thermodynamic approach**: In this, the process is studied from a consideration of the energy changes involved. The literal meaning of the term **thermodynamics is the study of flow of heat** but actually it deals not only with the relation between heat and work but includes all kinds of interconversion of one kind of energy into another.

Almost all process in nature are accompanied by energy changes. This energy manifests itself in various forms, e.g., mechanical, electrical, heat, radiation, etc. These different forms of energy are interconvertible under suitable conditions, and of all forms of energy, heat energy occupies a unique position as all other forms of energy tend to be ultimately converted into heat.

The science of thermodynamics attempts to study the laws underlying the interconvertibility of the different forms of energy into heat and to find out the relationships between the different properties of matter which are deducible from these laws.

3.1.2 Scope

1. The subject matter of thermodynamics is based on three generalizations acquired from human experience with energy. These have been subjected to rigorous mathematical treatment and have yielded logical deduction correlating the various properties of matter. For instance, generalisation of physical chemistry such as Gas Equation, Laws of Chemical Equilibrium, Phase Rule, Distribution Law, Vant Hoff's Laws of Dilute solutions, Raults Law of Vapour

Pressure Lowering and the Laws of Thermochemistry can be derived from thermodynamics consideration.

2. Thermodynamics also provides the basis for forecasting whether a certain transformation, physical or chemical, is possible or not under a given set of conditions of temperature, pressure and concentration.

3.1.3 Limitation of thermodynamics

1. While it gives the relationships between various properties experimentally observed, it is unable to give the actual values of these properties and does not offer any explanation as to why these properties arise in a system.

   Whereas, according to Kinetic molecular theory, the pressure of a gas is due to the impacts of the molecules on the walls of the vessels and is a measure of these impacts per second, in thermodynamics, pressure is regarded only as a property of the state of the system regardless of its constitution.

2. Another limitation of thermodynamics is in respect of time factor in various transformation. It fails to give any information regarding the time taken for the completion of a change or the rate at which it is brought about.

3. The laws of thermodynamics are based on human experience about the behaviour of macroscopic systems, i.e., comparatively large systems which are an assemblage of a large number of molecules, and hence these laws hold good rigidly.

4. Since human experience is limited to grossly palpable quantities of matter which are far from being anything on a molecular scale, the deductions of thermodynamics are independent of any hypothesis about the molecular structure of matter.

3.2 Glossary

1. **Work:** When a force acting on a body, moves it through a certain distance, work is said to be done.

   If a force F acting on a body, moves it through a distance S, the work done is $F \times S$. This type of work is called **mechanical work.**

   Other types of work are electrical work, work of expansion. Each form of work is the product of two factors:

   (a) **Intensity factor**: It is the measure of force or resistance overcome.

   (b) **Capacity factor**: It represents the extent to which it has been overcome when an operation is carried out.

   For example, in raising an object of mass (m) against gravity (g) through a height (h), the force exerted (intensity factor) in opposition to gravity is mg and the capacity factor is the height (h) so that the work done which is the product of these two factors is $mg \times h$. When a gas expands against pressure, the mechanical work is obtained by the product of pressure P (intensity factor) and volume change ΔV (capacity factor) or $W = P \times \Delta V$.

2. **System**: A system is defined as the specific portion of the universe, big or small, on which the effect of certain properties such as temperature and pressure is to be studied.

3. **Surroundings**: Regions outside the boundaries of the system are termed surroundings. A system is thus separated from the rest of the universe, i.e., surroundings, by real or imaginary boundary so that heat or other forms of energy may pass through it.

4. **Open, closed and isolated systems**

   (a) **Open system**: In an open system, mass as well as energy can enter into or leave the system and thus there can be exchange of these with its surroundings.

   (b) **Closed system**: In a closed system, exchange of energy in between the system and the surroundings is possible but mass can neither enter into nor leave the system.

   (c) **Isolated or insulated system**: A system in which there is no exchange of either mass or energy in between the system and the surroundings is called isolated or insulated system.

5. **Homogeneous and heterogeneous system**

   (a) **Homogeneous**: A system is said to be homogeneous when it has same chemical composition throughout, e.g., mixture of gases.

(b) **Heterogeneous**: A system consists of two or more different phases which are homogeneous in themselves and are separate from one another by definite bounding surfaces, e.g., ice in contact with water.

6. **State of a system**: A system is said to be in a particular state when specific values of the properties of the system called as **variables of state** are known. These variables of state are of four in number namely:

(a) Temperature

(b) Pressure

(c) Volume

(d) Composition.

If the values of these four variables of a system are known, all other properties such as mass, viscosity, density etc. are thereby definitely, fixed. Hence these thermodynamic properties completely define the state of the system.

If we consider a homogenous system consisting of a single substance, the composition is automatically fixed. Thus, we require only the values of temperature, pressure and volume to completely define the state of the system. But pressure, volume and temperature are interrelated and if the values of any two of them are known, the third can be calculated with the help of the Gas Equation $PV = RT$. The state of a simple homogenous system is, therefore, completely defined, if any two of these three properties are known.

7. **Extensitve and intensive properties**: The physical properties of the system are of two types:

(a) **Extensive properties**: Firstly, the extensive properties which depend on the quantity of matter in the system under consideration. Mass, volume and energy are the examples of this type.

(b) **Intensive properties**: Secondly, the intensive properties which depend upon the nature of the substance and are independent of its amount, e.g., temperature, pressure, viscosity, refractive index etc. are intensive properties.

8. **Thermodynamic equilibrium**: A system is said to be in thermodynamic equilibrium when any of its observable properties such as temperature, pressure, volumes etc. do not change with time. For thermodynamic studies, a system must be in three types of equilibria which must exist simultaneously. The system must be in

(a) **Thermal equilibrium**: Its temperature should not change with time.

(b) **Chemical equilibrium**: Its chemical composition should not change with time.

(c) **Mechanical equilibrium**: There should not be any movement of particles of the constituents of the system in itself and in between itself and surroundings. The system in which diffusion or fast chemical reactions are taking place not in thermodynamic equilibrium and as such cannot be studied by thermodynamic methods.

9. **Isothermal process**: A process is said to be isothermal if the temperature of the system remains the same throughout the process.

For an exothermic process to be isothermal, the heat evolved is given up instantaneously to the surroundings so that the temperature of the system does not rise at all.

Similarly, in an endothermic process, the required amount of heat enters the system from the surroundings instantaneously so as to maintain a constant temperature of the system at any stage during the process.

10. **Adiabatic process**: A process is said to be adiabatic if no heat enters or leaves the system during any stage of the process. In such a process, the system is in complete isolation from the surroundings. Thus, if heat is evolved or absorbed by the system during adiabatic process, its temperature rises or falls accordingly.

11. **Energy**: Energy is defined as the property which can be produced from or converted into work including work itself.

Energy manifests itself in many forms as represented by various forms of work. It is well known now that heat, which is a form of energy, cannot be completely converted into work but work can be converted into equivalent of heat energy. James Joule showed that the amount of mechanical work (W) is proportional to the quantity of heat produced (H) i.e.,

$$W \propto H \text{ or } W = JH$$

where J is Joule's mechanical equivalent of heat, when

$$H = 1;\ W = J.$$

Joule's mechanical equivalent of heat may, therefore, be defined as the amount of work done to produce one calorie of heat and is numerically equal to 4.185×10^7 ergs.

1 calorie = 4.185 joules = 4.185×10^7 ergs.

3.3 Laws of Thermodynamics and Living Organisms

Introduction: The molecular complexity and orderliness of structure of living organisms in contrast to the randomness of non-living matter, have profound implication to the physical scientist. The living system has been discussed under the purview of the laws of thermodynamics.

1. **The 1st Law of Thermodynamics**: It states that energy can be neither created nor destroyed. Living organism thus cannot consume or use up energy; they can only transform one form of energy into another.
2. **The 2nd Law of Thermodynamics**: It states that all physical and chemical processes always proceed with an increase in the disorderliness or randomness in the surrounding, that is its Entropy. How is it, then, that living organism can create and maintain their intricate orderliness in an environment that is relatively disordered and becoming more so with time?

Living organisms do not constitute exceptions to the laws of thermodynamics. Their high degree of molecular orderliness must be paid for in some way since it cannot arise spontaneously from disorder.

According to the laws of thermodynamics the living organisms have the following characteristics.

1. Useful "**Free energy**" utilization.
2. Living organism is an "**open system**".
3. They usually exist in a "**steady state**".
4. Living organism is a "**non-equilibrium**" system.
5. It is an "**isothermal system**".
6. It does not resemble heat or electrical engine, that is, it function as "**isothermal chemical engines**".

1. **Useful "free energy" utilization**: The living organisms absorb from their environment a form of energy that is useful to them under the special conditions of temperature and pressure in which they live and then return to the environment an equivalent amount of energy in some other less useful form. The **useful form of energy** that living cells take in is called 'free energy' or it is meant energy at high potential capable of performing work at constant temperature and pressure.

   The less useful type of energy that living cells return to the environment consist of heat and other forms that quickly become randomized in the environment and thus increase its disorder or entropy.
2. **Open system**: The environment of living organisms is absolutely essential to them not only as a source of free energy but also as a source of new material. In the language of thermodynamics living organisms are '**open system**' because they exchange both energy and matter with the environment and, in doing so, transform it. It is a characteristic of open system that they are not in equilibrium with the environment.
3. **Steady state**: Although living organisms may appear to be in equilibrium because they may not change visibly as we observe them when a period of time, actually they usually exist in a **steady state.** That condition of an open system in which the rate of transfer of matter and energy from the environment into the system is exactly balanced by the rate of transfer of matter and energy out of the system.

   It is, therefore, part of the molecular logic of the living state that the cell is a non-equilibrium open system, a machine for extracting 'free energy' from the environment which it causes to increase in randomness. Moreover, this is another reflection of the principle of maximum economy.

Living cells are highly efficient in handling energy and matter. They greatly exceed most man-made machines in the efficiency with which they convert input energy into work.

4. **Non-equilibrium**: The living system is a '**non-equilibrium**' open system which can perform work. If it is already in equilibrium, it cannot do work. The analysis of the magnitude and efficiency of energy exchange in steady state or open system is much more complex than in closed system but on extension of thermoyanmic theory, called 'non-equilibrium or irreversible thermodynamics' has been developed for the analysis of the open system. Three things can be attributed:
   (a) An open system in steady state is capable of doing work in non-equilibrium.
   (b) Process under non-equilibrium can be regulated.
   (c) In non-equilibrium thermodynamics, the steady state, which is a characteristic of all smoothly running machinery, may be considered to be the orderly state of an open system.

   The state in which the rate of entropy production is at a minimum, the system is operating with maximum efficiency under the prevailing conditions.

5. **Isothermal system**: The energy transforming machinery of living cells is built entirely of relatively fragile and unstable organic molecules that are unable to withstand high temperatures, strong electric currents or extremely acidic or basic conditions. The living cell is essentially 'isothermal', at any given time all parts of the cell have the same temperature. Furthermore, there are no significant differences in pressure between one part of the cell and another. For this reason cells are unable to use heat as a source of energy, since heat can do work at constant pressure only if it passes from a higher to a lower temperature zone.

6. **Isothermal chemical engines**: Living cells therefore do not resemble heat or electrical engines. Hence living cells function as isothermal chemical engines.

The energy that cells absorb from the environment is tranformed into chemical energy to carry out the chemical work for synthesis of cell components, osmotic work, transport of materials into the cell, nerve conduction, mechanical work of muscle contraction and locomotion which takes place at constant temperature.

Among familiar man-made machines, we see very few capable of using chemical energy to do work at constant temperature. In fact, engineering technology has yet to produce an useful engine that can convert chemical energy isothermally into mechanical energy, yet these types of energy conversion is familiar to all of us in the contraction of muscles.

Conclusion: Life is in a constant struggle against the tendency to produce entropy. The synthesis of large and information-rich macromolecules, the formation of intricately structural cells, the development of organization, all these are powerful anti-entropic doom imposed on all natural phenomena. Under the 2nd law of thermodynamics, living organisms choose the least evil—they produce entropy at a minimum rate by maintaining a steady state.

3.4 First Law of Thermodynamics

3.4.1 First law of thermodynamics

Failure to construct a perpetual motion machine, i.e., a machine which could produce more mechanical work than the equivalent energy used, lead to the universal acceptance of the principle of **conservation of energy**. The first law of thermodynamics is simply the law of conservation of energy. This law has been stated in various ways but the basic idea is that energy can neither be created nor destroyed; the only change which energy can undergo is a transformation from one form to another.

The first law of Thermodynamics was defined in the following ways:

1. The total energy of an isolated system always remains constant, although there may be a change from one form to another.

or, 2. The energy of an isolated system remains constant and whenever a quantity of some form of energy diasppears, an exactly equivalent quantity of some other form of energy must be produced.

or, 3. Any gain or loss of energy by the system must be exactly equivalent to the loss or gain, respectively, by the surroundings of the system.

or, 4. That whenever a certain quantity of energy is produced, an equivalent amount of other form of energy must be used up.

or, 5. Energy can neither be created nor destroyed, the only change which energy can undergo is a transformation from one form to another.

or, 6. The total energy of a system, plus its surroundings remains constant.

3.4.2 Discussion

The law of conservation of energy encountered more opposition than the second law, and numerous attempts were made to construct a perpetual motion machine which would produce work for eternity without any consumption of fuel.

The first law as expressed above needs qualification. Einstein has shown that matter and energy are equivalent and the interrelationship is given by the famous Einstein equation where E is the

$$E = mc^2$$

energy, m is the mass and c is the velocity of light. Hence 1 gm of matter is equal to 2.1×10^{13} calories, a stupendously large amount of energy. So in all chemical reactions which produce energy some loss of mass should take place, but normally, this would be too small to be detectable and so a chemist can as well forget about it. Any way, the first law is to be qualified by saying that matter is also interconvertible into energy, and with this provision the first law as stated above is rigorously correct.

3.4.3 Mathematical formulation of the first law

Suppose we put some amount of heat in a system. Since the heat energy cannot be lost, it must remain either wholly or partly as internal energy in the system, or can wholly or partly be used up by the system, in doing mechanical work. In the general case, when the heat absorbed goes both to increase the internal energy and to produce some mechanical external work, we must have Heat absorbed = Increase of internal energy + work done by the system or,

$$\begin{pmatrix}\text{Increase of}\\ \text{internal energy}\end{pmatrix} = \begin{pmatrix}\text{Heat}\\ \text{absorbed}\end{pmatrix} - \begin{pmatrix}\text{Work done by}\\ \text{the system}\end{pmatrix}$$

If the final and the initial internal energies of the above systems are E_2 and E_1 respectively, then the increase in internal energy is $E_2 - E_1 = \Delta E$ (the symbol Δ always signifies increase, i.e. Final – Initial, irrespective of whether it is positive or negative). If the heat absorbed is q and the work done by the system W, then by substituting these values in the foregoing equation, we get

$$E_2 - E_1 = \Delta E = q - W \qquad \text{... (1)}$$

which is the first law of thermodynamics. If the above change, as shown in equation (1) is very small we can write the above quantities in infinitesimals i.e.

$$dE = dq - dW. \qquad \text{... (2)}$$

If the system is under a pressure P and increases by a small volume dV, the work done is pressure multiplied by volume change, i.e., P.dV, and therefore the foregoing equation becomes

$$dE = dq - P.dV \qquad \text{... (3)}$$

This is the mathematical expression for the first law for infinitesimal changes where the work is exclusively of the pressure X volume type.

3.4.4 Mathematical implication of the first law equation

(Energy content or Internal energy)

The mathematical significance of this simple equation (1) $E_2 - E_1 = \Delta E = q - W$ is much deeper than which is apparent.

This equation means that if a system changes from a certain initial state to a certain final state, which it can do in an infinite number of ways, the increase in internal energy is always the same by whatever way the change is brought about. Although the heat absorbed or the work done by the system might vary with the path by which the change is effected. The quantity E, the internal energy, is thus a characteristic of the state of the system and is not dependent on the way in which that state has been reached. This is shown graphically in Fig. 3.1, where ΔE is the same and is equal to $E_B - E_A$, whether the change takes place by path (1) or by path (2) or by any other path.

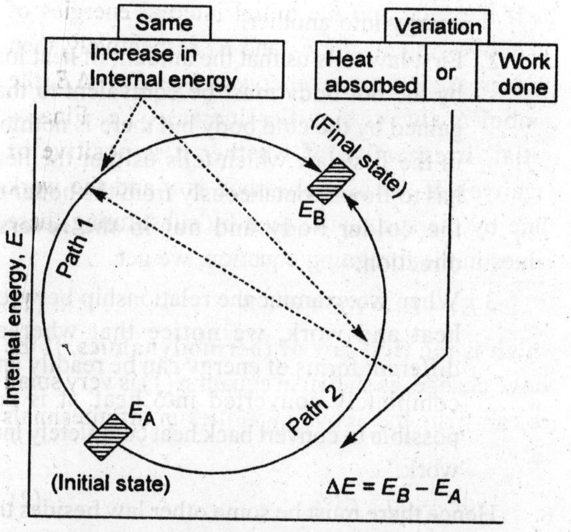

Fig. 3.1: Internal energy as a function of state

The above fact is expressed in mathematical paralance by saying that E is a **state function**, i.e., its value depends solely on the state of system and not the least on its previous history. So, E can be differentiated and dE can be integrated by standard mathematical techniques.

It may be pointed out that not all properties are state functions and so they do not admit of general differentiation and integration. For example, q and W in equation (1) are not functions of the state but depend on how that state has been arrived at. So q cannot be differentiated and dq cannot be integrated except under special conditions. (This is often expressed by saying that dq is not a perfect differential, and some even object to write dq and dW in equation (2).

3.5 Enthalpy (Heat Content)

(Heat changes at constant pressure)

Definition: To evalue the heat transferred in a **constant pressure** process, it is more convenient to make use of a function known as **enthalpy** or **heat content** denoted by the symbol H.

Constant pressure processes are more common in chemistry as a majority of the reactions are performed in open vessels.

Equation: Consider a system consisting of a gas in a cylinder filled with a frictionless piston, at constant pressure and volume. Let the system change from state A to state B by absorbing q calories of heat from surrounding at constant pressure P. Let EA and EB, V_A and V_B be the energy contents and volumes of the system in initial and final states of the system respectively. Let the work done due to expansion of the gas W. The system has gained energy equal to q and lost energy equal to W due to work done. According to first law of thermodynamics the difference of the two must be equal to increase in internal energy $(E_B - E_A)$,

or, $q - W = E_B - E_A$

$$\therefore q - W = \Delta E. \quad \text{... (2)}$$

The value of W, the work done due to expansion is the product of force displacement.

Force acting on the piston = pressure × area. Displacement is the distance through which the piston is raised = 1 (say).

$\therefore$ work done = pressure × area × 1 = $P \times a \times 1$.

But $(a \times 1)$ = change in volume = $(V_B - V_A)$.

$\therefore \quad W = P(E_B - E_A) = P\Delta V$

Substituting this value of W in equation (1) we have

$$\Delta E = q - P\Delta V$$

$$q = \Delta E + P\Delta V. \quad \text{... (3)}$$

At constant pressure, equation (3) may be written as

$$qP = \Delta E + P\Delta V \quad \text{... (4)}$$

Here $\Delta E = E_B - E_A$

and $\Delta V = E_B - E_A$

$$qp = (E_B - E_A) + P(V_B - V_A)$$
$$= (E_B + PV_B) - (E_A + PA_A). \quad \text{... (5)}$$

Just as the internal energy is characterised by the state of a system only and is independent of the process by which it is achieved, so is the product of pressure and volume, i.e., PV. The quantity, $(E + PV)$, therefore, is also dependent only on the sate of the system is known as "**enthalpy**" or "**heat content**" and is denoted by the symbol H.

or, $$H = E + PV \quad \text{... (6)}$$

Using this value in equation (5), we get

$$qP = H_B - H_A = \Delta H. \quad \text{... (7)}$$

Combining equation (4) and (7), we have

$$H = \Delta E + P\Delta V. \quad \text{... (8)}$$

i.e., the change in heat content at constant pressure is equal to the sum of the increase in internal energy and the external work of expansion.

Combining equation (1) which is according to the first law of thermodynamics,

$$\Delta E + W = q$$
$$\Delta E = q - W \quad \text{... (1)}$$

With (4), we have $qP = \Delta H$. ... (9)

i.e., enthalpy measures heat content of constant pressure and the heat of reaction of a process at constant pressure is equal to the change in enthalpy ΔH.

A reaction which gives off heat to the surroundings is said to be "**exothermic**" and H_B in such a case is less than H_A, so that ΔH has a negative value. This is the case for the reaction between hydrogen and oxygen:

$$2H_2(g) + O_2(g) \rightarrow 2H_2O(1); \Delta H = -136.6 \text{ cal.}$$

A reaction which absorbs heat from the surroundings is said to be "**endothermic**" and H_B in such a case is greater than H_A, so that H has a positive value and there is increase in enthalpy.

The heat of vaporisation and the heat of fusion of a substance are enthalpy changes and can be called "**enthalpy of vaporisation**" and "**enthalpy of fusion**" respectively. Enthalpy changes accopanying phase changes can be treated in exactly the same way as enthalpy changes accompanying any chemical change.

3.6 Second Law of Thermodynamics

3.6.1 Introduction

(Demerits or limitation of 1st law)

1. While the first law establishes the equivalence between heat and work, it imposes no condition on their mutual convertibility. In other words, it never tells us under what circumstances and to what extent it is possible to convert one form of energy into another.
2. First law tells us that the amount of heat lost by the hot body must be equivalent to that gained by the cold body but there is nothing in the first law which tells us that the heat has to flow spontaneously from the hotter to the colder body and not in the reverse direction.
3. When we examine the relationship between heat and work, we notice that whereas different forms of energy can be readily and completely converted into heat, it is not possible to convert back heat completely into work.

Hence there must be some other law besides the first law, which governs the direction of flow of heat and the extent of its convertibility into work. This limitation forms the basis of the second law of thermodynamics.

3.6.2 Second law of thermodynamic

1. "It is impossible for a self-acting machine unaided by any external agency to convey heat from a lower to a higher temperature."

or, 2. "It is impossible to lift a weight and cool a body without leaving any other change."

or, 3. "It is impossible to convert heat into an equivalent amount of work without producing other changes in some parts of the system."

or, 4. "The total entropy of a system must increase if a process is to occur spontaneously."

3.7 Comparison of 1st and 2nd Law of Thermodynamics

The modus operandi of the first law and that of the second law are basically similar.

1st Law	2nd Law
1. It introduces a new concept, a quantity called **"internal energy."**	1. It introduces a new function, called '**entropy**'.
2. We measure the change of internal energy by the energy we put into the system minus the energy we take out of it.	2. Similarly the entropy change is measured by the heat energy we put into the system divided by the corresponding temperature, summed up over the whole path of the change, provided the whole change is conducted reversibly.
3. The concept of internal energy is as much a logical invention of the 1st law.	3. The concept of entropy is as much a logical invention of the 2nd law.
4. That the energy is indestructible and so easily appeals to our material comprehension.	4. Entropy always increases in any natural process, which is unlike any material change we are familiar with.
5. It tells us without any bias towards any particular direction of change, which processes are energetically possible.	5. It points out the direction in which these processes would go.
6. The first law enforces a **"necessary"** condition.	6. The second law prescribes a "sufficient" condition for a change to take place.

3.8 Entropy

3.8.1 Definition

Entropy represents the extent of disorder or randomness of the system and becomes maximum in a system as it approaches true equilibrium.

3.8.2 Concept of entropy

The second law of thermodynamics states that "the total entropy of a system must increase if a process is to occur spontaneously." According to the second law of thermodynamics the expression we get:

$$\frac{\text{Heat absorbed at } T_2}{T_2} + \frac{\text{Heat absorbed at } T_1}{T_1} = 0$$

Since the equation derived for Carnot Cycle is valid for any other reversible cycle and in fact since any reversible cycle can be shown to be equivalent in the limit to a large number of Carnot Cycles, the above expression is valid for any reversible cycle whatsoever. Now, if we define a function S, to be called "entropy", such that:

$$\Delta S = \frac{\text{Heat absorbed reversibly}}{\text{Temperature at which the heat is absorbed}}$$

$$= \frac{qrev}{T}$$

We can immediately write equation as follows:

$$(\Delta S)\ T_2 + (\Delta S)\ T_1 = 0 \text{ (for a complete cycle).}$$

This equation is quite a general one and it asserts that the sum of all ΔS terms over any reversible complete cycle is zero. In other words, if we pass from a state A to a state B by any reversible path and again come back by any other reversible path to the original state A, the ΔS by the first path is equal to the ΔS by the second path or any other reversible path. Thus, S depends only on the state of the system and not in the least on how that state has been arrived at; i.e., S is a state function. That S (entropy) as defined above is a state function is undoubtedly the most fundamental and the most important consequence of the second law. This property S, defined by equation (1) is called the entropy of the system.

3.8.3 Molecular Interpretation or basis of entropy

Though thermodynamics is based on human experience and is not concerned with structure of matter or molecular concepts, it is very instructive to consider the molecular basis of entropy. Since all natural process leads to a net increase of entropy and also since there exists a natural tendency for an ordered set of things to get disordered through interplay of natural forces, we can identify increase of entropy (ΔS) as increase of disorder or randomness of a system on a molecular level. Thus, we can say that entropy is a measure of the disorder or randomness of a system.

3.8.4 Critical remarks on definition of entropy

1. It should be realised that this definition of entropy has nothing to do with the 2nd law,

but latter's existence only makes the entropy a state function, which means that entropy has a definite value for each state of a system independent of the way in which that state has been attained. Hence, the entropy change depends on the initial and the final state, and not at all on how the change of state has been brought about.

2. The second point and a very important one about entropy needs clarification at this stage. In the definition of entropy, in the equation, it should be noted that the numerator is the amount of **heat absorbed reversibly**. Suppose we take in a vessel a gram of atom of mercury and combine it with a gram of atom of chlorine to form a gram mole of mercurous chloride at 25°C (298 K) and 1 atm. The system, will liberate 31,300 calories of heat, i.e., the system absorb – 31,300 calories. The entropy change ΔS of the system, however, will not be equal to – 31,300/298 K (entropy unit) because the above reaction has not been conducted reversibly. If the same reaction is conducted reversibly, in a reversible galvanic cell, this cell will not evolve only 6,160 calories and so the entropy increase is only –6,160/298 = –20.5 (i.e., entropy unit, i.e., calories per K per mole of calomel formed).

3.8.5 Salient features of Entropy

1. Entropy is a measure of the degree of randomness or disorder of a system.
2. The entropy of a system increases when it becomes more disordered.
3. The second law of thermodynamics states that a process can occur spontaneously only if the sum of the entropies of the system and its surroundigs increases.
4. Entropy changes of chemical reactions are not readily measured.

3.9 Second Law of Thermodynamics Using the Concept of Entropy

3.9.1 Definition

Entropy represents the extent of disorder or randomness of the system and becomes maximum in a system as it approaches true equilibrium.

3.9.2 Second law of thermodynamics

"The total entropy of a system must increase if a process is to occur spontaneously."

Under the conditions of constant temperature and pressure, the relationship between the free energy change (ΔG) of a reacting system and the change in entropy (ΔS) is given by the following equation which combines the 2 laws of Thermodynamics:

$$\Delta G = \Delta H - T\Delta S,$$

where ΔH is the change in enthalpy of the system and T is the absolute temperature, ΔS is the change in entropy of the system and ΔG is the change in free energy of a system.

Under the conditions of biochemical reactions, because ΔH is approximately equal to ΔE, the total change in internal energy of the reaction, the above relationship may be expressed in the following way:

$$\Delta G = \Delta E - T\Delta S$$

1. If Δ-is negative in sign, the reaction proceeds spontaneously with loss of free energy, i.e., it is **exergonic**. If, in addition, ΔG is of great magnitude, the reaction goes virtually to completion and is essentially irreversible.
2. On the other hand, if ΔG is positive, the reaction cannot occur spontaneously and proceeds only if free energy can be gained, i.e., it is **endergonic**. If, in addition, the magnitude of ΔG is great, the system is stable with little or no tendency for a reaction to occur.
3. If ΔG is zero, the system is at **equilibrium and no net change takes place.**

3.9.3 Relationship between equilibrium constant and standard free energy change

In a model reaction:

$$A + B \rightleftharpoons C + D$$

$$\Delta G = \Delta G^{\circ} + RT \text{ in } \frac{[C]\,[D]}{[A]\,[B]},$$

where R is the gas constant and T is the absolute temperature. When the reactants $[A]$, $[B]$, $[C]$ and $[D]$ are present in concentrations of 1.0 M, ΔG° is known as the "**standard free energy change.**"

At equilibrium, $\Delta G = 0$, i.e.,

$$0 = \Delta G^{\circ} + RT \text{ in } \frac{[C]\,[D]}{[A]\,[B]}$$

and, therefore, at equilibrium,

$$\Delta G^{\circ} - RT \text{ in } \frac{[C]\,[D]}{[A]\,[B]}$$

For biochemical reactions, a standard state is defined as having a pH of 7. The standard free energy change at this standard state is denoted by ΔG^{o}. Since the equilibrium constant under standard condition is:

$$K' \text{ eq} = \frac{[C]\,[D]}{[A]\,[B]},$$

substitution gives

$$\Delta G^{o\prime} = -RT \text{ in } K' \text{ eq}$$

or $\quad \Delta G^{o\prime} = -2.303 RT \log K' \text{ eq.}$

Thus, the standard free energy change can be calculated from the equilibrium constant K' eq. It is important to note that ΔG may be larger or smaller than $\Delta G^{o\prime}$ depending on the concentration of the various reactants.

3.10 Physiological Steady State

(Maintenance of physiological steady state inspite of entropy or steady state in open system or energetic in open system).

The environment of living organism is absolutely essential to them, not only as a source of free energy but also as a source of raw materials. In the language of thermodynamics, living organisms are **open systems** because they exchange both energy and matter with their environment and, in doing so, transform it. It is a **characteristic of open system** that they **are not in equilibrium** with their environment. Although living organisms may appear to be in equilibrium because they may not change visibly as we observe them over a period of time, actually they usually exist in a steady state, that condition of an open system in which the rate of transfer of matter and energy from the environment into the system is exactly balanced by the rate of transfer of matter and energy out of the system (Fig. 3.2). It is, therefore, part of the molecular logic of the living state that the **cell is a nonequilibrium system,** a machine for extracting free energy from the environment, which it causes to increase in randomness. Moreover, this is another reflection of the principle of maximum economy, living cells are highly efficient in handling energy and matter. They greatly exceed most man-made machines in the efficiency with which they convert input energy into work performed.

The analysis of the magnitude and the efficiency of energy exchanges in **steady-state or open system (Equilibrium thermodynamics** may be applicable only to isolated chemical reaction in **closed system)**, but an extension of thermodynamic theory, called "**nonequilibrium or irreversible thermodynamics**" has been developed for the analysis of the open system.

Three things can be attributed to open system in a steady state:

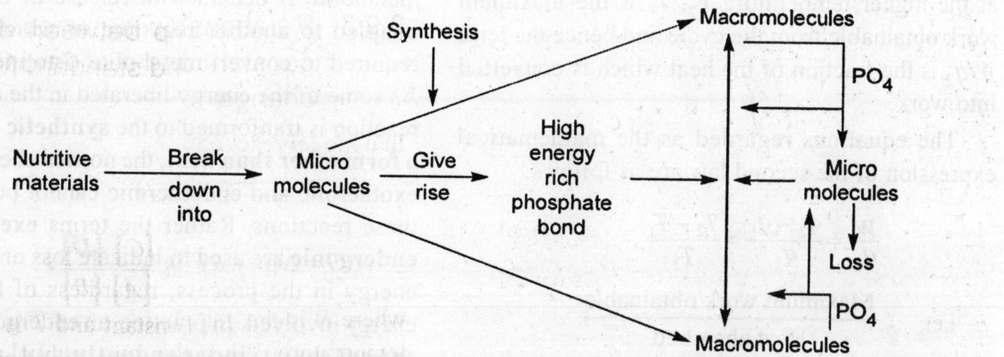

Fig. 3.2: A cell in a steady state

1. An open system in steady state is capable of doing work in nonequilibrium, because systems already at equilibrium can do no work.
2. Process under nonequilibrium can be regulated.
3. In nonequilibrium thermodynamics, the steady state, which is a characteristic of all smoothly running machinery, may be considered to be the orderly state of an open system. The state in which the rate of entropy production is at minimum, the system is operating with maximum efficiency under the prevailing conditions.

Life is in a constant struggle against the tendency to produce entropy. The synthesis of large and information-rich macromolecules, the formation of intricately structural cells, the development of organization, all these are powerful antientropic doom imposed on all natural phenomena. Under the 2nd law of thermodynamics, living organisms choose the least evil they produce entropy at a minimum rate by **maintaining a steady state.**

3.11 Efficiency and Free Energy in Thermodynamic System

3.11.1 Efficiency

Maximum fraction of the heat which is convertible into work from a Carnot Cycle depends only on the temperature of the two heat reservoirs and not only nature of the working substance, or anything else. The quantity W/q_2 is called the **Thermodynamic efficiency** of the cycle where q_2 is the heat absorbed at the higher temperature T_2, W is the maximum work obtainable from the cycle and hence the term W/q_2 is the fraction of the heat which is converted into work.

The equations regarded as the mathematical expression of the second law are as follows:

$$\frac{W}{q_2} = \frac{q_2 - q_1}{q_2} = \frac{T_2 - T_1}{T_2},$$

i.e., $$\frac{\text{Maximum work obtainable}}{\text{Heat absorbed}}$$

$$= \frac{\text{Temp. Difference}}{\text{Temp. at which heat is absorbed}}.$$

i.e., Efficiency (of any reversible engine) $= \frac{\Delta T}{T}$.

From the above equation for the efficiency of a Carnot cycle or any reversible cycle whatsoever, the interesting fact emerges that heat energy at higher temperature is more precious than heat energy at lower temperature, because more of the former can be converted into useful work. So, in the sense of convertibility to work, heat and work are not completely equivalent, and they can be regarded as completely equivalent only at an infinitely high temperature.

3.11.2 Free energy (Biochemical thermodynamics or bioenergetic)

Definition: By "free energy" is meant energy at high potential capable of performing work at constant temperature and pressure. It is designated as ΔG (or Δ*F*).

3.11.3 Discussion

The biochemical reactions are accompanied by liberation of energy, as the reacting system moves from a higher to lower energy level. The energy is liberated in the form of heat. In non-biologic system, heat energy may be tranformed into mechanical or electrical energy. Since biological systems are essentially isothermic, no direct use of heat can be made to drive vital processes requiring energy, e.g., synthetic reaction, muscular contraction, nerve conduction, active transport, etc. But these processes obtain energy by chemical linkage or "coupling" to oxidative reactions as shown in (Fig. 3.3). The conversion of metabolite *A* to metabolite *B* occurs with release of energy. It is coupled to another reaction, in which energy is required to convert metabolite *C* to metabolite *D*. As some of the energy liberated in the degradative reaction is tranformed to the **synthetic reaction in a form other than heat,** the normal chemical terms exothermic and endothermic cannot be applied to these reactions. Rather the terms **exergonic** and **endergonic** are used to indicate loss or gain of free energy in the process, regardless of the form of energy involved. In practice, an endergonic process cannot exist independently but must be a **component of a coupled exergonic/enedergonic system where the overall net change is exergonic.**

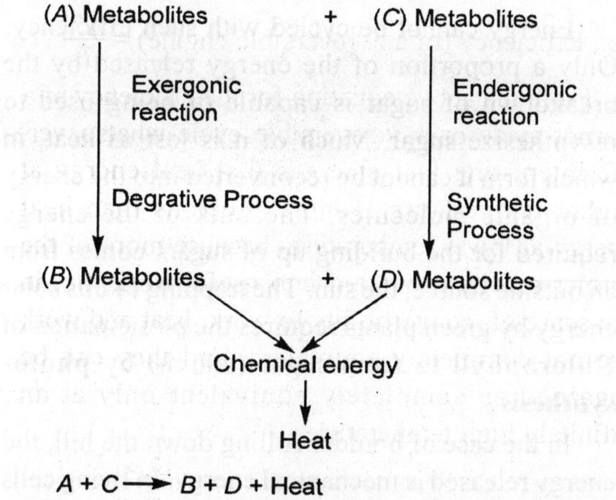

Fig. 3.3: Coupling of an exergonic to endergonic reaction

If the reaction is to go from left to right, then the overall process must be accopanied by loss of free energy as heat (Fig. 3.3).

1. One possible mechanism of coupling could be envisaged if a common obligatory intermediate, (i) took part in both reactions i.e., $A + C \longrightarrow (i) \longrightarrow B + D$. Some exergonic and endergonic reactions in a biologic system are coupled in this way. This type of system has a built-in mechanism for biologic control of the rate at which the oxidative processes are allowed to occur. The existence of a common obligatory intermediate for both the exergonic and endergonic reactions allows the rate of utilization of the product of the synthetic path (*D*) to determine by mass action the rate at which *A* is oxidized. Indeed, these relationships supply a basis for the concept of "**respiratory control**", the process that prevents an organism from burning out of control, is provided by dehydrogenation reaction.

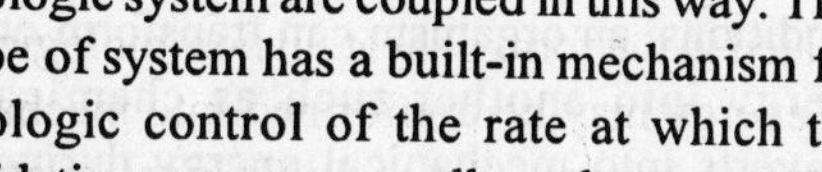

(Coupling of dehydrogenation and hydrogenation reactions by an intermediate carrier).

2. An alternative method of coupling an exergonic to endergonic process is to synthesize a compound of high-energy potential in the exergonic reaction and to incorporate this new compound into the endergonic reaction, thus effecting a transference of free energy from the exergonic to the endergonic pathway. $\sim E$ is a compound of high potential energy and E is low potential energy. The biologic advantage of this mechanism is that (a) E need not structurally related to A, B, C, D (b) This would allow E to serve as transducer of energy from a wide range of exergonic reactions (Fig. 3.3A).

3.12 Concept of Energy in the Biological System in the Light of Thermodynamics

3.12.1 Definition of energy

Energy is defined as the property which can be produced from or converted into work including work itself.

3.12.2 Concept of energy from the biological point of view

Let us look at energy from the biological point of view. One of the most important metabolic

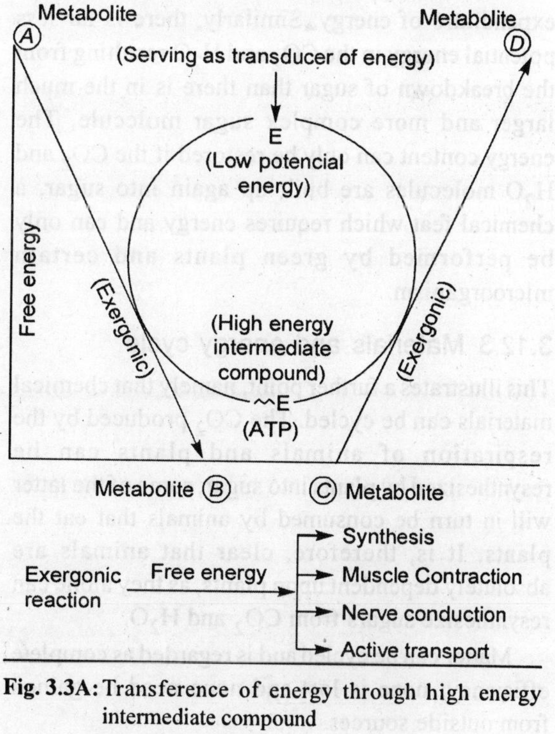

Fig. 3.3A: Transference of energy through high energy intermediate compound

pathways yielding energy is the breakdown of sugar: $C_6H_{12}O_6 + 6O_2 \rightarrow 6CO_2 + 6H_2O$ + Energy. This equation tells us the breakdown of sugar is an **oxidation** process: the sugar reacts with O_2 in some way and eventually breaks down into CO_2 and H_2O with the liberation of energy. The sugar molecule can be looked upon as containing **potential energy** which is freed only when the molecule is split. The potential energy is locked in the chemical bonds which join the atoms in the sugar molecule together, and only when these bonds are broken, the energy is released. The energy released is capable of doing useful work.

As an analogy, think of a boulder sitting at the top of a hill. While at rest, the boulder is full of potential energy. If, however, it is pushed so that it rolls down the hill, its potential energy is converted into the **Kinetic energy** of motion. The energy, formerly locked up in the boulder, is released.

The boulder analogy can usefully be taken a step further. When the boulder reaches the bottom of the hill and comes to rest, it contains less potential energy than it did at the top of the hill. In order to restore its potential energy it must be pushed back to its former position, and this requires the expenditure of energy. Similarly, there is far less potential energy in the CO_2 and H_2O resulting from the breakdown of sugar than there is in the much larger and more complex sugar molecule. The energy content can only be restored if the CO_2 and H_2O molecules are built up again into sugar, a chemical feat which requires energy and can only be performed by green plants and certain microorganism.

3.12.3 Materials and energy cycle

This illustrates a further point, namely that chemical materials can be cycled. The CO_2 produced by the respiration of animals and plants can be resynthesized by plants into sugar; some of the latter will in turn be consumed by animals that eat the plants. It is, therefore, clear that animals are absolutely dependent upon plants, as they alone can resynthesize sugars from CO_2 and H_2O.

Matter can be cycled and is regarded as complete efficiency; none is lost and none need be gained from outside sources.

Energy cannot be cycled with such efficiency. Only a proportion of the energy released by the breakdown of sugar is capable of being used to resynthesize sugar. Much of it is lost as heat, in which form it cannot be reconverted into the energy of organic molecules. The bulk of the energy required for the building up of sugars comes from an outside source, the sun. The trapping of this solar energy by green plants requires the participation of **Chlorophyll** to form sugar (starch) by **photosynthesis**.

In the case of boulder rolling down the hill, the energy released is mechanical energy. In living cells useful energy is contained in chemical bonds and is released when these bonds are broken. Anabolic processes like photosynthesis are concerned with building up bonds which can subsequently be broken for energy release.

3.12.4 Energy conversions

It is known from the laws of thermodynamics that different forms of energy (chemical, light, heat, etc.) are interconvertible. This is also important biologically because it means that, given the right conditions, an organism can transform one type of energy into another such as chemical energy converts into mechanical energy during muscles contraction. Other examples are green plant which converts light energy into chemical energy; in luminescent chemical energy into light energy; in electric fishes, chemical energy into electrical energy, etc.

The **First Law of Thermodynamics** states that energy can neither be created nor destroyed during conversion of energy. This does not mean that all the energy is converted into the mechanical energy of motion; the rest is used to overcome various frictional forces and is dissipated as heat. This is summed in the **2nd Law of Thermodynamics** which states: when one form of energy is converted into another, a proportion of it is turned into heat. This, too is an obvious fact of everyday life. Electricity is used to drive the motor of an electric train, but inevitably some of the energy is lost as heat. This unavoidable fact is of great importance in biology. It means that when sugar is broken down not all the energy liberated can be used for establishing new chemical bonds and so forth. A proportion of it is dissipated as heat. Of course, it does not follow that this heat is of no value to the organism, indeed in warm-blooded animals it contributes to the maintenance of a constant temperature, but from the thermodynamic standpoint the energy is changed into a from in which it has no direct metabolic use.

3.12.5 Activation energy

The boulder analogy illustrates another important concept. Imagine that the boulder resides in a slight depression at the top of the hill. It is obviously necessary to push it over the hump before it can start rolling down the hill and this requires the expenditure of energy. The hump represents an energy barrier, and the energy required to push the boulder over this barrier can be called the **activation energy**. Much the same applies to our sugar molecule, a small amount of activation energy must be supplied before the process of oxidative breakdown can get under way. Clearly any factor, physical or chemical, that helps the sugar molecule over its energy barrier will facilitate the process of oxidative metabolism.

3.13 Living Body as a Thermodynamic System

In engineering and technology, i.e., in case of non-living system, the thermodynamic system is simple. Here a few substances interacting according to known chemical formula under specified conditions in terms of composition, pressure, volume and temperature. Whereas in case of living system, there are large number of components whose nature and interactions are not known and not according to specified conditions like those mentioned in case of technology. The consideration of biological systems as analogous to thermodynamic system is not clear. Because in living systems, irreversible reactions are occurring continuously whereas in case of non-living thermodynamic systems, it is reversible one in equilibrium. One of the observables in living system is the rate of irreversible reactions indicated by the heat output of the reactions. When this rate is a studied as a function of the temperature of the system, some information sometimes can be obtained about the reactions. To know the type of reaction which may be recognised in this kind of experiment, which can be considered as **Arrheniun type**. In these reactions, the rate varies exponentially as the reciprocal of the absolute temperature. Thus the rate R is

$$R = R_0 e^{-\mu/T}$$

given by where R_0 and μ are constant. The rate R will change by a factor of about 2 when the temperature changes by 10° C according to **"Van't Hoff's rule"**. The plot of the log of the reaction rate R as a function of the reciprocal of the absolute temperature which is called **Arrhenius plot** which is s straight line. The Arrhenius type of reaction can proceed generally only after "activation". Molecules which have been activated by getting random motion higher than average speed may move from a higher to a lower energy level by penetrating a barrier between the level. Arrhenius form of equation for the reaction rate is the same as the equation giving the relative number of gas molecules having a given energy known as **Boltzman distribution,** valid for energies above the mean thermal energy. Some processes occur in living organisms that always require large activation energies whenever they occur in vitro, e.g., oxidation of carbon to produce heat. It can proceed in absence of large activiation energies by enzymes in vivo which form intermediate complexes with activation energies low enough so that each reaction may still follow the Arrhenius law. In all such reactions the fundamental law of thermodynamics still apply to each intermediate step in turn. It has been proposed that in a chain of reactions with different velocity is determined by the velocity constant of the slowest reactions which is known as **'Blackman-Putter principle'**. Moreover, at certain temperature, one reaction may predominate and determine the velocity, whereas at another temperature, a different reaction may predominate. This causes a "discontinuous Arrhenius plot" in which the slope of the plot would change suddenly at some "critical temperature". The reaction which determines the rate of reaction at the prevailing temperature is known as **master reaction.** The heats of reaction in living system may be intermediate in the biochemical chain but ultimate reaction is oxidation which is an irreversible reaction. The heat evolved in the reaction is the "heat of oxidation" in general sense; and in detail, which are probably

3.14 Comparison of Living and Non-living System as a Thermodynamic System

Living system	Non-Living system
1. First law applicable and application of second law is difficult.	1. Second law applicable.
2. Exchange matter with the surrounding.	2. Do not exchange matter to the surrounding.
3. Interaction of large number of substances of unknown nature.	3. Interaction of substances of known chemical formula.
4. Interaction is not known and not under specified conditions.	4. Interaction according to the specific conditions or observables e.g., pressure, volume, temperature etc.
5. Irreversible reactions are occurring continuously not in equilibrium.	5. Reversible one in equilibrium.
6. Rate of irreversible reaction is indicated by heat output of the reaction.	6. No.
7. Open System.	7. Closed system.
8. Exists in a steady state in which the rate of input of matter equals the rate of output of matter.	8. No.
9. Concentration of all components of the cell including ATP, remains constant under these conditions.	9. No.
10. The rate of formation of each component is exactly equals to its rate of utilization.	10. No.
11. The rate of entropy production is at a minimum and in which the system is operating with maximum efficiency under the prevailing conditions.	11. Rate of entropy production is maximum.
12. It has tendency to produce entropy but has antientropic forces for which they produce minimum entropy.	12. Entropy increases or constant but never decreases.

involved are heats of neutralization of acids and bases, heats of formation of compounds, heats of formation of ions and heats of solution and dilution. According to Hess's law, which states that the heat absorbed or evolved in a given reaction must always be constant and independent of the manner in which the reaction takes place which is a restatement of the first law of thermodynamics. So, we can infer that the biological reactions are temperature dependent. Application of **second law of thermodynamics** to living systems is difficult because second law states that in a closed system i.e., non-living system, entropy is either increasing or constant, never decreasing. This means that the degree of organization of the non-living system should be constant or decreasing. But in growing biological system materials in a low energy state (Nutrients) are re-formed into high energy substances with a high degree of organisation which contradicts the second law of thermodynamics. Moreover, the living system, is an orderly state of an open system, the state in which the rate of entropy production is at a minimum and in which the system is operating with maximum efficiency would be the prevailing conditions.

The living system is constantly struggling against entropy because this system has the tendency to produce entropy. The antientropic forces are:

1. The synthesis of large and information-rich molecule,
2. The formation of complicated structured cells,
3. The development of organization, etc.

In spite of that, there is no escape from entropy according to the second law for which living system produces entropy at a minimum rate to maintain a steady state.

Therefore, it may be corrected that second law applies only to closed thermodynamic systems (e.g., non-living) because the biological system is open and capable of exchanging materials and energy with the environment.

3.15 Carnot Cycle

(Transformation of heat into work)

If we are to find out the transformation of heat into work, i.e., amount of work which is obtainable from a given quantity of heat, it may be available by an imaginary cycle devised by Carnot known as **Carnot cycle** which consists of a series of operations on a perfect gas so that at the conclusion of the operations, the gas is restored to its original state.

According to **Kelvin's postulate**, it is impossible to transform heat into work if heat is taken from a source at an uniform temperature. We need at least two sources at different temperatures t_1 and t_2 in order to perform such transformations.

Consider a fluid whose state can be presented on a diagram (V, P) and consider two adiabatics and two isothermals corresponding to temperatures t_1 and t_2. These four curves interesect each other in the four points A, B, C and D as shown in Figure 3.4.

Let AB and CD be the 2 isothermal lines having the temperatures t_2 and t_1 respectively. AD and BC are the two adiabatic lines. This reversible cycle transformation ($ABCD$) is called **Carnot cycle.**

3.15.1 Performance of Carnot cycle

Enclose the fluid in a cylindrical container fitted with a piston having non-heat conductivity lateral wall and so that heat loss will be only through the base of the cylinder. Let t_1 and t_2 be the two sources of heat that are so large that their temperature remain sensibly unaltered.

Let t_2 be greater than t_1. Now assume V_A and P_A are initial volume and pressure of the fluid inside the cylinder corresponding to the point A in the figure. Since this point lies on the isothermal corresponding to t_2, temperature of the fluid is equal to t_2 initially. Keeping the cylinder on the source t_2 (no transfer of heat will occur), raise the piston very slowly and then increase the volume V_B, e.g., point B in figure. This part of the transformation is represented by the segment AB of the isothermal t_2. Now the state of the system is represented by B in the figure. Similarly, the volume may be increased to V_D. Keeping the system thermally insulated the process is represented by the adiabatic segment AD and the temperature of the fluid is decreased from t_2 to t_1. The state of the fluid is now given by the point D in the figure.

Again placing the cylinder on the source t_1, the piston is lowered compressing the fluid very slowly along the isothermal DC and the volume is V_c and finally the cylinder, keeping thermally insulated, is very slowly compressed, the fluid adiabatically gives out heat along the segment CB until the temperature is raised from t_1 to t_2.

The system will now be at its initial state again. In other words, the system gives up the amount of heat q_1 to the source t_1.

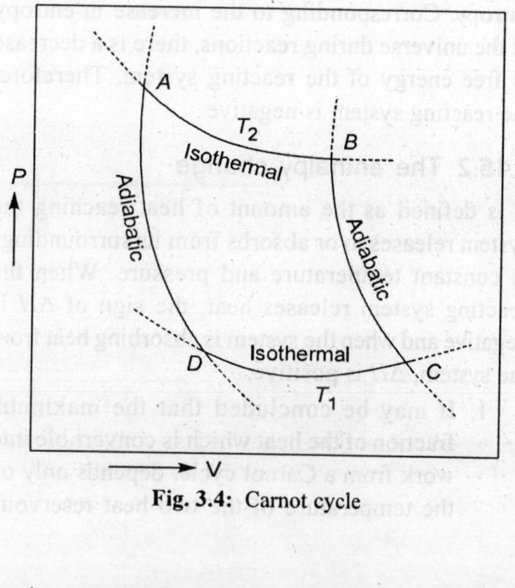

Fig. 3.4: Carnot cycle

Thus the total amount of heat absorbed by the system during the cycle is $q_2 - q_1$.

There is another aspect of second law that is necessary for understanding of its operations particularly in biological system.

First, the reaction system must t_1 be identified, the collection of matter that undergoing a particular chemical, or physical process, this may be an animal, a cell or two reacting compounds. In the real world and particularly in the biological world the reacting systems do exchange energy and often exchange matters with their surrounding.

Change in free energy heat and entropy in chemical reaction at constant temperature and pressure, the conditions existing in the biological systems are related to each other quantitatively by the equations,

$$\Delta G = \Delta H - T\Delta S$$

where ΔG = change in free energy of the reacting system.

ΔH = change in heat content or enthalpy (means warming within).

T = absolute temperature at which process is taking place.

ΔS = change in entropy of the universe (system and surrounding).

When a chemical process proceeds to its equilibrium point, the entropy of the universe increases and, therefore, always has a positive sign in the real world. Only theoretically in an ideal system, reaction can proceed with an increase in entropy. Corresponding to the increase in entropy of the universe during reactions, there is a decrease in free energy of the reacting system. Therefore, the reacting system is negative.

3.15.2 The enthalpy change

It is defined as the amount of heat reaching the system releases to or absorbs from its surroundings at constant temperature and pressure. When the reacting system releases heat, the sign of ΔH is negative and when the system is absorbing heat from the system, ΔH is positive.

1. It may be concluded that the maximum fraction of the heat which is convertible into work from a Carnot cycle, depends only on the temperature of the two heat reservoirs and not on the nature of the working substance on anything else.
2. Correlating the amount of work (W) obtainable from a given amount of heat q_2,

$$\frac{W}{q_2} = \frac{\text{Maximum work obtainable}}{\text{Heat absorbed}}$$

$$= \frac{\text{Temp. difference}}{\text{Temp. at which heat is absorbed}}$$

i.e., Efficiency (of any reversible engine)

$$= \frac{\Delta T}{T}.$$

The quantity $\frac{W}{q_2}$ is called the "*thermodynamic efficiency*" of the Carnot cycle and Carnot equations $\frac{W}{q_2} = \frac{\Delta T}{T}$ may be regarded as a mathematical expression for the second law.

3. From the equation of the efficiency of a Carnot cycle or any reversible cycle, the fact emerges that heat energy as higher temperature is more precious than heat energy at lower temperature because more of the former can be converted into useful work. Therefore, for convertibility to work, heat and work are not completely equivalent. And they can be regarded as completely equivalent only at an infinitely high temperature.

3.15.3 Carnot equation and Kelvin scale of temperature

From Carnot equation,

$$\frac{W}{q_2} = \frac{q_2 - q_1}{q_2} = \frac{t_2 - t_1}{t_2}, \quad \text{.... (i)}$$

it is evident that efficiency of the process becomes 'one' if the temperature $T_1 = 0$. This temperature at which the efficiency of the process is one, i.e., when heat converts completely into work is known as the 'absolute zero', of the thermodynamic scale of temperature or the **"Kelvin Scale of temperature."** It is found to be identical with the scale of temperature based on the ideal gas laws for which the symbol 'K' (degrees Kelvin) is used for absolute temperature based on the hypothetical ideal gas thermometer.

3.15.4 Carnot cycle and entropy

A new function of the state of the system was introduced by Claussius and designated by 'S'. It is numerically defined by the equation dS = dq/t where q = heat transferred reversibly and dS is the entropy change for an infinitesimal reversible process. In a Carnot cycle, q_2 has a positive value and q_1 has a negative value, because the former is taken up at the higher temperature and the latter is given out at the lower temperature. The Carnot equation may, therefore, be rewritten as:

$$\frac{q_2+q_1}{q_2}=\frac{t_2-t_1}{t_2} \quad \text{or} \quad \Sigma\frac{q}{t}=0$$

The reversible cycle may, therefore, be regarded as made up of a number of Carnot cycles, for each of the sum of the two q/t terms involved is zero, i.e.,

$$\frac{q_1}{t_1}+\frac{q_2}{t_2}=0\cdot$$

Energy

Definition: Energy means ability to do work.

FORMS OF ENERGY. Energy has many forms like matter:

1. Potential Energy
2. Kinetic Energy
3. Chemical Energy
4. Electrical Energy
5. Thermal (heat) Energy
6. Nuclear Energy
7. Electromagnetic Energy

1. **Potential Energy**: Ability to do work due to position. As for example a heavy iron ball is held 30 ft in the air by a rope and pulley is the example of potential energy possessed by the iron ball.

   Work is required to keep it in high position and because of this high position the iron ball is said to possess potential energy.
2. **Kinetic Energy**: Kinetic energy is the energy of motion. All matters possess this energy while in motion, e.g., automobiles. This energy can perform work because of their motion.
3. **Chemical Energy**: This energy is released due to chemical reaction. For example, this energy is provided by the biochemical reactions in the body involving the biochemical compounds like carbohydrates, lipids and proteins from food.
4. **Electrical Energy**: The work can be done when an electron or an electronic charge moves through an electrical potential. The most common form of electrical energy domestic electricity where the movement of electrons through a copper wire under an electric power e.g. moltor.
5. **Thermal Energy**: It is the energy of motion at the atom or molecule level. It is the kinetic energy of atom and molecules. This energy is measured by temperature. In a substance, the faster the atoms and molecules move more thermal energy is present in the substance and its temperature is higher.
6. **Nuclear Energy**: This energy is present in the nucleus of an atom. The release of energy and its use is controlled by electric power plants.
7. **Electromagnetic Energy**: This energy is most important because it is present in the X-ray, and other examples are visible light, radiowaves and microwaves.

Just like matter which can be transformed from size, shape and form to another, the energy also can be transformed from one type to another such as electric energy in the x-ray machine acting on tungsten filament produces electromagnetic energy, i.e. X-ray which is converted into chemical energy in X-ray film. Energy emitted and transferred through matter is known as *radiation.* Visible light is a form of electromagnetic energy. Energy radiated by the sun is also known as *electromagnetic radiation.* Usually electromagnetic energy is called electromagnetic radiation or simply radiation.

4

True Solutions

1. Definition, 2. Characteristics, 3. Comparison of true and colloidal solution, 4. Solvent for solutions, 5. Limitation of solubility, 6. Different types of solutions, 7. Methods of expression of concentration of solution, 8. Biological importance.

4.1 True Solutions

Many substances are soluble in water or other liquids. The word "soluble" means that when the substances are placed in selected liquids either all or portions of the substances will in time become evenly distributed throughout the liquid to make the solutions, e.g., NaCl in water, cane sugar in water, iodine in alcohol and sulphur in carbon disulphide.

4.1.1 Definition

True solution is the system in which a substance is evenly distributed throughout a liquid without any tendency towards separation. The disposing liquid is called the "solvent" and the dispersed substance the "solute". Usually, the solute is a solid substance but it may be a liquid (e.g., alcohol in water) or a gaseous substance (e.g., hydrogen chloride in water).

4.1.2 Characteristics

1. **Homogeneous**: Every solution is homogeneous, i.e., the solute or dispersed substance is evenly distributed throughout the solvent or the dissolving medium.
2. **Tendency of separation**: There is no tendency toward separation of the solute from solvent though there may be significant difference in their densities, provided the same temperature is maintained and no solvent is allowed to evaporate.
3. **Passage through a filter**: Both liquid and dissolved solute are able to pass through a filter very readily.
4. **Phase**: It is a single phase system.
5. **Boundary zone**: No boundary zone between the solvent and the solute.
6. **Size:** Passage through filter reveals that the dispersed particles of solid solute are exceedingly small and has been computed to be less than 1 millimicron (1 mμ) in diameter.
7. **Microscopic observation**: Indistinguishable even with ultramicroscope.

Now the question arises why the solvent and solute portions of a solution effect and maintain an even distribution regardless of the difference in their densities. The exact answer is not known but possible explanations are:

(a) Within a true solution the dispersed particles of both solute and solvent are single molecule of the substances concerned.

(b) These molecules are moving about with speeds which vary inversely with their relative weights, i.e., the heavier the molecule the less is the speed with which it travels while the lighter the molecule the greater is its speed.

(c) Despite the difference in relative weights of the molecules of solvent and solute they all move about among each other with the same kinetic energy.

(d) There are frequent collisions among the molecules and against sides of the container followed by rebounds which is an attempt to reduce these collisions to a minimum. The heavier and slowly moving molecules and the lighter and more rapidly moving molecules contribute for an even distribution in respect to each other.

(e) The even distribution is maintained for indefinite period because although the heavier molecules have a tendency to settle, their collision with other molecules moving with equal kinetic energy results compensating reboundings on their part.

8. **Boiling point**: If the solute is non-volatile, it has the effect of raising the boiling point of the solution above that of the pure solvent. If the solute is more volatile than the solvent, the boiling point of the solution is depressed below that of the pure solvent. As for example, water has a boiling point of 100 °C but when non-volatile substances are dissolved in water the boiling point of the solution is above 100 °C, the increase being proportional to the solution concentration. Suppose the increase is of 0.6 °C over that of water alone which does not follow from this, that water solutions of all non-volatile substances in molal concentrations have the same rise in boiling point. If the solute is an acid, base or salt the boiling point increase is even greater than 0.6 °C, the increase above this value varying with the specific acid, base or salt.

9. **Freezing point**: The presence of a solute in solution in a liquid solvent has the effect of

4.2 Table: 4.1 Comparison of True Solutions and Colloidal Solutions

Colloidal solutions	True solutions
1. Heterogenous system	1. Homogenous system
2. Two-phase system: (a) Continuous dispersion medium, (b) Discontinuous disperse phase.	2. Single phase system
3. Have distinct boundaries between the solvent and solute.	3. No boundary between the solute and solvent.
4. Particles are submicroscopic and can be observed under ultramicroscope.	4. Indistinguishable even with ultramicroscope.
5. Particle size 1 mμ to 0.1μ in diameter.	5. Less than 1 mμ in diameter.
6. Particles can be separated by the process of dialysis, electrodialysis, ultracentrifugation, ultrafiltration, salting out, isoelectric precipitation, adsorption, electrophoresis and chromatography.	6. The solute particles are dispersed among the solvent in the form of molecules or ions, cannot be separated from the solvent by passing through an ordinary filter paper, cellophane, parchment or fish bladder membrane. For separation of solute and solvent some processes like evaporation, distillation, etc. are employed.

Contd.

Colloidal solutions	True solutions
7. Dispersed particles are much smaller than the particles of a true solution so they form a stabler dispersion.	7. May not form a stabler dispersion.
8. Do not sediment under gravity.	8. May sediment under gravity.
9. Tendency toward separation of dispersed substance more or less delayed.	9. Dispersion being parmanent there is no tendency toward separation of dispersed substance.
10. **Example** (a) **Chemical**: Solutions of agar, starch, gum acacia, fine divided gold in water, fine divided sulfur in water, fine fat droplets in water, carbon particles in steel, water droplet in air, carbon particles in gas. (b) **Biological**: Proteins, lipoproteins, lipids in extracellular fluid, cytosol fraction of animal cell.	10. **Example**: NaCl, glucose, cane sugar in water, iodine in ethyl alcohol, etc.
11. **Light penetration**: Some are visibly translucent because light rays, upon effecting a partial penetration into the solution, are not permitted to get very far before they are reflected from surfaces of the numerous and sizable dispersed particles. Some may appear clear and transparent to the eye since the distributed particles are not visible but light ray makes it opalescence due to the presence of particles too small to be seen but large enough to act as reflectors of light.	11. True solutions with transparent liquid as solvent are transparent since light passing directly through them saves from some diffraction.
12. Can pass through ordinary filter paper more or less readily. But they are either absolutely incapable of diffusing through a membrane (e.g. cellophane or animal fat) or in case of smallest colloidal particles, diffuse to only a very slight extent.	12. Can pass through filter paper and diffuse through membranes. A mixture of true solution and colloidal solution separated by membranes called dialysis.
13. **Boiling and freezing point**: Solutions of albumen or starch with water have the same boiling and freezing points as does water alone.	13. Solution formed from sugar and water has a boiling point and freezing point different from that of water.

depression of freezing point of the solution below that of the solvent alone. As for example, water has a freezing point of 0°C but water solutions of various solutes have freezing points below this temperature, the extent of the depression being proportional to the concentration of the solution.

4.3 Solvents for Solutions

Liquid solvents

(a) **Water:** Water is the best solvent which can dissolve more substances to an appreciable extent than any other solvents.

The solvent property is sometimes disadvantageous such as chemically pure water can give desired result but naturally occuring water has greater or less quantities of many substances dissolved in it. It is advantageous to use water as a cleansing agent, for sugar and some fruit strains, as a carrier of substances in cookery, drugs, disinfectants and laboratory chemicals.

(b) **Other solvents:** The substances which are insoluble or slightly soluble in water, are readily soluble in other solvents. Such as Iodine is slightly soluble in water but readily soluble in ethyl alcohol and ethyl and methyl alcohol are used for substances which are insoluble in water e.g. drugs, flavours, perfumes etc. Fats are insoluble in water and soluble in fat solvents like chloroform, ether, alcohol, benzene etc.

4.4 Limitation of Solubility

(a) **Limited solubility:** In selected solvents, there is no limit to the solubility of solutes, e.g., ethyl alcohol in water and glycerol with water are soluble in each other in all proportions. But as a rule certain quantity of water or any solvents can hold in solution varying amounts of substances up to a certain limit under given temperature, e.g., 100 ml of water can hold 36.0 gm of NaCl at 20 °C but 39.8m g at 100 °C.

(b) **Unsaturated and Saturated solution**

i. **Unsaturated Solution:** When the solution can take more solute into the solution is known as unsaturated solution.

ii. **Saturated solution:** But when the solution is unable to dissolve more of the same solute it is known as saturated. Cane sugar solution having less than 203.9 gms of cane sugar in 100 cc of water at 20 °C is unsaturated but beyond this limit, the solution is saturated at this temperature. The quantity of solutes in saturated solution varies with the substance and with the solvent and the temperature of the solution.

(c) **Supersaturated solution:** When a solution has been saturated at a high temperature with a solid solute, the entire amount of solute will remain in solution as the temperature of the solution is lowered but there should not be any undissolved solute in contact with the solution. This type of solution at this lower temperature is called supersaturated solution. It is very difficult to maintain a supersaturated solution because slight shaking or stirring or introduction of solid solute will give rise to precipitation of the solid solute from the solution in excess of saturation at the lower temperature.

(d) **Dilute and Concentrated Solutions:** Dilute or concentrated solutions depend upon whether little or much of the solute is present in a given volume of the solution. Substances like sodium hydroxide and silver nitrate are very soluble in water and yield very concentrated solution in this solvent whereas mercuric chloride and calcium hydroxide are slightly soluble, hence produce dilute solution.

Table 4.2

Solutes	Solubility in gm. in 100 ml of water	
	At 20 °C	At 100 °C
Calcium Chloride	59.5 (at 0 °C)	159.0
Calcium Oxalate	0.00068 (at 25 °C)	0.0014 (at 95 °C)
Iodine	0.029	0.078 (at 50 °C)
Silver nitrate	222.0	952.0
Sodium bicarbonate	9.6	16.4 (at 60 °C)
Sodium Chloride	36.0	39.8
Cane sugar	203.9	487.0

Gaseous substance	In 100 ml of water at given temperature under standard pressure			
Oxygen	4.89 cc	at 0°C	2.61 cc	at 30°C
Carbon dioxide	179.7 cc	at 0°C	90.1 cc	at 20°C
Nitrogen	2.35 cc	at 0°C	1.55 cc	at 20°C

4.5 Different Types of Solutions

Substances in different physical states like solid, liquid and gas may form following types of solutions:

1. Solution of Solids in Solids: Frozen solution of iodine in benzene.
2. Solution of Solids in Liquids: Cane sugar in water.
3. Solution of Liquids in Solid: Liquid metallic mercury and gold.
4. Solution of Liquids in Liquids: Alcohol in water, glycerol in water.
5. Solution of Gases in Solids: Adsorption of H_2 by palladium.
6. Solution of Gases in Liquid: Mineral water with CO_2 gas.
7. Solution of Gases in Gases: Air.

4.6 Methods of Expression of Exact Concentration of a Solution

For physicochemical purposes, the various standards are used for solutions of more or less exact quantitative concentrations. The following standards are commonly used.

I. Percentage basis

(a) **Per cent by weight**: It is expressed in terms of gm of the solute in 100 gm of solvent. Both the solute and solvent must be measured on a common weight basis. Thus, 1 per cent solution of NaCl is one which contains 1 gm of NaCl in 100 gm of the solution. In this way it can be expressed as 5 per cent, 10 per cent etc.

(b) **Per cent by volume**: Sometimes the solvent, especially when water is used, is measured by volume rather than by weight. But the solute is given by weight in gm and the solutions are approximate, the degree towards accuracy depending upon the density of the solvent. It means grams of solute per 100 ml of solution, such as commercial hydrochloric acid contains 38 per cent HCl by weight and has a density of 1.19 g cc. Then each ml of HCl contains = $1.19 \times 0.38 = 0.45$ g of HCl, so 100 ml contains 45 g i.e., the solution is 45 per cent HCl by volume.

II. Ratio basis: Solutions are sometimes made up on a ratio basis of weight of solute to solvent as 1:5, 1:10, 1:20 etc. If volume be used for measuring the solvent, the accuracy of solutions is considered on the basis as to those prepared on a percentage basis.

Conclusion: Solutions on a percentage or on a ratio basis have no chemical significance but they are useful for many purposes such as therapeutic, industrial and laboratory uses.

III. Solution concentrations having chemical significances: (a) Molar solutions and (b) Normal solutions.

(a) **Molar solutions**

i. **Molarity**: One mole of substance is one gram molecular weight of the substances. Molarity means a solution of molar concentration which contains one mole of the solute in one litre of solution. As for example, a molar solution of NaCl contains 58.454 g of NaCl in one litre of solution. Molecular weight of NaCl is 58.454. A molar solution of silver nitrate (MW 169.888) contains 169.888 g of the substances in one litre (1 dm^3) of solution.

More concentrated or more dilute solutions can be prepared on the basis as multiples or fractional parts of a molar solution. Thus, a two molar solution contains two moles or two gram-molecular weights of substances in one litre of solution whereas fifth molar solution contains one-fifth of the gram molecular weight of the substance (i.e., solute) in one litre of solution.

Concentrations of these are solutions represented in the following ways. A solution of molar concentration is represented by 1 M, a two-molar solution by 2 M, etc. In case of mole dilute solutions like one-half mole of solute per litre is represented either by 0.5 M or by M/2, a tenth-molar solution by 0.1 M or M/10, a fifth-molar solution by 0.2 M or M/5, etc. For example:

M (Molar) solution
Sodium Chloride ... 58.454 g in 1 litre of solution.

2M (Two molar) solution
Sodium Chloride ... 116.908 g in 1 litre of solution.

0.1 M (Tenth molar) solution
Sodium Chloride ... 5.8454 g in 1 litre of solution.

Solution concentration on molar basis have chemical significance because molecular weights serve as measuring units for solutes as they are involved in chemical reactions. If the number of molecular weights of substances involved in a chemical reactions are known, it is easy to measure out the solution volumes, made up on a molar basis, as it contains the exact reacting weights of said substances. As for example:

$$H_2SO_4 + 2NaOH \longrightarrow 2H_2O + Na_2SO_4$$

(Sulphuric acid)	(Sodium Hydroxide)	(Water + Sodium sulphate)
1 Molecular weight	2 Molecular weight	

Therefore, solution volumes which contain such weights of H_2SO_4 and NaOH will exactly react with each other to produce H_2O and Na_2SO_4 in the following way:

	H_2SO_4	NaOH
	1 litre of M solution to	2 litres of M solution
or,	10 ml of M solution to	20 ml of M solution
or,	10 ml of 0.1 M solution to	20 ml. of 0.1 M solution.

- **Molecular concentrations of these solutions**

Since one mole of each substance contains 6.023 × 10^{23} molecules of the substance, then one litre of a molar solution must contain 6.023 × 10^{23} molecules of the substance. One litre of a half-molar must contain one-half of this number of molecules of substance. So it is obvious that equal volumes of solutions of various substances of the same molarity must contain equal number of molecules of the substances.

ii. **Molality**: Molality means moles or a gram-molecular weight of solute per 1000 g of solvent. Thus, one molal solution of NaCl contains 1 mole or 58.454 g of NaCl (MW 58.454) in 1000 g of water. This unit is helpful for calculation of freezing and boiling points as it gives a specific ratio between the molecules of solute and solvent.

iii. **Mole fraction**: The solution contains the mole fractions of all components, i.e., solute and solvent. The mole fraction of one constituent is equal to the number of moles of the constituent divided by the total number of moles present:

$$A_1 = \frac{a_1}{a_1 + a_2} \text{ and } A_2 = \frac{a_2}{a_1 + a_2},$$

where A_1 and A_2 are the mole fractions of the two constituents in the solution, a_1 and a_2 are the numbers of moles of the respective constituents present. Thus, when 180 g of glucose are dissolved in 1000 g of water, the mole fraction of glucose is:

$$A_1 = \frac{\frac{180}{180}}{\frac{1000}{18} + \frac{180}{180}} = 0.0177,$$

The more fraction of water in the solution is:

$$A_1 = \frac{\frac{1000}{18}}{\frac{1000}{18} + \frac{180}{180}} = 0.982$$

(b) **Normal solutions, normal system, normality:** It is a special case of the molar system ordinarily applied to solutions of acids, bases and oxidizing solutions. They are really concerned with concentration of radicals provided within this compounds rather than concentrations of the compounds in themselves. The normality unit is expressed as (N) and means number of gram equivalent (*g-eq*) of the solute in one litre of solution.

Solutions of greater or lesser concentration on a normal basis are also used and represented as N, 2N, 5N in respect to normal, two-normal and five-normal concentrations; $\frac{1}{2}$ N or 0.5 N and $\frac{1}{10}$ N or 0.1 N in respect of solutions of half-normal and tenth-normal concentrations.

The normal solutions used in the chemical laboratories differ from "Physiological normal solutions" used in biological experiments and medicinal therapy. There is no chemical significance in the term "normal" but it indicates the normal concentration present in the blood and isotonic to the body cells. Such as 0.9 per cent of NaCl known as "normal saline" is used for mammalian experiment and for transfusion in human beings.

Thus, one litre of normal solutions of hydrochloric acid has concentration as follows:

Formula for molecular weight of the substance and valences of radicals	Portion of molecular weight of substance to contain one-valent quantities of radicals	One litre of normal solution contains
HCl → H^+ + Cl^-; 1.008, 35.457; Molecular weight	1 Molecular weight	1 gram-molecular weight of Hydrochloric acid H^+ –0.008 g Hydrogen radical, Cl^- –35.457 g Chloride radical

- **Normal solutions of acids, bases and oxidizing agents**

Normal solution of an acid is defined on the basis of the concentration of the hydrogen radical which is common to all solutions of acids, while normal solution of a base is defined on the basis of the concentration of the hydroxide radical which is common to all solutions of bases.

A normal solution of an acid contains a formula weight of H^+, i.e., 6.023×10^{23} H^+ per litre of solution. Similarly, a normal solution of alkali contains a formula weight of OH^-, i.e., 6.023×10^{23} OH^- per litre. A normal oxidizing solution contains enough oxidizing agent per litre to increase the positive valence or reduction of the negative valence of a gram formula weight of reducing agent by one unit. A normal reducing solution is the opposite of a normal oxidizing solution. It can be explained that a litre of normal oxidizing solution has the capacity to take up 6.023×10^{23} electrons to an oxidizing agent and oxidation and reduction should proceed simultaneously.

4.7 Biological Importance of Solution

1. The water of the sap plants and of the blood of animals holds some inorganic (NaCl, KCl, Mg, Ca, etc.) and organic (urea, uric acid, creatinine, glucose etc.) substances in true solution which can enter and come out of the cells.
2. The intracellular, interstitial and extra-cellular fluids including blood plasma, contains electrolytes, non-electrolytes and respiratory gases in true solution (aqueous).
3. Water part of blood plasma carries different nutritional materials, hormones and respiratory gases in aqueous solution.
4. Different waste products like urea, uric acid, creatinine, etc. are transported in aqueous solutions to the kidney, sweat glands of the skin for excretion.
5. Urine contains metabolic end products, which usually are in aqueous solution.
6. Digestive juices, tear, etc. are aqueous solutions.
7. Sebum is a solution having only solvent in which lipids and sterols are present.
8. Blood plasma contains proteins as well as many organic and inorganic substances in solutions—nutritive and excretory materials, antibodies and hormones.
9. Amoeba, an unicellular organism, takes food through the cell membrane either in solution or as particulate matter, and waste products pass into the surroundings medium.
10. Saline (solution of NaCl) and glucose solution or gluco-saline solution are used in the dehydration, shock, etc. to correct salt depletion, haemoconcentration and fluid loss.
11. C.S.F., aqueous humour, synovial fluid are aqueous solutions.
12. Haemoglobin in R.B.C. remains in solution in cytosol.
13. Nutrients are absorbed in aqueous solutions. In case of fats, bile salts help to form micelles which are composed of 20 to 50 molecules of bile salts. Micelles develop because each bile salt molecule consists of sterol nucleus which is highly fat soluble and a polar group which is highly water soluble. The sterol nuclei of 20-50 bile salts molecules of the micelle aggregate together to form a small fat globule in the middle of micelle. This aggregation causes the polar groups to project outward for covering the surface of the micelle which are negatively charged. These negatively charged polar groups allow the middle globules to become dissolved in the water of the digestive juices and remain in stable solution in spite of the big size.

5

Biochemical Analytical Methods

PART I

BIOCHEMICAL ANALYSIS

Biochemical analysis can be done by the following methods:

I. **Qualitative method**: Detection of substance in the unknown biological material.

II. **Quantitative analysis: Determines the quantity of substance present in the unknown material. In quantitative analysis several types of procedure are employed. Such as:**

(a) **Gravimetric analysis**: In this method the substance to be determined is converted in an insoluble mass, filtered, dried and weighed.

(b) **Volumetric analysis**: In this analysis, solution of known concentration called **standard** solutions are employed. Here one adds from a burette, which is an apparatus for measuring volume of solution accurately, a standard solution of a compound which reacts with the substance to be determined. The standard solution is added until the reaction is completed as indicated by the change in colour of an **indicator**. This process is known as titration.

Volumetric analysis is commonly used in titrating with acid and bases or vice versa, this type of analysis known as:

1. Acidimetry
2. Alkalimetry

III. **Colorimetry**

A. **Visual colorimeter**

B. **Photoelectric colorimeter**

1. Single cell type,
2. Double cell type

IV. **Spectrophotometry**

V. **Electrochemical Analysis**

(a) Potentiometric method

(b) Voltammetric method

(c) Coulometric method

(d) Conductimetric method

VI. Electrophoresis

VII. Chromatography

VIII. Automated biochemical analysis

(a) Autoanalyser

(b) Semi auto analyser

IX. Flame photometry

X. Thermo-Analytical method

PHOTOMETRY

Photometry is the method to study the **concentration of substance** by applying the property of **absorption of light having definite wavelength** by the molecules of particles. When monochromatic light is passed through the solution of substance, it comes out with diminished intensity because of absorption of some portion of the light. This absorption of light by the solution containing the substance obeys **Beer-Lambert's law**.

Types of Photometry: Photometry (absorptiometry) are of 3 types:

I. Visual colorimetry (coloured)

II. Photoelectric colorimetry (coloured)

III. Spectroscopies i.e. spectrophotometry (Coloured and colourless)

Application of Photometry: Enzyme Immuno Assay of both types:

A. Homogeneous Enzyme Immuno-Assay (HEIA)
(Having only liquid phase which couples immunology, enzymology and photometry, applicable for T_3, T_4, oestriol, cortisol etc.)

B. Heterogeneous Enzyme Immuno-Assay (HJEIA) or Enzyme-linked Immuno-sorbent Assay (ELISA)

I. COLORIMETRY

A. **Visual Colorimeter**

B. **Photoelectric Colorimeter**
(a) Single cell type, (b) Double cell type

II. SPECTROSCOPIES

Spectroscopies is the **measurement and interpretation of electromagnetic radiation** (light, heat etc.) **absorbed, scattered, or emitted by atoms, molecules or other chemical substances.** This absorption of emission is associated with **changes in the energy states of the interacting chemical species** and, since each species has characteristic energy states, spectroscopies can be used to identify the interacting species and quantitative information may also be obtained.

Electromagnetic radiation consists of **electromagnetic waves** and also contains **energy** called **quanta or photons.** Energy content of each quanta determines the wavelength of radiation. An atom may exist:

1. **Ground state**: Electrons with lowest energy level.
2. **Excited state**: Electrons with high energy by absorption of some quantum of energy. The excited state of atom has a tendency to come back to its ground state by emission of radiation of a specific wavelength exhibit the spectroscopies phenomena of fluorescence and emission flame photometry. Molecules with **ground or excited state** can be grouped **into different energy levels showing specific band spectra.**

The energy levels change when the molecules absorb or emit other energy so these spectra are called:

(a) Absorption spectra

(b) Emission spectra

Different regions in the spectra have been characterized in terms of wavelengths and different instruments are used to measure the intensity of band of different regions.

ABSORPTION SPECTROSCOPY

Molecular absorption spectroscopy, based upon visible, ultraviolet and infrared radiation. The technique is generally used in physical, biological and medical sciences. In association with absorption spectroscopy, molecular fluorescence methods are preferable due to their selectivity and sensitivity. The type of spectroscopy which is suitable depends on the energy region of the electromagnetic spectrum and its interaction with matter

Table 5.1: Spectroscopies, electromagnetic radiation and interaction with method

Spectroscopy	Wave length Å	Type of interaction
1. Gamma ray (Mösshauver)	<0.15 nm	Nuclear
2. X-ray	0.01-10 nm	Core electrons
3. Ultraviolet (UV)	10-250 nm	Bonding electrons
4. Visible	300-750 nm	Bonding electrons
5. Infrared (IR)	0.75-300 µm	Molecular vibration/ rotation
6. Microwave	0.75-4 mm	Molecular rotation/ orientation
7. Electron spin Resonance (ESR)	–30 mm	Electron spin
8. Nuclear Magnetic Resonance (NMR)	0.5-10 m	Nuclear spin

The spectra in the visible and ultraviolet range have been measured with the help of the instrument called **spectrophotometer**. The techniques and instrumentation required for different regions of electromagnetic spectrum are so diverse that each region has given rise to seemingly quite different methods, although all are related by the fundamental principle discussed above.

SPECTROPHOTOMETER

The spectrophotometer has got superiority over the photoelectric colorimeter because it **functions at various range of wavelength. Ultraviolet spectrophotometers** work at wavelengths **below 400 mμ** whereas, **infrared spectrophotometer** works at the wavelengths **above 800 mμ e.g. the U.V. spectrophotometer is very much useful for biochemical estimations** such as many dehydrogenase enzymes requires the extinction to be taken at about 340 mμ and the purines and pyrimidines at the extinction of 260 mμ. The **superiority of photoelectric colorimeter** are:

1. The spectrophotometer uses prisms as monochromatic;
2. The wavelength of incident light can be adjusted precisely;
3. Maintains greater accuracy;
4. Requires smaller volumes of liquids;
5. Samples of solids and gases can not be used in photoelectric colorimeter.

- **Types**

1. Ultraviolet spectrometry
2. Fluorescence and phosphorescence spectrophotometry
3. Flame Emission Spectroscopy
4. Atomic Absorption Spectroscopy
5. Atomic Emission Spectroscopy
6. Infrared Spectrometry
7. Raman Spectroscopy
8. Auger Emission Spectroscopy
9. Nuclear magnetic Resonance Spectroscopy
10. Mass Spectrometry

ULTRAVIOLET AND VISIBLE SPECTROMETRY

Instrumental Components: Spectrophoto-metres are commonly used in various kinds of spectroscopies. The common components for any spectrophotometry are:

I. A radiation source
II. Sample holder
III. Wavelength selector/monochromator
IV. Cuvettes
V. Detector/recorder system.

I. Radiation source

1. Arc lamp, mercury vapour lamp. Special lamps using elements e.g., Zinc emitting of 472 nm for cortisol and thallium for oestrogens at 538 nm, Xenon arc (extending down to 220 nm).
2. Argon, Xenon lamps (150-600 nm).
3. Hydrogen/deuterium lamps (180-380 nm by electrical excitation under low pressure — for continuous radiation sources in UV regions.
4. Tungsten filaments, Halogen lamps, xenon lamp for continuous radiation in the visible region.
5. Nernst glower (350-20,000 nm) for continuous radiation in the visible and infrared regions.
6. Nichrome wire and glower for continuous radiation (750-40,000 nm) in infrared studies.
7. Organic dye lasers—for continuously-tunable radiation in the UV/visible regions.
8. Hollow Cathode lamps—for UV/visible (230-700 nm).
9. AN_2-laser (gives intense line at λ= 3370 nm) useful for exciting fluoresence in a variety of molecules.
10. Excimer laser provide spectral lines (193 nm for Ar F; 248 nm for KrF and 350 nm for Xe F) in the UV region.
11. Gas lasers have spectral lines in the visible regions 488 and 515 nm for Ar laser and 633 nm for He-Ne laser.
12. Solid state laser—For infrared studies.

II. Sample Holders

The variety of sample holders depends on the transmission ranges of the materials.

1. Lithium fluoride (LiF)

2. Fused silica or quartz—for UV and also for the visible regions (150-3000 nm).
3. In general glass is used in the visible region (300-500 nm)
4. NaCl and KBr Cells—For scanning UV to IR regions (150-30,000 nm)

III. Wavelength Selection

Wavelength selection is done by:

1. Absorption filters or,
2. Interference filters.
3. Fused silica or quartz in the UV region,
4. Glass in the visible region.
5. NaCl and KBr prisms in the IR region.
6. Interference filters and gratings provide a wide spectrum, from UV to IR regions.

(a) Absorption Filters

1. These are generally coloured glass.
2. They are restricted to the visible regions of the spectrum (400-700 nm).
3. They have low transmittance i.e. about < 20 per cent and wide pass width i.e., about 50 nm selection of the filter depends on the colour absorbed by the sample, as it is intensity of this colour of the spectrum that varies with concentration. A filter that allows least transmittance of that colour should be selected.

(b) Interference Filters

1. Provide better transmittance i.e. about > 50 per cent and narrower pass width i.e. about < 10 nm.
2. It works on the principle of optical interference.
3. The filter consists of transparent dielectric strip (magnesium fluoride or Calcium fluoride) with both sides coated with semitransparent silver film.
4. It is sandwiched between two glass plates to protect it from damage.
5. Incident radiation striking the first silver film is partly transmitted and partly reflected.
6. The transmitted rays further undergo transmission/reflection at the second metallic layer.
7. Then a particular wavelength is transmitted, excluding other wavelengths by destructive interference.
8. The thickness of the dielectric plate determines the wavelength to be transmitted.

MONOCHROMATORS

Monochromators are designed for spectral scanning, for proper selection of the wavelength in any spectrophotometry. A monochromator consists, in general of:

1. An entrance slit that provides a narrow optical image of the radiation source.
2. A collimator that renders the radiation emanating from the entrance slit parallel.
3. A grating or prism for dispersing the incident radiation.
4. A collimator to reform images of the entrance slit on the exit slit.
5. An exit slit to isolate the desired spectral band by blocking all of the dispersed radiation except that within the desired range. The primary function of a monochromator is to provide a beam of radiant energy of a given nominal wavelength and spectral bandwidth.

The basic requirements of a monochromator are:

(a) Design simplicity
(b) Resolution
(c) Spectral range
(d) Purity of exciting radiation
(e) Dispersion.

IV. Cuvettes

1. For below 320 nm, quartz or preferably synthetic silica is required and above this wavelength, glass cells or plastic cuvette may be used but these absorb light below 310 nm.
2. The faces of the cuvette should be clear for passing of light and for avoiding the excitating radiation.
3. Temperature of the cuvette should be maintained.
4. The cuvettes should have an optical length of 1 cm with a volume of the sample from 1.5 ml to 3.0 ml.

The cuvettes which are transparent cells in

which the sample to be measured is dissolved in the suitable solvent. A control cuvette is used which contains the same solvent as in the sample.

• **Radiation Detectors**: All radiation detectors are either photon or thermal radiation detecting instruments. A detector is a transducer that converts physical or chemical signals such as electromagnetic radiation into an electron flow and subsequently into a current flow or voltages in the read-out circuit which can be amplified and modulated when measuring low levels of radiant energy. All **photon detectors** are of the following type:

(a) Photoemission or
(b) Photoconduction

These are employed in the high-energy range, from X-rays to the visible region i.e. < 2000 nm.

Thermal detectors are:

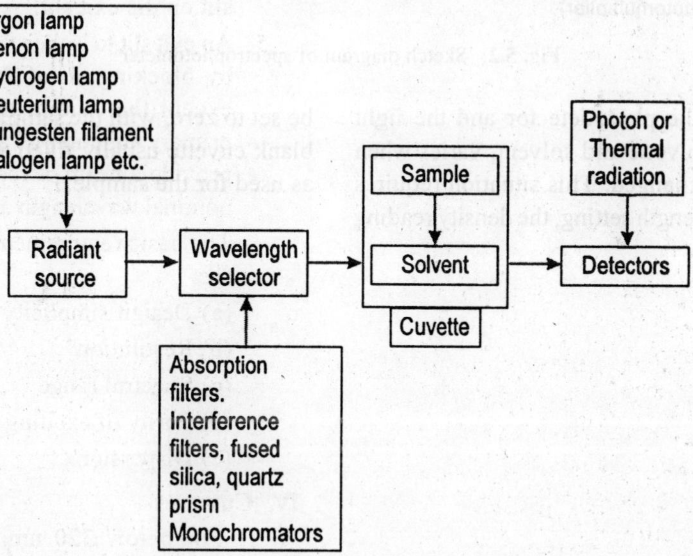

Fig. 5.1: Block diagram of measurement of absorption radiation

(a) Thermocouples
(b) Bolometer

and these are employed in the measurement of infrared radiation.

They are also grouped into:

A. **Single-element detectors**

1. Solid-state photodiodes
2. Photoemissive tubes
3. Photomultiplier tubes

B. **Multiple-element detectors**

Solid-state array detectors

The important characteristics of any type of detectors are spectral sensitivity, wavelength response, gain and response time (Fig. 5.1).

In the device of the spectrophotometer the monochromator is used in place of simple selection filter of colorimeter. A monochromator has a diffraction-grating to disperse light from a lamp. The light passes through the entrance slit into its spectral components. An exit slit selects a narrow band of the spectrum which is utilised to measure the absorption of a sample in the cuvette. The narrower the exit slit, the narrower the bandwidth of the light with smaller intensity. A sensitive photodetector (or a photomultiplier) is required with an amplifier and a meter which is calibrated in units of transmittance or absorbance. The wavelength of the light can be changed by rotating the grating. A mirror folds the light path to reduce the size of the instrument (Fig. 5.2).

The spectrophotometer allows the determination of the absorption of samples of various wavelengths. The light output of the lamp, however, as well as

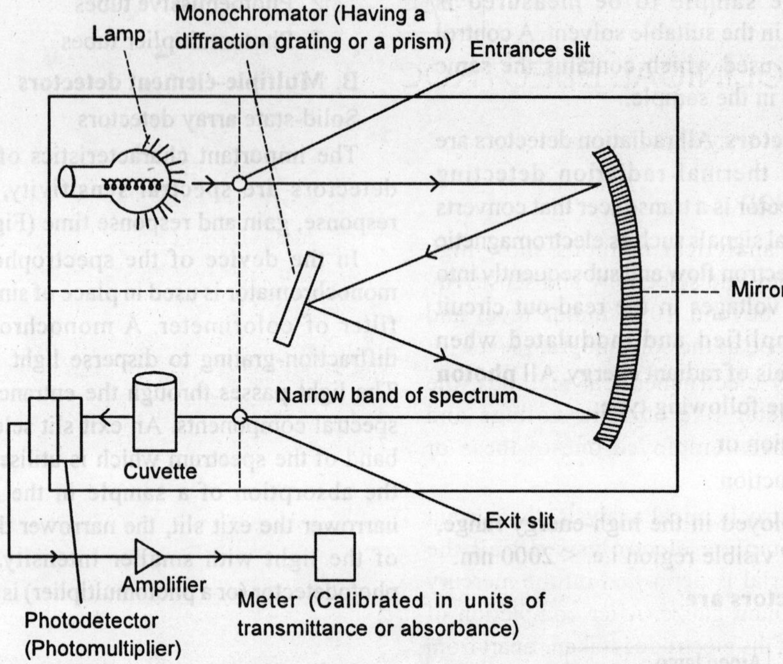

Fig. 5.2: Sketch diagram of spectrophotometer

the sensitivity of the photodetector and the light absorption of the cuvette and solvent, varies when the wavelength is changed. This situation requires that, for each wavelength setting, the density reading be set to zero, with the sample being replaced by a blank cuvette usually filled with the same solvent as used for the sample.

PART II

CLINICAL ELECTROCHEMICAL METHODS

5.1 Introduction

Electrochemical analytical methods show high sensitivity, accuracy and selectivity and sensitivity can be achieved beyond 10^{-10} molar level and analysis can be done at the subnanogram level.

In analysis by electrochemical methods, the electrical parameter like current, voltage and resistance have been employed one of these or jointly.

In simple electrochemical analysis, the current is passed for complete electrolysis till all the electrolysed material is deposited on the mercury electrode or platinum gauze. After completion of analysis, weight of the electrode is taken, apart from direct analysis, electrical measurements are good indicator for titrimetry. Reagents are added volumetrically up to end, detected electrometrically, sometimes reagents are used coulometrically. Electrochemical sensors are based on EMF i.e. potentiometric, resistive i.e., conductimetric, or current i.e., amperometric principles.

Analytical techniques based on electrochemical principle may be grouped into three parts. Each basic electrical measurement of current, resistance and voltage has been used alone or in combination for analytical purposes. If these electrical properties are measured as a function of time, many additional electroanalytical methods of analysis are possible. The electroanalytical methods of analysis most often encountered clinically are:

1. Potentiometry
2. Voltammetry
3. Coulometry
4. Conductimetry

5.2 Potentiometry

The most common electroanalytic method used in the clinical laboratory is *potentiometry* as exhibited by the wide use and acceptance of *ion selective electrodes* (ISEs). Two principles govern the manner in which ISEs work. The principles are:

1. Electrode-generated potential
2. Membrane-generated potential.

Potentiometry involves the measurement of the potential produced by an electrochemical (galvanic) cell at zero, or near zero current flow and relating the measured potential to the concentration of an ion or ions in the cell. *The type of ion or analyte measured depends on the particular electrode employed.*

5.3 Application

Potentiometry, and the application of ion selective electrodes, is the basis for the clinical measurements of:

(a) pH
(b) Sodium (Na^+)
(c) Potassium (K^+)
(d) Chloride (Cl^-)
(e) Calcium (Ca^{2+})
(f) Carbon dioxide (CO_2)
(g) Bicarbonate (HCO_3^-)

The first measurement technique employing potentiometry, was the measurement of pH (hydrogen ion activity).

Potentiometric techniques
(when used as an end-point detection system)

1. Constant current potentiometry
2. Zero current potentiometry
3. Null-point potentiometry
4. Cyclic chronopotentiometry

5.4 Voltammetry

Voltammetry involves the measurement of the current through an electrochemical (electrolytic) cell while the magnitude of voltage applied to the cell is changed. The applied voltage, called the *excitation voltage*, forces electrolysis of the reacting species to occur at the electrode, the result of an oxidation reduction reaction. Under appropriate conditions the current through the cell (called the

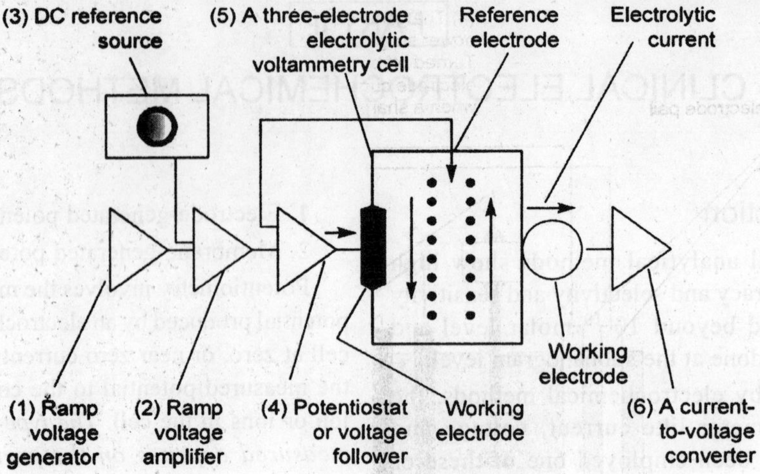

Fig. 5.3: Block diagram of an instrument for anodic stripping voltammetry

diffusion current or *limiting current*) becomes a measure of the concentration of the analyte ion and the identity of the analyte ion is defined by the applied potential (called the *decomposition potential*). The determination of *serum dissolved oxygen* and *serum glucose* by way of *oxygen* and *glucose electrodes* is a clinical application of this technique.

A variation of the process is employed in *anodic stripping voltammetry* for the determination of *serum lead*. In this technique the cell voltage is varied such that the lead ion solution is plated out onto an inert electrode. The voltage is then reversed and all of the lead is "stripped" from the electrode, with the resulting current pulse being proportional to the amount of lead present in the sample (Fig. 5.3).

- **Working Principle**

The ramp voltage signal generator (1) provides a linear increasing voltage that is then passed to the ramp/reference voltage amplifier (2) where there is signal variation of voltammetry. It is an electrochemical titration in which the *titrant* is *electrochemically generated*. The basis for this measurement is Faraday's law, which states that passing I Faraday of electricity through a solution will oxidize or reduce 1 g-equivalent of reactant. If the current is held constant, then the total charge transferred is proportional to the product of the cell current and time (Q = IT), and time becomes a measure of concentration.

In the clinical coulometric determination of *serum chloride*, two-electrode pairs are employed: a generator pair for production of the titrant (silver used to precipitate the chloride) and a second (or indicator) pair to measure the cell current and determine the end point. The end point is indicated by an increase in cell current, which occurs when all the chloride has been precipitated and the excess silver produces a current surge and is combined with a preselected DC reference voltage (3). This combined (X-axis) signal is then applied by the potentiostat (voltage follower) (4) to the electrolytic cell auxillary electrode (5) while the reference electrode voltage is held constant. Electrolytic current through the cell is due to the oxidation-reduction taking place at the auxillary working electrode pair. This current is sensed and converted into a voltage (γ-axis signal) by the current-to-voltage converter (6). The signals applied to a display device yield an X-Y graph of voltage versus current-the voltammogram or polarogram.

5.5 Coulometry

Coulometry: If a current (at a selected voltage) through an electrochemical cell is hold constant then the duration of the current flow will be directly proportional to the concentration of the reacting ion. Coulometry method of analysis depends on the exact measurement of the quantity of electricity that passes through a solution during the course of an

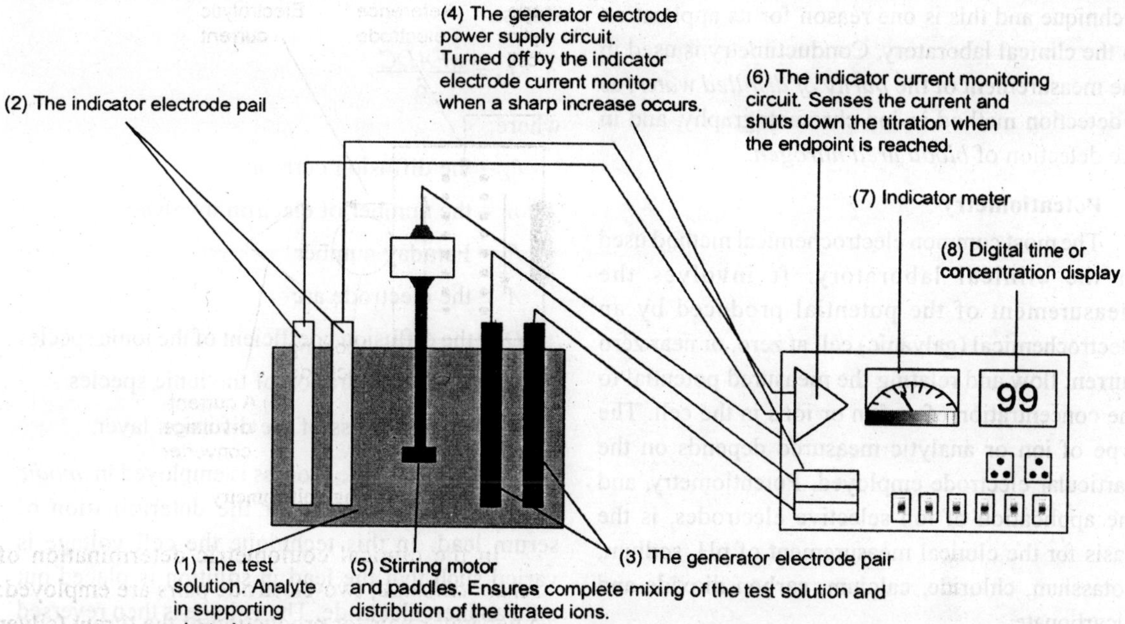

Fig. 5.4: Diagrammatic representation of a basic coulometric titrator

electrochemical reaction. Coulomety can be defined as **the analytical method based on the measurement of the quantity of electricity is called as a generic term coulometry** which is derived from **coulomb**. Coulometry is based on Faraday's law which relates electrical charge (Q), Current (I), and time (T) to concentration by Q = IT (Quality measured = I and/or T, controlled = V, I [when measuring T] [Solution is stirred]).

- **Working Principle**

The electrodes are immersed in the sample solution (1) containing chloride, and the indicator electrodes (2) are activated to monitor the current through the solution. While the residual current is stabilizing, the timer/display unit (8) is set to zero. The titration is then started by activating the generator electrode (3) so that silver ion is released into solution. The silver ion binds the chloride ion in the insoluble precipitate silver chloride. When all the chloride has been consumed, the silver ion is generated in excess, causing a rapid increase in current that is sensed by the current monitor circuit (6). The rapid rise in current represents the end point of the titration, and the silver ion generator circuit (4) along with the timer/display (7) & (8) are both stopped. The sample concentration is read directly or calculated from the time of the titration. The solution is stirred to improve reproduce ability by a motor-driven paddle system (5) (Fig. 5.4).

5.6 Conductimetry

Conductimetry involves measuring the electrical conductivity (conductance) of a 1 cm^3 volume of electrolyte solution. This is accomplished by applying an AC voltage of 2 to 10V, with a frequency of 10-1000 Hz across a set of 1cm^3 platinum black electrodes. The use of an AC excitation voltage presents polarization of the electrodes and improves reproduce ability of the technique. When the electrodes are placed in an electrolyte solution and the voltage applied, a current will flow that is proportional to the applied voltage and the total number and valence of ions in solution. Hence, the application of conductimetry is based on the fact that within a given range of concentrations current is proportional to ionic concentration. Although the conductivity of a given solution is nonspecific (all ions present carry current), it is a highly sensitive technique and this is one reason for its application in the clinical laboratory. Conductimetry is used in the measurement of the *purity of distilled water*, as a detection method in ion chromatography, and in the detection of *blood urea nitrogen.*

- **Potentiometry**

The most common electrochemical method used in the clinical laboratory. It involves the measurement of the potential produced by an electrochemical (galvanic) cell at zero, or near zero current flow and relating the measured potential to the concentration of an ion or ions in the cell. The type of ion or analytic measured depends on the particular electrode employed. Potentiometry, and the application of ion selective electrodes, is the basis for the clinical measurement of pH, sodium, potassium, chloride, calcium, carbon dioxide and bicarbonate.

Potentiometric methods encompasses two major types of analyses.

1. It involves the direct measurement of an electrode potential from which the activity or concentration of an active ion may be derived.
2. If involves measuring the changes in the electromotive force (emf) brought about by the addition of a titrant to the sample through the cell called *diffusion current* or *limiting current* becomes a measure of the concentration of the analyte ion, and the identity of the analyte ion is defined by the applied potential called the *decomposition potential.* The determination of serum-dissolved oxygen and serum glucose by way of oxygen and glucose electrodes is a clinical application of the technique.

The graph voltage (*v*) versus current (*I*) is referred to as a *polarogram* and the process classically referred to as *polarography*. The measurement of the diffusion current (I_d) in an electrochemical cell is the rate-dependant process that is related to the concentration of reactive ions in the cell by a modification of Fick's law, which states:

$$I_d = \frac{nFADC}{\delta}$$

where:

I_d = the diffusion current

n = the number of electron involved

F = Faraday number

A = the electrode arc

D = the diffusion coefficient of the ionic species

C = the concentration of the ionic species

δ = the thickness of the diffusion layer.

A variation of the process is employed in *anodic stripping voltammetry* for the determination of serum lead. In this technique the cell voltage is varied such that the lead in solution is placed out onto an inert electrode. The voltage is then reversed and all of the lead is *stripped* from the electrode with the resulting current pulse being proportional to the amount of lead present in the sample.

The two most commonly encountered clinically are *anodic stripping voltammetry*, used for the end point is indicated by an increase in cell current, which occurs when all the chloride has been precipitated and the excess silver produces a current surge.

- **Summary of the four common clinical electrochemical methods.**

1. Potentiometry	The potential developed by a galvanic cell, measured at zero current, is proportional to the ionic activity (concentration). The magnitude of the potential developed is predicted by the Nernst equation (quantity measured = *V*; controlled = *I* = zero)
2. Voltammetry	A voltage of specific magnitude, called the decomposition of reduction potential, is applied to a chemical cell, the ions within the cell respond. The magnitude of the voltage determines which ion will respond, while the dilution current through the cell indicates how much

of that ion (the concentration) is present. (quantity measured = *I*, controlled = *V*) solution not stirred).

3. Coulometry — If the current (at a selected voltage) through an electro-chemical cell is held constant then the duration of the current, flow will be directly proportional to the concentration of the reacting ion. Coulometry is based on Farady's law which relates electrical charge (*Q*), current (*I*), and time (*T*) to concentration by $Q = IT$ (Quality measured = *I* and/or *T*, controlled = *V*, *I* [when measuring T] [solution is stirred]).

4. Conductimetry — The conductance of a solution is proportional to the total number of ions in solution. It is a non-specific method and requires non-reactive electrodes of a specified (usually $1 cm^2$) surface area (quality measured = *I*, controlled = *V*) [solution is stirred].

6

Hydrogen Ion Concentration and pH

1. Dissociation (ionization) or protolysis of water, 2. Hydrogen ion concentration, 3. pH value, 4. Experimental determination of pH, 5. Acids, bases and salts, 6. Reaction of acids, bases and salts with each other, 7. Biological importance of acids, bases and salts, 8. Buffers and buffering system, 9. Henderson-Hasselbalch equation, uses and applications, 10. Indicators (acid-base)-Colorimetric determination of pH, 11. Acid-bases balance or maintenance of pH of the body.

6.1 Dissociation (Ionization) or Protolysis of Water (Water as an Electrolyte)

Water was regarded as the only medium in which ionic dissociation takes place. But, even though acid or base or salt or many acids or bases and salts may be present, water itself as a weak electrolyte establishes an equilibrium by two reactions:

(a) Dissociation to make hydrogen ions and hydroxyl ions and

(b) Combination between these two ions to make undissociated molecules of water.

The water molecules dissociate (ionize) with limited capacity into (H^+) and (OH^-) ions depending on temperature. This process is known as **dissociation or protolysis of water.**

6.1.1 Concentration of equilibrium

Since the water is an ultra-poor electrolyte, at equilibrium, in the presence of a relatively ultra-low concentration of the ionized part of the water, a condition represented as:

$$H^+ + OH^- \longrightarrow H_2O$$

$$H_2O \rightleftharpoons H^+ + OH^-$$

This phenomenon has been shown by electrical conductivity, the hydrolysis of compounds and other methods. The concentration of [H^+] and [OH^-] ions in pure water is about thirty times greater at 100 °C than at 0 °C. In superheated steam at 306 °C, the concentration is about 50 times as large as at 0 °C. The reactions involving the ions of water, the concentration is greatly increased by raising the temperature. Raising the temperature of water from 37 °C to 40 °C increases [H^+] and [OH^-] by about 10 per cent. In case of fever in human cases, there may be relation of increased dissociation of water. But the reactions of pure water remains neutral, irrespective of the extent of dissociation, because [H^+] equals [OH^-]. But the ions at the same time joining with each other to form water molecule and dissociate again into ions. Hence it is difficult to state whether individual ions or part of water molecule are present. At particular state it is present as an ion but again in the later part it is part of a molecule. We are concerned to know the "probability" that a hydrogen will be present as an ion or as part of a water molecule. The actual probability of a hydrogen atom in pure water, present as a hydrogen ion is about 1.8×10^{-9}. For every hydrogen and hydroxyl ions in pure water, there are 1.8×10^{9} water molecules.

6.1.2 Mechanism of ionization and characteristic

Ionization of water occurs because (1) the mass of hydrogen atom is so small and (2) the atom's single electron is tightly held by the oxygen atom, there is a tendency for a hydrogen ion to dissociate from the oxygen in which it is covalently bound in one molecule and jump to the oxygen atom of the adjacent water molecule (Fig. 6.1) which is hydrogen bounded, provided that the internal energy is favourable:

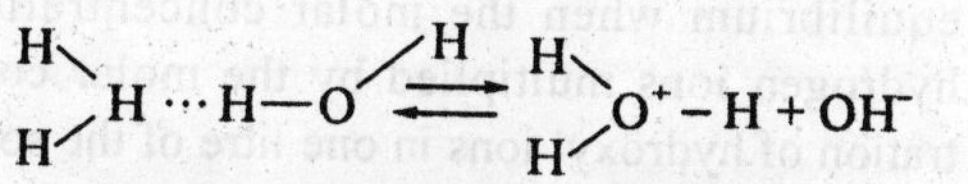

Fig. 6.1: Mechanism of ionization

Here, we see that two ions are produced, the hydronium ion $[H_3O^+]$ and the hydroxyl ion $[OH^-]$. Electrical conductivity has shown that one litre of pure water at 25 °C at any given time there is only 1.0×10^{-7} mol of $[H_3O^+]$ ions and an equal amount of $[OH^-]$ ions.

Though $[H^+]$ ion designates the hydronium ion $[H_3O^+]$, but it is a fact that protons or hydrogen ions do not exist in nonhydrated form in water. They exist only in hydrated form. Even the $[H_3O^+]$ (hydronium) ion is again hydrated through additional hydrogen bonding with water to form the $[H_9O_4^+]$ ion as well as more highly hydrated forms (Fig. 6.2). The hydroxide ion is also hydrated in liquid water.

Thus, we see that the water behaves as an "ampholyte' or "amphiprotic compound" as it can donate a proton like an acid and can also accept a proton like a base from an acid or from another molecule of water to form (H_3O^+) ions.

The rate of migration of (H_3O^+) ions in an electric field is many times (mobility, 36.3×10^{-4} $cm^2\ V^{-1}\ s^{-1}$) higher than that of univalent cations Na^+ (5.2×10^{-4} $cm^2\ V^{-1}\ s^{-1}$). Because, a proton can jump rapidly from a hydronium ion to a neighbouring water molecule to which it is hydrogen-bonded. Thus, a positive electric charge can move a given distance from one molecule of water to another with little movement of the water molecules themselves. A series of such proton jumps (Fig. 6.3) has the effect of translocating protons at a rate that is much higher than the rate of diffusive or bulk movement of (H_3O^+) ions. Conduction of protons through hydrogen-bonded water molecules is called "tunnelling". Hydrogen ion activity has been denoted by bracket (H^+) and hydrogen ion concentration by parenthesis $[H^+]$.

The dissociation of water can be expressed according to the "law of chemical equilibrium" for weak electrolytes.

$$K = \frac{[H^+]\,[OH^-]}{[H_2O]} \quad \text{and } [H^+]\,[OH^-] = K\,[H_2O]$$

where the bracketed terms represent the molar concentration (better to term it molar activity) of hydrogen ions, hydroxyl ions and undissociated water molecules and K is referred to as **dissociation constant.**

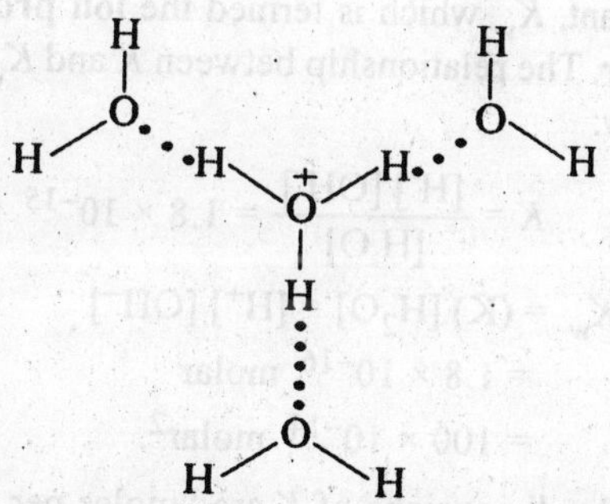

Fig. 6.2: Hydrated form of hydronium ions $(H_9O_4^+)$

Calculation of dissociation constant can be done in the following way:

1 mol of water weigh 18 g so one litre (1000 g) of water contains 1000/18 = 55.56 mole i.e., pure water in 55.56 molar. Hydrogen in pure water existing as an (H^+) ion is 1.8×10^{9}, then the molar concentration of (H^+) ions or of (OH^-) ions in pure water is $1.8 \times 10^{9} \times$ molar concentration of water, 55.56 molar $= 1.0 \times 10^{-7}$ molar.

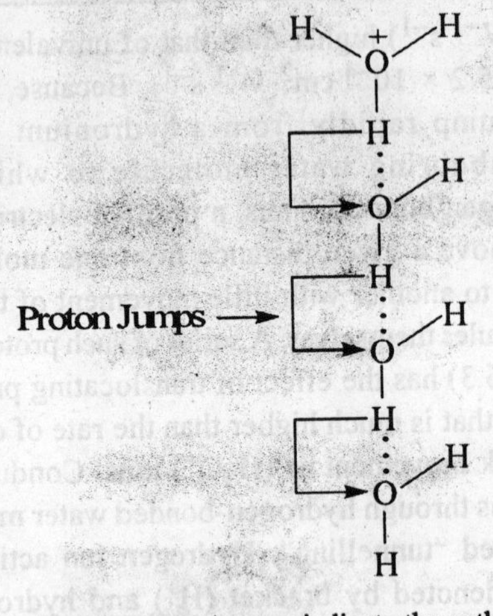

Fig. 6.3: Proton jumps-Arrows indicate the path taken by protons. Last H_2O molecule becomes a hydronium ion at the end of proton jump

Now calculation of K for water is

$$K = \frac{[H^+][OH^-]}{[H_2O]} = \frac{[10^{-7}][10^{-7}]}{[55.56]}$$

$$= 0.018 \times 10^{-14} = 1.8 \times 10^{-16} \text{ molar.}$$

As the high molar concentration of water is not affected by dissociation, and the constant is considered as essential, this constant may be applied into the dissociation constant, K, to get a new constant, K_w which is termed the **ion product for water**. The relationship between K and K_w is given below:

$$K = \frac{[H^+][OH^-]}{[H_2O]} = 1.8 \times 10^{-15} \text{ molar}$$

$$K_w = (K)[H_2O] = [H^+][OH^-]$$

$$= 1.8 \times 10^{-16} \text{ molar}$$

$$= 100 \times 10^{-14} \text{ molar}^2.$$

The dimensions of K are "moles per litre" and K_w "moles^2 per litre". The ion product K_w is numerically equal to the product of the molar concentration of H^+ and OH^-:

$$K_w = [H^+][OH^-]$$

At the human body temperature 37°C, the concentration of H^+ in pure water is slightly more than 10^{-7} molar. Within the limited temperature $K_w = 10^{-14}$ molar^2 for all aqueous solutions, even containing acid or bases.

The constant is used for the calculation of pH values for acidic and basic conditions.

6.2 Hydrogen Ion Concentration (cH)

Hydrogen and hydroxyl ions are measured in terms of molar quantities of these two substances in one litre of the solution. In case of H^+ ion it is expressed by cH and for OH^- ion cOH. As for example, within given solutions:

cH = 0.5 means that one litre of solution contains 0.5 mole of hydrogen ions.

cH = 0.002 means that one litre of solution contains 0.002 mole of hydrogen ions.

cOH = 0.150 means that one litre of solution contains 0.150 mole of hydroxyl ions.

cOH = 0.003 means that one litre of solution contains 0.003 mole of hydroxyl ions.

Hydrogen ion and hydroxyl ion mole-per-litre basis, the two reactions of water, as an electrolyte are at equilibrium whenever cH × cOH = 0.00000000000001 or 10^{-14}.

This shows that two reactions as to water are at equilibrium when the molar concentration of hydrogen ions multiplied by the molar concentration of hydroxyl ions in one litre of the solution is equal to this exceedingly small value 10^{-14}. Consequently, regardless of whether concentrations of hydrogen ions and hydroxyl ions are equal, or whether the concentration of hydrogen ions is greater than that of hydroxyl ions, or vice versa, if this relationship is obtained the two reactions as to the water system are at equilibrium. As for example,

1. If within a given one litre solution:

   cH = 0.0001 or 10^{-4}

   cOH = 0.0000000001 or 10^{-10}

   The water, as an electrolyte, is at equilibrium since

   10^{-4} mole of hydrogen ions × 10^{-10} mole of hydroxyl ions = 10^{-14}

2. If within a given one litre solution:

   cH = 0.0000001 or 10^{-7}

   cOH = 0.0000001 or 10^{-7}

   the water system is at equilibrium since

   10^{-7} mole of hydrogen ions × 10^{-7} mole of hydroxyl ions = 10^{-14}

3. If another one litre solution contains:

   cH = 0.00000001 or 10^{-8}

   cOH = 0.000001 or 10^{-6}

   water as an electrolyte, at an equilibrium since

   10^{-8} mole of hydrogen ions × 10^{-6} mole of hydroxyl ions = 10^{-14}

Here, in each case, the water, as an electrolyte, is at equilibrium, since each solution cH × cOH = 10^{-14}.

Such as:

when cH = 10^{-1} ... and cOH = 10^{-13}

when cH = 10^{-4} ... and cOH = 10^{-10}

when cH = 10^{-6} ... and cOH = 10^{-8}

6.2.1 Reaction of the solution and cH and cOH

1. **Neutral solutions**: When the solution contains 0.0000001 mole of hydrogen ions and 0.0000001 mole of hydroxyl ions per litre or when cH of the solution is 10^{-7} and the cOH is 10^{-7}, then the solution is neutral in reaction. This is always the case if water is the only electrolyte present because hydrogen ions and hydroxyl ions are equal in numbers.

   But a solution can also contain such proportions of acids, bases and selected salts that all of them, together with water, will attain their several equilibria in the presence of equal concentrations of hydrogen ions and hydroxyl ions.

2. **Acid solutions**: Solutions are acid when cH values are above 10^{-7} moles of hydrogen ions through 10^{-1} moles per litre and the cOH values are slightly below 10^{-7} down through 10^{-13} moles per litre.

   But in such solutions an acid or acids must be present along with the water, since the concentration of hydrogen ions is greater than that of hydroxyl ions, i.e., cH of the solution is greater than cOH. But so also based and salts, hence their ions, may be present.

3. **Basic solutions**: Solutions are basic in which cH values are slightly below 10^{-7} moles down through 10^{-13} moles per litre, whereas corresponding cOH values are slightly above 10^{-7} moles per litre through 10^{-1} moles per litre. In such solutions a base or bases must be present along with water, since the concentration of hydroxyl ions is greater than that of hydrogen ions. But so also acids and salts and their ions as well as other substances may be present.

6.2.2 cH or cOH of the solution—necessary to specify

The fixed relationship between the concentrations of hydrogen ions and hydroxyl ions must be met on the part of water at equilibrium, it is necessary to specify only the cH, or only the cOH of the solution, regardless of whether said solution is acid, basic or neutral, since a given molar concentration of hydrogen ions or the cH of a solution, means automatically a definite molar concentration of hydroxyl ions or a definite cOH within the solution.

6.2.3 Presence of substance in water

In extremely dilute solutions, the concentration in moles per litre of hydrogen ions and of other substances may be considered to approximate their "activity". The molar concentration of an ion [A⁻] is related to its "activity" by the equation [A⁻] fA⁻ = [A⁻], where fA⁻ is the "activity coefficient" of the ion.

Numerical values of "activity coefficients" for ions in dilute solutions of electrolytes may be calculated by means of the "Debye-Huckel limiting law" which has been deduced as follow:

$$-\log fi = 0.5\, Z_i^2 \sqrt{T/2}$$

This states that the negative logarithm of the "activity coefficient" of an ion (i) is directly proportional to the square of the number of charges on that ion (Z_i) and the square root of the molar "ionic strength" of the solution ($T/2$). The basic idea of Debye-Huckel theory is that the ions in a solution do not behave as independent entities, but interact with one another; thus an ion bearing a positive charge is encircled by a bunch of ions of opposite charge, and vice versa. The extent of this interaction is determined by the charge on the ion in question and the concentrations and charge of all the ions in the solution.

6.3 pH Value (Shorthand Notation of cH)

Most of the tissues in the human body contain about 80 per cent of water. In every reaction taking place in the organism, water plays an essential part. Pure water is only very slightly ionised:

$$H_2O \rightleftharpoons H^+ + OH^-,$$

the ionisation, however, markedly increases with rise of temperature. At a given temperature, K (law of mass action constant) = $[H^+][OH^-]$.

In a solution containing H^+ and OH^- ions the active mass in any reaction which may take place is equal to the product of the two concentrations $[H^+] \times [OH^-]$. Thus the more H^+ ions are present in a solution, the fewer will be OH^- ions and the shift is considerable, since it is the product of the two and not the sum which remains constant.

The behaviour of a living cell bathed in any solution in which H^+ ions are concerned will be determined, not by total titrable acid or base, but by the actual concentration or activity of the H^+ ions in the solution.

6.3.1 Definition

H^+ ion concentration means gram of hydrated H^+ ion present as H_3O^+ per litre of fluid denoted by "cH" or of OH^- ion acitivity by "cOH". As for example, water (H_2O) contains $\frac{1}{1000000}$ g of H^+ ion in one litre which is expressed as 10^{-7} cH or H^+ ion concentration.

Many times, molar concentrations of hydrogen ions are for convenience recorded on the basis of ten lowered to appropriate negative powers, but the scientific workers adopted to represent these values simply by the power notations concerned. The label given to such notation is pH. Usually pH notations are defined as the negative power on the basis of 10 or negative logarithm of the cH values of solutions. Sometimes pH is defined as the positive power on the basis of 10 or the positive logarithm of the reciprocal of the cH of the solution. But pH value by either definition is the same. Thus:

$$cH = 10^{-5} \quad \ldots \ pH = 5$$

$$\frac{1}{CH} = \frac{1}{10^{-5}} = 10^5 \quad \ldots \ pH = 5$$

The reason for usual concept is that computations for most cH values of solutions are more readily made from figures given on the basis of 10 to positive powers than from figures on the basis of 10 to negative powers. Therefore, the computation is made in terms of the reciprocal of the cH of the solution, then converted to the actual cH value.

Owning to the inconvenience of expression, Sorensen introduced the term pH. The pH of a solution is the logarithm of hydrogen ion concentration with a negative sign, i.e., pH = – log $[H^+]$. For example, in case of H_2O if H-ion concentration is 10^{-7}

$$pH = -\log [H^+]$$

$$= -\log 10^{-7} = -(-7) = 7$$

the pH is 7 at 25 °C which is neutral because water contains equal amount of H^+ and OH^- ions. Any solution having pH below 7 is acidic and above 7 is alkaline.

Thus:

A. pH 3 is the shorthand notation which means that within the solution the cH = 10^{-3} and the cOH = 10^{-11}.

B. pH 4.7 is the shorthand notation which means that within the solution the cH = $10^{-4.7}$ and the cOH = $10^{-9.3}$. pH scale is logarithmic but not arithmetic. pH scale ranges from 0 to 14 for dilute aqueous solutions inclusive biologic fluids. When two solutions differ in pH by 1 pH unit, it means only that one solution has 10 times the hydrogen ion concentration.

	pH notations
cH = 10^{-1}	... pH 1
cH = 10^{-2}	... pH 2
cH = $10^{-2.7}$	... pH 2.7
cH = 10^{-4}	... pH 4
cH = $10^{-8.3}$	... pH 8.3

Acids are "proton donors" and bases are "proton acceptor". Acids are differentiated into strong acids (HCl, H_2SO_4) which completely dissociate even in strongly acidic solutions and weak acids dissociate only in partially acidic solution. Similarly, there are strong bases (KOH, NaOH) and weak bases $Ca[OH]_2$. Only strong bases are dissociated at high pH. *Most of the biochemicals are weak acids except phosphorylated intermediate which contains strongly acidic primary phosphoric acid group.*

6.3.2 pH value of some body materials

Table 6.1: pH value of body materials

Materials	pH
pH above 7, i.e., alkaline	
Whole blood, plasma	7.4
Blood serum	7.3-7.4
Cerebrospinal fluid (CSF)	7.3-7.4
Aqueous humour of eye	7.4
Pancreatic juice	7.5-8.0
Intestinal juice (Succus entericus)	7.0-8.0
Hepatic duct bile	7.4-8.5
Faeces	7.0-7.5
Tears	7.4
Liver-peripheral cells	7.1-7.4
R.B.C.	7.25
Interstitial fluid	7.4
pH value below 7, i.e., acidic	
Materials	**pH**
Saliva	6.3-6.8
Pure gastric juice	0.9
Gastric juice	2.0-3.0
Gall bladder bile	5.4-6.9
Urine	4.8-7.5
Milk	6.6-6.9
Skin (intracellular, various layers)	6.2-7.5
Liver-Kupffer cell	6.4-6.5
Liver-Central cell	6.7-6.9
Intracellular fluid	7 or a bit below

6.3.3 Calculation for pH

Calculation of pH of a solution is done:

A. By calculating H-ion concentration $[H^+]$

B. By calculating the base 10 logarithm of $[H^+]$

C. pH is the negative of the value of "*B*".

Example of calculation for pH

1. **pH of a solution having 3.2×10^{-4} molar H-ion concentration**

$$pH = -\log H^+$$
$$= -\log (3.2 \times 10^{-4})$$
$$= -\log (3.2) - \log (10^{-4})$$
$$= -0.5 + 4.0 = 3.5$$

2. **pH of a solution having 4.0×10^{-4} molar OH-ion concentration**

$$[OH^-] = 4.0 \times 10^{-4}$$
$$pOH = -\log [OH^-]$$
$$= -\log (4.0 \times 10^{-4})$$
$$= -\log (4.0) - \log (10^{-4})$$
$$= -0.60 + 4.0 = -3.4$$

Hence, pH = 14 – pOH = 14 – 3.40 = 10.6

6.3.4 Biological importance of pH

1. **Enzyme activity:** Each enzyme shows maximum effect at particular pH known as "optimum pH". The activity depresses both above and below the optimum pH. Moderate pH change affects the substrate-binding and catalytic activities of the enzyme in the following ways:
   (a) It may affect ionic state of enzyme,
   (b) It may affect ionization of substrate,
   (c) It may dissociate the apoenzyme-prosthetic group linkage,
   (d) It may change in conformation of the enzyme,
   (e) It may denature the enzyme in extremely high or low pH by disturbing the ionic and hydrogen bonds holding its tertiary or quarternary structure.
2. **Effect on ammonia excretion in urine:** If the pH of plasma and urine differs greatly, there will be large differences in concentration of ionized base in the two fluids. In the acid region, the ratio of NH_3 to total ammonia is small, i.e., a difference of pH of one unit between plasma and urine or tubule cell and urine will lead to a concentration of NH_4^+ in the urine, some ten times that in the plasma or tubule cell.
3. **Action potential of myelinated nerve:** Heene (1962) examined the effects of pH on the resting and action potentials of myelinated nerve and found acid solutions have the same effects as those of a high concentration of Ca^{++} and may well be related to their influence on the activity of this ion at the membrane.
4. **Membrane permeability:** Variations of pH of the medium: It has been observed that the

permeability of Na^+ is maximum in the region of physiological pH while the permeability of K^+ showed no significant variation.

5. **Volume of erythrocyte**: Uneven distribution of negative ions is due to presence of indiffusible negative ion Hb^-. As for example, when erythrocyte is given on non-electrolyte medium, Cl^- and HCO_3^- can diffuse from cell to the medium but for electrical neutrality, negative ion should enter the cell. As the water does not have electrolytes, the only negative ion OH^- from dissociation of water enters the cells in exchange of Cl^- and HCO_3^-.

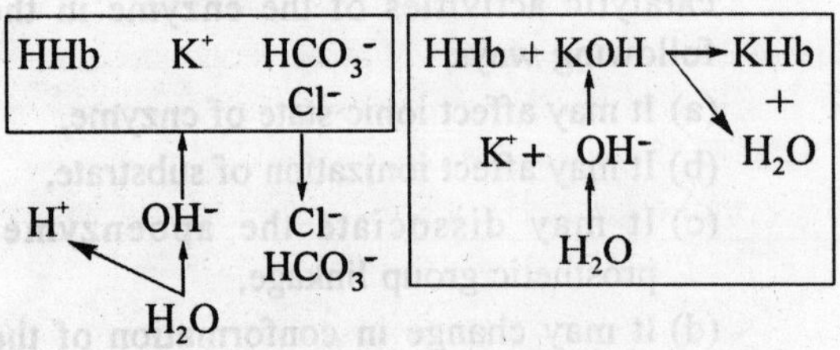

The condition for equlibrium is that:

$[Cl^-]$ Cell/$[Cl^-]$ Medium = $[OH^-]$ Cell/$[OH^-]$ Medium = $[H^+]$ Cell/$[H^+]$ Medium

In this period the following changes are seen:

(a) Shrinkage of cells,

(b) Medium becomes acid (less pH) due to loss of OH^-,

(c) High concentration gradient of Cl^- and HCO_3^- imposes a concentration gradient of OH^- and H^+.

6. **pH of different parts of gastric mucosa**: Bradford and Davi (1950) studied the secretion of a number of dyestuffs in the frog's gastric mucosa. These dyes were acid-base indicators, e.g., phenol red, methylene blue, toludine blue, acridine, etc. were secreted and were sensitive to different range of pH, so that the degree of acidity of the various regions of the mucosa could be estimated roughly, such as:

(a) with neutral red (1) the lower regions of mucosa and the cytoplasm of the parietal cells were slightly alkaline (pH 7.2-7.6), (2) the canaliculi and the lumen of some of the tubules and the secreting surface of the mucosa were acid (pH 6.8),

(b) with acridine—green fluorescence in ultra-violet light indicated that some tubules had a pH of less than 4.85.

7. **pH of kidney tubular fluid**: By introducing catheters at different levels in the collecting duct in golden hamster, it has been shown that concentration of inulin increased from the outer medullary zone towards the papilla where it is highest due to reabsorption of water and concentration occurred in the duct. Sodium analysis showed the decrease in concentration in spite of reabsorption of water and this happened due to active transport of sodium to the tubular cells. Since osmotic pressure rises, other solute must be supplied to replace sodium and this solute is ammonia and thereby increase in concentration of NH_4^+ in the tubular fluid. Since pH falls progressively towards papilla, that active transport of Na^+ outwards is associated with inward H^+ in this region.

8. **Free energy of reactions and pH**: In many biological reactions H^+ ions are involved and the free energy of reaction may vary greatly with variations of pH. And the pH at which the free energy is determined, must be specified if H^+ ions are involved.

9. **Decrease of intracellular pH by insulin**: The action of insulin in the presence of large amount of glucose may lead to a significant decrease of intracellular pH as a result of increasing the concentration of sugar phosphates, which are stronger acids than phosphoric acid.

10. **Oxidation-reduction potential and pH**: The potential of a reversible oxidation-reduction system (redox-system) in biological system varies in proportion to the logarithm of the ratio of oxidant to reductant just as the pH of a buffer system varies in proportion to the logarithm of the ratio of salt to acid. The equations for the two systems are similar:

$$E = E'_0 + \frac{RT}{nF}\log e\frac{[\text{Oxidant}]}{[\text{Reductant}]}$$

$$\text{pH} = P\kappa + \log 10\frac{[\text{Salt}]}{[\text{Acid}]}$$

Where E'_0 is specified at a given pH for systems in which it varies with pH.

11. **pH and three-dimensional forms of DNA and tRNA**: pH influences the maintenance of the natural three-dimensional forms of DNA and tRNA.

12. **Relation of pH to oxygenation of haemoglobin**: The marked effect of pH upon the oxygenation of purified haemoglobin has been observed. It has been found that as the pH is increased, the per cent oxygenation of the haemoglobin for a given O_2 tension increases, e.g., at pH 6 the oxygenation was 87.3 per cent when O_2 tension was 86.1 mm Hg, at pH 6.6, the oxygenation was 73.7 per cent when O_2 tension was 51.9 mm Hg and at pH 8.3, the oxygenation was 50 per cent when O_2 tension was only 5 mm Hg and 93.5 per cent at 18.1 mm Hg tension.

13. **Ionization of side chains of amino acids**: pH influences the ionization of side chains and free α-amino and α-carboxyl groups of amino acids.

14. **Plasma bicarbonate and pH**: Both total CO_2 and HCO_3^- vary directly with pCO_2 at any given pH, i.e., they are straight-line functions. At any given value of pCO_2, both total CO_2 and HCO_3^- increase pH, the rate of increase accelerating as the pH rises. There is variation in plasma $[HCO_3^-]$ with variations in pH at fixed pressures.

15. **pH and Gibbs-Donnan Effect**: pH affects the Gibbs-Donnan effect of non-diffusible protein ion because the ionic state of the protein changes with pH. At isoelectric point of the protein, the Gibbs-Donnan effect is minimum and the proteins exist as an anion. Their presence in one side of semipermeable membrane which is impermeable to the membrane, causes unequal distribution of cations and anions which are diffusible to the membrane and maintain electroneutrality of the both sides of the membrane (Gibbs-Donnan membrane equilibria). pH influences the Donnan effect by determining the quantity and net charge on the protein ions.

16. **pH and seperation of proteins and amino acids**: The pH-dependent charge of proteins and amino acids is employed for their separation from the biological materials by electrophoresis and ion-exchange chromatography.

17. **pH difference between cell and plasma**: Donnan effect causes the difference of pH between the cells and plasma. Intracellular pH is usually between 6-7.4 in different cells, averaging 7.0 whereas pH of plasma is 7.4. A rapid rate of metabolism in cells increases CO_2 production and consequently decreases pH. The cell membrane is not freely permeable to H^+, it is to OH^- and this assures the proportionality of pH values in cells and plasma.

18. **pH and charge of some biological organic compounds**: Some biological organic substances, e.g., nucleic acids, phosphoglycerides, sphingolipids and mucopolysaccharides may exist as zwitterions or cations or anions depending on pH.

19. **pH of intracellular fluid of RBC**: It is found that pH of intracellular fluid of RBC is lower than the plasma. This is partly due to high concentration of negative non-diffusible haemoglobin ions. This causes unequal distribution of H^+ ions on the opposite side of the RBC membrane with a higher concentration of H^+ ions inside the cells and pH falls and plasma becomes alkaline.

20. **pH and control of respiration**:The chemical control of respiration is mediated by:

(a) Directly or indirectly through the respiratory centre in the medulla oblongata which sends impulses to the respiratory apparatus. The neurons of the respiratory centre are stimulated direcly by increase in CO_2 pressure and H^+ ion concentration, i.e., decrease in pH.

(b) Chemoreceptors in the carotid and aortic bodies also play in the chemical regulation. These chemoreceptors are stimulated by a decrease in the O_2 pressure and increase in the CO_2 pressure and H^+ ion concentration i.e., decrease in pH of arterial blood.

21. **pH and saliva**: pH of saliva is between 6.35 to 6.85. Salivary pH depends primarily on the ratio of HCO_3^-: H_2CO_3 which in turn is governed to some degree by the ratio of these substances in blood. Saliva produced during or immediately after hypo-ventilation, which lowers the H_2CO_3 fraction of blood, has an increased pH.

22. **pH and optical rotation of amino acids**: Effect of pH on the optical activity of amino acids in aqueous solution has been investigated. If an L-amino acid is dissolved in water, and the pH is gradually decreased by the addition of acids, the solution becomes more dextrorotatory or less levorotatory. Thus L-Histidine is levorotatory in water and becomes dextrorotatory in acid solution.

23. **pH and light absorption spectrum of amino acids**: pH influences the light absorption in the near ultraviolet region of the spectrum which is important for behaviour of tyrosine. This amino acid has a maximum absorption at 275 mμ at acid and neutral pH values; but at pH values more alkaline than 10 the position of the maximum shifts to a longer wavelength of 290 mμ. The shift is associated with the dissociation of the phenolic hydroxyl group of the amino acid to form a phenolate ion.

24. **pH and heart activity**

(a) **Amphibian heart**: The pH of the perfusing fluid has significant effect on the frog's heart. If weak acids are added to the perfusion fluid, the heart dilates, the beats become smaller, slower and finally at about pH 2 or 3, all these irregularities disappear. If weak alkalines are added, the relaxation during diastole becomes more and more incomplete, so that the heart finally stops at pH 11 or 13 in a contracted condition. Some reactions are observed at pH 8.3 which represents the "optimum reaction" for the beat of the heart and it is different for different cavities of the heart and for the heart of different animals.

(b) **Mammalian heart**: The chief factor for the variation of the blood reaction is the CO_2 tension and any change in CO_2 tension causes changes in the heart activity due to alteration of pH. The optimum pH for mammalian heart is 7.5 to 8.0. Decrease in pH, i.e., reduction of alkalinity causes:

1. Slowing of the cardiac rhythm from its action on the S-A node,
2. Slower propagation and weak spread of the excitatory wave through the auricular muscle and the A-V bundle,
3. Dilation of the ventricles, which is greater resulting slowing of the heart and greater filling,
4. Dilation of coronary vessels. When the heart is perfused with fluid of very alkaline reaction having pH 8.4, the coronary flow is so reduced by constriction of the vessels that the heart fails due to lack of oxygen and up to pH 8.0 there is very much less effect.

25. **pH and plain muscle activity**: The plain muscle is very much sensitive to change in the pH. Lowering of hydrogen ion concentration, i.e., increase in pH causes contraction or tonus with increased rate of rhythmic contractions. Increasing the hydrogen ion concentration, i.e., decrease in pH causes the reverse action.

26. **pH and nerve excitability**: Changes in pH do not affect the spikes, but alter the excitability and the after-potentials. Increase in H^+ ion concentration decreases excitability, and decrease in H^+ ion concentration, increases the excitability until at pH 8.0, spontaneous discharges occur.

27. **pH and the affinity of haemoglobin for oxygen (oxygen haemoglobin dissociation**

curve): Three important factors affect the oxygen haemoglobin dissociation curve. They are the pH, temperature, and the 2, 3-diphosphoglycerate. A fall in pH or rise in temperature shifts the curve to the right. When shifted in this direction, a higher pO_2 is required for haemoglobin to bind a given amount of O_2. A rise in pH or fall in temperature shifts the curve to the left, and a lower pO_2 is required to bind a given amount of O_2.

The decrease in O_2 affinity of haemoglobin when the pH of blood falls is called "Bohr effect" and is closely related to the fact that deoxyhaemoglobin binds H^+ more actively than does oxyhaemoglobin. The pH of blood falls when its CO_2 content increases, so that when the pCO_2 rises, the curve shifts to the right.

28. **pH and carriage of CO_2 in the blood (Isohydric transport of CO_2)**: Oxyhaemoglobin is a stronger acid than the haemoglobin which has a great importance to CO_2 transport. When oxyhaemoglobin or arterial blood reaches the capillary bed of the tissues and gives up O_2 to the tissues and becomes less acidic, CO_2 enters the blood from tissues to form carbolic acid and decreases the blood pH. But the buffers of the blood react with this acid and prevents the fall of pH. Generally arterial blood has a pH of 7.4 and when CO_2 enters the blood, pH falls to 7.36, i.e., difference of 0.04 pH unit. The reverse occurs when the blood reaches the lungs, it releases CO_2 and oxyhaemoglobin formation helps to release CO_2 and pH rises to the arterial value, i.e., 7.4 again. The formation of this HCO_3 at the tissues and its conversion to CO_2 at the lungs take place because of the difference in acid strength of oxyhaemoglobin and haemoglobin and take place at a constant pH. This transport of CO_2 is called the "isohydric (constant pH) transport".

29. **pH of fluid for animal experiment**: Animal experiments like amphibian heart or capillary perfusion, mammalian heart perfusion, uterine or intestinal movement, liver or any organ perfusion etc. require Ringer or Dale solution whose pH is maintained usually at pH 7.4.

I. Bacteriology

Bacteriologists maintain certain pH for the culture media for bacteria. Different strains of bacteria are active and healthy only at specific pH. Moderate variations may result in significant modifications of their cultural characteristics, whereas more extreme variations may completely inhibit their growth. Most pathogenic bacteria find their optimum pH at about pH 7.4 which is dangerous because the body fluids, e.g., blood, lymph and cells of humans, have normal pH 7.4.

Table 6.2: Optimum pH of some bacteria

Bacteria	Optimum pH
Streptococci	pH 7.4
Meningococci	pH 7.4-7.6
Pneumococci	pH 7.6-7.8
Diptheria bacilli	pH 7.2
Bacillus Coli	pH 7.4 (for growth)
	pH 6 (for survival)
	pH 4.6 (cease to multiply)

II. Food Technology

1. **Length of the fermentation period of bread**: The pH of bread dough is one condition which determines the "length of the fermentation period"—whether it is too long, too short or just right. Consequently, whether the finished loaf will be light and tasty or heavy and undesirable in flavour.
2. **pH for maximum pectin precipitation**: Within the limits of a very small range of pH "maximum pectin precipitation" occurs. A jelly of desired flavour and fine texture is obtained and outside this range, either no pectin is precipitated or a jelly of poor flavour and texture will result.
3. **Length of sterilization period**: For canning of foods, the "length of the sterilization period" is determined in part by the pH of the food mass, and that, in general, the lower the pH, the shorter is the fermentation period.

- **The pH of some processes:**
  - (a) In bread making—for optimum fermentation period, pH 4.5-5.0.
  - (b) In jelly making—for maximum yield of pectin, most satisfactory flavour and texture-pH 3.0-3.55.
  - (c) In canning foods—the pH of tomatoes and some other foods is about 4.0; hence sterilization period is short.
  - (d) In canning carrots, beets, beans—pH 5.0-6.0, hence the sterilization period must be longer.
  - (e) In canning peas, corn and meat—pH below 6.0; hence the sterilization period must be still longer.

III. Plant Kingdom

pH and chloroplast membrane for ATP formation

Witt (1971)* used biophysical methods to establish the quantitative relationship between electron flow, H^+ movements and ATP formation and these observations also support the chemiosmotic hypothesis. When chloroplasts are illuminated with a very short light flash given by a single laser pulse of 20 ns, only a single "turnover" of electron carrier ensures. This is accompanied by absorption of H^+ ions by the chloroplasts.

Another evidence supporting the chemiosmotic hypothesis is work done by Gagendorf et al. A pH gradient artificially imposed across the chloroplast membrane which can drive the phosphorylation of ADP in the dark, without the input of light energy.

They lowered the internal pH of chloroplasts artificially by soaking them in an acid bath, a medium buffered at pH 4.0. The chloroplasts were then quickly mixed in the dark with an alkaline buffer at pH 8.5 in order to impose a monetary pH gradient across the membrane; the alkaline buffer also contained phosphate and ADP. Mixing was followed by a burst of ATP formation; simultaneously, the pH gradient disappeared. (Fig. 6.4)

6.4 Experimental Determination of H-Ion Concentration or pH

Determination of pH is very much useful for:

1. Analytical procedure in biochemistry,
2. Knowing the different aspect of the structures and activity of macromolecules,
3. Behaviour of cells,
4. Bacteriological and pathological procedure,
5. Determination of pH of biological fluids, e.g., blood.

6.4.1 Methods

I. **E.M.F. methods**
(Electrometric or potentiometric method)
   A. Hydrogen electrode,
   B. Glass electrode,
   C. Quinhydrone electrode.

II. **Potentiometric titration method**

III. **Acid-Base Indicators (Colorimetric method)**

1. Electrometric methods

The pH of solutions can usually be determined accurately by potential measurements of certain electrodes than by the indicators. The electrometric method requires expensive equipment, but provides quick and accurate measurements. The electrometric method depends upon the use of the (a) hydrogen electrodes, (b) glass electrodes, (c) quinhydrone electrodes.

Electrode Potential and Solution Tension of the Metal

When a metal strip or electrode e.g., zinc (or copper, silver), is dipped in water or aqueous solution, the reversible reaction takes place in the following way at the surface of metal.

Few atoms of Zinc on the surface of metal give up two electron, and produce Zn^{++} ions which pass into solution at the surface of the metal and give it a negative charge due to presence of electron and the process is reversible. After a short period, the rate of which Zn^{++} ions from the solution combine with electrons on the surface of the metal to form Zn atoms is equal to the rate at which Zn atoms lose electrons to form Zn^{++}.

When the rates of both processes are equal, the reaction is at equilibrium, and the zinc strip contains definite number of extra electrons, which give it a definite negative charge and also at equilibrium the concentration of Zn^{++} ions in the liquid has a fixed

*Witt, H.T.— Q. Rev. Biophysics 4, 365 1971.

value. The tendency of the neutral Zn atoms to lose electrons and pass into solution as Zn^{++} ion is known as the "solution tension" of the metal. At equilibrium the "solution tension" of the Zn atoms is balanced by the electrical attraction across the boundary of metal and solution.

Thus, surface of the zinc strip or electrode is covered with extra electrons or negative charges whose electrostatic attractions hold an adjacent layer of Zn^{++} ions of the solution. This combination of electron layer on the surface of the metal strip or electrode and ion layer in the liquid film is known

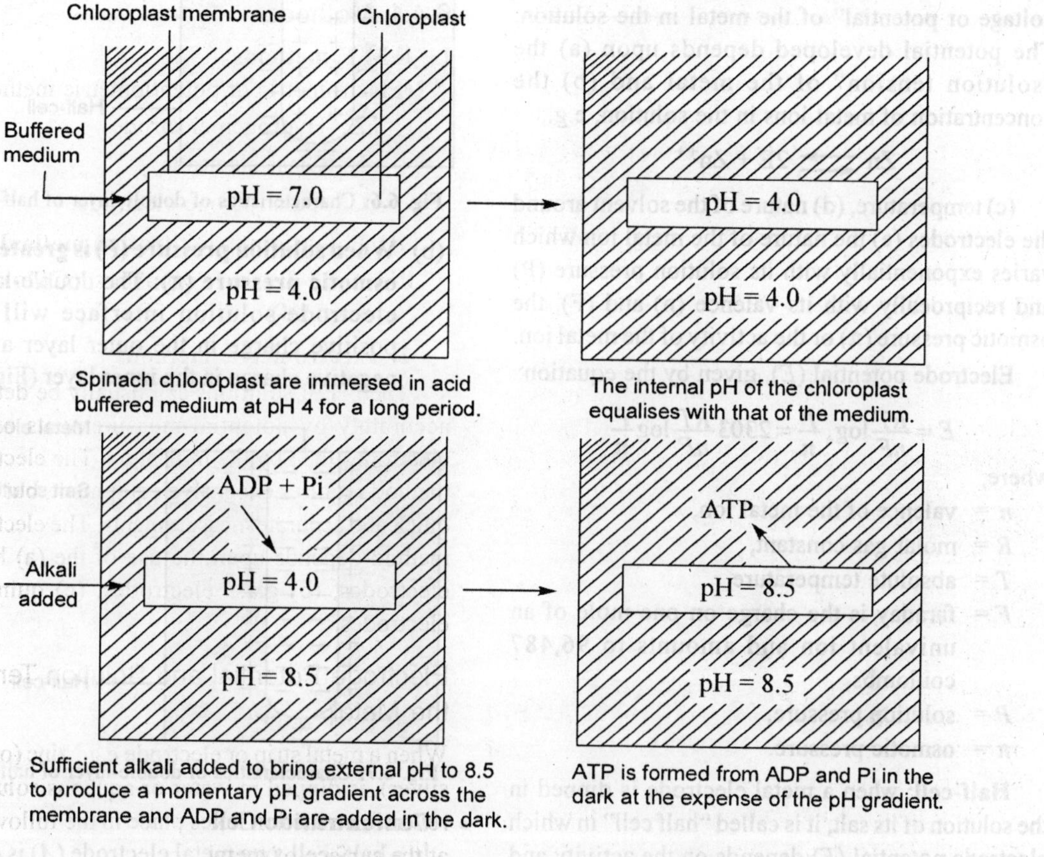

Fig. 6.4: pH and chloroplast membrane for ATP formation

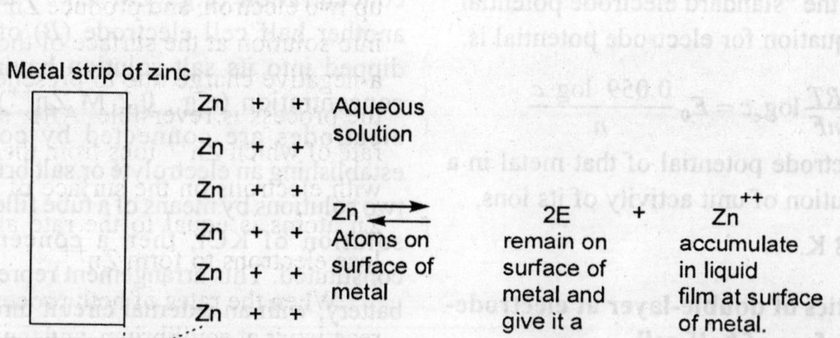

Fig. 6.5: Shows reversible reactions when metal strip is dipped in aqueous solution

as "electric double layer" (Fig. 6.5) at the electrode-solution interface and making the electrode negative relative to the solution (e.g., a Zn electrode in an IM $ZnSO_4$ solution). The magnitude of the electric attraction across this layer determines the potential difference between the electrode, and the solution is known as "electrode potential" (*E*) or "electrical voltage or potential" of the metal in the solution. The potential developed depends upon (a) the "solution tension" of the metal and (b) the concentration of metal ions in the solution, e.g.,

$$Zn \rightleftharpoons 2E + Zn^{++}$$

(c) temperature, (d) nature of the solvent around the electrodes (e) the nature of the metal ion which varies exponentially with its solution pressure (P) and reciprocally with its valence (n) and (F), the osmotic pressure (π) or the activity of the metal ion.

Electrode potential (*E*), given by the equation:

$$E = \frac{RT}{nF}\log_e \frac{P}{\pi} = 2303\frac{RT}{nF}\log\frac{P}{\pi}$$

where,

n = valence of the metal ion,

R = molar gas constant,

T = absolute temperature,

F = faraday is the charge on one mole of an univalent ion and amounts to 96,487 coulombs,

P = solution pressure,

π = osmotic pressure.

Half-cell: when a metal electrode is dipped in the solution of its salt, it is called "half cell" in which electrode potential (*E*) depends on the activity and on the molar concentration (c) of the metal ion in the solution, and the "standard electrode potential" (E_0). Then the equation for electrode potential is:

$$E = E_0 = \frac{RT}{nF}\log_e c = E_0 \frac{0.059 \ \log c}{n}$$

where E_0 = electrode potential of that metal in a solution of unit activity of its ions,

T = 298 K.

- **Characteristics of double-layer at electrode-solution interface of half-cell**

(a) **When solution pressure (*P*) is lesser than osmotic pressure (π)**: The double-layer at electrode-solution interface will show positive charge in the inner layer and negative charge in the outer layer (Fig. 6.6).

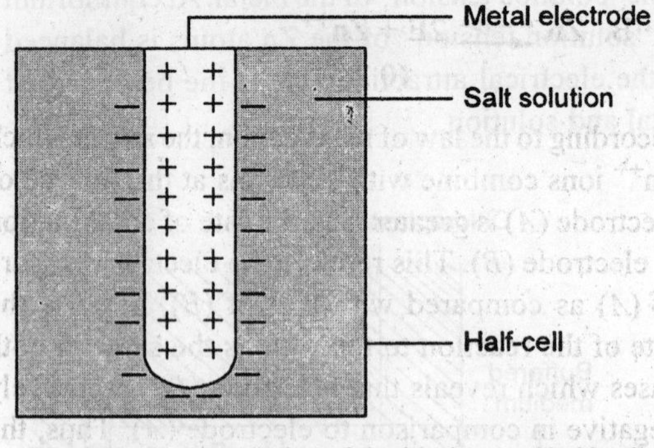

Fig. 6.6: Characteristics of double-layer of half-cell

(b) **When solution pressure (*P*) is greater than osmotic pressure (π)**: The double-layer at electrode-solution interface will show positive charge in the outer layer and the negative charge in the inner layer (Fig. 6.7).

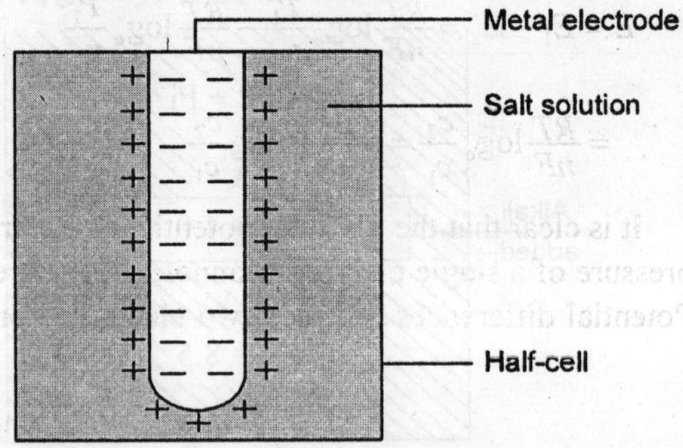

Fig. 6.7: Characteristics of double-layer of half cell

- **Concentration Cell**

In a half- cell, one metal electrode (*A*) is dipped into its salt solutions having different molar concentration (e.g., 2M Zn^{++}) and similarly in another half cell electrode (*B*) of same metal is dipped into its salt solution having lower molar concentration (e.g., 0.2 M Zn^{++}) than (*A*). The electrodes are connected by copper wire and establishing an electrolyte or salt bridge between the two solutions by means of a tube filled with saturated solution of KCl, then a concentration cell is constituted. This arrangement represents an electric battery, with an external circuit through the copper wire in which conduction is by the movement of electrons, and an internal circuit through the salt bridge in which production is by the movement of the ions (Fig. 6.8). The reaction occurs:

A. $Zn \rightleftharpoons 2E + Zn^{++}$
(2 M)

B. $Zn \rightleftharpoons 2E + Zn^{++}$
(0.2 M)

according to the law of mass action, the rate at which Zn^{++} ions combine with electrons at the surface of electrode (*A*) is greater than the rate of combination at electrode (*B*). This reduces the electron pressure of (*A*) as compared with that of (*B*), because the rate of the reaction to the right is the same in both cases which reveals that electrode (*B*) is relatively negative in comparison to electrode (*A*). Thus, the concentration cell consists of two half- cells which are inter-connected. An electromotive force (*E*) develops between the two electrodes due to difference of electrode potential ($E_1 - E_2$) and can be measured by potentiometer when placed in the circuit (Fig. 6.8). At 298 K,

$$E = E_1 - E_2 = \frac{RT}{nF}\log_e \frac{P_1}{\pi_2} - \frac{RT}{nF}\log_e \frac{P_1}{\pi_2}$$

$$= \frac{RT}{nF}\log_e \frac{c_2}{c_1} = \frac{0.059}{n}\log\frac{c_2}{c_1}$$

It is clear that the absolute potential or electron pressure of a single electrode cannot be measured. Potential differences between two electrodes only can be measured. One electrode should be used as a standard against which others are compared, e.g., normal hydrogen electrode is the standard for comparison.

II. Potentiometric Determination of pH

- **Common Potentiometric Indicator Electrodes (ion selective electrodes). For clinical laboratory**
  1. Glass membranes electrodes.
  2. Solid state membrane electrodes.
  3. Liquid and polymer membrane electrodes.
  4. Immobilized enzyme electrodes.
  5. Gas sensing electrodes.
  6. Disposable ISES, chemfets and biosensor.

Potentiometric determination is done by pH meter which consists of a concentration cell, i.e., two half-cells and a potentiometer is placed in the circuit to supply emf to oppose the emf of the concentration cell (Test cell). The half-cells of pH meter are having (a) Hydrogen electrode, (b) Calomel electrode or (c) Glass electrode.

The two electrodes are dipped in different solutions, one of which is the standard solution of known pH and another is the test solution. An electrode potential difference is generated

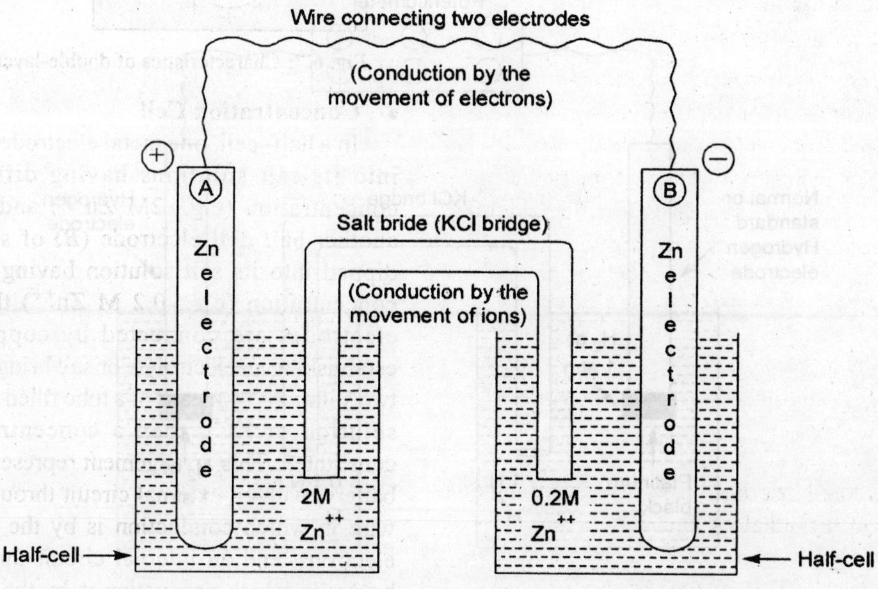

Fig. 6.8: Concentration cell

depending on H-ion concentration between the two solutions which is measured by the potentiometer.

(a) The hydrogen electrode

The principle of hydrogen electrode is similar to that of the zinc electrode. Since hydrogen is a gas, it cannot be used as electrode. If a piece of platinum foil, which is itself inert as an electrode, is covered with finely divided platinum black which is prepared by placing it in a solution containing H^+ ion concentration and by bubbling H_2 gas at 1 atmosphere pressure through the solution, it takes up hydrogen gas up to saturation and then functions as a hydrogen electrode. Inert electrodes prepared from gold, platinum or palladium and covered with platinum or palladium black may be used. Platinized platinum is generally used. If the electrode is placed in the aqueous solutions, the hydrogen molecules adsorbed on the platinum black surface, break into hydrogen atoms, and these possess a "solution tension" like metal atoms. The reactions at a hydrogen electrode can be given by equation:

Hydrogen adsorbed		Hydrogen atom	Lose	Electrons	Pass into liquid	
H_2	$\rightleftharpoons$	$2H$	$\rightleftharpoons$	$2E$	$\longrightarrow$	$2H^+$ ions
on the platinum surface				remain on platinum surface		at surface of electrode

The rate at which H atoms lose electrons to the platinum and pass into the liquid as H^+ ions is proportional to the concentration of H_2 molecules dissolved in, or adsorbed on the platinum black. The concentration of H_2 molecules in the platinum is in turn proportional to the pressure of gaseous hydrogen around the electrode. As this pressure goes up, more H_2 molecules are taken up by the platinum, and more H^+ ions pass into the solution. The hydrogen electrode with H_2 gas at one atmosphere pressure and dipped in a IN H^+ ion solution, i.e., 1g-equivalent of "active" H^+ per litre is called the normal or standard hydrogen electrode whose potential is zero and designated as Eho and is used as the standard of reference for other electrodes. If a concentration cell is (Fig. 6.9) constituted by joining the "normal or standard hydrogen electrode" with another hydrogen electrode placed in an unknown solution of different H^+ ion activity, a potential difference or electromotive force (emf) is developed by the cell depending on the difference between the two solutions in the H^+ ion activity, i.e., in H-ion concentration. The potential of hydrogen electrode in unknown solution is designated as Eh. Then

emf (volts) = Eh – Eho.

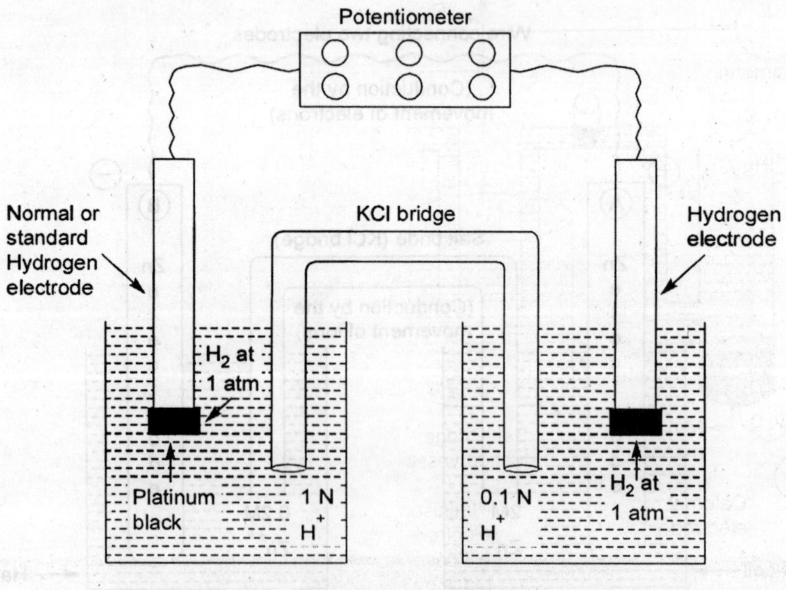

Fig. 6.9: Hydrogen concentration cell

Potential difference or emf between two hydrogen electrodes placed in solutions of H^+ ion concentrations $[H^+]$ and $[H^+]_2$ is expressed in the following way:

$$\text{emf} = \frac{RT}{nF}\log_e \frac{[H^+]_1 \text{ (unit } H^+ \text{ activity)}}{[H^+]_2 \text{ (activity of unknown solution)}}$$

$$\text{since } [H^+]_1 = 1 = \frac{RT}{nF}\log_e \frac{1}{[H^+]_2}$$

$$= 0.00019837\,T \log \frac{1}{[H^+]}.$$

$$\text{but } \log \frac{1}{[H^+]_2} = \log 1 - \log [H^+]_2$$

$$= -\log [H^+]_2 = pH$$

and we have = 0.00019837 T × pH

$$\text{or} \quad pH = \frac{\text{emf}}{0.00019387\,T}$$

where

R = the gas constant (8.315 joules per degree),

T = the absolute temperature,

n = the valence change,

F = faraday, representing the quantity of electricity per gram-ion equivalent (96,500 coulombs),

$\log_e$ = the natural logarithm.

- **Calomel and glass electrode for measurement of pH**

Due to difficulty of preperation and maintenance of a normal hydrogen electrode, a "calomel electrode" is substituted for it in making the potential difference. The calomel electrode is also combined with a glass electrode instead of hydrogen electrode for unknown solution.

(b) Calomel electrode

The calomel electrode is a standard half-cell consisting of metallic mercury beneath the layer of $Hg\text{-}Hg_2Cl_2$ paste in a little saturated KCl solution which is in contact with a layer of saturated Hg_2Cl_2 in KCl solution (0.1 M, 1.0 M or saturated KCl) above. This calomel electrode half-cell should be standardized against a normal hydrogern electrode (Fig. 6.10). The potential of the calomel electrode varies with the concentration of the KCl solution;

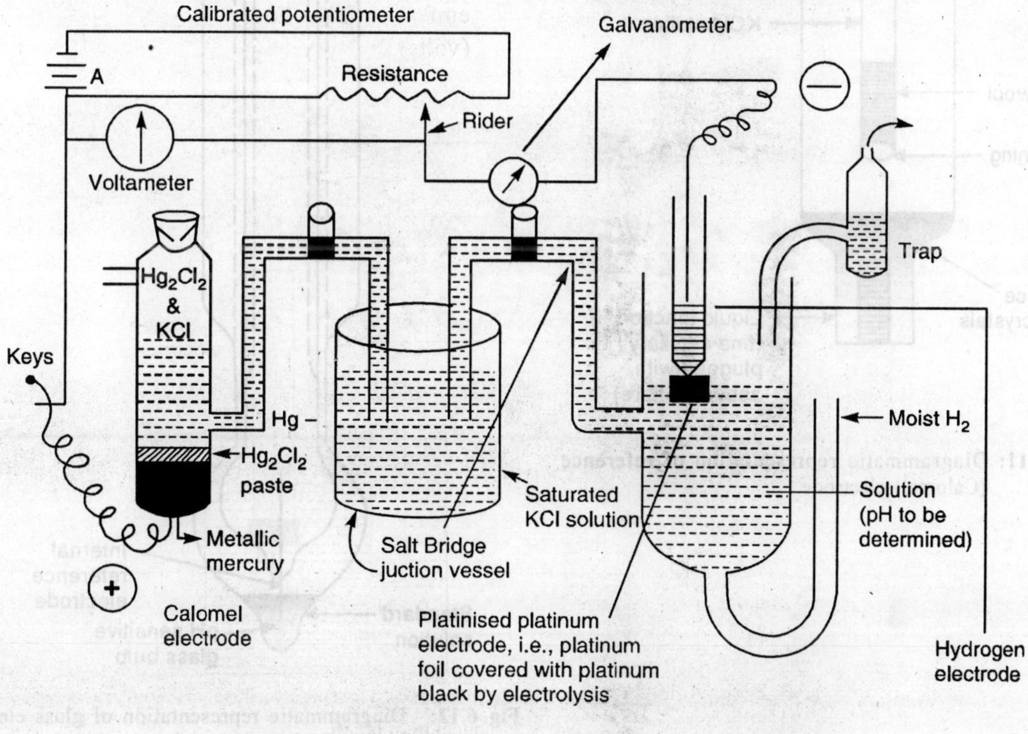

Fig. 6.10: Calomel-Hydrogen electrodes for pH measurement

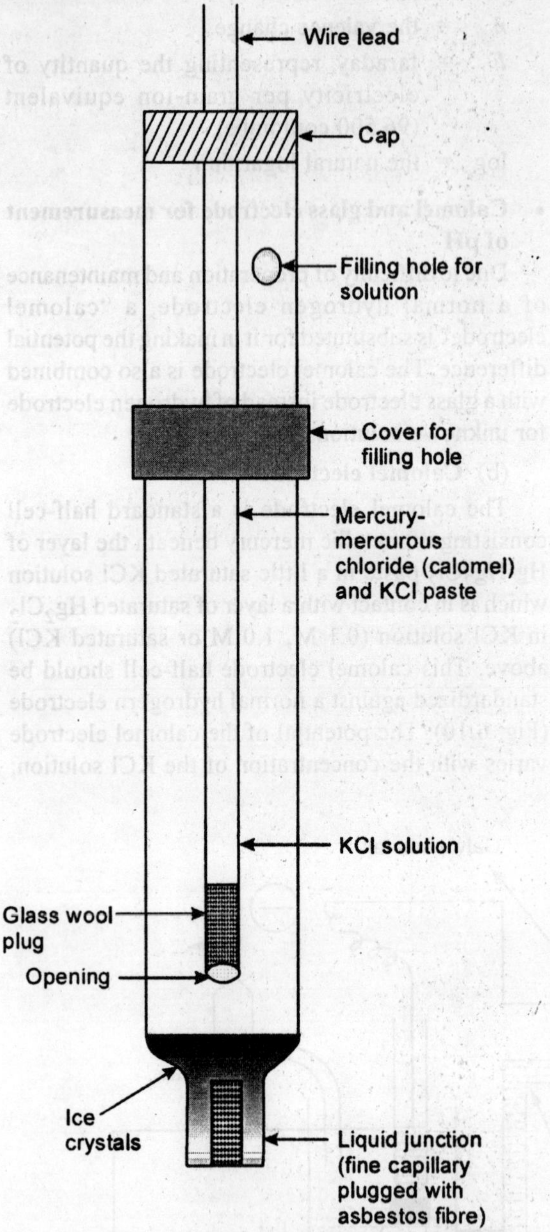

Fig. 6.11: Diagrammatic representation of reference (Calomel) electrode

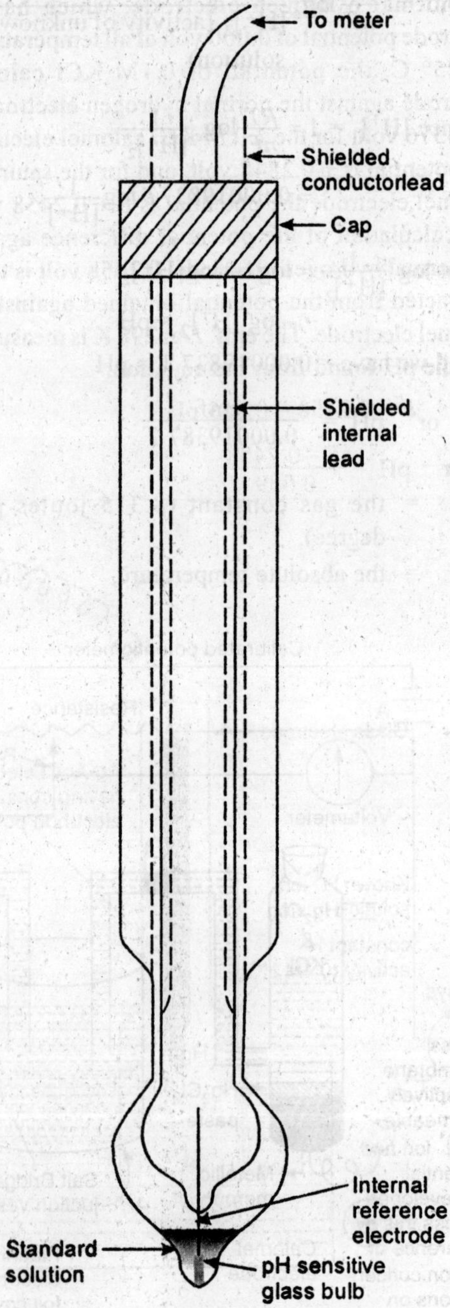

Fig. 6.12: Diagrammatic representation of glass electrodes (Beckman)

but for a given temperature and concentration of KCl the potential of a calomel electrode against the normal hydrogen electrode is constant. At 25° C the potential of calomel electrode will be greater by – 0.2458 volt than the potential measured against the normal hydrogen electrode, which has an electrode potential of 0.000 volt at all temperatures. At 25° C, the potential of 0.1M KCl calomel electrode against the normal hydrogen electrode is + 0.3376 volt, for the 1.0 M KCl calomel electrode the potential is + 0.2848 volt, and for the saturated calomel electrode the potential is 0 + 0.2458 volt. For calculation of the potential difference against the normal hydrogen electrode, 0.2458 volt is to be subtracted from the potential obtained against the calomel electrode. The emf, E at 298 K is measured, and the pH found from the equation:

$$E = 0.2458 = 0.059\text{ pH}$$

$$\text{or} \quad \text{pH} = \frac{E - 0.2458}{0.059}$$

(c) Glass Electrode and combination electrod

pH meter for determination of pH of biological fluids (e.g., blood) consists of saturated calomel electrode and a glass electrode. The glass electrode consists of a glass bulb filled with 0.1 M HCl having constant H^+ activity (e.g., known solution whose pH is known).

The combination electrode contains both pH sensing and reference electrodes, combined in a single probe body.

The thin membranes of the proper glass are selectively permeable to H^+ ions and potential is developed across this by difference of H^+ ion concentration on the opposite side having unknown solution which is present outside the glass bulb having variable H^+ ion activity. An Ag-AgCl electrode having constant electrode potential is dipped into the known solution of the glass bulb (Fig. 6.13).

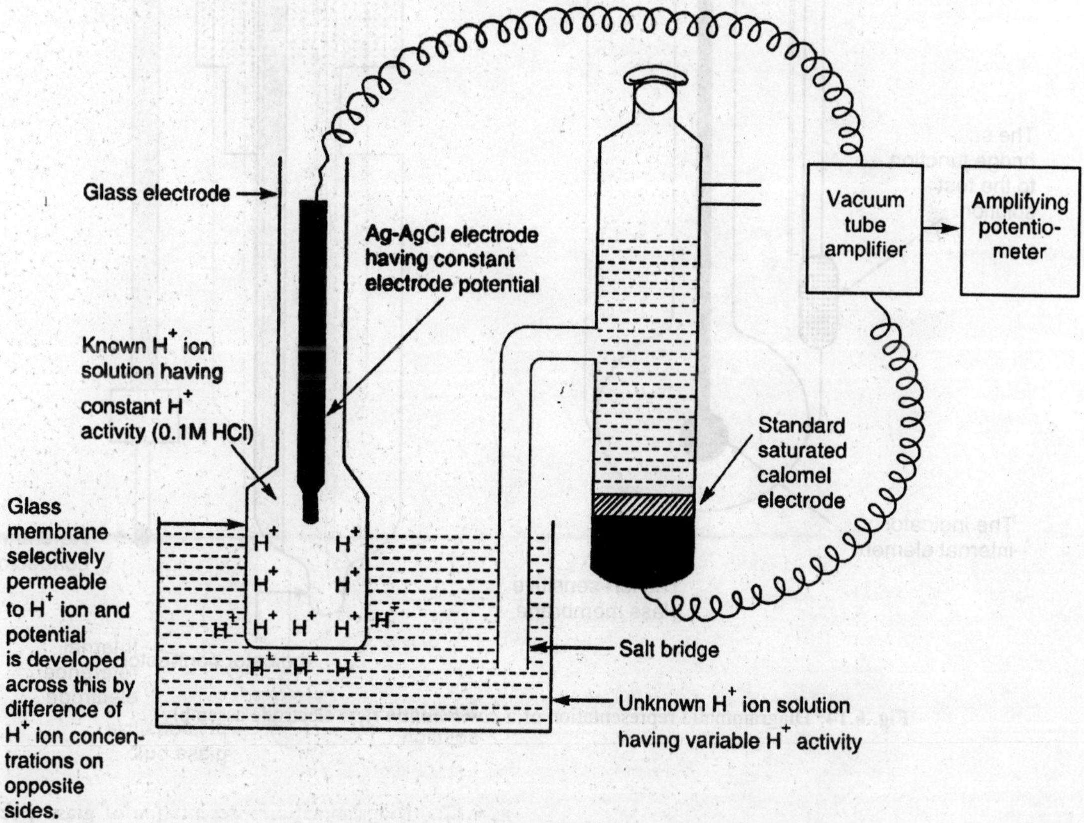

Fig. 6.13: Glass electrode assembly

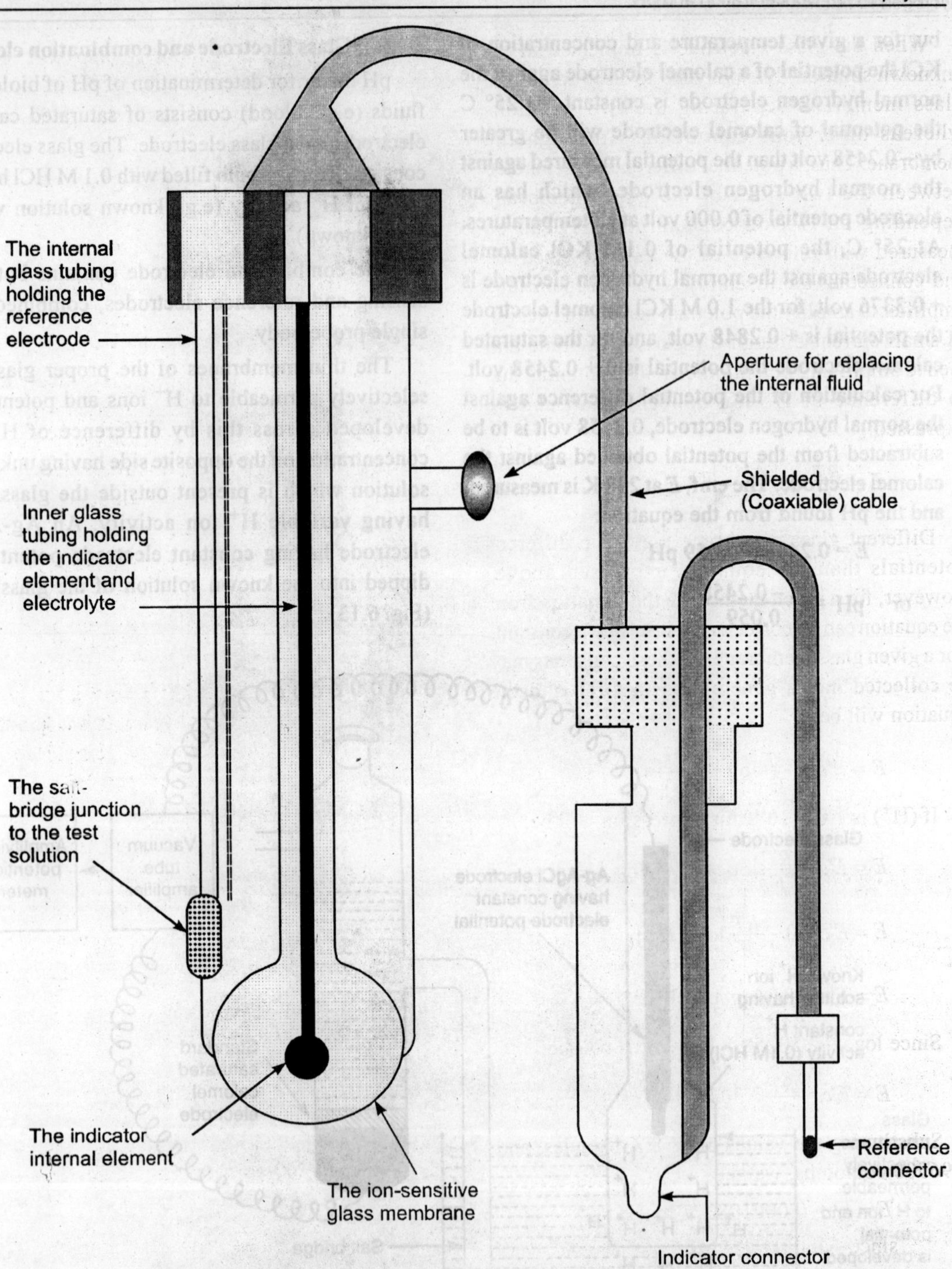

Fig. 6.14: Diagrammatic representation of a *combination type* electrode assembly

When the glass electrode is immersed in the unknown solution, H^+ ions tend to flow across the glass membrane seperating the two solutions whereas other ions cannot pass through the membrane. Thus, a potential difference is generated between the two sides of the glass membrane depending on H^+ ion concentration. It can be measured with reference to the calomel electrode and connecting the assembly with a vacuum tube amplifier to amplify the potential generated because of small current through the glass membrane and then to amplifying potentiometer. The potential due to difference of H^+ ion concentration can be expressed:

$$E = \frac{RT}{F}\log_e \frac{[H^+]_1}{[H^+]_2}$$

Different glass membrane may give different potentials than the equation expressed above. However, for a given membrane the variation from the equation can be corrected by inserting a constant. For a given glass membrane, all these constants may be collected into a general constant E′c and the equation will be:

$$E = E'c + \frac{RT}{F}\log_e \frac{[H^+]_1}{[H^+]_2}$$

If (H^+) is 1.0 = 10^0, this expression will be:

$$E = E'c + \frac{RT}{F}\log_e 1 - \frac{RT}{F}\log_e [H^+]_2$$

$$E = E'c + 0 - \frac{RT}{F}\log_e [H^+]_2$$

$$E = E'c - \frac{RT}{F}\log_e [H^+]_2$$

Since $\log_e = 2.30258 \times \log_{10}$, this will be:

$$E = E'c - \frac{RT}{F}2.30258\log[H^+]_2 .$$

Substituting the values of R, 8.315 joules, and F, 96500 coulombs, we get:

$$E = E_c' - 0.00019837T\log[\mathrm{H}^+]_2 ,$$

since $-\log[H^+] = pH$

$$E = E_c' - 0.00019837T\,\mathrm{pH}$$

$$\therefore \mathrm{pH} = \frac{E - E_c'}{0.00019837\,T}.$$

At 25° C: $\mathrm{pH} = \frac{E - E_c'}{0.05916}$.

If the instrument is calibrated against a series of buffers of known pH and which is plotted against E, there should be a smooth curve, straight between pH values of 1 and 9.

- **Important characteristics of Glass Electrode**

1. Low melting point
2. High hygroscopecity
3. Relatively high electrical conductivity

pH range: The useful pH range for glass electrode is generally from pH electrodes will as a rule perform to pH_{11}. Below pH_1 acid errors may be expected, requiring frequent standarisation against a buffer of pH close to the actual pH of the test solution. Above pH_{11} alkaline errors are observed. The alkaline error depends on the composition of the membrane glass.

Temperature for Measurement: Due to large change with temperature of the electrode resistance, glass-satisfactorily only within a temperature range of about 60°C. Better do work our 27°C.

Life time: The useful lifetime of a glass electrode is generally several years, if it is handled with care. The bad condition of the glass electrode shows slow response and reduced sensitivity. A glass electrode may be considered good:

If it has a fast response (< 1min) thickness of glass membranae is 0.05 mm to 0.15 mm.

Response to H^+ ion or any other cations depends on the composition of the glass membrane.

Stable response should be ≤ 1 min. for an good glass electrode.

Sensitivity should be 95%

Accuracy –0.1 pH

- **Merits of glass electrodes**

1. Applicable to all kinds of solutions including those containing strong oxidising and reducing agents.
2. It can be employed in case of semisolid materials like cheese.
3. Ideal assembly for biological fluids. Hydrogen electrode is not suitable for blood without special forms of electrode vessels, because hydrogen gas bubbling through the solution, evolves out carbon dioxide and changes the pH.

4. Solutions are not contaminated by glass electrode and can be used for other determinations.
5. Commercial glass-electrode assembly is calibrated to read directly the pH reading without any calculations.
6. The accuracy is within a few hundredths of a pH unit.
7. It is not necessary to deoxygenate the solution.

- **Demerits**

1. It should be calibrated against a series of buffers of known pH.
2. A vacuum-tube amplifying potentiometer is required because of small current through the glass membrane.
3. It does not function well below pH 1.
4. It does not function well in solutions of very high pH when appreciable quantities of certain metal ions are present. In these conditions it act as a Na^+ ion electrode because of the solubility of the sodium salts of the glass at high pH. The presence of Na^+ ions in solutions of high pH causes error. If the Na^+ ion concentration of the solution to be tested is known, a correction may be made and correct pH values are obtained.
5. The presence of Li^+, Ca^{++}, Ba^{++}, K^+, NH_4^+ ions in the solutions also gives incorrect pH values at high pH but not so much as do Na^+ ions. The order of decreasing interference is: Na^+, Li^+, K^+, NH_4^+, Ca^{++}, Ba^{++}.
6. With new glasses, the membranes give good result up to pH 11 or 12 and by making small corrections for Na^+ ions, these values may be extended to pH 13 or 14.

- **Combination Electrode**

A combination electrode is one in which the pH sensitive glass electrode are the silver/silver chloride type. Also, the inner solutions in the two parts of the electrode are identical and they both are symmetrical electrode chains. The two silver silver chloride reference electrodes are combined into the same glass or plastic body protected against light by means of ruby red glass, which absorbs light (Fig. 6.14).

A combination electrode can be used within the temperature range –10°C to + 100°C as they have silver.silver chloride reference electrodes. Also, the measuring error due to small temperature differences of the buffer solutions and the sample.

Combination electrodes are easily recognized by the type of a cable [shielded (coaxial) cable]. They are supplied with: it terminates in both a reference connector and an indicator connector.

- **Advantages of combination Electrode**

1. In many applications, the sample size of the vessel holding the sample may restrict the use of an electrode pair. In such cases the use of a combination electrode is required.
2. In this type of assembly the reference electrode and the indecator electrode are combined into the same glass or plastic body.
3. This electrode is probably the best all-around electrode for routine laboratory use unless the test solution is very viseous or is a colloidal solution (may cause blockage of the salf-bridge junction) in which case an electrode pair with an appropriate reference junction is used.
4. The main advantage is that temperature variations and electrical noise problems are somewhat reduced.

- **Electrodes Care and Storage**

1. Electrodes should not be handled roughly and when not in use should be stored correctly to maintain their life span. When storing glass pH electrodes overnight or for short periods of time, one should store them in a beaker containing a buffer of either pH 4 to 7, whereas reference electrodes are best stored in a solution of 0.1 M potassium chloride (to be sure the electrolyte level inside the reference electrode in higher then the level of the storage solution to prevent hydrostatic back flow into the electrode).
2. Because combination electrodes require that both the sensing membranes and the reference junction be maintained in a single solution, if should contain both pH buffer (4 or 7) and 0.1M potassium chloride in 50; 50 ratio.
3. For longer periods of storage, indicator electrodes should be moistened, capped, and stored.

4. New electrodes, or those that have been stored dry should be conditioned or activated before use by soaking the bulb for a period of 12 to 24 house in 0.1N hydrochloric acid.
5. The pH sensitive tip of the electrode should not become scratched or cracked through contact with the sample or container. Therefore, if should not be allowed to rest on the bottom of the sample container. The tip should dried by gentle rubbing with absorbent tissue.
6. Sometimes the presence of an inhibiting film can have a deleterious effect on pH response. Even the extraction of the soluble components of the glass from the surface of the membrane may cause a type of deterioration marked by sluggish behaviour. Sluggishness caused by emersion in alkali can often be eliminated by soaking in hydrochloric acid. and that caused by strong acids is best overcome by standing the electrode in water.
7. If this treatment fails to correct the difficulty, it is sometimes possible to rejuvenate the electrode by immersing it for 1 min. in a 20% solution of aminophylline bifluoride at room temperature. For the fluoride solution, a waxed paper cup or a waxed beaker should be used. This treatment should be carried out very carefully, since attack can cause puncture of the thin membranes.
8. Glass electrodes are detoriated for prolonged use but can be restored by etching the glass membrane in a 20% ammonium bifluoride solution.

Note: Combination electrodes can be essily identified by the type of cable they are supplied with; it terminates in both a references connector and an indicator connector.

I. Potentiometric Titration

Potentiometric titration method is used when:

1. The use of indicator is not feasible due to the intense colour or high dilution or dense turbidity of the solution,
2. The solution contains a polybasic acid or a mixture of weak and strong acids,
3. The solution contains interfering buffers.

This titration depends on the potential change (ΔE) of the concentration cell having a standard reference electrode and an indicator electrode and a potentiometer in the circuit when the indicator electrode is dipped in the unknown solution. Addition of each volume (ΔV) of titrant changes the electrode potential of the indicator electrode and the potential change (E) of the concentration cell. The rate of potential change ($\Delta E/\Delta V$) is highest at the end point. The end point is indicated graphically by plotting the potential change ($\Delta E/\Delta V$) against the total volume (V) of the titrant delivered during titration.

Use

1. Precipitation titrations, e.g., $AgNO_3$ with NaCl.
2. Oxidation-reduction titrations, e.g., Fe^{++} or dichromate.
3. Acid-base titrations.

pH Meter

Measurement of pH is undertaken to determine whether a particular solution is acidic or neutral or alkaline and to find out the hydrogen ion concentration of the solution. The electrode system consisting of a glass electrode and a reference electrode (Ag/AgCl electrode) provides the emf which is in turn amplified by means of high input impedance dc amplifier. The measuring system consists of an amplifier and microammeter and using the equation $E_0 = E_g{}^o + 0.05912\,(pH_2 - pH_i)$ the scale of the microammeter is calibrated in pH. To match with the high resistance of the electrode system and FET input, operational amplifier is used to amplify the signal voltage. The gain of the amplifier is made adjustable over a narrow range to provide for temperature compensation,

where,

$E_g{}^o$ is the equivalent membrane potential

E_0 the net potential difference

pH_x = pH of test solution

pH_i = pH of buffer solution

The instrument, which is measured the value of pH in a unknown solution by the indication or display system is called pH meter.

Electrodes for pH measurement: Three types of electrodes are used for the measurement of pH, these are (1) Hydrogen electrode; (2) Glass electrode; (3) Reference electrode.

Types of pH meter: There are 9 types but in general, commercial pH meters can be categorised broadly into two main types.

1. Potentiometer type (Null-Detector)
2. Direct reading type.

- **Types**

1. Null-detector type or potentiometric type
2. Direct reading type
3. Chopper amplifier type
4. Vibrating condenser amplifier type
5. Zero corrected dc amplifier type
6. Based on FET input stages
7. Digital pH meter
8. Industrieal pH meter
9. Microcomputer based pH meter.

Failure of pH Meter

1. Defective electrode
2. Defective input circuitry
3. Defective electronic circuitry.

- **Measurement Principle for pH**: Potentionmetry is the measurement of the potential developed at an indicator electrode. The electrode potential is a function of several variables and can be predicted by the Nernst equation developed by Nernst. If states that:

$$E_{cell} = E^{o} - 2.3\ \mathrm{RT} \times nF \times \log\ (\mathrm{H^{+}\ ion})$$

where,

E_{cell} = The electrode potential
E^{o} = The standard cell potential
R = The gas constant
T = The temperature in K
n = The valence of the ion measured
F = Faraday's constant.

pH METER: The measurement of hydrogen ion concentration (pH) in a test solution is made by measuring the potential developed in an electro chemical cell. The electrochemical pH cell consists of a measuring electrode and a reference electrode, both immersed in the solution under investigation. The two electrodes are connected to a measuring instrument and the emf between these two electrodes is measured. The measuring electrode is pH sensitive and its potential is proportional to the pH of the solution in which it is immersed while the reference electrode would always develop a constant electrical potential against which the potential of the glass electrode is measured.

This type pH measurements, glass electrode employing an ordinary potentiometric circuit with an electrometer tube is used for signal amplification. The glass electrode lead is connected to the grid of the electrometer, so that changes in the electrode potential cause changes in the anode current. A standard cell sends a constant current through the shield wire and establishes a voltage difference across it as in any potentiometer. The potentiometer circuit is connected to the cathode of an electrometer valve. A sensitive galvanometer is placed across the electrometer tube. The calibrated potentiometer dial is marked off by the known pH value of buffer solution.

It can be used for the estimation of chemical analysis, for electrophoresis and chromatography, estimation of pH of body fluid, e.g., blood and extraction of enzymes and other biological materials (Fig. 6.16).

- **Direction for use**

1. Shake the electrode gently, to ensure that the internal buffer solution covers the whole membrane and no air bubbles are entrapped.
2. Single rod assemblies should be filled with the appropriate electrolyte to a height of about 1 cm below the filling port.
3. Wash off any salt film present on the exterior of single rod assemblies, using distilled water.
4. To ensure pressure equalisation, the stopper of the filling port should be removed, or perforated with a pin.
5. Soak the electrode in water for some hours (preferably overnight).
6. Electrodes which have a slow response due to drying out of the membrane or use under extreme conditions, may be reactivated by soaking in 0.1 N hydrochloric acid for several hours.

Electrodes which fail to respond to the above treatment can be further activated by dipping them to 2% hydrofluoric acid for 5-10 seconds and immediately washing with

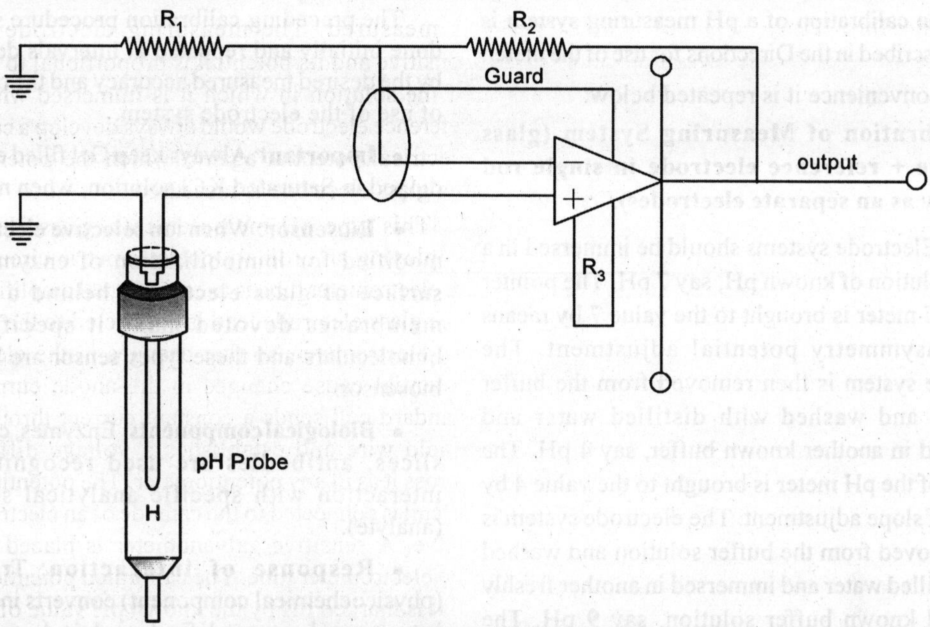

Fig. 6.15: Circuit diagram of pH meter

distilled water. The drastic treatment should, however, be applied only as a last resort as it reduces the life of the electrode.

7. After conditioning the electrode as in (6) above, if it is not to be used it should be kept in distilled water. To prevent the entry of water through the diaphragm of the reference electrode of single rod assemblies the stopper of the filling opening should be removed, or perforated with a pin.

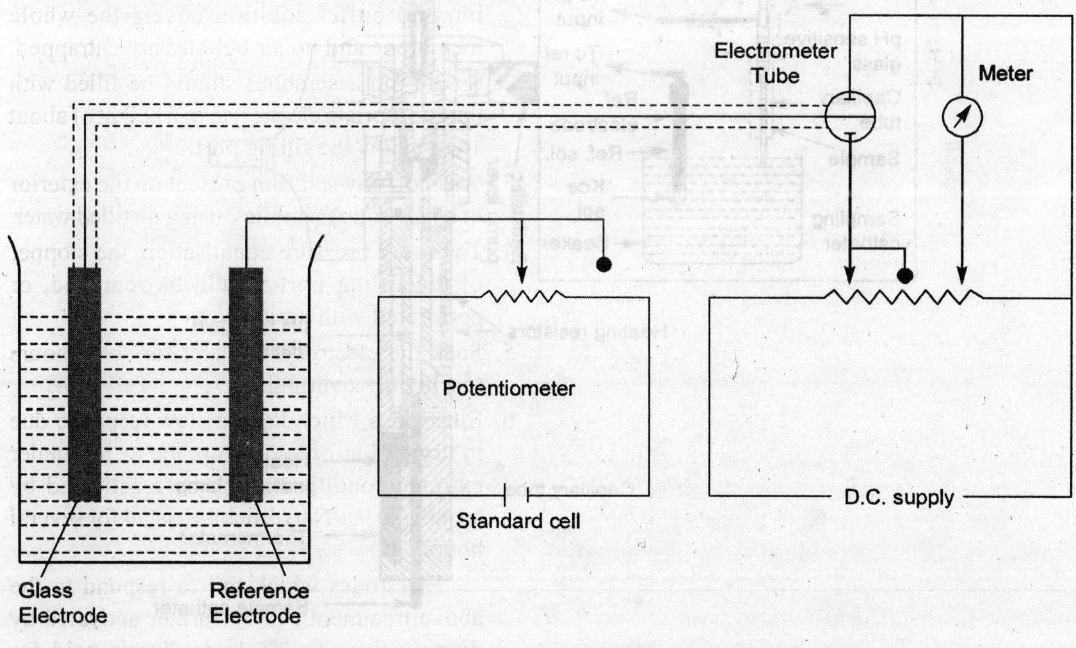

Fig. 6.16: Null Detector type or potentiometer type pH meter

8. The calibration of a pH measuring system is described in the Directions for use of the meter.

For convenience it is repeated below:

Calibration of Measuring System (glass electrode + reference electrode in single rod assembly as an separate electrodes).

The Electrode systems should be immersed in a buffer solution of known pH, say 7 pH. The pointer of the pH meter is brought to the value 7 by means of the asymmetry potential adjustment. The electrode system is then removed from the buffer solution and washed with distilled water and immersed in another known buffer, say 4 pH. The pointer of the pH meter is brought to the value 4 by means of slope adjustment. The electrode system is now removed from the buffer solution and washed with distilled water and immersed in another freshly prepared known buffer solution, say 9 pH. The pointer of the measuring instrument should now indicate this value and if it does so, the electrode system is functioning correctly. The effect of temperature on the pH value of the buffer solution must be considered.

The preceding calibration procedure should be done initially and repeated at intervals determined by the desired measured accuracy and the conditions of use of the electrode system.

- **Important**: Always keep Gel filled electrodes dipped in Saturated KCl solution, when not in use.
- **Biosensor**: When ion selective electrodes are modified for immobilization of enzyme on the surface of glass electrodes behind a polymer membranes devoted make it specifying for bimoleculars and these types sensor are known as biosensor.
- **Biological components**: Enzymes, cell, tissue slices, antibodies are used recognition and interaction with specific analytical substance (analyte).
- **Response of interaction**: Transducer (physicochemical component) converts inderaction into signal → amplified and it has a direct relationship with the concentration of the analytical material analyte, the transducer may show properties change in the forms of optical, calorimetric, magnetic potentiometric, ammperometre or conductance.

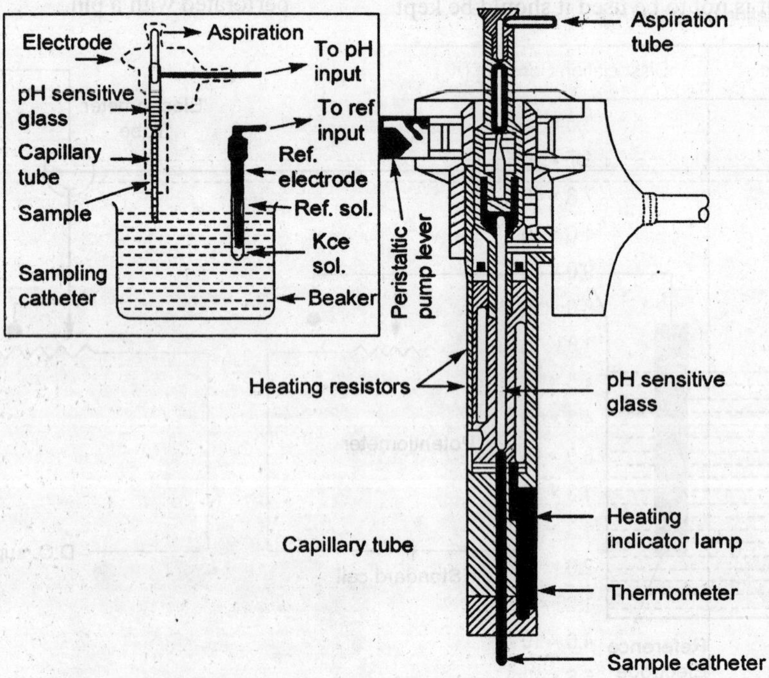

Fig. 6.17: Electrode for measurement of blood pH

- **Example of biosensors**
  1. Urea electrode contains an ammonium ion selective electrode coated with urease (an enzyme) behind a teflon membrane.
  2. Biosensors for measurement of glucose
  3. Biosensors used in food and beverage industries.

6.5 Acids, Bases and Salts

6.5.1 General concept of acids, bases and their dissociation constants

The general definition of acids as substances containing hydrogen ions replaceable by metals and having properties as colour change of indicator, corrosive action, etc. which are qualitatively acceptable but the problem arises when quantitative measurement of the strength of the acid is done. All acids in aqueous solution increase H^+ ion concentration of water and all bases decrease H^+ ion concentration, and therefore, increased OH^- concentration and quantitative measurement of acidity depend on this property of acids and bases.

Table 6.3: Some weak acids (and bases) and their dissociation constant (K_a).

Weak Acids	Dissociation Constant (K_a)
Oxalic acid	3.8×10^{-2}
Sulphurous acid	1.7×10^{-2}
Phosphoric acid	7.6×10^{-3}
Salicylic acid	1.0×10^{-3}
Hydrofluoric acid	9.0×10^{-4}
Citric acid	8.0×10^{-4}
Acetic acid	1.8×10^{-5}
Butyric acid	1.5×10^{-5}
Carbonic acid	4.3×10^{-7}
Hydrogen sulphide	8.0×10^{-8}
Hydrocyanic acid	1.3×10^{-9}
Phenol	1.3×10^{-10}
Hydrogen peroxide	2.0×10^{-12}
Diethylamine	1.26×10^{-3}
Ethylamine	4.6×10^{-4}
Trimethylamine	6.5×10^{-5}
Hyrazine	2.2×10^{-6}
Pyridine	2.3×10^{-9}

The H^+-ion concentration is related to the dissociation constant of Ostwald's dilution law, so the comparison of strength of acids and bases can be made using their dissociation constants.

$$\underset{\text{(Acid)}}{HA} + H_2O \rightleftharpoons H_3O^+ + A^-$$

$$\text{or} \quad K_a = \frac{[H_3O^+][A^-]}{[HA]}$$

$$B + H_2O \rightleftharpoons BH^+ + OH^-$$

$$\text{or} \quad K_b = \frac{[B^+H][OH^-]}{[B]}$$

But these equations are applicable only for weak acids and bases and not to strong acids and bases. So, the strength of only weak acids and bases are comparable in terms of their dissociation constants (or protolysis constants).

When the values of K_a i.e., dissociation constant of some weak acids and bases are arranged in decreasing order of strengths, it is observed that the acids and bases which are grouped as weak differ in strength amongst themselves.

6.5.2 Bronsted-Lowry theory of acids and bases

The classical concept of acids and bases has given much importance to H^+ and OH^- ions and is satisfactory for aqueous solution. But difficulties are encountered when attempts are made to apply them to non-aqueous solutions. Hydroxyl ion does not have importance pertaining to the basic properties, basic functions can be exhibited by non-hydroxylic bases in non-hydroxylic solvents (e.g., sodamide in liquid ammonia).

Another comprehensive concept was introduced by Lewis. According to Lewis, acid is a potential **electron-pair acceptor** and base is a potential **electron-pair donor**. However, the concept introduced by Bronsted and Lowry is known as **proton transfer theory** which is more widely used in describing acid-base reactions in all solvents. According to this concept, an acid is a **proton donor** and a base is a **proton acceptor** independent of any solvents.

Acid Proton donor		Base Proton acceptor
CH_3COOH	$\rightleftharpoons$	$CH_3COO^- + H^+$
H_2PO_4	$\rightleftharpoons$	$HPO_4{}^- + H^+$

$$NH_4 \rightleftharpoons NH_3^- + H^+$$
$$HOH \rightleftharpoons OH^- + H^+$$

The definition of an acid is equivalent to the older view, but the definition of a base is much broader in its implications.

6.5.3 Characteristics of the Bronsted-Lowry theory

(a) **Absoluteness** Substances are acids and bases by their right and independent of the solvent. For example, HCl is an acid because it can donate protons. But whether it donates or not depends on the conditions which does not exclude their right to be called an acid. OH^- ions are also bases because it tends to combine with protons.

(b) **Reciprocity-conjugate acid and base**: An acid-base reaction always involves a "conjugate acid-base pair" made up of a proton donor and the corresponding proton acceptor. For example, acetic acid (CH_3COOH) is a proton donor, and the acetate anion (CH_3COO^-) is the corresponding acceptor, together they constitute a conjugate acid-base pair. Thus, acetate ions and for that matter all anions, should be regarded as base. Such bases and the corresponding acids are said to be conjugate to each other. It is observed

$$\underset{\text{(Acid)}}{CH_3COOH} \rightleftharpoons \underset{\text{(Base)}}{CH_3COOH^-} + H^+$$

that the stronger, the acid, the weaker will be the conjugate base and vice versa, i.e., the conjugate of a strong species is weak and that of a weak species is strong. So, the acetate ion is a stronger base than the chloride ion because acetic acid is weaker than HCl.

(c) **Charge type**: This theory does not make any special charge type of the acid and acids of varied charge types are observed.

$$\underset{\textbf{Acid}}{HPO_4^{-2}} \rightleftharpoons \underset{\textbf{Base}}{H^+ + PO_4^{-3}}$$

$$HCO_3^- \rightleftharpoons H^+ + CO_3^{-2}$$
$$NH_4^- \rightleftharpoons H^+ + NH_3^-$$

(d) **Neutralisation**: All acid-base reactions are formation of the corresponding conjugate acid and base. The neutralisation of acetic acid by ammonia, $CH_3COOH + NH_3 \rightleftharpoons CH_3COO^- + NH_4^+$ causes production of acetate ion, i.e., the conjugate acid, acetic acid and the ammonium ion, i.e., the conjugate acid of ammonia. The general equation of all reactions between an acid and a base is:

$$Acid_1 + Base_2 \rightleftharpoons Acid_2 + Base_1 \text{... (A)}$$

where $acid_1$-$base_1$ and $acid_2$-$base_2$ are conjugate acid-base pairs.

If the strength of $acid_1$-$base_1$ and $acid_2$-$base_2$ are known, calculation can be done for the extent of neutralisation, i.e., the final position of equilibrium in the above reaction.

(e) **Effect of Solvent**: It is known that an acid is an acid or a base is a base independent of solvent. But most of the solvents have weak acidic or basic properties or sometimes both. So, proton transfer to or from the solvent takes place giving rise to many properties due to solvated proton.

$$\text{Acid + Solvent} \rightleftharpoons \text{Conjugate Base + Solvated } H^+$$

Thus, in water, hydrogen ions remain as H_3O^+ called oxonium or hydronium or hydroxonium in liquid ammonia, as NH_4^+, etc. The magnitude of acidity depends on the magnitude to which the solvolysis goes from left to right known as "Levelling Effect of Solvent". (Table 6.5).

(f) **Hydrolysis**: A solution of sodium acetate is alkaline because it contains the strong base, the acetate ions (CH_3COO^-) which contains strong base because its conjugate acid i.e., acetic acid, is weak. According to the Bronsted-Lowry theory, the different reactions which we know as dissociation (acid and base), hydrolysis, neutralisation, buffer action, etc. are basically the same reaction as expressed by the basic equation of Bronsted-Lowry theory (equation A) and are under the same mathematical calculation.

Table 6.4: Acid and solvent reaction

Acid$_1$	Base$_2$		Acid$_2$	Base$_1$
HCl	+ H_2O (solvent)	$\rightleftharpoons$	H_3O^+	+ Cl^-
HCl	+ NH_3 (solvent)	$\rightleftharpoons$	NH_4^+	+ Cl^-
HCO_3^-	+ H_2O (solvent)	$\rightleftharpoons$	H_3O^+	+ CO_3^{2-}
CH_3COOH (solvent)	+ NH_3	$\rightleftharpoons$	NH_4^+	+ CH_3COO^-

6.5.4 Acids

According to Bronsted-Lowry theory, acids are substances which tend to donate protons (H^+), thereby increasing the H^+ ion concentration of their solutions. For an acid to donate its protons, either a base or a solvent behaving as a base must be present to act as the proton-acceptor. Acids are of two types according to the number of protons, donated by its molecules.

1. Monoprotic (monobasic)
2. Polyprotic (polybasic) e.g., phosphoric acid, sulphuric acid etc. ionise stepwise and have separate dissociation constants for the dissociation of each proton.

Thus, $H_3PO_4 \rightleftharpoons H^+ + H_2PO_4^-$;

$$K_1 = \frac{[H^+][H_2PO_4^-]}{[H_3PO_4]}$$

$H_2PO_4^- \rightleftharpoons H^+ + HPO_4^{2-}$;

$$K_2 = \frac{[H^+][HPO_4^{2-}]}{[H_2PO_4^-]}$$

$HPO_4^{2-} \rightleftharpoons H^+ + PO_4^{3-}$;

$$K_3 = \frac{[H^+][PO_4^{3-}]}{[HPO_4^{2-}]}$$

It is observed that the successive steps in ionisation is difficult owing to increase in negative charge of the acid residue. The higher the negative charge on the acid residue, the more strongly is the proton attracted, which makes $K_1 > K_2 > K_3$. The conjugate bases, formed at the intermediate steps of ionization of a polyprotic acid (i.e., $H_2PO_4^-$ and HPO_4^{2-} ions from H_3PO_4) are "amphiprotic ions" which may act as proton-donors (acids) and as proton-acceptors (bases), the acid and basic strengths of the amphiprotic ion being given respectively by its own K_1 and the K_1 of the conjugate acid formed by its acceptance of proton. The equation for the dissociation or ionisation of an acid (HA) in dilute aqueous solution involves the transfer of a proton from the acid to water, which itself can act as a proton acceptor to yield the acid H_3O^+:

$$HA + H_2O \rightleftharpoons H_3O^+ + A^-$$

Each conjugate base has a characteristic affinity for a proton, relative to the proton affinity of OH^-. Acids that have only a slight tendency to give up protons to water the "weak acids", acids that readily give up their protons are "strong acids". The tendency of any given acid to dissociate is given by its dissociation constant at a given temperature.

6.5.5 *pK* values (expressions of acid and base strength)

At equilibrium, the equilibrium constant (K_{eq}) is:

$$K_{eq} = \frac{[H_3O^+][A^-]}{[H_2O][HA]} = \frac{[H^+][A^-]}{[H_2O][HA]}$$

because the concentration of H_3O^+ is equal to the concentration of H^+. The terms in the bracket indicate the concentrations of respective ions or molecules in moles per litre. To simplify the expression by elimination the water required for hydration of proton:

$$\frac{[H^+][A^-]}{[HA]} = K_{eq}[H_2O] = K_a$$

where K_a = temperature-dependent ionization constant called the "**acid dissociation constant**".

Though the acid dissociation constant expresses the acid strength accurately, it is inconvenient to express it in words. To avoid this inconvenience the modern method is to convert K_a as a power of 10 and express acid strength by the exponent i.e., power of 10 with sign changed and call this new unit pK called ionization exponent of an acid. Thus if K_a is $10^{-5}, pK_a = 4$.

$$pK_a = -\log K_a = \log (1/K_a)$$

pK is related to K as pH is to H^+ ion concentration. K_a is the measure of the strength of an acid, which is the magnitude of its proton-donating activity.

The strength of an acid in solution is dependent not only upon the chemical structure of the acid but also upon that of the solvent. An acid may be relatively weak in one solvent and strong in another. When the conjugate base of an acid is weak relative to the basic property of the solvent, then the acid is highly ionized in that solvent and conversely, when the conjugate base of the acid is strong relative to the solvent, the acid is poorly ionized. The strength of an acid is determined by (a) the strength of its conjugate base, (b) the basic strength of the solvent and the (c) dielectric constant.

The stronger acid means (a) the greater is the proportion of its molecules dissociated in solution, (b) the higher is its K_a, (c) the lower is its pK_a and (d) the lower is the pH of its solution. A weak acid has a high pK_a and dissociates little in solution. The pH of a solution of weak acid is related to its acid dissociation constant. The relationship can be stated in the convenient form of the Henderson-Hasselbalch equation, derived below at equilibrium.

$$HA \rightleftharpoons H^+ + A^-$$

$$K_a = \frac{[H^+][A^-]}{[HA]}$$

or $$H^+ = K_a \frac{[HA]}{[A^-]},$$

$$-\log [H^+] = -\log K_a - \log \frac{[HA]}{[A^-]}$$

Substituting pH and *pK* for $-\log [H^+]$ and $-\log K_a$, then

$$pH = pK_a - \log \frac{[HA]}{[A^-]}$$

removing minus sign and inverting the last term

$$pH = pK_a + \log \frac{[A^-]}{[HA]}$$

It follows from the equation that when $(A^-) = (HA)$, $pH = pK_a$ i.e., the *pK* of an acid group is that pH at which the protonated and unprotonated species are present at equal concentrations.

Thus, during the titration of an acid with a base or alkali, the pK_a of the acid is given by the pH of the titration mixture when exactly half of the molecules of the acid have been changed by titration to the conjugate base, because at this point (HA) equals (A) and this point is called **midpoint titration or half-equivalence point**.

6.5.6 Inductive effect of groups on acid strength

The electrons of covalent bonds between dissimilar atoms tend to associate with the more electro-negative or electron attracting atom forming a "dipole".

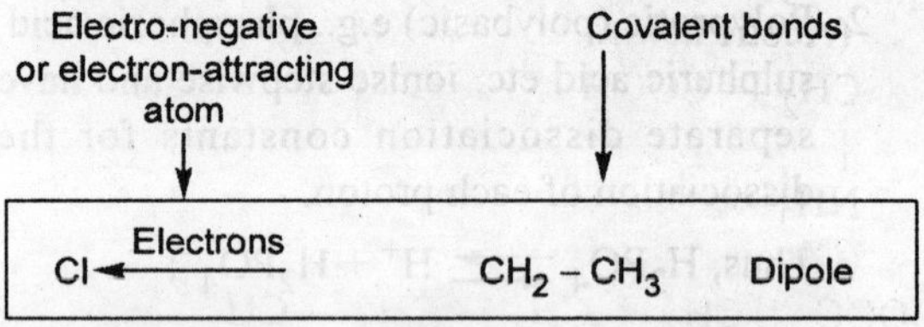

Fig. 6.18: Electrode for measurement of blood pH

In the figure (6.18), it is seen that the electrons are drifted to left hand side.

6.5.7 Influence of different groups

1. **Carboxyl group**: The factors that increase the electron density on the carboxyl group from which the positively charged proton must dissociate, resist its leaving, causing an "acid-weakening effect".
2. **Carbonyl group**: Any factor which decreases the electron density on the carbonyl group facilitates dissociation of the proton and have an "acid-strengthening effect".
3. **Distance between carboxyl group and electronegative atom**: When an electro-negative atom is closer to the carboxyl group the acid-strengthening effect is more

pronounced as we see in case of strong electro-negative atom chlorine:

	pK
$CH_3\,CH_2\,COOH$	4.9
$CH_2—CH_2—COOH$ (Cl on the terminal CH_2)	4.1
$CH_3\,CH—COOH$ (Cl on CH)	2.8

4. **Alkyl groups**: Alkyl groups supply electrons in the following manner:

	pK
$CH_3\,COOH$	4.7
$CH_3—CH_2COOH$	4.9
$(CH_3)_3\,C—COOH$	5.0

5. **Charged groups**: Charged groups may either supply or withdraw electrons:

	pK for carboxyl
$CH_3—COOH$ (Acetic acid)	4.7
$CH_2—COOH$ (NH_3^+ on CH_2) (Glycine)	2.3
$HOOC—CH_2—CH_2—CH_2—CH_2$ (NH_3^+ on the second carbon; COOH on the last CH_2) (Glutamic acid)	2.2

Dissociation of the second carboxyl of glutamic acid is intermediate in acid strength between that of acetic acid and glycine since the molecules have both positive and negative charge.

6. **Carbonyl group and hydroxyl group**: The carbonyl group and hydroxyl group exert inductive effects and are acid strengthening:

	pK
$CH_3\,CH_2COOH$ (Propionic acid)	4.9
$CH_3—CH—COOH$ (OH on CH) (Lactic acid)	2.9
$CH_3—CO—COOH$ (Pyruvic acid)	2.7

7. **Aromatic amines**: The aromatic amines, e.g., aniline and nitrogen atoms of cyclic amines, e.g., pyridine or purine, and pyrimidines are moderately strong acids. Aromatic amines, therefore, exist for the most part in the dissociated or uncharged form at pH 7.4. Their acidity is due to their aromatic electron which reduces the negative charge on the nitrogen and assists the dissociation of a proton.

8. **Drugs and pharmacologically active compounds**: Many drugs and pharmacologically active compounds are amines and their uncharged forms behave as bases, i.e., proton acceptors, whereas the charged forms behave as acids, i.e., proton donors. The relative strengths of various amines may be expressed by pK_a values for the dissociation. The pK_a values show that aliphatic amines are weaker acids (stronger bases) than ammonia and that at pH 7.4 all of an aliphatic amine is in the charged form. In body fluids, we see these amines are associated with an anion such as Cl^-.

6.5.8 Bases

The Bronsted theory defines the base as a proton acceptor in presence of proton donor or proton-donor solvent. This definition of a base is much broader in its implications. The following equations are given for illustration:

Acid		Proton		Base
CH_3COOH	$\rightleftharpoons$	H^+	+	CH_3COO^-
H_2CO_3	$\rightleftharpoons$	H^+	+	HCO_3^-
NH_4^+	$\rightleftharpoons$	H^+	+	NH_3^-
HOH	$\rightleftharpoons$	H^+	+	OH^-
H_3O^+	$\rightleftharpoons$	H^+	+	H_2O

All the acids mentioned above dissociate into protons (H^+ ions) and ions or molecules capable of combining with protons, i.e., bases. According to Bronsted theory, an acid dissociates into a proton and a base. An acid and its corresponding base are called "conjugate", i.e., joined in pair. According to this view, the weakest acid has the strongest conjugate base and the strongest acid has the weakest conjugate base.

According to the number of protons accepted by a molecule of the base, it may be grouped into (1) monoacidic or (2) polyacidic. In case of monoacidic base B ionizing in dilute aqueous solution into conjugate acid BH^+ and OH^-,

$$B + H_2O \rightleftharpoons BH^+ + OH^-$$

(Monoacidic base)

$$\therefore \quad K_{eq} = \frac{[BH^+][OH^-]}{[H_2O][B]}$$

$$\therefore \quad \frac{[BH^+][OH^-]}{[H_2O][B]} = K_{eq}\,[H_2O] = K_b$$

where K_b is temperature-dependent ionizing constant called the "basic dissociation constant" involving negative power of 10. K_b is conveniently expressed as pK_b known as the "ionization exponent of a base".

$$pK_b = -\log K_b = \log (1/K_b)$$

where K_b of a base is a measure of strength i.e., the magnitude of its proton acceptance. The stronger the base, (a) the greater is the proportion of its molecules changed to its conjugate acid, (b) the higher is the pH, (c) the lower is its pK_b, (d) the higher is its K_b.

According to Bronsted's theory, the strength of bases depends solely upon the (a) readiness of their combination with protons and the (b) strength of valence bond formed. If all acids were listed from the weakest to the strongest, then the conjugate bases of these acids would represent a series, in reverse order, according to the basic strength. Thus Cl^- is a very weak base and CN^- a very strong one.

The strength of a base can also be expressed by the pK_a of its conjugate acid.

Bronsted stated that the bases are proton acceptor, e.g., NH_3 (base) accepts a proton to form NH_4^+, OH^- ions (base accepts a proton to form H_2O). Here NH_4^+ and H_2O are called conjugate acid of the base NH_3 and OH^-. For expression of base strength as pK_a value of the conjugate acid i.e., to express the base strength of ammonia, instead of expressing it in terms of K_b for NH_3, better to express it in terms of pK_a of the acid NH_4^+, the conjugate acid of NH_3.

$$\therefore \quad \frac{[NH_4^+][OH^-]}{[NH_3]} = K_a = \frac{[NH_3][H^+]}{[NH_4^+]}$$

$$\therefore \quad pK_a + pK_b = pK_w$$

If ammonia at 25°C has a basic dissociation constant, K_b of 10^{-5}, its pK_b is 5 and its pK_a is:

$pK_w - 5 = 14 - 5 = 9$; so, ammonia has a pK_a of 9.

The higher the pK_a, the lower is the acid strength and, therefore, higher the base strength.

It is understood that we can express both acids and bases as acids and their strengths in term of the exponent (power) of their acid dissociation constant which is simple and rational.

Table 6.5: *pK* values of bases at 25° C

Base	pK_b	pK_a
Diethylamine	2.90	11.1
Ethylamine	3.34	10.66
Ammonia	4.74	9.26
Pyridine	8.64	5.36
Aniline	9.30	4.70
Diphenylamine	13.16	0.84

Bronsted's theory leads us to the broad concept of neutralization, which is ordinarily considered as the reaction of H^+ with OH^- to form H_2O. Since OH^- is only one of many bases capable of combining with protons, the process of neutralization may be more completely defined as the combination of base with protons.

6.5.9 Salts

A salt is a compound of a metal radical or metal-acting radical and an acid radical.

They can be made readily from reactions between bases, which provide the desired metal radicals, and acids which provide the acid radicals, e.g.,

Base	+ Acid		Salt	+ Water
2NaOH	+ H_2SO_4	⟶	Na_2SO_4	+ H_2O
(Sodium hydroxide)	(Sulphuric acid)		(Sodium Sulphate)	
↑	↑			
(Provides Sodium radical)	(Provides Sulphate radical)			

The nomenclature of the salts is done by naming the salts which comprises of specifying the two radicals that are present, the metal (or metal-acting) and the acid radicals carrying the same names as they do when they contribute to acids.

For example:

Metal .... ides- (or metal-acting)	Sodium chloride (NaCl) Ammonium Chloride (NH_4Cl)
,, ... ates-	Sodium chlorate (NH_4ClO_3) Ammonium Chlorate (NH_4ClO_3)
,, ... ites-	Sodium chlorite ($NaClO_2$) Ammonium Chlorite (NH_4ClO_2)
,, ... ites-	Sodium hypochlorite (NaClO) Ammonium Chlorite (NH_4ClO)
,, ... ates-	Potassium permanganate

Table 6.6: Names of some salts

Name of Salt	Formula
Sodium Chloride	NaCl
Calcium Chloride	$CaCl_2$
Sodium Carbonate	Na_2CO_3
Sodium bicarbonate	$NaHCO_3$
Calcium Carbonate	$CaCO_3$
Calcium Phosphate	$Ca_3(PO_4)_2$
Magnesium Sulfate	$MgSO_4$
Mercurous Chloride	HgCl
Mercuric Chloride	$HgCl_2$
Ethyl Chloride	C_2H_5Cl
Sodium Bromide	NaBr
Potassium Bromide	KBr
Ammonium Bromide	NH_4Br
Ferrous Carbonate	$FeCO_3$
Silver Nitrate	$AgNO_3$
Sodium Nitrite	$NaNO_2$
Calcium Sulphate	$CaSO_4$
Sodium Stearate	$NaC_{17}H_{35}CO_2$
Zinc Stearate	$Zn(C_{17}H_{35}CO_2)_2$
Glyceryl Stearate	$C_3H_5(C_{17}H_{35}CO_2)_3$

6.5.9.1 Some characteristics of salt solution

1. **Ionization:** In aqueous solution, the salt dissociates into its constituent as anion and a cation. Thus:

$$\underset{}{NaCl} \rightarrow \underset{(Cation)}{Na^+} + \underset{(Anion)}{Cl^-}$$

Therefore, salts are ionic compounds.

2. **Reaction of the salt solution**: The aqueous solution of salt may be neutral, acidic or alkaline, e.g., aqueous solution of NH_4Cl is acidic because NH_4Cl ionized into NH_4^+ (the strong conjugate acid of the weak base NH_3) and Cl^- (neutral or very weak conjugate base of the strong acid HCl).

3. **Water in hydrolysis with salts**: When water and a salt react in hydrolysis, the hydrogen radical from water combines with the acid radical from the salt to make an acid product; at the same time the hydroxyl radical combines with the metal radical to make a base product

$$\text{Water} + \text{Salt} \rightarrow \text{Acid} + \text{Base}$$

A portion of each salt only reacts with water and as a result, some salt always remain in solution together with the base and acid made. From the above reactions certain consequences are known.

(a) **Four compounds are present**: The relative quantity of the salt which reacts with the water depends upon the particular salt. With some salts, the reaction does not get very far, in which, there will be present within each solution a relatively high concentration of the remaining salt and a low concentration of the base and acid products. With some salts, the reaction is carried almost to completion until very little of the initial salt remains, while concentrations of the base and acid products will be relatively high. But in either cases, even four substances are always present:

i. The **remaining water**: Considerable in quantity because of the great excess water used to obtain the solution,

ii. The **remaining water**: More or less in relative quantity, depending upon the particular salt,

iii. The **base product** and
iv. The **acid product**
} in relative quantities that depend upon the extent of the reaction.

Within solution in which salt hydrolysis occurring, important consequences are seen from the comparative character of the acid and base products, i.e., whether the base is strong and the acid is weak or whether they are both strong.

(b) **Strong base and weak acid**: Within such a solution, although the acid is present, its effect is counteracted by the presence of the strong base. The solution is, therefore, "basic in reaction" (i.e., alkaline), if tested with pink and blue litmus, the presence of the base will be revealed but not that of acid and this base will react like a base regardless of the presence of the other three compounds. Because such a salt ionizes into a very weak conjugate acid cation or a neutral metal cations and a strong conjugate base of the parent weak acid, the strong conjugate base anion absorbs significant amounts of H^+ from water to make the solution alkaline. Some commonly used salts which react with water in hydrolysis to make such associated products:

Sodium carbonate	Potassium carbonate
Sodium phosphate	Potassium phosphate
Sodium bicarbonate	Potassium bicarbonate
Sodium stearate	Potassium stearate
Sodium palmitate	Potassium palmitate
Sodium oleate	Potassium oleate

- **Importance of strong base and weak acid in water**

i. **Physiologically**: Sodium bicarbonate salt is present in solution in lymph and blood in considerable amounts and reaction of these fluids is maintained a little on the basic side (pH 7.4) due to the hydrolysis reaction between water and this salt to make sodium hydroxide, a strong base, and carbonic acid, which is weak.

$$NaHCO_3 + H_2O \rightarrow NaOH + H_2CO_3$$

(Sodium bicarbonate) (Water) (Sodium hydroxide) (Carbonic acid)

ii. **Therapeutically**: Sodium bicarbonate is advocated for temporary correction of hyperacidity of the stomach. It reacts with water in hydrolysis to make sodium hydroxide, then this base reacts with some of the HCl of the gastric juice to reduce acidity of the gastric juice.

$$NaOH + HCl \rightarrow H_2O + NaCl$$

(Sodium hydroxide) (Hydrochloric acid) (Water) (Sodium Chloride)

iii. **Softening of hard water**: Sodium carbonate and Sodium phosphate are put into hard water for reacting with substances that render water hard. But if excess salts are added, this excess of either salts reacts now with the water to make strong base sodium hydroxide.

Softening salts

Sodium Carbonate + Water →
Sodium Hydroxide (strong base) + Carbonic acid (weak acid)

Sodium Phosphate + Water →
Sodium Hydroxide (Strong base) + Phosphoric acid (weak acid)

(c) **Strong acid and weak base**: Within such a solution, although base is present, it is the acid product which makes the solution acid in reaction due to the higher acid strength of the conjugate acid cation than the basic strength of the conjugate base anion.

Salt				**Weak base**		**Strong acid**
Ammonium Chloride (NH_4Cl)	+	Water	→	Ammonium Hydroxide	+	HCl

(d) **Strong acids and strong bases**: When these salts react with water in hydrolysis, they produce strong bases and strong acids. Such salts are called "neutral salts" since solutions in which the hydrolysis take place are "neutral in reaction" because such a salt ionizes into a very weak conjugate base anion of the parent acid and either a very weak conjugate acid cation or a neutral metal cation.

Salt				**Strong base**		**Strong acid**
Sodium chloride	+	Water	→	Sodium Hydroxide	+	HCl
Potassium chloride	+	Water	→	Potassium Hydroxide	+	HCl

(e) **Weak acids and weak bases:** The reaction between a weak acid and a weak base, produces a strong conjugate base and a strong conjugate acid on ionization; as both these ions can affect H^+ concentration, the solution of such salt may be neutral, acidic or alkaline.

4. **Salt hydrates:** Salt hydrates are made more or less readily by direct combination between water and the salts concerned.

Water	+	Salts	→	Hydrates of salts
$CuSO_4$ (Copper Sulphate)	+	$5H_2O$	→	$CuSO_4.5H_2O$ Copper Sulphate hydrate

Some common hydrates are:

Calcium Chloride hydrate	: $CaCl_2.6\ H_2O$
Copper Sulfate hydrate	: $CuSO_4.5\ H_2O$
Sodium Carbonate hydrate	: $Na_2CO_3.10\ H_2O$
Magnesium Salphate hydrate	: $MgSO_4.7\ H_2O$.

Salts in reaction with metals: Whether a salt in solution will or will not react with a metal substance depends on whether the metal element within the metal substance is more active or less active than the metal element within the salt. Thus:

Zinc (element) (less active) + Sodium Chloride → No reaction

Zinc (element) (more active) + Lead Nitrate → Lead + Zinc Nitrate

6.5.9.2 Importance of salts

I. Laboratory

Apparatus made of glass or porcelain should be used rather than of metals.

1. Aluminium beakers, bottles, etc. could be used for solutions of magnesium, calcium, sodium or potassium salts without any reactions.
2. Salts containing zinc, chromium, iron or any metal less active than aluminium, reactions with aluminium container will take place.
3. If the apparatus is made of chromium or of nickel, reaction would not take place between either of these substances and salts containing metal elements more active than the elements chromium or nickel.

II. Domestic

The salts and metal substances which we commonly encounter in daily life are:

Source		Salts
Meats and Vegetables.	→	Sodium Chloride, Potassium Chloride, Calcium Chloride, Magnesium Chloride.
Beans, eggs, oatmeal, oysters, spinach.	→	Iron Salts.

Metals	Presence
Aluminium, Zinc	Containers, spoons, tubs, garbage pails etc.
Iron	Cooking utensils.
Nickel, Silver	Utensils, cookery, cutlery.

1. Foods containing Sodium, Potassium, Calcium and Magnesium salts are cooked in containers made of any above metal and silverware is used for handling these foods, there will be no reactions.
2. When foods containing iron salts, are cooked in aluminium utensils, some reaction can take place between said salts and the aluminium. Thus:

Sodium Salt + Aluminium → No reaction.

Magnesium Salt + Aluminium → No reaction.

Iron Salt + Aluminium → iron + Aluminium Salt.

Iron Salt + Nickel → No reaction.

III. Clinic and hospital

In clinic and hospital practice, we handle some salts with metal substances. Some of the common salts and metals which are encountered in daily practice are given below.

1. Sodium chloride and Magnesium sulphate dressing and packs can be handled with chromium or nickel-plated instruments without any reaction.
2. Mercuric chloride dressing should be handled rapidly with these instruments and should not be kept for a long time in the mercury salt solution.

Table 6.7: Salts, their use and metal in instruments.

Salts	Use	Metals
Potassium Permanganate	Irrigation of wounds	Chromium plated instruments, table tops etc.
Magnesium Sulphate, Sodium Chloride	Dressing and packs	Nickel-plated instruments etc.
Mercuric Chloride,	Dressings	
Silver nitrate	Irrigation	

3. The same precaution should be taken in case of silver nitrate solution.

Magnesium sulphate + Nickel → No reaction.

Mercuric chloride + Nickel → Mercury + Nickel chloride

Silver nitrate + Chromium → Silver + Chromium nitrate.

IV. As electrolytes

Acid, bases and salts are the only compounds whose water solutions conduct the electric current appreciably; and for this reason, these compounds are called electrolytes. They are of two types:

1. Strong electrolytes, e.g., strong acids, strong bases and salts.
2. Weak electrolytes, e.g. solutions of weak organic acid and bases are feeble conductor of electricity. Nearly all salts are strong electrolytes, hence in water solution the salt molecules relatively ionized completely, the ions concerned being metal ions or ammonium ions and acid radical ions. Some of these ionization reactions are given below:

$$NaCl \rightarrow Na^+ + Cl^-$$
$$KCl \rightarrow K^+ + Cl^-$$

V. Absence of common ions

There is not one common ion substance that is supplied by all salts in water solution. Metal ions and acid radical ions must be present within solutions of all salts, but there can be different acid radical ions.

VI. Salts and soluble proteinates

Proteins, in their acid capacity, react with salts in the manner of all acids, to make new acids, and physiologically such reactions are constantly taking place between the proteins of cells, lymph and blood and the various salts are present; and the proteinate thus made are more soluble compounds than are proteins themselves. But in vitro, this type of reaction may result in insoluble salt of protein.

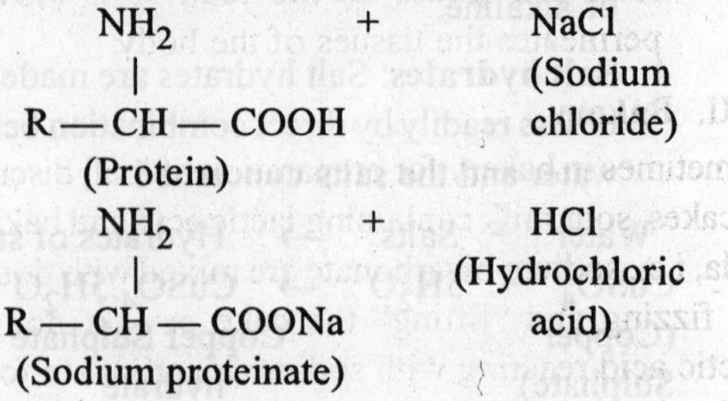

VII. Salts and insoluble proteinate

When silver nitrate solution reacts with solution of proteinate, a white insoluble precipitate of silver proteinate appears. Some salts are used for precipitation of proteins by full or half saturation.

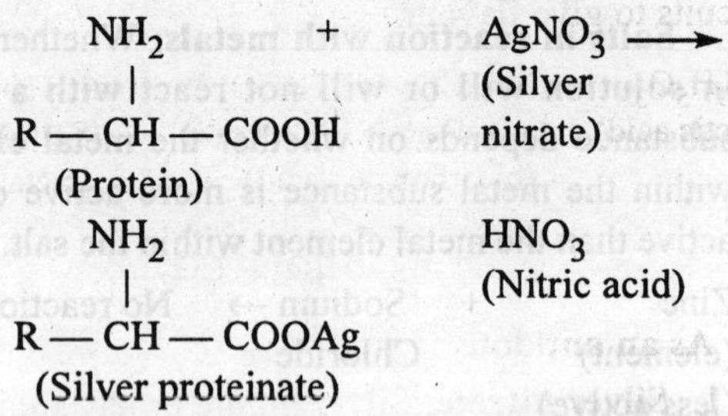

Consequences

(a) These insoluble protein salts are more or less caustic, so when they come into contact with albumins and globulins of cell protoplasm, they damage these proteins with the resulting death of the cells.

(b) Solution of silver nitrate and mercuric chloride and other salts are used as disinfectants and mode of action is to destroy the proteins of bacterial cells.

(c) Sometimes a silver nitrate pencil (lunar caustic) is used to destroy the excess granular tissue around the wound in similar manner.

(d) Accidental intake of such salts as mercuric chloride or silver nitrate causes poisonous effect due to reaction with protein of the digestive tract. The effect of plenty of egg albumen or milk as "antidote" provide proteins for the salts to react with instead of those of the tissue cells.

(e) The toxic effect due to absorption of lead salts, is due to in part production of insoluble lead proteinate as the lead salts slowly permeates the tissues of the body.

VIII. Bakery

Sometimes in bakery for preparation of loaf, biscuits or cakes, sour milk containing lactic acid and baking soda, i.e., sodium bicarbonate are mixed with dough for fizzing and "lifting" the cake or loaf dough. Lactic acid reacting with sodium bicarbonate form carbonic acid which is unstable and decomposes into water and carbon dioxide causing initial fizzing at ordinary temperature. But the loaf is put into oven at high temperature, the remaining carbonic acid decomposes and CO_2 make "lifts" or aerates the dough mass. Sodium lactate only remains in the biscuits to give flavour.

$$\underset{\text{(Lactic acid)}}{HC_3H_5O_3} + \underset{\text{(Sodium bicarbonate)}}{NaHCO_3} \rightarrow \underset{\text{(Carbonic acid)}}{H_2CO_3} + \underset{\text{(Sodium Lactate)}}{NaC_3H_5O_3}$$

$$H_2CO_3 \downarrow H_2O + CO_2$$

IX. As an antidote

1. **Silver nitrate**: Silver nitrate or any soluble silver salt is poison to man and prescribed antidote is a solution of sodium chloride to be given plentifully and rapidly to destroy the silver nitrate before entering the cells. Sodium chloride which is slightly soluble.

$$AgNO_3 + NaCl \rightarrow AgCl_2 + NaNO_3$$

2. In case of newborn baby, one per cent silver nitrate is dropped into the eyes to destroy gonococci. Then the eyes are irrigated with physiological normal saline (8.5 gm per litre) to prevent any injury to the cell by silver nitrate.

6.5.9.3 Type of salts

1. **Normal Salts**

Most salts are normal salts in that they have as constituents only metal radicals and acid radicals.

e.g. $NaCl \rightarrow Na^+ + Cl^-$

$KCl \rightarrow K^+ + Cl^-$

$Na_2SO_4 \rightarrow 2Na^+ + SO_4^{--}$

$CuSO_4 \rightarrow Cu^{++} + SO_4^{--}$

$NH_4NO_3 \rightarrow NH_4^+ + NO_3^-$

2. **Acid Salts**

Acid salts are those in which part of the hydrogen radical, originally in the acid from which the salt was made, remains in more or less close combination with the acid radical, and as long as this combination persists, this hydrogen constitutes a part of the acid radical, e.g., Sodium bicarbonate. The acid salts are:

Sodium bicarbonate	$NaHCO_3$
Sodium monohydrogen phosphate	Na_2HPO_4
Potassium monohydrogen phosphate	K_2HPO_4
Sodium dihydrogen phosphate	NaH_2PO_4
Potassium dihydrogen phosphate	KH_2PO_4

3. **Organic salts**

Salts that are composed of metal or ammonium radicals in combination with organic acid radicals are classed as organic salts, e.g.,

Sodium acetate	$NaC_2H_3O_2$	CH_3—COONa
Ammonium oxalate	$(NH_4)_2\ C_2O_4$	COO — NH_4 | COO — NH_4
Potassium tartarate	$K_2C_4H_4O_6$	CHOK | CHOH | CHOH | COOK

Organic salts react chemically as do all salts. They are weak electrolyes, e.g.,

$$\underset{\text{(Sodium acetate) (Undissociated)}}{NaC_2H_3O_2} \rightarrow \underset{\text{(Sodium ions)}}{Na^+} + \underset{\text{(Acetate ions)}}{C_2H_3O_2^-}$$

They react with acid, e.g.,

$$\underset{\text{(Sodium acetate)}}{2CH_3COONa} + \underset{\text{(Sulphuric acid)}}{H_2SO_4} \rightarrow \underset{\text{(Sodium Sulphate)}}{Na_2SO_4} + \underset{\text{(Acetic acid)}}{2CH_3COOH}$$

They readily undergo hydrolysis, e.g.,

$$\begin{array}{l} COO-NH_4 \\ | \\ COO-NH_4 \end{array} + 2H_2O \rightarrow 2NH_4OH + \begin{array}{l} COO-H \\ | \\ COO-H \end{array}$$

(Ammonium oxalate) (Water) (Ammonium hydroxide) (Oxalic acid)

4. Esters

Esters are organic salts that contain hydrocarbon as metal-acting radicals in combination with either inorganic or organic acid radicals, e.g.,

Methyl Chloride — CH_3Cl

Ethyl acetate — $C_2H_5C_2H_3O_2$

Glyceryl stearate — $C_3H_5(C_{17}H_{35}CO_2)_3$

Glyceryl nitrate — $C_3H_5(NO_3)_3$

They are not electrolytes. In water, hydrolysis occurs in the presence of a dilute acid or base functioning as catalyst.

$$\underset{(C_2H_5Cl)}{\text{Ethyl chloride}} + H_2O \xrightarrow{\text{Acid or base}} \underset{\text{(Ethyl alcohol)}}{C_2H_5OH} + \underset{\text{(Hydrochloric acid)}}{HCl}$$

Glyceryl tretyrate $C_3H_5(C_3H_7COO)_3$

Glyceryl palmitate $C_3H_5(C_{15}H_{31}COO)_3$

Glyceryl oleate $C_3H_5(C_{17}H_{33}COO)_3$

Thus, fats are esters in which the hydrocarbon radical is always the glyceryl radical ($C_3H_5^{+++}$), but the fatty acid radicals can vary. These esters are present in the natural fats, some are used as a local anaesthetic (ethyl chloride), in angina pectoralis (amyl nitrite), etc.

6.6 Reaction of Acids, Bases and Salts with Each Other

Due to significant similarities in reactions among acids, bases and salts, it is worthy to discuss it separately.

Acid and Base: When solutions of an acid and a base are mixed, due to reaction, water and salt are produced. The salt contains the metal radical of the base and the acid radical of the acid, thus:

$$\underset{\text{(Acid)}}{HCl} + \underset{\text{(Base)}}{NaOH} \rightleftharpoons \underset{\text{(Salt)}}{NaCl} + H_2O$$

The reaction of the acid and base can take place reversibly to make acid and base in which a smaller or larger portion of a salt reacts with water in double decomposition, or hydrolysis, to make acid and a base. This indicates that no sooner have a base and an acid reacted to make water and a salt, than some of the salt product, if soluble, begins to react with the water to remake portions of the acid and base. The extent of the reverse reaction depends on (1) the strength i.e., chemical activity of the acid and base, and (2) whether the water and salt are kept in contact.

Here the reaction is carried almost to completion toward the making of water and sodium chloride and when the two reactions attain equilibrium, concentrations of the water and NaCl products are relatively high, while those of the initial products (i.e., HCl and NaOH) are very low. But if the water is continuously evaporated, reaction is then carried entirely to completion in the direction of making of water and salt.

6.6.1 Importance

1. Acid poisoning

In case of acid poisoning, a base is advocated as an antidote and be given as early as possible. The bases are calcium hydroxide, a component of lime water, magnesium hydroxide, a component of milk of magnesia.

$$\underset{\text{(Oxalic acid)}}{H_2C_2O_4} + \underset{\text{(Calcium hydroxide)}}{Ca(OH)_2} \rightarrow \underset{\text{(Calcium oxalate)}}{CaC_2O_4} + \underset{\text{(Water)}}{2H_2O}$$

$$\underset{\text{(Hydrochloric acid)}}{2HCl} + \underset{\text{(Magnesium Hydroxide)}}{Mg(OH)_2} \rightarrow \underset{\text{(Magnesium Chloride)}}{MgCl_2} + \underset{\text{(Water)}}{2H_2O}$$

2. Strong base poisoning

In case of strong base poisoning, an acid is administered quickly as an antidote. For this purpose, acetic acid, the acid component of vinegar, or the weak acid of lemon juice is usually given.

3. Physiologically

This reaction is necessary in the body to prevent the excess acids or bases in the cells, blood and lymph. For example, hydrochloric acid, sodium hydroxide, potassium hydroxide and other acids and bases are always present in the body fluids; carbonic acid, phosphoric acid, sulphuric acid and ammonium hydroxide are produced in the body from the metabolic reactions. But none of these products attain high concentration because of various reactions among the acids and bases to make water and salt.

I. Acid and salt

When an acid and salt having different acid radicals react in water solution, another acid and salt are produced, such as:

$$H_2SO_4 + 2NaNO_3 \rightarrow Na_2SO_4 + 2HNO_3$$

H_2SO_4	$2NaNO_3$	Na_2SO_4	$2HNO_3$
(Sulphuric acid)	(Sodium Nitrate)	(Sodium sulphate)	(Nitric acid)

II. Base and Salt

If solutions of a base and a salt having different metal radicals are allowed to react, another base and salt are produced.

$$Ca(OH)_2 + 2NH_4Cl \rightarrow CaCl_2 + 2NH_4OH$$

$Ca(OH)_2$	$2NH_4Cl$	$CaCl_2$	$2NH_4OH$
(Calcium hydroxide)	(Ammonium chloride)	(Calcium chloride)	(Ammonium hydroxide)

III. Salt and salt

If a salt reacts with another salt having different metal radicals and acid radicals in water solution, two different salts are formed.

$$NaNH_3 + KCl \rightarrow NaCl + KNO_3$$

$NaNH_3$	KCl	$NaCl$	KNO_3
(Sodium nitrate)	(Potassium chloride)	(Sodium chloride)	(Potassium nitrate)

- **Reversible reactions**: In the above four groups, all reactions are reversible. In every reaction, when the products are kept in contact with each other, they react to a greater or lesser extent to remake certain amount of original compounds. Then we can infer that within such a solution there are four compounds and two opposing reactions to establish an equilibrium or an equality of reaction speed.

- **Importance of the reactions**

1. **Softening of the hard water**: Hard water contains some soluble salts, e.g.

Calcium chloride	$CaCl_2$
Calcium sulphate	$CaSO_4$
Calcium hydrogen carbonate	$Ca(HCO_3)_2$
Magnesium salts.	

When sodium bicarbonate is added, the salts of hard water become insoluble calcium or magnesium salts and make the water soft. Thus,

(a)

$$CaCl_2 + Na_2CO_3 \rightarrow CaCO_3 + 2NaCl$$

$CaCl_2$	Na_2CO_3	$CaCO_3$	$2NaCl$
(Calcium chloride)	(Sodium carbonate)	(Calcium carbonate)	(Sodium chloride)

(b)

$$Ca(HCO_3)_2 + Na_2CO_3 \rightarrow CaCO_3 + 2NaHCO_3$$

$Ca(HCO_3)_2$	Na_2CO_3	$CaCO_3$	$2NaHCO_3$
(Calcium hydrogen carbonate)	(Sodium carbonate)	(Calcium carbonate)	(Sodium hydrogen carbonate)

2. **In Bakery**: When sour milk and baking soda are mixed together in dough for preparation of cake or biscuit, at first there will be hissing sound and then swelling of cake or biscuit in the hot oven. This reaction is according to the reactions between acid and salts which is carbonate. The sour milk contains lactic acid which reacting with baking soda (sodium bicarbonate), carbonic acid is formed, which, being unstable, decomposes into water and carbon dioxide, for which there is hissing sound initially. At ordinary temperatures, the effervescence is completed, some carbonic acid is left in the dough but, when it is placed in the oven, the high temperature favours decomposition of the remaining carbonic acid and escape of CO_2, causing swelling of the dough. Eventually, there is no excess lactic acid or sodium bicarbonate in the baked products but sodium lactate is left which gives flavour to the baked products.

$$HC_5H_2O_3 + NaHCO_3 \rightarrow NaC_3H_5O_2 + H_2CO_3$$

$HC_5H_2O_3$	$NaHCO_3$	$NaC_3H_5O_2$	H_2CO_3
(Lactic acid)	(Sodium bicarbonate)	(Sodium Lactate)	(Carbonic acid)

$$H_2CO_3 \downarrow H_2O + CO_2$$

3. **Disinfecting function of salts**: When silver nitrate or mercuric chloride is used as a disinfectant, it destroys bacteria because like other proteins, bacterial proteins are acid (HPr, i.e., Hydrogen proteinate), it reacts

with salt to form another salt and acid. Since silver proteinates and mercuric proteinates are insoluble, the forward reactions continue; hence destruction of bacterial proteins is more active than the reverse reaction and continues as long as salt is there.

Silver nitrate + Hydrogen proteinate	→	Silver proteinate Hydrogen nitrate
Mercuric chloride + Hydrogen proteinate	→	Mercuric proteinate + Hydrogen chloride

4. **Reactions against antidote**
   (a) Silver nitrate is fatal to proteins of human cells as to those of bacteria; hence soluble silver salt is poison. In case of accidental poisoning, the prescribed antidote is a solution of sodium chloride in plenty and rapidly to make it silver chloride before entering the cells. Silver chloride is very slightly soluble which is precipitated outside the cell and may pass through cell membrane to attack proteins. Thus:

$$AgNO_3 + NaCl \rightarrow AgCl + NaNO_3$$

   (b) Antidote reaction in the eye: In another instance, solution of sodium chloride is used as antidote for silver nitrate. It is the usual procedure to instill silver nitrate (1 per cent) into the eyes of newborn babies as a prophylactic measure to destroy gonococci, if present. Then silver nitrate is irrigated with physiological normal saline (0.9 per cent) to destroy the silver nitrate and prevent tissue damage.

5. **Physiologically**: Various acids, bases and salts are present in the blood, lymph and interstitial fluid. These either get entrance in the blood from the food in the intestine and some are formed in the cells and poured in the circulation. In the body there are reversible reactions between acid and base, acid and salt, base and salt, and salt and salt.

6.7 Biological Importance of Acids, Bases and Salts

6.7.1 Acid

A. Inorganic acids

1. **Hydrochloric acid**: Gastric juice contains hydrochloric acid. It has several functions:
   (a) Antibacterial due to acid pH.
   (b) Maintenance of acid pH of gastric juice.
   (c) Activation of pepsinogen to pepsin.
   (d) Stimulates gastrin hormone secretion.
   (e) Stimulates secretion of intrinsic factor.
   (f) Maintenance of blood pH by mobilising H^+ ion.
   (g) Helps conversion of ferric iron to ferrous iron.
   (h) Break-down of cane sugar may take place though it is hydrolysed by invertase of intestine.
   (i) Denaturation of protein molecules. As a result of contact with gastric HCl, proteins are denatured, i.e., tertiary protein structure is lost as a result of the destruction of hydrogen bonds. This allows the polypeptide chain to unfold, making it more accessible to the actions of proteolytic enzymes (Proteases).
   (j) Causes secretion of secretin and cholecystokinin hormones from duodenum.

2. **Carbonic acid**
   (a) It is a weak diprotic acid and formed in the RBC, gastric parietal cells, renal tubular cells. CO_2 produced by the oxidation of metabolites, some amount is carried in but major part reacts with water to make carbonic acid with the help of carbonic anhydrase provided by RBC.

$$2CO_2 + 2H_2O \rightleftharpoons 2H_2CO_3$$

Carbonic anhydrase / $2H^+ + 2HCO_3^-$ (↑↓)

↓

Hb.2H

↓

$2H^+ + 2HCO_3^-$

Exhaled in the lungs ↑

$$2CO_2 + H_2O \rightleftharpoons 2H_2CO_3$$

Carbonic acid is dissociated into proton ($2H^+$) and bicarbonate ion. Deoxygenated haemoglobin acts as a buffer by binding protons and delivering in the lungs. In the lungs the binding of oxygen by Hb, forces the proton off from the Hb.

The protons combine with bicarbonate ion generating carbonic acid which is hydrolysed by carbonic anhydrase into H_2O and CO_2, CO_2 is exhaled in the lungs. This phenomenon is called **"Bohr effect"**.

(b) $H_2CO_3^-$ forms salt $NaHCO_3$ which acts as buffer and is secreted in bile and intestinal juices giving alkaline reaction which is due to amphiprotic nature of HCO_3^-.

3. **Sulphuric acid**: Sulphuric acid is produced in the body from the sulphur containing amino acid metabolism. It is strong but present in small quantities. It is used up in the following ways:

(a) Enters into the molecules of acid mucopoly-saccharides, e.g., chondroitin-4-sulphate, dermatan sulphate, keratan sulphate, and heparin which are distributed in the cartilage, skin, cornea and lung respectively.

(b) Detoxication of certain substance which are excreted in the urine as ethereal sulphate, e.g., indican.

(c) Certain hormones, e.g., oestrogen and other steroids and drugs are excreted with sulphuric acid.

(d) Formation of Heparin Chomdroition, Sulphatide etc.

4. **Phosphoric Acid: Present in Phospho**-proteins, phospholipids, nucleo-proteins, neucleotides, DNA, phosphate buffer Na_2HPO_4, NaH_2PO_4, K_2HPO_4, and KH_2PO_4 and Calcium phosphate to form bones.

B. Organic acids

1. **Acetic acid**: Acetic acid is a weak acid and formed from amino acid and fatty acid metabolism. It is utilized for formation of amino acid and fatty acid. In case of ketosis, it is converted into acetoacetic acid β-hydroxybutyric acid and acetone which are excreted in the urine.

Acetic acid + COA → Acetyl COA → Kreb's cycle

2. **Pyruvic acid and lactic acid**: These are weak acids and formed from anoerobic glycolysis and utilized in the Kreb's cycle to form CO_2, H_2O and energy. Lactic acid is a component of sour milk.

3. **Citric acid**: It is weak acid and absorbed from the intestine during ingestion of citrous fruits and also formed from carbohydrate metabolism via Kreb's cycle to form succinic acid, malic acid and formic acid. Citric acid after absorption reacts with potassium hydroxide which is a strong base in the blood.

Potassium citrate + water (in the intestine) ⟶ Citric acid (weak) + Potassium hydroxide (strong) (Blood)

4. **Oxalic acid**: Oxalic acid, together with its salts, is widely distributed in leaves and fruits of plants. The sour taste of some leaves is due to its presence. It is the strongest of the organic acid and is excreted in the urine as sodium oxalate after absorption.

5. **Malic acid**: It is widely distributed in fruits, e.g., apples, peers, grapes, straw-berries and other fruits and absorbed by intestine.

6. **Tartaric acid**: Tartaric acid occurs in fruits and is especially abundant in grapes. It is absorbed by the intestine and reacts with potassium hydroxide in the blood.

Potassium tartarate + water (In the intestine) ⟶ Tartaric acid (weak) + Potassium hydroxide (In the blood)

7. **Fatty acids**: Fatty acids are made from the hydrolysis of the fats butyrin, palmitin, stearin and olein. They are butyric acid, palmitic acid, stearic acid and oleic acid. They are made as digestive products of fats in the intestine and as hydrolysis products in the cells. Butyric acid is present in rancid butter, in cheese. But there are other fatty acids, e.g., valeric, caproic, linoleic and linolenic acids.

They are utilised for formation of fat in the body, formation of amino acids and oxidised to CO_2, H_2O and energy.

8. **Bile acids**: Bile acids are weak acids and they are cholic acid, chenodeoxycholic acid which are produced in the liver from cholesterol called "primary bile acids". These are poured in the intestine as salts with bile. In the intestine, cholic acid is converted into deoxycholic acid and chenodeoxycholic acid into lithocholic acid by the activity of intestinal bacteria. These two are called "secondary bile acids". The primary and secondary bile acids are absorbed almost exclusively in the ileum except lithocholic acid because of its insolubility. These circulate via portal circulation—go to liver and are again poured into the intestine and called "enterohepatic circulation". Bile acids are formed from cholesterol in the liver and present as salts, e.g., sodium taurocholate and sodium glycholate in the bile.

- **Importance of bile acids**

(A) **Hydrotropic action**: It keeps insoluble cholesterol in solution in bile. When there is alteration of ratio between cholesterol and bile salts (normal 1 : 20 to 1 : 30) to 1 : 13, gall-stone formation takes place.

(B) **Fat digestion and absorption**: Fat digestion and absorption take place by the following mechanism.

- **Emulsifying agent (or Stabilizer)**

(a) Bile salts with cholesterol and soluble proteins, reduce the surface tension of water and fat is emulsified making larger surface area for enzymatic acitivity of lipase.

(b) Activating the lipase.

(c) Acting as a good solvent for insoluble fat for making suitable medium for enzyme activity.

(d) Making insoluble fatty acid into soluble bile acid-fatty acid complex (choleic acid) which is permeable to intestinal epithelial membrane and thereby help better absorption.

(C) **Intestinal movement**: Increase peristalsis of both small and large intestine and thus act as a laxative.

(D) **Cholagogue**: Reabsorbed bile salts stimulate liver for further secretion of bile.

9. **Vitamins**: Several vitamins e.g., nicotinic acid, pantothenic acid, folic acid, ascorbic acid and biotin (monocarboxylic acid) are weak organic acids.

6.7.2 Bases

1. NH_3

(a) NH_3 is a weak base formed in the renal tubular cells from glutamine and oxidative deamination of amino acids. It enters the tubular fluid where it combines with H^+ from H_2CO_3 of tubular cells to form NH_4^+. This NH_4^+ then replaces Na^+ in the NaCl salt. The Na^+ is reabsorbed by the H^+—Na^+ exchange mechanism and re-renters the plasma as $Na^+HCO_3^-$. The NH_4 is excreted in the urine as NH_4Cl—after taking Cl^- from NaCl.

(b) Detoxication by urea formation in the liver.

2. **Choline**: This nitrogenous base is present in the lecithin, sphingomyelin, plasmalogen, phosphatidyl choline and in the neurotransmitter actylcholine. It also acts as lipotropic factor.

6.7.3 Salts

The inorganic foodstuffs comprise the mineral salts and water. The salts are essential substances for life. Depletion of salts in the food of the animal shows loss of appetite, becomes ill and dies. If the animal

Table 6.8: Sources of Salt radicals

Salt radicals	Sources
A. Cations	
1. Sodium (Na^+)	Sodium chloride in the food, water.
2. Potassium (K^+)	Meat, fruits and vegetable.
3. Calcium (Ca^{2+})	Milk, vegetables.
4. Magnesium (Mg^{2+})	Cells of plants and animals, i.e., vegetables and meat.
5. Iron (Fe^{2+} or Fe^{3+})	Meats, egg yolk, green vegetables.
6. Copper (Cu^{2+})	Liver, oysters.

Contd.

B. Anions	
1. Chloride	From NaCl through food.
2. Carbonate or bicarbonate (HCO_3^-)	Formed in the process of cell metabolism from CO_2 and H_2O, as end products of carbohydrate, fat and protein metabolism.
3. Phosphates ($H_2PO_4^-$) (a) Phosphate (b) Monohydrogen Phosphate (c) Dihydrogen phosphate	From metabolism of nucleoproteins and phospholipids, absorption from the intestine during digestion of nucleoproteins and phospholipids of the food.
4. Sulphate (SO_4^{2-})	From sulphur containing aminoacids, e.g., cystine, methionine, cysteine during metabolism.
5. Iodide (I^-)	Sea salts, water, vege-tables, marine fishes.
6. Proteinate⁻ (Na Proteinate salt)	Cellular proteins.
7. Haemoglobinate⁻ (KHb)	RBC.

is on a state of salt-hunger, it becomes very weak and shows signs of paralysis. Foster has shown that during salt-hunger, there is continually giving off inorganic constituents in the urine. The amount of these was reduced when the animal was given rich organic foodstuffs. It indicates that some inorganic ions are held in a state of combined tissue constitutents. If the amount of rich foodstuffs is inadequate, the animal lives on its own tissue constituents and sets free salts which excrete in the urine. In case of miner's cramp, the worker drinks water but due to sweating losses NaCl and there is retention of water in the tissue causing cramp but drinking of water containing 10 grams of salt per gallon removes cramps.

6.7.3.1 Salt radicals and sources

Most common solutes found dissolved in water are mineral salts, compounds of a metal with a non-metal or non-metallic radical, e.g., NaCl.

When the salt dissolves in water, it dissociates into its constituent ions.

$$\underset{\text{(Sodium chloride)}}{NaCl} \longrightarrow \underset{\text{(Sodium ion)}}{Na^+} + \underset{\text{(Chloride ion)}}{Cl^-}$$

Salts are, therefore, ionic compounds and are decomposed by electricity and the free ions are capable of carrying electric currents—they are called electrolytes. The salts are having two radicals, e.g. cations and anions (Table 6.8).

6.7.3.2 Requirement

A large number of elements are essential for the maintenance of health and normal growth in very different amounts. The daily intake for an adult may be approximately at :

Table 6.9. Daily intake of salts

Calcium	0.7 g
Sodium	6.0 g
Potassium	4.0 g
Iron	21 mg
Phosphorus	1.25 g
Iodine	0.05 mg
Chloride	1.5-5
Zinc	12.0 mg

6.7.3.3 Functions

1. **Constituents of different chemical components**
   (a) Nitrogen, sulphur taken as nitrates and sulphates enter into the composition of proteins, some of which also contain phosphorus.
   (b) Phosphorus enters into the phospholipid molecules and ATP, GTP, UDP.
   (c) Iodine enters into the molecule of thyroxine in thyroid and iodopsin in the retina.
   (d) Calcium enters into the molecule of calmodulin and calcium phosphates in the bones.
   (e) Sulphur in chordroitin sulphate, dermatan sulphate, keratan sulphate, heparin.
   (f) Caseins (milk) and ovovitellin (egg yolk) contain phosphorus.
   (g) Ferritin (intestinal mucosa) contains Fe^{3+}.
   (h) Ceruloplasmin (plasma) contains Cu^{2+} and Cu^+.
2. **Structural constituents**
   (a) Calcium and phosphorus as calcium phosphate in the bones, teeth.
   (b) Calcium in the plant cell wall.

(c) Connective tissue fibres, epithelial tissue containing proteins have nitrogen, sulphur.

(d) Nucleic acids in the chromosomes contain nitrogen and phosphorus.

(e) Phospholipid in the cell membrane contains phosphorus.

(f) Myelin sheath contains lipids having phosphorus.

3. **Constituents of enzyme (Metalloenzymes)**

(a) Zn^{2+} containing enzyme—Carbonic anhydrase (RBC, parietal cells, renal tubular cells).

(b) Cytochrome oxidase (mitochondria) contains Fe^{3+}.

(c) Catalase, peroxidase, cytochromes, tryptophan oxidase, homogentisicase contain iron.

4. **Metal-activated enzymes or activator of enzymes**: The metal is bound less tightly, easily dissociable complexes with the metal and can easily release the metal without denaturation.

(a) Mg^{++}, Mn^{++} or Zn^{++} activates enolase.

(b) Mg^{++} and K^{+} activate pyruvate phosphatase.

(c) Mn^{++} for arginase, phosphoglucomutase, dipeptidases.

(d) Co for peptidases.

(e) Mg^{++} for peptidases, phosphatases, ATP-enzymes, e.g., hexokinase.

(f) Ca^{2+} for lecithinases A and C, lipases.

(g) Cu for tyrosinase, phenolase, ascorbic oxidase.

(h) Mo for xanthin oxidase, etc.

5. **Constituents of pigment**

(a) Fe — Haemoglobin, myoglobin, cytochromes and in animals- chlorocruorin, haemorythrin.

(b) Mg — Chlorophyll of plant

(c) Iodine — Iodopsin of retina

(d) Cu — Haemocyanin (worms, molluscs)

6. **Contractility of muscles**: The relative concentrations of sodium, potassium, calcium, magnesium and chloride are necessary for contractility of muscles.

Sodium, potassium, calcium are essential for normal rhythmic beating of the heart. Sodium salt initiates contractions and maintains osmotic pressure of the cell, calcium maintains contractions and potassium favours relaxation.

7. **Transmission of nerve impulse**: All these sodium, potassium, calcium and magnesium are essential for the transmission of nerve impulse in both central and autonomic nervous system. High concentration of potassium delay the transmission of nerve impulse. A decrease in calcium ion concentration below normal or an increase in magnesium ion concentration above normal acts as a depressant on the CNS affecting both sensory and motor nerves. Decreased calcium ion concentration causes irritability of the nerve, e.g. excitability, muscle twitching and tetany which is corrected by administration of calcium salts or parathormone according to the condition. An increase in magnesium concentration in the blood depresses the CNS for which, in certain neural conditions, magnesium sulphate is sometimes given intravenously to produce an anesthetic effect.

8. **Bioelectric potentials**: Sodium and potassium are mainly (as cations) and proteinate and chloride ion (as anions) play a role for bioelectric potentials. Due to semipermeability of the cell membrane there is unequal distribution of ions which makes the cell negative inside and positive outside.

9. **Maintenance of osmotic pressure**

(a) RBC is isotonic with 0.9 per cent NaCl.

(b) Isotonic salines, e.g. Mammalian Ringer, Tyrode, Locke, Amphibian Ringer etc. are prepared with different salts (e.g., NaCl, KCl, $CaCl_2$, $MgCl_2$ etc.) in different concentration for biological experiments to maintain osmotic pressure and other functions.

(c) In blood, plasma proteins (colloids) exert much lower osmotic pressure in solutions compared to ionizable mineral salts because the colloid particles are much larger and non-ionized.

10. **Digestive processes**

I. **Salivary**

(a) **Control of H-ion concentration of saliva**: Saliva maintains pH 6.6 to 7.1. Acids from foods or from bacterial action on food particles, or bases from the use of sodium bicarbonate or basic tooth-pastes or mouth-washes produce slight variation but within a short period it becomes normal. For maintenance of constant pH, the compounds responsible are carbonates and phosphates present in saliva. They act as buffers, in pairs: potassium hydrogen carbonate with carbonic acid, dipotassium hydrogen phosphate with potassium dihydrogen phosphate.

$$\begin{cases} KHCO_3 + H^+ \rightarrow H_2CO_3 + K^+ \\ H_2CO_3 + OH^- + K^+ \rightarrow KHCO_3 + HOH \end{cases}$$

$$\begin{cases} K_2HPO4 + H^+ \rightarrow KH_2PO_4 + K^+ \\ KH_2PO_4 + OH^- + K \rightarrow K_2HPO_4 + HOH \end{cases}$$

(b) **Favourable pH for salivary digestion by ptyalin**: The enzyme ptyalin is very much active in the acid medium of around pH6. It is reasonable to think optimum pH for catalysing hydrolysis of starch by ptyalin takes place in the stomach. But because of the presence of salts like potassium hydrogen carbonate and dipotassium hydrogen phosphate in the food entering stomach combine with H^+ ion of HC1 and allow the enzyme to continue its action in the stomach for a longer period (30-40 minutes) before the ptyalin is destroyed by gastric juice and accumulated chloride activates its activity.

(c) **Salt as activator of ptyalin**: The ptyalin (salivary amylase) is inactive in absence of chloride or bromide ions. Removal of salts by dialysis, renders it almost or quite inactive, activity is restored after addition of salts.

II. **Gastric digestion**

The gastric juice has pH 1.2 to 3.5 for optimum action of pepsin due to presence of HC1. When stomach is secreting HC1, there would be an accompanying increase of sodium bicarbonate or disodium phosphate. But the evidence has provided that intake of food is often followed very soon by change in the reaction of urine from acid to alkaline which indicates the regulatory mechanism of the body to prevent excess of OH^- on concentration in the blood. This increase in hydroxide (OH^-) ion concentration of the blood and urine is called "alkaline tide".

$$H_2CO_3 + NaCl \rightarrow HC1 + NaHCO_3$$

$$NaH_2PO_4 + NaCl \rightarrow HC1 + Na_2HPO_4$$

(Blood stream) (Stomach) (Blood stream)

III. **Intestinal digestion**

Acid chyme reaches the small intestine where salts, chiefly sodium bicarbonate, help to achieve proper pH favourable to the various enzymes. Sodium bicarbonate is poured into the duodenum from the liver (via bile), pancreas (via pancreatic juice) and intestinal glands (via intestinal juice) and mixed with the intestinal contents, through reactions with water, it is indirectly supplied OH^- ions. These OH^- ions, coming into contact of H^+ ions from HC1 carried along from the stomach with the food, make water, thereby decreasing H^+ ion concentration.

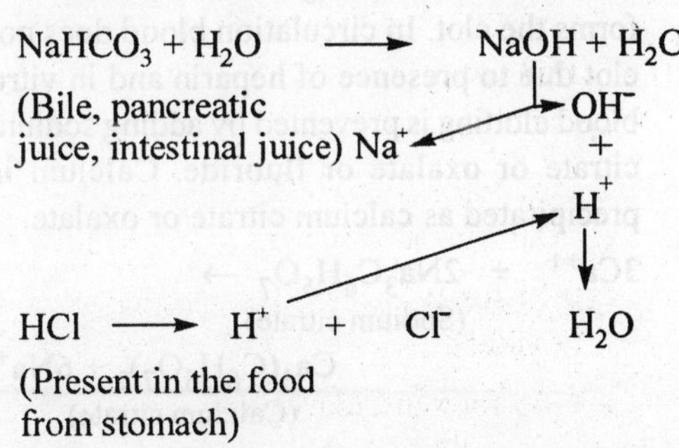

The process of mixing is gradual; hence decrease in H^+ ion concentration takes place gradually providing significant variations of pH at different parts of the intestine which favours the enzymatic activity at specific pH.

11. **Salts for building bones and teeth**: Bone and teeth cells collect calcium ions, phosphate ions and carbonate ions. About 99 per

cent of body calcium and 70 per cent of body phosphate are present in the bones and teeth. They are present as insoluble salts of calcium phosphate in larger proportion and calcium carbonate in smaller proportion to provide rigidity. These salts are not present separately nor in haphazard manner, but mutually assembled in a definite crystalline pattern.

12. **Haemoglobin and myoglobin formation:** Iron and copper salts are necessary for haemoglobin formation. Copper acts as a catalyst to speed up the formation of haemoglobin. Daily requirement of iron is about 21 mg and deficiency produces anemia. Ferrous salts are more readily utilisable than the ferric salts. The salts administered are:
   (a) **Inorganic salts**: ferrous carbonate, ferrous sulphate etc.
   (b) **Organic salts:** Ferrous succinate, ferrous fumerate. Apart from haemoglobin and myoglobin, iron enzymes, e.g. cytochromes, catalases, peroxidases, tryptophan 2, 3-deoxygenase contain iron.

13. **Clotting of blood**: Calcium salt is required for clotting of blood. Calcium salts act on prothrombin to form thrombin which acts on fibrinogen to form fibrin which is insoluble and entangles blood cells and forms the clot. In circulation blood does not clot due to presence of heparin and in vitro blood clotting is prevented by adding sodium citrate or oxalate or fluoride. Calcium is precipitated as calcium citrate or oxalate.

$$3Ca^{++} + \underset{\text{(Sodium citrate)}}{2Na_3C_6H_5O_7} \rightarrow \underset{\text{(Calcium citrate)}}{Ca_3(C_6H_5O_7)_2} + 6Na^+$$

14. **Iodides for Thyroid gland**: Thyroid gland makes use of iodides from the food and water or iodide salts given therapeutical-ly, as Lugol's solution, which is iodine dissolved in potassium iodide. Deficiency of iodide salts causes goitre. In regions where there is deficiency of iodine compounds, people take iodized salts as a protective measure. Iodide is required 100 μg to 140 ug daily for synthesis of thyroxine and triiodothyronine.

15. **Salts as "buffers" for blood, lymph and cells**: The blood, lymph and cells maintain a fairly constant pH despite the absorption of acids and bases, the entrance of strong and weak acids into the blood from the cells and continuous presence of salts of strong bases and weak acids. The buffer salts present in the blood, lymph and cells are chiefly bicarbonates, mono and dihydrogen phosphates and proteinates of sodium and potassium. The sodium salts predominate in the blood and lymph and potassium salts predominate in the cells.

   Various factors influences the efficacy of salt/salt and salt/acid buffer system such as (1) the relative concentration of the buffer pairs themselves and (2) the hydrogen ion concentrations of the medium. Buffer action of sodium bicarbonate/hydrogen carbonate is very much effective when the blood pH range is 7.3 – 7.5 and pair of phosphate salts is very much effective when the blood approaches the dangerous neutral pH 7.0.

Table 6.10: Salt of buffer pairs

Blood and lymph	Cells
1. Na_2HPO_4 and NaH_2PO_4	1. K_2HPO_4 and KH_2PO_4
2. $NaHCO_3$ and H_2CO_3	2. $KHCO_3$ and H_2CO_3
3. Sodium Proteinate (Na^+Pr^-) and Hydrogen Proteinate (H^+Pr^-)	3. Potassium proteinate (K^+Pr^-) and Hydrogen Proteinate (H^+Pr^-)
	4. Haemoglobin (K+Hb-and H^+Hb^-)
	5. $HbO_2^-/HHbO_2$

16. **Salts function in reversible reactions during transport of O_2 and CO_2**
   (A) **At the cells**: O_2 is released from oxyhaemoglobin to the cells which oxidises foodstuffs to CO_2 and H_2O. Due to higher concentration of CO_2 in the cells, it diffuses out to the lymph and then to the blood where major part is

converted to H_2CO_3 with the help of an enzyme carbonic anhydrase of RBC and minor amount of CO_2 is carried in solution and remaining part is carried as carbamino compound by combination of CO_2 with NH_2 of globin of haemoglo-bin.

$$CO_2 + H_2O \xrightarrow[\text{(from RBC)}]{\text{Carbonic anhydrase}} H_2CO_3$$

$$HbNH_2 + CO_2 \longrightarrow \underset{\text{(Carbamino compound)}}{HbNHCOOH}$$

The continuous production of H_2CO_3 (Carbonic acid) tends to react with salts in the blood, lymph and RBC to make bicarbonates chiefly sodium bicarbonate, in the blood and lymph and potassium bicarbonate in the RBC. Due to large quantities of CO_2 in the venous blood it causes slightly higher H-ion concentration than the arterial blood, and favours decomposition of oxyhaemoglobin. This facilitates escape of oxygen to the cells.

(B) **At the lungs**: When CO_2 - loaded blood reaches the lungs, its concentration is higher than CO_2 of the atmospheric air, so CO_2 evolves out of the blood in the lungs. At the same time, the concentration of O_2 entering the lungs being greater than that within the blood, O_2 from the air combines with haemoglobin to form oxyhaemoglobin which is a stronger acid and it reacts with potassium bicarbonate in the RBC to make more carbonic acid which decomposes to CO_2 and H_2O.

$$HHbO_2 + KHCO_3 \rightleftharpoons KHbO_2 + H_2CO_3 \xrightarrow{\text{Carbonic anhydrase}} CO_2 + H_2O$$

CO_2 goes out of the lungs

17. Selective reabsorption and elimination of salts by the kidneys

(a) **Phosphate mechanism**: Both Na_2HPO_4 and NaH_2PO_4 salts are brought to the kidneys by the blood and both are excreted in the urine. But in the blood Na_2HPO_4 predominates whereas in the urine NaH_2PO_4 predominates and the result is when the blood is in slightly basic side, the urine is acid in reaction (pH6). In the blood the salts are in the ratio of $Na_2HPO_4 : NaH_2PO_4 :: 4 : 1$ whereas in the urine the ratio is $Na_2HPO_4 : NaH_2PO_4 :: 1:9$. This difference is brought by reactions and reabsorption in the renal tubular cells in the following way:

$$\underset{\text{(Blood)}}{Na_2HPO_4 + H_2CO_3} \rightarrow \underset{\text{(Acid salt in the urine)}}{NaH_2PO_4} + \underset{\text{(Reabsorbed by the blood)}}{NaHCO_3}$$

When greater quantities of acids enter the blood, excessive amount NaH_2PO_4 salts appear in the urine. When the concentration of acid in blood is low, e.g., after taking food or persistent vomitting or when there is excess base making salts (e.g., organic sodium and potassium salts) in the food rich in fruits and vegetables or when large quantity of NaCl is taken medicinally, normal concentration of sodium bicarbonate in the blood is maintained without reabsorption by the kidney. During this period, formation of Na_2HPO_4 and $NaHCO_3$ is depressed or reversed and then Na_2HPO_4 appears in the urine and reaction is alkaline and the phenomenon is called "alkaline tide".

(b) **Ammonia mechanism**: Due to ion-exchange mechanism some urinary Na^+ ions are actively reabsorbed in renal tubules in exchange of H^+ ions secreted by tubular cells. The H^+ ions are mainly formed by the ionization of H_2CO_3 formed from $CO_2 + H_2O$. Sometime H^+ ions are buffered by the NH_3 synthesized and secreted by the tubular cells. NH_3 is derived from glutamine and asparagine with the help of the enzymes glutaminase and asparaginase. This NH_3 accepts H^+ ions to form NH_4^+ which is excreted as NH_4Cl or $(NH_4)_2SO_4$ taken from NaCl or Na_2SO_4 and Na^+ is returned to blood to form $NaHCO_3$.

18. **Excretion of salts**: NaCl is the salt that is excreted in the urine in largest quantity. It is also excreted through sweat. The salts in the normal urine may vary according to the type of salts and quantities. The following salts are normally present in greater or lesser quantities in the urine.

A. Inorganic

1. Sodium 2. Potassium 3. Ammonium	→ as →	1. Chloride 2. Dihydrogen phosphate 3. Monohydrogen phosphate 4. Sulphate 5. Carbonate (in alkaline urine)	C O M M O N

In traces

1. Calcium
2. Magnesium
3. Iron

B. Organic

1. Urates—Common
2. Oxalates
3. Citrates → more in alkaline urine.
4. Lactates

19. **Permeability of salts**: In aqueous solution, salts dissociate into ion, so the permeability of a membrane to a salt means the permeability to the individual ions of the salt. If the salt contains a weak base or weak acid, any penetration due to the migration of ions cannot be inferred, because it may be that the undissociated weak acid or base penetrates.

(a) $K^+Cl^- + Na^+NO_3^- \rightarrow Na^+Cl^- + K^+NO_3^-$
(cell) (Extra-cellular fluid)

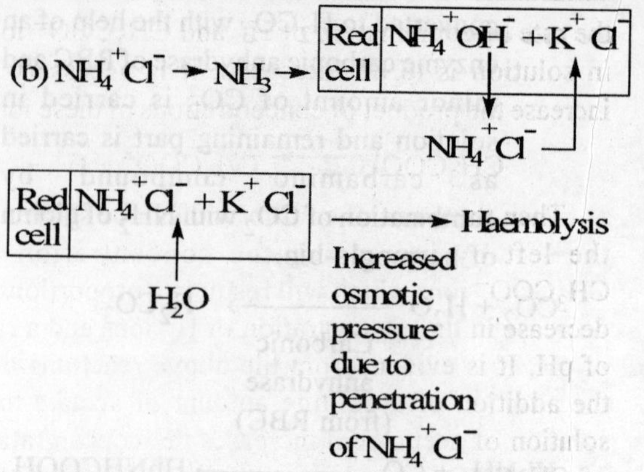

6.8 Buffers and Buffering System

It is necessary to have a stock solution of known pH for comparison, preparation and measurement of pH. When a solution is prepared by proper dilution of a standard acid or alkali it is seen that owing to impurities from the vessel or air, the value of pH of the solution slowly changes and never attains a steady value. So a solution of standard pH value is required provided:

1. Its pH value is accurately known.
2. It has some reserve power of neutralising acid or alkali so that it maintains a steady pH value i.e., long shelf life.

6.8.1 Principle (effect of salts upon the dissociation of acids)

When a salt of a weak acid is mixed with the weak acid in solution, the dissociation of the acid is reduced, and it shows less acidity and a higher pH than when the salt is absent. It is explained that salts of weak acids, as opposed to weak acids themselves, are generally highly ionized. For example, in a solution of acetic acid and sodium acetate, we have the equations for dissociation :

$$CH_3COOH \rightleftharpoons CH_3COO^- + H^+$$

$$CH_3COONa \rightleftharpoons CH_3COO^- + Na^+$$

The acetate ions from the dissociation of the acid are relatively few against the molecules of undissociated acid. Each hundred molecules of acetic acid produce less than two acetate and hydrogen ions. Again, the number of acetate ions from the sodium acetate is equal to the number of sodium acetate molecules dissolved. According to the Mass law, if the rate of combination of H^+ and CH_3COO^- ions in solution is to be increased, it is necessary to increase the product of concentrations of these ions.

$$CH_3COOH \rightleftharpoons CH_3COO^- + H^+$$

Then the point of equilibrium can be shifted to the left by increasing the concentration of CH_3COO^- ions which will result in a proportionate decrease in the concentration of H^+ ions and a rise of pH. It is evident from the above reactions that the addition of a definite amount of acetate to a solution of acetic acid increases the concentration of CH_3COO^- ions by a definite value, and this causes the concentration of hydrogen ions and pH of the solution to be fixed. Thus, it can be stated that the pH of a solution of a weak acid and its salt is determined by the ratio of salt to acid in the solution. The higher the salt concentration, the greater is the mass action in repressing the dissociation of the acid.

This principle is generally applied to solutions of weak acids in the presence of their salts and can also be applied in case of solutions of weak hydroxides and their salts.

6.8.2 Buffer solution

A buffer solution is a mixture of either a weak acid and its conjugate base or a weak base and its conjugate acid which can resist the change of pH when acid (a proton donor, such as hydrogen ion) or base (a proton acceptor) is added to it so that it maintains a steady value of pH without corresponding change in the reaction of the medium.

6.8.3 Example of buffer solution

1. Acetate buffer (Acetic acid : Na-acetate).
2. Bicarbonate buffer (H_2CO_3 : $NaHCO_3$). (Important for the fluids of the body)
3. Phosphate buffer (NaH_2PO_4 : Na_2HPO_4). (Important in the cell).
4. Borate buffer (Boric acid : Na-borate).
5. Ammonia buffer (NH_3 : NH_4CI).
6. Pyridine buffer (Pyridine : Pyridine).
7. Glycine buffer (Glycine : Glycine hydrochloride).
8. Glycine buffer (Glycine : NaOH).
9. Phosphate buffer (K_2HPO_4 : KH_2PO_4).
10. Pthalate buffer (Acid K. Pthalate : HC1).
11. Pthalate buffer (Acid K. Pthalate : NaOH).

6.8.4 Mechanism of buffer action

Buffer solutions generally contain a weak acid or base with its salts, e.g. acetic acid and sodium acetate solution. The mechanism of action of buffer is like Acetic acid and Sodium acetate (Acetate –Acetic buffer) which neutralises H^+ ion of acid when added in the following way :

Table 6.11: Range of Buffer Mixtures

Buffer Mixture	Range of pH
Glycine and HC1	1.15-3.68
Acid K. Pthalate and 1 HC1	2.2-3.8
Acid K. Pthalate and NaOH	4.0 - 6.2
Acetic acid and sodium acetate	3.24 - 6.55
$K_2HPO_4 + KH_2PO_4$	5.19-8.21
Boric acid, HC1 and NaOH	7.8 -10.0
Glycine + NaOH	8.58 - 12.97

$$H^+ + CH_3COO^- \rightarrow CH_3COOH$$

and counteracts alkalies (OH^- ion) when added as :

$$OH^- + CH_3OOH \rightarrow CH_3COO^- + H_2O$$

and maintains a steady pH. The dissociation reactions for members of a buffer pair are:

Weak acid HA : HA $\rightleftharpoons H^+ + A^-$

Conjugate base NaA : NaA $\longrightarrow Na^+ + A^-$

In such mixtures, all the anions A^- result from the conjugate base. According to the Law of Mass Action the equation is:

$$Ka = \frac{|H^+||A^-|}{|HA|}$$

where K_a is the apparent dissociation constant for the acid which can be rewritten as:

$$[H^+] = K_a \frac{|HA|}{|A|} \text{ (Henderson)}$$

or in its logarithmic form (Hasselbalch). Substituting pH for $-\log [H^+]$ and defining $pK_a = -\log K_a$, the so called Henderson-Hasselbalch equation is obtained for buffer solution:

$$pH = pK_a + \log \frac{|A^-|}{|HA|}$$

This equation can be used to illustrate how a weak acid functions as a buffer. By its use one may directly calculate the pH of a buffer solution if pK_a for the buffer acid and the molecular ratio of salt to acid in the solution are known.

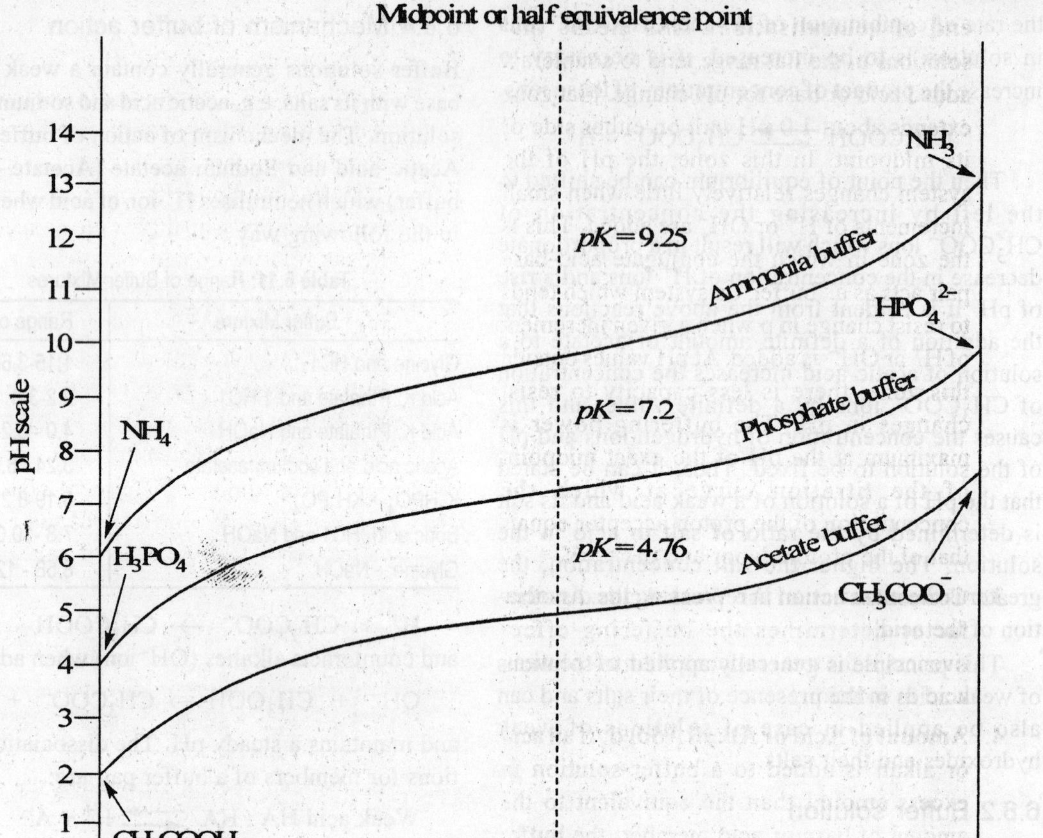

Fig. 6.19: Acid-base titration curves of some acids showing the major ionic species at the starting, midpoint and end of the titration

6.8.5 Buffer capacity and p^{H}

"Buffer capacity" means the magnitude of its resistance against pH changes. It is obtained by "Van Slyke's buffer value" which is the number of moles of a strong monoacidic base required to be added to 1 litre of the buffer solution to raise the pH of the latter by 1. Van Slyke proved that the maximal value for any single buffer system is 0.575 per pH unit, i.e. 0.575 moles of H^+ must be added to a molar solution of the buffer to induce a change of pH of one unit.

- **Factors influencing Buffer Capacity and pH.**

1. **When pH = *pK*:** The ability of a buffer system to resist changes of pH is maximal when pH of the solution is identical with that of the *pK* value. When pH = *pK* then from Henderson-Hasselbalch equation:

$$\log \frac{[A^-]}{[HA]} \text{ is } 0,$$

thus $[A^-] = [HA]$ for the log of unity is 0.

2. **At the middle of Titration Curve:** The buffer value of the solution is greatest in the middle range of the titration curve where the slope is virtually linear and this gives a measure of the *buffering power of the system* which means the number of moles of acid or base must be added to a molar solution in order to change the pH by one unit. It would be clear if we analyse the titration curve (Fig. 6.19). It is observed that the pH neutralization curve is flat at the midpoint or half equivalence point except at the beginning or

end of neutralisation. This means that solutions in the flat range tend to counteract added acid or base for pH change. Flat zone extends about 1.0 pH unit on either side of its midpoint. In this zone, the pH of the system changes relatively little when small increments of H^+ or OH^- are added. This is the zone in which the conjugate acid-base pair acts as a "buffer", a system which tends to resist change in p when a given increment of H^+ or OH^- is added. At pH values outside this zone, there is less capacity to resist changes in pH. The buffering power is maximum at the pH of the exact midpoint of the titration curve at which the concentration of the proton acceptor equals that of the proton donor and pH = pK.

3. **Concentration of the Weak Acids:** Another factor determines the buffering effectiveness is the total concentration of the weak acid in solution.
4. **Amount of Acid or Alkali Added:** If an acid or alkali is added to a buffer solution in excess amount than the equivalent to the amount of base or acid member; the buffer solution losses buffering capacity due to the removal of the whole amount of one of its members by the reactions with the added acid or alkali.
5. **Concentrated Buffer Solution:** More concentrated buffer solution has a higher buffering capacity and suffers from a smaller change in H concentration than a dilute buffer solution when a given amount of acid or base is added because the buffering capacity depends on the numbers of molecules or ions of the members of the buffer pair present in the solution.
6. **Value of pK_a:** According to the Henderson-Hasselbalch equa-tion, pH = pK_a + log [Salt]/[Acid], the pH of a buffer solution is determined by the value pK_a. The smaller the pK_a the stronger is the acid. pK_a represents the pH at which a weak acid is half-titrated or half-converted to its salts. If the pK_a of an acid is 4.0, the acid exists half as free acid and half as a salt at pH 4. Thus, the lower the value of pK_a in the buffer equation, the lower the pH of the solution.
7. **Ratio of Salt to Acid Concentration:** According to Henderson-Hasselbalch equation, another factor determining the pH of buffer solution is the ratio of salt to acid concentration. The higher the concentration of salt relative to acid, the higher is the pH of the solution. It is observed that, according to Henderson-Hasselbalch equation, the actual concentrations of salt and acid in a buffer solution may be varied widely, with no change in pH, so long as the ratio of concentration remains the same.

   The maximum efficiency of a buffer with a given total concentration of salt plus acid is greatest when the ratio of salt to acid is equal to 1.
8. **Addition of Strong Acid or Alkali:** Van Slyke has shown that the buffer value, or capacity of a solution, is determined by the pH change caused by the addition of increments of strong acid or alkali to it. The smaller the pH change caused by the addition of a given amount of acid or alkali, the greater is the buffer capacity.
9. **Limited Efficiency:** The efficient buffer range of a given buffer system is rather limited. It consists of about 1 pH unit on either side of the pK_a value of the buffer acid. As, for example, the acetate buffer is more or less efficient as a buffer from about pH 3.7 to 5.7, a total of 2 pH units. If one wants to prepare a buffer solution to cover the pH range 6-8, it would be well, if selection of a buffer acid is made with pK_a value close to 7, and similarly for other pH ranges. Increasing the concentration of the buffer solution increases its effective range (Fig. 6.19).

6.8.6 Illustration of phenomenon of buffering

1. The phenomenon of buffering is best illustrated by titrating a weak acid or base using a pH meter.
2. Alternatively, calculating the pH shift that accompanies addition of acid or base to a buffered solution.

6.8.7 Use of buffers

1. Standard buffer solutions are used with indicators for determination of pH.
2. Buffers are used to check the performance of electrodes used for determination of pH.
3. Many chemical reactions, including those catalyzed by enzymes, require pH control that is provided by appropriately buffered solutions.
4. Buffer are used in the pathological laboratory to control pH of culture media for bacteria and tissues.
5. Buffers are very important in regulating the pH of the body fluids, e.g., blood, interstitial fluid, lymph and intracellular fluid.

6.8.8 Buffer system in the body

In the presence of the need to protect the body from the over running of this cH/cOH relationship beyond the healthy normal limits, nature has provided control mechanism, one of which is through buffer system. Such a buffer system within the body is capable of preventing marked fluctuations in H^+ ion/ OH^- ion concentrations upon the addition, within limits, of either acids or bases.

The buffer salts provided in the blood, lymph and cells are chiefly :

1. Bicarbonates,
2. Mono and dihydrogen phosphate,
3. Proteinates of sodium or potassium.

The sodium salts are predominately present in the blood and lymph and the potassium salts predominately in the cells. The buffer acids provided are carbonic acid and hydrogen proteinates, i.e., proteins in their acid capacity.

- **Buffer Pairs Chiefly in Blood and Lymph**

(A) **In Plasma and Lymph**

[Sodium monohydrogen phosphate : Na_2HPO_4
Sodium dihydrogen phosphate : NaH_2PO_4

[Sodium Bicarbonate : $NaHCO_3$
Hydrogen Carbonate : H_2CO_3

[Sodium proteinate
Hydrogen proteinate
These are effective buffers because, both their free Carboxyl and Amino groups dissociate.

(B) **Buffer Pairs of Red Blood Cells**

[Potassium bicarbonate : $KHCO_3$
Hydrogen carbonate : H_2CO_3
(contribute much to the buffer capacity)

[Potassium monohydrogen phosphate : K_2HPO_4
Potassium dihydrogen phosphate : KH_2PO_4
(Relatively insignificant)

[Potassium proteinate (Haemoglobin) : KHb
Hydrogen proteinate (Haemoglobin) : HHb
(Play major role)

Haemoglobin is an important buffer system due to dissociation of the imidazole groups of the histidine residue in haemoglobin.

(C) **Buffers of Tissue Fluids:** The buffer systems in the lymph, spinal fluid and transudates are much the same as those of blood plasma, with the exception that these fluids contain much smaller quantities of protein buffers. The main buffer present in them is the Bicarbonate buffer system.

(D) **Buffer Pairs of the Tissue Cell**

[Potassium proteinate : Kpr^-
Hydrogen proteinate : Hpr^-
(chief role)

[Potassium bicarbonate : $KHCO_3$
Hydrogen carbonate : H_2CO_3
(chief role)

[Potassium monohydrogen phosphate: K_2HPO_4
Potassium dihydrogen phosphate : KH_2PO_4
(Relatively insignificant)

6.8.9 Reactions of buffer pairs in the blood and lymph

The following reactions show the function of three buffer pairs in the blood and lymph to prevent and increase in the hydrogen ions and hydroxyl ions.

1. **Phosphate Pair of Buffers**

$$Na_2HPO_4 + H^+ \rightarrow NaH_2PO_4 + Na^+$$

$$NaH_2PO_4 + OH^- + Na^+ \rightarrow Na_2HPO_4 + HOH$$

2. **Carbonate Pair of Buffers**

$NaHCO_3 + H^+ \rightarrow H_2CO_3 + Na^+$

$H_2CO_3 + OH^- + Na^+ \rightarrow NaHCO_3 + HOH$

3. **Protein Pair of Buffers**

Sodium Proteinate + $H^+ \rightarrow$ Hydrogen proteinate + Na^+

Hydrogen proteinate + OH^- + $Na^+ \rightarrow$ Sodium proteinate + HOH

In all three types of places—blood, lymph and cells—the proteins, because of their abundance, are the most important hydrogen ion control agents through buffer action. In this respect the haemoglobin salts of the red blood cells are very much efficient.

Various factors influence the effectiveness of these salt/salt and salt/acid buffer systems, among these:

1. The relative concentrations of the buffer pairs themselves, and
2. The hydrogen ion concentration of the medium.

The action of the bicarbonate buffer system is at its maximum efficiency when the H^+ ion concentration of the blood is within the normal range of pH 7.3 to 7.5; whereas the phosphate pairs become most effective as the H^+ ion concentration approaches the dangerous neutral pH7.0. The buffer systems can act within a fraction of a second to prevent excess changes in H^+ ion concentration.

6.9 Henderson-Hasselbalch Equation

Many biochemical products have functional groups which are weak acids or bases. One or more of these functional groups which may be Carboxyl or Amino or the secondary Phosphate group and dissociation of Phosphate esters are present in all proteins and nucleic acids, some coenzymes and some intermediary metabolites. The dissociation behaviour, i.e., protonic equilibria of weakly acidic and weakly basic functional group is, therefore, fundamental to an understanding of the influence of intracellular pH on the structure and biochemical activity of these compounds. Their separation and identification during investigation is also helped by knowledge of the dissociation behaviour of their functional groups. The relative strengths of weak acids and of weak bases are expressed quantitatively as their dissociation constants which express their tendency to ionize.

6.9.1 Henderson-Hasselbalch equation

The pH of a solution containing weak acid is related to its acid dissociation constant. The relationship can be calculated in the convenient form of the Henderson-Hasselbalch equation. This equation relates the pH of a mixture of a salt and a weak acid, i.e., buffer, e.g., $NaHCO_3$ and H_2CO_3 to the dissociation constant of the acid,

$$K' = \frac{[H^+][A^-]}{[HA]}$$

and the concentration of salt and acid as follows:

$$pH = pK' + \log\frac{[Salt]}{[Acid]}$$

pK' being the negative logarithm of K' (dissociation constant). The pK' values are less cumbersome to handle than K' values, just as the pH numbers are less cumbersome than actual H-ion molarities. This equation is known as the Henderson-Hasselbalch equation for buffer solution.

In a weak acid (HA) and its salts (BA), B is positive and A is negative ions. In the equilibrium reactions for the dissociation of HA and BA, a weak acid ionizes as follows:

$$\underset{\text{(undissociated acid)}}{HA} \rightleftharpoons \underset{\text{(weakly dissociated)}}{H^+ + A^-\ \text{(anion)}}$$

In case of its salts which ionize as follows:

$$\underset{\text{(undissociated salt)}}{BA} \rightleftharpoons \underset{\text{(Highly dissociated)}}{B^+ + A^-\ \text{(anion)}}$$

Now, the weak acid is slightly ionized, even when the salt is absent. Most of the acid is present as HA with a very small amount as H^+ and A^-. If the salt BA is added to the acid solution, all the A^- ions come from the highly dissociated salt. In dilute buffer solution, it may be assumed that the concentration of undissociated acid [HA] is equal to the total acid concentration and also that the concentration of acid ion $[A^-]$ is equal to the total salt concentration [BA]. According to the Law of Mass Action, the product of the concentrations of the products in a chemical reaction divided by the product of the concentration of the reactions at equilibrium is a constant. The equilibrium constant for this dissociation is written conveniently to simplify this expression by eliminating the water required for hydration of proton.

$$K' = \frac{[H^+][A^-]}{[HA]},$$

Cross-multiplying, $[H^+][A^-] = K'[HA]$

Brackets indicate concentration in moles per litre. K is constant. Such a constant is called "apparent or concentration dissociation constant" and is designated as K' to distinguish it from "true or theremodynamic dissociation constant" K' employed by the physical chemist.

Dividing both sides by $[A^-]$,

$$[H^+] = K'\frac{[HA]}{[A^-]}$$

If we know the dissociation constant, K', for the acid and the molecular ratio of acid and salt in the buffer solution, $[H^+]$ ion concentration of the solution can be calculated. This equation can be simply changed to calculate the pH of a buffer solution directly by taking the log of both sides

Taking the log of both sides,

$$\log [H^+] = \log K' \frac{[HA]}{[A^-]}$$

$$= \log K' + \log \frac{[HA]}{[A^-]}$$

Multiplying through by – 1,

$$-\log [H^+] = -\log K' - \log \frac{[HA]}{[A^-]}.$$

Substituting pH and pK for $-\log [H^+]$ and $-\log K$, respectively; then,

$$pH = pK' - \log \frac{[HA]}{[A^-]}$$

Then removing the minus sign and inverting the last term, we obtain, Henderson-Hasselbalch equation:

$$pH = pK' - \log \frac{[A^-]}{[HA]}$$

which in more general form is:

$$pH = pK' + \log \frac{[\text{Proton-acceptor}]}{[\text{Proton-donor}]}$$

6.9.2 Indication of equation

1. The Henderson-Hasselbalch equation has proved to be an expression of great predictive value in protonic equilibria.
2. It is apparent from these equations that the buffering capacity of a system is greatest when the amount of free anion is equal to the amount of undissociated HA, i.e., $[A^-]/[HA] = 1$, so that $\log [A^-]/[HA] = 0$ and $pH = pK$. This is why the most effective buffers in the body would be expected to be those with pK_s close to the pH in which they operate.
3. If the equation is evaluated at several ratios of $[A^-]/[HA]$ between the limits 10^3 and 10^{-3} and the calculated pH values plotted, the result obtained describes the titration curve for a weak acid.

The shape of the titration curve can be expressed by the Henderson-Hasselbalch equation, which is a logarithmic transformation of the expression for the dissociation constant.

6.9.3 Accuracy of equation

1. The equation applies more accurately when concentrations are converted to activities by multiplying by the appropriate activity coefficients, if these are available.
2. The equation does not apply with accuracy for the calculation of pH of buffer solutions when the proportion of weak acid in the form of its salt is less than about 10 per cent of the total buffer concentration, i.e., salt plus acid.
3. The H^+ and OH^- ion concentrations of the buffer solutions must not be sufficiently great to constitute a significant proportion of the total ions present if the equation is to apply with accuracy. Usually, these conditions are fulfilled by the buffer solutions present in body fluids and tissues.

6.9.4 Use of the equation

1. This equation is used for determination of pH of a buffer solution. One may directly calculate the pH of a buffer solution if *pK* for the buffer acid and the molecular ratio of salt to acid in the solution are known.
2. For determination of relative concentration of the salt and acid required for achieving the normal pH.
3. For calculation of *pK* of any acid from the molar ratio of proton-donor and proton-acceptor species at a given pH.
4. For calculation of molar ratio of proton-donor and proton-acceptor given the pH and *pK*.

6.9.5 Application

1. **Bicarbonate Buffer System and CO_2 Tension:** The buffering power of a buffer system is highest at a pH equal to its pK'_a. The pH of extracellular fluid is 7.4, while pK'_a of the bicarbonate $-CO_2$ system is 6.1. From the Henderson-Hasselbalch equation, it is seen that at pH 7.4, the concentration of bicarbonate ion is 20 times greater than that of dissolved CO_2. The Henderson-Hasselbalch equation indicates that $[HCO_3^-]/[H_2CO_3]$ ratio must be maintained at 20 for maintaining a pH of 7.4 in the blood. This is observed by the following calculation, applying the Henderson-hasselbalch equation to conditions in the blood.

7.4 = pH of blood

6.1 = pK'_a of H_2CO_3

$$pH = pK'_a + \log \frac{[Salt]}{[Acid]}$$

$$7.4 = 6.1 + \log \frac{[HCO_3^-]}{[H_2CO_3]}$$

$$1.3 = \log \frac{[HCO_3^-]}{[H_2CO_3]}$$

Antilog 1.3 = 20

$$\frac{[HCO_3^-]}{[H_2CO_3]} = \frac{20}{1} = 20.$$

This system is not a very powerful buffer system for (1). It operates on a portion of its buffering curve where the buffering power

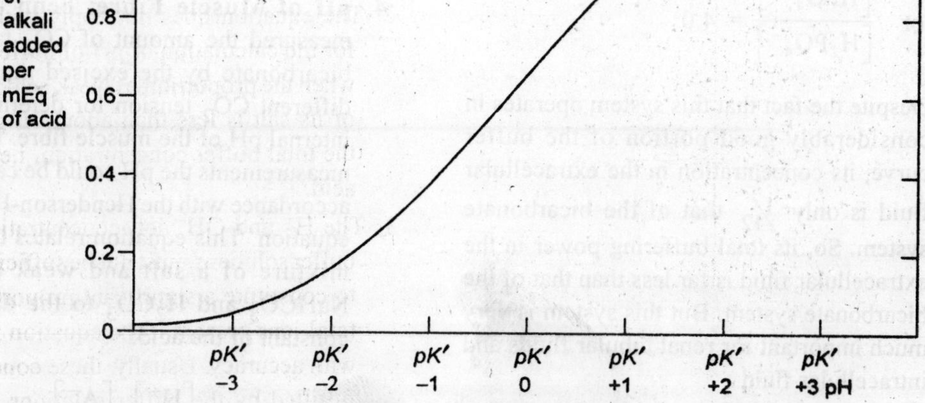

Fig. 6.20: Common form of titration curve calculated from the Henderson-Hasselbalch equation

is poor, (2). The concentrations of the two elements of the system, CO_2 and $H_2CO_3^-$ are not great. Yet, despite the fact that the concentration of each of these two components of the bicarbonate-carbonic acid system can be physiologically regulated by respiratory system and kidney, it provides a very powerful buffering system for the body. A fall in the ratio below 20 tends to depress the blood pH producing acidosis while a rise above 20 causes a threat to life raising pH which leads to alkalosis.

2. **Phosphate Buffer System**: The Phosphate buffer system (HPO_4^{2-}/H_2PO_4) is important next to bicarbonate system in the extracellular fluid. It has a *pK* of 6.8, which is closer to normal pH 7.4 in the body fluids, this allows the system to operate near its maximum buffering power. This system should maintain a $[HPO_4^{2-}]/[H_2PO_4^-]$ ratio of 4.0 for giving a pH of 7.4. According to Henderson-Hasselbalch equation:

$$pH = pK'_a + \log \frac{[HPO_4^{2-}]}{[H_2PO_4^-]}$$

$$\text{or} \quad 7.4 = 6.8 + \log \frac{[HPO_4^{2-}]}{[H_2PO_4^-]}$$

$$\text{or} \quad \log \frac{[HPO_4^{2-}]}{[H_2PO_4^-]} = 0.6$$

$$\text{or} \quad \frac{[HPO_4^{2-}]}{[H_2PO_4^-]} = 4.0.$$

Despite the fact that this system operates in considerably good portion of the buffer curve, its concentration in the extracellular fluid is only $\frac{1}{12}$ that of the bicarbonate system. So, its total buffering power in the extracellular fluid is far less than that of the bicarbonate system. But this system is very much important for renal tubular fluids and intracellular fluid.

3. **The Protein Buffer System**: The plentiful buffer of the body is the proteins of the cells and plasma, because of their high concentrations. it has been shown experimentally that about $^3/_4$ of all chemical buffering power of the body fluids is inside the cells and most of this results from the intracellular proteins. The buffer pairs are:

$$\frac{NA^+ \text{ Proteinate}^-}{\text{Hydrogen Proteinate (H. Proteinate)}} \quad \text{(Blood and lymph)}$$

Approximate quantities of protein buffer system is 13 in plasma and 55 in RBC in milliequivalents per litre of plasma or cells. The method by which this system operates is the same as that of the bicarbonate system. The protein is composed of amino acids bound by peptide linkages, but some of the different amino acids have free acidic radicals (free carboxyl and amino groups) that can dissociate into base plus H^+.

$$RCOOH \rightleftharpoons RCOO^- + H^+$$

$$pH = pK'RCOOH + \log \frac{[RCOO^-]}{[RCOOH]}$$

$$RNH_3 \rightleftharpoons RNH_2 + H^+$$

$$pH = pK'RNH_3^+ + \log \frac{[RNH_2]}{[RNH_3^+]}$$

And the *pK's* of a few of the protein buffering systems are not far from 7.4 which helps to make the protein buffering systems the most powerful of the body.

4. **pH of Muscle Fibre**: Fenn and Stella measured the amount of CO_2 taken up as bicarbonate by the excised muscle under different CO_2 tension for determination of internal pH of the muscle fibre. From these measurements the pH could be calculated in accordance with the Henderson-Hasselbalch equation. This equation relates the pH of a mixture of a salt and weak acid, e.g., $NaHCO_3$ and H_2CO_3 to the dissociation constant of the acid:

$$K'_a = \frac{[H^+]\,[A^-]}{[HA]}$$

and the concentrations of salt and acid as follows:

$$pH = pK'_a + \log \frac{[Salt]}{[Acid]}$$

pK'_a being the logarithm of K'_a. From the tension of CO_2 in the gas phase, the concentration of free acid may be obtained experimentally and from measurements of the total CO_2 liberated by a strong acid the amount of HCO_3^- may be obtained; hence with a knowledge of pK'_a, the pH of the muscle may be computed.

5. **Ammonia Buffer and Urine Acidity**: The dissociation reaction for the ammonia buffer system is as follows:

$$NH_4 \rightleftharpoons NH_3 + H^+$$

Since this reaction has a pK'_a of 9.3, in all the body fluids, the ammonia is present as the protonated ammonia ion $[NH_4^+]$. The plasma and glomerular filtrate having pH 7.4 the $[NH_3]/[NH_4]$ ratio is approximately 1 : 100. It is apparent to think that the amount of unprotonated NH_3 is insignificant but it is actually an efficient buffer for secreted H^+ in the tubular fluid because it is synthesized in renal tubular cells (60%) and for solubility characteristics, i.e., NH_3 is highly lipid-soluble and can passively diffuse across tubular cellular membrane whereas NH_4 is highly polar and crosses membrane poorly. The amount of ammonia in the urine vary directly with acidity. But there is continuous production of NH_3 by the tubular cell. It has been emphasized (Orloff and Berliner) that the passage of the undissociated base (NH_3) across the membrane is much more rapid than that of the ionized base (NH_4). Where the membrane is completely impermeable to the dissociated base, the condition for diffusion equilibrium will be that the concentration of undissociated base, i.e., NH_3, will be equal in plasma or tubule cell and urine. If the pH of plasma and urine are greatly different, this can lead to very large differences in concentration of ionized base in the two fluids, since in either fluid this concentration will be determined by the Henderson-Hasselbalch relationship:

$$pH = pK'_a + \log \frac{[NH_3]}{[NH_4^+]}$$

6. **Complete Titration of Monoamino Monocarboxylic acid** (e.g., Alanine): If a crystalline zwitterion amino acid, e.g.,

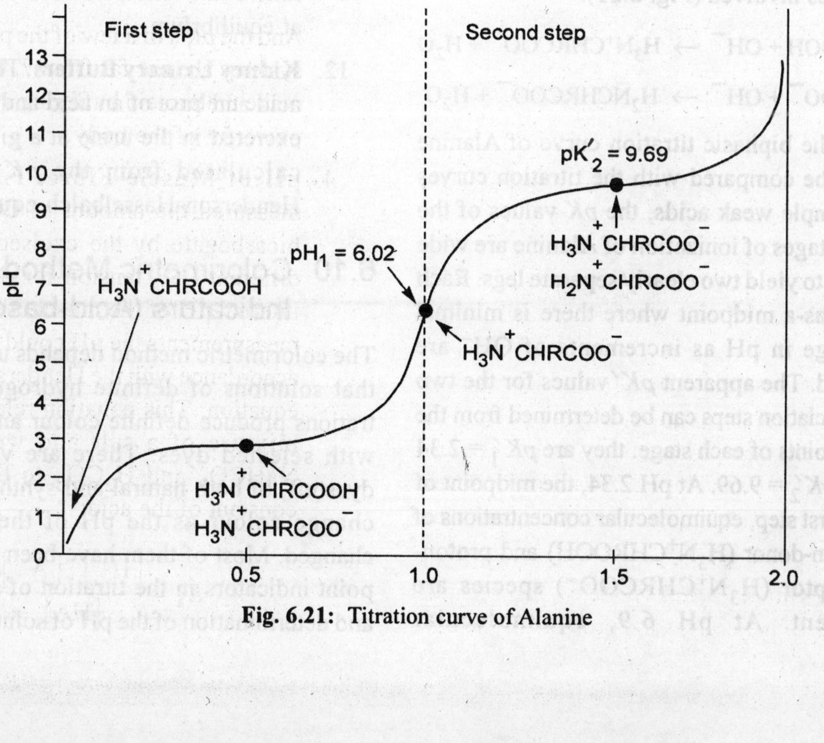

Fig. 6.21: Titration curve of Alanine

alanine is dissolved in water, it can function either as an acid or proton donor or as a base, i.e., proton acceptor:

$$H_3N^+CH(CH_3)COO \rightleftharpoons H^+ + H_3N^+CH(CH_3)COO^-$$

(As an acid)

$$H^+ + H_3N^+CH(CH_3)COO^- \rightleftharpoons H_3N^+CH(CH_3)COOH$$

(As a base)

The substances having this property is called "amphoteric" and the substance is known as "ampholytes" derived from "amphoteric electrolytes". This means, they have double nature (1) through their carboxyls they are acid, hence they react with bases (2) through their amino group they react with acids. For this reason, amino acids for their double reacting capacity, they regulate H^+ and OH^- ions in the cells, lymph and blood. This acid-base behaviour of ampholytes is explained by Bronsted-Lowry theory. The Alanine is a monoamino monocarboxylic acid and is considered as a dibasic acid in fully protonated form which donates two protons when titrated completely with a base. The two-stages of titration with NaOH is expressed below indicating each ionic species involved (Fig. 6.21).

$$H_3N^+CHRCOOH + OH^- \rightarrow H_3N^+CHRCOO^- + H_2O$$

$$H_3N^+CHRCOO^- + OH^- \rightarrow H_2NCHRCOO^- + H_2O$$

The biphasic titration curve of Alanine may be compared with the titration curves of simple weak acids, the *pK* values of the two stages of ionization of Alanine are wide apart to yield two clearly separate legs. Each leg has-a midpoint where there is minimal change in pH as increments of OH^- are added. The apparent pK' values for the two dissociation steps can be determined from the midpoints of each stage: they are $pK'_1 = 2.34$ and $pK'_2 = 9.69$. At pH 2.34, the midpoint of the first step, equimolecular concentrations of proton-donor ($H_3N^+CHROOH$) and proton-acceptor ($H_3N^+CHRCOO^-$) species are present. At pH 6.9, equimolecular concentrations of $H_3N^+CHRCOO$ and $H_2NCHRCOO^-$ arc present. Each of the two legs of the biphasic curve can be expressed mathematically by the Hcndcrson-Hasscl-balch equation i.e. we can calculate the ratios of ionic species of an amino acid at any pH given the values for pK'_1 (Fig. 6.21).

7. **Mechanism of Colour Change of Indicator:** (See mechanism of colour change of indicators).
8. **Buffer Salts to Buffer Acid Ratio**: The ratios of the buffer salts to buffer acids change in proportion to the change in H^+ ion concentration or pH and the pH alteration within a fluid is expressed by this equation for any one of the buffer system.
9. **Bicarbonate Buffer Systems of Blood**: (See bicarbonate buffer system of blood.)
10. **Mechanism of Buffer Action:**(See mechanism of buffer action.)
11. **pH of Solution of Weak Acid and Acid Dissociation Constant**: A weak acid has a high pK_a and dissociates little in solution. The pH of a solution of weak acid is related to its acid dissociation constant. The relationship can be stated in the convenient form of the Henderson-Hasselbalch equation at equilibrium.
12. **Kidney Urinary Buffers**: The proportion of acid and base of an acid and base of an acid excreted in the urine at a given pH may be calculated from the *pK* values using Henderson-Hasselbalch equation.

6.10 Colorimetric Method—Indicators (Acid-base)

The colorimetric method depends upon the finding that solutions of definite hydrogen ion concentrations produce definite colour and shade effects with selected dyes. There are various organic dyestuff, of both natural and synthetic origin, that change colour as the pH of their solutions is changed. Most of them have been applied as end-point indicators in the titration of acids and bases and determination of the pH of solutions. Such dyes when applied quantitatively, produce definite colour and shade effects. They function within solutions of specific H-ion concentrations, together with the respective cH ranges. The dyes are also called "range finder" because they function for measuring approximate molar concentration of H-ions provided in the solution which is not at exactly 4 or 5 or 6, etc.

6.10.1 Definition

Indicators are the substances which are used to indicate, usually by their colour change, the particular physico-chemical condition of the chemical system.

Acid-base indicators or pH indicators are weak acids or bases which change colour during change of pH of their aqueous solutions over a definite range of pH.

6.10.2 Chemical nature

They are chiefly salts of either a very weak acid or a weak base or the free acid or base itself. And the principles which apply to solutions of weak acids and weak bases also apply to indicators.

These may be regarded as very weak organic acids having capacity of dissociation like all acids into hydrogen ion and an anion expressed as HIn $\rightleftharpoons H^+ + In^-$. The main characteristic is that the colour of the undissociated acid, HIn, is different from that of the anion, In^-.

6.10.3 Types of indicators

1. Acid-base indicators (or neutralisation indicators) e.g. Methyl Orange, Phenolphthalin.
2. Oxidation-reduction indicators e.g. Diphenylamine.
3. Adsorption indicators e.g. Red dye eosin.
4. Starch paste as Iodine indicator.
5. Radioactive indicator.
6. Metalochrome indicators.

6.10.4 Mechanism of colour change

The two members of the conjugate pair of each indicator possess different colours. As for example, if acid indicator, Methyl red, represented by the general formula HIn and the dissociation is expressed as follows:

$$HIn \rightleftharpoons H^+ + In^-$$

The indicator acid HIn first undergoes a reversible "tautomeric transformation" into a different substance called "quinonoid form" which has a different colour (HIn*) and then ionises completely to form In^- and the whole equilibrium is represented as under:

$$\underset{\text{Tautomeric change}}{HIn} \rightleftharpoons \underset{\text{Quionoid form}}{HIn^*} \xrightleftharpoons{\text{Ionisation}} H^+ + In^*$$

But the concentration of HIn* is usually negligibly small and so the simple dissociation equation and consequently Henderson-Hasselbalch equation can be used.

The dissociation is expressed as:

$$HIn \rightleftharpoons H^+ + In^-$$

and the mass law equation for the equilibrium is:

$$\frac{[H^+]\,[In^-]}{HIn} = K_{in}$$

By rearrangement:

$$[H^+] = K_{in}\frac{[HIn]}{[In^-]}.$$

which is analogous to the equation of weak organic acid, acetic acid.

$$H^+ = K_a\frac{[CH_3COOH]}{[CH_3COO^-]}$$

Since the indicator, HIn is an acid with a dissociation constant K_{in}, it is titrated with alkalis through a definite pH range like other weak acid. If alkali is added to HIn, after reaction it forms BIn, salt of the indicator. Consequently the proportion of acid form of the indicator, HIn becomes lesser and the proportion of Bin, salt form becomes greater. After completion of the titration, whole indicator converts into salt form, BIn. Since HIn is a weak acid and BIn is the salt of weak acid, the mixture becomes a buffer pair to which the Henderson-Hasselbalch equation for buffers is applicable. For an acid indicator it is expressed:

$$pH = pK_{in} + \log\frac{[In^-]}{[HIn]}$$

$$\text{or} \quad pH = pK_{in} + \log \frac{[\text{Indicator salt}]}{[\text{Indicator acid}]}$$

pK_{in} is a constant called indicator exponent and pK_a is an ionization exponent of HIn. pK_{in} equals the pH of the solution when HIn and In^- are present in it in equal concentrations i.e. at that pH, half the original number of HIn molecules have been changed to In^- and this point is called "half-equivalence point".

From the Henderson-Hasselbalch equation it is certain that the colour of the indicator solution changes with pH. When the indicator is present as acid, the solution is red e.g. when a few drops of an acid are added to an indicator solution, the acid due to large H ion concentration produced by dissociation, checks the ionisation of the indicator acid. The solution would then contain a large amount of HIn molecules, whose colour will, therefore, be the "acid colour" of the solution. When, only salt is present, the solution is yellow, e.g., when small amount of alkali, NaOH, is added to the solution, sodium salt of the indicator will be formed which, being a salt, will ionize completely giving an excess of In^-, whose colour will be the "alkaline colour" of the indicator, i.e. yellow. And at all intermediate pH values, the colour will be a mixture of red and yellow and the solution contains an equal concentration of HIn molecules and In^- ions, the solution shows a "neutral colour" of the indicator.

(a) **Shade of the Colour**: The shade of the colour of Methyl red is fixed for any given pH value within its titration range, due to the ratio of salt to acid form being constant for a given pH.

(b) **Intensity of the Colour**: The intensity of the colour for a given pH and shade is determined by the concentrations of salt and acid forms present.

Table 6.12: Some common indicators, acid and alkaline colour, pH range and pK_{in}

Scientific name	Common name	Colours		pH range	pKin
		Acid	Alkali		
Thymolsulphonphthalein	Thymol blue	Red	Yellow	1.2-2.8	1.7
p-Benzenesulfonic acid-azo-dimethylaniline	Methyl orange	Red	Yellow	3.1-4.4	3.5
Tetrabromo-m-cresol-sulfon-phthalein	Bromcresol green	Yellow	Blue	3.8-5.4	4.7
Tetrabromophenol sulphonphthalein	Bromphenol blue	Yellow	Blue	3.0-4.6	4.0
Dimethylaminoazobenzene-o-carbo-Xylic acid	Methyl red	Red	Yellow	4.2-6.3	5.1
Dichlorophenolsulphonphthalein	Chlorphenol red	Yellow	Red	5.1-6.7	6.0
Dibromo-o-cresolsulphonphthalein	Bromcresol purple	Yellow	Purple	5.4-7.0	6.2
Dibromothymolsulphonphthalein	Bromthymol blue	Yellow	Blue	6.0-7.6	7.0
Phenolsulphonphthalein	Phenol red	Yellow	Red	6.8-8.4	7.9
O-cresolsulphonphthalein	Cresol red	Yellow	Red	7.2-8.8	8.3
Phenolphthalein	Phenolphthalein	Colourless	Red	8.3-10.0	9.7
Thymolsulphonphthalein	Thymol blue	Yellow	Blue	8.0-9.6	8.9
Thymolphthalein	Thymolphthalein	Yellow	Blue	9.3-10.5	9.9
Diphenyl-disazo-bis-α-napthylamine-4-sulphonic acid	Congo red	Blue	Violet red	3.0-5.2	—
Sodium alizarine sulphonate	Alizarine red	Yellow	Pink	3.7-4.2	
Azolitmin	Litmus	Red	Blue	5.0-8.0	—
Dimethyldiamino-phenozine'chloride	Neutral red	Yellow	Orange	6.8-8.0	6.85

6.10.5 Theories of colour change of indicator

1. **Ostwald Theory**: It is due to the presence of coloured ions which are produced when a salt is dissolved in water.
2. **The Quinoid Theory**: It is due to intramolecular change. It is believed that indicator is a mixture of two tautomeric forms, one exist in an acid solution and another in alkaline solution. The two forms possess different colours. The solution shows a change of colour due to the passing of one form to the other.

6.10.6 Criterion of a good indicator

The characteristics of a good indicator are:

1. Colour change should be sharp.
2. Colour should be stable and brilliant and contrasting.
3. Colour change should be at the end point of the reaction.

6.10.7 Indicator equilibrium

Indicators should obey Ostwald's dilution law because they are weak electrolytes and the equilibrium is expressed as:

$$\frac{[H^+][In^-]}{[HIn]} = K_{in} \text{ or } [H^+] = K_{in}\frac{[HIn]}{[In^-]}$$

where K_{in} = dissociation constant of the indicator.

For a bi-colour indicator, there is equal concentration of HIn molecules and In–ions and the solution shows the neutral colour, then we have

$$[HIn] = [In^-]$$

since HIn and In^- are responsible for acid and alkaline colour of the indicator.

For neutral colour,

$$[H^+] = K_{in} \quad \text{or } pH = pK_{in}$$

where K_{in} = indicator constants, which is different for different indicators, then the result is that the neutral colour pH of an indicator is constant for the same indicator and is different for different indicators.

6.10.8 Sensitivity range of indicators

The change of colour of an indicator does not take place exactly at the neutral pH of the indicator but within a pH range round it. Suppose, the sensitivity range i.e. colour change interval of an indicator is from [HIn] : $[In^-]$ = 8:1 to 1:8, then by substitution of these values in equation:

$$pH = pK + \log\frac{[Salt]}{[Acid]}$$

the sensitivity range covers roughly a pH interval of $pK_{in} - 1$ to $pK_{in} + 1$ i.e. about 2 units of pH with pK_{in} as the midpoint. It has been observed that phenolphthalin shows neutral colour in the alkaline region having pH = 8.6 but being one-colour indicator it does not follow the above equation and shows the onset of pink colour about one pH unit earlier than its $p^{pK}{}_{in}$ which is 9.6. Litmus shows the same with a neutral solution having pH = 7 and Methyl orange points to neutrality in an acid region having pH = 3.7.

6.10.9 Limitations of colorimetric determination

1. **Improper Buffered Solutions:** The indicators are weak acids or bases and when any indicator is added to pure water, there will be a change of pH of water and the pH value obtained is not correct. For which it is necessary to buffer the solution adequately before pH determination and addition of small amount of indicator will not cause any significant change of pH. For this reason, pH of water or of solution of strong acids and alkalis causes problem during deter-mination of pH. Very dilute solutions of bases and acids may cause a change in pH upon the addition of indicator.
2. **Other Factors**
   (a) Large variations of temperature.
   (b) The presence of proteins, organic solvents, heavy metal ions, oxidizing and reducing agents, high concentrations of neutral salts in the solutions.

6.10.10 Standard buffer solution and indicators for pH determination

An indicator can be used for pH determination in a limited range which is estimated by the pK_{in} of the indicator and also by the optical value of the colours of the indicator. For which one should select an indicator which exists as both salt and acid in the pH range of the solutions to be tested. The standard buffer solution also to be selected to use with indicators. As for example, if somebody wants to work in a pH range of 2.8-3.8, he should use Bromphenol blue as an indicator ($pK_{in} = 3.98$) and standard buffer solutions prepared by mixing Potassium acid phthalate and HCl solutions of proper concentrations.

Table 6.13: Indicators and Buffer Standards for pH 1.2-10

Indicator	pH range	pK_{in}	Buffer standards
Thymol blue (acid)	1.2-2.2	1.5	HCl + KCl
Thymol blue	2.2-2.8	1.5	Potassium acid phthalate + HCl
Bromphenol blue	2.8-3.8	3.98	-do-
Bromphenol blue	3.8-4.4	3.98	Potassium acid phthalate+NaOH
Methyl red	4.4-5.8	5.1	Acetic acid + NaOH
Bromcresol purple	5.8-6.8	6.3	KH_2PO_4 + NaOH
Phenol red	6.8-8.0	7.9	KH_2PO_4 + NaOH
Thymol blue	8.0-10.0	8.9	H_3BO_3 + KCl + NaOH

6.10.11 Ideal indicators for titrations

By plotting a titration curve we can study the pH change during titrations. The pH titration graphs given below show the pH change during titrations of a strong or weak base (Fig. 6.22). It is observed that near the equivalence point, the pH rapidly goes up, which reveals that an indicator whose neutral colour pH lies within the range of this ascent is ideal. By this method it has been found that for:

(a) Strong acid by a strong base, the rapid ascent is between pH 4 to 10, so, any indicator is applicable for titration.

(b) Strong acid and weak base, the indicator should be used whose pK_{in} i.e. neutral colour pH is in the acidic region i.e., about pH 4 and Methyl orange is ideal.

(c) Weak acid and strong base, e.g. acetic acid with caustic soda, an indicator should be selected showing its neutral colour in the alkaline region e.g., Phenolphthalein whose pH of colour change (pK) is 8.6.

(d) Weak acid and weak base, no indicator is suitable since such a solution on addition to alkali or acid during titration changes, its pH vary slowly like buffer solutions.

Table 6.14: Ideal indicators for titrations

Titrations indicator	Ideal colour	pH of change
Strong acid + strong base	Any indicator	4 to 9
Strong acid + weak base	Methyl orange	3.7
Weak acid + strong base	Phenolphthalein	8.6
Weak acid + weak base	No indicator is suitable	Slow-change of pH

- **Reason of Rapid Ascent:** The reason for rapid ascent can be explained in the following way. The percentage of acid neutralised for each addition of alkali determines the corresponding pH change, i.e. pH change at each step does not depend on the amount of acid neutralised but depends on the "fraction of the free acid neutralised". Suppose, at the initial point, 0.1 cc neutralises 1 per cent only but near the equivalence point, remaining 99 per cent has been neutralised by the same 0.1 cc, causing 100 per cent change of the free acid present in the solution. So, at the equivalence point the percentage change of acidity for a given volume of alkali added is maximum, for which the pH – titration curve is the steepest at the equivalence point and based on this principle, all potentiometric titrations are done.

6.10.12 Different ways of indicator method

(a) In most cases, the estimation of pH can be made rapidly by indicators. These are dyes which change colour at definite value of H-ion concentration.

(b) Alternatively, Henderson-Haesselbalch equation can be used if pK_a of the indicator is known.

(c) For rapid measurement, an "Universal Indicator", i.e., a mixture of indicators which

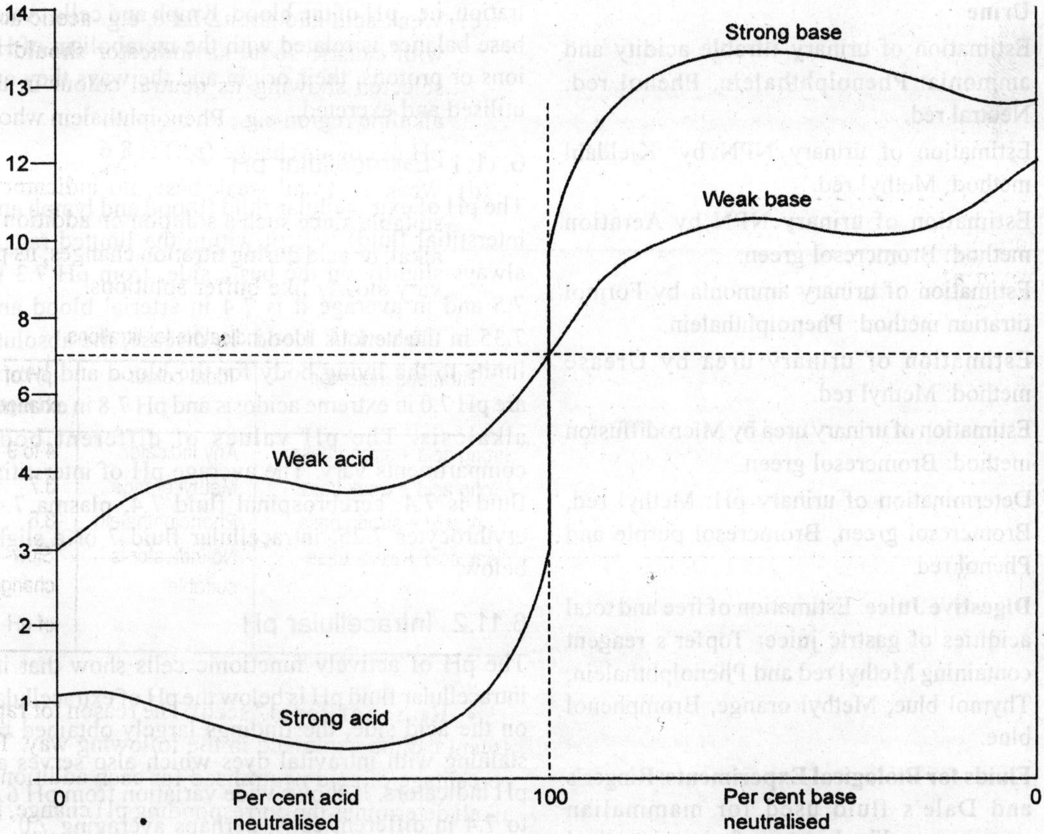

Fig. 6.22: pH change during titration of acids by alkali (Titration curves)

is sensitive over a very wide pH range, i.e., almost 0 to 14, can be used.

6.10.13 Determination of pH by indicators in case of coloured biological fluid

1. **Blood**: In determinings the H-ion concentration of fluid such as blood, the reaction may be tested in the dialysate obtained by placing the blood in a sac of collodin, which is immersed in a small quantity of 0.85% NaCl solution.
2. **Yellow Coloured Urine**: A series of standard buffer with selected indicator is prepared. Urine sample is treated with the above indicator solution. So the urine contains the colour for indicator plus its own yellow colour while standard buffer solutions contain only the colour of indicator. A tube containing only urine is placed behind the standard buffer and a tube containing only water is placed behind the tube of urine plus indicator and the colour of light passing through the pairs of tubes can be observed. In this way the correct pH values are recorded by making the colour of urine constant for both the standard tubes and the unknown sample.

Biological Application of Indicators

I. **Blood**

1. Total fatty acid of blood: Thymol blue.
2. Serum sodium estimation (Winbach's method): Phenolphthalein
3. Blood urea estimation by Van-Slyke method -Methyl red.
4. Estimation of blood NPN by Micro-Kjeldahl method: mixture of indicators called Tashirows indicator containing Methyl red and Thymol blue.

II. **Urine**

1. Estimation of urinary titrable acidity and ammonia: Phenolphthalein, Phenol red, Neutral red.
2. Estimation of urinary NPN by Kjeldahl method: Methyl red.
3. Estimation of urinary NPN by Aeration method: Bromcresol green.
4. Estimation of urinary ammonia by Formol titration method: Phenolphthalein.
5. Estimation of urinary urea by Urease method: Methyl red.
6. Estimation of urinary urea by Microdiffusion method: Bromcresol green.
7. Determination of urinary pH: Methyl red, Bromcresol green, Bromcresol purple and Phenol red.

III. **Digestive Juice**: Estimation of free and total acidities of gastric juice: Topfer's reagent containing Methyl red and Phenolphthalein, Thymol blue, Methyl orange, Bromphenol blue.

IV. **Fluids for Biological Experiments**: Ringer's and Dale's fluid used for mammalian experiments, like heart perfusion, intestinal or uterine movement, etc. are estimated for pH visually: Phenol red.

V. **For Spectrophotometric Estimation of pH of Solution**

VI. **Histochemical Studies of Gastric Mucosa for HCl Secretion**: Neutral red, Acridine, Toludine blue, Methylene blue, Phenol red etc.

VII. **Determination of Muscle Fibre pH**: Rous studied the changes in colour of injected indicators, the pH in the fibre which was acid (pH 5.6) but results based on intracellular use of indicators are generally not reliable because the dye concentrates at surfaces in the cell where the pH may be quite different from that in the bulk of the solution.

6.11 Acid-Base Balance or Maintenance of pH of the Body

One of the important phenomenon in the body is the maintenance of constant H^+/OH^-–ion concentration, i.e., pH of the blood, lymph and cells. Acid-base balance is related with the metabolism of H^+ ions or protons, their origin and the ways they are utilised and excreted.

6.11.1 Extracellular pH

The pH of extracellular fluid (blood and lymph and interstitial fluid) is kept within the limited range, always slightly on the basic side, from pH 7.3 to 7.5 and in average it is 7.4 in arterial blood and 7.35 in the venous blood. In disease, the absolute limits to the living body for the blood and lymph are pH 7.0 in extreme acidosis and pH 7.8 in extreme alkalosis. The pH values of different body compartments vary. The average pH of interstitial fluid is 7.4, cerebrospinal fluid 7.4, plasma 7.4, erythrocytes 7.25, intracellular fluid 7 or a slight below.

6.11.2 Intracellular pH

The pH of actively functionic cells show that its intracellular fluid pH is below the pH of extracellular on the acid side, the findings largely obtained by staining with intravital dyes which also serves as pH indicators. It shows wide variation from pH 6.4 to 7.4 in different cells, perhaps averaging 7.0. It has been observed in the rat tissues that pH range of the different cells are: bone cells 6.5-7.3, peripheral liver cells 7.1-7.5, central liver cells 6.7-7.0, Kupffcr cells 6.4-6.5, duodenal cells 6.7-6.8, stomach parietal cells 6.4-6.6.

A rapid metabolism in cells increases the rate of CO_2 production and consequently decreases pH. An inadequate blood flow to the cells causes accumulation of CO_2 and a decrease in pH.

6.11.3 Acids, bases and salts in the body

The constant pH range is maintained in blood, lymph and cells despite:

1. The absorption of variable quantities of acids and bases from the intestine,
2. The entrance into the blood from the cells of variable quantities of strong and weak acids,
3. The continuous presence of salts of strong bases and weak acids.

The chief acids of the body which are H^+ ion or proton donors are:

(a) **Inorganic**: Carbonic acid, phosphoric acid, sulphuric acid.

(b) **Organic**: Acetic acid, lactic acid, citric acid, fatty acids, organic phosphates, proteins, HHb, $HHbO_2$, tartaric acid, oxalic acid (Table 6.15).

The chief bases of the body which are H^+ ion or proton acceptors are the anions of the weak acids, e.g., HCO_3, Hb^-, HbO_2^-, protein, HPO_4^- and the anions of organic acids and organic phosphates.

6.11.4 Protection against changes in pH in the body

As a protective measure for acidosis or alkalosis, the body has the following regulatory mechanism.

I. **Buffer System**: The body fluids (blood plasma, lymph and cells) are provided with acid-base buffer system which immediately combines with acid or alkali and thereby prevent increase in H^+ ion concentration. It acts within a fraction of a second to adjust.

II. **Respiratory Mechanism**: Change in H^+ ion concentration causes stimulation of respiratory centre to alter the rate of respiration. As a result, the rate of CO_2 removal is also changed and causes the H^+ ion concentration to return to normal. It takes one to fifteen minutes to readjust.

III. **Kidney Mechanism**: When there is change of H^+ ion concentration, the kidneys excrete either an acid or alkaline urine, to regulate H^+ ion concentration of body fluids. It takes several hours to several days to adjust H^+ ion concentration.

- **Buffer System**: An acid-base buffer is a solution of two or more chemical compounds that prevents excessive changes in H^+ ion concentration when either an acid or a base is added to the solution. There are pairs of buffers, either salts with salts or salts with weak acids.

	B^+	B^+	B^+	B^+
Proton → acceptors	HCO_3^-	Hb^-	HbO_2^-	Protein⁻
Proton → donors	H_2CO_3	HHb^-	$HHbO_2^-$	H Protein

where B^+ = a cation which may be Na^+ or K^+.

Table 6.15: Sources of acids to blood

Intestine	Blood	Cell
From diet →	Acetic acid	Carbohydrate, fat and Proteins
↓	Lactic acid	↓
Acids →	Citric acid	
	Fatty acid	$H_2O + CO_2$
	Carbonic acid	←
	Lactic acid	← Glycogen breakdown
	Phosphoric acid	← Nucleoprotein, Phosphoprotein, Phospholipids metabolism
	Acetic acid	
From Food, Fruits, etc. ↓ Salts	Sulphuric acid	← Cysteine, Cystine and methionine metabolism
Pot. tartarate + water →	Tartaric acid + Pot. hydroxide	
Pot. citrate + water →	Citric acid + Pot. hydroxide	
Sodium bicarbonate + water →	Carbonic acid + Sod. hydroxide	

The buffer salts are distributed in the extracellular fluids (plasma, lymph) and in the cellular fluids.

If the pH of the body fluids is below normal, the ratio of proton acceptors, i.e., bases, to proton donors, i.e., acids, is below normal and if the pH is above normal, the ratio of proton acceptors to proton donors is above normal. If protons are added to the body fluids, buffer bases take up protons to form buffer acids; and if there is less protons, they are diverted to the body fluids by the buffer acids. In this way the process is going on to regulate H^+ ion concentration.

$$\text{Protons} + \text{Buffer base} \underset{\text{Low protons}}{\overset{\text{High protons}}{\rightleftharpoons}} \text{Buffer acids}$$

The ratios of the buffer salts to buffer acids change in proportion to the change in H^+–ion concentration or pH and the pH alteration within a fluid is expressed by the Henderson-Hasselbalch equation for any one of the buffer system.

The buffer salts in the extracellular fluids and cells are chiefly bicarbonates, mono and dihydrogen phosphate, proteinate of sodium or potassium. The sodium salts are predominantly present in the extracellular fluid, and potassium predominantly present in the cells.

A. **Extracellular Buffers**

1. Sodium monohydrogen phosphate: Na_2HPO_4
   Sodium dihydrogen phosphate: NaH_2PO_4
2. Potassium Bicarbonate: $KHCO_3$
   Hydrogen Carbonate: H_2CO_3
3. Sodium Proteinate: NaPr
   Hydrogen Proteinate: HPr
4. Na^+ organic $acid^-$
   H. organic acid

B. **Intracellular Buffers**

1. Potassium monohydrogen phosphate: K_2HPO_4
   Potassium dihydrogen phosphate: KH_2PO_4
2. Potassium Bicarbonate: $KHCO_3$, Hydrogen Carbonate: H_2CO_3
3. Potassium Proteinate: KPr
   Hydrogen Proteinate: HPr

   Besides these, RBC contains the following buffers:

4. $\frac{KHb}{HHb}$
5. $\frac{KHbO_2 \quad K^+ \text{ org. acid}}{HHbO_2 \quad \text{H. org. acid}}$

I. **Extracellular buffers (in the Blood, Lymph)**

A. Bicarbonate Buffer System.

B. Phosphate Buffer System.

C. Protein Buffer System.

(A) **Bicarbonate buffer System**

The bicarbonate buffer system HCO_3^-/H_2CO_3 is important for buffering strong acids of metabolic origin. But it is not so strong buffer for two reasons:

1. The pK' of the bicarbonate buffer system is 6.1 whereas pH of extracellular fluids is 7.4 which indicate that 20 times of the bicarbonate buffer is in the form of bicarbonate ions as in the form of dissolved CO_2 for which the system acts, where the buffering power is poor in the buffering curve.
2. Another reason is that concentration of the two elements of the system, CO_2 and HCO_3^- are not high.

In spite of these reasons, this system is important like other buffer systems because:

1. The concentrations of each of the two elements (i.e. the ratio $[HCO_3^-]/[H_2CO_3]$) can be regulated, CO_2 by the respiratory system and bicarbonate ions by the kidneys.
2. The amount of bicarbonate exceeds other buffers in amount.
3. The bicarbonate buffer operates jointly with haemoglobin buffer of RBC for buffering CO_2.
4. H_2CO_3 is in labile equilibrium with CO_2, so rise or fall of CO_2 causes reverse changes in the ratio $[HCO_3^-]/[H_2CO_3]$.

The ratio $[HCO_3^-]/[H_2CO_3]$ should be maintained at 20 for a given pH 7.4 according to Henderson-Hasselbalch equation. The bicarbonate buffer system acts in the blood and lymph in the following way for removal of H^+ ion or OH^- ion in excess of normal concentrations.

$$NaHCO_3 + H^+ \rightarrow H_2CO_3 + Na^+$$
$$H_2CO_3 + OH^- \rightarrow NaHCO_3 + H_2O$$

Plasma HCO_3^- is often called the "alkali reserve" because of its great role in buffering strong acids coming to blood due to metabolism.

(B) **Phosphate buffer system**

Considering the importance, the phosphate buffer system is next to bicarbonate system. It acts in almost identically the same manner as the bicarbonate system. It is composed of two elements HPO_4^{2-} and $H_2PO_4^-$. It has to maintain a $[HPO_4^{2-}]/[H_2PO_4^-]$ ratio of 4 for a given pH 7.4.

The phosphate buffer system has a pK of 6.8 which is near to normal pH 7.4 in the body fluids which favours it to operate near the maximum buffering power of the buffer curve, but its concentration in the extracellular fluid is 1/12 that of the bicarbonate buffer. Hence, its buffering power is less than that of the bicarbonate system.

This system is important in the renal tubular fluids because:

1. Phosphate becomes highly concentrated in the tubules, thereby increasing the buffering power of the system,
2. The tubular fluid becomes more acidic than the extracellular fluid, brings operation range of the system closer to the *pK* of the system. It is also important in case of intracellular fluids because (1) the concentration of phosphate in these fluids is far greater than the extracellular fluids, and (2) the pH of intracellular fluids is closer to the *pK* of the phosphate buffer system than is the pH of the extracellular fluids.

The phosphate buffer system acts in the following way:

$Na_2HPO_4 + H^+ \rightarrow NaH_2PO_4 + Na^+$

$NaH_2PO_4 + OH^- + Na^+ \rightarrow Na_2HPO_4 + H_2O$

(C) Protein buffer system

The most abundant buffer is the proteins of the plasma and cells. According to importance, it comes next to the bicarbonate system. CO_2 can diffuse readily through the cell membrane and bicarbonate ions to some extent whereas H^+ ion can diffuse slightly. The diffusion of the two elements of the bicarbonate buffer system causes alteration of pH in the intracellular fluid in proportion to the changes in pH in the extracellular fluids. In this way all the buffer system in the cells help to buffer the extracellular fluids and large amount of proteins is one of them. It has been found that 3/4 of the buffering power of the body fluids is due to intracellular proteins. The mechanism of action is the same as that of the bicarbonate buffer system such as:

1. Some of the amino acids have free acidic radicals which dissociate into base and H^+ ion,
2. The pKs of few protein buffering system are close to 7.4.

These two factors help to make the protein buffering systems very much effective.

Sodium Proteinate + $H^+ \rightarrow$ Hydrogen Proteinate + Na^+

Hydrogen Proteinate + OH^- + $Na^+ \rightarrow$ Sod. Proteinate + H_2O

• Isohydric Principle

All the above buffer systems act together for the H^+ ions which is common to the reactions of all the systems. If there is any change in H^+-ion concentration, it changes also the balance of all the buffer systems and the phenomenon is known, as *Isohydric Principle*, and the important feature of this Principle is that the buffer systems buffer each other.

(A) Intracellular Buffer System

(a) Tissue Cell Buffer System

1. Potassium Proteinate / Hydrogen Proteinate (Chief role)
2. Potassium Bicarbonate / Hydrogen Carbonate (Chief role)
3. Potassium Monohydrogen Phosphate / Potassium Dihydrogen Phosphate (Relatively insignificant)

(b) Red Blood Cell Buffer System

1. Potassium Bicarbonate / Hydrogen Carbonate (contribute much to the buffer capacity)
2. Potassium Monohydrogen Phosphate / Potassium Dihydrogen Phosphate (Relatively insignificant)
3. Potassium Proteinate (Haemoglobin) / Hydrogen Proteinate (Haemoglobin)

(B) Buffers of the Tissue Fluids

The buffer systems in the lymph, spinal fluid, transudates, are much the same as those of blood plasma, with the exception that these fluids contain much smaller quantities of protein buffers. The main buffer present in them is the bicarbonate buffer system.

I. Intracellular Buffer Mechanism (RBC)

The phosphate buffer is important in tissue cells but haemoglobin in the RBC plays a great role in buffer system for CO_2—buffering inside RBC. It is due to dissociation of the imidazole groups of the histidine residues in haemoglobin (Fig. 6.23).

In the pH 7.0-7.7 range, free carboxyl and amino groups of haemoglobin contribute relatively little to its buffering capacity. But the haemoglobin has 6 times the buffering capacity of the plasma proteins for the following reasons:

1. Haemoglobin molecule contains 38 histidine residues. The majority of the —NH_2 and —COOH groups of the amino acids in forming protein do not take part in the buffering but make peptide linkage — NH.CO — between the amino acids. The only intact —NH_2 and —COOH groups are at the ends of the peptide chains. However, certain amino acids have extra —NH_2 and —COOH groups in their molecules which are available for buffering as they do not participate in peptide link formation, and great example is Histidine. Haemoglobin molecule contains 38 histidine residues and, as a result, haemoglobin acts as a good buffer.

   The histidine molecule contains the equivalent of a second —NH_2 group in its imidazole ring. The ═N in this ring is a hydrogen acceptor and becomes ═N^+. It is thus basic and neutralizes acids. Thus, haemoglobin exists in two forms: Hb and HbH^+. During carriage of CO_2 in the blood and formation of bicarbonate in RBC will be:

   $Hb + H_2CO_3 \rightarrow HbH^+ + HCO_3^-$

   and carried as "haemoglobin-histidine bicarbonate".

   The ionization of imidazole ring depends upon the degree of oxygenation of the haemoglobin molecule.

2. The imidazole groups of deoxygenated haemoglobin dissociate less than those of oxyhaemoglobin, making Hb a weaker acid and, therefore, a better buffer than HbO_2.
3. Large concentration of Hb in RBC.
4. Reversible changes in its buffering capacity on oxygenation and deoxygenation.
5. Deoxygenated haemoglobin is a weaker acid and has much higher capacity for proton -acceptance and CO_2-buffering than oxyhaemoglobin which is a stronger acid and because HHb dissociates less readily than does $HHbO_2$ which is more effective in buffering against decreased CO_2.
6. Reversible interconversion between oxygenated haemoglobin (HbO_2^-/$HHbO_2$) buffer and deoxygenated haemoglobin buffer (Hb^-/HHb) help in CO_2 buffering.
7. Because of the operations and large capacity of the haemoglobin buffer systems, much CO_2 can be taken up by the blood as HCO_3^-, or can be given up by the blood through decomposition of HCO_3, with very little change in pH.
8. The buffer capacity of the haemoglobin buffers is so great that when the PCO_2 of blood is reduced to zero, the haemoglobin buffer acids decompose all of the HCO_3^- and all of the CO_2 is removed from the blood.
9. When the HCO_3^- is formed from the haemoglobin buffer systems, it is distributed between cells and plasma. But, if plasma buffers form small amount of HCO_3^- the supply of HCO_3^- in the plasma is maintained indirectly by the haemoglobin buffers.
10. Since the components of the haemoglobin and oxyhaemoglobin buffer systems interact with the components of the bicarbonate-carbonic acid buffer system, when other variables are constant, the total CO_2 and HCO_3^- of whole blood vary inversely as the haemoglobin content.

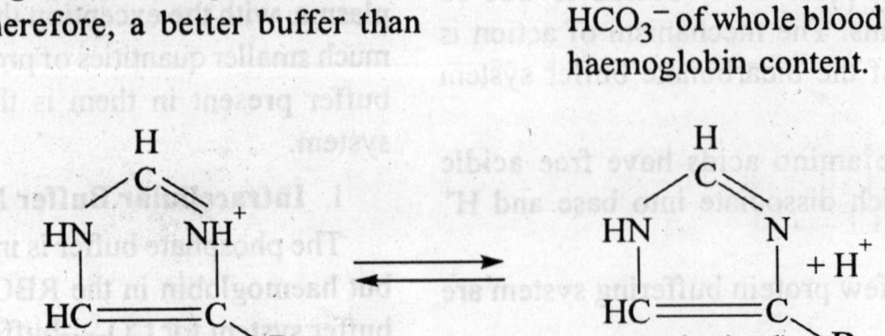

Fig. 6.23: Imidazole groups of histidine residues of Hb

11. Deoxygenated blood is much more efficient than oxygenated in buffering against CO_2, because Hb^- combines with H^+ more firmly than does HbO_2, and consequently Hb^- reacts with H_2CO_3 to form more HCO_3 than does HbO_2 at a given pH.

- **Chloride-Bicarbonate Shift (Buffering against CO_2) (Hamburger Phenomenon)**

When the CO_2 tension rises in the tissue cells, it diffuse out into the blood in gaseous form in the dissolved state but not in the bicarbonate form because it is less permeable to the membrane. On entering the capillaries, the dissolved CO_2 reacts with water to form carbonic acid (H_2CO_3) very slowly for which it is less important and the amount of protein to react with H^+ is relatively small and the interstitial fluid has less capacity for less protein to buffer CO_2. But inside the red blood cells this reaction is accelerated by the enzyme Carbonic anhydrase by 5000-fold and reaches almost complete equilibrium within a fraction of a second and allows large amounts of CO_2 to react with the red cell water before the blood leaves the tissue capillaries. In the next small fraction of a second the H_2CO_3 dissociates into H^+ and HCO_3^- i.e. bicarbonate ions. Then hydrogen ions combine with the haemoglobin (protonated Hb) in the RBC to form HHb. Since the rise of bicarbonate (HCO_3^-) ions content of RBC is much higher than that in the plasma as the blood passes through the capillaries, HCO_3^- diffuses into the plasma. The protein anions are not permeable to membrane and Na^+ and K^+ movement are regulated by Na-K pump, so, to maintain electrochemical neutrality, equimolecular quantities of Cl^- ions diffuse into the red cell in exchange of HCO_3^- and the phenomenon is called "Chloride shift" which occurs rapidly and is complete within one second (Fig. 6.23). For this, the Cl^- content of RBC in venous blood is higher (mosm/l in cells) than in the arterial blood. The conversion of CO_2 to carbonic acid to bicarbonate ion accounts for about 70 per cent of CO_2 transport (Fig. 6.24).

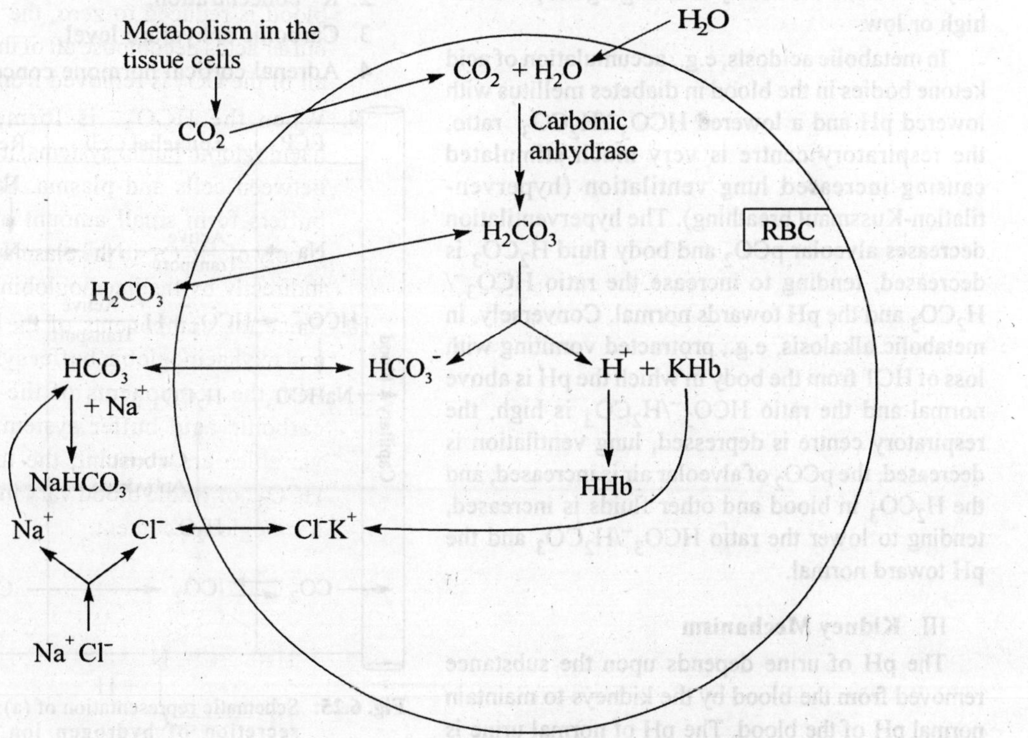

Fig. 6.24: Chloride Shift (Chloride-bicarbonate shift or Hamburger phenomenon)

Because there are many negative charges on each protein molecule but only one on each HCO_3^- and Cl^-, the number of osmotically active particles in the RBC increases as the H^+ ions are buffered and the HCO_3^- accumulates. The RBC takes up water causing increase in size of RBC for which the hematocrit of venous blood is 3 per cent higher than that of the arterial blood. In the lungs, the Cl^- moves out of the cells and the RBC shrinks.

This HHb is then oxygenated in the lungs to form $HHbO_2$ and immediately ionizes into H^+ and HbO_2^-. These H^+ ions combine with HCO_3^- to form H_2CO_3 which is dissociated by carbonic anhydrase into CO_2 and H_2O. CO_2 diffuses out of the RBC and escapes in alveolar air. Some HCO_3^- ions come back to RBC from plasma in exchange of Cl^- ions and are changed to CO_2 and H_2O.

II. Respiratory Mechanism

An increase in CO_2 concentration in the body fluids decreases the pH toward the acidic side, whereas a decrease in CO_2 raises the pH towards the alkaline side. Based on this effect, the respiratory system has the ability of changing the pH either high or low.

In metabolic acidosis, e.g., accumulation of acid ketone bodies in the blood in diabetes mellitus with lowered pH and a lowered HCO_3^-/H_2CO_3 ratio, the respiratory centre is very much stimulated causing increased lung ventilation (hyperventilation-Kussmaul breathing). The hyperventilation decreases alveolar pCO_2 and body fluid H_2CO_3 is decreased, tending to increase the ratio HCO_3^-/H_2CO_3 and the pH towards normal. Conversely, in metabolic alkalosis, e.g., protracted vomiting with loss of HCl from the body in which the pH is above normal and the ratio HCO_3^-/H_2CO_3 is high, the respiratory centre is depressed, lung ventilation is decreased, the pCO_2 of alveolar air is increased, and the H_2CO_3 in blood and other fluids is increased, tending to lower the ratio HCO_3^-/H_2CO_3 and the pH toward normal.

III. Kidney Mechanism

The pH of urine depends upon the substance removed from the blood by the kidneys to maintain normal pH of the blood. The pH of normal urine is usually on the acid side on normal diet. Acidification occurs in the proximal, distal and collecting ducts where H^+ secretion occurs. The amount of acid secreted depends upon the events in the tubular urine.

- **Secretion of H^+**

The H^+ secretion takes place in the cells of proximal and distal tubules and also in the collecting ducts like secretion of H^+ by the parietal cells of the stomach. The H^+ secretion in the proximal tubules is due to Na^+–H^+ exchange by secondary active transport. Efflux of Na^+ from the tubular cells into the interstitial fluid by Na^+–K^+ ATPase by the process of primary active transport lowers the intracellular Na^+, and this initiates entrance of Na^+ to the cell from tubular urine and extrusion of H^+ to the urine from the cell. Source of H^+ is from the dissociation of H_2CO_3 ($H^+ + HCO_3^-$) and HCO_3^- diffuses into the interstitial fluid. In the distal tubule, the H^+ secretion does not depend on Na^+ of tubular urine but it is secreted by the "ATP-driven proton pump", which is influenced by aldosterone to increase distal H^+ secretion (Fig. 6.25).

Renal acid (H^+) secretion is influenced by:

1. Changes in the intracellular pCO_2,
2. K^+ concentration,
3. Carbonic anhydrase level,
4. Adrenal cortical hormone concentration.

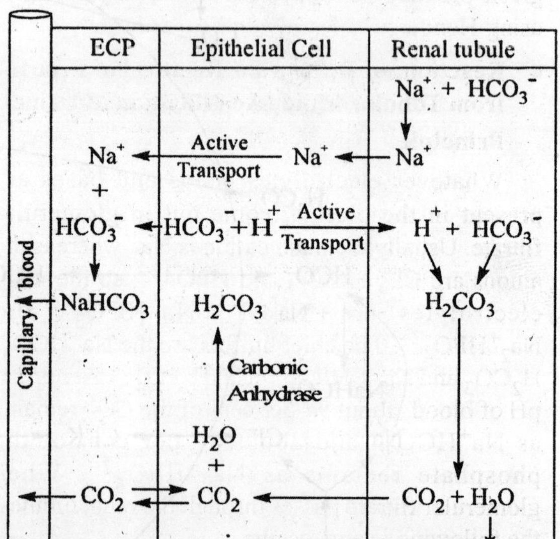

Fig. 6.25: Schematic representation of (a) Mechanism of secretion of hydrogen ion, (b) Sodium absorption in exchange of hydrogen ion, (c) Combination of hydrogen ions with bicarbonate ions in the renal tubule

- **Role of H^+ in the Urine**

The pH of urine normally varies between 4.8 to 7.4. The kidneys cannot excrete urine more acid than pH 4.5 and more alkaline than pH 8.2, i.e., an H^+ concentration in the urine which is 1,000 times the concentration in plasma and pH 4.5 is the "limiting pH" of the urine. If these H^+ ions are not taken up by any substances or buffered, the limiting pH, i.e., 4.5, will be reached very rapidly and H^+ ion secretion will be stopped. For this reason **three mechanism are involved for removal of H^+ ions.**

- **Urinary Buffers**

Urine is a strongly buffered solution, particularly because of large amounts of phosphates excreted in the urine which also contains little quantities of other buffer system. The phosphate buffer which is most important, composed of $HPO_4^-/H_2PO_4^-$. When the blood pH is 7.4, the ratio $B_2^+HPO_4^-/B^+H_2PO_4^-$ is 5, which means 83 per cent $B_2^+PO_4^-$ and 17 per cent $B^+H_2PO_4^-$. The excretion of $B^+H_2PO_4^-$ in the urine means removal of acid from the blood and plasma pH increases. At urinary pH 4.8 about 98 per cent of $B^+H_2PO_4^-$ and at 8.2 about 96 per cent $B_2^+HPO_4^-$ is excreted. The proportion of acid and base of an acid excreted in the urine at a given pH may be calculated from the *pK* values using Henderson-Hasselbalch equation.

- **Reactions of Buffers for Removal of Free H^+ from Tubular Fluid (Acidification of Urine)**

Principle

Whatever electrolytes, acids and bases are present in the plasma, come out in glomerular filtrate. Usually the main cation is Na^+ whereas the anions are Cl^-, HCO_3^- and $HPO4^{2-}$, so the main electrolytes are Na^+Cl^-, $Na^+HCO_3^-$ and $Na_2^+HPO_4^{2-}$. The chief buffers are the Na^+HCO_3/H_2CO_3 and $Na_2+HPO_4^{2-}/Na^+H_2PO_4^{2-}$. At normal pH of blood, about 94 per cent of the CO_2 remains as $Na^+HCO_3^-$ and about 84 per cent of the phosphate remains as $Na_2^+HPO_4^{2-}$. When glomerular filtrate passes through the renal tubules, the following events occur:

1. Reabsorption of water.
2. Secretion of H^+ by the tubular cells from H_2CO_3.
3. Large amount of Na^+ is taken up by the tubular cell in exchange of H^+.

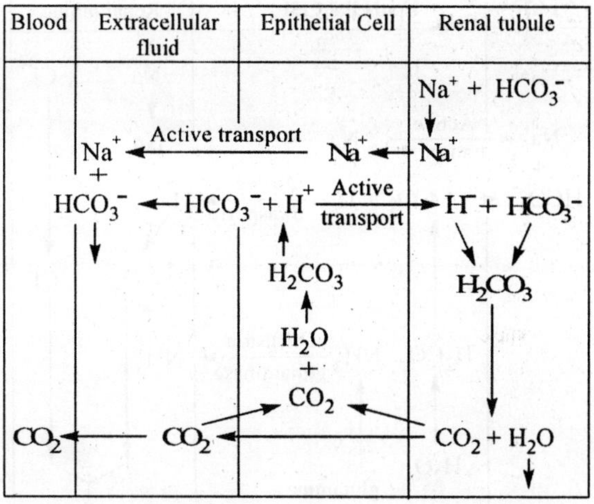

Fig. 6.26: Reaction of H^+ with bicarbonate in the renal tubule

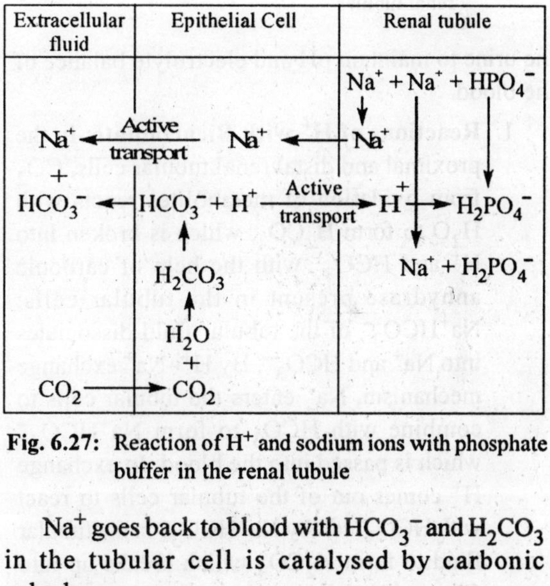

Fig. 6.27: Reaction of H^+ and sodium ions with phosphate buffer in the renal tubule

Na^+ goes back to blood with HCO_3^- and H_2CO_3 in the tubular cell is catalysed by carbonic anhydrase.

HPO_4^- is also another chief H^+ acceptor in the H^+–Na^+ exchange and secretion of acid mechanism which shows the following overall reaction:

$$H_2CO_3 \text{ (Tubular cell)} + Na_2^+HPO_4^- \text{ (Tubular fluid)} \rightarrow Na^+HCO_3 \text{ (Plasma)} + Na^+H_2PO_4^- \text{ (Urine)}$$

Here we see that half the sodium of the basic salt $Na_2^+HPO_4^-$ converts to Na^+HCO_3 which goes to blood and the remaining half is excreted in the urine as the acid salt Na^+–$H_2PO_4^-$. By this means, Na^+ and HCO_3^- go to blood and H^+ is secreted in

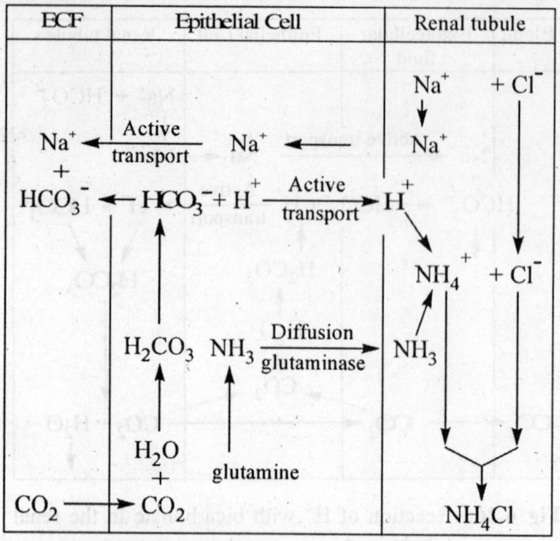

Fig. 6.28: Base conservation by ammonia secretion in the renal tubules

the urine to maintain pH and electrolyte balance of the blood.

1. **Reactions of H^+ with Bicarbonate**: In the proximal and distal renal tubular cells, CO_2 from oxidation of metabolites reacts with H_2O to form H_2CO_3, which is broken into H^+ and HCO_3^- with the help of carbonic anhydrase present in the tubular cells. $Na^+HCO_3^-$ of the tubular fluid dissociates into Na^+ and HCO_3^-. By H^+–Na^+ exchange mechanism, Na^+ enters the tubular cells to combine with HCO_3 to form $Na^+HCO_3^-$ which is passed into the blood. In exchange H^+ comes out of the tubular cells to react with HCO_3^- left by $Na^+HCO_3^-$ in the tubular fluid to form H_2CO_3 which breaks up into CO_2 and H_2O in the tubular fluid by carbonic anhydrase in the brush border of the cells of the proximal tubule but not in the distal tubule. The CO_2 enters the tubular cells to form equivalent amount of H_2CO_3 which provides further H^+ and HCO_3^-. Since H^+ is removed from the tubule, the pH of the fluid is not changed. HCO_3^- is reabsorbed for each mole of HCO_3^- removed from the tubular fluid (Fig. 6.26).
2. **Reactions of H^+ with Dibasic Phosphate**: Secreted H^+ ions are buffered in the urine by the phosphate buffer. H^+ reacts with dibasic phos-phate (Na_2HPO_4) to form monobasic phosphate i.e. acid phosphate (NaH_2PO_4) which is excreted in the urine. This reaction occurs in maximum extent in the distal tubules and collecting ducts because the amount of phosphate which is not rebsorbed in the proximal tubules is concentrated here due to reabsorption of water. Na^+ from Na_2HPO_4 enters that tubular cell by Na^+ - H^+ exchange mechanism to form $NaHCO_3$ which goes to be blood (Fig. 6.27).
3. **Reaction of H^+ with Base NH_3**: Ammonia is secreted in the proximal, distal tubules and collecting ducts. The ammonia is formed from the following reactions:
   1. Break-down of glutamine into glutamic acid and ammonia by the enzyme glutaminase,
   2. Deamination of glutamic acid,
   3. Deamination of other amino acids,
   4. Comes directly from the arterial blood.

After formation, ammonia diffuses (nonionic diffusion) into the tubular fluid easily because it is lipid-soluble and passes through the lipid containing membranes of the tubular cells. In the tubular fluid it reacts with H^+ from the tubular cells to form NH_4^+, i.e., ammonia ion. NH_4^+ is lipid-insoluble and remains in the tubular fluid. The amount of NH_4^+ formation depends upon the following factors:

1. The pH of the tubular fluid
2. The rate of NH_3 production. At any given rate of NH_3 formation, the amount of NH_4^+ formed is proportionate to the amount of H^+ available and, therefore, to the rate of H^+ secretion. Thus, the alkaline urine does not contain NH_4^+ but, in acid urine, NH_4^+ content is maximum.

This NH_4^+ then replaces Na^+ in a sodium salt like Na^+Cl^-. The Na^+ is reabsorbed by the H^+ – Na^+ exchange mechanism by the tubular cells and combines with HCO_3^- to form $Na^+HCO_3^-$ which enters the blood. The NH_4^+ is excreted in the urine as NH_4Cl after combination with Cl^- of Na^+Cl^- (Fig. 6.28).

7

Colloidal System

1. Colloidal system; Crystalloid and Colloid comparison; Phases of Colloids; classification; Lyophobic and Lyophilic comparison; Sol to Gel transformation; Dispersion Colloids; Association Colloids; Molecular Colloids; Settling down; Preparation, 2. Properties of Colloids, 3. Importance of Colloids, 4. Electrical charge of Colloidal particles, 5. Combination between Colloids, Protective Colloids, 6. Emulsion and Suspension, 7. Gel or Gel State, Gel and Sol comparison, 8. Concept of Macromolecules, 9. Physical Structure of Protoplasm, 10. Physical State of Protoplasm, 11. Biomolecules and Molecular organization of cell, 12. Microdissection (Micro-manipulator), 13. Ultramicroscope, 14. Birefringence, 15. Determination of the Size and Shape of the Colloidal Particles, 16. Haemodialysis; Artificial Kidney, 17. Biological membranes and Ultrafiltration, 18. Separation of Colloids, 19. Electrophoresis, 20. Chromatography. 21. Peritoneal dialysis.

7.1 Introduction

In course of investigation on diffusion, Thomas Graham (1861) found that soluble substances (inorganic or organic) could be divided into two classes— Crystalloids and Colloids according to the rates at which their solutions passed through a parchment membrane (i.e., dialyser). Substances such as salts, sugar, urea etc. which diffuses readily, were termed **Crystalloids** because they usually exist in crystalline form, whereas substances like gelatine, albumen, gum, silicic acid, starch, etc. which diffuse at a very slow rate, were termed **Colloids** (= Glue-like). This distinction, which is no longer tenable was realized by Graham and others because of the following facts:

1. Many crystalline substances can be obtained in colloidal solution, e.g., Sodium Chloride in Benzene.
2. X-ray studies have shown that the particles in colloidal system are often truly crystalline in character.
3. Molecular colloids (Macromolecules) can be crystallised, e.g., proteins, enzymes, polysaccharides and their derivatives starch, cellulose and high polymers, viruses, etc.
4. Other workers have shown that the Colloidal condition is primarily a state of matter and not a special kind of matter, for which, instead of the term colloids or colloidal substance, it is the general practice to refer to the **Colloidal state** or **Colloidal system**. Thus, the old idea of non-cystalline property of colloid has been discarded by experimental evidences.

Table 7.1: Differences between Crystalloids and Colloids

Crystalloids	Colloids
1. **Crystallisation** Can be crystallised with ease.	Non-crystallisable but some can be crystallised.
2. **Examples** Salt, sugar, urea.	Gelatin, gum, eggalbumin, starch, dextrin.

Contd.

Crystalloids	Colloids
3. **Diffusion** Diffuses rapidly	Diffuses with extreme slowness through water.
4. **Passage through animal membrane** Capable of diffusing through dead animal membranes.	Does not pass through animal membranes.
5. **Effect of evaporation** Does not form gummy mass on evaporation	Forms gummy mass when evaporated to dryness.
6. **Stickiness** Non–sticky.	Sticky.
7. **Molecular weight** Low molecular weight	High molecular weight.

7.2 Phases of Colloids

A Colloidal system is always heterogenous and consists of two phases, i.e., a two phase system:

(a) **Discontinuous dispersed phase or internal phase**: The phase constituting the colloidal particles.

(b) **Continuous dispersions medium or external phase**: The medium in which these colloidal particles are dispersed, i.e., it may be solid or liquid or gases.

7.3 Classification

I. According to Dispersion Medium

(a) **Hydrosol**: When dispersion medium is water.

(b) **Alcosols**: When dispersion medium is alcohol.

(c) **Benzosols**: When dispersion medium is Benzene.

(d) **Aerosols**: When dispersion medium is air.

(e) **Suspensoids**: Colloidal solutions consisting of large insoluble particles lying in a state of suspension in the dispersion medium are called "suspensoids".

(f) **Emulsion**: When two immiscible liquids (e.g., oil and water) are mixed up, they form "emulsion". Emulsion can be stabilized by adding a third substance called "emulsifier" such as casein (Protein) in milk stabilizes fat globules. Thus, milk is a suspension of oil droplet in a watery phase and butter is a Colloid consisting of water suspended in a continuous phase of oil. the conversion of milk into butter is an example of "phase reversal".

(g) **Gel**: When a colloidal solution resembles a solid and jelly-like substance, it is called "gel", e.g., gelatine gel, agar agar gel, pectin gel, silicic acid gel, methyl-cellulose gel.

(h) **Sol**: When a colloidal solution looks like a liquid with a liquid dispersion medium called a "sol".

II. Based on Solvent Affinity

Colloidal solutions with a liquid as dispersion medium may also be classified according to the existence of attractive forces between the dispersed particles and dispersion medium (Table 7.2).

A. Lyophobic (Solvent hating)

Insoluble substances like metals, metal sulphides, metal hydroxides which do not readily yield colloidal solutions with water are called *Lyophobic Colloids*. They are also called "irreversible colloids". If water is used as dispersion medium, then it is termed as "hydrophobic."

B. Lyophilic (Solvent loving)

Substances like gelatine, agar agar gum, silicic acid, various albumins, starch, soap and many dyes, etc. which directly pass into colloidal solutions in water are called *Lyophilic Colloids*. If water is the medium, then they are termed as "Hydrophilic Colloids". These are "Reversible Colloids."

Table 7.2: Comparison of Lyophobic and Lyophilic Colloids

Lyophobic	Lyophilic
1. **Nature of Substance** Mostly inorganic materials, e.g., metals, sulphides, oxides, etc. colloidal gold, colloidal sulphur.	1. Mostly organic materials (or of organic origin), e.g., starch (aqueous solution), proteins, gums, high polymers, plasma proteins in blood.
2. **Microscopic Observation** Well-defined under a microspcope or an ultramicroscope.	2. Invisible under a microscope or an ultramicroscope.
3. **Coagulation** Easily coagulated by electrolytes.	3. Very stable, not easily coagulated. Coagulated by adding hygroscopic substances, e.g., alcohol, acetone or by cooling.

Contd.

Lyophobic	Lyophilic
4. **Reversibility of Coagulation** Irreversible coagulation.	4. Reversible coagulation.
5. **Viscosity** Same viscosity like the solvent which is usually water.	5. More viscous than water.
6. **Gelatinisation** Do not gelatinise readily and are very unstable on reversal of charge.	6. Gelatinise rapidly and have an isoelectric point.
7. **Other physical properties** Density, refractive index, etc. follow the law of mixtures.	7. Do not follow the law of mixtures.
8. **Preparation method** Indirect method.	8. By a simple solution process.
9. **Protective action** Protective action on lyophilic colloids is not known.	9. Protective action on lyophobic colloids.
10. **Stabilisation** Stabilises due to charge on the surface.	10. Stabilises by forces of solution.

III. Based on the Nature of Dispersed Phase: General Classification

All colloidal solutions are dispersed system. Depending on the nature of the dispersed phase, modern classification of colloids is as follows:

(a) **Dispersion Colloids**: Mostly lyophobic except silicic acid, e.g., Colloidal sulphur, hydrated ferric hydroxide sol, gold sol, etc.

(b) **Association Colloids**: It means presence of lyophobic and lyophilic parts in the same molecule, e.g., dyes, soaps, surface active agents etc.

(c) **Molecular Colloids or Macromolecules**: Most of them are lyophilic, e.g., cellulose, proteins.

7.4 Transformation of Sol to Gel

Sol can be transformed to gel by:

1. Altering the temperature—Sol on cooling forms gel,
2. Changing pH.
3. Changing the electrolytic concentration,
4. Changing the concentration of solute.

7.5 Reversibility of Colloid

Gel can be transformed to sol by warming. Hence, a gel and a sol may be reversible and protoplasm is a reversible colloid.

7.6 Dispersion of Colloid

Some substances like sulphur, gold, sulphides, silver halides, heavy metal hydroxides are insoluble in water but in certain conditions they could be made apparently homogenous solution to the naked eye but under microscope or ultramicroscope reveals the non-homogeneity, these are colloidal solutions and called **dispersion colloids,** but in general they are lyophobic colloids or Sols. These are colloidal solutions and are evident by the experiments like (1) they have a very slow rate of diffusion through membranes like all colloidal solutions (2) they scatter light, e.g., Tyndal phenomenon.

7.7 Association Colloids

Certain substances like soap, when dissolved in water, forms aggregates of molecules having colloidal dimension due to reversible equilibrium between these aggregates of molecules; they are called "**Micelles**". As for example, when soap, e.g., Sodium Palmitate is dissolved in water, it dissociates into Sodium and Palmitate ions. These Palmitate ions, due to cohesion property, form groups of such ions to become colloidal dimension. These aggregate ions may contain a very few undissociated molecule and called **Micelles.** The colloidal property of soap solution is achieved for the presence of Micelles and the colloidal solution is called "**Association Colloids**". The soap micelles are good conductors of electricity owing to ionic nature and they are moderately strong electrolytes with colloidal properties.

7.8 Molecular Colloids (Intrinsic Colloids) or Macromolecules

The molecules of the substances are very big and show colloidal properties such as proteins, polysaccharides, high polymer and their molecular weight may be from 30,000 to a few millions and can form well-defined crystals due to high degree of symmetry.

7.9 Various Types of Colloidal Dispersion

The colloidal dispersion comprises of two phases viz:

1. **The solute phase**: The particles of the solute phase should be of colloidal dimension.
2. The solvent phase: It may be anything like solid, liquid or gas. Hence the colloidial dispersion may be:
   1. **Solid in Solid**: Colloidal gold particles of Colloidal dimension embedded in glass such as Ruby glass.
   2. **Solid in Liquid**: The two phases will be slightly repellant, e.g., NaCl in alcohol, rubber in water.
   3. **Solid in Gas**: Dust particles in air, smoke.
   4. **Liquid in Solid**: Pearl (it contains H_2O in $CaCO_3$), Opal (H_2O in Silicon dionide).
   5. **Liquid in Liquid (or Lyophilic or Emulsoids)**: Creams, jelly.
   6. **Liquid in Gas**: Water vapour, fog, aerosol.
   7. **Gas in Solid**: Solid foam, Pumice stone.

Among these, solid in liquid (Sols) and liquid in gas (Aerosols) have importance technically for hair spray, insecticide spray, etc.

7.10 Phases of Colloids

Colloids do not settle down on standing because of Brownian movement caused by the small size of the colloidal particles. But on prolonged standing the small particles may coalesce into bigger particles and the latter may settle down which usually does not occur due to: (a) The electrical charge of the particles repel each other and prevents formation of larger particles. (b) Solvation: each particle holds around itself a sheath-like layer of the solvent called solvation (If water is the dispersion medium, then it is known as "hydration") which prevents the particles from coalescing with each other. This solvation is the important factor for stability of lyophilic colloids by preventing aggregation and precipitation of the particles.

7.11 Preparation of Colloidal Solutions

A. Organic Substances

1. By simply dissolving in water: Gelatine, starch, soap, dye stuff.
2. By preparing alcoholic solution of the substance and pouring in warm or boiling water and the alcohol is removed by gentle warming to obtain coarse colloidal solution. e.g., palmitic acid, sulphur etc.

B. For Inorganic Colloids

1. **Condensation Method**: In this method the particles are allowed to attain colloidal size by aggregation of particles in the molecular state, e.g., for formation.
2. **Disintegration Method**: In this method, the substances in bulk is broken down by a colloid mill into colloidal size.

7.12 Properties of Colloids

7.12.1 General properties

A. Size: All matter may be classified according to the size of the particle. If we compare the size of the colloidal particle, it is found that its diameter is between the diameter of molecules and ions and matter in mass (Table 7.3).

B. Surface Forces and Surface Area: The fundamental properties of true solutions and colloidal solutions are qualitatively same but differ quantitatively due to difference in particle size and structure. Every particle, whether colloidal or non-colloidal:

1. has certain fields for force which are probably electromagnetic.
2. has property of coordination owing to certain groupings and unshared electron pairs.
3. is negatively or positively charged.

Table 7.3: Classification of matter according to particle size.

Molecules and Ions	Colloidal Particles	Matter in Mass
Particle Size: Less than 1 millimicron (1 mm) usually 0.1 mμ to 1 mμ (10^{-8} to 10^{-7} cm or less than 0.000001 mm).	**Particle Size:** 1 mm to 100 mμ, i.e., 10 Å (10^{-7} to 10^{-5} cm or 0.000001 –0.0001 mm)	**Particle Size:** Greater than 100 mμ (10^{-5} to 10^{-3} cm or greater than 0.0001 mm).
Visibility Under Microscope: Not Visible under ordinary micro-scope and ultramicroscope	Not visible under ordinary micros-cope but visible under ultramicro-scope.	Visible under ordinary micro-scope.

Table 7.4: Relative size of the particles

The average diameter of colloidal particle is 50 mμ and the lower limit is 1 to 5 mμ and the upper is 100 to 500 mμ. Relative sizes of the different particles compared to the colloidal particles are:

1 millimicron (1 mμ) = 0.000001 mm = 1 nano metre = 10 Å = 1^{-1} cm.

Microscopic limit of visibility of the particles having diameter	...	0.1 μ
H_2 molecule	...	0.1 mμ
Anthrax Bacilli	...	6 μ
Micrococcus	...	1 μ
Ordinary atom	...	0.2 – 0.5 mμ
Colloidal particles	...	1 – 100 mμ
Influenza Virus	...	100 mμ
Tobacco Mosaic Virus	...	300 mμ
Starch Molecule	...	8 mμ
Cane Sugar Molecule	...	0.7 mμ
NaCl Molecule	...	0.28 mμ
Egg Albumen	...	4 mμ
Cellulose	...	15 to 100 mμ
Cellobiose unit in cellulose	...	1.03 mμ
Human red blood cell	...	7.5 mμ
Wavelength of visible light	...	400 – 800 mμ

4. May be highly polarised and act as a dipole and form associations with like or different particles.
5. May have active chemical groups and undergo ordinary reactions characteristic of those groups.

But the forces of the particles of colloidal size are less per unit mass than the molecules of the same substance in true solution. Yet the colloidal state of matter has been considered separately because of:

1. visibility under ultramicroscope;
2. flocculation from solution,
3. separation from other molecules by dialysis,
4. adsorption and holding other molecules and ions to their surfaces which can be measured,
5. show relatively high activity of surface forces as compared with primary chemical forces.

• **Surface Forces**: The surfaces of all atoms and molecules are surrounded by electric or electromagnetic field of forces arising from the electrical particles forming the structure of atoms. In some atoms and molecules the surface fields of force are weak due to neutralization within the particles and in case of others have very strong surface fields of force due to weak internal neutralization. Whenever the particles of matter are surrounded by other particles, their fields of force do not mutually neutralize, resulting in a region of free fields of attraction. This happens at all surfaces or boundary phases (interfaces) between different kinds of matter.

• **Surface Area**: Colloidal state of matter possesses greater surface area per unit of weight than the matter of mass. The colloidal state of matter also possesses unneutralized surface forces to higher degree than the coarse particles of matter in mass. To understand the effect of particle size upon surface area: If one gram-mol of NaCl (58.454 gm) in a cube form has surface area of 54 sq. cm, but if it is crushed into smaller particles of colloidal size (10 mμ), the number of particles would be 27×10^{18} and the total surface area will be 16,000 sq. m.

1. The great example in physiology is the emulsification of fat by bile salts in the small intestine to increase the surface area of the fat particles for better contact with the enzyme lipase where the fat particles are dispersed in the aqueous phase. This function is done by any wetting agent. Bile salts, being a natural detergent, fulfil this purpose in the body.
2. The greater surface area of R.B.C. helps quick diffusion of O_2 in the lungs because blood remains only for a short period.

C. Colour

The colour of a colloidal solution may differ from the colour of the substance, e.g., Zsigmondy's Gold solution is red in colour or mixture of red and blue depending on the size of the colloidal particles, silver solution is grey or greenish grey, sulphur sols may be colourless.

D. Shape

According to the shape, the dispersed colloidal particles are classified into globular or spherical and fibrous or linear and there may be some intermediate shapes of the particles. (Table 7.5).

E. Density

Regarding density of colloidal solutions, it is insignificant for lyophobic solutions since it usually

Table 7.5: Classification of dispersed Colloidal particles according to shape

Globular Shaped Colloids	Fibrous Shaped Colloids
1. Swell less.	1. Swell more.
2. Have lesser streaming birefringence (double refraction).	2. Have higher streaming birefringence.
3. Lesser tendency to form jellies.	3. Stronger tendency to form jellies.
4. Example	4. Example
(a) Chemical: Silver Sol.	(a) Silicic Acid, Nitro-cellulose.
(b) Biological: Albumin, blood proteins, milk proteins.	(b) Biological: Structural proteins, e.g., myosin of muscle, Keratin of nail and hair, fibrin of clotted blood.

follows the law of mixture. But in case of lyophilic colloidal solutions, such as gelatine, they react strongly with water and there is contraction of volume and there is alteration of density.

F. Crystallization

Colloids can be crystallized such as molecular Colloids, e.g., proteins, enzymes, viruses, high polymers, etc. X-ray studies have shown that the particles in Colloidal system are often crystalline in character.

G. Osmotic Pressure and Molecular Weight

The osmotic pressure of the Lyophilic Colloids can be determined and one of the reliable methods for obtaining values of the molecular weight. But in case of Lyophobic Colloids it is difficult to determine osmotic pressure due to presence of electrolytes which cannot be removed completely and do not allow to give correct result of osmotic pressure of the Colloids. The osmotic pressure can be measured by Govaert's Osmometer (Verney's modification). Osmotic pressure of Colloidal solutions, e.g., haemoglobin, proteins, soap, colloidal dyes, etc. has been established. The osmotic pressure of haemoglobin corresponds to a molecular weight of about 68000. Sometimes the osmotic pressure is lower than that would be expected from the molecular weight of the substance due to molecular aggregation. The osmotic pressure being proportional to the number of separate molecules or groups of molecules in a given volume of solution, the larger the aggregate the smaller would be the osmotic pressure.

H. Freezing Point Depression and Boiling Point Elevation

Freezing point depression and boiling point elevation caused by colloidal solutions are quite small.

I. Diffusion

Like solute particles, colloidal particles diffuse at a slower rate from a region of higher to a region of lower concentration.

J. Sedimentation

Colloidal particles tend to settle down very slowly under the influence of gravity on prolonged standing. The rate of sedimentation can be increased to a large extent by the use of a high speed centrifuge known as "ultracentrifuge".

K. Solvation

Each colloidal particle holds around itself a sheath-like layer of the solvent called "Solvation". If water is the dispersion medium, then it is known as "Hydration". This mechanism prevents coalescing of the colloidal particles with each other and important for stability of Lyophilic Colloids by preventing aggregation and precipitation of the particles.

L. Non-Dialysable Property

The colloidal particles are held back and cannot pass through the pores of cellophane or parchment membranes. But these membranes allow the dispersion medium and electrolyte ions and molecules to pass through the pores. In this process, Colloids are separated from a mixed solution and the process is called **Dialysis**.

M. Gelation

A sol may set into a semisolid jelly-like mass called gel by standing, cooling, etc. It holds liquid, ions and molecules, e.g., gelatin gel, skin and eye ointments. By the process of mechanical agitation, the gel may change to the sol by lowering the viscosity and the mechanism is known as **Thixotropy**.

According to the change of volume and energy it may be:

1. **Isothermal gel**: When there is no change in volume or internal energy during gelation.
2. **Endothermic gel**: When there is increase in internal energy and volume during gelation, e.g., methyl-cellulose gel.
3. **Exothermic gel**: When there is liberation of heat and fall in internal energy, e.g., gelatin gel. Gelation occurs both in lyophillic and lyophobic sols. Though it contains significant portion of liquid, it possesses a greater or lesser degree of rigidity. Gelation is caused by the development of a very definite structure within the dispersed system which is brought about by the adherence of some of the colloidal particles to one another to form fibrillar network to enmesh a more dilute colloidal solution by capillary action and partly by hydration of the fibrils. Examples of gels are fruit jellies, cooked starch and heat-coagulated egg-white. Later on, a little volume of liquid diffuses out of the gel for shrinking and condensation of the gel mass due to the gradual shortening of the fibril threads; this is called **Syneresis** or **Weeping of the gel** or **Aging of gels** which can be regarded as the reverse of the absorption which previously took place in the gel formation.

N. Imbibition or Quellung

Many Colloids, such as starch or gelatin, which are insoluble in cold water, absorb water and exhibit a phenomenon called **Quellung** or **Imbibition** exhibited by swelling.

O. Viscosity

Regarding viscosity, Lyophobic Colloids do not show much difference from pure solvents. But Lyophilic Colloids have higher viscosity than the pure solvents because of:

(a) big size of the Colloidal particles,

(b) solvation around the particles and much of the liquid engaged for the process,

(c) zeta potential around the dispersed particles,

(d) elongated and coiled shape of the dispersed particles.

P. Protective Action

Certain Colloids may exercise a protective action to the other colloidal solution. The addition of Lyophilic Colloid which is highly stable, to a solution of Lyophobic Colloid solution, prevents or delays the coagulation and imparts stability to the Lyophobic Colloids. This is known as **Protection** and the colloid is called **Protective Colloid**. Colloidal Gold is much more stable in the presence of little gelatin. It is due to adsorption of protective Colloid by the Sol particles.

Q. Peptization

Specific substances in small amounts called **Peptizer** which change the insoluble solid of low molecular weight which may be a coagulum or precipitate or any other form into the Lyophobic Sol called **Extrinsic Colloid**. Many ions in low concentration act as peptizer such as AgCl precipitate is peptized by Ag^+ or Cl^- ions.

R. Thixotropy

Mechanical agitation may change a Gel to a Sol. It is due to lowering of viscosity. This mechanism is known as **Thixotropy.** When agitation ceases, it solidifies again. Practically all Colloidal solutions show thixotropy but the favourable result is seen in case of rod-shaped micelle. These particles link up with each other to give mesh-like structure but this weak linking cannot stand the stresses by mechanical agitation. Exposure of cell to magnetic field shows thixotropy.

S. Adsorption

Adsorption is another property of Colloids. Colloidal particles called **adsorbent**, adsorbate substances called **adsorbate** on their surfaces such as colloidal plasma proteins adsorb mineral ions, sugar, lipids, etc. and carry to other places.

7.12.2 Optical properties

(a) Brownian Movement

The English Botanist Robert Brown (1827) observed that pollen grains dispersed in water exhibited rapid movement in all directions, rendering impossible to count their number in the field under ultramicroscope. Later, this random zigzag motion was observed in all colloidal particles, including colloidal particles suspended in gaseous medium. Smaller particles exhibited more rapid and brisk motion than the larger ones. Such movements are called **Brownian movements**. Brownian movement cannot be inhibited by agitation, heat and light. The cause of Brownian movement is due to the moving molecules of the dispersion medium constantly colliding with the colloidal particles caused by thermal motions of the liquid molecules and impart momentum to the colloidal particles. This movement is a strong evidence in support of the Kinetic molecular theory of liquid and gases. Fig. 7.1 shows Brownian movement of Colloidal particles recorded by cinematography and ultramicroscope. Ultramicroscopic studies have shown that some of the granules present in the living cells move about within the cells and, in certain cases, display Brownian movement, e.g., in muscle cells lencocytes.

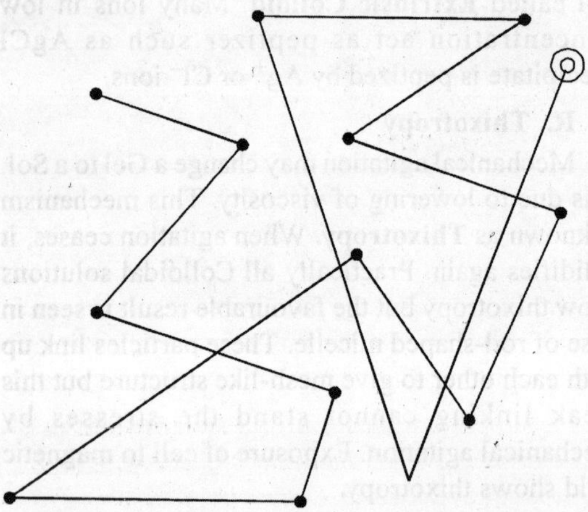

Fig. 7.1: Brownian movement of Colloidal particles recorded by Cinematography and ultramicroscope

(b) Tyndall Effect or Phenomenon or Faraday Tyndall Phenomenon

The presence of colloidal particles is not detectable under a microscope but can be made visible by optical means. When a beam of light is passed through an optically homogenous solution, the course of the beams is not visible. When a beam of sunlight is falling in a dark room, the dust particles in the path of light is visible due to scattering of light from their surfaces. Thus, if an intense beam of light be passed at right angle through a colloidal solution, the particles in it scatter the light and the beam is rendered visible and the individual particles can be counted. This phenomenon is known as **Tyndal effect** and the apparatus using the principle of dark ground illumination is called **Ultramicroscope** contructed and used by Zsigmondy (Fig. 7.2). The limit of resolution is 5 mμ or 60Å. The tyndal effect may be regarded as a test for the presence of ultramicroscopic particles having size between 5 mμ to 50 mμ. Tyndal showed that if the particles were sufficiently minute, the light dispersed by them at right angles to the beam was polarised which can be tested by looking at the beam through a Nicol's Prism. Not all colloidal solutions show these particles in ultramicroscope. Lyophobic sols only show this effect, but not the Lyophilic colloids. The reasons are:

1. The absence of optical heterogeneity which may be due to coincidence of refractive indices of hydrated Colloids and water.
2. Absence of any clear dividing surfaces between the two phases.

   It can be stated that it is not the Colloidal particles itself which is rendered visible but only the light scattered from it.
3. Small size of the particles and addition of any substances causing aggregation and, therefore, increase in size of colloidal particles, render them into view.

7.12.3 Electrical properties

(a) Electrical Charge

The Colloidal particles (both lyophilic and lyophobic) carry an electric charge either positive or negative. The charge upon the particles has importance such as stabilizing colloidal solutions because all the particles in a given sol are charged alike and repel each other and remain in suspension. The neutralization of these charges has importance in the precipitation or flocculation of Colloids. The electrical charges on colloidal particles arise:

1. From the ionization of groups in the particle surface, e.g., protein,
2. Through selective adsorption of ions from the medium upon surfaces.
3. Both the mechanisms may act concurrently.

(b) Electrical Charge at the surface

This theory explains the charge relations at the surface of the colloidal particles. According to this theory the surfaces of the colloidal particles are negatively charged and are surrounded by oppositely charged ions, e.g., positive ions. There are two layers of positively charged ions.

1. An immobile layer of ions from the liquid adsorbed on the surface.
2. More diffuse layers of mobile ions in liquid around the particles.

The mobility of mobile layer increases according to the distance from the negatively charged surface. But if the colloidal particle is positively charged, the relations of the double-layer are opposite and the surrounding liquid contains negative ions.

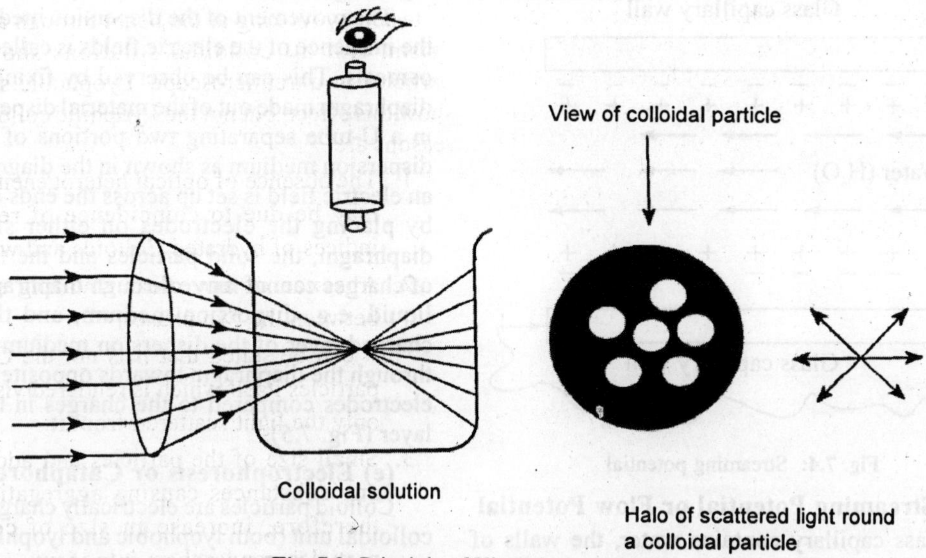

Fig. 7.2: Principle of Ultramicroscope

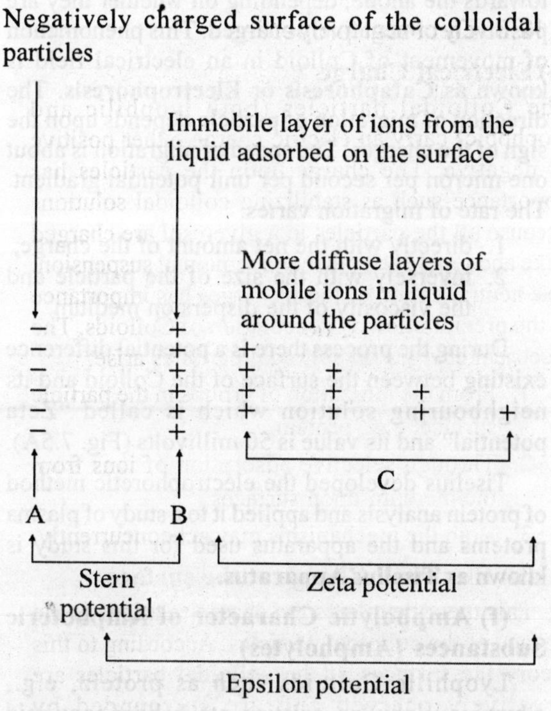

Fig. 7.3: Helmholtz-Gouy double layer and potential

Potential

1. **Epsilon (∈) Potential or Electro-chemical Potential:** The potential drop across all the ionic layers from the negatively charged surface of the particle (A) up to the mobile positive ions into the solution (C) (i.e., A to C) is the **Epsilon potential.**
2. **Zeta Potential (ξ) or Electrokinetic Potential:** The potential between the immobile positively charged layers of ions (B) and the mobile layer of positive ions (C) is called **Zeta potential (ξ) or electrokinetic potential** (i.e. B to C).
3. **Stern Potential:** The potential between the immobile positive layer of ions (B) and the negatively charged particle surface (A) is called **Stern potential** and is equal to Epsilon (Electrochemical) potential minus Zeta (Electrokinetic) potential (i.e., A to C – B to C). Stern Potential = Epsilon potential-Zeta potential.

Epsilon potential = Stern potential + Zeta potential.

Among these potentials, the Zeta potential is important pertaining to the properties of Colloids. By addition of electrolyts, there is drop of Zeta potential to a certain level called **Critical Potential** and the double layers of the colloidal particles do not persist and there is aggregation of the particles, and precipitation occurs (Fig. 7.3).

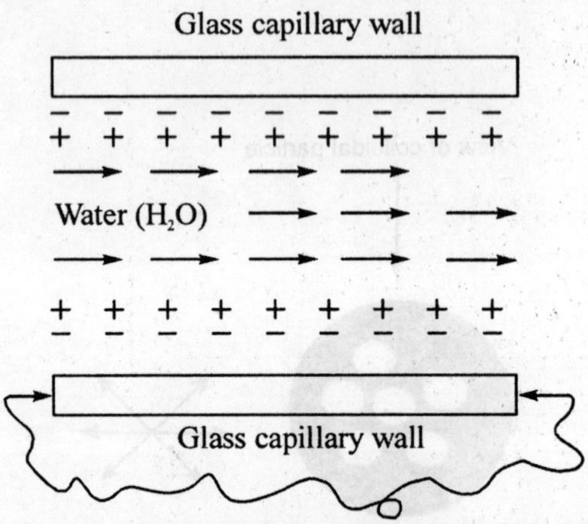

Fig. 7.4: Streaming potential

(c) Streaming Potential or Flow Potential

A glass capillary contains water, the walls of the capillary become negative and the water is positive. If the water is forced to flow through the glass capillary, a potential is developed between the two ends of the capillary and this potential is known as **Streaming Potential** or **Flow Potential.** The explanation is that the capillary walls adsorb OH^- ions from the water leaving H^+ ions which charges the water molecules positively. These positively charged water molecules are mobile and accumulate in excess at one end while flowing through the capillary leaving excess negatively charged molecules at the other. This develops a potential difference between the two ends of the glass capillary. Difficulty is encountered during filtration through filters with fine pores due to establishment of strong streaming potential in the membrane pores. This is observed particularly in case of solutions having low electrolyte concentration because electrolytes reduce the potentials at the pore surface (see reduction of Zeta potential on the surface of Colloids by electrolytes (Fig. 7.4).

- **Biological importance**

1. Probably the flow of fluids through the cell membranes of tissues produces streaming potential which is a contributory factor for the potential differences across the membranes.
2. The electrically charged pore surfaces of filters is important for assessing the impermeability of the filters to bacteria and other micro-organisms which are usually negatively charged.

(d) Electro-osmosis

The movement of the dispersion medium under the influence of the electric fields is called **Electro-osmosis.** This can be observed by fixing a porous diaphragm made out of the material dispersed phase in a U-tube separating two portions of the liquid dispersion medium as shown in the diagram. When an electric field is set up across the ends of the tube by placing the electrodes on either side of the diaphragm, the solid particles and the fixed layer of charges cannot move through diaphragm but the liquid, e.g. dispersion medium, and the diffuse charged layer of the dispersion medium can move through the diaphragm towards oppositely charged electrodes compared to the charges in the diffuse layer (Fig. 7.5).

(e) Electrophoresis or Cataphoresis

Colloid particles are electrically charged. So, the colloidal unit (both lyophobic and lyophillic) move in an electric field either towards the cathode or towards the anode, depending on whether they are positively or negatively charged. This phenomenon of movement of Colloid in an electrical field is known as **Cataphoresis** or **Electrophoresis.** The direction of migration of particle depends upon the sign of the charges. The speed of migration is about one micron per second per unit potential gradient. The rate of migration varies:

1. directly with the net amount of the charge,
2. inversely with the size of the particle and the viscosity of the dispersion medium.

During the process there is a potential difference existing between the surface of the Colloid and its neighbouring solution which is called "Zeta potential" and its value is 50 millivolts (Fig. 7.5A).

Tiselius developed the electrophoretic method of protein analysis and applied it to a study of plasma proteins and the apparatus used for this study is known as **Tiselius Apparatus.**

(f) Ampholytic Character of Amphoteric Substances (Ampholytes)

Lyophilic Colloids such as protein, e.g., albumins, globulins, gelatine, etc. behave as an acid

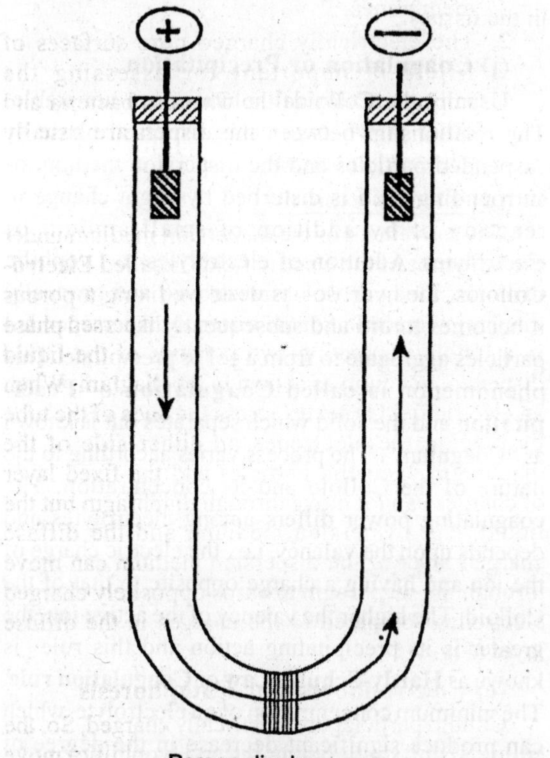

Porous diaphragm

Fig. 7.5: Electro-osmosis

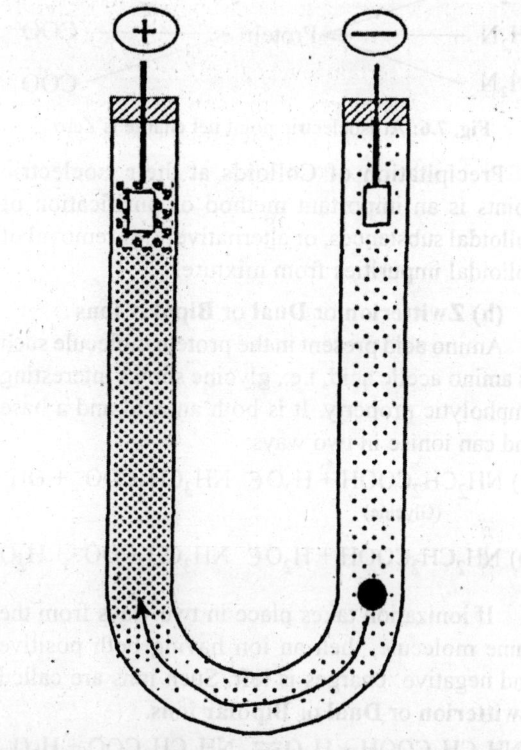

Negatively charged colloidal particles of the sol migrating to the positive electrode and after loosing charge, are deposited on it.

Fig. 7.5A: Electro-osmosis

in alkaline media and as a base in acid media. Hence they are called **Ampholytes** because they carry both positive and negative charges. It is due to ionisation of the molecules. The proteins contain amino acid and it has NH_2^+ group and COO^- (Carboxyl) group and they are positively charged in acid media and negatively charged in alkaline media.

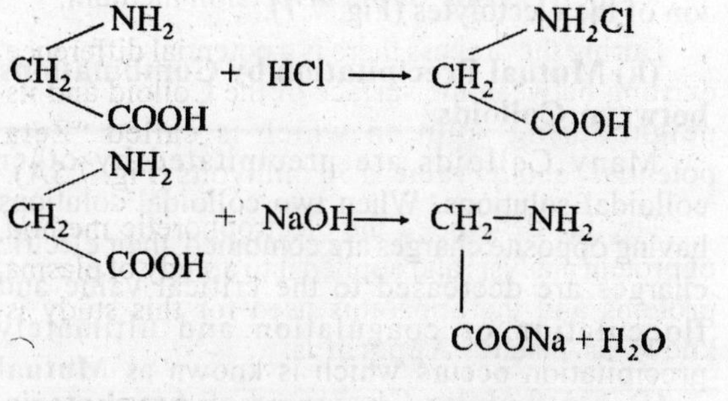

(g) Isoelectric pH or Isoelectric Point

Whether a portein has a positive or a negative charge depends cheifly on the hydrogen ion concentration or pH becase pH has the influence on the ionisation of the amino acids of protein. Due to the ampholytic character of Colloid, during cataphoresis, at particular pH, the colloidal particle does not move to the anode or cathode. At this point, no cataphoresis takes place. This pH of electrical neutrality of the molecule is known as **Isoelectric Point or Isoelectric pH**. At this point, following events are observed:

1. Complete ionisation of both groups according to the Zwitterion theory.
2. Ion carries equal number of positive and negative charges, hence has a net charge of zero, i.e., it carries no charge (Fig. 7.6.).
3. Often shows flocculation.
4. Exhibits maximum precipitability.
5. Minimum viscosity.
6. Minimum osmotic pressure.
7. Minimum imbibition.
8. No cataphoresis.

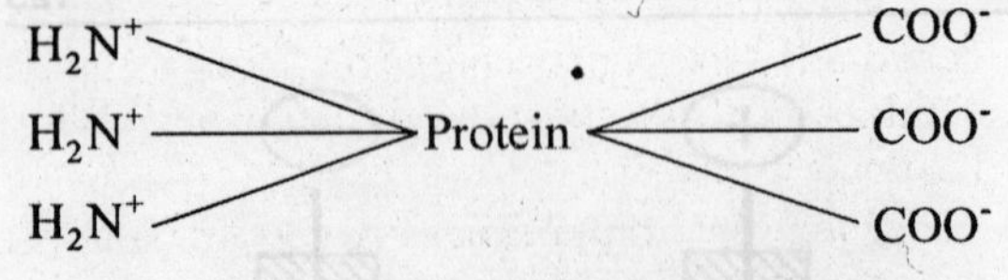

Fig. 7.6: At Isoelectric point net charge is Zero

Precipitation of Colloids at their isoelectric points is an important method of purification of colloidal substances, or alternatively, of removal of colloidal impurities from mixture.

(h) Zwitterion or **Dual** or **Bipolar ions**

Amino acid present in the protein molecule such as amino acetic acid, i.e., glycine shows interesting ampholytic property. It is both an acid and a base and can ionise in two ways:

(a) $NH_2CH_2COOH + H_2O$ € $NH_3CH_2COO^- + OH^-$
(Glycine)

(b) $NH_2CH_2COOH + H_2O$ € $NH_3CH_2COO^- + H_2O$

If ionization takes place in two ways from the same molecule, then an ion having both positive and negative charges is left. Such ions are called **Zwitterion** or **Dual** or **Bipolar ions.**

$$NH_2CH_2COOH + H_2O \rightleftharpoons {}^+NH_3CH_2COO + H_2O$$

Amino acid may exist in aqueous solution near its isoelectric point as Zwitterion. The Zwitterion theory can be applicable to polypeptides and proteins.

(i) The Hofmeister Series and Colloids

When gelatin is immersed in aqueous solutions, its swelling is hastened by NaCl, NaBr, KCl, NH_4Cl and $NaNO_3$ and retarded by Sodium Sulphate, Tartarate, Citrate and Acetate. The action of ions of Colloid (e.g., protein) is a complex phenomenon since both positive and negative ions exert an effect which is dependent on the nature of the ions. It can be arranged for certain ions in the order of their efficacies in affecting the phenomenon and other properties of Colloids. Arrangement in order of series is called the **Hofmeister series** which is as follows: Sulphate < Taratarate < Accetate < Cl < NO_3 < I < SCN. These have a tendency, in the order above, to precipitate positively charged, e.g., acidified protein, while the cations like Li > Na > K > NH_4 cause precipitation of positively charged protein in the above order. But in alkaline media there is reversal of the series. The Hofmeister series has importance in the physiological effects of ions in the tissues.

(j) Coagulation or Precipitation

Usually the Colloidal solutions are not stable. The relationship between the dispersed phase or suspended particles and the dispersion medium or surrounding fluid is disturbed by slight change in reaction or by addition of small amount of electrolytes. Addition of electrolytes to Lyophilic Colloids, the hydrosol is destroyed and gradually it becomes turbid and subsequently the suspended particles aggregate to from a gel or precipitate. This phenomenon is called **Coagulation** or **Precipitation** and the solid which separates out is known as "Coagulum". The process varies according to the nature of the Colloid and its concentration. The coagulating power differs among electrolytes and depends upon the valency, i.e., the electric charge of the ion and having a charge opposite to that of the Colloid, The higher the valency of the active ion, the greater is its precipitating action and this rule is known as **Hardy-Schulze Law** or **Coagulation rule.** The minimum concentration of an electrolyte which can produce significant decrease in the degree of dispersion is called the **Flocculation value**. The mechanism of coagulation of lyophobic colloidal solution may be due to neutralisation of the protective charge of the Colloids by the oppositely charged ions of the electrolytes and ultimately agglomeration of the colloidal particles. But the main reason for coagulation and precipitation is both dehydration of the colloidal particles by combining with layers of water surrounding them, i.e., hydrate shell and at the same time reduction of the charge of the colloidal particles by the positive or negative ion of the electolytes (Fig. 7.7).

(k) Mutual Precipitation by Combinations between Colloids

Many Colloids are precipitated by other colloidal solutions. When two colloidal solutions having opposite charges are combined, their electric charges are decreased to the critical value and flocculation or coagulation and ultimately precipitation occurs which is known as **Mutual Coagulation of Colloids** or **Mutual Precipitation of Colloids**. Thus, ferric hydrate in collodial solution is precipitated by colloidal silica or colloidal gold, both Colloids being thrown out of solution.

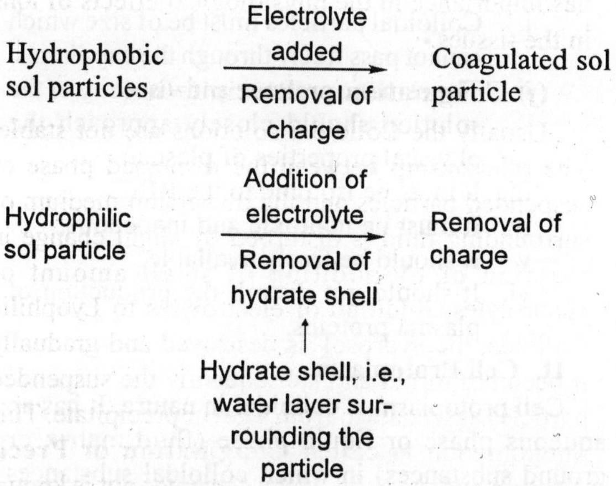

Fig: 7.7: Mechanism of coagulation or precipitation

(l) Salting Out

Emulsoids (Lyophilic Colloids) are not precipitated by addition of salts in amounts which is sufficient to precipitate Suspensoids (Lyophobic Colloids). In order to precipitate Emulsoids, higher amount of salts are required to dehydrate the colloidal particles and reduce the charge upon the particles by the positive or negative ion of the salt. The precipitation of Emulsoids (Lyophilic Colloids) by adding large amounts of soluble salts is known as **Salting Out**. For example, the colloidal protein particles in blood serum is precipitated by large amount of ammonium sulphate which is outstanding due to its high solubility in water (760 gm per litre at 20 °C) and relatively low temperature coefficient of solubility. The effectiveness of salting out depends upon the nature of both the cation and anion of the salt. Cations and anions arranged according to decreasing salting out power are as under:

Cation Series

$Li^+ > Na^+ > K^+ > Rb^+ > Cs^+$ and

$Mg^{++} > Ca^{++} > Sr^{++} > Ba^{++}$

Anion Series

$Sulphate^- < Cl^- < Br^- < NO_2^- < I^- < CNS^-$

These arrangements of cations and anions in series according to their salting power are called **Lytropic or Hofmeister or Hofmeister Lyotropic** series. These series of salting out may indicate the order of hydration of ions. More hydration capacity of ions means more efficiency for dehydration of colloidal particles causing precipitation. But in case of precipitation of Emulsoids of protein by heavy metal ions like Pb^{++}, Hg^{++} Fe^{+++}, Ag^+, a small amount is required for protein solution made alkaline to the isoelectric pH. In this case dehydration does not occur but the heavy metal ions combine with the negative protein ions to form insoluble complex.

Biological improtance: A good example of salting out is the fractionation of the mixture of proteins such as serum proteins by different saturation of ammonium sulphate.

(m) Salting in

Addition of minute amount of electrolyte ions especially divalent or multivalent ions in the colloidal solution increasing the stability of the solution is referred as **Salting In** process, e.g., the uglobulins are brought into solution and the solubility of albumins is increased. It may be due to either (1) some ions are adsorbed on the colloidal particles to increase the like charges of the particles or (2) some ions may interact with the surface ionic groups of the colloidal particles to depress the ionization into opposite ionized group on the colloidal particles surface.

As the ionic strength is increased further in a colloidal solution of mixed proteins, the effective concentration of water is decreased and ultimately a point is reached where protein-protein interaction exceeds the protein-water interaction and now the protein is being salted out.

(n) Coacervation

De Jong defines that, it is the process occurring in a condensed colloidal system, in which the particles with positive and negative electric charges are mutually attracted as it is seen in case of conversion of fibrinogen (Hydrosol) to fibrin (Hydrogel) a less rigid structure. Fibrinogen itself is composed of long fibre-like molecules. The fibrin forms ultramicroscpic crystal-like needles which are deposited and form tenuous, interlacing filaments within the structure of the protein and the immobilization of the molecules in this process is called **Coacervation.** According to Mommaertz, a profibrin is formed then by electrostatic attraction of its particles, a mesh-like interlacement causes formation of fibrin.

7.13 Importance of Colloids

7.13.1 Biological

I. Pharmaceutical or medicinal industry

1. Argyrol or protargol are protected colloidal solutions of silver used in eye infections.
2. Colloidal gold, manganese, calcium are used as intramuscular injection for tuberculosis and ricket.

3. Colloidal sulphur – used as germ killer.
4. Colloidal antimony – used for Kala-azar.

7.13.2 Physiological

I. Blood

1. **Blood Clotting or Coagulation**: When blood is collected in a test tube, it is observed that the blood has lost its fluidity and becomes jelly-like mass after 4-5 minutes which is known as clotting and it is a colloidal process. It is due to conversion of plasma protein fibrinogen which was in soluble form (hydrosol) into an insoluble fibrin (hydrogel).
2. **Stoppage of Bleeding (Styptic action)**: Blood is a colloidal solution and in case of bleeding, it can be arrested by forming clot by application of trivalent aluminium ions of alum or ferric ions of ferric chloride locally.
3. **Oxygen Carraige**: The greater part of O_2 taken up by the blood in the lungs is adsorbed on the surface of colloidal RBC and not dissolved in it.
4. **Carbon Monoxide Poisoning:** Carbon monoxide gas is a deadly poison to human system because once the RBC adsorbs the gas, they cease to carry O_2.
5. **Globulin of Plasma**: Globulins are held in solution by protective action of albumin.
6. **Calcium Phosphate of Plasma**: It is a colloidal suspension and protected by plasma proteins.
7. **Transport by Adsorption**: Many electrolytes and organic substances are transported by adsorption of the materials on the surface of plasma proteins.
8. **Colloids as Blood Substitutes**: In certain pathological conditions it is necessary to restore the blood volume to normal (e.g., burn, shock, severe haemorrhage, etc.) by transfusion of certain materials as blood substitutes. The following materials are usually employed: (i) Whole blood, (ii) Plasma or serum, (iii) solution of Colloids, e.g., gum acacia, isinglass, dextran, polyvinyl, e.g., pyrallidone, etc. and (iv) solution of crystalloids, saline, glucose solution or glucosaline etc. Solutions of Colloids are being used as blood substitutes due to limited availability of whole, blood, serum etc. These colloidal solutions fulfil apparently the physical properties of plasma as under:
   i. Colloidal particles must be of size which cannot pass freely through the capillaries.
   ii. The osmotic pressure and viscosity of the solution should closely approach the physical properties of plasma.
   iii. It must be isotonic to the RBC
   iv. It must be nontoxic and innocuous.
   v. It should be freely available.
   vi. It should not disturb the production of plasma proteins.

II. Cell Protoplasm

Cell protoplasm is colloidal in nature. It has an aqueous phase or liquid phase (fluid matrix or ground substances) in which colloidal substances (granules, globules of solid and semisolid particles), i.e., dispersed phase are suspended. Protoplasm is a **Sol** and commonly known as **Hydrosol** and passes readily to **Gel**, i.e., **Hydrogel** which is reversibly seen in case of pseudopodia.

III. Movement of Pseudopodia

Movement of pseudopodia in protozoa (e.g., amoeba), macrophages and phagocytic, leucocytes is due to tansformation of protoplasm from gel to sol reversibly causing increase in volume and fluidity.

IV. Cytokinesis and Protoplasm

During cytokinesis in a dividing cell, there is flow of protoplasm owing to reversible sol-gel transformation.

V. Plasma, Lymph and Milk

They are in a colloidal state and usually as Emulsoid. Blood is a colloidal solution where the RBCs are suspended.

Milk is an aqueous emulsion of fat droplets. The phosphoprotein casein in the milk acts as stabilizing agent for emulsion of fat in milk.

VI. Surface area

Colloidal state of matter possesses greater surface area per unit weight than the matter of mass. The Colloidal state of matter also possesses unneutralized surface forces to higher degree than the coarse particles of matter in mass. This characteristic is helpful for some processes like adsorption, enzyme-substrate reaction, surface tension, etc.

1. The emulsification of fat by bile salts in the small intestine is for better contact with the enzyme lipase where the fat particles are dispersed in the aqueous phase.

2. The individual RBC having surface area of 0.000128 mm^2 and total surface of Red Blood Corpuscles in the blood of a man of 70 kg is 1500 sq. metres is helpful for quick diffusion of O_2 in the lungs to combine with haemoglobin because the blood remains only for a short period in the lung capillaries.

VII. Adsorption

Adsorption is one of the characteristic property of Colloid. Various physiological processess show adsorption under as:

1. O_2 is adsorbed on the surface of RBC but not dissolved.
2. Enzyme-substrate reaction.
3. Histological staining with dyes is due to this process.
4. Transport of minerals, organic substances by the plasma proteins, etc.

VIII. Colloidal Osmotic Pressure

Water balance between blood and tissue fluids is maintained by the plasma proteins. Since the protein concentration of plasma is higher (7%) than the tissue fluids (1.5%), the colloidal osmotic pressure of plasma exceeds the osmotic pressure of tissue fluids and difference is about 22 mm Hg. and is know as **Effective Osmotic Pressure or Oncotic Pressure**. The distribution of fluid between blood and tissue fluids is regulated by the balance between the blood pressure tending to force fluid into tissue fluids and the effective protein osmotic pressure of plasma tending to draw fluid from the tissue fluids into the blood. The albumin fraction of plasma is mainly responsible for most of the effective osmotic pressure because of its high concentration, smaller molecular weight (69,000), lowest isoelectric pH, more binding capacity with cations which are held up and are not allowed to diffuse out (Donnan effect). This unequal distribution of diffusible ions due to Donnan effect is responsible for about 20 per cent of the osmotic pressure of plasma protein.

There will be excessive accumulation of water in the tissue spaces called edema or dropsy due to pathological conditions like nephritis (loss of albumin in urine), liver diseases, blood loss. In these cases either filtration increases or decreases or osmosis decreases due to prolonged, persistent elevation of the filtration pressure (i.e., blood pressure) in the capillaries and a reduced protein concentration of blood (Fig. 7.8).

7.13.3 Industrial application and others

1. Smoke precipitation.
2. Sewage disposal.
3. Purification of water.
4. Formation of delta.
5. Rubber industry.
6. Tanning of leather.

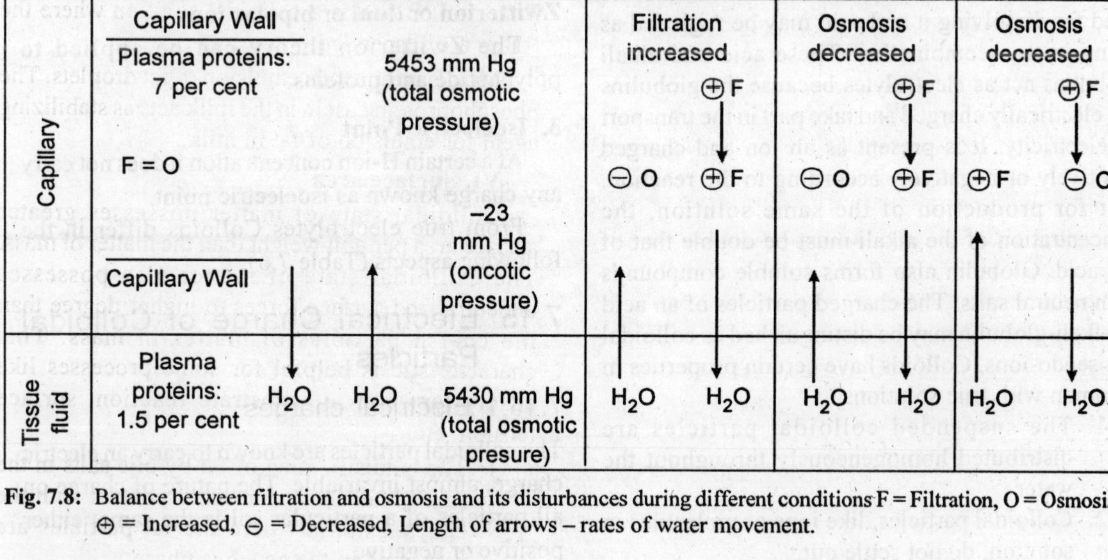

Fig. 7.8: Balance between filtration and osmosis and its disturbances during different conditions F = Filtration, O = Osmosis, ⊕ = Increased, ⊖ = Decreased, Length of arrows – rates of water movement.

7. Cleansing of soap.
8. Froth floatation process.
9. Chemical warfare.
10. Photographic plate.

7.13.4 Colloids in nature

1. Rain. 2. Blue colour of the sky.
3. Tail of comets. 4. Milk.
5. Soil.

7.14 Colloidal Ions

Colloids are characterized by (1) the chemical structure of the colloidal molecule and (2) electric charge carried by the colloidal molecule.

The chemical behaviour of the Colloids is usually determined by the surface phenomena. Hardy (1905) studied the behaviour of globulins and observed the combinations of both characteristics. Four states of globulins were identified:

A. Solid condition,

B. In solution such as:
1. Globulin itself,
2. Globulin compound with acid,
3. Globulin compound with alkali,
4. Globulin compound with neutral salt.

7.14.1 Addition of acid or alkali

When acid or alkali is added to the neutral globulin, there is gradual conversion of opaque suspension into a transparent solution. The amount of acid or alkali combining with the globulin cannot be determined. If the solution is dried up, the solid HCl-globulin is obtained and it contains all the chlorine used for dissolving it and acid may be regarded as being in true combination. These acid and alkali globulins act as electrolytes because the globulins are electrically charged and take part in the transport of electricity. It is present as an ion and charged positively or negatively according to the reaction. But for production of the same solution, the concentration of the alkali must be double that of the acid. Globulin also forms soluble compounds with neutral salts. The charged particles of an acid or alkali-globulin may be distinguished as colloidal or pseudo-ions. Colloids have certain properties in common with true solutions.

1. The suspended colloidal particles are distributed homogeneously throughout the water.
2. Colloidal particles, like ions or molecules in solution, do not settle out.

Colloids have shown the following properties:

1. Ampholyics

The relation of globulin to acids and alkalis is similar to that of ampholytes or amphoteric substances, e.g., amino acids. The amino acid can react as a basic anhydride with acids or as an acid with bases, thus:

As a Base with Acid

$$CH_2\begin{cases}NH_2\\COOH\end{cases} + HCl = CH_2\begin{cases}NH_2\ HCl\\COOH\end{cases}$$

As an Acid with Base

$$\begin{array}{l}CH_2.NH_2\\ |\\ COOH\end{array} + NaOH = \begin{array}{l}CH_2NH_2\\ |\\ COONa\end{array} + H_2O$$

2. Zwitterion

Amino acids have a Zwitterion structure: $^{+}NH_3.R.COO^{-}$.

It can ionise in two ways:

$$1.\ NH_2CH_2COOH + H_2O \rightleftharpoons NH_3CH_2COOH + OH^{-}$$

$$2.\ NH_2CH_2COOH + H_2O \rightleftharpoons NH_2CH_2COO^{-} + H_2O$$

When both types of ionisation occur from the same molecule, an ion carrying both a positive and a negative charge is left and these ions are called **Zwitterion or dual or bipolar ions.**

The Zwitterion theory can be applied to polypeptide and proteins.

3. Isoelectric Point

At a certain H-ion concentration it does not carry any charge known as isoelectric point.

From true electrolytes Colloids differ in the following aspects (Table 7.6).

7.15 Electrical Charge of Colloidal Particles

7.15.1 Electrical charges

The colloidal particles are known to carry an electric charge, almost invariable. The nature of charge on all particles of a particular sol is the same, either positive or negative.

Table 7.6: Comparison of true electrolyte and Colloids

Colloids	True electrolyte
1. Colloidal particles are much larger than ionic or molecular particles.	1. Smaller
2. Particles are of varying size according to the conditions in which they exist.	2. An ion has a mass which (e.g., Na^+, Cl^-) is constant.
3. Larger total surface area is required because most reactions occur on the surface.	3. Reaction occur at the point of contact or adjoining surfaces of those compounds.
4. Carry varying charges of electricity.	4. Always carries the electric charge.
5. Charge depends upon the H-ion.	5. Does not.
6. At isoelectric point there is no charge, i.e.. neutral.	6. Does not occur.

Because of the identical type of charge present on all the particles of a sol they repel one another. They carry varying charge of electricity (ions always carry the same charge) and whether it has a positive or negative charge depends chiefly on pH of the medium and at certain pH called isoelectric point, these do not carry any charge and do not migrate in an electric field. The particles usually move if placed in an electric field and the direction of their movement reveals the nature of charge, e.g., colloidal ferric hydrate is electro-positive and travels to the cathode. The presence of charge is not, however, a necessary condition for a stability of a colloidal solution. But any agency which will tend to discharge them will often cause precipitation and coagulation. Thus colloidal ferric hydrate, silica, gold or boiled albumin are thrown down by the addition of traces of neutral salts and it is found that in this process they carry with them some of the ion with the opposite charge to that of the colloidal particle. Thus, in the precipitation of the electro-positive ferric hydrate, the negative ion of the salt is the determining factor, the coagulating power increasing rapidly with the valency of the acid.

7.15.2 Origin of electric charge

Several explanations have been offered for it, but the origin of the charge depends upon the nature of the colloidal system.

1. **Due to Self-Dissociation or Ionisation:** Colloidal electrolytes such as soap (Sodium salt of fatty acid) dissociate, i.e., ionise in solution giving ions. The hydrocarbon parts of the ions have a marked affinity with one another and they are drawn together. Thus, the negative ions aggregate to form an ionic micelle which is of colloidal size. This accounts for the presence of negative charge on the colloidal soap particles.

2. **Due to Presence of Acid or Basic Groups:** Proteins readily yield sols. The molecule of a protein has a carboxyl group – COOH and also a basic amino group $-NH_2$. Thus, the complex protein particles may be represented as H_2N-X-COOH. In acid solution, the protein particle becomes positively charged. because of the ionization of the amino group.

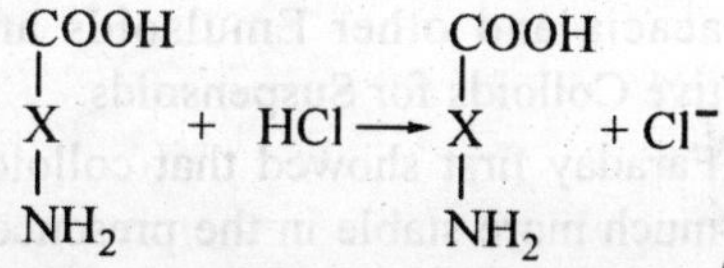

On the other hand, the protein molecule is negatively charged in alkaline solution due to the ionization of the carboxyl acid group. It may be clearly understood that the protein Colloidal solution particle will possess many of these carboxyl and amino groups on its surface and thus possess many units of positive or many units of negative charge on it.

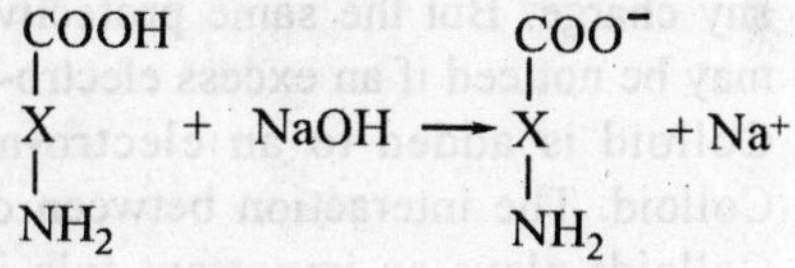

3. **Due to Selective Adsorption:** According to this, the origin of the charge is due to selective adsorption of a certain type of ions present in the dispersion medium. The particles of a particular sol are positive or negative according to the sol prepared in an acidic or alkaline medium which can be explained to be due to the selective adsorption of H^+ ions or OH^- ions, respectively, present in excess of the dispersion medium. According to Helmholtz and Gouy an electric double layer is formed on the surface of the particle known as Helmholtz-Gouy double layer theory (see, electrical properties of Colloid, Fig. 7.3).

7.16 Combination between Colloids

I. Mutual Precipitation or Mutual Coagulation of Colloids.

See (K) under electrical properties of Colloids.

II. Protective Colloids or Protective Action of Lyophilic Colloids

Certain Colloids may exercise a protective action on the other colloidal solution. The addition of Lyophilic Colloid which is highly stable to a solution of Lyophobic Colloid solution, prevents or delays the coagulation and imparts stability to the Lyophobic Colloids. Even addition of electrolytes does not produce coagulation. This is known as **Protection** and the Colloid which is added to achieve it is called **Protective Colloid**. Gelatin, Gum acacia and other Emulsoids are used as protective Colloids for Suspensoids.

1. Faraday first showed that colloidal gold is much more stable in the presence of a little gelatin. Here the particles of gelatin (emulsoid) are adsorbed by the particles of gold (suspensoid).
2. The Colloids of serum can dissolve a significant amount of purified globulin. But globulin alone shows a drift in the electric field and the resulting solution is not affected by the passage of a current through it and in these cases protective Colloid does not carry any charge. But the same protective effect may be noticed if an excess electro-positive Colloid is added to an electro-negative Colloid. The interaction between different Colloids plays an important role in many biological phenomena.

7.16.1 Mechanism

The mechanism is that the particles of the sol adsorb the protective Colloid particles which thus form a protective layer around them.

7.16.2 Power of protection

The Lyophilic Colloids differ with each other regarding power of protection. The protective power of certain Colloids is usually measured in terms of **Gold Number**. Protective power is also measured in terms of the **Congo Rubin Number**.

Gold number can be defined as "the least weight in milligrams of the hydrophilic Colloid which is just capable of preventing a change in colour, i.e. red to violet of 10 ml of a standard gold sol (0.0053-0.0058%) by addition of one ml of 10% NaCl solution."

Low Gold Number indicates higher protective action, e.g., gelatin and non-proteins have higher Gold Number. The importance of Gold Number has a great diagnostic value clinically for the diseases of central nervous system where cerebrospinal fluid (CSF) shows characterstic change in Gold Number.

7.16.3 Importance of protective colloid

A. Non-biological

1. In photographic plate, the Lyophobic Silver Bromide is protected by Lyophilic Gelatine.
2. In soil, the clay particles are protected by the Humic Acid Colloid present in the soil.
3. Gelatin and Carboxy Methyl Cellulose are used in ice-cream to prevent formation of coarse ice crystals.

B. Biological

1. Calcium phosphate of plasma in colloidal suspension is protected by proteins.
2. Aggregates of calcium phosphate suspended in the milk are protected by the milk proteins.
3. Less soluble cholesterol and the calcium salt of bilirubin are held in colloidal suspension by the bile salts and bile protein which act as protective Colloids. Gallstones may form by precipitation of cholestrol due to insufficiency of protective Colloids.
4. Urinary stone formation is prevented by the protective Colloids in urine.
5. The ass's milk is freely digestible than the cow's milk because of better protection from coagulation.
6. As high as 70 per cent solution of colloidal silver (Collargol) can be prepared in medicine due to protective action.
7. There may be protection between toxin-antitoxin reaction and enzymes-substrate reactions, as they are the complex formed due to adsorption combinations.
8. **Liver function test based on protein constituent of the Plasma**: The liver is the sole organ for albumin formation and normally albumin: globulin ratio is maintained. But in case of liver damage, albumin fraction is

reduced, and β and γ globulin fractions are increased. Usually albumin fraction tends to protect the globulin from the action of reagents of the test like Colloidal Gold Test, and inhibits the reaction but due to low albumin, globulins are exposed to pronounced reaction.

7.17 Emulsion and Suspension

In certain circumstances the dispersed particles are having diameter more than 100 nm and is visible under microscope. Hence sols are frequently subdivided into two classes:

1. **Suspension**: Dispersion of large solid particles in dispersion medium, e.g., metallic sols.
2. **Emulsion**: Dispersion of large liquid droplets in a liquid dispersion medium e.g. milk, protein solution, dilute egg white. The dispersed liquid particles in an emulsion may be of colloidal dimensions but frequently they are of suspension dimensions and large enough to be discernible to the naked eye or under an ordinary microscope.

Liquid substances which are not capable of forming true solutions in a given solvent can sometimes become dispersed throughout said solvent in the form of Emulsion. It, therefore, can be defined as a mixture of two mutually insoluble liquids, one of which is dispersed in the other in more or less finely divided particles.

7.17.1 Example

A. Non biological

(i) Liquid Hydrocarbons, (2) Mixtures of Liquid Hydrocarbons.

B. Biological

(1) Milky secretion of certain plants, e.g., milkweed, poppy. (2) Milk: Fine particles of fat suspended in an aqueous medium with the help of Casein as emulsifying agent and this type of emulsion is referred to as "oil-in-water" type. The conversion of milk to butter is an example of phase reversal. (3) Butter: This is the opposite type of emulsion, i.e., fine particles of water in a medium of fat, i.e., "water-in-oil" type. (4) Fats: Water with oily substance.

7.17.2 Emulsifying agents or emulsifier

Though an Emulsion can be prepared by mixing and shaking the two liquids together, it is difficult to maintain unless some substances are provided to check the marked tendency of the two liquids to seperate. These selected agents or substances, when added to the Emulsion, are capable of prolonging the period of Emulsion. These substances are called **Emulsifying Agent** or **Emulsifier** or **Stabilizer.**

7.17.3 Classification of emulsifying agents

A. For oil-in-water type Emulsion–Alkali metal soaps, Gum acacia, Gum Tragacanth, Egg-yolk, Casein, Irish moss, Saponin, Bile salts.

B. For water-in-oil type Emulsion–Heavy Metal Soaps.

7.17.4 Function of emulsifying agent

The function of the Emulsifier as an aid in prolonging the life of the Emulsion is believed to be:

1. due to its ability to collect in a thin film at the surface of the suspended particles, thereby preventing them from coalescing to make larger drops, or
2. due to marked adsorptive attraction in respect to one or other of the liquid components of the Emulsion, or
3. due to capability of Colloidal dispersion with one of the liquids involved in the Emulsion, e.g. Sodium and Potassium Soaps, Resins, Gum acacia, Irish moss, Albumin, Gelatin and Haemoglobin, Dextrin and Starches, Cholesterol and Lecithins–all form Colloidal solutions with water and all are stabilizer for water and oil Emulsions. This can be attributed to the semisolid films which they form at the interfaces between the liquid components of the Emulsion.
4. Sometimes finely divided and insoluble solids which are not capable of forming colloidal solutions with liquid of an Emulsion, function as emulsifier and the more finely divided, the greater is its stabilizing effect, e.g., Fuller's earth and Kaolin for water and oil emulsion is used in making water and mineral oil Emulsions for insecticidal spray. These agents must be attracted to one of the two liquids in emulsion and form a thin film of the finely divided particles around the suspended droplets,

thereby checking their tendency to coalesce and thus make a protective layer around the oil droplet.

5. When liquids are mutually insoluble as in case of Emulsion, they exhibit a high degree of interfacial or surface tension where they come in contact with each other. The cohesive force between particles of each of the liquids tends to make these particles coalesce and a gradual seperation of the two liquids takes place. To overcome the forces of surface tension to obtain a fine dispersion, it is necessary to prevent the operation of these forces. The addition of the agents causes concentration at the interfaces between the dispersed phase and continuous liquid phase and lowers the interfacial tension of either one of the liquids and functions to stabilize the emulsion.

7.17.5 Energy expenditure for lasting emulsion

For obtaining lasting emulsion, another essential condition is to break up the suspended liquid into very small droplets, the smaller the better. This requires the expenditure of energy to overcome the cohesive force operating between the particles of the same substance which tends to hold them together in a continuous mass.

This can be done in the following ways:

(1) Shaking, (2) Stirring, (3) Whipping, (4) Rubbing.

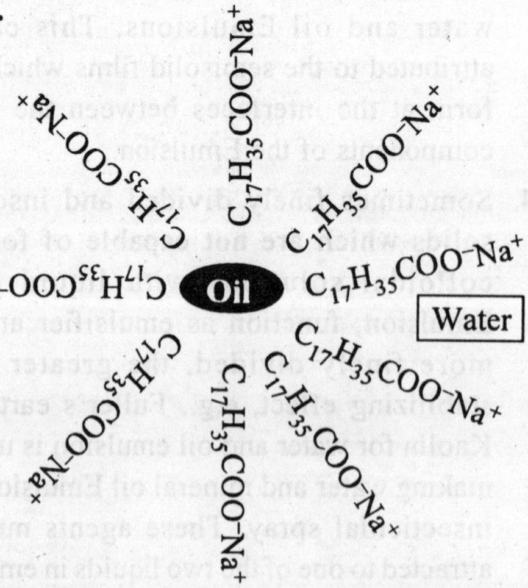

Fig. 7.9: Orientation of tonized soap molecules

In case of oil and water, the greater the proportion of oil to water, the greater the energy is expended to make a lasting Emulsion. It has been observed that intermittent shaking or stirring is more effective than continuous agitation. For this, electrically driven machine is very much helpful than the hand. The process of homogenization used in large-scale preparation of lasting Emulsion produces fine dispersion by means of forcing the mixture of the two liquids through very fine tubes under high pressure.

7.17.6 Orientation of molecules and Ions of emulsifier in emulsion

If oil is shaken with water vigorously, oil is broken into droplets and forms Emulsion which is not stable because the oil droplets coalesce and seperate as a layer on water. If small amount of soap, e.g., Sodium Stearate, is added before shaking, the Emulsion formed is stable.

The mechanism is that, the ionized soap molecules contain the hydrocarbon radical $C_{17}H_{35}$ which is soluble in oil and insoluble in water and the –COO–group which is souble in water and insoluble in oil. These Soap ions are oriented at the interface of oil and water with the hydrocarbon radicals ($C_{17}H_{35}$) in the oil and carboxyl (–COO–) ions in the water. This negative carboxyl groups give the droplets a negative charge at the surface and is balanced by the Na^+ ions in the water. The droplets thus possess Helmholtz-Gouy double layer, across which there is zeta potential. In this way soap stabilizes the Emulsion by (a) contributing a charge to the droplets. The droplets are charged alike and repel each other and cannot coalesce; (b) making the droplets hydrophilic. Because the Soap ions oriented, makes the surface of the droplets strongly attracted by water due to presence of –COO– groups for which water has tremendous attraction (Fig. 7.9).

7.17.7 Properties of emulsion

1. It possesses high viscosity like lyophilic sols.
2. It resists salting out by electrolyte ions.
3. It exhibits Tyndall effect and Brownian Movements like lyophobic sols because of the big size of the dispersed droplets.
4. These big droplets have smaller surface area per unit mass compared to the sol particles.
5. Multivalent ions and ions which can cause chemical changes in the emulsifying agents can precipitate the droplets.

6. The liquid droplets dispersed in an Emulsion may exhibit electrophoretic properties due to electrical double-layer which surrounds the droplets.

7.17.8 Importance of emulsion

A. Domestic

1. **Cleaning**: For removal of fats and oils from skin, garments and utensils with soap and water by way of an Emulsion. The hand rubbing or the agitation by washing machine is also very much effective to get desired dispersion of the grease.
2. **Cookery**:
   (a) Making of mayonnaise which is an oil and water mixture. The latter may be vinegar or lemon juice. The two liquids are mixed properly by beating to break the oil into finer droplets, and egg yolk is added, the proteins of which function as an emulsifying agent.
   (b) Preparation of gravies and food sauces, cooked starch provided by the flour or cornstarch to hold the fat and water in Emulsion dispersion.
   (c) Preparation of sweet milk and cream for food uses: When either milk or cream is desired, dried skim milk and butter fat from stock are assembled to obtain an emulsion from the mixture of water, fresh butter and dried milk through the aid of homogenizer.
3. **Pharmaceutical**: Certain oily medications, e.g., mineral oil and cod liver oil for Emulsion, Agar and Gum acacia are used as Emulsifier.
4. **Physiological**:
   (a) Bile salts stabilizes the intestinal Emulsion of fat for bringing fats, water and digestive enzymes into contact with each other favouring the digestion.
   (b) Egg albumin stabilizes the Emulsion of lipids in the egg.
   (c) Milk is an emulsion and fat in the milk is stabilized by the Phosphoprotein Casein.
   (d) Suspension of fats in the lymph and blood for transportation to the cells is an Emulsion phenomena.

7.18 Gel or Gel State

The peculiar characteristic of a colloidal suspension is the capability to undergo distinct changes in consistency or rigidity, i.e., gel formation. A gel is distinguished from colloidal solutions by a colloidal system although containing a considerable quantity of liquid, possesses a greater or lesser degree of rigidity. Gels may vary all the way from fluid jellies which flows on slight pressure to gels of rigid structure. The same colloidal solution may at one time be a fluid like water and at another time, has the consistency of jelly, e.g., gelatin is a protein which forms a colloidal suspension in water. When the system is warm, its consistency is that of water. When cooled, it becomes jelly. The fluid condition is called **Sol state** and the semisolid condition the **Gel state** and they are reversible (Fig. 7.10).

7.18.1 Definition

It is a Lyophilic Colloid system which is more or less rigid and makes up of fibrillar structures surrounded by the dispersion medium.

7.18.2 Mechanism of formation

There is no change in the chemical composition of the suspension. The change is of rearrangement of the relationship between the solid particles or solid phase and the watery phase of the colloidal system. In Sol state the solid phase is discontinuous and the waterly phase is continuous and solid particles are separated from one another by water throughout. In Gel state the watery phase is discontinuous and the solid phase is continuous. In both cases there has been no change either in the size or nature of the solid particles or in the number of particles per unit volume of the suspension. Although the gel contains a considerable amount of liquid, it shows rigidity which is due to development of a very definite structure within the dispersed system. it is brought by the adherence of some of the colloidal particles to one another to form fibrils and these fibrils serve to enmesh a more dilute colloidal solution by capillary action and hydration of water of fibrils.

7.18.3 Example of gels

Gelatin, gum acacia, agar, silicic acid gel, inorganic precipitates, fruit jellies, cooked starch, heat-coagulated egg white.

7.18.4 Classification of gel

A. According to physical state

1. Elastic gel, e.g., gelatin, agar.
2. Rigid gel, e.g., silicic acid gel, inorganic precipitates.

B. According to change of heat, internal energy or volume.

1. **Isothermic gel**: No change in volume or internal energy.
2. **Exothermic gel**: Fall in volume and internal energy and liberation of heat.
3. **Endothermic gel**: Increase in volume and internal energy.

7.18.5 Properties of gel

1. When a solution of Gelatin sets on cooling, the firm mass is a typical hydrogel.
2. It is clear, hyaline, apparently structureless and elastic.
3. It may be formed by the separation of the warm pseudosolution of Gelatin into two phases: a solid phase, rich in Gelatin, and forming a meshwork in which is embedded another phase of very weak solution of Gelatin.
4. Usually the fluid part is pure water.
5. The gel (e.g., Gelatin) loses water in dry atmosphere and becomes dry but, placing in water, it resumes its original size and characters by swelling, i.e. change is reversible. But Gels like silicic acid gel and ferric hydrate gel cannot swell up in water after drying, i.e., the change is irreversible.
6. A gel takes up last traces of water with great tenacity.
7. It produces a greater surface on which adsorption takes place.

   Osmotic pressure of dissolved substances and the vapour tension of fluids are diminished at this surface.
8. **Hysteresis**: For attaining a constant melting point, a Gel takes about 24 hours and this phenomenon is called **Hysteresis.**
9. **Imbibition (or quellung)**: Many Colloids (e.g. Starch, Gelatin) which are insoluble in cold water may swell up by absorbing water into its fibrillar network by adhesion and surface tension and the process is known as **Quellung or Imbibition.** In some cases it is difficult to distinguish from the process of solution.
10. **Syneresis or Aging of Gel**: After gel formation, a small amount of liquid diffuses out without any significant change of volume of gel as it is seen during blood clotting. After formation of clot, some amount of fluid squeezing out is called serum.
11. When immersed in aqueous solutions, the amount of water taken up varies inversely with the osmotic pressure of the solution.
12. Sometimes the insoluble Emulsoids, if soaked in aqueous solutions, behave as osmotic cells with boundaries capable of expansion. They many be regarded as cells filled with Colloid and surfaces are impermeable.
13. When dry Colloids are allowed to swell up by moistening they exert a significant force.

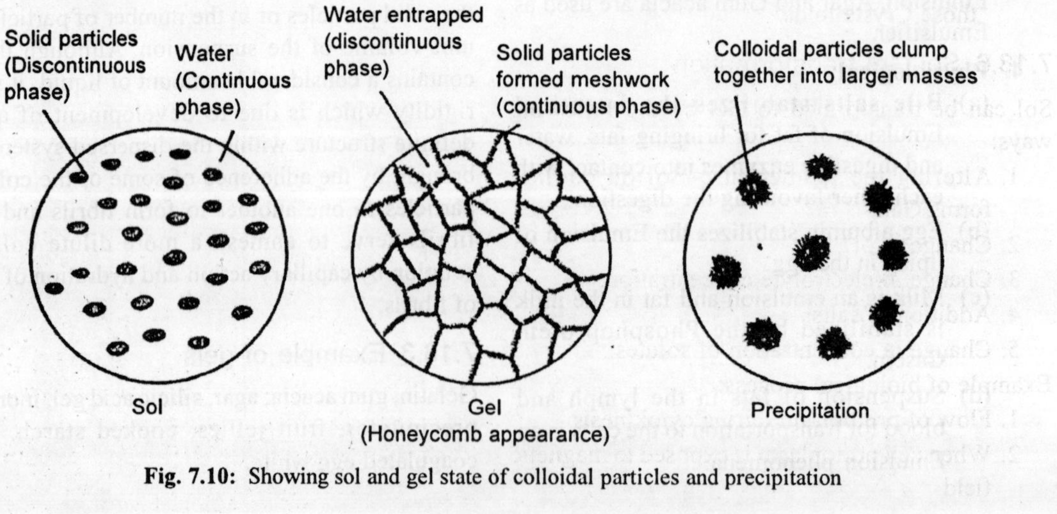

Fig. 7.10: Showing sol and gel state of colloidal particles and precipitation

14. A ten per cent solution of Gelatin sets at 21 °C but does not melt until 29.6 °C is reached. Similarly, agar solution in water sets at 25 °C but do not melt below 90 °C.
15. It is impossible to wash out the inorganic constituents from a Gel due to the extent of surface.
16. The lowering of osmotic pressure of dissolved substances at the surfaces causes greater concentration at the surfaces of the solid phase than that of the surrounding medium.
17. If dry Gelatin is placed in a salt solution, it swells up. But the solution absorbed by it is more concentrated than the solution in which it is immersed. So, the proportion of the salt of the remaining solution is diminished.
18. When an equilibrium is established between the Gel and the surrounding fluid, there is no significant resistance to the passage of dissolved Crystalloids.
19. If Gelatin is in pure water, salt or sugar can diffuse through the column of solid Gelatin.
20. Gels are impermeable to other Colloids in solution which is the basis of separation of Crystalloids from Colloids by dialysis and the membranes used in this process are generally irreversible gels (e.g., vegetable parchment, animal membranes, collodion or cellophane).
21. Some Gels, e.g., Tannate of Gelatin or Copper Ferocyanide, are impermeable to both Colloids and Crystalloids and these membranes can be used for the determination of the osmotic pressure of those Crystalloids.

7.18.6 Sol-Gel transformation

Sol can be transformed to Gel by the following ways:

1. Altering the temperature–Sol on cooling forms Gel.
2. Change in pH.
3. Change in electrolyte concentration.
4. Addition of salts.
5. Change in concentration of solutes.

Example of biological process:

1. Flow of protoplasm during cytokinesis.
2. When cell protoplasm is exposed to magnetic field.

7.18.7 Reversibility of gel to sol

Gel is transformed to Sol by warming, hence a Gel and a Sol may be reversible Colloid.

Example of biological process:

The pseudopodia of protozoa, macrophages and leucocytes form and show movement due to local transformation of protoplasmic Gel to Sol.

Therefore, the two states are readily reversible simply by warming and cooling.

7.18.8 Preparation of gel

Three methods are generally employed for prepation of Gel:

1. One method consists of allowing suitable substances, e.g., Gelatin, Starch or Soap to soak in contact with water or other desired dispersion medium until the colloidal material absorbs the fluid and swelling occurs with the production of jelly-like mass.
2. Another method is a colloidal solution of the material is prepared then altering the temperature and in some cases the concentration of acids, bases or salts present, finally the solution is allowed to stand undisturbed until the Gel gradually "sets", e.g., cooling of cooked starch.
3. Combination of the two methods are sometimes applied. The cold water is used first until the maximum "swelling" of the material takes place. Then hot water is used to secure a colloidal solution followed by cooling until the Gel "set".

7.18.9 Importance of gel

I. Cookery Procedures

Preparation of cornstarch, gelatin moulds, fruit jellies, custards, etc.

II. Pathological Laboratory

Preparation of culture media for the bacteria.

III. Industrial

Preparation of various mucilages, glues and pastes.

IV. Physiological

A. Cell structure

(a) Cytoplasm:

1. Gels also contribute to cell structure. Greater part of water is held within the cells of muscle, nerve tissues and other tissues. Despite this larger amount of water, the cell content is in a semifluid condition due to the presence of colloidal Gel.

2. It can fluctuate between the Sol and Gel as is seen in pseudopodia formation of protozoa, macrophages and leucocytes.
3. Peripheral zone of cytoplasm of many cells (not all cells) appears to be permanently gelated. Some of the chemical constituents of cytoplasm, the proteins and lipids are organized in the form of a **Cytoskeleton** consisting of membrane and microtubules which is a labile structure, and permits the streaming movements of the cytoplasm occurring in living fibroblasts, tips of growing nerve fibres and perhaps in all cells.

(b) Plasma membranes are colloidal gel.

(c) Network of membranes extending through the cytoplasm are colloidal gel.

(d) Micro-dissection method has helped to determine that astral configurations, e.g., division, spindle, etc. observed during cell-division are gelated parts of the cytoplasm. The mitochondria and the chromosomes appear to be Gels because they can be stretched, with micro-needles and on release restore their original size.

(e) Healing of "wound" of surface membrane by gel formation—if the surface membrane is mechanically disrupted by micro-dissection, a portion of interior fluid tends to escape. If the surrounding is favourable, the escaping Sol is converted to Gel and the membrane is restored and there is healing of cellular wound, and Ca^{++} is required for this process.

B. Clotting of blood

The clotting of blood is due to formation of gel. Suspended in the fluid portion of the blood are colloidal particles of a special protein called **fibrinogen** which is in sol state. But after bleeding, this protein converts into fibrin which is a Gel. The fluid blood changes into semisolid clot and subsequent exudation of serum, which has the characteristic of the hardening and drying of the clot which is known as **Syneresis.**

Table 7.7: Comparison of Sol and Gel

Sol	Gel
1. Liquid state of colloidal solution.	1. Jelly-like solid or semisolid state of colloidal solution.
2. Dispersion medium may be water (Hydrosol) or alcohol (Alcosol) etc.	2. Dispersion medium is formed by hydrated Colloid particles.
3. Easily dehydrated.	3. Cannot be dehydrated easily.
4. Diffusion and dialysis are not shown.	4. It may act as a semipermeable membrane showing diffusion (selective molecules) and dialysis.
5. Sol can be converted to Gel by cooling.	5. Gel can be converted to Sol by warming.
6. It takes the shape of a container.	6. It does not take the shape of a container.
7. No definite structure.	7. Assumed to have honeycomb structure.
8. Types: According to the type of dispersion medium, e.g., Hydrosol, Alcosol, Benzosol.	8. Type: (a) Elastic gel–gelatin, agar. (b) Rigid Gel–silicic acid gel, inorganic precipitate.
9. Syneresisi: absent.	9. Present.
10. Does not offer resistance to the diffusion of any substances dissolved.	10. The Gel structure offers resistance to the diffusion of any substances dissolved before setting.
11. Does not produce enromous surface area.	11. It produces enormous surface on which adsorption takes place.
12. Does not show the process of imbibition.	12. Shows the process of imbibition (Quellung).
13. Does not show the process of hysteresis.	13. Hysteresis process is seen.
14. Viscosity: low.	14. Viscosity: high.
15. Freely permeable to both non-colloidal ions and molecules and Colloids in solution.	15. Gels are freely permeable to non-colloidal ions and molecules. But Gels are impermeable to other Colloids in solution. This impermeability has been utilized to separate Crystalloids from Colloids.
16. No Micelle formation.	16. The large aggregate of colloidal particles formed in the process of Gel formation are called "Micelles".

C. Arteriosclerosis

It is claimed that changes in blood vessel walls during old age known as arteriosclerosis are the result of the **Aging of Gel or Syneresis.**

D. Tissue Gel

The intercellular fluid is an ultrafiltrate of plasma derived from the capillaries. This fluid is mainly entrapped in the minute spaces of proteoglycan filaments and this combination gives the characteristics of a Gel and, known as **Tissue Gel.**

7.19 Concept of Macromolecules

All colloidal solutions are dispersed system and the modern trend is to classify Colloids into following three general types depending on the nature of the dispersed phase:

1. Dispersed Colloids: Mostly lyophobic, e.g., gold sol.
2. Association Colloids: Lyophobic and lyophilic in the same molecules, e.g., soaps, dyes.
3. Molecular Colloids or Intrinsic Colloids or Macromolecules: Predominantly lyophilic, e.g., protein, starch, cellulose, nylon, polythylene, polystrene, etc.

7.19.1 Definition

One class of Colloids is called Macromolecules or Molecular Colloids where the molecules themselves are big enough to be of colloidal dimension and hence show colloidal properties having MW between 10,000 to 2,00,00,000.

7.19.2 Example

A. Synthetic : Polybutadine, nylon, polystrene, polythylene, lucite.

B. Plant : Rubber, resins, gums, polysaccharides, proteins, cellulose, starch etc.

C. Animal : Proteins, nucleic acids, lipoproteins, etc.

7.19.3 Formation

A Macromolecule is a giant molecule obtained by the polymerisation of a large number of simple, low molecular weight molecules of similar structure in a more or less repetitive manner. This polymerisation may take place either by condensation or addition reaction.

Hundreds of monomeric units are joined with each other by similar covalent bonds under as:

1. Proteins– Amino acids are joined with each other by peptide bonds.
2. DNA–Riboneucleotides are joined by phosphodiester bonds.
3. Cellulose–Glucose molecules are joined by glycosidic bonds.
4. Polybutadine–Butadine molecules with MW 54, polymerise to form polybutadine with molecular weight as high as 5 to 6 million.

7.19.4 Structure

The macromolecules are:

1. Long flexible fibrous molecules.
2. Remain helically or irregularly coiled such as α-helix of Keratin and double helix of DNA or,
3. In extended zigzag conformations such as β-pleated sheets of silk fibron because of noncovalent bonds cross linking different parts of the chain.

7.19.5 Characteristics or properties

The physical properties depend on the size, shape and conformation of the macromolecules.

1. **Size**: The particle size of macromolecular solute in a lyophilic sol is between 10 to 10^3 nm.
2. **Shape**: Spheroidal, asymmetric or elongated shape because they are highly folded.
3. **Flexibility**: They are very flexible and can take various shapes.
4. **Solubility**: Although they dissolve in solvents to form true solutions, they are treated as Colloids because of the large size of their molecules which come within the colloidal range.
5. **Behave as Reversible Colloid or Lyophilic Sol**: Solutions of macromolecules behave like reversible Colloids or lyophilic sols in many respects. The size of particle is between 10-10^3 nm. The dispersed particle may be:
   (a) Monodisperse: When the particles have same molecular weight.
   (b) Polydisperse: When the particles have different molecular weight. If a lyophilic sol of a macromolecular solute is added to a solvent in which the macromolecular solute is insoluble, the macromolecules are precipitated from the Sol. If it is a polydisperse macromolecular solution, precipitation occurs in descending order of mol.wt. of the dispersed particles, i.e., the particles of high molecular weight precipitate first and the particles of low molecular weight precipitate last. It does

not require any substance, i.e., peptizer, for its stability.

6. **Tyndall Effect**: They show the "Tyndall effect".
7. **Viscosity**: It possesses a high viscosity with coefficient of viscosity varying with rate of shear. High viscosity is due to extensive solvation. The viscosity rises with the rise of molecular weight and elongation of macromolecules. Viscosity of blood plasma depends upon plasma proteins. But major part is taken by asymmetric and elongated globulin molecule having mol.wt. of 1,67,000 than the symmetric and spheroidal albumin.
8. **Solvation**: Significant amount of solvent is held, (a) in the folds of the coils of the macromolecules and (b) by the surface of the hydrogen bonds.

   This solvation process by the macromolecules decreases the movement of solvent molecules and another factor also reduces the solvent movement that is due to entangling of the long coils of the macromolecules.
9. **Sedimentation**: Because of their large size, they can be sedimented and isolated by an ultracentrfuge.

   Using sedimentation velocity method and sedimentation equilibrium method, isolation, determination of shape, homogeneity, sedimentation coefficient, molecular weight can be done.
10. **Electric Charge**: Macromolecules in solution do not carry electric charge and as such no migration of the particles occurs under the influence of an electric field unless the molecules themselves produce ions. This happens when sulphuric, amino or carboxyl groups are present in the molecules. Macromolecules producing ions in solutions are called **Colloidal Electrolytes**. Such macromolecules producing ions in solution migrate towards oppositely charged electrodes in an electric field called **Electrophoretic mobility**. Such Macromolecular ions at particular pH, they neither go to the cathode or anode and called isoelectric pH or point.
11. **Number of Molecules and Mole Fraction**: Number of molecules and mole fraction of the macromolecular solute in a solution is always smaller than the solution of a small solute in the same per cent of mass concentration.
12. **Colligative Properties**: Colligative properties like elevation of boiling point, lowering of freezing point and lowering of vapour pressure are too small and difficult to measure with accuracy. These properties depend on the mole fraction of the macromolecular solute.
13. **Osmotic Pressure**: Osmotic pressure is measurable but gives very low value. Thus, plasma proteins have higher mass or per cent concentration than the electrolytes, but they exert lower osmotic pressure than the electrolytes.
14. **Crystallization**: Some of these "giant" molecules can form well-defined crystals owing to their high degree of symmetry. For example, many proteins and enzymes whose solutions have typical colloidal (lyophilic) properties, have been obtained in the crystalline form.
15. **Molecular Weight**: Because of the high molecular weight of macromolecules, the colligative properties are very small. Their molecular weight ranges from 10,000 to 2,00,00,000. The following are the methods used for determination of molecular weight of macromolecules:

    (a) Svedberg's sedimentation method,

    (b) Light scattering method,

    (c) Viscosity determination method,

    (d) Osmometric method.
16. **Gibbs-Donnan Effect**: Macromolecules are not diffusible through many membranes due to their large size. If present on one side of semipermeable membrane, they exert some influence on the membranous diffusion of diffusible ions, thereby changing their distribution on both the sides of the membrane producing Gibbs-Donnan effect. Due to this effect, there is also change in osmotic pressure across the membrane. The osmotic pressure of the macromolecular ions decreases due to presence of diffusible ions for which osmometric determination of molecular weight of macromolecules does not give correct result.
17. **Turbidity**: Solution of macromolecules show turbid appearance becase light is scattered by the macromolecules when beam of light passes through the solution. Turbidity is the fraction of the scattered light of a given wavelength while passing through the unit length of the macromolecular solution. Turbidity is determined from the difference in intensities between incident and transmitted

beams of light or from the intensity of the scattered light coming at right angle to the direction of incident light correcting the dissymmetry of the particles.

18. **Mie Effect**: The Mie effect is observed when light is scattered by macromolecules, whose size is commensurate with the light wavelength. It is a specific asymmetric scattering indicatrix (the angular dependence of the scattered light intensity), the intensity of the light scattered backward. Certain absorption bands characteristics of haemoglobin solution could not be observed in the blood spectrum for a long period. It was found that the absorption of light by haemoglobin was masked by the Mie effect–the enhanced scattering of light forward by RBC. The Mie effect is determined by the interference of light waves with different phases scattered by various parts of the molecules.

19. **Flow Birefringence–The Maxwell Dynamo-optical Effect**: The shape of macromolecules and their anisotropic polarizability can be determined by means of flow birefringence. In flow birefringences, two coaxial cylinders are used, one is fixed and the other rotates. The macromolecular solution is placed in the sapce between the two cylinders. The inner cylinder, called rotor, rotates at the common axis carrying off the liquid. A velocity gradient is set up in it. The layer of liquid adjacent to the wall of the inner cylinder (rotor) rotates rapidly. The layer adjacent to the wall of the stationary cylinder is immobile. The macromolecules become oriented in a solution and are acted on by a tensile force. The liquid becomes anisotropic, like a diaxial crystal.
Birefringence is observed in a direction parallel to the axis of the Dynamometer.

20. **The Kerr Effect**: Birefringence in an electrical field is known as Kerr effect which has no importance in case of biopolymers. Biopolymers are polyelectrolytes, the Macromolecules carry charges and, therefore, not only become oriented but also move in an electric field.

21. **Polyelectrolytes**: Many macromolecules, including proteins and nucleic acids, are macroions in aqueous solution, carrying a multitude of charged groups. Macromolecules (or biopolymers) are polyampholytes becase they have both anionic and cationic groups. Proteins have acidic residues like Glu, Asp, Tyr and basic, like Arg, His, Lys. Nucleic acids have acidic phosphate groups and the basic groups N≡, HN= and H_2N– of nitrogenous bases. There are existence of solid macroions like globular proteins etc. and flexible polyelectrolyte chains. Hence the properties of macromolecules depend on the interactions of the charges within them. The magnitude of charges depend on the degree of dissociation of ionogenic groups and the ion atmosphere that surrounds them. Polyelectrolyte properties of proteins enable their separation and investigation by electrophoresis.

22. **Diffuse Scattering of X-ray by Solution of Macromolecules**: The diffuse scattering of X-rays by the solutions of macromolecules yields direct information on the distribution of scattering objects. Diffuse scattering is the total scattering by randomly arranged individual macromolecules. For this reason, the scattered light intensities are averaged over the orientations of the macromolecules. The larger the scattering angles, the smaller are the sizes of structural details on which X-rays are diffracted. Small angle scattering is being used to study the gross structure of macromolecules in solution. The specific features of the internal structure of macromolecules are studied at medium and large scattering angles. The theory of small-angle diffraction is based on conception similar to those employed in the theory of light scattering by solutions of macromolecules.

7.20 Physical Structure of Protoplasm

Protoplasm after fragmentation, its constituents and molecules are separated from one another and analysed which reveals the properties and composition of each fragment. If a living cell or protoplasm is crushed and grinded and then extracted without losing any constituent, then we see the collection of non-living chemicals. If these chemicals are mixed in correct proportion we cannot get a living cell or protoplasm. Our knowledge is lacking pertaining to the precise orderliness of internal interelationship and the chemical, physical and structural integration of the constituent parts of the cell. Therefore, protoplasm can be regarded more than the sum total of the chemical properties and states.

7.20.1 Experimental observations

1. Microscopical

(a) **Living cytoplasm**: It is difficult to assess the structure of living cytoplasm even using the highest power of microscope. Though some inclusions and granules are present and are visible under microscope, these are not living materials and are not the essential components. It looks transparent and structureless under microscope.

(b) **Fixed, hardened and unstained cytoplasm**: If a cell is examined, after fixing in hardening fluid and in unstained condition, it shows some differentiation between regular framework or meshwork and a fluid portion filling up the interstices. The cytoplasm appears to be homogenous liquid mixture of Colloids.

(c) **Study of artificial cytoplasm**: Fischer and Hardy had shown that various structures and characteristics of cytoplasm can be imitated artificially by hardening a homogenous colloidal solution, e.g., egg-white, by different methods. All fixation methods also produce similar conditions causing precipitation of cell proteins. Though fixed preparations gives some information they do not give valuable picture of the structure in the cell cytoplasm.

(d) **Fixed preparation with differential staining**: After fixation, when studied with differential staining, the result cannot be interpreted confidently as it is not known which factor combines to give a specific result. So, there are various theories regarding structure of protoplasm based on fixed tissues.

Many histologists are of opinion that the cytoplasm may be structureless or certain structural modifications may be seen due to nutrition or presence of metabolic products. Though structural differentiation may be seen, it is not visible due to minute size of the constituents or identical refractive index between the wall and its contents.

Cytoplasm may show fibrillation under stress which disappears after removal of stress. The radiations of aster are probably due to strain. It originates from the vicinity of centrosome and disappears immediately if the cell is etherised. Such behaviour of the cell reveals that cytoplasm has more complicated structure than what we observe under the microscope.

(e) **Ultramicroscopic study**: Observations under ultramicroscope reveal that cytoplasm is full of granules of different sizes. These granules can move within the cell and in certain cases it shows Brownian movement, e.g., muscle cells, WBC, etc. In many animal cells, the cytoplasm appears to be viscous, having active movement. According to modern researches, the living cytoplasm is regarded as structureless at the microscopic level but it is highly organised at the ultramicroscopic level. Most part of the cytoplasm is fluid or semifluid within which there are various bodies having different refractive indices. When the refractive indices of an inclusion like mitochondria is the same or near to that of the cytoplasm, then it is not visible under ordinary microscope. But bodies like the nucleus, secretory granules having different refractive index are visible to different degrees. The above structures can be clearly visible under phase contrast microscopy.

(f) **Phase contrast microscopy**: The great advantage of this method is that it introduces contrast by optical means in unstained preparations of transparent objects like living cells.

7.20.2 Recent developments

Cytoplasm cannot be defined in chemical terms exclusively. The cytoplasm is filled with minute and large dispersed particles and organelles having diameters from a few nanometres to 3 microns.

Cytoplasm has the following phases:

1. Aqueous phase: It is a clear fluid in which the particles are dispersed and is called "Hyaloplasm" containing various inorganic ions, small organic molecules, e.g., glucose, amino acids and macromolecules, e.g., polypeptides, dissolved proteins, nucleic acid and small quantities of phospholipids, cholesterol and esterified fatty acids. These are submicroscopic but are often organised into microscopic organelles with clearly defined morphology and function.

   The portion of the cytoplasm beneath the cell membrane often contains large number of microfilaments consisting of fibrillar proteins which provide a semisolid support for the cell membrane. This portion of cytoplasm is known as "Cortex or Ectoplasm".

   Some chemical constituents, particularly proteins and lipids, appear to be organized in the form of "cytoskeleton" which consists of membranes and microtubules. This is regarded as a labile rather than a rigid structure in the mechanical sense and has to permit streaming movement of the cytoplasm which occurs in living fibroblasts, the tips of growing nerve fibres.

2. Dispersed particles: The large dispersed particles are neutral fat globules, glycogen granules, secretory granules, ribosomes, mitochondria and lysosomes.

7.21 Physical State of Protoplasm

Regarding the physical condition of the protoplasm, the question may arise whether it is a **Viscous fluid** or **Soft solid**. The following evidences have been forwarded in favour of **Viscous fluid character**.

7.21.1 Viscous fluid character

1. The mobile character of protoplasm of many cells, e.g., flow of protoplasm of an amoeba into its pseudopodia or the movements of granules or mitochondria within the living cells, reveals the fluid character of the protoplasm.
2. When small amount of water is injected into the protoplasm of a cell, the water is mixed freely with the protoplasm.
3. The basic qualities of solidity are shape, organization and power of resistance to deformation and the protoplasm may possess these qualities if it is solid.

The absence of resistance of deformation is a characteristic of the liquid and is applicable to the internal molecules. The surface of the liquid is under constant tension which limits deformation as well as offers great resistance to any enlargement of the surface. But the protoplasm exists continuously in a watery surrounding revealing that not only its composition, it is different from its surrounding but there must be a distinct surface between the two. The superficial layers of the protoplam must be under constant tension and exerts pressure on the internal portion of the protoplasm which brings the cytoplasm into spherical shape. If a fluid drop is suspended in a fluid in which it is not miscible, it offers greater rigidity than which is imparted to an emulsion. Here every droplet offers great resistance to deformation and every surface resists penetration and extension.

The resistance of the surface of a colloidal solution is further increased by the aggregation of the dissolved substances in the surface layer than in the internal portion. Therefore, considering the above facts, protoplasm may be regarded as fluid. The form and rigidity are acquired due to physical differentiation in the substances of the protoplasm as in the case of Colloid when it is stiff emulsion or when, as at the surface, there is the formation of a solid condensation layer.

7.21.2 Colloidal nature of protoplasm

Physically, protoplasm is a granular, semifluid or gelatinus substance. It has characteristic properties of both liquid and solid. Protoplasm is an aggregate of:

(a) **Crystalloids** which are of two groups:
   1. Electolytes: Which are in solution, conducts electricity, e.g., salts, acids, bases.
   2. Non electrolytes: Which cannot conduct electricity, e.g., sugar, urea, etc.

(b) **Colloids**: Proteins, nucleic acid, lipids. Colloidal nature of protoplasm was suggested by Fischer and later supported by Hardy.

Protoplasm has two phases:

A. Aqueous phase or liquid phase: (Fluid matrix or ground substances) In which minute and large particles are dispersed. It contains glucose, amino acids, macromolecules, dissolved proteins, etc.

B. Dispersed phase: Contains dispersed particles, e.g., neutral fat globules, glycogen granules, secretory granules, ribosomes, mitochondria and lysosomes.

These are colloidal substances and remain suspended.

The Colloidal particles are of diameter 0.001 micron to 0.1 micron. So they are too large in size to dissolve in the water to form a true solution or crystalloid and too small to settle down to form a suspension. Hence they remain suspended throughout the aqueous phase to form colloidal system.

7.21.3 Sol state of protoplasm

Protoplasm is a Sol since its fluidity is due to water, it is commonly known as hydrosol. It passes readily into a gel condition, and becomes a hydrogel. In living protoplasm this metamorphosis is known as **reversible process** because gel state again converts into sol state. Fixation by hardening fluid or killing of cells for study causes separation of the more solid part of colloidal protoplasm from a more liquid part.

7.21.4 Role of proteins in protoplasmic organization in colloidal suspension

Proteins have great role in protoplasmic organization and it is present in colloidal suspension. Proteins and other Colloids in protoplasm have certain properties in common with true solution as mentioned below:

(a) The suspended Colloids are distributed homogenously throughout the water;

(b) Colloidal particles like ions or molecules in solution, do not settle out. These colloidal particles–whether large lysosomes or mitochondria or small granules–are hydrophilic, i.e., attracted to water. Because of the ionic negative electrical charges on their surfaces, the particles remain dispersed in the cytoplasm mainly due to mutual repulsion of the like charges of the different particles. Thus, the cytoplasm is actually a colloidal solution. But in case of colloidal suspension of protein, there are certain important distinctions which reveal the organization and properties of protoplasm and cells. Colloidal particles have diameter of 0.001 μ–0.1 μ, the latter size just at the lower limit of visibility under the microscope and far bigger than a Na^+ ions or a glucose molecule.

But protein molecules are relatively large, having diameter of about 0.1 μ, and constitutes a discrete colloidal particle in protoplasm. Though the colloidal particles are large, which is a relative term in comparison to the size of a cell, they are very small. As a result, there is a large total surface area of colloidal particles in a cell. Intracellular reactions seem to occur at the surface of the Colloidal particles. Similarly, reactions between two inorganic reagents take place at the points of contact or adjoining surfaces of those compounds.

Eventually, the distribution of the cell proteins into these minute particles and the multiplication of reactive surface area play a significant role in determining the rapid rate of intercellular reactions. Protoplasmic poison produces the untoward effect by covering the highly reactive surfaces of the colloidal particles.

7.21.5 Sol and gel state

One peculiar characteristic of colloidal solution is the ability to undergo changes in consistency or rigidity. Protoplasm also converts into Gel and again becomes Sol. The two states are reversible. During reversal from one state to another the change is not due to change in chemical composition but the change is in the rearrangement of the relationships between the solid particles or solid phase and the aqueous phase. This Sol and Gel states and transformation from one to other are common in the colloidal state of protoplasm:

1. When a living cell is damaged by a microdissection needle, the homogenous fluid cytoplasm changes immediately to a jelly-like semisolid in which definite granules appear.
2. Bayliss showed that when an amoeba was stimulated electrically and observed under dark ground illumination, the fine movement of the granules was momentarily checked which may be due to a temporary gelation of the cytoplasm.
3. In some cell-like muscle fibres, the cytoplasm on microdissection behaves as a gel.
4. The cytoplasm of many cells have low viscosity, e.g., certain sea urchin eggs (Arbacia eggs) are centrifuged, the granules settle in layers within the cell.
5. The cortical protoplasmic layer of cell, called ectoplasm is an elastic gel, while the central portion called endoplasm, is more fluid in character and exists as an inelastic colloidal sol.
6. The pseudopodia of protozoans, e.g., amoeba, macrophages and leucocytes are formed due to local transformation of protoplasm into gel and to sol. For sol forma-tion there is an increase in volume and fluidity which causes protrusion of pseudopodia.
7. During cytokinesis in a dividing cell, there is a flow of protoplasm and is due to reversible Sol-Gel transformation of protoplasm.
8. If a cell is exposed to a magnetic field, the protoplasm in Sol state transforms into Gel state due to lowering of viscosity, called **thixotropy.**
9. Seifriz found that merely compressing a "Phizopus", with just sufficient pressure to make a hypha, caused it to solidify. Again, removal of pressure caused it to return to its original liquid state.
10. It has been shown experimentally that by forcing a marine egg through a tube narrower in diameter than that of the egg, the deformation caused solidification of it. After removal from the tube, it retained its abnormal shape.

7.22 Biomolecules and Molecular Organization of Cell

The molecules found in the living organisms not only conform to all the known physical and chemical principles governing the behaviour of all molecules but in addition, interact with each other in accordance with the principles which is referred as molecular logic of the living state. This molecular logic is maintained by biomolecules.

7.22.1 Definition

The macromolecules found in the living cells having definite structure and functions are called "Biomolecules".

7.22.2 Biomolecules of living cell

Proteins, nucleic acids, polysaccharides, lipids etc.

7.22.3 Characteristics

1. **Difference from physical environment**: The chemical composition of biomolecules is different from the physical environment in which they live. Chemical components are made of organic compounds of carbon (reduced or hydrogenatic), nitrogen but these are scarce in non-living.
2. **Type**: Biomolecules occur in extraordinary variety and they are complex, e.g., *E. coli* contains 5,000 compounds including 3,000 kinds of proteins and 1,000 kinds of nucleic acids.
3. **Molecular weight**: Most of them are "macromolecules" having large molecular weight.
4. **Nature**: They are usually molecular Colloids or macromolecules predominantly lyophilic.
5. **Species difference**: Although some of the proteins in cells of *E. coli* function in ways quite similar to some of the proteins in human cells, none of the protein molecules of *E. coli* is identical with any of the proteins found in man. Each species of organism has its own chemically distinct sets of protein molecules and of nucleic acid.
6. **Simplicity of diversification**: There are 300 amino acids present in nature but only 20 amino acids are found in protein molecule and they are arranged in different sequences to form thousands of different protein molecules because of genetic code which represent 20 amino acids and there are 20 + RNA for sequences. Similarly, thousands of nucleic acids of which there are two kinds DNA and RNA, are formed out of eight different building blocks "nucleotides", four for DNA and four for RNA.
7. **More than one function**: Each serves more than one function in living cells, some are extremely versatile and play a number of roles. The amino acids serve not only as building blocks of proteins but also as precursor of hormones, porphyrins, pigments, enzymes and many other biomolecules.
8. **Molecular organization of cell**: All organic biomolecules are ultimately derived from very simple, low molecular weight "precursors" obtained from the environment particularly CO_2, H_2O and atmospheric nitrogen. The schematic representation is given below:

Elements from the environment, e.g., CO_2, H_2O. Nitrogen, ammonia etc. having molecular weight between 18-44 as "precursors".

↓

These precursors are converted by living matter via sequences of "metabolic intermediates" to form pyruvate, citrate, malate, glyceraldehyde-3-PO_4 having molecular weight between 50-250 as "intermediate metabolic products".

↓

These intermediate metabolic products produce organic compounds of large molecular weight between 100-350, referred to as "building units", e.g., nucleotides, amino acids, monosaccharides, fatty acids, glycerol etc.

↓

The building units are linked to each other covalently to form macromolecules of the cells having molecular weight 10^3-10^6, e.g., nucleic acids, proteins, polysaccharides, lipids which are known as **Biomolecules.**

↓

At the next highest level of organization of cells, biomolecules of different classes associate with each other to from "supramolecular assemblies" having molecular weight between 10^6-10^9, e.g., ribosomes enzyme complexes, contractile system, microtubules.

↓

At the highest level of organization of cells, various supramolecular assemblies and systems are further assembled into cell organelles, e.g., nucleus, mitochondria, lysosomes, microbodies, Golgi boides, etc.

7.22.4 Functions of biomolecules

The major classes of biomolecules have identical functions in all species of cells.

1. Nucleic acid: To store and transmit genetic information.
2. Proteins: a. Direct products and effectors.
   b. Some have specific catalytic (enzyme) activity.
   c. Some have structural function.
   d. Hormonal function.
   e. Receptor or repositories of genetic information.
3. Polysaccharides: Storage forms of energy-yielding fuel for cell.
4. Lipids: Structural components of membranes or as a storage form of energy-rich fuel.

7.22.5 Fundamental differences in biomolecules

There is an important and fundamental difference between the biomolecules like nucleic acids and proteins on the one hand and the polysaccharides and lipids on the other.

1. **Nucleic acids and proteins**: These are "informational biomolecules". Each nucleic acid molecule contains four or more types of nucleotides arranged in a specific information rich sequence. Similarly, each protein molecule contains a specific information–rich sequence of some 20 different amino acids.
2. On the other hand, the polysaccharides and lipids do not have a information carrying function.

7.22.6 Protoplasmic streaming

The *protoplasmic streaming* can be best observed in the active state in the slime mould, the plasomodium of "Physarum polycephalus" which is a non-cellular protoplasmic mass. The active flow in it is found continuously and also the frequent shifting of channels of flow.

Kamiya* (1940) had grown small pieces of plasmodia on a cover-glass coated with agar. Then this was spread immediately into fine sheets which later developed into protoplasmic strands. He was able to make dumb-bell shape of protoplasm by causing two blobs of protoplasm to fuse with the two ends of a strand. He observed a regular ebb and flow between the blobs. He also observed that applying an uniform pressure on one blob caused stoppage of flow and he also measured the pressure which is required to stop flow is –20 cm H_2O. He also demonstrated rhythmic pressure changes in the protoplasmic bodies, by recording pressure developed in the protoplasmic bodies.

According to Seifriz, the flow is the result from an actual contraction of the wall of the protoplasm, so that streaming may be regarded as the result from "muscular contraction" of a very primitive type. The flow is also associated with a reversible conversion of solid cortex (ectoplasm) which is a gel to fluid endoplasm, i.e., sol (Fig. 7.11).

* Kamiya, Science, 92, 462, 1940.

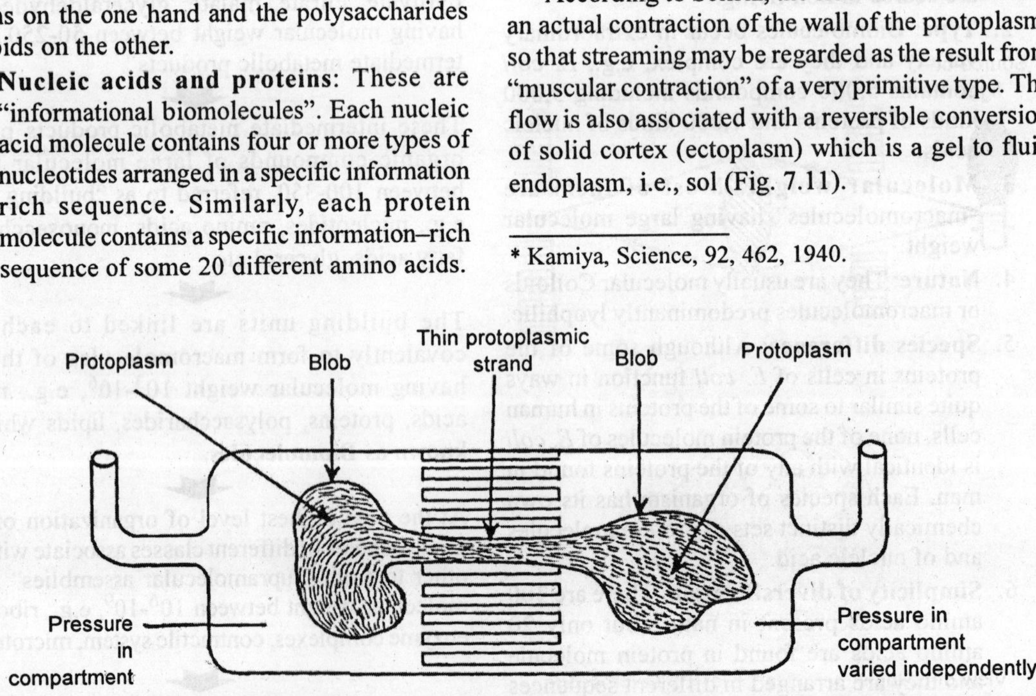

Fig. 7.11: Experiment for measurement of pressure necessary to check protoplasmic streaming

7.23 Micro-dissection (Micro-manipulation)

The dissection of living cells can be done microscopically with the help of fine quartz or glass needles, mounted in a special apparatus for manipulating them mechanically. The movements of the glass needle in all directions are controlled by micrometer adjustments.

The living cells are suspended, as a hanging drop preparation, from a cover-glass, and are manipulated and dissected by means of two needles. In dissection, one needle is brought up so as to hold the cell while the other needle is employed to dissect it and whole process can be observed under microscope. Injection is performed by fixating the cell with a needle and then perforating the cell at the selected place with a fine glass capillary through which the fluid can be pushed from a micrometer syringe.

7.23.1 Application

1. Fresh preparation of living cell can be studied.
2. Structure and behaviour of living cell can be studied.
3. Physical state of cytoplasm, i.e., sol or gel can be studied.
4. Individual living cell can be dissected.
5. By this process it is possible to determine the pH of the living cytoplasm.
6. Using hollow needle fluids, drugs, toxic substances, chemicals can be introduced into the living cytoplasm.

7.23.2 Experimental observation and importance

1. *Micro-dissection* has revealed that surface film of the protoplasm of a free cell like an amoeba can be stretched and ruptured and it will rejoin.
2. This method has shown the outer portion (ectoplasm) of the cytoplasm of a living cell such as the amoeba is more viscous, e.g., Gel than the endoplasm which is fluid, i.e., Sol. This is also true for leucocytes.
3. This method has revealed that the network of membranes extending through the cytoplasm are colloidal gel.
4. Healing of "wound" of surface membrane by gel formation (see importance of gel).
5. This method has helped to determine the astral configuration, e.g., division, spindle, etc. observed during cell division. (see importance of gel).

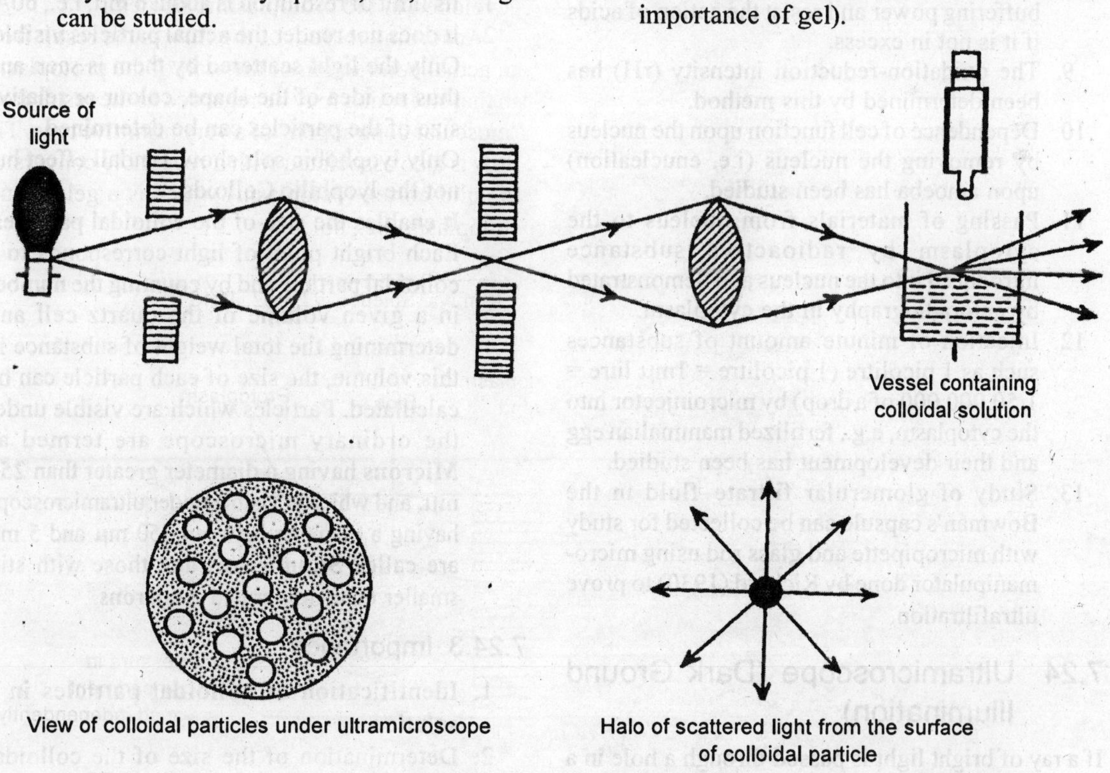

Fig. 7.12: Ultramicroscope

6. The injection of salts of monovalent electrolytes like Na^+ and K^+ increases the fluidity of the cytoplasm while the salts Ca^{++} and Mg^+ produce coagulation. It appears that electrolytes maintain a balanced condition of the Colloids in living cytoplasm. It is proportioned in such a way that the coagulating action of one group of salts is neutralised by the dispersive action of the other group.
7. By this method it has been shown that the nuclear membrane is not solid but consists of viscous material which allows itself to be drawn out by a micro-manipulator needle and on being released, regains its previous form and shape.
8. By using indicator dyes, the pH of living cell can be determined. Chamber found the pH of living nucleus in various cells of "Rana" and "Necturus" to be about 7.5, whether normal or injured. But the pH of cytoplasm was more acid, being about 6.9 in the normal state, and about 5.3 when injured and cytolysing. Needham found the pH of the cytoplasm of "Amoeba proteus" to be about 7.5 and Pollack found a lower range between 6.6-7.2 but the variety of amoeba may have been different. The cytoplasm has significant buffering power and resist the action of acids if it is not in excess.
9. The oxidation-reduction intensity (rH) has been determined by this method.
10. Dependence of cell function upon the nucleus by removing the nucleus (i.e. enucleation) upon amoeba has been studied.
11. Passing of materials from nucleus to the cytoplasm by radioactive substance introduced into the nucleus and demonstrated by autoradiography in the cytoplasm.
12. Injection of minute amount of substances such as 1 picolitre (1 picolitre = 1mμ litre = 1/50,000,000 of a drop) by microinjector into the cytoplasm, e.g., fertilized mammalian egg and their development has been studied.
13. Study of glomerular filtrate–fluid in the Bowman's capsule can be collected for study with micropipette and glass rod using micro-manipulator done by Richard (1930) to prove ultrafiltration.

7.24 Ultramicroscope (Dark Ground Illumination)

If a ray of bright light is passed through a hole in a dark room, the dust particles in the path of the light from their surfaces known as Tyndall effect. This principle is the basis of the ultramicroscope to observe the colloidal particles.

7.24.1 Ultramicroscope

The nature of colloidal solution was revealed as a result of the invention of ultramicroscope by Zsigmondy (1902). If a colloidal solution whose particle having smaller (below about 0.2 μ) size which is not visible under ordinary microscope is placed under the instrument against a dark background and when a strong beam of light is passed through paraboloid condenser at right angle to the line of view, the fild becomes filled with coloured specks of light due to scattered light from the colloidal particles. The individual particles, which were invisible, could be observed and counted. The above phenomenon is known as **Tyndall effects** and the bright track or cone of light is called **Tyndall cone** and the apparatus for observing the colloidal particles, utilizing the above principle of "dark background illumination", called "ultramicroscope" (Fig. 7.12).

7.24.2 Limitation

1. Its limit of resolution is about 6 mμ, i.e., 60Å.
2. It does not render the actual particles visible. Only the light scattered by them is seen and thus no idea of the shape, colour or relative size of the particles can be determined.
3. Only lyophobic sols show Tyndall effect but not the lyophilic Colloids.
4. It enables the size of the colloidal particles. Each bright point of light corresponds to a colloidal particle and by counting the number in a given volume of the quartz cell and determining the total weight of substance in this volume, the size of each particle can be calculated. Particles which are visible under the ordinary microscope are termed as **Microns** having a diameter greater than 250 mμ, and which is visible under ultramicroscope having a diameter between 250 mμ and 5 mμ are called **Submicrons** and those with still smaller diameter termed **Amicrons.**

7.24.3 Importance

1. Identification of colloidal particles in a solution.
2. Determination of the size of the colloidal particles.

3. Reveals the position of colloidal particles in living protoplasm and gives us some idea of their size.
4. Identification of causative organism of syphilis, treponema pallidum (Spirochetes).
5. Study of isolated living cell in fluid medium on a slide.
6. It reveals that protoplasm is highly organised.
7. **Chylomicrons during fat absorption:** Frazer's observation of blood lipid depends on chylomicron count. Ultramicroscopic examination of blood serum reveals bright particles in Brownian movement. These are chylomicrons and ranging in size from 35m μ to 1μ. They are covered with a protein layer which helps to maintain the colloidal state. A correlation was observed between chylomicron count and the level of fat in blood determined chemically.

7.25 Birefringence

Some changes that take place in the cell during its normal activity can be studied as under.

1. Polarization microscope.
2. X-ray diffraction analysis.

These methods have given valuable information regarding.

(a) Ultimate structure of living matter,
(b) Mechanism of some of the functional changes.

Analysis by this method depends upon the fact that most solid bodies are, in effect, crystalline, i.e., their constituent atoms or molecules are packed together in a definite regular three-dimensional pattern. **Polarization microscopy** depends on the anisotropy (doubly refracting) of crystalline matter. **Anisotropy** means the variation in some characteristic of the material with directions within it; thus its magnetic susceptibility, or its coefficient of linear expansion may vary according to the direction along which it is measured. This anisotropy follows from the ordered packing of its constituent elements or units. In a crystalline matter, the order in which the various atoms are encountered varies with the direction, so that light travelling in different directions may be differently affected. But in an isotropic material (singly refracting) due to random distribution of atoms or structural units, on the average one direction through the solid will be equivalent to any other.

The anisotropy or doubly refracting property under polarisation microscope is known as birefringence and the material is referred to as **Birefringent**.

The polarizing microscope is thus most useful in the analysis of the structure of bodies in terms of arrangements of comparatively large structural units.

- **Birefringence may be of four types**

1. **Form Birefringence**: In biological system, it frequently happens that the structure can be resolved into a fairly regular array of large "micelles" or "crystallite" (large on an atomic scale but submicroscopic); these micelles generally themselves possess a crystalline structure, so that the deserved birefringence of the body as a whole is due to both the orientation of the large micelles as units in a crystalline or semicrystalline form.
2. **Micellar Birefringence**: The birefringence is due to the crystalline arrangement of the atoms or molecules in the "micelles". A material made up in this way is called **Wiener mixed body**.
3. **Positive Birefringence**: With long particles or molecules, the refractive index for light vibrating in the direction of the long axis of the molecule is greater than for light vibrating across this axis. Consequently, for a crystal made up of long molecules we may say that the birefringence is positive.
4. **Negative Birefringence**: Flat molecules or lamellae show a greater refractivity for light vibrating in the plane of the molecules. The optic axis is perpendicular to the plane of the flat molecule so that the ellipse lies with its major axis at right angles to the optic axis and the birefringence is negative.

7.25.1 Birefringence in Biology

1. **Protoplasm**: Resting cytoplasm under polarizing microscope gives little evidence of submicroscopic structure probably due to random orientation of the micellar elements within it. Pfeiffer studied frog eggs with high speed ultracentrifuge designed to permit observation with a polarizing microscope; he found no evidence of orientation during centrifugation. But at 20000 r.p.m. some birefringence was observed at the centrifugal pole, it reveals the presence of fibrils with their axis parallel to the centrfugal force and

gives picture about normal resting condition of the cytoplasm.

Electron microscopical studies of the cortex region of cell show some birefringence. Analysis by immersion of the cell in solutions of varying refractive index indicate both form and micellar birefringence. The form bire-fringnce is negative in respect to the radius suggesting the presence of concentric lamellae while the intrinsic is positive which corresponds to radially oriented long molecules. The low cortical birefringence observed in most cells is presumably due to the partial or complete cancellation of the positive and negative components. The inference of this birefringence is probably that the cortex is made up of concentric leaflets of protein with layers of radially oriented lipid in between two layers of protein.

Mitchioson suggested that all birefringence is due to protein, the polypeptide chains of which are folded in a Chinese-cracker pattern. Such arrangement might provide a mechanical basis for the process of development of the cleavage furrow.

2. **Muscle**: Muscle fibrils under polarized light between crossed Nicols, i.e., polarizing microscope, show crossbanding in the sarcous elements. There is alternate:
   (a) Dark band or A-band which is doubly refracting (high refractive index) known as anisotropic, exhibits birefringence.
   (b) Light band or I-band which is singly refracting (low refractive index) known as isotropic, does not show birefringence.

   In contracted parts of the muscle fibrils the anisotropous sarcous elements are seen to have increased in bulk, while the isotropous substance of the light band has correspondingly diminished.

   The anistropy is due to two causes:

   1. The submicroscopic units are rod-shaped and lie with their axis more or less parallel with the axis of the fibre (form double refraction).
   2. These units are themselves built up of molecules in an orderly or crystalline fashion (micellar, double refraction).

   Both types of birefringence are positive. They are due to particles arranged with their long axis parallel to the axis of the fibre. So that a fundamental structure, composed of long rod-shaped bundles of polypeptide chains is consistent with the birefringence of the A-bands. This isotropy and less birefringence of I-bands may be due to lack of order in the arrangement of the submicroscopic units in this region or to the presence of some strongly negative birefringent material which cancelled the positive birefringence of the contractile units. Eletron microscopy has revealed that the characteristic difference between the two bands due to the presence of extra material other than Myosin is known as **A-substance**.

3. **Myelin sheath of the nerve**: X-ray diffraction pattern of nerve with those of the lipids isolated from its sheath suggested that the sheath was made up of concentric leaflets of lipids with a thickness of 63Å. The polarization microscopy showed that it was made up of long lipid molecules oriented radially, their long molecules being packed side by side, while the proteins would form flat laminae in a plane at right-angles. Birefringence seen in the sheath due to these lipid molecules is positive (Fig. 7.13).

4. **Nerve**: The birefringence of the nerve is negative, the axis of reference being the long axis of the nerve which is at right-angles to the true optic axis, i.e., the refractive index for vibrations parallel to the axis of the nerve is less than that for vibrations at right-angles, so that the birefringence is referred as negative.

7.26 Determination of the Size and Shape of the Colloidal Particles

The size and shape of biological materials may be observed with the help of electron microscope by negative staining. But the method depends upon the precipitation of the particles from solution. Again it does not give definite information pertaining to the condition of particles in solution so that we should rely on indirect methods mentioned below.

Indirect methods are:

1. Ultrafiltration
2. Diffusion rate
3. Tyndall effect (light scattering)
4. X-ray diffraction analysis
5. Osmotic pressure
6. Ultracentrifuge
7. Viscosity
8. Ultramicroscope.

7.26.1 Ultrafiltration

Colloidal particles can pass through the filter papers and collodion membranes without any difficulty. If collodion membranes are prepared having pores of different sizes, and the solution is allowed to pass through the collodion membranes under pressure, colloidal particles may pass or may not pass. The size of the particle is between the size of the pore through which it cannot pass and the next larger pore through which it can pass.

But determination by this method does not give correct result due to certain drawbacks.

7.26.2 Diffusion rate

The rate at which the colloidal particles diffuse in solution has been used for determination of the size and molecular weights. It is also used for smaller particles than the Colloids (subcolloidal size), e.g., sugars and amino acids. The rate of diffusion is measured by optical methods like light absorption. Tyndall effect, refractive index and fluorescence without any disturbance of the solution.

7.26.3 Tyndall effect or light scattering method

Tyndal effect, i.e., the scattering of light by colloidal particles is employed for determination of size and molecular weight for colloidal solution, the light transmits through solution and the certain part scatters. The fractional decrease of intensity of transmitted light has relationship with molecular weight M, as follows:

$$M = -\left(\frac{\log_e I/I_0}{H \times C \times l}\right)$$

in which

I = Intensity of transmitted light,

I_0 = Intensity of incident light,

H = Proportionality constant,

C = Concentration of the colloidal particles in gms per ml,

l = Length of the path through the scattering solution

The value of constant H is given by:

$$H = \frac{32\pi^3 n_0^{\,2}}{3N\lambda^4}\left(\frac{n - n_0}{c}\right)^2,$$

in which λ = wavelength of the incident light, n_0 and n = Indices of refraction of the solvent and solution respectively,

N = Avogadro's number,

The advantage of light scattering method over the osmotic pressure method in determination of molecular weight is that the lower limit of determination of molecular weight is 12000 but for high molecular weight, the light scattering effect is far better. Because, while the intensity of scattering goes high with increase in molecular weight, the osmotic pressure decreases.

7.26.4 X-ray diffraction analysis or X-ray crystallography

Electron microscopy can be highly informative about the structures of biological macromolecules, but the technique is not so effective delineating molecular architecture in atomic detail. Bragg

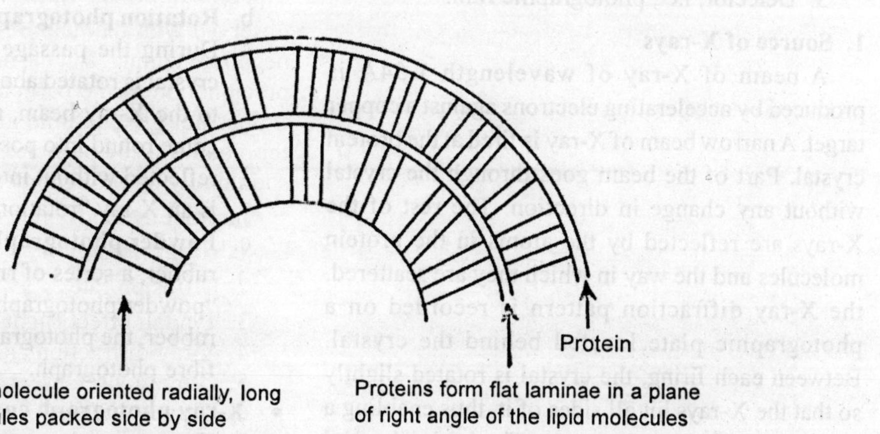

Fig. 7.13: Myelin sheath showing arrangement of lipid and protein

(1913) worked out the structure of solid sodium chloride by bombarding crystals of this salt with X-rays and analysed the way the X-ray scattered. Since then this technique of X-ray Crystallography has been used in the analysis of many different molecules including large molecules like proteins.

The spacing of regularly repeating atomic or molecular units in crystals can be determined by studying the angles and intensites at which X-rays of a given wavelength are scattered or diffracted, by the electrons that surround each atom. Atoms with highest electron density, such as heavy-metal atoms, diffract X-ray, most atoms with the lowest electron density, e.g., hydrogen atom, diffract X-ray least. X-ray analysis of NaCl is relatively simple, because only two different kinds of atoms are involved and they are regularly spaced. But when crystals of complex organic molecules such as proteins are analyzed, the mathematical analysis of the diffraction patterns is very complicated, because the large number of atoms in the molecule may yield thousands of diffraction spots. However, with the help of high-speed computers it is now possible to derive, from X-ray diffraction patterns, the precise positioning of the atoms and the shape of the whole molecule.

- **Method**

In X-ray Diffraction analysis or X-ray Crystallographic experiment consists of three components.

1. Source of X-rays.
2. Crystal of the substance to be examined.
3. Detector, i.e., photographic film.

1. Source of X-rays

A beam of X-ray of wavelength 1.54Å is produced by accelerating electrons against a copper target. A narrow beam of X-ray is fired at the protein crystal. Part of the beam goes through the crystal without any change in direction. The rest of the X-rays are reflected by the atoms in the protein molecules and the way in which they are scattered, the X-ray diffraction pattern is recorded on a photographic plate located behind the crystal. Between each firing, the crystal is rotated slightly so that the X-rays hit all sides of it, thus enabling a complete analysis to be made. The basic physical principles are:

a. Electrons scatter X-rays. The amplitude of the wave scattered by an atom is proportional to its number of electrons. Thus, a carbon atom scatters six times as strongly as a hydrogen atom.

b. The scattered waves recombine. Each atom contributes to each scattered beam. The scattered waves reinforce one another, if they are in phase and they cancel one another, if they are out of phase.

c. The way in which the scattered waves recombine depends only on the atomic arrangement (Fig. 7.14).

2. A crystal of substance (e.g., protein)

Crystals of the protein such as myolobin is crystallized by adding ammonium sulphate to a concentrated solution of the protein. Ammonium sulphate at a concentration of 3M markedly reduced the solubility of myoglobin and leads to crystallization. The solubility of most proteins is reduced on addition of high concentrations of any salt. This effect is known as "salting-out" and the crystal is several millimetres long. The reflection of X-rays occurs from the set of crystal planes of the crystal lattice (Fig. 7.15).

3. Detector i.e. photographic film

An X-ray diffraction pattern is usually recorded on photographic film, but ionization methods of recording are much more accurate and sensitive. Different types of photograph are seen:

a. **Laue photograph**: X-ray beam after passing through crystal may be made to impinge on a photographic plate to give a series of spots.

b. **Rotation photograph or fibre photograph**: During the passage of X-ray beam, if the crystal is rotated about an axis perpendicular to the X-ray beam, all the reflection planes come round into position where they may be reflected without interference, and the result is an X-ray "rotation photograph".

c. **Powder photograph**: In case of unstretched rubber, a series of rings are produced called "powder photograph". But on stretching the rubber, the photograph appears as one of the fibre photograph.

- **X-ray photograph enables us**

a. To measure the sequences in which different atoms are repeated along the chains.

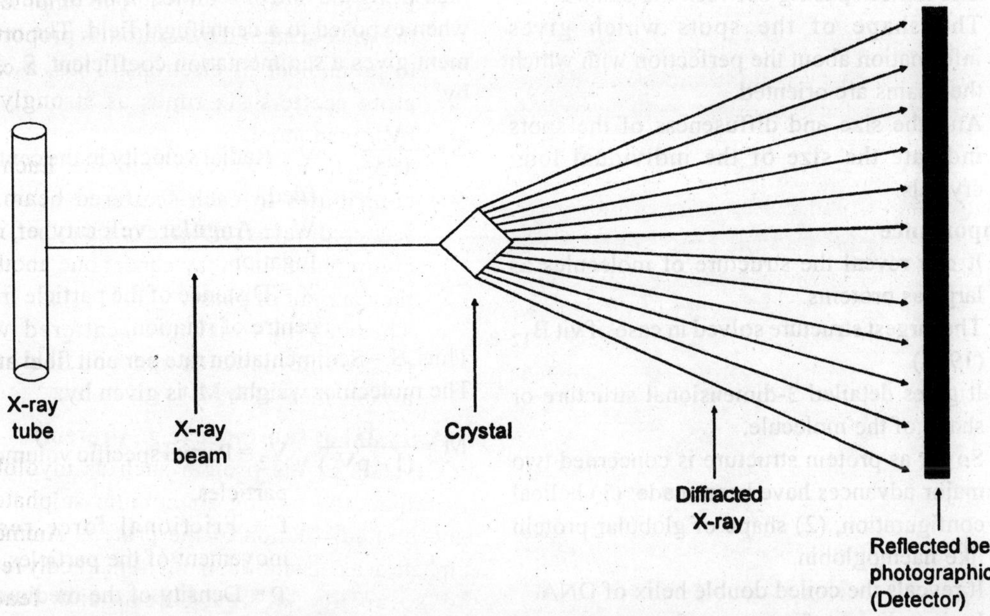

Fig. 7.14: Schematic representation of X-ray diffraction analysis

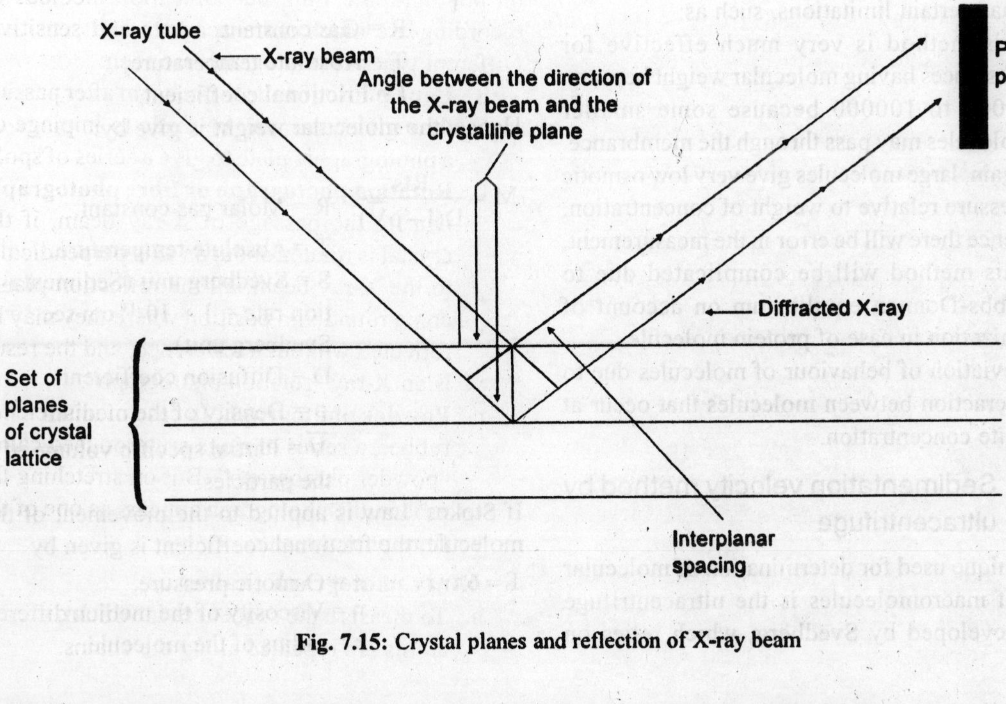

Fig. 7.15: Crystal planes and reflection of X-ray beam

b. The lattice spacing between the chains.
c. The shape of the spots which gives information about the perfection with which the chains are oriented.
d. And the size and diffuseness of the spots indicate the size of the individual long crystals.

- **Importance**

a. It can reveal the structure of molecules as large as proteins.
b. The largest structure solved in case of vit B_{12} (1957).
c. It gives detailed 3-dimensional structure or shape of the molecule.
d. So far as protein structure is concerned two major advances have been made: (1) helical configuration, (2) shape of globular protein like haemoglobin.
e. It reveals the coiled double helix of DNA.
f. It gives direct information of arrangement of atom and molecules and calculation of the size of the particles.

7.26.5 Osmotic pressure

Molecular weight can be determined by osmotic pressure measurement. This procedure was applied to many substances like proteins and was corroborated by the ultracentrifuge method. But this method has certain limitations, such as:

a. This method is very much effective for substances having molecular weight between 30000 to 100000 because some smaller molecules may pass through the membrance. Again, large molecules give very low osmotic pressure relative to weight of concentration. Hence there will be error in the measurement.
b. This method will be complicated due to Gibbs-Donnan equilibrium on account of ionization in case of protein molecule.
c. Deviation of behaviour of molecules due to interaction between molecules that occur at finite concentration.

7.26.6 Sedimentation velocity method by ultracentrifuge

The technique used for determination of molecular weight of macromolecules is the ultracentrifuge mehod developed by Svedberg, which relies on measuring the rate of sedimentation of the particles when exposed to a centrifugal field. The measurement gives a sedimentation coefficient, S, defined by:

$$S = \frac{V}{W^2X}$$

V = Radial velocity in the centrifugal filed,
W = Angular velocity of centrifugation,
X = Distance of the particle from the centre of rotation.

Thus, S = Sedimentation rate per unit field or force, The molecular weight, M, is given by:

$$M = \frac{fs}{(I - pV_2)}$$

V_2 = Partial specific volume of the particles,
f = Frictional force resisting movement of the particles,
ρ = Density of the medium.

To determine the molecular weight, besides the sedimentation rate, the partial specific volume and the frictional force are essential. The frictional resistance may be determined directly because it is simply related with the diffusion of the molecule. Einstein showed that the diffusion coefficient, D, was expressed by:

$$D = \frac{Rt}{t}$$

D = Diffusion coefficient,
R = Gas constant,
T = Absolute temperature,
f = Frictional coefficient.

Hence, the molecular weight is give by:

$$M = \frac{Rts}{D(1 - \rho\overline{V})}$$

R = Molar gas constant,
T = Absolute temperature,
S = Svedberg unit (Sedimentation rate = 1 × 10^{-13} cm sec^{-1} = Svedberg unit).
D = Diffusion coefficient,
ρ = Density of the medium
$\overline{V}$ = Partial specific volume of the particles.

If Stokes' Law is applied to the movement of the molecule, the frictional coefficient is given by

$$f = 6\pi\eta r$$

π = Osmotic pressure,
η = Viscosity of the medium,
r = radius of the molecule.

Hence, $D = \frac{Rt}{6\pi\eta r}$.

Thus, we can deduce the radius of the molecule from its measured diffusion coefficient provided Stokes' Law is applied. Applying Stokes' Law under limiting condition, we may deduce an estimate of molecular weight from a measurement of the diffusion coefficient and the partial specific volume.

7.26.7 Viscosity

The asymmetry of a molecule is indicated by the viscosity of its solution. According to Einstein, the theoretical equation related to viscosity to concentration is calculated assuming that the molecules are perfect spheres; deviations form spherical shape. Such asymmetrical particles turn and twist and appear to occupy a larger volume than they do actually. The twisting of the particle interferes the streamline flow and, therefore, more work is required to maintain a given velocity of flow, eventually the viscosity of the solution is greater than if the particles were spherical.

Simha developed equations relating the viscosity of a solution to the axial ratio of the particles. Axial ratio means the ratio of the long to the short axis, if the molecules are treated as ellipsoids. Measurements of the viscosity of solutions of large molecules may thus provide an information regarding the shape of the molecules, given their molecular weights.

7.26.8 Ultramicroscope

The size of the Colloidal particles can be determined with the help of ultramicroscope.

7.26.9 Molecular weight determination by ultracentrifuge

Smaller particles can be sedimented by the ordinary centrifuge having 4000-7000 revolutions per minute due to generation of less centrifugal force for sedimentation. But in case of macromolecules, e.g., protein, the molecular weight could be determined by sedimentation in a strong centrifugal field and observing the rate of movement of the Colloid particles outward from the centre of rotation. This method was developed by Svedberg (1925) who invented the instrument called **Ultracentrifuge** in which materials may be subjected to centrifugal forces which are many thousands of times stronger than the force of gravity (Fig. 7.16).

The principle of the method relies on measuring the rate of sedimentation of the particles under the action of centrifugal force whose rotor revolves at a speed of 70000 revolutions per minute. The centrifugal acceleration is much greater than the force g, which in modern centrifuge attains 3,50,000 g (70000 rpm).

The intensity of a centrifugal field is usually expressed in terms of the magnitude of the relative centrifugal force (RCF) which is related to the speed of rotation by the equation:

$$RCF = 1.118 \times 10^{-5} \times r \times (rpm)^2$$

Here RCF is expressed as the gravitational force acting on a 1 gram mass at a distance r (in centimetres) from the axis of rotation.

7.26.9.1 Procedure

The cell called sample cell, containing the solutions of Colloid is a cylinder with transparent windows made of quartz or glass which is placed in the rotor. A similar cell, called Ballast cell, having solvent or water is placed at the opposite side of the arm for counterbalance. The rotation is done maintaining vacuum environment inside the centrifuge to avoid thermal effect due to friction with air. The dispersed particles are gradually sedimented during rotation and their concentration changes in the different layer formed. If the rate of sedimentation exceeds the rate of diffusion of the particles, sedimentation occurs. Two regions are formed in the cell—the pure solvent and the solution, e.g., solute containing layer. There is also a transient zone, where the solute concentration varies from zero to a maximum value. As the sedimentation progresses this zone and its boundary are displaced. the velocity of travel of the boundary towards the base cell, e.g., from the axis of rotation of the rotor, is measured. The rate at which the particles move depends upon:

1. the force depending upon (a) angular velocity of the centrifuge (b) and the distance of the colloidal particles from the axis of rotation,
2. shape,
3. size,
4. density of the particles,
5. viscosity of the suspending medium.

In case of homogenous materials where the particles are all alike, the particles move under the centrifugal force as a sharp boundary in the medium.

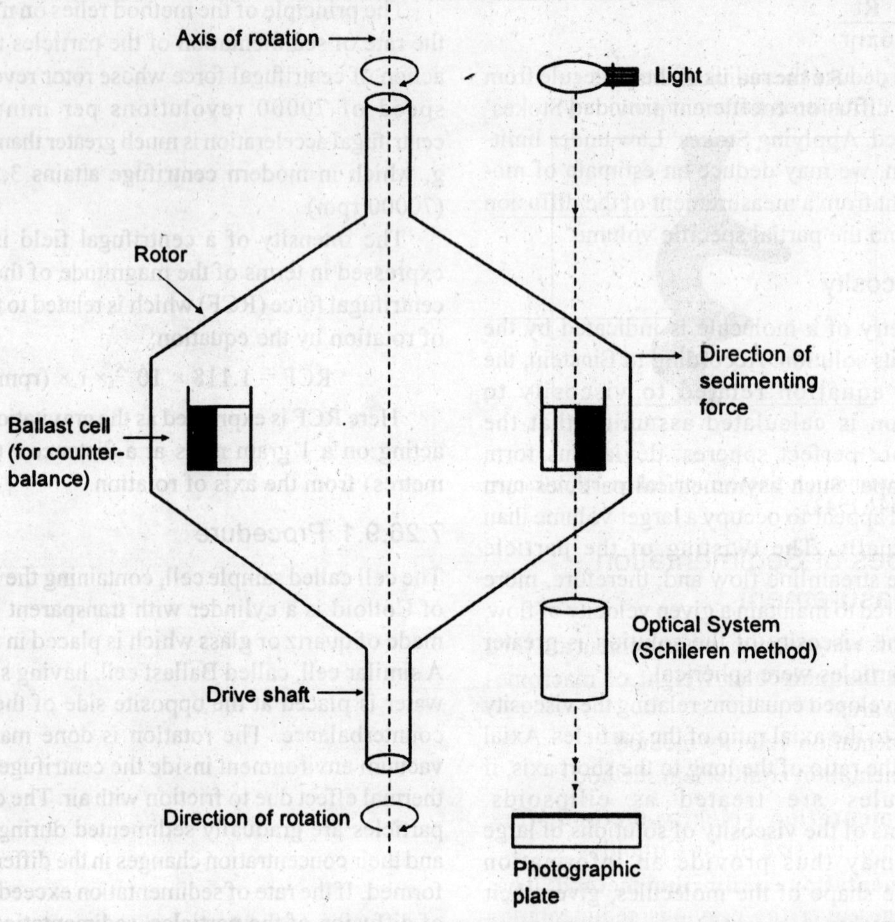

Fig. 7.16: Principle of analytical ultracentrifuge, while sample is undergoing sedimentation

- **Observation**

Since the particles absorb more light than the medium, their rate of movement may be detemined by photographing the position of the particle boundary in the sample cell at various time-intervals.

But the refractive index of the particles is different from that of the medium, there is every possibility to follow the movement of the boundary by the optical methods. These methods are based upon the fact that a light beam is bent most as it passes through the solution in the region of the boundary where the gradient of colloidal particles concentration is greatest. This refractive index effect is recorded photgographically and indicates the distance of the boundary from the axis of rotation at different time-intervals. The refractive index method gives:

1. **the position of the particle boundary in the solution,**
2. **an estimate of the concentration of a Colloid or a protein.**

The optical methods are:

1. **Schlieren: scanning method–Devised by Longsworth in connection with the study of the electrophoresis of proteins.**
2. **Schelieren method: Involves the use of a diagonal slit and a cylindrical lens. It is widely used in measurement of the sedimentation velocity of proteins as well as electrophoretic studies.**

In the photograph of the sedimentation cell, made by the refractive index method, the boundary appears as the top of a sharp peak. The area under the peak is a measure of the concentration of the Colloid or protein in question (Fig. 7.17).

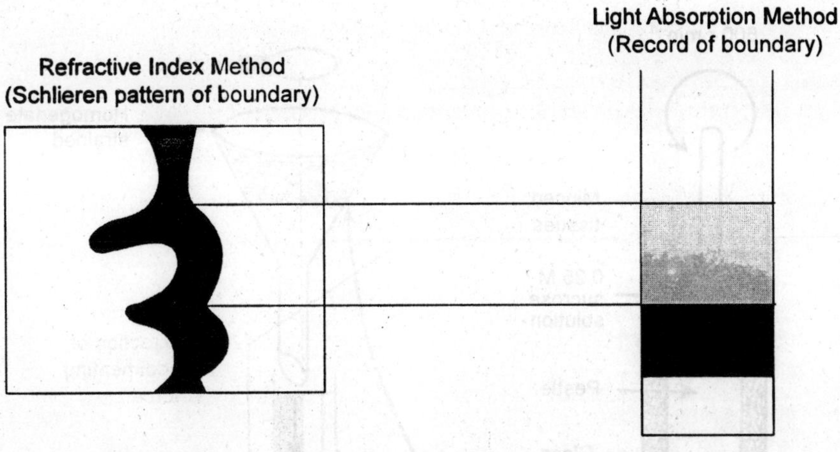

Fig. 7.17: Sedimentation photograph taken by refractive index and light absorption method (Pickels, Chem, Revs, 30,341,1942)

7.27 Types of Sedimentation Measurement

Three types of sedimentation measurement are used to determine the molecular weight of macromolecules. They are:

(a) Sedimentation velocity method,
(b) Sedimentation equilibrium method.

(a) Sedimentation velocity method: (See determination of the size of the Colloid).

(b) Sedimentation equilibrium method: It has two advantages over the previous sedimentation velocity method:

1. It does not require the knowledge of the diffusion coefficient or the shape of the macromolecules.
2. The ultracentrifuge is operated at a relatively low speed, just high enough for the system to come to an equilibrium state in which the rate of sedimentation of the macromolecules is exactly balanced by the opposing diffusion force. At equilibrium, there is no pure solvent of region at the surface meniscus; instead, a gradient of macromolecules is formed down the centrifuge tube. The bottom layer may contain macromolecule concentration about double that of the top layer. By measuring the concentration of the macromolecules as a function of the distance from the centre of rotation, data can be obtained to calculate molecular weight from the equation:

$$M = \frac{2RT\ \ln(C_2/C_1)}{\omega^2(1-\overline{V}\rho)(X_2^2 - X_1^2)}$$

where R = Gas constant (8.31×10^7 ergs mol^{-1} K^{-1}),
T = Absolute temperature (in kelvin),
C_1 and C_2 = Concentrations of substances (e.g., protein) at two points in the tube,
X_1 and X_2 = Distances from the centre of rotation, X_1 for C_1 and X_2 for C_2,
ω = Angular velocity,
ρ = Density of solvent,
$\overline{V}$ = Partical specific volume of the substance.

Though it is the most accurate method among the sedimentation methods, the disadvantage is that it may require several days of centrifugation to attain equilibrium. This problem can be solved partly by using cells with a very short column of protein solution (1-2mm) and the protein must be pure and homogenous.

(c) Approach-to-equilibrium method: It is more rapid measurement of molecular weight. The rotor is brought to the equilibrium speed over a 1- to 2h period by a series of adjustments. Each time, the protein concentration is measured near the bottom of the tube. From these, the molecular weight is determined.

7.27.1 Importance

1. Separation and isolation of the particles of different sizes and shapes.
2. Determination of molecular weight of the substance.
3. For estimating the sedimentation velocity of the substance.

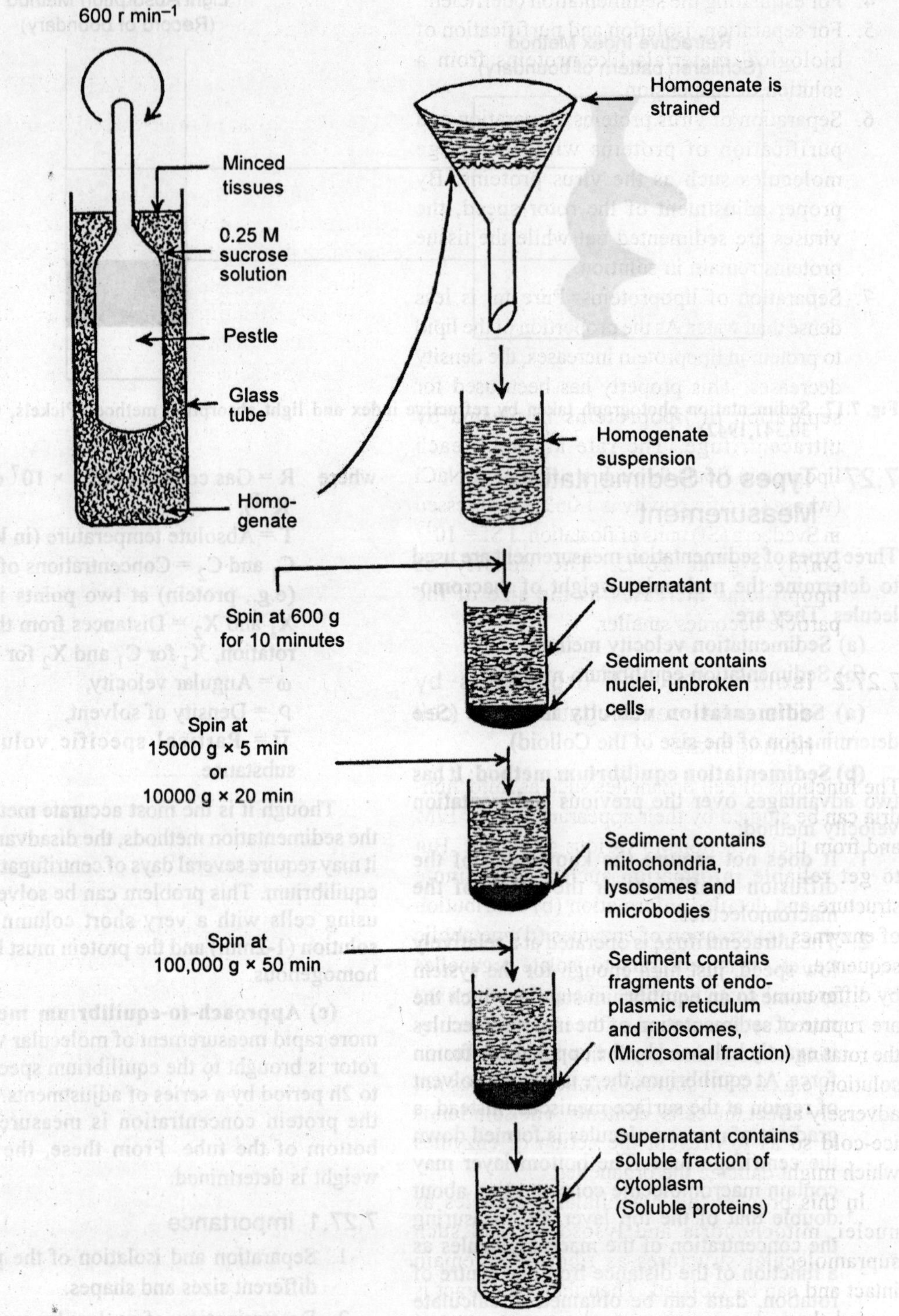

Fig. 7.18: Differential centrifugation (g is the unit of gravitational force, e.g., spining of 600 g means that the homogenate is subjected to a centrifugal force 600 times the force of gravity in preparative ultracentrifuge

4. For estimating the sedimentation coefficient.
5. For separation, isolation and purfification of biological materials like proteins from a solution or suspension.
6. Separation of virus proteins: Separation and purification of proteins with very large molecules such as the virus proteins. By proper adjustment of the rotor speed, the viruses are sedimented out while the tissue proteins remain in solution.
7. Separation of lipoproteins: Pure fat is less dense than water. As the proportion of the lipid to protein in lipoprotein increases, the density decreases. This property has been used for separation of lipoproteins in plasma by ultracentrifuge. The rate at which each lipoprotein floats through a solution of NaCl (whose specific gravity is 1.063) is expressed in Svedberg (Sf) units of floatation. 1 Sf = 10^{13} cm/dyne/g at 26°C. The density of lipoproteins increases as the size of the particle becomes smaller.

7.27.2 Isolation of cell organelles by differential centrifugation from cell homogenate

The functions of cell organelles such as mitochondria can be studied by their appearance under E/M and from their reactions to various stains, etc. But to get reliable information such as (a) minute structure and detailed information (b) distribution of enzymes (c) isolation of enzymes (d) metabolic sequence, etc. it is necessary to isolate organelles by differential centrifugation. The cell membranes are ruptured by the shearing forces developed by the rotating homogenizer pestle in a suitable isotonic solution, e.g., 0.25M sucrose solution that does not adversely affect the cells' contents. The solution is ice-cold so as to prevent the action of enzymes which might damage the organelles.

In this process the subcellular organelles as nuclei, mitochondria and lysosomes and such supramolecular structures as ribosomes remain intact and can be isolated. Then the cell extract is passed through a stainless steel sieve to remove fragments of blood vessels, connective tissues, etc. After staining, the homogenate is centrifuged at a series of increasing rotor speeds with the supernatant and taking out the sediment containing **Cell organelles** (Fig. 7.18).

7.28 Artificial Kidney (Haemodialyser)

For mechanism of urine formation in Kidney (See p. 161)

Definition: A method of extracorporeal blood purification in which blood is continuously circulated in contact with a permeable membrane whereas a large volume of balanced electrolyte solution circulates on the other side of the membrane. Diffusion of dissolved substances from one stream to the other removes molecules that are in excess in the blood and replaces these for which there is a deficiency.

7.28.1 Principle

The basic mechanism of the artificial kidney is to pass blood from a person having damaged kidneys through very narrow channels bounded by thin cellophane membrane. Outside the membrane, there is fresh dialyzing fluid which receives the waste products from the blood by diffusion.

7.28.2 Unit of artificial kidney

Blood from radial artery of the person flows continually through very narrow channels bounded by thin sheets of cellophane membrane. Outside the sheet of cellophane membrane there is fresh dialyzing fluid which comes from a container maintaining constant temperature. On both sides of the cellophane sheets there are plates of polypropylene. The fresh dialyzing fluid while flowing outside the cellophane sheets in the instrument, called Kill dialyzer, takes up unwanted materials from the blood and passes through the tube to a container placed for dialyzed fluid, i.e., used dialyzing fluid. The cellophane is porous enough to allow all constituents of the plasma except plasma proteins to diffuse freely in both directions, e.g., from plasma to the dialyzing fluid and from dialyzing fluid to the blood. If a substance present in the blood in higher concentration than in the dialyzing fluid, there will be net transfer of the substance from plasma to the dialyzing fluid. The transferable amount of the substance depends upon: (a) difference of concentration on both sides of the membrane, (b) size of the molecule, the smaller molecules diffuse rapidly than bigger ones, (c) the period of time that the blood and the fluid remains in contact with the membrane.

Blood comes from the radial artery of the person, passes through the Kill dialyzer to the superficial vein through a bubble trap to prevent any bubbles formed during the process. To prevent clotting of blood in the artificial kidney, an anticoagulant, heparin, is infused into blood. And to prevent bleeding in the patient as a result of heparin, an antiheparin substance, protamine, is infused into the blood.

Blood continually flows from the artery through the artifical kidney to the vein for 12 hours in single sitting every 3-4 days. The total amount of blood flowing at one time is usually less than 500 ml/ minute and the total diffusing surface is between 10000 to 20000 square centimetres (Fig. 7.19).

Biophysical Mechanism in artificial kidney: Removal of water and solutes from the blood takes place by the following ways:

a. Solute diffusion in response to concentration gradients,
b. Water ultrafiltration and solute convection in response to hydrostatic and osmotic pressure gradients,
c. Water migration in response to osmotic gradient.

Table 7.8: Composition of dialysate, i.e., dialyzing fluid

Ingradients	gm/l
1. Sodium	130.00
2. Potassium	1.34
3. Calcium	3.30
4. Magnesium	1.00
5. Chloride	101.00
6. Acetate	35.00
7. Lactate	1.30
8. Glucose	1.80

7.28.3 Dialyzing fluid

The dialyzing fluid contains all electrolytes, e.g., Na^+, K^+, Ca^{++}, Mg^{++}, Cl^-, HCO^-_3, lactate$^-$, in the same concentration like that of plasma but there is no HPO^-_3, urate–, sulphate$^=$. Among non-electrolytes it contains glucose but no creatinine and urea.

- **Membrane for haemodialyis**

1. **Cellulosic membrane**: Sieve type of membrane, hence allow the solutes to pass through the small pores according to size, shape and molecular weight of the solutes.
2. **Cuprophane membrane:** Most commonly used for haemodialysis, made of natural cellulose (cuproammonium regenerated collulose) having tenacity, elasticity, puncture-proof, allow to pass solution of different molecular weight not more than10000. It is moisture sensitive cellulose hydrate. Its formation depends upon the water content and diversity processing there is change in lengths for which glycerine is added to the membrane as humectant and plasticizer.

Type of Dialyser: The haemodialyser are classifies considering their design:

(1) Coil type, (2) Parallel plate type, (3) Hollow fibre type.

1. **Coil type**: If consists of a tubular membrane placed between flexible support wrapped around a cylindrical core. The coil is dipped in a dialysate i.e., dialysing fluid. It uses cellophane or cuprophane membrane.
2. **Parallel plate or parallel flow type**: It allows the blood flow at the pressure of the patient blood flow and does not require pump. The dialysing surface area is about 1 sq.m.
3. **Hollow Fibre Haemodialyser**: Most commonly used, having 10000 hollow capillaries made of deacetylated diacetate, packed in plastic cylinder of 1g cm in length and 7 cm in diameter. The capillaries are closed at both ends. The thickness of capillaries is 25-30 μm and diameter of 200-300 mm and dialysing area of 9000 cm^2/unit.

Table 7.9: Comparison of normal plasma, dialyzing fluid and uraemic plasma

Normal plasma	Dialyzing fluid	Uraemic Plasma (Damaged Kidney)
Sodium, potassium chloride present	Same as normal plasma	Same as normal plasma except potassium which is higher
Phosphate, urea (26 mg%), urate, sulphate, creatinine present	Absent	All are in higher concentration (urea-200 mg%)

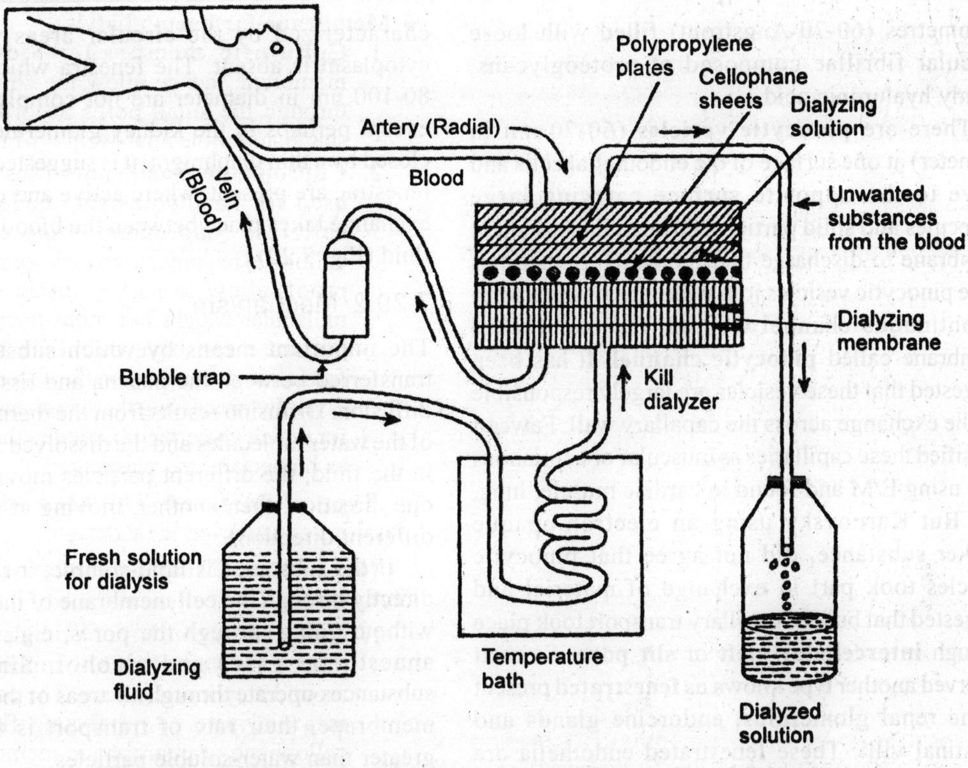

Fig. 7.19: Diagrammatic representation of artificial kidney (Haemodialyser)

Thus, the substance which is present in higher concentration in plasma such as urea (200 mg%) in uraemic patient can be removed rapidly while the normal electrolytes remain normal (Table: 7.9).

- **Factors to increase removal**

Increase removal can be achieved by increasing:
1. the duration of the process,
2. the overall membrane area,
3. the membrane permeability.

Efficacy: The effectiveness of the artificial kidney is expressed in terms of the amount of plasma that can be cleared of different substances each minute which is also applied for functional effectiveness of the normal kidneys. Artifical kidney can clear 100 to 200 ml of plasma per minute which indicates it can function about 2 times rapidly as the two normal kidneys together having urea clearance only 70 ml per minute.

Indication: It is used in case of (1) severe renal failure due to disease, (2) in acute renal failure due to mercury poisoning or circulatory shock, to tide over the crisis till the kidneys resume normal function, (3) permanent renal failure, (4) total kidney removal.

- **Complications of haemodialysis:**

1. Intra dialysic hypotension
2. Nausea
3. Vomiting
4. Cramps
5. Chest pain
6. Fever.

7.29 Biological Membranes and Ultrafiltration

7.29.1 Capillary membrane and ultrafiltration

The capillary wall is composed of a unicellular layer of endothelial cells which is surrounded by a basement membrane on the outside. The total thickness of the wall is 0.5 micron and the diameter of the capillary is 5 to 9 microns through which a blood cell can manage to squeeze through.

The "intercellular cleft" which is a thin "slit" or "pore" at the junction between the adjacent endothelial cells. The space between the endothelial cells, e.g., width of the intercellular slit pore is 6-7 nanometres (60-70 Angstrom) filled with loose reticular fibrillae composed of proteoglycans, mainly hyaluronic acid.

There are **pinocytic vesicles** (60-70 nm in diameter) at one surface of the endothelial cells and move to the opposite surface carrying large molecules and solid particles through the basement membrane to discharge their contents. Sometimes these pinocytic vesicles join with each other to form a continuous channel through the endothelial membrane called **pinocytic channel**. It has been suggested that these vesicles are largely responsible for the exchange across the capallary wall. Fawcett classified these capillaries as muscular or continuous type using E/M and found in cardiac muscle, lung, etc. But Karnovsky using an electron-opaque marker substance, did not agree that pinocytic vesicles took part in exchange of material and suggested that bulk of capillary transport took place through **intercellular cleft** or **slit pore.** Fawcett observed another type known as **fenestrated** present in the renal glomerulus, endorcine glands and intestinal villi. These fenestrated endothelia are characterized by the circular areas where the cytoplasm is absent. The fenestra which is about 80-100 nm in diameter are not completely open, except perhaps in the kidney glomerulus, but are closed by a thin diaphragm. it is suggested that these fenestrae are present, where active and unimpeded exchange takes place between the blood and tissue fluid (Fig. 7.20).

7.29.2 Mechanism

The important means by which substances are transferred between the plasma and tissue fluid is diffusion. Diffusion results from the thermal motion of the water molecules and the dissolved substances in the fluid, the different particles moving first in one direction, then another, moving at random in different directions.

If the substance is lipid-soluble, it can diffuse directly through the cell membrane of the capillary without going through the pores, e.g., O_2, CO_2, anaesthetic gases and alcohol. Since these substances operate through all areas of the capillary membrane, their rate of transport is two times greater than water-soluble particles.

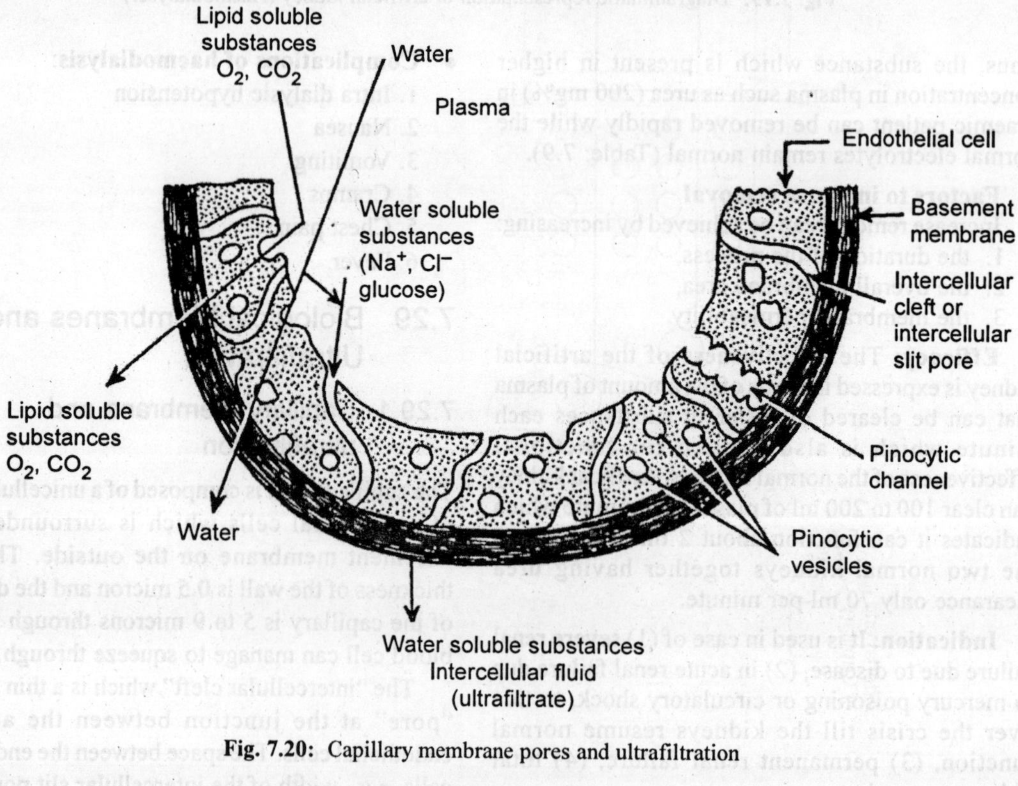

Fig. 7.20: Capillary membrane pores and ultrafiltration

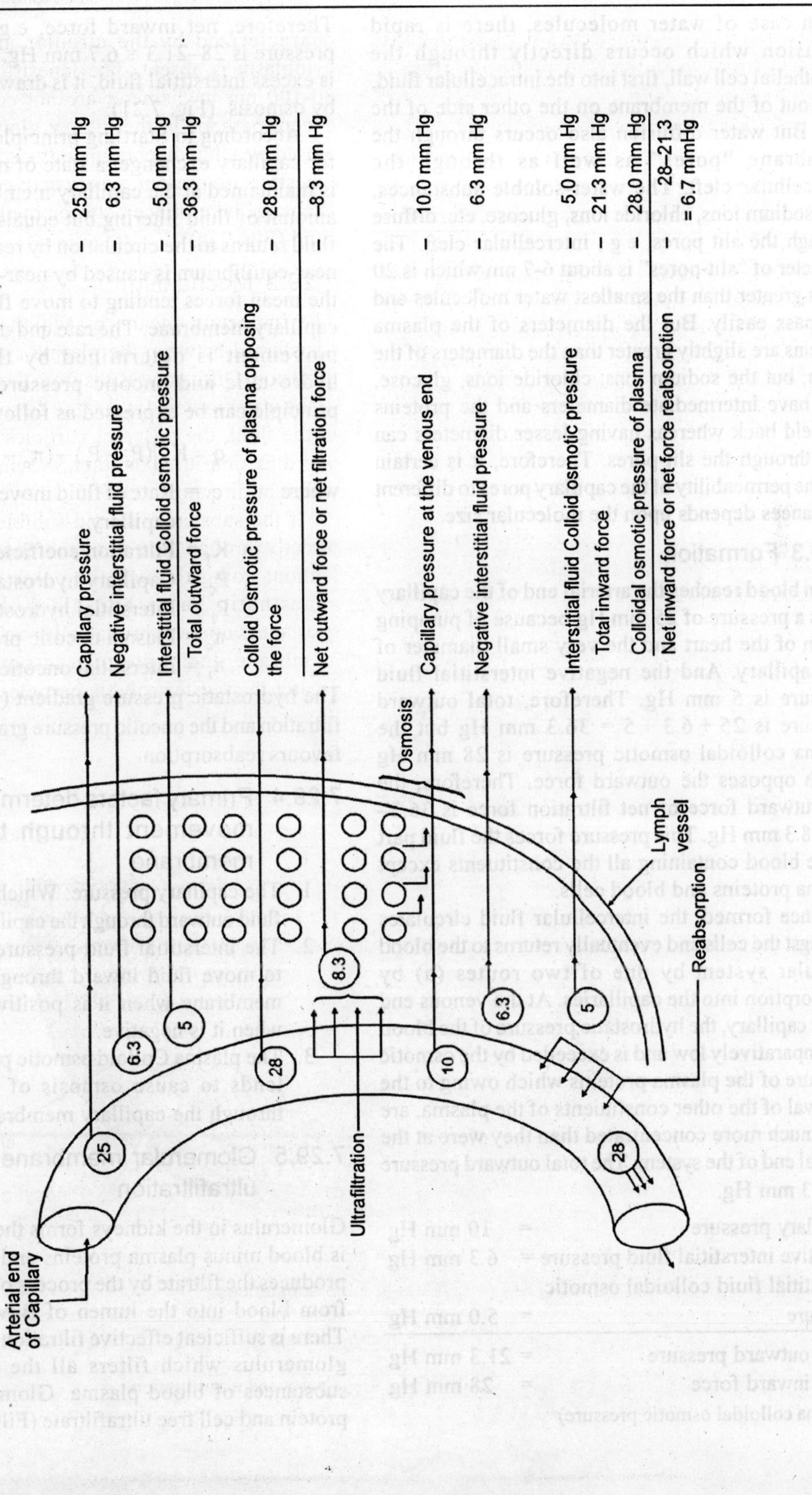

Fig. 7.21: Exchange of tissue fluid

In case of water molecules, there is rapid diffusion which occurs directly through the endothelial cell wall, first into the intracellular fluid, then out of the membrane on the other side of the cell. But water diffusion also occurs through the membrane "pores" as well as through the intercellular cleft. The water-soluble substances, e.g., sodium ions, chloride ions, glucose, etc. diffuse through the slit pores, e.g., intercellular cleft. The diameter of "slit-pores" is about 6-7 nm which is 20 times greater than the smallest water molecules and can pass easily. But the diameters of the plasma proteins are slightly greater than the diameters of the pores; but the sodium ions, chloride ions, glucose, urea have intermediate diameters and the proteins are held back whereas having lesser diameters can pass through the slit-pores. Therefore, it is certain that the permeability of the capillary pores to different substances depends upon the molecular size.

7.29.3 Formation

When blood reaches the arterial end of the capillary it has a pressure of 25 mm Hg because of pumping action of the heart and the very small diameter of the capillary. And the negative interstitial fluid pressure is 5 mm Hg. Therefore, total outward pressure is 25 + 6.3 + 5 = 36.3 mm Hg but the plasma colloidal osmotic pressure is 28 mm Hg which opposes the outward force. Therefore, the net outward force or net filtration force is 36.3–28 = 8.3 mm Hg. This pressure forces the fluid part of the blood containing all the constituents except plasma proteins and blood cells.

Once formed, the intercellular fluid circulates amongst the cells and eventually returns to the blood vascular system by one of two routes (a) by reabsorption into the capillaries. At the venous end of the capillary, the hydrostatic pressure of the blood is comparatively low and is exceeded by the osmotic pressure of the plasma proteins which owing to the removal of the other constituents of the plasma, are now much more concentrated than they were at the arterial end of the system. The total outward pressure is 21.3 mm Hg.

Capillary pressure	=	10 mm Hg
Negative interstitial fluid pressure	=	6.3 mm Hg
Interstitial fluid colloidal osmotic pressure	=	5.0 mm Hg
Total outward pressure	=	21.3 mm Hg
Total inward force (Plasma colloidal osmotic pressure)	=	28 mm Hg

Therefore, net inward force, e.g., reabsorption pressure is 28–21.3 = 6.7 mm Hg. (b) When there is excess interstitial fluid, it is drawn by lymphatics by osmosis. (Fig. 7.21).

According to Startling principle of equilibrium for capillary exchange, a state of near equilibrium is maintained at the capillary membrane where the amount of fluid filtering out equals the quantity of fluid returns to the circulation by reabsorption. This near-equilibrium is caused by near-equilibration of the mean forces tending to move fluid through the capillary membrane. The rate and direction of fluid movement is determined by the balance of hydrostatic and oncotic pressure. The Starling principle can be expressed as follows:

$$q = K_f \left|(P_c - P_i) - (\pi_c - \pi_i)\right|$$

where
- q = Rate of fluid movement across the capillary,
- K_f = Filtration coefficient,
- P_c = Capillary hydrostatic pressure,
- P_i = Interstitial hydrostatic pressure,
- π_c = Plasma oncotic pressure,
- π_i = Interstitial oncotic pressure.

The hydrostatic pressure gradient ($P_c - P_i$) favours filtration and the oncotic pressure gradient ($\pi_c - \pi_i$) favours reabsorption.

7.29.4 Primary factors determining the fluid movement through the capillary membrane

1. The capillary pressure: Which tends to move fluid outward through the capillary membrane.
2. The interstitial fluid pressure: Which tends to move fluid inward through the capillary membrane when it is positive but outward when it is negative.
3. The plasma Colloid osmotic pressure: Which tends to cause osmosis of fluid outward through the capillary membrane.

7.29.5 Glomerular membrane and ultrafiltration

Glomerulus in the kidneys forms the filtrate which is blood minus plasma proteins and blood cells. It produces the filtrate by the process of ultrafiltration from blood into the lumen of Bowman capsule. There is sufficient effective filtration pressure in the glomerulus which filters all the non-colloidal substances of blood plasma. Glomerular fluid is protein and cell free ultrafiltrate (Filtration process produces filtrate of blood excluding cell only) and glomerulus acts as an ultrafilter. As in the case of other capillary beds and in accord with Starling principle, the rate of filtration through the walls of the glomerular capillaries is determined by a balance of force favouring and opposing filtration. The main driving force is the hydrostatic pressure.

7.29.6 Effective filtration pressure

The afferent blood vessels of glomerulus have hydrostatic pressure of 75 mm Hg. This pressure is opposed by colloidal osmotic pressure of plasma (30 mmHg), Bowman capsular pressure (10 mm Hg) and interstitial fluid pressure 10 mm Hg. Therefore, effective glomerular filtration pressure is (75 mm Hg–30 mm Hg + 10 mm Hg + 10 mm Hg) 25 mm Hg (Fig. 7.22).

1. **The total surface area of glomerular** membrane helps in the glomerular filtration. 2. The permeability of the glomerular membrane helps ultrafiltration by diffusion 3. The Donnan effect of non-diffusible plasma proteins favours the passage of diffusible anions and prevents the passage of diffusible cations into the filtrate.

- **Evidence**: Richards inserted a micropipette into individual Bowman capsule and drew off very small quantities of fluid from the space of Bowman capsule after presssing the proximal renal tubule with the help of a glass rod. It was found that the fluid in the capsular space has exactly the same composition as blood plasma minus the plasma proteins (Fig. 7.23).

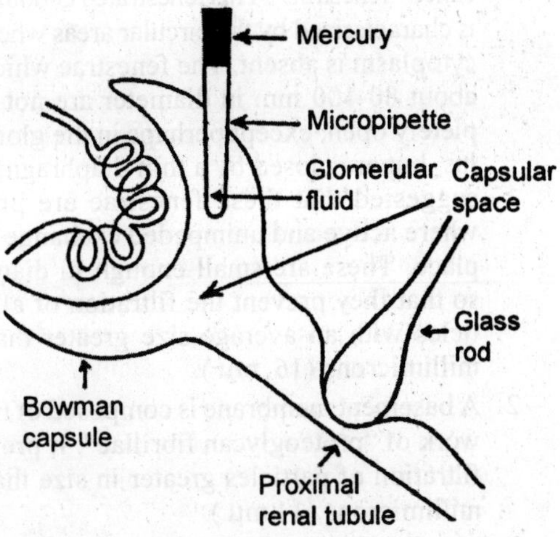

Fig. 7.23: Showing the method of collection of glomerular filtrate

7.29.7 Parameter for glomerular ultrafiltration

(a) The glomerular membrane must be so constructed as to retain molecules above a certain size whilst allowing smaller ones to pass through.

(b) There must be a head of pressure to force fluid through the filter.

7.29.8 Glomerular membrane

The membrane of the glomerular capillaries is called glomerular membrane (Fig. 7.24). In general, this membrane is similar to that of other capillaries but has several differences:

1. The **endothelial layer** of the capillary itself is perforated by thousands of small holes

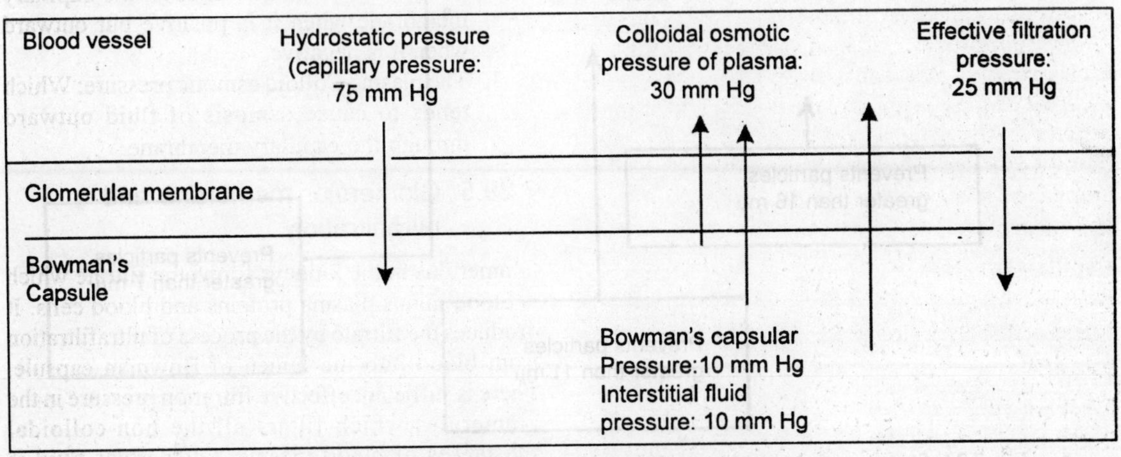

Fig. 7.22: Showing effective filtration pressure

called "fenestrae". This fenestrated endothelium is characterized by the circular areas where the cytoplasm is absent. The fenestrae which are about 80-100 mm in diameter are not completely open, except perhaps in the glomerulus, but are closed by a thin diaphragm. It is suggested that these fenestrae are present where active and unimpeded exchange takes place. These are small enough in diameter, so that they prevent the filtration of all particles with an average size greater than 16 millimicrons (16 mμ).

2. A basement membrane is composed of meshwork of "proteoglycan fibrillae". It prevents filtration of particles greater in size than 11 millimicrons (11 mμ).
3. The third layer is the **layer of epithelial cells** that line the outside of the glomerulus. These cells are not continuous sheet but instead consists of finger-like projections and each cell is known as a "podocyte" (Fig. 7.25). The finger-like projections join with each other to form a network having several pores, called "slitpore". It prevents filtration of all particles with diameter greater than 7 millimicrons (7 mμ).
4. **Channels or pores:** Less than 1% of serum albumin (MW 69000 daltons, radius 360Å) is filtered which suggests that the filtration barrier contains channels or "pores" up to approximately 60Å in diameter (Fig. 7.26). The location of these pores in the filtration barrier is not established. Recent evidence suggests that the pores may be hydrated channels between the glycoprotein chains of the basement membrane. Such channels probably are tortuous and may not be stable anatomical structures and indicate the failure to observe the pores under E/M.

B. The second difference is the **tremendous permeability** of the membrane for its special structure which is from 100 to 1000 times as great as that of the usual capillary.

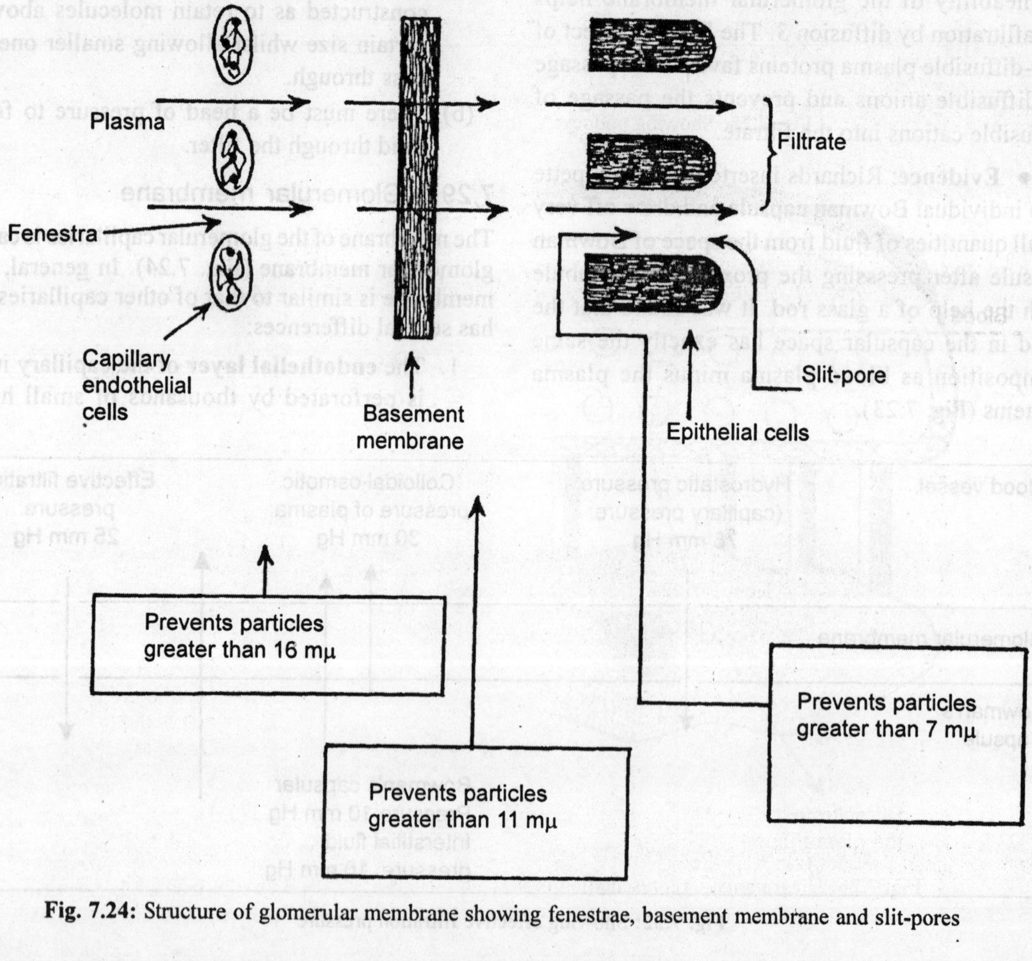

Fig. 7.24: Structure of glomerular membrane showing fenestrae, basement membrane and slit-pores

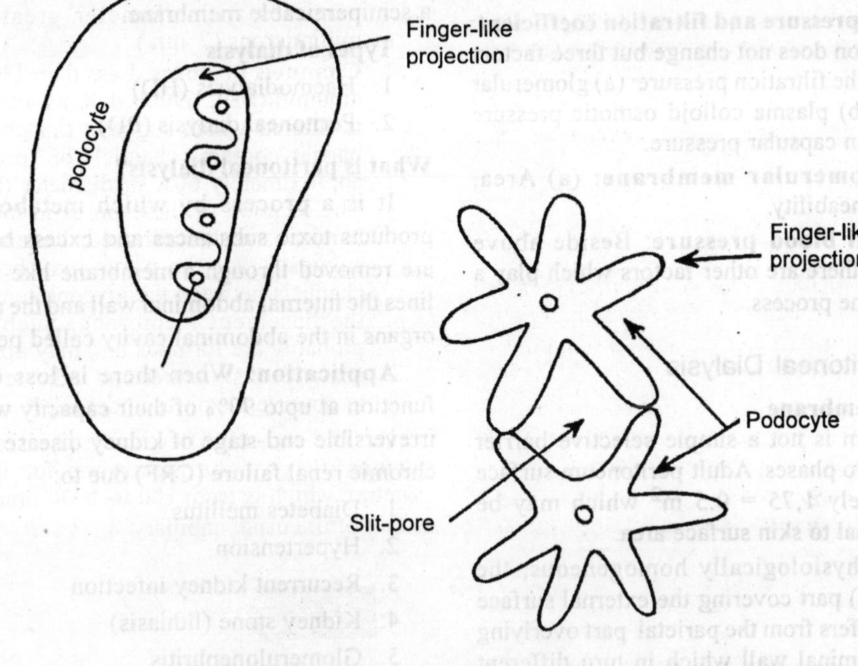

Fig. 7.25: Showing podocytes, finger-like projection and slit-pore

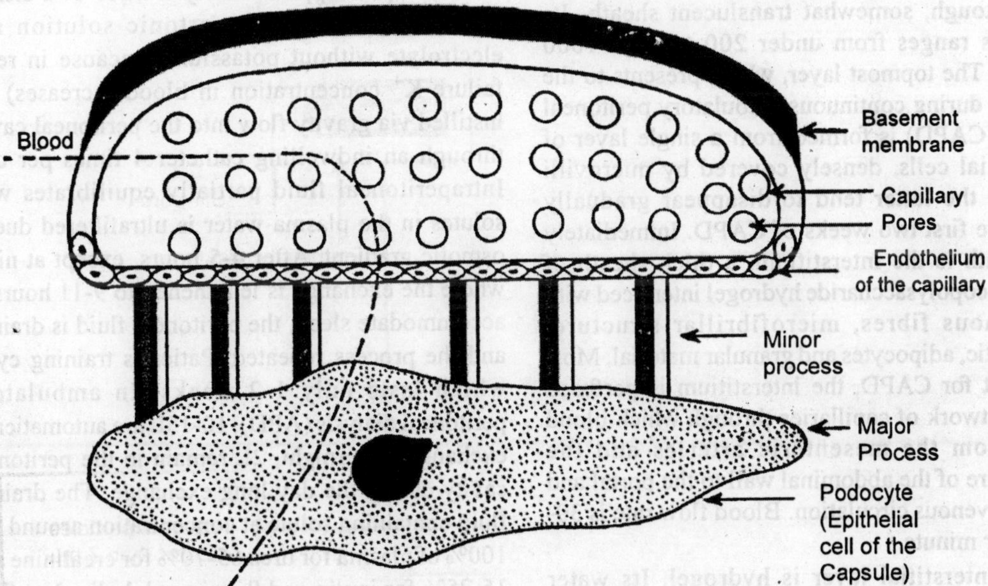

Fig. 7.26: Diagrammatic representation of structural basis of ultrafiltration in glomerulus

7.29.9 Factors affecting the glomerular filtration rate

1. **Filtration pressure and filtration coefficient:** The filtration does not change but three factors determine the filtration pressure: (a) glomerular pressure, (b) plasma colloid osmotic pressure (c) Bowman capsular pressure.
2. **The glomerular membrane:** (a) Area, (b) Permeability.
3. **Arterial blood pressure:** Beside above factors, there are other factors which play a role in the process.

7.29.10 Peritoneal Dialysis

Peritoneal Membrane

The peritoneum is not a simple selective barrier between the two phases. Adult peritoneum surface is approximately 1,75 = 0.5 m^2 which may be considered equal to skin surface area.

It is not physiologically homogeneous, the visceral (organ) part covering the external surface of the organ differs from the parietal part overlying the inner abdominal wall which in turn different from the folded or planted mesentery connecting the two. Morphologically, the peritoneum is a smooth tough, somewhat translucent sheath. Its thickness ranges from under 200 to over 1000 microns. The topmost layer, which presents to the dialysate during continuous ambulatory peritoneal dialysis (CAPD) is formed from a single layer of mesothelial cells, densely covered by microvilli although the latter tend to disappear gradually during the first two weeks of CAPD. Immediately underneath is the interstitium, a thick sheath of dense mucopolysaccharide hydrogel interlaced with collagenous fibres, microfibrillar structure, fibroplastic, adipocytes and granular material. Most important for CAPD, the interstitium is perfused with a network of capillaries through which blood flows from the mesenteric arteries and the vasculature of the abdominal wall to the portal and systemic venous circulation. Blood flow rate is 30-60 ml per minute.

The interstitial layer is hydrogel. Its water content and its transport properties will vary in response to the osmolarity of the peritoneal dialysis.

What is dialysis?

Dialysis is the process of seperation of non-diffusible colloidal substances from smaller diffusible molecules (crystalloid), e.g., salts, ion, sugar, urea, creatinine and water by passing through a semipermeable membrane.

Types of dialysis

1. Haemodialysis (HD)
2. Peritoneal dialysis (PD)

What is peritoneal dialysis?

It is a process by which metabolic waste products toxic substances and excess body water are removed through a membrane like tissue that lines the internal abdominal wall and the abdominal organs in the abdominal cavity called peritoneum.

Application: When there is loss of kidney function at upto 90% of their capacity which is in irreversible end-stage of kidney disease known as chromic renal failure (CRF) due to:

1. Diabetes mellitus
2. Hypertension
3. Recurrent kidney infection
4. Kidney stone (lithiasis)
5. Glomerulonephritis
6. Polycystic kidney disease etc.

Principle: Approximately 2 litres of a sterile, nonpathogenic and hypertonic solution and electrolyte without potassium (because in renal failure K^+ concentration in blood increases) are instilled via gravity flow into the peritoneal cavity through an indwelling catheter 4 times per day. Intraperitoneal fluid partially equilibrates with solutes in the plasma water is ultrafiltered due to osmotic gradient. After 4-5 hours, except at night where the exchange is lengthened to 9-11 hours to accommodate sleep, the peritoneal fluid is drained and the process repeated. Patient's training cycle which lasts only 1-2 weeks. In ambulatory peritoneal dialysis (APD); 10-15L are automatically exchanged overnight, 2L remain in the peritoneal cavity during the day long exchange. The drained fluid containing solute at concentration around 90-100% of plasma for urea 65-70% for creatinine and 15-25% for inulin and β_2 microglobulin. Net fluid ranges up to 100 ml per exchange. CAPD generally removes the same quantity of toxins and fluid as haemodialysis (HD). CAPD requires a higher plasma concentration as the driving force for this removal.

Procedure: Access to the peritoneum is done via a double cuff (Tenekhoff) catheter of 50-100 cm length of silicone tubing with side holes at the internal end, and connector fittings at the end of the exposed end. Most are implanted in a routine surgical procedure requiring 1 hr and are allowed to heal for 1-2 weeks prior to routine clinical use. Sterile and nonpathogenic fluid is supplied in a 2L container fabricated from dioctyle phathalate plasticized polyvinyl chloride. The fluid is K^+ free, lactated Ringers to which has been added 15-42.5 gm per litre of glucose (dextrose monohydrate) buffered to a pH of 5.5.

Since the glucose would caramelize during autoclaving at higher pH level, thus patient simply rolls up the empty bag after installation and drains of the three steps involved in a single CAPD exchange fluid infusion dwell and drain. Some administration sets require the bag to stay connected during dwell (it is rolled and fits in a girdle around the waist) other allow it to be disconnected. Drain and infusion take about 10 minutes each three days time dwells are 4-6 hours each, the overnight dwell lasts 8-10 hours. The bag filled with drained is each, disconnected and a fresh bag is reconnected. The patient removes on 500 ml in a single 6 hours exchange. The bag filled with drained fluid is disconnected and a fresh bag is reconnected, the patient removes 500 ml in a single 6 hours exchange.

Fluid and solute Removal: The rate at which solutes are removed depends primarily upon the rate of equilibrium between blood and instilled peritoneal fluid i.e., dialysed solution the patient removes 500 ml of fluid in a single hour exchange. Displacement of waste products and water from the blood ends after a certain period of time. For this reason, new solution has to introduce every 4-6 hours. The dialysis solution contains glucose which has the special characteristic of attracting or pulling water from one side to the other. After 4-6 hours the water stops leaving the blood. This is another reason why it is necessary to exchange the solution to prevent the water from reentering the blood. An exchange consists of 3 steps:

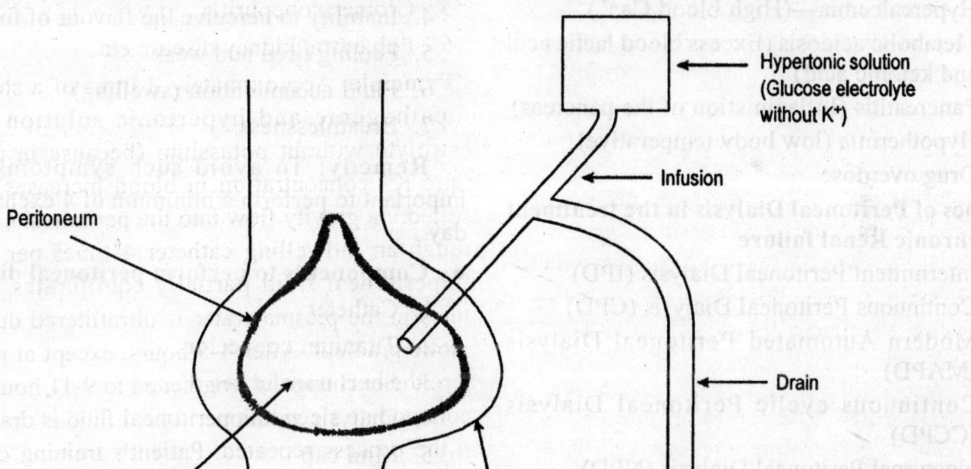

Fig. 7.26(A): Diagrammatic representation of peritoneal dialysis

1. **Drain**: Removal of the solution from the peritoneal cavity by gravity which takes approximately 20 minutes.
2. **Infusion**: Introducing dialysis solution into the peritoneal cavity through a catheter for less than 10 minutes.
3. **Dwell time**: This is the time in which the dialysis solution remains inside the peritoneal cavity for an average duration of 4-6 hours, depending on the attending physician's instruction. The exchange is generally repeated 4 or 6 hours and a minimum of four times a day, seven days a week. Exchanges can be performed at home, at workplace or at other location but preventing any possibility of infections. Each exchange lasts approximately 30 minutes.

- **Peritoneal dialysis in special Medical Conditions**

1. Diabetes mellitus (High blood sugar)
2. Edema (Swelling due to accumulation of fluid in tissue spaces)
3. Hyperkalaemia—(High blood K^+)
4. Hypercalcemia—(High blood Ca^{2+})
5. Metabolic acidosis (Excess blood lactic acid and ketonic acid)
6. Pancreatitis (Inflammation of the pancreas)
7. Hypothermia (low body temperative)
8. Drug overdose.

- **Types of Peritoneal Dialysis in the treatment of chronic Renal failure**

1. Intermittent Peritoneal Dialysis (IPD)
2. Continuous Peritoneal Dialysis (CPD)
3. Modern Automated Peritoneal Dialysis (MAPD)
4. Continuous cyclic Peritoneal Dialysis (CCPD)
5. Nocturnal Peritoneal Dialysis (NPD)
6. Tidal Peritoneal Dialysis (TPD).

- **Advantages of Peritonea (Dialysis)**

1. It frees the patient from the pervasive lifestyle invasions associated with thrice weekly in centre of HD.
2. It is popular to patient in rural areas distant from a HD centre.
3. The continuous nature of CAPD eliminates fluctuations in the concentrations of uraemic metabolites and avoids sawtooth pattern haemodialysis peak toxin concentration.
4. Fluid and dietary constrains are less restrictive for patients CAPD than those on HD.

- **Disadvantages of Peritoneal dialysis**

1. Complications like peritonitis. The rate of peritonitis was around 2 episode per patient per year which is now about less than 0.5 episode per patient year.
2. Morbidity of peritonitis has also decreased for improvement in the treatment without hospitalization.
3. Peritonitis caused by staphylococcus aureous, bacteria aureus, Pseudomonas and fungai remains a problem.
4. Daily transperitoneal administration of 100-150 gm of glucose providing 600 calories, and tedium of the exchanges.

- **Results of inadequate dialysis**

1. Nausea
2. Vomiting
3. Loss of appetite
4. Inability to perceive the flavour of foods
5. Feeling tired and weak
6. Fluid accumulation (swelling)
7. Breathlessness.

Remedy: To avoid such symptoms, it is important to perform a minimum of 4 exchanges a day.

- **Components to perform peritoneal dialysis**

1. Catheter
2. Titanium connector
3. Short transfer set
4. Dialysis solution
5. Mini cap
6. Outlet prop clamp.

7.30 Separation of Colloids

Colloids can be separated from Crystalloids, electrolytes, solvent or other colloidal materials by the following methods:

1. Dialysis
2. Electrodialysis
3. Ultrafiltration
4. Salting out
5. Isoelectric precipitation

6. Adsorption
7. Ultracentrifugation
8. Electrophoresis
9. Chromatography.

7.30.1 Dialysis

Colloidal substances do not diffuse through the membranes like parchment paper, cellophane, collodion or inert cellulose esters like acetate or nitrate but these membranes are permeable to small particles like sugar, amino acids, ions and water in solution. It was first observed by Graham and this property is the simplest way to observe the colloidal nature of a solution and is used for purification of Colloids. Though parchment paper is being used extensively for colloidal investigation, but sacks made of collodion or cellophane paper are favoured now-a-days. Dialysis is the process of separation of non-diffusible colloidal substances from smaller diffusible molecules like salts, ions and water by passing through a semipermeable membrane. The apparatus used for dialysis is called **dialyser** or **dialysing apparatus**.

7.30.1.1 Dialyser

It consists of a glass jar or bottle (A) having stopper (B). A collodion tube (C) is fastened to the glass stopper (B) which fits airtight into the neck of the glass jar. Sterile distilled water is introduced into the glass jar through (D) and colloidal solution mixed with crystalloids is placed inside the collodion tube (C). A partial vacuum is made in the bottle by exhaustion at (E), which opposes the osmotic pressure of the solution in (C) and so prevents its dilution or if great enough may even concentrate it by ultrafiltration (Fig. 7.27).

Most Gels, such as gelating and collodion, are of porous structure and collodion membranes may be obtained of various degrees of permeability according to their mode of preparation and the concentration of the collodion in the membrane.

Such membranes are used for the process of ultrafiltration. The membranes of loosest texture will allow proteins such as albumin to pass, in addition to Crystalloids. But the denser collodion membrane or cellophane will keep back all Colloids and allow only Crystalloids and water. All the smaller molecules and water and electrolytes diffuse freely from the collodion tube to the surrounding fluid while the colloidal particles are held inside the tube. The surrounding fluid is changed periodically by giving fresh fluid until the Colloid in the tube or sac is free from Crystalloids which are known by testing the colloidal solution for small solutes.

7.30.1.2 Influence of pressure on dialysis

If some pressure is exerted on the contents of the dialyser or suction is applied outside, fluid passes out by ultrafiltration. But there is certain limit of the pressure below which no filtration is observed. This limiting pressure is equivalent to the osmotic pressure of the Colloids in solution. This pressure varies according to the state of aggregation of the Colloid, e.g., to the number of molecules, simple or compound, of the Colloid present. As for example, serum pressure is equal to about 30 mm Hg and if the difference of pressure between the two sides of the membrane are less than 30 mm Hg, water and salts pass from outside through the membrane to the inside causing dilution of the serum.

7.30.1.3 Explanation for dialysis

1. Colloidal particles are larger and heavier than those in true solutions. As for example:
   (a) Cooked starch and most proteins of the food mass are not able to pass through the intestinal membrane into the blood.
   (b) Glycogen and cell proteins cannot diffuse through cell membranes into the tissue fluid; it is believed to be due to the inability of their relatively very large molecules or aggregation of molecules.

   But there are reasons to believe that molecules of egg albumin, lactalbumin and some intermediate products of protein digestion are able to diffuse through intestinal membrane to a slight extent causing sensitivity or allergy.
2. Most dispersed colloidal particles are electrically charged, with all particles of any given Colloid carry like charges. Consequently, both the relatively heavy weight and the effect of the electric charge must be considered in seeking an explanation for the behaviour of these colloidal particles in solution.

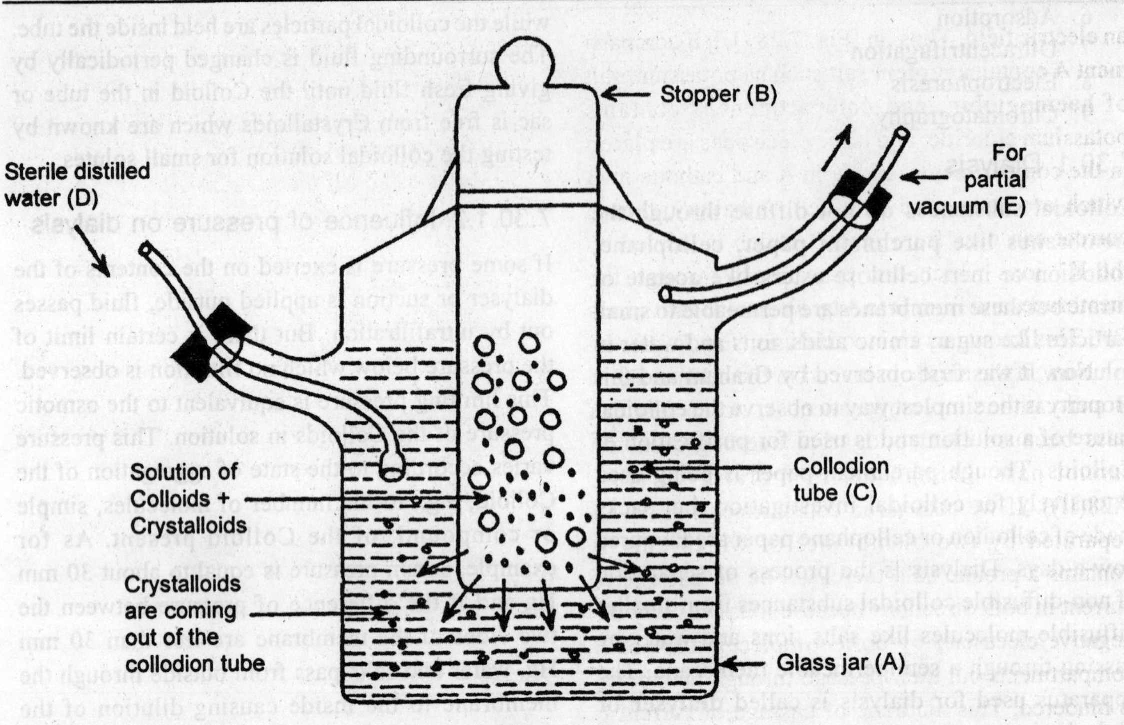

Fig. 7.27: Dialyser

7.30.1.4 Biological importance

This difference in diffusion regarding the substances in true solution and colloidal solution has importance in biology under as:

1. The water of the sap of plants and of the blood of the animals holds some substances in true solution and others in colloidal solution but only the substances in true solution are able to enter the cells.
2. Within the cells of both plants and animals there are substances held in colloidal solution which are essential for the cell activities and being colloidal these substances are not capable to escape from the cells.
3. During intestinal absorption the bigger food particles are held back whereas the smaller particles pass easily through the membrane. Besides dialysis, other factors also play a role in the absorption such as osmosis, diffusion, etc.
4. In the capillary bed, it helps the exchange of fluid due to presence of albumin and globulins.
5. The bigger particles of albumin and globulins are not allowed to pass to the tissue fluid through the capillary membrane.
6. Haemodialysis (Heamodialyser, artificial kidney): The principle of dialysis is employed for periodic dialysis of blood for person with damaged kidneys. It is used for 12 hours continuously at every 3 days interval. The blood is allowed to pass through a dialyser having two concentric cellophane tubes and water is passed through the annular space. The non-colloidal impurities (e.g., urea) in the blood is dialysed to dialysed fluid and this instrument acts as a kidney.
7. Separation of Colloid from the mixture of Colloid and Crystalloid.
8. Purification of enzymes.
9. Purification of proteins.

7.30.2 Electrodialysis

It is the process separating Colloids from electrolyte ions when the semipermeability of the dialyzing membrane and migration of electrolyte ions are in an electric field. Thus, in [Fig. 7.28 (I)], if compartment A contains protein salt such as potassium salt of haemoglobin, and compartment B contains potassium chloride, and if the electrodes are placed in the compartments–anode in A and cathode in B which are connected to a cell or battery, then a current can pass from A to B which is carried by the K^+ ions which can pass freely through the membrane, and from B to A by electro-negative ions Cl^-. This known as **electrodialysis** (Fig. 7.28).

Now, if the anode is placed in the compartment B and cathode in compartment A, no current can pass because electronegative haemoglobin ion cannot pass through the membrane (x) (Fig. 7.28(II)). If we take three compartments A, B, C, separated by two membranes of which only B contains a protein salt, there will be no passage of current in both directions because it has to carry the negative electricity by protein molecule to A or B compartments but their passage through membrane is hindered. This hindrane of passage of current is more marked if the membrane is impermeable to certain inorganic ions [Fig. 7.28(III)] . On account of this phenomenon, whole blood offers greater resistance to the passage of current than the serum. Each RBC serves as non-conductor and diminishes the total area through which the current pass.

The resistance to the migration of ions by the membrane prevents the passage of other associated ions which are permeable to the membrane. These ions travel for a little distance through the membrane till they are held back by electrostatic attraction of the oppositely charged which cannot pass. This electrostatic tension gives rise to a charge on the surface of the cell. In most cases, RBC carries negative charge and if a current is passed through blood, the RBC migrates to the positive pole.

7.30.2.1 Importance

1. Separation of Colloids from the electrolyte ions in a mixed sample.
2. This principle is utilised for production of pure water from sea water on a large scale and the process is known as "Desalination of sea water".
3. Polarisation of the membrane—If a current is passed across a membrane immersed in an electrolyte there will be a chemical difference between the two sides of the membrane producing a difference of potential. Breaking the current, ions move in the opposite direction to re-establish the equilibrium. This process of passage of current causes **polarisation** and after the passage of current, the membrane is **polarised,** i.e., it presents a difference of potential between its two sides.
4. Transference of dissolved substances—The difference of electrical charge between the two sides of the membrane probably plays in the transference of dissolved substances across membranes in the body. As the hydrostatic pressure differences determine the movement of the fluid, difference of electrical charge serve, as driving force for the charged particles of fluids.

7.30.3 Ultrafiltration

Colloidal solution can pass easily through ordinary filter paper because its pores are not small enough to retain the colloidal particles. But a semipermeable membrane made of collodion, cellophane, inert cellulose esters or hardened gelatine supported properly on a grid of wire mesh or deposited on a piece of porcelain or filter paper does not allow the colloidal particles to pass through these membranes. Membranes of different pore size can be made by varying the concentration of colloidion.

Ultrafiltration is the process of separation of particles of colloidal system from electrolytes, water, glucose, etc. by filtering the mixed solution through a membrane of different pore-sizes under pressure. The collodion filters of different pore-sizes are called **graded filters** having size from 10 nm to 5 μm and colloidal particles of different sizes can be separated. Ultrafiltration is the direct opposite of osmosis; osmosis causes migration of water from the lower to the higher concentration and is a "mixing" process whereas ultrafiltration causes migration of water from the higher to the lower concentration and is a "separation" process.

In order for ultrafiltration to occur two criteria must be fulfilled such, as (1) the barrier, i.e., membrane must be so constructed as to retain molecules above a certain size whilst allowing smaller

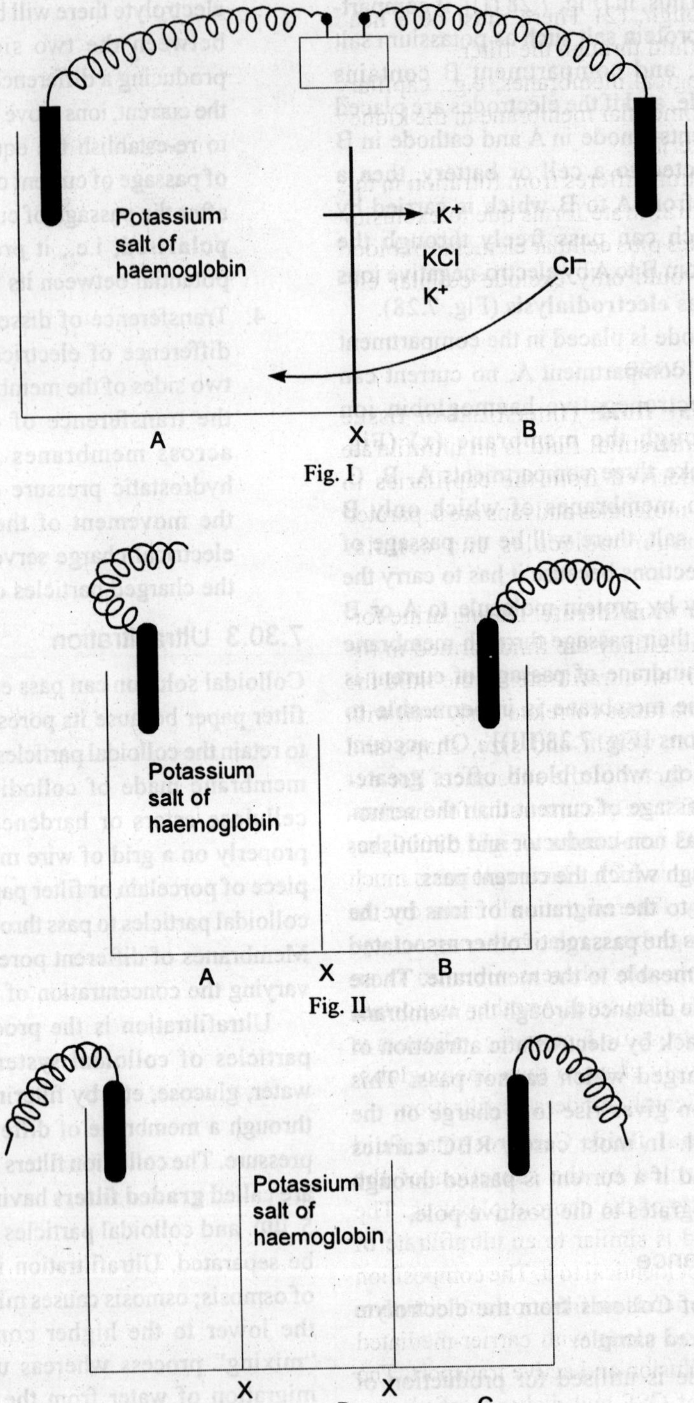

Fig. 7.28: Electrodialysis

ones to pass through. (2) There must be a head pressure to force fluid through the filter.

The two biological membranes, e.g., capillary membrane and glomerular membrane in the kidney fulfil both the criteria.

The ultrafiltration differes from filtration in this respect that the ultrafiltrate forms due to exclusion of colloidal particles plus cellular elements of blood whereas filtrate would only exclude cellular elements of blood.

7.30.3.1 Importance

(a) **Intercellular fluid**: (Interstitial or tissue fluid) The interstitial fluid is an ultrafiltrate of plasma derived from the capillaries in which small molecules and ions are separated from the larger molecules and cellular elements.

(b) **Glomerular ultrafiltrate**: During urine formation in the kidney the fluid formed in the glomerulus is an ultrafiltrate. In the fluid the filtrate/plasma ratios correlate fairly well with the molecular weight and size, shape and electrostatic factor of the substance. For example, the filtrate/plasma ratio for haemoglobin is 0.03 (molecular weight 68000, radius 32.5 Å) it is 0.01. Haemoglobin is much more readily filtered than albumin due to shape; haemoglobin is relatively compact, cylindrical molecule whereas albumin is an asymmetrical ellipsoid. Another important factor is electrostatic factors, i.e., albumin is more negatively charged than haemoglobin which significantly hinders its filtration.

(c) **Cerebrospinal fluid**: Cerebrospinal fluid (CSF) is formed by the secretion of the epithelial cells of the choroid plexuses. The fluid secreted is similar to an ultrafiltrate of plasma but not identical to it. The composition of the CSF depends on filtration and diffusion from the blood along with carrier-mediated facilitated diffusion and active transport. The comparison of CSF and dialysate of plasma shows the concentration of K^+, Ca^{++}, Cl^- and glucose are significantly lesser in CSF than the dialysate of the plasma. Therefore, CSF is not a protein-free transudate or dilysate of plasma.

7.30.4 Salting out

It is one kind of separation of Colloid, e.g., protein, by precipitation from its solution by adding certain amount of neutral salts. Solution of Colloids, e.g., protein, is saturated or partially saturated neutral salts of sodium chloride, magnesium sulphate or ammonium sulphate.

7.30.5 Isoelectric precipitation

At certain pH, the Colloid, e.g., from protein in an electric field, does not migrate to anode or cathode and the proteins are precipitated at the isoelectric pH, or point.

7.30.6 Adsorption

The Colloids, e.g., protein can be separated from a mixed solutions by differential adsorption on insoluble adsorbents like silica gel, alumina, calcium hydroxyapatite and charcoal powder in adsorption chromatography.

7.30.7 Ultracentrifugation

(See 7.26.6 determination of size, shape and molecular weight of Colloid).

7.30.8 Electrophoresis or Cataphoresis

7.30.8.1 Principle

The electrophoresis method is based upon the principle that Colloids, i.e., separation of mixture of electroactive colloidal particles, e.g., protein solution at pH, value above and below their isoelectric points migrate in an electric field towards the pole bearing a charge opposite to that of the protein. It is actually a separation method, not a measurement technique.

7.30.8.2 Mechanism

The electrokinetic potential is the potential difference between a particle and its surrounding medium, e.g., an ionized protein molecule, and is manifested by the migration of the particle in an electric filed, i.e., electrophoresis similar to the movement of ions in electrolysis. This is due to **ampholytic properties** of the protein molecules whose net charge depends on the pH, of the surrounding medium in an electric field. During the process there is a potential difference between the surface of the Colloid and its neighbouring solution which is called **zeta potential** and its value is 50 millivolts. Through the electric charge depends ultimately on their acid-base properties, which are largely determined by the number and types of ionizable groups.

7.30.8.3 Rate of migration

Basic Principle: **The rate of migration is about one micron per second per unit potential gradient which depends on:**

(a) the pH of the solution,

(b) the size and shape of the charged particles,

(c) the viscosity,

(d) the amount of net charge on the ion,

(e) the magnitude of the current,

(f) the dielectric constant of the liquid dispersion medium,

(g) nature of proteins,

(h) nature and concentration of dissolved salt,

(i) the magnitude of the current employed.

The rate of migration varies—

(a) Directly with the net amount of the charge,

(b) Inversely with the size of the particle and the viscosity of the dispersion medium.

7.30.8.4 Electrophoretic mobility

The electrophoretic mobility of such a particle is a function of its net charge and ionic radius.

This means that two particles having the same net charge can be separated based on their ionic radii, whereas two particles having the same ionic radii can be separated based on their net charge. This is the basis for the separation technique called electrophoresis.

The four basic parts of an electrophoresis system are as under:

1. A series of electro-active particles.

2. An electrolyte solution.

3. A physical support or stabilizing medium.

4. A high voltage DC power supply. Items (2) and (3) combined make up an electrophoretic chamber.

The rate of electrophoretic movement or electrophoretic mobility is expressed in terms of distance per time in a unit electric field (potential gradient, 1 volt per cm) and is the characteristic of the protein under the specified conditions of pH and ionic strength. The mobility (μ) equals cm per sec/ volt per cm and is given in units of $cm^2\ sec^{-1}\ volt^{-1}$, i.e., in square centimetres per volt-second of a molecule.

$$\text{(Mobility) (Square centimetres per volt-second) } \mu = \frac{\text{V (Velocity of migration) (centimetres per second)}}{\text{E (Electric field strength) (volts per centimetre)}}$$

By convention, μis given a plus sign if the protein moves towards the cathode and minus sign if the protein migrates towards the anode. When the pH is below the isoelectric pH, then the protein moves to the anode and when the pH is above, it moves to the cathode. If a solution contains 2 proteins of qualitatively similar, e.g., positive charge but of different electric mobility at a given pH, the two proteins will move toward the cathode at different rates, and the Schlieren diagram will show to peaks for which the electrophoresis is very much helpful for study of mixtures of proteins like plasma proteins.

For small ions, e.g., chloride, μ is between 4 and $9 \times 10^{-4}\ cm^2\ V^{-1}\ s^{-1}$ (25°C), for proteins, it is about 0.1 to $1.0 \times 10^{-4}\ cm^2 V^{-1}\ s^{-1}$. Proteins thus migrate slowly in an electric field than small ions like Na^+, Cl^-, becuase they have a smaller ratio of charge to mass.

7.30.8.5 Limitation of electrophoresis

The chief limitation of electrophoretic analysis is that the separation of protein components in a mixture of solution depends mainly upon their mobility in an electric field. And the mobility of the particle depends upon the size, shape and electric charge, the latter varies with pH.

7.30.8.6 Use of electrophoresis

1. Separation of protein from a mixture of proteins or polypeptides.

2. Separation of plasma proteins, glycoprotein.

3. Separation of Lipoproteins.

4. Separation of proteins extracted from tissues.

5. Separation of amino acids from protein hydrolysate.
6. Separation of immunoglobulins.
7. Separation of isoenzymes.

7.30.8.7 Different methods of electrophoresis

I. Free electrophoresis of Moving boundary electrophoresis.
II. Zone electrophoresis (Ionophoresis).
III. Disc electrophoresis.
IV. Electrofocusing or Isoelectric focusing electrophoresis.
V. Electrophoresis-convection.
VI. Immunoelectrophoresis.

I. Free Electrophoresis or Moving-Boundary Electrophoresis

The technique for studying the electrophoresis involves the use of an apparatus invented by Arne Tiselius (1933). This method was used by him for fractionating human plasma proteins. A buffered solution (e.g., Sodium diethyl barbiturate (veronal) at pH 8.6 in case of plasma proteins) of the protein mixture is placed in a U-tube observation cell. Same pure buffer is layered over the protein solution. The cell is kept immersed in a constant temperature bath insulated from vibrations. The temperature at which the density of water is maximal, reduces the influence of electrodes. Negatively charged proteins move toward the anode and positively charged proteins move toward the cathode. To get the actual picture of a mixture of protein solution the pH is selected so that all the proteins will have the same charge but different mobilities. The negatively charged protein molecules move toward the anode and they migrate from the protein solution into the zone of protein-free buffer and form a "boundary" or "front". The refractive index of the solution changes sharply at this boundary because the refractive index of protein molecule is different from that of the pure buffer. The migration of the boundary is observed by measurement of changes in the refractive index in the electrophoresis cell (Fig. 7.29).

To accomplish this, optical method known as "Schlieren method" is used (see ultracentrifuge).

The photograph of the Schlieren diagram after a given time interval will show that the boundary has moved upward into left channel known as "ascending boundary" and has moved downward in the right channel known as "descending boundary." Optical measurements of the refractive-index which change along the electrophoresis cell and

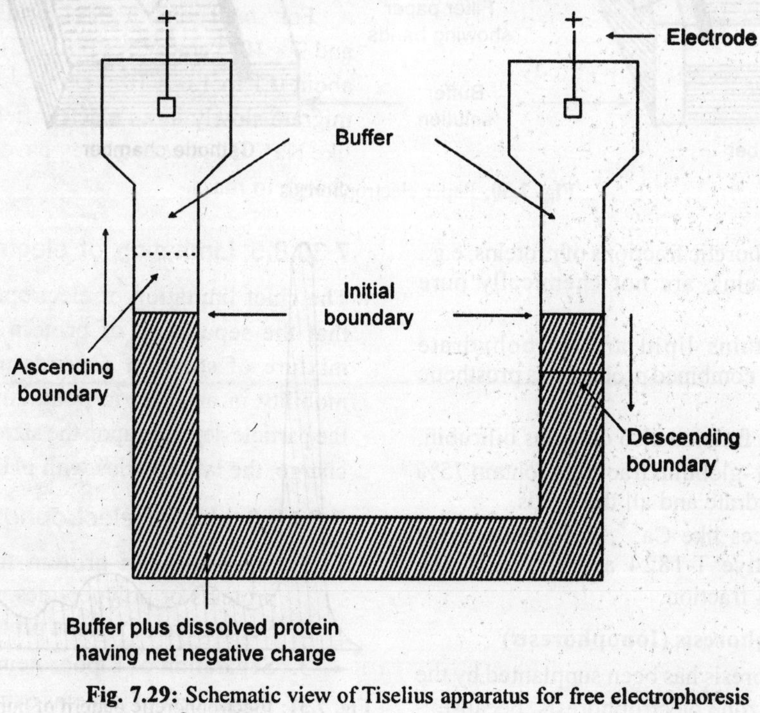

Fig. 7.29: Schematic view of Tiselius apparatus for free electrophoresis

yield electrophoretic pattern called "Schlieren patterns" show the direction and relative rate of migration of the major proteins in the mixture.

- **Disadvantages of free electrophoresis**

1. Slow process.
2. Very large samples of proteins are needed.
3. Optical system is required for analysis.
4. The moving boundaries are very prone to distrubances from vibrations and convection.
5. The method cannot be used for separating proteins in builk.

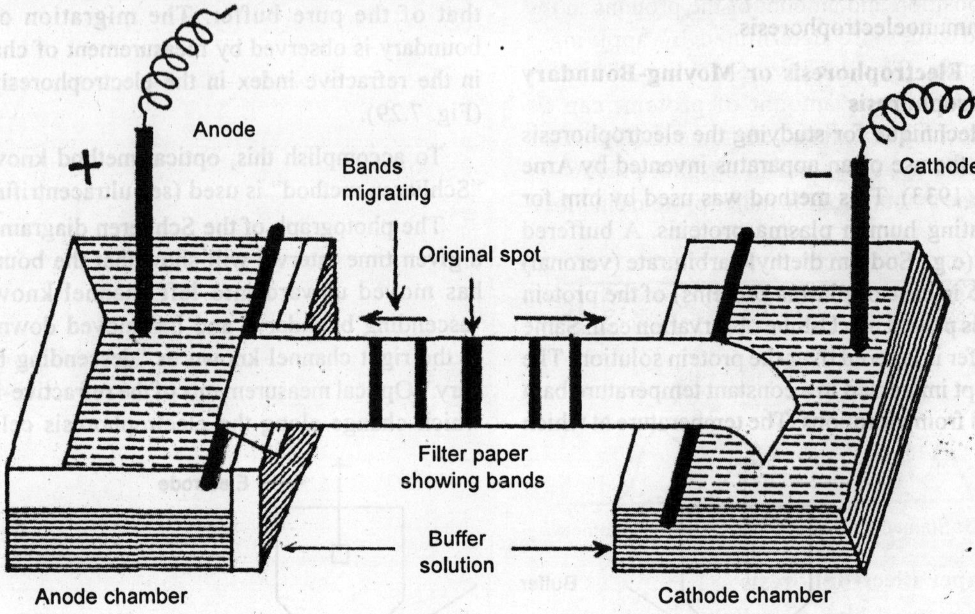

Fig. 7.30: Paper electrophoresis

6. The electrophoretic fractions of proteins, e.g., plasma proteins, are not chemically pure because:
   (a) all contains lipid and carbohydrate material combined probably as prosthetic group.
   (b) albumin fraction also contains bilirubin,
   (c) α and β -globulin fractions contain 75% carbohydrate and all the lipids.
   (d) substances like Ca^{++}, P, Sulphonamide drugs, dye T-1824 are bound to the albumin fraction.

II. Zone Electrophoresis (Ionophoresis)

Free electrophoresis has been supplanted by the different forms of zone electrophoresis, because:

(a) it is a simpler process,

(b) having much greater resolving power,

(c) requires smaller sample.

In this process the aqueous solution of protein is immobilized in a solid matrix or support, hydrated porous and inert material having mechanical rigidity, eliminating convention and vibration disturbances. The most widely used supports are moistened filter paper, cellulose acetate strips, agar gel, agarose gel, polyacrylamide gel or starch gel, relatively inert material and do not interact with

Fig. 7.31: Electrophoretic pattern of human blood plasma

migrating proteins or retard them. The electrophoretic process is allowed to continue until the major protein component separate into discrete zones; hence the name zone electrophoresis (Fig. 7.32).

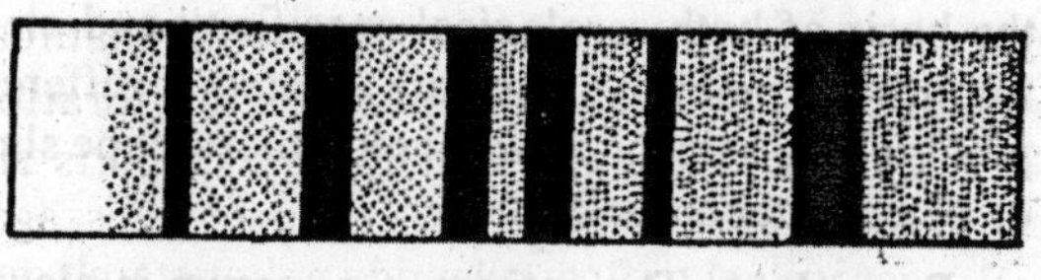

Fig. 7.32: Zone electrophoresis on a cellulose acetate strip which is stained after electrophoresis

The position and amount of the proteins in the separated zones are determined by applying a protein stain. The density of staining, which is proprotional to the amount of protein, can be estimated with a Scanning Densitometer (Fig. 7.33). This method is being used in hospital practice to measure the amounts of the major proteins in blood plasma.

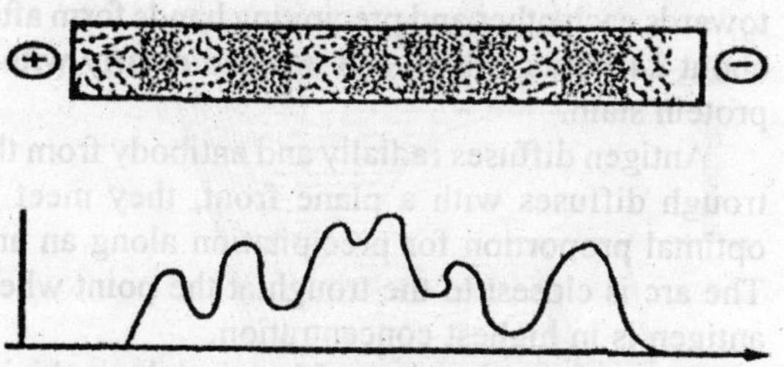

Fig. 7.33: Stained strip is scanned in the densitometre

(a) Paper Electrophoresis

In this method, the test solution is placed as a band transversely across the central part of the filter paper. The filter paper is saturated with a buffer solution for maintaining desired pH. The ends of the filter are dipped into buffer solution in the anode and cathode chambers. These chambers are fitted with electrodes. When the electric field is established, the charge particles migrate towards the opposite electrodes according to the size and charge of the particle. The paper strip is stained and measurements done by Scanning Densitometre (Fig. 7.33).

(b) Gel Electrophoresis

Much higher electrophoretic resolution is obtained if the support of the matrix material can retard or exclude protein molecules on the basis of both electric charge and molecular size. For this purpose, gels of potato starch or agarose or polyacrylamide are used commonly. By this method the protein components of blood plasma can be resolved into 15 or more bands, compared with about 5 or 6 observed with free or zone electrophoresis.

Polyacrylamide Gel Electrophoresis can be used to carry out isolation of larger quantities of purified proteins. The most common support is the cross-linked polymer of acrylamide ($CH_2 = CH.CONH_2$), i.e., polyacrylamide. Protein solutions are applied to buffered tubes or blocks of polyacrylamide cross-linked 2-10% by inclusion of methylene bisacrylamide of "bis" or similar crosslinking reagents. Direct electric current is applied. Result is observed by staining with coomassic blue for polypeptides, ethidium bromide for polynucleotides, etc. A variation of result is observed in this process when the substance is in denaturing conditions. When proteins are boiled in urea or sodium dodecyl sulphate (SDS) before separation based mainly on molecular size. This SDS-PAGE process is largely used to establish subunit molecular weights of proteins by comparison of mobilities with those of standards of known molecular weight. The polyacrylamide method is known as **Polyacrylamide Gel Electrophoresis (PAGE).**

III. Disc Electrophoresis

This is another variation of zone electrophoresis and refinement of gel electrophoresis. The protein mixutre is subjected to an electric field in a retarding gel support with is separated into two parts having different porosity and buffered at different pH. The protein mixture shows migration from the more porous gel into less porous gel accompanied by a change in pH. Now each protein becomes concentrated as a very thin, sharp band producing much higher resolution than can be achieved in a continuous buffer. This kind of zone electrophoresis is known as **Disc Electrophoresis** because of (a) the discontinuous buffer employed and (b) the discoid appearance of the protein zones.

IV. Electrofocussing or Isoelectric Focusssing Electrophoresis

This method was invented by Svensson and this is the most ingenious and effective method for separating proteins.

In this method the protein mixture is subjected to an electric field in a polyacrylamide or agar gel block in which a stable pH gradient has first been generated along its length by suitable ampholytes of low molecular weight.

Each protein migrates towards the opposite electrode and gradually loses its charges and the rate of migration is "focussed" at that portion of the pH gradient where the pH is equal to its isoelectric pH and forms a sharp stationery band there. All the components are separated as bands and finally the gel blocks are stained for study or

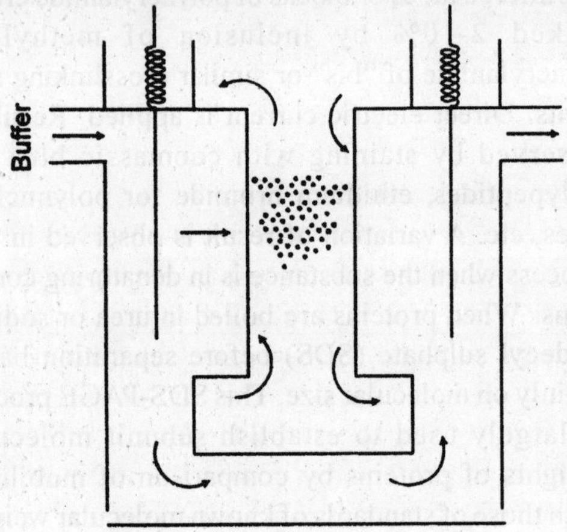

Fig. 7.34: Schematic representation of electrophoresis convection

gel block is sectioned and each section may be eluted with buffer for analysis. The power of this method is tremendous and can also be used on a large scale to prepare purified proteins.

V. Electrophoresis-Convection (Cann and Kirkwood)

This method involves a combination of horizontal electric transport and vertical convective transport. The construction of the apparatus is that the individual proteins themselves in a horizontal electric filed. The protein molecules set up a horizontal density gradient which leads to convection, and the proteins with a higher mobility move down into the vertical channel, where they descend under the force of gravity and are collected in the lower horizontal chamber (Fig. 7.34).

VI. Immunoelectrophoresis

The electrophoresis is one of the most powerful tools of biochemistry and permits the separation of macromolecules carrying different electrical charges which has already been established. Now combining this method with "agar gel double diffusion precipitin techniques", the sensitivity and discrimination of the method are enormously increased. It allows the differentiation of components of an antigen mixture or an antibody mixture on the basis of both serological specificity and electrophoretic mobility. They are located at different points along a line in agar gel on a microscope slide or other suitable surface.

Procedure: The antigen, e.g., serum is placed in a small hole cut in a layer of agar on a microscopic slide. A direct current is applied and differential migration of the serum components takes place. And it cannot be seen normally unless it is stained. After electrophoresis for an hour or so the current is stopped and a trough is cut longitudinally in the agar and an antiserum, i.e., antibody (prepared against the electrophoresed antigen) is placed in the trough. The slides are then kept at room temperature in a closed container to minimize evaporation. The two components, i.e., antigen and antibody diffuse towards each other and precipiting bands form after about 24 hours, which can be seen clearly with a protein stain.

Antigen diffuses radially and antibody from the trough diffuses with a plane front, they meet in optimal proportion for precipitation along an arc. The arc is closest to the trough at the point where antigen is in highest concentration.

There are developments of the principle combining electrophoresis with immuno-precipitation in which movement in an electric field drives the antigen is in highest concentration (Fig. 7.35).

- **Use of Immunoelectrophoresis**

1. Especially valuable for the identification and differentiation of serum proteins, some 75 of which have been distinguished.
2. Particular importance in the differentiation of the immunoglobulins.
3. The immunoglobulins differ in amino acid structure but every attempt to analyse the amino acid structure has been failed due to presence of incredible number of different molecules. This heterogeneity may be inferred from analysis by this method.
4. Particular importance in the differentiation of the components of complement.
5. Separation of a variety of bacterial and viral antigens.

6. Semiquantitative information regarding immunoglobulin concentration.
7. Identification of myeloma proteins (M-Protein) in the human disease known as multiple myeloma or myelomatosis.

1. Countercurrent Immunoelectrophoresis (C.I.E.) or Cross-over Immunoelectrophoresis

It is a rapid and very sensitive variant of radial gel-diffusion, in which an electric field is used to hasten the diffusion of the reactants towards one another.

It is applied to antigens which migrate towards the positive pole in agar. Antibody moves 'backwards' in the gel on electrophoresis due to endosmosis. An antigen which is negatively charged at the pH employed will move towards the positive pole and precipitate on contact with antibody (Fig. 7.35A).

- **Use**

(a) For detection of hepatitis B antigen or antibody.

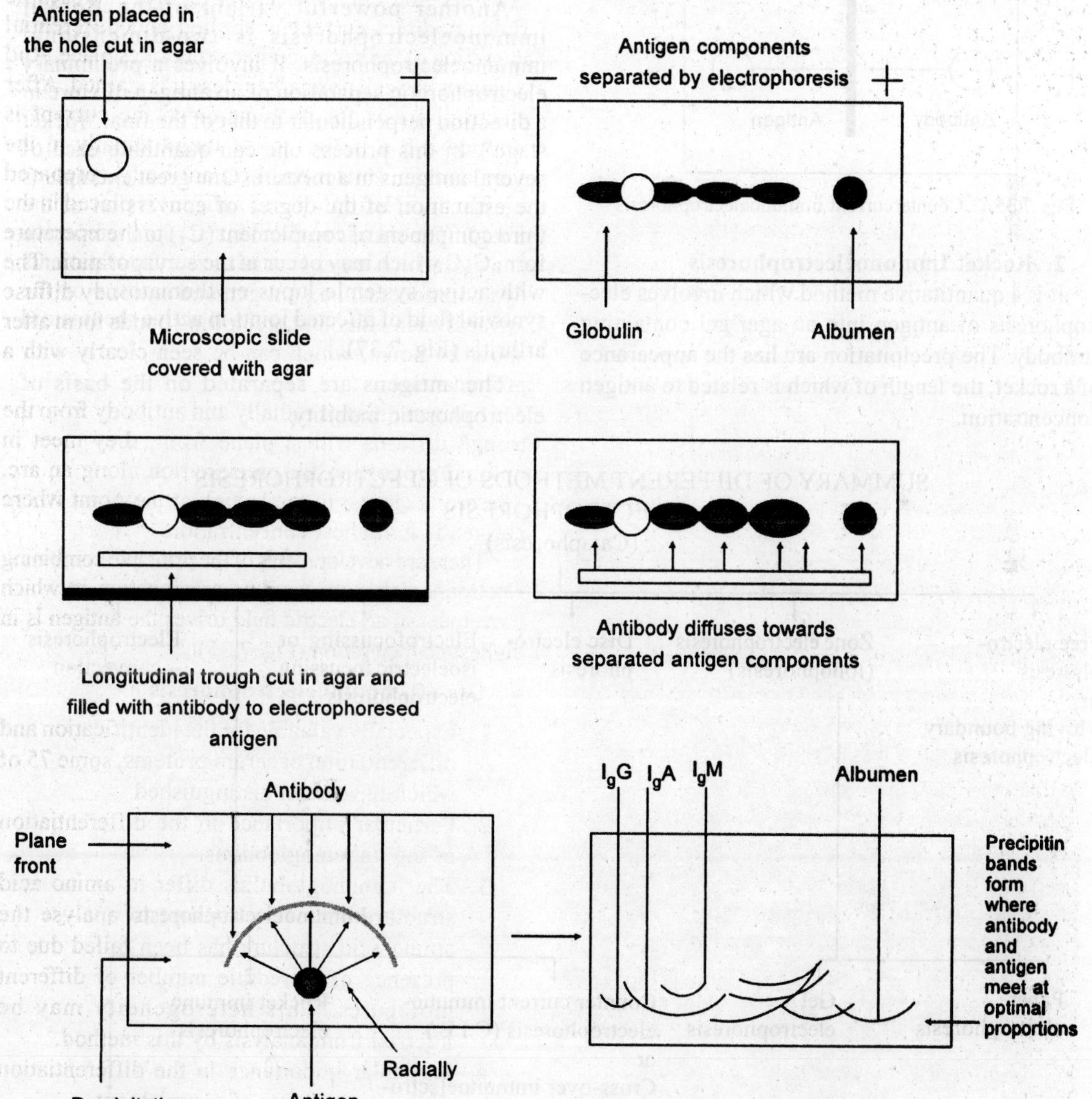

Fig. 7.35: Immunoelectrophoresis

(b) For detection of DNA antibodies in systemic lupus erythematosus.

(c) For detection of autoantibodies to soluble nuclear antigen in mixed connective tissue disease.

(d) For detection of Aspergillus precipitins in cases with allergic broncho-pulmonary aspergillosis.

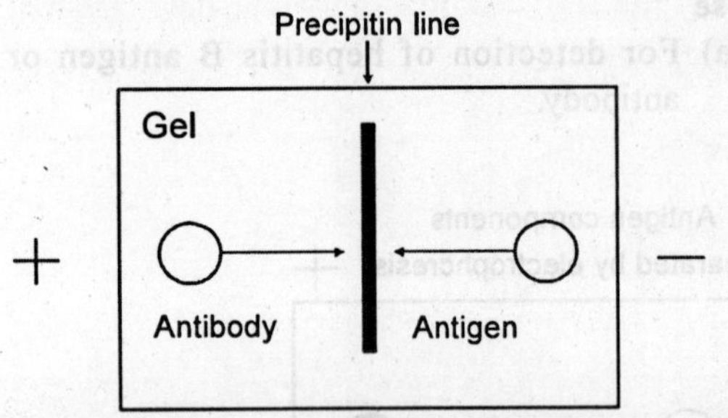

Fig. 7.35A: Countercurrent immunoelectrophoresis

2. Rocket Immunoelectrophoresis

It is a quantitative method which involves electrophoresis of antigen into an agar gel containing antibody. The precipitation arc has the appearance of a rocket, the length of which is related to antigen concentration.

Antigen is electrophoresed into gel containing antibody. The distance from the starting hole to the front of the rocket shaped arc is related to antigen concentration. Like counter-current electrophoresis, this is a rapid method but the antigen must move to the positive pole on electrophoresis; it is, therefore, suitable for proteins such as albumin, transferrin and ceruloplasmin but immunoglobulins are more conveniently quantitated by single radial immuno-diffusion (Fig. 7.36).

3. Two-Dimensional Immunoelectrophoresis (Laurell)

Another powerful variant of the Rocket immunoelectrophoresis is two-dimensional immunoelectrophoresis. It involves a preliminary electrophoretic separation of an antigen mixture in a direction perpendicular to that of the final "rocket stage". In this process one can quantitate each of several antigens in a mixture. One great example is the estimation of the degree of conversion of the third component of complement (C_3) to the inactive form C_3C which may occur in the serum of patients with active systemic lupus erythomatosus or the synovial fluid of affected joints in active rheumatoid arthritis (Fig. 7.37).

The antigens are separated on the basis of electrophoretic mobility.

SUMMARY OF DIFFERENT METHODS OF ELECTROPHORESIS

ELECTROPHORESIS
(Cataphoresis)

- Free electrophoresis or Moving boundary electrophoresis
- Zone electrophoresis (Ionophoresis)
  - Paper electrophoresis
  - Gel electrophoresis
- Disc electrophoresis
- Electrofocussing or isoelectric focussing electrophoresis
- Electrophoresis – convection
- Immunoelectrophoresis
  - Counter current immunoelectrophoresis (C.I.E.) or Cross-over immunoelectrophoresis
  - Rocket immunoelectrophoresis
  - Two dimensional immunoelectrophoresis

The second run at right angles to the first, drives the antigens into the antibody (antiserum) containing gel to form precipitin peaks; the area under the peak is related to the concentration of antigen.

7.30.9 Electrophoresis of plasma proteins

Plasma and sodium diethylbarbiturate (veronal) buffer solution are placed in the U-shaped glass vessel of Tiselius electrophoresis apparatus, the pH is maintained at 8.6 and the ionic strength at 0.1, two electrodes are placed in the two limbs of the glass vessels, and a potential gradient of 6 volts/cm^{-1} is established in the solution. As pH 8.6 is on the alkaline side of the isoelectric points of all plasma proteins, they will all exist as anions and migrate towards the positive electrode.

Each protein moves with a characteristic mobility which varies with the size and charge of the proteins, pH, viscosity of the solvent and the nature and concentration of the dissolved salts; hence in a solution of mixed proteins, the proteins move at different rates. The rate of migration being higher for smaller and more heavily charged protein anions (e.g., Albumin) than for larger and less charged ones. This causes separation of proteins into several fractions or zones marked by distinct boundaries in the limbs of the vessel. The rate of mobility is expressed in cm per sec when the protein is in the electric filed of one per cm^{-1}. After 4 hours, boundary of each separated fraction is located and photographically recorded. It will be observed that the patterns consist of series of peaks in a definite sequence of sizes and shapes. Depending on the difference of the refractive index which depends upon the concentration, their characteristics, and amount of each fraction is calculated from the areas under the peaks. Separation can be done by zone electrophoresis.

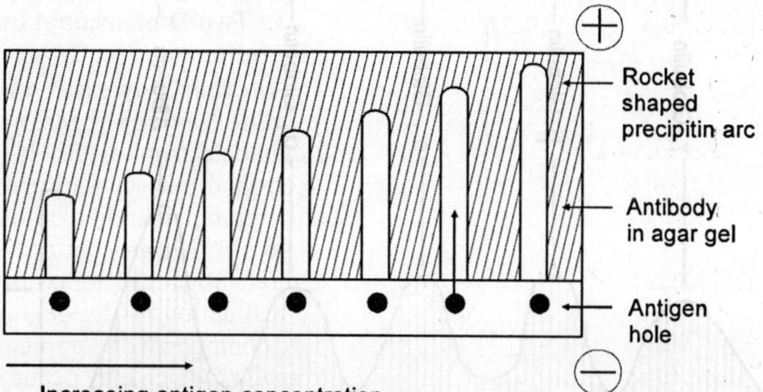

Fig. 7.36: Rocket immunoelectrophoresis

7.30.9.1 Result

1. Albumin which is smaller and highly charged is the fastest moving plasma protein and γ -globulin is the lowest.
2. By this method, four fractions have been isolated such as Albumin, α-β-γ globulins.

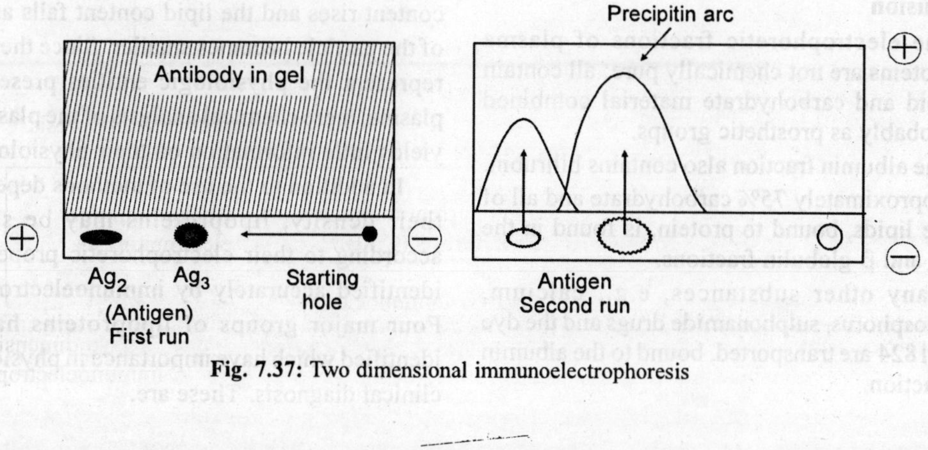

Fig. 7.37: Two dimensional immunoelectrophoresis

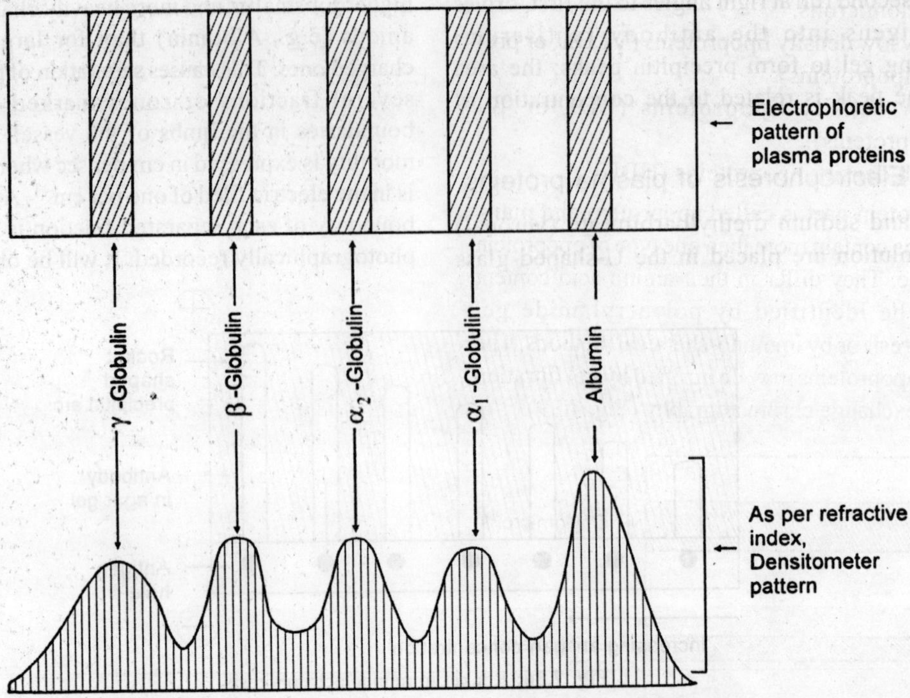

Fig. 7.38: Electrophoretic and Densitometer pattern of plasma proteins

3. Average figures for plasma proteins obtained by this method are:

Albumin	:	55%
α-globulin	:	13%
β-globulin	:	14%
Fibrinogen	:	7%
τ-globulin	:	11%

4. The albumin-globulin ratio obtained in this method is 1.2, a figure which is $\frac{2}{3}$ of that obtained by chemical fractionation because the two procedures isolate somewhat different entities (Fig. 7.38).

- **Conclusion**

1. The electrophoretic fractions of plasma proteins are not chemically pure; all contain lipid and carbohydrate material combined probably as prosthetic groups.
2. The albumin fraction also contains bilirubin.
3. Approximately 75% carbohydrate and all of the lipids, bound to protein, is found in the α- and β-globulin fractions.
4. Many other substances, e.g., calcium, phosphorus, sulphonamide drugs and the dye T-1824 are transported, bound to the albumin fraction.
5. The electrophoretic patterns of the plasma proteins of different animal species are different in characteristic and represent species variations in the various plasma proteins.
6. Electrophoretic patterns of human plasma may vary from normal in pathological conditions.
7. It contributes knowledge regarding plasma proteins in different clinical conditions.

7.30.10 Electrophoresis of lipoproteins

The density of lipoproteins increases as the protein content rises and the lipid content falls as the size of the particle becomes smaller. Since the fractions represent the physiologic entities present in the plasma, mere chemical analysis of the plasma lipids yields little information on their physiology.

In addition to the use of methods depending on their density, lipoproteins may be separated according to their electrophoretic properties and identified accurately by immunoelectrophoresis. Four major groups of lipoproteins have been identified which have importance in physiology and clinical diagnosis. These are:

1. Chylomicrons.
2. Very low density lipoproteins (VLDL, or pre-β-lipoproteins).
3. Low density lipoproteins (LDL or β-lipoproteins).
4. High density lipoproteins (HDL).

The protein part is called apoprotein and many lipoproteins contain more than one type of apoprotein polypeptide. They differ in their amino acid content and may be identified by polyacrylamide gel electrophoresis or by immuno-chemical methods. The lipid-free apoproteins may be purified by gel filtration or by ion exchange chromatography (Fig. 7.39).

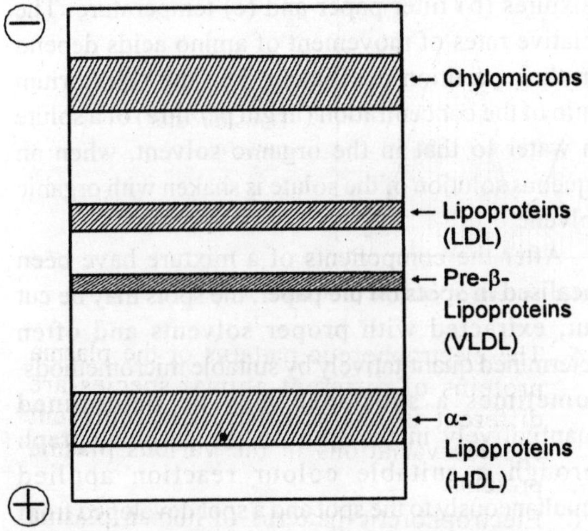

Fig. 7.39: Separation of plasma lipoproteins by electrophoresis

7.31 Chromatography

7.31.1 Introduction

Chromatography was first studied by Schoenbin in 1861. Then its development as a systemic method was due to work of the Russian botanist Michael Tswett, in 1906. He was interested to separate the leaf pigments including the chlorophyll. Further advancement in amino acid analysis came in 1941 by Martin and Synge. They introduced the technique of chromatography. The method devised by Tswett depends on the establishment of an equilibrium between a solid (calcium carbonate) and liquid (petroleum ether) phase which is usually referred to as adsorption chromatography. Martin and Synge established an equilibrium between two liquid phase.

7.31.2 Definition

Chromatography is the procedure to separate solutes by passing their solution through a solid porous support. Molecules are partitioned between a stationary and a mobile phase.

7.31.3 Chromatographic methods

I. (1) Partition chromatography, (2) Reversed phase partition chromatography.
   (a) Paper chromatography—(a) Ascending, (b) Descending, (c) 1-dimensional, (d) 2-dimensional, (e) Combination of Vertical paper Chromatography and Horizontal Electrophoresis.
   (b) Liquid-liquid chromatography
   (c) Thin-layer chromatography (TLC)
   (a) Partition TLC (b) Adsorption TLC.
   (d) Gas-liquid chromatography (GLC).

II. Adsorption chromatography or adsorbent or column chromatography.

III. Affinity chromatography.

IV. Ion-exchange chromatography.

V. Molecular-Exclusion chromatography or molecular-sieve chromatography or gel-filtration chromatography.

VI. Reversed Phase High-Pressure Liquid Chromatography (RPHPLC).

I. Partition Chromatography: In chromatography, molecules are partitioned between a stationary and a mobile phases. Separation of molecules in a mixture depends on the relative affinity of molecules to associate firmly either one of the phases.

Chromatography	Stationary phase	Mobile phase
Partition chromatography-paper, thin layer	Liquid	Liquid
Ion exchange	Solid	Liquid
Partition, gas-liquid		

As for example, for amino acid separation, solvents are polar binary, ternary or more complex mixtures of water, alcohol, and acid or bases. The more polar components of the solvent associate with the cellulose and form stationary phase. The less polar components are the stationary phases. This is referred to as **partition chromatography**. In case of "Reversed phase partition chromatography", the polarities of the mobile and stationary phase are reversed such as dipping the paper first in a solution of a silicone. Reversed phase partition chromatography is used for separation of nonpolar peptides or lipids but not amino acids.

1. Paper chromatography

This is a liquid-liquid chromatography and widely used for separation of mixture of amino acids, peptides, sugars and sugar derivatives.

(a) Ascending Paper Chromatography: When strip of filter paper is suspended vertically with the lower end dipping into a mixture of an organic solvent, e.g., **n-butyl alcohol, n-propyl alcohol, phenol, isobutyric acid** or **collidine**, the mixture of organic solvent moves up the filter paper. The water being held by the filter paper constitutes a stationary solvent phase, whereas the organic solvent for which the filter paper has little attraction constitute the mobile phase.

A line is drawn by a pencil about 5 cm from the end of a dry strip of filter paper (15 by 50 cm). A very small quantity of sample, e.g., amino acid solution about 5 to 15 μg of amino (0.005 ml) acids is applied to the pencil line. The paper strip is supported with the sample end dipping into the solvent. The paper and solvent is enclosed in a covered glass jar or other container, and the temperature is maintained constant, so that the solvent moves up the paper in an atmosphere saturated with the vapour of both water and organic solvent at that temperature. When a solute is spotted on the paper close to the end dipping in the solvent, as the solvent moves up the strip and over the solute, the solute is partitioned between the stationary water phase on the paper and the moving organic phase. As the moving organic phase reaches a portion of paper containing no solute, partition again takes place, and solute is transferred from the organic to the stationary water phase. As the solvent continually moves up the paper, the net effect of this partition between the two phases is the transfer of the solute from the point of its application to the filter paper to a point some distance along the paper in the direction of solvent flow. The solute is observed as a localised spot on the paper. After the solvent front has run sufficient distance in 15-20 hours depending on the solvent, paper and temperature, the paper strip is removed and the position of the solvent front is marked. The strip is dried and sprayed with a solution of **Ninhydrin** which with amino acids gives a colour, purple or orange. The amino acid is seen localised as a spot at some position up the strip. The ratio of the distance of the spot from the pencil line to the distance of the solvent front from the pencil line is R_F value. The R_F values are characteristic for the different amino acids when run under the same conditions. The R_F values and degree of separation depend on the (a) selection of proper solvent mixtures (b) filter paper and (c) temperature. The relative rates of movement of amino acids depend on their partition coefficient, i.e., the equilibrium ratio of the concentration (in gm per litre) of a solute in water to that in the organic solvent, when an aqueous solution of the solute is shaken with organic solvent.

After the components of a mixture have been localised in spots on the paper, the spots may be cut out, extracted with proper solvents and often determined quantitatively by suitable micromethods. Sometimes a substance may be determined quantitatively in the spot on the chromatograph through a suitable colour reaction applied simultaneously to the spot and a spot developed from a known amount of the same material (Fig. 7.40).

(b) Descending Paper Chromatograph: Sample and solvent may be applied at the top of the strip to give descending paper chromatography instead at the bottom to give ascending chromatography. The arrangement is to insert the end of the paper strip into the solvent mixture placed in a narrow trough mounted near the top of the cylinder; here the solvent descends into the paper (Fig. 7.40A).

(c) One-dimensional Paper Chromatography: The filter paper strip is stained with Ninhydrin solution in case of an amino acid mixture, so that separate fractions may be visualised as coloured spots.

(d) Two-dimensional Paper Chromatography: A sheet is used instead of a strip of filter paper, the solution containing an amino acid mixture

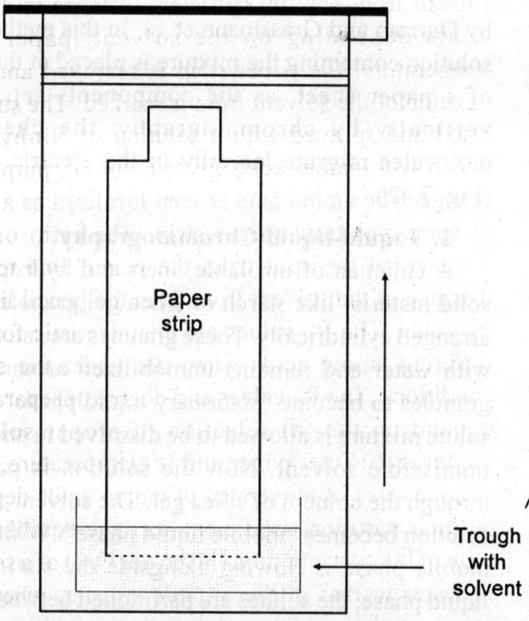

Fig. 7.40: Ascending Paper Chromatography

is placed at one corner of the sheet. The components of the mixture may then be separated chromatographically along one edge of the sheet by means of one pair of solvents (e.g., phenol-water). The paper is then dried and turned through 90°, and chromatography is effected with a different pair of solvents (e.g., Collidine-water), in a direction perpendicular to that used first. Upon treatment of the dried sheet with Ninhydrin, a "two-dimensional" chromatogram is obtained in which there is considerable separation of the spots corresponding to the individual amino acids.

Two-dimensional separation of electrolytes may be achieved by applying a direct current across the paper at right angles to the chromatographic flow.

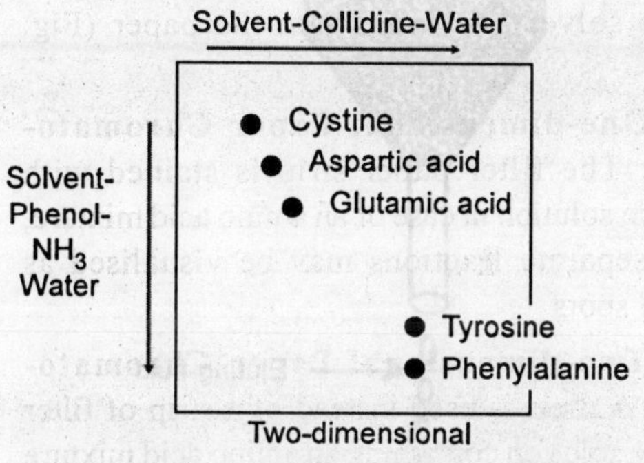

Fig. 7.40A: Ascending Paper Chromatography

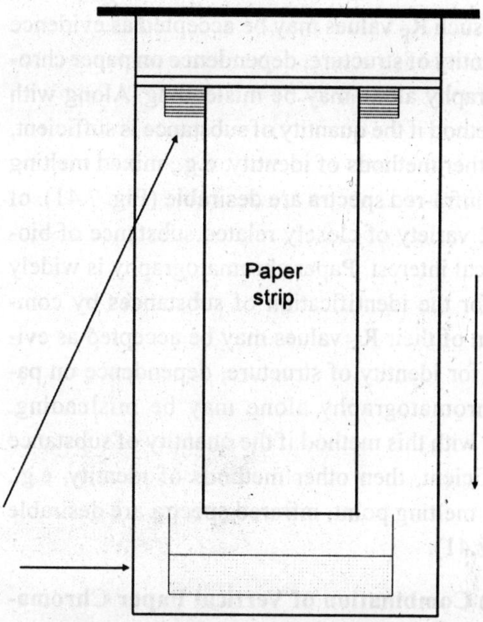

Fig. 7.41: Descending Paper Chromatography

This method is known as **Electrophoretic chromatography** employed for separation of blood serum proteins. One-dimensional and two-dimensional paper chromatography have been of great value not only in amino acids but also for the separation of a large variety of closely related substance of biochemical interest. Paper chromatography is widely used for the identification of substances by comparison of their R_F values in several solvents with those of known samples. Although the iden-

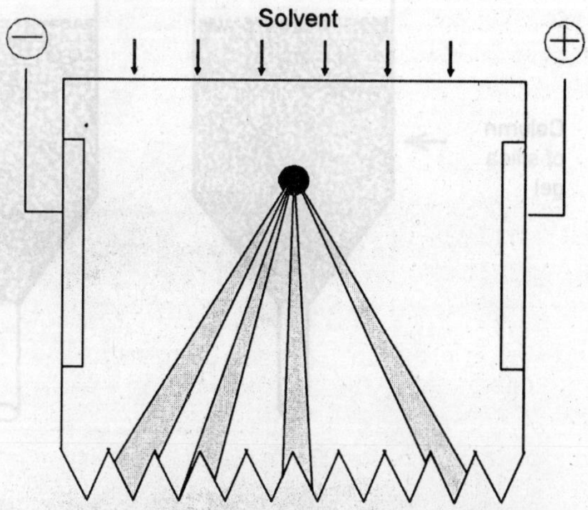

Fig. 7.42: Combine Paper Chromatography and Electrophoresis

tity of such R_F values may be accepted as evidence for identity of structure, dependence on paper chromatography alone may be misleading. Along with this method if the quantity of substance is sufficient, then other methods of identity, e.g., mixed melting point, infra-red spectra are desirable (Fig. 7.41). of a large variety of closely related substance of biochemical interest. Paper chromatography is widely used for the identification of substances by comparison of their R_F values may be accepted as evidence for identity of structure, dependence on paper chromatography alone may be misleading. Along with this method if the quantity of substance is sufficient, then other methods of identity, e.g., mixed melting point, infrared spectra are desirable (Fig. 7.41).

(e) Combination of Vertical Paper Chromatography and Horizontal Electrophoresis (Durrum and Grassmann et al.)

Combination of vertical paper chromatography and horizontal electrophoresis has been developed by Durrum and Grassmann et. al. In this method, a solution containing the mixture is placed at the top of a paper sheet, as the components separate vertically by chromatography, the charged molecules migrate laterally in the electric field (Fig. 7.42).

2. Liquid-liquid Chromatography

A columan of insoluble, inert and hydrophilic solid material like starch or silica gel granules are arranged cylindrically. These granules are saturated with water and remains immobilized around the granules to become "stationary liquid phase". The solute mixture is allowed to be dissolved in another immiscible solvent. Now the solution is passed through the column of silica gel. The solvent of the solution becomes "mobile liquid phase". While the mobile phase is flowing alongside the stationary liquid phase, the solutes are partitioned between the two liquid phases as per partition coefficient. In this way the solutes are separated as bands along the gel column. This method was originally used for separation of N-acetylamino acids when the

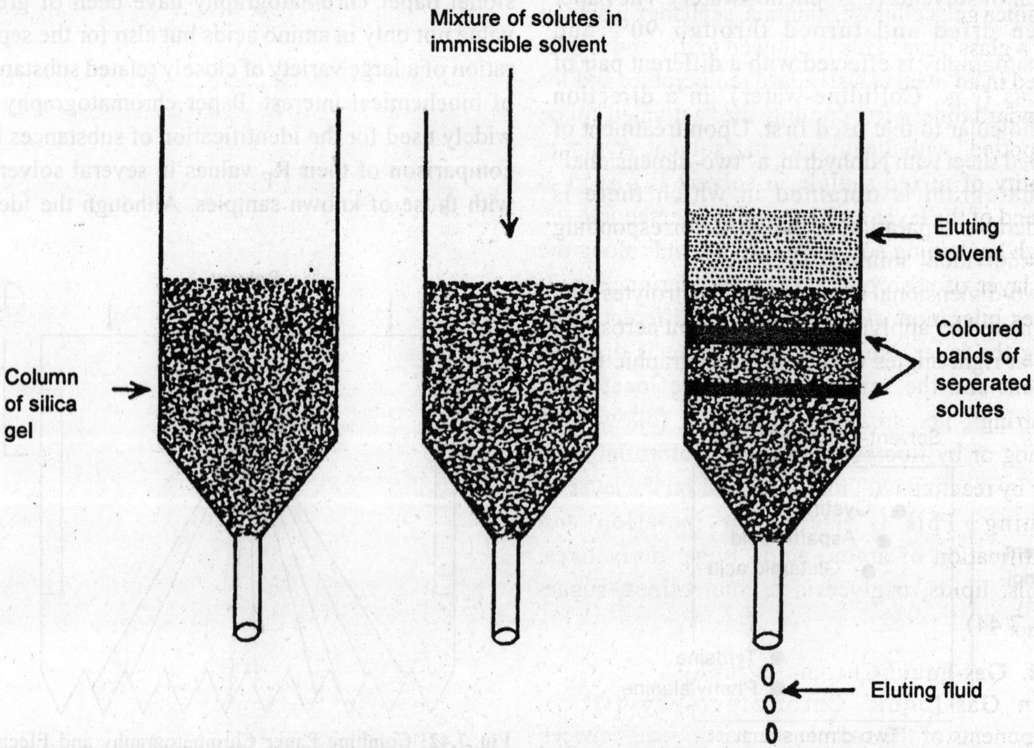

Fig. 7.43: Liquid-liquid Chromatography

stationary phase is coloured yellow with methyl orange which becomes red due to presence of N-acetylamino acids and is observed as red bands in the phases while moving. These can be separately collected in the different fractions of the outgoing fluid of the mobile phase (Fig. 7.43).

3. Thin-layer Chromatography (TLC)

There are two types of TLC:

(a) Partition TLC (PTLC): It resembles partition paper chromatography. For PTLC on powdered cellulose or other relatively inert supports, the solvent systems and detection reagents used for paper chromatography are also used for this method. Reversed phases PTLC can be done.

(b) Adsorption TLC (ATLC): Adsorption TLC (ATLC) does not have any similarity to paper chromatography except that both are conducted on thin sheets. ATLC chromatography depends on the ability of the solvent to elute sample components from adsorption sites on an activated sorbent such as heated silica gel. It is used for nonpolar materials such as lipids but not for amino acids.

In this method a thin layer of an inert adsorbent like silica gel, cellulose, alumina, sephadex is spread over a glass plate. This is allowed to dry and is then heated in an oven at a standard temperature and for a standard time. After cooling, the "activated" plate is "spotted" with the mixture in a solvent. A minute quantity of mixed sample in solvent is placed at one end of the layer and that end is imersed into the trough containing solvent which ascends along the thin layer of adsorbent due to capillary action. It causes migration of the solutes to different spots along the layer. Next, the glass plate is dried of solvent and the separated spots are located by "charring", i.e., spraying, of H_2SO_4, followed by heating or by fluorescence with dichlorofluorescein or by reacting with iodine vapour on the layer by staining. This is useful for isolation and identification of amino acids, sugar-derivatives, sterols, lipids, triglycerides, nucleotides, sugars (Fig. 7.44).

4. Gas-liquid Chromatography

In Gas-Liquid Chromatography (GLC), components of the sample are partitioned between a thin film of a relatively nonpolar, nonvolatile liquid like silicone, coated on a finely divided inert support like firebrick, diatomaceous earth. The mobile gas phase is usually argon, helium or nitrogen. This method is applicable only to substances having tendency to vaporize at temperatures below their decomposition points. Therefore, highly polar materials like amino acids and peptides must be derivatized first to make volatile. This method is highly successful for resolution of highly complex lipid mixtures but not favourable for amino acids and peptides because of difficulties in achieving quantitative derivatization and thermal decomposition on the column and moreover the detection technique destroys all or some of the sample. In this method the fatty acid mixtures are first converted into a more volatile state usually as methyl esters. An inert carrier gas, e.g., nitrogen, is used as the moving phase of the vaporized mixtures of methyl esters between the moving gas phase and a stationary phase of a high melting polyester of silicone polymer coated on particles of diatomaceous earth or on the inner surface of a long, heated capillary tube. The methyl esters of the various fatty acids are partitioned between the moving gas phase and the stationary liquid phase according to their gas-liquid partition coefficients. The separated methyl esters in the gas phase leaving the column can be measured by sensitive detectors such as the "flame-ionization detector" in which

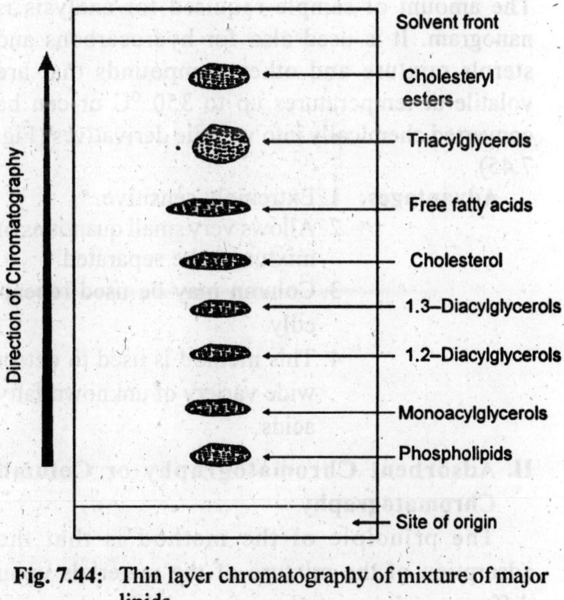

Fig. 7.44: Thin layer chromatography of mixture of major lipids

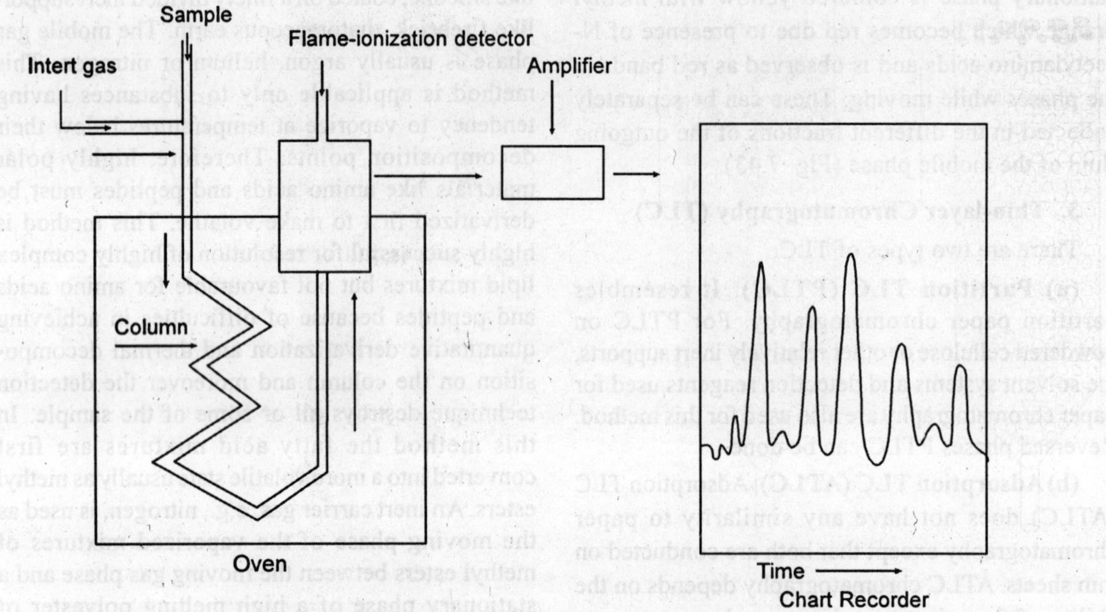

Fig. 7.45: Diagrammatic representation of gas-liquid Chromatography

the carrier gas stream containing the fatty acid esters is mixed with a stream of hydrogen and air and burned in a high voltage electric field. The current generated by the flow of ionized fragments of the fatty acid in the flame is automatically recorded on a chart, which shows a series of separate peaks. Each peak corresponds to a separate fatty acid, and the arc under the peak is proportional to the amount. The amount of sample required for analysis is nanogram. It is used also for hydrocarbons and sterols mixture and other compounds that are volatile at temperatures up to 350 °C or can be converted chemically into volatile derivatives (Fig. 7.45).

Advantages: 1. Extremely sensitive.

2. Allows very small quantities of mixture to be separated.
3. Column may be used repeatedly.
4. This method is used to detect wide variety of unknown fatty acids.

II. Adsorbent Chromatography or Column Chromatography

The principle of the method is that the adsorption of the mixture of the materials upon different solid particles of insoluble adsorbing agents, e.g., silica gel, alumina, calcium hydroxyapatite and charcoal powder arranged as columns in glass tubes, followed by differential elution of the components of the mixtures from the columns with appropriate solutions. This procedure is known as **Adsorbent Chromatography or Column Chromatography** and has proved valuable for many substances like amino acids, protein hydrolysate, carotenoids, proteins, porphyrins, etc. For this method, only few miligrams of sample are required. Moore and Stein used starch column initially, through which a solvent such as aqueous butanol was run for equilibration. Nowadays, column of finely divided relatively inert materials with a very large surface area in relation to particle in relation to particle size are being used. They include nonpolar substances like charcoal and polar substances like silica gel or alumina. The nature of forces binding the materials is not known but may be due to van der Waals' forces and hydrophobic interactions with nonpolar adsorbents, whereas ionic attractions and/or hydrogen bonding with polar adsorbents. Perhaps the most effective adsorbent for protein purification is a crystalline calcium phosphate or "hydroxyapatite", the same mineral present in bone. Probably negatively

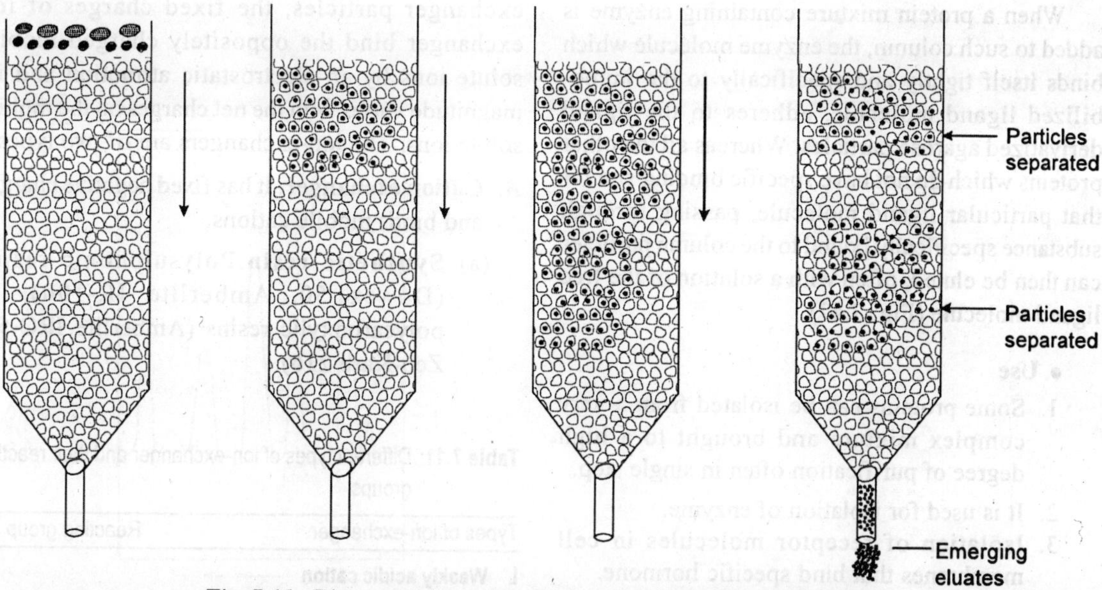

Fig. 7.46: Diagrammatic representation of Adsorption Chromatography

charged groups in protein molecules bind to the Ca^{++} ions in the hydroxyapatite crystal lattice. Proteins are eluted from hydroxyapatite columns with phosphate buffers.

The mixture of the sample is dissoved in the solvent; it is then poured on the column and more solvent is passed through the column. The individual component is adsorbed at the surfaces of the adsorbing agent from which they are eluted, i.e., extracted by dissolution by the solvent and move slowly down and from the columns at rates determined largely by their chemical natures and the kind of solvent used. When the conditions are proper, the individual component emerges from the column at different rates and each may be collected separately in a definite volume and fraction of the solution emerging from the column. A useful apparatus called "fraction collector" which automatically collects the column effluent as aliquots of definite volume has been developed. The aliquots are then analysed, e.g., amino acid content of protein hydrolysate by the "Ninhydrin colorimetric method" (Fig. 7.46).

III. Affinity Chromatography (Separations based on Ligand specificity)

This method is based on a biological property of some proteins, i.e., their capacity for specific, noncovalent binding of another molecule known as **ligand**. As for example, some enzymes have coenenzymes as prosthetic group. These enzymes bind the coenzymes tightly by noncovalent forces. If these enzymes are to be separated from other proteins by affinity chromatography, the coenzyme is covalently attached by chemical reaction with a functional group on the surface of large hydrated particles of a porous column material like the polysaccharide agarose or dextran which allows protein molecules to pass freely.

Table 7.10: Some commonly used ligands in affinity chromatography

Ligand	Affinity
1. 5' AMP	Certain kinases and NAD + dependent dehydrogenase
2. Poly U	mRNA containing poly (A) tail
3. Lysine	rRNA
4. Concavalin A	Glycoprotein, glycolipids, etc.
5. Cibacron Blue F_3GA	Nucleotide requiring enzymes, albumin, etc.
6. Lectin (from Triticum vulgare)	Cells and macromolecules containing N-Acetyl-B-glucosamine residues.

When a protein mixture containing enzyme is added to such column, the enzyme molecule which binds itself tightly and specifically to the immobilized ligand molecule, adheres to the ligand derivatized agarose particles. Whereas all the other proteins which do not have specific binding site for that particular ligand molecule, passing through substance specifically bound to the column particles can then be eluted, often with a solution of the free ligand molecule (Fig. 7.47).

- **Use**

1. Some proteins can be isolated from a very complex mixture and brought to a high degree of purification often in single step.
2. It is used for isolation of enzyme.
3. Isolation of receptor molecules in cell membranes that bind specific hormone.
4. It is also used for purification of enzyme. Rosenberry et al. (1972) have shown in a comparative study that ammonium sulphate precipitate of acetylcholinesterase showed specific activity of 470 μ/mol min^{-1} mg^{-1} whereas in only a single step of affinity chromatography yielded higher specific activity of 9750470 μ/mol min^{-1} mg^{-1}.

IV. Ion-exchange Chromatography

The principle of Ion-exchange Chromatography is to utilize the acid-base behaviour of proteins as a basis for separation having a number of variations of the method. The same basic principle was originally used for separation of amino acids or peptide mixtures on columns of ion-exchange resins but later on was used for protein mixtures by Sober et al. (1950).

The ion exchange materials are water-insoluble polymers, e.g., dextran, agarose, synthetic resins (Polysulphonic resins, polycarboxylic resins, polyamine resins), cellulose (diethylaminoethyl-cellulose, carboxymethylcellulose).

The surfaces of ion-exchanger have either acidic or basic groups which are fixed. These groups ionize at suitable pH to yield fixed negative or positive charges on the ion-exchanger material. If a solution of ionized solutes, e.g., amino acid or protein mixtures, is passed through a column of ion-exchanger particles, the fixed charges of ion exchanger bind the oppositely charged mobile solute ions by an electrostatic attraction and its magnitude varies with the net charges on the mobile solute ions. The ion-exchangers are of two types:

A. **Cation exchanger**: It has fixed negative charges and binds mobile cations.

(a) **Synthetic resin**: Polysulphonic resins (Dowex 50, Amberlite IR-100), or polycarboxylic resins (Amberlite IRC-50, Zeo-Karb 216).

Table 7.11: Different types of ion-exchanger and their reacting groups

Types of ion-exchanger	Reacting group
I. Weakly acidic cation exchanger	
a. Amberlite IRC 50	$-COO^-$
b. Bio-Rex 70	
c. Zeocarb 226	
d. CM–Sephadex	$-CH_2COO^-$
e. Cellex—CM	
f. Agarose (CM–Sepharose)	
II. Strongly acidic cation exchanger	
a. Amberlite IR 120	$-SO_3^-$
b. Bio-Rad AG 50	
c. Dowex 50	
d. SP–Sephadex	$-CH_2CH_2CH_2SO_3^-$
III. Weakly basic anion exchanger	
a. Amberlite IR 45	$-CH_2N^+HR_2$
b. DEAE–sephadex	$-CH_2CH_2N^+H(CH_2CH_3)_2$
c. Cellex-D	
d. DEAE–Sepharose	
IV. Strongly basic anion exchanger	
a. Amberlite IRA 410	$-CH_2N^+(CH_3)_2$ with CH_2CH_2OH
b. Bio Red AG_2	
c. Dowex 2	
d. QAE–Sephadex	$-CH_2CH_2N^+(CH_2CH_3)_2$ with $CH_2CH(OH)CH_3$

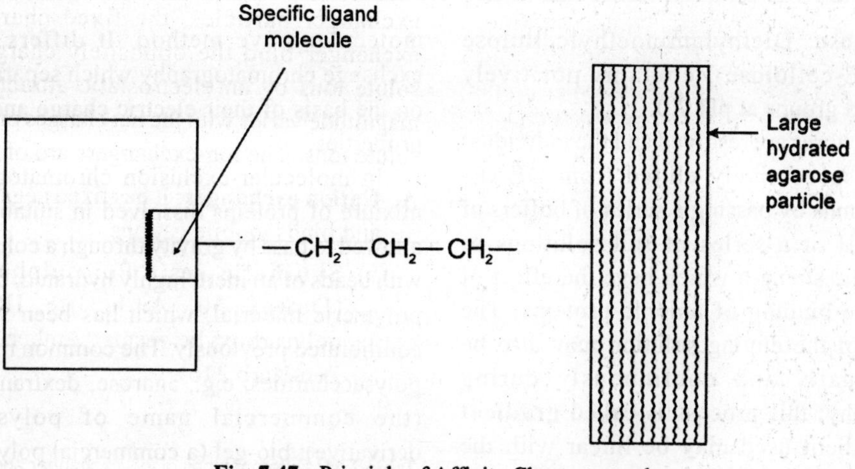

Fig. 7.47: Principle of Affinity Chromatography

The action of a cation exchanger is exchanging one ion (e.g., Na^+) for another (NH^+ R) is

Resin : $SO_3^- \ Na^+NH_3^+ R \rightarrow$

Resin: $SO_3^- \ NH_3^+ R + Na^+ \leftarrow$

(b) **Cellulose**: Carboxymethylcellulose (CM-cellulose) contains negatively charged groups at neutral pH.

B. Anion exchanger: It has fixed positive charges and binds mobile anions.

(a) **Synthetic resins:** Polyamine resins (Dowex 2, Amberlite IR-410).

The action of an anion exchanger is as follows:

Resin: $NR_3^+ + OH^- + R'COO^- \rightarrow$

Resin: $NR_3^+R'COO^- + OH^- \leftarrow$

Table 7.12: Different types of Column materials used in liquid chromatography

Column materials	Reacting group	Fractionated materials
I. For Ion-exchange chromatography		
a. Bio-Rex 70	$-COO^-Na^+$	Positively charged molecules
b. Dowex 50	$-O-SO_2-H^+$	Positively charged molecules
c. DEAE-Cellulose	$-O-C_2H_4-N^+H$ $(C_2H_5)_2d^-$	Very charged molecules
d. QAE–Sephadex	$-O-C_2H_4^-$, $N^+(C_2H_5)_2$ CH_2-CHOH^-, CH_2Br	Very charged molecules
II. For Gel filtration Chromatography		
a. Biogel P	Polacrylamide bead	Any molecules
b. Biogel A	Agarose bead	Any molecules
c. Sephadex G	Dextran bead	Any molecules
III. For Adsorption Chromatography		
a. Hydroxylapatite (hydroted Calcium phosphate)		Separation by single and double strand DNA Proteins
b. Alumina		Steroids

(b) **Cellulose**: Diethylaminoethylcellulose (DEAE-cellulose), contains positively charged groups at pH 7.0.

Protein mixtures are resolved and the individual components successively eluted from DEAE-cellulose columns by passing a series of buffers of decreasing pH or a series of salt solutions of increasing ionic strength which have the effect of decreasing the binding of anionic proteins. The composition of the eluting solution may also be varied gradually and continuously during chromatography; this process is called **gradient elution**. Gradient used may be linear with the volume of the eluting solution or varied in some other relationship by using devices of mixing liquids in varying ratios. The protein concentration in the elute, collected in fractions, is estimated optically by absorption of light in the ultraviolet region.

V. Molecular-Exclusion Chromatography

(Molecular-sieve or gel-filtration chromatography)

This is one of the powerful methods for separating protein mixtures on the basis of size or molecular weight. The columns of polysaccharide or polyacrylamide gels are used as molecular sieves; hence the method is also known as gel-filtration or molecular-sieve method. It differs from ion-exchange chromatography which separates solutes on the basis of their electric charge and acid-base properties.

In molecular-exclusion chromatography, the mixture of proteins dissolved in suitable buffer is allowed to pass by gravity through a column packed with beads of an inert, highly hydrated, cross-linked polymeric material, which has been washed and equilibrated previously. The common materials are polysaccharides, e.g., agarose, dextran, Sephadex (the commercial name of polysaccharide derivative); bio-gel (a commercial polyacrylamide derivative) all are prepared with different degrees of internal porosity. The proteins of different molecular size penetrate into the internal pores of the beads of the column at different degrees and they travel down through the column at different rates. Very large molecules cannot pass through the pores of the beads and are excluded and remain in the "excluded volume" of the column and referred to as "the volume of the aqueous phase" outside the beads and pass through rapidly. Molecules of intermediate size are also excluded from the beads to a degree depending on the size. Very small molecules can enter the pores of the beads freely.

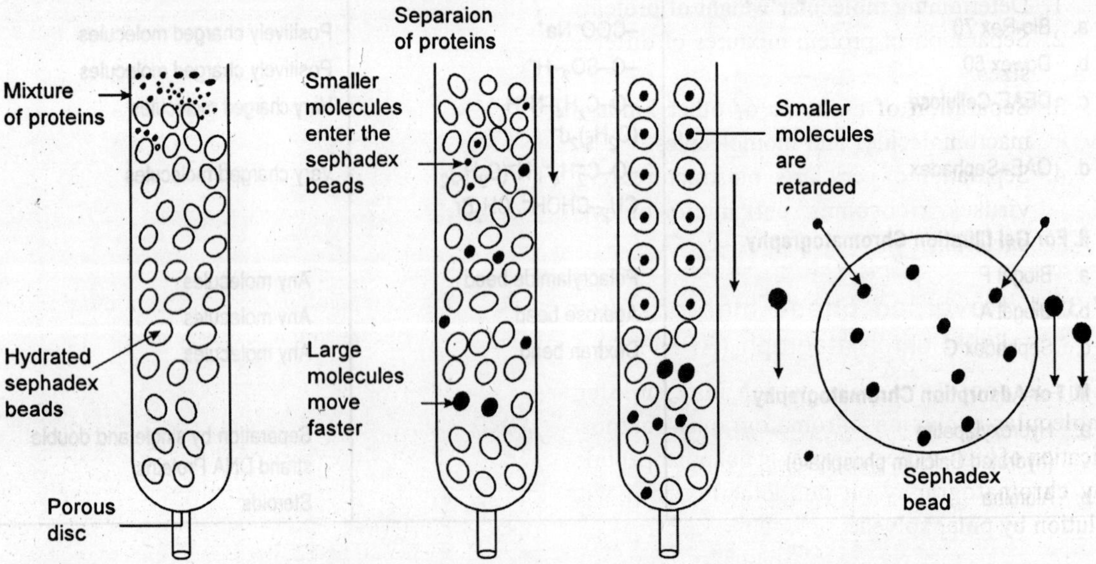

Fig. 7.48: Diagrammatic representation of Molecular-Exclusion Chromatography

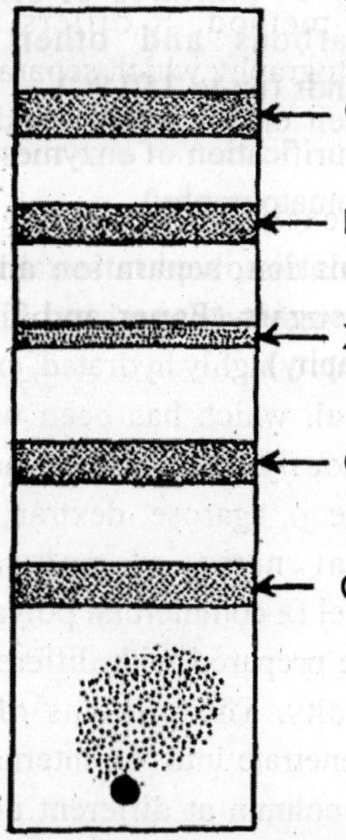

Fig. 7.49: Principle of Affinity Chromatography

Small molecules are retarded by the column. Exclusion of molecules according to size provide the term **Exclusion Chromatography**. From measurement of the protein concentration in small fractions of the eluate, an elution curve can be constructed (Fig. 7.48).

- **Use**
  1. Determining molecular weight of proteins.
  2. Separation of protein mixtures of different sizes.
  3. Separation of mixtures of other kinds of macromolecules and biomolecules.
  4. Separation of very large biostructures, e.g., viruses, ribosomes, cell nuclei or even bacteria.

7.31.4 Reversed phase high-pressure liquid chromatograph (RPHPLC)

This is another method which may supplant molecular-exclusion chromatography for purification of high-molecular weight nonpolar peptides by chromatography on nonpolar material with elution by polar solvents.

7.31.5 Use of chromatography

1. Separation of proteins, peptides, amino acids, enzymes, lipids from the mixture.
2. Purification of proteins, peptides, enzymes, etc.
3. Separation of proteins from protein hydrolysate, tissue extract.
4. Separation of other macromolecules and biomolecules.
5. Separation of very large biostructure like viruses, ribosomes, cell nuclei or even bacteria.
6. Identification of chemical compound into which the radioactive isotope has become incorporated. The tissue is ground up and the chemical constituents separated by chromatography. If paper chromatography is used the chromatogram is exposed against a photographic plate and the radioactive compound identified in the resulting autoradiograph.
7. Separation of chlorophyll: Chlorophyll, the green pigment of the leaf, is in fact a mixture of closely-related pigments of which chlorophyll is one. This is demonstrated by extracting the pigments with acetone and separating them by paper chromatography. The strip of paper containing a spot of concentrated chlorophyll extract is dipped in a solvent rises up the paper giving bands. Five pigments have been identified:
   (a) Chlorophyll a (blue-green)
   (b) Chlorophyll b (yellow-green)
   (c) Xanthophyll (yellow)
   (d) Carotene (orange)
   (e) Phacophytin (grey), a breakdown product of chlorophyll (Fig. 6.49).
8. **Lipids in chromatography**
   (a) Separation of major lipid classes in a mixture (Phospholipids, monoglycerols, 1,2-Diacylglycerols, cholesterol, free fatty acids, triglycerols, 1,3-Diacylgly-

cerols, cholesteryl esters) in a solvent [Hexane-diethylether-formic acid (80;20;2 vlvlv)'] is made by thin layer chromatography.

(b) For analytical purposes, thin layer chromatography can be used for purifying lipids.

(c) Separation of individual fatty acids (gas-liquid chromatography-GLC).

(d) Analysis of mixtures of sterols and hydrocarbons and other volatile compounds (up to 350 °C).

9. **Enzymes**: Purification of enzymes (Affinity column chromatography).

10. **Sugars**: Isolation, separation and identification of sugars (Paper and Thin-layer chromatography).

8

Nucleonics in Biology and Medicine

Part I: Elementary particles, Atom, atomic nucleus, protons, neutrons, electrons, accommodation of electrons. **Part II**: Radioactivity, energy of radiation, Detection of nuclear radiation, Radiation from radioactive materials or radioactive emission, α-rays, β-rays, γ-rays, detection of nuclear radiation, Radioactive decay, Radioactive decay series, Radioelements in the periodic table, Rate of radioactive decay and half-life, Types of radioactivity. **Part III**: Isotopes, Isobars, isotones characteristic, Varieties of isotopes. **Part IV**: Biological application of Radioisotope. **Part V**: Nuclear Radiation therapy. **Part VI**: Imaging System. **Part VII**: X-rays, nature, production, circuit, classification, application. **Part VIII**: Image intensifying tube. **Part IX**: X-ray machine, protective housing of the X-ray tube, factors influencing X-ray quantity and quality. **Part X**: Detection and measurement of radioactivity. **Part XI**: Nuclear radiations and interactions with living materials. **Part XII**: Biological effects of radiation, Factors influencing the biological effect. **Part XIII**: Radiation damage in the embryo and fetus during pregnancy. **Part XIV**: Demerits of different diagnostic and therapeutic methods of nuclear medicine during pregnancy. **Part XV**: Radiation hazards in human being. **Part XVI**: Fluoroscopy. **Part XVII**: Image contrast. **Part XVIII**: Pattern of X-ray beam and recording. **Part XIX**: Factors governing exposure. **Part XX**: Xero-Radiography. **Part XXI**: Contrast media. **Part XXII**: Digital Subtraction Angiography (DSA). **Part XXIII**: Radiotherapy. **Part XXIV**: Tomography. **Part XXV**: Magnetic Resonance imaging. **Part XXVI**: Mammography (Soft Tissue Radiography).

PART I. Elementary Particles

8.1 Atom

Dalton (1803-1808) introduced the concept that the matter is made up of minute individual mass, called atom. Later on discovery of "subatomic particle", the "electron" by Thomson (1898) encouraged the subsequent discoveries by Rutherford, Bohr, Mosley, Chadwick, etc. From the discoveries it was evident that the **atom of all elements** is composed of the smallest and elementary particles like **neutron, proton** and **electron**. In addition to these observable "particles" and quanta, there are other elementary particles like **positron, antiproton, antineutrino, neutrino** have been discovered and more are being discovered. Neutrino and antineutrino are supposed to permit energy and spin or angular momentum to be preserved in a nuclear "β-decay". When an electron or positron is emitted, the particle has a spin but leaves the nucleus with the same spin it had before the disintegration. Accordingly, the neutrino and antineutrino has a spin of $\frac{1}{2}$ and zero or very minute mass.

The neutron and proton occur in the centre of an atom where they form an **atomic nucleus**. Neutron and proton are called **nucleon** which has one unit of "atomic weight" or "mass" and is about 1800 times heavier than the electron. A proton's electric charge is positive, whereas a neutron is electrically neutral. The electron orbits the atomic nucleus at some distance from its centre where it is virtually weightless and carries negative electrical charge. Hence, the number of positive and negative charges are exactly balanced and so an atom is electrically neutral. The structure of atoms of some elements has a distinctive number of protons and electrons.

However, the chemical properties of the element depend upon the number and configuration of electrons. The number of such planetary electron, in the neutral atom of any elements, is equal to the number of protons in the nucleus of that atom (Fig. 8.1). An element is characterised by a fixed and definite atomic number of protons in the nuclei of its atom balanced by an equal number of planetary electrons.

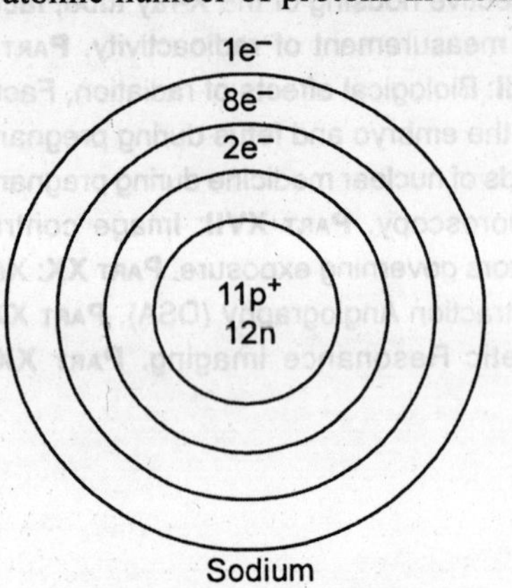

Fig. 8.1: Atom showing the number and disposition of the protons, neutrons and electrons

With the exception of the nucleus of the ordinary hydrogen, all atomic nuclei contain one or more neutrons in addition to proton.

The number of neutron in the nuclei of the atoms of an element are not fixed, but vary within certain limits. The atoms of most elements contain nuclei with different numbers of neutrons and consequently of different mass numbers. Each such atomic variety with a specific mass of an element constitutes **isotopes**.

8.1.1 Atomic nucleus

Rutherford (1911) opined that the atom consists of two parts, a very small, i.e., one-ten thousandth times smaller in diameter than the atom (whereas atoms are generally 2-5Å in diameter) but massive positively charged nucleus occupying a negligible fraction of the total volume of the atom and an extra-nuclear part containing electronegative electrons to make the atom electro-neutral.

Salient features are:

1. It is composed of dense packing of protons and neutrons.
2. The number of protons and neutrons is equal to the number of electrons outside the nucleus.
3. The protons are positively charged whereas neutrons are neutral for which nucleus is positively charged known as nuclear charge.
4. The number of protons in the nucleus of an atom is equal to its atomic number.
5. The total number of protons and neutrons is equal to the atomic mass number.
6. The space around the nucleus is empty.
7. **Electrons and nucleus**: It was originally postulated that the composition of nucleus to be wholly of electrons and protons but later discovery of Chadwick (1932) showed that there is neutron but no electron. But there is a controversy that the nucleus of an atom does not have electrons but it emits electrons, e.g., β-ray activity of radioactive materials. Again, the nucleus can absorb electrons, called K-capture. This can be attributed to the Fermi's theory according to which a neutron can be annihilated with simultaneous formation of a proton, an electron and a neutrino and the reverse change can also take place. It is obvious that electrons cannot be present in the nucleus because the volume of an electron is bigger than that of the nucleus.
8. **Radioactivity**: Emission of active radiation or radioactivity occurs from the nuclei of atoms which are unstable and break down into new atoms emitting active radiation and release enormous quantity of energy. The rate of radioactive change is unaffected by temperature, pressure, chemical combination, etc.

- **Proton**

Proton present in the nucleus is positively charged for which nucleus is positively charged, called nuclear charge, and it has a negative counterpart which is electron, present outside the nucleus and the atom is counterbalanced and is electroneutral. It was identified as the positively charged hydrogen atom, being the lightest positively charged stable material particle ever observed and was named "Proton". Proton is the universal constituent and was shown by the nuclear disintegration experiments of Rutherford (1919) who observed that by bombardment of some lighter elements with a α-particles, protons are set free. Thus, a proton has a mass practically equal to that of hydrogen atom but has a positive charge numerically equal to the negative charge on an electron. It has a mass equal to 1.00813. The number of protons in the nucleus of an atom is equal to its atomic number; this gives the nucleus a positive charge equal to its atomic number. The total number of protons and neutrons is equal to the atomic number.

- **Neutron**

Several workers observed that a very penetrative type of ray is emitted when boron, lithium or beryllium is bombarded by fast α-particles. Curie et al. (1931). Chadwick (1932) and others conducted extensive research and found them to be composed of particles having no electric charge and a mass equal to that of a hydrogen atom which were named neutrons. In these cases, formation of neutrons was due to nuclear reactions, e.g.,

$${}_5B^{10} + {}_2He^4 (\alpha\text{-particles}) \rightarrow {}_7N^{13} + {}_0n^1$$

These neutrons and protons constitute the atomic nucleus. Neutrons are considered as an element of zero atomic number because of zero charge. Neutrons are produced by nuclear reactors for experimental purposes. Its mass is equal to 1.00866.

Outside the nucleus

The atom contains electrons outside the nucleus.

- **Electron**

1. The number of electrons outside the nucleus is equal to the number of protons and neutrons in the nucleus.
2. Electrons are negatively charged and counterbalanced by the positively charged protons in the nucleus to make the atom electroneutral.
3. Electron has a mass equal to 0.000548. it is about 9.10×10^{-28} gm which is 1/1800 times the weight of a hydrogen atom.
4. It is known that the negative charge of an electron is equal to the positive charge of a hydrogen ion. Hence, an electron is about 1/1800 times as light as a hydrogen atom, but has a charge numerically equal to that of a hydrogen ion.
5. **Accommodation of electrons**
   (a) The electrons are arranged in the atom in different energy level called **electron-shells**. Each shell is designated by a quantum number called "principal quantum number" having symbol "n". This "n" has integral values 1,2,3 and so on. 1 for the first shell of K-shell; 2 for the second shell or L-shell, 3 for the third shell of M-shell, etc. The maximum capacity of each shell to contain is $2n^2$ number of electrons arranged in different sub-shells, i.e, different energy levels.

Table 8.1: Elementary particles of matter

Positive particles (Charge = + e)	Charge	Mass	Negative particles (Charge = – e)	Charge	Mass	Neutral particles (Charge = 0)	Charge	Mass
Proton	+ 1	1.00813	Electron	– 1	0.00054	Neutron	0	1.00897
* Positron	+ 1	0.00054	Negative Meson	• •	0.1	Neutrino	0	0.005
Alpha particle (He nucleus)	+ 2	4.0332	Anti-proton	• •	• •	Photons:		
						X-ray	0	• •
						τ-ray	0	• •
Deuteron	+ 1	2.0166						
Positive Meson	• •	0.1						

* Positron is positively charged particle and counterpart of negatively charged electron having very short life of millionth of a second, and radioisotopes emit positron.

(b) Each shell has **sub-shells**. Each subshell is designated by another quantum number called "azimuthal or orbital quantum number" having symbol "l" which defines the shape of the orbital (The space inside the atom which has a high probability of occupation by an electron is called an orbital).

(c) The sub-shells are further sub-divided into energy levels called "orbitals", each is associated with third quantum number called "magnetic quantum number" (m).

(d) Every electron has a spin quantum number which can be either $+\frac{1}{2}$ or $-\frac{1}{2}$ corresponding to the clockwise or the anticlockwise rotation of the electron.

(e) No two electrons in an atom can have all the four quantum numbers identical according to "Pauli's Exclusion Principle".

Table 8.2: Maximum number of electrons that can occupy each electron shell

Shell number	Shell symbol or designation	Number of electrons
1	K	2
2	L	8
3	M	18
4	N	32
5	O	50
6	P	72
7	Q	98

PART II. Radioactivity

8.2 Radioactivity

8.2.1 Becquerel (1896) observed that uranium minerals, e.g., pitchblende, emit a kind of radiation continuously which exhibited the following:

1. Penetration of thin sheets of matter;
2. Affecting photographic plate in the dark even if it is wrapped with paper;
3. Ionisation of a gas through which it passes, i.e., making the gas conductor of electricity;
4. Charged gold-leaf electroscope gets also quickly discharged.

Then it was followed by the discovery of radioactivity of other elements by different workers like Curies (1898) who discovered the strongly radioactive element known as radium. Later on, various investigations have established that these active radiations emit from the nuclei of atoms which are unstable and break down into new atoms with formation of these active radiations or radioactivity and there is release of large amount of energy which is about one million electron-volts, i.e., 23 million kilocalories. The radioactivity is also considered by the fact that the rate of radioactive change is not altered by any change of pressure, temperature and chemical combination. Nuclei of the atom are stable when the binding energy is highest. For an isotope, the stable condition is reached when the number of neutrons is at some particular level. Thus, if the nuclei contain less or more neutrons than the protons, they indicate stability but the nuclei are unstable and transformation of energy takes place which tends to reduce the nuclei to the stable state. The binding energy of a nucleus is proportional to the square of the number of excess protons. The stable condition may be represented by a curve of neutron number against proton number which is known as the **neutron-proton diagram**. If any isotope is not stable, it tends to transform in such a way that stability is achieved. The nucleus transforms either (1) by capturing an electron from one of the orbits of the atom called K-capture, or (2) by emission of beta particle or positron.

These changes either

1. increase the number of neutrons at the expense of protons by addition of an electron or emission of a positron from the photon, or
2. change some neutrons to protons by emission of electron.

8.2.2 Energy of radiation

When radiant energy is emitted from a nucleus, their equivalent in mass is lost from the original mass so that the disintegration products are lighter than the original. Thus, if the energy of the emitted radiation is added to that of the other products of disintegration, the total energy on the two sides are expressed by an equation of which both sides balance and the law of conservation of energy is applicable.

8.2.3 Radiation from radioactive materials or radioactive emission

When electric or magnetic field is applied, it reveals that the radiation emitted by radioactive materials is composite in nature and it consists of three types of emanations, e.g., α-rays, β-rays and γ-rays (Fig. 8.2).

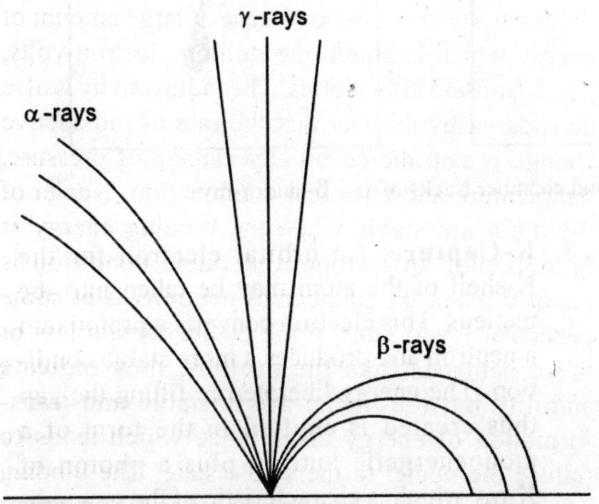

Fig. 8.2: Deflection of rays from radioactive material in a magnetic field

α-rays

1. These are positively charged particles (α-particles).
2. These are material particles of mass equal to 4, carrying 2 units of positive charge.
3. Have initial velocities of 1.4×10^9 to 2.2×10^9 cm.sec.$^{-1}$.
4. Have fixed initial kinetic energies of 3-9 million electron volts or MeV.
5. Have least penetrating power.
6. Have greatest ionisation power.
7. They are easily absorbed by the matter.
8. Have deflection towards cathode.
9. Each α-particle is a cluster of two neutrons and two protons.
10. Rutherford proved by spectrographic method that these rays are doubly charged helium atoms and so they are called **Helions**.

As the α-rays are continuously emitted from the radioactive materials (e.g., radium, uranium, etc.), it may be concluded that an element helium is produced from another element.

β-rays

1. More penetrating than α-rays.
2. Stream of very light negatively charged particles (β-particles).
3. Stream of electron coming out from the radioactive material.
4. Have a very high velocity like light.
5. Deflect towards anode.
6. Have viable initial kinetic energies about near 0 to 2 MeV.
7. Less easily absorbed by the matter.

γ-rays

1. These are electromagnetic radiations and are considered as non-material rays. These are hard X-rays with high penetrating power being identical with X-rays.
2. Have higher frequencies and shorter wavelengths than X-rays.
3. Not deflected in an electric field.
4. Very little absorbed by the matter.
5. Capable of penetrating thick layers of materials.
6. Unstable nucleus becomes excited after emitting α- or β-particles or positrons with less than the maximum kinetic energy and the excess energy is emitted as γ-rays from the nucleus.
7. Annihilation radiation: The mutual annihilation of a positron and an electron generates γ-rays, known as annihilation radiation.
8. Energy of γ-rays is about 0.5 MeV, 2 MeV and 0.5 MeV according to emission following α-decay, β-decay or positron-electron annihilation, respectively.

8.2.4 Detection of nuclear radiations

Radioactive emission is detected by an instrument, called Wilson CLOUD CHAMBER, where the radiations react with atoms within the instrument to produce charged ions and the ions are detectable by various processes. α-particles tracks appear as heavy lines having a few centimetre in length, β-particles tracks appear as lighter tracks having various lengths. γ-ray tracks are not usually visible except as a drop of water (Fig. 8.3) or possibly as a β-particle ejected by the γ-ray from the chamber wall. The sign of the charge on nuclear emissions and the ratio of charge to mass can be observed by exposing the cloud chamber in an uniform magnetic field. The direction and radius of track curvature can be seen.

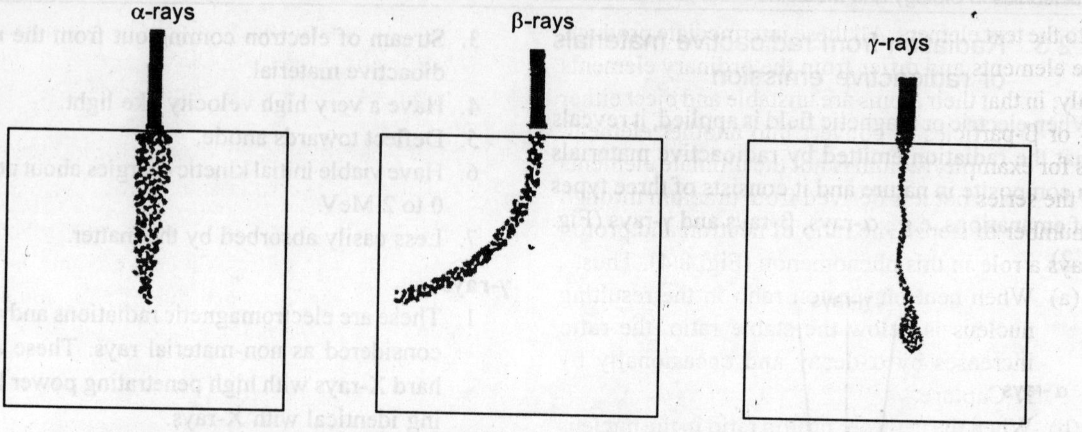

Fig. 8.3: Diagrammatic representations of cloud chamber tracks of α-, β- and γ-rays

8.2.5 Radioactive decay

It is the phenomenon of shifting from high-energy-level unstable radioactive nuclei to low-energy-level stable nuclei with the accompanying emission of energies or particles known as **radioactive decay**.

The process in radioactive decay are:

1. **α-decay**: The alpha particle consists of two neutrons and two protons. Alpha particles are monoenergetic for any one substance. Since the alpha particle is heavy, its expulsion is accompanied by recoil of the resulting nucleus. The total energy exchange is the rest mass of the alpha particle plus the kinetic energy of both fragments. Alpha decay causes lowering of the atomic number by 2 and the mass number by 4.
2. **β-decay**: High-energy electrons are also referred to as **negatrons.** These rays carry a negative charge and designated as β^-. Positrons having positive charge are referred to as **beta rays** and designated by β^+. These emissions are considered in the following ways:
   (a) During negation emission, there is also emission of neutrino. The actual energy of the negatron may vary from zero to the total energy represented by the mass defect. The difference is represented by the energy of the neutrino.
   (b) Positron emission also accompanies neutrino emission. The energy relations are the same as for the negatron.

   This process raises the atomic number by 1 without any change in the mass number.
3. **K-Capture**: An orbital electron for the K-shell of the atom may be taken into the nucleus. This electron converts a proton into a neutron and produces a more stable condition. The energy liberated in filling the gap thus created is emitted in the form of a monoenergetic nutrino plus a photon of X-ray which is characteristic of the new substance. This lowers the atomic number by 1 without any change in mass number.
4. **Gamma emission**: In an unstable nucleus, changes in the configuration of the nucleons may take place and the new configuration represents a high binding energy level. In such case, the excess energy is emitted as a photon of gamma radiation. From any one type of transition, the gamma-rays are monoenergetic, though the transition may consist of several steps. In such case, a line spectrum of gamma rays of several different energies may be emitted. *Internal conversion.* Sometimes it happens when a gamma ray is ejected from a nucleus, collision occurs with an electron in one of the orbits. The same electron is knocked from the orbit with an energy equal to the energy of the gamma photon minus the binding energy of the electron in the atom and the process is known is **internal conversion.**

8.2.6 Radioactive decay series—natural radioactive series

Each radioactive element always emits α- or β-particles from its nucleus and as a result converts into the text element. All these intermediate products are elements and differ from the ordinary elements only; in that their atoms are unstable and eject either α- or β-particles to convert into another element. As for example, radium is not the ultimate elements of the series but it is derived from uranium through a number of steps. The ratio of neutron and proton plays a role in this phenomenon (Fig. 8.4). Thus,

(a) When neutron: proton ratio in the resulting nucleus is below the stable ratio, the ratio increases by α-decay and occasionally by K-Capture;
(b) When the neutron: proton ratio in the nucleus is higher than the stable ratio, the ratio decreases by β-decay.
(c) When the nucleus has a stable neutron: proton ratio, then the product is a stable non-radioactive element, otherwise a new radio-active element from undergoing α- or β-decay through several steps until a stable ratio is achieved.

There are three natural radioactive series:

1. Uranium series: Radium is well-knwon member of this series or 4n + 2 series.
2. Thorium series: Starts with Th^{232} and is also known as 4n series because the mass numbers of member elements are multiples of 4.
3. Actinium series: Starts with $Uranium^{235}$ and also known as 4n + 3 series. The 4n + 1 series is not found in nature but it has been produced artificially. Many radio-isotopes do not exist in nature.

8.2.7 Radioelements in the periodic table (Radioactive displacement law)

Radium is Group II element in the periodic table. After emission of α-particles it converts into inert gas radon, a zero group element that means expulsion of an α-particles displaces the element two steps backward in the periodic table. In the manner if there is emission of a β-particle, the position of the new element is one step forward in the periodic table. The "radioactive displacement law" states that emission of an α-particle produces an atom two steps backward and emission of β-particle one step forward in the periodic table.

8.2.8 Rate of radioactive decay and half-life of radioactive element

The rate at which a radioactive element undergoes change to another element with emission of radiation is known as "rate of radioactive decay of the element". Radioactive decay is expressed in terms of the "half-life" of the substance. Half-life (T) means the time required for the initial number of radioactive atoms to be reduced to one-half. There is no practical importance of total life of radioactive element because in many cases it may approach

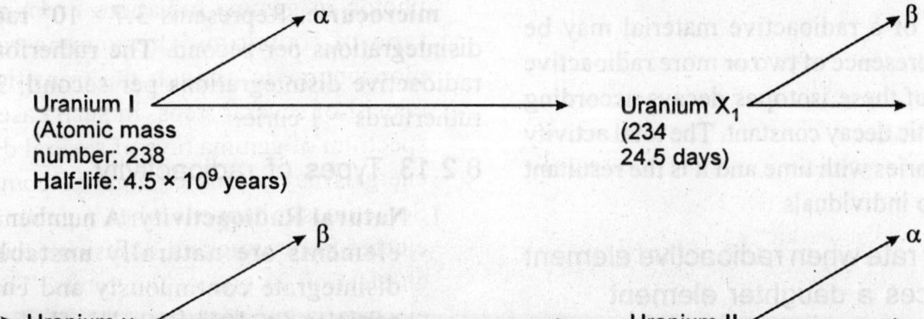

Fig. 8.4: Radioactive decay series of Uranium

infinity, but the half-life can be readily determined and is also very much useful in comparison to radioactive elements and in calculations. The shorter the half-life, the greater is the intensity of radiation per mol.

In view of kinetics, the radioactive decay is an actual first-order reaction. A given atom of a radioactive element will decay within a specific period and it is the intrinsic property of that particular atom and independent of other atoms of the element present, temperature, pressure, etc. In case of a homogeneous statistically large sample of the atoms, a constant fraction will decay in each succeeding unit of time. The "fundamental law of radioactive decay" may be calculated in the following way.

Let N = the number of atoms

ΔN = radioactive disintegration of change

ΔN = unit time is related to number of atoms

$\frac{\Delta N}{\Delta t} = -KN$ (the – sign represents decrease)

where K = the disintegration of "decay constant" and represents the fraction of all atoms undergoing decay per unit time.

Thus, from the decay constant, it is possible to calculate the fraction of radioactive atoms undergoing decay in any unit of time.

8.2.9 Decay rate when two or more radioactive isotopes present

The decay rate of a radioactive material may be complicated in presence of two or more radioactive isotopes. Each of these isotopes decays according to its characteristic decay constant. The total activity of the mixture varies with time and it is the resultant effect of the two individuals.

8.2.10 Decay rate when radioactive element produces a daughter element

When a daughter element is produced which is itself radioactive during the decay of a radioactive element, another complication of decay-rate arises. This occurs as under:

(a) If the half-life of the daughter element is short compared to that of the parent element, the activity of the daughter increases gradually over several half-lives of the daughter and after that is determined by the activity of the parent.

(b) If the daughter elements's half-life is of the same order of magnitude as that of the parent element or longer than that of the parent element, the daughter element's activity increases slowly and finally decays at the rate indicated by the longer half-life. In each case, the activity of the daughter element adds to the activity of the parent element,

(c) When the decay occurs in several steps, each of which produces radioactive substances, the situation is further complicated.

8.2.11 Disintegration schemes

Usually the disintegration of an unstable nucleus is not a simple matter of single event emission of radiant energies. The disintegration product, i.e., daughter nuclei may itself be radioactive and may decay with emission of an energy entirely different from that of the parent.

8.2.12 Unit of measurement of radioactivity

The primary unit of measurement of radioactivity is "curie" and "rutherford". Both are defined as some number of nuclear disintegrations per second.

curie: Is the number of nuclear disintegrations per second of the quantity of radon in equilibrium with a gram of radium, namely 3.70×10^{10} disintegration per second.

millicurie: Represents 3.7×10^4 radioactive disintegrations per second.

microcurie: Represents 3.7×10^4 radioactive disintegrations per second. The rutherford is 10^6 radioactive disintegrations per second; 3.7×10^4 rutherfords = 1 curie.

8.2.13 Types of radioactivity

I. **Natural Radioactivity**: A number of heavy elements are naturally unstable. They disintegrate continuously and emit alpha particles, i.e., helium nuclei of beta particles, i.e., electrons atom by atom. This process is known as **natural radioactive**. Most of the naturally radioactive elements have atomic number between 18 and 92.

II. **Artificial Radioactivity**: If the atoms some substances are bombarded with high-energy particles, and by the integration of the bombarding particles and the nuclear components, they create an unstable nucleus which will subsequently disintegrate with the emission of radiant energies. This process of production of radioactivity is known as **artificial radioactivity.**

PART III. Isotopes

8.3 Isotopes

8.3.1 Mass number

Mass number is equal to the sum of the number of its nucleons, i.e., neutrons and protons.

8.3.2 Atomic number

The atomic number of an element is equal to the free positive units of electric charge in the nucleus of its atom and hence, is also equal to its serial number in the periodic table.

8.3.3 Isotopes, Isobars, Isotones

1. **Isotopes**: Nuclides that have the same atomic number (e.g., thallium 201 and thallium 204).
2. **Isobars**: Nuclides that have the same mass number (e.g., iodine 131 and xenon 131)
3. **Isotones**: Nuclides that have the some neutron numbers (e.g., $^{201}T_{120}$ and $^{202}Pb_{120}$)
4. **Isomers**: Nuclides that have the same atomic number and the same mass number but different energy state.

Different specimens of the same element which have **different atomic** but **same number** and therefore, occupy the same place in the periodic table. (Greek, Isos = same; topes = place or position).

It differs from **isobars** and **isotones.** Isobars are the elements with the same atomic mass but having different atomic number, e.g., ^{40}Ar; ^{40}Ca. Isotones have the same number of neutrons.

8.3.4 Characteristic

Atomic structure

1. An isotope is one of a group of atomic species, all members of which have the same number of protons in the nucleus; it differs from the other members of the group only in the number of neutrons in the nucleus.
2. Two isotopes of an element have same atomic number but different atomic mass, e.g., nitrogen (atomic number 7) is found in nature both with a mass of 14 (designated N^{14}) and with a mass of 15 (designated N^{15}). These two naturally occurring nitrogen isotopes are "stable"; their nuclei do not undergo spontaneous disintegration.
3. Isotopes of the same element have same nuclear charge, i.e., the same atomic number.
4. Isotopes of the same element occupy the same place in the periodic table.
5. Isotopes of the same element share the same chemical properties (except for small differences in the isotopes of light atoms, such as hydrogen) but differ in physical properties.
6. Some isotopes are radioactive, emitting characteristic radiations such as alpha and beta particles and gamma rays. They are detected and the amount of radiation is measured by Geiger Counter or Scintillation Counter.

Other isotopes are stable and do not emit radiations and they have a greater mass. Such isotopes are usually detected and measured by Mass Spectrometer.

8.3.5 Varieties of isotopes

I. Natural isotopes:
   (a) Stable non-radioactive
   (b) Unstable radioactive

II. Artificial radioactive isotopes or radio-isotopes.

I. Natural isotopes

They remain as mixture in all natural sources and can be isolated by careful fractionation. The isotopes of biological importance are carbon, hydrogen, nitrogen, oxygen, sulphur, chlorine, etc. Carbon is an example of having isotopes of all the above varieties. Carbon has atomic weight 12. It has isotopes with atomic weight 11, 12, 13 and 14 and designated as ^{11}C, ^{12}C, ^{13}C, ^{14}C, or C^{11}, C^{12}, C^{13}, C^{14}. It occurs in nature as a mixture of two non-radioactive stable atoms, viz., C^{12} (98.9 per cent) and C^{13} (1.1 per cent) with minute trace of the radioactive C^{14}. C^{14} has also been artificially produced by nuclear bombardment. Radioactive C^{14}, the 14 indicates its atomic weight as compared with the normal carbon atom C^{12} having atomic weight 12.

Lead is an element having a number of stable naturally occurring isotopes as it is the final product of all the three natural radioactive series. These stable isotopes of lead have also a number of radioactive isotopes, e.g., Radium B, RaD, etc. as intermediate members of each of these three radioactive series. Another example is natural Uranium which is largely U^{238} with admixture of U^{235} (0.7%).

Hydrogen has the atomic weight 1 but isotopic "heavy hydrogen" has an atomic weight 2 (H^2). Oxygen with atomic weight, 16 has isotopes with atomic weight 17, 18 (O^{17}, O^{18}). One of the important isotopes are "hydrogen isotopes" (radio hydrogen, H^2) which have special names "deuterium" or D_2O (heavy water) and "tritium" or H^3. D_2O is present to some extent in all natural samples of water (e.g., tap water contains 1 part in 9000 parts) and traces of D_2O (deuterium) and THO (radioactive tritium oxide) are present throughout the body fluids. Electrolysis dissociates H_2O but not D_2O. Though D_2O has identical chemical properties with H_2O it differs in mass, which is recognised by Mass Spectrometer.

As a general rule, the elements of odd atomic number do not occur in nature in more than two isotopic forms; whereas elements of even atomic number may occur as many isotopes, those with an even number of neutrons generally being more abundant.

II. Radioactive isotopes (Radioisotopes, Radionuclide, Radioactive nuclide)

The radioactive elements are more useful as "tracers" than the stable isotopes. Some isotopes are stable and some are unstable due to their relative number of protons and neutrons in the nuclei. Major number of natural isotopes are not radioactive but they can be made radioactive artificially by bombardment of α-particles in a Cyclotron of Uranium Pile. The nuclear disintegration of these isotopes is accompanied by the emission of β-rays (negatron, mass 5.5×10^{-4}, charge 1) and γ-rays (photons, zero mass and charge). The emitting α-radiation from the disintegrating atom is not easily detected. β-radiation and γ-radiation are usually measured by Geiger-Muller Counter or in special cases by Scintillation Counter.

The energy of the radiation is designated as electron-volt (eV), which is defined as the kinetic energy acquired by an electron when it moves across a potential gradient of 1 volt. Usually the unit MeV (million electron-volts) is used. As for example, the β-radiation of C^{14} has an energy of 0.155 MeV, and that of P^{32} an energy of 1.7 MeV. C^{14} is usually referred to as a "weak" β-emitter and P^{32} as a "strong" β-emitter. The radioactive hydrogen isotope tritium (H^3, T) is a very weak B-emitter, the energy of radiation being 0.0176 MeV.

The radioactive decay of an isotope follows the kinetics of a first-order reaction and **the time required for the loss of half of its radioactivity, i.e., half-life is characteristics of each isotope.** Half-life period means the time required for the disappearance of one-half of the original amount of the radioactive substance, e.g., half-life period of radium is 1590 years. With short-lived isotopes, care must be taken for correction for the amount of radioactivity lost by decay during the experimental investigation but in case of C^{14} which is an important isotopic market for metabolic studies does require correction. Some biologically important radioactive isotopes are listed in Table 8.3.

Table 8.3: Showing biologically important isotopes with mass number, type of radiation and half-life

Element	Mass number	Type of Radiation	Half-life
Potassium	42	β^-, γ	12.5 hours
Sodium	24	β^-, γ	15 hours
Iodine	131	β, γ	8.1 days
Phosphorus	32	β, γ	14.3 days
Iron	59	β^-	45.1 days
Sulphur	35	β^-	87.1 days
Calcium	45	β^-	164 days
Cobalt	60	β^-, γ	5.2 years
Hydrogen	3	β^-	12.5 years
Carbon	14	β^-	5570 years

8.3.6 Nuclear reactions and the production of artificial radioactivity (induced radioactivity, nuclear bombardment)

Joliot and Curie (1934) found that bombardment of boron with fast α-particles caused nuclear reaction producing an isotope of nitrogen of atomic mass 13 with liberation of neutrons. The reaction can be expressed as

$$_5B^{10}(a, n)\ _7N^{13} \quad \text{or} \quad _5B^{10} + _2He^4$$

$$(\alpha\text{-particles}) \rightarrow\ _7N^{13} + _0n^1 (\text{neutron})$$

This experimental result established the following:

(a) Discovery of **neutron**,

(b) **Transmutation of element**, i.e., bombardment of some material with particles and results in the formation of some other isotope (transmutation),

(c) Nuclear bombardment could lead to products which are themselves radioactive and are isotopes of ordinary elements.

In this nuclear reaction, the nitrogen istope ($_7N^{13}$) produced is itself radioactive and decays according to the following reaction with a half-life of about 10 minutes with emission of the shortlived positron and converts into carbon having mass $_7N^{13} \rightarrow {}_6C^{13} + \beta^+$ (positron)

In this way, radioactive isotope of an ordinary element was produced and such radioactivity is called **induced radioactivity** and the radioactive element thus produced is called **artificial radioisotope** or better to refer it, as **radioisotope,** or radionuclide or radioactive nuclide.

In recent years, large number of radioactive and heavy and light isotopes of the elements have been produced artificially through bombardment of atomic nuclei with protons ($_1H^1$), deuterons ($_1H^2$) α-particles ($_2He^4$) and neutrons ($_0n^1$).

Protons and deuterons carry a single positive charge, α-particles and two positive charges while neutrons without any charge. Neutrons enter the nuclei of bombarded atoms most readily, because being electrically neutral, they are not repelled by the positively charged nuclei, as are protons, deuterons and α-particles.

Various types of apparatus have been used to provide the bombarding particles with the required energy. Initially workers used α-particles from natural radioactive substances, and neutrons by the action of radium upon beryllium. A great achievement in the production of artificial atoms was due to Cyclotron which may be used to provide any of the bombarding particles in quantities for use. Another apparatus called Nuclear Reactor (Atomic Pile, Reacting Uranium Pile) which is used for production of materials for atom bomb, is the best source for neutrons and constitutes the major source of isotopes for biological use.

The nuclear reaction for the preparation of isotopic C^{14} by neutron bombardment of N^{14} may be represented:

$$_7N^{14} + {}_0n^1 = {}_6C^{14} + {}_1H^1 \text{ or}$$

$$_7N^{14}\ (n, p)\ {}_6C^{14}$$

The equation states that neutron of atomic number 0 and mass 1 reacts with a nitrogen atom of atomic number 7 and mass 14, to form a carbon atom of atomic number 6 and mass 14 and a proton of atomic number 1 and mass 1.

There are certain advantages and disadvantages between the Cyclotron and Nuclear Reactor as stated below:

1. The Nuclear Reactor is adapted to bombardment of materials with neutrons whereas Cyclotron is for bombardment with protons, deuterons and alpha particles.
2. Most of the short-lived isotopes are produced by the Cyclotron and longer half-life isotopes are produced by the Nuclear Reactor.
3. It is possible to obtain higher specific activities by Cyclotron bombardment than by bombardment in the Nuclear Reactor.
4. Most of the isotopes are produced in the Reactor and there are a few isotopes which cannot be formed in the Reactor. But Cyclotron is more helpful for isotope-consuming laboratory and continued to be a significant source of radioactive isotopes. Some of the radioisotope produced by Cyclotron and Nuclear Reactor are given below (Table 8.4).

Table 8.4: Radioactive isotopes produced by Cyclotron and Nuclear Reactor

Cyclotron	Nuclear Reactors	Cyclotron and Nuclear Reactor
N^{13}	H^3	C^{11}, C^{14}
Mg^{27}	O^{15}	
Mn^{52}, Mn^{54}	S^{35}	Na^{22}, Na^{24}
Fe^{35}, Fe^{59}	Ca^{45}	
Co^{56}	Co^{60}	P^{32}
Cu^{61}	Mo^{99}	K^{42}
Zn^{65}	I^{131}	Cu^{64}

PART IV. Biological Application of Radioisotope

8.4 Radioactive Isotope

The radioactive isotope sometimes is introduced into a compound called "tagged" or "labelled" compound and the process is called "tagging" or "labelling". The radioactive isotope is called "tracer" or "tracer element" because it can be traced and observed by suitable instrument and other physico-chemical properties and the technique is called "tracer technique". Sometimes, when a suitable radioisotope is not available, molecules are tagged with nonradioactive, isotope, e.g., deuterium for hydrogen, and O^{18} for oxygen and the tracing is done with Mass Spectrometer. Common radioisotopes used as tracers in metabolic studies are C^{14}, Ca^{36}, Ca^{45}, H^3, I^{131}, Fe^{59}, P^{32}, S^{35}, Co^{60}, Na^{24}, K^{42}, etc.

8.4.1 By dilution of isotope technique

1. Measurement of total circulating red blood cell volume by labelled red blood corpuscle with Cr^{51}, Fe^{35}, P^{32}.
2. Determination of total plasma volume by labelled serum albumin with I^{131}.
3. Determination of sodium space with Na^{22} or Na^{24}, total exchangeable body K^+ by K^{42}.
4. Estimation of total body water with iodoantiphyrine labelled with I^{131}, D_2O.
5. Secretion rate of cortisol using C^{14} cortisol.
6. Determination of extracellular fluid volume by radioactive inulin prepared by substituting C^{14} for one of the carbon atoms of the molecules or by C^{36} and Cl^{38} or Br^{82}.

8.4.2 Membrane-transfer measurement

1. Absorption studies across the intestinal wall, e.g., iron (Fe^{59}), Vit B_{12} (Co^{60}), fatty acid (C^{13} oleic acid) etc.
2. Transport of Na^+ and K^+ across the intestinal wall.
3. Exchange of ions in the kidney tubules.
4. Transmembrane flux (unidirectional fluxes and net fluxes) across a membrane: The net flux across a membrane is the resultant of the two unidirectional fluxes. The unidirectional flux of a substance A, from one side to another side, can be determined by introducing an isotope A* as a tracer for A and measuring the rate of appearance at another side. Similarly, from other side to the former side can be measured by introducing a second isotope A** to the substance A and their appearance in the side.

8.4.3 Distribution studies

1. Uptake of I^{131} by thyroid gland in thyroid function text.
2. Uptake of P^{32} by malignant tumours for tracing the location of tumour.
3. Distribution of drugs tagged with radioisotope (Radiopharmaceuticals).

8.4.4 Synthesis of organic compounds

Synthesis of many organic compounds containing radioisotopes or stable isotopes has been done, e.g., various isotopic glycine are shown below:

Glycine-1-C^{14}	$NH_2CH_2C^{14}OOH$
Glycine-2-C^{14}	$NH_2C^{14}H_2COOH$
Glycine-N^{15}	$N^{15}H_2CH_2COOH$

8.4.5 Metabolic studies

This has the wide application in the biological sciences. Studies on synthesis, break-down, identification, steps of reactions and its variation of carbohydrate, protein, fat, steroids and mineral metabolism.

8.4.6 Disadvantages

1. **Isotopic effect**: A labelled compound may be metabolized at a different rate than the unlabelled substance, e.g., succinate containing 77 atom per cent excess deuterium in the methylene groups is oxidized by the heart muscle succinic oxidase at about 40 per cent of the rate for the unlabelled succinate (1) and known as "isotopic effects". These "isotopic effects" are roughly propor-tional to the difference in mass between two isotopes, and are especially marked for the hydrogen isotopes (H, H^2 H^3) and less evident for the carbon isotopes (C^{12}, C^{13}, C^{14}) (2).
2. **Exchange effect**: When labelled compounds are used in the metabolic studies, it should be noticed that isotope does not exchange with more abundant form present in the environment, e.g., an amino acid labelled with deuterium ions with the hydrogen ions of the biological fluid.

$$R-ND_3^+ + 3H^+ \rightleftharpoons R-NH_3 + 3D^+$$

For this disadvantage, amino acids labelled with deuterium in $\alpha-NH_2$ group are not useful in tracer studies of metabolism.
3. **Unphysiological effect**: The tracer technique involves the introduction into the organism of substances which may be unphysiological and may be harmful for the organism. This usually occurs when there is large amount of

1. Thorn, M.B. – Biochem. J., 49, 602 (1951)
2. Glascock, R.F et al. – Biochem. J., 58, 440 (1954), Rachelo, J. R. et al. – Am. Chem. Soc., 76, 4342 (1954).

isotopic substance or high isotopic concentration.

4. **Radiation effect**: Large amounts of radioactivity may cause radiation effects which in turn may alter the metabolic behaviour. For which in experiment, the investigators use very small quantity of labelled compound.
5. Laboratory should be properly equipped for handling and assaying the isotopes.
6. laboratory must have arrangement for disposal of waste products from the experiments, e.g., C^{14} is difficult to handle for its very long half-life. If large quantities of this material are distributed indiscriminately, a public hazard may be created for which very limited laboratories can handle C^{14}.

8.4.7 Advantages

Radioisotopes are useful in biological experiments for the following reasons:

1. The intensities of radiation are quantitatively measured easily.
2. The radiations are detectable and measurable when they are present in extremely dilute concentrations.
3. The radioactive materials may be incorporated into the chemical substances of the living tissues.
4. The radiations pass through the living tissues and layers of cells and can be detected *in vivo*.

8.4.8 Criteria of radioisotope for a tracer studies

The radioisotope must conform with the following criteria:

1. The half-life should be long enough that accurate determination of its transport and alterations can be made.
2. The isotope should be in highly specific activity because after dilution required in the experiment they can be assayed accurately.
3. The isotope should be capable of incorporation into the biological molecule.
4. The isotope should not leave the molecule easily after chemical incorporation.
5. The isotope should not alter the biological system during the presence through damage of the tissues.

8.4.9 Different type of metabolic studies

1. **Study of metabolic pathway**: The cellular organic substances may be converted from $A \rightarrow B \rightarrow C \rightarrow D$. This can be studied by inhibitors to identify the intermediate products formation and their precursors. But along with this, application of isotopes has confirmed the results obtained by above method. Compounds labelled with the radioisotopes of C^{14} are fed into the metabolising tissue and the radioactive product is subsequently identified which is chemically and biologically indistinguishable from its analog, In this way, some important metabolic pathways were studied as under:
   (a) Kreb's cycle or Tricarboxylic acid cycle: Acetate labelled with carbon in the carboxyl group (CH_3C^*OOH) was found in the γ-carboxyl group of α-ketoglutaric acid and subsequently in the cis-aconitic acid, isocitric acid, etc.
   (b) β-oxidation of fatty acid.
   (c) Elongation of fatty acid.
   (d) Urea cycle.
   (e) Cholesterol formation from acetate, etc.
2. **Metabolic conversion**: Rose indicated conversion of L-phenylalanine to L-tyrosine in growing rat. Definite evidence came out from the work of Moss and Schoenheimer. They prepared a sample of DL-phenylalanine in which the hydrogen atoms of the benzene ring had been replaced by deuterium, and fed the labelled amino acid to both growing the adult rats at various time intervals. After sacrificing the animals, the tissue hydrolysate was analyzed for deuterium and it was found that benzene ring of L-phynylalanine converted to the phenol ring of L-tyrosine. But this experiment does not give any information regarding metabolic transformation of phenylalanine into tyrosine.
3. **Metabolic synthesis and precursor**:
   (a) Discovery of carbon atoms of cholesterol derived from labelled acetate.
   (b) Labelled Glycine, acetate, succinyl-COA for synthesis of porphyrins for haemoglobin.
   (c) Labelled Glycine for synthesis of purines.

4. **Metabolic pool study**: The experimental results of Schoenheimer and others have led to the concept that, in the dynamic system involving dietary amino acids, tissue proteins and urinary nitrogen, there exists a "metabolic pool" of nitrogen. Using N^{15} it was observed that mixed proteins of different tissues differ greatly in the rate at which they become labelled with N^{15} and this result was confirmed using amino acids labelled with S^{35} or with C^{14}. It was seen that several animal tissues, e.g., intestinal mucosa, kidney, pancreas, liver, spleen, bone marrow, etc. are rapidly labelled after injection of an isotopic amino acid whereas other tissues, e.g., skin, muscle, brain, become labelled much more slowly. Again, different proteins, even from the same tissue, showed differences in the rate at which they are labelled.
5. **Metabolic turnover: The dynamic state of cell constituents**: Shoenheimer observed that when amino acids enriched with the heavy isotope N^{15} were fed to animals, the labelled amino acids became incorporated into the liver proteins at a high rate even though the total amount of protein in the liver did not change. When the labelled amino acids were removed from the diet and replaced by normal or N^{14} amino acids, the previously labelled liver proteins rapidly lost their isotopic nitrogen, with no change in the total amount of liver protein. It was concluded that the proteins of the liver cell are present in a "dynamic steady state", which a relatively high rate of synthesis is exactly counter-balanced by a relatively high rate of degradation. Thus, liver proteins undergo "metabolic turnover".
6. **Study of systematic degradation of isotopic compound and formation of ring compound**:
   (a) **Purine ring**: In systematic degradation of isotopic compounds isolated from a biological system, the labelled atoms may be identified, and their relation to a labelled atom in the administered precursor may be established. Thus, the administration of C^{14} labelled formate ($HC^{14}OO^{-}$) to a pigeon leads to the metabolic formation of labelled uric acid. Chemical degradation of this product has shown that the C^{14} is largely located in carbons 2 and 8 of the purine ring.

   Similarly, studies with isotopic glycine has shown that this amino acid contributes nitrogen 7 and carbons 4 and 5 derived from carboxyl and methylene groups and that the N-C-C skeleton of glycine is incorporated as a unit (Fig. 8.5).

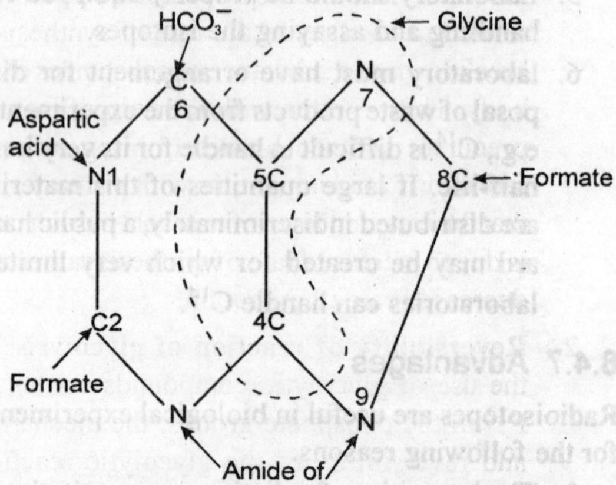

Fig. 8.5: Precursors for the biosynthesis of purine ring

   (b) **Pyrimidine ring**: Labelled aspartic acid has shown that it contributes N3, C4, C5, C6 of pyrimidine ring (Fig. 8.6).

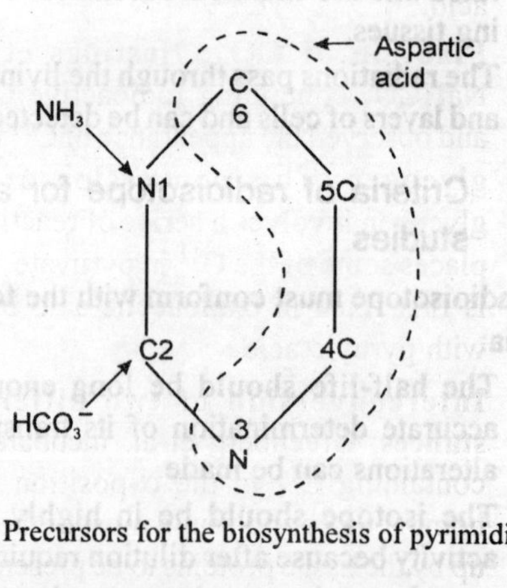

Fig. 8.6: Precursors for the biosynthesis of pyrimidine ring

8.4.10 Metabolic studies in microorganism

Microorganism is very much useful for metabolic studies with isotope because of the following reasons:

(a) Their nutritive requirements are simple,
(b) They can be grown in culture media,
(c) They can be grown in large quantities,

(d) Analysis of their metabolic products is convenient than in case of higher organisms. Thus, studies with isotopically labelled orotic acid on mutant strains of neurospora yielded labelled pyrimidines, indicating that orotate is an intermediate precursor of pyrimidines.

1. **Assessement of the cause of metabolic disease**: The metabolic disturbances are related to changes in the rate of synthesis or breakdown of a substance or substances of the body. Thus, Stetten with the help of isotopic precursor showed by determining turnover rate of uric acid in patients with primary gout, that the elevated level of uric acid in such cases is due to increased rate of synthesis rather than decreased rate of excretion.

2. **Reversibility of reaction of glycolysis**: By the use of glycolytic compounds containing P^{32} in the phosphate groups, the mechanism and reversibility of the glycolytic reactions have been demonstrated.

3. **Vitamin synthesis from carbohydrate**: The animal is capable of synthesizing ascorbic acid. Carbon – 1 in glucose –1-C^{14} becomes carbon –6 in ascorbic acid and carbon –6 in glucose –6-C^{14} becomes carbon –1 in ascorbic acid.

4. **Fixation of CO_2**: Hastings et al. gave $NaHCO_3$ containing radioactive C^{11} to rats and observed the appearance of C^{11} in the liver glycogen. This conversion of C^*C_2 to glycogen involves a series of reactions which places some of the C^{11} in pyruvate. The C^*C_2 is first fixed in oxaloacetic acid by reaction with pyruvic acid.

5. **Interconversion of two different substances**: Greenberg et al. incubated glycine containing C^{14} in the α-position with liver homogenates and found most of the C^{14} taken up by the tissue proteins to be present in serine, indicating conversion of glycine to serine. Sakami, gave glycine labelled with C^{13} in the carboxyl group and formate labelled with C^{14} to rats simultaneously. The serine isolated from liver proteins contained a large amount of C^{13} in the carboxyl group and C^{14} in the β-carbon position which indicates serine is formed by interaction of glycine and formate.

6. **Conversion of D-isomer to L-isomer**: Schayer gave rats D-tryptophan with N^{15} in the indole N and found L-tryptophan similarly labelled in the tissue proteins, proving the conversion of the D to the L-form.

7. **Breaking of ring structure during metabolism**: Rittenberg fed histidine labelled with N^{15} in the imidazole ring to rats and found N^{15} to be distributed among the tissue amino acids and other substances very much as found in the case of the nitrogen of ammonia and the α-amino nitrogen of other amino acids when they are fed. This revealed that the imidazole ring is broken up during metabolism.

8. **Enzymatic study as a rate limiting**: Measurements of the labelled intermediates following incubation of liver slices in the presence of Pi^{32} showed that phosphorylase, rather than phosphoglucomutase of glucose-6-phosphatase, was rate limiting.

9. **Synthesis of labile group**: Using isotopes it has been shown that the tissues can synthesize substances containing labile methyl groups (e.g. choline), if the diet contains folic acid and Vitamin B_{12}. These methyl groups take part in transmethylation process.

10. **Tissue metabolism**
    (a) Mucopolysaccharides of connective tissue exhibited very active metabolism. Radioactive S^{35} ($S^{35}O_4$ =) and glucose – C^{14} are rapidly incorporated into hyaluronic acid and chondroitin sulphate.
    (b) **Collagen fibrogenesis process**
    i. Jackson labelled the collagen of guinea pig skin and of carrageenin granulomas grown in the same animals with glycine –C^{14} and extracted the collagen in succession with increasing concentrations of NaCl– 0.14, 0.28, 0.45, 1.0., 2.0M at pH 7.4 and then with 0.2M citrate buffer at pH 3.6, and finally converted the insoluble collagen to gelatin with hot water. From the quantities of collagen and the specific activities (C^{14}) in each fraction, it was possible to formulate the process of collagen fibrogenesis.
    ii. Mitoma et al. (1957) demonstrated that fibroblasts incorporate C^{14} labelled hydroxyproline into collagen when

supplied with C^{14} labelled proline while C^{14} labelled hydroxyproline fed to rats is not incorporated into collagen (Stetten, 1949). This indicated that the hydroxyproline of collagen is formed from proline after the latter is bound in peptide linkage.

(c) **Mammary gland tissue**

Protein metabolism

i. Isotopic studies on cows and goats have shown that most amino acids for the synthesis of milk proteins are derived from blood and a small fraction of amino acids is synthesized by the mammary gland.

ii. Isotopic experiments have shown that α-lactalbumin is formed largely from amino acids of blood, the immune globulin and albumen of milk are derived unchanged from the blood. But in case of cow's colostrum, immune globulin is made by plasma cells of the gland from amino acids.

iii. **Fat metabolism**: With injection of glycerides containing labelled stearic acid into the cows, much of the labelled stearic acid appeared in milk fat, while very little of labelled stearic appeared in the fat after injection of labelled glucose or acetate. This indicated that the mammary tissue receives a large portion of long chain fatty acid from the blood. Folley et al. injected acetate into the goats and observed large proportion labelled shorter chain fatty acid in the milk fat and a small proportion of long chain fatty acid.

iv. **Carbohydrate metabolism**: When labelled C^{14} – glucose was injected into the goats or cow, both the glucose and galactose groups of the lactose become rapidly labelled. Barry observed that the C^{14} of injected carbon-1 labelled glucose appeared almost exclusively in carbon-1 of the glucose and galactose of lactose. The synthesis of lactose does not involve breaking the glucose chain of the carbon atoms.

(d) **Bone tissue**: Some radioactive isotopes may become incorporated into the bone known as **bone-seeking isotopes**, e.g., Ca^{45}, Sr^{89}, Y^{91}, Ba^{140}, Pu^{239}, Sr^{90}, P^{32}, Pu^{259}, etc. Such radioactive bone-seeking isotopes are very much useful for the study of bone metabolism and growth as their presence may be detected by Autoradiography.

The radioactive isotope Ca^{45} may substitute for the normal isotope or radioactive Sr^{90} may be found in place of calcium in the bone salt. P^{32} may also be seen incorporated into the mineral while Pu^{259} may be bound by the matrix. These radioactive isotopes irradiate the marrow and shoft parts and may eventually give rise to leukaemia or to bone tumours.

11. **Mineral Metabolism**

(a) **Iron metabolism**: Loftfield et al. administered Fe^{59} (as ferric salts) to rats, the isotope appeared rapidly in liver ferritin, and other iron-containing proteins, e.g., cytochrome b_5, catalase which were labelled more slowly.

(b) **Iodine metabolism**: The distribution, utilization and metabolism of iodine in the body have been studied after administration of tracer amounts of radioactive I^{131} (half life 8 days) or I^{132} (half-life 2.3 hours.), which mix with the unlabelled iodine but have no biological action. About one-third is taken up by the thyroid and about two-thirds excreted by the kidney. The thyroid contains 5000-8000 μg of iodine, i.e., 95 per cent of the total iodine content of the body, most of which is in the form of iodinated amino acids, i.e., iodotyrosine and iodothyronines. The large pool of iodine in the body turns over slowly, at a rate of about 1 per cent daily.

(c) **Copper metabolism**: Studies with isotopic Cu^{54} showed that ceruloplasmin copper and ionic copper exchanged only when ceruloplasmin was reduced by ascorbic acid.

8.4.11 Endocrine studies

1. **Thyroid Gland**

i. Use of radioactive iodine together with chromatography revealed the following:

(a) A unique ability to remove and concentrate blood iodine,

(b) Uptake of tracer iodine is dependent on TSH and amount of iodine used, and

(c) Iodine is held in bound form associated with protein.

ii. Comparative study of radioactivities of organic iodine compound. Taurog et al. found the time course of the formation of circulating hormone. They showed:

(a) 26 per cent of the total thyroid iodine is represented by thyroxine iodine,

(b) 15 minutes after injection of radioiodine, 95 per cent of the radioactivity was present in organic combination in the thyroid and of this about 80 per cent was found in the diiodotyrosine fraction and 10-15 per cent in the thyroxine fraction, and

(c) Progressive decrease in the diiodotyrosine fraction and an increase in the thyroxine fraction.

iii. Leblond et al. observed that one hour after the injection of radio-iodine. Radioautographs revealed the presence of protein-bound iodine in the apex of the follicular epithelia cells, with small amount in the periphery of the Colloid. After 24 hours, there was uniform distribution of organically bound radioiodine throughout the Colloid materials.

2. **Adrenal cortex**: Using C^{14} cortisol in isotope dilution method, the secretion rate of cortisol can be measured. Cortisol secretion in man ranges from 5-28 mg per day in normal condition with an average 15 mg per day. Little amount is stored in the adrenal cortex. There is continuous synthesis and release to maintain physiological concentration.

3. **Pancreas**

i. Experiments on diabetic patient, normal persons, intact animals, isolated perfused islets of Langerhan, cell and tissue cultures using labelled I^{131}, I^{125}, H^3 insulin or proinsulin or native isulin have revealed following information:

(a) Insulin is 10 to 20 times as active as proinsulin,

(b) Insulin does not produce effect on disrupted cells but it does become bound to separated cell membranes of responsive cells.

(c) All the actions of insulin are brought about by a reversible combination with specific glycoprotein receptors of cell membrane.

(d) Insulin does not penetrate into the cell, yet it exerts effects on the intracellular activities of muscle cells, liver cells and adipocytes.

(e) Its biological activity is proportional to the amount bound to the cell surface.

ii. Experimental studies of incubation of radioactive leucine or phenylalanine with islet tissue of rat pancereas or human pancreatic tumour (this type of tumour produces large quantity of insulin) showed that tumour produced two radioactive protein products, one was insulin and the other was proinsulin.

4. **Steroid hormone combination with specific receptors**

i. Jenson et al. studied the fate of very small amounts of highly radioactive estradiol injected into immature female animals. The labelled hormone became localised in the uterus and vagina, the targets of oestrogen action but without any chemical change. The uptake of labelled oestrogen closely paralleled the timing of the earliest detectable hormonal effects of oestrogen on these targets.

ii. Incubation of minute amounts of labelled oestradiol with intact uterine tissue or homogenates of the uterus revealed that the hormone became bound in two cells fractions, the cytosol fraction and nuclear fraction. The oestradiol in the soluble cytosol fraction binds to a specific oestrogen-binding protein. The labelled oestrogen found in the nuclear fraction has been localized in the chromatin.

8.4.12 Cellular studies

1. Cell respiration analysis using C^{14}.
2. DNA replication analysis using N^{14} and N^{15} known as Meselson-Stahl (1950) experiment giving unequivocal support to the semi-conservative (Zip-fastner) hypothesis. Cells of *E. coli* were grown for any generation on a medium in which normal nitrogen N^{14} was replaced by the heavy isotope N^{15}. Growth continued until it was certain that heavy nitrogen had been incorporated into all DNA.

After that, the heavy nitrogen N^{15} was replaced by normal nitrogen N^{14}. Samples of bacteria were taken at regular intervals. The relative amounts of two types of nitrogen were estimated by the technique depending on the heavier character of N^{15} than the lighter N^{14}. The light and heavy DNA were separated by ultracentrifuge and DNA detected by the absorption of ultraviolet light. The result revealed that DNA of the first generation contained equal amounts of heavy and light nitrogen. In the second generation, two types of DNA were detected, half was identical with the first generation and the other half contained no heavy nitrogen at all.

3. Location of synthesis and storage by different organelles, e.g., Golgi body, endoplasmic reticulum, mitochondria.

(a) **Golgi body and endoplasmic reticulum**: Experiments with distribution of radioisotope labelled substances taken up by the cell was followed by Autoradiography. The experiments revealed that cell secretions are glycoproteins. Newly synthesized proteins are found in the channels of the rough endoplasmic reticulum and then they move to the Golgi body where the carbohydrate molecule is added. Enveloped in part of the Golgi membrane, they leave the cell. Golgi body is the assembly centre where the carbohydrate is added to the protein and the finished product packaged before it leaves the cell.

(b) **Mitochondria**: Kreb's cycle (Tricarboxylic acid cycle) takes place within the inner compartment of the mitochondria. This cycle accounts quantitatively for the oxidation of carbohydrate, fatty acids and amino acids has been verified by isotopic test employing precursors and intermediates of the cycle in which specific carbon atoms were isotopically labelled with either the C^{13} or C^{14} isotopes.

4. **Movements of labelled molecules within the cell and labelled cells in tissues**: If tissue samples are taken at regular intervals after the injection of the labelled compound or labelled cells and Autoradiography made, movement of the labelled compound can be traced from one part of a cell or tissue to another part. For example, labelled amino acid is first seen in the basal region of a pancreatic cell, the principal sites of protein synthesis. After sometime, the radioactivity is predominantly seen in the apical region where the precursors of enzyme accumulate in the form of granules prior to their release to outisde. It may be concluded that enzyme synthesis first starts in the basal zone and then is transported to the apical pole of the cell.

5. **Life span of the Cell**: The life span of RBC can be determined by the method of "tagging" red blood cells with radioactive chromium. After withdrawal of few ml of blood, $Na_2Cr^{51}O_4$ is added to it and the cells are washed.

The cells are returned to the circulation and the radiations are counted at intervals. The radioactive salt is so strongly absorbed on the red blood cells that little elution occurs after injection. When a curve of loss of radioactivity is drawn, it is found to be linear falling to zero at about 120 days, the half-life of the cell being about 60 days.

- **Tissue Studies**

6. **Nature of tissue and histamine formation**: Kahlson et al. (1960) have demonstrated the histamine-forming capacity of rapidly growing embryonic tissues and in healing wounds. They have measured the amount of C^{14} – histamine produced from C^{14} -histidine both in *vitro* and *in vivo* and have shown a low content of histamine because it is being formed and disposed of continuously at a very high rate. This suggests that histamine has some role in initiating and sustaining rapid tissue growth.

7. **Brain tissue (visual cortex)**

(a) **Orientation columns of visual cortex**: The orientation columns in the visual cortex can be mapped with the help of a radioactive 2-deoxy-glucose. The uptake of this glucose derivative is proportional to the activity of neurons. When this method is employed in animals exposed to uniformly oriented visual stimuli, e.g., vertical lines, the visual cortex is shown as array of intricately curved but evenly spaced orientation columns over a large area.

(b) **Ocular dominance columns of the visual cortex**: The ocular dominance columns can be mapped by injecting a large amount of radioactive amino acid into one eye of monkey. The amino acid is incorporated into protein and transported by axoplasmic flow to the ganglion cell terminals, across the geniculate synapses, and along the geniculocalcrine fibres to the visual cortex. Layer IV becomes evenly labelled but above and below this layer in the cortex, labelled columns 0.8 mm in diameter alternate with unlabelled columns receiving input from the uninjected eye. The result is a vivid pattern of stripes that covers much of the visual cortex and is separate from and independent of the gird of orientation columns.

(c) **Brain areas activated by visual stimuli**: The brain areas activated by visual stimuli have been investigated in monkeys with the help of radioactive 2-deoxyglucose. Activation occurs in the occipital lobe, inferior temporal cortex, posteroinferior parietal cortex and portions of the frontal lobe. The subcortical structures activated are: the lateral geniculate body, superior colliculus, pulvinar, caudate nucleus, putamem, claustrum and amygdala. These findings support the other evidences of existence of occipito-temporo-prefrontal pathway concerned with object vision and an occipito-parieto-prefrontal pathway concerned with spatial vision.

(d) **Spatial pattern of increased metabolic activity in the olfactory bulb by radioactive 2-deoxyglucose method**: Glucose uptake is increased in active neurons, and 2-deocyglucose is taken up like glucose but is not metabolized. For this reason the radioactive 2-deoxyglucose is used for mapping the activity of neurons on a microscopic scale. By this method it has been shown that different odours produce different spatial patterns of incresed metabolic activity in the olfactory bulb. Further, it was shown that particular odours have produced localized increase in metabolism in particular portions of the olfactory cortex.

8.4.13 In plant physiology

1. Analysis of mineral salt uptake by plants.
2. Analysis of nitrogen-fixation.
3. Analysis of photosynthesis.
4. Analysis of translocation.
5. Analysis of virus life cycle.

8.4.14 Diagnostic use

1. **I^{131} for thyroid function**: Iodine is taken up by the thyroid gland. If the gland is nonfunctional, the uptake by the gland would be much less than normal and if the gland is hyper-functional, as in thyroid tumour, the uptake would be greatly increased. Iodine is administered as a solution of potassium iodide and time is allowed for its distribution. A small Geiger-Muller counter, shielded so that the tube occupies only a small angle of space, is then placed at various points near the surface of the neck over and around the thyroid gland.
2. **Localization of brain tumours using the dye fluorescein labelled with I^{131}.**
3. **Localization of breast and other tumours using P^{32}.**
4. **Radioactive clearance studies for renal function test**: Several substances containing iodine in their molecules e.g. diodrast, Hippuran and Iopax are prepared with radioactive iodine or other radionuclide, one can measure the radioactivity from both renal pelvis by placing radioactivity detection counter over the area of the kidneys. As very small amount of the substance is injected intravenously and the degree of radioactivity recorded during the following few minutes to determine roughly the renal clearances.
5. **Measurement of cerebral blood flow and effect of cerebral activity on the blood flow using radioactive xenon in conscious man (Xenon wash out technique).**

(a) Recently a method has been developed to record blood flow in 256 isolated segments of the conscious human cerebral cortex simultaneously. Radioactive Xenon (Xe^{133}) gas dissolved in saline is injected into the carotid artery, then the radioactivity of the cortex passing through the brain tissue is recorded. The arrival and clearance of the gas in various regions are then monitored by 256 small radioactive Scintillation Counters which are focussed on the same number of separate parts of the cortex. Each detector is collimated to scan about 1 cm^2 of brain surface. The rate of decay of the radioactivity after it once appears in each tissue segment is a direct measure of the rate of blood flow through the segment. The output from the detectors is processed in a computer and displayed on a colour television screen in such a way that the colour corresponding to the location of each detector is proportionate to the flow it is detecting. With this technique, only the radioactivity from the cerebral cortex is detected. Even by this method, it has been shown clearly the change in the blood flow in each individual segment in response to change in local neuronal activity, e.g., reading of book, clasping the hand. It is also used to localize the origin of epileptic attacks.

(b) **2-deoxyglucose method with Positron Emission Tomography in conscious man**: Another method for measuring blood flow in different parts of the brain in conscious man is the 2-deoxyglucose method combined with Positron Emission tomography. Alterations in blood flow are related to alterations in metabolism and is detected by 2-deoxyglucose method. And if the 2-deoxyglucose is labelled with the positron emitter F^{18}, its concentration in various parts of the brain can be determined through the intact skull by Tomography.

(c) **Xe^{133} technique in brain diseases**: Employing Xe^{133} method for the study of various brain diseases in man, the following results were obtained.

Epileptic foci: hyperaemic (high bloodflow).
Memory deficits: decreased temporal lobe flow.
Agnostic symptoms: decreased parieto-occipital flow.

Senile dementia: patchy decrease in flow, proportionate to the degree of dementia.

Chronic Schizophreenia: decreased blood flow at rest in the dominant hemisphere.

8.4.15 Cardiovascular studies

1. **Radioactive microsphere technique for measurement of coronary blood flow in experimental study**: The method is useful for the heart because it allows determination of "regional myocardial blood flow", including the "distribution" of blood flow across the ventricular wall. For coronary blood flow, a batch of radioactive microspheres measuring 9 to 15 μm in diameter is suspended in a saline-detergent solution and injected into the left atrium, they are mixed with blood in the left ventricle and ejected by the aorta and then enter the coronary arteries. The microspheres lodge in a few capillaries only without any damage or disturbance in blood flow. But the number of spheres trapped per unit of myocardial tissue is directly proportional to the myocardial blood flow to that region. After one or more injections the experiment is stopped, the heart is removed and the radioactivity in the samples of myocardium is directly measured in a Gamma Scintillation Counter.
2. **Dilution technique for determination of actual amount of blood in heart and lungs in experimental study**: Determination of the actual amount of blood in the heart and lungs by injecting cells tagged with Cr^{51} and then calculating the local blood volume from the total radioactivity in the blended homogenate of these organs.
3. **Isotopic clearance for blood flow in organ**: A small amount of radioactive ion, e.g., Na^{24} or I^{131} injected into a tissue like muscle and the rate of clearance from the area is then followed by a counting device. It is not a reliable one for measuring blood flow because the rate of clearance does not parallel the blood flow simultaneously measured.
4. **Cerebral blood flow in experimental animals**: By determination of the distribution of

an inert radioactive gas in frozen sections of the brain of experimental animal and comparison of these values with the level of the gas in the blood, the regional blood flow in the brain can be measured. This method reveals that the flow in cerebral and cerebellar cortex is high but the inferior colliculus has the highest blood flow per gram. The blood flow in gray matter is about 6 times the flow in white matter.

8.4.16 Organ scan

Certain substances have an affinity or specificity for certain organs. By tagging these substances with radioactive tracers, an idea of the anatomy and physiology of the organ, which is the principle of the organ scan. A variety of radiopharmaceuticals (e.g., 99 m technetium tagged to phytate for liver scan) is used for scanning depending upon the organ is to be investigated. Assessment of function of the organ has been improved due to introduction of computer technology. The organs which can be scanned are liver, heart, kidney, bone, brain, etc.

Tracer isotopes and radiopharmaceuticals for organ scan are:

1. Liver: 90 m technetium tagged to phytate.
2. Bone: Radiocerium, radiogallium, radiosulphur, radium, technetium phosphate.
3. Brain: 99 m T.C. pertechnetate.
4. Kidney: Technetium 99 m DTPA (Diethylene triamine penta acetic acid).

A. Positron Emission Tomography (PET)

Positron Emission Tomography (PET) method utilized the positron emission radioisotopes. Gamma ray-emitting radioisotopes of the elements which make up the organic molecules are required to label selected biochemical or pharmacological compounds. Radionuclides like oxygen 15(O^{15}), nitrogen 13 (N^{13}), carbon 11(C^{11}) and fluorine 18 (F^{18}) have very short half-lives and are used in this technique. This technique provides valuable information on regional tissue biochemistry, physiology and pharmacology; as for example, in the *in vivo* quantitative measurement of regional blood flow, oxygen consumption, glucose utilization and incorporation of selected amino acids in the tissues.

B. Cardiac imaging

The principle of organ imaging is to detect the presence, concentration and distribution of a radiopharmaceutical after administration into the system. Cardiac imaging gives information of the myocardial perfusion, myocardial ischaemia and the ventricular function. The isotope tracers most commonly used are:

(a) Myocardial perfusion imaging: Thallium,
(b) Myocardial infarct imaging: Thallium,
(c) Multiple Gated Aquisition Studies (MUGA): Technetium 99 m tagged to the patient's own RBCs or serum albumin.

C. Myocardial Scintigraphy

Radionuclide technique is used to detect coronary artery disease and to detect, localize and estimate the size of acute myocardial infarcts. Gamma-emiting radionuclides are used, since these are readily available and applicable, e.g., potassium and its analogues, Thallous Chloride ($Tl^{201}Cl$). Thallous^{201} Chloride is the imaging agent of choice because the myocardium actively and rapidly absorbs Thallium^{201} in proportion to its delivery. If there is significant stenosis limiting the coronary flow, the affected area of myocardium receives less thallium ^{201} and is detected as "cold spot" on scanning. More specific in detecting myocardial infarcts is Technetium TC 99 m pyrophosphate myocardial imaging, which shows increased uptake in areas of necrosis known as "hot spot".

8.4.17 Immunological

1. **Radioactive binding technique for immunoassay**

A. Measurement of antibody

These methods assess antibody level either by determining the capacity of an antiserum to complex with radioactive antigen or by measuring the amount of immunoglobulin binding to an insoluble antigen preparation.

(a) Using radioactive antigen, the methods (the Farr method, the antiglobulin coprecipitation method) involve the addition of excess radiolabelled antigen to the antiserum followed by assessment of the amount of antigen which has been complexed with antibody.
(b) Using insoluble antigen, the antibody content of a serum can be assessed by the capacity to bind to antigen which has been insolubilized either by coupling to an immunoadsorbent or by physical adsorption to a plastic tube, the bound estimated by

addition of a radio-labelled anit-Ig raised in another species.

B. Measurement of antigen

(a) **Radioimmunoassay**: A method in which immunological properties of a substance are utilized for its detection and quantification is known as immunoassay. When isotope tracers are used in this assay it is called Radioimmunoassay. This is the most sensitive method for estimation of antigen. It has been extensively used for the assay of hormones (Polypeptide). This method is capable of measuring as little as 10^{-12}m/ml of anitgen.

(b) **Immunoradiometric assay**: Immunoradiometric assay differs from radioimmunoassay in the sense that the labelled reagent is used in excess. For the estimation of antigen, antibodies are coated into a solid surface of plastic and the test antigen solution added; after washing, the amount of antigen bound to the plastic can be estimated by adding an excess radiolabelled antibody.

(c) **Radioimmunosorbent test (RIST)**: This is widely used for IgE. Rabbit anti-IgE is coupled by cyanogen bromide to microcrystalline cellulose and reacted with dilutions of an IgE–containing standard serum or the patient's serum under test. Bound IgE is, then measured by addition of labelled anti-IgE.

- **Bacteriological application (Disinfection):**

Ionizing radiation is one of the physical methods of sterilization. Gamma rays from radioactive isotope Co^{60}, which is available in substaintial quantities as by-products of atomic energy installation, are applied. For mechanism of action and use, see X-rays.

8.4.18 Radioautography

Radioautographic technique is considered as a development of histochemistry. If a radioisotope of iodine (I^{131}) is injected into an animal, it accumulates in the thyroid cells. It can be localized by making histological section of thyroid gland and covering it with a layer of photographic emulsion and subsequent development, fixation and staining show localized blackening of the emulsion due to radiations emitted by the I^{131} and it reveals the concentration of I^{131} in the colloid of thyroid vesicles. Similarly, P^{32} has been used for the studies of the demonstration of the movement of phosphate in the transpiration stream of plants and phosphate turnover in bone, etc. But the difficulty arises for obtaining good resolution in the technique because the radiations are not strictly confined to the point from which they are emitted, so poor resolution may be obtained due to thicker histological section and powerful radiation. For this reasons, ultra-thin section and weak radiating isotope like hydrogen isotope, tritium (H^3) are being used. As for example, when tritium labelled amino acid is injected into an animal, it is rapidly incorporated into cell's proteins and the location of this protein within the cell can be identified by the silver granules liberated by the radiation. The thinness of the sections allows examination by Electron-microscopy. Ocular dominance columns of the visual cortex have been demonstrated by Radioautography following injection 2-deoxy-glucose with one eye of the subjected closed.

PART V. Nuclear Radiation Therapy

A. Interstitial Therapy

1. Cancer therapy

Cancer can be treated by implanting into the cancer a piece of radioactive material which led to development of interstitial therapy.

(a) *Radium or radon gas*: Interstitial therapy with radium is available either as radium salt or as radon gas, enclosed in needles which may be inserted into the cancerous tissue. Radon has a relatively short half-life, so that radon implants may be left in permanently. Temporary implants may be either radon or radium.

(b) *Cobalt*: Cobalt (Co^{60}), having half-life 5.3 years, is available in large quantities from the Nuclear Reactors. Radium salts and radon are much more difficult to handle than cobalt which is malleables. The cobalt (Co^{60}) is used in the form of needles which can be inserted into the cancerous tissue area.

Co^{60} is also being used as a source of high energy gamma radiation for external use in the same manner as Supervoltage X-rays.

B. Distribution Therapy: Some cancerous tissue preferentially absorbs certain elements and makes it possible to irradiate these cancer tissues internally by administering large quantities of radioactive isotope to the patient. The dangers of damage to surrounding tissue are largely obviated by applying this method.

1. Radioiodine in toxic goitre.
2. Radiophosphorus in the treatment of polycythaemia rubra, vera and some chronic leukaemias.

C. Energy from radioactive isotope for heart pacemakers: A continuous powerful long acting but compact energy source is required for the batteries of heart "pacemakers". This energy comes from the decay of unstable radioactive isotopes. This type of energy is also used for scientific apparatus in space vehicles.

PART VI. IMAGING SYSTEM

8.6.1 What is medical imaging?

Medical imaging is the procedure to obtain information about the internal body organs or the bones and also to demonstrate pathological lesions within them to examine the physical state which can not externally observed.

8.6.2 Principle

Image is a representation of some object and its nature is indirect and its form is different from the original state, **usually two dimensional**. From the observation it can be evaluated the characteristics in the image, deviation from the normal and how they are derived.

8.6.3 Types of imaging modalities

Various imaging modalities are used:

1. **Photon**: X-ray, CT scan, DSA, mammography
2. **Ultrasound**: Scanner, echocardiograph, ultrasonic EEG, ultrasonic ophthalmograph, ultrasonic blood flow meter.
3. **Thermal**, 4. **Magnetic field**, 5. **Radionuclide**
6. **Electrical Impedance Tomography**.

These methods are used to obtain characteristics of the body being examined. Then the characteristics are evaluated and transformed into a form of picture which indirectly gives information about the pathology of the body.

8.6.4 Imaging techniques

There are different techniques to provide images of human internal organs.

The techniques are:

I. **Methods using ionizing radiation (invasive):** Ionizing radiation in large doses has dangers including cancer treatment and local tissue damage but the amounts used in modern imaging practice are minimum not so much hazardous.

II. **Non-ionizing methods (non-invasive)**

1. Magnetic resonance imaging (MRI)
2. Ultrasound Imaging
   (a) Ultrasound Scanner
   (b) Doppler Ultrasound
   (c) Colour Doppler ultrasound
   (d) Pulse wave Doppler
   (e) Duplex Doppler Scanner
   (f) Endosonography: (i) Transesophageal, (ii) Transvaginal
   (g) Echocardiography
   (h) Echoeneephalography
   (i) Echoophthalmography
3. Magnetic Resonance Microscopy (MRM)
4. Electrical Impedance Tomography
5. Positron Emission Tomography (PET)
6. Single Photon Emission Tomography
7. Gamma Camera.

8.6.5 Cardiac imaging

I. Invasive

1. Single Photon Emission tomography (SPECT)
2. Perfusion imaging
   (a) Exercise perfusion scintigraphy
   (b) Exercise Perfusion Imaging
   (c) Exercise Scintigraphy
3. Myocardial Perfusion Scintigraphy
4. Single-Photon Metabolic Imaging
5. Clot Detection by Imaging Technique
6. Indium III-Labeled Lymphocy Imaging to detect cardiac rejection
7. Atherosclerosis Imaging
8. The Multiwire Gamma camera
9. Myocardial Infact Imaging (using Thallium Technetium 99 m in stannous pyrophosphate)

10. Multiple gated Acquisition studies (MuGa) (using Technetium 99 m)

II. Non-invasive

1. 2-dimensional Echocardiography
2. Ultrafast Computed Tomography
3. Magnetic Resonance Imaging
4. Positron Emission Tomography.

8.6.6 Salient features of imaging process

1. **Imaging modalities**: Photon, ultrasonic, thermal, electrical impedance.
2. **In principle**: Is a representation of some objects.
3. **Nature**: Indirect.
4. **Form**: Different from original incarnation.
5. Images use in medical application—Two dimensionally represented.
6. Evaluation from examination of the representation—Certain characteristics in the image such as:
   (a) How they are derived or
   (b) How they deviate from the normal
7. **Use**: 1. To examine unusual growth which may be malignant
   2. To examine bone structure which has been physically damaged
   3. To find out damaged blood vessels in the body part
   4. To find out damaged brain tissue.
8. **Purpose**: For information about the internal organs or damaged structure which can not be visualised e.g., brain, bone, blood vessel etc. helps for diagnosis and correction.

PART VII. X-rays

8.7.1 Historical background

Rontgen (1895) found that some photographic plates wrapped with paper became fogged. He tried to find out the cause for fogging. He traced a gas-discharge tube which he was using with a low pressure and high voltage. This tube emitted a radiation that could penetrate the wrapping paper. Later on, it was found that it can even penetrate wood, glass, rubber and aluminium sheet having centimetre one and a half thickness. Rontgen could not find out the reason whether the radiation was due to stream of particles or waves and he referred it as X-rays.

8.7.2 Nature of X-rays

The X-rays are similar to the nuclear γ-radiation but regarded as electromagnetic waves, similar to light waves, but of much shorter wavelength; about 10^{-10}m, or 0.1–1 Angstrom unit and having less energy. These radiations come not from the nuclei of atoms but from the inner electron shells (Fig 8.7).

8.7.3 Production

X-rays are produced by bombarding matter with electrons. The production of X-rays is, therefore, the inverse process to the photoelectric effects, where electrons are ejected from metals by incident light waves. X-rays are produced when fast electrons or cathode rays are emitted from the heated cathode and are accelerated toward the anode. The electrons are provided by thermionic emission from a white-hot tungsten filament. The kinetic energy of electron is transferred into electromagnetic energy. The X-ray tube does not contain gas or as little as high-vacuum technique is present; the pressure is about 10^{-5} mm Hg. Because there is very little gas, the electrons do not lose energy in ionizing atoms on their way to the anode. From the A.C. mains, the transformers supply about 10 volts for heating the filament and about 100000 volts for accelerating the electrons. X-rays are produced in specially constructed glass tube which comprises the following:

(a) A source for the production of electrons (collision), (b) Energy source to accelerate the electrons (collision of electrons), (c) A free electron path (vacuum), (d) A means of focussing the electron (focussing step), (e) A device to stop the electron (Anode).

Most of the modern tubes are enclosed within metal casings in which the space in between the casing and the tube is filled with insulating oil. The anode is cooled by the radiation of heat from it to the oil and the outer casing from where the heat is dissipated due to radiation and convection to the air. The electrons strike a target such as the walls or anode. The anode is usually constructed of some metal which conducts heat from the bombarded area rapidly. The target in the X-ray tube may be tungsten which has a high melting point. The heat generated

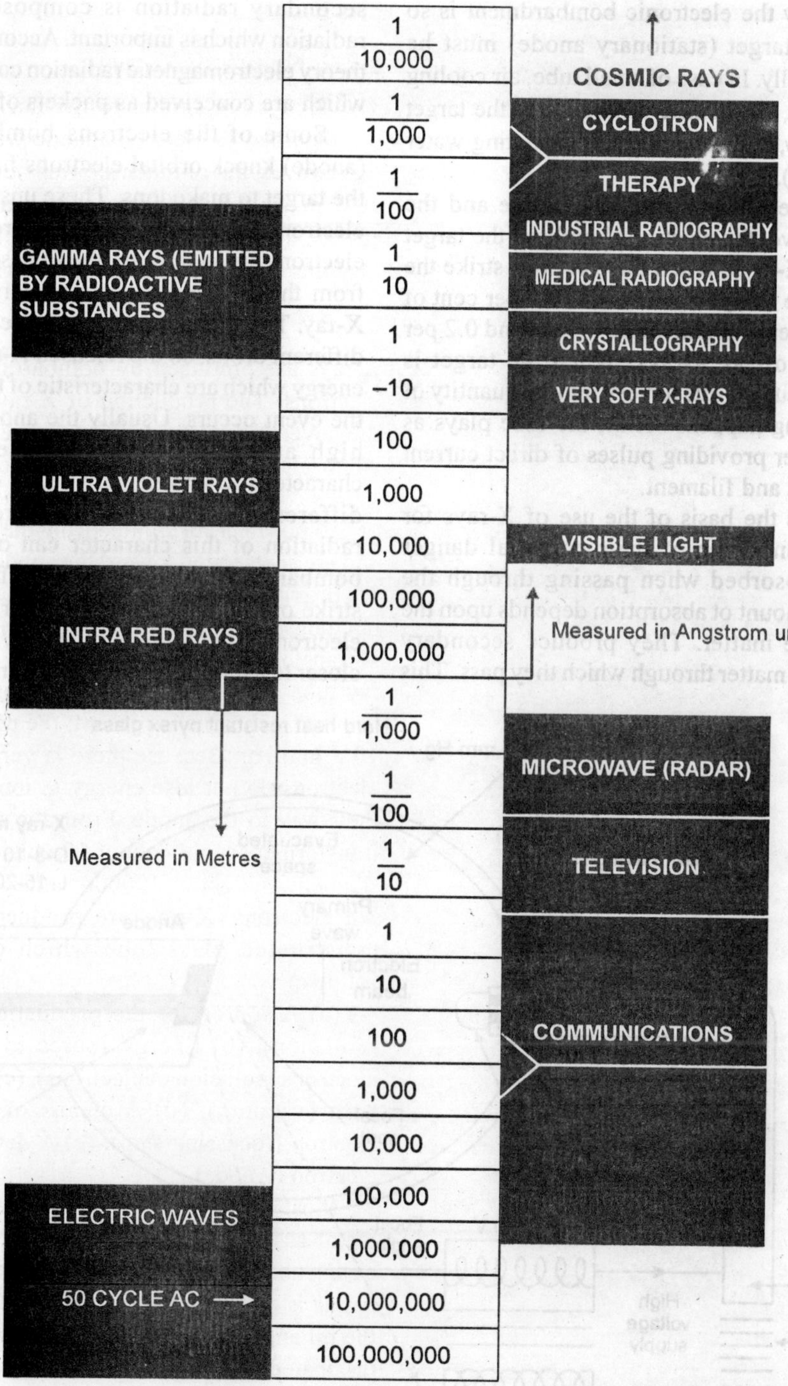

Fig. 8.7: The electromagnetic spectrum

at the target by the electronic bombardment is so high that the target (stationary anode) must be cooled artificially. In case of small tube, air cooling system is there, but in case of large tube the target is made hollow, and is cooled by circulating water or oil (Fig. 8.8).

When the electron strikes the anode and the target is positive, the electrons bombard the target and generate X-rays. When the electrons strike the anode, they are stopped and about 99.8 per cent of the electron energy is converted to heat and 0.2 per cent is given off as X-rays. When the target is negative, it is cut-off due to very minute quantity of gas and nothing happens. Thus, the tube plays as its own rectifier providing pulses of direct current between target and filament.

This forms the basis of the use of X-rays for radiography and even of their potential danger X-rays are absorbed when passing through the matter. The amount ot absorption depends upon the density of the matter. They produce secondary radiation in all matter through which they pass. This secondary radiation is composed of scattered radiation which is important. According to quantum theory electromagnetic radiation consists of photons which are conceived as packets of energy.

Some of the electrons bombard the target, (anode) knock orbital electrons from the atoms of the target to make ions. These unstable ions attract electrons to fill gaps which is created. When the electrons are in the gap, the release binding energy from the gap and the same energy is emitted as X-ray. The binding energy of electrons varies with different orbits, so the released X-ray possesses the energy which are characteristic of the orbit in which the event occurs. Usually the anode material is of high atomic number, and there are several characteristic wavelengths; each corresponds to a different atomic orbit. The production of the radiation of this character can occur only if the bombarding electrons have sufficient energy to strike orbital electrons from their positions. Other electrons lose their energy, partly when they come closer to a nucleus of orbital electrons, but continue

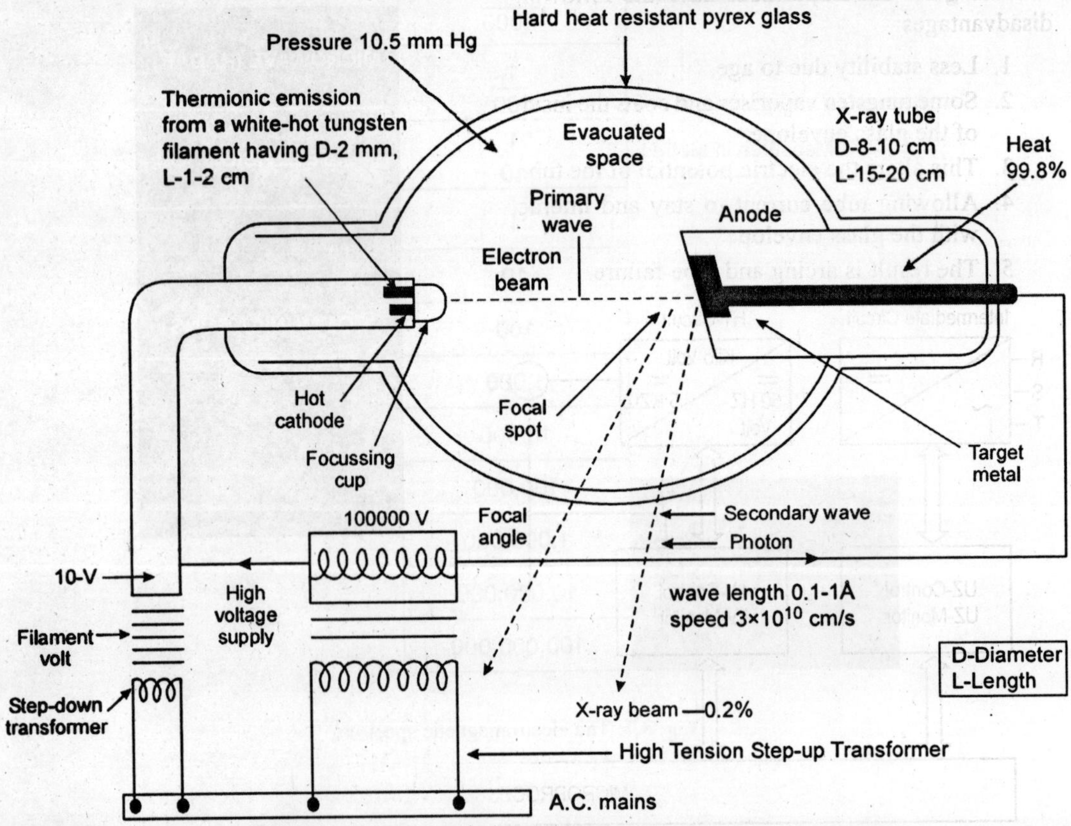

Fig. 8.8: Schematic diagram of a common X-ray tube, D-Diameter, L-Length

on with less energy. A photon of X-ray with the energy lost by the electron is ejected. The photon resulting from the reduction of energy has less energy than the full energy of the incident electron. The distribution of photon energies in a particular bombarding system is comparable to the energy spectrum of a beam from an X-ray tube. This is known as continuous, white **general radiation** from the X-ray tube. X-rays are photons of electro-magnetic energy and the wavelength associated with these photons is a function of the energy and the velocity according to relationship:

$$\lambda = \frac{hc}{E}$$

λ = the wavelength of the radiation in cm

h = Planck's constant,

c = the velocity of light,

E = the energy of the photon.

8.7.4 Metal X-ray tube

A recent improvement in tube design, incor-porates metal rather than glass as part or all of the envelope. because **glass envelope** tubes have the following disadvantages:

1. Less stability due to age.
2. Some tungsten vaporises and coats the inside of the glass envelope.
3. This alters the electric potential of the tube.
4. Allowing tube current to stay and interact with the glass envelope.
5. The result is arcing and tube failure.

Whereas **metal envelope tubes** have the following advantages:

1. Maintains a constant electric potential between electrons of the tube current and and the envelope.
2. They have longer life.
3. Less likely to fail.
4. Virtually all high-output X-ray tubes now use metal envelops.

8.7.5 Circuits

There are two Circuits:

(a) Tube circuit, (b) Filament circuits.

1. For producing high voltages, applied to the tubes anode and cathode and comprises a high voltage step-up transformer followed by rectification current passing through the tube is measured by mA meter. A kV selector switch enables to change voltage between the exposure. The voltages measured with a kV meter. The exposure which controls the timer and thus the duration of the application of kV to compensate the mains voltage variations, a voltage compensator is included in the circuit (Fig. 8.9).
2. The second circuit related to the control of heating of X-ray tube filament. The filament is heated with 10 volts of ac supply at a current of 3-5 amperes. The filament determines the tube current or mA. For temperature control mA selector is attached.

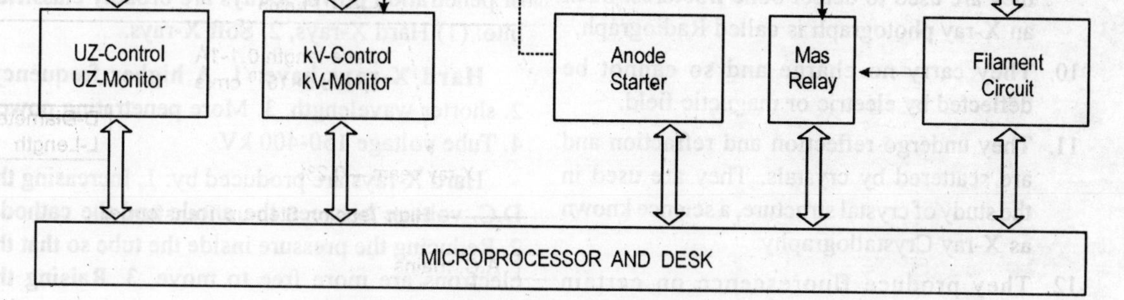

Fig. 8.9: Block diagram of circuit of X-ray machine

8.7.6 Characteristics of X-rays

1. There are differences in the penetration of X-rays of different energies so that low voltages are most convenient for some application of X-rays.
2. The penetrating power is inversely proportional to the wavelength of the ray.
3. In case of high-voltage X-rays, e.g., more than 200 kilovolts, some form of electron accelerator, e.g., betatron or Van de Graff generator is needed.
4. X-rays cannot be focussed since they are electromagnetic radiations of short wavelength similar to light rays and ultraviolet rays having a higher frequency. The wavelength is of the order of 10^{-10} metre.
5. The absorption of X-rays is similar to that of gamma rays, involving the production of photoelectrons, Compton scattering and pair production. Pair production takes place only in very high-voltage X-rays, since the minimum energy necessary for this is 1.02 MeV.
6. X-rays do possess enough energy to ionize absorbing materials and exceed the ionization potentials of ordinary materials.
7. They travel with a velocity of 10^8 m/s, and can pass through vacuum like all electromagnetic waves.
8. Penetrating power is greater than that of visible light. They can pass through many solid substances which are opaque to visible light, e.g., flesh, leather, wood, etc.
9. They affect photographic plates. Since they can pass through flesh but not through bones, they are used to detect bone fractures. Such an X-ray photograph is called Radiograph.
10. They carry no charge and so cannot be deflected by electric or magnetic field.
11. They undergo reflection and refraction and are scattered by crystals. They are used in the study of crystal structure, a science known as X-ray Crystallography.
12. They produce fluorescence on certain substances.
13. They can cause chemical change in the molecules of body tissues. Thus, an over-exposure to X-rays is harmful.
14. They are electrically neutral because it is electro-magnetic wave.
15. Wave length ranges from 0.1-0.5 Å.
16. Energy level 25-125 keV.
17. Travel in a straight line.
18. Travel or a speed of light i.e., 3×10^4 m/s.
19. Photographic effect on silver halide i.e., a chemical reaction which taken place in the X-ray film.
20. Can not be focussed by a lens.
21. Intensity of X-rays depends on the current through the tube.
22. The wavelength of the X-rays depends on the target material and velocity of the electrons holding the target.
23. X-ray equipment for diagnostic purposes uses target voltages in the range of 30-100 kV.
24. The diagnostic X-ray machine requires high mA current for heating the element and low exposure time whereas therapeutic X-ray machine requires low mA and high kV.
25. Special techniques are employed such as grid for absorption of scattered radiation, **contrast media** for ventriculography of brain, angiography of blood vessels, cardiac catheterisation for heart valves, cholecysto-graphy for gall bladder, cystography for urinery bladder, salpingography for Fallopian tube etc.

8.7.7 Classification of X-rays

According to the energy of X-rays they produce or penetration power X-rays are broadly classified into: (1) Hard X-rays, 2. Soft X-rays.

Hard X-rays have: 1. A higher frequency, 2. shorter wavelength, 3. More penetrating power, 4. Tube voltage 150-400 kV.

Hard X-rays are produced by: 1. Increasing the D.C. voltage between the anode and the cathode, 2. Reducing the pressure inside the tube so that the electrons are more free to move, 3. Raising the filament temperature.

- **Soft X-rays**

1. Lower penetrating power than hard X-rays.
2. Tube voltage 20-60 kV. Higher the tube voltage, higher would be energy and hardness of X-rays generated, so in practice the quality of X-rays is specified by tube voltage mentioned in the following table 8.5.

According to the application

A. **Diagnostic**

1. Head stand (Static), 2. Mobile, 3. Portable, 4. Dental, 5. Digital, 6. Tomographic, 7. Mammographic.

B. **Therapeutic** (Lower mA and high kV) (Fig. 8.9A)

8.7.8 Application of X-rays

A. **Diagnostic**: The field of application is known as **roentgenography**

1. If human body is placed between the X-ray tube and a fluorescent screen, the shadows of the bones can be observed on the screen, because of more absorption of X-rays by bones than the soft tissues.
2. To detect foreign materials inside the body, e.g., safety-pins, bullet injury, surgical scissors, scalpel, etc. left after the surgical operations.
3. To detect lesion in the lungs, e.g., tuberculosis, pneumothorax, pleurisy, tumours etc.

Table 8.5: Radiation quality specified by tube voltage

Tube Voltage	Radiation
1. Less than 20 kV	Very soft
2. 20-60 kV	Soft
3. 60-150 kV	Semi soft
4. 150-400 kV	Hard
5. 400-3000 kV	Very hard
6. Above 3000 kV	Ultra hard

4. **Spealised Dental X-ray equipmental: Dental Radiography:**
   (a) **Cephalostats (craniostats): It is an apparatus which persents correlation of position to be establised between an X-ray film, the head of the patient and the anode of the X-ray tube.**
   (b) **Pantomography: Pantomographic equipment protects the means to obtain panoramic tomographs of the jaws and face, i.e. all the teeth together with the mandible and maxilla are seen on a single 15 cm × 30 cm film.**
5. **Introducing radioopaque substances into the body and observation on different system such as:**
   (a) **Barium meal X-ray of oesophagus for constriction, obstruction.**
   (b) **Barium meal X-ray of stomach, duodenum for ulcer crater, filling defects etc. for peptic uncler, carcinoma.**

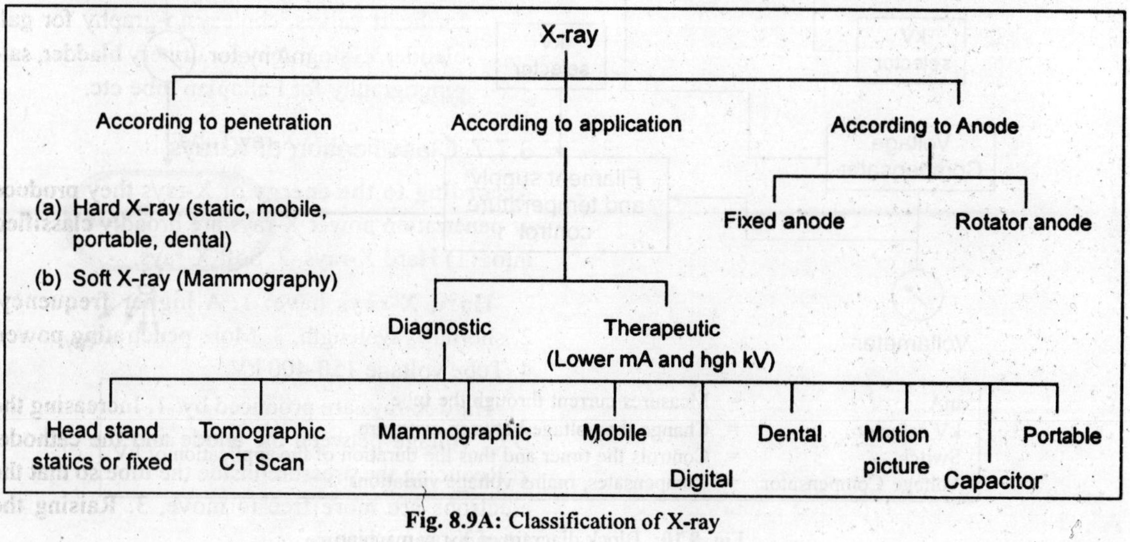

Fig. 8.9A: Classification of X-ray

(c) Oral cholecystography for functional or non-functional gall bladder, gall stones (cholelithiasis).

(d) Barium meal X-ray of whole G.I. tract for any tumour, obstruction, etc.

(e) Hystosalpingography: For any obstruction in the fallopian tube in case of infertility.

B. Therapeutic: Tumours are radiosensitive and tumours are divided into three major groups based on their sensitivity:

(a) **Radiosensitive group**: Lymphoma, Hodgkin's disease, leukaemia, embryonal carcinoma, endothelial myeloma, and some undifferentiated carcinoma.

(b) **Radioresponsive group**: Squamous cell carcinoma, basal cell carcinoma, adenocarcinoma of the uterus and breast.

(c) **Radioresistent group:** Adenocarcinoma of other regions except uterus and breast, sarcoma, teratoma and endothelioma.

Among these, only radioresponsive group is better for radiation therapy. Because, the radiosensitive group of tumours is generally diffused over the adjacent area or they have tendency to metastasize, and radioresistant group does not respond to radiation.

Low-energy radiations appear to be more effective than the high-energy radiations for reducing the tumour growth. Because of some recovery after radiation damage, it is better to administer radiation over several small doses for greater effect than administration in a single dose. The amount of X-ray required to destroy cancerous tissue varies from 1200 r (for medulloblastoma) to several times higher to this value for some skin cancers.

But very low-energy X-ray cannot penetrate the deeper tissues and can be used for superficial therapy. Low-energy Grenz rays having wavelength of about 2A produce electron tracks which are no more than a fraction of the total diameter of a cell. At energies of about 25 kV, X-rays have sufficient penetration and 99 per cent of the radiation is absorbed by the first 14 mm of tissue. Grenz rays may be used for skin lesions.

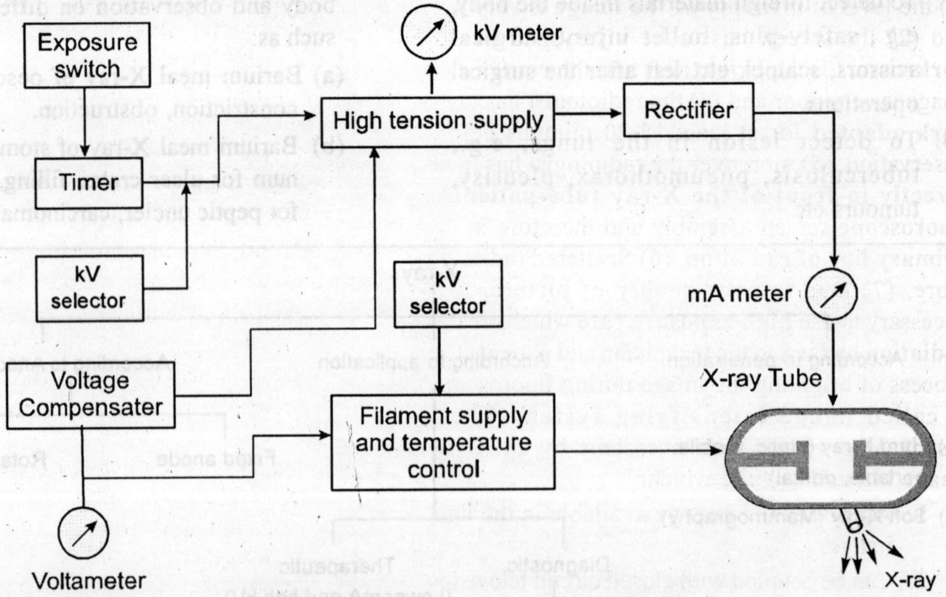

mA	=	Measures current through the tube
kV selector	=	Change the voltage between exposure
Switch	=	Controls the timer and thus the duration of the application of kV
Voltage Compensator	=	Compensates, mains voltage variations

Fig. 8.10: Block diagram of X-ray machine

Sometimes X-rays are to be applied to deeper lesions, in such cases it has to pass through the skin and tissues overlying the lesions and that causes damage of the tissue when a lethal dose to the tumour is to be administered. To avert this problem, two methods are followed:

1. Arranging the tube, several exposures are given in such a way that beams in the various exposures reach that tumour from different angles.
2. The patient is placed in a chair which can be rotated, with the lesion located on the axis of rotation. A beam of X-rays is directed through this axis, and the tumour gets a dose higher than the surrounding tissue continuously whereas the other tissue get interruptedly.

PART VIII

8.8.1 Image intensifier tube

The information given by the beam of X-ray converts into a visible image with either a fluorecent screen (radioscopy) or photographic film (radiography). But in usual method of fluoroscopy: (1) the level of **intensity of the image** is very low and (2) the screen has to be **visualized in the darkroom,** (3) the **brightness** and the **contrast** of image is very poor and (4) the radiologist has to be **dark adapted** for at least 15-20 minutes before observation, (5) moreover the radiologist has to be **directly in front of the X-ray tube-patient—fluoroscope** screen assembly and therefore in the **primary line of radiation.** (6) Scattered radiasion more, (7) Due to poor quality of picture, it is necessary to use **high exposure rate** which creates **radiation hazard** to the technician and patient. The process of brighting the image during fluoroscopy is called **image intensifying system.** These disadvantages have been overcome by an **X-ray image intensifier tube** which:

1. **Amplifies the energy** available in the light image.
2. Can be coupled with closed circuit television (CCTV, TV Cameras) and still or movie cameras, photo-spot camera and video recording. Potentially lowers radiation doses.
3. Permits observation with noticeable reduction in radiation doses.
4. Remote monitoring of X-ray images, the final image observed on the TV monitor depends on characteristic of TV chain.
5. Resulting in working conditions, better personnel safety and the facility of small-size image recording with 105 mm spot film which is cheaper and easier to pressure.

8.8.2 Components of image intensifier (Fig. 8.11)

The components of the image intensifier are housed in an evacuated glass envelope and include:

1. **Input screen** (input phosphor) made of vaporising sodium activated iodide on a special surface of aluminium which gives thin input screen (CsI) having diameter 15-32 cm which converts the X-ray image into a light image and it is transmitted through the glass of the tube to a **photocathode.**
2. **Photocathode:** A combination of antimony and caesium compounds. As X-rays from the patient are absorbed in the imput phosphor, light is produced and directed into the photocathodes which in turn, absorbs light and emits electrons. These electrons are attracted by the anode, which is kept at a positive potential of 25 kV with respect to photocathode.
3. **Anode:** These electrons are accelerated towards the anode and focussed by the electronically charged focussing electrodes through an opening of the anode centre.
4. **Electron Lens**
5. **Output Screen** (output phosphor): It is made of vary small crystals of CdS : ZnS : Ag.

Then the electrons are made to strike the output phosphor of the intensifier. As the electrons pass through a crossover point near the anode opening, the images on the output phosphor is inverted and reversed. The **light image** on the **output phosphor** is **much smaller** but also much **brighter** than the **input phosphor image.** The image intensification takes place due to very small size output screen and electron magnifi-cation in the tube.

8.8.3 Choice of intensifying screen film

The selection of screen depends on collective description of **rare earth** present there, are four types:

1. Gadolinium Oxysulphide, activated by terbium.
2. Lathanium oxysulphide, activated by terlieum.
3. Lanthanium oxybromide, activated by thulium and terbium.
4. Yttrium oxysulphide, activated by terbium.

From above list 1, 2, 3 are preferable.

8.8.4 Types of intensifying screen

1. Standard Intensifying Screen
2. High Speed Intensifying Screen
3. High Definition Intensifying Screen.

8.8.5 Advantages

1. Better for brightness and contrast
2. More rapid work
3. Low radiation exposure
4. No adaptation for eye in the darkroom
5. Automatic close regulation
6. Can be connected to X-ray television
7. Cinefluorography can be done.

8.8.6 Application

1. Fluoroscopic control during insertion of probes, canula, and catheter or injecting conduct media.
2. Intraoperative control during surgery of the gall blader, kidney and bone frac-tures
3. Locating tumours.

8.8.7 Viewing and display system

The final image is projected through a system of lenses and mirror on a viewing mirror for examination by fluoroscopies. Alternatively, the image on the output phosphor may be viewed by TV camera and 70 mm camera or cine camera (two channel attachment) and displayed on TV monitor. The intensifier tube envelope is contained within a metal housing to protect it from light and strong magnetic field (Fig. 8.11).

PART IX

8.9 X-ray Machine

The function of X-ray machine is to provide a sufficient intensity of electron flow in a controlled manner to produce an X-ray beam of desired quantity (Fig. 8.10).

8.9.1 Three types of X-ray machine

According to either

(a) the energy of the X-rays they produce i.e. a hard or soft.

(b) the purpose for which those X-rays are intended i.e., **Diagnostic or therapeutic.** Diagnostic X-ray machine come in different shapes, sizes e.g., 1. Tomo-graphy, 2. Head stand. 3. Mammo-graphic, 4. Mobile, 5. Portable, 6. Dental, 7. Digital, 8. DSA.

- **Voltage and Current**

They are usually operated at maximums

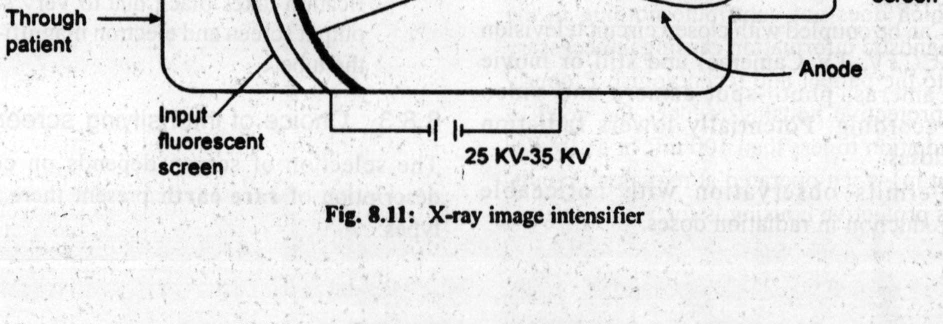

Fig. 8.11: X-ray image intensifier

voltages ranging from 25-250 KVP and tube currents from 25 to 1200 mA.

Therapeutic X-ray machines, on the other hand, can be operated at higher or lower voltages but at **tube current not exceeding 20 mA.**

The modern **general purpose X-ray examination room** usually contains

(a) a radiographic unit attached to overhead movable crane for positioning of the tube and aiming of the X-ray beam.

(b) a fluoroscopic unit under the examining table

(c) an electronic image intensifier.

8.9.2 Parts of X-ray machine

Every X-ray machine, regardless of its design has three principal parts:

(a) X-ray tube

(b) The control console

(c) the high-voltage section or generator.

In dental and portable machines, these **three components are housed compactly**. Most, however, have the head of the X-ray tube located in one room, the control console in an adjoing room, and a protective barrier separating the two. The protective barrier must have a window viewing the patient during examination. The high-voltage generator may be housed in a cubicle located in the corner of the room.

8.9.3 Protective housing of the X-ray tube

Mounted inside a **lead** lined protective housing designed to control serious hazards

(a) *excessive radiation exposure*

(b) *electric shock.*

When X-rays are produced, they are emitted *isotropically* (i.e. with equal intensity in all directions). Those emitted only through the window of the X-ray tube called *useful beam* and those escape through the protection housing called *leakage radiation* which does not contribute nothing in the way of diagnostic information causing unnecessary exposure to the patient and technologist. Properly designed protective housing reduces the level of leakage radiation to less than 100 mR/hr at 1m (26 μc/kg-hr at m) when operated at maximum conditions. This protective housing is for:

(a) Prevention of electric shock

(b) Mechanical support

(c) Electrical insulator

(d) Thermal caution

- **Glass envelope**

X-ray tube → electronic vacuums tube (like those contained in tube-type radios and television) → 20-35 cm and 15 cm in diameter → made of *pyrex glass for withstand tremendous heat* and *maintenance of vacuum.*

If the tube is filled up with gas the following may happen:

(a) electron flow from cathode to anode would be reduced.

(b) fewer X-rays would be produced and

(c) more heat would be produced earlier tube known as **crookes** tube were not vacuum but rather contained controlled amount of gas. The modern tube of X-ray called **Coolidge tube** is a vacuum tube, becomes gassy, X-ray production will fall off and the tube will fail.

- **Metal X-ray tube**

A recent improvement in tube design incorporated metal rather than glass as part or all of the envelope. Because glass envelope tubes have the following disadvantages:

1. Less stability due to age.
2. Some tungsten vaporizes and coats the inside of the glass envelope.
3. This alters the electric potential of the tube.
4. Allowing tube current to stray and interact with the glass envelope.
5. The result is arcing and tube failure.

Where metal envelope tubes have the following advantages:

(1) Maintain a constant electric potential between the electrons of the tube current and the envelope, (2) They have longer life, (3) Less likely to fail.

Virtually all high-output X-ray tubes now use metal envelopes.

Glass window: 5 cm square, this section of glass, serves to allow maximum emission of X-ray with minimum absorption in the glass envelope.

Cathode: It has two parts

1. Primary part—a filament
2. A focussing cup

Filament — Length— 1 to 2 cm
Diameter— 2 mm
Made of— Thoriated tungsten

When the current through the filament is sufficiently intense approximately 4A and above, **the outer shell electron of the filament atoms are** literally **boiled off and ejected from the filament** and the phenomenon is known as **thermionic emission.**

It has;

1. Higher thermionic emission than other metal
2. Melting point is 3410 °C and is not likely to burn out
3. Does not vaporise easily and if it happens the tube would be gassy and internal would be coated. Ultimately the tungsten does vaporise and deposit on the internal components which upsets some electrical characteristics of the tube and can lead to tube failure.
4. High atomic number –79 (higher efficiency X-ray production, higher energy X-rays.
5. Thermal conductivity.

The addition of 1% to 2% thorium to the tungsten filament serves the following:

(a) increases efficiency of thermionic emission and

(b) prolongs tube life.

- **Focussing Cup**

1. Filament is embedded in a metal shroud called focussion cup.
2. The electrons accelerated from cathode are electrically negative.
3. The beam tends to spread out owing to **electrostatic repulsion** and some electrons can even miss the anode completely.
4. The focussing cup is negatively charged so that it condenses the electron beam to a small area of the anode.
5. The effectiveness of the focussing cup is determined by:

   (a) Its size and shape

   (b) Its charge

   (c) The filament size and shapes.

   (d) Position of the filament within the focussing cups.

6. **Focussing cup is the grid**: Certain types of X-ray tubes called **grid controlled tubes** are designed to be **turned on and off very rapidly** and this type of tubes are used in DSA **(digital subtraction angiography)** and **cineradiography** in which precise exposure times are required. The grid in vacuum tube electronics means switch. In a grid-controlled X-ray tube the **focussing cup is the grid** and therefore **the exposure switch.**

8.9.4 Factors influencing X-ray quantity and quality (Table 8.6)

The factors are:

1. **Exposure Time**: Quantity and quality of X-ray generated depends on the power supply to the x-ray tube i.e., kilovoltage (kVp) and milliampere seconds (mAs).
2. **Kilovoltage (kVp)**: Two characteristics of X-ray beam, emitted from the X-ray tube i.e.,

Table 8.6: Factors influencing X-ray quantity and quality

Increasing the factor	Resulting change in X-ray	
	Quantity	Quality
Killovolt	Increase	Increase
Milliampere seconds	Increase	No change
Milliampers	Increase	No change
Exposure time	Increase	No change
Distance	Decrease	No change
High-Voltage generation	Increase	Increase
Filtration	Increase	Increase

quality and quantity are affected by changing tube kilovoltage.

3. **Quality**: The quantity of X-ray beam is an expression of the effective photon energy of the beam. Its importance in radiography is three fold.
   (a) Influences the penetrating power, increase in tube kVp causing more ponetrating power. It indicates that higher tube kVp is required for thick and dense tissue.
   (b) Increase in radiographic contrast, decrease of kVp which causes an increase in contrast.
   (c) Influencing radiation dose.
4. **Intensity**: The intensity of emitted radiation and the intensity of radiation reaching the film, both can be increased by increasing tube kilovoltage and causes an increase in image density.
5. **mAs (milliampere second)**: Changing mAs affects X-ray beam energy in tube during exposure.
6. **Focal spot**: Selection of mAs and the size of the tube are closely related. Most X-ray tube show two focal spots, one is fine and the other is broad considering spot size. The size combination differ in different X-ray tubes but common pair of focal spot is obtained which are 0.6 and 1.2 mm, 0.3 and 1.2 mm and 1.0 and 2.0 mm. Fine focus is selected where object-film distance is wider. Broad focus is selected because if permits higher tube loading and enables maximum mA to be most. It requires when a high mAs and minimum exposure time is necessary. Modern X-ray machines do not contain focal spot size selection.
7. **Filtration**: X-ray beam emerges from the X-ray tube is modified by filtration. This is only adjustable in mammography and directly under the control of radiographer. Increasing filtration there will be increase in quality and reduction of beam intensity of the beam emerging from the X-ray tube. The intensity only of the beam reaching the image receptor is affected by changing the filtration.
8. **Focus-film distance**: Increasing the distance, divergence of primary beam causes reduction of its intensity. Its energy is spread over an increasing area as the distance from the source increases.

Other biological effects

1. **Spontaneous mutations**: Spontaneous mutations are rare but its likelihood can be increased by exposure of the dividing organism to X-rays.
2. **Bactericidal activity of X-irradiation**: Ionizing radiations have disinfectant effect due to (i) much higher energy content than ultraviolet radiation and (ii) much greater capacity to induce lethal chemical changes and (iii) greater penetration. Of the various types of ionizing radiations only gamma-rays and X-rays have prospect of practical application. Generators and accelerators are now available for cathode-rays on a commercial scale (radioactive isotopes, e.g., Co^{60} which produces gamma radiation are available in sufficient quantities as a by-product of atomic energy installations used for disposable plastics). These have:
   (a) *Primary action*: Produce tracks of ionization in the cells. These initiate changes in the proteins and nucleic acids particularly in DNA.
   (b) *Secondary action*: Formation of free radicals – H, OH, HO_2 and hydrogen peroxide, H_2O_2 from water molecules, both extracellularly and intracellularly.

   These are used for sterilization of surgical materials made of rubber or plastic, disposable syringes, catgut, bone and tissue grafts and adhesive dressings.
3. **Sensitivity of Viruses**: All viruses are inactivated by ultraviolet light and X-rays though there are quite large variations in the dosage required by different viruses.
4. **Immunosuppression**: Among the three methods of immunosuppression, i.e., immuno-suppressive drugs, immunological methods using antilymphocytic serum (ALS), the X-irradiation method with

sublethal doses of X-rays have markedly destructive effect on lymphoid tissues and abolish the immune response. The response depends upon the animal and the radiation dose. But 3-4 weeks are required for regeneration of lymphoid tissues to appear histologically normal. Irradiation acts as a primary response and affects antibody production much more than cellular immune responses.

Whole body X-irradiation used in the early days of organ transplantation surgery to control allograft rejection is no longer employed for this purpose and has been replaced by immunosuppressive drugs and anti-lymphocyte serum (ALS).

5. **Secondary immunodeficiency**: Immunodeficiency may arise secondarily by many factors and X-rays are one of them. It causes B-Lymphoproliferative disorders like chronic lymphatic leukaemia, myeloma, etc. which are associated with varying degrees of hypo-γ-globulinaemia and impaired antibody responses.
6. **Removal of infected hair**: In some dermatophyte infections (Dermatomycoses), infected hairs were removed by X-irradiation in the past.

8.9.5 Instrument using x-ray energy

1. Computerised tomography (CT scan)
2. Mammography
3. Digital subtraction angiography (DSA)
4. X-ray microscopy
5. X-ray crystallography
6. Dental X-ray
7. Therapeutic
8. X-ray spectroscopy.

8.9.6 X-ray therapy (tube)

1. Low mA values of tube current to produce X-rays of four main KV energy ranges.

(a) Low voltage 50-120 kV—Treatment of skin diseases.

(b) Intermediate voltage 130-150 kV Treatment of interior part of the skin.

(c) Orthovoltage 160-300 kV—Deep X-ray therapy

Table 8.7: Comparison of Diagnostic and Therapeutic Machine

Diagnostic Machine	Therapeutic Machine
1. Low Voltage, (100-150 kV)	1. Mostly high voltage (more than 200 kV)
2. Short focus for short exposure	2. Large focus for larger exposure
3. Wide beam used	3. Narrow beam used so extra radiation avoided, more homogenous and less scattering effects
4. Exposed a larger area	4. Exposed a particular area because of more adjusted by diaphragm and scattering collimation
5. Long frequency so less penetrating power and thus give more skin and bone reactions	5. Short frequency so more penetrating power and thus less skin and bony reactions upto 600 kV. Above 1MeV no skin and bone absorption

(d) Megavoltage or super voltage 4MV or more.

Orthovoltage therapy tube has:

(a) A larger filament

(b) A stationary anode

(c) A larger focal spot

(d) Anode angle 30°

(e) Hollow anode.

(f) Cooling heat by circulating oil (45 lb pressure) between the tube and housing.

(g) A pump to transfer the hot oil to a special tank where it is cooled by water circulation through coils (See p. 278).

8.9.7 Factors governing the radiography

To maintain a consistent high radiographic quality such as the amount of blackening or diversity on an X-ray film depends upon the following factors:

1. (a) kilovoltage (kV) of X-ray, (b) Type of radioactive source (gamma-rays).
2. (a) milliamperage of X-ray, (b) millicuries (gamma-rays).
3. Time of exposure.
4. The distance of the source from the film.
5. The X-ray film used.
6. The intensifying screen used.
7. Processing conditions.
8. Filtration.

9. The X-ray apparatus in use and the circuit employed.
10. The object being radiographed.

8.9.8 Visualization or observation of X-ray

1. Radioscopy or fluroscopy: observation by fluroscopic screen (a) plain, (b) Big TV or cine camera.
2. Radiography: observation by X-ray film.
3. Image intensifier (to utilize actually wasted X-ray to activate).
4. Using contrast, media, e.g., Bariumm.

8.9.9 Special techniques of X-ray

1. Grids,
2. Contrast media,
3. Angiography,
4. Cardiac catheterisation,
5. 3-D visualisation (Tomography-Tomo-Slice or section),
6. Stereo-radiography.

8.9.10 Types of X-ray machine

1. Static or head stand or fixed
2. Mobile
3. Portable
4. Tomographic
5. Mammographic
6. Dental
7. Therapeutic
8. DSA
9. Digital.

PART X. Detection and Measurement of Radioactivity

8.10.1 Measurement of radioactivity

For measurement of radioactivity of radioactive materials, some units are employed which can be divided into two groups.

8.10.2 Units of source strength

(a) Curie, (b) Rutherford.

Both are defined as some number of nuclear disintegration per second.

(a) Curie (C): 3.7×10^{-10} disintegration per second. It is the number of disintegrations per second of the quantity of radon in equilibrium with a gram of radium.

One millicurie (mc) = 3.7×10^{7} disintegration per second or 10^{-3} curie.

One microcurie (μc) = 3.7×10^{4} disintegration per second or 10^{-6} curie.

(b) Rutherford: 10^{6} disintegration per second; 3.7×10^{4} rutherfords = 1 curie.

Source of strength is measured by Geiger-Muller Counter or Scintillation Counter.

8.10.3 Units of dosage

These units are used for:

1. Description of radiation effect,
2. Safety of the person handling or present in the surroundings of the radioactive materials. The dosage specification is a difficult task because of: (i) different types of radioactive emission, (ii) different depths of penetration, (iii) different effects which vary with their energy.

The units for dose are given below:

1. **Roentgen**: The unit is named after the German Physician Wilhelm Conrad Röntgen. Any unit named after any scientist person must be written starting with a small letter roentgen (r). This is the primary unit of dose. It is the amount of X-rays or gamma-rays such that the associated corpuscular emission per 0.001293 gm of air, 0 °C, 760 mm Hg produces, in air, ions carrying 1 electrostatic unit of quantity of electricity of either sign. The amount of air under the conditions has a volume of one cm^3. Quantities which are equivalent to this amount of ionization are:

1. electrostatic unit (esu) of ion pairs produced per cc air,
2.083×10^{9} ions pair produced per cc air,
1.61×10^{12} ion pairs produced per gm air,
6.77×10^{4} MeV absorbed per gm air,
83 ergs absorbed per gm air.

The roentgen unit is used pertaining to the ionization produced in the air by the incident radiation. If the same goes through a dense substance, more ionization would be produced where one roentgen of radiation goes through the soft tissue, 93 ergs may be absorbed per gm and in case of bone which is a very dense tissue, several hundred ergs may be absorbed, the roentgen unit gives the activity of a given quantity of radiation for comparison of its ionizing power in the exposed material with that in air. Radiation absorption varies with the electron density of the substance through which it passes but the living tissue is not a homogeneous material, so application of roentgen unit directly as a measure of effective radiation is a problem.

2. **Roentgen equivalent physical (rep)**: This dosage unit can be applied directly for living tissue than the roentgen unit. It is defined as the dose of any ionizing radiation which produces energy absorption of 83 ergs per cubic centimetre of tissue. One roentgen of radiation is equivalent to 93/83 = 1.1 rep, provided the tissue is soft tissue and it can absorb 93 ergs per gm of tissue. And if the tissue is dense, one roentgen is equivalent much more rep. The rep is also reffered to as tissue roentgen, equivalent roentgen and roentgen equivalent but not used.

3. **Roentgen equivalent man (rem)**: It is very difficult to define biological equiva-lence for which its application is very much restricted.

It is defined as the dose which is delivered to man (or mammal) exposed to any ionizing radiation, and is biologically equivalent to the dose of 1 roentgen of X-rays or gamma-rays. Generally the following equivalence obtains:

1 rem = 1 rep for β-particles,
1 rem = 0.25 rep for protons, deuterons and neutrons,
2 rem = 0.1 rep for α-particles.

4. **Gram roentgen:** It is defined as that amount of gamma-ray energy converted into kinetic energy of secondary electrons which is equal to the energy absorbed by 1 gram of air exposed to 1 roentgen. The gram-roentgen is an unit of radiation. It applies to absorbing material irrespective of the volume on which this quantity of radiation is distributed over. Thus, 1 rep = 1 gm-r per gram of tissue. The dimensions of the gram- roentgen differ from other units.

5. **Energy units:** It is defined as such dose delivered to tissue by ionizing radiations that the energy absorbed per gram of tissue is equal to the energy absorbed per gram of water exposed to 1 r of gamma radiation. This unit is usually not used because the tissue cannot be considered as water.

6. **J unit**: It is defined as such dose one J has been received at any point in a medium when the ionization which would have been observed in an infinitesimal cavity containing the point is 1.58×10^{12} ion paris per gram of air enclosed in the cavity.

Measurement of radiation is generally referred to ionization in air, so this unit may be applicable.

7. **The u unit**: This unit is for measurement of the ionizing power of fast neutrons. It is defined as that of dose of 1 n is delivered to tissue by fast neutrons when the ionization produced in the Victoreen 100 r thimble chamber equals the ionization produced by 1 r of gamma radiation.

8. **Radiation absorbed dose (rad)**: Radiation absorbed dose (rad) means the amount of ionizing radiation absorbed by the irradiated tissues which is expressed as "rad" and equivalent to 100 ergs of energy absorbed per gram of tissue.

9. **Relative biological effectiveness (RBE)**: Relative biological effectiveness is the ratio between the dose in rads of a radiation which is to be applied and the dose of γ-rays producing the same biological effect. It can be expressed as the ratio between rem and rad, i.e., RBE = rem/rad.

8.10.4 Interventional radiology

Recent radiodiagnostic techniques have been used for **diagnosis** as well as for **therapy**. This interventional radiology include a wide variety of procedures such as:

1. **Fine Needle Biopsies (FNB)**: Fine needle biopsy under imaging control is widely used.
   (a) In many deep seated tumours FNB can now be performed by simple percutaneous puncture by the radiologist with the help of image intensifier screening e.g., peripheral nodules in the lung, tumour of the lung and abdominal masses of all kinds.
   (b) Bone biopsies from the spine (replacing exploratory thoracotomy) the or limbs percutaneous puncture
   (c) Retroperitoneal nodes
   (d) The development of ultrasound and CT have enabled the method to extend further in case of inaccessibly organs.
   i. Ultrasound guided biopsy in case of kidneys and abdominal organs.
   ii. Greater accuracy and for quite small lesions, CT guidance is more reliable.

2. **Percutaneous catheterisation and embolisation**: For the treatment of tumours which is mainly to reduce their size and vascularity prior to operation in difficult cases or as a palliative treatment of internal haemorrhage and for the treatment of angioma and arteriovenous fistulae.
3. **Percutaneous Catheterisation with balloon Catheters**: It can be used to occlude arteries temporarily, either to stop haemorrhage or to obtain a bloodless field at operation.
4. **Percutaneous catheterisation**

   It is also used:

   (a) For the delivery of chemotherapeutic drugs to tumours.

   (b) For the delivery of vasospastic drugs in patient with internal haemorrhage.

   (c) For thrombolysis by delivering thrombolytic drugs directly to the clot.
5. **Percutaneous transluminal dilatation**: It is used for dilatation of arterial stenosis in the fermoral and iliac arteries and extended to many arteries including the renal artery and the coronary arteries.
6. **Transhepatic catheterisation of portal vein**: It is used for the portal vein and embolisation of varicose.
7. **Transhepatic catheterisation of the bile duct**

   It is used:

   (a) For drainage in obstructive jaundice

   (b) For dilatation of stenosing structures or insertion of prostheses.
8. **Needle puncture and drainage of cysts**: In kidney or other organs using X-ray or ultrasound.
9. **Percutaneous catheterisation of renal pelvis**: It is used in case of renal pelvis for antegrade pyelography or hydronephrosis drainage.
10. **Percutaneous removal of residual biliary duct stones** through the "T" tube tract.
11. **Percutaneous catheterisation and drainage** of intra-abdominal abscesses.

8.10.5 Intensity of radiation in space

For determination of the absolute activity of the source, measurement of the radiation at a point in surrounding, the source is required. Then one can calculate the number of ionizing events in the entire surface area of a particular sphere with a centre and correct the same for (a) the number of radiations emitted per single disintegration and (b) the efficiency of the counter. To accomplish this calculation, one must know the complete disintegration scheme. It is also necessary to know the efficiency of the working counter accurately because of (a) different types of radiation, (b) one disintegration may emit more than one type with different energies. For these reasons, it is difficult to determine the absolute efficiency of a counter. To calculate with accuracy the activity of any unknown source, it is reasonable to compare the activity of a particular source of radiation with that of a standard emitting the same type and energy of radiation. One should be careful during comparing the acitivity of an element with that of a standard having a different element because there is every possibility of error for difference in quantity of radiation.

8.10.6 Protection from X-ray radiation

Protective measure can be discussed under the following categories:

8.10.6.1 Designing the X-ray apparatus

1. *Proper collimation*: reduces patient dose and improves image contrast
2. *Filtration*: reduces patient dose
3. Protective housing of the X-ray tube of diagnostic tube
4. *Control panel*: to indicate the condition of the exposure and when the X-ray .... energized.

8.10.6.2 Radiographic equipment and operation Radiography

1. Source to image distance (SID) by SID indicator as a

   (a) Tube measure attached to the tube housing or, (b) Laser light
2. *Collimation:* By:

   (a) variable aperture rectangulars collimators, (b) cones, (c) Diaphragm.

3. *Positive beam limitation (PBL)*: To provide X-ray beam equal to the image receptor.
4. *Beam alignment*: The tube head is to be provided a mechanism by which a proper alignment of X-ray beam and the film will be obtained.
5. *Filtration*: Diagnostic X-ray beam must have a total filtration of at least 2.5 mm A.
6. *Reproducibility*: The output radiation intensity should be constant from first exposure to another exposure.
7. *Linearity*: The mA and exposure time should be adjusted to maintain constant mA.
8. *Operator shield*: The operator should be within a fixed protective barrier having control console. But should not use long cord.
9. *Protective Dress*: While operating, lead apron and gloves should be used and the operator should be at least of 180 cm away from the x-ray tube.

8.10.7 Inverse square law (distribution of energy in space)

If a radioactive mass is considered as a point in space, then radiation emits from it in all directions. The energy flows per unit area of a plane (A), placed perpendicular to a line streching from the sources (S) of radiation to the plane (B). This has the importance for consideration of the effects of radiation on material (M) present in the path of radiation. The total energy emitted from the source

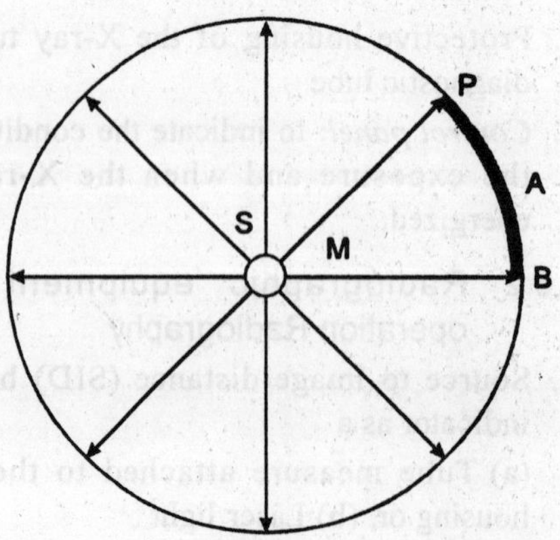

Fig. 8.12: Illustration showing inverse square law and geometric relationship necessary for it

(S) is distributed over an area equal to the surface area of a sphere of radius r, where r is the distance of the plane from the radiation source. Since the surface area of a sphere equals $4\pi r^2$, the energy flux per unit area of surface is inversely proportional to the square of the distance of the surface from the source known as the "inverse square law" (Fig. 8.12). Its main application is to calculate the intensity of radiation at one distance from a source if the intensity at another distance is known. The equation for calculation is:

$$I_x = I_o\left(\frac{r_o}{r_x}\right)^2$$

where I_x = the intensity of radiation at some distance r_x from the source,

I_o = a known intensity of radiation at a distance r_o from the source.

8.10.8 Instruments used for measurement of radiation intensities

The measurement of radiation intensities is based on the ability of the emitted radiation to ionise atoms in a suitable detection apparatus. We have to bear in mind that (1) measurements of ionizing radiation are of indirect nature, and (2) it is the measurement of ionization produced during the passage of radiations through the matter. These ionization causes to appear as a current flow due to application of a high voltage across the absorbing space. Then the ions migrate to the electrodes placed across the absorbing space which are attached to a condenser of electroscope. The amount of discharge occurring per unit of time is noted. Such device is generally known as "Ionizing Chamber".

The ionizing chambers are grouped under two headings:

I. **For accurate measurements**: Condenser discharge or electroscope discharge methods, e.g.
   1. Victoreen Condenser R-meter,
   2. Lauritsen Electroscope,
   3. Extrapolation Chamber.

II. **For survey meters**: These are current measuring devices, e.g,
   1. Fountain-pen Dosimeter,
   2. Geiger-Muller Counting Tube,
   3. Scintillation Counter.

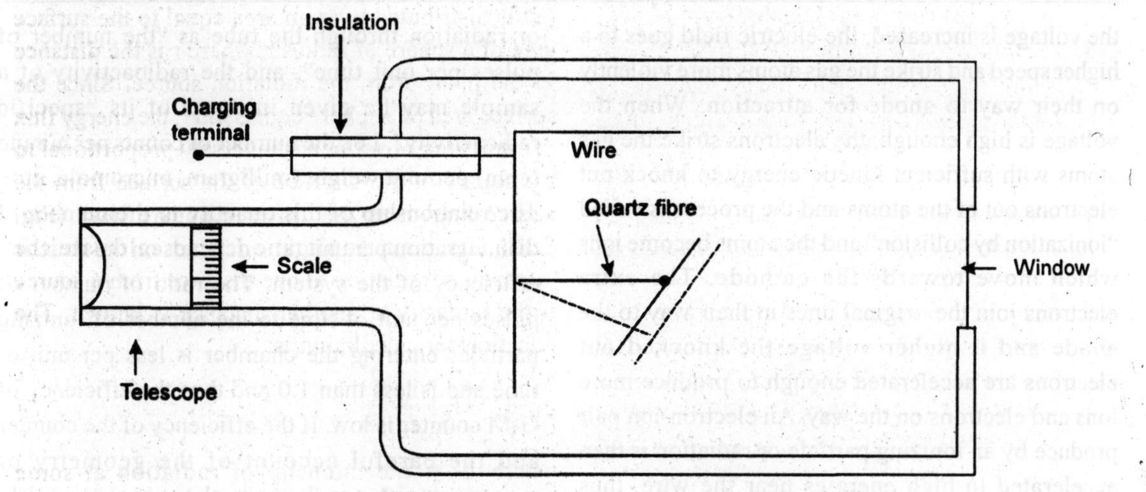

Fig. 8.13: Schematic diagram of Lauritsen electroscope

1. **Victoreen Condenser R-meter**: It is one of the useful instruments to measure exposures to X-ray or gamma radiation.
   It has a condenser chamber whose charge is determined by an auxiliary electroscope. The sensitivity of the instrument can be varied by varying the size of the chamber. Its construction is such that the scale gives the reading, the amount of charge of the condenser calibrated in roentgens for a particular type of chamber.
2. **Lauritsen Electroscope**: Lauritsen electroscope is used for comparison of the intensity of two sources (Fig. 8.13).
3. **Extrapolation Chamber**: Since the measurement of β-radiation intensity is very difficult for the short range of β-particles in air, the ionization associated with several volumes are measured and the plotted curve extrapolated to zero volume to get beta radiation intensity on the chamber and this modified chamber is called "Extrapolation Chamber". The distribution of ionization is not uniform throughout the volume of the chamber and it can be charged easily. Electrometer vacuum tube circuit is used for determining the ionization continuously as the current associated with ionizations in a chamber of reasonable size are very small.
4. **Geiger-Mueller Counter**: Principle—The principle in this instrument is that the ionizing particles produce ions in a gas-filled tube. These ions are accelerated toward a high-voltage electrode which in turn collide with gas molecules to form other ions. These shower of ions appear as a voltage pulse across a resistor placed in the high voltage circuit. This pulse is then passed for amplification, shaping and activating the counting device.

8.10.9 Instrument

The instrument consist of a cylindrical metal tube called Geiger-Muller (GM) Tube. One end is closed by a thin mica window protected by gauze through which the particles can enter. The other end is closed by insulated plug through which a thin metal wire, the anode, enters the tube and the wire is insulated from the surrounding cylinder B which is metal or graphite coated. The anode wire A must be thin, so that the charge on it produces an intense electric filed E close to its surface where E is inversely proportional to the radius of the wire. Anode is kept at a positive potential V of about +400 V relative to cathode B, served by the metal wall of the tube which may be earthed. The tube is filled with an inert gas like helium (He), argon (Ar), etc. which separates the two electrodes. The gas is present at low pressure along with a small volume of an easily dissociable "quenching" substance like ethanol or a polyatomic organic gas. When the radiation emitted by the radioactive substance enters the tube through the window, ionization of gas atoms produces few electrons and ions. When the voltage is increased the number of electrons and ions are multiplied enormously. The mechanism is that when the voltage is increased, the electric field goes to a higher speed and strike the gas atoms more violently on their way to anode for attraction. When the voltage is high enough, the electrons strike the gas atoms with sufficient kinetic energy to knock out electrons out of the atoms and the process is called "ionization by collision" and the atoms become ions which move towards the cathode. The extra electrons join the original ones in their way to the anode and it higher voltage the knocked out electrons are accelerated enough to produce more ions and electrons on the way. An electron-ion pair produce by an ionizing particle or radiation is then accelerated to high energies near the wire, thus producing by collision with gas molecules an "avalanche" of more electron-ion pairs. Eventually a point is reached, the shower of ions appears as voltage pulse, i.e., electric current flows across a resistor in the high resistance placed in the high-voltage circuit and the gas is said to break down. At a sufficiently high voltage (800 to 1500 volts) between the electrodes, a plateau is reached where the number of pulses of current is independent of the voltage and is proportional to the number of ionizing particles entering the tube. The current flowing in the high resistance produces a potential difference which is amplified, shaped and used to activate counter, e.g., dekatron counter or ratemeter, which registers the passage of an ionizing particles or radiation through the tube as "the number of pulses per unit time", and the radioactivity of a sample may be given in terms of its "specific radioactivity", i.e., the number of counts per minute (cpm) per unit weight (milligram, micromole, etc). The relationship of this quantity to the number of disintegration per unit time depends on the counting efficiency of the system. The ratio of number of pulses per unit of time to the number of ionizing particles entering the chamber is less per unit of time and is less than 1.0 and thus the efficiency of G-M counter is low. If the efficiency of the counter and the careful account of the geometry of measurement are known then a reasonable measurement of the absolute strength of a radiation source can be obtained. But in ordinary measurement, it is not necessary as the G-M counter is used only for comparative measurement.

The current flow persists for a short time as the secondary electrons are emitted from the cathode by the positive ions reaching there. This disturbs the recording of other ionizing particles. The air in the tube is, therefore, replaced by argon mixed with halogen vapour or other quenching material which has the property of quenching the pulse quickly. Sometimes electrical methods are also used for quenching. These quenching materials absorb the energy of electronic avalanche and terminate the current pulse (Fig. 8.14).

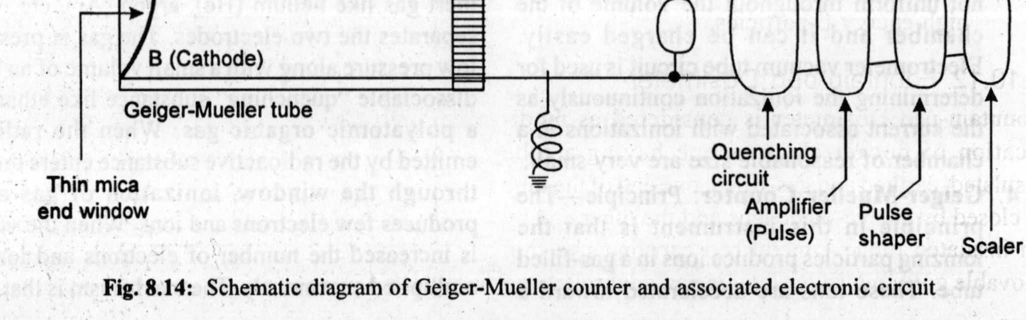

Fig. 8.14: Schematic diagram of Geiger-Mueller counter and associated electronic circuit

8.10.10 Measurement of uniformly layered solid material

Radioactivity measurement of uniformly layered sample of solid material like Ba $C^{14}O_3$ depends on many factors which must be considered if any corrections are necessary. These factors are:

1. The background count caused by radiation not emitting directly from the sample;
2. The geometry of the system, i.e., spatial relation between the sample and the detector;
3. The interaction of the radiation with the sample, as reflected in "self-absorption" and "backscatter" of the radiation, and varies with the thickness of the sample.

To overcome the problem, methods have been employed for stepwise degradation of many carbon compounds to convert each carbon atom to CO_2 which may be introduced directly into the GM tube and, in this process, the counting efficiency is much higher than with solid material.

8.10.11 Advantages

1. The best known of the radiation measuring instruments.
2. More sensitive than the other methods.
3. It can be used with quantities of radioactive material which are easy to handle and do not damage biological tissue.
4. It is sensitive to α-and β-particles and γ-rays.
5. It can detect particles with weak ionizing effects.
6. It can count individual particles at a high rate and suitable for enumeration of particles at high radiation level.
7. Modifications of the instrument may be done to eliminate low-ionizing γ-rays.
8. Very efficient for detection and counting of high-energy β-particles.

8.10.12 Fountain-pen Dosimeter

Fountain-pen Dosimeter is considered as modification of pocket electroscope having a tube insulated, earthed and gas-filled, one end of the tube is closed by an eye-piece lens and the other end by an insulated plug. A metal rod carrying two thin movable gold foils at one end is passed through the insulated plug and the other end remains outside the tube and terminates in the charging terminal. When the metal rod is charged, the gold foils carrying like charges diverge, making an angle proportional to the potential which can be observed through the eye-piece lens.

Radiation from radioactive materials, e.g., β-particles and γ-rays except α-particles penetrates the wall of the tube and ionizes the gas molecules inside the tube. These ions discharge the gold foils for making an angle of divergence. The alteration of angle of divergence is read from a scale present in it for measurement of the amount of ionizing radiation (Fig. 8.15).

8.10.13 Disadvantages

1. The instrument is sensitives to β-particles and γ-rays only.
2. It cannot detect neutrons or α-particles.

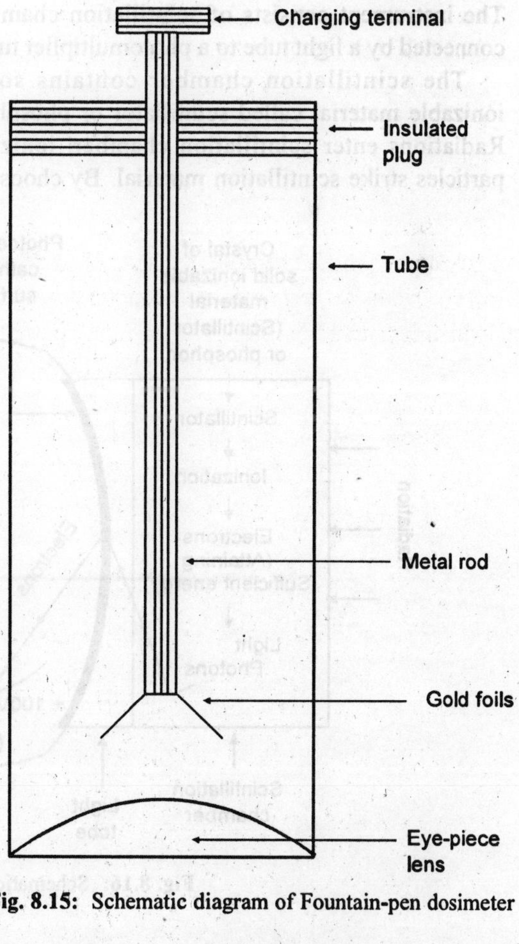

Fig. 8.15: Schematic diagram of Fountain-pen dosimeter

3. Individual emitted particles cannot be enumerated.
4. Requires recharging after use.
5. Used for short period only.

8.10.14 Scintillation counter

During various experiments on radioactivity, Rutherford observed that the scintillations are produced when α-particle was incident on a material like zinc sulphide. This phenomenon was utilized in the Scintillation Photomultiplier.

8.10.15 Principle

The instrument works on the principle that certain crystals give off flash of light when a photon of ionizing radiation is absorbed by crystal. The crystal is placed in such a way in front of the photomultipler that the flash of light produces a current in the photomultiplier tube.

8.10.16 Instrument

The instrument consists of scintillation chamber connected by a light tube to a photomultiplier tube.

The scintillation chamber contains solid ionizable material called scintillator or phosphor. Radiations enter scintillation chamber, ionizing particles strike scintillation material. By choosing a suitable phosphor, scintillation counters can detect electrons and gamma rays as well as fast neutrons. Ionization takes place in the phosphor to emit electrons. These electrons gain sufficient energy to give light photons. Photons pass through the light tube and fall on a photo-sensitive material, known as photoelectric cathode surface of photomultiplier tube. The electrons are emitted from the photoelectric cathode surface. These electrons are focussed toward and accelerated to an electrode A, coated with a material which emits secondary electrons four or five times as numerous as those incident on it. The secondary electrons then strike the electrode B after further acceleration, thus multiplying the number of electrons further, and so on along with the tube. There is a series of such electrode for acceleration and multiplication of electrons. A single ionizing particle can produce a million electrons in a photomultiplier tube. The enormous number of electrons strike the collector plate to generate a pulse of current which is amplified further and electronically recorded.

The counting of current pulses of scintillation counter can be integrated over a period of time by placing capacitors in the counter output circuits, and the integrated currents can be read on a meter for continuous indication of the number of counts

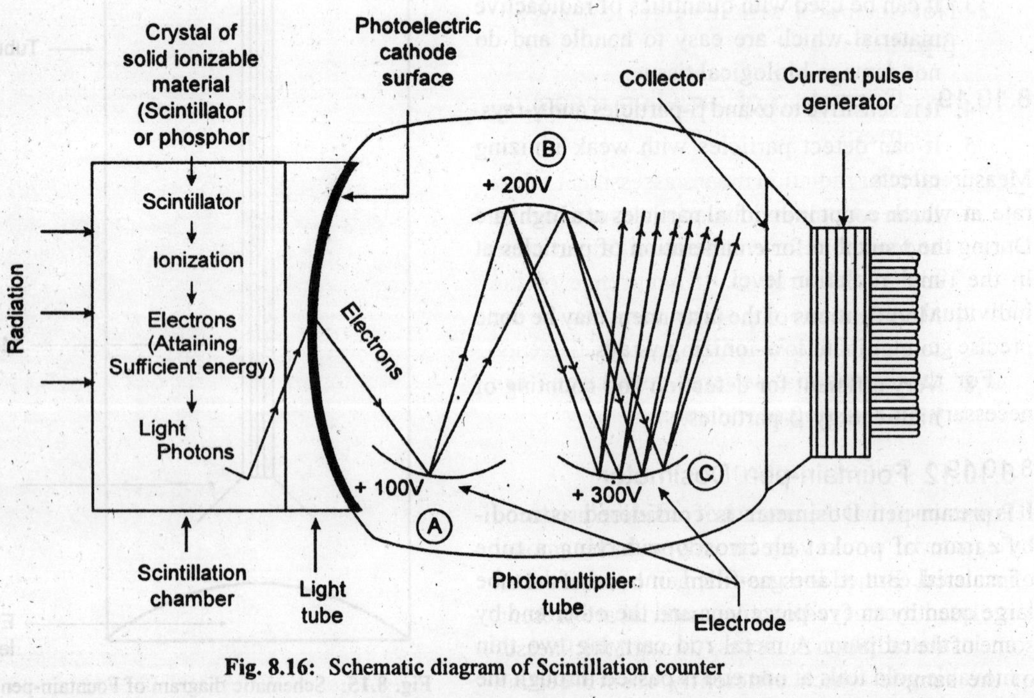

Fig. 8.16: Schematic diagram of Scintillation counter

per unit of time. A device called "count rate meter" performs the same function and is often used in making dynamic measurements of changing radiation sources (Fig. 8.16).

8.10.17 Advantages

1. More sensitive than the Geiger-Muller Counter.
2. The efficiency approaches 100 per cent.
3. Very minute quantities of radiation can be detected and measured.
4. It can detect α-, β-particles and γ-rays.
5. It can detect α- and β-particles at a faster rate and at higher radiation level than the G-M Counter.
6. Even very high-energy γ-rays absorbed by the phosphor produces detectable ionizations making it more sensitive than G-M Counter to γ-rays.
7. It distinguishes different types of emissions. It is due to height of the pulse of the current which is proportional to the energy contents of emitted particles.

8.10.18 Disadvantages

1. Can only be operated with higher voltage than the G-M Counter.
2. Not so much sensitive to low-energy β-particles.

8.10.19 Correction of radiation measurement

Measurement of radiation means assessment of the rate at which a number of random events occur. During the procedure, one can expect (a) variation in the time of distribution of occurence of the individual events and (b) some variation in the precise geometric location of the individual events.

For these reasons, several corrections are necessary for counting of radiation.

8.10.19.1 Self-absorption

It is presumed that the amount of radiation produced by a material would vary linearly with the amount of material. But it does not happen because when large quantities of radioactive material are present, some of the radiation must pass through thicknesses of the sample which efficiently absorb certain amount of radiation. When the thickness of the sample is great, all the radiation from the lower layers is absorbed. And the activity of the counter does not increase when more layer are added to the bottom of the sample. This self-absorption property of material limits the intensity of a source of radioactivity. When the source reaches a certain thickness, the intensity can be increased either by increasing the specific activity of the material from which it is made or the area of the source. If this samples cannot be prepared as in case of β-emitters, it is necessary to make correction for self-absorption in assaying activities of samples.

8.10.19.2 Coincidence correction

It is the general characteristic of all the counter to recording of nuclear events and to take very small significant time. In case two nuclear events occur in the same period, only one is recorded. When a counter records the radioactivity of successive increased areas of uniform source, the count will be proportional to the area in case of low counting rates. When the counting rates are high, the rate is less proportional to area due to coincidence of events. As the number of counts per unit of time increases, the error due to these coincidence events increases. If a curve is plotted for activity of sample and number of counts, there is deviation from linear increase of count rate as the absolute number of disintegration increases due to an increase in the number of simultaneous absorptions.

8.10.19.3 Background

Every point in space, there is little residual radioactivity, known as "background" which is due to radioactive disintegration from the environment, e.g., natural radioactivity and cosmic rays. Count should be corrected by subtracting background activity. Natural radioactivity can be prevented by shielding but cosmic rays can penetrate it, so that the background cannot be reduced to zero.

8.10.19.4 Scattering

(a) Radiations bombarding any material and their secondary radiations are scattered. So any material present close to the counter causes an increase in count.

(b) Sample containing the radioactive material may produce scattering.

(c) Any object between the sample and the counting system produce scattering.

Therefore, for correct counting, scattering should be considered.

PART XI. Nuclear Radiations and Interactions with Living Materials

8.11 Nuclear Radiations and Interactions

8.11.1 Introduction

Influence of nuclear radiations on living materials depends on the interaction of the radiations with the materials which again depends on (a) the absorption of the radiation by the living material, and (b) the physical events accompanying such absorption.

According to degree of absorption the different radiations can be arranged in decreasing order as alpha-particles, beta-particles and gamma-rays.

All radiations are attenuated in the last part during penetration when its energy changes to heat. During the process an important event is ionization, i.e., the removal of electrons from atoms.

When ionization is the only immediate mechanism of energy absorption, as occurs in case of charged high-energy particles and electro-magnetic radiation, the effectiveness of absorptive material (Absorber) varies directly with the number of electrons in it per unit volume. It is assumed that all materials absorb radiation equally when the radiation passes through the same weight of material. Hence the absorbing power is usually expressed in grams per cm^2 or an equivalent mass per unit area.

8.11.2 Absorption of alpha-particles

Alpha-particles, having large mass and charge, after traversing a short distance through the living matter are stopped very quickly. If the results are plotted, the absorption curve shows square shape which indicates that alpha particles are monoenergetic. They lose energy rapidly and are totally absorbed by the living matter due to large number of collisions with atoms and every time alpha particle ejects orbital electron to ionize atom. The distance of penetration is called the 'range' of the particles.

8.11.3 Absorption of beta-particles

Beta particles are having low mass and negative charge. These are not usually monoenergetic, so that their penetration produces the flat absorption curve. It shows finite range and beyond this range, beta-particles are not seen. It traverses a tortuous course through the matter due to deflections caused by orbital electrons and elastic collisions with atomic nuclei. Beta-particles generally do not react with atomic nuclei but influence the orbital electrons to ionize the atoms. Sometimes high-energy beta particles encounter a type of collision with nuclei resulting in conversion of the beta energy to a photon of soft gamma-rays, known as "brehmsstrahlung". The amount of photon of soft gamma-rays (Brehmsstrahlung) increases with increasing beta energy.

8.11.4 Absorption of gamma-rays

Gamma-rays have high penetrating power into the matter due to high energy and absence of mass and charge. Gamma-rays are absorbed in the matter by three mechanisms:

1. Photoelectric absorption
2. Compton scattering
3. Pair production.

1. **Photoelectric absorption**: Photoelectric mechanism of absorption predominantly occurs in case of soft gamma radiation and materials of low atomic weight. This low-energy gamma-ray is completely absorbed by an atom of the matter by losing total energy. And its energy causes ejection of orbital electron from the innermost shell of the atom and this scattered electron causes ionization of other atoms.

2. **Compton scattering**: Compton scattering predominantly occurs at higher gamma-ray energies. It is the most effective process of attenuation of the usual gamma energies produced in Cyclotron and Reactor. The mechanism is elastic scattering of gamma-ray quanta by electrons. These gamma-ray quanta and electron are considered as particles. In this period, some amount of energy is transferred from the scattered gamma-ray quantum to electron resulting in

Table 8.8: Comparison of G-M Counter and Scintillation Counter

G-M Counter	Scintillation Counter
1. Inert gas in the tube, e.g., argon, helium etc.	1. Crystal of phosphor or scintillator.
2. Quenching material is required, e.g., halogen gas, ethanol, etc.	2. Not required.
3. Electrons and ions are produced by ionization of gas by the process of "ionization by collision".	3. Electrons are produced and no ions.
4. There are two electrodes, anode and cathode, producing secondary electrons.	4. Series of electrodes produce secondary electrons and multiply it.
5. Operated with lower voltage.	5. Operated with higher voltage.
6. Less sensitive than scintillation counter.	6. More sensitive than G-M counter.
7. Efficiency is low.	7. Efficiency approaches 100 per cent.
8. Output does not indicate the type of emission.	8. Distinguishes different types of emission.
9. Less efficient than scintillation counter in detecting γ-rays.	9. More efficient and sensitive to γ-rays.
10. It can detect α-, β-particles and γ-rays.	10. It can detect α-, β-particles and γ-rays.
11. Very minute quantities of radiation cannot be detected.	11. Very minute quantities of radiation can be detected.
12. Poor efficiency for low energy β-particles.	12. Not so much sensitive to low energy β-particles.
13. Very efficienty for detection and counting of high-energy β-particles.	13. It can detect α- and β-particles at faster rate and higher radiation level than the G-M counter.
14. Modification can be done to eliminate low-ionizing γ-rays.	14. No such modification.
15. Smaller than scintillation counter.	15. Bigger than G-M Counter.

lower energy in the scattered gamma quantum and change of direction. Eventually the path through the matter may be lengthened. Later on, many scatterings may be absorbed photoelectrically.

3. **Pair production**: Pair production occurs when high-energy gamma-rays strike heavy nuclei. It loses its whole energy while creating and emitting a pair of electrons, positive and negative. And the total energy is equal to that of the quantum and this energy includes the kinetic energy of the electrons and their rest mass of roughly $\frac{1}{2}$ MeV apiece.

In the process the neutron looses its kinetic energy as heat and the positron follows the same but at the later part recombines with an electron. This combined particles are annihilated by emission of a pair of gamma-rays having intermediate energy (0.5 MeV gamma quanta) in opposite direction, known as "annihilation radiation". The pair production cannot occur unless the energy of the incident gamma photon is at least 1.02 MeV. A gamma photon produces little or no ionization. But the secondary particles produced by these three processes ionize matter through which they pass. Therefore, the gamma photon is, of necessity, detected only by the ionization produced by these secondary particles. The photoelectric effect accounts for only less than 20 per cent of energy exchange by an average gamma-ray, the pair production accounts approximately one-half of the energy loss if the material is heavy and in case of light material the loss is few per cent and by Compton scattering a large proportion of the energy loss takes place.

8.11.5 Absorption of neutrons

Neutrons do not have any charge and they can pass through electron layers and penetrate great thickness of the matter. Fast neutrons are absorbed in minute amount but scatter with loss of energy to any light scatterer like hydrogen. Neutrons having energies of a few volts are absorbed readily by a few atoms but at thermal energies, they are strongly absorbed by many atoms. After absorption there is emission of hard gamma-rays or other nuclear reaction occurs with total energy production of about 6 MeV.

PART XII. Biological Effects of Radiation

8.12 Biological Effects

8.12.1 Introduction

Biological effects from nuclear radiation were observed after the discovery of radioactivity. Becquerel while carrying a vial of radium in his pocket, noticed reddening of the skin. Pierre Curie observed the progress of healing of burn injury after exposing an arm to radium. Animal experiments in 1920 demonstrated the various effects of radiation, and relationship was observed between the dose of radiation and degree of damage.

In 1946, when radioisotopes were produced, knowledge was widened regarding biological effect and the permissible exposure.

8.12.2 Basic principle

The energy required for ionization of an absorber varies with the nature of the molecules present in the absorber, called "ionization potential" which is 10 electron volts for gas. Considering the wavelength, of radiations, the greater the wavelength, the smaller is the energy of the photon. The visible lights, like ultraviolet and infrared light do not have enough energy to ionize the materials. Absorption of radiation energy is a quantum phenomena and finite amounts of energy are transferred in discrete steps to individual atoms of the irradiated matter. For visible light, these quantum energies match the energy level of molecular bonds of the complex compounds that make up living matter.

X-rays, gamma-rays, beta-particles, alpha-particles, neutrons and cosmic rays have short wavelength of the electromagnetic spectrum and possess sufficient energy to exceed the ionization potential of ordinary matters. For these radiations, the quantum energies exceed more and more those of chemical bonds. As a consequence, conductive action ceases and destruction originates in removal of an electron. This process is called "ionization with disruption" of one or more valence bonds of the affected molecule. Therefore, ionizing radiations are basically a destructive phenomena for living matter.

8.12.3 Physical consequences of radiation absorption

The interaction of nuclear radiation with biological materials takes place through two processes:

1. Primarly: The process of ionization events,
2. Secondarily: Through excitation of molecules.

Ions are more reactive than excited nonionized molecules. As the mechanism of absorption of ionizing radiations has been discussed in the previous chapter, let us know the consequences of radiation absorption in the biological material producing the biological effects.

Three types of events might occur in the living body:

1. Photon may pass through the molecule and mechanically damage the molecule directly,
2. The energy of the ionizing particle may be degraded to heat producing high temperature causing damage,
3. While passing through the solvent of the material, the particle may ionize the solvent, producing highly reactive molecules acting chemically on the materials.

8.12.4 Theories of radiation effect after absorption

A. **Target Theory**: Ionizing radiations cause changes in hereditary characteristics in the offspring of irradiated animals. This is attributed to the specific damage to the chromosomes or genes. Ionizing events must have localized a single molecule and presumably the damage is due to direct hit. Ionizing event occurs within a specific sensitive area of the biological material of some cases. The sensitive molecule or senstitive area is known as "target" and ionizing event affecting the material is called "hit".

Demerits: (a) Applicable only in certain types of radiation damage. (b) Some phenomena are not considered for explanation by this theory.

B. **Point-heat theory**: All radiant energy after absorption in the tissue is degraded to heat. Apparently the rise of temperature is not significant (e.g. ,10^5 roentgen raises 0.25 °C) but within small spaces, the temperature is considered to be high and may cause abnormal chemical reactions such as breaking of molecules, coagulation of proteins etc.

C. **Chemical theory**: Tissue contains solvent and solutes, solvent is much more than the solute proportionately. Solvent is body water, which is 70 per cent of the total body weight. Hence water changes due to radiation absorption producing the effect which might be many times than the effect due to direct hit on the solute molecules. Indirect damage is mainly caused by radiolysis of body water which is similar to electrolysis.
When water is ionized, it dissociates immediately. The "free electron" removed from water reacts with other molecules to yield "free hydrogen" which is very reactive and can affect solute molecules. When water molecules are excited due to absorption of energy, which is small enough to produce excitation only, probably a reaction produces H_2O_2.
If ionization occurs in water and if the diffusion times of the ions are large, then two hydroxyl ions come together to form H_2O_2. These free H_2 and H_2O_2 products of water, by ionizing radiation, react with different types of biomolecules causing biological damage.

8.12.5 Phases or sequence of events of radiation damage

8.12.5.1 Physical phase

The physical phase consists of:

(a) **Absorption of ionizing radiation**: It is first event for development of radiation injury. The ionizing radiations having biological importance and absorbed by the tissues are the following:
Alpha-particles, beta-particles, gamma-rays, X-rays, protons, deuterons, heavy nuclei, neutrons negative pimesons (pions). These are called ionizing radiations because the main way of energy dissipation is the ejection of electrons from atoms of interacting materials. It takes about 10^{-16} seconds.

(b) **Ionization**: In the progress of ionization, an orbital electron is ejected from the atom. Chemical bonds holding a molecule together consists of electrons shared between atoms. Now the removal of binding electrons causes dissociation of atoms within the molecules which undergo chemical changes.

(c) **Excitation of atoms**: Another process of energy dissipation by radiation in tissue is excitation of atoms. Here the electrons are not removed completely from the atom but are raised to higher energy level and eventually the atom or molecules is in an excited state. For dissipation of this excess energy, the excited molecules undergo reorganization of the electrons which hold the atoms together. As a consequence, the chemical bonds rupture leading to destruction of the molecule releasing stable and unstable molecular fragments. The time taken by ionization and excitation process is 10^{-12} seconds.

Table 8.9: Change in some important biomolecules by radiation

Biomolecules	Changes due to radiation	Damaging effect in the cell
Enzymes	Changes in tertiary structure due to disruption of chemical bonds.	Enzyme inhibition causing changes in metabolic reaction.
DNA	Disruption of linear arrangement of bases.	1. Inhibition of DNA synthesis. 2. Synthesis of defective DNA. 3. Inhibition of mitosis. 4. Synthesis of defective protein.
Structural lipids in the cell membrane.	Disruption of molecular bonds.	Increased permeability of the cell membrane to Na^+ and K^+, etc.

8.12.5.2 Chemical phase

Absorption of radiation producing ions and excited atoms induces chemical changes in the cell. The chemical changes occur by two ways:

1. *Direct action:* Absorption of energy from radiation directly by the molecules resulting changes.
2. *Indirect action:* Absorption of energy by the molecules from another molecule which acts as an intermediary for transfer of energy. The intermediary in the cell is H_2O which by ionization produces highly reactive ions and free radicals like H^+, OH, H_2O and their products like H_2O_2 which act as powerful oxidising agents react with important biomolecules like DNA, RNA, and enzymes for alteration of their function. The chemical phase occurs within millisecond.

8.12.5.3 Biological phase

A. **Changes in the biomolecules, i.e. at the molecular level**: The important biomolecules are intimately related to cellular function. If they are damaged, the cellular function will be seriously hampered or there may be cellular death. Time involved in this process is from seconds to hours (Table 8.9).

B. **Changes at the different levels of organization** (Time involved hours to years)

(a) **Cellular damage**: Some important changes that occur in the cells due to radiation:

1. Aberration (breakage) of chromosomes.
2. Onset of mitosis is delayed followed by normal mitosis.
3. Complete inhibition of mitosis.
4. Death of cells after few subdivisions.
5. Death of cells without division.
6. Transformation to malignancy.

All types of mammalian cells are not equally sensitive to radiation. It has been observed experimentally that the cells which divide regularly, e.g., haemopoietic stem cells are highly sensitive to radiation. But differentiated cells like muscle and nerve which do not divide are resistant to radiation. But differentiated cells like muscle and nerve which do not divide are resistant to radiation. Based on these observations, Bergonic et al. (1906) formulated a law which states that "radio-sensitivity varies directly with the mitotic activity, and with the number of future divisions of the cells will still undergo and inversely with the degree of morphologic and functional differentiation known as "law of Bergonie and Tribondeau." But many exceptions to the law have been observed, such as small lymphocyte which is much radiosensitive but does not divide. Experimentally, it has been observed that, haemopoietic stem cells, spermatogonia are highly sensitive; endothelial cells of liver, kidney, etc. are moderately sensitive and neurons, RBC, muscle cells, etc. are low sensitive. Radiosensitivities according to law are consistent with the observation on relative sensitivities. In spite of exception, it is still used as a guide for predicting relative sensitivity of cells.

- **Morphological changes in the cell**

The following changes are observed in the cells due to radiation.

1. Staining characteristics of the cell.
2. Vacuolization.
3. Swelling.
4. Increased granularity.
5. Distorition of the intercellular structures.
6. Destruction of the cell boundaries.
7. Nucleus may be completely disintegrated or the chromatin material of the nucleus may be coagulated.
8. Chromosomes may be fragmented or clumped.
9. Viscosity of cellular substances may be decreased.

- **Functional changes**

1. Membrane permeability may be either decreased or increased or both in sequence.
2. May stop mitosis in the equatorial plate stage or it may cause abnormal mitosis.
3. Suppression or restriction of motility, growth, reproductive activity, metabolism and secretory activity.

8.12.6 Sensitivity at different level of organization in different conditions

I. **Cellular**

(a) **Differentiated Cells** (neurons, neuroglia): Are less sensitive to radiation than primitive cells.

(b) **Embryonic and immature cells**: Are more sensitive than the adult cells.

(c) **Cells having higher metabolism**: Shows more radiosensitivity.

(d) **Tissue having more blood supply**: Shows more sensitivity.

(e) **Dose of radiation**: The manner in which the radiation dose is delivered also contributes to the degree of biologic effect, e.g., biologic effect is more when the dose is at higher rate than the lower rate.

(f) **Ionizing effect of radiation**: The greater the ionizing effect, the higher is the injury.

(g) **Proliferating cell**: Rapidly proliferating cells, e.g., germ cells lymphoid tissue cells, erythropoietic cells of red bone marrow, skin epithelia and G.I. mucosa cells are more sensitive than differentiated cells (neurons, neuroglia).

(h) **Selective collection and concentration of radioisotope**: Throid cells selectively take up I^{131} and concentrate them resulting in destruction of thyroid cells. In case of bones, Ra^{226} and U^{238} known as bone seeker are concentrated and may cause cancer.

(i) **Cell component and variation in damage**: Cytoplasm shows less damage due to radiation than the cell membrane and membraned organalles, osteoblasts are more sensitive than osteocytes.

(j) **Age group and variation in damage**: Fetus, infant and children, are more susceptible to damage. In infants, the cells are rapidly proliferating and their tissue mass is smaller.

(k) **State of health**: Sickness, fatigue, etc. make the cells more sensitive to radiation.

II. Tissue damage

The haemopoietic tissue, lymphoid tissue, spermatogenic epithelium, intestinal epithelium are highly sensitive to radiation. The haemopoietic tissues show effect by lymphopenia after whole body doses with 25 rads. Whereas muscle and nerve tissue have low sensitivity and can withstand massive dose of 1000 rads without any injury. The interstitial connective tissue, fine vascular, growing cartilage and bone have moderate sensitivity. Variation of sensitivity may be due to:

1. Different sensitivity of the cells that constitute the tissue,
2. Normal turnover time of the cells,
3. Interaction of the various cell types in the tissue,
4. Capacity to repopulate.

III. Whole body damage

The response of whole body to radiation in man is divided into two groups:

A. **Acute effects**: Occurs within days of exposure.

B. **Late effects**: Requires several months to years for full manifestation.

A. **Acute effects**: When the tissues and organs are exposed to radiation, the changes produced in it cause effects which are collectively known as **The Radiation Syndrome**, some of which are lethal. The time interval between the exposure and the onset of syndromes, duration, probable survival depend on the radiation dose, uniformity of the dose over the body, age, condition of health. The dose-response curve of the whole body for acute radiation lethality shows sigmoid in shape. It reveals the relation between radiation dose and per cent lethality in exposed population. Low doses (below 200 rads) result in no lethality, intermediate doses (about 200 to 600 rads) result in lethality in a fraction of the population directly proportional to dose and high doses (excess of 600 rads) result in 100 per cent lethality in the exposed population. Acute radiation lethality is referred to as LD_{50} and it is not known in case of man but is believed to be 250-450 rads (X-ray or gamma radiation) and comparison of LD_{50} between mammals and man shows that the man is relatively sensitive.

The acute radiation syndrome:

(a) *CNS syndrome:* Lethargy, tremor, convulsions, ataxia.

(b) *G.I. syndrome:* Malaise, anorexia, nauseas, vomiting, G.I. malfunction, diarrhoea, dehydration, fever, loss of electrolytes, circulatory collapse.

(c) *Haemopoietic syndrome:* Malaise, dyspnoea on exertion, fatigue, leukopenia, thrombopenia, purpura.

B. Late effects: Late effects occur in man who survived the acute effects of radiation and are belived to be due to irrepairable damage to the genetic material of the cells which survive after radiation. It takes many months or years after exposure to radiation and the time interval between the exposure and onset of manifestations is called **latent period.**

Late effects are grouped into two:

(a) *Somatic tissues:* Induction of leukaemia, cancers, cataracts, non-specific life shortening.

(b) *Germinal tissues:* Sterility, mutation and genetic errors, foetal death and abortion.

8.12.7 Physiological mechanism of radiation damage

1. **Manifestation of general symptoms**: Symptoms of the radiation syndrome is of a general nature except of some specific damages like cataract, skin burns, etc. Most of the general symptoms can be considered in the light of general adaptation syndrome in which the main factor is exhaustion of adrenal cortical hormones. Though there is similarity between the two with a few differences but still the relationship is not clear.

2. **Direct or indirect radiation damage**: The evidence show that action is indirect but not direct one:

(a) Administration of large dose of radiation to a small part of the body causes radiation causes radiation syndrome.

(b) Irradiated animals show significant amount of histamine or histamine-like substance in the blood.

(c) Histamine administration can produce symptoms of radiation sickness.

In a cross circulation experiment taking two animals and exposing one of them to large dose such as 1500 rads of X-rays did not show any significant symptoms in non-irradiated animals. This experiment revealed the direct action.

Cellular damage may produce histamine and due to radiation, there is damage of the individual cells. Chemical agents which destroy histamine may have a protective actions on irradiated animals. From the above investigations it may be stated that both direct and indirect actions are involved in the radiation damage.

8.12.8 Theories for physiological mechanism in radiation damage

A. **Water Shift Theory**: During radiation syndrome, there is shift of large amount of water. The intracellular fluid volume and the blood volume are increased. There is swelling of nucleus and vacuoles form. These changes may be due to increase in the osmotic pressure of nuclear materials or to change of cell permeability, but the definite role of water shift in the the production of radiation sickness is not known.

B. **Enzyme Inactivation theory**: Barron was the person to study the enzyme inactivation. He observed the enzyme inactivation by breakdown products (H^+ and H_2O_2) of water. These products oxidize the SH group of enzymes to S-S causing inactivation. In case of excess doses, the protein molecules of the enzyme show breakage of chemical bonds. The prominent enzymes inactivated are phosphoglyceraldehyde dehydrogenase, adenosine triphosphatase and succino-oxidase of which the first one has shown 73 per cent inactivation by 300 rads alpha-particles which were partly reversible but reversibility was not observed in case of beta-particles and gamma-rays which may be due to alteration of protein molecule.

C. **Toxin Theory**: It has been postulated, some toxins are responsible and these are absorbed from G.I. tract due to increased permeability of the cells or from the damaged cells and one of the important toxins is histamine. Histamine may be produced either by the activation of the proteolytic enzymes or inactivation of their inhibitors.

8.12.9 Factors influencing the biological effects

There are many factors influencing the biological effect and among these physical factors are very important. The physical factors are given below:

1. **Radiation dose**: Increasing the radiation dose, there is increase in biological effects, for which minimum dose is used in nuclear medicine and radiological diagnosis. The minimum dose means when the radionuclide has an average effective time in the body equal to the time of the procedure of nuclear medicine. Minimum radiation dose has been very much advanced due to application of (a) metastable radionuclides e.g. ^{99m}Tc, which do not emit beta-particles, (b) radionuclides having short half-lives; (c) radiopharmaceuticals which facilitate rapid biological elimination.
2. **Way of delivery of radiation dose**
   i. Higher dose rate produces greater biological effect than the lower dose rate.
   ii. Single exposure produces greater effect than exposures delivered in a fractional way.
3. **Volume of tissue**: The greater the volume of tissue irradiated, the greater the biologic effect, e.g., severe skin reactions in therapeutic radiology because of wide area of skin irradiated.
4. **Nature of quality of the radiation**: The spatial distribution of ionizing events is different for different radiations. It is the general process that the more densely ionizing radiation, the greater the biological effects.

   The average rate of energy loss per unit per length is known as the "linear energy transfer or LET." LET through the tissue indicates the spatial distribution of ionizing events. Radiation with high LET, biologically more effective than low LET radiations which has low ionizing capacity. Gamma-rays and beta-particles have low LET (0.3 and 5.5. keV/μ) and less biological effect than neutrons and alpha particles (119 and 45 keV/μ).

PART XIII. Radiation Damage in the Embryo and Fetus During Pregnancy

8.13. Radiation Damage During Pregnancy

8.13.1 Introduction

Radiodiagnostic methods employed during the pregnancy period should be avoided unless it is essential and fetus are extremely sensitive to radiations. Irradiation of embryo or fetus generally occurs in two ways either from (1) radionuclides concentration in the mother's organs and (2) in the embryo itself when radionuclides cross the placental barrier.

The radiation damage during pregnancy can be grouped into four, such as:

1. Development anomalies,
2. Mutations,
3. Central nervous system damage,
4. Late effects.

8.13.2 Developmental anomalies

Radiation results disturbances in growth during organogenesis which is causative factor of developmental defects. Developing organ or tissue on exposure to radiation may produce absence of any structure, hypoplastic or may be severely distorted. The above effects may lead to congenital anomalies observed at birth which may be due to destruction of some cells and inhibition of mitosis in other cells. The common anomalies are:

(a) Small size at birth
(b) Stunted growth
(c) Microcephaly
(d) Cataract
(e) Microphthalmus
(f) Malformation of genital organs
(g) Malformation of skeletal system.

Studies on developmental defects are based on the experiments on rats, mice, etc. The relevant information supplied are:

(a) Irradiation during first 2 weeks of gestation cause prenatal death in a higher percentage.
(b) Irradiation during 2-7 weeks of gestation causes higher percentage of death rate and developmental defects in neonats.

(c) During radiotherapy of the pregnant women at 4-5 weeks of gestation, the organs are very much sensitive and limbs anomalies are observed.

(d) Malformation of human embryo occurs after exposures to doses of 10 rads or less.

8.13.3 Mutations

Mutations may occur both in somatic and germ cells of the developing embryo of fetus. In case of mutation, it is difficult to find out the safe dose for mutation induction. A condition known as "mosaicism", not detectable at birth but evident during post-natal development. It occurs when some cells in one or more organs contain the mutation(s) while the others are normal. Mutation effects depend on the function of affected genes and number of cells having the mutation.

8.13.4 Central nervous system effects

Whole gestational period is the critical one for central nervous system because of extreme susceptibility of CNS to radiation especially the neuroblasts which are present in the brain as long as 2 years after birth. Radiation causes gross structural defects changing the neurologic function.

8.13.5 Late effects

(a) Leukaemia, (b) Increased risk (50%) of developing childhood malignancies.

PART XIV. Demerits of Different Diagnostic and Therapeutic Methods of Nuclear Medicine During Pregnancy

8.14 Introduction

Methods are employed in nuclear medicine for diagnosis and therapeutic purposes which are hazardous to the embryo and fetus. The danger on damage depends on the dose of radiation. The determining factors for the radiation dose are:

(a) Type of radiation,

(b) Rate at which it passes through the placental barrier,

(c) Metabolic path of radiopharmaceutical in the maternal side,

(d) Nature of distribution of radionuclide in the fetal tissue.

Out of these factors, the potential dangerous factor is the permeability of the placenta to the radiation carrier. Because the placenta is permeable to small molecules like sugars, and amino acids and impermeable to large molecules like proteins, colloidal suspensions and small molecules attached to protein carriers. Stenberg[1] (1973) has shown that nuclear medicine procedures are hazardous to the embryo or fetus. Observations on the radionuclides in pregnancy show that the radionuclides can be grouped into two such as:

1. Permeable to placenta,
2. Impermeable to placenta.

8.14.1 Permeable to placenta

(a) **Thyroid imaging with radioiodine:** Radioiodine easily crosses the placenta and concentrates in the fetal thyroid which 6-7 times higher than the maternal thyroid. It has been observed that administration of millicurie radioiodine in unknown pregnancies caused complete destruction of foetal thyroid associated with severe exopthalmos and significant diminution of brain. Thyroid imaging and functional procedure during pregnancy is hazardous.

(b) **Scanning of pancreas**: Scanning of pancreas with Se^{75} selenomethione is also hazardous. Because this labelled amino acid passes easily through the placenta and can remain in the fetal tissue for a significant period.

(c) **Bone scanning**: Bone scanning is usually done by Ca^{47} and Sr^{85} which are known as "bone seeking isotopes". These radionuclides cross the placenta easily and are taken up by fetal skeleton if there is development of bone. These radio-nuclides irradiate the marrow and soft parts.

8.14.2 Impermeable to placenta

(a) Radiocolloids-Au^{198}, Tc^{99m} for liver scanning.

(b) Hg^{197} Chlormerodrin for kidney scanning.

(c) Macroaggregates of labelled albumin for lung scanning.

---

[1] *Stenberg, J. : Clin. Obstel. gynaeocol., 16, 235, 1973*

(d) Tc^{99m}-HSA, Tc^{99m}-DTPA for brain scanning.

Diagnostic procedures in nuclear medicine produce the same effects due to radionuclide exposure. Some examples are given below:
1. Genetic effects, 2. Late effects.

8.14.3 Genetic effects

(a) Standard dose of I^{131} used for treatment of hyperthyroidism caused chromosomal defects (Cantolino et al., 1996).

(b) Chromosomal aberration were observed in case of radiumdial painters (Boyd et al., 1996).

(c) Aberrations in the peripheral blood cells and bone marrow were observed. Clones of cytogenetically abnormal cells in the bone marrow have been observed. These observations were done during radioactive fall out (Lisco and Couard, 1967).

(d) After receiving colloidal thorium dioxide (thorotrast) 19-27 years earlier, 19 patients out of 20 showed peripheral blood cell aberrations with detectable radioactivity in the liver or spleen. (Fischer et al. 1967).

8.14.4 Late effects

(a) Patients treated with P^{32} for polycythae-mia showed higher incidence of leukaemia (Krauss et al., 1970).

(b) Osteosarcomas were significantly observed in radium-dial painters.

(c) Lung cancer is prevalent due to inhalation of radioactive dust and is common among uranium miners (Miller, 1972).

(d) In diagnostic radiology, colloidal thorium dioxide (Thorotrast) is used. It causes cholangiomas and haemangio-endotheliomas of the liver (Swarm, 1970).

8.14.5 Precautions

1. Amount of radionuclide should be low.
2. Procedure is mainly based on detection of gamma-rays outside the body but most of the dose of gama emitting radionuclides is not due to the gamma-rays only, but other radiations such as beta-particles are also present which does not give any information in diagnosis. Some metastable radionuclides such as Tc^{99m}, reduce the radiation doses as they emit gamma-rays only but not beta-particles.
3. Average time of effectiveness of the radionuclide should be equal to the total time of procedure. Because, after the completion of the procedure the remaining radioactivity is contributed to the dose unnecessarily.
4. Sometimes radiopharmaceuticals are concentrated in the tissues other than target tissue and giving unnecessary radiation. This concentrations can be minimized by:
   (a) Diluting the radionuclide with large amount of nonradioactive isotopes and
   (b) Decreasing uptake of radiopharmaceuticals by administering nonradioactive substances chemically identical or similar to the radiopharmaceuticals.

PART XV. Radiation Hazards in Human Being

8.15 Introduction

Man is constantly exposing himself to ionizing radiations either from natural sources or accidentally.

The radiation sources are:

A. Natural

1. Naturally occuring radioactive elements like U, Th, Ra, etc. in soil, atmosphere and rocks.
2. Radioactive isotopes like K^{40}, Rn^{222} present even normally in small amounts in tissues.
3. Therapeutic and diagnostic procedures of X-rays and radioactive isotopes.

- **Exposure**

A. Chronic exposure

Chronic exposure means, the persons are exposed for a long period, e.g., uranium mine workers, radium-dial maker, etc.

It leads to:

(a) Acceleration of ageing (shortening of life span);
(b) Neoplastic diseases (leukaemia, cancer, etc.)
(c) Genetic effects.

B. Acute exposure

It occurs in atomic explosion, atomic reactor accident.

Onset of manifestations:

(a) Acute effect, (b) Late effect.

(a) **Acute effect**: The symptoms produced are known as "acute radiation syndrome". These are:

1. Haemopoietic syndrome: Occurs at doses up to 800 rads.
2. G.I. syndrome: Occurs at doses 800-1200 rads. Death mainly due to changes in the G.I. system.
3. C.N.S. syndrome: Occurs at doses higher than 1200 rads. See biological effects of radiation.

(b) **Late effect**: See biological effects of radiation.

8.15.1 Damages at different levels of organization

A. Somatic effect, B. Genetic effect.

I. Changes at the biomolecular level (see biological effect of radiation).

II. Changes at the cellular level

(a) Morphological changes, (b) Functional changes (see biological effects of radiation).

III. **Changes in the major organ system**

1. **Blood and haemopoietic tissue:** Radiation of the whole body in the dose range of 500-1000 rads causes decreased concentration of all cellular elements due to:

(a) Direct destruction of the mature circulating cells,

(b) Loss of blood cells from the circulation by haemorrhage or leakage through the capillary walls and

(c) Loss of production of cells.

After exposure of the whole body to moderate doses of radiation, within fifteen minutes the following changes are observed:

(a) Decrease in circulating lymphocytes.

(b) Minimum number of platelets are seen in the second and third weeks which is responsible for haemorrhage and defective coagulation.

Matured RBC is relatively radioresistant.

2. **Bone marrow**: Bone marrow tissue cells may be radioresistant or radiosensitive. Radioresistant cells are fat cells, macrophages, reticular cells, connective tissue cells, etc. Blood forming cells are radiosensitive. The most radiosensitive cells are erythroblasts. The less radiosensitive cells are the myeloblasts but mitosis is inhibited in these cells. Megakaryocytes are reduced in numbers about after 3 days of exposure.

3. **Lymph node**: It lowers the lymphocyte number in the node in very low dose as 10 rads. The stroma is relatively radiosensitive and it is damaged. Regeneration of the lymph node may be restored in low doses and compensate the lymphocyte loss.

4. **Thymus and spleen**: The stroma and Hassal's Corpuscles of the thymus are resistant to radiation. If regeneration occurs, that occurs later and is complete in four weeks after exposure. The white pulp of spleen does not show lymphocytes after a moderate dose. Spleen becomes half the normal size in 24 hours.

5. **Digestive system**

(a) Oral mucosa becomes eroded.

(b) Pharynx and oesophagus are relatively radioresistant.

(c) Stomach is prone to ulceration (1-2 weeks or 1-2 months after irradiation) along with increased mucous secretion.

(d) Small intestine is the most radiosensitive in the G.I. tract. After 500-1000 rads of exposure, within 30 minutes, crypt cells of duodenum show nuclear fragmentations, swelling, etc. Within first few hours mitotic activity of crypt cells stops. At lower doses, recovery of the small intestine is rapid.

(e) Liver is radioresistant due to its tremendous regenerative power. After large doses of radiation, there is damage of vascular systems which eventually lead to degenerative changes in the liver cells; necrosis may occur 1-2 weeks after exposure.

(f) Pancreatic secretion decreases after large doses of radiation.

(g) Large intestine is less radiosensitives. This part is prone to ulceration (1-2 weeks or 1-2 months after irradiation) with increased mucous secretion. Haemorrhages are seen 10 days after exposure due to early changes in blood vessels of the gut wall. Capillary leakage, localized oedema, vacuolerization of muscle fibres, nuclear irregularities appear in the G.I. tract after exposures.

(h) Anorexia, prolonged stomach emptying time, decrease in metabolic transport and absorption, dearranged segmental and propulsive movements are functional changes observed after exposure.

6. **Cardiovascular system**: The heart and large arteries are relatively radioresistant. Endothelial cells of the capillaries and small arteries swell and block the capillary lumen. Destruction of cells and clot formation are additional factors for blockage. Venous pressure, and plasma volume of tissues increase in 0-2 days after exposure. DNA content and lipid of the heart are decreased by the second day. DNA, actomysin concentration, K^+ reaches minimum level at sixth day, ultimately results into cardiac arrythmia. Abnormalities in T-wave, decrease of S-T segment are also associated changes.

7. **Bone**: Non-growing parts of the bone are relatively radioresistant. After 30 days of moderate radiation exposure mitosis of the epiphyseal plate and diaphyseal plate ceases, cells die or degenerate. After two weeks, there is a complete separation of cartilage plate from marrows. At large doses, most cartilage cells and osteoblast degenerate, and acellular, avascular bone-like substances may be formed from the cartilage, resulting in shortening of bones. Large doses to growing and developing teeth produce malformation.

8. **Skin and Hair**: Low doses as 35 rads decrease the rate of cell dimension in germinal layer. The differentiated cells in the more superficial epidermis are radioresistant. At moderate doses, tight rigid structures of the dermis change to a looser tissue. Large doses destroy the cells in germinal layers leading to a condition called "dry desquamation", massive doses lead to a "wet desquamation" where fluid exudate is present. Regeneration occurs if the area of exposure is small. First change is seen in reddening of skin known as "radiation erythema" due to increased blood flow.

9. **Respiratory system**: All proteins of the lungs are relatively radioresistant. Radiation pneumonitis occurs frequently after large doses which is an inflammatory reaction. Oedema, fibrosis of alveolar walls and haemorrhage may occur in the lungs.

10. **Urinary system**: Secondary ischaemic kidney may occur after months or years of the radiation exposure due to changes in the blood vessels. Diminished kidney functions, retention of nitrogen, proteinuria, haematuria and diuresis are observed with large doses after several months of the exposure of the kidney.

11. **Muscles and connective tissue**: Muscles are the most radioresistant tissue. Slight muscular atrophy may result after localised massive doses. Connective tissue has been referred as morphologically radioresistant.

12. **Endocrine glands**: The endocrine glands are generally considered as radioresistant. The endocrine imbalances occur in case of thyroid, adrenals and gonads which may be due to pituitary reactions. After sublethal X-ray doses (580γ), the thyroid gland of rats showed hyperactivity and after lethal doses (800γ), hypoactivity. The activity changes are not constant. The parathyroid glands are considered to be the most resistant of all endocrine glands. After large doses, hypaeremia and an increase in the connective tissue of the gland are seen. Anterior pitutitary is most radiosensitive. Degenerative changes in the adrenals have been observed after heavy local irradiation (more than 5000 γ), but after 1000 γ there was only

minimal morphological alterations in the cortex and medulla of adrenals.

13. **Gonads**: The gonads and their products, especially the spermatogonia and the egg cells are sensitive to ionizing radiation. After low doses exposure, changes in germinal epithelium and ultimately sterilization may result. After irradiation with large doses the organ may atrophy and a temporary or permanent sterility results.
14. **Nervous system**: Neurons of adult animals are radioresistant. But transient functional changes have been reported at low doses. Large doses (more than 600 rads) of whole body irradiation cause increased neuron excitability. Delayed necrosis of brain cells and haemorrhage in brain have been observed. Peripheral nerves are more resistant to ionizing radiation than the brain and spinal cord.

B. Genetic effect

In the chromosomes and genetic material the following changes may occur:

(a) Chromosomes breakage, union and exchange of species,

(b) Mutation with no distinguished change in chromosome.

Induced mutation may give rise to: (1) dominant lethal mutance—causing abortion, neonatal death, even in heterozygous state, (2) sub-lethal mutance– is lethal only in homozygous state and causes only congenial malformation in the succeeding generation, (3) recessive mutance—damage is invisible, expresses only in homozygous state and arrives only after passing through generations.

8.15.2 Protection against radiation hazards

1. Mitotic inhibition with colchicine
2. Anorexia at tissue level, producing by:
   (a) hypothermia,
   (b) hypothyroidism,
   (c) other means.
3. **Scavenging agents**: Minimising the indirect action by neutralizing the toxic radicals formed by indirect action of radiation on water.
4. **Sulphydryl (-SH) donors**: Enzyme containing sulphydryl groups (-SH) are oxidised to -S-S- linkage. To protect, sulphydryl donors are given. Sulphydryl donors are:

Cysteine, methionine acts as -SH donors and thus are radioprotectors.

8.15.3 Radiation hazards in the atmosphere and space

One of the natural source of radiation in the atmosphere is Cosmic rays. Cosmic particles originate from the sun and some from the outer space. Large amount of cosmic particles are continually bombarding the upper atmosphere of the earth. Cosmic particles are high energy electrons, protons, small amount of alpha-particles, carbon, nitrogen, oxygen and iron. Many of these Cosmic particles are trapped by the magnetic field of the earth in two major belts known as **Van Allen radiation belts**. The inner belt extends from the altitude of 300 miles to 3000 miles and the outer belt starts from 3000 miles and extends up to 20000 miles. These belts exist only in the tropical and temperate zones of the earth but not the two poles because of the magnetic field of the earth. The type of radiation in these two belts are usually high energy electrons and protons. If a person traverses these two belts during interplanetary space travel, they may get about 10 roentgens of radiation even with proper shielding. This much radiation is one-fortieth of the lethal dose. If a person is in a spacecraft and orbits the earth within these two belts, he receives enough radiation which is sufficient to cause death.

It is better to orbit in a spacecraft below the altitude of 200-300 miles where radiation hazard is not significant. The radiation hazard can be minimised if the spacecraft leaves or returns to the earth near one of the poles than near the equator.

Part XVI. Fluoroscopy

8.16.1 What is fluoroscopy?

The term fluoroscopy means the use of a fluorescent screen, that is, a sheet of material which fluoresces when X-rays strike it. When a patient is placed between a source of X-rays and the screen the X-ray image becomes a visible light image as long as X-ray present which can be observed with naked eyes.

8.16.2 Advantages

1. Speedy and easy method.
2. Makes possible the study of movement.
3. Can be watched and elucidate among dynamic function as it occurs such as swallowing, breathing and the opening and closing of the valves in the heart.
4. Movement of diaphragm whether it is paralysed or not
5. Pt. can be examined by changing the body in different positions. Such as feet downwards, head downwards, horizontal and other position.
6. Able to make a film of image.
7. Mass screening of population: school, college students, factory workers or office workers etc.
8. To assess the relationship of an opacity to other structures more accurately by observing it with different degrees of rotation on the screen.
9. To confirm valvular calcification.
10. To confirm other intra-cardiac calcification, because this is much better visualized by screening with an image intensifier than on a simple film.

8.16.3 Component

In order to perform fluoroscopy the basic components are:

1. X-ray tube
2. Patient
3. Fluorscopic screen8: Nowadays it has become advantageous for:
   (a) Capable to make a radiographic film (fluorography) or other record of screen appearance.
   (b) Requirements of radiation protection.
   (c) Image intensifier can be placed behind the screen for brightness and amplication of the image.
   (d) Image can be observed by television camera, television monitor called television fluoroscopy or intensifier fluoroscopy or recorded by cine camera.

8.16.4 Type of fluoroscopy

There are two forms of fluoroscopy:

I. **Direct Fluoroscopy** (Fig. 8.18)
II. **Indirect Fluoroscopy**

I. Direct Fluoroscopy

In this procedure, X-ray beam from X-ray tube penetrating the body, the image is visualized on the screen and the radiologist or observer can observe. An X-ray film can be placed behind the fluoroscopic screen to obtain a permanent radio-graphic image called **fluoro-graphy**.

II. Indirect Fluoroscopy (Figs. 8.19, 20, 21, 22)

Apart from direct visualisation of fluorogram i.e., fluoroscopic image, there are other methods of fluroscopy to make the image 1. Sharp, 2. Bright, visualization recording and storage the indirect methods are employed such as:

1. Television fluoroscopy or intensifier fluoroscopy.
2. Closed-circuit TV fluoroscopy.
3. Image intensifier cinefluroscopy.
4. Video fluoroscopy (for recording image, storage or play back).
5. Marker fluoroscopy: Using this procedure the aortic valve of the heart can be studied. Aortic leaflets were surgically tagged with radio opeque markers and imaged with high speed X-ray.

1. X-ray tube.
2. Spot film device (serial changer).
3. Fluorographic camera.
4. Iris diaphragram (to reduce light for TV camera tube avoid over exposure).
5. Television camera tube.
6. Video tube Recorder.

8.16.5 Fluorescent screen

The fluorescent screen consists of a **plastic base** coated with a thin layer of fluorescent material **Zinc Cadmium Sulphide** is bonded to a **lead glass plate**. Through which the image is **visualized** and **provide protection from radiation**. Zinc cadmium sulphide emits light of 558 nm and it is used because it produces yellow-green light and the eye is most sensitive in the green part of the spectrum and a colour which affords us the greatest visual acuity (ability to separate or recognise the small details of an image) at low levels of illumination. [Materials which change invisible radiation to luminous radiation which is visible to naked eye are known

Fig. 8.17: Plate 1 X-ray table used for fluoroscopic examination

as phosphor and is luminescence. When the material is luminous only so long as it is irradiated and it is known as fluorescence but differs from phosphorescence where the material continuous to emit light for certain period after withdrawal of radiation.]

8.16.6 Disadvantage of fluoroscopy

1. **Radiation exposure more than the radiography causing radiation hazard.**
2. **The image is poor compared to X-ray film such as:**
   (a) Sharpness
   (b) Brightness
   (c) Contrast
   (d) Colour
3. **Radiation exposure is to be reduced with the help of accessories such as image intensifier etc.**
4. **More secondary radiation causing hazard.**
5. **Eyes of Radiologist should be accustomed for 10-25 seconds in the dark room before fluoroscopy.**

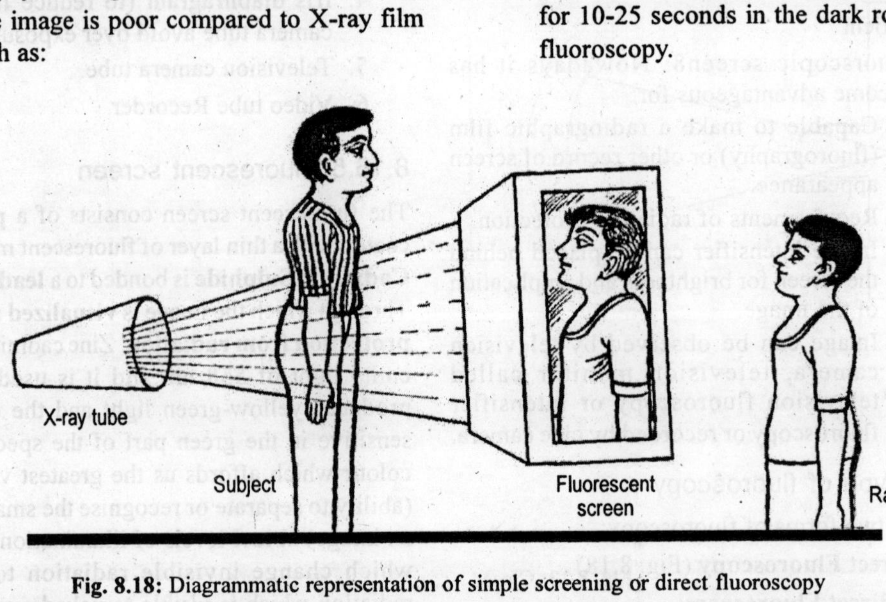

Fig. 8.18: Diagrammatic representation of simple screening or direct fluoroscopy

- **Clinical use of fluoroscopy (Screening)**
  1. Encysted pleural effusion.
  2. Suspected obstruction emphysema due to inhalation of foreign body.
  3. Assessment of diaphragmatic function.
  4. Study of mechanical properties of aortic valve which anisotropic, has shown variation of thickness across the leaflets.
  5. Marker fluoroscopy has shown changes of leaflets of aortic valve during cardiac defect cycle.
  6. Checking each structure in turn noting any deviation from the normal or difference from other side. Thus an irregular bony defect in a rib with an adjacent soft tissue swelling may represent the evidence of **metastasis.**

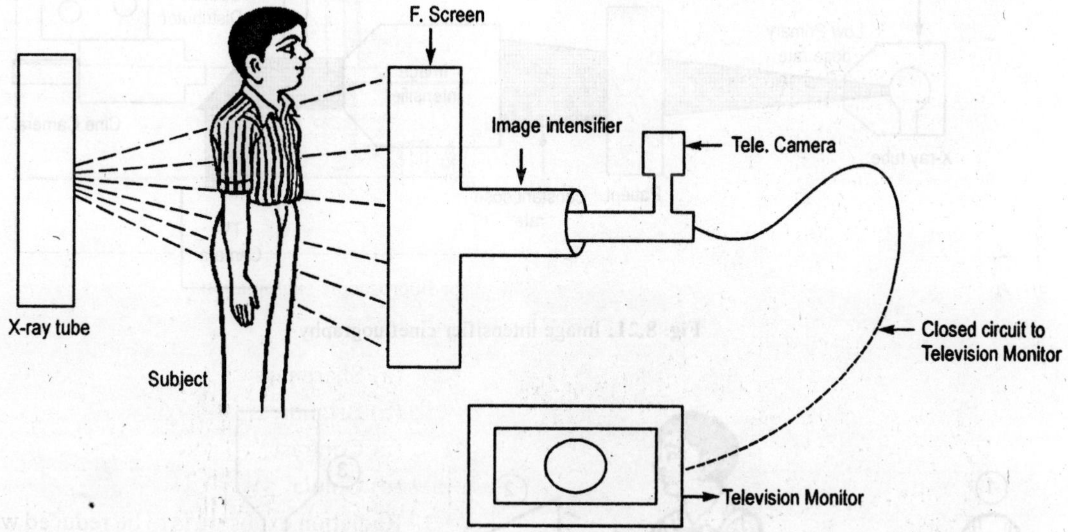

Fig. 8.19: Television fluoroscopy or intensifier fluoroscopy

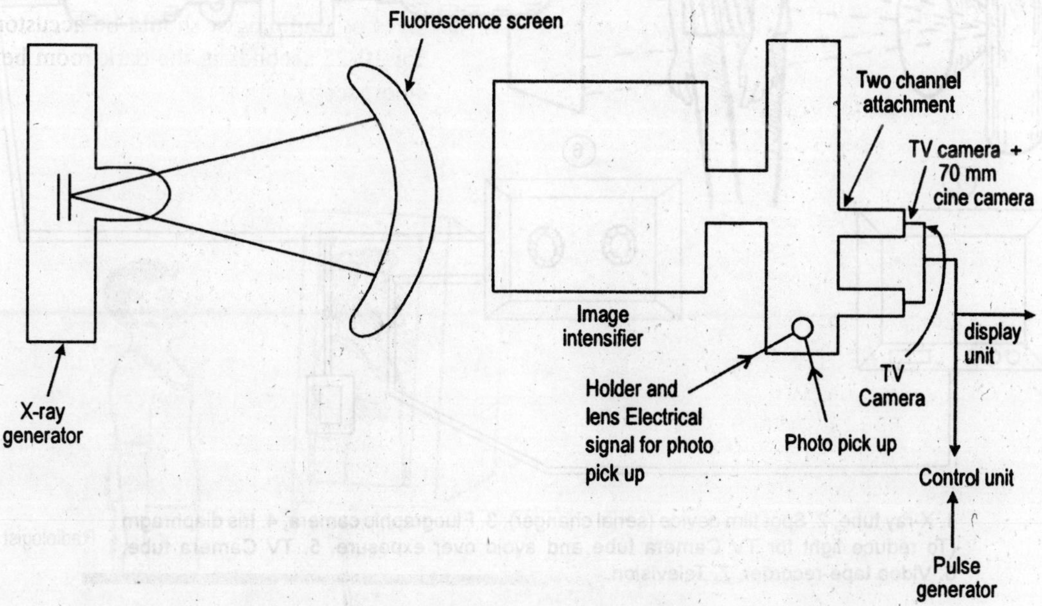

Fig. 8.20: Closed circuit TV fluorography

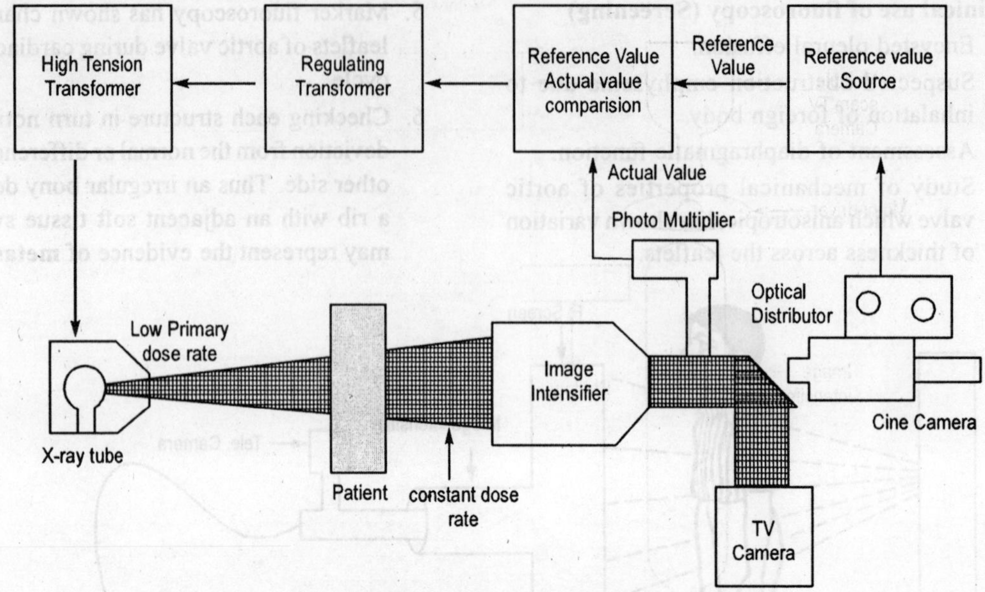

Fig. 8.21: Image intensifier cinefluography

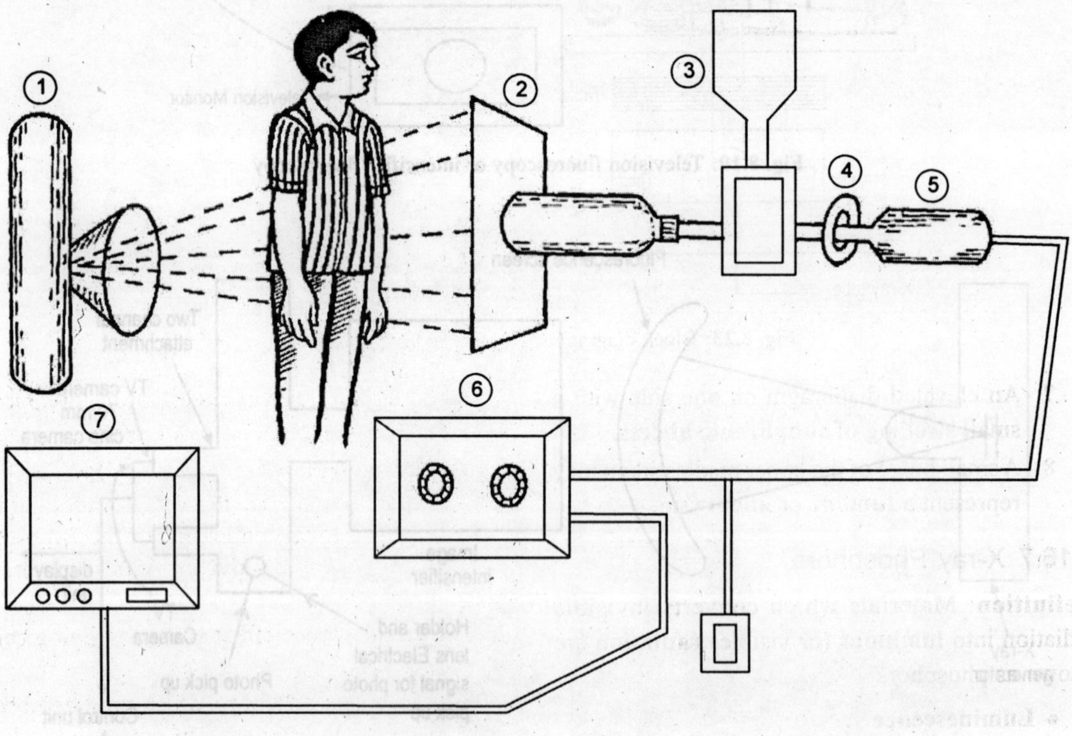

1. X-ray tube, 2. Spot film device (serial changer), 3. Fluographic camera, 4. Iris diaphragm (To reduce light for TV Camera tube and avoid over exposure, 5. TV Camera tube, 6. Video tape-recorder, 7. Television.

Fig. 8.22: Schematic diagram of video fluorography

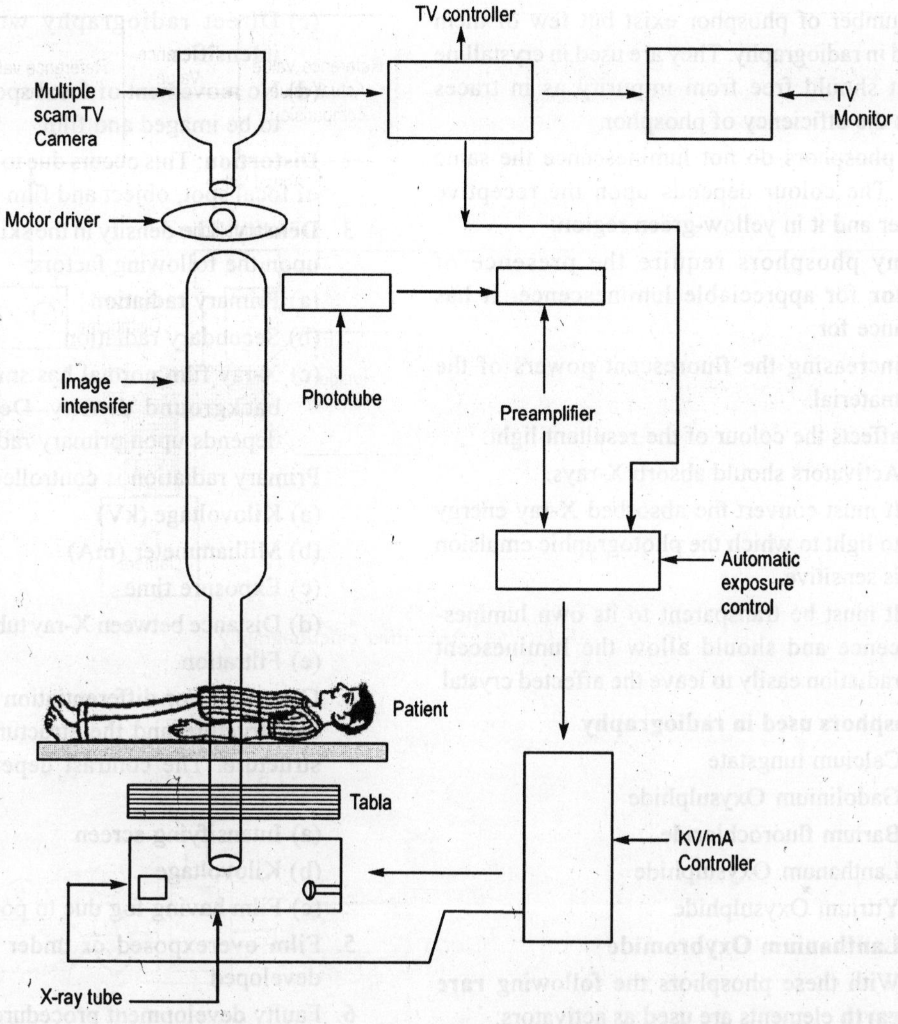

Fig. 8.23: Block diagram of Digital Fluoroscopic system

7. An elevated diaphragm on one side with a small swelling of **subphrenic abcess.**
8. A small bulge of the upper mediastinum may represent a **tumour or aneurysm.**

8.16.7 X-ray Phosphors

Definition: Materials which converts invisible radiation into **luminous** (or visible) **radiation** are known as phosphors.

• **Luminescence**

The emission of light from a substance bombarded by radiation is termed **luminescence.** It consists of two effects:

(a) **Fluorescence**: Luminescence is excited only during the period of irradiation and disappears on termination of X-ray exposure.

(b) **Phosphorescence**: It is after glow i.e., the irradiated material continue to emit light for a period after termination of the exposure. After glow, is disadvantageous as it blurs to moving images.

The luminescence effect is used radiologically in two ways:

(a) To obtain an image which may be observed on the **fluorescent screen** e.g., fluoroscopy.

(b) To increase the photographic response of silver halide in radiographic film (fluorography).

A number of phosphor exist but few of them are used in radiography. They are used in crystalline form. It should free from impurity as in traces reduces the efficiency of phosphor.

All phosphors do not luminescence the same colour. The colour depends upon the receptive character and it in yellow-green region.

Many phosphors require the presence of **activator** for appreciable luminescence. It has importance for:

(a) increasing the fluorescent powers of the material.
(b) affects the colour of the resultant light.

1. Activators should absorb X-rays.
2. It must convert the absorbed X-ray energy to light to which the photographic emulsion is sensitive.
3. It must be transparent to its own luminescence and should allow the luminescent radiation easily to leave the affected crystal.

- **Phosphors used in radiography**

1. Calcium tungstate
2. Gadolinium Oxysulphide
3. Barium fluorochloride
4. Lanthanum Oxysulphide
5. Yttrium Oxysulphide
6. **Lanthanium Oxybromide**

With these phosphors the following **rare earth** elements are used as activators:

1. Terbium, 2. Thulium, 3. Europium.

8.16.8 Factors for best X-ray image (Skiagram)

Good image should fulfill the following criteria:

1. **Definition** (maximum)
2. **Distortion** (minimum)
3. **Density** (optimum)
4. **Contrast** (optimum)

1. **Definition**: Means degree of sharpness, contour of lines of the object would be visible in the image. Sharpness depends upon the following:
   (a) Smallest focal spot.
   (b) Long distance between the focal spot and the subject.
   (c) Direct radiography without image intensifier.
   (d) No movement of focal spot and the part to be imaged and film.
2. **Distortion**: This occurs due to nonalignment of focal spot, object and film parallel.
3. **Density**: The density in the skiagram depend upon the following factors:
   (a) Primary radiation
   (b) Secondary radiation
   (c) X-ray film normal has small amount of background density. Density mainly depends upon primary radiation.

   Primary radiation is controlled by:
   (a) Kilovoltage (kV)
   (b) Milliammeter (mA)
   (c) Exposure time
   (d) Distance between X-ray tube and the film
   (e) Filtration.
4. **Contrast**: For differentiation of shadow of the structure and the structures within the structure. The contrast depends upon the factors are:
   (a) Intensifying screen
   (b) Kilovoltage
   (c) Film having fog due to poor storage
5. Film overexposed or under exposed and developed
6. Faulty development procedure
7. Proper illumination of the video box screen.

PART XVII

8.17 Image Contrast

Contrast

When a high-density radiograph is placed in an illuminator, the differences in optical density and result in the image is called contrast.

These optical density differences are called **radiographic contrast**.

(a) **High contrast radiograph**: A radiograph that has sharp differences in optical density to called a high-contrast radiograph.
(b) **Low-contrast radiograph**: If the optical density differences are small and not distinct.

Types of Radiographic contrast or image contrast:

I. Radiographic (objective) Contrast

Difference in optical density recorded in different parts of processed film.

(a) **Subject Contrast:** Differences in radiation intensities emerging from different parts of the patient i.e., determined by the size, shape, and X-ray attenuating characteristics of the subject.

(b) **Film Contrast:** It is inherent in the film and influenced somewhat by the processing of the film.

II. Subjective Contrast

It is due to personal appreciation of the differences in optical density on image being viewed.

Subject Contrast

1. **The region of patient being examined** – thickness difference: density difference. The greater the density, e.g., mass per unit volume of a tissue, the greater is its ability to attenuate X-rays. It may be due to atomic number difference. Photoelectric absorption is higher in substances with higher atomic numbers when KVp (Kilovolts peak) X-rays are used. Use of low KVp (below 30) X-rays is essential to produce greatest possible difference in photoelectric X-ray absorption between muscle and fat, both of which have low atomic number elements unlike bone, e.g., mammography.
2. **Kilovoltage**: At lower KVp there is high subject contrast.
3. **Contrast media**: Both positive (high density) and negative (low density) contrast media increase subject contrast.
4. **Pathology of the affected part**: Pathological part changes the density of the part, e.g., osteoporosis of the bones decreases the bone density.
5. **Scattered Radiation:** It reduces the subject contrast.

- **Film Contrast**

It is inherent in the film and influenced somewhat by the processing of the film.

1. **Intensifying screen**: It increases contrast. The same film exposed without screens has lower average gradient thereby less contrast.
2. **Type of film**: Double coated films give better contrast than single coated.
3. **Film Fog**: It may be due to faulty dark room, safelight or poor film storage.
4. **Film Density**: The film contrast changes with density, e.g., toc and shoulder regions.
5. **Development**: It can be controlled by careful control of developer, temperature, activity and development time. Increasing the time or temperature of development will:
   (a) Increase the average gradient (Decrease film contrast)
   (b) Increase film speed (Increase density for a given exposure).
   (c) Increase fog thereby decrease film contrast.
6. **Fog**: These are silver halide grains in the film emulsion which are developed though they are not exposed to light of X-rays. It causes unwanted film density which lower radiographic contrast. True fog is increased by:
   (a) Improper film storage, high temperature or humidity;
   (b) Contamination;
   (c) Exhausted developer solution;
   (d) Use of highly sensitized grains i.e., high speed film.

- **Subjective Contrast**

This factors on which subjective contrast depends:

(a) Radiographic contrast;
(b) The observer;
(c) Viewing box: Brightness, uniformity and colour of illumination;
(d) Ambient light.

PART XVIII.

8.18 Pattern of X-ray Beam and Recording

Pattern of X-ray beam which has been transmitted through the patient i.e., radiographic examination of a patient the following feature occurs.

I. Incident X-ray beams

Beam of uniform intensity: The X-ray beam of uniform intensity passes from the X-ray tube into the patient. As the beam penetrates through the body

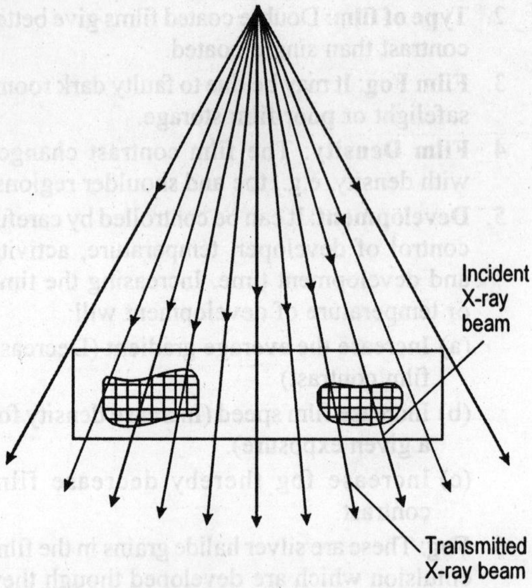

Fig. 8.24: Incident X-ray beam

tissues it gets **modified** or is **modulated**, each part of the beam being **attenuated** to a degree which depends on the particular tissue structure and thickness through which it has passed and this beam is known as **incident X-ray beam** (Fig. 8.24).

II. Transmitted X-ray beam

When the X-ray beam finally emerges from the patient, it no longer uniform intensity.

The beam shows **variation in intensity** depending upon the different tissues it has penetrated. It displays a pattern of intensities which carries information about the structures the X-ray beam has penetrated. This pattern of intensifies can be referred to an **invisible X-ray image**. The modified X-ray beam after penetration of body tissue called **transmitted X-ray beam.**

Conversion of Invisible X-ray image into visible high image

1. Employing **photographic effect of X-rays** on a sheet of **sensitive film**.
2. Using direct exposure X-ray film and then obtain chemically processing the visible image is obtained. The film is called **direct exposure X-ray film.**
3. Using fluorescent screen the X-ray beam produces visible image for a short time but to make it **permanent**, a X-ray film is placed behind screen called screen type X-ray film.

• Types of Radiation

There are three types of radiation emerging out from the X-ray tube:

1. Primary radiation or beam or useful beam
2. Composite beam or secondary radiation or non-useful beam two types:
   A. Leakage radiation
   B. Scattered radiation (Fig. 8.25)

1. Primary Radiation

It is called useful beam:

(a) Short wave length

(b) More penetrating

(c) Most intense

(d) Most hazardous

(e) Most difficult to protect against it.

It come out from the **focal spot.** It enters the body and after absorption, some amount comes out of the body in different direction called scattered radiation.

2. Secondary Radiation

Secondary radiation is of two types:

A. **Leakage radiation:** This radiation comes out from the **edge of the focal spot.** It is:
   1. Non useful beam
   2. Long wave length
   3. Non penetrating

It is emitted from the X-ray tube housing assembly in all directions other than that of the useful beam. If the tube housing is properly designed, the leakage radiation will never exceed the regulatory limit of 100 mR/hr (26 μc/kg/hr) at 1 m. Although in practice leakage radiation levels are much less than this limit, 100 mr/hr at 1 m is used for barriers (Fig. 8.25).

• Secondary Radiation Barriers

Barriers prepared to shield areas from secondary radiation are called **secondary radiation barriers.**

1. Lead is not so necessary for secondary barrier because compensation usually occurs in less than 0-4 mm lead.
2. This radiation barriers are less thicker than primary radiation barrier.

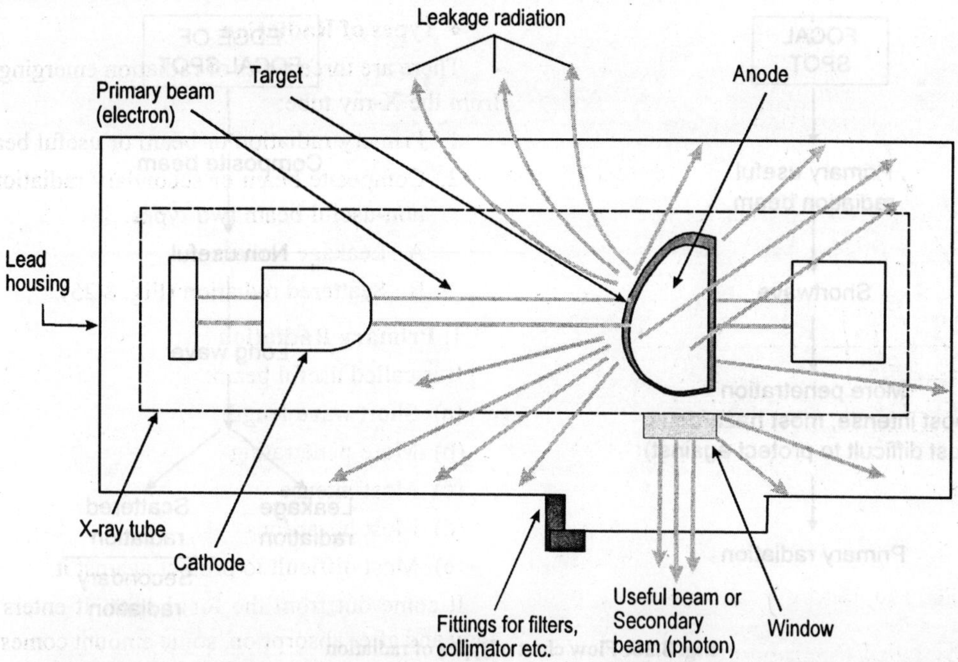

Fig. 8.25: Lead protective housing reduces intensity of leakage radiation to less than 100 mR/hr at m

3. Conventional gypsum board, glass or lead acrylic is sufficient.
4. Walls can be protected with four thickness of 5/8 inches gypsum board.
5. Usual control booth is protected from secondary protective barrier and primary radiation is never directed to control booth. Fourth thickness of gypsum board and 1/2 inch plate glass may be applied. Sometimes glass 1/2 inch thick can be used in control booth barriers.
6. Composition of the barrier and the additional distance will reduce more exposure to the operater.

B. Scattered radiation (Figs. 8.26, 27)

For the purpose of protective shielding calculations, the scattering object can be considered as a new source of radiation. During radiography and fluoroscopy, the patient is the **single most important scattering object**. As a general rule of thumb, the intensity of scatter radiation 1 m from the patient is 0.1% of the intensity of the useful beam (primary radiation) at the beam.

- **Factors affecting Scattered radiation**
  1. Kilovoltage,
  2. Field size or reduction
- **Control or reduction of Scattered Radiation:** The scattered radiation can be controlled by the introduction of the following:
  1. Aperture diaphragm
  2. Cones and cylinders
  3. Variable: Aperture collimator
  4. Collimator: filtration
  5. Orientation of the patient (To place the region of exposure as close to the film as possible).

8.18.1 Means and ways to control or minimizing scattered radiation

1. **Aperture diaphragm**: It is noted that while emitting radiation from the focal spot, the beam i.e., primary beam or useful beam are more penetrating having short wave but the radiation emitting from the edge of the focal spot are more composite beam i.e., useful and non-useful beams are mixer of short wave and long wave beams. By using an

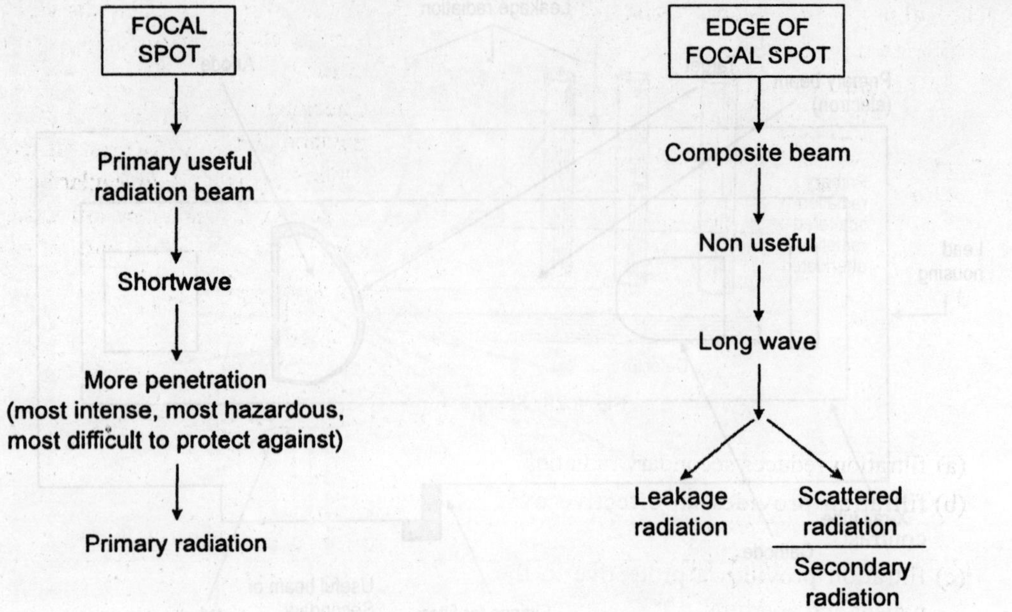

Fig. 8.26: Flow chart of types of radiation

aperture diaphragm at the X-ray tube opening considerable amount of composite beam which contains more long waves can be short waves. The shorter the opening more control rays of useful beam are obtained.

2. **Cone:** Another way to reduce the long waves by using different sizes of cone which is metal cylindrical cone of different height use of aperture diaphragm and cylindrical cone of different size and height should be in relation to object and sizes of the film. So it is to be observed that by judicial use of aperture diaphragm and cone, secondary scattered radiation can be reduced and thus contrast density and sharpness of the radiography image can be obtained.

3. **Aluminium filtration:** While cone diaphragm reduces the area of radiation which eliminates fogging of the film by reducing the scattered radiation. Another method is used which is the addition of simple aluminium filter at the opening of the X-ray tube which will act upon the beam itself and change it to a more desirable and useful quality through this filtration acts in three ways to improve the beam of X-ray:

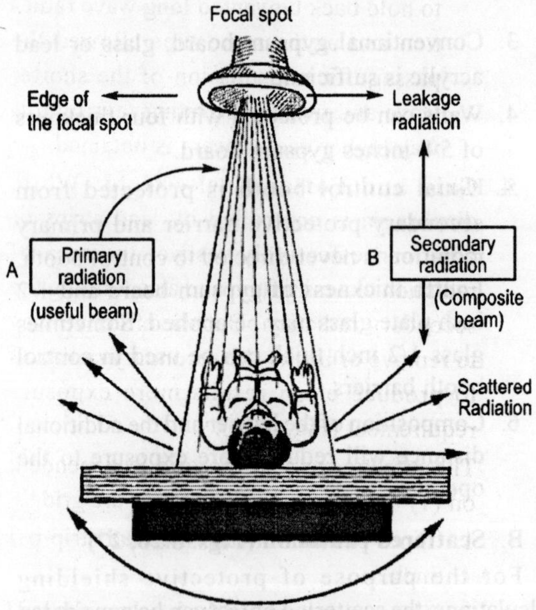

Fig. Showing two types of radiation—
A. Primary radiation (useful beam)
B. Secondary radiation (Leakage radiation and scattered radiation)

Fig. 8.27: Schematic representation of origin of different radiation

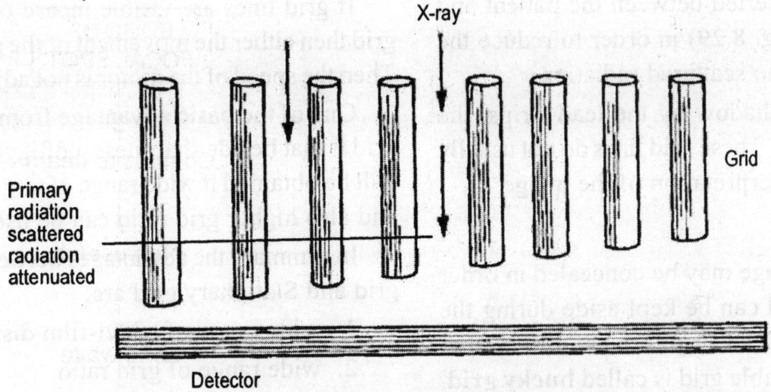

Fig. 8.28: Scatter reduction grid

(a) filtration reduces secondary radiation
(b) filtration provides an effective over contrast.
(c) filtration provides a protective to the patient and aperture.

The usefulness of the filtration principle is based upon the ability of the material to hold back unwanted long wave radiation and at the same time allowed the passage or transmission of the shorter wave and more penetrating rays by this way higher useful beam is obtained.

4. **Grid and Potter Bucky Grid**: While diaphragm is controlled by lead strips or cone can effectively restrict the area of radiation to a minimum and also limiting the secondary radiation. But the effective way to remove or bucky the secondary radiation to produce clean results more exposure requirement (Fig. 8.28).

The effectiveness of the grid is dependent on (1) lines per inch (2) ratio of the grid.

1. Lines per inch means the lead strip per inch.
2. The ratio is between the height of the lead strip is eight times the gap or spaces, then the grid ratio is 8 : 1.

Higher the grid ratio more kVp is required, generally upto 10 : 1 grid ratio, the effective kVp is upto 100 kV.

If the grid ratio is 16 : the effective kVp is between 90 to 140 kVp.

GRID

A grid made of strips of lead, separated by another material, for application of with low energy beam. This is normally low attenuated material, (Cotton) whereas with high energy beam aluminium is used. The grid may be left stationary, but will then cause the display of the grid lines. Alternatively it may be moved in same or reciprocating motion. The grid chosen is dependent on the application. Densities of 80 to 150 lines cm^{-1} are employed. Moving grid with a line density of 30 cm^{-1} and cotton or proper interspace are used where minimising the dose is essential.

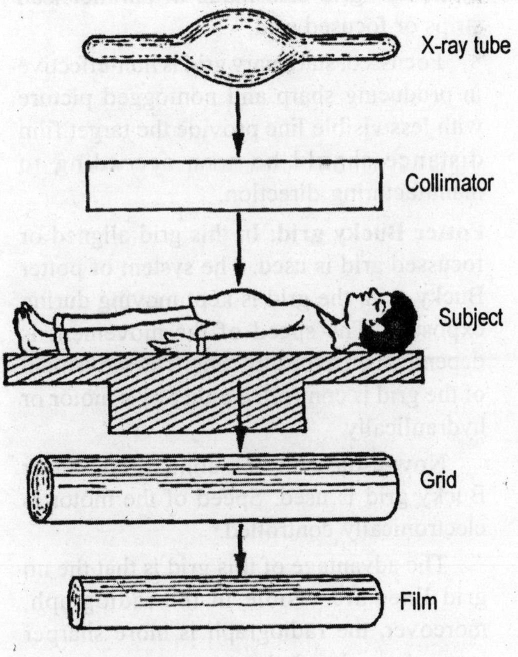

Fig. 8.29: Position and function of grid

The grids are inserted between the patient and the film cassette (Fig. 8.29) in order to reduce the loss of contrast due to scattered radiation.

Because of the shadow by the lead strips, the final image is striped. These grid lines do not usually interfare with the interpretation of the image.

- **Disadvantage**

Details in the image may be concealed in order so avert this the grid can be kept aside during the exposure so that the lead strips are not observed in the image. This movable grid is called **bucky grid**.

- **Use of grid**

There are two types grid:

1. Stationary
2. Potter Bucky grid

1. **Stationary grid**: This grid is used as a fixed grid. Generally, in this grid, thick grid lines with another short lead strip lines are visible. So finer the lines the visibility is less or to a minimum. Here the factor of grid ratio is very important, because if the lines are finer or more in an inch the cut off of the secondary rays is minimum due to the grid ratio. For a time, the ratio can not be made higher, due to the manufacturing hazard. Stationary grid also made of parallel lead strips or focused grid.

   Focussed, stationary grid is non-effective in producing sharp and nonfogged picture with less visible line provide the target film distance should be scen according to manufacturing direction.

2. **Potter Bucky grid**: In this grid aligned or focussed grid is used. The system of potter Bucky grid, the grid is kept moving during exposure. The speed of the movement is dependent on exposure time. The movement of the grid is controlled either by a motor or hydraulically.

   Now-a-days motor controlled Potter Bucky grid is used. Speed of the motor is electronically controlled.

   The advantage of this grid is that the no grid lines are visible in the radiograph, moreover, the radiograph is more sharper less unfogged and clear.

If grid lines are visible incase of Potter Bucky grid then either the movement of the grid is too slow. Then the speed of the motor is not adjusted properly.

One of the basic advantage from this stationary grid is that beside sharpness, unfogged clear picture will be obtained if wide range of target film distance and also higher grid ratio can be used.

In summary the advantages between Potter Buck grid and Stationary grid are:

1. wide range of target-film distance
2. wide range of grid ratio
3. unfogged clear, sharper image

COLLIMATORS

To increase the contrast image and so reduce the dose of X-ray beam, it is necessary to restrict the beam to the particular are to be visualised. This can be done by two ways:

(a) Collimator or

(b) Grid

The **collimator** is placed between the X-ray tube and the subject. It connects a sheet of lead with circular or rectangular slit (hole) of suitable size. Alternatively it may consists of the adjustable lead strips which can be moved relatively to each other.

In practice it is advisable to use smallest possible field size. This results in a low dose to the subject and simultaneously increase the image contrast because less scattered radiation reaches the image plane. The scattered radiation produces diffuse illumination and fogging of the image without increasing its information content and therefore, by choosing the smallest possible field size and using a collimator, the loss of contrast due to scattered radiation is reduced, collimators are usually provided with an optical device by which the X-ray field size can be exactly calculated by a light field. This can be projected on the patient to ensure proper positioning of the opparaters.

- **Use**

1. X-ray
2. Gamma-Camera
3. SPECT (Parallel hole collimator as used for imaging organs).
4. brain for special fan beam collimator are used.

5. Cobalt-60 machine
6. Medical Linear accelerator machine.

PART XIX

8.19 Factors Governing Exposure

The number of factors which determined the good radiography i.e., high radiographic quanity are:

1. (a) Kilovoltage (X-rays)
   (b) Type of radioactive source (gamma rays)
2. (a) Milliamperage (X-rays)
   (b) Millicuries (gamma rays)
3. Time of exposure
4. The distance from the source to the film
5. Type of X-ray film used
6. The intensifying screens used
7. Processing conditions
8. Filtration
9. The X-ray apparatus in use and the circuits employed
10. The object being radiographed.

8.19.1 Exposure timer

For radiographic examination the number of X-ray reaching the image receptor is very important because it is related to:

1. X-ray tube current
2. The time that the X-ray tube is energized.

X-ray operating console contains wide selection of beam on times and if it used along with mA then a wrider selection of mA value can be obtained.

For this purpose a timer circuit has been provided in X-ray machine which is separate from the other main circuit. It has either

(a) Mechanical device or
(b) Electronic device

The function is to 'make' and 'break' of the high voltage across the X-ray tube which is done on the primary circuit of the high voltage section.

There are five basic type of timing circuits of which four are controlled by technologist and one is automatic.

8.19.2 Types of Timers

1. **Mechanical timers**
   (a) Simple device
   (b) Used in dental and portable
   (c) Operated by clock work
   (d) Inexpensive, not very accurate
   (e) There is dial by turning a knob which winds a spring, when exposure button is depressed the spring is released and unwinds.
2. **Synchronous timers**
   1. When electric current of 50-60 Hz, a special motor called synchronous motor having revolution of 60 rpm.
   2. In some X-ray machine synchronous motors are used as timing mechanism.
   3. Minimum time is usually 1/60 second.
   4. Can not be used for serial exposure, must be reset after each exposure.
3. **Electronic timers**
   (a) Most sophisticated, complicated, and accurate.
   (b) Allow wide range of time intervals
   (c) Used for rapid serial exposure upto 60 per second.
   (d) Modern X-ray machine is provided electronic timers.
4. **mAs timers**
   (a) This is also electronic timer and for control of tube current and exposure time
   (b) If monitors the exposure when the product of mA and time and terminates the exposure for any mA selected.
   (c) It determines the number of X-rays emitted and therefore the density on the film.
   (d) Provides highest safe tube current and therefore exposure for shortest time of any mAs selected.
5. **Phototimers**
   (a) It measures the quantity of radiation reacting the image receptor.
   (b) Automatically terminates the exposure when sufficient radiation to provide the required optical density has reached.
   (c) Most common type is provide a flat, parallel plate ionization chamber between the patient and the film.
   (d) The chamber is made radiolucent so that it does not interfere with the radiographic image.
   (e) Ionization within the chamber creates a charge proportional to the film density.

When the appropriate charge is reached, the exposure is terminated.

8.19.3 Special techniques of X-ray

(1) Grids, contrast media, (2) Angiography, (3) Cardiac Catheterization, (4) 3-D visualization Tomography (Tomo = Slice or section), axeroradiography).

8.19.4 Visualization of X-rays

1. Fluoroscopy (Screen) → Radioscopy
2. Radiography (X-ray film)
3. Image intensifier.

PART XX

8.20 Xero-Radiography

Definition: Xero-radiography uses dry paper and reversed images rather than film used for mammography.

8.20.1 Principle

Xero-radiography is a method of imagine using a special equipment in which X-rays are made to discharge on **already charged selenium plate,** instead of making a photoimage on ordinary radiographic films. The sharp-image is then transferred onto a **thermoplast paper** with the help of blue powder (Toner) and is fixed by heats. The procedure of patient positioning and imaging is just like routine radiography except for the receptor and film processing. It is cheaper than X-ray films and finds special use in imaging of breast, soft tissue of neck, bone lesions.

8.20.2 Procedure

A selenium coated, aluminium plate is charged in a special charge unit and is ready for use. It is placed under the part to be imaged just like a film screen cassette. X-ray exposure is made. The plate gets discharged in proportion to the X-ray energy received by it after passing through the body part and a latteral image is formed on the plate.

The toner i.e., blue powder which is also charged, is sprinkled on the plate in a specialised processing unit, and a visible image is then formed. It is then transferred onto the thermoplast paper specifically provided for xero-radiography.

The selenium plate once used, has to be passed through the charging unit again to recharge before it is fit to be used.

The Xero-radiographs are superior to radiographs due to the edge enhancing effect by virtue of which the images are sharp and stand out better in a background of homogeneous density. The chest and retro-mammary space is well demonstrated as skin and subcutaneous fat due to wide latitude. Thus vascular margins and micro-calcifications in breast imaging are better visualised. Trabecular detail of bone lesions are better imaged and soft tissue masses of larynx are better detected (Fig. 8.30).

8.20.3 Methods of X-ray examination

1. **Simple Radiography**: With photographic plate i.e., X-ray film with exposure of 0.1 of a second a less (Fig. 8.31).
2. **Tomography**: Tomography is a variation of simple X-ray film method. In tomography, the **tissue section radiography** is obtained. In this method **X-ray tube moves in one direction** and the **film** moves in the **opposite direction**. The two are connected by a rod which can be made to a pivot at variable point A. Since A remains stationary during the whole procedure the part of the body in line with A is the only part which will clearly shown several films can be taken at the same time by **multi-section box**. In this way **multiple body sections** can be obtained with **a single exposure**. Thickness of section depends upon the angle of X-ray tube. Smaller the angle, thiner the section.
3. **Screening (fluoroscopy) and the image intensifier**: Screening or fluoroscopy mean passing of X-ray beam through the patient on the fluorescent screen containing **phosphor.** The visible image can be observed from the opposite of the screen the fluorescent image.

8.20.4 Angiography

An angiography is the radiological investigation of the blood vassel by making blood vassel visible with the help of iodine containing contrast medium.

- **Type of contrast**: It should be low osmolar agents because high osmolar agents cause pain and complications:

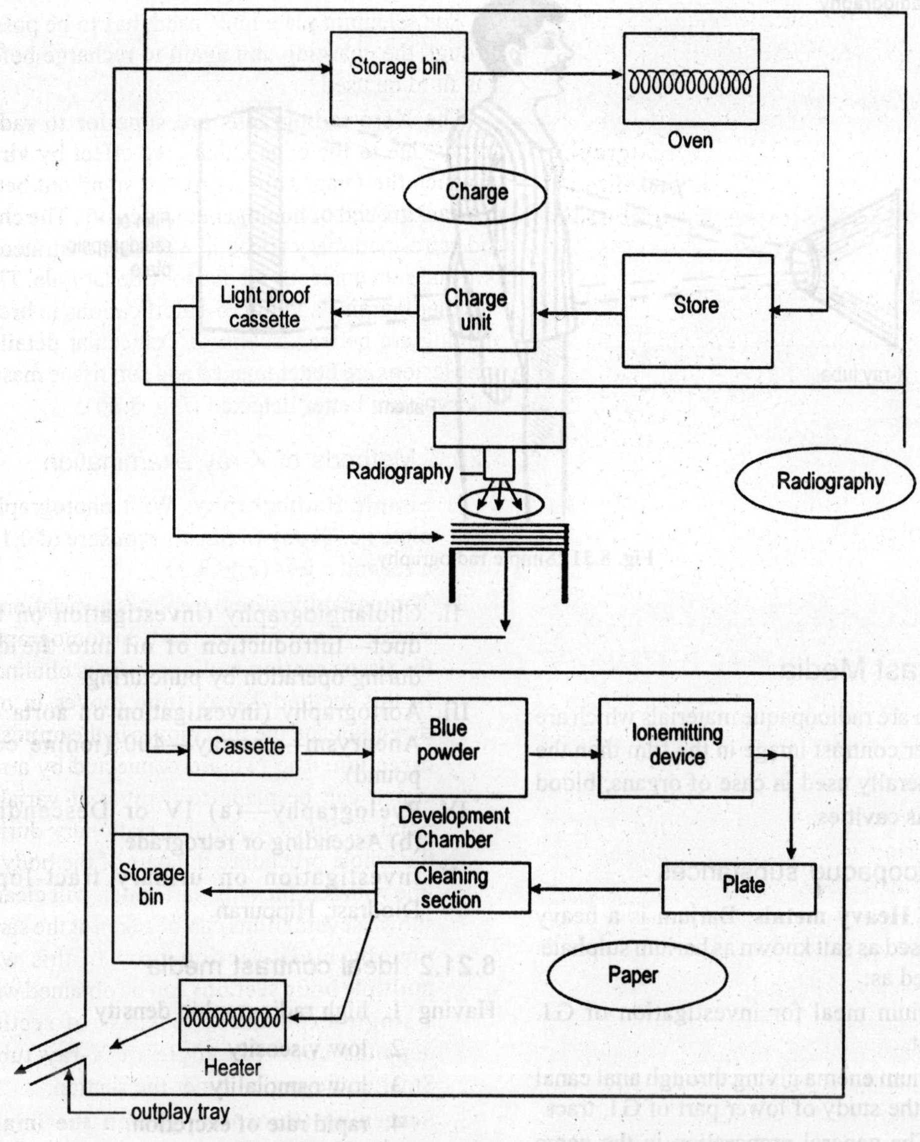

Fig. 8.30: Block diagram of xero-radiography processing unit

8.20.4.1 Division of angiography

1. Arteriography: Investigation of artery
2. Venography (Phlebography): Investigation of vein.

8.20.4.2 Information

The factors require for optimal information are:

1. High quality imaging
2. Recording in cine film
3. Opacity of the blood vessel
4. Reviewing the film carefully

8.20.4.3 Major indication

1. Disorders of blood vessel: Occlusion, stenosis, throumbi, aneurysm, vascular, malformation.
2. Anatomical information before surgery.
3. Tumours, cyst, having abnormal vascularity.
4. Application of DSA (digital subtraction angiography).

Simple Radiography

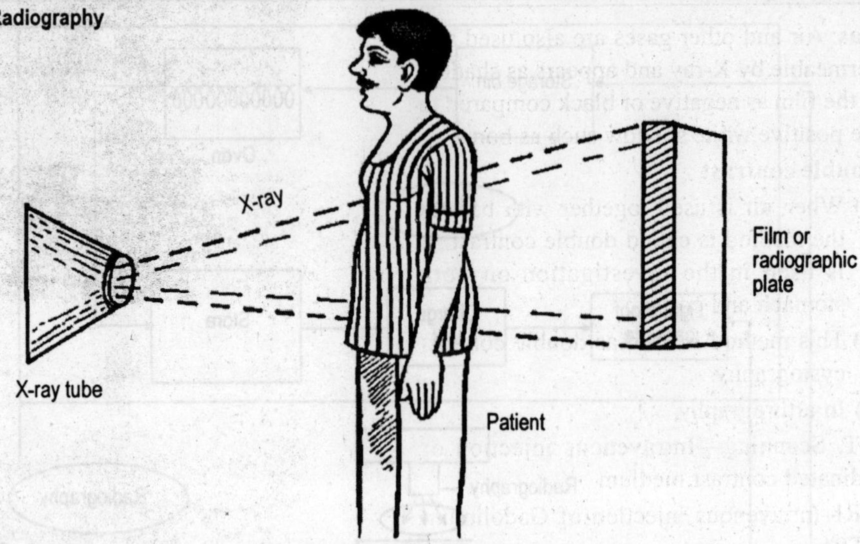

Fig. 8.31: Simple radiography

PART XXI

8.21 Contrast Media

Contrast media are radioopaque materials which are used for greater contrast image in the film than the body. It is generally used in case of organs, blood vessels, various cavities.

8.21.1 Radioopaque substances

1. **Salt of Heavy metals**: Barium is a heavy metal used as salt known as barium sulphate. It is used as:
   (a) Barium meal for investigation of G.I. tract,
   (b) Barium enema giving through anal canal for the study of lower part of G.I. tract.

   The general preparation in the name of microopaque is available. Barium causes better coating of the mucous membrane of the G.I. tract. Usually the parts are examined by barium meal are: oesophagus, stomach, duodenum, in follow through cases, whole G.I. tract, colon by enema.

Some radioopaque substances i.e., contrast media used in different systems:

I. Cholecystography (investigation on gall bladder)—Iodine compound—Teleopaque (Sodium tetra iodophenolphthalein) Biloptin, Solusiloptin

II. Cholangiography (investigation on bile duct—Introduction of oil into the duct during operation by puncturing.

III. Aortography (investigation on aorta e.g. Ancurysm—cowray—400 (iodine compound).

IV. Pyelography—(a) IV or Descanding, (b) Ascending or retrograde.

V. Investigation on urinary tract-Iopas, Diodrast, Hippuran.

8.21.2 Ideal contrast media

Having 1. high radiographic density
2. low viscosity
3. low osmolality
4. rapid rate of excretion.

8.21.3 Cases and dose

1. In aortography—Contrast media containing 350-370 mg iodine/ml and for selective angiography the requirement of contrast medium is 300 mg iodine per ml.
2. The contrast media should be warm to the body temperature before injection for reduction of its viscosity.
3. The total volume of contrast media should not exceed 4-5 ml/kg of body weight.
4. Water soluble organic iodide—Are also used for myelography.

5. **Gas**: Air and other gases are also used and permeable by X-ray and appears as shadow in the film as negative or black compared to the positive white shadow such as bone.
6. **Double contrast**
   (a) When air is used together with barium the method is called double contrast. It is used in the investigation on both stomach and colon.
   (b) This method is used as double contrast cystography.
   (c) In arthrography.
7. C.T. Scanning—Intravenous injection of iodinated contrast medium.
8. MRI–Intravenous injection of Gadolinium DTPA.

PART XXII

8.22 Digital Subtraction Angiography (DSA) or Digital Vascular Imaging (DVI)

Definition: It is the method in which digitized angiographic images are subtracted in order to provide contrast enhancement of the opacified vasculature. Clinically temporal subtraction in the algorithm used, though energy subtraction methods also fall under the generic term DSA. It is also known as DVI or Digital Vascular Imaging.

8.22.1 X-ray projection angiography

Angiography is a diagnostic and therapeutic modalities related to the diseases of the vascular system. There are many procedures e.g.,

(a) Ultrasound
(b) Computed Tomography
(c) Magnetic Resonance Imaging
(d) Angiography

are available either clinically or for research for vascular study but there is another procedure in where clinical angiographic study can be done by projection angiography.

8.22.2 Principle

In this procedure, the vessel under investigation is opacified by injection of a radiopaque contrast agent.

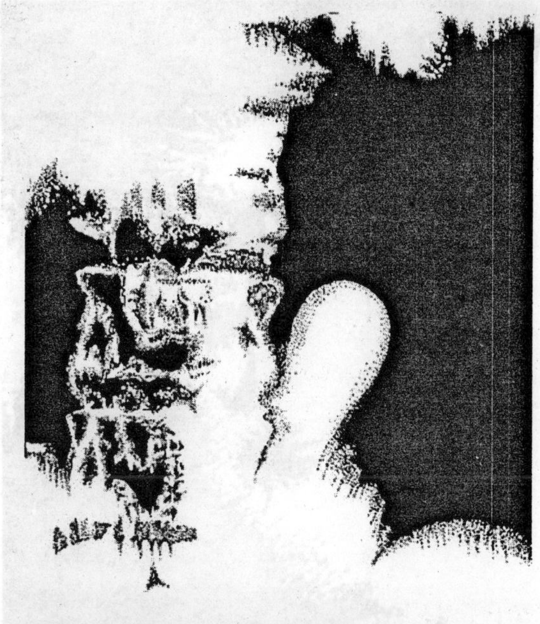

Fig. 8.32: An X-ray showing the normal gall bladder filled with an opaque media

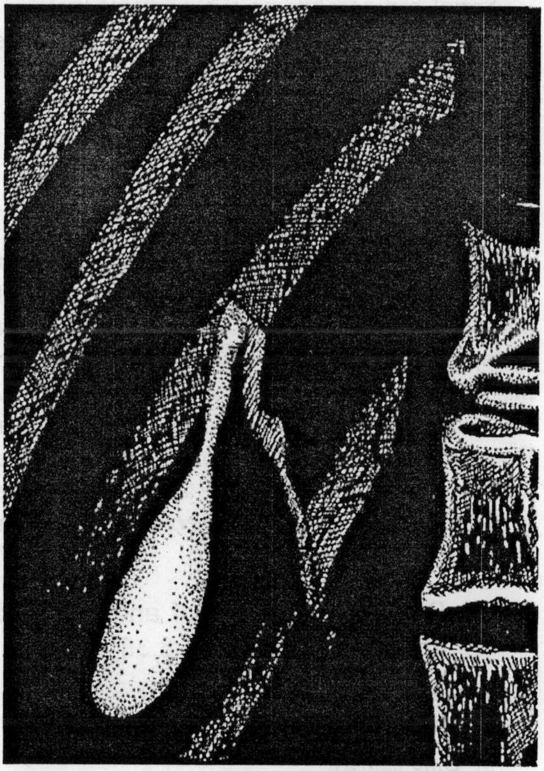

Fig. 8.33: The gall bladder, emptying after a fatty meal showing the bile ducts

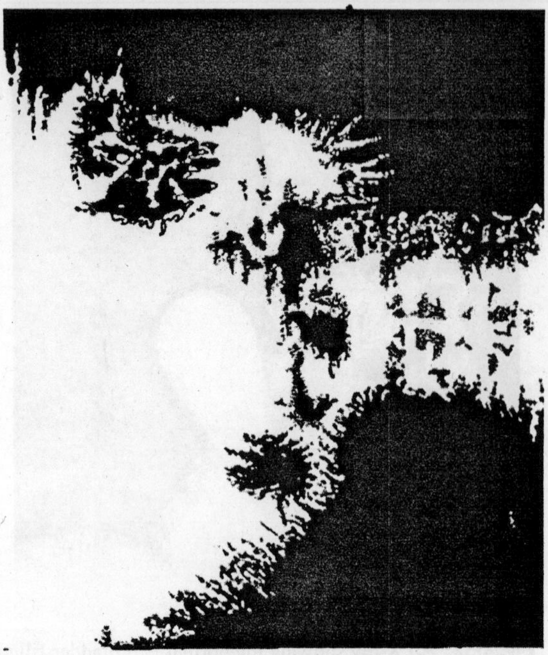

Fig. 8.34: Renal Angiogram

Serial radiographs of the contrast material flowing through the vessel are taken. This is done in either of the following ways:

(a) In an angiographic suite
(b) In a special procedure Laboratories
(c) In a Cardiac Catheterization Laboratory.

- **Contrast Material**

Contrast material is to opacify vascular structures because the radiographic contrast of blood is the same as that soft tissue. Contrast material consists of an iodine compound with maximum iodine concentration of about 350 mg/ cm^3. Contrast material is injected through a catheter having diameter from 1 to 3 mm, depending on the injection flow rate to be used. Radiographic image of the contrast filled vessel are recorded using either film or video.

8.22.3 Basic components of angiographic imaging system (Fig. 8.35)

1. Image intensifier
2. Video camera
3. Cine camera
4. Digital image Processor (optional)
5. Film changer (Optional)
6. X-ray tube (see X-ray)
7. **X-ray generator**: High power, sophisticated X-ray generation system is required for production of short, intense X-ray pulses to produce clear images of moving vessls. The exposure time range from 100-200 ms for cerebral studies, 1-10 ms for cardiac studies.
8. **Image Intensifier**: It is essential for modern angiographic investigation. The function is:
   (a) to produce a light image with sufficient brightness to allow the use of video and film camera.
   (b) to produce an output image of small enough size to allow convenient coupling to video and film cameras.
9. **Optical Distributor**: The image present at the image intensifier output phosphor is coupled to the video camera and cine camera, by the optical distributor. The components of the optical distributor are:
   (a) Lens
   (b) Beam splitter
   (c) Cine aperture
   (d) Cine lens
   (e) Video aperture
   (f) Video lens

- **Video System**

The video system consists of:
(a) Camera head
(b) Camera control unit (CCU)
(c) Videomonitors
(d) Video recording device
(e) Digital Image processor

Hardware is structured as a pipeline processor to Digital image processor.

- **Performance of Digital Image Processor**

Real time processing at Video rates. It meets requirements:
(a) Image subtraction
(b) Integration
(c) Spatial filtration
(d) Temporal fillration algorithm

Look up tables (LUTs) are used to perform intensity transformation e.g. contrast enhancement and logarithmic transformation.

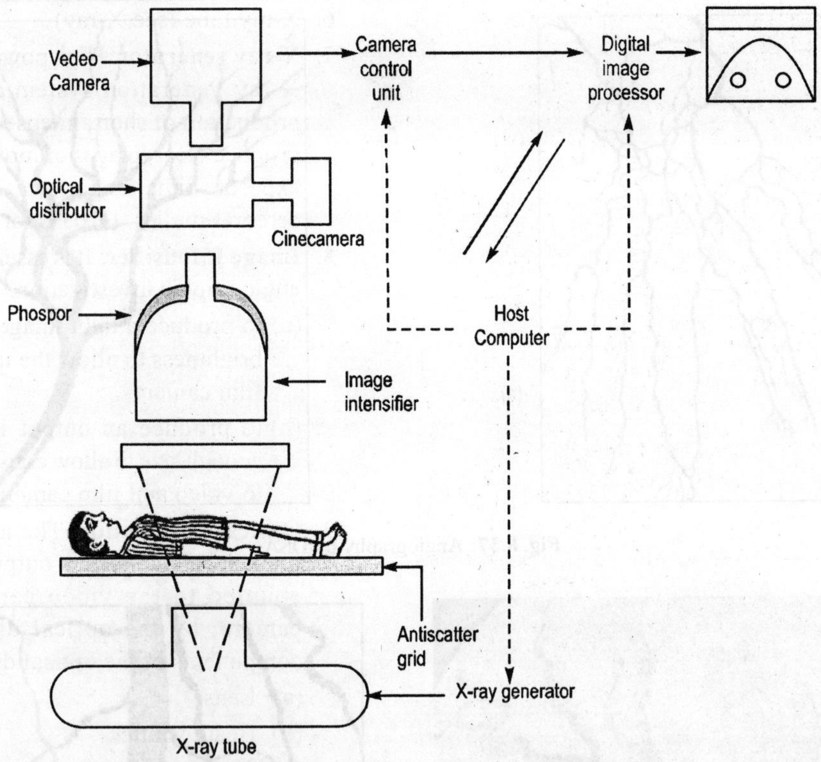

Fig. 8.35: Schematic representation of an image intensifier-based digital angiographic and cine imaging system. Solid arrow = image signals, Interrupted arrow = Control signals

Host computer: It is used to control pipe line processor and X-ray generator respond to user input and perform non-real-image manipulations.

Image storage: It should provide both:

(a) Online—provided by video rate digital diks

(b) Archival storage—long term storage

The image access to image provided by real-time disk technology and advantegeous over film.

- **Cine**

Cardiac imaging retains a significant dependence on film. Cine angiography is performed with 35 mm motion picture camera optically coupled to the image intensifier output phosphor.

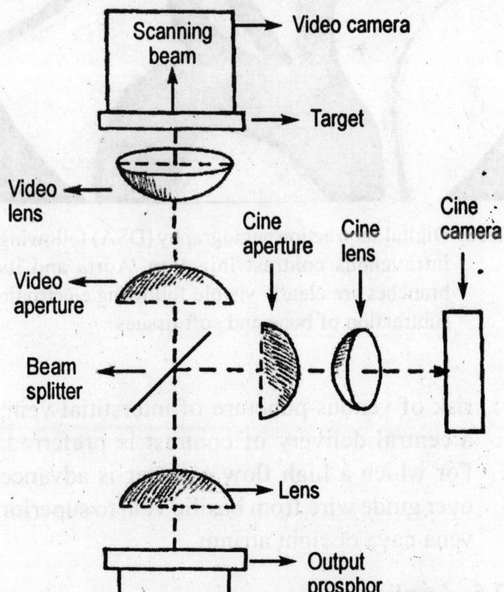

Fig. 8.36: Schematic representation of optical distributor coupling image intensifier output phosphor to a video and cine camera

8.22.4 Methods of DSA

I. **Intravenous DSA**: It is the usual procedure on outpatients basis, contrast medium is injected into the venous system and images of arterial system are taken in the area of interest as the contrast passes through it. About 30-40 ml of contrast medium is given at the rate of 20-25 ml per second. There is

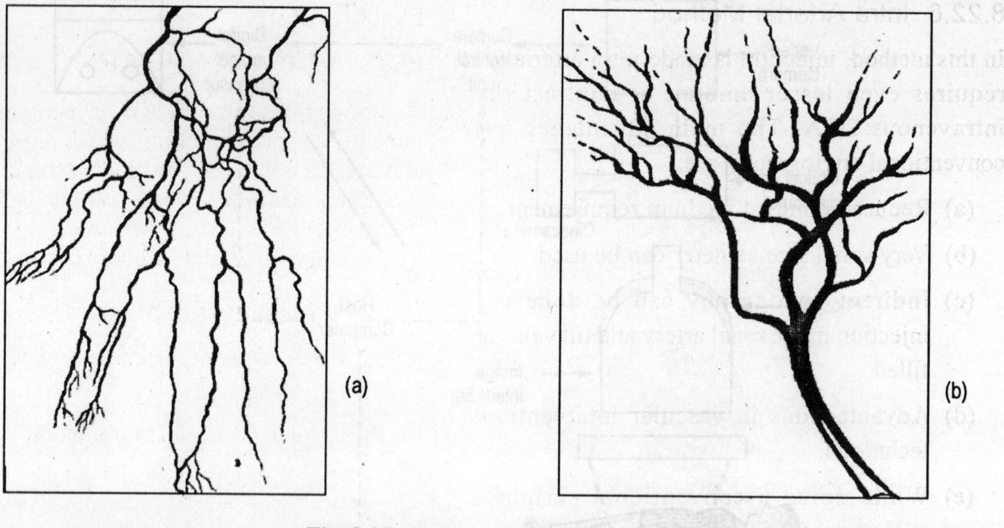

Fig. 8.37: Angiography by DSA

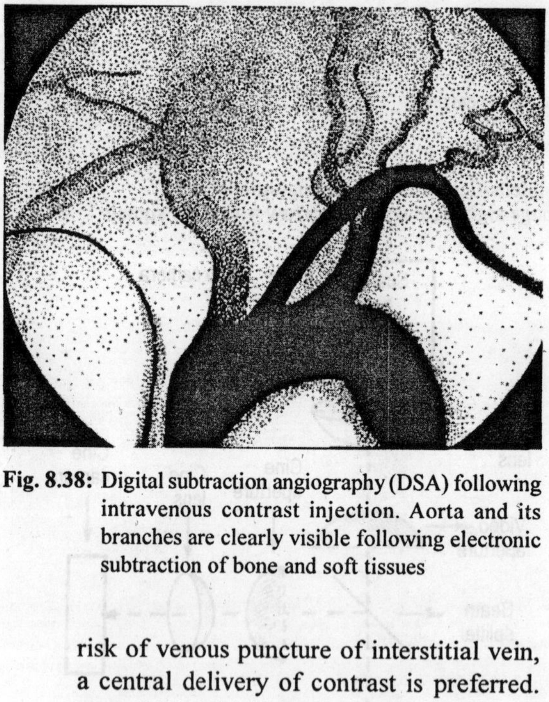

Fig. 8.38: Digital subtraction angiography (DSA) following intravenous contrast injection. Aorta and its branches are clearly visible following electronic subtraction of bone and soft tissues

risk of venous puncture of interstitial vein, a central delivery of contrast is preferred. For which a high flow catheter is advance over guide wire from basilic vein to superior vena cava or right atrium.

8.22.5 Application

1. Extracranial carotid arterial disease
2. Subclavian artery disease
3. Aortic arch anomalies
4. Pulmonary embolic disease

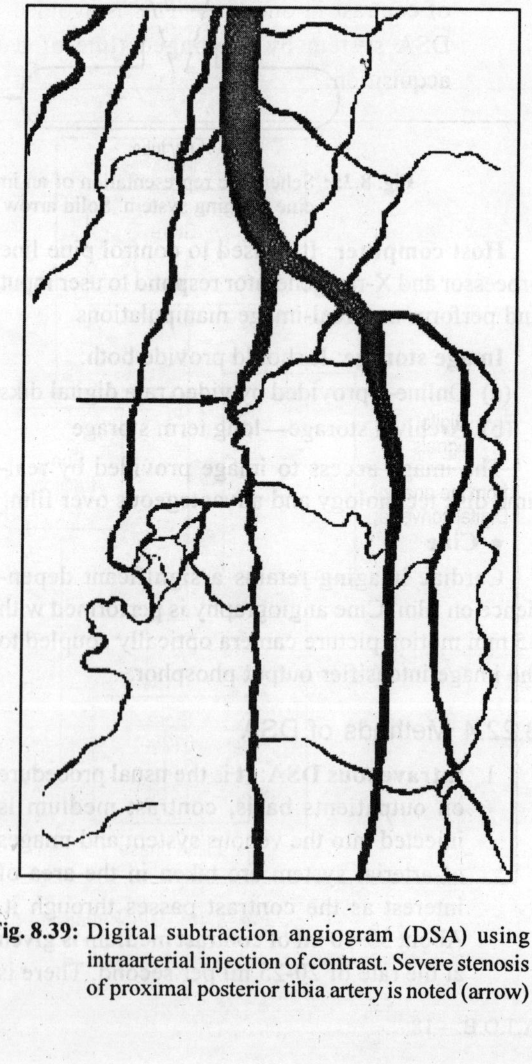

Fig. 8.39: Digital subtraction angiogram (DSA) using intraarterial injection of contrast. Severe stenosis of proximal posterior tibia artery is noted (arrow)

8.22.6 Intra Arterial Method

In this method, injection is made intra arterially if requires even lesser amount of contrast than intravenous DSA. The main advantages over conventional angiography are:

(a) Reduced contrast medium requirement.

(b) Very small size catheter can be used.

(c) Indirect venography can be done e.g., injection in the renal artery and till veins are filled.

(d) Advantageous in vascular interventional technique.

(e) While doing a conventional peripheral angiography, if there is differential flow in two limbs, one might not capture the flow of contrast in one limb. This is avoided in DSA system by prolonged time of data acquisition.

PART XXIII

8.23 RADIOTHERAPY

Radiotherapy is the branch of medicine which is done by ionizing radiations which produce specific and predictable biological changes in diseases e.g., malignant tumours. Radiotherapy includes both X-ray beam as well as radiation from radium and various other artificial radium substitutes.

8.23.1 Ionizing radiation

I. Electromagnetic radiation

(a) X-ray, (b) γ-ray

II. Particle (perticulate) radiation

(a) Electrons (β particle), (b) Neutrons, (c) Protons, (d) Pimesons, (e) Heavy ions, (f) Other particles.

8.23.2 Types of radiation application

(a) External beam radiotherapy-application of beams from outside the body.

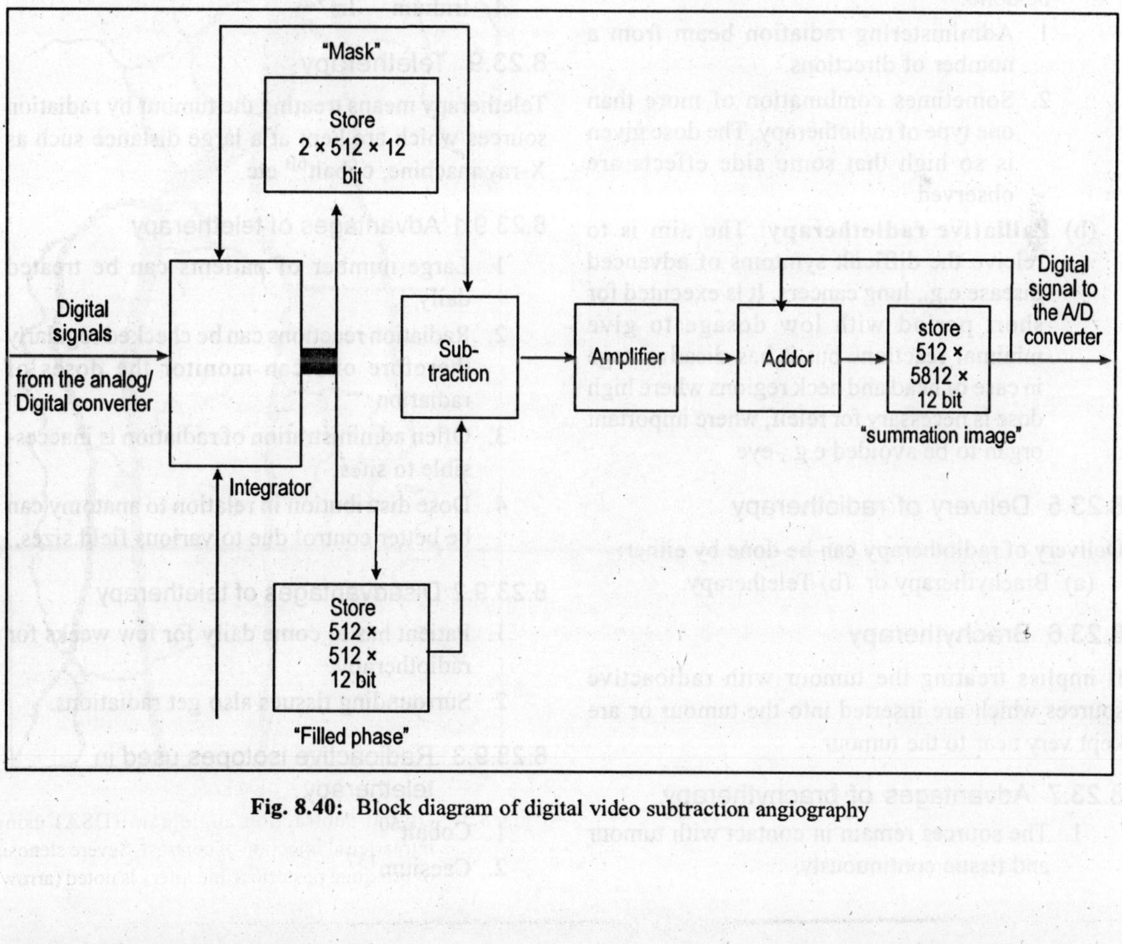

Fig. 8.40: Block diagram of digital video subtraction angiography

(b) Intra cavitary or intraluminal radiotherapy—by introducing radioactive, sources into the body cavities.

8.23.3 Sources and mode of administration

1. Implanted into the tissue is called **interstitial radiotherapy,** e.g., cheek, tongue.
2. Occasionally radioactive fluids are introduced into the body either via a vein or into a cavity called intracavitary brachytherapy, e.g., uterus, vagina, oesophagus.

8.23.4 Radiotherapy treatment

Radiotherapy is prescribed as:

(a) Radical radiotherapy
(b) Palliative radiotherapy.

(a) **Radical radiotherapy**: Treatment is given to 'cure' the patient's disease is called radical radiotherapy. In such cases the area to be treated includes the tumour and the area affected by the tumour. The treatment may be done:

1. Administering radiation beam from a number of directions.
2. Sometimes combination of more than one type of radiotherapy. The dose given is so high that some side effects are observed.

(b) **Palliative radiotherapy**: The aim is to releive the difficult symtoms of advanced disease e.g., lung cancers. It is executed for short period with low dosage to give minimal reactions but it has disadvantage in case of head and neck regions where high dose is necessary for releif, where important organ to be avoided e.g., eye.

8.23.5 Delivery of radiotherapy

Delivery of radiotherapy can be done by either:
(a) Brachytherapy or (b) Teletherapy.

8.23.6 Brachytherapy

It implies treating the tumour with radioactive sources which are inserted into the tumour or are kept very near to the tumour.

8.23.7 Advantages of brachytherapy

1. The sources remain in contact with tumour and tissue continuously.
2. It delivers a particular dose in a very short time.
3. Surrounding normal tissue can be spared.
4. Dose distribution can be controlled.

8.23.8 Disadvantages of brachytherapy

1. Does not allow the reaction of radiations to be checked.
2. Calculations of dose of the tumour volume and normal areas have to be very exact and accurate.
3. Difficult to apply in inaccessible sites.
4. Usually require admission in the hospital or nursing home.
5. Very view patients can be treated at a given time.

Isotopes used in Brachytherapy

1. Radium—Ra^{226}
2. Caesium—Cs^{137}
3. Cobalt—Co^{60}
4. Iridium— Ir^{192}

8.23.9 Teletherapy

Teletherapy means treating the tumour by radiation sources which are kept at a large distance such as X-ray machine, cobalt^{60} etc.

8.23.9.1 Advantages of teletherapy

1. Large number of patients can be treated daily.
2. Radiation reactions can be checked regularly therefore one can monitor the doses of radiation.
3. Often administration of radiation is inaccessible to sites.
4. Dose distribution in relation to anatomy can be better control due to various field sizes.

8.23.9.2 Disadvantages of teletherapy

1. Patient has to come daily for few weeks for radiotherapy.
2. Surrounding tissues also get radiations.

8.23.9.3 Radioactive isotopes used in teletherapy

1. Cobalt^{60}
2. Caesium^{137}

Table 8.10: Radiation dose indifferent organs

Organs	Radiation dose (Gy)
1. Whole brain	50-55 Gy; damage to optic nerve beyond this.
2. Spinal cord (10 cm at neck level)	44 Gy
3. Brachial plexus	50 Gy threshold for neuritis, arm function loss
4. Whole lung	20 Gy, 25-30 Gy, leads to pneumonitis
5. Part of the lung	40-50 Gy; irreversible fibrosis, al and beyond this level.
6. Pericardium	40 Gy.
7. Whole liver	20-30 Gy. (upto half liver higher dose tolerated)
8. Thyroid gland	30 Gy before functional problems occur
9. Skim	55 Gy. alopecia reversible at this level,
10. Parofid gland	10 Gy; temporary dryness
11. Both kidneys upto one third of each Kidney	20 Gy. 40-50 Gy
12. Ovary	2-6 Gy; permanent sterility
13. Testes	3-4 Gy; permanent sterility
14. Eye lens	5-10 Gy; cataract formation Sterility and cataract formation occurs after low doses.

8.23.10 Characteristics of various radioactive isotopes for radiotherapy

I. RADIUM—Ra^{226}

(a) Used for Brachytherapy purposes
(b) Energy— 0.19-2.43 MeV
(c) half life— 1620 yrs.
(d) Rate of decay— 1 per cent per 25 yrs.
(e) K factor— (R/h— mci at 1 cm) 8.25

Disadvantages: It is being discarded these days from clinical practice. The reasons are:

1. Potential leakage of **Radon gas** which is radioactive. In the event of a crack or hole in the tube/needle while in use it would leak out.
2. Low specific activity of approximately one curie per gramme. Therefore, the size of the source will be bigger.
3. Very high photon energy make the protection of the source housing difficult.
4. Since it emits all types of radiations (α, β and γ) and if α particles are absorbed into the body as a result of leaking sources, they would remain active during a very long physical and biological half life.

II. CAESIUM—Cs^{137}

It can be used for brachytherapy and teletherapy. But now a days maximally used for brachytherapy purposes and not favoured for teletherapy machines because it has a low specific activity like radium as well as low K factor which together result in relatively low output exposure rate even from large sources caesium units. Therefore, operate of source—skin—distance of less than 40 cm and the useful beam has a wide pneumbra (it is the dose gradient produces outside the useful radiation beam. There are two components, the transmission and the geometric pneumbra) and there are severe skin reactions when used in teletherapy.

- **Characteristics**

(a) Energy— 0.662 MeV
(b) Rate of decay— 1 per cent per 6 months
2 per cent every year, source can be replaced after 20 yrs.
(c) K factor (R/h— mci at 1 cm)— 3.3
(d) Typical SSD 20-40 cm

For teletherapy purposes the energy is lower and for brachytherapy purposes it is ideal (through little higher) therefore, necessary protection is cheaper and higher. It has an advantage that it can reproduce the dosimetry of radium which was used previously and to which radiotherapist is accoustomed.

III. COBALT^{60}

Cobalt^{60} used for both teletherapy and brachytherapy purposes. The effective photon energy of cobalt^{60} is only slightly lower than that of radium, and its half life is relatively short, but despite this, Cobalt^{60} is the most widely used isotope in teletherapy unit. It can be made to a higher specific activity which combined with K factor of R/m- mci at 1 cm, produce a source with a useful beam exposure rate. Because short half-life, treatment time have to be increased by approxi-mately 1 per cent per month.

- **Characteristics**

(a) Gamma ray energy 1.17 MeV, 1.33 MeV (average 1.25 MeV).

(b) Half life—5.3 years, therefore in practice have to replaced every 3-4 yrs.
(c) Rate of decay—1 per cent per month.
(d) K factor—(R/ mci at 1 cm) 13.0.

- **Normal Tissue Tolerance Dose Level in Radiotherapy**

A quantity of normal tissue is inevitably irradiated during radiotherapy. It is difficult to assess the highest doses tolerated by particular organs but widely accepted levels using 2Gy daily fractions are given below (Table 8.10).

NOTE: Gy—There various ways of defining radiation dose. The dose prescribed for and received by the patient, defines the required dose of energy actually absorbed by the patient at the tumour site, from the radiation beam. The SI unit of absorbed dose is the "gray" (Gy).

8.23.11 X-ray therapy tube

X-ray therapy tubes operate at low mA values of tube current to provide X-rays of four main kV energy changes.

1. **Low voltage** 50-120 kV for treatment, skin lesions.
2. **Intermediate voltage** 130-150 kV for treating lesions a few centimeters below the skin.
3. **Orthovoltage** 160-300 kV provides X-rays for deep X-rays therapy.
4. **Mega voltage** supervoltage 4 mV or more.

A typical orthovoltage therapy tube has a **large filament, a stationary anode,** and a **larger focal spot** as variations of radiography tube. **Anode angle** is usually 30°.

Special means of cooling the tube must be provided. The therapy tubus have hallow anodes cool by circulating oil (45 lb pressure) between the tube and housing. A pump then transfers the heated oil in a special tank where it is cooled by water circulating in coils.

PART XXIV

8.24 Tomography

8.24.1 What is tomography?

In a conventional radiography a 3D subject is displayed in 2D image for which there is super imposition of structures in successive layers of the object.

But tomography technique in which images of structure in selected plane is observed sharp whereas images of structures outside the selected layer are unsharp.

8.24.2 General features

During tomography the unwanted layers are made more relative to the film for which produces unsharp images whereas desired layer is kept stationary for a sharp image.

Out of three parts i.e., X-ray tube, film and patient, the patient remains stationary for synchronized movement of the other two.

1. If there is movement of patient only, then it is known as auto tomography e.g., movement of mandible for AP views of odontoid process.
2. Linear tomography machine causes the movements of the X-ray tube and the film across a adjustable pivot and is kept at a plane (object plane) requires for imaging.
3. The thickness of the section depends upon the angle through which X-ray tube rotater. The smaller the angle greater the thickness of the section of the angle is less than 10°, the thickness of section is enough is called **zonography.**
4. It enables a view to be obtained which is spatial distribution of structures including the depth. Apart from linear movement, are, rotational and pleuri-direction movement scan can be used such as elliptical, circular, hypocyclodal, figure of eight, spiral etc.
5. Simultaneous Multisection Tomography is another technique by which several sections of the patient can be taken using single tomographic exposure. In this method multilayer cassette in a light proof box having multiple layers of screen of difference speed are used to record different layers of object. All films are recorded in same position and phase of respiration in one exposure, raducing time and radiation dose to the patient.

8.24.3 Basic principle

In radiography the 3-D object is represented in 2-D image causing super-imposition of images is not a radiographic image but the image displayed is reconstructed from a large number of absorption profiles taken at regular angular intervals around a slice. Each profile being made up from a parallel set of absorption values through the object.

8.24.4 Importance of tomography technique

(a) Providing a visualisation of a section of the body.

(b) It allows to record on a radiograph for selected structure only, free from the superimposition of shadows of other organs and tissues.

(c) It enables a view to be obtained of the spatial distribution of structures including their depth.

(d) By taking views through a patient the tomographic techniques make a possible view of the plane through which the radiation has passed. The image then is as if the patient had been cut right through so that a picture is obtained looking perpendicularly onto the cut.

(e) The advantage of the procedure is to produce clearer visualisation immediately, e.g., the gall bladder may be separated from intestinal gas; the lung hila distinguished the palmonary vasculature.

8.24.5 Reasons for employing tomography

1. In case of radiography of abdomen, there may be visualisation of pelvic bone, gas in the colon and opaque stone in urinary or biliary tract, the renal outline and the vertebral bodies.
2. In plain X-ray film imaging, the three dimensional structure of the patient is presented as a 2 dimensional projection.
3. The gray level of the image on the film at point represents the sum of X-ray absorption of all tissues on the path of the X-ray of projection from the source to their point on the film e.g., in case of chest radiography, the image of ribs, lungs and the heart overlapping in many areas.

8.24.6 Tomographic imaging

Tomographic imaging solved the problem by taking a large number of different views at different angles.

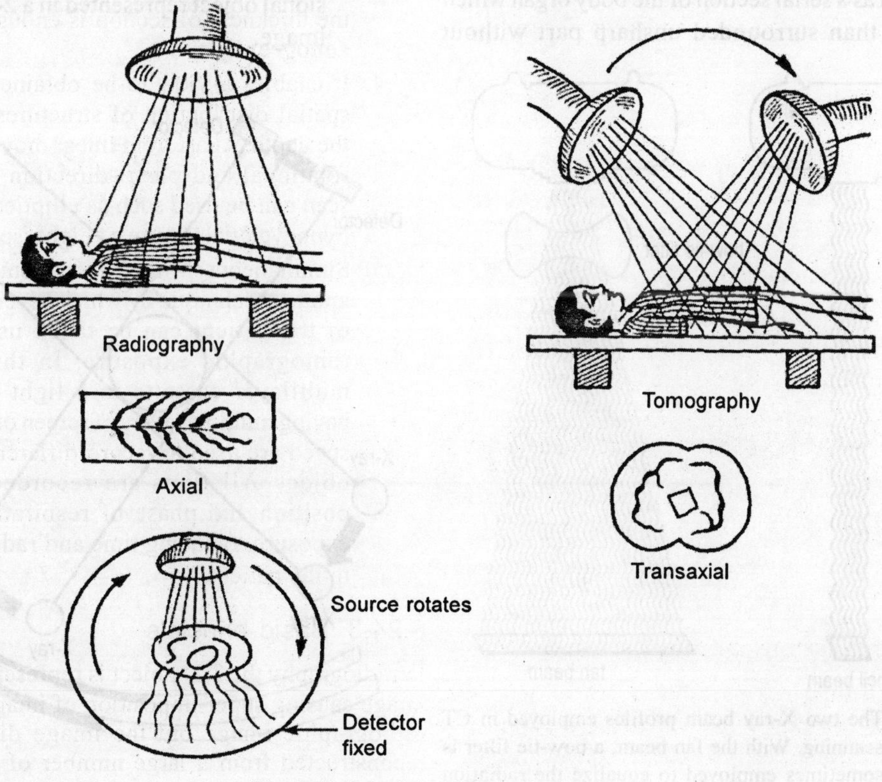

Fig. 8.41: Principle of Tomography

Information from all the images are computed with the help of computer, to obtain an image, e.g., one transverse plane of the body without overlapping of images of the adjacent organs. But the conventional X-ray images give global view of an entire segment of the body, computed tomographic (CT) scanning gives images of a plane. The latest development is **spiral CT** makes an assumbly of images which provide a 3-dimensional presuntation. In spite of these differences, the basis is that final image is the representation of gray levels proportion to the X-ray absorption of tissue is the same in the both methods (Fig. 8.41).

8.24.7 Computed tomography

What is Computed tomography?

A collimated X-ray beam is directed on the patient, and the attenuated radiation beam is measured by a detector whose response is transmitted to a computer which analyses the signal from the detector reconstruction image and displays the image on a television monitor. Then image can be photographed for future evaluation and filing. The image is visualised as a filing. The image is visualised as a serial section of the body organ which is bright than surrounded unsharp part without superimposition of the other tissue. The computer reconstruction of the cross sectional structure is accomplished with mathematical equation adapted for computer presentation called algorithm. Computed tomographic scanning have designated in various ways such as:

1. Computed axial tomography (CAT)
2. Computerised transaxial tomography (CTAT)
3. Computarized reconstruction tomography (CRT)
4. Digital axial tomography (DAT)
5. Computed tomography (CT)
   This nomenclature is widely used now a days.
6. X-ray computed tomography (CT Scanner)
7. Computed aided tomography
8. Transmission tomography (TCAT)
9. Emission tomography (ECAT)
10. Computer Assisted tomography.

8.24.8 Salient features of computed tomography

1. In conventional radiography a 3-Dimensional object represented in a 2-Dimensional image.

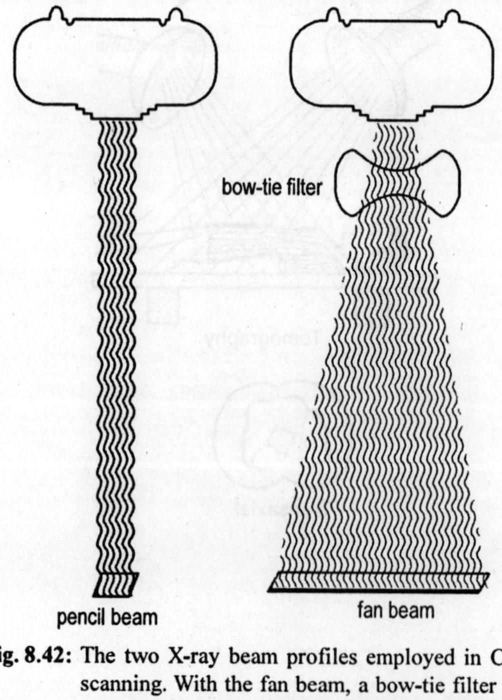

Fig. 8.42: The two X-ray beam profiles employed in CT scanning. With the fan beam, a bow-tie filter is sometimes employed to equalize the radiation intensity reaching the detector array

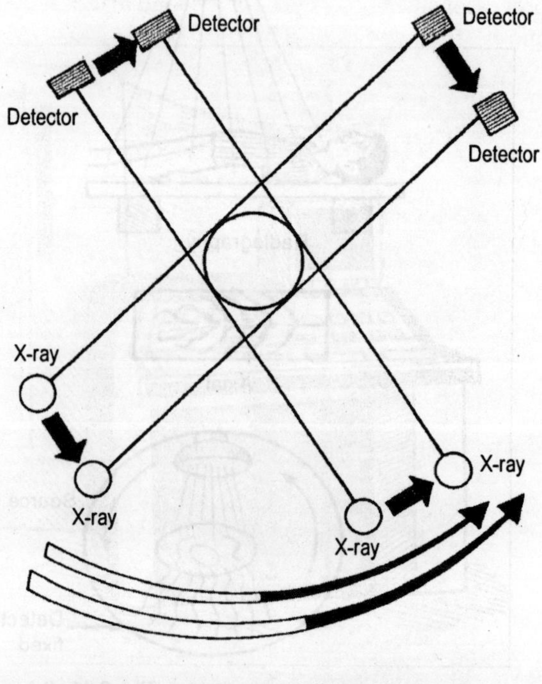

Fig. 8.43: Rotation of X-ray tube and detector

2. Conventional X-rays cause superimposition of images of structures in successive layers of the object.
3. CT technology where images of structures in any selected plane are recorded with sharpness while images of structures outside the selected layer remains unsharp.
4. Throught the procedure, the unwanted layers are made to more relative to the film and therefore make unsharp images while the wanted layer is kept stationary to be recorded sharp.
5. Synchronised movement of two (X-ray and dectors) of the three (Subject + detector + X-ray tube) elements, the X-ray tube and the film and the subject, the third element, i.e., the subject remains stationary.
6. But in case of movement in patient only then it is called **autotomography**, e.g., movement of mandible for AP view of adontoid process.
7. In linear tomography machine, the movement of the X-ray tube and the film across a pivot which is adjustable and is kept at the required plane to be imaged sharphly, i.e., object plane (Fig. 8.46).
8. Beside the linear movement, the other rotational and pleuridirectional such as are, ellipjical, circular, hypocyclodal, figure of eight and spiral. They have specific uses and indications (Fig. 8.52).
9. The thickness of the section depends on the angle through which the X-ray tube moves. The section will be thicker if angle is smaller. If the angle is less than 10^0 degree, the section will be enough thicker and is called **zonograph.**
10. **Multisection Tomography** is the technique by which several layers of the body can be taken simultaneously using one tomographic exposure. In this form process multilayer cassette in the light proof box having multiple layers of screen of different speed are used to record the different layers of object. All the films are recorded in same position and phase of respiration in single exposure reducing time and radiation dose to the patient.

8.24.9 System components computed tomography

Computed tomography system includes the following sub-system (Figs. 8.44, 45):

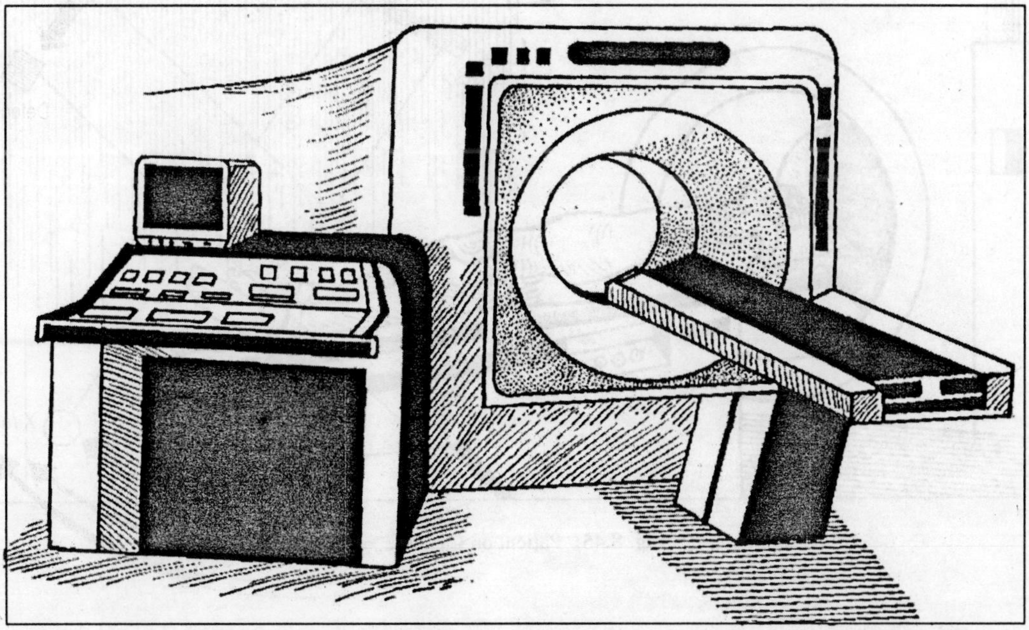

Fig. 8.44: CT equipment

1. **Scanning System** consists of
   (a) X-ray source
   (b) Detectors
   For taking readings for the picture to be reconstructed.
2. **Processing light**: For conversion of above readings into picutre information.
3. **Viewing unit**:
   (a) It gives information in visual form
   (b) Includes manipulative aids to assist diagnosis.
4. **Storage unit**: Information to be stored for analysis.

8.24.10 Computed tomography

Computed tomography was developed by a British Physicist Godfray Hounsfield in 1972. It consists of the following system components:

1. **Scanning system**: Responsible for taking suitable reading for an image to be reconstructed with includes X-ray tube and detectors.
2. **Processing unit**: Responsible for conversion of these readings into intelligible image information.
3. **Viewing part**: Responsible for presentation of this information in visual form and includes other manipulation accesseries to asset diagnosis.
4. **Storage unit**: This unit equables the information to be stored for subsequent analysis.

8.24.11 Principle

The **principle** is by using multi-directional scanning of the objects, multiple data are collected from desirable area containing all tissues in path of transmitted X-ray beam. The beam of X-ray does not fall on the film directly like conventional radiography but so detectors which converts photons of X-ray beam into scintillation. The response of the detector is directly related to the number of photons impinging on it and so to tissue density since more photons of X-ray are absorbed by denser tissues. The scintillation thus produced are

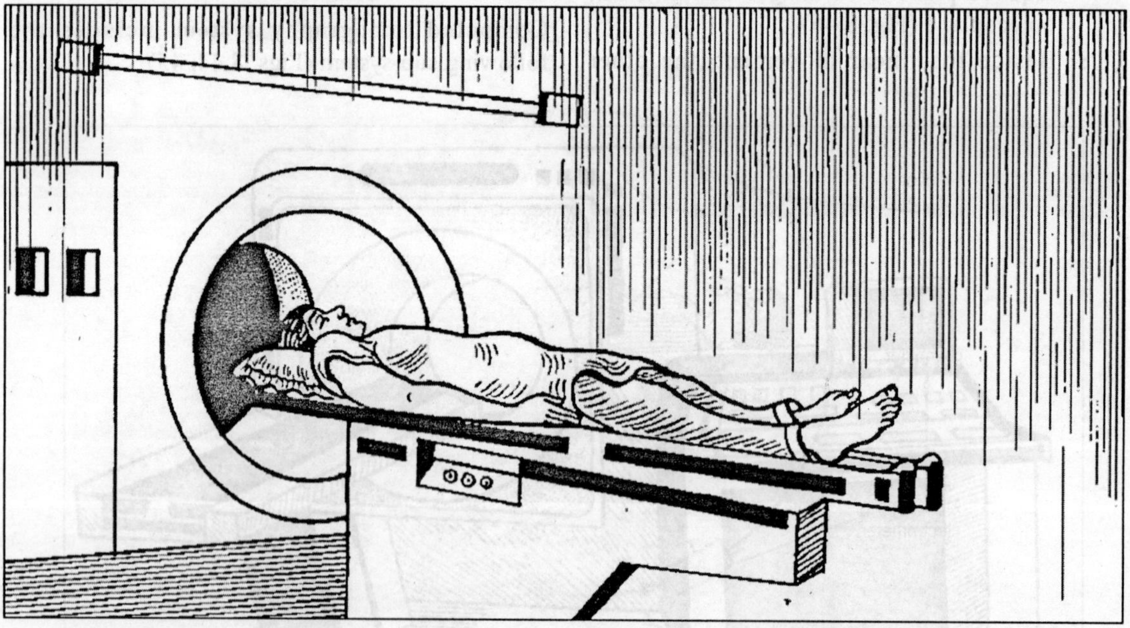

Fig. 8.45: Patient on CT table

quantified and recorded digitally. The information is fed into the computer which producers different readings when the X-ray beam rotates round the patient. The computer is necessary for dealing with large number of digital readings. Then readings can be presented as a numerical record out representing X-ray absorption in each small segment of the tissue transmitted by X-ray beam.

The information can also be represented in analogue form as a 2-D display of the **matrix** on a screen where each numerical value is represented as a single picture element called **pixel**.

The modern machine is provided with fan beam of X-ray and multiple detectors (Figs. 8.46, 8.47).

The earlier machine was provides two detectors only and housed a sharply collimated X-ray beam. The earlier machine was used for scanning head only but the modern machines are used to scan all parts of the body including head region. The earlier machine used to take 4.5 minutes to take single tomographic section whereas the modern machine takes scanning time from one to ten seconds according to the type.

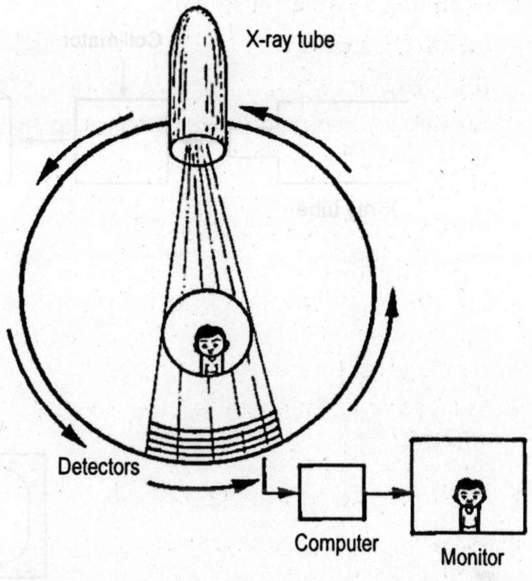

Fig. 8.46: Principle of computerised tomography (CT)

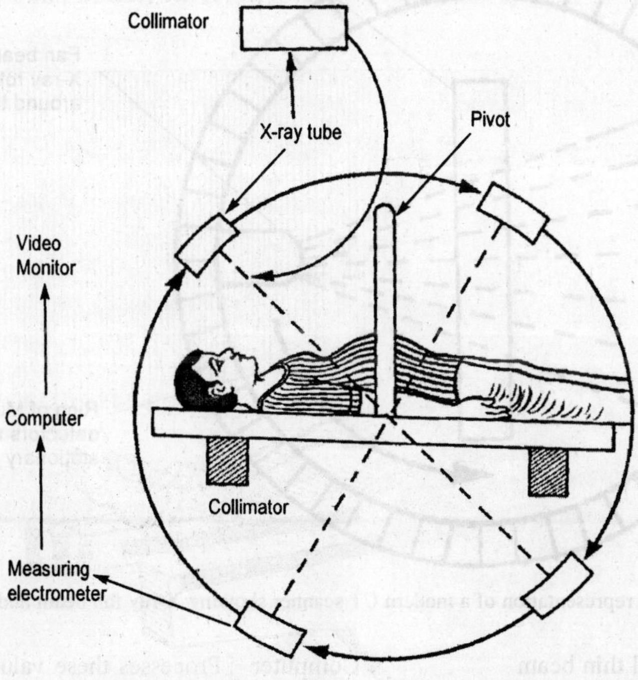

Fig. 8.47: Schematic diagram of principle of computed tomography

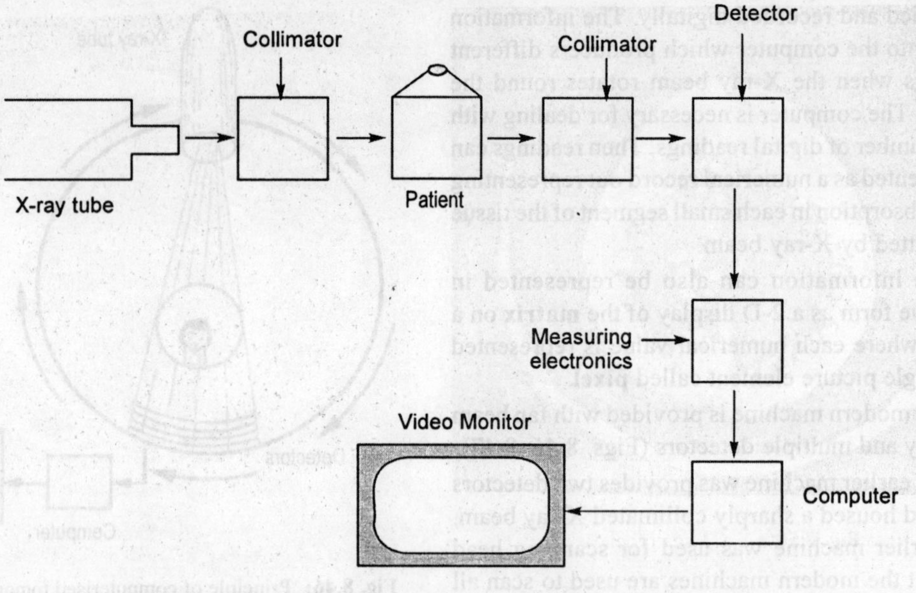

Fig. 8.48: Block diagram of computerised Tomography (CT Scan)

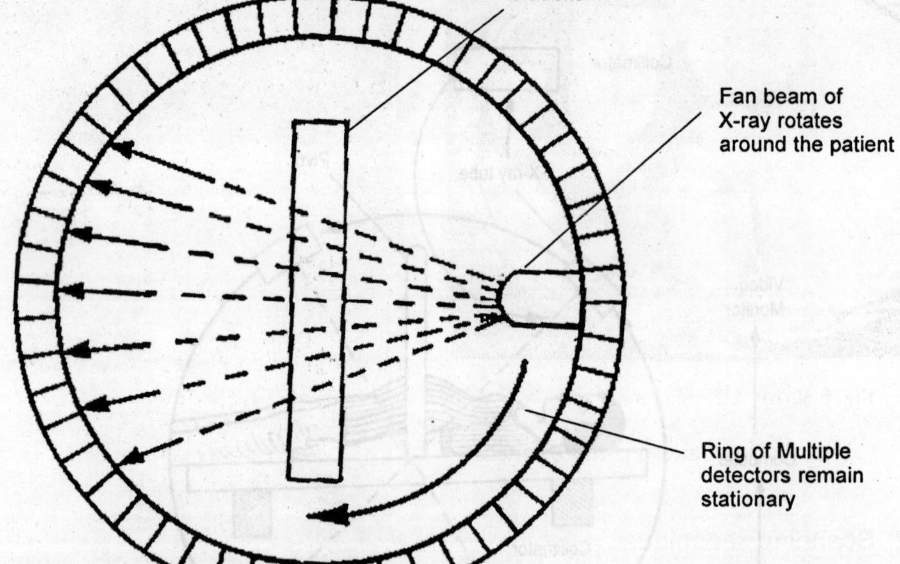

Fig. 8.49: Schematics representation of a modern CT scanner showing X-ray fan beam and detectors

Collimator : Produces pencil thin beam

Detector : Converts X-ray radiation into electrical signal

Electronic : Converts radiation and amplifying electrical signal and converts into digital value.

Computer : Processes these values and computed them into a matrix line density distribution pattern which is reproduced in the Video Monitor (mathematical equation adapted for computer processing called algorithms).

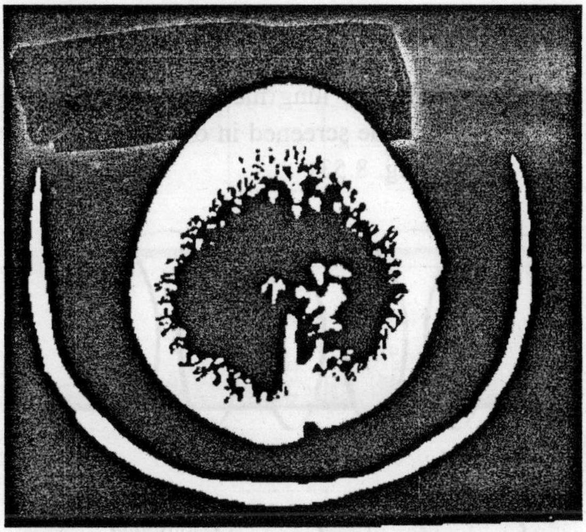

Fig. 8.50(a): CT scan (head)

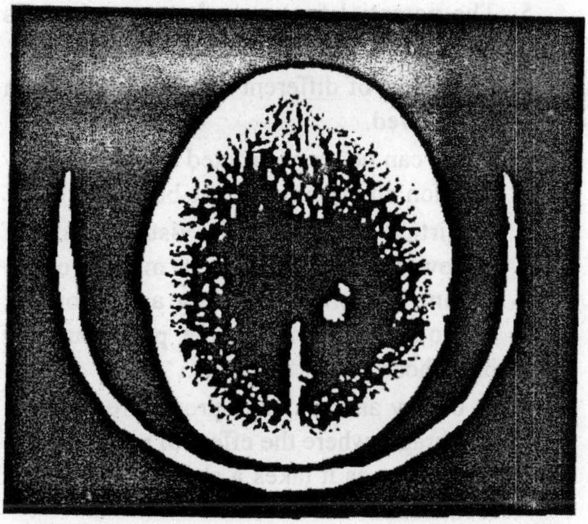

Fig. 8.51(a): CT scan (head)

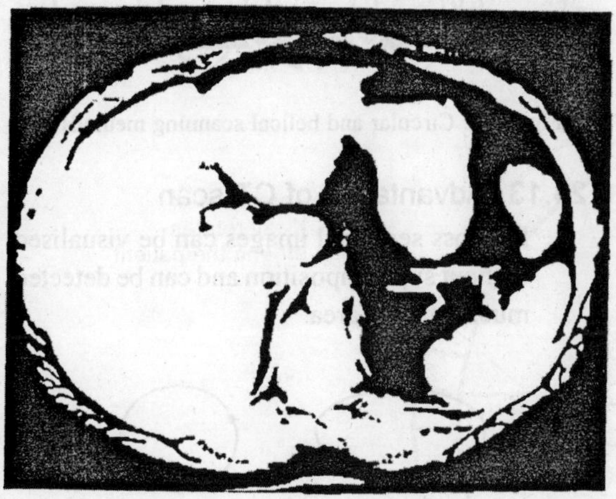

Fig. 8.50(b): CT scan (abdomen)

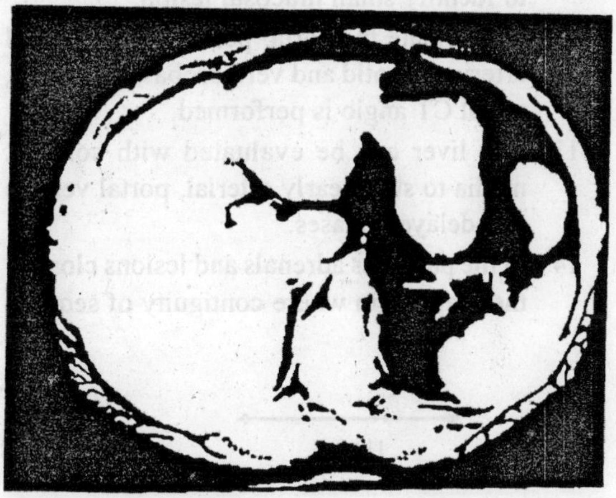

Fig. 8.51(b): CT scan (abdomen)

Video Monitor : Distribution pattern is represented

8.24.12 Scanning technique

There are several scanning techniques such as linear, arc, elliptical, circular, hypocyclodal, figure of eight, spiral which are used for tomography for specific uses and indications. Special Scanning was developed in 1990 by Kelendar. Here patient translocation data acquisition occur simultaneously through the use of a continuously rotating gantry and a high heat capacity X-ray tube.

But **spiral or helical CT has more advantages for the** following reasons:

1. The examination time is reduced by eliminating an interscan delay so that whole organ can be scanned on a single breath hold.
2. Volume of tissue is scanned rather than slice by slice obtained in other techniques hence a gap between slices is removed.
3. Lesions can be localised accurate as there is no shifting of anatomical structures. So the effect of different depth of respiration is removed.
4. Slices can be reconstructed at any arbitary position within the volume after the data have beam obtained.

5. The potential for a gap between slices is removed.
6. The effect of different depth of respiration is removed.
7. Slices can be reconstructed at any arbitrary position after the data have been required.
8. In virtual endoscopy, in visualization of hollow organs such as inner mucosa of the colon, intestines, trachea and bronchi, without the cumbersome procedure of bronchoscopy and hazards.
9. In biopsy and drainage procedure near the diaphragm where the effect of respiration is maximum, as it takes a short time to exact localisation.
10. It is useful in uncooperative and disoriented patient with trauma.
11. Due to excellent spiral, CT becomes useful to identify small mucosal lesion.
12. To evaluate the aorta, pelvic vessels, renal arteries, carotid and vertebrobasilar system, spiral CT angio is performed.
13. The liver can be evaluated with contrast media to study early arterial, portal venous and delayed phases.
14. In the pancreas adrenals and lesions close to the diaphragm where contiguity of sequential images is crucial, spiral CT scan is suitable.
15. Screening for lung metastases. The entire chest can be screened in one or two breath holds (Fig. 8.53).

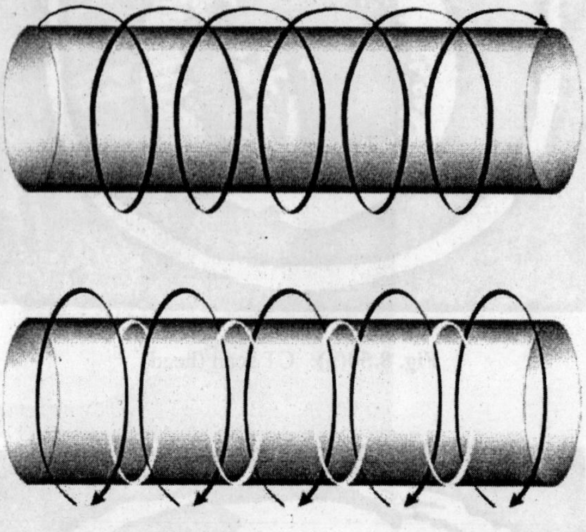

Fig. 8.53: Circular and helical scanning methods

8.24.13 Advantages of CT scan

1. Its cross sectional images can be visualised without superimposition and can be detected much smaller area.

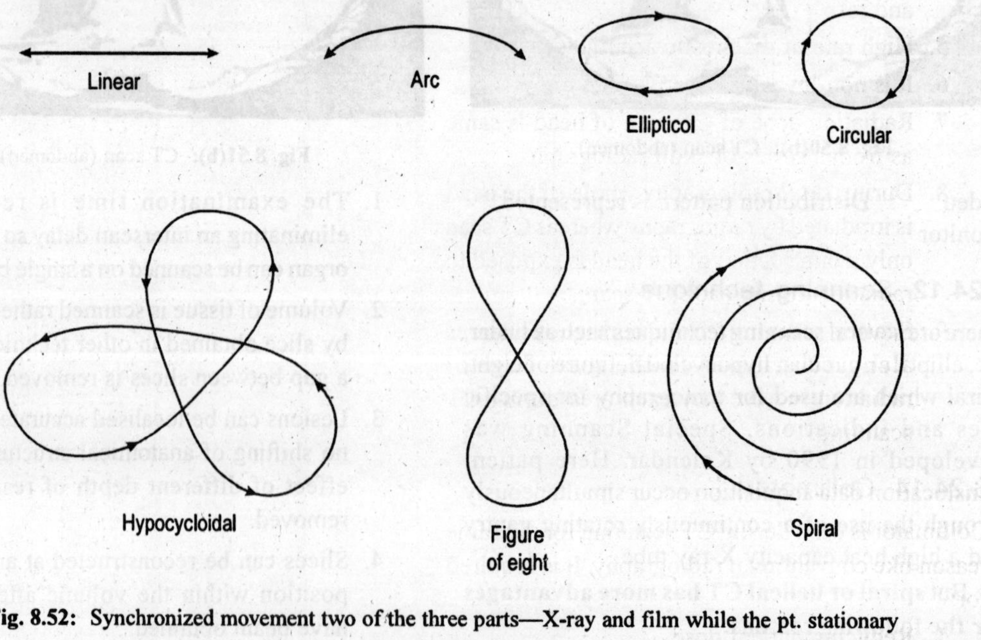

Fig. 8.52: Synchronized movement two of the three parts—X-ray and film while the pt. stationary

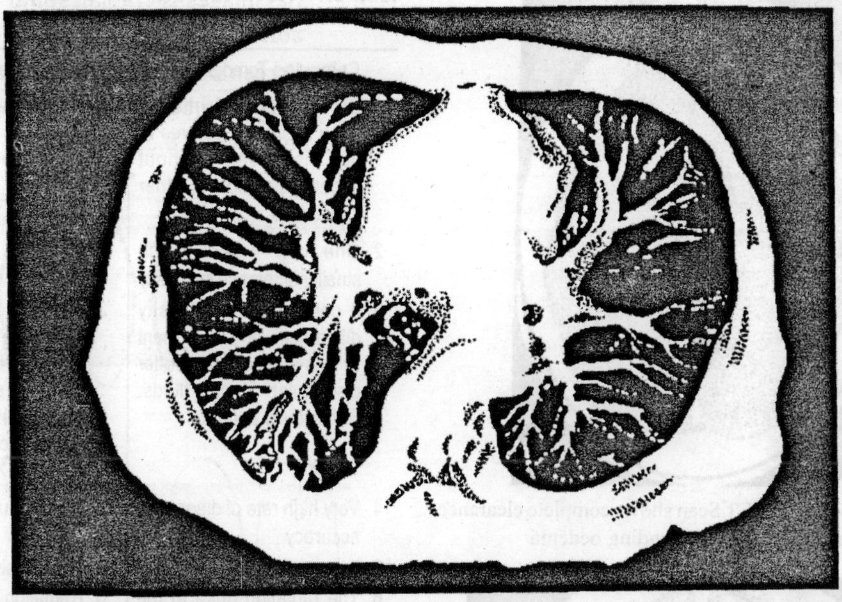

Fig. 8.54: Helical scan of a lung

2. Much smaller difference in density can be detected.
3. It can show density difference under 10 per cent whereas radiography is limited to 4 basic densities such as air, fat, water and bone.
4. It better distinguishes among fluid, soft tissue and fat.
5. High rate of diagnostic accuracy.
6. It is non-invasive.
7. Radiation dose of CT scan of head is same as from a series of skull radiograph.
8. During skull radiography whole of the head is irradiated by radiography whereas CT scan only a thin section of the head is exposed to radiation during each CT scan.
9. Scatter radiation from contiguous slices along contributes to the total dose of radiation received by each area during CT scanning.

8.24.14 Collimator

Collimator is used during CT scanning for the same reason like conventional radiography. It is required for:

1. Reducing patient dose
2. Restricting the volume of tissue irradiated
3. Enhances image contrast by timuting scatter radiation.

- **Number of Collimator**

(A) In conventional radiography there is only one collimator which is mounted on the tube housing, i.e., between the X-ray tube and the subject.

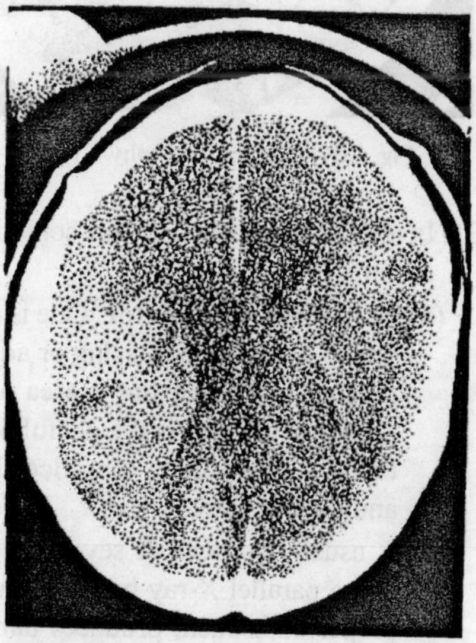

Fig. 8.55: Normal head scan, showing axial sections cut of different levels

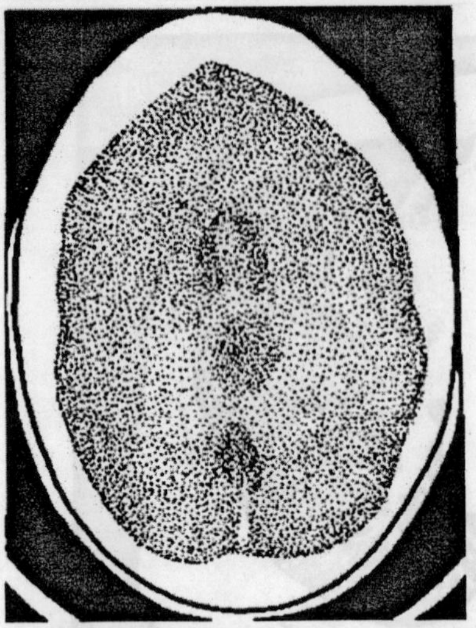

Fig. 8.56: Post-operative CT Scan shows complete clearance of the lesion and surrounding oedema

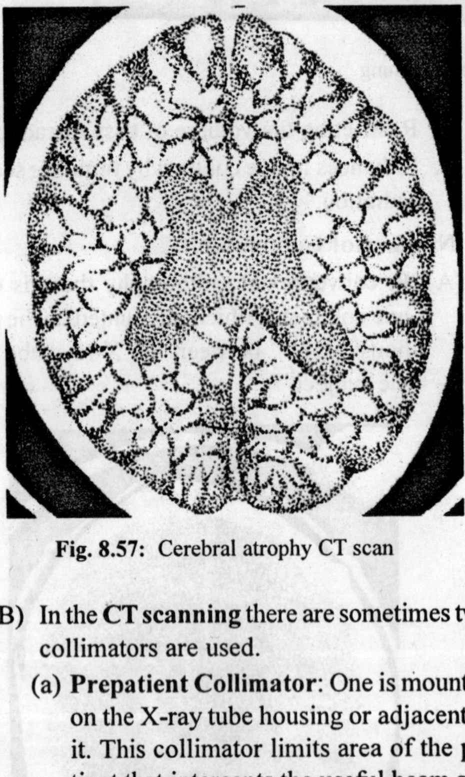

Fig. 8.57: Cerebral atrophy CT scan

(B) In the **CT scanning** there are sometimes two collimators are used.

(a) **Prepatient Collimator**: One is mounted on the X-ray tube housing or adjacent to it. This collimator limits area of the patient that intercepts the useful beam and thereby determines the slice thickness and patient dose.

It usually consists of several sections so that a parallel X-ray beam produces. Improper adjustment produces unnecessary radiation dose during CT scanning.

Table 8.11: Advantages of CT scan over conventional Radiography

Computed Tomography	Conventional Radiography
1. Cross-sectional images allow internal structures to be visualised without superim-position of other structures.	1. No cross-sectional image, Image is visualised with super-imposition.
2. One can detect much smaller difference in density	2. No.
3. It can determine density differences of 0.5 per cent or less and so it can better distinguish among fluids. Soft tissues and fat.	3. Can not show a density difference under 10 per cent, they are limited to 4 basic densities e.g., air, fat, water all soft tissues structures appear to have density of water on X-ray.
4. Very high rate of diagnostic accuracy	4. Not so much accurate
5. It is non-invasive	5. It is invasive
6. The radiation dose to the patient for the head by about the same as from a standard series of skull radiographs.	6. Radiation dose is higher.
7. Irradiation: when the entire head is irradiated during each exposure for a skull film, only a thin section of the head is directly exposed so radiation during each scan.	7. Total head is exposed
8. Scattered radiation from contiguous slices also contributes to the total dose of radiation received by each area during scanning.	8. No.
9. Uses more sensitive X-ray detection system (e.g. scintillation crystal, xenon gas, scientillare) than photographic film.	9. No such detector system
10. Manipulates the data from the detectors using a computer.	10. No such system
11. Both methods use X-rays generated in the same way	11. Same
12. In this method: (a) X-ray tube rotates around the patient in some system. (b) In some detectors also move along with X-ray tube fixed to opposite to it (c) In modern CT a stationary ring with detectors is arranged around the patient in which fan beam of X-ray rotates around the patients.	12. No such system

(b) **Post patient or protective collimator**

1. It restricts the X-ray field viewed by the detector.
2. Reduces the scatter radiation
3. If properly coupled with the prepatient collimator helps to define the slice thickness.

- **Ways of Imaging using lead Collimator**

1. Moving probe rectilinear scanner.
2. Array of multiple crystals autofluoroscope, "the eye" camera.
3. Two moving probes dual head reactilinear scanner.
4. Large single crystal system, Anger Camera.
5. Two crystals on opposite sides of patient for two views using anger logic.
6. Large multiple crystal systems using anger Logic—SPECT.

- **Pixel**

The conductivity distribution is usually represented as a set of connected piecewise uniform patches. Each of these patches is a pixel. The pixel may take any shape, but square or triangular shapes are most common.

X-ray computed tomography (CT scan)

X-ray → Detector → Converts X-ray Photons into scintillations

- **Limitation of CT scan**

1. Superimposition of the 3-dimensional information.
2. Limited area
3. Brain—Skull and Soft Tissue
4. Observation—various angles, mathematically reconstructing the detached structured and displaying the reconstruction on video monitor.
5. To distinguish different types of brain tissue even between normal and coagulase blood, easily observe ventricles of the brain, the repositories of the CSF.

8.24.15 Gantry assembly

It includes:

1. X-ray tube
2. Detectors
3. High voltage generator
4. The patient support
5. Positioning couch
6. Mechanical support for each

These subsystem → receive electronic commands from the operating console → transmits data to the → computer for image production and analysis.

A. **X-ray tube**

1. It has special requirements such as low tube current less than 100 mA.
2. The anode heating capacity must be at least 500000 HU. Some tube is designed specifically for CT have 2 million HU capacity.
3. Focal spot size is relatively is not important in some design, since the CT scanner is not based on the principle on direct geometric imaging and usually require small focal spot.
4. X-ray tubes are energized differently depending on the scanner design.

B. **Detectors**

Early scanner used only one detectors, contemporary scanner uses multiple detectors upto 2400.

Three types of detectors are employed such as:

1. Scintillation crystals and photomultiplier

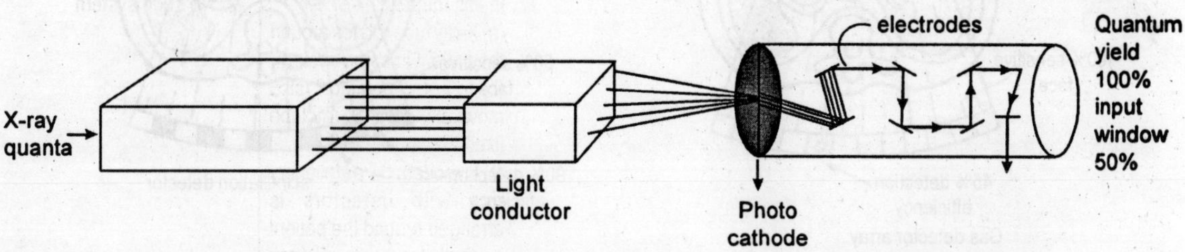

Fig. 8.58: Showing transformation of Kinetic energy of secondary into flashes of light

2. Xenon gas detector
3. Scintillare

The X-ray tube and the detector are rigidly coupled to each other. The system executes translations and rotational movement and transradiates the patient from various angular projections.

I. Scintillation crystal photomultiplier

1. These detectors could not be packed with crystals very tightly together and they require a power supply for each photomultiplier tube.
2. Later on, they have been replaced crystal-photodiode assemblies. Photodiodes are smaller and are efficient equally as a CT radiation detector.
3. Sodium Iodide, Na crystal has been replaced by Bismuth Germanate (Bi_4Ge_{12}) or BGO. Cesium Iodide (CsI) and cadmium tungstate ($CdWO_4$) are used currently.
4. Spacing of detectors are usually one to eight detectors per cm or one to five detectors per degree.
5. It has high intrinsic detection efficiency. Approximately 90% of the X-rays incident on the detectors will be absorbed and contribute to the output signal.
6. The detector interspace may occupy 50% of the total area intercepting the X-ray beam. Consequently the overall detection efficiency only about 45% (Fig. 8.59).
7. Approximately 55% of the remnant of X-rays exciting the patient which contributes to patient dose without contributing to the image.
8. The crystals transform the kinetic energy of the secondary electrons into flashes of light which can be detected by a photomultiplier. An example of a commercial unit using scintillation detector in the EMI 7000 system which employs 1088 cesium iodide detectors. showing transformation of Kinetic energy of secondar electrons into flashes of light (Fig. 8.58).

II. Xenon Gas Detector

1. They are constructed of a large metalic chamber with baffles spaced only at approximately 1 mm interval. The baffles are like grid strips and divide the chamber into many small chambers.
2. Each small chamber acts as a separate radiation detector.
3. The entire detector is hermetically sealed and filled under pressure with a high atomic number inert gas such as xenon or a xenon—krypton mixture. Ionisation of gas in each chamber is proportional to the radiation

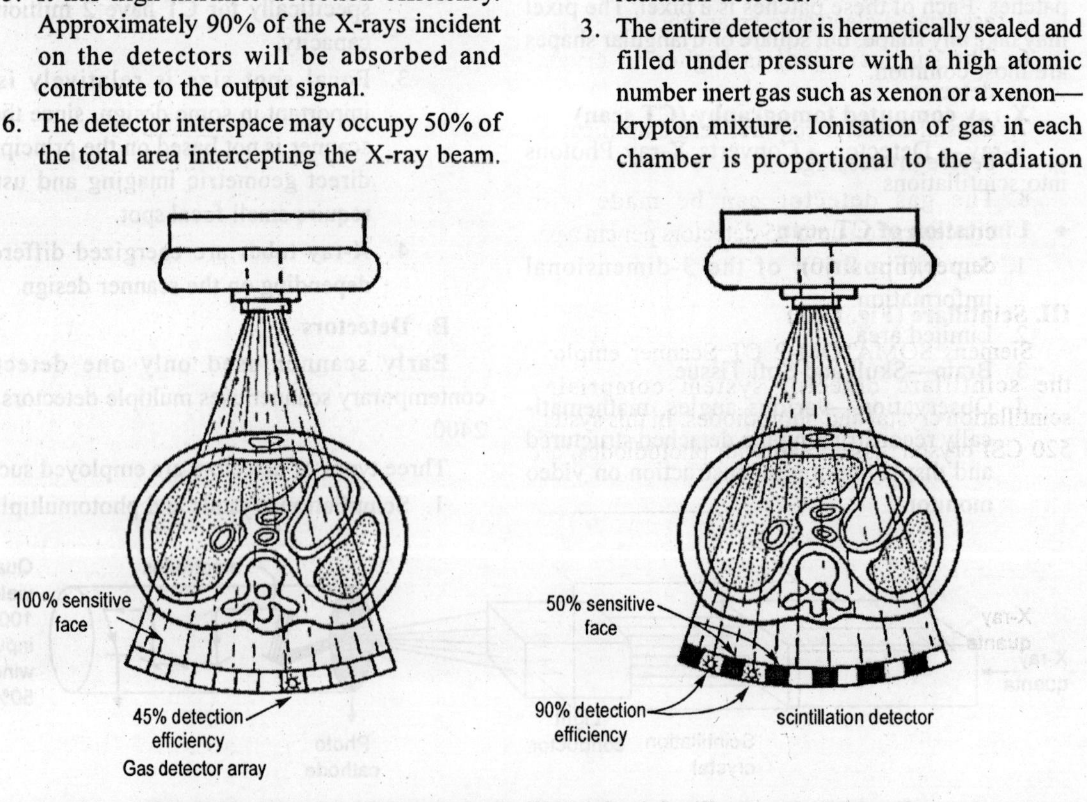

Fig. 8.59: Detection efficiency of scientillation detector and xenon gas detector

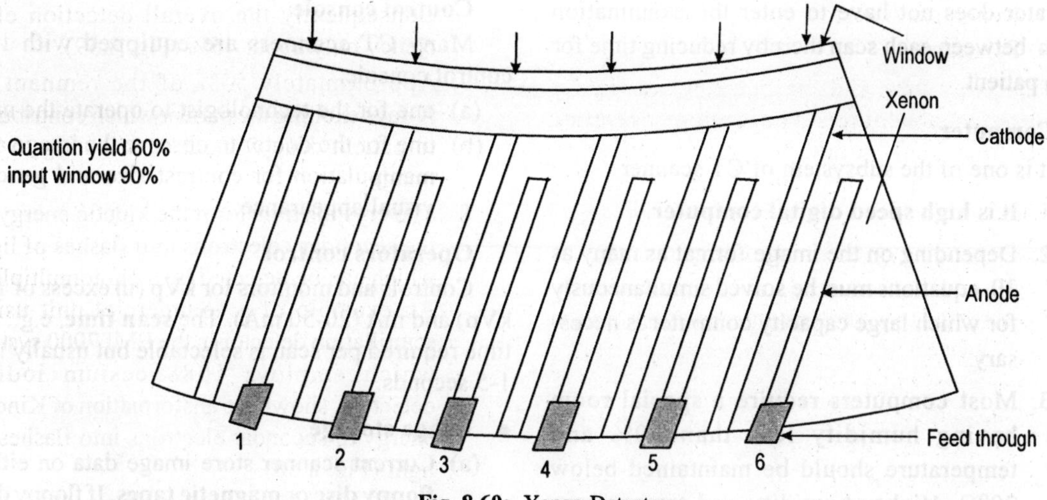

Fig. 8.60: Xenon Detector

incident on the chamber and is detected in much the same way as the ideal gas filled detector.

4. Detection efficiency is only about 45%
5. The gas detector is cheaper than crystal detector. crystal detector is costly due to associated electronic assembly.
6. The output signal is higher in crystal detector.
7. Crystal detector requires less amplification before processing.
8. The gas detector can be made with concentration upto 15 detectors per cm a per degree (Fig. 8.60).

III. Scintillarc (Fig. 8.61)

Siemens SOMATOM-2 CT Scanner employs the scintillarc detector system comprising scintillation crystal and photodiodes. In this system 520 CSI crystal, assembled with photodiodes, are arranged on a 42°arc. In the radiation entrance plane, the detectors have very small dimension of only 1-2 mm × 13.5 mm, thus permitting a good resolution.

- **High Tension generator**

All CT scanners operate on three-phase power. This helps X-ray tube rotor speeds and instantaneous power surges characteristics of pulsed system. **Patient positioning and support couch.** Apart from supporting the patient comfortably, it must be constructed of low-Z material so that it does not interfere with the X-ray beam transmission and patient imaging. The newer couches are fabricated from carbon fibre material. This is low Z-very thin, adequately motor driven so that perfect patient positioning is possible.

If positioning is not proper the same tissue can be scanned twice, doubling dose, or missed alltogether. It helps automatic indexing so that the

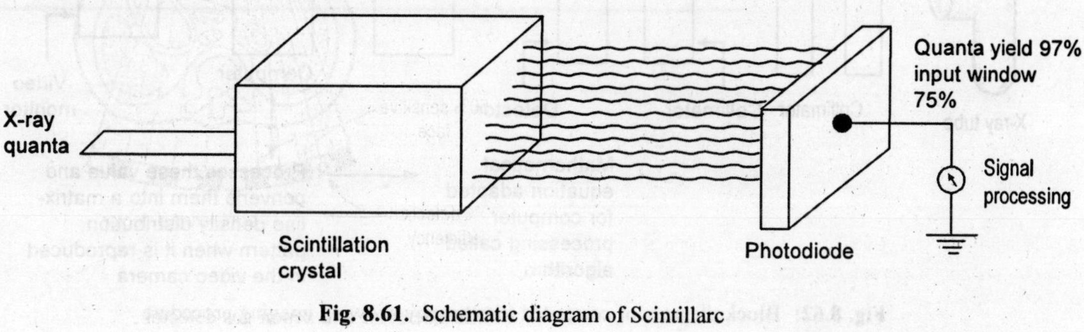

Fig. 8.61. Schematic diagram of Scintillarc

operator does not have to enter the examination room between each scan thereby reducing time for each patient.

- **Computer**

It is one of the subsystem of CT scanner

1. It is **high speed digital computer**.
2. Depending on the image format as many as 30, equations must be solved simultaneously for which large capacity computer is necessary.
3. Most computers require a special room having humidity less than 30% and temperature should be maintained below 20°C. Higher humidity and temperature cause computer failure.
4. At the heart of the computer is the microprocessor and primary memory. These determine the time between the end of a scan and the appearance of an image, the reconstruction time.
5. Some CT scanner use an array processor instead of microprocessor. Because this processor does many calculations faster and takes image is reconstructed in less than 1 sec.

Control console

Many CT scanners are equipped with two control console.

(a) one for the technologist to operate the unit.
(b) one for the doctor to observe the image and manipulation for contrast size and general visual appearance.

Operators control

Controls and monitors for kVp (in excess of 100 kVp) and mA (20-50 mA). The **scan time**, e.g., the time required per scan is selectable but usually it is 1-5 seconds.

- **Image storage**

(a) Current scanner store image data on either **floppy disc or magnetic tapes**. If floppy disc is used, data from a single patient is transformed to single disk that can be stored in a jacket with other patients record.
(b) If magnetic tape storage is used, data from many patients can be transformed into a single tape. Each tape accommodates 150s or 5 to 10 patients. For future observation and filing the scan images is recorded in a film in **multiformal camera** or **laser camera**. Using 8″ × 10″, or 14″ × 17″ film. Video-monitor from which the image is photographed. For quality images high resolution monitor should be used.

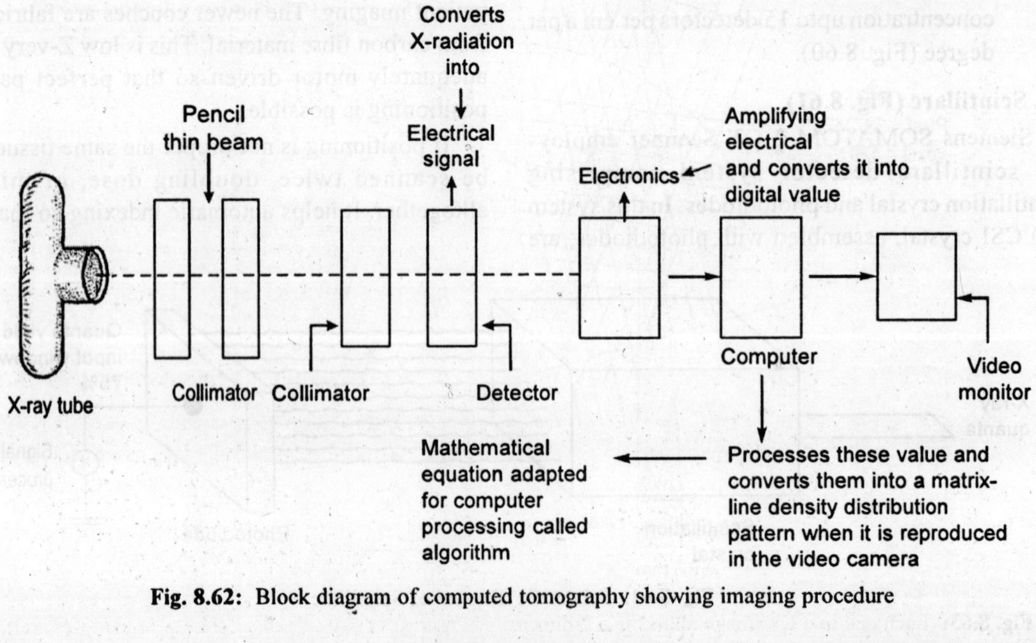

Fig. 8.62: Block diagram of computed tomography showing imaging procedure

8.24.16 Image features or characteristics

In the CT scanner image is not obtained like conventional radiography which requires image receptor, i.e., film. In CT scanner the X-ray form a stored electronic image that is displayed as a **matrix** of intensities.

1. In CT can the image consists of many cells each having a number and displayed as a density or brightness level on the video monitor.
2. The original EMI format consisted of an 80 × 80 matrix, for a total of 6400 individual cells of information.
3. Current scanners have 512 × 512 matrices providing 262,144 cells of information.
4. Each cell of information is a **pixel** (picture element) and numerical information contained in each pixel is a **CT number** or **Housfield unit**.
5. The pixel is a 2-D representation of a corresponding **tissue volume.**
6. The tissue volume is called **Voxel** (volume element) and it is determined by the product of the pixel size and thickness of CT scan slice section or slice.
7. The larger the diameter of reconstruction, the larger will be each pixel for a given matrix size or larger the reconstruction matrix, the smaller will be each pixel (Fig. 8.63).

Progressive scanning: Videoraster scan method in which all horizontal lines are read out in a single vertical scan of the video camera target.

Upscanning: Scan conversion method in which the number of pixels or video lines displayed higher (usually by a factor of 2). Then those actually acquired from the video camera. Extra display data are produced by either replication or interpolation of the acquired data.

8.24.17 Types of films used in radiography

1. Screen Film
2. Direct Exposure or non-screning Film
3. Mammographic Film
4. Duplicating Film
5. Digital Subtraction Angiography Film

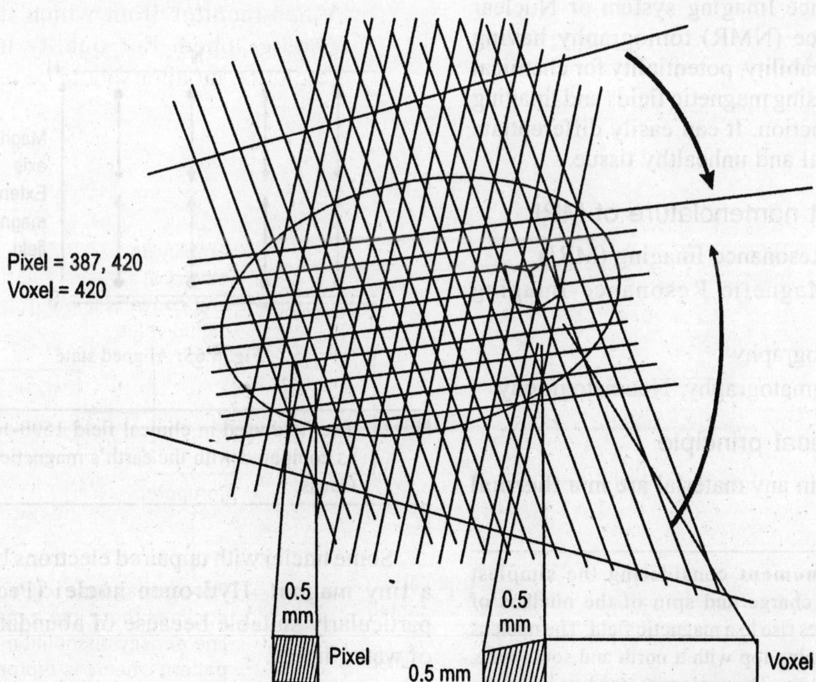

Fig. 8.63: Each cell in a CT image matrix is a 2-dimensional representation (pixel) of volume of tissue (voxel)

6. Dental X-ray Film
7. Medical Imaging Film (C.T Scan, MRI, USG, Digital X-ray, PET, etc.)

PART XXV

8.25 Magnetic Resonance Imaging

- **MRI**
  1. Why MRI is advantageous.
  2. Biophysical principle
  3. MRI Components and function
  4. Block diagram
  5. Preparation of the RP
  6. Scanning procedure
  7. Summary of MRI Imaging
  8. Imaging
  9. Advantages
  10. Advantages over other techniques
  11. Indication/application
  12. Why MRI is preferred
  13. Contrast media of MRI
  14. CT Scan Vs MRI.

8.25.1 Definition

Magnetic Resonance Imaging system or Nuclear magnetic resonance (NMR) tomography having high resolution capability, potentiality for chemical specific imaging, using magnetic fields and imaging in any desired direction. It can easily differentiate between the normal and unhealthy tissue.

8.25.2 Different nomenclature of MRI

1. Magnetic Resonance Imaging (MRI)
2. Nuclear Magnetic Resonance Imaging (NMRI)
3. NMR Tomography
4. NMR Zeugmatography, Neumatography.

8.25.3 Biophysical principle

1. The atoms in any material are in a **thermal** motion.

NOTE: **Magnetic moment** considering the simplest nucleis, the charge and spin of the nucleus of hydrogen gives rise to a magnetic field. The nucleus is a tiny spinning top with a north and south pole, just like a magnet. The nucleus said to be a magnetic dipole, and the name for its magnetism is magnetic moment.

2. All materials have atomic nuclei show three types of movements within the atoms
   (a) electron spinning on their own axis
   (b) electrons orbiting the nuclei.
   (c) The nucleus itself spinning about its own axis.

The nuclei of the materials contain either protons or neutrons or a combination of both. MRI is based on the:

(a) spinning motion of the nuclei and
(b) a magnetic moment which has both magnitude and direction.

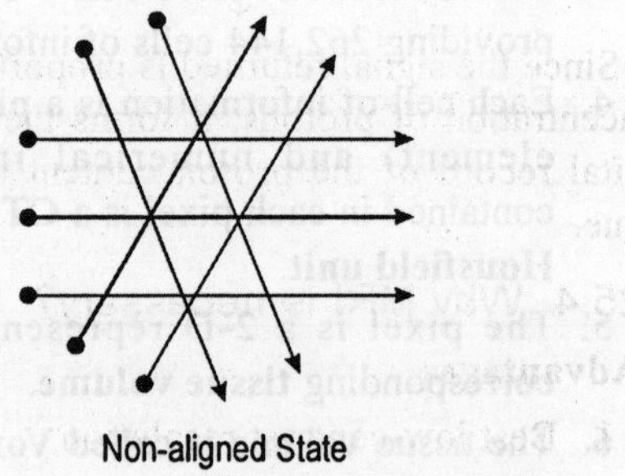

Fig. 8.64: Atomic nuclei non-aligned state

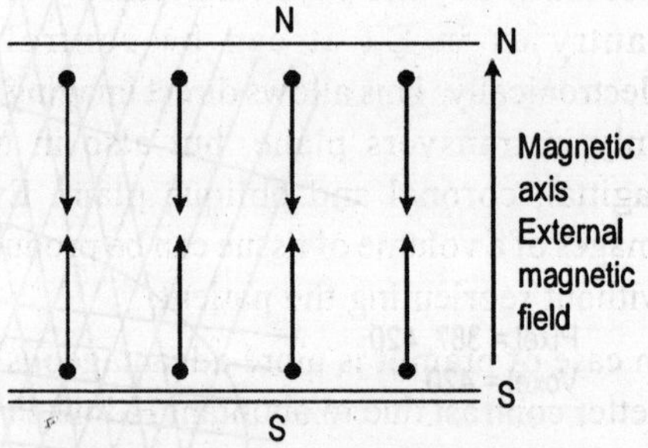

Fig. 8.65: Aligned state

NOTE: The field used in clinical field 1500-15000 Gauss as compared with the earth's magnetic field of 0.5 Gauss.

Some nuclei with unpaired electrons behave like a tiny magnet. Hydrogen nuclei (Protons) are particularly suitable because of abundant quantity of water, lipids.

↓

These are in non-aligned state (Fig. 8.64).

↓

A strong external magnetic field forces them (aproportion) to aligns (Fig. 8.65) in a new magnetic axis.

↓

Displaces nuclei from their new alignment.

↓

They return to this immediately after the pulses stops.

↓

All the same they release energy absorbed as a radiosignal of the same frequency (MR signal).

↓

Since the signal returned is proportional to the concentration of protons, it forms the basis for a digital record of the proton content of the body tissue.

8.25.4 Why MRI is necessary?

- **Advantages**

1. Best low-contrast resolution.
2. No ioninizing radiation
3. Direct multiplanar images
4. Not limited by the physical geometry of the gantry as in CT it can be controlled electronically. This allows direct imaging not only in transvers plane, but also in true sagittal, coronal and oblique plain. Even images of a volume of tissue can be produced without reorienting the patient.
5. In case of brain it is more advantageous for better contrast due to abuntant H-ions in the lipids of brain tissue.

NOTE: **MR Signal**: 'Relaxation' is the term used for the process by which nuclei give up energy.

8.25.5 Hazard

1. There is also a danger from the field such as it can generate an induced *emf* in the blood stream due to passage of ionic material in the blood which in the right sort of range to induce depolarisation voltages in the heart.
2. The hazard of the static field is less than the sudden removal of the field.
3. Similar adverse effect due to sudden removal during emergency reason causing massive rate of change of flux.
4. The magnetic is sufficiently strong for which it requires super conducting magnets where

Table 8.12: Comparison of Radiography and MRI

MRI	Radiography
1. An image can be obtained by the use of magnetic field and radio waves to get mathematically reconstructed image	1. Electromagnetic waves
2. The image represents differences among various tissues of the patient in the number of nuclei and in the rate at which these nuclei recovered from stimulation by radiowaves in the presence of magnetic field.	2. Described in terms of their wavelength, frequency and the amount of energy that each "wave packet" or photon carries.
3. 0.0000001 (eV) energy wave length 10^3 to 10^{-2} meters. Frequency 10^5 to 10^{10}	3. A typital X-ray photon used in medical imaging have a wavelength of 10^{-9} centimeters, a, frequency of 10^{19} hertz (cycles/sec) and an energy of 60,000 electron Volts (eV).
4. It is possible to obtain an image of the body through the use of electro-magnetic waves having energies far below that requires to ionize atoms thereby reducing if not eliminating the threat of biological harm to the patient.	4. Imaging with X-ray is possible because the photon has enough energy to ionize atoms. Thus, some photons are removed from the total number sent into the patient. The pattern of photons transmitted through the patient constitutes a radiographic image which may then be captured by an image receptor, i.e., film.
5. MRI makes use of the radioportion of the electromagnetic spectrum where-in. Photons have relatively long wavelength of 10^3 to 10^{-2} meters with frequencies of only 10^5 to 1-^{10} hertz. A typical photon used in MRI has an energy of only 10^{-2} electron volts (eV) (a tenth of a millionth of an electron volt)	5. The fact that X-ray photons have enough energy to ionize atom implies that spectrum some small biological hazard is associated with a radiographic examination.

field strength greater than 0.3T are required (compared to earth's field at about 50 μT).

5. At the higher fields than the superconducting magnet requires liquid helium temperature.

8.25.6 Advantages of MRI

1. MRI technique is non-invasive
2. No irradiation problem for which it is safe
3. No adverse biological effect
4. Images can be taken in any plane
5. Having high image contrast.

8.25.7 Disadvantages of MRI

1. Costly technique
2. There is limitation of taking image of bone and stone (calculus) in any parts of the body.
3. It is not suitable for a person having any ferromagnetic implants such as surgical clips, orthopedic appliances or pacemaker.
4. Time is long, so there is every possibility of artifacts due to movement.
5. Technically high skillful operator is required.

8.25.8 Application of MRI

1. Observation of the function of the myocardium, to assess irregularities, heart diseases, the effect of infarction and to monitor the response to therapy.
2. It enables identification of regions of excessive fat deposition in the various organs or in the major blood vessels.
3. Measurement of blood flow.
4. Localised phosphours (P^{31}) chemical shift imaging.
5. **In sodium NMR Imaging**: Ejection fraction measurement of working heart. Based on the presence of higher sodium concentration in the blood than in healthy cardiac tissue.
6. Imaging anatomical section (spin imaging), blood flow, fluid flow, mass transport and other dynamic activity.
7. Imaging of skull, skeletal frame, and spinal cord. These are better by MRI than with C.T scan because:
   (a) bones obstructs X-ray
   (b) Structures are static in patient, imaging is faster and simpler
8. Diagnosis of infarction, tumour and other damage to the spinal cord, brain.
9. Other damge of the organs which are not easily identificable by C.T. scan can be done easily.
10. NMR Angiography—can perform imaging of arteries, veins and capillaries and find out defects in circulation.
11. Injection of paramagnetic contrast agents in the blood enhances the contrast between blood vessels and surroundings.
12. NMR signals from tissues are determined by water content but exception in case of lipids. This is the basis of diagnosis of lymph from the surrounding fat tissues.
13. Comparison of healthy tissue with the cancer biology alteration of metabolites level in the tumour cells can be monitored.
14. **Cardiac Application of MRI**
   (a) Assessment of left ventricutlar function.
   (b) Assessment of left ventricular hypertrophy
   (c) Assessment of pericardial diseases

Table 8.13: MRI components and functions

Component	Use	Function
1. Magnet	Application for static megnetic field	Nuclei align and process.
2. Gradient coil	Selection of slice by applying gradient magnetic flied, i.e., variation of magnetic field strength over patient	Nuclei process as a particular frequency.
3. RF sending coil or antenna	Application—RF pulses	Nuclei in the slice area process in phase at a greater angle
4. RF receiving coil or antenna	Receive of RF signal	Electrical signal is received from nuclei and to computer.
5. Computer	Converts signal to image	Reconstructed image is displayed.

Table 8.14: Summary of MRI Imaging process

The following process is a summary of five steps·

Procedure	Component involved	Function
1. Apply static Magnetic field ↓	Magnet	Nuclei align and process
2. Select slice by applying gradient magnetic field (variations of magnetic field strength over patient) ↓	Gradient coil	Nuclei process at a particular frequency
3. Application of RF pulses ↓	RF sending coil or antenna	Nuclei in the slice area process in phase at a greater angle
4. Receive RF signal ↓	RF receiving coil or antenna	Electrial signal is received from nuclei and to computer
5. Converts signal to image	Computer	Reconstructed image is displayed

(d) Assessment of intracardiac thrombi and masses
(e) Detection of valvular regurgitation
(f) Evaluation of congenitat heart disease
(g) Detection of aortic dissection
(h) Visualisation of coronary anatomy
(i) Measurement of coronary blood flow.

8.25.9 Sequence of the imaging process of MRI

The steps involved in the production of an MRI study is summarised below:

1. A powerful, uniform, external magnetic field is employed to align the normally randomly oriented water nuclei contained in the tissue being examined.
2. The alignment or magnetization is next disrupted by **introduction of external RF energy at an appropriate frequency** so as to induce **resonance.** Spatial localization is obtained through application of a spatially dependent magnetic field (gradient) during the same time that RF energy is introduced into the tissue. The gradient field selectively modulates the resonant frequency of the patient in accordance to the Larmor equation.
3. The nuclei return to their resting alignment through various relaxation process and in so doing emit RF energy proportional to the magnitude to their initial alignment or magnetization.
4. After an appropriate period following initial RF deposition, the emitted signals are measured or read out.
5. A mathematical process called Fourier transformation is used to convert the frequency information contained in the signal from each location in the imaged plane to corresponding intensity levels, which are then displayed as shades of gray in a matrix arrangement of, for example, 256 × 256 pixels (picture elements).
6. Protons in the various tissues in the imaged slice realign with the magnetic field at different rates, so that at any given moment there is a difference in signal strength between various tissues. This difference in signal strength from region to region constitutes the basis of tissue.

8.25.10 Equipment for MRI

1. Patient couch
2. Magnet
3. Radio frequency coil
4. Gradient coil
5. Electronic supporter
6. Computer
7. Display unit
8. Digital storage facilities

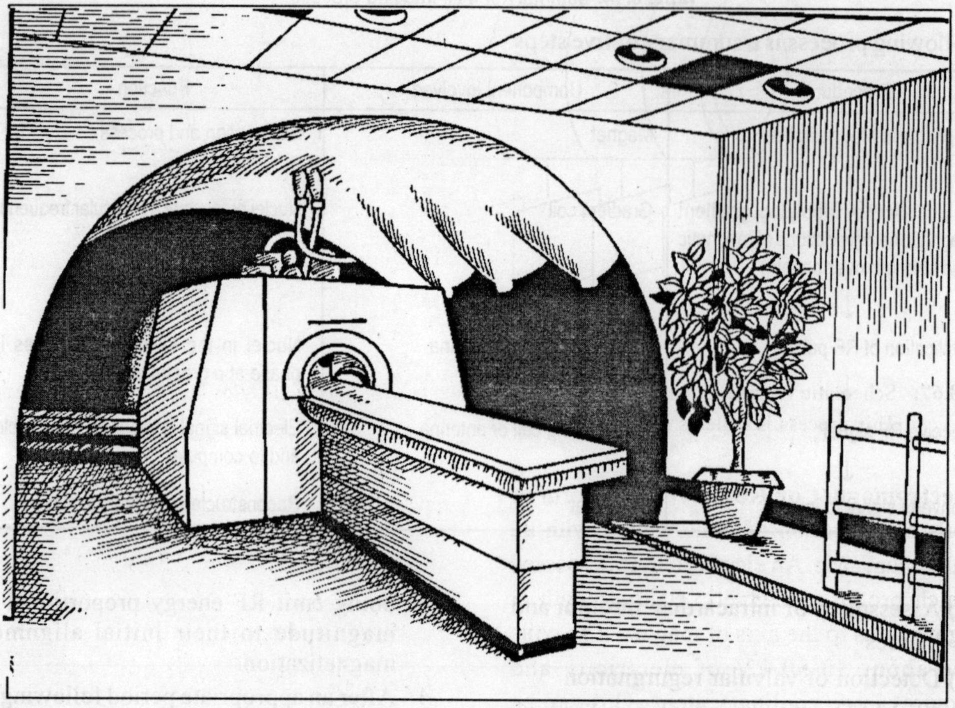

Fig. 8.66: Typical MR imager and patient positioning couch

- **Patient Couch**

There is a large gantry-type assembly with an attached movable patient couch to transport and position the patient in the aperture of the gantry. It is made of a nonmagneted material to prevent disturbing the uniformity of the magnetic field in the scanning region. The top plate is controlled from a control panel for raising or lowering it, moving back and forth accurately. It can also be driven by remote control from the controller console and set to the scanning or can be converted to automatic or manual feed.

- **Magnet**

Magnetic resouance requires static stable, homogeneous magnetic field. The magnetic field exists inside the magnet as well as outside the magnet which known as stray or fringe field.

Magnets are of three types:

A. **Permanent Magnet:** These are made of certain alloys with ferromagnetic properties and have a north pole and south poles separately by an air gap. They require no power and no cooling system. They have small fringe field.

Table 8.15: Advantages and disadvantages of the three main types of MRS magnets

Types of magnet	Advantages	Disadvantages
1. Permanent	Low cost. Low operating cost. Negligible fringe field.	Limited field strength. Fixed field strength, very heavy.
2. Electromagnet Resistive	Low cost coil, maintenance easy. Negligible fringe field.	High power consumption, water cooling required, significant fringe field.
3. Superconductive	High field strength, High field homogeneity, Low power consumption.	High cost. High cryogin cost. Intense fringe field. More than 95% imagers use this type.

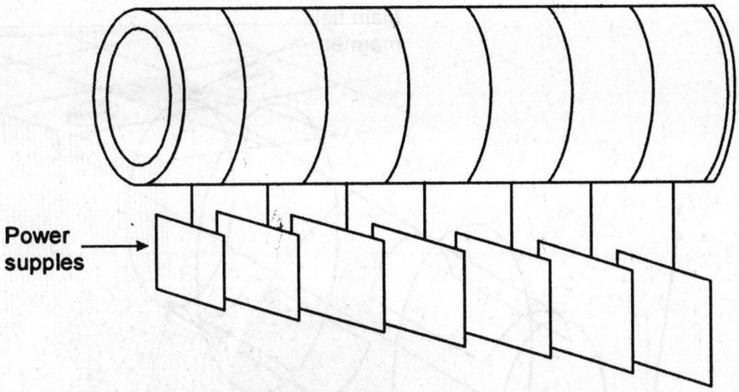

Fig. 8.67: Schematic representation of shim coils for improving the homogeneity of the main magnetic field. This process is called shimming the magnet

B. **Electromagnet or Resistive**: This form of coils or collection of coils wound with an aluminium strip. An electric current is passed which produces magnetic field strength at right angles to the axis of magnet. The coils use about 40-50 kV of electricity and requires large cooling system. This can be switched off when not required.

C. **Super conducting Magnet**: The core is made of a special super conducting alloy (Niobium/Titanium). When these are cooled down to temperature close to absolute temperature zero, they become super conductors i.e., they show virtually no electric resistance and the cell is therefore coiled with liquid helium. These coils are insulated with liquid nitrogen and the entire system is in a vacuum jacket. The helium and nitrogen should be replaced at regular intervals.

8.25.11 Secondary coils

I. SHIM Coils

A principla requirement of the imaging magnetic field homogenity. A1T magnet for instance should produce a field in the imaging volume varying by not more than ± 50μT or 50 ppm. If the homogenity is worse than this image quality will be degraded. Just inside the aperture of the main magnet is positioned a drum with upto 30 individual windings called shim coils each with its own power supply (Fig. 8.67). After the main magnet has been brought upto field, the current and polarity of each shim coil will be adjusted to produce maximum homogeneity in the magnetic field. This process is called **shimming the magnet.**

II. Gradient Coils

To obtain spatial information about the tissue from which the MRI signal is emitted, it is necessary that the primary magnetic field be slightly varied by using a gradient magnetic field. The gradient fields is produced by electric coils called gradient coils.

To obtain projections from variety of directions one must be able to orient the gradient magnetic field along either the X, Y, Z axes, or along any oblique plane. Three parts of coils exist (Fig. 8.69) and they identified as the X, Y, Z gradient coils respectively.

Normally the Z gradient coils are used for selection of a transverse slice. Where the Z gradient magnetic field is on, the RF pulse can be precisely tuned so that only the hydrogen nuclei in a given slice of the patient are energized. The strength of the gradient magnetic field and the shape of the RF pulse determine the width of the slice selected. If a coronal slice is desired, the X gradient coils will be energized. Energizing the Y gradient coils will produce a sagittal slice, and energizing all three parts of coils simultaneously will result in an oblique slice. For 2-dimensional Fourier transformation (2FT) imaging of transeverse anatomy, the Z

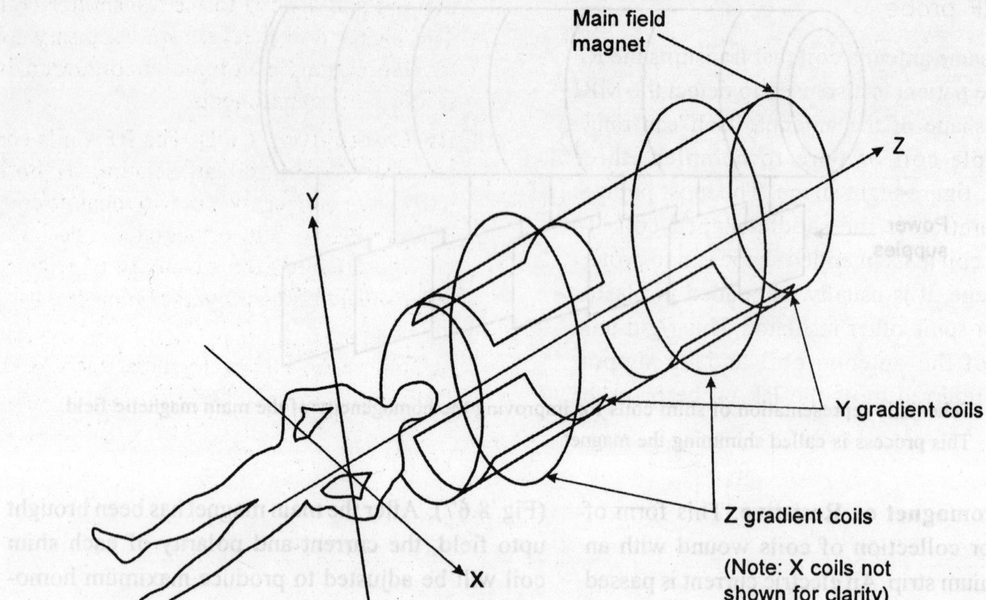

Fig. 8.68: Schematic representation of NMR scanner

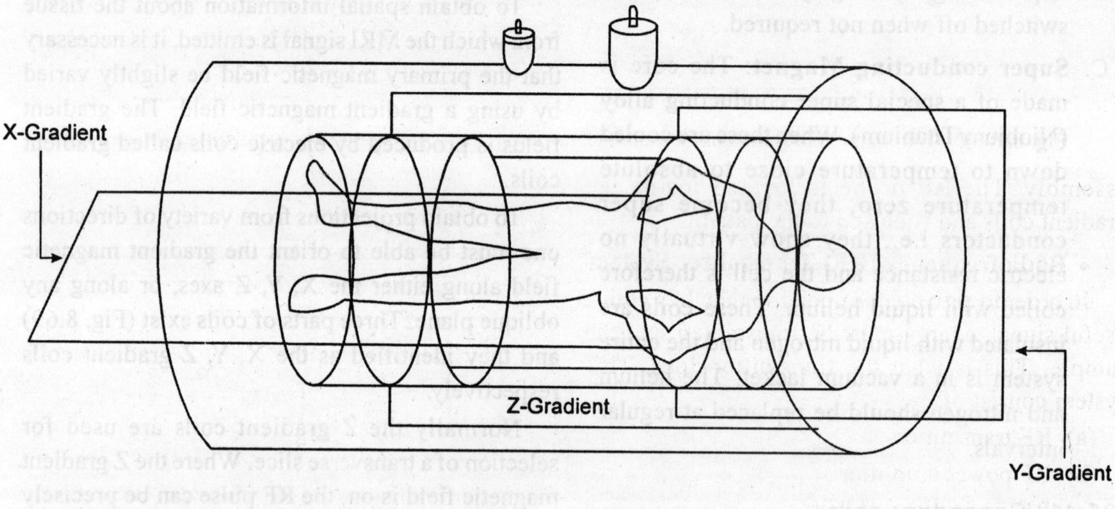

Fig. 8.69: In relation to the patient, position of the three pairs of gradient coils are diagrammatically represented

gradient is on during the excitation RF pulse to select the appropriate slice. While receiving the MRI signals, the X and Y gradient coils will be turned on in the sequence.

The X-gradient conventionally is termed the **frequency—encoding gradient** and the Y gradient the **phase-encoding gradient**.

Therefore, unlike, CT, the plane of the scan can be determined totally electronically. The MRI images contains no moving parts and can produce not only coronal and induced any oblique image of the volume of tissue lying within the gradient coils. Physically the gradient coils are usually embedded in a ring fitting snugly in side the shim coils within the patient aperture.

8.25.12 RF probe

Usually the same antenna coil that transmits the RF pulse into the patient is also used to detect the MRI signal. The shape of the antenna itself can range from a simple coil of wire to complex, three dimensional, figure-eight shape. The most popular coil configuration is the saddle-shaped coil. To maintain the coil in its intended shape and to protect it from damage, it is usually embedded in plastic, fibre glass or some other insulator. This rigid unit, consisting of the antenna coil and its support material, is referred to as the RF probe or probe assembly. The RF probe is located inside the gradient coils and is closed to the patient.

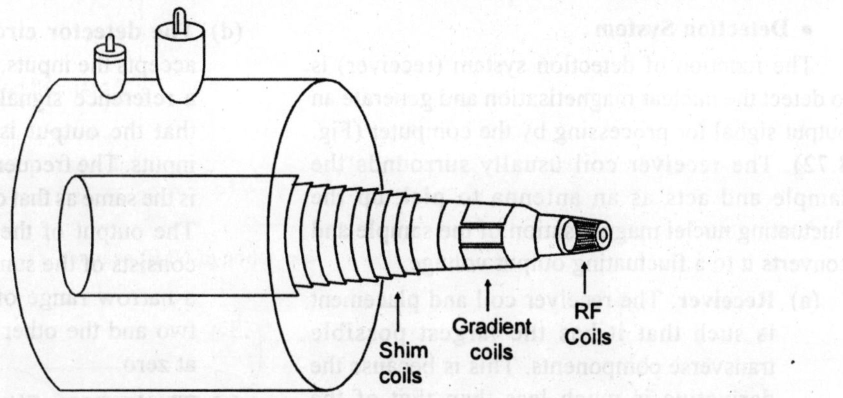

Fig. 8.70: Diagrammatic representation of main magnet and relative position of the secondary coils inside it

- **Radiofrequency (RF) Transmitter System**

In order to activate the nuclei so that they emit a useful signal, energy must be transmitted into the sample. This is what the transmitter does. This system consist of:

(a) RF transmitter
(b) RF power amplifier
(c) RF Transmitting Coils

(a) **RF Transmitter**: The RF transmitter consists of an **RF crystal oscillator** at the Larmour frequency. The RF Voltage is gated with the pulse envelops from the computer interface to generate RF pulses that excite the resonance.

(b) **RF Power Transmitter**: These pulses are amplified to levels varying from 100 W to several KW depending on the imaging method and are fed to the transmitter coil. The higher power levels are necessary for the large sample volume encountered in whole body experiments.

(c) **RF Transmitting Coils**: The RF Coils can be either a single coil serving as both transmitter and receiver or two separate coils that are electrically orthogonal. The latter configuration has the advantage of reduced pulse break through into the receiver during the pulse.

In both cases, all coils generate RF fields orthogonal to the direction of the main

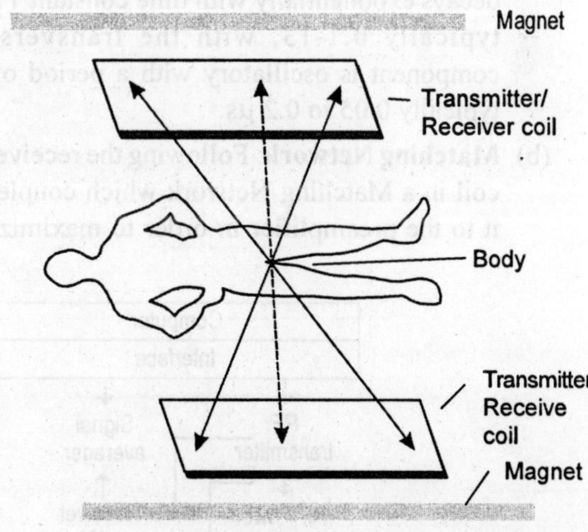

Fig. 8.71: Schematic diagram of MRI machine. The patient's body is within a strong magnetic field. The radio-frequency transmitter coils send radiowaves into the body. The same coils receive signals from within the body. The intensity and source of these signals is converted into an image and displayed.

magnetic field. Sadde and Solemoidal shapped RF coils, are typical geometries for the RF coils. The coils are tuned to the NMR frequency and usually isolated from the remainder of the system of enclosure in an RF the system of enclosure in an RF sheilding cage.

For magnetic field in the ranging 0.05 to 2T used for immaging of the human body. The resonance frequency fall in the radio-frequency band. Usually the resonance is extremely sharp. Width in the range of 10 Hz are typical of biological systems.

- **Detection System**

The function of detection system (receiver) is to detect the nuclear magnetisation and generate an output signal for processing by the computer (Fig. 8.72). The receiver coil usually surrounds the sample and acts as an antenna to pick up the fluctuating nuclei magnetisation of the sample and converts it to a fluctuating output voltage.

(a) **Receiver**: The receiver coil and placement is such that it has the largest possible transverse components. This is because the derivative is much less than that of the transverse component. Time derivative decays exponentially with time constant T_1, typically 0.1-13, with the transverse component is oscillatory with a period of, typically 0.05 to 0.2 μs.

(b) **Matching Network**: Following the receiver coil in a Matching Network which couples it to the preamplifier in order to maximize energy transfer into the amplifier. This network introduces a phase shift to the phase of the signal.

(c) The preamplifier is a low noise amplifier which amplifies the signal and feeds it to a quadrature phase detector. The detector accepts the RF NMR signal which consists of a distribution of frequencies centered around or near the transmitted frequency ω and shifts the signal down in frequency ω. By this process, the distribution of frequencies is unchanged except that it is now centered about zero.

(d) **The detector circuit**: The detector circuit accepts the inputs, the NMR signal V(t) and a reference signal and multiplies them, so that the output is the product of the two inputs. The frequency of the reference signal is the same as that of the irradiating RF pulse. The output of the phase sensitive detector consists of the sum of two components, one a narrow range of frequencies centered at two and the other a narrow range centered at zero.

(e) **The low pass filter**: The low pass filter following the phase sensitive detector reverses all components except those centered at zero from the signal. It is necessary to convert the complex (two channel) signal to two strings of digital numbers by analog to digital converters. The A-D converter output is passed, in serial data form to the computer for processing.

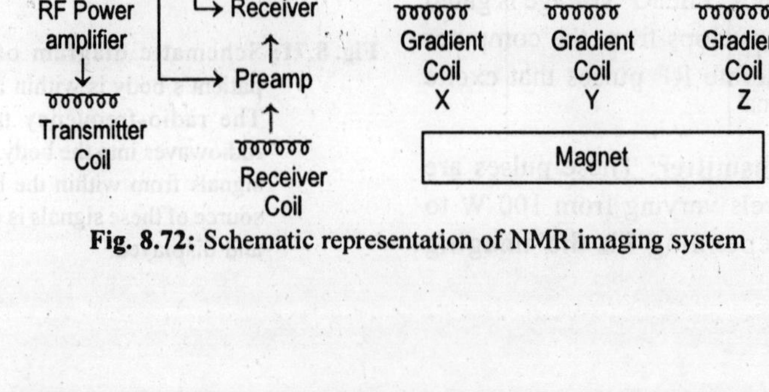

Fig. 8.72: Schematic representation of NMR imaging system

8.25.13 IMAGE System

The image system has the following components:

(a) Computer Image processing

(b) Display system

(c) Controle console

(a) **Computer Image System**: The computer is the source of both the voltage waveforms of all gradient pulses and the envelopes of the RF pulses. A general purpose minicomputer of the type used for the CAT scanner is adequate for those purposes. The computer system collects the nuclear magnetic resonant signal after A/D conversion, correct, recomposes, the ability to perform comprehensive image processing. The image reconstruction software is used to rebuild an image, register images, display and position scanned image. Displays, and stores it. High speed data are sent from the system controller to the computer. An exclusive high seed computer is used to reduce the circulating time of these data. The reconstructed image data are transmitted to the display console by a high capacity image memory disk.

(b) **Display Console or system**: The display console is usually an intelligent console that can be used as an independent image processing unit in an interactive system. The screen of the high resolution monitor can be divided into 4 parts, if desired. It is possible to show simultaneously, display of proton density and T_1, T_2 distribution image. A multiformal camera is used for making hard copies of the image. The desirable features of the software are superior operating characteristics, high speed image reconstruction and the ability to perform comprehensive image processing. The image reconstruction software is used to rebuild an image, register images, display software transfer data registered in the image data file to the display console for display ends other image data to the floppy disk or magnetic tape and also ensures data protection and editing.

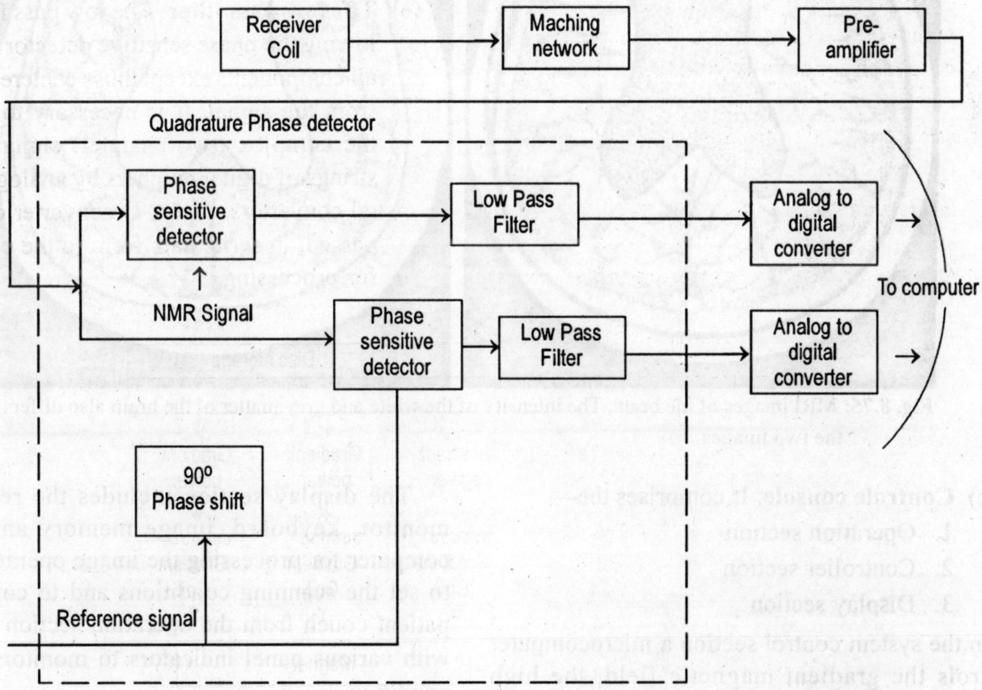

Fig. 8.73: Block diagram of NMR system

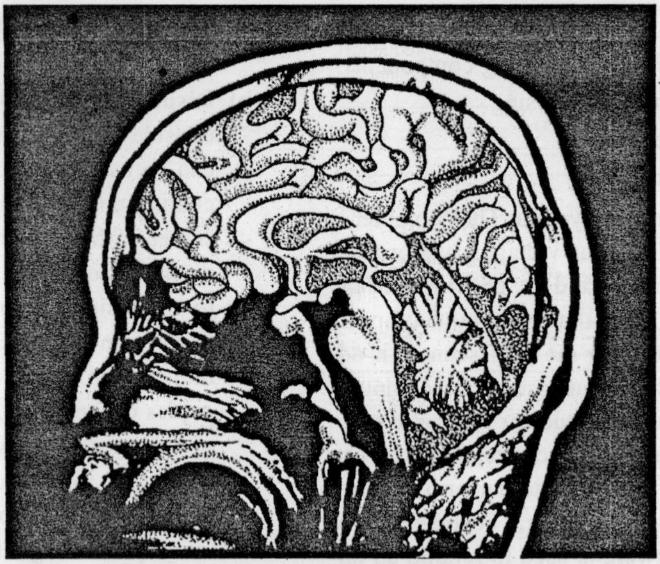

Fig. 8.74: Longitudinal section of the brain

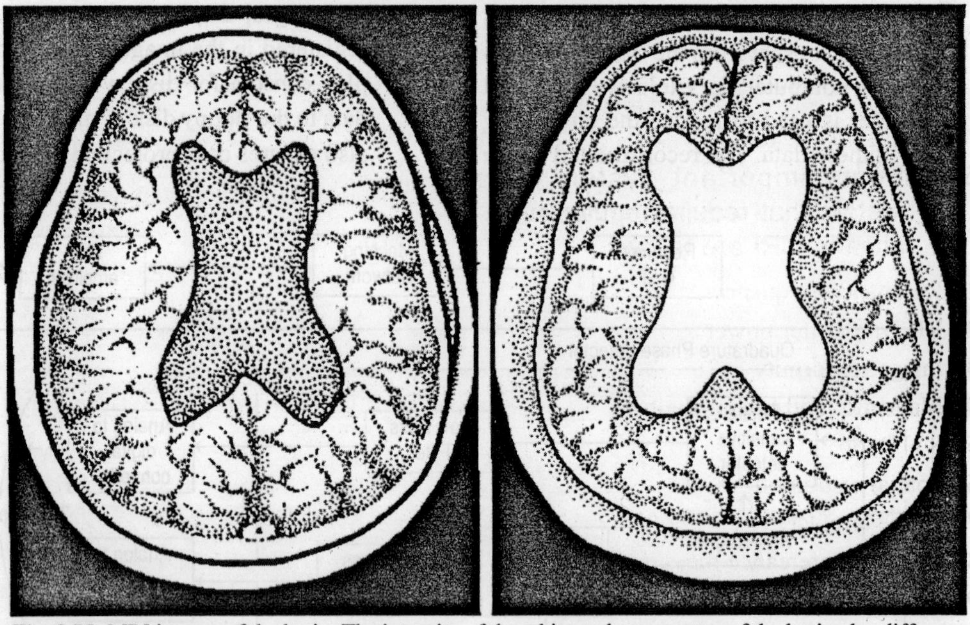

Fig. 8.75: MRI images of the brain. The intensity of the white and grey matter of the brain also differs on the two images

(c) **Controle console**: It comprises the—

1. Operation section
2. Controller section
3. Display section

In the system control section a microcomputer controls the gradient magnetic field, the high frequency pulse train (RF pulses) and the timing A/D conversion of the signals received.

The display section includes the resolution monitor, keyboard, image memory and micro computer for processing the image operating keys to set the scanning conditions and to control the patient couch from the operation section together with various panel indicators to monitors system condition.

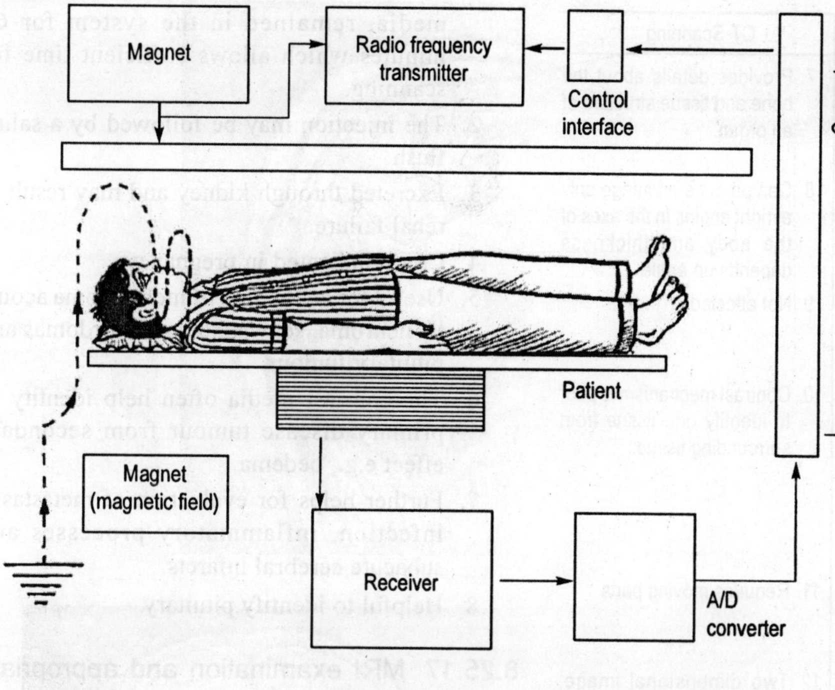

Fig. 8.76: Schematic representation of MRI

8.25.14 Some important system para meters that require regular evaluation in MRI are follows

1. Resonance frequency
2. RF Transmitter power
3. Image uniformity
4. Magnetic field homogeneity
5. S/N Ratio
6. Spatial resolution
7. Image artefact
8. Standard refereance measurement
9. Spatial linearity
10. Phantom T_1 and T_2 values

8.25.15 Biological effect of NMR

1. **Heating**—Due to the RF power (0.3°C)
2. **Static Magnetic Field**
   (a) Slight decrease in cognitive skills
   (b) Mitotic delay in slime molds
   (c) Delayed wound healing
   (d) Elevated serum triglyceride

Table 8.16: Comparison of MRI and CT Scanning

MRI	CT Scanning
1. Powerful imaging technique because of its high resolution capability and potential for chemical specific imaging	1. It is also similar to MRI but has no potential for chemical imaging
2. It uses magnetic field and radio frequency signal	2. Uses electron beam
3. Obtain anatomical information about the human body as cross-sectional images in any desired direction.	3. Information as slice section at different angle and total image is obtained by algorithm
4. Easily discriminate healthy and diseased tissue	4. Not discriminated easily
5. Imager are a map of the distribution density of hydrogen nuclei and parameters reflecting their motion in cellular water and lipids	5. Density of the tissue
6. It is useful non-invasive immaging technique for totals avoidance of ionizing radiation, lack of known hazard, penetration of bone and air without attenuation	6. Invasive, ionizing radiation hazardous, does not penetrate and air, attenuation during penetration.

Contd.

MRI	CT Scanning
7. It highlights the liquid-like areas on those organs and can also be used to detect flowing liquids like blood.	7. Provides details about the bone and tissue structure of an organ.
8. Can produce any desired cross-section, a distinct advantage	8. Can produce an image only at right angles in the axes of the body and thickness depends on angle
9. Imaging parameters are affected by chemical bonding	9. Not affected
10. Very good contrast and high resolution provides promising possibilities of new diagnostic for pathologists, which are difficult or impossible for other present technique.	10. Contrast mechanism is poor to identify one tissue from surrounding tissue.
11. Does not require moving parts, gantries or sophisticated component.	11. Requires moving parts
12. Permits imaging of entire three dimensional volumes simultaneously instead of slice by slice employed in other imaging systems.	12. Two dimensional image, slice by slice section
13. Power supply is not a high generator like X-ray but a high current power supply for the magnet and or precision power supply for the secondary coils	13. Require high generator

3. **Electric current induction**
   (a) It can cause ventricular fibrillation
   (b) It could generate an induced emf in the blood which as a result of the passage of ionic material in the blood which is in the right sort of range to induce depolarisation voltages in the heart.
4. There is also hazardous as a result its sudden removal during an emergency shutdown causing a massive rate of change of flux.

8.25.16 Contrast agents of MRI and advantages

1. **Gadolinium**: DTPA (GD-DTPA) most popular, dose of 2 ml/kg with the injection, not to exceeds 10 ml/min, lower toxicity, fewer side effects than iodinated contrast media, remained in the system for 60 minutes which allows sufficient time for scanning.
2. The injection may be followed by a saline flush.
3. Excreted through kidney and may result in renal failure.
4. Contraindicated in pregnancy.
5. Useful for evaluation of meningioma acoustic neuroma, schwannomas, chordomas and pituitary tumour.
6. The contrast media often help identity of primary disease tumour from secondary effect e.g., oedema.
7. Further helps for evaluation of metastasis, infection, inflammatory processes and subacute cerebral infarcts.
8. Helpful to identify pituitary.

8.25.17 MRI examination and appropriate imaging sectional view of different parts of the body

1. **Brain Imaging** (Routine Section)
   (a) Sagittal
   (b) Coronal
   (c) Axial
2. **Spinal Imaging** (Routine Section)
   (a) Sagittal
   (b) Axial
3. **Abdomen and Pelvis Imaging** (Routine Section)
   (a) Sagittal
   (b) Coronal
   (c) Axial
4. **Cardiovascular system** (Routine Section)
   (a) Sagittal
   (b) Coronal
   (c) Oblique
   (d) Axial
5. **Limb and Joint Imaging** (Routine Sections)
   (a) Sagittal
   (b) Coronal
   (c) Axial

Relative Indications of CT scan and MRI in case of Central Nervous System (CNS)

1. **Head Trauma**	
Acute	CT
Subacute	MRI
2. **Vascular stroke**	
Acute stroke	CT
Subacute	MRI
3. **Neoplasm**	
Extra cerebral	CT or MRI
Intra cerebral	MRI or CT
4. **Convulsive disorders**	MRI
5. Headache: Vagus Symptoms	MRI
6. Suspected Multiple Sclerosis	MRI
7. Degenerative diseases	MRI
8. Posterior fossa lesion	MRI
9. Cranio-vertebral junction	MRI
10. Suspected artero-venous malformation	MRI or CT
11. Suspected saccular ancurysm	MRI
12. **Spine**	
Spinal cord	MRI
Spinal Trauma	CT or MRI
13. **Emergencies**	
In rapid progressive paraparesis	MRI

PART XXVI

8.26 (Soft Tissue Radiography) Mammography

Radiographic examination of the soft tissue, called soft tissue radiography. But this technique differs from the conventional radiography because of substancial difference in the anatomical structures which are radiographed.

In conventional radiography the subject contrast is much due to large differences in **mass density** and effective **atomic number** among the different structures like bone, muscle, fat and lung tissue. But in soft tissue radiography only muscle and fat structures are imaged and these structures have similar mass densities and atomic numbers. And the soft tissue radiography techniques are designed to enhance the **differential absorption** in these two tissue.

The great example of soft tissue radiography is called **mammography**.

8.26.1 Physical basis

Because mass density and effective atomic number of soft tissue components of the breast are so similar, conventional radiography technique is not effective. In the 70-100 kVp range, **compton scattering** predominates with soft tissue differential absorption among tissues of similar composition is minium low kVp must be employed to maximize the **Photoelectric effect** and therefore enhance **differential absorption.**

X-ray absorption in tissue occurs principally **by photoelectric effect and Compton effect**. The degree of absorption is determined by the mass density and the effective atomic number.

Absorption caused by differences in the effective atomic number, however, is directly proportional for Compton interactions and proportional to the cube of the effective atomic number for photoelectric interaction.

Further more, at low X-ray energy, photoelectric absorption becomes increasingly more frequent than compton scattering.

Therefore, X-ray mammography requires a low kVp technique.

As kVp is reduced, however the penetrability of the X-ray beam is also reduced which in turn requires an increase in the mAs. If the kVp is to low, an inordinately high mAs may be required and this could be unacceptable because of the increased patient dose. Technique factors of approximately 28 kVp for screen film and 45 kVp for xeroradiography are employed as an effective compromise between the increasing dose at the low kVp range and reduced image quality at the high kVp range.

The X-ray tube for Mammography different from other X-ray tube is specifically designed for mammography.

Features of an X-ray tube of mammography are:

1. Closer spacing of electrods, allow the tube to be used with lower filament heat for a given milliamperage.
2. The anode is made of molybdenum causing X-ray beam of narrow band of long wavelengths which are intense.

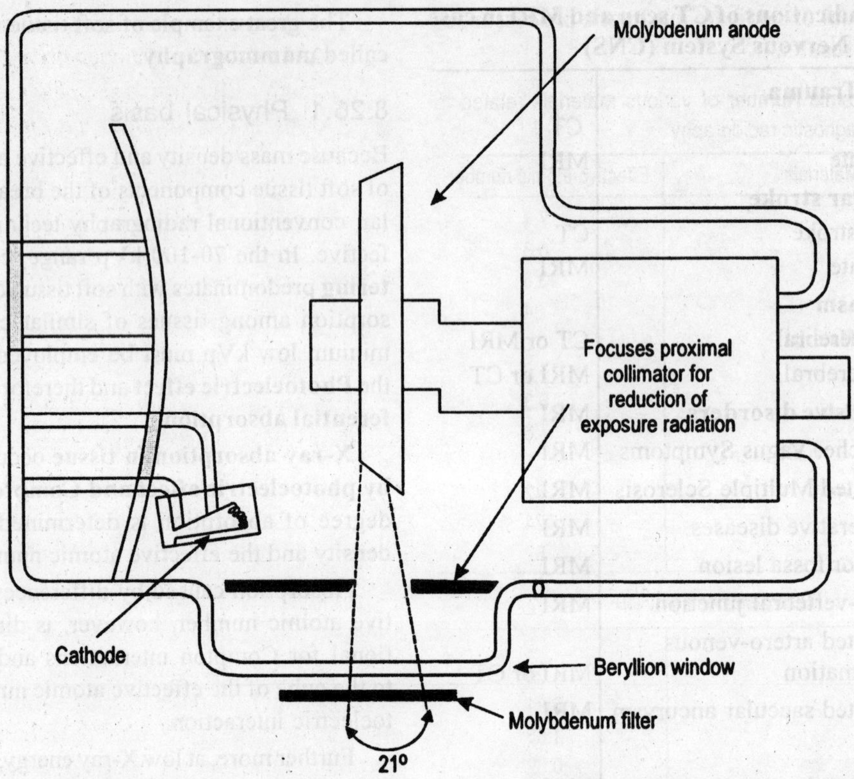

Fig. 8.77: Diagrammatic representation of an X-ray tube for mammography

3. The window of the X-ray tube is made of thinned glass or berrillium for less filtering effect than the thickness of borosilicate glass used for the rest the tube envelope and for the windows and envelops of conventional X-ray tubes. Thus the inherent filtration of the mammographic tube is reduced.
4. The tube has a molybdenum filter which by selective absorption removes certain wavelengths from the beam which are not intended for mammography.
5. The lower filament temperature causes less deposition of evaporated tungsten on the tube wall which may cause further reduction of inherent filtration.

8.26.2 Differences from conventional X-ray

1. The anode is made of molybdenum, provides monoenergetic beam of X-ray (energy level 17.9-19.5 keV).
2. Use of Molybdenum-lead filters—cut off soft X-ray which are otherwise absorbed by the skin and may produce cancer but not to image production.
3. Focal point should be as small as possible to prevent distortion of image.
4. Uniform compression of the breast should be imaged for which the following are necessary:
   (a) Breast immobilisation
   (b) Reduction of geometric sharpness
   (c) Reduction of scatter radiation
   (d) Homogeneity of the image density
5. Perfect breast film contact
6. Single side coated film should be used.
7. Fine crystal single screen helps to reduce the radiation.
8. Vacuum sealed film screen cassettes having single emulsion film and single intensifying screen for image sharpness. Lower radiation dose, shorter exposure time, better contrast at lower film densities.
9. Fine grid used—permits degradation of the image by scattered radiation.

10. Manual or dedicate automatic processor for best results.

Table 8.17: Atomic Number of various materials related to diagnostic radiography

Materials	Effective atomic number
I. Muscle	7.4
Fab	6.3
Bone	13.8
Lung	7.4
II. Contrast Material	
Barium	5.6
Iodine	6.3
Air	7.6
III. Other	
Molybdenum	42
Tungsten	74
Lead	82
	Mass densities kg/m^2/10^3
I. Tissue	
Muscle	1.00
Fat	0.91
Bone	1.85
Lung	0.32
II. Contrast material	
Barium	9.5
Iodine	4.93
Air	0.001298
III. Other	
Tungsten	19.3
Lead	11.35

8.26.3 Factors to be considered during radiographic examination of the breast

1. Breast is soft tissue structure and only muscle and fat structures are imaged.
2. The soft tissue of breast and lesions in the breast have nearly the same degree of radiolucency.
3. An X-ray beam of soft radiation is required and hence a low kilovoltage must be used.
4. The method must be capable of recording very small structures like pin-head size calcification.
5. The low intensity beam for low kilovoltage along with slow imaging system make it necessary to use a high milliampere seconds value in the exposure technique in order to obtain satisfactory density in the image.
6. High milliampere/seconds are achieved by long exposure times e.g., may be up to 10 seconds.
7. Careful immobilisation in necessary because movement may occur during long exposure.

8.26.4 Mammographic equipment

The mammographic equipments are:

I. Mammographic stand

II. Complete mammographic units.

I. Mammographic stands

Mammographic stands are designed for flexibility in operation and easily manoeverable so that patient may be examined standing, seated or lying down and required projections can be conveniently and reproducibly obtained. The movements of the tube unit and the film-holder/ breast-supporl are controlled by electromagnetic breakes which hold them firmly when applied. The stand has the special X-ray tube mounted on a tube carriage. The tube carraige holds the X-ray tube rigidly coupled and aligned to a little table which serves as a holder for the film and support for the breast under examination.

(a) Some stands have mounted above the table a radiolucent compressing plate which can be lowered to compress the breast between the table and the plate.

(b) In other stands there is no breast compressor is achieved by the use of a long beam-limiting cone on the tube unit the cone having a radioluscent plastic cover over its wide lower aperture. The tube unit and the cone are brought down so that the cone compresses the breast against the film. The carraige which holds the tube unit and the film-holder/breast—support is mounted on a cross-arm which itself is carried on a tube stand. This may be a heavy steel vertical columm which moves on a floor track.

8.26.5 Complete mammographic units

Such units are often made mobile, so that they can be wheeled into position for use. Mobility makes them capable of being used to examinea patient being down on an ordinary X-ray table. But a mammographic unit need not always to be mobile and may be a fixed installation. The equipment has the following components:

1. **A breast support/film-holder** which may be designed so that a special cassette for xeroradiography may be put into the film holder as alternative to a conventional cassette or film pack.
2. **A breast compressor** either as a separate compression plate or integral in the beam-limiting cone by virtual of a radiolucent plate covering its open.
3. Some **breast cones;**
4. A **tube support** by means of which the tube unit is rigidly coupled and aligned to the breast support/film holder.
5. A full range of movements for the tube assembly and the breast—support/film holder so that the required projections may be conveniently obtained and the patient may be examined easily seated or lying down;
6. A special high tension generator for the unit which will provide a range of low kilovoltage (e.g., 20-40 KVp);
7. An **automatic timer** to control the exposure;
8. A **special X-Ray tube.**

8.26.6 The X-ray tube for Mammography

1. **Closer spacing of electrodes,** allow the tube to be used with lower filament heat for a given milliamperage.
2. The anode is made of **molybdenum** (MO) causing X-ray beam of narrow band of long wavelength which are intense.
3. The window of the X-ray tube is made of thinned glass or of **beryllium** for less filtering effect than the thickness of **borosilicate** glass used for the rest of the tube envelope and for the window and envelops of conventional X-ray tubes. Thus the inherent filtration of the mammographic tube is reduced.
4. The tube has a **molybdenum filter** which by selective absorption removes certain wavelengths from the beam which are not intended for mammography.
5. Lower filament temperature causes less deposition of evaporated tungsten on the tube wall which may cause further reduction of inherent filtration.

8.26.7 Filtraton of X-ray beam

Conventional radiology uses filters made of **aluminium or copper** to provide selective removal of low X-ray energies from the beam before it is incident on the patient. Whereas in mammography when a molybdenum anode X-ray tube is employed a **molybdenum filter** 20-30 μm thick is generally used. This filter attenuates X-rays both at low energies and at those above its own K-absorption edge allowing the molybdenum characteristic X-rays from the target to pass through spectrum enriched with X-ray energies in the range of 17 to 20 keV.

8.26.8 Compression device and usefulness

Compression of breast during examination without causing pain is necessary for the following reasons:

1. Causes different tissues to be spread out.
2. Minimizing superposition from different planes.
3. Scattered radiation can degrade contrast.
4. Decreases the ratio of scattered radiation to directly transmitted radiation reaching the image receptor.
5. Decreases the distance from any plane within the breast to the image receptor and reduces unsharpness.
6. Provides lowering overall attenuation to the incident X-rays beam allowing radiation dose to be reduced.
7. Provides more uniform attenuation over the image, which reduces the exposure range.
8. Provides clamping action.

8.26.9 Importance of breast compression

Breast compression has several advantages

1. Causes more uniform thickness, and therefore the optical density of the image will be more uniform.
2. Tissue near the chest wall are less apt to be underexposed and tissue near the nipple are less apt to be overexposed.

9

Biophysical Principle

1. Diffusion, 2. Facilitated diffusion (carrier-mediated diffusion), 3. Physicochemical criteria for diffusion across membranes, 4. Active transport, 5. Osmosis, 6. Osmotic pressure, 7. Osmoregulation, 8. Endocytosis-phagocytosis and pinocytosis, 9. Receptor-mediated endocytosis, 10. Secretion and exocytosis, 11. Absorption, 12. Reabsorption, 13. Adsorption, 14. Hydrotropy, 15. Viscosity, 16. Surface tension, 17. Donnan membrane equilibria, 18. Membrane hydrolysis, 19. Ultrafiltration, 20. Dialysis, Electrodialysis, 21. Haemodialysis, 22. Peritoned dialysis.

9.1 Diffusion

The particles, ions or molecules (solute) are in the body fluids (solvent), both solute and solvent are in constant, rapid random thermal movement and frequent collisions occur between the molecules. Each particle moves its own separate way, and the greater the motion, the higher is the temperature and it never ceases under any condition except absolute zero temperature.

The continual movements of molecules among each other in liquids or gases is called **diffusion** or **simple passive transport**. The movements of the frequently colliding molecules cause an equilized distribution of molecules throughout the solution. During the process there is migration of molecules from regions of higher concentration, where collisions are frequent to regions of lower concentration.

9.1.1 Causes of diffusion

1. Motion of solute molecules from a concentrated solution into the solvent.
2. Motion of solvent molecules into the concentrated solution.

9.1.2 Kinetics of diffusion

When large amount of solute is placed in the solvent in one side of the container, it diffuses to the opposite side of the container and again, if the above amount is added to the opposite side of the container, it starts diffusion to the first side, the same amount diffuses in each direction; hence the **net rate of diffusion** from one side to other is zero.

(a) **Concentration difference**: When the concentration of solute is higher in one side of the container than the other side, the net rate of diffusion from the side of the high concentration to low concentration is directly proportional to the larger concentration minus the lower concentration. The total concentration change along the axis of the container is called **concentration difference**.

(b) **Concentration or diffusion gradient**: The concentration difference divided by the distance is called "concentration or diffusion gradient".

(c) **Rate of diffusion and factors affecting it:**

1. **Molecular size**: If the molecular size is greater, the rapidity with which a molecule diffuses from one side to another is less, because the large particles are not forced so intensely by collisions with other molecules. Diffusion is inversely related to the diameter of diffusion molecules or ions.
2. **Molecular weight:** The rate of diffusion is approximately inversely proportional of the square root of the molecular weight. The less the square root of the molecular weight, the greater is the rate of diffusion.
3. **Shape of the molecule:** Elongated molecules encounter higher frictional resistance in a solution and consequently has a low rate of diffusion than a spehrical molecule having same molecular weight.
4. **Concentration difference**: The greater the concentration difference, the greater is the rate of diffusion.
5. **Diffusion distance**: The rate of diffusion is inversely proportional to the distance over which it has to take place. It indicates that the thicker the tissue the slower is the rate at which O_2 reaches the cell furthest from the surface. In smaller organisms, the diffusion distance is short and, therefore, there is no problem. In larger organisms the problem has been overcome either by developing a circulatory system with an O_2 carrying pigment or by reducing the diffusion distance so that none of the cells are far removed from the surface. The latter has been achieved in various ways:
   (a) The body may be flattened, thus reducing the distance between the surfaces (e.g., flatworms, leaves of plants).
   (b) The body may be constructed in such a way that tissues are thin (e.g., hydra, sea anemones).
   (c) There may be a system whereby the external medium is brought into the body so that it comes into intimate association with all tissues (e.g. tracheal system of insects).

   The shorter the distance, the greater is the rate of diffusion.
6. **Cross-section of the container**: The greater the cross-section of the container in which diffusion takes place, the greater is the rate of diffusion.
7. **Temperature**: The greater the temperature, the greater is the molecular motion and also, the greater is the rate of diffusion.

$$\text{Diffusion rate} \propto \frac{\text{Concentration difference} \times \text{Cross-sectional area} \times \text{Temperature}}{\sqrt{\text{Molecular weight}} \times \text{Distance}}$$

8. **Viscosity of the solvent**: The more viscous is the solvent, the greater is its frictional resistance against free movements of solute, hence the lower is the rate of diffusion and mean free path.
9. **Mean molecular velocity**: In cases of solutes and also for gases, diffusion varies directly with the mean molecular velocity. All the molecules do not move at the same velocity, they may possess an average velocity, called the mean molecular velocity and mean free path (average linear distance traversed by each molecule between two successive collisions is called the mean free path).
10. **Graham's law of diffusion of gases:** Graham (1833) arrived at the following law for the rate of diffusion of different gases: The rate of diffusion of any gas is inversely proportional to the square root of its density (Graham's law of diffusion).

    i.e., $\text{Rate of diffusion} \propto \frac{1}{\sqrt{\text{Density}}}$.
11. **Fick's first law of diffusion**: The rate of diffusion is given by Fick's first law of diffusion which states that the amount of solute diffusion across the area A in a period of time dt is proportional to the concentration gradient de/dx at that point:

$$\frac{ds}{dt} = -DA\frac{dc}{dx}.$$

The proportionality constant D is the **"diffusion coefficient"**.

Diffusion coefficient: It is the quantity of solute diffusing per second across a surface area of 1.0 cm^2 when there is a concentration gradient of unity. The sign of the expression of D is negative because the diffusion is in the direction of lower concentration.

The difusion coefficient depends upon the following factors:

(a) Size and shape of the molecule,
(b) Frictional resistance offered by the viscosity of the solvent,
(c) For spherical macromolecules, it is inversely proportional to the cube root of the MW.

9.1.3 Types of diffusion

1. Simple diffusion
2. Non-ionic diffusion
3. Restricted diffusion
4. Exchange diffusion
5. Facilitated diffusion
(see the factors governing the passage of materials across the cell membrane.)
6. Translational diffusion

It has been demonstrated that phospholipids undergo rapid redistribution in the plane of the membrane. This diffusion within the plane of membrane is called **translational diffusion** which is rapid for a phospholipid. Within a plane of membrane, one molecule of phospholipid can move several macrometres per second.

9.1.4 Diffusion across cell membranes

Diffusion occurs when a membrane or boundary separates two fluids differing in concentration at the same hydrostatic pressure. When the membrane is **freely permeable** to all the solutes, then all the solutes are freely diffusible through the membrane and an equilibrium is reached with equal concentrations on each side and concentration gradient disappears. If one solution contains substance which is not diffusible through the membrane, then passive diffusion to equilibrium may result in an unequal distribution of freely diffusible solutes producing Donnan membrane equilibrium of ions.

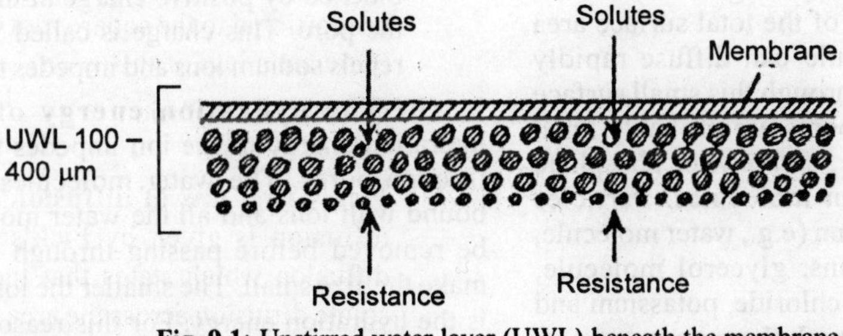

Fig. 9.1: Unstirred water layer (UWL) beneath the membrane

The time for equilibration by diffusion is proportional to the square of the diffusion distance.

(a) **Limiting boundary in the membrane:** The cell membrane is a lipid matrix in which globular proteins are embedded in the matrix. Some of these protein molecules penetrate through the membrane water and dissolved substances (small) which pass through the protein part of the membrane freely but the lipid layer is limiting boundary for the fluids between the two sides of the membrane.

(b) **Unstirred water layer (UWL):** In living animals, beneath the membrane, there is a layer of relatively unstirred water of 100-400 μm thick. Solutes pass through this UWL by diffusion, which is considered during transport across membrane because it offers a significant resistance during the movement of the solute across the membrane (Fig. 9.1).

9.1.5 Methods of membranous diffusion

There are two ways by which diffusion across membrane takes place.

A. Diffusion in the dissolved state in the lipid matrix.
B. Diffusion through minute pores in the membrane.

A. Diffusion in the dissolved state in the lipid matrix

Oxygen, CO_2, alcohol, fatty acids are soluble in lipids of the cell membrane as well as in water. Any one of these substances, when in contact with the membrane, is dissolved in the lipid matrix and diffuses like diffusion in watery medium. In spite of lipid solubility, the diffusion process is slowed down due to high viscosity of lipid matrix (300 times that of water).

- **Lipid solubility and diffusion**

The main factor to determine the rapid diffusion through the lipid matrix is the lipid solubility.

(a) *Very soluble*: If it is very soluble, it becomes dissolved in the lipid matrix easily and passes on rapidly.

(b) *300 times soluble*: If it is 300 times more soluble in lipid matrix than in water, it diffuses more rapidly through the membrane than in the water.

(c) *Very poorly dissolved*: Water, glucose and the electrolytes cannot pass through the lipid matrix.

(d) *Very insoluble*: The substances which are very insoluble in the lipid matrix, e.g., sugar and amino acids, can pass through the lipid matrix by a process called carrier-mediated or facilitated diffusion".

In general, although there are exceptions, according to **Overton's rule**, the greater the oil-water partition coefficient for a substance, the greater is its rate of transfer through the cell membrane.

B. Diffusion through minute pores in the membrane

Some substances like water and dissolved ions pass through holes in the membrane called **membrane pores**. The pores pass directly through the membrane at wide intervals over its surface. These pores are intramolecular spaces within the structures of the protein molecules that penetrate all the way through the membrane. The diametre of pore is about 0.8 nanometer (8 Angstroms) and the total area is about 1/5000 of the total surface area of the cell. Water molecule can diffuse rapidly within 1/100 of a second through this small surface area. The surface of the pores are probably lined with negative charges, e.g., carboxyl or phosphate groups within the protein molecules. Particles having diameter up to 0.8 nm (e.g., water molecule, urea molecule, lactate ions, glycerol molecule, ribose molecule, hydrated chloride, potassium and sodium ions) can pass through these pores.

9.1.6 Factors influencing diffusion through pores

1. **Permeability of pores to substances**: Permeability is the rate of transport through the membrane for a given concentration difference. Permeability of pore depends upon the diameter of the substances in comparison with the diameter of the pores. Water molecule, urea molecule and chloride ion are smaller in diameter than the pores and pass through easily whereas sugar, e.g., galactose, glucose, lactose, mannitol cannot pass due to bigger diameter and they enter the cell through facilitated diffusion. Excess calcium in the external fluid causes decrease in permeability and diminished calcium internally increases permeability producing excessive diffusion of ions. ADH increases permeability of cells lining the collecting duct of the kidney by increasing the diameter of the pores.

2. **Electrical charge of ions**: The electrical charges of ions affect their diffusion through pores which may cause slowing or accelerating effect. They are affected by three ways:
   (a) The intensity of electric charge of ions lining the pores.
   (b) The hydration energy of the ions, and
   (c) The potential difference across the membrane.

(a) The intensity of electrical charge of ions lining the pores

i. Pores are lined either by positive or negative charges. One group of pores is called "sodium channel" which is lined with "negative" charges and attract sodium ions and shuttle sodium from one negative to the next allowing the rapid diffusion. But the larger sized positive, e.g., potassium ions cannot pass.

ii. Sometimes the end of sodium channel is blocked by positive charge at the opening of the pore. This charge is called "gate" which repels sodium ions and impedes their passage.

(b) The hydration energy of ions: The hydration energy of the ion impedes the diffusion through pores. The water molecules are loosely bound with ions and all the water molecules must be removed before passing through the pores to make the size small. The smaller the ion, the greater is the hydration energy. For this reason, hydration of sodium is greater and this factor retards the sodium transport through some pores far more than it does potassium transport because the water molecules can be removed easily than the sodium. The hydration energy of sodium can be overcome if the pores are strongly negatively charged. But most cell membranes are having uncharged or weakly charged pores, causing impermeability to sodium ions and relatively permeable to potassium ions for lower hydration energy; these pores are called **potassium channels**.

(c) The potential difference across the membrane: The fluids on either side of the membrane show different electrical potential, i.e., the potential is more positive on one side and more negative on the other. The positive fluid attracts negative ions through the pores and the negative fluid attracts positive ions.

9.1.7 Net diffusion through membrane and factors influencing it

It is obvious that substances which diffuse from outside to the inside of the cell can also diffuse from inside to outside of the cell. The total quantity of substances diffusing in both directions is not so important for the cell but the "net quantity" diffusing in both directions is very important

The rate of net diffusion of a substance depends upon:

(A) The concentration difference of the substance across the membrane.

(B) The electrical potential difference across the membrane.

(C) The pressure difference across the membrane.

A. Influence of concentration differences

Suppose a substance is in high concentration outside the cell, (C_0) and low concentration on the

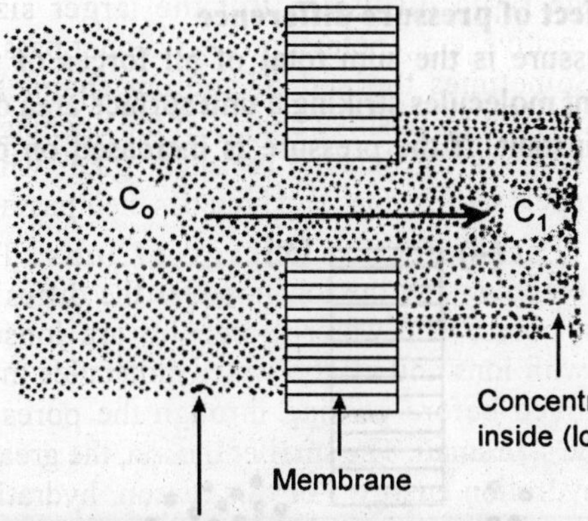

Fig. 9.2: Effect of concentration difference on diffusion of molecules and ions across a membrane

inside (C_i). The rate at which the substance diffuses inward is proportional to the concentration of molecules present outside (Fig 9.2).

This concentration determines the number of molecules knocking the outer part of the pore each second. Again, the rate at which the molecules diffuse outward is proportional to their concentration inside the cell membrane. Therefore, the rate of net diffusion into the cell is proportional to the concentration on the outside minus the concentration on the inside, i.e.,

$$\text{Net diffusion} \propto P(C_o - C_i),$$

where C_o = Concentration on the outside

C_i = Concentration on the inside

P = Permeability of the membrane for the substance

While discussing diffusion of molecules through the cell membrane, we should define "permeability constant" and "permeability coefficient".

Permeability Constant: The ratio of the diffusion constant to the unknown constant which has the dimensions of a velocity in centimeters per second.

It can be defined quantitatively as the ease with which a given substance can permeate a given cell membrane, and it is a property of the membrane.

Permeability coefficient of the membrane for the given substance: Fick's equation can be applied to describe the passage of substances across the cell membrane by passive diffusion. And the **concentration gradient** would be the difference between the concentrations of substance just within its outer and inner borders of the membrane divided by the thickness of the membrane i.e.,

$$\left(C_{0_m} - C_{i_m}\right) / X_m$$

Where, C_{0_m} = Concentration of substance just within its outer border of the membrane.

C_{i_m} = Concentration of the substance just within its inner border of the membrane.

X_m = Thickness of the membrane.

Thus, $J = D_m A_m (C_{o_m} - C_{i_m}) / X_m$,

Where, D_m = Diffusion coefficient for the molecule in the material of the membrane.

A_m = Area of the cell membrane.

Being the lipid nature of the membrane, the concentrations of a substance just within the outer and inner borders of the membrane are not the same as the concentration in the immediately adjacent aqueous phases of the extracellular fluid and cytoplasm. The "partition coefficient" (K_m) is the ratio of the concentration when the solute is distributed at equilibrium between water and the material of the cell membrane, i.e., $K_m = C_m/C$ water. Now using this K_m value, the Fick equation for the diffusion across the membrane can be expressed in terms of the concentrations of solute in the extracellular and intracellular fluid (C_o and C_i).

So we can rewrite the equation: $J = D_m A_m K_m (C_o - C_i)/X_m$. The terms D_m, A_m and K_m are combined and replaced by P, i.e., **permeability coefficient** of the membrane for the given solute and membrane, i.e. $P = D_m K_m / X_m$ and to simplify the equation it is expressed as:

$$J = PA_m (C_o - C_i).$$

From the above equation it is evident that the transport across the membrane is a "passive" transport.

Permeability coefficient (*P*) defines the flow rate of solute per unit membrane area per unit concentration gradient maintained across the membrane. The greater the *P*, the greater the rate of which the solute diffuses across the membrane. *P* may include the factors, e.g., the solute diffusion coefficient, the membrane thickness, the geometry of aqueous channels in the case of polar solutes, etc.

The relation with molecular size is that the greater the MW, the smaller the permeability coefficient, e.g., proteins, nucleic acids and viruses have small permeability coefficient.

B. Effect of an electrical potential difference

When an electrical potential difference is applied across the membrane, the ions for their electrical charges move through the membrane though there is no concentration difference between two sides of the membrane (Fig 9.3).

In (A), the negative ions concentrations are same on both sides, but when a negative charge is applied to the left and positive charge to the right an electrical gradient is developed across the membrane. The positive charge attracts the negative ions while negative charge repels them. So the net diffusion occurs from left to right. After some time, large amount of negative ions move to the right creating a concentration difference of the same ions in the direction opposite to the electrical potential difference. At normal body temperature, the electrical difference which balances a given concentration difference of univalent ions is determined from the formula known as Nernst equation:

$$\text{EMF (millivolts)} = \pm 61 \log \frac{C_1}{C_2}$$

(Electromotive force voltage)

C. Effect of pressure difference

Pressure is the sum total of all forces of the different molecules striking a unit surface area of a given instant. If the pressure is increased on one

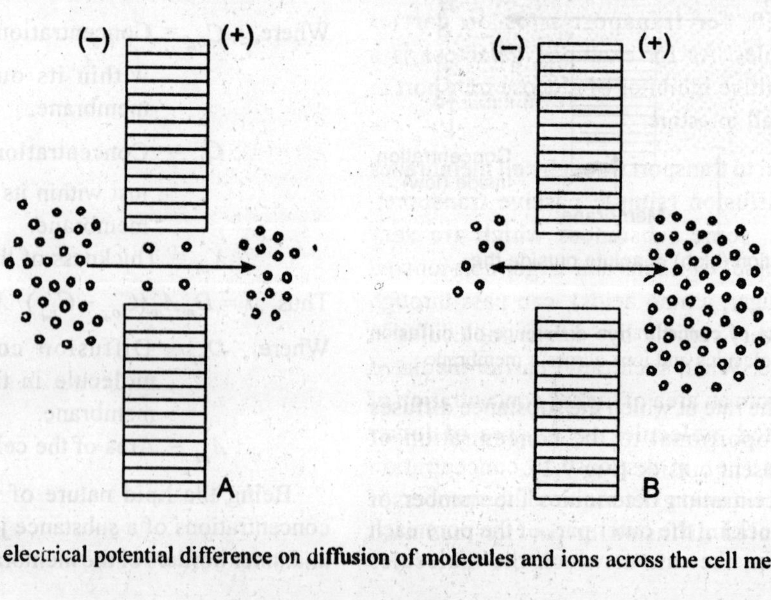

Fig. 9.3: Effect of electrical potential difference on diffusion of molecules and ions across the cell membrane

side of the membrane, the sum total of all forces of the molecules striking the membrane pores on the same side is greater than the other side. This happens either from large number of molecules striking the membrane per second or from greater kinetic energy of the average molecules striking the membrane. However, in either case, increased amount of energy is available for the net movement of molecules from the high pressure side to the low pressure side.

9.2. Facilitated Diffusion

(Carrier-Mediated Transport)

Characteristics

1. **Carrier-mediated transport comprises**
   (a) Facilitated diffusion
   (b) Primary active transport
   (c) Secondary active transport.
2. **Stereospecificity**: D-glucose which is natural isomer is transported by facilitated diffusion, but the 1-isomer is not. Simple diffusion, however would not distinguish between the two isomers because it does not involve carrier.
3. **Saturation**: The transport rate increases as the concentration of the solute increases, until the carriers are saturated. The **transport maximum** (T_m) is analogous to the maximum velocity (V_{max}) in enzyme kinetic.
4. **Competition**: Structurally related solutes compete for transport sites on carrier molecules. As for example, galactose is a competitive inhibitor of glucose transport in the small intestine.

In addition to transport through cell membranes by simple diffusion (simple passive transport), osmosis, etc. some substances which are very insoluble in lipid, e.g., ions and larger non-ionized molecules (sugar, amino acids), can pass through lipid matrix very rapidly by carrier molecules in the membrane. When such rapid carrier-mediated transport is from an area of higher concentration of the transported molecules to an area of lower concentration (i.e., in response to concentration gradient) and no energy is required, the process is called **facilitated diffusion or carrier-mediated diffusion.**

9.2.1 Carrier-mediated model for glucose

Facilitated diffusion depends upon the presence in the cell membrane of a small number of "carrier" molecules. At first glucose (or amino acid) combines with specific carrier at the outer border of the cell membrane at which the glucose concentration is higher to form sugar-carrier complex. The sugar-carrier complex transfers the glucose to the other border of the membrane, where it is dissociated to deliver the sugar into the fluid on that side and the carrier molecule moves back to the original site again, to pick up still more sugar and transport it to the inside and act as ferry-boat. Thus, the carrier molecules show their effect in the following ways:

(a) To make the glucose soluble in the membrane;

(b) To act as ferry-boat for sugar molecules to increase the rate of transport across a barrier;

(c) To reduce the thermal energy needed by the substrate molecule to leave the aqueous phase and to enter the membrane phase.

(d) Another view regarding carrier transport is that the carrier rotates within the membrane exposing binding sites first to the compartment, origin of the transport process and then to the compartment of destination for the transported entity (Fig. 9.4).

9.2.2 Nature of carrier

Knowledge regarding carrier substance is meagre. A protein having MW 45000 and having the property of combining specifically and reversibly with the monosaccharides, e.g., glucose, mannose, galactose, xylose and arabionse, has been discovered in the cell membrane using radioactive labelled sugars which was attached with the protein molecule. Probably this protein is the carrier substance for transport of glucose.

9.2.3 Factors influencing the rate of transport by facilitated diffusion

Facilitated diffusion depends upon the following:

(a) Difference in concentration of the substance on the two sides of the membrane.

(b) Amount of carrier molecules available for binding.

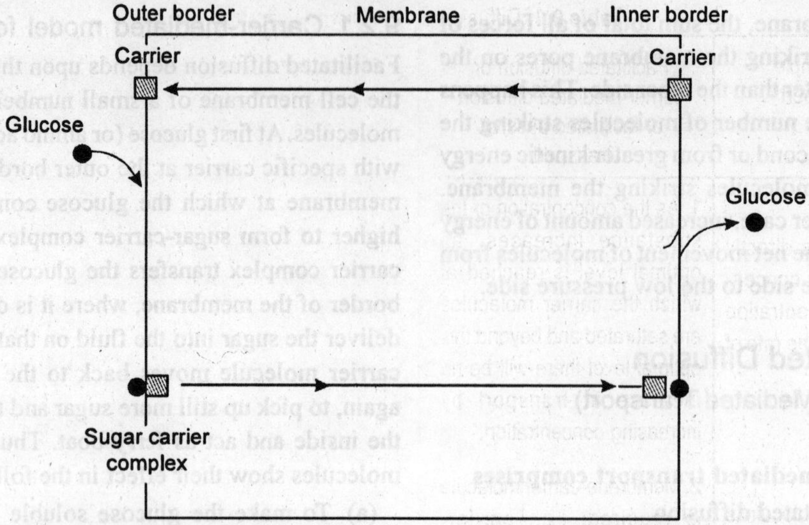

Fig. 9.4: Carrier model for sugar transport across cell membrane (Facilitated diffusion)

(c) Speed of chemical or physical reactions,

(d) Hormone: Insulin in large quantities can increase the rate of glucose transport 7-10 fold, though it is known that insulin increases the quantity of carrier in the membrane or increases the rate at which the chemical reactions take place between the carrier and glucose.

1. **Kinetics**: The rate of glucose entery is always much greater than the simple solubility-diffusion mechanism. When the external glucose concentration increases, the faster rate of entry of glucose into the cell continues to increase linearly at the initial period. But as the glucose does not pass through pores, even on further increase in glucose concentration, the linear increase does not continue indicating saturation of the carrier molecules responsible for active uptake of glucose.

2. **Substrate specificity**: Facilitated diffusion exhibits substrate specificity. The entry of glucose is stereospecific, such as isomer D-glucose is rapidly transported while optical isomer L-glucose cannot. This can be attributed to the highly specific structure of the substance. The carrier molecule for D-glucose can also transport other sugars like D-galactose, D-mannose, D-xylose and D-arabinose, which are hexoses or pentoses having six-membered pyranose ring in the same chair conformation as in D-glucose. But their relative affinities for the carrier molecule for sugar transport are lesser than that of D-glucose. It may be related to the number of OH groups in the equatorial plane of the pyranose ring.

3. **Influence of poisons**: Due to similar kinetics of enzymes and facilitated diffusion, the poisons like $HgCl_2$ and Dinitroflurobenzene react with proteins in the cell membrane producing inhibition of facilitated diffusion. Cyanide and Dinitrophenol inhibit production of metabolic energy but does not disturb the facilitated diffusion, revealing that no metabolic energy is required for this process. This process is always in the direction of the glucose concentration gradient.

- **Sites of facilitated diffusion**
  1. **Sugars:** Muscle, adipose tissue, choroid plexus, capillaries in the brain, red and white blood cells.
  2. **Amino acids:** Intestinal epithelium, red and white blood cell.
  3. **Sugars, fatty acids and amino acids in the small cell**: When in one cell the three substances—sugars, fatty acids and amino acid are to be transported, there are also several types of carriers present in the cell membrane and there is substrate competition with each other. In case of amino acids, it is grouped into neutral, basic and acidic. When there is substrate competition, it occurs between the members of the same group but not with different groups.

Table 9.1: Diffusion vs. Facilitated Diffusion

Diffusion or simple passive transport	Facilitated diffusion or carrier-mediated diffusion or facilitated passive transport
1. **Saturation** Rate of transport is directly proportional to the concentration. If the concentration difference doubles, the rate of transport doubles.	1. As the concentration of the substance increases, an optimal level is reached at which the carrier molecules are saturated and beyond this optimal level there will be no increase in transport by increasing concentration.
2. Carrier Membrane carrier molecule is not required, i.e. not carrier-mediated.	2. Membrane-carrier molecule is required, i.e., carrier-mediated.
3. Specificity No specificity	3. One type of carrier molecule may transport only one kind of substance. In actuality, specificity is never absolute and a given carrier may transport several chemically related substances but with a different efficiency, *e.g.* one carrier for several sugars and another for a number of amino acids.
4. Competition There is no competition for diffusion between two substances A and B in a solution.	4. If two substances A and B use the same carrier molecule in the cell membrane, then addition of B to the extracellular fluid will depress the net transport of A. Since the total number of carrier molecules in the membrane is limited, B will occupy some of the carrier sites previously available to substance A. Thus the substances complete for the same loci on the carrier molecules; likewise, the presence of A will reduce the transport B, as competition is a mutual effect.
5. Inhibition Inhibitory substance does not exert any influence on diffusion.	5. During inhibition an inhibitory substance (I) combines with the carrier molecule strongly and irreversibly. Consequently, the transport of substance A is inhibited completely, even though (I) is not transported but remains bound to the active sites on the carrier molecule.
6. Energy Not requried.	6. Not required.
7. Concentration Depends upon concentration gradient.	7. Diffuses down its own concentration of the substance.
8. Temperatures Coefficient Does not show.	8. Generally shows the high temperature co-efficient characteristic of chemical reaction.

Contd.

9.2.4 Sodium co-transport (secondary active transport)

Sodium co-transport system is the combination of two processes, i.e., pure diffusion and active transport. In brush-border epithelial cells of the mucosa of the small intestine and renal tubules, there is a membrane mechanism which can transport glucose and amino acids from the surroundings to the interior of the cell against a concentration gradient. This mechanism is called sodium co-transport system or secondary active tranport.

- **Characteristics of Secondary Active Transport (sodium co-transport)**
  1. **Coupling**: The transport of two or more solutes is coupled.
  2. **Na^+ Provides Energy**: One of the solutes, usually Na^+ is transported "downhill" and provides energy for the "uphill" transport of the other solute(s).
  3. **Metabolic Energy**: Metabolic energy is not provided directly, but indirectly from the Na^+ gradient, which is maintained across cell membranes. Thus, inhibition of Na^+, K^+-ATPase will decrease transport of Na^+ out

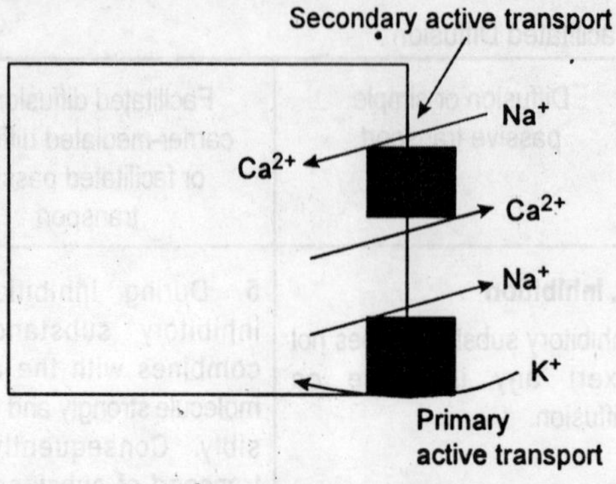

Fig. 9.5: $Na^+–Ca^{2+}$ Countertransport (antiport)

of the cell, decrease the transmembrane Na^+ gradient, and eventually inhibit secondary active transport.

4. **Co-transport or Symport**: If the solutes move in the same direction across the cell membrane, it is called **co-transport** or **symport,** e.g., Na^+ -glucose co-transport in the small intestine and $Na^+–K^+–2Cl$ co-transport in the renal thick ascending limb.
5. If the solutes move in opposite directions across the cell membranes, it is called **countertransport, exchange,** or **antiport.** For example, $Na^+–Ca^{2+}$ countertransport or exchange, Na^+–H exchange.

9.2.4.1 Site for sodium co-transport

1. *Sodium and sugars*: Proximal renal tubule, gall bladder, small intestine.
2. *Sodium and amino acids*: Intestine, proximal renal tubule, brain cells, choroid plexus of cerebral ventricle, white blood cells, tumour cells.
3. *Catecholamines*: Cells of the nervous system.
4. Choline and para-aminohippuric acid—Kidney.

9.2.4.2 Disorders of sugar and amino acid co-transport system

Heriditary diseases like:

(a) *Glucose*: galactose malabsorption by the intestine,

(b) Malabsorption of neutral amino acids by the intestine—Hartnup's disease,

(c) Cystinuria.

9.2.4.3 Mechanism of sodium co-transport

The association of sugar transport with sodium transport has led to term the process as sodium co-transport system.

- **Evidence**

1. Replacing sodium by lithium or potassium in the lumen of the intestive, the transport of sugar through brush-border epithelium was markedly reduced.
2. Increasing the luminal concentration of sodium, stimulated the sugar transport.
3. Increasing the luminal concentration of glucose, stimulated uptake of sodium through the brush-border membrane of the epithelial cell. At physiological sodium concentration, one sodium ion enters the cell for each glucose molecule transported.
4. Ouabin causes blockage of transport of sugars which is the inhibitor of the membrane-bound sodium and potassium dependent ATPase which is responsible for active sodium transport.

9.2.4.4 Model for co-transport of glucose and sodium

There is a carrier in the membrane which is not active in absence of sodium, i.e., it shows very low affinity for glucose. When sodium is present, the carrier is activated by allosteric modification and its affinity for glucose also increases. Thus, the binding capacity of glucose with the carrier depends upon the concentration of sodium and glucose. The brush-border side is impermeable to sodium but it becomes permeable when combines with a carrier of the membrane known as **sodium-glucose carrier.** But the carrier has the peculiarity not to carry sodium alone but also the glucose molecule at the same time. At the brush-border side sodium concentration is high, so if the glucose concentration is low, the sodium activated carrier has high affinity for glucose which results in a high concentration of **sodium-sugar-carrier-complex**. It is the sodium gradient across the brush-border membrane that provides energy (not metabolic energy) to promote the transport of sodium and glucose. Since energy is provided from sodium gradient, glucose can be transported against a concentration gradient. Thus, if there is low concentration of glucose in the human intestine or renal tubules it will be transported to the interior of the cell where glucose concentration is high. At the inner border of the membrane (cytoplasmic side), the sodium ion concentration is kept low by Na^+/K^+ exchange pump (Fig. 9.6).

Therefore, sodium leaves the sodium-glucose-carrier complex and reducing the affinity of the

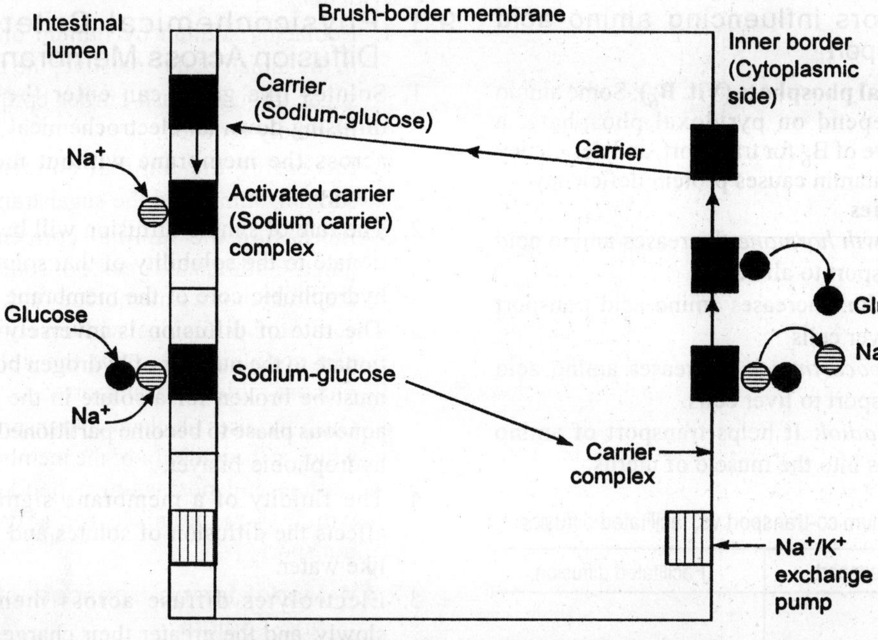

Fig. 9.6: Model for co-transport of sodium and glucose

carrier for glucose, glucose is released from the carrier to the cytoplasm, even the glucose concentration is higher in the cytoplasm than the lumen of intestine. Glucose crosses the basal side of the cell by the usual process of facilitated diffusion, in the manner it crosses all the membranes of the body. There are different types of sodium co-transport carriers. One is sodium-glucose carrier responsible for transport of glucose and galactose and four others are responsible for transport of different types of amino acids.

Almost all monosaccharides are transported to the body, e.g., glucose, galactose, fructose, mannaose, xylose, arabinose and sorbose. Fructose is transported by a separate carrier system. The nature of carrier system and chemical reactions that take place for sodium co-transport are not known. But for glucose and galactose, an intact –OH group on the C_2 carbon of the monosaccharide molecule is essential and it is presumed that the sugar attaches to the carrier at this point.

9.2.4.5 Sodium co-transport of amino acids

Most of the amino acids are transported to the interior of the cell by facilitated diffusion mechanism, and insulin accelerates this process like glucose.

But in certain region, the membranes transport amino acids by sodium co-transport like glucose. These are bursh-border membrane of the intestinal mucosa, renal tubules and some exocrine glands:

1. There are at least four different carrier systems for transport of amino acids:
   (a) For neutral amino acids
   (b) For basic amino acids
   (c) For acidic amino acids
   (d) For imino acids (Proline and hydroxyproline).
2. The carrier transport mechanisms have much affinity for L-stereoisomers of amino acids than D-stereoisomers.
3. Amino acids transport occurs only in presence of simultaneous transport of sodium like glucose transport.
4. It occurs in the brush-border of the epithelial cell.
5. Amino acids are transported by the same sodium co-transport mechanism as for glucose.
6. Carrier has receptor sites for amino acid and a sodium ion and when both the sites are filled the carrier moves to the interior of the cell.

Due to sodium gradient on the brush-border side, diffusion of sodium inside cell drags the carrier along with amino acid to the inner side of the membrane, then it is released to the cytoplasm. Here the concentration of amino acid increases and by diffusion it passes through the basal side of the cell into the portal blood.

9.2.4.6 Factors influencing amino acid transport

1. **Pyridoxal phosphate (Vit. B_6)**: Some amino acids depend on pyridoxal phosphate, a derivative of B_6 for transport, and deficiency of this vitamin causes protein deficiency.
2. **Hormones**
   (a) *Growth hormone*: Increases amino acid transport to all cells.
   (b) *Insulin:* Increases amino acid transport to liver cells.
   (c) *Glucocorticoids*: Increases amino acid transport to liver cells.
   (d) *Estradiol*: It helps transport of amino acids into the muscle of uterus.

Table 9.2: Sodium co-transport vs. facilitated diffusion

Sodium co-transport	Facilitated diffusion
I. Similarity	
1. The rate of sugar accumulation exhibits saturation kinetics.	1. Same
2. Stereospecific process. Only D-glucose but not L-glucose accumulates in the intestinal epithelium.	2. Same
3. Several sugars share the same transport system.	3. Same
4. Only hexoses with an equitorial OH group on carbon number.	4. Same
5. Mutual competition among the sugars that are transported.	5. Same
II. Dissimilarity	
1. It is related to the concentration of sodium.	1. No.
2. It is the sodium gradient across the brush border and provides energy to promote the transport of glucose and sodium.	2. No.
3. It requires energy for transport (Not metabolic).	3. No.
4. Since energy is provided from sodium gradient, glucose can be transported against a concentration gradient.	4. In the direction of concentration gradient of glucose
5. Carrier undergoes allosteric modification by sodium and increases affinity for glucose.	5. No.

9.3 Physicochemical Criteria for Diffusion Across Membranes

1. Solutes like gases can enter the cell by diffusing down an electrochemical gradient across the membrane without metabolic energy.
2. The rate of simple diffusion will be proportionate to the solubility of that solute in the hydrophobic core of the membrane bilayer.
3. The rate of diffusion is inversely proportionate to the number of hydrogen bonds and must be broken for a solute in the external aqueous phase to become partitioned into the hydrophobic bilayer.
4. The fluidity of a membrane significantly affects the diffusion of solutes and solvents like water.
5. Electrolytes diffuse across membranes slowly, and the greater their charge density, the slower the diffusion rate.
6. In the membranes, there are transmembrane channels or pore-like structures generating ion-conductive pathways.
7. The nerve cell membranes contain ion channels which are responsible for the action potentials generated in the membrane.
8. *Ionophores and ionophoresis*: A large number of antibiotic substances capable of inducing passage of specific ions across biological membranes called ionophores or carriers of cations and the process is called ionophoresis. These are valinomycin, gramicidin, nonactin, nigericin. These function as ion shuttles for the movement of ions across membranes. These agents form lipid soluble complexes with specific cations, which are carried through the mitochondrial membrane, e.g., valinomycin forms valinomycin-K^+ complex readily passes through the membrane but in the absence of valinomycin, K^+ passes only very slowly. Ionophores prevent also the oxidative phosphorylation.
9. *Metabolic uncoupler*: Dinitrophenol, a metabolic inhibitor increases the permeability of mitochondria to protons which is impermeable and provides shuttles for protons across membranes, causes collapse

of proton gradient necessary for the generation of ATP according to chemiosmotic theory of Mitchell.

10. Some specific solutes diffuse down electrochemical gradients across membranes rapidly, irrespective of their size, charge, and partition coefficients. This diffusion is carried by a carrier and hence is called carrier mediated or facilitated.
11. The rate of diffusion can be saturated, i.e. the number of sites involved in diffusion of the specific solutes appears to be finite.
12. Most of the facilitated diffusion systems are stereospecific and no metabolic energy is required.
13. *Lipid solubility*: The greater the oil-water partition coefficient for a substance, the greater its rate of transfer through the cell membrane.
14. *Hydrophilicity*: The permeability of the cell membrane to a substance is inversely related to the hydrophilic nature of the substance.
15. *Molecular size*: For a series of homologous substances, the greater the molecular weight, the smaller the permeability coefficient. The larger the particle, the slower its thermal motion and consequently, the smaller its rate of diffusion.

- **Biological importance of diffusion**

1. **External respiration**: Exchange of respiratory gases (O_2 and CO_2) between the lungs and atmosphere.
2. **Internal respiration**: Exchange of respiratory gases between the blood through the capillary membrane to the tissues.
3. **Absorption of food**: Passage of food materials (pentoses, some minerals and most of the water soluble vitamins) from intenstine to the blood.
4. **Elimination in the kidney**: Elimination of some of waste products (NH_3) in the renal tubules (non-ionic or trapping diffusion).
5. **Reabsorption in the kidney**: Reabsorption of glucose, water, and urea in the renal tubules.
6. **Transcapillary exchange**: Diffusion is the major process by which nutrients and metabolic products cross the capillary membrane.
7. **Placental diffusion**: The placenta serves as a fetal lung and respiratory gases are transferred between fetal and maternal compartments by simple diffusion.
8. **Return of CSF**: Cerebrospinal fluid is returned to the circulation at a rate equal to its rate of production. It is returned via arachnoid villi, invaginations of arachnoid covered blood vessels into the space of venus sinus. Transport at this site is passive in response to hydrostatic pressure and diffusion. The endothelial cells covering the villi shows under EM, formation of large vesicular holes directly through the bodies of the cells. These are big enough to allow free flow of CSF, protein molecules, even RBC into the venous blood.
9. **Passage of aqueous humor flow**: Aqueous humor from ciliary body, some protein is diffused to the vitreous humor.
10. **Return of aqueous humour to the blood**: Some portion of aqueous humour after passing through vitreous humour, return to the retinal vessels by diffusion and filtration.
11. **Plant Cell:** Molecules or ions enter plant cells and move from one to the other by following the simple law of diffusion.
12. **Animals:** In planarian worms, the gut, instead of being a straight tube, as it is in most animals, gives off numerous blindly-ending branches which ramify throughout the body penetrating deep into the tissues. Digested food is forced into these branches by muscular action and is thus conveyed directly to the cells. Planarians have no circulation and this extraordinary alternative means that they do not have to depend on diffusion alone for distributing their digested food.
13. **Carbohydrate absorption**: Little carbohydrate absorption results from simple diffusion for the pores of the mucosa through which diffusion occurs are essentially impermeable to water soluble solutes with molecular weights greater than 100.
14. **Water absorption**: Water is transported through the intestinal mucosa entirely by the process of diffusion. Thus diffusion obeys the usual laws of osmosis.

15. **Diffusion of ATP into the cilia**: The ATP required for the ciliary movements is provided by mitochondria near the base of the cilia from which ATP diffuses into the cilia.

9.4 Active Transport

9.4.1 Introduction

It is seen that concentration of a substance is high in extracellular fluid but it is present in low concentration in intracellular fluid such as sodium. Again, some substance is present in low concentration in the intracellular fluid but in high concentration in the intracellular fluid, such as potassium. It is also observed that diffusion takes place through a cell membrane from a high concentration region to low concentration region, i.e., not against the concentration gradient, hence no energy is required for the molecules but reverse process also occurs in the system.

9.4.2 Definition

When the movement of the substance takes place through cells membrane from low concentration region to higher concentration region i.e., against concentration gradient, and energy is required for the molecules, the process is called active transport.

- **General features of Na^+/K^+ in the active transport**
  1. Most cells have a high concentration of K^+ ions and low concentration of Na^+ ions inside the cell.
  2. Na^+ must be inside and K^+ must be outside the cell for active transport and to activate ATPase. That is, asymmetric distribution of cations is coupled to the functions of the enzyme.
  3. ATP is not hydrolyzed unless Na^+/K^+ are transported.
  4. The face energy required for the active transport is provided by the hydrolysis of ATP.

- **Characteristics of Primary Active Transport**
  1. **Electrochemical gradient:** It occurs against an electro-chemical gradient, i.e., "uphill".
  2. **Metabolic Energy:** It requires direct input of metabolic energy in the form of adenosine triphosphate (ATP) and, therefore, is active.
  3. **Carrier-mediated:** It is carrieer-mediated and, therefore, exhibits stereospecificity, saturation and competiton.

9.4.3 Difference between mediated diffusion and active transport

In case of diffusion taking through a "carrier", it is called mediated diffusion but does not require energy against concentration gradient, whereas in active transport, *"carrier" system is associated with energy consumption* which is about 30-40% of the total energy produced by the cell and against the concentration gradient.

9.4.4 Substances for active transport

1. **Na^+, K^+, Cl^-, Ca^+, Fe^+, Iodide, Urate, Hydrogen ions.**
2. **Sugars.**
3. **Amino acids.**

9.4.5 Examples of active transport

1. **Cation 'Pumps' of RBC.**
2. **Movement of Na^+, K^+ (Na^+-K^+ pump) Ca^+ in nerve.**
3. **Ca^+ in a sarcotubular system of muscle (Ca^{2+} pump)**
4. **Secretion of H^+ by gastric oxyntic cells Inhibited by omeprazole (proton pump).**
5. **Alkaline secretion from pancreas.**
6. **Reabsorption and secretion by kidney tubules.**
7. **Secretion of hormones by endocrine glands.**
8. **Large amount of Na^+ in aqueous humour during formation in ciliary body.**

9.4.6 Various forms of active transport

1. **Acceleration of movements along the gradient**: **When a glucose solution of higher concentration is placed in the intestine, it is more rapidly absorbed through living intestinal mucosa than through dead mucosa, owing to the active intervention of the mucosal cells.**
2. **Prevention of movement with the gradient: Accumulation of Na^+ inside the cell may be prevented when the concentration is higher outside the cell. At this stage Na^+ can enter the cell through the membrane but a steady state is maintained inside the cell by pumping out Na^+ by continuous expenditure of chemical energy.**

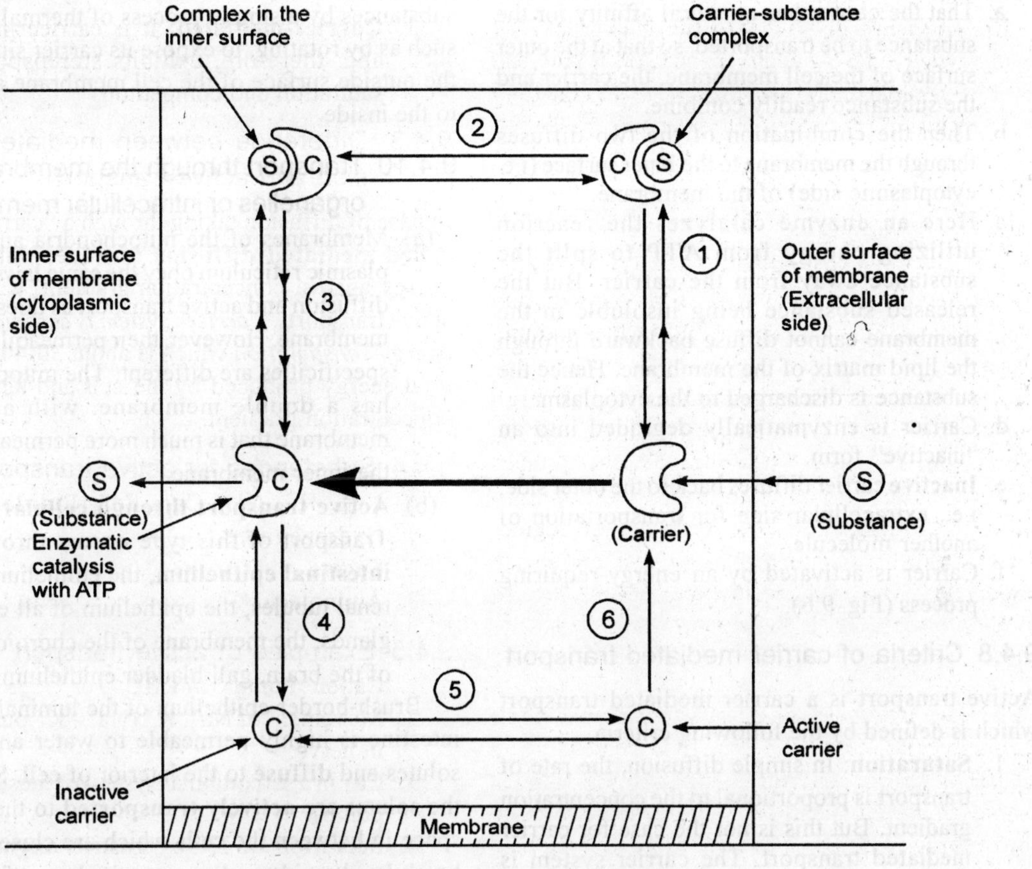

Fig. 9.7: Sequence of events in active transport

3. **Movements of substance against the gradient:** Thus K^+ can be taken up by the cell even when the intracellular K^+ is higher. The cell maintains equilibrium of higher K^+ concentration in the cells which actively maintain concentration gradient by the expenditure of energy, are said to be **selectively permeable.**

9.4.7 Basic mechanism of active transport

The mechanism of active transport is similar to most substances and has to depend on transport meterials called **carriers**. There is similarity between "active transport" and "facilitated diffusion" but having difference, i.e.,

(a) Active transport works against concentration gradient but not in case of facilitated diffusion.

(b) Active transport requires metabolic energy whereas facilitated one does not require energy.

Though the mechanism of energy consumption by active transport is not known, some features of this process is known:

1. First, the energy is supplied to the inner surface of cell membrane from high energy substances, e.g., ATP from cytoplasm.
2. Second, active transport obeys the usual laws for chemical combination of one substance (which is to be transported) with another (the carrier).
3. Third, a specific "carrier" molecule is required to transport each type or each class of substance.
4. Fourth, a specific enzyme (or enzymes) is required to promote active transport. From the foreign features we can deduce the following sequence of events:

a. That the carrier has a natural affinity for the substance to be transported, so that at the outer surface of the cell membrane, the carrier and the substance readily combine.
b. Then the combination of the two diffuses through the membrane to the inner surface (i.e. cytoplasmic side) of the membrane.
c. Here an enzyme catalyzes the reaction utilizing energy from ATP to split the substance away from the carrier. But the released substance being insoluble in the membrane cannot diffuse backward through the lipid matrix of the membrane. Hence the substance is discharged to the cytoplasm.
d. Carrier is enzymatically degraded into an "inactive" form.
e. **Inactive** carrier diffuses back to the outer side, i.e., extracellular side for transportation of another molecule.
f. Carrier is activated by an energy-requiring process (Fig. 9.6).

9.4.8 Criteria of carrier mediated transport

Active transport is a carrier mediated transport which is defined by the following criteria.

1. **Saturation**: In simple diffusion, the rate of transport is proportional to the concentration gradient. But this is not the case for carrier mediated transport. The carrier system is saturated and the carrier-mediated transport operates at the maximum possible rate.
2. **Competition**: Two substances A and B which are attached to the same carrier compete for the carrier system.
3. **Specificity**: One carrier molecule may transport one amino acid or group of amino acids, while another transports one sugar or group of sugars.
4. **Inhibition**: This is a special case of competition. When a substance "I" may combine with a carrier strongly and irrever-sibly so that the carrier cannot combine with its usual substrate.
5. **Requirement for energy**: Since this is an energy-coupled process, active transport can propel molecules against an electrochemical gradient.

9.4.9 Chemical nature of carrier

It is believed to be proteins, conjugated proteins or loose physical combinations of more than one protein molecule. The carrier might transport substances by a simple process of thermal motion, such as by rotating, to expose its carrier site first to the outside surface of the cell membrane and then to the inside.

9.4.10 Transport through the membranes of organelles or intracellular membranes

(a) Membranes of the mitochondria and endoplasmic reticulum obey the same laws of both diffusion and active transport as does the cell membrane. However, their permeabilities and specificities are different. The mitochondria has a double membrane, with an outer membrane that is much more permeable than the inner membrane.

(b) **Active transport through cellular sheets**: Transport of this type occurs through the **intestinal epithelium**, the epithelium of the renal tubules, the epithelium of all exocrine glands, the membrane of the choroid plexus of the brain, gall bladder epithelium, etc.

Brush-border epithelium of the luminal side of intestine is highly permeable to water and some solutes and **diffuse** to the interior of cell. Some of the solutes are **actively transported** to the lateral space in between the cells which are closed at the brush-border side and open at the base of the cell and rest on the basement membrane which is extremely permeable and the substances flow towards connective tissues. The most important substance, Na^+, is actively transported in this manner. When Na^+ ions are in the lateral space, their positive electrostatic charges pull negatively charged Cl^- ions to follow electroneutrality. When concentration of both ions increases in the lateral space, which causes **osmosis** of water out of cell Na^+ and Cl^- flow along with water to the connective tissue where water and ions diffuse into the capillaries (Fig. 9.8).

9.4.11 Secondary active transport: Sodium co-transport of glucose and amino acids

Glucose and amino acids are transported through the epithelial cells of intestine or renal tubules by a mechanism that is neither pure diffusion nor active transport but a mixture of the two called **Secondary active transport** or **Sodium co-transport**, Na^+

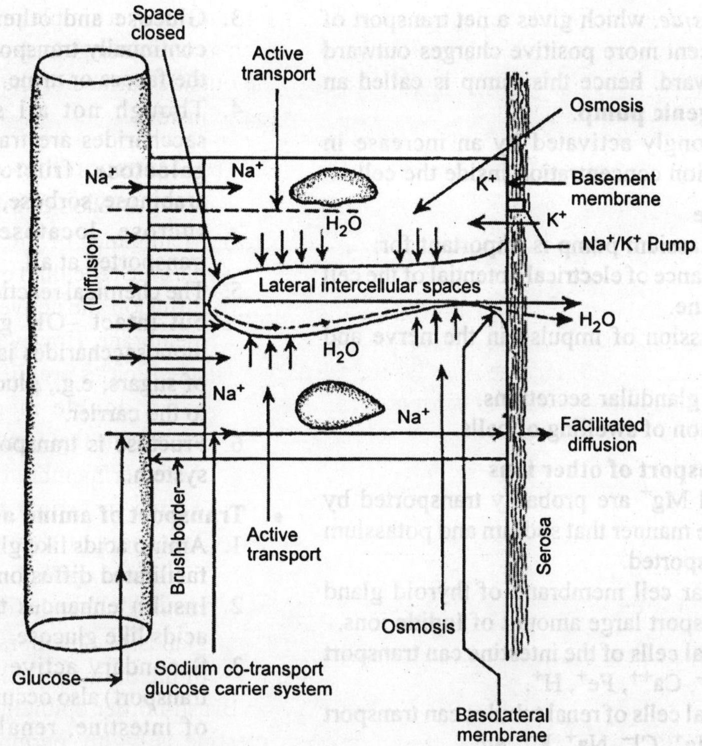

Fig. 9.8: Active transport through a layer of cells or cellular sheets, e.g., intestinal epithelium

enters through brush-border side of the epithelium, passes to the basement membrane blood capillaries which occurs by active transport and marked depletion of Na^+ in the cell, creates diffusion gradient for Na^+. Na^+ attempts to enter the cell through brush-border epithelium which is impermeable to Na^+. But it is permeable when shares with glucose for combination with a common "carrier" (symport) called **sodium-glucose carrier** or which transport both Na^+ and glucose. There are several different types of "sodium co-transport carriers" such as sodium-glucose, sodium-galactose, sodium-amino acids (4 types), etc. Glucose crosses the basal side of the cell by facilitated diffusion.

9.4.12 Active transport of different substances

A. Active transport of Sodium and Potassium

Sodium concentration is very high and potassium is low outside the cell and reverse concentration is present inside the cell. But some minute quantities of Na^+ and K^+ can diffuse through the pores of the cell membrane and if it is allowed to continue, there will be equal distribution between the two sides of the membrane. But there should be unequal distribution to maintain electrial potential difference between outside (i.e., positive) and inside (i.e., negative) the cell and which is maintained by the mechanism of active transport called **Sodium-Potassium Pump**. The carrier for this mechanism is called **Sodium-Potassium ATPase** which can also split up ATP to provide energy. This ATPase is composed of two protein molecules, globulin and glycoprotein (Fig. 9.9)

- **Mechanism**

Magnesium-ATP binds with "Sodium-Potassium ATPase" at the inner surface of membrane. Energy released from ATP causes K^+ to split away from "carrier" and causes Na^+ ions to bind. At the outer surface of the membrane, Na^+ splits away from the carrier and K^+ ions bind without any energy (at the outer surface) as both release of Na^+ and binding of K^+ are exothermic process.

- **Peculiarities**

1. Sodium-Potassium pump normally *transport 3 sodium ions inside, for each 2 potassium*

ions outside, which gives a net transport of 50 per cent more positive charges outward than inward. hence this pump is called an **electrogenic pump.**

2. It is strongly activated by an increase in sodium ion concentration inside the cell.

- **Importance**

Sodium-Potassium pump is important for:

1. Maintenance of electrical potential of the cell membrane.
2. Transmission of impulse in the nerve and muscle.
3. Various glandular secretions.
4. Prevention of swelling of cells.

- **Active transport of other ions**

1. Ca^+ and Mg^+ are probably transported by the same manner that sodium and potassium are transported.
2. Glandular cell membrane of thyroid gland can transport large amount of Iodide ions.
3. Epithelial cells of the intestine can transport Na^+, Cl^-, Ca^{++}, Fe^+, H^+.
4. Epithelial cells of renal tubules can transport Ca^{++}, Mg^+, Cl^-, Na^+, K^+, etc.

- **Transport of sugars**

1. Facilitated diffusion of glucose and certain other sugars occur in all cells.
2. But "secondary active transport" (sodium co-transport) occurs **only in the renal tubules and intestinal epithelium.**
3. Glucose and other monosaccharides are continually transported without any loss in the faeces or urine.
4. Though not all sugars but all monosaccharides are transported, e.g., glucose, galactose, fructose, mannose, xylose, arabinose, sorbose, etc. Disaccharides, e.g., sucrose, locatose and maltose are not transported at all.
5. The chemical reaction involved is not known but intact –OH group at C_2 carbon of monosaccharides is necessary in one group of sugars, e.g., glucose, galactose to attach to the carrier.
6. Fructose is transported by separate carrier system.

- **Transport of amino acids**

1. Amino acids like glucose are transported by facilitated diffusion.
2. Insulin enhances this transport of amino acids like glucose.
3. Secondary active transport (Sodium co-transport) also occurs through the membrane of intestine, renal tubules and exocrine glands.
4. There are four different carrier systems for four groups of amino acids:
   (a) Neutral amino acids.
   (b) Dibasic amino acids (basic).
   (c) Imino acids.
   (d) Dicarboxylic acids (acidic).

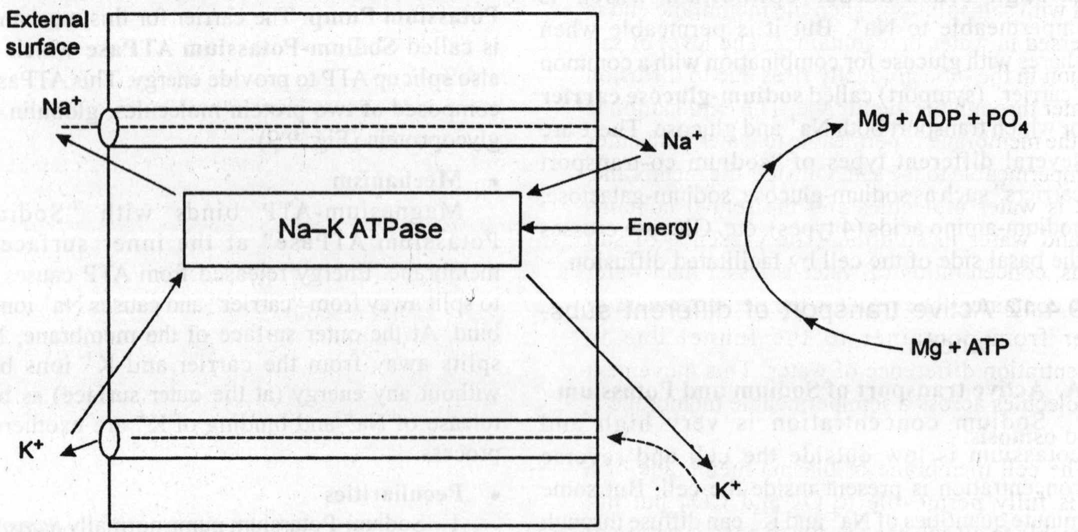

Fig. 9.9: Active transport of Na^+ and K^+ (Sodium-Potassium pump)

5. Some amino acids depend upon Vit B_6 (Pyridoxine).
6. Growth hormone, insulin, glucocorticoids, estradiol control amino acid transport.

9.5 Osmosis

(Net movement of water across cell membrane)

9.5.1 What is osmosis?

Among all the substances which diffuse through the cell membrane is the water in large quantity. But inward and outward movements are perfectly balanced to avert the swelling or shrinkage of the cell and there is no net movement of water. But under certain conditions, there is net movement of water across cell membrane due to concentration difference of water causing alteration of cell volume depending on the direction of the net movement. This mechanism of net movement of water due to concentration difference is called **Osmosis** (Greek – push), i.e., **diffusion of water through the membrane**. Abbe Nollet (1748) placed pure water and solution of alcohol in water between the sides of the animal membranes which is permeable to the solvent but not to the solute. He observed that pure water passes through the membrane to the side of the solution of alcohol. The process is called **osmosis**.

Vierordt and Dutrochet placed animal membrane over the end of a glass tube which is filled with salt solution. Then the membrane end is immersed in water in a container. The level of salt solution in the tube gradually rises due to diffusion of water through the membrane. The explanation is that, the membrane is permeable to water molecules but impermeable to the salt. Outside the membrane, there is water molecules and the funnel contains salt and water in solution. The presence of salt means, concentration of water is lesser than water in the container. The result is the net movement of water from container to the funnel due to concentration difference of water. This movement of molecules across a semipermeable membrane is called **osmosis**.

The cell membrane or plasma membrane of a cell is fully permeable to O_2 and CO_2 but not permeable to all substances due to porous nature of the plasma membrane. Only small molecules diffuse through it but larger molecules either penetrate slowly or not at all. The plasma membrane is thus **semipermeable or differentially permeable.** If the membrane allows all substances to pass through, then the membrane is called **permeable**. If the membrane does not allow any substance to pass through it, then it is called **impermeable**.

9.5.2 Semipermeable membrane

When a membrane is inserted between a solution and a solvent, the membrane does not allow the solute molecules to pass through but allows free passage to the solvent molecules. Such a membrane which allows the solvent to pass through preferentially but does not allow the passage of solute molecules through it is called **semipermeable membrane.**

Membranes may be of different degrees of permeability and have been classified according to the sizes of their structural pores.

I. **Permeable**: Cellulose wall of plant cell.

II. **Impermeable**: (a) Surrounding of certain eggs. (b) Protecting covering of seeds.

III. **Semipermeable**

A. Biological or natural

1. Cell membrane
2. Fish bladder
3. Dead pig's bladder
4. Egg membrane or lining of egg shells
5. Ectoplasm of plant cell (but not cell wall).

B. Artificial

1. Parchment
2. Cellophane paper
3. Collodion
4. Many inorganic precipitate (jelly), e.g., copper ferrocyanide, calcium phosphate, copper silicate, etc. No completely semipermeable membrane is known, which allows the passage of water only but is impermeable to all dissolved substances because there is slight leakage of dissolved substances.

Trauble (1867) observed that the gelatinous precipitate of copper ferrocyanide was very effective as a semipermeable membrane. Pfetter, a botanist, gave the idea of making the ferrocyanide membrane strong and capable of withstanding pressure by depositing them in the pores of clay cups.

More working with this, improved the method making of these membranes by placing potassium ferrocyanide solution inside the clay cup and a solution of copper sulphate outside the cup. An electric current, when passed through the solutions, so that Cu^{++} ions and $Fe(CN)_6^{4-}$ ions migrated towards each other in the walls of the cup and due to contact, membrane of copper ferrocyanide precipitated.

The best known artificial semipermeable membrane is that composed of copper ferrocyanide, supported in the meshes of clay cell. This membrane is impermeable to copper salts, ferrocyanides, sugar and to certain dissolved substances. It allows KCl to pass freely. Its semipermeability may be increased by depositing AgCl, when it becomes impermeable to potassium chloride. A semipermeable membrane may not have to be a solid, it may be a layer of liquid or even a gas can serve as a semipermeable membrane. The better consideration for perfect semipermeable membrane is air or any gaseous phase. In fact, for osmotic pressure measurement, a short air space in between a solution and a solvent has been used successfully as a perfect semipermeable membrane.

- **Living semipermeable membrane vs. dead semipermeable membrane**

The living membranes differ from dead membranes, in that they are able to select certain soluble substances for diffusion through them and to prevent others. Thus, monosaccharides like glucose, fructose and galactose, which are in true solution, can pass through the intestinal membrane while the disaccharides like sucrose and lactose are impermeable though they are also in true solution.

- **Living membrane vs. artificial semipermeable membrane**

Variation of permeability

Living membrane shows variation of permeability according to the state of the cell whereas variation is not seen in case of artificial semipermeable membranes.

(a) **Structure and previous history of the membrane**: Raid found that fresh frog's skin allows water to pass from without inwards. Wertheimer found that in frog's skin, sodium salts and acid dyes only pass from without inward while glucose and basic dyes only pass outwards.

(b) **Activity of the cell**: Activity of a cell is always related with permeability of its surface membrane. The surfaces of all cells show in greater or lesser degree, the phenomenon of semipermeability. Thus RBC are impermeable to most cations, e.g., Na^+, K^+, Ca^+ but allow the passage of urea, CO_2, chlorine, OH^- ions, NH_4 and glucose very slowly.

(c) **Composition of cell and surrounding**: RBC of man, rabbit and rat contains more K^+ and less Na^+, while the blood plasma contains more Na^+ and less K^+, in dog and cat, Na^+ is the main salt of the RBC as of the plasma.

(d) **Alteration in the external medium:** Many cells, e.g., muscle cells entirely after their permeability if immersed in a fluid without Ca^+ salts and eventually it becomes permeable to KCl which is not ordinarily allowed to pass.

(e) **Lipid solubility** of the substance in the membrane.

9.5.3 Mechanism of semipermeable membrane

There are several theories regarding the mechanism of semipermeability of the membrane:

1. Theory of sieve action.
2. Theory of preferential solubility of the solvent in the membrane material.
3. Theory of the formation of a loose chemical compound with the membrane.
4. Theory of the adsorption of the solvent on the walls of the membrane.
5. Thermodynamically it can be interpreted that the vapour pressure of solution is lower than that of the vapour pressure of the solvent, so the vapour distils through the semipermeable membrane from the high vapour pressure side, i.e., solvent side to the low vapour pressure side, i.e., solution side, thus causing the solvent molecules to diffuse across the membrane from the solvent side to the solution side, until the vapour pressure of the solvent becomes identical on the two sides of the membrane to establish the osmotic equilibrium. Application of pressure increases the vapour pressure of any liquid, osmotic pressure is equal to the applied pressure and it equalises the vapour pressure of the solvent and the solution. The pressure is considered in osmotic pressure determination.

- **Definition of osmosis**

Osmosis is defined as the spontaneous passage of solvent to the solution when the two are separated by a semipermeable membrane.

Table 9.3: Comparison of diffusion and osmosis

Diffusion	Osmosis
1. **Solute** diffuses from higher concentration to lower concentration.	1. **Solvent** passes from lower concentration to higher concentation.
2. Flow of particles in all directions.	2. Flow of particles in one direction only.
3. Permeable membrane is required.	3. Semipermeable membrane is required.
4. It is the basis of osmosis and imbibition.	4. Based on diffusion phenomenon through semipermeable membrane.
5. Equal diffusion.	5. Unequal diffusion.
6. Hydrostatic pressure not developed.	6. Development of hydrostatic pressure with solution to equalise the hydrostatic pressure of the solvent, to prevent the inflow of solvent and outflow of solvent from the solution.
7. Similarity: Equalisation of concentration throughout all accessible regions.	7. Equalisation of concentration.

9.5.4 Abbe Nollet's experiments

The instrument used for the study of osmosis is called osmometer. The experiment was done by Abbe Nollets (1784). The mouth of a thistle funnel is covered with a semipermeable membrane such as dead pig's bladder. The funnel is then filled with a strong, i.e., concentrated sugar (sucrose) solution and immersed in a beaker of pure water. After some times, the level of the solution of the funnel shows that it gradually becomes more dilute indicating that the water is passing into the funnel from the beaker (Fig. 9.10).

- **Theoretical explanation**

The explanation is that the membrane is permeable to water molecules of both solvent and solution (i.e., in both direction) but not to the solute (sugar). In the beaker there are nothing but water molecules only. In the funnel there are water and sugar molecules. The presence of sugar molecules decreases the concentration of water molecules in the funnel, whereas the concentration of water molecules is higher in the beaker. Since the concentration of water in contact with the membrane is greater outside than inside the funnel, more water passes into the solution than from the solution to the outside, i.e., the result is a net movement of water molecules from beaker. This movement of water molecules across a semipermeable membrane is called **osmosis**. As a result the volume of solution in the funnel increases and in the course of a few days, the level of the solution will attain a definite maximum value.

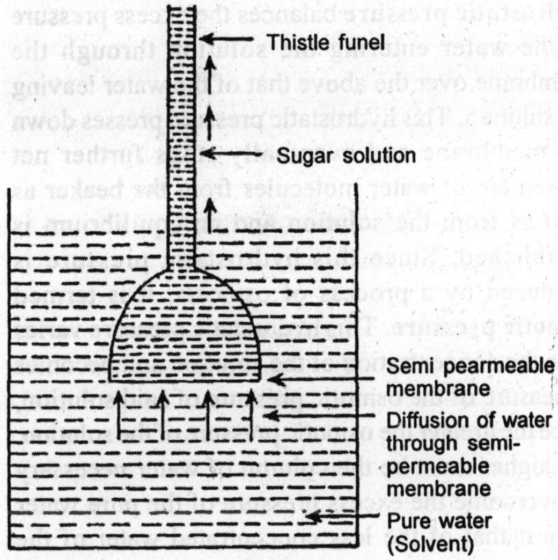

Fig. 9.10: Simple osmometer (Abbe Nollet's Experiments)

When a solution is separated from the water (solvent) by a membrane, a potential difference develops between the two phases and equilibrium can be established by:

(a) Free distribution of solute on both sides of the membrane with diffusible solute and permeable membrane.

(b) Passage of water through the membrane to the solution for dilution to remove the potential difference.

(c) Exerting pressure to the solution to increase the potential to the pure water (solvent) to prevent net movement of water called **osmotic pressure.**

This unequal diffusion through the membrane will continue accompanied by a continued rise of the volume of solution in the funnel until the pressure of the solution column in the funnel called

hydrostatic pressure balances the excess pressure of the water entering the solution through the membrane over the above that of the water leaving the solution. This hydrostatic pressure presses down the membrane and eventually stops further net movement of water molecules from the beaker as well as from the solution and an equilibrium is established. Since this hydrostatic pressure is produced by a process of osmosis, it is termed **osmotic pressure**. This hydrostatic pressure varies with the concentration of the solution and becomes a measure of the osmotic pressure of said solution, since the greater the osmotic pressure of the solution, the higher must be the column of water necessary to overcome the excess pressure of the pure water against that of the less concentrated water of the solution. The condition can be reversed if the solvent, i.e., pure water, is kept inside the funnel and the solution in the beaker, the inequality of water pressure against the membrane would be reversed, hence more water would pass from funnel through the membrane to the beaker.

9.6 Osmotic Pressure

9.6.1 Definition

If a solution is separated from the solvent by means of a semipermeable membrane, the excess hydrostatic pressure which must be applied to the solution to check inflow of solvent and establish an equilibrium is called the **osmotic pressure** of the solution.

9.6.2 Reverse osmosis

If the applied pressure is higher than the osmotic pressure of the solution, solvent will flow from the solution through the membrane to the solvent side. This is called **reverse osmosis**. This process has been applied for obtaining pure water from sea water.

9.6.3 Mechanism of osmotic pressure

In Abble Nollet's experiment it is seen that water molecules can pass through the membrane in both directions, but sugar molecules cannot pass through it. The rate at which water molecules are allowed to pass through the membrane is determined by the concentration, or activity of the water molecules which are in the contact with the membrane. It is a fact that there are more water molecules per unit volume in the pure water (solvent) than the water molecules present in the sugar solution. As a result, the diffusion pressure of water molecules from the pure water through the membrane is higher than the diffusion pressure of water molecules from the sugar solution rises above the level of the pure water. The level of solution continues to rise until the hydrostatic pressure equalizes the diffusion pressure of the water molecules in the sugar solution and in the pure water (solvent), at which point water passes through the membrane in opposite directions at the same rate and the system is in equilibrium. In this process, the solute molecules decrease the diffusion pressure of the solvent in direct proportion to their concentration, or number and independently of the kind of molecules. The higher the osmotic pressure shown by a solution, the lower is the diffusion pressure of its solvent molecules. Thus the osmotic pressure is the representation of the difference between the diffusion pressure of pure water (solvent) molecules and the diffusion pressure of solvent molecules present in the sugar solution. The more concentrated the sugar solution, the greater is the difference in diffusion pressure and the greater is the osmotic pressure.

The vapour pressure or diffusion pressure of the solvent molecules in the more dilute solution is greater than that of the solvent molecules in the more concentrated solution, so the net result is that the solvent molecules distil from the more concentrated solution until the concentrations and vapour pressure of the two solutions are equal. The use of a semipermeable membrane in determining the osmotic pressure is the only means of estimating the diffusion or vapour of solvent molecules. When we insert a semipermeable membrane between the solutions, only water molecules can pass and there is the phenomenon of osmotic pressure. But when a membrane is permeable to sugar molecules and not the water molecules, we get an osmotic pressure due to diffusion of sugar molecules and in a direction oppsoite to the osmotic pressure of water molecules. If there is a membrane which separates a solution from the solvent, and is readily permeable to the solvent and slowly permeable to the solute, a transitory rise of osmotic pressure, followed by a fall, is seen. When the membrane is equally readily permeable to both solute and solvent, no osmotic effect is seen, though osmotic pressure is there (Fig. 9.11).

9.6.4 Salient features of osmotic pressure

1. **Number of particles**: Osmotic pressure being a fundamental colligative property of solutions like vapour pressure lowering, freezing point depression and boiling point elevation, depends upon the number rather than the type of particles in a solution.
2. **Reduction of the activity of the solvent**: Osmotic pressure is due to a reduction in the activity of the solvent molecules in a solution. The activity of a substance is its effective concentration as manifested by its behaviour in solution.
3. **Solution must contact more dilute solution**: The solution exhibits an effective osmotic pressure only when the solution is in contact with more dilute solution across a membrane permeable to solvent but not to the solute.
4. **Temperature and Volume**: Osmotic pressure (P) is related to temperature and volume in the same way as the pressure of a gas:

$$P = \frac{nRT}{V}.$$

where n = Number of particles
R = Gas constants
T = Absolute temperature
V = Volume.

5. **Nonionized and ionized solute**: When the solute is a nonionizing like glucose, the osmotic pressure is a function of the number of glucose molecules present in the solution. And when the solute is ionizing and forms an "ideal" solution each ion is an osmotically active particle, e.g., NaCl dissociates into Na^+ and Cl^- ions and each mole in solution will supply 2 osm, Na_2SO_4 dissociates into Na^+, Na^+ and SO_4^{2-} which provides 3 osm.
If we consider the body fluids, it is not an ideal solution, because the dissociation of strong electrolytes is complete but the number of particles are free to exert an osmotic effect but it is reduced due to interactions between the ions. Thus, it is the actual concentration rather than number of equivalents of an electrolyte in solution which determines the osmotic effect in case of body fluid.
6. **Rate of diffusion of water**: The rate at which water molecules pass through the membrane is determined by the concentration or activity

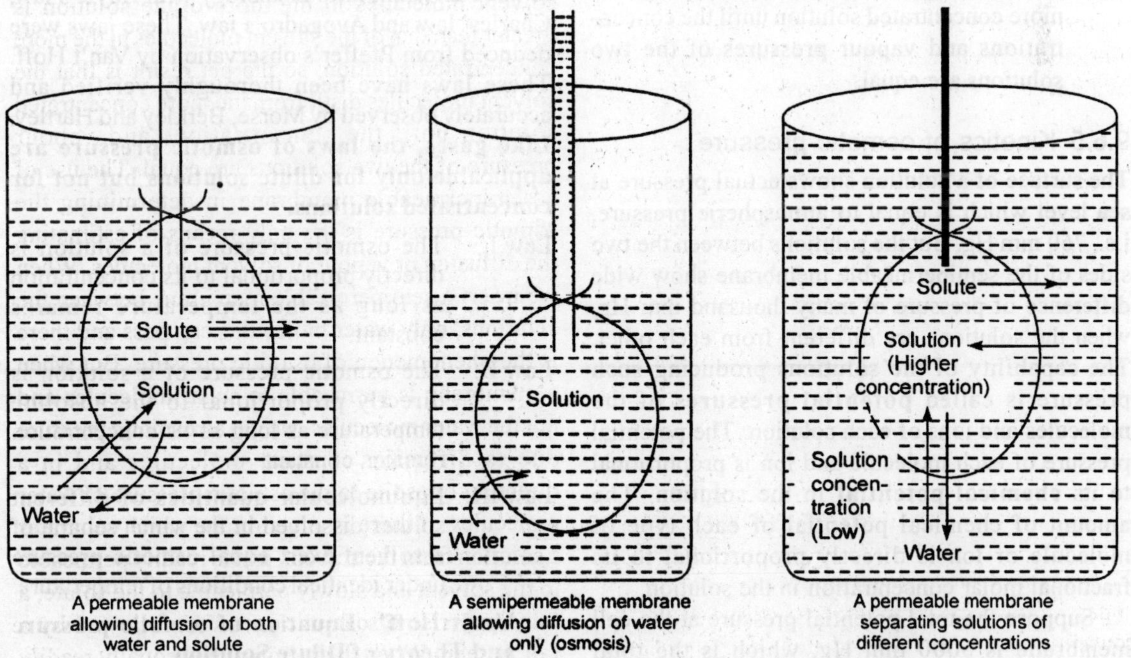

Fig. 9.11: Diagrammatic representation of diffusion and osmosis across a membrane

of the water molecules in contact with the membrane.

7. **Solute molecules and diffusion pressure**: Solute molecules decreases the diffusion pressure of the solvent in direct proportion to their concentration, number and independently of the kind of molecules.
8. **Diffusion pressure of solvent in solution**: The higher the osmotic pressure exhibited by a solution, the lower is the diffusion pressure of its solvent molecules.
9. **Difference of diffusion pressure of two solvents is the osmotic pressure**: The osmotic pressur is the manifestation of the difference between the diffusion pressure of pure solvent and the solvent present in the solution.
10. **Concentration of solution and osmotic pressure**: The more concentrated the solution, the greater is the difference in diffusion pressures and the greater is the osmotic pressure.
11. **Vapour pressure**: The vapour pressure or diffusion pressure of the solvent molecules in the more dilute solution is higher than that of the solvent molecules in the more concentrated solution, so the net result is that solvent molecules distil from the more dilute into the more concentrated solution until the concentrations and vapour pressures of the two solutions are equal.

9.6.5 Kinetics of osmotic pressure

The surface of a solution shows actual pressure at sea level which is equal to atmospheric pressure, i.e., 760 mm Hg. But the solutions between the two sides of the semipermeable membrane show wide difference of pressure of many thousand mm Hg, when the solutions are different from each other. The capability of the solutions producing such pressure is called **potential pressures** of the molecules and ions of each solution. The potential pressure of each molecule and ion is proportional. to its **chemical potential** in the solution. The amount of chemical potential of each type of molecule or ion is directly proportional to its fractional molar concentration in the solution.

Suppose the total potential pressure at the cell membrane is 5000 mm Hg, which is the total potential pressure of non diffusible particles in the extracellular fluid, e.g., sodium, potassium, calcium, magnesium, chloride, bicarbonate, protein ions and non ionic substance like glucose.

Suppose the pure water having total potential pressure of 1006800 mm Hg is separated by a semipermeable membrane from the solution of electrolyte of 1006300 mm Hg, i.e., 5000 mm Hg less than the pure water present on the lower, because of the presence of nondiffusible particles has diluted the concentration of water molecules than the concentration of pure water molecules. Due to this greater potential pressure of water molecules of pure water below the membrane than above, water diffuses from below upward than in the opposite direction causing **osmosis**. The hydrostatic pressure which is necessary on the top of the membrane to check inflow of water by osmosis is 5000 mm Hg which is equal to the potential pressure of the non-diffusible particles in the solution. This potential pressure of non-diffusible particles caused by the chemical potential of the particles is called the **osmotic pressure of the solution**. It leads to osmosis of water in the opposite direction to the osmotic pressure for which **the term "osmotic pressure" is not proper because it is not an actual pressure but a potential pressure only** (Fig. 9.12).

9.6.6 Van't Hoff's laws of osmotic pressure

Van't Hoff's laws of osmotic pressure are expressed in three different laws in analogy with Boyle's law, Charles' law and Avogadro's law. These laws were deduced from Pfeffer's observation by Van't Hoff. These laws have been thoroughly verified and accurately observed by Morse, Berkley and Hartley. **Like gases, the laws of osmotic pressure are applicable only for dilute solutions but not for concentrated solutions.**

Law I: The osmotic pressure of a solution is directly proportional to its concentration as long as the temperature remains constant.

Law II: The osmotic pressure of a solution is directly proportional to the absolute temperature as long as its concentration remains constant.

Law III: Equimolecular quantities of different solutes dissolved in the same volume of a solvent exert equal osmotic pressure under identical conditions of temperature.

- **Van't Hoff's Equation of osmotic pressure and Theory of Dilute Solution**

Van't Hoff showed that all the phenomenon

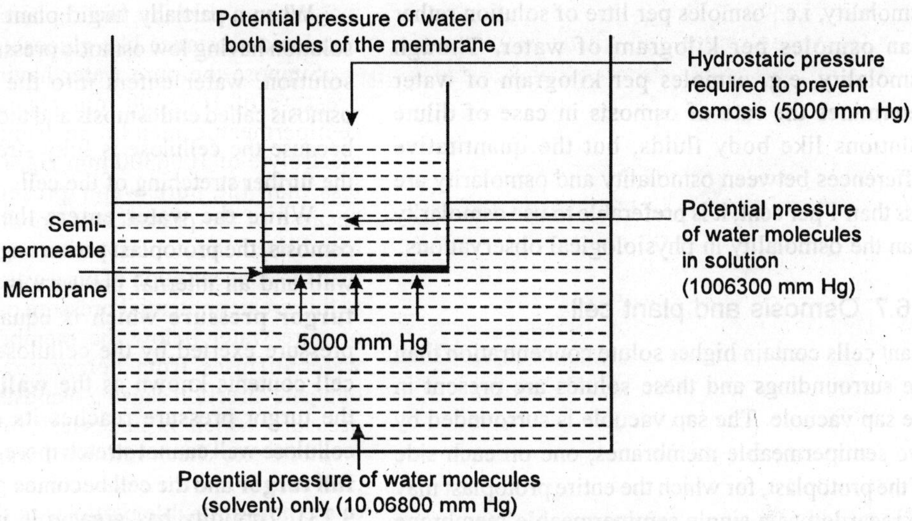

Fig. 9.12: Potential pressure of water on both sides of the membrane

regarding osmotic pressure of dilute solutions of non-electrolytes are encompassed by the equation mentioned below:

$\pi = c\,RT$ or $\pi V = n\,RT$,

where π = Osmotic pressure,
R = Universal gas constant,
T = Absolute temperature
V = Volume containing n moles of solutes
n = Moles of solute.
c = Molar concentration.

This equation is identical with ideal gas equation PV = RT. It can be concluded that the ideal gas equation is also applicable for substances in dilute solution, provided p stands for osmotic pressure π.

• Deviation from Vont Hoff's equation

The laws of osmotic pressure incorporated in the Vont Hoff equation $\pi = c\,RT$ are only applicable for dilute solutions of non-electrolytes, which are known as ideal solutions and identical to the definition of ideal gas. But deviation may occur due to any one of three reasons:

(a) During high concentration of solute,
(b) During association in solution,
(c) During dissociation in solution.

• Osmole

The amount of osmotic pressure exerted by a solute is proportional to the concentration of the solute in numbers of molecules or ions and the concentration is expressed in terms of number of particles, the unit is called **osmole**. One osmole is the number of particles (molecules) in gram molecular weight of undissociated solute.

Undissociated substance like glucose of 180 grams is equal to 1 osmole of glucose. When the solute dissociates into two ions, 1 gram molecular weight of the solute is equal to 2 osmole, e.g., 1 gram molecular weight of NaCl (58.5 gm) is equal to 2 osmoles.

• Osmolality

A solution that has 1 osmole of solute dissolved in each kilogram of water is called **osmolality** of 1 osmole per kilogram. A solution that has 1/1000 osmole dissolved per kg, has an osmolality of 1 milliosmol per kg.

The extracellular fluid (e.g., plasma) and intracellular fluid have normal osmolality about 300 milliosmoles per kg. A concentration of 1 osmole per litre causes 19300 mm Hg. osmotic pressure in the solution at normal body temperature of 38 ^{o}C. Similarly, 1 milliosmole per litre concentration is equivalent to 19.3 mm Hg osmotic pressure.

• Osmolarity

Osmolarity means when the solute concentration of solution is expressed as osmoles per litre of solution.

While considering the osmotic characteristics of body fluids, it is difficult for measuring kilogram of water in a solution, it is preferable to use the term osmolarity, i.e., osmoles per litre of solution rather than osmoles per kilogram of water. Though osmolality, i.e., osmoles per litre of solution rather than osmoles per kilogram of water. Though osmolality, e.g, osmoles per kilogram of water determines the rate of osmosis in case of dilute solutions like body fluids, but the quantitative differences between osmolality and osmolarity are less than 1 per cent, it is preferable to use osmolarity than the osmolality in physiological observations.

9.6.7 Osmosis and plant cell

Plant cells contain higher solute concentration than the surroundings and these solutes are present in the sap vacuole. The sap vacuole is surrounded by two semipermeable membranes, one on each side of the protoplast, for which the entire protoplast may be regarded as a single semipermeable membrane but the cellulose wall is fully permeable.

I. Turgidity in hypotonic solution

When a partially turgid plant cell is placed in a solution having low osmotic pressure, i.e., hypotonic solution, water enters into the sap vacuoles by osmosis called **endosmosis** and the cell does not burst because the cellulose is fully stretched, and resists the further stretching of the cell.

While the water enters the sap vacuole by osmosis, the protoplast presses against the cellulose wall and an internal pressure is developed called **turgor pressure** which is equal to the opposite pressure exerted by the cellulose wall against the cell contents known as the **wall pressure**. When the turgor pressure reaches its optimum and the cellulose wall cannot stretch more, the point is called **full turgor** and the cell becomes "fully turgid" (Fig. 9.13). Turgidity has great role in plant kingdom, such as:

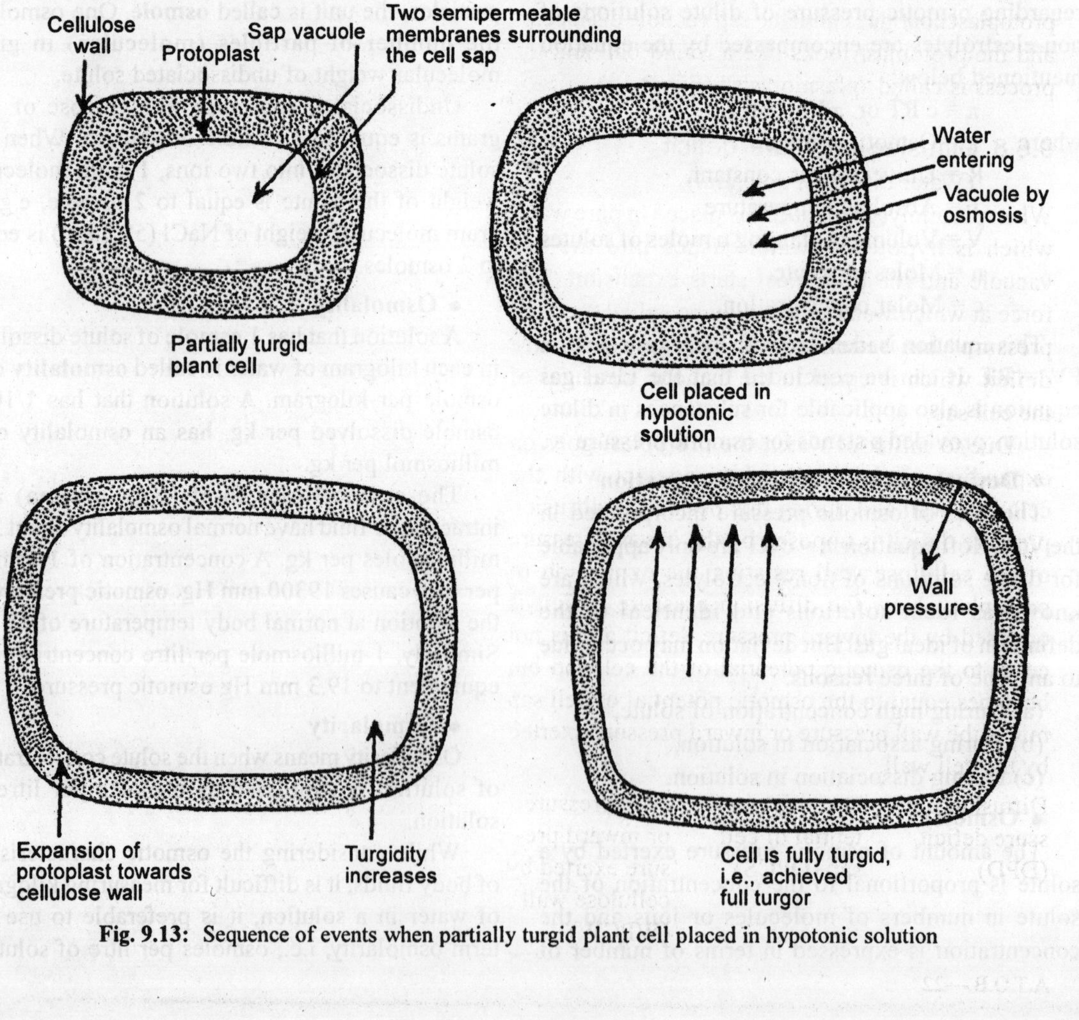

Fig. 9.13: Sequence of events when partially turgid plant cell placed in hypotonic solution

1. Supporting the plant,
2. Maintaining the shape and form of the plant,
3. Keeping erect position of stems of herbaceous plants,
4. Holding leaves in a flat and horizontal position.
5. Preventing drooping of plants or "wilting": When turgidity is reduced due to loss of water in hot dry days, the plant droops and the phenomenon is called "wilting".

II. Plasmolysis in hypertonic solution

When a plant cell is placed in a hypertonic solution, i.e., solution having high osmotic pressure (e.g., strong sugar solution), water comes out of the sap vacuole (exosmosis) and the cell shows gradual decrease in cell volume. Within a few minutes, the protoplast shows shrinkage in such a way that it pulls away from the cellulose wall, leaving a gap between the wall and the outer membrane of the protoplast continues until the sap vacuole disappears and the protoplast looks like a round ball and the process is called "plasmolysis" (Fig. 9.14).

9.6.8 Diffusion pressure deficit

(Water diffusion potential)

When plasmolysed plant cell is placed in pure water which is hypotonic, water rushes into the sap vacuole and the protoplast starts expansion. The force at which water enters may be called "suction pressure" but better term is **diffusion pressure deficit** which is equal to the osmotic potential of the cell sap.

Due to influx of water, the protoplast goes on expanding until it comes into contact with the cellulose wall and during this period the diffusion pressure deficit is opposed by the inward pressure of the cellulose wall resisting the expansion of protoplast. Now the diffusion pressure deficit is opposed by the inward pressure deficit and is not equal to the osmotic potential of the cell sap but becomes equal to the osmotic potential of cell sap minus the wall pressure or inward pressure exerted by the cell wall.

Diffusion pressure deficit (DPD)	= Osmotic potential of cell sap (OPCS)	— Wall pressure or inward presure exerted by cellulose wall (WPCW)

As the cell goes on expanding the following events occur:

1. Resistance offered by cellulose wall becomes higher.
2. Diffusion pressure deficit reduces,
3. Full turgor is achieved,
4. Cell expansion ceases,
5. Diffusion pressure deficit becomes zero,
6. Osmotic potential of the cell sap is counterbalanced by wall pressure of the cellulose wall.

The diffusion pressure deficit is also termed as "water diffusion potential" because of net tendency of the cell for influx of water from the surroundings.

1. In plasmolysed cell, wall pressure = 0
   Diffusion pressure deficit = Osmotic potential of cell sap.
2. In case of incipient plasmolysis (zero turgor), wall pressure = 0.
   Diffusion pressure deficit = osmotic potential
3. In case of full turgidity – wall pressure = osmotic potential.
   Diffusion pressure deficit = 0.

- **Measurement of diffusion pressure deficit**

Pieces of tissue having known weight or volume are placed in a series of solutions of different solute (sucrose) concentration. The solution in which there is no change in weight or volume, has a solute concentration and osmotic potential equal to the diffusion pressure deficit of the tissue.

9.6.9 Osmosis and human cell (cell as osmometers)

Human or animal cells show their properties due to semipermeable membrane but it does not show perfect semipermeability. It selectively allows diffusion of water and some solutes and prevents the diffusion of other solutes of extracellular or intracellular fluid. Thus, the extracellular fluid separated from the intracellular fluid by the cell membrane does not exhibit the actual osmotic pressure but a fraction of the total osmotic pressure, which is known as **tonicity** of the solution. The 'tonicity' of the solution is proportional to the sum of the molar concentration of the solutes only which are impermeable to the cell membrane. The term tonicity is used to describe the effective osmotic pressure of a solution related to plasma. Solutions having same effective osmotic pressure as plasma are known as **isotonic**, those with greater osmotic

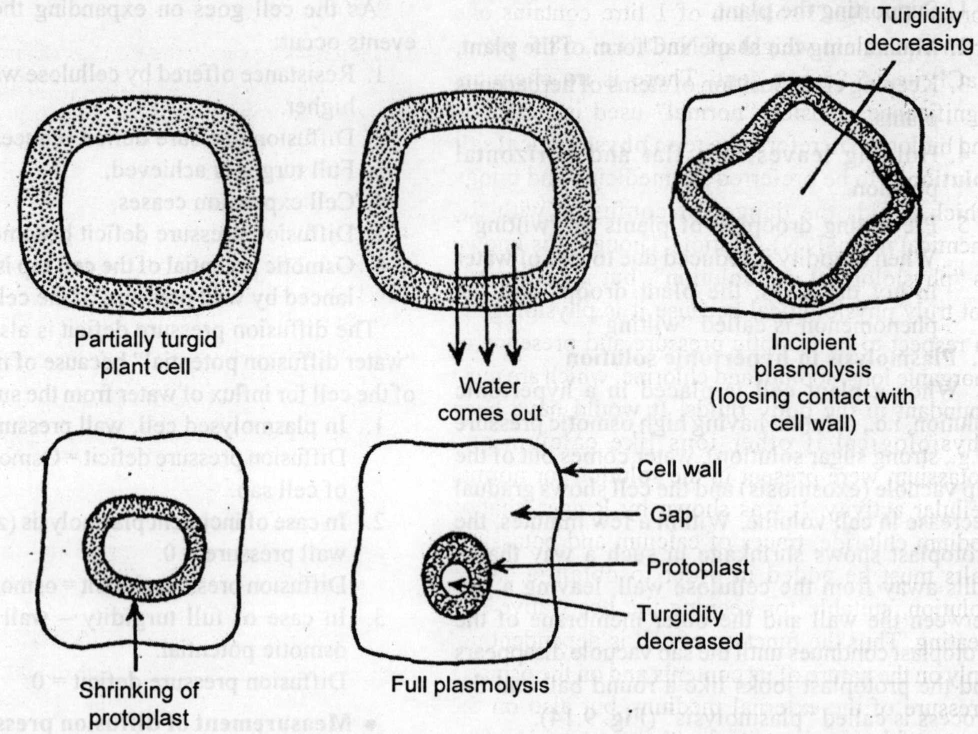

Fig. 9.14: Sequence of events when partially turgid plant cell is placed in hypertonic solution

pressure are **hypertonic** and those with lesser pressure are **hypotonic**.

When a solution is separated from pure solvent (water) by perfect semipermeable membrane, water enters into the solution through the membrane but not the solute, then the solution exhibits a total osmotic pressure which is proportional to the sum of the molar concentrations of all the solutes. Solutions having identical osmotic pressure are called **isosmotic** with identical vapour pressure, boiling point and freezing point. When the solution has higher osmotic pressure called **hyperosmotic** compared to other solutions.

A. Hypotonic solution

If human RBC is placed in pure water which is hypotonic or in a solution having lower solute concentration (e.g., 0.3 per cent NaCl solution) than that of the cell content, water molecules enter the cell by diffusion (**endosmosis**) and the cell gradually swells up and consequently may burst in the solution and the phenomenon is called **haemolysis**. In this case, the osmotic pressure (or osmotic potential) is lower than that of the cell and hence the solution, and accordingly the solution is said to be hypotonic. In case of RBC, 0.34 per cent NaCl solution is hypotonic and causes haemolysis. The osmotic fragility test of RBC is done with different hypotonic NaCl solutions having different concentration below 0.9 per cent.

B. Hypertonic solution

If RBC is placed in a solution having higher solute concentration (more than 0.9 per cent and higher osmotic pressure than that of the cell's content, water passing out of the cell (**exosmosis**) causes shrinkage of the cell known as **laking or crenation**. This process usually occurs in case of 1.2 per cent NaCl solution.

C. Isotonic solution

If the solute concentration of the cell and its surrounding medium, i.e., the osmotic pressure in the cell and outside are the same, there will be no osmosis in either direction and the external solution is termed **isotonic solution**. A solution of 0.9 per cent NaCl is isotonic to the human cell (0.65 per cent for amphibians) and is known as **physiological salt or normal solution** or **normal saline**, which is used in medical and biological (mammalian) purposes and differs from "chemical normal solution of NaCl" designated as N. The chemical normal (N) NaCl solution of 1 litre contains one gram equivalent weight of NaCl, i.e., 58.5 gm of NaCl, i.e., 5.85 per cent. There is no chemical significance of using "normal" used in medicine and biology. Therefore, the term **physiological salt solution** is to be preferred for medicine and bilogy which avoids the danger of confusion with the chemical normal (N) solution. Though it is known as "physiological salt solution", but actually, it is not truly physiological because it is physiological in respect to its osmotic pressure and presence of inorganic ions, sodium and chlorine, which are most abundant in the body fluids. It would have been physiological if other ions like calcium and potassium were present in the solution for normal cellular activity. It was shown by Ringer, that to sodium chloride, traces of calcium and potassium salts must be added in order to obtain a saline solution, suitable for keeping the heart alive and beating. Thus the function of cell is dependent not only on the nature of its contents and on the osmotic pressure of the external medium, but also on the nature of the ions in content with the external surface of the cell membrane. It is not significant what salts of calcium or potassium are used because it is metallic ions that are important and responsible for the activity. For keeping a tissue alive "Ringer's fluid" is used in which sodium, potassium and calcium ions are added in proper amount to make it isotonic and to perform proper activity. Sodium bicarbonate and sometimes sodium phosphate are added to make the reaction of the body fluids and to neutralize the acid if it is produced during the activity of the tissue. The following table is given to show the compositon of "physiological saline solutions" which are isotonic (Table 9.4).

9.6.10 Isotonic vs. isosmotic

Isotonic is not exactly the same as isosmotic because most naturally occurring membranes are not perfectly semipermeable. Two solutions having same tonicity is called isotonic, but they are not necessarily isosmotic.

Tonicity is referred to the action of the solution with regard to some cell membrane, and, therefore implies the impermeability of the membrane to the substance or substances in solution. In fact, the membranes are often selectivity permeable to certain solutes, whereas they are impermeable to other solutes present in the cell or outside the cell. For example, human RBC membrane is permeable to most types of ions present in blood but is impermeable to sodium and potassium. In such cases, two types of mechanism may occur.

1. When two solutions are isotonic as determined by plasmolysis, their osmotic pressure with respect to the impermeable ions are the same but if these solutions contain solutes permeable to the cell membrane in different proportions, the osmotic pressure of these two isotonic solutions may be different.
2. Conversely, it is possible to have unequal concentration of a certain solute between the inside of a cell and its surrounding medium even when the total osmotic pressure inside and outside is the same. So "tone" is not quite the same as osmotic pressure.
3. Equimolecular solutions of urea and glucose are isosmotic but when tested by the RBC method, the urea solution would be seen to be isotonic with distilled water, because the cell membrane is freely permeable to urea but not to sugar. Therefore, urea is unable to exert any osmotic pressure in relation to RBC.
4. Solutions with identical osmotic pressure, i.e. isosmotic pressure, will show the same

Table 9.4: Physiological saline solutions which are isotonic

	Water (ml)	NaCl (gm)	KCl (gm)	$CaCl_2$ (gm)	$NaHCO_3$ (gm)	NaH_2PO_4 (mg)	MgCl (gm)	Glucose (gm)	Na_2HPO_4 (gm)	$\frac{M}{1}$ H_3PO_4 (cc)
Ringer (Apmphibian)	100	0.65	0.014	0.012	0.02	0.001	—	(0.2)	—	—
Ringer-Locke (Mamamlian)	100	0.9	0.042	0.024	0.01-0.03	—	—	(0.1-0.25)	—	—
Ringer-Tyrode	100	0.8	0.02	0.02	0.1	0.005	0.01	0.1	—	—
Ringer-Dale	100	0.9	0.042	0.024	0.05	—	0.0005	0.05	—	—
Buffered Ringer	100	0.85	0.042	0.024	—	—	—	0.1	0.06	0.02-0.06

boiling point and depression of freezing point. **Isotonic** solutions will not show the same properties like isosomotic solutions.

5. Equimolecular solutions of non-electrolytes have the same osmotic pressure, i.e., isosmotic. Since the osmotic pressure of such a solution is dependent on the number of molecules of solute in unit volume, the properties of solutions which are functions of the number of molecules present are vapour pressure, boiling point and freezing point. Isotonic solutions may not show the above properties of isosmotic solutions due to unequal distribution of molecules between the two sides of the membrane.
6. All solution that are isosmotic with plasma would also be isotonic if the solutes do not diffuse into the cells and others are not metabolized. Thus 0.9 per cent NaCl solution is isotonic because there is no net movement of the osmotically active particles in the solution into the cells and the particles are not metabolized. But if the cells are suspended in aqueous solutions which initially contains 290 osm of urea per litre, urea diffuses rapidly into the cells, so that the effective osmotic pressure drops. It a 5 per cent glucose which is isotonic initially and infused intravenously, glucose is metabolized, so the net effect is the infusion of a hypotonic solution.

9.6.11 Methods of determination

A. Mechanical methods

1. **By putting weights**: The simplest way is to apply adequate pressure (i.e., weight) upon the stronger solution to prevent any rise of volume. That pressure which is just needed to stop the increase of volume of a particular solution is the measure of its O.P. (Fig. 9.15).

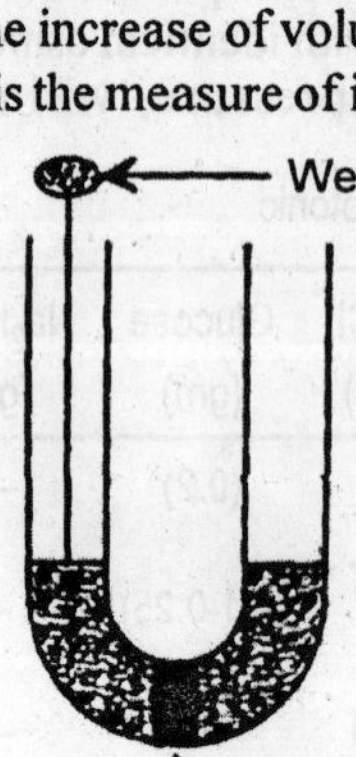

Fig. 9.15: Determination of osmotic pressure by weight method

2. **By manometer (Pfeffer's method)**: The same thing can be done by connecting the apparatus with a suitable manometer, in which the pressure will gradually rise till it equalizes with the osmotic pressure of the solution, at which point further rise will stop (Fig. 9.16).

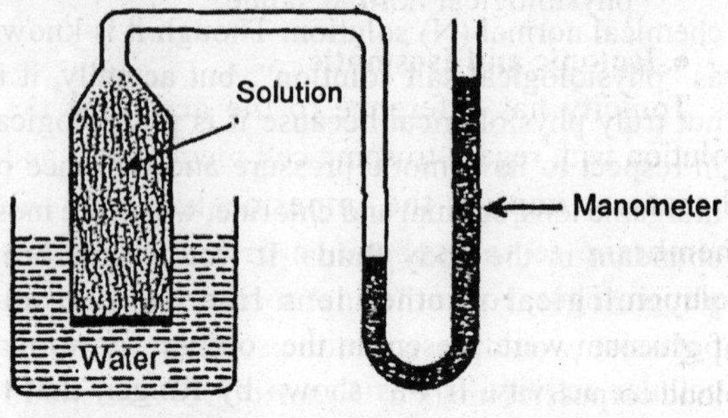

Fig. 9.16: Pfeffer's method

3. **Morse's method**: Morse and others modified the Pfeffer's method in two respects: (a) devising a method of making porous cells, (b) deposition of membrane of copper ferrocyanide.

B. Biological methods

1. **De Vries plasmolysis method**: In this method plant cells are used. These cells are bounded by more or less firm cellulosic wall lined by a membrane permeable to water but impermeable to substances dissolved in the cell sap, e.g., glucose, K-salts, etc. If such a cell is immersed in solution of lower concentration and lower osmotic pressure than the solution inside the cells, no appreciable change in the size of the cell takes place because of rigidity of cell wall; water is not allowed to come in. If the cell is immersed in solution having high osmotic pressure, the water will be diffused out from inside of the cell to the exterior. This causes collapse of the cell membrane and the process is called "plasmolysis" (Fig. 9.17).
2. **Humberger blood corpuscle method**: In this method red blood corpuscles are used instead of plant cell. When RBC are placed in high osmotic pressure solution, water comes out of the cell by exosmosis, hence cells are shrunken. Again, when the cell is placed in low osmotic pressure solution, the

cell swells up by endosmosis and ultimately it bursts called **haemolysis.** The solution in which the cell neither swells up nor bursts is called **isotonic solution** 0.85 per cent NaCl is isotonic, hence it is used in transfusion is known as "normal saline", better to say "physiological normal saline".

- **Isotonic and Isosmotic**

Tonicity has reference to the action of the solution with regard to some cell membranes, and therefore, implies the impermeability of the membrane to the substance or substances in solution. Thus, equimolecular solution of urea and of glucose would be isosmotic; yet tested by the blood corpuscles method, the solution of urea would be found to be isotonic with distilled water, since the outer layer of a blood corpuscle is freely permeable to urea but not to sugar. Urea is, therefore, unable to exert any osmotic pressure in relation to blood corpuscles.

C. Indirect method or physical method

I. By depression of freezing point: (Generally represented by Beckman's apparatus).

Higher the concentration, lower will be the freezing point, and therefore, higher will be osmotic pressure.

II. By depression of vapour pressure

(a) **Hill's Method:** Using Thermopile. Higher the rate of evaporation, more will be the fall of temperature and less will be the osmotic pressure. Comparison is made with a solution of known osmotic pressure.

(b) **Barger's Capillary Method:** alternative drops of known and unknown solution separated by air bubbles are drawn in a standard capillary glass tube and after some time the edges of the solutions are to be noted. The edges will shift according to the rate of evaporation. From these data, osmotic pressure can be calculated.

D. Berkeley and Hartley's Method

Both the difficulties met with in the Pfeffer's method, were overcome by Berkeley and Hartley.

The apparatus consists of two concentric tubes, the inner one, MM being of fine uniform-textured porcelain, in the pores of which is electrolytically

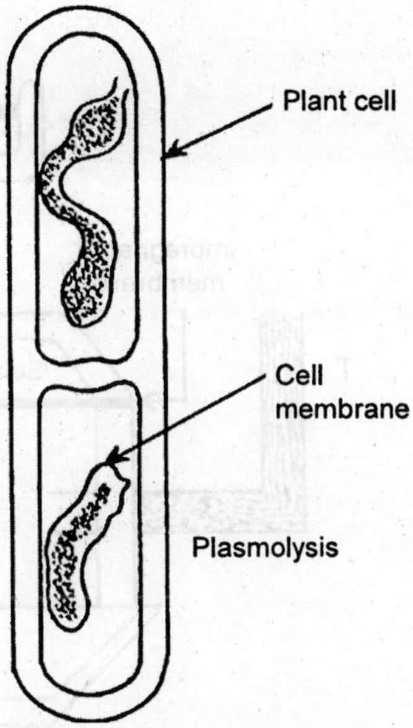

Fig. 9.17: Pfeffer's method

deposited the membrane of copper ferroycyanide. The ends of the tube are connected to a fine capillary at one end and a funnel provided with a stopcock at the other. The outer tube made of a gun-metal is solidly fixed to the porcelain tube and is provided with an arrangement for applying definite pressures by means of a piston. In the annular space between the two tubes is introduced the solution, while the inner porcelain tube contains water up to definite level in the capillary tube T. Water from inside diffuses into the solution and the level of water falls in the capillary tube T. By applying a suitable pressure on the solution, the water in the capillary tube is forced to the original level, and this pressure is the osmotic pressure of the solution. The advantages of the method are that, it is very simple and rapid and one can record very high pressures, and that the concentration of the solution is not appreciably changed by dilution with the solvent (Fig. 9.18).

9.6.12 Importance of osmosis in medicine and biology

I. Medicine

1. **Transfusion**: Isotonic solutions of crystalloid, e.g., NaCl (0.9 per cent) and glucose (5 per cent) are used for **transfusion** in medicine and surgery for dehydration, burn, shock, etc.

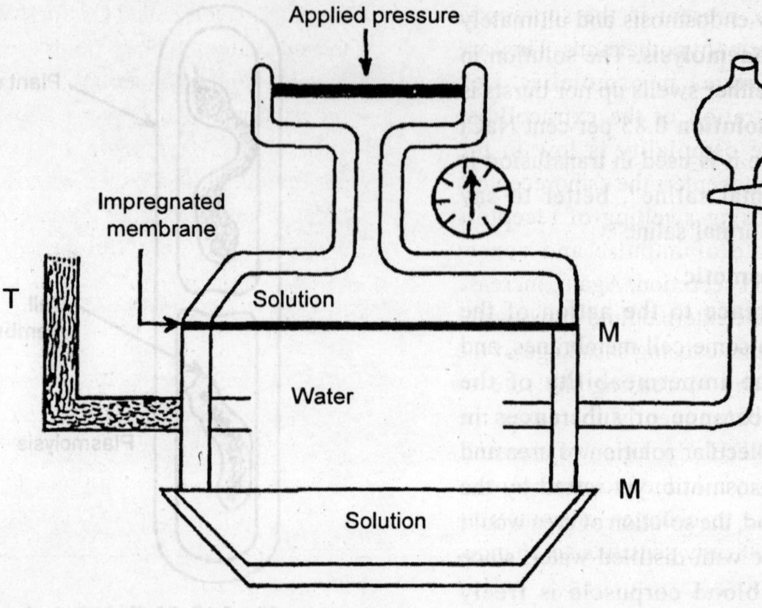

Fig. 9.18: Berkeley and Hartley's method

2. **Purgative:** Hypertonic saline **purgative** or other purgative draws more water in the large intestine causing purgation.
3. **Irrigation of Wounds**: Isotonic solutions are used for **irrigation of wounds**, etc.
4. **Osmotic Diuresis**: It is produced by the administration of compounds like mannitol and related polysaccharides that are filtered but not reabsorbed. It can be produced by excess amount of glucose administration which are present in amounts exceeding the capacity of renal tubules for reabsorption (e.g., in diabetes mellitus causing polyuria). Osmotic diuresis can be produced by the infusion of large amounts of sodium chloride or urea.
5. **Red Blood Cell Osmotic Fragility Test:** The resistance which RBC offer to the haemolytic action of hypotonic solution is used as an index of the fragility of the RBC. In solution of low osmotic pressure, RBC swells up, becoming spherical rather than disc-shaped and eventually lose their haemoglobin called **Heamolysis.** When osmotic fragility is normal, RBC begins to haemolyse in 0.48 per cent NaCl solution and haemolysis is complete in 0.33 per cent NaCl. In **hereditary spherocytosis** (congential haemolytic icterus), the RBCs are spherocytic and haemolyse more readily than normal cells because of abnormalities of the protein lattice which maintains cell shape and flexibility.
6. **Moistening or Immersion of Living Tissue**: When living tissues are removed from the body, (e.g., biopsy) of man or experimental animal for pathological, toxicological or physiological studies, it is moistened or immersed in a fluid having the proper salt osmotic pressure.
7. **Observation of Living Cells**: If the cells in living state (cells of the blood, isolated cell of tissue or organ) are to be observed under microscope, it is moistened with isotonic salt solution (0.9 per cent in mammals, 0.65 per cent in amphibians) calls "Physiological Salt Solution".
8. **Injection of Solid substance:** Solid substance, e.g., drug, blood substitutes, etc. are dissolved in physiological salt solution for administration in the body.

II. Man and Animals

1. **Osmoreceptors or Osmo-Sodium Receptor**: There are specialised neuronal cells called

Bosmoreceptors, present in the supraoptic nuclei of the anterior hypothalamus. They are sensitive to changes in osmolality, i.e., sodium concentration of the extracellular fluid. When the osmolality is low in the external fluid, water enters the osmoreceptors by osmosis, causing swelling of receptors reducing the rate of impulse and consequently less ADH secretion. Again, increase in osmolality in the external fluid draws water out of the receptors causing shrinkage of the receptors, thereby increasing the rate of discharge and consequently increases the secretion of ADH. In this way water, regulation takes via kidney. osmoreceptors are not sensitive to changes in potassium concentration but only sensitive to changes in sodium concentration, hence called **Osmosodium Receptors.**

2. **Water Balance Between Blood and Tissue Fluid**: Water balance existing in the body between the blood and the tissue fluid is maintained during equilibrium when osmosis and filtration become equalized. The filtration of water out of the blood is, on the average, approximately equal to the osmotic pull of water into the blood vessels. But this is not balanced dynamic equilibrium because the filtered fluid through capillaries re-enters the blood by the lymph vessles.
3. **Absorption of Food and Water**: Osmosis is one of the factors by which water and soluble substances of digestion are absorbed from the intestine into the capillaries of the villi and pass from blood through the successive membranes into the cells.
4. **Osmotic Pressure of Digestive Juices and CSF**: The osmotic pressures of blood plasma, gastric juice, pancreatic juice, intestinal juice, liver bile and cerebrospinal fluid are approximately equal. These fluids are roughly 0.3M in total concentration of dissolved particles including nonelectrolytes and ions.
5. **Osmotic Work and Fluid Secretion**: The work done to maintain or increase concentration differences across a membrane by water transfer is called **Osmotic Work**. When the substances in body fluids are raised from lower to higher concentration, Osmotic Work is performed but the reverse also occurs when substances pass from higher to lower concentration. The Osmotic Work done by the kidneys for concentrating a urea solution from 5 mMl^{-1} in blood filtrate to 330 mMl^{-1} in urine is about 10.5 kJ mol^{-1} irrespective of direct transfer of water or a transfer secondary to the movement of sodium ions. When kidneys do the Work concentrating Cl in urine passing from plasma, the work sign is positive. When the concentration of water is higher in plasma than in urine, the Work passing from plasma to urine and the sign of the Work is negative.

   Marine teleosts (bony fishes), e.g., Cod, Whiting, have hypotonic body fluids but in sea water they perform Osmotic Work by extruding salts through chloride secretory cells in the gills moving against a concentration gradient from the blood to the sea water. Marine birds and marine reptiles also get rid of excess salt by performing Osmotic Work. Marine birds, e.g., cormorant, penguine, gulls, pelicans perform Osmotic Work by eleminating salt through special nasal glands; marine reptiles, e.g., turtles perform Osmotic Work by eliminating salt from glands close to the eye, hence the turtle "shedding tears".

   The relation between Osmotic Work and concentration change for a given substance is represented by the equation:

$$W_{min} = NRT \ln \frac{C_2}{C_1} - 2.3\, NRT \log \frac{C_2}{C_1},$$

where W_{min} = Minimum osmotic work in small calories, for transfer of N mol of substance from a molal concentration of C_1 to C_2.

N = Mols of substance,

R = Gas constant = 1.987 calories per mol per degree,

T = Absolute temperature,

C_1 = Molal concentration,

C_2 = Molal concentration,
In = Natural logarithim,
2.3 = Factor for the conversion of natural logarithms to logartithms of base 10.

6. **Physiological Saline Solution for Experimental Observation**: Different physiological saline solution, e.g., amphibian or mammalian Ringer are used for moistening, perfusion of heart, organs, etc.
7. **Colloidal Osmotic Pressure of the Plasma**: The total proteins content in the plasma determines, the collodial osmotic pressure or oncotic pressure. Out of the total osmotic pressure of 7.6 atmospheres, only 25-30 mm Hg, i.e., about 1.30 of an atmosphere, is due to the presence of colloids in the plasma. This negligible osmotic pressure of plasma proteins are compared to ionized mineral salts, because of high molecular weight and nonionized properties of colloidal plasma proteins. This amount of total osmotic pressure has great importance because, unlike crystalloids, they do not diffuse through cell membranes and counteracts the filtering effect of blood pressure and thereby helps to retain water in plasma.
8. **Normal Osmotic Distensibility of RBC**: Osmotic pressure inside the RBC is slightly higher than plasma by 1.5 atm, due to a higher intracellular concentration of electrolyte ions than in the plasma, yet the cells remain well-distended with the fluid from the blood maintaining normal volumes.
9. **Reabsorption of Water by Osmosis through the Tubular Epithelium**: During transportation of the different solutes from the lumen of renal tubules, through the epithelium, their total concentration decreases in the lumen and increases in the epithelium. This produces a concentration difference causing osmosis of water in the same direction that the solutes are transported. In the highly permeable parts of the renal tubules, e.g., proximal tubule, where osmosis of water is very rapid, the osmolar concentration of solutes on the peritubular side of the membrane is more than a few milliosmoles higher than on the intratubular side. Since the membrane of the proximal tubule is highly permeable to water, osmosis occurs instantaneously into the peritubular fluid to re-equilibrate the **osmolar** concentrations on the two sides of the membrane.
10. **Cytolysis, Haemolysis, and Crenation**

(a) **Cytolysis**: When the cells are placed in a solution having lower salt concentration (hypotonic) than that within the cell, water passes from the solution into the cell by osmosis (**endosmosis**). Under microscope, it is seen that the cells swell up and increase in size. Due to greater concentration difference between inside and outside the cell, the water influx causes disruption of the cell membrane and appears to burst and destroyed. The process of such cellular disintegration is called **Cytolysis**.

(b) **Haemolysis**: Separation of stroma and haemoglobin of RBC by any procedure is called **Haemolysis** and it is the other kind of general cellular phenomenon of cytolysis. Haemolysis is regarded as cytolysis of RBC. After placing RBC in 0.6 per cent NaCl, it is observed under microscope—it shows a bigger size and swelling of the cell because of water influx into the cell (**endosmosis**). In further diluted salt solution, the bigger swollen cells disappear, observed under microscope. At this stage, the cells lose their haemoglobin and become invisible. But there is no actual disintegration because on staining with methylene blue, the cell is visible, due to presence of stroma without haemoglobin. The invisible, unstained stroma is called "Ghost". The shape of the ghost looks like the intact cell. The mechanism is that during osmotic haemolysis, the loosely bound stroma-haemoglobin complex is disrupted. The character of the cell membrane is changed by distension of the cell due to influx of water allowing the escape of haemoglobin from the cell. In 0.4 per cent salt solution, all the cells are haemolyzed, and the haemoglobin as a colloidal particle disperse throughout the solution. If the solution is centrifused, the invisible ghosts settle at the bottom leaving transparent red coloured solution.

(c) **Crenation**: If the cells are placed in a solution having higher salt concentration (hypertonic) than that within the cells, water comes out **(exosmosis)** of the cells, producing shrinkage of the cells called **Crenations**. Again, if it is placed in isotonic solution, they fill out and resume their normal size and appearance.

Similarly, RBC is seen to decrease in volume and to shrink and the surface appears wrinkled called "**Crenation**" which remains in creanated condition as long as the cells are kept in solution, without loosing haemoglobin. If the cells are placed again in the isotonic solution, they resume their normal shape and appearance due to "endosmosis" of water.

11. **Osmoregulation in Animals**: Osmoregulation is the process by which the osmotic pressure of the blood and tissue fluids is kept constant. Osmoregulating animals are grouped into two classes:

(a) Not capable to regulate their osmotic pressure (osmoconfirmers), e.g., all marine animals, e.g., sea anemones, spider crabs, etc.

(b) Capable to regulate to some degree (osmoregulators), e.g., amoeba, crayfish, freshwater teleosts.

- **Structures for Osmoregulation**

(a) **Contractile Vacuole of Amoeba and Freshwater Protozoons**: Contractile vacuole is a small sac, lined with unit membrane present in the cytoplasm. The cell membrane is semipermeable and internal osmotic pressure; is higher than external osmotic pressure; water enters the cell by osmosis. To prevent accumulation of water, water is secreted into the contractile vacuole rapidly, which expands and migrates towards the cell membrane, where it discharges water through the pores of the cell membrane to the exterior.

(b) **Antenal glands** of the shore crab and crayfish participate in the process.

(c) **Chloride secretory cells** of marine teleosts (bony fishes) and freshwater teleosts.

(d) **Nasal glands** of marine birds, e.g., cormorants, penguine, gulls.

(e) **Glands close to the eye**, marine reptiles, e.g., turtles.

III. Plant Life

1. **Uptake of Water by the Roots**: Root hairs absorb water from the soil. Water is drawn into root hair, mainly by osmosis. Due to presence of sugars and other metabolites, the concentration of solutes in the sap vacuole is greater than that of the surrounding soil water. Water molecules pass through the permeable cellulose wall and semipermeable protoplast into the vacuole.

2. **From Root Hair, Water to the Vascular Tissue (Xylem Vessel)**: From root hair, water passes to the vascular tissue in the centre of the root via the parenchymal cell (cortex). There are three possible mechanisms: (a) One is through the sap vacuoles, water is drawn from one vacuole to the next by osmosis (osmotic gradient). The other two: (b) One is through cytoplasm from cell to cell, and the (c) Other along the cell walls through the cellulose of adjacent cells and through the small intercellular spaces between them, water diffuses (diffusion gradient) (Fig. 9.19).

3. **Transport of Water from Parenchyma Cells into the Vascular Tissues of the Root**: Transport of water from parenchyma cells into the vascular tissues of the root is due to high osmotic pressure developed in the leaf cells. As a result of transpiration the water is drawn into and along the vascular tissues, i.e., pull exerted by the leaves. Beside that there is a force, pushing water up the stem from the roots is called **Root pressure**.

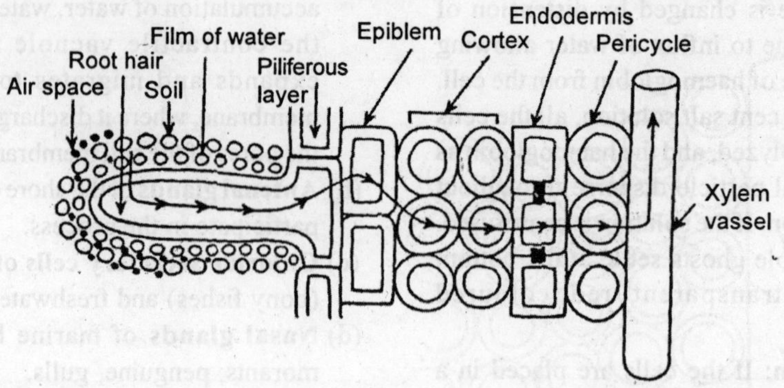

Fig. 9.19: Uptake of water by the root. One of the routes showing water and mineral salts may be drawn from the soil into the vascular tissues

4. **Transport of Water to the living Cells Surrounding Xylem Vessel**: The living cells surrounding xylem draw water from it by the process of Osmosis.
5. **Turgidity of the Cells**: Osmosis makes the cell turgid.
6. **Various processes of plant**: Various movements, turgor movements particularly, sleep movement, opening and closing of stomata, bursting of many fruits are due to osmotic pressure.
7. **Plasmolysis**: By the process of plasmolysis, it is possible to determine the osmotic pressure of a cell.
8. **Mechanism of Opening and closure of Stomata**: There are two explanations—
(a) The guard cells have chloroplasts, start photosynthesis during daytime, which leads to an accumulation of sugar in the guard cells, whose osmotic pressure increases. This, in turn, causes water to be drawn into them from the surrounding epidermal cells, resulting in the opening of the pore.
(b) The enzymatic conversion of starch to sugar takes place readily, when little acid is present whereas conversion of sugar to starch is favoured in higher acid media. In the night, CO_2 accumulates in the intercellular spaces of leaf and this increases the concentration of Carbonic acid. As a result, there is decrease in pH, which favours the conversion of sugar to starch in the guard cells, thereby decreasing their osmotic pressure and causing the closure of stomata. In the daytime, during photosynthesis, CO_2 concentration is lowerd, the level of Carbonic acid falls, the pH rises, starch converts into sugar, the osmotic pressure of the guard cell increases and the *stomata opens*.
9. **Movement of Water in the Leaf**: During evaporation from the walls of the spongy mesophyll cells into the sub-stomatal air chambers, water molecules diffuse out of the sap vacuoles to replace the amount lost. The mechanism is that, this causes more concentration of the solutes, so the osmotic pressure is increased in the outermost cell than the inner cell. Thus, water is drawn into the surface cells from the neighbouring cells. The latter gets a higher osmotic pressure and draws in water from the neighbours and so on. There is osmotic gradient in the leaf, the cells at the periphery have the highest osmotic pressure and those at the centre, adjacent to the xylem tissues, have the lowest osmotic pressure. This osmotic gradient remains, as long as water evaporates from the cells at the surface. The osmotic gradient also provides the way, by which water is drawn into the mesophyll cells from the vascular tissues. This process offers high resistance to the movement of water, than process of diffusion along the cellulose walls of adjacent mesophyll cells (Fig. 9.20).
10. **Mechanism of Translocation**: According to the **Mass flow hypothesis**, there may be passive mass flow of solutes through the phloem resulting a turgor gradient, highest in the leaf, where the sugar is formed and lowest in the root. The high turgor pressure in the leaves due to high sugar concentration, and therefore, high osmotic pressure in the leaf cell. But the different substances move at different speeds and even in opposite directions, within the sieve tube, which can

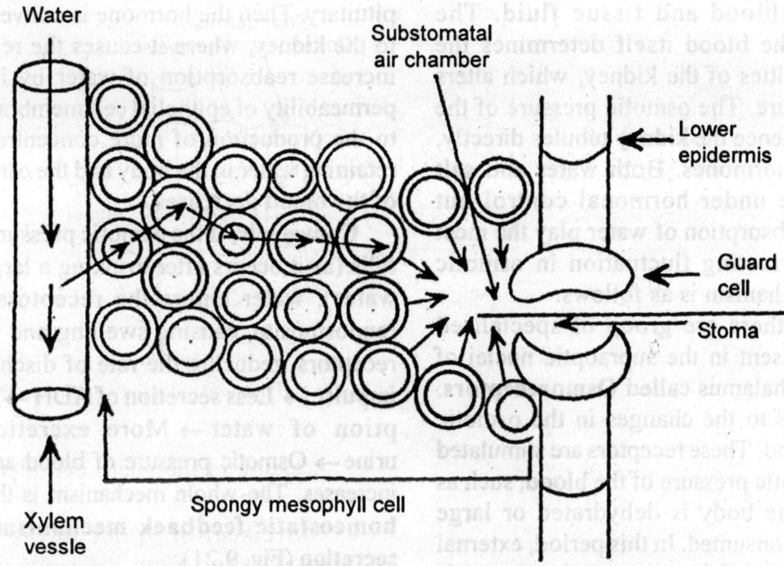

Fig. 9.20: Showing one of the routes by which water through the leaf, from the vascular tissue (xylem vessel) to the substomatal air chamber

be explained by their differing molecular properties or the walls of the sieve tubes are not equally impermeable to all solutes. What is the explanation for bidirectional movement in one and the same sieve tube?

It has been suggested that the mass flow is helped at the sieve plates by "electro-osmosis". Electroosmosis is the passage of water across a charged membrane but experimental evidence are lacking.

IV. In Food Preparation

1. The swelling of dried fruits, beans, etc. when soaked for some time in water.
2. Shrivelling of fresh fruits, e.g., peaches and prunes, when preserved in a concentrated sugar solution.
3. When cucumber is soaked in brine.

All these processes involve osmosis phenomenon, the skins of the fruits and seeds serving as the permeable membrane.

9.7 Osmoregulation

When a freshwater fish whose body fluid has low osmotic pressure is placed in sea water having high salt concentration and higher osmotic pressure, the fish dies. The reason is that, water is drawn out from the body by osmosis. Though the skin of the fish is impermeable but the gills and lining membrane of the mouth act as a semipermeable membrane. If the freshwater fish is placed in hypotonic solution, water enters the body of fish by osmosis. Thus, aquatic animals cannot tolerate significant fluctuation of salinity of external medium, and it is also not expected that the external medium will maintain same osmotic pressure like the body fluid. So, there must be some adaptation mechanism to fight against the fluctuation of osmotic pressure in the external medium.

If a cell has to maintain its proper volume and shape, it should either exist permanently in an isotonic surrounding or have some special adaptive mechanism, enabling it to survive in a hypotonic or hypertonic surrounding, known as **osmoregulation**. According to the osmoregulatory mechanism, the animals have been grouped into two classes:

1. Not capable to regulate their osmotic pressure, i.e., does not have osmoregulatory mechanism, e.g., all marine animals, called **osmo-confirmers**.
2. Capable to regulate the osmotic pressure, i.e., having osmoregulatory; mechanism called **osmoregulators, e.g., amoeba, crayfish, freshwater teleosts, etc.**

- **Osmoregulation in Mammals and Human Beings**

The kidney has two functions: 1. Excretory, and 2. Regulatory. The regulatory function is to regulate water and salts by reabsorption to maintain osmotic pressure of the blood and tissue fluid. The composition of the blood itself determines the reabsorptive activities of the kidney, which alters the osmotic pressure. The osmotic pressure of the blood cannot influence the kidney tubules directly, but does so via hormones. Both water and salt reabsorption, are under hormonal control but changes in the reabsorption of water play the most direct part in correcting fluctuation in osmotic pressure. The mechanism is as follows:

In the brain, there are group of specialised neuronal cells present in the supraoptic nuclei of the anterior hypothalamus called **Osmoreceptors.** They are sensitive to the changes in the osmotic pressure of the blood. These receptors are stimulated by rise in the osmotic pressure of the blood, such as might occur, if the body is dehydrated or large quantity of salt is consumed. In this period, external fluid draws water out of the receptors (exosmosis), causing shrinkage of the receptors, thereby stimulating the receptors to the hypothalamus, which flows along the wall of axolemma of the supraoptico-hypophyseal tract to the posterior lobe of the pituitary. Then the hormone is conveyed via blood to the kidney, where it causes the renal tubules to increase reabsorption of water by increasing the permeability of epithelial cell membrane. This leads to the production of more concentrated urine, by retaining water in the body and the osmotic pressure of the blood decreases.

Conversely, if the osmotic pressure of the blood falls (as it occurs after drinking a large quantity of water), water enters the receptors by osmosis (endosmosis), causing swelling and depression of receptors, reducing the rate of discharge of nerve impulse → Less secretion of ADH → Less reabsorption of water → More excretion of dilute urine → Osmotic pressure of blood and tissue fluid increases. The whole mechanism is the example of **homeostatic feedback mechanism** and **neurosecretion** (Fig. 9.21).

I. Osmo-Confirmers

A. Life of many animals begins in the sea, such as marine invertebrates, e.g., sea anemones, spider crabs and star fishes. Their body fluids are isotonic

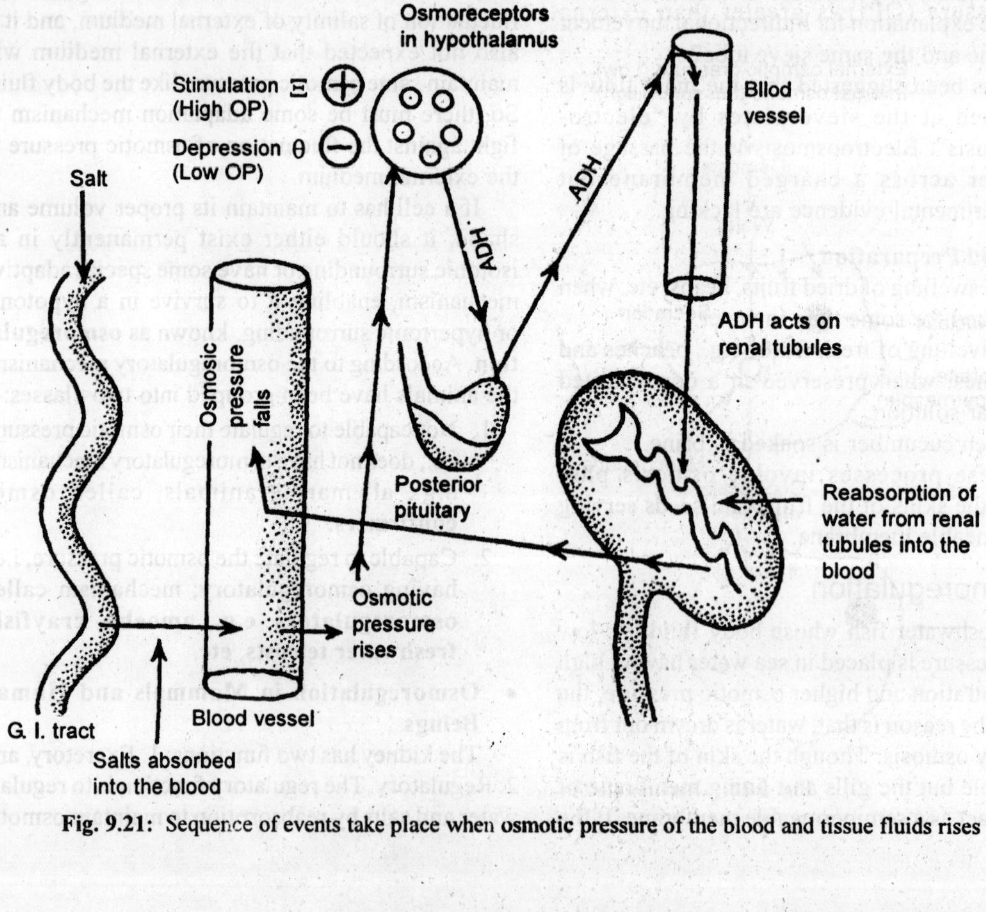

Fig. 9.21: Sequence of events take place when osmotic pressure of the blood and tissue fluids rises

with sea, such as marine invertebrates, e.g., sea **anemones, spider crabs and starfishes. Their body fluids are isotonic with sea water and in starfishes the tissues are perfused with sea water. Since, their** internal osmotic pressure (OPi) is equal to the **external osmotic pressure (OPe), these animals do** not need osmoregulation due to lack of any **mechanism.**

B. Migratory Animals from Sea to Fresh Water

The marine animals, e.g., Maia, Carcinus, Eriocheir, Arenicola, Hydrobia and Neresis dilversicolour can withstand dilution of its surrounding medium by two possible mechanisms:

1. May have tissues that can tolerate a greater range of salinities, e.g., Arenicola, neresis diversicolour and called **Eusyhaline.**
2. May avoid the effects of dilution by behavioural means, e.g., Hydrobia, which burrows into the mud when the tide is going out, thus avoiding the periodical dilution of its external medium. In case of freshwater animals (estuarine), the internal osmotic pressure (OPi) is greater than external osmotic pressure (OPe); so water will be drawn into the body and to overcome this problem, two possible mechanisms may occur:
   (a) Water eliminates rapidly by means of the kidney or some equivalent structure. Salts being reabsorbed from the water by the structure before leaving the body, a hypotonic urine is produced.
   (b) Taking up salt actively from the external medium compensates the diluting effect due to influx of water.

Both mechanisms involve the movement of ions against a concentration gradient and requires the **expenditure** of energy, i.e., **active transport.**

II. Osmo-Regulators

A. Contractile Vacuoles in Amoeba and Freshwater Protozoans

These animals contain contractile vacuole which is a small sac, lined with unit membrane lying freely in the cytoplasm. The cell membrane surrounding the animal is semipermeable. Since internal osmotic pressure is greater than external osmotic pressure,

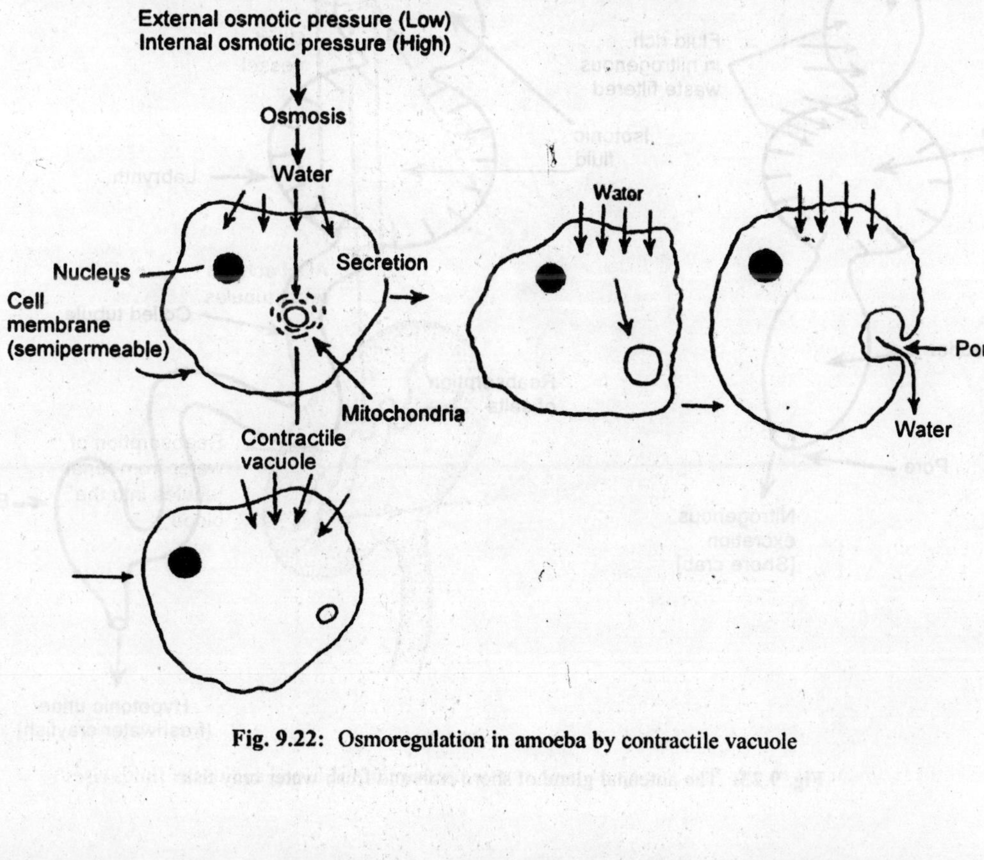

Fig. 9.22: Osmoregulation in amoeba by contractile vacuole

water enters the cell by osmosis (endosmosis). To minimize this, water is secreted into the contractile vacuole rapidly. This mechanism requires energy expenditure, as evident by presence of several mitochondia, close to the vacuolar membrane. The contractile vacuole expands and eventually discharges its contents to the exterior, through a small pore in the cell membrane, after which the whole process is repeated. Frequency of discharge depends upon salinities because increase in salt concentration in the external fluid medium, decreases frequency of discharge (Fig. 9.22).

B. Antennal Gland of Shore Crab and Fresh-Water Crayfish

(a) In the shore crab "carcinus", water is eliminated by a pair of antennal glands, because they are at the base of the antenna in the head. Each gland consists of a small end sac called "Coelomic sac", which is derived from coelom, which is connected to a larger sponge-like cavity called "Labrynth". This connects with a "bladder", which opens to the exterior by a small "pore" at the base of the antenna. The antennal glands are for nitrogeneous excretion of fluid, rich in nitrogeneous waste, which is filtered into the end sac and labrynth from the surrounding blood. The urine formed is isotonic with the blood, because the antennal gland is unable to retain salts. It eliminates both water and salts. The loss of salts in the urine is compensated by the gills, which absorb salts from the surrounding medium and secrete them into the blood against a concentration gradient. Thus, internal osmotic pressure is maintained at a higher level than external osmotic pressure.

In freshwater crayfish, a pair of antennal glands are capable of producing a hypotonic urine. This is because the antennal glands are able to reabsorb salts, a process which takes place in a coiled tubule connecting the labrynth with the bladder. Fluid collected from the antennal gland at various points,

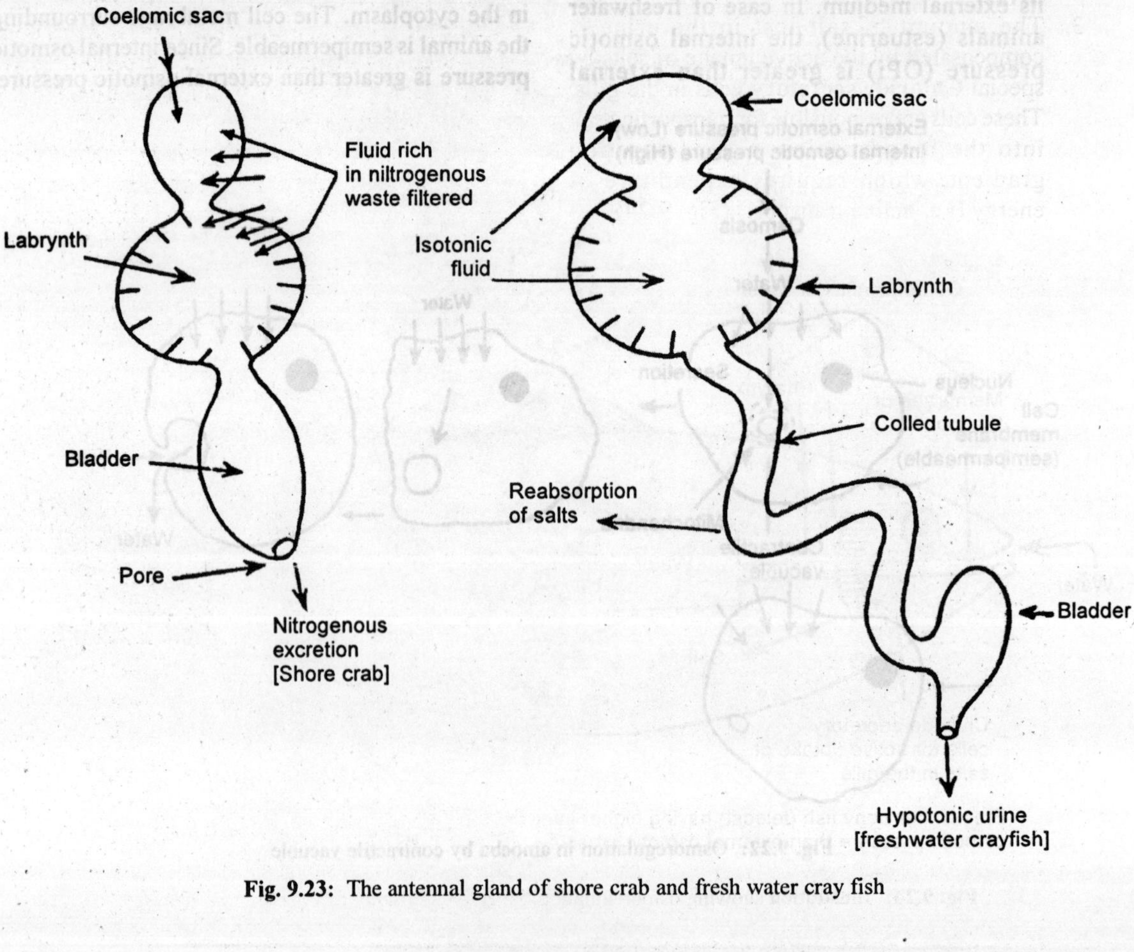

Fig. 9.23: The antennal gland of shore crab and fresh water cray fish

along its length has shown, that the contents of the "end sac" and labrynth are isotonic with the blood, but when the urine flows along the coiled tubule, salts are reabsorbed from it. In this way, the antennal gland of the crayfish gets rid of excess water and retains salts, resulting in higher internal osmotic pressure than external osmotic pressure (Fig. 9.23).

C. Freshwater Teleosts (Bony Fishes e.g., Stickleback)

Freshwater animals are the bony fishes (teleosts) and there is osmotic influx of water across the gills and lining of the mouth cavity and pharynx. This is countered by the following processes:

1. The rate of glomerular filtrate in the kidney is very high. The cortex of the fish kidney is structurally similar to that of the mammals and the high filtration rate is achieved by the numerous large glomeruli in the kidney.
2. When renal fluid flows along the kidney tubules, large amount of salts are reabsorbed into the blood stream, producing **hypotonic urine**.
3. The amount of salt lost in the urine is compensated by the active uptake of salts by special **Chloride secretory cells** in the gills. These cells are responsible for transporting salt into the blood against a concentration gradient, which requires expenditure of energy (i.e. active transport) (Fig. 9.24).

D. Marine Teleost (Bony Fishes)–From Freshwater to the Sea, e.g., Cod, Whiting etc.

Certain fishes return from freshwater to the sea, during evolutionary period, e.g., marine teleosts like cod and whiting. They have body fluids, which are hypotonic to their surrounding, i.e., internal osmotic pressure is lesser than external osmotic pressure. As a result, there is outflux of water (exosmosis) from the body causing dehydration of the tissue. This is countered by the following processes:

1. Swallowing sea water and having a kidney, with a relatively low filtration rate.
2. Removing salts by means of chloride secretory cells in the gills against a concentration gradient from blood to the external sea water.
3. Eliminating nitrogenous waste which is solute but nontoxic, is trimethylamine oxide, which requires comparatively little water for its elimination and isotonic urine (Fig. 9.25).

E. Marine Elasmobranch

Marine elasmobranches, e.g., shark, rays and dogfish have their bodyfluids slightly hypertonic to sea water due to retention of nitrogenous waste product **urea**, which raises osmotic pressure of body fluids and slightly exceeds the surrounding sea water, causing influx of water, which is readily eliminated by the kidney.

The retention of urea, is made by reabsorption in the kidney tubules together with the impermeability of gills to urea and the tissues can

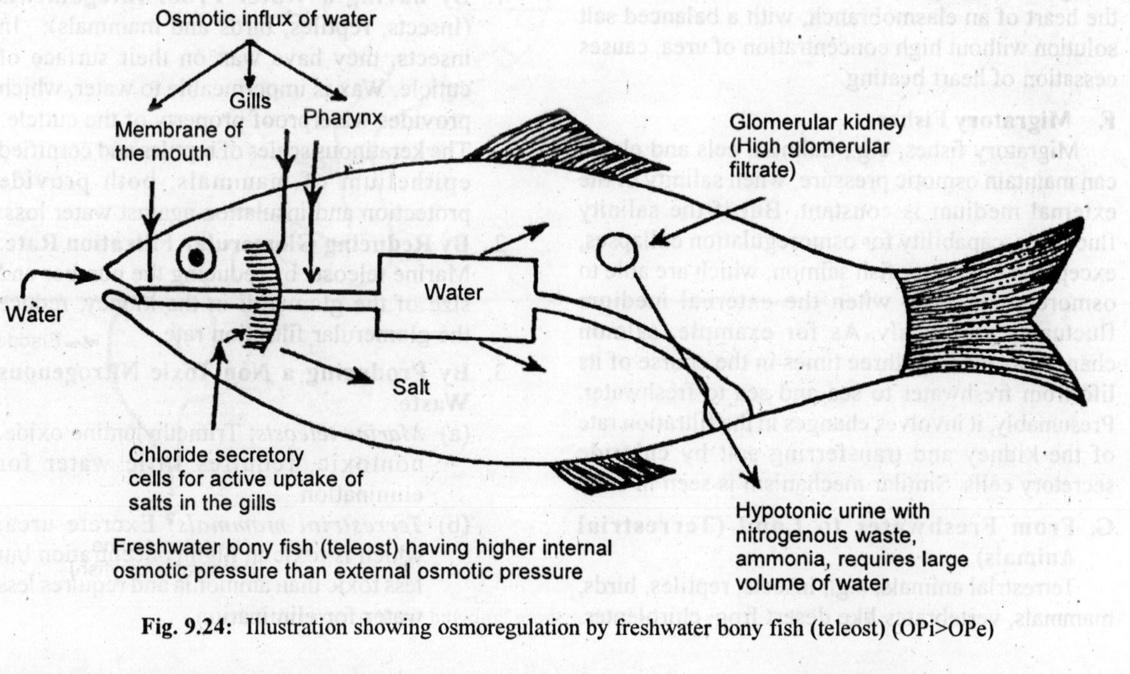

Fig. 9.24: Illustration showing osmoregulation by freshwater bony fish (teleost) (OPi>OPe)

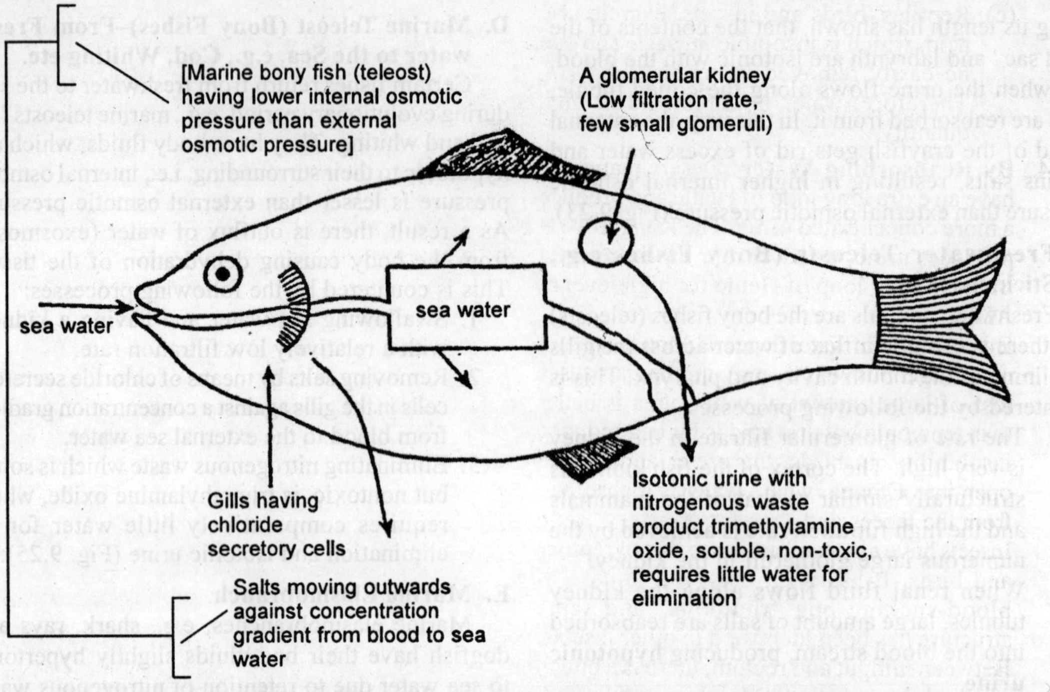

Fig. 9.25: Illustration shown osmoregulation of marine bony fish (teleost) OPi < Ope

tolerate urea. This high concentration of urea alters the shape of proteins by breaking the hydrogen bonds, linking adjacent polypeptide chains, disrupting the smooth functioning of the cells, particularly, when the enzymes are affected. Elasmobranch proteins are adapted to these effects, though the tissues try to thrive urea. Perfusion of the heart of an elasmobranch, with a balanced salt solution without high concentration of urea, causes cessation of heart beating.

F. Migratory Fishes

Migratory fishes, e.g., Salmon, eels and elvers can maintain osmotic pressure, when salinity of the external medium is constant. But if the salinity fluctuates, capability for osmoregulation collapses, except the migratory fish salmon, which are able to osmoregulate even when the external medium fluctuates vigorously. As for example, salmon changes its medium three times in the course of its life from freshwater to sea and sea to freshwater. Presumably, it involves changes in the filtration rate of the kidney and transferring salt by chloride secretory cells. Similar mechanism is seen in eels.

G. From Freshwater to Land (Terrestrial Animals)

Terrestrial animals, e.g., insects, reptiles, birds, mammals, vertebrates like desert frog, chiroleptes, desert mammal, kangaroo rat, etc. Earthworm, African and South American lungfishes, camel, etc. experience the water loss by evaporation from permeable surfaces exposed to the atmosphere. These animals solve this problem in the following ways.

1. **By having a Water Proof Integuments** (Insects, reptiles, birds and mammals): In insects, they have wax on their surface of cuticle. Wax is impermeable to water, which provides waterproof property of the cuticle. The keratinous scales of reptiles and cornified epithelium of mammals, both provide protection and insulation against water loss.
2. **By Reducing Glomerular Filtration Rate**: Marine teleosts by reducing the number and size of the glomeruli in the kidney, reduce the glomerular filtration rate.
3. **By Producing a Non-Toxic Nitrogenous Waste:**
   (a) *Marine teleosts*: Trimethylamine oxide, nontoxic, requires little water for elimination.
   (b) *Terrestrial mammals*: Excrete urea, which is toxic in high concentration but less toxic than ammonia and requires less water for elimination.

(c) Reptiles, birds and insects, excrete uric acid, which is insoluble in water and can be easily removed in semi-solid form. Therefore, water is eliminated in high quantity.

4. **By Reabsorbing Water**: Desert mammals have an extra long loop of Henle and produce a more concentrated urine. The Kangaroo rat produces urine, which is four times concentrated in long loop of Henle for high level of ADH.

But retention of water depends upon the nature of nitrogenous waste, e.g., ammonia causes loss of large amount of water, urea is better than ammonia but uric acid is far better due to insolubility. In birds, the ureter and rectum open into **Cloaca**. where water is reabsorbed from the faeces and excretory waste.

Insects have a system similar to that of reptiles and birds. It has been observed in case of blood-sucking bug "rhodnius" and same structure has been identified in other insects. Between midgut and rectum, there is a bunch of blind ducts called "Malphigian tubules". One end of these open into midgut and the other end is free and floats in the blood-filled body cavity. Body tissues produce nitrogenous waste which is soluble and is called "Potassium urate" which liberates into the blood and takes up by the cells of the Malphigian tubules. The muscular structure of tubules causes within movements to stir the blood for better absorption of urate. In the tubular cells, the potassium urate reacts with water and CO_2 from respiration to form potassium bicarbonate and uric acid. The potassium bicarbonate is absorbed back into the blood, and the uric acid is deposited in the lumen of the tubule. When the uric acid moves towards the gut water, it is reabsorbed in large quantity into the blood, and the proximal end of the malphigian tubules is filled with solid crystals of uric acid. Epithelial cells are having numerous microvilli, which increase the surface area for water reabsorption. At the proximal end of the tubule, the microvilli are separate but at the distal end, they are close. Water is reabsorbed also by the folded walls of the rectal gland and when the urine leaves the body, it is very much concentrated and thus they conserve water for which they are found in the hottest and driest places (Fig. 9.26).

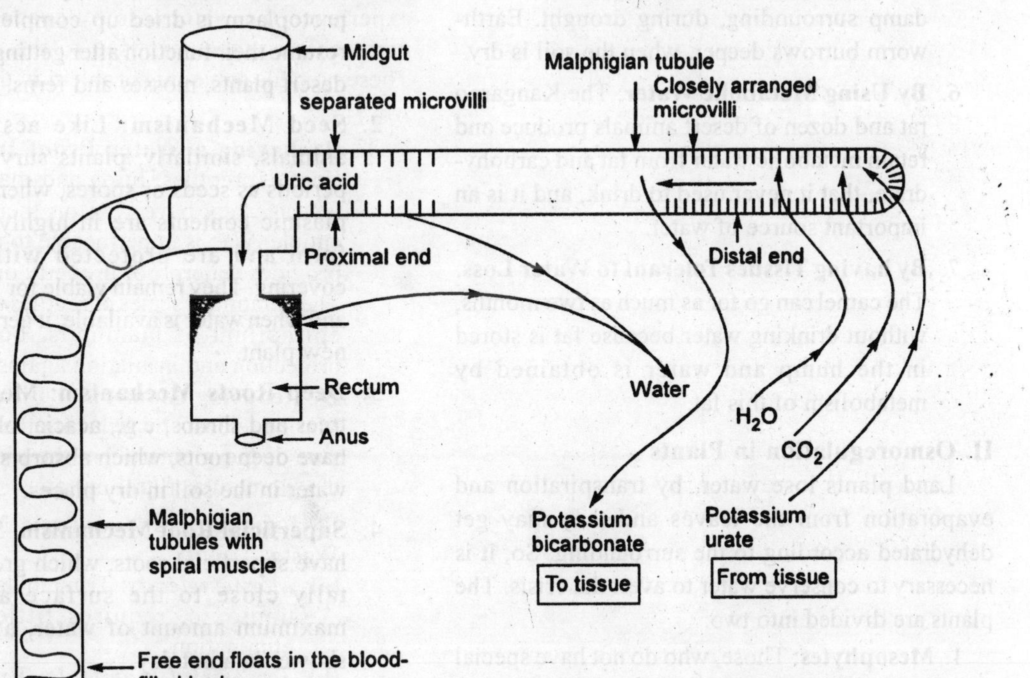

Fig. 9.26: Diagrammatic representation of structure and function of Malphigian tubules of an insect (Rhodnius)

Table 9.5: Comparison of the process of osmoregulation in elasmobranches, marine teleosts fishes, marine birds, marine reptiles and mammals

Animals	Water loss by osmosis	Water loss by evaporation	Osmotic condition of blood compared to ext. medium	Osmotic condition of urine compared to the blood	Type of drinking water	Type of excretion (urine)	Organs for Osmoregulation
Elasmobranch	No	No	Isotonic	isotonic	Freshwater	Isotonic	Hypertonic NaCl from rectal gland
Teleost	Yes	No	Hypotonic	Isotonic	Sea water	Isotonic	Secretes salt from gills
Reptiles	No	Yes	Hypotonic	Isotonic	Sea water	Isotonic	Hypertonic tears.
Mammals	No	Yes	Hypotonic	Hypertonic	Freshwater	Strongly hypertonic urine	Hypertonic urine through kidney
Birds	No	Yes	Hypotonic	Hypertonic	Sea water	Weakly hypertonic urine	Hypertonic nasal secretion

5. **By Means of Behaviour**: Many animals reduce the water loss by modifying their behaviour or changing their habitat, e.g., the Kangaroo rat cuts down evaporation by the lungs (25 per cent) remaining in the burrow during the heat of the day. Many animals seek damp surrounding, during drought. Earthworm burrows deeper, when the soil is dry.
6. **By Using Metabolic Water**: The Kangaroo rat and dozen of desert animals produce and retain metabolic water from fat and carbohydrate, that it never used to drink, and it is an important source of water.
7. **By having Tissues Tolerant to Water Loss**: The camel can go for as much as two months, without drinking water because fat is stored in the hump and water is obtained by metabolism of this fat.

II. Osmoregulation in Plants

Land plants lose water, by transpiration and evaporation from the leaves and they may get dehydrated according to the surrounding. So, it is necessary to conserve water to avert the crisis. The plants are divided into two:

1. **Mesophytes**: Those, who do not have special mechanism of conserving water.
2. **Xerophytes**: Those, who have special mechanism to conserve water.

A. Structural Mechanism

1. **Protoplasm**: Plants usually are more efficient than animals to withstand the fluctuations in water content of the tissues and some have tissues which are tolerant to dessication. Some plants are so adaptable that, their protoplasm is dried up completely but yet resume their function after getting water, e.g., desert plants, mosses and ferns.
2. **Seed Mechanism**: Like aestivation in animals, similarly, plants survive the dry periods as seeds or spores, where the protoplasmic contents are in highly dessicated form and are protected within a hard covering. They remain viable for long periods and when water is available, it germinates into new plant.
3. **Deep Roots Mechanism**: Mediterranean trees and shrubs, e.g., acacia, oleander, etc. have deep roots, which absorb subterranean water in the soil in dry places.
4. **Superficial Root Mechanism**: Some plants have superficial roots, which grow horizontally close to the surface and absorb maximum amount of water, after a short shower in a hot day.
5. **Succulent Mechanism**: Some plants store water in large parenchyma cells within swollen stems and leaves and make the tissue

wet and juicy and such plants are called **Succulents**. The stem succulents in the desert, e.g., Giant Saguaro Cactus of Arizona, in which water is stored in the thick stem, and side branches and the leaves are reduced to slender spikes. Loss of water from the succulents is prevented, by reducing the number of stomata. In some plants (Shrub hakea of Australian desert) the stomata are sunk down into "pits", in which humid air accumulates and thereby, reduction of rate of evaporation takes place from the leaves. Sinking of stomata and folding of the leaves are seen in some plants (e.g., certain grasses) in dry weather.

B. Physiological Mechanism

1. **By Revising the Normal Stomatal Rhythm**: Opening the stomata at night and closing by day, is an ideal mechanism to prevent water loss, but stomata must be open for a certain period in day-light for photosynthesis.
2. **By Thickening the Waxy Cuticle**: In some ordinary plants, water evaporates through the cuticle. This cuticular transpiration is reduced by thickening of the "waxy cuticle" which becomes impermeable to water (e.g., desert cacti, prickly pear). These plants get water shortage in winter due to frosts and snow cutting off supply from the soil and the phenomenon is known **Physiological draught**.
3. **Shedding the Leaves**: Trees shed the leaves before the winter to avoid the loss of water by transpiration, e.g., Ocotillo ("vine cactus" of the North American desert), which comes with leaf every time during raining and sheds the leaves immediately after the rain. This happens 6-7 times in a year.
4. **Halophytes and the Mechanism of Water Balance**: Some plants grown in mud flats and salt marshes are known as **Halophytes** which absorb water, having higher osmotic pressure than ordinary soil water. This is countered by two mechanisms:
   (a) The root cells exert a much greater osmotic pressure, than other ordinary plants, which enable them to receive water by osmosis.
   (b) They store enough water in the root cells to avoid shortage of water, if the external pressure exceeds that of the root cells.
5. **Guttation**: By this mechanism, the plants get rid of excess water, which has been stored. During summer, on the cool humid evening, the leaves of some trees exude water and the phenomenon is called **Guttation**. This is due to removal of water by evaporation, which has been taken in excess.
6. **Water Secretion by Hydathodes**: Some plants of tropical rain forests, where the humidity is high, water is actively secreted by the structures called **Hydathodes**, present at the terminal ends of the veins round the edges of the leaves.

9.8 Filtration

Filtration is a physical process by which undissolved substances are separated from a liquid, through a membrane or pores, due to difference of mechanical pressure on the two sides of the membrane. The mechanical pressure producing filtering force, is either hydrostatic pressure or gravity or osmotic pressure.

- **Experiment**

Filtration process can be done with filter paper or linen. If a closed bag of linen having fairly fine mesh and containing water with dissolved ions or molecules is immersed in water and the bag is compressed by mechanical squeezing, then the water with dissolved substances come out to the surrounding water.

- **Characteristics**

1. **The amount of fluid filtered in a given interval, is proportionate to the difference in pressure, the surface area and permeability of the membrane.**
2. **Molecules that are smaller in diameter than the pores of the membrane, pass through with fluid and larger molecules are retained.**
3. **The denser the membrane, the greater will be the difference of pressure required to effect a passage at a certain rate.**
4. **In pore-restricted filtration, the hydrodynamic flow obeys Poiseulle's Law, provided that the radius of solvent and solute particles is less than 1/20th of that of the pores.**

9.8.1 Capillary filtration

The physical phenomenon of filtration occurs in mammals, vertebrates and some invertebrates.

Filtration process is seen with the fluid blood present in the blood vessels, at a pressure above the atmospheric pressure partitioned by the cellular layer of capillary walls from the tissue fluid, bathing the tissue cells at an atmospheric pressure. The cellular layer of the capillary walls allows to pass the water along with the dissolved ions and other substances. Beside that, the continuous squeezing of the blood permit water and dissolved substances out of the capillary wall. In this way, tissue gets nutrition and oxygen from the blood. Along with filtration, diffusion also plays a role for transport of nutrition and oxygen from the blood to the tissues. This shift of nutritional products, when there is difference of pressure between the two sides of the capillary membrane, i.e., pressure in the blood and the tissue fluid and produces unidirectional movement, but there is also a reverse movement of materials from tissues to the blood, which is not under the process of the difference of hydrostatic pressure. It can be explained by the fact that, the fluid is filtered from the blood due to difference of hydrostatic pressure. There is a continuous influx of fluid from the tissue to the blood, by osmotic pull due to difference of protein osmotic pressure between the two sides. By virtue of higher protein osmotic pressure, owing to high protein concentration in the blood, it draws fluid from the tissue fluid for its low protein osmotic pressure due to low concentration of protein. Apparently, the two forces (filtering force and osmotic pull) are equal but cannot be regarded as a perfectly balanced, dynamic equilibrium because some amount of fluid, filtered through the capillaries again enter the circulation via lymph. If filtration and absorption are balanced in any organ, conditions are called **Isovolumetric or Isogravimetric.**

9.8.1.1 Factors influencing the rate of filtration across the capillary membrane

1. **The Filtration Coefficient**: The filtration coefficient (K_f) is the flow, crossing the membrane in unit time under unit difference of pressure and is expressed as rate of filtration/min/mm Hg/100 gm of tissue. For average tissues, it is 0.008 to 0.015 ml/mn/mm Hg/100 gm of tissues. The coefficient varies from one tissue to another due to extreme differences in permeabilities of the capillary systems in different tissues.
2. **Capillary Hydrostatic Pressure**: There is a pressure gradient between the arterial to venous end of the capillary and this pressure is influenced by changes in the calibre of the precapillary and the postcapillary resistance vessels.
3. **The Interstitial Fluid Hydrostatic Pressure**: The actual pressure is not known but under normal conditions, it may be positive.
4. **The Colloidal Osmotic Pressure of the Interstitial Fluid**: The proteins (0.7–3 gm/100 ml) and hyaluronic acid in solution in the interstitial space, produce colloidal osmotic pressure (oncotic pressure) which is determined primarily, by the number of molecules in solution E. H. Starling developed an equation to apply for filtration across the capillary membrane, known as Starling's Law of Ultrafiltration. The law states that:

$$FM = K(P_c + \pi_i - P_i - \pi_c),$$

where FM = Fluid movement (outward +, inward –),

K = Filtration coefficient of capillary wall,

P_c = Hydrostatic pressure in the capillaries,

P_i = Hydrostatic pressure in the interstitial fluid.

π_c = Oncotic pressure in the plasma in the capillary,

π_i = Oncotic pressure in the interstitial fluid.

It has been found that the capillary beds in different organs differ in their filtration properties, e.g., kidney capillaries in the glomeruli, filter through the whole length whereas absorption predominantly occurs in some capillary beds in the intestinal wall. Usually the filtration occurs in the arterial end

Table 9.6: Comparison of filtration and diffusion in biological processes

Filtration	Diffusion
1. It is the separation of undissolved substances from a liquid through a membrane or pores.	1. It occurs due to continual movement of molecules among each other in liquid or gases. Substances can diffuse through the membrane by: (a) Becoming dissolved in lipid matrix, or (b) Diffusing through minute pores.
2. It depends on: (a) Pressure difference across the filtering surface. (b) Pore size of the surface. (c) Osmotic pressure of the fluid to be filtered.	2. **Concentration Factor** – The net rate of diffusion from area of higher concentration is directly proportional to the larger concentration minus molecular sizes.
3. The amount of fluid filtered in a given interval is proportionate to the pressure difference, surface area and permeability of the membrane.	3. The **rapidity** with which a molecule diffuses from one point to another is less, the greater the molecular sizes.
4. The denser the membrane, the greater will be the difference of pressure required to effect a passage at a certain rate.	4. The **rate of diffusion** is approximately proportional to the square root of the MW but it is affected by the shape of the molecule.
5. **Mechanical force,** i.e., difference of hydrostatic pressure or gravity or osmotic pressure is required.	5. The movement of molecules is due to concentration gradient.
6. **Net movement** of liquid in one direction, e.g., out of the capillaries at the arterial end.	6. The **net movement** of molecules is away from the solute in any direction.
7. **Rate of filtration** at the arterial end of the capillary is 16 ml/min.	7. **Rate of diffusion** through the cell of capillary membrane is 240000 ml/min.

of the capillaries and absorption in the venous end.

5. **Capillary Surface Area.**
6. **Permeability of the Capillary Membrane.**

9.8.2 Glomerular filtration

Filtration in the Glomerular Capillary Membrane

The average renal plasma flow (FPF) exceeds 650 ml/min or 900 l/day. Approximately, 125 ml of plasma are filtered per minute at the glomerulus, i.e., the "glomerular filtration rate" (GFR) is 125 ml/min or 180 L/day. About 20 per cent of the RPF is filtered into Bowman's capsular space. The rates of GFR to RPF is termed as **Filtration fraction,** which is normally 0.16-0.20.

The glomerular filtrate is often termed as **ultra-filtrate** because of elimination of cellular element plus colloid particle whereas a **filtrate** means elimination of only the cellular elements of blood.

9.8.2.1 Glomerular capillaries

Glomerular filtrate comes out of the glomerular capillaries into Bowman's Capsule, through the membrane of the glomerular capillaries called **Glomerular membrane.** The glomerular capillaries differ from the other capillaries in the following respect:

1. **Structure**
   (a) Glomerular capillaries are invaginated into and covered on their outer surface by the visceral layer of the epithelium of Bowman's capsule. This layer of flattened cells is continuous with the parietal layer of epithelium and with the living cells of the renal tubules.
   (b) Glomerular capillaries are interposed between two arterioles.
2. **Filtering Surface**: Total filtering surface area of the glomerular capillaries is to be about 1.5 sq. metres (Body capillaries – 6300 m^2).
3. **Hydrostatic Pressure**: The glomerular capillaries have higher hydrostatic pressure because it arises directly from straight interlobular arteries, which are short afferent arterioles. Two major areas of resistance to blood flow through nephron are: (a) afferent

arterioles and (b) efferent arterioles. Hydrostatic pressure remains constant along the glomerular capillary whereas it declines in the other capillaries.

4. **Diameter:** Diameter of the efferent capillaries is lesser than afferent capillaries, which is advantageous for effective filtration.
5. **Permeability:** Higher permeability of the glomerular membrane which is 100 to 1000 times that of muscle capillaries due to presence of pore of different sizes.

   Recent evidence suggests that the pores may be hydrated channels between the glycopro-tein chains of the basement membrane. These channels may be tortuous and may not be stable anatomical structures for which its identification under EM has not been established.
6. **Presence of Fenestrae, Slit Pores:** Many large fenestrae, which are small oval windows develop through the very middle of the endothelial cells and small in diameter and prevent the filtration of all particles with an average size greater than 16 millimicorns (160Å). The mesh-work of basement membrane prevents filtration of particles in size greater than 11 millimicrons. Slit pores prevents the particles with diameter greater than 7 millimicrons.
7. **Glomerular Membrane:** The glomerular membrane (filtration barrier), in general, is similar to that of other capillaries of the body but there are some differences:

   A. It consists of three layers:

   (a) Endothelial layer of the capillary,

   (b) A basement membrane,

   (c) A layer of epithelial cells on the outer surface of the capillaries.

   Inspite of that, it has greater permeability than the body capillaries.

   B. The membrane has pores of different sizes, which allow neutral substance with effective molecular diameters of less than 4 nm to be freely filtered and the filtration of neutral substances having diameters of more than 8 nm are not filtered. Between these values, filtration is inversely proportionate to diameter.

   C. Sialoproteins in the endothelium and basement membrane, are negatively charged and repel negatively charged particles of blood, with the result that filtration of anionic substances 4 nm in diameter is less than half that of neutral substances. Perhaps this explains the concentration of 0.2 per cent albumin in the filtrate with a diameter of 7 nm, instead of higher concentration on the basis of the diameter. The charge of the albumin is negative. In nephritis, there is albuminura, where the negative charges in the glomerular wall are dissipated and leakage of albumin occurs, without increase in the size of the pores in the membrane.
8. **Plasma Oncotic Pressure**: The plasma oncotic pressure rises along the length of the capillaries because of filtration of large volume of water whereas it remains constant in other capillaries.
9. **Net Filtration Pressure**: The filtration occurs at the afferent end where net filtration pressure is 10 mn Hg, as in other capillaries but no fluid movement at the efferent end, where filtration pressure is zero. Hence, filtration equilibrium exists at the efferent end of glomerular capillaries.

9.8.2.2 Glomerular filtrate

The glomerular filtrate has the same composition as the fluid that filters at the arterial ends of the capillaries without any blood cells and about 0.02 per cent of protein (albumin) of the plasma. The filtration-barrier is freely permeable to water and to small solutes, i.e., crystalloids, e.g., ions, glucose and urea. It has limited permeability to larger solutes, i.e., colloids, e.g., large proteins and lipids and impermeable to the blood cells. The glomerular filtrate is identical to plasma except proteins and lipids. The electrolytes composition is also similar to that of the tissue fluid. But the concentration of the other negative ions, e.g., chloride, bicarbonate ions are 5 per cent higher in the glomerular filtrate than in plasma due to paucity of negatively charged protein ions in the filtrate, a Gibbs-Donnan equilibrium effect occurs and the concentration of positive ions is about 5 per cent lower.

9.8.2.3 Dynamics of glomerular filtration

Glomerular filtration occurs by the effective filtration pressure in the glomerulus with the help of the following factors:

1. Glomerular pressure,
2. Pressure in the Bowman's capsule,
3. Colloidal osmotic pressure in the glomerular capillaries.

1. **Glomerular Pressure**: It is the average pressure in glomerular capillaries which is in average 45 mm Hg in rat, 55-70 mm Hg in dog. In man, a reasonable average value can be considered to be 60 mm Hg.
2. **Bowman's Capsular Pressure**: Pressure in the Bowman's capsule, has been measured in the animals and on the basis of these findings, the capsular pressure in human has been estimated to be around 18 mm Hg.
3. **Colloidal Osmotic Pressure in the Glomerular Capillaries**: The normal colloidal osmotic pressure when the blood enters the glomerular capillaries is 28 mm Hg and it rises to 36 mm Hg when it reaches the venous end of the glomerular capillaries because of increase in concentration of plasma protein by 20 per cent for filtration of large amount of fluid. So, the average colloidal osmotic pressure is 32 mm Hg.

9.8.2.4 Filtration pressure

The filtration pressure is the net pressure, forcing fluid through the glomerular membrane. This pressure is equal to the glomerular pressure minus the total of glomerular colloidal osmotic pressure and capsular pressure, i.e., – 60 mm Hg = (28 mm Hg 60 mm Hg – (28 mm Hg + 18 mm Hg) = + 14 mm Hg + 18 mm Hg) = + 14 mm Hg.

9.8.2.5 Filtration coefficient

Filtration coefficient is expressed as K_f, which is constant. It is the glomerular filtration rate for both kidneys per mm Hg of filtration pressure, e.g., glomerular filtration rate is equal to the filtration pressure times the filtration coefficient or GFR = Filtration pressure K_f, which is 12.5 ml/min/mm Hg. of filtration presssure.

Characteristics of Glomerular Membrane or Filtration Barrier
(See ultrafiltration)

9.8.2.6 Factors affecting the glomerular filtration rate (GFR)

The glomerular filtration rate is determined by:

1. Net filtration pressure,
2. Filtration coefficient.

The filtration pressure is determined by:

1. Renal blood flow,
2. Afferent arteriolar constriction,
3. Efferent arteriolar constriction,
4. Effect of sympathetic stimulation,
5. Effect of arterial pressure,
6. Change in hydrostatic pressure in Bowman's Capsule (ureteral obstruction),
7. Changes in plasma protein concentration (dehydration, hypoproteinaemia).
8. Changes in permeability of glomerular membrane,
9. Decrease in total area of capillary bed.

9.8.2.7 Biological importance of filtration

1. Passage of water, salts, foodstuffs, etc. through the capillary membrane and formation of tissue fluid.
2. Formation of glomerular filtrate for urine production.
3. Formation of certain portion of cerebrospinal fluid.
4. Absorption of foodstuffs from the intestine.

Table 9.7: Comparison of capillary and glomerular filtration

Capillary Filtration	Glomerular Filtration
1. Structure-Filter Barrier (a) Single layer of endothelium. (b) Capillaries not interposed between the two arterioles. (c) Diameter of the capillary remains constant along the entire length.	1. It has three layers: (a) Endothelial layer of the capillary wall. (b) A basement membrane. (c) A layer of epithelial cells. (d) Capillaries are interposed between the two arterioles (e) Diameters of the efferent capillaries are lesser than the afferent arteriole, which is advantageous for effective filtration.
2. Filtering Surface To be about 6300 m^2.	2. Total filtering surface area is to be about 1.5 sq. meters.
3. Hydrostatic Pressure Lesser than glomerular capillaries and declines while approaching the venous end.	3. It is higher because of: (a) Afferent arterioles are short. (b) Straight branch of interlobular arteries. (c) Efferent arterioles have high resistance and remain relatively constant along the capillary.
4. Osmotic Pressure It remains relatively constant in the capillary.	4. Because of large volume of fluid filtered, the plasma oncotic pressure rises along the length of a capillary.
5. Fluid Movement Filtration pressure varies along the length of systemic capillary. At the arterial end the balance of pressure favours filtration having net filtration pressure + 16 mm Hg. At the venous end reabsorption occurs where net filtration pressure is –14 mm Hg. The result is that filtration and reabsorption is in balance.	5. Filtration occurs at the afferent end where net filtration pressure is about 14 mm Hg. But at efferent end fluid movement stops and net filtration pressure is zero. So, filtration equilibrium exists at the efferent end of capillaries.
6. Filtration Rate Filtration rate average than 2 ml/min due to difference in filtration coefficient (K_f).	6. Glomerular filtration rate (GFR) averages 125 ml/min due to larger filtration coefficient (K_f), probably for large number of fenestrae in the capillary endothelium.
7. Propelling the Filtrate No such mechanism.	7. Hydrostatic pressure in Bowman's capsule is not only helpful for net filtration pressure but also to propel filtrate to the proximal tubule and the remainder of the nephron.
8. Region of Filtration Capillary beds of different organs, differ in their filtration properties, e.g., absorption occurs predominently in some capillary beds in the intestinal wall.	8. Glomerular capillaries filter throughout its entire length.
9. Permeability Lesser	9. 100 times more than muscle capillaries.
10. Mechanism of Movement of Fluid (a) Fluid 1. Through the membrane, or 2. Through the pores. (b) Proteins 1. Leakage through pores in the capillary walls or 2. Moved by vesicular transport. (c) Large Molecules 1. By vesicular transport (micropinocytosis), or 2. Directly through capillary fenestrations.	10. Through pores of different size or hydrated channels in the glycoprotein chains of the basement membrane.
11. Flow-Limited or Diffusion-Limited It is flow-limited rather than diffusion-limited.	11. Exchange across the glomerular capillaries is flowlimited.

9.9 Endocytosis

(Phagocytosis and Pinocytosis)

Cells take up their nutrients from the surrounding fluid for growth and survival in three ways:

I. Diffusion
II. Active transport
III. Endocytosis.

Endocytosis is a process by which the cell membrane engulfs particulate matter, e.g., bacteria, other cell, degenerated tissue, etc. or extracellular fluid with dissolved substances.

Endocytosis are of three types:

I. Not receptor-mediated
   A. Phagocytosis (cell-eating)
   B. Pinocytosis (cell-drinking)
II. Receptor-mediated
   C. Receptor-mediated endocytosis.

9.9.1 Phagocytosis

Phagocytosis is the process by which a cell engulfs foreign particle, i.e., large particulate matter which is visible under ordinary microscope called **Phagocytosis** (cell-eating). The particulate matters are bacteria, dead tissues, cell fragments, worn out red blood cells, etc. It differs from pinocytosis (cell-drinkings). **Pinocytosis** means engulfing the extracellular fluid with dissolved substance in the form of minute vesicle, which is not visible under ordinary microscope but is visible under Electron Microscope. The dissolved particles are usually proteins and strong electrolyte solutions, Both mechanisms are identical except for the sizes and natures of the ingested vesicles.

- **Phagocytes**

The cells which are invloved in phagocytosis are called phagocytes. They can be grouped into two:

(a) **Mobile**
1. WBC – Neutrophil, Eosinophil (weak), Lymphocytes, Monocytes.
2. Platelets.
3. Amoeba.

(b) **Stationary**
1. Tissue macrophage system (Reticuloendothelial system)—Phagocytic cells lining the blood capillaries of liver, spleen, bone marrow, skin, alveoli, etc. The tissue macrophages in various tissues, differ in appearance and are known by different names, e.g., kupffer cells (liver), reticulum cell (lymph nodes, spleen, bone marrow), alveolar machrophages (alveoli of the lungs), tissue histiocytes, clasmatocytes or fixed macrophages (subcutaneous tissue), microglia (brain).
2. Lymphoid tissue.

9.9.1.1 Factors for phagocytosis

Phagocytosis occurs when:

1. The objects contact the cell membrane,
2. The surface of the particle is rough,
3. The objects having electronegative charges are repelled, while those having electropositive charges are susceptible to phagocytosis. The phagocytic cells are electronegative for which the different reactions are observed. Most natural substances of the body have electronegative surface charges and repel the phagocytes, whereas dead tissues and foreign particles are frequently electropositive and are prepared for phagocytosis, by attaching an antibody **Opsonin** and the process called **Opsonization** and acquire positive charge.
4. The body has the specific power to recognize certain foreign materials with the help of immune system. The immune system develops antibodies opsonin against infections agents like bacteria. These antibodies adhere to the bacterial membranes, to make them susceptible to phagocytosis. After adherence to the bacteria, opsonin combines with an immune factor present in the plasma called **Complement**.
5. Phagocytosis of bacteria requires the presence of divalent cations and is prevented by calcium chelating agents.
6. Phagocytosis occurs in the pH range 6.0–8.0.
7. Phagocytosis is not dependent on oxygen.

9.9.1.2 Sequence of events in phagocytosis

1. **Chemotaxis**: When bacteria enter the body, the bone marrow is stimulated to produce and release more neutrophils. Bacterial products react with plasma factors and of complement

system (C5a), leukotrienes and polypeptide from lymphocytes, mast cells and basophil attract neutrophils towards the infected area and the phenomenon is called **Chemotaxis**.

2. **Opsonization**: The bacteria is coated with other plasma factors called Opsonin, an immunoglobulin (IgG) and complement proteins to make them sensitive for phagocytosis and the process is known as **Opsonization**.
3. **Adsorption on the Membrane Receptor**: The immunoglobulin or antibody coated bacteria adsorbed on the receptors on membrane of phagocytic cells.
4. **Polarization of Cell Membrane**: Hyper-polarization of cell membrane, followed by depolarization and slow repolarization in 5-10 seconds and there is increased free calcium in cytoplasm.
5. **Projection of Pseudopodia**: Then the cell projects pseudopodia around the particle and the pseudopodia meet with each other on the opposite side and fuse. This creates an enclosed chamber containing the phagocytized material.
6. **Phagosome**: Then the enclosed chamber invaginates inside the cytoplasm and the portion of the cell membrane that surrounds the phagocytized particle, breaks away from the outer cell membrane to form a free-floating **Phagocytic vesicle** or **Phagosome** inside the cytoplasm. This process of endocytosis of antigen-antibody complex with release of superoxide ions takes place in 30 seconds.
7. **Digestion of Material**: The phagosome migrates towards the centre of the cell where they fuse with lysosome, permitting the contents of the vacuole and the lysosome to mix up with common membrane. Some of the products of the digestion of the engulfed material are absorbed through the walls of the vacuole and the remnants (indigestible product) are rejected and are expelled by the vesicle moving to the surface of the cell and fuses with the plasma membrane and comes out of **Exocytosis** (Fig. 9.27).

9.9.1.3 Biological importance

1. Phagocytosis is important for elimination of bacteria as well as removal of inanimate particles like antigen-antibody complex.
2. Carbon particles, sodium urate crystals are taken up by neutrophils (microphage), which act as **Scavengers.**
3. Eosinophil is very weak in phagocytic action but they attack some parasites.
4. Monocytes (macrophages) like neutrophils are active phagocytic.
5. Platelets are phagocytic. They engulf carbon particles, immune complexes, and virus particles.
6. The lymphoid tissue and reticulo-endothelial cells (macrophages) participate in defence reactions against toxins (diptheria, tetanus), bacteria and larger parasites by the formation of antibodies and by phagocytosis.

9.9.1.4 Disorders of phagocytes

1. **Neutrophil hypomotility syndrome**: Action in neutrophils does not polymerize normally and neutrophils move slowly.
2. **Chronic granulomatous disease**: There is a failure to generate O_2 in both neutrophils and monocytes and consequent inability to destroy many phagocytosed bacteria.
3. **Severe congenital glucose: 6-phosphate dehydrogenase deficiency**: There are multiple infections because of failure to generate the NADPH necessary for O_2 production.
4. **Myeloperoxidase deficiency**: Microbial killing power is decreased but not absent because most of the oxygen-dependent bacteriocidal mechanisms remain intact.

9.9.1.5 Discrimination of phagocytosis

Phagocytosis is a selective process because the cells select the material which is suitable for them and reject the others.

the plasmalemma into two layers, but high-resolution preparations revealed that the particles lining the vesicle had never penetrated into the more superficial membrane, but the plasmalemma enmeshed in the filamentous extensions to this membrane. The whole membrane complex had been turned inside-out; so, the vesicle contained the innermost layer, the filamentous extension with the particle enmeshed while the outermost layer of the vesicle was the plasma membrane, which could be resolved into two dense lines separated by a less dense space.

Brandt and Pappas (1962) studied the formation of vesicles during absorption of ferritin. They observed that the funnels were formed from the surface by invagination and their linings were of filamentous extensions, which were closed to form vesicles.

Chapman, Anderson and Holter (1955) when incorporated C^{14}-glucose into the vesicles by placing this in the medium in addition to protein, though glucose penetrates into amoeba very slowly without pinocytosis but radioautography showed that the glucose left the vesicles soon after they were formed which suggest that the plasma membrane becomes more permeable to sugar, after pinching off to form vesicle. Policard and Besis (1958) observed that, when ferritin particles (ion oxide particles coated with protein) were absorbed by the erythroblasts, there were similar vacuolization and they called it **Ropheocytosis**.

9.9.2.3 Mechanism

Protein molecules or other macromolecules attach to the membrane by the process of adsorption. presence of the protein changes the surface properties of the cell membrane for invagination forming tiny **Pinocytic channel**. From the inner end of each channel the small vacuoles are pinched off immediately after closing the membrane over the proteins and form the **Pinocytic vesicles** or **Pinosomes** or **Endosomes** and move towards the centre of the cell. The content of pinocytic vesicle faces the "outside" surface of the cell membrane. For formation of pinocytic vesicles as well as for phagocytic vesicle the following factors are involved.

1. It requires energy from ATP in the cell.
2. Presence of calcium ions in the extracellular fluid is necessary.
3. Contractile function of the microfilaments beneath the cell membrane.
4. Contact of the macromolecules may change the surface tension of the membrane for invagination.
5. Microfilaments stimulated by calcium may provide force for pinching the vesicle away from the cell membrane.

As soon as pinocytic vesicle appears inside the cell, one or more primary lysosomes become attached to the vesicle and fuses with it and empty their hydrolases into it to form **Digestive vesicle**. In the digestive vesicle, hydrolases start hydrolizing the proteins, glycogen, lipids, nucleic acids, mucopolysaccharides. The product of digestion, e.g., small molecules of amino acids, glucose, fatty acids, phosphates diffuse through the membrane into the cytoplasm. The remaining structure of digestive vesicle is called **Residual body** which contains undigestible products which is excreted out of the cell by exocytosis (Fig. 9.28).

9.9.2.4 Advantage of pinocytosis

Pinocytic channels provide a means by which extracellular fluid with dissolved substances having high molecular weight, which cannot diffuse through the cell membrane, can be brought into the body of the cell and their breaking up into numerous vacuoles helps distribution and increases the surface area across which, absorption can take place.

9.9.2.5 Biological importance of pinocytosis

1. **Kidney Tubular Reabsorption**: Reabsorp-tion of peptides and proteins from the proximal tubular fluid occurs predominantly by pinocytosis.
2. **Genetic Transformation of Bacteria**: The genetic transformation of bacteria by transforming DNA.
3. **Virus Infection**: The infection of cells by viruses.
4. **Penetration of Antibodies**: The penetration of cells by antibodies.

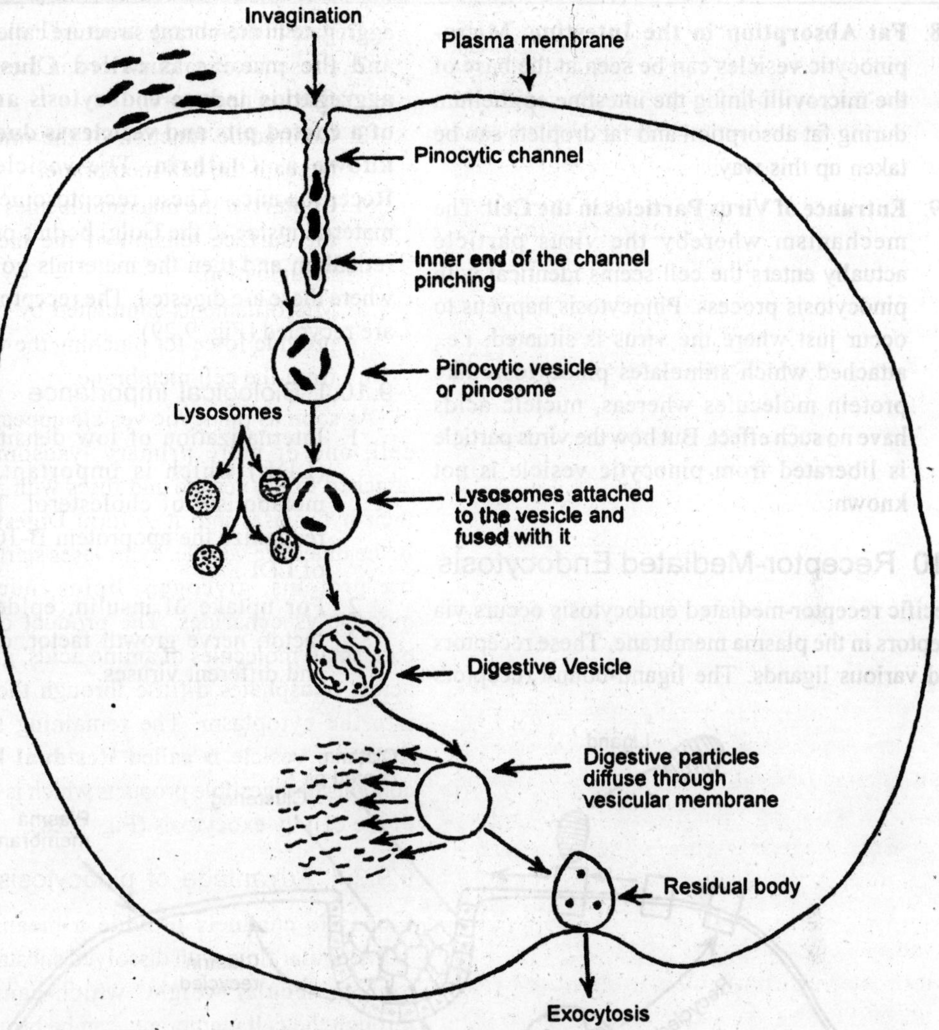

Fig. 9.28: Schematic diagram of pinocytosis

5. **Proteins and Strong Electrolytes Transport**: The two most important substances, i.e., proteins and strong electrolytes, are transported by this process because it is the only mechanism by which proteins can pass through the membrane.

6. **Transport Through Capillaries**: The intercellular cleft of the capillaries has a width of about 6-7 nanometres, while plasma proteins have minimum diametre of about 8 nanometres and cannot pass through intercellular cleft. Other larger substance in the plasma, e.g., lipoprotein molecules, dextran and proteoglycan molecules also cannot pass through capillary membrane by pincytosis revealed by electron microscope. The mechanism is that the endothelial cells of the capillary wall take up small amount of plasma and form small intracellular pinocytic vesicles which migrate from the inner surface of the endothelial cells to the outer surface where the plasma and its contents are released. Sometimes these vesicles move from one side of the endothelial membrane to the other side of the membrane in a few seconds.

7. **Other Macromolecule Transport**: Other macromolecules other than proteins transported by pinocytosis are peptides, glocygen, lipids nucleic acids, mucopoly-saccharides, etc.

8. **Fat Absorption in the Intestine**: Micropinocytic vesicles can be seen at the base of the microvilli lining the intestine epithelium during fat absorption and fat droplets can be taken up this way.

9. **Entrance of Virus Particles in the Cell**: The mechanism whereby the virus particle actually enters the cell seems identical with pinocytosis process. Pinocytosis happens to occur just where the virus is situated, i.e., attached which stimulates pinocytosis like protein molecules whereas, nucleic acids have no such effect. But how the virus particle is liberated from pinocytic vesicle is not known.

9.10 Receptor-Mediated Endocytosis

Specific receptor-mediated endocytosis occurs via receptors in the plasma membrane. These receptors bind various ligands. The ligand-bound receptors aggregate in membrane structure called **Coated pits** and the process is called **Clustering. The aggregation induce endocytosis and formation of a coated pits and vesicles is due to a protein known as Clathrin.** The vesicles are called **Receptosomes.** These receptosomes transfer the material inside to the Golgi bodies or endoplasmic reticulum and then the materials go to lysosomes where these are digested. The receptors and clathrin are recycled (Fig. 9.29).

9.10.1 Biological importance

1. Internalization of low density lipoprotein (LDL) which is important for cellular metabolism of cholesterol. The receptors recognize the apoprotein B-100 component of LDL.
2. For uptake of insulin, epidermal growth factor, nerve growth factor, diptheria toxin and different viruses.

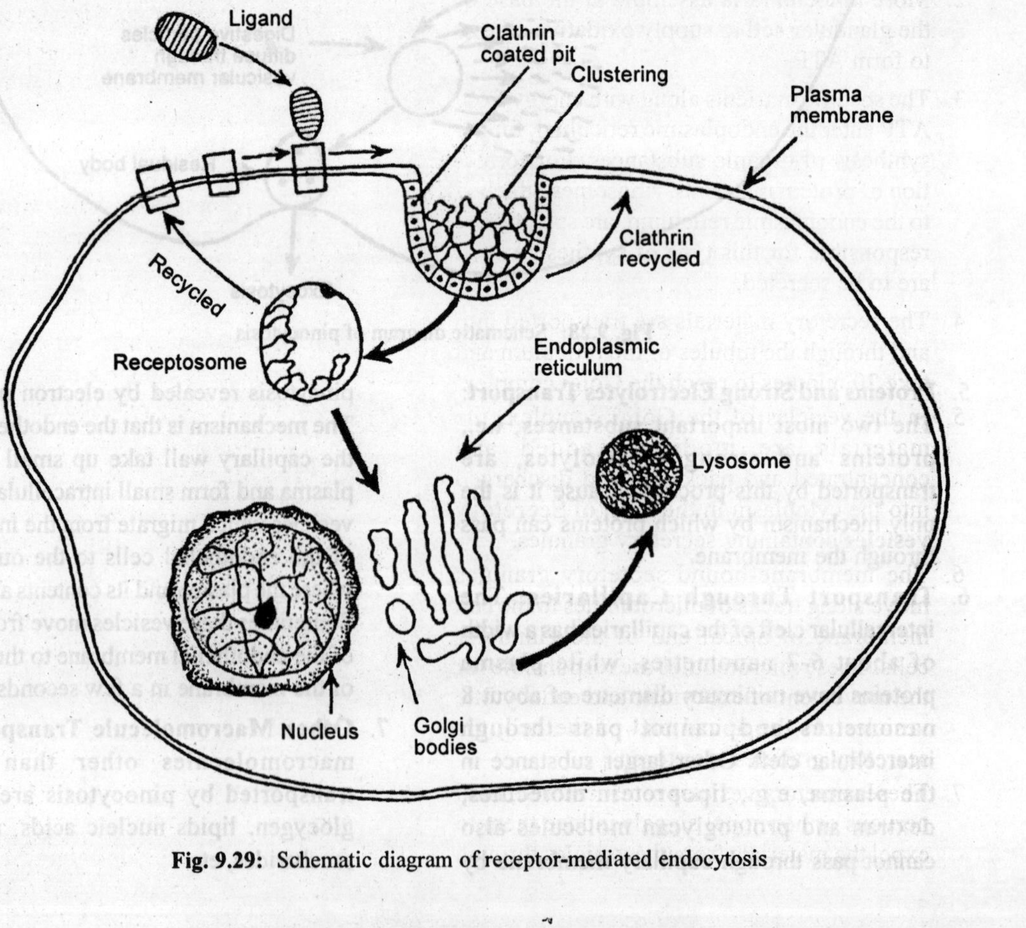

Fig. 9.29: Schematic diagram of receptor-mediated endocytosis

3. **Cytopempsis or Vesicular Transport**: It is important for transport of substances across epithelial cells. There are numerous vesicles in the capillary endothelium. Tagged protein molecules have been found in these vesicles, indicating that proteins of plasma are transported out of the capillaries across endothelial cells by endocytosis and then by exocytosis. This transport mechanism is called **Vesicular transport** or **Cytopempsis**.

9.11 Secretion and Exocytosis

The secretory granules which are expelled outside the cells, the steps in this process is called **Exocytosis** or **emeiocytosis** (cell vomitting) or **Reverse endocytosis**.

9.11.1 Mechanism

1. The nutrients necessary for secretion, enter the cell by diffusion or active transport from the plasma in the capillary into the base of the glandular cells.
2. More mitochondria assemble at the base of the glandular cell to supply oxidative energy to form ATP.
3. The specific nutrients along with energy from ATP enter the endoplasmic reticulum, for the synthesis of organic substances. For formation of protein materials, ribosomes attached to the endoplasmic reticulum, are specifically responsible for this type of synthesis which are to be secreted.
4. The secretory materials are transported into and through the tubules of the reticulum and take 20 minutes to reach the Golgi complex.
5. In the vesicles of the Golgi complex, the materials are modified, added and concentrated and packaged and discharged into the cytoplasm in the form of **Secretory vesicles** containing secretory granules.
6. The membrane-bound secretory granules move along tracks of microtubules to the cell membrane of apical end of the glandular cells. These microtubules are responsible for the movement of exocytotic and endocytotic vesicles. At the apical end of the cell, the secretory granules are stored.
7. These secretory vesicles remain stored, until nervous or hormonal signals induce them to expel the materials from the vesicles through the cell surface. The sequence of events are, the signal increases the permeability of the cell membrane to calcium for entering the cell which, in turn, initiates the secretory vesicle to fuse with the cell membrane.
8. The secretory vesicles attach to the cell membrane fuse (fusion), the area of fusion breaks down (elimination of layers and tissues), leaving the contents of the secretory vesicle, i.e., secretory granules outside the cell and the cell membrane remains intact, this process is called **Exocytosis**. The secretory granules are extruded onto a surface either directly or by means of a duct (exocrine glands) while others, the internally secreting or endocrine glands, pass their secretion directly into the blood or lymph.
9. The mechanism of break-down of the membrane is not known. However, according to chemiosmotic hypothesis, an anionic channel opens into exocytotic vacuole and that the influx of anions produces an increase in osmotic pressure, leading to dissolution of the area of membrane fusion (Fig. 9.30).
10. The method of release of the secretory product from the gland cell is sometimes used to classify glands.
    (a) **Merocrine Gland**: In the merocrine type of secretion, the product is released by the process of exocytosis without the loss of any cellular cytoplasm. This type of secretion is found in the majority of glands, e.g., pancreas, salivary gland.
    (b) **Apocrine Gland**: When the secretion accumulates in the apical part of the cell which is later pinched off to release the product, then mode of secretion is called apocrine e.g., lactating mammary gland.
    (c) **Holocrine Gland**: When the whole of cell together with its accumulated secretion is shed, then the gland is called Holocrine type (e.g., sebaceous glands) having high mitotic rate to replace the gland.
11. **Calcium Dependent Exocytosis at the Nerve Endings**: The hormones of the posterior pituitary gland (vassopressin and oxytocin) are synthesized in the cell bodies of neurons in the supra-optic nucleus and

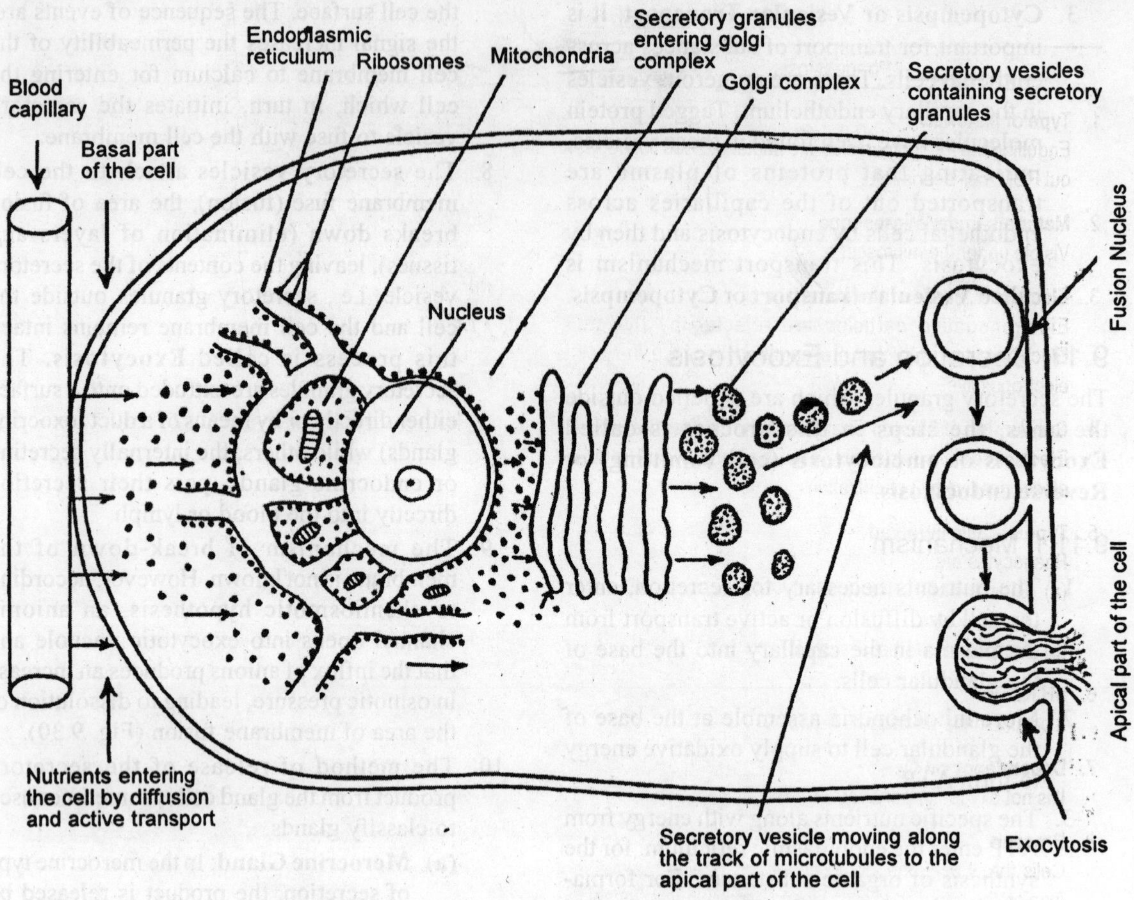

Fig. 9.30: Schematic diagram of secretory pathway and exocytosis

paraventricular nucleus. Some of the neurons make oxytocin and other make vasopressin. Oxytocin-containing and vasopressin-containing cells are found in both nuclei. The precursors molecules are synthesized in the ribosomes of the neurons. They are packaged into secretory granules in the Golgi apparatus and are transported down the axons by axoplasmic flow to the nerve endings in the posterior pituitary and these **secretory** granules are called **Haering bodies**, and are easy to stain. These cleavage precursors as secretory granules reach the nerve endings called **Storage granules** and are stored as free oxytocin and vasopressin loosely bound with corresponding neurophysin. The neuron also conduct action potentials and action-potential reaching the axon endings trigger release of the hormones by **Calcium-dependent exocytosis.**

9.12 Absorption

Absorption is the passage of various substances through a layer of cells into the blood.

Passage into the blood may be from:

1. Various parts of alimentary tract,
2. Internal organs, e.g., bladder, uterus, bronchi, lungs,
3. Body surface, e.g., skin,
4. Body cavities, e.g., pleural cavity, peritoneal cavity,
5. The subcutaneous cellular tissue,
6. Muscle,
7. Lung alveoli,
8. CSF by arachnoid villi.

I. Absorption in the G.I. Tract

Absorption of substances from the intestines is of greatest **Physiological importance** for:

(a) Nutrition,
(b) Oral method of administering medicine,

Table 9.8: Comparison between phagocytosis and pinocytosis

Phagocytosis	Pinocytosis
1. **Type of Materials** Engulfing the foreign particles like bacteria, dead tissue, worn out RBC, cell debris, etc.	1. Engulfing the extracellular fluid with dissolved substance, e.g., proteins, strong electrolytes.
2. **Materials under Microscope** Visible under the microscope.	2. Materials are dissolved in solution hence, not visible under microscope but are visible under electron microscope.
3. **Electrical Charge of the Materials** Electronegative particles are rejected by the cells. Electropositive particles are engulfed because the cell is electropositive.	3. Electrical charge is not involved.
4. **Conversion of Electric Charge of Materials for the Process** Electronegative materials, e.g., bacteria which is made electropositive by opsonization.	4. No such process occurs.
5. **Type of Cells involved** Phagocytes are: (a) Blood cells: Neutrophil, Monocytes, Platelets, Lymphocytes. (b) Reticulo: Endothelial system.	5. No classified cell is involved.
6. **pH Range** It occurs in the pH range 6.0-8.0.	6. No specific pH range and occurs at cellular pH.
7. **Dependency on O_2** It is not oxygen dependent.	7. Oxygen dependent.
8. **Process** Cells throw pseudopodia to engulf the foreign material and fuse on the opposite to form closed chamber, then invaginate and pinched off to form phagosomes or phagocytic vesicles.	8. Materials, e.g., proteins adhere to the membrane by adsorption and by invagination of cell membrane, pinocytic channel is formed. The inner end of the channel is pinched off to form pinocytic vesicle or pinosome.
9. **Fusion with Lysosome** Phagosome fuses with lysosome for digestion.	9. Pinosome also fuses with lysosome for digestion.
10. **Fate of Digested Material** Absorbed by the cytoplasm.	10. Absorbed by the cytoplasm.
11. **Fate of Undigested Material** Expelled by the vesicle by exocytosis.	11. Expelled through residual body by exocytosis.
12. **Calcium** There is increase in free calcium in the cytoplasm.	12. Calcium is required.
13. **Advantage** One of the defence mechanism for fighting bacteria, elimination of other foreign materials.	13. Helping metabolism of macromolecules which cannot diffuse cell membrane and distribution and increases the surface area, across which absorption takes place.

The rate of absorption depends on:

(a) Histological structurue of intestinal part, e.g. villi having greater surface area,

(b) The length of time, the substances are retained in the intestine,

(c) Composition of the substances,

(d) State of the substances, e.g. diffusible or not, soluble or insoluble.

- **Mouth and Oesophagus**

Absorption from the mouth and oesophagus is

practically negligible, as regards food. Because food is not broken by enzyme, the mucosa is not suitable, the food is not retained for sufficient time.

Atropine, nitroglycerine (drug for heart attack given sublingually) and epinephrine are absorbed for pharmacological action. The best site is below the tongue (sublingually) and between gums and cheek.

- **Stomach**

Substances are not absorbed by the stomach due to absence of villi-like structure and the junction between the epithelium is very tight. But these simple substances which do not require digestion at all may pass through the gastric epithelium. Hence, water, alcohol, saline, glucose and certain easily diffusible drugs are absorbed by stomach to some extent.

- **Small Intestine**

It is the chief place for absorption. From 5 to 8 hours after eating most of the chyme (semi-digested food) passing from the small into the large intestine hardly contain any absorbable substances except water, which is vigorously absorbed from large intestine.

Different portions of the small intestine vary in their ability to absorb difrerent substances. For example, in the upper position, sugar is absorbed faster than water, in the lower position water is absorbed faster than sugar. NaCl is absorbed in the lower portion even faster than water. Fat and bile salts are absorbed in the ileum. The facility of absorption is due to:

1. Greater length of the small intestine which has provided big surface area,
2. Presence of villi, which has increased the absorptive surface area, further. (In man there are about 18-40 villi per sq. mm. and each villus is 0.2 to 1mm in height. The total absorptive area is about 60 sq. feet, i.e., 1.5 time the surface area of the body.)
3. Presence of brush-border epithelium which is nothing but microvilli, has further increased the surface area and the property of epithelium is absorption.
4. The villus has both arterial and lymphatic vessel and its tremendous ramification at the tip of villus, help absorption of non-lipid and lipid substances.
5. *Lymphatic absorption*: The lymph vessel which drain the intestinal walls serve another special function. Fats after digestion and absorption pass mainly into the lymphatics.

- **Large Intestine or Colon**

The colon absorbs water, salts and glucose, for which, in medicine, rectal infusion of saline, glucose solution or gluco-saline are given in certain cases. The faeces contains only about 1/3rd of the water, which enters the large intestine. The human colon absorbs 80 cc of water per hour. Water soluble drugs including anaesthetics, even ether are absorbed from the colon. Absorption of glucose is only about 1/20th, as rapid as in the small intestine. Certain drugs are absorbed from the rectum and are given as suppositories or small retention enema (e.g., sedatives such as barbiturates and chlora hydrate).

9.12.1 Factors or mechanism or forces for absorption

A. Physical factors
B. Biological factors
C. Chemical factors
D. Anatomical and histological factors

A. Physical Factors

1. **Diffusion**: Diffusion is a passive or physical process by which small molecules or ions pass through the pores of a permeable membrane, from a higher concentration to a lower concentration, i.e., along a concentration gradient. For intestinal absorption by diffusion, the concentration of the substance must be higher in the intestinal lumen, than in intestinal cells or blood—greater is the difference between these two concentrations, higher is the rate of absorption by diffusion. Moreover, particles of the substance must have smaller diameters than the pores of intestinal cell membranes. Pentose and D-amino acids are absorbed by diffusion. Diffusion does not require energy from cell.
2. **Carrier Mediated Diffusion or Facilitated Diffusion**: In this process of diffusion, a specific carrier molecule present in the cell membrane which is protein in nature, participates in the transport of the food material across a membrane. It does not act along concentration gradient like ordinary diffusion and again it does not require energy from ATP in the cell like active transport, which also requires a carrier molecule (see Art. 9.2).
3. **Active Transport or Active Absorption**: Active transport or absorption always

requires cellular work and energy expenditure and takes place irrespective of concentration and electrical gradient. It requires a carrier molecule in the membrane for transport of food substance, e.g. Na^+, K^+, sugar, amino acids. Substances combine with carrier to form substance-carrier complex, which at the base of cell releases substance and carrier is set free which is again utilized (Fig. 9.31) [See Art. 9.4].

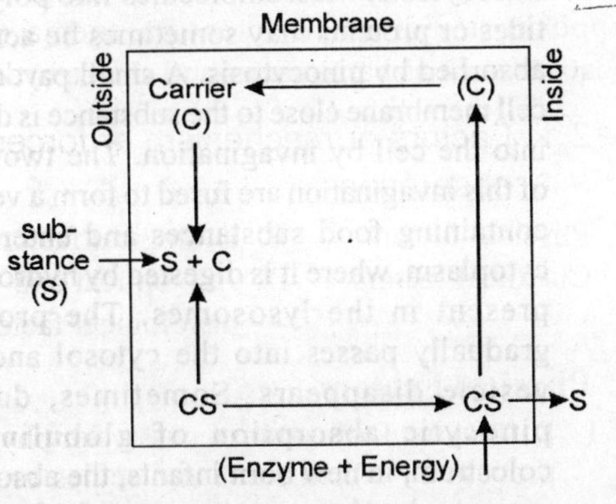

Fig. 9.31: Basic mechanism of active transport or active absorption

4. **Hydrostatic Pressure**: In the intestine, there are forces capable of exerting a certain mechanical (hydrostatic) pressure. These include:
   (a) Action of abdominal wall,
   (b) Intestinal peristalsis,
   (c) Suction movements of the villi,
   (d) **Intestinal contents at the height of digestion.**

   Humberger found that, amount of absorption is directly proportional to the hydrostatic pressure. This higher pressure causes filtration of water and dissolved substances across intestinal epithelium. The intra-intestinal pressure is 4 to 6 mm Hg during digestion and blood pressure in the capillaries of the villi is higher (8-15 mm Hg), hence Starling concluded hydrostatic pressure does not ensure absorption. However, there is no constant hydrostatic pressure in the small intestine and the mechanism of filtration is rather a supplementary factor.

5. **Filtration**: The mechanism of filtration consists in the passage of a solution through pores under the action of some hydrostatic (mechanical) pressure. The rate of filtration depends on the amount of this pressure and the size of the pores. If the dissolved particles are larger than the pores, they do not filter. If filtration occurs, it results in similar concentrations of the solute and the solvent on both sides of the membrane.

6. **Osmosis and Osmotic Pressure**: Osmosis is the process of diffusion through a semi-permeable membrane which is permeable to the solvent and impermeable to solutes and depends upon the concentration of a solute on the two sides of the semipermeable membrane. The pressure exerted on the membrane during osmosis is called Osmotic pressure. The colloidal osmotic pressure in the capillaries of the villi is 2-3 times higher than the blood pressure, helps in drawing fluid from the lumen of intestine like the venous capillaries absorbing water and metabolites from tissues.

7. **Adsorption**: This process is involved during absorption where certain substance is carried along with a carrier where the substance is adsorbed on the surface of the carrier.

8. **Hydrotrophy**: Hydrotropy is the process where the insoluble substances in water are converted into soluble forms. The agent which renders the insoluble into soluble is called **Hydrotropic agent**. e.g. bile salts help the fatty acid to be in soluble form. Diffusion and osmosis can only act if substances are in a dissolved state. The hydrotropic action of a substance is not general and it may be a specific, e.g., bile salts only for fatty acid and Ca salts and sterols.

9. **Electrical Effects (Electrical Gradient)**: Heber suggested that a process of cataphoresis, i.e., the transfer of a solute across a membrane from anode to cathode may be set up during absorption. The cations will move towards having negative potential charge and the anions towards having the positive potential charge. In addition to this, the fluid exerts effect on the solute particles which is known as **Solvent drag phenomenon.**

10. **Electrical Difference of Potential**: All asymmetric membranes, including the intestinal mucosa show a difference of electrical potential between the two sides. This factor probably plays an important part

in the transference of dissolved substances across membranes in the body. Just as differences of hydrostatic pressure, determine movements of a fluid en masse, differences of electric potential will serve as a driving force for the charged constituents of fluids.

11. **The Influence of Surface Activity**: Substances with high surface activity and substances capable of changing the surface activity of cell membranes, may profoundly affect absorption. Thus saponin greatly increases the absorption of substances from intestine. The substances are called **Wetting agents** such as **Aryl** and **Alkyl sulphates**, also increase intestinal permeability to certain substances.
12. **Swelling**: The permeability of membrane is influenced by the degree of swelling, i.e., by the hydration of the colloids in the membrane. The sulfates are among those electrolytes which dehydrate colloids thereby, decreasing the absorption of other salts. This may be the mechanism of the laxative action of sulphate.
13. **Solubility in the Cell Membrane**: The membrane is a bimolecular layers of lipids sandwiched between the two layers of proteins. The lipid layers has "hydrophilic group" and "hydrophobic group". The substance whether lipids soluble or water soluble and the solubility in the cell membrane is a factor in absorption. Substances insoluble in cellular lipoids, are absorbed only through the intercellular spaces while the lipid soluble substances (e.g., urea, glycerol, alcohol), are absorbed by the epithelial cells.
14. **Surface Tension**: Surface tension is also another factor for absorption. During fat digestion and absorption bile salts emulsify fat (i.e., detergent function) by lowering surface tension which helps better lipase action and micelle formation for absorption).
15. **Donnan Membrane Equilibria**: This factor also plays an important role in absorption.

B. Biological Factors

1. **Permeability of Membrane**: The nature of the membrane cannot add to the force concerned in the transference of water and dissolved substances across it, it can only serve as a limiting condition for the action of the forces acting for absorption. How these forces will act in regard to different kinds of membrane will depend on the permeability of the membrane, and this again on the physical and chemical structure.
2. **Phagocytosis**: Phagocytosis by leukocytes, may play a part in the absorption of particulate matter such as dyestuffs, carbon, etc. It is very doubtful whether this is important or not.
3. **Pinocytosis**: Macromolecules like polypeptides or proteins may sometimes be actively absorbed by pinocytosis. A small part of the cell membrane close to the substance is drawn into the cell by invagination. The two ends of this invagination are fused to form a vesicle containing food substances and enters the cytoplasm, where it is digested by hydrolases present in the lysosomes. The product gradually passes into the cytosol and the vesicle disappears. Sometimes, during pinocytic absorption of globulins of colostrum, in new born infants, the absorbed macromolecules may pass as such, from the pinocytic vesicle into the blood without an intracellular enzymatic intervention (See 8.9.2).
4. **Influence of Nerves**: Impulses from the autonomic nervous system can alter the rate of absorption from the intestine. How much these effects depend on vasomotor changes and changes in the movement of the villi is not known.
5. **Influence of Blood Circulation**: The rate at which substances can diffuse from the intestine will be directly affected by the rate at which they are carried away by the blood in the vessels of the intestinal wall. Furthermore, the very sensitive mucosa alters its physical and biological properties in response to changes in blood flow.
6. **Movement of the Villi**: Rythmic contractions of the villi, i.e., pumping actions occur during absorption. Not all the villi move at once; some remain quiet for long periods while their neighbours are steadily contracting and relaxing. The normal rate of rhythmic, contraction, i.e., pumping movements is about 6 a minute. The villi shorten quickly and lengthen more slowy. The villi become shorter but not thicker on contractions. The

epithelium can be seen to become wrinkled as the villus contracts.

The factors responsible for villi contraction are:

(a) **Mechanical**: A villus can be stimulated to contract, by touching a horse-hair to its base but stimulation of the tip is ineffective.

(b) **Chemical**: Chemical stimuli found in the intestinal cavity, act on the villi with the aid of the nervous structures in the intestine. The chemical stimuli are yeast extracts, peptone solutions, histamine, bile acids, a number of amino acids, several pharmacological substances, e.g., calomel. The movements are inhibited by the salts of calcium and potassium.

(c) **Hormonal or Humoral**: Mucous membrane of the intestines produce a hormone which influences the movement of villi is known as **Villikinin**. This is neither liek **Histamine** nor like **Secretin**.

(d) **Nervous:** Villi move under the influence of nerve impulses, which may arise not only in the central nervous system but also in Meissner's plexus. Touching the top of villus with a hair produce no effect, whereas touching its base causes contraction. The movement of villi causes acceleration of absorption because absorption takes place even after paralysing the villi.

By contraction and expansion, the villi press blood and lymph out of themselves, now suck in the substances dissolved in the intestines. It is probable that the absorption of substances, such as fat, which enter mainly through the lacteals, is more greatly influenced by villi movements than water and crystalloids.

7. **Concentration of Substances in the Blood**: Iron absorption is regulated by the concentration of it in the blood. If the concentration in the blood is normal and there is no deficiency in the body, no absorption takes place. Again, if there is deficiency and low concentration in the blood, there is absorption. The high concentration of iron in the intestine also inhibits iron absorption.

8. **Presence of Carrier Substance in the Cell and Membrane**
   (a) Presence of carrier substance which is a protein in nature helps absorption of Na^+, K^+, Ca^+, sugar, amino acids during "carrier mediated diffusion" and "active transport". This substance is present in the membrane.
   (b) Presence of apoferritin and its concentrates in the cell help iron absorption.
   (c) *Adenly cyclase*: Its presence in the membrane helps in absorption by hormones.

9. **Selective Absorption**: When diffusibility of various sugars are compared through non-living membrane, e.g., cellophane, there is no difference but in the intestine it is seen that glucose is faster than fructose and galactose and the rate of absorption is glucose > fructose > galactose > mannose. This indicates that, some special mechanism is provided in the gut epithelium for the absorption of glucose in faster rate, than the other sugars is referred to as **Selective absorption.**

10. **Dependency on other Substance**
    1. Vit. A, D, E, K are absorbed in presence of fat.
    2. Fat is absorbed in presence of bile salts.
    3. Calcium is absorbed in presence of PO_4, fat, etc.
    4. Glucose and Sodium relationship during absorption.

C. Chemical Factors

Beside other factors, there is chemical factor which plays a role inside the epithelial cells during absorption. This may be explained from the following:

1. Rapid absorption.
2. Selective absorption,
3. Absorption of glucose and NaCl from hypotonic solutions,
4. Oxygen consumption during absorption,
5. Expenditure of energy, etc.

It is evident by the fact that, when the temperature of a loop of intestine is reduced to 0°C, preferential absorption of glucose is inhibited. All these facts indicate that beside other factors, there are some chemical factors involved during absorption.

D. Anatomical and Histological Factors

1. **Length of the Small Intestine**: The length of the intestine, helps a lot for absorption as the substances take time for passage and to allow the enzymes to act for complete digestion.
2. **Presence of Villi**: The folding of mucus membranes called Villi, increase the surface area for absorption.
3. **Presence of Columnar or Brush-Border Epithelium**: The cells of the absorption epithelium are found to be longitudinally striated in their free border. This type of bursh-border epithelium is also found in the kidney tubules whose function is also absorption. Under Electron Microscope, it is seen that, brush-border is villi like projection called **Microvilli** which increases effective absorbing surface.
4. **Double Vascular Arrangement**: The villi have enormous capillary complex or net-work, which help absorption of water and crystalloid whereas lymphatic arrangement i.e., lacteals) is for fat absorption.
5. **Presence of Pores, Gates, Pumps and Recptors**: The structures present in the membrane like pores, gates, pumps and receptors, play some essential role during absorption.

- **Interelationship of Absorption with Digestion**

Digestion should be adequate in all respect, for suitable absorption. Digestion makes the substances smaller, soluble and permeable for absorption. Again, absorption should start effectively unless there will be disturbances, e.g.,

1. Increase in osmotic pressure in the lumen of intestine due to accumulation of particles which can draw fluid from intestinal wall,
2. Enzymes, which hydrolyze the particles may proceed reversibly for which end products should be absorbed quickly. So it can be concluded that, digestion and absorption are interdependent.

- **Intestinal Absorption Vs. Absorption in the Alveoli or Capillaries**

Intestinal absorption and absorption in the alveoli or capillaries are not similar. Though some processes are common there are some differences.

I. Intestinal Absorption

Intestinal mucosa containing the epithelial cells, have the capability to take up very dilute substances from the lumen of intestine and to pass these to blood of the capillaries, where the concentration of the substances may be higher than the luminal concentration, i.e., absorp-tion takes place against the diffusion gradient or water against osmosis.

II. Absorption in the Alveoli of the Lungs

The oxygen concentration in the alveoli is higher than the blood in the lung capillaries. So, there is diffusion of oxygen, from alveoli to blood of the capillaries, i.e., a diffusion gradient is there for continuous transport of oxygen in the same direction. In this process the alveolar cells and the capillary cells are passive and their structures are also adapted for the process e.g. flat, very thin and differ from intestinal epithelial cells (brush-border) or kidney tubular epithelium, which participate actively in the absorption.

Blood Capillaries of Intestine	Blood Capillaries of Alveoli of the Lungs
Movement of materials and absorption against diffusion gradient in the intestine.	Movement of gases and absorption in the direction of diffusion gradient in the alveoli of the lungs.

Similarly, diffusion takes place in case of CO_2, from blood capillaries to the alveolar cavities of the lungs. Because the concentration of CO_2 in the blood capillaries is higher than the alveolar cavities, which is also passive process, the cells of alveoli and capillaries do not have active role.

III. Capillaries of the Tissue of the Body

Diffusion takes place in the reverse direction in the capillaries of the tissue of the body. Oxygen goes out of the capillaries (due to high concentration) to the tissues where carbon dioxide goes to the capillaries from tissues (due to higher concentration of CO_2). Thus, in the body, between lungs and tissues, the respiratory gases move in the direction of decreasing diffusion gradient. The oxygen gradient decreases from alveoli to blood, to tissue cells and the decreasing diffusion gradient of CO_2 occurs in reverse direction, i.e., tissue cells to blood to alveoli of the lungs.

9.13 Reabsorption

Reabsorption means, once the material has been absorbed by the intestine or by any structure and again is absorbed in another site, e.g., glucose, water, etc., absorbed by the intestine, again absorbed by the renal tubules and goes back to circulation.

- **Sites of Reabsorption**
  1. Renal tubules, e.g., proximal tubule.
  2. Venous end of capillaries.
  3. Lymphatics.
  4. Arachnoid villi in brain.
  5. Retinal blood vessle.
  6. Reabsorption lacunae of thyroid.

I. Renal Tubules

The substances reabsorbed are:

(a) Absorbed by the intestine go to blood then filtered in the kidney for reabsorption in the renal tubules, e.g., glucose, water, Na^+, K^+, Cl^-, HCO_3^-, ascorbic acid, amino acids, etc.

(b) Some are produced in the tissue and absorbed in the blood and then filtered in the kidney for reabsorption in the renal tubules, e.g., urea, uric acid, creatinine, creatine, sulphate, ketone bodies, lactate, citrate, phosphate, etc.

Substance	Source	Percentage of Reabsorption
Water	Intenstine	99.4
Glucose	"	100
Na^+	"	99.4
K^+	"	100
Cl^-	"	99.2
HCO_3^-	"	100
Urea	Tissue	53
Creatinine	"	7
Uric acid	"	98

According to the mechanism of reabsorption in the renal tubules the substances can be grouped under three headings.

1. **Reabsorption by Active Transport**: Glucose, amino acids, bicarbonate along with Na^+, creatine, sulphate, uric acid, etc.
2. **Reabsorption by Secondary Active Transport**: Lactate, citrate, phosphate, Cl^-, etc.
3. **Passive Diffusion**: Na^+, Cl^-, etc.

II. Venous End of the Capillaries

At the venous end of the capillary, the pressure is low, which alters the balance of forces in favour of reabsorption. The driving force of 28 mm Hg. causes fluid to move into the capillary and the pressure opposing the reabsorption is 21.3 mm Hg. The difference of 6.7 mm Hg. is the **Reabsorption pressure**. This reabsorption pressure cause reabsorption of 9/10 of the fluid which is filtered at the arterial end of the capillaries and the remaining 1/10 enters into the lymph vessels.

III. Lymphatics

The one-tenth of the fluid which enters the lymphatic capillaries and returns to the blood through the lymphatic system has special significance. Because the substances of high molecular weight (e.g., proteins) cannot pass easily through the pores of venous end of the capillaries they can enter the lymphatic capillaries easily without any hindrance because of special structure.

The endothelial cells of the lymphatic capillaries are attached by filamentous structures called **Anchoring filaments**, to the connective tissue between the surrounding tissue cells.

At the junctions of adjacent endothelial cells, there are very loose connections between the two cells. The edge of one cell overlaps the edge of adjacent cell in such a way that, the overlapping edge is free to flap inward and forms a minute valve that opens to the interior of the lymphatic capillary.

Interstitial fluid containing its suspended particles pushes the valve open and flows in the lymphatic capillary. This fluid cannot leave the capillary after entering because backflow causes closure of the flap valve. Thus, the tips of terminal lymphatic capillaries have valves and they have also the valves along the whole vessel (Fig. 9.32).

IV. Arachnoid Villi of Brain

The arachnoid villi or arachnoid granulations are finger-like projections of the arachnoid membrane through the walls of the venous sinuses. These villi, passes through minute valve-like openings, directly into the blood. When the meningeal pressure is sufficiently greater than that in the sinus, the villus is open and when the meningeal pressure is below that in the sinus, reflux of blood is prevented by collapse of the villus (Wetch and Friedman, 1960). Electron microscopy has revealed, that the endothelial cells covering the

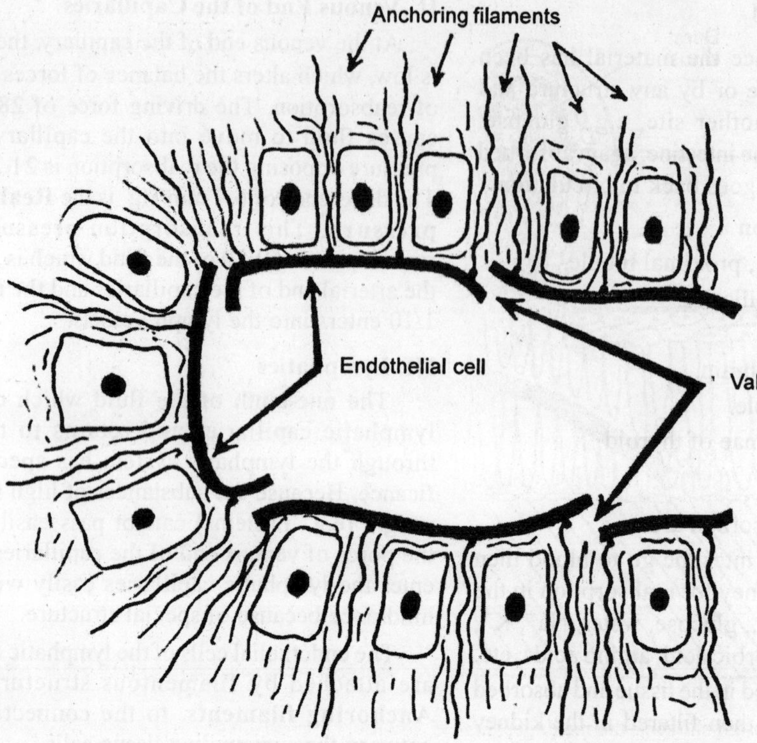

Fig. 9.32: **Structure of lymphatic capillaries showing the mechanism of allowing the macromolecules of high molecular weight (e.g., protein)**

villi, form large vesicular holes directly through the bodies of the endothelial cells. These are having enough size to pass CSF freely, protein molecules and even RBC into the venous sinuses and then to the general circulation (Fig. 9.33).

V. Intraocular Fluid (Aqeous Humour)

Aqueous humour, a clear liquid, produced in the ciliary body by diffusion and active transport and flows through the pupil to fill the anterior chamber of the eye. It is normally reabsorbed through a network of trabeculae into the Schlemm's canal, a venous channel at the junction of the sclera and the cornea, i.e., anterior chamber angle. The endothelial membrane is so porous that even large protein molecules, as well as small particulate matter, can pass from the anterior chamber into the Schelmm's. Though the canal of Schlemm's canal is a venous blood vessel, much aqueous humour normally flows into it so that it is filled only with aqueous humour rather than with blood. The small veins that lead from the Schlemm's canal to the larger veins of the eye, usually contain only aqueous humour, and these are called **Aqueous veins**. Some amount of aqueous humour diffuses through the vitreous humour and is reabsorbed by the retinal vessels.

9.14 Adsorption

9.14.1 Introduction

If a piece of charcoal is introduced into a closed space containing ammonia gas or carbon monoxide, it is observed that a good quantity of gas is taken up by the charcoal. It is proved that the gas remains on the surface of the charcoal and does not go into the interior of the solid. If the same charcoal is finely divided to produce greater surface area per unit mass, it can take up more gas. This process of accumulation of any substance on the surface of another substance is called **ADSORPTION'**.

9.14.2 Definition

Any particle such as colloidal particles possess the property of attracting and holding other substances on their surface. It is called **Adsorption.** The substance that sticks or adheres on the surface of

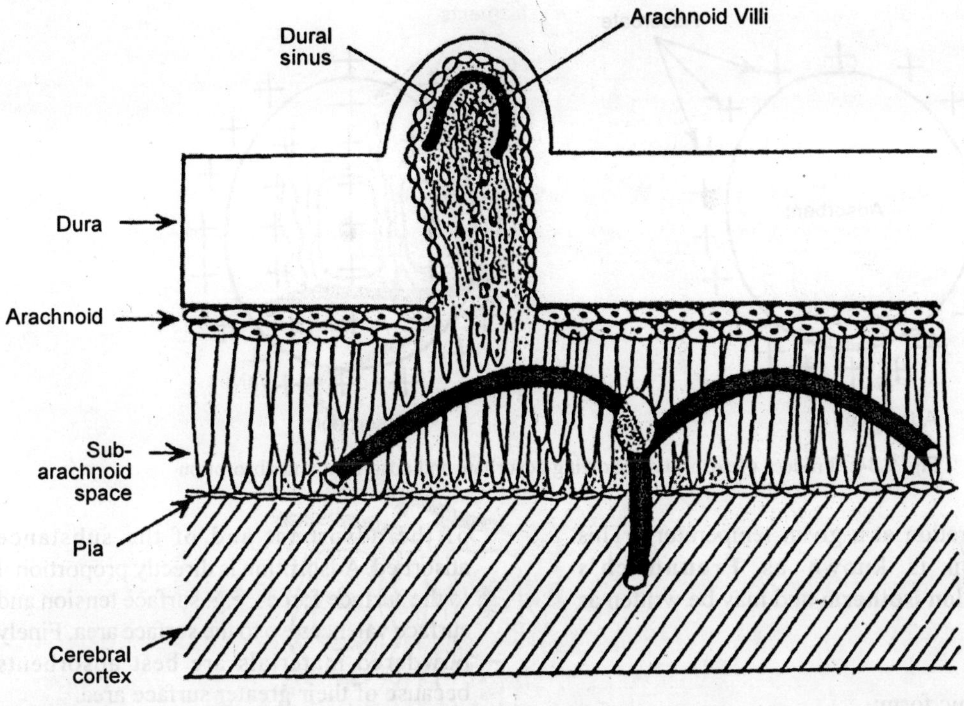

Fig. 9.33: Section of brain showing the meninges, arachnoid villi and venous sinuses for reabsorption of CSF

colloid is called **Adsorbate** and the surface on which the adsorbate settles is called **Adsorbent**. And the common surface between the two phases where the adsorbed molecules concentrate is called the **Interface**. The "SORPTION" is also used to indicate all kinds of adsorption and related phenomena (Fig. 9.34).

9.14.3 Adsorption vs. absorption

Adsorption is a fairly rapid process and surface phenomena, whereas absorption is a slow process and involves diffusion into the interior of the material, i.e., penetration into the body of the solid. Thus a chalk, when dipped in ink, adsorbs the latter and when broken it is found to be white from within. On the other hand, water is adsorbed by $CaCl_2$ crystals. Both adsorption and absorption, often go side by side and it is difficult to distinguish between the two experimentally. McBain introduced the term "Sorption" to describe the general phenomenon of gas uptake by solids (Fig. 9.34).

9.14.4 Langmuir's theory of adsorption on solids

Langmuir's contribution to the adsorption is especially important for adsorption on solids. According to him, each solid has a characteristic crystal lattice, or structure and the surface is like a checkboard with atoms arranged in orderly fashion. Each atom at the surface, has free valences or attracting forces which are capable of attaching one, and only one molecule from a solution or gas. A dynamic condition exists at the surface with molecules continually passing into and away from it. At an equilibrium, the rate of adsorption is equal to the rate of desorption.

9.14.5 Principles governing adsorption

1. **Gibbs-Thomson's Principle**: The principle states that those substances which decrease surface energy, tend to concentrate at a liquid-air interface, and those substances which decrease interfacial energy, tend to concentrate at a liquid-solid or liquid-liquid interface.

   This principle indicates that, adsorption is best from solvents of high surface tension, because here the solute causes the greatest decrease in surface energy at the interface of adsorbing solid and liquid solvent.
2. **Freundlich Equation**: An empirical equation was proposed to describe the relative amount of substance adsorbed and its equilibrium

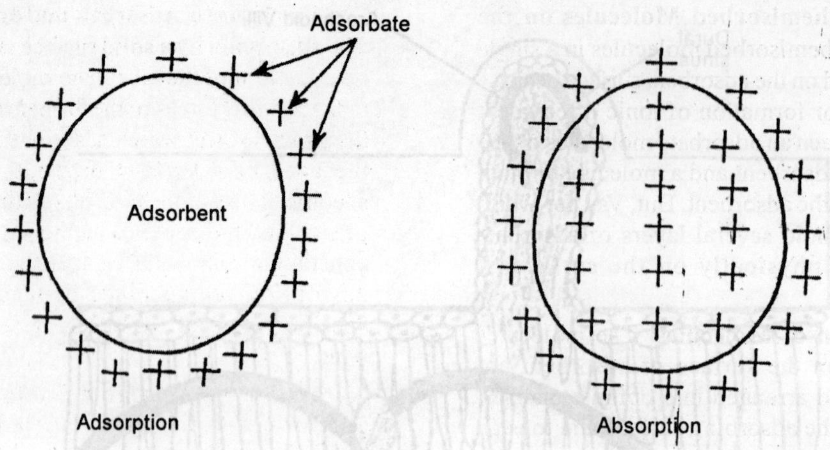

Fig. 9.34: Process of adsorption showing adsorbate and adsorbent and absorption

concentration at a given temperature. This equation is known as **Freundlich's adsorption isotherm** and may be written as below:

$$a = kC^{1/n}$$

or in logarithmic form:

$$\log a = \log k + \frac{1}{n}\log C$$

where, a = the amount of substance adsorbed per gram of adsorbent,

C = the equilibrium concentration of substance in solution (or pressure of gas in case of gas adsorption),

k and n = constants depending upon the natures of adsorbent and the substance adsorbed and the temperature.

3. **Reversible Process**: Adsorption is a reversible process and a condition of equilibrium is reached for any given concentration of solute in contact with the adsorbing agent.
4. **Temperature Rise Decreases Adsorption**: Rise in temperature decreases adsorption, which is due to the increased kinetic energy of the adsorbed molecules, which causes them to escape more readily from the adsorbing surface.
5. **Time Period for Adsorption**: Adsorption takes place quickly and an equilibrium is usually reached within one hour.
6. **Surface Area and Adsorption**: The amount of adsorption is proportional to the surface area and varies with the nature of the surface of the adsorbent and of the substance adsorbed. Adsorption is directly proportional to the surface forces, e.g. surface tension and surface valences, so to the surface area. Finely powdered materials are best absorbents because of their greater surface area.
7. **Ratio of Adsorbent to Substance Adsorbed**: Adsorption is nearly complete when the ratio of adsorbent to substance adsorbed is high, i.e., adsorption proceeds more efficiently from dilute solution.
8. **Shape of Pores of Adsorbing Agent**: In the surface, narrow pores of the adsorbing agent are more effective than globular openings.
9. **Adsorption and Temperature**: Adsorption is exothermic and involves a reduction in enthalpy; so, it varies inversely with temperature, i.e., heat releases the adsorbed molecules by enhancing their thermal movements. In chemisorption there is more heat of adsorption than physical adsorption. For reversal of chemisorption, a higher temperature is required than that of physical adsorption because the ionic and covalent bonds have greater bond energies than hydrophobic bonds and van der Waals forces.
10. **Heat of Adsorption**: The heat of adsorption may be great because it has been determined that, the heat of adsorption of oxygen on charcoal varies from 50000 to 224000 calories while heats of combustion of 94000 to 97000 calories, are for the carbon which adsorbs this amount of oxygen.

11. **Holding Chemisorbed Molecules on the Surface**: Chemisorbed molecules in a single layer, is held on the adsorbent surface because of failure for formation of ionic or covalent bonds between an adsorbate molecules on the surface of adsorbent and a molecule or atom deep inside the adsorbent. But, Van der Waals forces can hold several layers of adsorbate molecules physically on the surface of adsorbent.

12. **Orientation of Molecules**: The molecules adsorbed on the surface of adsorbent are oriented and arranged in a definite manner, relative to the adsorbing surface and to each other.

13. **Adsorption of Solute from Solution:** A solid adsorbent can adsorb a solute from the solution, if
    1. the solution is very dilute,
    2. the solvent has high surface tension,
    3. the interfacial tension is lowered by a higher concentration of the solute on the solvent surface than the interior.

9.14.6 Mechanism of adsorption

Atoms or molecules of a solid surface behave like the surface molecules of a liquid. These are not surrounded on all sides by atoms or molecules of their kind and hence, posses unbalanced attractive forces (residual forces) similar to valency forces. These forces attract the molecules of the adsorbent A, that come in contact with the solid (say material M). Due to adsorption, the residual forces decrease and, therefore, the surface energy gets decreased considerably. This energy is lost as heat energy. Thus adsorption is always accompanied by evolution of heat and the energy liberated, when 1 mg-mole of gas adsorbed on the surface of the solid, called **Heat of adsorption** (Fig. 9.35).

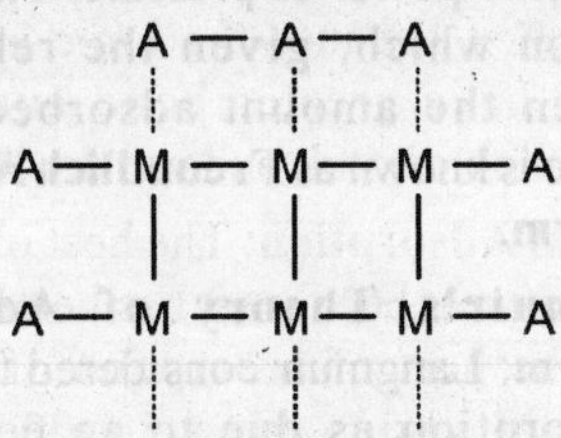

Fig. 9.35: Unbalanced residual forces on the material (M) surface hold the adsorbent (A)

1. **Specific Nature of Adsorbate and Adsor-bent:** The adsorption by a solid surface takes place by attraction of the adsorbate molecules due to the residual forces of the former. Therefore, the firmness with which adsorbate is held by the adsorbent surface depends upon the specific nature of the two, just as the stability of a compound depends on the nature of the constituent elements. Hence, the extent of adsorption taking place depends on the specific nature of the adsorbate and adsor-bent.

2. **Accessible or Available Surface Area:** Greater the available surface area of the adsorbent, greater will be the adsorption. Since Colloid particles have large surface area for a given mass of the solid, they possess high adsorption power. Such a power increases by increasing the surface area and also the residual force by subdivision.

3. **Experimental Conditions**: Adsorption being akin to an exothermic reaction, is governed by the experimental conditions. It is greatly decreased by an increase of temperature (Le Chatelier's principle).

9.14.7 Types of adsorption

Two types:
(a) Physical adsorption.
(b) Chemisorption.

(a) **Physical Adsorption**: When the adsorbed molecules are held by intermolecular attracting forces, often called Van der Waals forces, the adsorption is called **Physical adsorption of Van der Waals adsorption** (or physisorption). It is a **weaker** force.

(b) **Chemisorption**: When the adsorbed molecules are held by sort of chemical bonds to the molecules of the surface, the phenomenon is called **Chemisorption or Activated adsorption**, as it has a rather high energy of activation. It is a **stronger** force.

9.14.8 Characteristics of physical adsorption

1. **Lack of Specificity**

It is due to interplay of weak Van der Waals forces between molecules. Van der Waals forces are universal for which adsorption does not show specificity, i.e., a given surface does not exhibit strong attraction for a few particular gases.

2. **Influence of Temperature**

This process occurs even at a very low temperature because of:

(a) **Zero-energy of activation** and the rate is very little affected by lowering temperature,

(b) **Occurrence by forces,** which cause liquefaction and becomes stronger near the critical temperature or below the critical temperature.

(c) **Influence of Pressure**: Adsorption increases with pressure. Multilayer adsorption starts at near saturation pressure, so the adsorption tends to increase rapidly with pressure in this zone. Since vapour pressure inside capillaries is lower than the normal vapour pressure due to concavity of the liquid surface, strong condensation inside the pores called **Capillary condensation,** which is a kind of physical adsorption, takes place in this region.

A characteristic of physical adsorption is complete **Reversibility** because adsorbed gas can be taken out by a lowering of pressure.

(d) **Nature of Gas**: The easier a gas is condensable, i.e. the higher the boiling point, the more is the physical adsorption.

(e) **Heat of Adsorption:** Physical adsorption is exothermic (chemisorptions are also exothermic) but heat is much smaller than that of chemisorption, i.e., about 5000 calories/mole, similar to heat of evaporation.

9.14.9 Characteristics of chemisorption

1. **High Specificity**: Chemisorption is highly specific.
2. **Influence of Temperature**: Since chemisorption is exothermic, according to Le Chatelier's principle, it should show higher adsorption at lower temperature. The energy activation being high, the rate falls off rapidly, making it slow at low temperature.

   Adsorption is always associated with a loss of entropy and a decrease of free energy. So it must be exothermic. So, accordingly to Le Chatelier's principle, all types of adsorption should decrease with increasing temperature.
3. **Influence of Pressure**: Increase in pressure causes increase of chemisorption for which high pressure is applied in industrial catalysis
4. **Nature of Gas**: Gases which have the capacity of formation of compound, with the adsorbent are chemisorbed.
5. **Heat of Adsorption**: Heat of adsorption is high and it is about 20000-60000 calories similar to the values of heat of reaction.

9.14.10 Adsorption isotherm

When a finely divided solid is shaken with the dilute solution of a dye, the intensity of the colour decreases, as the dye is adsorbed by the solid particles. It is noticed that the decrease in the intensity of the colour takes place to a certain minimum beyond which there is no change at the temperature. In such case of adsorption, equilibrium is said to have been reached. And at this stage, the amount of the substance adsorbed and the concentration of the substance in solution becomes constant.

At adsorption equlibrium stage, the magnitude of the adsorption depends upon:

(a) the temperature,

(b) the nature of the adsorbate and its concentration,

(c) the nature and subdivision (or surface area) of the adsorbent.

The relationship at a given temperature between the equilibrium amount of the substance adsorbed and the concentration of the adsorbent (pressure in case of gaseous adsorbents) is known as the **Adsorption Isotherm.**

(a) **Freundlich Theory of Adsorption Isotherm:** (Effect of pressure and concentration on adsorption)—An increased pressure of a gas or vapour or an increase in the concentration of a solute causing increased pressure, is experimentally found not to be proportional to pressure but is somewhat less; and so Freundlich puts the adsorption proportional to a fractional power of pressure. The resulting equation which, given the relationship between the amount adsorbed and the pressure is known as **Freundlich Adsorption Isotherm.**

(b) **Langmuir's Theory of Adsorption Isotherm**: Langmuir considered the process of adsorption as due to an equilibrium between the adsorbed and the unadsorbed gases. Langmuir imagined that, a gas molecule on striking a surface gets condensed there (this

Table 9.9: Comparison of physical adsorption and chemisorption

Physical Adsorption	Chemisorption
1. Caused by non-specific Van der Waal's forces and hence, strong near condensation temperature.	1. Caused by the same kind of forces as are responsible for chemical union.
2. Highly reversible.	2. Not quite reversible.
3. Tends to form multilayer, with increase of pressure.	3. The adsorbed layer is one molecular thick.
4. Almost zero energy of activation, i.e. rate of adsorption hardly changes with temperature.	4. High energy of activation.
5. Low heat of adsorption.	5. High heat of adsorption comparable to usual heats of reaction.
6. Lack of specificity.	6. High specificity.
7. Occurs at low temperature.	7. Occurs at higher temperature.

is called Inelastic Collision) and evaporates off only after an appreciable time, and this time-lag between condensation and evaporation causes adsorption. His another premise which evoked great controversy and discussion is that the adsorbed layer is normally only one molecule thick because attractive forces on the surface are limited to distances of about one molecular thickness, though he conceded that under some circumstances polymolecular layer (i.e., a layer more than one molecule thick) could be built up.

9.14.11 Adsorption complex and law of adsorption

The charged condition of a **Colloidal Particle** makes it behave in an electric field in the same way as a charged ion of an electrolyte. This similarity extends also to its chemical behaviour. For this we have a class of combinations formed resembling in many respect, chemical compounds. But it differs from the chemical compound in the absense of definite quantitative relations between the reacting substances. This class of compounds of varying composition has been designated **Adsorption complexes**. The factors involved in the formation of complexes are:

1. Extent of surface,
2. Chemical nature of particle, residual valencies, etc.
3. Electric charge on the surface.

The formation of adsorption complex can be studied by the adsorption of dyestuffs by filter paper. If we take a series of solutions of a dye, such as congo red, in progressively diminishing concentration, and place in each solution the same amount of filter paper, we find that part of the dye is taken up by the paper and the proportion taken up is larger than the more dilute solution. This relation is in accordance with what Bayliss called the **Law of Adsorption.**

9.14.12 Langmuir theory on orientation of adsorbed molecules at interface

Langmuir states that the adsorbed molecules get attached to the surface by a bond, which is chemical in origin. For this reason, they do not remain adsorbed in a random way but get oriented in an orderly manner at the interface.

Langmuir also showed that when palmitic acid was spread on water surface, the molecules instead of lying flat on the surface, remained standing erect in a closely packed monomolecular film. The Carboxyl ends were downwards and the Hydrocarbon ends upwards owing to the attraction of the Carboxyl groups for the polar water molecules.

The thickness of the film is 24 Å and area per molecule is about 21 sq. Å and length of C-C bond is 1.5 Å. In all adsorption processes, the adsorbed molecules show more or less this type of "directional orientation due to interactions of physiochemical forces acting at the interface" (Fig. 9.36).

9.14.13 Biological importance of adsorption

1. **Enzyme-Substrate Combination**: During enzyme-substrate reaction, the substrate at first is adsorbed on the surface of enzyme molecule to form enzyme-substrate complex.

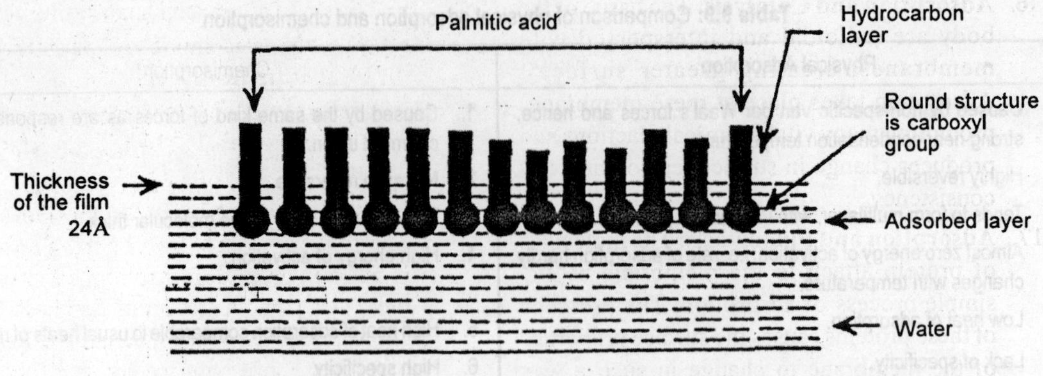

Fig. 9.36: Orientation of adsorbed molecules at interface

2. **Toxin-Antitoxin Reaction**: Toxin-antitoxin interaction is under the process of adsorption. The exact adsorption which takes place, must be a function of the chemical configuration of the substance forming the surface, since otherwise, it would be impossible to account for the extremely specific character of the interaction between toxins and their corresponding antitoxins. It probably depends upon the operation of other forces as electronic Van der Waals attraction, Coulomb attraction, electrical dipoles or multipoles, formation of hydrogen bonds and other forces.
3. **Separation and Purification of Enzymes**: Separation of enzymes from a mixture of enzymes and impurities, can be done by selective adsorption.
4. **Histological Staining of Tissue by Dyes**: Staining of histological section of tissue by dyes like eosin, haematoxylin, etc. is an adsorption process. The dyes are adsorbed on the surface of the Colloids present in the cell.
5. **Lipoid-Protein Structure of Cell Membrane**: The cell membrane consists of two layers of lipids, sandwitched between two layers of proteins. This protein layer is adsorbed on the surface of the lipid layer.
6. **Adsorption of O_2 on the Surface of RBC**: Though certain amount of O_2 combines with Fe^{++} of haemoglobin but greater part of O_2 taken up by the blood is adsorbed on the surface of colloidal RBC and not dissolved, in ordinary sense.
7. **Carbon Monoxide (CO) Poisoning**: CO is deadly poisonous to human system, because once the corpuscles adsorb this gas, they cease to function as "O_2 carrier".
8. **Pulversised Charcoal in Digestive Tablet**: Often the digestive tablet contains powdered charcoal to adsorb intestinal gases.
9. **Pectin-Kaolin in Anti-diarrhoeal Preparation**: The anti-diarrhoeal preparation often contains Pectin and Kaolin, which adsorb bacterial toxins.
10. **Aluminium Hydroxide in Antacid Preparation**: Some antacid (which is used for gastric acidity) contains some salts such as Aluminium hydroxide which adsorbs HCl of gastric juice.
11. **Nourishment of Plant from Soil**: Nourishment of plant from soil takes place by the process of adsorption.
12. **Adsorption Indicator**: During Chloride estimation, Silver nitrate ($AGNO_3$) is used. For this purpose adsorption indicators such as cosin, dichlorofluorescein, etc. are applied.
13. **Electrical Charge of Colloidal Ions**: The colloidal particles develop electrical charge by the adsorption of ions, which exist in the surrounding medium and help to develop zeta potential.
14. **Protection of Colloid**: Protection of Lyophobic Colloid with Lyophilic Colloids is an adsorption phenomenon.
15. **Adsorption Chromatography**: Adsorption chromatography is based on the difference in adsorption of various components on a solid.

16. **Adsorption and Catalysis**: The tissue of the body are colloidal and interspersed with membranes, creating greater surfaces. Adsorption takes place at these membranes promoting many vital chemical reactions and produces change in surface tension and cell consistency.

17. **Adsorption and Pinocytosis**: The molecules of protein attach to the membrane by the simple process of adsorption. The presence of these proteins causes the surface properties of the membrane to change in such a way that it invaginates and then rapidly closes over the proteins.

18. **Protoplasmic Organization**: There are joined complex compounds in protoplasm, e.g., combinations of proteins and lipids, proteins and salts, proteins and carbohydrates, etc. These compounds are being continually formed and broken down and enter into the vital structure and function of living protoplasm. Adsorption plays a role in the formation of some of these complexes.

19. **Action of Drugs**: Certain drugs and poisons adsorb on the cell surfaces and exert their action from the surfaces. Certain specific action are seen due to selective adsorption.

20. **Transportation and Adsorption**: Many mineral ions, carbohydrate or lipids may be transported in the plasma by forming adsorption complex with colloid plasma proteins.

21. **Ion Exchange Resins and Adsorption**: A group of substances is called Ion Exchange Resins which are used for the adsorption of positive and negative ions from solutions in the laboratory, in industrial processes and in medicine. These substances are insoluble synthetic polymers having acidic or basic groups.

A. The acidic or cation exchange resins contain:
(a) Sulphonic acid groups (—SO_3H), e.g., Dowex 50, Amberlite IR - 105.
(b) Carboxyl groups (—COOH), e.g., Amberlite IRC-50, Zeo-Karb 216,
(c) Phenol groups (—OH),

B. Basic or anion exchange resins contain:
(a) Weakly basic amine groups (NH_2), e.g., Amberlite IR-4B, or
(b) Strongly basic quaternary ammonium groups $(-NH_3)^+ OH^-$, e.g.; Amberlite IRA-400, Dowex 1).

22. **Multiplication of Virus in the Host Cell**: The first step of invasion of fresh host cells is attachment and adsorption of viruses to their surfaces. Adsorption depends on an antigen-antibody-like mechanism between parts of the virus surface and receptor areas on the surfaces of the host cells.

23. **Multiplication of Virus in Tissue Culture (Haemadsorption)**: If a cell-sheet in which a haemagglutinating virus is multiplying, is exposed to a dilute suspension of the appropriate red cells, these may adhere to the tissue culture cells. This phenomenon is known as **Haemadsorption**. It is of considerable value in the detection of haemagglutinating viruses, especially myxoviruses, paramyxoviruses and poxviruses.

24. **Preparation of Pure Suspension of the Myxoviruses**: For preparation of pure suspension of the myxoviruses, the virus is adsorbed onto red blood cell in the cold and allowed to elute, through the action of the virus enzyme, at 37°C. Adsorption-elution procedures employing ion exchange resins, such as DEAE cellulose, and inorganic adsorbents, such as calcium phosphate and diatomaceuos silica, are also extensively used.

25. **Precipitation of Virus from Mixtures of Protein and Virus**: Isoelectric precipitation in which the pH is adjusted to the isoelectric point of the virus is also valuable. The virus may, however, be precipitated at the isoelectric point of protein impurities, to which it is adsorbed on precipitation.

26. **Bacteriophages**: From one of the ends, emerges a thin hollow tubular tail, with and end-plate and terminal fibres that are the means by which the virus becomes attached and adsorbed to the cell wall of its bacterial host. Such adsorption is possible only if the phage, which is non-motile, happens to come into contact with a bacteria of a strain with

the appropriate specific receptors on its surface, and if various other conditions are fulfilled. Once, adsorbed, the phage digests a small area of bacterial cell wall and then contracts, injecting its DNA into the bacterial body.

- **Industrial Importance**

1. **Exchange Adsorption-Permutit**: When a salt solution (e.g., calcium) is percolated through clay, salt ions are adsorbed by the clay and the clay also gives out an equivalent quantity of salt (e.g., sodium). In case of Kaolin, the reaction will be:
   Sodium Kaolin + ($Ca^{2+} + 2Cl^-$) → Calcium Kaolin + 2($Na^+ + Cl^-$)
   This process of adsorption of one of the ions by a solid material in exchange for another similarly charged ion, present in the solid is called **Exchange Adsorption** and is applied in case of softening and de-ionisation of water.
2. **Surface-Active Agents:** Soap and various synthetic detergents in small concentration lower the surface tension. Such substances which cause depression of surface tension of water is called **Surface active agents.** It has higher concentration on the surface layer where they are oriented (Fig. 9.36). These compounds are used for emulsification, formation of spray, etc. (Phase-subdivision processes) and wetting, orefloatation, cleaning, etc. (Phase-displacement processes) are related with surface activity and adsorption at an interface.
3. **Molecular Sieves**: The adsorbent called **Molecular Sieves,** are highly porous synthetic zeolites. They are powerful, selective and are used for drying gases and for separation of components of preferential absorption.
4. **Adsorption and Catalysis**
   (a) Chemical reactions are accelerated by the presence of adsorptive surfaces. Oxygen and hydrogen are adsorbed together upon platinum black and combine rapidly at ordinary temperatures to form water.
   (b) Charcoal, suspended in a solution of glycine enables gaseous oxygen to oxidize the glycine at ordinary temperatures.
5. **Solid Adsorbent in Industry**
   1. *Fuller's earth*: For refining petroleum and vegetable oil.
   2. *Activated charcoal*: For decolourising sugar solutions, for recovery of volatile solvents from air in gas mask deodorising purposes. When charcoal is heated in steam and then at temperature of 700° C to 800°C in a closed container, it becomes "activated" and is the most powerful adsorbing agent or adsorbent. The activation process affects the surface of the charcoal and greatly increases it adsorbing power.
   3. *Silica Gel and activated alumina*: For refining petroleum distilates, in gas mask, separation of gases by fractional adsorption, vapour phase partition chromatography, corrosion protection, etc.
   4. Kaolin.

9.15 Hydrotropy

(Hydrotropic Action)

9.15.1 Definition

Hydrotrophy is the property of certain substances to render other insoluble substances soluble in water.

9.15.2 Example

1. 100-fold increase in the solubility of sulphonal by sodium salicylate.
2. 5-fold increase in the solubility of uric acid by atophan.
3. Bile salts for digestion and absorption of fat.

Hydropic substance are usually salts of certain organic acids which powerfully lower surface tension. They are:

1. Bile salts, e.g. Sodium glycocholate, Sodium taurocholate,
2. Sodium salicylate, Atophan, Phenylacetic acid salt, Benzoic acid salt, Hippuric acid salt.

9.15.3 Insoluble substances rendered soluble

Fats, phospholipids, higher alcohols, sterols, e.g., cholesterol, uric acid, sulphonal, insoluble soaps, Ca, Mg, Fe, Cu, Mn.

9.15.4 Mechanism

The mechanisms of action is not know. Probably the hydrotropic substances form loose compounds with the insoluble substances and subsequently make them soluble and diffusible through membranes.

9.15.5 Evidences

I. Experimental Evidence

Verzar and Kutthy (1931) showed experimentally that, various substances especially fatty acids are dissolved and rendered it dialysable by the bile salts. Oleic acid was emulsified in a phosphate buffer-solution, and then 5 per cent Sodium glycocholate added. The mixture was placed in a dialyser, outside which was the same mixture without any fatty acid. After 16 hours dialysis, the dialysate, i.e., the outer fluid and the contents of the dialyser were analysed for presence of fatty acid. The experiment revealed that, the more dilute solutions, the fatty acid was almost diffusible; but with concentrated ones, only about two-thirds of it was really dissolved.

II. Biological Process of Fat Digestions and Absorption by Hydrotropic Action

Bile salts reduce the surface tension of water and fat is emulsified and rendering larger area for the enzyme pancreatic lipase to act. Fats are finely emulsified in the small intenstine by the detergent action of bile salts, lecithin and monoglycerides because bile salts are relatively poor emulsifying agents but in the presence of phospholipids and monoglycerides, cholesterol and soluble proteins, particles of 200-5000 nm in diameter are formed.

Lipase cannot penetrate fat droplets covered by emulsifying agents without **Collipase,** a protein with a MW 11000 secreted by the exocrine part of pancreas. Collipase binds to the surface of the fat droplets at the oil-water interphase, displacing the emulsifying agents and anchoring lipase to the droplet which hydrolyzes triglycerides into free fatty acids and 2-monoglycerides.

Bile salts and lipids interact spontaneously to form "micelles" which are spherical globules about 3-10 nm (25 Angstroms) in diameter and composed of 20-50 molecules of bile salts.

These are formed because each bile salt molecule is composed of the nonpolar sterol nucleus, which are highly fat soluble forming hydrophobic centre and the polar group (hydroxy, peptide bond and carbonyl or sulphate portion of the bile salts) which is highly water-soluble facing outward. The primary bile acids form micelles above a critical micellar concentration. Micelles are aggregates of bile acids and phospholipids on the outside with cholesterol or monoglycerides in the centre. Depending on the lecithin and bile acid ratio, micelles may be small and simple or large and mixed.

The sterol nuclei of the 20-25 bile salts molecules of the micelle aggregate together to form a small fat globule in the middle of the micelle. This polymolecular aggregation causes the polar groups to project outward to cover the surface of the micelle. Since these polar groups are negatively charged and they allow the entire micelle globule to become dissolved in the water of the digestive fluid or chyme and remain in stable solution despite the very large size of the micelle. Micellar formation further solubilizes the lipids and provides a transport mechanism to the intestinal mucosal cells. Then the micelles move down their concentration gradient through the **Unstirred water layer (UWL)** to the micelles, and a saturated aqueous solution of the lipids is maintained in contact with the brush-border of the mucosal cells. The lipids enter the cells by passive diffusion through the epithelial membrane, leaving the bile acid micelles still in the chyme. The micelles then diffuse back into the chyme and absorb still more monoglycerides and fatty acids and transport them to the epithelial cells, thus performing as "ferrying agent".

After entering the cell, the lipids are esterified rapidly inside the cells, maintaining a favourable concentration gradient from the lumen into the cells.

A low pH facilitates micellar diffusion, as well as micellar dissociation. Converting a nonprotonated fatty acid to its protonated form in a low pH microclimate, further facilitates diffusion across the lipid membrane of the cell crossing. The limiting factors for passing to the cells are:

1. Crossing the **Unstirred water layer** is the rate-limiting step in the uptake of medium and long chain fatty acids.
2. Mucus coat overlying the brush-border membrane which must be penetrated.
3. Passive diffusion is dependent upon lipid solubility, but there are evidences of binding or carrier mediated transport.

9.16 Viscosity

9.16.1 Definition

Viscosity means the internal resistance against the flow of gas or liquid due to frictional forces between the layers while moving one layer over another at different velocities.

I. Viscosity of Gases

9.16.2 Molecular theory of viscosity

Viscous forces are observed in gases as well as in liquids. The viscosity of gases is explained by the transfer of momentum which takes place between adjacent layers of the gas, as it flows in a particular tube. Fast moving molecules in a layer A (Fast) cross with their own velocity to the layer B (Slow), where molecules are moving with as slower velocity.

Molecules of gas layer B, likewise, move to A (Fast). The net effect is an increase in momentum in B and corresponding decrease in A (Fast), although on the average the total number of molecules in the two layers is unaltered. Thus, the layer B speeds up and the layer A slows down, i.e., a force acts on the layers of the gas while they move known as Viscous force (Fig. 9.37).

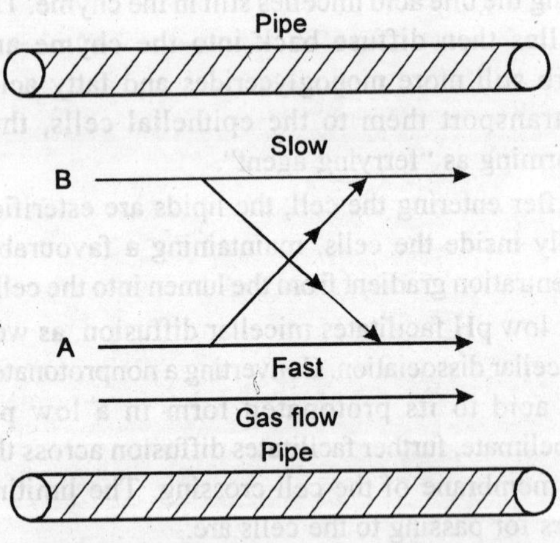

Fig. 9.37: Viscosity of gas – momentum effect

Though, there is transfer of momentum as in the gas, the viscosity of liquid is mainly due to the molecular attraction between molecules in the adjacent layers. Energy is needed to drag on layer over the other, against the force of attraction. Thus, a shear stress is required to make the liquid move in laminar flow.

9.16.3 Factors influencing the viscosity of gases

1. **Density**: Viscosity of a gas is directly proportional to the density of the gas.
2. **Temperature**: Viscosity of a gas is directly proportional to the square root of its absolute temperature.
3. **Mean Molecular Velocity and Mean Free Path**: Viscosity of a gas varies directly with the mean velocity and the mean free path of the molecules.

9.16.4 Viscosity of liquids

9.16.4.1 Definition

When a liquid flows, each portion of the liquid experiences a resistance, to flow or movement of one layer over another. This property of liquids, which determines their rate of flow is termed **Viscosity**. Viscosity of a liquid is a measure of this **internal friction of a liquid or lack of slipperiness** and it determines the **rate of flow** of a liquid. Isaac Newton (1713) described it as the "internal friction" which we now call viscosity.

9.16.4.2 Coefficient of viscosity

When a liquid flows in contact with a plane surface S. we can see two events in this flow process.

1. The plane surface would experience a tangential force parallel to the direction of flow and would tend to move along the line of flow.
2. There is a "gradient of velocity" or "rate of shear" of flow at right angles to the direction of flow. In other words, the velocity of flow is zero on the plane surface S, and increases continuously, as we go up into the liquid at right angles to the directions of flow. The tangential force on S depends on the: (1) Area of the surface, and on the (2) Velocity gradient. This "tangential force per unit area per unit velocity gradient" is called the **Coefficient of viscosity** of the liquid or viscosity and is designated by η. Viscosity governs the flow property of a liquid, the higher the viscosity the lower is its tendency to flow (Fig. 9.38).

The above concept of viscosity applies even when the stationary surface is absent. Viscosity is generally expressed as follows,

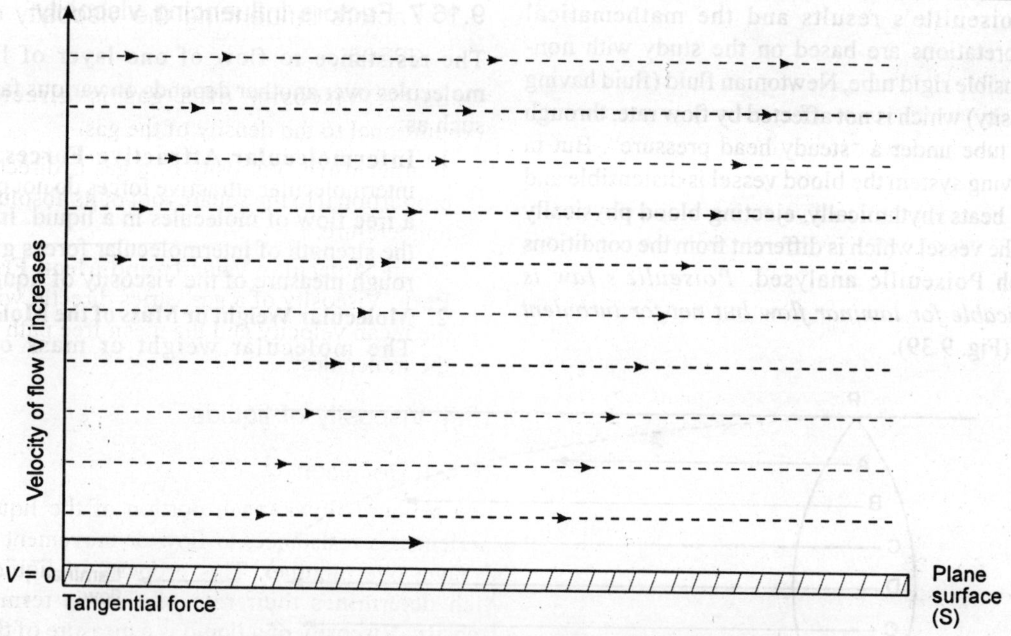

Fig. 9.38: Viscous flow on a plane surface by a flowing liquid

Coefficient of viscosity, η

$$= \frac{\text{Tangential force}}{\text{Area} \times \text{Velocity gradient or rate of shear}}$$

- **Newton's Formula-Coefficient of Viscosity**

The basic formula for the frictional force, F, in a liquid gas was first suggested by Newton. He found that the larger the "area" of the surface of liquid, the greater was the frictional force, F. He also stated that F was directly proportional to the velocity gradient at the part of the liquid. This is the case for most common liquids, called **Newtonian liquids**, e.g., water. A simple viscous liquid is known as a **Newtonian fluid** and it is defined as one whose viscosity does not vary with **rate of shear** and remains constant at different rates of laminar flow or uniform flow.

If V_1, V_2 are the velocities of D and B, respectively, and h is their distance apart, the velocity gradient between two liquids, is defined as $(V_1 - V_2)/h$ and is expressed in (m/s)/m or "S^{-1}".

Thus, if A is the area of the liquid surface, the frictional force F on the surface is given by $F \propto A \times$ velocity gradient, $F = A \times \eta \times$ velocity gradient, where η is a constant of the liquid, known as the **Coefficient of viscosity.**

9.16.4.3 Poiseuille's law

Poiseuilles discovered the law governing the rate of flow of liquids from a study of laminar flow by a Newtonian liquid in narrow rigid glass tubes of capillary diameter. It is expressed by equation as follows:

$$\eta = \frac{\pi p r^4}{8vl} t,$$

where η (eta) = Viscosity,

V = Volume of liquid flowing through a capillary tube,

l = Length of the tube,

r = Radius of the tube,

p = Pressure,

t = Time.

If the times of flow of equal volumes of two liquids through the same capillary, are measured under the same head of liquid, the ratio of viscosities is given by:

$$\frac{\eta_1}{\eta_2} = \frac{d_1 t_1}{d_2 t_2},$$

where η_1 and η_2 = Coefficient of viscosity of the two liquids;

d_1 and d_2 = Densities of two liquids,

t_1 and t_2 = Times of flow of the two liquids.

Poiseuille's results and the mathematical interpretations are based on the study with non-distensible rigid tube, Newtonian fluid (fluid having viscosity) which is not affected by flow rate, through rigid tube under a "steady head pressure". But in the living system the blood vessel is distensible and heart beats rhythmically, ejecting blood physically into the vessel which is different from the conditions which Poiseuille analysed. *Poiseulle's law is applicable for laminar flow but not for turbulent flow* (Fig. 9.39).

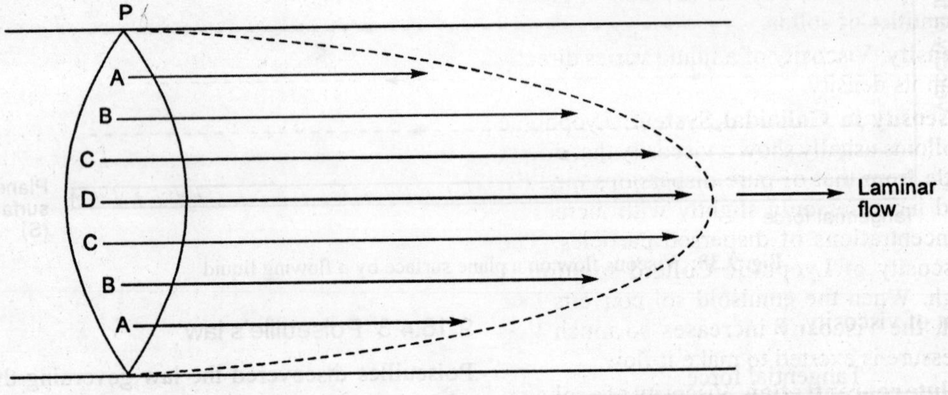

Fig. 9.39: Laminar or uniform flow of liquid through pipe

9.16.5 Unit of viscosity

Since the unit of force is dyne, that of area is square cm (cm^2) and that of velocity gradient is cm per second per cm (sec^{-1}), the unit of viscosity is "dyne-second per sq. cm^2". This unit is called a **poise**, after the name of Poiseuille, who first made systematic study of liquid flow through capillaries.

The viscosity of water at room temperature is about 1 centipoise whereas, that of glycerine is about 800 centipoise and water is 800 centipoise at room temperature. The value of viscosity as defined above is called **Absolute viscosity** and also less frequently **dynamic viscosity**. In industry, however, **Kinematic viscosity** defined as absolute viscosity divided by density, is more commonly used.

9.16.6 Fluidity

The reciprocal of the coefficient of viscosity is called **Fluidity**. Fludity $= \frac{1}{\eta}$. Fluidity expressess the tendency of a liquid to flow, whereas viscosity is a measure of resistance which a liquid offers to flowing.

9.16.7 Factors influencing viscosity

The resistance to flow of one layer of liquid molecules over another depends on various factors, such as:

1. **Intermolecular Attractive Forces**: The intermolecular attractive forces do not permit a free flow of molecules in a liquid. In fact, the strength of intermolecular forces gives a rough measure of the viscosity of liquids.
2. **Molecular Weight or Mass of the Molecule**: The molecular weight or mass of the molecule of a liquid also determines the flow of a liquid. Thus, heavier the molecule of a given liquid, the greater will be its viscosity.
3. **The Structure and Shape of the Molecule**: The structure and shape of the molecules of a liquid plays an important role in influencing its viscosity. Liquids with large, irregularly shaped molecules are generally known to be more viscous than those with small and symmetrical molecules. Since only hard, symmetrical molecules have perfectly elastic collisions, the large and irregular molecules will have less elastic collisions amongst themselves. Thus, the collisions between large molecules involve the loss of kinetic energy and as a consequence the intermolecular forces dominating the molecules tend to stick together. This increases the viscosity of the liquid.
4. **Temperature**: An increase in temperature decreases the viscosity of liquids. As the temperature increases, the molecular motion increases at the expenses of cohesive forces causing resistance to flow. Therefore, the

viscosity of liquids is found to decrease by 1-2 per cent for each degree rise of temperature.

5. **Increase in Pressure**: The increase of pressure goes to strengthen the cohesive forces between molecules. That is why, with the increase of pressure, the viscosity of a given liquid increases to some extent.
6. **Dissolved Substances**: Little amount of nonelectrolytes, e.g., sugar, alcohol increases the viscosity, whereas, small quantities of highly ionized salts decrease it. Viscosity is usually increased by the solution of greater quantities of solids.
7. **Density**: Viscosity of a liquid varies directly with its density.
8. **Viscosity in Colloidal System:** Lyophobic Colloids usually show a viscosity that differs little from that of pure dispersions medium and increases only slightly with increasing concentrations of dispersed particles. The viscosity of Lyophilic Colloid solution is high. When the emulsoid sol converts to a gel, the viscosity increases so much that pressure is exerted to make it flow.
9. **Solute concentration**: Viscosity of a solution is directly proportional to the solute concentration.
10. **Velocity of Flow**: Viscosity of a solution during flow is independent of velocity of flow. Viscosity of a suspension, (e.g., blood) during flow shows a progressive rise at low velocities.
11. **Influence of Suspended Particles**: Suspended particles cause an increase in viscosity, in proportion to the volume of suspended material relative to the total volume.

(a) **Viscosity of Blood**: The viscosity of blood is very much important in relation to the resistance offered to the heart for blood circulation. The viscosity of blood is mainly due to the emulsoid colloidal system, e.g., plasma protein and to a greater extent for the suspended red blood cells.

(b) **Viscosity of Protoplasmic Gel and Effect of Metallic Ions**: The study of the effect of metallic ions upon the viscosity of protoplasmic gel by Micro-manipulation Apparatus has been made by Chambers and Rezuikoff. When solution of chloride of sodium, potas-sium, calcium and magnesium were injected into amoebae they showed the following results.

i. Sodium and potassium chloride cause liquefaction of the surrounding area of injection and the protoplasm tended to convert from the gel to sol colloidal state.

ii. Calcium chloride caused a local solidification of the protoplasm, which was "pinched off" by the amoebae and rejected. The remaining part of the amoebae appeared normal.

iii. Injection of magnesium chloride caused local solidification and there was no "pinching off". The solidification process was gradually spread throughout the protoplasm and death of the cell occurs.

iv. When solution of sodium, potassium and calcium chlorides in correct proportion, there was no effect. The ions exerted an antagonistic effect upon the protoplasm. These ions probably change the interfacial tension and the type of protoplasmic emulsions.

(c) **Viscosity of Protoplasm**

i. Viscosity of protoplasm was determined by Heilrunn et al. in eggs of the sea urchin Arbacia, Clam Cumingia and Amoebae by estimating the rate of fall of granules through the protoplasm under centrifugal force.

ii. Viscosity of protoplasm was determined from the Brownian movement of cellular granules and equation of Einstein was applied:

$$\eta = \frac{RT_1}{D_x^2 \times N_3 \pi a}$$

where η = The viscosity,

R = The gas constant,

T = The absolute temperature,

N = The Avogadro number,

D_x = The displacement of a particle in Brownian movement along one axis,

t = The time,

a = The radius of the granule.

The above methods show that hayline protoplasm has viscosities from 2-20 Centipoise and varies with species and different conditions. The whole protoplasm has higher viscosity due to presence of inclusion of granules and bodies.

9.16.8 Experimental observation

It was shown experimentally (Reynolds, 1993) and proved mathematically, that a Newtonian fluid moving parallel to the axis of a tube with a constant velocity possesses a paraboloid profile. Reynolds used a long cylindrical tube through which fluid flowed at a pressure head provided by a raised reservoir. He introduced a thin stream of dye into the axial stream and noted that the dye remained in the centre of the stream, showing this paraboloid profile. Only if the velocity of flow became high did flow change from laminar flow or streamlined flow to turbulent (Fig. 9.40).

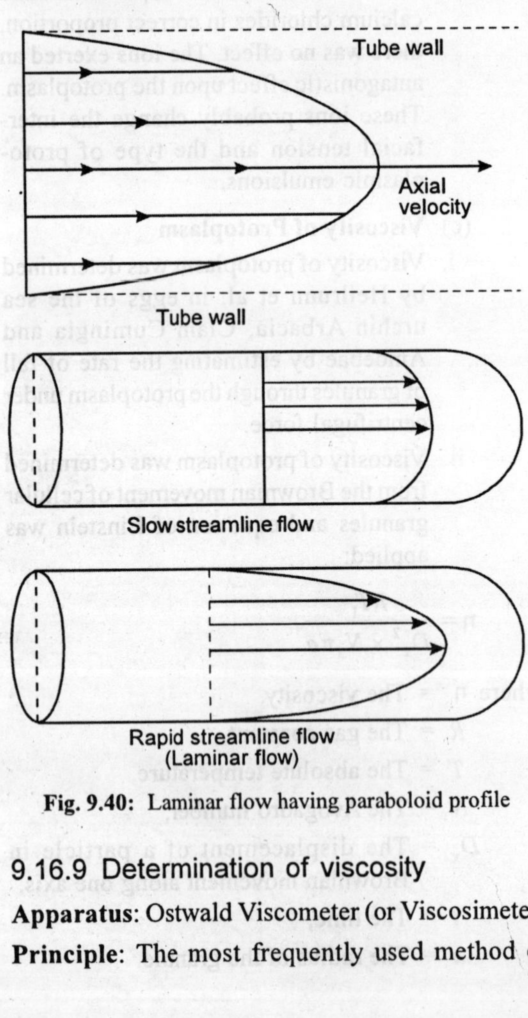

Fig. 9.40: Laminar flow having paraboloid profile

9.16.9 Determination of viscosity

Apparatus: Ostwald Viscometer (or Viscosimeter)

Principle: The most frequently used method of determining liquid viscosity, is based on the determination of time of flow, of a given volume of a liquid, through a length of capillary tubing and to compare this time of flow with the time of flow of the same volume of another liquid of known viscosity.

Method: A definite quantity of the liquid under examination is put into the wide limb and sucked up the right limb *C*, slightly above the etched mark *A*. The quantity of the liquid should be enough to fill the viscometer between *C* and *C′* approximately. The liquid is now allowed to flow back and the time taken for the meniscus to fall from *A* to *B* is noted. The liquid passes through the capillary tube *BD* and it is clear that the time of flow *t*, will be directly proportional to the coefficient of viscosity. The time of flow is also determined for a liquid of known viscosity, say, water. If all the requisite precautions are taken, the absolute viscosity is proportional to the time of flow, multiplied by the density, i.e.,

$$\frac{\eta_1}{\eta_2} = \frac{d_1 t_1}{d_2 t_2}$$

and so the viscosity of the experimental liquid is easily computed (Fig. 9.41).

Conclusion: The direct experimental measurement of absolute viscosity offers considerable difficulty. The relative viscosity of liquid with this apparatus can be determined easily.

9.16.10 Viscosity and chemical constitution

Since viscosity depends on resistance offered when a molecular layer moves over another, some relationship between viscosity and chemical constitution is to be anticipated. The following general rules have been discovered.

1. In a homologous series, the increase in viscosity per CH_3 group is approximately constant.
2. An interesting relationship between viscosity and molecular volume discovered by Dunstan may be stated as:

   $$\frac{d}{M} \times \eta \times 10^6 = 40 \text{ to } 60,$$

   where d is the density and M_l is the molecular weight.
3. **Rheochor**: It has been shown by Newton Friend, that if molecular volume be multiplied by the eighth root of the coefficient

of viscosity, it leads to a constant, to which he has given the name Rheochor.

$$R = \frac{M}{D} \times \eta^{1/8}.$$

9.16.10 Biological Importance

1. **Viscosity Giving Information About the Shape of Protein Molecule:** In addition to diffusion methods, the study of the viscosity of protein solution has given information about the shape of protein molecules. Einstein showed that a dilute solution of spherical particles should obey the equation $\eta sp = \eta r - 1 = 2.5\Phi$,
   where ηsp = specific viscosity,
   Φ = volume fraction of the solute.

   Protein solution give value of $\eta sp / \Phi$ greater than 2.5 as Φ approaches zero. And the magnitude of this viscosity increment may be used to calculate the ratio of the axes of an assumed ellipsoid. Elongated or spheroids will increase the viscosity of a solvent to a greater extent than, will an equal number of equivalent spherical molecules.
2. **Viscosity of Blood Plasma**: Plasma proteins are responsible in large part for the viscosity of the plasma, which markedly affects the arterial blood pressure. Increase or decrease in the blood viscosity impairs the circulation proportionally to the degree of deviation from the normal. Its normal viscosity is 15-20 poises at 20°C is mainly due to the large size and elongated molecular shapes of plasma proteins. In haemorrhage, transfusion of fluids (e.g., saline, glucosaline, etc.) fail to restore an adequate circulation, partly due to viscosity, which does not approximate that of the the plasma, which is fulfilled by the blood transfusion.
3. **Viscosity of Blood**: Blood is more viscous than water but since it is not a homogenous liquid, its apparent viscosity varies according to the method used for the measurement. If the viscosity is determined by the rate of flow along a capillary tube, the apparent viscosity is increased.
   (a) By increasing the relative number of RBC,
   (b) By diminution of the diameter of the capillary,

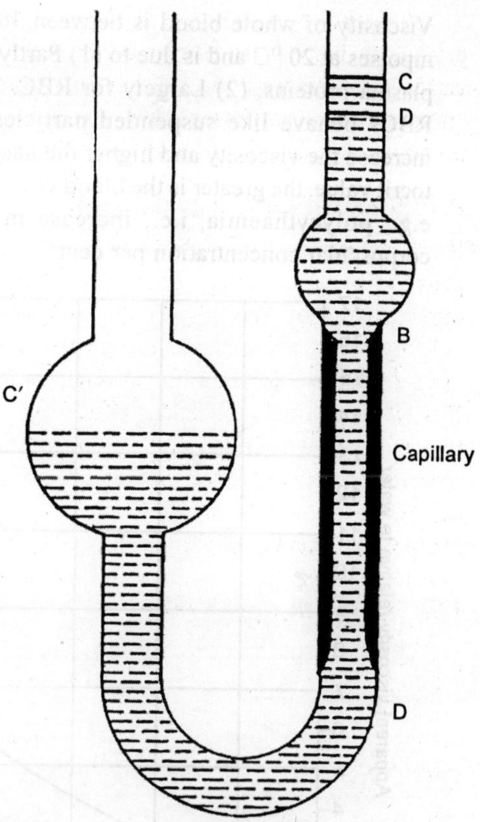

Fig. 9.41: Determination of viscosity

   (c) By reduction of driving pressure and the result is five times that of water.

   Experiment with the capillaries of the vascular system, blood shows twice the viscosity of water but the glass capillary viscometers, generally show a viscosity double of this. The reason for low viscosity in the blood capillaries and compensatory mechanism during rise in viscosity due to increased haematocrit and reduced blood velocity in the blood capillary vessels having diameter less than 150 µm (Fahraeus-Lindquist effect) is that; (1) RBC's tend to move like a single row in the middle of the stream, leaving the plasma at the periphery, to encounter the frictional resistance of the vascular wall, and (2) they do not move randomly and collide with one another. But water does not show Fahraeus-Lindquist effect on viscosity while flowing through narrow vessels.

Viscosity of whole blood is between 30-40 mpoises at 20 °C and is due to (1) Partly for plasma proteins, (2) Largely for RBC. The RBCs behave like suspended particles to increase the viscosity and higher the haematocrit value, the greater is the blood viscosity, e.g., polycythaemia, i.e., increase in the corpuscular concentration per cent.

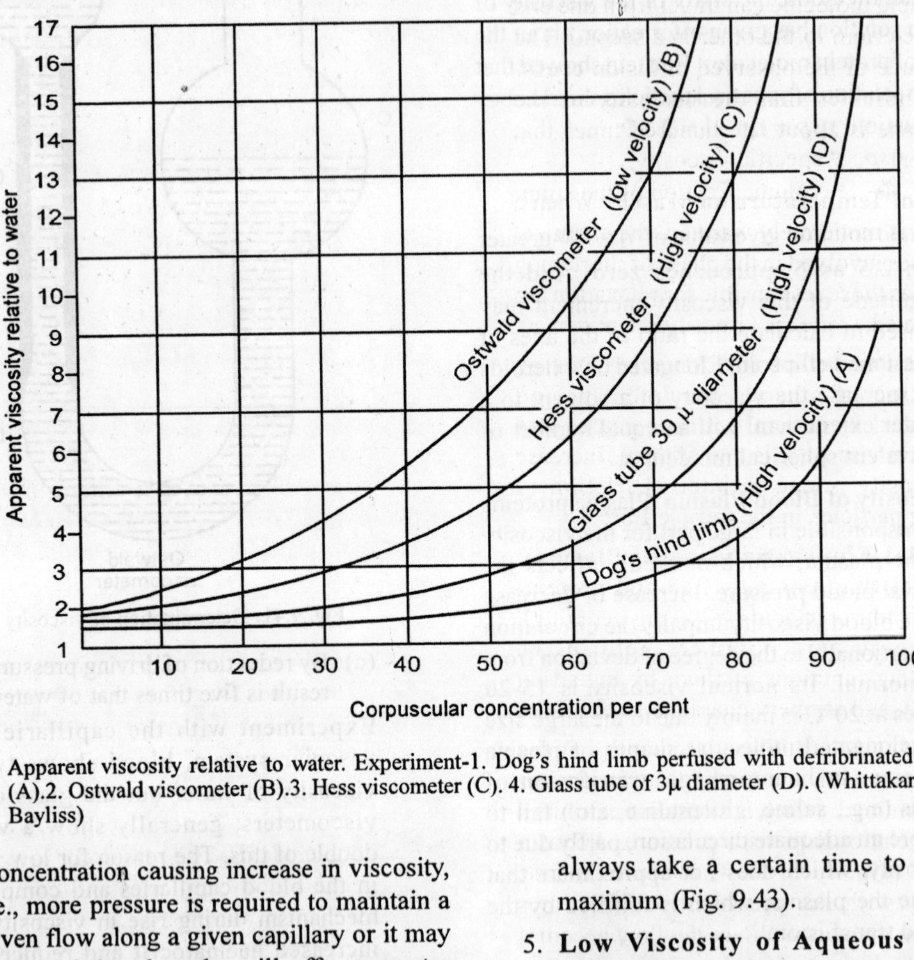

Fig. 9.42: Apparent viscosity relative to water. Experiment-1. Dog's hind limb perfused with defribrinated dog's blood (A).2. Ostwald viscometer (B).3. Hess viscometer (C). 4. Glass tube of 3μ diameter (D). (Whittakar, Winton and Bayliss)

Concentration causing increase in viscosity, i.e., more pressure is required to maintain a given flow along a given capillary or it may be stated that the tube will offer a greater resistance (Fig. 9.42).

Increase in rigidity of RBC, e.g., genetic spherocytosis, and fall in blood velocity cause increase in viscosity. Increase in viscosity due to fall of velocity is due to rulleaux formation and random orientation of the elongated plasma protein molecules. Viscosity decrease in chronic anaemia.

4. **Viscosity of Changes in Muscle**: The changes in the muscle on excitation involve: (a) A sudden increase in the viscosity of the muscle fibre, and (b) a sudden rise of tension. Both causes delay of both the changes in form and the changes in tension of the contracting muscle. If it were not for this increase in viscosity, the contraction might begin at the moment of stimulation, though it would always take a certain time to rise to its maximum (Fig. 9.43).

5. **Low Viscosity of Aqueous Humour**: Aqueous humour, a thin watery fluid present in the anterior chamber of the eyes, contains Crystalloids in concentrations, similar to that of plasma. It also contains hyaluronic acid, which is kept in the depolymerized state by hyaluronidase present in the ciliary body. Hence the viscosity of aqueous humor is low.

6. **High viscosity of Vitreous Body**: The vitreous body present in the posterior chamber of the eyes is an amorphous

transparent gel. It contains albumin and hyaluronic acid; the latter is responsible for the high viscosity of the vitreous body.

7. **Viscosity of the Membrane**: The diffusion coefficient of lipids in a variety of membranes is about 10^{-8} cm^2 sec^{-1}. Thus a phospholipid molecule, diffuses an average distance of 2×10^{-4} cm or 2 μm in 1 second. This means that a lipid molecule can travel from one end of a bacterium to the other in a second. The magnitude of the observed diffusion coefficient indicates that the viscosity of the membrane is about one hundred times that of water.
8. **Effect of Temperature on Muscle**: When a muscle is moderately warmed, the chemical processes involved in the phase of shortening as well as those responsible for relaxation are accelerated, while the viscosity is diminished.
8. **Latent Period of the Muscle**: During excitation of muscle, the latent period manifested by several intrinsic causes and change in viscosity, is one of them. Increase in viscosity is due to internal friction of moving particles in the muscle fibre.
9. **Synovial Fluid (Dialysate of Plasma)**: Synovial fluid contains free and protein-bound hyaluronic acid (4-300 mg per 100g). Hyaluronic acid polymers help in the lubricating functions of fluid, by imparting viscosity and slippery character to the fluid.
10. **Turbulent Blood Flow**: Generally, in the vascular system, the blood flow is laminar or streamlined, which is silent and exhibits a linear relationship with pressure. About a certain velocity of flow, it ceases to rise linearly with the presssure and becomes turbulent gradually. When the flow becomes fully turbulent, it is approximately proportional to the square root of the pressure head.

Reynolds showed that the critical average velocity of flow at which the flow became turbulent is dependent on the radius of the tube and on the viscosity and density of the fluid.

$$V = \frac{R_e \eta}{\rho r}$$

where V = Critical mean velocity,

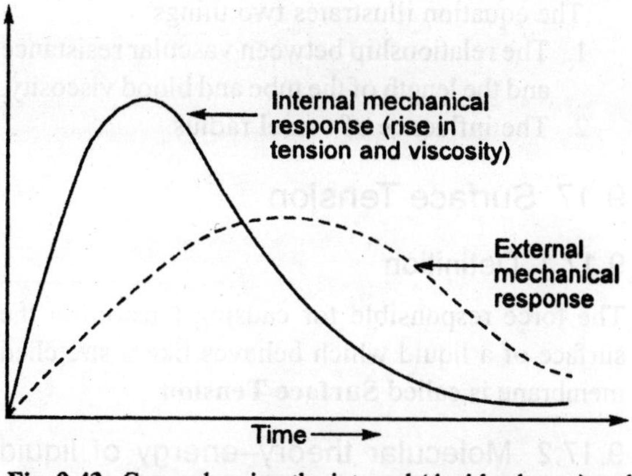

Fig. 9.43: Curve showing the internal (double change) and external mechanical response on excitation of muscle

R_e = Reynod's number,
ρ = Density,
η = Viscosity of the fluid,
r = Radius of the tube,

Hence $R_e = \dfrac{V \rho r}{\eta}$.

If V is measured in cm s^{-1}, r in cm, density in unity, and viscosity in poises, then, in a long straight tube, turbulence develops when R_e exceeds 1000.

11. **Haemic murmurs of the heart**: Haemic murmurs develop in anaemic patients as a result of decreased viscosity and increased blood velocity in the presence of an augmented cardiac output.
12. **Relationship between Vascular Resistance, Length of the Tube and Blood Viscosity-Poiseuille-Hagen Equation**: Viscosity is the important factor in the resistance term of the Poiseuille equation. The factors influencing resistance revealed by the analysis of the Poieseuille-Hagen equation, from the study of laminar flow by a Newtonian fluid in narrow rigid glass tube. It states that:

$$Q = \pi \frac{(P_1 - P_2) r^4}{8 \eta L}$$

where Q = Flow rate,
P_1-P_2 = Pressure difference across the circuit,
r = Radius of the tube,
η = Viscosity of the liquid,
L = Length of the tube.

The equation illustrates two things:

1. The relationship between vascular resistance and the length of the tube and blood viscosity.
2. The influence of vessel radius.

9.17 Surface Tension

9.17.1 Definition

The force responsible for causing tension on the surface of a liquid which behaves like a stretched membrane is called **Surface Tension.**

9.17.2 Molecular theory–energy of liquid surface

A molecule lying **in the surface** of the homogenous liquid is pulled laterally and downwards due to the attractive forces of other adjacent molecules (shown by arrows). The pull operative on this molecule towards the right and the left is counterbalanced but the downward pull remains unbalanced, since there is no molecule above to balance it.

On the other hand, a molecule placed in the body of the homogeneous liquid, is subject to molecular attractive forces operating equally in all directions, as indicated by arrows in the (Fig 9.44). These are completely balanced and there is no resultant force left on the interior molecules.

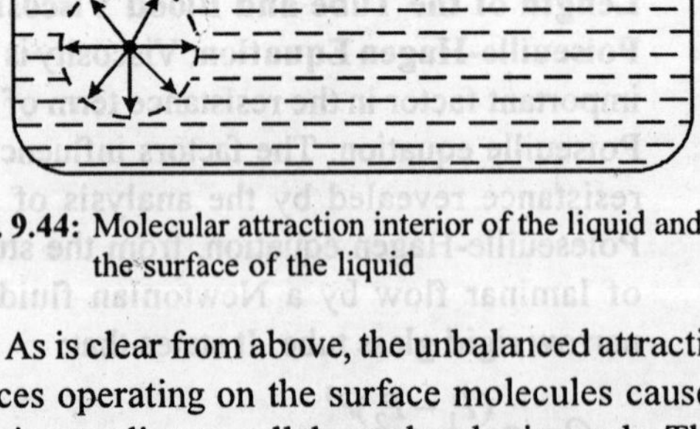

Fig. 9.44: Molecular attraction interior of the liquid and in the surface of the liquid

As is clear from above, the unbalanced attractive forces operating on the surface molecules cause a tension tending to pull the molecules inwards. Thus the surface of a liquid is a seat of special force as a result of which molecules in the surface are in a way bound together to form something like a stretched membrane tending to compress the molecules below to the smaller volumes. In other words, the inward pull operating on the surface molecules tends to give the minimum surface area to a liquid. Many in a capillary tube, the spherical shape of rain drops and the movement of water in the soil are adequately explained with the help of this concept. The force responsible for causing tension on the surface of a liquid is called **Surface tension.**

The surface tension γ of a liquid is precisely defined as "the force acting per cm length on the surface of the liquid, which tends to compress the surface area. This force acts tangentially to the surface and perpendicular to the boundary line of the liquid. It is measured in dynes/cm.

Existence: (1) Wherever a liquid touches the walls of the container, (2) Surface of solid particles in the liquid, (3) At the boundary between immiscible liquids, e.g., oil drops emulsified in water.

9.17.3 Factors influencing surface tension

1. **Nature of the Liquid**: The surface tension is different for different liquids due to varying magnitude of the attractive forces operating between their molecules.
2. **Temperature**: The surface tension differs appreciably with temperature. A rise in temperature, causes an increase in the kinetic energy of the liquid molecules, thereby lessening the strength of inter-molecular attraction. This decreases the inward pull that were operating on the molecules in the surface. Thus, the surface tension of a liquid is diminished with a rise of temperature.

- **Ramsay-Sheild's Equation**

Eotvos (1885), deduced theoretically the "molar surface energy", $\gamma = \left[\frac{M}{d}\right]^{2/3}$ of aliquid should be a linear function of temperature, where M = molecular weight, d = density of the substance, and γ = surface tension. Later, Ramsay and Shields (1898) experimentally verified the same and deduced an equation known as Ramsay-Shield's Equation, which is valid for many liquids within a few degrees of the critical temperature.

$$\gamma = \left[\frac{M}{d}\right]^{2/3} = K(T_c - T - 6)$$ –Ramsay-Shield's equation,

where K = Constant (Eotvos' constant),
= 2.12 (for many liquids, but differs for other liquids)
T_c = Critical temperature.

- **McLeod Equation**

McLeod equation expresses more accurately the variation of surface tension with temperature,

$$\gamma = K(D-d)^4$$

where D = densities of the liquid,
d = its vapour,
γ = surface tension,
K = constant, McLeod constant, which is characteristic of the liquid and is independent of the temperature.

3. **Effect of Inorganic Ions**: Inorganic salts cause a slight rise of surface tension.
4. **Effect of Organic Substances**: Most organic substances cause a diminution of surface tension. It was shown by Gibbs and Thomson that, dissolved substances, which lower surface tension, tend to concentrate at the surface, on account of the fact that the free energy of the system is thereby lessened. In many cases this concentration at the surface may be so marked that the substance forms a solid pellicle.
5. **Effect of Electrical Charge**: Many surfaces such as glass, paper or wool in water are negatively charged. This charge affects the type of adsorption at the surface of such particles and will tend to favour the concentration of electro-positive, as against electro-negative molecules of the interface. A charge on a surface implies, a diminution of the surface tension, and this is the principle of the capillary electrometer.
6. **Concentration at the Surface and the Adsorption Phenomena**: While dealing with an interface between liquid and liquid or between liquid and solid, this concentration of solute or dissolved substance at the surface is equivalent to a deposition on the other substance with which the solution is in contact. It is responsible for many cases of concentration at surfaces for which, the name "adsorption" is used. Adsorption of gases by charcoat is due to the very large surface. In the staining of histological section, adsorption is the main factor invloved on chemically active solute. This process of adsorption will have a two-fold effect:
   (a) That part, which is adsorbed at the surface, is taken out of the main solution and loses, for instance, its osmotic effect.
   (b) On the other hand, the concentration of two reactive substances in the same surface may bring about their interaction, so that chemical changes are furthered which, without this concentration, would have gone on much more slowly.

9.17.4 Surface energy

According to thermodynamic principle, a system, whether it is physical or chemical, tends to move towards the minimum possible free energy under the circumstances. Free surface energy of a liquid is the surface tension of a liquid multiplied by surface area of the liquid. If the surface area is large, the free surface energy is large. According to thermodynamic principle, to decrease surface energy the surface area of the liquid tends to minimize as much as possible. It is expressed in the following way:

S = γs, where S = free surface energy,
γ = surface tension in dynes per cm,
s = surface area in square cm.

9.17.5 The Gibbs-Thomsom principle of surface tension

Gibbs-Thomsom's principle is based on the determination of the effect of substances upon surface and interfacial tension. It states that "substances which lower the surface tension be- come concentrated in the surface tension, are more concentrated within the interior of the liquid than in the surface". This principle has importance pertaining to adsorption process. In colloidal system, many properties, e.g., precipitation, electrical charges, surface tension and viscosity are related to the adsorption process.

9.17.6 Capillarity (capillary action, capillary attraction)

When a capillary tube is immersed in water and placed in vertical position with one end in the liquid which wets the glass, the liquid spreads as a film over the glass surface energy. This film of liquid tends to contract due to its free surface energy and pulls a column of liquid up into the capillary to a height above the surface of the outer liquid. The height of the liquid is proportional to the surface tension of the liquid and to the bore of the capillary tube. The narrower the tube, the greater is the height. This phenomenon is known as **Capillarity**. The liquid rises up to a level when the force of surface tension counterbalances the weight of the liquid in the tube. The surface of liquid in the capillary tube is in concave form.

When the capillary tube is placed inside the mercury, the liquid does not wet the walls of the capillary, and, instead of rising, is depressed below the outside level and the surface of the liquid is convex in form. The depression increases as the diameter of the capillary tube decreases (Fig. 9.45).

Both the concave and convex surfaces are expression of the tendency for the free surface energy to reduce surface area to minimum.

Explanation: The force of attraction between molecules of the same substance is called **Cohesion** and the force between molecules of different substances is called **Adhesion**. The adhesion of water to glass is stronger than the cohesion of water. Hence, when water is spilled on a clean glass surface, it wets the glass and spreads out in a thin film. Consequently, the meniscus curves upwards (i.e., concavity) and the liquid rises in the tube.

Conversely, the cohesion of mercury is greater than its adhesion to glass. Therefore, when mercury is spilled on glass, it forms small spherical droplets or larger flattened drops. The meniscus, therefore, curves downwards (i.e., convexity), and this is accompanied by a capillary depression.

A liquid rises in a capillary tube if the angle of contact is acute, and a liquid (e.g., mercury) is depressed in the tube, if the angle of contact is obtuse.

9.17.7 Colloidal solutions and surface tension

a. **Suspensions and Suspensoids**: Coarse suspensions and suspensoids have little effect upon the air-liquid surface tension of the liquid, in which the particles are suspended. But an interfacial tension exists between the surfaces of the particles and surrounding liquids.

b. **Emulsions and Emulsoids**: These decrease the air-liquid surface tension of liquids in which they are suspended. These solutions,

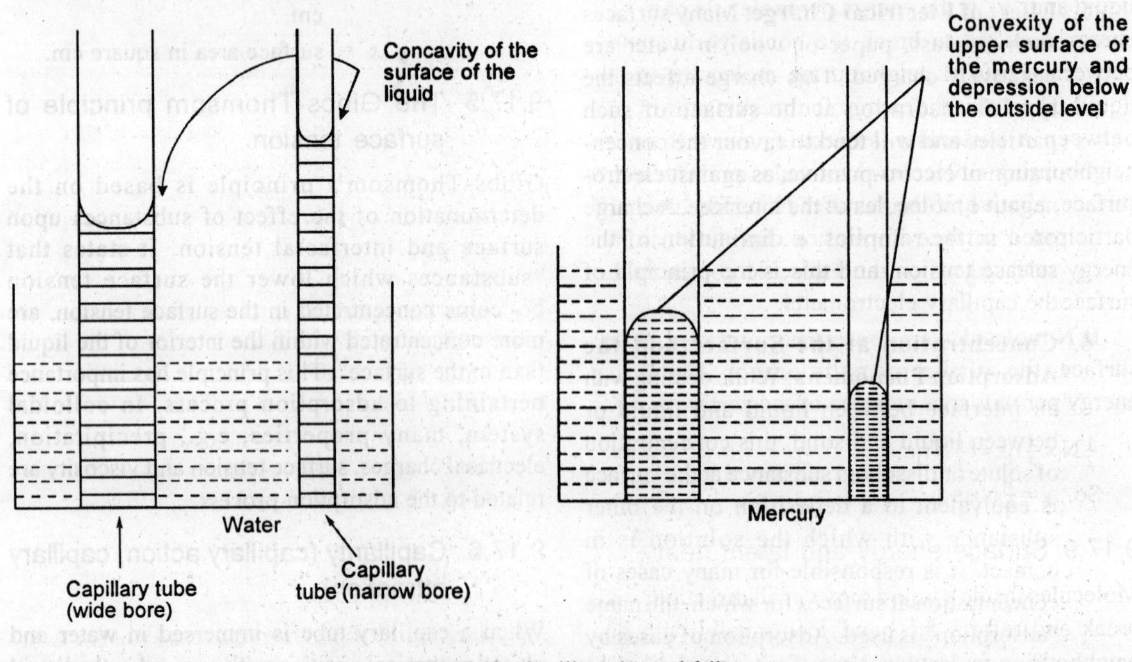

Fig. 9.45: Capillary rise and fall

when shaken, foam greatly due to lowering of the surface tension of the suspending liquid but suspensoids solutions do not foam significantly.

c. **Temporary Emulsion**: When olive oil is shaken with water, a temporary emulsion forms. The fine droplets produced during shaking coalesce and the emulsion breaks. The interfacial tension between olive oil and water exists which is 23 dynes at ordinary temperature.

d. **Permanent Emulsion**: When soap is added to the above mixture and shaken, a permanent emulsion is formed. Soap molecules intrudes at the interface between oil and water and decreases the interfacial tension and favours emulsion formation.

9.17.8 Molecular bonds and surface tension

Molecules reach the surface from the interior part, it breakes and reforms the bonds with the neighbouring molecules along the path while ascending. At the surface, half the bonds are not reformed and bond energy released becomes greater in the surface than in the total fluid. If η is the number of nearest neighbours per molecule in the liquid and "E" is the energy per molecule pair, i.e., energy required to break the bonds permanently between a pair of neighbouring molecules in the liquid. The $\eta\varepsilon$ is the energy for breaking the bonds between a particular molecule, as well as its neighbouring molecules. Molecules present in the surface, about half the number of bonds have not participated in the reformation. So $\frac{1}{2}\eta\varepsilon$ is the energy released, when the molecule comes to the surface.

If N molecules are present per unit area in the surface, the number of pairs is $\frac{1}{2}N$. Then, the energy per unit area increase of the surface is:

$$\frac{1}{2}N \times \frac{1}{2}\eta\varepsilon = \frac{1}{4}N\eta\varepsilon.$$

So, $\gamma = \frac{1}{4}N\eta\varepsilon$

9.17.9 Surface energy and latent heat

Molecules inside a liquid move in all directions and break and reform the bonds continually with the neighbouring molecules. Now if a molecule in the surface transforms into vapour outside, certain amount of energy is required to break the bonds with the neighbouring molecules in the liquid permanently. The energy required to evaporate a liquid has relation to its surface energy or surface tension. The latent heat of vaporisation which requires energy to change liquid to vapour at the boiling point, is related to the free surface energy.

9.17.10 Biological importance

1. **Hays Sulphur Tension Test**: It is under the principle of surface tension. When sulphur powder is sprinkled on the surface of the urine containing bile salts, it reduces suface tension. Bile salts concentrate on the liquid surface to lower the surface tension of the solution; so sulphur powder penetrates easily through the liquid surface. Sulphur powder sinks if there is no bile salt, the sulphur powder floats on the surface it bile salts are present

2. **Emulsification of Fat by Bile Salts**: Fats are finely emulsified in the small intestine by the detergent action of bile salts, (i.e., reducing surface tension) lecithin and monoglycerides because bile salts alone are poor emulsifying agents but in the presence of phospholipids and monoglycerides, particles 200-5,000 nm diameter are formed—preparatory to fat digestion and absorption. The mechanism is that the carboxyl or polar groups of bile salt is highly soluble in fat. Therefore, the sterol group dissolves in the surface layer of the fat globule and projecting the carboxyl group outward and soluble in the surrounding fluids which causes great decrease of the interfacial tension of the fat. When the interfacial tension of a globule of immiscible fluid is low, this fluid on agitation can be broken up into minute particles easily than, it can when the interfacial tension is great (See 9.15.5 II).

   Bile salts help fat digestion and adsorption by reducing surface tension of the solution of fat.

3. **Invagination During Pinocytosis**: Initial contact of the cell membrane with the material to be engulfed changes the surface tension of the cell membrane causing it to invaginate and draw the material inward.

4. **Alveolar Surface Tension and Lung Compliance**: An important factor affecting the compliance (stretehability) of the lungs is the surface tension of the film of fluid that lines the alveoli. The low surface tension, when the alveoli are small is due to presence of **Surfactant** in the fluid lining the alveoli, a lipid, surface tension – lowering agent.

Surfactant is produced by large epithelial cells lining the alveoli called **type II alveolar epithelial cells** or **pneumonocytes,** which contain osmophilic inclusion bodies which are the secretory products and secretion takes place by exocytosis and removed by alveolar marcophages. These cells show microvilli on their free alveolar surface for transport functions. The surfactant molecules are spread apart when alveolar size increases, i.e., in inspiration but come closer when alveolar size decreases, i.e., in expiration and adjust surface tension during respiration.

Surfactant is a complex 2-phase mixture of protein and lipids, but the major component is dipalmitoyphosphatidylcholine (DPPC). This phospholipid has a hydrophilic "head" and 2 parallel hydrophobic fatty acid "tails" like phospholipids in cell membranes. Surfactant acts by forming a monomolecular layer and the molecules are parallely oriented at the interface between the fluid lining the alveoli and the air in the alveoli. This prevents the development of a water-air interface, which has 1-14 times as much surface tension as the surfactant-air interface. Surface tension is inversely proportional to the concentration of surfactant.

- **Mechanism of Action**

(a) **Prevention of Collapse of Alveoli**: The surfactant molecules are more concentrated on the surface of the alveolar fluid film and reduce the surface tension of the fluid film according to Gibbs-Thomsom's principle and prevent the collapse of the lung alveoli (atelectasis). If the surface is not kept low of the fluid when the alveoli become smaller during expiration, they collapse in accordance to the Laplace law. In spherical structure like the alveoli, the distending pressure equals 2 times the tension divided by the radius, $R = 2R/T$, if T is not reduced as R is reduced, the tension overcomes the distending pressure.

During expiration, an alveolus becomes smaller and reduces the diameter, the surfactant molecules become more concentrated at the surface of the lining fluid film of the alveolus (i.e., at the air-liquid interface) and reduce surface tension progressively and prevent collapse of alveolus. Again, during inspiration, the alveoli are inflated and become larger and the surface fluid film is expanded. The surfactant molecules are spread more thinly and become widely separated on the fluid surface and their effect is reduced which increases surface tension of the fluid film, and help in the elastic recoil of lungs.

Thus, this special characteristic of surfactant helps to "stabilize" the sizes of the alveoli, causing the larger alveoli to contract more and the smaller ones to contract less. This effect helps to ensure that the alveoli in any one area of the lung, remain approximately the same size.

(b) **Preventing Accumulation of Edema Fluid in the Alveoli**: Surfactant also helps to prevent pulmonary edema. It tends to pull fluid into the alveoli from the alveolar wall. If surfactant were not present, the unopposed surface tension in the alveoli would produce a 20 mm Hg force favouring transudation of fluid from the blood into the alveoli.

(c) **Fetal Lung**: Surfactant is important at birth. The fetus makes respiratory movements in the uterus but the lungs remain in collapsed state until birth. After birth, the baby makes several strong inspiratory movements and the lungs expand. Surfactant keeps them from collapsing again. Surfactant deficiency causes hyaline membrane disease or respiratory distress syndrome, which develops in baby born before the surfactant system is functional. Surface tension in the lungs of those babies is high and there are many areas in which, the alveoli are collapsed (atelectasis).

Surfactant deficiency occurs in:

1. Hyaline membrane disease of respiratory distress syndrome (RDS),
2. Undergone cardiac surgery during which pump oxygenator was used,
3. Following occlusion of a main bronchus,
4. Occlusion of one pulmonary artery,
5. Long-term inhalation of 100 per cent oxygen,
6. Cigarette smoker.

5. **In Living Cell**: Since electrolytes may augment or diminish the charge at a surface, we see how important may be the constituent of a solution to the living cell with which it has to come in contact.

The fact that protoplasm consists largely of colloids involves a heterogeniety of structure and hence, large areas of surfaces. The tension at the cell surface has been measured in various ways and found to be usually between 1.0 and 0.1 dyne/cm. This value is very much lower than that would be expected, if the surface layer consisted of lipoid alone, and is explained as being due to the presence of an adsorbed layer of protein covering the lipoid membrane (Harvey and Danielli).

9.17.11 Method of determination of surface tension

1. **The Capillary Height Method**: When the end of a capillary tube is placed vertically in a liquid, the liquid rises in the capillary. It continues to do so until, the force of surface tension tending to pull the liquid upwards is exactly counterbalanced by the weight of the liquid in the capillary. At this equilibrium point, the liquid rises in the capillary to a height (say, h), which is the characteristic of the liquid. The surface of the liquid in the capillary tube, often called Meniscus, is generally curved. It may be concave (∪) or convex (∩) accordingly as the liquid wets the tube (e.g., turpentine) or, does not wet the tube (e.g., mercury). The curved surface makes an angle ∅ with the wall of the capillary which is called the **angle of contact**. For a concave meniscus the angle of contact is less than 90° while for the convex meniscus, it is more than 90°. It has been mentioned earlier that, the surface tension acts along the tangential line to the surface. The capillary rise of a liquid method are of two types,

(a) Single capillary method,

(b) Double capillary method.

It is given by the equation:

γ = surface tension in dynes per centimetre = $\frac{1}{2}$ hdgr

where h = height to which liquid rises in the capillary tube.

d = density of the liquid,

g = acceleration due to gravity,

r = capillary radius.

2. **The Drop-Weight Method:** It is based upon the weight of liquid drops, delivered from the tip of a tube calibrated against known liquid. When a liquid is allowed to flow very slowly through a capillary tube, a drop will form, which will increase up to a certain point and then fall. The apparatus is used called **Drop pipette**. The weight of the drop is proportional to the surface tension of the liquid.

3. **Torsion Balance Method**: A torsion balance was devised, in which, a platinum loop is dipped into the liquid and the force, in dynes, required to separate the loop from the surface is determined. The calibrated scale in the instrument gives a direct reading of the surface tension.

Beside these methods, there are Ring Method, Maximum Rubber Pressure Method, Microscope Slide Methods, etc.

- **Stalagmometer**

Stalagmometer is an instrument for measurement of surface tension of liquid. The principle of the method based on the direct proportionality between the surface tension and the weights of liquid, droped from the vertical capillary tube. It is made of glass and looks like: (a) Pipette having a capillary tube end, (b) Bulbous part, (c) Stem on the both sides of the bulbous part which is calibrated (Fig. 9.46).

The liquid whose surface tension is to be determined, sucked in the capillary tube (γ) and placed in the thermostat bath vertically. A known volume (v) of the liquid is read from the calibration and is dropped out slowly in regular drops from the capillary end and number of drops (n) is counted.

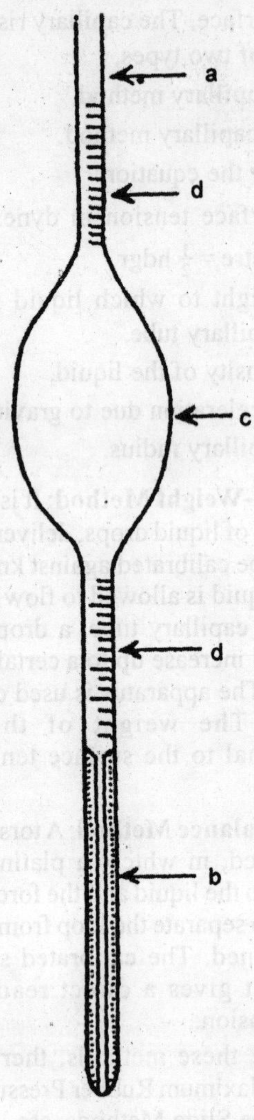

Fig. 9.46: Stalagmometer

The same procedure is followed with a liquid of known surface tension (γ') with the same volume (v) and counting the number of its drops (n'). Then the equation for determination of surface tension would be:

$$\frac{\gamma}{\gamma'} = \frac{n'\rho}{n\rho'} \text{ or, } \gamma = \frac{\gamma' n' \rho}{n\rho'},$$

where ρ and ρ' - known densities of the two liquids.

9.18 Donnan Membrane Equilibria

When a solution of NaCl is placed in a collodian sac and is immersed in distilled water, the final equal concentrations of salt reaches on both sides of the permeable membrane by simple law of diffusion.

I. Two Ionized and Diffusible Solution (Symmetrical Distribution)

It two ionised solutions (1) and (2), are placed in a container of constant volume and separated by a semipermeable membrane, an equilibrium is reached and phenomenon occurs at equilibrium. This was shown theoretically by Gibbs and confirmed experimentally by Donnan (1911). Hence, it is called Gibbs-Donnan Effect or Donnan Membrane Equilibria.

The phenomenon occurs at equilibria:

1. Solutions of both sides are electrically neutral because total charges on cations are equal to those charges on anions,
   $(\text{Cations})_1 = (\text{Anions})_1$, and
   $(\text{Cations})_2 = (\text{Anions})_2$.
2. The product of diffusible ions of one side of the membrane is equal to the product of the diffusible ions on the other side.
   $(\text{Diffusible cations})_1 \times (\text{Diffusible anions})_1 = (\text{Diffusible cations})_2 \times (\text{Diffusible anions})_2$
3. For diffusible ions (e.g., NaCl), at equilibrium, both sides of the membrane have the same concentration of both ions (Na^+ and Cl^-) and it is called **Symmetrical distribution**. When nondiffusible ions are present, at equilibrium, the distribution of diffusible ions are present, at equilibrium, the distribution of diffusible ions is **Asymmetrical.**

II. Non-Diffusible Ion and Diffusible Ion

A. Non-Diffusible Ion and Water (Change of pH)

When a sodium salt of non-diffusible anion (R^-), e.g., Sodium proteinate (NaR) (R^- may be other non-diffusible ion like congored or organic phosphate) is placed on one side of the membrane and water on the other side to consider H^+ as the diffusible cation, then the Gibbs-Donnan effect will lead to establish a pH difference on either side of the membrane.

(a) The Na will tend to pass through the membrane provided an equivalent amount of OH from the water diffuses in the same direction.

(b) The water will become alkaline while the solution of NaR will become less alkaline.

(c) If a chloride salt of non-diffusible cation is placed on the left-hand side, then the water outside will become acid and the solution on the left side will be less acid.

(d) At the isoelectric point of the protein, the Gibbs-Donnan effect will be at its minimum (potential difference).

- **B. Na R and Sodium Chloride**

One of the property of proteins, which is influenced by their net electric charge, is their effect in causing an unequal distribution of diffusible ions on either side of a membrane, through which proteins cannot pass. If on one side of a membrane, there is sodium chloride (Na^+ and Cl^-) solution (2) of diffusible ions Na^+ and Cl^- which penetrate in pairs and in another side (1) there is sodium proteinate (NaR) and NaCl containing indiffusible protein molecules (anions) of net negative charge (P^-) and an equivalent concentration of diffusible cations (Na^+).

Since there is a concentration gradient of Cl^-, this ion will diffuse into the protein solution in order to maintain electric neutrality, an equivalent quantity of Na^+ must also pass into the protein solution. However, this will tend to establish a concentration gradient of Na^+ and to promote the diffusion of Na^+ from the protein solution. A steady state will be attained at which the concentration gradient of Cl^- toward the protein solution is balanced by the concentration gradient of Na^+ outward from the protein solution.

This can only occur, if:

$(Na^+)_1 > (Na^+)_2$ and $(Cl^-)_1 < (Cl^-)_2$.

This difference in the concentration of the diffusible ions on opposite sides of the membrane, gives rise to a potentional difference which is equal to $58 \log \dfrac{(Cl^-)_1}{(Cl^-)_2}$ millivolts.

The two phenomenons are seen during equilibrium, e.g.,

(1) Electrical neutrality

$$(Na^+)_1 = (Cl^-)_1 + (R^-)_1$$

$$(Na^+)_2 = (Cl^-)_2$$

(2) Products and ratios of diffusible ions

$$(Na^+)_1 \times (Cl^-)_1 = (Na^+)_2 \times (Cl^-)_2.$$

Hence at equilibrium, the cation (Na^+) concentration on the side of the membrane containing the non-penetrating anion R^- is greater than the cation concentration on the other side; the opposite is true of the diffusible anion (Cl^-) concentration, which will be greater on the side without the non-diffusible anion. Donnan derived his theory based on thermodynamic consideration and his derivation was used based upon kinetics which is known as the **Gibbs-Donnan Membrane Equilibria.** Assuming that the volumes of the two compartments are maintained constant, i.e., during the distribution of the ions in accordance with the Gibbs-Donnan equilibrium, osmotic exchanges are excluded (Table: 9.1).

The ratios $(Na^+)_1/(Na^+)_2$ and $(Cl^-)_2/(Cl^-)_1$ are termed as "**Donnan ratios**". The equality of these ratios for all types of diffusible ions can be explained by the **Nernst Equation.** At equlibrium,

$$Em = E_{Na} = 60 \log_{10} \frac{(Na^+)_1}{(Na^+)_2}$$

Similarly,

$$Em = Ecl = 60 \log_{10} \frac{(Cl^-)_2}{(Cl^-)_1}$$

$$\text{Thus } \frac{(Na^+)_1}{(Na^+)_2} = \frac{(Cl^-)_2}{(Cl^-)_1}.$$

- **Salient Features**

1. There is greater number of ions in (1) than in (2).
2. The greater the equivalent concentration of protein, the more uneven will be the final distribution of the diffusible ions.
3. The greater the disparity in the total concentration of ions on either side of the membrane, the greater will be the osmotic pressure at the steady state.
4. The Gibbs-donnan effect always increases the osmotic pressure and the magnitude of the effect is decreased by raising the salt/protein ratio. For this reason, the determination of molecular weight by osmotic pressure measurements is best done in the presence of relatively high salt concentration.
5. pH affects the Gibbs-Donnan effect of non-diffusible protein ion because the ionic state of the protein changes with pH. At the

Table 9.10: Donnan membrane equilibria

Initial Sate		Final State		Initial State		Final State	
Na'		Na'	Na'	Na'	Na'	Na'	Na'
R'	H_2O	H	OH'	R'	Cl'	Cl'	Cl'
		R'				R'	
				(1)	(2)	(1)	(2)
Soldium salt of non-diffusible anion R' and water				NaR and NaCl			

isoelectric point of the protein, the Gibbs-Donnan effect will be at its minimum.

6. Donnaneffect may lead to a difference of pH between the two sides of a semipermeable membrane.
7. Gibbs-Donnan effect can give rise to potential difference due to the differential distribution of diffusible ions on both sides.
8. Donnan effect occurs by differential permeability and passive transport but the effect can be magnified or impeded by active transport which maintains a constant concentration of an intracellular ion, though it is impermeant.
9. Living cells contains impermeant anions, e.g., organic phosphates and proteins. These intracellular ions do exert some influence in the following way:

(a) The distribution of diffusible anion (e.g., Cl^-) and cations (e.g., Na^+, K^+) is unequal between cells and extracellular fluid even excluding the active transport and consequently a permanent potential difference develops across cell membrane.

(b) The concentration of cation (e.g., Na^+, K^+) inside cell is higher than the extracellular fluid e.g., RBC has excess cations of about 10 mM per litre of cells fluid but complication arises when there is active intake of K^+ and efflux of Na^+. As a result, this intracellular H^+ concentration is higher than the outside, e.g., RBC's pH is 7.2 and in plasma it is 7.4.

(c) The diffusible anion (e.g., Cl^-) concentration is lower inside the cells than external medium. In case of human RBC, the Donnan ratio is 0.7, in case of ratio between intracellular chloride to plasma chloride, as we see in case of "chloride shift" mechanism and in stomach, there is HCl secretion but lower Donnan ratio is observed in nerve due to high membrane potential and other reasons.

9.18.1 Relationship between Donnan effect, osmosis and active transport

If two compartments are separated by semi-permeable membrane of which one compartment contains solution of NaCl and non-diffusible anion NaR^- [Sodium Proteinate (R^-)] and in another compartment there is diffusible ions only (NaCl). i.e., Na^+ and Cl^-, an equilibrium is established and it is observed that at this state, the concentration of diffusible cations and anions is higher in the compartment of solution having non-diffusible ions (R^-), i.e.,

$$(Na^+)_1 + (Cl^-)_1 > (Na^+)_2 + (Cl^-)_2$$

1. **Exertion of Greater Osmotic Effect:** Due to increase in concentration of diffusible cations and anions in the compartment of non-diffusible ions, the total concentration of ions (i.e., diffusible ions + non-diffusible ions) in this compartment is higher, so, this compartment (1) exerts higher osmotic effect.
2. **True Equilibrium and Transfer of Water and Ions:** In the experiment the wall of the compartment is rigid like plant cell wall and both the solutions are within these walls. If the wall is elastic like living animal cell membrane, then water will flow from (2) to (1) having higher osmotic pressure. When it happens, there is dilution of solution in (1),

and (2) becomes concentrated, which cause disturbance of Donnan equilibrium but it can be re-established by diffusible ions passing with water from (2) to (1). With no limitation on volume, true equilibrium cannot be established until the total water and ions in (2) have passed to (1) compartment.

3. **Active Transport, Donnan Effect and Excess of Cation, Water Influx**: Animal cells having elastic walls, cannot resist the influx of water due to osmotic effects and these may rupture. But the active transport by the Na^+/K^+ pump in the membrane, opposes the Donnan effect and prevents, excess of cations and therefore, of water influx in the cells.

4. **Tonicity, Hydrostatic Pressure and Metabolic Work**: Though there is a large difference in tonicity across the natural membrane of metabolizing cell, but does not create any great difference of hydrostatic pressure because:
   (a) The living system is not in true equilibrium, and
   (b) Metabolic work is going on continuously.

5. **Passive Diffusion of Water**: When isolated tissue slices of kidney or brain are incubated in physiological saline solution with α-Ketoglutarate, the cells show gradual swelling with little loss of K^+ and gain of Na^+. When depression of cellular metabolism is done by reducing temperature to zero, or addition of cyanide, dinitrophenol or exclusion of substrate, cells show rapid swelling with loss of K^+ and gain of Na^+. It is evident that the cellular swelling is not for failure of pump of active transport of water out of the cell but for the failure of a metabolic pump which causes extrustion of Na^+.

- **Conclusion**

(a) Water always diffuse passively.

(b) Water movement in and out of the cell is commanded by the active transport of solute, usually Na^+. The work done to maintain or increase concentration differences across a membrane by water transfer is called **Osmotic Work**, e.g., for kidney is about 10.5 K J mol^{-1}.

9.18.2 Relationship of Donnan effect with osmotic pressure (colloid)

At the state of Donnan equilibrium, there is unequal distribution of diffusible ions across membrane, leading to an osmotic pressure difference between the solutions across the membrane. This is due to greater concentration of the diffusible ions (Anion + Cations) in the side having non-diffusible ions. Suppose the concentrations of the ions between the two sides of the membrane are as under:

(1)	(2)
Na^+	Na^+
$a + x$ moles	$(b - x)$ moles
Cl^-	Cl^-
x moles	$(b - x)$ moles
R^-	
a moles	

The value of x, the amount of NaCl diffusing from (2) to (1), is:

$x = \frac{b^2}{a+2b}$. For illustration, let $a = 1$ mole and $b = 2$ moles; then $x = \frac{4}{1+4} = \frac{4}{5} = 0.8$.

After adding up the concentrations in (1) and (2), we have:

(1)		(2)	
Na^+	1.8	Na^+	1.2
Cl^-	0.8		
R^-	1.0	Cl^-	1.2
3.6 moles		2.4 moles	

Since osmotic pressure is expressed by the equation:

$$P = cRT,$$

where c = concentration, then the osmotic pressure at the state of Donnan effect will be:

$$P = (3.6 - 2.4)\, RT = 1.2\ \text{RT},$$

At 27°C (300Å) the osmotic pressure would be:

$$P = 1.2 \times 0.082 \times 300 = 29.52\ \text{atm}.$$

When a colloid in the from of a compound with acid or base undergoes ionic dissociation and it is used for Donnan effect experiment, certain phenomenon can be observed. Congo red is a sodium salt of a colloid acid and if it is placed in an osmometer and then immersed in water, a certain osmotic pressure is developed. Sodium ions of Congo red pass easily through the membrane of osmometer and carry a positive charge and the fluid inside becomes negative. Due to negativity of non-diffusible ions containing solution, the positively charged sodium ions are arranged in the immediate neighbourhood of the membrane by the electrostatic attraction between Congo red acid ions and the sodium ions. Now, if sodium chloride is added to either, the inner or outer, fluid, the osmotic pressure falls, if time is allowed for equilibrium between the two sides of the membrane. At this state, the outer fluid without Congo red, contains more chloride than the inner fluid and the difference is permanent and is more marked if the concentration of dye is greater. The general features are:

(a) The greater the equivalent concentration of non-diffusible ions (Congo red, Protein, etc.), the more uneven will be the final distribution of the diffusible ions.

(b) The Gibbs-Donnan effect always increases the osmotic pressure, and the magnitude of the effect is decreased by raising the salt/protein ratio. For this, the determination of molecular weight by osmotic pressure measurements is best conducted in the presence of salt concentrations.

(c) It follows also, that the greater the dispartiy in the total concentration of ions on either side of the membrane, the greater will be the osmotic pressure.

When osmotic pressure of a solution is determined by using membrane, the presence of non-diffusible ions (e.g., Protein), may distrub the result, if precautions are not taken to minimize or correct the error for the osmotic pressure due to Donnan effect. This error is particularly prevalent in experimental Protein solutions (e.g., Serum), in which protein present as nondiffusible negative or positive ions and salts are present. It would be more effective if, a membrane is used which is freely permeable and impermeable to protein, and to make the Donnan effect minimum, measurement is made at the isoelectric point of the protein by using high salt concentration.

When gelatin, a protein gel, is placed in water solution at a pH above or below its isoelectric pH, electrolytes in the solution induces the gel to take up more water and swell. The swelling is due to partly by some specific effect of the ions and partly by the **Donnan osmotic effect** between the gel and the outside solution.

The mechanism of swelling can be explained that the gelating surface acts as a semipermeable membrane to the surrounding fluid containing diffusible electrolytes. Some of the proteins in the gelatin present as P^+ or P^- ions, depending on pH which cannot diffuse through the gel surface into the surrounding solution. As a result, a Donnan equilibrium is established between the gelatin and the elctrolytes in the solution which leads to higher concentration of diffusible ions in the solution having non-diffusible ions; hence, water diffuses from the surrounding solution into the gel causing swelling. The amount of swelling depends upon:

(a) the concentration of electrolyte,

(b) the pH,

(c) the valence,

(d) the nature of the diffusible ions.

It, therefore, can be stated that, the Donnan osmotic effect is one of the reasons for the swelling of the proteins due to electrolytes. Sometimes, in certain pathological conditions, abnormal swelling of tissues are observed which is known as "Cloudy Swelling" by the histopathologist and is due in part to the **Donnan osmotic effect.**

9.18.3 Biological importance of Donnan principle

1. **Ionic Concentration Between Plasma and Interstitial Fluid and Lymph**: The intersti-tial fluid and lymph have lower concen-trations of inorganic cations (Na^+, K^+) and higher concentrations of inorganic anion (Cl^-), than the blood plasma. Starling in his "Starling principle", described these fluids as a protein-free ultrafiltrates of blood plasma. While analysing the fluid movement through capillaries, Starling principle stated that rate and direction of fluid movement is determined by the balance of hydrostatic and oncotic pressure. But the relationship of blood plasma to these fluids with respect to their diffusible ions can be explained by the higher content of proteins in plasma separated by the semipermeable capillary membrane and the resultant

distribution according to the Gibbs-Donnan Law. Higher protein (anion) concentration in the plasma, then in the interstitial fluid which is impermeable to the capillary membrane causes enhancement of efflux of inorganic anion like Cl^- from the plasma and decreases efflux of diffusible cations like Na^+. The equilibrium state between the two compartments is charac-terised by the following features:

(a) Higher concentration of diffusible cations Na^+ in the high protein containing plasma and higher diffusible anion Cl^- in the interstitial fluid,

(b) The total concentration of equivalent of charge is higher in the protein-containing plasma.

(c) Electrical neutrality is maintained in both compartments in spite of differences in ion concentration and total concentration of equivalents of charge. Electrical neutrality = Total number of anionic charges equals the total number of cationic charges.

In spite of the above features, the process is not a simple one because of presence of small amount of proteins (less than 0.1 meq/l H_2O) in the interstitial fluid.

2. **Ionic Concentration between Intracellular Fluid (ICF) and Extracellular Fluid (ECF):** ICF contains cations Na^+ in low and K^+ in high concentrations and anions like Cl^-, HCO_3^- in low concentration and predominantly organic phosphates, e.g., ATP and proteins in contrast to plasma and interstitial fluid (ECF). ICF is not a continuous fluid phase and composition differs in different tissues. The differences between ICF and ECF can be explained by the following factors:

(a) The Na-K ATPase in cell membranes is responsible for active transports of Na^+ from the ECF and K^+ into the cells.

(b) The cell membrane is limited permeable to organic phosphates and proteins, and establishes Donnan membrane equilibrium across the cell membrane.

Ionic composition differences between ICF and interstitial fluid for Donnan equilibria are manifested by the following features:

(i) Higher diffusible cations (Na^+, K^+) in protein containing ICF and higher concentration of diffusible anions (Cl^-) in the interstitial fluid. The transport of cations is not mainly dependent on Donnan effect but due to presence of Na–K ATPase.

(ii) Protein-containing ICF contains greater total concentration of equivalents of charge.

(iii) Maintenance of electrical neutrality in each compartment but an electrical potential difference is seen between ICF and interstitial fluid due to excesses or deficits of these small number of ions.

3. **Chloride-Bicarbonate Shift in RBC and Plasma**: During respiration, there is constant influx of CO_2 into the blood of the blood vessels and a corresponding loss of CO_2 from the blood vessel in the lungs. As a result of high CO_2 pressure in the tissues, CO_2 passes rapidly into the plasma, then to erythrocytes. After entering erythrocyte, CO_2 undergoes **Hydration** by the enzyme carbonic anhydrase into carbonic acid and then subsequent **Ionization** by simple dissociation and by interaction with the potassium salts of oxyhaemoglobin, gives rise to bicarbonate.

$$CO_2 + H_2O \xrightarrow[\text{anhydrase}]{\text{Carbonic}} H_2CO_3 \rightarrow H^+ + HCO_3 \rightarrow$$

$$H_2CO_3 + K^+ + \begin{bmatrix} (HbO_2)^- \\ \text{or} \\ Hb \end{bmatrix} \rightarrow \begin{bmatrix} HHbO_2 \\ \text{or} \\ HHb \end{bmatrix} + K^+HCO_3^-$$

H^+ cannot move out easily because erythrocyte cell membrane is relatively impermeable to cations and is taken up by the histidine group of haemoglobin and becomes HHb. This reaction is necessary because reduced Hb is better "proton acceptor", i.e., less acid than the oxygenated form and favours loading of CO_2 while oxygenation in the lungs favours unloading of CO_2.

When bicarbonate concentration in RBC increases, it may cause imbalance of ionic distribution between RBC and plasma and as a compensatory mechanism, there occurs diffusion of 70 per cent of the formed HCO_3^- from cell to plasma but the corpuscular membrane is impermeable to the anion Hb, which imposes Donnan principle. As bicarbonate anions come out of the cell, chloride (Cl^-) anions enter the cell to maintain electrochemical neutrality in order to restore the Gibbs-Donnan equilibrium ratio involving these two anions and the phenomenon is known as **Chloride-bicarbonate shift** or **chloride-shift.** The Cl^- content of RBC in venous blood is, therefore, significantly greater than in arterial blood. The chloride shift occurs rapidly and is complete in one second.

In such system the concentration of diffusible negative ions in the plasma will be greater than their concentration in the cell. In other words,

(HCO_3^-) plasma > (HCO_3^-) cell,
(Cl^-) plasma > (Cl^-) cell, and
(OH^-) plasma > (OH^-) cell.

According to Donnan principle, the following relative ratio of diffusible ion concentration in cells and plasma is given by:

$$\frac{(HCO_3^-)\text{ cell}}{(HCO_3^-)\text{ plasma}} = \frac{(Cl^-)\text{ cell}}{(Cl^-)\text{ plasma}} =$$

$$\frac{(OH)\text{ cell}}{(OH^-)\text{ plasma}} = \gamma = 0.7 .$$

The magnitude of γ depends on the concentration of the impermeable ion.

- **Reverse Chloride Shift**

In the lungs, the chloride-bicarbonate shift is reversed. As HbO_2 reacts with HCO_3^- to form H_2CO_3 and liberates CO_2, which is blown off, HCO_3 in the cell goes down, again disturbing the ionic ratio, but in the reverse direction. In this state.

$$\frac{(HCO_3^-)\text{ cell}}{(HCO_3^-)\text{ plasma}} < \frac{(Cl^-)\text{ cell}}{(Cl^-)\text{ plasma}} =$$

$$\frac{(OH^-)\text{ cell}}{(OH^-)\text{ plasma}} = \frac{(H^+)\text{ plasma}}{(H^+)\text{ cell}} .$$

Consequently, HCO_3^- ions pass from plasma into cells and Cl^- and OH^- ions from cell to plasma, until the ratios are restored.

- **Water Balance in the Cell during Chloride-Shift**

During the exchange reaction in chloride-shift, the cells are still having a relatively higher concentration of osmotically active ions, so some water enters the cells to restore osmotic equilibrium between RBC and plasma. The passage of blood through the lungs is accompanied by the release of CO_2 from the blood to the alveolar air and the above reactions are reversed and RBC loses water. This results in venous erythrocytes being a bit larger than arterial erythrocytes and causes 3 per cent increase in haematocrit value in venous blood, than that of the arterial blood.

4. **Difference in pH, between the Cells and Plasma**: Donnan effect causes the difference in pH between the cells and plasma. Intracellular pH is usually between 6-7.4 in different cells, averaging 7.0 whereas, pH of plasma is 7.4. A rapid rate of metabolism in cells increases CO_2 production and consequently decreases pH.

   The cell membrane is not freely permeable to H^+, it is to OH^- and thus assures the proportionality of pH values in cells and plasma.

   OH^- ions move into the cells, as HCO_3^- ions move out and vice versa. The plasma is very poorly buffered against it. The HCO_3^- shift from cells to plasma and reverse as CO_2 is taken up and given off, it is very much important in reducing pH changes in plasma during CO_2 transport.

5. **Colloidal Osmotic Pressure**: There is an increase of 50 per cent colloidal osmotic pressure than the predicted value obtained by protein alone during Donnan effect. This can be explained: The proteins are negative ions and to maintain a balance of these negative ions, a greater number of positively charged ions, i.e., cations, especially sodium ions, are attracted towards electronegative charges of the proteins. This excess cations increase the number of osmotically active particles and increase the total colloidal osmotic pressure of plasma. The colloidal osmotic pressure of normal human plasma is on average 28 mm Hg., 19 mm Hg caused by the dissolved protein, and 9 mm Hg by the cations held in

the plasma by the Donnan effect of the proteins. If the concentration of protein is higher, the Donnan effect becomes more significant.

6. **Osmolality of Body Fluids**: Osmolality of plasma exceeds that of the interstitial fluid by 1-2 mosmol/kg H_2O. This small difference is attributable both to the osmotic contribution of the plasma proteins and to the larger concentration of diffusible ions in plasma, as a result of Donnan effect. This difference is not so significant (for which, the major body fluids can be considered isosmotic for clinical purposes).

7. **Glomerular Filtrate**: The ratio between plasma/filtrate for small nonelctrolytes is 1, but there is slight difference in ratio for small electrolytes due to different concentrations in plasma and the glomerular filtrate, which is attributable to the Donnan effect. The proteins in plasma which is absent in the glomerular filtrate establish the Donnan effect between plasma and the filtrate. Proteins in the plasma have negative charge and cations, e.g., Na^+, K^+ are attracted for which plasma contains higher concentration, while anions, e.g., Cl^-, HCO_3^- are in higher concentration (5 per cent) in the filtrate and the concentration of cations is about 5 per cent lower than in plasma.

8. **Electronegativity of Plasma**: The non-diffusible plasma protein anions, according to the Gibbs-Donnan effect, decreases the potential inside the blood vessel relative to the extra-vascular tissue and interstitial fluid, making the intravascular plasma electro-negative.

9. **Determination of MW by Osmotic Pressure Method**: The Gibbs-Donnan effect always increases, the osmotic pressure and the magnitude of the effect is decreased by raising the salt/protein ratio. For example, determination of MW of protein by osmotic pressure method gives error due to contaminated diffusible salts. For this reason, measurement is best done:

   (a) by adding relatively high concentration of diffusible salt in the solution, which impedes the Gibbs-Donnan effect of the protein.

   (b) by maintaining the solution at the isoelectric pH of the protein to change the protein ions to Zwitterions with the minimum Gibbs-Donnan effect.

10. **Maintenance of Constant Intracellular Concentration of Ions**: Donnan effect occurs by differential permeability and passive transport but the effect can be magnified or impeded by active transport, which maintains a constant concentration of an intracellular ion though it is impermeant.

11. **Unequal Distribution of Anions and Cations in the Cell**: Living cells contain impermeable anions, e.g., organic phosphates and proteins. These intracellular anions do exert some effect in the following way according to Gibbs-Donnan principle:

    (a) The distribution of diffusible anions (e.g., Cl^-) and cations (e.g. Na^+, K^+) is unequal between cells and extracellular fluid.

    (b) The concentration of cations (e.g., K^+) inside cell is higher than the extracellular fluid.

    (c) The diffusible anion (e.g., Cl^-) concentration is lower inside the cells than external medium.

12. **Relationship between Donnan Effect, Active Transport, Osmosis, Excess Cations**: Animal cells having elastic walls, cannot resist influx of water due to osmotic effect caused by Gibbs-Donnan principle and these may rupture. But the active transport by $Na^+ - K^+$ pump in the membrane opposes the osmosis caused by Donnan effect and prevent excess cation transport and, therefore, of water influx in the cells.

13. **Swelling of the Protein (Cloudy Swelling)**: Donnan osmotic effect is one of the reasons for the swelling of the proteins due to electrolytes. Sometimes in certain pathological conditions, abnormal swelling of tissues are observed, which is known as **Cloudy swelling**, by the histopathologist and is due partly to the Donnan osmotic effect.

14. **Intracellular pH of RBC Fluid:** It is found that, pH of intracellular fluid of RBC is lower

than the plasma. This is partly due to high concentration of negative non-diffusible haemoglobin ions. This causes unequal distribution of H^+ ions on the opposite side of the RBC membrane with a higher concentration of H^+ ions inside the cells and pH falls and plasma becomes alkaline.

15. **Impermeability of Living Cell Membrane to Certain Electrolytes**: The impermeability of living cell membrane to certain electrolytes is due partly to the Donnan effect. The cytoplasmic proteins exist partly as negative ions, because they have isoelectric pH values below the pH of the cells. Whenever there is impermeability of diffusible ions to cell membrane, it will cause Donnan effect across the cell membrane.

16. **Incerease in RBC Size and Haemetocrit Value**: During chloride-bicarbonate shift, the cells are still having a relatively higher concentration of osmotically active ions, so some water enters the cells to restore osmotic equilibrium between RBC and plasma. The passage of blood through the lungs, is accompanied by the release of CO_2 from the blood to the alveolar air and the above reactions are reversed and RBC lose water. This results in venous erythrocytes being a bit larger than arterial erythrocytes and causes 3 per cent increase in haematocrit value in venous blood than that of the arterial blood.

17. **Diffusion Potential**: Electrolytes diffuse across cell membrane, like in free solution. Due to their polar nature, they cannot cross lipid membrane easily. They have mobility and move in solution under the influence of electrical potential and their diffusion establishes an electrical potential. Cl^- ions are permeable to cell membrane and their concentration is lower inside the cell, relative to higher concentration in the interstitial fluid. So, Cl^- tends to diffuse across the cell membrane along its concentration gradient. The non-diffusible anion is higher in the cell making it electronegative which causes an outflow of Cl^-, so the diffusion influx of concentration gradient is counterbalanced by the diffusion influx by electrical, gradient, i.e., due to presence of non-diffusible anion in the cell produces Gibbs-Donnan effect manifested by higher concentration of Cl^- outside and low concentration in the cell. These diffusion potentials occur spontaneously due to kinetic forces when one ion species diffuses at a different rate than its counterions(s). The size of the potential is proportional to the differences in ionic mobilites. Concentration gradient is the source of energy of these potentials. If diffusion proceeds for longer period to abolish the concentration gradient, the diffusion potential will also be abolished.

18. **Equilibrium Potential**: Chloride ion is higher in ECF than in the cell's interior. Chloride ions diffuse along the "concentration gradient" into the cell. The interior of the cell is negative relative to external medium and chloride ions are driven out of the cell along this "electric gradient" and an equilibrium is reached at which Cl^- influx and Cl^- efflux are equal in spite of its unequal distribution. It is for any ion whose unequal distribution across the membrane is mainly for Gibbs-Donnan principle, equilibrium potential E depends solely on the concentration difference on the two sides of the membrane resulting from Gibbs-Donnan principle only. The membrane potential at which this equilibrium exists is **Equilibrium potential**. The magnitude can be calculated from the "Nernst equation" and denoted by E ion. Three main equilibrium potentials are important for cell, i.e., sodium (E_{Na}), poitassium (E_K) and chloride (E_{Cl}).

$$E_{Cl} = \frac{RT}{FZ_{Cl}} \text{ in } \frac{(Cl_o^-)}{(Cl_i)}$$

where E_{Cl} = equilibrium potential for Cl^-,
R = gas constant,
T = absolute temperature,
F = (Faraday number of coulumbs per mole of charge),
Z_{Cl} = valance of Cl^- (–1),
(Cl_o^-) = Cl^- concentration outside the cell,
(Cl_i^-) = concentration inside the cell.

From the natural log to the base 10 log and replacing some constants with numerical values, the equation is:

$$E_{Cl} = 61.5 \log \frac{(Cl_i^{o})}{(Cl_o^{-})} \text{ at } 37^{o}\text{C}.$$

When it is converted to the simple expression, the concentration ratio is reversed because the minus 1 valence of Cl^- has been removed from the expression. E_{Cl} value is –70 mV by calculation and identical to measure resting membrane potential of –70 mV. The equilibrium potential for an ion is found by equating the diffusion force on this ion to the electric force on this ion. The result is the "Nernst potential" which is taken to be the intracellular potential minus the extracellular potential. If the membrane potential of a cell equals the equilibrium potential of an ion, there will be no net movement of that ion into or out of the cell. All ions do not have the same equilibrium potential and, therefore, they cannot all be in equilibrium at any given cell membrane potential.

19. **Membrane Potential**: The potential difference across the membrane of most cells (not all cells), with the inside negative and outside positive, known as **Resting membrane potential** or **Steady potential.** It is written with a minus sign, signifying that the inside is negative relative to the outside. Its magnitude varies from tissue to tissue, ranging from –9 to –100 mV.

- **Genesis**

Though the ions like K^+ and Cl^- may sometimes be in equilibrium across cell membranes, they are not the main cause for the membrane potential, even their diffusion would set up potentials.

The advent of intracellular recording techniques, i.e., Micro-Electrode Technique made it easy to measure membrane potentials directly and test their dependence upon K^+ and other ions. It has been observed that, (a) the distribution of ions across the cell membrane and (b) nature of the membrane give the explanation for the membrane potential. Several factors should be discussed such as:

(a) K^+ diffuses out of the cell along its concentrations gradient, while the non-diffusible anion remains in the cell, creates Gibbs-Donnan effect, and sets up a potential difference across the membrane. For this, there is slight excess of cations outside the membrane and slight excess of anions inside the membrane. But the number of ions responsible for the membrane potential is minute compared to the total number present.

(b) Resting membrane is much less permeable to Na^+ and K^+. So, Na^+ influx cannot compensate for the K^+ efflux.

(c) In spite of Gibbs-Donnan effect of non-diffusible anions (organic phosphates, proteins), the Na^+ concentration is higher outside the membrane than inside the membrane. And passive influx of K^+ for electrical gradient, causing higher concentration in the cell, leads to passive outflow for concentration gradient, ultimately excess cations (Na^+, K^+) make the outer membrane positive and negative inside due to higher anions (non-diffusible ions, Cl^-). The separation of charges by the membrane, known as **Polarization** of the membrane.

(d) Cl^- diffuses into the cell along its concentration gradient, but it is pushed out by the electrical gradient to balance its movement.

(e) It is electrogenic sodium-potssium pump which set up the membrane potential, where upon all permeable ions would move until they come into equilibrium with this potential. It has long been known that increase in the potassium ions can depolarize nerve and muscle, and this leads to the belief that the diffusion of potassium ions cause, the membrane potential. But this pumps functions to maintain the concentration gradients which is responsible for membrane potential. If the pumps are made inactive by metabolic poisons or inhibitors (e.g., azide, cyanide, dinitrophenol), Na^+ influx and K^+ efflux occur and the membrane potential drops. The rate of fall varies with size of the cell. In large cells it takes hours but in nerve fibres with diameter less than 1 μm, complete depolarization can occur in less than 4 minutes. The magnitude of the membrane potential at any given time depends upon the distribution of Na^+, K^+ and Cl^- and the permeability of the membrane to each of these ions. An equation which describes the relationship of these ions are known as **Goldman Constant-field Equation**:

$$V = \frac{RT}{F} \ln \left[\frac{P_{K^+}(K_o^+) + P_{Na^+}(Na_o^+) + P_{Cl^-}(Cl_i^-)}{P_K^+(K_i^+) + P_{Na^+}(Na_i^+) + P_{Cl^-}(Cl_o^-)} \right]$$

where V = membrane potential,

R = gas constant,

T = absolute temperature,

F = Faraday number

P_{K^+}, P_{Na^+}, P_{Cl^-} = permeability of the membrane to K^+, Na^+, Cl^- respectively,

( ) = concentration,

i = inner,

o = outer.

Since P_{Na^+} is low relative to P_{K^+} in the resting cells, Na^+ contributes little to the value of membrane potential (V). It can be predicted from the Goldman equation: changes in external Na^+ creates little change in the resting membrane potential, whereas increase in external K^+ decreases it.

20. **Edema Fluid**: Edema fluid may be taken as representation of interstitial fluid so far as the Gibbs-Donnan principle is concerned. It has been observed that the distribution of ions between edema fluid and plasma, so far as it has been determined, corresponds approximately with the requirements of the Donnan effect, though its abnormal conditions of formation may sometimes contains smaller concentration of protein than that in the true interstitial fluid.

21. **Erythrocyte Potential**: Due to impermeability of the cell membrane to the negative Hb-ion, the diffusible negative ions are unevenly distributed between the cell and plasma fluid:

$$\frac{(OH^-)\text{ Cell}}{(OH^-)\text{ Plasma}} = \frac{(Cl^-)\text{ Cell}}{(Cl^-)\text{ Plasma}} = 0.7$$

This uneven distribution with greater concentration of negative ions in the plasma can be maintained only by Gibbs-Donnan potential which slows down the rate of diffusion of negative ions from plasma to erythrocyte, i.e., there will be a Gibbs-Donnan potential across the membrane, outside being positive and the inside negative. With Gibbs-Donnan ratio 0.7, the potential is about 4.9 mV and the plasma and tissue fluid having Donnan ratio 0.96, the potential is 1 mV only.

22. **Erythrocytes, Cation Impermeability and Haemolysis**: Stored human blood for a period of weeks at low temperatures has revealed some important facts, e.g., consideration of the consequences on the basis of simple equilibrium theory of a normal permeability of the erythrocyte to both anions and cations relative to a virtual impermeability to cations (except H^+ ions) and rapid exchange of univalent anions.

Significant differences between:

(a) the system of plasma separated by its ultrafiltrate, and

(b) the system of the erythrocyte, separated from plasma are discussed.

Although there is a Gibbs-Donnan principle in both systems, in the first system osmotic equilibrium is only involved by the exertion of hydrostatic pressure on the fluid containing the non-diffusible anion, whereas, in the latter, osmotic equilibrium may be achieved without any such hydrostatic pressure. The explanation for the differences of the two is the effective impermeability to the erythrocyte membrane to cations, beside the Hb^- ion; with this impermeability, it is possible to balance the osmotic pressure of the non-diffusible protein in erythrocyte. Without such an impermeability, no such balance can be attained. Thus, from the point of osmosis, the integrity of the erythrocyte would appear to depend on its permeability to cations; any agent that destroys the impermeability should cause haemolysis.

9.19 Membrane Hydrolysis

Donnan principle states that:

(a) The presence of a non-diffusible ion on one side of a membrane causes membrane hydrolysis, and

(b) Difference in pH on the two sides of the membrane.

A. When a salt, sodium proteinate (NaR), is placed in water in one side of the compartment separated by a semipermeable membrane and water in the other side only, then at equilibrium the result will be like the final stage. Since R^- causes a greater concentration of diffusible negative ions on the opposite side of the membrane, OH^- negative ions from water move through the membrane along with positive Na^+ ions to maintain electrical neutrality, the solution No. 2 becomes alkaline and solution No. 1 shows fall of pH. The membrane allows the hydrolysis reaction which is:

$$NaR + HOH \rightarrow NaOH + HR$$

and there is separation of NaOH and HR across the membrane (Table 9.11A).

Table 9.11A: Membrane hydrolysis

(1)	(2)
Na^+ R^- H^+ OH^-	Na^+ OH^- H^+
pH falls	Alkaline

B. If protein hydrochloride (RCl), is taken in the compartment (1) instead of NaR, then positive H^+ ions along with negative Cl^- ion will move from compartment (1) to (2), resulting in greater concentration of H^+ ions in compartment (2) for non-diffusible R^+ ions in (1). Ultimately, solution (2) becomes acid in reaction and solution (1) shows rise in p^H (Table: 9.11B)

Table 9.11B: Membrane hydrolysis

(1)	(2)
R^+ Cl^- H^+ OH^-	H^+ Cl^- OH^+
pH rises	Acid

The membrane allows the hydrolytic reaction in this way:

$$RCl + HOH \rightarrow ROH + HCl$$

As a result, there is retention of ROH in one side of the membrane and secretion of HCl through the membrane to the opposite side.

1. It is found that pH of intracellular fluid of RBC is lower than the plasma. This is partly due to high concentration of negative non-diffusible haemoglobin ions. This causes unequal distribution of H^+ ions on the opposite side of the RBC membrane with a higher concentration of H^+ inside the cells and pH falls and plasma becomes alkaline.

   The tissue cells have lower pH than the outer fluid which is partly due to presence of high concentration of negative protein ions in the cells.
2. The impermeability of living cell membrane to certain electrolytes is due, partly, to the Donnan effect. The cytoplasmic proteins exist partly as negative ions because they have isoelectric pH values below the pH of cells. Whenever there is impermeability of diffusible ions (K^+, Ca^{2+}) to cell membrane, it will cause Donnan effect across the membrane.
3. Regulation of distribution of electrolyte ions across the cell membrane in living organism is operated by the Donnan effect having secondary importance because of the limited permeability to certain ions. Na^+, K^+, Ca^{2+} and Mg^{2+} have low permeability to RBC membrane which causes maintenance of high concentration Na^+ and Ca^{2+} in the plasma compared to the inner fluid of RBC and high concentration of K^+ and Mg^{2+} inside the RBC, than their concentration in the plasma. These electrolytes are responsible for the Donnan effect across the cell membrane.
4. The distribution of HCO_3^- and Cl^- ions between plasma and RBC as per Donnan effect, the ratio should be:

$$\frac{\text{Cells } (HCO_3^-)}{\text{Plasma } (HCO_3^-)} = \frac{\text{Cells } (Cl^-)}{\text{Plasma } (Cl^-)},$$

when the ionic activity coefficients are the same in the plasma and RBC. But, these are not equal and the above equation will be valid, if proportionality factor, f, equal to 0.8 to 0.9 is applied:

$$\frac{\text{Cells } (HCO_3^-)}{\text{Plasma } (HCO_3^-)} = f\frac{\text{Cells } (Cl^-)}{\text{Plasma } (Cl^-)},$$

According to Donnan principle, the distribution of cations Na^+, K^+, Ca^{2+} and Mg^{2+} between blood plasma and lymph fluid, is higher in plasma due to higher negative protein ion than the lymph. The ratios of these concentrations are:

$$\frac{Na^+ (P)}{Na^+ (L)} = 1.01 \qquad \frac{K^+ (P)}{K^+ (L)} = 1.5$$

$$\frac{Ca^{2+} (P)}{Ca^{2+} (L)} = 1.25 \qquad \frac{Mg^{2+} (P)}{Mg^{2+} (L)} = 1.4$$

(Concentrations in millimols per kg of water)

(P) = Plasma, (L) = Lymph

These are not equal according to the Donnan effect.

- **Discussion**

Concentration gradients of diffusible ions are essential for the normal physiological process of the living system. Though Donnan principle may exert some effect in some living system, the living system is a dynamic one in which energy is required for maintenance of concentration gradient which is provided by the intracellular chemical reactions. We see in case of concentration of Cl^- is higher in lymph than in plasma according to Donnan effect, the HCO_3^- ions in plasma is higher than in lymph contrary to the Donnan effect. Though relative concentration of anions in RBC and plasma are in accordance with the Donnan equilibrium but the relative concentration of cations cannot be explained by the Donnan principle. Because cations like Na^+ and K^+ are "actively transported" and need not conform the Donnan effect because the active transport requires:

1. energy and dependent on the oxidative metabolism of cells, and
2. acts against the concentration gradient.

So, application of the Donnan effect to the living system is a complicated problem.

9.20 Ultrafiltration

A colloidal solution can pass easily through ordinary filter paper and porcelain filter because the size of the pores are not so small to retain the colloid particles. However, filters can be prepared which will allow the colloidal particles of different sizes of and retain the larger size colloidal particles. Thus colloidal particles of different sizes may be separated by filtration under a pressure, the process is called **Ultrafiltration and Colloidal filters** of different pore-sizes are called **Graded filters.**

- **Preparation of Graded Filters or Ultra-filters (Semipermeable Membrane)**

1. Filter papers are soaked in gelatin solution of different strengths and is hardened the gelatin in formaldehyde.
2. Impregnation of filter paper with collodion solution of a suitable concentration.
3. Cellophane membranes are best ultrafilters.
4. Intercellulose esters are either supported on a grid of wire mesh or deposited on a piece of porous porcelain or filter papers.

- **Ultrafiltration vs. Osmosis**

During ultrafiltration, a pressure greater than osmotic pressure is exerted causing the movement of water from a higher concentration to a lower one and this condition is represented as ultrafiltration. Ultrafiltration is the opposite of osmosis, which represents the movement of water from the lower to the higher concentration and it is a "mixing process". Ultrafiltration is represented by the movement of water from the higher to the lower concentration and it is a "separation process".

- **Biological Importance**

1. **Determination of the Size of Colloidal Particles:** If the colloidal particles pass through pores of a certain size, it may be assumed that their diameter is lesser than that of the pores in which they cannot pass through. A series of collodian membranes with graded pore sizes may be prepared and used for determination of colloidal particle size. The colloidal solution to be tested is filtered under pressure through series of filters. The method has a number of limitations.
2. **Separation of plasma proteins**: From electrolytes and small organic substances.
3. **Isolation of protein of specific molecular sizes:** From contaminated proteins in which pressure or centrifugal force is applied to filter the watery medium and small solute molecules through a semipermeable membrane,

which retains the protein molecules (Fig. 9.47).

4. **Biological Fluids (Ultrafiltrate)**
   (a) **Bicarbonate Secretion in the Gastric Mucosa**: Flemstrom (1981) showed that some ultrafiltration of bicarbonate across the mucosa may also occur during gastric secretion.
   (b) **Glomerular Filtrate**: The glomerular membrane is freely permeable to water, crystalloids (ions, glucose, urea). It has limited permeability to larger solutes, i.e., colloidal particles (large proteins, lipids), and completely impermeable to the cellular elements of blood. Without proteins, lipids and cells, the glomerular filtrate is identical to plasma for which it is called **ultrafiltrate** compared to simple filtrate which excludes only the cellular elements.

   For ultrafiltration, two things must be fulfilled:

   1. Presence of a barrier to retain molecules above a certain size while, allowing smaller one to pass through.
   2. There must be a head of pressure to force fluid through the barrier.

   These two criteria have been fulfilled by the glomerular membrane (See. 6.29.8).

   (c) **Intercellular Fluid**: Intercellular fluid is formed from blood by the process of ultrafiltration. The capillary pressure at the arterial ends of the capillaries is 15-20 mm Hg greater than at the venous ends. Because of this difference, fluid filters out through the capillary walls into intercellular space. It contains all the constituents of blood plasma, less the proteins. The wall of the capillaries act as filter holding back the plasma protein molecules together with the cellular components of the blood but allowing everything else to go through. Small amounts of fluid circulates amongst the cells and return to the circulation by two routes, one is venous end of the capillaries and excess fluid is drained into the lymph vessles.

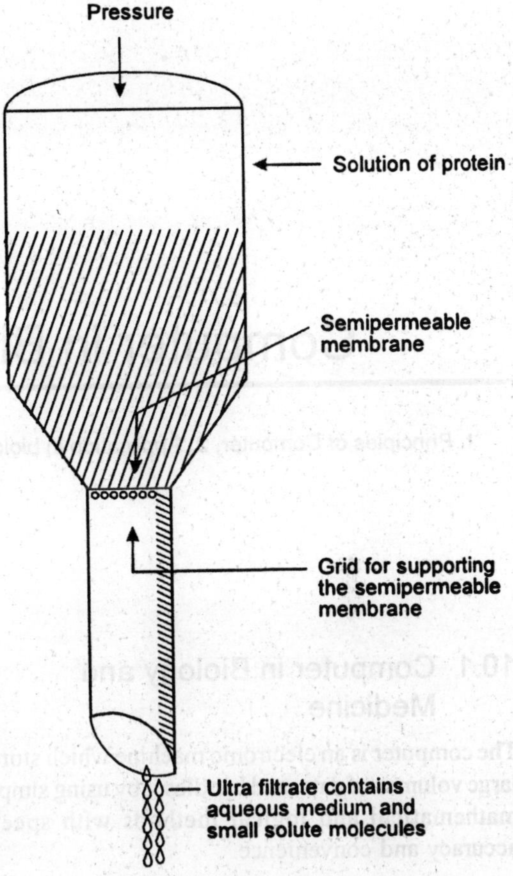

Fig. 9.47: Ultrafiltration of a protein solution by applying positive pressure above the membrane

   (d) **Artificial Kidney**: The artificial kidney is used for severe renal failure and in certain types of acute renal failure, following shock, poisoning, etc. to eliminate the excess waste products from the blood retaining plasma proteins and blood cells. In case of normal person, the dialyzing fluid does not contain HPO_4^-, urate, sulphate, urea and creatinine but dialyzing fluid of uraemic patient, contain all these substances due to large quantities in the plasma and are eliminated in the fluid.

10

Computer in Biology and Medicine

1. Principles of Computer, 2. Application in biology and medicine, 3. Computer versus human brain.

10.1 Computer in Biology and Medicine

The computer is an electronic machine which stores large volumes of data, and handles it by using simple mathematical and logical methods with speed, accuracy and convenience.

10.1.1 Historical background

1. **Ancient period**: Ancient China used the abacus which was probably the ancestor of the modern computers.
2. **1642**: Pascal Blaise developed adding machine worked on wheels and gears while the figures were displayed on dials. Perhaps that was the earliest version of a computer.
3. **1694:** A machine was devised by Liebniz, G. W., which had the capability of addition, multiplication and division.
4. **1883**: The "analytical engine", designed by Charles Babbage, was a fully automatic general purpose digital computer but it was not developed for use.
5. **1880**: Hermal Hollerith, devised a punch card machine and completed 1890 census in America in less than 3 years, which was to take 9 years.
6. **1946**: Presper Eckert, J and John Mauchly were to devise first electronic computer in 1946. 30 tonnes, occupied 15000 square feet spreading over two storeys and carried out 300 multiplications per second.

Subsequently the development of transistors and silicon chips accelerated further development in the next decades.

10.1.2 Computer "generations"

1. **Vacuum tube device**: The computers developed in 1956 using vacuum tubes as the main logical units were bulky, carried out 5000 basic mathematical manipulation per second.
2. **Semiconductor device**: Then in 1954 transistors, i.e., semiconductor devices replaced the vacuum tube, made the size of computer smaller, consumed less power became faster and more accurate and carried out 100000 instructions per second.
3. **Silicon chip device (Integrated circuits)**: Silicon chip device was introduced in 1970 which made the computer smaller and faster. Each silicon chip could accommodate hundreds of transistors.
4. **Reduction in size of silicon chip (Large scale integrated circuits)**: Further develop-

ment of computer observed when silicon chip was used having size of 0.5 square millimetre which can accommodate hundreds of thousands of parts. A silicon chip can hold 2,000 transistors. This chip is now the headquarter of a computer having the control processing unit and arithmetic and logical unit.

10.1.3 Units of a computer

The main units which function in a computer are:

1. Input unit,
2. Storage unit,
3. Control unit,
4. Arithmetic and logical unit,
5. Output unit.

10.1.4 Functional process

Problems or information are fed to the input units. The central processor unit consists of three units, viz., Central unit, Storage unit and Arithmetic and logical unit. The input data is transmitted by the control unit along with the necessary information or data from the storage which is recalled. The input data may be stored as memory. The storage unit holds the programme on which one is working and the data to be analysed. The stored data along with input data are transmitted to arithmetic and logical unit by the control unit. The function of arithmetic and logical unit is just like a calculator such as addition, subtraction, multiplication, division, preparation of list, selection and comparison based on the stored data. After that the data is transmitted by central processor unit to the output unit to give final answer (Fig. 10.1).

Whatever information or data is fed to computer through the input by our alphabet is not understood by the computer. Hence the data has to be converted into "yes" or "no", or "zero" and "one" which is known as "binary system". Now the input unit has an "user" to type out whatever data is to be fed on a normal typewriter keyboard. The matter which is being fed is displayed on a "visual display units". Inside, there is a converter which converts all this normal data to the binary system. Output unit also having a converter which give a visual display or a print out in human readable form. The function of this converter is to convert binary numerals to letters or figures as needed.

Simple computer: In simple computer, the output signals are controlled directly by the input signals.

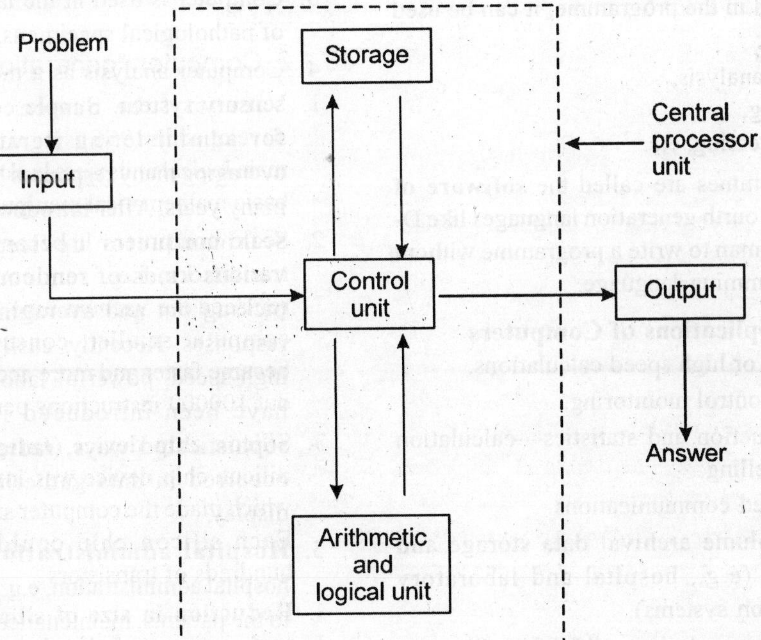

Fig. 10.1: Block diagram of an electronic computer showing the basic functional units

Complex computer: In this complex compter, the output is determined both by the input signals and by information that has already been stored in memory in the computer.

More Complex computer: In addition of the units present in the complex computer, in this computer it is necessary to add still another unit known as "central programming unit" which determines the sequence of all operations.

10.1.5 Languages of the computer

We cannot command or order the computer in our spoken language which is to be converted into machine language or binary system for which specific short forms have been deviced to order the computer. These languages can be different for different applications such as:

1. **Fortran**: Formula Translator is used for scientific purposes.
2. **Cobol**: Common Business Oriented Language used for general purposes.

These computer languages are helpful for a programmer to command the computer how to proceed for doing a calculation in stepwise and logical manner. These orders or instructions constitute a programme. Depending upon the instruction given in the programme, it can be used in the following:

(a) Scientific analysis,
(b) Accounting,
(c) Word processing, etc.

These programmes are called the **software of the computer**. Fourth generation languages like D-base, allow a layman to write a programme without knowing programming language.

- **General Applications of Computers**

1. Complex or high speed calculations.
2. Quality control monitoring.
3. Data reduction and statistics—calculation and modelling.
4. High speed communications.
5. Large volume archival data storage and retrieval (e.g., hospital and laboratory information systems).
6. Interpretive analysis—artificial intelligence.
7. Education—self-instructional.

- **Application of Computer in Medicine**

1. Hospital Administration
2. Laboratory tests.
3. Clinical tests.
4. Medical teaching.
5. Surgical procedure display.
6. Computer assisted imaging.
7. Classical diagnosis and decision making.
8. Patient Monitoring.
9. Storage of patients, history and clinical data for future use.
10. Computerise aid of the physically handicapper.
11. Patient management.
12. Radiotherapy administrations.

10.2 Application in Biology and Medicine

1. Assessment, diagnosis and treatment of acid-base balance disorder.
2. It is being used widely in case of Siggard-Anderson nomogram. The nomogram is used with pH determination of the blood in the Astrup method of measurement of P_{CO_2}.
3. Computer is used in the laboratory analysis of pathological specimens, e.g., blood, etc.
4. Computer analysis as a method of studying sensory system. Simple computer methods for administering iterative stimuli and averaging many responses have been used for many years. After introduction of laboratory scale computers it became easier to give various kinds of randomized stimuli and picking out and averaging the individual responses. Recently, easily programmable, high-speed, powerful laboratory computers have been introduced into research in sophisticated ways, radical improvement occurred in data gathering, analysis and display.
5. **Hospital administration**: It is used in hospital administration, e.g., allocation of bed to the patients, maintenance of inventories of drugs, surgical equipment, linen, etc. payroll, hospital accounting, etc.

6. **Storage of information for patient**: Information regarding the condition of patients, details of laboratory investigations done and clinical reports can be stored in the memory of the computer system. This combined information can provide instructions and reminder regarding the care of individual patients.
7. **Intensive care unit**: It is used for monitoring a patient's condition. Scanning instruments attached to the patient are connected to the computer system so that the duty staff can be informed regarding patient's condition.
8. **Alarming signal**: It can print out or display a log of the patient's condition, drawing attention to measurements that fall outside the critical level set by the physician or the computer itself may trigger directly the necessary corrective action.
9. **Physician's clinic**: In clinic, it is used to collect information during the interview of the patient with the physician for the patient's records and to assist with the diagonostic process.
10. **Assisting medical diagnosis in particular analysis**: It assists the medical diagnosis in case of analysis like electro-cardiogram to determine both normal and abnormal heart conditions.
11. **Acting as medical encyclopaedia**: It acts as a store of vast medical knowledge which is not possible to keep in one's brain. Hence it can be considered as medical encyclopaedia. In case of complicated or critical patients where the diagnosis is doubtful, the physician can verify or correct the diagnosis by opening a dialogue with the computer.
12. **Prescription of correct dosage and pattern of treatment**: The computer helps in prescribing the correct dose and pattern of treatment, e.g., for cancer treatment by radiography, where correct dose of radium is required to avoid radiation hazards and to locate the actual area of treatment. Computers are pressed into the service for such delicate calculations.
13. **Treatment timetable**: It is being used to prepare a treatment timetable for individual patients. It is done by using the data provided by the consultant, the computer prepares a timetable complete with the calculated doses for the individual patient.
14. **Medical research**: The computer plays an important role in medical research. It stores information, data, reference in large scale and detailed records of case histories of particular illness are available for scrutiny and ready references.
15. **Teaching:** The computer is being used for teaching the doctors and nursing staff.
16. **Computer "models"**: The different "models" of the parts of the body, e.g., lungs and heart, can be constructed with the help of the computer. The models simulate the behaviour of the parts of the body.
17. **Testing the effect of treatment**: It can be used to launch programmes for testing the effect of a kind of treatment, might have on a patient, before administration.
18. **Computer in renal scanning**: The radioactivity over the kidney can be detected by the Gamma Camera and fed to the computer. In most cases, the radiopharmaceutical flows to the kidneys as well as to other organs present in the surrounding. This background radioactivity is automatically eliminated by the computer to give radioactivity over the kidneys only.
19. **Multiple Gated Aquisition Studies (MUGA) for ventricular function**: Ventricular function is scanned by Multiple Gated Acquisition Studies. The cardiac cycle recorded in ECG is divided from R wave to next into 16, 32, 64 different portions. In the computer memory, the same number of frames or pigeon holes are made. Information picking up by Gamma Camera from a particular portion of the cardiac cycle is fed to its corresponding frame or pigeon hole of the computer. Then the information can be presented either as a curve or as a picture. Thus, the computer can supply us the information of different parameters for ventricular function.
20. **Positron emission tomography (PET)**: A complete Positron Emission Tomography unit consists of:

1. A data acquisition system,
2. A computer-data from the acquisition system is fed into a computer with a large memory.

This method supplies useful information regarding in *vivo* quantitative measurement of:

(a) Regional blood flow,
(b) Glucose utilisation,
(c) Oxygen utilisation,
(d) Incorporation of selected amino acids in tissue.

(For details, see chapter 11 on Nuclear Medicine.)

21. **Single phtoton Emission Computed Tomography for myocardial imaging**: Recently this technique is being used for myocardial imaging for detection of coronary artery disease. It is another form of Positorn Emission Tomography which uses single photon emitters, e.g., technetium-99 m or iodine-123.

It is a major advance in medical technique for myocardial imaging. By this method, multiple view images are obtained circumferentially and then reconstructed by computer as a single three dimensional image.

22. **Magnetic Resonance Imaging (MRI) for cross-sectional images of body**: The nuclei of some atoms (e.g., Hydrogen nucleus) behave as small bar magnets. If it is placed in a magnetic field and excited by pulse of radiowaves at certain frequency, the nuclei emit a radio signal (MR signal) and detected by a receiver. Analysis of this signal provides information about the chemical nature and environment of the nuclei. The associated computer transforms the received MR signals into:

(a) *Anatomical information*: e.g., cross-sectional images of body.
(b) *Physiological*: three dimensional description in health and diseases.
(c) *Biochemical*: three dimensional description in health and diseases.
(d) *Characterisation of pathology*: e.g., tumours, inflammation and infection.

23. **Storing and retrieving data**: While storing data (e.g. reference of journals, abstract, review or excerpt with references) the source and the abstract is fed to the computer. In this way thousand pages of printed matter can be stored on a single floppy disc. After getting order, the computer scans through the huge amount of data and within minutes locates the matter wanted. A print out of the data can also be obtained if necessary. This facility can be available from far distant places, through telex and satellite. Even the reprints of abstracts are also available by telex from large libraries on a mainframe computer. The physician can book a personal computer. The connection between the physician's computer with the library computer in other place is through telephone lines. At both ends there are modem which converts computer data to electrical signals compatible with the telephone lines.

The general mode of transmission is shown below (Fig 10.2).

24. **Computer as artificial intelligence**: Recently the computer is being used as artificial intelligence to assist in decision making at the bed side. The computer simulates the actual behaviour of an experienced clinician. The computer is able to "think" like a human and to "learn" from the past experiences. Though the fastest computer carries about 50000000 instructions per second but still it is too slow to approximate higher forms of human thought involving rapid association and analysis of ideas. **The human brain still remains far superior of any computer.**

25. **Computer in X-ray Diffraction or X-ray Crystallography**: High computer is employed in X-ray Diffraction or X-ray Crystallography to study the precise positioning of the atoms and the shape of the whole molecule.

26. **Computer in Radiotherapy**: Computer systems are very much useful for:

(a) Radiotherapy administration,
(b) Information generation,
(c) Workload management,
(d) Treatment planning and delivery.

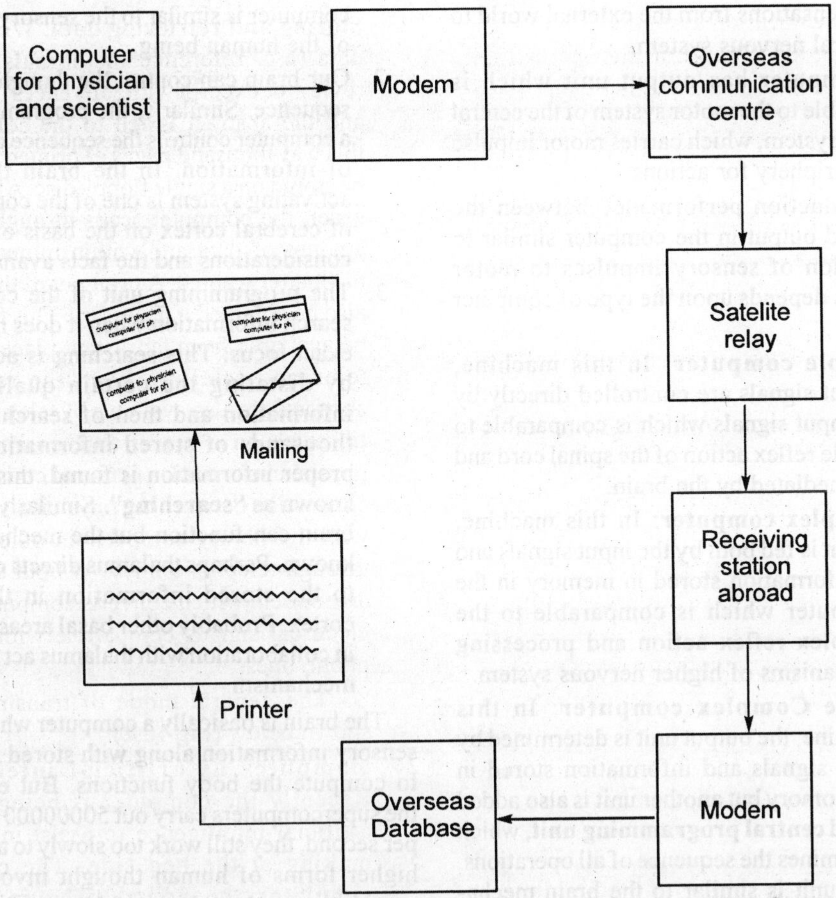

Fig. 10.2: Schematic representation showing data storage and retrieval

27. **Computer in ECG Analysis:** Computers are used in medical diagnosis for analysis and evaluation of electrocardiograms such as:
    1. Reduces the time required for interpretation of electrocardiograms.
    2. Ensures better reliability and consistency of interpretation.
    3. Lowers cost of electrocardiographic interpretation because it permits better use of the cardiologist's time by releasing him from routine burdensome tasks of reviewing large number of recordings.
    4. It is also possible to increase the diagnostic potential of the electrocardiogram.
    5. It assists the physician in making the diagnosis with much greater accuracy than by manual method of visually inspecting the analog tracing.
    6. It assists the physician to meet the ever increasing ECG load with greater efficiency.
    7. The recorded electrocardiogram which is analog in character is digitized by taking several hundred samples per second, the sampling rate is usually, but not always, 500 samples per second per load.
28. **Computer in Scanning:** Computer is essential for scanning different parts of the body for image reconstruction.

10.3 Computer Versus Human Brain

The electronic computer machine is comparable to the human central nervous system. They have many features in common with the central nervous system. These are as under:

1. Computer has input which is comparable to the sensor unit of the nervous system, which

carries sensations from the external world to the central nervous system.

2. The computer has output unit which is comparable to the motor system of the central nervous system, which carries motor impulse to the periphery for actions.
3. For conduction performance between the input and output in the computer similar to conversion of sensory impulses to motor impulses depends upon the type of computer such as:
   (a) **Simple computer**: In this machine, output signals are controlled directly by the input signals which is comparable to simple reflex action of the spinal cord and not mediated by the brain.
   (b) **Complex computer**: In this machine, output is fed both by the input signals and by information stored in memory in the computer which is comparable to the complex reflex action and processing mechanisms of higher nervous system.
   (c) **More Complex computer**: In this machine, the output unit is determined by input signals and information stored in the momory but another unit is also added called **central programming unit**, which determines the sequence of all operations. This unit is similar to the brain mechanism which gives directions to focus first to one sensation, thought or motor activity, then to another and so forth until we get a complex sequence of thought, or motor activity takes place.

- **Programming unit of brain computer**

The reticular activating system of the brain probably acts as a programming unit of a complex computer. The functions of programming unit and reasons for considering activating system as programming unit are given below.

1. The programming unit of the computer receives input signals and recalls the information stored in memory of computer and directs the attention of different parts of the computer to specific information stored in other parts of the computer. Similarly, the reticular activating system of the brain probably directs one's attention to specific mental activities. The information in the computer is similar to the sensory information of the human being.
2. Our brain can control the thoughts in orderly sequence. Similarly, the programming unit of a computer controls the sequence of processing of information. In the brain the reticular activating system is one of the controlling unit of cerebral cortex on the basis of anatomical considerations and the facts available.
3. The programming unit of the computer can search information when it does not know the exact locus. This searching is accomplished by distating the certain qualities of the information and then of searching through thousands of stored information until the proper information is found, this function is known as "**searching**". Similarly, the human brain can function but the mechanism is not known. Perhaps thalamus directs our attention to the stored information in the cerebral cortex. Probably other basal areas of the brain in collaboration with thalamus act in searching mechanism.

The brain is basically a computer which receives sensory information along with stored information to compute the body functions. But even though the supercomputers carry out 50000000 instructions per second, they still work too slowly to approximate higher forms of human thought involving rapid association and analysis of ideas. Moreover, the brain of an expert in any subject uses a parallel processing of the information whereas the computers use sequential processing. When a human brain receives various sensory impulses, it analyses the same, both simultaneously and independently. The resultant output is linked continuously with each other through the various neural pathways. The final interpretation is made by the cerebral cortex and then decision is taken. This decision is stored in the memory to be used for future reference. Thus, human brain learns from past experience.

A different approach is required to simulate a human brain. A separate computer having different design is necessary which can learn from past experiences. Present technique of sequential processing used in conventional programmes does not suit for this purpose. **The human brain still remains far superior to any computer in the world.**

11

Nuclear Medicine (Diagnostic and Therapeutic Technology)

1. Radioimmunoassay (RIA), 2. Organ Scan, 3. Liver (Imaging) Scan, 4. Bone Scan, 5. Brain Scan, 6. Renal Imaging, 7. Radioactive renal clearance studies, 8. Cardiac Imaging, 9. Myocardial Scintigraphy, 10. Positron Emission tomography, 11. Myocardial Imaging, 12. Magnetic Resonance Imaging, 13. Radio-iodine for diagnosis in thyroid disorder, 14. Radio-iodine therapy in thyroid disorder, 15. Isotopic Method of determination of blood volume, 16. Laser Beam in biology, medicine and surgery, 17. Nuclear Medicine for Therapy.

11.1 Radioimmunoassay Method (RIA) (Radioactive competitive binding assay)

Radioimmunoassay was described first by Yalow and Berson (1959) for serum insulin assay. But it was first used by Ekins (1960).

- **Introduction**

The technique of RIA pursuits measurement of antigens up to picogram (10^{12}g) quantities. RIAs have proved to the very useful for the quantification of hormones etc. in very small volumes of biological samples and eliminated often, a need for purification or extraction of the principle to be analysed from the samples.

- **Principle**

It is basically a competitive binding assay in which fixed amounts of antibody and labelled antigen react in the presence of known (standards) or unknown (samples) amounts of the antigen. The amounts by which the binding of labelled antigen to its antibody are competitively inhibited by increasing amounts of standard antigen preparation are recorded (dose response curve or standard curve). The amount of antigen present in an unknown sample can then be calculated from standard curve by interpolation of the binding of labelled antigen. The assay in which both labelled and unlabelled antigens (present in standard sample) are allowed to compete simultaneously for the limited number of binding sites (limited amount of antibody) is called equilibrium type of RIA. The one in which the unlabelled antigen is first allowed to bind the antibody (first incubation) and then further incubated with the labelled antigen is known as non-equilibrium type of RIA.

11.1.1 Definition

Immunoassay is a method utilizing the immunological properties of a substance for its detection and quantification. When isotopes are used as tracers

in this assay it is called **radioimmunoassay**, also known as **radioactive competitive binding assays** because both labelled and unlabelled substance compete for the binding sites on the antibody. This method can detect immuno-competent proteins like antigen, hormone and also can detect smaller molecules and even ions. This is the most sensitive method for antigen but now it is used extensively for hormone assay.

11.1.2 Principle

The basis of the assay is the antigen-antibody reaction. The antigen hormone is the substance being measured. The antibody to this antigen or hormone is raised by injecting small quantities of antigen or hormone in animals like rabbits, guinea pigs and sheep. This antibody can reversibly bind to the specific antigen or hormone for which it is produced. When this antigen and antibody are combined at equilibrium, some of the antigen or hormone will be free or some antibody receptor sites will be utilized depending upon what is excess, the hormone or antibody.

Ag (Antigen or hormone)	+	Ab (Antibody)	⇄	AgAb Antigen-antibody complex (or hormone-antibody complex)

A. Antigen Measurement

The procedure is to mix a known amount of isotopically labelled antigen, with specific antibody. Then either the antigen-antibody complex is precipitated with ammonium sulphate or antiglobulin sera or the free uncombined antigen is removed by adsorption on charcoal. The ratio of free to bound radioactivity is then estimated. The labelled and unlabelled antigens will compete with each other for combining sites on the antibody in proportion to their relative concentration. From the ratio of free to bound radioactive antigen the concentration of unlabelled antigen can then be calculated. This method is only useful in the presence of antigen excess but yet it is capable of measuring as little as 10^{-12} g/ml of antigen.

B. Hormone assay

• **Principle**: The hormone assay method is also based on antigen-antibody reaction. The hormone to be measured acts as an antigen. The antibody is produced by injecting small quantities of hormone in animals like sheep, rabbits and guinea pigs. The anti-hormone antibody can reversibly bind to the antigen hormone. When hormone and anti-hormone antibody combine, at equilibrium, some of the hormone will be free or some antibody receptor sites will be unutilized depending upon what is excess, the hormone or the antibody.

H Hormone	+	Ab Antibody	⇄	H Ab Hormone-antibody complex

Now if the hormone is labelled with a radioisotope, this helps to measure the bound and free amount of hormone. The amount of hormone that binds to a unit quantity of antibody can also be measured. If some labelled and some unlabelled hormone are added to a unit quantity of antibody, these two hormones (labelled and unlabelled) compete with each other for receptor site on the antibody. The amount labelled hormone bound to the antibody varies according to the concentration of the unlabelled hormone and is inversely proportional to it.

H* Labelled hormone	+	H Unlabelled hormone	+	Ab Specific antibody of hormone	⇄
HAb Unlabelled hormone-antibody complex	+	H* Ab Labelled hormone-antibody complex	+	H + H* Free hormones-labelled and unlabelled	

By adding different known concentrations of unlabelled hormone to a fixed quantity of labelled hormone and antibody, the amount of bound labelled hormone can be measured for different concentrations of unlabelled hormone and the results are plotted and a standard curve is prepared (Fig. 11.1). If serum or fluid containing an unknown quantity of labelled hormone is added to the fixed quantity of labelled hormone and antibody, the amount of bound labelled hormone can be measured and the amount of hormone present in the serum can be observed from the curve.

11.1.3 Method

Suppose if we are to estimate gonadotrophin, e.g., FSH level in blood, the following method is employed. Nowadays composite Kits for different

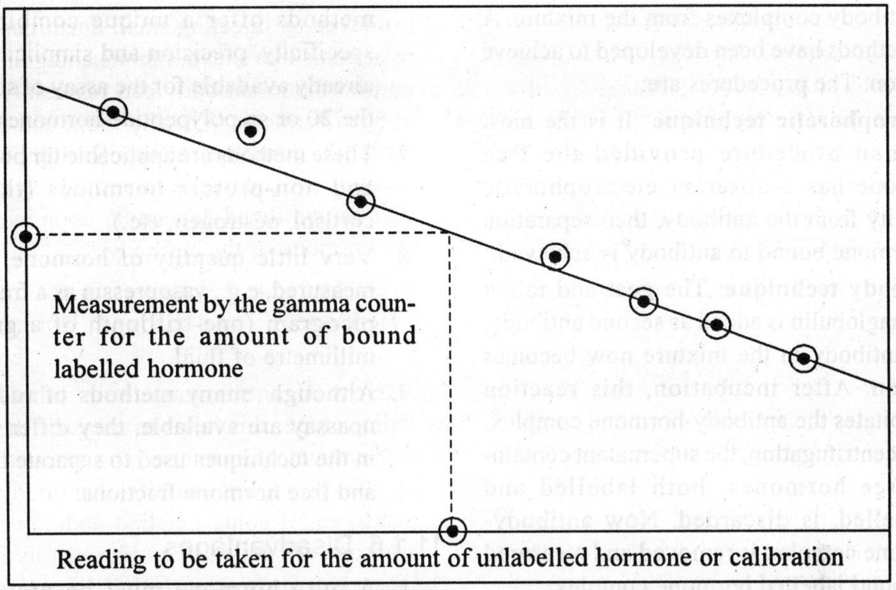

Fig. 11.1: Standard or calibration curve for assay of hormone by radioimmunoassay (RIA)

substances are available. If it is FSH Kit, then it consists of (1) I^{125} labelled FSH, (2) Rabbit anti-FSH serum (Antibody raised against FSH in rabbit), (3) Precipitating antiserum reagent (Goat-anti-rabbit gammaglobulin, (4) Standard FSH-6 vials containing increasing quantities of unlabelled FSH. After taking blood from the patient, it is allowed to clot at room temperature and serum is separated by centrifugation.

↓

• **Mixing:** Then in a test tube a mixture is prepared by mixing three different substances:

1. Patient's serum containing unknown quantity of FSH to be assayed.
2. Fixed amount of I^{125} labelled (purified) hormone, i.e., FSH.
3. A fixed amount of antibody, i.e., anti-FSH serum.

Monoclonal antibodies are now extensively in the tests. It would be noticed that one specific condition is to be maintained, i.e., there must be too little antibody for both of the labelled and unlabelled FSH hormones to combine completely. Therefore, the unlabelled FSH of serum and the labelled radioactive hormone compete for the binding sites on the antibody.

↓

Incubation for the period specified on the kit.

↓

• **Binding**: The test tube contains the complex of antibody combined with both unlabelled and labelled FSH plus some free labelled and unlabelled FSH.

The quantity of the bound radioactive, i.e., labelled FSH, is inversely proportional to the concentration of the natural unlabelled FSH of patient's serum. After binding is complete, the antibody-hormone (FSH) complex is separated from the remainder of the solution and the quantity of radioactive FSH that has bound with the antibody is measured by means of radioactive counting equipment called Gamma Camera. The result obtained with the test sample is to be plotted and its concentration is obtained against the standard curve (Fig. 11.1).

If a large amount of radioactive hormone (FSH) has bound, there was only a small amount of test hormone (i.e., patient's FSH) to compete. Conversely, if only a small amount of labelled hormone (FSH) has bound, then there was a very large amount of test hormone (i.e., patient's FSH) to compete for the binding sites of antibody.

11.1.4 Separation technique

In order to measure the amount of labelled hormone attached to antibody it is necessary to separate the hormone-antibody complexes from the mixture. A variety of methods have been developed to achieve this separation. The procedures are:

(a) **Electrophoretic technique**: It is the most common procedure provided the free hormone has a different electrophoretic mobility from the antibody, then separation of hormone bound to antibody is achieved.

(b) **Antibody technique**: The goat and rabbit gammaglobulin is added as second antibody. The antibody in the mixture now becomes antigen. After incubation, this reaction precipitates the antibody-hormone complex. After centrifugation, the supernatant containing free hormones, both labelled and unlabelled, is discarded. Now antibody-hormone complex is removed and measured for bound labelled hormone complex.

(c) **Antibody-insoluble support technique**: Though a second antibody is provided in the kit for separation of bound and free hormone, sometimes antibody being linked to an insoluble support like cellulose, charcoal, ion exchange resin or silicates. The insoluble complex is then mixed with labelled and test hormone (tube A, Fig. 11.2), and allowed to interact (tube B, Fig. 11.2), and later the unreacted hormone is removed and the amount of labelled hormone attached to the insoluble complex is measured (tube C, Fig. 11.2).

11.1.5 Advantages

Methods based on radioimmunoassay have the following advantages:

1. Facilities are more economical and easily available.
2. Methods are highly specific and convenient.
3. These methods can assay 20-40 samples per hour without any human intervention.
4. These methods are more sensitive than bioassay.
5. These are the most sensitive methods for measurement of antigen and capable of measuring as little as 10^{-12} g/ml. of antigen.
6. The methods offer an accurate quantitative estimation of polypeptide hormones. These methods offer a unique combination of specificity, precision and simplicity and are already available for the assay of some 14 of the 20 or so polypeptide hormones in man.
7. These methods are applicable for both protein and non-protein hormones (thyroxine, cortisol, oestrogen, etc.).
8. Very little quantity of hormone has been measured, e.g., vasopressin as a fraction of a picogram (one-trillionth of a gram) per millimetre of fluid.
9. Although many methods of radioimmunoassay are available, they differ primarily in the techniques used to separate the bound and free hormone fractions.

11.1.6 Disadvantages

1. A pure hormone must be prepared for standard.
2. These methods can measure degradation fragments or precursors of the hormones that retain some binding activity but can vary in their biologic activity.
3. Appropriate species of animal is required.
4. Each reading has to be in duplicate since microlitre quantities are involved and even a small human error could be magnified.
5. Along with each assay a standard curve has to be plotted.
6. Calibration is necessary on every occasion of assay as the radiolabelled standards will decay over time.

- **Types of RIA**: 1. Equilibrium type, 2. Non-equilibrium type.

11.1.7 Conclusion

If possible, both chemical and biological assays should be performed on identical sample.

11.1.8 Competitive binding technique using specific carrier globulin of plasma (for cortisol and thyroxine)

Other competitive binding technique have been used in place of antibody for assay of minute quantities of hormones. The specific carrier globulins of plasma which are natural binding agents for some hormones such as transcortin for cortisol and

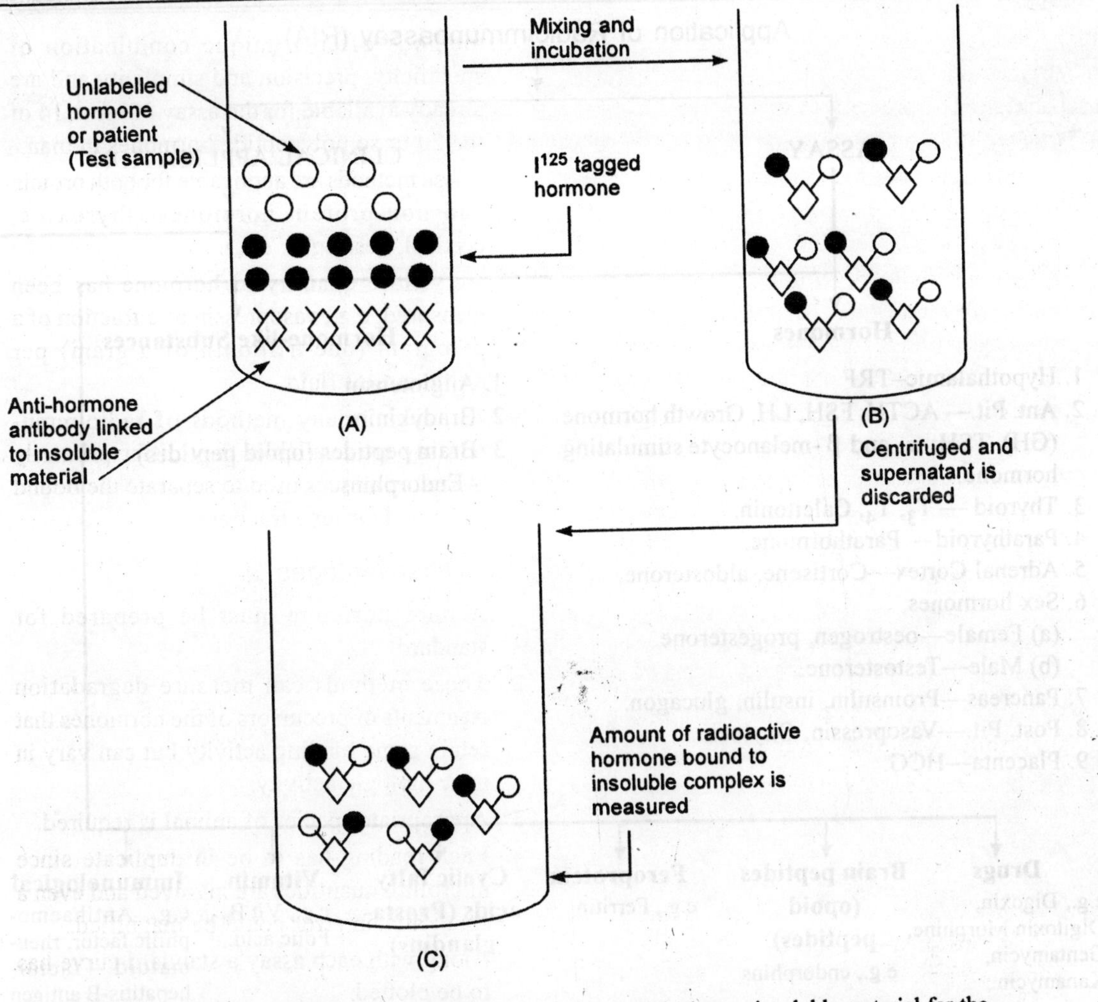

Fig. 11.2: Illustration of the use of antihormone antibody linked to an insoluble material for the assay of hormone in a sample

thyroxine binding globulin for thyroxine can be used for estimation of cortisol and thyroxine. Then the assay is carried out in exactly the same way as the radioimmunoassay procedure. This technique is mainly used for estimation of cortisol and thyroxine.

- **Application of radioimmunoassay**

I. Immunological

1. Assay of carcinoembryonic antigen.
2. Assay of hepatitis B (Australia) antigen in human serum but not suitable for large scale routine use.
3. Assay of 3gE levels in serum using iodine labelled IgE and anti-IgE linked to cellulose.

II. Hormonal

1. Polypeptide hormones, e.g., FSH, L.H., vasopressin, oxytocin, insulin, calcitonin, growth hormone, glucagon TSH, ACTH, parathormone α and β-melanocyte stimulating hormone.
2. Non-protein hormone, e.g., steroids thyroxine. **Brain peptides (opoid peptide):** Endorphins

III. Cyclic fatty acid

Prostaglandins (local hormone).

IV. Pharmaceutical

Morphine-related drugs (appropriate antibodies are raised by coupling to an immunogenic carrier).

Application of Radioimmunoassay (RIA)

ASSAY | **CLINICAL APPLICATION**

Hormones

1. Hypothalamic–TRF
2. Ant. Pit.—ACTH, FSH, LH, Growth hormone (GH), TSH, α- and β-melanocyte stimulating hormone.
3. Thyroid — T_3, T_4, Calcitonin.
4. Parathyroid— Parathormone.
5. Adrenal Cortex—Cortisone, aldosterone.
6. Sex hormones,
   (a) Female—oestrogen, progesterone.
   (b) Male—Testosterone.
7. Pancreas—Proinsulin, insulin, glucagon.
8. Post. Pit.—Vasopressin, Oxytocin.
9. Placenta—HCG.

Hormone-like Substances

1. Angiotensin
2. Bradykinin
3. Brain peptides (opoid peptides) –Endorphins.

Drugs e.g., Digoxin, Digitoxin Morphine, Gentamycin, Kanamycin.

Brain peptides (opoid peptides) e.g., endorphins

Feroprotein e.g., Ferritin

Cyclic fatty acids (Prostaglandins)

Vitamin e.g., Vit.B_{12}, Folic acid.

Immunological e.g., Antihaemophilic factor, rheumatoid factor, hepatitis-B antigen (australia antigen), Carcino-embryonic antigen, IgE.

Clinical Application

I. Diagnosis e.g., selection of treatment and follow-up

II. Oncology e.g., Tumour markers

III. Therapeutic e.g., drug monitoring

IV. Research

11.1.9 Clinical application

I. Diagnosis, selection of treatment and follow up

1. **Male Infertility**: Measuring serum FSH, LH and testosterone levels, helps to categorise the patient into pregerminal or germinal infertility.
2. **Diagnosis of Hypothyroidism**: Measurement of TSH level has become an important tool in the diagnosis of hypothyroidism distinguishing between hypothyroidism of thyroid origin and pituitary of hypothalamic hypothyroidism.
3. **Parathyroid Hyperfunction**: It helps to find the cause of hypercalcemia. If it is due to

parathyroid disease, parathormone levels are raised or detectable, while if the cause is not a parathyroid disease, parathormone levels are undetectable.

4. **Immunological**: To detect rheumatoid factor, hepatitis-B antibodies and rubella antibodies, etc.
5. **Management of Hypertension**: To evaluate the renin activity, angiotensin and aldosterone levels are assayed.

II. Oncology–Tumour Marker

Certain tumours are known to produce some hormones, proteins or some antigens. These substances are called **tumour markers**. RIA helps to diagnose certain tumours by identification of these tumour markers, e.g., Alfa fetoproteins to detect liver or testes cancers.

III. Therapeutic drug monitoring

RIA helps to adjust doses of drugs, measurement of drug levels in serum, to detect toxicity (e.g., digoxin) while the patient is on therapy.

IV. Research

1. Sympathomimetics have been used for many years for bronchial asthma. Clinician had often felt the need to increase the dose of the drug used over a period of time, as the patient seemed to have developed a tolerance to the drugs.

   For RIA we now know the mechanism of this phenomenon.
2. Radiolabelled adrenergic receptor antagonists have been utilised as ligands to study the properties of adrenergic receptors. This diminished response to adrenergic agonists has been explained on the basis of a reduction in the density of adrenergic receptors on the cell surface.
3. RIA studies are also being used in the field of basic research on cell physiology.

11.2 Organ Scan

Some substances have affinity for particular organ or some organs specifically take up the substances. If these substances are tagged with radioactive tracer, one can get the idea of the anatomy of the organ as well as its function to some extent. This is the basic principle of organ scan. Various radiopharmaceuticals are being used according to the organ which is to be studied. Computer is also included in procedure to get dynamic studies like renal excretion or heart condition. The organ scan can be done in case or kidney, liver, bone, brain, etc.

11.3 Liver Scan

Liver imaging is a simple non-invasive technique. It gives us information as follows:

1. Shape, size and function of the liver.
2. Idea of distortion or displacement of the liver due to surrounding structures.
3. Helps to identify the space occupied by any lesion.
4. Can differentiate between extrahepatic and intrahepatic causes in case of severe jaundice.
5. Helps to identify shrunken cirrhotic liver along with proof of its functional deterioration.
6. Liver scan gives an idea of the functional capacity to some extent and along with other studies we get additional information for diagnosis and management.

11.3.1 Radiopharmaceuticals used for liver scan

1. 99 m technetium tagged to phytate.
2. BULIDA tagged with technetium.

11.3.2 Radionuclide scanning vs. ultrasound

Ultrasound is quicker and easier method than Radionuclide Scanning. Moreover, Ultrasound helps to diagnose all above cases and in addition it may help as a visual monitor during surgical procedure.

- **Advantage**: Gas in the abdomen does not interfere with the procedure.

11.3.3 Observation on some pathological conditions of the Liver

I. Hepato-splenomegaly with cirrhosis

Liver scan shows the following:

1. Enlarged liver and spleen.
2. Uptake of tracer by the spleen is more than that of the liver.

3. The picture taken after 35 minutes of I.V. injection of tracer 99 m technetium tagged to phytate shows accumulation of radiopharmaceutical in the reticuloendothelial system of the body. The Kupffer cells form 20 per cent of the liver mass and 90 per cent of the reticuloendothelial cells in the body.
4. In case of cirrhosis, the tracer is taken up by spleen and bone marrow, thus showing them clearly in the scan.

II. Hepatomegaly with amoebic liver abcess

Technetium scan shows enlarged liver with cold area. Cold area means a space occupied by lesion. Destruction of liver tissue results in lack of tracer and hence a cold area. The cold area is due to amoebic liver abcess but may be cysts, hepatomas or metastasis.

III. Diagnosed case of Obstructive Jaundice

After one hour of a BULIDA scan, the picture showed enlarged gall bladder. But after a fatty meal the gall bladder is emptied and bile has drained into the duodenum. Thus, obstruction was not present.

IV. Chronic Cholecystitis

Liver scan shows activity in the intestines but the gall bladder picture is not seen which is suggestive of chronic cholecystitis.

11.4 Bone Scan

Bone mineral is a hydroxyapatite crystal which consists of calcium and phosphate. These can be exchanged for other phosphates or for other elements. Now, if these exchangeable substances are tagged to a tracer isotope, the bone will be highly exhibited on a scan.

As the bone contains osteoid and chondroitin sulphuric acid, some of these can be exchanged for a suitable radiopharmaceutical.

11.4.1 Radioactive substances for bone scan

1. Radiocerium and Radiogallium are deposited in the osteoid tissue.
2. Radiosulphur labels Chondroitin sulphuric acid.
3. Radium can be exchanged with calcium.
4. Technetium phosphate is the common radiopharmaceutical used nowadays.

11.4.2 Factors for the positivity of bone scan

Bone scan depends upon the following factors:

1. **Increased vascularity,**
2. **Increased osteoblastic activity,**
3. **Increased new collagen deposition.**

These factors attract the radiopharmaceutical right from the onset of bone involvement for which bone scan shows lesions several weeks ahead of the radiograph.

11.4.3 Some clinical observations

1. **Stress fractures: Stress fractures can be diagnosed within 1 to 2 days of pain by bone scan and it is difficult to ascertain by radiograph.**
2. **Avascular necrosis of femoral head: An early diagnosis can be done by bone scan whereas radiologically it can be diagnosed after a long time.**
3. **Bone metastasis: Bone scan is ideal for early detection of bone metastasis due to cancer of the breast and prostate.**
4. **Bone tumours and arthritis: These cases are seen as hot areas, i.e., areas of increased intake of tracer.**
5. **Breast cancer with chest pain: Multiple hot areas are seen in the ribs and vertebral column.**
6. **Arthritis: Bone scan shows hot knee and ankle. When compared to the normal joint, the tracer uptake is definitely higher in these joints.**

11.4.4 Advantages

1. **Bone scan gives earlier diagnosis than Radiography.**
2. **Bone scan pin-points the specific areas of lesions.**
3. **Difficult cases (e.g., stress fracture) can be ascertained easily and earlier than Radiograph.**

11.4.5 Disadvantages

During bone scans, kidneys are visualised. Radiopharmaceutical should be excreted via kidneys. If the excretion is delayed, the radiopharmaceutical

is accumulated and the scans show hot areas with greater intensity. Some normal tissues sometimes might be interpreted as a hot area.

11.5 Brain Scan

The basic principle of brain scan is that most intracranial lesions alter the blood-brain barrier. This initiates accumulation of radio-pharmaceuticals around the lesions. The localised accumulation of radioactive material is detected as hot area by Gamma Camera.

11.5.1 Procedure

The patient is given 99 m Tc per technetate intravenously and scan can either be (a) dynamic or (b) static.

For doing a dynamic study, serial pictures are taken for about forty seconds. The pictures show the arterial capillary and venous phase of circulation. For static study, subsequently brain scan is done.

11.5.2 Diagnosis by scanning

1. Tuberculoma.
2. Infarct.
3. Abscesses.
4. Neoplasm.
5. Vascular malformations.
6. Extra cerebral haematomas.

11.5.3 Radionuclide imaging vs. CT scan

Though CT scan is much superior to radionuclide scanning, still it is very much helpful in many clinical conditions.

11.6 Renal Imaging

11.6.1 Basic principle

The basic principle is to use radio-pharmaceuticals which specifically concentrate in a particular organ for scanning of the organ is also employed for renal scanning.

11.6.2 Information from imaging

1. This gives us: (a) static picture of the kidneys or urinary bladder or (b) dynamic observation—when it is fed to computer to give a graph of the radioactivity of each kidney in relation to time.
2. We get idea of the functional capacity of the kidneys and their shape, size and position.
3. In case of kidney stone (urolithiasis) though the stone is not visualised it helps to make decision to operate and assess the functioning capacity of other kidney, the response to medical treatment and to follow up after surgical treatment.
4. This is a simple, safe and noninvasive method for serial assessment of renal function after kidney transplantation. This serial assessment is very much important to detect acute rejection of the transplant. Because, IVP is contraindicated for first 48 hours after transplantation.
5. Renograms can quantify the disturbances of renal function better than the IVP as it is an advancement over the IVP. This procedure causes minimal radiation exposure.

11.6.3 Procedure

The patient is administered a dose of radiopharmaceutical known as technetium 99 m DTPA (Diethylene triamine penta-acetic acid). Pictures are taken every second for one minute and then every minute for 28 minutes.

At the same time the radioactivity over each kidney can be detected by the Gamma Camera and fed to the computer. In most cases the radiopharmaceutical flows to the kidneys as well as to other organs present in the surrounding. This background radioactivity is automatically eliminated by the computer to give radioactivity over the kidneys only.

11.6.4 Observations

1. Anatomical defects like stricture or presence of renal stone cannot be visualised but their impact on the renal function can be diagnosed.
2. If the radioactivity continues to remain high over a kidney it indicates an obstruction to excretion.
3. If radioactivity reaches baseline it indicates only the functional obstruction like stone.

4. If the radioactivity continues to remain high, it is highly suggestive of organic obstruction or parenchymal damage, e.g., polycystic kidneys. The findings are:
   (a) Sluggish vascular flows to both kidneys,
   (b) The kidneys are enlarged and show multiple, well circumscribed, cold areas or cysts,
   (c) The graphs show good uptake of tracers by both the kidneys. The peak time is normal but the excretion is very slow. Even this excretion cannot be accelerated after giving diuretic.
5. In case of renal calculi (stone) which has been diagnosed by X-ray and IVP but sometimes some persons may show hypersensitivity to iodine; in such cases renogram is done. The renogram shows:
   (a) A normal uptake of tracers but a delay in excretion by the kidneys,
   (b) The excretion of radiopharmaceutical improved after giving diuretic,
   (c) The graph exhibits a composite nature having tracing of the radioactivity on both kidneys in relation to time and the urinary bladder,
   (d) If the excretion by the kidneys are improved, the radioactivity in the urinary bladder is observed. This result, along with scan, does suggest passive obstruction.

11.6.5 Normal excretion pattern of the kidneys and the urinary bladder

1. As the radioactivity in the kidneys fall, that in the urinary bladder increases.
2. The peak radioactivity in the kidneys observed within a minute and then immediately show fall of radioactivity.
3. Compared to the kidneys, the urinary bladder shows peak almost at the same time but there is no fall and the radioactivity in the urinary bladder is at its minimal.
4. After administration of diuretic the radioactivity over the kidneys is reduced and that over the urinary bladder is increased. But this rise is lesser than the normal renogram (Fig. 11.3).

11.7 Radioactive Renal Clearance Studies

(Radioactive Renography, Radioactive Pyelography)

The renal function tests can be grouped into three:

1. Physical and chemical examination of the urine.
2. Determination of substances in the blood which are excreted in the urine.
3. Measurement of renal clearances.

Renal clearance studies with inulin, para-aminohippuric acid, Diodrast, mannitol are used for renal function tests. These substances give information pertaining to glomerular filtration rate, renal blood flow per minute, the filtration fraction and other functions but most of the tests are very difficult to perform. So two types of pyelography are used for renal clearance studies.

1. **Intravenous Pyelography (IVP)**: Substances having large quantities of iodine, e.g., Diodrast, Iopax and Hippuran, are excreted in the urine by glomerular filtration and tubular secretion and concentration rises within 5 minutes of injection. These compounds become opaque to X-ray due to presence of iodine. Therefore, X-ray gives shadow of pelvis, ureters and urinary bladder. Failure to show distinct shadow of the pelvis within 5 minutes indicates depressed clearance.
2. **Radioactive Pyelography or Radioactive Renography or Radioactive Clearance Studies**: If any of the above substances is prepared with radioactive iodine (I^{131}) or other radioactive nuclide, radioactivity can be measured by placing Radioactivity Counters over the kidneys. A trace amount is injected intravenously and degree of radioactivity is recorded for few minutes which gives a rough estimate of renal clearance. This procedure of radioactive pyelography is known as **radioactive renography** and the result plotted in curve is known as **radioisotope renogram.**

Continuous external measurement of the renal uptake and excretion of a labelled contrast medium

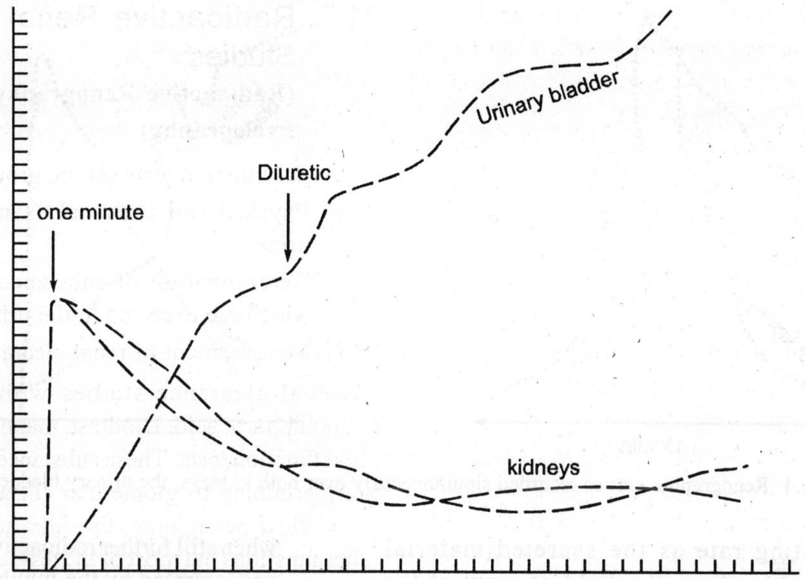

Fig. 11.3: Normal excretion pattern of the kidneys and the urinary bladder by detecting radioactivity

was performed and reported by Taplin et al. (1955). Nordyke et al. (1960) introduced I^{131}-Hippuran, which is now the substance almost exclusively used for renographic studies.

11.7.1 Use of other radioisotope substance

1. **Radiomercury-labelled Chlormerodrin**: Chlormerodrin is usually retained by the renal parenchyma longer than Hippuran. After the administration, an evacuation phase is not observed and the secretion phase can be followed over a longer period. A slight impairment of tubular function can thus be recognizable more easily.
2. **Radioiodine-labelled Albumin**: To know the degree of vascularization of each kidney in which the counting rates after I.V. injection of I^{131} labelled albumin may be recorded. But it has limited clinical value because lack of information regarding its concentration and excretory function of the kidney.

11.7.2 Renographic curve and its interpretation

Patients with normal kidney function exhibits a renogram as shown in (Fig. 11.4) and (Fig. 11.5). The renogram show three phases which are discussed below (Fig. 11.5).

1. **The initial or vascular phase**: It begins about 15-20 seconds after the injection with a steep increase in the counting rate, it lasts about 20 seconds and is caused by the appearance of the injected radioactive material in both renal and extrarenal tissue. In arterial stenosis there is less blood resulting in the vascular phase slope becoming less.
2. **The second or secretion phase**: It corresponds to glomerular filtration and tubular secretion. This processes commence as soon as the material is entering the renal parenchyma. It is, therefore, impossible to distinguish clearly between the end of the initial and the beginning of the second phase. During the latter, the counting rate rises more slowly than during the first phase. The radioactive material goes from the renal vascular bed through the renal parenchyma to the lumina of the nephrons, about 80 per cent being transported by tubular secretion and about 20 per cent by glomerular filtration.
3. **The third or excretory or evacuation phase**: It is characterised by a decrease in

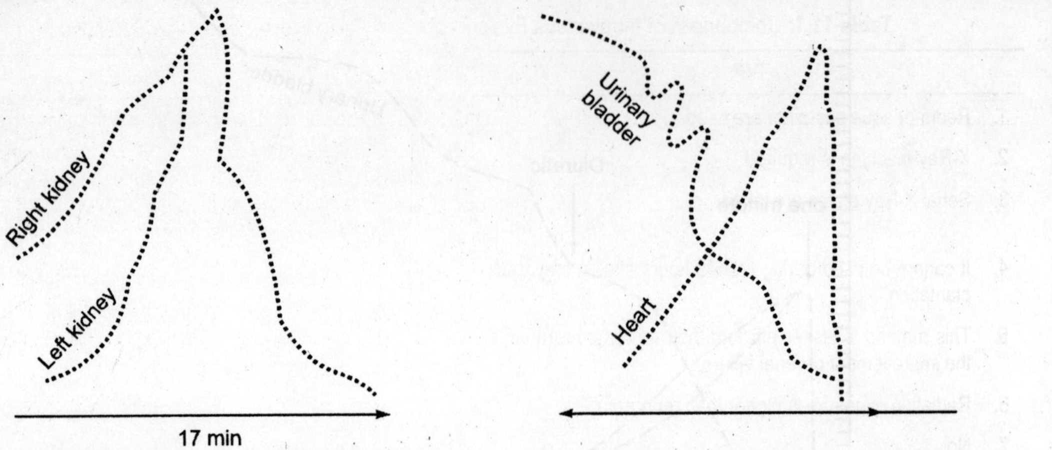

Fig. 11.4: Renographic curves recorded simultaneously over both kidneys, the urinary bladder and the heart

counting rate as the secreted material disappears from the field of view of the detection through the upper urinary tract. It is accompanied by a simultaneous increase of the counting rate recorded over the urinary bladder (Fig. 11.4). This phase, again, cannot be clearly separated from the secretion phase since evacuation already begins at a time when still further radioactive material is being concentrated by the tubular cells.

Although a separate evaluation of the three phases of the renogram is, thus, not possible, typical disorders of renal function are observed to be associated with characteristic changes in different parts of the recorded curve.

Fig. 11.5: Renogram showing three phases

Table 11.1: Comparison of Introvenous Pyelography and Radioactive Renography (Renogram)

IVP	Renogram
1. Radio-opaque sustainer are used.	1. Radioactive material is tagged with substance.
2. X-Ray machine is required.	2. Radioactive counter.
3. Serial X-Ray plates are taken.	3. Radioactivity is detected by placing detecting instrument on the kidney region.
4. It cannot be used during first 48 hours after kidney transplantation.	4. It is used and safe, simple and non invasive after transplant.
5. This method is lesser efficient than renogram to quantify the improvement of renal failure.	5. It can quantify the impairment of renal function better than IVP.
6. Radiation exposure is higher than renogram.	6. Radiation exposure is minimal.
7. No.	7. It gives qualitative idea of the blood flow to both kidneys.
8. No.	8. Pictures in the first minute gives an idea of the potency of the renal vessels.
9. The urinary bladder is also visualised.	9. The bladder is also visualised.
10. No.	10. X-ray and radioactivity measures function of each kidney independently of the other rather than total function of both kidneys together compared to other functional tests like IVP.
11. Iodine sensitivity may occur.	11. Anatomical defects as stone, stricture, etc. cannot be identified but their effects on function can be diagnosed. In case iodine sensitivity IVP, renogram is an advantage over IVP.

11.7.3 Renogram in different kidney conditions

I. Reduction of renal blood supply

(a) **Bilateral stenosis**: It does not result in a typical change in renogram though in most cases the curve rises initially to a lower level.

(b) **Unilateral stenosis**: The two curves differ. The curve of affected kidney rises initially and ceases at a lower level. The rise is during the secretion phase and delay in the attainment of peak. The latter effect is caused by the reduction in the urine flow and the consequent delay of transport of the radioactive material to the collecting system.

II. Imparied tubular function

(a) Pyelonephritis, glomerulonephritis, interstitial nephritis, nephrosclerosis the rise in the counting rate during the secretion phase is slower, attainment of the peak is delayed and the decline during the evaluation phase is also slower.

(b) **Severe impairment**: The height to which the curve rises during the vascular phase is also reduced, no further increase in the counting rate occurs during the secretion phase. So the curve may remain horizontal or may even show a decline.

(c) **Non-functioning kidney**: The height of the curve is further reduced during the vascular phase, which is then immediately followed by a decrease in counting rate, which runs parallel to the decrease in the counting rate observed over the heart.

(d) **Disorder of evacuation of urine**

1. **Slight reduction in urinary drainage:** The only change in the renogram is a slower decrease of the counting rate during the evacuation phase. With increasing obstruction, the counting rate tends to remain high.

2. **Complete obstruction of urinary flow:** The counting rate continues to increase beyond the end of the secretory phase provided that tubular function is not yet impaired. If the obstruction persists over more than a few days, the secretory

capacity of the tubular cells will be affected and blood supply will be reduced. In case of acute failure this sequence proceeds much more rapidly than in patients with uretral obstruction.

11.7.4 Modification of Procedure-Repetition of test

In certain diseases, e.g., pyelonephritis, frequently normal curve is seen because sufficient volume of functioning tubules eliminate labelled Hippuran, for which it is advisable to repeat the test with a modified technique.

1. **PAH-loading**: This test may be repeated by giving I.V. infusion of 75 mg PAH/kg. body weight over a period of 12 minutes immediately followed by injection of Hippuran. In normal kidney, there is slight depression in the curve in the second phase and in imparied kidney tubular function, there is significant alteration of the curve.
2. **Antidiuresis**: The patient is kept on a dry diet for 12-18 hours and the examination is carried out in the usual way. Compared with the curve obtained in diuresis, alteration in the second phase is evident, particularly in patients with renal artery stenosis or pyelonephritis. During antidiuresis, the peak of the curve is always reached later than during diuresis.

11.7.5 Advantages

1. A trace amount of the substance is to be injected.
2. After few minutes, determine roughly the renal clearance.
3. It can be used in iodine sensitive cases.
4. It measures function of each kidney independently of the other, rather than total function of both kidneys together given by other renal function tests.

11.7.6 Disadvantages

1. Several attempts have been made to analyse the renogram quantitatively. However, it is impossible to evaluate the three phases of the renogram separately.
2. Moreover, the geometry of the *in vivo* measurement is influenced by so many factors that counting rates over the kidney vary widely from patient to patient even when renal function is normal and administered doses are normalized for body weight.

11.8 Cardiac Imaging

11.8.1 Basic principle

The basic principle of cardiac imaging is to detect the presence, concentration and distribution of radiopharmaceutical in the heart after administration.

11.8.2 Information

Cardiac imaging is a convenient non-invasive method and it helps to give information regarding

(a) Myocardial perfusion,
(b) Myocardial ischaemia,
(c) Ventricular function.

(a) **Myocardial perfusion imaging:** This helps to get information about the blood supply to the myocardium. Thallium is commonly used isotope tracer which enters the cell like potassium. Then the scan is done at different positions (anterior, left anterior, oblique, etc.) which gives a picture having ring shape or a 'W' shape. The body of the latter depicts the ventricular mass and inside the ventricular cavity. The high or low activity recorded depends on the perfusion.

- **Exercise-ECG Test vs. Thallium imaging**

When the diagnosis is difficult after routine treadmill test or post-exercise ECG is doubtful, thallium imaging is very much helpful. At the peak of exercise, if thallium is administered intravenously and then the imaging is done it shows clearly the ischaemic area.

(b) **Myocardial infarct imaging**

1. **Thallium imaging:** For myocardial infarct imaging, thallium can be used. It can detect an infarct as early as 6 hours when sometimes ECG cannot give information properly. Infarction is detected as cold area by thallium imaging and if the area is more than 16 mm, then poor prognosis is expected.
2. **Technetium 99 m stannous pyrophosphate imaging:** If there is injury in fresh tissue, it gives rise to leakage of intracellular calcium which is necessary for pyrophosphate binding. Pyrophosphate

binds to calcium and organic macromolecules in irreversibly damaged myocardial cells. If there is sufficient coronary flow to necrotic myocardium, technetium 99 m stannous pyrophosphate binds there. This is visualised as a hot spot in the scan.

- **Advantages**
  1. Infarct can be visualised after 3 days.
  2. Differentiate between a fresh infarct and an old scar.

(c) **Ventricular functions**: Ventricular function is scanned by Multiple Gated Aquisition Studies (MUGA).

11.9 Multiple Gated Aquisition Studies (MUGA)

11.9.1 Method

Patient's RBCs or albumin is tagged with technetium 99 m and administered intravenously. Scanning is usually done after 15-20 minutes when state of equilibrium is reached. The cardiac cycle recorded in ECG is divided from one R wave to the next into different portions like 16, 32 or 64. In the computer memory, the same number of frames or pigeon holes are made (Fig. 11.6). Gamma Camera picks up information from a particular portion of the cardiac cycle which is fed to its corresponding frame or pigeon hole of the computer. In the same procedure, information for 200-400 cardiac cycles are collected because the more number of cycles are scanned, higher the accuracy. Then the information can be presented either as a curve or as a picture. Thus, the computer can supply us the information of different parameters for ventricular function, e.g.,

1. Ventricular ejection fraction,
2. Ventricular ejection velocity,
3. Ventricular diastolic volume,
4. Ventricular systolic volume.

11.9.2 MUGA studies in case of normal person

MUGA studies in a normal person depict the following:

(a) **Regional Wall Motion Histogram**: It represents two concentric circles of which outer one is the outline of the ventricle at the end of the diastole and inner one is the outline of the ventricle at the end of the systole.

Divisions are made in the circles into different segments and the percentage radial motion of each segment is plotted by the computer. This is known as "regional wall

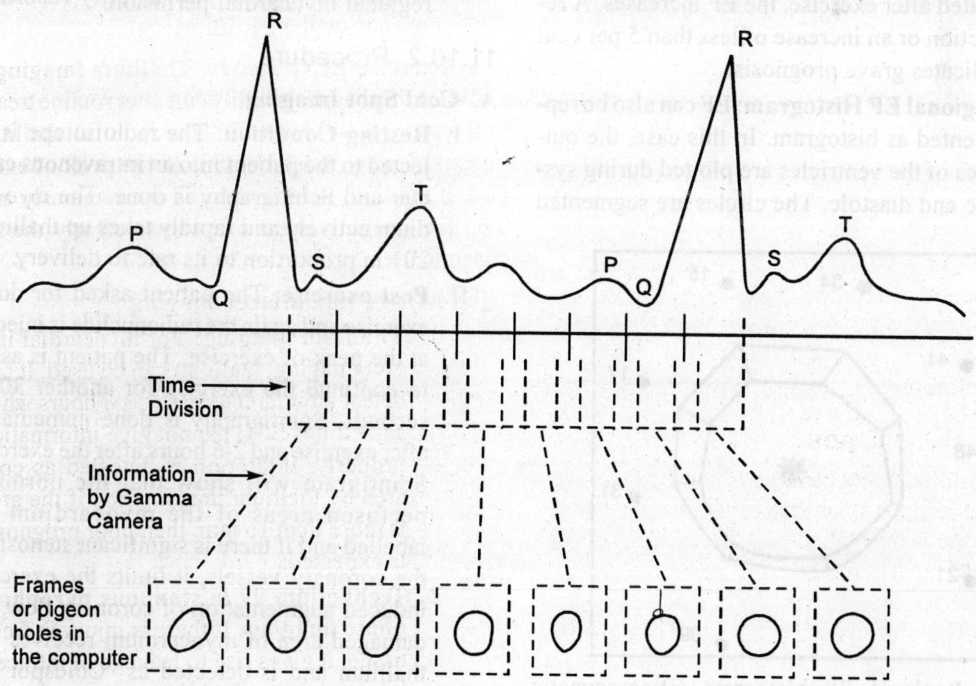

Fig. 11.6: Diagrammatic representation of MUGA or ECG-gated cardiac imaging in frame code

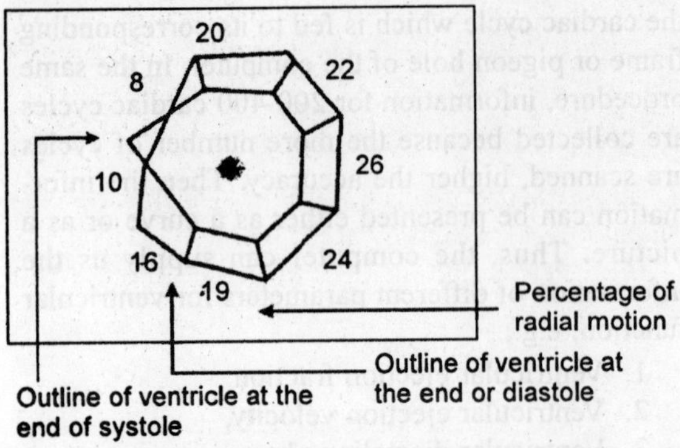

Fig. 11.7: Diagrammatic representation of regional wall motion histogram

motion histogram" (Fig. 11.7)—an ischaemic area of the ventricle or infarcted area of the ventricle which does not exhibit adequate regional wall motion:

(b) **The ventricular change**: The ventricular change in the volume in relation to time can be shown graphically. The drop in volume between diastole and systole is known as "Ejection Fraction (EF)". From the cardiac parameter displayed by the computer, the EF can be calculated as the ratio of stroke volume to end diastolic volume. The normal EF is 55-60 per cent. When MUGA study is repeated after exercise, the EF increases. A reduction or an increase of less than 5 per cent indicates grave prognosis.

(c) **Regional EF Histogram**: EF can also be represented as histogram. In this case, the outlines of the ventricles are plotted during systole and diastole. The circles are segmented and the EF for each segment is separately represented. A low EF in certain segments indicates that part is affected. This method is known as **regional EF histogram** (Fig. 11.8).

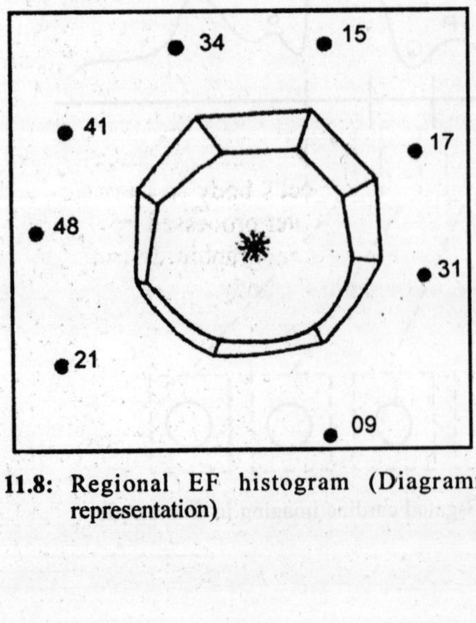

Fig. 11.8: Regional EF histogram (Diagrammatic representation)

11.10 Myocardial Scintigraphy

Recently progress has been made in nuclear cardiology due to introduction of new radiopharmaceuticals, and more sophisticated imaging instruments. These have made the diagnosis and therapy easier in cardiovascular medicine than the past. The relatively low sensitivity and specificity of exercise electrocardiography has opened the new era of radionuclide technique for detection of coronary artery disease, localisation and assessment of the size of myocardial infarcts and measurements of the cardiac response to increased oxygen demands.

11.10.1 Radionuclides for imaging

The radioistopes used for this technique are usually gamma-emitting. These are readily available and generally applicable. The radionuclides for measuring myocardial perfusion are:

1. Potassium and its analogues,
2. Thallous chloride ($^{201}TlCl$)—It is imaging agent of choice for qualitative estimation of regional myocardial perfusion.

11.10.2 Procedure

A. Cold Spot Imaging

I. **Resting Condition**: The radioisotope is injected to the patient into an intravenous catheter and Scintigraphy is done. The myocardium actively and rapidly takes up thallium-201 in proportion to its rate fo delivery.

II. **Post exercise**: The patient asked for doing exercise and again the radionuclide is injected at the peak of exercise. The patient is asked to continue the exercise for another 30-60 seconds. Scintigraphy is done immediately after exercise and 2-6 hours after the exercise. Scintigram will show that the normally perfused areas of the myocardium are labelled and if there is significant stenosis of the coronary vessels, it limits the exercise-induced augmentation of coronary flow, the damaged area of myocardium receives less thallium and is detected as "Coldspot" on postexercise scanning. Scintigraphy is done immediately after exercise and after 2-6 hours shows areas of decreased perfusion during exercise and reperfusion later. The normalisation of thallium uptake several hours after exercise observed on scanning helps to differentiate between ischaemic areas from the scar areas.

- **Advantages and disadvantages**: Thallium scanning is indicated when the diagnostic results are not obtained properly in routine treadmill testing or when the exercise-ECG cannot be interpreted. Though the method is accurate and specific it sometimes fails to detect existing diseases and gives very little information regarding prognosis.

B. Hot spot imaging: It is very difficult to detect acute myocardial infarcts by the conventional methods. Several hot-spot myocardial Scintigraphic methods are being used for this purpose. But it has been found that technetium Tc99m (^{99m}Tc) pyrophosphate myocardial imaging is the agent of choice. It is more specific in detecting myocardial infarcts because of increased uptake of radionuclide in regions in which coronary blood flow is decreased. Increased accumulation of radionuclide in areas of necrosis leads to formation of "hot-spot" which helps to identify the size and location of the infarct accurately.

- **Application of ^{201} thallous chloride Scintigraphy**
  1. Useful in the detection of coronary artery disease and its specificity in detection is 90 per cent.
  2. Useful for detection of other myocardial diseases, e.g., infarcts, septal hypertrophy and right ventricular hypertrophy.
  3. Useful in detecting the presence of one cardiac chamber as in congenital heart disease.
  4. Thallium Scintigraphy is being used for follow up after coronary revascularization and it is done before and after surgery. Increase in regional uptake of radionuclide with exercise during postoperative investigation indicates good revascularization.

11.11 Positron Emission Tomography (PET)

Positron Emission Tomography (PET) is an ideal scanning technique which utilizes positron emitting radioisotopes making cross-sectional images by detecting the distribution of radioactive isotopes. This technique provides valuable information regarding regional tissue biochemistry, physiology and pharmacology.

11.11.1 Type of information

It supplies useful information regarding *in vivo* quantitative measurement of:

(a) Regional blood flow of human brain, heart muscles.
(b) Glucose utilization,
(c) Oxygen utilization,
(d) Incorporation of selected amino acids in tissues.

11.11.2 Basic principle

Gamma-ray-emitting radioisotopes photons or positron (pos. electron) of the elements which constitute the organic substances in the body are taken up for labelling the selected biochemical or pharmacological compounds. The radionuclides selected for the technique are:

1. Oxygen : 15 (O^{15})
2. Nitrogen : 13 (N^{13})
3. Carbon : 11 (C^{11})
4. Fluorine : 18 (F^{18})

These radioisotopes have **very short half lives** for which they have to be prepared near the site of the application.

These radioisotopes are manufactured by the **bombardment of the stable elements with high energy ion beams in a Cyclotron**. During disintegration, they **emit a positron** which is positively charged electron-like particle. **Positrons are captured by the electrons during travelling a few milimetres away from the nucleus. Both positron and electron are destroyed and the resulting energy is released as two gamma-rays** (511 keV), which **travel in opposite direction.** Correlated pairs of gamma-rays emerge from a body containing positron-emitting radioisotopes. These gamma-rays are recorded at specific angles throughout the subject's body to detectors and the information is computer processed by algorithm for reconstructing the tomographic distribution of the isotope throughout the body.

11.11.3 Procedure

Technique or Procedure: The technique is analogous to that of computerised axial tomography. A chemical compound with desired biological activity is labelled with a radioactive isotope that

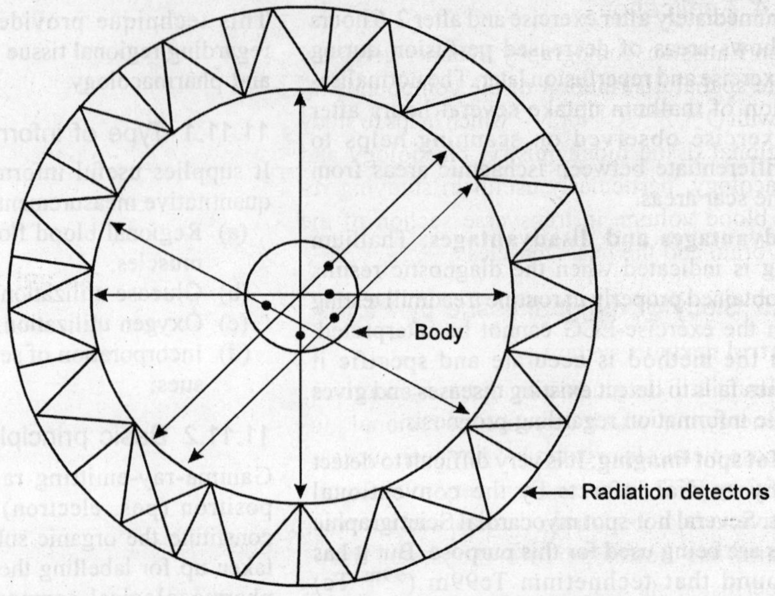

Fig. 11.9: Diagrammatic representation of the principle of PET

decays by emitting positron, or positive electrons. The **emitted positron** almost immediately **combines with an electron** and the two are mutually annihilated with the **emission of two gamma-rays.** The two gamma-ray photons travel in almost **opposite directions**, penetrate the surrounding tissue and are recorded outside the subject by array of **detectors.** A **mathematical algorithm** applied by **computer** rapidly **reconstructs the spatial distribution of the** radioactivity within the body for a **selected plane** and displays the resulting image on a CRT (cathode ray tube).

The patient is injected radiopharmaceutical with radioisotope having short half life which are positron emitters and then scanned in the PET scanner. The image-forming process in Positron Emission Tomography has three steps:

1. The detection of the gamma-rays emitted by the positron + electron combination.
2. The identification of the radiation's direction of travel, and
3. The reconstruction of the distribution of radiation into an accurate geometric image.

This procedure not only exhibits transaxial tomographic images throughout the body, but also provides absolute values of the biologic entities being traced.

- **IMAGE or SCAN**: From the above examples, it is clear that radioisotopes or radionuclide are used to absolute the integrity of physiological systems and the anatomical integrity of organs. To assess the anatomy or an organ, the **distribution of a given radionuclide in the organ is mapped.**

The maps obtained are called **images or scans**, of the organs and are **produced by the gamma-rays emitted by the radionuclide. The image represents the distribution** of the radio nuclide and hence reflects the structural integrity of the organ. Initially only two-dimensional images of organs were possible—no depth information was provided. However, due to progress in the field of **emission tomography,** it is now possible to obtain three-dimensional images of organs. Two approaches are followed in one, single photon emitters (radionuclides) such as $^{99}TC_m$ are used, and this approach is, therefore, called **single photon emission computerised tomography** (SPECT). In the second approach, a positron-ejecting radioisotope such as ^{11}C, ^{13}N, O^{15} or, ^{18}P, is used, and this with a half-life roughly 20 (C), 2 (O), 10 (N) minutes approach is, therefore, called **positron emission tomography** (PET). The use of computerised image reconstruction techniques for displaying the concentration of radioactively labelled substances in the living body has led to the development of positron emission tomography (PET).

11.11.4 Application

Positron Emission Tomography Scanner gives the accurate spatial distribution of the compounds labelled with O^{15}, F^{18}, C^{11} or N^{13} which helps to study the regional tissue biochemistry, physiology and pharmacology, particularly useful in studying regional blood volume in transverse section of the human brain and the heart muscles.

11.11.5 Study of regional tissue by PET

I. Central nervous system

1. Brain utilizes glucose as a primary source of energy, for which detection of regional glucose metabolism using F^{18} has been successful by PET scanning. It is very much sensitive indicator of brain damage and recovery and has better performance than tissue perfusion studies.
2. PET scanning is very much helpful for detection of epileptic patient because the areas show less metabolism and the visualised areas get activated during seizures.
3. PET imaging accurately select the patients for brain surgery and determine the areas of the brain for surgery.

II. Cardiovascular disorders

1. PET is useful for detection of ischaemic areas of myocardium.
2. It is helpful for assessment of functional significance of coronary stenosis in a more quantitative way.

III. Pulmonary disorders

1. This technique is useful for accurate measurement of regional ventilation and ventilation-perfusion ratio.
2. Recently PET studies in lung tumours have shown a relatively high glucose uptake by tumour tissue.

11.11.6 PET unit

A complete PET unit consists of:

1. **A Data Acquisition System:** It consists of radiation detectors either sodium iodide or bismuth germanate crystals. In most cases a mechanical system is present which imparts a small motion to the detectors in order to get better sampling. Most recent scanners use smaller detectors for higher resolution and more than one plane of data can be recorded, simultaneously.
2. **Computer:** Data from the acquisition system is then fed into a computer with a large memory.

11.12 Myocardial Imaging (SPECT)

Single Photon Emission Computed Tomography (SPECT)

11.12.1 Historical background

Recently the technique is being used for myocardial imaging for detection of coronary artery disease is known as SINGLE PHOTON EMISSION COMPUTED TOMOGRAPHY (SPECT). It is another form of Positron Emission Tomography which uses single photon emitters, e.g., technetium 99 m or iodine 123. It is a major advance in medical technology for myocardial imaging. By this method, multiple view images are obtained circumferentially and then reconstructed by computer as a single three dimensional image.

11.12.2 Application

1. Detection of coronary artery disease, especially occult coronary artery disease.
2. Detection of involvement of the left anterior descending and left circumflex cornoary arteries.

11.12.3 Advantages

1. It is more improved and accurate method than the planar imaging, stress Electrocardiography and Scintigraphy.
2. It is more sensitive and specific for detection of cornonary artery disease.
3. It gives single three-dimensional image of the lesion.
4. It is more accurate for detection and exclusion of coronary artery disease.
5. It has greater advantage in patients with milder form of coronary diease, i.e., less than 60 per cent lesion without any prior myocardial infarction.
6. It has greater advantage in the detection of single vessel coronary artery disease.

7. It has greater diagnostic value in doubtful cases with exercise Electrocardiogram, abnormal baseline Electrocardiograms or suspected false-positive or false-negative conventional exercise tests.
8. Exercise imaging studies offers greater value to assess prognosis following infarction and evaluate patients after Coronary Angioplasty.
9. SPECT offers greater advantage to detect abnormalities limited to the subendocardium, the part of myocardium highly susceptible to regional ischaemia.
10. It is very much helpful for grouping high risk and low risk patients after uncomplicated myocardial infarction before discharging from the hospital.
11. When patients are reluctant for the technique Coronary Angiography, this method is more acceptable to them.
12. It is more advantageous than planar imaging for achieving high detection rates for infarct related vascular territory and non-infarcted vascular territories with stenosis or a regional wall abnormality.

11.12.4 Disadvantage

This procedure is not available in all centres, very few centres can offer these facilities.

- **Conclusion**: It is the diagnostic imaging procedure of choice in myocardial perfusion studies.

11.13 Radiopharmaceuticals for the Study

1. **Thallium-201**: Thallium-201 is the radio-pharmaceutical of choice for myocardial perfusion imaging. This radiopharmaceutical is very much useful in (a) assessing myocardial ischaemia, (b) identifying patients at high risks for cardiac problems, (c) evaluating the efficacy of interventional therapy.
2. **Technetium-99 m methoxy-butyl isonitrile**: Technetium-99 m methoxy-butyl isonitrile is an alternative imaging agent. It has some advantages over thallium-201 such as:

   (a) Localising perfusion abnormalities,

   (b) Showing regional wall motion abnormalities,

   (c) Estimating left ventricular function.

11.13.1 Procedure

1. SPECT is done with a Gamma Camera which rotates around the patient on a specially devised arrangement. It takes up images of the thorax, usually 32 images each of 45 seconds duration through a 180^{0} circular orbit.
2. The patient is asked to exercise on the bicycle ergometer. Thallium-201 is administered intravenously at the peak exercise when 85 per cent of age-predicted heart rate. Thallium-201 is rapidly distributed within the myocardium depending on the regional myocardial blood flow and regional cellular viability.
3. Then the patient is taken to the tomography table and imaging is done with the rotating unit, 10 -15 minutes after stopping exercise. another image is done after 3-5 hours. At first transaxial images are taken from the original planar views. Images in other planes are then computed from the transaxial images namely vertical long axis, horizontal long axis and short axis. Images obtained after stress testing are then reconstructed and displayed.

11.13.2 Interpretation of images

After thallium-201 administration, the normal image shows homogenous distribution within the myocardium. But in normal myocardium, mild defects are seen in the basal septum and the inferior wall. Areas with decreased thallium-201 uptake which are observed in more than one tomographic slice are considered as abnormal regions.

Using a single functional image known as "BULL'S DISPLAY" has been introduced as new quantitative methods and display techniques to project a series of tomographic data at rest and/or at exercise.

11.14 Magnetic Resonance Imaging (MRI)

Magnetic Resonance Imaging (MRI) is the greatest advance in medical technology. It is a new approach in medical imaging among the various techniques available for visualization. This new technique of medical imaging has been developed from new concepts and development of medical science and applied mathematics.

11.14.1 Importance of MRI

MRI reveals the following:

1. Anatomical: Three-dimensional image of the body structure in detail.
2. Physiological: Three-dimensional description of physiology of the parts of the body.
3. Biochemical: Three-dimensional description of biochemical process.
4. Pathological: Highly sensitive for characterization of pathological conditions, e.g., tumours, inflammation, infection, etc.

11.14.2 Principle

Magnetic Resonance Imaging (MRI) produces tomographic images very similar to CT scan but it is based on different physical principle. The nuclei of some atoms have properties to behave like small bar magnets or neutron or both having nucleus with odd number of proton atoms having nucleus with odd number of protons or neutrons in orbit. When these nuclei are placed in a static magnetic field and if they are excited by a pulse of radio waves at certain frequency, the nuclei start emitting a radio signal called "magnetic resonance signal (MR signal)" which is detected by a receiver. Analysis of the signal supplies information about the chemical nature and environment of the nuclei. For clinical purposes, MRI uses the hydrogen nucleus with its single proton. Hydrogen emits strong MR signals and hydrogen is present in large amount in the body water, fat, protein and carbohydrates.

11.14.3 Procedure

The whole body of the patient is placed within a tunnel inside the coil of a powerful electromagnet. The MR signals are fed to the associated computer which transforms these signals into cross-sectional images of the body. Components of normal and pathological structures of biologic tissues have different MR signals. This method can detect the differences of MR signals.

11.14.4 Application

I. Central nervous system

1. The cerebellum, brain stem, tentorium and cerebrospinal fluid spaces are routinely demonstrated.
2. It is the method of choice in case of suspected posterior fossa lesion.
3. Identification is easily done in case of lesions within the spinal cord, protrusions of disc and infections like encephalitis and cerebral abscesses.
4. It is very much sensitive in detecting white matter disease and shows changes due to multiple sclerosis at an early stage than CT scan.
5. Probably this imaging technique is the only choice for the assessment of pituitary tumours.

II. Skeletal system

MRI has superseded CT scan in the imaging of primary bone and soft tissue tumours.

1. The calcified part of bone does not give MR signal and so bone marrow is observed unobscured by calcium.
2. Haemopoitic malignancy produces changes in the bone marrow which can be monitored during treatment and this may prove useful in assessing responses.

III. Muscle and other soft tissue

Muscle and other soft tissue structures, e.g., cartilage, ligaments, nerve, fat, blood vessels, etc. are observed clearly. The contrast between tumours and normal tissue clearly exhibits the extent of lesion and its relationship to neighbouring structures.

IV. Cardiovascular system

1. In ischaemic heart disease, the regions of acutely damaged myocardium, the progress of repair, the complications of myocardial infarction and coronary artery bypass graft patency, can be detected with great accuracy.
2. High intrinsic tissue contrast between flowing blood and adjacent chamber or vessel wall can be visualised without contrast media because protons in motion do not generate signals.

11.14.5 Advantages

It has many advantages over other imaging techniques.

1. It exhibits higher sensitivity.
2. It has better contrast resolution.
3. It offers greater facility to scan in any orthogonal plane.

4. There is no image degradation from bone or air artifacts.
5. No exposure to ionizing radiation or contrast injection.
6. No clinical evidence of biologic hazard from exposure to magnetic waves.
7. Accurate detection of the differences between normal and abnormal tissues.
8. Very useful for localizing targets and guiding biopsy needles and fibre optics.

11.14.6 Disadvantages

1. High cost.
2. Limited availability.
3. Procedure may not be favourable to some patients because of long scan time.
4. Limitation of investigation of the abdomen and thorax due to motion artifacts from cardiac and respiratory movements, intestinal peristalsis and aortic pulsation.

11.14.7 Precautions

Because of strong magnetic field, the following cases are excluded.

1. Pregnancy.
2. Epileptic patient.
3. Patients with metallic implants or surgical clips.
4. Patients with cardiac pacemakers because the pacemaker may be displaced which may be fatal.
5. Prosthesis.

11.15 Radioiodine (I^{131}) for Diagnosis in Thyroid Disorder

11.15.1 Thyroid gland

The thyroid is one of the largest endocrine gland having two lobes connected by a thin band of tissue called isthmus. Microscopically it consists of follicles filled with clear, protinaceous material called Colloid containing iodine.

11.15.2 Intake of iodine

Iodine is an essential raw material for the synthesis of thyroid hormone. The richest source of dietary iodine is sea fish but it also occurs in bread, milk and vegetables. All iodine is converted to iodide in the intestine and in this form is completely and rapidly absorbed from the upper G.I. tract into the blood and extracellular fluid. The normal dietary intake of iodine per day is 100-200 μg. Iodine deficiency can develop with a daily intake of less than 50 μg over a long period of time, but if thyroid hormone production falls, TSH secretion is increased and may restore normal hormone output with or without enlargement of the thyroid (simple goitre). Japanese take large amount of sea fish, may assimilate several mg of iodine daily, but in spite of this huge iodine intake, thyroid is usually kept in normal limits (euthyroid). Thus, within the range of 50 μg to several mg of iodine daily, the body can adapt thyroid hormone production so as to maintain euthyroid state. This homeostasis is maintained by the following mechanism:

(a) Changes in renal excretion of iodine,
(b) Autoregulatory process within thyroid,
(c) High sensitivity of the anterior pituitary thyrotropic cell to changes in blood concentration of thyroid hormones so that there is a fall in hormone level secretion of TSH and vice versa.

11.15.3 Biosynthesis of thyroid hormones

Thyroid hormones are tri-iodothyronine (T_3) and thyroxine (T_4). Iodide is trapped by the thyroid gland from blood which is specific and inherent character of the thyroid gland. The mechanism is active transport using iodide pump. This uptake of iodide is blocked by anti-thyroid agents (e.g., thiocyanate, perchlorate).

The next step is iodination of the amino acid tyrosine to form:

1. Monoiodotyrosine (MIT) containing one molecule of iodine, and
2. Diiodotyrosine (DIT) containing two molecules of iodine.

The next step is combination of two-molecule DIT to form thyroxine (T_4) and condensation of molecule of MIT and one molecule of DIT to form tri-iodothyronine (T_3). This T_3 may form from intra-thyroidal deiodination. Deiodination in the thyroid, some MIT and DIT cause release of iodide which is accumulated by the thyroid for further use in the synthesis of hormones.

The next step is the storage of thyroid hormones as thyroglobulin, the main iodoprotein of the gland present in the follicles. It is glycoprotein having molecular weight 64000. It contains 10 per cent carbohydrate.

The next step is the cleavage of the stored thyroglobulin by proteolysis and pour into circulation. About 20 per cent of the total circulatory T_3 comes from thyroid and remaining 80 per cent comes from peripheral deiodination of T_4. About 0.05 per cent of T_4 and 0.5 per cent of T_3 are free in the circulation which are considered only metabolically active (Figure 11.10). The remaining part circulates in binding form with thyroxine binding globulin (TBG) and albumin. T_4 is also bound to prealbumin. T_3 is about four times more potent than T_4. The normal blood range of T_4 of healthy adults is 11mcg/dl and T_3 is between 80-180 ng/dl.

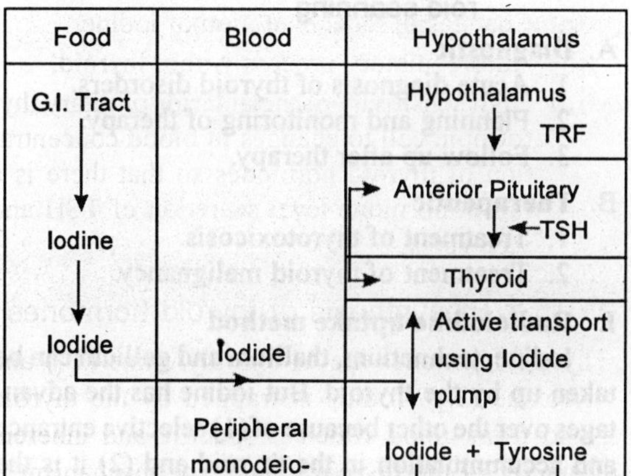

Fig. 11.10: Biosynthesis of thyroid hormones

- **Secretion**: Secretion of thyroid hormone is under the control of anterior pituitary hormone TSH. TSH regulates:
  1. Iodine uptake by thyroid gland,
  2. Growth and maintenance of the thyroid gland,
  3. Synthesis of thyroid hormones,
  4. Secretion of thyroid hormones T_3 and T_4 in turn regulates the secretion of TSH by negative feedback mechanism. The secretion of TSH is under the influence of hypothalamus through TRF (Thyrotropic releasing factor). Thus, the function of thyroid is regulated by:
     - i. Hypothalamus—TRF,
     - ii. Anterior pituitary—TSH,
     - iii. Thyroid gland—By autoregulation, i.e., controls iodide uptake which depends on blood iodide and thyroxine level pool.

11.15.4 Iodine turnover in the body

The distribution, utilization, thyroid take up, metabolism of iodide in the body have been studied after the administration of tracer amounts of radioactive I^{131} (half life 8 days) or I^{132} (half life 2-3 hours), which mix with the unlabelled iodide but have no biological action. The rate of uptake of a standard injected dose of radioactive iodine by the normal thyroid gland when measured by a calibrated radioactive detector placed over the neck is about 4 per cent per hour. In the hyperthyroidism this can rise to as high as 20-25 per cent per hour. About 1/3 of ingested iodine (120 μg) is taken up by thyroid out of about 500 μg of iodine in diet, 2/3 is excreted by the urine (480 μg). The thyroid contains about 5000-8000 μg of iodine which is 95 per cent of total iodine content of the body, most of which is in the form of iodinated amino acids (iodothyrosine and iodothyronine). Out of which only 80 μg comes out of the thyroid as T_3 and T_4. About 20 μg of iodine is excreted in the bile via liver. Extracellular fluid gets 40 μg from thyroid and 60 μg from the peripheral tissue. The large iodine pool in the body turns over slowly at a rate of about 1 per cent daily. The iodine pool and overall turnover is shown in the (Fig. 11.11).

11.15.5 Thyroid function test

There are several tests to determine the function of thyroid. They are grouped under five categorises:

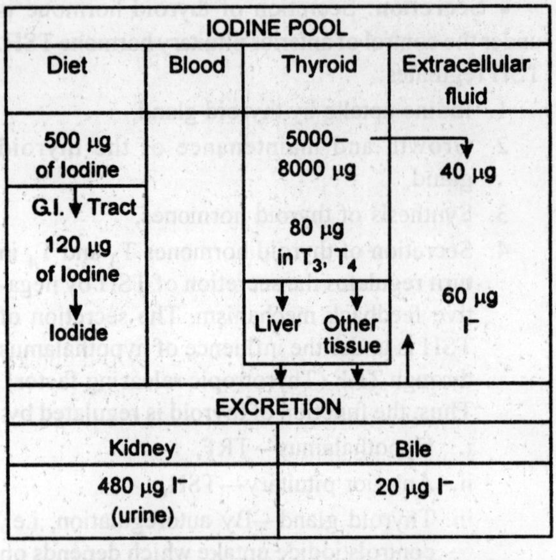

Fig. 11.11: Iodine pool and overall turnover of iodine

I. Tests depending on the blood level of thyroid hormones: Measurements of total T_3 and T_4 levels.

II. Tests depending on alternations of thyroxine binding proteins:
   1. T_3 resin uptake,
   2. Radioimmunoassay (RIA) of thyroxine binding globulin.

III. Test for functional integrity of hypothalamic-pituitary axis.

IV. Tests to establish the etiology of thyroid dysfunction.

11.15.6 Dynamic tests of thyroid function

11.15.6.1 Radioisotope uptake and scan for diagnosis

Application of nuclear medicine in case of thyroid was first made in 1939. Initial studies of thyroid function with I^{131} were concerned with *in vivo* measurements at varying intervals following oral or intravenous administration of the isotope. These were combined with the estimation of urinary excretion, but due to the difficulty of accurately collecting specimens and the advent of more reliable measurement of function, the urinary excretion test has been discarded. It was soon learned that the rate of uptake of I^{131} by the thyroid is as important as the total uptake at a given time. Thus, the most widely employed study is the 24-hour thyroid uptake measurement combined with an earlier determination at 2,4,6 or 8 hours. These earlier intervals are necessary to aid in demonstrating the rate of iodine metabolism and to avoid erroneous results due to rapid turnover in the gland.

- **Methods:** The methods for investigations for diagnosis and follow-up of thyroid disoreders are:

A. In Vivo

I. Radioiodine uptake.

II. Thyroid scanning.

B. In Vitro: Estimation of T_3, T_4, protein bound iodine (BI) of blood, radioimmunoassay (RIA), etc.

11.15.6.2 Use of radioiodine uptake and thyroid scanning

A. Diagnostic

1. Acute diagnosis of thyroid disorders.
2. Planning and monitoring of therapy.
3. Follow-up after therapy.

B. Therapeutic

1. Treatment of thyrotoxicosis.
2. Treatment of thyroid malignancy.

I. Radioiodine uptake method

Iodine, technetium, thallium and gallium can be taken up by the thyroid. But iodine has the advantages over the other because of (1) selective entrance and accummulation in the thyroid and (2) it is the only agent which is not only taken up but processed and released from the thyroid of labelled products occurs, (3) I^{131}, rather than I^{123} has relatively long half-life permits prolonged observation. Thus, iodine provides valuable information regarding functional status of the thyroid. For these reasons, uptake studies are done with radioactive isotope, i.e., radioiodine, I^{131}.

- **Procedure**

1. The patient should not take iodine containing substances at least 8 days before the investigation, e.g., cough syrup containing iodine salts, halogenated oxyquinolines (for amoebiasis treatment), local application of iodine containing ointment, tincture of benzoin and tincture of iodine.
2. The patient should not take cabbages and fish which contain iodine.
3. Radio-opaque dyes contain iodine, hence the uptake study should be done 3 months after

a myelogram and at least 15 days after an intravenous pyelogram.

4. The patient should not take any antithyroid agent, e.g., potassium perchlorate, Lugol's iodine for at least 15 days prior to the investigation.
5. The investigation is absolutely contraindicated in pregnancy (I^{131} may produce irreversible hypothyroidism in fetus) and lactation.

Administration of I^{131}

In this technique, a moderately large dose of I^{131}, about 50 μ Ci or more, is administered orally and allowed to accumulate in the thyroid. Serial epithyroid counts are done by placing the detector over the thyroid on the neck at 2, 14, 48 and 96 hours which reveal the rate of release of accumulated I^{131}.

Instrumentation

Sensitive Scintillation Detector is placed over the thyroid on the neck for detection. The development of Radiation Analyzers, Pulse-height Selectors and Spectrometers for use with the Scintillation Counter, has further improved the statistical aspects of these procedures. Quantitative measurements by external counting require a Rate Meter or Scaler in conjunction with the Detector.

- **Information**

The investigation reveals the following:

1. Amount and rate of uptake of I^{131},
2. The time required to reach peak, i.e., peak hours,
3. The biological half-life of I^{131},
4. Helps to calculate the therapeutic dose of I^{131} for thyrotoxicosis in the patient.

If the result is plotted in a curve (Fig. 11.12) it shows the following:

1. Euthyroid (Normal thyroid): Thyroid uptake of I^{131} reaches a plateau in 24 hours.
2. Hypothyroid: At this interval hypothyroid subject uptake may continue to increase.
3. Hyperthyroid: Uptake may have passed the peak at the same interval.
4. The normal rate of uptake is 5-30 per cent of administered dose in 24 hours.
5. In hyperthyroid case, the rate of uptake is 60-70 per cent of the administered dose.
6. In hypothyroid cases it is about 5 -10 per cent.
7. About 1/3 of ingested iodine is taken up by thyroid and 2/3 is excreted by urine.
8. The uptake is substantially affected by alteration of the total body iodine pool and also by drugs that interfere with the trapping mechanism.

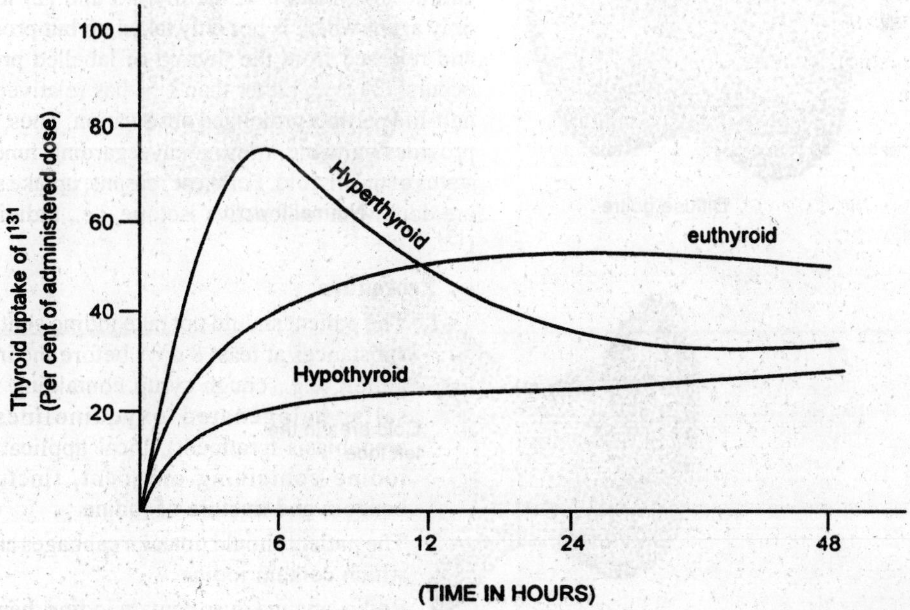

Fig. 11.12: Typical iodine uptake curve in euthyroid, hyperthyroid and hypothyroid subjects

1A. Measurement of thyroid radioiodine release or turnover

Measurement of thyroid release or turnover of I^{131} is done by administering a moderately large doses of I^{131}, i.e., about 50 μ Ci or more. Serial epithyroid counts reveal the rate of release of accumulated I^{131}, I^{131} is preferred to I^{123} because of long half-life which allows prolonged observation. The result is plotted in a curve which indicates the release of initial accumulation of I^{131}. But this curve may be obscured by the reaccumulation of I^{131} already released by the thyroid. For this reason, completely blocking doses of an antithyroid substance are generally administered. After this, the curve of epithyroid counts closely conforms to a single exponential function of time, indicating the release of a constant fraction of thyroid I^{131} unit time. After correction of physical decay of the isotope, this investigation reveals that the rate of turnover of accumulated I^{131} is 1 per cent daily in normal humans.

- **Advantage**

Very effective for physiological study.

- **Disadvantages**

It is rarely used for clinical diagnosis because of:

1. Multiple observation,
2. Indirect methods of measurement of hormonal radioiodine available.

- **Discussion**

After accumulation of I^{131}, labelled hormone releases from thyroid. Fractional rate of release or turnover can be interpreted either from:

1. Direct study of the rate of disappearance from the thyroid, or
2. The rate of appearance of radioiodinated products in the blood.

But several difficulties are encountered to give conclusion about the absolute rate of release of thyroid hormone from the above observations. These are:

(a) Inference regarding the quantity of hormone released is not possible because it requires the knowledge of the iodine content of the pool or pools whose turnover is being measured which is rarely available.

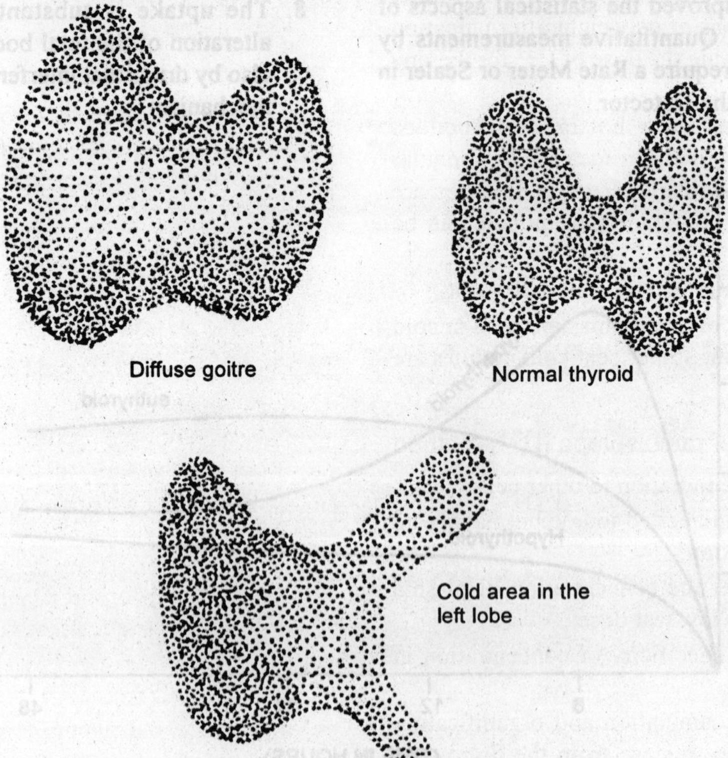

Fig. 11.13: Thyroid scan showing different disorders

(b) The concept of activity of entire thyroid as a homogeneous entity has been discarded. Both anatomic and functional heterogeneity exists in the thyroid. Hence the radioiodine released is likely drawn from several iodine pools of varying iodine content and turnover rate.

(c) Along with radioiodinated hormone, other substances like iodide, iodiotyrosine and thyroglobulin or other iodoproteins may be released and it has some impact in the loss of I^{131} from the gland or the appearance of radioiodinated products in the blood.

11.15.6.3 II Thyroid scanning

Thyroid scanning is done with radioiodine, I^{131} which is commonly used. Technetium ^{99}TC radioisotope scans can also be done. These isotopes give an estimate of the functional anatomy of the thyroid. I^{131} is given intravenously and imaging of the organ is done on a Gamma Camera.

- **Observations**
  1. Thyroid scanner gives an actual size picture and shape of the gland. Along with these, approximations of its weight after palpation is used for fixing the dose of I^{131} for therapy of disorder.
  2. The scan may show hot or cold nodules. When I^{131} is used, an idea of the modular function is available. If other isotopes are used, only a nonfunctioning nodule can be seen.
  3. Hot functioning nodules are malignant.
  4. The majority of carcinomas are seen in cold nodules. About 35 per cent cold nodules are suspicious.

11.15.6.4 Merits of radioisotope (I^{131}) method

1. Delivers little radiation to other cells.
2. Mixes with unlabelled iodide but having no biological action.
3. Dose not alter the concentration of plasma iodide due to low test dose.
4. Having high specificity of concentration in thyroid tissue.
5. Following accumulation and organification of radioiodine, release from the thyroid of labelled products occur.
6. Easy to detect from external arrangement.
7. Having very desirable biochemical, physiological and physical characteristics of isotope.
8. It is a measure of thyroid function supplements rather than supplants the BMR, serum cholesterol and serum PBI (Protein-bound iodine).
9. It has particular value in certain nervous disorders, cardiac disorders where BMR is inconclusive.
10. Reliability is comparable to that of PBI and both are significantly more accurate than the BMR and cholesterol estimation.

11.15.6.5 Demerits

1. It is a poor test to determine hypo- or hyperthyroidism.
2. Due to radiation from I^{131}, there may be tumour or thyroid-like thyroid adenoma.
3. Production of late hypothyroidism which is about 20 per cent.
4. Genetic exposure.

11.16 Radioiodine (I^{131}) Therapy in Thyroid Disorders

Radioactive iodine is used for thereapy in thyroid disorders like hyperthyroidism, metastatic thyroid carcinoma and ablative effects of surgery, etc. It is relatively simple, effective and economical means of treatment.

11.16.1 I Hyperthyroidism

- **Type of patients**

In the treatment of thyrotoxicosis with I^{131} the following patients are selected for therapy:

(a) Elderly patient refusing surgery.

(b) Where surgery is contraindicated.

(c) Who has undergone thyroidectomy, the anatomical landmarks are changed. If the patient again suffers from hyperthyroidism, surgical intervention may cause laryngeal palsy or parathyroid tetany and in this case I^{131} is the choice of therapy.

(d) Patients having relapse or recurrence following drug therapy.

(e) Patients who cannot tolerate drug therapy.

(f) All age groups since there is no evidence or carcinogenic or leukemogenic effect of treatment.

- **Procedure**

Antithyroid drugs should be given to the patient before I^{131} therapy which induces resistance to the radioiodine therapy. Antithyroid therapy should be interrupted 1-2 weeks before the radioiodine therapy which helps adequate accumulation and retention of I^{131}.

- **Precautions**

1. Flushing the toilet properly since I^{131} is excreted.
2. No breast feeding for one month.
3. No conception for about 2-3 years.
4. Pregnant women should not be treated.

Evaluation of the patient should be done at intervals of 3 months, 6 months and 12 months. A routine TSH estimation is necessary to detect hypothyroidism.

11.16.2 II Metastatic carcinoma

I^{131} is not used as primary therapy for thyroid carcinoma but only it can be used in case of matastasis. All surgical excision of thyroid lesion along with surrounding tissue, the metastasis cannot be prevented by the surgery and in these cases I^{131} is used for identification and ablation of residual thyroid tissue and metastatic lesions.

- **Post-surgical care**

Thyroid replacement therapy is not started after surgical thereapy. After 6 months scanning of the whole body is done for detection of thyroid tissue left during surgery and the functioning of metastatic lesions. Nonfunctioning metastases are not located. If remnants and functional metastases are observed. I^{131} therapy should be started.

11.16.3 III Ablative effects of surgery

I^{131} can produce the ablative effects of surgery without the immediate operative and post-operative complications. The ablative dose of I^{131} is much more than a therapeutic dose for which the patient should be hospitalised. If the patient is on thyroid replacement therapy, it should be stopped 1-10 weeks prior to the procedure to prevent TSH suppression and improve uptake. Eltroxin is not stopped if the metastases are in the spinal column as they might flare up and cause neurological defect.

- **Procedure**

The patient is given I^{131} in a few millitires of water and should be isolated in a separate leadlined room. The ceiling of the room has the detector to detect radiation. The patient's excreta should be disposed off separately.

Another method is to measure the radiation in the patient's urine. Then the patient is discharged only when the amount of radiation decreases to acceptable levels.

The precautions are the same like therapy.

After 3-6 weeks, scanning of the whole body is repeated to detect metastasis. If there is metastases, then therapy should be repeated. If the patient's scan does not show any metastasis or thyroid remnants, then the patient should be followed up at 6 months interval for 3 years and yearly for 2 years. The patient should be advised to have an yearly X-ray of chest and thyroglobulin levels taken.

11.16.4 Complication

1. Radiation sickness: Anoresia, nausea, vomiting.
2. Radiation sialoadenitis: Metallictaste, swelling and tenderness of the salivary glands.
3. Pelvic examination in women: To rule out cystocele due to retention of urine containing I^{131} producing radiation cystitis and haematuria.

11.17 Blood Volume Determination by Isotopic Method

11.17.1 Introduction

The principle of blood volume determination is the dilution of a known amount of marker substance in the circulating blood. The ideal marker substance satisfies the following criteria:

1. Substance should be very small for administration into the blood.
2. Substance can be measured accurately in the blood or plasma.
3. Substance should remain in the circulating blood at least for 30 minutes after injection or it should leave the blood at a constant rate,
4. Substance should not be toxic to the body.

The substances are of two groups:

1. For plasma volume: Albumin tagged with radioactive I^{131}, I^{132}, I^{125}.
2. For red cell volume: Radioactive Fe^{55}, Cr^{51}, P^{32}.

11.17.2.1 Single isotopic method

A. Plasma volume.

B. Red cell volume.

11.17.2.2 A. Determination of plasma volume

The dye method of estimating plasma volume has the following advantages:

1. No radioactivity,
2. Can be done in pregnancy,
3. Eliminated very slowly (4.8% per hour) from the circulation.
4. Remains in circulation binding with proteins.

But the disadvantages are:

1. Stains blood and tissue,
2. Affected by haemolysis or lipaemia,
3. Does not give accurate result.

The most accurate and reliable serial estimation of plasma volume is done with protein tagged with radioactive isotope. Human serum albumin tagged with radioactive I^{131} is commonly used for this estimation. In special study γ-globulin and fibrinogen are also used. In many cases I^{125} or I^{132} are more convenient for tagging. But albumin tagged with I^{131} or I^{125} gives accurate result. The advantages of I^{131} are the following:

1. Gives accurate result,
2. Serial estimation can be done,
3. Not affected by haemolysis or lipaemia,
4. Does not stain the tissue,
5. Minute radioactivity is used in each investigation, repeated determination can be made accurately and rapidly.
6. This method can be used at the bed side with the development of sophisticated instrument.

The disadvantages are the following:

1. Radioctivity is involved,
2. Requires information regarding its disappearance rate from the blood,
3. Cannot be used in pregnancy and lactation.

- **Procedure**

A. **Preparation**: Approximately 5 μCi of an aqueous solution of labelled albumin are added to 10 ml of isotonic saline solution and 1 ml of isotonic sodium iodide solution to make a volume of 11.5 ml and the solution is mixed thoroughly. Out of this solution 11 ml is taken into a syringe calibrated by gravimetric means to a known mark of the barrel. The syringe is filled up with 1 ml. mole solution above the calibration mark and the bubbles are removed. Excess labelled solution is removed into a tube for standard. 1 ml of labelled solution is removed with a grade A pipette to a 100 ml graduated grade A flask containing 5 ml of a 10 per cent solution of carrier unlabelled human albumin. The contents of the flask is diluted to 100 ml with water containing 1g/litre sodium or potassium iodide.

The contents of the syringe are injected intravenously taking care of the needle which must be inside the vein initially as well as at the end of the injection by sucking a little volume of blood into the syringe. Blood samples of 10 ml are taken at 15, 30 and 45 minutes from the different vein, and preferably a different limb other than that has been used for injection for elimination of contamination from the site of injection. Heparin or dry oxalate is used as an anticoagulant. 2 ml of blood from each sample are used for duplicate haematocrit and the remaining amount of blood is centrifuged to separate the plasma.

The radioactivity of the duplicate aliquots of the diluted standard and the plasma samples are measured by the well type Scintillation Counter.

- **Observations**

Radioactivity levels of the plasma is plotted on a semilogarithmic paper against time and the resulting straight line is extrapolated to zero time. This is done for correction for radioactivity lost during initial mixing period. The slope of the plotted line shows the rate of disappearance which is between 6-10 per cent per hour in normal person and 18 per cen in case of burn-patient.

If one blood sample is taken and extrapolation method is not possible, then this sample should be taken 15 min after injection. A factor of 10.2 is used to correct for loss of radioactivity in normal persons and 1.045 in case of burn-patients.

- **Calculation**

The plasma volume (PV) is calculated in the following way:

Extrapolation method: $PV \frac{VDn_s}{n_0}$

Single sample method: $PV\frac{VDn_s}{n_{15}f}$,

where V = volume of labelled albumin solution injected,

D = dilution factor of the standard,

n_s = counting rate of the standard,

n_0 = counting rate of the plasma sample at time 0,

n_{15} = counting rate of the plasma sample taken 15 minutes after injection and

f = correction factor for a 15 minute sample.

- **Discussions**

1. Variation of the disappearance rate of labelled albumin from the circulation makes this approximation less accurate than the estimate obtained from extrapolation. This uncertainty is encountered in abnormal circulatory conditions, e.g., heart failure after severe haemorrhages.
2. Addition of solution of carrier unlabelled human albumin during preparation of the standard reduces the loss of labelled albumin from solution by its absorption onto glass surfaces.
3. A final protein concentration should be more than 2 mg/ml. If the stock solution of labelled albumin having less than 20 mg/ml of protein, unlabelled albumin should be added to minimise the hazards of irradiation.
4. Errors in dilution during the preparation of the standard and samples for assay are decreased by using either grade A glassware or a gravimetric procedure.
5. Measurement of plasma volume can be done with Cr^{15} chromic chloride but it gives an overestimate of plasma volume and contributes a larger radiation dose to the subject than that from iodine isotope.
6. Slightly smaller values for plasma volume are obtained with substances having high molecular weight than albumin, e.g., γ-globulin, fibrinogen, high molecular weight dextran because of slower rate of transcapillary exchange.
7. Increased rates of capillary permeability as in case of burn-patient, plasma measured with albumin or γ-globulin are high.

In order to obtain the value for the total blood volume, the ratio between the plasma and RBC is determined by means of the haematocrit. If the packed cell volume is 45 per cent, the plasma volume is, therefore, 55 per cent. Thus, the total blood volume can be calculated from the plasma volume and the haematocrit reading.

11.17.2.3 B. Determination of red cell volume by isotopic method

For calculation of the blood volume from the plasma volume and the haematocrit reading it is assumed that the proportion of red cells to plasma is the same throughout all parts of the vascular system but experimental evidences show that the ratio of red cells to plasma in capillaries significantly less or more than in the large vessels. To exclude the error of unequal distribution of red cells throughout the vascular system, methods have been devised for measurment of the total circulating red cells volume.

- **Isotopes for determination**

The radioactive labels of red cells are Fe^{55}, Fe^{59}, P^{32}, K^{42} and Cr^{51} are used. Among these radioactive labels, isotopes of iron are rarely used because a donor for labelled red cells is required and the heterologous cells may not be completely compatible. Since subject's own red cells are used *in vitro*, no problem arises in case of P^{32} or K^{42} are used as red cell lables. It has now been largely replaced by Cr^{51}.

(a) Isotope Fe^{55} method

According to the method of Hahn and modified by Gibson et al. radioactive iron Fe^{55} is incorporated into Ferric ammonium citrate which is injected into a donor belonging to group O. The new cells take up the radioiron and appear in the circulation in 24 hours and reaches a maximum in 24 days.

70-100 ml of donor's blood having about 30-40 ml or red cells with a radioactivity of 2500-3000 counts per minute measured by Geiger Counter are injected intravenously or intraarterially into subject. 10-20 minutes are allowed to mix up the donor's and recipient's blood. 15 ml sample of recipient's blood is taken and followed at 10 and 20 minutes intervals. 20 ml of donor's blood is diluted to 100 ml, 10 aliquots are then prepared. The recipient's blood samples are centrifuged at 3000 r.p.m. for 30 minutes. Donor and recipient samples are wet-ashed and the iron deposited electrolytically on copper. The radioactive iron in the two samples is determined by a Geiger Counter.

Calculation

$$Vrr = \frac{CD \times VaD}{VaR},$$

where Vrr = red cell volume,
CD = number of ml of donor cells injected,
VaD = radioactivity of the donor's cells,
VaR = radioactivity of the recipient's cell.

Modification

Simplification can be done by eliminating the wet-ashing and electroplating. The whole blood or its dilution can be counted directly in a well-type Scintillation Counter.

Calculation

For determination of whole blood volume, the plasma volume is measured by the dye method and the result is added to the result obtained for red cell volume.

Discussion

This method is applied considering the following facts:

1. That radioactive iron does not escape again once it has been taken up by red cells
2. That all the tagged cells are mixed up properly with the recipient's cells throughout the circulation.
3. Radioactive iron is used because the investigation can be continued for a long time as the radioactivity remains in the red cells and the blood for much longer period than the radioactive chromium or phosphorus.

(b) Determination of red cell volume using Cr^{51}

The method employing Cr^{51} is simpler to use than Fe^{55} because the labelling of red cells can be done *in vitro* with the subject's own blood, hence the problem of compatibility does not arise. Radioactive Chromium (Cr^{51}) is the isotope of choice which is being used frequently.

• **Procedure**

5-10 ml blood is taken out from the subject and added to a bottle containing 4 ml of acid-citrate-dextrose (ACD) solution. The blood is properly mixed and centrifuged at 1000 r.p.m. 3-5 ml of plasma is removed and added to about 75 ml isotonic solution to provide plasma/saline washing solution. Approximately 20 μCi of Cr^{51} as sterile isotonic chromate solution is added to the red cell suspension and the whole is well mixed. The mixture is left at room temperature at least for 10-30 minutes. The labelled red cells are washed twice with 30 ml Plasma saline for decreasing the extracellular Cr^{51} level to less than 1 per cent of the intracellular level. The result of the washing should be confirmed for each red cell volume estimation. It is done by assay of known volume of supernatant taken from part of the final labelled red cell suspension. The labelled red cell suspension is made up to 11 ml with plasma saline. A known volume (10 ml) in a calibrated syringe is injected into the subject.

The standard is prepared by the transfer of 1 ml of the remaining labelled red cell suspension with a grade A pipette to a 100 ml grade A graduated flask and dilution with 0.1 per cent unlabelled sodium chromate solution.

After the injection of the Cr^{51}-labelled red cells, sample of blood (12 ml) is taken from a vein at 15 min. The sample blood is added to a tube containing anticoagulant (heparin or dry balanced oxalate). Other samples may be taken at 25 and 35 minutes. The radioactivity in duplicate 5 ml aliquots of the diluted standard and the blood sample is determined in well-type Scintillation Counter. Duplicate haematocrit estimations are made on the remaining blood.

Calculation

The red cell volume (RCV) is calculated in the following way:

$$RCV = \frac{VDn_s H}{n_{15} 100},$$

where V = volume of labelled red cells injected,
D = dilution factor of standard,
n_s = counting rate of the standard,
n_{15} = counting rate of venous blood sample taken at 15 minutes after injection.

H = haematocrit of venous blood sample corrected for trapped plasma.

Discussions

1. Considering the problem of incompatibility, the patient's own red cells should be labelled with Cr^{51}.
2. Satisfactory result can be obtained using red cells stored at 4°C in ACD solution for transfusion purposes. But the period of storage at 4°C should not be more than 12 days.
3. The cells injected should be preferably of the same ABO and Rh groups.
4. Though both ACD solution and radioactive sodium chromate solution are required for labelling red cells, they cannot be mixed together and stored as a mixture, since the sodium chromate is thereby converted into a chemical form which will not label red cells.
5. Since there is no significant loss of Cr^{51} from labelled red cells during the first hour after their injection, no correction factor need be applied to the observed counting rate of the blood sample.
6. The firmness of the Cr^{51} label leads to shorter preparation time.
7. In case of burn, spherocytosis where the red cells are fragile, accurate result can be obtained after washing three times.

- **Isotope haematocrit**

Apart from the haematocrit, which is generally measured by centrifugation of blood in Wintrobe haematocrit tube, an isotopic haematocrit is being used at present and a new electronic instrument has been devised to give instant readings.

It is done by the use of labelled albumin in plasma volume measurements. The difference between the counting rate per ml of whole blood and that per ml of plasma depends upon the proportion of red cells in the blood. Thus,

$$H = \frac{\text{C/min per ml whole blood} \times 100}{\text{C/min per ml plasma}}$$

With this method no correction for trapped plasma is required. A number of studies by this method have shown that the result agrees with that measured by centrifugation appropriately corrected for trapped plasma.

11.17.3 Double isotopic method

(Simultaneous measurement of plasma and red cell volumes)

Most accurate measurement of the blood volume is the sum of the plasma and the red cell volume measured separately and simultaneously when I^{125} labelled albumin and Cr^{51} lebelled red cells are used to measure plasma volume and red cell volume simultaneously. The two labelled preparations may be either mixed in the syringe prior to injection or administered separately from two syringes preferably using a three-way tab attached to the injection needle so that the time interval between the injections is minimal.

11.17.3.1 A. Single injection method

If a mixture of the two labelled preparations is injected, a diluted standard of the mixture should be prepared. An aliquot of this together with the samples are measured for I^{125} and Cr^{51} under the appropriate conditions. Since some Cr^{51} γ radiation is detected under the optimum condition for I^{125}, a pure Cr^{51} source should be measured under both conditions. The I^{125} measurement can then be subsequently corrected for the contributions from Cr^{51}. No I^{125} radiation should be detected under the optimum condition for Cr^{51}, but a pure I^{125} source should be measured under both conditions to confirm that this is so. If a pulse-height analyser is not available to separate I^{125} and Cr^{51} radiations, a cylindrical liner of brass or steel with a wall and base thickness of 1 mm may be placed inside the well of the NaI (T1) crystal to absorb the I^{125} radiation. Such a liner does not significantly reduce the counting rate due to Cr^{51}, nor unfortunately that due to any I^{125} present as a contaminant the I^{125}.

Blood samples are taken at 15, 30 and 45 minutes after injection. Aliquots of whole blood may then be measured for both I^{125} and Cr^{51} as for the injection mixture.

Alternatively, plasma be separated and measured for I^{125} alone and whole blood measured for Cr^{51} alone; this avoids the need to correct the I^{125} measurements on the samples for contributions from Cr^{51}.

11.17.3.2 B. Sequential injections method

With sequential injections, separate individual diluted standards are prepared from each labelled preparation. Measurements of I^{125} are made directly on plasma separated from the post-injection blood samples. Measurements of Cr^{51} are similarly made in whole blood. The methods of calculation are then identical with those discussed previously.

- **Precautions**

1. Whatever method is used, determinations are made with the subject in recumbent position and under basal conditions, 12-14 hours after a meal.
2. The loss of dye into the tissue at the site of injection should be avoided unless there will be error.
3. Care should be taken to prevent evaporation of fluid from the samples after they are drawn.

11.17.4 Merits and demerits of single and double isotopic method

I. Single isotopic method

Plasma volume (Dye) + Red cell volume (Cr^{51}) combination:

1. Double radioisotope measurement is not required.
2. No overlapping of the radiation.
3. Accuracy depends upon the accurate measurement of dye T-1824.
4. Infiltration of the dye into the tissue at the site of injection may produce error in the result.
5. Cr^{51} count may be taken for red cell volume at least after 24 hours.

II. Double isotopic method

(a) Plasma volume (I^{125}-albumin) + Red cell volume (Cr^{51}) combination:

1. Ideal combination.
2. I^{125} can be measured alone or with Cr^{51} in whole blood in well-type Scintillation Counter.
3. Cr^{51} alone can be measured in whole blood by the same Counter.
4. Procedure is more complex if used with "closed" red cell volume measurement technique.

(b) Plasma volume (I^{131}) + Red cell volume (P^{32}) combination:

Two separate instruments are required, I^{131} by Scintillation Counter, P^{32} by thick walled jacketed G-M Counter or end-window G-M Counter and absorbers.

(c) Plasma volume (I^{132} -albumin) + Red cell volume (Cr^{51}) combination:

1. Plasma volume is measured from I^{132} soon after injection.
2. Red cell volume is measured from Cr^{51} in whole blood at least 24 hours after injection.
3. Slight correction is necessary for I^{121} contamination of I^{132}.
4. The disadvantage is that: I^{131} albumin should be prepared just before use.

(d) Plasma volume (I^{131} albumin) + Red cell volume (Cr^{51}) combination:

1. Gamma-ray energies of I^{131} and Cr^{51} make the measurement difficult.
2. It is possible to measure I^{131} alone in plasma.
3. It is possible to measure mixed radioisotopes in whole blood.
4. It is possible to indicate plasma content of whole blood using haematocrit.
5. I^{131} can be measured alone in whole blood for its weak 0.64 Me Vr radiation.

(e) Plasma volume (I^{125} albumin) + Red cell volume (P^{32}) combination:

1. Satisfactory combination.
2. I^{125}-albumin can be measured alone in plasma by well-type scintillation Counter.
3. P^{32} can be measured in whole blood by jacketed G-M Counter.

11.18 LASER Beam in Biology, Diagnosis and Therapy

11.18.1 Definition

L = light, A = amplification, S = stimulated, E = emission of R = radiation.

Laser is an acronym, formed from the initial letters of the words mentioned above.

A laser is an amplifier which emits electromagnetic energy in the form of a concentrated light beam. The word LASER stands for Light Amplification by Stimulated Emission of Radiation, it is based on the principle of Stimulated Radiation which was recognised by Einstein in 1917.

11.18.2 Development

Based on the principle of Einstein (1917), methods of utilising the principle were found in 1950. C.H. Townes and his co-workers (1953) doing research on this, developed a microwave amplifier called MASER or Microwave Amplification by Stimulated Emission of Radiation. Townes, Prokhorov and Basov (1958) doing independent research, showed that it was possible to build a "MASER" emitting visible light and shared the Nobel Prize in 1964. Theodore Maiman (1960) developed the first working laser. Bell Laboratories (1961) built the first continuously operating model.

11.18.3 Spontaneous and stimulated emission

An atom may undergo a transition between energy states if it emits or absorbs a photon of the appropriate energy. If an atom is in the ground state, it absorbs photon and makes a transition to a higher energy state or level, the photon is referred to have "stimulated" the absorption. The atom cannot increase its energy "spontaneously" in absence of a photon. Then it may be expected that when the atom is at higher level, it would emit a photon, which means, the emission would be independent of the number of photons present in the environment of the atom. But Einstein disproved it in 1917. According to him, the probability per unit time is that an atom will decay to a lower energy state and emits a photon in the sum of two terms, one is a "spontaneous emission" and the other is a "stimulated emission", and is proportional to the number of photons present in the environment of the atom. But Einstein disproved it in 1917. According to him, the probability per unit time is that an atom will decay to a lower energy state and emits a photon in the sum of two terms, one is a "stimulated emission", and is proportional to the number of photons of the relevant energy already present in the environment of the atom. The photon produced by stimulated emission is always "in phase" with stimulating photon.

11.18.4 Laser principle

If a few photons of energy E (excited state energy)—E_o (ground state energy) are introduced, these immediately stimulate the emission of a number of photons of the same kind. There is increase in number of photons and these again stimulate the emission of more photons. In this way a "chain-reaction" or "avalanche" is formed and as a result all the atoms present give up their photons rapidly. This process is called "laser action" or "light amplification" by stimulated emission of radiation as the original "pulse" of photons has been amplified into a more powerful pulse of light.

11.18.5 Nature of laser light

The light pulse from a laser shows the following characteristics and differs from ordinary light in the following ways:

(a) Monochromatic (single wavelength) light, because:
   1. All the photons have the same energy above the ground level,
   2. Same frequency,
   3. Single wavelength.

(b) Coherent light, because all photons or waves in phase, i.e., all electromagnetic waves are in step.

(c) Unidirectional light: because all the emitted waves are parallel.

(d) High intensities: Laser light achieves very high intensities through concentrations of energy in a small area.

11.18.6 Ruby laser beam

First successful development of laser was due to an American scientist in 1960 by using ruby crystal (aluminium oxide with a chromium impurity). Subsequently other substances like organic dyes, gases (e.g., CO_2), various rare elements, e.g., Neodymium, were used to produce laser beam. But the lasers produced by these substances differ in their effects for all types of cases. Application of lasers in biology and medicine is based on the interaction with biological substances. Laser section is the light amplification by stimulated emission of radiation. Laser beam is produced by exciting the molecules of ruby crystal. The molecules of ruby crystal then emit a characteristic red light and excite the neighbouring molecules and the process continues like a chainreaction. The action is enhanced by silvering both the ends (and sides) of ruby crystal which keep on reflecting and thus amplifying the intensity of light. The sides of the ruby crystal are kept free to allow the light from the pumpin lamp which then excites the molecules of the ruby crystal. A very powerful and narrow beam of red light having wavelength 694.3 nm emerges through a section at one end of ruby crystal which is only partially silvered (Figs. 11.14, 11.15).

Depending on the medium used, the emitted light has a wavelength which can be anywhere from infrared through the visible spectrum to ultraviolet. Methods of excitation of the medium can range from gas discharge to photo-dissociation.

Lasers which emit energy intermittently are said to be operating in the pulse mode, and can provide output energy levels as high as several thousand kilowatts. Continuous laser, on the other hand, typically emit energies in the 1 to 10 kilowatt range.

11.18.7 Types of lasers

I. Solid, Gaseous, Semi-conductor

II. According to mode of operation
   A. Pulsed operation.
   B. Continuous wave operation.

Main types of lasers in medical use are:

1. Argon,
2. Argon/Krypton,
3. Tunable Dye,
4. Yttrium-Aluminium-Garnet (YAG).

In an Argon laser, Argon atoms are excited by an electric discharge to produce an infrared beam which is readily absorbed by Hb but not by water and other transparent tissues. Carbon dioxide and neodymium, YAG lasers also produce light in the infrared region.

11.18.8 Mechanism of laser effect

The mechanism of laser effects are of four types such as:

A. **Photocoagulation**: The principlal lasers used for this process are *thermal lasers*. They depend upon absorption of the laser light by tissue pigments. This absorbed light then converts into heat which raises the temperature of the target tissue high enough to coagulate and denature the cellular elements. This mechanism is used for opthalmic therapy.

B. **Photodisruption**: The lasers for this mechanism release a giant pulse of energy with a pulse duration of a few nanoseconds. When this pulse is focussed to a small spot of 15-25 micrometre, "optical breakdown" occurs in which temperature rises to high level such as about 10000°C. During this period, the electrons are stripped from atoms, resulting in a physical state called "Plasma".

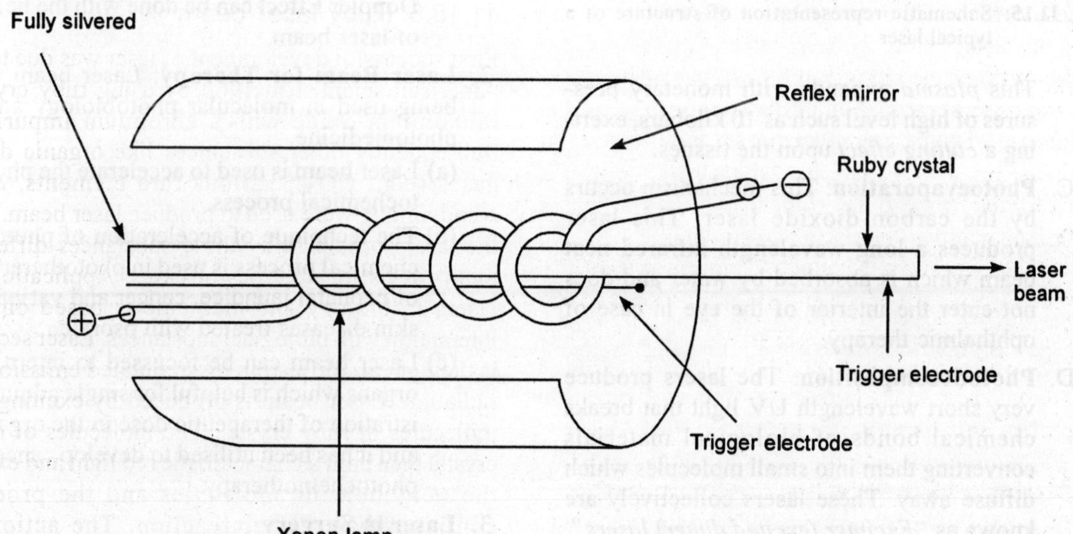

Fig. 11.14: Diagrammatic representation of mechanism of ruby laser beam

[Excited atoms spontaneously return to original or ground energy by shedding their excess energy through the emission of photons. The net result is intense and concentrated beam of light].

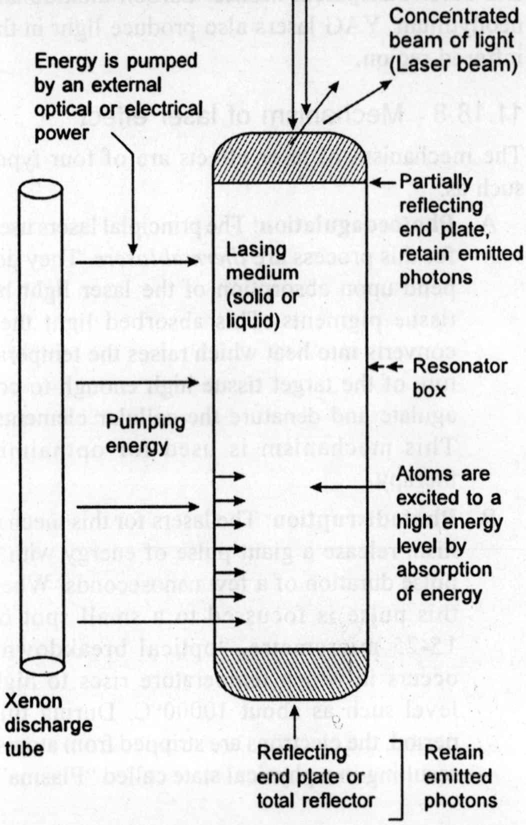

Fig. 11.15: Schematic representation of structure of a typical laser

This *plasma* expands with monetary pressures of high level such as 10 kilobars, exerting a *cutting effect* upon the tissues.

C. **Photoevaporation**: This mechanism occurs by the carbon dioxide laser. This laser produces a long wavelength infrared neat beam which is absorbed by water and does not enter the interior of the eye in case of ophthalmic therapy.

D. **Photodecomposition**: The lasers produce very short wavelength UV light that breaks chemical bonds of biological materials converting them into small molecules which diffuse away. These lasers collectively are known as *"Excimer (excited dimer) lasers"*. Because the cavity contains argon and fluorine, that react into unstable molecules which then emit the laser light. In case of ophthalmic therapy, they help to recontour the cornea with precise lesions which are controlled by a computer.

11.18.9 Applications of laser beam

Laser beam is applied in biology and medicine for the following purposes:

1. Diagnositc,
2. Therapeutic,
3. Surgical.

1. **Laser for diagnosis**

(a) Laser beam has helped to develop ultrasensitive analytical methods to detect even a single atom of molecule. This is being used for analysis of trace elements in metabolic study and toxicology.

(b) Laser fluorescence helps to localise early lung cancer.

(c) Laser beam exhibited the fluorescence pattern difference between normal and atherosclerotic vessels which can be of use in vascular surgery.

(d) Laser beam can be focussed to internal organs through a Fibreptic Endoscope for diagnostic and therapeutic purposes.

(e) Three dimensional photography of internal organs known as Holography can be done with the help of laser beam.

(f) Measurement of capillary blood flow of Doppler Effect can be done with the help of laser beam.

2. **Laser Beam for Therapy**: Laser beam is being used in molecular photobiology and photomedicine.

(a) Laser beam is used to accelerate the photochemical process.

(b) The technique of acceleration of photochemical process is used in phototherapy of neonatal jaundice, cancer and various skin diseases treated with psoralen.

(c) Laser beam can be focussed to internal organs which is helpful for single administration of therapeutic dose in the organ and it has been utilised to develop cancer photochemotherapy.

3. **Laser in Surgery**

Surgical Laser or Lasers Surgery: A laser is an instrument equipped to produce a fine parallel beam of high intensity light which can be focussed on a very small area. It is used with an operating microscope or fluoroscope. By this microsurgical procedures have been established e.g., sutureless microvascular anastomosis, excision of posterior capsule of lens through the transparent cornea, using a flexible gastroscope, stopping gastric bleeding by craguletion. In laser surgery the hearts the target tissue causing the following reactions:

(a) Thermal tissue destruction
(b) Contraction of tissue
(c) Coagulation in blood vessel
(d) Boiling of cell water and vaporisation
(e) Incision by vaporisation of the tissue occurs by higher energy exposure.

(a) **Photocoagulation in Diabetic Retinopathy**: Focussed laser beam on a tissue causes intense and localized heat. This localized heat produces tissue damage which is propotional to the intensity of the laser beam. This photo-destruction happens in aseptic condition without any physical contact. Hence laser beam is being used in proliferative diabetic retinopathy by photocoagulation.

(b) **Treatment of Glucoma**: Laser beam can be focussed on a few micrometre small spot, hence it has been utilized to open up blocked canals of Schlem during treatment of glucoma which is referred to as *Microsurgery.*

(c) **Microsurgical Technique for Alteration of Structure at Subcellular Level**: Microsurgical technique is utilized (at subcellular level) for alteration of the structure of an organelle or a chromosomes.

(d) **Microsurgical Technique for genetic study**: This microsurgical technique helps to introduce a new gene through a submicron hole drilled in a cell membrane in case of genetic engineering.

11.18.10 Use of laser beam

1. Treatment of glucoma.
2. Treatment of retinal detachment.
3. Surgical renoval of tumours in selected cases.
4. Endoscopic diagnosis of early malignancy.
5. A laser "knife" for reduction of blood loss and post-operative pain.
6. Photochemotherapy of malignancy.
7. Cosmetic surgery for facial wrinkles, deeply pitted acne scars.
8. Microsurgery in the subcellular level, e.g., organelles, chromosomes.
9. Genetic engineering.
10. **Cardiac**
    (a) Removal of coronary occlusion. [Ultraviolet excimer (excited dimer) laser]
    (b) As an alternative to bypass surgery and balloon angioplasty for removing plaque from coronary arteries (Holmium cold laser).
    (c) As an adjunt to coronary artery bypass grafting to form small channels within the heart while the heart continues to beat. [Mid–infrared thulium-holmiumchromium Yttrium-aluminium-garnet (ΓHC : YAG) laser]
    (d) For channel formation with an arrested heart during cardiopulmonary bypass. [Far-infrared carbon dioxide lasers]
    (e) Treatment of vascular disorders.
    [Copper vapour laser]
11. **Orthopaedics**: In orthopaedic surgery to minimise pain and swelling. [Holmium laser].
12. **Dental**
    (a) For approaching difficult recesses of the mouth, remaining painless and requiring no local anaesthesia.
    [Neodymium : Yttrium-aluminium-garnet (Nd : YAG) laser]
    (b) Treatment of root canal.
    [Ultraviolet excimer lasers]
13. **Spinal disc surgery**
    [Holmium : Yttrium-aluminium-garnet (H : YAG) laser]
14. **Eye therapy**: Laser are being used in opthalmology for every anatomical structure of the eye, e.g., eyelids (Haemangiomas), conjunctiva (pterygium), cornea (Neovas-

cularisation), ciliary body (partial destruction), iris (Iridotomy), pupil (photomyriasis), retina (detachment of retina, diabetic retinopathy), sclera (sclerotomy).

Laser Scalpel

The laser beam acts as a scalpel which can perform **bloodless** surgery and has welding capacity for blood vessels that are being cut. Beside that, this type of surgery high, sterility and is painless. Ulcers turnovers and modules are removed by thermal effects and immuno suppressive effects usually hard lasers (Na–YAG, CO_2 and excimer lasers) are used for surgical procedures on hard tissue line, bone and teeth.

11.18.11 Advantages

1. The unique properties of laser light allow the beam to be focussed down to extremely small spot resulting in very high energy densities that are the basis of the tissue altering effects
2. Patient compliance increases because of reduced hospital stay.
3. A safer and quicker alternative to surgery.
4. Serious ocular diseases can be treated.
5. Effective therapeutic modality is offered by experienced surgeon.

11.19 Nuclear Medicine for Therapy

11.19.1 Introduction

Beta-rays have the property to ionize and destroy the tissue-close to their sources. This property is being utilized in the **therapeutic nuclear medicine,** whereas gamma rays cannot cause any internal radiation injury and can be detected externally for which it is being used in **diagnostic nuclear medicine.** The therapy concerns only open radioactive sources and the substances are introduced into the patient's body and circulates in the body and come in contact with the body fluids.

Lawrence (1933) used ^{32}P for treating leukaemia and Hertz and Roberts (1942) used radioiodine for treatment of Graves' disease.

11.19.2 Common conditions for therapy

1. Bone metastasis in cancer of the breast or prostate.
2. Polycythaemia vera.
3. Hyperthyroidism.
4. Thyroid cancer.
5. Malignant pleural effusion.
6. Maligant ascites.
7. Rheumatoid arthritis.

11.19.3 Bone metastasis in breast and prostatic cancer

1. **Bone Metastasis in Breast and Prostatic Cancer**: When chemotherapy is not successul, radioisotopes treatment is ideal. The common isotope is ^{32}P which relieves pain due to shrinkage of tumour metastasis and reduction of periosteal stretching.

   Prior administration of androgen, in case of breast cancer metastasis, there is increase in uptake of ^{32}P. In case of prostatic cancer metastasis, ^{32}P uptake increases for prior administration of parathyroid hormone.
2. **Malignant Pleural Effusion and Malignant Ascites**: Tapping of fluid does not prevent accumulation of fluid but colloidal gold ^{198}Au or colloidal chromic phosphate ^{32}P instillation causes irradiation of free floating malignant cells, the surface seedlings and mesothelial cells.
3. **Polycythemia Vera**: Polycythemia Vera shows overproduction of red blood cells, pan myclosis, having increased red cells in the bone marrow and increased red blood cells, white blood cells and platelets in the peripheral blood. ^{32}P is used for therapy which is readily diffusible and tissues utilise phosphorus in their metabolism. The nucleus contains highest amount of phosphorus. Nucleus is larger and needs more phosphorus during division. Therefore, ^{32}P concentrates heavily in the cancer cells. In case of Polycythaemia Vera, ^{32}P concentrates in the marrow and acts directly on the dividing parent cells and also concentrates in the reticular system.
4. **Rheumatoid arthritis**: In rheumatoid arthritis (intractable) radio-nuclides are used as an alternative to synovectomy. The radionuclides are colloidal gold ^{198}Au, Ittrium ^{90}Y, ^{32}P or Erbium ^{169}Er.

12

Concept of Valency and Chemical Bonding

1. Valency or Combining Capacity, 2. Electronic Theory of Valency, 3. Types of Bonds, 4. Chemical Bonds, 5. Energy Concept of Chemical Bonding, 6. Bond Energy, 7. Strong and Weak Bonds, 8. Strong Bonds, A. Covalent Bonds or Shared Electron Pair Bond: (a) Peptide Bond; (b) Glycosidic Bond; (c) Ester Bond; (d) Phosphodiester Bond, (e) Disulphide Bond; 9. Weak Bonds: (a) Electrovalent or Ionic Bond; (b) Hydrogen Bond; (c) Hydrophobic Bond; (d) Van der Waals Force; (e) Ion-Dipole Interaction; (f) Orientation Force; (g) Inductive Force; (h) Dispersion Force (London Force).

12.1 Valency or Combining Capacity

Valency is the general term to explain the power of the atoms to combine with another atom to form molecules by chemical bonds. It is usually defined as "the number of Hydrogen atoms or twice the number of Oxygen atoms with which one atom of that element can combine". It is usual procedure to compare the combining capacities of different elements with that of Hydrogen. The reason is that, there is no compound of Hydrogen (exception, Hydrazoic acid, N_3H) which contains more than one atom of an element combined with one atom of Hydrogen, for which chemists have considered an atom of Hydrogen as the standard for measuring the valency or combining capacity of the atoms of different elements. But it is difficult to agree that the term valency is appropriate. Because,

1. **In water (H_2O), Ammonia (NH_3) and Hydrochloric acid, etc. molecules, it is observed that in water (H_2O) one atom of Oxygen combines with two atoms of Hydrogen and in ammonia (NH_3) one atom of Nitrogen combines with three atoms of Hydrogen. From this we can infer that atoms of different elements have different capacities for combining with Hydrogen.**
2. **Some elements do not combine with Hydrogen but they combine with other elements like Chlorine, Fluorine etc.**
3. **Again, Carbon forms many compounds in series with Hydrogen, e.g., C_2H_3, C_2H_4, C_2H_6 and CH_4, etc. in which the valency of Carbon is 1, 2, 3, 4. Now, if we consider the previous definition, valency of Carbon varies from 1-4 in these compounds whereas Carbon is known to be tetravalent in these compounds.**

Therefore, the conception of valency as a number is being discarded gradully and efforts are being made to explain the "combining capacity" of elements in the light of the modern concept of atomic structure through the discovery of electronic

mechanism by which atoms are held together in molecules–known as Electronic Theory of Valency. It has been shown that there are at least three different ways in which an atom can form links with other atoms.

12.2 Electronic Theory of Valency (Lewis)

After the discovery of the electron, further knowledge on the electronic structure of the atom made it possible to know the chemical combinations in terms of the ultimate units of matter and the electronic theory of valency was propounded.

The theory of valency was originated by Kossel (1916) for ionic, i.e., electrovalent compounds, then Lewis (1916) succeeded in developing a theory of bonding. Because it was based on the electronic conception, it is known as Electronic Theory of Valency (or bonding). According to him, bonding of an atom depends on the electrons in the outermost shell and that the number of electrons in the last orbits of the atoms of inert gases (e.g., Helium having 2 and the rest 8) formed a most **stable grouping**. All other elements except the inert gases have atoms with **unstable and incomplete outer shell**. These atoms with unstable and incomplete outer shells, tend to lose or gain electrons to possess an electronic configuration identical with that of the nearest inert gas in the periodic table.

This tendency of atoms to complete and stabilise their outermost shell of electrons causes the formation of valence forces or bonds to combine chemically to form molecules. In this way molecules are formed by the redistribution of electrons in two or more atoms in such a way that outermost shell or orbit of each atom acquires complete group of two electrons (electron duplet) or eight electrons (electron octet). Almost all atoms tend to attain octet except H, Li and Be which tend to attain duplet and for this reason this theory is also known as the Octet Theory.

12.2.1 Variation of the valency

Sometimes variation of the valency of the element occurs as it is not a constant property of the element for which some elements have more than one valency. As for example, Sulphur is divalent in H_2S, tetravalent in SO_2 and hexavalent in SO_3. It is a problem to assess the exact valency of an element. To avoid this problem, it is the general rule to take the highest number of monovalent atoms with which one atom of an element is capable of combining as the valency of the element.

12.2.2 Active valency, saturated compounds

When the valency of an element exists in a particular compound it is called **active valency**. The saturated compound means the element exhibits its maximum valency in the compound and its chemical affinities are completely satisfied; the compound is known as saturated and in this case the maximum valency of the element is the active valency of the element.

12.2.3 Latent valency, unsaturated compound

When the element in a compound is present in a lower valency than its maximum valency, then the compound is called unsaturated and the difference between the maximum valency and the lower valency is called the **latent valency**. In an unsaturated compound, the combining capacity of the element is not lost but remains in depressed state and this latent power becomes active in favourable conditions when the unsaturated compound converts into saturated state.

12.3 Types of Bonds

The forces which operate to keep the atoms and molecules in biological system are of the following types:

I. **Strong Bonds**
1. Covalent bond
2. Coordinate bond

II. **Weak Bonds**
3. Electrovalent or ionic bond
4. Hydrogen bond
5. Hydrophobic bond
6. Van der Waals forces
7. Ion-dipole interactions
8. Orientation forces
9. Inductive forces
10. Dispersion force (or London force)

The chemical compounds formed out of these bonds or attractive forces are of two types:

A. Covalent Compounds,
B. Ionic Compounds.

12.4 Chemical Bonds

It is seen that the molecule of a compound is formed by the union of two or more atoms which hold atoms together by interacting force of attraction, i.e., chemical force known as chemical bonds. It may cause union of two parts of the same molecule to form a specific folded conformation (Conformation refers to the relative positions in space of each of the constituent atoms of a molecule) to the latter.

1. This chemical bond does not take place in irregular manner, but it follows a definite rule. As for example, one atom of Bromine always combines with one atom of Hydrogen yielding a molecule of the compound of Bromine and Hydrogen and one atom of Calcium always combines with two atoms of Hydrogen yielding one molecule of the compound of Calcium and Hydrogen.
2. The forces playing role for the chemical bonds act in definite directions in the space from the concerned atom. As for example, the carbon is tetravalent which takes position in the centre of the tetrahedron and each of the four bonds of the atom directing towards one of the corners of the tetrahedron. The capacity of the atom of an element of combination with the atoms of another by chemical bonds is known as **Combining Capacity** or **Valency** of that element.

12.5 Energy Concept of Chemical Bonding

The behaviour of atoms and their atomic orbitals interaction to form chemical bonds can be discussed in the light of electronic structure of atoms and their orbitals.

It is the general property of the atoms to have random motion and hold energy, and if the atoms are farther, more will be the energy of the system which renders it lesser stable. When a molecule forms by the union of two or more atoms, the energy of the system is lowered and the molecule becomes stable in comparison to separate atoms. It can be stated that "when the atoms come closely and if the energy is lowered, then a stable chemical union or bond between the two or more atoms can be achieved."

The term energy means the sum of the kinetic and potential energies. The kinetic energy originates from the motion of the electrons and the potential energy of a system when atoms X and Y approach each other and is determined by the following factors:

i. Nucleus-electron attractions,
ii. Nucleus-nucleus repulsions, and
iii. Electron-electron repulsions.

Both kinetic and potential energies are related by the virial Theorem–according to it the potential energy (P.E.) is double the kinetic energy (K.E.) and opposite in sign. It can be expressed mathematically.

$$\text{P.E.} = -2\text{K.E.}$$

The total energy (E) of the system is given as

$$E = \text{P.E.} + \text{K.E.}$$

$$\text{or, } E = -2\text{K.E.} + \text{K.E.}$$

$$= -\text{K.E.}$$

Regarding the change in energy of the system, the net potential energy of the system must decrease as a result of the formation of a chemical bond by electronic rearrangements. Out of the three determining factors for potential energy the second and the third are repulsing and unable to lower the potential energy but the first one, i.e., nucleus-electron attractions is responsible for lowering the potential energy.

12.5.1 Atomic orbital theory for bonding

Chemical bonding between atoms to form chemical compounds can be explained in the light of Orbital theory of atomic structure. Heitler and London opined that electron wave of the valency orbital of one atom overlaps the electron wave of the other bonding atom for forming a covalent linkage.

But for the formation of electrovalent bond, a physical transfer of the electron is involved for which the orbital concept does not gain importance. The "Maximum Overlap Theory" gives valuable explanation of covalent bond formation.

The parameters for the overlap of electron waves are:

(a) The orbitals coming into combination should possess only electron because of having a pair of electrons which render if unable for combination.

(b) The atoms with bonding orbitals should come close to one another with the axis of their orbitals in proper alignment.

When these two parameters are fulfilled, such orbitals merge into one bigger orbital called **Molecular Orbital** (M.O.) and allows the electrons greater freedom of movement and this causes lowering its energy.

The amount of energy given off or released per mole at the time of overlapping of orbitals for formation of a bond, is known as **bond energy** or **Stabilisation energy**.

The term is used by the Physical Chemist which should not be confused with the term phosphate-bond energy used by the Biochemists which denotes specifically the differences in the free energy of the products and the free energy of the reactants when a phosphorylated compound undergoes hydrolysis.

Bond energy also implies that the mean energy required for breaking the bond between two atoms in a molecule expressed in KCl $mole^{-1}$. According to the bond energy, bond can be grouped into two:

(a) Bond with high bond energy known as "strong bond" which is more stable, e.g., covalent and coordinate bonds,

(b) Bond with low energy bond known as "weak bond" which is dissociable, e.g., hydrogen and hydrophobic bonds.

The merger or overlapping of atomic orbitals causes stabilising effect on the system. But this effect stops at certain stage when the atomic nuclei come close at a distance to exhibit a repulsive force on one another for balancing the force of merger; this distance is known as *Bond Length* and is expressed in Å.

During bond formation, if overlap is greater, more is the energy set free and higher is the bond strength and shortening of the distance between the nuclei leading to shorter bond length.

Table 12.1: Different bonds, bond energy and bond length

Bond	Bond energy ($KCl.mole^{-1}$)	Bond length (Å)
H – H	103.2	0.74
C – N	98.7	1.10
C ≡ N	213.0	1.15
C ≡ C	200.0	1.20
C = O	177.0	1.24
C – N	148.0	1.29
C = C	146.0	1.34
C – O	84.0	1.43
C = N	70.3	1.52
C – C	83.0	1.54

12.6 Strong and Weak Bond Relationship in Biological System

In biological molecules (also in the synthetic organic compounds), the interactions of atoms are, first of all, chemical covalent bonds which are strong interactions and is based on quantum mechanics or quantum chemistry.

Strong bonds determine the following:

1. Chain structure of biopolymer,
2. Binding of monomers, e.g., amino acid residues, hexoses, nucleotides.

But only strong interactions may form:

1. Too rigid structures,
2. Too stable structures,
3. Lack of mobility in the molecules to accomplish various functions.

Weak interactions determine the following:

1. Molecular flexibility, i.e., the conformational properties of biopolymers and small molecules.
2. Molecular recognition in enzyme catalysis,
3. Molecular recognition in biosynthesis, e.g., protein.
4. Existence and function of cells and organelles which are heterogeneous system are determined by intermolecular interactions of nonvalent character.
5. The molecular functions perfomed by the protein in the cells interact with carbohydrates, lipids, nucleic acids and other small molecules.
6. Molecular mobility required for biological molecules for various functions like delicate control of chemical reactions, compartmentation, setting-up concentration gradient, etc.

Since strong interactions form too rigid and stable structures lacking molecular mobility, these weak interactions are required in addition to strong interactions. Because biological molecules cannot function and life cannot exist without nonvalent, nonchemical and weak forces which operate inside the biological molecules. Weak interactions are essential in addition to strong interactions.

12.7 Strong Bonds

12.7.1 Covalent bonds or shared electron pair bond

Electrovalent or ionised bonds can operate only in case of inorganic compounds where the union is between two atoms dissimilar in character but incapable of accounting for non-ionised bonds present in organic substances, non-metal compounds which are formed by the union of atoms of equal or nearly equal electronegativity. Lewis (1919) solved the problem by stating that the chemical combination is due to the tendency of electrons to redistribute themselves among the atoms to form more stable arrangements like inert gases. He further pointed out that not only by transfer of electrons but also by an electron to be shared between two atoms, i.e., sharing of electron pair, which causes the stability of both; hence it is also known as **Shared Electron Pair Bond**.

The bonds are formed by the sharing of one or more pairs of electrons by two atoms to make a complete and stable electron octate in the outermost orbit of each atom. One electron of each electron pair comes from each atom to make electron orbits of the two atoms overlap to hold them together as a molecule. For stability of the bond, two electrons of the overlapping orbital must spin in opposite direction.

As for example, there are two atoms X and Y, each having only seven valency electrons. Both of them are short of one electron to fill up their "octet" to form a stable structure. This cannot be achieved by donating of electrons. But it is possible if they have electrons in common, one of which is contributed by each atom. This "sharing of a pair of electrons" makes each atom surrounded by an outermost orbit of eight electrons in spite of having total fourteen electrons. This "shared electron pair" is called **covalent bond** and this type of bonding formed by sharing of an electron pair is called **"Covalency"** (Fig. 12.1). Since the linkage comes into existence by the joint combination of the two atoms, it is known as "covalent linkage" and is expressed by a line (–).

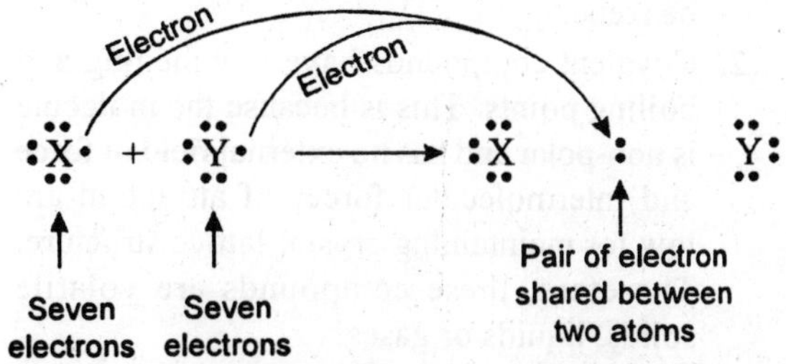

Fig. 12.1: Covalent bond formation

1. Covalent linkage may connect two similar atoms when both are a few electrons short of a stable number.
2. Bonding electron pair between two dissimilar atoms is more under the control of the more electronegative atom which is the origin of polarity of bonds.
3. In a covalent linkage, electrons are not transferred from one atom to the other, the molecule cannot get polarity and no ions are formed for which covalent linkage may be called **"non-polar"** and **"non-ionised"**.
4. Covalency is caused due to tendency of atoms to get the outermost shell of eight electrons to form molecules which have a state of lower energy than the individual atoms.
5. Covalency is exhibited in almost all organic compounds and also in some inorganic compounds.
6. It is not the only process by which one atom can be linked to another by one covalent linkage but polycovalent elements may form molecules in which two atoms are linked by 2 or more pairs of shared electrons.
7. The bond length, bond angle and bond energy are specific for a particular bond.
8. It has a "directional character" and is maximum in a specific direction.
9. It does not dissociate spontaneously or by change in H^+ concentration in a biological system. For lowering the activation of energy, it needs catalysts of enzymes.
10. Sufficient stability of the bond is due to low rates of dissociation and acquire longer half-lives.

12.7.1.1 Characteristics of covalent compounds

1. Covalent compounds are insoluble in water and soluble in nonpolar organic solvents, e.g., benzene.
2. Covalent compounds have low melting and boiling points. This is because the molecule is non-polar and has no external field of force and intermolecular forces of attraction are low for maintaining crystal-lattice structure. Therefore, these compounds are volatile solids, liquids or gases.
3. In covalent compounds, the atoms are firmly held by electrons and cannot be separated without chemical decomposition. Therefore, when they are in water or melted, they do not dissociate or conduct electricity.
4. In covalent compounds, the molecules have the opportunity for different spatial arrangements.

12.7.1.2 Biological application

Biomolecules related to the covalent bonds are:

1. Peptide bond
2. Glycosidic bond
3. Ester bond
4. Phosphodiester bond
5. Disulphide bond

12.7.1.2.1 Peptide bond

When a molecule of water is splitted out between a carboxylic acid and ammonia or an amine it forms an amide. Important amides are peptides. When α-amino group of one amino acid and the α-carboxyl of another amino acid are united with elimination of one molecule of water, they form an "**Amide linkage**". The product of such a condensation reaction is known as "peptide" and an amide linkage joining two amino acids is called "**Peptide bond**" or "**Peptide linkage**" (—CO—NH—) (Fig. 12.2).

The peptide bond itself has a specific "**planar structure**" found by X-ray structure analysis and has some "**double bond character**", thus free rotation about this bond does not occur. This semi-rigidity has important impact on protein structure above the primary level.

All the four atoms (—CO—NH—) in the peptide bond lie in one plane (Fig. 12.3). The N—C bond

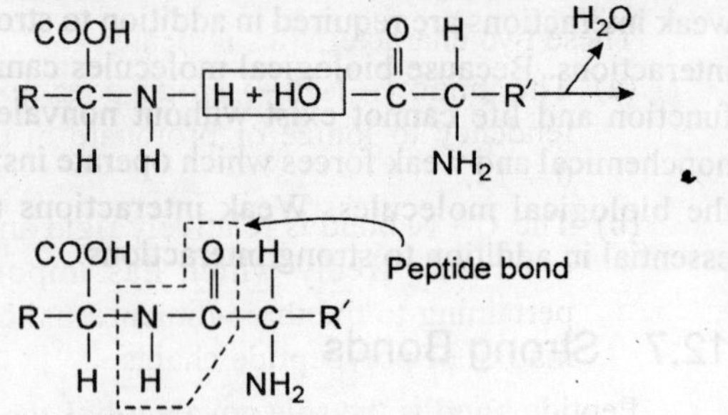

Fig. 12.2: Formation of peptide bond

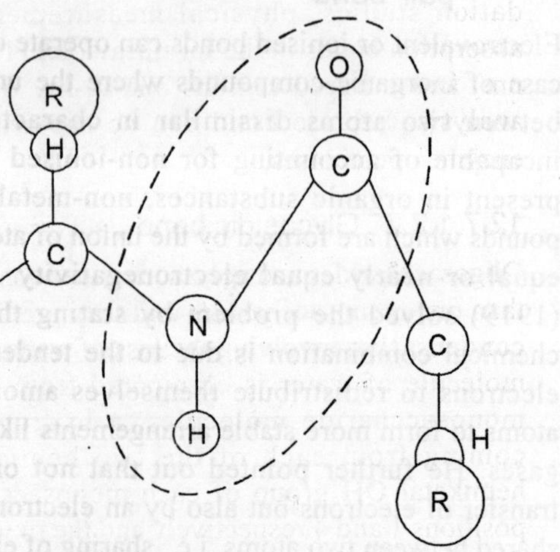

Fig. 12.3: Planar structure of peptide bond

is shorter than the same bond in R—NH_2 (Aliphatic amines), where the length is 0.147 nm. This shortened length of the N—C bond and the planar arrangement of the bonds which is rigid and is the result of resonance stabilization give evidence of the union of the N—C and C=O bonds and hence merger or overlap of their electron shells. This is being accompanied by a shift of electron density from N to C which can be seen by the following structure (Fig. 12.4):

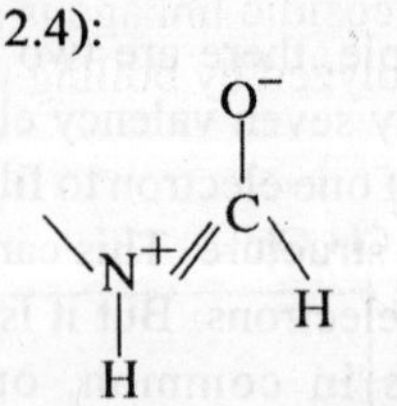

Fig. 12.4: Shift of electron density from N to C

This indicates that the N—C single bond is partially double bond character (40 per cent) and C=O double bond is partially singlebond character (40 per cent).

These two characteristics indicate:

(a) The imino (—NH—) group does not have tendency to ionize or protonate in the pH 0—14.

(b) The C—N bond is relatively rigid and cannot rotate freely which has importance pertaining to the three-dimensional conformation of polypeptide chains.

Peptide bond is the sole covalent linkage and it has been supported by chemical, enzymatic degradation studies, physical measurements, e.g., absorption bands in the far ultraviolet (180 to 220 nm) and infrared regions and X-ray diffraction analysis. Peptide bond can be ruptured by the enzymes peptidases.

12.7.1.2.2 Glycosidic bond

Oligosaccharides and polysaccharides contain more than one monosaccharides and they are formed by **condensation** involving the loss of water. A single molecule of water is eliminated from a pair of monosaccharide molecules, a hydrogen atom coming from each of the two hemiacetal or hemiketal OH group of each monosaccharide at positions 1 and 4 respectively and the oxygen atom from one of the two OH groups. As a result **covalent bonds** known as **glycosidic bonds** are established joining the two monosaecharides molecules at positions 1 and 4 and gives 1, 4 linked disaccharide—the component is called **glycosides.** The glycosidic bond is **acetal link** because it results from a reaction between a hemiacetal group and another —OH group. If the hemiacetal part is glucose, the resulting compound is **glucoside**; if galactose, a galactoside, etc.

The glycosidic bond may be $\alpha-1, 4$ or $\alpha-1, 6$ glycosidic bond as these are present in polysaccharides. The glycosidic linkage is stable to bases but it can be hydrolyzed by boiling with acid to get the free monosaccharides and free alcohol. Glycosides can be hydrolyzed by enzymes known as glycosidases which differ in their specificity according to the (1) types of glycosidic bond, i.e., α or β (2) structure of monosaccharides unit (s) and structure of the alcohol (Fig. 12.5).

(a) The participation of the reducing group of a sugar in a glycosidic bond abolishes its capacity to react with phenyl-hydrazine, Fehling's solution and the configuration of the ring compound is stabilized in either the $\alpha[-]$ or the $\beta[-]$ form. The glycosidic bond can be ruptured by the enzymes glycosidases.

(b) β-glycosidic linkage is present in most other nucleotides whereas ribonucleotide in vit B_{12} is exceptional having base 5, 6-dimethyl-benzimidazole in the unusual α-glycosidic linkage with D-ribose.

12.7.1.2.3 Ester bond

An ester linkage is formed when an alcohol reacts with acid splitting out water. The acid may be organic or inorganic.

In case of lipid, alcohols are glycerol, cholesterol, etc, the acids are fatty acids. The triglycerols (triglycerides) or neutral fats are esters of the alcohol glycerol where three hydroxyl groups are esterified with fatty acids. In naturally occurring fats, the proportion of triacylglycerol molecules containing the same fatty acid residue in all 3 ester position is very small and most of them are mixed acylglycerols where more than one fatty acid is involved.

Esters of H_3PO_4, e.g., phosphorylated sugar, phospholipids and H_2SO_4 are also present in the body. Ester linkage also presents between the 3-phosphate group of the ribonucleotide and a side

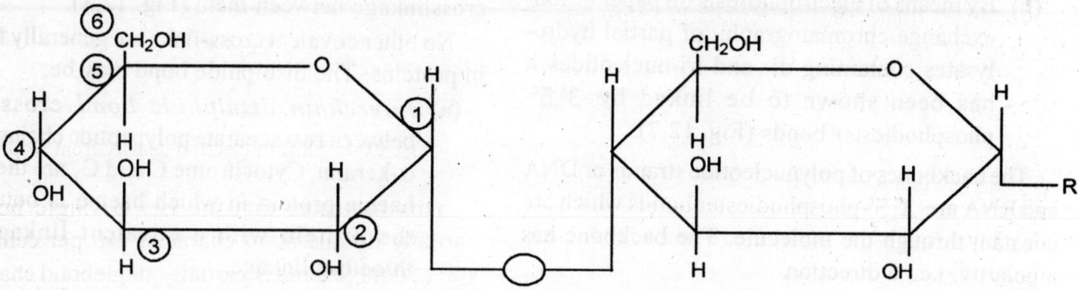

Fig. 12.5: $\alpha-1, 4$ Glycosidic linkage between C_1 and C_4 of two glucose residue of portion of polysaccharides

chain of the Corrin ring of Vit B_{12}. Many lipids contain carboxylic ester linkages. The ester bond can be ruptured by the enzymes esterases (Fig. 12.6).

In Chlorophyll *a*, the linkage between the porphyrin and phytol is an ester bond.

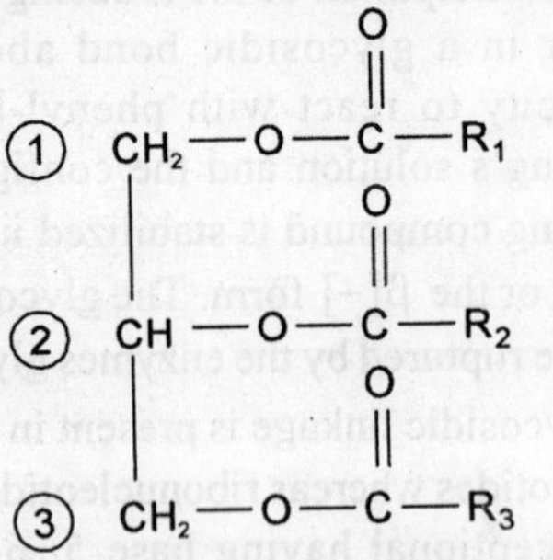

Fig. 12.6: Triglycerols showing ester bonds

12.7.1.2.4 Phosphodiester bond

The nucleic acids are polynucleotides in which one of the two acidic groups of the phosphoric acid residue of a mononucleotide is esterified by one of the sugar hydroxyl of another mononucleotide evident by the acid-base titration curves of nucleic acid. In the nucleotides of DNA the only sugar hydroxyl available is at the 3′ and 5′ positions of deoxyribose for formation of a phosphoric acid ester. Thus, the nucleotides must be joined by phosphate ester linkages involving the 3′ position of one deoxyribose unit and 5′ position of another sugar and known as 3′,5′-phosphodiester bonds.

The principal mode of internucleotide linkage is through 3′,5′-phosphodiester bonds as is evident by the following studies:

(a) Partial degradation of PNA preparations by ribonuclease or by mild acid treatment.

(b) By means of electrophoresis on paper or ion-exchange chromatography of partial hydrolysates containing di- and tri-nucleotides it has been shown to be linked by 3′,5′-phosphodiester bonds (Fig. 12.7).

The backbones of polynucleotide strands or DNA and RNA are 3′,5′-phosphodiester bonds which are constant through the molecule. The backbone has a polarity, i.e., a direction.

DNA ligase, an enzyme, catalyses the formation of phosphodiester bond between two DNA chains.

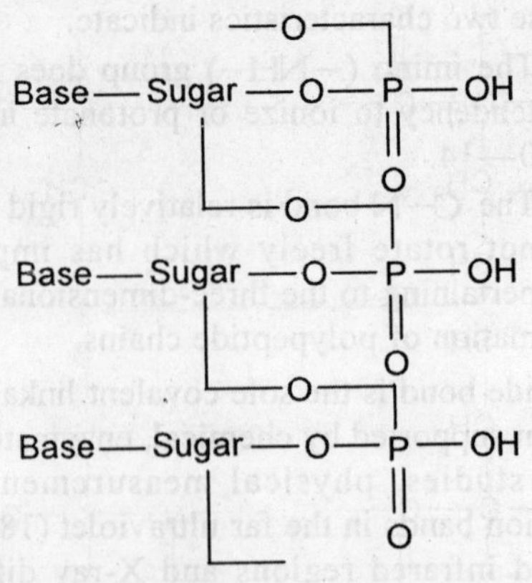

Fig. 12.7: Portion of polynucleotide of DNA showing 3′,5′-phosphodiester bonds

This enzyme requires a **free OH group** at the 3′ end of one DNA chain and a phosphate group at the 5′ end of the other. Formation of the bond is an endergonic reaction and an energy source, i.e., ATP or NAD^+ is required for the joining reaction. This joining reaction is necessary for:

1. The normal synthesis of DNA,
2. Repair of damaged DNA,
3. Splicing of DNA chains in genetic information.

This enzyme cannot link two molecules of single-stranded DNA but DNA chains joined must be part of a double-helical DNA. The bond can be ruptured by the enzymes phosphodiesterases.

12.7.1.2.5 Disulphide bond

In some proteins, a few side chains are cross-linked by disulphide (S—S) bonds. These cross-links are formed when cysteine (SH) occurs in oxidized form cystine (S—S), in which thiol groups (SH) of two molecules of cysteine have been oxidized to a disulphide (S—S) group to provide a covalent crosslinkage between them (Fig. 12.8).

No other covalent cross-links are generally found in proteins. The disulphide bond may be:

(a) *Interchain disulphide bond*–cross-link between two separate polypeptide chains, e.g., α-keratin, Cytochrome C and C_1 are the only haeme protein in which haeme is bound to the protein with a covalent linkage by thioether linkage.

(b) *Intrachain disulphide bond*–cross-link between loops of a single chain, e.g., ribonuclease.

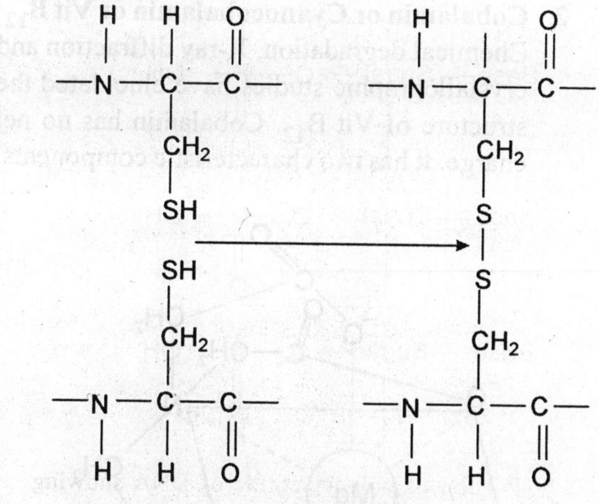

Fig. 12.8: Formation of disulphide bond (S-S)

Disulphide bonds can be irreversibly oxidized by **performic acid** to give cysteic acid residues and it can be cleaved reversibly by reducing them with a reagent such as β-mercaptoethanol which forms mixed disulphides with cysteine side chains. The —S—S— linkages of proteins may be reduced to two —SH groups by treatment with reducing agents, e.g., sodium cyanide, cysteine, glutathione and glycolic acid.

All processes of denaturation of protein cause an increase in the number of reactive —S—S— bonds. The —SH is not the outcome of reduction —S—S— groups but may involve release from a coordinate type of linkage. Intrachain disulphide cross-linkages are formed enzymatically in the microsome by oxidation of —SH groups. This occurs once a polypeptide chain is formed by ribosome, it may need to undergo covalent modification to yield its biologically active form. The bond can be ruptured by the enzymes oxidoreductases. The presence of certain —S—S groups is necessary for full activity of insulin, and reduction to —SH groups causes loss of about 50 per cent activity.

12.8 II Coordinate bond

Coordinate bond (coordinate = to bring into relation) is a special type of stable covalent bond in which both of the shared pair of electrons are coming from one atom only and not one by each atom as in covalent bond.

The shared pair electrons belonging to X can be counted as one electron to each atom. It may be

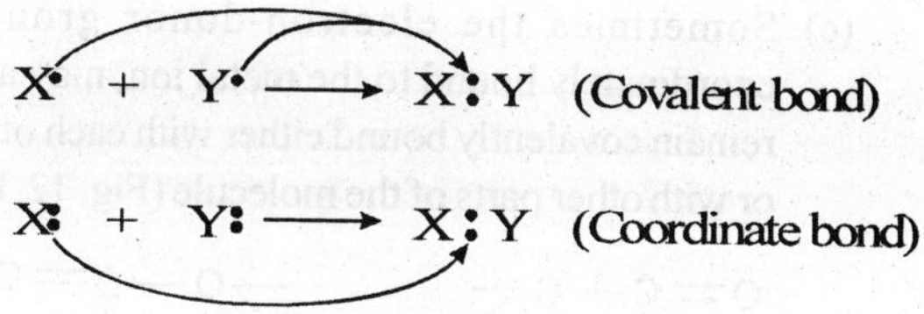

Fig. 12.9: Covalent and coordinate bond

considered that one atom is associated to X and another to Y atom. It can be stated that the formation of this bond is performed by transfer of electron from X to Y and the molecule XY acquires polarity and is represented as $X^+–Y^-$, and this bond is often referred as to **semipolar bond** (Fig. 12.9).

When a coordinate bond is formed between the two atoms, one atom possesses a complete octet having a pair of free valency electrons, and the other is short of two electrons, while the atom X donates the free pair of electrons, which can be called **donor** and atom Y accepts the electrons in order to complete its octet and it can be called **acceptor**. The pair of valency electrons possessed by the donor atom (X) is called **long pair**. This can be illustrated in the following way (Fig. 12.10).

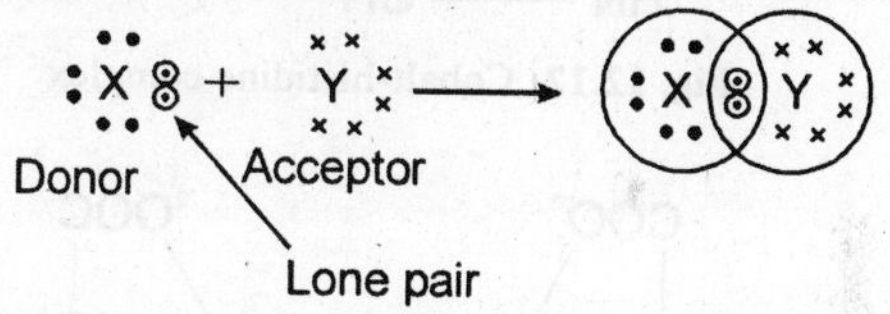

Fig. 12.10: Schematic representation of lone pair

12.8.1 Metal Chelate

Several heavy metal ions, e.g., Cu^{2+}, Co^{2+}, Mn^{2+}, Fe^{2+}, etc. form chelated complexes having both coordinate and covalent bonds with proteins, amino acids, porphyrins, etc. called metal chelate (Greek chela–claw) in which both carboxyl and other groups such as —NH_2 and —SH, are involved. This can be formed in the following ways:

(a) The metal chelate forms coordinate bond by sharing of a single pair of electrons by a charged metal ion like Fe^{3+} and Co^{3+} and an electron-rich molecule or ions like NH_3, Cl^-, etc.

(b) The metal ion of the chelate remains covalently bound with other groups beside the coordinately bound to one or more electron-donor electronegative ion or group.

(c) Sometimes the electron-donor groups, coordinately bound to the metal ion, may also remain covalently bound either with each other or with other parts of the molecule (Fig. 12.11).

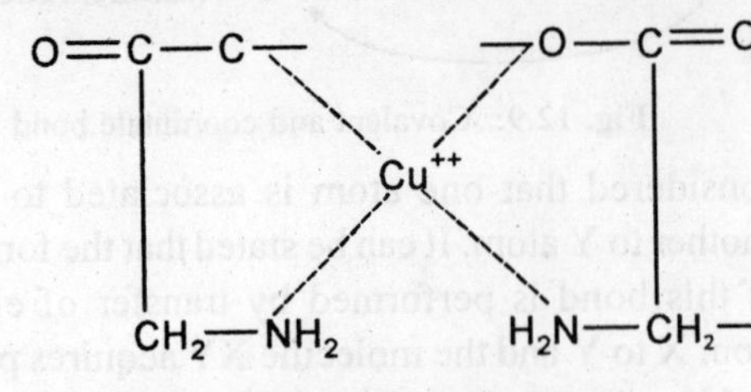

Fig. 12.11: Copper diglycinate

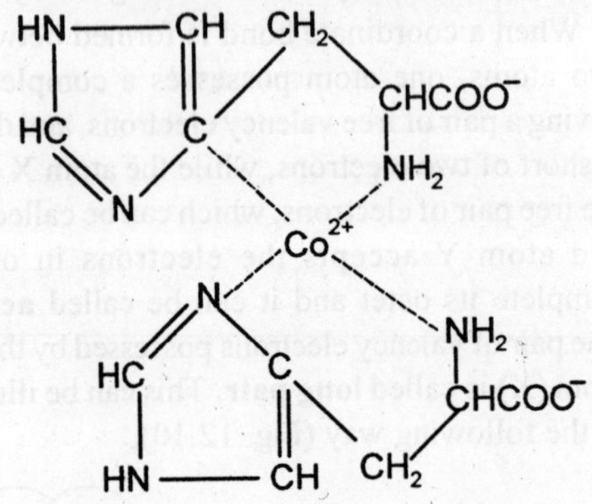

Fig. 12.12: Cobalt-histidine complex

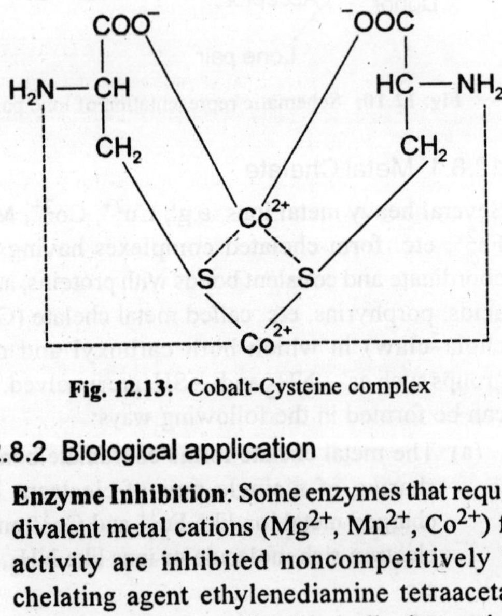

Fig. 12.13: Cobalt-Cysteine complex

12.8.2 Biological application

1. **Enzyme Inhibition**: Some enzymes that require divalent metal cations (Mg^{2+}, Mn^{2+}, Co^{2+}) for activity are inhibited noncompetitively by chelating agent ethylenediamine tetraacetate (EDTA) which reversibly binds divalent cations by coordinate bonds to form metal chelates. The plane of the coordination bonds is shown by the shaded part in the (Fig. 12.14).

2. **Cobalamin or Cyanocobalamin or Vit B_{12}:** Chemical degradation, X-ray diffraction and crystallographic studies have elucidated the structure of Vit B_{12}. Cobalamin has no net charge. It has two characteristic components:

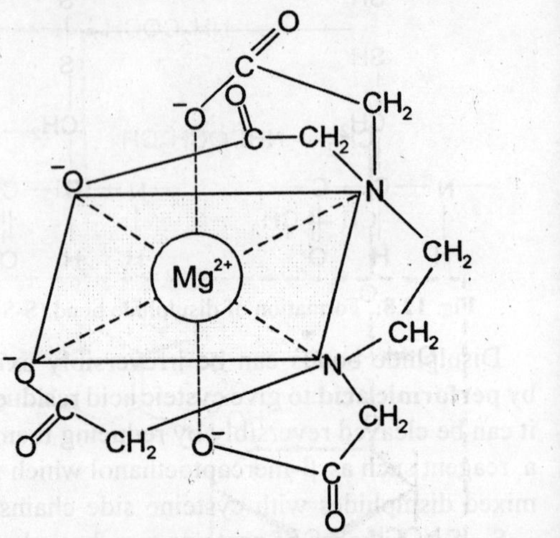

Fig. 12.14: A metal chelate of EDTA with a divalent metal cation (Mg^{2+}). The shaded part shows the plane of the coordination bonds

(a) The larger is the corrin ring which is a highly substituted and partially hydrogenated tetrapyrrole but in which a pair of these five numbered rings is joined directly rather than through a methane bridge in haemoglobin. There are 6 coordinated valences of the cobalt atom (Co^{2+}) of which 3 are satisfied by coordinated bond with the three inner nitrogen atoms of the corrin ring and one covalently with other nitrogen of corrin ring.

(b) The second major component is a substituted and charged ribonucleotide which is exceptional due to presence of a base 5, 6-dimethylbenzimidazole in the unusual α-glycosidic linkage in other nucleotides. This ribonucleotide is joined to the corrin ring by coordination between the other nitrogen atom of the nucleotide and the fifth coordination valence of the cobalt atom and by an ester linkage between the 3′-phosphate group of the ribonucleotide and a side chain of the corrin ring.

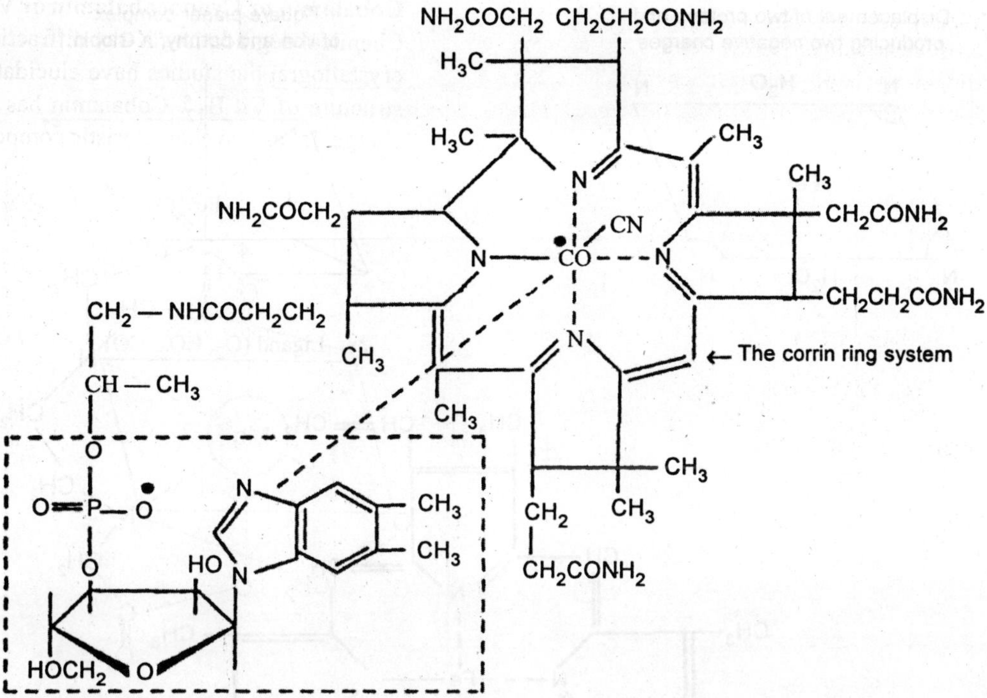

Fig. 12.15: Structure of cobaltmine showing cobalt ion chelated to five nitrogen coordinated bond (—) and covalent bond (—)

The sixth coordination valence is satisfied by the cyanide (CN—) group (Fig. 12.15) covalently.

3. **Haem of haemoglobin**: Protoporphyrin nucleus present in the haemoglobin is the derivative of the cyclic tetrapyrrole nucleus called prophin in which 4 pyrrole rings are linked by methene bridges. Protoporphyrin, like other prophyrins, forms quadridentate chelate complexes not only with iron but also with copper, manganese, magnesium, zinc, nickel and cobalt. Such a chelate complex of protoporphyrin with ferrous ions (Fe^{2+}) is termed haem or protohaem. In haem, Fe^{2+} is coordinately bound with the tertiary nitrogens of the 2 pyrrole rings and forms square-planar complex. For formation of haemoglobin, the fifth coordination bond of iron directed perpendicularly to the ring plane is also coordinately bound to imidazole nitrogens of two histidine residues of globin and the sixth valence is occupied by a ligand (O_2, CO_2, CO). The resulting structure is called haemochrome. The iron atom is paramagnetic, and magnetic moment of Fe^{2+} is retained and has a magnetic susceptibility corresponding to 4 unpaired electrons. When iron enters the protoporphyrin ring, it occupies the centre of the ring and forms bonds with the nitrogen atoms and displaces two protons (H^+) from two nitrogen atoms. This leaves two net negative charges on the two N atoms of the porphyrin which are equally distributed by resonance among the nitrogen atoms of the 4 pyrrole rings. These are balanced by positive charges of the iron. In case of ferrous iron the compound formed possesses no net charge (Fig. 12.16).
4. **Haemin**: It is another chelate complex of protoporphyrin in which Fe^{3+} is coordinately bound.
5. **Chlorophyll *a***: It is a magnesium-porphyrin compound in which magnesium is bound by two coordinate bonds and two covalent bonds with the four nitrogens of 4 pyrrole rings of porphyrin.
6. **Myoglobin**: It has single porphyrin ring and possesses only one iron atom.

Displacement of two protons and producing two negative charges

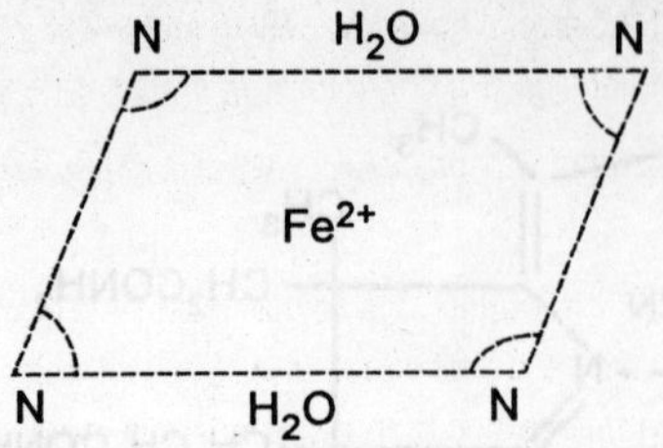

Square-planar complex of iron and porphyrin Globin

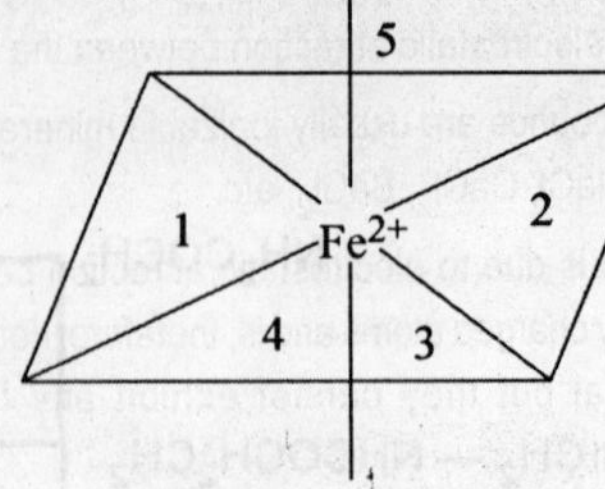

Ligand (O_2, CO_2, CO)

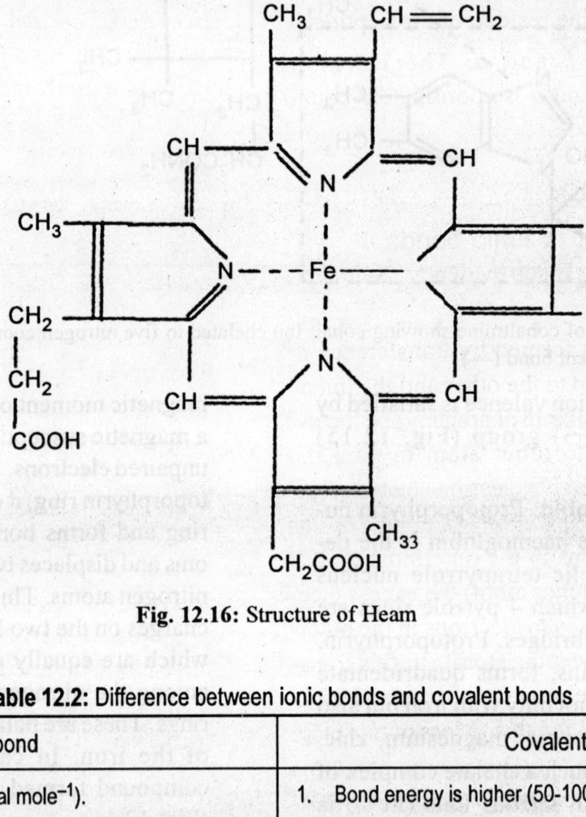

Fig. 12.16: Structure of Heam

Table 12.2: Difference between ionic bonds and covalent bonds

Ionic bond	Covalent bond
1. Bond energy is lower (3-7k cal $mole^{-1}$).	1. Bond energy is higher (50-100 k cal $mole^{-1}$).
2. Change in H^+ ion concentration destroys the bond by altering the charges.	2. Does not affect the bond.
3. The bond lengths and the angles are more variable.	3. The bond angle and bond length are specific for a particular bond.
4. Does not have directional character and have equal strength in all directions.	4. It has a directional character and is maximum in a specific direction.
5. Ionic compounds conduct electricity when dissolved in water or in the molten state.	5. It is non-conductor in the molten state.
6. Ionic compounds are soluble in water and insoluble in organic solvents.	6. Insoluble in water and soluble in organic solvents.

Contd.

Ionic bond	Covalent bond
7. Ionic compounds have high melting and boiling points because of strong electrostatic attraction between the ions.	7. The compounds have low boiling and melting points because the molecules are nonpolar and has no external force.
8. The compounds are usually ionizable mineral salts like KCl, $BaSO_4$, NaCl, $CaCl_2$, $BaCl_2$, etc.	8. The compounds are usually volatile solids, liquids or gases.
9. The bond is due to electrostatic attraction between the two oppositely charged atoms and is, therefore, non-rigid and non-directional but they cannot exhibit any type of space isomerism.	9. The bond is due to shared electrons not by lines of force for which it is rigid and directional. It can exhibit different spatial arrangements and stereoisomerism of molecules.

This is coordinately bound with the imidazole-N of histidine residue of the peptide.

7. **Cytochromes** *b* and *a*: The Haem in cytochromes *b* and *a* is coordinately bound with the protein.

12.9 Weak Bonds

12.9.1 Electrovalent or ionic bonds

(Electrostatic, Electrovalency, Salt linkages)

This type of linkage is formed by transference of electrons from one atom to the other and the linked atoms should be dissimilar in character, one having more electrons loses to other atom to gain the electrons to acquire the stable electron octet arrangement in its outermost shell. For example if X has more than eight electrons in the outermost shell it may donate (Electron-donor atom) the excess electrons to Y atom having seven electrons in the outermost shell (Electron-acceptor atom). In this transference, X acquires one positive charge and turns into a positively charged ion due to loss of one electron and Y acquires one negative charge and turns into a negative charged ion due to gain of one electron, i.e., the molecule X^+Y^- acquires polarity and the linkage established between the atoms may be called **Polar bond.** Since these oppositely charged two ions are held together by electrostatic lines of force, the linkage is often called **Electrovalent** or **Electrostatic** or **ionic bond** or **salt linkages** and the compound thus produced in called **Electrovalent compound.**

This type of linkage is seen in ionisable compouns (e.g., acids, bases and salts). When a salt (e.g., NaCl) is dissolved in water, the electrostatic lines of force are cut off between the oppositely charged ions Na^+ and Cl^- due to the high dielectric constant of the solvent, and ions are separated; hydration of the ions is also another factor. The salts in which electrovalency is present and linked ions even in solid state are separated when dissolved in water; hence electrovalent linkage is also known as **Ionised** or **Ionisable bond.**

12.9.1.1 Characteristics of ionised compounds

1. It is non-rigid and non-directional simply due to the electrostatic attraction between the two oppositely charged atoms. They cannot exhibit any type of space isomerism.
2. It conducts electricity due to dissociation of the compounds when dissolved in water or in molten state.
3. It neither vaporises nor dissolves in organic solvents. Their boiling points and melting points are high because tremendous work is required to separate the molecules from each other.
4. It is usually soluble in water and insoluble in organic solvents.

12.9.1.2 Biological applications

1. Ionic bonds are formed between ionogenic groups, e.g., Glu^- and Lys^+. These are salts bonds formed between oppositely charged groups in the side chains of amino acids. The epsilon-amino group of lysine bears a net charge of +1 at physiological pH and the non-α-Carboxyl of glutamate and aspartate a net charge of −1 and may interact electrostatically.
2. Ionic bonds are formed between the ionogesic groups and small counterions.

3. Ionic bonds are formed between phosphate groups in nucleic acids and cations.
4. **Stabilization of Tertiary Structure of Proteins**: The tertiary structure globular proteins and in addition a quarternary structure of oligomeric proteins (contains two or more separate polypeptide chains or subunits) require four major types of weak bonds for stabilising the structure, e.g., ionic bonds, hydrogen bonds, hydrophobic bonds.
5. **Mineral Salts**: The ionic compounds are mineral salts, e.g., NaCl, KCl are present in the living system.
6. **Dissociation of ionic bonds**: Ionic bonds may dissociate in a living system due to change of pH and does not require enzyme for its dissociation.
7. **Antigen-Antibody Reaction**: Antigen-antibody reactions are due to some bindings of the molecules which are due to ionic bonds between charged polar groups of antigen and antibody.

12.9.2 Hydrogen bonds (H-Bond or H-Bridge)

Definition

The electrostatic interaction, i.e., strong intermolecular attractive forces which exist in a com-pound having a hydrogen atom covalently bonded highly between hydrogen nucleus of one group and the electronegative atom of another group (e.g., unshared electron pair of another) is termed a **hydrogen bond** or **H-bond or H-bridge.**

Hydrogen possesses only one electron, and is monovalent, therefore capable of a true valency of unity only but, under certain circumstances, hydrogen can exhibit a kind of weak bivalency. For this to happen, the hydrogen must be between two atoms of strongly **electron attracting** character, such as oxygen, nitrogen, fluorine. The specific donor-acceptor interactions are induced by the hydrogen atom in groups O—H, N—H, F—H and Cl—H, and sometimes S—H groups. The hydrogen atom binds these groups with valency-saturated oxygen, nitrogen and fluorine atoms. As for example:

1. **H-bond with Oxygen**

Water molecules are strongly polar. The dipolar character of water molecules favours their mutual association. Hydrogen bonding is very pronounced in them.

Association of two dipolar water molecules is shown in Fig. 12.17A. Association of a central water molecule with four other molecules also occur in ice and liquid water (to a lesser extent) shown in Fig. 12.17B.

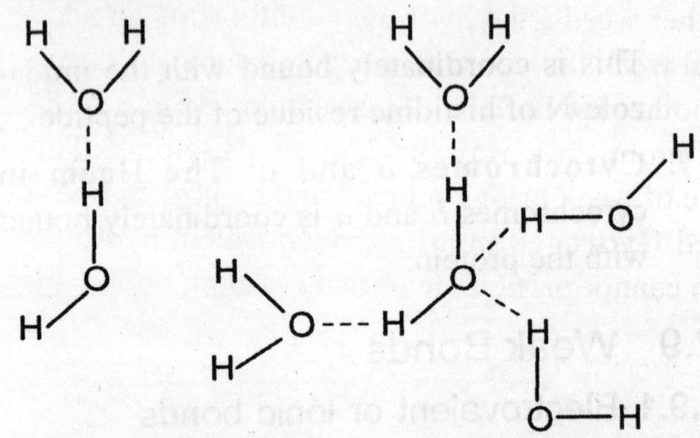

A. Association of two dipolar water molecules. The dotted line shows hydrogen bond

B. Association of central water molecule with four other water molecules by hydrogen bonding exist in ice and liquid water (to a lesser extent)

Fig. 12.17: Hydrogen bonding in water molecules

This crystalline structure is open and occupies more space. That is why water expands as it freezes to ice which floats on the surface. As the temperature increases, the kinetic energy of water molecules increases, the ice melts and many of the hydrogen bonds are broken, the crystal lattice collapses and the water molecules again move closer together. Thus, water is more dense than ice. As the water is heated alone above 0°C, more of the hydrogen bonds are broken, and the molecules continue to move closer together until, finally, at 4°C, almost all hydrogen bonds have been broken. Hence, above 4°C, the normal expansion of water with increased temperature becomes effective. At this temperature, the density of water starts to decrease for which water has its maximum density at 4°C.

2. **H-bond with Fluroine**

If a hydrogen atom is linked with a strongly electro-negative atom F, it will form the positive end of the polar bond in case of HF molecule

$$H \rightarrow — F \text{ or } H\delta^{+} — F\delta^{-}$$

The arrow in the middle of the bond indicates the direction of displacement of the electrons. The hydrogen atom develops a significant positive charge. Such a positive charge between bearing H-atoms in this molecule will exercise an attractive force on the negative end of the other molecules of Hydrogen Fluoride (HF). Since the intensity of the charges is not very large, this attractive force is quite small. In other words, the hydrogen atom forms an additional weak linkage with a second atom of fluorine of another HF molecule shown below (Fig. 12.18).

H-atom, in a way, bridges two F-atoms. This mode of weak bridge represented by broken line is called **Hydrogen bonding.** Monovalent hydrogen atom cannot participate in two covalent bonds, the

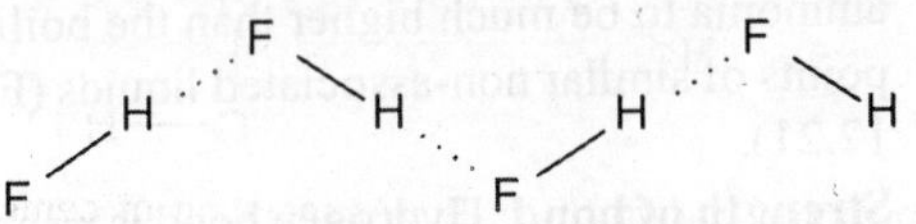

Fig. 12.18: H-bond with Fluroine

second bond is merely **electrical in nature** and **very weak**. The hydrogen bond energy lies in the range of 2-8 kcal per mole as against a normal covalent bond energy of 120 kcal per mole. As a result of such weak bonds, a large number of HF molecules get associated to form a big molecule like ammonia and water (Fig. 12.19).

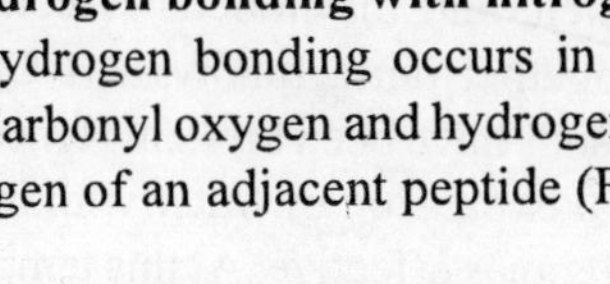

Fig. 12.19: Bonding of ammonia

3. **Hydrogen bonding with nitrogen**

This hydrogen bonding occurs in between the peptide Carbonyl oxygen and hydrogen on the peptide nitrogen of an adjacent peptide (Fig. 12.20).

Fig. 12.20. Hydrogen bonding with nitrogen

Mirsky and Pauling (1936) suggested that a major factor in conferring upon an extended peptide chain of a protein its characteristics folding is the presence of **hydrogen bonds**. The importance of hydrogen atom in serving as a bridge between 2 atoms (e.g., 2 oxygen atoms or an oxygen and nitrogen) was demonstrated by Pauling in formic acid. These bonds are very weak individually and the conformation is not stable. In order to find out the most stable conformation of the polypeptide chain, it is necessary to minimize its total energy, including the energy of the intramolecular hydrogen bond. It happens when a molecule has many hydrogen bonds, and they reinforce one another and thus produce a stable structure. In case of proteins and polypeptides, the hydrogen bond hypothesis has observed the sharing of hydrogen atoms between the nitrogen and the Carbonyl oxygen of different peptide bonds to form linkages of the type NH — H ... O. Evidence for the existence of such linkages in polypeptides and proteins has been observed by Infrared Spectroscopy.

12.9.2.1 Type of hydrogen bond

There are two types of hydrogen bond

1. **Inter-molecular H-bond:** (H-bond between two different molecules).
2. **Intra-molecular H-bond** (H-bond between atoms contained in the same molecule).

1. **Inter-Molecular H-Bond Formation:** Such H-bond occurs between two different molecules and is responsible for the association of HF, H_2O, alcohols etc. Many of the peculiar properties of water are due to this ability of inter-molecular hydrogen bond formation of the H_2O molecules. Hydrogen bond formation and association is the reason for higher boiling point and heat of vaporization of many N and O compounds. Hydrogen bond is very much important for the life processes, e.g., helical structure of proteins and double helix structure of nucleic acids, having hydrogen bond mainly of the N—H···O type.

2. **Intra-Molecular H-Bond Formation:** Physical properties of orthocompounds can be explained based on H-bond formation. Such as, O-nitrophenol and salicylaldehyde

are easily volatile in steam but the para-isomers are not, and this is due to formation of six-membered ring through H-bond formation.

- **Mechanism of Hydrogen Bond**

The mechanism of hydrogen bond is electrostatic in origin. The bare proton is due to its positive charge attracting the unshared electrons of strongly electronegative elements. The intermolecular forces of the polar compounds are stronger than ordinary dipole-dipole interactions. In all these substances one hydrogen atom is covalently bonded to a strongly electronegative atom N, I or F. Under the conditions, the electronegative atom attracts towards it almost completely, and the electrons constituting the covalent bond. Thus, the electronegative atom takes almost full possession of the electrons shared with the hydrogen atom. This leaves hydrogen with a large partial positive charge. The molecule is therefore, highly polar and the hydrogen end of it is nearly bare hydrogen nucleus or proton (H^+).

In a substance made of such polar molecules, the positive hydrogen atom is attracted to the negative pole of another molecule. Because the hydrogen atom has been reduced to a proton almost devoid of electrons, the attractive force is strong, stronger than ordinary dipole-dipole attraction. The very small hydrogen atom bridging the electronegative atoms of two molecules brings the interacting molecules sufficiently close. Thus attractive force binding the two molecules is strong enough to be called a bond, the **Hydrogen bond.**

12.9.2.2 General features of H-Bond

1. **Association:** Association may not always be taking place between molecules of the same species, but molecules of different substances may also get associated through hydrogen bonding.
2. **Change in physical properties:** As a result of hydrogen-bonding, the physical properties such as boiling point, solubility and ionisation undergo a significant change. For instance, the boiling points of water and ammonia are plotted in the figure below. Extra energy required to break apart associated molecules causes the boiling points of water and ammonia to be much higher than the boiling points of similar non-associated liquids (Fig. 12.21).
3. **Strength of bond:** Hydrogen bond is weaker than ionic or covalent bonds, but stronger than the oridinary van der Waals forces of intermolecular attraction.
4. **Effect of higher electro-negativity:** Hydrogen fluoride forms stronger hydrogen bonds than water or ammonia because of the greater electro-negativity of fluorine (Fig. 12.22).

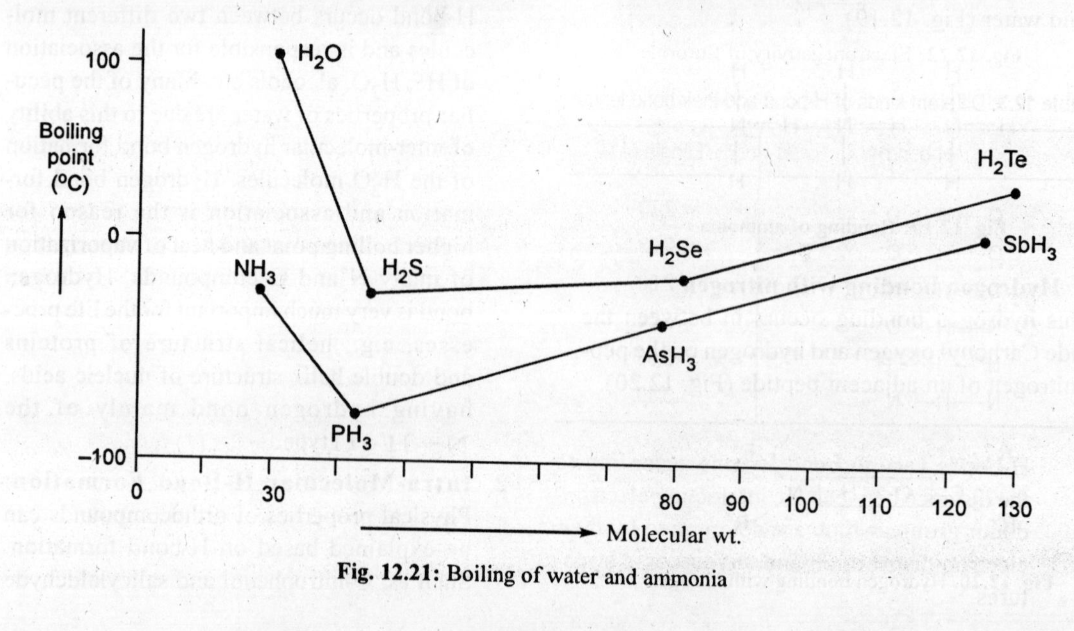

Fig. 12.21: Boiling of water and ammonia

5. **Bond energy**: Hydrogen bond is quite a weak type of bond having a bond energy of only 4-10 kcal as against 50-100 kcal for ordi-nary covalent bonds.
6. **Presence of many hydrogen bonds and stability**: Although these bonds are very weak individually, if a molecule has many hydrogen bonds, they will reinforce one another and produce a stable structure.
7. **Energy requirement for breaking H-bond**: To break a hydrogen bond in liquid water requires almost 4.5 kcal of energy per mole, about 4 per cent of the energy required to rupture the O—H bond in water (110 kcl/mol).
8. **Specificity**: Hydrogen bond shows considerable sepcificity for involvement of covalently bond atoms of either F, N or O.
9. **Bond length**: The kinds of hydrogen bonds and thier bond lengths are given below (Table : 12.3). The length of hydrogen bond is intermediate between that of a covalent bond and van der Waals bond.

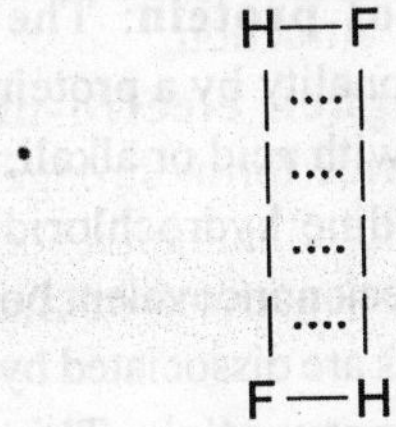

Fig. 12.22: Electronegativity of flurorine

Table 12.3: Different kinds of H-bond and their bond length

H-bond	Length (Å)
O—H···· O	2.70
O—H···· O⁻	2.63
O—H···· N	2.88
N—H···· O	3.04
N⁺—H···· O	2.93
N—H···· N	3.10

H-bonds vary in bond lengths according to the distance between the interacting electron-donor groups which are determined by their electron distribution and stereoscopic structures.

10. **Directional character and strongest H-bond**: An important feature of H-bond is one in which the donor, hydrogen and acceptor atoms are **colinear**.

The strength of H-bond varies inversely with the distance between the interacting groups; it also depends on the spatial orientation of the attracting groups and reaches a maximum when the bond is linear with the acceptor electronegative atom.

If the acceptor atom is at an angle to the line joining the donor and hydrogen atoms, the bond becomes weaker with increasing angle (Fig. 12.23).

11. **Dielectric constant and H-bond**: Substances associated by hydrogen bonds have large values of dielectric constants (ε). As for example, at 20°C the dielectric constants of the following substances are:

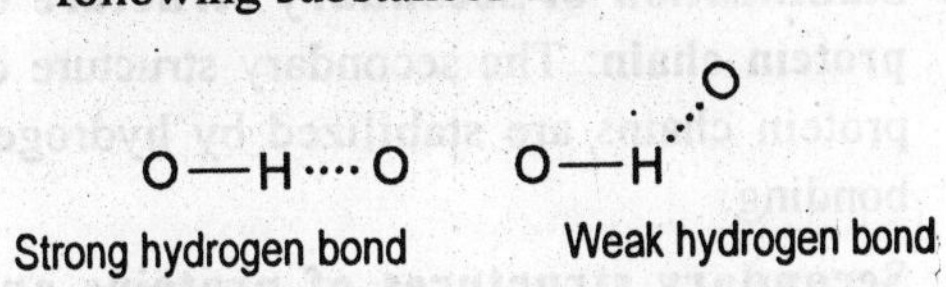

Fig. 12.23: Strong and weak Hydrogen bond

1. H_2O—80
2. HCN—95
3. $HCONH_2$—84 (Formaldehyde)
4. NH_3—15.5
5. $(C_2H_5)_2O$—4.3 (Diethyl ether)

12.9.2.3 Methods of study of H-bond

1. **Optical spectra**: The hydrogen bond reveals itself distinctly in optical spectra and in NMR spectra. The characteristic vibration frequencies of groups containing a hydrogen atom, say O—H and N—H groups, decrease if hydrogen forms a hydrogen bond.
2. **Infrared absorption bonds**: The infrared absorption bands of O—H groups are strongly broadened when a hydrogen bond is formed, and their intensities are enhanced.
3. **Electronic spectra**: Hydrogen bonding affects electronic spectra.

4. **Proton-magnetic-resonance (PMR)**: In PMR, there are observed chemical shifts caused by a hydrogen bond. ThePMR spectroscopy is the most efficient method of studying hydrogen bonds.

12.9.2.4 Biological application

1. **Structure and properties of water**: H-bonds determine the structure and properties of water (see Ch. 15).
2. **Structure of biopolymers**: They play an important role in the formation of the structure of biopolymers and in their interactions with small molecules. They are responsible for folding and overall three-dimensional conformation and intermolecular bindings of biomolecules.
3. **Stabilization of secondary structure of protein chain**: The secondary structure of protein chains are stabilized by hydrogen bonding.
4. **Secondary structures of proteins and nucleic acids**: The spiral or regularly folded arrangements of three-dimensional **secondary structures** of proteins and nucleic acids are produced by H-bonds. H-bonds between complementary pairs of purine and pyrimidine bases of two separate polynucleotide chains of DNA hold them together in the form of a helically coiled pair of DNA strands called **DNA double-helix** and constitute a DNA molecule.
5. **Folded secondary structure of tRNA**: Hydrogen bonding of complimentary bases in different parts of the same polynucleotide strand of tRNA gives the latter a folded **clover-leaf form** of secondary structure.
6. **Fibrous proteins (e.g., -Keratin) and secondary structure**: The peptide chain of a few fibrous proteins, e.g., α-Keratins have regular, helically coiled secondary structure (α-helix) which is due to regularly occuring hydrogen bonds. The oxygen of each α-carbonyl (CO) group involving peptide bond remains hydrogen-bonded with the α-amide (NH) group of another peptide bond in the same chain.
7. **Triple-stranded helix of collagen fibre**: The **triple-stranded helix or triple-helix secondary structure** of collagen fibre consists of three polypeptide strands winding around a common long axis and held together by hydrogen bonds between α-amide and α-carbonyl groups of separate strands.
8. **Tertiary structures**: Folding or twisting of proteins and nucleic acids into somewhat irregular or globular **tertiary structures** are due to some extent by hydrogen bonds though other noncovalent bonds including van der Waal's forces contribute some share at this stage.
9. **Quarternary structure of oligomeric protein**: Sometimes the hydrogen bonds join more than one polypeptide chain together to form the quaternary structure of an oligomeric protein.
10. **Denaturation of protein**: The loss of biological functionality by a protein when it is heated, reacts with acid or alkali, addition of urea or guanidine hydrochloride causes dissociation of weak noncovalent bonds such as hydrogen bonds are dissociated by changes in pH called **denaturation**. This process causes destruction of the secondary, tertiary and quaternary structures of protein and nucleic acids.
11. **Hydrogen bonds in myoglobin and haemoglobin**: Hydrogen bonds are present in the structure of myoglobin and haemoglobin. In the α-helix, the peptide —NH and —CO groups are hydrogen bonded to each other. The nitrogen atom is the hydrogen donor, whereas the oxygen atom is the hydrogen acceptor. The distance between the nitrogen and oxygen atoms is 2.9Å. The hydrogen atom is closer to the nitrogen than to the oxygen atom by 0.9 Å.
12. **Substrates and enzymes interaction**: The role of hydrogen bonding in the interaction

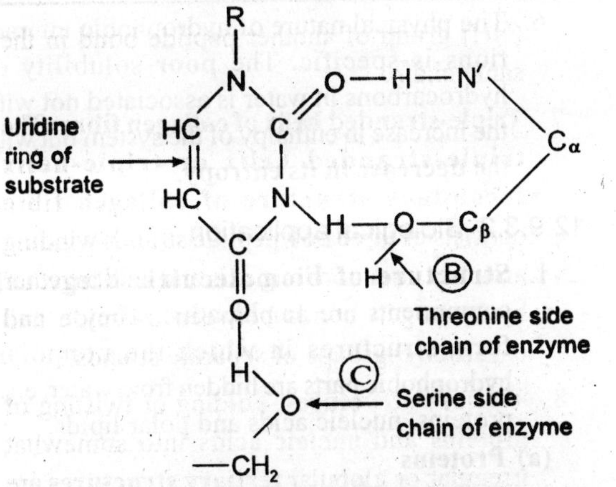

Fig. 12.24: Hydrogen bond interactions in the binding between a substrate and ribonuclease

of substrates with enzymes is illustrated above to show the binding of the uridine portion of the substrate to pancreatic ribonuclease, the enzyme causing cleavage of ribonucleic acid (Fig. 12.24).

In this reaction, three hydrogen bonds are invloved:

(a) One C═O group of uridine ring is hydrogen bonded to a peptide N—H(A).

(b) The N—H group of the uridine ring is hydrogen bonded to the —OH of a threonine residue (B).

(c) The other ring C═O is hydrogen bonded to the —OH group of a serine residue (C).

13. Antigen-antibody reactions: Hydrophilic groups such as OH, NH_2 COOH, etc., of antigen and antibody form hydrogen bond between them in a few antigen-antibody reactions.

12.9.3 Hydrophobic interactions or bonds

Biopolymers proteins and nucleic acids function mainly in aqueous medium. These and many organic compounds are dispersed and solubilized in water in the form of micelle which contains both strongly polar and strongly nonpolar groups. Such molecules are called amphipathic. The example of amphipathic biomolecules that tends to form micelle is the sodium salt of the long-chain fatty acid oleic acid. This organic molecule has a single carboxyl group, which is polar and thus tends to hydrate readily, hydrophilic (water attracting) groups, exposed to the water medium, negatively charged and form hydrogen bond with water and this group is known as head.

The other part is a long hydrocarbon chain, nonpolar and intrinsically insoluble in water and hydrophobic (water repellant) is called tail (Fig. 12.25).

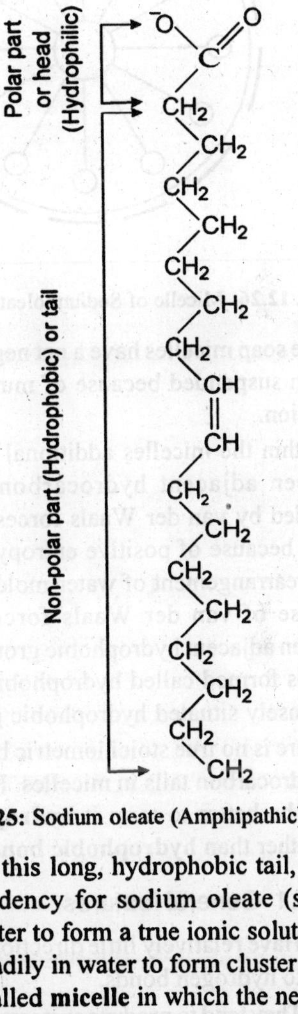

Fig. 12.25: Sodium oleate (Amphipathic)

Because of this long, hydrophobic tail, there is very little tendency for sodium oleate (soap) to dissolve in water to form a true ionic solution. But it disperses readily in water to form cluster close to one another called micelle in which the negatively charged carboxylate groups are exposed and form hydrogen bonds with water molecules and the nonpolar, insoluble hydrocarbon chains, which do not form hydrogen bond with water are hidden in the interior of the folds of the organic molecule (Fig. 12.26).

The micelle squeezes out the water molecules to reduce the number of water molecules.

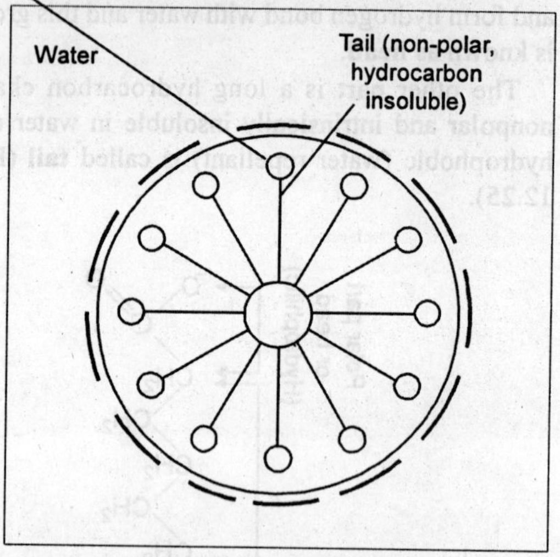

Fig. 12.26: Micelle of Sodium oleate (cross section)

The soap micelles have a net negative charge and remain suspended because of mutual electrostatic repulsion.

Within the micelles additional attractive forces between adjacent hydrocarbon structures are provided by van der Waals forces or interactions. Partly because of positive entropy changes due to local rearrangement of water molecules and partly because of van der Waals forces of attraction between adjacent hydrophobic groups, an attractive force is formed called hydrophobic bond between two closely situated hydrophobic groups.

There is no true stoichiometric bonding between the hydrocarbon tails in micelles. For this reason it would be better to term it **hydrophobic interaction** rather than **hydrophobic bond.**

12.9.3.1 General features

1. Have relatively little directionality compared to hydrogen bonds.
2. They tend to produce systems of high stability.
3. They are not influenced by H^+ concentrations unlike hydrogen bonds.
4. They are weakened by a fall of temperature.
5. Though they are weak individually give collective strength to five stable folded informations to molecules when they are present in large numbers at frequent intervals.
6. The physical nature of hydrophobic interactions is specific. The **poor** solubility of hydrocarbons in water is associated not with the increase in enthalpy of the system but with the **decrease in its entropy.**

12.9.3.2 Biological application

1. **Structure of biomolecules**: Many cell components are amphipathic. They usually form structures in which the nonpolar, hydrophobic parts are hidden from water, e.g., proteins, nucleic acids and polar lipids.

(a) **Proteins**

(i) **Tertiary structures of globular proteins**: Hydrophobic interactions may be formed between nonpolar sidechains of different amino acids residues of a polypeptide strand, resulting further folding and twisting of the coiled secondary structures of proteins into globular **tertiary struc-tures.**

(ii) **Stabilization of tertiary structures of globular proteins**: After formation of native tertiary structures of globular proteins, four major types of weak bonds or interactions cooperate for stabilization of it:

1. Hydrophobic interactions between nonpolar R group;
2. Hydrogen bonds between peptide groups are in α-helical or β-pleated sheets;
3. Hydrogen bonds between R groups;
4. Ionic bonds between positively charged and negatively charged groups, such as the —COO^- of aspartate or glutamate R groups and the —NH_3^+ of lysine R groups. It has been observed by different studies regarding relative contribution to the total conformational stability of native protein molecule, the hydrophobic interactions between the nonpolar R groups are most important.

5. **Nucleic Acids**: The Watson-Crick double helix having diameter 2.0 nm has one shallow and one deep groove. The relatively hydrophobic bases are within the double helix, shielded from water; the

polar sugar residues and negatively charged phosphate groups are oriented. The amphipathic molecules in aqueous medium may remain in the following forms:

6. **Surface monolayer**: In a very **dilute water solution**, amphipathic molecules get concentrated on the surface of water in the form of single layer of molecules as a monolayer showed by Langmuir in which the polar head groups which are lyophilic are oriented towards the water while the nonpolar groups are oriented towards the vapour phase. Similarly, the amphipathic molecules may also form a monolayer at an oil-water interface, with the polar groups of the molecules directed towards the water phase and the nonpolar groups towards the oil phase (Fig. 12.27). The periphery exposes to the water. The double helix is stabilized not only by the hydrogen bonding of complementary base pairs but also by electronic interactions between the stacked bases, as well as by hydrophobic interactions.

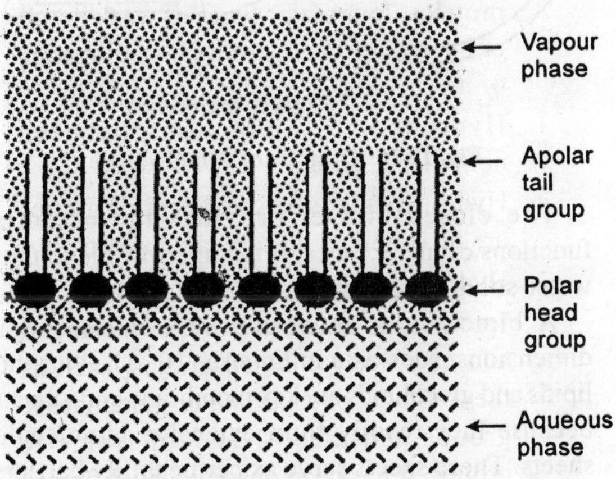

Fig. 12.27: Surface monolayer at the water-vapour interface

7. **Polar lipids—Phosphoglycerides**: The phosphoglycerides possess a polar, hydrophilic head and nonpolar hydrocarbon tails, they are called **amphipathic** or **polar lipids**. The amphipathic character of phospholipids suggests that the 2 regions of molecule have incompatibilities. If we consider water as a solvent, phospholipids organizes themselves into a form that thermodynamically satisfies both ends (Fig. 12.28).

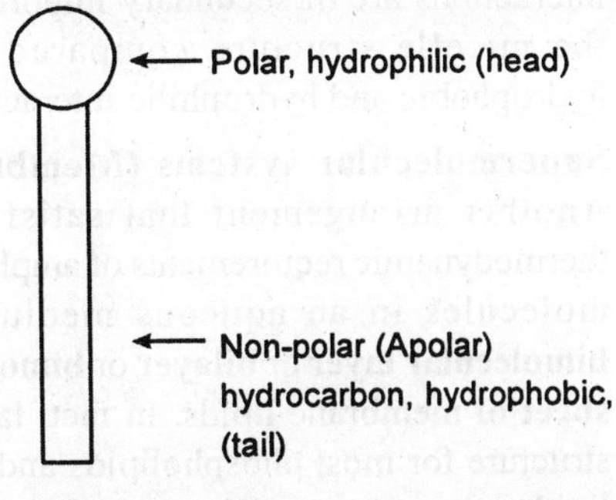

Fig. 12.28: Diagrammatic representation of phospholipid or glycolipids or other membrane lipid

II. **Micelle**: When a polar lipid, like phosphoglyceride, is added to water, only a small fraction dissolves to form a true molecular solution. Above a specific minimum concentration called **Critical micelle concentration**, the polar lipids associate into various types of aggregates resembling the **micelle** formed from surfactant. Arrangement of phospholipids and glycolipids in an aqueous medium is that their polar head groups have affinity for water, i.e., hydrophilic whereas their hydrocarbon tails have no affinity for water and shielded or hidden from water; hence called hydrophobic. This is accomplished by forming **micelle**, depicted in (Fig. 12.26). These micelles are small, spherical, droplet-like aggregates, stable, having 50-500 molecules of a surfactant, diameter of 3-10 nm and being polymolecular have high molecular weights of 2000-2000000 in which the polar head groups are on the surface and the hydrocarbon tails are sequentered inside. Thus, the nonaqueous interior of the micelles, may influence other nonpolar and lipid-solute substances to keep them in solution in the central part of the micelle. The stability of this structure is based on the fact that significant **free energy** is required to transfer a nonpolar molecule from a nonpolar medium to water. At the other end, much energy is required to transfer a polar moiety from water to a nonpolar medium. Thus, micelle provides a minimal energy configuration and accordingly is **thermodynamically stable.** Hydrogen bonding and salt or electrostatic interactions are of secondary importance to the micelle structure compared to the hydrophobic and hydrophilic interactions.

III. Supermolecular systems (Membranes): Another arrangement that satisfies the thermodynamic requirements of amphipathic molecules in an aqueous medium is a **bimolecular layer or bilayer or bimolecular sheet** of membrane lipids. In fact, favoured structure for most phospholipids and glycolipids in aqueous medium is a bimolecular sheet rather than a micelle (Fig. 12.29).

Table 12.4: Hydrophobic and hydrophilic unit of membrane lipid

Membrane lipids	Nonpolar Hydrocarbon tail unit or hydrobhobic unit	Polar head unit or hydrophilic unit
Phosphoglycerides	Fatty acid chains	Phosphorylated alcohol
Sphingomyelin	Fatty acid chain and Hydrocarbon chain of sphingosine	Phosphoryl choline
Glycolipid	Fatty acid chain and Hydrocarbon chain of sphingosine	One or more sugar residues
Cholesterol	Entire molecule except for OH group	OH group of C-3

A bilayer exists as a sheet in which the hydrophobic regions of the phospholipids are protected from aqueous medium while the hydrophilic regions are dipped in water. Only the ends or edges of the bilayer sheet are exposed to an unfavourable environment, but even those exposed edges can be eliminated by folding the sheet back upon itself to form and **enclosed vesicle with no edges** or **liposome** which is formed by exposing phosphoglyceride water suspensions to sonic oscillation, i.e., agitated by high-frequency sound waves (Fig. 12.30).

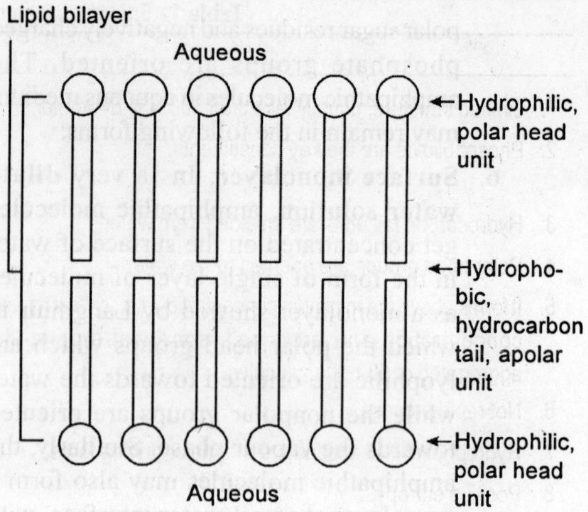

Fig. 12.29: Diagram of a section of a bilayer membrane formed from phospholipid molecules

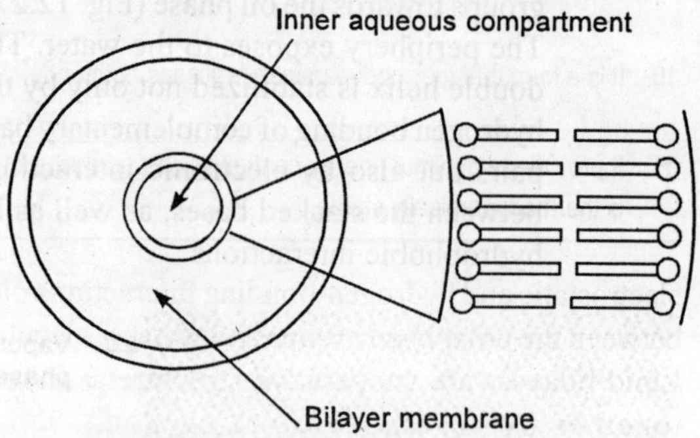

Fig. 12.30: Diagram of a lipid vesicle

The closed bilayer provides the essential functions of membranes. It is impermeable to most water-soluble molecules.

A bimolecular sheet can have macroscopic dimensions, such as a milimetre (10^7 Å). Phospholipids and glycolipids are key membrane constituents because they readily form extensive bimolecular sheets. These sheets serve as permeability barriers, yet they are quite fluid. The formation of lipid bilayer is a **self-assembly process** and it is a rapid and spontaneous process in water. Hydrotrophic interaction is the major driving force for the formation of bilayers, due to sequestration of hydrocarbon tail to release water resulting increase in entropy.

There are van der Waals forces between the hydrocarbon tails which favours close packing of the hydrocarbon tails. Apart from hydrophobic interactions, van der Waals forces, there are

Table 12.5: Comparison of Micelle and Bimolecular Sheet

Micelle	Bimolecular Sheet
1. Limited structure, usually less than 200 Å in di-ameter.	1. Macroscopic dimension such as a millimeter (10^7Å).
2. Phospholipids are the key constituents.	2. Phospholipids and glycolipids are key constituents because they readily form extensive bimolecular sheets
3. Hydrocarbon tail units are shielded from water.	3. Same.
4. Polar head groups are oriented to the water.	4. Same.
5. It forms above the minimum concentration, i.e., critical micelle concentration and are small, spherical droplet-like aggregated of 50-500 molecules.	5. Two tail-like hydrocarbon chains which cannot be accommodated in the globular micelle, instead they orient themselves in a regular manner to form a bilayer sheet.
6. Not rapid process.	6. It is a rapid and spontaneous process in water.
7. Hydrophobic interactions are the major driving force.	7. Same.
8. Positive entropy.	8. During sequestration of the tail, release of water causes large increase in entropy.
9. Other forces—Van der Waal's force, Hydrogen bonding and salt or electrostatic bonds are of only secondary importance.	9. van der waals' force operates close packing of the hydrocarbon tails. There are electrostatic and hydrogen bond interactions between polar head group and water.
10. It is a form that thermodynamically satisfies both ends.	10. It can also satisfy the thermodynamic requirements of amphipathic molecules.
11. It provides a minimal energy configuration and accordingly is thermodynamically stable.	11. Thermodynamically stable.

electrostatic and hydrogen-bonding interactions play between the polar head groups and water molecules. Lipid bilayers are cooperative structures and held together by several reinforcing noncovalent interactions. Phospholipids and glycolipids cluster together in aqueous medium to minimize the number of exposed hydrocarbon chains and this process is favoured by the van der Waals' forces between adjacent hydrocarbon chains. These energetic factors have biological significances for lipid bilayers:

(a) They have an inherent tendency to be **extensive;**

(b) They tend to **close on themselves;**

(c) They are **self-sealing** because a hole is energetically not favourable.

2. **Membrane lipoproteins:** Animal and plant cells consist of lipids and proteins which are held together by hydrophobic bonds between nonpolar hydrocarbon tail of lipids and the nonpolar side chains of amino acid residue of proteins.
3. **Antigen-antibody reaction:** The binding of nonpolar groups of an antigen with the nonpolar groups of an antibody is due mainly to hydrophobic bonds for antigen-antibody reaction.

- **Biological Importance of Micelle and Bimolecular Sheet**

I. Micelle

A. Insolubility of fat and micelle: All fats are usually soluble in fat solvent e.g., ether, chloroform, benzene and hot ethanol. Glycerides not containing hydroxy fatty acids are soluble in petroleum ether and glycerides with lower fatty acids only may dissolve in water. The lower fatty acids only may dissolve in water. The insolubility of fat and lipids in water is due to water repulsive (hydrophobic), nonpolar, hydrocarbon chains of the fatty acid residues. When emulsifying agents of detergents, e.g., soaps, bile salts and lecithins are added to a mixture of fat and water, a very fine thin layer of the emulsifying agent cover each fat droplet, in such a way that the polar hydrophilic groups direct to the surrounding water phase and the nonpolar hydrophobic groups orient towards the nonaqueous centre of the droplet. The fat molecules are aggregated in the interior of the droplet which is nonaqueous between the hydrophobic group of the emulsifying agent. This molecular aggregation is called **micelle**. It remains stably dispersed in water because hydrophobic groups in the interior are shielded by the surface hydrophilic group.

B. Absorption of lipids

1. **Absorption of triglycerides**: The hydrolysis product of triglycerides, i.e., free fatty acids, glycerol, monoglycerides and diglycerides of which only lower fatty acids and glycerol are fairly soluble in water and easily absorbed by the small intestine.

   But higher fatty acids, monoglycerides and diglycerides are relatively insoluble in water and absorbed with the help of hydrotropic action of bile salts. Bile salts, soaps (sodium salts of fatty acids formed in the intestine), bicarbonate (from pancreatic and intestinal juice) help the molecules of higher fatty acids, monoglycerides, diglycrides, lecithins, cholesterol, etc. to form the water-soluble molecular aggregates called **mixed micelles** which are much smaller than the droplet of emulsified fat and are absorbed by the small intestine.

2. **Cholesterol absorption**: Cholesterol esters after hydrolysis yields free cholesterol which is incorporated into the interior of the water-soluble mixed micelle by the action of bile salts and remain in dissolved state and is absorbed by the small intestine.

3. **Absorption of other fat-soluble substances**: Carotenes, Vits A, D, E, K, higher fatty acid and other nonpolar substances remain in dissolved state in the interior of hydrophobic region of the mixed micelle. The mixed micelles remain finely dispersed in the aqueous medium of the intestinal contents due to association or hydrogen-bonding of the polar groups of the surface with the water molecules. It is also due to mutual electrostatic repulsions between the micelles due to the charges of their ionic polar groups. The micelles are absorbed easily from the small intestine.

4. **Solubility of cholesterol in bile. Formation of Gallstone**: Free cholesterol is completely insoluble in an aqueous medium like bile. But it remains dissolved in the hydrophobic interior due to incorporation into the lecithinobile salt micelle. Lecithin is the predominant phospholipid in the bile, is itself insoluble in aqueous medium and can be dissolved by bile salts in micelles. The large quantities of cholesterol in bile are solubilized in these water-soluble mixed micelles, allowing the cholesterol to be carried in bile to the intestine. These micelles are first formed by bile salts and lecithin gets incorporated afterwards.

   The actual solubility of cholesterol in bile depends on the relative proportions of bile salts, lecithin and cholesterol. The solubility also depends on the water content of the bile.

   Using **triangular coordiante**, Redinger and Small were able to determine the maximum solubility of cholesterol in human bile. If the relative proportion of the concentration of bile salts and lecithin in the micelle is below the optimum ratio of about 5 : 1, the hydrophobic central micellar fluid can dissolve the cholesterol and can carry it in the aqueous medium of bile. If there is failure of micelle formation in the bile or of the incorporation of cholesterol in the micelle it may cause precipitation of the nonpolar cholesterol molecules as crystals in the gall bladder and bile ducts leading to gallstones and stones in the bile duct.

5. **Excretion of cholesterol and lecithin in bile**: Bile salts help in the excretion of lecithins and cholesterol in bile by forming water-soluble micelles with them.

II. Surface Monolayers–Surfactants

(a) **Lung Surfactant**: A lipoprotein mixture called **Surfactant** is secreted by special **surfactant secreting cells** which are component parts of the alveolar epithelium. It is the mixture containing especially the phospholipid "dipalmithoyl lecithin". Other surfactants are lecithins, soaps and detergents. These are amphipathic substances or amphiphiles which carry both strong polar and nonpolar groups. Surfactants concentrate on the surface of the liquid medium in the form of monolayer to lower the surface tension.

Surfactant is very much important to minimize the effect of surface tension in causing collapse of the lungs. In hyaline membrane disease or respiratory distress syndrome of new born premature babies inadequate secretion of surfactant cause disturbed ventilation resulting in death soon after birth.

Surfactant acts by forming a monomolecular layer at the interface between the fluid lining the alveoli and the air in the alveoli. This prevents the development of a water-air interface which has two to fourteen times as much surface tension as the surfactant-air interface.

(b) **Bile salts and soap as surfactants in the digestion and absorption of fact**: Bile salts of bile and soaps (the metallic salts of the fatty acids above C_6 are known as soaps, e.g., sodium salt of palmitic, stearic and oleic acids. The alkali metal soaps are soluble in water and act as efficient emulsifying and cleansing agents) take active part as surfactants in the small intestine during digestion and absorption of fats. These surfactants concentrate on water-air or water-liquid interfaces because of amphipathic nature and form monolayers. For their nonpolar hydrophobic groups, they decrease interfacial tensions by reduction of intermolecular attractions between water molecules.

III. Lipid bilayers of the cell membrane

The basic structure of the membranes of cells and organelles is a lipid bilayer, which is a thin film of lipids only 2 molecules thick and continuous over the entire cell surface, 2-dimensional solutions of large globular integral proteins are dipersed in a fluid of phospholipid matrix depicted in the (Fig. 12.31) known as Fluid Mosaic Model. The lipid bilayer consists of phospholipids and cholesterol. The **phosphate part of the phospholipid and hydroxyl radical of cholesterol** are watersoluble hence called **hydrophilic**, and orient themselves towards the two surfaces of the bilayer in contact with the aquoues phase on each side of the latter.

The fatty acid radicals of phospholipid and steriod nucleus of cholesterol are insoluble in water and fat soluble—they are called hydrophobic. Because the hydrophobic part of both these molecules are repelled by water but are mutually attracted to each other, they have a tendency to line up with the hydrophobic part in the centre of the membrane and the hydrophilic portions are directed towards the two surfaces. The helically coiled nonpolar group of some amphipathic protein molecules is placed deeply in the hydrophobic central part of the lipid bilayer, whereas, the nonhelical random coils of peptide chains of the protein molecules having ionic polar groups orient on the two surfaces of the bilayer.

- **Characteristics of lipid bilayer**

1. It is a lipid fluid and not a solid.
2. Portions of the membrane can flow from one point to another.
3. Proteins or other substances dissolved in or floating in the lipid bilayer tend to diffuse to all areas of the membrane.
4. The lipid bilayer is totally impermeable to water and water-soluble substances, e.g. ions, glucose, urea etc.
5. Fat-soluble substances can penetrate the bilayer membrane, e.g. oxygen, carbon dioxide, alcohols etc.

12.9.4 Van der Waal's Forces/Bonds

The concept of Van der Waal's forces or bonds was introduced in 1873 for explanation of the deviation of gases from ideal behaviour. In the molecules of N_2, O_2, Cl_2, P_4, etc. in which all the valency orbitals are either used in normal bonding or are occupied by non-bonding electrons, they are rather firmly held in the liquid and solid states by Van der Waal's forces. The intermolecular forces of different types, when all other forces are absent, are referred to as Van der Waal's forces.

The orientation forces, inductive forces and dispersion forces are also usually termed Van der Waal's forces.

These are very weak attractive forces between groups of atoms in different molecules or in different

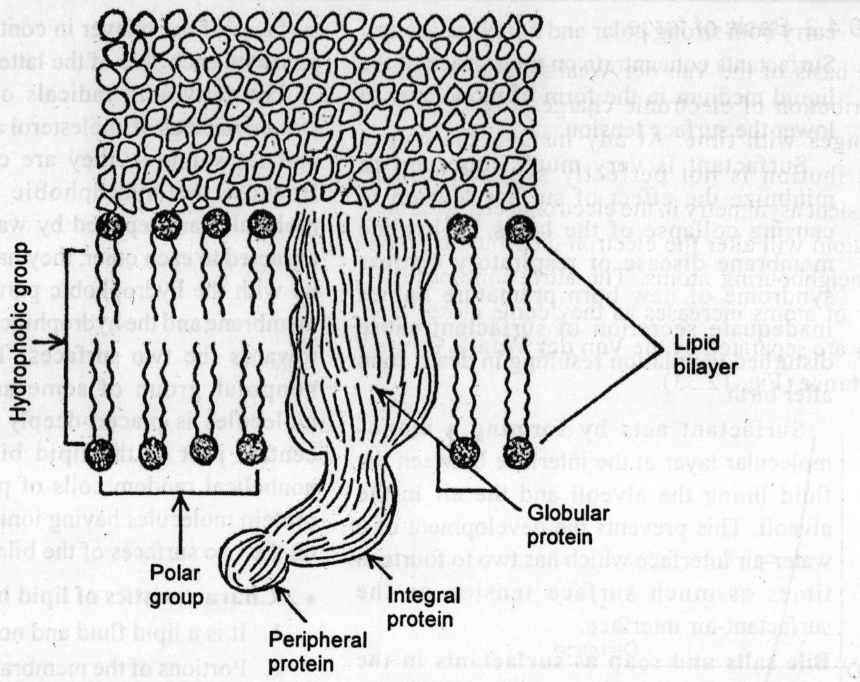

Fig. 12.31: Fluid Mosaic Model showing structure of cell membrane

parts of the same molecule, irrespective of their polar or nonpolar nature.

12.9.4.1 Distance of atoms and force

These forces are absent when the molecules in gases are far apart and in rapid kinetic motion. But when the molecules or atoms of a gas are brought closer under high pressure and the kinetic energy is withdrawn by cooling, the Van der Waal's forces are capable to hold the molecules together to form the liquid or the solid state.

There is a nonspecific attractive force between any two atoms when they are about 3 to 4 Å apart. This interaction, i.e., Van der Waal's bond, is less specific than electrostatic and hydrogen bond (Fig. 12.32).

In Fig. A, when the distance between atoms or molecules is long, there is neither attraction nor replusion and the force is zero.

In Fig. B, when the two atoms or molecules are close together at a few Å apart, the intermolecular attraction due to instantaneous dipole-induced dipoles, i.e., London forces, become appreciable.

In Fig. C, when the atoms are at a very close range, 3-4Å apart, there is overlapping of electron clouds, and the electrical repulsions between the clouds and the positive nuclei of the two atoms become pronounced..

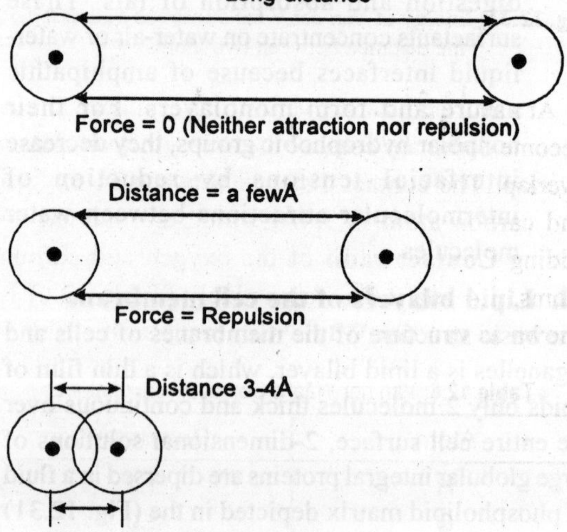

Fig. 12.32: Distance between atoms (or molecules) and the force

12.9.4.2 Basis of force

The basis of the Van der Waal's force is that the distribution of electronic charge around an atom changes with time. At any instant, the charge distribution is not perfectly symmetric. This transient asymmetry in the electronic charge around an atom will alter the electron distribution around its neighbouring atoms. The attraction between a pair of atoms increases as they come closer, until they are separated by the **Van der Waal's Contact Distance** (Fig. 12.33).

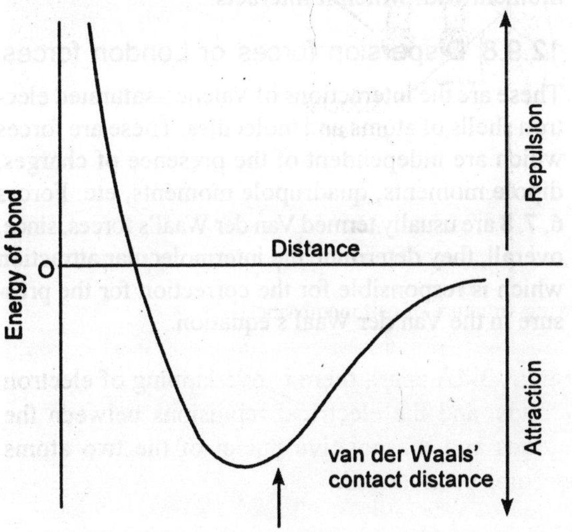

Fig. 12.33: Energy of Van der Waal's force as a function of the distance between two atoms

At a shorter distance, very strong repulsive forces become dominant because the outer electron clouds overlap. The contact distance between an oxygen and carbon atom is 3.4Å, which is obtained by adding **Contact radii** of the oxygen and carbon atoms, 1.4 and 2.0Å, respectively (Table 12.6) is known as **van der Waals' Contact Radii.**

Table 12.6: Van der Waal's Contact Radii of atoms

Atom	Radius (Å)
H	1.2
C	2.0
N	1.5
O	1.4
S	1.85
P	1.9

The Van der Waal's force fades rapidly when the distance between a pair of atoms is even 1 Å greater than their contact distance.

12.9.4.3 Bond energy

The Van der Waal's bond energy for a pair of atoms is about 1 kCal/mol. This is comparably weaker than a hydrogen or electrostatic bond, which is in the 3-7 kCal/mol range. Thus, a single Van der Waal's bond has little importance. Its strength is only a little more than the average thermal energy of molecules at room temperature, i.e., 0.6 kCal/mol.

12.9.4.4 Specificity

Though there is no specificity in a single Van der Waal's bond, **specificity arises when there is a large number of Van der Waal's bonds present simultaneously.**

12.9.4.5 Factors holding molecules in Van der Waal's effect

The effect of van der Waals' which holds the molecules together in liquids or solids is a combination of two factors:

1. Appreciable instantaneous dipole-induced dipole attraction when the atoms or molecules come very colse;
2. The repulsion resulting from the overlaps of electron clouds of the atoms or molecules and the nearness of the positive nuclei of the interacting particles.

12.9.4.6 Characteristics of Van der Waal's force

1. This force is unaffected by changes in the H^+ concentrations.
2. Weaker than hydrogen bonds.
3. Bond energies being slightly higher than the average kinetic energy of the molecules.
4. Presence of Van der Waal's forces in regular and close sequence can generate sufficient collective strength to produce stable folds in the molecular chain.
5. Van der Waal's forces (F) are inversely proportional to the seventh power of the distance (d) between the dipoles: $F \propto 1/d^7$.

12.9.4.7 Biological importance

A. Proteins

(a) Foldings of the polypeptide chain to produce tertiary structure.

(b) Participation in stabilizing the higher order of structure.

B. Enzymes

Numerous atoms of a substrate can interact with many atoms of an enzyme only if their shapes match. In other words, an effective Van der Waal's force between a substrate and an enzyme can occur only if they are sterically complementary.

C. Lipoproteins

The bond between lipids and proteins in the lipoprotein molecules is often due to Van der Waal's forces. For the weak bonds joining lipids and proteins in the lipoproteins, most of the lipids can be removed by extraction with fat solvent.

D. Antigen-antibody reaction

Van der Waal's forces frequently make the non-covalent bond of the antigen-antidboy reaction.

12.9.5 Ion-dipole interactions

Interactions between ions and polar groups of the molecules are called Ion-dipole interactions. They are determined by the ionic charge and the dipole moment of the atomic group that interacts with it.

12.9.6 Orientation forces

The electrostatic interactions between dipoles.

12.9.7 Inductive forces

The permanent dipole of a molecule or atomic group induces in another molecule or atomic group a dipole moment with which it interacts.

12.9.8 Dispersion forces or London forces

These are the interactions of Valence-saturated electron shells of atoms and molecules. These are forces which are independent of the presence of charges, dipole moments, quadrupole moments, etc. Forces 6, 7, 8 are usually termed Van der Waal's forces, since, overall, they determine the intermolecular attraction which is responsible for the correction for the pressure in the Van der Waal's equation.

13

Light and Associated Phenomena

PART I : 1. Stereoisomerism, optical isomerism, mutarotation, 2. Polarimeter, 3. Colorimeter, 4. Photometer

PART II : 1. Photochemical Reactions, 2. Photosynthesis, 3. Photochemistry of vision, 4. Transmitter in vision

PART III : **Biological effect of light**: 1. Lethal effects on small animals and plants, 2. Effect of long-continued exposure in large animals and man, 3. Effect on growth patterns in plants-Phytochrome system, 4. Synthesis of materials by plants and a few bacteria, 5. Photochemical processes, 6. Photoperiodism, 7. Photodynamic action, 8. Solarization (Photo-oxidation)

PART IV : **Biophysics of Vision:** 1. Light and its attenuation for vision, 2. Light and field of view, 3. Light and photopigment, 4. Illumination of retina, 5. The eye as an optical instrument, 6. Critical fusion frequency (CFF), 7. Changes in the retina on exposure to light, 8. Bioelectric potential from the retina

PART V : 1. Light and effect on animals, 2. Effect of photoperiods on gonads of animals, 3. Effect of light on secretion of anterior pituitary hormones other than gonadotrophins

PART VI : 1. Ultraviolet light on living system, 2. Photoreactivation

PART VII : 1. Light and application in therapy

PART VIII : 1. Biological light or animal light or bioluminescence

PART IX : 1. Interaction of light with biological materials

PART X : 1. Light and pigmentary response

PART I

13.1 Stereoisomerism, Optical Isomerism, Mutarotation

13.1.1 Definition of isomerism

Various organic compounds represented by the same molecular formula are called isomerides while the phenomenon is known as *isomerism.*

13.1.1.1 Types of isomerism

13.1.1.2 A. Structural isomerism

It is simple type of isomerism due to the difference in the arrangement of the atoms within the molecule i.e., the isomerides have different structural formulae. Structural isomerism are of different kinds:

1. Chain or Nuclear Isomerism
2. Position Isomerism
3. Functional Isomerism
4. Metamerism
5. Tautomerism

13.1.1.2 B. Stereoisomerism

Stereoisomerism is the phenomenon of existence of the same compound in more than one form due to different spatial arrangement. It is of two types:

I. Geometrical Isomerism

It is also caused by the difference in arrangement of the various groups in space. Two *Carbon atoms attached by a "double bond"* are incapable of rotation around the double bond and the position of various groups attached to those atoms is fixed in space and causes Geometrical Isomerism. For example, the formula HOOC—CH═CH—COOH represents two acids, Maleic acid and Fumaric acid. The molecule *does not contain any asymmetric carbon* atom, hence does not show optical isomerism. The isomerism is caused by the presence of a double bond which fixes the position of two hydrogen atoms and the two carboxyl groups in space. The two possible space models represents the two isomeric acids as given below:

Fumaric acid	Maleic acid
Two — COOH groups on opposite sides (Trans-variety)	Two — COOH groups on the same side (Cis-variety)

Fig. 13.1: Showing trans- and cis- variety

The isomer in which the H and OH groups are on the same side is termed Cis (or Syn) (Latin = this side) while the other in which H and OH groups are on the opposite sides is termed Trans (or Anti) (Latin = across). Since *the double bond is rigid, the atoms in it are not free to rotate* as are *those attached to a single bond.* Thus, the above structures are not equivalent and *have different chemical and physiological properties.* Fumaric acid, but not maleic acid is physiologically active.

Introduction of "trans" double bonds in an otherwise saturated hydrocarbon chain deforms the shape of the molecule relatively little. A "cis" double bond, by contrast, entirely changes the shape of the molecule. It can thus be appreciated why "*cis*" *and* "*trans*" *isomers of a compound are not interchangeable in cells.* Membranes composed of "trans" and "cis" isomers would have entirely different shapes (Fig. 13.1). Enzymes acting on one isomer might be expected to be entirely inert with the other.

13.1.1.3 Optical Isomerism and Optical Activity

Certain spatial relationships are readily visualized using ball-and-stick atomic models (Fig. 13.2). A compound having asymmetric carbon atoms exhibits "*optical isomerism*" and the product is called "*optical isomer*". An asymmetric carbon has four different groups or atoms satisfying its four valencies. Due to presence of asymmetric carbon atom, optical activity is shown by all compounds capable of existing in *two forms* that *are non-superimposable mirror image of each other,* such compounds which can exist in *right-handed* and *left-handed* forms are called "**Chiral**" compound (Greek for "hand"). The phenomena of stereoisomerism is also called "**Chirality**" ("handedness"). The four different substituent groups can occupy two different arrangements in space around the carbon atom to yield two different "**stereoisomers**" or "**enantiomers**". Thus, lactic acid has 2 nonequivalent optical isomers, one being the mirror image or enantiomer of the other. Stereoisomerism in respect of the groups connected to a single asymmetric carbon atom of a molecule

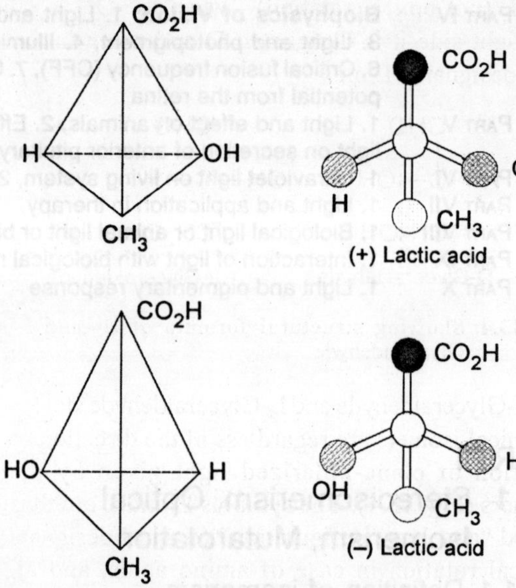

Fig. 13.2: Tetrahedral and ball-and-stick model representation of lactic acid enantiomers

containing more than one asymmetric carbon atom is called "*epimerization*". Ribose and xylose are epimers of one another while galactose and manose are two different epimers of glucose (Fig. 13.3).

The number of possible stereoisomers of any given compound is 2^n, where "n" is the number of asymmetric carbon atoms. All optically active centres can be related stereochemically to the

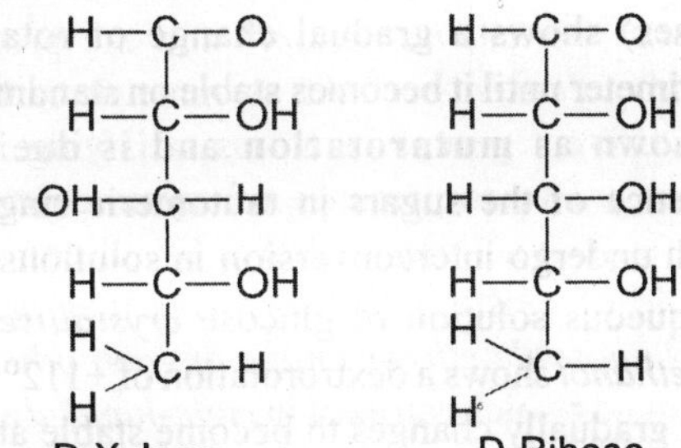

D-Xylose D-Ribose

Fig. 13.3: Showing structural formula of D-Xylose and D-Ribose

optically active centre of a single parent compound arbitarily chosen to serve as a standard of reference for stereoisomers. This is the three-carbon sugar glyceraldehyde, the smallest sugar to have an asymmetric carbon atom.

By convention, the mirror image of one of its isomers (i.e., enantiomers), i.e., two possible stereoisomers of glyceraldehyde are designated D and L (not optical rotation)—when the OH is on the right side it is designated D and when in the left, designated L, i.e.,

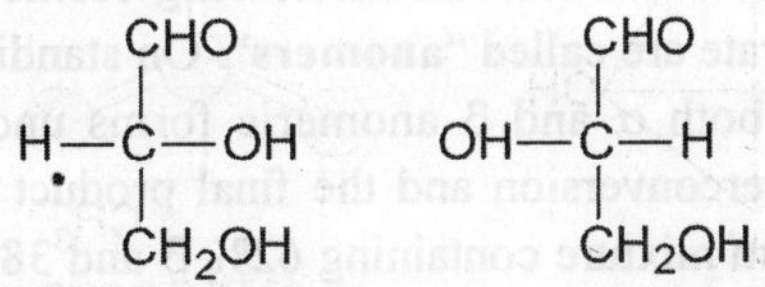

D-Glyceraldehyde L-Glyceraldehyde

Fig. 13.4: Showing structural formula of D- and L-Glyceraldehyde

D-Glyceraldehyde and L-Glyceraldehyde stereochemical isomers are regardless of the direction of rotation of plane-polarized light given by the isomers (Fig. 13.4). The symbols D and L are thus called "absolute configuration" (d and l designate optical rotation in case of amino acids, and all natural amino acids are l, i.e., levo variety and L stereochemical series).

The D and L stereoisomers of any given compound have identical physical properties and identical chemical reactivities with two exceptions:

1. They rotate the plane of plane-polarized light equally but in opposite directions;
2. They react at different rates with reagents that are themselves asymmetric.

Most enzymes acting upon amino acids have asymmetric binding sites and are thus capable of discriminating completely between D and L forms of amino acids.

An optically active substance may exist as two isomers in a polarimeter, one rotating the plane of polarized light to the right i.e., dextro-rotatory or d-isomer and designated by prefix (+) and angle of rotation 52.5°; when expressed it is indicated as +52.5° and if the substance is of stereo-isomer of D series then the substance is written as D-glucose + 52.5°, (all natural glucose is D-glucose +52.5°). Another rotates towards the left and is called levorotatory and designated as (–) or l-isomer. The optical activity is expressed quantitatively as the "specific rotation" $[a]^{25}$ D.

$$[a]^{25}\,D = \frac{\text{observed rotation. deg} \times 100}{\text{optical path length} \times \text{concentration g/100 ml}}$$

The optical activity of substances is the algebraic sum of the optical activities due to all the asymmetric carbon atoms in its molecule; so sometimes a substance may not have any optical activity inspite of possessing several asymmetric carbon atoms. An equimolecular mixture of the D- and L-isomers of an optically active substance is also optically inactive which is known as **"racemic mixture"** and the substance is called **"racemate"** and is designated as DL but it can be easily resolved or separated into two optically active isomers. Except the diose sugar glyceraldehyde, all other monosaccharides possess optical activities and all amino acids except glycine (it does not have asymmetric carbon atom) show optical activity.

"Allo" form or "diastereoisomers"—The amino acids with two asymmetric carbon atoms, threonine and isoleucine have four stereoisomers. The form of threonine isolated from protein hydrolyzates is by convention designated L; its mirror image is the D form. The other two stereoisomers are "diastereoisomers" or "allo" forms, also mirror images of each other. It is the configuration about the L carbon atom that is the basis for configurational assignment.

"Meso" form—Cystine, the oxidized form of cysteine which contains two asymmetric carbon atoms, one in each half of the molecule, may occur not only in L and D forms but also as an isomer in which the asymmetric carbon atoms are mirror images of each other. This *internally compensated* isomer, which does not occur biologically is called a **"meso"** form.

The correct configuration about a single asymmetric carbon is indicated by a projection formula, that is when the carbon atom is shown with four single bonds joining it and the four dissimilar substituents (Fig. 13.5). Whenever the four single bonds are not explicity shown, the stereochemistry is not specified.

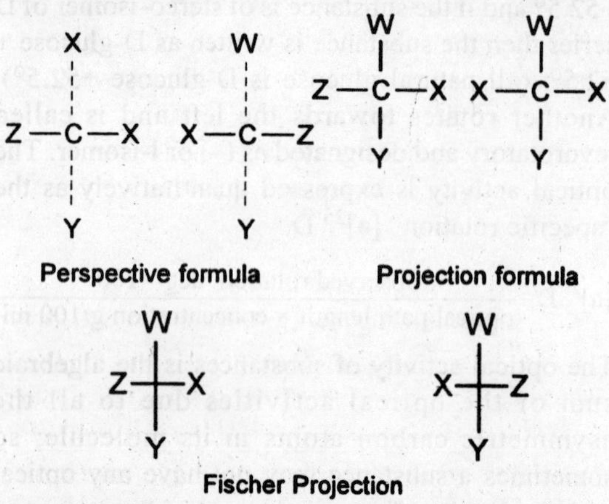

Fig. 13.5: Showing perspective and projective formula

Naming Stereoisomers of compounds with more than one asymmetric carbon atom can lead to ambiguities. These and other difficulties are avoided by a new system of designating stereoisomers, the *Cahn–Ingold–Prelog convention* usually called the *RS System.* Although it is not yet widely used in biochemistry, it provides the only unambiguous way of designating absolute configurations of molecules having two or more asymmetric centres.

13.1.1.4 Mutarotation

The optical activity of freshly prepared solution of a monosaccharide, e.g., glucose (except some ketoses) shows a gradual change of rotation in polarimeter until it becomes stable on standing. This is known as **mutarotation** and is due to the existence of the sugars in tautomeric ring forms which undergo interconversion in solutions.

Aqueous solution of glucose *crystallised from cold ethanol* shows a dextrorotation of +112° at first, but it gradually changes to become stable at 52.5°; on the other hand, an aqueous solution of glucose *crystallised from hot pyridine* shows a dextrorotation of +19° initially but this also ultimately comes to +52.5°. The two forms of glucose are known as α —D— glucose (mp 146°C) containing *pyran ring*—a six membered ring (5 carbon and 1 oxygen), hence also called α —D *glucopyranose.* The other is known as β —D— glucose (mp 148-150°C) containing *pyran ring,* called β —D—*glucopyranose.* The two isomeric forms α and β being due to the difference of arrangements of the H and OH groups attached to the first carbon (rendered asymmetric due to the ring formation) (Fig. 13.6).

Such pairs of stereoisomers ring forms of a carbohydrate are called "**anomers**". On standing in solution, both α and β anomeric forms undergo partial interconversion and the final product is an equilibrium mixture containing 62% β and 38% α-glucose, shows the stable optical rotation +52.5°. This is the reason for mutarotation. All reducing sugars including fructose (although fructose is a ketose) exhibit such mutarotation for identical reasons. Besides the pyranose forms, sugars may also exist in *furanose* ring forms having five-membered ring (4 carbon and 1 oxygen) in the molecule; furanose forms also may cause mutarotation (Fig. 13.7).

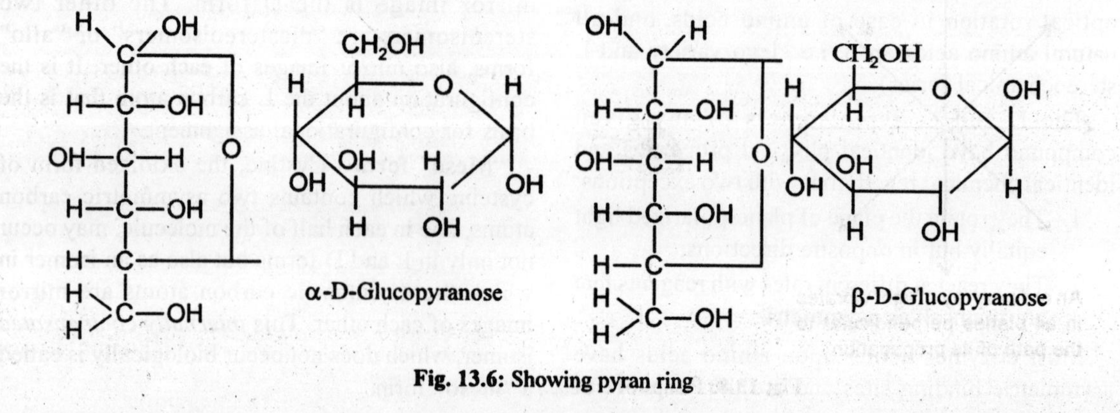

Fig. 13.6: Showing pyran ring

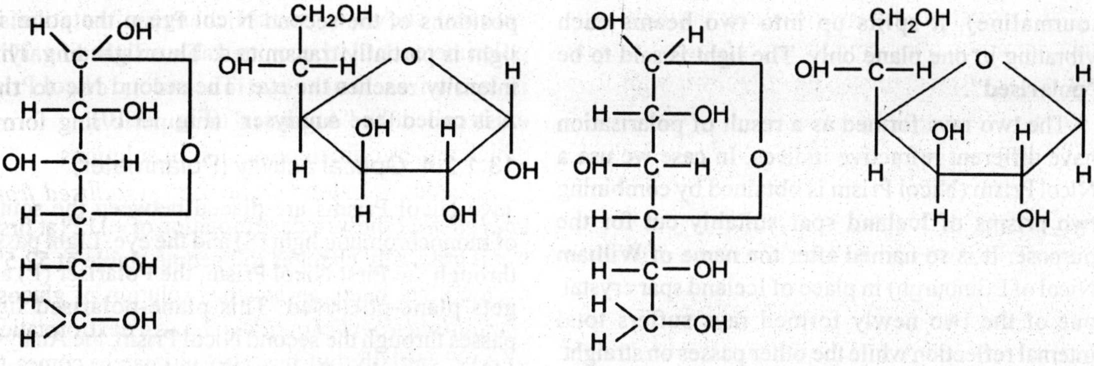

Fig. 13.7: Showing furanose forms of glucose

In a solution of a hexose or a pentose, comparatively large amounts of two pyranose anomers and much smaller amounts of two furanose anomers, as also of the open-chain form of the sugar, may ordinarily remain in equilibrium with each other.

13.1.2 Polarimeter (Determination of optical activity)

13.1.2.1 Introduction

Optical activity of a substance depends upon the presence of asymmetric carbon atom, i.e., four valencies of carbon are satisfied by four different univalent groups or atoms. Optical activity can be determined by polariscope, or polarimeter. For polarimeter, the polarised light is used, hence it is necessary to discuss polarised light first.

13.1.2.2 Source of light

Because different wavelengths of polarised light are rotated differently, it is necessary to use incident unpolarised monochromatic, i.e., a single wavelength light. Ordinarily, the light from incandescent Na or Hg is used. Special electric lamps as sources of these lights are available. Polariscopes often have special filters designed to purify the beam of light as it passes through. In practice, D line of the Na flame is really composed of the wavelengths D_1 (5896Å) and D_2 (5890Å) and except for work of high precision is commonly used. The most satisfactory is green line of a Hg discharge lamp (5416Å) since only one wavelngth is involved and is less frequently used. Because of the shorter wavelength, the Hg line gives slightly greater rotation than Na-D light.

13.1.2.3 Polarised light

According to the wave theory an ordinary ray of light is considered vibrating in all planes perpendicular to its path of propagation (Fig. 13.8). On passing a beam of monochromatic light (e.g., yellow light from a sodium flame) through crystal of Iceland spar (or

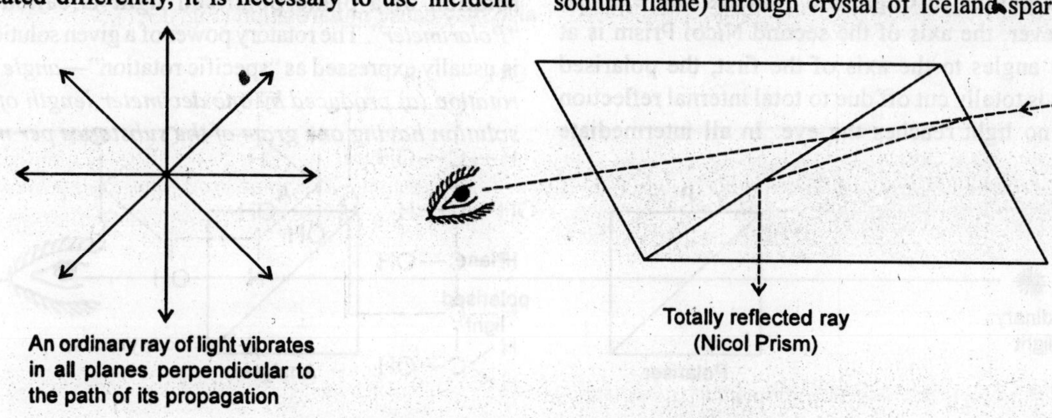

Fig. 13.8: Planes of ordinary and polarised light

tourmaline), it splits up into two beams each vibrating in one plane only. The light is said to be **"polarised"**.

The two rays formed as a result of polarisation have different refractive indices. In case we use a Nicol Prism (Nicol Prism is obtained by combining two prisms of Iceland spar suitably cut for the purpose. It is so named after the name of William Nicol of Edinburgh) in place of Iceland spar crystal, one of the two newly formed rays suffers total internal reflection while the other passes on straight. The Nicol Prism used for polarisation is called **"Polariser"**.

The light on passing through a Nicol Prism is polarised. In a plane polarised light vibrations take place in one plane only, say horizontal, as shown in (Fig. 13.9). The plane perpendicular to the plane to which the vibrations are restricted is called the **"plane of polarisation"**. In the present case the vibrations are restricted to the horizontal plane and the vertical plane is the plane of polarisation.

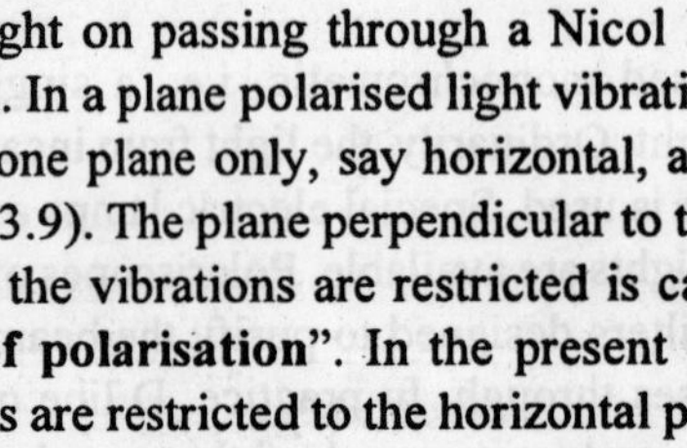

Fig. 13.9: Plane polarised light in which vibrations take place only in the horizontal plane

It is interesting to note, what happens when a plane-polarised light is passed through another Nicol Prism. If the axis of the second Nicol Prism is parallel to that of the first, the polarised beam of light passes through and reaches the eye. If, however, the axis of the second Nicol Prism is at right angles to the axis of the first, the polarised light is totally cut off due to total internal reflection and no light reaches the eye. In all intermediate positions of the second Nicol Prism the polarised light is partially transmitted. Thus light of varying intensity reaches the eye. The second Nicol Prism A is called the **"Analyser"** (Fig. 13.10).

13.1.2.4 Optical activity (Polarimeter)

Two Nicol Prisms are placed between the source of monochromatic light (S) and the eye. Light passes through the First Nicol Prism, the Polariser (P) and gets plane-polarised. This plane-polarised light passes through the second Nicol Prism, the Analyser (A) and reaches the eye. No light reaches the eye when the axis of the analyser is perpendicular to the axis of the polariser.

Now the analyser and the polariser are arranged for no light reaching the eye. Tube (T) filled with cane sugar solution or any other solution is interposed between the polariser and the analyser, some light is visible. The analyser has to be rotated a bit in order to get the original position of no light reaching the eye. What has happened which made light visible on the right? Neither the polariser nor the analyser has moved. This is possible only if the plane of polarisation is rotated. This shows that cane sugar solution has rotated the plane of polarisation (Fig. 13.11).

Substances of the type of cane-sugar which can rotate the plane of polarisation of light are called *optically active* while the phenomenon is called *optical activity*. Substances which rotate the plane of polarisation of light towards the right are called *dextrorotatory* whereas those which do so towards the left are termed *levorotatory*.

The instrument used for measuring the angle of rotation of a plane polarised light is called a "*Polarimeter*". The rotatory power of a given solution is usually expressed as "specific rotation"—*angle of rotation (a) produced by one decimeter length of a solution having one gram of the substance per ml.*

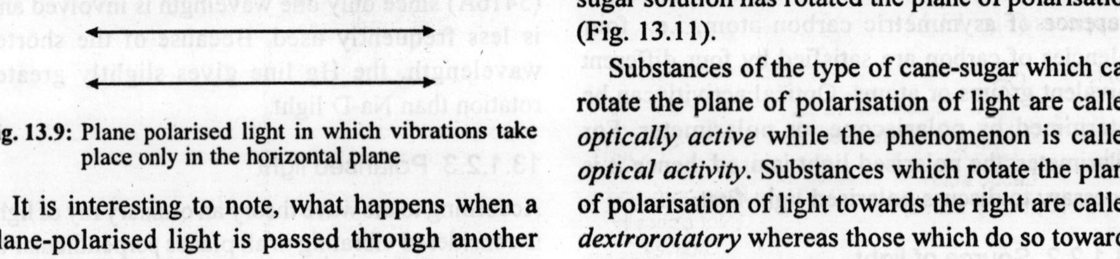

Fig. 13.10: Passing plane-polarised light through another Nicol Prism

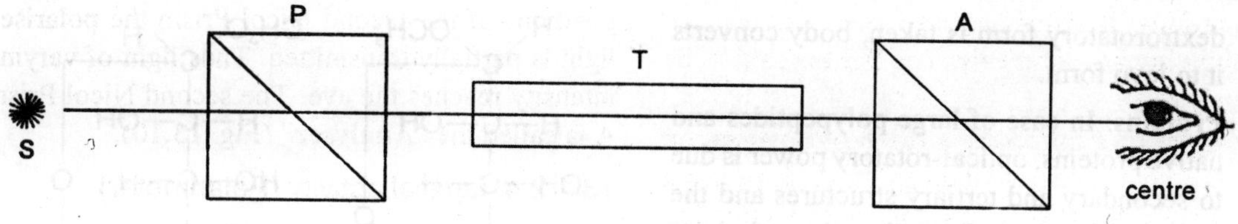

Fig. 13.11: Principle of polarimeter

$$\text{Specific Rotation} = \frac{100 \times \text{observed angle of rotation}}{\text{length in decimetre} \times \text{grams of substance present in 100 ml of soln.}}$$

$$\text{or, } [\alpha] = \frac{100 \times \alpha \text{ observation}}{1 \times C}$$

In some cases the optical activity is caused by crystalline structure, e.g., in sodium chlorate or bromate crystals, and is destroyed by fusion. In organic compounds the property is due to the presence of one or more carbon atoms whose four valencies have been satisfied by four different univalent groups or atoms. Such carbon atoms are called *asymmetric carbon* atoms. For example, lactic acid is optically active and possesses one asymmetric carbon atom. On reduction of lactic acid to give propionic acid, the asymmetric carbon atom disappears and with that optical activity also disappears (Fig. 13.12).

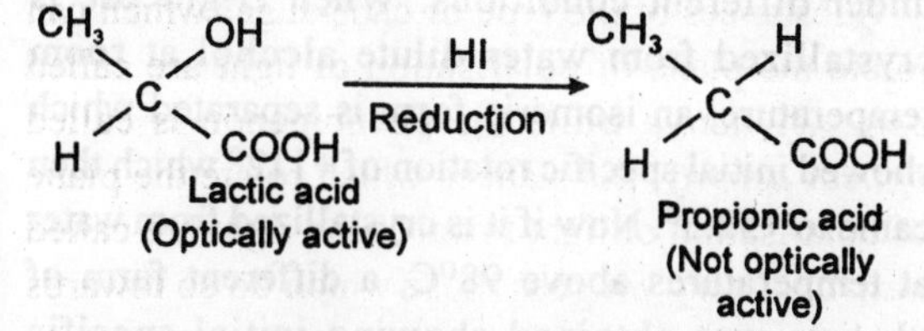

Fig. 13.12: Showing asymmetric carbon atom of lactic acid

13.1.2.5 Magnitude of the optical rotation

Practically all the principles of optical rotation were formulated by Biot. Among these, the followings are of direct applications which contributes to the magnitude of the optical rotation:

1. An optically active substance has a definite rotation for a given wave length.
2. Rotation is proportional to the concentration of optically active substance.
3. Rotation is proportional to the thickness of solution through which the light passes.
4. Rotation varies with temperatures, some substances showing an increase and others a decrease with rising temperature.
5. Rotation varies with wavelength of light. Higher the rotation, shorter the wavelength.
6. Rotation varies with the nature of the solvent in which the substance is dissolved.
7. Rotation varies with the length of the column of liquid or solvent.

In view of these principles, it is necessary to express the rotation of all substances according to a single definition in order that the values may be comparable. This is expressed as *specific rotation* (designated α) which is defined as the rotation in degrees of gram of substance per millilitre in a tube in dm (10 cm in length). The specific rotation (α) is given by the formula:

$$[\alpha] = \frac{\text{observed rotation in degrees}}{\text{length of tube (dm)} \times \text{concentration (g/cc)}}$$

13.1.2.6 Application

1. **Biomolecules:** Most of the biomolecules are optically active such as hexoses, pentoses, polysaccharides, amino acids, proteins and some vitamins. These substances may be either dextrorotatory or levorotatory.
2. **Carbohydrates:** Glucose, galactose, mannose, ribose, sucrose, maltose, dextrins, starch and glycogen are dextrorotatory. Fructose is levorotatory. They are designated as [+] or [–]. The body utilizes dextrorotatory form. But if levorotatory form is taken, it is converted into dextro form by the body.

   Glucose in the blood is present in the form of α D [+55°]-glucose.
3. **Amino acids:** Optical activity depends upon the side-chain on its α-carbon and also on the pH of the solution. Some are dextro-rotatory designated d not [+] and some are levorotatory designated 1, not [–]. The body utilizes levorotatory form of amino acids. If

dextrorotatory form is taken, body converts it to levo form.

4. **Proteins**: In case of large polypeptides and native proteins, optical-rotatory power is due to secondary and tertiary structures and the asymmetry of a α-helical coils and show dextrorotation.
5. **Vitamins**: Pantothenic acid and ascorbic acid are dextrorotatory.
6. **DNA**: Shows dextrorotation due to asymmetry of the double-stranded helical coils.
7. **Hormone**
   (a) Adrenaline isolated from adrenal medulla is levorotatory. Adrenaline can be artificially prepared, both dextro and levorotatory. Levo form is 12 times stronger than dextro form in physiological action.
   (b) The naturally occurring forms of thyroxine and its congeners with an asymmetric carbon atom are the L-isomers. D-thyroxinc has only a small fraction of the activity of L-form.

Inversion of Cane Sugar (Sucrose)

Sucrose is dextrorotatory (αD + 66.5°) but boiling with dilute mineral acids yields levorotation equilibrium mixture of hydrolytic products, glucose and fructose. The mixture is called **invert sugar** and the change in the direction of the optical rotation from dextrorotation to levorotation is known as "inversion". This levorotation is due to levorotatory power of fructose and is higher than dextrorotatory power of glucose. This process is utilized in a "polarimetric method" of estimating sucrose and glucose in the same solution.

The same hydrolysis occurs by the small intestinal enzyme called *sucrase* or *invertase*.

13.1.2.7 Mutarotation

Fischer obtained a product by treating glucose with methyl alcohol and hydrogen chloride while Ekenstein showed that glucose consists of two isomeric methyl derivatives of glucose. These derivatives had specific rotations of +159° and –34° and were termed as "α-methyl-D-glucoside and β-methyl-D-glucoside", respectively (Fig. 13.13). They had the following ring structures:

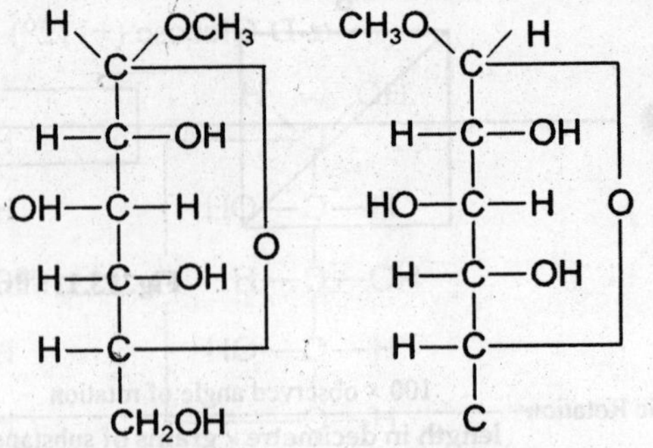

α-Methyl-D-Glucoside (α) D +159° β-Methyl-D-Glucoside (α) D –34°

Fig. 13.13: Showing methyl derivatives of glucose.

It is observed that carbon number one has become asymmetric which accounts for the two isomeric forms. Dubrunfaut was the first to observe that the optical rotation of a freshly prepared glucose solution gradually decreases, and finally becomes constant. This change of rotation of sugar solutions upon standing is a general property of reducing sugars except some Ketoses and is known as mutarotation. The authentic interpretation of mutarotation was given by Tanret. He prepared two isomeric forms of D-glucose by crystallization under different conditions. When D-glucose is crystallized from water dilute alcohol at room temperature, an isomeric form is separated which showed initial specific rotation of +112° which then came to +52.5°. Now if it is crystallized from water at temperatures above 98°C, a different form of glucose was obtained showing initial specific rotation of +19° which changed to +52.5°. The first isomers of glucose was termed "α-D-glucose" and the second "β-D-glucose" (Fig. 13.14). Tarnet showed that glucose exists in isomeric forms which in solution change into the same equilibrium mixture regardless of which form is dissolved:

α-D-Glucose (+112°) → +52.5° ← β-D-Glucose (+90°)

In glucose solution approximately two-thirds of the sugar exists as the β-form at equilibrium. About few hours may be required for the completion of mutarotation and the attainment of equilibrium in pure solution. The α and β forms of the isomers are also called *anomers*. The asymmetric carbon atom gives rise to the α and β forms, carbon number one is known as the *anomeric carbon atom*.

α-D-Glucose (+112°) → + 52.5° ← β-D-Glucsoe (+19°)

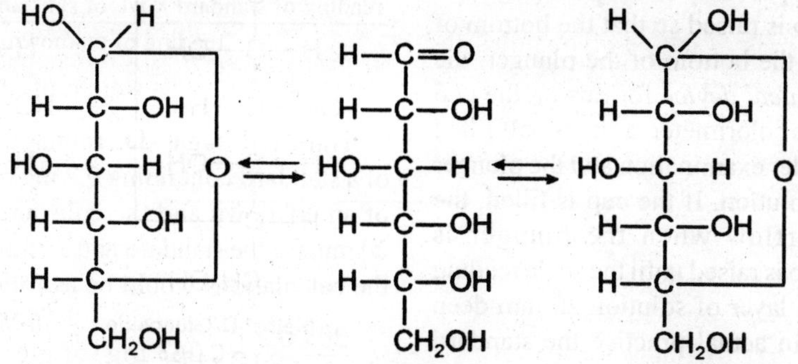

β-D-Glucose + 19°   only trace of chain form of D-Glucose   α-D-Glucose + 112°

Fig. 13.14: Mutarotation of glucose

13.1.2.8 Estimation of proportion of α-helical segments in protein

(a) **From reduction of its dextrorotatory activity:** Proportions of α-helical and random coils in a globular protein molecule is estimated from the reduction in its dextrorotatory activity on its denaturation to a randomly arranged peptide bond.

(b) **From the optical rotatory dispersion curves:** The proportion of α-helix in a protein in native and desaturated states may be estimated from the "optical rotatory dispersion curves". When the results are plotted, the curves represent the variations of change of optical rotation with the change of wavelength of monochromatic light. These show the difference in the helical and non-helical forms of the same polypeptide strand.

13.1.3 Colorimeter and photometer

13.1.3.1 Introduction

In quantitative analysis several types of procedure are employed such as:

1. **Gravimetric analysis**: Substance is converted into an insoluble compound, filtered, dried and weighed.
2. **Volumetric analysis**: This is titration process using standard solution and indicator.
3. **Colorimetry**: Another type of analytical procedure of estimating of the concentration of coloured substances used extensively in biochemical work is known as *Colorimetry*.

Principle: Colorimeter is a simple instrument used for reading of colour, against a blank in the visible range (400-720 nm) by the use of various filters. It gives fairly accurate results of *optical density or percentage transmission of a coloured solution.*

In colorimetry the sample is treated with a reagent which produces a colour with the constituent to be determined. A standard solution of the substance to be determined is treated in the same way and the colours of the standard and of the unknown are compared in an instrument called *colorimeter*. The *intensity of the colour is directly proportional to the concentration of the coloured substances.*

According to Beer's Law, light passing through a coloured solution is absorbed in direct proportion to the concentration of the coloured substance in solution and in inverse proportion to the depth of the observed layer. Thus a 20 mm layer of a solution of a coloured compound will give the same intensity of colour as a 10 mm layer of a solution of the same compound of twice the concentration.

4. **Photometry**: Estimation of the concentration of a light absorbing substance.

13.1.3.2 Visual colorimeter instrument

A colorimeter is an instrument with *cups filled* on to a stage which may be moved up and down. Above each cup is a *glass rod* or *plunger* which fits into each cup as the cup is elevated. By means of *an optical system* it is possible to look through the plungers when a cup is raised so that the bottom of it is in contact with the bottom of the plunger, the *scale on the mechanical device* for raising the cup reads zero. In using a colorimeter, a cup is half filled with the solution under examination and the plunger is dipped into the solution. If the cup is filled, the solution will overflow when the plunger is introduced. If the cup is raised until the scale reading is 20 mm, there is a layer of solution 20 mm deep under the plunger. In actual practice the standard solution is placed in the left-hand cup and the scale set at 20 mm. The unknown solution is then placed in the right-hand cup, which is moved up and down until the colour of the unknown matches the colour of the standard, as indicated by a field of uniform intensity of colour in the eyepiece. The depth in millimetres of the unknown solution is then determined by reading the scale, and the data obtained are used in calculating result (Fig. 13.15).

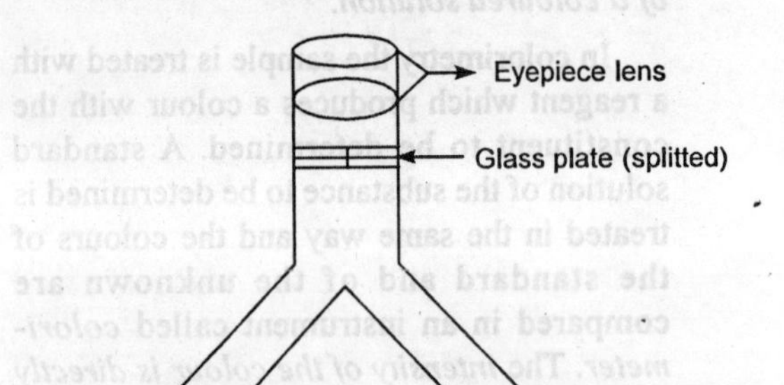

Fig. 13.15: Diagram of visual colorimeter

13.1.3.3 Calculation

$$\frac{\text{reading of standard} \times \text{wt. of substance in standard}}{\text{reading of unknown}}$$

= weight of unknown in sample

Thus, in a sugar determination, if we used 2 cc of a standard containing 0.2 mg of glucose and 2 cc of an unknown and the colorimeter readings were 20 mm for the standard and 15 mm for the unknown, the calculations would be as follows:

$\frac{20 \times 0.2}{15} = 0.267$ mg of glucose in 2 cc of the unknown solution.

13.1.3.4 Importance

Colorimetry is of special value in determining substance which are present in only small amounts. Since traces of highly coloured compounds give highly coloured solution, methods of analysis based on colour comparisons are of great value in determining many compounds present in low concentrations.

13.1.3.5 Precautions

1. Cups should never be filled to the extent that there is danger of overflowing. If overflowing does occur, the instrument should be throughly cleaned immediately.
2. Many of the solutions compared in a colorimeter are corrosive and will destroy the mirror which reflects the light into the base of the cups or will corrode the cup supports.
3. After a colorimeter is used the cups should be rinsed throughly with distilled water.
4. The plungers showed also be rinsed by dipping them in several changes of distilled water in the clearance cups.
5. Finally the plungers should be carefully dried with a clean cloth.
6. For accurate results, it is important that the instrument be in proper adjustment.
7. Another test which should be applied to a colorimeter before using it is to make sure that both cups read zero when they are elevated so that their bottoms are in contact with the plungers.
8. Also, the mirror and the light should be adjusted so that the two halves of the field

obscured through the eyepiece appear uniform and identical.

13.1.3.6 Modification

Advanced models of colorimeters are available in which there is an arrangement of automatic zero facility, i.e., the blanking is done automatically with the pressing of a button. The filter selection is also done automatically. These may have built-in incubators and temperature control. The times are also in built so that the reaction takes place inside the colorimeter itself. The digital readings are given directly in concentration unit. They consume on an average 2.5 to 3 ml of reagent. The major drawback of a colorimeter is its less accuracy and repeatability.

13.1.3.7 Beer's law

Beer's law states that, if I_o is the intensity of the light incident on a solution, I, the intensity of the transmitted light, is given by:

$$I = I_o 10^{-ecd},$$

where c = concentration of the solution in moles/l,

d = thickness through which the light is transmitted,

e = molar extinction coefficient for the solute for a particular wavelength.

If the law is used in the form:

$$I = I_o e^{\alpha cd}$$

α is called the "molar absorption coefficient"; for a pure liquid the law takes the form:

$$I = I_o 10^{-kd}$$

when k is called the extinction coefficient". By plotting extinction coefficient against wavelength an "absorption spectrum" is obtained, the regions of peak absorption are called "absorption bands".

The law is not obeyed by:

(a) Concentrated solutions

(b) Dilute solutions.

13.1.3.8 Lambert's Law

Lambert's law states that each successive layer of identical thickness of a pure light-absorbing substance absorbs the proportion of the light entering it. Lambert's Law **is obeyed by concentrated** and **dilute solutions.**

13.1.3.9 Lambert-Beer Law

The fraction of the incident light absorbed by a solution at a given wavelength is **related** to the **thickness** of the absorbing layer and the **concentration** of the absorbing substance. These two relationships are combined into the Lambert-Beer Law, given in integrated form as:

$$\log \frac{I_o}{I} = eci$$

where I_o = Intensity of the incident light,

I = Intensity of the transmitted light,

e = Molar absorption coefficient (in units of litres per mole centimetre),

c = The concentration of the absorbing substance in moles per litre,

I = The thickness of the light-absorbing sample in centimetres.

The Lambert-Beer Law assumes that the incident light is parallel and monochromatic and that the solvent and solute molecules oriented. The expression log (I_o/I) is called the "absorbance", designated A. It is necessary to know that each millimetre thickness of absorbing solution in a 1.0 cm cell absorbs not a constant amount but a constant fraction of the incident light. However, with an absorbing layer of fixed thickness, the absorbance A is directly proportional to the concentration of the absorbing solute.

The molar absorption coefficient varies with (a) nature of the absorbing compound (b) the solvent (c) the wavelength and also (c) with pH if the light absorbing substance is one that is in equilibrium with another substance having a different spectrum through gain or loss of protons.

13.1.4 Photoelectric colorimeter and photometry

The photoelectric colorimeter is used for measurement of colour by electrical devices. It is always preferred to the visual colorimetry. It does not cover full colour band but none the less has a smaller and narrower band of wavelength components such as 500 to 510 nm.

The advantages of this method are:

1. Measurements are recorded electrically and eliminate the errors due to procedure,
2. Measurement can be done within short time,

3. Blank readings, which are not available in a visual colorimeter, eleminate a definite source of error,
4. Smaller volumes are required for estimation,
5. Standard curves can be prepared and the concentration of the unknown can be read directly off the curves. So preparation of standard and blank is not necessary every time.

• **Use of Photometry**: Photometry is used in Enzyme Immunoassay of both types, e.g., **homogenous enzyme** immunoassay (HEIA) and having only liquid phase which couples immunology, enzymology and photometry, applicable for T_3, T_4, oestriol, cortisol, etc. and heterogenous enzyme immunoassay **(HIELA)** or enzyme-linked immuno-sorbent assay (ELISA).

Many types of photoelectric colorimeter with various modification are available, generally two types are used:

1. **Single cell type**: In this instrument, the light transmitted through solution which is interposed between the light source and the photo-cell or photo-emissive cells (because in them light caused electrons to be emitted) is measured directly by the current output of cell and compared with that obtained with blank and standard.
2. **Double cell type**: In this instrument there are two photo-cells, one is called Working photoelectric cell and another is Reference photoelectric cell having the same light source and balanced against each other by means of a null point instrument like galvanometer. The solution to be measured is placed between the light source through coloured filter and one cell. The extent to which the two cells are thrown out of balance with respect to each other is a measure of the light transmitted by the solution. A monochromatic light is allowed to pass through a filter and it is adjusted to give 100 per cent transmission against distilled water or a blank. The percentage transmissions given by the unknown, standard and the blank are then recorded in the similar way. The working principle of the instrument depends on the *Beer's law*. The concentration of the substance is directly proportional to the optical density, i.e., "Extinction".

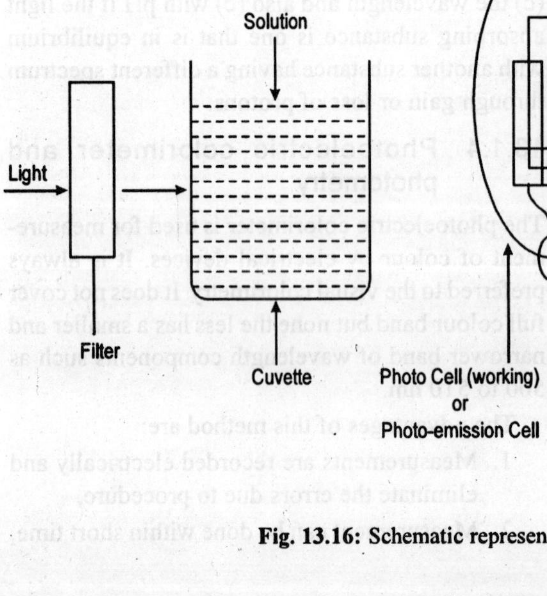

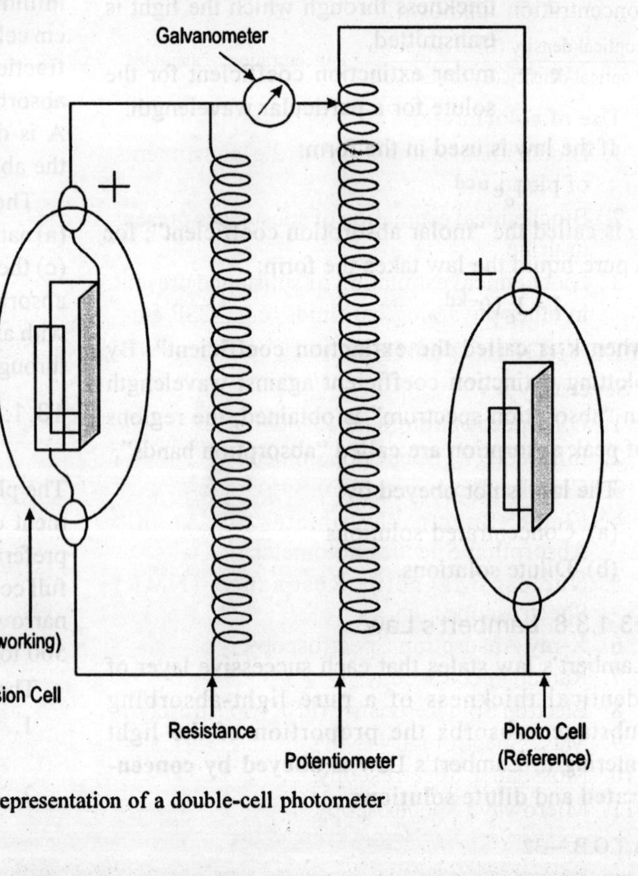

Fig. 13.16: Schematic representation of a double-cell photometer

The extinction is defined as the logarithm of the ratio of the incident to transmitted lights. Thus,

$E = \log 100 - \log T = 2 - \log T$,

where T = percentage of transmission given by a solution.

Suppose a solution is having 70 per cent transmission, then the optical density (E) is:

$E = 2 - \log 70 = 0.1549.$

Some instruments are available with graduation in logarithmic scales which give the reading of optical density directly. For convenience of expression, the values of E which comprise the logarithmic scale are multiplied by 100 and are plotted as whole numbers.

Now the optical density is directly proportional to the concentrations, it follows that:

$$\frac{E_1}{E_2} = \frac{C_1}{C_2}$$

where C_1 and C_2 = concentrations

E_1 and E_2 = their optical densities

From the readings of the standard and unknown and the known concentration of the standard, the concentration of the unknown can be calculated as follows:

Concentration unknown

$$= \frac{\text{optical density (E) of unknown}}{\text{optical density (E) of standard}} \times \text{concentration of standard}$$

- **Use of colorimeter**

1. Biochemical estimation of the constituents of blood.
2. Biochemical estimation of substances present in urine.
3. Biochemical estimation of substance present in other physiological fluids, e.g., CSF etc.

SPECTROPHOTOMETRY

- **Several types**

1. UV Spectrophotometer.
2. Infrared (IV) Spectrophotometer.
3. Circular Dichroism (CD).
4. Flame Spectrophotometer or Atomic Absorption Spectrophotometer.
5. Nuclear Magnetic Resonance (NMR) Spectroscope.
6. X-ray Absorption Spectroscopy.
7. Raman spectroscopy.
8. Electron Spin Resonance (ESR) Spectroscopy.
9. Mass Spectrometry (MS).
10. Mossbauer Spectroscopy.
11. Microwave Spectroscopy.

PART II

13.2.1 Photochemical reactions

Before discussing the photochemical reactions it is necessary to know the fundamentals of reaction Kinetics. Chemical reactions involve a redistribution of energy, which generally manifests in the absorption or liberation of heart. It is not necessarily the exothermy or endothermy of a reaction that determines whether it will occur spontaneously at a measurable rate or not. An exothermic reaction proceeds spontaneously at a measurable speed if there are suitable number of molecules with the necessary "activation energy". The form of activation energy may be:

1. *Kinetic or translational energy*
2. *Internal energy* consisting of:
   (a) *Rotation* about various axis,
   (b) *Vibrations of the constituent atoms.* Transition of the orbital electrons of the constituent atoms from one energy state to another.
3. *Electronic energy*
   Energy associated with the rays or light:
   1. For infrared rays: For emission or absorption of infrared rays, rotational and vibrational energy are associated.
   2. For ultraviolet or visible light: For emission or absorption of ultraviolet or visible light, changes in electronic energy is associated.

13.2.1.1 Definition

Since any molecule emitting energy of a given wavelength must be able to absorb it, it follows that activation energy can be supplied in the form of visible or ultraviolet light. Reactions activated in this way are called **Photochemical reactions.** The energy may be acquired directly by the reacting molecules or vicariously from some light-absorbing molecule. The latter reaction is said to be **Photosensitized.**

13.2.1.2 Effect of light on reactions

The effect of light on the reactions may be as follows:

1. The reaction may be faster, i.e., catalytic effect than without light,

2. The normal equilibrium point may get shifted,
3. An alternative reaction path may be taken to yield a product which is different from that of the product without light,
4. A novel reaction may occur.

13.2.1.3 Nature of light

Light is a form of electromagnetic radiation or waves and it has definite wavelength (400-700 nm) which depends on its colour or its position in the spectra. When wavelength has a value, there are electromagnetic waves of a wide range of wavelengths measured in Angstroms. Starting from x-rays of very short wavelength of an Angstrom to radiowaves of very long wavelengths of hundreds of metres are known.

Ordinary visible light has wavelength of a fraction of wide range which is about 4000-8000 Å and below this wavelength, i.e., 2000 Å is called Ultraviolet and above the wavelength of the visible light, i.e., 10000 Å is called Infrared rays. The whole wavelength pattern is shown schematically below (Table 13.2).

Table 13.1: Comparison of Thermal Reaction and Photochemical Reaction

Thermal Reaction	Photochemical Reaction
1. Reaction can occur in absence of light.	1. Reaction occurs in presence of light.
2. It is due to kinetic energy of the reactant molecules.	2. It is due to activation of the reactants through their absorption of light of specific wavelengths.
3. Depends solely on the temperature because it increases kinetic energy of the reactant molecule.	3. Depends solely on the intensity and wavelength of the light, not on the temperature.
4. Spontaneous reaction shows decrease in free energy.	4. Initially there is increase in the free energy of the reactants to a very high level and drop at the end of the reaction.
5. Not driven against entropy and may have a negative G^o.	5. Driven against entropy and may have a positive G^o.

Table 13.2

About 1Å	2000 Å	4000-8000 Å	10000 Å	Few hundred Meters
X-ray	Ultraviolet	Visible light	Infrared	Heat waves	Radio-waves

Light has properties suggesting that it is propagated in a discontinuous or corpuscular manner in the form of **photons or quanta.**

Absorption of a light quantum is, in an activated state, favourable for undergoing a reaction.

13.2.1.4 Energy content of light

The total amount of solar energy falling on the surface of the earth in the form of **photon** is enormous which is estimated to exceed 2×10^{25} cal/year. Only 12 per cent of this energy is available to plant life; the rest is either:

(a) Outside the visible range, or
(b) Absorbed by the atmosphere, or
(c) Absorbed by the non-living portion of the earth's surface.

The energy content of a photon is represented by hν, in which h is the frequency of the radiation, E, the energy in Planck's constant (1.58×10^{-34} cal-s) and ν kilocalories of 1.0 einstein, that is, 1.0 mol of light, containing 6.023×10^{23} (Avogadro's number) quanta, and is most simply given by the formula.

$$E = \frac{28600}{\text{wavelength (nm)}}$$

In the visible range, 1 einstein carries from 40 to 72 kcal of energy, depending on the wavelength of the light (Table 13.3).

Table 13.3: Energies equivalent of 1 einstein at different wavelengths

Wavelength (nm)	Colour	kilocalories	kilojoules
700	Far-red Red	40.9	171
600	Orange Yellow	47.7	199
500	Green Blue	57.2	239
400	Violet	71.5	299

Photons of short wavelength, at the violet end of the spectrum, have the greatest energy content. It is found that the energy content of a "mole" of photons, regardless of wavelength, is considerably greater than the amount of energy required to synthesize 1 mol of ATP from ADP and phosphate, which is 7.3 kcal under standard thermodynamic conditions. The ability of a compound to absorb photons depends on the atomic structure, particularly on the arrangement of electrons surrounding its atomic nuclei. The **absorption spectrum** of a compound indicates its capacity to absorb light as a function of wavelength.

13.2.1.5 Change in molecule after light absorption

When a molecule in ordinary state absorbs a quantum of light, then this energy rich molecule is known as **excited molecule**. The change of electron distribution in the excited molecule makes it different from the ordinary state molecule in its physical and chemical properties, e.g., in size and sometimes shape for which it is regarded as a new molecule. This excited molecule with electronic and vibrational changes may undergo the following changes:

1. It may emit light immediately as fluoresence and drop to a lower energy level,
2. It may pass to a different state which may emit light for a longer time as phosphorescence,
3. It may get the whole or part of its excess energy converted to heat, i.e., kinetic energy of the surrounding molecules,
4. It may react chemically as it is in an energetically activated state,
5. It is not only electronically excited but also have excess vibrational energy, i.e., nuclear motions, bounding and also nonbounding electrons are energised.

13.2.1.6 Primary photochemical reaction

In the above discussion, the processes (1), (2), (3) are photophysical process whereas (4), (5) are under photochemical reaction. In photochemical reaction, the excited molecule may show either of the two pathways: It may either (a) directly undergo a chemical change e.g.,

$$X + h\nu \longrightarrow X \longrightarrow \text{Products}$$

(Ordinary state molecule) (Energy in each quantum of light) (Excited molecule)

or, (b) It may go into active intermediates, usually free radicals

$$Cl_2 + h\nu \longrightarrow Cl + Cl$$

These two reactions are caused by direct result of light absorption; hence called **primary photochemical reactions**. All subsequent reactions of these free radicals are **secondary reactions.**

13.2.1.7 Basic laws of photochemical reactions

(a) **Grothus-Drapar Photochemical Law or First Law of Photochemical Reaction:** The law states that only light which is absorbed is photochemically active. Each light-absorbing substance has one or more specific band or wavelengths or absorption maxima (λ max) at which it absorbs light maximally and only then this light may produce maximal activation of the substance and produces maximum photochemical reaction. The absorbed light is not always utilised or fully utilised in photochemical reactions and may re-emit partly or wholly or it may be transformed into heat.

(b) **Einstein's Law of Photochemical Equivalance or Second Law of Photo-chemical Reaction:** The law states that the primary process in photochemical reaction is absorption of one quantum of light of energy hν by one molecule or atom to form an excited molecule or atom. The proportional ratio between the number of quanta and the number of excited molecules is 1 : 1. This is deviation from the physics of wave nature of light according to which the light energy should have been spread out over a number of molecules and not one quantum per molecule.

From Einstein's law, each mole of light absorbing substance must absorb the Avogadro number of photons or quanta of light, this amount of energy is known as an "einstein (E)".

$$E = Nhc/\lambda,$$

where hc/λ = Energy of photon or quantum.

N = Avogadro number or the number of molecules per mole.

The quantum is given by $h\nu$; ν, the frequency of the light, is related to the wavelength, λ, by $\nu\lambda = c$, where c is the velocity of light, so that the quantum is hc/λ. If each molecule absorbs one quantum, the activation energy for one mole is Nhc/λ, where N is Avogadro's number.

$N = 6.0273 \times 10^{23}$, $h = 6.624 \times 10^{-27}$ and $c = 2.997.7 \times 10^{10}$ cm/sec.

The energy will be given in ergs.

Thus $E = 1.96 \times 10^{8}/\lambda$ ergs $mole^{-1}$, when λ is given in cm.

Again, on dividing by 4.184×10^{10}, the result comes out in calories (kcal).

Thus, $E = 2.859.1\text{-}^{10}/\lambda$ Cal/mole.

In the visible range, 1 einstein carries from 40 to 72 kcal of energy, depending on the wavelength of the light.

The observed photochemical change is the result of secondary reactions of the excited molecules of atoms of free radicals produced by the primary reaction as governed by Einstein's law.

13.2.1.8 Quantum efficiency

The interpretation of the Einstein equivalence law indicates that the "Quantum efficiency", i.e., the number of molecules reacting per quantum absorbed, would be equal to unity.

$$\text{Quantum efficiency, } \varnothing = \frac{\text{number of molecule decomposed}}{\text{number of quanta absorbed}}$$

If the sole reaction is the decomposition of this excited molecule, we should get unity for the quantum efficiency for all photochemical reactions. Actually, very few reactions exhibit this efficiency, e.g., decomposition of HI by ultraviolet light shows an efficiency of 2, whereas many others have values less than unity. The value ranges from very low to very high values. Low values are due to some kind of nonutilisation of excited molecules for the reaction and high efficiency is due to chain reaction. The high efficiency in the decomposition of HI is due to secondary reactions following the primary reaction, which consists in the splitting of the molecule into two atoms:

$$HI + h\nu \longrightarrow H + I$$

Following this, the atoms may react in a variety of ways, the principal ones being:

$$H + HI \longrightarrow H_2 + I$$
$$I + I \longrightarrow I_2$$

Thus, one quantum causes the splitting of two molecules. A quantum efficiency of less than unity is given by **inelastic collisions** of the excited molecule with normal molecules, the absorbed light energy being dissipated as heat; alternatively, some of the light may be lost by **fluorescence**, the excited molecule retaining its energy for some 10^{-7} to 10^{-8} sec and the re-emitting most of it at a longer wavelength.

13.2.1.9 Photosensitization and Photosensitizer

According to the first law of photo-chemistry only the absorbed light is photochemically reactive. But sometimes, an external substance makes a system sensitive to some region of spectrum to which the system is insensitive becuase the system does not have absorption in this region. The external substance, i.e., sensitizer, absorbs the light and makes it available to the system. Such a phenomenon is called **photosensitization** and the substance responsible for photosensitization is called **photosensitizer or a sensitizer**. Thus, ozone does not decompose in visible light, but if a trace of chlorine is present, it is rapidly decomposed by visible light which is absorbed by chlorine. The great example in biology is **photosynthesis** in which *chlorophyll acts as photosensitizer*.

13.2.1.10 Example of photochemical reaction

1. Photosynthesis in plants.
2. Bleaching of rhodopsin in the retina of the eye when exposed to light ,i.e., isomerization of II-(Cis)-retinal to all(trans)-retinal.
3. Biosynthesis of Vit D_3, i.e., Cholecalciferol from 7-dehydrocholesterol in the skin.
4. Photolysis of riboflavin (vit B_2) to lumiflaving and lumichrome.
5. In histological preparation of cornea or urinary bladder of toad, if it is treated with silver nirate and exposed of sunlight, photochemical reaction of silver halides to metallic silver is observed.

6. Photoperiodism in plants.
7. Photographic reaction.

13.2.2 Photosynthesis

13.2.2.1 Definition

It is the photochemical reaction causing production of starch in green leaves from carbon dioxide and water using sunlight as the source of energy and chlorophyll (or some other closely related pigments) for trapping energy which acts as photosensitizer. The overall equation is as follows:

$$nCO_2 + nH_2O + \text{Chlorophyll}\ (CH_2O)n + nO_2 \times h\nu\ (\text{Light energy})$$

- **Occurrence**

The capacity to carry out the process is found in wide range of organisms in both prokaryotic and eukaryotic groups.

(a) Eukaryotes : 1. Higher green plants
2. Unicellular and multicellular green, brown and red algae, euglenoids, dinoflagellates and diatoms.

(b) Prokaryotes: Blue-green algae, the green sulphur bacteria and the purple bacteria.

(Microscopic algae, diatoms and dinoflagellates together constitute phytoplankton.)

13.2.2.2 Pigments for photosynthesis

All photosynthetic cells contain one or more of the pigments known as **Chlorophylls** most of which are green. All photosynthetic cells are not green, for example, photosynthetic algae and bacteria may be green, brown, red or purple depending on the species. These colours are produced due to presence of one or more accessary pigments in the photosynthetic cells which include the yellow, red or purple **Carotenoids** and the blue or red **phycobilins.**

What we call chlorophyll is in fact a mixture of closely-related pigments of which chlorophyll is one. When acetone extract was separated by paper chromatography, five pigments were identified and showed different absorption spectra such as chlorophyll a (blue-green), chlorophyll b (yellow-green), xanthophyll (yellow), carotene (orange) and phaeophytin (grey) which is a breakdown product of chlorophyll.

13.2.2.3 Photosynthetic system

In eukaryotic cells the photosynthetic apparatus is localised in the *chloroplasts*, one of several different kinds of plastides, membrane bounded organelles, self-replicating like mitochondria contain DNA, having size 1-10 μm in diameter, globular or discoid in form. The photosynthetic pigments are present in the chloroplast and they are organized into two functional sets or assemblies, which in turn are connected with characteristic electrontransport chains. These functional units are called **Photosystems I and II.**

13.2.2.4 Facts about photosynthesis

For photosynthesis, the materials required can be shown by the different experiments and facts are given below:

1. The oxygen which is given out, comes from water and not from CO_2.
2. By using radioactive CO_2 ($^{14}CO_2$) as tracer, it has been found that incorporation of labelled carbon occurred into a large number of compounds like formaldehyde, glyceraldehyde and indicates the formation of some intermediate products during the process.
3. Water or CO_2 cannot absorb light and it is evident that chlorophyll acts as a photosensitizer.
   Chlorophyll has absorption in the red and green and both these absorption are photochemically active. 5 to 6 quantas are necessary to fix one molecule of CO_2 under favourable condition, which is very close to the limits of efficiency calculated from energy consideration.
4. Experimental study has revealed that CO_2 is not the universal electron hydrogen acceptor in all photosynthetic cells.

13.2.2.5 Leaf as photosynthetic organ and relation of respiration with photosynthesis

Carbon dioxide from the atmosphere diffuses through the stomata into the sub-stomatal air chambers and then via the intercellular spaces to the chloroplasts in the mesophyll and palisade cells. The stomata have a vital role in photosynthesis. When the stomata are open, the rate of photosynthesis may be 16-20 times as fast as the maximum rate of respiration. Under these conditions, the plant uses CO_2 from its respiration for photosynthesis but the bulk of its CO_2 is taken from the atmosphere.

If the stomata are closed, photosynthesis can still continue, using the CO_2 from respiration. In fact, an equilibrium can be reached between photosynthesis and respiration, photosynthesis using CO_2 from respiration, and respiration using oxygen from photosynthesis. The rate of photosynthesis under these conditions will be much slower than when external CO_2 from atmosphere is available. The stomata cannot remain closed for indefinite period. Open stomata are necessary in order to maintain the transpiration stream which is the only way the leaf can obtain water.

13.2.2.6 Photosynthetic process or the chemistry of photosynthesis

In the process of photosynthesis, the salient features are:

1. Energy from sunlight is trapped by chlorophyll and used in the manufacture of carbohydrate from carbon dioxide and water.
2. The oxygen evolved does not come from CO_2.
3. This photochemical reaction is temperature insensitive. But, in fact, the rate of photosynthesis is strongly influenced by temperature, a 10°C rise in temperature, approximately doubling the rate.
4. This process proceeds in two stages each consisting of many steps.

(a) **Light stage—requiring light**

i. Light induced noncyclic electron transport or noncyclic photosynthetic phosphorylation.

ii. Cyclic electron transport or cyclic photosynthetic phosphorylation.

(b) **Dark Stage:** Does not require light. Formation of carbohydrate through **Calvin Cycle**.

5. The amount of carbohydrate formed is greater if a plant is provided with alternate light and dark period than if it is given the same amount of light continuously. It is inferred that the light stage results in a product which is then fed into the dark stage.

$$X \longrightarrow Y \text{ (Light stage)}$$

$$Y \longrightarrow Z \text{ (Dark stage)}.$$

It is assumed that Y build up during a period of light cannot be used up by the dark reactions as quickly as it is formed. Intermittent light ensures the complete conversion of Y into Z, and prevents Y from accumulating in the plant.

6. In the light stage, chlorophyll traps the radiant energy of sunlight and converts into the chemical energy. This is used from splitting water molecules into hydrogen and oxygen.
7. All photosynthetic organisms except bacteria use water as electron or hydrogen donor to reduce various electron acceptors and from the water they evolve molecular oxygen.
8. The photosynthetic bacteria normally neither produce nor use molecular oxygen. In fact, they are anaerobes and are poisoned by oxygen. Instead of water these organisers use other compounds as electron donor. The green and purple sulphur bacteria use *hydrogen sulphide* (H_2S).

$$2H_2S + CO_2 \xrightarrow[\text{light}]{} (CH_2O) + H_2O + 2S$$

9. Some nonsulphur purple bacteria use an organic hydrogen donor such as *isopropanol*, which is oxidised to acetone.

$$\underset{\text{(Isopropanol)}}{2CH_3CHOHCH_3} + CO_2 \xrightarrow{\text{light}} (CH_2O) + \underset{\text{(Acetone)}}{2CH_3COCH_3} + H_2O$$

10. Most higher plants also use *nitrate* as electron acceptor, which they reduce to ammonia. In nitrogen-fixing photosynthetic organisms, molecular nitrogen as well as CO_2 may be used as electron acceptors during photosynthesis, the nitrogen is reduced to ammonia.
11. Many photosynthetic organisms can use hydrogen ions as the ultimate electron acceptor, from which they form molecular hydrogen; still others can use sulphate as electron acceptor.

Equations for photosynthesis with different electron acceptors are the following:

$$\underset{\text{Donor}}{\overset{\text{Electron}}{2H_2D}} + \underset{\text{Acceptor}}{\overset{\text{Electron}}{CO_2}} \longrightarrow (CH_2O) + H_2O + 2D$$

$$9H_2D + 2NO_3 \longrightarrow 2NH_3 + 6H_2O + 9D$$
$$3H_2D + N_2 \longrightarrow 2NH_3 + 3D$$
$$H_2D + 2H^+ \longrightarrow 2H_2 + D$$

12. The direction of net light induced electron flow in photosynthesis is toward the more electronegative system, i.e., against the normal gradient of the standard oxidation-reduction potentials. The energy for such reverse electron flow is provided by the absorbed light. Photosynthesis has two stages:

I. Light reaction stage: The light stage involving splitting of water has been ascertained by the use of isotope. Besides that, it has been shown that isolated chloroplast-as well as splitting water into hydrogen and oxygen are capable of producing ATP.

The function of light stage is twofold:

1. *Photochemical splitting of water*: It provides hydrogen atoms for the reduction of CO_2 and NADP.
2. *Production of ATP*: It provides a source of chemical energy for the subsequent synthesis of carbohydrates in the dark reaction, i.e., phosphorylation, ADP $\rightarrow$ ATP.

II. Dark reaction or stage: It involves the reduction of carbon dioxide to form carbohydrate.

III. Light stage: In eukaryotic cells photosynthesis takes place in the *thylakoids,* flattened membranous vesicles within *Chloroplasts*: they are stacked into structures called *grana.* By ultracentrifuge, smallest component of chloroplast has been isolated and it has been proved that *quantasome* is the smallest unit for photosynthesis in the light stage. Photosynthetic cells contain three types of light capturing pigments—*chlorophylls, carotenoids* and *phycobilins* arranged in two sets called photosystems I and II (Table 13.4).

Table 13.4: Comparison between Photosystem I and Photosystem II

Photosystem I	Photosystem II
1. It contains: (a) Chlorophyll a (200 molecules), (b) Chlorophyll b (50 molecules), (c) B-Carotene (50-200 molecules), (d) P-700 (Pigment 700, chlorophyll a + Protein).	1. It contains: (a) Chlorophyll a, (b) Chlorophyll b, (c) P 680 (Pigment 680, chlorophyll-protein complex).
2. Longer wavelength (710 nm) is required for reaction.	2. Shorter wavelength (670 nm) is required for reaction.
3. Not responsible for O_2 evolution.	3. Responsible for O_2 evolution.
4. P-700 is the "exciton" agent, i.e., a trap for collecting *excitons*, quanta of *excitation* energy from the other chlorophyll molecules. When an exciton is trapped, a high-energy electron is lost from it.	4. P 680 – *exciton factor.*
5. The ratio of Chlorophylls a and b is higher.	5. Lower.
6. Large amounts of chlorophyll a-protein complexes absorbing at long wavelength are present.	6. Absent.
7. Red drop–Above 680 nm, the red region of the spectrum, the efficiency of monochromatic light in supporting oxygen evolution decreases greatly in relation to the absorption spectrum, which is largely due to chlorophyll. This deficit is called red drop. It is absent in this system.	7. Red drop – present.

Contd.

Photosystems I and II are the energy delivering components in a continous electron-transport chain extending from water, the electron donor, to $NADP^+$, the electron acceptor. It has similarity with electron-chain of mitochondria having some differences:

1. Photosynthetic chain from water to $NADP^+$ is opposite in direction to mitochondrial electron transport chain, contains a large number of electron carriers which is also true for mitochondrial electron transports but the main difference is the presence of the two pigment systems, which act as electron boosters utilizing light energy to push electrons via a series of carriers from H_2O to $NADP^+$. The photosynthetic electron-transport has three functional units:

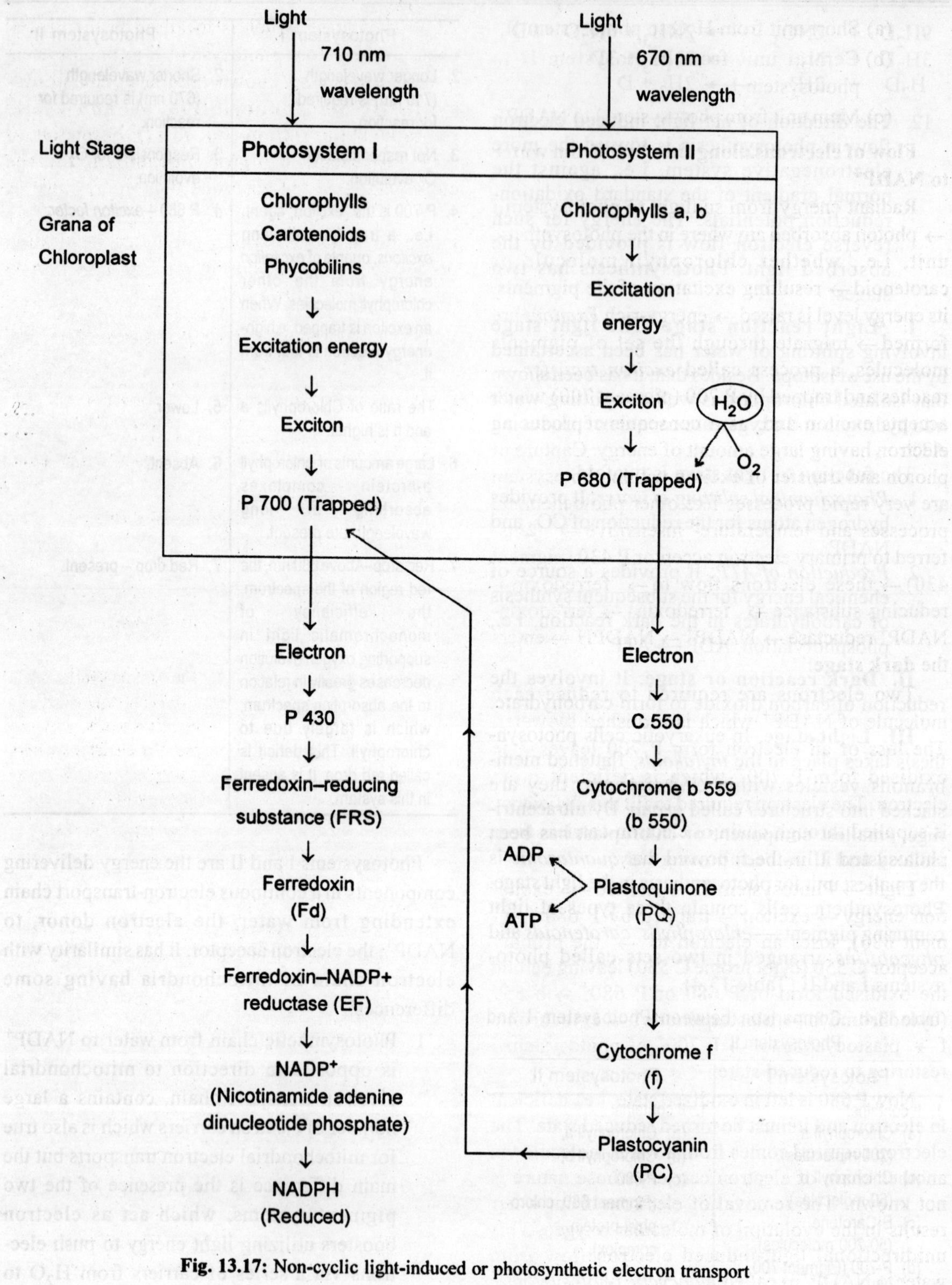

Fig. 13.17: Non-cyclic light-induced or photosynthetic electron transport

(a) Short unit from H_2O to photosystem II,

(b) Central unit from photosystem II to photosystem I,

(c) Main unit from photosystem I to $NADP^+$.

Flow of electrons along the chain from water to $NADP^+$

Radiant energy from sunlight to photosystem I → photon absorbed anywhere in the photosynthetic unit, i.e., whether chlorophyll molecule or carotenoid → resulting excitation of the pigments, its energy level is raised → energy-rich *Excitons* are formed → migrate through the set of pigments molecules, a process called *exciton transfer* → reaches and trapped by P 700 (pigment 700) which accepts exciton and, as a consequence, loses an electron having large amount of energy. Capture of photon and transfer of exciton within photosystem are very rapid processes like other photochemical processes and temperature- insensitive → transferred to primary electron acceptor P 430 (pigment 430) → these electrons flow to → ferredoxin – reducing substance → ferredoxin → ferredoxin-$NADP^+$ reductase → $NADP^+$ → NADPH → enters the **dark stage**.

Two electrons are required to reduce each molecule of $NADP^+$ which is furnished by water. The loss of an electron form P 700 leaves it in oxidised form P 700^+ which is deficient in an electron. The electron required to fill this deficiency is supplied through chain of electron carriers from photosystem II in the following way:

Light → Photosystem II → excited → excitation energy → exciton → trapped by P 680 (pigment 680), loses an electron to → the primary acceptor C 550 (cytochrome C 550) leaving behind the oxidised form of P 680 or P 680^+ → b 559 (cytochrome b → plastoquinone) → cytochrome f → plastocyanin to → P 700^+ of photosystem I restoring to reduced state.

Now P 680 is left in oxidized state, i.e., deficient in electron and it must be turned reduced state. The electron required comes from a water molecule via another chain of electron carriers whose nature is not known. The removal of electrons from water results in the evolution of molecular oxygen. This unidirectional light-induced electron flow from water to $NADP^+$ is called, *noncyclic photosynthetic electron transport*. During this electron transport from water, formation of one molecule of ATP from ADP takes place which is localised in the central chain leading from photosystem II to photosystem I; hence it is also known as **non-cyclic photosynthetic phosphorylation** or **non-cyclic photophosphorylation** (Fig 13.17). Formation of ATP represents a major mechanism for the conservation of the absorbed light energy just as ATP is the major process, i.e., oxidative phosphorylation by which energy of respiration process is conserved during electron transport in mitochondria (Table 13.5).

Table 13.5: Comparison between Photophosphorylation and Oxidative Phosphorylation

Photophosphorylation	Oxidative Phosphorylation
1. **Occurrence** During Photosynthesis.	1. During Cellular respira-tion.
2. **Illumination** Presence of light is necessary.	2. No light is required.
3. **Source of Energy** Radiant energy from sunlight via chlorophyll.	3. From oxidation of orga-nic compounds.
4. **Energy level** Energy level is raised due to stimulation of chlorophyll by light.	4. The electron carriers are at different energy levels and as the electron is transferred from one carrier to the next, energy is removed from it for synthesis of ATP.
5. **Production of ATP** Two molecules of ATP (one from non-cyclic and one from cyclic) are formed. At least one molecule of ATP is generated per pair of electrons passing form H_2O to $NADP^+$.	5. Three molecules of ATP.
6. **Electron donor and acceptor** Chlorophyll functions as the donor and acceptor of electrons.	6. The chain of respiratory enzymes act as donor and acceptor of electron.
7. **Uncoupling or inhibition** It may be uncoupled or inhibited by specific agents.	7. It may be inhibited by uncoupling agents without affecting electron transport (separate agents inhibit both).
8. **Coupling factor** Coupling factor isolated having ATPase activity,	8. Electrochemical gradient of H^+ ions is the coupling agent.

Contd.

Photophosphorylation	Oxidative Phosphyorylation
reconstituted in depleted chloroplast by adding chloroplast ATPase.	
9. **H⁺ absorption** Chloroplast absorbs H⁺ during illumination.	9. H⁺ ion is absorbed by NAD enzyme and no illumination is required.
10. **Mechanism** Chemiosmotic coupling may occur.	10. Chemical or conformational or Chemiosmotic coupling may occur.
11. **Reversibility** Irreversible process.	11. Appears to be reversible process. ATP can stimulate reverse electron flow.
12. **Site** Quantosome in the chloroplast.	12. Mitochondria in the Cell.

13.2.2.7 Splitting of water

Though we do not know the details but the main idea is as follows: Water dissociates spontaneously into hydrogen (H^+) and hydroxyl (OH^-) ions. When light stimulates chlorophyll, an electron is produced. All the electrons produced do not return to chlorophyll via electron carrier, so an electron may combine with a hydrogen ion to form hydrogen atom. Hydrogen atoms formed are taken up by a hydrogen acceptor $NADP^+$ to form NADPH (reduced). The NADPH enters the dark stage, hand over the hydrogen which is used for reduction of CO_2 and NADPH becomes $NADP^+$.

During this period, the chlorophyll molecule has been deficient of an electron and is in an unstable state. To restore its stability, the hydroxyl ion is required. The hydroxyl ion donates an electron to chlorophyll, and the OH resulting from this produces water and oxygen. The electron derived from the hydroxyl ion is transported to the chlorophyll via an electron carrier system as in cyclic photophosphorylation.

$$OH^- + \text{Chlorophyll} \rightarrow \text{Chlorophyll} + OH$$

$$4(OH) \rightarrow 2H_2O + O_2.$$

In noncyclic electron transport, apparently 8 light quanta, 4 absorbed by photosystems 1 and 4 by photosystem II, are required one molecule of oxygen and to reduce the molecule of CO_2, for which two molecules NADPH and three molecules of ATP are required.

13.2.2.8 Cyclic photosynthetic phosphorylation

Cyclic photoinduced electron flow or cyclic photosynthetic phosphorylation occur in photosystem I alone, through a shunt mechanism or bypass when $NADP^+$ is not available to act as electron acceptor and in which no net reduction of an electron acceptor takes place. In this process chlorophyll functions as the donor and ultimate acceptor of electrons. When light stimulates the chlorophyll molecule, the energy level of an electron is raised. In this excited state, the electron is emitted and transferred to an electron acceptor.

Then the electron is transported through a series of electron carriers, the last one sends it back to the chlorophyll molecule which was positively charged and unstable due to loss of electron, becomes neutral and stable when the electron is returned as unchanged and energy being released as heat and flourescence. When the electron is transported from one carrier which includes cytochromes and then to the next, energy is removed from it for the synthesis of ATP. This process is known as cyclic photosynthetic phosphorylation (Fig. 13.18) which differs from non-cyclic photosynthetic phosphorylation (Table 13.6).

Table 13.6: Comparison of cyclic photosynthetic phosphorylation and non-cyclic photosynthetic phosphorylation

Cyclic Photosynthetic Phosphorylation	Non-cyclic Photosynthetic Phosphorylation
1. Performed by Photosy-tem I only.	1. Performed jointly by photosystems I and II.
2. It is a shunt or bypass mechanism when $NADP^+$ is not available to act as an electron acceptor.	2. Regular and usual pro-cess.
3. No evolution of O_2.	3. Evolution of O_2.
4. Help from H_2O is not re-quired.	4. Help from H_2O is requi-red.
5. Presence of $NADP^+$ is not required.	5. $NADP^+$ is essential.
6. No net reduction of electron acceptor.	6. Net reduction of electron or electron acceptor.
7. Electron does not go to $NADP^+$.	7. Electron goes to $NADP^+$.

Contd.

Cyclic Photosynthetic Phosphorylation	Non-cyclic Photosynthetic Phosphorylation
8. Electron returns to Chlorophyll, unchanged to make it neutral and stable.	8. Electron comes from H_2O via photosystem II to P 700 to restore electron deficiency.
9. During transportation of electron, energy release produces ATP from ADP + Pi.	9. During Electron transport, ATP is formed.

II. Dark Stage

1. The light reaction of photosynthesis is primarily responsible for converting light energy into chemical energy in the form of ATP and NADPH whereas the dark stage involves the utilization of the chemical of ATP and NADPH to bring about the reduction of carbon dioxide to hexose and other products.
2. The dark stage does not mean it occurs in the dark or in the night but it makes place together with the light stage.

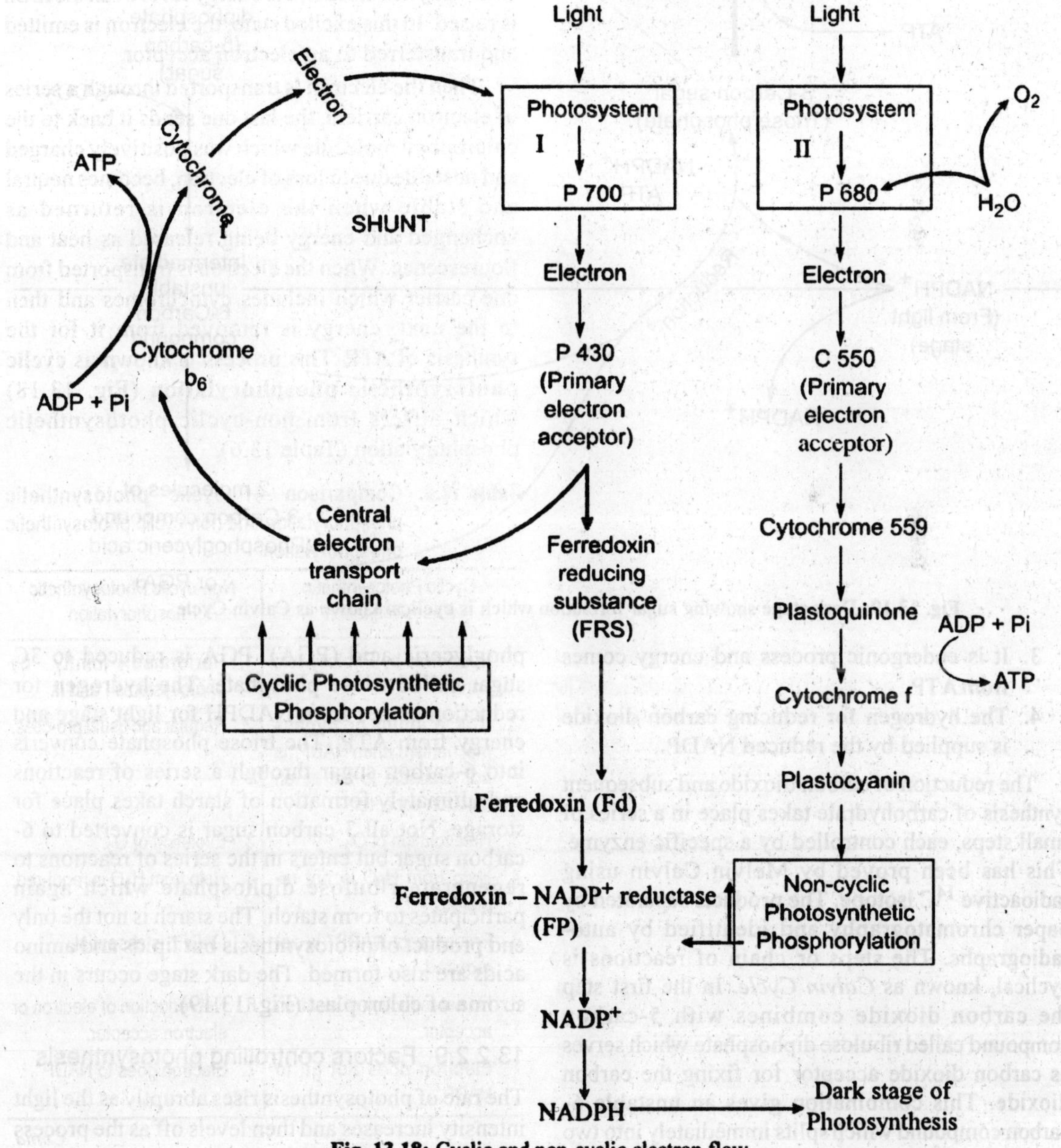

Fig. 13.18: Cyclic and non-cyclic electron flow

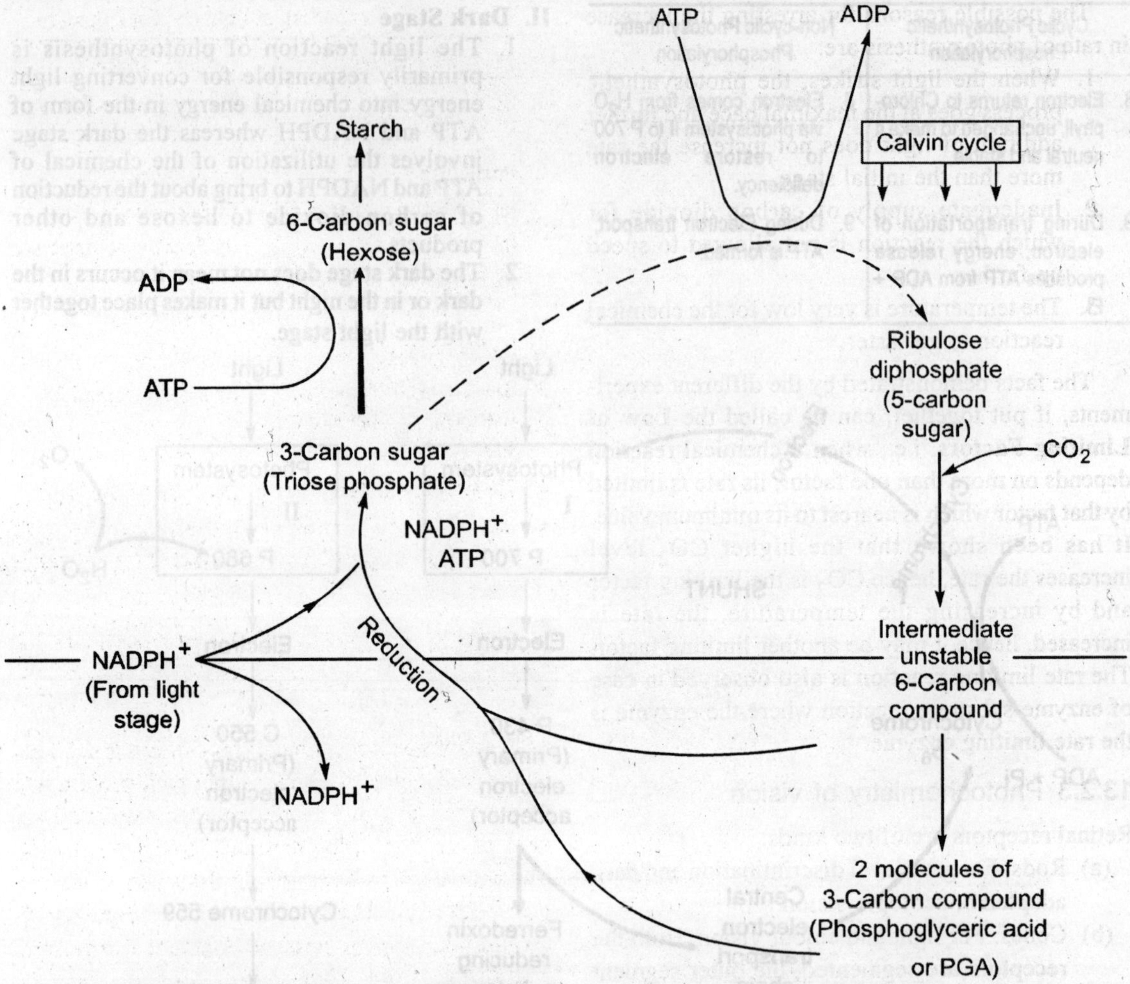

Fig. 13.19: Dark stage showing sugar formation which is cyclical known as Calvin Cycle

3. It is endergonic process and energy comes from ATP.
4. The hydrogen for reducing carbon dioxide is supplied by the reduced NADP.

The reduction of carbon dioxide and subsequent synthesis of carbohydrate takes place in a series of small steps, each controlled by a specific enzyme. This has been proved by Melvin Calvin using radioactive ^{14}C isotope. The products separated by paper chromatography and identified by autoradiographs. The steps or chain of reactions is cyclical, known as *Calvin Cycle*. In the first step the carbon dioxide combines with 5-carbon compound called ribulose diphosphate which serves as carbon dioxide acceptor for fixing the carbon dioxide. This combination gives an unstable 6-carbon compound which splits immediately into two molecules of 3-carbon compound called Phosphoglyceric acid (PGA). PGA is reduced to 3C sugar called triose phosphate. The hydrogen for reduction comes from NADPH for light stage and energy from ATP. The triose phosphate converts into 6-carbon sugar through a series of reactions and ultimately formation of starch takes place for storage. Not all 3-carbon sugar is converted to 6-carbon sugar but enters in the series of reactions to regenerate ribulose diphosphate which again participates to form starch. The starch is not the only end product of photosynthesis but lipids and amino acids are also formed. The dark stage occurs in the stroma of chloroplast (Fig. 13.19).

13.2.2.9 Factors controlling photosynthesis

The rate of photosynthesis rises abruptly as the light intensity increases and then levels off as the process reaches its maximum rate.

The possible reasons for arresting the increase in rate of photosynthesis are:

1. When the light strikes, the photosynthetic process goes at the maximal pace and further addition of light does not increase the rate more than the initial stage.
2. Inadequate supply of carbon dioxide for which the reaction is not allowed to speed up further.
3. The temperature is very low for the chemical reaction to go faster.

The facts demonstrated by the different experiments, if put together, can be called the **Law of Limiting Factors**, i.e., when a chemical reaction depends on more than one factor, its rate is limited by that factor which is nearest to its minimum value. It has been shown that the higher CO_2 level increases the rate, hence CO_2 is the limiting factor and by increasing the temperature, the rate is increased, hence it may be another limiting factor. The rate limiting reaction is also observed in case of enzyme-substrate reaction where the enzyme is the rate limiting enzyme.

13.2.3 Photochemistry of vision

Retinal receptors are of two kinds:

(a) Rods: For grey level discrimination and dark adapted or scotopic vision.

(b) Cones: For light and colour vision. Both the receptors are segmented, the outer segment being a modified cilium which contains many photoreceptive discs. They contain photochemicals which decompose on exposures to light and in the process excite the nerve fibres leading from the retina. Thus, rods and cones are *photoelectric transducers*.

13.2.3.1 Photochemistry of rod vision or the rhodopsin-retinal visual cycle and excitation of rods

A typical rod has a long, slender outer segment which is specialized for photoreceptor and projects into the pigment layer of the retina having cross marks under E/M which represents "discs" or "shelves" formed in foldings of the cell membrane. These are double membrane discs or membrane vesicles are arranged in parallel to the light surface of the retina.

Table 13.7: Comparison of Light and Dark Stages

Light Stage	Dark Stage
1. Light is necessary.	1. Occurs in absence of light along with light stage.
2. **Phenomena** (a) Splitting of H_2O. (b) Evolution of O_2. (c) Reduction of $NADP^+$ $\rightarrow$ NADPH. (d) Formation of ATP.	2. (a) Fixation of CO_2. (b) Formation of sugar. (c) Formation of lipids and amino acids. (d) Formation of ATP.
3. Site–Grana of Chloro-plast, specifically the smallest unit is quanto-some.	3. Stroma of Chloroplast.
4. Calvin Cycle – absent.	4. Calvin Cycle–present.
5. Hydrogen acceptor $NADP^+$, reduced to NADPH $\rightarrow$ enters dark stage.	5. NADPH from light stage reduces CO_2 and $NADP^+$ is formed.
6. Energy from sunlight.	6. Energy from ATP and it is endergonic reaction.

The photoreceptor membrane of outer segment of the rods contains about 60 per cent protein and 40 per cent lipid and rhodopsin makes up to 80 per cent of its total composition. Lipids are mainly phospholipids, e.g., phosphatydyl choline, phosphatydyl ethanolamine and phosphatydyl serine and 5-10 per cent glycolipids are also present. The discs of the outer segments of the rods are floating freely as intracellular organelles like mitochondria. The disc membrane occupies some space separated from the intracellular space of the photo-receptor where these discs are suspended. The disc membrane is impermeable to Na^+, K^+, Ca^{2+}, Mg^{2+}, Cl^- and PO_4^{3-}. Experimental study has shown that the isolated discs undergo swelling and contract in response to changes in osmotic pressure of the medium. But during exposure to light, permeability of the membrane increases. The conformational change of rhodopsin alters the state of membrane and increases passive permeability to K^+ and Na^+ ions.

Experimental studies with X-ray Structural Analysis and Electron Microscopy have shown that exposure to light causes structural changes of the disc membrane which is proportional to the time exposure. The main change is the **translocation** of rhodopsin. Exposure to light causes transference of rhodopsin from the interdisc hydrophilic surface into the internal hydrophobic phase of the membrane. These translocations are equivalent to phase transitions and these phase transitions have been detected by Differential Colorimetry. Direct evidence of the conformational changes of the membrane on exposure to light was available by Electron Paramagnetic Resonance (EPR) using a spin probe inserted into the photoreceptor membrane. When the suspension of the membrane was exposed to light, there were changes in EPR spectra. It showed that photolysis of the pigment is coupled with the change of the conformational state of the membrane and with an increase in microviscosity of its hydrophobic regions.

Gamma-Resonance Spectroscopy (the Mossbauer effect) using photoreceptor membrane has revealed the similar results. Iron ascorbate labelled with isotope Fe^{57} was used and source of gamma-radiation was Co^{57}.

It has been proved that the rods adapted to the dark exhibit dichroism. Hence, chromophore of rhodopsin is oriented in parallel to the plane of the disc membrane. But chromophores are oriented at random, in this plane and no dichroism occurs. Usually no dischroism occurs which can be accounted for by the Brownian rotational motion of rhodopsin molecules in the retina.

Receptors contain hundreds to thousands of discs, with a disc surface of hundreds to thousands of square microns. The disc are stacked so that their broad surfaces present normal to incoming light having half-life of only 10 days. They undergo rapid turnover, being replaced at a rate of about 30 discs per day from specialized plasma membrane, but remain structurally and functionally connected with the plasma membrane. They are connected to the cytoplasmic surface by narrow tube-like processes. The photosensitive pigments called "Rhodopsin" or "Visual purple" are incorporated into these membrane vesicles or discs as a part of their structure. They are concentrated at each surface both inside and outside of the membrane and concentration is about 40 per cent. The inner segments of the rods continually synthesize new rhodopsin. This in turn migrates to the base of the outer segments where the cell membrane is continually forming new folds becomes the discs of the rod and to which the rhodopsin attaches. The sugar units of rhodopsin are located inside the discs, whereas they face extracellular space while still on the plasma membrane. These discs or membrane vesicles serve as receptors and increase surface area for incorporation of rhodopsin, the biochemical machinery for both transduction and light adaptation (Fig. 13.20).

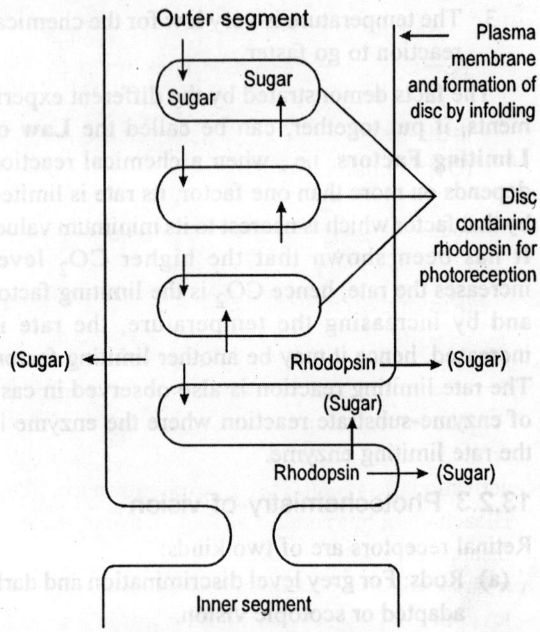

Fig. 13.20: Rod showing disc

13.2.3.2 Mechanism of reaction

Rhodopsin is a globular molecule and a combination of the protein called "Scotopsin" (lipoproteins) and the carotenoid pigment called "retinal" or "retinene" (chromophore group) (Fig. 13.20). The membranes contain an ordinary lipid bilayer. The rhodopsin molecules reside on the inner hydrophilic surface of the membrane. The 11-Cis-retinal only can combine with scotopsin to form rhodopsin. Due to exposure to light, rhodopsin begins to decompose which is shown below (Fig. 13.21). This photo-activation of electrons in the rhodopsin leads to isomeric change of the "Cis" to "all-trans" form which has the same chemical structure but has a different physical structure—a straight molecule rather than a curved molecule. Because the three dimensional orientation of the reactive sites of the "all-trans" retinal no longer fits with that of the reactive sites on the protein scotopsin, it begins to pull away from the scotopsin. Bleaching occurs between Metarhodopsin I to Metarhodopsin II.

At the stage Schiff base linkage is deprotonated. The retinal forms a Schiff base as a result of the interaction of the aldehyde group of retinal with the amino group of opsin, namely with the amino group of lysine (fig. 13.21).

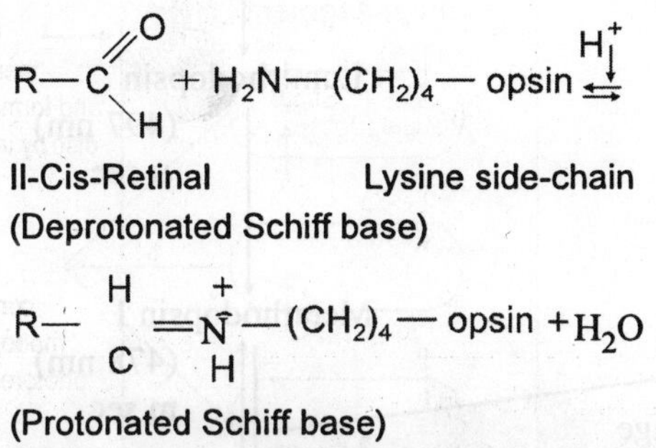

Fig. 13.21: Showing Schiff base linkage formation and deprotonation

The immediate product is prelumirhodopsin (543 nm). The main action of light appears to be a 10-fold activation of a cyclic GMP phosphodiesterase (DPE). Bleaching of one rhodopsin molecule can lead to the disappearance of 5×10^4 molecules of cyclic GMP, a large amplification of the light signal. The dark level of GMP concentration is restored in less than 1 min after light is extinguished.

13.2.3.3 Regeneration of Rhodopsin

Regeneration takes place by reconversion of all-trans-retinal into 11-cis-retinal catalyzed by retinal isomerase. This requires matabolic energy from rods and cones. Once the 11-cis-retinal is formed, it automatically recombines with scotospin to form rhodopsin, an exergonic process. The product is stable compound until its decomposition by light (Fig. 13.22).

13.2.3.4 Excitation of the rods during rhodopsin decomposition

Mechanism of excitation of the rods by rhodopsin is not known but it is believed to occur in the following way:

The retinal molecule of rhodopsin is photo-sensitive and activated by light. Light causes photoexcitation of electrons to higher energy states. In this period, conversion of II-cis-retinal in physical structure (not chemical) within nano-seconds or microseconds. And at the same time rhodopsin molecule undergoes conformational changes, from an angulated molecular structure of "cis" form to straight molecular structure of the "trans" form. But how photoexcitation and decomposition of rhodopsin excite the rods? It occurs at the very instant when rhodopsin becomes excited by light causing a sudden decrease in electrical conductance of the rod membrane which leads to a change in the membrane potential. Decrease in membrane conductance may take place in either of the following ways:

1. Excited electrons themselves directly change the membrane conductance,
2. Excited electrons cause a change in a second chemical substance which in turn affects membrane conductance,
3. Rhodopsin molecule itself undergoes an instantaneous conformational change that alters the actual conductance. It is based on the fact that the rhodopsin molecule is the physical part of the membrane structure, i.e., membrane folds or vesicles of shelf-like discs.

13.2.3.5 Generation of the rod receptor potential

The function of rhodopsin is like a photogenerator which produces photocurrent. Rhodopsins are proton pumps and its operation is explained by Mitchell theory. Proton pumps operation is coupled with the phosphorylation of ATP and with the operation of the sodium pump.

In unexcited condition, i.e., in absence of light, the inner segment is pumping out through Na^+ pump creating a negative potential for the entire cell. But the outer segment is leaky, so Na^+ leakes inside, neutralizing much of the negativity on the inside of the entire cell and there is reduced amount of electronegativity inside membrane of the rod. It is presumed that Na^+ permeability is maintained in the dark by GMP-mediated protein phosphorylation. When light strikes rhodopsin in a disc membrane, i.e., in excited condition, there is transient decrease of sodium permeability of the rod plasma mem-brane, increased of further negativity which is

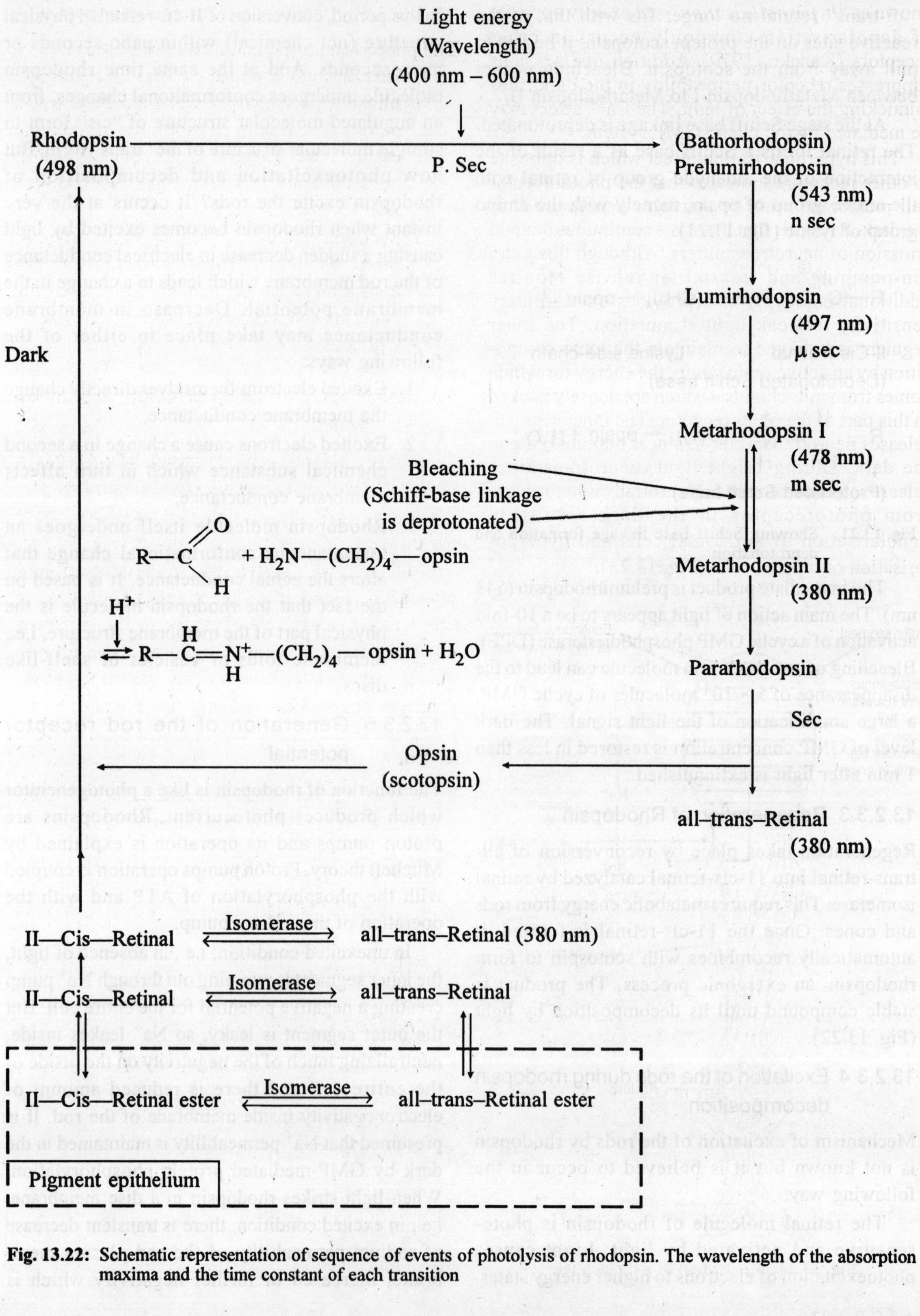

Fig. 13.22: Schematic representation of sequence of events of photolysis of rhodopsin. The wavelength of the absorption maxima and the time constant of each transition

known as "*hyperpolarisation*" and exactly opposite of depolarisation commonly occurs in other receptors (Kaneko, 1979). Addition of Ca^{2+} ions suppresses receptor permeability in the dark, thus mimicking the effect of light. The Ca^{2+} effect may be mediated by the cyclic GMP system.

This hyperpolarisation effect which is opposite to other receptors indicates that the photoreceptor cell membrane must be actively "*depolarised*" in the dark. It reveals that there is a continuous quantal emission of neurotransmitters. Although this extra ion-pumping and transmitter release requires additional energy, it also provides extraordinary sensitivity to weak light stimulation. The inner segment is believed to maintain the ionic composition by an active ionic pump, the energy for which comes from mitochondria which are densely packed in this part of the photoreceptor. The inner segment releases neurotransmitter at a near maximal rate in the dark. During bright light, neurotransmitter release is almost stopped. Neurotransmitter release from photoreceptor in the dark results in depolarisation of horizontal cells and hyperpolarisation of bipolar cells (Fig. 13.23).

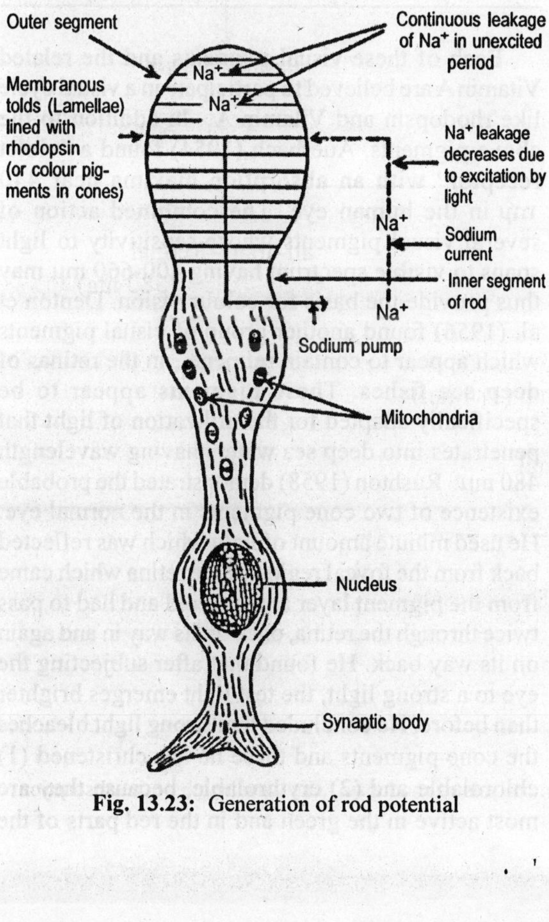

Fig. 13.23: Generation of rod potential

13.2.3.6 Characteristic of rod receptor potential

The photoreceptor potential has two phases:

1. **Instantaneous phase or very low voltage phase**: In instantaneous phase the electrical potential is directly proportional to the light energy. The potential is the result from the immediate interaction of light with rhodopsin.
2. **Secondary phase**: Having greater voltage, the potential of which is proportional to the logarithm of the light energy. It is the result from the sodium current mechanism described above.

13.2.3.7 Photochemistry of colour vision by cones

The cones contain the pigment having same composition like rhodopsin except the protein part called photopsin (i.e., retinal + photopsin). There are different types of photochemicals (Fig. 13.24), e.g., iodopsin, porphyrinopsin (porphyropsin) and cyanopsin present in different cones, thus making these cones selectively sensitive to the different colours—blue, green and red and the photochemicals are called blue-sensitive pigment, green-sensitive pigment and red-sensitive pigment.

Photopigment	Absorption maxima (nm)
1. Rhodopsin $\xrightleftharpoons{\text{Light}}$ Rod opsin +	
Retinal$_1$ $\xrightleftharpoons[\text{DPNH}]{\text{DPN}^+}$ Retinol$_1$	500
2. Iodopsin $\xrightleftharpoons{\text{Light}}$ cone opsin +	
Retinal$_1$ $\xrightleftharpoons[\text{DPNH}]{\text{DPN}^+}$ Retinol$_1$	562
(Alcohol dehydrogenase)	
3. Porphyrinopsin	
(Porphyropsin) $\xrightleftharpoons{\text{Light}}$ Rod opsin +	
Retinal$_2$ $\xrightleftharpoons[\text{DPNH}]{\text{DPN}^+}$ Retinol$_2$	522
4. Cyanopsin $\xrightleftharpoons{\text{Light}}$ cone opsin +	
Retinal$_2$ $\xrightleftharpoons[\text{DPNH}]{\text{DPN}^+}$ Retinol$_2$	620
(Alcohol dehydrogenase)	

Fig. 13.24: Showing different photopigments, reactions and absorption maxima

13.2.3.8 Other photovisual pigments

I. Prophyropsin

Rhodopsin is not the only visual pigment. Wald found that fresh water fishes, lamp rays and certain larval and adult amphibia contained another pigment absorbed in maximally at 523 mμ and is known as *porphyropsin*. It is bleached in the similar manner as that of rhodopsin on illumination. But retinene and Vitamin A differ from those found in mammals. He also observed that photosensitivity of those animals shifted towards the long-wavelength of the spectrum. This pigment produces retinene$_2$ differing from retenence$_1$ by the presence of an extra double bond in the ionone ring and produce Vitamin A$_2$ on further bleaching. The significance of this variation in structure is not yet fully understood but probably it is connected to ionic content of the animals environment. It is also pointed out that it has some advantages to fishes living in deeper water of the sea to have a pigment that is more sensitive to the blue rays which is the only ray penetrating through any depth of water.

In other fishes, those migrating from saline to freshwater (Euryhaline fishes), like eel and salmon, Wald found that there was usually a mixture of two systems, i.e., porphyropsin system and rhodopsin system as identified by the presence of both retinenes and Vit. A. Datnall observed that there is a wide range of scotopic visual pigments of eyes of different fishes, each responding to a particular wavelength of light.

II. Iodopsin

Wald extracted from the chicken retina a violet pigment, which had an absorption maximum at 560 mμ. Subsequent studies showed it to a chromoprotein containing Retinene$_1$ linked to a protein which Wald called "photopsin", to distinguish it from "scotopsin". It seems it undergoes the same series of reactions as that undergone by rhodopsin, in the sense that neoretinene-b is isomerized during bleaching to the all-trans form, which forms Vitamin A (Hubbard et al.).

III. Cyanopsin

Retinene$_2$ combined with "opsin", a new pigment from chick retina gives cyanopsin with its maximum at about 600 mμ.

In addition to rhodopsin which is found in the retinal rods of frogs, birds and mammals, other related visual pigments have also been found in these and other species. It has also been observed that these pigments are conjugated proteins that differ with respect of their carotenoid retinene$_1$ or retinene$_2$ (Vitamin A$_2$ aldehyde) or their protein, rod "opsin" or cone "opsin". Thus rhodopsin and porphyropsin are associated with the retinal of rods and responsible for vision at low illumination. Iodopsin and cyanopsin are associated with the retinal of cones and responsible for colour vision (Table 13.8).

Table 13.8: Visual pigments of vertebrates

Source	Visual pigment		Carotenoid	Protein
	Name	Absorption maximum		
Land & marine animals	Rhodopsin	500 mμ	Retinene$_1$	Rod opsin
Land animals	Iodopsin	562 mμ	Retinene$_1$	Cone opsin
Fresh water animals	Porphyropsin	522 mμ	Retinene$_2$	Rod opsin
Fresh-water fish	Cyanopsin	620 mμ	Retinene$_2$	Cone opsin

Each of these visual pigments and the related Vitamin A are believed to participate in a visual cycle like rhodopsin and Vitamin A$_1$. In addition to the above pigments, Auerbach (1954) found a "violet receptor" with an absorption maxima near 440 mμ in the human eye. The combined action of several visual pigments whose sensitivity to light spans to visible spectrum having 400-660 mμ may thus provide the basis for colour vision. Denton et al. (1956) found another group of visual pigments which appear to contain retinene$_1$, in the retinas of deep sea fishes. These pigments appear to be specifically adapted for the utilization of light that penetrates into deep sea waters having wavelength 480 mμ. Rushton (1958) demonstrated the probable existence of two cone pigments in the normal eye. He used minute amount of light which was reflected back from the foveal region of the retina which came from the pigment layer and choroid and had to pass twice through the retina, once on its way in and again on its way back. He found that after subjecting the eye to a strong light, the test light emerges brighter than before. He concluded that strong light bleaches the cone pigments and these he has christened (1) chlorolable and (2) erythrolable, because they are most active in the green and in the red parts of the spectrum, respectively. He further studied the pigments in the fovea of the normal eye analysed by bleaching with coloured lights. In the particular form of red-green colour blindness known as protanopia, only one pigment chlorolabe appears to be present in the fovea. When the fovea was tested with red-green light, it had shown different spectrum indicated by the curve called "protanope curve" or "chlorolabe curve" or "P-curve". But fovea was tested by bleaching with deep red light and with white light it had shown different curve known as "D-curve" or "denteronope curve" which indicates that the cones may contain a mixture of erythrolabe and chlorolabe, so the sensitivity curve is likely to depend upon both these pigments.

13.2.4 Transmitters in vision

Mechanism of Transmission of Signal from Photolyzed Rhodopsin to the Plasma Membrane of Rod by Transmitter

The photoreceptor cell in the retina converts **light into atomic motion** and then into **nerve impulse**. The electrical signals generated by the photoreceptors are processed by nerve cells within the retina by the optic nerve for visual analysis. Thus, the retina serves two functions:

1. To transform light into nerve impulses and
2. To integrate visual information.

The outer segment of rod is photosensitive and Hecht (1938) discovered that human rod cell can be excited by a single photon and a photon coverts into atomic motion. Light somehow blocks the Na^+ channels in the plasma membrane of the outer segment of the rod. Consequently, the inward flow of Na^+ decreases, and the membrane becomes more negative inside. In other words, the membrane potential of an illuminated rod shifts towards the K^+ equilibrium potential causing hyperpolarization transiently. This light-induced hyperpolarization in the vicinity of the illuminated discs is then passively transmitted by the plasma membrane to the synaptic body.

This change in permeability of the plasma membrane to Na^+ and consequent hyperpolarization are "highly amplified responses" of the outer segment of the rod. A dark adapted rod by absorption of a "single" photon blocks and flow of more than a "million" sodium ions. This is a remarkable amplification but the actual mechanism is not clear because:

1. The disc membranes containing photochemical pigment rhodopsin are not continuous with the plasma membrane of rod;
2. These disc membranes are not coupled electrically;
3. A rhodopsin molecule which absorbs a photon may be far away (several thousand angstroms) from sodium channels of the plasma membrane of the rod, so a direct interaction between these sites is not expected. Therefore, the signal must be carried from photolyzed rhodopsin in the disc membranes to the plasma membrane by diffusible transmitters. For this work, a large number of transitters must be formed or destroyed by the action of single photolyzed rhodopsin for production of high degree of amplification.

But the transmitter has not yet been identified but investigations have focussed on two substances i.e., calcium ion and cyclic GMP.

(A) Evidences in favour of Ca^{2+}

1. When Ca^{2+} level in cytosol is raised, sodium channels in the plasma membrane are closed and they are opened when the Ca^{2+} level is lowered.
2. Chelating agents which specifically bind Ca^{2+} lower the sensitivity of the rod to light when they are introduced into the cytosol. This desensitization indicates that photolysis of a single rhodopsin induces the release of several hundred Ca^{2+} into cytosol.
3. When the outer segment of the rod is illuminated following a light, there is extrusion of more $Ca_2{}^+$ from the outer segment.

(B) Evidences in favour of cyclic GMP

1. When the level of cyclic GMP in the cytosol is raised, sodium channels in the plasma membrane are opened and they are closed when level of cyclic GMP is lowered.
2. Light regulates the level of cyclic GMP. Light activates a phosphodiesterase that hydrolyzes cyclic GMP which is shown below (Fig. 13.25). The free energy for driving the amplification cycles comes from the hydrolysis of GTP and~P is used to amplify a signal.

3. Photolysis of a single rhodopsin molecule causes rapid hydrolysis of more than 10^5 molecules of cyclic GMP.

- **Neurotransmitter in the retinal synapses:**

The specific neurotransmitters that mediate impulse conduction at retinal synapses are not known with complete certainty. There is some evidence that acetylcholine is mediator at some junctions in the retina. This structure also contains fairly high concentrations of dopamine, 5-hydroxytryptamine, τ-amino butyric acid (GABA) and substance P (a polypeptide).

13.2.4.1 Bacteriorhodopsin

Halobacteria (Halobacterium halobium) are external halophiles which grow in saline lakes and have a positive phototaxis to visible light and a negative phototoxis to ultraviolet hight. A certain region of the membrane surface of red halobacteria contains the protein called "bacteriorhodopsin".

Light ↓

Rhodopsin → Photolyzed rhodopsin

GDP-Transducin → GTP-Transducin

(Regulatory protein-stimulation of photolyzed rhodopsin mediated by this)

Inactive phosphodiesterase

GTP – Transducin – Active phosphodiesterase

Cyclic GMP + H_2O → GMP + H^+

(10^5 cyclic GMP are hydrolyzed per sec per photolyzed rhodopsin)

(A large amplification of the light signal)

↓

Na^+ channel closes

Na^+ channel opens

Fig. 13.25: Cyclic nucleotide cascade in vision

This protein has a molecular mass 26000 having one protein molecule attached to a "retinal" molecule by an aldimine bond. This photopigment has absorption maxima at 570 nm.

- **Characteristics**

1. The purple membrane of bacteria is crystalline rather than liquid in case of rods and cones membranes.
2. The bacteriorhodopsin molecules are present in clusters, 3 molecules in one cluster and packed into a hexagonal lattice which remains in a compact lipid matrix.
3. In the middle of the cluster there is a channel 1-2 nm in diameter.
4. The photopigment molecule is 70-80 per cent α-helical.
5. Each molecule contains 7 rod-like α-helical, 3.5-4 nm long which are oriented perpendicular to the membrane plane.
6. The size of the protein $2.5 \times 3.5 \times 4.5$ nm^3.
7. The photopigments runs through the entire membrane and comes into contact with aqueous phases on both sides of the membrane.
8. This pigment molecule is asymmetrical.

- **Reactions**: The photon induced proton transport causes changes in the conformation of the photopigment and its spectrum. Bacteriorhodopsin in light-adapted form is known as BR_{570} which enters the cycle of photoinduced reactions, forming several intermediate compounds (e.g., K_{630}, L_{550}, M_{412}, N_{520}, O_{640}, etc.) and accompanied by the transport of H^+ across the membrane and regeneration of BR_{570} takes place. One revolution of cycle, absorbing one quantum light, produces proton passing through the membrane. In certain phase of the bacteriorhodopsin (BR) cycle, BR releases a proton into the acidic (pH = 3) medium on one side of the membrane and in another phase, BR absorbs the proton from the alkaline (pH = 12) medium on the other sides of the membrane.

- **Function**: Bacteriorhodopsin with lipids makes a system which provides the transport of protons across the membrane against the pH gradient. BR functions as a photon induced proton pump (Oesterhelt et al. 1971) like rhodopsin. The proton pump is operated coupling with the phosphorylation of ATP and with the operation of sodium pump providing desalting of the internal medium of the cell of bacteria. The Mitchell theory may be applied to this system.

PART III

13.3 Biological Effects of Light

The biological effects can be grouped under six headings. These are:

I. Lethal effects on small animals and plants.

II. Effect of long continued exposure in large animals and man.

III. Effect on growth patterns in plants and behaviour patterns in animal phytochrome system.

IV. Synthesis of materials by plants and a few bacteria, i.e., photosynthesis.

V. Photochemical processes.

VI. Photoperiodism.

VII. Photodynamic action.

VIII. Solarization (Photo-oxidation).

- **Light Reaching the Living Organism, its Effects and Limitations**

(a) The sun radiates like a black body at a temperature of 6000°F. The intensity of sun's radiation reaching the earth varies as the distance between the earth and the sun changes seasonally. As the sunlight passes through the atmosphere, some of it is absorbed by the normal gases present, ozone in particular, by water vapour and some energy is lost due to scattering by dust particles. As a result of these factors the lower limit of wavelength of the sunrays actually reaching the earth is about 2950 Å. Therefore, we cannot expect biological action from sunlight if the reactions require light of even shorter wavelength. The absorption spectrum of water has an even greater influence in limiting the quantity of light reaching marine organisms. Light which actually reaching marine organisms. Light which actually reaches the surface of the organisms is greatly attenuated by absorption in the outer layers of living material. Single celled organisms like protozoa are specially susceptible to light because they are so thin that light penetrates vital areas. In larger mammals the penetration of light is superficial only. In man visible light can penetrate the skin few millimetres only, ultraviolet light penetrates a fraction of a milimetre. The reflection of visible light from the skin which further reduces the effectiveness of sunlight varies from about 17 per cent in the Negro to about 81 per cent in the lightest bond.

The depth of penetration of light of different wavelengths are the following:

1. Cornea ... 2800 Å
2. Epidermis ... 3000Å
3. Epidermis + bloodvessels ... 5000Å
4. Epidermis + bloodvessels + subcutaneous tissue ... 6000Å

Light of longer wavelengths penetrates the skin to a greater depth than that of shorter wavelengths.

(b) The human lens absorbs light of lower wave lengths only from the spectrum thus limiting the availability of higher wavelengths for rod action.

(c) Bacteria and protozoa are usually immersed in a liquid culture medium which has its own specific absorption bands.

13.3.1 Lethal effects on small animals and plants

The lethal effect of light on viruses, a large number of species of bacteria, fungi, etc. has been known for a long time. Stuhlman (1943) used different wavelengths to study maximum effect of killing of various microorganisms and observed that the maximum lethal effect is of about 2650 Å in the ultraviolet region. The energies necessary to kill may be estimated from the minimum values (those doses killing at the wavelength of maximum effect). These values vary according to the method of measurement. The explanation of lethal action of light on small organisms is that some vital protein is affected since the action spectrum so closely resembles the absorption spectrum of proteins. This effect is not secondary to chemical changes in the surrounding medium, but rather is direct. There is a quite linear relationship between the total effect and the total exposure, regardless of whether the exposure is in a short, high intensity dose or a long, low intensity dose; that is to say, the lethal effect of light follows the "law of reciprocity". This fact implies that the lethal effect is exerted through a single reaction. The temperature at which the organism is exposed has no effect on the size of dose necessary to kill.

13.3.2 Effect of long-continued exposure in large animals and human (Harmful Effect)

The great example of harmful effects of light on large animals is the sunburn produced by prolonged exposure to high intensity sunlight. Sunburn is largely restricted to man because the animals possess protective hair layer which prevents the sunrays from reaching sensitive skin areas.

• **Skin effects**: Depending on the extent of exposure, the symptoms associated with sunburn are:

Reddening (erythema), peeling of the external layers of skin, pain, blistering, and the appearance of pigmentation (suntan). The reddening of the skin is due to dilation of the blood vessels of the skin which in turn seems to be due to the liberation of some dilating substance from the cell layers external to the blood vessels. This substance may be identical with histamine. The release of such a substance from the superficial cells implies some damage to those cells. The outer layers of cells finally die; and are sloughed off in the familiar peeling process. If the damage is extensive enough, fluid accumulate in the epidermal layers and blisters form. After some hours or days, the erythema disappears and a brownish colour takes its place. This brown colour is the first phase of tanning, is due to the formation of brown pigments in the skin. When the first phase is over, exposure to large quantities of light may cause the second phase of tanning, which is the deposition of large amounts of pigment in the skin. The exact nature of tanning is not known, though the formation of the pigment melanin is certainly involved. These changes are produced primarily by light of wavelengths less than 3200 Å except the second phase of tanning which has maximum action spectrum at about 3500 Å.

Other harmful effects of light on larger animals include skin effects other than sunburn, possible increase of incidence of cancer of skin, photosensitivity to wavelengths longer than 3200Å (urticaria solaris), damage to the eye, photoophthalmia (damage to eye, is an inflammation of cornea, quite similar in origin and effects to sunburn) which results from overexposure to sunlight on water or snow or exposure to extremely bright artificial light source. Eye damage due to looking to directly into the sun includes loss of sight in the area of focus of the image of the sun, and is due to the "burning glass" action of the optical system of the eye. The cataract which is an occupational disease of glass blowers also result from the infrared exposure.

• **Miscellaneous effects of light on living matter**: Both visible light and ultraviolet light increase the permeability of animal and plant cell membranes. There is complex changes in the visocsity of cell materials. Using ultraviolet light, cell contents show varying degree of coagulation which may be preceded by liquefaction. Animal tropisms are primarily due to reactions brought about through photoreceptors on the surface of the animals, both larger and smaller animals, even protozoa, show **phototropism**. In these small organisms, the tropism must be the result of the effect of light on cellular contents. Both positive and negative phototropisms exist in the animal kingdom.

Tropism is also exhibited by the plants and plant phototropism has been explained on the basis of the fact that plant hormone auxin concentrations are less in an illuminated area of the plant. This results in slower growth on the illuminated side, which bends the plant toward the source of light (positive tropism). This may or may not be the true explanation of phototropic effects. The roots of plants bend away from the source of light (negative tropism).

13.3.3 Phytochrome system (Light and plant growth)

Light has a profound effect on the growth of the plants. It affects:

(a) Photosynthesis,
(b) The synthesis of chlorophyll,
(c) Phototropic movements,
(d) Stem and root elongation,
(e) The opening of stomata and
(f) Flowering.

These processes are related to growth directly or indirectly. If a plant is grown in the dark, it shows pale yellow colour due to lack of chlorophyll, the stem is elongated and distorted and the leaves do not expand, and the results are known as *etiolation*.

Before showing any manifestations due to light, it must be absorbed by photoreceptor substance present in the plants as light does to animals.

13.3.3.1 Experiments

It has been shown experimentally that seeds of many plants germinate only if they are exposed to light, at least for a brief period. It was discovered that red light in the range 580-660 nm was the most effective, whereas far-red light between 700-730 nm inhibited germination. Again, if a flash of red light is followed immediately by exposure to far-red, the stimulatory effect of the red light is removed and germination is inhibited. If seeds are exposed to alternating red and far-red flashes, the response is determined by the last flash, if it is red, the seeds germinate, they fail to do so if it is far-red.

13.3.3.2 Photochemical mechanism

The light is absorbed by a single substance called **phytochrome**. It is present in very small amounts, about 1 part in 10 million in the tips of the growing shoots. It is a pale blue compound consisting of a pigment molecule attached to a protein (chromoprotein). It exists in two forms which are interconvertible, and are called P 665 and P 725, respectively. One absorbs red light and has its absorption peak at 665 nm and other absorbs far-red light with its peak at 725 nm. When P 665 absorbs red light, it converts rapidly into P 725, and when P 725 absorbs far-red light, it converts slowly into P 665. In sunlight P 665 converts into P 725 and P 725 into P 665, but the first reaction predominates as sunlight contains more red than far-red light. So P 725 accumulates during daylight whereas at night it converts slowly into P 665 (Fig. 13.26).

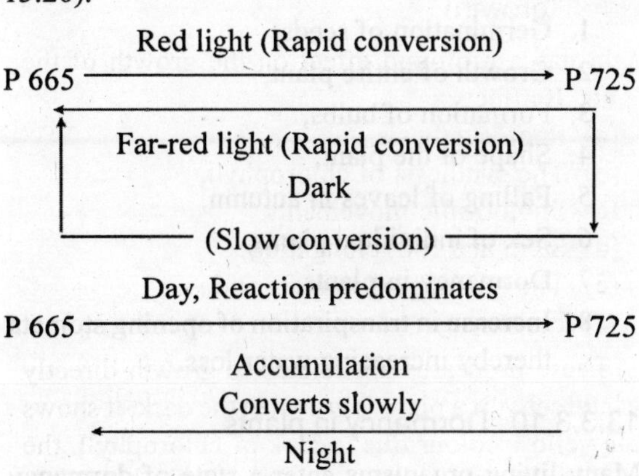

Fig. 13.26: Showing conversion of phytochrome pigments

The significance of day and night conversion is that P 725 initiates enzymatic action and it is biologically active, whereas P 665 is inactive. This does not indicate P 725 always stimulates growth but it sometimes inhibits it.

13.3.3.3 Function of phytochrome system

The phytochrome system is very important for many other growth processes. In some cases red light is the effective stimulus, in other cases far-red. Thus:

1. Elongation of the stem is stimulated by far-red light and inhibited by red light,
2. Leaf expansion is stimulated by red light and inhibited by far-red, but brief exposure to very dim light is sufficient for leaf expansion but this effect is removed by far-red light,
3. The growth of lateral roots is stimulated by far-red and inhibited by red,
4. Flowering depends on alternate periods of light and dark.

13.3.3.4 Mechanism of action

Though the actual mechanism is not known it appears that after absorption of particular wavelength of light → activation of phytochrome system → causes production of hormone from an inactive precursor → affects growth by influencing either cell division or cell expansion. The involvement of hormone was suggested from several evidences, such as the amount of light required to bring a response can be reduced by previous treatment with Gibberellin (Gibberellic acid, GA_3).

13.3.3.5 Phototropism (Response to light)

It is known that the stems grow towards light, hence they are "positively phototropic" and the roots are either "negatively phototropic" or do not respond at all. Growth is slower on the illuminated side compared to the dark side resulting of bending of shoots towards light. It has been proved experimentally that response is proportional to the total amount of light energy received by the shoot. Thus, strong light of short duration generally produces the same degree of curvature as weak light of long duration. Regarding mechanism of phototropism there are two opinions:

1. Light might cause an unequal distribution of the hormone in the shoot, more on the dark side than the light side,
2. The cells responsible for growth in the region of elongation might be prevented from responding to the hormone on the illuminated side of the shoot.

It has been proved experimentally that phototropic effect is due to unequal distribution of the hormone rather than a failure of the cells to respond to it.

13.3.3.6 Photoperiodism in plants

Photoperiodism means influence of the relative lengths of the day and night on the activities on the living organism. This was first reported in plants by Garner and Allard (1920).

Relationship between photoperiod and flowering in plants was demonstrated by exposing the plants to a brief flash of light in the middle of night. If the procedure is continued for a number of days, flowering is inhibited. If the plants are to be made to flower early then extra exposure to darkness is required and in some cases on long night is sufficient.

But reverse result may be observed in some plants, such as light treatment induces early flowering whereas dark treatment delays it. By careful adjustment of the photoperiods, early and late varieties of plants can be made to flower simultaneously.

13.3.3.7 Classification of flowering plants based on responses to light and dark

1. **Long day plants**: Those require long days and bloom with 12 hrs of light/day and short nights. These plants flower only when the light period exceeds a certain critical length in each 24-hour cycle which varies on an average about 10 hours, e.g., radishes and lettuce.
2. **Short-day plants**: Those require short days, i.e., bloom with less than 12 hours of light/day and long nights. These plants flower only when the light period is shorter than a critical length in each 24-hour cycle, e.g., cocklebur.
3. **Day-neutral plants or plants of wide tolerance**: Those are indifferent to day-length, i.e., light not a limiting factor.

There is definite demarcation between long and short-day plants. Long day plants tend to flower in the summer or to inhabit temperate regions where days are long and nights are short. Short day plants, on the other hand, tend to flower in the winter and spring or to live nearer the equator.

13.3.3.8 Factors involved in the process

Various experiments indicate that phytochrome system is involved in the photoperiodic control of flowering. It has been observed that only red light inhibits the flowering of short day plants and this inhibitory can be removed by following the red treatment with far-red light. Far-red light reconverts P 725 back into P 665, it seems that short day plant will only flower if a sufficient proportion of its phytochrome is in the P 665 form. The stimulation to flowering could by either a high accumulation of P 665 or a low concentration of P 725. Recent view is that P 725 inhibits flowering and its conversion back to P 665 removes the inhibition and thus allow flowering. In long-day plants, the reverse process is observed, i.e., accumulation of P 725, due to long exposure to light, stimulates flowering (Fig. 13.27).

13.3.3.9 Mechanism of action of phytochrome system

Photoperiodic stimulus is perceived by the leaves. Message is transmitted from the leaves to the buds for flowering. The message is carried in the form of hormone.

Apart from the flowering of the plants, light also control other phenomena in plants, which are mentioned below:

1. Germination of seeds,
2. Growth of entire plant,
3. Formation of bulbs,
4. Shape of the plant,
5. Falling of leaves in autumn,
6. Sex of individual plants,
7. Dormancy in plants,
8. Increase in transpiration of opening stomata thereby increasing water loss.

13.3.3.10 Dormancy in plants

Many living organisms enter a state of dormancy either as adults or sometimes during their life pe-

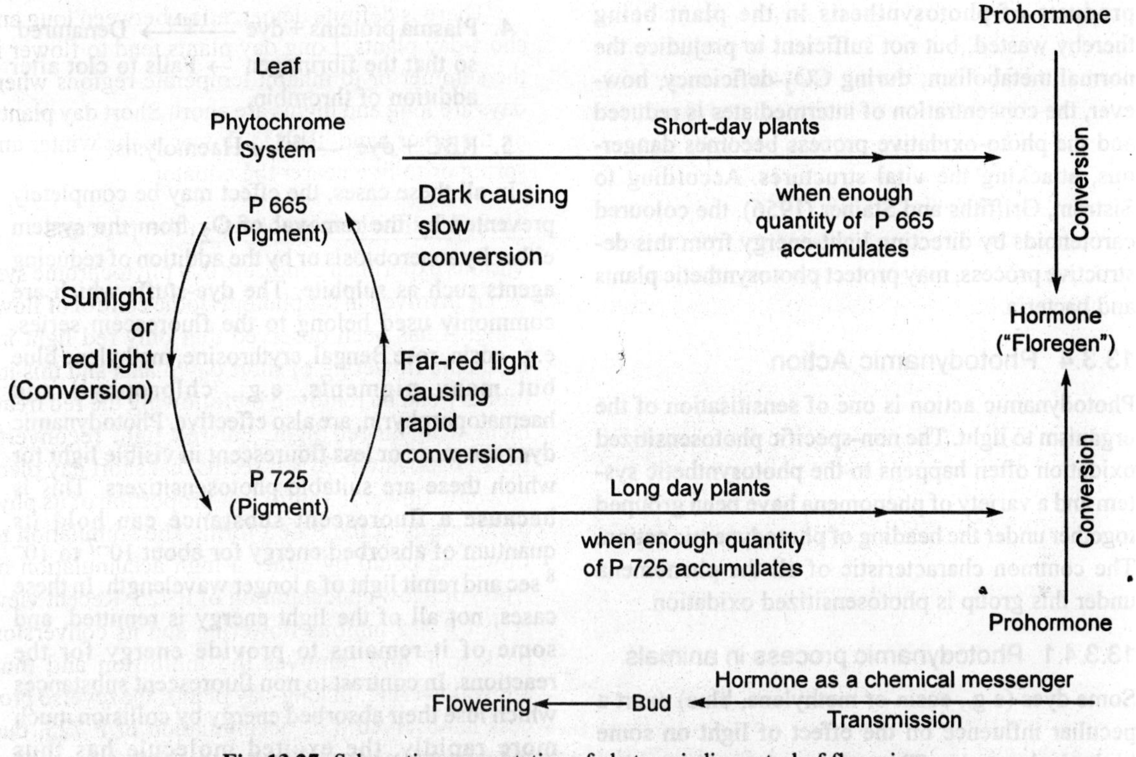

Fig. 13.27: Schematic representation of photoperiodic control of flowering

riod. Growth and development cease, and metabolic rate may drop to a level that it is just sufficient to keep the cells alive. In this way the organism can survive for months or years without exhausting its food reserves. Dormancy enables the organism to withstand unfavourable conditions, e.g., drought, food shortage and winter cold and takes necessary time for internal distribution. Dormant structure can survive for variable length of time.

The dormancy is brought on—and subsequently ended—by environmental factors, particularly light and temperature acting via hormone. This process is applicable to both plants and animals. Dormancy of seeds is caused by the presence of high concentration of a germination inhibitor. Dormancy often breaks by Gibberellic acid suggesting that this germination may be inactivated by Gibberellins.

13.3.3.11 Solarization (Photo-Oxidation)

It is known for a long time that excessive light causes injury to the green plant and this damaging effect is known as "solarization" or "photo-oxidation". Experimental observation has shown that it may be considered as the result of a transition from one type of photochemical reaction to another. It has been observed that the damage is more pronounced in CO_2 deficiency atmosphere. It has also been observed that excessive illumination causes stoppage of O_2 evolution and allows absorption of O_2 and the amount of O_2 absorption depends on the partial pressure of O_2 in the atmosphere. Similar photo-oxidation can be observed if plants are killed by boiling or in plant juices. The important fact is that the concentration of CO_2 is critical in determining capability of a given light-intensity to cause photo-oxidation. This suggests that, in the absence of suitable concentrations of substrate, the light-energy, absorbed by chlorophyll, is used in a less specific manner to oxidize the constituents of cells. If this oxidation proceeds further without any inhibition, the structural components of the leaf, i.e., protein, carbohydrate will be involved in the process and ultimately the damage will be permanent. According to Frank and French (1941), the non-specific photo-oxidation normally runs parallel with photosynthesis, a certain amount of the intermediate products of photosynthesis in the plant being thereby wasted, but not sufficient to prejudice the normal metabolism; during CO_2-deficiency, however, the concentration of intermediates is reduced and the photo-oxidative process becomes dangerous, attacking the vital structures. According to Sistrom, Griffiths and Stainer (1956), the coloured carotenoids by directing light-energy from this destructive process, may protect photosynthetic plants and bacteria.

13.3.4 Photodynamic Action

Photodynamic action is one of sensitisation of the organism to light. The non-specific photosensitized oxidation often happens to the photosynthetic system and a variety of phenomena have been grouped together under the heading of photodynamic action. The common characteristic of all the phenomena under this group is photosensitized oxidation.

13.3.4.1 Photodynamic process in animals

Some dyes (e.g., eosin or methylene, blue) exert a peculiar influence on the effect of light on some biological systems. These dyes, when subjected to radiation, undergo the sensitization reaction:

$$D + h\nu = D^*$$

where D = Dye molecule before exposure,
$h\nu$ = Energy of quantum of light,
D^* = Molecule of dye after it has been activated.

The activated dye (D^*) molecule may now react with a molecule of its substrate X to form a reactive substrate molecule Xr thus:

$$D^* + X \rightarrow Xr + D$$

This reactive molecule (Xr) may now be oxidized to the products X_{ox}.

$$Xr + O_2 \rightarrow X_{ox}$$

In many cases this abnormal type of oxidation may be harmful to the organisms, such as:

1. Unicellular organisms and viruses + dyes (eosin) $\xrightarrow{\text{Light}}$ Killed.
2. A frog muscle + dye (eosin) $\xrightarrow{\text{Light}}$ series of twitches → Contracture.
3. Enzymes, e.g., invertase, peroxidase and catalase + dye $\xrightarrow{\text{Light}}$ Inactivation.
4. Plasma proteins + dye $\xrightarrow{\text{Light}}$ Denatured, so that the fibrinogen → Fails to clot after addition of thrombin,
5. RBC + dye $\xrightarrow{\text{Light}}$ Haemolysis.

In all these cases, the effect may be completely prevented by the removal of O_2 from the system either by anaerobiosis or by the addition of reducing agents such as sulphite. The dye-stuffs which are commonly used belong to the fluorescein series, e.g., eosin, rose Bengal, erythrosine, methylene blue but many pigments, e.g., chlorophyll and haematoporphyrin, are also effective. Photodynamic dyes are more or less flourescent in visible light for which these are suitable photosensitizers. This is because a fluorescent substance can hold its quantum of absorbed energy for about 10^{-7} to 10^{-8} sec and remit light of a longer wavelength. In these cases, not all of the light energy is remitted, and some of it remains to provide energy for the reactions. In contrast to non fluorescent substances which lose their absorbed energy by collision much more rapidly, the excited molecule has thus sufficient time to pass its quantum on to energy-absorbing reactants.

13.3.4.2 Activated molecules

In photodynamic action, the problem arises regarding (1) the nature of the oxidized substance i.e., whether it is carbohydrate, protein or fat and (2) the nature of the activated molecule, i.e., whether O_2 is activated and subsequently oxidizes substrate or (3) whether the substrate is activated and reacts with normal O_2. The former opinion has been supported by Kautsky (1936) and he suggested that O_2 is activated with a relatively long life of 10^{-3} to 10^{-2} sec. The energy of this state requires a wavelength of less than 7623 Å. Quenching of fluorescence by substrates indicates that O_2 need not be the energy acceptor and it favours the activation of the oxidizable substrate as the primary process.

13.3.4.3 Oxidized substrate

Protein nature of the oxidizable substrate is favoured by the major group of workers. Thus, when blood plasma is treated with haematoporphyrin and irradiated with visible light, it exhibited a rapid consumption of O_2. Smetana (1939) observed that this

O_2 consumption is due mainly to the albumin and globulin. Fats, glucose and urea have negligible contribution. Howell (1921) had shown that the clotting of plasma can be inhibited by photodynamic action due to denaturation of fibrinogen molecules. Thus, during cell damage by photodynamic action, the structural proteins are attacked and the amino acids with aromatic nuclei are most vulnerable to attack, e.g., histidine, tryptophan and after their disappearance, tyrosine begins to be destroyed.

13.3.4.4 Photodynamic haemolysis

When suspension of R.B.C. in saline containing low concentration of rose Bengal is exposed to light from a 100 watt lamp it causes haemolysis after an induction period. The induction period varies from a matter of seconds to few minutes depending on the concentration of dye and the intensity of the light. If the light is switched off before the haemolysis is complete, the process of haemolysis does not stop immediately but continues for sometime in the dark. Photodynamic haemolysis is a primary damage to the cells followed by a dark or "after-light" haemolytic process. Blum (1941) showed that dye is absorbed on the surface of the cell and efficacy depends on the extent to which it is absorbed. He also showed that the number of quanta of light required to haemolyse a single cell was approximately 1.10^{10} and it does not depend on dye or its concentration. Therefore, it may be stated that the first step in haemolysis is the photosensitized destruction of a constituent of the cell membrane. The presence of reducing agents completely inhibits the light haemolysis; consquently the destructive action is oxidative.

13.3.4.5 Colloid-osmotic haemolysis

The haemolysis after irradiation, the after-light haemolysis, was not due to a continuation of the oxidative process initiated in the light and was unaffected by reducing agents, this was observed by Davson and Ponder (1940). Further they found that it was not due to the presence of lytic substance formed during the light-phase because dilution of the suspension of irradiated cells with isotonic saline had not effect on the course of after-light haemolysis. It was observed that before any haemolysis occurred, the cells lost potassium ions and this loss continued during the after-light period. It was concluded that the after-light haemolysis due to the cation permeability induced during the light-phase gives rise to colloid-osmotic haemolysis. Therefore, the persistence of the haemolytic process after irradiation is due to secondary changes, colloid-osmotic swelling, resulting from the primary oxidative damage to cell membrane.

13.3.4.6 Absence of the intermediate substance

From the study it was concluded that photodynamic action results from the transfer of light energy by an absorbed molecule of dye to the structural proteins and as a result of this activation, they combine with molecular O_2 and the damage is not due to formation of toxic materials in solution which was proved in perfused heart-experiment. The turtle hearts were perfused with ringer solution containing haematoporphyrin. Irradiation of the first heart, when the second heart was in darkness, caused irregularity in the heart beat and eventually stoppage, whereas the second was unaffected.

13.3.4.7 Evidence for absorption of the dye to the cell (Inhibition by plasma)

It was observed that addition of quite small quantities of plasma can reduce or inhibit the photodynamic action. Now to produce the photodynamic effect, a higher concentration of dye is required. The active molecule in the plasma is the albumin. It may be stated that the plasma prevents the dye from adsorbing to the cell surface mainly because the dye is adsorbed on the plasma albumin (Haemolysis due to saponin and many other lysins are also inhibited by plasma).

13.3.4.8 Intracellular action

Hyman and Howland (1940) studied the effect of injection of eosin and rose Bengal directly into the cytoplasm of amoeba followed by irradiation which produced a series of changes leading to the bursting of the cell. This damage was attributed to increased fluidity of the cytoplasm with marked increase in permeability to water leading to overwork by the contractile vacuole.

The dye Janus green has a specific affinity for mitochondria and is stained when the living

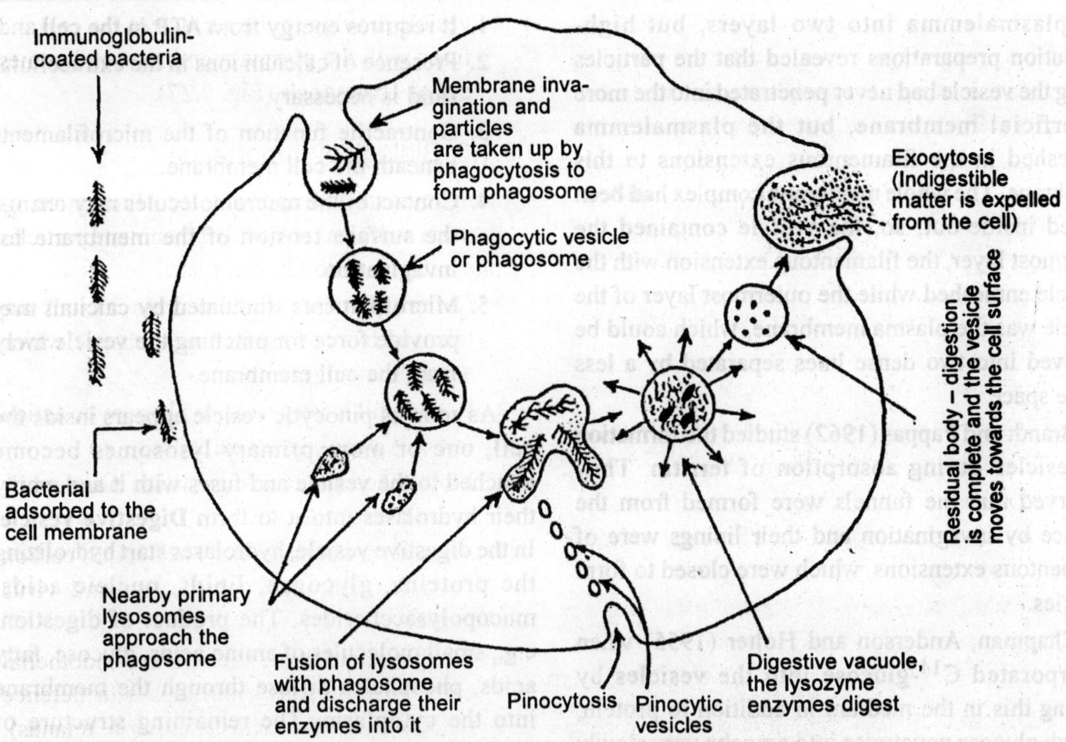

Fig. 9.27: Schematic diagram of sequence of events in phagocytosis and pinocytosis

9.9.2 Pinocytosis

Certain substances, i.e., macromolecules of high molecular weight, e.g., proteins cannot diffuse through cell membrane, are taken into cytoplasm by invagination of cell membrane and are called Pinocytosis which literally means "cell drinking". It is one type of endocytosis and essentially similar to phagocytosis except the following:

1. The size and nature of vesicle.
2. Not selective process.
3. Vesicular content – Excreta, cellular fluid and dissolved substances.
4. Content of vesicle is not visible under ordinary microscope but visible under electron microscope.

9.9.2.1 Reverse pinocytosis or emeiocytosis or exocytosis

Proteins secreted by the cells, packaged into secretory granules. The membrane-bound granules move along tracks of microtubules of the cells membrane. The granule membrane and cell membrane fuse. The area of fusion breaks down and contents of the granules come outside the cell and the cell membrane remains intact. This process is called **Reverse Pinocytosis** (cell vomitting) or **exocytosis** or **Emeiocytosis.**

9.9.2.2 Experimental observation

Leivis (1931) observed that white blood cells formed large numbers of vesicles in a medium containing protein and it was explained that these were formed due to engulfment of protein—containing external medium by invagination of the cell membrane.

Holter and Marshall (1954) showed that the vesicles contained labelled proteins, when the WBC placed in the medium.

Brandt and Schumakar (1958) after using labelled proteins, indicated that the primary event is adsorption to the cell wall and it was clearly exhibited under electron microscope, when ThO_2 particles were engulfed.

Brandt (1962) stated that, the formation of the vesicle appeared to be associated with splitting of organism is treated with it. If the living organism like spermatozoa is kept in the dark the mobility is unaffected after exposure to the dye but after illumination their mean velocity is decreased. Van Duijn (1961) opined that this effect is due to attack on the mitochondria, although the influence on the cell membrane was not excluded. Norman and Goldberg (1959) showed that bull spermatozoa are actually sensitive to light in the absence of dyes; this is due to the presence of a natural photosensitizer and is a true photodynamic effect since adding catalase, which breaks up peroxide as soon as formed, inhibits it. The toxicity of Janus green to cells in tissue culture may be due entirely to this photosensitizing effect since the cells are exposed to high intensity of illumination during examination. Rosenblum (1960) demonstrated that muscle can produce twitch after photosensitizing with Rose Bengal and end-plate blocking agents can inhibit this, it indicates that the primary action is on the nerve or nerve-muscle junction. The contracture of the muscle observed after depolarizing by KCl reveals an action on the contractile machinery.

13.3.4.9 Photodynamic action in living animals

Visible light cannot penetrate the deeper tissues, therefore, photodynamic action can be produced in animals by injection of dye and subsequent irradiation Rats and mice exhibited symptoms like generalized stimulation of the sensory nerve endings. Later, necrotic lesions developed with loss of hair. But it is known for long time that cattle and sheep develop skin diseases after eating certain weeds (e.g., Hypericum) due to presence of photodynamically active pigment named *hypericin*. Other plants, e.g., *Lippia-Sp.*, *Tribulus. Sp.* having toxic substance are not photodynamically active but they damage liver and cause various breakdown products of chlorophyll which appear in the blood stream, and *phylloerythrin* is one of them and photodynamically active. Another plant called *buckwheat* (Fagopyrum esculentum) causes sensitization of animals and the condition is known as "fagopyism".

13.3.4.10 Characteristics of Photodynamic Action

1. Photodynamic action follows the law of reciprocity.
2. Photodynamic action is thermodynamically irreversible.
3. Photodynamic action does not need any cellular or protoplasmic device like cellular metabolism.
4. Photodynamic action is a non-enzymatic process.
5. O_2 metabolism being an enzymatic process is inhibited by heat but boiling does not abolish photo-sensitized O_2 uptake.
6. Cyanide stimulates photodynamic action but inhibits cellular metabolism.
7. Photodynamic action has $Q_{10} = 1$.
8. R.Q $\left(\frac{CO_2}{O_2}\right)$ of photodynamic action is 0.047 but that of cellular metabolism is near unity.
9. Photodynamic action requires dye or pigment.

PART IV

13.4 Biophysics of Vision

13.4.1 Light and its attenuation for vision

(Regulation of the amount of light which falls on the retina)

Most important photobiological process is "photoreception". The eye is a self-regulatory system and this regulatory mechanism optimize the operation of the eye. These are essential for (1) focussing the image on the retina and (2) regulation of the amount of light which falls on the retina.

Before the actual absorption of light by the photochemical pigments, there are several steps of attenuation of the light incident on the cornea. Therefore, not all of the 54-148 quanta actually are involved in vision.

13.4.1.1 Role of cornea

About 4 per cent of the light falling on the cornea is reflected from it, at the angle of incidence represented by a 20° to 30° deviation of the beam from the optic axis of the eye. Approximately 50 per cent light passing through the corneal surface is absorbed by the ocular media (e.g., aqueous humour, lens, vitreous humour) before it reaches the retina. The cornea transmits rays from 295 mμ to 2,500 mμ, but above 1800 mμ the transmission is slight. Maximum transmission occurs at around 1000 mμ and continues high at lower wavelengths until the ultraviolet is reached. Absorption then becomes pronouncd and is complete at about 295 mμ.

13.4.1.2 Role of media

The aqueous humour and vitreous humor are somewhat less transparent than the cornea. All rays above 2700 mμ are absorbed by the aqueous humor and all above about 1600 mμ by the vitreous humour. The lens absorbs all rays below 300 mμ or above 2500 mμ ; most of the rays longer than 1,300 mμ are absorbed. Maximal transparency is between 400 and 1200 mμ. That is, most of the rays outside this range are absorbed by the lens. Wavelengths between 350 mμ and 400 mμ cause flourescence in the lens. The phenomenon of fluorescene, in general, is attributed to the transference of the energy of the incident radiations to particles of the substance absorbing them. The particles then act as independent light sources, emitting waves which are, for the most part, longer than the original radiation. Thus, the lens converts the harmful shorter waves to longer one which are permitted to reach retina.

13.4.1.3 Threshold energy of light entering the eye

Radiant energy is expressed in "ergs" when the necessary calibrations of the light source, neutral filters, etc. have been made, a curve is plotted to relate the threshold energy in ergs of the light entering the eye to wavelength. This "scotopic visibility curve" shows that under these conditions the threshold energy is least, i.e., the eye is most sensitive, at a wavelength of about 507 mμ. At this wavelength the minimal energy entering the eye was found to be between (2.1 to 5.7) × 10^{-10} ergs as measured at the cornea. On the other hand, the curve indicates that the threshold energy rises enormously beyond about 400 mμ and 650 mμ, at the end of the visible spectrum.

13.4.1.4 Role of retina

Upon reaching retina, a large proportion of light must pass through the transparent entire retina (i.e., through all layers) before it is absorbed by photopigment rhodopsin in outer segments of rods. This probably limits visual activity in the peripheral retina. Stimulating light must also pass through the retinal blood vessels. Not all the light which falls on the rods is absorbed in the rods by the rhodopsin. Most of it, 80 per cent, passes through the rods to be absorbed by the pigment layers of the retina and the choroid coat; a small fraction is supposed to be reflected back. If this light were reflected back, it would degrade the retinal image. However, this light is largely absorbed by melanin, a black pigment contained in granules in the pigment epithelium and choroid. By greatly reducing reflected light, the optical quality of the image at the photoreceptors is thus improved. It can be stated that the maximum sensitivity of the retina is such that only from 2-14 quanta of light need to be absorbed by the rods in a given area and within a given time for a sensation of light to be evoked. This means that a stimulated rod must give its signal to the bipolar cell when only a single quantum of light has been absorbed and as a result of the breakdown of a single molecule. Several such discharges must be combined to cause sufficient impulses for the stimulus to be perceived. It is different to assay exactly the quantity of light absorbed by the visual purple. Hecht et. al have done it and concluded that about 20 per cent is a reasonable figure for absorption by rhodopsin of light 5100 Å. It is apparent that the actual number of quanta absorbed by photosensitive substance at absolute threshold at the eye (i.e., 2-14 quanta) is an upper limit, and the actual sensitivity is something less than these values as this sensitivity is considered high.

13.4.2 Light and field of view

Determination of the absolute threshold of various wavelengths, the field of view is considered to be large. With small fields and with fixed positions on the retina the results in terms of threshold may be different. But the process of summation go on extensively in the retina, e.g., when the field is small, all the light falling within a certain area, the "summation area" within a given time, the "summation time" is summed in order to give the response. With very small fields the eye cannot distinguish between an intense point-source and a less intense but larger area of light, provided it falls within the limits of the summation area up to a certain size of field which depends on the wavelength of the light used and on the position on the retina, IA is constant, where I = light intensity and A = area of the field. This is known as *Ricco's Law*. With larger fields and in the retinal periphery the relationship is much more nearly $I \times \sqrt{A} = K$ and this is known as *Piper's Law* which are only approximations.

13.4.3 Light and photopigment

While discussing the characteristic of vision as mediated by the rods, the essential visual process is the photochemical breakdown of rhodopsin which are photosensitive pigments where light energy is transferred into chemical energy.

When the absorption of light by the pigment at different wavelengths is plotted against wavelength, a curve is obtained which, when suitably corrected for absorption by break-down products, etc., is similar to the "visibility curve" for the dark-adapted eye, althouth there are slight differences in shape and in the position of the maxima between the "absorption spectrum" and the "visibility curve". This indicates that vision is a photochemical reaction and the vision up to 0.03 sec obeys *Bunsen-Roscoe Law*. If the law is valid, then various differences between the "Scotopic visibility curve" and "absorption spectrum" should be removed. First of all, the matter which is to be considered is the energy distribution of the light source used in each case and of its photochemical efficiency at different wavelengths because different source of light have not the same spectral composition. It is customary, therefore, to express all visibility and absorption curves in terms of *equal energy spectrum*, which is a hypothetical concept. It means that the light source must be calibrated in terms of the energy which it emits different wavelengths so that the sensitivity on absorption can be standardized for each wavelength and so made independent of the particular form of illuminant. Next, the photochemical effects on the pigment depend not so directly on the energy absorbed as on the number of quanta absorbed, so that the energy values at each wavelength have to be converted into quanta. Comparisons of the curves are only legitimate when the data are fully corrected to an equal quantum spectrum. After doing this, the two curves become coincident over a large part of the spectrum, but still they are not identical as they diverge widely and inconstantly in the short wave part of the spectrum (Blue). This difference can be explained partly on physiological grounds. The actual amount of light, which actually reaches the photoreceptors, is only that part which is not absorbed by the structural media on its way.

13.4.4 Illumination of retina

The unit of retinal illumination is called the **photon**, or, preferably, the **troland**. The amount of light which illuminates the retina depends not only upon the brightness of the object but also upon the size of the pupil. A troland, is therefore, defined as the illumination of the retina by light from a surface having the brightness of one standard candle per square metre, as seen through a pupil having an area of 1 sq. mm. The pupils play a great role in limiting the light entering the lens to a relatively small area of the lens surface near the optical axis of the system. This results in the three phenomena of interest in image formation in the eye such as:

1. The total quantity of light entering the eye is reduced.
2. Amounts of spherical and chromatic aberrations are reduced.
3. The depth of focus is increased.

13.4.5 The eye as an optical instrument

The eye is optically equivalent to the usual photographic camera. It has (1) a lens system (2) a variable aperature system, e.g., the pupil and (3) a retina which corresponds to the film.

The lens system consists of:

(a) A concavo-convex lens comprised of the cornea and aqueous humour.

(b) A biconvex lens—the crystalline lens.

(c) Single surface concave lens—vitreous humour.

Between the first two refracting elements is the iris, which acts as a diaphragm.

Light rays are bent, i.e., refracted when they pass from one medium into a medium of a different density, except when they strike perpendicular to the interface. Parallel light rays striking a biconvex lens are refracted to a point called principal focus behind the lens. The principal focus is on a line passing through the centres of curvature of the lens called the principal axis. The distance between the lens and the principal focus is the principal focus distance (Fig 13.28).

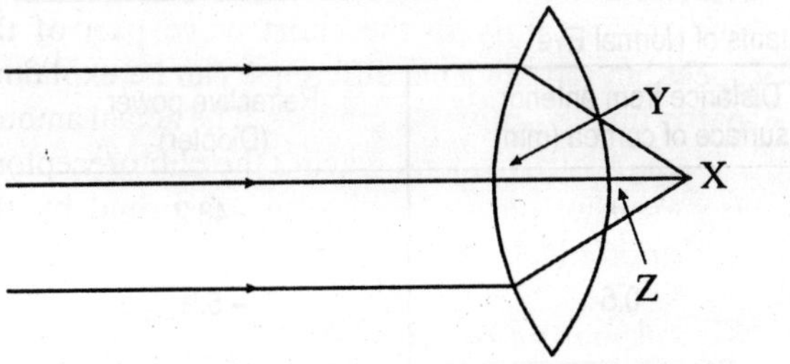

Fig. 13.28: Refraction of light rays by a biconvex lens

X = principal focus,

Y = principal axis,

Z = principal focal distance.

To produce an image of the outside world on the retina, the eye has a refracting apparatus consisting cornea and the crystalline lens. The optical properties of refracting surfaces are determined by:

(a) Their radius of curvature

Refracting surfaces	Radius of curvature (mm)
Anterior cornea	7.8
Posterior cornea	6.8
Anterior lens	10.0 (during maximum accommodation)
Posterior lens	6.0

(b) Index of refraction of the media that they separate and

(c) The characteristics of the optical apparatus as a whole depend on those of the surfaces and their separation.

Total refractive power = 59 diopters.

The different refractive media of the eye are cornea, the aqueous humour, the crystalline lens and the vitreous humour. The refraction occurs at the following interfaces:

1. The interface between air and the anterior surface of cornea.
2. The interface between the posterior surface of the cornea and the aqueous humour,
3. The interface between the aqueous humour and the anterior surface of the crystalline lens and
4. The interface between the posterior surface of the lens and the vitreous humour. The refractive indices of air = 1.0, the cornea = 1.38, the aqueous humour = 1.33, the lens (on the average) = 1.40 and the vitreous humour = 1.34 (Fig. 13.29).

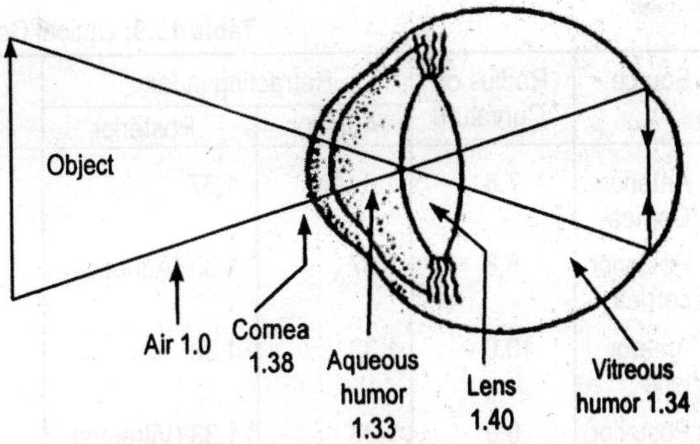

Fig. 13.29: Refractive indices of different structures of the eye

The refractive indices of the cornea, aqueous humour and vitreous humour are approximately same, hence for practical purposes they may be considered as identical and the eye then taken having two refracting surfaces:

(a) The anterior surface of the cornea in contact with air, and

(b) The lens surrounded by a common medium.

The greatest refraction occurs at the corneal surface (42 diopters), hence any irregularities in it are of major significance in image formation in the eye. The refraction at the surfaces of the lens (19 diopters with accommodation relaxed and 36 diopters in full accomodation) is of less importance. Refracting power of the whole eye is approximately between 59-65 diopters. The value of refracting power of lens is calculated as a whole though it is a homogenous structure. But the lens consists of an almost spherical "nucleus" with high refractivity (1.41) surrounded by a zone called "cortex" of lower optical density (1.38). The cortex is composed of a series of concave meniscus lens and this peculiar structure accounts for the mean value of the refractive indices of this part which is 1.39 and lesser than the refracting power of the whole (1.42) (Fig. 13.30). This peculiar structure is advantageous for the following reasons:

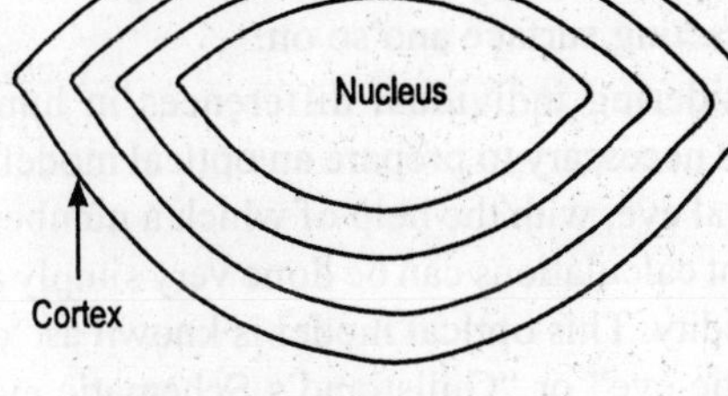

Fig. 13.30: Structure of lens

Table 13.9: Optical Constants of Normal Eye

Source	Radius of Curvature	Refracting index Anterior	Refracting index Posterior	Distance from anterior surface of cornea (mm)	Refractive power (Diopter)
1. Anterior Cornea	7.8	1.0 (Air)	1.37	0	+ 48.2
2. Posterior cornea	6.8	1.37	1.33 (Aqueous)	0.5	– 5.9
3. Anteror lens	10.0	1.33	1.38	3.6	+ 5.0
4. Posterior lens	6.0	1.38	1.33 (Vitreous)	7.2	+ 8.3
5. Retina				24.0	

1. Diminishes spherical aberration,
2. Tends to prevent the scattering of light within the eye and
3. Enhances the power of the lens to alter its converging power during accommodation.

13.4.5.1 The constant of the eye

If we know the values of (1) the curvatures of the refractive surfaces of the eye and (2) the distances between them and (3) refractive indices of the media which are known as "constant of the eye" given above (Table 13.9), then the path taken by the rays of light can be determined and the image constructed.

The total refracting power of the eye cannot be obtained by simple addition of the values from the constants of the eye for two reasons:

1. The separation of the surfaces affects their cumulative refractive effect and
2. The crystalline lens behaves in quite an unusual ways as discussed above.

Again, construction of an image from the above data is a difficult procedure. To start with the image formed by the first refracting surface is constructed; this image now serves as the object for the next refracting surface and so on.

Considering individual differences in human eyes, it is necessary to prepare an optical model for the typical eye, with the help of which a number of important calculations can be done very simply and with validity. This optical model is known as "diagrammatic eye" or "Gullstrand's Schematic eyes" by the application of the *Theorem of Gauss.* This states that every optical system composed of spherical surfaces with their centres on the principal axis has three pairs of cardinal points. These are:

1. Two principal points,
2. An anterior and a posterior focal points, and
3. Two nodal points.

13.4.5.2 The reduced eye

The compound optical system of the schematic eye can be further simplified by the so-called *reduced schematic eye* of listing or simply "reduced eye" when we consider the refracting system of the eye as a single surface lens, i.e., if all the refractive surfaces of the eye are algebraically added together and then considered to be one single lens. This is useful in simple calculations. In the reduced eye, a single lens is considered to exist with the central point 17 mm in front of the retina and to have a total refractive power of approximately 59 diopters when the lens is accommodated for distant vision.

The optical constants of the reduced eye have been chosen because it operates as image-forming mechanism most like the real eye. There are some important differences—the reduced eye cannot accommodate in the way the real eye does but the optical properties are sufficiently similar and so very much simpler that it is an ideal model on which to illustrate the major optical effects and defects occurring in a normal human eye.

By means of X-ray beam without refraction projected into eye, Goldman and Hagen measured the length of the eye globe in the living human subject. The value 23.4 mm obtained agrees closely with that of the schematic eye. The value for the total regracting power of the normal human eye obtained by these workers is also in close agreement, namely 59.22 diopters.

13.4.5.3 Formation of image

The image, e.g., the replica of the outside world, formed by the optical lens system on the retina is two dimensional. The geometric relationship between the outer environment and the retinal image is based on the projection system. The centre of the pupil entrance acts as centre of projection. An object sends out light energy in many directions but only a small proportion (i.e., chief ray) which passes through the pupil, contributes to the formation of the retinal image. The optical lens system focuses the image on the retina is inverted and reversed way with respect to the object. Scheiner furnished the proof by observing the back of an excised eye from which the sclerotic and choroid coats had been removed. The inverted image of an object was clearly visible upon the translucent retina. The mind perceives the object in the upright, despite the upside-down orientation on the retina because the brain is trained to consider an inverted image as the normal.

The size of the retinal image depends on the angle subtended at the nodal point of the eye by an object in the visual field is called the *Visual angle*. The size of the image can be calculated if the following values are known:

1. The distance of the nodal point of the eye from the object.
2. The distance of the nodal point of the eye from the retina.
3. The size of the object.

An object which subtends one minute of angle at the eye forms an image on the retina of size approximately 4.5 microns. The full moon subtends about $\frac{1^{\circ}}{2}$ and its image on the retina would therefore, be about 0.135 mm diameter.

13.4.6 Critical fusion frequency (CFF)

(Intermittent stimulation of retina, Flicker, resolution of stimuli in time by eye)

The eye can resolve stimuli that are separated in time. The limit of this ability is indicated by measuring the minimum frequency at which repetitive stimuli appear to fuse together into a continuous stimulus. This minimum frequency causing fusion is called *critical fusion frequency* (CFF). CFF is a measure of time resolving power of the eye.

13.4.6.1 Mechanism of CFF

An instantaneous flash of light excites the visual receptors for as long as $\frac{1}{10}$ to $\frac{1}{5}$ second, and because of the persistence of excitation, rapidly successive flashes of light become fused together to give the appearance of being continuous. The effect can be observed during motion pictures or television. The images on the motion pictures screen are flashed at a rate of 24 frames per second where as those of the television screen are flashed at a rate of 60 frames per second. As a result, the images fuse together, and continuous motion is observed.

This is due to each light stimulus falling upon the retina during the time of the positive after image of its predecessor. The suppression of the after image by the second stimulus causes the first sensation to end more abruptly, and, through contrast, to render the succeeding one more brilliant.

If at the instant of fusion the illumination of the bright patch in milliambers be designated (a), its area designated (b), and the total area of the disc, (c), then the sensation produced is equal to that which would result from a continuous stimulus having the values (a × b)/c. This is known as *Talbot-Plateau Law*. The law holds true only for moderate light intensities.

13.4.6.2 Role of light intensity

The value of the CFF is variable, depending upon several conditions of which most powerful is light intensity, the value rises as the intensity increases. That is a higher rate of stimulation is required for fusion as the intensity of illumination is increased. The influence of light intensity obeys *Ferry-Porter Law*, which states that the critical fusion frequency is directly proportional to the logarithm of the light intensity. Thus, $n = K \log I + K'$ where n = flashes per second at the instant of fusion, I = the light intensity; K and K′ are constants involving the size of the stimulated area and the sensitivity of the observer's eye. The *Ferry-Porter Law* is valid, however, only under certain special conditions; it holds over moderate ranges of illumination of the test object when the image is restricted to the fovea.

If the results of relationship between the

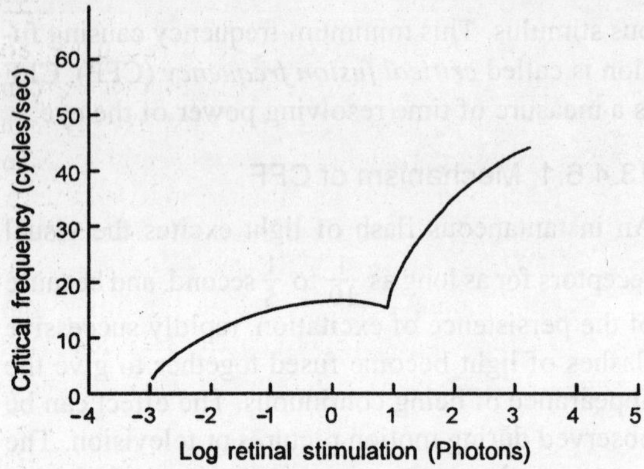

Fig. 13.31: Relationship between light intensity and critical frequency for fusion

intensity of light and critical frequency is plotted, it would be like given illustration (Fig. 13.31). At a low intensity, fusion results even when the rate of flicker is as low as 2 to 6 per second. In bright light, the critical frequency for fusion rises to as great as 60 flashes per second. This difference is due partly from the fact that the cones, which operate mainly at high levels of illumination, can detect much more rapid alterations in illumination than can the rods, which are the important receptors in dim light.

13.4.6.3 Retinal potentials during intermittent stimulation

A study of retinal potentials during intermittent stimulation shows that when a light flash falls upon the retina during the "off effect" of a preceding stimulus, the (d) wave is interrupted and a pronounced negative dip occurs. This is an exaggerated (a) deflection. The negative deflection is followed by a large positive swing which, if the second stimulus occurs soon after the first, is simply the return of the momentarily interrupted "off effect" of the first. If the interval between the two is greater, the upswing is higher, and is then due to (b) deflecton of the after-coming stimulus. Thus, if the light flashes are so timed that each interrupts the "off effect" of its predecessor, a series of regular ripples appear in the ERG which apparently are the cause of the flickering sensation. No negative dip occurs, nor do the characteristic ripples appear if a flash falls upon the retina before the "off effect" of the preceding one—a continuous sensation should, therefore, result.

13.4.6.4 Changes in the retina on exposure to light

A light stimulus falling on the retina causes a number of changes to occur which may be classed into four:

I. Structural changes.
II. Physical changes (Electrical changes).
III. Chemical changes.
IV. Physiological changes.

I. Structural changes

1. Firstly, movement of the pigment from the outer epithelial layer into the spaces between the rods and cones.
2. Secondly, shortening of the cones themselves.
3. Swelling of the outer limbs of the rods.
4. Disappearance of chromatin granules from the ganglion cells. The (3) and (4) are said to occur more rapidly under the action of rays of short wavelength.

The (1) movement of pigment and (2) shortening of cones occur only:

(a) When the connections of the eye with the brain are intact,
(b) When the rate of movement appears to vary with intensity of light,
(c) When violet light is said to be more effective than red,
(d) When electrical stimulation of the optic nerve or the falling of light on the other retina also causes these cone movements. It is supposed that it is due to descending nerve fibres from the brain to the retina called "retinomotor" fibres.

II. Physical Changes (Electrical changes)

When retina is stimulated by light there is an electrical response somewhat similar to the current of action in the nerve.

The typical curve is shown and called *Electroretiogram* (Fig. 13.32).

Some results are obtained in the response.

(a) With light of any one colour, a geometric rise of intensity causes an arithmetic increase in the current,

(b) With coloured rays of apparent equal intensity, yellow rays are said to give a larger current in the light adapted eye and green in the dark adapted eye,

(c) Current commences after a latent period.

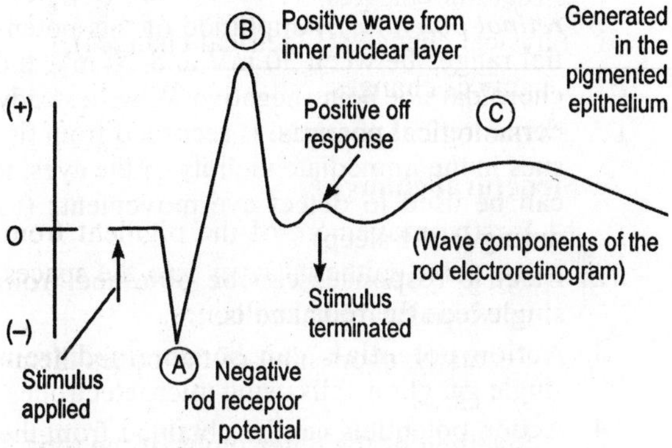

Fig. 13.32: Electroretinogram

III. Chemical changes

1. Firstly a tendency of the retina as a whole to become acid in reaction, as shown in the change in its behaviour to certain stain.
2. Bleaching of three pigments:
   (a) Rhodopsin or visual purple,
   (b) Iodopsin and,
   (c) Fuscin.

(a) **Rhodopsin or visual purple**: This pigment present in the rods of retina is a chromoprotein containing a *protein part* "scotopsin" and a *carotenoid pigment "retinal"* or "retinene" and it is 11-cis retinal. This "cis" form is important because only this form can combine with scotopsin to synthesize rhodopsin. When light energy is absorbed by rhodopsin, it immediately begins to decompose in the following way (Fig. 13.33). The cause of decomposition of photoactivation of electrons in the rhodopsin is to change *cis* form into *all-trans* form.

(b) **Photochemistry of Cones**: This pigment is present in cones and has similar composition except the protein part which is called "*photopsin*". Three different types of photochemicals are present in different cones (e.g., iodopsin, porphyrinopsin, cyanopsin) selectively sensitive to three different colours, e.g., blue-sensitive pigment, green-sensitive pigment and red-sensitive pigment. Ruston concluded that strong light bleaches the cone pigments and he proposed the existence of two nonbleached pigments (1) Chlorolabe and (2) Erythrolabe in the normal human eye as they have maximum sensitivity at green and red part of the spectrum, respectively (see photo-chemistry of vision).

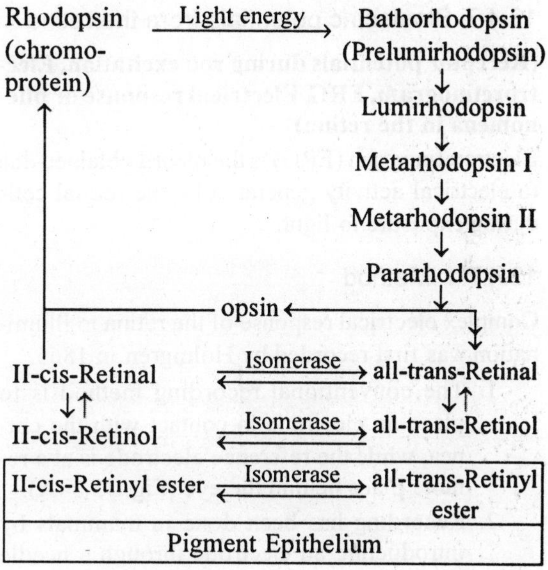

Fig. 13.33: Photochemistry of the rhodopsin-retinal-Vit A visual cycle

(c) **Fuscin and Melanin**: This pigment found in the form of needles, plates or prism in the **stratum** pigment layer. It absorbs light and prevents **spreading** of light but bleaching by light does not have any visual function. Melanin is the black pigment which prevents light reflection like camera.

IV. **Physiological changes**: Physiological changes produced by light depend on the region of the retina on which it falls since this may contain (1) rods only (2) cones only (3) both rods and cones or (4) neither rods nor cones.

1. **Rods only**: The peripheral part of retina contains numerous rods and very few cones. This part is extremely sensitive to dim light, i.e., for night vision and colour blind—no colour is perceived.
2. **Cones only**: The fovea centralis contains cones only and it is extremely sensitive to

bright light (i.e., day vision) and colour perception).

3. **Both Rods and Cones:** Parts of the retina around the macula lutea contain both rods and cones, perceive colour and react to light of low intensity, i.e., dim light.
4. **Neither Rods nor Cones:** At the papilla where the optic nerve comes out is the "blind spot" where rods and cones are absent. Here perception of light and colour is absent.

13.4.7 Bioelectric potentials from the retina

(Receptor potentials during rod excitation, Electroretinogram, ERG, Electrical response or phenomena in the retina)

Electroretinogram (ERG) is the record obtained due to electrical activity generated by the retinal cells during exposure to light.

13.4.7.1 Method

Complex electrical response of the retina to illumination was first recorded by Holmgren in 1865.

1. The conventional recording method is to place an electrode in contact with the cornea, while the reference electrode is at a remote place behind the eye (Fig. 13.34A).
2. Recording has been done in mammals by introducting an electrode through a needle into the vitreous humour.

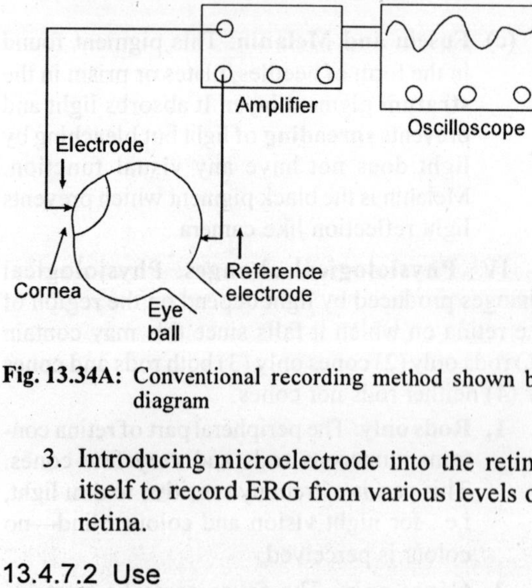

Fig. 13.34A: Conventional recording method shown by diagram

3. Introducing microelectrode into the retina itself to record ERG from various levels of retina.

13.4.7.2 Use

1. To study retinal physiology.
2. To diagnose retinal disorders.

Four types of bioelectric potential can be obtained from the retina, depending on size and location of the electrode used to record such activity.

1. A steady potential can be recorded between the front and back of the eyeball (*Corneo-retinal potential*). Amplitude of this potential ranges between 30 mV and 60 mV, the choroidal side being negative. When a steady corneoretinal potential is recorded from tissues in the immediate vicinity of the eyes, it can be used to detect eye movements (as during REM sleep).
2. Electric responses can be obtained from single bipolar retinal cells.
3. Action potentials can be obtained from single ganglion cells using microelectrodes.
4. Action potentials can be obtained from the optic nerve fibres using microelectrodes.
5. The sequence of compound, rapid potential changes recorded extracellularly following excitation of the retina by light is called the *Electroretinogram or ERG*.

13.4.7.3 Rod electroretinogram—wave component

It has been studied by use of drugs as well as microelectrodes inserted into the retina:

A-wave: It is the negative rod receptor potential abrupt in onset and this potential is manifested initially as the A-wave. It reflects receptor cell responses to light.

B-wave: The electropositive B-wave produced in the inner nuclear layer gives rise to the "off response" a characteristic finding in rod ERGs. It presumably arises from glial cells in deep layers of the retina and it also corresponds closely to ganglion cell activity.

C-wave: A large, slow C-wave is triggered by receptors and sustained by potassium leakage which prolongedly hyperpolarizes the pigmented epithelium. It is generated in the pigmented epithelium revealed by intracellular recording from single pigment epithelia cell of cat retina.

DC component—The origin of DC component of the ERG is not known. Maximum amplitudes of both B-wave and DC components were in the inner nuclear layer. The B-wave and DC components were then isolated by two methods. Xylocain, local anesthetic, abolishes the B-wave without disturbing the DC component Again, reduction of intensity of stimulus can separate the DC component to determine its shape but the type of cell that generates the DC component has not yet been identified (Fig. 13.34B I).

13.4.7.4 Cone electroretinogram-wave component

If the ERG is recorded from an area of the eye containing only cones (i.e., fovea), the compound pattern of bioelectric potential changes is like that of rod receptor potential, except (1) that the cones receptor potential decays more rapidly than that of the rods (2) a small negative "of response" (D) wave is seen following termination of the stimulus (Fig. 13.34B II).

13.4.7.5 Receptor potentials of rods and cones

Receptor potentials of both rods and cones are unique in comparison with all other types of receptor potential. Following generation of receptor potential

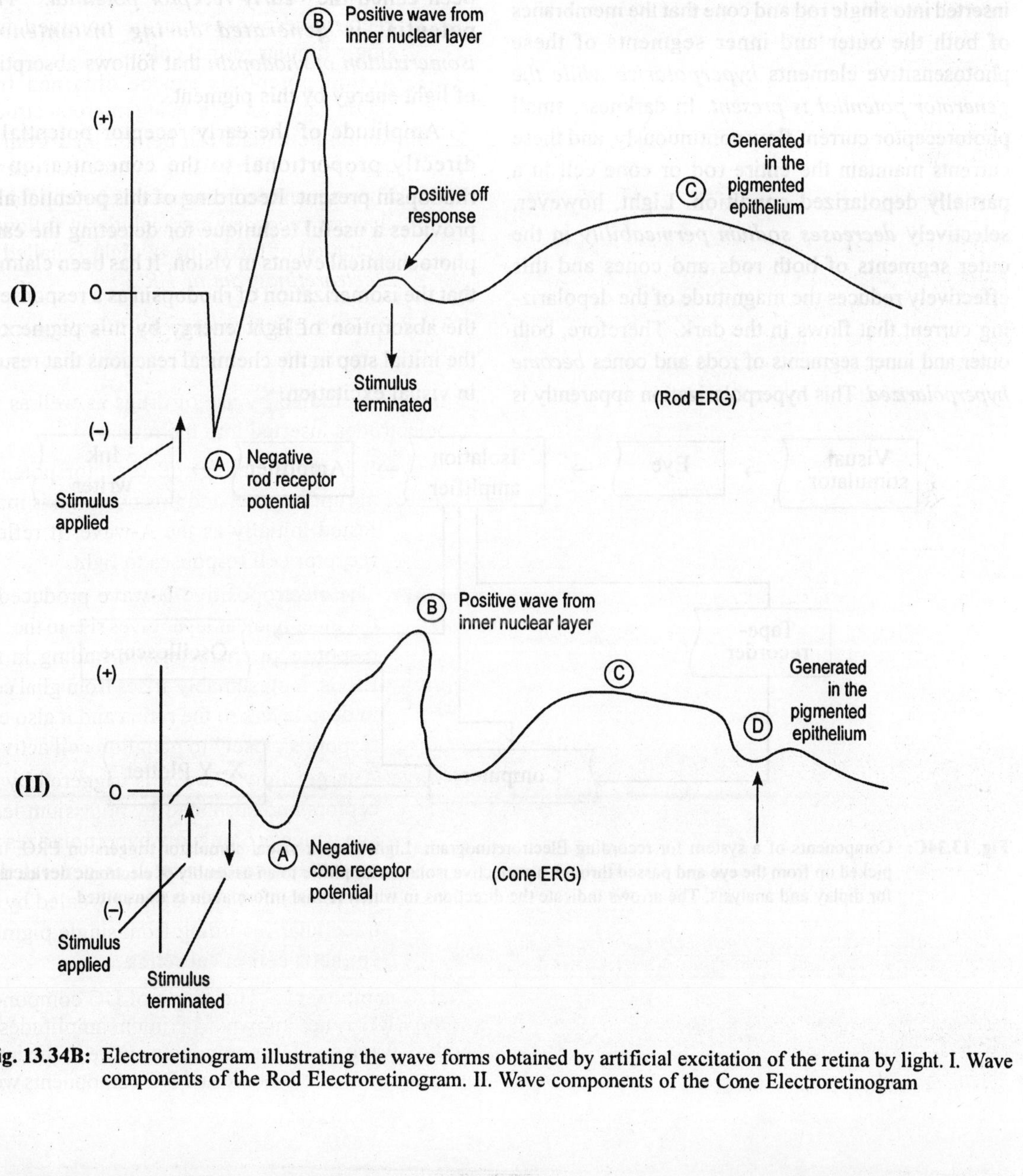

Fig. 13.34B: Electroretinogram illustrating the wave forms obtained by artificial excitation of the retina by light. I. Wave components of the Rod Electroretinogram. II. Wave components of the Cone Electroretinogram

in these retinal elements, the outer and inner portions of the rods and cones, are *positive* with respect to the axon terminals. Current flows in an extracellular pathway from the outer and inner segments of rods and cones toward the axon terminals, a route diametrically opposite to that followed by current flow in other receptors. This current flow serves as a *generator potential* to produce nerve impulses.

13.4.7.6 Hyperpolarization of rods and cones

It has been demonstrated by use of microelectrodes inserted into single rod and cone that the membranes of both the outer and inner segments of these photosensitive elements *hyperpolarize while the generator potential is present.* In darkness, small photoreceptor current flow continuously, and these currents maintain the entire rod or cone cell in a partially depolarized condition. Light, however, selectively *decreases sodium permeability* in the outer segments of both rods and cones and this effectively reduces the magnitude of the depolarizing current that flows in the dark. Therefore, both outer and inner segments of rods and cones *become hyperpolarized.* This hyperpolarization apparently is conducted along the rods and cones by an "electrotonic process", so the bipolar cells, which are contiguous with the photoreceptor cells, are activated by neurotransmitters released by this current flow.

13.4.7.7 The early receptor potential

In addition to the receptor potential, both rods and cones exhibit a minute but measurable biphasic electric response that precedes the A-wave of the ERG without any demonstrable latency. This response has been called the *"early receptor potential."* This potential is *generated during instantenous isomerization of rhodopsin* that follows absorption of light energy by this pigment.

Amplitude of the early receptor potential is directly proportional to the concentration of rhodopsin present. Recording of this potential also provides a useful technique for detecting the early photochemical events in vision. It has been claimed that the isomerization of rhodopsin as a response to the absorption of light energy by this pigment is the initial step in the chemical reactions that results in visual excitation.

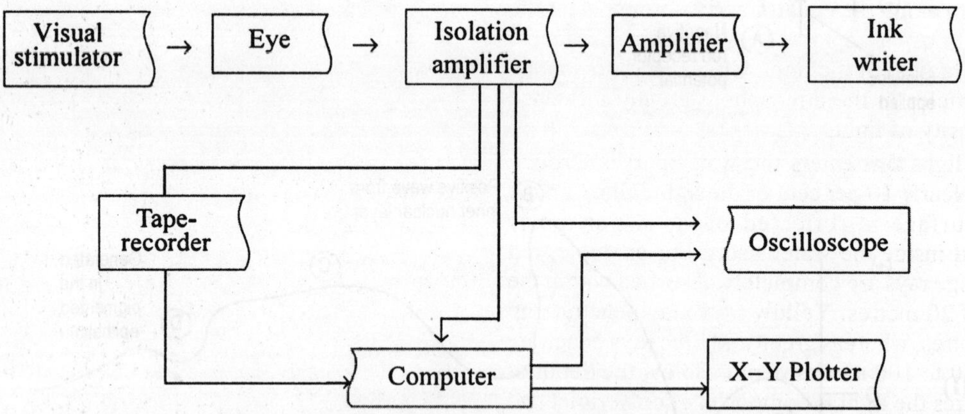

Fig. 13.34C: Components of a system for recording Electroretinogram. Light from a visual stimulator triggers on ERG. It is picked up from the eye and passed through a protective isolation amplifier to an assembly of electronic device used for diplay and analysis. The arrows indicate the directions in which retinal information is transmitted

PART V

13.5 Light and Effect on Animals

Light as an environmental factor is of great ecological value. On light depends the manufacture of food by green plants which in turn support the animal life on the earth. The main source of light is the sun, although the moon, stars and luminiscent organisms exert some amount of light energy. Practically, all the energy required for life on the earth is derived from solar energy. Light is a highly directional factor, its intensity (energy), quality (wavelength) and duration (length of the day) have influence on plants and animals. It also acts as an important limiting factor in the manifold activities of the living organisms.

13.5.1 Distribution of light

The intensity of light (energy) reaching the earth's surface depends upon various factors. The variation in intensity is due to the amount of absorption and dispersion in the atmosphere and also a number of climatic and topographical factors. When the angle of incidence is similar, the light rays have to travel a longer distance through the atmosphere which causes reduction in its intensity. The light intensity decreases with the increase of latitude. It is maximum along the equator and minimum near the poles. Atmospheric gases also absorb and disperse a small fraction of the light. Clouds and suspended dust particles in the atmosphere greatly influence the intensity of light.

The light that enters the water passes through the air. Nearly 10 per cent of the light falling on the water surface is reflected back. A portion is absorbed inside the water itself. For example, red and orange rays are completely absorbed within the depth of 20 metres. Yellow rays may penetrate up to 50 metres, whereas green and blue rays penetrate nearly 50 to 100 metres depth. Below the depth of 100 metres the light is composed of only violet and ultraviolet rays. No light can penetrate below 200 metres inside water and in this zone there is total darkness. Depending upon the light penetration in the ocean, the zone up to 50 metres is called a **euphoric**, between 50 to 200 metres **diphotic** and below 200 as **ophotic** zones.

13.5.2 Absorption of light by matter

In passing through matter, light intensity is cut down by two processes:

(a) By absorption.

(b) By scattering.

Absorption of light may be (a) general or (b) specific (i.e., selective absorption at only certain wavelengths).

The energy transferred to absorbing molecules may cause:

(a) Vibration of atoms along axis of connection,

(b) Rotation around the axis of connection or

(c) Changes in the energy levels of elctrons within the atoms.

In each of these three types of changes, molecular energy is altered in integral amounts, that is as an integral multiple of the quantum of energy. Because of this quantum nature of energy transfer, only those wavelengths which possess energy of a particular range of frequencies will be absorbed by a substance.

Absorption of energy raises molecular energy above ground level or base level. This extra energy may be liberated as heat by collision of the excited molecules with others or it may be re-emitted as light or the energy may be utilized to participate in a chemical reaction. Some reaction require activation by light excited molecules and cannot be excited in any other way. Such reactions fall under the *Grothus-Draper Category*. Usually there is a linear relationship between the quanta of light absorbed and the amount of end product produced. In other words, the intensity × time is a constant. This is known as the *Bunsen-Roscoe Law* or "the law or reciprocity". The law was formulated for photochemical processes in general, namely, that for the production of a given photochemical effect, a constant quantity of energy is required which can be distributed within certain limit by varying either the illumination or its duration. In other words, the product of illumination (I) and the time of exposure (T) is constant, Energy = KIT.

13.5.3 Effect of light on animals

Light acts as a limiting factor for the multi-sided activities of living organisms. It regulates a number of biological processes and also the structural as well as behavioural characteristics of animals. The important effects of light on animals are as follows:

1. **Effect on Metabolism**: Light has a profound effect on the metabolism of animals. Besides heating the tissues, the absorbed radiant

energy brings about ionisation of protoplasm. This increases the rate of enzyme activity and general metabolic rate. Animals living in caves, where there is little light, show slow rate of metabolism.

2. **Effect on Locomotion**: Light controls the speed of locomotion in many lower organisms. This phenomenon is known as **photokinesis.** Experiments on the blind larvae of mussel crab (Pinnothers maculatus) have shown that their swimming rate increases with increased illumination. Light often plays an important role in orientating locomotion. Movement of animals in response to light is known as **phototaxis.** If an animal moves towards the source of light it is said to be **positively phototectic** (e.g., Euglena, Ranatra) and if it moves away from the light source, it is termed **negatively phototectic** (e.g., Earthworms, slugs). Orientation by growth or **phototropism** is also seen in many tubiculous worms.
3. **Effect on Development**: Light in some animals accelerates the development and in some cases retards the process. For example, the larvae of Salmon undergoes usual development when sufficient light is available. In the absence of light the development of egg becomes abnormal and there is a heavy mortality of larvae. The larvae of "Mytilus" grows larger in darkness rather than in the light.
4. **Effect on Eye**: The degree of development of eye is dependent on the intensity of light available in the environment. In many cave dwelling animals like "Proteus", the eyes are either absent or rudimentary. Nocturnal animals and deep sea fishes have generally large eyes.
5. **Effect on Vision:** The most significant aspect of light in the vision of higher animals lies in its visibility. Man is able to see only in the presence of some form of light which has been observed by "Scotopic Visibility Curve". The human eye is most sensitive at a wavelength of about 507 mμ . At this wavelength the minimal energy entering the eye was found to be between 2.1×10^{-10} and 5.7×10^{-10} ergs as measured at the cornea. On the other hand, the curve indicates that the threshold energy rises enormously beyond about 400 mμ and 650 mμ at the end of the visible spectrum. Many fishes depend upon their eyes for locating the food organisms. This limit of visibility varies in different animals.
6. **Effect on Pigmentation**: Light may affect pigmentation in animals and plants in various ways. The cave animals mostly lack pigments due to the absence of light. The pigmentation of many animals form a colouration that appears to give it protection from enemies. Such coloration is called "protective coloration". For example, the body colour of leaf insects is like that of a leaf. Some animals like Gecko are able to change their colour abruptly to match with the surroundings. Such changes are said to be associated with the stimulus of sight.
7. **Photoperiodism or Photoperiodicity**: The response of an organism to the day length is termed as *photoperiodism*. Various organisms exhibit a 24 hour cycle in their activities which is called *diurnal periodicity* or *Circadian rhythm*. The most fundamental diurnal rhythm is photosynthesis which fluctuates due to daily changes of light. A large number of Copepods and other planktonic animals move towards the surface of water during the night but go downwards during the daytime. Some animals are active in the light but less active in the darkness and vice versa. Many defenceless animals like mice come out in the night for search of food, when they are less likely to be detected by the enemies. The development of fur and feather coats of animals appear to be influenced by photoperiodicity. The activities of many organisms in relation to their reproduction cycle show a close relation to the moon. This is known as *lunar periodicity*. In this case biological rhythm appears once or twice in every lunar month. If the rhythm occurs in 15 days, it is termed *semilunar*, but when it happens in 30 days it is called *lunar* (see chapter on chronobiology).

8. **Effect on Reproduction or Gonads**: In many animals like birds, light initiates the annual breeding activities. Gonads of birds become active during summer with increased illumination. Animals have been classified as short-day, long-day and indifferent day length animals:
   (a) Deer, sheep and goats are **short day animals** and can be brought to sexual activity with decrease in the length of exposure to daylight.
   (b) Birds breed during spring with lengthening of day and are, therefore, called **long-day animals.**
   (c) Guinea pigs and squirrels are not at all affected by daylight, hence they are **indifferent-day length animals.**
   (d) Fee and Parkes (1933) have shown that animals, e.g., ferret, having an anoestrous season in winter can be brought into oestrous by prolong exposure to light.
   (e) There is much evidence that light has an important influence on sexual maturity and it has been shown that blind girls have an acceleration of menarche compared with normally sighted girls.
9. **Effect on Endocrines**
   (a) **ACTH Secretion**: There is "diurnal" or "circadian rhythm" of glucocorticoids secretion due to rhythmic variations of ACTH secretion from anterior pituitary and corticotrophin releasing factor from hypothalamus.
   (b) **TSH Secretion**: There is some evidences for a "circadian rhythm" of Thyrotrophin releasing factor secretion from hypothalamus.
   (c) **Pineal Gland**: If rats are exposed constantly to light for prolonged period, the cyclic pattern of gonadotrophins discharge is abolished and the normal oestrus cycle ceases and is replaced by state of *constant oestrus*. It has been suggested that the constant light stimulus acts by inhibiting the activity of the pineal gland.
10. **Diapause**: In animal kingdom, a type of dormancy is observed—like in plants—known as diapause which can occur at any stage in the life cycles such as egg, larva, pupa or adult. The causative mechanism of diapause is the same like plants, i.e., hormones regulated by the environmental stimuli. These hormonal changes are related to day-length, as in plants. As soon as the hours of daylight fall below a certain critical level, diapause sets in. The critical amount of daylight varies from insect to insect, e.g., for Cabbage white butterfly it is about 12 hours. When the light falls below this level, the pupae enter diapause and do not resume development, until the following spring. It has been shown experiementally that the light does not stimulate eyes because covering them does not abolish the diapause response. Probably the light stimulates the brain directly and produces another substance or substances which inhibits growth-stimulating hormone. The end of diapause in some species, the change to longer days may be the effective stimulus. But in many insects a particular period of cold is necessary before growth is resumed which is comparable to stratification in seeds and buds.

13.5.4 Effect of photoperiods on gonads of animals

Among all the seasonal changes in the environment, the length of the day is the most regular, being directly related to the fixed tilt of the earth and fixed rotation of the earth round the sun. There is much evidence that variation in the length of the day—either day-length or the rate of change in the length of day—trigger off the physiological preparations for breeding in many species of both invertebrates and vertebrates. However, the sensing of day-length or of the rate of change of day-length is a complex biological function which remains largely unexplained.

13.5.4.1 Experimental observation

1. Scharrer suggests that in transparent organisms, light can exert a direct effect on the gonads. Organisms with opaque integuments may have

a transparent membrane or window which admits light to a part of the brain, in which photic stimuli are converted into chemical messages by neurosecretory cells which reach organs such as the gonads whose functions are light-dependent, but to which light can no longer penetrate (Figs. 13.35A and 35B).

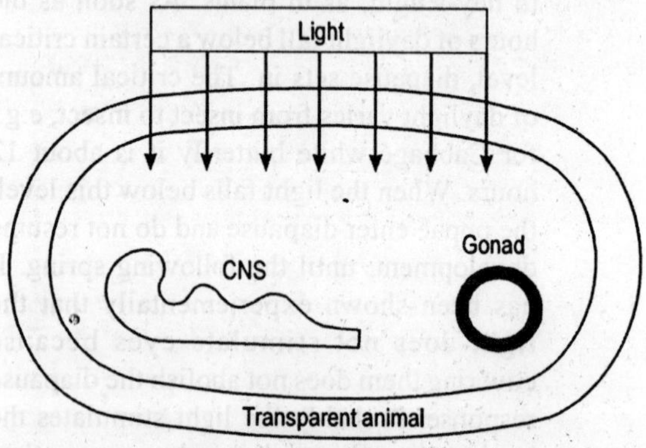

Fig. 13.35A: In transparent animals light can exert a direct effect on internal organs including the CNS and gonads

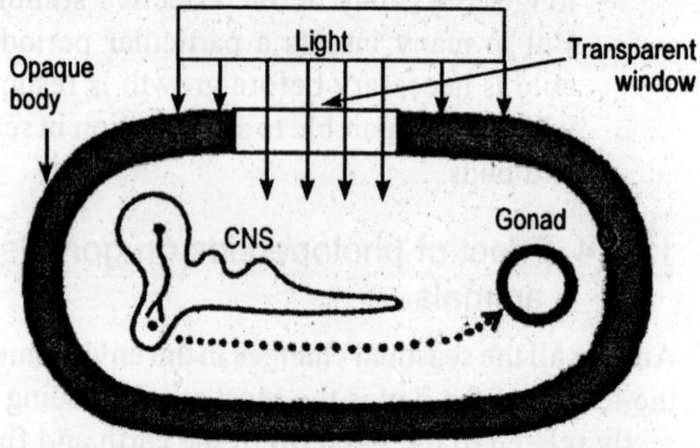

Fig. 13.35B: Where an opaque integument excludes light from internal structure, a transparent window may admit light to photosensitive parts of the brain, where the photic stimuli are converted by neurosecretory cells into chemical messages which reach organs whose functions are light dependent but to which light can no longer penetrate

- **Scharrer's representation of influence of light on gonads**

2. Embryologically speaking, the retina is a specialised light sensitive tissue of CNS origin and in vertebrates (except fishes) there is a pathway from the retina which is distinct from that of the visual system. This goes to the hypothalamus, reaching the median eminence from which releasing factors (RF) flow via the hypophyseal-portal vessels to the anterior pituitary which releases gonad stimulating hormones. The current evidence is that the influence of light on mammalian reproduction is wholly or principally via this pathway from the retina to the hypothalamus (Fig. 13.35C).

3. In some mammalian species pinealectomy changes the effects of light on the gonads, and it has been suggested that light and pinealectomy operate independently to reduce the availability of the same inhibitory factor. In hamsters there is good evidence that the pineal gland is involved in the gonadal atrophy that occurs after blinding or exposure to short light periods and in pinealectomized hamsters, the normal decrease in spermatogenesis during December and January did not occur. On the other hand, Roche found no effect on oestrus, ovulation and luteinizing hormone (LH) secretion in owls after pinealectomy. But there is no evidence that the pineal gland functions as a primary light-sensor in mammals.

4. Although the seasonal gonadal stimulation of blinded ducks is less than that of ducks with undamaged eyes and optic tracts, retinal photoreceptivity is not essential for photoperiodic control of gonadal activity. Ducks blinded by enucleation or transection of the optic nerve will still respond to long days by gonadal development. It has also been shown that light applied to the hypothalamic region by means of implanted light-conducting mineral (e.g., quartz) stimulated the development of the testes of the ducks. This suggests that some light may still penetrate to skull of the duck and act on intracranial sensors.

Although day-length exerts a powerful influence on the breeding patterns of perhaps the majority of both birds and mammals, and especially those at high latitudes, the effect of light varies not only between species but also between breeds.

Conclusion: The conclusion is that there is no essential difference in the photosensitive mechanism of short-day animals. In both there is an endogenous refractory period which is longer in breeds, genetically adapted to high latitudes. The implication of this finding is that the length of daylight acts as the trigger for sexual preparation and is fixed by some physiological cycle of responsiveness and refractoriness to light.

13.5.4.2 Light and gonadotrophin secretion

1. Fee and Parkes (1933) have shown that animals, e.g., ferret, having an anoestrous season in winter and ovaries remain silent, can be brought into oestrus by prolonged exposure to light. This responses to light cannot be brought about in hypophysectomised animals and is impaired by section of the optic nerves. It is, therefore, concluded that the awakening of the gonads is brought about by stimulation of the anterior pituitary from the retina. In this reflex, the visual cortex and superior colliculi appear not to be involved, and the path seems to be either to the ventral nucleus of the lateral geniculate body or to the subthalamus by way of the accessory optic tracts. But how the impulses then reach the anterior pituitary is doubtful; it may be via the hypothalamus (Le Gros Clark et al. 1939).
2. The phenomenon of light influencing structure and function through the mediation of a retino-hypophyseal mechanism is not peculiar to cold-blooded animals. The work of number of experimenters indicates that the well-known association of the seasonal periods with morphological changes (e.g., colour and texture of hair or plumage) and the sexual cycles of certain mammals and birds as well as with the migration of birds, is due in part at least to light acting upon the anterior pituitary through the medium of retinal impulses. Rown has able to induce sexual activity in crows and canaries at any desired time of the year by varying the periods of their exposure to artificial light.
3. It is well known that the domestic fowl will lay regularly in winter if its period of exposure to light is lengthened by artificial illumination. Similar experiments have been carried out upon mammals, hedgehogs and racoons.

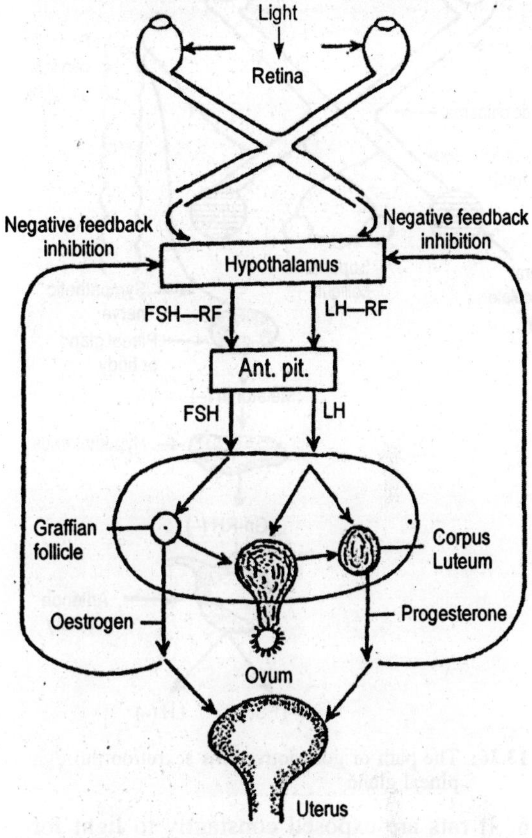

Fig. 13.35C: Photoperiodic effects on gonads by neuroendocrine mechanism

4. The extreme sensitivity of light of the pigeon with respect to the activity to the time of egg laying is extraordinary. The pigeon lays a pair of eggs in the morning and the second of the pair being laid, with the regularity of clock-work, half an hour after the first. Lengthening or shortening the period of exposure to light alters the duration of the interval between the laying of the first and second eggs of the pair (see effect of light on animals).
5. The menstrual rhythm is frequently disturbed in nurses on night duty or air hostesses travelling on intercontinental flight.

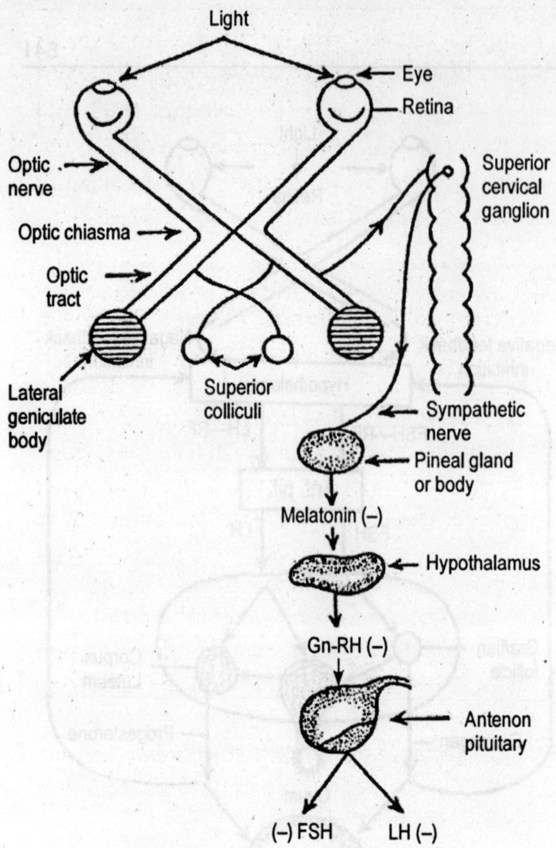

Fig. 13.36: The path of gonadotrophins secretion through pineal gland

6. If rats are exposed constantly to light for prolonged periods the cyclic pattern of gonadotrophin discharge is abolished and the normal oestrus cycle ceases and is replaced by state of "constant oestrus". It has been suggested that the constant light stimulus acts by inhibiting the activity of pineal gland. Darkness stimulates the secretion of hormone melatonin from the pineal body whereas exposure to light inhibits melatonin secretion. These effects are mediated by the retina → optic tract → brain stem → fibre from optic tract → superior cervical ganglion → sympathetic nerve → pineal gland → melatonin → hypothalamus → GnRH (Gonadotrophin releasing hormone) → anterior pituitary → FSH and LH serection. In darkness due to melatonin secretion, gonadotrophin serection is depressed whereas in the light due to inhibition of melatonin secretion, gonadotrophin release is stimulated.

Physiologically, the pineal body may be concerned with regulation of time-dependent effects such as onset of puberty and rhythmic activities such as hormonal regulation of ovulation.

In lower vertebrates, e.g. amphibians and dogfish, the pineal contains light receptor cells and nerve cells which conduct impulses to brain. In higher vertebrates including man the photoreceptor elements have been completely replaced by parenchymal cells called "pinealocyte" and also contain neuroglial cells and numerous postganglionic sympathetic nerve fibres from cells in the superior cervical ganglion. These fibres contain noradrenaline and mostly terminate on the pinealocytes which secrete hormone melatonin. Melatonin reduces gonadal activity by decreasing the rate of discharge of hypothalamic GNRH and thus reducing the secretion of FSH and LH from the anteior pituitary (Fig. 13.36). At least there is no evidence that pineal gland functions as a primary light-sensor organ.

In mammals and men, the neurosecretory path for light stimuli for secretion of gonadotrophines has been illustrated in the previous discussion (Fig. 13.35C).

13.5.5 Effect of light on secretion of other anterior pituitary hormones

1. **ACTH (Adrenocorticotrophic hormone)**: There is "diurnal" or "circadian rhythm" of glucocorticoids secretion due to rhythmic variations of ACTH from anterior pituitary and CR-H (corticotrophin releasing factor) from hypothalamus. The circadian rhythm of glucocorticoids secretion depends upon cyclic variation of CR-H secretion from hypothalamus due to **biological clock** in the limbic system.

The activity of this intrinsic "biological clock" in determining the circadian rhythm of CR-H release is supplemented by the effects of the rhythmically alternating cycles of sleep and activity and the rhythmically alternating cycle of darkness and light during each 24-hour period. If human subjects are exposed to constant light and the length of the sleep/activity cycle varied from 12 to 33 hours, the length of the glucocorticoids cycle is adjusted in such a way that a peak is still attained first before waking. In addition, subjects maintaining a normal sleep/activity cycle, but exposed to abnormal cycles of light and darkness, showed two peaks in plasma glucocorticoids levels. One peak corresponds to the transition from sleep to activity and the other coincides with the change from darkness to light.

2. **TSH (Thyroid stimulating hormone):** There is some evidence for a "circadian pattern" of TR-H (Thyrotrophic releasing hormone) release from hypothalamus. Thyroid hormone levels show a peak at about 4 a.m. and less at about 5 p.m. Whether this is due to the same factors that determine the circadian rhythm of glucocorticoids secretion or whether it is a secondary consequence of the changes in corticosteroid level is not clear.

PART VI

13.6 Ultraviolet Light on Living System

Light is an electromagnetic vibration. It has a dual nature; in some cases light acts like a wave motion similar to sound. In other instances, it behaves more like a particle, a packet of energy, with many of the properties commonly associated with mass particles. The phenomena occur in the eye such as reflection, refraction, diffraction and interference, light can be considred as a wave motion. Again, whenever light is absorbed, with transformation of its energy into chemical energy or heat, it is treated as a particle or quantum. Source of ultraviolet light is sunlight. The electromagnetic spectrum shows that ultraviolet rays cover the wavelength between 1000 Å to 4000 Å, visible light between 4000 Å to 8000 Å and beyond 8000 Å is infrared region. Ultraviolet is also available artificially from mercury vapour lamp (Fig. 13.37).

13.6.1 Effect on living system

I. **Nuclear and Cytoplasmic Damage**: Study with cells shows that characteristics of absorption of the nucleic acid is in the region of 2600 Å whereas proteins absorbed most strongly at the longer wavelngth region of 2800 Å. The strong absorption of nucleic acids in the ultraviolet indicates that the lethal action of ultraviolet rays is due to nuclear damage as is seen in case of bacteria, yeast and dermatophytes. Sensitization to haematopyrin in case of humans in photodynamic action is very similar to sunburn due to ultraviolet light but with the following differences:

1. Sunburn due to ultraviolet light is a direct effect whereas photodynamic action is indirect effect,
2. Sunburn due to ultraviolet does not require photosensitizer whereas photodynamic action requires photosensitizer,
3. Effect of ultraviolet light is not oxidation reaction but the haemolysis, stimulation of muscle, destruction of bacteria etc. in photodynamic action being effected in presence of O_2.

II. **Transforming Principle**: The influence of ultraviolet light on nuclear material can be revealed and studied quantitatively by measuring the behaviour of transforming DNA.

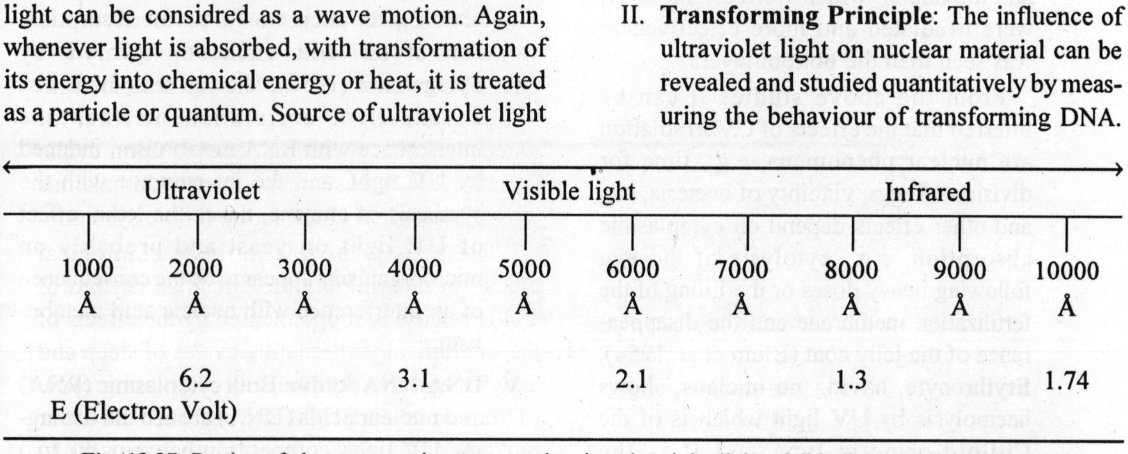

Fig. 13.37: Portion of electromagnetic spectrum showing ultraviolet light, visible light and infrared rays

Extract of DNA from streptomycin resistant "Haemophilus influenzae" is measured for its power by transforming wild strains into resistant organisms. The effectiveness may be measured in terms of the number of "transforms" in unit time. After irradiation with ultraviolet light, the transforming effectiveness decreases in proportion.

III. **Cytoplasmic Effects**: Apart from nuclear damage, cytoplasmic absorption also takes place.

1. Giesa showed that irradiation of "Paramoecium" delayed its cleavage, and the amount of delay was proportional to the dose of irradiation. The action spectrum was closely to the absorption spectrum of nucleo-protein, the maximal effect being at 2650 Å. If "Paramoecia" were well fed before irradiation, their susceptibility to UV light was reduced and the action-spectrum was not seen like nucleic acid or protein absorption.
2. Marshak et al. (1949) studied the delay of cleavage in "Arbacia" and opined that nuclear absorption is the main factor for cleavage delay. He also observed that the dose 2537 Å radiation is sufficient to delay cleavage and it was very much less if the sperm, rather than the egg, was irradiated, indicating that absorption by the egg's cytoplasm was ineffective in causing delay.
3. Harding and Thomas centrifuged eggs and the nuclei, which were on the layer, were irradiated and more effectiveness was seen than the bottom layers.

From the above studies it can be inferred that the effects of UV irradiation are nuclear phenomena, e.g., time for division of eggs, viability of bacteria, etc. and other effects depend on cytoplasmic absorption, e.g., cytolysis of the egg following heavy doses or the lifting of the fertilization membrane and the disappearance of the jelly coat (Blum et al. 1954). Erythrocyte, having no nucleus, shows haemolysis by UV light which is of the Colloid-osmotic type and it is the consequence of a direct attack on the cell membrane. The escape of organic phosphates from the yeast cell irradiated with UV light is a complex phenomenon, involving interference with nucleic acid metabolism as well as permeability changes (Swenson, 1960).

IV. **RNA Absorption**: Goldman and Setlow (1956) studied the superficial cleavage of "Drosophilia" eggs to differentiate the nuclear and cytoplasmic damage. After fertilization, the zygote nucleus migrates to the centre of the egg immediately where it divides several times; at this stage if it is irradiated by UV light, the nucleus is protected from the damage by surrounding cytoplasm. At a later stage, most of the daughter nuclei migrate to the cortical cytoplasm, where subsequent development occurs. At this stage nuclei are subject to damage. Cytoplasmic damage due to UV light causes defects of morphological development such as herniation of the intestine and damage of nuclei causes complete arrest of developments and to produce maximum effects, wavelength of 2500-2700 Å was sufficient. This corresponds to the absorption of purines, hence it can be presumed that cytoplasmic damage is due to absorption by RNA. This view was substantiated by studies on the strength of yeast to synthesize enzymes. Swenson and Giesa (1950) showed that galactozymase formation was inhibited by UV light and the action spectrum is that for purine absorption. Halvarson and Jackson, quoted by Spiegelman (1956), showed that there was an excellent correlation between the interference with RNA metabolism, induced by UV light, and the interference with the synthesis of enzyme, hence the lethal effect of UV light on yeast and probably on microorganisms appear to be the consequence of an interference with nucleic acid metabolism.

V. **DNA/RNA Radio**: Both cytoplasmic (RNA) and nuclear acids (DNA) absorb the damaging UV light, competition between the two may be expected. Since nuclear damage is more serious in a growing cells, the damage caused by a given dose will be greater.

The proportion of DNA to RNA will be greater. This may be attributed to:

1. Resistance of starved yeast, starved "Shigella sonnei",
2. If the DNA/RNA ratio of *E. coli* is doubled—induced by copper—the resistance to UV light damage is increased (Weed and Longfellow, 1954).

VI. **Formation of Thymine Dimers**: The most common form of damage from UV irradiation is the formation of "**thymine dimers**" in which two adjacent thymine residues become covalently joined by a cyclobutane ring (Fig. 13.38).

VII. **Growth of Cancer Cells**: A cancer cell results from a permanent genetic change in a normal somatic cell. Such a change called the "malignant transformation", may be triggered by some external physical agent, such as X-ray or excess ultraviolet irradiation.

VIII. **Xeroderma Pigmentosum**: Xeroderma pigmentosum is a rare human genetic disease in which skin is abnormally sensitive to the ultraviolet component of sunlight. Such patients have a tendency to develop skin cancer. Cleaver (1969) discovered that this disease is due to genetic deficiency of an enzyme normally participating in excision repair of ultraviolet induced thymine dimers probably the endonuclease required for the first incision or cut.

IX. **Ultraviolet Light in Microscopy:** Most of the methods used can yield little information regarding the location and function of nucleoprotein in the nucleus and cytoplasm. Apart from Feulgen method and specific enzyme method, the Casperson's technique is being used for examining the tissues directly and without injurious chemical treatment with the help of ultraviolet light through microscope known as *Ultraviolet microscope* based on ultraviolet light absorption of nucleic acid. By his technique, the presence of as little as 10^{-11} mg. of nucleic acid may be determined. But this technique does not differentiate DNA and RNA. Casperson has shown that U-V absorption spectrum can distinguish not only the presence of nucleic acid but also that of certain types of protein. Such as, the tyrosine and tryptophan residues absorb strongly at 2750 Å and 2800 Å, respectively, whereas the aliphatic types have no specific absorption in the ultraviolet.

X. **Biosynthesis of Vit D_3 in the Skin:** Vit D is a prohormone of a sterol type that occurs in nature chiefly in animals (e.g., fish liver oil) but also in plants and yeast. D vitamins are generated from the provitamins (1) ergosterol of plants and (2) 7-dehydrocholesterol of animals and plants. Ergosterol and 7-dehydrocholesterol differ chemically only in the side-chains at 21 position. Ultraviolet irradiation (λ256-313 nm) from sunlight sponteneously cleaves B ring of ergosterol or 7-dehydrocholesterol. In plants, the UV irradiation of ergosterol leads to the production of ergocalciferol (Vit. D_2). In animals the 7-dehydrocholesterol converts into cholecalciferol (Vit D_3) by irradiation of the skin. Ergocalciferol and cholecalci-

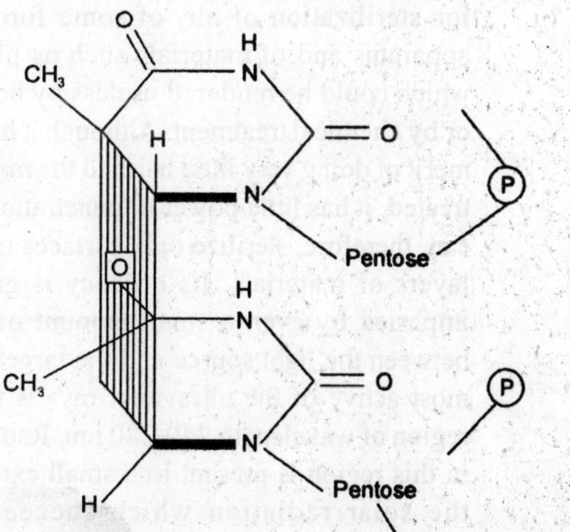

Fig. 13.38: Structure of thymine dimer. Cyclobutane ring between successive thymine residues. The shaded portion is the plane of the cyclobutane ring

ferol are of equal biologic potency Vit D preparations available commercially are products of the ultraviolet irradiation from yeast.

XI. **Effect on Enzymes**: Urease is irreversibly inactivated by ultraviolet light. Kauffman (1944) showed that papain which is one of the sulphydryl enzymes, was inactivated to the extent of 90 per cent in 53 sec by UV light whereas trypsin, not a sulphydryl enzyme, which required the same degree of inactivation. From the observation, it may be interpreted that—SH group is not sensitive to UV light. Purified acetylcholinesterase was found to be very resistant to ultraviolet radiation. Hargreaves (1955) is of opinion that the inactivation after very large doses of ultraviolet light is not due to the effect on –SH groups and this group does not play any role.

XII. **Biosynthesis of Melanin in the Skin**: Sunlight through its ultraviolet rays accelerates the conversion of tyrosine to dopa by photochemical reaction during biosynthesis of black pigment melanin in the skin.

XIII. **Antimicrobial Effect**: Sunlight has an antimicrobial effect by virtue of its content of ultraviolet light. Artificial UV light is used for sterilization of air, of some forms of apparatus and of materials such as plasma which could be rendered useless by heating or by chemical treatment. Although it has the merit of doing very little harm to the material treated, it has little power of penetration and can, therefore, sterilize only surfaces or thin layers of materials. Its efficacy is greatly imparied by even a small amount of dust between the light source and the target. The most active of the ultraviolet rays is in the region of wavelength 240-280 nm. Radiation in this region is present to a small extent in the solar radiation which succeeds in penetrating the earth's atmosphere, which contains little radiation with wavelength below 290 nm. But artificial UV rays from "low pressure mercury vapour lamp" emits 95% of the radiation of wavelength 253.7 nm. The lethal effect of this light having bacteriocidal range causes damage to nuclear DNA producing **thymine dimers**.

XIV. **Sensitive to Viruses**: All viruses are inactivated by ultraviolet light and by X-rays though there are quite large variations in the dosage required by different viruses.

XV. **Lysogeny and UV Light**: Infection of a bacteria by phage is not necessarily followed by phage multiplication and bacteriolysis. Sometimes when phage DNA enters a host cell it does not derange its synthetic activities but is integrated into the bacterial chromosome. Such a phage is described as **temperate**, as distinct from virulent or lytic, and the integrated non-lytic form in which it persists in the cell is called **prophage**. In this form it is reproduced synchronously with the bacteria and handed down to all the progeny of the original cell. This relationship is known as **lysogeny** and the host bacteria as **lysogenic** because in certain circumstances propagation and release of the phage may occur and it may then cause lysis of other bacterial strains. Such propagation may occur spontaneously, and it can also be induced by using ultraviolet light or various chemicals.

XVI. **Identification of Dermatomycoses**: Fungi of the genera Trichophyton, Epidermophyton and Microsporum, collectively known as "dermatophytes", cause superficial infections of the Keratinized layers of the skin, hair and nail but never attack deeper tissues. Scales collected from skin lesions and recognition of hairs infected with "Microsporum" species is helped by the fact that they fluorescein UV light.

13.6.2 Photoreactivation

Kelner (1949) observed that suspension of spores of "Streptomyces griseus" showed 100000-fold recovery by exposure to visible light after ultraviolet irradiation injury—the phenomenon is known as **photoreactivation**. Later it was shown that the destructive effect of ultraviolet light could be largely reversed by subsequent illumination of the organisms and the effect is due to the visible portion of the spectrum (not to the infrared) and increasing with the duration of radiation. Irradiation of the organisms with visible light before treatment with ultraviolet was ineffective to prevent damage. Dulbecco (1950) carried out experiments on photoreactivation of bacteriophages inactivated with ultraviolet radiation. He incubated equal quantities of ultraviolet inactivated phage with sensitive bacteria, the one in the light and the other in darkness. The light one showed 1000 times as many plaques as those in the dark. He further observed that wavelengths longer than 4400 Å were ineffective in reactivation.

13.6.2.1 Role of cytoplasm

Studies on delayed division of "Arbacia" or "Paramoecium" have shown that photoreactivation is a general phenomenon but not universal because Jagger (1958) has shown the instances of non-photoreactiveness in many bacteria, e.g., in strains of B. Cereus, B. megatherium, S. faecalis, Diplococcus pneumonae, etc.

Although ultraviolet effect is mainly a nuclear phenomenon, reactivation seems to depend on the cytoplasm. Marshak (1949) observed that the delayed division of the zygote, obtained by fertilizing a normal egg with an ultraviolet irradiated sperm, could not be reversed by visible light. Irradiation of the zygote, however, caused photo-reversal. Dulbecco observed that exposure of the ultraviolet inactivated phage to visible light in the absence of bacteria would be ineffective. Cook and Rieck (1962) showed that in the sand-dollar the following systems could be photoreactivated:

1. Eggs exposed to ultraviolet and fertilized with normal sperm,
2. Eggs exposed to ultraviolet and fertilized with irradiated sperm,
3. Zygotes exposed to ultraviolet.

The other system examined—namely normal eggs fertilized with ultraviolet irradiated sperm that had been treated with longer wavelength light—showed more delayed cleavages than if no attempt had been made to reactivate the sperm. Thus, all the above experiments reveal that presence of cytoplasm is essential for successful photoreactivation.

13.6.2.2 Cytoplasmic agent (Reactivating enzyme)

Rupert (1960) extracted a material from baker's yeast and, when added to inactivated transforming DNA, showed protoreactivation. In absence of this agent, even the long-wavelength light was not effective. Not all bacterial extracts were effective such as *E. coli* extract was effective whereas Haemophilus influenzae was ineffective. The extracted material had the characteristics of an enzyme known as **reactivating enzyme** being heatlabile, non-dialysable and forms complex with inactivated DNA but not with normal DNA. The mechanism of reactivation seems to be as under:

Formation of enzyme-inactivated DNA complex → DNA repairs itself → light releases the repaired molecule from the complex. The photoreactivation process may be competitively inhibited by adding another DNA so long as this has been inactivated by ultraviolet light. When an intact cell has been photoreactivated, the cause of this reactivation is the restoration of the DNA to normal.

13.6.2.3 Changes in DNA in Photoreactivation

Studies of the physico-chemical properties of the photoreactivated DNA revealed that the repair consists of reversal of stable cross-linkages between complementary DNA strands. Photoreactivation is never complete, and the residual damage probably due to failure to repair the broken hydrogen-bond links (Marmur and Grossman, 1961).

PART VII

13.7 Light and Application in Therapy

Penetrating power of ultraviolet light and visible light are proportional to the wavelength of the light. The most active of the ultraviolet light are those with wavelengths in the region 240 to 280 nm. Radiation in this region is, however, present to only a small extent in the sunlight which succeeds in penetrating the earth's atmosphere, the latter containing little radiation with wavelength below 290 nm. UV light has the serious disadvantage that is penetration capacity is very slight and active only on the surfaces. Ultraviolet light is, therefore, therapeutically useful only to produce actions at or very near the body surface. UV light penetrates only a fraction of a millimetre of the skin. UV light of 2800-3000 Å is absorbed practically completely in the corneum of the skin.

The energy of photon of light is inversely proportional to the wavelength, and many potentially useful photochemical reactions cannot occur with photons of less energy than those of UV light. Therefore, the therapeutic applications are limited to those, actions in which UV light is effective and which can occur at the surface.

13.7.1 Source of UV light and visible light

(A) Natural: Sunlight is a source of visible and near ultraviolet light (i.e., light at λ = greater than 2900Å) in addition to infrared rays. The atmospheric envelope absorbs light of shorter wavelength than this.

(B) Artificial: Artificial sources of ultraviolet light can be produced by special lamps. With these lamps, over 95% of the radiation emitted is of wavelength 253.7 nm. These lamps are:

1. Carbon arc (particularly those with metals work into the carbon electrodes).
2. Mercury vapour lamp (Low pressure).

Artificial visible lamp (Low pressure).

1. Incandescent lamps (in reflectors or in light-bathing cabinet).
2. Carbon arc lamps.

Relative distribution of wavelengths of different sources of light are shown in (Fig. 13.39). These filters have been used to limit the spectral distribution of light from various sources. And with these limitations one vary the spectral quality of the light uses therapeutically by selecting the correct sources.

13.7.2 Physiological effects

Penetrability of various radiations is considered by scanning the entire electromagnetic spectrum in terms of absorption by tissues. The pattern shows

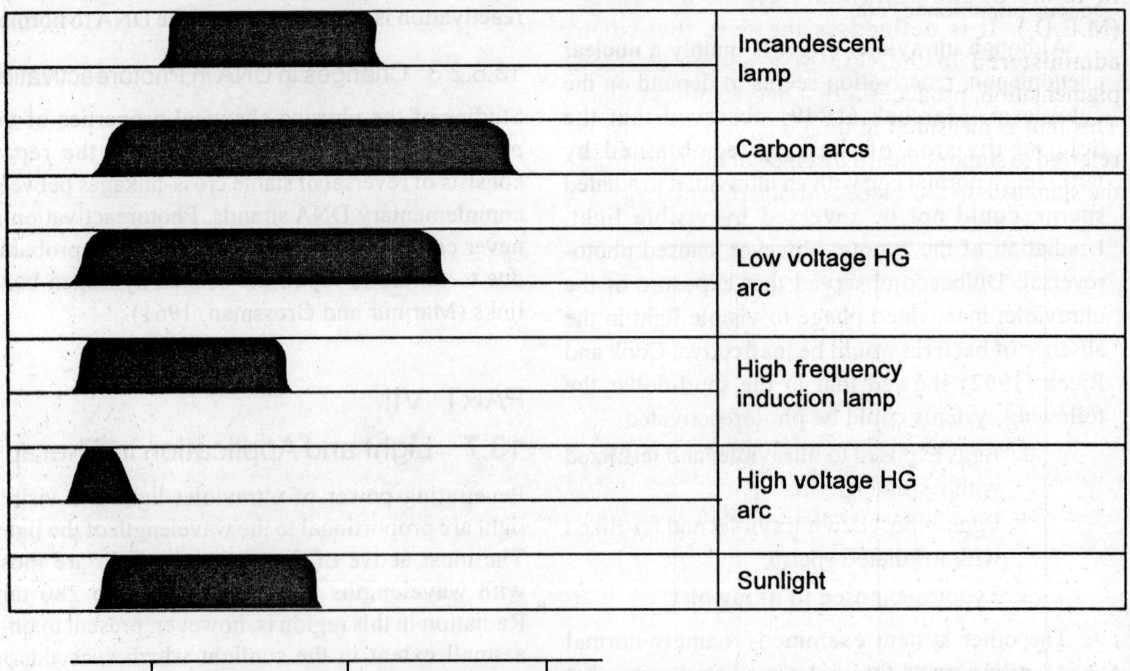

Fig. 13.39: Various sources of light used in light therapy and their spectral distribution

two "humps" in the spectral absorption curve (Fig. 13.40), one in the ultraviolet region and the other in the far infrared region. These absorption maxima are closely related with wavelengths of atomic dimensions and molecular dimensions, respectively.

Most ultraviolet sources emit only small amount of far ultraviolet light less than 2900 Å. Many of the **injurious effects** of UV rays and many beneficial effects are produced by the near UV light. The biological effects have been discussed previously such as:

1. The photochemical effects,
2. The metabolic effects, e.g., formation of Vit. D,
3. The bactericidal effect.

13.7.3 Clinical effect

- **Dosage Determination**

In clinical practice, the dose of UV light is measured in terms of the "minimum erythemal dose" (M.E.D.). It is defined as the dose that when administered to the skin of a person average pigmentation, produces a first degree of erythema. This unit is measured in time, i.e., minutes, and is referred to some standard distance of the lamp from the skin, usually 30 inches (about 76 cm).

1. **Treatment of Rickets**: Gerstenberger Hartman (1929) recommended that dose of one M.E.D., given once per week for five to eleven weeks, prevent or correct rickets. Luce-Clausen (1938) recommended doses of one M.E.D. every other day for four weeks and stated that in some cases eight to ten weeks may be required. The results of treatment are in general and especially when combined with proper nutrition and medication programme.
2. **In Tuberculosis**: In certain tuberculosis, there is beneficial effect of UV light, particularly of sunlight. Treatment by exposure to sunlight (Heliotherapy) is used in bone and joint tuberculosis, tuberculous lymphadenitis, genito-urinary tuberculosis without active pulmonary tuberculosis. Low intensities of sunlight are used when the tuberculosis is active and accompanied by fever. Sometimes heliotherapy is supple-mented with artificial light therapy. Advancement in chemotherapy gradually reduces the importance of the light therapy in pulmonary tuberculosis.
3. **Other Diseases**: UV light may be used in neuritis, lumbago, chronic asthma, conjunctivits, trachoma, corneal ulcer, surface ulcers,

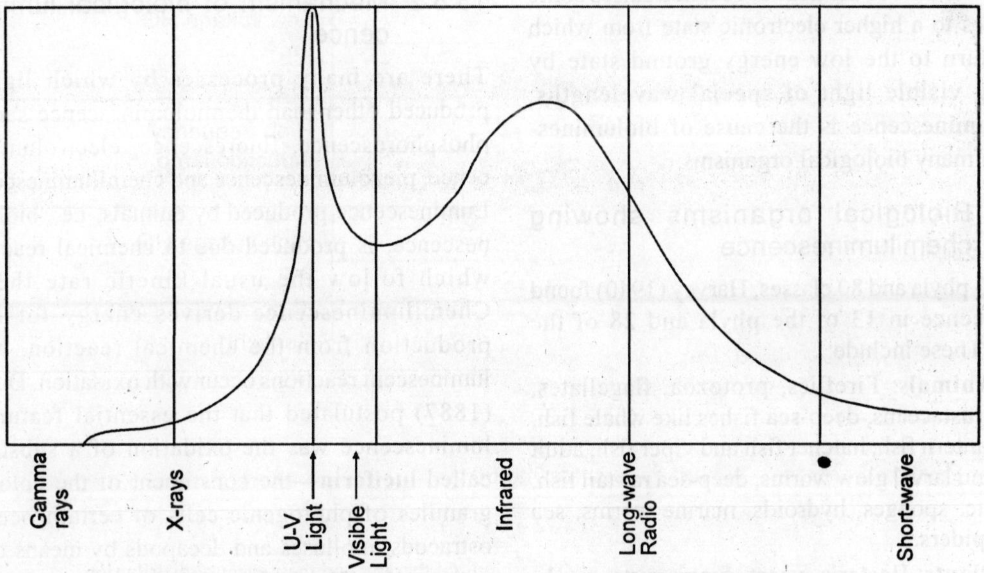

Fig. 13.40: Spectral absorption curve in semiquantitative fashion of tissues

less serious skin burn or abrasions, psoriasis, X-ray burns, acne vulgaris, erysipelas.

- **Precautions in UV Therapy**

1. UV light is to be avoided in active and progressive tuberculosis, extreme wasting, advanced heart disease, advanced kidney or liver diseases, hyperthyroidism, diabetes, severe generalised dermatitis.
2. Therapist should be careful not to burn the patient severely over a large skin area and not to over-expose the eyes of the patient.
3. Patient should be advised not to take the following drugs and chemical, before starting and during therapy such as quinine, methylene blue, insulin, epinephrine, thyroid extract, certain heavy metals, e.g., gold, silver, calcium, mercury, etc. because photo-sensitisation can occur due to these agents.

PART VIII

13.8 Biological Light or Production of Light by Living Organisms or Animal Light (Bioluminescence)

Chemiluminescence is the emission of visible light at ordinary temperature due to exergonic chemical reactions and represents a reversal of photochemical reactions. A part of the heat of reaction raises some molecules to a higher electronic state from which they return to the low energy ground state by emitting visible light of special wavelengths. Chemiluminescence is the cause of bioluminescence in many biological organisms.

13.8.1 Biological organisms showing chemiluminescence

Out of 33 phyla and 80 classes, Harvey (1940) found luminescence in 13 of the phyla and 28 of the classes. These include:

(a) **Animals**: Fireflies, protozoa, flagellates, crustaceans, deep-sea fishes like whale fish, lantern fish, hatchet fish and viper fish, adult and larval glow worms, deep-sea rat-tail fish, etc. sponges, hydroids, marine worms, sea spiders.

(b) **Plants**: Bacteria, yeast, fungi, rusts, mushrooms.

- **Site of Biological Light**

In animals, the light is emitted from the following structures:

1. Specialised glandular organs from which secretory material comes out for the production of light and this secretion is squeezed out on stimulation of contracting muscles;
2. Glandular cells present on the surface of the body which secretes material on stimulation;
3. Intracellular luminescence which is produced due to nervous or hormonal stimulation; or
4. Small particles scattered throughout the cell contents which may be broken by any kind of stimulation.

The light may be turned off and on in all of these animals. The special organs which emit light continuously may have some mechanism for turning the organ so that its luminous surfaces is alternatively open or closed or may cover with flaps drawn down over the organ. In some species, the luminescene may occur only when two or more substances are brought into contact and producing a "flash". Bacteria are generally continuously luminescent and in some cases the luminescence of animals is the result of the symbiotic relationship between the bacteria and animals.

13.8.2 Mechanism of biological luminescence

There are many processes by which light is produced other than thermoluminescence such as phosphorescence, fluorescence, electroluminescence, piezoluminescence and chemiluminescence. Luminescence produced by animals, i.e., bioluminescence, is produced due to chemical reactions which follow the usual kinetic rate theory. Chemiluminescence derives energy for light production from the chemical reaction. Most luminescent reactions occur with oxidation. Dubois (1887) postulated that the essential feature of luminescence was the oxidation of a substance called **luciferin**—the constituent of the coloured granules of photogenic cells of certain beetles, ostracods, molluscs and decapods by means of an enzyme, **luciferase** found in the water extract of the same organ. Thus, on boiling a luminous organ to destroy the enzyme luciferase but not the luciferin, then adding fresh organ to it, the boiled organ was able to luminescence. Luciferase catalyzes the breakdown of luciferin. There is species difference of these materials, so that luciferin from one species does not react with luciferase of another species if the species are not closely related. Secretions of **cypridina** and of many other organs, only show luminescence when extruded into sea-water and the extracts always require the presence of O_2 to produce luminescence, suggest that luciferin and luciferase are secreted as "granules", their dissolution in water permitting the oxidative reaction. That breakdown of the granules is an important element in luminescence is shown by the fact that many cytolytic agents, e.g., saponin, increase luminescence.

Dubois showed that the lanterns of fireflies contained material that could be resolved into "thermolabile enzyme-luciferase" and a "thermostatic-luciferin". McElroy et al. (1955) showed that luciferin consisted several substances; a faintly yellowish compound with a brilliant yellow-green fluorescence—luciferin proper—ATP, and an inorganic divalent cation, Mg^{++} or Mn^{++}. In some glands, the enzyme and substrate are secreted by separate types of cells, but only the combined secretions can exhibit potentiality for luminescence. In others it may be that both enzyme and substrate are secreted by the same cell, in which the dissolution of the granules and the aerobic conditions in sea-water allow the luminescence to proceed. In intracellular luminescence, as in photophores and bacteria, both enzyme and substrate must be present in the same cell; the bacteria luminesce continuously but in case of the photophores some mechanism is necessary to prevent the reaction from proceeding.

- **Structure of Luciferin**

Luciferin is a thermostable, heterocyclic phenol. It has an asymmetrical C-atom and may be synthesized in D-or L-form, only the natural D-form participates in the luminescence reaction with luciferase and ATP (Fig. 13.41).

- **Luciferase**

It is euglobulin having molecular weight 100000. It is thermolabile enzyme, catalyzes the adenylation of luciferin with the help of ATP and Mg^{++} to produce luciferyl adenylate which remains bound to the catalytic site of the enzyme as luciferase—luciferyl adenylate.

13.8.3 Chemical reaction

McElroy et al. clarified the reactions in case of fireflies. All the bioluminescence reactions appear

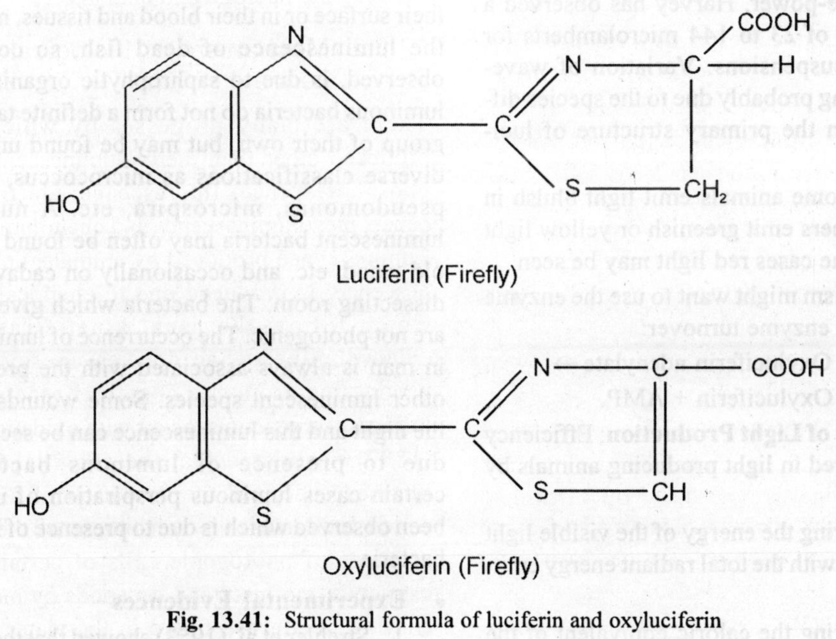

Fig. 13.41: Structural formula of luciferin and oxyluciferin

to be the enzyme substrate type but of differring complexity.

Luciferase + Luciferin – ATP, Mg^{++}
↓
Luciferase – Luciferyl adenylate + PP (AMP)
(Enzyme – Substrate complex)
↓ Aerobic Oxidation
Luciferase – Oxyluciferin adenylate + PP

(Oxyluciferin produced at a higher electronic or activated state with the help of the free energy change during the oxidation.)
↓
The activated oxyluciferin molecule reverts back to the low energy ground state by emitting the excess energy as visible light (Luminescence).

(a) As much as 90 per cent of the free energy change is transformed to light in fireflies.

(b) The aerobic oxidation of each luciferin molecule generates 1 photon of light.

(c) The emitted light flashes about 10-20 times per second, every time for a fraction of a second.

(d) Bioluminescence is dependent on the presence of oxygen, pH, temperature and salt concentrations.

(e) **Intensity**: It has intensity of less than 0.025 of a candle-power. Harvey has observed a luminance of 23 to 144 microlamberts for bacterial suspensions. Variation of wavelength being probably due to the species differences in the primary structure of luciferase.

(f) **Colour**: Some animals emit light bluish in colour; others emit greenish or yellow light and in some cases red light may be seen.

(g) The organism might want to use the enzyme again, i.e., enzyme turnover:
Enzyme + Oxyluciferin adenylate →
Enzyme + Oxyluciferin + AMP.

(h) **Efficiency of Light Production**: Efficiency is considered in light producing animals by two ways:

1. Comparing the energy of the visible light emitted with the total radiant energy emitted and
2. Comparing the caloric equivalent of the visible light emitted with the caloric equivalent of the oxygen, which goes into the photoemissive reaction. Regarding the first point, it may be observed that there is very little radiation from the observed that there is very little radiation from the emitted light that visible from a bioluminescent organ. Based on this, then, efficiency of the firefly is nearly 90 per cent. "Cypridina" gives blue light, much of it is invisible, the efficiency is more than 20 per cent. These values are more than the efficiencies calculated on a similar basis for artificial system (e.g., carbon filament lamp 5%). Based on total energy input, the efficiency of bacteria luminescence is less than 1 per cent. This is greater than the usual efficiency of artifical chemiluminescence (e.g., phosphorus oxidation: 0.017%).

Quantum efficiencies of the luminescence may be expressed in terms of the number of molecules of oxygen required to give 1 quantum of light (Bacteria: 33 and 193, "cypridina": 50, luminol oxidized with hydrogen peroxide: 350).

13.8.4 Bacterial luminescence

The luminescence of many living marine forms is due to the presence of luminous bacteria, either on their surface or in their blood and tissues, moreover, the luminescence of dead fish, so commonly observed, is due to saphrophytic organisms. The luminous bacteria do not form a definite taxonomic group of their own, but may be found under such diverse classifications as micrococcus, bacillus, pseudomonas, microspira, etc. A number of luminescent bacteria may often be found on meat, old wood, etc. and occasionally on cadaver in the dissecting room. The bacteria which give off light are not photogenic. The occurrence of luminescence in man is always associated with the presence of other luminescent species. Some wounds glow in the night and this luminescence can be seen usually due to presence of luminous bacteria. In certain cases luminous perspiration of urine has been observed which is due to presence of luminous bacteria.

- **Experimental Evidences**

1. Strehler et al. (1953) showed that the acetone powders obtained by precipitating the

bacteria with acetone, washing with acetic acid, and drying it, when added to distilled water, gave a weak luminescence. They found that when DPN was added to an aqueous extract of the powder a bright luminescence was observed with spectral characteristics similar to those of bacterial luminescence. If the extract is dialyzed, DPN was only effecting in the reduced form—$DPNH_2$ unless a substrate, such as malate, was added; and it was concluded that $DPNH_2$ was the actual luciferin in the sense that it was the first substance to become rate-limiting. Other factors favouring emission were shown to be a long-chain plasmal or palmitic aldehyde, extracted from animal tissues such as kidney cortex and FMN (Flavin mononucleotide).

2. McElroy et al. showed the absolute importance of FMN. A thermolabile luciferase preperation was obtained by extracting bacteria with water and fractional precipitation with ammonium sulphate; this was irradiated with ultraviolet light to remove flavines. Addition of $DPNH_2$ had a negligible effect whereas FMN gave a strong luminescence. It was concluded that the function of $DPNH_2$ was to reduce FMN, which subsequently reacted with oxygen to give the chemiluminescent reaction. These authors also noted that some other factors was necessary since addition of bacterial extracts without luciferase activity would allow more luminescence to occur, in spite of the system's containing plenty of FMN; this additional factor they called bacterial luciferin and it was subsequently shown by Strehler et al. to be identical with the long chain aldehyde that they had already denoted as "Kidney-Cortex factor".

Thus, the various factors that favour bacterial luminescence are:

1. Malate, 2 DPN, 3. FMN,
4. Aldehyde, 5. Oxygen and 6. Luciferase.

By studying the rate of development of luminescence on adding one component to the remaining five, some idea of the order in which the various components were used, could be obtained. For example, addition of Malate to a mixture of DPN, FMN, aldehyde and O_2 gave a "half rise time." (i.e., the time required for the luminescence to reach half its maximum intensity) of 108 sec. This was the same as when DPN was added to a mixture of Malate, FMN, Aldehyde and O_2, so that it could be presumed that DPN and Malate reacted with each other. If $DPNH_2$ was added to a mixture of FMN, aldehyde and O_2, the time was now only 2.7 sec, indicating that the Malate-DPN reaction occurred first on the basis of these studies, the suggested order of reaction was:

$$DPN + Malate \longrightarrow DPNH_2$$
$$DPNH_2 + FMN \longrightarrow FMNH_2$$
$$FMNH_2 + Aldehyde + O_2 \xrightarrow{\text{Luciferase}}$$
$$FMN + h\nu \text{ (Luminescence)}$$

13.8.5 Light organ (Photuris, Photopores) of Firefly, Control and Significance of Light Emission

13.8.5.1 Bacterial luminescence

The firefly (Photinus pyralis) is an example of an organism containing a definite *light organ* which may be activated by nervous influences to emit a bright flash of light, the luminescence being intracellular. In the male, the organ occupies the whole *ventral surface of the 6th and 7th segments* whilst in the female only about $\frac{2}{3}$ of these segments are occupied. The structure of the organ of "Photurus" is shown below (Fig. 13.42). "Photurus" consists of two layers (a) Reflecting layer (b) Photogenic layer.

The cells of former contain *crystals of some urate* which give the layer of reflecting property, whilst the latter is made up of large cells filled with *yellow granules.* An adequate supply of O_2 is ensured by the numerous *trachae* which penerate both layers and branch profusely in the photogenic layer; each branch ends in trachae profusely in the photogenic layer; each branch ends in a *tracheal end-cell* which gives rise to *tracheal capillaries or tracheoles* which are considered to enter the *photogenic cells.* However, Beams and Anderson, in their EM study, were never able to see a tracheole enter the cell. The end cell is peculiar to the tracheal system of the light organ, and is remarkable for the

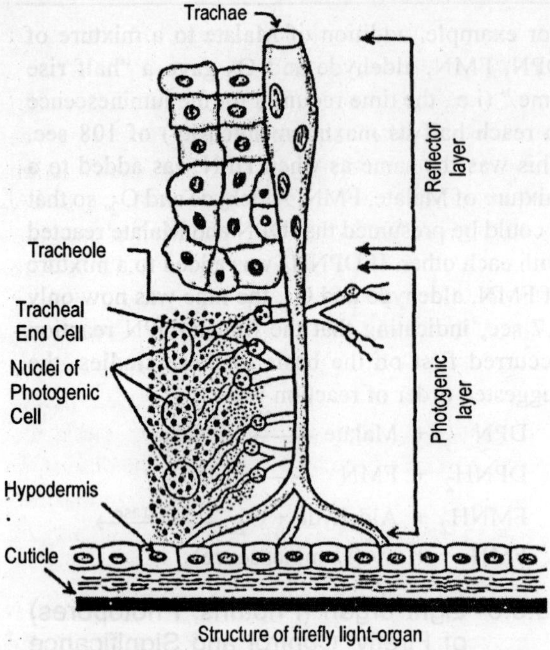

Fig: 13.42: Showing structure of firefly light-organ

large number of mitochondria it contains. The tracheolar twig is surrounded by the end-cell and, more distally, by a rounded body that has been called the *end-bulb* by Beams & Anderson (the rounded body by Dahlgren). Before emerging from this, it breaks into two tracheoles which pass between the photogenic cells, the membrane and cytoplasm of the end-bulb investing the outer layer of the tracheole. The outer cytoplasmic zones of the photogenic cells are differentiated from the inner regions by being free of photogenic granules and containing numerous small mitochondria. The photogenic granules have a laminar basis.

13.8.5.2 Control of emission of light

1. **Oxygen**: It was believed that the control of emission of light was regulated by O_2 supply to the photogenic cells through the tracheo-les, the end-body acting as a muscular valve. But EM study has shown that the structures that had been considered to be muscular elements were mitochondria. However, the end-cell is concerned in some way to emit a discrete flash.
2. **Nervous Control:** The "Porichthys notatus", California singing fish, a luminous vertebrate contains 840 small organs, "photophores" which emit light. The luminescence takes place within the photophores. Histological study failed to reveal any nerve supply to the organ and Green concluded that it was controlled by humoral mechanism. Nicol (1957) had identified nerve fibres supplying the organ and his experiments on the effects of cord section would suggest that both humoral and nervous (adrenergic) mechanisms are involved.
3. **Humoral**: Mechanical stimulation does not cause luminesence but powerful electrical stimulation causes all the photophores to light up. It may be due to that the nerve stimulation causes liberation of hormone which acts on the photophores. Green found that a subcutaneous injection of adrenaline was followed by luminescence, the number of activated photophores increasing progressively until, in about 10 minutes, the whole animal was alight and remained thus for longer than hour.

13.8.5.3 Significance of light emission

Significance of light emission in photinus pyralis (Firefly) has been elucidated by Buck. The female is wingless and in the evening perches on a blade of grass whilst the males, who during the day remain quiescent, fly about emitting flashes. The female never flashes spontaneously, but only in response to the flash of a male within some 3-4 metres, if the female makes this response, the male immediately turns directly towards her and flashes again, the female flashes in response and after about five of these exchanges the insect mate. The ability of the male fly to distinguish a flash emitted by a female, from one emitted by a male is not due to any difference in the spectral quality of the light emitted by the two sexes, but depends entirely on the interval elapsing between his own flash and the answering one. If this is precisely 2 seconds, the male responds to it, whereas, he ignores any flashes not occurring at other intervals after his own. Buck was able to attract males to a torch by flashing it exactly two seconds after the appearance of a flash, or alternatively by squeezing a male (which makes him emit a flash) just two seconds after seeing a flash.

PART IX

13.9 Light Interaction with Biological Materials

Light associated phenomenon involving *absorption of light*, i.e., some photo-chemical reactions, e.g., photo-synthesis, bleaching of rhodopsin, etc. have been discussed. Now let us consider the *interaction of light with biological matter* known as optical activity of the matter" which encompasses (1) the alteration of the plane of polarization of the light waves (e.g., carbohydrate, amino acids, etc.) and (2) birefringence.

- **Physical basis for optical activity**

Light consists of electric field vibrating in one plane and a magnetic field vibrating in a plane at right angles. Both planes intersect normally in the line of propagation. Light is emitted from a body due to vibration of electrons. These electrons may vibrate in one plane and in such case the emitted light from that electron will also vibrate in one plane only. If the atoms are randomly arranged in the light-emitting material, the total light emitted will contain waves with their plane of vibration oriented in random directions. Light is said to be unpolarized if the electric field oscillates in all planes including the axis of propagation. If all planes of electric vibration are cut off except a single plane, including the axis of propagation, then the beam is said to be "polarized light".

Light wave exhibits two basic characteristics: the velocity of propagation and the frequency. A derived characteristic is the wavelength but frequency of light is an absolute factor. Both *the velocity and the wavelength can be varied in the interaction of light with matter,* hence it is customary to indicate it by the "plane of vibration of a wave".

13.9.1 Isotrophy and anisotropy

Solutions of ordinary crystalloids contain molecules which are oriented in random directions in the solution and there is arrangement of the direction of vibration of the chemical bonds whereas in the colloidal suspensions, the molecules are arranged in orderly fasion. Long molecules may be lying parallel to each other and their chemical bonds may assume parallel directions. Materials in which there is random orientation of molecules are said to be **isotropic** and those with a definite orientation of molecules are **anisotropic.**

Biological materials exhibit no preferential direction for their chemical bonds. Therefore, we may expect that the velocity of light in one direction should equal the velocity of light vibrating in any other direction. When the chemical bonds are oriented in the anisotropic material, the bonds in the long chains react in one way to light with its electric field polarized parallel to the molecules and in a different manner with its electric field polarized perpendicular to the molcules. The effect of the reaction of the light and the electrons is that the polarized beam parallel to the long chains is transmitted more slowly than the beam polarised perpendicular to the chains. For any frequency of vibration, there would be two different indices of the material. Such materials are said to be **doubly refracting or birefringent.**

13.9.2 Different type of birefringence shown by different materials

1. **Crystal Birefringence or Intrinsic Birefringence**: Materials in which molecules are arranged in a sort of crystal lattice, is called **crystal birefringence** or **intrinsic birefringence.** It is independent of the refractive index of the medium in which it is suspen-ded.
2. **Form Birefringence:** Materials whose particles are not symmetrical but are oriented to an orderly fashion exhibit this type of birefringence. These individual particles must be of such a size that at least one of the dimensions is small with respect to the wavelength of light. Such materials are not birefringent and only become so when suspended in a medium which has a different refractive index. The birefringence can be abolished by placing the materials in a medium with the same refractive index as the materials.
3. **Strain Birefringence**: When a dielectric material is stressed, the oscillating bonds in it are distorted and become oriented regularly. This results in a type of birefringence called **strain birefringence.**
4. **Flow Birefringence**: Sometimes a case of strain birefringence is seen when a fluid is

caused to flow, thus creating shearing stresses, such birefringence is called "**flow birefringence.**"

13.9.3 Biological material showing birefringence

1. **Lipid and Protein**: Birefringence in living matter is due to presence of oriented lipid or protein molecules. Though the lipid molecules are short they are strongly birefringent. The protein molecules are long and less strongly birefringent. Protein fibres are made of long polypeptide chains, so oriented that their long axes are parallel. These molecules are connected laterally through linkages of their terminal groups of side-chains. The physical and chemical properties of these materials are largely determined by the stability of such bonds. According to Schmitt (1939), if the bonds are strong and the fibres are long, then tissue is strong and elastic as is seen in tendon and connective tissue, whereas if the chains are short, a compact tissue-like hair results. If the proteins are highly solvated, i.e., com-bined with large quantity of solvent, an active metabolizing form like muscle or intracellular fibril is formed. So far the evidences are available, they indicate the great importance of the orientation of molecules in biological materials, and reveal the nature of factors contributing to the life of the materials and also a number of fundamental concept of protoplasmic orga-nization can be implied from birefringence phenomenon.

2. **Skeletal and Smooth Muscle**: Skeletal muscle exhibits positive uniaxial birefringence, with the optical axis in the long axis of the muscle. In case of smooth muscle, the birefringence is uniform throughout the length of the muscle fibre, but in skeletal muscle, the regions of alternate weak and strong birefringence are present regularly along the length of the fibre. These are called **anisotropic** and **isotropic bands**, respectively. Muscle birefringence shows both crystalline and form birefringence and reveals that the structure of it is the long aggregates in which individual protein chains are at least partially oriented. The birefringence of smooth muscle varies with the stretch of the muscle, becoming greater as the muscle is stretched further. Bozler et al. (1937) have shown that the birefringence is proportional to the square root of the length. The change on stretching implies more perfect orientation of the muscle substructure.

   Contraction of skeletal muscle from the initial length produces a decrease of about 40 per cent in the birefringence, whereas shortening from a stretched length may result in an increase in birefringence. Thus, it appears that the molecules of protein making up the muscle substance are optically oriented at the resting length. The change occurring in contraction or stretching may be attributed to the "photoelastic effect" of the strain produced in the muscle. The birefringence of muscle can be associated from the development of tension.

3. **Cilia, Flagella and other Structures**: Cilia, flagella, axopodia, filopodia, myonemes of protozoa, sperm tails etc. are positively birefringent. Spindle and astral fibres and chromosomes of dividing cells also show birefringence. These materials exhibit positive crystal and form birefringence.

4. **Sperm Heads**: Sperm heads contain highly condensed chromosomes materials. The birefringence of these bodies varies from weak positive to strong negative, according to the degree of solvation. Chromosomes of dividing cells show birefringence.

5. **Nervous Material**: Nerve containing fibrils are birefringent. Myeling sheath of nerve shows positive birefringence and its optic axis are directed radially. It show form birefringence. Most of the birefringence is due to the presence of lipid molecules. Myelin sheath consists of radially oriented lipid (positively birefringent) alternating with protein layers oriented parallel to the surface (negatively birefringent). X-ray diffraction studies strongly support the impli-cations of birefringence information. No significant changes of birefringence occur during passage of nerve impulses.

6. **Red Cell Membrane:** The red cell membrane shows the same polarization phenomenon as the myelin sheath. The red cell membrane is about 160Å thick of which 100Å is occupied by lipid molecules. 60Å by protein molecules. Birefringence of this nature has also been observed in the membranes of fresh nerve cells. Nuclear membranes of nerve cells show negative birefringence revealing the protein composition mainly.

13.9.4 Dichroism of biological materials

Apart from two different refractive indices, biological materials may also possess two absorption coefficients in different optical axes. This phenomenon is known as "**dichroism or pleochroism**". It can be demonstrated with a single polarizing element by rotating the element and observing the changes in transmission. Dichrosim indicates anisotropy and in some cases it will be observed when birefringence is not. Retinal rods are highly dichroic, because of the presence of rhodopsin. Some dyes are highly dichroic, and can be used to stain a tissue preferentially.

PART X

13.10 Light and Pigmentary Response

The pigmentary response to light means colour changes in chameleon, the paling and darkening of frogs, crabs and fishes and certain adaptive colouration of some insect larvae. But winter dress of many mammals and the breeding dress of some fishes are not under the purview of this light response because they are the result of the breakdown or build-up of pigment in the tissues, not a redistribution which occur in response to light.

13.10.1 Chromatophores

The colours of animals depend on the nature and variety of the pigmented cells present immediately under the epidermis. In insects, the pigment is extracellular and distributed in granular form throughout the cuticle.

The changes in colour is due to alterations in the degree of dispersion of the pigment within the coloured cells known as **Chromatophore** and these cells are capable of dispersing or concentrating their pigment. There are four groups of chromatophore or pigment cells.

1. Melanophores: Most common, containing the brownish-black pigment called melanin derived from amino acid tyrosine.
2. Xanthophores: Containing a yellow pigment.
3. Erythrophores: Containing a reddish one.
4. Guanophores: Containing highly refractile guanine-type crystals. They are generally classified as leucophores and irridophores according to the state of the guanine crystals within them. Thus,
   (a) Leucophores: Containing very fine crystals and
   (b) Irridophores: Containing large reflecting platelets. Irridophores usually are not grouped under chromatophores becuase they do not alter their distribution of reflecting material, although they contribute greatly to the colour of animals due to highly reflecting qualities.

13.10.2 Organization of chromatophores

According to the nature of variety of the colour changes, the animal organize the chromatophores and it varies from simple arrangement of melanophores to a complex arrangement. Regarding complex organization, Schmidt (1920) studied the tree-frog (Hyla arborea) which may change from a pale green through dark green, lemon-yellow to grey. He described the findings of different stages. Such as:

(a) In bright green stage:
   1. A layer of yellow xanthopores occupies the most superficial position,
   2. Behind them are the reflecting guanophores,
   3. Whereas behind these, and partially enveloping them with their processes, are the melanophores:

(b) In the lemon-yellow stage:
   1. Concentration of the melanophores so that the dark pigment does not surround the guanophores.

(c) In grey condition with black spot:
   1. The reflecting guanophores are surrounded by the processes of the melanophores and

2. The xanthophores may be squeezed out of the surface.

In simple cases, changes of colour are brought about by changes in the degree of dispersion of pigments in the melanophores alone.

13.10.3 Mechanism of dispersion and concentration

The mechanism of dispersion and concentration of the pigments may be due to:

1. Change in shape of the cells or
2. The pigment diffuses without any change in form of the cell. The diffusion of pigment may be from the centre outwards during dispersion and back again during the concentration. To know the mechanism, experiments were done on the melanophores of vertebrates and cephalopods and the erythropores of crustacea and the results indicated the same mechanism.

(A) **Cephalopods**: In the cephalopodes, e.g., the squid or octopus, there is change in shape of melanophore brought by the contraction of radially arranged muscle fibres. The malanophore in relaxed condition is a small speherical cell and the pigment is concentrated (Fig: 13.43A); when all the melanophores in the skin are in this "concentrated" condition, the animal becomes pale. When the muscle fibres pull radially and contract, they draw the sphere into a large thin disc; the pigment is dispersed and the animal becomes dark (Fig. 13.43B).

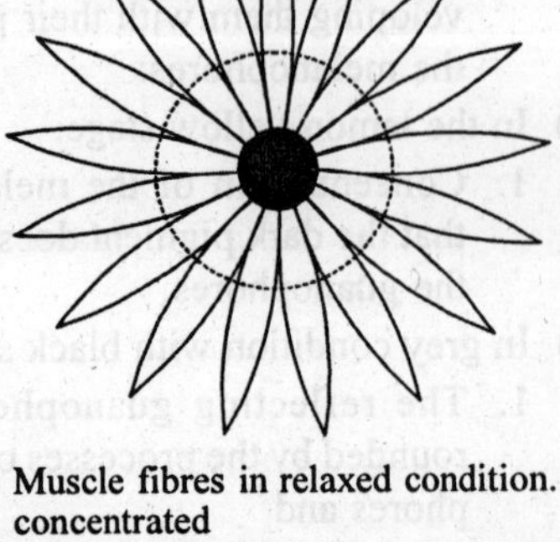

Fig: 13.43A: Muscle fibres in relaxed condition. Pigment concentrated

(B) **Vertebrates**: In these animals, the melanophores are with many branching processes. Different states are characterized by the varying degree of dispersion of the pigment along these branching processes. It is not clearly understood whether the changes in shape of the melanophore are due to the migration of pigment in amoeboid contractions and expansions of the cell and pigment limits the boundaries.

- **Tissue-Culture Studies:** Tissue-culture studies have shown that a given melanohpore retains its identical shape, on concentration and subsequent dispersion of its pigment, which favours the view that the pigment diffuses into the branching processes with fixed boundaries.

Fig: 13.43B: Muscle fibres in contracted condition. Dispersion of pigment

(C) **Arthropoda**: The dispersion of pigment is a filling up of collapsed tubes. Dispersion may be accompanied by an increase in cell volume brought about by osmosis from outside.

13.10.4 Control of pigment migration

Changes in pigment distribution is a result from a change in the condition of illumination such as frog becomes black in darkness and becomes pale in bright light. "Palaemonetes", becomes pale in darkness and black in daylight. But stimuli other than light play a role, e.g., humidity, temperature, 24-hourly rhythm of light and darkness. But the light stimulus is the most important one.

13.10.5 Response to light

There are two types of response to light:

(a) **The Primary Response**: It is a direct reaction for light on the skin and occurs in blinded animals which indicates it is not due to visual stimuli. It is manifested by dispersion of the pigment in the chromatophores which mainly determine the colour of the animal. This response is more prominent in chameleon. It has been shown that by illuminating a portion of a fully dispersed chromatophore, concentration of pigment was prevented that occurred in neighbouring chromatophores when the animal was in darkness. The illuminated part of the chromatophore thus retained its outline. When the pigment-free part of a concentrated chromatophore was illuminated, the pigment migrated into it, showing that migration occurred in the direction of the light and suggesting that the chromatophore was directly sensitive to the action of light.

(b) **Secondary Response**: The secondary responses are mediated by the eye i.e., visual stimuli to the effectors, the chromatophores.

1. Hogben and Winton showed that posterior pituitary extract caused the pigment granules in the melanophore of the frog to become dispersed throughout the bodies and branching processes of these cells.
2. Injection of a small dose of posterior pituitary extract into a frog causes its skin to become almost dark black as a result of melanophore reaction.
3. After hypophysectomy, owing to disappearance of the pituitary hormone from the circulation, the pigment granules gather near the centre of the cells and the skin becomes pale.
4. Hypophysectomized tadpoles have, instead of the usual dark brown or green colour, a silvery appearance.
5. The changes in colour which certain amphibia, reptiles, and fish undergo in order to blend into the colour of their surroundings is largely due to variations in the concentrations of the pituitary hormone in the blood or to the balance between the concentrations of the melanophore expanding principle and adrenaline.

Nervous impulses arising in the retina due to light through the mediation of a retino-hypophyseal mechanism causes liberation of melanophore expanding principle from parts intermedia of the pituitary. The production of "melanin" pigment in the skin of the frog is due to stimulation of this principle. Zondek has given the name "intermedin" or the "Chromatophorotropic" hormone to this principle.

The phenomenon of light influencing structure and function through the mediation of a retinohypophysieal mechanism usually occurs in cold-blooded animals.

14

Microscopy in Biology and Medicine

1. Phase-Contrast Microscope, 2. Interference Microscope, 3. Fluorescence Microscope, 4. Ultraviolet Microscope, 5. Electron Microscope, 6. Scanning Electron Microscope, 7. Cryo-electron Microscopy, 8. High Voltage Electron Microscopy, 9. Transmission Electron Microscopy, 10. Scanning-Transmission Electron Microscopy, 11. Immuno-electron Microscopy, 12. Scanning Tunnelling Electron Microscopy, 13. Scanning Electron-Probe Micro-analyses, 14. Low temperature Electron Microscope, 15. Centrifuge Microscope, 16. Confocal Microscopy, 17. Video Microscopy and digital imaging, 18. Atomic force Microscopy, 19. Interfocial force Microscopy, 20. X-ray Microscopy, 21. Nuclear Magnetic resonance (NMR) Microscopy, 22. Laser Microscopy, 23. Acoustic Microscopy, 24. Operating Microscopy.

14.1 Phase-Contrast Microscope

Study of living cells is gaining importance day by day. Conventional microscopes are of no use because of limitations such as:

1. Most living cells are transparent and look like structureless objects.
2. Constituent parts differ but little from one another by refractive index for which the different structures cannot be differentiated.
3. Refractive index of most living cells is very close to that of the aqueous media. For differentiation and identification, the living cells were chemically treated for fixation and coloured with different dyes but the controversial point is whether the structures are artifacts for the techniques rather than the structures actually present. To overcome the difficulties encountered for the study of transparent, and unstained objects, e.g., living cells, Zernike developed "Phase-Contrast Microscope", which is an optical system whereby the incident and diffracted waves are separated and the phase difference between them increases by $\frac{1}{4}\lambda$ (quarter of a wavelength) because original beam passing through transparent object is retarded in phase by about $\frac{1}{4}\lambda$ (quarter of a wavelength) but amplitude is unchanged (Fig. 14.1).

14.1.1 Light microscope vs. phase-contrast microscope

1. In the image plane of the light microscope, light arriving from areas corresponding to transparent regions of the specimen with an index of refraction different from that of their surroundings will be out of phase with the surrounding light. However, the ordinary Light Microscope

is not sensitive to such phase difference in intensity between these various regions and, hence, no contrast. The Phase Contrast Microscope is equipped with special apertures and with absorbing and phase-shifting plates.

2. Light Microscope gives contrast by differences of capacity in the object which are very small for which staining is necessary whereas Phase-Contrast Microscope gives contrast by differences of refractive index in the structure. These differences are such as to cause diffraction of the light passing through similar to that taking place at the classical diffraction-grating.
3. Living cell cannot be studied by Light Microscope unless it is chemically treated (Fixation) and stained, whereas it can be studied by Phase-Contrast Microscope without fixation and staining.
4. Phase-Contrast Microscope does not increase the resolving power of the optical system but it does render visible some micrscope structures, whereas Light Microscope depends on resolving power for visibility.

14.1.2 Principle

It was originally developed by Zernike as a method for testing telescopic mirror, the small phase-differences are amplified so that they are detected by the eye or the photographic plate.

When light passes through a transparent object in a medium having different refractive index, there is not change in the amplitude of the light waves. The retina of the eye is sensitive to amplitude changes, the object is, therefore, not visible. Though there is no change in amplitude of the transmitted light but it is affected during the passage through the object. Different protoplasmic constituents of the cell produces phase change as they vary in thickness and refractive index. This microscope converts phase variations into optical intensity variations. Diffraction of light generates a set of new waves which differ in phase from the direct light which has not passed through the object. The waves are retarded in phase of the diffracted light by about a quarter of a wavelength, i.e., $\frac{1}{4}\lambda$ for a cell (Fig. 14.1).

The microscope image is formed due to interference between. If the phase difference between these two sets of waves is further increased to $\frac{1}{2}\lambda$, then maximal destructive interference will result and thus formation of amplitude differences will take place. This can recognized by the eye and the transparent object on the stage of the microscope is seen as an absorbing object and appears dark to a greater or lesser degree depending on the amount of phase change which it has introduced into the diffracted light. The problem was to convert these differences of phase of the light, caused by variations in the refractive index of the object—which cannot be recognised by the eye—into differences of light intensity.

The problem was solved by allowing the undiffracted light to pass through a "phase-strip", i.e., equipped with special apertures and absorbing and phase-shifting plates. This device is to increase the effective path of the rays by a definite amount and the undiffracted light provides a background for the diffracted light.

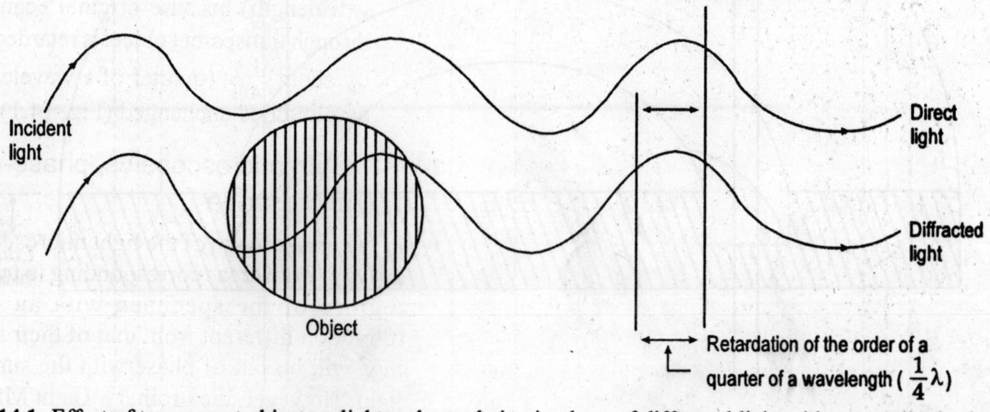

Fig. 14.1: Effect of transparent object on light and retardation in phase of diffracted light without amplitude change

36/F+B

retardation are converted into intensity difference.

3. In commonly used Phase Microscope, each phase contrast objective is designed to work optimally for only a limited range of retardation.

14.1.6 Advantages

1. The Phase Microscope utilizes two beams but no beam splitters.
2. It takes advantage of the same principle of obtaining interference between two sets of waves that are out of phase with one another.
3. Introduces contrast by optical means.
4. In this microscope there is no change of amplitude of light during transmission through the object. But there is change of light intensity.
5. Does not increase the resolving power of the optical system but it does not render the structures visible.
6. It helps to study the living cell and no chemical treatment or staining is required.

14.1.7 Importance

1. It has great advantage to give high contrast images of living cells and tissues.
2. It is valuable for observing cells cultured *in vivo.*
3. Using Cinematography, one can easily study and record the different nuclear and cytoplasmic changes occurring during cell division and cell movement, the continuous flow of mitochondria, water vacuoles, inclusion bodies, pinocytosis process, the formation of fine membranes and fibrillar expansion, etc.

14.2 Interference Microscope

With an ordinary Light Microscope it is very difficult to observe any detail in preparations of fresh unstained tissue. The reason for this is that the amount of light that comes through a thin preparation of unstained tissue depends on the optical density of the components and it is so much alike that they do not stand out in contrast to one another and hence cannot readily be distinguished.

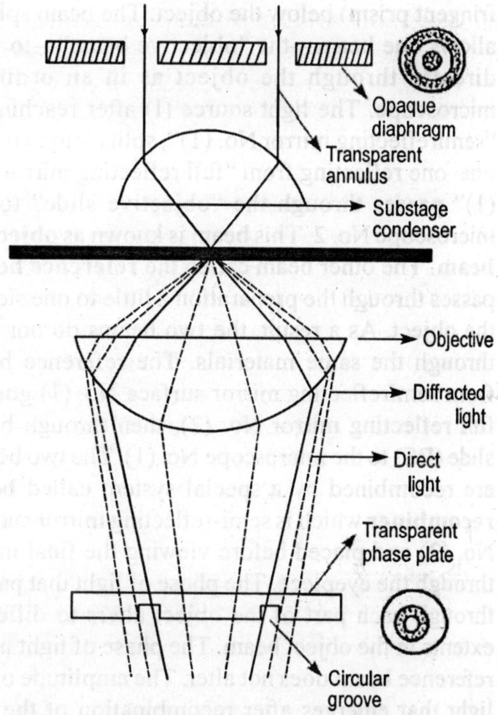

Fig. 14.3: Principle of phase-contrast microscope

The amplitude of light, which is responsible for its intensity, is reduced more or less equally if it passes through a nucleus of a cell or only through its cytoplasm. In order to see, for example, the nucleus of a cell distinct from its cytoplasm, it would be necessary for the amplitude of light waves to be reduced more or less by passing through a nucleus than through cytoplasm alone. If this could be accomplished, the nucleus would be either darker or lighter than the cytoplasm, and hence the two could be easily distinguished from one another. This can be done with either the Interference or the Phase-Contrast Microscope.

14.2.1 Interference microscope

It is known that the waves can interfere with one another. The basic idea, then, of the Interference and Phase Microscopes is to arrange matters to take advantage of light waves interfering with each other so that they cause differences in amplitude that can be observed by the eye.

To get interference there have to be two sets of waves. These are obtained in the Interference Microscope by inserting a "beam splitter" (birefringent prism) below the object. The beam splitter allows one beam—the "objective beam"—to pass directly through the object as in an ordinary microscope. The light source (I) after reaching to "semireflecting mirror No. (1)", splits into two two, one one reflecting from "full reflecting mirror No. (1)" passes through the "objective slide" to the microscope No. 2. This beam is known as **objective beam**. The other beam called the **reference beam,** passes through the preparation a little to one side of the object. As a result, the two beams do not pass through the same materials. The reference beam from semireflecting mirror surface No. (1) goes to full reflecting mirror No. (2), then through blank slide (BS) to the microscope No. (1). The two beams are recombined by a special system called **beam recombiner** which is semi-reflecting mirror surface No. (2) and placed before viewing the final image through the eyepiece. The phase of light that passes through each part of the object alters to different extents in the object beam. The phase of light in the reference beam does not alter. The amplitude of the light that emerges after recombination of the two beams is affected to different extents. It depends on how much each tissue ingredient altered the phase of the light waves (in the object beam) that passed through it. The combined beam will interfere, so that variations in refractive index in the intensity in the image. The different components is the object appear to be lighter or darker than one another (Fig. 14.4).

14.2.2 Retardation or phase change of objective beam

The beam that has crossed the object is retarded, i.e., it has undergone a phase change. As in the Polarising Microscope, the retardation (T) is determined by the thickness of the object (t) and the difference between the refractive indices of the object (n_0) and of the surrounding medium (nm) $n_0 - nm = \frac{T}{t}$, if nm is known, n_0 can be determined.

14.2.3 Nomarski interference microscope

Nomarski used a principle of beam separation and introduced another type of Interference Microscope, Linearly polarized light is divided into reference and object beams by a double quartz wedge, the Wollaston prism, placed above the microscope objective causing the recombination of the two beams. Images produced by the Nomarski Interference Microscope have a pseudopstereoscopic or "3-D" effect in addition to the usual interference colour contrast. The "3-D" effect is exhibited by a dark shadow on one side of refractile objects, structures of greater optical path difference than their surroundings appear to be higher and elevated above the background. Such an image may, on occasion, be of great value interpreting the micro-anatomy of transparent objects.

14.2.4 Interference microscope vs. phase-contrast microscope

1. The Phase-Contrast Microscope also utilizes two beams like Interference Microscope but without beam splitters. However, Phase-Contrast Microscope takes advantage of the same principle, that of obtaining Interference Microscope between two sets of waves that are out of phase with one another.
2. Phase-Contrast has the great advantage that high contrast images of living cells can be achieved easily and with relatively simple equipment; because of the incomplete separation of direct and diffracted light, however, the image suffers from certain defects, especially from the presence of a marked "halo" around the edges where there is a sharp difference in refractive indices. But Interference Microscope does not produce "halo".
3. Phase-Contrast does not require beam recom-biner whereas the Interference requires the same.
4. Phase-Contrast gives qualitative data whereas Interference provides both qualitative and quantitative data.
5. Interference Microscope helps to determine the optical phase difference for the various cellular structures and as a consequence to measure their dry weight, whereas Phase-Contrast does not provide the measurement of dry weight.
6. Furthermore, Interference Microscope permits detection of small continuous changes in refractive index, whereas the Phase Microscope reveals only sharp discontinuity.

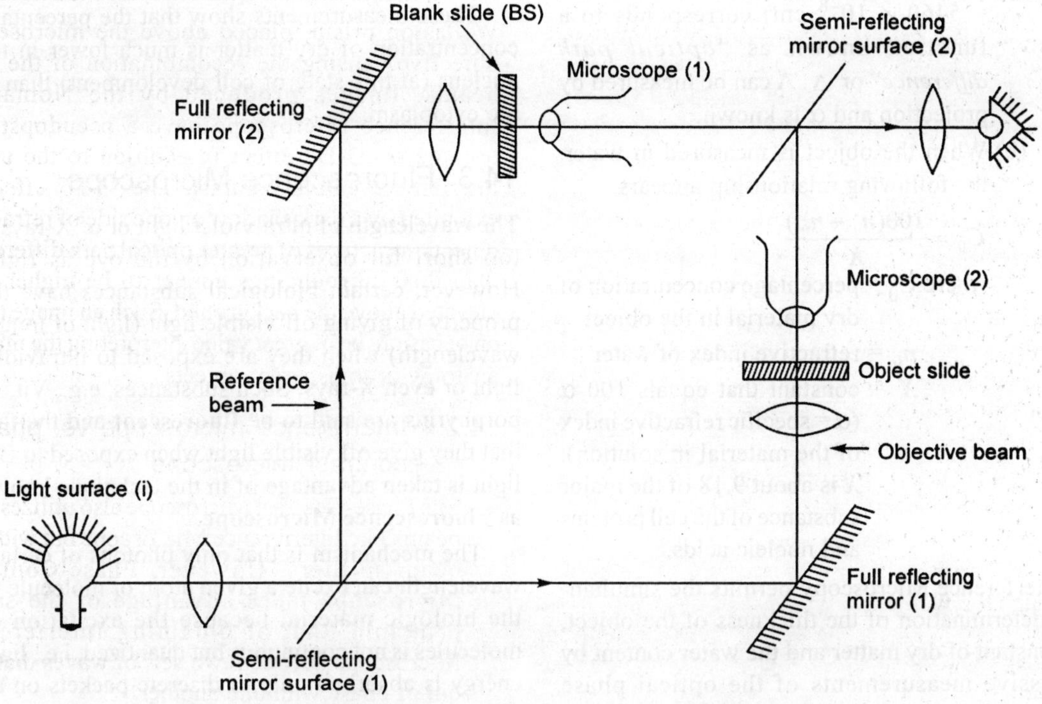

Fig. 14.4: Schematic diagram of interference microscope

14.2.5 Disadvantages

1. This system is expensive.
2. It is difficult to adjust.
3. A single objective and condenser lens are employed, i.e., an overlap of the reference and object beams but this is made as small as possible by making the two beams come to different foci as in (Fig. 14.5) where the object-beam is brought to a focus at A while the reference-beam comes to a focus at A_1. If the object at A is small compared with XX, the overlap will be very small.

14.2.6 Importance

1. It may be used qualitatively to provide contrast in transparent objects.
2. It may be used quantitatively to obtain information about the concentration of susbtance in cells.
3. It may be used to measure cell dry mass.
   (a) When there is no object in the field, the interference bands are evenly spaced and are separated by intervals representing one wavelength (λ). But when the object, e.g., cell, is in the field, the bands are displaced by varying amounts (Fig. 14.6). The degree of displacement can be measured at different points in the object and an average value obtained. If the displacement is D, and the distance between the two bands outside the object is d, then $\frac{D}{d\lambda}$ (where λ for green light

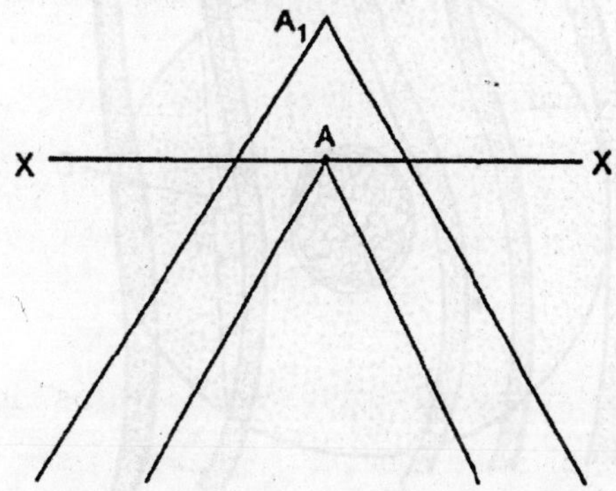

Fig. 14.5: Illustration of the principle of the interference microscope

= 5460 × 10^{-8} cm) corresponds to a function known as "*optical path difference*" or Δ, A can be measured by projection and α is known.

(b) When the object is measured in water, the following relationship appears:

$$C_0 = \frac{100(n_0 - n_w)}{X}$$

where C_0 = percentage concentration of dry material in the object

n_w = refractive index of water

X = constant that equals 100 α (α = specific refractive index of the material in solution). X is about 9.18 of the major substance of the cell proteins and nucleic acids.

Interference Microscopy permits the simultaneous determination of the thickness of the object, the constant of dry matter and the water content by successive measurements of the optical phase difference in two media of known refractive indices. The example of the application of this method to study of the sea urchin oocyte is illustrated in the table below (Table 14.1).

Table 14.1: Measurement of dry matter (C_0) in sea urchin oocytes.

	$n_0 - n_w$	C_0 + for X = 0.165
Cytoplasm	0.036	25.5
Nucleus	0.021	16.4

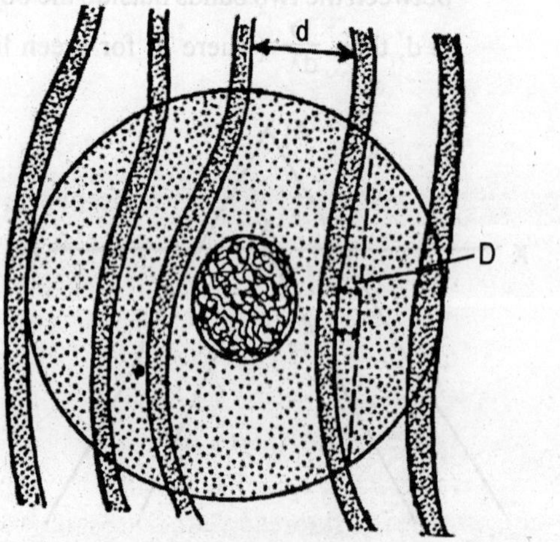

Fig. 14.6: Measurement of displacement of bands and the value of D/d

These measurements show that the percentage concentration of dry matter is much lower in the nucleus (at this state of cell development) than in the cytoplasm.

14.3 Fluorescence Microscope

The wavelength of ultra-violet light or of X-rays is too short for observation by the eye as light. However, certain biological substances have the property of giving off visible light (light of longer wavelength) when they are exposed to ultraviolet light or even X-rays. Such substances, e.g., Vit. A, porphyrins are said to be **fluorescent** and the fact that they give off visible light when exposed to UV light is taken advantage of in the technique known as **Fluorescence Microscope.**

The mechanism is that only photons of certain wavelength can excite a given atom or molecule of the biologic material because the excitation of molecules is not continuous but quantized, i.e., light energy is absorbed only in discrete packets on an all-or-none basis leading to the term "quantum". Excitation of a molecule by light is very rapid, taking less than 10^{-5}s. The excited molecule has two fates such as (1) it may return to its original low-energy state, the "*ground state*", with simultaneous emission of the energy originally absorbed during excitation, which reappears either as light or as heat or both. The emission of light by excited molecules is called "**fluorescence**". Fluorescent decay of excited molecules, which is complete in less than 10^{-8}s, occurs at a longer wavelength than the exciting wavelength, (2) the excited molecules react more readily with some other molecule and in the photo-chemical reaction, the excited molecule may lose an electron to the other reacting molecule and in the photo-chemical reaction, the excited molecule amylose an electron to the other reacting molecule. The above principle is known as **primary fluorescence** and has little application for cell components.

14.3.1 Equipment required

Sections that contain either naturally occurring fluorescent substance or substances made fluorescent by experimental procedure can be examined in a usual type of microscope equipped with an ultraviolet light source by high pressure mercury arc which is rich in UV light and a darkfield condenser. Those wavelengths in either the UV or the blue regions of the spectrum which are required for excitation of the specimen are **isolated** by means of suitable glass filters and focused by the substage condenser (or dark-field condenser) onto the slide. The exciting radiation is absorbed by fluorescent substance and then the energy is reemitted at longer wavelengths.

The microscope, unlike a true ultraviolet microscope, does not require special quartz or other types of objectives. The visible light from sites that fluoresce is transmitted by odinary objectives and eyepieces. The light source employed must be that emits light in the ultraviolet range. Any light that the source emits of wavelengths longer than those in the ultraviolet range must be filtered out. The wavelengths in the ultraviolet spectrum that are used in Fluorescence Microscopy are generally long enough to pass through an ordinary condenser in appreciable amounts; hence a condenser of the dark-field type generally is employed so that fluorescent sites can be seen easily against a dark background. Light of long wavelength in the ultraviolet spectrum is usually employed and that may continue through nonfluorescent part, objective and eyepieces. Accordingly, to prevent eye injury from this invisible light, filters—that exclude ultraviolet rays, but permit the passage of light in the visible spectrum—must be ensured in the eyepiece.

14.3.2 Secondary fluorescence

Though the "primary fluorescence" technique permits naturally occurring fluorescent (e.g., Vit A, porphyrin) substances that are present in the cell in histologic sections. A more important use of this technique having greater value is the practice of introducing "secondary fluorescence". The fact that treatment of the tissues or proteins with certain fluorescent dyes known as **fluochromes** which combines selectively with certain specific body components so that these can be identified in section because they fluoresce. The absorption of UV light between 290 nm and 495 nm by fluorescing and its emission on longer wavelength green light (525 nm) is used to visualise protein labelled with this dye shown by Coon and Kaplan (1950). These, although often not coloured in the visible regions of the spectrum, fluoresce vividly. Secondary fluorescence has the advantage of being much more intense than the primary form and by careful choice of fluochromes very valuable diagnostic information may be obtained. The fluorescent dyes which are generally used are: (1) fluorescin iso-thiocyanate giving green fluorescene, (2) lissamine-rhodamine giving orange fluorescence and (3) auramine for tubercle bacilli.

14.3.3 Importance

1. Study of distribution of Vit. A. porphyrin in the cell by primary fluorescence.
2. Demonstration of biogenic amines, e.g., 5-hydroxytryptamine of 5HT, noradrenaline and adrenaline by the technique of Falk et al. The method involves freezing and drying the tissue before treating it with hot formaldehyde vapour at about 70°C. The conditions of this treatment must be determined empirically. Afterwards, the tissue is embedded in wax, sectioned and the sections are mounted on microscope slides. The wax is removed and the sections are then mounted in liquid paraffin for examination with a Fluorescence Microscope.
3. Differentiation of end products of amines. The formaldehyde treatment induces the amines to become strongly fluorescent and by using suitable exciting wavelengths and combinations of filters in the microscope it is possible to differentiate between the fluorescent end products from different amines.
4. Demonstration of distribution of 5HT in Mast cell.
5. Demonstration of distribution of noradrenaline in the terminals of sympathetic nerve endings and in this way innervation of the muscle layers of a small vessel in the iris diaphragm of a rat has been observed.
6. **Fluorescent-Antibody Technique or Immunofluorescence**
   (a) If the antibody is labelled by coupling it with fluorescin, the presence of anti-body-antigen reaction will be revealed in the Fluorescence Microscope. Micro-

organisms and other antigens can be located and identified in sections and smears by staining these preparations with dye-conjugated specific antibodies.

(b) The Gonococcus is indistinguishable from the Meningococcus by ordinary microscopy, but can be identified by the Fluorescent-Antibody Technique.

(c) **Virus in Tissue Culture**: The multiplication of virus in tissue culture can be detected by attachment of virus-specific fluorescent antibodies to virus particles or virus antigens within the cells.

(d) **Virus in Host Cell**: Ordinary Light Microscopy is rarely used for virological diagnosis; examination of stained smears or tissue sections showing characteristics histological changes, or the presence of pathognomic inclusion bodies or even that of individual particles (elementary bodies) in pox virus infections. Electron Microscopy—though it is more infor-mative because a much wider range of viruses can be seen and recognized—is more expensive and relatively rare. Fluorescence Microscopy and Immunofluorescence Techniques are applied to the rapid identification of viruses and virus antigens inside host cells.

(e) **Bacteriological Identification in CSF**: CSF is centrifuged for bacteriological examination and the deposit is used for making smears and for inoculating cultures. Identification is done by Immunofluorescence and other methods.

(f) **Fluorescent Treponmal Antibody Test**: The Fluorescent Treponemal Antibody (F.T.A.). Test depends on the same source of T. pallidum, but they do not have to be alive, and so it can be carried out in routine laboratory. It is performed by applying dilute serum from a patient to a fixed smear of dead T. pallidum on a microscope slide, so that any antitreponemal antibodies present attach themselves to the treponemas during a period of incubation; the serum is then washed off and antibody-coated treponemas are detected by means of rabbit serum containing fluorescent antihuman-gammaglobulin antibodies using Fluorescence Microscopy.

(g) **Detection of Antinuclear Factor (ANF)**: The fluorescent antibody test is most common method for the detection of the "antinuclear factor (NAF)" found in the majority cases of systemic lupus erythematosus. Sections of tissue, e.g., rat liver, are used as the substrate and are layered with patient's serum. Binding of ANF to the liver cell nuclei is detected with fluorescein-labelled antihuman immunoglobulin.

(h) **Distribution of Microfilaments**: The cellular distribution of microfilaments can be displayed by Immuno-fluorescence Microscopy. Antibodies specific for action are covalently labelled with a fluorescent dye such as fluorescin and then added to cultured cells. The resulting Fluorescence Micrograph shows array of filaments. Some of these fluorescent bundles span the entire length of the cell. In quiescent cells attached to a solid support, many of these acting filaments are aligned parallel to the long axis of the cells. In contrast, ruffled borders exhibit a diffuse fluorescence, which comes from action monomers or short acting filaments. The distribution of other contractile proteins also has been observed by Immunofluorescence Microscopy.

(i) **Localization of Cellular Protein**: By this method, proteins can be identified and located in the cell or tissue. As for example, when a protein in muscle is exposed such as myosin, then the myosin is exposed in a section of tissue to antibody which is specific for this protein. Then the antibody precipitates on the protein and by labelling the antibody, the distribution of the protein is observed by Fluorescence Microscope. The antibody is obtained by injecting a pure preparation of myosin of an animal (i.e., rabbit) into another animal (e.g., guineapig). The serum from guineapig contains antibody that precipitates on rabbits myosin.

(j) **Localization of Hormones:** Immunohistochemical Technique have also been used for hormone localization. Initially a fluorescent compound was conjugated to the hormone antibody but more recently an enzyme, horseradish peroxidase, is being used as conjugate. Using this method, it has been found that STH<ACTH, LTH and TSH were localized in separate cells, FSH and LH were frequently found in the same cell. This method confirms that secretory granule size serves as a useful indicator for the identification of pituitary hormone-producing cell types.

7. **Identification of Tubercle Bacillia:** Tubercle bacilli can be selectively stained with one fluorescent dye, auramine, by a Ziehl-Neelsen like procedure in which the auramine replaces the carbol fuchsin. They can be recognized in smears of sputum or other material stained in this way and examined by Fluorescence Microscope.

14.4 Ultraviolet Microscope

Quantitative study of cell structures, i.e., the amount of constituents present in a cell and their variations during activity can be made by measuring the amount of light absorbed by a constituent when a beam of light passes through a microscopical section and the method is known as "**cytophotometry**". This procedure represents a development of the Ultraviolet Microscopy of originally devised by Kohler and later modified by Caspersson (Fig. 14.7).

The visible region of the spectrum extends from 6500 Å (deep red) to 4500 Å (violet). Wavelengths shorter than 4500 Å are called **ultraviolet light**. Since the wavelength of light proved to be the limiting factor in ordinary microscopy, it was only necessary that microscopes would be built to take advantage of the shorter wavelength of ultraviolet light. For example, nucleic acid absorbs most in-

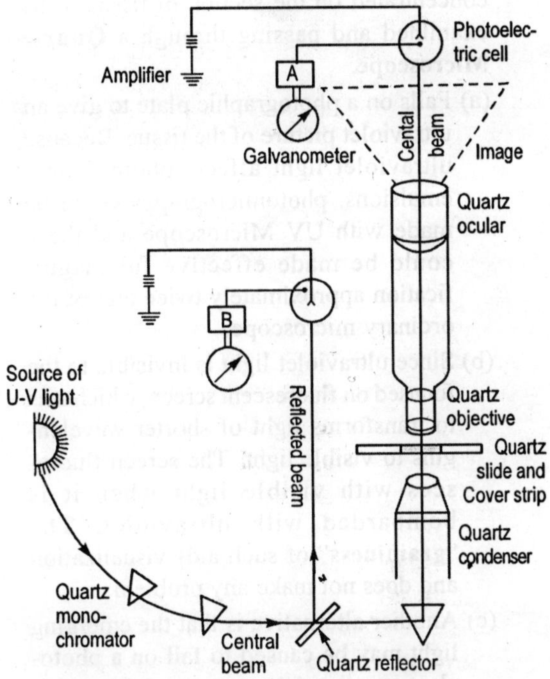

Fig. 14.7: Sketch diagram of UV Microscope

tensely at 2600 Å, and proteins generally absorb most intensely in the region of 2800 Å.

To make shorter wavelength of UV light it is necessary of making lenses out of some materials other than glass which does not transmit UV light, i.e., it absorbs UV light for which quartz was fitted with the microscope. The limitation imposed by the spectral absorption characteristics of such elements narrows the wavelength band to the region between 2200 Å and 4000 Å. The average wavelength is 2700 Å. This is approximately half the average of white light. Thus, the minimum separation with UV Microscope is about 0.1 micron (white light of average wavelength 5300 Å gives minimum separation of two points which is about 0.2 micron). The refractive index of a material increases with decreasing wavelength of light. Therefore, Ultraviolet Microsceope has a twofold advantage over the Optical Microscope:

1. The decrease in minimum separation.
2. The increased ability to discern objects, which results from the increase in refractive index.
3. Provides about twice the resolving power of the Optical Microscope. Ultraviolet light is

concentrated on the section of tissue to be examined and passing through a **Quartz Microscope.**

(a) Falls on a photographic plate to give an ultraviolet picture of the tissue. Because ultraviolet light affects photographic emulsions, photomicrographs could be made with UV Microscope and these could be made effective for magnification approximately twice that of the ordinary microscope.

(b) Since ultraviolet light is invisible to the focused on fluorescent screen, which acts to transforms light of shorter wavelengths to visible light. The screen fluoresces with visible light when it is bombarded with ultraviolet. The "**graininess**" of such aids visualization and does not make any pfoblem.

(c) Another alternative is that the emerging light may be caused to fall on a photoelectric cell and the amount of absorption by very small particles as small as 4000 Å may be determined; by varying the wavelength of UV light, the degree of absorption at different wavelengths may be measured to give an ultraviolet absorption spectrum of the particle observed.

14.4.1 Importance

1. Indetification of nucleic acids, proteins. Nucleic acid absorbs most intensity at 2600 Å on a account of the presence of pyrimidine rings while protein in the region of 2800 Å.
2. To study quantitative aspects of cell structures, i.e., how much of a certain constituent is present in a cell and variations in this amount with the functional state of the latter. In this way, the presence of as little as 10^{-11} mg of nucleic acid may be determined.
3. This method does not differentiate between the two types of nucleic acid, but when the results are compared with those gives by the Feulgen technique it is possible to deduce the presence or absence of the RNA. Thus the nucleoli absorb strongly at 2600 Å, and, therefore, contain a high concentration of nucleic acid; since they are Feulgen-negative, the acid must be of the RNA variety.
4. With the help of this microscope, rapid synthesis of nucleic acid in the developing embryonic tissue was observed, most of which was RNA of the cytoplasm.
5. It was observed by this microscope that the protein associated with cytoplasmic nucleic acids of embryonic cells to differ from that of adult cells. The embryonic protein was of the basic histone type, whereas the adult protein was of the more complex tryptophan and tyrosine containing type.

14.4.2 Precaution

1. UV rays have profound effects on living biological materials for which there is limitation for its use. Therefore, care must be taken in interpreting observations made on living material.
2. It is necessary to use filters to protect the eyes, which can be damaged by UV light.

14.5 Electron Microscope

Abbe showed that resolving power of the Light Microscope is irrevocably limited, not by the quality or the nature of lenses that can be made, but by the wavelength of light. With ordinary illumination this means that 2 objects separated by less than 0.2 microns can be resolved clearly as distinct entities by the Light Microscope.

The Electron Microscope which was built by Knoll and Ruska (1931) depends primarily on the scattering of a beam of electrons having extremely short wavelengths, less than 1Å (4000 to 7600Å of visible light). If an apparatus could be designed to form a picture with electrons as the source of illumination, details of less than 1Å (Atomic dimensions), should be resolvable. Electron rays may be made to deviate in a regular manner by the application of electric and magnetic fields, consequently it is possible to construct electron lenses capable of bringing a bundle of electron rays to a focus, and thus to construct an Electron Microscope.

When it was discovered that suitable shaped magnetic or electrostatic fields could be used as true

14.1.3 Optical system of the phase-contrast microscope

The optical system of Phase-Contrast Microscope consists of an opaque diaphragm containing transparent annulus placed at the back focal plane of the substage condenser. Direct light without any interference by the object placed on the state forms an image of the annulus in the back focal plane of the objective. The diffracted light, emanating from the object, evenly spreads over the same area. Inside the objective groove in it and it is mounted in such a way that the direct image of annulus coincides exactly with the groove in the plate whereas the diffracted light suffers a extra thickness and introduces a further $\frac{1}{4}\lambda$ phase change relative to the direct light and details of structures, e.g., nucleus, chromosomes, mitochondria, lipid droplet can be observed (Figs. 14.2, 14.3).

14.1.4 Cell refractometry

If a living cell is immersed in a medium whose refractive index is same with that of the cell, it shows minimum contrast and the cell is not visible in the Phase Microscope. It is known that the cell survives in osmotically-balanced aqueous solutions of proteins such as bovine plasma albumin and, if an effort is made, it is not difficult to find out the exact concentration of a solution for a particular type cell to make the cell invisible when immersed in it. Under this circumstance, the protein concentration in the mounting medium should be equal to the concentration of proteins in the cell. Since, in some cells, apart from proteins, there are significant amounts of other substances like carbohydrates, lipids, etc., the value obtained indicates the concentration of protoplasmic solids. This method of **Cell refractometry** gives the results which are in good agreement with those obtained by using other methods.

14.1.5 Drawbacks

1. There are halos around the images because of incomplete separation of the background light from the diffracted light.
2. Only phase differences from small objects or from border regions of rapidly changing

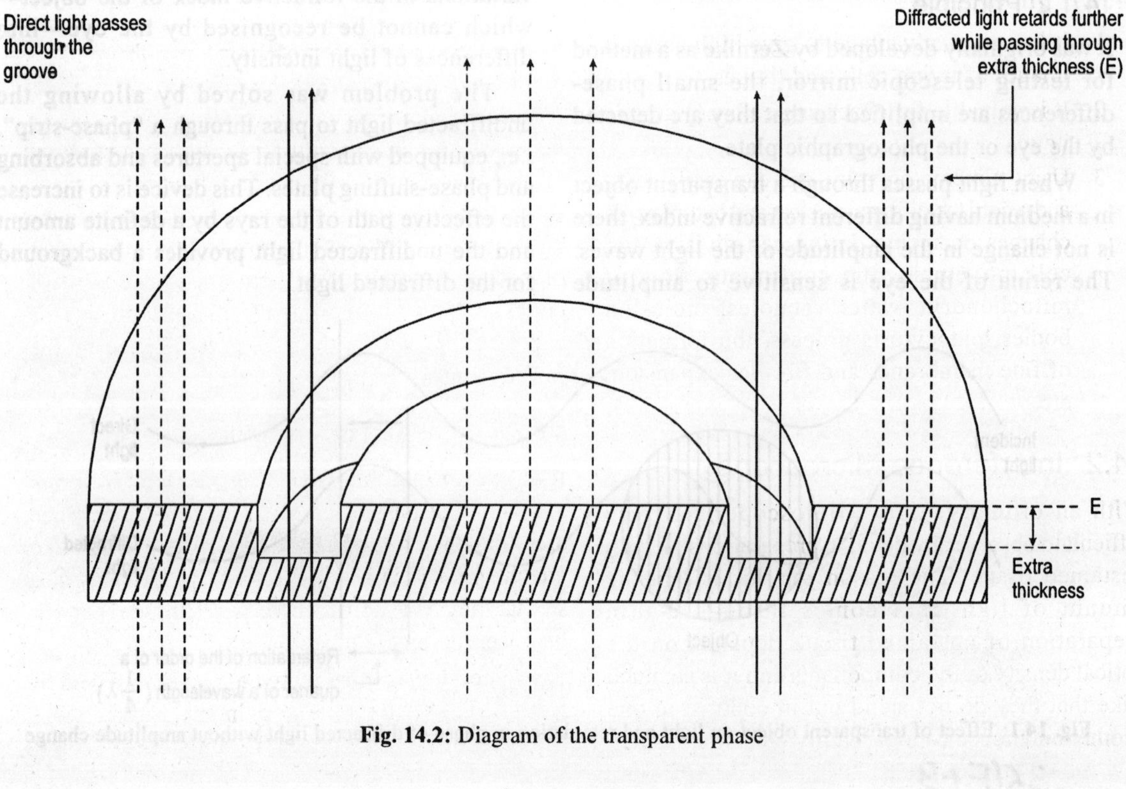

Fig. 14.2: Diagram of the transparent phase

lenses for an electron beam and that electrons could be accelerated to speed at which their wavelength would be an almost negligible factor in resolving objects with them. These were sufficient to stimulate to use electrons instead of light.

14.5.1 The electron lens

There are two types of electron lens:

1. **Electrostatic type**: The lens consists of a pair of electrodes of different voltages. The result of the action of this lens is the focussing of an electron beam in exactly the same fashion as an optical lens focuses light. There will be a focal distance at which a parallel beam will diverge; and an image will form in exactly the same as in geometrical optics.
2. **Magnetic Electron Lens**: This type of lens is widely used. It is an uniform magnetic field and can be formed inside a solenoid. It is the characteristic of this field that an electron moving in it exhibits a helical path around the magnetic lines of force. This helix shows a radius which is proportional to the component of electron velocity transverse to the magnetic lines of force: an electron moving strictly along those lines will not be affected by the field. To make a complete spiral circuit in the helical path, the time required is the same for all electrons of the same total speed, no matter what the initial divergence of the electron path. This means that a slightly diverging beam of electrons can be caused to converge by passing it into a uniform magnetic field having a direction parallel to the exis of symmetry of the beam. Instead of just one focal point as in the electrostatic lens, there will be a series of them corresponding to successive turns in the helical path of the electrons in the beam. Electron Microscopes with electrostatic or magnetic lens requires extreme control of potential or current, respectively.

14.5.2 General principle

The energy source in the E/M is a beam of electrons emitted from an electrically-heated tungsten cathode situated at the top end a tube in which a vacuum is constantly maintained by means of a series of pumps. The beam of electrons is accelerated, by a potential difference, towards the anode which has a circular aparture at its centre through which the electrons pass. Beneath the anode are placed, at appropriate intervals, a series of magnetic lenses which are, in fact, magnetic fields and correspond to the condenser, objective and eyepiece lenses of the optical microscope. The E/M has a two lens system like an optical system. The first lens is a "**Collimator**" which means it brings electrons to a proper convergence for object illumination and that corresponds to the condensing lens system of the optical microscope. its function is to illuminate the specimen uniformly with electrons from a point source (Fig. 14.8).

The object is placed between the condenser and the objective in the focal plane of the latter. The electron beam behaves in many ways like a beam of light and its path through the magnetic lenses is similar to that taken by light waves in the light microscope. The object under examination must be extremely thin and is mounted on a thin film of collodion or Formvar (methyl cellulose) or, polyvinyl formol or carbon which is supported on perforated metal plates or grid. The examination is carried out in a high vacuum which inevitably produces some distortion. Ultra-thin sections of tissues or of suitably embedded microbial or other cells can be examined, and so can films made from suspension of very small particles, such as the smaller viruses. To be clearly visible these objects must differ from their surroundings in their opacity to the electron beam. In general, the scattering of electrons depends on the mass of the atoms in their path, so that regions of high density—will appear black in contrast with regions of lower density. Contrast of opacity to the electron beam can be achieved artificially in the following ways:

1. **Negative Staining**: This is very much useful for particular material. Particles are suspended in a solution of electron-dense stain, e.g., phosphotungstate and allowed to dry on the grid; under appropriate conditions this will lead to the formation of an electron dense glass in which the particles are embedded and show up as areas of relatively low electron-density.

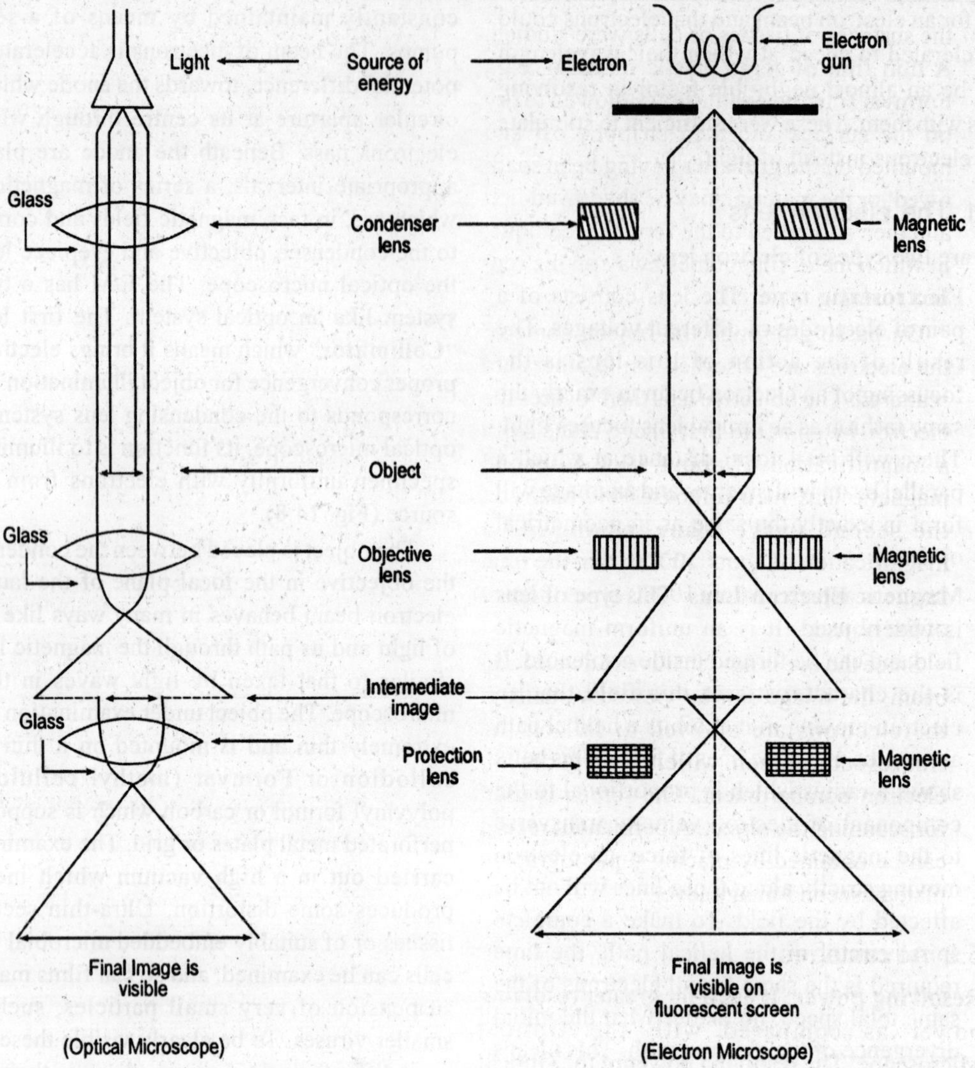

Fig. 14.8: Sketch diagram of electron microscope and optical microscope, and comparison thereof

2. **Positive staining**: This is done by "electron stains" with some electron-opaque material containing elements of high atomic number which react specifically will certain parts of the structure, e.g., Osmic acid or bichromate fixatives "staining" individual cell for E/M.

3. **Shadowcasting**: By this process the contours of the structure are exhibited. The specimen is exposed to a stream of evaporated heavy-metal atoms, e.g., Cr, Ni, Pd, Ag, Au and U impinging over the film obliquely, so that a thin layer of metal is formed all over it except in the "shadows" of particles. These shadows transmit electrons, and other regions are opaque; on the fluorescent screen of the instrument they appear bright, but are dark in the negative of the shadow gives a measure of the height of a projection such as the size of a protein molecule, if the angle of incidence of the stream of atoms is known.

4. **Shadowed replica method**: This is combination of two methods, i.e., replica technique and shadow technique. In replica technique,

the surfaces of tissues or cells were studied. A thin film of some plastic material, e.g., Formvar (methyl cellulose) is allowed to dry on the surface and then stripped off and mounted on the grid after having been shadowed; or the material may be shadowed first and then submitted to the replica technique, in which metal film comes away on the replica forming material.

On passing through the object, some of the electrons are absorbed while others are scattered. The objective lens then focuses the electron beam at the first image plane, with a magnification of about 100. A second magnetic lens then refocuses the beam on the second image plane, again with a magnification of about 100, so that the final image is approximately 10000 times the size of the object.

Since the eye is not sensitive to an electron beam, the image can be viewed visually if the electrons are allowed to fall on the fluorescent screen which fluoresces on electron bombardment. The screen is used for scanning the object. A permanent record can be obtained by placing a photographic plate at second image level.

14.5.3 Advantages

1. **Resolving power**: It has a far greater resolving power as compared with the Optical Microscope. The resolving power of the Optical Microscope is a function both of the numerical aperture of the objective and of the wavelength of light; in practice, the minimum resolvable distance between two objects is approximately 0.2 μm. The wavelength of the electron beam in the E/M, however, is very much less, so that its resolving power, and hence its ability to reveal detail, is greatly increased by at least 100 times, hence the photographic negative obtained can be greatly enlarged to reveal a vast amount of detail.
2. **Magnification**: An Optical Microscope can only magnify an object without loss of detail about 1500 times. The E/M can magnify 10000 and can achieve magnification of over 500000 times.

14.5.4 Disadvantages

1. The tissue only can be examined.
2. Material for examination has to be mounted in a vacuum, and is, therefore, dead before it can be viewed. The reason is that electrons are strongly absorbed and scattered by the evacuated, so that only dried specimens of biological material are suitable for study.
3. An electron lens has the same errors as an optical lens: chromatic and spherical aberration, astigmatism, etc. However, the two errors are reduced because the lens apertuers are small, focal depths are relatively large, and all rays are near the axis and nearly parallel with it.
4. The E/M is sensitive to external magnetic fields.
5. Some errors in image formation may result from electrical disturbances coming from traffic, nearby electric equipments, etc.
6. It is also sensitive to mechanical vibration.
7. In order to reduce the difference of the electron beam by collision with air molecules, the pressure within the E/M must be kept low, less than 10^{-4} mm Hg.
8. Extreme care must be taken that no volatile meterial is introduced with the sample.
9. Two fundamental limitations are imposed by the apparatus itself. These are ultimate resolving power of the microscope and image contrast. Actually, in practice, resolutions of about 10 Å have been attained though in many specimens less resolution may be required. Biological material is composed of carbon, nitrogen, oxygen and hydrogen, all of which absorb or scatter electrons partly; therefore, it is difficult to see significant details near the resolution limit. Hence, metal compounds like osmic acid and phosphotungstic acid—which absorb electrons well—are being used for staining biological materials.
10. In the process of viewing, the sample is subjected to considerable electron bombardment, which may significantly alter the structure.
11. Preliminary treatment of sample—which has to be subjected to—may cause distortion of

the delicate structure inside the cell and some artifacts may be observed.

12. Because of the very poor penetrating power of electrons, very thin slices of tissue have to be used. Because the loss of energy by the electrons on passing through a section changes their wavelength, they thus give rise to a form of chromatic aberration which sets the limit to the degree of resolution.

    The thin slices are prepared by taking small pieces of tissue which are fixed in a 1 per cent solution of osmium tetroxide buffered to pH 7.4 or in solutions of aldehydes. The tissues are dehydrated in a graded series of alcohols of increasing strength before embedding in a plastic; many different resins are in current use, but the majority are epoxy resins such as "Araldite" or "Epon". Sections are cut about 0.05 μm thick on a special microtome using glass or diamond knives and then mounted on a copper grid which is placed in the E/M.

13. It is difficult to prepare a support for the specimen on an ordinary glass slide. Therefore, the specimen is mounted on a very thin film of plastic, about 900 Å thick, which itself is mounted on a very fine wire mesh; the beam of electrons thus passes through the holes in the "gird". Movement of the specimen on the grid that takes place when it is bombarded with electrons is largely avoided by evaporating a fine film of carbon onto the plastic-coated grid.

14. Difficult to make sample adequately if the range of structure presented by cells is in a large heterogeneous organ.

15. Information regarding chemical composition of tissues is not available by the standard E/M technique.

16. Difficult to examine surface in the standard E/M technique.

17. Whereas ordinary E/M can show outlines and structural details in a flat plane, the more sophisticated technique—Scanning Electron Microscopy—can produce clear "aerial view" of particles such as microbial and other cells, showing the contours, irregularities and texture of their surfaces.

14.6 Scanning Electron Microscope

A new type of Electron Microscope introduced recently is Scanning Electron Microscope. It differs in principle from optical or Transmission Electron Microscope. In these latter types, the image is formed directly on an image plane. In the Scanning Microscope, the image is formed indirectly to accumulation of information from the specimen point by point.

14.6.1 Principle

A specimen in the column of the Scanning Microscope is bombarded with a focused beam of electrons 100 Å or less diameter. The electron beam is scanned over an area of the surface of the specimen in a raster fashion by varying the electrical currents in a pair of beam deflector coils placed in the microscope column. The specimen is placed at the lower end of the column. When the electron beam strikes the specimen surface, secondary electrons are emitted from its surface. The secondary electrons are emitted from its surface. The secondary electrons are collected by means of a positively-charged collector placed above and to one side of the specimen. The effect of these collected electrons on a phosphor causes flashes of light which stimulate a photomultiplier tube. If a biological specimen is given with an irregularly contoured surface, the intensity of the emission of secondary electrons will vary with the angle at which the electron probe strikes each point on the surface. A cathode ray tube on the display panel of the instrument is scanned in synchronization with the electron beam in the column, so that each point on the cathode ray tube corresponds to a point on the specimen.

The amplified output is used to modulate the intensity of the cathode ray tube in proportion to the intensity of the secondary electron emission and an image of the specimen is built-up point by point on the fluorescent screen of the cathode ray tube. A camera directed at the screen records the varying intensity of the signal along each line of scan and thus a magnified picture of the specimen is built-up.

14.6.2 Some salient features

1. The image is formed indirectly by accumulation of information from the specimen, point by point.

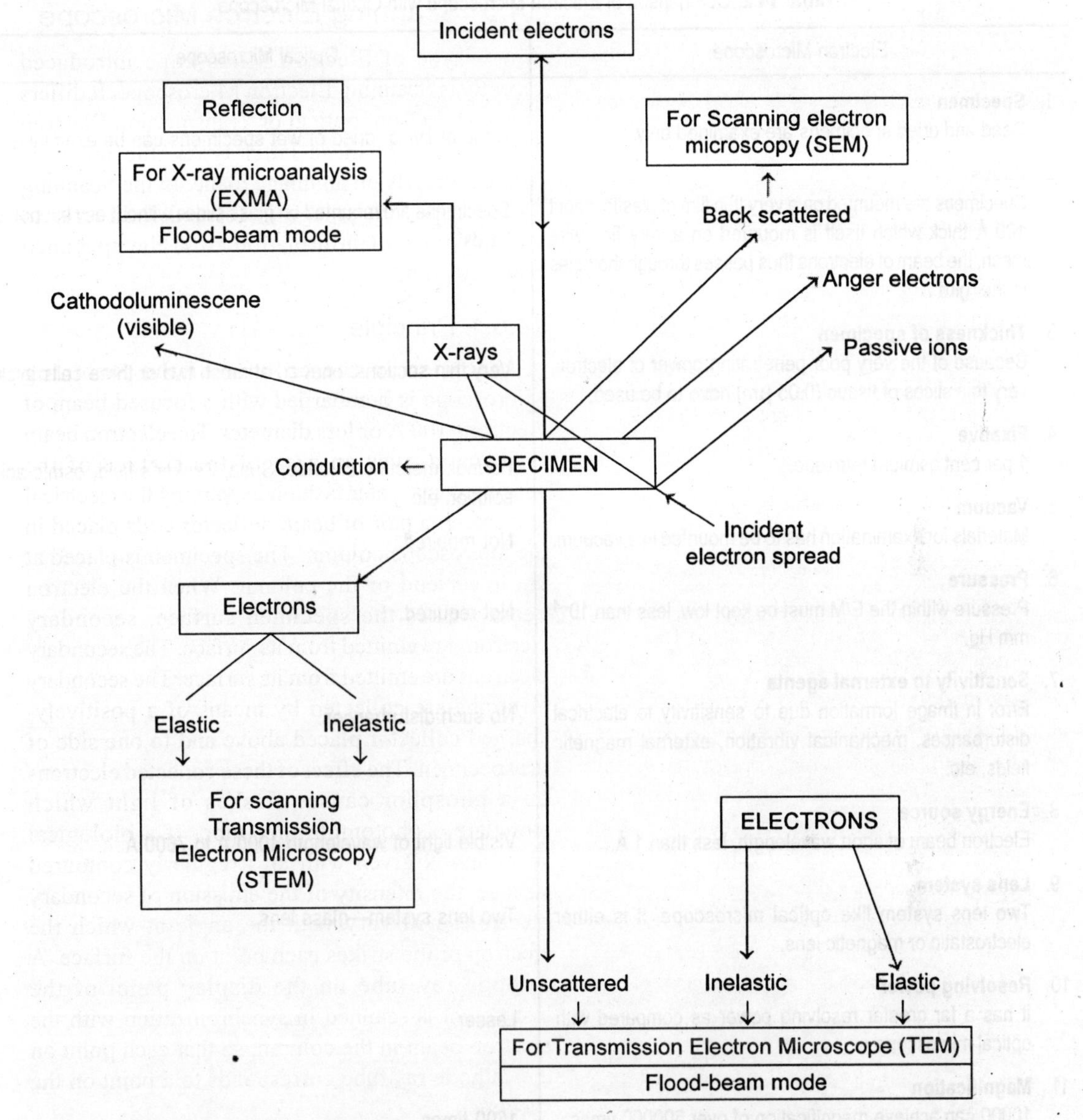

Fig. 14.9: Relationship between high energy electrons and matter and types of interaction

2. The electron beam does not pass through, hence no need to cut ultra-thin sections.
3. The resolution is only about 100 Å. At present the resolution of the Scanning Electron Microscope is not equal to that of the Transmission Electron Microscope.
4. Advantages for great depth of field (7-10 times that of the Compound Light Microscope at comparable magnification) which facilitates to obtain images with a remarkable three dimensional quality.

14.6.3 Advantages

1. Has a great depth of field.
2. Can operate at low magnification.
3. Can provide stereoscopic photographs very easily.
4. As the electron does not pass through the specimen, ultra-thin specimen is not required.

14.6.4 Importance

1. Study of the surface of the cell.

Table 14.2: Comparison of Electron Microscope with Optical Microscope

Electron Microscope	Optical Microscope
1. **Specimen** Dead and dried specimens are examined only.	Dead or living, dried or wet specimens can be examined.
2. **Slides** Specimens are mounted on a very thin film of plastic, about 100 Å thick which itself is mounted on a very fine wire mesh, the beam of electrons thus passes through the holes in the "gird".	Specimens are mounted on glass slides without any support of "girds".
3. **Thickness of specimen** Because of the very poor penetrating power of electron, very thin slices of tissue (0.05 μm) have to be used.	Very thin sections; one, or, at most, two or three cells thick.
4. **Fixative** 1 per cent osmium tetroxide.	Formol, mercuric chloride, Susa, Zenker's fluid, osmic acid solution etc.
5. **Vacuum** Materials for examination has to be mounted in a vacuum.	Not required.
6. **Pressure** Pressure within the E/M must be kept low, less than 10^{-4} mm Hg.	Not required.
7. **Sensitivity to external agents** Error in image formation due to sensitivity to electrical disturbances, mechanical vibration, external magnetic fields, etc.	No such disturbances.
8. **Energy source** Electron beam of short wavelength, less than 1 Å.	Visible light of wavelength 4000 Å to 7600 Å.
9. **Lens system** Two lens system like optical microscope, it is either electrostatic or magnetic lens.	Two lens system—glass lens.
10. **Resolving power** It has a far greater resolving power as compared with optical microscope.	Lesser.
11. **Magnification** 10000 can achieve magnification of over 500000 times.	1500 times.
12. **Visibility of image** Electron beam is not visible for which a fluorescence screen is required.	No fluorescence screen is required.

2. Study of hard materials such as cartilage, bone and dentine.

14.7 Scanning Electron-probe Micro-analysis

An electron beam is generated, focussed on the specimen and scanned regularly across its surface by means of the coils and magnetic lenses in the column of the instrument.

The secondary electron beam, i.e., the X-rays emitted from the specimen are collected. These rays are analysed by X-ray Crystal Spectro-Meter. When the angle of the spectrometer crystal is altered. X-rays of different wavelength are separated and

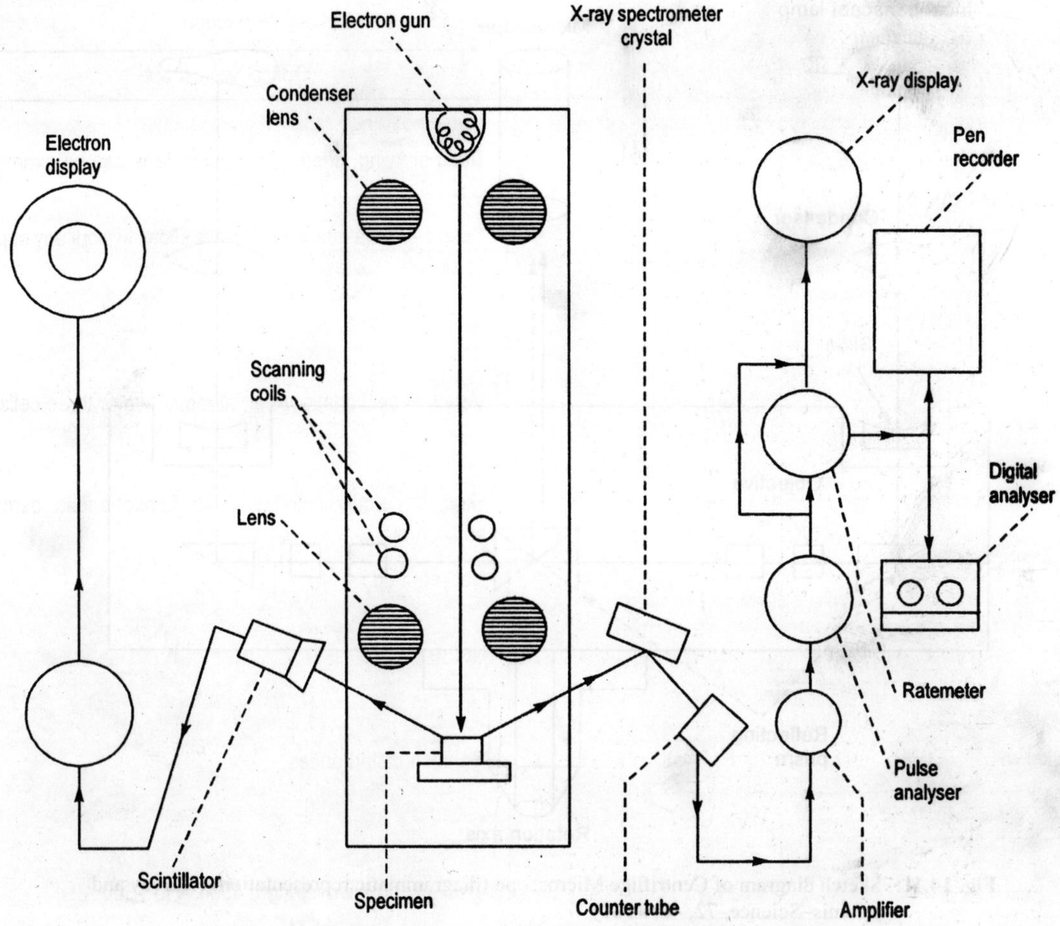

Fig. 14.10: Sketch diagram of scanning electron-probe micro-analysis

passed into the gas filled tube of the counter. The resulting signals are amplified and displayed on a cathode ray tube where the spot is scanned in synchrony with the original electron beam. A picture of the distribution of the particular element is built-up. A **Scintillation Phosphor** is used to pick up the secondary electrons which are emitted from the specimen at the same time when X-rays are emitted. This is used to assist the analysis.

Just as in the Scanning Electron Microscope, this allows a scanning electron image to form and a detailed morphological picture may be obtained from the area of the specimen that is emitting X-rays. Sometimes the Optical Microscope is introduced into the instrument, so that visual examination can be done simultaneously.

Performance of the instrument can be improved by incorporating the following:

1. Stationary probe of electrons.
2. Counting facilities.
3. A pen recorder.
4. Any process for varying the crystal angle continuously to get a wide range of X-ray wavelengths for which the elements present can be recognized rapidly and easily (Fig. 14.9).

14.7.1 Importance

1. To identify the chemical elements present in a specimen and also to get some idea of their location and amount. Each chemical element produces X-rays of a few specific wavelengths.
2. Studies have been made of the deposition of calcium in arterial walls.

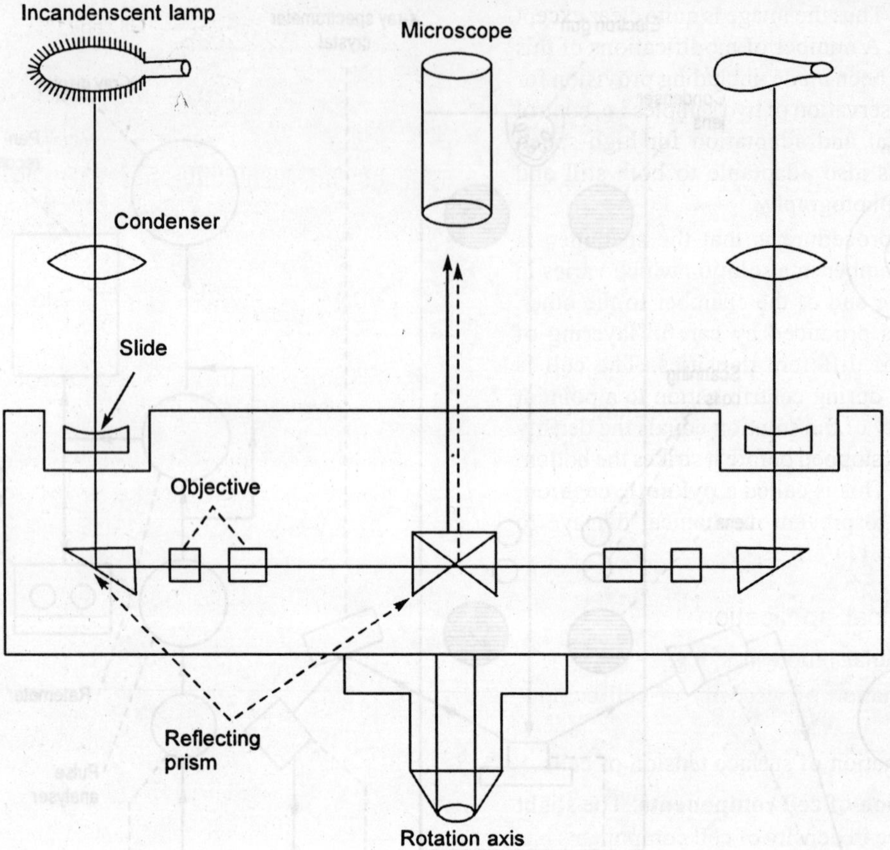

Fig. 14.11: Sketch diagram of Centrifuge Microscope (diagrammatic representation) (Harvey and Loomis–Science, 72, 42, 1930)

LOW-TEMPERATURE ELECTRON MICROSCOPE

Fernandez-Moran (1962) improved the negative staining technique for production of areas of low electron-density. These include cryofixation based on rapid freezing of glycerinated specimens with liquid helium at 1-2K, followed by freeze-substitution and embedding in plastics. In this way, damage by the electron-beam is minimized with these preparations by cooling the specimen—support with liquid nitrogen whilst a low energy microbeam is employed. This technique is known as *low-temperature* electron microscopy. Drying the specimen can be avoided by examination of ultra-thin frozen sections of fresh tissue directly without thawing, by using a liquid-nitrogen cooling device.

Fernandez-Moran, H—Circulation 26, 1039-1065, 1962

14.8 Centrifuge Microscope

Harvey and Loomis (1930) constructed a centrifuge with microscope optical system so that effect of centrifugation on single cell could be observed. The objective of the microscope is fitted into the rotating centrifuge head. The specimen is placed in a specially constructed slide mounting. The specimen is illuminated by light from an incandescent lamp mounted in a stationary position with the filament parallel to the radius of rotation of the centrifuge head.

The light beam passing through the objective is reflected by a prism into the microscope which is stationary. The optical system is placed in such that the material under study is illuminated only when the centrifuge rotates to such a position that the entire system is aligned. The assembly is done in such that the movement of the specimen and macroscope objective of 100 microns is manifested by an apparent movement in the ocular of only

1 micron or less. Thus the image is quite clear except at the periphery. A number of modifications of this instrument have been made, including provision for simultaneous observation of two samples i.e. control and experimental and adaptation for high speed centrifuges. It is also adaptable to both still and motion picture photography.

The actual procedure is that the specimen is placed in the chamber in a solution which varies in density from one end of the chamber to the other. This situation is produced by careful layering of two solutions of different densities. The cell is thrown outward during centrifugation to a point at which the density of the solution equals the density of the cell, and is stopped before it strikes the bottom of the chamber. This is called a **pyknotic cushion**, as is necessary to prevent mechanical damage to the cells (Fig. 14.11)

14.8.1 Biological application

1. Study in cellular biophysics, e.g.,
   (a) Determination of viscosity of cell components.
   (b) Determination of surface tension of cells.
2. **Separation of cell components**: The slight difference in density of cell components has made it possible to separate these components by centrifugation for their subsequent study separately. If gravitational forces produced are 4000 or more times gravity, and if the centrifugation is prolonged, the cell may be pulled apart into a light half and a heavy half. And if the forces are further increased, further fractionation may be accomplished.
3. **Layering of the cell components**: If the centrifugal forces are not great enough to break the cell into parts, the components will layer. Each layer represents a segment of the cell volume.

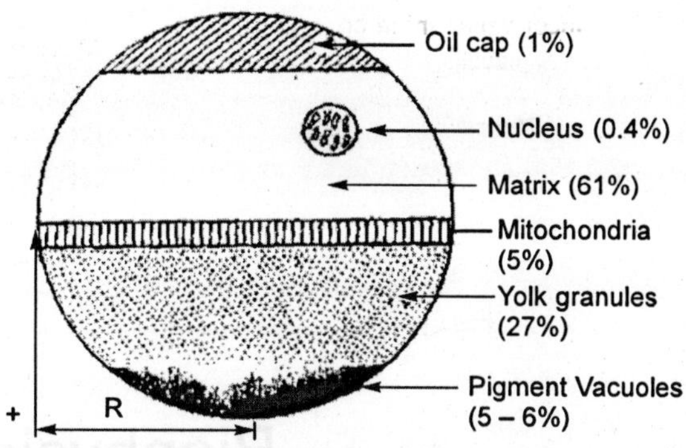

Fig. 14.12: Centrifuged Arbacia when the component are completely layered

14.9 Polarising Microscope

(See chapter on Colloid, Chp. 7).

14.10 Ultra Microscope

(See chapter on Colloid, Chp. 7).

14.11 Micro-manipulator

(See chapter on Colloid, Chp. 7).

14.12 Other Microscopes for Study

Confocal Microscopy, video Microscope and digital imaging, Scanning tunnelling Microscope, Atomic force Microscope, Interfacial force microscope, Laser Microscope, Acoustic Microscope, NMR Microscope, Surgical Microscope, X-ray Microscope, Immunoelectron Microscope, cryoelectron Microscope, High voltage electron Microscope etc. Operating Microscope.

15

Biophysics of Water

1. Introduction, 2. Molecular structure of water, 3. Macromolecular structure of water, 4. Hydrogen bonds and structure of water, 5. Physical properties of water and its biological applications.

15.1 Introduction

Water is the most abundant component and a cell contains approximately 80 p.c. of water, and water constitutes about 60 p.c. of the total body weight. All the biomolecules are present in the water medium and most of the biochemical reactions take place in an aqueous environment and is an important determinant of the properties of macromolecules e.g., proteins. It is, therefore, necessary to consider some physical features of significance in biological events.

15.2 Molecular Structure of Water

The water molecule is an irregular tetrahedron with oxygen at its centre. In tetrahedron, two bonds having hydrogen are directed towards 2 corners, whereas the unshared electrons occupy the two remaining corners. The angle between the two hydrogen atoms is 105 degrees and is slightly shorter than the tetrahedral angle which is 109.5 degrees forming a slightly skewed tetrahedron. Electrical charge is not uniformly distributed in the water molecule. Opposite the hydrogens, the side of oxygen is rich in electrons whereas the other side, the unshielded hydrogen nuclei, form a local region of positive charge and the oxygen becomes negative charge. For this reason, water is referred to as "dipole" which means it has electric charge, i.e., electrons unequally distributed in the structure.

The structure of water and ice is determined by hydrogen bonds. A water molecule can form 4 hydrogen bonds with neighbouring molecules with a tetrahedral coordination. Ordinary ice has a hexagonal crystalline lattice, each oxygen atom in the lattice residing in the centre of the tetrahedron at the apexes of which there are located the neighbouring oxygen atoms. The distances between the oxygen atoms are equal to 0.276 nm (2.76 Å) and between oxygen and hydrogen is 0.99 Å (Fig. 15.1).

The elementary unit cell has four molecules. The molecular lattice of ice is very loose, with large open channels, since its "coordination number" is very small which explains why ice is less dense than water if we consider the time scale in which water structure is recorded after experiencing vibrations, rotations and occasional translational displacement by the water molecules. Three types of structural display are seen, e.g.

(a) Instantaneous or M-structure,
(b) Vibration: Averaged or K-structure,
(c) Diffusion: Averaged or D-structure.

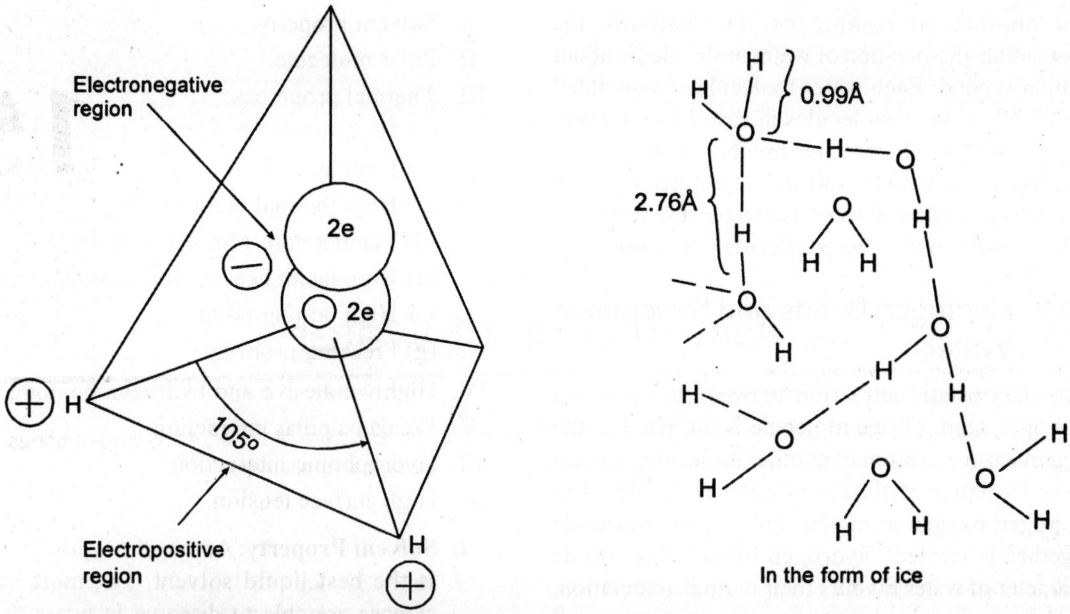

Fig. 15.1: The structure of water

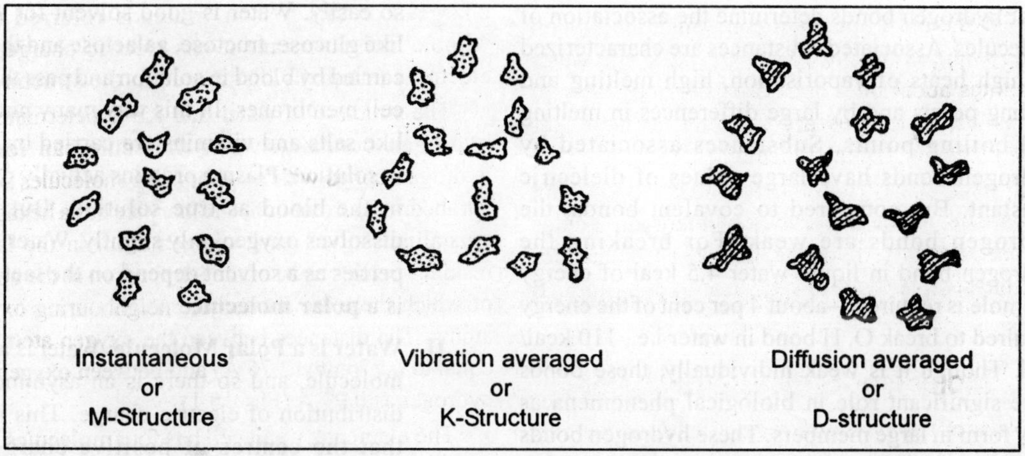

Fig. 15.2: Structural display of structures of water

Using various methods of investigations, e.g., X-ray diffraction, thermodynamics, light scattering, nuclear magnetic resonance, dielectric relaxation, inelastic magentic resonance, dielectric relaxation, inelastic neutron scattering, absorption of ultrasound and vibrational spectra with different durations of experiments the above structures are obtained. The thermodynamic properties naturally characterize the D structure (Fig. 15.2).

15.3 Macromolecular Structure of Water

Molecular ordering of water molecules is not always restricted to ice, Liquid water shows macromolecular structure which parallels the geometric disposition of water molecules in ice. Association of water molecules with one another in both solid and liquid states is due to dipolar character of water. It remains liquid than solid due to transient nature of these macromolecular complexes. The half-life for association-dissociation of water molecules is about 1 microsecond. Each water molecule is associated with 4 other water molecules in case of solid state, whereas in liquid state the number is about 3.5. Except the transient nature of intermolecular interactions in liquid water, the water resembles ice more closely in its macromolecular structure.

15.4 Hydrogen Bonds and Structure of Water

The electrostatic attraction between the positive hydrogen atom of one molecule is attracted to the negative oxygen atom of another molecule, leading to the formation of clusters of water molecules. The hydrogen-oxygen attraction holding the molecule together is termed "hydrogen bond". The dipole character of water favours their mutual association. Hydrogen bonds determine the structure and properties of water. The presence of hydrogen bonds exerts a strong effect on the physical and physico-chemical properties of substances. The intermolecular hydrogen bonds determine the association of molecules. Associated substances are characterized by high heats of vaporisation, high melting and boiling points and by large differences in melting and boiling points. Substances associated by hydrogen bonds have large values of dielectric constant. But compared to covalent bonds, the hydrogen bonds are weak. For breaking the hydrogen bond in liquid water 4.5 kcal of energy per mole is required—about 4 per cent of the energy required to break O, H bond in water i.e., 110 kcal/mol. Though it is weak individually, these bonds have significant role in biological phenomena as they form in large members. These hydrogen bonds contribute to the structure, not only for water but also for other dipolar molecules like alcohol, DNA and proteins. Thus the secondary structures of protein chains are stabilized by hydrogen bonding. These bonds play a key role in the conformational structure of nucleic acids and carbohydrates.

15.5 Physical Properties of Water and its Biological Applications

Water is a medium and an active participant in molecular interaction in biology. The following properties are important for biological processes:

I. Solvent property.
II. Polar molecule.
III. Thermal properties:
   (a) High specific heat or heat capacity.
   (b) High latent heat of vaporisation.
   (c) High thermal conductivity.
   (d) Temperature of maximum density.
   (e) High latent heat of fusion or melting.
   (f) High boiling point.
   (g) Freezing property.
IV. Highly cohesive and hydrogen bonding.
V. Weakens polar interaction.
VI. Hydrophobic interaction.
VII. High surface tension.

I. Solvent Property: Among the liquids, water is the best liquid solvent, i.e., more substances are able to dissolve in water to an appreciable extent than any other liquid. It acts as a solvent in varying degrees, i.e., some substances dissolve readily, others not so easily. Water is good solvent for sugars like glucose, fructose, galactose and they are carried by blood in solution and pass through cell membranes. In this way many nutrients like salts and vitamins are carried by water in solution. Plasma proteins actually present in the blood as true solution. But water dissolves oxygen only slightly. Water's properties as a solvent depend on the fact that it is a **polar molecule.**

II. Water is a Polar Molecule: Water is a polar molecule, and so there is an asymmetrical distribution of electric charge. This means that the centres of positive charge and negative charge are separated by a short distance due to configuration of the water molecule. Instead of being arranged in a straight line, the hydrogen and oxygen atoms are arranged asymmetrically. The hydrogen atom has net positive charge whereas oxygen has a net negative charge and the whole molecule thus shows **polarity**. Therefore, water is a good solvent, ionic solids and polar molecules readily dissolve in it. Polarity contributes towards solvent and cohesive properties of water molecule.

III. Thermal Properties

(a) High Specific Heat or Heat Capacity: Specific heat or heat capacity of a material is the number of calories of heat required for 1 gm of substance to raise its temperature through 1°C. If we say specific heat of water is 1.0 it means 1 gm of water requires 1 calorie heat for raising its temperature through 1°C. Water has a very high specific heat compared to other liquids, e.g., a high increase in heat causes a comparatively small rise in temperature. It means that water is good to maintain its temperature irrespective of fluctuations in the surrounding temperature. This has biological importance such as:

1. **Ideal for biochemical reaction**: The range of temperatures in which biochemical reactions take place is short, hence this property of water is ideal for the reaction.
2. **Regulation of body temperature**: Water acts as the heat transportation medium. Cell produces heat due to metabolism and blood and lymph collect it and carry it to the body surface for loss of heat to the surrounding with less rise in temperature in water due to high specific heat property. On account of this great heat storage power of water, sudden changes of body temperature are avoided.

(b) High Latent Heat of Vaporisation or Evaporation: Latent heat of vaporisation of a substance is the number of calories (small) of heat which 1 gram of the substance must absorb in order to change from liquid to gaseous state wihout undergoing any temperature change. If we say latent heat of vaporisation of water at 100°C is 539.5 calories it means 539.5 calories of heat must be absorbed to change 1 gram of water liquid at 100°C 1 gram of water gas at 100°C. Thus 1 c.c. of water requires about 0.6 calories of heat for its vaporisation. This figure is 50 per cent higher than that of water's closest competitors. Biologically, this property has vital significance for regulation of body temperature. The latent heat of vaporisation of water serves as one of the means to take care of the considerable quanity of heat, i.e., from 2000 to 3500 calories (large) per day produced in the body cells from kinetic or other energies. This heat is carried by the water in the blood to the lungs, skin areas and deeper parts of the body where nature disposes in the following ways:

1. Fluid is being constantly lost from the body through evaporation from the lungs and skin. By far the larger portion is used for vaporising the water and approximately 574 calories of heat for vaporising one gram of this water at natural body temperature immediately becomes latent as heat, hence is responsible for no temperature increase in these areas or the adjoining air.
2. A smaller portion of heat is brought by blood from deeper to superficial regions and spread out in fine vessels over a broad area just beneath the skin and is radiated as appreciable heat which, of course, keeps these areas warm and also warms the surrounding air which is a major control mechanism.

If conservation of heat is required, the fluid is drained from the surface areas and collected in the deeper parts of the body and internal organs, e.g., muscles etc.

(c) High Thermal Conductivity: The thermal conductivity of water is higher than that of any other liquid. The advantage of this property is the dissipation of heat from deeper regions of the body.

(d) Temperature of Maximum Density: When water reaches 4°C, it behaves in a peculiar manner. Further decrease in temperature causes change in volume, hence its density.

Most substances contract continuously as their temperatures go down further and further, therefore, they meet with increase in densities as they get colder and colder. If the substance is liquid and mobile and if it is not at the same temperature throughout, the lowest-temperature portion of the liquid drops down to the bottom and the highest rises at the top. Such liquids meet with their freezing temperatures at the bottom of their mass and freeze from the bottom up. In case of water it behaves as do other liquids down to 4°C; in that as its temperature from any level above 4°C is lowered, it contracts continuously and become continuously denser until it attains this temperature. But from 4°C, if it is lowered, it behaves peculiarly by showing expansion as it gets colder and colder instead of contraction. Therefore, whether the temperature of water goes either up or down from this 4°C level, it becomes lighter or less dense at both the higher and lower temperatures and water attains maximum density at 4°C.

(e) **High Latent Heat of Fusion or Melting**: Latent heat of fusion or melting of a substance is the number of calories of heat which the substance must absorb in order to change 1 gram of solid substance from the solid state to the liquid state without effecting any change in its temperature. If we say latent heat of fusion of water is 79.71, it means 79.71 calories of heat must be absorbed to change 1 gram of water solid at 0°C to 1 gram of water liquid at 0°C. When a substance changes from solid to liquid it must absorb heat. But the heat which is essential for melting disappears as heat since it does not produce any temperature increase to the substance. The initial solid and the resulting liquid are at the same temperature, if only just enough heat for the melting is provided. It is for this reason that this disappearing heat has been referred to as **latent heat of fusion or melting**.

(f) **High Boiling Point**: Water has a remarkably high boiling point having such low molecular weight. Water boils at 100°C or somewhat above or below this temperature, depending upon the pressure upon its surface.

(g) **Freezing Properties**: Water freezes at 0°C or somewhat below this temperature. Most liquids decrease in volume and increase in density as the temperature decreases. When such liquid freezes, molecules become densely packed together and the resulting ice sinks. With the water the reverse phenomena occurs. As water is cooled below 4°C its volume increases and its density decreases which means that ice tends to float rather than sink. Moreover, when the temperature drops, the coldest water is at the surface and slightly warmer water goes down. So ice forms at the surface first and the bottom later and organisms which live at the bottom of fresh water lakes are protected from freezing.

IV. **Highly Cohesive or High Affinity for Each Other**: The polarity of molecules is causative factor for their association with each other. A positive hydrogen atom of one molecule of water may be attached to the negative oxygen atom of another leading to the formation of small clusters of water molecules. This hydrogen-oxygen attraction for a link to hold the other molecules together is termed "hydrogen bonding" which is a weak one but plays an important role in holding organic molecules together.

V. **Weakens Polar Interaction**: Sodium chloride in crystal form, the sodium and chloride ions are held together by strong attraction with each other. But in water this attraction is reduced and sodium and chloride ions go into solution. Water weakens the attraction between sodium and chloride ions having opposite charge because:

1. Water having positive and negative charges, attract both the ions of sodium chloride, therefore, diminishes electrostatic interaction.
2. Moreover, once the ions have separated they are prevented from reunion by water molecules clustering round each of them. The oriented solvent shell around an ion and produce electric field of their own, which opposes the fields produced by the ions. Eventually electrostatic interactions of these

ions are markedly weakened. Water has an unusually high dielectric constant which is 80 at 20°C (Fig 15.3).

VI. Hydrophobic Interaction: As the oil droplets come together in water to form a single large oil drop, similar phenomenon occurs at the atomic level, e.g., nonpolar molecules or groups tend to cluster together in water. These association of nonpolar molecules in water are known as **hydrophobic interaction**. Hydrophobic interactions are the major driving force for biological processes like:

1. Folding of macromolecules,
2. Binding of substrates to the enzyme.

The basis of hydrophobic interaction is increased in entropy as a result from the enhanced freedom of released water molecules. Nonpolar molecules come together in water, not because of high affinity for each other but due to water bonds strongly itself.

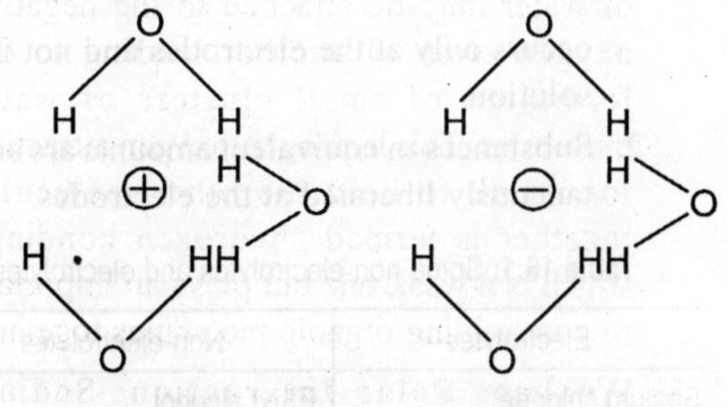

Fig. 15.3: Water surrounds the electric charged ions and diminishes electrostatic attraction

VII. High Surface Tension: Surface tension is the force that causes the surface of a liquid to contract and occupies the least possible area. It is due to inward-acting cohesive forces between the molecules at the surface for polarity of water molecule. Water has the highest surface tension among the liquids and this property has great importance.

If a substance is dissolved in water, the molecules lower the surface tension of water and tend to collect at the interface between the liquid phase and other phases. This may play a role in the (1) development of plasma membrane, (2) movement of molecules across the plasma membrane, (3) movement of water up the capillary vessels and tracheids in the stems of plants.

15.6 Summary

1. Reversible molecular interactions in biological systems are due to electrostatic bonding, hydrogen bonding and van der Waals' bonding. These interactions are enormously affected by the presence of water.
2. Water is polar molecule and has the capacity to serve as both a hydrogen donor and a hydrogen acceptor, it has unusually high dielectric constant and cohesiveness.
3. Water diminishes the strength of electrostatic and hydrogen bonding between other molecules and ions.
4. Water accentuates the interactions of nonpolar molecules.
5. Substrates are bound to enzymes at active-site from which water is largely excluded when the substrate is bound. The exclusion of water strengthens electrostatic and hydrogen bonds between enzyme and substrate.
6. Water is a good solvent and carries the products as true solution in the blood.
7. High specific heat is important for biochemical reactions and body temperature regulation of the body.
8. High latent heat of vaporisation plays a role in body temperature regulations.
9. High thermal conductivity plays a role in dissipation heat from deeper regions of the body.
10. It weakens polar interactions.
11. Hydrophobic interactions are the major driving force for:
    (a) Folding of macromolecules,
    (b) Binding of substrates to the enzymes.
12. Increase in surface tension may play a role in the:
    (a) Development of plasma membrane,
    (b) Movement of molecules across the plasma membrane,
    (c) Movement of water up the capillary vessels and tracheids in the stems of plants.

16

Electrochemistry

1. Ionization, 2. Electrolytes, 3. Electrolysis, 4. Conductance, 5. Solubility Product, 6. Common Ion Effect, 7. Ostwald's Dilution Law, 8. Dielectric Constant.

16.1 Ionization and Electrolysis Conduction

16.1.1 Electrolytes and non-electrolytes

Pure water is a very poor conductor of electricity. Water solution of sugar, urea, alcohols do not conduct electricity, hence these substances are called **non-electrolytes**. But water solution of acids, bases and salts are good conductors of electricity when electric current is passed. These substances are called **electrolytes** (Table 16.1). The presence of water is not essential because many of them are typical **electrolytes** even in fused state or in non-aqueous solution.

The acidulated water is decomposed into Hydrogen and Oxygen by electric current. The bases and salts also are decomposed when dissolved in water during passing electric current through such solutions. This decomposition by means of electric current is called **electrolysis**. And all the substances which conduct electric current and at the same time undergo decomposition are called **electrolytes**.

16.1.2 Salient features of the process of electrolysis

1. Though electric current flows through the solution during electrolysis, decomposition occurs only at the electrodes and not in the solution.
2. Substances in equivalent amounts are simultaneously liberated at the electrodes.

Table 16.1: Some non-electrolytes and electrolytes

Electrolytes	Non-electrolytes
Sodium chloride	Ethyl alcohol
Calcium chloride	Methyl alcohol
Barium sulphate	Glucose
Hydrochloric acid	Sucrose
Acetic acid	
Sodium hydroxide	
Ammonium hydroxide	Urea
Cupric sulphate	etc.
Potassium hydroxide	
Nitric acid, etc.	

Table 16.2: Electrolyte solution vs. metal regarding passage of electric current

Electrolytes solution	Metal
1. Far less conduction.	Greater conduction.
2. Chemical changes take place during conduction (e.g., salts) on the	There is no change during conduction except the heating effect.

Contd.

Electrolytes solution	Metal
electrodes, the process is called electrolysis.	
3. During conduction ions carry the current.	During metallic conduction, electrons carry the current.

16.1.3 Conductivity and chemical activity of electrolytes

It has been observed that in solutions of electrolytes having the same normality, the range of electrical conductivities is the same as the range of chemical activities of the specific acids, bases and salts. The following table shows the electrolytes arranged from the highest to the lowest of chemical activity of acids, bases and salts and it also shows the order of the electrical conductivities of their solutions of the same normality (Table 16.3).

The numbers mentioned in the (Table 16.3) in relation to electrolytes illustrate the degree of relationships specified.

Table 16.3: Chemical activity and conductivity of tenth-normal solution

Hydrochloric acid	...	92
Nitric acid	...	92
Sodium hydroxide	...	91
Potassium hydroxide	...	91
N/64 Barium hydroxide	...	92
N/64 Calcium hydroxide	...	90
(Not sufficiently soluble to make tenth-normal solution)		
Potassium chloride	...	86
Sodium chloride	...	84
Silver nitrate	...	81
Sodium nitrate	...	79
Barium hydroxide	...	77
Sulphuric acid	...	61
Oxalic acid	...	50
Copper sulphate	...	40
Zinc sulphate	...	40
Phosphoric acid	...	27
Tartaric acid	...	8.5
Acetic acid	...	1.3
Ammonium hydroxide	...	1.3
Mercuric chloride	...	1.0
Carbonic acid	...	0.17
Hydrosulphuric acid	...	0.07
Boric acid	...	0.01

16.1.4 Strong electrolytes and weak electrolytes

A strong electrolyte is an acid, base or salt which is very active chemically and where water solution conducts the electric current very well; whereas a weak electrolyte is an acid, base or salt which is not very active chemically and whose water solution does not conduct the electric current very well (Table 16.4).

Table 16.4: List of some strong and weak electrolytes

Strong electrolytes	Weak electrolytes
Strong acids (HCl, HNO_3).	Carbonic acid
Strong bases (NaOH, KOH)	Weak acids
All salts except a few Hg-Salts	(Most organic acids)
	Ammonia
	Weak bases
	(Most organic bases)
	Phenols
	$HgCl_2$
	$Hg(NH)_2$
	Water
	Amines

16.1.5 Theories of electrolytic dissociation (ionization)

1. Grothus (1806) was of the opinion that the electrolytic molecules contain two poles, i.e., positive and negative which align themselves along the lines of force and conduct the current by definite order or mechanism.

- **Demerits**

(a) According to this theory, dissociation of molecules require a definite voltage which varies with the type of electrolytes. But it is a fact that even small E.M.F. can produce electrolysis for a short period.

(b) Verification by Ohm's law shows that no part of the current is used up for dissociation.

He made an error by giving opinion that even strong electrolytes are incompletely dissociated into ions and their degree of dissociation changes with dilution which is valid for weak electrolytes like acetic acid but not for strong electrolytes.

2. Arrhenius (1887) was the person to postulate the theory of electrolytic dissociation on quantitative basis. He postulated that:
   (a) The extensive spontaneous dissociation (i.e., ionization) of electrolytes in a solution (ionizing solvent) into specific number of charged particles called ions and act as the transporter of electricity.
   (b) Ionization is brought on by the agency of the solvent.
   (c) The electrolyte molecule gives two types of ions having equal and opposite charges.
   (d) These ions attract each other constantly and form neutral molecules.
   (e) Ionization increases with dilution. At a particular dilution the equilibrium is reached when the rate of ionization is equal to the rate of formation of neutral molecules.
   (f) Properties (physical and chemical) of the electrolytes depend upon the nature and concentration of the ions coming from the electrolytes. He also first showed methods of calculations of the extent of ionization by different methods.

According to modern concept, strong electrolytes are completely ionized at all dilutions. Ionization proposed by Arrhenius are:

$$NaCl \rightleftarrows Na^+ + Cl^-$$

$$CH_3COOH \rightleftarrows H^+ + CH_3COO^-$$

$$H_2SO_4 \rightleftarrows H^+HSO_4^- ; HSO_4 \rightleftarrows H^+ + SO_4^=$$

16.1.6 Evidence for the theory of electrolytic dissociation

1. While studying for per cent dissociation, it has been observed by X-ray studies that crystals of salts, e.g., NaCl, do not contain molecules of the salts, but the rystals of NaCl composed of Na^+ and Cl^- ions are arranged in perfect order which is known as "space or ionic lattice". This lattice structure is separated when the salt crystal is placed in water and distributed homogeneously throughout the water molecules. Hence, NaCl is completely ionized in solutions of any concentration.

   The strong lattice energy of crystals is overcome by the hydration energy of the ions and also helped by the high dielectric constant of water which reduces the attractive force between negative and positive ions. The ionic concept is strongly supported by this evidence. But as the electrolytes exist wholly as ions, so it may not be termed as dissociation of the electrolytes but ionization only and formation and dissociation of ion-pairs.

2. Metallic iodide solutions in tubes, on rapid centrifugation, show negatively charged at the extreme end of the tube due to high sedimentation rate of the large heavy iodide ion.

3. Based on ionization concept facts have been simply explained, e.g.
   (a) High abnormal values of osmotic pressure and related colligative properties of the electrolytes.
   (b) The same value of heat of neutralisation of all strong acids and bases.

4. Based on ionic concept, there are theories which have been successful, e.g.
   (a) It explains precipitation through the theory of the solubility product, such as:
   (b) Ostwald's dilution law,
   (c) Kohlrausch's law,
   (d) Debye-Huckel equation, etc.

Ions are atoms or groups of atoms having charges due to acceptance or donation of the electrons. The number of unit charges on an ion equals its number of valency. Cations (positively charged) migrate towards the cathode in the electric field and anions (negatively charged) move towards the anode. These ionic motility conduct electric current through electrolyte solution. The extent of the current conductivity depends in part upon the number of ions per unit volume of the solution, but even more upon the motility of the ions. The total number of positive charges of cations of an electrolyte molecule equals the total negative charges of its anions, e.g.

$$NaCl \rightleftarrows Na^+ + Cl^-$$

Table 16.5: Electrolytic Dissociation vs. Gaseous Dissociation

Electrolytic dissociation	Gaseous dissociation
1. Reversible process.	Reversible process.
2. Leads to abnormally low values of molecular weight.	Same.
3. Products are charged ions.	Products are neutral molecules.
4. Ions cannot be easily separated due to the strong attraction between the oppositely charged ions.	Products can be easily separated by mechanical means, e.g., diffusion, partial solution.
5. Needs presence of medium, i.e., solvent.	No medium is required.

16.2 According to the Degree of Ionization

(a) **Strong Electrolytes**: Ionize completely in solution at all dilutions. There is no existence of neutral molecules. In water solution of strong electrolyte, two types of reacting particles are present; namely two types of ions, i.e., positively and negatively charged. The variation of conductivity with dilution is attributed to strong **interionic attraction.** A positive ion is related in its motion by the negative ions and vice versa and this interionic attraction decreases with dilution. Hence the equivalent conductivity increases with dilution and achieves a limiting value in sufficiently dilute solution, when the effect of interionic attraction becomes negligible. Thus the term "dissociation of strong electrolyte" does not carry any meaning, hence better to say **ionization** of NaCl. However, complete ionization does not mean that there is no incomplete dissociation due to **ion-pair formation**, particularly for high valent ions and at higher concentrations. However, as ion-pair, e.g., $Na^+ Cl^-$ is not the same as the molecule NaCl and the latter does not as the molecule NaCl and the latter does not exist. The examples of strong electrolytes are KOH, NaOH, H_2SO_4, HNO_3, HCl, NaCl, KCl, Na_2SO_4, $CuSO_4$, NH_4NO_3.

(b) **Weak Electrolytes**: Weak electrolytes are ionized incompletely. In the solution of weak electrolyte, there are three kinds of reacting particles; the two varieties of ion molecule and the undissociated molecules of the acid, base or salt with the undissociated molecules in predominance.

16.2.1 Opposite reactions and equilibrium

When a weak electrolyte is in a solution, the dissociation to form ions is accompanied by a combination between ions to remake some undissociated molecules of the electrolyte, and two oppsoite reactions take place simultaneously. Besides that, these two reactions tend to establish an equilibrium, i.e., a condition in which the speed of dissociation or the number of molecules that ionize during any given time is equal to the speed of association, or the number of the combinations between ions to remake undissociated molecules.

16.2.2 Relationship between concentrations of the Un-ionized and ionized weak electrolytes

When the reactions are at equilibrium, the concentration of the unionized particles is always higher than that of the ionized particles. Apart from this relationship, the percentage of ionisation at equilibrium differs with each electrolyte in different concentration in solution, e.g., 0.1 normal solution of acetic acid yields 1.3 per cent ionized and 98.7 per cent unionized acetic acid whereas normal solution of acetic acid gives 0.4 per cent ionized and 99.6 per cent unionized acetic acid.

16.2.3 Response to interfering factor at equilibrium

At equilibrium, two opposite reactions take place at equal speed and remain indefinitely until an interfering factor enters the field of reaction.

1. **Removal of Water**: When water is removed by evaporation which makes the electrolyte solution more concentrated. Though the two opposite reactions continue, the speed of combination between ions becomes more than the speed of dissociation as long as the evaporation continues.

   The explanation is that, due to increase in concentration of the weak electrolytes as a whole, there must be an increase in its un-ionized part in relation to the ionized part when the two reactions are at equilibrium.

2. **Addition of More Water**: When an electrolyte is diluted the speed of dissociation or ionization becomes more than that of the combination of ions until the dilution is stopped and equilibrium is again established between the two reactions.
3. **Removal of one of the Ions**: At this instance, the ionization reaction predominates over the combinations of the two ions as long as the removal continues.
4. **Addition of one of the Ions**: Addition of one of the ions from another electrolyte but not any one of the partner ion, combination between the ions predominates over the dissociation reaction as long as the disturbing factor is present.

16.2.4 Degree of ionization (α) of weak electrolytes

The degree of ionization (α) is the proportion of the total number of solute molecules which is ionized in the relevant solution.

$$\alpha = \frac{(i-1)}{(n-1)},$$

where i = Van't Hoff factor for the solution,
n = number of ions from each solute molecule.

Ionization constant (K_i)

Ionization constant (K_i) of a weak electrolyte is the equilibrium constant of ionizaton at a given temperature.

$$K_i = \frac{(B^+)(A^-)}{(BA)}$$

Ostwald's Dilution Law

Ostwald's dilution law helps to correlate ionization constant (K_i) with the molar concentration (c) and degree of ionization (α) of the weak electrolyte.

According to the Origin of Ions

(a) **Electrovalent Electrolytes**: These are ionic compounds. They exist in the crystal lattice structures as cations or anions. They are held together either by ionic or electrostatic bonds. They disperse readily in a solution as independent ions because of weakening of interionic electrostatic attractions. This weakening of interionic force is due to encirclement of ion by solvent molecules known as *solvation effect* because of exchange of electrons between the ions and the solvent molecules (NaCl, $CaCl_2$, $BaSO_4$).

(b) **Covalent Electrolytes:** They do not ionize in the pure state due to stability of covalent bonds in the molecules. They may ionize in a solution either by receiving or donating electrons from or to the solvent molecules (HCl, NaOH, H_2SO_4).

16.2.5 Activity of electrolytes

Lewis developed the theory of activity to note the behaviour of strong electrolytes. The activity theory as applied to strong electrolytes considers these electrolytes to be cent per cent ionized. Each ion has an activity to function as an ion according to the concentraton of ions surrounding it having like and unlike charge causing attractions and repulsions and will have varying activities according to variations in the concentration of surrounding ions. Usually, the activity of an ion decreases as the ionic concentration increases except in case of solvation of ions and the activity of solvent molecules change, with changing concentration to alter ionic activities. The term "activity" is used to indicate the apparent or effective concentration of a substance as shown by its properties in solution and designated by "a".

16.2.6 Activity and ionic strength

Activities depend upon:

1. The concentrations of the ions,
2. The total ion concentration in the solution. The activity of NaCl mixed with KNO_3 in solution is the function of all ions but not only of the Na^+ and Cl^- concentration. Charges of the ions are the determining factor, because of ionic effect upon activity is electrostatic in nature. Two divalent ions attract or repel each other four times as strongly as two univalent ions. The effect

on any given ions is half the total electrostatic effect because the ions are in equivalent plus-minus pairs.

Ionic strength concept was used by Lewis to include whole ionic effect and designated as μ and it is obtained by multiplying the concentration of each ion by its valence squared and adding the products together and taking half of the sum.

The ionic strength of an uni-univalent compound, e.g., NaCl or KNO_3, is:

$$\mu = \frac{C(1)^2 + C(1)^2}{2} = C$$

Therefore, ionic strength of uni-univalent compound is eqaul to the concentration C. A divalent compound, e.g., $ZnSO_4$ has an ionic strength equal to 4C:

$$\mu = \frac{C(2)^2 + C(2)^2}{2} = 4C$$

16.2.7 Activity coefficient

Lewis defined the activity coefficient as the ratio of activity "a" to molal concentration "m" and designated as " τ ":

$$\tau = \frac{\text{a (activity)}}{\text{m (molal concentration)}}$$

If a 0.4 molal electrolyte solution shows an activity of 0.24, then the activity coefficient is:

$$\tau = \frac{0.24}{0.4} = 0.60$$

The activity coefficients of most electrolytes decrease with increasing concentration. The activities of electrolytes may be obtained from a variety of measurements. Measurement of any of the colligative properties of electrolyte solutions, e.g., osmotic pressure, freezing point, boiling point or vapour pressure may be used. Activity coefficient may be obtained by measurements of the electromotive force of appropriate galvanic cells involving the electrolytic under consideration. Solubility measurements may also be used in certain cases. The mean **activity coefficient** ($\tau^{\pm}$) of an electrolyte in solution to ionic strengths of about 0.3 may be calculated from the ionic strength (μ) by using *Debye-Huckel equation*:

$$-\log \tau^{\pm} = \frac{(z_+ \times z_-) 0.5\sqrt{\mu}}{1+\sqrt{\mu}}$$

where z_+, z_- = number of charges on the positive and negative ions.

16.2.8 Solvent, ionization and activities

The calculations and discussion relative to ionization and activities have been applied to water solution. But other solvents are also very good ionizing solvents because they have high dielectric constants like water. These are formic acid, liquid HCN and hydrogen peroxide. Usually the ionizing power of a solvent is higher than higher its dielectric constant. Very poor ionizing solvents have very low dielectric constant, e.g., organic solvents like chloroform, ether and benzene. The activity and activity coefficient concept has been applied to solutions of electrolytes, it is also applicable to all types of substances in all three states of matter though it is difficult to determine the activity in some cases.

16.2.9 Biological importance

1. **Electrolytes in Blood**: In the blood there are many ions supplied by acid, bases and salts. Being strong electrolytes, they are completely ionized. But blood contains mainly weak acids, e.g., carbonic acid, lactic acid, phosphoric acid and fatty acids. As weak electrolytes these are incompletely ionized and two reactions, e.g., ionisation and combination, take place with an attempt to establish an equilibrium. Though there are disturbing factors in the blood such as:

   (a) The entrance of ions and their undissociated molecules from the G.I. tract,

   (b) Entrance and return through membranes to and from the lymph and cells,

   (c) Removal of ions and undissociated molecules by the kidney and reabsorption of some of them.

It is, therefore, expected that weak electrolytes of the blood, lymph and cell should maintain their equilibria but it does not, if there is equilibria, then it is perhaps temporary because if all the weak electroytes maintain their equilibrium, the animal will die.

2. **Osmotic Pressure and Electrolytes**: Each substance in solution possesses a certain amount of potential energy in the form of osmotic pressure. This pressure is independent of the nature of substance dissolved and is determined by its molecular concentration and degree of ionic dissociation. Electrolytes ionize into ions which exert the same osmotic pressure as a single undissociated nonelectrolytic molecule. Electrolytes exert more osmotic pressure than the pressure caused by equimolecular concentration of nonelectrolyte which is observed in case of plasma where electrolytes having lower concentration than the proteins in plasma exert higher osmotic pressure than the plasma proteins. The potential energy in the form of osmotic pressure can be used:
   (a) As a force for the attraction of water, or
   (b) By the use of appropriate mechanism such as:
   (i) For the performance of mechanical work or
   (ii) For the production of electrical differences of potential.

3. **Electrolytes in the Cell**: The important electrolytes in the cells are potassium, phosphates, sulphate, bicarbonate, magnesium and little quantities of sodium, chloride and calcium. They are dissolved in the water of cell and help in the following ways:
   (a) Providing inorganic chemicals for cellular reactions,
   (b) For operation of some of the cellular control mechanisms e.g.
   (i) Acting on cell membrane, allow transmission of electrochemical impulses in the nerve and muscle fibres,
   (ii) Determining the activity of enzymatic reactions for cellular metabolism.

4. **Electrolyte Secretion in the Glands**: Exocrine glands secrete sufficient quantity of water and electrolytes along with organic substances. The mechanism is postulated in the following way. Nerve stimulates the membrane of the basal part of the cell causing active transport of chloride ions into the cell → increase in electronegativity in the cell attracting the positive ions inside the cell → excess of those ions in the cell producing osmotic force, allowing water to move inside, thereby increasing the hydrostatic pressure inside the cell→ causing swelling of the cell→ producing minute ruptures of the secretory border of the cell→ flow of water, electrolytes and orga-nic substances. The evidences in favour of the theory are:
   (a) Histological studies has shown distribution of nerves mainly at the base of the cell,
   (b) Inserting microelectrodes, it has been observed that electrical potential at the base of the cell across the membrane is 30-40 mV, negatively inside and positivity outside.

5. **HCl Secretion in the Gastric Juices**: HCl is a strong and covalent electrolyte present in the gastric juice. The parietal cells of the stomach secrete an electrolyte solution having 160 millimoles of HCl per litre. HCl is formed at the membrane of the canaliculi of the parietal cells (Oxyntic cells). Chloride ion is actively transported from the cytoplasm of the parietal cell to the canaliculi which produces negative potential of –40 to –70 mV in the canaliculi. This in turn causes a passive diffusion of positive potassium ions from the cell cytoplasm into the canaliculi. Water in the cytoplasm dissociates into hydrogen ions and hydroxyl ions. Hydrogen ion is actively secreted into the canaliculi in exchange of potassium catalyzed by $H^+ - K^+$ ATPase. Hydroxyl ions combine with hydrogen ions of H_2CO_2 to form water. Water passes through the cells to the canaliculi by osmosis. So the final secretion contains 160 millimoles of HCl per litre (Fig. 16.1).

6. **Electrolyte Distribution in the Body Fluids**: Electrolytes are distributed in solution in all body fluids. The chief cations

of these electrolytes are Na^+, K^+, Ca^{++} and Mg^{++} with Na^+ and K^+ predominating. The chief anions of the body are Cl^-, HCO_3^-, $H_2PO_4^-$, $HPO_4^=$, $SO_4^=$, protein and small amounts of anions of many organic acids (Table 16.6).

Table 16.6: Chief cations and anions in body fluids

Fluid	Chief cations	Chief anions
(a) Plasma and other extracellular fluids	Na^+	Cl^-, HCO_3
(b) Intracellular fluid	K^+	Inorganic and organic phosphates, protein⁻, $SO_4^=$, HCO_3^-, very little Cl^-.

7. **Electrolytes and Parenteral Fluid Therapy:** In some pathological conditions (dehydration, acidosis, alkalosis, anoxia, periodic paralysis, diarrhoea, vomiting, lead to (a) changes in the K^+ concentration of extracellular and intracellular fluids and (b) the distribution of K^+ between these fluids. The administration of excess water or glucose solution—orally or parenterally—dilutes both the extracellular and intracellular fluids with a drop in K^+ concentration without any change of total body K^+. The fluid therapy is usually done by oral administration of water, NaCl solution, isotonic NaCl solution with KCl or electrolyte solutions with glucose. Glucose provides energy and by deposition as glycogen in cells increases transfer of K^+ from extracellular fluid into the cells.

8. **Electrolyte Balance through Kidneys and Hormone:** The reabsorption of electrolyte takes in the distal tubular cell with the help of cortical hormones, e.g., aldosterone and deoxycorticosterone. If electrolyte concentration and osmotic pressure of plasma decrease below a certain level, the adrenal cortex secrete more mineralocorticoids which increases reabsorption of sodium salts and recovers the original concentration of plasma

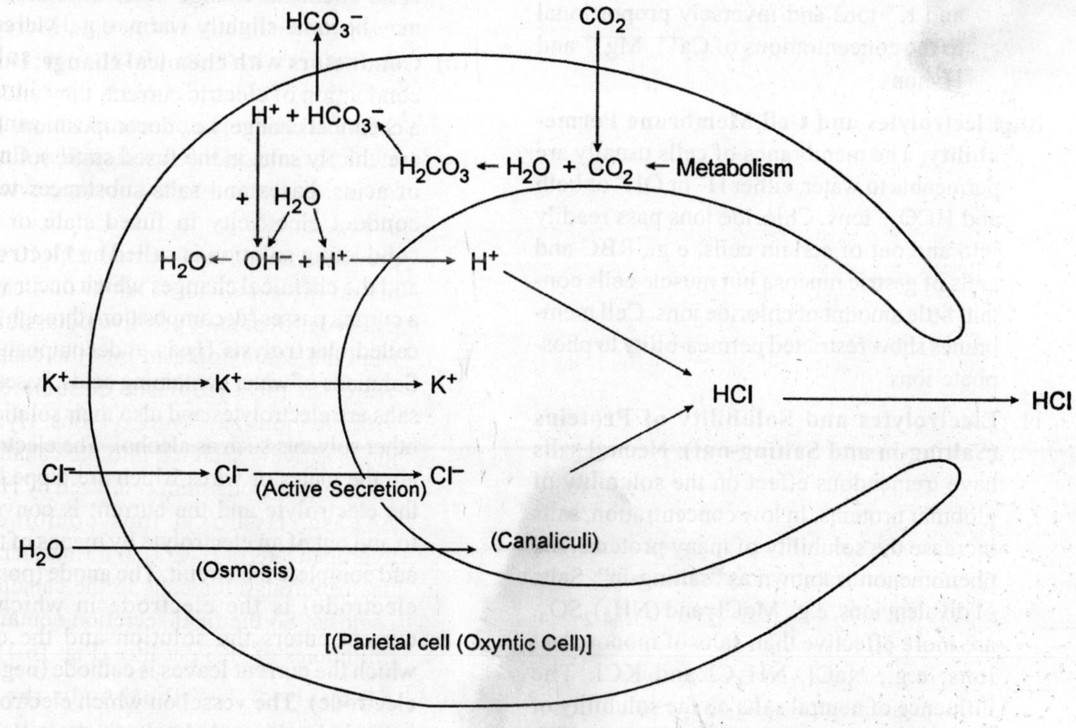

Fig. 16.1: Mechanism for the secretion of HCl in the canaliculi in the parietal cell

and osmotic pressure. If there is the rise of electrolyte concentration and osmotic pressure above normal, the adrenal cortex secretes less hormones allowing excretion of more sodium salt for lowering the electrolyte concentration and osmotic pressure of plasma.

9. **Tissue Irritability and Electrolyte**: The electrolytes of body fluids perform several functions, the main being:
   (a) To provide buffer system,
   (b) To participate in the acid-base balance,
   (c) To contribute most of the osmotically active particles,
   (d) To maintain proper ionic balance for normal neuromuscular irritability and tissue function. An approximate general expression of the relation between irritability and the ions of the body fluids is:

$$\text{Irritability } \alpha = \frac{(Na^+) + (K^+)}{(Ca^{++}) + (Mg^{++}) + (H^+)}$$

which shows that irritability is directly proportional to the concentrations of Na^+ and K^+ ions and inversely proportional to the concentrations of Ca^{++}, Mg^{++} and H^+ ions.

10. **Electrolytes and Cell Membrane Permeability**: The membranes of cells usually are permeable to water, either H^+ or OH^- or both and HCO_3^- ions. Chloride ions pass readily into and out of certain cells, e.g., RBC and cells of gastric mucosa but muscle cells contain little amount of chloride ions. Cell membranes show restricted permea-bility to phosphate ions.

11. **Electrolytes and Solubility of Proteins (Salting-in and Salting-out)**: Neutral salts have tremendous effect on the solubility of globular proteins. In low concentration, salts increase the solubility of many proteins, the phenomenon is known as "salting-in". Salts of divalent ions, e.g., $MgCl_2$ and $(NH_4)_2SO_4$, are more effective than salts of monovalent ions, e.g., NaCl, NH_4Cl and KCl. The influence of neutral salts on the solubility of protein is a function of their "ionic strength", a measure of both the concentration and the number of electric charges on the cations and anions contributed by the salt. The salting-in effects are due to changes in the tendency of dissociable R groups on the protein to ionize. When the ionic strength is increased, the solubility of a protein starts to decrease. At a sufficient high ionic strength a protein is precipitated from solution—the phenomenon is called "salting-out". The mechanism of salting-out can be explained—high concentration of salt may remove water of hydration from the protein molecules, reducing their solubility. Besides this factor, other mechanisms are also involved.

16.3 Electrolysis

The substances do not allow current to pass through them and are called **non-conductors**, e.g., liquids like turpentine, petroleum, etc. The substances which allow current to pass freely through them are called **conductors**, e.g., mercury. The conductors are of two types:

(a) **Conductors without chemical change**: During the passage of electric current there is no chemical change other than that they may become slightly warm, e.g., Mercury.

(b) **Conductors with chemical change**: During conduction of electric current, they undergo a chemical change, i.e., decomposition. They are chiefly salts in the fused state, solutions of acids, bases and salts substances which conduct electricity in fused state or in a conducting solution is called an **electrolyte** and the chemical changes which occur when a current passes (decomposition) through it are called **electrolysis** (lysis = decomposition). Solutions of water containing acids, bases and salts are electrolytes and also their solution in other solvents such as alcohol. The electrodes are the plates or wires which are dipped into the electrolyte and the current is conveyed to and out of an electrolyte by means of these and complete the circuit. The anode (positive electrode) is the electrode in which the current enters the solution and the other which the current leaves is cathode (negative electrode). The vessel on which electrolysis is carried out is called voltameter or electrolytic cell.

16.3.1 Faraday's laws of electrolysis

Faraday (1832) discovered the relationship between the amount of chemical action occurring on the electrodes with the quantity of electrolytic current passing. He expressed the results in the form of two laws which are known as Faraday's laws of electrolysis.

If a current is passed through copper sulphate solution with copper electrodes, copper is deposited on the cathode and lost from the anode. Faraday showed that:

(a) The mass dissolved off the anode by a given current in a given time is equal to the mass deposited on the cathode,

(b) The mass is proportional to the product of the current, and the time for which it flows.

Law I: The mass of any substance liberated at any electrode in electrolysis is proportional to the quantity of electric charge passing ($w \propto q \propto Ct$).

Law II: The mass of different substances liberated in electrolysis by the same quantity of electric charge are proportional to their chemical equivalent or equivalent weight.

Consequence of Law I

If W = the mass in gm of ion liberated

C = the strength of current in ampere

t = the time in second

Q = the quantity of electricity in coulomb

$W \propto Q$ or $W = Z \times Q$

Again, $W \propto Q$, but $Q = C \times t$

$\therefore \quad W \propto C \times t$

or, $\quad W = Z \times C \times t$, where Z is constant.

Now, if Q = 1, i.e., C = 1 and t = 1, Z = W = mass of ion liberated by unit quantity of electricity, it is known as "Electrochemical Equivalent" of the substance and designated E.C.E.

Consequence of Law II

From law II, it follows that Electrochemical equivalent is directly proportional to its chemical equivalent.

$$\frac{\text{Electrochemical equivalent of an element A}}{\text{Electrochemical equivalent of an element B}} = \frac{\text{Chemical equivalent of A}}{\text{Chemical equivalent of B}}$$

Similarly,

$$\frac{\text{Electrochemical equivalent of any element}}{\text{Electrochemical equivalent of Hydrogen}} = \frac{\text{Chemical equivalent of the element}}{\text{Chemical equivalent of Hydrogen}}$$

$= \text{Chemical equivalent of the element.}$

($\because$ Chemical equivalent of Hydrogen = 1)

Electrochemical equivalent of any element = Its chemical equivalent × Electrochemical equivalent of Hydrogen.

Faraday's second law implies that the same quantity of electric charge is required to liberate one gram-equivalent of substance. It is called one Faraday of electricity—denoted as F—and its value in units is 96500 coulombs or 26.8 ampere-hours per gm-equivalent. A Faraday contains the same number of unit charges as the number of molecules in a mole. Hence, a Faraday (F) may be called a **"Mole of Electricity"**.

The significance of Faraday's law is that the amount of electrolysis depends only on the quantity of electricity passing through but not on the size and shape of the vessel or electrodes, the pressure, the temperature, the voltage or current strength, etc. and the chemical equivalent of the product. In some electrolytic processes the yield of material is always less than the calculated quantity which is not due to deviation of Faraday's law but for various side reactions or disturbances on the electrodes. The ratio of the yield in an electrolysis to the theoretical yield calculated based on Faraday's Laws is called **Current efficiency** which varies within wide limits.

16.3.2 Mechanism of conduction: ions

The theory of electrolytic conduction is attributed to Arrhenius (1859-1927), though Faraday had stated this in 1834. It was suggested by Faraday that the current through an electrolyte was carried by charged particles, which he called ions. He stated that in case of silver nitrate solution in water, which contains silver ions, i.e., silver atoms with a positive charge, they are positive because silver was deposited at the cathode or negative electrode and the nitrate ions—which are group of atoms—NO_3 group with a negative charge because they moved towards the anode or positive electrode; when silver electrodes are used silver nitrate is formed. The present concept is that a sliver ion is a silver atom which donates an electron which transfers itself to the NO_3 group when the silver nitrate molecule is formed and gives the nitrate ion its negative charge and these are designated as Ag^+ and NO_3^-. When the ions come at the electrodes, they are discharged. The conventional current flowing in the external circuit brings electrons to the cathode and takes them away from the anode. At the anode, silver atoms lose electrons and go into solution as positive ions. As a result, the negative charges carried across the cell by the NO_3^- ions flow away through the external circuit. At the cathode, each silver ion gains an electron and becomes a silver atom which is deposited upon the electrode (Fig. 16.2).

16.4 Conductance

Electrolytic Conductivity or Conductance

Electrolytic conductivity (C) means the power of conduction of an electrolyte for electric currents and depends on the ionization of the electrolyte which is the reciprocal of the resistance R against the current flow. Electrolytic conductivity (C) is expressed in terms of a (a) specific conductivity or specific conductance and (b) equivalent conductance—equivalent and molar conductivity.

(a) **Specific Conductance:** The reciprocal of resistance is called **conductivity** and the reciprocal of specific resistance is called **specific conductance or specific conductivity**. Therefore, the specific conductance can be defined as the electrical conductivity between the opposite faces of a unit cube. It is denoted by κ (Kappa) and its unit is reciprocal ohms per cm and generally known as ohms per cm whereas units of specific resistance is ohms per cm.

(b) **Equivalent Conductance (Equivalent and molar conductivity):** Equivalent conductance of an electrolyte is the product of the specific conductivity of the solution and volume in cc which contains 1 gram equivalent of the electrolyte of conductivity of a solution, containing 1 gm-equivalent of the electrolyte, when placed between two electrodes one centimetre apart and it is denoted as Δ. Often, the term "Molecular conductivity" is applied for comparison among electrolytes denoted by μ. Molecular conductivity is equal to specific conductivity multiplied by that volume of solution in cc which contains one gram-molecule of the electrolyte. If a solution is having concentration C, then the volume in litres which

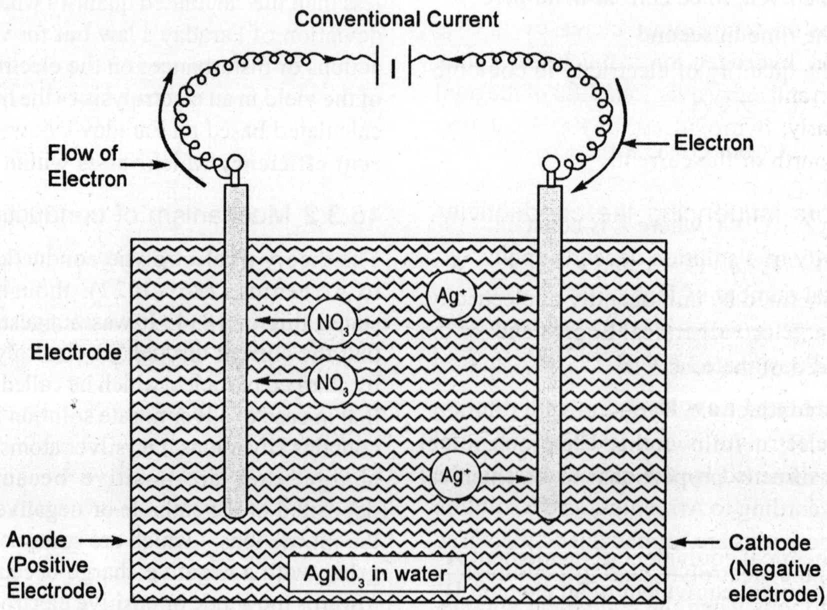

Fig. 16.2: Ion movements in electrolysis and electron movements in the external circuit

contains one gram equivalent of solute is $\frac{1}{C}$. Now, volume (V) in cc, containing 1 gram-equivalent of solute = $\frac{1}{C} \times 1000$. Therefore

Equivalent conductivity (Δ) =

$$\kappa V = \frac{\kappa \times 1000}{C},$$

where κ = specific conductivity of a solution of concentration C. Equivalent conductance depends on the specific conductance (κ) and the concentration of the solution.

16.4.1 Relative speed of the ions

The quantities of anions and cations liberated at the electrodes are in equivalent quantities but the fractions of the current carried by both the charged ions are not equal due to their unequal speed. Hittorf (1853) has done experiments on this and the results were observed extensively.

16.4.2 Transport numbers

The fraction of the total quantity of current carried across by a particular type of ion is known as **transport number** or **Hittorf's number** of that ion. The transport number for any ion is proportional to its speed. It gives an idea of the relative speed of the cations and anions. It tends to become equal for both the ions as the temperature increases. In case of HCl solution, hydrogen ion is much faster than the chloride ion and carry over four-fifth of the total current. Similarly, hydroxyl ion of KOH solution carries three-fourth of the current.

16.4.3 Factors influencing the conductivity

The conductivity of a solution depends upon:

(a) The total number of ions,

(b) The charge or valency of the ions and

(c) The speed of the ions,

Equivalent conductivity variation with dilution for the same electrolyte is due to change in total number of ions formed by the dissociation of the electrolyte. According to Arrhenius, as the solution gets diluted, more ions are formed due to increased dissociation of the electrolyte until it is completely dissociated into ions when the equivalent conductivity achieves its limiting value and there is no further change with dilution.

16.4.4 Kohlrausch's law

Kohlrausch (1876) observed experimentally which is known as "Kohlrausch's law of independent migration of ions" and states that the equivalent conductivity at infinite dilution of an electrolyte is equal to the sum of the ionic conductances of its constituent ions. Therefore,

$$A_o = \lambda_c + \lambda_s \text{... Kohlrausch's law,}$$

where A_o = equivalent conductance at infinite dilution, λ_c = cation, λ_s = anion.

Kohlrausch's law is very much useful for calculation of equivalent conductance at infinite dilution (A_o) for weak electrolytes.

16.4.5 Ionic conductance

Ionic conductance is the product of the transport number and equivalent conductivity at infinite solution. The values of ionic conductance (or limiting ionic mobility or ion mobility) are proportional to the speed of the ions.

16.4.6 Ionic mobility or absolute velocity of ions

Ionic mobility is the velocity of an ion in cm per second when it moves unit potential gradient. The unit of absolute velocity is cm per sec per volt per cm, i.e., cm^2 sec per volt.

16.4.7 Importance of conductance

1. Conductometric titrations are indicated where indicator-based titrations are difficult. These are:
   (a) Titration of very weak strength of the base or the acid,
   (b) Solution of a mixture of strong and weak acids,
   (c) Difficulty to find out the intensity of colour of an indicator,
   (d) Very dilute solution.

When an acid is titrated with a base, fast moving H^+ ions are replaced by the cation of the base. When a strong acid like HCl is titrated with a strong alkali like NaOH, at the beginning the acid solution shows high conductance because HCl is completely ionized giving a large number of H^+ ions having high mobility exhibiting high ionic conductance. During the progress of titration, the H^+ ions are removed by combination with OH^- ions of the alkali and are progressively replaced by the much slower metal cations Na^+ ions from alkali having two ionic conductance. As a result of this, the conductivity of the solution gradually decreases until it attains the minimum value at the equivalence point. Beyond the equivalence point, the conductivity increases progressively with the addition of more alkali because of high concentration of OH^- ions of high ionic conductance. If the conductivity of the solution is plotted against the volume of alkali added the graph would be like two straight lines meeting at the equivalence point. This type of behaviour is used to locate the end point in volumetric analysis to determine the conductivity of the solutions before and beyond the equivalence point and drawing two intersecting straight lines. The volume corresponding to the point of intersection indicates equivalence point. Again, during titration of a weak acid with a strong alkali, at the beginning the acid solution shows, low conductance due to less ionization a yielding a few H^+ ions. As a result, the conductivity of the solution rises linearly with the addition of more alkali until the equivalence point is reached. Addition of more alkali shows a steeper and linear rise in conductivity because of high conductance of excess OH^- ions of the alkali. In this case also, the equivalence point is seen by the point of intersection of the two straight lines which indicate the rises of conductance before and after the equivalence point.

The conductometric titrations method may be applied for:

(a) Finding the end-point for the titration of coloured or turbid solutions or

(b) Very weak acids or bases,

(c) Precipitation reactions.

2. Other applications

(a) Determination of hydrolysis constant of salts.

(b) Determination of the rate of reaction involving at least one electrolyte.

(c) Determination of the ionisation constant of water.

(d) Control of the purity of the water.

16.5 Solubility Product

In a saturated solution of NaCl in water, the undissolved solid of the same salt is in equilibrium with the solution. The solution contains both Na^+ ions and Cl^- ions of the salt and molecules of NaCl and this NaCl molecules again are in equilibrium with Na^+ ions and Cl^- ions. The active mass of the solid of the salt is constant at a given temperature. The active mass of the dissolved NaCl molecules must be constant as well. According to the law of mass action to these ions, the equation for equilibrium is

$$[Na^+] \times [Cl^-] = k \times [NaCl]$$

$$\text{or} \quad [Na^+] \times [Cl^-] = K = \text{constant},$$

for the active mass of the dissolved molecules is constant. The equation $[Na^+] \times [Cl^-]$ is called the *solubility product* of the salt NaCl and it is temperature dependable.

The law of mass action does not apply to the process of electrolytic dissociation except in the case of weak electrolytes. It applies in case of saturated solution of sparingly soluble salts, e.g., AgCl. Accordingly, dissociation of a saturated solution of silver chloride may be expressed by the equation:

$$\underset{\text{(Solid)}}{AgCl} \rightleftarrows \underset{\substack{\text{(In solution)}\\ \text{(undissociated}\\ \text{molecule)}}}{AgCl} \rightleftarrows \underset{\substack{\text{(In solution)}\\ \text{(ions)}}}{Ag^+ + Cl^-}$$

When the Law of Mass Action is applied, then

$$\frac{[Ag^+]\times[Cl^-]}{[AgCl]} = K \text{ (constant)}$$

$$\text{or,} \quad [Ag^+] \times [Cl^-] = [AgCl] \times K \text{ (Constant)}$$

$$= K_s = \text{constant} = \text{solubility product.}$$

The constant K_s is known as the *solubility product*. It is defined as the product of the active masses of its ions in a saturated solution, i.e., a solution in equilibrium with some of the undissolved solid. The significance of the above equation is that in a saturated solution of AgCl, the product of concentration of Ag^+ ions and Cl^- ions is constant and independent of the individual concentration of Ag^+ ion or Cl^- ion, provided the temperature is constant.

In case of sparingly soluble salt, e.g., BA whose solubility is S moles per litre and the concentration of $[B^+]$ and $[A^-]$ exhibiting complete dissociation are equal to S gm ion per litre.

Solubility product, $K_s = [B^+] \times [A^-] = S^2$. In case of sparingly soluble binary electrolyte, the solubility product is the square of its solubility in pure water expressed in gram-moles per litre,

When salts ionise into more than two ions, i.e., the salt is of the type $A_x B_y$, $(A_xB_y \rightleftharpoons A_x^+ + B_y^-)$, then the solubility product $K_s = (A^+)^x \times (B^-)^y$.

16.5.1 True solubility product

The solubility product of saturated solution of sparingly soluble salt, e.g., AgCl, is equal to the product of the concentrations of Ag^+ and Cl^- ions, i.e., $[Ag^+] \times [Cl^-] = K_s$, which as defined is an approximation and the correct definition uses activity in place of concentration such as:

$$^aAg^+ \times {}^sCl^- = K_s,$$

where a = activity of ion since

a = τc,

where τ = activity coefficient and

c = concentration

then the equation is

$$K_s = [Ag^+]\,\tau Ag^+[Cl^-]\,\gamma Cl^-$$

16.5.2 Effect of salt on solubility

Addition of any salt without a common ion to a saturated solution of sparingly soluble salt like AgCl—more AgCl will be dissolved. Because, according to Debye-Huckel equation, the mean activity coefficient of a salt decreases with increase of ionic strength, so the τ-terms in the last equation decrease on addition of salt to the solution of AgCl and the concentration $[Ag^+] \times [Cl^-]$ increases for maintaining K_s constant. It can be stated that solubility of a sparingly soluble salt increases on addition of any salt having no common ion and the phenomenon is known as "SALTING-IN" or **salt effect** or **ionic strength effect**. As for example, many proteins are sparingly soluble in water but dissolve in salt solution due to this salt effect.

16.5.3 Effect of a common ion

If a salt with a common ion is added to a sparingly soluble salt it reduces the solubility of salt due to the influence of the Law of Mass Action, e.g., if a little amount of $AgNO_3$ or Na-acetate is added to a saturated solution of Ag-acetate, Ag-acetate is precipitated.

16.5.4 Precipitation and solubility product

When the product of the active masses (number of gm-ions per litre) of any pair of ions present in the solution exceeds the solubility product of its salt (due to evaporative concentration or a common ion effect) precipitation of ions occurs from the solution. At this stage, the undissociated salt joins the solid phase until the ion product is decreased to the level of solubility product. This is know as **rule for precipitation.**

16.5.5 Application of solubility product

1. **Analytical Chemistry**: The precipitation process is governed by the principle of solubility product for which its application has gained great importance in the analytical chemistry.

   If increasing amount of chloride is gradually added to a solution of cuprous salt, precipitation begins as soon as the product $[Cu^+] \times [Cl^-]$ exceeds the solubility product of cuprous chloride. Addition of more chloride ion causes continuation of precipitation till the product $[Cu^+] \times [Cl^-]$ in solutions becomes equal to the solubility product of Cuprous Chloride. When the chloride ion concentration increases consequently the residual cuprous ion concentration decreases and on adding more chloride ion the residual copper in solution can be decreased to a very small value.

   This explains the reason of using excess precipitant in quantitative analysis and the process of precipitation helps to understand the different procedures of the analytical chemistry.

- **Demerits of the Application of the solubility Product**

  Some complications are encountered during application of the solubility product, e.g.

  (a) Salt effect or Salting-in or ionic strength effect,

  (b) Existance of other ionic equilibria, e.g., hydrolysis, complex formation. These two factors cause higher solubility value than the expected value. The effect of first factor has been discussed (effect of salt on solubility). As for example, solubility product, K_s, of lead chloride ($PbCl_2$) is much higher than the other salts like Hg_2Cl_2 and AgCl because the precipitation of $PbCl_2$ is not complete in comparison to the other two. It is due to the fact that certain portion of the Lead exists as $Pb(OH)^+$ ions instead of Lead ions in solution for hydrolysis and $Pb(OH)^+$ does not have any contribution to the solubility product of $PbCl_2$. In many cases the solubility is more than the theoretically expected value, i.e., the square root of solubility product ($4S^2 > K_s$).

2. **Precipitation of Ferric Phosphate in the Intestine:** Presence of high concentration of free phosphate ions in the intestine causes precipitation of ferric ions (Fe^{3+}) as ferric phosphate resulting depression of iron absorption. The explanation is that free phosphate ions increases the ionic product of ferric and phosphate ions above the solubility product of ferric phosphate causing precipitation of ferric and phosphate ions as ferric phosphate.
3. **Factor underlying the Calcification Process:** Many theories have been advanced to explain the mechanism of deposition of Calcium salts during calcification.
   (a) The saturation of the solution with the salt like Calcium Phosphate and precipitation of the same in solid form depends upon the product of the concentration of the [Ca^{2+}] and [PO_4^{3-}] ions in the solution, i.e., upon the 'ion-product' [Ca^{2+}] × [PO_4^{3-}] and not upon the total amount of Calcium and Phosphorus present in the solution. The ion product of which the solution is just saturated is called the *solubility product.*
   (b) The enzyme phosphates is present in the osteoblast, the proliferating cartilage cells and the cells of the inner layer of the periosteum which liberates inorganic phosphate from phosphoric esters and raises the concentration of [PO_4^{3-}] locally. The product of the concentrations of the [Ca^{2+}] and [PO_4^{3-}] ions then exceeds the solubility product of calcium phosphate, resulting in deposition of calcium phosphate in the cartilagenous matrix.
4. **Chloride Estimation (urine and CSF):** Chlorides in urine and CSF are estimated by titration with standard silver nitrate ($AgNO_3$) solution in presence of K_2CrO_4. A part of the $AgNO_3$ is used up in precipitating Chlorides and is called combined $AgNO_3$. Addition of $AgNO_3$ gradually increases the Ag^+ concentration, and ionic product of Ag^+ and Cl^- exceeds the solubility product of AgCl resulting in formation of a white precipitate of AgCl. The excess of $AgNO_3$ (uncombined) left over is determined by titrating against a standard solution of K_2CrO_4 required is equal to the uncombined $AgNO_3$ can be found out. But the formed Ag_2CrO_4 is not precipitated due to the ionic product of Ag^+ and CrO_4^{2-} does not surpass the solubility product of Ag_2CrO_4. When all the Cl^- ions have been precipitated as AgCl, further addition of $AgNO_3$ raises the ion product of Ag^+ and CrO_4^{2-}, surpassing the solubility product of Ag_2CrO_4 which is then precipitated showing the end-point by red colour.
5. **Calcium and Phosphate Level in Plasma in Certain Pathological Conditions:** In certain pathological conditions (e.g., genetic disorders—Milkman's syndrome, vitamin D-resistance cases) there is high excretion of phosphate in the urine causing reduction of plasma phosphate ion concentration and consequently decrease of the ion product of Ca^{2+} and phosphate ions below their solubility product in the plasma. As a result of this, more $Ca_3(PO_4)_2$ enters the blood plasma as Ca^{2+} and phosphate ions from the bones by demineralisation and produces high blood level of Ca^{2+} (hypercalcemia).
6. **Amino Acid Analysis of Proteins:** Max Bergmann and his associates have worked on the selective precipitation of amino acids from a protein hydrolysate. They used an extensive series of Aromatic Sulphonic

Acids which could be used for the selective precipitation of Amino Acids. Among these works one method was developed *solubility product method.*

If a known quantity of L-Lysine salt of a sulphonic acid (β-naphthalenesulphonic acid) is added to a sample of a hydrolysate, some amount of salt will dissolve; that amount may be expressed as S_1. According to the solubility product law,

$S_1 (A + S_1) = K_1$,

where A = Amount of amino acid, i.e., L-Lysine in the hydrolysate,

K_1 = constant, characteristic of L-lysine napthalenesulphonate under the particular conditions employed.

If an amount R of the sulphonic acid is added to a second sample of the hydrolysate which is insufficient to precipitate and again a known quantity of salt is added, a certain amount of the salt S_2 will dissolve. In this condition, the equation $(R + S_2) (A + S_2) = K_2$ will be applicable. If the solubility product law is followed, then $K_1 = K_2$ and $S_1 (A + S_1) = (R + S_2) (A + S_2)$ and in this equation, only A is unknown, hence the amount of lysine in the hydrolysate can be calculated. The disadvantages limiting the applicability are:

1. K_1 does not equal K_2,
2. Laborious procedure for which it is not used widely.

16.6 Common Ion Effect

(Change of equilibrium between electrolytes in solution)

When two electrolytic solutions are mixed together, there is an alteration of degree of dissociation which may be increased sufficiently if there is an ion in common and known as *common ion effect*, e.g., weak acid or base with ion of its salt.

As for example, when a solution of an ionizable salt AB is mixed with a salt CB, then the ionization produces a common ion B^-. This common ion B^- causes increase in the ion product of A^+ and B^- in the solution. When ion product reaches above the solubility product of AB, some amount of AB is precipitated by the common ion effect.

This can be further illustrated with the help of example of Acetic Acid and Sodium Acetate. If Acetic Acid solution is mixed up with Sodium Acetate, there will be change in equilibrium between electrolytes in solution. In case of only Acetic Acid solution, it ionizes into $[H^+]$ and $[CH_3COO^-]$ ions to maintain an equilibrium,

$$[H^+] \times [CH_3COO^-] = K_s \times [CH_3COOH]$$

But when Sodium Acetate is added, it ionises completely, concentration of acetate increases. In this period, the dissociation of Acetic Acid is small which is a weak electrolyte, so the concentration of unionised Acetic Acid does not increase very much. Because of depressed dissociation, the $[H^+]$ ion concentration decreases proportionately to establish the same value of the equilibrium constant. It is due to addition of salt having a common ion, and reduces the degree of dissociation of weak electrolytes which can be based on Ostwald's Dilution Law. The addition of a strong electrolyte to a weak electrolyte whose degree of dissociation is approximately inversely proportional to the concentration of the strong electrolytes.

Again, if two weak electrolytic solutions having a common ion are mixed together, there will be no change of degree of dissociation of any of them and these solutions are called *isohydric solutions.* Applying Ostwald's Dilution Law, it is observed that it does not depend on the total volumes of the mixed solution and the isohydricity condition is brought by the common ion which should be in the same concentration in both the solutions before mixing.

16.6.1 Biological application

1. **Absorption of Iron**: Absorption of iron is lowered due to precipitation as ferric phosphate for common ion effect where there is phosphate ion from the inorganic and organic phosphates of food which precipitate ferric iron of food.
2. **Bone Mineralisation**: Alkaline phosphatase hydrolyses the Phosphate Esters during bone mineralisation and releases Phosphate ion

whose common ion effect increases the ion product of calcium and phosphate above their solubility product causing precipitation of them.

3. **Formation of Renal Calculi**: When there is increased excretion of soluble sodium or potassium oxalate in urine, there is rise in urinary oxalate ion concentration. Common ion effect of oxalate ion increases the ion product of calcium and oxalate ions above the solubility product of the Ca-oxalate (slightly soluble) which is precipitated to from renal calculi.

4. **Absorption of Calcium**: When diet contains high phosphate, the common ion effect of phosphate ions increases the ion product of calcium and phosphates thereby exceeding the solubility product of calcium and phosphate and precipitate as calcium phosphate and its absorption is lowered. It occurs when diet contains more organic phosphates and hydrolyzed by intestinal phosphatase. Phosphate ion also comes from phosphates of alkali metals by ionization.

16.7 Ostwald's Dilution Law

The dissociation equilibrium of weak electrolytes can be decreased with the help of the Law of Mass Action. As for example, if one gram of Acetic Acid is added to "l" litres of water, then a little amount of it, "α", is dissociated into ions at equilibrium. Due to ionisation, "α" moles of CH_3COO^- and α moles of H^+ and 1-α moles of undissociated CH_3COOH will remain such as:

$$\underset{(1-\alpha)\text{ moles}}{CH_3COOH} \rightleftarrows \underset{(\alpha\text{ moles})}{CH_3COO^-} + \underset{(\alpha\text{ moles})}{H^+}$$

When concentration = No. of moles/volumes,

then $[CH_3COOH] = \dfrac{(1-\alpha)}{l}$ and

$[H^+] = [CH_3COO^-] = \dfrac{\alpha}{l}$ and

according to the law of mass action and using

$C = \dfrac{1}{l}$

$$K_a = \frac{[H^+]\times[CH_3COO^-]}{[CH_3COOH]} = \frac{\alpha^2}{(1-\alpha)l} = \frac{\alpha^2 C}{(1-\alpha)}$$

where K_a = ionisation constant or the "dissociation constant" of the acid at a constant temperature and it alters with ionic strength and temperature. The above equation signifies the variation of degree of dissociation with dilution and is knwon as *ostwald's dilution law*. This law was tested by conductivity measurements and seen to be applicable to weak acids and bases. In case of electrolytes at ordinary concentration, the degree of dissociation is very small and the Dilution Law will be:

$$\frac{\alpha^2}{l} = K \text{ or, } \alpha = \sqrt{Kl} = \sqrt{\frac{K}{C}}.$$

i.e., the degree of dissociation is inversely proprotional to the square root of concentration, and directly proportional to the square root of dissociation constant for weak electrolytes.

In another way, the degree of dissociation of a normal solution of a weak electrolyte is approximately equal to the square root of ionisation constant. Again, it may be stated that the concentration at which the electrolyte is half dissociated is twice the value of the dissociation constant.

16.7.1 Biological application

A. Strong Electrolytes

1. Body fluids contain strong and electro-valent electrolytes, e.g., mineral salts—Sodium Chloride, Sodium or Potassium Sulphates, Potassium Chloride etc.
2. HCl is a strong and covalent electrolyte present in the gastric juice.

B. Weak Electrolytes

Weak electrolytes, e.g., Amines, Amino acids, Fatty Acids, Ammonia—are present in the body fluids.

Each molecule of electrolyte ionizes into two or more ions and these ions exert the same osmotic pressure as a single undissociated nonelectrolyte molecule. It has been found that an electroyte can exert more osmotic pressure than that produced by an equimole-cular concentration of a nonelectrolyte. For these reasons, the mineral ions present in plasma exert more osmotic pressure than the plasma protein though the concentration of mineral ions is much lower than the plasma proteins.

16.8 Dielectric Constant

(Specific Inductive Capacity)

Dielectric constant (D) is the measure of the capacity of substance to oppose the electrostatic attraction between the positive and negative charged ions separated by it causing either lowering the electrostatic force of attraction or repulsion. It is also a measure of the insulating power of a material. Material having a high dielectric constant is a good insulator between the electric charges because it lowers the electric field intensity. In case of ionized polar molecules, charged polar amino acids and electrolyte salts, the electrostatic attractions between the positive and negative charged ions are reduced by the high dielectric constant of water. And each type of polar ion is surrounded by water molecules because of electrostatic attraction between this ion and the opposite charge in the water dipole. These two factors exert their influence to reduce the tendency of the solute ions to join oppositely charged ions to prevent formation of a crystal-lattice structure and thereby keep them in solution. In such cases, ions of an electrolyte move freely in a solvent of high dielectric constant and show higher conductance of the solution.

As for example, crystalline sodium chloride is dissolved in water because the strong electrostatic attraction between water dipoles and the Na^+ and Cl^- ions leads to the stable hydrated Na^+ and Cl^- ions which exceeds the tendency of Na^+ and Cl^- to attract each other and water also tends to oppose the electrostatic attraction between the positive and negative ions due to high dielectric constant (D) of water.

Again, the case of nonpolar solvents like benzene or chloroform having low dielectric constant than the water, Na^+ and Cl^- are held together by very strong electrostatic attractions between positive and negative ions of an electrolyte and reduce the solubility of the same and also favour the formation of crystal lattice. The attractive force between Na^+ and Cl^- ions of a given distance in water is only $\frac{1}{40}$ that in benzene. In benzene and chloroform solvent, considerable energy is required to pull these ions away from each other.

The nature of solvent affects ionization to a marked degree and cuts the electrostatic attraction force which binds the two ions and separates them and this effect is measured by its dielectric constant (D). The dielectric constant(D) of vacuum is the minimum and is taken as the standard which is 1 and dielectric constant of solvents is evaluated against the vacuum as unity. A comparative table of (D) of different solvents is given below (Table 16.7):

Dielectric constant (D) is inversely proportional to the magnitude of the force F between two charges e_1 and e_2 separated by a distance τ in the medium which can be expressed in terms of Coulomb's law.

Table 16.7: Comparison of dielectric constant of different solvents

Solvent at 20°C	Dielectric constant (D)
Vacuum	1
Water	80
Methanol	33
Ethanol	24
Acetone	21.4
Ether	4.1
Benzene	2.3
Hexane	1.9

$$F = \frac{e_1 e_2}{D\tau^2} \quad or, \quad D = \frac{e_1 e_2}{F\tau^2}$$

When

e_1 and e_2 = having like charges,

F = is of positive sign and represents a repulsive force.

When,

e_1 and e_2 = have opposite charges,

F = having a negative sign and represents an attractive force between two ions of opposite charges.

16.8.1 Biological importance

1. **Measurement of Thickness of the Cell Membrane**: Accurate measurement of the thickness of the cell membrane is much important for the study of various biological phenomena. Various methods were tried but did not reveal any reliable data due to dielectric constant of the membrane lipid. Danielli pointed

out that the lipid in an electric field which determines its dielectric constant depends on the dipole of the molecules and in the membrane, it is the polar region where they are anchored in the interface. The dielectric constant for the solid state is due to this anchorage.

2. **Electrical Resistance in the Cell**: If a current is passed through a membrane by placing electrodes inside and outside of a cell, the current flow encounters as ohmic resistance (a resistance where the current is always a definite fraction of the applied votage) due to the:
   (a) Resistance of the solution in the cell,
   (b) Resistance of the membrane,
   (c) Resistance of the extracellular fluids,
   (d) Dielectric resistance (because the membrane consists of poorly conducting lipoid material in the layers).

   The resistance to flow of the alternating current thus consists of ohmic resistance plus a "capacitative" resistance which produces what is known as **impedance** of the system.

3. **Amino Acids as Dipolar Ions**: The amino acids exist as dipolar ions in neutral aqueous solutions and is indicated by their (a) high dielectric constant and (b) large dipolar moments along with other evidences.

4. **Separation Procedures of Proteins based on Solubility**: Proteins in solution exhibit solubility differences due to four factors such as (a) the dielectric constant (b) pH (c) ionic strength and (d) temperature.

   (A) Solvent Fractionation: Water-miscible neutral organic solvents like acetone or ethanol decreases the solubility of most globular proteins in water and they precipitate due to dielectric constant of the medium.

   (B) Cohn's Fractionation of Plasma Proteins: In case of fractionation procedure based on the use of neutral salts and organic solvents as precipitants, the basic parameters of protein solubility such as dielectric constant, ionic strength, temperature, pH and protein concentration should be controlled. Cohn et al. applied successfully these parameters for fractionation of human plasma proteins.

   (C) Protein Solubility in Water: Albumins and pseudoglobulins are soluble in salt-free water because the protein-solvent interaction is stronger than the protein-protein interaction. Water has high dielectric constant, and in water solution, the electrostatic attraction between the charged group of separate protein molecules are depressed and the interaction of these groups with water molecules increases and favours the solubility.

5. **Insulators and Dielectric Constant**: Insulators prevent the flow of a current through them due to absence of free charge and a dielectric is the insulating material which does not ionize or does not have loosely bound electrons in the electronic shells of the atoms, so the charges cannot transmit freely through it to conduct the current. The insulator has a high dielectric constant and opposes the electrostatic force between the charges separated by the dielectric and decreases the electric field in the dielectric. Myelin sheath acts as an insulator of the axon which prevents the transmission of electric current from one axon to the other and lipids are present as dielectric. Sphingomyelin is an excellent insulator preventing all flow or ions. It also increases the resistance to ion flow through the membrane 5000 fold and decreases the membrane capacitance 1000 fold.

6. **Impulse Conduction by Axolema**: Though axoplasm (specific resistance: 10 to 100 ohms-cm) is a solution of electrolytes, it is not impulse conductor which is served by the axon membrane. The axon membrane has a dielectric constant of $\varepsilon = 5$, resistance of 1000 ohm-cm^2, a capacitance of 1 μF/cm^2 corresponding to a lipid bilayer having a thickness of 5 nm and a specific resistance of 2×10^9 ohms-cm. Myelinated portions are well insulated and myelin sheath has a low capacitance which accounts for the high rate of transmission of the nerve impulse.

7. **Cell Membrane as Insulator:** Cell membrane is the partition between the intracellular and extracellular fluids which contain electrolytes and function as electrolytic conductor. But the cell membrane is not a good conductor for the presence of lipid matrix in it and acts as insulator having dielectric constant of 3-10. In facts, the cell membrane is not a good insulator due to diffusion of anions and cations through pores.

8. **Electrokinetic Properties of Colloids**

   (A) Streaming Potential: The streaming potential is directly proportional to the:

   (a) Zeta potential at the solid liquid interface,

   (b) Pressure applied for movement of liquid,

   (c) Dielectric constant of the dispersion medium.

   But it is inversely proportional to the specific conductance and viscosity of the dispersion medium.

   (B) Electrophoretic Mobility: Electrophoretic mobility (μ) depends on:

   (a) The amount of net charge on the ion, i.e., Zeta potential (ζ),

   (b) The dielectric constant of the dispersion medium (D),

   (c) The size and shape of the charged particles,

   (d) The viscosity of the liquid dispersion medium (η),

   (e) The magnitude of the current and the fall of potential (E),

   (f) The pH of the solution.

   The electrophoretic mobility (μ) is a measure of velocity (cm) of migration of the particles for a fall of potential by 1 volt-cm and directly proportional to the Zeta potential (ζ) and the dielectric constant (D).

$$\zeta = \frac{4\pi\eta\mu}{D} \quad \text{or,} \quad \mu = \frac{D\zeta}{4\pi\eta}.$$

9. **Electrolytic Conductance of the Body Fluids:** Manifestations of many electrical properties of the body tissue are due to high dielectric constant of the aqueous body fluids. Because the ions of an electrolyte have much freedom of movement in watery solution having high D, it exhibits a higher conductance in such solution.

10. **High Solubility of Mineral Salts in the Body Fluids:** High solubility of mineral salts, e.g., NaCl, KCl and many ionic compounds is partly due to high dielectric constant. It is required to maintain ionic compounds (e.g., mineral salts) in solution in ionized forms by dispersing the ions in the aqueous biological fluids.

17

Chronobiology

1. Biological rhythm, 2. Significance of biorhythm, 3. Circadian rhythm, 4. Influence of environmental synchronization (Zeitgeber or synchronizer) on circadian rhythm, 5. Biological clock, 6. Mechanism of biological clock, 7. Concept of master clock.

17.1 Biological Rhythm or Biological Oscillatory Rhythm or Biological Oscillation

17.1.1 Introduction

The environment of continuous light or continuous darkness would be considered extreme. It demands physiological adjustments no matter whether a mammal experiences them naturally or in an environmental chamber. We must also realize that the light cycle of day and night results in two sets of environmental extremes. The sensitivity of the mammalian retina to light need hardly be stressed. Yet the eye in day-light must tolerate or protect itself from direct sunlight. Some mammals alleviate it by being nocturnal or staying in forests. Regardless of habitat the powerful daily cycle has a influence on animals and plants. All mammals at some time have had to experience the light-darkness extreme. The result has been that in some way a phenomenon common to plants and animals has been coupled with the *light-darkness cycle.* This phenomenon is a composite which we can call "the 24 hour" solar or circadian biological rhythm. Since the daily biological event does not usually begin exactly every 24-hour a day, its timing is called circadian, meaning "about 24 hours". Because biological systems change rhythmically, it follows that living organism must be biochemically different during different phases of the day and night. Consequently their reaction to the same stimulus will vary at different time.

17.1.2 Homeostasis

The maintenance of constancy in the internal environment of living organism represents one of the major evolutionary achievement of adaptive physiology known as **homeostasis** or **internal environment** by Cluade Bernard. It is a process of counter-regulation when a function approaches a limit. It always seeks an equilibrium through feedback mechanism. If the process comes back through a series of irregular variations then it is called "*irregular* (*non-specific*) *chronobiological variation*" and in orderly regular way is termed "*systemic or regular chronobiological variation*". Feedback mechanisms are of two types—positive and negative.

17.1.3 Definition of biorhythm

Biorhythm or biological rhythm is the regularly repeated fluctuations or oscillations in some biological activity due to environmental factors. Biorhythm governed by the physiological mechanism is known as **biological clock.**

17.1.4 Type of biorhythm

Biological rhythms or oscillations are usually classified into 3 types *according to origin or source.*

I. **External Rhythm**: Which are usually rhythm in response to the external environment like light, temperature, sound which act as **synchronizers.**

II. **Exogenous Rhythm**: Which are in direct response to physical change in the environment and *do not persist when conditions are kept constant.*

III. **Endogenous or Spontaneous Rhythm**: It is self-sustained, i.e., rhythm persisting under constant conditions.

Majority of species in their natural environment appear to show a combination of both exogenous and endogenous type and such rhythms are termed *composite rhythms.*

The endogenous rhythm may be considered in relation to time intervals ranging in deviation from milliseconds to millions of year such as:

1. **Circadian Rhythm**: An endogenous oscillation with a natural period of about 24 hours.
2. **Tidal or Circa-tidal Rhythm**: Endogenous oscillation with a natural period of 12 hours.
3. **Semilunar (or circasyzygic) Rhythm**: Endogenous oscillation with a natural period of 15 days.
4. **Lunar or Circa-lunar Rhythm**: Endogenous rhythm with a natural period of 30 days.
5. **Circannual Rhythm**: Endogenous rhythm with a natural period of 1 year.

Regarding organizations of endogenous rhythm, it may be at subcellular, cellular organ system, individual or social levels. All organisms, plants, invertebrates alike abound in endogenous or spontaneous rhythms with duration ranging from milliseconds to year. They run with frequency from 1000 cycles per second to duration of 1 year. The examples are EEG, biological tremor, walking, tapping, respiratory rhythm, nervous rhythm, urinary rhythm, temperature rhythm, eosinophil count, enzyme rhythms, mitotic rhythm, adrenal rhythm, pineal gland, metabolic rhythm, sexual cycles, protoplasmic rhythm, myogenic rhythm, motor performance rhythm, etc.

1. **Enzymic rhythm**: The oscillation rhythm in enzymic activity are due to feedback loops resulting in some cases from effects of metabolites on particular enzymes or in other cases from inhibition of enzyme production along the well-known chains from gene via mRNA.
2. **Neural or nervous rhythm**: They are concerned either with generation of impulses in single neurone or constitute oscillatory system built of several rhythms each with a frequency ranging from 20 cycles per second or even longer.
3. **Motor performance rhythm**: There is a limit to the rate at which we can perform voluntary rhythmic activity, i.e., tapping, chewing. The highest frequency in the human is always about 7 cycles/second.
4. **Myogenic rhythm**: They are seen in heart musculature and in invertebrate flight muscles (2200 cycles/second). In man, the pulse is one of the oldest spontaneous rhythm having frequency of 72 beats/minute.
5. **Respiratory rhythm**: It is an example of neural frequency. Normal frequency is from 20/minute.
6. **Protoplasmic streaming in cell**: Circular streaming or periodic reversal of direction has been noticed in some cells. The energy comes from ATP.
7. **Sexual cycle**: In many invertebrates, fish, reptiles and birds, the sexual cycle is definitely seasonal. It is mediated by the primitive neuroendocrine system.
8. **Metabolic rhythm**: A glycogen rhythm is seen not only in the liver but also in the heart, skeletal muscles, spleen and the pancreas.
9. **Urinary rhythm**: Variation in volume may be affected by cardiac insufficiency when the main peak of production is changed from day to night.

10. **Temperature rhythm in the body**: There is evidence of endogenous rhythm in the variation of balance of body temperature.
11. **Eosinophil count**: A morning fall in eosinophil count is observed and there is variation of count throughout the day. It may be due to circadian rhythmic pattern of ACTH and glucocorticoid secretion.
12. **Mitosis**: There is evidence that the adrenocortical rhythm may mediate the rhythm in mitosis in epidermis.
13. **Urinary K^+ and Na^+ rhythm**: Disturbances in Na^+ excretory rhythm have been reported in patients with adrenal disease.
14. **Adrenal rhythm**: It occurs in circadian rhythm pattern through ACTH which stimulates secretion of glucocorticoids.
15. **Pineal gland:** Circadian rhythm in pineal gland has been studied. It stores biological transmitter which influences the motor activity and reproductive activity in lower species. The materials stored are serotonin and melatonin. Melatonin acts on hypothalamus to secrete GnRH which acts on anterior pituitary to secrete FSH, LH. These hormones act on ovary for secretion of oestrogen and progesterone.
16. **Chronopharmacology and chronotherapeutics**: Most drugs act upon rhythmic processes hence efficacy depends upon the time of administration.

    i. The sensitivity of certain tissues to the drug concentration in blood:

    (a) Fixed dose of the drug has shown mortality of higher rate at particular time and the same dose produced lesser mortality.

    (b) Induction of abortion by intra-amniotic prostaglandin showed optimum effect at 8 A.M.

    ii. Particular dose of drug does not achieve same concentration of blood at different times. This may be due to: (a) rapid excretion of drug, (b) differences in the rate of uptake into the blood or (c) removal from blood by metabolism.

    iii. When a drug is given therapeutically, sometimes it shows multiple effect and in certain cases there is untoward effect. This can be averted if a suitable time is adjusted.

    iv. Evidences show that malignant cells show rhythms which are out of phase with those of the host. It is useful for the diagnosis but difficulties are encountered for therapy. If the rhythm of the malignant cell is known, at the time of therapy the malignant cells will be more sensitive and the host cells least sensitive.

17.2 Significance of Biorhythm

Significance

1. **Measure the passage of time**: The biorhythm or innate periodicities are synchronized with the daily, lunar and seasonal changes that take place around all. All living organisms (except bacteria and algae) possess biological clock by which they can measure the passage of time, and in *some rhythm activities result.*

   As for example, circadian rhythm is used by animals and plants to "time" daily events and thus allow the organisms to perform functions at the "*right time of the day*" or to attain synchrony with other individuals of the population.
2. **Reaction of biological system**: Because biological systems change rhythmically, it follows that living organisms must be biochemically different during different phases of the day and night. Consequently, their reaction to the same stimulus will vary at different times. This differential response to an identical stimulus has been confirmed repeatedly with respect to a variety of stimuli including drugs, poisons and other chemical substances; physical agents such as noise and radiation and biological agents such as endotoxins.
3. **Important for survival**: This rhythm may be important for survival, as is the timing of birth one of the most critical events in the life of an individual.

4. **Variation in rate of metabolism**
   (a) Diurnal variation in rate of alcohol metabolism has been demonstrated.
   (b) Seizure in mice is most common at the time of maximal locomotory activity in darkness when glycogen content of the liver reaches its lowest level.
5. **Heat resistance in frogs**: A daily rhythm of heat resistance in frogs, for example, could be of importance during the summer. After periods of heavy rain the animals are often found in temporary poots.
6. **Adaptive specialization**
   (a) A diurnal rhythm of sensitivity to toxic or other harmful environmental variables is not necessarily advantageous but, when the strength of such variables alters rhythmically, the diurnal rhythm can be seen to be an adaptive specialization. Circadian rhythm, *as an adaptation to the environment,* involves biotic as well as abiotic factors.
   (b) Timing of activity, with regard to microclimatic conditions, can be as

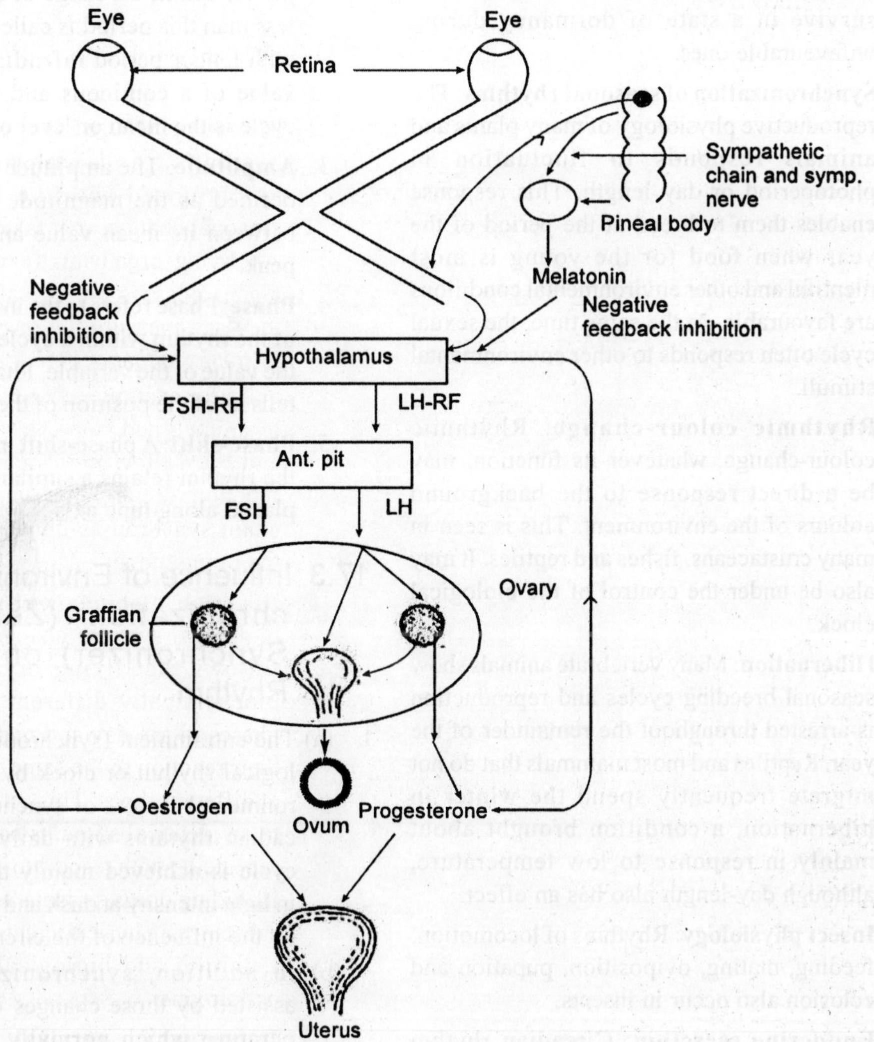

Fig. 17.1: Neuro-endocrine control of biorhythm

important as is the correct phase relationship between predator and prey, or the synchronization of sexual partners.

7. **Diapause**: (A state of suspended animation in which physiological processes are much reduced).

In most regions of the world, the physical and biological conditions suitable for growth, development and reproduction prevail only during particular seasons of the year. Some animals avoid inclement periods by migration while others synchronize their activities with favourable seasons and survive in a state of dormancy during unfavourable ones.

8. **Synchronization of seasonal rhythms**: The reproductive physiology of many plants and animals responds to fluctuation in photoperiod or day length. This response enables them to breed at the period of the year when food for the young is most plentiful and other environmental conditions are favourable. At the same time, the sexual cycle often responds to other environmental stimuli.

9. **Rhythmic colour-change**: Rhythmic colour-change, whatever its function, may be a direct response to the background colours of the environment. This is seen in many crustaceans, fishes and reptiles. It may also be under the control of the biological clock.

10. **Hibernation**: Many vertebrate animals show seasonal breeding cycles and reproduction is arrested throughout the remainder of the year. Reptiles and most mammals that do not migrate frequently spend the winter in hibernation, a condition brought about mainly in response to low temperature, although day-length also has an effect.

11. **Insect physiology**: Rhythms of locomotion, feeding, mating, oviposition, pupation and eclosion also occur in insects.

12. **Endocrine secretion**: Circadian rhythm influences ACTH, glucocorticoid, TSH secretion, regulation of pineal body for time dependent effects such as onset of puberty and rhythmic activities of hormonal regulation of ovulation (Fig. 17.1).

17.2.1 Biorhythmological terminology

1. **Rhythm (or oscillation, cycle, periodicity)**: A rhythm is a sequence of events which repeat themselves through tune in the same order and at the same interval.

2. **Period of rhythm**: The period of rhythm is the reciprocal of the frequency and is the time taken to complete one cycle. According to Halberg, a rhythm is circadian if its period lies in the range of 20-28 hours and less than this period is called **ultradian** and with longer period **infradian**. The average value of a continous and variable over a cycle is the mean or level of the rhythm.

3. **Amplitude**: The amplitude of the rhythm is defined as the magnitude of the variable between its mean value and the trough or peak.

4. **Phase**: Phase refers to the instantaneous state of the rhythm within a cycle, represented by the value of the variable. Phase of the rhythm tells us of the position of the rhythm in time.

5. **Phase-shift**: A phase-shift implies that while the rhythm retains a similar shape, it is displaced along tune axis.

17.3 Influence of Environmental Synchronization (Zeitgeber or Synchronizer) on Circadian Rhythm

I. (a) The entrainment (synchronization of a biological rhythm or clock by means of environmental factors or synchronizers) or circadian rhythms with daily environmental cycle is achieved mainly through changes in light intensity at dusk and dawn, enhanced by the influence of the circadian rule.

(b) In addition, synchronization may be assisted by those changes in ambient temperature which normally occur at these times. This is suggested by the fact when the activity rhythms of various insects, including spider-beetles have been gradually lost in constant conditions in case of exogenous rhythms. They can sometimes be reinitiated by periodic exposure to high and low temperatures. Regular thermal fluctuations also appear to maintain the activity rhythm of the cockroach in constant darkness, although they have no power to initiate a rhythm once it has died out. The effects of temperature perturbations depend on the phase at which they are administered; they are greatest when operating in cooperation with the cycle of daylight and darkness.

(c) Changes in relative humidity, earth's magnetic field, barometric pressure, food supply and other environmental variables may sometimes support or inhibit the primary Zeitgeber (light) and its chief subordinate (temperature).

II. In order to be effective a synchronizer must be reliable. No doubt the biological clocks are synchronized by light more often, and to a greater extent, than by any alternative environmental variable. Though, other factors such as temperature, relative humidity or predation may, at times, have greater ecological significance. Baker has distinguished between two groups of factors which influence the reproductive cycles of higher vertebrates:

(a) "Proximate" factors in response to which the reproductive organs undergo their physiological development from a seasonally quiescent state to functional breeding condition and

(b) "Ultimate" factor which exerts a selective pressure, ensuring that animals breed at the optimal seasons. In many animals the most important ultimate factor is the availability of suitable food supplies for the young. Other selective influences, however, impose numerous compromises and adaptations of this basic principle.

III. From the above account it will be seen that all the complex harmony of natural rhythm and counterpoint could be based on the operation of simple metronomes—cellular circadian clocks—coupled directly and indirectly with one another, and synchronized primarily by the natural periodicity of daylight and darkness. At the same time, the circalunadian clock is probably coupled to these circadian clocks in some complex manner, while by using them to measure photoperiod, seasonal cycles of reproduction, diapause and other physiological processes are synchronized with environmental influence resulting from the movement of the earth in solar orbit.

V. Biological clocks have evolved through the influence of natural selection, and this selection is responsible for their efficiency. But what are the uses of efficient clocks, and why should living organisms have evolved them at all?

17.4 Circadian Rhythms

In certain organisms, certain disorders are first manifested by changes in diurnal rhythm. Diurnal rhythm means variations of any parameters "during" 24 hours, whereas circadian rhythm means variations with a period of "about" 24 hours. When the animals are exposed for different periods to either continuous illumination or darkness, after some time, a rhythm with a period of approximately 24 hours was observed in the animals. The period was longer or shorter than 24 hours but never equal to exactly 24 hours. Hence, its name **circadian** which is from the Latin *circum* means around, approximately. Halberg has defined a rhythm as being "circadian" if its period lies in the range of 20-28 hours. Rhythms with a period less than this are called **ultradian** and those with a longer period **infradian**.

17.4.1 Explanation of circadian rhythm

There are two explanations of circadian rhythm:

I. **Endogenic**: The function of the living organism seems to evolve from (1) constant variations of geophysical parameters caused by the rotation of the earth round its axis which is believed to be the main factor in the development, and (2) genetic continuation of the mechanism of diurnal rhythms and in determining the diurnal oscillation of the organism's parameters, (3) and being very much active depends upon the biological clock and the timing can be varied by the environmental factors or Zeitgeber or synchronizers like an alarm clock to set at a particular hour and minute for the desired timing. The biological clock is situated in the hypothalamus and this internal clock is so synchronized with the exogenous influence that these two function in a concert.

II. **Exogenic**: Exogenic mechanism is passive. The exogenous factors influencing the circadian rhythm are:

1. Light, 2. Darkness, 3. Temperature, 4. Humidity, 5. Barometric pressure, 6. Food, 7. Water, 8. Earth's magnetic field.

Exogenous factors, particularly light and darkness, influence the internal environment greatly. Since often rhythms are measured in individuals exposed to all the rhythmic 24-hour influences of the solar day—such as the alteration between day and night, day-to-night environmental temperature changes and rhythms of social behaviour—it is more convenient to use a single adjective to describe such conditions. The word which has come into general usage is **nychthemeral**. Thus, if a rhythm is measured in nychthem-eral conditions, this implies that at least some of the normal periodicities of the solar day are present. Similarly, a nychthemeral rhythm refers to a bodily rhythm measured in nychthemeral conditions.

17.4.2 Experimental evidences

1. The animals were kept for a long time in deep caves and mines. Circadian rhythms were observed in the same period. This refuted all arguments against the endogenic interpretation of circadian rhythms.
2. Experimental conditions are not the only factor affecting the period of circadian rhythms. Nocturnal-habit animals were found to have a period of circadian rhythms, at constant illumination, of over 24 hours, while diurnal animals manifest shorter periods. Conversely, in constant darkness, the former have a period shorter than 24 hours, while the latter longer than that, hence the explanation is not convincing.

17.4.3 Salient features regarding circadian rhythms

1. Circadian rhythms are multioscillatory rhythms.
2. Well-adapted and healthy animals do not show circadian rhythms. A reduction in the animal's adaptability caused by some stressor or disturbed diurnal stereotype leads to the generation of circadian rhythm.
3. Circadian rhythms are seen in poorly adapted and unhealthy animals, and an increase in the circadian rhythms is seen in animals not well-adapted to certain conditions and in those subjected to a stressor a day earlier. The more intensive the stressor, the greater the amplitude of circadian rhythms.
4. A higher amplitude of circadian rhythms becomes clear during transition process. When well-adapted animals are placed in a new environment, the diurnal rhythms start to be gradually restructured till a new diurnal rhythm which meets the new conditions is formed, i.e., a new diurnal stereotype is generated.
5. The amplitude of circadian variations gradually grows, reaches its maximum and starts dropping during transition processes. Along with the increase of the amplitude of circadian variations, the variations of smaller periods gradually fade.
6. The growth rate and the maximal amplitude of circadian rhythms depend on a number of factors, and in particular, on the intensity of external factors. With a higher intensity, the amplitude grows more quickly.
7. In the response of the organism, each type of stressors has a respective combination of amplitudes of the circadian variation.

17.4.4 Remarks

Based on the experimental evidences available so far, it may be stated that

1. Circadian rhythm is not a normal rhythm; on the contrary, an evidence of the abnormal state of the organism.
2. A circadian rhythm characterizes the degree of the animal's inadaptability which can be caused not only by exogenic stressor but also by endogenic factors.

17.4.5 Example of circadian rhythms in human

1. Sleep-wakefulness cycle.
2. Deep body temperature.
3. Cardiovascular system—Blood pressure, pulse rate.
4. Respiratory rhythm.
5. Metabolic rhythms—Enzyme synthesis, e.g., Glutaryl CoA.
6. Hormonal rhythm:
   (a) Pituitary,
   (b) Hypothalamus,
   (c) Adrenal cortex,
   (d) Pineal gland,
   (e) Thyroid.
7. Kidney excretion rhythm.
8. Rhythm in mental performance and emotional state.
9. Prenatal to birth synchronization.

17.5 Biological Clock (Physiological Clock)

Ever since life first appeared on this planet it has been subjected to daily cycles of light and darkness and to seasonal cycles of climatic change, caused by the rotation of the earth around it axis and the revolution around the sun, respectively. Marine and intertidal organisms have, in addition, been subjected to tidal and lunar periodicities. Only those animals which have invaded the depths of the ocean, or underground caves and rivers, have avoided this fluctuating environment. Other species—especially those on the land, where daily and seasonal changes may include violent fluctuations in temperature and humidity—have developed strategies to counteract or to exploit this periodicity. Majority of organisms show daily and annual cycles of activity and development. They may be nocturnal (Night), diurnal (Day) or crepuscular (Twilight action). They may hibernate or aestivate.

17.5.1 Biological clock (exogenous and endogenous)

Some of these phenomena are direct responses to *environmental change* (exogenous) but many more are overt manifestation of an endogenous periodicity. These innate rhythms or oscillations underlying such phenomena are now known to provide a temporal organization for physiological and behavioural activities in practically every group of organisms apart from the prokaryotes. Of particular interest are those endogenous oscillations which have evolved with a periodicity close to 24 hours (circadian rhythms) and are used by animals and plants to "time" daily events and thus allow the organism to perform functions at the "*right time of the day*", or to attain synchrony with other individuals of the population. It is clear that these circadian oscillation in the cells and the organisms have evolved to match almost exactly the oscillations in the physical environment. These rhythms, therefore, are not "imposed" on the organisms by the environment, neither are they "learned". The natural cycles of light and temperature, however, do serve to entrain and phase-control those endogenous oscillators so that under natural conditions their periods become exactly 24 hours. In the absence of temporal clues from the environment (i.e., in darkness and constant temperature) the rhythms "free-run" and reveal their own natural period which is close to, but significantly different from, that of the solar day. The observations that this period is temperature-compensated, and that the rhythms are used by the organisms to measure the passage of time, justify the use of the term *Biological Clock.*

Apart from *Circadian rhythms* which have evolved as a match to the 24-hour periodicity of the earth's rotation around its axis, endogenous oscillations with:

(a) Tidal or circa-tidal rhythm (~ 12 hours),
(b) Semilunar (or circasyzygic rhythm) (~ 15 days),
(c) Lunar or circa-lunar rhythm (~ 30 days),
(d) Circannual rhythm (~ 1 year).

Periods are also to be found in organisms.

In many cases the endogenous nature of these rhythms has been demonstrated by allowing them to "free-run" in the absence of the environmental clues (Zeitgebers) which normally entrain them.

17.5.2 Classification of biological clocks:

A. **Endogenous**
B. **Exogenous**

Biological clocks have been classified in a number of ways. The endogenous clocks are divided into different ways:

I. According to Pittendrigh (1958):
   (a) "Pure" rhythms, such as colour-change in the crab.
   (b) "Internal timers", in which a particular event occurs (e.g., pupal eclosion) at a particular time of the day.
   (c) *Continuously consulted clock* in which time may be "recognized" at any time of the day.

II. According to Truman (1971):
   (a) Type A: These clocks are generally associated with *developmental rhythms,* e.g., hatching, moulding, eclosion or release of brain hormone.
   (b) Type B: Those controlling *locomotor activity rhythms,* the compound eyes are the principal and sometimes the only photoreceptors involved and the brain itself is insensitive to light (Fig. 17.2).

17.5.3 Functions of biological clocks (advantages)

I. **Insects**: Biological clocks control a wide variety of behavioural and physiological activities in insects. These include daily rhythms of locomotion, feeding, mating, oviposition, pupation and eclosion. These rhythms may be operational either in individual insects or in populations which behave in this respect like "super organisms." Clocks also control cuticle deposition, metabolism and the seasonal control of alternate developmental pathways (photoperiodism).

II. **In animals and human beings**: Biological clocks influence ACTH and glucocorticoid secretion, TSH secretion, regulation of pineal body for time-dependent effects such as onset of puberty and rhythmic activities of hormonal regulation of ovulation.

III. **Stand in for time-givers**: In free environment, time-givers are not regular and there is competition between them. During disturbed condition in the surrounding, the biological clocks maintain habitual activity and persistence of this regularity is of great advantage to the animal.

IV. **Assisting orientation**: During sunny day, navigation in birds, fishes, various reptiles and invertebrates is advantageous due to biological clock.

V. For adjustment of diurnal cycle in the environment.

VI. Finding direction.

VII. For adjustment of tidal and lunar cycle.

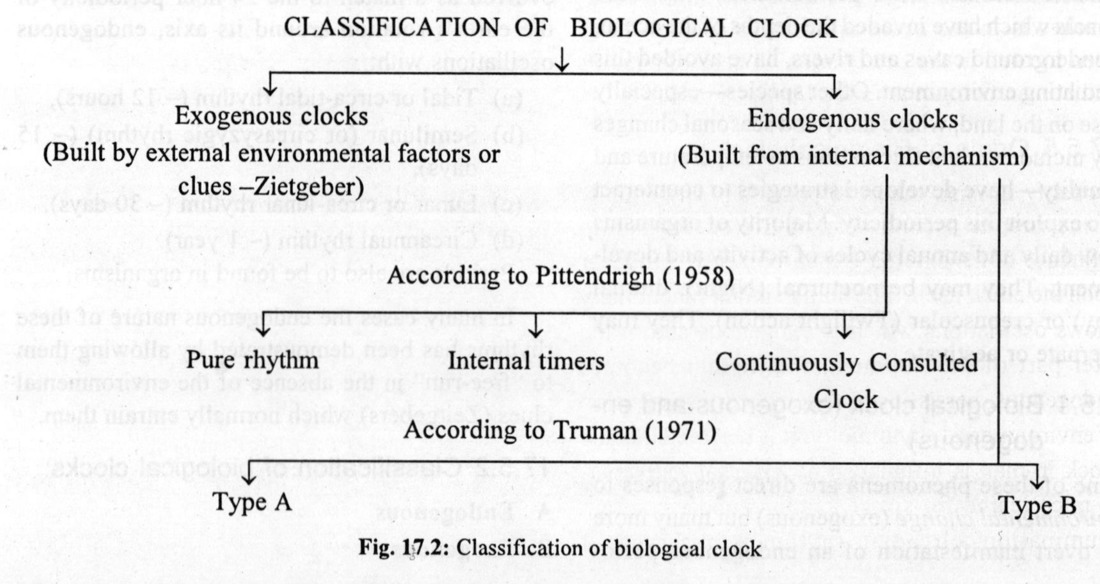

Fig. 17.2: Classification of biological clock

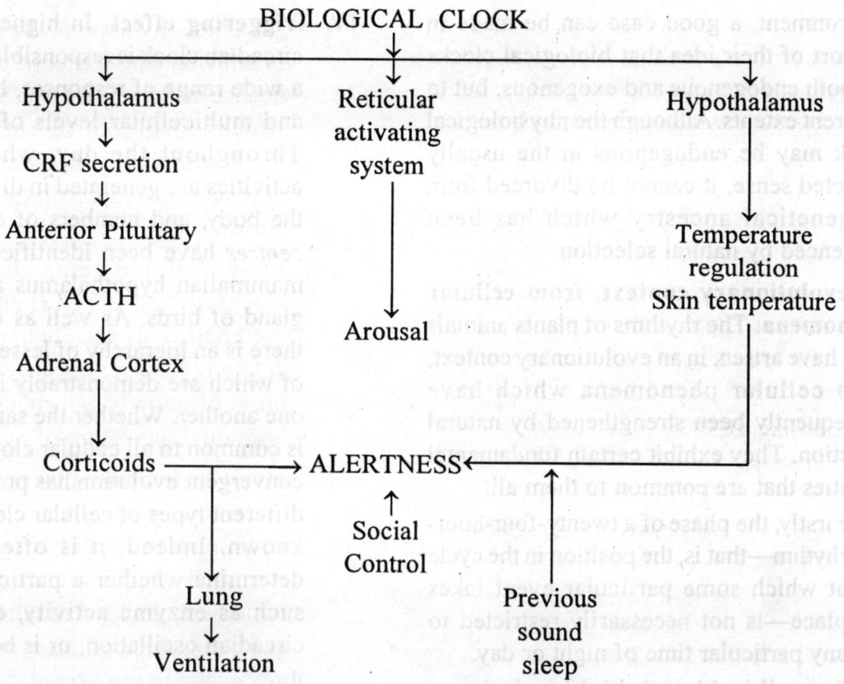

Fig. 17.3: Schematic representation of role of biological clock showing the mechanism of keeping the body alert to do other activities

VIII. Diagnosis of pathological conditions, e.g., blood sugar level and relation with glucocorticoid and ACTH secretion.

IX. For keeping the body alert to do other activities in the following way (Fig. 17.3).

17.5.4 Location of biological clocks

The clock controlling locomotory rhythm in cockroaches is in the *brain* and in two species it has been located in the *optic lobes.* In case of animals and human being it is in the *limbic system.*

17.5.5 Origin of Biological clock

Probably in man, the development of rhythmicity in the first or second year of life occurs due to influence of the external synchronizer or Zeitgebar upon the clock itself. The human infant functionally shows oscillations which is exogenous and in the later part of development, the rhythm becomes endogenous, persisting with temporary suspension of environmental rhythmicity. It is believed that the clock in man is influenced by external Zeitgeber. In mice reared for six generations under constant illumination, circadian rhythm is observed. It appears that in these animals, the clock is genetically determined and required an external Zeitgeber to adjust it precisely. In humans also, the clock is similarly genetically built, requiring an external Zeitgeber to develop the genetical potential.

17.6 The Mechanism of Biological Clock

17.6.1 The clock mechanism

Few biologists today doubt that organisms possess biological clocks, but there is a degree of controversy regarding the fundamental mechanism involved.

I. **Exogenous or Endogenous?** Brown and coworkers have maintained that the clocks of living organisms are timed by subtle, rhythmic, geophysical *forces*. Most other investigators, however, now hold that biological rhythms are endogenous (i.e. internal in their control), and completely independent of the environment for their fundamental timing. All biological clocks are part and parcel of their evolving rhythmic environment, a good case can be made in support of their idea that biological clocks are both endogenous and exogenous, but to different extents. Although the physiological clock may be endogenous in the usually accepted sense, it cannot be divorced from its genetical ancestry which has been influenced by natural selection.

II. **An evolutionary context, from cellular phenomena**: The rhythms of plants animals must have arisen, in an evolutionary context, from cellular phenomena which have subsequently been strengthened by natural selection. They exhibit certain fundamental qualities that are common to them all:

(a) Firstly, the phase of a twenty-four-hour-rhythm—that is, the position in the cycle at which some particular event takes place—is not necessarily restricted to any particular time of night or day.

(b) Secondly, although biological phenomena are usually extremely sensitive to thermal influences, the periods of biological rhythms are relatively independent of changes in the ambient temperature. If this were not so, of course, biological clocks could only function in constant environments.

(c) Finally, rhythms are not normally learnt, because apparently a rhythmic organisms, raised under constant conditions in the laboratory, may become rhythmic after experiencing only a single non-periodic stimulus, such as flash of light.

III. **Self-sustained oscillations**: Biological clocks are thought to be self-sustained oscillations whose phase can be reset or entrained by an external synchronizer or Zeitgeber. Holdberg has coined the word "Circadian" (from Latin "Circa", about, and "dies", a day) to describe them. Persistent lunar and tidal rhythms are likewise called "circalunadian" (about a lunar day) or "circalunar", because the period of the bimodal lunar-day rhythm is usually longer or shorter than the period displayed in nature, while endogenous yearly rhythms are called "circannual" for similar reasons.

IV. **Triggering effect**: In higher animals, the circadian clock is responsible for triggering a wide range of responses, both at cellular and multicellular levels, of organization. Throughout the day, whole series of activities are generated in different parts of the body, and numbers of different clock *centres* have been identified, such as the mammalian hypothalamus and the pineal gland of birds. As well as central clocks, there is an hierarchy of lesser clocks, some of which are demonstrably independent of one another. Whether the same mechanism is common to all cellular clocks or whether convergent evolution has produced several different types of cellular clocks, is not yet known. Indeed, it is often difficult to determine whether a particular function, such as enzyme activity, engenders the circadian oscillation, or is being driven by it.

17.6.2 Conclusion

Although the ability to measure time is apparently an innate property of the cell, the *fundamental nature of biological clocks* remains obscure. Feedback regulations of enzyme activation and inhibition; RNA hypothesis, ionic diffusion across cellular membrane and a chronon model which proposes a circadian transcription of genes (the rhythmic transfer of genetic information in DNA into an intermediate RNA through which it is then translated into proteins), have all been proposed as possible mechanisms.

17.7 Master Clock or Central Clock

(Interactions between cellular clocks within multicellular organisms)

17.7.1 Master clock

Single cell organisms have one characteristic endogenóus rhythm. In a multicellular organism, rhythms are usually coordinated and consequently synchronized. In animals such as man and rat in which large number or rhythm manifestations can be observed in the rhythms, the rhythms are commonly in phase with one another or, at least, show some phase relationship (regular). The concept of "*master clock*"—usually supposed to reside in the brain—has arisen. It controls other rhythms by one or another form of mediation. Individual cells, tissues, organs may have their own inherent rhythmicity, which may continue after their period set and locked in phase by the master clock while within the body.

The overt rhythm of an organism represents the sum of the reactions of its constituents cells to their biological clocks with, in turn, may be synchronized by a group or groups of cellular clocks forming a "**master clock**".

There may be no need for a central or master clock if all cells posses clocks which can be synchronized like mitotic rhythm. Nevertheless, certain groups of cells, such as those in the optilobes of the brain, are better placed than others to entrain with environmental cycles of light and darkness. These might, therefore, logically be expected to become specialized as master clocks, although few have yet been positively identified.

17.7.2 Types of master clock

In general, the master clocks which have so far been studied in animals appear to fall into one of two groups:

1. **Group I**—Those which stop in continuous light, such as the eclosion clocks of fruit-flies and other insects and those involved with activity rhythms. Clocks of the first type seem to be associated mainly with developmental rhythms and to be under neuroendocrine control.
2. **Group II**—These are the direct product of a complex nervous system. Their phase-response curves, in general, are more stable against interrupting light perturbations since the maximum phase-shifts are usually small. It is, at present, not known by which type of clock the time-sense and photoperiodic response of animals are controlled.

17.7.3 Mechanism of action

Circadian clocks of the second type have been implicated in many diverse kinds of behaviour. Generalised activities—such as locomotion—are influenced by an array of factors so complex that any clock system must interact with a network of other internal and external stimuli. In many instances the coupling between the master clock and effector organs may be hormonal, for example testosterones can desynchronize the circadian rhythms of birds and mammals. For fifteen years it was accepted that the circadian rhythms of insects were under the control of a hormone secreted by the subesophageal ganglion. Subsequent research having failed to confirm the central importance of this ganglion, it now seems more probable that hormonal rhythms are not primary oscillators but are driven. Indeed, while the physiology of the oscillator mechanism (which represents the escapement of the biological clock), is not yet understood, its coupling now appears to be primarily nervous (electrical) and only secondarily hormonal.

18

Cybernetics (Control Engineering)

1. Introduction, 2. Definition, 3. Examples of Homeostatic or Control Mechanism, 4. Principles or characteristics of Homeostasis or Control System, 5. Requirements for Homeostasis, 6. Amplification or Gain of Control system, 7. Positive feedback, 8. Oscillation and Damping of Control System, 9. Conclusion, 10. Regulation of Homeostasis, 11. Operation of Homeostatic Mechanism.

Homeostasis or Feedback Mechanism

18.1 Introduction

In the body a small fluctuation of osmotic pressure, temperature or the amounts of certain chemical substances may disrupt the biochemical activities or, in extreme cases, may destroy the cells. But the cellular activity is not disturbed due to constant maintenance of the normal condition of the surrounding called **internal environment**. This basic fact was first recognized by the French physiologist **Claude Bernard**. In 1857 he wrote "The Constancy of the Internal Environment". By internal environment is meant the immediate surroundings of the cells which are generally tiny channels and spaces filled with fluid called intercellular or interstitial or tissue fluid. The vital function of cell depends upon the tissue fluid constancy (Cannon).

18.2 Definition

Maintenance of a constant internal environment is known as **homeostasis**, a Greek word meaning "**staying the same**". In engineering, the terms feedback is to control a process. When this concept is used in biological process within the body, it is known as **biological feedback or biofeedback.**

- **Factors to be kept constant**

Many physiological processes are homeostatic and they are responsible directly or indirectly for regulating the internal environment. The most important features of the internal environment that must be kept constant are:

1. Its chemical composition e.g., glucose, ions, etc.
2. Its osmotic pressure which depends on the amount of water and solutes,
3. Its CO_2 and O_2 level,
4. Its temperature,
5. Its pH,
6. Elimination of waste products and toxin, etc.

18.3 Examples of Homeostatic or Control Mechanism

1. **Regulation of blood sugar**: When there is rise in blood sugar it stimulates insulin secretion and sugar level decreases. Again, low level inhibits insulin secretion and stimulates adrenaline, and by other hormone secretion sugar level rises.

2. **Regulation of O_2 and CO_2 concentration:** By buffering action and respiratory centre.
3. **Regulation blood pH, electrolytes,** etc.
4. **Regulation of water balance.**
5. **Regulation of osmotic pressure, temperature,** etc.

18.4 Principles or Characteristics of Homeostasis or Control System

Most of the Homeostatic Mechanisms or Control Systems have certain principles or characteristics in common. The Control System involves a "**self-adjusting mechanism**", the Control Process, being built into the system. The characteristics are:

(A) **Negative Feedback Nature of Control System**: The best example is homeostasis of blood sugar level. An increase in level sets into motion processes which decrease it, e.g., the response is negative to initiate stimulus. Conversely, a decrease in level causes increase in sugar where the response is negative to initiate stimulus. A change in the sugar level of the blood automatically brings about the opposite effect and keeps the level constant.

(B) **It has a "norm" or "set-point" for fluctuations**: Homeostasis must involve fluctuations, though these may be small. There must be periodical increase or decrease in the sugar level, then control mechanism of sugar can work. The level does not and cannot remain absolutely constant without fluctuations within narrow limit on either side of an optimum value which is known as "**norm of set-point**". Only by deviating from this set-point or norm, homeostasis can be brought into play which restores the normal level of sugar.

18.5 Requirements for Homeostasis

For physiological homeostatics mechanism the following units are required:

1. **Receptor**: Capable of detecting the change.
2. **Control mechanism**: Capable of initiating the appropriate corrective measure.
3. **Effectors:** Can carry out these corrective measures.

18.6 "Amplification" or "Gain" of Control System

The degree of effectiveness with which a Control System maintains constant conditions is called the "**amplification**" or "**gain**" of the system. For instance, a person is in a room for 24 hours at 60°F. If the temperature is suddenly raised to 110°F then the body temperature will increase from 98° to 99°F i.e., increase in °F for every 50° rise i.e., gain for the Control System is –49. The gain of negative feedback Control System is calculated by the formula,

$$\text{Feedback gain} = \frac{\text{Amount of correction of abnormality}}{\text{Amount of abnormality that still persists}}$$

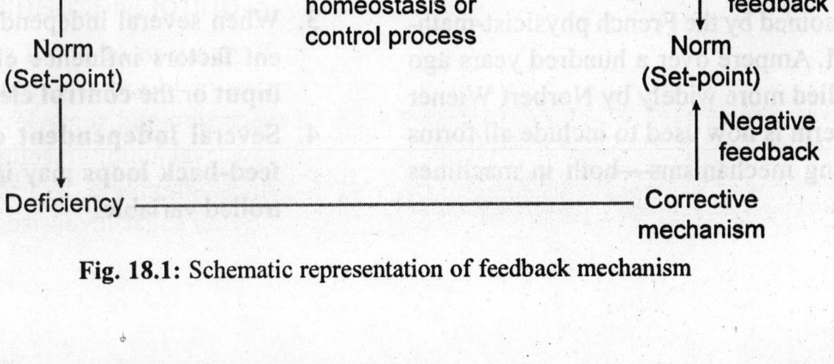

Fig. 18.1: Schematic representation of feedback mechanism

18.7 Positive Feedback

Sometimes corrective mechanism leading to negative feedback breaks down with the result that a deviation from the "norm" or "set-point" initiates further deviation known as **"positive feedback"** which is unstable and leads often to death. It is better known as **"vicious cycle"** because a mild degree of positive feedback can be overcome by the negative feedback control mechanism of the body. It is irreversible, self-perpetuating and self-amplifying, under physiologic conditions, e.g., parturition, micturition leading to severe dysfunction, e.g., cardiac failure, irreversible shock etc.

18.8 Oscillation and Damping of Control System

Even Feedback Control System sometimes become unstable and oscillate as we see in case of blood pressure waves caused by oscillation of baroceptors. But this oscillation does not occur very often but only under certain peculiar conditions due to **"damping"** which results from basic organization of the system. The oscillation may be **waxing oscillation, driving oscillation and damped oscillation.**

18.9 Conclusion

Homeostasis is not confined to physiological conditions but the size of population of animals or plants is kept constant and the same mechanism maintains the constancy of species over long periods of time. This mechanism is also not confined to biology e.g., thermostat operates on this basis. Many machines involve similar principles and even the industrial processes, economic system and stabilisation of prices can be analysed in similar terms. And the term now used to encompass all these concepts is **"Cybernetics"**, the science of communication and control. A Greek word meaning **"steersman"**, the term was first coined by the French physicist-mathematician A.M. Ampere over a hundred years ago but it was applied more widely by Norbert Wiener in 1948. The term is now used to include all forms of self-adjusting mechanisms—both in machines and organisms.

18.10 Regulation of Homeostasis

I. At the Molecular Level

II. At the Organ System (organismal) Level

I. **At the Molecular Level**: There are four general mechanisms:

1. **Enzymes themselves towards specific** substrate,
2. **Kinetic properties of "systems of enzymes"**,
3. **Compartmentalization of enzymes** or enzyme system in the cell or certain organelles,
4. (a) **Rate at which an enzyme is synthesized**
   (b) **Structure of the enzymes.**
   Both the factors are controlled by gene.

III. **At the Organ System (organismal) Level**

1. **The nervous system** is responsible for rapid transmission and integration of information among all regions of the body.
2. The **endocrine system** transmits information by chemicals, i.e., hormones, which is a slower process.
3. The **cardiovascular system** helps to convey the hormones in the blood to the appropriate sites of action (Target tissue).

They usually operate in a smoothly coordinated, integrated and interdependent fashion to achieve a homeostatic balance.

18.11 Operation of Homeostatic Mechanism

Operation is mediated by negative feedback control system of various types. It becomes more complex in the following situations:

1. When it exhibits **non-linear response,**
2. When a **time lag (delay)** exists in the control element or feedback loop,
3. When several independent or interdependent factors influence either the **reference input** or the **control element,**
4. Several **independent** or **interdependent feed-back loops** may impinge on the controlled variable.

19

Biomagnetism and Magneto-Biology

1. Definition of Biomagnetism, 2. Definition of magnetobiology, 3. Generation and nature of Biomagnetic Fields, 4. Ferromagnetic particles in the body, 5. Magnetic fields of skin, muscle, etc. 6. Neuromagnetic fields, 7. Magnetometer (SQUID), 8. Magnetography, 9. Magnetocardiography (MCG), 10. Magneto-encephalography (MEG).

19.1 Biomagnetism

Magnetic fields generated by living organisms and the analysis of these fields known as **biomagnetism.** Biomagnetic signals are very weak and their measurement is difficult mainly because of the high level of magnetic noise in our environment.

19.2 Magnetobiology

Magnetobiology is concerned with the effect of strong magnetic fields on biological processes. But no reliable data have yet been available on the effect of a magnetic field on biological phenomenon.

The iron atom is paramagnetic and iron containing compounds like haemoglobin and myoglobin are also paramagnetic. Hence, it could have an effect on the behaviour of haemoglobin and myoglobin, cells having these compounds, e.g., RBC and muscle cell and on the kinetic behaviour of the biochemical reactions involving free radicals.

Recently, a bacteria has been discovered in peat bogs in USA which contains ferrimagnetic magnetite, $FeO.Fe_2O_3$. These bacteria can orient themselves in a magnetic field and they can be magnetized.

19.3 Generation and Nature of Bio-Magnetic Fields

In living organisms, the magnetic fields are generated mainly by three different ways:

I. **Accompanied with an electric field:** Ionic currents originated from the electrical activity of the cell membranes like muscle cells and nerve cells. These generate their inherent fields.

II. **Not accompanied with an electric field**

1. Minute ferromagnetic particles enter the body or are introduced into the organisms. These also generate their inherent fields.
2. The superimposition of an external magnetic field.

As the above cases are not accompanied with an electric field, Magnetometric Methods are suitable for the studies of:

(a) Magnetic particles in the organism and

(b) The magnetic properties of various organs.

Biocurrents redistribute electrical potential on the surface of the body and these can be picked up and graphically recorded as we see in ECG, EEG. Similarly, we can expect that the methods like Magnetocardiography (MCG) and Magnetoencephalography, which record signals from the same electrical processes in the body, should provide the same information on the organs. But, according to the theory of electromagnetic phenomena,

(a) The shape of the source of current in a conducting medium of the organism and

(b) The inhomogeneity of this medium have different effects on the distribution of magnetic and electric fields. Some kinds of bioelectric activity manifest themselves mainly in the electric field and produce a weak magnetic field while for the others the picture is reversed. Thus, there are so many processes which can be studied Magnetographically.

19.4 Ferromagnetic Particles in the Body

Those are engaged in metal processing industries, contain small amount of ferromagnetic particles in their skin and the body and their magnetic fields pose problem to highly-sensitive biomagnetic measurements. To eliminate the disturbance, demagnetization of the particles in the external field is necessary. Again, the fields of ferromagnetic particles can be enhanced by magnetization in a large constant field. Thus (ferroprobe) Magnetometers are employed for measurement of the amount of iron dust in the lungs of welders. SQUID Magnetometers can detect minute amounts of paramagnetic whose magnetization is very weak and ferromagnetic impurities. The high sensitivity makes it useful for diagnostic purposes.

19.5 The Magnetic Fields of Skin Muscles, Eyes and Other Internal Organs

Magnetic manifestations of biological activities have been observed in many organs of the body:

1. **Stomach**: Fields, either continuous or oscillating for a period of several minutes, are typical manifestation of the human stomach, the shape of the signal depends on the state of stomach which is applicable for the diagnosis of pathological condition of the stomach.
2. **Skin**: Magnetic fields of D.C. were observed to be generated in the skin when the hair is touched.
3. **Skeletal muscles:** The magnetic fields produced from contraction of human contracting muscles, which have been measured and the record obtained from these fields as a function of time is called **Magnetomyogram** (MMG). Apart from the high frequency components of 10-150 Hz which can also be recorded Electromyographically there is a slowly varying MMG component which is seen to be generated by contraction or slight massaging of the muscle which is characteristic of leg muscle and persists for about an hour. The current which is generated in these fields seems to play a significant role in the regeneration of limbs, in particular, in curing fracture of bones.
4. **Eye:** The eye is a source of strong electrical field because the action of retina induces a potential of upto 0.01 V between the front and rear parts. For this, an electric current flows in the surrounding tissues; its magnetic field can be recorded as **Magnetooculogram** (MOG) due to eye motion and as a **Magnetoretionogram** (MRG) due to variation of the illumination. It has been observed that in the eye the magnetic field induction is significantly higher than the brain.

19.6 Neuromagnetic Fields

1. The brain produces both electric and magnetic fields. The strong signals are generated by spontaneous rhythms of the brain and can

be recorded as Electroencephalogram (EEG) which classifies the rhythms and their relationship during sleep, wakefulness, epilepsy, etc. Magnetoence-phalogram (MEG) differs from EEG, such as certain rhythms are seen only in the EEG and some in the MEG whereas both are very much alike. For this reason, SQUID Magnetometers are promising in brain studies.

2. The differences we see between EEG and MEG are not obvious. Such as, in the alpha rhythm (in 8 to 12 Hz oscillations), which is the characteristic of wakefulness in a relaxed state with eye closed, the magnetic and electric field are recorded simultaneously and their amplitudes are also correlated. If larger electrical alpha-rhythm signal is produced there is also larger magnetic signal.

3. While making differences between Electro and Magnetoencephalogram, it is seen that the pattern of potentials on the surface of the scalp of the head is a complex superposition of spatial distributions of signals from distant sources in the brain. The reasons of such nature of pattern are:

   (a) The brain is surrounded by bony tissue, i.e., skull whose electrical conductance is much lower than that of the skin and the brain itself.

   (b) And the natural orifices of the skull make the path of electric currents more complicated. Whereas in case of Magnetographic methods, (i) the magnetic sensor mainly responds to stronger currents in the region of bioelectrical activity which is important for Magnetometer input coil of SQUID. (ii) and the recording signals are undisturbed by other activities and coming from a specific source inside the brain. Hence, in case of epilepsy, MEG can reveal the source of pathological condition where EEG sometimes fails to show.

4. Magnetic Techniques are essential for studies of brain responses to tactile, visual and acoustic signals. These cause **"evoked magnetic fields"** (EFs) which is used for localization of the source of biological activity in the brain cortex. Some EF sources can be modelled by a current dipole. The dipole position, its depth inside the brain, direction and magnitude (the product of the current density by the active volume) can be determined Magnetographically. In response to visual stimuli a current dipole is formed in the occipital lobe. Stimulation of the little finger on the right hand causes the formation of dipole, normal to the central fissure in the left hemisphere. This dipole is in the projection zone of sensory receptors associated with various parts of the body, exactly where the neurosurgeons detect the "representation" of the little finger. Magnetography detects without surgery the exact spot in the cerebral cortex where the data from sensory organs arrive to be processed.

5. Simple shape of some EF makes the information reliable neurophysiologically. Such as the brain magnetic fields generated in response to a periodic pattern of dark and light on the Oscilloscope screen were studied. The amplitude of the magnetic signal in response to such stimulation has been found to be twenty times that in response to a single flash. In this way, the time of signal propagation along nervous paths from the eye to a specific point in the cortex can be determined by measuring the delay time of the magnetic responses.

6. The measurements of the active EF delay time make it possible to compare the response time of brain hemispheres exposed to visual stimulus.

7. By magnetic method, the active area of the cortex can be localized such as the active area in the cortex was found to move 1 cm into the so-called Sylvian fissure as the pitch of the sound increased from 200 to 5000 Hz., when the frequency changed smoothly, the gradualness of the movement was seen.

8. Accurate measurement of the nervous activity region accompanying the stimulation of a specific sensory organ, makes it possible to prepare maps, "**somatotopic**" for tactile stimulation, "**tonotopic**" for audio stimulation and "**retinotopic**" for vision.
9. Magnetography can penetrate beyond the cortex to deeper brain structure. Magnetic fields have been measured that appear when man is about to act.

Measurement

19.7 Measurement of Magnetic Field Generated by Living Organisms-Magnetometers

The Semiconducting quantum Interference Device (SQUID) is an instrument whose principle is based on purely quantum such as the Josephson effect and interference of the wave functions of electron pairs in a superconducting ring containing a Josephson junction. To maintain the superconducting state, which remains only at a very low temperature, the SQUID is placed in a Dewar flask with liquid helium. If the walls of the flask are metallic, the currents generated there distort the magnetic fields of sources outside the vessel. Special composite glass-plastic Dewar flasks have been recently developed where the SQUID or its input unit, made of superconducting wire (Magnetic Flux Transformer), are within one centimetre from the outer wall and record undistorted magnetic field produced by outside sources at room temperature.

The SQUID Magnetometers are being used in measurements of the magnetic fields generated by living organisms, above all—man. Biomagnetic fields are very weak and their measurement is not a simple procedure because of high level of magnetic noise in our environment.

Megnetic noise is mainly generated by (a) chaotic oscillations (fluctuations) of the earth magnetic field and (b) industrial electrical units. The noise can be eleminated by using a large room by magnetic shielding using multilayer walls of magnetically soft ferromagnetic alloy such as Permalloy.

Use

1. Measurement of magnetic field.
2. Measurement of physical quantities of electric current.
3. Magnetic susceptibility.
4. Displacement of magnetics, etc.

Advantage

The sensitivity of these instruments is at least 1000 times better than that of the best non-superconducting Magnetometers.

19.8 Magnetography

(a) The magnetic fields of living organism may be generated by three ways of which in two cases the magnetic fields are not accompanied with an electric field. For this reason only Magneto-metric Methods can be used for:

1. The studies of magnetic particles in the living organism, and,
2. The magnetic properties of various organs.

(b) Like the biocurrents which are redistributed, the electrical potentials which are recorded by EEG, ECG, etc. similarly, their magnetic analog-Magnetocardiography (MCG) and Magnetoencephalography—which record signals from the same electrical processes in the organism, all might be expected to provide the same information on the organ. But, according to the theory of electromagnetic phenomenon, the shape of the source of current in a conducting media like living organism and the inhomogeneity of this medium have different effects on the distribution of magnetic and electrical fields. Some kinds of bioelectric activity manifest themselves mainly in the electric field and produce a weak magnetic field while for others the picture is reversed. Thus, there are many processes which are better observed Magnetographically.

(c) Magnetography does not require any contact with a person because measurement is done through a bandage. This is convenient and more advantageous over electrical methods because the spots where the electrodes are placed on the skin may become the source of slowly changing contact potentials. These do not occur in Magnetographic Methods, so it is reliable in the study of slow processes.

(d) Magnetic fields fall off steeply with the distance from the source because they are generated by relatively strong currents in the surface potentials by weaker and widely spread currents in the skin. For this reason Magnetography gives us more accurate localization of the bioelectrically active spot.

(e) Additional useful information can be obtained because the inductance of the magnetic field as a vector is characterised by the direction, as well as the magnitude.

19.9 Magnetocardiography (MCG)

The heart is the most powerful source of electric and magnetic fields in the living organism. Magnetocardiography was being used long before the development of SQUID. Before the SQUID, MCG equalled the ECG. They are very much alike but heart troubles have different effects on electrical and magnetic measurements. Studies are going on to classify the changes in MCG during heart disturbances. The advantage of MCG are obvious when slowly changing and constant signals are to be recorded.

Observations

1. It first detected continuous "alarm currents" signalling the clot of the coronary artery in dogs.
2. In case of foetus, MCG shows the clear localization of the magnetic field in the vicinity of the source, makes it possible to differentiate the foetus signals from the stronger mother's heart signals whereas electrical signals are largely mixed up because of the spatial spreading of the surface ECG currents.
3. It measures the characteristics of the blood flow in the heart. In a small external magnetic field the periodic ejection of blood generates an alternating magnetic signal which is indicative of the flow rate and velocity of the moving fluid.
4. High Resolution Magnetocardiography—which is similar to neuromagnetic measurements—ensures a more convincing result of those intervals in the heart cycles during which the heart muscle is quiet and weak magnetic signals can be measured which accompany the nervous impulses propagating in the heart.

The technical difference between ECG and MCG (or technical reason to use the MCG)

Better ability to record:

1. Static sources,
2. Sources on the posterior side of the heart,
3. Monitor the fetal heart,
4. Without electrodes.

19.10 Magnetoencephalography (MEG)

(sec.19.6 Neuromagnetic fields (p.622.)

20

Biophysics of the Membrane

1. Salient features of the membrane, 2. Chemical composition of the membrane: (a) Lipids, types, organization of lipids, characteristics of bilayer membrane arrangement, (b) Membrane proteins, (c) Carbohydrates. 3. Molecular structure of the cell membrane (unit membrane), 4. Modification of cell membrane and biophysical importance, 5. Nuclear membrane or envelope, 6. Nuclear membrane vs. cell membrane, 7. Cell membrane (animal) vs. cell wall and membrane (Plant), 8. Functions of the cell membrane, 9. Physico-chemical properties of the membrane, 10. Conformational properties of the membrane (Dynamical properties), 11. Permeability of the membrane and the factors affecting it, 12. Membrane receptor (signal transmission), 13. Hormone receptor, 14. Peptide hormone receptor on the surface of the membrane, 15. Transport of substances across epithelia or cellular sheet, 16. Mitochondrial membrane, 17. Factors governing the passage of materials across the cell membrane.

20.1 Biophysics of the Plasma Membrane (Cell Membrane)

Every living cell has a definite boundary, i.e., one side of the boundary has intracellular material and other side has extracellular material. Every cell maintains its boundary by a membrane but boundary and membrane are not synonyms. Because gelating has a definite boundary but is not bounded by a membrane. Permeability and electrical studies have suggested that the cells are bounded by "**cell membrane**" or "**plasma membrane**". Recently, EM studies have shown that not only the cell but other organelles, e.g., nucleus, mitochondria, lysosomes, endoplasmic reticulum, microbodies, etc. are also bounded by the membrane. The plasma membrane separates the cell from its surroundings and it is important in controlling what enters and leaves the cell. It is actually a part of the cytoplasm which has become strengthened due to its functions and situations. It is elastic and semipermeable.

20.1.1 Salient features of cell membranes

Though the structure of membranes differs like their functions, it has certain characteristics which are common to all.

1. Membranes are *supermolecular (ultramolecular)* and *dynamical organised* system.
2. Membranes are 2-*dimensional oriented viscous solution* surrounding all living cells.
3. Two-dimensions considerably exceed their thickness which is in order of 5-10 nm (60-100 Å). All the mechanisms for biological functions of the membrane are *localized just in its thickness*.
4. Membranes form a *closed compartment* around the protoplasm, which separate one

cell from the other and the inside from outer environment.

5. Membranes are *sheetlike structures* which form *closed boundaries* between compartments of diffrent composition.
6. Membranes consist of *lipids* and *proteins*, having weight ratio of proteins to lipid ranging from 1 : 4 to 4 : 1. These also contains *carbohydrates* which are linked with lipids and proteins.
7. Membrane lipids are small molecules having both a hydrophilic and hydrophobic groups. These lipids spontaneously form *closed bimolecular sheets* in aqueous media which are barriers to the flow of polar molecules.
8. *Specific proteins* serve specific functions. These proteins act as pumps, gates, channels, receptors, energy transducers and enzymes.
9. Specific proteins in the membranes are anchored into lipid bilayers creating a favourable environment for the action of these proteins.
10. Membranes are *asymmetric sheet* like enclosed structure with inner and outer surface.
11. These sheet-like structures are *noncovalent assmeblies* that are thermodynamically stable but metabolically active. The constituent protein and lipid molecules are held together by many noncovalent interactions, which are cooperative in character.
12. Membrane are *heterogeneous.*
13. Membranes *act as a barrier with selective permeability.* Selective permeability are provided by gates, pumps, channels receptors.
14. Membranes *provide sites of energy transductions* (e.g., photosynthesis, oxidative phosphorylation).
15. They are not *passive,* semipermeable shells. They take a direct and important part in all functions of cells.
16. Water, small molecules and lipid-soluble substances pass easily through cell membrane; charged molecules and large molecules do not.
17. Portions of membranes are specialized for transport, ion excitation and conduction, metabolism, etc.
18. The other properties are given in Table 20.1.
19. Membranes are electric insulators or dielectrics with capacitance 1 $\mu F/cm^2$.

Table 20.1: Different properties of membrane

Property	Membranes at 25°C
1. Electronmicroscopy	Trilaminar
2. Thickness	6-10 nm (60-100 Å)
3. Electric capacitance ($\mu F/C\ mm^2$).	0.5-1.3
4. Electric resistance (ohm-cm^2).	10^2-10^5
5. Dielectric breakdown or breakdown voltage (mV).	100-200
6. Surface tension (dynes/cm).	0.03-1.0
7. Water permeability (μsec).	0.37-400
8. Activation energy of transport of water (kJ/mole).	40.3

20.2 Chemical Composition of the Membrane

Most membranes are composed of:

(A) Lipids,

(B) Proteins and,

(C) Carbohydrates.

(A) **Lipids**: The lipids of the membranes are:

I. **Phospholipids (25%)**: Lecithin, Sphingomyelin.

II. **Glycolipids**: Cerebroside, Gangliosides (other lipids 4 per cent)

III. **Sterol**: The common sterol which is more abundant in the membranes is cholesterol. It is also found in lesser quantity in the membranes of the mitochondria, golgi apparatus, nuclear membrane, etc.

Membrane lipids are *amphipathic* molecules containing both a *hydrophobic* and a *hydrophilic* region (Table 20.2).

The *hydrophilic unit* also called the *polar head group* whereas hydrophobic unit is called *apolar or nonpolar group* or *hydrocarbon tails* (Fig. 20.1).

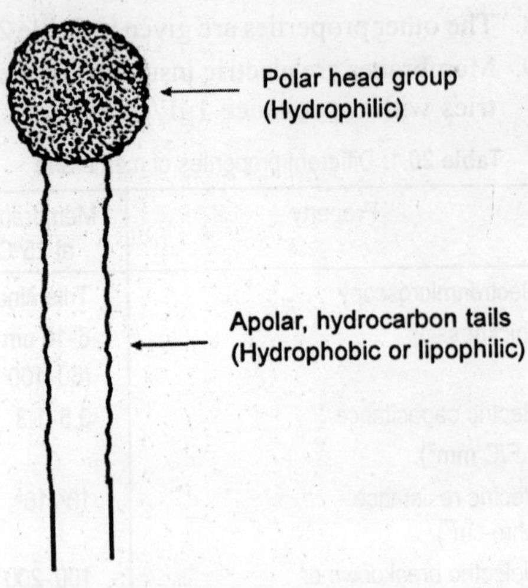

Fig. 20.1: Diagrammatic representations of a phospholipid or other membrane lipid

20.2.1 Organization of lipids in the membrane

The amphipathic character of phospholipids make the two regions of the molecule incompatible solubilities. However, in a solvent like water, phospholipids arrange themselves into a form which thermodynamically satisfies both regions. Their polar head groups (hydrophilic) have affinity to water and their hydrocarbon tails (hydrophobic) avoid water. In aqueous medium, this can be accomplished by forming a spherical structure called "micelle" where the hydrophilic polar head groups are on the surface immersed in the aqueous environment and the hydrophobic, apolar, hydrocarbon tails are sequentered inside (Fig. 20.2). The stability of this structure is based on the fact that significant "free energy" is required to transfer a nonpolar molecule from a nonpolar medium of water. The micelle provides a minimal energy configuration and is thermodynamically stable.

Hydrophobic and hydrophilic interactions, hydrogen bonding and salt or electrostatic bonds are of secondary importance to the micelle structure.

In aqueous medium, another arrangement was recognized by Gorter and Grendel known as "bimolecular sheet" or "lipid bilayer" (or bimolecular layer or bilayer) which can satisfy the thermodynamic requirements of amphipathic molecules. In fact, the favoured structure in aqueous media is a bimolecular sheet rather than a micelle (Fig. 20.3).

Table 20.2: Hydrophobic and hydrophilic unit of membrane lipids

Membrane lipid	Hydrophobic unit	Hydrophilic unit
Phospho-glycerides	Fatty acid chains	Phosphory-lated alcohol
Sphingo-myelin	Fatty acid chain and hydrocarbon chain of sphingosine	Phosphoryl choline
Glycolipids	Fatty acid chain and hydrocarbon chain of sphingosine	One or more sugar residues
Cholesterol	Entire molecule except for OH group	OH group at C-3.

20.2.2 Characteristics of bilayer membrane arrangement

The bilayer membrane is preferable for critical biological importance for the following characteristics

1. They have macroscopic dimensions such as a millimetre (10^7 Å) whereas micelle is a limited structure usually less than 200 Å in diameter.
2. Phospholipids and glycolipids are main constituents and can readily form extensive bimolecular sheets.
3. These sheets serve as permeability barrier and yet they are quite fluid.
4. Formation of lipid bilayer is self-assembly process.
5. The formation of lipid bilayers from phospholipids and glycolipids is a rapid spontaneous process in water.
6. Hydrophobic interactions are the major driving force for the formation of lipid bilayer which also plays a role in the folding of proteins in watery solutions.
7. Water molecules are released from the hydrocarbon tails which results in a large increase in entropy.
8. There are "van der Waals' attractive forces between the hydrocarbon tails which favor close packing of the hydrocarbon tails.

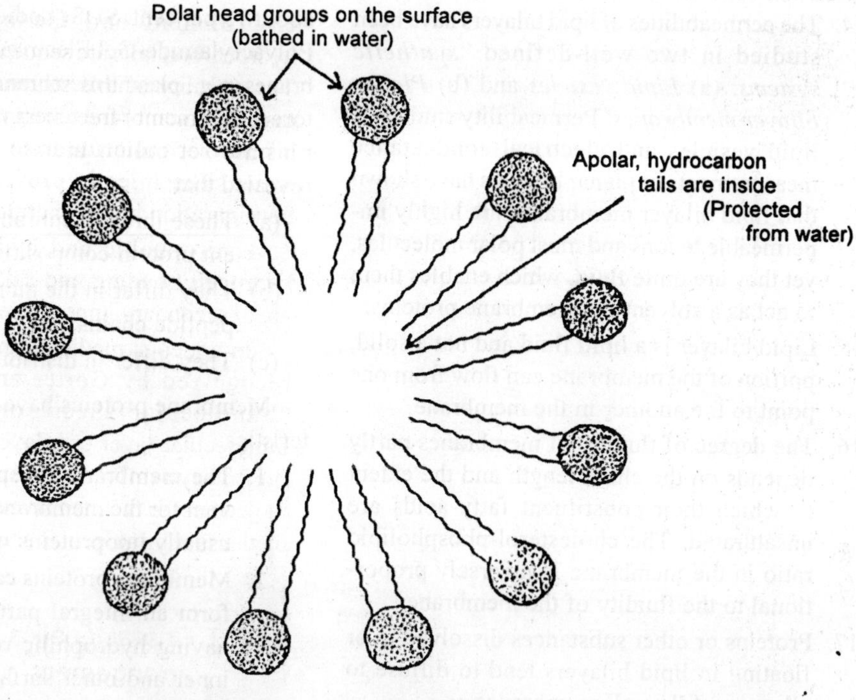

Fig. 20.2: Cross section of micelle

9. There are "electrostatics" and "hydrogen-bounding" interactions between the polar head groups and water molecules.
10. There are "electrostatic" and "hydrogen-bounding" interactions between the polar head groups and water molecules.
11. They are "cooperative structures" and are held together by many "reinforcing, non-covalent interactions".
12. Phospholipids and glycolipids cluster together in water to minimize the number of exposed hydrocarbon chains which is favoured by the van der Waals' attractive forces between the adjacent hydrocarbon chains. These energetic factors have significant biological consequences:
    (a) Lipid bilayers tend to "close on themselves" so that there are no ends with exposed hydrocarbon chains, which results in the formation of a compartment.
    (b) Lipid bilayers have an inherent tendency to be "extensive".
    (c) Lipid bilayers are "self-sealing" because a hole in a bilayer is energetically unfavourable.
13. Only the ends or edges of the bilayer sheet are exposed to an unfavourable surrounding but the exposed ends or edges can be eliminated by folding the sheet back upon itself to form an "*enclosed vesicle with no edges*". This closed bilayer provides the essential functions of membranes and is impermeable to most water-soluble molecules, since they would be insoluble in the hydrophobic core of the bilayer.

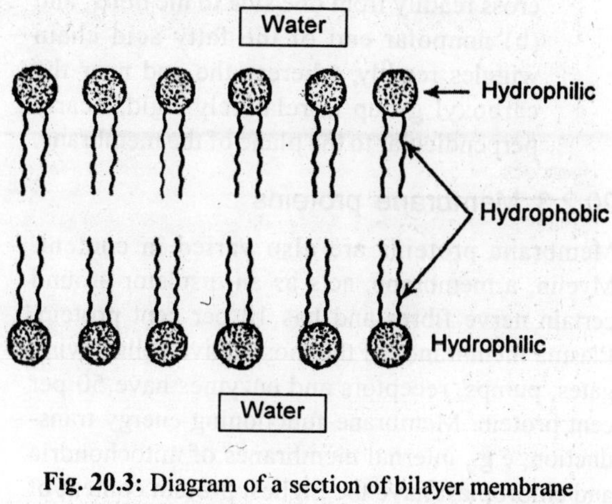

Fig. 20.3: Diagram of a section of bilayer membrane

14. The permeabilities of lipid bilayers have been studied in two well-defined "*synthetic systems*: (a) *Lipid vesicles* and (b) *Planar bilayer membranes.* Permeability studies of lipid vesicles and electrical conductance measurements of planar bilayers have shown that lipid bilayer membranes are highly impermeable to ions and most polar molecules, yet they are quite fluid, which enables them to act as a solvent for membrane proteins.
15. Lipid bilayer is a lipid fluid and not a solid, portion of the membrane can flow from one point to the another in the membrane.
16. The degree of fluidity of membranes partly depends on the chain length and the extent to which their constituent fatty acids are unsaturated. The cholesterol-phospholipid ratio in the membrane is inversely proportional to the fluidity of the membrane.
17. Proteins or other substances dissolved in or floating in lipid bilayers tend to diffuse to all areas of the cell membranes.
18. The lipid composition of RBC membranes of different mammals have shown that the proportion of cholesterol and phosphatidylethanol amine are approximately constant but that the ratio of phosphatidylcholine to sphingomyelin varies greatly with the species.
19. Electron Spin Resonance (ESR), Spectroscopy also called Electron Paramagnetic-Resonancy Spectroscopy (EPR), has shown that: (a) phospholipids in membranes move freely in the plane of membrane but do not cross readily from one side to the other, and (b) nonpolar end of the fatty acid chain wiggles readily, whereas the end near the carboxyl group is relatively rigid, nearly perpendicular to the plane of the membrane.

20.2.3 Membrane proteins

Membrane proteins are also varied in content. Myelin, a membrane, acts as an insulator around certain nerve fibres and has 18 per cent protein. Plasma membranes of the most active cells having gates, pumps, receptors and enzymes have 50 per cent protein. Membrane functioning energy transduction, e.g., internal membranes of mitochondria and chloroplast have the highest protein content of about 75 per cent. SDS (Sodium dodecyl sulphate)-Polyacrylamide-Gel Electrophoresis of three membranes (e.g., plasma membrane of erythrocytes, photoreceptor membrane of retinal rod cells and sarcoplasmic reticulum membrane of muscle), has revealed that:

(a) These three membranes have a very different protein composition,

(b) They differ in the number of kinds of polypeptide chains, and

(c) They differ in distribution mass.

Membrane proteins have the following characteristics:

1. The membrane phospholipids act as a solvent for the membrane proteins and they are usually lipoproteins or glycoproteins.
2. Membrane proteins can be amphipathic and form an integral part of the membrane by having hydrophilic regions protruding the inner and outer surfaces of the membrane. These hydrophilic regions are connected by a hydrophobic region which passes through the hydrophobic coat of the bilayer.
3. The asymmetry of the membranes can be partly attributed to:
   (a) The asymmetric distribution of proteins within the membranes,
   (b) Inside-outside asymmetry is also provided by the outside location of the carbohydrates attached to the membrane proteins,
   (c) Specific enzymes are located predominantly either on the outside or inside of the membranes as in the mitochondria and the cell membrane.
4. Regional asymmetries occur in the membranes, e.g., villous borders of mucosal cells are almost macroscopically visible, gap junctions, tight junctions and synapses occupy smaller regions of the membranes and produce local asymmetries.
5. On the weight basis, membranes contain larger amount of proteins than lipid (4 : 1) but due to high molecular weight of proteins, the ratio is reversed on a molar basis; protein : lipid ratio is 1 : 100 to 10 : 100.

6. Some of the proteins must be present in globular form rather than extended form, unfolding of enzymic protein result in loss of activity.
7. Membranes are dynamic structures in which proteins and lipids diffuse rapidly in the plane of the membrane, if it is not restricted by special interactions (lateral diffusion), whereas the rotation of proteins and lipids from one side of a membrane to the other is a very slow process (transverse diffusion).
8. Optical-Rotatory-Dispersion and Circular-Dichroism Measurements indicate that the protein molecules in the membranes have a rather high content of right handed α-helix.

20.2.4 Classification of membrane proteins

I. **Structural**: Membrane proteins are classified on the basis of the difference in dissociability as under:

(A) Integral proteins (Intrinsic proteins) –70 per cent or more, tightly bound.

(B) Peripheral proteins (Extrinsic proteins), loosely attached to the membrane surface.

(A) **Integral Proteins**: They are usually globular in shape and are amphipathic. They consist of 2 hydrophilic ends separated by a hydrophobic region which traverses through the hydrophobic core of the bilayer demonstrated by Freeze-Fracture Technique. These proteins are asymmetrically distributed across the membrane bilayer and during insertion in the bilayer some proteins show asymmetric orientation. The hydrophilic external region of an amphipathic protein, which is synthesized inside the cell, traverses through the hydrophobic core of the membrane and seen external to the membrane (Fig. 20.4). Erythrocyte membranes contain 17 different polypeptide chains having MW from 27000 to 220000 and among them are:

1. *Glycophorin*—A glycoprotein that extends completely across the membrane.

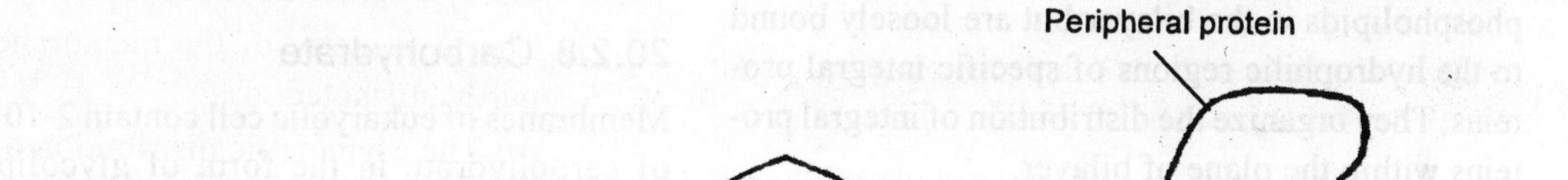

Fig. 20.4: Bilayer and integral and peripheral proteins

The inner mitochondrial membrane probably contains over 100 different kinds of polypeptide chains.

2. *Band-3 protein*—Anion channel is dimer of the band-3 protein and is located on both sides of the RBC membrane and that all molecules of this protein point in the same direction.

20.2.5 Role of integral proteins

(a) The integral proteins provide structural path ways through which water and water-soluble substances, especially the ions, can diffuse between extracellular and intracellular fluid.

(b) These proteins have selective properties that cause preferential diffusion of some substances more than others.

(c) Some of the proteins can also act as an enzyme.

20.2.6 Peripheral proteins

These proteins do not interact directly with the phospholipids in the bilayer but are loosely bound to the hydrophilic regions of specific integral proteins. They organize the distribution of integral proteins within the plane of bilayer.

(a) **Ankyrin:** A peripheral protein is bound to the integral protein "band-3" of the erythrocyte membrane.

(b) **Spectrin:** Composed of two components with MW 255000 and 220000 bound to ankyrin, a cytoskeletal structure within RBC, maintains biconcave shape of RBC.

(c) Series of proteins with MW 90000.

(d) Another group of proteins with MW 9000-15000.

(e) Membrane of simple composition is the inner membranes of the rods of retina having only one protein rhodopsin.

(f) Immunoglobulin proteins on the membrane of lymphocytes.

(g) Hormone receptor proteins.

(h) Specific polypeptide hormones that bind to the receptors role—these function entirely as enzymes.

20.2.7 II Physiological

In addition to structural proteins, there are a few proteins:

(a) **Pumps:** Actively transporting ions across the membrane.

(b) **Channels:** Passive channels for ions that can be opened or closed by changes in the conformation of the protein which is glycoprotein.

(c) **Receptors:** That bind neurotransmitter, and hormones initiating changes in the cell.

(d) **Enzymes**: Catalyzing reactions at the surface of the membrane. The enzyme content of membrane varies not only from cell to cell but also within the same cell.

(e) **Gap junctions**: These are composed of subunits, which called "connexons", have been studied by X-ray crystallography and each connexon consists of 6 protein subunits that span the membranes and connect with the analogous structure of the adjacent cells from passage ions and small molecules.

20.2.8 Carbohydrate

Membranes of eukaryotic cell contain 2-10 per cent of carbohydrate in the form of glycolipids and glycoproteins.

1. Sugar residues are always located on the external surface of the membrane. Because sugars are hydrophilic, the sugar residues of a glycoprotein or of a glycolipid will tend to be located at a membrane surface (Fig. 20.5).
2. Very high free energy is required for inserting oligosaccharide chain into the hydrocarbon core of the membrane.
3. There is high barrier to the rotation of a glycoprotein from one side of a membrane to the other.

20.2.9 Role of carbohydrate residue

1. It helps to orient glycoproteins in the membrane.
2. The sugar moieties of membrane glycoproteins help to maintain the asymmetric character of biological membrane.
3. Carbohydrate on cell surfaces may be important in intercellular recognition.

4. These sugar residues are the portions of the cell membrane that enters into immune reactions.
5. They often act as receptor substances for binding hormones, e.g. insulin.
6. The carbohydrate residues serve to modify the electrical charge at its surface and provide specific binding sites.
7. Polysaccharide: Protein complex demonstrated Cytochemically is associated with many cell surfaces as an extracellular layer. In certain places (e.g., luminal surfaces), this layer forms a coat called "glycocalyx" which is mucoprotein in nature and act as a crude filter or facilitate the attachment of molecules for endocytic transport across the cell membrane.

20.3 Molecular Structure of Cell Membrane (Unit Membrane)

Before the advent of Electron Microscopy, cell membranes could not be visualized by the ordinary Light Microscope as they were below the limits of resolution of the Light Microscope. The presence of this structure was inferred by osmotic, electrical and cytosurgical experiments. It was first visualized after the advent of Electron Microscopy in 1950, as very thin lines at the outer edges of the cells in a stained section.

Robertson (1964), from the studies with myelin formation, found that the membrane of which myelin is built is a "trilaminar or three-layered membrane" or "unit membrane" composed of two "*electron-dense layers*" each 25 Å thick on either side of a single "*electron lucent layer*" 25 Å thick. The former is considered to be the protein; the latter the lipid. The unit membrane hypothesis was supported by the "*Low Angle X-ray Diffraction*" experiment. However, unit membranes are not only present in the Schwann cell membrane but are also present in all plasma membranes.

Regarding molecular organization that might underlie such an appearance had been first proposed by Gorter and Grendel (1925). He concluded from the study of the RBC membrane that RBCs are covered by a layer of fatty substances, that is two molecules thick. Later on various workers proposed various models for molecular organization.

I. **Danielli-Davson Model (1935)**: ("Paucimolecular" model or "paucimoral membrane" model). According to them the unit membrane is considered to be a bilayer of mixed polar lipids with their hydrophobic tails extending inward and their hydrophilic heads exposed on the surface where they are bound to the outer nonmolecular proteins layers (Fig. 20.6).

II. **Robertson's Modification of Unit Membrane Model (1964)**: Some years after Danielli-Davson unit membrane model hypothesis, it was modified and refined by Robertson (1964).

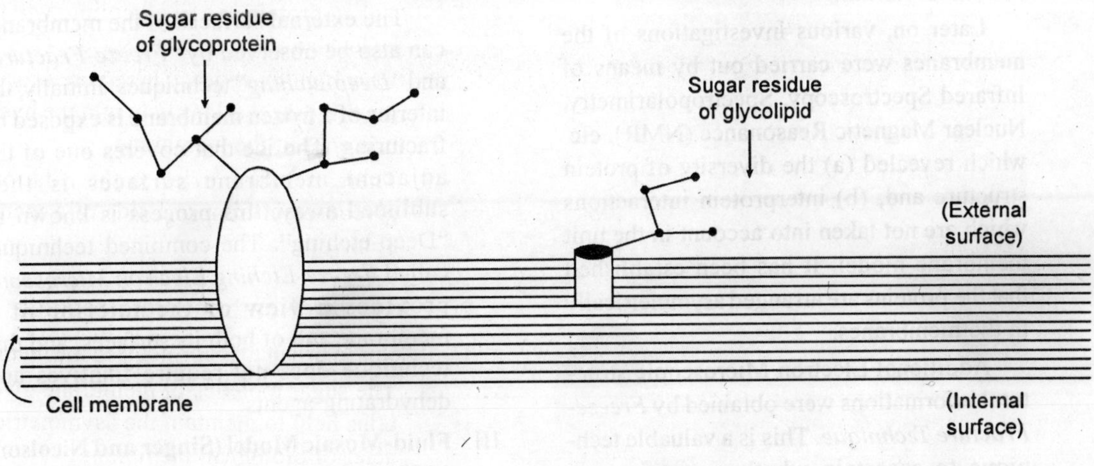

Fig. 20.5: Sugar residues of glycoprotein are located on the external surface of the membrane

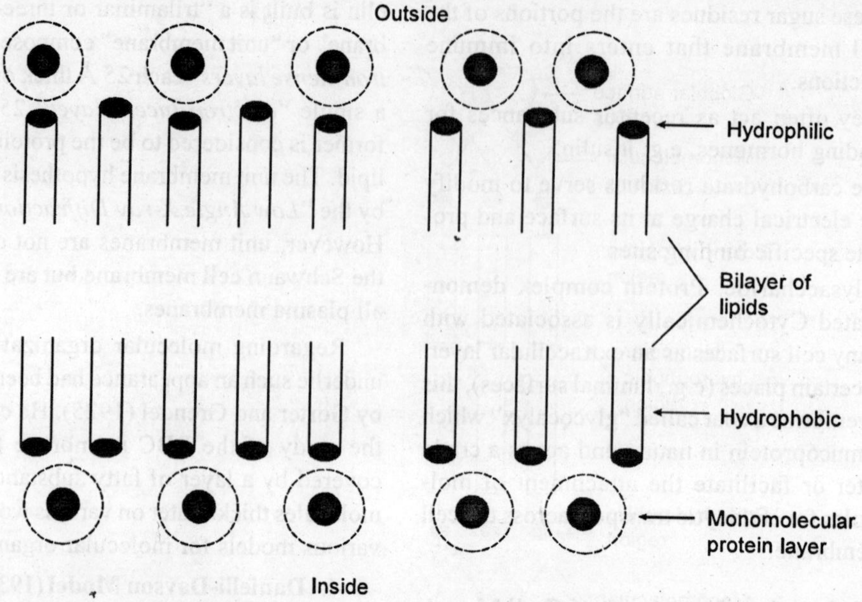

Fig. 20.6: Danielli-Davson's original paucimolar membrane model

The unit membrane was proved to consist of a bilayer of a mixed polar lipids, with their hydrocarbon chains oriented inward to form a continuous hydrocarbon phase and their hydrophilic heads oriented outward. Each surface was thought to be coated with a monomolecular layer of protein molecules, with the polypeptide chains in extended form (Fig. 20.7). The total thickness of the unit membrane was suggested to be about 7.5-9.0 nm, the thickness of the lipid bilayer was about 3.5-7.0 nm.

Later on, various investigations of the membranes were carried out by means of Infrared Spectroscopy, Spectropolarimetry, Nuclear Magnetic Reasonance (NMR), etc. which revealed (a) the diversity of protein structure and, (b) interprotein interactions which are not taken into account in the unit membrane model. It has been established that the proteins are arranged asymmetrically in the membranes.

Additional Electron Microscopic structural informations were obtained by *Freeze-Fracture Technique*. This is a valuable technique to ascertain whether proteins are located in the interior of biological membranes. This provided the evidence that substantial amounts of protein are deeply embedded in many membranes. Cells or membrane fragments are rapidly frozen at liquid nitrogen temperatures. The frozen membrane is then fractured by the microtome knife. Cleavage usually occurs along a plane in the middle of the bilayer. Extensive regions within the lipid bilayer are exposed. These exposed regions can be shadowed with carbon and platinum which yields a replica of the interior of the bilayer.

The external surfaces of the membranes can also be observed by "*Freeze-Fracture*" and "*Deep-etching*" techniques. Initially, the interior of a frozen membrane is exposed by fracturing. The ice that coveres one of the adjacent membrane surfaces is then sublimed away; this process is known as "Deep-etching". The combined technique, called *Freeze-Etching Electron Microscopy* provides a view of the interior of a membrane and of both its surfaces, and this technique does not require fixatives and dehydrating agents.

III. **Fluid-Mosaic Model (Singer and Nicolson, 1972)**: After studying membrane with different techniques, several modification of the

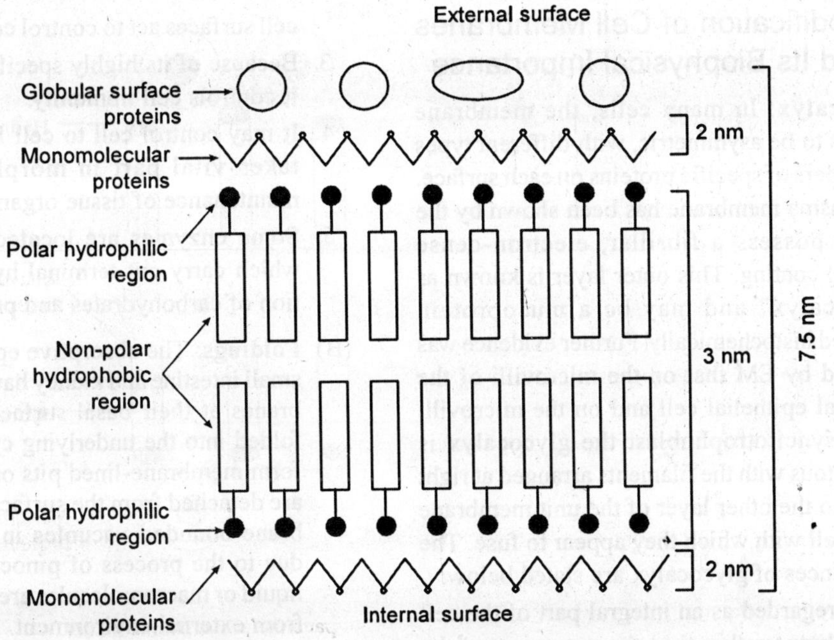

Fig. 20.7: Robertson's unit membrane model

unit membrane structure appears to be the *Fluid-Mosaic Model* postulated by Singer and Nicolson in (1972) (Fig. 20.8).

The model postulates:

1. The phospholipids of unit membranes are arranged in a bilayer to form a fluid, liquid crystalline matrix or core.
2. In this bilayer individual lipid molecules can move laterally, endowing the bilayer with fluidity, flexibility and a characteristically high electrical resistance and relative impermeability to highly polar molecules.
3. This model postulates that the membrane proteins are globular, to account for their high content of α-helix.
4. Some of the proteins are embedded in the membrane, penetrating into the lipid phase from either side, and others completely span the membrane.
5. The extent of penetration of globular protein into the lipid phase is determined by the amino acid sequence of the protein and the location on its surface of the nonpolar amino acid R groups.
6. Thus the various membrane proteins form a mosaic-like structure in the fluid phospholipid bilayer.
7. This mosaic is not fixed or static, since the proteins are free to diffuse laterally in two dimensions, at least in some membranes.
8. The relative viscosity of the lipid bilayer is about 100-1000 times than that of water. The fluid-mosaic model fulfills many features and properties of biological membranes.

(a) It provides for membranes with widely different protein content, depending on the number of different protein molecules per unit area of membrane.

(b) It provides for the varying thickness of different types of membranes.

(c) It can account for the asymmetry of natural membranes, since it permits proteins of different types to be arranged on the two surfaces of the lipid bilayer.

(d) It accounts for the electrical properties and permeability of membranes.

(e) It also accounts for the observation that some protein components of cell membranes move in the plane of the membrane at a rather high rate.

20.4 Modification of Cell Membranes and Its Biophysical Importance

(A) **Glycocalyx**: In many cells, the membrane appears to be asymmetric, with different types or numbers of specific proteins on each surface. The plasma membrane has been shown by the EM to possess a fibrillar, electron-dense external coating. This outer layer is known as "glycocalyx" and may be a mucoprotein observed histochemically. Further evidence was obtained by EM that on the microvilli of the intestinal epithelial cell and on the microvilli of the syncitotrophoblast the glycocalyx is filamentous with the filaments arranged at right angles to the other layer of the unit membrane of the cell with which they appear to fuse. The importances of glycocalyx are stated below:

1. It is regarded as an integral part of the cell membrane rather than the outer coat of the cell and plays an important role in cell life..
2. It is suggested that the glycocalyx on free cell surfaces act to control cell permeability.
3. Because of its highly specific composition, it controls cell immunity.
4. It may control cell to cell interactions and takes vital part in morphogenesis and maintenance of tissue organization.
5. Some enzymes are located in the region which carry out terminal hydrolytic digestion of carbohydrates and proteins.

(B) **Foldings**: The absorptive epithelial cells of small intestine and kidney have plasma membranes at their basal surfaces and become folded into the underlying cytoplasm. They form membrane-lined pits or tubules which are detached from the surface to form membrane-bounded vacuoles in the cytoplasm due to the process of pinocytosis whereby liquid or macromolecules are taken into cells from external environment. Small fat droplets may enter into intestinal epithelia by pinocytosis.

(C) **Interdigitation between the Plasma Mem-**

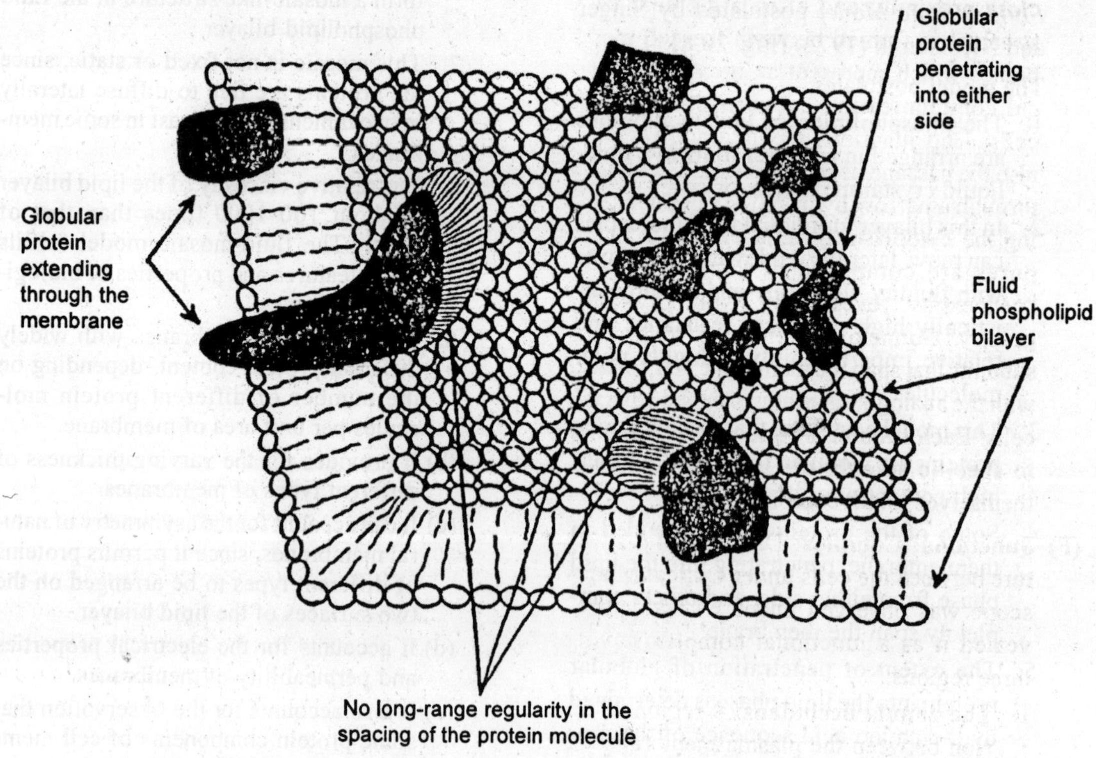

Fig. 20.8: The fluid-mosaic model of membrane structure

branes: In certain places, where cells are grouped together into sheets (e.g., in epithelia) due to interdigitation of plasma membrane between the adjacent cells which increases the surface of contact as well as subserving a mechanical function.

(D) **Desmosomes**: With the help of EM, some differentiated areas are seen in the plasma membrane between adjacent cells which are known as "desmosomes" or "**maculer adhaerens**". These represent a functional adaptation for the maintenance of a high degree of strength and structural integrity in the cells and probably enhancement of intracellular transport mechanism.

(E) **Nexus or Gap Junction**: By the addition of Electron-Opaque Marker substances to the fixative fluids used in Electron Microscopy, another type of specialized cell to cell contact has been observed which is known as a "**nexus or gap junction**", where the plasma membranes run parallel to one another in close proximity and often a globular substructure is formed between the cell membranes which represent an area for electrical communication between the adjacent cells. Gap junctions mediate and regulate also the passage of ions and small molecules through a narrow hydrophilic pore connecting the cytoplasm of adjacent cells. These pores are composed of subunits called "connexons" revealed by X-ray Crystallography. Connexons consist of 6 protein subunits that span the membrane and connect with the analogous structures on the adjacent cells. Each subunit is rigid and in response to specific stimuli, the subunits rearrange themselves relative to one another.

(F) **Junctional Complex**: Dark staining structure between the cells under Optical Microscope was observed under EM which revealed it as a junctional complex having three regions:

1. **The zonula occludens**: A region of fusion between the plasma membranes of adjacent cells.
2. **The zona adhaerens**: Regions of close apposition of the plasma membranes.
3. **The macula adhaerens or desmosome:** The junctional complex is important for maintaining the integrity of the surface layer or epithelium and in preventing any substances passing diffusion from the lumen into the tissues via the intercellular space. Recent opinion is that with nexus, it is an important means for ensuring rapid conduction of excitation from cell to cell.

(G) **Pores**: Pores in the plasma membrane are present at regular intervals. The substances which are lipid insoluble can still penetrate the membrane through the pores. The pores are believed to be lined with polar hydrophilic groups, e.g. polar groups of the protein molecules and are, therefore, penetrated readily by water-soluble substances. They are less than 1.0 nm and not visible by EM and their existence has been supported by the experimental evidences (Fig. 20.9).

(H) **Microvilli**: In some cells the plasma membrane is folded to form minute, hair-like structures which project (evagination) from the surface in the form of finger-like extensions called **microvilli**. These are 1.0 μm long and 0.08 μm wide. It is lined with plasma membranes and filled with cytoplasm which is continuous with the main body are seen under EM. The individual villi are not

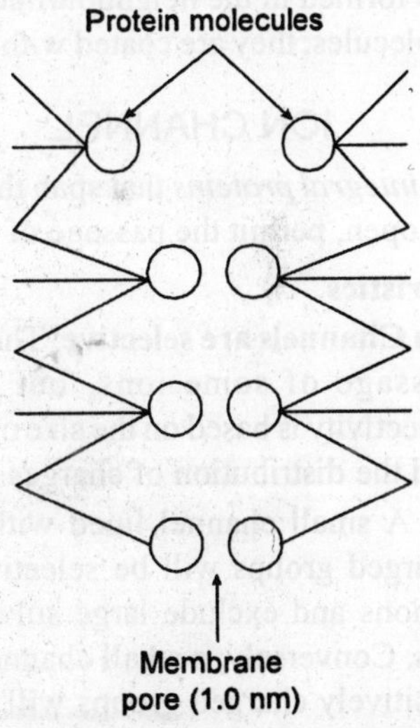

Fig. 20.9: Schematic representation of membrane pores

separately resolvable by the Light Microscope, but in mass they constitute the so-called **brush or striated** borders of intestinal epithelial cells and the cells of the proximal convoluted tubules of the kidney. It is the device for increasing the surface area thereof aiding absorption of materials from the enclosed cavity. In thyroid follicle cells and certain others, the length and frequency of microvilli appear to vary with the activity of the cell.

(I) **Invagination**: In striated muscle, there is an invagination of the cell membrane, the transverse tubule, associated with each sarcomere. Since the transverse tubules are continuous with the surface membrane, they provide a direct route for the communication of changes at the cell surface to the contractile system deep within the muscle fibre.

(J) **Membrane Channels**: There are channels in membranes through which small ions and molecules are transported. These channels are formed in the neighbourhood of protein molecules; they are coated with polar group.

ION CHANNEL

These are *integral proteins* that span the membrane and when open, permit the passage of certain ions.

Characteristics

1. **Ion Channels are selective**: They permit the passage of some ions, but not others. Selectivity is based on the size of the channel and the distribution of charges that line it.

   A small channel lined with negatively charged groups will be selective for small cations and exclude large solutes and anions. Conversely; a small channel lined with positively charged groups will be selective for small anions and exclude large solutes and cations.
2. **Ion channels may be open or close:** When the channel is open, the ion (s) for which it is selective can flow through. When the channel is closed, ions cannot flow through it.
3. **Conductance:** Conductance of a channel depends on the probability that the channel is open. The highest the probability that a channel is open, the **higher the conductance, or permeability**. Opening and closing of channels are controlled by **gates.**
   (a) **Voltage-gated channels** are opened or closed by **changes in membrane potential**. Such as, the **activation gate of the Na^+ channel** in nerve is opened by depolarization; when open, the nerve membrane is permeable to Na^+ (e.g. during the upstroke of the nerve action potential). The **inactivation gate of the Na^+ channel** in nerve is closed by depolarization; when closed, the nerve membrane is impermeable to Na^+ (e.g., during repolarization phase of the nerve action potential).

Advantages

1. Current of a few picoamp of single ion channel can be recorded directly.
2. Recording is 100 terms magnified.
3. Activity of individual molecules can be watched which is difficult.
4. Technique showed that channels are specific to particular ions, e.g., calcium or chloride ions.
5. It shows that the channels can be either open or closed.
6. This method can get rid of membrane "noise" by most of the cell membranes.

Application

1. To investigate the mechenisms of the nicotine acetyl receptors.
2. Study of fusion of secretary vesicles with the cell membrane in most cells.
3. Study of neurosecretion.
4. Development of synopsis.
5. Study of the relation between protein structure of ion channels, and their function.
6. Study of nature of synaptic transmission in the brain.
7. To investigate new drugs.
8. To investigate normal physiological and pathological processes.
9. To study of insulin secretion.
10. To understand cellular mechanisms underlying diseases like diabetes and cystic fibrosis.

11. The method opened the way to develop new and more specific drugs.
12. To study the effect on the channel of different transmitter.
13. To study the mechanisms of blocking certain druges and anaesthetics on the channels.
14. For electro-physical study of all types of cells which were not studid previously.
15. For better understanding of the propagation of the diseases.

20.5 Nuclear Membrane or Envelope

The nucleus is surrounded by a membrane called nuclear membrane or envelope. This envelope is a double membrane complex and each membrane is about 7-8 nm thick. This envelope, a flattened sac which is separated by (distance 150 Å) (space between the two folds) called **perinuclear space or perinuclear cisterns** resembles the rough endoplasmic reticulum.

Because: (a) The outer surface of the outer nuclear membrane (cytoplasmic) has granules which resemble ribosome, (b) Direct continuities has been observed between the outer nuclear membrane and the EF and, (c) Presence of certain enzymes in both the perinuclear space and the cisternae of the ER.

Each membrane is almost indentical to the cell membrane, having a basic lipid bilayer structure with globular proteins floating on the lipid fluid.

The inner part of the nuclear membrane is often associated with chromatin and an "internal dense lamella" and the latter provides some rigidity to the structure. The inner and outer membranes join at certain intervals to form **pores**, 50-80 nm in diameter. These **pores are not holes** through the membrane; instead, they are areas where the two membranes fuse with each other. The pores are often referred to as **annuli** to indicate that it is not a simple hole. Often pores are seen with a **diaphragm or plug**. The transnuclear transport (e.g., nuclear RNA gains access to the cytoplasm) occurs either across membranes of the perinuclear sac or through pores. The transport through the pores is supported by the facts:

1. The movement of electron dense material which may be RNA-containing granules through the nuclear pores in the insect salivary gland and some other tissues and,
2. The movement of a marker, e.g., colloidal gold into the nucleus when it is injected into the cytoplasm of amoeba. However, the fused membrane is so permeable in these pore areas that almost all dissolved or suspended substances, including even very large, newly formed ribosomes, can move between nucleoplasm and cytoplasm.

20.5.1 Membrane having trilaminar structure similar to cell membrane

(A) **Plant**

1. Membrane surrounding the sap vacuoles of plant.
2. Membrane surrounding plastids of plant.

(B) **Animal**

1. Membrane of the endoplasmic reticulum.
2. Nuclear membrane.
3. Membrane surrounding mitochondria.
4. Other organelle membranes.

20.5.2 Distribution of single and double membrane

(A) Structures having a **single layer** unit membrane: Membranes of cells, lysome, endoplasmic reticulum, golgi apparatus, microbodies.

(B) Structures having **two layers** unit membrane: nucleus, mitochondria.

(C) Structures having **multilayer membrane**, e.g., myelin sheath (up to 250 layers).

20.6 Nuclear Membrane vs. Cell Membrane

Table 20.3: Comparison of cell and nuclear membrane

Nuclear membrane or envelope	Cell membrane (plasma membrane)
1. Nuclear membranes are double membranes separated by a space having thickness of 7-8 nm each.	1. It is a single membrane having a thickness of 7.5-9.0 nm.
2. The space between the two membranes are called perinuclear space of 150 Å wide.	2. No such space.

Contd.

Nuclear membrane or envelope	Cell membrane (plasma membrane)
3. The outer membrane is continuous with rough endoplasmic reticulum.	3. No.
4. Two membranes fuse at certain intervals to make pores.	4. Fusion may occur with the membrane of the adjacent cells to form junctional structure.
5. There are pores, 50-80 nm in diameter and they are not holes but have diaphragm or plug.	5. There are pores without diaphragm, or plug.
6. The fused membrane is so permeable in these pores that all dissolved or suspended particles, even very large new formed ribosomes can move between nucleoplasm and cytoplasm.	6. Transport of ions, water and food material occur through pores, channels, gates, pumps, etc.
7. The presence of pores causes lack of osmotic activity and low electrical resistance of many nuclei.	7. Maintains osmotic activity and high electrical resistance.
8. It is not semipermeable membrane and the permeability regulated by the pore diaphragm.	8. It is semipermeable membrane.

20.7 Cell Membrane vs. Cell Wall and Thylakoid Membrane (Plant)

Table 20.4: Cell membrane (animal) and Cell wall and membrane (plant)

Cell membrane (animal)	Cell wall and membrane (plant)
1. Thin, elastic.	1. Thick, rigid.
2. Living.	2. Non-living.
3. Composed of lipids and proteins.	3. Cellulose.
4. Semipermeable.	4. Permeable.
5. Does not give mechanical support.	5. Gives mechanical support.
6. Phagocytosis and pinocytosis occur.	6. Does not occur.
7. Generation of electrical potential. (Biopotential).	7. No.
8. Propagation of nerve impulse is a membrane process.	8. No.
9. Reception—whether by touch, smell, taste, hearing or vision—occurs with the invariable participation of the membranes.	9. Some receptions like touch, light.
10. Provides an active transport of matter moving against the opposing chemical or electrochemical potential gradient.	10. Provides osmosis and diffusion.
11. Serve as the site of the main bioenergetic process—oxidative phosphorylation.	11. Bioenergetic processes—photosynthesis in the thylakoid membranes.
12. Synthesis of ATP in the mitochondiral membrane.	12. Synthesis of ATP in the thylakoid membranes of the chloroplast of green plants.
13. Insulation—The membranes of Schwann cell form a myelin sheath around axon and act as an insulator.	13. No.

20.8 Function of Cell Membrane

1. **Shape**: It provides external shape of the cytoplasm.
2. **Boundary**: It makes boundary between the cells and environment.
3. **Mechanical Support**: It checks the cytoplasm from flowing out.
4. **Protection**: It saves the cytoplasm from external damage.
5. **Compartmentation**: The appearance of a

cell membrane was evidently important in the origin of life—the compartmentation, the isolation of intracellular space from the external environment, one of the early stages of prebiological and biological evolution.

6. **Transportation**: In open nonequilibrium system biosynthesis and bioenergetic are interrelated which are linked with transportation of the materials from the surrounding by the semipermeable partition.
7. **Semipermeability**: Cell membrane is a semipermeable membrane, e.g., it exhibits selective permeability or permits the passage of selective materials.
8. **Biophysical Processes for Movement of Materials**: Various biophysical phenomena occur through membrane for passage of materials into the cell and out of the cell, e.g., diffusion, facilitated diffusion, nonionic diffusion, solvent drag or bulk flow, osmosis, filtration, active transport, pinocytosis, exocytosis, phagocytosis, secretion, active transport, absorption, adsorption, Donnan-membrane equilibria, dialysis, vesicular transport (micropinocytosis for large molecules), ultrafiltration (glomeruli, capillary).
9. **Electrical Properties (Biopotential)**: It can generate electrical potential (biopotential), e.g., electromotive forces.
10. **Propagation**: Propagation of nerve impulse is the membranal function (axon membrane).
11. **Insulation**: The membranes of Schwann cells form a myelin sheath when covers myelinated axon. This sheath is interrupted at regular intervals (every 1-2 nm of length of the axon) at the nodes of Ranvier where the axon membrane comes into contact with the environment which is 1μm long. This myelin sheath consists of a multiple helical wrapping of the cell membrane of Schwann cells around the axon. Thus, it is multilayer (up to 250 layers) membrane and molecular arrangement of each layers turns out similar to that of the cell membrane studied by Polarized Light, Biochemical Analysis, X-ray Diffraction and High Resolution Electron Micrograph.

    The myelin sheath acts as:

    (a) Insulator—Prevents the conduction of nerve impulse from one axon to another and almost all the flow of ions.

    It increases the resistance to ion flow through the membrane approximately 5000 fold and also decreases the membrane capacitance as much as 1000 fold.

    (b) Accelerator of fibre conduction—such as in very small unmyelinated fibre, the velocity of conduction is 0.5 metre per second up to as high as 130 metres per second in very large myelinated fibres.
12. **Molecular Reception, Visual and Mechanical Reception**: Various receptor cells have fine, hair-like projections-antennae whose plasma membranes contain specific sites, e.g., smell, taste.

    The visual reception occurs by single protein rhodopsin present in the rod membrane. The mechanical reception occurs by various receptors in the skin, muscles, etc.
13. **Membrane Receptors**: The membrane of the effector cell contains specific sites for specific substances, e.g., enzymes, substrates, signals—called hormones, for acting receptor. They can act in two ways:

    (a) Combining with receptor substances,

    (b) Activating the enzymes in the cell membrane.
14. **Energy Transduction**: It also provides site for energy transduction such as oxidative phosphorylation and photosynthesis (plant).
15. **Maintenance of Equilibrium**: Due to selective permeabilities provided by (a) Gates; (b) Pumps, e.g., Na^+, K^+, Ca^+; (c) Channels, etc., it maintains equilibrium inside the cell.

20.9 Physico-Chemical Properties of the Membrane

Living vs. Non-living Membranes

Lifeless membranes like parchment and collodion are permeable due to the presence of pores through which materials can pass. Another membrane of rubber is permeable to benzene and allied materials which passes through the membrane by solution in it.

But the cell membrane is a living one and the protoplasm is bounded by it and has the properties which are lacking in lifeless membranes.

- **Composition Differs from Protoplasm**

The membrane of the cell is formed from the constituents of the protoplasm (proteins, phospholipids, cholesterol, some neutral fat, free fatty acids, soaps, carbohydrates, various inorganic ions, etc.) in different proportions than the protoplasm. The explanation of the difference in composition is that the cell membrane represents an interface, according to the principle of Gibbs-Thomson. The substances which lower surface tension tend to concentrate at the interface. Hence, the membranes should contain more phospholipids, soaps, cholesterol and other lipids than the interior protoplasm.

1. **Molecular Orientation**: According to the principle of molecular orientation, the various molecules and ions in the surface of cell membrane more or less oriented producing characteristic type of structure which significantly influences the permeability.
2. **Molecular Structure**: The cell membranes are generally composed of organized and oriented bilayer of lipid sandwitched between two layers of proteins. The protein layers and lipid layers are held together by forces of adsorption and by formation of complex between them. The lipid bilayer is about 3.5 nm thick and the protein layer is 2 nm thick and the whole membrane is 7.5-9.0 nm thick.
3. **Pores in the Membrane**: There are holes or pores in the cell membranes through which small moleculles, water and certain ions may pass. The walls of the pores are usually negatively charged and as a result, the pores may exhibit the phenomenon of streaming or flow potential and electroen-dosmosis.
4. **Electric Charge**: The membrane surface is having an electric charge due to selective adsorption of ions and the ionization of compounds in the membrane surface. Consequently, the membrane possesses a Helmholtz double ionic layer.
5. **Potential Difference**: There is a potential difference between the outside and inside of the cell membrane which is about +0.1 volt. This potential difference is due to unequal distribution of K^+ ions in the intracellular and extracellular fluids. Cell membrane is freely permeable to K^+ and its permeability is 50-100 times greater than Na^+ permeability. There is generally 30-40 times more K^+ in the intracellular fluid than the extracellular fluid. According to the Nernst equation for a concentration cell, the potential is close to +0.1 volt at 37°C:

$$E = 0.061 \log C_2/C_1 = 0.061 \log 40/1$$
$$= 0.061 \times 1.6 = 0.097 \text{ volt},$$

where $C =$ the concentration of K^+ in the intracellular fluid.

$C_1 =$ the concentration in extracellular fluid.

Therefore, it can be concluded that the potential across the membrane is due to the maintenance of unequal distribution of K^+ between the two sides of the membrane.

6. **Potential Drop**: There is a potential drop across the membrane from the external surface to the protoplasmic layer inside. Flexner and Stiethler found that there was a potential drop of 0.23 volt between the epithelium and stroma of the secreting choroid plexus of the fetal pig. The potential in the epithelium was + 0.100 volt, and in the stroma –0.130 volt under aerobic condition. During anaerobiosis, the potential difference disappeared and both have a potential of about – 0.290 volt. The cellular oxidation processes play a significant role in regulating the membrane potential.

- **Other Electrical Properties**

(a) The capacitance per unit area of the cell membrane is about 1μ F/cm^2.

(b) The cell membrane can be considered to be the dielectric material of a charged condenser.

(c) The cell membrane material considered as an electric insulator, having a very high dielectric strength, i.e., 8,000000 V/m.

20.9.1 Physical driving force for transport of materials

Physical forces to drive the ions and molecules through the membrane are those arising from differences in concentration and activity coefficients of substances on the opposite side of the membrane. The electrical potential difference between the two sides of the membrane and solvent drag by which the materials are carried along with the flow of solvent through the membrane; which may arise from differences in osmotic pressure, electro-endosmosis, etc.

1. **Permeability**: The permeability of the membrane depends upon the size of the molecules, size of the pores or channels, lipid solubility, electric charges on the cell surface. The cell membranes are semipermeable, i.e., certain substances are allowed to pass through and some are prevented.
2. **Chemical Changes**: During the passage through membranes, chemical changes may occur. Galactose may be converted to glucose when it passes through intestinal membranes and sucrose splits into glucose + fructose when it passes through the membrane.
3. **Active Transport**: Membranes have the property to pass substances through it from a lower to a higher concentration, that is against a concentration gradient at the expense of energy which is referred to as "active transport", e.g., glucose and NaCl absorption from the intestine.
4. **Sodium Pump**: Cell membranes are permeable to both K^+ and Na^+. Intracellular K^+ is 40 times greater than the extracellular fluid while Na^+ is one-twelfth in the intracellular fluid than that of the extracellular. The distribution of the intracellular and extracellular K^+ is in accord with the Donnan membrane equilibrium principle and the potential across the membrane. But the distribution of Na^+ is opposite to the Donnan effect. Though Na^+ enters the cell freely but its intracellular concentration is kept low due to active transport of the Na^+ from the cell to the extracellular fluid across the membrane, i.e., extrusion of Na^+ from cell to maintain proper ionic balance and sodium pump operates, though there is no conclusive evidence of constituents of sodium pump and nature of pump substance or carrier. The operation of pumping requires expenditure of energy.
5. **Membrane Receptor**: The membranes of the effector cells contain specific sites for specific substances for acting called **membrane receptor**. The substances act by combination with the receptor in the cell membrane. The receptor in most instances is in the cell membrane and is a **protein molecule**.
6. **Viscosity**: A close value (0.1-0.4P) by the visocity of the cell membrane of mammals are determined from translational diffusion and also by the membranes of mitochondria and nerve axons (axolemma). Thus the membrane viscosity is two or three times higher than the viscosity of water and corresponds to the viscosity of vegetable oil.
7. **Floating of Globular Proteins**: The globular proteins in the membranes are "floating" in an oily layer of lipid. The protein molecules are embedded on both sides of the membrane to a varying degree in the double bilayer of mobile hydrocarbon tails of the lipid. The others completely span the membrane. A significant part of the membrane surface is free from proteins, e.g., in RBC membrane, 70 per cent of the surface and in microsomal membrane, 80 per cent of the surface are occupied by the proteins.
8. **Channels (Pores)**: There are channels in the membrane through which small ions and molecules are transported. They are formed in the neighbourhood of protein molecules and coated with polar hydrophilic group. Recent evidence suggests that pores may be hydrated channels between the glycoprotein chains of the membrane which are tortuous and might not be stable anatomical structures, which indicates the failure of visualization under EM.

20.10 Conformational Properties of Membranes (Dynamical Properties)

Facts about High Mobility of Lipid Bilayer

The lipid bilayer of the membrane shows *high mobility* and lipids in the membrane behave like *liquid crystal*. When a liquid crystal combines with fluidity and lability, it makes possible for membranes to carry out the important functions. The liquid-crystalline properties of membranes are determined by the fact that the lipids present in the membranes are in a molten state at the physiological temperature. The melting temperature of a hydrocarbon is low when there is larger number of double bonds present in it. The lipids having double bonds in the hydrocarbon chains melt at lower temperature than the physiological temperature and the percentage of such lipids is high in the plasma membrane of mammals. Many facts have proved the liquid properties of plasma membranes. The mobile property of the membrane structure has been revealed by Paramagnetic and Fluorescent Labels and Nuclear Magnetic Resonance.

The liquid-crystalline properties of photoreceptor membranes of rods containing one molecule of protein rhodopsin per 60-90 molecules of lipids of which 80 per cent contain an unsaturated fatty acid.

The Flash Photometric Technique has revealed that rhodopsin molecule rotates rapidly about the axis perpendicular to the membrane plane. The time of rotational diffusion is 20 μs at 20°C. The Microspectrophotometric has revealed that the bleaching rhodopsin in light undergoes translational lateral diffusion in the membrane.

20.10.1 Facts about conformational transitions

1. **Structural Changes**: Structural changes have been observed with the help of Fluorescent, Paramagnetic Labels, Measurement of Birefringence, Light Scattering and Circular Dichroism.
2. **Phase Transition–The Melting of Lipids**: Phase transition occurs near 0°C, when the mitochondrial and microsomal membranes are heated from −40°C. The outer monomolecular lipid layer shows transitions at 15°C and 31°C and inner monomolecular lipid layer at 21°C and 37°C.

The melting of lipids occurs through the rotational isomerization of hydrocarbon chains—which is conformational melting. The hydrocarbon tails of the lipids contains saturated hydrocarbons crystallize (like paraffins) in the form of continuous series of "transrotamers". Besides the melting gives rise to "gauche-rotamers", whose fracton is about 10 per cent in the hydrocarbon tails.

Trauble (1971) advanced the theory that the passage of molecules through the lipid membrane is *due to rotamerization* of hydrocarbon chains.

20.10.2 Role of conformational changes

1. **Interaction with Ligands**: Interactions of membranes with various ligands, which are essential to physiology and pharmacology.
2. **Amplifying Properties—Trigger**: The bacteria—*Escherichia coli* is colicinogenic. It produces a macromolecular antibiotic "Colicins" which can destroy the other sensitive starins of *E. coli* in a very minute quantity. This reveals that sensitive membrane of a sensitive cells has amplifying properties, i.e., reception of one molecule serves as a "trigger" for magnified events.
3. **Membrane Receptor action:** In the case of the membrane receptors, probably the trigger process acts similarly.
4. **Effect of Drugs:** Many drugs have an effect on the conformations of membranes and membrane lipids. Changeux et al. considered the triggering properties of the membranes which was based on the theory analogous to the theory of indirect cooperativity of enzymes developed by Manod, Wymax and Chanywx.

Receptor site of each subunit is for a specific ligand (biological property of some proteins, namely, their capacity for specific, noncovalent binding of another molcule, called the ligand), the affinity for which it is changed when there is alteration or conformation of it. The subunits or protomers interact with the neighbours and give rise to cooperative properties. Depending on the ligand activity and interaction energy of protomers, the membrane response to ligand may be gradual or sigmoidal becoming a phase transition.

This provides explanation for the action of Colicins for the following facts:

(a) Various related drugs cause different maximum responses of the membranes.
(b) Primary action of many drugs is localized in membranes and is cooperative.
(c) Many drugs are effective in very small amount and are highly specific.
(d) The effect of a drug on the membrane receptor is determined by molecular recognition but the knowledge about the exact nature of the receptor is lacking.

Membrane receptors are known to be proteins as rhodopsin in the membrane of rods, change their conformation when bound to the ligand, resulting to a change in depth of immersion and mobility of proteins in the lipid fluid. The factor responsible for cooperativity may be the interaction of floating proteins upon their collision. The fluid mosaic level may provide the basis for the molecular physics of the membranes.

The properties of the membranes are largely determined by **electronic conformational interactions** (ECI). [ECI—The transfer of electrons in molecular systems is accompanied by the movement of atomic nuclei. In biopolymeric systems these are ECI. The movement of electrons and charged densities in a biopolymeric systems in particular in a membrane coupled with conformational motions, may be regarded as the movement of a quasiparticle called *conformons*, consisting of carriers of electronic charge or as an electron density shift and conformational displacement of an electron or electron density in macromolecule causes a conformational rearrangement which may be treated as the excitation of *longwave phonons* and the system electron added to the deformation of the macromolecules is considered to be similar to the "polaron" and such system is called the "conformon".]

The local shift of electron density arising from the interaction of the membrane receptor with the ligand resulting in conformational changes in the biological molecules. The transport of electrons and ions across the membrane may be considered as the propagation of "conformons"—quasiparticles consisting of carriers of electronic charge and as an electron-density shift and conformational displacement of the surrounding medium.

20.11 Permeability of Membrane and Factors Affecting It

The permeability is defined as the rate of transport through the membrane for a given concentration difference. When dissolved particles can freely pass through the cell membranes without any hindrance, then the membrane is called *permeable* as we see in case of diffusion and filtration. In other cases the particles are not allowed to pass due to hindrance, the membrane is called *impermeable*. When a membrane allows certain materials to pass and hold back others, the phenomenon is known as *selective permeability* and the membrane is a called *semipermeable membrane.*

In case of nonliving membranes, parchment and collodion are permeable due to the presence of pores through which the materials can pass. Another membrane of rubber is permeable to benzene which passes through the membrane by solution in it. But the living cell membrane which surrounds the protoplasm has some properties which are absent in the non-living membrane. By virtue of these properties, the cell membrane permits the passage of materials.

1. Cell membranes are impermeable to intracellular protein and other organic anions and are usually represented by the symbol A^-.

   Though the cell membrane is not permeable to proteins, it is not true in case of certain specific proteins and certain type of cells, e.g.

   (a) Absorption of certain protein causing allergy to some persons,
   (b) Leaking of albumin through glomerular membrane,
   (c) Leaking of plasma proteins through capillary wall,
   (d) Maternal protein antibodies permeable to placental membrane,

(e) Access of foreign protein antigen to antibody producing cells,

(f) Secretion of digestive protein enzymes in the digestive tract,

(g) Liver cells take up protein and release proteins to blood and lymph,

(h) Penetration of enzyme ribonuclease and destruction of RNA.

2. O_2 and CO_2 are freely permeable.
3. Cl^- and K^+ are freely permeable. K^+ permeability is 50-100 times greater than the Na^+ permeability.
4. Na^+ is moderately permeable to cell membrane.
5. Cell membranes are more permeable to substances soluble in fat than they are to substances which are not fat-soluble, because the cell membranes contain a large proportion of lipid materials.
6. The surfaces of cell membranes and pore walls of the membranes are electrically charged, so the neutral or uncharged molecules pass through better than the ions of electrolytes. Thus, weakly dissociated organic substances pass into and out of cells more readily than their salts which are highly ionized.
7. The intestine absorbs Na^+, K^+ and Cl^- ions much readily than Ca^{++}, Mg^{++} and SO_4^-.
8. Small molecules pass through living membranes than the larger molecules of the same general type.
9. A large molecule can pass readily than a smaller molecule of different types of chemical.
10. Cholesterol molecules are little absorbed in the intestine but it is readily absorbed when combine with bile salts to form a complex, which is larger than cholesterol molecules.
11. The intestinal membrane absorbs Fe^{++} readily than Fe^{+++}.
12. The intestine absorbs glucose readily than the smaller xylose molecule.

20.11.1 Factors affecting permeability

1. **Size of the Particles**: The size of the particles may affect the permeability. There are very small pores in the semipermeable membrane which is large enough to permit water but too small to permit larger particls above a certain size. For example, the protein, which is the largest molecule, cannot pass through the membrane while the amino acid, which is a smaller molecule, can pass easily.

The size of the particle affects the ion permeability of the cell membrane (Table 20.5). In the body the ions are hydrated. For example, the atomic weight of K^+ is 39 while atomic weight of Na^+ is 23. But the hydrated Na^+ is larger than the hydrated K^+. These ions cross the membranes through the "ion channels" than the simple pores. These channels are formed through protein molecules and charge configurations around them and make them specific for particular ions, e.g., separate channels for Na^+, K^+ and Cl^-. The passage of ions are controlled or "gated" by voltage or by agents like neurotransmitters. The **gated ion channels** are of two types:

(a) **Voltage-gated ion channels**: In excitable tissue like muscle and nerve, passage of Na^+ through the Na^+ channels is increased by a decrease in membrane potential.

(b) **Chemically-gated ion channels**: At synaptic junction the ion channels are chemically gated which means capacity of channels to pass ions is increased by the binding of a given neurotransmitter or hormone to receptors associated with them.

Table 20.5: Relationship of diameter of substances, pore diameter and relative permeabilities.

Substances	Diametre Å	Ratio to pore diameter	Approximate relative permeability
Water molecule	3.0	0.38	50000000
Urea molecule	3.60	0.45	1500000
Chloride ion (Hydrated) (RBC)	3.86	0.48	500000
Potassium ion (Hydrated) (RBC)	3.96	0.49	1.1
Sodium ion (Hydrated) (RBC)	5.12	0.64	1.0
Glucose	8.60	1.04	0.4

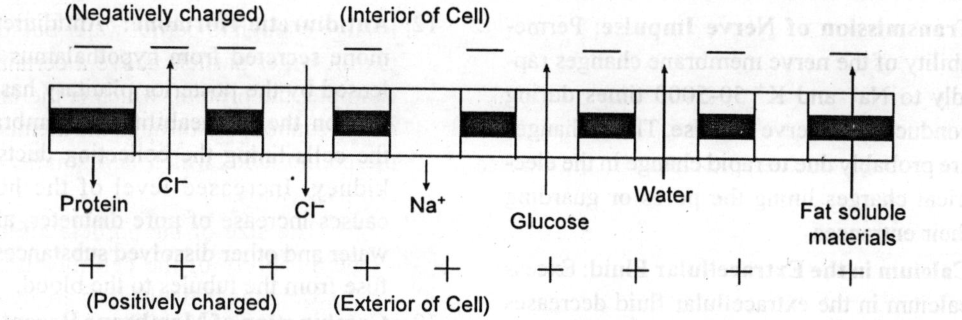

Fig. 20.10: Diagram showing the permeability of membrane

2. **Fat Solubility**: Another factor is the solubility of the molecule in fat. Those molecules are readily soluble in fat and they enter or leave cells more readily. A large fat-soluble molecule may pass through the membrane more readily and quickly than a smaller molecule which is not soluble in fat. The cell membrane consists of bilayer of lipids. The fat-soluble molecules enter the cell by first dissolving in the fatty portions of the membrane.

   Such fat-soluble molecules pass in excess. They may dissolve out sufficient fats of the cell membrane to disrupt the cell as it happens in case of general anaesthetics and narcotics, e.g., chloroform and ether (Fig. 20.10).

3. **Electrical Charge**: Electrical charges on the surface also determine the permeability of membrane to ions which also possess electrical charges. Cell membrane with a positively charged surface repels positively charged particles and thereby prevents their passage, while negatively charged particles or uncharged particles even larger than ions might readily enter or leave the cell. Migration of negatively charged ions into the cell could not occur unless there was simultaneous migration of equal numbers of negatively charged particles. Three types of electrical forces affect the ion permeability:
   (a) The intensity of electrical charges lining the ionic channel,
   (b) The hydration energy of ions,
   (c) The difference of potential across the membrane.

4. **Orientation of Molecules and Ions in the Membrane**: According to the principle of molecular orientation, the various molecules and ions in the surface of the membrane must be more or less oriented, leading to a characteristic type of structure and influencing the permeability of the membrane.

5. **Cellular Oxidation**: Flexner and Stiehler, Cattell and Civin have proved that cellular oxidation processes play an important role in regulating the permeability, structure and functioning of the cell membrane.
   (a) Flexner and Stiehler observed potential drop of 0.23 volt between the epithelium and stroma of the choroid plexus of the foetal pig under aerobic conditions. Under anoerobiosis which prevents cellular oxidation, the potential difference disappears and both show a potential of −0.290 volt which reveals that cellular oxidation is essential for maintenance of potential difference.
   (b) Cattell and Civin observed that when a cat is asphyxiated for a certain time, the K^+ content of blood increases greatly due to the excess passage of K^+ ions from the cell to the blood across the cell membrane. After restoration of respiration, K^+ goes back to the cell and normal ionic distribution is restored.

6. **Stimulation**: Stimulation of membranes, especially in excitable tissue like muscles and nerves, cause increase in permeability of membranes and exchange of ions between insides cause depolarisation of the membrane.

7. **Transmission of Nerve Impulse**: Permeability of the nerve membrane changes rapidly to Na^+ and K^+ 50-5000 times during conduction of nerve impulse. These changes are probably due to rapid change in the electrical charges lining the pores or guarding their entrances.

8. **Calcium in the Extracellular Fluid**: Excess calcium in the extracellular fluid decreases the permeability and decreases excitability, and diminished calcium causes increased permeability of nerve membrane and increases excitability. Calcium ion normally decreases the permeability of the membrane to sodium. In case of low calcium (tetany), nerve excitability increases. Therefore, high calcium level is called "membrane stabilizing factor" or "stabilizer".

9. **Potassium in the Extracellular Fluid**: Low potassium ion concentration in the extracellular fluids acts as "stabilizer". K^+ increases the negativity of the resting membrane potential, acts as "stabilizer" and reduces membrane excitability. In familial periodic paralysis there is excessesive reduction of K^+ in the extracellular fluid causing paralysis and is reverted by intravenous administration of K^+.

10. **Local Anesthetics**: Another group of important stabilizers are local anesthetics, e.g., cocaine, procaine, tetracaine which act directly on the membrane, decreasing the permeability to sodium and also reduces membrane excitability. If the excitability reduces so low, the ratio of action potential strength to excitability threshold known as "safety factor" reduces below unity, then nerve impulse fails to pass through the anaesthetized area.

11. **Veratrine**: Veratrine acts directly on the membrane and causes increase in permeability to sodium. Consequently, the strength of stimulus to elicit a response reduces greatly or, sometimes, the nerve fibre becomes so excitable that spontaneous impulses generate without any external stimulus.

12. **Antidiuretic Hormone:** Antidiuretic hormone secreted from hypothalamus and released by the posterior pituitary has the effect on the permeability of membranes of the cells lining the collecting ducts of the kidney. Increased level of the hormone causes increase of pore diameter, allowing water and other dissolved substances to diffuse from the tubules to the blood.

13. **Combination of Membrane Receptor with Specific Substance:** The membrane receptor located on the outer surface of the cell membrane binds the specific substance which causes basic change in the molecular structure of the receptor compound. This structural change often alters the permeability of the cell membrane to various ions.

20.12 Membrane Receptor

(Signal Transmission)

20.12.1 Definition

The effector cells contain specific sites to bind specific substances to transmit signal or information, called **cell receptors.**

20.12.2 Type

These are of three types:

(a) Membrane receptor,

(b) Intracellular receptor (for steroid hormones),

(c) Nuclear receptor (for thyroid hormones).

20.12.3 Membrane receptor

The membranes of the effector cell contain specific sites, for specific biochemical signals such as hormones, neurotransmitters and immunoglobulins to transmit information through the membrane to the cytoplasm.

20.12.4 Mode of action

The specific substances react at the membrane receptors in two ways:

I. Combination with receptor substance in the cell membrane,

II. Activation of enzymes in the cell membranes.

I. **Combination with receptor substances:** The acetylcholine, norepinephrine and epinephrine stimulate the effector cell by first reacting with receptor substances in the membranes. The receptor in most instances are exposed to the outside of the cell membrane and is an *integral protein* molecule. These receptor protein molecules are *not static components* of the cell but their numbers increase or decrease in response to various stimuli and their properties change with changes in physiological conditions. When a hormone or neurotransmitter is present in excess, the number of active receptors usually decreases which is known as **down regulation**, while in case of deficiency, there is an increase in the number of active receptors called **up regulation**. The most likely mechanism for function of receptors is that the receptor *located asymmetrically* on the outer surface of the target cell membrane which *stereospecifically* binds the substance and this causes a basic change in the molecular structure of the receptor substance. The receptor is an integral part of the cell membrane for which this structural change alters the permeability of the cell membrane resulting in rapid influx of Na^+, Cl^- or Ca^{++} into the cell or to allow rapid efflux of K^+ ions out of the cell. These ionic changes usually alter the membrane potential, sometimes eliciting action potentials (e.g., smooth muscles) and at other times causing *electrotonic effects* on the cells (e.g., glandular cells) to elicit responses. The ions have also direct effect e.g., Ca^{++} ions to induce smooth muscle contraction.

II. **Activation of enzymes in the cell membrane**: As for example, the catecholamines like epinephrine increases the catalytic activity of Adenylate cyclse or Adenyl cyclase, *asymmetrically* located on the inside of the membrane.

The mechanism is that the stimulating hormone (first messenger) combined with a specific receptor on the surface of the target cell. The specificity of the receptor determines which hormone will affect the target cell. After hormone binding with the receptor, the hormone-receptor complex activates the enzyme Adenyl cyclase in the membrane. The portion of adenyl cyclase exposed to the cytoplasm causes rapid conversion of cytoplasmic ATP into cyclic-AMP or cRMP (second messenger) which elicits number of cellular responses before destruction, e.g.

(a) Alteration of cell permeability,
(b) Muscle contraction or relaxation,
(c) Activation of enzymes,
(d) Synthesis of proteins and
(e) Cellular secretion.

The type of effect that occurs inside the cell are determined by the character of the cell itself.

(a) **Cyclic guanosine Monophosphate or cyclic GMP**: Another intracellular mediator or second messenger is cyclic Guanosine Monophosphate or cyclic GMP which is exactly like cAMP except that it contains the base guanosine. It usually promotes intracellular reactions that are different from those promoted by cAMP.

(b) **Protein Kinase and Ca^{++} translocation**: Alpha receptor effect is not yet established. In some cases, they result in less cAMP producing effect and some are unrelated to cAMP but act directly at the level of Ca^{++} translocation in the cell. It leads to activation of protein kinase and resulting in metabolic effect.

Rasmussen (1978) suggested that Ca^{++} in cytosol fraction is the important signal which may come from extracellular fluid or may be mobilized from intracellular tissue-bound calcium, i.e., **Calmodulin**.

(c) **Phospholipid Methyl Transferese I**: When the catecholamine binds to the beta receptor, the receptor undergoes *conformational change* and activates Phospholipid Methyltransferase I, an integral enzyme *oriented* towards the cytoplasmic side of the membrane. It acts on Phosphatidyl-ethanolamine and converts Phosphatidyl-N-monomethylethanolamine which flip-flops towards the outside of the membrane. It is further methylated to Phosphatidyl-N-dimethylethanolamine and then to Phosphatidylcholine by the integral enzyme Phospholipid Methyltransferase II, located towards the outside of the membrane. The increased local concentration of Phsophatidylcholine accelerates the local *membrane fluidity*.

The local enhanced membrane fluidity:

1. Allows the occupied beta receptors to react with the GTP-dependent coupling factor and Adenyl Cyclase to activate adenyl cyclase on the cytoplasmic side of the membrane.
2. Increased fluidity exposes more pre-existing beta receptors to the outside of the membrane surface.
3. Increased fluidity may allow rapid influx of Ca^{++} which binds to "Calmodulin" and activate specific enzyme Phospholipases. These Phospholipases form Arachidonic Acid, the precursor of Prostaglandin which is involved in signal transmission.

20.13 Hormone Receptor

20.13.1 Nature

Protein structure (e.g., receptor protein) can be regulated by small molecule effectors (e.g., hormone). These small molecule effectors are bound to one site of the protein and alter the *tertiary configuration.* In this way, proteins can exist in various structural conformation and states of activity. The activity of the protein can be modified by the regulatory effector molecules. Proteins which are under this type of regulation are called **Allosteric Proteins and the molecules bound to the** protein are called **Allosteric Effectors**. The allosteric proteins to which hormones bind are called **Hormone Receptors.**

20.13.2 Binding linkages

The binding of hormone to receptor protein requires:

1. Hydrophobic interactions,
2. Hydrogen bonds,
3. van der Waals' forces and
4. Salt bridges,

But not covalent linkages. The regulation of allosteric proteins can be terminated rapidly by reducing concentration of the allosteric effector (hormone). According to the mass action equation, active homone-receptor complex formation depends on (1) the concentration of both the hormone and the receptor, and (2) on the intrinsic affinity of the receptor for the hormone.

20.13.3 Parts of hormone receptor

Hormone receptor proteins are having two main sites:

(A) Binding site.
(B) Activity site.

(A) **Hormone Binding site**: It exhibits two properties:

(a) **High specificity**: It distinguishes between two hormone molecules with minor structural modifications e.g., presence or absence of a hydroxyl group on a steroid molecule or an amino acid change in a peptide.

(b) **High affinity for specific hormone ligand**: Affinity is high with equilibrium dissociation constants for hormone-receptor interactions of 10^{-8}-10^{-12} M.

Specificity and affinity of hormone receptors are sufficient for recognition of hormone from nanogram to picogram present per ml of blood.

(B) **Activity site**: Activity site couples it directly or indirectly to enzyme systems and biological responses. The activity site can distinguish receptors from other proteins which have without biological activity. Hormone receptors are identified by their binding capacity with natural and synthetic hormones with high specificity and affinity in the same concentration at which these act on target cells.

20.13.4 Types of hormone receptors

There are three types of hormone receptors:

1. **Peptide Hormone Receptors**: These are present in the cell membrane. The hormone

binding site is exposed to the outer cell surface.

2. **Steroid Hormone Receptors**: These are present inside the cell. Hydrophobic, lipid-soluble steroid hormone molecules diffuse easily into the cells to bind to the receptors. The hormone-receptor complex acts at a nuclear level for biochemical reactions for biological responses.
3. **Thyroid Hormone Receptor**: These are present in the nucleus.

20.13.5 Distribution of hormone receptors

Receptors may be present to a few cell types for certain hormones, and receptors for other hormones distributed widely, such as for:

ACTH : Adrenal cortex, fat cells (lesser extent).
Estrogen : Uterus, ovary, breast.
Progesterone : Uterus, ovary, breast.

20.13.6 Reaction with receptor

Reaction with specific receptor causes → induced conformational change → altered activity of the receptor → series of biochemical changes: (a) regulation of enzyme activity, (b) regulation of the biosynthesis and degradation of nucleic acids and proteins, (c) regulation of the capacity of the largest cell to grow and replicate.

Thus hormones control the differentiated function and growth potential of the target cells. Degree of expression of the genetic material of a differentiated target cell is regulated by the hormone but do not alter the type of genetic information expressed.

20.14 Peptide Hormone Receptor on the Surface of Cell Membrane

Peptide hormones (Pituitary, parathyroid hormones, pancreatic, neurohormones, gastrointestinal hormones, etc.) are recognized by specific and distinct cell surface receptors and then translated the hormone recognition into biological responses.

20.14.1 Nature of peptide hormone receptor

1. The receptor proteins for peptide hormones are asymmetric.
2. It has (a) a hormone binding site exposed to the outer surface of the cell, (b) a hydrophobic site which anchors them within the lipid bilayer, (c) an active site which interacts with other membrane proteins or with the interior of the cell.
3. These receptors, glycosylated, contain more than one subunit, and require phospholipid or an environment in the membrane for biological response.
4. The receptors are mobile within the plane of the membrane. Mobility was demonstrated using hormone conjugated to Fluorescent Indicators and cytoskeletal element anchored at the cell membrane may contribute to receptor mobility.
5. A target cell contains approximately 10^4-10^5 receptors which are turning over constantly. Half life of a receptor, e.g., insulin receptor has ~7h.
6. Receptors are diffusely lodged into the cell membrane after synthesis.

20.14.2 Mechanism

When the hormones are bound with receptors → they cluster → localized in the membrane invagination. The bound receptor-hormone are then internalized by adsorptive endocytosis → endocytic vesicles fuse with lysosomes → receptor and peptide hormone are degraded by lysosomal hydrolases. Some receptors escape degradation and are recycled to lodge in the cell membrane along with synthesized receptors. Available evidence indicates that the hormonal signal transmission takes place at the cell membrane though clustering may be required for generation of signal in some cases.

20.14.3 Regulation

These receptors are dynamic molecules and regulated by their homologous hormone resulting down regulation by decreasing the number of receptors on each cell. Chronic exposure to hormone leads to decrease in concentration of receptor which causes less sensitivity to hormone. Down regulation causes conformational changes in receptors and internalization.

20.14.4 Biological information

The active peptide Hormone-receptor complex is transmitted into the cell via second messengers cAMP.

Hormones using cAMP as intracellular meadiators are ACTH, ADH, calcitonin, FSH, LH, Glucagon, HCG, MSH, PTH, TSH, and those not using cAMP are Angiotensin II, GnRH, Growth hormone, Insulin, Nerve growth factor, Oxytocin, Placental lactogen, Prolactin and Somatostatin.

20.15 Transport of Substances Across Epithelia or Through Cellular Sheet

Epithelia present as a covering of the surfaces of the organs or lining the lumen or cavities. They are present as continuous sheets of cells. Their important function is of transport water and solutes, e.g., (a) formation of urine by the kidney (epithelium of renal tubules), (b) absorption of foodstuffs in the digestive tract (intestine epithelium), (c) epithelium of exocrine gland, (d) membrane of choroid plexus. Here the transport of substances occurs through an entire cellular layer instead of membrane of the cell only.

20.15.1 Transport mechanism

1. Transport of substances is by passive diffusion.
2. Passage through "tight junctions" or "zone occludens".

   The substances may not travel from apical to basal surface for crossing the epithelial layer. The membranes of the adjacent cells come in contact for a short length of their lateral border and form seals known as "tight junctions". Between the remaining part of the lateral border of adjacent cells, there are narrow clefts of different dimensiosn known as **lteral intercellular spaces** which are extensions of the extracellular fluid. The epithelia of gall bladder, proximal renal tubule, small intestine, choroid plexus, etc. the passive flux of Na^+ and Cl^- is across the tight junctions and through the lateral intercellular spaces but not across the epithelial cells, and rate of transport is higher.
3. Epithelia present in the distal renal tubule and urinary bladder, called "tight" epithelia, where transport is through the cell.
4. Some epithelia do not form tight junctions but form aqueous channels or pores which is prevalent in muscle capillaries and ependyma which allow solute to by-pass the cells and becomes highly permeable to polar molecules like sucrose, inulin and haemoglobin. The pores are 3-5 nm in diameter and occupying 0.2 per cent of the area of the muscle capillary wall.
5. **Active transport:** Epithelia in the stomach, gall-bladder, intestine, renal tubules and choroid plexus transport water and solutes like ions, sugar and amino acid across epithelia without external driving force like hydrostatic pressure gradient, osmotic pressure or solute concentration. In certain cases, the transport occurs against uphill gradients of electric potential or solute concentration.

   (a) **Glucose**: In the brush-border side (apical) of the "epithelia, glucose is transported by the sodium co-transport glucose carrier system" which raises local glucose concentration higher than the serosal border (basal) producing a concentration gradient and diffuses to the basal side of the cell. Glucose accumulation inside the epithelial cell causes higher concentration than the extracellular fluid, so that glucose diffuses out of the cell which is a passive process but accelerated by the facilitated diffusion. This glucose transport requires metabolic energy for continued operation of Na^+ and K^+ exchange pump.

   (b) **Sodium**: There is unequal distribution of Na^+ across the membrane of the intestinal epithelium. Na^+ enters the cell continuously by:

   1. Diffusion along their electrical and chemical gradient through the brush-border side due to high permeability than the basolateral membrane (basal side).
   2. The co-transport system through glucose and amino acids.

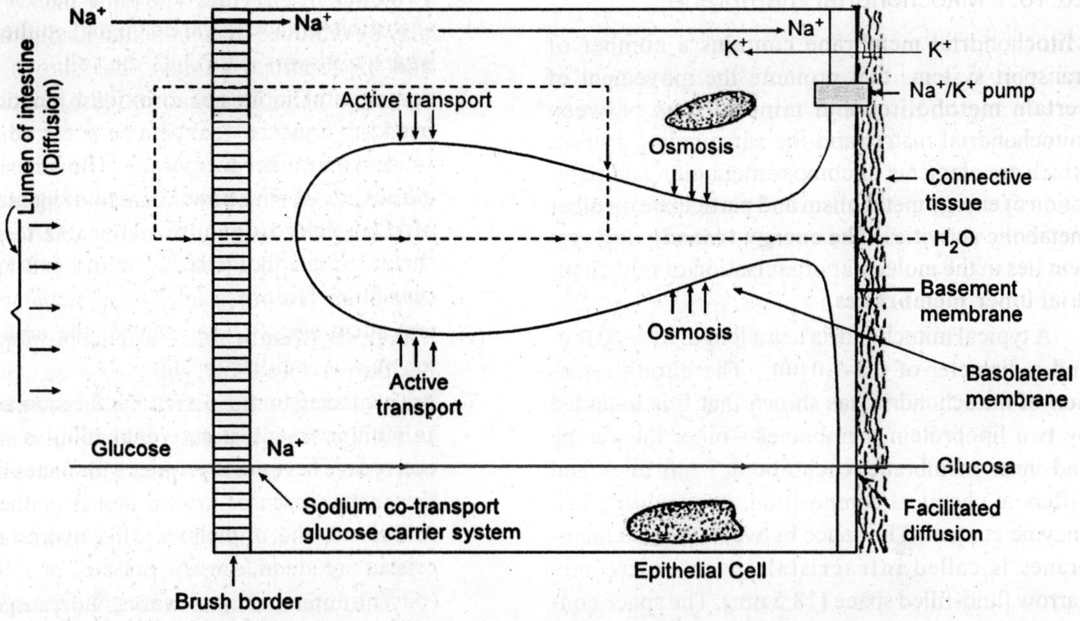

Fig. 20.11: Mechanism of active transport through cellular sheet

When Na^+ ions are transported into the cleft, their positive electrostatic charges pull negatively charged Cl^- ions through the membrane, showing the principle of electroneutrality.

But a low concentration state is maintained by Na^+/K^+ exchange pumps situated in the basolateral part of the cell membrane which is evident by the following:

1. Comparing biochemical analysis of brush border and basolateral membrane, Na/K-ATPase system was located in the basolateral membrane, and
2. Radioactive labelled ouabin caused inhibition of Na-pump and Autoradiography exhibited ouabin at the basolateral border. Na^+ transport in the intestinal epithelia is a two-stage system where Na^+ enters the brush border side by passive force and is actively pumped out across the basolateral membrane.

(c) **Water:** Recently water transport across the epithelia was observed by Diamond and Bossert. According to them, Na^+ transport is restricted mainly to the cleft (lateral intercellular space) close to the tight junction producing hypersomolality in this part causing flow of water both from the cell and more directly from the lumen across the tight junction. Water accumulation in this part of the cleft creates a **hydrostatic pressure** inducing flow of water along the cleft towards the basal end. When the fluid flows along the cleft its **osmolarity** differs from the fluid present inside the epithelium. The lateral membranes of clefts show much **infolding** and **interdigitation** which increase the **surface area** which is exposed to the hyperosmolar fluid in the cleft and thus increases the rate of diffusion of water through the membrane. Na/K exchange pumps play a great role in this process (Fig. 20.11).

20.16 Mitochondrial Membrane

The membranes of mitochondria and endoplasmic reticulum obey the laws of diffusion and active transport like cell membrane but having differences in permeabilities and specificities for transport.

20.16.1 Mitochondrial membrane

Mitochondrial membrane contains a number of transport systems that promote the movement of certain metobolites and mineral ions between mitochondrial matrix and the surrounding cytosol which are elements of compartmentation and regulation of energy metabolism and participate on other metabolic activities. The **energy transducing system** lies in the molecular organisation of mitochondrial **inner membranes.**

A typical mitochondria has a length of 5-10 µm and a diameter of 0.5-1.0 µm . The ultrathin section of mitochondria has shown that it is bounded by two lipoprotein membranes—outer membrane and inner membrane, each about 7 nm thick and differs in chemical composition, permeability and enzyme content. The space between the two membranes is called **intracristal space**—extremely narrow fluid-filled space (18.5 nm). The space contains enzyme, i.e., Adenylate Kinase. The inner membrane has been folded inwards or invaginations, called **cristae,** which project inside space, called **matrix,** a gel-like phase containing about 50 per cent proteins which is organized into a reticular network attached to the inner surface of the inner membrane. The matrix undergoes changes in volume and state of organization during changes in the respiratory system. The matrix also contains DNA and ribosomes and ribosomes are sometimes located where inner membrane closely opposes to the outer membrane. The matrix contains most of the tricarboxylic acid cycle enzymes, Fatty-acyl–COA oxidation enzymes, aspartate transminase, etc.

20.16.2 Structure

(A) **The Outer Membrane**: The outer membrane is smooth and elastic. The outer membrane of liver mitochondria contains 50 per cent lipids. The molecular core of the outer membrane showed hexagonal arrays of small particles, presumably protein molecules under Electron Micrograph following Freeze-Etching Method.

Function

1. **Permeability**: It is much more permeable than the inner membrane. It is permeable to most metabolites simple sugars, small molecules and ions.
2. **Presence of enzyme:** The outer membrane may be removed by treatment with digitonin and biochemical analysis has shown the presence of monoamino oxidase which is used as "marker" enzyme to indicate the presence of outer membrane. The other enzymes are Kynurenine-3-monooxygenase, NADH dehydrogenase, Acyl-COA synthetase, Phospholipase A_2, Nucleoside diphosphate Kinase.

(B) **The Inner Membrane**: The inner membrane has thrown into folds, called **cristae**, which are projected to the matrix. The **cristae** vary in number and structure depending on the cell type. These are the devices for increasing the surface area of the inner membrane in relation to the mitochondrial volume. The cristae are studded with enzymes.

The outer and inner mitochondrial membranes differ in ultrastructure, as revealed by Negative-Contrast Electron Microscopy. The inner surface of the inner membrane is covered with regularly spaced *spherical particles* (F_1 subunit) around *submitochondrial vesicle called inner-membrane spheres* having diameter 8-9 nm connected to the inner membrane by a **narrow stalk** (F_0 subunit).

20.16.3 Biochemical composition of inner membrane

The inner membrane contains only about 20 per cent lipids and about 80 per cent proteins. It is rich in **cardiolipin** which makes up about 20 per cent of the inner membrane lipids. The inner membrane of liver mitochondria contains about 60 different biological active proteins embedded in the phospholipid bilayer. These are:

1. Electron-Transferring enzymes and proteins,
2. Enzymes of ATP synthesis,
3. Different dehydrogenases,
4. Protein components of transport system for metabolites.

20.16.4 Organization of the inner membrane

Organization of the inner membrane has been studied by Freeze Etch Technique of Electron

Microscopy. This method reveals that the faces of the inner membrane has many pits and mounds corresponding to the position of the enzyme molecules. Many of the inner membrane enzymes and cytochromes have a fixed sidedness in the membrane and cannot rotate.

20.16.5 Function of inner membrane

1. **Permeability**
   (a) The inner membrane is intrinsically impermeable to nearly all ions and most unchanged molecules but selectively permeable.
   (b) Mitochondria are generally impermeable to protons and other ions but there is specific transport systems that enable ions to penetrate the inner membrane.
   (c) **NADH impermeability**: The inner membrane is impermeable to NAD, NADH, $NADP^+$, NADPH, AMP, CTP, GTP, CDP, CoA, Acyl-CoA, produced in the cytosol continually but it is not accumulated and is oxidized by the respiratory chain of mitochondria by **glycerophosphate shuttle** and **malate shuttle** system.
   (d) The inner membrane is freely permeable to oxygen, water, CO_2, monocarboxylate ions, e.g., β-hydroxybutyrate, acetoacetate and acetate.
   (e) **Long chain fatty acids**: Long chain fatty acids cannot penetrate inner membrane but are transported into mitochondria via **Carnitine system.**
   (f) **Pyruvate**: There is special carrier for pyruvate.
   (g) **Dicarboxylic, tricarboxylic anions and amino acids**: They require specific carrier systems to facilitate their transport across the membrane.
   (h) **Monocarboxylate anions**: They penetrate more easily because of lesser dissociation of these acids.
   (i) **Undissociated and lipid soluble acid**: They penetrate the lipid membrane.
   (j) **Neucleotides and metabolites**: The inner membrane of liver mitochondria does allow certain specific nucleotides (ATP, ADP) and metabolites (inorganic phosphate, pyruvate, citrate, succinate, α-Ketoglutarate, malate, amino acids, glutamate and aspartate) to pass readily. The passage of this group of solutes through the inner membrane takes place by the action of specific membrane transport systems.

20.16.6 Metabolite transport mechanisms of inner membrane

1. **Metabolite transport system**: The metabolite transport systems in the inner membrane are specialized proteins or groups of proteins and those are specific for particular molecules but do not act for closely similar molecules, e.g., transport system of ATP, act only for ATP, ADP, dADP and dATP but not for AMP, GTP, CTP and CDP.
2. **Major Transport System**: The major transport systems are also called "carriers", "translocases" or "porters". These are:
   (a) **ATP-ADP Carrier**: Promotes reversible equimolecular exchange of an external molecule of ADP for an internal ATP molecule formed by oxidative phosphorylation.
   (b) **Phosphate Carrier**: Promotes an exchange of H_2PO_4 ions and OH^- ions or its equivalent, the contrasport of both $H_2PO_4^-$ and H^+ in the same direction.
   (c) **Dicarboxylate Carrier**: Promotes equimolecular exchange of malate, succinate and fumerate with each other or phosphate.
   (d) **Tricarboxylate Carrier**: Promotes equimolecular exchange of citrate and isocitrate with each other and equimolecular exchange of a tricarboxylate with a dicarboxylate such as malate.
3. **Passive systems**: The metabolite-transport systems are passive system. In the absence of respiration, they transport metabolities down concentration gradients in the direction that result in thermodynamic equilibrium across the membrane. These systems

Table 20.6: Comparison between Inner and Outer Membranes of Mitochondria

Outer Membrane	Inner Membrane
1. Smooth, elastic,	1. Inwards fold or invagination called cristae to increase the surface area.
2. It does not contain F_1 and F_0 subunits.	2. It contains the subunits.
3. 50 per cent lipids, 50 per cent protein present in the membrane.	3. 20 per cent lipid, 80 per cent protein.
4. Enzyme content differs from inner membrane.	4. Enzyme content differs from outer membrane.
5. Cardiolipin-trace.	5. Rich in cardiolipin
6. Freely permeable to most solutes of low molecular weight.	6. It has selective permeability. (a) Permeable by specific transport system, ADP, ATP, Phosphate, Pyruvate, Citrate, Succinate, α-Ketoglutarate, Malate, Glutamic acid. (b) Impermeable: NAD, NADA, $NADP^+$, NADPH, AMP, CTP, GTP, CDP, GDP, CoA, Acyl CoA.

can transport their substrates against gradients of concentration when they are coupled to electron transport as a source of energy.

4. **Species-Specific:** The inner-membrane transport systems of mitochondria are species-specific and genetically determined.

20.16.7 Structural Distribution of Enzymes in Inner Membrane for Electron Transfer and Oxidative Phosphorylation

Respiratory assembly is an integral part of the inner membrane. The enzymes responsible for elec-

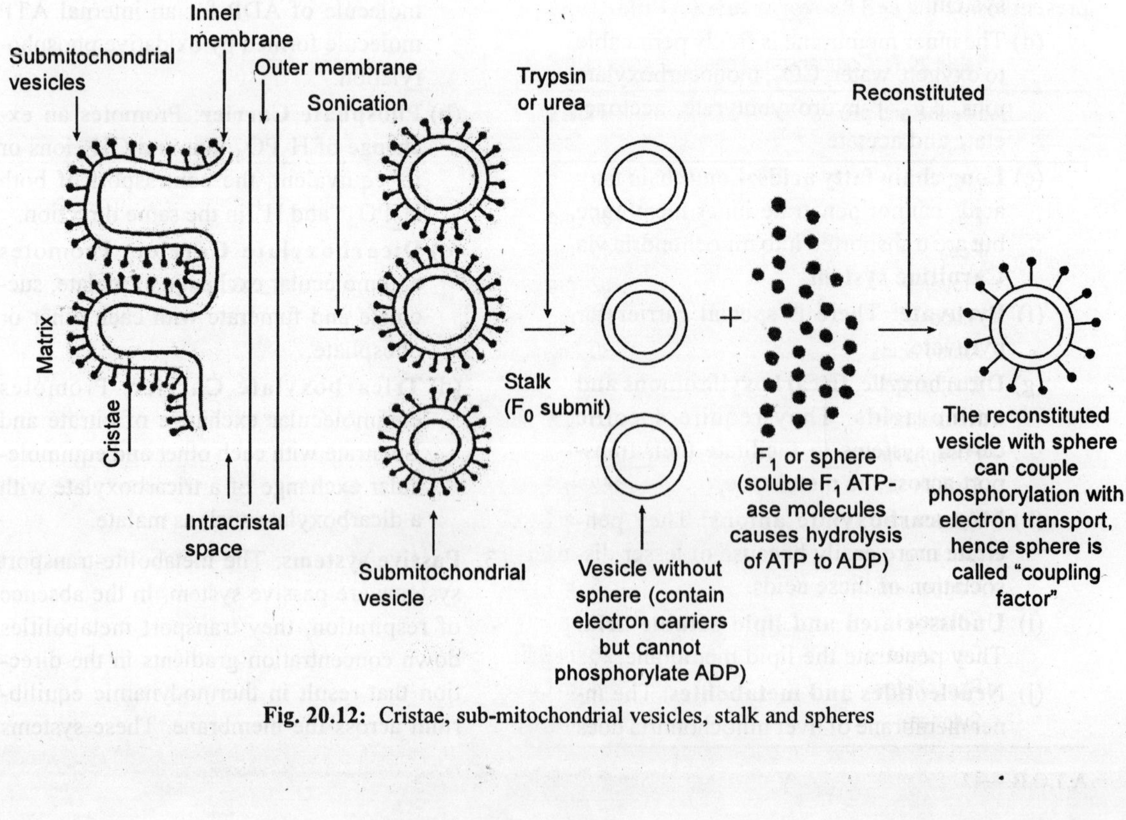

Fig. 20.12: Cristae, sub-mitochondrial vesicles, stalk and spheres

tron transfer and oxidative phosphorylation are embedded in the inner membrane and they form characteristic repeating unit visible under E.M. Each unit is called **submitochondrial vesicles** having:

1. **Vesicle or base piece**: Contains the enzymes of the electron transfer chain,
2. **Stalk or F_0 subunit** which probably extends through the membrane and
3. **Head piece or sphere or F_1 subunits** which projects into the matrix consist of several proteins collectively known as an F_1 subunits and contain the ATPase (Adenosine triphosphatase) which causes hydrolysis of ATP to ADP and other enzymes concerned with synthesis and metabolism of ATP like ATP synthetase in the stalk or F_0 subunit. For every proton pair passing through F_0-F_1 complex, one ATP molecule is formed from ADP and Pi (Fig. 20.12).

20.17 Factors Governing the Passage of Materials Across Cell Membrane

The cell bounded by the membrane and the fluid present inside the cell known as **intracellular fluid** which differs from the fluid present outside the cell called **extracellular fluid** for which it is necessary that water, salts and metabolized substrates should enter the cells. It is equally necessary that metabolic waste products come out of the cells through the membrane, for which there is difference in composition between the intracellular and extracellular compartmental fluids (Table 20.7). Solutes can enter the cell through the membrane by the process of diffusion and is dependent on thermal agitation and the magnitude of which is proportional to the concentration gradient. When a structure like a membrane is placed in the pathway of diffusion, the movement of solutes may be restricted or accelerated in some cases.

The forces which govern the passage of water and solutes are given below:

1. Solvent drag,
2. Simple diffusion,
3. Non-ionic diffusion,
4. Restricted diffusion,
5. Facilitated diffusion,
6. Exchange diffusion,
7. Endocytosis,
8. Vesicular transport (micropinocytosis),

Table 20.7: Comparison of chemical composition of fluid of extracellular and intracellular compartments

Substance and pH	Fluid of extracellular compartment	Fluid of intracellular compartment
pH	7.4	7.0
Po_2	35 mm Hg	20 mm Hg
Pco_2	46 mm Hg	50 mm Hg
Proteins	2 gm per cent	16 gm per cent
Glucose	90 mgm per cent	0.-20 mgm per cent
Amino acids	30 mgm per cent	200 mgm per cent
Cholesterol, Neutral fat, Phospholipids	0.5 gm. per cent	2-95 gm per cent
Sodium	142 m Eq/l	10 m Eq/l
Potassium	4 m Eq/l	140 m Eq/l
Calcium	5 m Eq/l	< 1 m Eq/l
Magnesium	3 m Eq/l	58 m Eq/l
Chloride	103 m Eq/l	4 m Dq/l
Carbonate	28 m Dq/l	10 m Eq/l
Phosphates	4 m Dq/l	75 m Eq/l
Sulphate	1 m Eq/l	2 m Eq/l

9. Exocytosis,
10. Permeability of membrane,
11. Filtration,
12. Ultrafiltration,
13. Dialysis,
14. Osmosis and Osmotic pressure,
15. Donnan membrane equilibria,
16. Active transport.

1. **Solvent Drag (Bulk flow)**: When water moves in one direction it tends to drag along some solutes through the membrane and this force is known as **solvent drag**. Such drag forces may arise from differences in osmotic pressure, electroendosmosis, etc. Its effect in the body is very small.
2. **Simple Diffusion**: All particles, molecules or ions in the body fluids including water and dissolved substances are in constant thermal motion resulting in frequent collision between the molecules. By virtue of their motion, they will tend to spread themselves throughout the solvent from regions of high concentration to regions of low concentration until the concentration is uniform throughout the solution. This process is called diffusion. In the process, movement of solute occurs when the two fluids of different concentrations are separated by a mechanical partition, i.e., a membrane which is permeable to the dissolved substances maintaining same hydrostatic pressure. If the hydrostatic pressure on the two sides of the membrane are not equal, a flow of water and dissolved solutes will occur by filtrations.
3. **Non-ionic Diffusion**: It has been found that some is soluble in cell membranes but in ionic form they cross membranes with great difficulty. If molecules of undissociated substance diffuse through membrane and then dissociate in other side, their movement from one side of the membrane to another is very much effective, e.g., gastrointestinal tract, kidneys.
4. **Restricted Diffusion**
   (a) Charged sites on or in the membrane may repel or attract ions and so mediate their movement.
   (b) Lipid-water partitioning of the solute may cause an enhanced movement through the lipid phase of the membrane or a repulsion of the solute.
5. **Facilitated Diffusion**: Ions and larger nonionized molecules (e.g., glucose) are transported by a mobile structure, i.e., carrier molecules present in the membrane which can combine reversibly with the solute, may be mediated by directional movements of the complex. When such carrier-mediated transport is from an area of greater concentration of the transported molecules to an area of lesser concentration, energy is not required (energy is required in active transport), and the process is called **facilitated diffusion**, e.g., glucose transport in RBC.
6. **Exchange Diffusion**: Solute movement through membranes without net movement can be observed if the solute exchanges with a like solute on the opposite side of the membrane. Exchange diffusion can be explained in case of isotope accumulation without net changes in total solute.
7. **Endocytosis** (Pinocytosis + Phagocytosis): Endocytosis is a transport process that allows cells to internalize extracellular material and involves the formation of endocytic vesicles. Endocytosis are two types:
   (a) **Phagocytosis**: It is a process by which the cell engulfs bacteria or foreign materials by throwing out pseudopodia and causes lysis with the help of lytic enzymes.
   (b) **Pinocytosis:** It is the property of all cells and leads to the cellular uptake of fluid and fluid containing dissolved nutrient particles.
   i. **Adsorptive Pinocytosis**: It is a receptor-mediated, selective process primarily responsible for the uptake of macromolecules for which there is a finite number of binding sites on the cell membrane. In this process the vesicles are coated (cytoplasmic side) with filamentous material called **clathrin**, which is probably a peripheral membrane protein. The process is seen in case of LDL and hormone.

ii. **Fluid-phage Pinocytosis**: It is non-selective fluid phase process in which the uptake of a solute is simply proportionate to its concentration in the surrounding medium.

8. **Vesicular Transport (Micro Pinocytosis)**: Large molecular movement of some proteins can occur directly through the membrane or by vesicular transport. In vesicular transport, enclosure of plasma proteins within vesicles occur and the vesicle then moves from one side of the cell membrane surface and the process is also called **micropinocytosis.**

9. **Exocytosis**: Exocytosis is also known as **reverse pinocytosis** or **emeiocytosis** (cell vomiting). Proteins secreted by cells move from the endoplasmic reticulum to the Golgi complex for packaging into secretory granules membrane and cell membrane fuse and the area breaks down and the contents come out of the cell.

10. **Permeability of Membrane**: When the particles can move through the membrane easily without any hindrance, the membrane is permeable and if the particles are not allowed to pass it is called "impermeable". But the cell membrane has the property of "selective permeability" in which membranes allow passage to certain substances and others are held back for which the cell membrane is called "semipermeable membrane". Semipermeability accounts for size of the molecules, size of the pores, lipid solubility, electric charges on the cell surface, orientation of molecules and ions in the membrane, cellular oxidation, stimulation Ca^{2+} and K^{+} in extracellular fluid, local anaesthetics, veratrine, ADH, combination of membrane receptors with specific substance, etc.

11. **Filtration**: It is a physical process by which undissolved substances are separated from a liquid through a membrane or pore due to difference of mechanical pressure, e.g., hydrostatic pressure, gravity or osmotic pressure. This process is seen during intestinal absorption of foodstuffs, formation of glomerular filtrate, passage of water, salts, foodstuffs, etc. from blood to the tissue fluid.

12. **Ultrafiltration**: When a solution of crystalloid and colloid are filtered through a membrane having pores of different sizes, the crystalloids pass through easily but the colloidal particles up to a certain size can pass but larger particles are retained and the process is called **ultrafiltration** and the fluid is called **ultrafiltrate.**

    The glomerular filtrate is actually an **ultrafiltrate** of plasma where blood cell and plasma proteins are retained, otherwise its composition is identical to that of plasma. Tissue fluid formation through capillary membrane is also ultrafiltration.

13. **Dialysis**: When the membrane allows to pass water, salts, urea and other crystalloids under pressure, it will keep back all colloidal constituents. The process is called **dialysis** and the fluid obtained in called **dialysate** or **dialysing fluid**. This process occurs in the intestine during absorption, exchange of body fluid through capillary membrane and formation of aqueous humour.

14. **Osmosis and Osmotic Pressure**: Osmosis is the spontaneous flow of solvent molecules through a membrane from a solution of lower concentration to higher concentration of solutes to which the membrane is impermeable. Osmosis is really a special case of diffusion (Greek; osmos = a push). The tendency for flow of solvent molecules to a region of higher concentration can be prevented by applying hydrostatic pressure to the higher concentrated solution. The pressure required to prevent solvent migration is called **effective osmotic pressure**.

15. **Donnan Membrane Equilibria**: Gibbs had shown theoretically and Donnan confirmed experiementally that the presence of non-diffusible anions like proteins and organic phosphate inside the cell, affect the distribution of diffusible, i.e., penetrating ions. The negative charge of a non-diffusible anion resists diffusion of diffusible cations and

favours diffusion of the diffusible anions. Donan and Gibbs had shown that in the presence of a non-diffusible ion the diffusible ions distribute themselves and at equilibrium their concentration ratios are equal. In the case of single cations and anions of the same valence, the product of the concentration of the diffusible ions on one side equal that on the other side.

16. **Active Transport**: When the carrier-mediated transport of solute is from an area of lower to an area of higher concentration, i.e., against concentration gradient. It is an energy-dependent process. It is carried out by protein "pumps" in cell membranes and the energy is generally supplied by the cell through ATP.

Molecular Transport Across Cell Membrane

Table 20.8: Categories of solute transport across cell membranes

Type of transport	Transport mechanism	Schematic representation
I. A. Uniport	Single molecule transport	in | out → glucose
B. Coupled transport		
(a) symport	Parallel transport of several molecules simultaneously	→ H^+ → H^+ Lactose
(b) Antiport (exchange transport)	Transport of a molecule linked with the transport of another molecule in the opposite direction	Na^+ K^+ ←

Contd.

Type of transport	Transport mechanism	Schematic representation
II. Electroneutral	No charge transfer across the cell membrane	→ Na^+ H^+ ←
Electrogenic	Net charge transfer across the membrane	→ $3Na^+$ $2K^+$ ←
Energy-coupled	Solutes transport against the gradient	nH^+ ← O_2 nH^+ ← ATP ADP+ Pi

- **Summary of Functions of Plasma Membrane**

I. **Biophysical**

1. Transport
2. Pinocytosis
3. Reverse Pinocytosis
4. Rhopheocytosis—This is a process by which larger fluid molecules are taken in
5. Podocytosis or cytopemphis
6. Phagocytosis
7. Reverse Phagocytosis.

II. **Protective functions**

(a) Cell protection

(b) Response stimuli.

III. **Modified Structural function**

1. Sodium pump
2. Potassium pump
3. Receptors
4. Signalling
5. Carrier protein.

21

Bioacoustic

1. Sound and its characteristics, 2. Function of ear as a hearing organ, 3. Physical basis of hearing, 4. Unit of intensity of sound, 5. Physical organisation of the ear, 6. Physical aspects of transmission of sound, 7. Transmission of sound waves in the inner ear–travelling waves, 8. Electrical responses of cochlea, 9. Pitch perception, 10. Perception of loudness, 11. Mechanism of hearing, 12. Beats and dissonance, 13. Combination of tones, 14. Consonance and dissonance, 15. Sound analysis, 16. Sound localization, 17. Physical basis of voice (speech), 18. Infrasound or subsonic sound, 19. Ultrasound or ultrasonics, 20. (a) Ultrasonography, (b) Ultrasonic therapy, (c) Ultrasonic surgery, (d) Ultrasound transducer, (e) Echocardiography, (f) Vascular Ultrasound, (g) P-Z materials, 21. Shock waves and Extracorporeal Shock Wave Lithotripsy, 22. Noise, 23. Effect of noise in the body, (a) Sound measurement, (b) Audiometer, (c) Hearing Aid, 24. Protection against noise or noise control, 25. Speech Interference Level (SIL) and measurement of level of noise.

21.1 Sound and its Characteristics

Sound signifies the sensation received by the ear. It is also referred to as the physical cause which stimulates the auditory nerves to produce the sensation of hearing.

Sound is produced by the vibrations of material bodies. The vibrations transmit in the surrounding medium and is carried by the medium to the ear. The ear drum is thrown into similar vibrations and produces the sensation of wound. Sound energy is the mechanical energy of the vibrating medium developed from the mechanical energy of the vibrating source. The vibrations must lie within a certain limit to make sound audible and it is known as **limits of audibility**, the lower limit being about 20 vibrations per second and the upper limit being about 20000 vibrations per second. The audible sound is known as *sonic sound* and waves carrying this sound are called *sonic waves*. The waves of vibrations above the upper limit are called **ultrasonic waves or ultrasound** and those below the lower limit are called **infrasonic waves or subsonic waves or infrasound**. Sound cannot travel through vacuum but a material medium is required for the propagation which may be solid, liquid or gas. It is the mechanical energy of the vibrating body that is transmitted by the medium in a peculiar way known as *wave motion* or *sound wave* which travels through the medium and reaches the ear

There are two distinct types of vibration in a medium:

(a) Transverse vibrations,

(b) Longitudinal vibrations.

The waves arising out of these vibrations are called:

(a) Transverse waves,

(b) Longitudinal waves.

The sound has the following properties:

1. Travels with finite velocity.
2. Reflection,
3. Refraction,
4. Dispersion,
5. Interference phenomenon,
6. Diffraction.

Sound may be divided into two groups:

1. *Noise*: Explosion of bombs, thunder of lightning, flying jet plane, whistle of train–which are unpleasant to the ear.
2. *Musical sound*: Sound from sitar, harmonium, etc. But there is no actual line of demarcation, between the two.

21.2 Functions of Ear as a Hearing Organ

1. **As a Mechano-Receptor**: The human ear as a special sense organ for hearing is highly differentiated and specialized mechano-receptor which extracts and transmits information to the central nervous system about the acoustic environment.
2. **As a Detector**: As a detector, the ear is capable of recording almost incredibly minute amounts of mechanical energy in the form of vibrations of the air molecules.
3. **As an Analyser**: As an analyser, it furnishes detailed information about the intensity and frequency of these airborne vibrations and thus gives rise to the sensations of loudness, pitch and timbre.
4. **Temporal Pattern of Sound**: At the same time, it provides the faithful, fine-grained record of the temporal pattern of sound stimulation that is essential for the understanding of speech and the appreciation of music.
5. **Monitoring Action**: It continuously monitors the speaker's own voice and thus helps him to maintain a standard of speech that he regards as acceptable.
6. **Alerting Organ**: It has a value as an alerting or "early warning" mechanism, which is of great importance for survival as it is seen in animals, e.g., elephant, deer, etc.
7. **As Direction Finder**: Acting together, the two ears serve as a "direction finder", providing the brain with data for localizing the source of a sound for deciding whether the sound should be attended to or may safely be ignored.
8. **As an Acoustic Transducer**: It can be regarded as a transducer, having the specialized function of converting the mechanical energy of sound into the form of electrochemical energy that is capable of activating the endings of the auditory nerve.

21.3 Physical Basis of Hearing

21.3.1 Sound sensation

Sound is the sensation produced by the air pressure change in the surrounding which travels through air as waves and in the form of longitudinal vibrations of the molecules in the surrounding, i.e., waves of alternate phase of condensation and rarefaction of the molecules in an elastic medium such as air, water, etc. and which strike the tympanic membrane and perceived by the auditory mechanism. These movement of changes in pressure on the tympanic membrane is called **sound waves**. The simplest sound waves is the sine wave produced by the vibrating tuning fork. The wave frequency determines the pitch of the sound. The wave form of any periodic vibration can be resolved into a series of simple waves by Fourier's Theorem. This theorem has unique role to analyse or synthesize a complex type of vibration. It states that any single valued periodic function can be analyzed into a sum of infinite number of simple harmonic vibrations, having frequencies which are multiples of the frequency of the given function) of which: (i) One is called fundamental having the lowest frequency and the (2) others are called harmonics having frequencies which are multiples of this fundamental. The sound waves travel through air at a rate of 344 m/s. or 770 mi/h at 20°C at sea level. The speed increases with temperature and with altitude. Other media also conduct sound waves such as speed in freshwater it is 1450 m/s at 20°C and greater in salt water.

• **Quality**: The quality of a sound depends on its waveform. Some sounds are of very complicated waveform but they can be resolved by mathematical analysis into fundamental and many harmonics if they are periodic. But if the sound contains many components having frequencies which are not simple multiple of the fundamental, then it is called "noise".

Sound waves that have repeating patterns, even though the individual waves are complex, are perceived as musical sounds; whereas aperiodic nonrepeating vibrations cause a noise. Most musical sounds are made up of a wave with a primary frequency that determines its pitch plus a number of harmonic vibrations (overtones) that give the sound its characteristic "timbre" or "quality". Variations in timbre permit of identify the sounds of the various musical instruments even though they are playing notes of the same pitch.

- **Loudness**: The loudness of a sound is correlated with the amplitude of a sound wave and its pitch with the frequency, i.e., number of waves per unit of time. The greater the amplitude, the louder the sound and the greater the frequency, the higher the pitch.

Loudness is not synonymous with intensity and it is a psychological term. Intensity of a sound refers to a purely physical aspect of sound whereas loudness of sound is a subjective phenomenon and it refers to the sensation it produces in the ear. The sensation of loudness depends on the intensity of sound but loudness is, however, not proportional to the intensity. A relation between physical stimulus and corresponding response was investigated by Weber and Fechner, known as **Weber-Fechner law**.

21.4 Unit of Intensity of Sound

The intensity of a sound is measured in terms of energy or power, e.g., in microwatts, or ergs per second passing through an area of 1 cm^2 placed at right angles to the direction of sound propagation. Alternatively it may be expressed in dynes cm^{-2} in terms of pressure. But because of great range of intensities involved, for convenience, a scale of change of intensity is employed to describe them. The scale is "decibel" (dB) scale. The unit of intensity of sound is "bels" which is the logarithm of the ratio of the intensities of the sound and a standard sound:

$$\text{bel} = \log \frac{\text{Intensity of sound}}{\text{Intensity of standard sound}}$$

The intensity is proportionate to the square of the sound pressure. Therefore,

$$\text{bel} = 2\log \frac{\text{Pressure of sound}}{\text{Pressure of standard sound}}$$

A decibel is 0.1 bel. The standard sound reference corresponds to 0 decibels at a pressure level of 0.000204 dyne/cm^2, a value which is just at the auditory threshold for the average human. The decibel scale is a log scale and the value of 0 decibels does mean the absence of sound but a sound level of an intensity equals to that of the standard.

21.4.1 Audible sound frequency or sensitivity of intensity of sound by human ear

The human ear is sensitive to sound energy and to pitch variations in the range of sound frequencies in the lower limit 20 cycles per second (cps, Hertz or Hz). This least intensity for the sound to be audible is called **threshold of audibility or hearing**. which varies greatly with frequency or pitch of the sound. The maximum limit is 20000 cycles per second (cps, Hz.) or 20 kHz. Young age group can hear up to 25000 cycles per second (25 kHz) and it is lost gradually as the age advances. In mid-life the limit is about 4-5 kHz. The most valuable range for speech and hearing is from 200 Hz to 4.5 kHz. Very large intensities cause is painful sensation and this upper limit is called **threshold of feeling**, which also varies greatly with frequency. These values vary to some extent from person to person.

In animals like bats and dogs, much higher frequencies are audible.

The greatest sensitivity of human ear is in between 1000-4000 Hz. The pitch of the average male voice in conversation is 120 Hz. and in the female it is about 250 Hz. The number of pitches that can be distinguished by an average individual is about 2000 but it is much improved in trained musicians. But pitch discrimination is seen between 1000-3000 Hz and is poor in high and low pitches.

An audibility curve can be plotted by applying sound of different frequencies for diagnosis and prescription or suitable hearing aid.

21.4.2 Masking

It is our common knowledge that the voice is raised during conversation in a noisy surrounding. If the conversation goes on at an intensity level of 50 dB, it will be inaudible if the noise in the surrounding exceeds this intensity level. This phenomenon is known as **masking**. It represents the ability of the ear to separate the total stimulation into the separate components. It may be due to the relative or absolute refractoriness of previously stimulated auditory receptors and nerve fibres to other sound stimuli.

21.5 Physical Organization of the Ear

The ear is divided into three parts:

I. External Ear: It consists of the following:

1. **Pinna**: It is made up of fibrocartilage covered by the skin. Its function is to receive sound waves and transmit it to external auditory meatus or canal. *It is also important for localization of sound.*
2. **External auditory meatus or Canal**: It is a "S"-shaped canal. A depression called *concha* forms the orifice of the external auditory canal which extends to end at

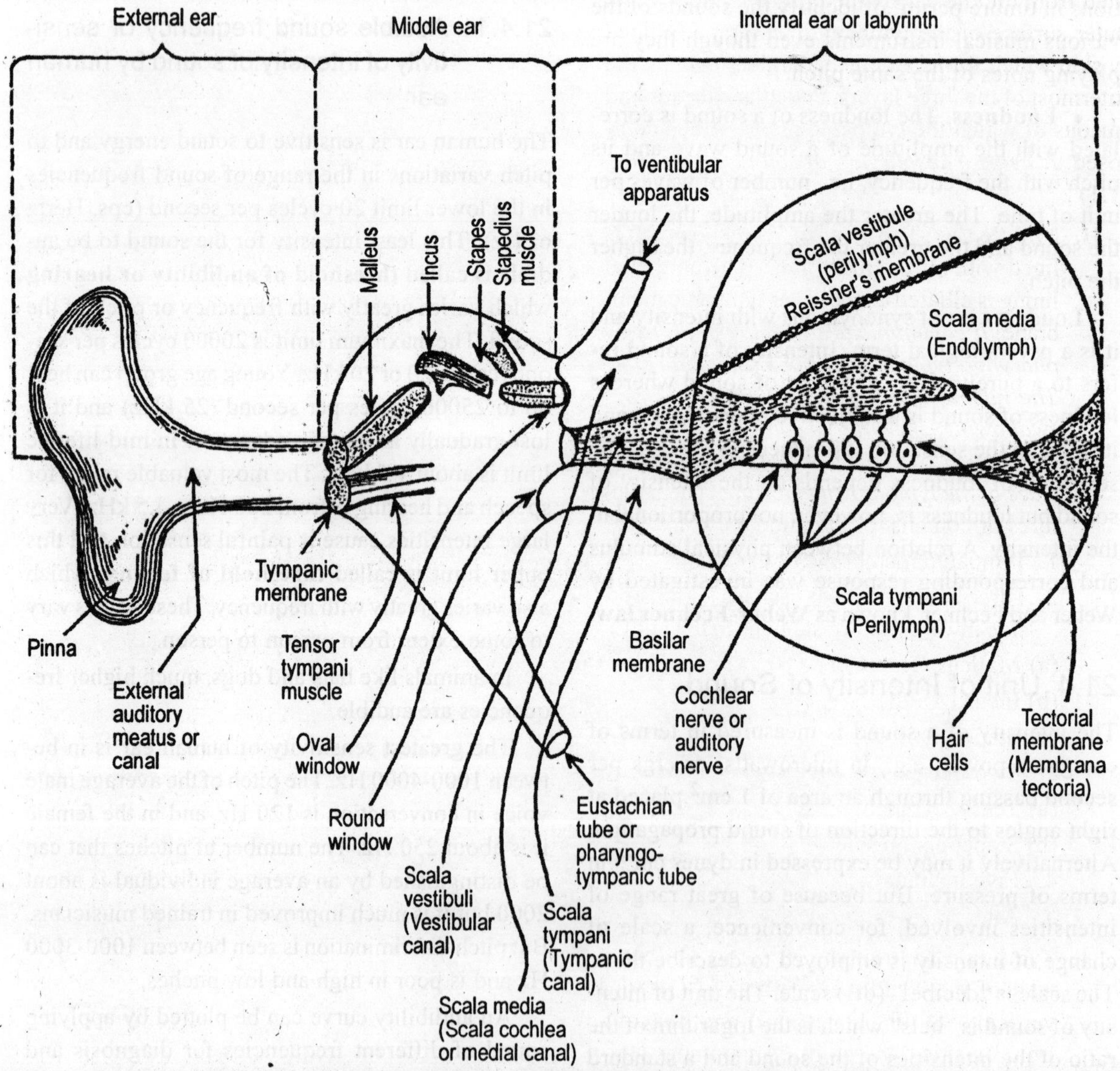

Fig. 21.1: Schematic diagram of the basic structure of the mammalian ear

the *tympanic membrane* or drum membrane. It is composed of two parts: *cartilaginous outer part,* 8 mm long and *osseous inner part* made up of temporal bone, 16 mm long.

The *skin lining* of the cartilaginous outer part has many *ceruminous gland* producing cerumen and laterally directed hairs which prevents entry of insects and foreign particles. Sometimes cerumen accumulates and blocks the passage of sound. It lubricates the canal. The skin lining of the osseous inner part is thin, closely adherent to the periosteum, and very much sensitive. The canal is separated from middle ear by the tympanic membrane. Outer surface of the tympanic membrane is lined by skin which forms a delicate lamina forming the outermost of the three layers, cuticular, fibrous and mucous of which the tympanic membrane is composed. *It helps transmission of sound waves* (Fig. 21.1).

II. Middle Ear: The area is 15 × 6 × 4 mm. It is lined with mucus membrane. The epithelial lining is ciliated except over tympanic membrane. *It communicates with the pharynx by pharyngo-tympanic tube or Eustachian tube. The tube remains closed but opens during swallowing, chewing, yawning and sneezing. Its function is to equalise pressure on the two sides of the tympanic membrane.* The structures or contents of the middle ear are:

1. Tympanic membrane or ear drum.
2. Three auditory or ear ossicles (From outside inwards, shortest bones in the body).

(a) Malleus (Hammer)
(b) Incus (Anvil)
(c) Stapes (Stirrup)

3. Two muscles

(a) Tensor tympani muscle
(b) Stapedius muscle

4. Five ligaments
5. Air.

1. **Tympanic Membrane**: It forms the outer wall of the middle ear, elliptical (10 mm vertical × 9 mm horizontal) and concave from outside. The point of maximum concavity is called *Umbo*, where *handle* of the malleus is attached on the inner side.

The tympanic membrane has three layers:

(a) Outer layer: Formed by the skin
(b) Middle layer: Is fibrous, called *membrane propria* containing *circular* and *radial set of fibres.*
(c) Inner layer: Formed of mucus membrane.

In between the middle and inner layer the handle of malleus is inserted. At the upper part of the membrane there is a small white projection, due to lateral process of malleus. From this white projection, 2 folds project upwards and forwards and are called *anterior* and *posterior malleolar folds.* The membrane in between the two folds is loose and triangular and is called **pars flaccida.** Other part of the membrane is called **pars tensa**. A ridge is noticed in the membrane which is due to the handle of malleus (Fig. 21.2).

2. **Three Auditory Ossicles:** The three ossicles—outside to inwards—are:

(a) Malleus, (b) Incus, (c) Stapes.

Ear ossicles are kept in position by five ligaments. They remain articulated together forming a *complex system of levers.* The *mechanical joint (malleus + incus) and bearing system* (incus + stapes) of the *single unit* of the bony ossicles are such that displacement of the tympanic membrane by air will be three times higher in the oval window and thus it acts like a **hydraulic press.** And the pressure changes in the liquid of the inner ear just inside the oval window will be greater than outside the tympanic membrane. Thus the three ossicles transmit sound wave from tympanic membrane to the internal ear and also **magnify the force of vibration ten times** for following reasons:

i. The surface area of the tympanic membrane is 90 sq mm whereas that of oval window is 3.2 sq mm. Hence the force of vibration is concentrated to a much smaller area,

ii. The ossicles constitute a "bent lever" of first class. But the actual magnification is *thirty times* dur to:

(a) Damping action of the ossicles, and

(b) Loss of energy in setting the ossicles into vibration, the *actual magnification is ten times.*

3. **Two Muscles**: They are

(a) **Tensor tympanic muscle**: Attached to the handle of the malleus. It has a protective function, i.e., when sound stimulus is of high amplitude, this muscle contracts and the tympanic membrane becomes tense and the rate of vibration of the membrane is diminished.

(b) **Stapedius Muscle**: Attached with the stapes ossicle (neck). Regarding its functions there are two views:

1. **Protecting theory**: When the sound is heard, by its contraction, the rate of vibration of tympanic membrane is diminished.

2. **Accommodative theory**: They tune up the transmitting mechanism and impart increased sensitiveness to any given vibration frequency.

- **Oval Window, Round Window, Mastoid Antrum**

In the medial wall of the middle ear there are two windows:

(a) **Oval window** (or fenestra ovalis): Fitted with the *footplate of the stapes* and sealed-in by a membrane.

(b) **Round window** (or fenestra cochleae or rotunda): Is closed by a membrane.

The posterior wall communicates with air cavities in the mastoid process called *mastoid antrum.*

III. Internal Ear or Labyrinth: The internal ear or labyrinth consists of bony labyrinth with which *membrane labyrinth* is present having the following structures:

A. **Cochlea**: For hearing,

B. **Vestibular apparatus**: For regulation of posture and equilibrium. It consists of two types of structures:

I. Vestibule Consists of two parts: (a) Utricle, (b) Saccule (for static equilibrium).

II. 3 Semicircular canals (For kinetic or dynamic equilibrium).

III. Iternal Ear or Labyrinth

A. **Cochlea**: It is embedded in a body cavity in the temporal bone. Outer part is made of bone and inner side is made up of bony canal arranged spirally (two and three-fourth turns) like the shell of a snail, hence the name is cochlea (Greek *kochlias*-snail). The spirals wind round a central bony pillar called *modiolus,* through which passes the cochlear nerve. A bony projection winding round the modiolus is called *lamina spiralis ossea,* it makes an incomplete partition which is completed by *basilar membrane* attached to the outer side of lamina. Three different tubes are coiled side by side:

1. Scala vestibuli,
2. Scala media (Scala cochlea or cochlear duct or cochlear partition),
3. Scala tympani.

The *basilar membrane* separates the scala tympani and scala media. Another membrane called *reissner's membrane* separates scala vestibuli and scala media (or Scala Cochlea). Thus, three *compartments are formed.* The scala vestibuli and scala tympani contain a fluid called *perilymph* and scala media (or scala cochlea) contains *endolymph.* Perilymph is identical with cerebrospinal fluid. Endolymph is identical with the composition of lymph. The Reissner's membrane probably has the role for production of endolymph. Scala vestibuli and scala tympani communicate with each other at the apex of the cochlea called *helicotrema.* When vibration is produced in the tympanic membrane it will be transmitted. It is believed that endolymph is secreted by stria vascularis and vestibular neuroepithelia. Site of absorption is not known.

Cochlear Partition of Cochlear Duct or Scala Media or Scala cochlea: The triangular endolymph fluid filled space (Space of Nuel) covered by Reissner's membrane, Stria vascularis and Basilar membrane is called *cochlear partition* or

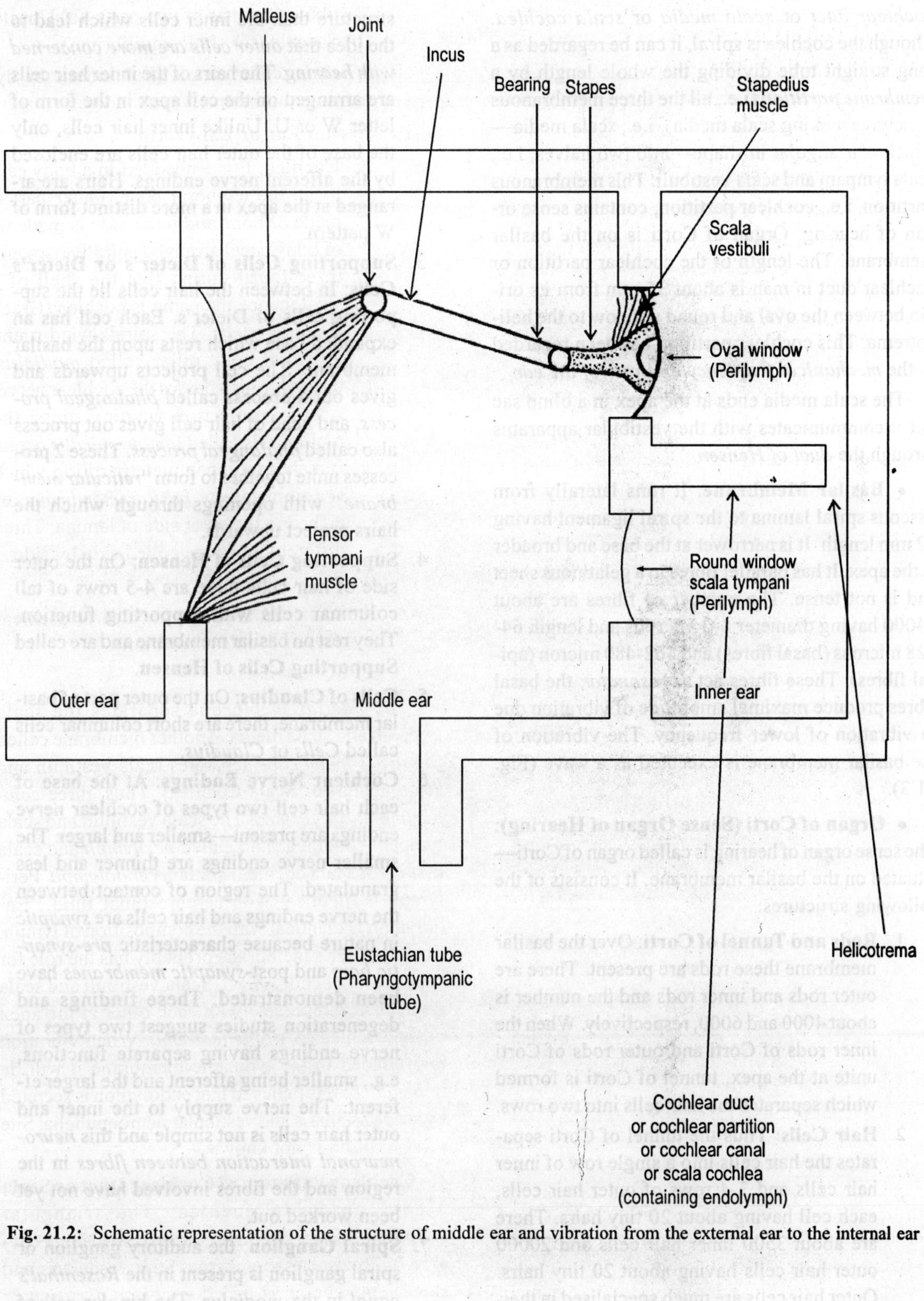

Fig. 21.2: Schematic representation of the structure of middle ear and vibration from the external ear to the internal ear

cochlear duct or *scala media* or *scala cochlea.* Though the cochlea is spiral, it can be regarded as a long straight tube dividing the whole length by a *membrane partition* (i.e., all the three membranous structures making scala media), i.e., scala media—which is triangular in shape—into two halves, i.e., scala tympani and scala vestibuli. This membranous partition, i.e., cochlear partition, contains sense organ of hearing. Organ of Corti is on the basilar membrane. The length of the cochlear partition or cochlear duct in man is about 35 mm from its origin between the oval and round window to the helicotrema. This cochlear partition has been regarded as the *mechanical frequency analyser of the ear.*

The scala media ends at the apex in a blind sac but it communicates with the vestibular apparatus through the *duct of Hensen.*

- **Basilar Membrane**: It runs laterally from osseous spiral lamina to the spiral ligament having 32 mm length. It is narrower at the base and broader at the apex. It has parallel fibres in a gelatinous sheet and is not tense. The number of fibres are about 24000 having diameter 1-2 microns and length 64-128 microns (basal fibres) and 362-480 micron (apical fibres). These fibres act as *resonator,* the basal fibres produce maximal amplitude of vibration due to vibration of lower frequency. The vibration of the basilar membrane is executed as a wave (Fig. 21.3).

- **Organ of Corti (Sense Organ of Hearing)**: The sense organ of hearing is called organ of Corti—situated on the basilar membrane. It consists of the following structures:

1. **Rods and Tunnel of Corti**: Over the basilar membrane these rods are present. There are outer rods and inner rods and the number is about 4000 and 6000, respectively. When the inner rods of Corti and outer rods of Corti unite at the apex, tunnel of Corti is formed which separates the hair cells into two rows.
2. **Hair Cells**: Thus the tunnel of Corti separates the hair cells into a single row of inner hair cells and 3-4 rows of outer hair cells, each cell having about 20 tiny hairs. There are about 3500 inner hair cells and 20000 outer hair cells having about 20 tiny hairs. Outer hair cells are much specialised in their structure than the inner cells which lead to the idea that *outer cells are more concerned with hearing.* The hairs of the inner hair cells are arranged on the cell apex in the form of letter W or U. Unlike inner hair cells, only the base of the outer hair cells are enclosed by the afferent nerve endings. Hairs are arranged at the apex in a more distinct form of W pattern.
3. **Supporting Cells of Dieter's or Dieter's Cells**: In between the hair cells lie the supporting cells of Dieter's. Each cell has an expanded base which rests upon the basilar membrane. The cell projects upwards and gives out a process called *phalangeal process,* and head of hair cell gives out process also called *phalangeal process.* These 2 processes unite together to form "*reticular membrane*" with openings through which the hairs project upwards.
4. **Supporting Cells of Hensen**: On the outer side of hair cells there are 4-5 rows of tall columnar cells with supporting function. They rest on basilar membrane and are called **Supporting Cells of Hensen.**
5. **Cells of Claudius**: On the outer part of basilar membrane, there are short columnar cells called *Cells* or *Claudius.*
6. **Cochlear Nerve Endings**: At the base of each hair cell two types of cochlear nerve endings are present—smaller and larger. The smaller nerve endings are thinner and less granulated. The region of contact between the nerve endings and hair cells are *synaptic* in nature because characteristic *pre-synaptic bars* and post-*synaptic membranes* have been demonstrated. These findings and degeneration studies suggest two types of nerve endings having separate functions, e.g., smaller being afferent and the larger efferent. The nerve supply to the inner and outer hair cells is not simple and this *neuro-neuronal interaction between fibres* in the region and the fibres involved have not yet been worked out.
7. **Spiral Ganglion**: the auditory ganglion or spiral ganglion is present in the *Rosenthal's canal* in the *modiolus.* The bipolar cell of

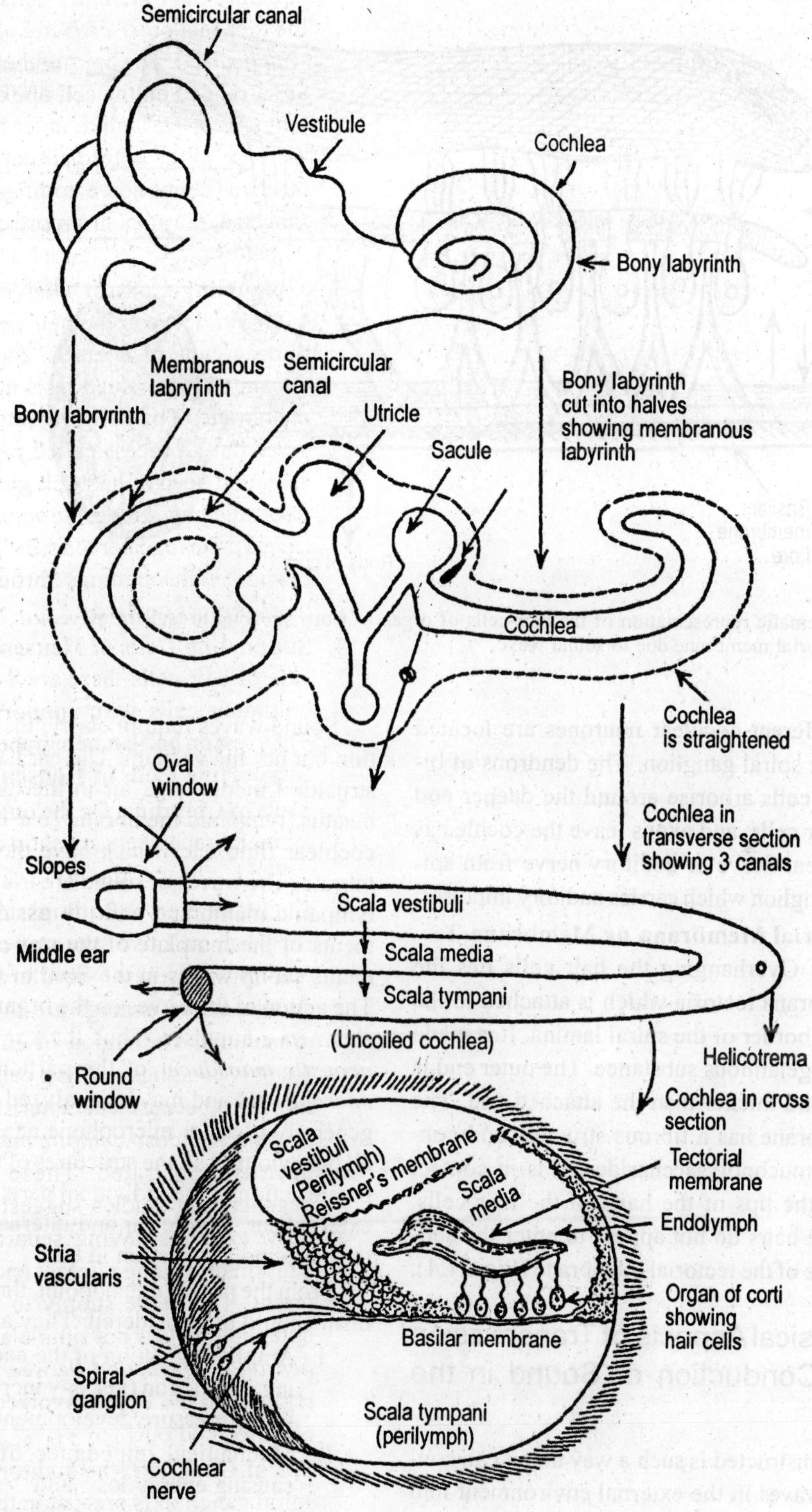

Fig. 21.3: Schematic representation of structure of labyrinth or internal ear by making different sections

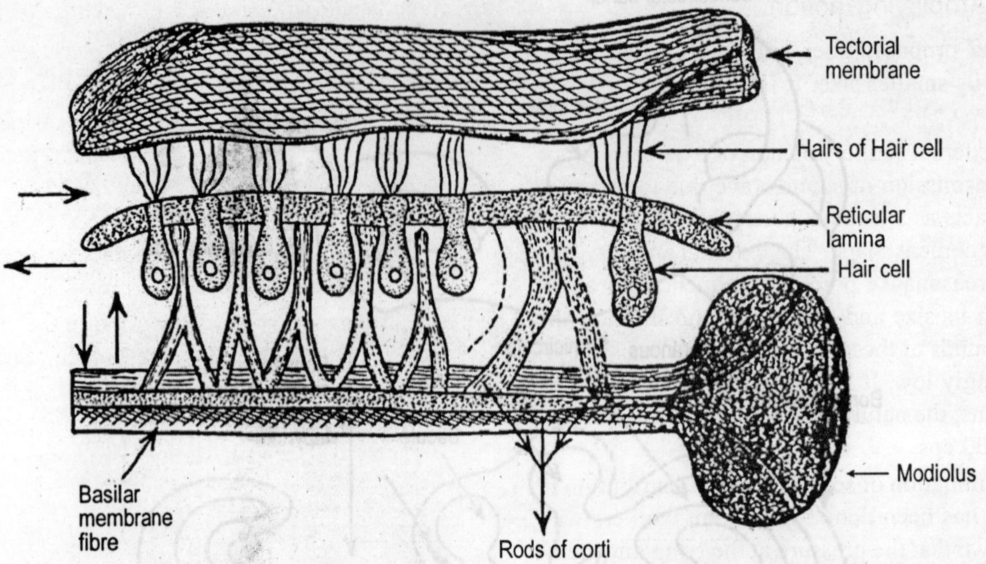

Fig. 21.4: Schematic representation of the hair-cells of organ of Corti showing to-and-fro movement of the hairs in the tectorial membrane due to sound wave

the afferent cochlear neurones are located in this spiral ganglion. The dendrons of bipolar cells arborise around the deeper end of hair cells, and axons leave the cochlea as cochlear nerve or auditory nerve from spiral ganglion which carries auditory impulses.

8. **Tectorial Membrana or Membrana Tectoria**: Overhanging the hair cells lies the membrana tectoria which is attached to the outer border of the spiral lamina. It is made up of gelatinous substance. The outer end is free and thicker than the attached end. The membrane has a fibrous structure and contains mucopolysaccharides. It is in contact with the tips of the hairs of the hair cells. These hairs do not appear to enter the substance of the tectorial membrane (Fig. 21.4).

21.6 Physical Aspects of Transmission or Conduction of Sound in the Ear

The ear is constructed is such a way that it can convert sound waves in the external environment into nerve impulses in the auditory nerves which go to brain for analysis.

Sound waves require elastic media for conduction but not the vacuum. The ear has all the elastic structural media. e.g., air in the external auditory meatus, tympanic membrane (ear drum), ossicles, cochlear fluid, etc. which have the capabilities to vibrate. The waves of sound are transformed by the tympanic membrane and the ossicles into movements of the footplate of the stapes. These movements set up waves in the fluid of the internal ear. The action of the waves on the organ of corti generates nerve impulses. Thus the *ear is a biological acoustic transducer* of the particular type called *microphones* and may be analyzed by the methods generally used for microphone analysis. The overall behaviour of all the structures of the ear is a composite action of the individual parts. In this way the external ear, middle ear and internal ear contribute to the sound conduction in ear.

From the physical standpoint, three basic mechanism should be considered. They are:

I. Amplifying action of the ear structures during conduction (i.e., any increase or decrease in the pressure development),
II. Mechanical impedance of ear structures causing energy loss, and
III. Resonance phenomenon causing frequency variation in behaviour of the ear structures.

21.6.1 Amplifying action

The pinna probably does not act as reflector because of its smaller size. It is important for localization.

The external auditory meatus or external ear tube helps transmission of sound wave due to presence of air. It acts as a trumpet concentrating the sound due to its conical shape. The external auditory meatus has reasonance points at frequencies predictable from its size and shape. The threshold of the ear for sounds of these resonant frequencies is correspondingly low. It has been calculated from the dimensions, the natural frequencies of this canal is about 2800 cps.

Determination of sound pressure distribution in the canal has been done by different workers who have shown that the pressure at the tympanic membrane side was greater than that at the opening of the canal. The maximum gain by both amplification and reasonance was 17-22 dB and occurred at frequency of 300 cps. At this frequency, the length of the canal is nearly one-quarter of wevelength.

21.6.2 Tympanic membrane

The middle ear is 15 × 6 × 4 mm, having air inside. It communicates with the pharynx by pharyngotympanic tube or Eustachian tube. The tube remains closed but opens during swallowing, chewing, yawning and sneezing. Its function is to equalise pressure on the two sides of the tympanic membrane. The tympanic membrane is elliptical in shape and has maximum concavity called "umbo" pointing inwards where the handle of malleus is attached. The ear drum probably is aperiodic and it moves in and out in response to the pressure variations produced by the sound waves. The membrane, therefore, functions as a "resonator". Helmholtz observed that cones vibrate in different frequencies of forcing, and he worked out the patterns of vibrations at different frequencies in the eardrum. At low frequencies the ear drum moves at a stiff cone, while at high frequencies only the centre portion of the tympanic membrane made with magnification and stroboscopic light have verified the existence of this phenomenon. Vigorous analysis is difficult since the eardrum is made of material whose properties are not the same over the entire membrane, and since the membrane is not symmetrical, it stops vibrating immediately when the sound wave stops, so it is very **critically damped.**

- **The Ossicles**: Movements of the tympanic membrane are relayed to the chain of ossicles—malleus, incus and stapes. The ossicles increase the efficiency of conduction of the sound waves. There are three types of sound conduction in the ear:

1. Tympnic: Ossicular conduction—usually occurs during hearing.
2. Air conduction: usually occurs in the external auditory canal due to presence of air.
3. Bone conduction: Transmission of vibration of the skull bones to the fluid of the inner ear as we feel by placing vibrating tuning fork directly to the skull.

Ear ossicles are kept in position by fine ligaments. They remain articulated together forming a complex system of levers. The mechanical joint (malleus + incus) and bearing system (incus + stapes) of the single unit of the bony ossicles are such that displacement at the tympanic membrane by air will be three times higher in the oval window and thus it acts like a hydraulic press. And the pressure changes in the liquid of the inner ear just inside the oval window will be greater than outside the tympanic membrane. The ossicles are suspended by the ligaments in such a way that the malleus and incus together act as a single lever having its fulcrum at the border of the ear drum. The large head of the malleus—which is on the opposite side of the fulcrum from the handle—balances the other end of the lever, so that changes in position of the body will not increase or decrease the tension of the tympanic membrane.

The handle of the malleus is constantly pulled inward by ligaments and by tensor tympanic muscle and keeps the tympanic membrane tensed. This allows the sound vibrations on any portions of the eardrum to be transmitted to the malleus.

The malleus rocks on an axis and through the junction of its long and short processes transmit the vibrations of the manubrium to the incus. The incus moves in such a way that the vibrations are transmitted to the head of the stapes. Movements of the head of stapes swing its footplate to and fro at the posterior edge of the oval window. Thus the three ossicles transmit the resonant sound wave from tympanic membrane into the movements of the stapes against the perilymph-filled scala vestibuli of the internal ear and also magnify the force of vibration 10 times for the following reasons:

i. The surface area of the tympanic membrane is 90 sq. mm whereas that of oval window is 3.2 sq mm. Hence the force of vibration is concentrated to a much smaller area.
ii. The ossicles constitute a "bent lever" of first class.

The actual magnification is thirty times but due to

(a) Damping action of the ossicles,
(b) Loss of energy in setting the ossicles into vibration, the *actual magnification is ten times.*

Thus, in the middle and external ear there is an overall transformer gain of pressure of about 20-30 dB.

- **Muscles of Middle Ear**: These two muscles in the middle ear—tensor tympani and stapedius—modify the transmission of sound waves:

(a) Tensor tympani muscle: Attached to the handle of the malleus. It keeps the eardrum taut. The muscle contraction pulls the handle of the malleus and the eardrum becomes tense. It has a protective function, i.e., when loud sound is transmitted, this muscle contracts reflexly and the ear drum becomes tense and the rate of vibration is diminished.
(b) Stapedius muscle inserted into the neck of stapes and tends to pull the stapes out from the oval window. It also contracts reflexly due to loud sound, diminishes the rate of vibration as a protective measure.

Attenuation of sound by contraction of the two muscles occurs in the following way:

During transmission of loud sound a reflex occurs after a latent period of 40-80 milliseconds to cause contraction of both muscles known as **tympanic reflex** or **attenuation reflex**. The tensor tympani muscle pulls the handle of maleus inward while the stapedius muscle pulls the stapes outward. These two forces oppose each other and causes high degree of rigidity in the ossicles. Thus it reduces the ossicular transmission to low frequency sound below 1000 cycles per second. This reflex can reduce the intensity of sound transmision by 30-40 dB. The functions of the mechanism are:

i. Protection of the cochlea from the damage by excessive loud sound but the latent period is 40-80 dB, so it does not protect against force of intense stimulation like bomb explosion,
ii. Masking low frequency sounds in loud environment.

21.6.3 Mechanical impedance of the ear structures

The low frequency response is determined by the elastic structures of the ear, e.g., eardrum, the air of the middle ear, ligamentous suspensions of the ossicles, membranes separating the scala of the cochlea and the membrane covering the round window. At very low frequencies the excursion of the eardrum is larger. Thus, the cushion of air in the middle ear is an important factor at these frequencies. In medium frequency region, the vibrations are controlled by viscosity of friction. Friction in the ossicle articulations and viscosity of the cochlear fluid (density is 1034 $gm.cm^2$ at 37°C and viscosity is 0.0197 cgs units) contribute in the frequency region and friction in the muscle in the middle ear may contribute. At higher frequencies the vibration is controlled by the mass of the eardrum, the ossicles and the cochlear fluid. The ossicles are so arranged that their centres of gravity are correlated with the axes of revolution and their contributions are kept to a minimum.

The amplitude of movement of the stapes with each vibration of sound is 3/4 of the amplitude of the handle of the malleus. Therefore, the ossicular system does not amplify the movement of the stapes. This system actually reduces the amplitude but increases the force of movement about 1.3 times. The surface area of the eardrum is 55 sq. mm and 3.2 sq. mm for stapes. This 17-fold difference makes 1.3-fold ratio of the lever system. Thus this ratio of the lever system allows all the energy of a sound wave striking the ear drum to be applied to the plate of the stapes. Consequently, this causes 22 times pressure on the fluid of the inner ear than which is exerted on the eardrum. It is known that fluid has greater inertia than air for which this amount of pressure is necessary to cause vibration in the fluid.

Therefore, the eardrum and the ossicular system proved *impedance matching* between the sound waves in the air and fluid in the inner ear of which 50-75 per cent is suitable for sound frequencies between 300-3000 cycles per second and permits full utilization of the energy in the receiving sound waves.

21.6.4 Resonance phenomenon

Frequency variation in behaviour of ear structures resulting from resonance phenomenon is observed.

The external auditory canal's natural frequency is about 2800 cps. The middle-ear shows about 800-1500 cps. The other structures may show different natural frequency but the overall natural frequency includes all these values. The overall resonance curve is about the same as that of the middle ear cavity. Bekesy (1936) determined the motion of the eardrum at different frequencies by fastening a small mirror to the eardrum and recording the deflection of a beam of light reflected by this mirror during the application of sound forcing. In such experiment, the natural frequency stopped completely within 0.004 second after stopping of the forcing. Thus, the ear structures are highly damped. The basilar membrane has 20000 to 30000 basilar fibres which are stiff, elastic, reed-like structures fixed at the basal ends, but not at the distal ends except that the distal ends are embedded in the loose basilar membrane. Because the fibres are stiff and also free at one end, they can vibrate like reeds of a harmonica. This type of behaviour is the basis for the Helmholtz Resonance Theory of hearing.

High frequency resonance of the basilar membrane occurs near the base and low frequency resonance occurs near the apex because of (a) difference in stiffness of the basilar fibres and (b) difference in "loading".

21.7 Transmission of Sound Waves in the Inner Ear—Travelling Waves

When the footplate of the stapes is pushed into the oval window, pressure of the cochlear fluid increases and as the stapes is withdrawn it decreases. There is no overall volume changes but this reciprocal action is necessary as the cochlear fluid is incompressible.

The movements of the footplate of stapes produces a series of "travelling waves" in the perilymph of the scala vestibuli. The height of the waves increases to a maximum as the wave travels in the cochlea and then drops within a very short time. The distance between the stapes and the maximum height point varies with the frequency of the vibrations initiating the waves. High-pitched sounds produce waves which reach maximum height near the base of the cochlea, and low-pitched sounds produce waves which reach maximum height near the base of the cochlea and low-pitched sounds produce wave and reach maximum height at the apex. It is not clear about the precise manner in which the structures of the cochlea are affected by the waves thus created. The walls of the cochlea are rigid and the Reissner's membrane is flexible. Then waves must be transmitted to all cochlear structures instantaneously. But the only means by which we can get differential displacement of different cells of the organ of Corti if there is vibration of some structure which varies in natural frequency according to its position along the cochlea. Such differential displacement is the only way for frequency discrimination. The basilar membrane can play a role in this respect. It is not under tension and it can be pressed readily into the scala tympani by the peak of waves in the scala vestibuli. Displacement of the fluid in the scala tympani are dissipated into the air at the round window. Therefore, sound waves produces distortion of the basilar membrane, and the site at which this distortion is maximal is determined by the frequency of the sound wave. The upper end of the hair cells are held rigid by the reticular lamina, the hairs are embedded in the tectorial membrane. With the movement of the stapes, both membranes move in the same direction, but they are hinged on different axes, so there is shearing motion which bends the hairs (Fig. 21.5).

21.7.1 Hair cells to nerve endings

Hair cells act as *mechanoelectric transducers*. The movement of hairs causes depolarization of the hair cells producing electrical potential across the hair cell membrane known as "*receptor potential*". This in turn stimulates the nerve endings at the base of hair cells. Regarding origin of receptor potential and mechanism of stimulation is that there is difference in composition of the two fluids (perilymph and endolymph), causing marked difference of different species of ions on the outside of the plasma membrane, in different regions of the same hair cell. It indicates that the membrane is depolarized (–60 mv to –40 mv) due to influx of K^+ ions (as the endolymph is rich in K^+ and poor in sodium) for increased permeability of the membrane causing receptor potential. Depolarization changes the receptor so that it rapidly admits the Ca^{2+} of the basal region of the hair cells to mobilize synaptic vesicles near the presynaptic membrane, induces fusion of vesicles to the plasma membrane, and releases neurotransmitter molecules into the synaptic cleft.

A reverse potential shift may occur when hairs are shifted to the opposite direction. The open ion channels are closed and by quick pumping of K^+ ions by the cell membrane, potential becomes hyperpolarised (from –60mV to –65 mV). Hyperpolarization reduces the calcium ion influx into the base of the cell and reduces the amount of neurotransmitter being released which reduces spontaneous activity in the primary neurones. The receptor function of the hair cells have the following features:

1. Degree of bending of hair: About 30.
2. Distance of bending: About 1 μm.
3. Amplifying function: Small shift of membrane potential has great amplifying action.
4. Activation: Probabilistic but not threshold process which permits detection of small displacement.
5. Process: Rapid movement of stereocilia and onset of membrane potential takes place within a few tenth microseconds.
6. Adaptation of stereocilia: Very rapid. If they are displaced by one micrometer, membrane potential is restored within a few tenths of a second before another stimulation. This facilitates the cells to be more excitable in the next displacement.

21.8 Electrical Responses of Cochlea or Organ of Corti or Cochlear Partition

21.8.1 Introduction

Wever and Bray discovered that the electrical potential picked up from the auditory nerve reproduced the exact frequencies and wave-forms of the sound reaching the ear. This finding supports Rutherford's Telephone Theory.

Davis and Saul demonstrated two distinct types of potentials:

1. **The Cochlear Microphonic Potential (CM)**
   (a) It precedes the action potential (AP) of the auditory nerve,
   (b) Having no all-or-one properties,
   (c) Closely reproduced the waveform and polarity of the stimulus.
2. **The Action Potentials (AP) of the Auditory Nerve Fibres**
   (a) Having typical threshold and refractory period of other nerves.
   (b) Does not differ from action potentials of other nerves.

Bakesy, Davis and others have shown that five different types of electrical potentials can be recorded from the cochlea by means of five electrodes inserted into the scalae.

(a) **Two** of these are the **resting potentials.**
(b) **Three** others are responses to sound stimuli.

- **Different potentials of cochlea**

1. Cortilymphatic potential.
2. Endocochlear potential.
3. Cochlear microphonic potential.
4. Summating potential.
5. Action potential.

1. **Cortilymphatic Potential**: Like other cells, the hair cells and supporting cells of the cochlea presumably have a negative intracellular potential with respect to the fluid that bathes them. After thrusting microelectrode into the organ of Corti, potentials of –20 to –80 mV is recorded which is referred to *perilymph.* Since this negative potential is found regardless of the postion of the electrode so long as it is in the organ of Corti and no one thus far has shown that it is recorded only from cells and not from the extensive fluid

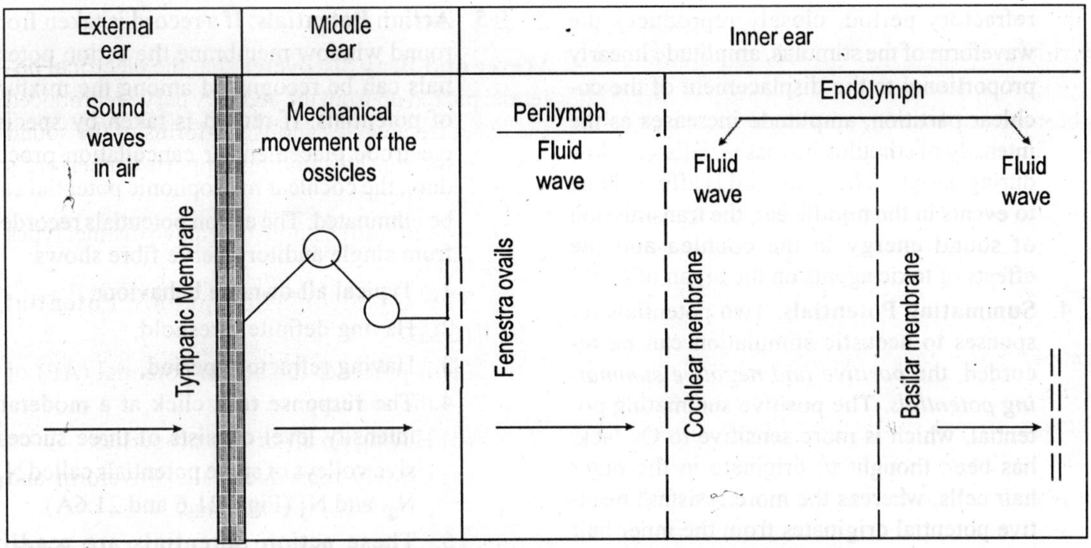

Fig. 21.5: Showing transmission of sound through the ear

spaces, it seems preferable to refer to it as the *cortilymphatic potential.*

2. **Endocochlear Potential**: The endolymph in the cochlear duct shows a remarkable *positive* potential of 80 mV. The high positive potential appears to be a unique property of the cochlea whereas the endolymph of vestibular apparatus shows only +1 to +5 mV. The endocochlear potential show the following characteristics:
   - (a) It does not depend upon the ionic composition of the endolymph,
   - (b) It depends upon the metabolic activity of the stria vascularis,
   - (c) It can be recorded from the surface of the stria even when the endolymph is removed.
   - (d) It decreases:
     - i. During asphyxia and returns to normal when O_2 is readmitted,
     - ii. When outward movement of the stapes causing upward movement of basilar membrane.
   - (e) It increases:
     - i. When the membrane is pushed downwards toward scala tympani,
     - ii. At the inward movement of the stapes,
     - iii. By injecting fluid into the cochlear duct into the scala media.
   - (f) It is modified by movements of the basilar membrane.
   - (g) It is abolished by injection of azide or cyanide into the scala media.
3. **Cochlear Microphonic Potential**: Cochlear microphonic potential is an immediate response to acoustic (hearing) stimulation. It was thought previously that elongation and compression of hair cells were causing *piezoelectric potentials* like crystal of a microphone. (Piezoelectric effect: When pressure is applied to certain nonconductive materials so that deformation takes place and a charge separation occurs in the materials and an electric voltage can be measured across the material. By this effect pressure energy is converted into electric energy. This principle is one kind of active transducer.) Recent studies have focused attention on the region of the upper end of the hair cells (transducer) as the site of *mechanoelectrical transduction* process. But the contributions of the tectorial membrane, stereocilia, cuticular plate and basal body to this process is a matter of speculation.

   Various authors suggested that it may be due to tectorial membrane alone for rapid ion fluxes. But it may be regarded as *receptor potential* having no true threshold or refractory period, closely reproduces the waveform of the stimulus, amplitude linearly proportional to the displacement of the cochlear partition, amplitude increases as the intensity of stimulus increases, falls very low during anoxia. This potential is affected due to events in the middle ear, the transmission of sound energy in the cochlea and the effects of toxic agents on the organ of Corti.

4. **Summating Potentials**: Two potentials responses to acoustic stimulation can be recorded, the *positive and negative summating potentials.* The positive summating potential, which is more sensitive to O_2 lack, has been thought to originate in the outer hair cells, whereas the more resistant negative potential originates from the inner hair cells.

5. **Action Potentials**: If a record is taken from round window membrane the action potentials can be recognized among the mixture of potentials. If record is taken by special electrode placement or cancellation procedure, the cochlear microphonic potential can be eliminated. The action potentials recorded from single auditory nerve fibre shows:
   1. Typical all-or-none behaviour,
   2. Having definite threshold,
   3. Having refractory period,
   4. The response to a click at a moderate intensity level consists of three successive volleys of spike potentials called N_1, N_2, and N_3 (Figs. 21.6 and 21.6A).
   5. These action potentials are readily "masked" by noise (Fig. 21.6 B).

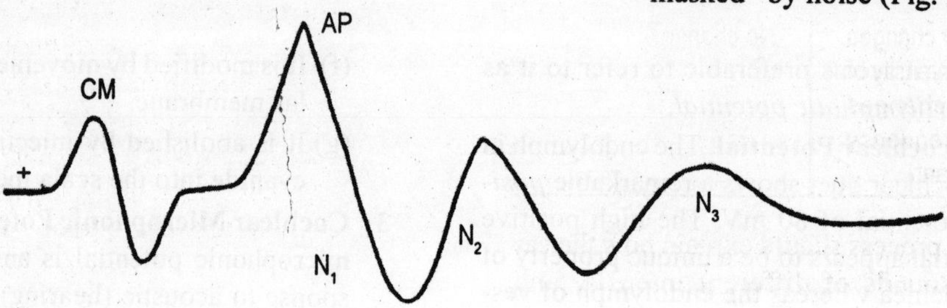

Fig. 21.6: Responses to clicks. CM-Cochlear microphonic potentials, AP-Action of auditory nerve with the spikes N_1, N_2 and N_3

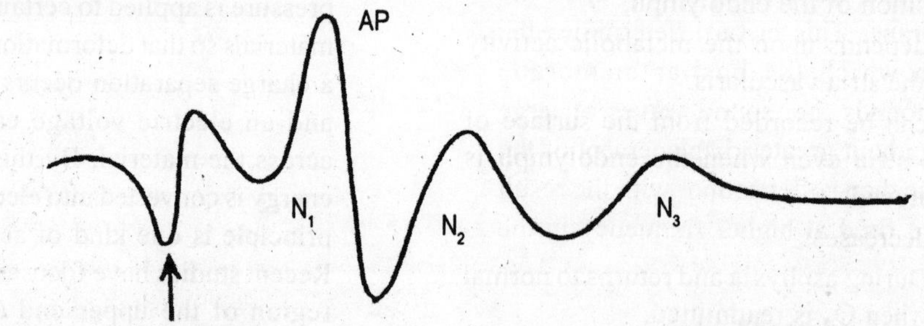

Fig. 21.6A: Polarity of stimulus reversed. CM shows complete reversal and AP shows only slight shift of latency

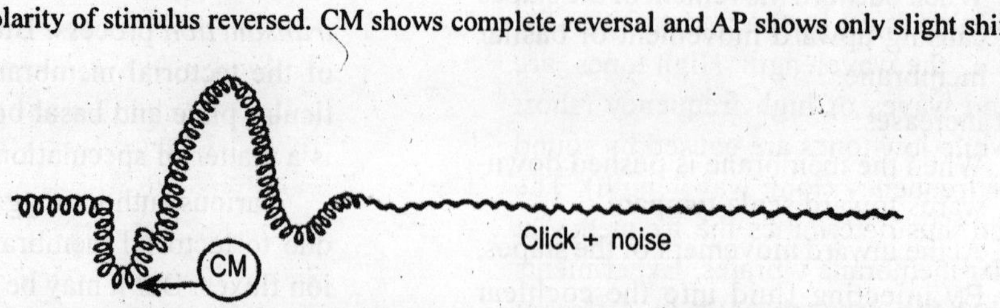

Fig. 21.6B: Masking by noise. CM-remains unchanged, AP-temporary disappearance

6. Frequency of action potentials proportionate to the loudness of the sound stimuli.

21.9 Pitch Perception

(Discrimination between sounds of different intensity and pitch, frequency analysis)

Sound pitch and sound frequency have little difference though apparently they are considered the same. The differences are:

Sound pitch	Sound frequency
1. Conscious perception of the sound frequency.	No
2. Deviates from the true sound frequency both at very high and low frequency.	No deviation
3. Pitch generally changes slightly as the intensity of the sound changes though the frequency remains constant.	No change.

The hearing process should explain how the ear discriminates sounds of different intensity and sounds of different pitch (frequency).

21.9.1 Sounds of different intensity

The intensity of a sound is determined by the amplitude of the sound waves, impinging upon the tympanic membrane. This in turn determines the amplitude with which the basilar membrane vibrates. Loud sounds, i.e. sound waves of large amplitude, bring about greater displacement of the basilar membrane than softer sound, with the result that impulses are fired at higher frequency in the auditory nerve.

21.9.2 Sounds of different pitch

The pitch of a sound is set by the frequency of the sound waves, i.e., the wavelength. High tones are caused by sound waves of high frequency (short wavelength), while low tones are caused by sound waves of lower frequency (long wavelength). The pitch of a sound thus determines the frequency at which the basilar membrane vibrates. Experiments have shown that the sensory cells in different regions of the cochlea responds to different frequencies. Those towards the apex respond to low tones (down to about 65 hertz in man) while those towards the base respond to high tones (up to about 16000 hertz in man). It has been estimated that these high frequency sounds vibrate the basilar membrane by a mere one-billionth of a centimetre.

21.9.3 Pitch discrimination and perception

Regarding pitch discrimation, how is localization of response in the cochlea accomplished? The answer can be given in the following way:

1. **Resonance and Basilar Membrane**: Each frequency of sound causes a different "pattern" of vibration in the basilar membrane and that this is by one of the important means by which sound frequencies are discriminated from the other.

   (a) **Basilar Fibres: Stiffness**: The mechanism seems to depend on the "stiffness" of the fibres of the basilar membrane and also free at one end, they can vibrate like reeds of the harmonica. The length of the fibres increases progressively from oval and round windows toward the helicotrema. The diameter of the fibres decreases from the base to the helicotrema and their overall stiffness decreases. As a result, the stiff, short fibres near the base of the cochlea vibrates at a high frequency while the long fibres near the helicotremma vibrates at a low frequency.

   (b) **Basilar Fibres: "Loaded"**: The basilar fibres are "loaded". Fibres vibrating near the base of the cochlea (near the two windows), cause high frequenc vibration because the total mass of the moving cochlear fluid is slight, whereas the fibres near the helicotrema near the tip of the cochlea having higher total mass of moving fluid cause low frequency.

   (c) **Travelling Wave**: The movement of ossicles set up a "travelling wave" in the fluid which passes along the basilar membrane for a certain distance, and then dies out. Travelling distance depends upon the frequency. High frequency waves travel only a short distance, low frequency waves tavel much further.

i. **Pattern of Vibration:** There are different patterns of transmission for sound waves of different frequencies. The general features are:

1. Each wave is relatively weak at the onset but becomes strong when it reaches at a point of basilar membrane that has a natural resonant frequency equal to the respective sound frequency.
2. At this point, the energy in the wave is completely dissipitated.
3. The wave ceases at this point and fails to travel further.
4. High frequency sound wave travels only a short distance and dies out at resonant point. Medium frequency waves travel the entire distance (Fig. 21.7).
5. It travels fast initially, becomes slow progressively as it goes into the cochlea which is due to high coefficient to elasticity of the fibres.

ii. **Pattern of Amplitude**: Maximum amplitude for 8000 cycles per sec occurs at the base of the cochlea, whereas less than 400-500 cycles/sec near the helicotrema. Another feature is that the basal end, the fibres vibrate weakly for all frequencies but beyond the resonant point for each given frequency, the vibration of the basilar membrane cuts off sharply. Sound frequencies above 400-500 cycles/sec are discriminated from each other and is based on the "place" of maximum stimulation of the nerve fibres from the organ of Corti lying on the basilar membrane (Fig. 21.8).

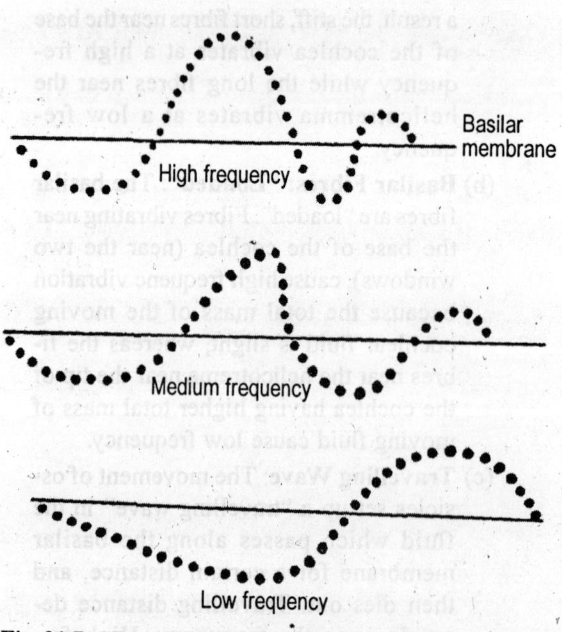

Fig. 21.7: "Travelling waves" along the basilar membrane for high, medium and low frequency sounds

(d) **Differential Sensitivity**: These coupled with differential sensitivity of the receptor cells, could result in different parts of the cochlea responding to different frequencies.

(e) **Spatial Organization and "Place Principle or Theory"**: The sensation received is determined by the pattern of impulses generated in many fibres of the auditory, i.e., cochlear nerve. Furthermore, there is spatial organization of the cochlear nerve fibres from the cochlea to the cochlear nuclei in the brain stem, the fibres from each respective area on the basilar membrane terminating in a corresponding area in the cochlear nuclei. The spatial organization continues all the way up the brain stem to the cerebral cortex. The recording of signals from the auditory tracts shows that specific neurons are activated by specific pitches. This is called the **"place principle"** for determination of pitch.

(f) **Frequency Principle or Law of Frequency**: But the distal end of the basilar membrane at the helicotrema is stimulated by all sound frequencies below 400-500 cps. Therefore, it is not properly understood from the **"place principle"**. How one can differentiate between very low sound frequencies. It is postulated that these low frequencies are discriminated by "frequency principle". i.e., the low frequency sounds cause volleys of impulses, at the same low frequencies impulses are transmitted by the cochlear nerve into the cochlear nuclei which distinguishes the different frequencies. In fact, destruction of basilar membrane where low frequency sound are produced does not eliminate the discrimination of low frequencies.

21.9.4 Theories of pitch perception (or theories of hearing) or frequency analysis

Many theories have been postulated to explain the role of cochlea in the *process of hearing* and particularly in the *perception of pitch.* Some workers have held that the cochlea performs a frequency analysis of sound, whereas others have reserved that function to the brain. The theories are:

1. **Resonance Theory of Helmholtz**: Helmholtz postulated a series of resonators in the cochlea each tuned to a different frequency. The rods of Corti with varying heights might serve as the resonantors. Later he proposed the fibres of the basilar membrane, which he compared to the strings of a harp or pianoforte, postulating that the shorter fibres near the oval window respond to higher frequencies and the longer fibres near the apex to lower frequencies. Demerits: No individual resonator is detected.
2. **Telephone Theory of Rutherford**: In opposition to resonance Theory, Rutherford ascribed cochlea as a telephone transmitter. He thought of it as transforming sounds into nerve impulses of the same frequency, which were then analyzed in the CNS. Wever and Bray (1930) discovered electrical potentials from the auditory nerve reproduced almost perfectly the frequencies and waveforms of the sounds reaching the ear which supports telephone theory. Demerits: 1. Auditory nerve can transmit 1000 impulses per second like other nerve fibre of large diameter. 2. Did not explain the frequency analysis by brain.
3. **Place Theory**: It holds that the entire cochlea is a tuned structure with different portions of the basilar membrane and organ of Corti responding to sounds of different frequencies. The corresponding nerve fibres then report to the brain which portions of the organ of Corti are being stimulated.
4. **Resonance-Volley Theory**: The theory uses the (1) Telephone or volley principle to explain the response of the whole cochlea to low frequencies and the (2) Place or resonance principle to explain the cochlear analysis of higher frequencies. Both principles now have a solid basis of experimental fact derived from two types of investigation:

   (a) Mechanics of cochlear partition.

   (b) Electrical potential of the organ of Corti.

(a) **Mechanics of Cochlear Partition:** Von Bekesy (1956) studied the physical properties of the various membranes and tissues that make up the cochlear partition by direct microscopic observation of their patterns of movement under *stroboscropic illumination,* and stated that the *place princile* is correct. The cochlea is a *tuned structure* which performs a frequency analysis of sound.

(b) **Electrical Potentials of the Organ of Corti (or Cochlea):** There are five different types of electrical potentials:

I. **Resting Potentials**

(a) Cortilymphatic Potential.

(b) Endocochlear Potential.

II. **Responses to sound stimuli**

(a) Cochlear Microphonic Potentials (CM)

(b) Summating Potentials

(c) Action Potentials.

(a) **Cochlear Microphonic Potentials**: It is an immediated response to sound stimulation and regarded as a receptor potential having no true threshold or refractory period. It closely reproduces the waveform of the stimulus and its amplitude is linearly pro-

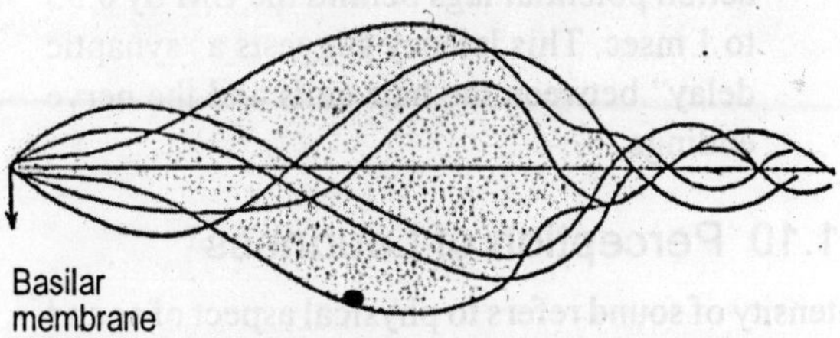

Fig. 21.8: Amplitude pattern of vibration of basilar membrane for a medium frequency sound. Shaded area is the maximum extent of vibration of basilar membrane during a complete vibration cycle

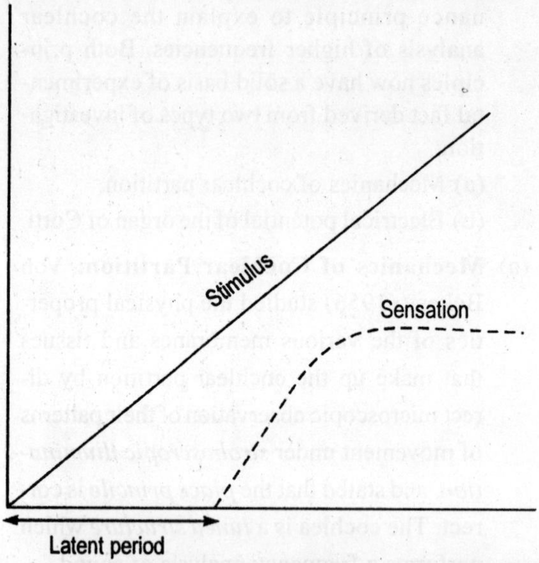

Fig. 21.9: Graph showing relationship between stimulus and sensation (Weber-Fechner Law)

portional to the displacement of the cochlear partition. This is further evidence of the broad tuning of the basilar membrane.

(b) **Summating Potentials**: The more sensitive positive summating potential originates in the outer hair cells and more resistant negative summating potential originates from the inner hair cells.

(c) **Action Potentials**: The action potentials show typical all-or-none behaviour of axon-spike potentials of myelinated nerve fibres having definite threshold and refractory period. Synchronous volleys of action potentials occur in response to sound waves of a steady tone of low or middle frequency. The action potential lags behind the CM by 0.55 to 1 msec. This latency suggests a "synaptic delay" between the hair cells and the nerve endings.

21.10 Perception of Loudness

Intensity of sound refers to physical aspect of sound whereas loudness of sound is a subjective phenomenon and it refers to the sensation it produces in the air. The sensation of loudness depends on the intensity of sound but loudness is not proportional to the intensity.

21.10.1 Loudness determination

Loudness is determined by three different ways:

1. As the sound progressively becomes louder, the amplitude of vibration of the basilar membrane and hair cells of organ of Corti also progressively increases resulting in excitation of the nerve endings at rapid rate.
2. As the amplitude of vibration increases, more and more hair cells are involved at the vibrating region of the basilar membrane for excitation resulting in "spatial summation" of the nerve impulses which means impulses pass through many nerve fibres than through a few fibres.
3. A few hair cells are not stimulated until the vibration of the basilar membrane attains high intensity and excitation of these cells by any means makes the CNS understand that sound is loud.

21.10.2 Ear perception

The ear responds to sound in a logarithmic manner in accordance with Weber-Fechner's law. Weber's law states that the increase of stimulus, which is necessary to produce an appreciable increase in sensation, always bears the same ratio to the whole stimulus. Fechner's law gives the result of an attempt to state Weber's law in mathematical terms. It states that the sensation varies as the natural logarithm of the stimulus (Fig. 21.9).

The lowest intensity that a 1000 cycle tone can have and still be audible to the normal ear is called **basic unit of sound**. The scale used for sound intensity proceeds logarithmically from this basal level. The difference of sound intensity cannot be expressed conveniently in terms of decibels (dB) change because of the logarithmic nature of the decibel scale. Therefore, for description of loudness of a tone, another unit—the "phon" is used. For standardization, simple harmonic sound with a frequency of 1000 cps at an intensity of 40 dB over the hearing threshold and listened to with two normal ears has been counted a value of 1000 phons. Measurement of the sensation indicate that a sound of 33 dB intensity has a loudness level of 500 phone, 49 dB one of 2000 phons, 73 dB one of 10000 phons, etc. Thus there is not a straight-line relation between the loudness experienced and the intensity applied (Stuhlman, 1943 shown in the Fig. 20.10).

Fletcher (1929) observed that the loudness function is different for each level of frequency. For this reason there is a family curve of loudness versus intensity, each of which corresponds to a different frequency. These lines are called **unequal loudness contours** (Figs. 21.11 and 21.11A).

Licklider (1951) plotted auditory curve and showed loudness depends upon both the amplitude and the frequency of sinusoid. The curve shows the dependency of the human ear. At each frequency two threshold intensities can be found, one for hearing and one for feeling.

Between these two, extreme loudness varies as logarithmic of intensities. The striking feature of the curve is the degree to which it depends upon the frequency. Hearing is the keenest in the 1000-4000 cps range, decreasing sharply for higher and lower frequencies. The curve is determined to a large extent by the energy losses in the middle and inner ear and to a small extent by the resonance properties of the external ear (Fig. 21.12).

21.10.3 Detection of changes in loudness (or interpretation or discrimination—power law)

A person interprets changes in intensity of sensory stimuli approximately in proportion to a power function of the actual intensity, i.e., Power Law.

Interpreted signal strength = K (stimulus-*K*)y. K and *K* constant, y = power to which the stimulus strength is raised, exponent y and constants K and *K* are different for each type of sensation. The plotted curve shows a linear one. But Power Law does not hold at either very weak or very strong stimulus strength. In the case of sound, the interpreted sensation changes approximately in proportion to the cube root of the actual sound intensity, i.e., ear can discriminate difference in sound intensity from low to the loudest sound showing several times increase in sound energy. Still the ear can discriminate this wide difference in sound level, having 10000 fold changes. Thus, the scale of intensity is "compressed" by the perception mechanism of the ear and this permits one to discriminate differences in sound intensities in wide range.

21.11 Mechanism of Hearing

Sound is the sensation produced by the air pressure changes in the environment which travels as waves in the form of longitudinal vibrations of the molecules in the surroundings, i.e., waves of alternate phase of condensation and rarefaction of the molecules in an elastic medium such as air which enters or being collected by the:

↓

Pinna and is directed into the:

↓

External auditory meatus or canal which contains elastic media like air and transmits the wave towards the:

↓

Tympanic membrane (Eardrum) on which sound waves impinge upon and set into vibrations. It is probably aperiodic and it moves in and out in response to the pressure variations produced by the sound waves. The **membrane**, therefore, functions as **resonators**. It stops vibrating immediately when the sound wave stops. So it is very **critically damped.** The vibrations are communicated through the:

↓

Auditory ossicles (malleus, incus and stapes). The ossicles increase the efficiency of conduction of the sound waves. They are kept in position by fine ligaments and remain articulated together forming a **complex system of lever**. The mechanical joint and bearing system of the ossicles make the displacement three times higher in the oval window than the tympanic membrane and thus act like a **hydraulic press**. The ossicles magnify the force of vibration 10 times. They communicate sound waves to the:

↓

Membranes covering the oval window (fenestra ovalis). The vibrations of this membrane in their turn set into motion

↓

The perilymph of scala vestibuli and scala tympani of the inner ear. The vibrations of the perilymph are communicated through the Reissner's membrane and basilar membrane to:

↓

The endolymph of the scala media (or scala cochlea, cochlear duct or cochlear canal or cochlear partition). Difference of surface area between eardrum and stapes makes 1.3 fold ratio which allows all the energy of sound waves from ear drum to stape

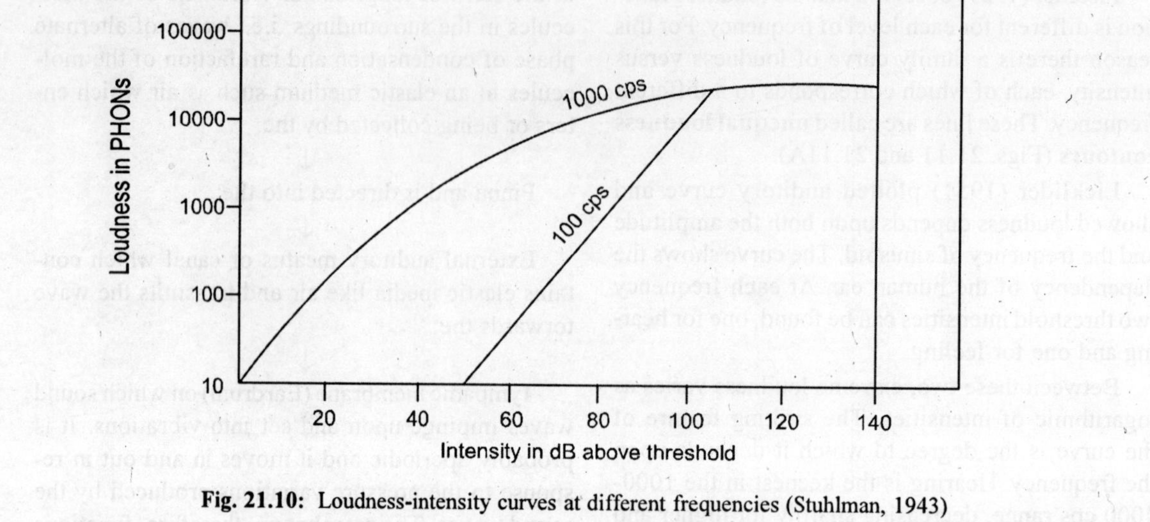

Fig. 21.10: Loudness-intensity curves at different frequencies (Stuhlman, 1943)

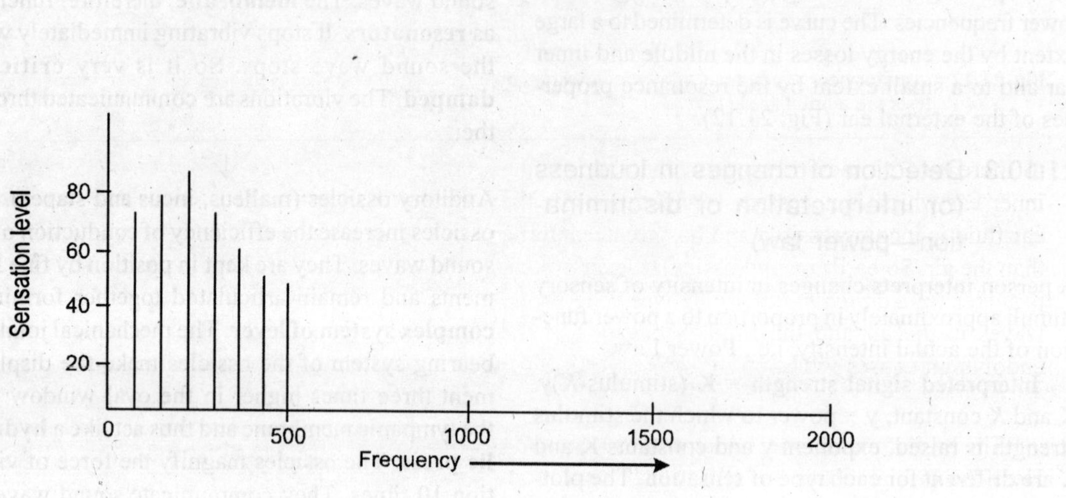

Fig. 21.11: Loudness spectrum (outer ear spectrum)

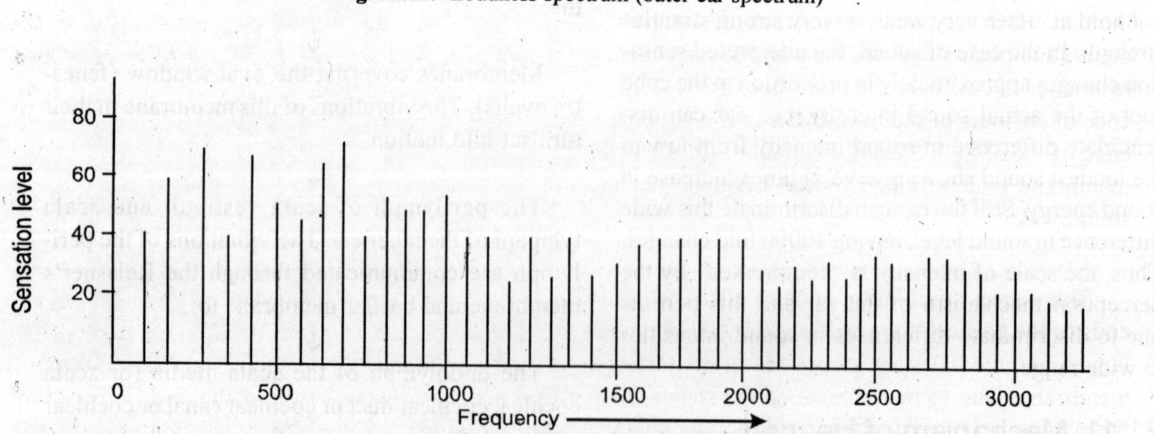

Fig. 21.11A: Loudness spectrum [Fletrum (Fletcher, 1929) (inner ear spectrum)]

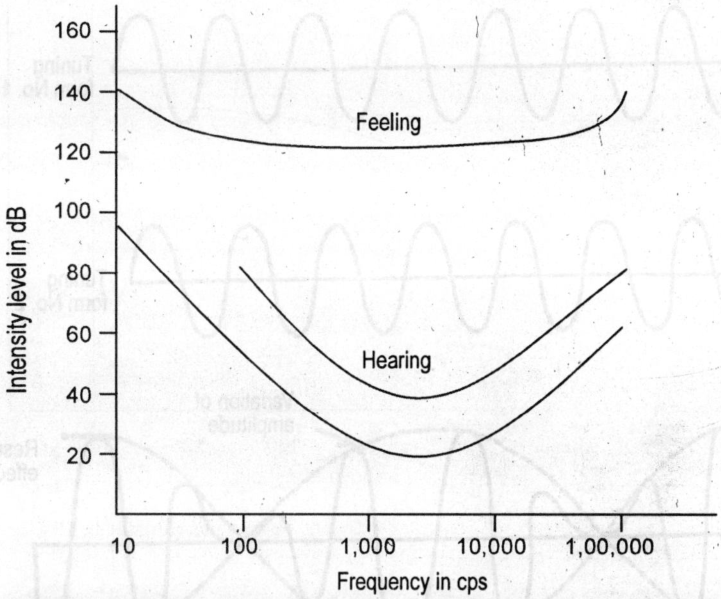

Fig. 21.12: Auditory curve of a man; showing threshold under ideal conditions and under conditions in which audiometer tests are administred (Licklider, 1951)

to increase 22 times pressure in the fluid of the inner ear which is necessary for vibrations as the ear fluid is incompressible and has greater intertia than the air. So eardrum and ossicular lever system provides *impedance matching* between the sound waves in the air and the fluid. Vibrations of the endolymph causes vibration of:

↓

The basilar membrane, which acts as *resonator.* Here the basilar fibres are stiff, elastic and reed-like structure which can vibrate like reeds of harmonica. High frequency resonance of the basilar membrane occurs near the base and low frequency resonance occurs near the apex, because (a) difference of stiffness of the basilar fibres and (b) difference of "loading". Site of distortion of the basilar membrane is determined by the frequency of the sound wave.

↓

This causes stimulation of hair cells. The upper ends of the hair cells are held rigid by the reticular lamina and the hairs are embedded in the tectorial membrane. Due to movements of the stapes, both the membranes move in the same direction, but they are hinged on different axes, so there is shearing motion which bends the hairs in one direction and depolarizes the hair cells producing electrical potential across the hair cell membrane knwon as **receptor potential** of the hair cells. Thus, hair cells act as *mechanoelectric transducer*. This in turn stimulates.

↓

Nerve endings arborise the base of hair cells which are the dendrons of the bipolar cells of the spiral ganglion.

↓

Impulses enter the bipolar cells of the spiral ganglion (situated in the Rosenthal canal in the modilous). These are the first order of the neurones.

Axons from the spiral ganglion unite to form cochlear nerve (nerve of hearing), i.e., cochlear division of eighth cranial (auditory) nerve.

21.12 Beats and Dissonance

When two tuning forks are vibrated, the vibrations differ by one or two per second, the phenomenon is called **beats**. The phenomenon is due to the summation and interference of the waves from the tuning forks. As for example, when two tuning forks are vibrated together having vibration 100 and 101 times per second, at first there is coincidence of waves of compression of each tuning fork and the

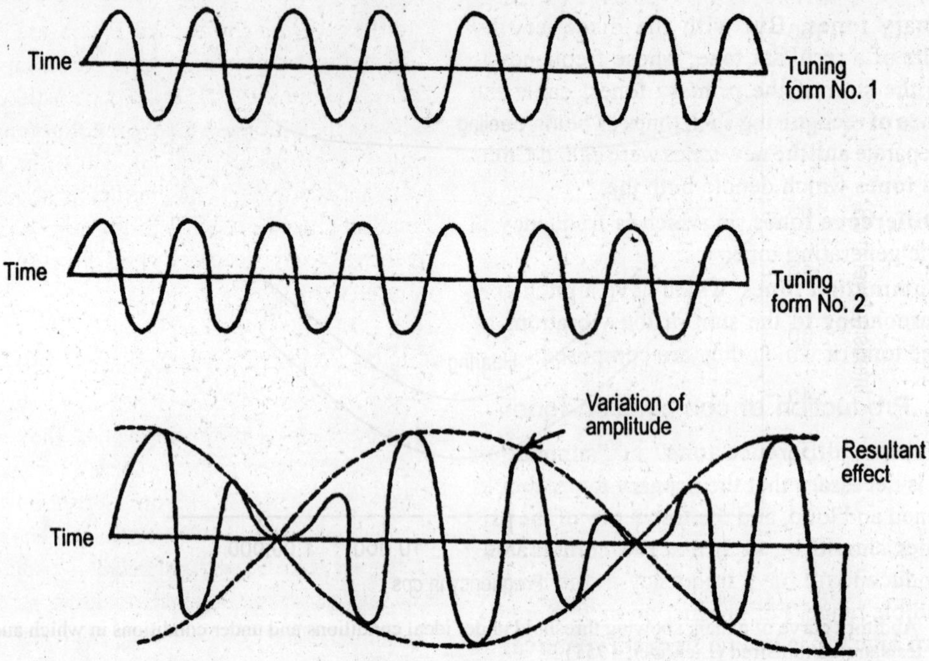

Fig. 21.13: Phenomenon of beats

total compression of the air at each beat is the resultant effect of the compression produced by each tuning fork. After half a second, the tuning forks show different phases so that the tuning fork of 100 vibrations moves in one direction while the fork having 101 in the other direction, the compression of one fork coincides with the expansion of the air produced by the backward movement of the other fork. The sound produced by one fork is reduced by the sound of the other fork, so the total sound is lesser than the sound produced by either of the tuning forks.

Again, at the end of one second, the phases of the two tuning forks coincide and the sound increases in loudness and there is alternate waxing and waning of the sound which occurs once in a second and known as *beat* (Fig. 21.13).

The number of beats may be used to determine the differences in the vibration frequencies of two tuning forks. Beats can be used to measure the unknown frequency of a note, to "tune" an instrument to a given note. For example, two tuning forks vibrating at 100 and 104 will give four beats per second. As the number of beats increases, the perception by the air is unpleasant, and it is more marked when the beats occur at about 33 times per second and the individual beats are not discriminated and the sound is referred as **discordant or dissonant.** It is observed that the loud sound is heard after every 1/8 second, and thus the beat frequency is 8 cycles per second. This is the difference between the frequencies 48 and 56 of the two notes. The beat frequency is always equal to the difference of the two nearly equal frequencies.

Time	Tuning fork No. 1
Time	Tuning fork No. 2
Time	Resultant effect

21.13 Combination of Tones

When two tuning forks are vibrated together at an interval of one-fifth between them, a weak lower tone may be heard, the pitch of which is an octave below that of the lower fork, i.e., when two tones are sounded together, a quite new tone is produced. The frequency of this low-pitched new tone is equal to the difference of the two high-pitched tones. The resultant tone was initially considered as a **beat tone**, the frequency being equal to the difference of the primary tones. But with the discovery by **Helmholtz of a resultant tone,** whose frequency is **equal to the sum of the primary** tones, came the **importance of recognising such** tones as being completely separate and the new tones were called **Combination tones** which denote both the:

1. **Difference tones,** in which is frequency is the generating tone;
2. **Summation tones,** which have a pitch corresponding to the sum of the vibrations of the tone of which they are composed.

21.13.1 Production of combination tones

To produce a "difference tone" or "summation tone", it is necessary that the primary tones should be sustained and loud, and the frequency of the primary tones should be such that the combination tones produced will have frequency in the middle of the audible range. For production of "difference tone", two high notes having frequency interval of 400-500 cps should be sounded loudly and it can be heard by sounding two tuning forks of moderately high frequency, differing in frequency by about 50 cps. It is difficult to produce "summation tone"; for its production low tones are preferred.

21.13.2 Objective and subjective combination tones

The subjective combination tones are those which are realised only by the ear and have no existence outside the ear. It was first thought and beat tone theory suggests that the "difference tone" which is identical with beat tone is only subjective. Later it was found that the "difference tone" can exist outside the ear, which can be demonstrated. By means of appropriate resonators these tones can be reinforced showing that they have an objective existence. They are, however, reinforced in the ear by the asymmetry of the tympanic membrane and auditory ossicles. As a result of these combination tones the sound spectrum of the motion conveyed to the cochlea is much richer in higher frequencies than is the wave motion incident on the eardrum.

Several theories, e.g., Beat Tone Theory and Helmholtz Intensity Theory have been proposed to explain the different aspects of combination tones, but most satisfactory among them is Wactzman's General Asymmetric theory. According to this theory, the eardrum with its load of ossicles on one side behaves as the loaded membrane producing "subjective combination tone". Similar rectifying properties may be associated with other systems also, so that "objective combination tone" may be expected in the external air. Thus, the theory explain satisfactorily the salient facts about the combination tones as regards to the amplitude and frequency; the existence of subjective and objective combination tones.

21.14 Consonance and Dissonance

When the combination of two or more notes produces a pleasing effect to the ear, they are said to produce a *consonance* or *concord.* When the combination gives an unpleasant sensation to the ear, then they are said to produce *dissonance* or *discord.* If two notes are selected at random, usually the sensation in the ear is unpleasant, and in the first place, there is absence of beats, e.g., of rhythmic oscillations of amplitude of sound waves which reach the ear. The complete consonance is attained when the two notes sounded together are identical in frequency and more complete consonance when a note is sounded together with its octave (freuqncy ratio 2 : 1). If we take two notes in unison (freuqney ratio 1 : 1) and maintain the pitch of one of them constant while that of the other is gradually increased, we find that the degree of dissonance goes through a series of maxima and minima. It is found that when the intervals between two notes are expressible in simple ratio as 1 : 2, 2 : 3 or 3 : 4, etc. The consonance is then attained. The octave and other consonant intervals which are employed in music are as follows:

Frequency ratio	Interval
1 : 1	Unison
1 : 2	Octave
2 : 3	Fifth
3 : 4	Fourth
4 : 5	Major third
5 : 6	Minor third
5 : 8	Minor sixth
3 : 5	Major sixth

It is observed that in all these consonant combinations, the vibration frequencies of the notes are in proportion to small whole numbers.

The physical explanation of dissonance was given by Helmholtz Theory of Dissonance which states that it is due to unpleasant beats produced by component notes that the degree of dissonance depends both on the beat frequency and on the interval between the notes. If the beat frequency is up to five or six per second, the effect is pleasing to the ear. As the beat frequency increases, the sensation becomes less pleasant and the beats become too rapid, they are perceived as a roughness in the notes and as the interval between two notes are still further increases, the sensation of harshness diminishes and finally disappears.

Beats are periodic variation of loudness of sound and they produce a jerking effect on the tympanic membrane which are sometimes unpleasant. During waning of sound, the auditory nerve becomes fatigued and during waxing of sound it comes to normal. If the beats are too rapid so that the next waxing pulse reaches the ear before it comes to its normal states, the effect becomes too tiring for the nerve and an unpleasant sensation is produced in the ear. But the effect does not only depend upon the number of beats. The number of beat that produces a dissonace is seen to vary in the different parts of the scale. The beat causing dissonance may also be produced due to the overtones. Thus dissonance depends to a certain extent on the quality of the notes. Beats may also be produced by the components of the combination tones. It is observed that the unison, the fifth and the octave are completely free from dissonance.

21.15 Sound Analysis

The ear can perceive appreciably the differences between different musical instruments having overtones present in the sound from each instrument. But when a number of these instruments are sounded simultaneously, the ear can pick out from the compound sound the notes due to an individual instrument. And a trained person can easily name the notes composing any chord struck on an instrument, e.g., piano, harmonium, etc.

This analytical power acquired by the auditory apparatus may be stated in the form of law, known as Ohm's Law:

"Every motion which the air corresponds to a composite mass of musical tones is capable of being analysed into a sum of simple pendular vibrations, and to each single vibration corresponds a simple tone, sensible to the ear and having a pitch determined by the periodic time of the corresponding motion of the ear."

21.16 Sound Localisation

Man and animal determine accurately the direction from which the sound is coming. Though the subject is not aware about it, it can obtain the information, and the following factors play a major role to perform the sound localisation:

1. The intensity of the sounds coming to both the ears. The ear gets more powerful stimulus when the sound comes to the same side of the ear. The relative intensities of the components of high and low pitches vary with the direction of the sound. Because the notes of low pitch (long wavelength) are diffracted to ear which is away from the sound than those of high pitch (short wavelength). The intensity factor is important for frequencies above 1000 vibrations/second.
2. The sound coming to the ear nearer to the sound reaches earlier than the other ear because of the time taken to travel round the head. So the nerve impulses from the two ears do not reach at the same time and by appreciation of the difference in time, the approximate position of an external object can be determined. The time factor is important when the frequencies are below 1000 vibrations/second.
3. The ability to turn the ears in different directions to find the direction of maximum intensity is seen in the animals, is another important factor in sound localization.

21.17 Physical Basis of Voice (Speech)

Among the stimuli which act on the auditory analyser of man, some are very essential for articulation of sound constituting speech. The speech consists of two separate mechanical functions:

A. Phonation: Performed by the larynx.

B. Articulation: Performed by the structures present in the mouth.

A. **Phonation**: The ability to produce sounds is performed by a special vocal apparatus which is also present in many animals. The human vocal apparatus consists of:

(a) *Larynx*: Specially adapted to act as vibrator.

(b) *Vocal cords present in the larynx*: A revibrating element. These are symmetrically situated folds of the mucous membrane of the larynx containing elastic fibres, along the lateral wall of the larynx, stretched and positioned by several muscles, within the larynx. These are set into vibrations by currents of air emitted by the lungs. These produce sound not only by the expiratory, but also by the inspiratory current of air.

21.17.1 Vibration of the vocal cords

Voice is produced not so much by the vibration of the vocal cords, as by the condensation and rarefaction of the column of the air, caused by these vibrations, mainly above the vocal cords,

Vocal cords may be set into vibrations provided they are in state of tension and the aperture of the glottis narrowed so as to afford resistance to the current of air.

In reed wind instruments, a column of air is caused to vibrate by the vibrations of an elastic reed in the direction of current of air, i.e., longitudinal direction, the vocal cords vibrate mainly in the direction transverse to the current of air. Experimental observations on tracheotomy patients reveal that the pressure of air in the trachea, necessary to cause production of voice, is, for a tone of ordinary loudness and pitch, between 140-120 mm of water and with loud shouting the pressure rises to 945 mm of water. The pressure is provided by the contraction of expiratory muscle, i.e., abdomen and thorax. The higher the pitch of the sound, the tenser the vocal cords, the shorter their vibrating part the higher the pressure of the air in the larynx. The intensity of the sound is approximately proportional to the force.

The cause of vibration can be explained in the following way: When the vocal cords are apposed and the air is expired, pressure of air below the cords makes the vocal cord apart, allowing rapid flow of air between their margins. The blast of air produces a partial vacuum immediately between the vocal cords which pull them close to each other. This causes stoppage of flow of air, pressure generates behind the vocal cords and the cords open again and continues in a vibratory way. The vibrating part of the vocal cords is shortened because the cords approach each other to the point of contact in the posterior part of the glottis, as a result of which only the anterior part of the vocal cords vibrates.

21.17.2 Frequency of vibration

For variations in the pitch of the sound, the following means are probably employed by the larynx:

1. By changing the tension of the vocal cords either by stretching or relaxing.
2. By changing the length of their vibrating part that is free to vibrate.
3. By varying the pressure of the blast of air passing through the glottis.
4. By changing the shape and mass of the vocal cord edges.

21.17.3 Characteristic features of human voice

The voice, like the sound produced on any musical instruments, may vary either in pitch, loudness or in quality. The scale of frequencies reproduced by human voices ranges between 80-1024 cycles. The range of any individual voice is generally about two octaves, i.e., the ratio between the extreme frequencies is approximately 1 : 4. In some cases, an individual voice may be of four octaves whereas piano has seven octaves and the organ eight octaves.

Since the pitch of the note produced rises with the increasing force of the blast of air, while the tension of the cords remains constant, the phenomenon is known as the act of "selling" on a note. At that time, tension of the cord should have been diminished but due to failure to achieve this muscular relaxation, often causes a vocalist to sing sharp when "swelling" on any note.

The pitch of the voice usually employed is determined mainly by the length and mass of the vocal cords and thus in children the voice is high-pitched. During puberty, there is significant development in the size of the larynx in both sexes, which is usually marked in the male causing rapid drop in pitch called "breaking" of the voice. In the female the greater size of the larynx is chiefly perceptible in the increase in fulness and richness of the voice which occurs at this age.

In singing, the voice may be produced in various ways, i.e., in different "registers" and thus it is distinguished into "the chest register", "the middle register" and the "the head register". The deeper notes of any individual voice are always produced in "the chest register".

According to the pitch of phonation, the following "vocal registers" are distinguished: bass –80 to 341 cycles, tenor –128 to 518 cycles, alto –170 to 683 cycles and soprano –246 to 1024 cycles. The vocal register is determined mainly by the length of the vocal cords. In females, the vocal cords are shorter (15-18 mm) than man (20-24 mm) and their voices have a higher pitch than those of men. In boys, the size of the larynx significantly increases at puberty, for which the voice passes to a lower register.

Depending on the frequency of vibration, the sound may be audible or not. Audible range covers frequencies of vibration between 16-16000 cps. According to the vibration, sound may be called:

1. **Infrasound, subsonic vibration or sound:** Sound of frequencies less than the lower limit, i.e., below the threshold of hearing (20 Hz).
2. **Sonic vibration or audio vibration:** Sound of frequencies between 20-20000 Hz is perceived by human ear.
3. **Ultrosonic vibration of sound or ultrasound:** Sound of frequencies above the upper limit of hearing (20000 Hz).
4. **Supersonic which is not audible vibration or sound:** Frequencies of sound that is greater than the upper limit (50-50000 kilocycles per second). Subsonic and supersonic, both belong to regions beyond human perception (The vibration of a lower frequency has a "bass effect" and that with a higher frequency a "shrill effect".)

(Hz = hertz is the unit of frequency and denotes the number of vibration produced by a body in one second).

21.18 Infrasound or Subsonic Vibration of Sound

21.18.1 Production

Infrasound or subsonic vibration is produced by chipping and riveting hammers, road drills and chain saws, etc.

21.18.2 Effects

A. **Acute Effects**: There are no immediate actue effects from this source.

B. **Chronic Effect**: These effects are related to the length of exposure rather than the physical characteristics of vibration:

1. Reduced sensitivity to pain and extremes of temperature in fingers and hands. Severe injuries may only be recognised when blood is seen rather than pain felt.
2. Pain of joints of the affected limbs.
3. Thickening of the skin over the fingers and palms of the hands.
4. Changes in the circulation of hands and fingers leading to cyanosis and dead white fingers when exposed to cold air or water. Reynaud's phenomenon or traumatic vasospastic disease may also occur.

21.18.3 Precautions

1. Limit the length of time spent operating these tools.
2. Provide padded gloves, particularly for the left or steadying hand.
3. Warning or working conditions.
4. Design of the tool to reduce vibration including silencers for the exhaust air.

21.19 Ultrasonics or Ultrasound

What is Ultrasound?

Ultradound is the name for mechanical oscillations (e.g., density waves pressure waves, etc.) with frequency (f) between 20 kHz and 1 GH_z. It lies beyond the audible range of human. The human audible range is about 20 Hz to 20000 Hz (20 kHz). Sound of frequency less than the lower limit is called **subsonic** sound and that greater than the upper limit of audible range is called **supersonic.**

Ultrasonics and sound waves of higher frequency (1.9-7.5 MHz) at or above the upper limit of hearing (20,000 Hz) which are inaudible to a human being and since velocity = wavelength × frequency, ultrasonics have short wavelengths compared with sound waves in the audio-frequency range.

21.19.1 Generation of ultrasonic vibration

The pulse generator produces a short electrical pulse at a required frequency such as 7.9 MHz which is fed to the transducer (Probe). The piezoelectric is the most effective method of generation of ultrasound waves.

There are several generators of ultrasonic vibrations and they are grouped according to the mode of excitation. They are:

1. Mechanical generator,
2. Thermal generator,
3. Piezoelectric generator,
4. Magnetostrictive generator.

21.19.2 Source of ultrasonics

Ultrasound is deliberately produced either as a continuous or pulsed beam which has a variety of uses in industry such as drilling, welding plastics, maxing liquids, washing machines.

21.19.3 Difference with audible sound

Ultrasound waves have same physical properties as the audible waves but they are preferred in medical diagnosis due to the following reasons:

1. The waves can be **focused easily** because of **directional and less spreading.**
2. They are **inaudible** and advantageous for medical application.
3. **High frequency waves associated with shorter wavelengths** have advantage to investigate the properties of small structures because the **wavelengths** utilized is of the **same order as the dimensions of the defect.**

Clinical Application	Depth of focused field	Ultrasound Frequency
1. Obstetric	2-100 mm	2 MHz
2. Cardiovascular	20-100 mm	4 MHz
3. Peripheral vascular	8-15 mm	8 MHz
4. Ophthalmic/ Peripheral	7-10 mm	10 MHz

4. Information received in dynamic studies cannot be obtained by any other convenient technique.

21.20 Properties of Ultrasound

Velocity: Like other forms of sonic energy, ultrasound exists as a sequence of alternate compression and rarefactions of a suitable medium i.e. air, water, bone, tissue etc. and propagated through the medium at some velocity.

Frequency i.e. wavelength of the sonic energy: The behaviour also depends on the frequency i.e. wavelength of the sonic energy and the density and the mechanical compliance of the medium.

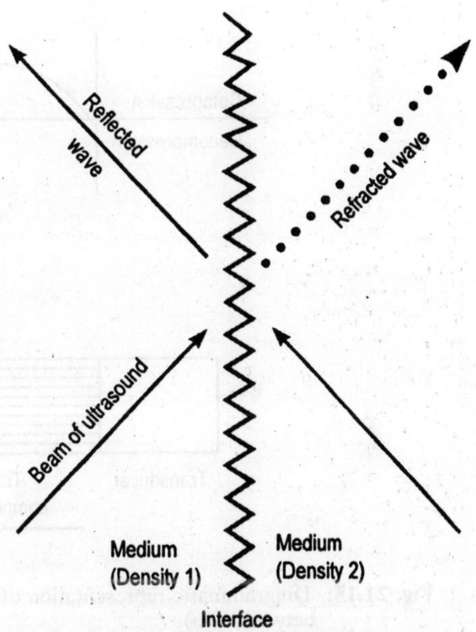

Fig. 21.14: Reflection and refraction of ultrasound at an interface of two media

Focuses into a beam: It can be focussed into a beam and obeys the laws of reflection and refraction which is used in diagnostic application.

Reflection and refraction: Whenever a beam of ultrasound passes from one medium to another a portion of the sonic energy is reflected and the remainder is refracted.

1. **Propagation of wave**

(a) Electromagnetic waves such as light and X-rays can propagate in a vacuum, but the propagation of ultrasound is related to matter. They can propagate in solids, liquids and to a limited in gases.

(b) Regarding sound waves propagation the *human body is considered as a loaded fluid due to lack of homogeneity.*

(c) When ultrasound waves are introduced into the human body, the *particles in the tissue produce longitudinal oscillation* (longitudinal waves are in which the oscillation direction of the particles is the same as the direction of wave of propagation) about their points of rest. The propagation is characterized by periodic compression i.e. pressure rise and decompression i.e. pressure reduction in the direction of propagation (Fig. 21.15).

(d) The ultrasound propagation speed is dependent on the compressibility of the medium and its density at constant temperature.

(e) The speed of sound in the human body is 1540 m/s for the transmitted pulse and echo. The table 21.1 given shows the actual speed of sound in human tissue range between 1440-1700 m/s according to the density.

2. **Doppler Effect**: Ultrasound frequently utilized in biomedical instrumentation is Doppler effect in which the frequently of the reflected ultrasonic energy is increased or decreased by a moving interface.

Ultrasound is sonic or soner energy at frequencies above 20 kHz. Ultrasound waves

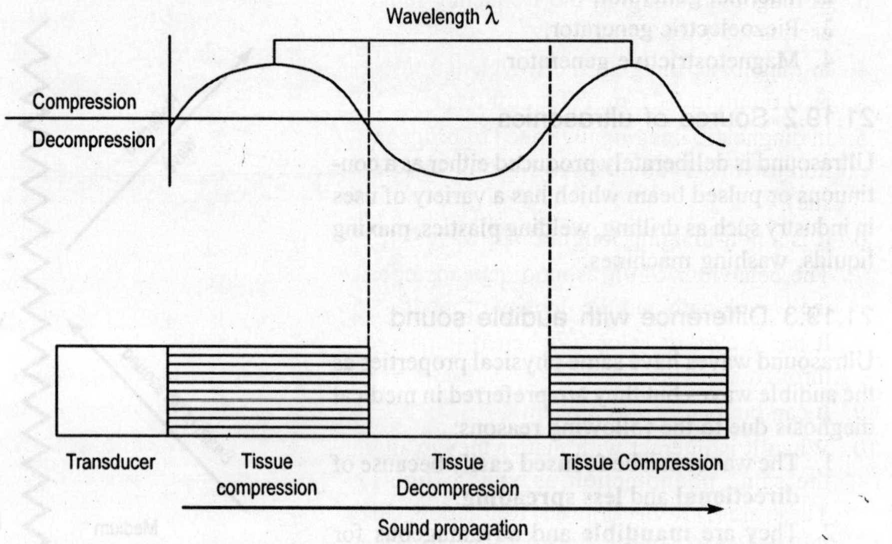

Fig. 21.15: Diagrammatic representation of longitudinal wave (wavelength is defined as the last separation between two).

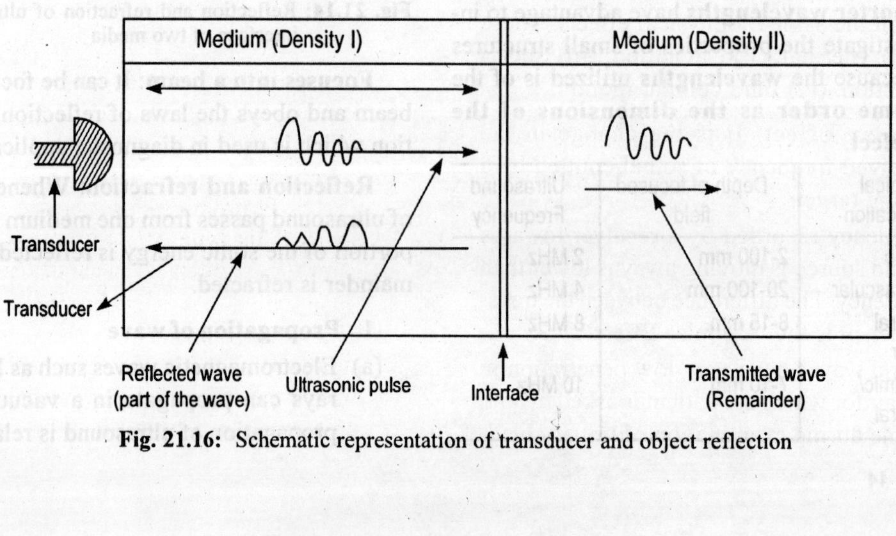

Fig. 21.16: Schematic representation of transducer and object reflection

are longitudinal waves. They are transmitted mechanical vibrations along the axis of transmission, forming bands of compression and rarefraction. The direction of ultrasound oscillation of the particles is longitudinal to the direction of transmission of the resultant waveform.

The waveform ■ λ ■ of a longitudinal wave is the distance between two consecutive bands of equal compression or rarefactions.

$\lambda = c/f$

where, c = speed of propagation
f = frequency of ultrasound.

3. In diagnostic imaging the frequency range used is 2.5 MHz to 10 MHz.
4. In diagnostic imaging the wavelength used is: 0.6 mm to 0.15 mm (for soft tissue).
5. In diagnostic imaging the speed of propagation used is: 1480 m/s (Average for soft tissue).
6. It is a non-imaging longitudinal wave.
7. The behaviour of ultrasound depends upon the wavelength and the density of media.
8. It has a certain velocity when it is passed through a media.
9. It can be focussed as a beam.
10. When it is interact with matter they exhibit the same phenomenon as visible lights. These are (i) Reflection, (ii) Refraction, (iii) Diffraction (scattering), (iv) Attenuation or absorption.
11. When it interacts with biological matter, they exhibit the same biophysical effects. These are (i) Thermal, (ii) Cavitation, (iii) Shearing action, (iv) Intra cellular motion.
12. **Dopplar Effect**: It is the change in the preceived frequency of sound emitted by a moving target.

The frequency recorded by a receiver towards which a sound source is moving away is lower than the true frequency. The difference between the transmitted frequency is the "Dopplar Effect".

The sound waves have very low penetration in presence of air for which the patient is asked to hold back the urine during examination of urinary bladder which serves as an *acoustic window*. Because sound waves have:

1. Very low penetration power in air.
2. Rapid dissipation in air for which it has little value for investigation of:
   i. Air filled structures like lungs
   ii. Structure surrounded by bones e.g. brain, spinal cord.

21.21 General Features of the Diagnostic Ultrasound

Ultrasound is usually a combination of the following consideration:

1. Tissue structures can be imaged in detail. This is particularly true for soft tissue, which are difficult to depict by conventional X-ray techniques.
2. Movement of tissues within the body can be presented on display screens and chart recorders.
3. Rapid examinations are possible with relatively inexpensive equipment.
4. Portable instruments are available.
5. Blood flow can be recorded using Doppler effect and in some instances, measured quantitively.
6. Regions of blood flow can be visualised and related to the vascular structure.
7. No harmful effects have been shown to the associated with diagnostic ultrasound. There is no risk to the operator.
8. Many examinations are non invasive and do not distress the patient. Repeated examinations can therefore be made.
9. Treatment can be monitored.
10. Transducer can be made of small and suitable for invasive applications.
11. Needles catheters can be directed under ultrasonic guidance.
12. Information can be obtained about the nature of a tissue from its effect on the ultrasonic beam. This type of information is being gained more often as techniques are refined.
13. Quantitative measurements can be made of structures within the body. For example, the biparietal diameter of a fetal head, the volume of a liver, or the movement of cardiac structures can be quantified.

21.21.1 Diagnostic application of various Ultuasonic techniques

B mode

1. Liver
2. Gall bladder
3. Spleen
4. Pancreas
5. Lymph nodes
6. Kidneys
7. Urinary bladder
8. Prostate
9. Uterus and pregnancy diagnosis
10. Testes and penis
11. Abdominal vessels
12. Breast
13. Nasal cavities
14. Neck vessels
15. Thyroid
16. Neonatal brain
17. Ophthalmology
18. Joints and muscles
19. Extremity vessels
20. Cardiac diagnosis

A mode

1. Echoencephalogram.
2. Ophthalmology (Echophthalmogram)

M mode

1. Cardiac diagnosis (Echocardiography)

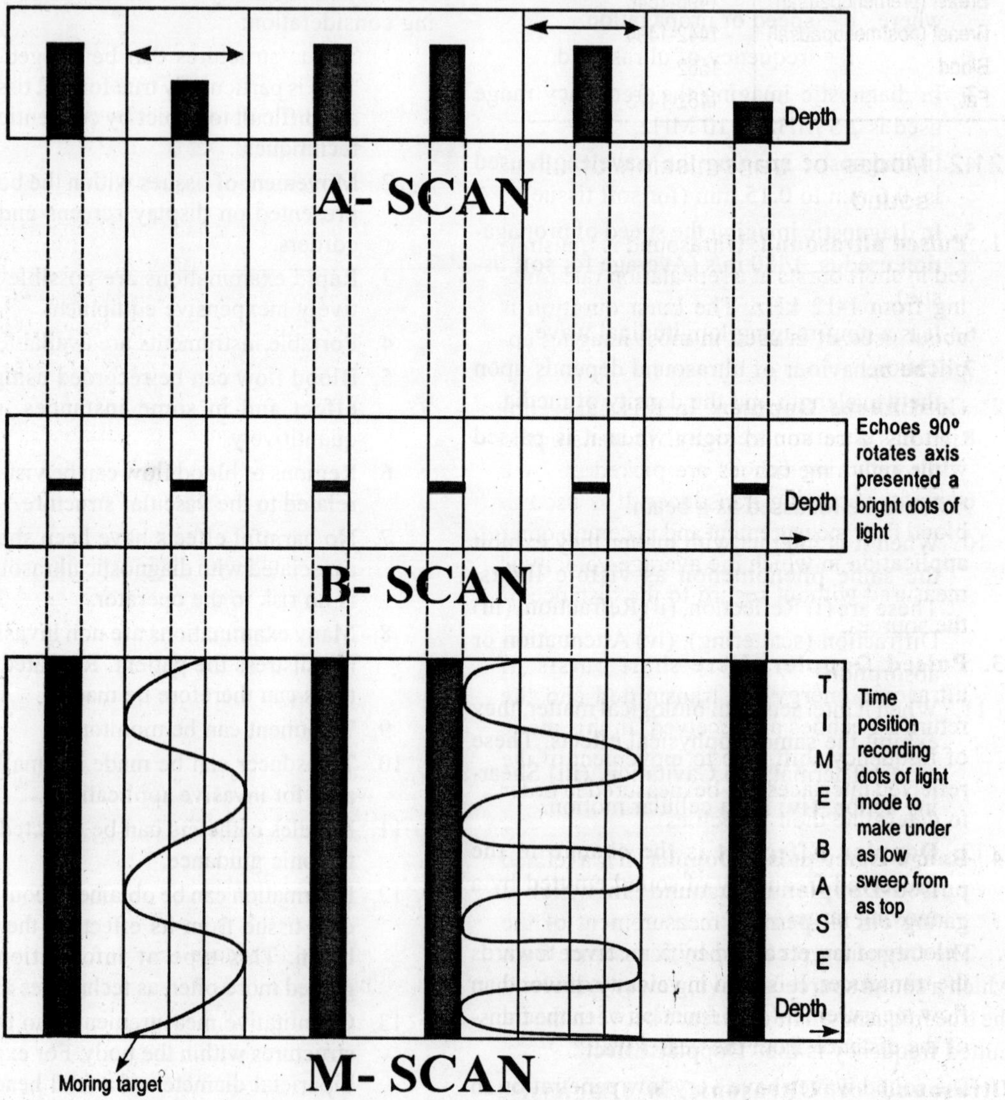

Fig. 21.17: Modes of display of ultrasound

PROPAGATION OF ULTRASOUND

Table 21.1: Speed of sound in various media of the body

Body Media	Speed of sound (m/s)
1. Air	331
2. Water	1485-1626
3. Bone	3000
4. Nerve tissue	1508-1667
5. Muscle	1548-1632
6. Liver	1538-1580
7. Kidney	1564
8. Spleen	1556-1577
9. Breast (premenopausal)	1490-1546
10. Breast (postmenopausal)	1442-1494
11. Blood	1562
12. Fat	1462-1478

21.21.2. Modes of transmission of ultrasound

1. **Pulsed ultrasound**: Ultrasound is transmitted in short bursts at a repeatation rate ranging from 1-12 kHz. The burst duration is about μsec. It is used in most imaging application.
2. **Continuous Doppler**: In this case continuous ultra sound signal is transmitted while returning echoes are picked up by a separate receiving transducer. It is used in blood flow measurement and in certain other application in which the average velocity is measured without regard to the distance of the sources.
3. **Pulsed Doppler**: Here short bursts of ultrasonic energy are transmitted and the returning echoes are received. In this mode of frequency shift, due to movement of the reflected interfaces can be measured in order to determine their velocities.
4. **Range-Gated pulsed Doppler**: It is a refined pulsed-Doppler ultrasound, in which a gating circuit permits measurement of the velocity of targets at a specific distance from the transducer. It is used in velocity of blood flow measurement as a function of time and of the distance from the vessel wall.

• **Ultrasound or Ultrasonic in Diagnosis**: Ultrasonic or Ultrasound is used clinically for diagnosis by different methods. The different methods are in different branches of medical science such:

1. Cardiology e.g. Echocardiography Phonocardiography, ultrasonic blood flow measurement.
2. Abdominal Imaging e.g. liver, kidney, prostate, testes, pancreas, gall bladder.
3. Brain studies e.g. Echoencephalography.
4. Eye investigation e.g. Echophthalmography.
5. Reproductive system e.g. testes, ovary, prostate.
6. Gynaecology and obstetrics e.g. pregnancy, fetus, uterus.

For diagnostic records have different names with the words "echo" or "sound" e.g.

(a) Echocardiogram—For heart
(b) Echoencephalogram—For brain
(c) Ultrasonogram—For eye

The instruments used have different names such as:

(a) Ultrasonograph, (b) Ultrasonoscope
(c) Gonofluroscope (used for moving structure) and other devices for static images.

A. **Phased Array Ultrasonography**: It represents a class of instruments recently developed in which whole medical ultrasound laboratory in one mobile unit. This V-3000 machine consists of:

1. A large video display
2. Video monitoring facility
3. Two separate sets of gain controls
4. A set of seven slides control
5. M-mode and A-mode are available on the V-3000.

21.22 Clinical Applications of Ultrasound and Differences

I. Therapeutic application
II. Diagnostic application

The main differences between the two applications in the

I. **Ultrasonic power level**: At which the equipment operates such as:

(a) *Therapeutic systems:* Operate at ultrasonic power level of upto several watts per square centimeter.

(b) *Diagnostic system:* Equipment operates at power levels of well below 100mW/cm

II. **Energy Level**

(a) *Therapeutic system*: Equipment is designed to agitate the tissue to the level where thermal heating occurs e.g. treatment of lumbago.

(b) *Diagnostic system*: As long as sufficient amount of signal has returned for electronic processing no additional energy is necessary. For which lower ultrasonic power levels are employed.

I. **Diagnostic**: Diagnostic ultrasound are used as:

(a) *Continuous waves*: It depends for their action on Doppler effect e.g. instruments based on this effect are the foetal heart detector, blood flow meter.

(b) *Pulsed waves*: Modern ultrasonic diagnostic instrument is based on this pulse technique. This pulse echo based instrument is used for direction of abnormalities or defects in the structure at various depths in the body. This achievement is due to the time of travel of a short pulse which can be measured easily as compared to continuous waves e.g. ultrasonic scanners for imaging, echocardiography and echoencephalography.

21.23 Biological Effect of Ultrasonics

I. Physical Changes due to Absorption of Ultrasonics (Acoustic Beams): When the acoustic energies like ultrasonics are propagated through a medium of living material, there are three major events observed which may be considered as physiological effects of the beam.

1. **Temperature Effect**: Two types of temprature changes occur in the absence of cavitation:

   (a) A periodic temperature change associated with condensation and rarefaction,

   (b) A steady temperature rise associated with the absorption of acoustic energy.

   The temperature changes are very small and less significant for production of biological effects.

2. **Pressure Variation**: Experimental studies reveal that a beam of 35 W/cm^2 produces pressure changes upto 10.0 atmosphere. But it is doubtful whether this pressure is sufficient to produce any change in the properties of protoplasm.

   Other factors which are influencing the response of living tissue are radiation pressure and viscous flow of medium particles. The forces produce spatial disorientation of parts of the driven system. Such disorientation may result in alteration of biological function.

3. **Cavitation**: Cavitation means physical separation of liquid layers which occurs on drastic reduction of hydrostatic pressure providing gas molecules. Expansion of gas nuclei occurs under negative pressure allowing other molecules in the space to form bubbles. Cavitation is necessary for occuring the effects of ultrasonics, e.g.,

   (a) The liquefaction of thixotropic gels,

   (b) Reduction of visocity of the structure of some materials.

   (c) Disruptive effects on single cells, etc.

II. Biological Changes: The biological effects of ultrasonics are the same as those of sonics; with the difference that production of large intensities are much easier in the ultrasonic range than the sonic range. The biologic effects are:

A. **Heating Effects**: These are primary effects with low intensity beams.

1. Heating effect of high intensity ultrasonics is partly due to absorption of energy and partly to cavitation mechanism. The cavitation causes very high local temperatures.
2. Lipid and protein materials exhibit increases of temperature in ultrasonic beams associated with acoustic absorption.
3. Other less viscous materials may show large cavitation temperature changes.

B. **Paralytic Effect**: The paralytic effects of ultrasonics with moderate intensity beams are similar to be produced by excessively intense audible sound.

1. Men exposed chronically to ultrasonics lose the ability to concentrate and may suffer loss of equilibrium.
2. Fry et al. (1950) exposed frogs to beams of about 35 W/cm^2 intensity (f = 1 megacycle/sec) and observed that 4.3 sec of exposure produced paralysis of the hind legs.
3. Isolated peripheral nerves were not affected by this beam with prolonged exposure.

The above experiments indicate that temperature increase is slightly responsible for paralytic effect. Again, hydrostatic pressure of 13 atmosphere decreased the paralytic effect (Fry et al. 1957) by ultrasonics which does not clarify the physical basis of paralytic effect.

C. **Lethal Effects**: Lethal effects are produced by high intensity beams. These effects are associated with cavitation for the most part.

1. **Protozoa**: Lethal action on protozoa by ultrasonics has been manifested by paralysis or death depending on the exposure. Large protozoa were much susceptible than small ones because the small one did not undergo large internal pressure changes.
2. Arbacia eggs are destroyed; tobacco mosaic viruses are inactivated RBCs and some bacteria are destroyed by high intensity ultrasonic beams.
3. Yeast cells loose their reproductive powers.
4. Luminous bacteria lose their luminosity (Gregg. 1950). The evidence indicates that the lethal action of ultrasonics are direct or indirect result of cavitation. The destruction of protozoa can be reduced significantly by application of high hydrostatic pressures. The reason of lethal action is due to rapid motion of cell materials subsequent to cavitation and formation of minute bubbles.
5. *Sterilizing agent*: It is not so much effective. It disintegrates only a few bacteria and only some of these are fully destroyed. The factors influencing the efficacy of ultrasonics as a sterilizer are as under:
   (a) Viscosity of suspension,
   (b) Shape of the container, etc.
   (c) Larger animals like frogs, fishes, etc, may be disabled or killed by acting on some vital cell or group of cells.

Ultrasonics are not considered because of the following:

(a) Short range in air,

(b) Tremendous quantities of power necessary to make injury,

(c) Low surface absorption coefficient limits the power which can be absorbed.

21.23.1 Minor effects of ultrasonics

1. Electrical charges are produced at the air-gas interface when cavitation occurs in presence of dissolved gases. These electrical charges cause formation of H_2O_2 in the presence of dissolved air and may cause secondary chemical effects.
2. Dispersion and emulsification of mixtures in liquid suspension.
3. Coagulation or agglutination of materials in gas suspension.
4. Pain in the parts of the body exposed to the direct beam.
5. The ears: Tinnitus, dizziness, noises.
6. Nausea, headache, fatigue, irritability.

CHRONIC EFFECT: There is no evidence regarding chronic effect.

21.23.2 Precautions

1. Avoidance of exposing the body to the direct beam as in Radiography or LASER.
2. Guarding the operator from the source of wave emission by screening.
3. It might be necessary in some applications to provide ear protection (Glass down wool).

21.24 Importance

I. Biological, II. Clinical.

I. Biological

1. **Echo Depth-Sounding**: Ultrasonic waves: because of their extreme short wave-lengths

—can be obtained in narrow beam and hence, can be conveniently used in Echo Depth-Sounding. The time taken by the wave to reach the bottom of the sea from the surface and back being determined are used in coastal vessels and submarines.

This directive property is used for the flight of bats in darkness.

2. **Night Fly of Bat**: The bats emit squeaks of ultrasonic waves from the larynx and hear echoes as a result of reflection from an obstacle in front by the ear having big pinna and changes its course accordingly.

The mechanism is: they make a series of twittering sound from the larynx so high pitched that the human ear cannot hear the ultrasound. The sound wave strikes against the object in their path and send back echoes to the ear, and the bat can fly skillfully in the night.

Some insect-eating bats that feed on the wings of insects, have a keen sense of hearing which guides them in the dark. It is believed that the highly sensitive nose of a bat acts as a recorder and picks up air vibrations set in motion by the movements of other animals and also the vibrations, reflected from the stationary objects through a continuous vocal utterance made by the bat themselves. These supersonic utterances are pitched in a key too high for the human ear to hear. The vocal sounds made by the bats have frequencies ranging from 50 to 50000 kilocycles per second. The bats have also a relatively deeper voice which we hear as high shrill squeak. It appears that the nose and ears of the bats are both important factors in this radar operation of bats.

3. **Dolphin**: Dolphins are aquatic animals who make all sorts of sounds—grunts, chirps and squeaks—with which they communicate with each other. Dolphins also emit high frequency sounds and are able to detect fishes at night or even in murky water, through which it is not possible to see. This process is called **echolocation**.

4. **Fish Trawlers**: The trawlerman is able to locate shoals of fish and their distance from him by a similar echosounding arrangement. The transducer keeps on turning and sending ultrasonic pulses along different directions. When the pulses falls on a shoal of fish, it is reflected and the direction and distance of the shoal from the trawler can be determined.

5. **Dogs:** Dogs can hear up to a frequency of 40000 Hz. Thus, dogs can be trained for special purposes to an ultrasonic whistle (Galton whistle) which can not be heard by human beings.

6. **Dispersion:** Ultrasonic wave cause mixing of immiscible liquids such as water and oil to form a homogeneous solution. This is known as dispersion and emulsification or homogenization.
   (a) *Homogenising milk*: Milk is agitated with ultrasonic vibrations so that the larger fat particles break to form smaller particles.
   (b) Ultrasonic vibration used for homogenising cosmetics.
   (c) Dissipating fogs on runway.

Table 2.2: Difference of Power level

	Diagnostics	Therapeutics
Ultrasonic Power level at which equipments operate	1. Below mW/cm²	1. Several watts per sq. cm.
	2. As long as sufficient amount of signal has, returned for electronic processing, no additional energy is necessary.	2. Agitate the tissue for thermal heating, e.g. lumbago.

7. **Coagulation or Agglutination**: Coagulation or agglutination may be done under the influence of ultrasonic vibrations to form heavier particles.

8. **Disruption of Organic Molecules**: High polymers and molecule chains may be disrupted by such high frequency vibration.

9. **Killing of Small Animals**: Ultrasonic vibrations can kill fish in water, protozoa, and red blood corpuscles.

10. **Sterilizing Effect**: Ultrasonics is not so much effective because it can disintegrate only certain bacteria. Only some of these

bacteria show full effect and depend upon the shape of the container, viscosity of suspension, etc.

11. **Disintegration and Fraction of Organelles**: The submitochondrial vesicles are isolated by sonication from mitochondria for obtaining and purification of F_1ATPase (Racker, 1970).
12. **Breaking The Tough Cell Wall**: In bacteria, yeasts and many plant cells much more vigorous procedures are necessary to break the tough cell walls by sonic radiation, grinding with sand or fragmentation in a high-pressure.

21.25 Clinical Ultrasonography

The basic components

1. **Transducer**: It transmits the ultrasound wave to the subject and receives the corresponding echoes. The transducer comprises the same transmitting and receiving section for pulsed wave ultrasound and seperate transmitting and receiving section for continuous wave of ultrasound.

   In selecting a transducer for investigation, the following *factors must be considered for optimum results.*

   (a) The type of investigation to be performed
   (b) Physical size of the patient.
   (c) The anatomical area involved
   (d) The type of tissue to be encountered
   (e) The depth of the structured to be studied.
2. **Start pulse generator**: This sets pulse repeatition frequency which is determined at the time of manufacture.
3. **High voltage pulse generator**: This provides the voltage to apply to the transducer electrode.
4. **T.G.C. unit**: It amplifies the signal proportional to the time delay between the transmission and detection of the ultrasound pulses. There are two components in this control:

   (a) Near gain: which is used to diminish near eachoes.
   (b) Slope: which sets a fixed rate of enhancement of echos with time.

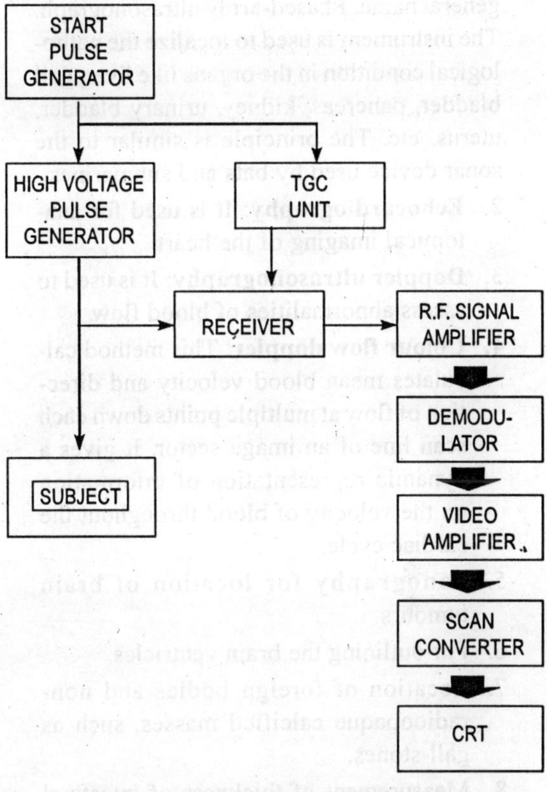

Fig. 21.18: Block diagram of USG machine

5. **Receiver and Pre amplifier**: These detect the echoes received by the transducer and amplify it for demodulation.
6. **Demodulator**: These rectifies the received echoes and smooth the resultant waveform.
7. **Video amplifier**: The demodulated signal is amplified by the video amplifier.
8. **Scan Converter**: It stores signals in specific location in a micro computer memory. These stored digital signals are then read electronically and used to drive the electron gun of a CRT monitor.
9. **CRT display:** The final image is displayed in the CRT monitor.

II. Clinical

Ultrasound is used clinically for:

A. Diagnostic, B. Therapeutic

A. Diagnostic

1. **Ultrasonography**: General name: Ultrasonograph, Ultrasonscope, Sonofluoroscope

general name: Phased-array ultrasonograph. The instrument is used to localize the pathological condition in the organs like liver, gall bladder, pancreas, kidney, urinary bladder, uterus, etc. The principle is similar to the sonar device used by bats and submarines.

2. **Echocardiography**: It is used for anatomical imaging of the heart.
3. **Doppler ultrasonography**: It is used to assess abnormalities of blood flow.
4. **Colour flow doppler**: This method calculates mean blood velocity and direction of flow at multiple points down each scan line of an image sector. It gives a dynamic representation of information on the velocity of blood throughout the cardiac cycle.
5. **Sonography** for location of brain tumours.
6. For outlining the brain ventricles.
7. Location of foreign bodies and non-radioopaque calcified masses, such as gall-stones.
8. Measurement of thickness of intestinal mucosa.
9. In obstetrics and gynaecology, it is used for detection of pelvic mass, tumours, pregnancy, ovulation, site of placenta and texture, etc. During labour it helps to assess the foetal conditions, foetal distress or detection of a cord loop around the neck, post-partum placental lobe retention. It can detect foetal anomalies.

I. Abnormalities of early pregnancy diagnosed by sonography

1. Missed abortion
2. Anembryonic pregnancy (blighted ovum)
3. Live abortion
4. Hydaliform mole
5. Ectopic pregnancy

II. Mid and late pregnancy (second and third trimesters)

Ultrasound investigation at this stage, 18–20 weeks is a good time the ultrasound investigations are done for the following:

1. Fetal number
2. Fetal life (heart beat)
3. Fetal position
4. Gestational age (fetal age)
   (a) Biparietal diameter of the skull (BPD)
   (b) Femoral length (FL)
   (c) Head circumference (HC)
   (d) Abdominal circumference (AC)
5. Abnormalities of fetus
   (a) Neural tube defect
   (b) Thorax, heart
   (c) Congenital diaphragmatic hernia (CDH)
   (d) Abdomen
   (e) Hydropic fetalis
6. Polyhydramnios
7. Increased placental thickness
8. Skin thickening
9. Ascites, pleural and pericardial effusion.
10. Localisation of placenta is necessary particularly in:
    (a) Antepartum haemorrhage (placenta previa, abruptio placentae)
    (b) Amniocentesis.

III. Gynaecology: Ultrasound of the female pelvic is the primary imaging investigation such as:

(a) Congenital anomalies
(b) Pelvic masses (e.g. uterine fibroid, uterine carcinoma, hydatiform mole, choriocarcinoma, ovarian masses such as cystadenoma, cystadenocarcinoma, dermoid, teratoma etc).
(c) Pelvic inflammatory disease
(d) Monitoring of ovarian function
(e) Ovarian pathology
(f) Assessment of uterine pathology
(g) Control of intrauterine contraceptive devices (IUCD).

• Ultrasonography in obstetrics and gynaecology

I. Early pregnancy (first trimester) (5-6 weeks after the first day of the last menstrual period but transvaginal scanning can identify the sac in the uterus as early as 4 weeks).

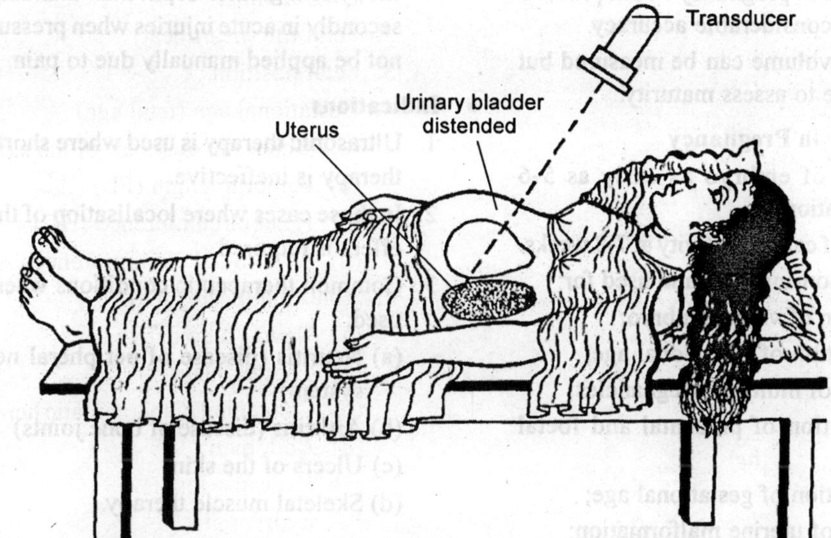

Fig. 21.19: Diagramatic representation of the position of the ultrasound transducer, urinary bladder and uterus, longitudinal and transverse scan

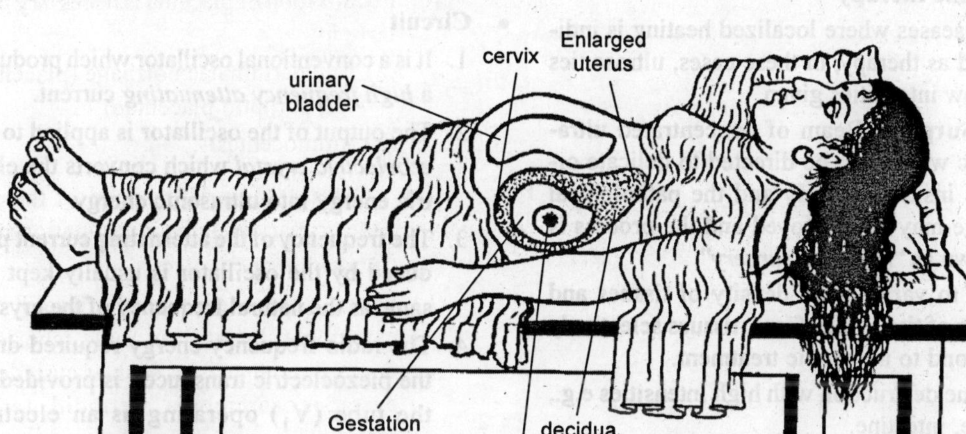

Fig. 21.20: Schematic diagram of early pregnancy in longitudinal scan at 9 weeks. The gestational uterus is large and the fetus (fetal node) is seen as an oval collection of mass within it

Observation

1. A cystic area with a rim of high intensity echoes. The sac enlarges by 10 wks to occupy most of the uterine cavity. The developing embryo or 'fetal node' can be identified from the 6th or 7th week and the fetal head at 14 weeks.
2. Presence of two or more gestation sacs indicate multiple pregnancy.
3. The fetal heart beat can be identified soon after the fetal node is seen and considered as evidence of live fetus.
4. The fetal node length can be measured accurately by electronic calipers between the

7th-14th weeks of pregnancy which pradicts maturity with considerable accuracy.

5. Gestation sac volume can be measured but is less accurate to assess maturity.

- **Ultrasonography in Pregnancy**

1. Visualisation of embryo as early as 5-6 weeks of gestation.
2. Visualisation of cardiac activity at 7-8 weeks.
3. Because it is non-invasive and used for:
   (a) Verification of viable embryo;
   (b) Determination of gestational age;
   (c) Diagnosis of multiple pregnancies;
   (d) Determination of placental and foetal position;
   (e) Determination of gestational age;
   (f) Detection of uterine malformation;
   (g) Diagnosis of foetal anomalies.

21.26 Therapeutic

- **Ultrasonic therapy**

1. In diseases where localized heating is indicated as therapy, in these cases, ultrasonics of low intensities given.

   Surgery: Beam of concentrated ultrasonic waves can be directed to delicate organs inside the body and the pathological tissue may be destroyed and the process is known as "*knifeless surgery*".

2. Due to variation in density of tissues and types of tissue interfaces, tissues selectively respond to ultrasonic treatment.
3. Tissue destruction with high intensities e.g., bone, intestine.
4. Treatment of pain syndrome, e.g., postoperative neurofibroma.
5. Ultrasonics are bing used for dissolution of painful scars.

- **Reason of heating**

1. The heating effect is produced due to *ultrasonic energy absorption* by the tissue.
2. The property of *specific heat distribution* in the tissue plus additional effect of a *mechanical component* has opened a number of special application.
3. The effect on the tissue is a *high speed vibration of micromassage.* This massage is firstly to a greater depth than manually and secondly in acute injuries when pressure can not be applied manually due to pain.

- **Indications**

1. Ultrasonic therapy is used where short wave therapy is ineffective.
2. In those cases where localisation of the heat effect is required.
3. Common therapeutic conditions where it is used:
   (a) Neuritis (disease of peripheral nervous system).
   (b) Arthritis (disease of bone joints)
   (c) Ulcers of the skin.
   (d) Skeletal muscle therapy.

Electric Power: Electric power required in most of the application is usually less than $3 W/cm^2$ of the transducer area that is in contact with the part of the body to be treated.

- **Circuit**

1. It is a conventional oscillator which produces a *high frequency attenuating* current.
2. The output of the oscillator is applied to *piezoelectric crystal* which converts the electric energy into ultrasonic energy.
3. The frequency of the atternating current produced by the oscillator is usually kept the same as the natural frequency of the crystal.
4. The radio frequency energy required drive the piezoelectric transducer, is provided by the tube (V_1) operating as an electron coupled oscillator.
5. In this circuit R_1 provides the class C operating bias for this oscillator by virtue of the voltage drop produced across it and as a result of grid current flowing in the oscillator.
6. Inductance L_2 and capacitance C_7 and C_8 in parallel, provide the tank circuit and are tuned to the resonance frequency by an adjustable iron core.
7. C_8 is a temperature compensator capacitor.
8. Valves of capacitor C_5 and the inductance L_1 are adjusted to achieve matching from the high output impedance of the oscillator

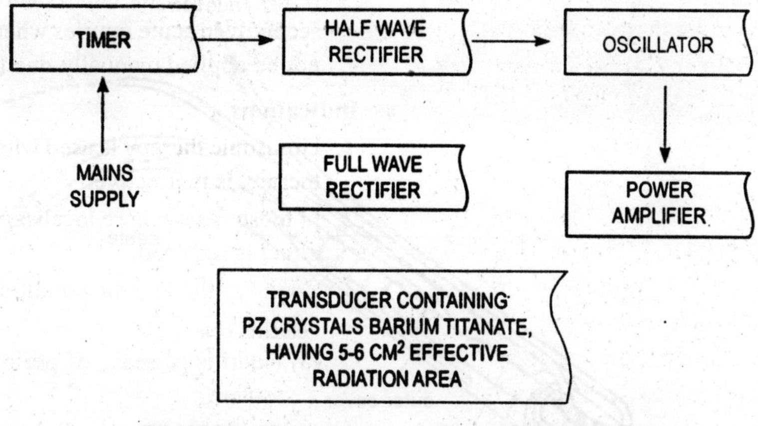

Fig. 21.21: Block diagram of ultrasonic therapy machine

tube V_1 transducer coaxial cable paralleling the piezoelectric crystal. A portion of the plate current V_1 flows through the power output meter M. The meter calibration is achieved by adjusting the meter shunt R_2.

9. The transducer may be barium titanate or lead zirconate titanate crystal, having 5-6 cm^2 effective radiating area.
10. The length of the cable connecting the transducer with oscillator is of a critical dimension should not be altered.
11. The intensity control varies the output power from approximately 0-15 W by varrying the voltage drop across the tube V_1.
12. In front of the crystal lies a *metal plate* which is made to vibrate by the oscillations of the crystals. Ultrasonic waves are emitted from this plate. The crystal has a metal electrode pressed against its back surface by a coiled spring. Voltage is applied to the crystals via this electrode. The front diaphragm is grounded and provides a return path for the excitation voltage.

Dosage control: The dosage can be controlled by varying any one of the following:

1. Frequency of ultrasound
2. Intensity of ultrasound
3. Duration of the exposure

Exposure Time

1-15 minutes

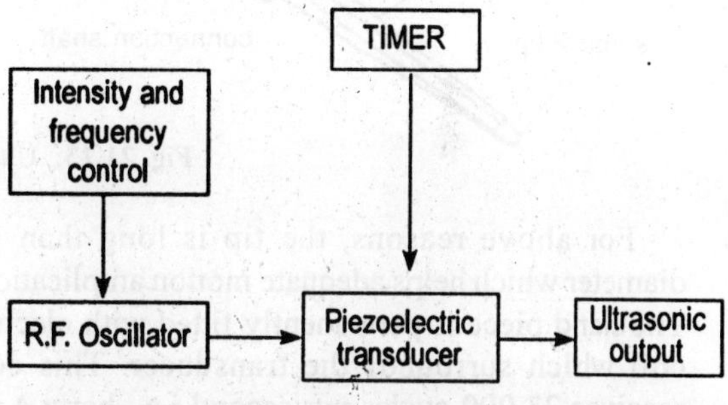

Fig. 21.22: Block diagram of ultrasonic diathermy

21.27 Ultrasonic Surgery

Ultrasonic surgery is done with the help of a special instrument called CUSAtm (cavitron TM Ultrasonic surgical Aspirator). This acoustic vibrator has the following components which are:

1. **Transducer**: It is magnetic device which causes conversion of electromagnetic energy into mechanical vibration. Transducer has stack of nickel alloy plates. Production of magnetic field occurs by a coil placed around the nickel alloy plates and produces mechanical motion of around 100 microns.
2. **Connecting body**
   (a) It has the function to convey motion mechanically to the surgical tip.
   (b) It has other function, amplification of vibration motion of the transducer.
3. **Surgical tip**
   (a) It plays for completion of amplification of the motion.
   (b) It conducts the tissue.

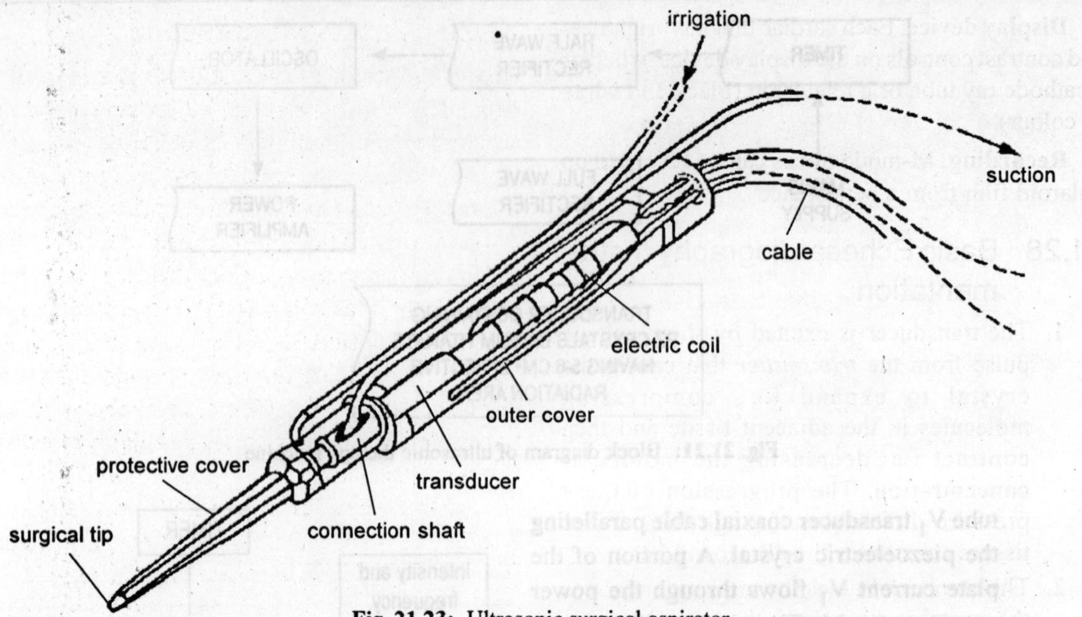

Fig. 21.23: Ultrasonic surgical aspirator

For above reasons, the tip is long than its diameter which helps adequate motion amplication. The hand piece is permanently fitted with electric coil which surrounds the transducer. This coil receives 23,000 cycles per second i.e., hertz A.C. current from the console for activation of the transducer.

The hand piace has connection to the console with a cable which is inside a tube for circulation of fluid between the cooling water can in the console and the hand piece. Because the electric current flows in the electric coil, heat is produced for which fluid circulation is necessary. It keeps the hand piece cool which is very much helpful for the surgeon.

Applications

1. *Neurosurgery*: Rapid removal of intracranial and spinal cord tumours.
2. Ressection and dissection of liver, kidney, and spleen.
3. Pancreatic and urological surgery.

21.27.1 Summary: basic ultrasonic

1. A generator for electric signal
2. A transducer
3. Necessary amplifiers
4. Other electronic processing devices
5. Display unit.

Transducer converts → electrical signals → into the mechanical vibration → thus the acoustic waves → from transducer → enters body through skin → propagated in predetermined beam pattern (wide or narrow depending on requirements) towards the structure to be examined → ultrasonic waves strikes an acoustic interface such as boundary of an organ → some energy is reflected → reflected energy is picked up.

Refraction: There are only minor differences CRT or on continuous-roll 35 mm film. Film development is time consuming and expensive the most common method uses the strip chart recorder, which provides a continuous reproduction of the trace. The M-mode trace is recorded in a line-by-line sequence as the recording paper moves past a *specialised CRT.*

Reflection: Reflection redirects some of the ultrasonic energy back toward the transducer in the form an echo. The returning pressure wave induces an electrical signal in the transducer crystal that is *amplified* from a microvolt or millivolt range to a few volts. One component of this amplification process is *time gain compensation* (TGC), sometimes referred to as *depth gain compensation (DGC).* After further processing, the scan data are presented on a display device normally in different shades of gray, that shows the relative amplitudes of the echoes.

Display device: Each cardiac unit has brightness and contrast controls on the display device, whether a cathode ray tube or a television (black and white or colour).

Recording: M-mode traces can be recorded on polaroid film from a persistence.

21.28 Basic Echocardiography Instrumentation

1. The transducer is excited by short voltage pulse from the *transmitter* that causes the crystal to expand i.e., compressing molecules in the adjacent tissue and then contract i.e. decreasing the molecular concentration. The progression of these pressure differences through the cardiac tissues constitutes the ultrasound beam.
2. The number of times the transmitter excites the crystal each second is the pulse repetition frequency (PRF). Typically the PRF ranges from 200 to 9000 per/s and is set by the manufacturer. For some scanner the operator can adjust the PRF in relation to the transducer, type of study and greatest depth of interest (R). The maximum PRF is given by $PRF_{(max)} = \frac{C}{2R}$

   where C = velocity of ultrasound in tissue.

   The depth control on some units automatically reduces the PRF as the depth of penetration increases.
3. When ultrasound beam enters the patient, the various interactions take place.

20.28.1 Advantages

1. Noninvasive procedure,
2. Quick procedure,
3. No radiation hazards,
4. Helps to narrow down the provisional diagnosis and when used with the clinical data, it may help to come to an exact diagnosis.
5. It not only visualises the organ but also indicates the shape, size and texture.
6. It detects physiological and pathological collections of fluids, both generalised and localised.
7. It also informs whether a mass is cystic, solid or mixed.

21.29 Vascular Ultrasound

One of the leading cause of death is stroke. It is caused generally due to atherosclerosis i.e. accumulation of lipid (e.g. cholesterol) on arterial walls. Pulmonary emboli caused by venous thrombosis also another cause of mortality which can be attributed to vascular disease. Disruption of blood flood in the peripheral vascular system may cause of loss of an extremity. With the development of high-resolution X-ray imaging angiography (DSA), Computed tomography and magnetic resonance imaging imaging of the vascular system has become very sophistacated and highly effective. But these techniques often require introduction of contrast agents and are invasive. Whereas ultrasonic imaging does not require contrast media and non-invasive, lower costs than the costs associated with other imaging modalities. Moreover, ultrasonography provides

(a) Flow dynamics,

(b) Vascular wall anatomy,

(c) Tissue characteristics which can not be achieved by other modalities.

- **Instrumentation**

  I. **Real-time imaging**

  It provides:

  (a) Anatomical information regarding the vessel and surrounding tissues

  (b) Vessel size

  (c) Perivascular masses and luminal thrombi visualisation

  (d) Does not depict blood flow because the echoes from RBCs are weak.

  II. **Doppler imaging**: Doppler imaging (e.g. Duplex and Color Droppler scanners) are used extensively in vascular diseases. This method is useful for investigation of blood flow dynamics e.g.

  (a) Presence of flow

  (b) Its direction

  (c) Other physiological aspects of flow e.g. pulsatility, resistivity, high velocity gets and turbulence are assessed.

- **Intravascular Imaging**: A phased array or mechanical transducer is mounted at the fit of a catheter. The catheter is placed in the artery 1-and a 2-D image of the vessel wall is obtained at each point of interest. The scanning range is 1-2 cm. A very high frequency extending from 20-40 MHz is used. The axial images are often reconstructed in 3-D format to help in the presentation of scan data.

(a) This technique allows examination of the vessel wall not immediately adjacent to the lumen.

(b) Information regarding the histological layers is now achieved which cannot be available by angiography.

(c) The severity of disease may be more accurately assessed with intravascular imaging.

(d) Information regarding extent of plaque formation, measurement of luminal.

21.30 Ultrasonic Transducer

Introduction: The ultrasonic transducer (also called probe) is one of the most important component of an ultrasound pulse echo system. It converts electric signals into acoustic pulses during transmission and during reception it transforms acoustic pulses into electric signals. Shape of scanning beam is also regulated by transducer and related electronic components of the ultrasonic machine. Development regarding new piezoelectric materials, resolution, arrays both linear and annular have been established.

Pz-Transducer: Transducer produces an electrical phenomena called *piezoelectricity.* It is due to deformation of Pz materials by a mechanical pressure exerted on it producing voltage. On the other hand application of this voltage, the Pz-materials are deformed mechanically by lengthening or shortening depending on the polarity the applied voltage.

The transducer consists of piezoelectric crystals which *generates* and *detects* ultrasound pulses piezoelectric properties are due to the chemical morphology of the material. In a piezoelectric material the constituent molecules are electrically polarised i.e. one end of the molecule is electrically negative and the other end positive.

The polarization is due to the unequal sharing of electrons within the molecules. The electrically polarized molecules are reffered to as dipoles. In naturally occurring Pz-materials e.g. quarts, the

Different sectional view of Ultrasonic Transducer (Figs. 21.23, 21.24, 21.25, 21.26, 21.27)

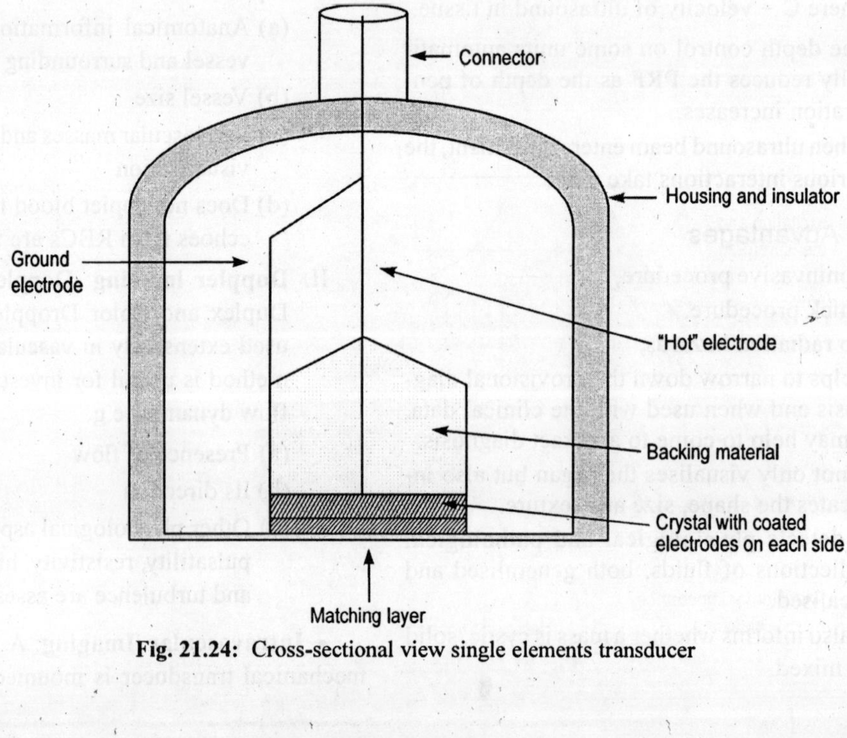

Fig. 21.24: Cross-sectional view single elements transducer

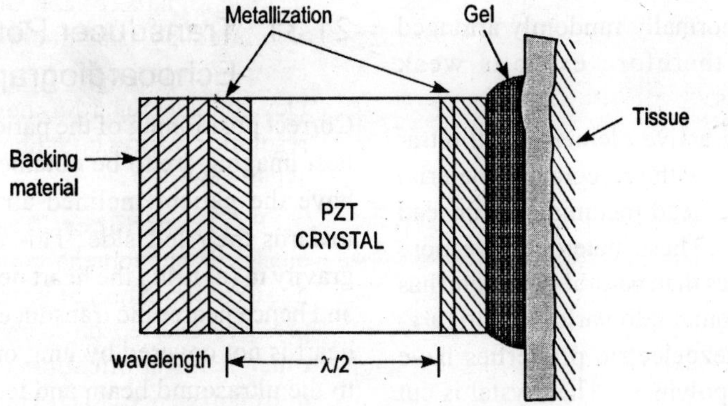

Fig. 21.25: A PZT ultrasonic transducer

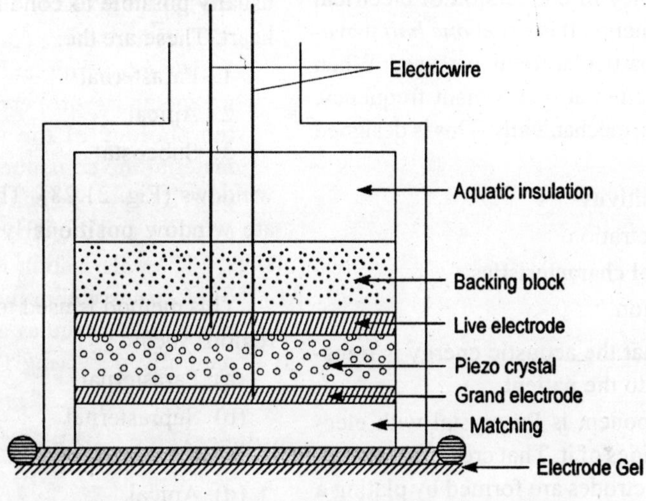

Fig. 21.26: Diagram of basic ultrasound transducer

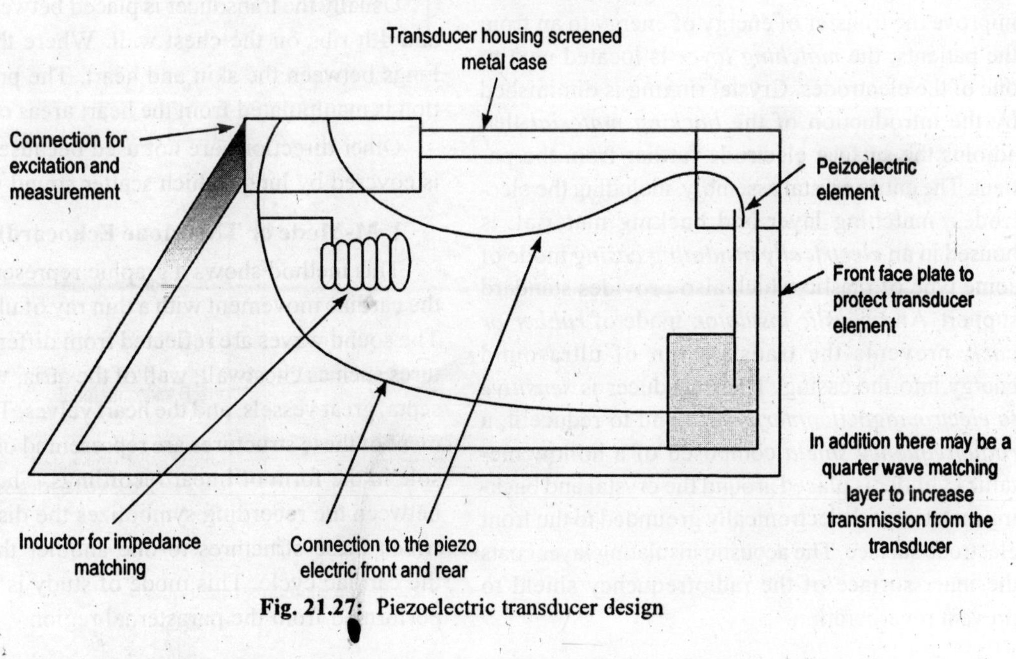

Fig. 21.27: Piezoelectric transducer design

electrical dipoles are normally randomly arranged and the materials therefore exhibits weak piezoelectric properties.

But now a days the active element in all ultrasonic transducers is a synthetic ceramic material such as barium titanate, lead metaniobate, or lead zirconate titanate (PZT). These materials have more controlled characteristics than natural quarts and has ability to be cut and ground into various configurations. But recently, piezoelectric properties have been shown in plastic polymers. The crystal is cut *hexagonally* so that it is mechanically resonant for an increased efficiency of conversion of electrical energy to acoustic energy. It is usual *one half wavelength thich* for the particular frequency used. When the transducer is excited at its resonant frequency, it continues to vibrate mechanically. This is designed to achieve the

(a) highest sensitivity

(b) highest penetration

(c) optimal local characteristics

(d) best resolution.

This requires, that the acoustic energy is transmitted efficiently into the patient.

The major component is Pz crystal with electrodes on opposite sides of it. That create the changing polarity. The electrodes are formed by plating a thin film of gold or silver on the crystal surface. To improve the transfer of energy of energy to an from the patients, the *matching layer* is located next to one of the electrodes. Crystal ringing is diminished by the introduction of the *backing material* that adjoins the surface electrode further from the patient. The entire crystal assembly, including the electrodes, matching layer and backing material, is housed in an *electrically insulating casing* mode of some type of plastic which also provides standard support. An *acoustic insulator,* made of *rubber or cork,* prevents the transmission of ultrasound energy into the casing. The transducer is *sensitive to electromagnetic interference* and to reduce it, a *radiofrequency shield* composed of a hollow metallic cylinder is placed around the crystal and backing material and electronically grounded to the front electrode surface. The acoustic insulating layer coats the inner surface of the radiofrequency shield to prevent reverberation.

21.31 Transducer Positioning During Echocardiography

Correct positioning of the patient is necessary if the best imaging are to be obtained. It is customary to have the patient inclined about 45° and rotated towards their left side. This utilizes the effect of gravity to (a) bring the heart nearer to the chest wall and hence nearer the transducer (b) areas where the heart is not covered by lung or bones allow access to the ultrasound beam and termed '*windows*'.

There are three main windows from where it is usually possible to conduct an examination of the heart. These are the:

1. Parasternal
2. Apical
3. Subcostal

windows (Fig. 21.28). There are many intermediate window positions lying between these main sites.

This method is used to study the heart from the regions such:

(a) Parasternal

(b) Suprasternal

(c) Substernal

(d) Apical.

Usually the transducer is placed between the 3rd and 4th ribs on the chest wall. Where there is no lungs between the skin and heart. The probe position is manipulated from the heart areas of interest.

Other directions are not used because the heart is covered by lungs which scatter sound waves.

I. M-Mode or TM-Mode Echocardiography

This method shows a graphic representation of the cardiac movement with a thin ray of ultrasound. The sound waves are reflected from different structures such as chestwall, wall of the atria, ventricles, septa, great vessels, and the heart valves. The movement of these structures are represented on the console in the form of linear recordings. The distance between the recording symbolizes the distance between these structures to one another throughout the cardiac cycle. This mode of study is generally performed from the parasternal region.

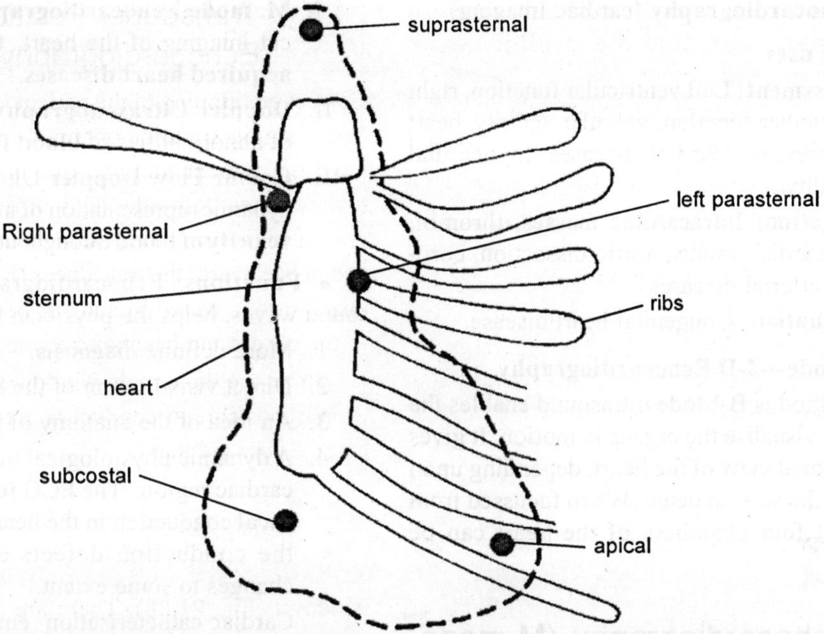

Fig. 21.28: Illustrating anatomical sites commonly used in echocardiogram window

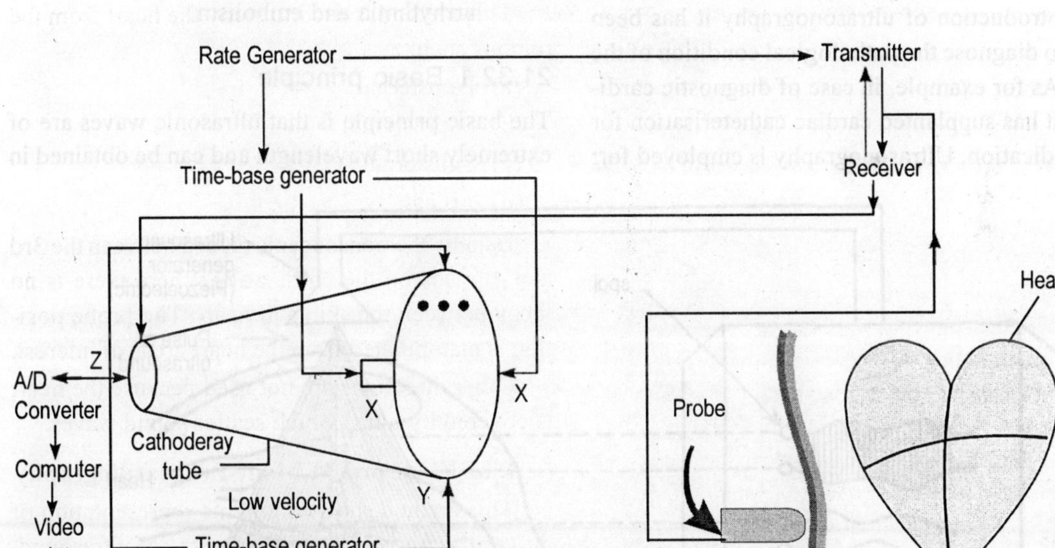

Fig. 21.29: Block diagram of a Echocardiography

2-D Echocardiography (cardiac Imaging)

Cardiac uses

I. **Assessment:** Left ventricular function, right ventricular function, valvular activity, heart diseases, pericardial diseases, myocardial viability.

II. **Detection:** Intracardiac masses, thrombi, intracardiac shunts, aortic dissection, coronary arterial diseases.

III. **Evaluation:** Congenital heart disease.

II. B-Mode—2-D Echocardiography

This method is B-Mode ultrasound enables the physician to visualise the organs in motion. It gives a cross sectional view of the heart, depending upon the plane of the section beam. When focussed from the apex all four chambers of the heart can be visualised.

21.32 Echocardiography (M-mode, Echocardiography) 2-D

After introduction of ultrasonography it has been easier to diagnose the pathological condition of the organ. As for example, in case of diagnostic cardiology, it has supplanted cardiac catheterisation for most indication. Ultrasonography is employed for:

I. **M. mode Echocardiography** for anatomical imaging of the heart. **Congenital and acquired heart diseases.**

II. **Doppler Ultrasonography** for assessment of abnormalities of **blood flow.**

III. **Colour Flow Doppler** Ultrasonography for dynamic representation of information on the **velocity of blood** throughout the cardiac cycle

- **Functions**: Echocardiography uses ultrasound waves, helps the physician to come to for

1. More definite diagnosis.
2. Direct visualisation of the heart in motion.
3. An idea of the anatomy of the heart.
4. A dynamic physiological information of the cardiac region. The ECG records only electrical conduction in the heart giving class to the conduction defects and anatomical changes to some extent.

Cardiac catheterisation employed to lesser number of patients is excellent method for diagnosis but it is invasive, carries risks like arrhythmia and embolism.

21.32.1 Basic principle

The basic principle is that ultrasonic waves are of extremely short wavelength and can be obtained in

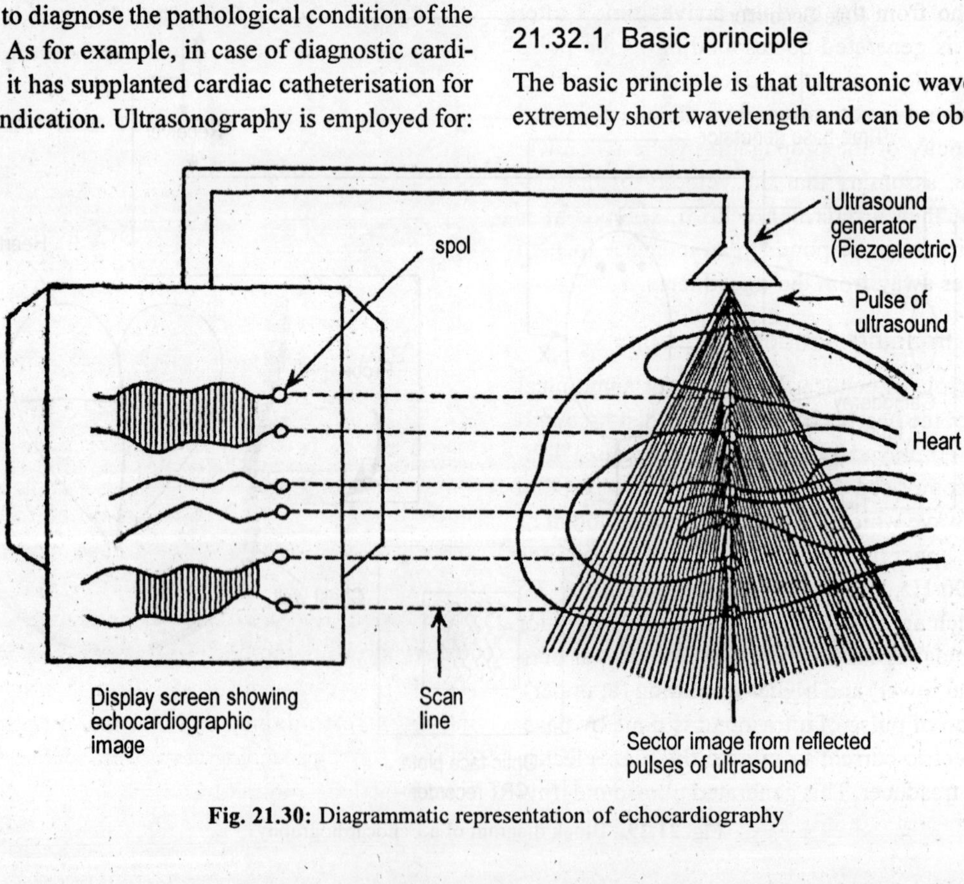

Fig. 21.30: Diagrammatic representation of echocardiography

Types of Echocardiography

- Echocardiography
  - M-mode
  - 2-Dimensional realform imaging (2DE)
  - Contrast echocardiography
  - Doppler technique
    - Continuous wave (CW)
    - Pulsed wave (PW)
    - Colour code of doppler imaging or colour flow mapping technique (CFM)

a very narrow beam and hence, can be used in echo depth-soundings. This principle is being used by submarines and coastal vessels for ocean depth. This principle is also used by the bats during their flight in the darkness. The bats send out ultrasonic waves and hear echoes as a result of reflection from an obstacle in front and change their course. The reflected wave travels back through the medium and is detected by the crystal transducer. Therefore, a return echo from the medium arrives time t after the pulse is generated corresponding to the pulse having travelled a distance 2d (to and fro). In imaging situation d is not known, but can be calculated if the velocity of the propagating wave is known. Therefore, assuming that the velocity of the medium is v, then a returning echo ti, seconds after pulse emission corresponds to an interface located vt/2 metres away from the transducers.

21.32.2 Imaging echocardiography

The principle of echocardiography for anatomical imaging of the heart is based on echo-depth-sounding devic employed by submarines and bats. Ultrasound is comprised of mechanically induced sound pressure wave which is high frequency sound at or above the upper limit of hearing of man in the range of 50-2000 Hz, and having frequencies of 1.9-7.5 MHz which are used in the Echocardiography for cardiac imaging. At this range, there is better penetration (at lower) and higher resolution (at upper). Generation of pulse of ultrasound is done by passing an electric current to vibrate the piezoelectric crystal in trasducer. This generated ultrasound travels at 1540 m/sec, through the body and most of the sound energy is either scattered or absorbed during transmission. But a little amount is reflected at the interfaces between the tissues having different acoustic impedance like pericardium and myocardium traveling back through this underarms and the reflected pulses are detected by a receiving crystal. Density of the spot on the screen is controlled by the magnitude of the received signal. The time difference between transmission through the body and the reflection of the pulse determines the position of the spot. Formation of a two dimensional sector, scan pulses are transmitted along 120 lines over a 90° arc, 30 times per second. An M. mode image is formed due to transmission and recording at only one scan line out of 120 scan lines. The returning echoes are displayed as a graph of depth against time.

21.32.2.1 Application

In various cardiac diseases like valvular lesions (identification of site, type and severity of lesion, assessment of dilation of chamber size, for deciding surgery, for diagnosis and follow-up of infective endocarditis), congenital heart disease, myocardial function and disease, prosthetic valves, pericardial disease, detection of effusion as small as 15-20 ml, follow up after tapping, detection of abnormal structures (between 3 to above 5 mm) like intracardiac tumours, thrombi and vegetation, dilation of the ascending aorta, ventricular studies (identification of mural thrombi).

21.32.2.2 Version of echocardiography

1. **Endoscopic Transoesophageal Echocardiography**: Advantageous for optical window and views that will be complementary to those obtained with transthoracic echo-cardiography. Advantages are that the whole of the thoracic aorta can be imaged and the atria, the interatrial septum, and the proximal left-cooronary are visualised properly, minor abnormalities, e.g., small thrombi, vegetations are detected easily than transthoracic echo-cardiography, monitoring changes in left ventricular function in intensive care unit or operation theatres or for assessing the surgical repair of mitral regurgitation or intracardiac shunts, investigation of choice in aortic dissection. This method can be done when transthoracic imaging quality is suboptimal.

21.32.2.3 Transthoracic echocardiography

Disadvantages

1. Dependent on the technical person because of 10-20 per cent error in quality of images.
2. Good pictures are essential for diagnosis of vegetations.
3. Image of root of the pulmonary artery is not well.
4. Artefacts may cause over-interpretation.
5. There is every possibility to miss abnormalities lateral to the field of view.
6. Tissue characterisation is poor.

21.32.3 Doppler ultrasonography

Basic Principal

It is our everyday experience that the pitch of a note appears to change if there is a relative motion between the source and the observer. The pitch increases as its moving source approaches the observer and decreases as the source recedes from the observer. As for example, if a motor car sounds its horn, the pitch increases as it approaches the observer and the pitch decreases when it recedes from the observer. This apparent change in the pitch of a note when there is relative motion between the source and the observer is known as *Doppler effect*. Doppler effect is nothing but change in frequency due to relative motion between the source of sound and the observer and this apparent change of frequency was first predicted by Doppler in 1845. He stated that a change of frequency of the wave-motion should be observed when a source of sound or light was moving and this is known as the **Doppler effect.**

21.32.3.1 Method

Information about velocity and direction of blood flow and turbulence is derived from the change of wavelength and hence frequency occuring between transmitted and reflected ultrasound. It is also useful in the study of altered direction of flow blood in valvular defects like stenosis and regurgitation. Ultrasound is transmitted at a constant frequency. If RBCs move directly towards the transducer A, it reflects a wave of ultrasound and decreases its wavelength. When they move away directly from transducer B, it reflects a wave ultrasound and increases its wavelength. The change in frequency is in the audible range and the audiosignal is used for the guidance of the transducer to get visual display. Fig. 21.30 illustrates how the change of wavelengths, and hence frequency—occurs when a source of sound is moving towards a stationary observer.

The change in wavelength depends on the velocity of the RBC and on its direction relative to the ultrasound.

The returning ultrasound signal analysed by the computer, forms a plot of velocity against time. In this the density of a point which is displayed is in proportion to the number of red cells moving at the velocity.

As for instance, in normal mitral valve, there is normal flow across the mitral valve, and the flow is laminar having a narrow range of blood velocities. But in case of mitral stenosis, peak velocity rises, the rate of decay of peak velocity falls, and the flow becomes turbulent causing a greater range of velocities which may be used to measure the degree of stenosis.

When there is drop in pressure at any time during diastole and across any other valve or shunt, it may be calculated using modified Bernouilli equation (Holen, J. et al. Acta. Med. Scand. 199, 455-60, 1976):

Pressure drop = 4 × (velocity).

The rate of decay of peak transmitral velocity is expressed as the pressure half time $T(\frac{1}{2})$ and this measurement was initially used during cardiac catheterisation. And it is the time taken for peak pressure across the mitral valve to half. For Ultrasonography, the pressure half time is the time taken for peak velocity to fall by the square root of half, i.e., 0.7 × the peak velocity. The pressure half time is used to calculate the area of the orifice of the mitral valve in stenosis, provided the area is less than 1.8 cm^2 with the empirical formula:

Estimated area of orifice = 220 / pressure half time.

The pressure gradient across the valves: (1) is highly dependent on flow and (2) large differences may occur from day to day and even (3) from beat to beat for which difficulty arises in case of aortic stenosis and correction is made for flow by calculating the estimated area of the orifice with the continuity equation:

Area of the orifice + Subaortic area × Subaortic velocity/Aortic velocity.

21.32.3.2 Why echocardiography is advantageous for cardiovascular imaging over the other imaging methods?

The following reasons have made ultrasound to be more preferable in the cardiovascular imaging.

1. It is possible to distinguish between different soft tissues.
2. It is possible to measure the motion of structures of the heart.
3. Another factor is that there is actually no interference by echoses from other body structures, because the heart is surrounded by the lungs which are nothing but air bags structures, because the heart is surrounded by the lungs which are nothing but air bags which is helpful for interpretation.
4. The heart has a number of acoustic interfaces, e.g., atrial and ventricular walls the septum and various valves. The position and movement of each interface can be measured by the reflected ultrasound.
   The echoes from these walls and the valves are predictable since the components of the heart move in a known manner.
5. Another advantage of echocardiography is in the detection of fluids in pericarditis (inflammation of pericardium, the sac surrounding the heart).
6. The temporal and spatial imaging properties and lack of demonstrated adverse effects make ultrasound (i.e., echocardiography) valuable for cardiac imaging.

(a) Temporal resolution is the ability of an imaging system to interrogate the structures of importance repeatedly in a very short time.

Speed of ultrasound is 1540 m/s or 1 cm every 6.5 μs. The time required for passage to a depth of 15 cm which is the most distal cardiac structure and back is 195 μs.

Even an ultrasound system can interrogate the heart at least 5000 times each second for one line of sight which provides the opportunity to simple numerous times during one cardiac cycle.

(b) *Spatial resolution*: It is the efficiency of an imaging system to detect and display as separate entities, anatomical structures, that are close together.

21.32.3.3 Type of technique of doppler ultrasonography

There are two basic types of Dopplar techniques:

1. **Continuous Wave Doppler**: In this, one crystal transmits and the second crystal receives continuously. It is very much sensitive and permits measurement of high maximal velocities. But the demerit is that all velocities along with the length and width of the ultrasound beam are recorded so that regions of abnormal flow cannot be localised correctly.
2. **Pulsed Doppler**: In this technique, a single crystal transmits ultrasound in pulses and is set to receive echoes after a time delay. The reflected signals are recorded from a particular depth and can be localised within an echocardiographic image. But the demerit of this is that if there is high frequency shift, sampling error occurs and, above a certain maximum blood velocity, measurements project error. This problem has been reduced by Colour Flow Doppler which is a two-dimensional version of a Pulsed Doppler.

• **Merit:** It is useful for hospital because it provides direct haemodynamic data which are complementary to imaging.

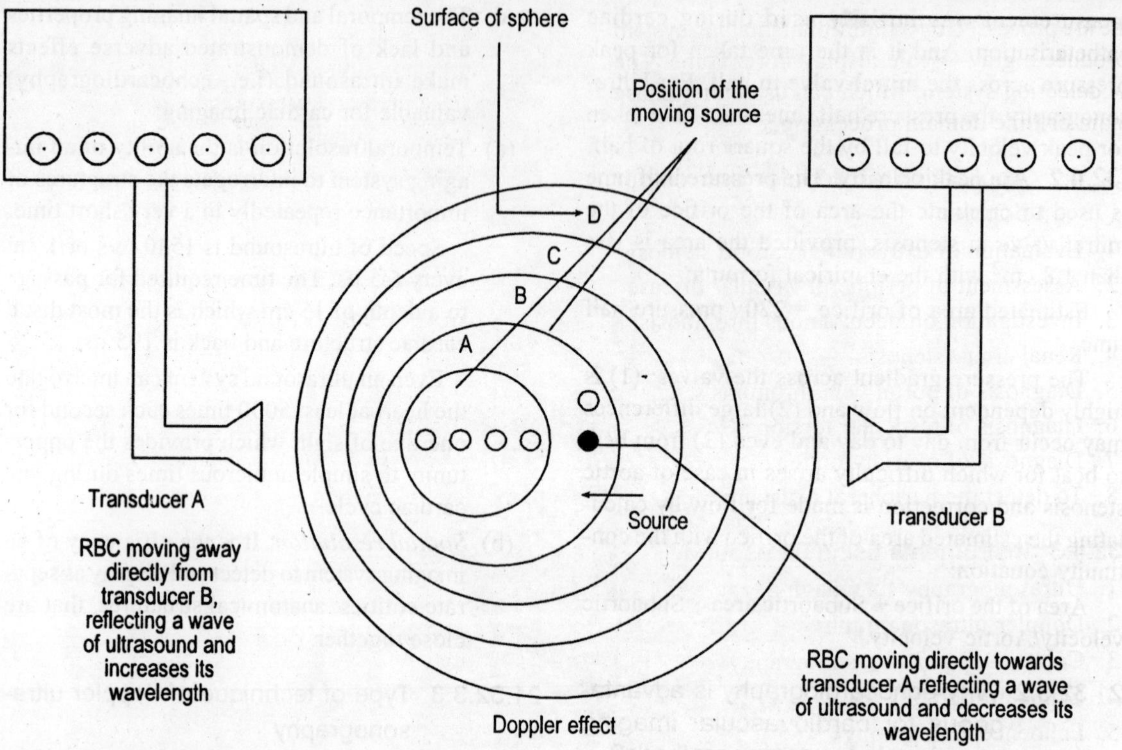

Fig. 21.31: The Doppler effect illustrates the change of wavelengths, and hence of frequency occurs when a source of sound is moving downwards a stationary observer

21.32.4 Colour flow doppler

The Colour Flow Doppler is a two-dimensional version of Pulsed Doppler. The demerit of measurement error above a certain maximum velocity is greatly reduced by this method. This colour flow system measures mean blood velocity and the direction of flow at multiple points down each scan line of an image sector. This velocity observation is then superimposed on to the echocardiographic image with colour coding. The velocities having direction towards the transducer are displayed in red and those velocities that are away from the transducer are in blue. Initially, the increasing velocities are seen in progressively lighter shades or changes of hue and afterwards as a change in colour from blue to red or vice versa and this type of display is known as "aliasing", which helps the visual detection of abnormal flow.

21.32.4.1 Information

The Colour Flow Doppler gives us a dynamic representation of information on the velocity of blood throughout the cardiac cycle. The display looks superficially an image produced by Cine-angiography but gives information on the volumetric of blood rather than on spatial velocity and is not strictly comparable.

Advantages of colour flow imaging

1. Arteries can be differentiated from the veins.
2. Differentiation between vascular and avascular structure.
3. Visualisation of pattern of blood flow and intravascular filling defect.
4. Identification of terbulenec and high velocity of blood flow.
5. Visualisation of intravascular filling defect.
6. Visualisation of small arterial plaques and non-occlusive venous thrombosis.
7. Visualization of the contrast between vessel wall and lumen.
8. Helps identification of branches, coils, kinks and aneurysm.

Colour imaging system in this example an ultrasonic probe containing a phased-array transducer produces a sectal scan Echoes from within the scan plane are processed by a tràditional imaging system to produce a gray—scale display in real time. The flow detection system, which can use either phase domain or time domain processing.

21.32.4.2 Application of doppler and colour doppler

1. Evaluation of extracranial carotid stemosis
2. Assessment of peripheral vascular disease
3. Investigation on deep venous thrombosis
4. Renal artery stenosis
5. Diagnosis of portal vein thrombosis
6. Diagnosis of testicular torsion
7. Diagnosis of ovarian torsion
8. To detect the peripheral resistance.

21.32.4.3 Instruments using ultrasound

1. Ultrasonography for imaging
2. Doppler ultrasonography
3. Colour Doppler ultrasonography
4. Echocardiography
5. Echoencephalography
6. Echoophthalmography
7. Phonocardiography
8. Ultrasonic blood flow meter
9. Endosonography
10. Ultrasonic microscopy
11. Ultrasonic blood pressure method
12. Ultrasonic camera
13. Ultrasonic therapy machine
14. Ultrasonic surgery machine.

21.33 Classification of PZ materials

I. **Natural**: Quarts or Rochelle salt (sodium potassium tartarate. $KNaC_4H_4$ 06-$4H_2O$)

II. **Biological**—Bone,

III. **Artificial**

A. *Crystalline*—Lithium Niodate

B. *Ceramic*—Lead Zirconate Tannate.

C. *Polymer*—Polyvinyl difluoride → heated (PVDT) curic point or temperature → *P-z* domain disappears → reappears on cooling → when close to curie point, domains can move and therefore a potential may be applied to attain domain alignment as the material is cooled. Polymer materials are streteted to increase their P-Z properties.

D. *Copolymer*—Powdered P-Z ceranic within polymer material.

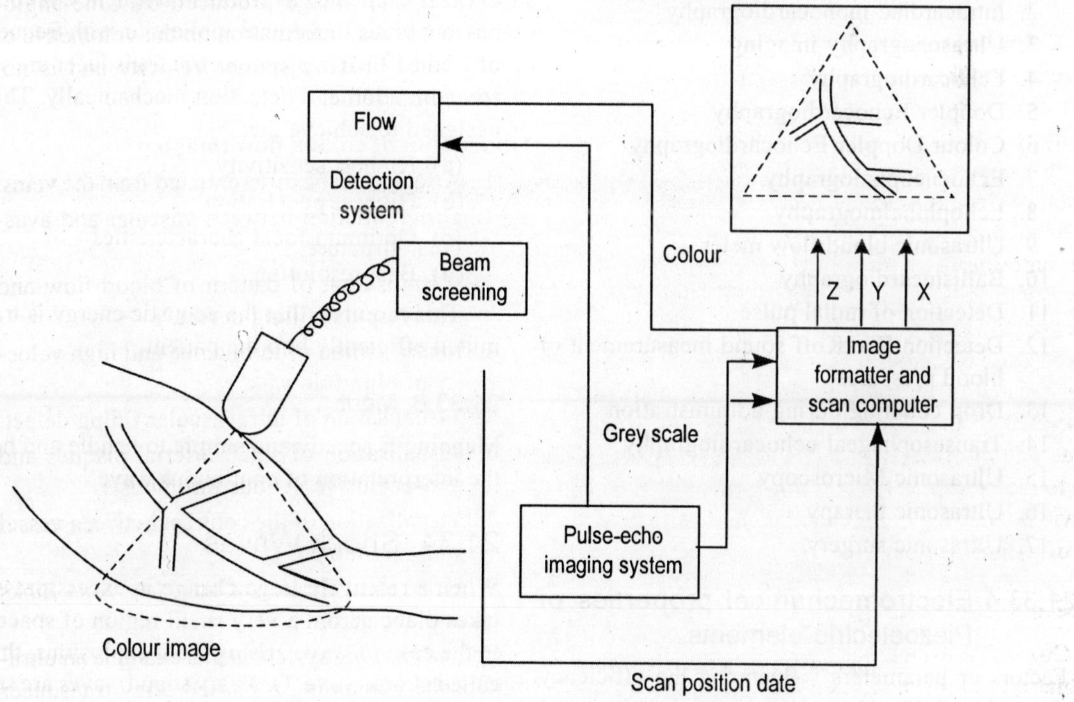

Fig. 21.32: Block diagram of an ultrasonic colour imaging system

1. Good P-Z electrical properties
2. Low acoustic impedance.

21.33.1 Properties

1. High acoustic impedance.
2. Produces high mechanical and electrical activity when used as generator or detector of USG.
3. Poor transmission of USG through the tissue.
4. Has an acoustic impedance well matched to the acoustic impedance to body tissue.
5. Has low P-Z electric factor.

21.33.2 Some characteristis

1. P-Z materials are non-conductive.
2. P-Z transducer is active transducer because self generating active transducer.
3. P-Z transducer can be considered as fixed capacitor because fixed capacitor carrying charge proportional to the applied force.
4. Magnitude of the P-Z electric effect is dependent on the axis of the cutting of the crystal.

21.33.3 Application of P-Z crystal

1. Phonocardiography
2. Intracardiac monocardiography
3. Ultrasonography imaging
4. Echocardiography
5. Doppler Echocardiography
6. Colour Doppler Echocardiography
7. Echoencophalography
8. Echophthalmography
9. Ultrasonic blood flow meter
10. Ballistocardiography
11. Detection of radial pulse
12. Detection KrotKoff sound measurement of blood pressures
13. Drop counting during administration
14. Transesophageal echocardiography
15. Ultrasonic Microscopy
16. Ultrasonic therapy
17. Ultrasonic surgery.

21.33.4 Electromechanical properties of Piezoelectric elements

(Factors or parameters influencing the efficiency of transducer). The characteristics of piezoelectrical elements are considered by a number of coefficients such as:

1. Electromechanical coupling coefficient (K)
2. Piezoelectric voltage constant (g)
3. Piezoelectric charge constant (d)
4. Transmission coefficient (h)
5. Dielectric constant (E)
6. Acoustic Impedance (X)
7. Mechanical coefficient (Q).

21.33.5 Production of ultrasound by P-Z materials

The ultrosonograph contains *pulse generator* which produces a *short electrical pulse* at a required frequency such as 7.5 MHz which is fed to the transducer (probe). The Piezoelectric generator is the most effective method of generating ultrasonic waves. The transducer (probe) contains *one or series of piezoelectric elements,* in which the electric *oscillation* is converted to a mechanical one i.e., ultrasound. The P-Z crystal is cut hexagonally so that it is mechanically resonant for an increased efficiency of conversion of *electrical energy to acoustic energy*. It is usually one half wavelength thick for particular frequency used when the transducer is excited of its resonant frequency, it continues to vibrate. In medical application with frequency of 1 and 12 MHz piezoelectric transducer are used for generation and detection mechanically. This is designed to achieve the:

(a) Highest sensitivity
(b) Highest penetration
(c) Optimum focal characteristics
(d) Best resolution

This requires, that the acoustic energy is transmitted efficiently into the patient.

21.33.6 Merit

Mapping is speedy and simple to handle and helps the interpretation of continuous wave.

21.34 Shock Waves

When a relatively large change in excess pressure takes place across a very small region of space, as in the case of wave arising from explosions, this is called **shock wave**. Ordinary sound waves are small amplitude waves and such waves propagate with-

Table 21.3: Comparison of the devices for Medical Diagnosis by X-ray and ultrasound

Diagnostic X-ray	Diagnostic Ultrasound
1. Imaging Modalities Invasive	1. Imaging Modalities non-invasive.
2. Responds to atomic weight differences and often require the injection of more dense contrast medium for visualisation of a non-bony tissues.	2. Distinguishes interfaces between soft tissues
3. Soft tissue X-ray is the only method to detect breast cancer e.g. mammography	3. Externally applied and non-traumatic but also apparently safe at the acoustic intensities and cycles presently used in diagnostic equipment
4. It can not be used in dynamic blood flow in cardiovascular system.	4. Applied to obtain images of the entire range of internal organs in the abdomen e.g., kidney, liver, splean, pancreas, bladder, major blood vessels, the fetus of pregnancy, thyroid gland, eyes, breast etc.
5. It can propagate through medium.	5. Detection of cysts, tumours or cancer in the organs.
	6. Ideal in differentiation solid from cystic structures although the nature of fluid is difficult to identify.
	7. This applicable when other diagnostic methods inapplicable, insufficient and hazardous.
	8. Extensively used for dynamics of blood flow in cardiovascular system which was not previously possible.
	9. *Limitations:* Its propagation is related to the matter can not propagate through the medium e.g. hollow organs e.g. lungs, stomach, empty urinary bladder if it can hold water then USG is possible.

out change in form. But for waves of large amplitude the region of higher pressure (crest) move faster than regions of lower pressure (trough). As a result the wave changes form and acquires a saw-tooth shape. When a macrosonic sound—i.e., sound of large amplitude—acquires this shape, it becomes a shock wave. Explosion produces such shock wave sounds. The low wave of projectile moving with supersonic speed is also an example of shock wave. The shock wave arising from a high speed bullet or aircraft is transmitted laterally but not in a forward direction. Hence, sound in such cases is heard only when the projectile passes the observer. The study of shock waves is of great importance in aerodynamics of aricraft and missiles and in connexion with explosions and ballistics.

Later on, when it was discovered that shock waves can pass through body tissue without causing any damage, it was employed for the treatment of stones in the gall-bladder (cholelithiasis) and kidney (urolithiasis) and the method is known as "Extracorporeal Shock Wave Lithotripsy (ESWL)".

21.35 Extracorporeal Shock Wave Lithotripsy (ESWL)

21.35.1 Principle

Lithotripsy means "stone crushing" (Litho = stone, tripsy = crushing). Extracorporeal Shock Wave Lithotripsy utilizes shock waves to crush the stones present in the gall-bladder or kidney. The shock waves are high-pressure waves which are generated in 3 ways:

A. By an imersion spark-gap generator or thermal generator,
B. By an electromagnetic acoustic generator, or
C. By a piezoelectric generator.

• **The Spark-Gap Generator**: It uses an electrode under water in the geometric focus of an ellipsoidal reflector. This electrode emits high-voltage spark which causes sudden evaporation and expansion of surrounding water producing shock waves. The shock waves then reflect and focus on the stone, the point of highest energy density. The shock waves can penetrate. The human body without any hindrance because the acoustic impedance at the stone leads to absorption of energy from the shock waves resulting in disintegration of the stone (Fig. 21.33).

21.35.2 Method

The patient is under epidural or general anaesthesia. The patient is placed in a reclined position on a chair-like support and is partially immersed in a tube containing full amount of degassified and demineralized water.

Two-axes X-ray systems are used with the axes crossing at the second focal point. To position the patient correctly, there is a fluoroscopic display on a television monitor.

The whole procedure takes 30-45 minutes and within the time the stone is crushed into tiny sand-like particles and passes out of the body (Fig. 21.31).

21.35.3 Indication

I. Kidney or Ureteric Stone: For treatment by ESWL some criteria should be fulfilled:

1. Radio-opaque stone in the kidney or ureter should be above the bony pelvis.
2. The passage should be clear for passing stone fragmentation.
3. The urine should be sterile, i.e., absence of infection, if there is infection, that must be controlled before starting Litho-tripsy.
4. Absence of significant aortic or renal artery calcification.

II. Gall Stones: ESWL is also being used for disintegration of gall stones using the Biliary Lithotripter. There are also certain limitations;

1. Stone should be within certain size and number.
2. Gall bladder should not be infective.
3. Cystic duct should be without any obstruction.

21.35.4 Post treatment period

1. Most patients do not require any further ESWL treatment.
2. Very small number of patients may require another ESWL treatment for residual large fragments.

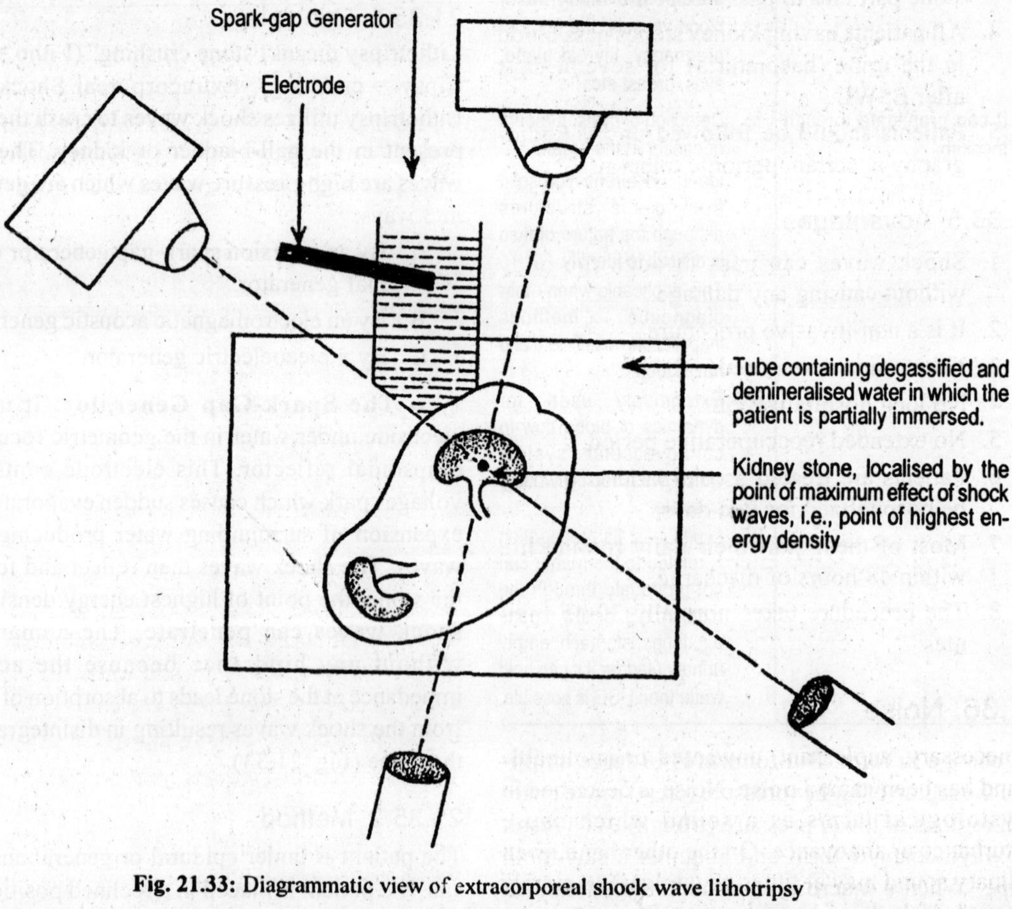

Fig. 21.33: Diagrammatic view of extracorporeal shock wave lithotripsy

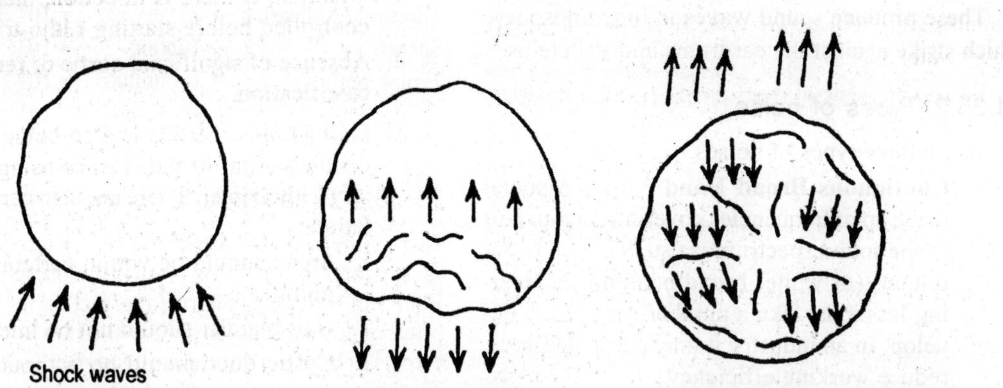

Fig. 21.34: Diagrammatic representation showing sequence of events of fragments and disintegration of stone

3. 2-4 weeks are required for disintegrated stone particles to pass out of the body.
4. All patients having kidney stones pass blood in the urine (haematuria) for several days after ESWL.
5. Patients should be followed up by Radiography at certain periodic interval.

21.35.5 Advantages

1. Shock waves can pass through the body without causing any damage.
2. It is a non-invasive procedure.
3. No incision is required.
4. No long hospitalization.
5. No extended reoccuperative period.
6. Patients are treated as out-patients or may be hospitalized for two days.
7. Most of them start their daily routine life within 48 hours of discharge.
8. The procedure takes normally 30-45 minutes.

21.36 Noise

Unnecessary, unpleasant, unwanted or intolerable sound has been named **noise**. Noise is described in physiological terms as a sound which cause disturbance or annoyance. On the other hand, even ordinary sound may at times be unpleasant, simply because of lack of conditioning. Noise can be defined in engineering terms as any disturbance in an integrated equilibrium. We human beings are in an exquisitely delicate integrated equilibrium. This can be disturbed in many ways. The disturbance is usually called noise. Noise is one of the stress factors that can produce widespread disturbances in various biochemical and physiological activities of the human body. Hawel defined it more precisely: "sound only becomes burdensome if the person affected feels that it is discordant, i.e., if it does not harmonise with his intentions at that particular moment"—this is apt when applied to noise at work.

21.36.1 Source of noise

Disturbing noise may be either external or internal.

(a) **Internal**

1. Playing of transistor, record player, radio and T.V. at full volume.
2. Band and rock music.
3. Barking of dog.
4. Noise in factory, machines, rotors, compressed air, milling machines, looms, sawmills, domestic water pumps.

(b) **External**

1. Roaring of jet planes.
2. Rattling of cars, motorcycles, trucks.
3. Industry.
4. Loudspeaker.
5. Whistle of train.
6. Siren.

These produce sound waves in the atmosphere which strike against the eardrums and irritate us.

21.36.2 Types of noise

There are three types of noises

1. **Continuous Broad Band Noise**: A sound made up of frequencies covering a major part of the sound spectrum is usually called continuous (or white) broad band noise. Hearing loss must take a number of years to develop. In an industry it may cause deafness, reduce working efficiency.
2. **Intermittent Broad Band Noise**: Noise with these impulsing nature of sound is more damaging than with continuous noise. If it is regular, it may have effects which differ little from those of continuous noise, but if it is irregular and unexpected, it may cause a startle reaction which can be most disturbing.
3. **Meaningful Noise**: The sound which conveys an unwanted meaning to the subject and its effect may have no relationship to its loudness.

21.36.3 Unit of noise

The intensity of noise is expressed in "decibels" (dB), i.e., it is the unit for measuring the pitch of noise. The intensity of sound of a jet plane at the time of take off is 120 dB, and in metropolitan cities where cars, trucks, horse-carts are continuously plying, it is about 90 dB. The bearable limit of noise is about 140 dB.

Several evidence reveal that if a man stays continuously for long periods in an area where the sound intensity is 85 dB, his hearing power is affected.

21.36.4 Magnitude or loudness of sound

The ear is capable of hearing a very wide scale of loudness of sound from the softest sound, i.e., the threshold of hearing to the intense and distressing roar of a jet aircraft which is 10 million million times as loud (10^{13}). This range is quite unmanageable in an arithmetical sense and a logarithmic scale has to be used—decibel scale—which indicates how many louder a noise is than the basic reference point, which is usually 0.0002 dynes/cm^2.

Threshold of hearing	= 1	= 10^9	= 0 dB
Speech	= 100000	= 10^5	= 50 dB
Typing pool	= 10000000	= 10^7	= 70 dB
Jet aircraft	= 10 million million	= 10^{13}	= 130 dB

21.36.5 Frequency or pitch

Just as the sunrays are made up of many types of rays, sounds of different pitches are present in noise. Each pitch produces harmful effect on the ears. The pitch of sound is governed by the number of waves emitted by the sound source per second, for example, middle C on the piano is produced by a frequency of 256 cycles per second. The range of sound detectable by the human ear lies between 16 cycles/second and 16000 cycles/second. Other animals detect a different frequency range e.g., dogs, birds, and bats. The normal speech range is upto about 2000 cycles/second.

21.36.6 Measurement of noise load

The "noise load" is the extent of the noise, as measured in physical units, taking account of all the acoustical factors over a given time. Research work has shown that noise level is not the only factor operating, but that the frequency with which the noise occurs and other quantities contribute to the total noise load.

21.36.7 Sound measurement

Sound is measured and recorded in decibels for each octave wave band. This measurement is not suitable for translating the effects of sound on the human ear, however, since the response of the ear varies not only with the loudness but also with the pitch of the sound than it is to high frequency sounds of the same loudness. Sound produced by heart—12-13 dB, movement of levels—18-19 dB, normal conversation—60-62 dB, barking—68-70 dB, aircraft take off—more than 120 dB, cracker—more than 120 dB have been observed.

In order to express "sound pressure level" (SPL) in meaningful human terms the sound pressure, matter is weighed to take account of this phenomenon and levels of noise are expressed on the "A" weighted scale as dBA.

21.37 Effects of Noise in the Body

1. **On Auditory System (Ear)**: (a) Partial loss of hearing (deafness) and interference with

understanding, (b) Displacement of bones (ossicles) in the middle ear, (c) Disruption of some parts of organ of Corti, e.g., loss of hair cells in patches, (d) Hearing loss is proportional to duration of exposure, (e) Excessive noise interfers with reception of audible information and thus irritation is aroused.

2. **Emotional Disturbance**: Perception of sound takes place in the cerebral cortex. But it first passes through midbrain and it is then directed to the cortex and it is here that noise may produce emotional disturbances of annoyance, irritation, quarrelsomeness, loss of attention and other conscious factors which are disturbing. The impulse may spread to autonomic nervous system to produce complicated effects on the cardiovascular system, renal system, endocrine system, reproductive system, etc.

3. **Susceptibility to Infection and Audiogenic Stress**

(a) Subjects inoculated intranasally with virus just before exposure to sound were most susceptible to the infection. If inoculated after the exposure to sound they were more resistant.

(b) The inflammatory and interferon responses were also impaired in subjects exposed to noise.

(c) Animals exposed to sound intensities about 120 dB for 3 hours each day suffer increased stimulation of pituitary adrenocortical axis causing increase in ACTH secretion.

(d) During exposure, leucopenia followed by leucocytosis, after a stress period were observed.

4. **Endocrine System**

(a) **Ant. Pituitary**: Loud sound stimulates pituitary-adrenocortical axis and there is increased secretion of ACTH. It is characterised by leukopenia followed by leukocytosis after stress period.

(b) **Adrenal Cortex**: Elevation of ACTH level in blood causes more secretion of corticoids and aldosterone. Subjects also develop hypertension due to repeated exposure to sound and have an elevation of urinary catecholamines. Hypertension may be due to renin-angiotensin system for hyper activity of endocrines.

(c) **Thyroid**: Sound reduces the secretion of TSH from anterior pituitary depressing thyroid activity.

(d) **Adrenal medulla**: Increased secretion of adrenaline and noradrenaline takes place due to overactivity of sympathetic nervous system associated with emotional disturbances, e.g., anger, irritation, etc. It is characterised by initial increase in urinary catecholamines.

5. **Digestive System**: Those people who work in noise level become hypersensitive. Their digestive glands in the stomach and duodenum may develop ulcers.

6. **On Reproductive System**

(a) It has been proved experimentally that girls in cities who are exposed to intensive noise develop menstruation earlier.

(b) There is an increased secretion of oestrogen which interfers with the development of fertilized ova and the maintenance of pregnancy.

(c) The breasts enlarge and there is excess milk secretion during pregnancy.

(d) Increased weight of ovaries, rapid follicle maturation and increase in number of corpus luteum are seen.

(e) There is also a secretion of oxytocin from post-pituitary.

7. **On Pregnancy**

(a) A pregnant woman suffers from its unwholesome effects. The foetus develops defects in the nervous system. Later on, its ill-effects are reflected in the behaviour of the boy or girl.

(b) The foetus is capable of preceiving sounds and responding to them by motor activity and cardiac rate.

(c) Maternal anguish produces increased motor activity of the foetus.

(d) Emotion-inducing music causes change in foetal heart rate.

(e) In later months, severe emotion causes hyperkinetic, uncomfortable poor functioning infants.

8. **Psychological Effects**: The schizophrenics exposed to noise have high secretion of norepinephrine and are similar to psychotics during extreme aggressive episodes.

9. **Cardio-Vascular System**

(a) Causes high blood pressure, either systolic or diastolic, or both.

(b) Degenerative changes in the myocardium due to increase in cholesterol and triglyceride release from adipose tissue by catecholamines.

(c) Vasoconstriction.

(d) Left ventricular hypertrophy.

10. **Brain**

(a) Morphological changes in the neurone.

(b) High frequency oscillation electrically recorded from medulla and subcortical areas.

(c) Less pronounced oscillation in the cortex.

(d) Different types of seizures.

Audiogenic seizure has been widely studied in animals which have considerable resemblance to human epilepsy. Mice, rabbits and chicken are susceptible to such seizure but not the cats and monkeys. Pretreatment with semicarbazide of metheonine, sulphoxine in cats and monkeys produce seizure and lower the threshold in case of susceptible animals. During the tonic phase of seizure, due to asphyxia or cerebral haemorrhage, the animal could die. Lockbett (1970) suggested that genetic factors resulting in biochemical deviations from normal, determine the susceptibility to seizure. Because urinary excretion of epinephrine and norepinephrine and plasma cholesterol and cortisol increase significantly during seizure; patients with essential hypertension or with some mental disease are more susceptible to seizure. There is significant decrease of 17 OH-Corticoids in seizure cases.

11. **Eye**

(a) Diminished night vision which is transitory.

(b) Narrowing of vision after 6 months.

(c) The concentration of melanin in the stria vascularis of cochlea is directly proportional to that of the iris and inversely proportional to dehydrogenase in the cochlea.

12. **Skin**: Changes in the electrical potential.

13. **Pain sensation**: Patients undergoing dental treatment do not require anaesthesia if they listen to controlled white band noise while undergoing the operation. It can, therefore, be said that noise reduces pain sensation.

14. **Physical and Mental Health, Sleep**: Noise causes disturbance in sleep and it affects man's physical and mental health. Victor Green, the internationally noted city planning expert, holds that "noise is the agent that brings death closer bit by bit!" Experimental evidences have shown that noisy environment:

(a) Seriously curtails the total time asleep,

(b) Cuts down the amount of deep sleep,

(c) Increases the time spent awake or in light sleep,

(d) Prolongs the time of falling asleep.

15. **Electrolyte and Water Excretion**

(a) Rats exposed to sound having 150 cycles/sec show increased excretion by sodium and potassium at a slow rate and stay at plateau level and then, by seventeenth day, the rate of excretion returns back to normal level. Adrenalectomy or adrenal demodullation caused failure to produce any such effect.

(b) No marked effect on water excretion (Lockett, 1970).

16. **Vegetative Effect**: Auditory tracts are linked with the reticular activating system and alarm-sensitive structures of the brain. For this connection many of the effects of noise and the spread of a state of activation or alarm occur and may result in:

(a) Impaired alertness,

(b) Disturbance of sleep,

(c) A feeling of stress.

At the same time this activation affects the autonomic centres and produces the so-called "vegetative effects" in the internal organs (Fig. 21.35). Finally a special problem of noise is that it makes it difficult to understand what people say. Due to long exposure the vegetative irritations are manifested by:

(a) Rise of blood pressure,
(b) Acceleration of heart rate,
(c) Contraction of the blood vessels of the skin,
(d) Increase in metabolism,
(e) Slowing down of the digestive system,
(f) Increased muscular tension,
(g) Gastric ulcer,
(h) Colitis,
(i) Migraine.
(j) Various mental diseases known as "neurohumoral stress response".

17. **Understanding of speech**: The sensitivity of the ear to one particular sound becomes less and less as the ambient noise increases. The ability to pick out one particular sound from the rest depends on its auditory threshold with the sound intensity up to 80 dB.

18. **Human Performance**

(a) No general criteria exists to assess man's complex behavioural response to noise levels below 140 dB when speech communication is not involved. Although significant decrements in the performance of mental or motor tasks have not been reported, increased fatigue and changes in motivation towards the job—at least in some groups of individuals—have been observed which indicates that noise control is required.

(b) Noise principally affects human performance in tasks requiring communication. Recent investigations have shown that aircraft noise influences speech communication in office in the same predictable way as industrial noise.

(c) Background noise often interferes with complex mental activities, as well as certain kinds of performance that make heavy demands on skill and on the interpretation of information.

(d) Noise can make it more difficult to learn certain kinds of dexterity.

(e) Many studies have shown that high levels of noise (over 90 dB as a rule)—either discontinuous or unexpected—can impair mental performances.

19. **Mental Work**

(a) An intermittent noise, especially if unexpected, is more disturbing than a continuous one.

(b) High frequencies are more disturbing than low ones.

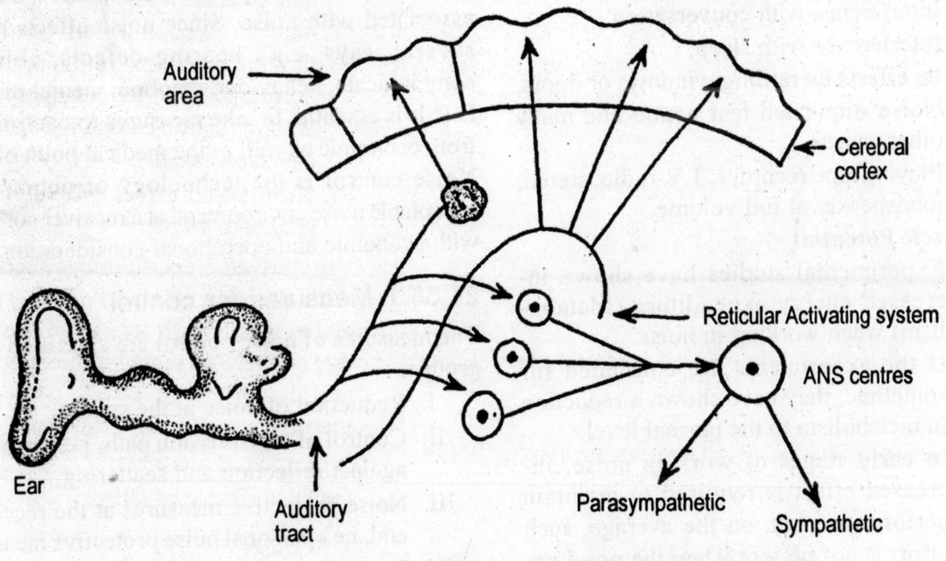

Fig. 21.35: Diagrammatic representation of vegetative effect of noise

(c) Work which calls for sustained alertness over a long period and without pause is particularly sensitive to noise.

(d) Disturbance is greater during a learning stage than when the work has become more automatic.

(e) Sounds which have a certain information content are more disturbing than meaningless noise.

20. **Behaviour response**: Behaviour in response to noise can normally be measured in three ways:

(a) A man may be asked to inform on his own feeling or sensation.

(b) Physiological measurements may be employed such as metabolism, rate of breathing, tension in the muscles, etc.

(c) The man may be asked to do some task and his efficiency on that task is measured.

21. **Annoyance**: Some noises are annoying to almost all people and probably any particular noise is annoying to some person. But two points must be considered—first, the annoyance produced by some sounds are not bad for health, it does not mean that it can be ignored. But it is possible to reduce annoyance either by training or by compensating. In residential areas, annoyance of people with the presence of noise can arise from:

(a) Interference with conversation,

(b) Interference with sleep,

(c) Its effects on rattling windows or doors,

(d) Noise stimulated fear action and many other causes,

(e) Playing tape recorder, T.V. radio, stereo, loudspeaker at full volume.

22. **Muscle Potential**

(a) Experimental studies have shown increased energy expenditure (Metabolism) when working in noise.

(b) If the experiments are continued for sometime, they have shown a reduction in metabolism to the normal level.

(c) In early stages of work in noise, increased effort is required to maintain performance but, on the average, such effort is not present when the noise becomes familiar.

(d) Further studies have shown a relation between muscle-tension changes and efficiency of work in high noise. Persons who maintained their level of work in the distracting conditions showed a larger increase in muscle potentials than those whose performances were less efficient.

23. **Other Physiological Effects**

(a) Feeling of vibration in the head.

(b) Movement of the air in the nose.

(c) Loss of equilibrium, probably due to vibration of the labyrinthine sense organs which control posture.

(d) Vibration of eyeballs may cause disturbances of vision.

(e) There may be some heating of the skin. Animals with hair on their skins (e.g., rats, guinea pigs) may suffer from burns because of the energy absorbed. A hairless skin seems to lessen this effect considerably.

21.38 Protection against Noise (Noise Control)

Revolution in industry has caused the employment of new machines for manufacturing products, for high speed transportation, for furnishing additional comfort, for speeding up our daily routines to provide additional leisure hours, etc. These are associated with noise. Since noise affects man in several ways, e.g., hearing defects, ability to communicate, behaviour, emotion, mental diseases, B.P. It is essential to take measures to control noise from economic as well as the medical point of view. Noise control is the technology of obtaining an acceptable noise environment, at a receiver consistent with economic and operational consideration.

21.38.1 Measures for control

The measures of noise control are classified into 4 groups:

I. Reduction of noise at the source,

II. Control of transmission path, i.e., insulation against reflection and scattering,

III. Noise protective measures at the receiver's end, i.e., personal noise protective measures,

IV. Protective planning or hearing conservation programme.

I. **Reduction of Noise at the Source**: The most effective and rational means to deal with noise. The important methods of controlling noise at its source include:

i. The reduction of the amplitude of the exercising forces,

ii. The reduction of the response of various components of the system to these exciting forces (components that generate noise when excited or that transmit vibratory energy to surfaces that will radiate sound).

iii. Changes in operating procedure.

iv. **New Plant and Process**: Where these items are to be purchased, a noise specification should accompany the enquiry.

v. **Existing Plant and Process**: The noise from small items of plant can sometimes be reduced by enclosure and even large items can benefit from acoustic treatment. Much work has been done on the silent handling of billets and impact noise of these sort should not be accepted as inevitable.

vi. **Noise Refuges**: Control cabins and pulpits serve as noise refuges to which men can retreat to escape general noise level. Consideration should always be given to building, other refuges are solely for the purpose of noise protection in pump houses, generating room, etc.

The windows and doors of such refuges must be kept shut as even a small opening will offset their value. There must be some form of air conditioning in hot environments. Noise refuges are of great value in the reduction of total noise exposure, with the additional benefit of reducing the periods when ear protection needs to be worn in the open shops.

i. **Reduction of the Amplitude**

(a) The noise creating by heavy, hard surfaces grating together can be reduced by replacing the hard material with something softer, e.g., rubber or felt. Transport vehicles are, therefore, quieter with rubber types than with metal wheels.

(b) All machinery should be examined to see if they are still in good condition, since old and worn components create unnecessary noise.

(c) An especially effective way of reducing noise is to enclose the source. The inside wall of a housing of noise should be lined with sound-absorbing material, while the wall itself should be as heavy and airtight as possible. The housing should enclose the source of noise with as few gaps as possible to avoid the transmission of structural vibrations.

ii. **Reduction of the Response of Various Components of the System to the Exciting Force**: When components in a system are set into vibration, they will radiate sound. Application of the following techniques can reduce this noise at its source.

(a) **Alteration of the Natural Frequency of a Resonant Element**: A panel may be set into strong vibration when the frequency of the exciting force corresponds with or is near the natural frequency of the panel, as for example, this may be done by increasing the mass of the panel (which will lower the natural frequency) by increasing its stiffness (which will raise the natural frequency) or by changing the dimensions of the panel.

(b) **Increasing the Energy Dissipation:** Vibration-damping materials may be applied to the radiating surfaces. In this way, very substantial reduction in noise output may be obtained.

iii. **Change in Operating Procedure:** Changes in the usual procedure of operation may be effective as a noise control technique. Thus, some factories adjacent to the residential areas suspend or reduce noise operation at night, when the normal activity in a community diminishes and the general background noise is decreased.

II. **Control of Transmission Path**: Technique of noise reduction is that of controlling the transmission path so as to reduce the energy that is communicated to the receiver. This may be done as follows:

(a) **Sitting:** In the open air, maximum attenuation should be provided by increasing the distance between the source and the receiver. Since many noise sources do not radiate uniformly in all directions, by altering the relative orientation of the source and the receiver, a considerable reduction in noise level at the receiver may be possible. As for example, the orientation of an airport runway may be an important consideration in reducing noise in an adjacent community.

(b) **Design and Layout of Buildings**: The careful planning of the location of rooms within a building with respect to the relative position of the noise sources and those areas where quiet conditions are desired, may result in a considerable economy by reducing the extent of the noise-control measures that would otherwise be required.

(c) **Path Deflection**: Barriers in the open air can be effective when they are large in size compared with the wavelength of the noise to be deflected. As for example, deflecting surfaces—which make an angle of 45° to the horizontal—have been used in the noise field of jet aircarft engines to reflect the high frequencies upward toward the sky.

(d) **Enclosures:** Considerable attenuation may be provided by the use of a properly designed enclosure around a noise source or around the receiver.

(e) **Absorption:** One of the most effective measures of attenuating sound in its transmission path is by means of absorption. As for example, suppose 10 machines are in operation in a large office. Most of the noise from these sources that reaches the workers on the opposite side of the room will have been reflected by the ceiling, walls and floor.

The use of sound absorption in the form of acoustical materials on the ceiling, on carpet, and on the floor, will provide attenuation in the path between the source and the receiver. If noise is communicated by a ventilating duct, attenuation along this path may be placed in the form of sound-absorbing lining.

(f) **Magnetic Shielding**: Radical elimination of noise can be done by taking a large room where the noise is reduced by magnetic shielding. For biomagnetic brain research the noise must be reduced to one-millionth of its usual value, for which a room of multilayer walls of a magnetically soft ferromagnetic alloy, e.g., Permalloy, is constructed. But this type of shielded room is costly and very few are present in the world.

(g) **Impedance Mismatch (Acoustic Filters, Mufflers, etc.)**: The flow of acoustic energy along the path from source to receiver can be impeded which reflects the energy back towards the source (i.e., by an impedance mismatch). In dwellings, this may be provided by a break in the building construction. Sound transmission in the open air can be similarly impeded. Acoustic filters and mufflers operate on this principle.

III. Personal Protection of Ear

(a) Ear plugs to block the outer ear passage: These may be of cotton wool, wax or synthetic material.

Ear Plugs: A simple plug of cotton wool or wax in each outer ear passage is an old device that is often used. Properly used ear-plugs can reduce the noise level by up to 30 dB. Instead of cotton wool, conical plugs of a synthetic material (Selectone) are available, which cut more of the higher frequencies than the cover ones and so interfere less with conversation.

Drawback: They need to be used carefully and correctly if they are to give sufficient protection, and, if used regularly, they may set up irritations, such as aural eczema.

(b) **Protective Caps, which Cover the Entire External Ear**: Protective caps which cover the whole external ear are called "ear muffs". It usually gives the greatest protection even for the highest noise levels and if correctly fitted, can reduce noise levels by about 40-50 dB in the frequencies between 1000-8000 Hz. There is considerable variation, however, in the weight of muff, the material used for damping sound, and the choice of whether there is a need for apparatus to amplify signals and communication.

The insulating pad must fit closely round the ear, on to the skull so that it does not exert any pressure on the ear itself, otherwise it may set up pains and headache.

Drawback: Unfortunately, many workers object to any kind of ear protectors. Their main ground of complaint is "acoustic isolation". The worker thinks that he is missing some vital information from his surroundings. He also thinks that his output, and hence his earnings, will be adversely affected. In spite of these objections the wearing of ear protectors should be encouraged in the interests of the workers themselves.

It has been suggested that:

(a) For levels 85-100 dBA, ear plugs,

(b) For levels above 100 dBA, ear muffs.

(c) **A Sound-Proof Helmet**: A sound-proof helmet encloses the head including the ears.

(d) **Acoustic Glass Wool:** Acoustic glass wool is different from cotton wool. This gives a far degree of protection if properly fitted, but the degree of fitness is very dependent on the wearer's skill in fastening the plug and inserting it properly. It is comfortable to wear, cheap and can be made readily available on site from a dispenser.

IV. Protective Programme

(a) **Education and Public Relation**: In some cities where noise has been a serious problem, both the government and the industrialists should get in touch for factory, machine installation, and proper scientific advice from the government.

Educational programmes should be undertaken to educate workers and to make them understand the abuse of noise.

(b) **Exposure Control:** Under some circumstances it is impracticable to reduce extremely intense noise levels in areas where people must work to levels which are considered acceptable for the usual working period, for which authority should divide the city into zones, e.g.,

1. Commercial zone: Where sound should be between 55-65 dB throughout the whole day and night,
2. Industrial zone: Where sound should be between 70-75 zone: Where sound should be between 70-75 dB,
3. Residential zone: Where sound should be between 40-50 dB,
4. Silence zone or sensitive zone: As for example, hospital, college, school, court, etc.—about 100 metres around each building is called "silence zone", where sound should be between 40-45 dB like residential areas. In the silence or sensitive zone, blowing horns, crackers, loudspeakers, etc. should be prohibited.

V. Protective Planning or Hearing Conservation Programme

1. The measurement of noise (Industrial Hygiene),
2. The control of noise (Design and Engineering),
3. Identification of noise: susceptible people (Medical),
4. Issue of appropriate personal protective devices (Safety and Medical),
5. Implementation of the programme.

21.39 Speech Interference Level (SIL) and Method of Measurement of the Level of Noise

The audible sound spectrum is divided into 20 bands. Those are not harmonically related. A method has been employed for determining the maximum tolerable noise levels for satisfactory speech intelligibility. In this, acoustical measures of the noise are made by means of **Octave-Band Filters** and a **Sound-Level Meter**. In this method, the arithmetic average of the decibel level over all of the noise in each of the octave bands (600-1200 cps 1200-2400 cps and 2400-4800 cps) is found. It has been observed that, when the arithmetic average of the octave-band levels of wide-band noise is 12 dB below the overall rms level of undistorted speech at the listener's ear, in general, nearly 100 per cent of sentences will be heard correctly. This level of noise (12 dB below the average overall levels of undistorted speech) is called the **speech Interference Level** (SIL) for reliable speech communications. The SIL gives us information about the intensity of the speech signal at the listener's ear, which should be for a given noise condition in order to be heard correctly.

21.39.1 Measurement of Noise

Because of the variety of information that is needed and the variety of conditions encountered, different types of instruments are employed in noise measurement. All these instruments have to depend on a microphone. The instrumentation set up comprises of (Fig. 21.38):

1. Microphone,
2. Sound-Level Meter,
3. Magnetic-Tape Recorder,
4. Graphic Level Recorder,
5. Oscillograph,
6. Spectrum Analyzer,

1. **Microphone**: Microphone is connected either to a Sound-Level Meter or to a Magnetic Tape recorder. Its function is to transform acoustical noise into an electrical signal so that electronic instruments can be used to measure some characteristics of the signal or to store the signal for later measurements. The electrical wave should be an exact replica of the sound wave with a certain scaling factor called **response sensitivity**. In a given measurement—e.g., if an instantaneous sound pressure of 1 microbar is reproduced as 1mv in an electric circuit, then an instantaneous sound pressure of minus $\frac{1}{2}$ microbar must be reproduced as minus $\frac{1}{2}$ mV. In addition, measuring system should not vary with time or with ambient conditions.

• **Direction of Arrival of Sound at the Microphone:** Since the typical Sound-Level Meter is adjusted so that its frequency response most approaches the design objective for sounds of random incidence, it follows that no corrections are necessary for the directional characteristics of the microphone when the noise which impinges upon the microphone is randomly distributed with respect to angle of incidence.

A simple check by rotating the microphone in a plane through its normal axis will generally indicate whether or not such a condition exists; if it does, the reading will not vary. This condition generally obtains when measurements are made in a large scale reverberant room at a distance from the source so that the sound field is relatively diffuse.

Unless otherwise specified, Sound-Level Meters are so adjusted that the overall frequency response is closest to the design objective for sounds of random incidence. For the average microphone that random response closely approximates the response of the microphone at an angle of incidence of about 70°.

2. **Sound-Level Meter**: The Sound-Level Meter is used if the sound pressure level of a noise is to be measured directly.

   A Sound-Level Meter is normally equipped with a particular Microphone, a Sound-Level Meter and a Spectrum Analyzer. Even in these combined instruments, provision is commonly made for the use of different types of Microphones, and output circuits are often included so that other measuring instruments can be connected to yield supplementary information about a noise.

   Sometimes a Sound-level Meter is not used and a Microphone is connected directly to a sensitive Analyzer. A particular measurement may be made by a number of different set-ups and the one chosen depends in many instances on the equipment available. But since the Sound-Level Meter has been specifically designed for acoustic measurement, it is usually found to be the most satisfactory basic instrument for noise measurement.

• **Different Parts of Sound-Level Meter and its Function**: The Sound-Level Meter is a sensitive electronic voltmeter used to measure the electrical signal from a Microphone. It is calibrated in sound pressure level for use with a particular Microphone, which is ordinarily attached to the instrument. It is constructed according to the principle shown in the simplified diagram (Fig. 21.36). The electrical signal from the Microphone is amplified sufficiently so that, after conversion by means of a rectifier to direct current, it can deflect an indicating meter. An output connection that can supply the amplified signal to another instrument for analysis is also generally provided. An attenuator controls the overall amplification of the instrument, and the response versus frequency characteristics is controlled by electrical circuits called *weighting network*.

• **Amplifier**: The Amplifier in a Sound-Level Meter must satisfy many criteria:

1. It must have a large available gain in order to measure the signal from a Microphone in a quiet location.
2. It should have a wide frequency range, usually it is made to the order of 20 to 10000 or 20000 cps, although the frequency range from about 50 to 6000 cps is the range of the greatest interest in noise measurements.
3. The background noise level should be low.
4. It should be stable in gain.

• **Attenuator:** Sound-Level Meters are used for measuring noises that are of widely different level. Some range of level is covered by the relative deflection of the indicating meter. The rest of the range is covered by an adjustable attenuator, which is an electric resistance network inserted in the Electronic Amplifier to produce a certain decrease in the signal level. To simplify the calculation of level, it is customary to have the attenuator adjustable in steps of 10 dB. Some instruments have been built with less than 10 dB steps in order to reduce the range to be handled by the indicating meter.

• **Weighting Networks**: The response versus frequency characteristic of a Sound-Level Meter is controlled by electrical weighting networks. One of these networks is called "C" weighting, and it is intended to provide uniform response over the frequency range from 25 cps to 8000 cps. The networks called "A" and "B" weighting have a response that decreases with decreasing frequency. The responses approximate two of the equal loudness contours for pure tones; the "A" weighting corresponds

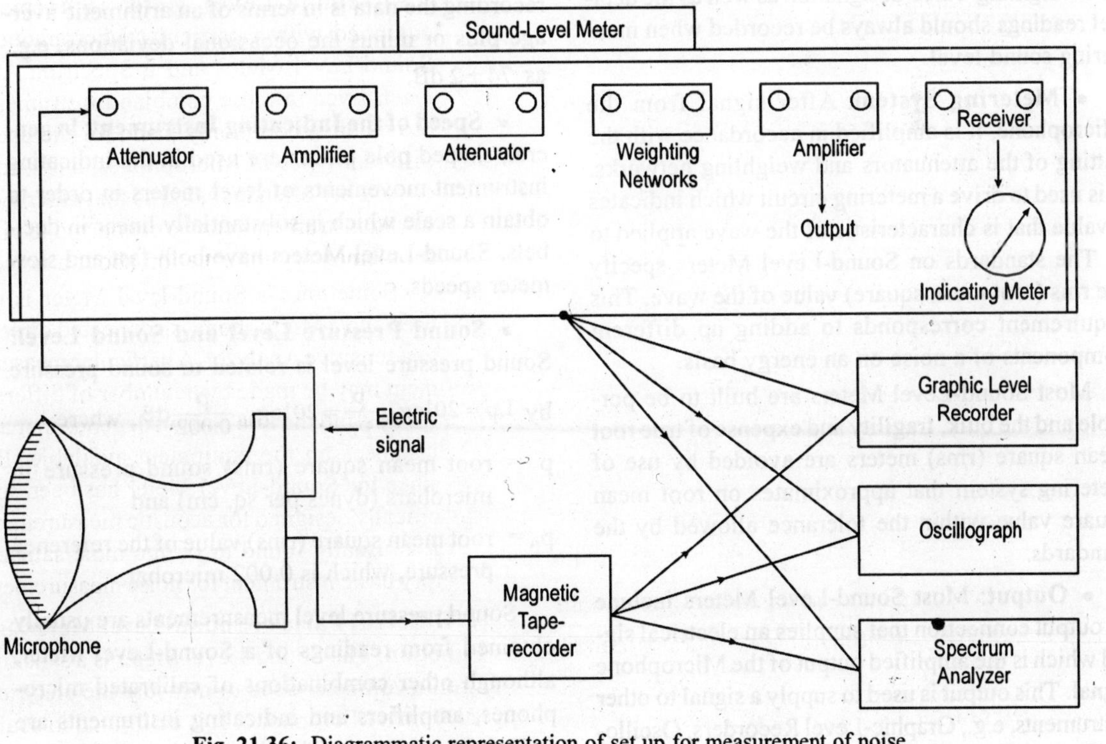

Fig. 21.36: Diagrammatic representation of set up for measurement of noise

to the 40 dB contour and the "B" weighting to the 70 dB contour. Readings taken using these responses are weighted levels in decibels, not sound pressure levels.

• **Use of Weighting Networks**: Sound-Level Meters are generally equipped with weighting networks which adjust the frequency response mainly by attenuating the lower frequencies.

The "C" network of the Sound-Level Meter provides the closest practical approach to overall flat response, i.e., equal response of the Sound-Level Meter to sounds of all frequencies within its frequency range. The "C" network is used for all measurements at high sound levels (85 dB or higher).

Since a reading with the "C" weighting represents a measurement of the actual physical sound pressure level, this network is used in all cases where the sound is to be analysed.

The greatest low-frequency attenuation is provided in the "A" network, while the "B" network provides an intermediate degree. These networks have been known, respectively as the "40 and 70 dB networks" because of their resemblance to the equal-loudness contours for the ear at these levels. The weighting-curve designation as well as the decibel readings should always be recorded when measuring sound level.

• **Metering System**: After signal from the Microphone, it is amplified in accordance with the setting of the attenuators and weighting networks. It is used to drive a metering circuit which indicates a value that is characteristic of the wave applied to it. The standards on Sound-Level Meters specify the rms (root mean square) value of the wave. This requirement corresponds to adding up different components of a noise on an energy basis.

Most Sound-Level Meters are built to be portable and the bulk, fragility and expense of true root mean square (rms) meters are avoided by use of metering system that approximates on root mean square value within the tolerance allowed by the standards.

• **Output**: Most Sound-Level Meters include an output connection that supplies an electrical signal which is the amplified output of the Microphone signal. This output is used to supply a signal to other instruments, e.g., Graphic-Level Recorders, Oscillograph or Spectrum Analyzers.

• **Commercially Available Sound-Level Meters:** Sound-Level Meters are made by many different companies. Most of the instruments are similar in many electrical characteristics as outline, because they are generally built to conform closely with existing standards. The portable battery-operated instrument is ordinarly supplied with a Rochelle salt diaphragm type microphone. A typical range of sound level covered directly is 40-130 dB or more.

• **Reading the Sound-Level Meter:** The Sound-Level Meter reading is the sum of the attenuator settings and the reading of the indicating instruments. In some Sound-Level Meters there is more than one attenuator and the reading of all must be added together. The instructions of the manufacturer should be followed closely to obtain the correct overall reading.

If the meter readings fluctuate widely in a reasonably regular manner the readings may be recorded giving both minimum and maximum values, e.g., 72-76 dB. On the other hand, if the readings remain relatively steady with only occasional increases or decreases, the preferred method of recording the data is in terms of an arithmetic average plus or minus the occasional deviations, e.g., as 74 ± 2 dB.

• **Speed of the Indicating Instrument**: In general, shaped pole pieces are used in the indicating instrument movements of level meters in order to obtain a scale which is substantially linear in decibels. Sound-Level Meters have both fast and slow meter speeds.

• **Sound Pressure Level and Sound Level:** Sound pressure level is related to sound pressure by $L_p = 20\log_{10}\frac{p}{p_0} = 20\log_{10}\frac{p}{0.002}$ dB, where

p = root mean square (rms) sound pressure in microbars (dynes per sq. cm) and

p_0 = root mean square (rms) value of the reference pressure, which is 0.002 microbar.

Sound pressure level measurements are usually obtained from readings of a Sound-Level Meter, although other combinations of calibrated microphones, amplifiers and indicating instruments are sometimes employed. The characteristics of the rectifier in the indicating instrument are important in determining the indicating level. As for example, if the indicating instrument on the Sound Level Meter provides a true root mean square (rms) reading, the measurement is one of rms sound pressure. Sound-Level Meters are equipped with weighting networks which alter the frequency response, mainly by attenuating the lower frequencies. Measurements made with these weighting networks do not, therefore, indicate true sound pressure levels, but weighted sound pressure levels, reported simply as sound level.

DEAFNESS

This is defined as the inability to hear wanted sounds, i.e., conversation.

It may be brought on by a variety of causes which include:

1. Disease of the ear,
2. Old age (presbycusis),
3. An assault on the ears by loud noise, i.e., gunfire,
4. Industry.

There are difficulties in assessing the proportion of deafness due to these various causes and indeed deciding on the degree of disability caused by any form of deafness. The deafness is temporary at first, but—after being "deafened" repeatedly—some permanent damages may occur. This is called **noise deafness** and is brought about by slow but progressive degeneration of the sound sensitive cells of the organ of Corti. The louder the noise, and the more often it is repeated, the greater the damage to hearing. Noise consisting of predominantly high frequencies is more harmful than low frequency noise. Intermittent noise, such as hammering, is more harmful than continuous noise, and a singular loud noise, a detonation, or an explosion, can damage the ears immediately.

Individual sensitivity to noise varies greatly from one person to another. Some—who are particularly sensitive—may suffer permanent deafness after only a few months, whereas less sensitive persons may not show the first symptoms until after many years of exposure. Noise deafness starts with the higher frequencies of about 4000 Hz. and extends only gradually to the lower frequencies. At first the worker is unaware of it and only gradually notices his loss of hearing when it begins to involve the lower frequencies. Noise deafness is progressive, and often combines with the deafness of increasing age, or is mistaken for the early onset of the latter.

The extent of loss of hearing is usually measured by means of so-called "pure tone audiometry", which determines the threshold of hearing of pure tones of various frequencies.

The risk of damage increases both with the sound intensity and the duration of exposure, the damaging intensity being those above approximately 90 dBA.

In a survey it has been observed that in a metropolitan city 8 persons out of 1000 are weak in hearing and out of 8 persons most of them are having deafness due to persistent noise called **steady state noise**, due to traffic, loud speaker, cassette shops, louder conversation among the persons, slogans of demonstration, etc. Again, **instant noise.**—e.g., sudden explosion of cracker or gunfire—may cause deafness.

21.40 Pure Tone Audiometer

Pure tone audiometers are usually used for investigation of loss of hearing function.

Function of the instrument: They generate tones of audible frequency with adjustable audio output.

Components

1. Tone generator
2. Noise generator
3. Tone Amplifier
4. Noise Amplifier
5. Tone Attenuator
6. Masking attenuation
7. Output selector
8. Head phones
9. Bone vibrator
10. Microphone
11. Audio amplifier.

Function of components

1. **Tone Generator:** It is a LC oscillator and generates tone of frequency between 125 Hz to 10 kHz in eleven steps.
2. **Noise Generator**: It is used to inject certain amount of noise or masking in another ear during measurement of air conduction

threshold. This is wide band noise and often it is referred as white noise. Noise is usually generated using semiconductor diode. A diode is connected across the D.C. power supply with reverse bias through a resistor and the noise generated across the diode is amplified by a pre-amplifier.

Tone and noise amplify these signals to the expected level. The acoustic output reaching the ears then it is set between 10 dB 1o 100 dB through driver amplifier and independent ladder type attenuator.

Attenuator: An attenuator is generally rotary switch or electronically controlled up and down electronic switch.

Output selector: The output selector blocks switches either headphones or bone vibrator as per the test to be performed. It also helps to select the ears for testing and masking.

Headphones: Most of the headphones in audiometers are dynamic type have that frequency response over the testing bandwith between 125 Hz to 10 KHz. Headphones and bone vibrators are used to measure air and bone conduction thresholds respectively.

Microphone and audioamplifier: These are employed for communication between the operator and the patient.

- **LED digital displays**

Seven segments LED digital displays are used for continuously indicate the setting of frequency, tone and masking attenuators (Fig. 21.37).

Advantages

1. This is specific for making threshold sensitive measurement.
2. Simplest type of auditory stimulus.
3. It is specific and accurate for frequency and intensity.
4. These frequency and intensity can be controlled with a high degree of precision.
5. Patients having impaired high frequency response due to occupational noise such as 4000 or 6000 Hz, this method is more accurate than spech audiometry.

Types of Audiometer: Based on the type of stimulus they provide to evoke auditory response. They are divided into groups.

1. Pure tone Audiometer
2. Speech Audiometer

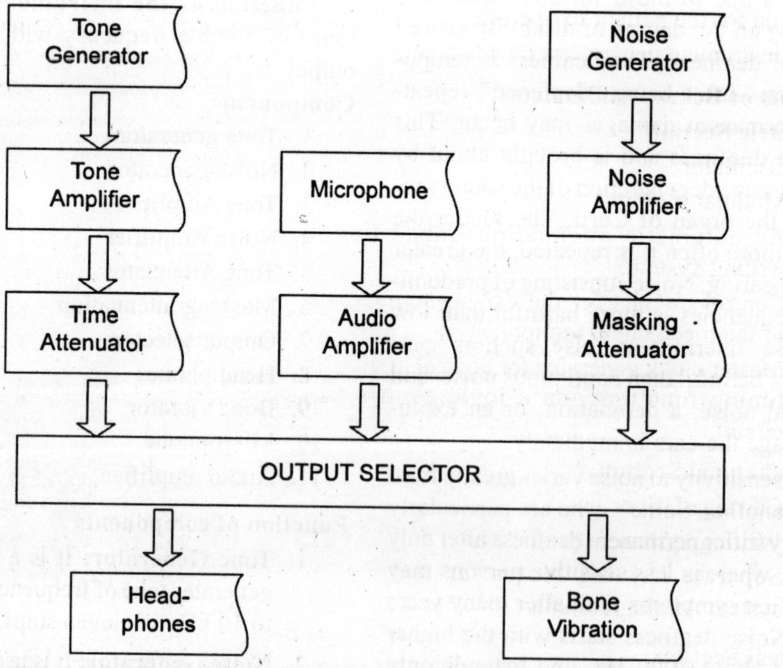

Fig. 21.37: Block diagram of pure tone audiometer

3. Screening Audiometer
4. Audiometer (Bekesy).

Types of Transducer in Audiometery

1. Earphones
2. Microphone
3. Bone vibrator
4. Loud speaker.

21.40.1 Audiometer (BEKESY)

1. It is automatic audiometric testing method by plotting of the hearing threshold based on the patient's signal.
2. It differs from pure tone method is that the interdependence of the patients response and stimulus intensity. Responses govern intensity and are affected by changes they introduce in it.
3. The audiogram recorded represents the obsolute threshold value in all frequencies in the tested range.
4. It also shows the difference in dB between the levels at which the patient just hears a signal of increasing intensity and those at which he just ceases to hear the signal of decreasing intensity.
5. Simple for the patient to operate
6. The instrument generates a pure tone signal.

Components of Bekesy Audiometer

I. Electric Section

1. Oscillator
2. Modulator circuit
3. An automatic attenuation linked to writing system.
4. Control circuits for the driving motors of the mechanical section.
5. Master clock generator for control of all timing functions via a logic control circuit.

II. Mechanical Section

1. Carraige Driver
2. Writing system
3. Separate Driving Motor for writing system Audiogram chart.

- **Electrical Section**

1. **Sine wave oscillator**: It generates test signal with frequencies of 125, 250, 500, 1000, 1500, 2000, 3000, 4000, 6000 and 8000 Hz. This sequence is first presented to the left ear automatically, each tone for 30s, and then to the right ear, the shift between the frequencies being noiseless. After both ears have been tested, a kHz tone is presented to the right ear to provide a useful indication to test reliability.
2. **Modulator**: From the oscillator, the test signal is fed to the modulator, where the mode of operation is selected by the 'Tone' switch, via the logic control circuit. Two models, 'Pulse' or 'Cont' are available. In the 'Pulse' mode the test signal is modulated giving a signal which is easily recognised by the subject. In the 'cont' mode no modulation is applied, giving a signal suitable for use, when calibrating the audiometer.
3. **Automatic attenuator**: The signal from the modulator feeds to the automatic attenuator situated on the carraige together with the writing system. The attenuator consists of logarithmic potentiometer which has its wiper attached to the pen drive so that the attenuation of the potentiometer corresponds to the position (X-axis) of the pen on the audiogram chart. The potentiometer has infinite resolution. The attenuation range is 100 dB, thereby covering the range of hearing levels from –10 to + 90 dB. When the test signal switches between the ears and when retesting is 1 kHz, the attenuator decreases the signal level to –10 dB to ensure that the right ear does not receive a tone at the elevated level possibly required at 8 kHz.
4. **Handswitch**: The pen drive is controlled are the logic control circuit by means of the handswitch operated by the patient. Pressing the switch decreases the output from the potentiometer and thereby the level in the earphones, while releasing the switch increases the output both ways with a speed of 5 dB/s.

Buffer Aplifier and Calibration Circuit

From the attenuator the signal is fed via a buffer amplifier to the hearing level calibration circuit. The

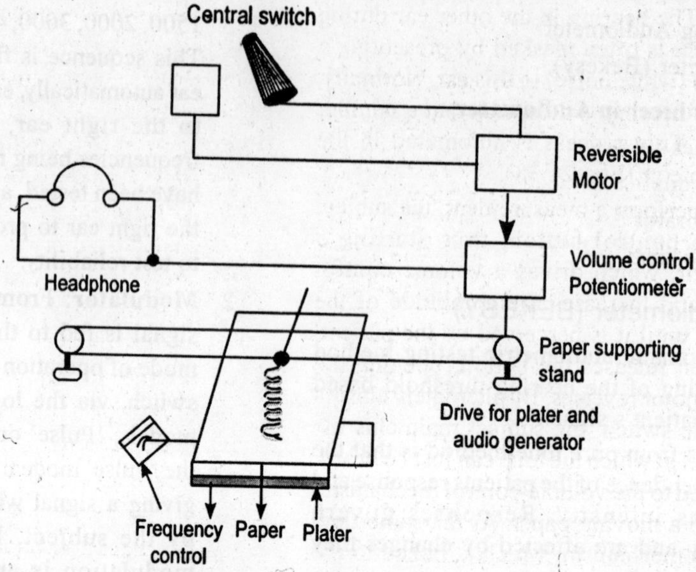

Fig. 21.38: Bekesy audiometer

buffer amplifier isolates the attenuator from the calibration circuit in order not to affect it output. The calibration circuit consists of seven potentiometers, one for each test frequency. During calibration, the potentiometers are adjusted one at a time until the correct level measured in a coupler, is obtained in the earphones.

Earphones: The earphones are a matched pair with distortion which is typically less than 1%.

Master Clock Generator: A stable clock generator supplies the necessary signals for control of motor speed, attenuator speed, frequency shift, modulation and other timing functions. This makes the system independent of variations in line voltage and frequency.

- **Mechanical**

1. **Carriage**: The Carriage with the writing system is driven by a stepping motor via a toothed belt. The speed and directions of rotation of the motor are automatically controlled via the logic control system. When the test is initiated and the patient indicates that he hears the signal by pressing the handswitch, the carraige moves along the X-axis (frequency axis) of the audiogram in agreement with the frequency of the test signal. When the frequency shifts, the carriage stops until the patient again, by pressing the handswitch, indicates that he bears the signal. Hence, avoiding waste of recording space on the audiogram if a patients hearing threshold varies from frequency or from left to right ear. When the complete test is finished the carriage and writing system return to the start position. To prevent carriage over-run, two limit switches are included in the carriage drive circuit.

Writing system: The writing system, is operated by the pen drive, which is driven by a stepping motor. The pen drive moves the pen, and with it the wiper of the automatic attenuator, along the Y-axis (hearing level axis) with a constant speed corresponding to the change in attenuation of 5 dB/s. The direction of movement of the pen is determined by the position of the handswitch operated by the patient. Limit switches are also included with the pen drive.

Audiogram Chart: The audiogram is printed in standard A5 format (148 × 210 mm). The recording space is large, 0.8 dB/mm, to enable easy reading. Space is provided on the audiogram side for registration of information on the patient, audiometer, operator etc., while the other side has space for recording the patient's medical and occupational history. Four holes in the chart give precise and automatic location of the audiogram on the chart bed.

Procedure: The sound intensity in an earphone is gradually increased until the sound is perceived by the subject. The hearing in the other ear during this measurement is often masked by presenting a neutral stimulus (white noise) to this ear. Normally, the threshold of hearing is determined at a number of frequencies. This process is automated in the Bekeásy audiometer (Fig. 21.38).

In order to perform a measurement, the subject first presses a control button, thus starting a reversible motor, which drives a volume control potentiometer and increases the amplitude of the stimulus signal until it is perceived by the patient. The patient then releases the button, opening the switch and the motor reverses. By alternately closing and opening the switch, the subject maintains the volume at a level at which the tone can just be heard. A pen, connected to the volume-control mechanism, draws a line on a moving paper. At the same time the paper-drive mechanism, which is linked to the instruments frequency control, slowly changes the frequency of the tone within about 15 minutes, a recording, called an audiogram, is obtained. The audiogram is often calibrated, not in absolute values of the perception threshold but in relative values referred to the acuity of normal subjects (which is stored in the instrument in a mechanical can, not shown in the diagram. The resultant graph corresponds directly to the hearing loss as a function of frequency. The diagram shows a simplified version of the original Bekesy audiometer, which changes the frequency of the stimulus in steps instead of continuously.

21.41 Hearing Aid

Conventional Analog Hearing Aid

(a) The system works on single pen battery or button cell.

(b) It is available as
   1. Pocket conventional model
   2. Spectacle model
   3. Behind the ear model
   4. All in the ear model

(c) Different types of models are available to use for different types of losses.

(d) The standard instrument uses pencil cell and other works on button cell.

(e) The instruments are having an *electrat* or condens microphone for conversion of sound into electrical signal.

Because electret microphones have very sensitivity to weak sound signal. At the same time, they are small in size and have light weight.

(f) Now a days, dedicated *integrated circuits* (ICs) are incorporated in hearing aid circuit *as a signal processing device.* It consists of an audio amplifier and filter. The instrument has high amplification power and in case of deaf and dumb sound may be amplified one million times.

(g) A two position tone control is usually used in hearing aid to suit hearing of individual one's problem.

(h) The amplified audio signal is fed to the electromagnetic earphone.

(i) In ordinary pocket unit, earphone is attached to the instrument by a flexible wire but in other units it is fixed to the main body of the instrument and an audio is coupled to the ear by hollow flexible rubber a plastic tube.

21.41.1 Technical information

I. Power: 1.Battery, 1.5 AA type pencil cell in standard unit.

   2. Battery Mercury, 1.35V, or Silver Oxide 1.5V or Manganese 1.5V in behind.

II. Electronics, IC the ear units based

III. Frequency response—100 Hz to 8000 Hz (200 Hz to 5000 Hz sufficient)

IV. Peak output—130 to 140 dB

V. Gain—50 to 70 dB.

Control Components

1. ON/OFF control switch
2. Volume control mechanism
3. Tone control mechanism.

Principle of Operation: The principle of operation can be summarised in the following ways.

1. By the components
2. By sequence of events by making block diagram.

1. **Components (Fig. 21.39)**
   (a) Microphone
   (b) Amplifier and Filter
   (c) Small earphone which fits into ear.

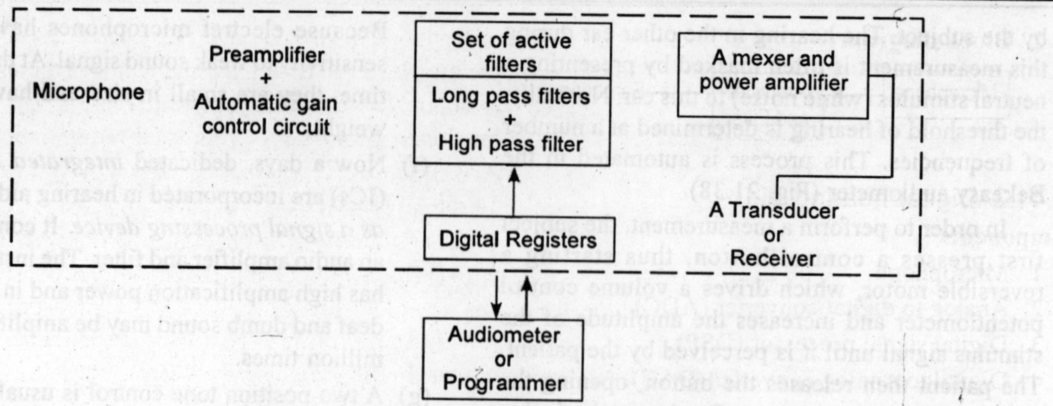

Fig. 21.39: Conventional hearing aid-analog type

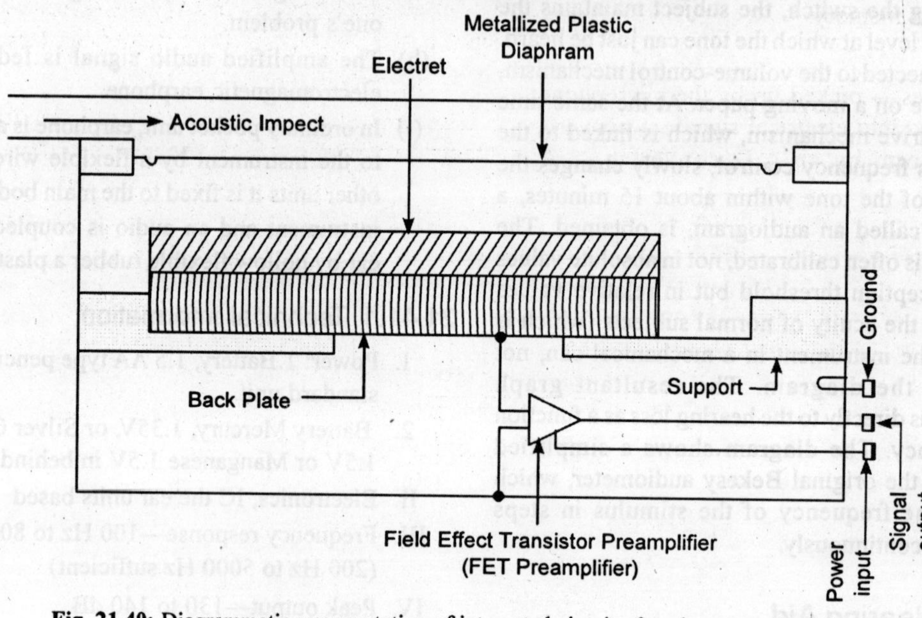

Fig. 21.40: Diagrammatic representation of integrated circuit of analog type of hearing aid

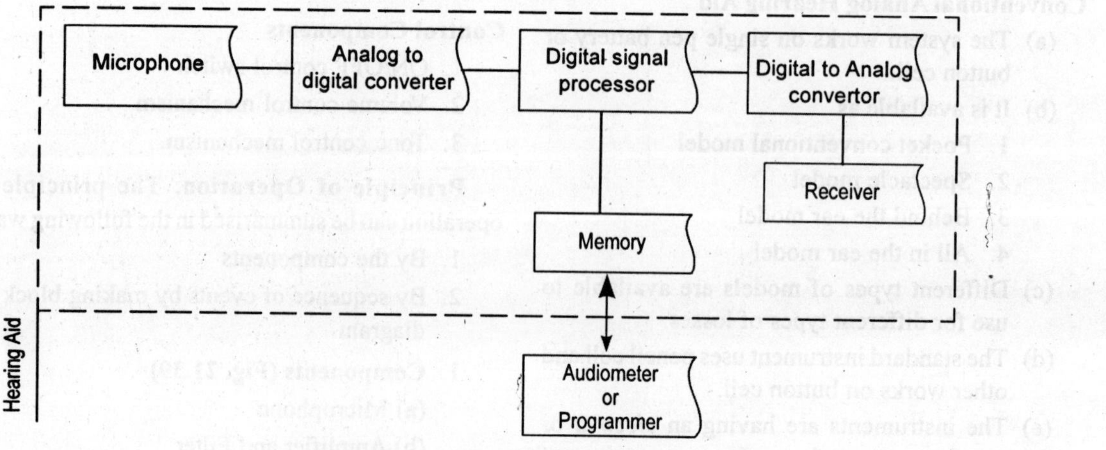

Fig. 21.41: Block diagram of digital hearing aid

2. **Block diagram**

21.41.2 Digital hearing aid

Components

1. Microphone
2. Analog to digital converter (ADC)
3. Digital signal processor (DSP)
4. Digital to Analog converter (DAC)
5. Receiver
6. Two port memory.

Process

Sound waves → picked up by the microphone → Transformed into electrical signal → converts into Digital form by an A/D converter → A microphone having internal noise of 20 dB sound pressure level (SPL) of input and output of the signal is about 90 dB sound pressure level (SPL). For maximum performance, the total requirement of ABC is 80 dB. This requirement is achieved with a 14 bit A-D converter (Fig. 21.41).

The Digital signal processor (DSP) is a fixed wired program digital processing device containing an array of adders, multipliers, registers → provide the fundamentals operation necessary for implementing various digital algorithms.

For general purpose DSP, consumes significant power in executing the programms instruction→ since power consumption is considerable factor in the design of hearing aid.

Advantage: It provides (1) proper fitting (2) stable performance, (3) superior performance.

22

Biophysics of Sense of Balance and Rotation

1. Physical organization of vestibular apparatus as mechanical sensors: (a) Semicircular canals, (b) Utricle and saccule (vestibule), (c) Hair cells as mechanical transducer, Kinocilium, directional sensitivity, 2. Fluid dynamics of semicircular canals (angular rate sensors), 3. Artificial stimulation of semicircular canals for experiments and clinical diagnosis, 4. Utricle and saccule operation, otolith organs as linear accelerometer, Clinical test for function of utricle, 5. Schematic representation showing mode of stimulation and role of vestibular apparatus, 6. Comparison of semicircular canals with utricle and saccule, 7. Function of vestibular apparatus.

22.1 Physical Organization of Vestibular Apparatus as Mechanical Sensors

Internal ear or labyrinth consists of:

I. Cochlea: For hearing

II. Vestibular apparatus: For posture and equilibrium.

It consists of two parts:

(A) Semicircular canals: For dynamic equilibrium.

(B) Vestibule having two structures: (a) Utricle, (b) Saccule: For static equilibrium.

(A) **Semicircular canals:** The semicircular canals are arranged in three planes, each of which is at right angles to the other two, so that the three planes of space are represented. The semicircular canals are arragned in space so that they complement each other. On each side of the head there are two canals lying vertically at right angles to each other, with the apex of the right angle pointing to the centre of the skull. The arrange-

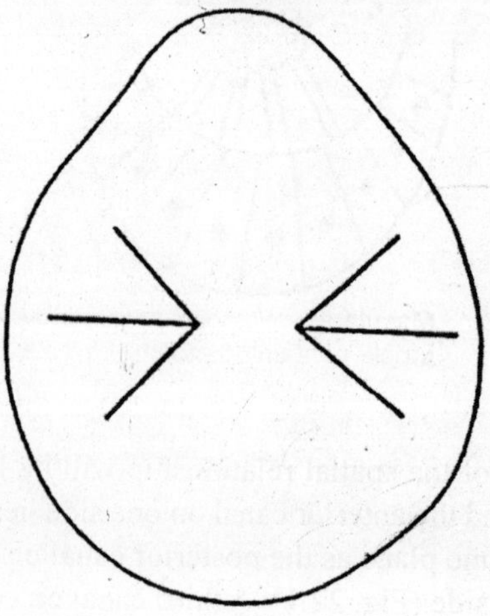

Fig. 22.1: Showing spatial relationships of the semicircular canals

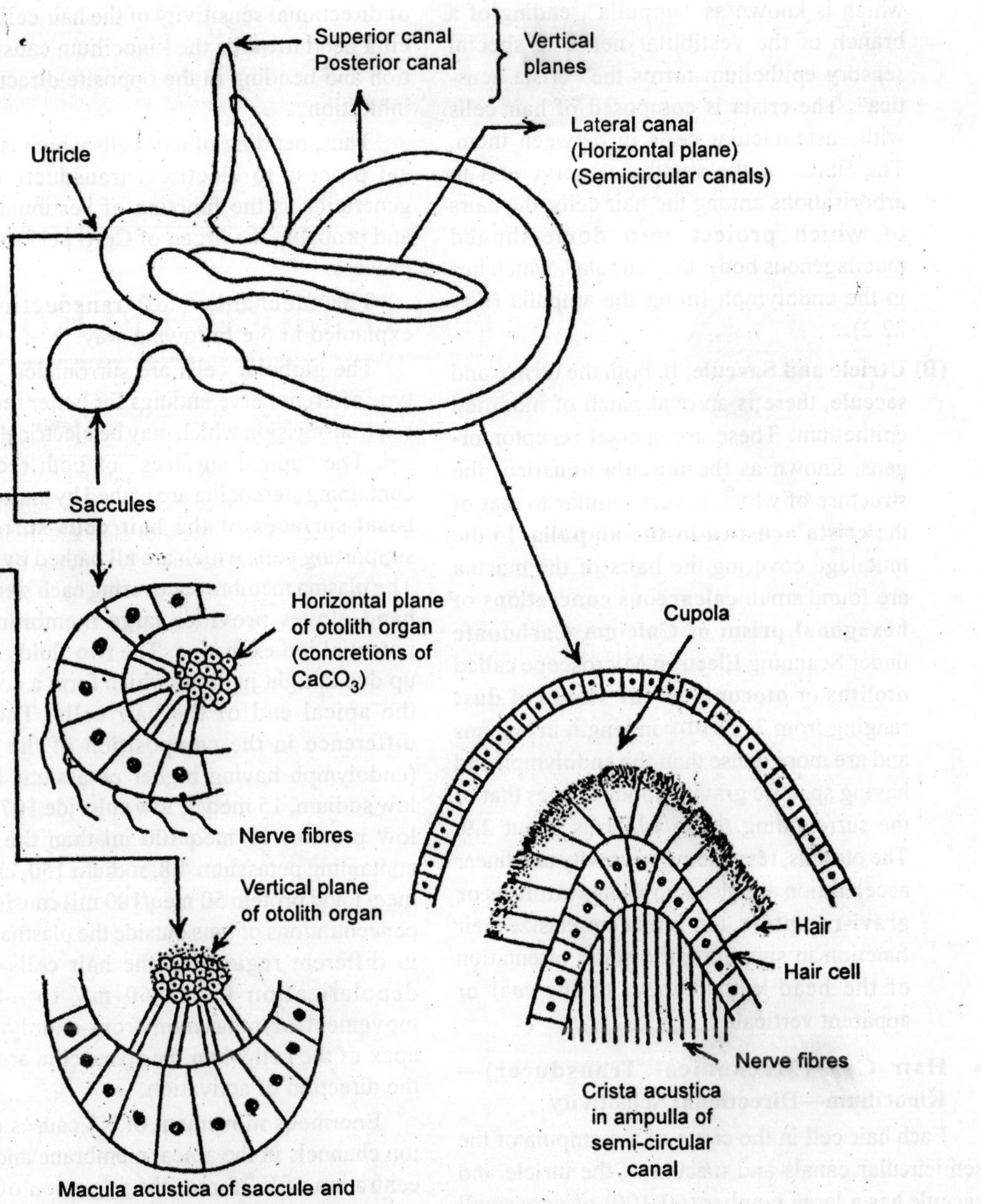

Fig. 22.2: Structure of semicircular canal

ment of the spatial relationship will be like >< and the anterior canal on one side lies in the same plane as the posterior canal on the other side (Fig. 22.1). A third canal on each side lies in the horizontal plane with its arc outwards. Thus, these canals form together three planes, one nearly horizontal and two vertical. These six semicircular canals respond to the movement of the skull in any plane or combination of planes and the resulting patterns of nerve impulses provide information.

At one end each of the osseous canals is dilated and the membranous canal undergoes a corresponding dilatation so as to fill up the whole bony canal. In this dilatation, which is known as "ampulla", ending of a branch of the vestibular nerve in special sensory epithelium forms the "crista acustica". The crista is composed of hair cells with sustentacular cells in between them. The fibres of the vestibular nerve end in arborisations among the hair cells, the hairs of which project into dome-shpaed mucilagenous body, the "cupola", which lies in the endolymph filling the ampulla (Fig. 22.2).

(B) Utricle and Saccule: In both the utricle and saccule, there is an oval patch of modified epithelium. These are special receptor organs, known as the **macula acustica**, the structure of which is very similar to that of the **crista acustica in the ampulla.** In the mucilage covering the hairs of the macula are found small **calcareous concretions** of **hexagonal prism of Calcium Carbonate** under Scanning Electron Microscope called **otoliths** or **otoconia** or **ear stones of dust** ranging from 3-19 μm in length in humans and are more dense than the endolymph and having specific gravity about 3 times that of the surrounding tissue which is about 2.9. The otoliths, responding to gravity and linear acceleration are also known as **statoliths or gravi-receptors**, terms that emphasize their function in signalling the static orientation of the head with respect to the real or apparent vertical.

- **Hair Cell (Mechanical Transducer)—Kinocilium—Directional Sensitivity**

Each hair cell in the crista of the ampula of the semicircular canals and macula of the utricle and saccule has a large number (60-100) of very small club-like cilia called **stereocilia** whose lengths increase regularly with their proximity to the **Kinocilium** and they are graded in length, the longest being closest to the Kinocilium. One very large cilium known as **Kinocilium** is located to one side of the hair cell. The kinocilium contains 11 filaments, 9 of which are double, which projects from a centriole on one side of the hair bundle. This is always seen on the same side of the cell related to its orientation on the ampulla. This is the reason of directional sensitivity of the hair cells, i.e., when cilia bend towards the kinocilium causing stimulation and bending in the opposite direction causing inhibition.

Thus, bending of hair cells which is a mechanical process to electrical transduction or signal generation is the function of vestibular apparatus and probably the organ of Corti performs a similar process.

The mechanism of transduction may be explanied in the following way.

The globular cells are surrounded by a calyx-type afferent nerve endings for better receptor-neuron transmission which may be electrical and chemical. The "apical surfaces" of both receptor cells containing stereocilia are bathed by endolymph. The basal surfaces of the hair cells surrounded by supporting cells which are all bathed by perilymph. The plasma membrane covering each stereocilia like hand gloves provides large membrane area for potential ion exchange. The two fluids cannot mix up due to tight juctions which form a collar around the apical end of the hair cells. Thus there is difference in the composition of the two fluids (endolymph having higher potassium 144 meq/1, low sodium, 15 meq/1, low chloride 107 meq/1 and low protein, 15 meq/100 ml than the perilymph containing potassium 4.8, sodium 150, chloride 121 meq/1 and protein 50 meq/100 ml) causing different concentrations of ions outside the plasma membrane in different regions of the hair cells producing depolarization from –60 mV to –40 mV by movement of potassium from endolymph at the apex of the cell when the stereocilia are pushed in the direction of activation.

Enormous movement of K^+ causes opening of ion channels in the apical membrane and high concentration of K^+ forces the migration of K^+ across the membrane and plasma membrane of stereocilia producing the highest amplitude of potential at the apical part.

Due to depolarization there is a change in the hair cells which permits calcium at the basal part, i.e., presynaptic region which mobilizes the synaptic vesicles near the presynaptic region, inducting fusion of vesicles to the plasma membrane and releases neurotransmitter in the synaptic clefts.

Reversability of potential, i.e., hyperpolarization may occur when hair bundle moves in opposite direction, away from the Kinocilium. Ion channels which were opened are closed and then the potassium pumped out of the cell by the cell membrane and the potential becomes from –60 mV to –65 mV. Hyperpolarization diminishes calcium influx into the base of the cell resulting in reduction of neuro-transmitter release.

The salient features regarding function of hair cell receptor are:

1. Bending of stereocilia: 1 μm distance.
2. Angle of bending –3°.
3. Amplifying response: Very small, shift of potential has larger effect.
4. Receptor activation: Not a threshold process, which is advantageous to detect small displacement.
5. Onset of process: Very rapid process, movement of stereocilia and initiation of cell membrane potential take millionths of a second,
6. Adaptation: They adapt quickly, after displacement by one micrometre, within a few tenths of a second, the membrane potential is restored to the initial state which permits the cell for maximum excitation when next displacement occurs.

22.2 Fluid Dynamics of Semicircular Canals (Angular Rate Sensors)

If the head suddenly starts to rotate in any direction, i.e., rotational acceleration in the plane of a given semicircular canal, it causes stimulation of crista. The endolymph in the membranous semicircular canals, because of inertia, remains stationary while the canals themselves turn,causing displacement of endolymph in the direction opposite to the direction of rotation (Fig. 22.3). This can be explained by a model known as "Torsion Pendulum Model". Such motion stimulates the end organs and produces not only consciousness of the rotation but reflex adjustments to it, such as eye and head movement in a direction opposite to the rotation. As a rule, rotation is followed by "postrotatory nystagmus", i.e., quick rhythmic jerking of the eyeballs in the direction opposite to rotation. Evidence for this concept comes from experiments and this is also used clinically in which the fluid moves without rotation. It is done by irrigating the external auditory canal by warm (about 43°C) or cold (12 to 30°C) water. Convection currents are set up in the endolymph of the canal. Nystagmus produced is known "Coloric nystagmus" and the test is called "Caloric test". The physical basis for the Caloric test rests on the density gradient in endolymph.

This movement of endolymph causes bending of cilia of hair cells. Bending in the direction of the kinocilium results in "depolarization" and increases neural activity. Bending in the opposite direction results in "hyperpolarization".

When a constant speed of rotation is achieved, the fluid spins at the same rate as the body and the cupula swings back into the upright position. When rotation is stopped, deceleration produces displacement of the endolymph in the direction of the rotation and the cupula is deformed in a direction opposite to that during acceleration. It returns to mid- position in 25-30 seconds. The movement of cupula in one direction causes increase in nerve impulse in the single nerve fibre whereas movement in the opposite direction inhibits nerve activity.

Experimental studies with animals rotating for 40 seconds and recording from a single hair cell showed that:

1. Cupula in resting, still hair cell produces tonic discharge of 15 impulses per second, and
2. During continued rotation, the increases in discharge of the hair cells diminishes gradually and resting level is attained in 20 seconds.

The explanation of adaptation of hair cells is that within a second or more of rotation, friction in the canal causes the endolymph to rotate as rapidly as the canal itself. Then in another 15-20 seconds the cupula slowly returns to its initial position for its own elastic recoil.

When the rotation is stopped suddenly, the opposite effects take place and cupula returns gradually to its original position and the discharge of hair cells comes back to its initial tonic level. Rotation causes maximal stimulation of the canal, in the plane of rotation. Since the canal's one side is the mirror image of those on the other side, the endolymph is

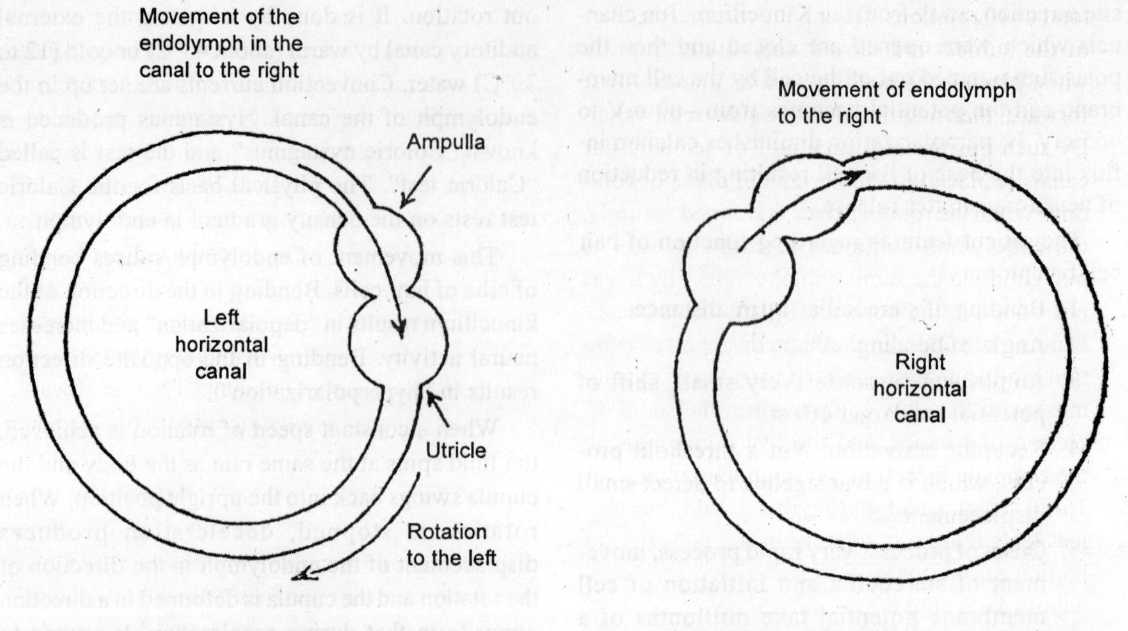

Fig. 22.3: Left and horizontal semicircular canals showing movement of endolymph to the right during rotation to the left due to inertia of the fluid

displaced in the plane of rotation. Since the canal's one side is the mirror image of those on the other side, the endolymph is displaced towards the ampulla on one side and away from it on the other. The pattern of stimulation projected in the brain varies with the direction as well as the plane of rotation. Linear acceleration cannot displace the cupula, so there is no stimulation of the hair cells.

Thus, the canals are responsible for transmission of positive signals when the head starts to rotate and negative signals when rotation is stopped. It has been further observed that some hair cells always respond to rotation in any plane because of fluid movement that always occurs in at least one canal.

22.2.1 Role in angular velocity and angular acceleration

Angular velocity is the rate of rotation in a given period of time and angular acceleration is the rate of change of angular velocity. The canals are mainly a detector of angular acceleration (angular rate sensors) but their actual role is to detect a combination of angular acceleration and angular velocity where angular velocity persists only for 5-15 seconds. The angular acceleration required to stimulate the canals is about 1 degree per second per second.

22.2.2 Semicircular canals and maintenance of equilibrium

They dectect when a person's head is rotating or starting to rotate or stopping rotation in one direction or another. Though it does not participate in static equilibrium, equilibrium in linear acceleration or exposure to centrifugal force, there is very weak response in case of rapid and complicated body movements. The explanation is that the semicircular canal mechanism sends information well ahead of time before occurrence of disturbed equilibrium for preventive adjustment.

22.3 Artificial Stimulation of Semicircular Canal for Experiments and Clinical Diagnosis

1. **Barany Chair Test (Rotational Stimulation):** This method is usually applied in the clinic and experimental research for adequate stimulation of the semicircular canals which consists in rotation on the vertical axis on the Barany's chair at the rate of one rotation per two seconds for a

total of 10 rotations. During the rotation the eyes of the subject are closed and rotated rapidly while placing the head in various planes—first forward, then angulated to one side or the other. By such positioning, each pair of semicircular canals is placed in the horizontal plane of rotation. When the rotating chair is stopped abruptly, the endolymph continues to rotate due to its moventum in the semicircular canals which was positioned at the horizontal plane. This causes bending of cupula in the same direction of rotation producing nystagmus (post-rotatory nystagmus) with the slow component in the same direction of rotation and the fast component in the opposite direction. Nystagmus remains for 15-20 seconds and, during this period, the subject feels that he is rotating in the opposite direction of rotation of the chair. Sometimes nystagmus lasts upto 80 seconds and over, depending on the excitability of the vestibular apparatus. The duration of post-rotatory nystagmus diminishes by training.

The advantage of this test is that it checks the semicircular canals on both sides of the head at the same time.

2. **Caloric Stimulation (Hallpike Method)**: The common clinical method for unilateral testing of the semicircular canals in man is the caloric test. According to Hallpike method, the head is elevated to 30 degree position from the supine position for placing the horizontal semicircular canals in the vertical plane. External auditory meatus of each ear is irrigated for 40 seconds, in turn with warm (about 43°C) or cold water (12 to 30°C), i.e., water 7°C above and below body temperature. Nystagmus produced by this procedure is known as 'Caloric nystagmus'. Different measures for caloric response are used, e.g. measurement of nystagmus duration, frequency and peak slow-phase velocity with eyes open or closed. A decrease in nystagmus response to both hot or cold stimulations in one ear reveals the reduced activity. The physical basis for the caloric test is the density gradient in endolymph produced by heating or cooling.
3. **Galvanic Stimulation**: Horizontal nystgmus and sensation of rotation can be produced by moderate galvanic stimulation. Direct current passing between each mastoid or between a single ear and a second electrode on the back of the neck produces nystagmus called "Galvanic nystagmus" with a fast phase towards the cathode and a sensation of rotation towards the anode. The intensity of response is approximately proportional to current intensity with a threshold for sensation of about 600 μamp. The site of stimulation is not known but the vestibular nerve should be intact.
4. **Movement of Objects Before the Eyes (Optokinetic Nystagmus)**: By the movement of objects in a given direction before the eyes, nystagmus can be produced—called "optokinetic nystagmus". If the eyes are open during rotation, the optokinetic nystagmus will be in the same direction.
5. **Ice Water Test**: This clinical test is only for one vestibular apparatus of one side. Ice water is introduced in one ear. Cooling the ear causes transference of sufficient heat from the external semicircular canal to cool the endolymph. This increases the density of the endolymph resulting in sinking it downwards and producing slight movement of fluid around the semicircular canal. This movement stimulates the canal giving sensation of rotation and nystagmus (cold nystagmus). This test reveals that if semi-circular canals are normal, then utricle and saccule are also normal.

22.3.1 Function of semicircular canals

1. Maintenance of dynamic equilibrium.
2. Detection and measurment of angular acceleration around any axis in three-dimensional space.
3. Stimulation develops a number of vegetative reflexes; giddiness, sweating, reddening or pallor of the skin, nausea, vomiting, changes in the cardiovascular system.
4. Detection of angular velocity.

5. Alerting function in maintenance of equilibrium when malequilibrium is going to occur.
6. It gives information about the movement of the head.

22.4 Utricle and Saccule Operation (Otolith Organs as Linear Accelerometer)

22.4.1 Mechanical characteristic

The otoliths are more dense than the endolymph and having specific gravity about 3 times that of the surrounding tissue which is about 2.9. The otoliths responding to gravity and linear acceleration are also known as **statoliths** or **graviceptors**, terms that emphasize their function in signalling the static orientation of the head with respect to the real or apparent vertical. As the otoliths are concerned with the linear acceleration, they are also regarded as **linear accelerometers**. Though we cannot differentiate between gravity and linear acceleration by instrument the otolith fully responds to orientation with respect to the resultant gravitoinertial force or direction of the vector $\overline{g} - \overline{a}$, i.e., gravity minus linear acceleration (Fig. 22.4). But human beings can differentiate between linear acceleration and tilt with respect to the vertical by using additional information such as sensation of rotation from the semicircular canals and orientation to the visual verticle.

The otolith and otolith membrane show the non-rigid characteristics of the system. The different areas of the irregular macula are covered by otoconia to different degrees and there is dynamic coupling among the different portions.

22.4.2 Stimulation

The adequate stimuli of the otoliths are: acceleration or deceleration of rectilinear motion, jolting, tossing and any changes in the position of the head and of the whole body, which displace the otolith membrane. Here, they are stimulated by tension of the sensitive hairs or the pressure of the otolith membrane upon them due to pull of gravity. It may hang vertically from the hair or pull sideways or press down on them according to the position of the head in space (Fig. 22.5). That receptors are stimulated by changes in the gravitational pull was experimentally shown by using crayfish having similar but less complex organ. In these animals, the calcium carbonate crystals were replaced by iron particles which can be pressed against the hair cells of any part of the sphere by placing a magnet near the animal. If the magnet is placed below the animal, the iron particles are pulled against the lower hair cells like normal otolith shows for gravity. The animal can maintain upright position. Similarly, any disturbance by an altered position of the head, the normal gravitational pressure directly down upon the utricular hair cells or of the shearing pull of the saccular otoliths.

Stimulation of the otolith depends upon the differences in density of the otolith (otoconia) with respect to the gelatinous substrate (otolith membrane) which supports them. Having higher specific gravity of the otolith, the action of a component of linear acceleration on the otolith results in the exertion of a force for deformation of the otolithic membrane and, consequently, bending of the hair cells. The direction of bending will depend upon the orientation of the resultant acceleration force (mechanical transducer).

Hair cells do not all have the same orientation with respect to location of their kinocilia. The pattern of polarization is in the different directions and directions of bending of stereocilia are anterior, medial, lateral and posterior. In case of saccula, macula is medial and otolith is laterlal and the pat-

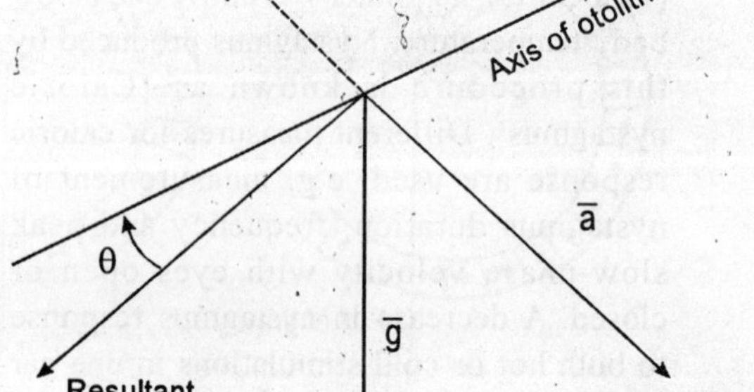

Fig. 22.4: Schematic representation of otolith action

tern of polarization is reversed in the region of the middle of the structure, the direction of bending of stereocilia is reversed. In case of utricle the macula is in horizontal plane.

22.4.3 Sensory neurophysiology of cells

Electrophysiological studies with the animals in tilted position, recording form the utricular unit of the restibular nerve reveals that there are several types of receptors (e.g., flask-shaped type I or HC I surrounded by a nerve calyx, cylindrical type II or HC II). The responses are:

1. Some to the angle of tilt mainly about one particular horizontal axis, continuous variation of firing rate with tilt angle for maximum or minimum response or sharp localized response increasing activity near the normal head position.
2. Other fibres to the rate of change of the tilt angle in other direction and also to the angle itself.
3. Remaining to the mechanical vibrations only, independent of head position.

Such nerve provides a complex parallel information flow where the signals are formed from the concert of many sensory cells responses. It reveals that: (1) The movement of the head and direction of the apparent vertical and (2) the structure to linear acceleration and angular acceleration to some extent.

Further studies reveals that afferents from the utricular macula show tonic, phasictonic and phasic responses to changes in head position (Precht, 1979).

Information delivered by the vestibular afferent nerve fibres from the utricle and saccula may be compared to that of "multidirectional linear accelerator" in each of the two major planes whereas information from the semicircular canals may be compared to that of "bidirectional angular accelerometer". These two information together may be compared to that of "intertial guidance system."

22.4.4 Function of utricle and saccule

1. To monitor gravity.
2. To detect and measure linear acceleration.
3. To detect and measure vibrations to some extent in any plane.
4. To maintain static equilibrium.

22.4.5 Clinical test for function of utricle

Balancing test: It is the simple clinical test for integrity of equilibrium mechanism. The subject is asked to stand properly with eyes closed. If the static equilibrium system of the utricles does not function, the subject will waver to one side or the other or may even fall. But some of the propioceptive mechanism of equilibrium is sometimes sufficiently active to maintain balance even with the eyes closed.

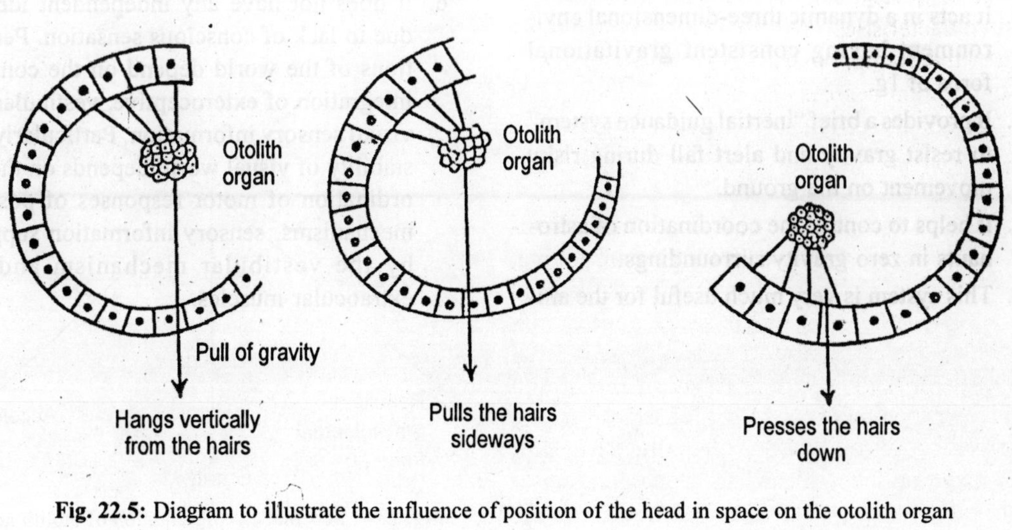

Fig. 22.5: Diagram to illustrate the influence of position of the head in space on the otolith organ

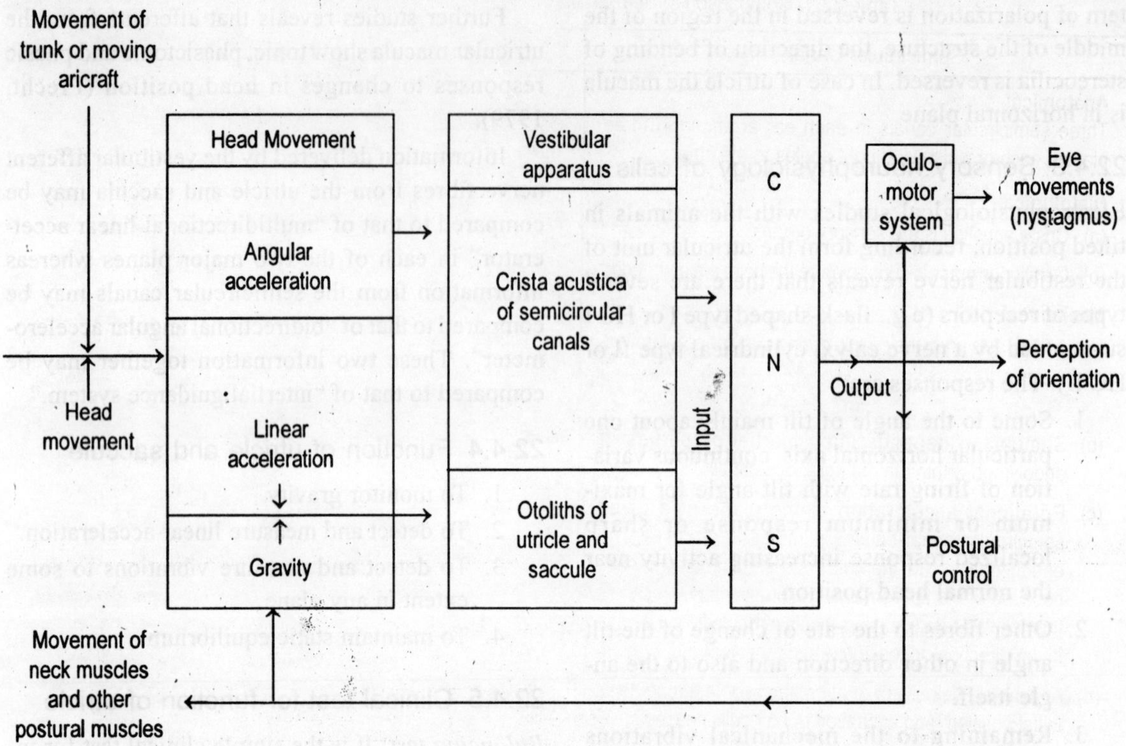

Fig. 22.6: Schematic representation showing mode of stimulation and role of vestibular apparatus

22.4.6 Function of vestibular apparatus (see Fig. 22.6)

1. **Vestibular apparatus plays a role in controlling posture reflexly to stabilize vision and to coordinate movements and balance of head, body and limbs.**
2. **It acts in a dynamic three-dimensional environment having consistent gravitational force of 1g.**
3. **It provides a brief "inertial guidance system" to resist gravity and alert fall during risky movement on the ground.**
4. **It helps to control the coordination of astronauts in zero gravity surroundings.**
5. **This system is very much useful for the animals which are using three-dimensional space, e.g., birds, marine mammals, fishes, amphibia and in case of human beings using handgliders and in case of SCUBA divers for effective maintenance of flight and swimming.**
6. **It does not have any independent identity due to lack of conscious sensation. Perceptions of the world depend on the complex integration of exteroceptive, vestibular and visual sensory information. Particularly, the stability of visual world depends on the coordination of motor responses of postural mechanisms, sensory information supplied by the vestibular mechanism and the extraocular muscles.**

Table 22.1: Comparison of Semicircular Canal with Utricle and Saccule

Semicircular Canal	Utricle and Saccule
1. **Anatomical** Three semicircular canals in each ear arranged in three planes, each of which is at right angles to the other.	One utricle and one saccule in each ear.
2. **Histological** (a) Crista acustica is the sensory organ.	Macula with otolith is the sensory organ.
(b) Gelatinous mass is cupola where the hairs are embedded	Gelatinous mass is the otolith membrane in which hairs are extended.
3. **Mechanical** (a) A force acts upon the cupola during angular acceleration.	For a given position of the head, otolith may exert a force, which will maintain hair cells in a bent position, even though the head is maintained stationary.
(b) Stimulation depends upon inertial properties of endolymph within the membranous canals.	Depends upon the difference in density of the otoconia with respect to gelatinous substrate which supports them.
(c) Rotational acceleration in the plane of a given semicircular canal causes stimulation of crista. The endolymph in the membranous semicircular canals, because of inertia, remains stationary while the canals themselves turn, causing displacement of endolymph in the direction opposite to the direction of the rotation.	The action of a component of linear acceleration on the otolith, exerts a force for deformation of the otolithic membrane leading to bending of hair cells, and direction of bending depend, upon the orientation of the resultant force.
(d) At constant velocity of rotation, the cupola returns to its initial position, and excitation of hair cells ceases.	Some cells of the otolith organs respond continuously for a given head position.
(e) Hair cells have the same orientation with respect to location of their kinocilium.	Hair cells do not have the same orientation with respect to location of their kinocilium.
(f) **Pattern of Polarization** Pattern of polarization responses for bending in the opposite direction.	The pattern of polarization is in the different directions for bending.
(g) Information delivered from these afferent nerve fibres may be compared to that of a "bidirectional angular accelerometer."	Information delivered by the afferent nerve fibres may be compared to that of "multidirectional linear accelerator" in each of the two major planes.
4. **Functional** (a) It evolves and developes to detect and measure angular acceleration in any direction.	It us evolved and developed to monitor gravity and detect and measure linear acceleration and to some extent vibrations in any plane.
(b) It is for maintenance of dynamic equilibrium.	These are for maintenance of static equilibrium.
(c) It gives information about movements of the head.	It gives information about position of the head in space.
(d) It detects angular velocity.	It does not.
(e) It does not monitor gravity.	It monitors gravity.

23

Influence of Gravity, G-Force, Acceleration, Deceleration

1. Influence of gravity, 2. G-Force, positive and negative G, 3. Influence of linear acceleration on the body, 4. Effects of brief acceleration, 5. Deceleration.

23.1 Influence of Gravity

The force which we are encountering constantly in our daily lives is that which pulls all towards the earth and is known as **gravitational force**. By Newton's second law, the gravitational force F acting on a body near the earth is equal to the product of the mass 'm' of the body and its acceleration 'a' is due to gravity:

$$F = ma.$$

23.1.1 Maintenance of erect posture of the body against gravity

The erect standing posture of a person or an animal against the gravity is maintained by the continuous discharge of impulses from the vestibular nuclei and the nuclei in the reticular formation into the spinal cord and then to the extensor muscles to stiffen the limbs. This allows the limbs to support the body against the gravity. These are mediated reflexly known as **vestibuloreticular reflexes** and the impulses are transmitted mainly by the vestibulo-spinal tract and reticulospinal tract. In turn the vestibular nuclei and nuclei of the reticular formation receive impulses from the exteroceptors, propioceptors. As a result, the antigravity extensor muscle contractions fix the hip, knee, ankle and vertebral joints to stiffen the limbs to support the body weight, while on the opposite side relax.

I. **Cardiovascular System and Homeostasis** The physical factor which alters the pressure in the large arteries or large veins is the hydrostatic pressure due to gravity.

(A) **Arterial Blood Pressure:** A maintained high arterial blood pressure is required to force the blood through the small vessels, where the resistance is great. It must be high enough to force blood against gravity to the brain whose blood requirements are always high. Here, both frictional resistance encountered in the smaller vessels and gravitation must be overcome. In case of arterial pressure, it is usually high, the effects of gravity are relatively less than in the venous system and is capable to compensate in case of alteration of position.

The pressure in any vessel below heart level is increased (large artery in the foot, 180 mm Hg) and in any vessel above heart level is decreased (large artery in the head, 62 mm Hg) and at the heart level it is 100 mm Hg. The magnitude of the gravitational effect is the product of the density of the blood, the acceleration due to gravity (980 m/s/s) and the vertical distance above or below the heart is 0.77 mm Hg/cm at the density of normal blood. As for example, if the pressure at the heart level is 100 mm Hg, the pressure 50 cm above the heart level should be 62 mm Hg (100 mm Hg— [0.77 × 50]).

When a person lies down, changing the axis of the body from the vertical to the horizontal position, gravity assists the heart to send blood to the brain. Here, gravity no longer opposes the upward pumping of blood by the heart. When a person lying in horizontal position suddenly stands upright, gravity opposes the action of the heart to pump in the brain. The pressure in the carotid artery falls and inadequate blood is supplied to the brain, and "darkness before the eyes", giddiness or even fainting, may occur. Now fainting acts as a compensatory mechanism for restoration of brain circulation. If fainting mechanism is not furnished, the person should be allowed to lie down in horizontal position or to a crawling position. But usually there is a compensatory vasoconstriction and cardio-acceleration to prevent sudden fall of pressure, especially in the brain vessels. Beside that, the arteries are thick-walled and except animals like giraffe, the differences of arterial pressures between head and foot usually is not a problem. There is also prompt increase in the circulating levels of renin and aldosterone.

(B) Venous System: The veins are thin-walled and are often of considerable capacity, and the pressure within them is so low as often to be outweighed by quite moderate hydrostatic pressure. If a length of vein is filled with fluid and held vertically, the upper part collapses and lower part distends. In erect position, blood tends to accumulate in the lower parts of the body, the veins become distended, while those of the head and neck become empty. This phenomenon is evident in the following cases:

1. If the arm hangs down at the side, the veins of the hand distends with blood. Again, raising the arm above the head, the veins collapse.
2. Cut finger bleeds more when the hand hangs at the side than the elevated hand because of gravitational effect.
3. During brain surgery, the surgeon applies the influence of gravity by performing operations in the sitting position. Because, gravity helps to decrease haemorrhage by its effect on the pressure in the brain vessels.
4. **Muscle Pump and Varicose Veins**: When a skeletal muscle contracts, it compresses the veins inside the muscle. This forces the blood from the veins in the direction of the heart. As the muscle relaxes, the veins are filled up again with the blood coming from the capillaries but does not flow in reverse direction due to the valves. This intermittent muscular contraction pumps the blood mechanically—which is essential for return of blood to the heart from the lower limbs against the action of gravity. During standing, when the effect of gravity is fully manifested, venous pressure at the ankel is 85-90 mm Hg. Rhythmic contractions of the leg muscles—while the subject is standing—serve to lower the venous pressure in the leg to less than 30 mm Hg. Sometimes the movement of blood towards heart is decreased and produces engorged, tortuous veins due

to incompetent valves known as "varicose veins" and such persons may show venous stasis and ankle oedema.

This type of upset is usually compensated in different ways:

1. Physical, 2. Physiological

1. **Physical**

(a) When the blood flows along the veins in the same direction of the pull of gravity, the collapse of the higher parts prevents the venous pressure in those parts of the vein still further up from falling below a certain level.

(b) In many parts which are normally dependent, the adventitial coat is thickened, and the walls supported by surrounding structures cause limitation to the degree of distension.

(c) Another factor is the continuous massage action moving the blood onwards due to muscular movement and valvular activity.

2. **Physiological**

(i) **Vasomotor system**: When blood begins to accumulate in the large veins, the venous return is reduced, and with it the cardiac output, so that arterial pressure begins to fall and it is further aggravated by the direct effect of gravity on the arterial pressure. This causes stimulation of carotid sinus and aortic arch mechanism and there is also reduction of blood to the medulla and both mechanisms cause splanchnic vasoconstriction and raises cardiac output again.

(ii) When the above mechanism is inadequate, the respiratory centre is brought into action due to increased excitability, thereby increasing activity of the diaphragm, the thoracic and abdominal walls to hasten the venous return.

(C) **Blood Pressure:** It is observed in some persons that sudden standing from horizontal position causes fall of blood pressure, dizziness, dimness of vision and sometimes fainting and the condition is known as **orthostatic or postural hypotension.** This manifestation is common when there is:

1. Damage of sympathetic nervous system (surgical sympathectomy, sympatholytic drugs, diabetes and syphilis), where the compensatory mechanism due to vasoconstrictor fibres is not available.
2. In "autonomic insufficiency", postural hypotension occurs due to defect in the CNS, though there is normal epinephrine in the blood, but fails to compensate on standing.
3. In one form of this disease, the defect is peripheral norepinephrine level is low with little or no response to baroceptor stimulation. Though baroceptor reflexes are abnormal in primary hyperaldosterism, these persons do not show postural hypotension due to expansion of blood volume to maintain cardiac output even in changing the position. For this reason, mineralocorticoids are given to the patients having postural hypotension.

(D) **Cardiac Output**: The decrease in cardiac output is adjusted partially—by 30 per cent—by cardiac acceleration on standing from supine position.

(E) **Cerebral Circulation**: The decrease in cardiac output may reduce the cerebral blood flow causing temporary fainting on prolonged standing—the condition is called **postural syncope.** But there are compensatory mechanisms called **autoregulatory mechanisms** which permit only 20 per cent decline of cerebral blood flow. The arterial pressure at the head level drops to 20-30 mm Hg but jugular venous pressure falls to 5-8 mm Hg, reducing the drop in perfusion pressure, i.e., arterial pressure minus venous pressure. The pressure on the cerebral vessels is decreased due to reduction of cerebral vascular resistance because intracranial pressure falls as venous pressure. The reduction in cerebral blood flow increases CO_2 partial pressure, decreases PO_2 and pH in brain tissue causing dilatation of the cerebral vessels, and the net cerebral O_2 consumption is about the same in the supine and upright position.

(F) **Blood Volume:** The fall in circulating blood volume is adjusted by 25 per cent rise in peripheral resistance through arteriolar constriction due to prompt release of renin from kidney and aldosterone secretion from adrenal cortex.

(G) **Pulse Rate:** The pulse rate increases during change of position from horizontal to vertical.

II. Respiratory System

(a) **Expiration:** In expiratory phase, as soon as the inspiratory muscles relax, several factors contribute to restore the thoracic cavity to its initial smaller size and shape. The elasticity of the thoracic walls, in case of man, the action of gravity lowering the front end of the ribs, tends to reduce the size of the chest cavity.

(b) **Intrapleural Pressure:** Due to gravitational forces, intrapleural pressure is about 5 mm Hg higher at the bases than the apexes. At the same time the transmular pressure, i.e., the difference between intrapulmonary and intrapleural pressure may become negative at the end of forced expiration, resulting in closure of airways. For this reason, more inspired gas during the first part of inspiration goes to the apexes than to the bases.

(c) **Pressure in the Pulmonary Vessels:** Gravity affects the pressure of the pulmonary blood vessels. In erect position, blood flow at the apexes is lesser than at the bases.

(d) **Gas Exchanges in the Apex and Base:** Gas exchanges in the apex and base of the lungs in the upright position are shown below (Table 23.1):

Table 23.1: Showing ventilation-perfusion relationship

	Apex	Base
PO_2	132 mm Hg	89 mm Hg
PCO_2	28 mm Hg	42 mm Hg
O_2 content	20 ml/dl	19.2 ml/dl
CO_2 content	42 ml/dl	49 ml/dl
pH	7.51	7.39
O_2 (In)	4 ml/min	60 ml/min
CO_2 (Out)	8 ml/min	39 ml/min

23.1.2 Tolerance for gravity

The tolerance for G forces (force acting on the body as a result of acceleration is usually expressed in G units, 1 G being the force of gravity on the earth's surface) exerted across the body is far greater than it is for axial G. Man can tolerate 11 G acting in a back-to-chest direction for 3 minutes and 17 G acting in a chest-to-back direction for 4 minutes. Therefore, the astronauts are positioned to take G forces of space flight in the chest-to-back direction. In this position, tolerance is greater to allow acceleration to orbital or escape velocity and deceleration back into the earth's atmosphere without any adverse effect.

23.1.3 Precaution against gravity reaction—antigravity "G-suits"

The effects of positive g are effectively averted by the use of antigravity "G-suits". These are double-walled pressure suits containing either water or compressed air and regulated in such manner that they compress the abdomen and legs with a force proportionate to the positive G. This decreases venous pooling and helps to maintain venous return.

23.2 Influence of G-Force

In modern civilization, due to high speed vehicles like aircraft and spacecraft, man is exposed to several types of acceleratory forces which affect the body during flight. Simple linear acceleration occurs at the starting of flight whereas linear deceleration occurs at the termination of the flight. Angular and centrifugal accelerations occur during every turn of the aircraft or spacecraft or during loops or dives of aircraft. But in aviation, centrifugal acceleration is more important for consideration because the modern aircraft can withstand more acceleration than man. Centrifugal acceleration can be experienced in the earth's field or it can be artificially created by the human centrifuge machine.

23.2.1 Terminology in aviation medicine

1. Positive acceleration or positive G or G^{+G} is the acceleration in which the intertial forces are directed from head-to-foot with respect to the aviator.
2. Negative acceleration or negative G or G^{-G} is the acceleration in which the intertial forces are directed from foot-to-head with respect to the aviator.
3. Transverse G is the acceleratory force along the antero-posterior or ventro-dorsal axis across the body.

23.2.2 Expression of magnitude of acceleratory force or measurement of acceleratory force "G"

The G-unit is used to express the magnitude of the acceleratory forces. One G is equal to the force exerted by one time the normal gravitational force. Thus, 5G is the magnitude of an acceleratory force, i.e., 5 times the force of gravity.

23.2.3 Centrifugal acceleration—effect on the body

Angular and centrifugal accelerations occur during every turn of the aircraft or spacecraft or during loops or dives of aircraft. It can be experienced in the earth's field or it can be artificially created by use of the Human Centrifuge. Centrifugal acceleration has gained importance because the modern aircraft can withstand more acceleration than the human body. Moreover, unlike an aircraft, the spacecraft cannot take rapid turn where centrifugal acceleration is of little value except in abnormal gyration of the spacecraft, In case of spacecraft, blast-off acceleration and landing deceleration might be vigorous and both are types of linear acceleration.

23.2.4 Determination of force of centrifugal acceleration

When an aircraft makes a turn, the force of centrifugal acceleration is determined by the following equation:

$$f = \frac{mv^2}{r}$$

in which f = centrifugal acceleratory force,
m = mass of the object,
v = velocity of travel,
r = radius of curvature of the turn.

Form this formula, it can be inferred that, as the velocity increases, the force of centrifugal acceleration increases in proportion to the square of the velocity. It can also be inferred that the force of acceleration is directly proportional to the sharpness of the turn.

23.2.5 Effect of G-forces on the human body

In empty space, a human being never feels the sensation of weight, but he feels weightlessness due to the absence of gravitational force. He may get the feeling of fall when he approaches the earth or other body with rapid speed.

But when a person is steadily supported by solid or liquid object or medium, some force exerts upwards on the body, muscles, bones, and transmits the force upwards on each organ to a certain degree which is necessary to maintain it fixed with reference to the supporting system. The strains produced by stresses of these internal organs stimulate various sensory receptors of the body to make the person feel the sensation of weight. When a person is steadily supported by a solid object, the reaction to the supporting forces is interpreted as weight, whether its origin is the attraction of gravitation or acceleration of the body by the supporting force. It is a fact that the origin of the G-field cannot be differentiated in any way between gravitation or acceleration. The sensation of weight (the G-force) is the resultant of the gravitational attraction and the d'Alembert G-force or "centrifugal force". During the turns of the aircraft, it is so oriented that the G-forces are exerted in a head-to-foot direction, similar to the experience gained in loops and pullouts from dives. But the difference is that, in vertical angular acceleration, the forces of natural gravity must be taken into account in the effects, where as in horizontal angular acceleration, the natural gravity forces are exerted along a plane which is sideways to the long axis of the aviator.

23.2.6 Importance of G-forces in aviation

The importance of G-forces arises from the fact that the hazards in modern aircraft which are encountered by the pilots for G-forces are to be prevented to maintain his efficiency.

23.2.7 Production of artificial G-forces-human centrifuge

Influence of angular acceleration on the animal body has been studied for long years with the help of a large centrifuge. The human centrifuge was first constructed during the Second World War (1939-1945) for the development of aircraft. The instrument gained more importance due to jet aircraft and rocket and it is applied primarily for problems in military aviation for different manoeuvers.

The centrifuge machine is driven by a powerful electric motor and produces rotational speed providing up to 20G within a few seconds after starting. The machine consists of a cab or chamber for the subject mounted at the end of a horizontal rotor beam which may be hung loosely so that rotating forces cause the bottom to swing outward during rotation, or it may be fastened securely in such a position that the head of the subject is directed outward. These two conditions produce positive and negative G-forces, respectively. The modern human centrifuge is provided with a number of channels for electrical transmission of physiological changes from the subject through various leads fixed on the body and their recording is done in the control room. These channels may utilize slip ring commutators with Hg contacts or telemetering apparatus (Fig. 23.1).

23.2.8 Effect of positive G-forces

The main effect of centrifugal acceleration is on the cardiovascular system because the blood is mobile and loose and can be translocated. Some tissues are loose with respect to the remainder of the organism. The centrifugal acceleration tends to displace them but due to solid structure they sag; hence it does not produce abnormal functions.

The sequence of events due to gradual increase of G-forces are:

1. Up to 3G: Increased pressure on the seat.
2. Between 3G to 4G: Difficulty in the use of muscles, tissues of the face are dragged downwards.
3. At 5G: Movement of the body is difficult and respiration is stopped.
4. Between 5G to 9G: The legs become congested, calf muscles cramp, vision is lost, sharpness of hearing is decreased and consciousness is lost in majority of cases.

The symptoms of positive G-forces are understood if it is considered in the light of "laws of hydrostatics" with respect to the cardiovascular system. The pressure gradient in the arteries between the head and the foot of the seated aviator is about

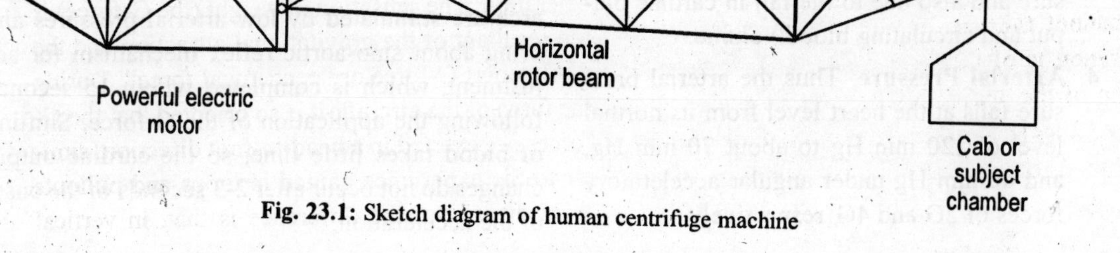

Fig. 23.1: Sketch diagram of human centrifuge machine

1 metre (height of the blood column) times the density of the blood (slightly greater than 1.0). When the body is exposed to acceleratory forces which are 5 times than that of normal gravity, the apparent density of the blood is 5.0, and the pressure gradient is consequently 5 times as great. This causes migration of the blood towards the lower limbs. If the acceleration is prolonged, the blood causes expansion of the venous reservoir of the lower part of the body until the elastic force of the reservoir equals the pressure head tending to expand it.

23.2.9 Different cardiovascular phenomenon with mechanism of production

1. **Heaviness of the Lower Limbs and Calf Muscle Cramp**: Positive G force raises the pressure in abdominal and leg veins by 5 times or more by translocating large volumes of blood to the veins → consequent passive dilatation of those veins leads to the pooling of considerable volumes of blood in the vein → the excessive pressure in these lower vessels causes diffusion of large volume of fluid → into the tissue spaces of lower limbs. The pooling of blood and the rise in interstitial fluid volume may produce congestion and heaviness of legs at about 3G, restrict locomotory movements at 5G, and reduce circulation in leg muscles leading to gastrocnemius cramps.
2. **Venous Return and Cardiac Output**: The pooling of blood reduces the venous return and the cardiac output; these are partly compensated by a reflex tachycardia (increase in heart rate) and some addition to the circulating blood volume from the pulmonary venous reserve.
3. **Pressure in Thoracic and Cranial Vessels**: Rise in +G depresses progressively the pressure in cranial and thoracic vessels due to the direct effect of positive G on their pressure and also due to the fall in cardiac output and circulating blood volume.
4. **Arterial Pressure**: Thus the arterial pressure falls at the heart level from its normal level of 120 mm Hg to about 70 mm Hg, and 40 mm Hg under angular acceleratory forces of 2G and 4G, respectively.
5. **Vasoconstriction and Renin Secretion**: Lowering of arterial pressure is partly compensated by baroceptor reflexes initiated within 15 seconds and by peripheral vasoconstriction due to renin secreted from kidney in consequence of the fall in circulating blood volume.
6. **Cerebral Blood Flow**: The fall in arterial pressure in the head region may still lower the cerebral blood flow; however, the cerebral flow may be maintained to some extent by a simultaneous fall in intracranial pressure due to reduction in cerebral venous pressure as well as CSF pressure and also by vasodilatation due to the rise in brain PCO_2 and a consequent rise in H^+ concentration in the brain.
7. **Temporary Loss of Vision**: Within a second of onset of an acceleratory force of 4G to 6G, vision is temporarily lost before the loss of consciousness, and is due to failure of the retina because of insufficient blood supply to it, and this loss of vision is termed "blackout."
8. **Loss of Consciousness:** At 5G to 9G, severe failure in cerebral circulation renders the brain hypoxic, power of audition is impaired and consciousness is lost shortly after blackout.
9. **Fracture of Vertebrae:** Loose facial tissues are dragged downward even at 3G, producing a sagging facial appearance. Very high G like 20G may fracture vertebrae and tear softer tissues within a fraction of a second.

23.2.10 Compensatory mechanism

All these symptoms vanish quickly when G-forces are removed. Even some physiological adjustments occur during short duration of high acceleration—such as, baroceptors in the carotid sinus and aortic arch are stimulated by low arterial pressures and bring about sino-aortic reflex mechanism for adjustment, which is completed within 15 seconds following the application of the G-force. Shifting of blood takes little time, so the cardiac output changes do not occur after 2-3 seconds of the onset of the acceleration.

23.2.11 Effect of negative G

1. In negative G-forces (foot-to-head), the movement of large volume of blood towards the head takes place.
2. Up to –3G, the aviator feels a gritty sensation and pressure in the eye socket. Headaches may occur due to congestion and intense rise of pressure in cranial vessles.
3. Above –3G, the retinal vessels are engorged and vision is lost because of obstruction of the light path to the retina by excess blood. The light passing through the blood gives a sensation of redness, thus the phenomenon is termed "red out."
4. At high negative G, the cerebral blood vessels are not ruptured in spite of the rise in cerebral arterial pressure, because it is prevented by the simultaneous rises in the intracranial, cerebral venous and CSF pressures which act as cushioning buffer on the outside of the brain.
5. At very high negative G (–20G), retinal or cerebral haemorrhage may occur.
6. The arterial pressure at the heart level increases greatly and the pressure in the head is about 400 mm Hg, which elicits baroceptor reflex causing severe vagal response of the heart and sometimes may stop the heart for 5-10 seconds.
7. Animals exposed to –20G to –40G acceleratory force have developed subarachnoid haemorrhages (Fig. 23.2).

23.2.12 Characteristics of negative G effects

(a) Much more lasting than those of +G.
(b) Headaches may persist for hours following the exposure.
(c) Less acute.
(d) Slower.
(e) Less tolerable than the effects of +G.
(f) Symptoms may continue for even some hours.
(g) More permanent damage than the effects of +G.

23.2.13 Transverse G

Transverse acceleratory forces act along the anterio-posterior axis of the body, i.e., lying down in the aircraft.

Transverse G is better tolerated than +G and –G because the various effects on the cardiovascular system like motion of blood, blood pressure, etc. are very much lower. If the transverse acceleratory forces are exerted uniformly on the body, +15G to + 25G can be withstood for many seconds without any serious effects other than occasional collapse of a lung, which is not lethal. For this reason, when there are large acceleratory forces to be encoun-

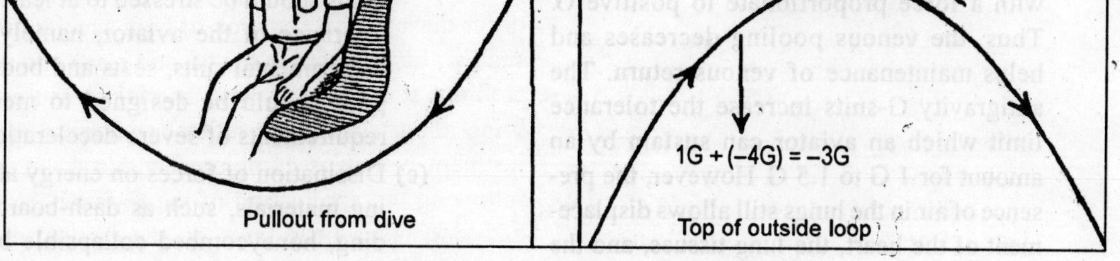

Fig. 23.2: Illustrating positive and negative G-forces

tered, the pilot or astronaut flies in the semireclining or lying position.

23.2.14 Tolerance of G-forces

1. The tolerance of man to G-forces is a time-dependent function.
2. The tolerance for G-foves applied across the body is much greater than it is for axial (head-to-foot and foot-to-head).
3. Men tolerate +G 11 acting in a back-to-chest direction for 3 minutes and +G 17 acting in a chest-to-back direction for 4 minutes. For this reason astronauts are positioned in the chest-to-back direction to take G-forces during rocket flight. The advantages are:
   (a) Better tolerance in this position to greater acceleration to orbital, or
   (b) Escape velocity, and
   (c) Deceleration back into the earth's atmosphere without ill effects.
4. If a final curve of tolerance, i.e. G-forces vs. maximum duration of exposure is drawn, it shows a composite of several curves. The individual curves are hyperbolic; the composite curve also resembles a hyperbola.

23.2.15 Protection against G-forces

It is very difficult to prevent the migration of blood into the lower limbs during negative G-forces. But in case of positive G-forces some steps can be taken to prevent the pooling of blood in the lower limbs by application of external pressure to these areas of the body.

1. **Antigravity "G-suits"**: The effects of positive G can be cushioned effectively by antigravity "G-suits". It is a double-walled pressure suits containing water or compressed air and arranged in such a way that they compress abdomen and lower limbs with a force proportionate to positive G. Thus, the venous pooling decreases and helps maintenance of venous return. The antigravity G-suits increase the tolerance limit which an aviator can sustain by an amount for 1 G to 1.5 G. However, the presence of air in the lungs still allows displacement of the heart, the lung tissues, and the diaphragm into seriously abnormal positions despite submersion in water. Even when this procedure is applied, the limits of safety are about 15 G to 20 G.
2. **Tightening the Abdominal Muscles**: If the pilot tightens the abdominal muscles to an extreme degree and bends forward to compress the abdomen, some of the pooling of blood in the large vessels of the abdomen can be prevented, thereby delaying the onset of blackout.
3. **Position of the Aviator and Astronaut**: It is known that transverse G is better tolerated than positive and negative G. Again, chest-to-back (prone) position tolerance (+17G) is far better than (+11G) back-to-chest (supine) position. Due to technical difficulties for lying down position in the cabin, when very large acceleratory forces are to be involved, the aviator or astronaut flies in the semi-reclining position.
4. For a reduction of the effects of landing deceleration, the spacecraft is decelerated at a much slower rate than the rate of subsonic aircraft.
5. **Protecting Equipment and Materials**: The designer should design the aircraft or spacecraft to prevent or reduce the occurence of accidents after following the recommendations based on the experiments. Because, injury is a function of design and not the inevitable result of accident. For the experiments, the equipment required are dummies, crash protection items (helmets, padding), pressure suits, ejection seats, all types of vehicles including space vehicles.
   (a) Absorption of force on substances other than the human body, such as chassis, seat, etc. which prevents the occupant from colliding with solid objects.
   (b) Seats, seat-belts, tie-downs and similar items should be stressed to at least the G tolerance of the aviator, namely 50G. Experimental suits, seats and body mopeds should be designed to meet the requirements of severe deceleration.
   (c) Dissipation of forces on energy absorbing materials, such as dash-board padding, honeycombed collapsible buffers and immersion media for attenuating forces before they reach the occupant,

light weight metals or honeycombed paper have been shown experimentally to absorb rather than to transmit impact forces.

(d) Exposure of less vulnerable portions of the body.

(e) The experiment has shown that there is little difference between forward and backward seated tolerance to linear deceleration assuming proper restraint in the forward facing position.

23.3 Influence of Linear Acceleration on the Body

In case of spacecraft, the centrifugal acceleration is not important because it cannot take rapid turn as aircraft except during abnormal gyration. But during take-off of a spacecraft there is blast-off acceleration and landing deceleration, both are under the purview of the linear acceleratory forces.

During blast-off in case of three-stage spacecraft, an astronaut experiences acceleratory forces of 9G in the first stage booster and second stage booster shows 8G and the acceleratory forces continue for 5 munutes. The astronaut cannot withstand so much acceleratory force on standing but he can in semi-reclining position—transverse to the axis of the acceleration—for which the astronaut is provided with a reclining seat. Whereas an aircraft during take-off has only about +0.5 G for 10 seconds.

Main problem is during decleration when the spacecraft re-enters the atmosphere producing –G while the aircraft produces –45G for a second during crash-landing.

If the pilot travels at the speed of sound (331.7 m/sec at sea-lvel) or a fast aircraft they can be safely decelerated in a distance of 0.12 mile (0.19 km) whereas an astronaut travelling at a speed of interplanatory space travel takes a distance of 10000 miles (16000 km) for safe deceleration.

The reason for the difference is that the total amount of energy must be dispelled during deceleration. It is proportional to the square root of the velocity, which alone increases the distance 10000 fold. Further, the person can withstand more deceleration if it lasts for a long time than for a short time. For this, deceleration must be accomplished more slowly from very high velocities than from slower velocities.

23.3.1 Parachute jump

The opening of a parachute causes deceleratory forces of 8G for 1 sec depending on distance of free fall. A normal parachute lands with a velocity of 6 m/sec, the parachutist encounters a linear deceleratory force of 3-4G for a fraction of a second.

When the parachutist leaves the aircraft, the velocity of fall is 0 feet per second initially, 32 feet (9.6 m) per second within one second due to acceleratory force of gravity without any air resistance and 64 feet (19.2 m) per second within 2 seconds and so on. As the rate of fall increases, the air-resistance also increases and ultimately the deceleratory force of the air resistance is equal to the acceleratory force of gravity so that after falling for 12 seconds and a distance of 1400 feet (427 m) the parachutist will be falling at a "terminal velocity" of (the body increases velocity during free fall until the wind resistance equals the pull of gravity. The velocity is called the terminal velocity) 175 feet (53 m) per second. If the parachutist leaves the aircraft at a very high altitude where the resistance is very little in the atmosphere, the terminal velocity is higher than 175 feet (53 m)/second but when the body reaches the higher density atmosphere close to earth, the terminal velocity gradually comes down to 175 feet (53 m)/second.

The parachutist reaching the terminal velocity without opening the parachute, gets "opening shock load" of about 1200 pounds (544 kg) on the parachute.

23.3.2 Landing

The normal size of the parachute decreases the fall of the parachutist to about one-ninth the terminal velocity, i.e., the speed of landing is about 20 feet (6 m)/second and the force of impact against the earth is about 1/81 the impact force without a parachute. In spite of that, this impact is sufficient to cause extensive injury to the body if the parachutist is not properly trained. If he lands stretching the legs and using pelvis to strike against the earth, there is great deceleratory forces along the skeletal axis of the body, resulting in the fracture of the pelvis, vertebrae or a leg. But if he is trained to strike the earth bent knee with taut muscles it acts as cushion against the shock of landing.

23.4 Effects of Brief Acceleration

In accelerations which last less than about 1 second, the tolerance will be considered including the factors other than those in connection with longer accelerations such as inclusion of the third derivative of distance with respect to time.

The relations of the various derivatives are:

ds/dt = velocity
d^2s/dt^2 = acceleration
d^3s/dt^3 = jolt (the rate of change of accelerative forces).

In addition to jolt, factors influencing the tolerance to brief accelerations are:

(a) Peack acceleration,
(b) Time of exposure,
(c) The momentum (the time integral of accelerative forces, and is measured as the area under the acceleration-time curve).
(d) Geometric factors:
1. Direction of acceleration
2. Area of pressure.

23.4.1 Occurence

Brief accelerations occur in aviation under such conditions as:

1. Catapult take-offs of aircraft,
2. Explosive ejection of the pilot from the aircraft in emergencies,
3. Crash landings,
4. Parachute openings.

For the most part, these are linear accelerations, and they usually are accelerations directed either from seat-to-head or from front-to-back.

23.4.2 Factors determining tolerance

1. **Sitting Posture**: When a person is sitting upright and forward velocity is suddenly decreased, the effect is that the person will be thrown forward into contact with some hard object producing injury to the head or the chest. This can be reduced by supplying belts and shoulder harness to the flier to limit his travel away from the seat. In automobilies, the common incident is head injury due to strike against the dashboard or rear of front seat and chest injuries due to strike against the steering wheel for which modern automobiles are provided with sponge rubber crash pads for the dashboard to reduce the number and severity of the injuries.

2. **Area for Absorption of Shock**: If the person takes seat with his back turned to the direction of original velocity and his back is supported by high panel or seat back, he can withstand brief G-forces up to 30G without any injury because of large area of contact of the back with the seat causing less contact pressure and local tissue stresses. If the person is protected with belt and shouler straps, the area for absorption of shock is approximately 300 cm^2, whereas when the person is seated backward, the area is approximately 1300 cm^2. For this reason the modern aircraft and projected automobiles are designed to provide backward riding by the passengers.

3. **Strength of the Spinal Column**: If the acceleration is from foot-to-head, the force is usually applied to the lower end of the spinal column because of the sitting position, so the ultimate limits of tolerance are offered by the structural strength of the spinal column. This type of acceleration occurs in explosive ejection from aircrafts. Deceleration is by a harness in which the main force is carried by a gluteal strap.

4. **Dynamic Impacts**: Bones are more susceptible to fracture by dynamic impacts than by steady forces, which is also applicable to vertebral column. If a curve is drawn for tolerance to seat-to-head accelerations, three general regions are seen:

(a) Dynamic strength of the vertebral column is the limiting factor.
(b) A slower region in which the static strength of the vertebral column is important.
(c) The third region in which the circulatory disturbances in slow acceleration limit the tolerance.

5. **Deceleration Distance and Deceleration Forces**: During sudden deceleration, the impact of the heart against the chest wall produces injury to the heart muscle resulting in delayed death. Development in this respect is the use of undrawn nylon ropes which stretch when tension is mounted, thus increasing the deceleration distance and decreasing the deceleration force.

24

Problems and Solutions in Aviation and Space Voyage

1. Weightlessness (zero gravity) in Space Vehicle, 2. Temperature in Aviation and Space, 3. Radiation in Aviation and Space, 4. Atmosphere and Climate in the Spacecraft.

24.1 Weightlessness (Zero Gravity in Space Vehicles

If a person stands on a spring weighing machine and the earth's rotational speed continues to increase, then at certain critical speed, the centripetal force would be just equal to the total gravitational force and no resultant would be left over to provide weight and so the machine would give zero reading. At this stage, the person becomes weightless although the full gravitational force still continues to act.

24.1.1 Weightlessness in the spacecraft

It is known that the astronaut in the orbiting satellite or in spacecraft experiences weightlessness. The astronaut feels absence of gravitational force and he is not drawn toward the bottom, sides or top of the spacecraft but simply floats inside the cabin (zero gravity). In this time, the earth's gravitational pull is just sufficient to provide the centripetal force required for their particular mass, speed and orbital radius. Mechanically, the astronauts are in a similar position to the person standing on weighing machine on a fast rotating earth.

24.1.2 Time of occurrence and duration

The effects of weightlessness usually occurs and reach their maximum within the first few weeks after the astronaut enters space and does not progress further; it persists for a short time after the termination of the space voyage.

24.1.3 Reason

The reason is not due to failure of gravity to pull on the body because the gravity is still active from any nearest heavenly body. The gravity acts on both the astronaut and the spacecraft in the same time in absence of any resistance to movements in space. Both astronaut and the spacecraft are pulled with the same acceleratory forces and in the same direction, and the astronaut is not attracted towards any wall of the spacecraft.

24.1.4 Effects

The main physiological effects of weightlessness are transient and not so severe. They are related to two effects of weightlessness:

(a) Translocation of fluids within the body because of absence of gravity to cause hydrostatic pressures, and

(b) Diminution of physical activity because no strength is required to contract muscles to oppose the force of gravity.

(A) **Effects on Cardiovascular System**

1. **Onset of Effects**: Weightlessness for up to 237 days causes only transient untoward effects on the cardiovascular system which are similar to the effects if a person lies in bed in a recumbent position at sea level for an extended period of time.
2. **Translocation of Body Fluids**: In zero gravity, the pressures in the vessels of different parts of the body are devoid of gravitational force. So the usual gradients of hydrostatic pressure are not maintained between the blood and the interstitial fluids and fluid is translocated in the body in quantity.
3. **Cardiac output**: Decrease in cardiac output.
4. **Blood Volume**: Decrease in blood volume.
5. **Red Cell Mass**: Decrease in red cell volume reduces O_2 carrying capacity.
6. **Heart Rate**: Depression of heart rate by 50%. This, with the diminution of venous return, bradycardia depresses both the stroke volume and the minute volume of the heart.
7. **Blood Pressure**: Transient postural hypotension may result from circulatory failure after return to the earth from the orbital flights and full readaptation to normal gravity after return from space flights takes 4-7 weeks.
8. **Myocardium**: Due to less cardiac output, there is some atrophy of the myocardium. In prolonged weightlessness, this may lead to more severe disuse atrophy of the heart.

(B) **Physical Work**: Work capacity decreases against gravity in zero gravity condition when the objects to be moved are weightless. The decrease in the normal proprioceptive impulses due to action of gravity on the body reduces the normal tension of the extensor, antigravity muscles lead to weakness, flaccidity and atrophy of skeletal muscles. Due to less circulation, muscles cramps may appear. For this reason, a training programme of exercise against resistance is required.

(C) **Other Effects**

1. Motion sickness,
2. Diuresis,
3. Loss of bone mineral,
4. Increased Ca^{2+} excretion,
5. Decrease in blood lymphocytes.

24.1.5 Prevention

1. **Handholds or Other Stabilizing Means**: Weightlessness is actually for the design of the spacecraft. The astronaut cannot pour liquid into a cup, neither can he drink from it, food and water do not stay on open plates, etc. The astronaut should be provided special technique for eating and drinking, special waste disposal systems, handrails or handholds or other means for stabilization of the astronaut for control of spacecraft.
2. **Exercise Programme**: With an appropriate exercise programme the physiological effects of zero gravity can be minimised and the muscular activity for physical work can be increased.
3. **Direction of Walking**: If the astronaut walks round the space-station in the direction of its rotation, his weight increases whereas walking in the opposite direction causes decrease in weight.
4. **Design of the spacecraft**: It has been proposed by the scientists that the spacecraft might be built in the form of large wheels with hollow rims. When this would be placed in rotation, the outer rim—acting as the floor—has to exert a radial centripetal force to the astronaut inside to keep him moving in a circle. The equal and opposite reactions to this centripetal force which the astronaut exerts on the floor, would act as an artificial

weight, which allows eating, drinking and physical works comfortably. The value of this weight would be made equal to or less than the normal earth weight by adjustment to the speed of rotation (Fig. 24.1).

24.2 Temperature in Aviation and Space

The physiological problems due to cold temperature on the earth are similar to the cold temperature in the upper atmosphere. When the temperature on the surface of the earth is 20°C, the temperature at 10000 feet (3050 m) is 0°C and –55°C at 40000 feet (12200 m). For this reason special clothing and heating arrangement are required.

But in space travel, the temperature several miles above the earth rises high. In the ionosphere, several hundred miles above the earth surface and in space, the kinetic energy of the few molecules, atoms and ions is very high. Because any particle that escapes this far away from the gravitational pull of the celestial bodies must have a great velocity. As a result, when the spacecraft reaches above 350 miles (560 km), the temperature of the surrounding particles is approximately 3000°C. This amount of temperature does not have any effect on the spacecraft because of the presence of very few molecules, atoms and ion particles which do not produce any significant amount of heat. The temperature of the spacecraft is maintained by keeping a balance between relative absorption of radiant energy from the sun and re-radiation of energy from the spacecraft. The spacecraft has different coatings for different absorptive and radiation characteristics. With the help of proper coating and orientation of the spacecraft toward the sun, the temperature inside the aircraft is controlled at proper level.

24.3 Radiation in Aviation and Space

24.3.1 Aviation at high altitude

There is considerable difference of electromagnetic radiation between the earth's surface and at high

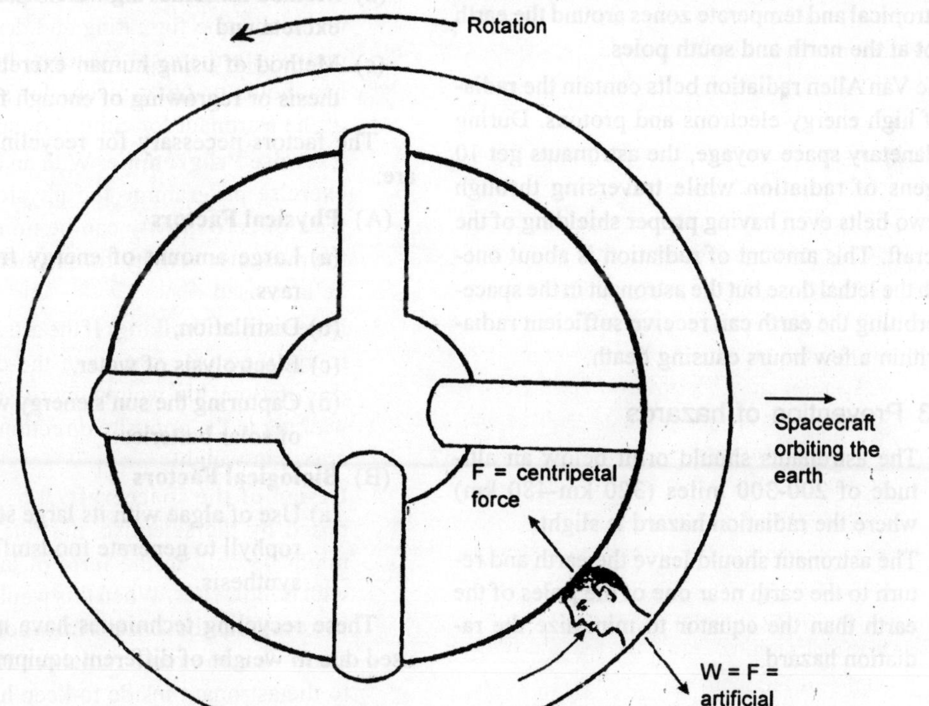

Fig. 24.1: Proposed design and production of artificial weight in spacecraft

altitude. This is because the radiations from the sun rays is filtered by the atmosphere before reaching the surface of the earth—such as most of the ultra-violet light is absorbed before reaching the earth's surface. It has been observed that 18 per cent of the visible light is absorbed by the atmosphere before reaching the earth's surface for which upper atmosphere is the brighter to the aviator than the earth's surface and makes earth's surface less distinct. No problem arises to the aviator from the gamma rays and X-rays which are hazardous in space flight.

24.3.2 Space radiation hazard

The upper atmosphere of the earth is continually bombarded by a large amount of cosmic particles which originate from the sun and outer space. Many of these cosmic particles are trapped by the magnetic field of the earth into two main belts called "Van Allen radiation belts". The inner belt starts at an altitude above 300 miles (480 km) and extends up to about 3000 miles (48000 km) and the outer belt starts from 3000 miles (4800 km) and extends up to 20000 miles (3200 km). These two belts occur in tropical and temperate zones around the earth and not at the north and south poles.

The Van Allen radiation belts contain the radiation of high energy electrons and protons. During interplanetary space voyage, the astronauts get 10 roentgens of radiation while traversing through these two belts even having proper shielding of the spacecraft. This amount of radiation is about one-fortieth the lethal dose but the astronaut in the spacecraft orbiting the earth can receive sufficient radiation within a few hours causing death.

24.3.3 Prevention of hazards

1. The astronauts should orbit below an altitude of 200-300 miles (320 km-480 km) where the radiation hazard is slight.
2. The astronaut should leave the earth and return to the earth near one of the poles of the earth than the equator to minimize the radiation hazard.

24.4 Atmosphere and Climate in the Spacecraft

In outer space there is no atmosphere for which artificial atmosphere and climate are created in the spacecraft.

It is essential to keep high concentration of oxygen and low concentration of carbon dioxide. In most of the spacecraft, the atmosphere has been maintained between 260-270 mm Hg pressure. Nitrogen has been used in the mixture to avert the likelihood of fire and explosion.

But during space travel for several months, it is very difficult to maintain the atmosphere having arrangement for adequate oxygen supply and sufficient carbon dioxide absorbent. For these difficulties a techniques has been developed known as **recycling technique** for using over and over again the same oxygen and also re-using the same food and water. The technique comprises:

(a) Method of removing oxygen from carbon dioxide.

(b) Method for removing water from the human excreta and

(c) Method of using human excreta for resynthesis or regrowing of enough food supply.

The factors necessary for recycling technique are:

(A) **Physical Factors**

(a) Large amount of energy from the sun rays.

(b) Distillation,

(c) Electrolysis of water,

(d) Capturing the sun's energy with the help of solar batteries.

(B) **Biological Factors**

(a) Use of algae with its large store of chlorophyll to generate foodstuffs by photosynthesis.

These recycling techniques have not yet been used due to weight of different equipments.

25

Influence of Barometric Pressure

1. Low Barometric Pressure, High Altitude, 2. High Barometric Pressure, Sea Diving, 3. Decompression, 4. Physical Factors involved in Diving.

When a person ascends gradually—such as in mountain climbing—or ascends speedily—like aviation or space flight—there is an impact of altitude on the human body due to changes in barometric pressure (low) and gas pressure (low). Initially, during aircraft flying, the following problems were considered with great care:

(A) Low partial pressure of oxygen (PO_2) producing hypoxic hypoxia.

(B) Effect of physical factors like temperature, radiations, etc. (discussed in Chapter 23).

(C) Working capacity.

(D) Different acceleratory forces which has been focused due to space flight (discussed in Chapter 23).

25.1 High Altitude

25.1.1 Effect of reduced atmospheric pressure (low barometric pressure) on respiration, circulation, etc. and compensatory mechanism

1. **Altitudes and Barometric Pressure:** During ascent to a progressive higher altitude the barometric pressure decreases and the total pressure of all gases also diminishes with a distance above the earth's surface in an approximately exponential manner (Table 25.1). The fall in barometric pressure with increasing altitude is not linear, because air is compressible — barometric pressure at sea level is 760 mm Hg, at 10000 ft (3 km) is 523 mm Hg and at 20000 ft (6 km) is 349 and is only one-half the normal 760 mm Hg and at 50000 ft (15 km) it is 87 mm Hg.

The physiology of respiration at reduced atmospheric pressure is of great importance for mountaineering, aviation and space travel. The chief difference between mountaineering and aviation being that, in the former, physical exertion is coupled with reduced atmospheric pressure and adverse climate while, in the latter, the reduction of pressure is rapid and onset of symptoms are sudden.

Table 25.1: Barometric pressure (P_B) at different altitudes and inspired PO_2 calculated from (Barometric pressure (P_B) − 47) × 0.2093 = 63 mm Hg.

Height above sea level (ft)	Barometric Pressure (mm Hg)	PO_2 in Air (mm Hg)	PO_2 in alveoli (mm Hg)
0	760	159	104
7350 (2.2 km)	580	112	
10000 (3 km)	523	110	67
11000 (3.3 km)	500	95	
20000 (6 Km)	349	63	40
30000 (9 km)	226	47	21
40000 (12 km)	141	29	8
50000 (15 km)	87	18	1
63000 (19 km)	47	0	

25.1.2 Altitudes and partial pressure of oxygen (PO_2) in inspired air in the lungs

The decrease of barometric pressure is the main cause of hypoxic hypoxia at high altitudes. As the barometric pressure decreases, the oxygen pressure decreases proportionately—slightly less than 21 per cent of the total barometric pressure. The oxygen pressure at sea level in dry air is 159 and can be decreased by 10 mm Hg, if there is large amount of water vapour. The barometric pressure at 20000 ft (6 km) is 349 mm Hg, so the PO_2 of moist inspired O_2 (349 − 47) × 0.2093 = 63 mm Hg (Table 25.1), while 47 mm Hg is the partial pressure of water vapour at body temperature. At the peak of Mount Everest the altitude is 29028 ft (8848 mt) and the barometric presure is 253 mm Hg, the inspired PO_2 is 43 mm Hg. At 63000 ft (19 km), the barometric pressure is 47 mm Hg, and PO_2 of inspired air in the trachea is zero and if the pilot bails out from the aircraft, the tissue fluids will be vapourised.

25.1.3 Altitudes and alveolar PO_2 arterial O_2 saturation

When PO_2 in the atmosphere reduces at higher altitude, a decrease in alveolar PO_2 also occurs. At low altitude alveolar PO_2 does not decrease like atmospheric O_2 due to compensatory mechanism by hyperventilation. But at higher altitudes the alveolar PO_2 decreases more than atmospheric PO_2 for the following:

(a) At high altitudes, CO_2 is excreted continuously from the pulmonary blood into the alveoli, and

(b) Water vaporizes from the lung surface to the inspired air and both these dilute the oxygen and nitrogen in the lungs, thus lowering the oxygen concentration.

Still the CO_2 and water vapour in the alveoli have importance at high altitude because the barometric pressure falls but the pressure of CO_2 and water do not fall comparably. Moreover, a greater volume of the alveolar air must consist of CO_2 since the tension of this gas tends to be kept nearly constant. The lowest limit at which human life is possible corresponds to an oxygen tension in the alveoli of 27 to 30 mm Hg. Table 25.1 shows alveolar PO_2 at different altitudes.

25.1.4 Altitudes and arterial oxygen saturation

At sea level, arterial oxygen saturation is 97 per cent. At an altitude of 10000 ft (3 km) even breathing air, the oxygen saturation is still as high as 90 per cent, at 20000 ft (6 km) it is 70 per cent and at 50000 ft (15 km) it is 1 per cent.

25.1.5 Effects of hypoxia caused by low barometric pressure

Hypoxia may be explained under three conditions:

(A) Sudden exposure to low barometric pressure.

(B) Gradual exposure for a few weeks during mountaineering when the person acclimatizes.

(C) Life-long expsoure of a person residing at high altitude.

(A) Sudden exposure

When a pressurized aircraft loses cabin pressure or a person is detached from special oxygen equipment while flying at a high altitude, the alveolar PO_2 falls within a few seconds but due to storage of some amount of oxygen (in Hb, myoglobin, and other O_2 carriers), a short time passes before the onset of fatal result. It has been observed that the time of fatal result depends upon the altitude, such as at 28000 ft (8.5 km) decrease in consciousness starts in one minute and coma occurs in three minutes. At 38000 feet (11.6 km) decrease in consciousness starts in thirty seconds and coma occurs in one minute.

(B) Gradual exposure during mountaineering

At a height below 8000 ft (2.4 km) there is no appreciable pulmonary ventilation but at 8000 ft the pulmonary ventilation increases because of the fall of arterial oxygen saturation to 93 per cent causing stimulation of chemoreceptor. The progressive increase of chemoreceptor mechanisms causing increasing in pulmonary ventilation continues till the person reaches 20000 ft (6 km). Pulmonary ventilation reaches maximum of 65 per cent above normal.

Other effects of hypoxia occur when the person reaches 12000 ft (3.6 km) and these are drowsiness, lassitude, mental fatigue, sometimes headache, occasionally nausea and euphoria and, at still higher altitude, twitching or convulsions occur.

The mental proficiency like judgement, memory, performance of discrete motor movements remain normal up to 9000 ft (2.7 km) but, beyond that height, if the person is exposed to hypoxia for a long time, mental proficiency decreases.

Acute Mountain Sickness: A person climbing high altitude for the first time often complains of headache, fatigue, dizziness, palpitations, nausea, anorexia (loss of appetite) and insomnia which is known as acute mountain sickness. Though cause is not known it has been suggested that hypoxemia and respiratory alkalosis or mild cerebral oedema may be the reason. Retinal haemorrhages frequently occur and heal up usually without any visual defects. Periodic or cheyne-stokes breathing occurs usually.

(C) Chronic mountain sickness or Monge's disease

When a person remains for a long time at high altitude he develops chronic mountain sickness or Monge's disease which is manifested by:

1. Very high red cell mass and haematocrit,
2. The pulmonary arterial pressure increases,
3. Enlargement of right heart,
4. Fall of peripheral arterial pressure,
5. Congestive heart failure,
6. Death may occur if the person is not removed to a lower altitude.

The reasons for the above manifestations are:

1. Very high cell mass → increase in viscosity of blood → decrease of blood flow to the tissues → O_2 supply seriously disturbed.
2. Vasoconstriction of pulmonary vessels due to hypoxia → diverson of blood flow from unaerated to aerated area of lungs → increase in pulmonary arterial pressure → causing right heart failure.

- **Working Capacity**: Working capacity of all the muscles, i.e., skeletal muscles and cardiac muscles, are decreased due to hypoxia. Eventually the maximum level of cardiac output is also decreased. Reduction of work capacity is directly proportional to the decrease in maximum rate of oxygen uptake by the body.

25.2 High Barometric Pressure Effect on the Body

When the air pressure is greatly increased as in deep sea diving, digging underwater tunnels in the chambers (Caissons) or with those using submarine escape apparatus (within the submarine the pressure is normal), air also must be supplied under high pressure to prevent the collapse of his lungs. thus, his blood in the lungs is subjected to high alveolar gas pressure. Adverse reactions occur due to these high pressures when beyond the limit.

25.2.1 Depth of water and pressure

A column of 33 ft (10 m) high of sea water exerts the same pressure at its bottom which is exerted by 34 feet (10.3 m) high of freshwater. There is an increase of 1 atmosphere for every 33 ft (10 m) depth of sea water. Therefore, a person is exposed to a pressure of 2 atmosphere, 1 atmosphere due to air above the sea water and another 1 atmosphere by the weight of the water itself. The atmospheric pressure at 66 ft (20 m) would be 3, at 100 ft (30 m) 4 and so on.

25.2.2 Depth of water and the volume of gases

Depth causes compression of gases. As the depth increases, compression of gases to smaller volumes also occur. At sea level, if a container contains 1 litre of air, it compresses to one-half litre at 33 ft (10 m) depth where atmospheric pressure is 2. At 100 ft (30 m) it becomes one-fourth litre and so on. This has gained importance in diving, because it has an effect on the chamber of the diver. The lungs of the diver may become small that might pose some danger to the diver.

25.2.3 High partial pressure of gases

A diver takes breathing air having three gases.

1. Nitrogen, 2. Oxygen, 3. Carbon dioxide, Helium is often substituted for nitrogen.

25.2.3.1 Nitrogen at high pressure–nitrogen narcosis

A diver breathes air and other gases at high pressure to equilize the high pressure on the chest wall and abdomen. CO_2 is removed as a routine procedure to prevent its accumulation. At high pressure, 100 per cent oxygen causes oxygen toxicity producing central nervous system symptoms. This can be prevented by decreasing the concentration of oxygen to 20 per cent or less by the mixing nitrogen or helium. Nitrogen at sea-level pressure is about 79 per cent and does not have any effect on bodily function. Because the nitrogen of air is normally a purely inert gas, the very inertness of nitrogen may be of danger at a higher pressure. It produces varying degress of nitrogen narcosis. At 120 ft (37 m) depth, first symptoms of mild narcosis drowsy, at 250 ft (76 m) his strength reduces, becomes clumsy. Beyond 300 ft (90 m) at 10 atmospheric pressure, the diver is completely out of intellectual functions. Nitrogen narcosis has similarity with alcohol intoxication for which it is called "rapture of the deep or depths".

The mechanism of the narcotic effect may be like gas anaesthetics. Nitrogen is especially soluble in the fats of adipose tissues and nervous tissues. The body normally contains about 1 litre of dissolved nitrogen at sea level, one-fifth of which can be washed out in a few minutes by breathing oxygen. But if a man is exposed, as in a Caisson, to a pressure of three to six atmospheres, the nitrogen dissolved in the body increases in direct proportion to the pressure. Like other anaesthetic gases, it is dissolved in the membranes or other lipid structures of the nerve cells producing physical effect like alteration of electric charge transfer, thereby reducing their excitability.

25.2.3.2 Oxygen at high pressure (hyperbaric oxygen) oxygen toxicity

(A) **High PO_2 on Blood Oxygen Transport**: When the PO_2 increases in the blood above 100 mm Hg, oxygen is dissolved in the water of the blood in more quantity. In normal alveolar PO_2, total oxygen dissolved in the blood is not accounted for but as the pressure rises, a large quantity of oxygen is dissolved rather than bound with haemoglobin.

(B) **High Alveolar PO_2 and Tissue O_2**: If PO_2 in the lungs is 3000 mm Hg at 4 atmospheric pressure (hyperbaric oxygen—oxygen under more than one atmosphere pressure) which contains 29 volumes per cent of total oxygen in 100 ml blood. After reaching tissues which take up 5 ml of oxygen per 100 ml the blood returns with 24 volumes per cent of oxygen and PO_2 is still 1200 mm Hg instead of normal range –40 mm Hg. Thus, due to rise of alveolar PO_2, the haemoglobin-oxygen buffer mechanism fails to keep the tissue PO_2 in the normal range.

(C) **Acute Oxygen Poisoning**: When the breathing oxygen at a very high alveolar oxygen pressure, high PO_2 occurs in the tissues. This causes adverse effects in the body tissues, especially in the brain. Exposure to 3 atmospheric pressure of oxygen ($PO_2 = 2280$ mm Hg) causes convulsions followed by coma after one hour. Other symptoms are nausea, muscle twitchings, dizzingess, distrubrance of vision and irritability.

High PO_2 is dangerous for neuronal metabolism. The rapid oxygen usage of the brain is due to the grey matter. White matter uses only 0.3 ml per 100 gm per minute (which is 60 per cent of the brain weight). Grey matter (40 per cent of total brain weight) uses 3 ml per minute when to total

3.3 ml per 100 gm per minute is used by brain.

The reasons for acute oxygen toxicity is known to be based on experiemental observations and opinion has been given.

(a) During oxygen toxicity, concentration of certain oxidative enzymes in the tissue is reduced, and it decreases the ability to form high energy rich phosphate bonds.

(b) It has been further observed that high pressure oxygen cuases inactivation of sulphydryl groups of enzymes and coenzymes in the Kreb's Cycle, such as two reactions like:

(i) Pyruvic acid-acetyl coenzyme A and

(ii) L-Ketoglutarate-Succinyl coenzyme A which requires a cofactor lipoic acid which contains reduced sulphur at 6 and 8 positions which are oxidized and the above reactions are incomplete. As a result the turning of Kreb's Cycle is stopped and formation of 15 mols of ATP from one mol of pyruvic acid is disturbed and the cells suffer from deficiency of ATP. Toxic action of high pressure oxygen can be protected by giving reduced glutathione, chelating metallic ions such as CU^{2+}, EDTA. Nowadays hyperbaric oxygen chambers having 2-3.5 atmospheres are being used, such as:

(i) In surgical operations on the heart and lungs,

(ii) In the treatment of CO poisoning, and

(iii) In infections by anaerobic organisms.

(c) It has been shown experimentally that there is a production of large quantity of oxidizing radicals causing destruction of some essential elements of the cells, thereby damaging the metabolic systems.

(d) **Chronic Oxygen Poisoning**: When a person is exposed to 100 per cent oxygen at normal atmospheric pressure, there is no acute oxygen poisoning but after 12 hours he develops congestion and oedema of lung passage due to damage to the linings of the bronchi and alveoli, as a result of oxidative destruction of the essential elements of the cells. This damaging effect does not occur in the tissues because the lungs are exposed directly to the high oxygen pressure, while oxygen is delivered to the tissue at normal PO_2 because of haemoglobin-oxygen buffer system. If haemoglobin-oxygen buffer system is no longer effective, which allows the rise of PO_2 in the tissues it produces acute oxygen poisoning.

25.2.3.3 Carbon dioxide toxicity

Depth alone does not increase the carbon dioxide toxicity if diving gear is constructed and functions properly. But in certain cases, in the diving gear and the rebreathing apparatus, carbon dioxide can accumulate in the dead space air of the apparatus and rebreathed by the diver. If PCO_2 is 80 mm Hg, the diver can tolerate by increasing minute respiratory volume upto 6-10 folds but if it is beyond 80 mm Hg, diver cannot tolerate due to depression of the respiratory centre and his respiration begins to fail. The diver also develops severe respiratory acidosis, lethargy, narcosis, and finally anaesthesia occurs.

25.2.3.4 Helium and high-pressure nervous syndrome (HPNS)

Sometimes helium is used in the place of nitrogen to prevent nitrogen narcosis. Helium has one-fourth to one-fifth the narcotic effect of nitrogen and no delecterious effect is seen up to 500 ft (150 m) dive. But beyond this level, the divers begin to feel "high-pressure nervous syndrome" which is manifested by tremor at 700 ft (210 m), drowsiness from 1000 ft (300 m) onwards and lack of coordination at deeper levels.

Other advantages over the nitrogen are:

(i) Low atomic weight, slight density reducing the airway resistance,

(ii) Diffuses through the tissues rapidly due to low atomic weight and more rapid removal from the body fluids,

(iii) Less soluble in the body rapidly due to low atomic weight and more rapid removal from the body fluids,

The cause of HPNS is not known but it may be explained that physiologically inert gases at atmospheric pressure are anaesthetics at increased pressure. These are applicable to nitrogen, xenon, krypton, argon, neon and helium. The anaesthetic activity is for their lipid solubility and action on nerve cell membranes.

25.3 Decompression

When a diver is exposed to air for several hours under high pressure, more nitrogen is dissolved in the body fluids and the amount of dissolved nitrogen increases. The cause is that the pulmonary capillaries are saturated with nitrogen. And during the long period of exposure the dissolved nitrogen makes the tissues saturated. As the nitrogen is not a metabolizer, it remains in the tissues till the pressure in the lungs decreases when nitrogen evolves out by respiration.

25.3.1 Amount of nitrogen dissolved

One litre of nitrogen is normally dissolved in the body fluids at sea level. Out of this, less than half of this is dissolved in water and more than half in the fat of adipose tissue and brain. Nitrogen is five times as soluble in fat as in water. The nitrogen dissolved in the body is increased in direct proportion to the pressure. The volume of nitrogen dissolved at different depths is 2 litres at 33 ft (10 m), 4 litres at 100 ft (30 m), 7 litres at 200 ft (60 m) and so on. For a short period (a few minutes) at deep level, nitrogen does not dissolve so much in the body fluids and tissues but if the diver remains for a long time at deep level, his body fluids and tissues become completely saturated with nitrogen.

25.3.2 Decompression sickness (compressed air sickness, bends, caisson disease, diver's paralysis, dysbarism)

If a diver breathing 80 per cent nitrogen ascends from a dive, the increased alveolar PN_2 falls. Nitrogen diffuses from the tissues into the circulation and then to the lungs along the partial pressure gradient. When he ascends gradually to atmospheric pressure (decompression), no adverse effects are seen, but if the diver ascends rapidly and suddenly comes back to the surface of the sea, significant quantities of nitrogen form in the blood and tissues causing the symptoms of decompression sickness. The degree of damage depends on the amount of bubbles formed. Exercise hastens the bubble formation because of increased agitation of the fluids and tissues.

25.3.3 Mechanism of bubble formation

During the period when the diver remains at deep level of the sea, the pressure against outside the body, about 3000 mm Hg compresses all body tissues to hold the dissolved nitrogen in the fluids. When he ascends rapidly and reaches at sea level, the pressure outside the body becomes 1 atmosphere (760 mm Hg) whereas the pressure inside the body fluids is about 4065 mm Hg which is the sum of pressures of water vapour, carbon dioxide, oxygen and nitrogen which is higher than the outside. Hence, the gasses escape from the dissolved state and forms bubbles inside the tissues. Bubbles may not appear from minutes to hours because the gases can remain in the supersaturated state for sometimes before bubbling.

- **Symptoms of Mountain Sickness**: Severe pains around joints, neurologic symptoms like paresthesia, itching, major paralysis and respiratory failure due to obstruction of the arteries to the brain, dyspnoea (the chokes) due to bubbles in the pulmonary capillaries, myocardial damage due to bubbles in the coronary arteries, collapse with unconsciousness.

25.3.4 Treatment of decompression sickness

Decompression in a pressure chamber as early as possible followed by slow decompression. Recovery is often complete but there may be residual neurological symptoms due to irreversible damage of the CNS.

25.3.5 Role of helium-oxygen mixture in diving

Helium has the advantage over nitrogen for mixture with oxygen during diving. The advantages are:

(a) Reduction of decompression time due to:
1. 40 per cent dissolving capacity in the body,
2. Small atomic size and, for this property, it diffuses through the tissues at a velocity about 2.5 times that of nitrogen.

(b) Absence of narcotic effect.

(c) Reduced airway resistance in the lungs.

But helium also has disadvantages, such as, during short and shallow dives, it diffuses to the tissue rapidly and forms more dissolved helium in the tissue fluids than the nitrogen which has slower rate of diffusion. Therefore, nitrogen is preferable for depth below 100 feet (30 m) and helium for greater depths.

25.4 Physical Factors Involved in Diving

The following physical factors are involved in diving:

I. High gas pressures.
II. Relationship between volume of air pumped to the diver and rate of carbon dioxide elimination.
III. Relationship between density of air and maximum breathing capacity.
IV. Rapid descent and its effect.
V. Air embolism—over-expansion of the lungs due to rapid ascent.

I. **High Gas Pressures** (Previously discussed)

II. **Relationship between Volume of Air Pumped to the Diver and Rate of Carbon Dioxide Elimination**: To eliminate the carbon dioxide from the lungs of the diver, the tidal volume of air going in and out of the lungs with each breath must be the same regardless of the depth. The amount of air that should be pumped to the diver to maintain his alveolar carbon dioxide normal is directly proportional to the pressure under which he is operating.

For adequate elemination of carbon dioxide from the diver's helmet, the diver requires about 1.5 cubic feet (42.5 dm^3) of air per minute at sea level. This sea level volumes of air should be pumped each at different depths such as at 33 ft (10 m) depth, 3 cubic feet (8.5 dm^3) of air is required, at 66 ft (20 m), 4.5 cubic feet (127.5 dm^3) air, at 100 ft (30 m), 6 cubic feet (1170 dm^3) air and so on.

III. **Relationship between Density of Air and Maximum Breathing Capacity**: Increased density of air increases the work of breathing which eventually decreases maximum breathing capacity, i.e., the amount of air that a person can breath per minute.

The density of air increases proportionally to the pressure, such as density is 4 times as great at 100 ft (30 m) than the sea level. The resistance to air flow in the respiratory tract increases directly in proportion to the density of the breathing mixture.

IV. **Rapid Descent and its Effect**: During rapid descent, the volume of all gases decreases due to increasing pressures exerted outside the body. Hence, an extra amount of air is supplied to thr air cavities to prevent any harmful effect. But, if the diver continues to descent rapidly and no extra air is supplied to the air cavities, the volume of all gases greatly reduce and there is an occurence of serious physical damage known as **the squeeze.**

1. The great damage of squeeze occurs in the lungs and, to prevent this, the diver should inspire additional air when he descends.
2. If air is trapped in the middle ear during descent, the squeeze causes rupture of tympanic membrane.
3. If air is trapped in the nasal sinuses, there will be intense pain.

V. **Air Embolism—Over Expansion of the Lungs due to Rapid Ascent**: During rapid ascent, if the diver fails to expel air from the lungs on the way, opposite effect of squeeze, i.e., over expansion of the lungs occur. During expansion of the lungs to the limit, the pressure continues to rise and, above an excess alveolar pressure of 80-100 mm Hg, air is forced into the lung capillaries resulting in air embolism in the circulation—often death occurs. Increased pressure in the lungs produces large blebs on the surfaces of the lung or lungs and may rupture producing pneumothorax.

Rapid ascent may cause serious damage to the gastrointestinal tract when the diver had stayed for a long time (under water) and a large amount of gas accumulates in the abdomen. Air embolism also occurs during escape from submarine.

Air embolism also occurs as a result of expansion of the gas in the lungs when the wall of the pressurised cabin of an aircaft or rocket at high altitude is breached (explosive decompression).

26 Vibrations

1. Physical basis of vibrations, 2. The human body is a mechanical system (oscillatory characteristics), 3. Physiological effects, 4. Raynaud's phenomenon.

Vibration is the term describing oscillatory conditions within a mechanical system, i.e., mechanical oscillations produced by either regular or irregular periodic movements of a body about its resting position and must have three components:

1. **The storage of potential energy,**
2. **The storage of kinetic energy and**
3. **Some means for dissipating this energy.**

26.1 Physical Basis of Vibrations

Five physical quantities are necessary to understand vibrations:

1. **Application Point to the Body**
   - **(a) Two points at which vibrations enter the body are important ergonomically, e.g., feet and buttocks in case of driving or riding in a vehicle and hands when working with vibrating tools or a machine.**
   - **(b) The direction of oscillation is important, usually it lies in the vertical plane such as head to foot or along the line of hand and arm.**
2. **Frequency Dependent Oscillation**
   - **(a) The degree of physiological and pathological manifestation of vibrations depends on frequency.**
   - **(b) The important frequencies are those which fall into the range of natural frequencies of the body causing resonance. Low and high range of frequencies can be distinguished.**
   - **(c) The threshold of frquency lies between 30 to 50 Hz.**
3. **Acceleration of the Oscillation: Within the physiologically important frequency range, the acceleration of the oscillations is usually taken as a measure of the vibrational load. The unit is the acceleration due to gravity (g) = 9.8 m per sec^2.**
4. **Duration of Effect: The vibration effect depends greatly on their duration.**
5. **Individual Frequency and Resonance**
   - **(a) Every mechanical system having the elementary properties of mass and elasticity exhibits oscillation.**
   - **(b) The force required to set the system in motion is called the "exciting force" and the resultant oscillations are called the "forced vibrations".**

(c) Each system has its own natural frequency and nearer the frequency of the exciting force comes to this, greater will be the amplitude of the forced vibrations.
(d) When the amplitude of the forced vibrations exceeds that of the exciting force, the system is in "resonance".
(e) Oscillations of any system are subject to "damping" which reduces their amplitude.
(f) Frequencies above 30 Hz are particularly highly damped by the tissues of the body.

26.2 The Human Body is a Mechanical System (Oscillatory Characteristics).

The human body does not vibrate as a simple mass, with its own natural frequency. Natural frequencies are different in different parts of the body. Man's sensitivity of external vibrations is highest in frequency between 0.5 and 20 Hz, since the human system absorbs most of the vibrating energy which is applied within this range. The resonant or natural frequency reacting to vertical vibrations for other part of the body has been determined experimentally such as cervical vertibrae (3-4 Hz), girdle (5 Hz), head and shoulder (20-30 Hz), eyeballs (60-90 Hz), and lower jaw (100-200 Hz). The most effective exciting frequency for vertical vibrations lies between 4 and 8 Hz.

Studies of mechanical impedance have been used to define the mechanical properties of the human body and is calculated by the ratio of transmitted force to the angular velocity at the point of applied force.

The natural frequencies of the smaller components of the body, such as group of muscles, eyes, etc. lie in higher ranges of frequency. Hence the operation of machines with frequencies above 30 Hz are likely to set up resonances in the fingers, hands and arms. On the other hand, the damping effect of body tissues are greater for these higher frequencies, and tends to confine them to the vicinity of the point of application.

26.3 Subjective Response

Animal studies with high intensity vibrations have shown tissue damage, with maximal damage occurring at resonant frequency of the organs of the body members involved. The higher frequencies (14-20 Hz) primarily affect muscle tone, head sensations, speech and the urge to defecate and micturate. The chest pain, voluntary muscle contraction, abdominal pain, lumbosacral pain and general discomfort tends to occur in the range between 5 Hz and 10 Hz, whereas the respiration and dysponea were primarily observed at very low frequencies, around 2 Hz. In general, it appears that frequencies below 10 Hz affect organ tissue complexes of the chest and adbomen, whereas frequencies above 10 Hz primarily affect the organ tissue complex located in the periphery.

26.4 Physiological Effects

Physiological effects to tolerable levels of vibrations have been reported by several workers.

1. **Heart and B.P.**: Heart rate and blood pressure both increase when exposed to a low level of frequency vibrations but the B.P. decreases at a frequency of 10-20 Hz.
2. **Respiration**: Tidal volume increases but is maximum at 5-7 Hz. Both mild hyperventilation and respiratory rate show increase, and vital capacity characteristically shows slight decrease at 5-7 Hz.
3. **Blood, kidney or endocrine**: There has been little or no evidence of blood, kidney or endocrine effects under moderate vibration levels.
4. **Muscular reflexes**: Vibrations seems to generate muscular reflexes having protective function causing extensors to shorten. Increase in fatigue causes these reflexes to diminish or to disappear.
5. **Vision**: Effect of vibration on eyesight has importance, because it impairs the efficiency of drivers and increases the risk of accident. Visual acuity is poorer and the image in the visual field becomes blurred and unsteady. Visual powers are not affected by vibrations of less than 2 Hz. Optical aberrations appear from 4 Hz upwards and are greatest in the range 10-30 Hz. Vibrations of 50 Hz and an oscillatory acceleration of 2 m per sec^2 cause reduction of visual acquity by half.

6. **Skills:** Strong vibration impairs performance in various psychomotor tests, marked decline in manual skill.
7. **Driving Tests:** The psychophysiological effects on driving tests have shown that efficiency imparied at around 4 Hz, errors increase when the seat is accelerated in the order of 0.5 m per sec^2 and when the accelerations reach 2.5 m per sec^2 it makes the error positively dangerous.
8. **Equilibrium ability:** Vibration is also shown to interfere with man's equilibrium ability though individual variations are extreme.

26.5 Vibration as a Nuisance

Vibration is subjectively felt as nuisance. The extent of nuisances depends on:

(a) The exciting frequency,
(b) The rate of acceleration of the oscillations and
(c) The length of time they continue.

26.6 Complaints

In addition to annoyance or nuisance, sometimes there are complaints, some of them frequency-dependent (1-4 Hz), pain in chest and abdomen (4-10 Hz), backache (8-12 Hz), muscular tension, headache, eyestrain, throat pain, speech disturbance, irritation of intestine and urinary bladder (10-20 Hz).

26.7 Health Damage

Exposure to vibration causes morbid changes in the organs affected. The effects are different in the two parts of the body depending on vibration.

(a) Vertical oscillations in standing or sitting condition, caused by vibration from underneath (e.g., vehicle), producing degenerative changes in the spine, whereas.
(b) Vibration of power tools affect the hands and arms.
   1. Spinal trouble: Disc troubles, arthritis.
   2. Intestinal ailments: Haemorrhoids.
   3. Prostate troubles.
   4. Hand and arm troubles.
   5. Arthritis: Due to degenerative symptoms in the bones, joints and tendons of hand and arms leading to arthritis in the wrist, elbow and shoulder.
   6. Atrophy: Bone atrophy and fracture.
   7. "Dead fingers" (Raynaud's disease): Power tools with frequencies between 40 Hz and 300 Hz have a very small amplitude of oscillation (from 0.2-5 mm.) and their vibrations are damped rapidly in the tissues. Such vibrations may cause ill-effects on the blood vessels and nerves of the hands, resulting in one or more of the fingers going "dead". Usually the middle finger is most affected, becoming white or bluish, cold and numb. After a short period the finger turns pink again, it is painful and due to a cramp-like condition of the blood vessels, known as Raynaud's disease. This is also observed among miners using pneumatic drill, among forestry workers using power saw. This disease occurs at an earliest, 6 months after starting the work and cold is an important factor for the onset of this diease. It is common in northern countries than in the warmer latitudes.

26.8 Raynaud's Phenomenon

26.8.1 Localised vibration

This type of vibration due to hand-held tools and equipments which is transmitted through the body members represents a related vibration problem. It is encountered commonly in the use of hand operated devices and in numerous other work situations in which imbalanced rotary equipment or oscillatory hand tools are used.

26.8.2 Etiology

Localised vibrations for prolonged exposure (not less than 2-5 years) of the hands to vibrations in the 90-300 Hz frequency range can generate damage to the peripheral vascular tissues, to other soft tssues and under higher intensity levels and exposure, damage to the bones and joints.

26.8.3 Symptoms

Symptoms that have been obscured include: a stiffness or numbness of the hands, a loss of muscle control, inability to hold, grab or manipulate, finger swelling, reduced sensory preception, parasthesia, cyanosis, etc. Due to similarities of symptoms to that of Raynaud's disease—a vascular disease affecting the arteriole, these vibration problems commonly are referred to "Raynaud's Phenomenon". During recent years this disorder has been regarded as one of the characteristic signs of vibration disease—an independent nosological entity of occupational pathology.

26.8.4 Objective findings

Osteophytic alterations at the radius head, inferior humerus, carpals, etc; areas of decalcification in the bones of hands and forearms; compression and loss of elasticity of cartilage, etc. Symptoms are most likely with vibrations between 25 Hz and 150 Hz.

26.8.5 Types of machine

Vibration of this frequency range may be produced by machines and various power-driven hand tools (pneumatic or electric drills, chisels and other tools).

26.8.6 Factors influencing the effect

The effect produced by this vibration depends on:

(a) The frequency spectrum,
(b) Speed of oscillation,
(c) Position of the tools,
(d) Weight of the tool,
(e) The type of material being worked, and
(f) The work posture.

The harmful effects of vibration may be aggravated by muscular strain and by cooling of the engaged hand by the tool.

26.8.7 Pathogenesis

The pathogenesis of vibration disease consists in a microtraumatic overstimulation of the peripheral neuromotor system and impaired microcirculation, which are followed by neuro-reflectory disturbances and neural dystrophy. The disease evolves, in general, slowly over several years. The speed of evolution depends on the degree of vibration exposure, aggravations by other harmful factors and individual neuromotor and cardiovascular particularities of the exposed worker.

26.8.8 Stages of vibration disease

There are 3 stages in vibration disease:

1. The initial stage,
2. The moderately pronounced stage, and
3. The pronounced stage.

27

Microwave

1. Source and victims to exposure, 2. Effect of interaction on living system.

The term microwave radiation is used to designate that portion of the electromagnetic radiation spectrum which lies between frequencies 300-300000 MHz. When corresponds to wavelengths in air of 1 metre to 1 millimetre (UK), 10-100000 MHz. (USA), whereas according to Amercian St. Inst. UV light ranges from 10-200 MHz or 30-1.5 m. wavelength in the air.

27.1 Source and Victims to Exposure

(A) **Wavelength**: 300 to 30000 MHz.

(a) **Source**: Radar, T.V., telecommunication, telemetry, medicine, microwave ovens, food industry.

(b) **Occupational Exposure**: Broadcasting transmitter and T.V. personnel, microwave testers, microwave diathermy operators, and maintenance workers, medical personnel, from micro-oven.
(Housewives and children are specifically exposed to micro-oven, medicine, T.V., telecommunication)

(B) **Wavelength**: 3000 to 30000 MHz.

(a) **Source:** Altimeters, air born radar, navigation, satellite, communication, microwave point to point.

(b) **Occupational Exposure**: Marine and coast guard personnel, sailors, fishermen, persons professionally present on the board of the ship.

(C) **Wavelength**: 30000 to 300000 MHz.

(a) **Source:** Radio astronomy, radio meteriology, space research, nuclear physics technique, radiospectroscopy.

(b) **Occupational Exposure**: Operators and users.

27.2 Effect of Interaction on Living System

The effects of interaction of microwave on living system are as follows:

I. The penetration and propagation of microwaves within a biological target organ or tissue.

II. The primary interaction of microwaves with living system.

III. The secondary effects induced by primary interacton.

27.2.1 Penetration and propagation of microwaves within a biological target organ or tissues

1. When microwaves arrive at the surface of a biological target, they enter into a medium of different dielectric properties, e.g., skin, fat, muscle, bone, etc., thus affecting the condition of propagation.
2. Reflection and scattering occur at the surface. Because of the different properties of the biological medium as compared to air, the field intensity, wavelengths, direction of propagation and plane of polarization may be affected by such phenomena as reflection, defraction, scattering, absorption, interference and polarization.
3. The biological object consists usually of several layers of complex geometry with various dielectric properties causing alteration in propagation condition at the boundary of each consecutive layer. Thus, the final picture is extremely complex. The water content of different tissues are different and the water content of the same tissue may alter conditions. The velocity of propagation of microwave through a medium depends on its electric and magnetic properties which may be determined by using several parameters. One of such is the dieletric constant. Johnson et al. (1972) observed that the dielectric constant is the highest in high water content tissue with lower frequency microwave. But in tissue with low water content with higher frequency, the dielectric constant is the highest and, as a whole, the power absorbed in tissues with high water content may be many times greater than that with low water content (17.7 : 7.45) due to difference in conductivity. It was also observed by shapira et al. (1971) that in the primate head structure, the pattern of microwave absorption with 3 giga Hz frequency (1 giga Hz = 10^9 Hz. 1 Mega Hz = 10^6 Hz.), has maximum dielectric constant and refraction index and minimum dielectric constant and refraction index in bone and in fat (low water content).

27.2.2 Primary interaction of microwave with living system

Primary interaction of microwave with living system causes:

1. Thermal effect, II. Non-thermal effect

These effects are produced due to:

(a) Absorption of microwave radiation,
(b) Induction of microwave radiation.
(a) Absorption of microwave causes:
1. Relaxation of the molecules,
2. Dimensional excitation of molecules.
(b) Induction process causes:
1. Induced dipoles,
2. Conductor rectification,
3. Ionic currents.

27.2.3 Interaction of microwave with living system at different level of organisation

(A) **Molecular Level**

Primary Effect: Primary heating effect, segmental rotation of biopolymer, conformational changes in biopolymers, excitation of molecules, ionic currents and changes in ionic distribution.

Secondary Effect: Temperature rises and metabolic rate increases due to effect on biochemical reactions through conformational changes, excitation, changes in ion distribution, etc. Structural changes in subcellular elements like chromosomes and mitochondria.

(B) **Subcellular and Cellular**

Primary Effect: Interference with biomembranes or secondary effects through ionic currents and changes in ionic effect, semiconductor effect, changes in bound water.

Secondary Effect: Chromosomal effect, lymphoblastoid transformation, interference with mitosis, genetic effects, carcinogenic or leukemogenic effect.

(C) **Organs and System**

Primary Effect: Focal thermal stimulation and lesion.

Secondary Effect: Cardiovascular effect may be local or generalised, nervous system stimulated by peripheral receptors and disorganisation of functions (focal lesion of stimulation), is same for the endocrine system. Changes in metabolic rate and/or in abnormal metabolites production.

(D) **Highly Organised System**

Primary Effect: Interference with electromagnetic wave transmission and/or reception of biological information.

Secondary Effect: Stress effect (accumulation of chronic microstress and micro-trauma), adaptive responses, disada-ptation, interference with biorhythms and their synchronization.

27.2.4 Secondary effect induced by primary interaction

(A) **Thermal Effect**

1. About 40 per cent of microwave incident energy (may be up to 100 per cent) is absorbed and transformed into heat.
2. Power density above 100/cm square (ms) should be considered as high level of exposure causing death for over-heating in laboratory animals within a relatively short time.
3. In principle, exposure at power density levels below 10 miliwatt/sq. cm do not lead to rise in rectal temperature.

(B) **Nervous System**

1. In rats, repeated irradiation for 75 days (at high power-density 40-100 microwave/sq. cm) causes brain hypermea, pycnosis and vacuolisation of nerve cells. Proliferation of microglial cells was observed. Similar but less pronouned changes were observed in exposures at 10-20 microwave/sq.cm.
2. In EEG patterns desynchronization of dielectric brain function, i.e., basic amplitude decreases and frequency increases.
3. Motor excitation initially but decrease in activity occurs later. At higher power density flaccid paralysis takes place and this paralysis disappears with withdrawal of exposure.
4. Animal experiments show they become emotional and the behaviour is altered.

(C) **Cardiovascular System**

1. High rise in B.P., decreased peripheral resistance.
2. Cardiac stroke volume and rate increased. Sometimes arythmia may occur.
3. Haemodynamic response related to thermoregulatory reaction.

(D) **Endocrine and Metabolic Effects**

1. Hypothalamic region of the brain is affected, discrete changes in cells and synaptic connections are observed and affects the function of the hypophysis.
2. Gonadotrophic hormone in hypophysis increases, ACTH remains same.
3. Decrease in lipid and Vit. C content in adrenal gland.
4. Liver glycogen content altered, pyruvic and lactic acids and muscle glycogen decreased. EMG (Electromyogram) shows prolongation of duration of single multiphasic potentials indicating altered metabolic effect in skeletal muscles, disturbances in carbohydrate metabolism.
5. An increase in calcium level in blood and urine. In severe conditions sodium and potassium levels are also altered.
6. Thyroid activity increases. It should be kept in mind that temperature effects influence the metabolism and lead to displacement in water and mineral equilibrium.
7. O_2 consumption and CO_2 production are increased.
8. Respiratory distress observed.

(E) **Testes, Female Genital System, Pregnancy, etc.**

1. Testicular activity decreased, abnormal maturation of sperm, acid and alkaline phosphatase activity decreased, succinyl acid dehydrogenase and 5-nucleotidase activity decreased.

2. Disturbance in sexual cycle in female, vaginal smear altered.
3. Foetal development—Induced metabolic disturbance would cause death at early stage. At later stage, the effect is less.

(F) **Chromosomal and Genetic Effect**

1. Chromosomal and mitotic aberration could occur. Fragmentation and chromosome tagging occur.
2. Initial depression in DNA synthesis, followed by rise.
3. Down's syndrome (Mongolism) occurs to offspring due to father who is frequently exposed to microwave.

(G) **Cataractogenesis**: Produces cataract. Rise in temperature in the eye, changes in ascorbic acid concentration or content, DNA synthesis and mitosis of lenticular cells cause change in membrane permeability.

(H) **Blood and Blood Forming System**

1. No significant alteration in RBC count, slight hypochromic anaemia.
2. A marked increase in WBC total count after single exposure (or two or more), mainly due to increase in lymphocyte count.
3. Erythrocyte production rate decreases significantly.

(I) **Chest, Abdominal Cavity, G.I. Tract**

1. Morphologic sign of liver injury.
2. Various macro- and microscopic lesions in liver, kidney, spleen, heart muscle, nervous system and adrenal glands are observed.

28

Haemodynamics

PART I: RHEOLOGY

1. Biophysical properties of plasma, 2. Viscosity of blood, 3. Viscous fluid vs. engineering fluid, 4. General concept of viscosity for Newtonian and non-Newtonain fluid, 5. Non-Newtonian fluids—blood plasma, whole blood, 6. Comparison of Newtonian and non-Newtonian fluids.

PART II: DYNAMICS OF PERIPHERAL CIRCULATION

(Dynamics of peripheral circulation pertaining to the patterns of blood flow, pressure and resistance during the flow through the blood vessels.

7. Different blood vessel in the vescular system, 8. Modification of phasic blood flow to continuous flow and pressure in the capillary bed (Windkessel function), 9. Poiseuille's law (Physical factors influencing vascular flow), 10. Difference between Poiseulle's experimental preparation and circulation, 11. Application of Poiseuille's law to the living system, 12. Blood flow—laminar flow, turbulent flow, 13. Measurement of blood flow (flowmeter), 14. Velocity of blood flow, Bernoulli's principle, 15. Comparison between blood velocity and flow, 16. Turbulent flow and Reynold's number, 17. Critical closing pressure (Effect of pressure on vascular resistance), 18. Tension in vascular wall (Laplace law), 19. Vascular resistance, 20. Pressure in the vascular system, 21. Physical laws determining the blood pressure, 22. Arterial blood pressure, 23. Measurement and recording of arterial blood pressure, 24. Effect of hydrostatic pressure on venous and arterial pressure.

HAEMODYNAMICS

The blood circulates in a closed system of branching tubes. Haemodynamics means the study of blood flow and related forces in moving the blood through the circulatory system. It discusses the physical principles of blood flow through the blood vessels with special reference to the interrelationships among pressure, flow and resistance. Under haemodynamics, it is essential to discuss two matters:

I. **Rheology**: The physical properties of blood to understand the flow and pressure characteristics of the circulation. This branch is known as *rheology* which deals, with investigation of the deformation and flow of matter (liquid or gas).

II. **Dynamics of Circulation**: Dynamics of peripheral circulation pertaining to the patterns of pressure and blood flow through the blood vessels.

PART I: Rheology

28.1 Biophysical Properties of Plasma Proteins

(A) **Blood**: 6-8 per cent of body weight, 66-78 ml/kg., 2000-2900 ml/sq. m. specific gravity 1000 average, viscosit 3.6-5.3, pH about 7.4 average, range 7.33-7.51, freezing point about –0.55°C.

(B) **Plasma**: 50-60 per cent by volume, solids 8-9 per cent, plasma proteins 6.5-7.5 per cent, serum proteins 6.0-6.9 per cent, albumin (serum) 1.3-2.5 per cent, A/G ratio 1.2-1.8, specific gravity 1.026, viscosity 1.7-2 (Water = 1), freezing point about –0.55 °C, total molal concentration about 0.30, osmotic pressure at 37°C about 7.6 atm., pH about 7.4 average, range 7.33-7.51, molecular weight, 69000 (albumin) to 400000 (fibrinogen).

I. **Precipitation, Separation**

(A) **Precipitation by Salts**: Different proteins are precipitated from the solution by addition of different concentrations of salts:

1. Albumin by saturation with ammonium sulphate.
2. Globulin by half-saturation with ammonium sulphate.
3. Albumin-globulin mixture with 22.2 per cent of sodium sulphate. Albumin remains in solution.
4. Fibrinogen with one-fifth saturation with ammonium sulphate and half saturation with NaCl.

(B) **Salting Out**

1. Euglobulin fraction by saturation with NaCl, half-saturation with $MgSO_4$ or one-third saturation ammonium sulphate. Insoluble in water.
2. Pseudoglobulin by $MgSO_4$ or half-saturation with ammonium sulphate. Soluble in water.

(C) **Cohn's Fractionation**: Fractionation with low salt concentration at low temperatures varying the pH and ethanol concentraton. Six main fractions have been isolated by this method.

(D) **Electrophoretic Mobility**: Plasma proteins in an electric field move either to the anode or cathode according to the nature of the protein and different surface charges. They migrate at different rates: prealbumin and albumin moves the fastest and gamma globulins the slowest, with the other plasma proteins at intermediate rates. Due to its low isoelectric pH, albumin has 18 negative charges at pH 7.4 and low MW causes rapid electrophoretic mobility. Their concentrations are measured by Densitometric Scan and staining.

(E) **Isoelectric Point**: Proteins can ionize either as acids or as bases owing to the presence of side chain of amino group and carboxyl group of amino acids in the protein molecule. In alkaline solution at pH 7.4, the proteins ionize as acid, and in acid solution they ionize as alkali. In acid solutions, the protein amino groups act as base, taking up H^+, while ionization of the carboxyl groups is suppressed. The protein carries a net positive charge.

At certain pH, which is specific for each protein, the protein carries equal numbers of positive and negative charges, i.e., net charge is zero. This pH value for electrical neutrality is known as the "isoelectric point or pH (pI)" which depends on the relative number and relative dissociation strengths of its acidic and basic groups and at this point protein precipitates. It reveals the following:

1. Characterization of protein,
2. As an aid to their isolation and purification,
3. Minimum solubility,
4. Precipitation of protein (exception chymotrypsinogen),
5. The Gibbs-Donnan effect will be at its minimum,
6. Protein would not move toward the electrode in an electric field either,
7. Protein exists as a dipolar ion or zwitterion carrying equal numbers, of positive and negative charges on its ionizable groups and the net charge is zero.

(F) Immunoelectrophoresis: It is a valuable quantitative method for special puposes which utilizes the pattern formed by precipitation at the site of local antigen-antibody reaction.

(G) Ultracentrifugation (Svederg): Sedimentation of the different plasma proteins at different rates when solution of them are spun at a very high speed in the ultracentrifuge. It separates protein fractions on the basis of differences in density. It is of value in determining the molecular weight of the proteins.

(H) Molecular Weight and Shape: The molecular weight of albumin is 69,000, globulins vary from 90000 to 156000 (β_1-globulin 90000, γ-globulin 156000), fibrinogen 4000000.

Passing through the capillary wall depends not only on the size but also on the shape of the molecules. Albumin passes more readily than the others because of low molecular weight, small size and ellipsoid shape (15 nm × 3.8 nm), made up of a single folded, low α-helical polypeptide chain of about 580 amino acids residue and more compact than other proteins. But all plasma proteins pass through in small amounts and appear in the lymph. If capillary permeability is increased all the proteins pass more readily than normal.

Serum globulins are having lower degree of dispersiveness and an unequal molecular weight which is higher than 10,000 and the molecules are more or less oblong.

Fibrinogen has large ellipsoid molecules (45 nm × 9 nm). Though it has a large fibrous structure, the molecule is kept in solution in plasma by intermolecular repulsions due to strong negative charges at its terminals, contributed by a large number of aspartate and glutamate. These negatively charged terminals of the fibrinogen molecules not only contribute to its water solubility but also repulse the terminals of other fibrinogen molecules, thereby preventing aggregation (Fig. 28.1).

II. **Buffer Action**: Plasma proteins play a significant role among the various buffer systems. According to the situation, they act as proton acceptor. They may also act as acids or as bases and they are called "amphoteric" because the amino acid in the protein molecules contains both basic —NH_2 groups and acidic —COOH groups.

As for example, glycine NH_2CH_2COOH in which carboxyl group —COOH is thc acidic group or proton dónor, and —NH_2

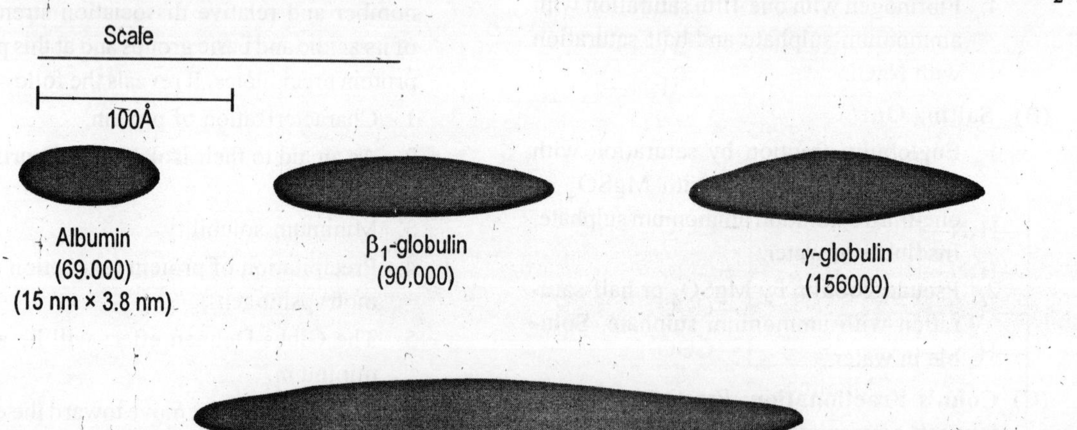

Fig. 28.1: Dimensions of various plasma proteins

group is the basic group or proton acceptor, hence amino acids are called "**hybrid**" or "**Zwitterions**" as a result of transfer of a proton from the carboxyl group to the amino group, thus:

$$NH_2CH_2COOH \rightleftarrows NH_3^+ CH_2COO^-$$

Glycine Zwitterion of glycine

Though one part of the glycine molecule is positive and other part negative, the molecule of amino acid as a whole remains neutral.

If H^+ ions (protons) are added to the amino acid solution they are taken up by the —COO-part of the zwitterion to form an undissociated —COOH group and a positively charged glycine cation, such as:

$$NH_3^+CH_2COO^- + H^+ \rightleftharpoons NH_3^+CH_2COOH$$

If a base such as OH^- is added, the NH_3^+ donates a proton which unites with the hydroxyl group to form water and a negatively charged glycine anion, such as:

$$NH_3^+CH_2COO^- + OH \rightleftharpoons NH_2CH_2COO^- + H_2O$$

The glycine or other amino acids can act as buffer by changing pH when acids or bases are added.

But in the protein molecule, though the amino and carboxyl groups are present in united peptide bond they cannot participate in the buffer system, for which other acidic or basic groups are present and free in the molecule, e.g., NH_2 groups of lysine, the guanidine group of arginine, the imidazole group of histidine, free carboxyl groups of the dicarboxylic acids, phenolic hydroxyl groups of tyrosine and the sulphydryl group of cystein which permit the protein to act as zwitterion and to act as buffer system, such as:

Sodium proteinate + $H^+ \rightarrow$ Hydrogen proteinate + Na^+

Hydrogen proteinate + OH^- + $N^+ \rightarrow$ Sodium proteinate + HOH.

It can be represented graphically (Fig. 28.2).

III. **Osmotic Pressure or Oncotic Pressure:** Osmotic pressure or oncotic pressure of the blood is maintained by the three proteins of the plasma. Of the total osmotic pressure of

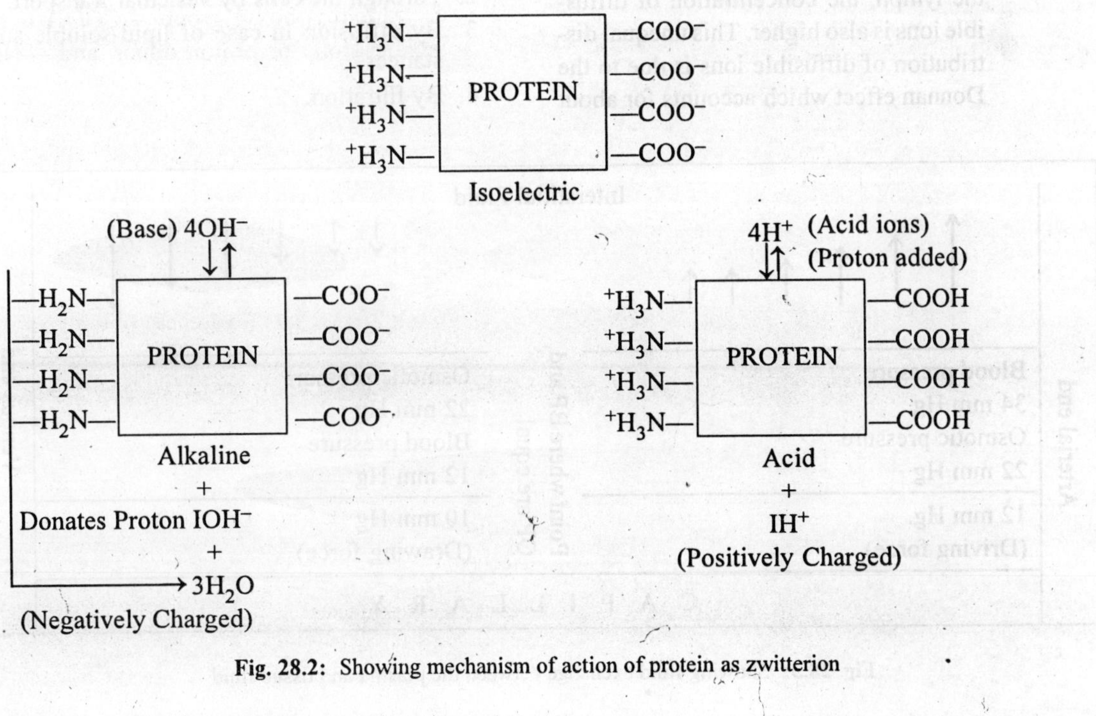

Fig. 28.2: Showing mechanism of action of protein as zwitterion

7.6 atmosphere, about 1/3 of an atmosphere, i.e., 25-30 mm Hg, is due to the presence of colloids. The pressure which each fraction of plasma proteins exerts is inversely related to the size of its molecule and directly related to its concentration in the plasma. 80 per cent of the osmotic pressure depends on the albumin fraction because of:

1. Least molecular weight (68000).
2. Higher concentration.
3. Attraction exerted by its negatively charged hydrophilic groups on the neighbouring water molecules.
4. Lowest isoelectric pH.
5. Highest fluid holding capacity: It has been found that each gram of albumin holds about 18 ml of fluid in the blood through its osmotic effect.
6. It binds more cations at the pH 7.4 of plasma than the other plasma proteins. Those bound ions are restràined from passing through the capillary walls by the non-diffusible negative protein ions (Donnan effect).
7. Since the concentration of non-diffusible protein ions in plasma is higher than in the lymph, the concentration of diffusible ions is also higher. This unequal distribution of diffusible ions is due to the Donnan effect which accounts for about 20 per cent osmotic pressure of plasma proteins.
8. The number of molecules of solutes, albumin/globulin ratio (A/G) which is 1.2-1.8 is one of the important factors to maintain osmotic pressure. The globulins are present in lower concentration and MW varies 150000 to 1000000 and contributes about 20 per cent of osmotic pressure.

Fibrinogen concentration is very low and MW is about 400000 and contribute very little osmotic pressure.

28.1.1 Equilibration of plasma with tissue fluid

The tissue fluid is in equilibrium with the blood plasma through the capillary walls and the chemical composition of the two is almost the same except that the protein content of plasma is much higher. There is continuous exchange of water and chemicals between tissue fluid and blood plasma and the tissue fluid is completely renewed every four or five minutes.

Substance from plasma pass through the capillary walls by the following ways:

1. Passing through the junctions of the endothelial cells.
2. Through the cells by vesicular transport.
3. By diffusion in case of lipid-soluble substances.
4. By filtration.

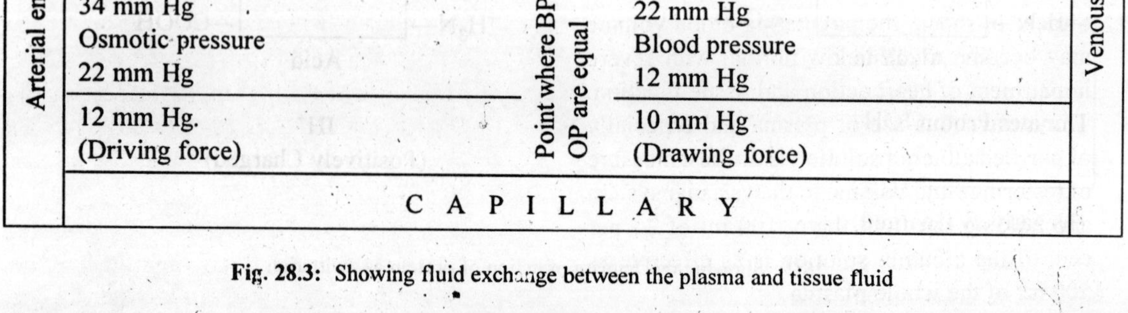

Fig. 28.3: Showing fluid exchange between the plasma and tissue fluid

The rate of filtration in a capillary depends upon a balance of forces at that point which is sometimes called **Starling forces**, after the name of Starling who described the phenomenon.

According to "Starling Hypotheses":

1. "Hydrostatic pressure gradient" (that hydrostatic pressure in the capillary minus the hydrostatic pressure of the interstitial fluid) is one of the forces.
2. "Osmotic pressure gradient" across the capillary wall (colloid osmotic pressure of plasma minus colloid osmotic pressure of interestitial fluid) is the other force. The osmotic pressure of plasma exceeds the osmotic pressure of lymph due to higher concentration of plasma protein than the lymph.

This difference of pressure is about 22 mm Hg and known as **effective osmotic pressure or oncotic pressure** of plasma. The distribution of fluid between the plasma and tissue is regulated by the balance between the blood pressure (34 mm Hg) tending to force fluid out of the capillary into the tissue spaces and the effective protein osmotic pressure of plasma (22 mm Hg) tending to draw fluid from the tissue spaces into the plasma. Therefore, the driving force is 12 mm Hg (34-22 mm Hg) which forces the fluid from the plasma to the tissues.

As the blood flows along the capillary, at certain point the blood pressure and protein osmotic pressure are equal and fluid exchange takes place. But at the venous end of the capillary, the blood pressure falls to 12 mm Hg and plasma protein osmotic pressure exceeds (22 mm Hg) the blood pressure by 10 mm Hg (22-12 mm Hg) of protein osmotic pressure which draws fluid from the tissue spaces into the plsama (Fig. 28.3).

28.1.2 Disturbance of fluid balance

1. **Shock**: In shock, the circulating blood volume may become much below normal with severe impairment of heart action and tissue function. The intravenous use of plasma and especially of purified albumin solution is the best measure of restoring fluid volume to the vascular system and holding the fluid there. 100 ml of 25 per cent of the albumin solution is as effective as 500 ml of the whole plasma.
2. **Oedema:** When plasma osmotic pressure decreases due to less plasma protein concentration and there is alteration of A/G ratio in the plasma, there is accumulation of excess fluid in the tissue spaces producing oedema. Such reduction of protein osmotic pressure increases the pressure to force fluid to the tissue and decreases the osmotic pressure for drawing fluid from the tissues.
3. **Nephrosis**: In nephrosis there is leakage of plasma protein through the kidneys, especially the albumin in higher concentration due to least MW which leads to severe oedema. It usually appears in man when the plasma albumin falls below 2.5 per cent of the total protein below 5.5 per cent and plasma specific gravity below 1.023.
4. **Malnutrition**: Protein deficiency diseases of protein synthesis as in hepatic diseases may also lead to oedema.
5. **Burns and Toxins**: In these conditions, capillary permeability increases and allows leakage of plasma proteins in the tissues producing severe oedema.
6. **Right Heart Failure and Obstruction of Large Veins**: In these cases, the flow of blood is obstructed from the venous end of the capillary and the blood pressure within the capillary rises. This increases the flow of fluid from the capillary to the tissues and decreases the return of fluid to the plasma producing oedema.

IV. **Viscosity**: The plasma proteins contribute to the viscosity of blood which is a helping factor to offer resistance to blood flow in the vascular system for maintenance of blood pressure and efficient cardiac activity (For details, see Physical Properties of Blood—Blood Viscosity).

V. **Suspension Stability of the Blood (E.S.R.)**: The physical changes in the blood may induce the rapid rate of red cell settling in the blood. Four factors are responsible for this process:

1. **Clumping of Red Cells**: The globulin and fibrinogen fractions influence the tendency of the red cells to adhere to one

another and form aggregation of the cells called rouleaux or clumps which cause rapid sedimentation.

2. **Specific Gravity of the Plasma**: Normal red cells sediment more rapidly in plasma with a low specific gravity. Corpuscles of high specific gravity sediment quickly in normal plasma.
3. **Lowered Viscosity of the Plasma**: This is another factor which may cause an increase in the rate of sedimentation.
4. **Size of the Red Cells**: Increase in size of the red cells increases their mass disproportionately to their surface and consequently accelerates the rate of sedimentation.

VI. **Fibrinogen and Clotting**: Blood present in sol form converts to gel form during clotting, and fibrinogen plays a great role in this process by conversion into fibrin by the following ways:

1. Alteration of structure from globular into fibrous and
2. Polymerization.

The soluble fibrinogen converts into fibrin by thrombin. Fibrin separates as long fibres or threads in a three-dimensional network that holds the blood cells. Fibrin threads consist of needle-like micelles packed lengthwise into bundles. These threads, when fresh, are very much adhesive and adhere to each other, blood cells, tissues and foreign substances. This adhesiveness of the fibrin threads causes the clot to hold together and stick firmly to the injured tissues to stop haemorrhage. The steps are given in (Fig. 28.3A).

- **Mechanism**

1. The platelets attach themselves to the fibrin threads which then contract and, drawing the fibrin compactly together, squeeze out the serum.
2. Contraction of contractile protein *thrombosthenin* produced by the fused platelets.

28.2 Viscosity of Blood

When a person moves through a pool of water, he experiences a resistance of his motion. This shows that there is frictional force in the liquids which is due to viscosity of the liquid. If the frictional force is low, as in water, the viscosity of the liquid is low, if the frictional force is high, as in glue or glycerine, the viscosity of the liquid is high.

28.2.1 Definition

Viscosity is the internal resistance against the free flow of a liquid (or gas) due to the frictional forces between the fluid layers moving over one another at different velocities.

Newton defined it as the ratio of the sheer stress (force resisting or pushing flow) to the sheer strain rate of a moving fluid.

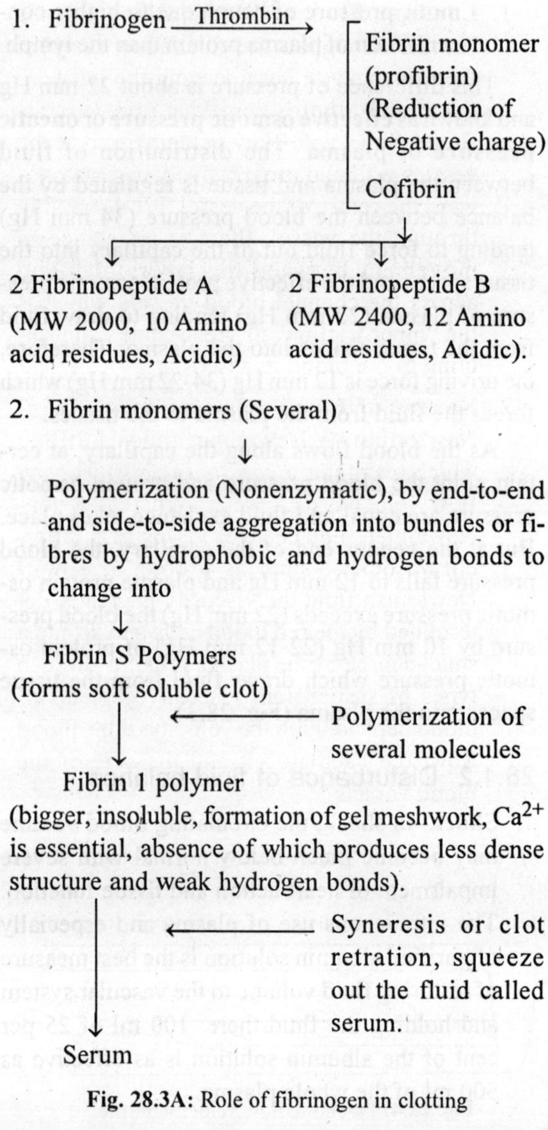

Fig. 28.3A: Role of fibrinogen in clotting

Blood is a viscous fluid having cells and plasma in a ratio of 45 : 55 and the percentage of blood cells, i.e., 45 per cent, is called haematocrit. More than 99 per cent of the cells are red blood cells which indicates that white blood cells do not have any role in determining the physical characteristics of the blood. The normal count is 5 million/mm^3 and 40 per cent of the volume of blood. The red blood cells are suspended on the viscous fluid plasma and their characteristics have important influence on the viscosity.

28.2.2 Rheological behaviour of blood (Non-Newtonian fluid)

1. **Laminar Flow of Blood**: Blood flow is laminar throughout the circulation except at the heart valves, aorta leaving the left ventricle and atherosclerotic blood vessels. Blood flows with flow velocity zero at the vessles wall and progressively increases toward the centre of the vessel. While moving through the vessel, there is resistance to movement between the layers of blood across the channel blood vessels which is the "inner friction" or "resistance" contributing to the blood its viscosity.
2. **Nature of Fluid—Newtonian or Non-Newtonian**: A Newtonian fluid is a fluid which maintains a *constant viscosity* at any flow velocity, e.g., water. Whereas a non-Newtonian fluid changes viscosity at different velocity—such as blood. In a non-Newtonian fluid the viscosity is greatly determined by the red blood cells and also to a small degree by the plasma proteins. In such types of suspension, viscosity changes of blood depend upon the velocity of the blood flow, the haematocrit and the size of the blood vessels.

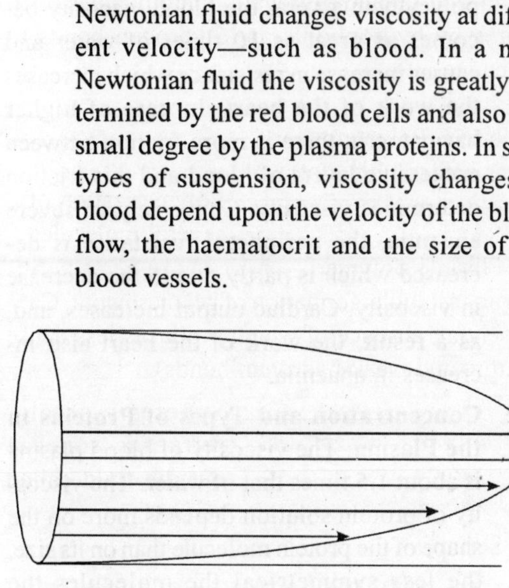

Fig. 28.4: Parabolic flow velocity profile

3. **Parabolic Flow Velocity Profile**: A Newtonian fluid flowing through a *parabolic velocity profile* across the channel. The fastest velocity occurs in the centre and a zero velocity at the boundary wall. Whereas, in blood showing laminar flow in arteries and veins, the profile is not parabolic. But in case of wider vessels having diameter more than the red blood cells, blood acts as Newtonian fluid and the prófile resembles parabolic (Fig. 28.4).
4. **Nonlinear with Alteration in Flow Velocity**: The plasma behaves like Newtonian fluid but due to the presence of suspended red cells behaves like non-Newtonian and alteration in blood viscosity are non-linear with the changes in flow velocity. When flow velocity decreases, viscosity increases and vice versa. The shapes of this viscosity-velocity relationship depend on the behaviour of the red blood cells. During decrease in the flow velocity, the red cells form "rouleaux" (increases aggregate of stacked red cells helped by fibrinogen and globulin), which increases viscosity. Whereas, during increase in velocities when shear rates over 1, viscosity decreases due to deformation, i.e., elongation of the red cells with improvement of flow characteristics. This progressive decrease in viscosity can be checked by making red cells nondeformable. In case of sickle-cells anaemia, the red blood cells are less deformable and cause increase in blood viscosity. For most pure liquids (e.g., water), the relation between applied force and rate of flow is linear. Such liquids are said to be Newtonian. On the other hand, when materials are suspended in the liquid, particularly if these materials have structural continuity, the flow is not linearly proportional to force and the material is non-Newtonian.
5. **Resistance to Blood Flow**: The viscosity of blood is important in relation to the resistance offered to the heart in circulating the blood. The heart muscle functions best when working against a certain resistance.

According to the Law of Ohm (1=E/R); as I is represented by Q and E by (P_1–P_2), the resistance R is represented by $R = 8L\eta / \pi r^4$. The resistance (R) against blood flow is directly proportional to the blood viscosity (η) and the vessel length (L) but is inversely proportional to the fourth power of the vessel radius (r). It is the fourth power of the radius of the tube which influence the resistance. The longer or the higher the viscosity or the smaller the radius, the higher is the resistance to flow. So, a rise in blood viscosity increases this resistance.

Poiseuille showed the results of flow of fluid based on Newtonian fluid using rigid tube under a steady head of pressure. This is known as Poiseuille's Law and the mathematical interpretation is known as Poiseuille-Hagen Equation $Q = \Delta p \pi r^4 / 8\eta L$, where Q = flow in a narrow rigid tube.

Δp = pressure difference between the two ends of the tube.

But in the actual situation in life, it does not hold properly because the heart beats rhythmically, ejecting blood physically into a system of elastic tubes [blood is not a pure fluid (i.e. Newtonian) but a 2-phase system of liquid and cells (i.e., non-Newtonian]. Therefore, the behaviour of the circulatory system sometimes deviates markedly from this principle. But it indicates the mechanism of maintenance of unaltered blood flow in spite of a rise in blood viscosity by increasing the work of heart or the vessel radius which should be increased by vasodilatation.

6. **Turbulence in Blood Flow and Viscosity**: Blood viscosity helps streamline or laminar blood flow by reducing the turbulence. Reynolds number (Re) indicates the occurrence of turbulence. So, if there is fall in blood viscosity (hypoproteinaemia, anaemia) it may increase the probability of turbulence.

28.2.3 Factors influencing anomalous viscosity of blood

In vitro, blood has a relative viscosity (the term relative viscosity being the viscosity of the fluid is expressed relative to water. Water has a viscosity of 0.695 cP at 37°C, so plasma which has a viscosity of 1.2 cP at 37°C has a relative viscosity of 1.7) of 4-5 when measured at moderate or high shear rates. When the in vitro measurements are made at very low shear rates the viscosity is greatly increased. These facts reveal that blood in vitro shows anomalous viscosity, i.e., it does not manifest the properties of a Newtonian fluid whose viscosity is independent of shear rate.

1. **Haematocrit**: Blood viscosity depends on the haematocrit, i.e., the percentage volume of blood is occupied by red blood cells. The relative viscosity of blood increases when the value increases above 45 per cent in vitro. But in vivo, in large vessels, increase in haematocrit causes significant increase in viscosity whereas in smaller vessels having a diameter of less than 100 μm (arterioles, capillaries and venules) the viscosity change per unit change in haematocrit is much less than the wide-bore vessels. This is because the RBC is very deformable and can squeeze themselves and pass through the smaller vessels. So the net change in viscosity per unit change in haematocrit is significantly less in the body than it is in vitro. This is the reason for haematocrit changes that have little effect on the peripheral resistance except when the changes are large. But when there is an enormous increase in the number of RBC (haematocrit above 50 per cent), e.g. polycythemia vera, the blood viscosity becomes as great as 10 times of water and causes increase in resistance which increases the work of the heart. In case of higher haematocrit, there is more friction between successive layers of blood and this friction determines viscosity. Conversely, in severe anaemia, the peripheral resistance is decreased which is partly due to the decrease in viscosity. Cardiac output increases, and, as a result, the work of the heart also increases in anaemia.

2. **Concentration and Types of Proteins in the Plasma**: The viscosity of blood plasma is about 1.5 times that of water. The viscosity of protein solution depends more on the shape of the protein molecule than on its size, the less symmetrical the molecules the

greater is its viscosity. Normal range of plasma viscosity is 15 to 20 millipoises at 20°C and is mainly due to large elongated plasma protein molecule. Albumin with smaller and more globular molecules contributes less viscosity than globulins and fibrinogens which have large and more ellipsoid molecules. Clinically significant increases in viscosity are observed in diseases in which plasma proteins—such as the immunoglobulins—are markedly elevated.

The viscosity of whole blood is only that of twice concentrated plasma which indicates that the red cells and the plasma contribute equally to the total viscosity of the blood.

3. **Shear Rate**: The viscosity of blood rises at low shear rate. When the shear rate is low, the tendency of the red cells to occupy the central axis of the stream is minimal and increase in collisions between the suspended particles influence increase in internal friction and viscosity. Blood behaves as non-Newtonian fluid when the shear rate is high, the red cells occupy the central axis of the vessel, moving with their long axes parallel to the direction of flow where the flow rate is the highest, levaing cell-free zone of plasma at the periphery of the vessel. The phenomenon is known as "*plasma skimming*" and hence the friction between cells and plasma is the lowest. In this stage, blood behaves as a Newtonian fluid with a constant viscosity.

   The blood along the side of the tube is usually directed to the branch vessel leaving the large vessel at right angles, takes disproportionate amount of red cells in the blood and has a low haematocrit. Plasma skimming is not so significant as is the case of branches from large vessels but may be of appreciable significance in case of branches from smaller vessels.

   Plasma skimming may be the cause of lower haematocrit in capillary blood than the whole body haematocrit.

4. **Fahraeus-Linqvist Effect: Tube Diameter and Blood Viscosity**: Fahraeus and Linqvist (1981) used different tubes having 575 and 40 μm diameters for measurement of apparent viscosity of human blood under a constant pressure head of 100 mm Hg. The length of the tube was proportional to its diameter. It was shown that the relative viscosity of blood decreased when the tube diameter was less than 300 μm. It has been observed by them that the blood flow in minute blood vessels shows less viscous effect than in the large vessels. This effect is known as *Fahraeus-Linqvist effect*. Its appearance is seen when the diameter of the vessels is less than 1.5 mm. In the capillaries, this effect is very much prominent and the viscosity of blood is half than in large vessels. When water, serum or plasma is used, there is no change in its relative viscosity when flowing through tubes of different sizes and it revealed that this effect is certainly due to the presence of RBC. The cause is probably due to the red cell that does not move at random but moves through the vessels in rows as a single plug. Thus they avoid viscous resistance which occurs within the blood itself.

5. **Velocity of Flow**: When the velocity of blood flow decreases, the viscosity of blood increases. Since the velocity of flow in small vessels is less than 1 mm per second, blood viscosity increases 10-fold. This effect is due to the formation of "rouleaux" and larger aggregates.

6. **Constriction of Small Blood Vessels**: Cells aggregate at constriction in the small blood vessels like capillaries where the nuclei of the endothelial cells protrude into the lumen resulting in blockage of blood flow to a short or longer periods of time causing an increase in viscosity.

7. **Temperature**: The superficial vessels—as well as the deeper vessels—show certain effects due to the alteration of temperature. Thus, in cooling the hand in ice water, there is three-fold increase in the regional viscosity of blood. But the capillaries are not influenced by the temperature and blood viscosity is unaltered, i.e., similar to plasma. And the blood behaves as a Newtonian fluid.

28.3 Viscous Fluids in Biological System and Comparison with Engineering Fluid

Table 28.1: Comparison of engineering and biological fluid

Engineering Fluid	Biological Fluid
1. Flow of liquid through tubes requires the action of a force. This force is to: (a) Accelerate the fluid from rest to its flow speed, (b) To overcome the viscous friction of the fluid.	1. Flow is slow, inertia effect is less important than the viscous effect.
2. These are handled at high speed to make the flow turbulent.	2. Being viscous, has a low speed of flow in the body and not turbulent except blood flow through aorta.
3. These are Newtonian fluids having viscosity which is constant with respect to the flow of speed.	3. Non-Newtonian fluid and it is not a simple fluid. It changes viscosity at different velocities.
4. In most pure liquids, the relation between applied force and rate of flow is linear.	4. Materials are suspended, the flow is not linearly porportional to force.
5. Example—water, glycerine.	5. Example—Plasma, blood.

28.4 General Concept of Viscosity for Newtonian and Non-Newtonian Fluid

1. **Velocity Gradient and Frictional Force:** If a fluid flows across a plane stationary surface, the fluid layers nearest to the surface will be at rest and the layers away from the surface moves faster, so that a velocity gradient exists in the fluid and is expressed as:

$$\frac{dv}{dx}$$

where v = local speed parallel to the wall,

x = distance from the wall.

But the fluid flow is slowed down by a force which is proportional to the velocity gradient $\frac{dv}{dx}$, known as *frictional force,* F, given by

$$F = \eta\frac{dv}{dx}, \text{ when } \eta = \text{viscosity}$$

which is for Newtonian fluids and is constant with flow conditions but varies slowly with fluid speed, temperature and detailed shape of the body of which the wall is a part, whereas, non-Newtonian fluids show change in viscosity at different velocities and depend on the velocity of blood flow, the haematocrit, size of the blood vessels.

2. **Turbulence and Reynolds Number:** When the rate of flow is low, the fluid flows in an orderly fashion and there are no vortices or turbulence. But when the rate of flow increases, the smooth laminar flow begins to change in character, i.e. it becomes turbulent and the friction force tends to increase. The condition at which it occurs is known, i.e., the ratio of the intertial forces to viscous forces at which this alteration occurs is known as *Reynolds number* (Re).

$$Re = \frac{\rho v l}{\eta}$$

where ρ = fluid density,

v = local speed,

l = length of the solid experiencing the drag of the flowing fluid.

Reynolds number greater than about 1000 is called "*critical Reynolds number*". Most homogenous fluids, i.e., Newtonian fluids, are having Reynolds number about 1000. For blood (non-Newtonian) the apparent critical Reynolds number averaged 970 ± 80. This value is independent of tube size, RBC concentration and temprature. When turbulence occurs (Reynolds number is exceeded) a part of the driving energy of the system goes into motion in other than the forward direction. It is important to measure

Reynolds number to determine whether turbulence plays a significant role in the vascular system. When the Reynolds number is high, flow tends to become more turbulent in its movement down the tube; when the Reynolds number is low, turbulence tends to damp itself out.

Whereas, in case of blood (non-Newtonian), it being a viscous fluid has a low speed of flow in the body and not turbulent except blood flow through aorta.

3. **Steady Viscous Flow in a Tube (Poiseulle's Law):** The Newtonian fluid obeys Poiseulle's Law which states that when a volume of Newtonian fluid flows through a rigid tube in a given period of time, i.e., per unit of time, it is directly proportional to the pressure head (P) between the two ends of the tube, the fourth power of the radius (r^4) and is inversely proportional to the tube length (L).

$$Q = K\frac{Pr^4}{L},$$

where K is constant for the liquid concerned at any one temperature.

Whereas non-Newtonian fluid, e.g., blood does not obey this law because the tube, i.e., blood vessel, is distensible or elastic, the fluid contains suspended materials, pressure is non-steady and flow-pressure relation is non-linear.

4. **Flow-Pressure Curve Characteristic for Various Deformable Substance:** Variation of slope of the pressure-flow curve occurs for one or more reasons.

(a) **Shape of Suspended Particle:** When a liquid consists of small particles having different shapes (spherical, ellipsoid, disc-like, etc.) which break up the neat laminar arrangement and the resistance causes the variation of the slope of the pressure-flow curve, the fluid exhibiting such variations is called *"generalized Newtonian liquid"*.

(b) **Alteration in the Consistency of Fluid Layers:** Variation of the pressure-flow curve occurs during alterations in the consistency of fluid layers immediately adjacent to the wall of the tube. This factor causes variation in the resistance to flow with varying velocity and its effect is greater in smaller tubes than in larger ones. Then the slope of the pressure-flow curve is not proportional to the minus fourth power of the radius of the tube.

(c) **Liquid having both Plastic and Viscous Materials:** Variation also occurs when a liquid consists of both plastic and viscous materials.

The truly plastic material forms *St. Venant body* which shows that when there is increase in stress applied to the body, no flow occurs until some minimum stress, i.e., the *yield value*, is reached. Beyond the yield value, a very small increment of stress produces great increments of flow.

But in liquids having combined properties of plastic and viscous materials, there is no flow up to a yield value as they are plastic but, beyond this yield value, flow varies as the applied stress varies. This combined body is known as "*Bingham body*". In case of non-Newtonian liquid (e.g., blood), there is no substantial report on pressure-flow effect in vivo except on artificial system done by Bayliss (1952) with rigid tubes. It indicates that the flow pressure relation is more or less linear over a considerable pressure range. At high pressures, i.e., high rates of shear, the apparent viscosity increases gradually which may be due to the occurrence of turbulence in the flow. From this flow-pressure curve, it is difficult to interpret exactly what happens during low shear rate. Pappenheimer (1949) showed by in vitro experiments that, in these low shear rates, the blood shows nonlinearity when the concentration of RBC is in the order of magnitude found in actual blood.

5. **Einstein Equation:** The presence of suspended particles in a liquid causes a change in the apparent viscosity as expressed by the Einstein equation:

$$\eta_s = \eta(1+K.\phi),$$

where η_s = viscosity of the suspension,
η = viscosity of the suspending medium,
K = proportionality constant,
ϕ = fraction of the total volume occupied by the particles.

It was shown that when the particles are spheres, the value of K is 2.5, if these are thin rods or discs, the value of K is 2.00 or 2.061, respectively. The Einstein equation was expressed based on suspension containing less than 3 per cent of the volume in suspended particles. This equation does not hold for the viscosity of liquids containing 40-60 per cent of the suspended particles of non-Newtonian fluid like blood.

28.4.1 Newton's equation and velocity profile

The fundamental equation expressed by Newton which allowed the definition of coefficient of viscosity is:

$$F = A\eta \times \frac{\Delta V}{\Delta X},$$

where, F = tangential force as "drag" or resistance exerted between the fluid laminae (the tendency of a moving layer travelling at a faster velocity to pull a slower layer along),
A = area between the two layers,
η = coefficient of viscosity,
$\Delta V / \Delta X$ = the velocity gradient between the two layers, i.e., the increase in velocity ($cm\ s^{-1}$) for each centimetre distance at right angles to the flow movement.

Then, rearranging the equation,

$$\eta = \frac{F/A}{\Delta V/\Delta X} = \frac{\text{shear stress}}{\text{shear strain rate}}$$

In case of Newtonian fluid flowing through cylindrical tube having laminar flow, application of this equation for each multiple layer yields a "parabolic velocity profile" across the channel with the highest velocity in the centre and a velocity to be 0 at the boundary wall.

Whereas, in non-Newtonian fluid, i.e., blood, having laminar flow in arteries and veins, the velocity profile is not parabolic though, in large vessels having a diameter much greater than those of RBC, blood behaves as a Newtonian fluid and the profile is parabolic.

28.4.2 Wall-effects

When a fluid containing suspended materials (e.g., rubber spheroids in glycerine) flows through a tube, the suspended materials migrate toward the axis flow leaving behind a particle-free layer adjacent to the tube wall. This causes disruption in the continuity of the velocity gradient producing changes in the pressure-flow relations. This wall effect may be due to deformation of the suspended material and more pronounced when they are plastic or may be due to lubrication effect when they are nearly solid.

In case of blood (non-Newtonian) when the shear rate is high, the red cells occupy the central axis of the vessel where the flow rate is high leaving cell-free zone of plasma at the periphery of the vessel. In this stage, blood behaves as a Newtonian fluid with a constant viscosity.

28.5 Non-Newtonian Fluids

I. **Blood Plasma**: Plasma itself is a water solution of salts and heavily hydrated macromolecules, i.e., plasma proteins. Plasma viscosity is 1.5 times that of water and in unit 15 to 20 m poises at 20°C. Plasma behaves like Newtonian fluid but because of the suspended RBC which changes the viscosity and non-linear with alteration in flow velocity. Viscosity of plasma depends on the shape of the protein molecules than on the size, the less symmetrical the molecule the greater is its viscosity. It mainly depends on the large elongated plasma protein molecules. Albumin with smaller and more globular molecules contributes less viscosity than globulins and fibrinogen which have larger and more ellipsoid molecules. For the above facts, the solutions of 250 gl^{-1} albumin, 150 gl^{-1} r-globulin, 20 gl^{-1} fibrinogen have equal viscosities. Each of the above solutions has a viscosity equal to that of twice the concentration of plasma, i.e.,

50 g of "mixed" proteins per litre. The viscosity of whole blood is that of twice concentrated plasma, this indicates that the RBC and the plasma contribute equally to the viscosity of the blood.

Rise in plasma proteins concentration increases viscosity (e.g., immunoglobulins). Plasma viscosity contributes to haemodynamics by:

(a) Offering a resistance against blood fluid,
(b) Reducing the tendency of turbulence inside vessels.

II. **Whole Blood**: Whole blood is several times viscous as water. The viscosity of whole blood is 5. In healthy people the viscosity of the blood is between 4 and 5. In unit, the viscosity is 30-40 m poises at 20°C. The viscosity of blood depends on (1) the emulsoid colloidal systems present in the plasma, i.e., mainly plasma proteins, and (2) to the great proportion of the suspended particles, i.e., concentration of RBC or haematocrit.

When there is abnormal increase in number of RBC as in polycythemia, if one stays in high altitude for some time, e.g., in mountain, shock after extensive burns and poisoning with carbon monoxide, viscosity increases and may reach 8. In diseases such as hereditary spherocytosis—in which the RBCs are abnormally rigid—viscosity increases. Whereas, in case of anaemia—when haematocrit is low-viscosity decreases and may drop below 4.

It has been observed that if water or saline is allowed to pass through the tube, the viscosity increases progressively with introduction of plasma and then RBC. Thus increased viscosity and the fluidity (q) for plasma was about half (85) than for water $\left(144\frac{\text{cm}}{\text{sec}}\Big/\frac{\text{dyne}}{\text{cm}}\right)$ and increasing the concentration of RBC further increased the viscosity and reduced the fluidity still further.

Blood viscosity also increases at low velocity in capillaries which may be due to: (1) rouleaux formation and large aggregates, and (2) random orientation of the elongated plasma protein molecules. The increase in viscosity due to high haematocrit and reduced velocities in capillaries is partly compensated by a fall in viscosity due to Fahraeus-Lindqist effect while flowing through narrow vessels diameter below 150 µm. This effect may be due to alignment of the RBC when they pass through the vessels. Viscosity of blood also increases due to constriction in small blood vessels because RBC's become stuck at the constriction region. Blood viscosity also increases in cold.

Most real liquids are at least slightly "non-Newtonian" and the most complex example of this behaviour is whole blood. Yet, to a first approximation within the physiological range of operation, both plasma and

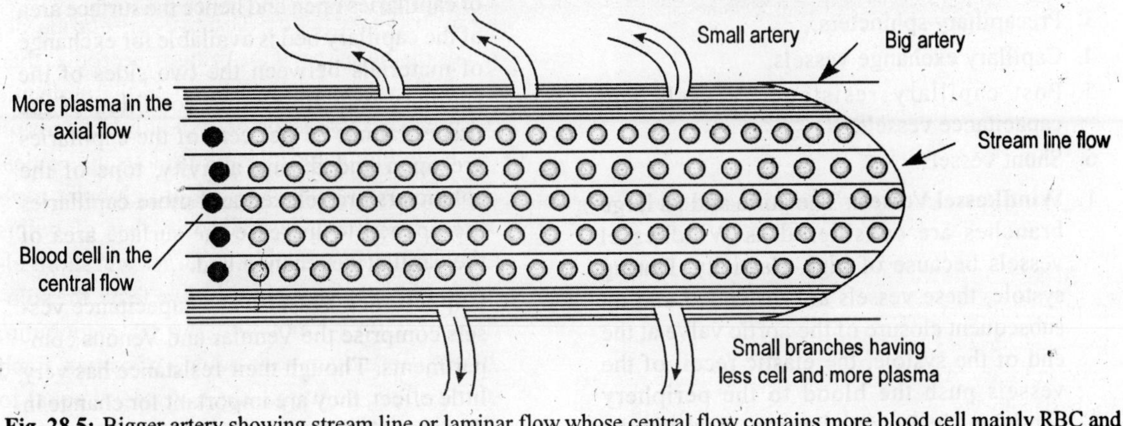

Fig. 28.5: Bigger artery showing stream line or laminar flow whose central flow contains more blood cell mainly RBC and the axial flow and side branches having more plasma and less RBC

whole blood are essentially Newtonian, i.e., their curves are linear, Poisecuille's law of laminar flow is obeyed.

The blood viscosity is thus faster in maintaining the peripheral resistance and thereby the arterial pressure.

PLASMA SKIMMING

After studying perfusion pressure and flow measurement have shown during normal blood flow that the blood viscosity is lower in vivo study than the in vitro using rotational viscometers. The cause is due to variation of ratio between plasma and blood cells in different blood vessels varying in diameter in the different region of the body. RBC are accustomed to deform themselves to hydrodynamic forces and tend to move from axial region to central region producing plasma-rich zone in the axial zone resulting low haematocrits value. The side branches of these blood vessels also get this plasma rich axial blood flow giving low haematocrits value. This phenomena is known as *plasma skimming* (Fig. 28.5).

PART II: Dynamics of Peripheral Circulation

28.6 Different Blood Vessels in the Vascular System

The vascular system may be considered consisting of the following:

1. Windkessel vessels,
2. Precapillary resistance vessels,
3. Precapillary sphincters,
4. Capillary exchange vessels,
5. Post capillary resistance vessels and capacitance vessels,
6. Shunt vessels.

1. **Windkessel Vessels**: The aorta and its large branches are considered as Windkessel vessels because of high elasticity. During systole, these vessels are distended and on subsequent closure of the aortic valve at the end of the systole, the elastic recoil of the vessels push the blood to the periphery continuously. Potential energy stored in the elastic tissue of these vessels during heart contraction converts into kinetic energy for the blood flow during diastole. (for further discussion, see Windkessel effect and Windkessel vessels).
2. **Precapillary Resistance Vessels**: The precapillary resistance vessels are the arterioles which offer greater resistance to the blood flow. Changes in radius of the arterioles determine the blood flow to the region. The radius is controlled by local myogenic tone of the vessel itself and is further activated by the extrinsic neural control.
3. **Precapillary Sphincters**: These are the part of the precapillary resistance vessels. Increase in potency of the precapillary sphincters causes an increase in the number of capillaries open. The sphincters are important in determining size of the capillary exchange area which is perfused in the tissues at any one time. The radius of the sphincters is regulated by the local metabolites concentration and the neural factors.
4. **The Capillary Exchange Vessels**: The capillary exchange vessels are tube-like structures having single layer of epithelium. The capillaries are the important part of the cardiovascular system because the solutes pass to the tissues through the capillary walls. They are not regulated by either metabolites or neural factor. Changes of tone of precapillary sphincter determine the number of capillaries open and hence the surface area of the capillary bed is available for exchange of materials between the two sides of the capillary wall. In resting condition of the tissues, about 25 per cent of the capillaries are open and during activity, tone of the sphincters are relaxed and more capillaries are opened to increase the surface area of the capillary exchange bed.
5. **Capacitance Vessels**: The capacitance vessels comprise the Venular and Venous compartments. Though their resistance has very little effect, they are important for change in the capacity. The capacity adjustment in

28.7 Comparison of Newtonian and Non-Newtonian Fluids

Table 28.2: Comparison of Newtonian and Non-Newtonian fluids

Newtonian Fluid	Non-Newtonian Fluid
1. **Action of Force** Flow of fluid through tube requires the action of a force in the nature of pressure gradient. This force is: (a) To accelerate the fluid from rest to its flow speed, (b) To overcome the viscous friction of the fluid.	1. Flow is slow, inertia effect is less important than the viscous effect.
2. **Action of High Speed** High speed is required to make it turbulent.	2. Being viscous, has a low speed of flow in the body and not turbulent except blood flow through aorta.
3. **Constant Viscosity** Newtonian—having viscosity which is constant with respect to flow of speed.	3. Non-Newtonian fluid—as it is not simple fluid, the viscosity is not constant and changes depend on different velocities of blood flow, haematocrit, size of the blood vessel, etc.
4. **Viscosity and Rate of Shear** The viscosity is independent of the rate of shear.	4. The viscosity is not independent of the rate of shear.
5. **Reynolds Number** About 1000	5. Apparent critical Reynolds number average 970 ± 80. This value is independent of tube size, RBC concentration and temperature.
6. **Poiseuille's Law** It obeys the law because the fluid is simple, does not contain suspended materials and flowing in a rigid tube.	6. It does not obey the law because the blood vessel is distensible or elastic, the fluid contains suspended materials, pressure is non-steady.
7. **Flow-Pressure Curve** Flow-pressure relationship is linear.	7. Non-linear.
8. **Newton's Equation** Applicable.	8. This equation does not hold for the viscosity of the fluids containing 40-60 per cent of the suspended particles like blood.
9. **Linear** It is linear with alterations in flow velocity.	9. It is nonlinear with alterations in flow velocity.
10. **Fluidity** It has fluidity and water is the most simple liquid having fluidity (Q) $144 \frac{cm}{sec} \Big/ \frac{dyne}{cm}$	10. Plasma is about half that of water, i.e., 85, addition of 20% RBC reduces to 20. The slope of the curve for rate of fluid flow is proportional to the fluidity (Q).
11. **Parabolic Flow Velocity Profile** Flowing through tube shows laminar flow and exhibits parabolic velocity profile. The fastest velocity occurs in the centre and zero velocity at the boundary wall.	11. Blood shows laminar flow in arteries and veins, the profile is not parabolic except in wider vessels having a diameter more than RBC, blood acts as Newtonian and the profile resembles parabolic.

postural changes is greatly influenced by alteration of luminal configuration, i.e., from normal elliptical or oval to circular cross-sectional appearance together with myogenic tone of the veins activated by the sympathetic constrictor nerves. When a person is in upright posture, veins are distended and the hydrostatic pressure of the column of the blood in veins below the heart level tends to poll blood in the legs and feet. But this hydrostatic effect is compensated by the myogenic tone of the veins influenced by the sympathetic nerves and external compressive effect by the contraction of the skeletal muscle.

The shape of the pressure-volume curves are compared between the aorta and vena cava. This is done by filling and distending one segment of the vessel with increasing volume of fluid.

In case of aorta, the pressure in the segment initially rises in a linear fashion. In case of vena cava or any other large distensible vein, the pressure does not rise rapidly until large volumes of fluid are injected. The veins act as a good reservoir in the body and remain in partially collapsed state and oval in cross-sectional profiles. A large quantity of blood can be added to veins before it is distended to the point after which further addition of more volume of blood causes a great rise in venous pressure. For this reason veins are known as *Capacitance vessels*.

About 54 per cent of the circulating blood is in the systemic veins. 12 per cent is in the heart cavities and 18 per cent is in the low-pressure pulmonary circulation. In arterial system, only 2 per cent is in the aorta, 8 per cent in the arteries, 1 per cent in the arterioles and 5 per cent in the capillaries. In administration of blood by transfusion, less than 1 per cent is distributed in the arterial system, i.e., in the systemic circulation known as *high-pressure system* comprising left heart, Windkessel vessels (aorta, large arteries), precapillary resistance vessels having blood pressure between 100 and 20 mm Hg, according to the region whose function is control of systemic pressure and flow distribution. And the remaining portion is seen in the systemic veins, pulmonary circulation and cardiac cavities except left ventricle which is known as *low-pressure system* having pressure between 25 to 0 mm Hg whose function is to control blood volume and venous return.

6. **Shunt Vessels**: A few tissues are having shunt vessels, of which the most important is the skin. The shunt vessels bypass the capillaries and, when patent, allow rapid flow of blood to the tissues without any nutritive function of the tissues. During thermal stress, inhibition of sympathetic vasoconstrictor action via central nervous system causes high increases of skin blood flow resulting in dissipation of heat from the skin surface.

28.8 Modification of Phasic Blood Flow to Continuous Flow and Pressure in the Capillary Bed (Windkessel Function)

28.8.1 Phasic blood flow

Modification of phasic character of blood flow to continuous and fairly steady flow pattern against pressure (continuous flow and pressure in the capillary or vascular bed) is a basic problem in haemodynamics, not explained by Poiseuille as to how to represent and analyse pulsatile pressure and flow. An effort to solve this problem was made by Frank (1895, 1927) who described the circulation as a system in which the heart pumps blood into an elastic reservoir which was called Windkessel—which is the German word for an elastic reservoir (literally, air chamber or compression chamber). This chamber was considered to include (1) aorta, and (2) the proximal part of its larger branches with the smaller arteries taken to be essentially nonelastic tubes. McDonal (1947) has shown that the Windkessel hypothesis has limited value in describing the circulation and that a better model is to view the entire arterial bed as a system put into a steady-state oscillation by the repetitive beat of the heart. The pressure pulse is described as a set of component sinusoidal waves of frequencies determined by the harmonic or Fourier series (Taylor, 1996).

28.8.2 Windkessel vessels

Windkessel vessles are represented by the aorta and its large branches—vessels which are highly elastic. Systolic ejection of the blood by the heart distends these vessels and subsequent closure of the aortic valve at the termination of systole, the elastic recoil of the vessels sustains the pressure head better and renders the blood flow to the periphery steadier than it would be otherwise.

28.8.3 Windkessel function

The mean velocity of the blood in the proximal part of the aorta is 40 cm/s and the flow is phasic. Usually the velocity ranges from 120 cm/s during systole to a negative value at the time of the transient back flow before the aortic valves close in a diastole. In the distal parts of the aorta and in the large arteries, velocity is also greater in systole than it is in diastole. Large phasic pressure changes are generated by the left ventricle when it delivers the stroke volume into the aorta. The elastic properties of the large arteries and aorta and the vessel store part of the energy produced by the phasic cardiac ejection as potential energy which permits them to rapidly distend or expand their walls as the volume of blood enters the arterial bed during systole. This distension of the large arteries helps to store blood transiently. In diastole, the recoil of the walls of the Windkessel vessels converts this stored potential energy into kinetic energy and, together with the resistance offered by the arteries, these two together influence the transient stored blood to pass onward. Here the arteries act like an accessory pump and thereby maintains a fairly steady flow through the systemic capillaries during diastole. This process is known as *Windkessel effect*. The combination of elasticity and resistance converting the pulsative ejection of the heart into a steady flow was shown experimentally by Borelli in the seventeenth century.

The distension of the smaller resistance vessels at a given pressure increases the internal radius and eventually causes a raised flow. But these arteriolar vessels have smooth muscle in the middle coat, i.e., media which are distensible or elastic and responds sharply to stretch by contraction like cardiac muscle.

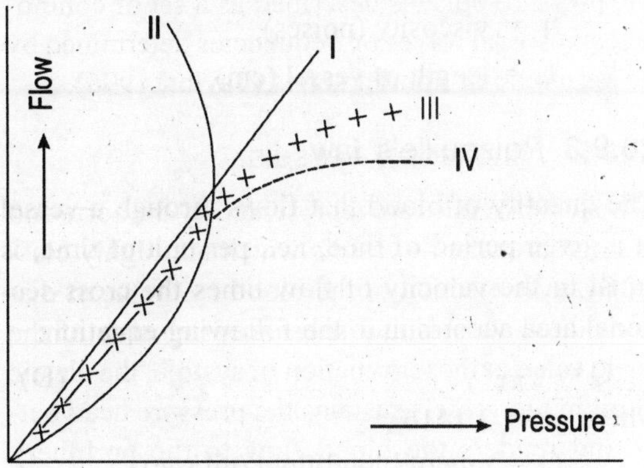

Fig. 28.6: Flow-pressure curve

This myogenic property limits or may compensate or exceed the initial distension response of the elastic vessel to transmural pressure.

28.8.4 Flow-pressure curve

If the graphs are drawn for flow against the pressure in different conditions, such as with: (1) rigid or nondistensible tube, (2) distensible or elastic vessel, (3) distensible vessle having active myogenic element, (4) distensible vessel having myogenic contractile element, the results will be like Fig. 28.6.

Graph I: Rigid tube shows that the relationship is linear as defined by a Poiseuille formula.

Graph II: Distensible vessel shows the increase in flow due to rise in pressure head which is greater than rigid tube (Graph I). The collapsed vessels are distensible and limited by inextensible fibrocollagenous elements in the adventita coat, so, at increased pressures, the flow will again be linear.

Graph III: Distensible vessel containing active myogenic elements shows myogenic contractile response to stretch and offsets the distensible or elastic effects exerted by the increased pressure. The curve shows concavity to the axis of the pressure.

Graph IV: Distensible vessel containing myogenic contractile elements which can stabilize flow between the pressures of 80 to 200 mm Hg. It shows that myogenic elements of the vascular wall can exceed the distensible or elastic effects of increased pressure and these vessels show the process of "autoregulation".

28.8.5 Physiological features

Besides the physical features, the physiological features are the effects produced by high sympathetic vasoconstrictor discharge to the resistance vessel which constricts the lumen and thereby decreases the flow against a given pressure. Artificially induced pressure do not increase the flow due to maximum sympathetic vasoconstriction. In natural conditions this happens, e.g., muscular exercise reflexly reduces sympathetic vasoconstrictor effect through baroceptors and causes relaxation of the smooth muscle of the arterial wall and hence vasodilation with increase in radius. The distension due to increased pressure is thus minimized by reflex vasodilation lowering the resistance.

28.8.6 Autoregulation

The capacity of the tissues to maintain a constant blood flow or to regulate blood flow during the changes in blood pressure is referred to as *autoregulation.*

Most vascular beds have an intrinsic property to offset for moderate changes in perfusion pressure by changing in vascular resistance, so that blood flow remains constant. This mechanism is well developed in the following tissues:

1. Kidney: Only cortical blood flow but not the medullary flow.
2. Myocardium: Intrinsic autoregulation of myocardium according to Frank-Starling law.
   Two types:
   (a) Heterometric autoregulation of the heart,
   (b) Homeometric auto-regulation.
3. Brain, 4. Liver, 5. Skeletal muscle, 6. Mesentery,
7. Vasomotion: Blood usually does not flow at a continuous rate through the capillaries. Instead, it flows intermittently due to intermittent or cyclic opening and closing of precapillary sphincter and metarterioles, the blood flow to the tissues is autoregulated. This is called *vasomotion.* The mechanism of autoregulation is partly due to intrinsic contractile response of smooth muscle to stretch (myogenic theory of autoregulation).

28.9 Poiseuille's Law

Physical factors which influence vascular flow

Hagen (1839) and Poiseuille (1842, 1846) showed experimentally the factors which influenced the volume flow of flow rate of fluid flowing steadily (laminar flow) through rigid glass tube of capillary diameter using Newtonian fluid in vitro at constant temperature. Analysis of the law gives us the information regarding various factors influencing resistance.

28.9.1 Conductance and resistance relationship

Conductance is the measure of blood flow through a vessel at a given pressure difference and expressed in ml/sec/mm Hg pressure or litres/sec/mm Hg. The conductance is the reciprocal of resistance according to the following equation:

$$\text{Conductance} = \frac{1}{\text{resistance}} \qquad \text{(A)}$$

28.9.2 Conductance and vessel diameter

If the blood flow is streamline, slight changes in diameter of the vessel cause great change in the conductance with the same pressure difference between the two ends of the vessels. If the increase in diameter of the vessel is four-fold, respective increase in flow is 256 fold. Thus, the conductance of vessel increases in proportion to the "fourth power of the diameter" according to the following equation:

$$\text{Conductance} \propto \text{Diameter}^4 \qquad \text{(B)}$$

This is due to the velocity of flow of the blood touching the wall of the vessel being practically slow because of adherence to the endothelium, next layer away from the wall flows at a slightly rapid rate but the layers further away from the wall flow with progressive rapid velocities. In small vessels, all the blood is near the wall, the rapid flowing central axial stream does not exist.

By integrating the velocities of all the layers of flow, the formula is derived related to "mean velocity" or blood flow to vascular radius:

$$V = \frac{(P_1 - P_2).r^2}{8\eta L}, \qquad \text{(C)}$$

where V = velocity (cm/sec),

$P_1 - P_2$ = pressure difference (dynes/sq. cm),

r = radius of the vessel (cm),

η = viscosity (poises),

L = length of vessel (cm).

28.9.3 Poiseuille's law

The quantity of blood that flows through a vessel in a given period of time, i.e., per unit of time, is equal to the velocity of flow times the cross-sectional area according to the following equation:

$$Q = V\pi r^2, \qquad \text{(D)}$$

where π = 3.1416

Q = rate of blood flow (ml/sec)

r^2 = cross-sectional area (square centimetres)

Now, substituting the value for velocity of blood from equation (C) into equation (D) gives the following equation which is known as *Poiseuille's law.* The equation is also known as **Hagen-Poiseuille's equation.**

$$Q = \frac{\pi(P_1 - P_2)r^4}{8\eta L} \qquad \text{(E)}$$

The equation shows that the rate of blood flow is directly proportional to the fourth power of the radius of the vessel and also illustrates that the diameter of a vessel plays a great role in determining the rate of blood flow. The law states that volume of blood flowing through a circulating system increases with the resistance to flow. It shows that the flow is accelerated by an increase in any of the factors in the numerator and decreased by any factor in the denominator.

28.9.4 Liquid flow at different conditions

1. **Tube Length**: If the tube length is doubled or if the blood becomes twice as viscous, the blood flow would be decreased by half.
2. **Pressure:** Blood is allowed to flow from one point to another in a tube by the pressure difference between the two ends of the tube. Hence, doubling the pressure gradient along the tube (i.e., pressure difference between the ends of the tube, P_1–P_2), will double the flow.
3. **Radius:** Change of radius is not an effective means in changing the flow. Doubling the radius will increase flow 16 times.
4. **Temperature:** If the fluid is warmed, its flow is faster than when it is cold but it still flows at a much slower rate than water.
5. **Viscosity:** The term viscosity describes the behaviour of these different fluids (Newtonian, non-Newtonian). It is already known that temperature affects viscosity.

What does Poiseuille's equation reveal?

1. Important relationship between vascular resistance and the length of the tube and blood viscosity. If one doubles, the flow will be halved. If flow remain constant, the pressure will be double.
2. Great influences of radius of the vessel. If the radius is doubled, the flow will be increased 16-fold.

28.9.5 Resistance and Poiseuille's equation

Poiseuille's experiments were performed with a Newtonian fluid. If the radius and length of the tube remain constant, the viscosity will not vary when the flow is changed and the resistance remains constant.

In the equation of Poiseuille's law, Q = flow and $P_1 - P_2$ = pressure difference. The remaining part of the equation represents the conductance by the equation:

$$C = \frac{\pi r^4}{8\eta L}.$$

Since conductance is the reciprocal of resistance, the following equation illustrates the factors influencing resistance. From the Ohm's law (1 = E/R) as 1 is represented by Q and E by ($P_1 - P_2$), the resistence R is represented by $8\,\eta\,L/\pi r^4$. The longer the tube or the higher the viscosity or the smaller the radius, the higher is the resistance to flow.

$$R = \frac{8\eta L}{\pi r^4}$$

If those factors are considered separately, than the resistance of a vessel is directly proportional to the blood viscosity (η) and length (L) of the tube and is inversely proportional to π and the fourth power of the radius (r^4) of the tube.

But this equation contains three components:

1. A constant $\left(\frac{8}{\pi}\right)$.
2. A geometric component $\left(\frac{L}{r^3}\right)$.
3. The viscosity component (η).

Most of these factors cannot be measured in vivo, i.e., in the intact circulation. Therefore, a simple equation of resistance has been employed. This is done by substituting R for the above factors into Poiseuille's equation by the following formula:

$$Q = P_1 - P_2 \text{ or } R = \frac{P_1 - P_2}{Q},$$

where R = resistance in distensible (i.e., elastic) tubes (e.g., blood vessels).

28.10 Difference between Poiseuille's Experimental Preparation and the Circulation

Table 28.3: Difference between Poiseuille's Experimental Preparation and the Circulation

Poiseuille's Experimental Preparation	Circulation
1. Tube: Rigid or nondistensible.	1. Tube: Blood vessel, distensible or elastic.
2. Fluid: Newtonian.	2. Non-Newtonian.
3. Pressure: Steady head pressure.	3. Non-steady pressure.
4. Temperature: at one temperature, extremely temperature dependent, flow is higher as the temperature rises.	4. Not at one temperature, not temperature dependent.
5. Resistance: Remains constant.	5. As the pressure increases, resistance falls.
6. Viscosity: Remains constant.	6. Not constant.
7. Flow-pressure curve: As pressure increases, flow increases linearly.	7. Flow and pressure relationship is non-linear. As the pressure difference increases the flow is augmented in a nearly exponential manner as the blood vessel calibre is augmented.
8. Study: In vitro	8. In vivo.
9. Vasomotor tone: Nil.	9. With higher degree of vasomotor tone, the pressure flow curves are progressively shifted to the right and flattened and intercept at zero flow (critical closing pressure) and becomes progressively higher.

28.11 Application of Poiseuille's Law to the Living System

McDonald (1947) described the conditions under which Poiseuille's equation applies to fluid flow and compared each condition with circumstance actually present in the circulation.

1. **The fluid is homogeneous and its viscosity is the same at all rates of laminar flow (Shear)**: In larger arteries and veins, blood may be considered to be a homogeneous fluid with viscosity independent of the velocity gradient. In vessels having an internal radius below 0.5 mm, however, changes in apparent viscosity occur. Thus blood behaves as a Newtonian fluid in a vessel with a diameter of red blood cells.
2. **The liquid does not slip at the wall**: There is no indisputable evidence that blood plasma in the narrow cell-free layer at the vessel wall slips at the endothelium.
3. **The flow is laminar, i.e., the liquid at all points moves parallel to the vessel wall**: This is not true at blood flow rates above a critical value (Reynolds number), where flow becomes turbulent. It appears, however, that laminar blood flow occurs in vessels where flow is sufficiently steady.
4. **The rate of flow is steady and no subject to cyclical acceleration or deceleration**: Since flow is pulsatile in large arteries and intrathoracic veins, it is evident that this condition of Poiseuille's experiment cannot be applicable.
5. **The tube is long compared to the section in which flow is studied**: Close to the inlet of a tube, laminar flow may not be established. The distance required to establish the parabolic velocity profile is called the "inlet (or entrance) length" and in this length, Poiseuille's findings are not applicable. For example, at the beginning of the aorta, although flow is laminar, the velocity profile is flat rather than parabolic.
6. **The tube is non-distensible (rigid)**: Blood vessels are distensible or elastic and diameter varies with distending pressure.

From these considerations it is clear that the Poiseuille's equation describing steady flow in narrow, rigid tubes can only approximate the conditions in the living circulation. Nevertheless, understanding the relationships, the description of Poiseuille is fundamental to the study of haemodynamics.

The applicability of Poiseuille's law to the circulation depends upon whether resistance is dependent on pressure and flow or not. Because, in case of circulation, peripheral resistance is the indicator of the total influence of all the factors affecting blood flow, such as:

(a) Change in apparent blood viscosity,
(b) Existence of laminar versus turbulent flow,
(c) The length of the vessels,
(d) The cross-sectional areas of the blood vessel which is determined by:

1. External pressure from the surrounding tissue,
2. Mechanical dilatation with rising perfusion pressure,
3. Opening of new capillaries and vessels due to:
   i. Changes in metabolism,
   ii. Rise in perfusion pressure.
4. Active changes in the contraction of muscular wall due to the following factors:
   i. Vasomotor nerves,
   ii. Humoral substances, and
   iii. Metabolic products.

The flow-pressure curves with blood in vascular system is different from the curve obtained in the artificial experiment done by Poiseuille. The differences are:

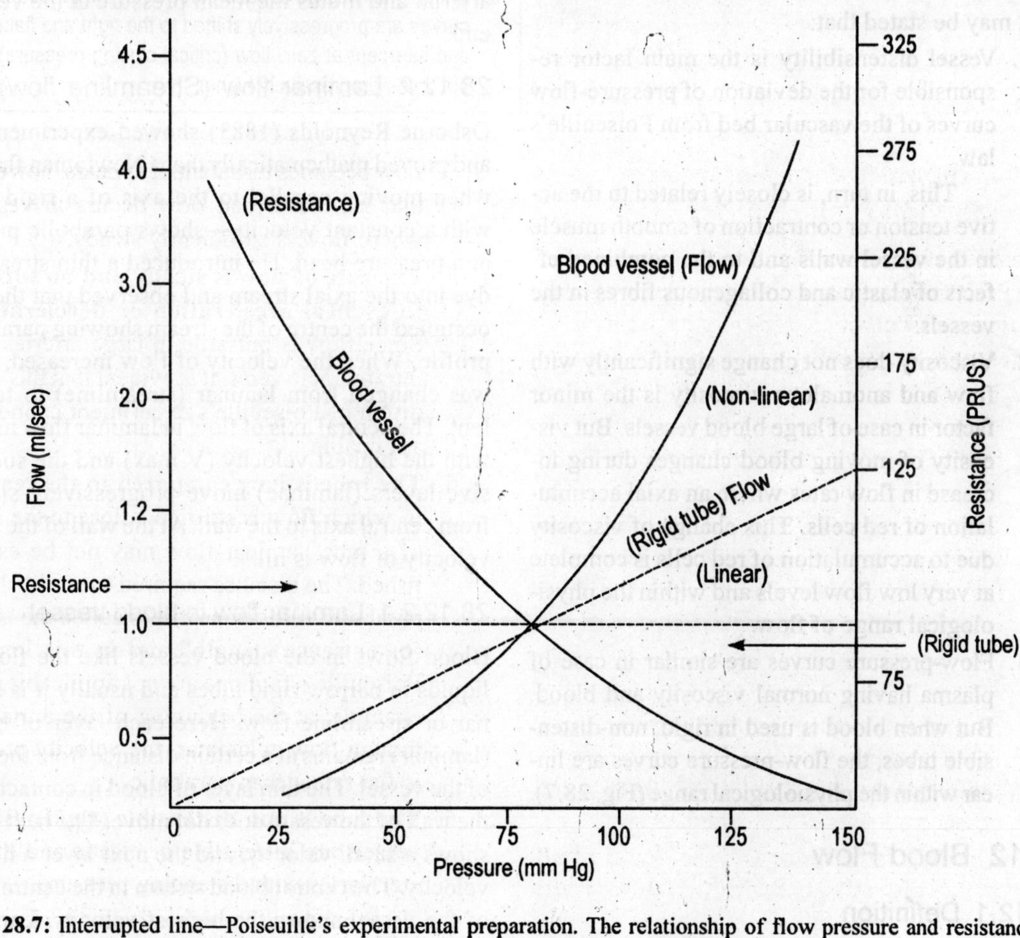

Fig. 28.7: Interrupted line—Poiseuille's experimental preparation. The relationship of flow pressure and resistance in a non-distensible (rigid) tube using Newtonian fluid. Solid line—blood in vessel

1. The curves do not pass through the origin,
2. The flow is not significant until ciritical closing pressure is attained,
3. For dilated vessels, the curves are linear. But the curves become sigmoid with increasing tone and then convex to the flow axis.
4. The curves tend to become straight lines at sufficient high pressures is decreased to the values near the critical closing pressure. Such deviations from constant resistance to increased resistance could occur when:
   i. The geometric factor of Poiseuille's law is not constant but varies with the distensibility of the blood vessels, and
   ii. Blood has anomalous viscosity so that the viscosity factor of Poiseuille's equation varies with the rate of flow and hence, the driving pressure.

It may be stated that:

1. Vessel distensibility is the main factor responsible for the deviation of pressure-flow curves of the vascular bed from Poiseuille's law.

   This, in turn, is closely related to the active tension or contraction of smooth muscle in the vessel walls and to the combined effects of elastic and collagenous fibres in the vessels.
2. Viscosity does not change significantly with flow and anomalous viscosity is the minor factor in case of large blood vessels. But viscosity of moving blood changes during increase in flow rates where an axial accumulation of red cells. This change of viscosity due to accumulation of red cells is complete at very low flow levels and within the physiological range of flow.
3. Flow-pressure curves are similar in case of plasma having normal viscosity and blood. But when blood is used in rigid, non-distensible tubes, the flow-pressure curves are linear within the physiological range (Fig. 28.7).

28.12 Blood Flow

28.12.1 Definition

Blood flow is the quantity of blood that passes a given point in the circulation in a given period of time. Usually, blood flow is expressed in millilitres/minute or second or litres/minute. The total blood flow in the circulation of an adult person at rest is about 5 litres/minute which is equivalent to cardiac output.

Blood flows only from a region of higher pressure to a region of lower pressure except in certain cases when momentum transiently sustains flow. The relationship between the mean flow (F), mean pressure (P) and resistance (R) is expressed in Ohm's law:

$$\text{Mean flow (F)} = \frac{\text{mean pressure (P)}}{\text{resistance (R)}}$$

Flow in a part of vascular system is equal to the "effective perfusion pressure" in that part divided by the resistance. The effective perfusion pressure is defined as the mean intraluminal pressure at the arterial and minus the mean pressure at the venous end.

28.12.2 Laminar flow (Streamline flow)

Osborne Reynolds (1883) showed experimentally and proved mathematically that a Newtonian fluid—when moving parallel to the axis of a rigid tube with a constant velocity—shows parabolic profile at a pressure head. He introduced a thin stream of dye into the axial stream and observed that the dye occupied the centre of the stream showing parabolic profile. When the velocity of flow increased, flow was changed from laminar (streamline) to turbulent. The central axis of flow in laminar flow moves with the highest velocity (V max) and the successive layers (laminae) move progressively slowly from central axis to the wall. At the wall of the tube, velocity of flow is nil.

28.12.2.1 Laminar flow in blood vessel

Blood flows in the blood vessels like the flow of liquids in narrow rigid tubes and usually it is laminar or streamline flow. Here each layer of blood (laminar) remains at a certain distance from the wall of the vessel. The thin layer of blood in contact with the wall of the vessel does not move. The next layer shows a small velocity, and the next layer a higher velocity. The central blood stream in the central part of the vessel shows the highest velocity. Laminar flow occurs at velocities up to a certain "critical velocity". At or above this velocity, laminar flow changes into turbulent flow in which blood flows in all directions in the vessel and continually mixing within the vessel. Laminar flow is silent and turbulent flow makes a sounds. The resistance met by the blood moving in laminar flow through the vessel has nothing to do with friction between the fluid and the vessel wall, but is due to the friction between the adjacent laminae of the fluid.

The bigger this internal friction, the greater is the difference of velocity between the laminae and the greater the coefficient of viscosity.

28.12.2.2 Parabolic velocity profile of laminar flow

In laminar flow, each layer of the stream has different velocity and the central stream shows the highest velocity. During the flow of fluid, a "parabolic interface" develops in the fluid. The fluid portion adjacent to the vessel wall does not move at all, the next portion slightly away from the wall moves at a smaller distance and the central portion of the fluid in the centre of the vessel moves a longer distance. This effect is known as *parabolic velocity profile* of the blood flow (Fig. 28.7A).

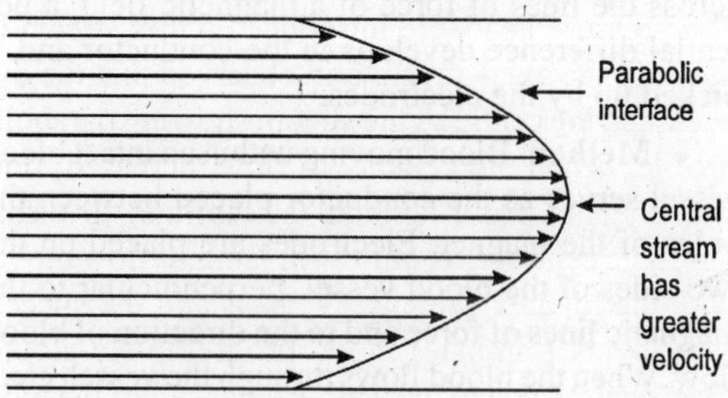

Fig. 28.7A: Laminar (streamline) flow showing parabolic velocity profile

28.12.2.3 Cause of parabolic velocity profile

The molecules of the fluid adjacent to the wall of the vessel cannot move due to adherence to the vessel wall. The molecules of the next layer slips over the first layer, and the third layer slips over the second layer and so forth. Thus, the fluid in the centre of the vessel can move fast because of the existence of many layers of molecules between the centre of the vessel and the vessel wall and all these molecules have the capacity of slipping over each other but the fluid adjacent to the wall does not have this advantage.

28.12.3 Turbulent flow

Normal laminar blood flow becomes turbulent flow when:

1. The rate of blood flow, i.e., the velocity of flow, is high,
2. It passes by an obstruction in a vessel (as in case of compression by cuff of sphygmomanometer),
3. It makes a sharp turn,
4. It passes over a rough surface.

The characteristic of the flow is that the blood flows crosswise in the vessel as well as along the vessel forming whorls in the blood similar to whirlpools in a rapidly flowing river at a point of obstruction and it is known as *eddy currents*. During eddy current, the blood flow encounters greater resistance because eddies add vigorously to the overall friction of flow in the vessel (Fig. 28.7B).

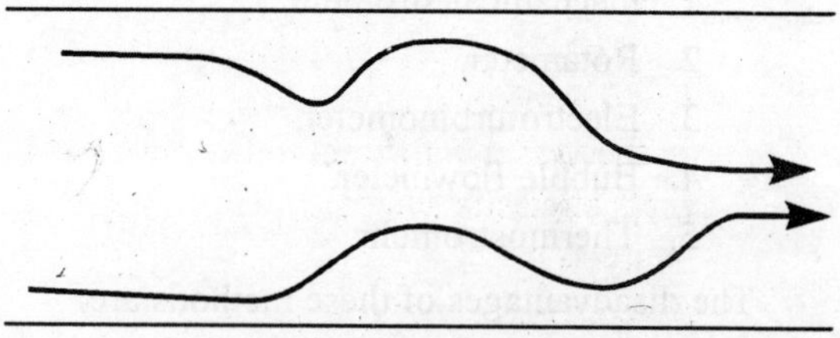

Fig. 28.7B: Turbulent flow

Turbulence is also related to the diameter of the vessel and the viscosity of the blood which can be expressed by the ratio of inertia to viscous forces as follows:

$$R = \frac{\rho DV}{\eta},$$

where R = reynolds number,
ρ = density of the fluid,
D = diameter of the tube,
V = velocity of the flow,
η = viscosity of the fluid.

If R is higher, there is a greater probability of turbulence. (See, for details, turbulent flow, Reynolds number, 28.16).

28.13 Methods for Measurement of Blood Flow (Flowmeter)

Many methods have been introduced to measure blood flow, some of which are applied inserting in a series with a blood vessel or are applied outside the vessel. These mechanical or mechanoelectrical devices are called *flowmeters*. Most of them are not suitable for venous flow, because resistance to flow but can be used for arterial flow because slight drop in pressure across the meter is insignificant.

28.13.1 Types of flowmeter

Based on Physical Principle

(a) Electromagnetic induction
(b) Ultrasound reflection of transmission
(c) Thermal convection
(d) Radiographic Principle
(e) Indicator dilution (dye or thermal)

Generally all blood flowmeters currently used in clinical and research applications are as under:

I. **Mechanical**

1. mechanical stromuhr.
2. Rotameter.
3. Electroturbinometer.
4. Bubble flowmeter.
5. Thermostromuhr.

The disadvantages of these methods are:

1. Operation is extensive.
2. Require anaesthesia.
3. Require insertion between the cut ends of the vessel in which flow is being measured.
4. Response time is slow.

Despite the above disadvantages, it is very much useful in case of arterial flow which is highly phasic.

II. **Phasic Flowmeters**: These types of methods can record the phasic or moment-to-moment changes in arterial and venous flow during a cardiac cycle. These flowmeters are:

ELECTROMAGNETIC AND ULTRASONIC DOPPLER FLOWMETERS

1. Both electromagnetic and ultrasonic flowmeters are applied for measurement of blood flow in intact vessels in unanaesthetised active animals. Electromagnetic flowmeter is used in anaesthetised humans.
2. These can be mounted chronically on the external surface of intact blood vessels.
3. These record instantaneous flow for a long period, weeks or months without changing calibration.
4. These are much sensitive and have frequency response to monitor flow in arteries and veins.

• **Electromagnetic Flowmeter**: The electromagnetic flowmerter was developed by Kolin and Wetter. Recently several devices have been developed to measure blood flow in a blood vessel without opening it. But electromagnetic flowmeter is very useful for this purpose.

• **Principle**: An electromotive force or voltage is generated or induced in a wire, i.e., the conductor moving rapidly through a magnetic field at right angles to the lines of force which is proportional to the velocity of its motion.

Same principle is applied for generation of electromotive force in blood when it moves through a magnetic field. Here the column of blood in the vessel serves as an electric conductor when it moves across the lines of force of a magnetic field, a potential difference develops in the conductor and is picked up by the electrodes.

• **Method**: Blood moving within an intact blood vessel serves as the conductor placed between the poles of the magnet. Electrodes are placed on the two sides of the blood vessel, perpendicular to the magnetic lines of force and to the direction of blood flow. When the blood flows through the vessel, electrical voltage, i.e., potential difference, proportional to the rate of flow is generated and is drawn from the walls of the vessel by the non-polarizable electrodes. As the potential difference is small (one-millionth volt), it must be amplified before being recorded with a galvanometer or oscillograph.

The flowmeter is encased in plastic. It consists of

1. A miniature electromagnet made up of coils of wire mounted on steel laminations,
2. A slotted plastic tube into which the vessel is inserted,
3. Wires of energizing the electromagnet and

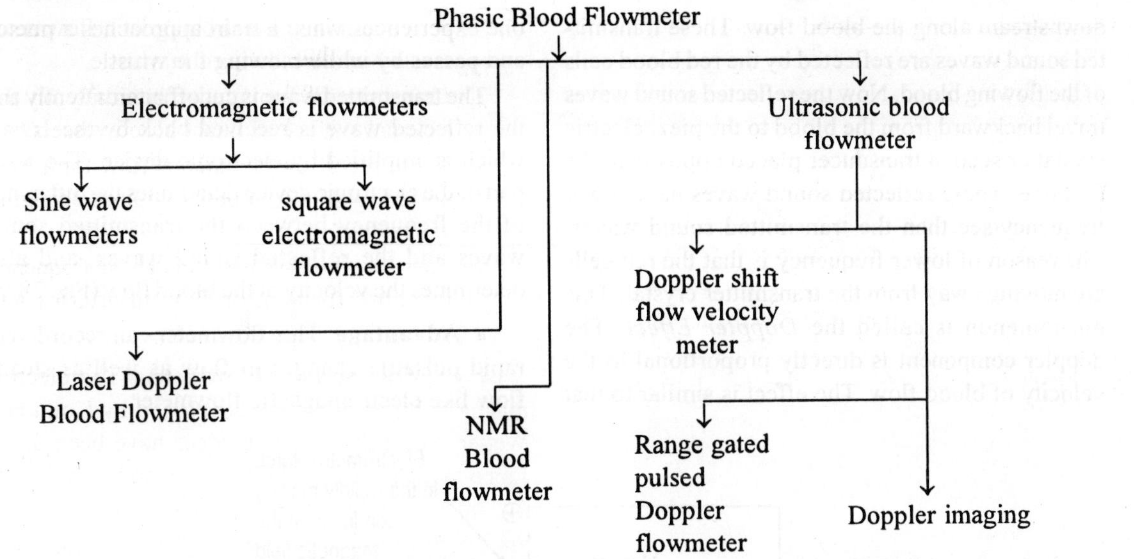

4. Wires for picking off the generated voltage, i.e., potential difference (one-millionth volt) (Fig. 28.8A; 28.8B).

An acutal "probe" (Fig. 28.8C) is placed around the intact blood vessel by temporarily decompressing and forcing the vessel through the longitudinal slit in the probe to record its blood flow. This probe contains both the strong magnent and the electrodes.

After amplification of the composite signal, the Doppler frequency can be obtained at the output of the detector as the difference between the direct and the scattered signal components. In case of normal blood flow, the Doppler signal is in the low audio frequency range. Due to velocity profile of the blood flow, the Doppler signal is not a pure sine wave, but has a form of narrow-based noise and in a loudspeaker or earphone, the Doppler signal of the pulsating blood flow will be heard as "swish—swrish". When the transducers are placed in a suitable mount, a frequency-meter used to measure the Doppler frequency can be calibrated directly in flow-rate units. Transducers for ultrasonic flowmeters can be implanted for chronic use and some instruments includes telemetry system to measure the blood flow in unrestrained animals.

- **Advantages**

1. Can be used to record pulsatile flow in the aorta and arteries of dog and man in intact vessel.
2. Can be used to measure mean and phasic flow in unopened large and small blood vessels of chronic animals.
3. It can record changes in flow that occurs in less than 0.01 second.
4. It allows accurate recording of pulsatile changes in flow as well as steady flow.
5. Its linearity of calibration and its high frequency response are advantageous.
6. The flow beam can be made to approximate the zero flow position by "electronic zeroing" or repeated mechanical flow zeros can be easily obtained by temporary (3-4 sec) inflation of a rubber pneumatic cuff distal to the flowmeter.

ULTRASONIC DOPPLER FLOWMETER

Another flowmeter which can be applied outside the blood vessel is called *Ultrasonic Doppler Flowmeter.* It has the same advantages as the electromagnetic flowmeter.

- **Principle**

A very small piezoelectric crystal made of barium Titanate (transducer) is placed in the wall of the instrument.

The crystal is energized by an electronic device, i.e., an oscillator, Operating at a frequency of several megahertz and excites to transmit sound waves at a frequency of several million cycles/second downstream along the blood flow. These transmitted sound waves are reflected by the red blood cells of the flowing blood. Now the reflected sound waves travel backward from the blood to the piezoelectric crystal or second transducer placed opposite to the first one. These reflected sound waves have lower frequency/sec than the transmitted sound waves. The reason of lower frequency is that the red cells are moving away from the transmitter crystal. This phenomenon is called the *Doppler Effect*. The doppler component is directly proportional to the velocity of blood flow. The effect is similar to that one experiences when a train approaches a person and passes by while blowing the whistle.

The transmitted wave is cut off intermittently and the reflected wave is received back by the crystal which is amplified by electronic device. The other part of the electronic device determines the difference of the frequency between the transmitted sound waves and the reflected sound waves, and also determines the velocity of the blood flow (Fig. 28.9).

- **Advantage**: This flowmeter can record very rapid pulsatile changes in flow as well as steady flow like electromagnetic flowmeter.

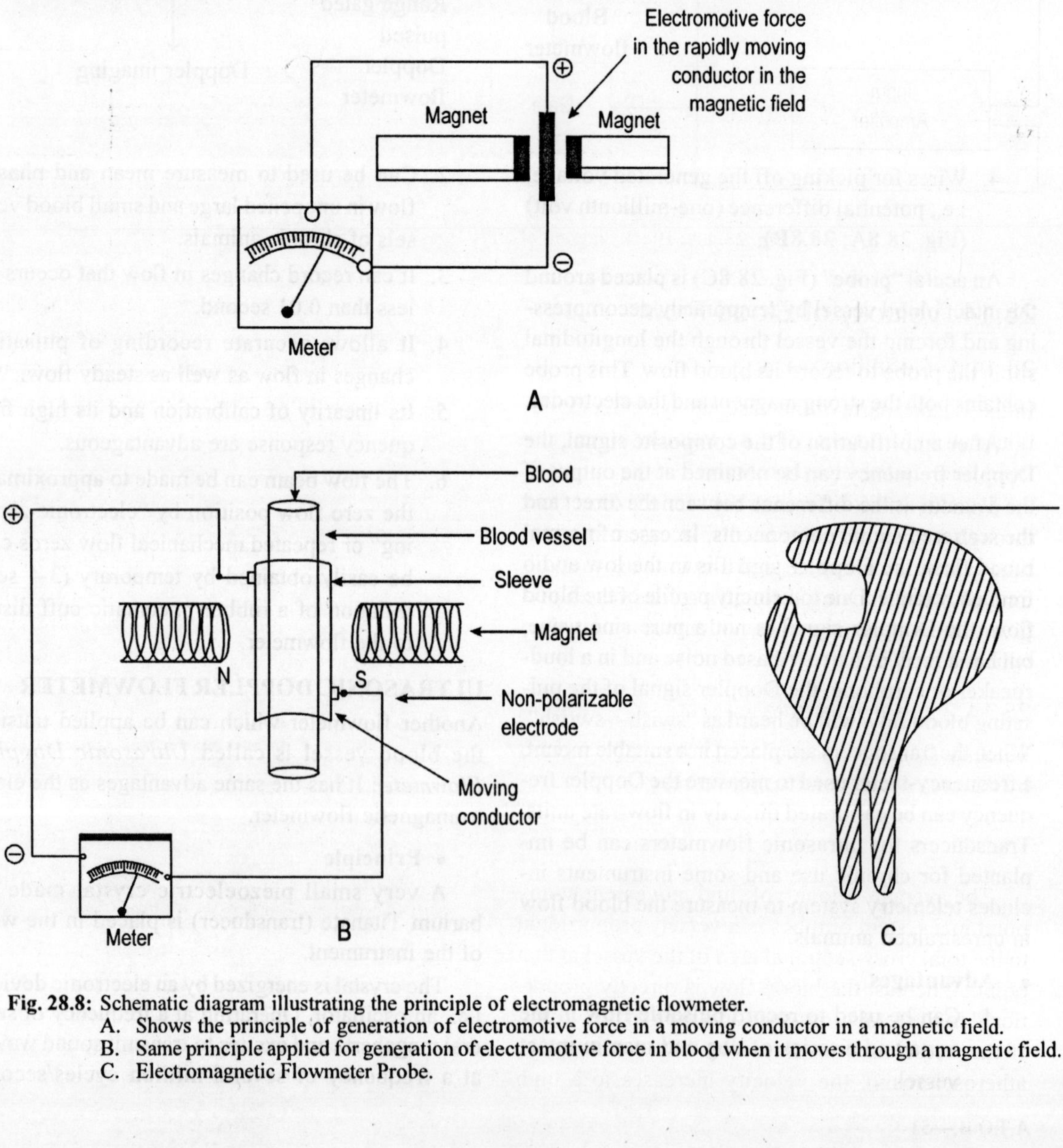

Fig. 28.8: Schematic diagram illustrating the principle of electromagnetic flowmeter.
A. Shows the principle of generation of electromotive force in a moving conductor in a magnetic field.
B. Same principle applied for generation of electromotive force in blood when it moves through a magnetic field.
C. Electromagnetic Flowmeter Probe.

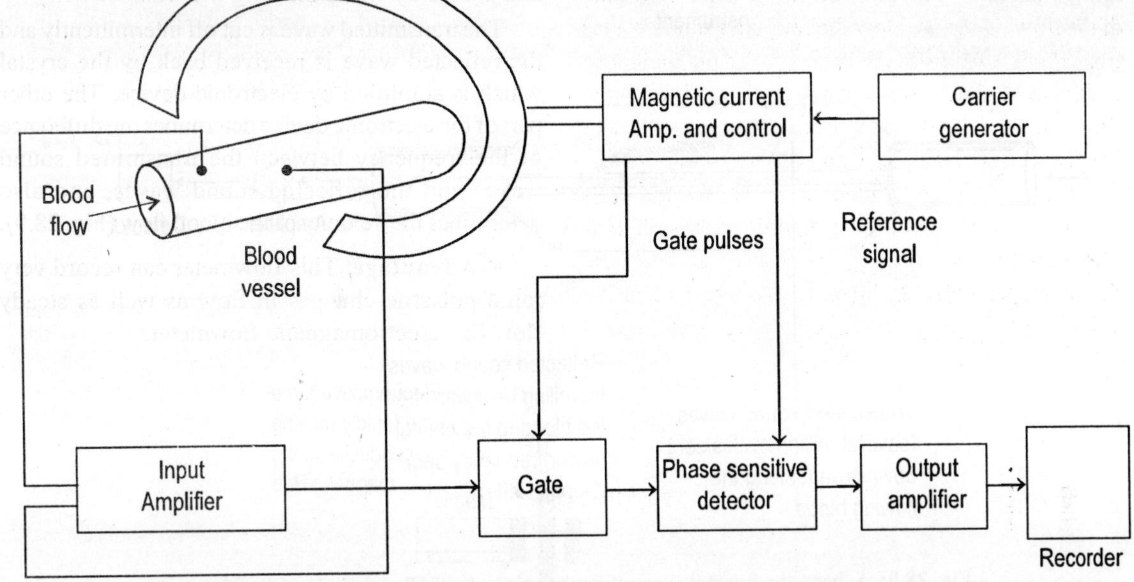

Fig. 28. 8A: Block diagram of electromagnetic blood flowmeter

28.14 Velocity of Blood Flow

28.14.1 Definition

In the various parts of the vascular system, circulation is also characterized by the velocity of the flow in addition to the blood pressure.

Velocity of blood means the rate of blood flow through a given vessel, i.e., displacement per unit time which is cm/sec. Blood flow means the volume of blood flowing through a given vessel in a given interval of time, i.e., volume per unit time which is cm^3/sec.

28.14.2 Characteristic

Velocity (V) is proportionate to flow (Q) divided by the area of the conducting vessel (A):

$$V = \frac{Q}{A}.$$

The average velocity of fluid movement at any point in a system of tubes is inversely proportional to the total cross-sectional area of the vessel at that point. Whereas, the blood flow is directly proportional to the total cross-sectional area of the vessel. If the vessel is constricted by any means (e.g., atherosclerosis), the velocity increases to a high level known as "critical velocity" and laminar flow becomes turbulent at or above this velocity and creates sounds. Although the mean velocity of the blood in the proximal part of the aorta is 40 cm/s, the flow is phasic, and velocity ranges from 120 cm/s during systole to a negative value at the time of the transient backflow before the aortic valves close in diastole. In the distal part of the aorta and in the large arteries, velocity is also greater in systole than it is in diastole. But the forward flow is continuous because the vessels are elastic and store blood temporarily during systole by distension during systole. Due to recoil activity during diastole the blood is pushed forward to maintain continuous flow. The effect is know as *Windkessel effect* and the vessel is called *Windkessel vessel.*

28.14.3 Factors influencing velocity of blood

1. **Kinetic Energy for Flow, Lateral Pressure and Velocity (Bernoulli's Principle)**

- **Bernoulli's Principle**: According to Burton (1965), fluids flow from one point (1), to another point (2), if the total fluid energy at (1) exceeds the total fluid energy at (2). Out of the 3 components of the total fluid en-

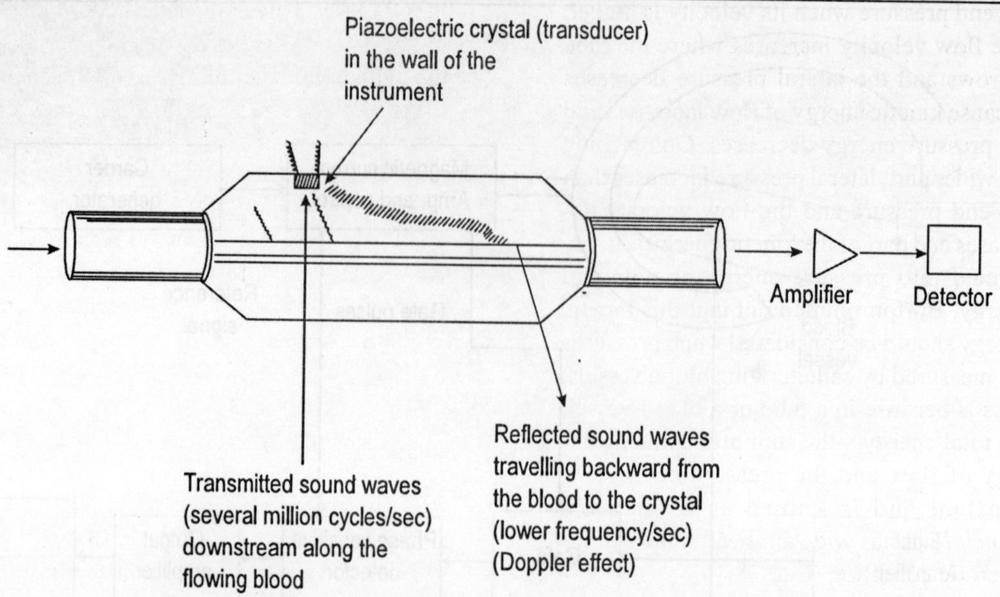

Fig. 28.9: Schematic diagram illustrating the principle of Ultrasonic Doppler Flowmeter

ergy (pressure or potential energy (P), gravitational potential energy (Ogh) and Kinetic energy (1/2 ρV^2), the kinetic energy per unit volume plays a greater role.

The kinetic energy which is proportional to the square of the velocity is sometimes ignored but it is important when blood normally flows from a higher pressure to a lower pressure. If the kinetic energy component (velocity) is sufficiently high, blood occasionally can flow from a lower to a higher pressure.

Bernoulli showed that when fluid flows along a tube, its lateral pressure is lower than

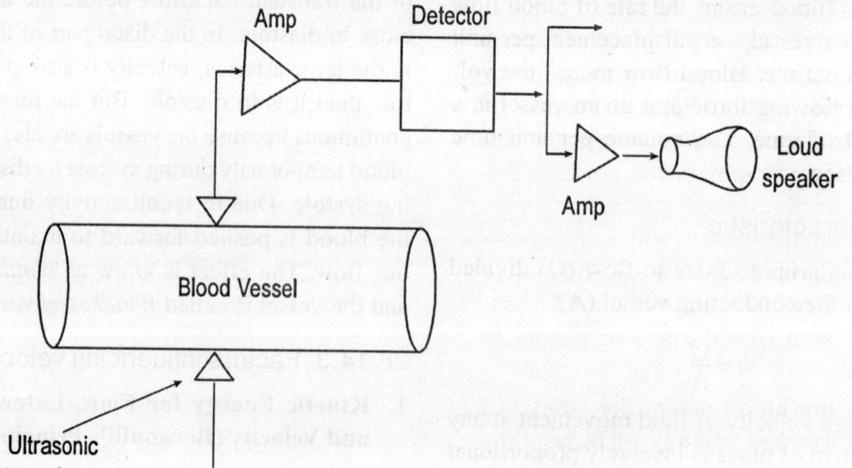

Fig. 28.9A: Block diagram of blood flow measurement (Doppler type)

the end pressure when its velocity is higher. The flow velocity increases where the tube narrows and the lateral pressure decreases because kinetic energy of flow increases and the pressure energy decreases. On reaching the wider part, lateral pressure increases than the end pressure and the flow velocity decreases and part of the kinetic energy is transformed into pressure energy or potential energy. Burton pointed out that this kinetic energy should be considered when pressures are measured by catheterizing blood vessels. This is because in a tube or a blood vessel, the total energy—the sum of the kinetic energy of flow and the pressure energy—is constant and is known as *Bernoulli's principle.* It has wide application in arteriosclerotic condition (Fig. 28.10).

In a fluid flowing at one level and ignoring the gravitational energy as factor, it can be expressed:

$$E = P + \frac{1}{2}\rho V^2,$$

where E = total fluid energy,
ρ = density,
V = velocity,
P = pressure or potential energy.

2. **Total Cross-sectional Area of the Vascular Bed**: The average velocity of blood movement at the site in the vessel of cardiovascular system is inversely proportional to the total cross-sectional area at that site of the vessel. Therefore, the average velocity of the blood is rapid in the aorta (0.5-1 metres/s), declines steadily in the larger arteries and smaller vessels, and is the slowest in the capillaries (0.5-1 mm/s), though the capillaries have 1000 times the total cross-sectional area (4500 cm^2) of the aorta (4.5 cm^2).

When the blood enters the veins (40 cm^2), the average velocity of blood flow increases again and is rapid in the vena cava (18 cm^2) but not so rapid as in the aorta.

3. **Different Portions of the Cross-sectional Area of the Vessel:** The velocity varies in different parts of the cross-sectional area of blood vessel, especially in the larger vessels. In the larger vessels, during laminar flow, the cylindrical layer of blood which adheres to the blood vessel is motionless having zero velocity. The layer in the central flow (central axial core) has the maximum velocity. In arterioles and capillary red cells are in the axial part and plasma flows along the walls. As a result, red cells move faster than the plasma and haematocrit is lesser than that of larger vessels. This effect is to change the distribution in the various parts of a cross section of vessel.

When the velocity of flow reaches a critical velocity, turbulence occurs. The formula RVD/v = a critical constant for turbulence or Reynolds number expresses the conditions in which turbulence is produced in fluid through tubes having constant diameter. The formula in which:

v = viscosity, D = density, V = mean velocity, R = tube radius.

The formula indicates that turbulence develops when the fluid of low viscosity flows through the tubes having a large diameter at quite high velocity.

In case of cardiovascular system, application of this formula indicates that the blood velocity is the main variable factor since the vessel diameter and blood viscosity do not alter significantly.

4. **Cardiac Output**: It is evident that an increased output of the heart in unit time will accelerate the rate of blood flow in all parts of the vascular system. At any given rate of cardiac output, however, the velocity of the blood varies as it passes along the vessel. Velocity of blood is directly proportional to the force with which blood is ejected. Therefore, velocity depends upon the minute output of the heart.

5. **Resistance to Flow and Velocity**

(a) **Resistance in the Layers of Flow (Internal Resistance to Flow or Viscosity):** In the laminar flow, the layer in contact with the walls of the tube is motionless, the next layer slides over it slowly with low velocity, the next layer over that and

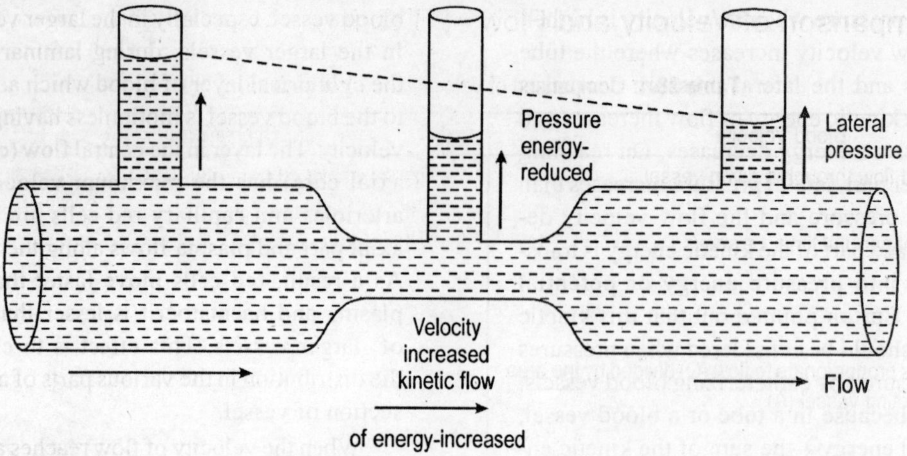

Fig. 28.10: Bernoulli's principle

so on until the axis of the tube where the axial flow shows maximum velocity. In such cases the resistance is due to the friction of the molecules of the layers over one another, varies directly as the mean velocity. In case of turbulent flow, the resistance varies approximately as the square of the mean velocity.

(b) **Resistance in the Arterioles (External Resistance)**: The greatest difference of pressure exists between the arteries on the one side and the small arterioles. The distribution of pressure points to the chief gradient in the vascular system being situated in the arterioles. The resistance offered by these arterioles is due to the small calibre, since they are maintained in a state of vasomotor tone. The total bed of the flow in the arterioles, though greater than that of the arteries, is lesser than that of the capillary network. Again the difference between the diameters of the arterioles and dilated capillaries is not so big and for this reason the velocity of the blood in the arterioles is greater than that in the capillaries, since friction, and, therefore, the resistance, probably varies as the square of the velocity. The resistance to the flow of blood through the arterioles must be greater than that offered by the capillaries. During constriction of the capillaries, however, they also add largely to the peripheral resistance.

(c) **Peripheral Resistance**: The resistance encountered by the arteriolar and capillary regions is peripheral resistance. The peripheral resistance depends largely upon the calibre of the arterioles. Peripheral resistance is a matter of the factors which control arteriolar vaso-constriction and vaso-dilation. Blood flow is inversely proportional to the peripheral resistance. Vascular dilatation reduces the resistance and increases blood flow while vascular constriction increases the resistance and reduces blood flow. But in case of velocity of blood flow we get the reversal effect. If the blood vessel is dilated, the velocity of blood flow is decreased and if it is constricted, the velocity is increased.

28.15 Turbulent Flow of Blood, Reynolds Number

28.15.1 Turbulent flow

When blood flows at a steady rate through a long, smooth vessel, it flows in streamlines with each layer of blood remaining at the same distance from the wall. Also, the central portion of the blood stays in the centre of the vessel. This type of flow is called

28.16 Comparison of Velocity and Flow of Blood

Table 28.4: Comparison of Velocity and Flow of Blood

Blood Velocity	Circulation or flow
1. Rate of blood flow through a given vessel.	1. Volume of blood flowing through a given vessel in given interval of time.
2. Displacement per unit time, i.e., cm/s.	2. Volume per unit time, i.e., cm^3/s.
3. Velocity (V) is proportionate to flow (Q) divided by the area of the conducting vessel (A). $V=\frac{Q}{A}$	3. $V=\frac{Q}{A}$.
4. Velocity is inversely proportional to the total cross-sectional area of the vessel.	4. Directly proportional to the total cross-sectional area of the vessel.
5. If vessel is constricted, velocity is increased. If the velocity reaches critical level, turbulence occurs.	5. If vessel is constricted, flow is decreased.
6. If the vessel is dilated, velocity is decreased.	6. In case of dilatation, flow is increased.

laminar flow or streamline flow and it is opposite to turbulent flow where the blood flows in all directions in the vessels (Fig. 28.11). When Poiseuille's experiment on fluids flowing through rigid tubes are extended to very high rates of flow, the relation between the driving pressure and the resulting rate of flow changes suddenly. Up to this part, the relation is linear, silent and streamline as predictable from Poiseuille's law, but, when a critical velocity of flow is reached, the slope of the curve changes, and thereafter, the velocity does not increase. An English engineer, Osborne Reynolds, in 1883, showed that at this point of transition, the flow changes from streamline flow to turbulent flow which is noisy (Fig. 28.12).

Critical velocity of flow occurs:

1. When the rate of blood flow becomes too great,
2. When it passes by an obstruction in a vessel e.g. measurement of blood pressure,
3. When it makes a sharp turn,
4. When it passes over a rough surface, the flow may then become turbulent rather than streamline.

Turbulent flow means that the blood flows crosswise in the vessel as well as along the vessel usually forming whorls in the blood called *eddy currents*.

These are seen in a rapidly flowing river at a point of obstruction. When eddy currents are present, blood flows with much greater resistance than when the flow is streamline—because the eddies add tremendously to the overall friction of flow in vessel.

28.16.1 Reynolds number-equation

It may be defined as the measure of the tendency for turbulence to occur. Osborne Reynolds (1883) made a classic investigation of this phenomenon which Poiseuille noted but did not analyse. Reynolds showed that a thin stream of dye—which he injected into fluid flowing under a high pressure head through a long glass tube—ceased to have an axial distribution when the velocity of flow became high and show vortex-lke eddies. The dye then fills the tube. The critical mean or average velocity of flow at which the motion became turbulent was dependent on:

(a) Radius or diameter of the tube,

(b) Viscosity of the fluid,

(c) Density of the fluid,

$$V = \frac{Re\,\eta}{\rho r}$$

Then, $Re = \frac{V\rho r}{\eta}$,

where V = critical average velocity (cms^{-1}) of the flow

Re = Reynolds number,

η = Viscosity (poises) (not centi-poises) of the fluid,

ρ = density of the fluid,

r = radius of the tube (cm).

In a long tube, turbulence develops when Re exceeds 1000. Turbulent flow entails a greater energy loss than the streamline (laminar flow) because more energy is dissipated in creating the kinetic energy of the eddies.

28.16.2 Significance of Reynolds number

1. **Blood**: The viscosity of blood is normally 1/30 poise and the density is only slightly greater than 1. Reynolds number of blood and other fluid is near 1000. The critical Re number for blood is 1000 i.e., turbulent occurs when its Re exceeds 1000. Since the viscosity of blood is relatively high, the Re for turbulent flow does not exceed in most parts of the circulation.
2. **Blood Vessel and Turbulent Flow**: None of the small "resistance" vessels of the vascular system shows turbulent flow but streamline flow only. The ventricles and the aorta, however, are normal sites of turbulence.

   When the Re rises above 200 to 400, turbulent flow occurs at some of the vessel. However, when Re rises above, approximately 2000, turbulence will usually occur even in a straight, smooth vessel. Reynolds number for flow in the vascular system even normally rises to 200 to 400 in larger arteries; as a result there is almost always some turbulence of flow at the branches of these vessels. Re can rise to several thousands, during the rapid phase of ejection by the ventricle in the proximal aorta and pulmonary artery and causes considerable changes in the flow in these two vessels where many conditions are appropriate for turbulence:

   (a) High velocity of blood flow,
   (b) Pulsatile nature of the flow,
   (c) Sudden change in vessel diameter,
   (d) Large vessel diameter.

   However, in small vessels, Re is almost never high enough to cause turbulence. Reynolds number of different vessels are given below (Table 28.5).
3. **Stenosis or Atherosclerosis, i.e., Pathological Obstruction of Blood Vessel**: The critical value of 1000 for turbulent flow applies only to long straight tubes. If a small area of narrowing occurs in a tube or in a vessel affected by the development of an atheromatous plaque (i.e., atherosclerosis), the mean velocity through this reduced lumen is greatly increased and turbulence develops at and beyond the site of constriction. This causes a vibration which can be felt and a murmur or bruit which can be heard with stethoscope. Turbulence due to a stenosis occurs at much lower Re than 1000.
4. **Heart Sounds**: The heart sounds are due to the closure of the valves and do not necessarily involve turbulence. However, if the flow becomes very high, as in heavy exercise, quite normal systolic murmurs can be heard with the stethoscope.
5. **Systolic Ejection**: Aortic turbulence occurs during systolic ejection. During systolic ejection alone, however, the Re exceeds 1000 and turbulence results. In heavy exercise, aortic turbulence becomes very marked and extends into the large arterial branches.
6. **Korotkow Sounds during Blood Pressure Measurement by Sphygmomanometer**: The Korotkow sounds heard in auscultatory method by sphygmomanometer originate from turbulence as the systolic blood pressure pushes blood though the semi-collapsed vessel beyond the partially occluded branchial artery. Once the cuff pressure is below that of the diastolic pressure the downstream vessel is wide open and the sound disappears.
7. **Blood Speeds through a Narrowed Region to Wider Lumen (Vascular Stenosis)**: Turbulent flow itself need not cause murmurs,

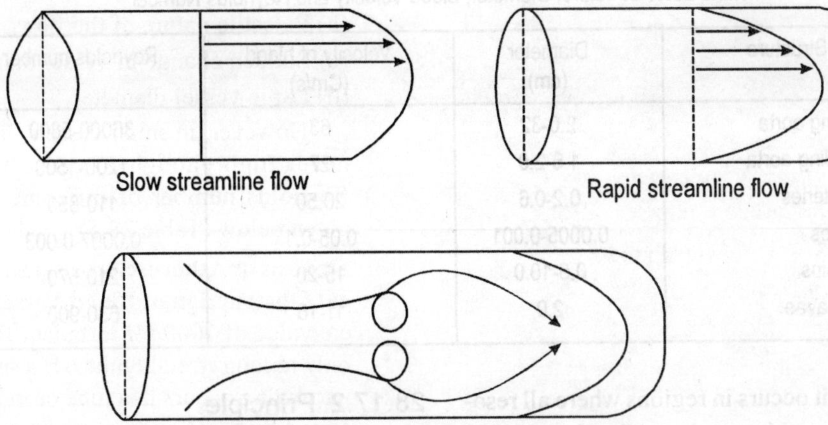

Fig. 28.11: Different types of blood flow in vessel

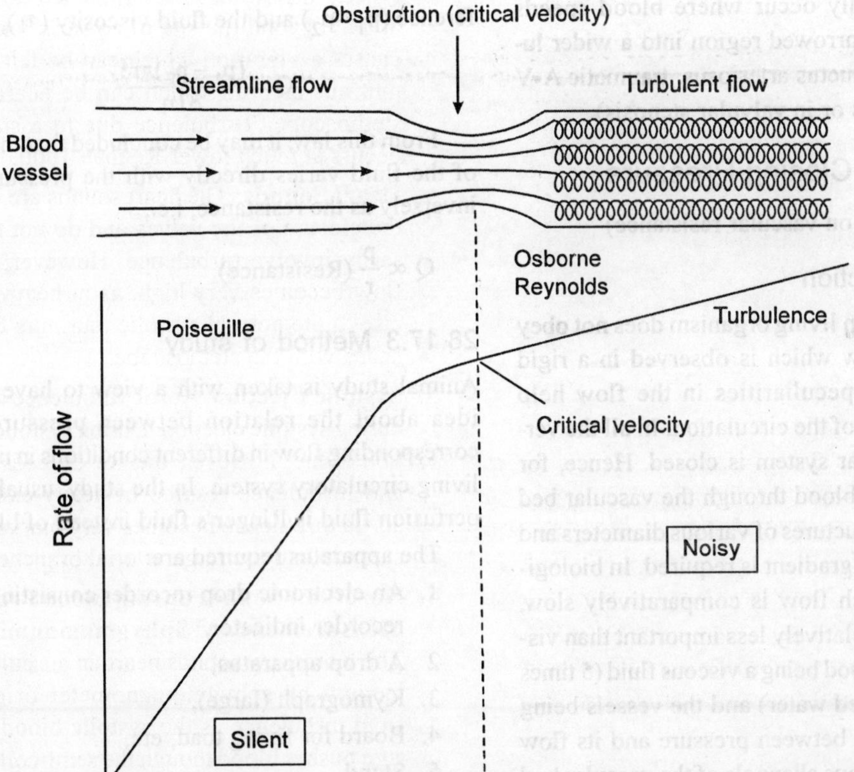

Fig. 28.12: Changes in type of flow from streamline to turbulent

Table 28.5: Structure, Diameter, Blood Velocity and Reynolds Number

Structure	Diameter (cm)	Velocity of blood (Cm/s)	Reynolds number
Ascending aorta	2.0-32	63	36000-5800
Descending aorta	1.6-2.0	27	1200-1500
Large arteries	0.2-0.6	20.50	110-850
Capillaries	0.0005-0.001	0.05-0.1	0.0007-0.003
Large veins	0.5-10.0	15-20	210-570
Venae Cavae	2.0	11-16	630-900

but when it occurs in regions where all resonance is prominent, then vibrations of the wall are aroused and these are responsible for the murmur heard. Such vibrations characteristically occur where blood speeds through a narrowed region into a wider lumen (e.g., ductus arteriosus, traumatic A-V anastomosis or in valvular stenosis).

28.17 Critical Closing Pressure

(Effect of pressure on vascular resistance)

28.17.1 Introduction

The flow of blood in living organism does not obey the Newtonian flow which is observed in a rigid system and these peculiarities in the flow help proper functioning of the circulation. In all the vertebrates the vascular system is closed. Hence, for the proper flow of blood through the vascular bed having tube-like structures of various diameters and calibres, a pressure gradient is required. In biological system in which flow is comparatively slow, inertia effects are relatively less important than viscous effects. The blood being a viscous fluid (5 times viscous than distilled water) and the vessels being elastic, the relation between pressure and its flow throughout the various channels of the vascular bed will not be as identical as that will be observed in a rigid tube.

28.17.2 Principle

From Poiseuille's law, we know that when total volume of fluid (Q) is flowing steadily through a tube of radius (r), length (1), there is a pressure difference ($p_1 - p_2$) and the fluid viscosity (η), etc.

$$Q = \frac{(p_1 - p_2)\pi r^4}{81\eta}$$

From this law, it may be concluded that the flow of the fluid varies directly with the pressure and inversely as the resistance, i.e.,

$$Q \propto \frac{p}{r} \text{ (Resistance)}.$$

28.17.3 Method of study

Animal study is taken with a view to have some idea about the relation between pressure and corresponding flow in different conditions in normal living circulatory system. In the study, usually the perfusion fluid is Ringer's fluid instead of blood.

The apparatus required are:

1. An electronic drop recorder consisting of a recorder indicator,
2. A drop apparatus,
3. Kymograph (large),
4. Board for fixing toad, etc,
5. Stand.

Toads are used in this study. Animals are pithed and chest is opened. Amphibian Ringer from bu-

Records of flow under various changes of pressure are tabulated in the following way.

Initial height of the fluid in burette (in cm.)	Height of perfusion fluid (in cm) in burette	No. of drops (Total)	Time in seconds	No. of drops per minute

rette is passed through the canula into the sinus venosus and the great arteries are cut to allow the fluid come out and the drops recorded. The records of flow are taken under various pressure in the burette. Experiments are also performed with the perfusion fluid containing drugs of different dilution.

28.17.4 Discussion

From physics it is known that pressure and flow have got a linear relationship and a straight line may pass through the origin when plotted on a graph paper. But in biological system this rule is not followed. The line is not a perfect straight line and does not pass through the orgin. The flow of fluid ceased at a certain pressure of the perfusion fluid and this pressure is called "critical closing pressure", i.e., the pressure below lower limit at which no flow ensues. Lee (1977) in his experiments with mammals and perfusing the vascular bed with plasma, whole blood and the influence of sympathetic stimulation and inhibition show the results in the flow-pressure curves (Fig. 28.13).

The curve illustrates normal effect on blood flow through a small vascular bed caused by changing the arterial pressure. As the arterial pressure falls from 130 mm Hg the flow decreases rapidly at first because of two factors:

(a) The decreased pressure difference between the artery and the vein of the tissue, and

(b) Decreasing diameter of the vessels.

At 20 mm Hg, blood flow ceases entirely. The pressure at which the blood flow stops flowing is called the "Critical Closing Pressure" because at this point, at this pressure the small vessels—the arterioles in particular—close so completely that all flow ceases.

28.17.5 Mechanism of critical closing pressure

(A) Biological Mechanism

1. **Resistance Exerted by the Capillaries:** Blood vessel diameter increases when the pressure inside increases because of distensibility of the blood vessel. As the diameter increases, it reduces the resistance of the vessel. Conversely, decrease in pressure increases the resistance.
2. **Tissue Pressure:** The blood vessels are surrounded by tissues and exert some amount of pressure on the vessels. The tissue pressure is certainly lower than intraluminal pressure. But when the intraluminal pressure falls below the tissue pressure, the blood vessels collapse.

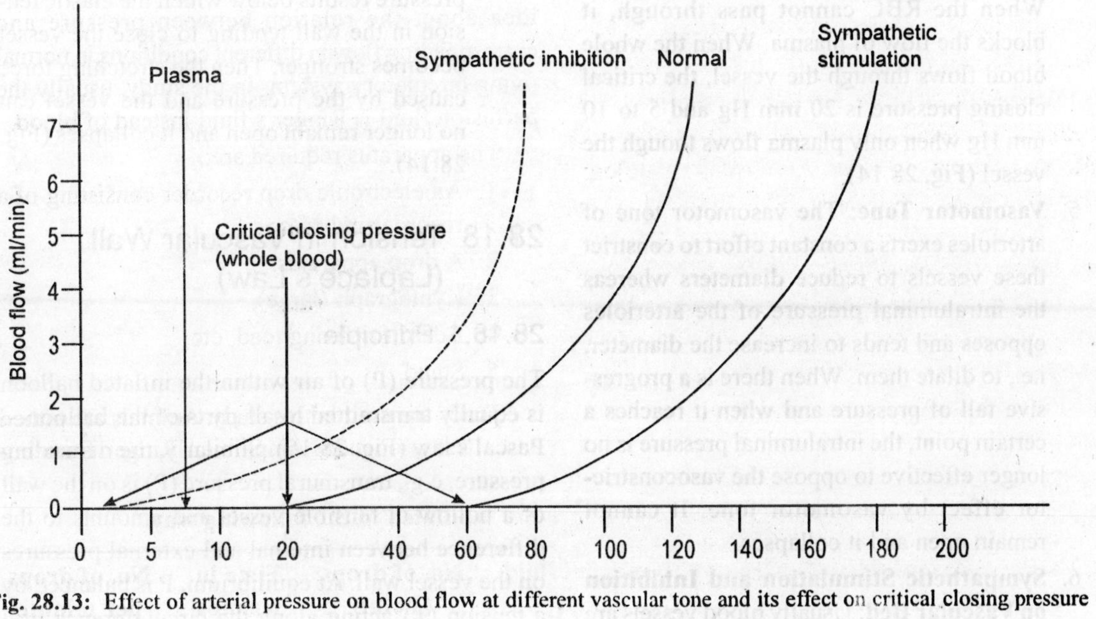

Fig. 28.13: Effect of arterial pressure on blood flow at different vascular tone and its effect on critical closing pressure

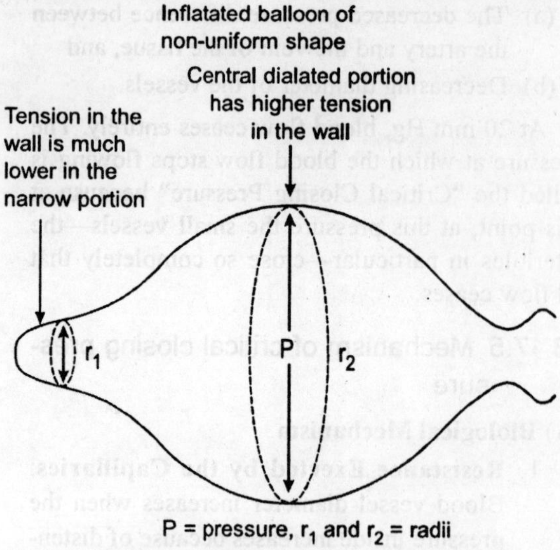

Fig. 28.14: Pressure-tension relation in a balloon. The pressure (P) is equally exerted to all sides of the air in the inflated balloon (Pascal's law). The tension (T) varies with the radius in that portion (Laplace's law)

3. **Inactive Tissues**: In inactive tissues, the pressure in many capillaries is low because the precapillary sphincters and metarterioles are constricted and most of these capillaries collapse.
4. **Size of the Red Blood Cell**: Some pressure is required to force RBC through capillaries which have diameters less than the red cells. When the RBC cannot pass through, it blocks the flow of plasma. When the whole blood flows through the vessel, the critical closing pressure is 20 mm Hg and 5 to 10 mm Hg when only plasma flows though the vessel (Fig. 28.14).
5. **Vasomotor Tone**: The vasomotor tone of arterioles exerts a constant effort to constrict these vessels to reduce diameters whereas the intraluminal pressure of the arterioles opposes and tends to increase the diameter, i.e., to dilate them. When there is a progressive fall of pressure and when it reaches a certain point, the intraluminal pressure is no longer effective to oppose the vasoconstrictor effect by vasomotor tone. It cannot remain open and it collapses.
6. **Sympathetic Stimulation and Inhibition on Vascular Bed**: Usually blood vessels are stimulated by sympathetic nervous system. Vascular reflexes may cause stimulation or inhibition of blood vessels. Stimulation decreases rate of blood flow whereas inhibition allows more blood flow and decreases the critical closing pressure to as low as 5 mm Hg. Whereas stimulation increases the critical closing pressure to as high as 100 or more mm Hg, it indicates that strong stimulation can stop blood flow through some tissues even when the arterial pressure is at its normal level of 100 mm Hg.
7. **Adrenaline and the Flow**: Adrenaline causes diminution of flow which is due to the resistance offered by the blood vessel by increasing its vasomotor tone. In this, critical closing pressure is higher than that was observed in the case of perfusion with plain Ringer's fluid (Dilution factor).

(B) Physical Mechanism—Laplace's Law: Laplace's law states that the circumferential force tending to stretch the muscle fibres in the vascular wall is proportional to the diameter of the vessel times the pressure or $P \propto D.P$. Therefore, when the pressure falls in the vessels, the vascular diameter decreases, the force tending to keep the vascular wall stretched decreases rapidly. Then a pressure results below which the elastic tension in the wall tending to close the vessel becomes stronger. Then the stretching force caused by the pressure and the vessel can no longer remain open and it collapses (Fig. 28.14).

28.18 Tension in Vascular Wall (Laplace's Law)

28.18.1 Principle

The pressure (P) of air within the inflated balloon is equally transmitted to all parts of the balloon—Pascal's law (Fig. 28.15). Similarly, the distending pressure, e.g., transmural pressure (P) is on the wall of a hollow distensible vessel and amounts to the difference between internal and external pressures on the vessel wall. At equilibrium, P is balanced by a tension (T) acting along the circumference of a cross-section of the vessel and tending to collapse it (Fig. 28.15).

28.18.2 Laplace's law

Laplace's law states that the distending pressure (P) in a distensible hollow object is equal at equilibrium to the tension in the wall (T) divided by the 2 principal radii of curvature of the object (r_1 and r_2):

$$P = T\,(1/r_1 + 1/r_2)$$

In the equation, P is actually transmural pressure, the pressure on one side of the wall minus that on the other. T is expressed in dynes/cm and r_1 and r_2 in cm, so P is expressed in dynes/cm^2. In a sphere, $r_1 = r_2$, so $P = 2T/r$.

In a cylinder such as a blood vessel, one radius is infinite, so

$$P = T/r.$$

• **Example:** The capillaries which are thin-walled structures and delicate are not more vulnerable to rupture due to distending pressure because of their small diameter. The smaller the radius of a blood vessel, the less the tension in the wall necessary to balance the distending pressure. This protective small size is the operation of Laplace's law which is one of the important physical principles which is applicable in case of blood vessels. It has several applications in case of other hollow structures.

28.18.3 Indication

This law indicates that the smaller the radius, the greater the mechanical advantages in term of the tension that must be borne by the vessel wall.

28.18.4 General feature of the law

(a) **Radius of the Vessel:** According to Laplace's law, the smaller the radius of the vessel, greater the mechanical advantage in terms of tension. Thus a capillary having the diameter of 7 µm can withstand an intravascular pressure of 25 mm Hg with a tension of only 16 dynes, whereas the aorta having a large radius of 2.5 cm carries a tension in the wall of 170000 dynes

(b) **Thick-walled Structure**: In case of arteries, which are thick-walled structures, the equation is modified for wall thickness, so that

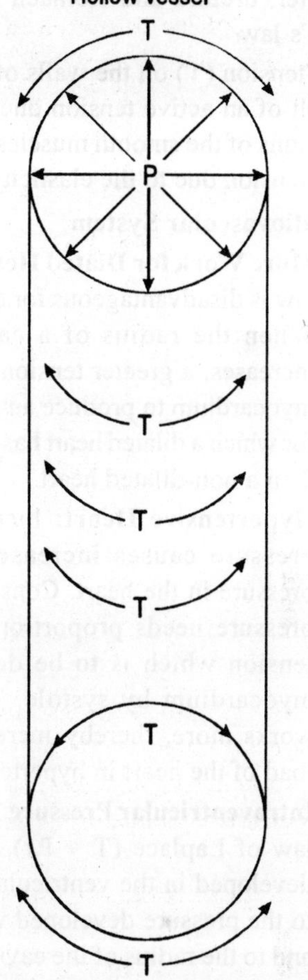

Fig. 28.15: Transmural pressure (distending pressure) (P) and tension (T) on the wall of hollow vessel

$$T = P.r/h,$$

where h = wall thickness.

Thus, the thick wall of the vessels, e.g., arterioles, exhibits lower wall tension and the smooth muscle layer which is thicker than the aorta (elastic and fibrous layer are thinner than aorta) receives smaller load in spite of high intravascular pressure. This provides a mechanical advantage for alteration of blood vessel diameter in small vessels. This equation is applied in considering aortic aneurysm (bulging area for the disease in the aortic wall).

28.18.5 Biological application of Laplace's law, structures obeys Laplace's law

Besides hollow blood vessels, the hollow visceral organs like heart, urinary organs like urinary bladder, ureter, urethra and stomach and lungs obey Laplace's law.

The tension (T) on the walls of these organs is the result of an active tension due to the tone and contractions of the smooth muscles on the wall and passive tension due to the elasticity of the wall.

I. Cardiovascular System

1. **More Work for Diated Heart**: The Laplace law is disadvantageous for the dilated heart. When the radius of a cardiac chamber increases, a greater tension develops in the myocardium to produce any given pressure, for which a dilated heart has to do more work than a non-dilated heart.
2. **Hypertensive Heart:** Increase in arterial pressure causes increase in distending pressure in the heart. Consequently, rise in pressure needs proportionate increase in tension which is to be developed in the myocardium by systole. Thus, the heart works more, thereby increasing the work load of the heart in hypertension.
3. **Intraventricular Pressure**: According to the law of Laplace (T = P.r), the tension (T) developed in the ventricular wall is related to the pressure developed within the cavity and to the radius of the cavity. For the active ventricle, this indicates that the greater the ventricular size, the greater the wall tension needed to maintain a given intraventricular pressure.
4. **Thickness of the Blood Vessel Wall**
   (a) A large artery (e.g., aorta) requires generation of a high tension (170000 dynes/cm^2) for balancing the distending force of a mean blood pressure of 100 mm Hg. So, a large artery is provided with very thick coats of elastic nature having more elastic and collagen fibres.
   (b) In case of arterioles where the wall is relatively thicker, a lower wall tension (1200 dynes/cm^2) is required. The smooth muscle coat (smooth muscle coat is thicker than large artery like aorta and elastic and collagen fibre coats are thinner than aorta) receives smaller load for balancing a relatively high intravascular pressure which provides a mechanical advantage for changing blood vessel diameter in these small vessels.
   (c) A minute capillary requires a small tension (16 dynes/cm^2) for balancing a pressure of 30 mm Hg. The capillary having thin endothelium layer only—without an elastic layer—can provide this tension (Fig. 28.16).

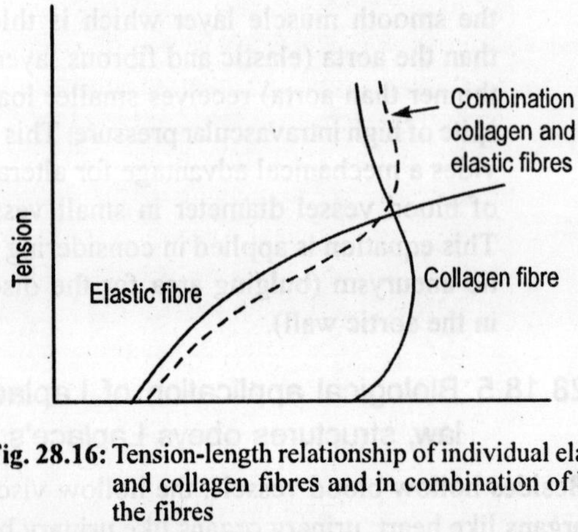

Fig. 28.16: Tension-length relationship of individual elastic and collagen fibres and in combination of both the fibres

5. **Aortic Aneurysm:** The equation T = P.r/h has a relation for development of aortic aneurysm (a large saccular bulging area in the aortic wall). In aneurysm, rise in blood pressure during any disease causes increases in radius with further stretching and thinning (because of lesser muscle layer and greater amount of elastic and fibrous layer than the arterioles) of the aortic wall, thereby increasing the wall tension; this sequence lead to further stretching and thinning, etc. Thus, an aneurysm becomes unstable, and expansion leads to rupture of the aorta in some cases.
6. **Closure of Vessels at Critical Pressure Level:** The closure of vessels at an exact critical pressure level can be explained with the help of Laplace's law. Ths law states that the circumferential force tending to stretch the muscle fibres in the vascular wall is proportional to the diameter of the vessel times the pressure, i.e., $F \propto D.P.$ (where F = force, D = diameter of the vessel, P = pressure). Therefore, as the pressure in the vessel falls—thus allowing the vascular diameter

also to decrease—the force tending to keep the vascular wall stretched decreases rapidly. Then, at a certain pressure below which the elastic tension in the wall—which tends to close the vessel—becomes stronger than the stretching force produced by the pressure, and the vessel collapses.

II. Urinary System

The urinary system consists of a series of tube-like structures transmitting the fluid and reservoirs for storage of fluid and is obviously under the laws of fluid mechanics. Among these laws, the most appropriate for this system is Laplace's law which states:

P = 2T/r in a sphere,

and P = T/r in a tube,

where P = intraluminal pressure in case of sphere, resistance to flow in case of tube, provided external pressure is constant.

T = tension on the wall of the sphere or tube, i.e., the sum of the force of active tension developed by muscular contraction plus the passive tension exerted by the elastic properties of the wall.

r = radius of the tube or sphere.

If the system is at equilibrium, these relationships are very much helpful respec-tive of the characteristics of the wall of the viscus.

1. **Urinary Flow in the Renal Pelvis**: The renal pelvis is relatively indistensible. Like calyces, the pelvis contracts and dilates rhythmically—systole and diastole. Renal pelvis fills passively from the calyces under a very low pressure during diastole and then contracts to press only a small percentage of its volume into the upper end of the ureter. The normal intrapelvic pressure in a very few instances exceeds more than a few cm of water, even in a systole, but tends to be higher in the narrow ureter-like pelvis than in the large ampullary-type pelvis. This is a direct effect of Laplace's law.
2. **Filling up the Urinary Bladder and Micturition**: Urine enters the urinary bladder without producing much intravesical pressure until it is well filled. Besides that, smooth muscle of urinary bladder has the property of plasticity and, when it is stretched, the tension produced initially is not maintained. Study with cystometrogram for relationship between intravesical pressure and volume and the curve shows three components:
   (a) Initial slight rise in pressure when the first increments in volume are produced (segment 1a),
   (b) A long, nearly flat segment as further increments are produced (between intravesical volume of 50 ml to 400 ml) (segment 1b), and
   (c) A sudden, sharp rise in pressure as the micturition reflex is triggered at 400 ml (segment 1c).

   When the bladder volume is about 150 ml, the first urge to void is felt and marked sense of fullness is felt at about 400 ml.

   The flatness of segment 1b between 50 ml to 400 ml is the manifestation of Laplace's law which states P = 2T.r in a sphere. In case of urinary bladder, the tension increases as the organ fills and also the radius increases until the bladder is full, the intravesical pressure increase is slight.
3. **Continence of Urine**: In spite of periodical discharge of urine under voluntary control, the urinary bladder must also retain urine. Continence entirely depends upon detrusor muscle. The neck of the bladder, a tubular structure having muscularis, which is continuous with the body of the detrusor. Except micturition, the inherent tonicity of the smooth muscle and elastic tissue of the bladder neck holds its walls in firm apposition. According to Laplace's law, the resistance to flow through a tube varies inversely to its radius, the tone in its walls being constant, it follows that the intravesical pressure, helped by accommodation, never exceeds the intraurethral pressure under resting conditions.
4. **Prevention of Reflux of Urine During Bladder Contraction**: During micturition,

the urinary bladder contractions increase the intravesical pressure and at the same time the lower ends of the ureters—which pass obliquely through the wall of the bladder—collapse. Under these conditions, the radius of ureter (r) is very small or may be 0. In a cylindrical tube, when r = 0 (e.g., ureter), for any given tension on the uretheral wall, the pressure (P) becomes very high in the collapsed ureter, certainly higher than the intravesical pressure and prevents any reflux or backflow of urine into the ureters from the urinary bladder. This is one of the corollaries of this law.

5. **Urinary Flow through Urethra**: According to Laplace's law, P = T/r in case of cylindrical tube. When the proximal part of urethra remains constricted where radius is 0 (r = 0), pressure (P) in the urethra is extremely high and prevents entry of urine into urethra from urinary bladder. But, during micturition, longitudinal muscles of the urethral wall contract for dilatation of urethra and increase the radius (r) and lower the tension (T), thus reducing the elasticity and constrictor muscle tones of urethral wall. Thus, increase of radius (r) and fall of tension (T) decrease the intraurethral pressure (P), thereby allowing urinary flow through the urethra.

III. Respiratory System

1. **Role of Surfactant to "Stabilize" the Alveoli to Prevent Instability of Alveoli of Lungs**: The effect of surface tension in causing collapse of an alveolus becomes greater when the diameter of the alveolus decreases. This effect can be explained by Laplace's law. It states that the transalveolar pressure (P) required to keep the alveolus expanded in directly proportional to the tension (T) in the alveolar wall (in this case, the surface tension of the fluid), divided by the diameter (r). Therefore, as the diameter becomes less, the required pressure becomes proportionaly greater.

   Surfactant plays an important role to minimize the effect of surface tension in causing collapse of the lungs and thereby preventing the phenomenon of "instability of alveoli" to "stabilize" alveoli in any area of the lung for remaining approximately the same size.

2. **Air Entry into the Fetal Lungs with the First Breath at Birth**: The lungs of the fetus at birth are filled with a fluid of pulmonary origin produced by type II pulmonary alveolar cells or pneumocytes called surfactant. The first breath of fetus at birth causes entry of air-liquid interface in the alveoli. At this interface, surfactant acts by forming a monomolecular layer between the fluid lining the alveoli and the air in the alveoli. This prevents the development of a water-air interface, which has 2-14 times as much surface tension as the surfactant-air interface. There is surface tension at the air-liquid interface and because of the Laplace's law, P = 2T/r, the reduction of radius of the alveoli during expiration, the surface tension increases resulting in collapse of the lung but is prevented by surface active material called surfactant which reduces surface tension.

IV. Digestive System

Storage Function of Stomach: The muscular wall of the body of stomach has little tone, so that it can bulge progressively outward to accommodate more and more food up to a limit of one litre. The pressure in the stomach is low until the limit is approached for these reasons:

(a) Plasticity of smooth muscle of stomach (plasticity means increase in length without significant change in tone),

(b) Laplace's law: The stretching force on the walls increases in direct proportion to this radius of curvature, i.e., pressure increases slightly despite distension.

(c) Vagal reflex due to stretch inhibits the muscle activity.

28.19 Vascular Resistance (Resistance of Blood Flow)

28.19.1 Principle

The resistance to flow through a hydraulic circuit can be expressed according to Ohm's law of electric circuit. The analogy applied in case of vascular bed is known as "general resistance equation" which is:

Blood flow =

$$\frac{\text{pressure difference across a vascular bed}}{\text{vascular resistance}}$$

i.e., $Q = \frac{P_A - P_V}{R}$,

where, Q = blood flow,

$P_A - P_V$ = pressure difference between artery and vein,

R = resistance in arbitrary units.

The blood flow to an organ (Q) is largely controlled by metabolism in the organ, the products of which affect local vascular resistance (local VR). Application of the above equation for defining the control of blood flow to an organ is:

$$Q = \frac{P_A - V_V}{\text{local VR}},$$

where local VR = vascular resistance in the organ or tissue. This equation is applied for describing the total vascular resistance across the systemic organ beds of the body which causes maintenance of blood pressure. After rearranging the above equation:

SAP = CO × TPVR,

where, SAP = systemic arterial pressure,

CO = cardiac output,

TRVR = total peripheral vascular resistance.

28.19.2 Resistance, blood flow, series vessels resistance and series vessels

Suppose three vessels (a, b, c) are connected in series flow as in across an organ bed, or when vascular lesions occur in two portions of the same vessel and their resistances are R_1, R_2 and R_3. If we calculate the total resistance of R (total), then it would be the sum of the three R_1, R_2 and R_3, according to Ohm's law.

$$R\text{ (total)} = R_1 + R_2 + R_3.$$

Any number of resistance in a series must be added together, i.e, total peripheral resistance is equal to the total resistances of the arteries, arterioles, capillaries and veins together (Fig. 28.17A).

- **Resistance and Parallel Vessel**: If four vessels are in a parallel system having resistances of R_1, R_2, R_3 and R_4—the diameters of the vessels being slightly different from each other—greater amounts of blood will flow through this system than through any one of the vessels at a given pressure difference. Therefore, the total resistance is lesser than the resistance of any single vessel (Fig. 28.17B).

For calculation of the total resistance, determination of the conductance of an individual vessel is done (expressed as millilitre per second per mm Hg), which is equal to the reciprocal of the resistance $\frac{1}{R_1}, \frac{1}{R_2}, \frac{1}{R_3}$, and $\frac{1}{R_4}$. The total conductance of all the vessels is equal to the sum of the conductances of each vessel, such as:

$$C\text{ (total)} = \frac{1}{R_1} + \frac{1}{R_2} + \frac{1}{R_3} + \frac{1}{R_4}.$$

And the resistance through this parallel system is the reciprocal of the total conductance, i.e.,

$$R_{(\text{total})} = \frac{1}{\frac{1}{R_1} + \frac{1}{R_2} + \frac{1}{R_3} + \frac{1}{R_4}}.$$

28.19.3 Characteristics of parallel circuits

1. In the parallel circuit, adding resistance which is even higher than the individual resistances in

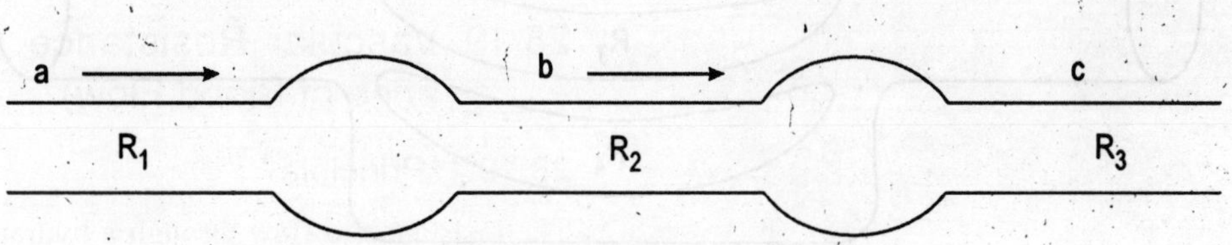

Fig. 28.17A: Vascular resistances in series vessels

the circuit will add another conductance raising the total conductance and lowering the total conductance as well as the total resistance. This phenomenon explains how the capillary bed having large number of high resistance units with larger cross-sectional area shows a lower total resistance than the arteriolar bed.

2. If there is drop of one resistance below the others in a parallel circuit, the total conductance will rise and the total resistance will fall.
3. In a circuit like the peripheral circulation where many organsare placed in parallel, moderate increases or decreases of resistance in one organ do not have any great effect on total vascular resistance or blood pressure.

28.19.4 Importance of equations

1. Clarification of interrelationships among pressure, flow and resistance.
2. Measurement of vascular resistance and the "vascular resistance unit" are very much helpful in studying human disease because these resistance relations may be altered in diseased states:
   (a) The systemic vascular resistance is elevated in hypertension.
   (b) In ventricular septal defect, there is rise in resistance in the pulmonary arterioles.
   (c) To solve the diagnosis, such as normal pulmonary resistance is 2RU and if it is elevated to 24 RU, then it indicates the pulmonary disturbances.

28.19.5 Factors responsible for vascular resistance in the circulation

1. Characteristics of blood flow,
2. The diameter of the vessels,
3. The distending pressure in the vessels,
4. Viscosity of the blood.

28.19.6 Units of resistance

Resistance is the hindrance to blood flow in a blood vessel and cannot be measured directly. Therefore, resistance is calculated by measuring the blood flow and pressure difference in the vessel. If the pressure difference between two points in a vessel is 1 mm Hg and the flow is 1 ml/sec, then the resistance is 1 "peripheral resistance unit", abbreviated by PRU. The units of resistance, i.e., pressure divided by flow, are expressed in dyne -sec-cm^{-5}. But this is a complex unit for which the cardiovascular system resistance is expressed in R units by dividing pressure in mm Hg by flow in ml/s. Thus, for example, when the aortic pressure is 100 mm Hg and the left ventricular output is 100 ml/s, the "total peripheral resistance" is

$$\frac{100 \text{ mm Hg}}{100 \text{ ml/s}} = 1 \text{ R unit}$$

28.19.7 Expression of resistance in CGS units

Sometimes, a physical unit called the CGS (centimetres, grams, seconds) unit is applied for expression of resistance. This is expressed in dyne-sec-

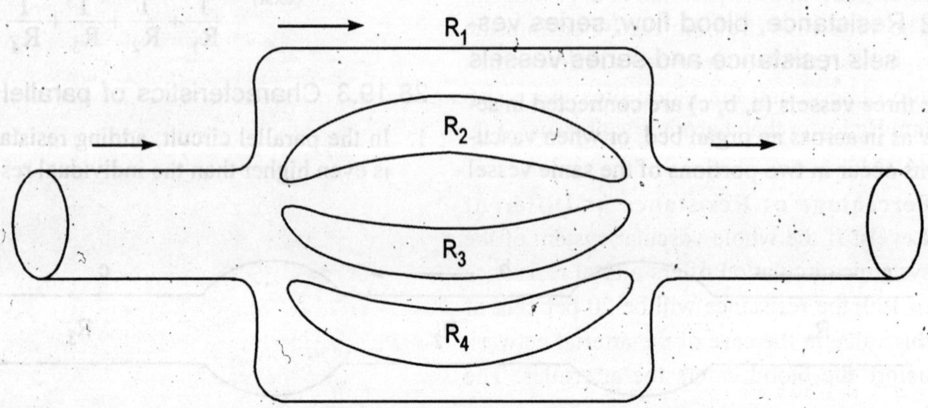

Fig. 28.17B: Vascular resistances in parallel vessels

ond/centimetre^{-5} (dyne-sec-cm^{-5}). The resistance unit is calculated in the following way:

$$\frac{\text{Pressure}}{\text{Flow}} \text{ i.e., } R\left(\text{in}\frac{\text{dyne-sec}}{\text{cm}^{-5}}\right)$$

$$= \frac{1333 \times \text{mm Hg}}{\text{ml/sec}}$$

- **Resistance in the Distensible Tube**

(Sec comparison of Poiseuille's experimental material and circulation).

28.19.8 Vascular resistance at different levels of circulatory system

1. **Slight Resistance in the Large Arteries**: By measuring blood pressure in the various vessels at different levels of the circulatory system, it has been shown that, in all large arteries, the pressure is slightly lower than in the aorta. It is due to the resistance offered to the flow of blood in the large arteries which is slight and a minor amount of potential energy contributed to the blood which comes out of the aorta and is utilized to the blood flow along the arteries.

   The normal pressure in the aorta during systole is 120-130 mm Hg and pressure at the beginning of the arterioles is 70-80 mm Hg. While the blood is flowing through the arterioles, the pressure falls sharply and the blood flows into the capillaries under a pressure of 20-30 mm Hg. After flowing through the capillaries, the blood enters the small veins under a pressure of 8-10 mm Hg. The vena cava draining the blood into the right heart flows under a pressure lower than atmospheric and the blood pressure is 2-6 mm Hg lower than barometric pressure (Fig. 28.18).

2. **Percentage of Resistance at Different Levels**: If the whole vascular system of the systemic circulation offers a total resistance of 100, the resistance will be 20 per cent of this value in the case of the arterial network before the blood enters the arterioles. The entire venous system will be about 10 per cent of this value. The arterioles and capillaries will offer 70 per cent of the total resistance of the entire blood stream and $\frac{3}{4}$ of this value is due to friction of the blood against the wall of the arterioles and $\frac{1}{4}$ to the resistance to the blood flow in the capillaries.

3. **Highest Resistance in the Arterioles**: The highest resistance is offered by the arterioles. It is due to the summary value of the internal surface of all the arterioles, i.e. their summary surface is many times as great as the summary surface of all the larger arteries. But the velocity of the blood flow in the arterioles is still high which is about $\frac{1}{2}$ to $\frac{1}{3}$ of the velocity of the blood flow in the aorta due to the summary section of all the arterioles which is about three times that of the aorta. The resistance overcome by a liquid in its friction against the walls of blood vessels is proportional to the square of its velocity. The factors responsible for the resistance to the blood flow in the arterioles which exceeds that in the system of all the large arteries are:

   (a) Relative high velocity of the blood flow,

   (b) Large number of long arterioles,

   (c) Very large surface against which the blood rubs (Fig. 28.18).

4. **Less Resistance in the Capillaries**: More number of capillaries (several thousand million) are present in body than the number of arterioles (several hundred thousand). The sum of the lumens of all the capillaries is, therefore, 200-400 times than that of the summary lumen of all the arterioles and about 800 times as great as the cross-section of the aorta. Due to this, the blood in the capillaries flows very slowly. Since the blood flows through them slowly and the capillaries are short, they offer less resistance than the arterioles.

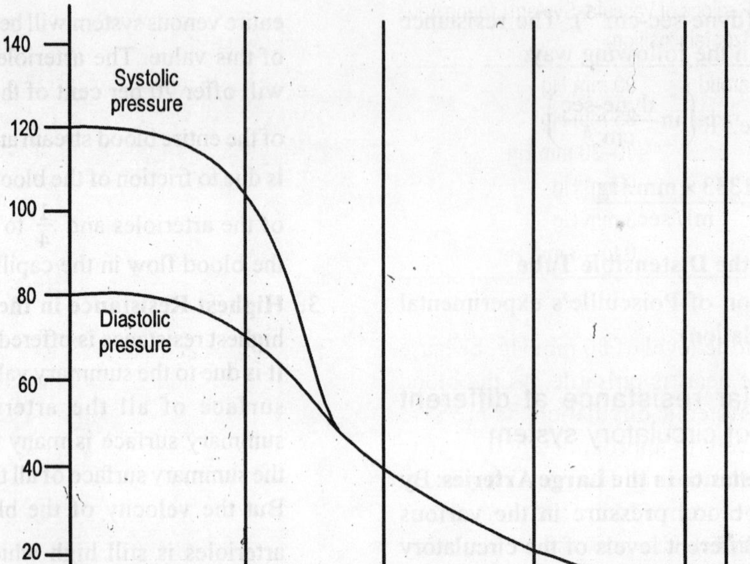

Fig. 28.18: Graph showing gradient of pressure into the large and small arteries, capillaries and veins due to resistance

5. **Resistance in the Pulmonary Circulation:** The resistance in the pulmonary circulation is about $\frac{1}{3}$ to $\frac{1}{6}$ of the systemic circulation. The pressure in the pulmonary artery is, therefore, $\frac{1}{5}$ to $\frac{1}{6}$ the pressure in the aorta.

28.20 Pressure in the Vascular System

28.20.1 Pressure variation in vascular system

The veins are more easily collapsible by external applied pressure than the arteries. The reason is not for the thinner wall but for low pressure inside the vein. Before the blood travels from arteries to veins, if we measure of the average pressure of the blood in successive regions of the vascular system, we find that the pressure which drives the blood onwards declines from the heart through the arterial and capillary and venous regions as it passes along. In the initial part of the aorta close to the heart, the average pressure is the highest. In the large arteries (e.g., carotid, brachial, femoral, etc.) the pressure is high—about 120 mm Hg and about 2-10 mm Hg lower than the aorta, because the resistance is very weak. It is somewhat lower in the smaller arteries. The chief function of the large arteries is to serve as elastic conduits, whereas the small arteries and arterioles leading from the arteries to the capillaries have in addition the function of regulating the amount of blood flowing through the capillary area of the organs which they supply. The pressure is lower in the arterioles and capillaries (30-50 mm Hg) and still lower in the veins—not greater than 1 mm Hg. The lowest pressure is in the veins closest to the right atrium and where the pressure is about at atmospheric pressure or even lower in mammals. The veins have the function of conducting blood at low pressure from capillaries to heart and of holding any excess of blood which is not immediately taken up by the heart (Table 28.6).

Because the heart pumps blood continually into the aorta, the pressure in the aorta is definitely high, averaging approximately 100 mm Hg. The arterial pressure fluctuates betwen a systolic level of 120

Table 28.6: Average Height of the Mean Blood Pressure at Different Regions of Vascular System (Young Adult Man, Horizontal Position)

Large arteries, e.g., carotid	90 mm Hg
Medium arteries, e.g., radial	85 mm Hg
Capillaries	10–30 mm Hg
Small Veins (arm)	9 mm Hg
Inferior vena cava	3 mm Hg
Large Veins (neck)	0 to –8 mm Hg

mm Hg and a diastolic level of 80 mm Hg, because the pumping of the heart is pulsatile. As the blood flows through the systemic circulation, its pressure declines progressively to approximately 0 mm Hg by the time it reaches the right atrium. The rate at which blood escapes from a blood vassel indicates the differences between the pressure in the blood vessel and the atmospheric pressure. On cutting through an artery, blood escapes from the central end, which is nearest to the heart, with great force and a series of spurts rapidly indicates: (1) The pressure of the blood is greater than atmospheric pressure, and (2) The flow from the heart is pulsatile (Fig. 28.19).

The rate and force at which blood escapes is a rough indication of this pressure and this varies considerably in different parts of the circulation. If a vein is cut across, blood escapes chiefly from the peripheral end and instead of spurting out with each heart beat, it flows steadily and with every little force. Even the large veins in the neck near the heart may bleed very little though the small cut portion in the wall of the vessel indicates the pressure is about atmospheric pressure or lower. If a vein is ligatured, it swells up at the peripheral part of the ligature. If the periphereal end of the vein is connected to a mercurial manometer, the pressure is seen to be a few mm Hg for which it is customary to use manometer with an aqueous solution, e.g, sodium citrate.

28.20.2 Reasons

1. This gradual decrease in pressure is a physical phenomenon. This is due to the resistance to flow, i.e., friction while passing through the vessel. The energy of pressure is gradually dissipated as the heat of friction. Thus, in the aorta, the resistance is almost zero and in the large arteries it is very weak. The resistance increases rapidly in the very small arteries and it is very much high in the arterioles and pressure is 30 mm Hg. The pressure at the arterial end of capillaries is 30 mm Hg but at the venous end it is about 10 mm Hg. The decrease in arterial pressure is directly proportional to the vascular resistance.

   The pressure at the beginning of the venous system (i.e., venules) is about 10 mm Hg and at the right atrium it is almost 9 mm Hg. This large decrease in pressure in the veins indicates that the veins have more resistance. Much of ths resistance is caused by compression of the veins from the external region.

2. **The Gradient of Pressure:** Along the length of the horizonatal tube, if friction becomes greater at any point, the fall of pressure will be greater at that region as the fluid flows through the region. And this fall is greater than the fall in any other tube. The fall of pressure is greater because of the increase in frictional surface and dissipation of energy in overcoming the friction. Though the fall of pressure is continuous, the rate of flow is not. The speed of flow decreases while crossing the resistance of different diameters, then the rate accelerates again due to the effect of changing the cross-sectional area. Similarly, the downward gradient of pressure

Table 28.7: Average mean pressure

Blood vessels		Pressure
1. Aorta		100 mm Hg
2. Large arteries (carotid)		95 mm Hg
3. Medium arteries (radial)		85 mm Hg
4. Smaller arteries		80 mm Hg
5. Arterioles		55 mm Hg
6. Capillaries	ARTERIAL	30 mm Hg
7. Venules		10 mm Hg
8. Small veins		2 mm Hg
9. Large veins		1 mm Hg
10. Venae cavae		0 mm Hg
11. Right atrium	VENOUS	0 mm Hg

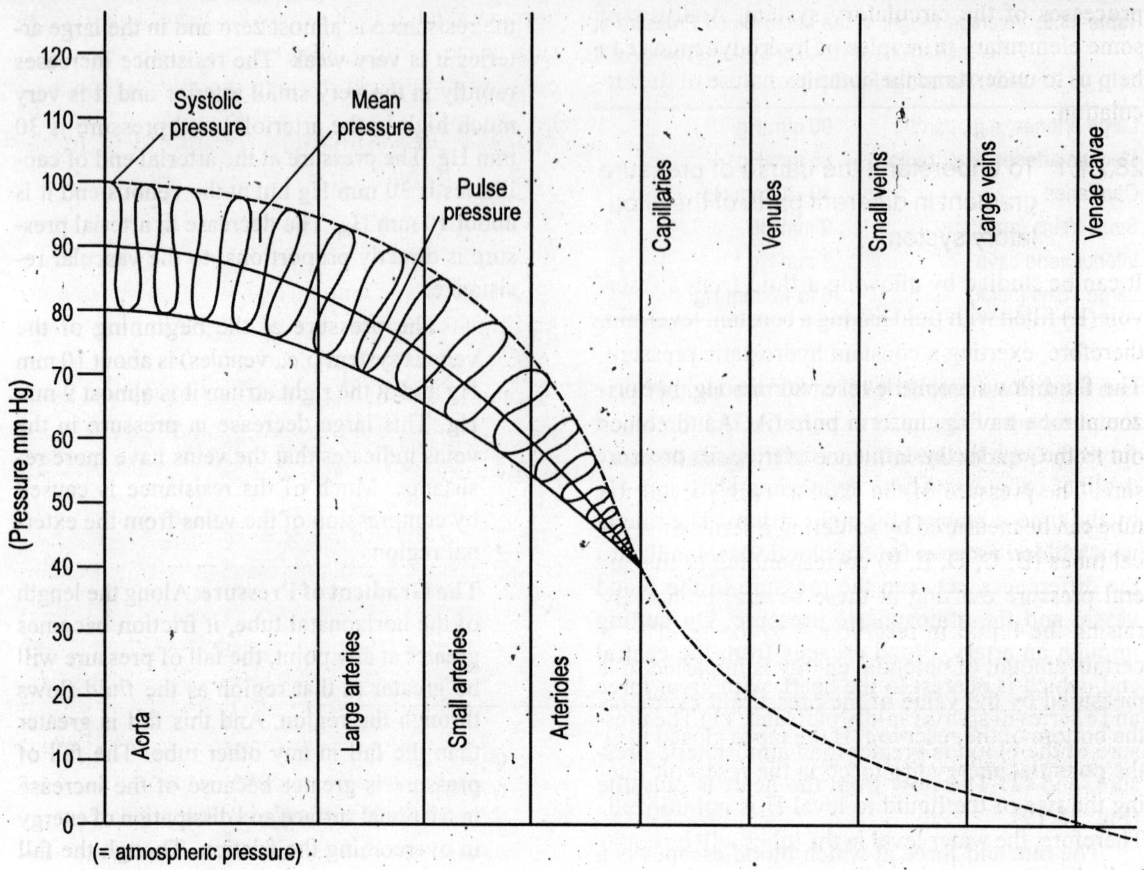

Fig. 28.19: Graph showing average mean blood pressure in the different portions of the systemic circulation

in the circulatory system is not uniform throughout. A sustained high arterial pressure is necessary to push the blood through the small blood vessels, where the resistance is high. The blood pressure must be high enough to force blood against the gravity to the brain, whose blood requirements are always high. Blood pressure must overcome both frictional resistance encountered in the smaller vessels and gravitation. Sufficient high pressure is also required, because the residual pressure which still remains beyond the capillary bed should drive the blood through the veins to the right heart. Some of the venous flow must also be against gravity as it occurs in case of blood flow towards the heart from the leg veins.

When there is high pressure in the arteries, more rapid adjustments of blood flow takes place, such as generalized vasoconstriction in organs drives the blood to flow rapidly into unobstructed channels, into unconstricted vessels. If there is local vasodilation, the high pressure ensures an immediate inrush of blood into an organ.

28.21 Physical Laws Determining the Blood Pressure

The circulatory system through which blood flows has the following characteristics:

(a) Branched.
(b) Closed system of tubes.
(c) Elastic properties.
(d) Variable diameter.
(e) Intermittent pumping to the arteries.
(f) Heterogenous (non-Newtonian) liquid.

Considering the above factors, it is difficult to assess the physical principles governing the processes of the circulatory system. Analysis of some elementary principles of hydrodynamics can help us to understand the complex nature of the circulation.

28.21.1 To understand the cause of pressure gradient in different parts of the circulatory system

It can be studied by allowing a fluid from a reservoir (R) filled with fluid having a constant level and, therefore, exerting a constant hydrostatic pressure. The fluid flows from the reservoir through a horizontal tube having uniform bore (AG) and comes out from G under the influence of hydrostatic pressure. The pressure of the fluid at each part of the tube can be measured by soldering a series of vertical tubes (B, C, D, E, F) corresponding to the lateral pressure existing at these several points. By raising the liquid in reservoir A to the level H, a certain amount of potential energy is imparted to it measured by the value of the pressure it exerts on the bottom of the reservoir. If the tap is closed at G, the potential energy produced in the reservoir during the rise of the liquid to level H is not utilized. Therefore, the water level in the tubes will be equal to that in reservoir A. If the tap G is open, water flows out of the tube through G, then the water level in the vertical tubes shows a continuous drop of fluid level, so that the line joining the tops of the fluid in the tubes is a straight one. Comparing the fluid level tube B and in the reservoir, it is observed that the fluid level, i.e., lateral pressure of the fluid, is lower than the fluid pressure in the reservoir. The movement of the fluid from A to B can be regarded as due to the differences of pressure $P_1 - P_2$. It is also observed that the straight line joining the tops of the fluid in the tubes does not touch the surface of the liquid in reservoir R but passes just a little below it. Of the total potential energy represented by the pressure H in the reservoir R, the large portion of h_1 is employed in overcoming the resistance to the flow in the tube AG while the remaining $H - h_1 = h_0$, energy is spent in the very beginning of the system in imparting a certain velocity to the flow of the liquid. The potential energy (pressure energy) of the flowing liquid decreases as it flows along the horizontal tube, as shown by the fall of lateral pressure in the vertical tubes but the kinetic energy is constant throughout the tube of uniform bore.

28.21.2 To understand the effect of resistance

If the velocity of the flow of the liquid is diminished by friction against the walls of horizontal tube by: (a) partially clamping at G or (b) increasing the length of the horizontal tube up to H or (c) narrowing certain parts of the horizontal tube or (d) making tube of wider bore, or (e) placing a stopcock in between the two vertical tubes, i.e., between D and E, the rate of fall of lateral pressure will diminish. The reason is that, in such flow, velocity increases and velocity varies inversely as the frictional resistance and vice versa. A smaller portion of h_0 of the hydrostatic pressure is used to impart velocity to the fluid and a large portion of h_1 is used to overcome the resistance. The same result can be seen by raising the levels of G or by lowering the level of reservoir—and so the pressure at A—and it is due to the velocity—and hence, the frictional resistance—which depends on the pressure difference.

The difference of pressure between the two points such as D and E may be regarded as the pressure which is essential to maintain a certain volume-flow of the fluid against the resistance offered by the friction of the fluid with the walls of the tube. The friction, and, therefore, the resistance to the flow, can be increased by narrowing, i.e. reducing, the diameter of the tube, when a greater difference of pressure will be required to maintain the same flow. This can be done by producing resistance between D and E by partially clamping the tube at this point. There will be no more uniformity of fall of pressure in the vertical tubes; instead, a continuous but slower fall between A and D and a steep fall between D and E and again a gradual fall between E and G. This can be explained by Poiseuille's law which states that the volume of a fluid flowing along a straight tube in unit time under constant pressure varies as the fourth power of the radius of the tube and the pressure difference between the two ends, and inversely as the length of the tube and the viscosity of the liquid (Fig. 28.20).

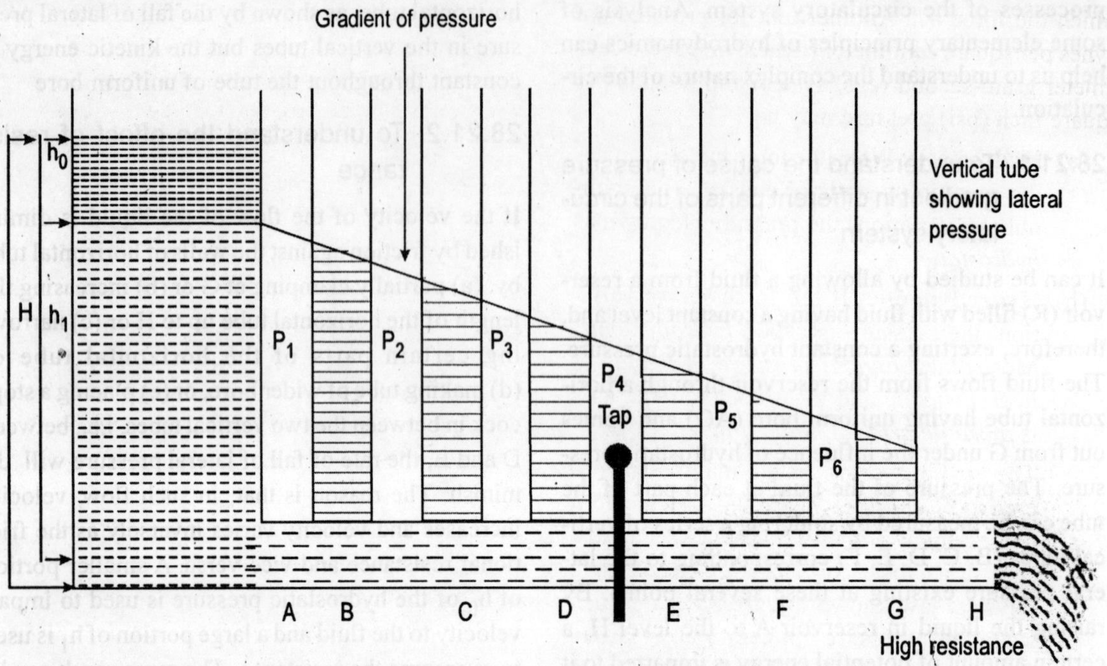

Fig. 28.20: Diagram illustrating significance of physical factors affecting blood pressure

28.21.3 Discussion

Thus the energy communicated from the reservoir to the moving fluid is spent on imparting:

(a) A certain velocity to the flow of liquid, and

(b) On overcoming the friction against the walls of the tubes, along with the liquid flows. In the cardiovascular system, the total amount of energy produced by the work of the heart—and communicated to the blood—physically corresponds to the potential energy which the liquid had in the reservoir. The pressure exerted by the blood on the wall of the blood vessels is analogous to the lateral pressure of the liquid in the vertical tubes.

In the circulatory system the amount of energy spent on imparting velocity of blood is comparatively very small than the potential energy communicated to it; the final pressure, therefore, approximates the lateral pressure. When we consider the value of blood pressure, we are not aware of the fact that whether the value includes the kinetic factor of the work of the heart or it is the lateral pressure exerted by the flowing blood on the wall of the vessels. But the amount of the total potential energy communicated by the work of the heart to a unit of blood volume is the value of the blood pressure in the initial part of the aorta. The pressure in all the large arteries (e.g., brachial, carotid, femoral) is 2-10 mm Hg lower than in the aorta. Because the resistance in the large arteries is very weak, blood hardly spends any of the potential energy imparted to it by the heart. Therefore, blood pressure in any large artery, i.e., arterial blood pressure—may be considered the means of the potential energy of the blood pumped by the heart into the aorta.

Blood Pressure

What is pressure?

Pressure is force per unit area.

$$P = \frac{F}{A}$$

where P = the pressure in newton per square meter (N/m²) or pascals (Pa); 1 N/m² = 1 Pa

F = the force in newton (N)

A = the area in square meter (m²)

The pressure can be increased either by increasing the applied force or by decreasing the area on

which the force act. Another unit for pressure are dynes per square centimeter (dyne/c m^2) in the centimeter gram-second (cgs) system and pounds per square inch (psi) pressure may be:

(a) **Hydrostatic**: When the force in a system under pressure is static or constant (i.e. without any variation) the pressure is known as *hydrostatic.*

(b) **Hydrodynamic**: If the force shows variation, the pressure is called *hydrodynamic.* Most *physiological pressure in human is hydrodynamic* and most common example is *pulsatile arterial blood flow.*

The cardiovascular system may be considered as *closed-loap hydraulic* system and it possess many properties of this system. Pressure in a *closed system* is governed by a physical law known as *Pascals law* which state that "Pressure applied to an enclosed fluid is transmitted undiminished to every portion of the fluid and the walls of the containing vessel."

Pascal's law holds true in *hydrostatic system* and also in *quasistatic* system in which a small change is made and the turbulence is allowed to die down before subsequent measurements are made, Pascal's law holds approximately true in hydrodynamic system in which the flow is nonturbulent and the vessel lumen is small except the vessel wall boundaries.

The study of pressure in turbulent or large human system or near the vessel wall is the subject of engineering mechanic and physics discipline. Anyhow, it can be assumed that Pascal's law hold generally true but recognize that it results only in approximations when applied in the turbulent human cardiovascular system. Pressure in the cardiovascular system is exerted on the fluid blood produced by the force of the pumping heart and is transmitted through the blood to the wall of the blood vessel.

The cardiovascular system is under the control of autonomic nervous system which causes constriction (sympathetic nervous system) and dilatation (parasympathetic nervous system) alters the surface area of the blood vessel by altering the diameter of the blood vessel. For which the pressure is not constant and as a result is obtained as average result. Pressure applied to an enclosed fluid is transmitted undiminished to every portion of the fluid and the walls of the containing vessel.

The heart generates flow and pressure under the influence of four major control mechanism such as

(a) end diastolic volume (preload)
(b) ejection pressure (afterload)
(c) heart rate and (d) contractility.

The control mechanism of each factors has been clarified by the investigation which opened the hydraulic loop of circulation and studied the heart in a more or less isolated condition and applied the knowledge in the physiological closed-loop condition. Cardiac surgeons utilize the knowledge using infusion scheme for post surgical patients combining starling's law of the heart with the role of blood volume for venous return. Realising the beneficial effect of reducing systolic pressure afterload has led cardiologist to carefully synchronized intra aortic balloon counter pulsation and more recently, to an extensive use of vasodilator therapy, ventricular tracing appropriately delayed from sensed atrial excitation, or even an automatic defibrillation, is now incorporated in an implantable pacer. This indicates the bright prospect of use of closed-loop physiological control of cardiac contraction including pharmacological closed-loop control of the contractility

28.22 Arterial Blood Pressure

Blood pressure (BP) is the lateral pressure exerted by the blood on the wall of the blood vessel, i.e., artery while flowing through it. Blood pressure means the force exerted by the blood against any unit area of the vessel wall.

28.22.1 Lateral pressure and end pressure

lateral pressure is the pressure when force is exerted at right angles to the direction of flow at any point within a tube filled with a circulating fluid. When a canula is inserted into the artery and if the artery is tied off beyond the site of insertion of canula, the record shows "end pressure". The blood flow in the artery shows interruption and all the kinetic energy of flow converts into potential energy or pressure energy.

In a T-tube whose long arm is inserted into the artery and the pressure is measured from the side arm which is at right angle to the long arm, the record shows "lateral pressure" or "side pressure" which is lesser than the "end pressure" by the kinetic energy of flow (some amount of pressure energy has been diverted to the kinetic energy of flow) and drop of pressure due to resistance is negligible. The reason is that in a tube or blood vessel, the total energy is the sum of the kinetic energy of flow and the pressure energy which is constant according to Bernoulli's principle. The drop of pressure in any part of the artery is due both to resistance and conversion of pressure energy or potential energy into kinetic energy. If the drop of pressure is due to loss of energy to overcome the resistance, then the process is irreversible because it dissipates as heat. But if the vessel is narrow at some portion, the pressure drops due to conversion of pressure energy (or potential energy) to kinetic energy, the process is reversible when the vessel widens out again.

28.22.2 Systolic, diastolic, pulse and mean pressure

(A) **Systolic (or Maximal) Pressure**: Arterial pressure rises to a peak value, i.e., maximum pressure during ventricular systole (120 mm Hg). It establishes in the arteries for a brief period, 0.1-0.12 second after the systole. It undergoes considerable fluctuations which increases during excitement, exercise, meals, etc. and decreases during sleep, rest, etc. The height of systolic pressure indicates:

(a) Extent of work done by the heart,

(b) Force of the heart,

(c) Withstanding power of the wall to the degree of blood pressure.

(B) **Diastolic (or Minimal) Pressure**: Minimum pressure is at the very end of ventricular diastole (70-80 mm Hg). It is a measure of constant stretch to which the walls of the arteries are subjected. It is more important than systolic pressure because:

(a) It undergoes much less fluctuations in health and remains within a limited range,

(b) It is the constant load, against which the heart has to work,

(c) It is the measure of peripheral resistance and depends mainly on the tone of the arterioles.

In the treatment of high blood pressure (hypertension), the drug is considered to be ideal if it can lower the diastolic pressure efficiently.

- **Expression of Systolic and Diastolic Pressure**: Clinically, after measuring blood pressure, it is expressed in the following way:

$$\frac{\text{Systolic pressure}}{\text{Diastolic pressure}} = \frac{120 \text{ mm Hg}}{80 \text{ mm Hg}}$$

1 mm Hg equals 0.133 KP_a so in SI units this value is 16.0/10.6 KP_a.

(C) **Pulse Pressure**: The pulse pressure is the difference between the systolic and diastolic pressures (i.e., 120 mm Hg and 80 mm Hg) is normally about 40 mm Hg. The variations in the systolic and diastolic pressures, equal in magnitude and direction, do not affect the value of the pulse pressure. During each cardiac cycle the arterial pressure fluctuates between the systolic and diastolic pressure and this fluctuation is the pulse pressures. The elasticity of the arteries and the action of the semilunar valves contribute significantly to the phenomenon of pulse pressure. Passing from aorta to the arterioles, the pulse pressure becomes less and less, the effect of the arterial elasticity reduces progressively in both the rise of systolic and the fall of diastolic pressures. In the capillary bed, the pulse pressure disappears, so there is constant flow in the capillaries but not the intermittent flow as in the aorta. In case of hardening of arteries—where elasticity of the blood vessel decreases—the pulse pressure increases. Damage of the semilunar valves causes increase in pulse pressure.

(D) **Mean Pressure:** The mean pressure is the average pressure throughout the cardiac cycle. The mean pressure is slightly less than the average of systolic and diastolic pressure because systole is shorter than diastole. The mean arterial pressure is about 96 mm Hg which is slightly less than the average 100 mm Hg (systolic 120 mm Hg + Diastolic 80 mm Hg). It signifies the average pressure to push blood through the vascular system and is important for tissue blood flow.

Alteration of the mean pressure occurs by:

(a) Changing the capacity of the total system either by contraction of the wall of the vessel or by exerting pressure on them from outside.

(b) Changing the total volume of the circulating blood,

(c) Changing the output of the heart,

(d) Changing the peripheral resistance.

28.22.3 Physical factors influencing the arterial pressure

1. **Contractile Force of the Heart**: The driving force of blood pressure is mainly created by the pumping action of the heart beat. The efficiency of the heart is considered upon how much amount of blood is driven out by the heart into the aorta in each beat for controlling cardiac output, blood pressure and flow within the blood vessel.
2. **Cardiac Output**: The greater the cardiac output, the greater is the blood pressure provided other factors remain constant. Alterations of cardiac output will alter the blood pressure. Cardiac output depends upon venous return, force and frequency of heart beat.
3. **Peripheral Resistance in the Arterioles:** Maximum peripheral resistance is offered by the arterioles and, to a smaller extent, by the capillaries. The arterioles act as "stopcocks" of the vascular system. If the lumen of the vascular system is constricted and primarily in the arterioles, the amount of blood pumped by heart remains the same, the same amount of blood flows as a narrow stream. As a result, the linear rate of flow of the blood through the arterioles increases. Since the expenditure of energy on overcoming friction is proportional to the square of the velocity of the flow of liquid, the flow of the blood through the arterioles can be accelerated only by increasing the expenditure of energy. The energy is communicated to the blood by the work of the heart, hence an increase in peripheral resistance causes an increase in the potential energy imparted to the blood by the heart, this increase in energy is responsible for the rise in arterial pressure.
4. **Volume of Blood**: An increase in the volume of blood in the circulation raises both the systolic and diastolic pressures because an increase in the mass of blood moved by the heart leads to an acceleration of flow of blood and consequently to an increase in the resistance to the flow of blood. A decrease in volume causes a drop in blood pressure, everything else being equal.
5. **Size of Lumen of Capillaries and Veins:** Peripheral resistance is inversely proportional to the lumen of the vessel. But the resistance to the blood flow in the capillaries and veins is relatively low compared to the arterioles. Though the capillaries have the smallest lumen, they should have shown higher pressure. But it does not happen because, in the capillaries, the velocity of blood flow is very slow and the frictional effect is also very low; hence, the pressure is low. Dilation in the capillaries and veins leads to congestion in them. The flow of blood decreases and the blood pressure drops. Again, a constriction of the capillaries and veins over a large area leads to a flow of the blood away from them into the large veins and to the heart. The work of the heart increases and consequently leads to rise in arterial blood pressure until the blood flow to the heart decreases.

6. **Viscosity of the Blood**: The greater viscosity of the blood leads to higher friction against the walls of the vessels and, consequently, the higher resistance is offered to the flow in the blood vessels. An increase in viscosity of the blood causes rise in the arterial blood pressure when other factors remain constant. Changes in the viscosity of blood play a minor role among the other factors because in normal condition the viscosity of the blood does not change significantly, except some pathological conditions. A rise or fall in blood viscosity tends to affect the diastolic pressure in a way similar to changes in the calibres of the peripheral resistance.
7. **Elasticity of the Arterial Wall**: Due to elastic properties of the arterial wall, it can accommodate considerable amount of blood by stretching or distending of the wall during ejection of blood by the heart during systole and there is less rise of blood pressure. The elasticity does not play any role when the pressure is below 30-40 mm Hg and the vascular wall behaves as a rigid tube. Elasticity is important for origin and maintenance of the diastolic pressure and sustaining mean pressure at a higher level.
8. **Elastic Recoil of the Arterial Wall**: By virtue of the pressure of the elastic tissue in the arterial walls, the stretched walls tend to recoil against the distending force at the existing diastolic pressure. Due to elasticity the flow is pulsatile in the arteries but not beyond the arterioles, i.e., in the capillaries and venules, where flow is a continuous stream. The conversion of pulsatile flow to continuous flow depends upon the existence of diastolic pressure. Thus, elastic recoil acts as a subsidiary pump to drive the temporary stored blood in the distending arteries onwards in a continuous stream between the heart beats and without this function the pressure would fall to zero after each systole.
9. **Gravity:** The pressure in the blood vessel below heart level increases and that in any vessel above heart level decreases due to the effect of gravity which has hydrostatic effect on the circulation. Changes in the bodily posture and in the relationship of the body axis to the direction of gravitational pull alter the pressure of the blood in the various parts of the body. In case of arterial pressure, when a person is in a standing position, if this blood pressure is 100 mm Hg at the level of heart, he has a pressure in the feet about 190 mm Hg. If we say a blood pressure of 120 mm Hg, it usually means that this is the *pressure at the hydrostatic level of the heart.*
10. **Effect of Branching (Piezometric Circulation)**: When a tube has branches with different angles, the pressure exerted in any branch by the fluid flowing along the tube depends upon the angle of the branch with the main tube.
    1. Branch I leaves the main vessel at right angle, then the hydrostatic pressure in the branch is equal to the lateral pressure on the wall of the main vessel (Fig. 28.22A).
    2. Branch II leaves the main vessel at an acute angle to the direction of flow, then the pressure will be greater than the lateral pressure of the main vessel due to the kinetic energy of the moving fluid, the lateral pressure is more than the hydrostatic pressure.
    3. Branch III leaves the main vessel at an obtuse angle, the pressure will be less due to more hydrostatic pressure than the lateral pressure of the main vessel.

    It follows the principle of "piezometer" (Fig. 28.21A). When the branches II and III are connected, there will be a subsidiary circulation of fluid along the tubes and branches in the direction of Branch II to Branch III, known as "piezometric circulation". The mechanism in Fig. 28.21B occurs in may parts of the circulation predominantly in the smaller arterial anastomotic circulations, e.g., from each of aorta to internal mammary artery, then via ascending aortic intercostal arteries to the descending aorta. Most of the branches are coming out from the arteries at an acute angle and

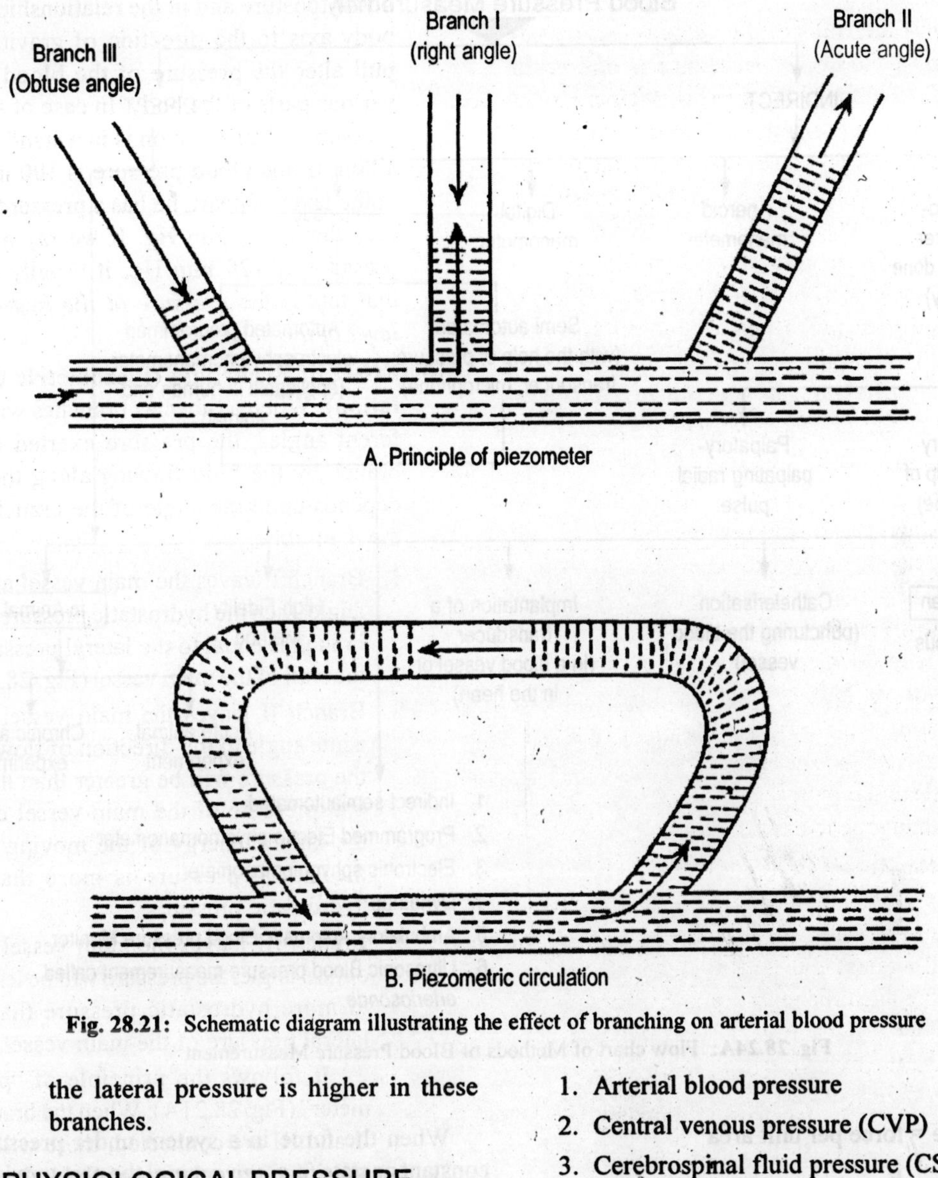

Fig. 28.21: Schematic diagram illustrating the effect of branching on arterial blood pressure

the lateral pressure is higher in these branches.

PHYSIOLOGICAL PRESSURE

Under physiological pressure usually we are interested in fluid and gas related pressures. The gases are compressible but the liquid is not. This compressibility affects the measurement technique. The most common physiological fluid pressure is encountered by the physician in daily practice and in intensive care unit (ICU), intensive coronary care unit (ICCU), intensive therapeutic unit (ITU).

Besides these there are other physiological pressures are considered in the medical profession are:

1. Arterial blood pressure
2. Central venous pressure (CVP)
3. Cerebrospinal fluid pressure (CSF)
4. Intraocular pressure (IOP)
5. Intra ventricular pressure (in brain)
6. Pulmonary arterial pressure
7. Rt. intra arterial pressure
8. Lt. intraventricular pressure
9. Intrathoracic pressure
10. Intrapleural pressure
11. Glomerular filtration pressure
12. Pulmonary capillary pressure.

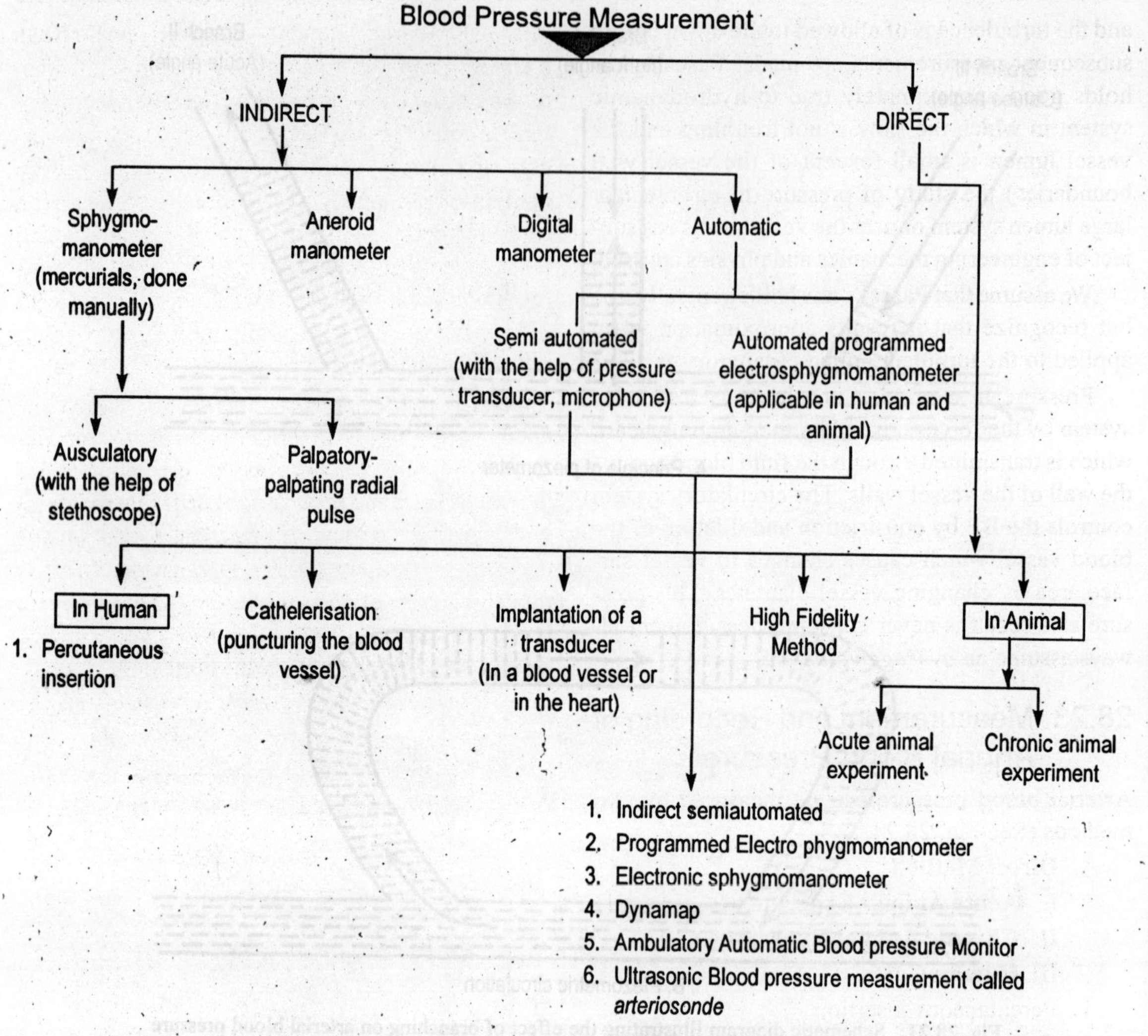

Fig. 28.24A: Flow chart of Methods of Blood Pressure Measurement

Pressure = force per unit area

$$P = \frac{F}{A}$$

P = pressure in newtons per square meter (N/m^2) in pascals (Pa) $=1N/m^2$

F = is the force in newton (N)

A = area in square meter (m^2).

Pressure can be increased either by *increasing the applied force* or by *decreasing the area on which the force acts alternate units* for pressure are *dyne per square centimeter* (dyn/cm^2) in the centimetergram second (cg/s) system and *pounds per square inch* (psi) in the Brits Eng. Sys.

When the force in a system under pressure is constant or state (e.g. *unvarying)* the pressure is said to be *hydrostatic.* If the force is *varying,* the pressure is said to be hydrodynamic. Most human physiologically pressure are hydrodynamic of which the most easily recognized is the pulsatile flow of arterial blood.

Pressure in a closed system obeys a physical law known as Pascal's principle which states that pressure applied to an enclosed fluid is transmitted undiminished to every portion of the fluid and the walls of the containing vessel.

Pascal's law holds true in hydrostatic system and also in *quasistatics system* in a small change is made and the turbulence is of allowed to die down before subsequent measurements are made. Pascal's law holds good approximately true to hydrodynamic system in which the flow is not troubling and the vessel lumen is small (except of the vessel wall boundaries) the study of pressure in turbulent or large lumen system or near the vessel walls is a subject of engineering mechanics and physics coursed.

We assume that Pascal's law holds generally true but recognize that it results approximation when applied to the turbulent lumen circulatory system.

Pressure in exerted in the human circulatory system by the forces created by the pumping heart, which is transmitted through the fluid blood against the wall of the vessel walls. The circulatory system controls the BP by constriction and dilation of the blood vessel which causes changes in vessel surface area by changing vessel diameter. This pressure as a result is never constant measurement always assume an *average value.*

28.23 Measurement and Recording of Arterial Blood Pressure

Arterial blood pressure can be measured by two methods (See Fig. 28.24A):

A. **Direct Method**
   I. **Acute Animal Experiment,**
   II. **Chronic Animal Experiment,**
   III. **Human Cases.**
   1. Percutaneous insertion
   2. Catheterisation (puncturing blood vessel)
   3. Implantation of transducer (in a blood vessel a in the heart
   4. High Fidelity Method.

B. **Indirect Method** (Human cases)
   I. Mercury manometer
      Instrument—Sphygmomanometer
   II. Aneroid manometer.
   III. Digital manometer
   IV. Automatic apparatus.
      (a) Semi-automated, (b) Automated

28.23.1A Direct Method

28.23.1.1.1 Acute Animal Experiment

The animal is anaesthetized and the carotid artery is exposed by dissection. An arterial glass canula (also called carotid canula)—of which one limb has a tapering end—is inserted directly into the lumen of the central end of the artery. The other end of the limb is connected to the U-shaped mercury manometer of suitably calibrated strain gauge and an osillograph arranged to write directly on a moving strip of paper. The other side-limb is the outlet which is kept clamped and used for washing out the clotted blood from the canula which is filled up with 3-4 per cent sodium citrate solution (to prevent clotting of arterial blood in the canula) (Fig. 28.22).

The U-shaped mercury manometer shows the actual recording of the blood pressure in mm Hg. For recording the blood pressure, a floating stylus (a float of vulcanite from which a stiff fine rod rises) with a writing pointer is used that marks on the moving smoked paper of the kymograph, which is rotated electrically. As the mercury column in one limb, which has direct contact with the artery, descends and the mercury in the other limb ascends, *the value recorded by the scale (attached with the*

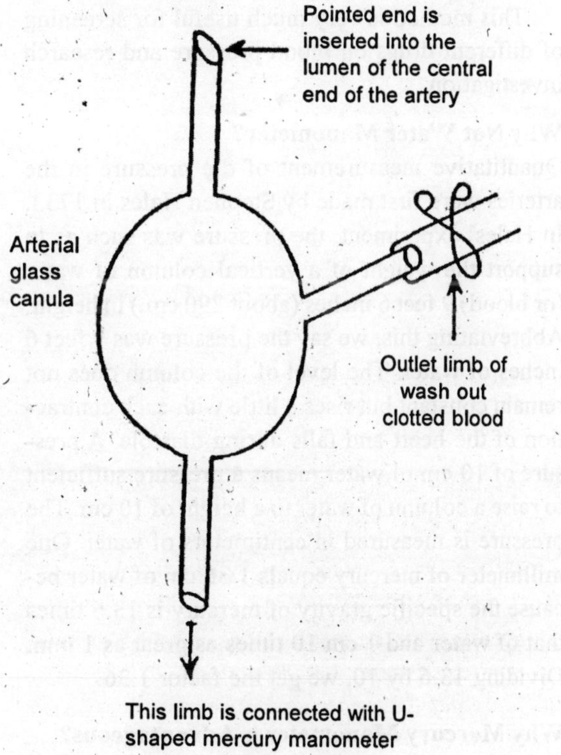

Fig. 28.22: Diagram showing arterial glass canula and different limbs

U-*tube) will be multiplied by two so as to get the actual blood pressure.* For convenience it is generally considered to be 1 mm in the scale equivalent to 2 mm of Hg of pressure. Before recording the blood pressure, the mercury levels in both the limbs of the U-tube must be adjusted to the zero mark of the scale. *It is ideally measured as* dynes *per square* centimetre *but physiologically it is measured as* millimetres *of mercury* (1 mm Hg = 1332 dynes/cm^2).

- **Disadvantages**
  1. Suitable only for acute experiment in anaesthetized animals.
  2. Surgical intervention is required.
  3. Gives the *record of mean arteral pressure* and *heart* beats only.
  4. Owing to high inertia of the mercury, the blood pressure changes associated with cardiac cycle are damped.
  5. Respiratory undulation of mean arterial pressure waves are clearly seen in this direct method.

- **Advantages**

This method is very much useful for screening of different drugs on blood pressure and research investigation.

Why Not Water Manometer?

Quantitative measurement of the pressure in the arteries were first made by Stephen Hales in 1733. In Hales' experiment, the pressure was such as to support the weight of a vertical column of water (or blood) 9 feet 6 inches (about 290 cm.) in height. Abbreviating this, we say the pressure was 9 feet 6 inches of water. The level of the column does not remain constant but rises a little with each contraction of the heart and falls during diastole. A pressure of 10 cm of water means a pressure sufficient to raise a column of water to a height of 10 cm. The pressure is measured in centimeters of water. One millimeter of mercury equals 1.36 cm of water because the specific gravity of mercury is 13.6 times that of water and 1 cm 10 times as great as 1 mm. Dividing 13.6 by 10, we get the factor 1.36.

Why Mercury Manometer is Advantageous?

Hales' device of water manometer is cumbersome and cannot be used extensively in experiment procedures. At present a modified procedure is employed in which—instead of supporting a vertical column of blood—the pressure supports a column of mercury.

The height of the column of mercury would be only a fraction of that of an equal weight of water. As mercury is 13.6 times the weight of the water, a mercury column 1.0 cm in height would be the equivalent of a column of water of 13.6 cm.

This mercury manometer consists of a glass U-tube having two vertical limbs about a foot in height, the bottom of which is half-filled with mercury. The mercury columns remain at the same level when both arms are open to the air because the surface of the mercury of both sides are influenced equally by the pressure of air (Fig. 28.23A). If one arm of the manometer is connected with an artery (e.g., carotid) there will be an extra downward pressure P which leads to all of mercury level and pushes the mercury upwards into the open arm of the manometer resulting in a rise in the level of mercury from Q to R. Equilibrium is reached when the downward pressure of the blood at P is equal to the downward pressure of mercury column RQ at point Q. The height of the column RQ in mm represents a quanitative measure of pressure exerterd by the blood at P, i.e. mean arterial blood pressure expressed in mm Hg; average or mean blood pressure in mammals is 110-120 mm Hg. *Poiseuille* used the mercury manometer for blood pressure measurement purpose and Ludwig placed a float on the manometer some years after Poiseuille used the plain mercury tube.

Fluctuations in arterial blood pressure occur with each heart beat. The arterial blood pressure as recorded by a mercurial manometer also exhibits a series of pulsations, each corresponding to a heart beat. These pulsatons are due to the fact that at each systole the blood (about 60 c.c. in man) is pumped out through the aortic valve in about one-fifth of a second, so that the artery becomes distended. During diastole, the aortic valves are closed and blood is pushed onwards to the capillaries.from the distended arteries for elastic recoil action, so the pressure falls. But these rapid changes of pressure in the artery cannot be accurately registered by the mercurial manometer owing to the inertia of the column of the mercury. The inertia of the heavy

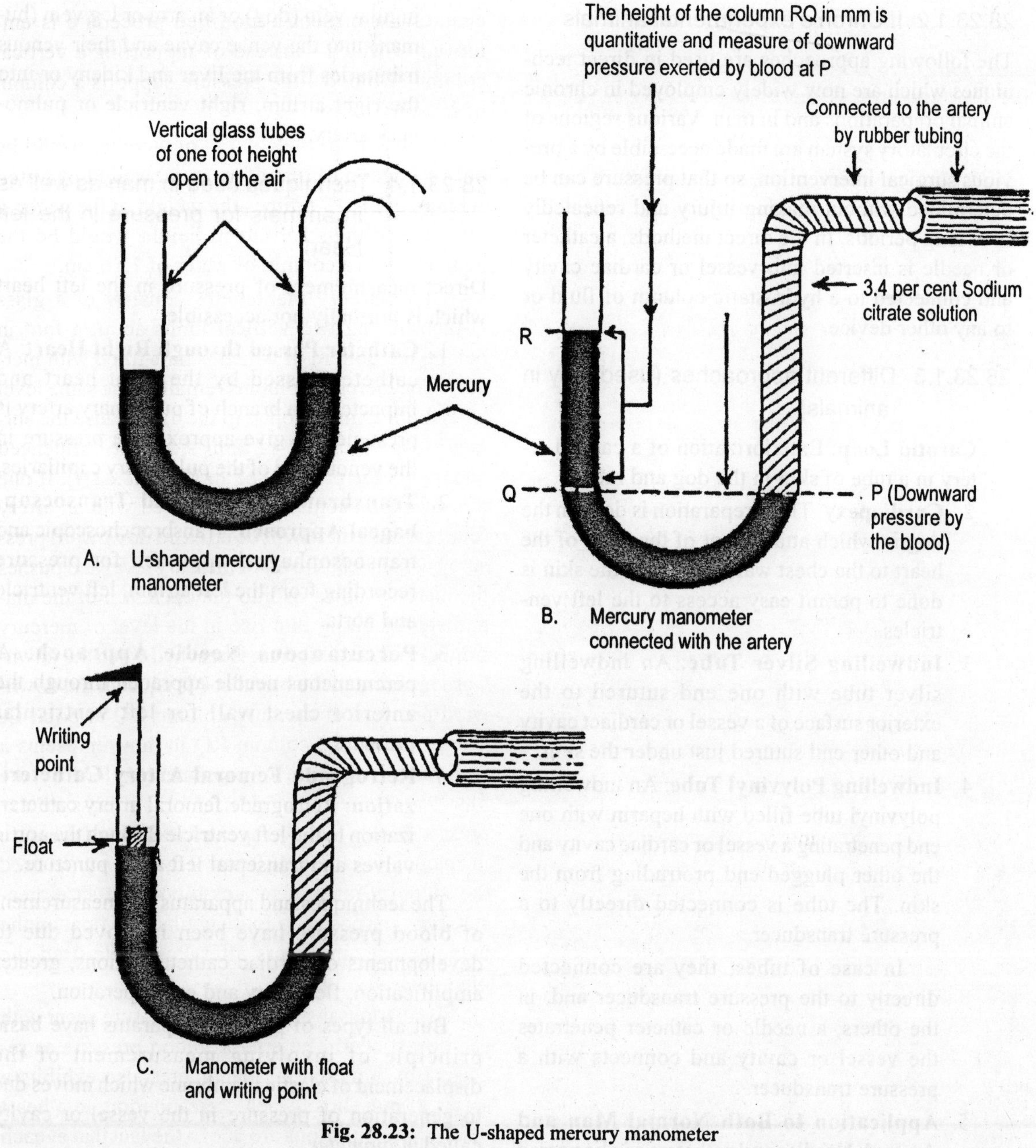

Fig. 28.23: The U-shaped mercury manometer

liquid is such that when it is pushed up to reach the actual systolic level, the heart has already gone to diastole. Again, before reaching the actual diastolic level by the mercury, the systole has already started. As a result, the excursion of mercury manometer with each cardiac cycle are a few mm instead of actual pulse pressure (about 40 mm is normal condition in mammals). In case of slow heart beat, the excursions of the manometer are greater.

To record a true pressure change in arteries, it is essential to reduce the instrumental intertia with the help of Hurthle's or Frank's membrane manometer or one of the forms of manometer which can record the pressure changes photographically. Frank's membrane manometer is more accurate than the mercury manometer because its period of vibration is short, so that it readily follows rapid changes of pressure.

28.23.1.2 II Chronic experimental animals

The following approaches are used in direct techniques which are now widely employed in chronic animal preparations and in man. Various regions of the circulatory system are made accessible by a previous surgical intervention, so that pressure can be determined without making injury and repeatedly over long periods. In the direct methods, a catheter or needle is inserted into vessel or cardiac cavity and connected to a hydrostatic column of fluid or to any other device.

28.23.1.3 Different approaches (used only in animals)

1. **Carotid Loop**: Exteriorization of a carotid artery in a tube of skin in the dog and rabbit.
2. **Cardiopexy**: This preparation is done in the dog in which attachment of the apex of the heart to the chest wall just under the skin is done to permit easy access to the left ventricles.
3. **Indwelling Silver Tube**: An indwelling silver tube with one end sutured to the exterior surface of a vessel or cardiact cavity and other end sutured just under the skin.
4. **Indwelling Polyvinyl Tube**: An indwelling polyvinyl tube filled with heparin with one end penetrating a vessel or cardiac cavity and the other plugged end protruding from the skin. The tube is connected directly to a pressure transducer.

   In case of tubes, they are connected directly to the pressure transducer and, in the others, a needle or catheter penetrates the vessel or cavity and connects with a pressure transducer.
5. **Application to Both Normal Man and Animal**: Needles and catheters can be passed into many accessible arteries and veins without previous preparation in man and animals and without fluoroscopy or under fluoroscopy. For the right-sided circulation, one method is to "float" a polyvinyl catheter with a balloon tip down the jugular vein into the right ventricle or pulmonary artery without fluoroscopy. But the standard method is to pass a catheter under flouro-scopy via the jugular vein (dog) or an arm or leg vein (human) into the venae cavae and their venous tributaries from the liver and kideny or into the right atrium, right ventricle or pulmonary artery.

28.23.1.4 Techniques used in man as well as in animals for pressure in the left Heart

Direct measurement of pressure in the left heart which is normally not accessible.

1. **Catheter Passed through Right Heart**: A catheter passed by the right heart and impacted in a branch of pulmonary artery is presumed to give approximate pressure in the venous end of the pulmonary capillaries.
2. **Transbronchoscopic and Transoesophageal Approach**: Transbronchoscopic and transoesophageal approach for pressure recording from the left atrium, left ventricle and aorta.
3. **Percutaneous Needle Appraoch**: A percutaneous needle appraoch through the anterior chest wall for left ventricular pressure.
4. **Retrograde Femoral Artery Catheterization**: Retrograde femoral artery catheterization to the left ventricle through the aortic valves and transeptal left atrial puncture.

The techniques and apparatus for measurement of blood pressure have been improved due to developments of cardiac catheterizations, greater amplification, flexibility and ease operation.

But all types of pressure apparatus have basic principle of involving measurement of the displacement of elastic membrane which moves due to generation of pressure in the vessel or cavity called *manometer*.

There are two groups of manometer:

(a) **External**: This group of manometers are nondistensible conduit, e.g., needle, lead tube, cannula or catheter filled with fluid and connected to the membrane outside the body, which, in turn, are connected to a device for amplification and recording.

(b) **Internal**: In this group a micromanometer or similar device is placed in the tip of a catheter at the pressure source. The catheter acts only as a conduit for wire to the amplifying and recording device. The detection and transmission of the membrane movements take place by:

(a) Mechanical means—obsolete nowadays.
(b) Optical means.
(c) Electrical means.
(d) Combination of these.

The manometers combining the optical and electrical means are also in the same category.

28.23.1.5 Manometers with fluid-filled conduit

(A) **Optical Manometer**: This type of manometer is largely used. It consists of a stiff rubber or metal membrane of minimum diameter connected to a non-distensible, fluid-filled conduit of minimal volume, which is inserted into the source of pressure. Movements of membrane caused by pressure change are recorded by reflecting from an attached mirror or a beam of light to give a rapid response.

(B) **Electrical Transducer**: Movement of the manometer membrane is transformed into electrical energy which is then transferred to a recorder. Transformation of the pressure impulse is effected by the piezoelectric effect or by a change in capacitance, resistance and inductance. The piezoelectrical phenomenon depends on the potential difference produced in certain crystals exposed to mechanical pressure. It is not so much used due to disadvantages like:

1. Stationary difference of crystal potential cannot be maintained,
2. Constant pressure cannot be measured accurately.

(C) **Electronic Transducer**: Recently, for determination of blood pressure, the fluctuations of pressure in a vessel or cavity are transmitted to a membrane through a long non-distinsible catherter filled with fluid. This activates an electronic pressure transducer in which movements of the membrane with minimal fluid displacement and high frequency of response—produce a change in capacitance, resistance or inductance. The electronic transducer which is commonly used is the resistance wire strain gauge.

- **Amplification and Recording**: The signal from the various pressure transducers (except the optical manometer) is amplified electronically to drive a high frequency galavanometer or oscilloscope. The devices used to record the pressure pulses are:

(a) Optical manometer with camera,
(b) Direct pen writing galvanometer,
(c) Optical galvanometer, and
(d) Cathode-ray oscilloscope.

28.23.1.6 III Human cases

High-Fidelity Method: Direct method is also used in human beings without anaesthesia, both at rest and during physical activity. This instrument is an electronic transducer for recording rapidly changing blood pressure in man by direct method.

- **Rationale**: Direct method is also used in human beings without anaesthesia, both at rest and during physical activity. This instrument is an electronic transducer for recording rapidly. It is excellent for recording steady pressure but cannot respond rapid changes of blood pressure than approximately one cycle every 2-3 seconds. High-fidelity is used only in special circumstances like management of shock and cardiac surgery.

Pressure cycles up to 500 cycles per second have been recorded accurately with some of these high-fidelity types of instruments. These are commonly used and can register pressure changes as rapidly as 20-100 cycles per second.

- **Principle**: The basic principle is that—commonly used in three electronic pressure transducers—they convert pressure into electrical signals and the pressure is recorded on a high speed electrical recorder.

- **Instruments**: The method comprises of:
  I. Electronic pressure transducers.
  II. High speed electrical recorder.

I. **Electronic-Pressure Transducers**: The electronic pressure transducers have the following parts:

(a) Needle or catheter,

(b) Fluid chamber,

(c) Highly stretched thin metal membrane,

(d) Metal plate above the metal membrane or small iron slug or very thin stretched resistance wire attached to the thin metal membrane.

There are three types of electronic pressure transducers:

1. **Type I**: It consists of a very thin and highly stretched membrane wall of the fluid chamber. The fluid chamber is connected with blood vessel in which the pressure is to be measured through the needle or catheter. A simple metal plate is placed a few thousands of an inch above the thin membrane. When the pressure rises, the membrane bulges outward, the capacitance between the metal plate and thin metal membrane increases. This change in capacitance can be recorded by an electronic device (Fig. 28.24).

Type II: This type of transducer has a small iron slug placed on the metal membrane and there is a movement upward, this is displaced upward into a coil. Movement of iron inside the coil causes changes of inductance in the coil which can be recorded electronically (Fig. 28.24).

Type III: In this type a very thin, stretched resistance wire is connected to the metal membrane. Due to stretching of resistance wire, resistance increases and when there is less stretching, there is decrease in resistance. The variations of changes in resistance can also be recorded electronically (Fig. 28.24).

These pressure transducers are used for converting pressure into electrical signal and the pressure is recorded by high-speed electrical recorder.

More recently, miniature pressure transducers have been developed of inductance, strain gauge or piezoelectric types which are small enough to be attached at the tip of a cardiac catheter, and these devices provide much higher fidelity pressure wave forms up to 100 Hz; but these are expensive.

- **Disadvantages**: Some artifacts may be observed due to:

1. Microbubbles in the needle or catheter or transducer which must be removed before use,
2. Movement of the catheter within the heart due to heart motion can produce oscillation of the fluid column contained within the catheter.

28.23.2.B Indirect method

The blood pressure can be determined in man by inserting a hollow, wide-bore needle into an artery and connecting it with a suitable manometer but such a method is generally used for cardiopulmonary diagnostic work. In general clinical practice, for convenience, safety and frequent check-up, indirect method is used.

28.23.2.1 Principle

The principle employed involves external pressure exerted required just to collapse an artery. This is done by balancing air pressure against the pressure of the blood in the brachial artery. The air pressure is determined by means of:

1. Mercury manometer—the instrument is called sphygmomanometer.
2. Aneroid manometer (clock-like).
3. Digital manometer
4. Semiautomated
5. Automated—Programmed electrosphygmomanometer.

Pressure of the cardiovascular system are conveniently recorded using *atmospheric pressure as the zero reference level* and *further referenced to the level of the heart to eliminate gravitational level.*

28.23.2.2 Advantages of indirect method

The indirect method is advantageous for the following reasons:

1. Convenient for routine pressure check-up in humans,
2. Giving systolic, diastolic, pulse pressure,
3. Without surgical intervention or anaesthesia, and
4. Gives accurate results when compared with the result of direct measurements by means of a needle inserted into the artery.

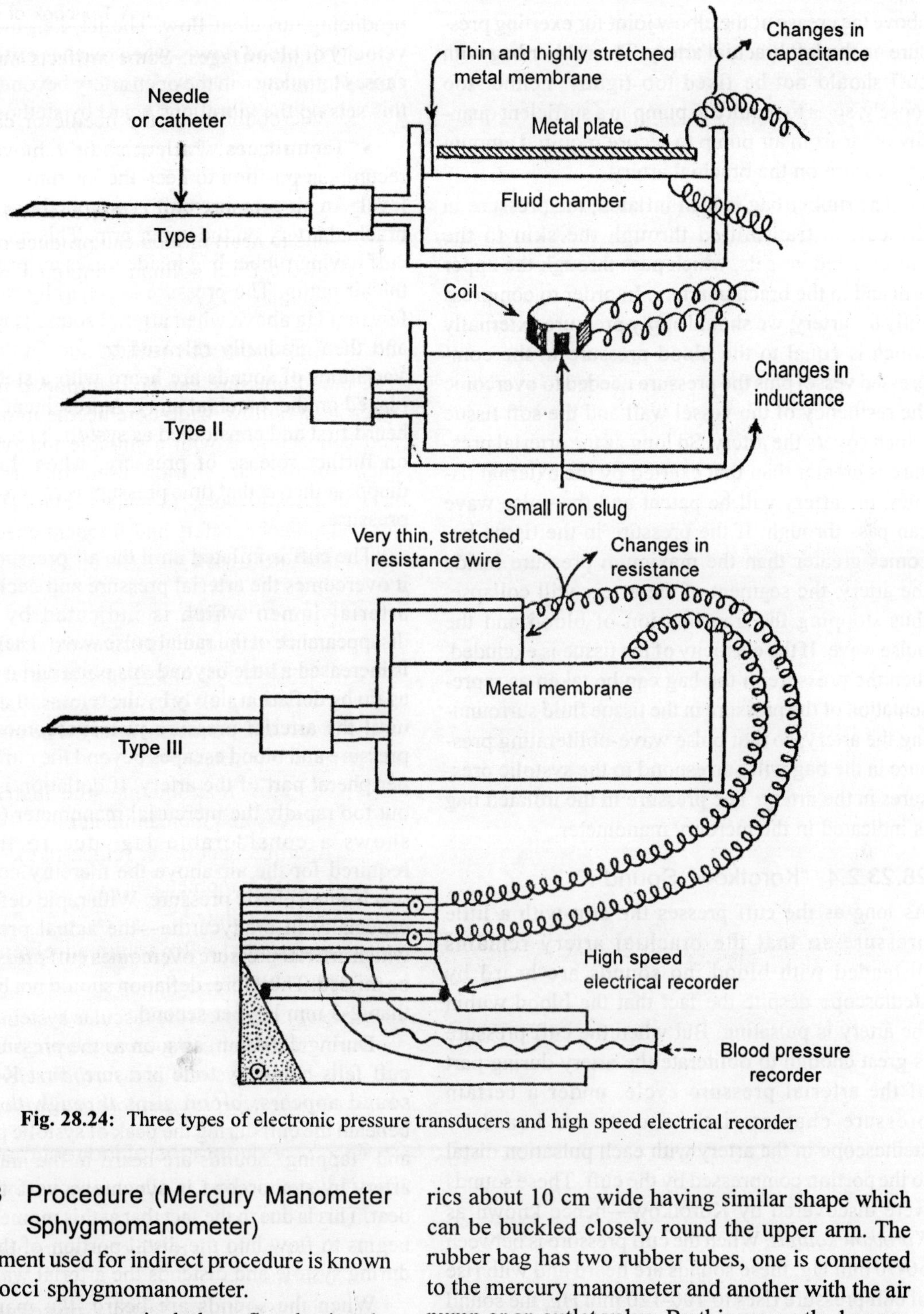

Fig. 28.24: Three types of electronic pressure transducers and high speed electrical recorder

28.23.2.3 Procedure (Mercury Manometer Sphygmomanometer)

The instrument used for indirect procedure is known as Riva-Rocci sphygmomanometer.

Sphygmomanometer includes a flat elastic rubber bag (9″ × 4.5″, i.e., 23 × 11.5 cm) which is designed to wrap around to encircle the arm which is covered by a nondistensible canvas or cotton fabrics about 10 cm wide having similar shape which can be buckled closely round the upper arm. The rubber bag has two rubber tubes, one is connected to the mercury manometer and another with the air pump or an outlet valve. By this means the bag can be inflated to any desired pressure. The uninflated rubber bag with the canvas cover (cuff) is wrapped around the upper left arm just 1-1.5″ (2.5-3.8 cm) above the crease of the elbow joint for exerting pressure on the left brachial artery. The rubber bag with cuff should not be fixed too tightly, neither too loosely, so as to require to pump in a sufficient quantity of air from air pump to secure required amount of pressure on the brachial artery.

The rubber bag is then inflated, the pressure in the cuff is transmitted through the skin to the muscles and vessels, which pass through the upper arm and to the brachial artery. In order to compress fully the artery, we should apply pressure externally which is equal to the blood pressure in the compressed vessel plus the pressure needed to overcome the resiliency of the vessel wall and the soft tissue which covers the artery. So long as the arterial pressure is greater than that exerted by the external tissues, the artery will be patent and the pulse wave can pass through. If the pressure in the tissue becomes greater than the maximum pressure inside the artery, the segment of the artery will collapse, thus stopping the transmission of blood and the pulse wave. If the elasticity of the tissue is excluded, then the pressure in the bag can be taken as representation of the pressure in the tissue fluid surrounding the artery, so that pulse wave-obliterating pressure in the bag will correspond to the systolic pressures in the artery. The pressure in the inflated bag is indicated in the mercury manometer.

28.23.2.4 "Korotkow" Sound

As long as the cuff presses the arm with a little pressure so that the brachial artery remains distended with blood, no sounds are heard by stethoscope despite the fact that the blood within the artery is pulsating. But when the cuff pressure is great enough to obliterate the artery during part of the arterial pressure cycle, under a certain pressure characteristic sounds are heard by stethoscope in the artery with each pulsation distal to the portion compressed by the cuff. These sounds were discovered by Korotkow—hence known as *Korotkow sounds*. When the cuff pressure is between 50-70 mm Hg, these sounds are heard and with rise of cuff pressure rises to 100-120 mm Hg, the sound suddenly disappears.

The exact cause of these sounds are not known but it is believed to be caused by a jet of blood passing through the partially occluded artery producing turbulent flow. The jet, i.e., increase in velocity of blood flow—when reaches critical level causes turbulence in the open artery beyond the cuff, this sets up the vibrations heard by stethoscope.

• **Techniques**: Patient should lie down on recumbent position to keep the instrument at heart level. An air pressure cuff is wrapped over the left brachial artery on the upper arm. This air pressure cuff having rubber bag inside raises air pressure by the air pump. The pressure is generally raised to a few mm Hg above when arterial sound is not heard and then gradually released by the Outlet valve. Variations of sounds are heard with a stethoscope placed on the brachial atery. A prominent sound is heard first and considered as systolic pressure. But on further release of pressure, when this sound diappear then at that time pressure is called diastolic pressure.

The cuff is inflated until the air pressure within it overcomes the arterial pressure and occludes the arterial lumen which is indicated by feeling disappearance of the radial pulse wave. The pressure is increased a little beyond this point and is reduced again by deflation slowly by the release of the valve, until the arterial pressure just overcomes the air pressure and blood escapes beyond the cuff into the peripheral part of the artery. If deflation is carried out too rapidly the mercurial manometer (clinical) shows a considerable lag, due to the time required for the air above the mercury column to reach atmospheric pressure. With rapid deflation—especially in bradycardia—the actual pressure at which arterial pressure overcomes cuff pressure may be missed. Therefore, deflation should not be higher than 2-3 mm Hg per second.

During deflation, as soon as the pressure in the cuff falls below systolic pressure, first Korotkow sound appears, blood slips through the artery beneath the cuff during the peak of systolic pressure, and "tapping" sounds are heard in the antecubital artery by stethoscope in synchrony with the heart beat. This is due to the fact that at this moment blood begins to flow into the distal portion of the artery during systole and distends the arterial walls.

When the sounds are heard, the manometer reading indicates the systolic pressure. The mercury manometer should denote, with its upper meniscus, a reading on the scale (after use, the mercury level should coincide the zero level).

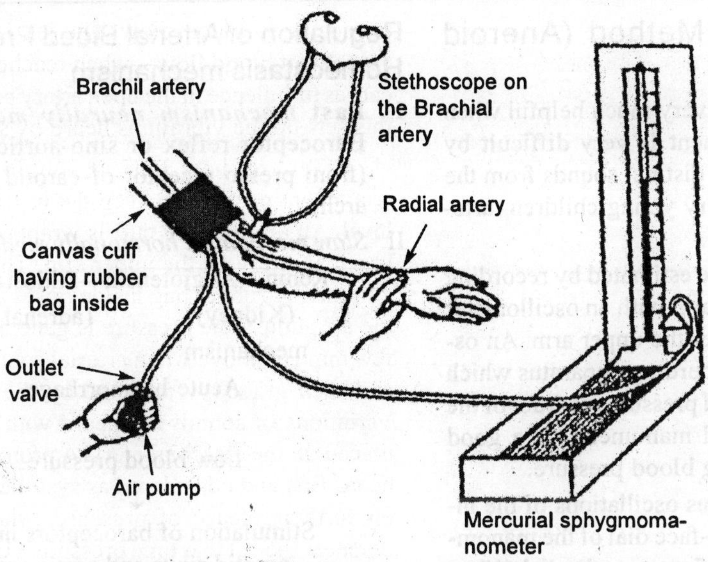

Fig. 28.25: Ausculatory method for measurement of blood pressure

While releasing the air pressure gradually, blood just begins to flow through the narrowed vessel and the pattern on flow is changed from streamline (laminar) flow to turbulent (noisy) and tapping sounds are heard. When the pressure is further released, normal streamline flow sets in and Korotkow sounds change in quality with less of the tapping quality having more of a rhythmic, harsh quality. The sounds are heard in 4 phases:

Phase I : "Tapping" sound, persists during fall of 15 mm Hg.

Phase II : "Murmur" sound, lasting for 15 mm Hg.

Phase III : "Gong" sound, clear, loud, lasting for 15 mm Hg.

Phase IV : "Muffled" sound and begins to fade away, indicates diastolic pressure.

Phase V : No sound.

When the pressure in the cuff falls equal to diastolic pressure, the artery is open and there is no jet flow of blood in the artery. Therefore, the sounds suddenly change to a muffled quality from Korotkow, the manometeric pressure is approximately equal to the diastolic pressure. The sounds usually disappear totally after another 5-10 mm Hg drop in the cuff pressure. This is probably due to the fact that when the pressure in the cuff drops below diastolic pressure, the artery over which the sounds are ausculated during diastole is not empty because the blood penetrates into the artery under the cuff also during diastole (Fig. 28.25).

By exerting cuff pressure, the blood pressure measurement can be done in three ways:

1. Ausculatory method (described above).
2. Palpatory method.
3. Oscillatory or oscillometric method.

Palpatory Method: Cuff pressure on the arm is raised to certain higher level than the systolic pressure and the valve is released gradually till the radial pulse wave is felt by pressing the radial artery against the radius bone at the wrist and the pressure is noted in the manometer which is systolic pressure.

- **Disadvantages**

1. It is rarely used because of less accuracy.
2. Diastolic pressure cannot be measured.
3. It is known that the first release of blood beneath the cuff causes pulsation in the peripheral artery but due to lack of evidence, we do not know that the amount of blood first released beneath cuff is sufficient to produce pulsation wave which can be felt by the finger.
4. Definite pulsation may not occur until the cuff pressure has been lowered 5-10 mm Hg below the point when the artery first becomes pervious.
5. Gives low reading of pressure.

28.24 Oscillatory Method (Aneroid Manometer)

The oscillatory method is very much helpful when blood pressure measurement is very difficult by ausculatory method to get distinct sounds from the antecubital artery, e.g., very young children, arterial spasm in shock.

Arterial pressure can be estimated by recording the pulsation in the lower arm with an oscillometer while a cuff is inflated over the upper arm. An oscillometer consists of a recording apparatus which registers pulsation. A blood pressure recorder of the aneroid type, i.e., aneroid manometer, is a good oscillometer for estimating blood pressure.

The pulsations appear as oscillations of the indicator needle on the clock-face dial of the manometer. Pressures are noted by figures on the dial. When the cuff is inflated or deflated, the needle moves to indicate the applied pressure according to the condition.

The pressure in the cuff on the upper arm is raised above the systolic pressure, the oscillations are minimal but when the cuff pressure is slowly lowered, and the pulsations pass beneath the cuff distinction oscillations will be recorded by the oscillometer with a rapid increase in amplitude and duration. This is the indication of systolic pressure. Further decrease in the cuff pressure causes a progressive increase in intensity of oscillometer pulsation and then decreases at approximately diastolic level and suddenly becomes smaller and at this point the figure on the dial to which the indicator points is noted is the diastolic pressure.

Automatic Apparatus: This is an electronic equipment which can pick up Korotkow sounds and is audible by a microphone. The blood pressure is digitally displayed on the equipment. The equipment can record blood pressure continuously by deflecting the pulsatile flow in the artery by pizoelectric crystal gauze. It is very much advantageous for 24 hours monitoring of blood pressure.

Regulation of Arterial Blood Pressure and Homeostasis mechanism

I. **Fast** *mechanism neurally mediated*: By Baroceptor reflex or sino-aortic mechanism (from presso receptor of carotid sinus, aortic arch)

II. *Slow mechanism hormonally mediated*

Renin—Angiotensin—Aldosterone mechanism
(Kidney) (adrenal cortex)

Acute haemorrhage
↓
Low blood pressure
↓
Stimulation of baroceptors in the carotid sinus and aortic arch
↓
Firing of nerve impulses through sino-aortic nerves (Herring's nerve)

↓ Depression of parasympathetic nerve to the heart
↓ Increase in heart rate

↓ Increase in sympathetic nerve activity of the heart and blood vessel

Decreases unstressed blood volume

Increases
1. Heart rate
2. Contractility of the heart
3. Contraction of arterioles (total peripheral resistance)
4. Contraction of veins
5. Increased various pressure
6. Increase mean systemic pressure

Increased mean arterial pressure (MAP) to restore normal stage

28.25 Indirect Automated and Semiautomated B.P. Instrument

1. **Indirect Semiautomatic**: The method consists of a *transducer* connected to the sphygmomano-meter cuff, a *microphone* placed under the cuff over the brachial artery above the elbow joint and a *recording system* to record cuff pressure and Korotkow sound. The procedure is similar to manual method. The cuff is automatically inflated to a certain high level such as about 180-200 mm Hg then it is allowed to deflate slowly. The microphone picks up Korotkoff sound of the artery from the surface. Depending on the type of instrument either superimposes the signal of the Korotkoff sound on the voltage recording the fall of blood pressure or records the two separately during the gradually fall of pressure, the first sound indicates systolic blood pressure, and when the last signal represents last sound that indicates diastolic blood pressure.
   It is called semiautomatic because the available record is *interpreted-by the operator*. The *disadvantage* is the false indication caused by motion artifacts.
   It differs from fully automated instrument which uses some kind of signal—detecting circuit to detect the first and last Korotkoff sound either electronically or with mercury manometer which is cut off by solenoid valves.

2. **Programmed Electrosphygmomanometer**: The instrument consists of:
   1. Occluding cuff
   2. Microphone of pulse transducer
   3. A recorder

   The instrument incorporates:
   (a) A transducer—preamplifier—provides two output signals
   (b) A voltage proportional to the cuff pressure
   (c) Amplified Korotkoff sounds or pulses.

   These signals can be monitored individually or with the sounds or pulses superimposed on the calibrated cuff recorder. The combination signal can be recorded on a graphic pen recorder.

   The self contained cuff inflated system is programmed to inflate and deflate the cuff at various rates and time intervals. Equal and linear rates of cuff inflation accomplishes two phases of B.P. i.e., systolic and diastolic pressure measurement in each cycle. The instrument can be programmed for each cycle after adjusting the time interval for monitoring the B.P. for a certain period of time. Single cycle may be started by pressing a panel mounted on switch. Arrangement is also made for remote control via by external contact closure.

   The maximum pressure of the cuff can be adjusted and the front panel meter gives a continuous vascular observation of the B.P. through cuff (Fig. 28.26).

3. **Electronic Sphygmomanometer**: It is a device which incorporates a large gauge for taking reading easily. The scale measures 6 inches (over 15 cm) in diameter. The cuff is inflated manually for which this device is also called semiautomatic. The Korotkoff automatically detected by microphone and a light flashes at 1st Korotkoff sound i.e., systolic blood pressure which stops at diastolic pressure.

4. **Dynamap** (for Mean arterial pressure): Mean arterial pressure (MAP) is generally calculated by taking average of systolic plus diastolic blood pressure. A simple formula is followed for MAP such as $\frac{1}{3}$ of the systolic + diastolic pressure such as:

   MAP = $\frac{1}{3}$ systolic BP + Diastolic BP + diastolic BP

   MAP has the importance for the following reasons:
   (a) Indicates the condition of tissue perfusion.
   (b) Continuous flactuation of MAP ultimating leads to hypertension or hypotension. Dynamap follows the principle of oscillatory method.

   Pulsation or oscillation is introduced within the cuff bladder which is sensed

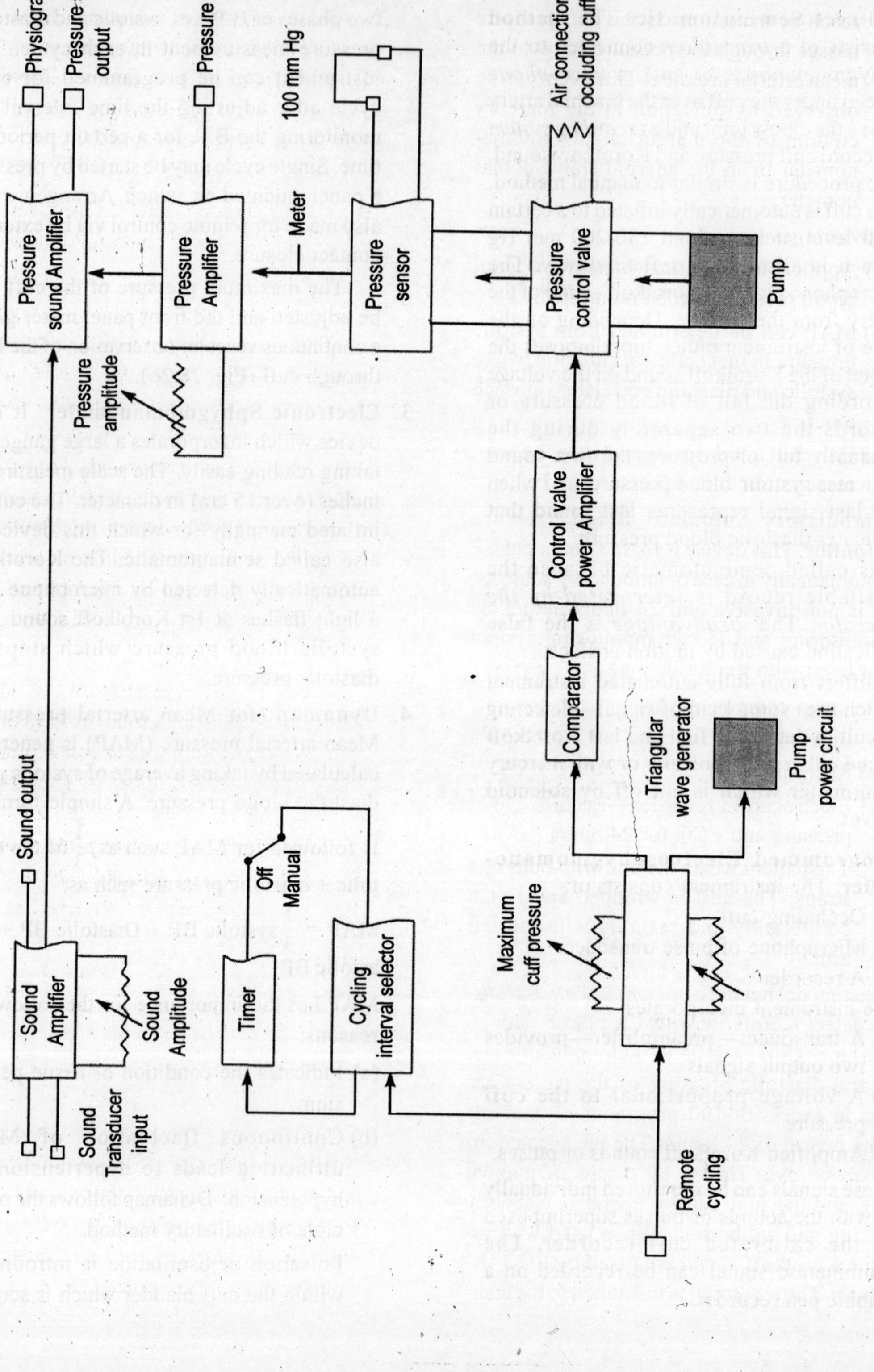

Fig. 28.26: Block diagram of programmed electrosphygmomanometer

by a solid-state pressure transducer located within the enclosure when the air pressure in the cuff is reduced, pressure oscillations increase in amplitude and reach or maximum as the cuff pressure passes through the pressure equal to the mean arterial pressure. This event can be observed by pulsation in the mercury column of the mercurial sphygmomanometer or in the aneroid sphygmomanometer.

- **Advantages**
  (a) Applicable in critical patient's arterial blood pressure can be monitored,
  (b) MAP can be determined from systolic pressure in most clinical situation where systolic and diastolic correlate with each other,
  (c) Can be used in operation theatre, resucitation unit, ICU etc.

5. **Ambulatory Automatic Blood Pressure Monitor**: This device is for measurement BP automatically in case of ambulatory subjects. It is non invasive and 24 hours automatic monitoring and Hotter monitoring system for heart with trend ride outs.
   The main components are:
   (a) Pressurometer
   (b) Recording unit
   (c) Electrocardio recorder—Records blood pressure and ECG for 24 hours.
   The patient wears the unit with belts and straps. The cuff is wrapped and fasten comfortably and securely with the left arm.

Another model having a *pneumatic cuff.* A *transducer* to detect Korotkow sound is placed with an adhesive tape. Cuff is inflated automatically and systolic and diastolic pressure of the patient are measured in excess of 100 programmed intervals during the period of 24 hours. The preselected intervals can be over ridden by means of manual switching and the unit can be operated manually for checking and calibration. When used with recorder, the Korotkoff sound are gated by ECG 'R' wave signals. The recording is trasmitted to the companion trend *computer,* which gives digital read out of the data on the recording. The whole system is powered by a portable rechargeable *nickel-Cadmium battery* which can continue upto 26 hours recording which is important because it can be used with electrocardio-scanner which can scan all the data at a rate 120 times as fast as they were recorded. The whole record of 24 hours can be scanned in 12 minutes which includes the record of:

(a) total heart beat
(b) printing hourly trends on heart rate
(c) operator selected abnormalities when it rewinds itself

6. **Ultrasound**: Ultrasound is employed to measure the pulsatile motion of the brachial artery wall. High frequency ultrasound energy is transmitted into the patients arm and reflected back from the arterial wall. Using Doppler effect the movement of the arterial wall is detected when they snap opens and closed with each pulsation of blood. The advantage is that the result is closer to the direct method. It can be used in patient is under shock and in ICU when the direct method is not feasible.

   One of the instrument is called arteriosonde having cloth cuff and air bag with an electric air pump to create pressure. The pump can be regulated by a front-panel control. The cuff is placed on the left arm in the same manner as the sphygmomanometer except that there is a transducer under the cuff. The transducers are arranged as alternate transmitter and receiver. The motion of the artery produces a Dopple shift which identifies the artery is opened and close with each heart beat between the systolic and diastolic pressure.

   The first signal is to stop the mercury fall in the first manometer out of the two which indicates systolic pressure. Stopping the second manometer at the moment of disappearance of the pulses is the indication of diastolic pressure and both the readings can be observed simultaneously.

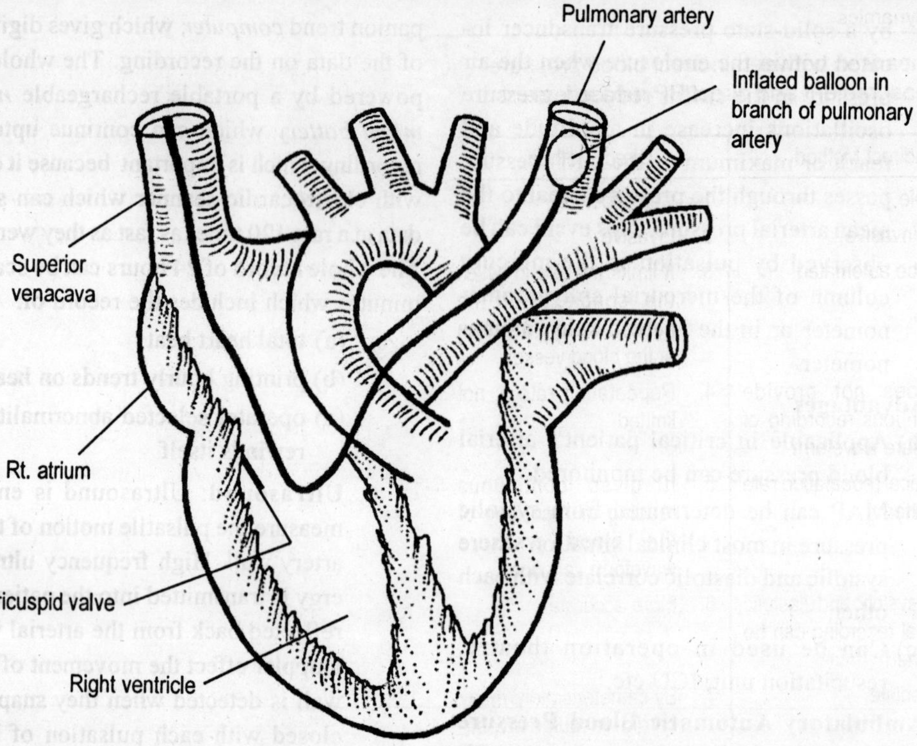

Fig. 28.27: Diagrammatic representation showing the course of pulmonary artery catheter through the heart

28.26 Direct Blood Pressure measurement

28.26.1 Percutaneous insertion

A local anaesthesia is given near the site of invasion after occlusion of the vessel, a hollow needle is inserted at an angle toward the vessel. When the needle is in place, a catheter is introduced through the wide bore hollow needle with guide when the catheter is in place in the vessel, the needle and guide are withdrawn.

For some measurements, a type of needle attached to an airtight tube is used, so that the needle can be left in the blood vessel and the *blood pressure* sensed directly by attaching a transducer to the tube other types have the transducer built into the tip of the catheter. This later type is used in both percutaneous and in catheterisation.

28.26.2 Cardiac catheterisation

Insertion of a cardiac catheter with transducer at the tip is passed through brachial artery under television screen. It is used for:

1. The pressure measurement in the different chambers of the heart.
2. Pressure measurement in the great blood vessel.
3. Measurement of cardiac output.
4. Detection and localisation of shunt.
5. Calculation of area of an oritice.
6. Angiocardiography.
7. Coronary arteriography.
8. To obtain blood samples from the heart for oxygen content analysis.
9. To detect and localise the abnormal blood flow path.
10. For X-ray studies by injection of radiopaque diges.
11. Injection of colour dyes for indicator dilution studies.
12. For injection of vasoactive drugs directly into the heart and certain vessels.

Catheter is a long narrow tube is passed into the heart or great vessel by way of a superficial vein e.g. basilic vein or artery e.g. brachial artery.

Comparison between indirect blood pressure measurement and direct BP measurement

Indirect Method	Direct Method
1. Simple	1. Complicated
2. Non-invasive	2. Invasive
3. Can be automated	3. It limits the use in which the condition of the patient warrant invasion of the blood vessel
4. It does not provide continuous recording of pressure waveform	4. Repeatation rate is not limited
5. Practical repeatation rate is limited	5. It gives continuous reading or recording of the blood pressure waveform variations.
6. Only systolic and diastolic arterial recording can be obtained	6. More accurate
7. Subjective	7. By catheterisation, many cardio-vascular investigation such as pressure of individual chambers, great vessels X-ray study etc.
8. It often fails in subject when BP is very low e.g. shock	8. MAP can not be done.
9. Not so accurate	
10. Can not be used during activity, ambulatory patient	
11. It gives general blood pressure of the C.V. system	
12. Mean arterial pressure (MAP) can be measured	

Measurement can be done in two processes:

1. Introducing a sterile saline solution into the catheter so the fluid pressure can be transmitted to a transducer outside the body to prepare a complete fluid pressure system with provision for checking against atmospheric pressure and for establishing a reference point. The frequency response of this system is a combination of the frequency response of the transducer and the fluid column in the catheter.
2. The second process, the transducer is introduced in the catheter and pushed to the point of measurement or the transducer is mounted at the tip of the catheter and this method is called the catheter tip blood pressure transducer. For mounting the transducer at the tip two types of transducer are used:
   (a) Unbonded resistance strain gauge in the transducer
   (b) Variable inductance transducer.

28.26.3 Implantation Technique

This technique is usually employed in the research experiments. The advantages of this technique are

(a) the transducer can be kept fixed at the site of the desired vessel

(b) can be maintained for a long period.

Transducer can be selected by the type of circuit element used to sense the pressure variation such as capacitive, inductive and resistive but resistive type is frequently used. In the capacitance manometer, a change in the distance between the plates capacitor changes its capacitance. In usual application one of the plates is a metal membrane separated from a fixed plate by some $\frac{1}{1000}$th of an inch of air. Changes in pressure change the distance between the plates, thereby change the capacitance. If this element is placed in high-frequency resonant circuit, the changes in capacitance vary the frequency of the resonant circuit to produce a form of frequency modulation with a proper circuit the BP can be obtained and record in time.

28.27 Effect of Hydrostatic Pressure on Venous and Arterial Pressure

The pressure at the surface of the water is equal to atmospheric pressure. But the pressure rises 1 mm Hg for every 13.6 mm depth from the surface which is the result from the weight of the water and known as hydrostatic pressure.

Similarly, hydrostatic pressure occurs in the vascular system due to weight of the blood in the blood vessel, i.e., gravity of blood column.

28.27.1.1 Effect on venous pressure

1. **Right atrium**: When a person is standing, the pressure in the right atrium is approximately 0 mm Hg. The reason is that the heart pumps into the arteries excess amount of blood which tends to accumulate at these points.

2. **Veins of the Feet, Thigh and Abdomen**: On standing, the pressure in the veins of the feet in man is +90 mm Hg for greater height between the heart and the fet and weight of the blood column in the veins between the heart and feet. Similarly, according to distance from the heart and weight of the blood column, venous presssure at high level is about + 40 mm Hg and the abdominal veins shows about + 22 mm Hg. Leg veins are not rigid tubes and are responsive to elevation of intramural pressure both geometrically—by changing in luminal configuration from oval to circular cross-sectional profiles—and distensibly—by changing myogenic tone. They accommodate more blood and there is "venous pooling" resulting low venous return which in turn induces a fall of cardiac output and ultimately tends to decrease the arterial pressure head. Peripheral venous pressure, like arterial pressure, is increased by 0.77 mm Hg. for each cm below the right atrium and decreased a like amount for each cm above the right atrium.

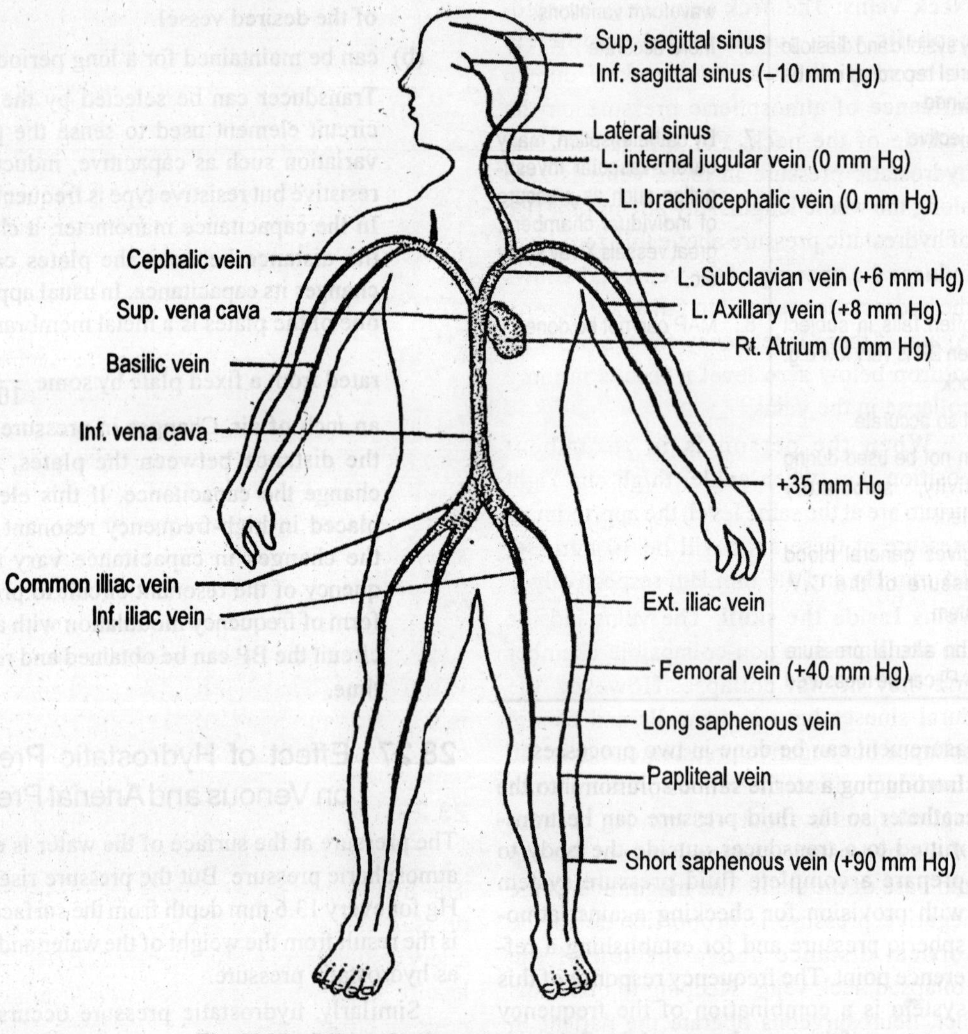

Fig. 28.28: Effect of hydrostatic pressure on venous pressure

3. **Veins of the Arm and Hand**: The pressure in the arm veins (e.g., subclavian vein) at the level of the first rib is about +6 mm Hg due to compression of the subclavian vein by sharp angulation over the first rib. The hydrostatic pressure along the arm and forearm is determined by the distance below the level of the first rib. If the hydrostatic pressure difference between the level of the first rib and the hand is 29 mm Hg, then hydrostatic pressure of 6 mm Hg—caused by compression of the vein when it crosses the rib—is added to 29 mm Hg, making a total hydrostatic pressure of 35 mm Hg, in the veins of the hand.
4. **Neck Veins**: The neck veins (e.g. brachio-cephalic vein, jugular vein) completely collapse all the way to the skull due to influence of atmospheric pressure on the outside of the neck. For this reason, the hydrostatic pressure in these veins is zero along the whole length. Any attempt for rise of hydrostatic pressure above this zero level induces the opening of the veins and allows the hydrostatic pressure to fall again to zero. Again, if there is tendency for the pressure to drop below zero level it results in more collapse in the veins.

   When the person is in recumbent position, i.e., when ankle, thigh and right atrium are at the same level, the approximate pressure at these sites will be 10 mm Hg, 7.5 mm Hg and 4.6 mm Hg, respectively.
5. **Veins Inside the skull**: The veins indside the skull are in a non-collapsible chamber and they do not collapse. However, the dural sinuses have rigid walls and cannot collapse but negative pressure can exist. In standing position, the venous pressure in the sagittal sinus is about –10 mm Hg because of the hydrostatic suction between the top and base of the skull. The magnitude of the negative pressure is proportionate to the vertical distance above the top of the collapsed neck veins. Keeping in mind this fact, neurosurgeons operate the patient in sitting position. If one of the sinuses is opened during neurosurgical procedures, air is sucked in causing air embolism (Fig. 28.28).

28.27.2 Role of venous valves and "venous pump"

The venous pressure in the feet is always +90 mm Hg in standing position in adult because of hydrostatic pressure. But if the person moves the legs, contraction of the muscles compresses the veins and squeezes the blood towards the heart. Due to arrangement of the valves, the direction of flow is only toward the heart but not the reverse flow and the pressure in the vein is lowered. This pumping system is known as the "venous pump" or "muscle pump" and the pressure in the veins of the feet remains +25 mm Hg during walking.

If persons stand without movement of the muscles, the full hydrostatic pressure value at the ankle is +90 within 30 seconds. In this condition capillary pressure also increases and fluids leaks from the circulation into the tissue spaces resulting in oedema of the legs.

28.27.3 Varicose veins

Sometimes the venous valves become incompetent due to overstretching of veins for high venous pressure for a prolonged period (e.g., pregnancy, duty on standing—as in case of traffic police). Thus, stretching the veins causes increase in cross-sectional areas, but the valves do not increase in size. Hence the veins cannot close completely and prevent the reverse flow. Owing to high rise of venous pressure for failure of venous pump there is further distension of the veins and finally due to venous stasis, the veins become swollen and there is bulbous protrusion under the skin—of the leg called *varicose vein.*

28.27.4 Effect of hydrostatic factor on arterial pressure

The hydrostatic pressure also affects the arterial pressure like venous pressure. The pressure in any artery below heart level is increased and above heart level is decreased. Suppose, a person in standing position having mean arterial pressure of 100 mm Hg at heart level has mean arterial pressure in the

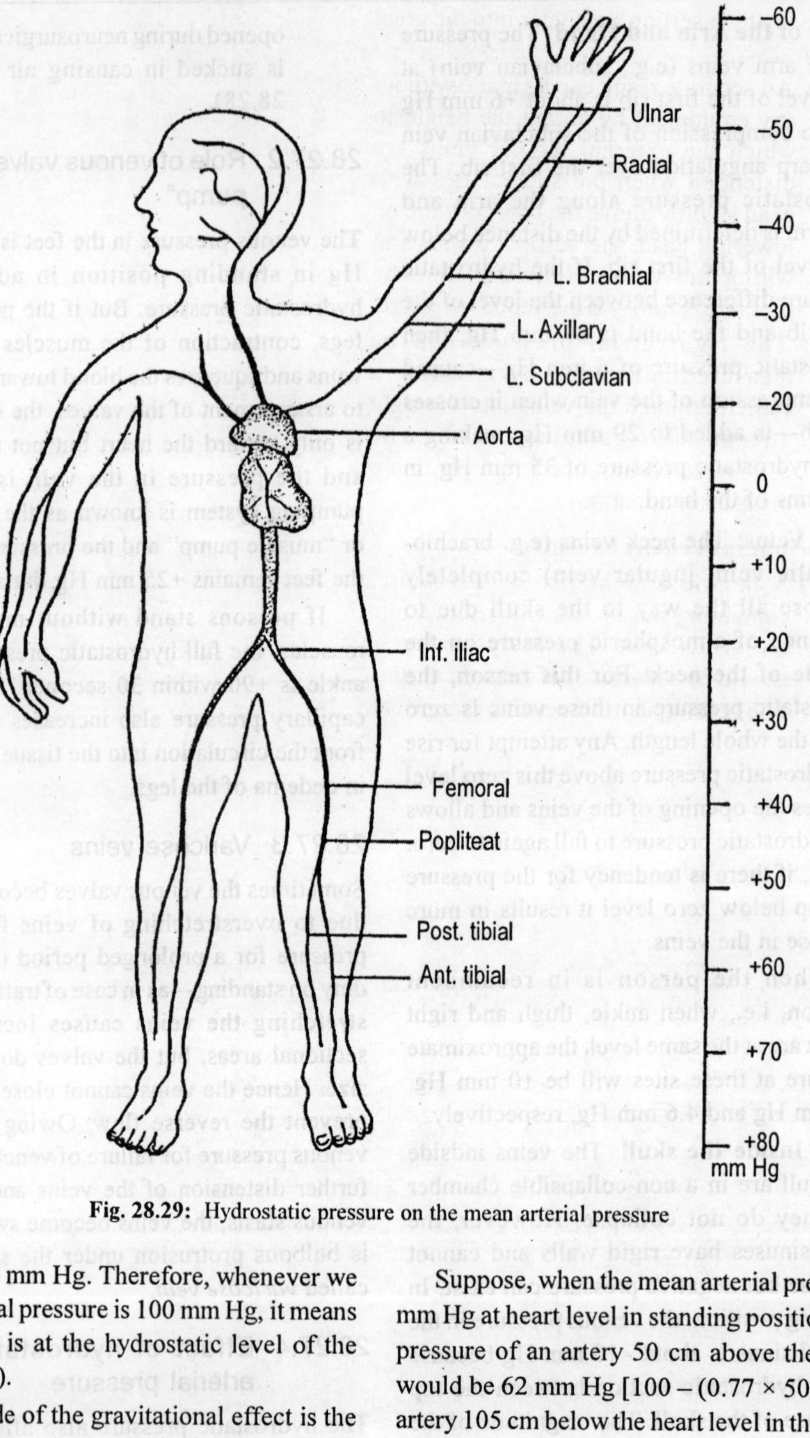

Fig. 28.29: Hydrostatic pressure on the mean arterial pressure

feet of about 190 mm Hg. Therefore, whenever we say that the arterial pressure is 100 mm Hg, it means that the pressure is at the hydrostatic level of the heart (Fig. 28.29).

The magnitude of the gravitational effect is the product of the density of the blood, the acceleration due to gravity (980 cm/s/s), and the vertical distance above or below the heart is 0.77 mm Hg/cm, at the density of normal blood.

Suppose, when the mean arterial pressure is 100 mm Hg at heart level in standing position, the mean pressure of an artery 50 cm above the heart level would be 62 mm Hg [100 – (0.77 × 50)] and in the artery 105 cm below the heart level in the foot would be 180 mm Hg [100 + (0.77 × 105)].

28.27.5 Measurement of venous pressure

1. **Central Venous Pressure**

(A) Central venous pressure can be measured

directly by inserting a catheter into the great vein.

(B) An accurate result can be made without any equipment by noting the height to which the external jugular veins are distended when the subject lies with head slightly above the heart. Again, on lowering head, the vertical distance at which the vein collapses—i.e., the pressure is zero—is the venous pressure.

(C) The right atrial pressure can be measured accurately by inserting a catheter though the veins into the right atrium. This catheter can then be connected to an appropriate pressure measuring apparatus.

II. Peripheral Venous Pressure

(A) A needle attached to a manometer containing sterite saline is inserted into the arm's vein. The peripheral vein should be at the level of the right atrium. The reading in millimetres of saline can be converted into millimetres of mercury by dividing by 13.6 (the density of mercury).

(B) Roughly, anyone can measure the venous pressure by raising or lowering an arm of a reclining person while noting the degree of distension of the antecubital or hand veins. When this is progressively raised, the veins collapse and the level at which it occurs from heart reference the level is a rough measure of the peripheral venous pressure.

(C) Venous pressure can be recorded by inserting a syringe needle connected to a water manometer or pressure recorded directly into a vein. The measurement is expressed in height in centimetres of water above the level of the tricuspid valve (reference level for pressure measurement). This can be converted to mm Hg by dividing by a factor of 1.36.

29

Tissue Engineering

PART I: CELL

1. Human cell, 2. Organelles of cell, 3. General features of cell, 4. Cell as a living unit, 5. Some specialised cell and function, 6. Cells of lymphoid system, 7. Cells of mucosa; liver; bone, 8. Cell membrane (see chapter part III), 9. Stem cell, 10. Stromal cell, 11. Cell differentiation, 12. Cancer cell, 13. Apoptosis or programmed cell death, 14. Junctional complex, 15. Growth factor, 16. Soluble Growth factor; Growth inhibitor, 17. Measurement in cell, 18. Flow cytometer, 19. Cell fractionation, 20. Cell; tissue; organ culture, 21. Fine Needle Aspiration Cytology (FNAC), 22. Growth and Development, 23. Exfoliative Cytopathology.

PART II: CYTOKINETICS

1. Cell adhesions, 2. Cell migration or cell motion, 3. Signal reception, 4. Cell communication, 5, Cell cycle, 6. Cell signalling, 7. Cell recognition, 8. Cell receptors, 9. Stem cell, 10. Biologic modifiers, 11. Cell therapy, 12. Biomaterials, 13. Cytokines, Bioreactors, 14. Chemical mediators of inflammation.

PART III: CELL MEMBRANE

(For details see chapter 9.)

1. Dynamics of cell membrane (Fluidity of cell membrane), 2. Study of cell membrane, 3. Membrane recycling, 4. Functions of membrane proteins, 5. Ion channels, 6. Gap junction, 7. Membrane proteins for dynamic processes, 8. Techniques for study of cell membrane.

PART IV: TISSUE DYNAMICS

1. Definition; classification; origin of tissue, 2. Connective tissue, 3. Comparison of different tissue, 4. Component of connective tissue, 5. Tissue, cell spaces, 6. Extracellular Matrix, components, 7. Fibrous proteins, adhesive fibrous proteins, 8. Organisation of tissue into different functional system and subunits, 9. Functions of proteins, 10. Collagen-rich tissue, 11. Tissue dynamics, 12. Collagen, 13. Micro-environment, 14. Tissue repair, healing of wounds.

PART V: TISSUE GRAFT, TRANSPLANTATION, CRYOTECHNIQUE

1. Transplantation, Grafting, 2. Rejection of grafts, 3. Prosthetic heart valve, 4. Cryo-technique or preservation technique.

PART VI: BONE MARROW AND TRANSPLANTATION

1. Bone Marrow-Nature, amount, Type, composition, Origin, Ratio of cells, Function, Distribution of marrow, Microscopical observation, Components, average percentage of different cells, Significance of study of bone marrow, Myeloid-erythroid ratio, Indications for bone marrow study. 2. Transplantation of bone marrow.

PART I

CELL

1. Human cell, 2. Organelles of cell, 3. General features of cell, 4. Cell as a living unit, 5. Some specialised cell and function, 6. Cells of lymphoid system, 7. Cells of mucosa; liver; bone, 8. Cell membrane (see chapter part III), 9. Stem cell, 10. Stromal cell, 11. Cell differentiation, 12. Cancer cell, 13. Apoptosis or programmed cell death, 14. Junctional complex, 15. Growth factor, 16. Soluble Growth factor; Growth inhibitor, 17. Measurement in cell, 18. Flow cytometer, 19. Cell fractionation, 20. Cell; tissue; organ culture, 21. Fine Needle Aspiration Cytology (FNAC), 22. Growth and Development, 23. Exfoliative Cytopathology.

29.1 Human Cell

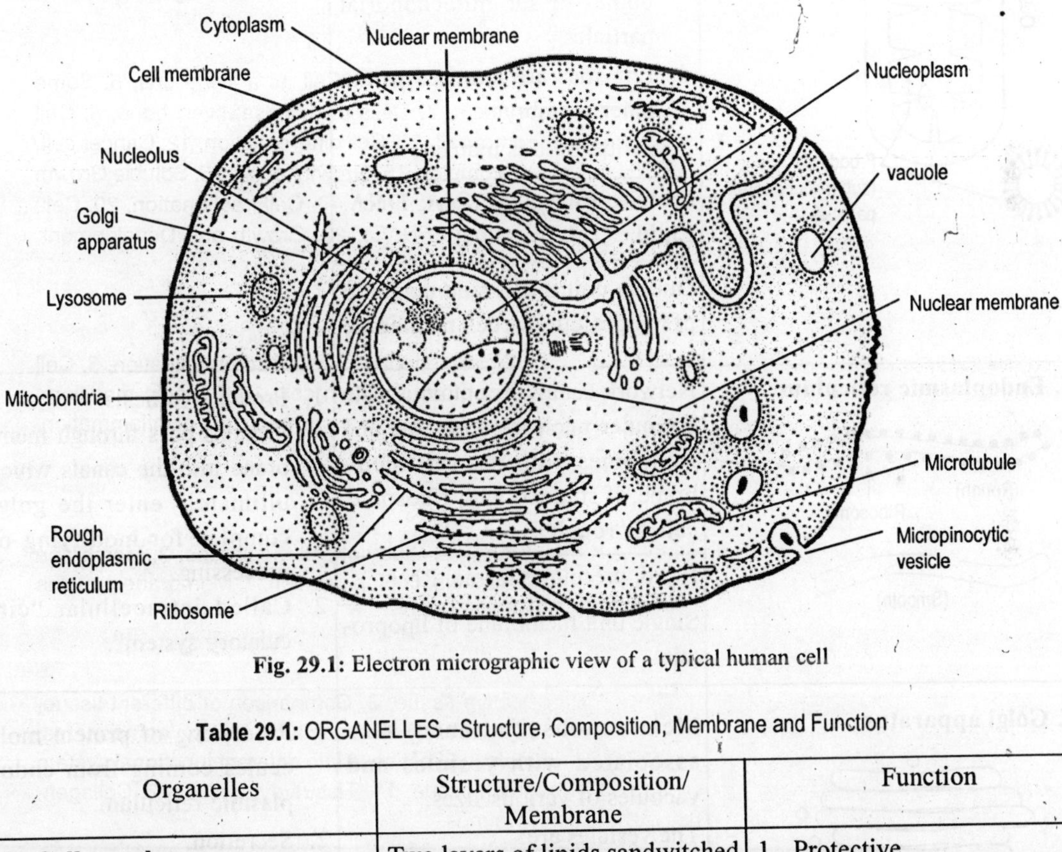

Fig. 29.1: Electron micrographic view of a typical human cell

Table 29.1: ORGANELLES—Structure, Composition, Membrane and Function

Organelles	Structure/Composition/ Membrane	Function
I. Cell membrane 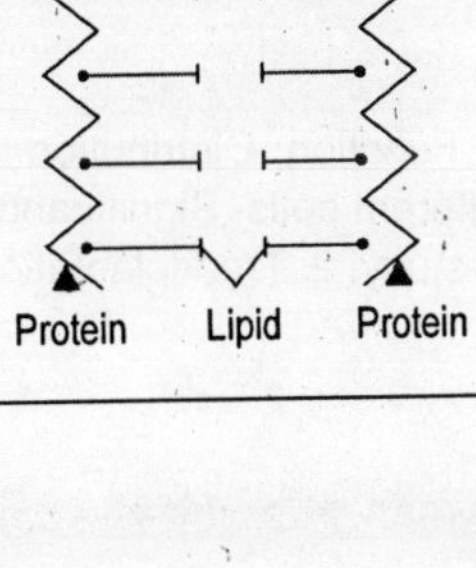	Two layers of lipids sandwitched between two layers of protein. Phospholipids, glycoproteins. Single unit membrane.	1. Protective 2. Transport of materials by diffusion, carrier mediated diffusion, active transport, pinocytosis.

Contd.

Organelles	Structure/Composition/ Membrane	Function
II. Mitochondria 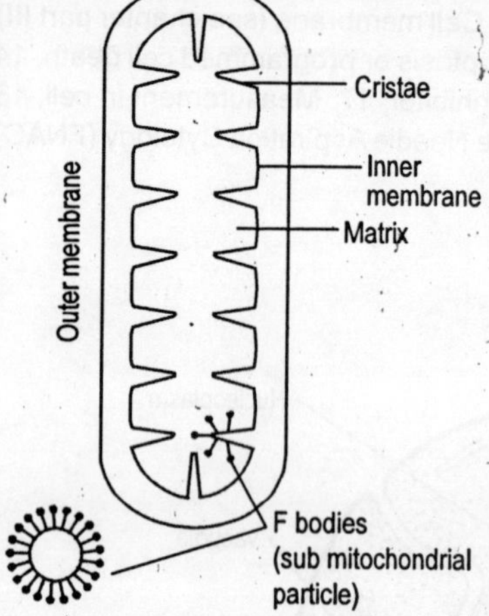	1. Two layers of membrane and having space between the two. 2. Inner membrane has been folded to form cristae to increase surface area. 3. Granular round bodies attached by stalk called F bodies or sub-mitochondrial particles. 4. Enzymes are attached to the inner membrane. 5. Inside fluid is matrix. Protein, Enzymes, RNA, DNA, Lipid. Double unit membrane made up of lipoprotein like cell membrane	Enzymes present for 1. Kreb's cycle. 2. Urea cycle (Partly). 3. Fatty acid oxidation. 4. Amino acid oxidation. 5. Electron transfer. 6. Regeneration of ATP (95%). 7. Protein synthesis.
III. Endoplasmic reticulum 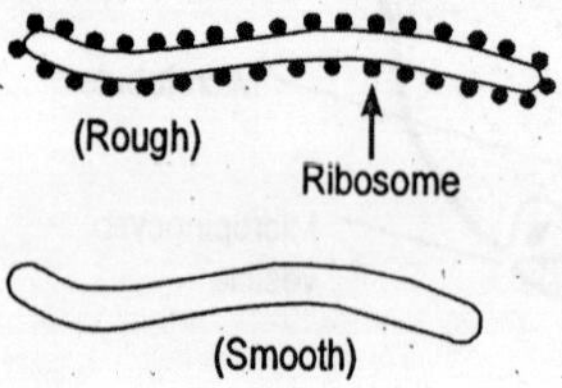	Network canals, continuous with the outer nuclear membrane and sometimes with the cell membrane, if studded with ribosome, called "rough". Lipid, protein, enzyme, RNA. Single unit membrane of lipoprotein.	1. Protein synthesis—Proteins pass through membrane into the canals which ultimately enter the golgi complex for modifying or processing. 2. Called intracellular "circulatory system".
IV. Golgi apparatus 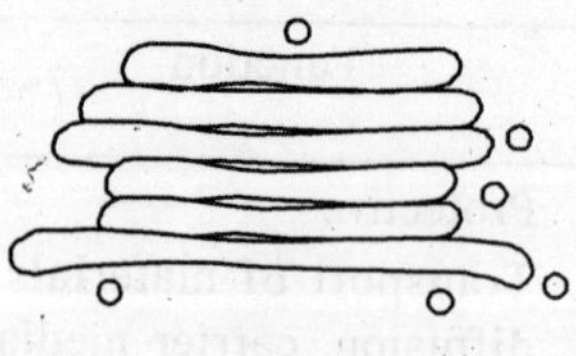	Membranous flattened sacs associated with vesicles and vacoules of various sizes. The vesicles are: 1. Flattened vesicles. 2. Secretory vesicles.	1. Processing of protein molecules coming from endoplasmic reticulum. 2. Secretion. 3. Synthesis of polysaccharides for glycoprotein.

Contd.

Organelles	Structure/Composition/ Membrane	Function
	3. Microvesicles. Protein, lipids, polysaccharide Bounded by single unit membrane.	4. Helps to form the cell membrane and the membrane of lysosomes.
V. Lysosomes 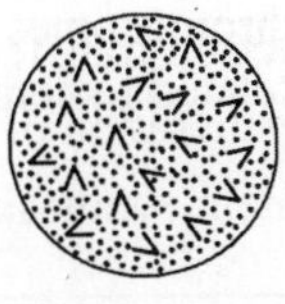	Membranous sac containing many hydrolytic enzymes. Lipid, protein, enzymes. Bounded by single unit membrane.	1. Digestive. 2. Autolysis. 3. Phagocytosis. 4. Wound healing. 5. Acrosome of spermatozoa appears to be a lysosome.
VI. Nucleus 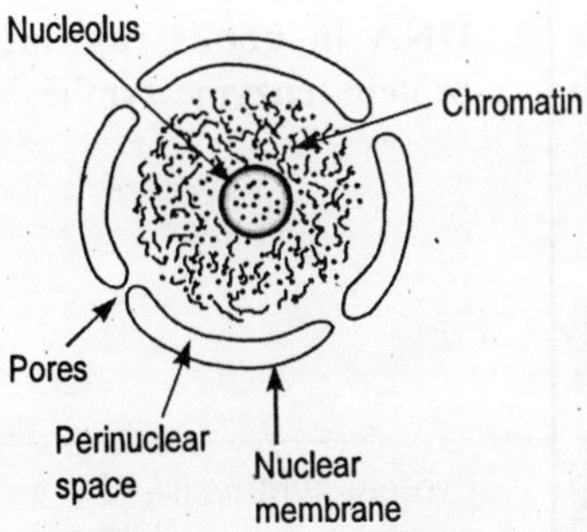	1. Surrounded by membrane called perinuclear envelope having perinuclear space, and spore. 2.. Inside nucleoplasm containing chromatin threads and nucleolus. Protein, DNA, RNA (2%). Bounded by double unit membrane.	During mitosis, chromosomes undergo replication of their DNA and separation into daughter chromosomes.
VII. Nucleolus (one or more per cell) 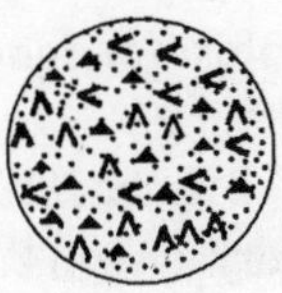	1. Organelles having the highest density. 2. Composed of a reticular network of helically coiled fibrils. 3. Rich in RNA 4. Contains still smaller nucleus known as nucleolus. 5. No unit membrane.	Formation of ribosomes.
VIII. Ribosome 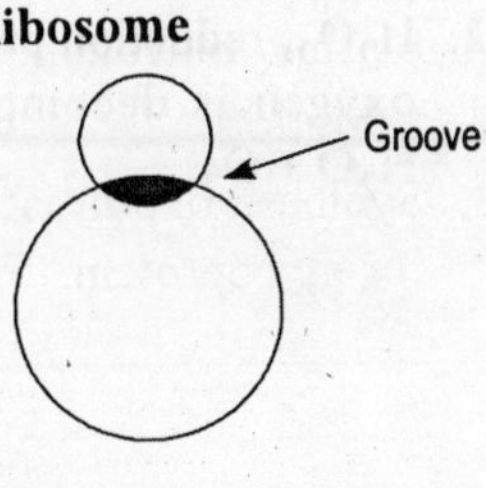	Ribonucleoprotein in nature, attached to the endoplasmic reticulum (rough) or scattered in the cytoplasm singly or in groups (polyribosome or polysome). RNA—50% Protein—50% No unit membrane	Protein synthesis as per genetic code present in mRNA.

Contd.

Organelles	Structure/Composition/ Membrane	Function
IX. Centrosome and Centriole 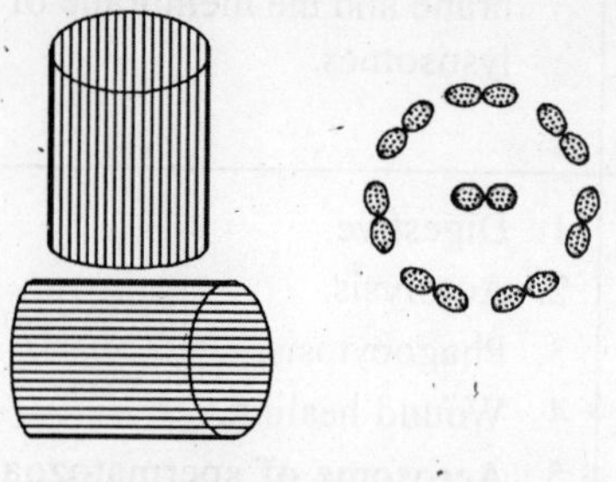	1. Always found in pairs (diplosome) inside the centrosome. 2. Wall of centriole consists of nine tubules arranged longitudinally. 3. Each tubule consists of three subunits, called "triplets". 4. Satellite or pericentriolar bodies are attached to tubules. No unit membrance.	1. Involved in cell motion. 2. Found at centre of mitotic potes. 3. Found at base of all cilia and flagella. 4. Centromere (Kinetochore) of chromosomes believed to be related to centriole.
X. Chromosome (48 per human diploid cell) 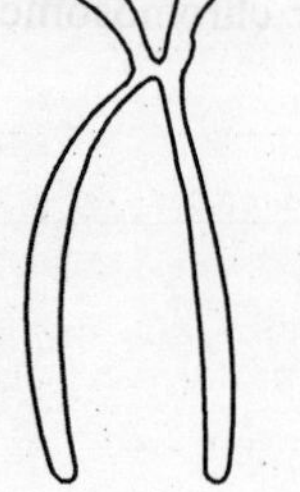	1. Having one to four coiled threads called chromonema. 2. Contains minute particles called "**Chromomeres**" rich in DNA. 3. Two constrictions { Primary, Secondary Primary constriction called kinetochore. 4. Three types of chromosomes based on position of kinetochore: (a) Acrocentric chromosome (b) Metacentric chromosome (c) Sub-metacentric chromosome. DNA, RNA, Protein (Histone). No unit membrane.	1. Having genes carrying hereditary characters. 2. DNA in genes regulates protein, enzyme synthesis.
XI. Microbody (Peroxisome) 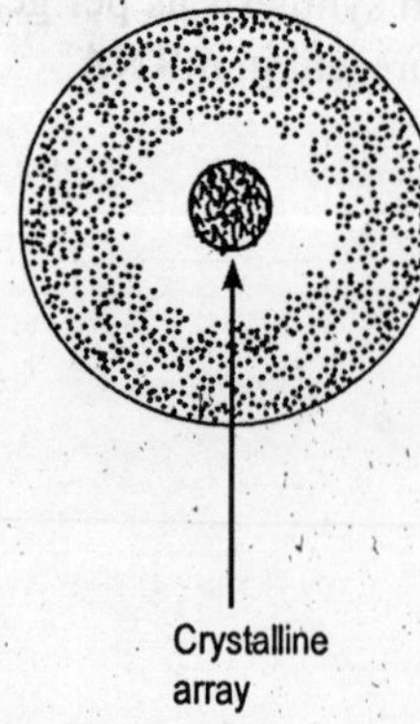	1. Single membranous vesicles (0.5 μm). 2. Contains enzymes: Catalase, D-amino acid oxidase, urate oxidase and other oxidative enzymes present in crystalline arrays. Enzyme proteins Single unit membrane.	1. Oxidation of nutrients. 2. H_2O_2, reduction product of oxygen is decomposed to $H_2O + O_2$.

29.2 Size, Shape, Organization, Type

• **Size**: Cells of the body vary in size. They vary from a few microns (one micron = $\frac{1}{1000}$ of a mm) to few centimetre long. Most of the body cells are within 5μ (micron) to 15μ (micron). Variation in size depends upon the parts of the cell. This is perhaps the adaptation of the cell. Because, in case of small cell, there may be lack of space to accommodate the structure such as RBC to accommodate (having 7.5μ diameter) respiratory pigment haemoglobin, it does not have nucleus. Again, if the cell is too big, it would be difficult to maintain. Some example of size of a cell:

(a) Individual voluntary muscle fibre is an individual cell, so the longest muscle sartorius extending from hip bone to knee joint—is the longest cell in the body, i.e., about 46 cm.

(b) Some nerve cells of animals are about a metre long.

(c) Egg cell of ostrich is about 15 cms.

• **Shape**

The shape of a cell has relation with the nature of its function, such as RBC is biconcave circular disc, which has facilitated to release oxygen from it equally through all surfaces. Muscle cells are elongated and adapted for contraction and relaxation. A nerve cell having long process (axon and dendrons), i.e., nerve fibres, because the nerve cell is situated in nervous system but has to carry message to or from the effective organ which is situated far away.

• **Organization**

Body cells of individual differ in their structural organization. Such as RBC does not have nucleus, WBC has nucleus, platelets without nucleus, voluntary muscle with number of nuclei, involuntary (plain) muscle with one nucleus, cardiac muscle with one nucleus, nerve cell with one nucleus but without centrosome. Though there are variations between the various types of cells, basically there is similarity between them.

Different Types of Somatic or Body Cells (Fig. 29.3).

A. **Without nucleus**

1. RBC
2. Platelet

B. **With nucleus**

3. WBC
   (a) Having one nucleus
      i. Lymphocyte
      ii. Monocyte
   (b) Having more than one nucleus
      i. Neutrophil
      ii. Eosinophil
      iii. Basophil
4. Skeletal muscle fibre
5. Plain or smooth muscle fibre
6. Cardiac muscle fibre
7. Nerve cells
   (a) Pyramidal cell
   (b) Purkinje cell
   (c) Apolar cell
   (d) Bipolar cell
   (e) Golgi type II
8. Neuroglia
   (a) Oligodendrocyte
   (b) Astrocyte
   (c) Microglia
9. Osteogenic cells
   (a) Osteoblast
   (b) Osteoclast
   (c) Osteocyte
10. Cartilage cell—Chondrocyte.

Classification:

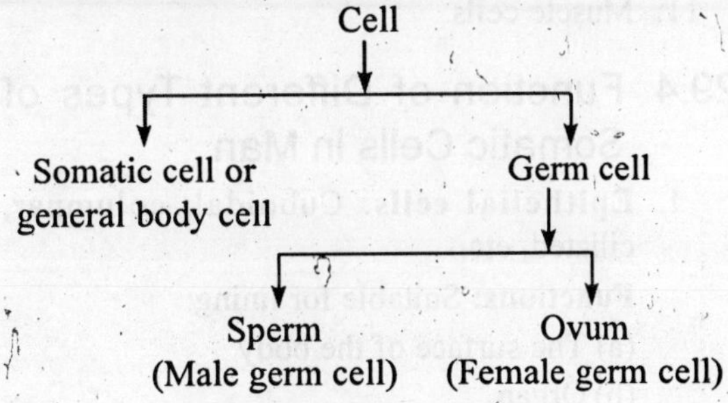

Fig. 29.2: Classification of Cell

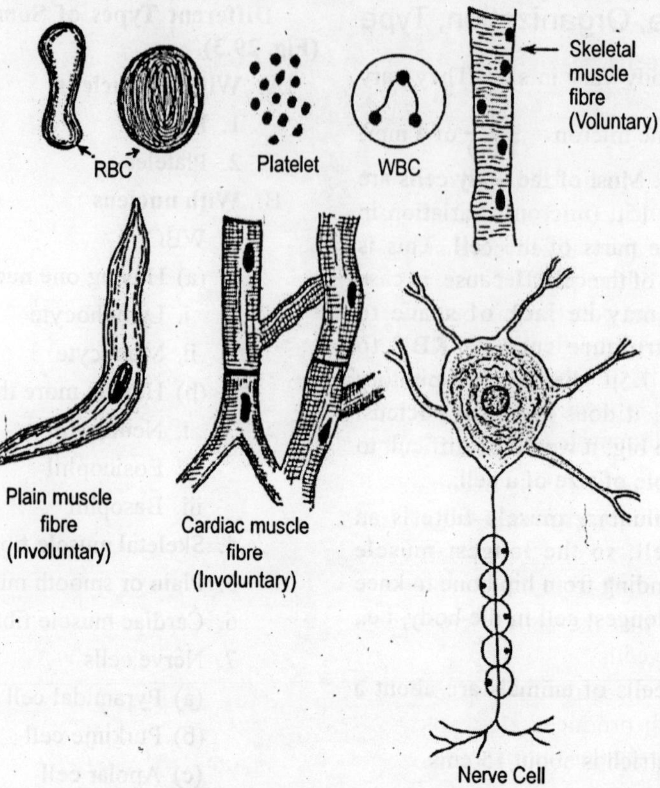

Fig. 29.3: Different types of somatic cells

29.3 Somatic Cell (General Body Cell)

1. Epithelial cells
2. Glandular cells
3. Fibroblasts
4. Chondroblasts
5. Osteoblasts
6. Erythrocytes
7. Leucocytes
8. Nerve cells
9. Sensor cells
10. Receptors
11. Muscle cells.

29.4 Function of Different Types of Somatic Cells in Man

1. **Epithelial cells**: Cuboidal, columnar, ciliated, etc.

   Functions: Suitable for lining:

   (a) The surface of the body

   (b) Organ

   (c) Cavities within the organs.
2. **Glandular cells**: For secretion, e.g., mucus
3. **Fibroblasts**: Secretes protein fibres that make up connective tissue, binding tissues and organs together.
4. **Chondroblasts**: Produces cartilage.
5. **Osteoblasts**: Produces bone.
6. **Erythrocytes**: Biconcave, disc-shaped and loaded with haemoglobin.
7. **Leucocytes**: Amoeboid, important for defence against infection.
8. **Nerve cell**: Slender processes transmit electrical impulses.
9. **Receptor or sensory cells**: Can receive stimulation like touch, sound, light, heat etc.
10. **Muscle fibres or muscle cells**: Shows contractility.

- **Are the somatic cells idential? How do they differ?**

Structures like chromosomes, mitochondria, endoplasmic reticulum and ribosomes are common to virtually all cells, but the shape, form and contents of individual cells show much variations.

- **Cells which show movement**
  1. Certain white blood cells.
  2. Muscle fibre.
  3. Spermatazoon.

- **Change in structurally specialized cells to perform particular function**

In case of structurally specialized cells, to perform particular function, there may be loss of protoplasm or some important constituents but in majority of cases it involves modification of the shape and form of the cells. Its basic features (nucleus, mitochondria, endoplasmic reticulum) remain same, e.g., nerve cell, muscle cell.

- **In some cells, the shape of the cells become modified (e.g. free cells leucocytes, ova)—what are the reasons?**

This may be the result of:

1. Mutual pressure, so that the cells become polygonal (e.g., liver cells).
2. It may be the result of active growth and differentiation which produces elongated forms or cells with irregular outlines (e.g., many neurons).
3. During life, cells may change their shape, e.g., amoeboid movements of leucocytes when examined under microscope.

- **Relation between the size of the cell and size of the animal**

The size of an individual animal depends on the number of cells rather than on their size, small animals do not necessarily have small cells. The cells of the human body vary greatly in size, the average tissue cell is between 10 μm and 20 μm in diameter, e.g.,

(a) Small lymphocytes—6.5 μm
(b) Large nerve cell—120-160 μm
(c) Ovum—250 μm

- **Cell without nucleus**

RBC, Platelet.

- **Cell having more than one nucleus**

Neutrophil, Megakaryocyte, striated muscle fibre.

- **Cell without centrosome**

Nerve cell, multiplication absent.

- **Cell having half number of chromosomes**

Germ cells.

- **Longest cell in the body**

Muscle cell—may be upto 130 mm.

- **True cell**

Cell having nucleus.

- **Anatomical structures without blood vessels**

Cornea, lens of the eye, because it disturbs the vision.

- **Osteogenic cells and Functions**

Osteoblast—Bone forming cell.
Osteoclast—Bone destructive cell.

- **Plasma cell and Function**

Present in bone marrow (1-3 per cent) as isolated cells, occasionally in the connective tissues common in the serous membrane (particularly the omentum), in the alimentary mucous membrane, in the lymphoid tissue.

Function—Formation of antibodies

- **Mast Cell and Functions**

Present in the capillaries.

Functions

1. Heparin production
2. Histamine release
3. Slow-reacting substance–A release, (SRS-A, where A = Anaphylaxis)
4. Eosinophil chemotactic factor (ECF-A) release.
5. Bradykinin release.
6. Serotonin release.

- **Lacis cell or mesangial cell and Function**

At the site of division of glomerular capillaries into afferent and efferent arterioles, there is thickening of wall in the tunica media called juxtaglomerular apparatus complex. This complex contains lacis cell.

Function

1. Production of renin
2. Production of erythropoietin.

- **Oxyntic cell or parietal cell-situation and Function**

In the fundic part of the stomach.

Functions: Secretion of HCl, mucus, less pepsinogen, intrinsic factor.

- **Glandular cells present in the fundus of the stomach**

1. Peptic (chief) cells.
2. Oxyntic (parietal) cells.

3. Mucoid cells.
4. Argentaffin cells at the base of the glands.

Microvilli present in the oxyntic cell and enzyme in this cell

1. Yes, to increase the surface area of the membrane.
2. Carbonic anhydrase.

Functions of peptic cells

Secretion of gastrin hormone, more pepsinogen, mucus.

- **Goblet cell and Function**

Most areas of the G.I. tract on the surface epithelium there are billions of **single cell mucous glands called goblet cells.** These function independently and extrude mucus directly into the lumen of G.I. tract providing lubricating coat of mucus to facilitate peristaltic movement of semi-solid mass of food residues.

- **α-cell and Function**

Pancreas. Secretion of glucagon hormone.

- **Schwann cell and Function**

Present in nerve fibre. The membrane of Schwann cell wound around the axon to form myelin sheath due to rotation of the Schwann cell.

- **Name of neuroglia cell of the posterior pituitary**

Pituicytes.

- **Different types of nerve cells**

1. Neuroglia cell
2. Pyramidal cell
3. Purkinje cell
4. Unipolar cell
5. Bipolar cell
6. Multipolar cell
7. Golgi cell Type II.

- **Neuroglia cells**

The neuroglia is the supporting tissue of the nervous system. The cells of this tissue are called glial or neuroglial cells. They are:

1. Protoplasmic astrocyte
2. Fibrous astrocyte
3. Oligodendrocytes
4. Microglia

Functions

1. Supporting the CNS (Astrocyte)
2. Formation of myelin sheath (oligodendrocyte).

- **Phagocytic cells**

1. Blood cells—Neutrophil,
2. Blood vessels—Reticular cells, endothelial cells of liver, bone marrow, spleen.
3. CNS—Microglia.

Function—Engulf foreign particles.

- **Paneth cell and Function**

Intestinal glands (crypts of Lieberkühn) contains paneth cells near the bottom of crypt having granules, appear to be precursors of the intestinal enzymes.

- **Zymogenic cells**

1. Salivary glands
2. Stomach
3. Pancreas
4. Small intestine.

- **Argentaffin cells**

The cells are having granules which have affinity for silver due to presence of serotonin. Present in stomach, crypts of Lieberkühn.

- **Amoeboid cell**

Cell having amoeboid movement, e.g., neutrophil.

- **Mitral Cell**

Olfactory bulb in nose.

- **Chromophore and chromophobe cells**

Anterior pituitary has three parts:

1. Pars distalis
2. Pars intermedia
3. Pars tuberalis.

Pars distalis contains:

1. Chromophobe—Empty-looking without granules.
2. Chromophil—Containing granules.

These cells are subdivided

A. Acidophil cells—Take acid dyes, orange
   (a) Mammotrophs—Produce prolactin
   (b) Somatotrophs—Produce growth hormone (STH)

B. Basophil cells—Megenta colour with PAS technique.
   (a) Thyrotrophs—Produce TSH
   (b) Gonadotrophs—Produce FSH, LH.

- **Where are chromaffin cells present? Why are they so called?**

Adrenal Medulla, sympathetic ganglia.

When the adrenal gland is fixed in potassium dichromate, the medulla shows yellow-brown due to presence of adrenaline and the reaction is known as chromaffin reaction.

29.5 Cell as the Living Unit

I. Physiologically

1. Movement e.g. rhythmicity
2. Ecitability, irritability
3. Contractibility
4. Conductivity
5. Reception of nutrients material and excretion of end products.
6. Building cellular materials
7. Metabolism — Anabolic / Catabolic
8. Energy production
9. Circulation
10. Respiration
11. Reproduction
12. Many cells differ in structural appearance (e.g., muscel cell, nerve cells, etc.) but there is production of energy CO_2, water etc.

II. Biophysically

1. Semipermeable membrane
2. Materials can pass through the membrane according to the size of the pores.
3. The membrane is electrically positive outside and electrically negative inner-side of the membrane.
4. Production of bioelectricity
5. Conduction of impulse.

- **Cells for lymphoid system**

I. Lymphocytes

1. B-lymphocyte (B-cells)
2. T-lymphocyte (T-cells)
3. Natural Killer Cells (NK).

II. Macrophages

III. Plasma Cells

1. Mesenchymal reticular cells
2. Epithelial reticular cells.

IV. Antigen Presenting Cells

1. Lymph nodes—(a) Follicular dendritic cells (b) Dendritic cells

Table 29.2: Some specialised somatic cells and functions

Specialised Cells	Function
1. Muscle cell	Contraction, relaxation—movement
2. Nerve cell	Conductivity
3. Pancreatic acinar cells	Synthesis and secretion of enzymes
4. Mucous-gland cells	Synthesis and secretion of mucous substances
5. Adrenal gland, testis and ovary cells	Synthesis and secretion of steroids, adrenaline
6. Cells of the kidney and salivary gland ducts	Ion and water transport
7. Macrophages and some white blood cells	Intracellular digestion
8. Sensory cells	Transformation of physical and chemical stimuli into nerve impulses
9. Cells of the intestine, kidney, etc.	Metabolic absorption

2. Skin—Langerhan's cells
3. Liver—Kupffer's cells.

- **Structures present in the mucosa of the small intestine**

1. Villi
2. Intestinal glands (crypts of Lieberkuhin)
3. Enterocytes (absorptive cells)

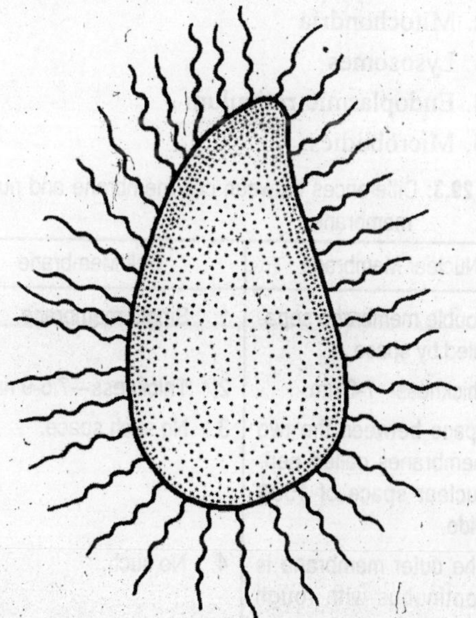

Fig. 29.4: Diagram of osteocyte cell

4. Goblet cells
5. M-cells
6. Paneth cells
7. Enteroendocrine cells
8. Undifferentiated cells.

- **What are the types of Cells present in the liver**
  1. Hepatocytes
  2. Kupffer's cells
  3. Fat storing cells.
- **What are the bone cells?**
  1. Osteoblast (bone forming cell)
  2. Osteoclast (bone destructive cell)
  3. Osteocyte (matured bone cell) (Fig. 29.4).

29.6 Cell Membrane

What is cell membrane or plasma membrane?

Every living cell has a definite boundary, i.e., one side of the boundary has intracellular material and the other side has extracellular material, and the boundary is maintained by a membrane called cell membrane or plasma membrane (or unit membrane).

Other cellular structures bounded by membrane

1. Nucleus
2. Mitochondria
3. Lysosomes
4. Endoplasmic reticulum
5. Microbodies.

Table 29.3: Differences between cell membrane and nuclear membrane

Nuclear Membrane	Cell Membrane
1. Double membrane separated by space.	1. Single membrane.
2. Thickness—7-8 nm.	2. Thickness—7.5-9 nm.
3. Space between the two membranes called perinuclear space of 150Å wide.	3. No such space.
4. The outer membrane is continuous with rough endoplasmic reticulum.	4. No such.
5. Two membranes fuse at certain intervals to make pores.	5. Fusion may occur with the membrane of the adjacent cell to form junctional structure.
6. There are pores with diaphragm or plug.	6. There are pores without diaphragm or plug.
7. Pores are permeable to dissolved or suspended particles and ribosome.	7. Transport of ions, water and food materials.
8. Not semipermeable.	8. Semipermeable.

Composition of cell membrane

It consists of lipids, carbohydrates and proteins.

Specific proteins in the cell membrane

Specific proteins serve specific functions such as pumps, gates, channels, receptors, energy transducers and enzymes.

Structure of cell membrane

It is bimolecular layer of lipid sandwiched in between an outer and inner layer of protein molecules.

What is "brush" or "striated" border epithelium? How is it formed? What are the functions and distribution?

In some cells (e.g., intestinal epithelial cells, proximal convoluted tubules) the plasma membrane may project from the sufrace, numerous finger-like extensions called "microvilli which cannot be resolved separately under optic microscope but in mass they constitute the so-called "brush" or "striated border" which increases the cell surface area for better absorption.

Structures bounded by single and double membrane

Double—Mitochondria, Nuclear membrane.

Single—Cell membrane, Endoplasmic reticulum, Golgi bodies, Lysosome, Microbody (peroxisome).

Structures having microvilli

1. Columnar epithelium of small intestine.
2. Oxyntic cell of gastric mucosa.
3. Mesothelium, non-secreting.
4. Goblet cell.
5. Epithelium of apocrine gland.

29.6A CYTOPLASM (Protoplasm–Nucleus)

Physical characteristic of protoplasm

Physically, it is a granular, semifluid or gelati-nous substance. It has characteristic properties of both liquid and solid.

Composition of protoplasm

It is an aggregate of:

(a) **Crystalloids**

1. Electrolytes—which are in solution—salts, acids, bases.
2. Non-electrolytes—sugar, urea, etc.

(b) **Colloids**—Proteins, lipids, nucleic acids.

Streaming movements of the cytoplasm

1. Living fibroblast
2. Tips of growing nerve fibres.

Microscopically visible structure in cytoplasm of living and fixed cells

The visible structures are divided into two groups:

1. **Organelles:** These are specialised parts of the living substance, often able to divide, and probably always present, e.g., mitochondria, golgi apparatus, centrosphere, lysosomes, microtubules, membranes, etc.
2. **Inclusions:** These are very commonly the products of cellular activity and are often of a temporary nature, e.g., fat droplets, glycogen particles, pigment.

29.7 Organelles

What are organelles?

The cell contains highly organized physical structure called organelles (organoids).

1. These are formed by combination of macro-molecules with smaller molecules.
2. They have individual important functions which are contributory to the total function of the cell.
3. They can be considered as small internal organs of the cells—hence called organelles or organoids.

How do you classify organelles with examples?

I. **According to their position**

A. **Cytoplasmic organelles:** Cell membrane, mitochondria, endoplasmic reticulum, ribosome, lysosome, golgi apparatus, microbodies, microfilament and micro-tubule, centrosome or centriole, cilia and flagella.

B. **Non-cytoplasmic:** Nucleus, Nucleolus, Chromosomes.

II. **According to the presence of membrane**

A. **Membranous:** Cell membrane, mito-chondria, endoplasmic reticulum, lysosome, golgi complex, nucleus, microbodies.

B. **Non-membranous:** Ribosome, micro-tubule, centrosome, cilia, nucleolus, chromosomes (Table 29.1).

Do you find any organelles in red blood cells?

The mature RBC is devoid of nucleus, endo-plasmic reticulum, golgi apparatus, centriole and mitochondria which have been lost during devel-opment.

29.8 Stem Cell

1. **Haemopoietic stem cells** are most primitive cells having extensive proliferative capacity.
2. They have two functional properties:
   (a) Capacity of producing new stem cells, i.e., self-renewal to maintain haemo-poietic, stem cell pool means to replenish.
   (b) Pluripotent stem cell: Ability to differentiate into blood cells. Animal studies and studies of chromosomal and enzyme markers in the blood cells of human with chronic granulocytic leukaemia and the subject *receiving bone marrow transplant* suggests origin from this primitive cells.
3. Stem cells cannot be distinguished from lymphocytes.
4. The frequency of origin of stem cells in human bone marrow is not known but does not exceed 1-2/1000 nucleated cells. Small numbers of stem cells also circulate in the blood.
5. In case of acute demand for increased blood cell production, stem cells are recruited into cell cycle divide and temporarily increase

the member. In chronic need, both pluripotent stem cells and cells committed to the production of one or a limited number of cell types are increased in number.

6. At the beginning of differentiation the stem cell loses its ability to produce all blood cells and becomes committed to give rise to one or more cell lines.
7. It is not known whether the process of commitment is gradual or sudden.

• **Characteristic**

1. *Renewal*: Haemopoietic stem cell possesses the property of ability of cell division to give a new stem cell that is known as *self-renewal.*
2. *Frequency*: Blood stem cell in human bone marrow is 1 to 2 per thousand nucleated cells.
3. *Formation of Clone*: When multitude of cell arises from a single cell that multitude of cell is called *Clone.* When mature blood cell having origin from multiple stem cells, hence normal hemopoiesis is *poly cloner.*
4. *Production of one or more cell lines*: Stem cell differenciates and that time it loses its ability to give rise of all blood cells and becomes committed to the production of one or more cell line.
5. *Diminution of self-renewal*: As committment progresses in the capability of a stem cell for cell renewal diminishes. When it is markedly limited or lost, their cell is no longer called a stem cell, but it turns to *progenitor cell.*

• **Identification:** It can be identified by conventional procedure or technique but it can be done by a special technique called *cell shorting technique.*

• **Mechanism of conversion of stem cell to progenitor cell:** Probably it may occur by the following ways:

1. Intracellular interaction
2. Local hormones
3. Intracellular interaction such as:
   (a) Inter-leukines no. 1, 6, 4, 3 from T-cell.
   (b) The peptide called BPA (Burst Promoting Activity).
   (c) Protein coded by the (gene) AEL cell GM-CSF (Granulocyte Macrophage Colony Stimulating Factor) secreted from the macrophages, T-cells, endothelial cell and fibroblast act as a local hormone for production of committed stem cell.
   (e) Committed stem cells are also regulated by individual growth proteins.

• **Production of embryonic stem cell**: From a patient, somatic cell is taken by *biopsy* transfer. genetic material called enuclear transfer. Then it is transplanted to enucleated (when nucleus has been removed) donors oocyte (ovum) then blastocyst (6 day old embryo) and cell formed called *embryonic stem cell.* These are used for *therapeutic cloning* (Fig. 29.5).

• **Types of stem cell**: All tissues consist of stem cell compartment, that produces cellular offspring that mature into all tissue phenotype, such as embryonic stem cell, bone stem cell, which are used for clinical research or therapy. The maturation process are

(a) **Commitment cells,** in which the pluripotent stem cell produce *daughter* cell with restricted genetic potential appropriate for single cell activity (unipotent)

(b) **Differentiation**: *In which sets of genes are activated* or alter in their labels of expression.

29.9 Stromal Cells

Tissue consists of parenchymal cells and stroma:

Parenchymal cells are functional cells such as:

1. Hepatocyte in the liver.
2. Hematopoietic cell in the red bone marrow.

Stroma consists of primarily connective tissue elements which together with their products form structural framework of the tissue. *Parenchymal cells* are developed from any of the germ layers e.g., ectoderm, mesoderm, endoderm and during development they grow in the area populated by stromal cells or their progenitor cells (Fig. 29.6).

It is understood that *stromal cells* are derivative of mesenchyme, e.g.,

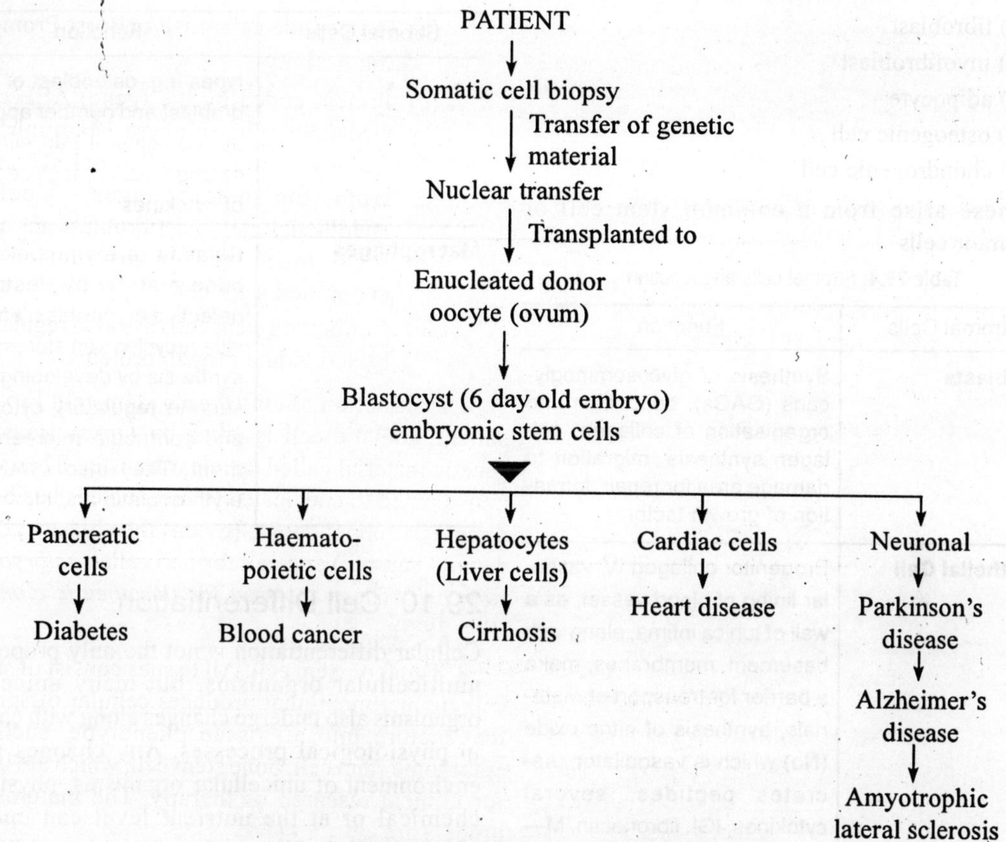

Fig. 29.5: Flow chart of production of embryonic stem cell

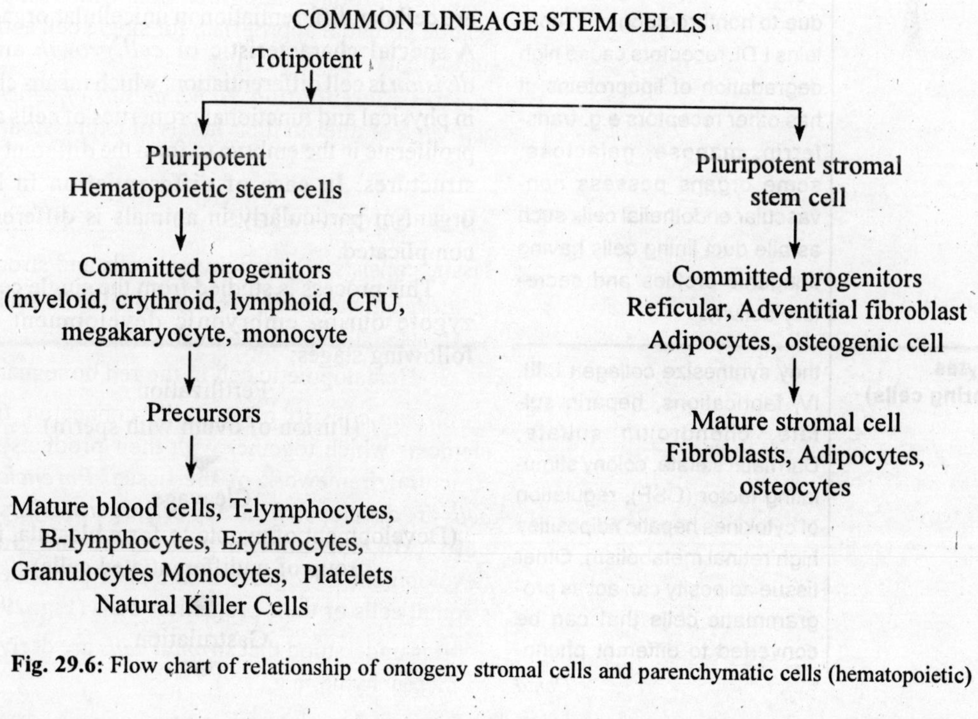

Fig. 29.6: Flow chart of relationship of ontogeny stromal cells and parenchymatic cells (hematopoietic)

(a) fibroblast
(b) myofibroblast
(c) adipocyte
(d) osteogenic cell
(e) chondrogenic cell

These arise from a common stem cell or progenitor cells.

Table 29.4: Stromal cells and function

Stromal Cells	Function
Fibroblasts	Synthesis of glycosominoglycans (GAGs), deposition and organisation of collagen, collagen synthesis, migration to damage area for repair, formation of growth factor.
Endothelial Cell	Progenitor collagen IV, vascular lining of blood vessel, as a wall of tunica intima, along with basement, membranes, make a barrier for transport of materials, synthesis of nitric oxide (No) which is vasodilator, secretes peptides, several cytokines, IGL fibronectin, M—CSF, GM-CSF, platelet activating factor, helps angiogenesis due to nonthrombogenic, contains LDL receptors cause high degradation of lipoproteins, it has other receptors e.g. transferrin, manose, galactose, some organs possess nonvascular endothelial cells such as bile duct lining cells having antigenic profiles and secretory activity.
Adipocytes (Fat storing cells)	they synthesize collagen I, III, IV, fabrications, heparin sulfate, chondroitin sulfate, Darmatin sulfate, colony stimulating factor (CSF), regulation of cytokines hepatic adiposities high retinal metabolism. Other tissue adiposity can act as programmatic cells that can be converted to different phenotypes e.g. osteoblast or chondroblast and number appropriate condition if can stimulate osteogenesis by their secretion of cytokines.
Macrophages	Relation to erythropoiesis in bone marrow by destroying defective er,throblast and provide recycled iron stores for Hb synthesis by developing RBC, secrete regulatory cytoplasm and contribute microenvironment. They also synthesize erythropoietin and store it.

29.10 Cell Differentiation

Cellular differentiation is not the only property of multicellular organisms, but many unicellular organisms also undergo changes along with changes in physiological processes. Any changes in the environment of unicellular organisms, physical or chemical or at the nutrient level can undergo physical cellular changes like formation of different types of spores, sporulation in bacteria, fungi, etc. are cellular differentiation in unicellular organisms. A special characteristic of *cell growth* and *cell division* is cell differentiation, which means changes in physical and functional properties of cells as they proliferate in the embryo to form the different bodily structures. In case of differentiation in higher organism particularly in animals is different and complicated.

This process is studied from the single cell, i.e., zygote during embryonic development in the following stages:

Fertilization
(Fusion of ovum with sperm)

▼

Cleavage
(Development of zygote to form blastula, i.e., a group of undifferentiated cells)

▼

Gastrulation

▼

(Differentiation and movement of cells to form specialized cell layers, i.e., ectoderm, mesoderm and endoderm)

▼

Differentiation

(Development of specialized cells into tissues, organs and growth of the embryo)

▼

Growth

(Maintenance and regeneration of some cells)

29.10.1 Molecular changes during oogenesis

1. Increase of RNA synthesis
2. Increase in size of the nucleus
3. Amplification of ribosomal granules
4. Accumulation of protein, carbohydrate, lipid

29.10.2 Role of cytoplasm in cellular differentiation

The cytoplasm has a great role in cellular differentiation has been established by several experiments.

1. During differentiation of snail egg, snail produces a lobe at the vegetal end known as *Polar lobe* if it is removed then defect in the embryo appears.
2. In Amphibian egg a particular area is produced which is coloured during cell division after fertilization. This is known as *Grey crescent.* If this, is damaged, then there is abnormality in the nervous system.
3. In case of amphibian egg, if the 1st cleavage is *perpendicular* to the grey crescent and the resultant blastomere is separated, each blastomere produces normal animal. But if the cleavage takes *parallel* to the grey crescent and if the two blastomeres are separated, one having grey crescent will produce normal animal. Thus the differentiation of cells depends on the participating substance in the cytoplasm.
4. During embryonic stage of *Drosophila*, the primordial grown cells arise from the posterior end of the egg. In an experiment by removing some amount of cytoplasm from the posterior end the egg of drosophila and injected into the anterior end of the another egg. If has been observed that the germ cells are produced from the anterior end of the injected egg. Therefore, it can be concluded that the cytoplasm has some significant role for inducing differentiation of cells.
5. Another experiment has shown that during embryonic development of the egg of round worm (*Ascaris lumbricoides*) that during development of the embryo, the 1st cleavage occurs perpendicular to the animal vegetal axis. When the animal pole of the new blastomere starts dividing, the heterochromatic portion of the chromosome showed degeneration. The euchromatic portion of the chromosome is fragmented into numerous small chromosomes by the process known as *chromosome diminution* and this occurs during embryonic development.
6. In another experiment in which eggs are centrifused before cleavage to disturb the polarity of the cell. By centrifugation the mixture of both animal and vegetal cytoplasm occurs and the 1st cleavage occurs along animal vegetal axis not perpendicular to it. No chromosome diminution occurs and thus signifies the role of cytoplasm in chromosomal behaviour.

29.11 Cancer Cell

Introduction

Cell division in a multicellular organisms is the normal process and it is necessary for:

1. Growth
2. Development of organs
3. Healing of wounds
4. Replacement of older and damaged cells.

Cell division is controlled by regulatory mechanism at the:

1. Molecular level
2. Cellular level.

By these process acquire its specification and specialisation is known as *Cellular differentiation.* This cellular differentiation starts at the embryonic gastrulation stage which has a genetic basis.

In multicellular organism, cell division, differentiation and survival of cell are regulated to cater the need of the organisms. In case of *failure* or regulation for any reason, the cell's abnormal behaviour, the cells grow and divide in *uncontrolled* manner and spread throughout the body (metastasis) disturbing the normal function of the tissue and organ and this condition leads to malignancy or cancer.

29.11.1 What is cancer or malignancy?

It is *uncontrolled* multiplication or proliferation of cells *without having differentiation.*

It is a non-infectious disease which starts at molecular level of the cell and ultimately affects the cellular activity.

29.11.2 What are the types of tumour?

Types of Tumours: The uncontrolled cell growth and multiplication of any different kinds of cells produces a mass of cells called neoplasm or tumour. Tumours are of two types:

I. Benign tumour

1. Slow growing
2. Remains confined to its original site
3. Does not spread from origin to distant places (metastasis)
4. Does not invade surrounding tissue
5. Consists of closely resemble normal cells and may function like normal cells
6. Generally they are not harmful and can be removed surgically
7. Sometimes it becomes harmful if present in brain or liver
8. Sometimes some of these tumours may become malignant
9. Example—Lipoma.

II. Malignant tumour or Carcinoma

1. Does not remain confined to it original site
2. Rapid growing
3. Invades surrounding tissue
4. Spreads to distant places via lymphatic or circulatory system
5. It is life threatening
6. The cells of malignancy are derived from a single cell, i.e., monoclonal in character
7. It contains aberrant cells. They behave like embryonic cells, undifferentiated having irregular, large nucleus and deficient of cytoplasm.

29.11.3 What are the characteristics of cancer cells?

1. Cell transformation
2. Cellular changes in the cancer cell
3. Immortality
4. Loss of contact inhibition
5. Irregular masses of several layers
6. Structural changes in the cell
7. Reduced adhesiveness
8. Specificity of adhesiveness
9. Invasiveness
10. Angiogenesis
11. Failure of differentiation
12. Autostimulation of cell division
13. Apoptosis.

1. **Cell transformation**: The process of cell alteration in which a normal cell loses the ability to control. When normal cell is cultured, it shows that normal cells are surrounded by other cells and these normal cells form *a monolayer.* There is initiation of growth and number of cells in the culture media remains constant.

   In case of cancer cells, continue to mutiply and form *irregular masses of several layers.*

2. **Cellular changes in cancer cells**

   (a) Change in property of the cell membranes and cell coat, such as:

   (i) Disappearance of gap junction

   (ii) Loss of coupling changes in glycolipid and glycoprotein

   (iii) Reduction in gangliosides

   (iv) In the cell coat, a large glycoprotein called fibronectin found in foot prints of moving cultured cells is reduced in cancerous cells.

   These changes cause dissociation of cancer cells from neighbouring cells and show loss of contact inhibition.

3. **Immortality**: Normal cells do not survive for definite period in culture, and die after about 50 generations. Whereas cancer cells can survive indefinitely and remain immortal if the nutrition is provided and overcrowding is avoided.

4. **Loss of Contact Inhibition**: Normal cells grown in tissue culture tend to make cells contact by adhesion to the neighbouring cells. At the point of adhesion there is some kind of electron dense plaque in both contacting cells. At the same time there is a slowing down of the amoeboid movement which results in contact inhibition.

   But cancer cells do not form adhesive junctions and do not show this type of contact inhibition.

5. **Formation of mass of several layers**: Cancer cells continue to multiply forming irregular masses of several layers deep.

6. **Cell membrane–change in properties**: Cancer cells undergo a change in property of their cell membranes and cell coat such as disappearance of gap functions, loss of coupling changes in glycolipid and glycoprotein and reduction in gangliosides.

7. **Reduced Adhesiveness**: (a) cancer cells are less adhesive than the normal cell because of reduction of cell surface adhesive molecule. When normal cells are transformed into cancer cells, there is change of stickiness resulting a less stickiness or adhesiveness.

8. **Specific of Adhesiveness**: Adhesiveness shows specificity such as liver cell tends to stick with another liver cell but not the kidney cell. Cancerous cells do not show this property. They are able to mix and stick to any type of normal cells.

9. **Invasiveness**: It is the most important property of cancer cells. This is the ability to invade other tissues. Malignant cells generally secrete *proteases* that digest extra cellular matrix components allowing the cancer cells to invade adjacent normal tissues, e.g., *Collagenase* of the cancer cells helps to digest and penetrate through basal laminae to invade the underlying connective tissue.

10. **Secretion of growth factor**: Cancer cells also secrete *growth factors*. That promote the formation of new blood vessels which is known as *angiogenesis* which is necessary to support the growth of tumour beyond the size of about a million cells at which point new blood vessels are needed to supply oxygen and nutrition for multiplication of tumour cells. As a result metastasis process begins.

11. **Failure of differentiation**: Cancer cells usually fail to differentiate which is closely related to abnormal proliferation. Normal cells are fully differentiated, at this time cell division ceases. But in cancer cells, normal differentiation programme is blocked at the early stage of differentiation. Defective differentiation and rapid proliferation is noticed in leukaemia.

12. **Autostimulation of cell division**: Cancer cells produce *growth factor* that stimulates their own cell division known as *autocrine growth stimulation*.

13. **Apoptosis**: Many normal cell types including blood cells undergo *apoptosis* i.e. the programmed cell death. But cancer cells fail to undergo apoptosis which leads to tumour development. The survival of normal cells is dependent on signals from growth factors or from the extracellular matrix which inhibits apoptosis. But cancer cells survive in absence of growth factors. Normal cells also undergo apoptosis following DNA damage but cancer cells fail to do so. Cancer cells are even resistant to irradiation and many toxic or lethal chemicals which act by damaging DNA.

29.11.4 Development of cancer cell

It is a multistep process. It consists of mutation → multiple mutation → transformation of cells → progressively increasing capacity for cell division → survival → invasion metastasis, i.e., spread of cancer cells through the blood or lymphatic system to other organs (Fig. 29.7).

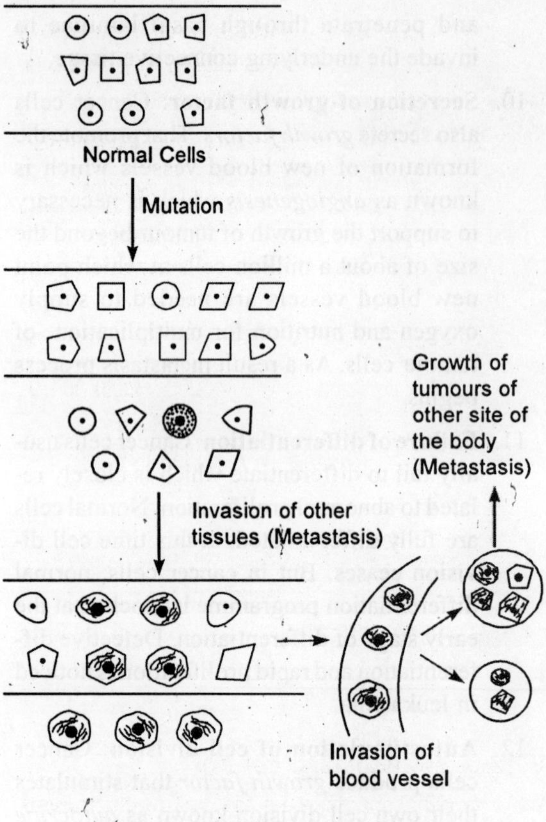

Fig. 29.7: Development of Cancer

Microscopical Observation

1. High nucleus to cytoplasm ratio
2. Prominent nuclei
3. Many mitosis
4. Relatively little specialised structures

Chromosomal Changes

1. Both structural and numerical changes.
2. Parent cell contains 46 chromosomes. Later a series of abnormal division, the cancer cells contain a series of chromosome number.

29.12. Apoptosis or Programmed Cell Death

Definition: Process of programmed cell deaths of cortical lymphocytes and fragmentation death called apoptosis followed by phagocytosis by macrophages.

- **Examples**

1. Most T-lymphocytes originating in the thymus have the ability to attack and destroy body components and would cause serious damage if they enter the blood circulation. Inside the thymus, T-lymphocytes receive signals that activate the apoptic progam encoded in their chromosomes.
2. Embryologists observed that during morphogenesis, cells and tissues were being deleted in a predictable fashion.
3. During human development as ontogeny recapitulates phylogeny, there is the loss of branchial arches, the tail, the cloaca, and webbing between fingers.
4. Pediatric surgery is replete with examples where there has been failure of this process, with double aortic arch, sacro-coceygeal, teratomas, extrophy of the bladder, and syndactyly, etc.
5. In vertebrate nervous system, during development the death of neurons is a regular feature. Neuronal death (50%) is due to a nerve growth factor (NGF) produced by the target cell. It is required for the sustenance of the neuron responsible for making contact with that target cell. During development and maturation many types of neurons are being produced in excess. Neurons which find their way to the wrong target cell do not survive and the result is that a balance is achieved between the surviving neurons and target cells.

29.12.1 Significance

This may eliminate prematurely activated cells or targeted toward "self" antigens.

T-lymphocyte production is the primary function of the thymus. T-lymphocyte precursors formed in the bone marrow enter the thymic cortex for proliferation and acquire T-lymphocyte (Thymocytes) characteristics. Thymic programming comprises two important development:

1. Acquiring antigen-recognition capacity by reordering antigen-receptor genes;
2. Acquiring T-cell characteristics through *expression* of the cell-surface molecules and *signal transduction* complexes that regulate:
   (a) Interaction with antigenic and accessory cells and

(b) The homing capabilities that direct virgin T-cells to secondary sites for further programming. Most cortical thymocytes undergo cell death and fragmentation, i.e., apoptosis.

Maturing survivors move in the medulla, where they enter the circulation through postcapillary venules or efferent lymphatic vessels. They enter the T-dependent regions of secondary lymphoid organs, e.g., lymph nodes, spleen etc. where they further differentiate into functional T-lymphocytes.

3. Most cells of the body can activate their apoptosis programme when major changes occur in their DNA, e.g., just before a tumour appears. In this way, apoptosis prevents the proliferation of malignant cells that develop as a result of accumulated mutations in the DNA. To form a clone and develop into a tumour, the malignant cell needs to deactivate the genes that control the apoptotic process.
4. Apoptosis was first observed in developing embryos, where programmed cell death is an essential process for shaping the embryo (morphogenesis).
5. Later investigations revealed that apoptosis was also a common event in the tissues of normal adults.

29.12.2 Changes in the cell in apoptosis

1. The cell and its nucleus become compact, decreasing in size.
2. The cell shows a dark-stained nucleus (pyknotic nucleus), easily identified with the light microscope.
3. The chromatin is cut into pieces by DNA endonucleases.
4. The cell shows cytoplasmic large vesicles (blebs) that detach from the cell surface. These detached fragments are contained within the plasma membrane, which is changed in such a way that all cell remnants are readily engulfed or phagocytosed by macrophages. But the apoptotic fragment do not stimulate the synthesis of molecules in macrophages which cause inflammatory reactions.

It differs from *necrosis* which is accidental death of cells a pathological process caused by infections, viruses, chemicals, etc. (Table 29.5).

Table 29.5: Difference between necrosis and apoptosis

	Necrosis	Apoptosis
1. Process	Pathological	Physiologic
2. Energy	Passive	Active
3. Stimulus	External	Internal
4. Host response	Inflammation	None
5. Cell membrane	Ruptured	Intact Blebs 36 + 9
6. Mitochondria	Loss	Intact (release Cytochrome C)
7. Duration	Days	Hours
8. Relation with growth	Scar	None
factor (NGF) and target cell	No	Yes

29.12.3 Homeostatic control

Two opposing mechanisms that maintain homeostatic control of apoptosis are:

1. Fas ligand
2. Bcl-2 gene-protein expressed by this gene.

These two important mechanisms modulate or control apoptosis to a large degree and help to maintain the equilibrium in normal cell populations.

29.12.4 Apoptosis and diseases

1. In patients with *acquired immune-deficiency syndrome* (AIDS) the human immuno deficiency virus causes the death of T-helper cells which prevents apoptosis of cytotoxic-T-cell by elaborating kinins that are necessary for maintenance of the cytotoxic T-cells.
2. *Hashimoto thyroiditis* shows abnormal rates of apoptosis. Increased quantities of interleukin 1B produced by infiltrating macrophages induce Fas expression by the thyrocytes.
3. *In myocardial infection* and *strokes* though there is a loss of tissue from ischaemic

necrosis, the larger loss of tissue results from the extension of the original vascular injury by apoptosis.

4. *Cancer* represents failure in the processes that regulate cell survival and/or growth. Cells with DNA defects are defected by intranuclear monitoring processes that activate the p53 gene, whose protein in turn initiates apoptosis.

29.13 Junctions between Cells (Cell Junction) Junctional Complex in Epithelial Cells, Morphological Specialisation of the Junction between the Cells

Introduction: The tissues of higher organisms are composed of enormous numbers of individual cells that function as harmonious units. Therefore, it is essential that appropriate intercellular *communication be established and mechanical contact is maintained between adjacent cells. Such functional and structural integrity is accomplished by lateral surface specializations of the epithelial cells (Intercellular junctions)* comprising many tissues and organs. There are several types of junction present in the same tissue, they are known collectivity as a "*junctional complex*".

• **Types of intercellular junction**: There are four major kinds of intercellular junction, each having particular physiologic properties:

1. **Tight junction (Zonulae occludeues)**: It is the intimate lateral contact between the cell membranes. The intercellular space is abolished in tight junctions by fusion of the outer leaflets of the two opposed cell membranes. These regions are essentially *impermeable*, so the passage of solutes in a plane vertical to the surface of the epithelial sheet, although *solutes can be transported* in this direction *through the apical plasmalemma* of the epithelial cells.

Tight junctions enable tissues and organs to maintain a chemical environment distinct from their surroundings. This is accomplished by the cells actively transporting solutes via their lateral membranes, directly

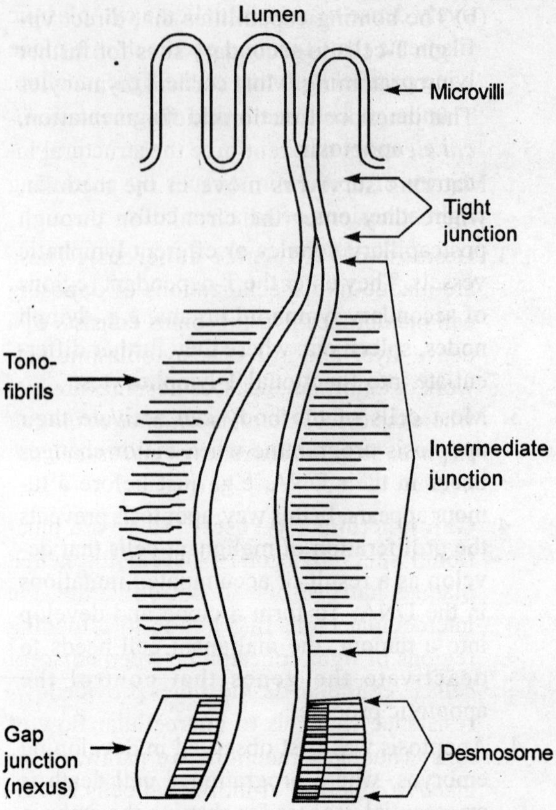

Fig. 29.8: Morphologic specializations present at the junction between two columnar epithelial cells. A gap junction (nexus) is indicated by dark shading within the desmosome)

into the intercellular space below the tight junctions; thus an osmotic gradient is developed. The osmotic gradient in turn enables water to pass across the epithelial sheet to the intercellular space, thereby concentrating the luminal contents.

Owing to their strategic location and structural characteristics, tight junctions are important physiologically for e.g. transporting epithelia such as the gall bladder, intestine, proximal renal tubules and choroid plexus of the brain. Tight functions also have a mechanical function, as they help to maintain the structure of epithelial tissues (Fig. 29.8).

2. **Intermediate junction (Zonulae adherens)**: These are located immediately beneath the tight function. These circumferential regions, where the cell membranes diverge somewhat, are strengthened by rela-

tively dense masses of delicate cytoplasmic filaments that surround each cell in a plane lying parallel with that of the tight junction. This junctions function chiefly as *mechanical elements* that reinforce the structural integrity of epithelium tissues by promoting tight adhesions of adjacent cells.

3. **Desmosomes (Maculae adherens)**: These are plaque-like specializations of opposing cell membranes. Desmosomes consists of a dense layer and a deeper cytoplasmic network of tonofilaments. Desmosomes are loci where the cytoskeleton is *attached to the cell surface.* They are also sites of *cell-to-cell adhesions.*
4. **Gap junctions (Nexuses)**: These are junctional specializations wherein thickened plaques narrow, but do not obliterate, the intercellular cleft. In addition to providing regions of tight cohesion between adjacent cells in epithelia, desmosomes provide low-resistance channels to intercellular flow of ions. Thus, gap junctions are pathways for *electric coupling* of individual cells in epithelial structures. These are also found in smooth and cardiac muscle. In both of these contractile tissues, gap junctions provide low-resistance channels that permit rapid cell-to-cell excitation of the entire tissue mass. Hence these types of muscle cell (fibre) behave as *functional syncytium* because of the rapid conduction of signals via gap junctions (Fig. 29.8).

29.14 Growth Factor

In addition to well characterized hormones, a number of *new polypeptide hormone* growth factors have been identified. The growth factors are:

1. Epidermal growth factor (EGF)
2. Transforming growth factor (TGF)
3. Platelet derived growth factor (PDGF)
4. Fibroblast growth factor (FGF)
5. Nerve growth factor (NGF)
6. Somatomedin; insulin-like growth factor (IGF)
7. Tumour angiogenesis factor (TAF).

29.15 Soluble Growth Factors

Soluble specific growth factors are an **essential requirement for proliferation and differentiation of developing cells.**

Table 29.6: Growth factor differs from hormones

Hormone	Growth Factor
1. Hormones are secretory products of endocrine organs	1. Secreted by submaxillary gland, megakaryocyte and platelet etc.
2. Hormones like growth hormone, anabolic steroids are necessary for general cell function and tissue formation.	2. If drives the developmental programming of differentiating cells.
3. Hormone carried by blood and act at distant organ to target cell and maintenance of homeostasis.	
4. Mediated by messenger.	4. Function is inductive action and direct effect on gene expression and hence cell proliferation and differentiation revealed by demonstrating linkages between receptor ligand and interaction and gene activation.

29.15.1 Study of growth factors

For in vitro study of cell growth, animal serum has been used for years. These appreciate mixture of hormones have been used. Later on to obtain optimal growth in culture, combination of hormones along with other factors such as nutrients, the extracellular matrix, the concentration of hormone receptors and concentration of antigrowth factors have been used.

Table 29.7: Growth factor source and action

Growth factor	Source	Action
1. Epidermal growth factor (EGF)	submaxillary glands, platelets, human urine	1. Eye opening in new born mice 2. Inhibit gastric secretion 3. Stimulates growth of endodermal, ectodermal, mesodermal cells.
2. Platelet derived growth factor (PDGF)	Mega karyocyte carried in alpha granule in platelet	Acts via transmembrane receptor with intrinsic protein tyrosine kinase
3. Fibroblast growth factor (FGF)	Many cells, brain, retina	Variety of cells derived from mesoderm
4. Insuline like growth factor (IGF)	Synthesis depends on growth hormone, insulin, thyroid	Homologous to insulin, IGF—acts via a receptor containing intrinsic protein, tyrosine kinase IGF_2—via a receptor for manose 6-phosphate
5. Nerve growth factor (NGF)	Synaptic junction	Stimulation, replication, embryonic narrow neural outgrowth.

29.16 Growth Inhibitor

In addition to growth factors, there are other factors, nutrients, presence of surrounding cells, extracellular matrix, blood supply are important for helping the growth of cells. Some other factors are produced locally to limit cell tissue growth called *growth inhibitors chalones*. These are:

1. *Cartilage produced factor*: inhibit growth of both cartilage and capillaries.
2. *Epithelial produced factors*: inhibit epithelial cell growth
3. *Transforming growth factors*: B Turnover growth factor, B (TGF r) or Polyergen: inhibits proliferation of epithalial, messenchymal and haemopoietic cells.
4. *Tumour necrosis factor or GTNE or Cachectin*: couses necrosis of turnover cells.
5. *Interferon*: antiviral activity, inhibit growth of variety of cells.

29.17 Measurement in Cell Study (Cytophotometry)

Methods for qualitative and quantitative determination of the chemical composition of cells and tissues.

These methods may be divided roughly into:

I. Physical Methods

1. Fluorescence Microscopy
2. Scanning electron-probe microanalyser
3. Autoradiography

II. Chemical Methods (By histochemistry): **Cytophotometry**: Cytophotometry involves the *optical quantification* of some chemical substance in the cells. Optical quantification can in principle be based on any of the optical phenomena of:

(a) Absorption, (b) Luminescence
 1. Fluorescence, 2. Phosphorescence

(c) Retardation
 1. Interference, 2. Polarisation

(d) Reflection
 1. Specular, 2. Scattered

The most commonly used methods are:

1. Absorptiometry
2. Fluorimetry
3. Interferometry
4. Reflectionometry

Instruments

1. Microdensitometer
2. Microscope photometer

Topographical measurement

I. Measurement of Section thickness

1. Focussing on lower and upper surface
2. Stereomicroradiography
3. Mechanical Methods
4. Interferometry

II. Measurement of areas giving a particular histochemical reaction

1. Planimetry
2. Photography
3. Line and point sampling
4. Automatic scanning.

29.18 Flow Cytometer

Principle: Flow cytometer is based on the principle that single cell in suspensions are passed through a field of illumination and each cell is quantitatively assayed by staining cells with fluorochromes or noting the scattering of light by each cell.

29.18.1 Procedure

Cell suspensions are placed at first in a flow cell fitted with a device to form a liquid jet. Cells will travel through the centre of a liquid jet at the rate of 5-10 m/sec. Cells are then passed through the area of intense light. If the cells contain some fluorochromes, fluorescence emission can be detected using lens, beam splitters and photomultiplier tubes (Fig. 29.9).

Different fluorochromes used will show different types of fluorescence emission. These signals are then quantified and are kept in the memory of computers as histograms. During the flow, droplet formation is done using piezoelectric crystal in the flow cell. At the end of the instrument there is a droplet deflector assembly. Giving a definite quantitative value to the computer for collecting a specific cell desirable will then be collected in a microliter plate or tube and undesirable cells will be collected in the waste. Different fluorochromes are used for different types of cells which will then help to sort the cells having two types of nuclei (hetero-karyons).

29.19 Cell Fractionation or Isolation of Cell Component

Cell fractions is the physical process (mechanically breaking) by which centrifugal tree is used to separate organelles and cellular components as a function of their sedimentation coefficient or EM for biochemical analysis (Fig. 29.10).

Note: *Sedimentation coefficient* is defined as the sedimentation velocity in cm per unit time and per unit field of force generated by centrifugation.

The sedimentation coefficient of a particle depends on its size, form and density, and on the viscosity of the medium.

The method comprises

1. **Homogenization**: The mechanical homogenization means preparation of a suspension of cells and tissues components by disrupting the cell membrane and release the cell components (Fig. 29.8A).

II. **Differential Centrifugation**: The components in the suspension (e.g., individual organelles) are then separated by their characteristic sedimentation rates, using different amounts of centrifugal force for various periods. There are two centrifugation methods:

Density gradient centrifugation: Components in the homogeneities is layered on top of a gradient of solute and centrifuged until its components come to rest in portions of the gradient with densities similar to their own (Fig. 29.8A).

29.19.1 Importance

1. Isolation and collection of cellular components in quantity to study their chemical composition and contribution to the cell is determined.
2. Electron micrographic study of individual organelles.

29.19.2 Density gradient centrifugation

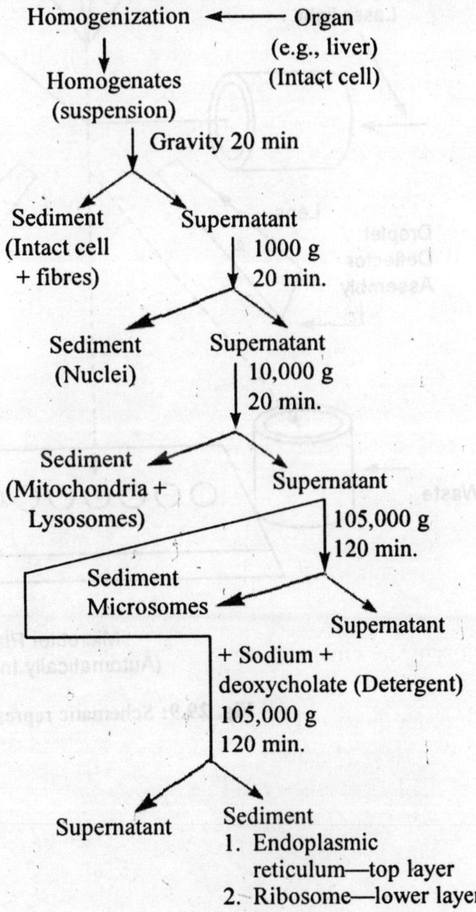

Fig. 29.8A: Schematic representation of Density gradient Centrifugation

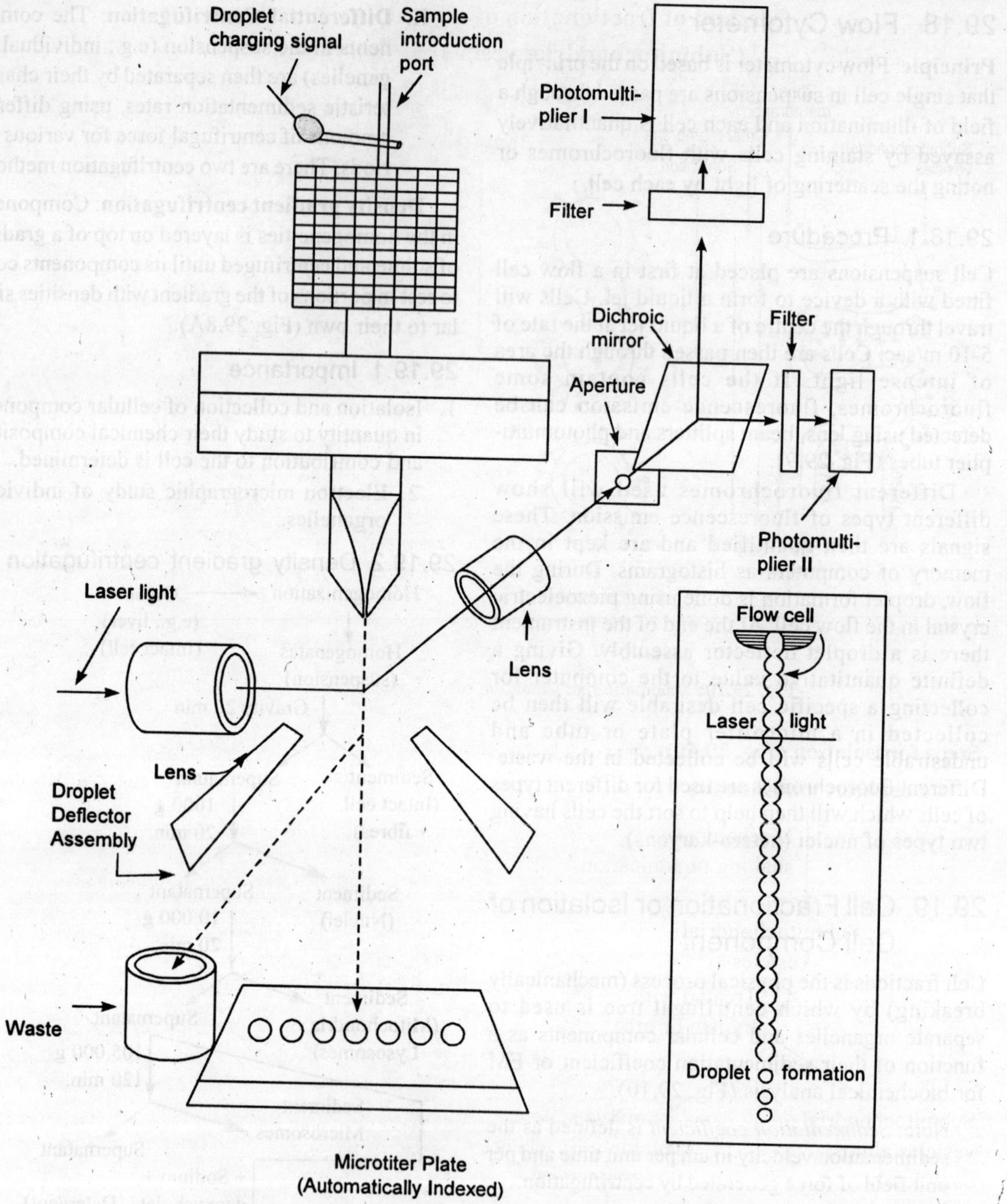

Fig. 29.9: Schematic representation of flow cytometer

Method of fractionation of mitochondrial components (Submitochondrial vesicle) for enzymatic study

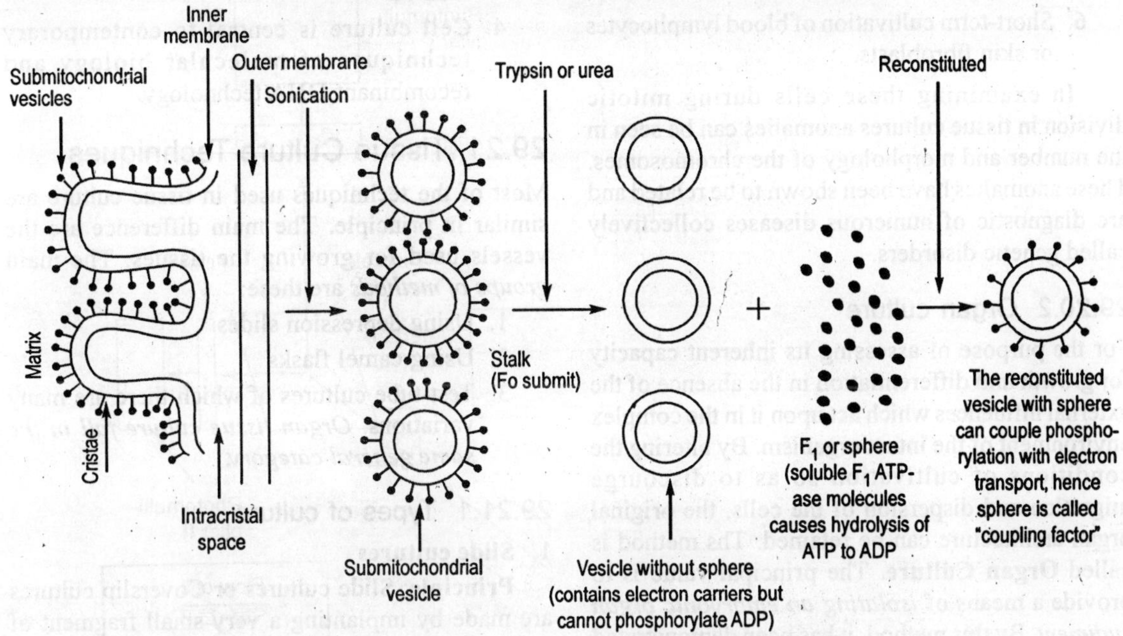

Fig. 29.10: Schematic representation of vesicles and spheres of mitochondria

Steps for isolation (Fig. 29.10)

Beef heart mitochondria

↓ Mechanical shaking or sonication

Submitochondrial vesicles

↓ Treated with trypsin or urea

Submitochondrial vesicles almost completely lacking spheres observed by negative-contrast electron microscopy → Soluble sphere containing F_1 ATPase molecules

↓

Reconstituted vesicle with shperes

Observation from the study

1. Mitochondrial membrane contributes the enzymes of electron transport.
2. Soluble spheres represents the enzyme complex required to make ATP.

29.20 Cell, Tissue and Organ Culture

29.20.1 Tissue culture

Tissue Culture: In this method a small portion of embryonic tissue is taken and kept at suitable temperature in nutrient fluid, e.g., mixture of embryo extract, blood plasma and saline solution. The whole preparation must be aseptic. The cells grow and multiply and reaches a certain size. A subculture can be made by taking small piece of tissue and can be repeated indefinitely.

Use:

1. To learn structure
2. To learn growth
3. To learn differentiation of embryonic cells.
4. In cytogenetic research, tissue culture are used to study mitosis and chromosomes in human cells.

5. Determination of karyotype (the number and morphology of an individual's chromosomes).
6. Short-term cultivation of blood lymphocytes or skin fibroblasts.

In examining these cells during mitotic division in tissue cultures anomalies can be seen in the number and morphology of the chromosomes. These anomalies have been shown to be related and are diagnostic of numerous diseases collectively called genetic disorders.

29.20.2 Organ culture

For the purpose of assessing its inherent capacity for growth and differentiation in the absence of the external influences which act upon it in the complex environment of the intact organism. By altering the conditions of cultivation so as to discourge migration and dispersion of the cells, the original organ architecture can be retained. Ths method is called **Organ Culture**. The principal value is to provide a means of *isolating an embryonic organ rudiment*. By this method, it has been demonstrated that bone rudiments from early embryos would grow in length and undergo calcification outside the body.

29.20.3 Cell culture

In preparing cell culture, the cells are usually dispersed mechanically or by previous treatment with enzymes, such as trypsin or collagenase. Once isolated the cells can be cultivated either in a suspension called *suspension culture* or spread out on a culture plate to which they can adhere to as a single layer of cells called *plate culture*. The use of extracelluar matrix such as collagen and laminin and addition of growth factor and hormone increase the survival of cells in vitro.

29.20.4 Use of cell culture

1. For the study of the metabolism of normal and cancerous cells.
2. For the study of parasites that grow only within cells, such as viruses, mycoplasma and some protozoa.
3. *Transformation*: Most cells from normal tissues have a finite, genetically programmed life span. Certain changes mainly related to oncogens can promote cell immortality, a process called transformation, which may be a first step in a normal cell's becoming a cancer cell and can be studied by this method.
4. Cell culture is central to contemporary techniques of molecular biology and recombinant DNA technology.

29.21 Tissue Culture Techniques

Most of the techniques used in tissue culture are similar in principle. The main difference are the vessels used for growing the tissues. The main *groups of methods* are these:

1. Using depression slides
2. Using camel flasks
3. Test-tube cultures of which there are many variations. *Organ tissue culture fall in the same general category.*

29.21.1 Types of culture

1. **Slide cultures**

Principle: Slide cultures or Coverslip cultures are made by implanting a very small fragment of tissue on a coverslip which is subsequently inverted over a small cavity of a *depression slide.* There are *six* general types of slide cultures:

(a) Single coverslip with plasma clot (Harrison)
(b) Double coverslip with plasma clot (Maximum)
(c) Single coverslip with fluid hanging drop (Lewis & Lewis)
(d) Double coverslip with perforated cellophane (Schilling & Esle)
(e) Thin drilled metal or glass slide with plasma & fluid medium (Yey)
(f) Special slide with a circulating fluid medium-perfusion chamber (Pomenat).

29.21.2 Methods

(a) Single coverslip with plasma clot:

i. All the instruments required, include a rack with sterile test tubes for holding Pasteur pipettes. Constituents of the medium are prepared in two parts—

(a) One part containing plasma.
(b) The other part containing embryo extract. Such as one solution contains 50% plasma in balanced salt solution and the other contains 50% embryo extract in serum.

ii. With a pair sterile forceps, one or two coverslips (22 mm) placed on a clean sterile surface. One drop of the plasma containing solution is placed in the centre of each coverslip by a capillary pipette.

iii. One or two explants to this drop either by means of knife, blades or by a fine pair of forceps, taking care not to crush the tissue, are transferred.

iv. A drop of the embryo extract containing solution is added to this. Immediately it is mixed thoroughly, before clotting starts, and spread out into an area of 15 to 20 mm diameter. The explant in this is located.

v. With a glass rod, two small spots of vaseline, are placed, near the concavity of a depression slide, in such a position that they will be covered by the coverslip. The slide over the coverslip preparation is inverted and pressed down in such a way that the vaseline sticks the coverslip to the slide. The culture, thus prepared, is placed aside to permit the medium to clot.

vi. The cultures are then turned over and the margin sealed with hot paraffin and labelled and incubated at 37°C.

(b) Maximised double coverslip method with plasma clot

This technique is very much similar to that already described. One or two large (40 mm) square coverslips are placed on a sterile surface and a small drop of a balanced salt solution is put on each of them. A square or round 22 mm coverslip is then placed on the centre of each large one, being held in position by the surface tension of the balanced salt solution. The culture is then prepared on the small coverslip exactly as before. A large depression slide is used and the entire preparation is attached to it by vaseline and wax in such a way that the small covership is not in contact with the slide at any point (Fig. 29.11).

Single coverslip with liquid medium (Hanging drop cultures): This is very simple but its application is very limited. The prepared explants are placed in culture medium in a water glass. Coverslips are laid out on a sterile surface, the explants drawn into dip of a capillary pipette and deposited one in the centre of each coverslip. Fluid can be drawn off and carefully remeasured (1 drop for a 1 inch slide, 2 to 3 drops for a large slide). The fluid must then be spread out in a very thin circular film with the explant protruding above its surface. A vaselinated depression slide must be applied immediately and the preparation turned over with a quick flick to prevent the fluids from running into the crevice between the slide and the coverslip. After labelling and paraffining, slides must be incubated in either the upright or inverted position. The cells grow directly in the glass of the coverslip.

29.21.3 Advantages of slide cultures

1. It is simple and relatively inexpensive.
2. The cells spread out in a manner which is favourably for microscopic examination and photography in living state.
3. The cells grow directly on the coverslip so that they are easily fixed and stained and made into permanent preparation.

- **Disadvantages**

1. Small supply of O_2 and nutrients is very rapidly exhausted so that the medium quickly becomes acid and hence rapidly grown tissues must be transferred frequently.
2. It is difficult to maintain sterility for a long period.

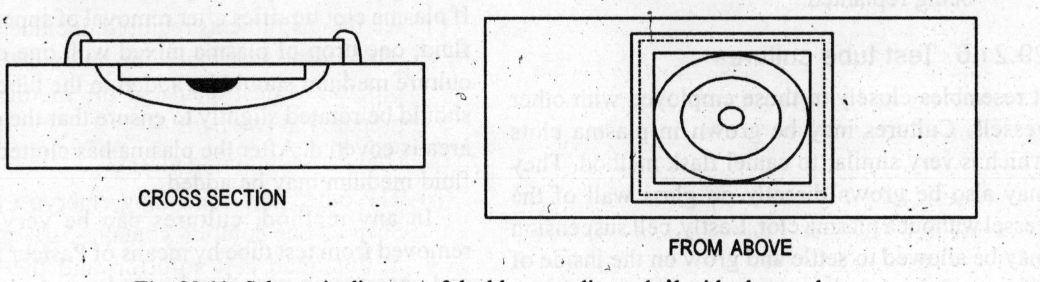

Fig. 29.11: Schematic diagram of double coverslip method with plasma clot

3. Only very small amounts of tisssues can be cultured and hence the application of the method is limited.

29.21.4 Camel flask. Technique (Earle)

Its main use is the establishment of strains from fresh explants of tissue. There are two types of camel flask techniques—

(a) *Thick clot*: Encourages rapid growth and is suitable for short term cultures or for the maintenance of strains.

(b) *Thin clot*: These cultures are best for maintaining cultures over a considerable period of time and they are better for testing the effects of the materials added to the medium.

29.21.5 Procedure

i. One drop of plasma is placed on the floor of each flask using a platinum spatula, spread this plasma out in a circle extending to the edge of the floor of the flask.

ii. With the spatula, the desired number of explants to the plasma are transferred and allowed for clotting.

iii. After clot, the explants are fixed in position and extra medium can be added. In case of *thick clot cultures* 1-2 ml of dilute ($\frac{1}{4}$) plasma are placed on top of the explants and the whole is then left to clot.

In case of *thin cultures,* 1-2 ml of dilute ($\frac{1}{4}$) serum is added instead of the plasma.

iv. The flasks should be gassed with 5% CO_2 in air.

v. Medium can be renewed.

vi. Cultured tissues can be transferred to other vessels and put into suitable pieces before being replanted.

29.21.6 Test tube cultures

It resembles closely to those employed with other vessels. Cultures may be grown in plasma clots which is very similar to camel flask method. They may also be grown directly on glass wall of the vessel without a plasma clot. Lastly, cell suspension may be allowed to settle and grow on the inside of the test tubes.

There are various types of tissue cultures. They are:

1. Plasma clot technique.
2. Feeding test tube cultures.
3. Patching test tube cultures.
4. Flying coverslips in test tubes.

29.21.7 Plasma clot technique

(a) With a Pasteur pipette, a drop of plasma, near the bottom of each tube, is placed and spread over the lower part of the tube.

(b) The explants are transferred, usually about four, to the plasma and left to clot.

(c) After they are fixed in a position, add medium, usually 0.5 to 1 ml/test tube.

(d) Stopper and label the tubes and place them in a stationary rack or in a roller drum in the incubator.

29.21.8 Feeding test tube culture

It is simply achieved by removing the supernatant fluid and replacing it with fresh medium.

29.21.9 Flying coverslip in test tube

Flying coverslips are small round coverslips (11 mm × 400r, 20 mm long) which can be inserted into ordinary 6″×5/8″ test tube. cultures are prepared on them in the standard manner, but instead of mounting them on depression slides, they are slid into test tubes.

A fluid medium is added to the test tubes (102 ml each) which may then be placed in a stationary rack or in a roller drum. Feeding is performed by removing the fluid medium and replacing it with fresh material. Large number of cultures can be prepared in this way.

29.21.10 Patching test tube cultures

If plasma clot liquifies after removal of supernatant fluid, one drop of plasma mixed with one drop of culture medium should be added to the tube and it should be rotated slightly to ensure that the culture area is covered. After the plasma has clotted, fresh fluid medium may be added.

In any method, cultures can be very much removed from test tube by means of Pasteur pipette and observation may then be made.

For preparing large number of cells in culture, the monolayer. method is exten-sively used. The chapped tissue is treated with dilute trypsin, an enzyme which breaks up the intercellular cement and produces a suspension of separated cells. The trypsin is then removed by washing in a centrifuge and the cell suspension mixed with the culture medium. The cells set on the glass, attach to the surface of the vessel, and grow to form a monolayer. When the culture has covered the glass surface, the cells may be harvested by removal with tayking. They can then be used to prepare fresh monolayers. By this method, a large number of cells can be grown from a small mass of starting material.

Although monolayer cultures are useful for study of cell growth, the cells tend to loose many of their characteristic properties under these conditions. To preserve the *in vivo* character of cells and tissues, the organ culture method is generally used. This is now not used so much for culture of whole organs but fragments of tissue appx. 1.5 mm in size. (With larger fragments, there are problems in obtaining adequate gas exchange). These fragments are placed on a soft or permeable sheet such as millipore material.

For studies of movements and behaviour of cells by the time-lapse method, the cells are grown on glass coverslips which are then sealed in a chamber.

29.22 Fine Needle Aspiration Cytology (FNAC)

Fine needle aspiration cytology (FNAC) is the method of study of taking cells from any site of the body taken out by fine needle aspiration. This is done in case of superficial nodule, and thyroid has become routine diagnostic procedure. Now-a-day, it is use, for the diagnosis of the bone lesions, breast lesions and in lesions of internal organ like prostrate, salivary gland, pancreas, etc. The material which is aspirated is taken on a glass slide and a smear is prepared and stain with Leishman stain, Giemsa or Papanicolaou's stained (for cancer cell) and also by Hematoxylin-eosin stain. The advantages of this procedure are:

1. Easy method
2. Less damage than by biopsy
3. Rapid and accurate
4. Treatment can be done earlier
5. This is done without excision, mass can be investigated
6. Pre-operative information of deep seated disease.

The disadvantages are:

1. Tumour cell can be disseminated in the needle tract and natural cavity.
2. Bleeding may occur.
3. Complication like, air emboli, pneumothorax may arise.
4. Point of needle may not take the mass from proper site.
5. It is indicated—
   (a) When patient refuses surgery
   (b) If the patient cannot tolerate anaesthesia.
   (c) Dangerous site of mass.
   (d) When malignancy already diagnosed clinically
   (e) When the mass is cyst or brain tumour.
   (f) Very old patient.

29.23 Exfoliative Cytopathology

Exfoliative cytopathology is the study of normal cell and diseased cells where desquamation has taken place from the sites. The type of cell which are desquamated are mainly epithelial cell but some are mesodermal in origin, such as connective tissue. These are labile cells, which are shedding spontaneously and again replaced by a new cell. The rate of shedding varies according to type of tissue. Some of the sheded cell may accumulate in cavities like peritoneal cavity and pleural cavity, most of the exfoliated cells are lost from the surface or may pass through the GI tract.

The types of exfoliation are—

1. Natural spontaneous exfoliation, e.g., vagina cervical cells, mesothelial cells.
2. Artificial exfoliation—It occurs due to scraping of musosal surface, by spatula or aspiration by fine needle. e.g., cervical carcinoma.

These is done by—

i. Early detection of cancer
ii. Confirmation of suspected disease
iii. To know the cause of the disease.

29.24 Growth and Development

Growth: The term *growth* means increasing of size. In human beings or in living organism this growth comprises three processes:

1. Increase in cell number (mitosis)
2. Increase in size of cells
3. Increase in secretion. Different organs and systems do not grow and mature at the same rate—prenatally and postnatally, e.g., thymus and heart are functionally mature prenatally whereas nervous system matures only some years after birth.

29.24.1 Developments

Development refers to the continuous process by which an individual changes from one life phase to another. Two main periods in human's development are:

I. Prenatal period
II. Postnatal period

29.24.2 Prenatal period

This period begins with fertilization of ovum and ends at birth at about 40 weeks.

In this period of growth and development the new individual is at different stages of its development, such as:

Fertilised ovum → Zygote → Morula → Blastula → Embryo → Foetus at the *second week* of prenatal life. The embryonic cells or embryoblasts become organised in two layers of cells:

(a) Upper (outer) layer known as ectoderm
(b) Lower (inner) layer known as endoderm
(c) At the *third week* a third layer of cells forms between the two layers by mitosis of the existing two layers known as **Mesoderm:** All the body structures are developed from these three layers such as:
   1. *Ectoderm*—skin, glands, epidermis, nervous system, oral and nasal epithelium.
   2. *Mesoderm*—Muscle tissue, connective tissues, kidneys, gonads, endothelium.
   3. *Endoderm*—Thyroid, lungs, digestive tract, urinary bladder.

Rate of increase of foetal mass

1. Slowly—First trimester of pregnancy
2. Very rapidly—Second trimester
3. Moderately—Third trimester.

29.24.3 Postnatal period

This period begins from birth and continues until death in old age. This period is distinguished into several periods such as:

1. Neonatal
2. Infancy
3. Childhood
4. Puberty
5. Adolescence
6. Adulthood
7. Old age or sensescence.

Phases of postnatal Period

1. Neonatal Phase

Period—First four weeks after birth.

Changes—Unfolding of the lungs, normalisation of pulmonary circulation, rhythmic breathing, closure of ductus arteriosus and foramen ovale normalisation of RBC count.

2. Infancy

Period—Second month to 1 years. Rapid growth in this phase, the body mass may be tripled, the height may increase by 25-30 cm. Then the growth continues at slower rate.

Changes—Decidous teeth begin to erupt through the gums at the age of 6 months, muscular and nervous systems mature in such a way that coordinated muscle movement is possible, e.g., sitting, crawling and standing, learns smile and lough, responds to sound, tries to express or to say some syllables.

3. Childhood

Period—From first year and ends at puberty, growth at slower rate (50-60 mm per year) until puberty.

Changes—Deciduous teeth erupt by the end of second year. Permanent teeth begin to replace the deciduous ones by the age of 6 years. Development of high degree of voluntary muscular control, learns to walk, run, balance, jump, climb and wash hands. Urinary bladder and bowel controls are established.

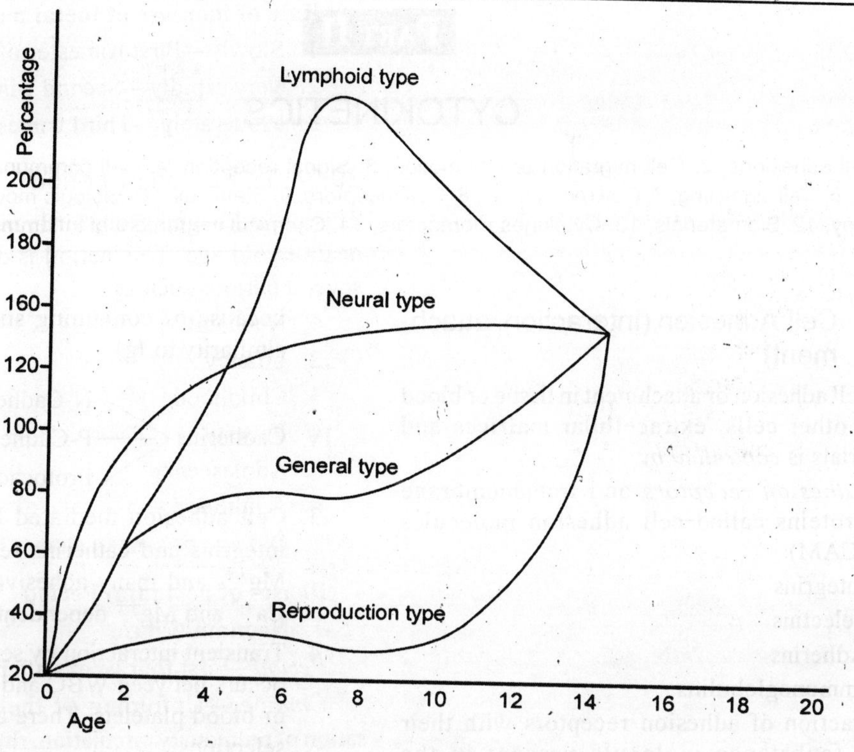

Fig. 29.12: Growth curve

The child recognises persons, can write and read and develops emotion. Apart from nutrients, several growth factors play roles in prenatal and postnatal growth such as:

(a) The genes inherited from the parents

(b) Hormones, e.g., growth hormone, insulin, thyroxine, glucocorticoids, sex hormones, etc.

(c) Non-hormonal chemical compounds, e.g., growth factors (thrombocyte growth factor, insulin-like growth factor, etc.).

(d) Disease processes—slow down and even suppress growth. Gain in weight (body mass) and height, and increase in head circumference are used to assess normality of growth (Fig. 29.12).

4. Puberty: It is the stage where the young reaches sexual maturity. There is anatomical and physicological changes controlled by the gonadotrophins and sex hormones. The general growth curve applied to skeleton.

PART II

CYTOKINETICS

1. Cell adhesions, 2. Cell migration or cell motion, 3. Signal reception, 4. Cell communication, 5, Cell cycle, 6. Cell signalling, 7. Cell recognition, 8. Cell receptors, 9. Stem cell, 10. Biologic modifiers, 11. Cell therapy, 12. Biomaterials, 13. Cytokines Bioreactors, 14. Chemical medeators of inflammation.

29.25 Cell Adhesion (Interaction, attachment)

Cell to cell adhesion or attachment in tissue or blood cells to other cells, extracellular matrices and biomaterials is *controlled by*

1. *Adhesion receptors* and transmembrane proteins called cell adhesion molecules (CAM)
2. Integrins
3. Selectins
4. Adherins
5. Immunoglobulins

Interaction of adhesion receptors with their ligands (substance molecule present in the substratum which combines with receptor of the cell membrane responsible for *physical attachment* and *biochemical signalling.*

29.25.1 Type

Cell to cell interaction is essential for development in multicellular organism. These type cell-cell interaction may be of two types:

Transient: 1. Interaction between cells of the immune system.
2. Direct-WBC to the sites of inflammation.

Stable: Cell-cell interaction in organisation of cells for tissue, e.g., membrane and function of epithalial sheet.

29.25.2 Characteristics

Cell to cell adhesion is a *selective process.* Cells adhers only to other cell of the same tissue e.g. liver cell will adhere to liver cell specifically but not to the cell of brain (Fig. 29.13).

Selective cell to cell adhesion mediated by *transmembrane proteins* called *cell adhesion molecules (CAMs).* They are of 4 major groups:

I. Selectins, II. Integrins

III. Immunoglobulins superfamily (named because of containing structural domains similarity to Ig).

IV. Cadherins — N-Cadherin (neural); P-Cadherin (Placontal); Protocadherin (Synapse)

3. Cell adhesion mediated by the selectins, integrins and catherins requires Ca^{++} and Mg^{++}, and many adhesive interaction, are Ca^{++} and Mg^{++} dependent.
4. Transient interaction by selectins mediation occurs between WBC and endothelial cells or blood platelets. There are three types of selections:
   (a) L-selectin is expressed on leukocytes
   (b) E-selectin is expressed on endothelial cell
   (c) P-selectin is expressed on platelets.

These cell adhesion materials of adjacent cells bind togather giving stability in the tissue. These may be:

(a) *Homophilic*—bind to same type of molecules.

(b) *Heterophilic*—bind to different types of molecules.

Stable adhesion junction involving the *cytoskeleton* of adjacent cell are instead mediated by the cadherins, these cell-cell junctions are of two types:

(a) *Adherens junction,* (b) *Desmosomes.*

In which cadherins or related proteins, i.e., *desmogleins* and *desmocollins* are linked to *actin bundles* and *intermediate filaments.* The role of cadherins in linking the cytoskeleton of adjacent cells is thus analogous to that of integrins in forming stable junction between the cells and the extracellular matrix (ECM) in addition to adhesion junction mediated by the *cadherins,* two other types of specialized junctions

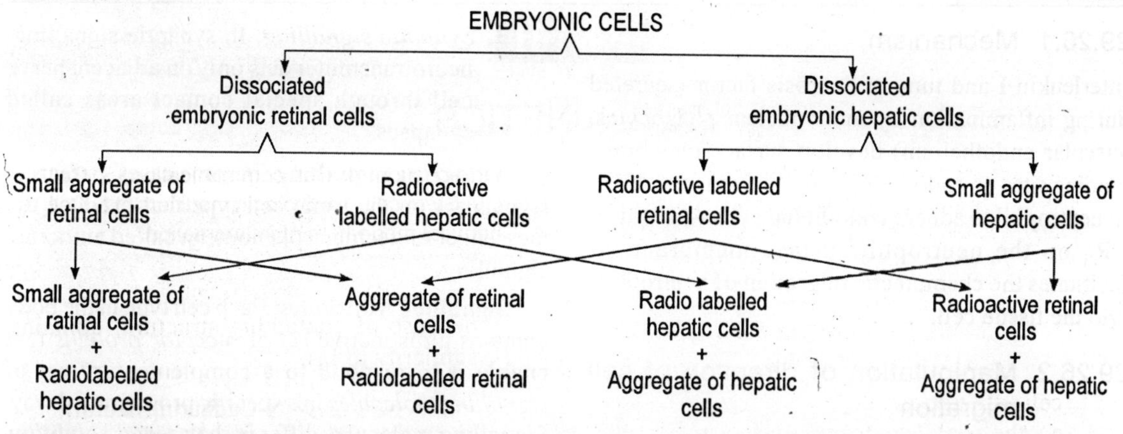

Fig. 29.13: Adhesion of specific tissue dissociated embryonic cells studied by radioisotope labelled cells binding assay

(a) *Tight junction*—which are usually associated with adherens junction and (b) Desmosomes in a junctional complex.

29.25.3 Functions of CAMs

1. Stability of the tissue
2. Development of the tissue
3. In inflammation
4. In wound healing
5. In metastasis of tumours.

29.26 Cell Migration (Cell motility, cell movement)

Cell movement through two or three dimensional substrate occurs by the process of receptor-ligand interaction in two:

1. *Intracellular motile force* initiates active migration. Motile force requires that these forces should be transmitted significantly to the substratum as differential traction.
2. *Adhesion receptors* mainly of the *integrin* superfamily are involved in this interaction. Integrin interaction with extracellular matrix (ECM) ligands influence the migration speed. *Intracellular motile forces* must be stimulated under specific condition for desired movement. The ligands stimulating cytoskeleton *generate motile force* which can induce cells to extend *membrane lamellipodia* to the direction of higher concentration of ligands leading to a ligand concentrated gradient. These ligands are called *chemotactic attractants*—source of them can modulate speed of movement, hence are termed *chemokinetic agents.* These ligands may be growth factor. Endothelial vascular growth factor (EVGF) and some other factor e.g., chemotactic factors seen in inflammation and phagocytosis done by WBC and the process is called chemotaxis (Table 29.8).

Table 29.8: Chemoattractants

Source	Materials
1. Breakdown of bacterial protein	N-Formyl Methionyl oligopeptides
2. Breakdown of mitochondrial proteins from damaged tissue	do
3. Complement cleavage product	C5a, C5a—des—Arg
4. Platelet and granule secretion	Platelet factor 4 Platelet derived growth factor
5. Neutrophilic granulocyte activation	Metabolites of lipoxygenation of arachidonic acid (lukotriene B4, 5-HETE, etc) AGEPC (Acetyl glycerol ether phosphorylcholine crystal induced chemotactic factor (CF).
6. Tissue mast cells degranulation	Eosinotactic tetrapeptide (eosinophil chemotactic factor of anaphylaxis, High molecular weight neutrophil chemotactic factor of anaphylaxis.
7. T-Lymphocyte antigen or motigen stimulation	Lymphokine chemoattaraction for granulocytes

29.26.1 Mechanism

Interleukin I and tumour necrosis factor (secreted during inflammation by mononuclear phagocytes, vascular endothelium) develop surface membrane binding sites on endothelial cells → enhances ability of neutrophil to adhere endothelial cells → requires CR_3 on the neutrophil surface membrane → facitiates the chemotactic migration of neutrophils into the tissue cell.

29.26.2 Manipulation of direction of cell migration

It can be considered under three categories:

1. Soluble chemotactic ligands
2. Immobilized hepatactic ligands
3. Contact guidance.

29.26.3 Significance of cell migration or motions

1. During organogenesis and embryonic development.
2. Tissue repair response both wounded healing and angiogenesis.
3. Immune response rely on cells migration.
4. Pathological situation such as cancer metastasis are characterized by cell motion.

29.27 Signal Reception

Multicellular organism, the cells need to communicate with each other to regulate (a) development, (b) growth, (c) division, (d) coordination of functions for tissues. Many cells from communicating junctions with the adjacent cells for exchange of ions and small molecules. Through these channels, called gap junctions, signals pass directly from cell to cell without reacting the extracellular fluid. In other cases, cells display membrane-bound signalling molecules that influence other cells in direct physical contact.

Extracellular signalling molecules, or messengers, mediate *three kinds of communication* between cells.

1. *Endocrine signalling*: In endocrine signalling, hormones are carried in the blood to target cells throughout the body.
2. *Paracrine signalling*: In paracrine signalling, chemical mediators are rapidly metabolized so that they act on local cells only.
3. *Synaptic signalling*: In synaptic signalling, neurotransmitter acts only on adjacent nerve cell through special contact areas called synapses.

Autocrine Signalling: In some cases, paracrine signals act on the same cell type that produce the messenger molecule, a phenomena called autocrine signalling.

Signalling Molecules: Each cell type in the body contains a distinctive set of *receptor proteins* that enable it to respond to a complementary set of *signalling molecules* in a specific programmed way. Signalling molecules differ in their *water solubility.*

Hydrophobic signalling molecules: Small hydrophobic signalling molecules, e.g., steroid and thyroid hormones diffuse through the plasma membrane of the target cell and activate receptor proteins inside the cell.

Hydrophilic signalling molecules: Includes neurotransmitters, most hormones, and local chemical mediators (paracrine signals), activate receptor protein on the surface of target cells. These receptors which span the cell membrane, relay information to a series of intracellular intermediaries that ultimately pass the signal to its final destination in either the cytoplasm or the nucleus.

G-protein—The numerous intercellular hydrophilic messengers rely on membrane proteins that direct the flow of information from the receptor to the rest of the cell.

The best studied of these proteins are the G-proteins, so named because they bind to *guanine nucleotide.*

One *first messenger* (hormone, neurotransmitter, paracrine signal) → binds to a receptor → conformational change occurs in the receptor → activates the G-protein-guanine diphosphate (GDP) complex → GDP-GTP (guanine triphosphate) exchange releases the subunits of the G-protein → which acts on other membrane-bound intermediaries called *effectors* → often, the effector is an enzyme that converts an inactive precursor molecule into an active *second messenger* → which can diffuse through the cell membrane → second messenger trigger a cascade of molecular reactions that lead to change in cell behaviour. e.g., epinephrin, glucagon (stimulus) → liver cells (effected cells) → Gs protein → adenyl cyclase (effector) → breakdown of glycogen (effect).

Applied: For defective receptor, pseudo-hypoparathyroidism and dwarfism are observed for non-functioning receptors. The blood levels of these hormones are normal but the target cells do not respond for lacking normal receptors.

29.28 Cellular Communication

The cells within the tissue communicate with another cells by three processes:

1. With the help of cytokines or chemokines which are secreted as soluble signals
2. Cells touch other, i.e., direct cell-cell contact
3. Proteins made by cells and change the chemical microenvironment.

29.28.1 Features

1. These communication processes differ regarding their characteristics time and length scale.
2. Differ in their specificity.
3. Each carries a particular type of message.

29.28.2 Soluble signals (growth factors)

1. These are small proteins having size of 15-20 KDa (1Da $\approx$ wt. of hydrogen atom)
2. Chemically stable
3. Having long half-lives
4. Can be cloned directly as ligands for known receptors e.g. thrombopoietin as C-mpl ligand, stem cell factor as C-kit ligand.
5. Produced by a signalling cells and secreted to react a target cell.

29.28.3 Mechanism

Growth factors bind with receptor which triggers a complex signal transduction process, as a result there is change in the receptor complex leads to catalytic activity of the extracellular components. These activities are due to Jonus Kinases (JKs). These kinases then act on transducers and activates the transcription process which transmit the signal into the nucleus.

29.28.4 Direct cell-cell contact (insoluble factor I)

Surface of the cell containing, proteins called cell adhesion molecules (CAMs), includes cadherins (adhesion belt, desmosome and connexins (gap junction). These are involved direct cell-cell communication and are often termed as cell junction molecules because junctions are formed between the cells through which cytoplasmic communication takes place as well as in the intercalated disc of heart muscle. These junctions having diameter of 1.5 nm and allow the molecules below 1KD to pass through cytoskeleton having a mechanical role, signals may be transmitted to the nucleus by cellular stretching and the integrins acts as mechanical transducer of the signals. Long distance information from our cells to another is transmitted by weak mechanical interaction. This mechanism is smilar to fluid mechanical shear forces influence on cell and tissues functions.

29.29 Cell-Matrix Interaction

The cells in tissue and their cytoskeleton are interconnected by complex weaves, gels and glues of extracellular matrix.

ECM has many functions such as mechanical support, adhesion of cell, migration of cell, place to locate signals for communication, because:

1. ECM consists of many components having various structures and functions.
2. On the cell surface there are many adhesion and ECM receptors molecules which help to cell-ECM communication.
3. These signals contain instructions for replication, migration, differentiation and apoptosis.
4. The nature of signals is governed by the component of ECM which in turn can be changed by the cell in the tissue. Thus all the cellular processes are governed by ECM and thereby help in the communication.
5. The signals in ECM is stable
6. More specific
7. Stronger than the soluble growth factors.

29.30 Signalling of Cell

Biological cells in the tissue interact with environment. For this interactions *three mechanisms* play role to bind cells together. These are:

I. *The glycocalyx of the cell* membranes contain glycoprotein having proteoglycan molecule which is glycosylaminoglycan, help in adhesion.

II. *The plasma membrane of the adjacent cells* are wavy, i.e., there is elevation and depression and the cell is lodged in the socket like tooth in the socket.

III. Formation membrane junction: This interaction, adhesion, etc. necessary for receiving and sending information from cell to cell. But the questions are how interaction takes place and how the information are triggers to transmit to the environment.

Apart from interaction between the cell and environment, the cell responds to extracellular chemicals like hormones (e.g., insulin, thyroxine, etc.) neurotransmitter (e.g. acetylcholine, adrenaline, etc.) cells also interact with molecules present in the substratum which acts as guide to cell migration and repair.

It is apparent that glycocalyx is involved and it may be direct or indirectly. It is known that glycocalyx of the cell membrane has *two participating molecules* which take part in the above mechanism: They are *membrane receptors* (e.g., hormone *receptors,* etc.) and *adhesive molecules* (CAMs) such as *adhesin, integrin, filbronectin, cadherin.*

29.30.1 Function of adhesion molecules (CAMs)

1. Adhesion molecules are very much essential for cell adhesion.
2. It is an useful in migrating function of the cell and is involved such as embryonic development, wound repair, immunity, inflammation and infection, etc.
3. The molecules anchor themselves as the molecules of the extra cellular space and to such other.
4. In case of inflammation or infection the migration of WBC (Phagocytes) from blood vessel by emergency signal in the form of *selectin.*

29.31 Function of Membrane Receptors (Signalling)

There are several membrane receptors which are integral proteins (e.g., glycoprotein) participate in signalling after binding with *ligands,* e.g., (signalling chemicals present in the substratum which combines with membrane receptors).

The signalling are of *three* types: (1) *Contact signalling,* (2) *Chemical signalling,* (3) *Electrical signalling.*

29.31.1 Contact signalling

Contact signalling is the process by which the cell can recognize the other cells. The importance of this signalling are:

(a) For development of cells and tissue applied.
(b) Immune process
(c) Few bacteria and some infectious agents prefer this signalling to identify suitable tissue or organs.

29.31.2 Chemical signalling

Major group of membrane receptors bind to chemical ligands. Among these chemical ligands most of them are:

A. *Neurotransmitters* which are of neural origin, e.g., acetylcholine, adrenaline, signalling of the nervous system)
B. *Hormones,* e.g., insulin, thyroxine, etc signalling of the endocrine system.
C. *Paracrines*—The chemicals of this system *act locally* and is destroyed rapidly.

Some chemical ligands act differently on different cells such as acetylcholine in case of skeletal muscle is released from motor nerve into the neuromuscular junction which acts on the acetylcholine receptor in the postsynaptic membrane creating depolarisation causing muscle contraction whereas same chemical ligand causes depression in heart muscle.

29.31.3 Response to receptor binding

There is variation of receptor binding but basically similarity is observed such as:

I. When *ligand binds* with receptor, there is *conformational change in* the receptor protein and the cell protein also changes, e.g., (1) muscle contraction causes change in shape for creating force, (2) activation or inactivation of enzymes, (3) opening and closure of ion channel by channel proteins, etc.

II. Some membrane receptor proteins act as enzyme, i.e., catalytic function.

III. Chemically gated channel linked receptors are found in muscle and nerve cell, when bond to ligands causes opening and closing the ion channels which causes excitability of the cell.

IV. Some type of receptors are coupled to an *enzyme or ion channels* by a regulatory protein called G-protein.

V. *G-protein* linked receptors produce the effect indirectly through G-protein which acts as a middle man or relay to activate or inactivate a membrane bound enzyme or ion channel. As a result a intracellular chemical signal is produced known as *second messenger* which makes a linkage between cell membrane and internal metabolic system.

There are two mediators or messengers are present inside the cell such as cyclic AMP (cAMP) and calcium which activate the enzyme protein kinase which transfer the high enzyme rich phosphate bond (~ph) to other proteins, protein kinases, in turn activate whole series of enzymes including other kinases. This activated phosphorylated protein kinase causes cellular reaction like metabolic and structural changes in cascade mechanism.

VI. *Nitric oxide* (NO) *signalling agent*: Another signalling molecule which is environmental pollutant and does not act like other signalling molecule is *Nitric Oxide* (NO), a biological messenger. It is very small in size and can easily enters and comes out of the cell. Due to presence of the unpaired electron it is a highly reactive free radical which reacts speedily with other molecule so trigger the cell for activities.

29.32 Cell to Cell Recognition

Cell in a tissue can recognize same type of cell. After this phenomenon, there is cell adhesion. This process has been studied in sponge and chick embryo using sea water lacking Ca^{++}, Mg^{++} does not show the process. For another study it has been observed that for a growth factor normal *aggregation factor* for aggregation of cell is required. This factor are proteoglycans which seems to be two extra cellular bridging materials which are joined by Ca^{++}. The free terminals of the fused aggressive factor has carbohydrate molecule which are recognized by the receptors on the surface of cell membrane. The receptors of the two cells bind with the free terminal of the aggressive factor by Ca^{++}.

In case of immunological process, the different investigations have shown that several cell glycoproteins are involved on the surface of cell membrane, all proteins are not involved in the process of adhesion.

But antibodies are produced against all proteins of the membrane which are divalent. These antibodies bind with membrane proteins or antigen. If there is fragmentation into monovalent, there will be inhibition of cross-linkage and adhesion of these cells because of blocking recognition site by monovaleny antibody. Then bound antibody is made free and others intact and ultimately will adhere with each other. Two groups of glycoproteins are involved in the cell-cell adhesion. They are:

(a) This group operates initially, does not require Ca^{++}. The first group are glycoproteins known as *neural cell adhesion molecule*
   1. Operates in the early embryonic development.
   2. Present in the nerve and neuroglial cell membrane.
   3. Help to bind together without Ca^{++}.

(b) This group acts later, requires Ca^{++}. The second group are glycoproteins and all are *cadherins*. These are of three types:
   1. E-cadherins—They are bound to most of the epithelial cells.
   2. N-Cadherins—They are in heart, lens and nerve cells.
   3. P-Cadherins—In the cell of placenta and dermis

It is understood that cell-cell adhesion followed by *recognition*. The steps are as below:

1. Adhesion takes place in the presense of glycoprotein in the cell membrane surface.
2. Attachment of one cell to other cell by making high affinity non-covalent bond with the complementory molecules.

This type of binding revealed that there are three processes during cell adhesion:

1. Complementary molecules interact with a homophilic bond or a heterophilic bond or with extracellular agent making linkage.
2. N-CAM usually binds by a hemophilic bond.
3. Cadherin may interact by any one of the above processes but how signal is given to recognise other cell is not clear.

In experiments with Amoeba revealed that

1. It secretes cyclic AMP (C-AMP) which acts as a chemotactic signal which attracts other amoeba.
2. In this process C-AMP serve as an extra cellular signalling molecule.
3. When aggregation formation takes place, aggregation cells responding only to the C-AMP signal and transmit from the cell to other cell.
4. Apart from C-AMP signalling system, during starvation for first 8 hours cells, adhere by Ca^{++} dependent mechanisms involving cell-cell adhesion known to *contact site* B and after another 8 hrs cells adhere by Ca^{++} independent membrane called contact site A. Both A and B contact site are integral proteins of the cell membrane glycoprotein.

   The above procedure has been investigated but result is known. But beside adhesion extracellular matrix (ECM) also play role in adhesions and signalling.

29.33 Biologic Modifiers or Modulators of Biologic Response

Genetic engineering and recombinant DNA technique have achieved a great success to produce immunologic cell products for use clinically and dramatic results have been observed in tumours in the general treatment (Fig. 29.14).

29.33.1 What are the modulators?

1. Interferons— α, β, γ
2. Colony stimulating factors
   (a) G-CSF (Granulocyte colony stimulating factor).
   (b) GM-CSF (Granulocyte-monocyte colony stimulating factor.
   (c) Interleukins.

I. Interferons: These are one family of proteins which are produced and secreted by: (a) peripheral white blood cells (b) fibroblasts (c) epithelial cells.

29.33.2 Types

Based on antigenicity, biologic and chemical properties they are:

(a) Interferon alpha (α)
(b) Interferon beta (β)
(c) Interferon gamma (γ).

Commercially available preparations are:

(a) Interferon alfa 2a
(b) Interferon alfa 2b
(c) Interferon alfa nc
(d) G-CSF
(e) GM-CSF
(f) Interleukin-3

29.33.3 Indications

1. Viral infection
2. Malignancies
3. Bone marrow transplantation
4. Failure of other treatment
5. Immunotherapy.

29.33.4 Action

Appears to be a complex cascade of biological modulations and pharmacologic effect by influencing many cell functions such as:

(a) Augmentation
(b) Restoration
(c) Modulation of host's immune system
(d) Antineoplastic
(e) Antiproliferative
(f) Modulation of cell differentiation
(g) Modulation of cell translation
(h) Modulation of cell transcription
(i) Reduction in oncogene expression.

29.33.5 Clinical Application

1. Viral infection, e.g., viral hepatitis
2. Malignancies
3. Others e.g. exophytic genital and wart, multiple sclerosis, AIDS (without kaposis sarcoma), angiomatous disorders.

29.33.6 Colony stimulating factors

During cell culture technique for leucocytes certain substances were identified to help growth of cell colonies known as colony stimulating factors (CSFs). They act as intracellular signals to activate their target cells . It has been identified also that which factor causes stimulation of growth of which cell.

29.33.7 Chemical nature and production

These are glycoprotein in nature. Now-a-days these are produced by recombinant DNA technology and genetic engineering. These are

1. Granulocyte-Monocyte growth stimulating factor—produced in the body by (a) T-lymphocyte, (b) monocyte, (c) fibroblast, (d) endothelial cell.
2. Granulocyte colony stimulating factor (G-CSF)—Produced by monocytic fibroblast and endothelial cells. The gene has been located on the chromosome 17.

29.33.8 Clinical application

1. Chemotherapy-induced neutropenia
2. Bone marrow transplantation
3. Myelodysplastic syndrome
4. Acute myeloid leukemia
5. Leukopenia associated with AIDS
6. GM-CSF associated in combination with zidovudine and gansiclovir in AIDS patients counteracts neutropenia frequently encountered with these agents.

29.33.9 Interleukins

These are a family of proteins called lymphokines produced by lymphocytes which have diverse effect on the function of blood cells in the immune system

1. Interleukin-II is widely studied protein which is previously known as T-cell growth factor because it regulates growth and differentiation of T-lymphocytes.

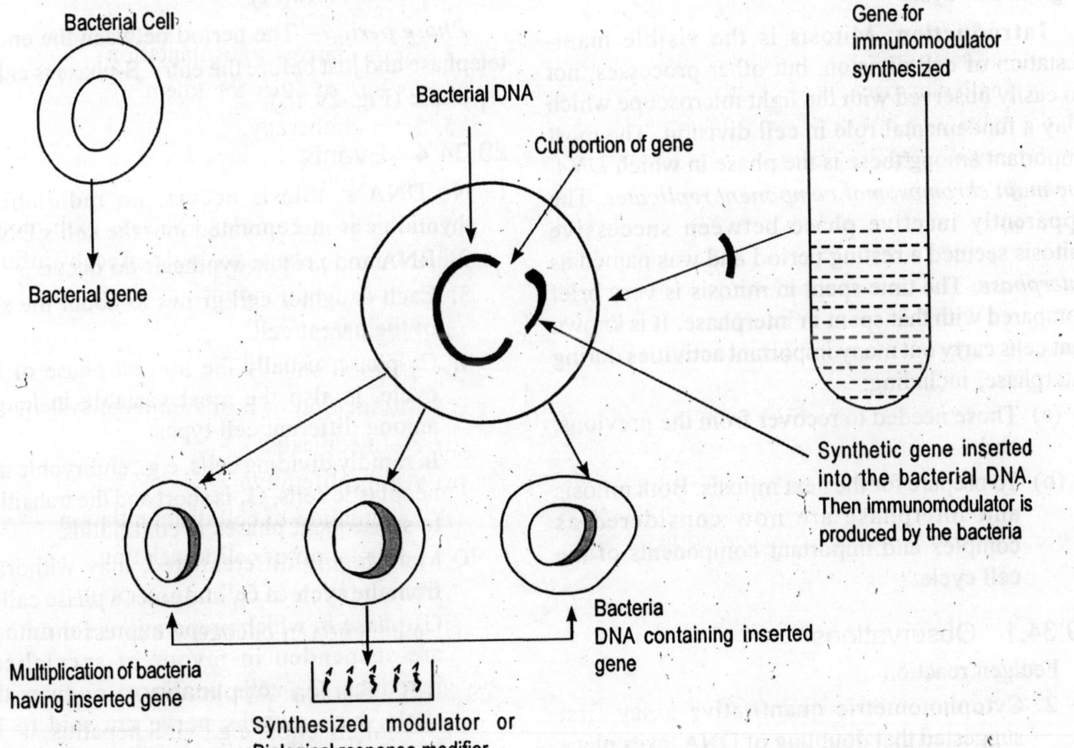

Fig. 29.14: Recombinant DNA genetic engineering and synthesis of biologic modifiers

2. It can stimulate growth of certain cells called lymphokine activated killer (LAK) cells.
3. Some killer cell types are affected by SL-II such as natural killer cells, T-lymphocytes, and tumour-infiltrating lymphocytes within the immune system.
4. They have the ability to recognise and destroy tumour cells without changing normal cells.
5. IL-II has shown promising result on the tumour cells which are resistant to other conventional treatment such as renal cell carcinoma and malignant melanoma.

Genetic engineering and recombinant DNA technology can induce bacteria to produce immune modulator or biologic response modifiers. The gene is obtained by removing it from human DNA and then purified and spliced into the DNA (Fig. 29.11).

29.34 Cell Cycle

Cell cycle, Phases of cell cycle, Factors regulating the cell cycle

Introduction: Mitosis is the visible manifestation of cell division, but other processes, not so easily observed with the light microscope which play a fundamental role in cell division. The most important among these is the phase in which *DNA, the main chromosomal component replicates.* The apparently inactive phase between successive mitosis seemed a resting period and was named as *interphase.* The time spent in mitosis is very brief compared with that spent in interphase. It is known that cells carry out many important activities during interphase, including:

(a) Those needed to recover from the previous, and
(b) To prepare for the next mitosis. Both mitosis and interphase are now considered as complex and important components of the cell cycle.

29.34.1 Observations

1. Feulgen reaction
2. Cytophotometric quantitative assay first suggested that doubling of DNA takes place during interphase.
3. Autoradiographic study showed by introducing labelled radioactive DNA precursors, e.g., H^3-thymidine into the cell and tracing them by means of biochemical and autoradiographic method.

29.34.2 Definition

The reproductive cycle of a cell is termed cell cycle, i.e., alternation between the mitosis and interphase, known as the cell cycle.

29.34.3 Phases of cell cycle

Autoradiographic studies have shown that doubling of DNA, i.e., replication or synthesis of DNA did not take place throughout the entire interphase. It occurs only in a restricted phase of the interphase such as:

I G_1-Phase (Gap I period)
II S-phase (Synthetic period)
III G_2-phase (Gap II period)

The gap phases (G, G_2) of interphase in which there is no DNA synthesis. These three comprise 90% or more of the total cell cycle.

I G_1-Phase (Gap I)

Phase period—The period between the end of telephase and just before the entry. S-phase is called G_1-phase (Fig. 29.15).

29.34.4 Events

1. No DNA synthesis occurs, no radiolablled thymidine is incorporated into the cell's DNA.
2. RNA and protein synthesis do occur.
3. Each daughter cell grows to about the size of the parent cell.
4. G_1-phase, usually the longest phase of the cycle, is also the most variable in length among different cell types.
5. In rapidly dividing cells, e.g., embryonic and neoplastic cells, G_1 is short and the transition to subsequent phases is continuous.
6. More highly differentiated, may withdraw from the cycle in G_1 and enter a phase called G_0 phase in which preparations for mitosis are suspended in favour of specialised functions. G_0 cells unable to re-enter the cycle, e.g., muscle, nerve are said to be terminally differentiated. Other cells in G_0, e.g., hepatocytes, fibroblasts can re-enter the cell cycle in response to growth factors encountered during or after an injury.

S-Phase (Synthesis phase)

1. It is intermediate phase between G_1 and G_2.
2. Synthesis of DNA and replication occur.
3. The centrioles often self-duplicate during this stage.

G-2 (Gap II phase)

1. The period from the end of S-phase until mitosis is called G_2 phase
2. It is the shortest phase.
3. The final preparations for cell division occur, these includes:
   (a) Repair of damaged DNA
   (b) Synthesis of tubulin for the spindle apparatus
   (c) ATP accumulation for the energy-expensive mitosis.
4. Very little synthesis occur during mitosis.

29.34.5 Proteins controlling cell-cycle

The transitions between the phases of the cell-cycle require for continuous cycling are:

(a) Permissive proteins (b) Inhibitory proteins

These are the following:

1. *Cyclin-dependent kinases (cdk)*: These are *enzymes* (e.g., cdc 2, cdk 2, cdk 4, cdk 6) that phosphorylate serine and threonine residues on other proteins, initiating or blocking activities crucial progression through the cycle.
2. *Cyclins*: *cdk*s are active only when bound to certain *cyclins* (e.g. A, B, C, D and E) to form *cdk-cyclin complexes.* Concentrations of cyclins rise and fall as the cell-cycle progresses.
3. *Growth factors*: It can induce cyclin and cdk synthesis and thus reentry from G_0 into active cycling.
4. *Cyclin-dependent kinase Inhibitors (CKI)*: These are protein families (e.g., Kip/Cip and INK4) whose members are capable of binding to cdk-cyclin complexes and inhibiting their activity.

29.34.6 Cell-cycle checkpoint

Certain checkpoints in the cell cycle are susceptible to control by cdk-cyclin checkpoints. During progress of the cycles enough cdk-cyclin complexes

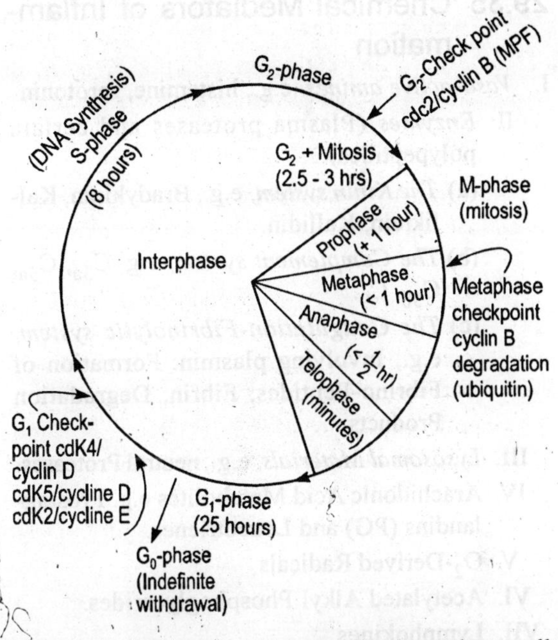

Fig. 29.15: Cell cycle of a generalised vertebrate cell showing different phases of interphase, check points and proteins associated with these

must be present to overcome inhibition. Its importances are:

(a) Progression may be halted or even may continue for extended period.
(b) Some aspects of cell-cycle control involve feedback from specific steps in the cycle.
(c) Due to these checkpoints, progression occur only when conditions indicate that the previous phase has been successfully completed.
(d) Involves signals from the cell's environment.

There are two main checkpoints:

I. G_1 checkpoint
II. G_2 checkpoint

Beside these there is metaphase checkpoint (M)

I. *G_1 checkpoint*: Controls the onset of DNA synthesis (S-phase).

II. *G_2 checkpoint*: Controls the onset of mitosis (M). Each of these checkpoints requires the presence of a different class of cyclin i.e., G_1 cyclin or mitotic cyclines. The different complexes phosphorylate different target proteins. In mammalian cells there appear to be various cdk proteins that function at different checkpoints.

29.35 Chemical Mediators of Inflammation

I. *Vaso-active amines*, e.g., histamine, serotonin

II. *Enzymes* (Plasma proteases and certain polypeptides)

(a) *The Kinin system*, e.g., Bradykinin, Kallikrein, Kallidin.

(b) *The Complement system*, e.g. C_{3a}, C_{5a}, C_{5b}, C_9.

(c) *The Coagulation-Fibrinolytic system*, e.g., Involving plasmin: Formation of Fibrino-Peptides, Fibrin, Degradation Products.

III. *Lysosomal Materials*, e.g., neutral Proteases.

IV. Arachidonic Acid Metabolites e.g. Prostaglandins (PG) and Leucotrienes.

V. O_2-Derived Radicals

VI. Acetylated Alkyl Phosphoglycerides.

VII. Lymphokines

VIII. Other Mediators of inflammatory Response.

29.36 Interleukins

Cells give origin to interleukins, e.g., Macrophages, Keratinocytes, T-cells, mast cells, fibroblasts, bone marrow stroma, monocytes, possibly B-cells.

29.36.1 Functions of interleukins produced by the cells

1. Proinflammatory endogenous pyrogen, activates fibroblasts, granulocytes, osteoclasts, makes T-cells responsive to signals (Macrophages, Keratocytes cells, etc.).
2. Proliferation of T, B and natural killer cells, Proliferation of early haematopoietic cells (multi-CSF) [T-cells].
3. Governs B-cell isotype switching to IgG and IgE (T-cell, mast cell).
4. Eosinophil differentiation and proliferation IgA production (T-cells, mast cell, possibly B-cell).
5. Proinflammatory, B-Cell differentiation, thymocyte growth.
6. B-cell differentiation and maturation (Bone marrow stroma).
7. Neutrophil chemotaxis and activation (keratinocytes, fibroblasts, monocytes).
8. Proliferation of T-cell, thymocytes, mast cells (T-cells).
9. Inhibition of cytokine synthesis in various cells, proliferation of mast cells, (T-cells, mast cells, possibly B-cells). (See page 891)

29.37 Cytokines

What are Cytokines?

These are soluble factors—collectively called *cytokines*—that are made by, or act on, elements of the immune system and modulate the activities of this system.

29.37.1 Chemical nature of cytokines

Cytokines are small protein molecules that are not antigen-specific, although their production is often antigen-driven.

29.37.2 Different types of cytokines

1. *Lymphokines*: Cytokines produced by lymphocytes.
2. *Interleukins*: Mediators that act between leucocytes. There are about 10 kinds of interleukins.
3. *Colony-stimulating factors*: Made by T-cells to stimulate lymphoid and haematopoietic organs.
4. *Tumour necrosis factors* (α *and* β).
5. *Transforming growth factor*-(β): Both (4 + 5) are associated with inflammation, tumour defense, cell growth and wound healing.

29.38 Bioreactors

Bioreactor system has great role in tissue engineering for their reproducible and controlled activity in particular environmental condition.

29.38.1 Salient feature

1. For cultivation of microorganisms
2. For cultivation of mammalian cell under controlled environment e.g. pH, temperature, O_2 supply etc.
3. It is useful for every type of tissue like bone, skin, cartilage, blood vessels etc.
4. Every tissue requires individual bioreactor.

29.38.2 Use

1. Cell proliferation

2. Generation of tissue from isolated cell or tissue *in vitro*.
3. Direct organ support devices.

29.38.3 Bioreactor and cell division

Cell devision of different cells

1. Stem cells of bone marrow division takes place continuously.
2. Blood cell e.g., neutrophils stop division after maturation.
3. But some cells e.g., lymphocyte divides even after maturation.

29.38.4 Control of cell division

Cell division is under the control different protein factors such as:

(a) growth factor

(b) cyclin cdc kinase

Growth factor acts first properly then cyclin activates the cdc kinase which then activates cell division. Most engineered tissue having limited growth because of lacking blood supply but the cell layer of 100–200 μm thick where O_2 supply is taken place by diffusion only.

(c) Mechanical stimulations have effect on tissue formation such as:

(1) mechanical compression

(2) hydrostatic pressure

(3) fluid flow.

29.38.5 Bioreactor processes

Bioreactor helps the different processes such as:

(a) batch

(b) fed batch

(c) continuous cultivation

Continuous perfusion enables cultivation in controlled environment and the bioreactor system is to be involved in the cultivation process encompassing the following:

(a) biopsy

(b) proliferation

(c) cell seeding

(d) tissue formation

(e) delivery to the proper site

The engineering of 3D tissue structure formation involves the following:

(a) cell maintenance

(b) proliferation

(c) tissue formation

(d) cell maintenance

These generally occur in culture process followed for monolayer culture of adherent cells as it has been seen in small well issued bioreactors such as shaken flasks, stirred vessels and super spinner may be preferable for cell proliferation.

In fixed bed and fluidized bed bioreactors, cells are immobilized in macroporous carriers which are arranged in a column either packed (fixed bed) or floating (fluidized bed). The column is perfused permanently from a reservoir of medium in a circulation loop. These kinds of bioreactors are very much useful for long-term cultivation for production of monoclonal antibodies, recombinant drugs etc.

29.38.6 Membrane bioreactors

In case of membrane bioreceptors including hollow fibre reactor, miniperm system or technansuse cells are cultivated in a compartment having several types of membanes for nutrient and O_2 supply and removal of toxic products.

29.38.7 Exception

1. They are used very limited way in case of generation of implantable tissue e.g. skin, heart valve, blood vessel, cartilage etc. which require specific bioreactor especially mechanical type stimulation.
2. Engineered tissue should be transferred from the culture to patient after cultivation.
3. For these reasons several tissue specific culture and bioreactors are to be prepared.

29.38.8 Cell surface markers

Cell surface markers are proteins present on the external surface of the cell membrane. The most significant cell markers which are responsible for transplant acceptance or rejection called **major histo-compatibility complex** (MHC).

These are expressed on cell surface and display as **self-antigen** and **non-self antigen**. The MHC region is divided into three groups such as:

1. MHC-Class I: binding proteins and antigen processing molecules

2. MHC-Class II: Modulation of antigen loading, expansion on antigen processing cells.
3. MHC-Class III: Encoded for immune component e.g. complement component, cytokinase.

29.39 Biomaterials

29.39.1 Definition

Biomaterials are materials which are used clinically for the treatment of diseases, injury, joining the tissue, implantation, transplantation, grafing transfusion, etc.

29.39.2 Clinical Applications

1. Ankle implant
2. Artificial leg
3. Total hip-joint replacement
4. Knee-joint replacement
5. Shoulder-joint replacement
6. Dental implant
7. Artificial ear
8. Hydrocephalous shunt
9. Testicular prosthesis
10. Temporary Tendon
11. Cochlear implant
12. Nasal implant
13. Shoulder prosthesis
14. Breast prosthesis
15. Artificial heart
16. Artificial heart valves
17. Blood substitute
18. Artificial chin
19. Ocular lens
20. Contact lens
21. Cartilage replacement
22. Artificial liver
23. Spinal fixation.

29.39.3 Structure and properties of Biomaterial

Properties of biomaterials depend upon the nature of bonding, such as:

1. **Metallic bond**: These bonds allow high electrical and heat conductivity because of free valance electrons which act as the medium.
2. **Ionic bond**: Ionic bond containing materials are insulators of heat and electricity because their electrons are tightly held by the ions. Ionic materials have a limited number of slip planes due to repulsion of like-charged ions making these materials brittle.
3. **Covalent bond**: Covalent bonds share valence electrons with neighbouring atoms. These bonds containing compound show poor electrical and thermal conductivity.
4. **Other bonds (secondary bonds)**: They play a great role in determining the properties. According to the strength they are: Ionic > metallic > covalent > hydrogen > Van derwalls.

A. **Crystal structure:** According to the structure, materials are divided into two:

(a) Crystalline
(b) Non crystalline or amorphous.

B. **Non crystalline structure**: In this type of structure, in solids the atoms, ions or molecules comprising the individual crystal are arranged in three dimensional lattices following a periodic patterns with minor defects disturbing the general order of pattern.

Surface Properties: The efficacy of a biomaterial depends upon two factors i.e. bulk and surface properties. A biomaterial must display a surface behaviour in addition to the bulk properties for reduction of interfacial trouble encountered with host tissue and fluids.

These are:

I. **Surface energy**: It is a measure of the extent at which the bonds are not satisfied on the surface of biomaterial.

II. **Contact angle:** This can be observed when one drop of liquid is placed on the surface of a solid or another liquid, two things may be observed alignment such as:

(a) The liquid may be present as a droplet or
(b) It spread out throughout the whole surface.

These two events depends upon the interfacial free energies of the two substances.

III. **Critical surface tension**: The critical surface tension of a biomaterial is measured by the different values of contact angle formed by the values of solid or liquid.

IV. **Electrokinetic Phenomena**: When a biomaterial with a surface having charge placed in solution with ions, a diffused layer of oppositly charged ions is formed close to the surface.

• **Mechanical Properties:** Deformation of metals initially result in streching of the bonds that hold atoms together and this produces a response between stress and strains, i.e., it follows the Hook's law which states that the force is proportional to the displacement. The proportionality constant is known as Elastic modulus or Young's modulus.

29.39.4 Different biomaterials and their applications in the body systems

1. Muscular system—Sutures, muscle stimulator.
2. Vision—Viscoelastic lotion, contact lenses, Intraocular lenses (Table 29.10).
3. Skeletal system—Bone plate, serews, intramedullary rod, total joint replacement. (Table 29.9).
4. Cardio-vascular system—Artificial heart, Heart valves, Blood vessel graft.
5. Respiratory system—Heart-lung machine, oxygenators.
6. Urinary system—Catheter, Haemodialyser machine, stent.
7. Nervous system—Hydrocephalus drain, Nerve stimulator, Cardiac pacemaker.

Table 29.9: Biomaterials used in total joint replacement

Type	Biomaterials
A. **Metal**	Stainless steel 3161, cobalt-based alloys, cast cobalt-chromium—molybdenum, Wrought cobalt—chromium-W-Nickel Titanium based materials C.P.—Ti T1—6Al—4V Ti—5AL–2.5Fe Ti—Al—Mb
B. **Ceramic**	Bioinert carbon, Alumina Zirconia, Bioactive calcium phosphates, Bioglasses
C. **Polymers**	PMMA, UHMWPE/HDPE Polysulfolane, PTFE
D. **Composite**	Polymer based Polysufolane—Carbon Polycarbonate—Carbon Polysulfone—Kevlar Polycarbonate—Kevlar

Table 29.10: Biomaterials used in ophthalmology

Ophthalmic	Purpose
1. Contact lens	Correction of defect in vision
2. Intraocular lens	Replacing lens having cataract
3. Epikeratoplasty	For changing corneal curvature
4. Scleral buckling materials	For detached retina
5. Viscous Polymer materials	Insertion in the intraocular lenses, removal of cataract for maintaining position of the retina

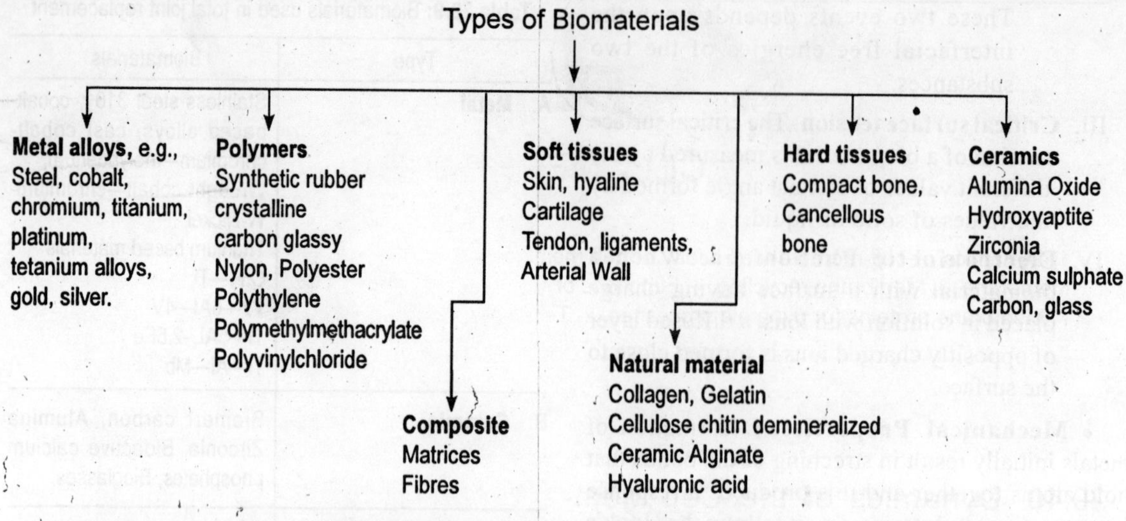

Fig. 29.15A: Flow chart showing types of biomaterials

8. Endocrine system—Insulin delivery system, microencapsulated pancreatic islet cells.
9. Reproductive system—Birth control devices, cosmetic replacement, Augmentation mammoplasty.
10. Integumentary system—Burn dressing, suture, artificial skin.

29.39.5 Collagen materials used in dentistry

1. Collagen sponges
2. Collagen membranes
3. Collagen gel allogenic bone
4. Collagen tricalcium phosphate
5. Collagen coated root transplant
6. Collagen solution
7. Collagen graft
8. Collagen allogenic bone
9. Collagen film.

PART III

CELL MEMBRANE

1. Dynamics of cell membrane (Fluidity of cell membrane) For details see chapter 9), 2. Study of cell membrane, 3. Membrane recycling, 4. Functions of membrane proteins, 5. Ion channels, 6. Gap junction, 7. Membrane proteins for dynamic processes, 8. Techniques for study of cell membrane.

29.40 Dynamics of the Cell Membrane (Fluidity of the cell membrane)

The technique of Electron spin resonance spectroscopy (ESRS) has established the dynamics of lipid molecules, e.g., fluidity of the cell membrane. The scientists labelled the tail end of the fatty acid of tepid layer of the cell membrane with a nitroxide group having an unpaired electron. When nitroxide group is exposed to external magnetic field it emits the energy. Maintaining control of phospholipid compound using ESRS, and the spectrum is observed. It is observed that the spectrum of the membrane is *intermediate* between the completely *mobile and immobile molecules.*

The intermediate state of the membrane is known as *liquid crystal state.* The lipid molecules can shift laterally within the bilayer maintaining this orientation in the same state keeping the hydrophilic group (solvent loving) towards the outer side and the hydrophobic (solvent hating) group towards inner side of the membrane.

It usually generates heat during transformation from one state to another, such as solid to liquid or liquid to gas.

Hence *Differential colorimetry* has demonstrated the change of physical state in the cell membrane like solid to liquid state. It has also been domonstrated that lipid layer changes to solid or gel state at low temperature and it melts to fluid or liquid crystalline state at higher temperature.

This transitional temperature at which the change of state occurs is below the temperature at which the most physiological functions occur.

29.40.1 Factors controlling the membrane fluidity

1. *Increase in fluidity*—Due to increase in unsaturated fatty acid and decrease in fatty acid chain length.
2. *Increase in fluidity* due to decrease in cholesterol content.
3. *Increase in fluidity* is inversely proportional to the transition temperature.
4. Alteration of physical state the membrane has been observed by the *fluorescent-labelled antibodies* regarding mobility of the protein by rapid lateral migration in fluid lipid layer.
5. Random distribution of membrane, proteins has been observed.
6. The behaviour of membrane proteins depends on the membrane fluidity.

29.40.2 Technique for study of cell membrane

1. *Transmission electron microscopy*: This method has revealed the structure of cell membrane as well as there are membranes in the cytoplasm dividing it into several intracellular compartments. It also revealed the structure of plasma membrane, i.e., cell membrane and membrane of intracellular compartments are similar.
2. *Freeze-fracture electron microscopy*: In this method the frozen cell membranes are fractured at the temperature of liquid nitrogen causing splitting of the membrane into half through the bilayer. This method is used to study integral proteins of the membrane embedded in the bilayer.

3. *X-ray diffraction study*: It is the only quantitative technique to get complete 3-dimensional structure of the material of molecular and atomic levels. It is used to study the arrangement of lipid molecules in the bilayer and its fatty acid chains.
4. *Nuclear magnetic resonance or electron spin resonance*: This method is employed to study motion of lipid molecules in the bimolecular layer of phospholipid and asymmetry in the layer.
5. *Sonification.*
6. *Extraction method*: Extraction with chaotic agents and detergent have revealed the hydrophobicit property of membrane proteins in purified state.
7. Electron spin resonance spectroscopy (ESR).
8. Polyacrylamide—Gel electrophoresis
9. Optical Rotatory Dispersion.
10. Circular Dichroism Measurements
11. Low Angle X-ray diffraction
12. Intra Red spectroscopy
13. Spectropolarimetry
14. X-ray crystallography

29.41 Study of Cell Membrane Proteins (Freeze-etching Electron micrography)

29.41.1 Freeze-fractures electron microscopy

This technique has been employed to know *whether membrane proteins are present in the interior of the cell membrane.* At first the cells or membrane are fragmented and then rapidly frozen at liquid-nitrogen temperatures. The frozen membrane is fractured by the impact of microsome knife. Cleavage forms along a plane in the middle of the bilayer and causes exposure of the regions within the bilayers and the regions are then shadowed by carbon and platinum which makes a replica of the *interior of the bilayer.* The *external surface* of the cell membrane can be studied by combining freeze-fracture technique and deep-etching technique. Initially, the interior of the frozen membrane is exposed by fracturing. The ice covers one of the adjacent membrane surfaces which is then sublimed away. This process is known as *deep etching.* The combined technique called *freeze-etching electron micrography* reveals a view of the interior of a membrane and of both its surfaces. The peculiarity of this technique is that there is no fixative and dehydrating agent.

Freeze-etching technique has revealed the following:

1. Presence of integral proteins in many biological membranes.
2. RBC membrane contains a high density of globular particles having diameter 75Å approximately in its interior.
3. The inside of the endoplasmic reticulum is rich in globular particles.
4. In case of myelin membrane the fracture faces of large areas are smooth which is ideal for insulation.
5. RBC the peripheral proteins has been located and several proteins are rich in carbohydrate. They have been located in the cytoplasm face. Band 6 is glyceraldehyde-3-phosphate dehydrogenase enzymes, bond 5 is actin (cell motility) Band 1 and 2 cells spection cells filamentous network and with other proteins stabilises and regulate the shape of cell membrane while passing through smaller capillaries. The RBC membrane cell tolerate great mechanical stress when pumped by the heart glyceraldehyde-3-phosphate dehydrogenase.

29.41.2 Functions of membrane proteins

There are seven types of proteins in addition to structural proteins such as:

1. **Pumps**: Active transport of ions across the membrane.
2. **Channels**: Passive channels for ions that can be opened or closed by changes in the conformation of the protein.
3. **Receptors**: It helps binding neuro-transmitters and hormones.
4. **Enzymes**: It catalyzes reactions at the surface of the membrane.
5. **Gates**
6. **Energy transducer**
7. **Pores.**

29.41.3 Cell membrane coat

Cell membrane is not naked but it is coated outside by a protective layer. But it is not *cell wall* made of cellulose present in plant. The external surface of cell membrane is surrounded by a protective layer of polysaccharide called *cell coat,* also known as *glycocalyx,* containing *glycoprotein* and *oligosaccharide.* Beyond the cell coat there is a separate layer made of carbohydrates secreted by the cell. Cell coat can be observed under light microscope by staining with PAS (Periodic Acid Schiff) or Alcian blue. Having thickness of 10-20 nm, resistant to mechanical and chemical agents but washed away by urea solution and enzyme action and renewed by cell production. It has the following functions.

Depending on the attachment of the glycocalyx on cell surface, this layer is divided into two kinds, i.e., attached glycocalyx and unattached glycocalyx. The first one is attached to the cell membrane in such a way that it cannot be separated from the plasma membranes under EM it shows fibrillar structure containing glycoproteins and glycolipids. The second one, i.e., unattached layer is external and can be easily removed from the cell surface.

29.42 Ion Channels

These are membrane proteins. These are very important for the cell activity. *Ion channels* membrane proteins give passage of different ions such as Na^+, K^+ out of the cell or inwards of the cell. The membrane contains layers of lipids which are impermeable to ions, so ionic current, are carried through the membrane by specialised protein molecules called *ion channels* which span the membrane on sensing the stimulus—electrical or chemical. Ion channels change their shape forming pore in the membrane through which it can flow easily. With the help of Patch clamp technique, Neher and Sakmanm succeeded observing the current through single ion channels. They may be *leak channel,* i.e., continuously open to the ion and others are *gated channel,* i.e., may open either due to change in the membrane voltage called *Voltage-gated,* or may open due to binding with chemicals or ligands called *ligand-gated channel.* These proteins are selective and specific which allow only one ion such as Na^+ but not the K^+ ion. Some of the channels allow more than one ion. If the channel is opened, ions pass along their concentration or electrical gradient (Fig. 29.16).

29.42.1 Different ion channels

1. Na^+ Channels

These are of different types of proteins—one type is composed of large molecules with several membrane spanning segments. There are four units of segments which encircle the central pore.

These are gated and open when there is low membrane voltage. It facilitates the easy flow of Na^+ ions. These channels can be blocked by tetrodotoxin (TTX) and saxitoxin (STX). The channels are opened and closed automatically. These voltage gated Na^+ channels are responsible for origin of action potential (Fig. 29.16).

2. K^+ Channles

K^+ channels are of many types such as:

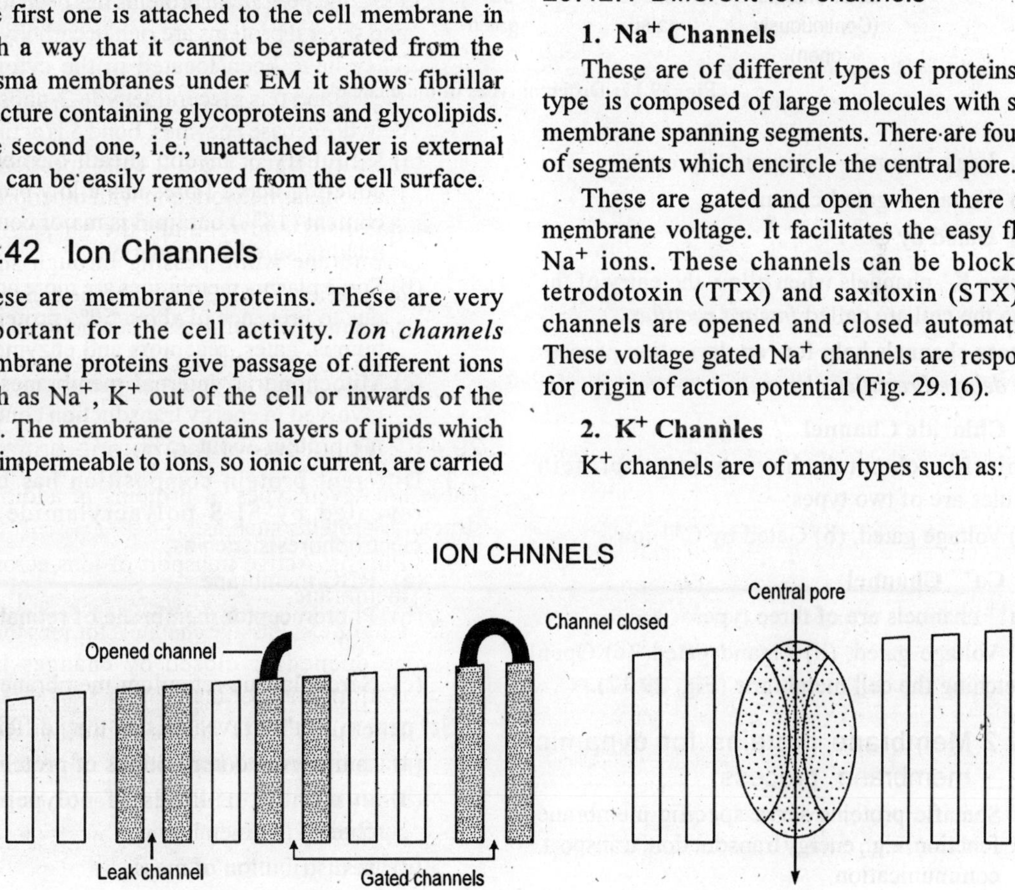

Fig. 29.16: Schematic representation of ion channels and Na^+ channel

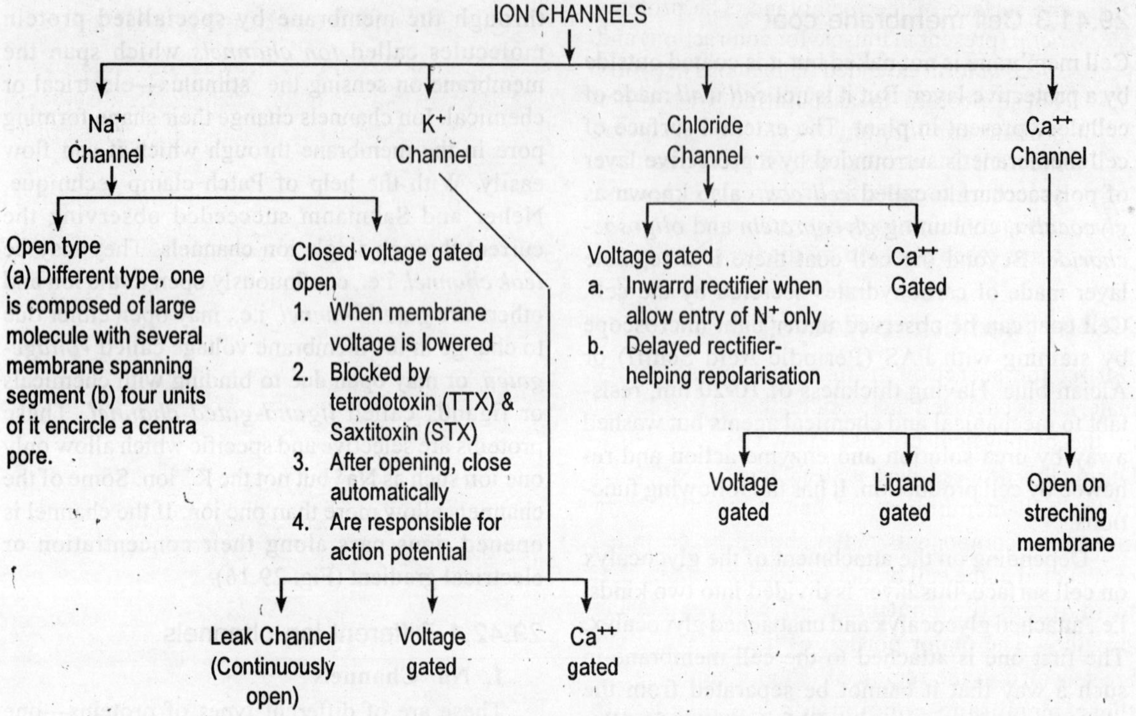

Fig. 29.17: Different types of ion channels

(a) Leak channel—continuously open

(b) Voltage—gated channel

(c) Gated by C^{++}

These K^+ channels when allow the entry of the K^+ into the cell are called *inward rectifier.*

Those channels help for repolarisation process called *delayed rectifier channels.*

3. Chloride Channel

Chloride channels having large protein molecules are of two types:

(a) Voltage gated, (b) Gated by C^{++} ion.

4. Ca^{++} Channel

Ca^{++} channels are of three types:

(a) Voltage gated, (b) Ligand gated, (c) Open on stretching the cell membrane (Fig. 29.17).

29.42.2 Membrane proteins for dynamic membrane process

1. Specific proteins serve specific membrane function, e.g., energy transduction, transport, communication.
2. Different membranes differ in their protein content, Such as
   (a) Medullary or myelin sheath membrane covering nerve fibres has a low protein content (18%) but lipid is major content making suitable insulation.
   (b) Some plasma membranes are more active due to presence of about 50% protein as pumps, gates, receptors and enzymes.
   (c) Mitochondrial internal membranes are involved in energy transduction containing protein about 75%.
3. Different protein composition has been revealed by SDS polyacrylamide gel electrophoresis such as
   (a) RBC membrane
   (b) Photoreceptor membrane of retinal rod cell.
   (c) Sarcoplasmic reticulum membrane.

Gel pattern further revealed that they differ

(a) in different compositions of proteins

(b) in number of kinds of polypeptide chains.

(c) in distribution of mass

Membranes performing different functions due to having different proteins.

4. RBC membrane has peripheral proteins in the surface of the membrane. The proteins *actin* (present in muscle for contraction) and *spectrin* form extended filamentous network. These protein in association with other proteins *stabilize* and *regulate* the shape of the RBC membrane which changes according to the size of the capillaries and probably helps to resist ground mechanical stress when pumped by the heart.

29.43 Membrane Recycling

During the secretion from a cell the secretory materials is enclosed in a vesicle, then it is secreted by the biophysical process known as *exocytosis.* During this process there is stretching or expansion of the cell membrane and hence that the extra membrane components either should be degraded or recycled in the cell by the process of *endocytosis* to maintain the continuity of cell membrane structure. The major portion of these membranes are recycled to the region of Golgi complex when these membrane components are fused with associated membrane of Golgi complex.

29.43.1 Evidences

1. By analysis of the binding of IgG (immunoglobulin) to the cell membrane using radioisotope or fluorescent as a labelling agent.
2. By comparison of internalisation of epidermal growth factor (EGF) and transferrin. Both the substances were inserted into the cytoplasm through endocytosis forming vesicle. The EG is transferred to the lysosome through some degradation. Transferrin is transferred to the Golgi complex → both transferrin and transferrin receptor are transferred → to the cell membrane through the extension (dumbbell shaped) of the Golgi complex.

29.43.2 Function of membrane recycling

i. Helpful to deliver materials inside the cell.
ii. Help in translocating material from one compartment to another.
iii. Translocate receptor from one extra cellular compartment to a cell surface or another cellular compartment.

29.44 Carriers of Membrane (Carrier membrane proteins)

This group of membrane proteins also known as *transporters* because they transport substances across the cell membrane. The transport of materials by carriers may be:

A. *Active transport*—when the transport requires energy and the carrier is called *pump.*
B. *Passive transport*—when the process does not require energy.

29.44.1 Mode of action

It works in three ways:

(a) *Uniport*—when it transports one substance in one direction.
(b) *Symport*—when it transports two substances in one direction.
(c) *Antiport* (Counter-transport)—when it is responsible for moving few materials in opposite direction.

These transport proteins bind with substance to be carried and it is specific like membrane receptors.

Examples of carrier proteins

Na^+K^+ ATPase: This enzyme removes Na^+ from interior of the cell to outside of the membrane against concentration gradient and electrical gradient with utilization of energy and simultaneously allow K^+ to enter the cell. It has two units i.e., α and β units.

29.44.2 Mechanism of Action

$3Na^+$ binds to the intracellular portion of α unit and this α -unit binds to ATP → ATP is then hydrolysed to ADP → α-unit is phosphorylated → phosphorylated α -unit undergoes conformational changes → removes Na^+ from the cell → in the next stage $2K^+$ binds to the extracellular part of the α-unit → dephos-phorylated → comes back to initial position → bringing K^+ inside the cell.

1. This is antiport process activated to Na^+ inside the cell and K^+ outside the cell.
2. This pump inhibited by ovabin and cardiac glycoside digoxin.
3. This pump is responsible for maintenance of *membrane volume and action potential.*

29.45 Gap Junction

Gap junctions are found between the cells in tissues which serve as *direct connection* between the cytoplasms of adjacent cells. They provide open channel through the cell membrane allowing ions and other small biomolecules less than one thousand dalton which diffuse freely from one cell to the adjacent cell preventing the passage of proteins and nucleic acids (Fig. 29.18).

Tissue having gap junctions: In electrically excitable tissue like heart, skeletal muscle, smooth muscle.

29.45.1 Structure

The adjacent cell membranes are very close and the cells are connected by hollow cylinders known as *connexions* composed of transmembrane proteins *connexins,* ions, sugars and other small biomolecules pass through these water filled channels from one cell to the next cell.

29.45.2 Functions

1. Gap junctions between the embryonic cells are essential for distribution of nutrients before the development of circulation.
2. It exists in electrically excitable tissues where passage of ions from cell to cell helps synchronize electrical activity and contraction.
3. It couples both metabolic activities and electrical response of the cell. And they connect cells in the tissues e.g., epithelial cells, cardiac cell, endothelial cells, smooth muscle (single unit). They communicate with each other.
4. In electrically excitable cell in tissue (e.g., cardiac, smooth muscle) direct passage of ions through this junction and also allow ions and couples and synthesizes the connection of adjacent cell.
5. It also allow the passage of some intracellular signalling molecules, e.g., C-AMP and Ca^{++} between the neighbouring cells and potentially coordinating the response of the cells in the tissue.
6. Skeletal muscle, it is electrically excitable tissue but the individual muscle fibre does not have gap junction. They are stimulated through myoneural junction in the motor unit
7. Cardiac cell membrane is connected through dark staining junction called *intercalated disc.* It contains anchoring *desmosome and gap junction.* The desmosomes prevent separation of cardiac cell during contraction. Whereas gap functions allow the passage of ions and free cytoplasmic flow carrying electrical stimulus directly from cell to cell transmission of depolarizing current across the whole heart. Because the cardiacs muscles are small, branched and connected with each other by intercalated disc and electrically coupled by the gap function and whole myocardium acts as a single unit in coordinated way called syncytial function, gap junction contains 6 cylindrical channel of which two are joined to make a pipe like channel revealed by EM, X-ray diffraction studies. Through this channel cytoplasm, ions, depolarising current pass.

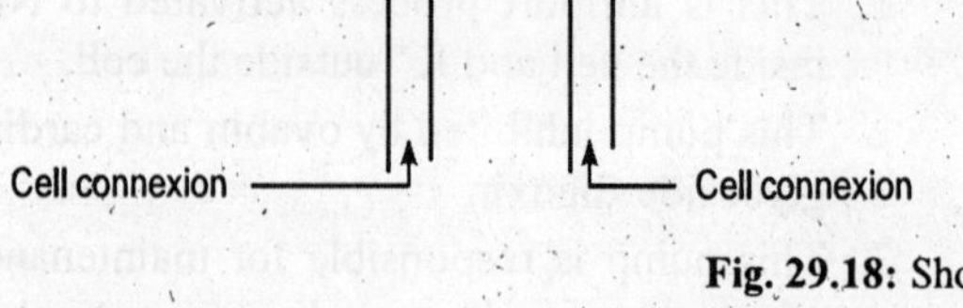

Fig. 29.18: Showing gap junction

PART IV

TISSUE DYNAMICS

1. Definition; classification; origin of tissue, 2. Connective tissue, 3. Comparison of different tissue, 4. Component of connective tissue, 5. Tissue, cell spaces, 6. Extracellular Matrix, components, 7. Fibrous proteins, adhesive fibrous proteins, 8. Organisation of tissue into different functional syntem and subunits, 9. Functions of proteins, 10. Collagen-rich tissue, 11. Tissue dynamics, 12. Collagen, 13. Micro-environment, 14. Tissue repair, healing of wounds.

29.46 TISSUE

Protoplasm in a cell, differs in character depending upon the function, the cell is earmarked to perform. The four types of tissues namely (i) Epithelial, (ii) Connective, (iii) Muscle, (iv) Nervous can be further subdivided according to their location and function in the body. It can be considered as collection of several cells constitutes a tissue.

Table 29.11: Epithelial Tissue

Location	Function
i. Surface of skin ii. Surface layer lining body cavities	Protection from mechanical and bacterial invasion.
iii. Lining glands.	Secretion and excretion.

Table 29.12: Muscle Tissue

Location	Function
Skeletal, voluntary or straited	
1. Muscles in arm, legs	Movements of limbs
2. Eyeball muscles	Eyeballs movements
3. Upper third of oesophagus, tongue, & mouth.	Swallowing
Visceral, (involuntary or smooth).	
1. Walls of tubular viscera	Movements as in intestines
2. Walls of blood vessels	Control the size of vessels
3. Ducts of glands	Movements of secretions.

Cardiac muscle is faintly striated, but at the same time is an involuntary type. Owing to the branching arrangement, when viewed under a microscope, a lattice-work-like arrangement is observed due to branching of muscle fibres joining with each other forming a mass called **syncytium**.

Table 29.13: Connective Tissue

	Location	Function
1. Aerolar	Between other tissues Between two organs and superficial fascia	Supports and cements parts together.
2. Adipose (fat)	Subcutaneously. Padding at various sites like the shoulder, abdominal wall.	Affords protection, insulation, support and/or source of fat for emergency.
3. White fibrous	Tendons, ligaments scars. Forms capsules over various organs.	Furnishes flexible but strong connection.
4. Dense fibrous	Walls of blood vessels Ligaments of spine	Furnishes elasticity
5. Lymphoid which is also called as "adenoid" tissue.	Tonsils, Adenoids, Spleen	Localisation and prevention of infection.
6. Osseous tissue	Skeleton	Forms strong framework for body.
7. Cartilage—3 types		Forms strong framework for body.
1. Hyaline (Articular surface of bones)	Nasal septum Larynx	Furnishes firm but flexible support.
2. Fibrous	Discs between vertebrae	Flexibility Binding
3. Elastic	External ear.	Elasticity

29.46.1 Origin of different tissue

Fertilisation

[Union of ovum (female germ cell) with spermatozoa (male germ cell)]

⇓

Cleavage

(Development of *zygote* to form *blastula,* i.e., group of undifferentiated cells)

⇓

Gastrulation

(Differentiation and movement of cells to from specialised *cell layers*)

⇓

Differentiation

(Development of specialised cells into tissues, organs and growth of the embryo)

⇓

Growth

(Maintenance and regeneration of some cells into three layers)

⇓

Ectoderm gives origin to	Mesoderm gives origin to	Endoderm gives origin to
↓		↓
Skin, nervous system	1. Connective tissue 2. Muscular tissue, etc.	Endocrine organs, etc.

Fig. 29.19: Origin of different tissue

29.46.2 Connective Tissue

29.46.2.1 Definition

The tissue which forms framework or support or binds the different parts together and consists of cells which are held together by a large quantity of intercellular substance called "matrix" or "ground substance" formed by the cells is called connective tissue (Table 29.13).

29.46.2.2 Characteristics

1. Developed from mesoderm
2. Presence of intercellular matrix in large amount.
3. The matrix contains many types of fibrous proteins.
4. Gives tensile strength, elasticity, etc.

29.46.2.3 Types of connective tissue

1. Adipose tissue
2. Areolar tissue
3. Yellow elastic tissue
4. White fibrous tissue
5. Haemopoietic tissue
6. Reticular tissue
7. Jelly-like tissue
8. Cartilage
9. Bones
10. Reticulo-endothelial tissue.

29.46.3 Cartilage

It is specialized connective tissue in which cells are embedded in slightly hardened matrix due to deposition of calcium. The rounded cells called "chondrocytes" present in groups within a space called "lacunae". The matrix consists of gel like

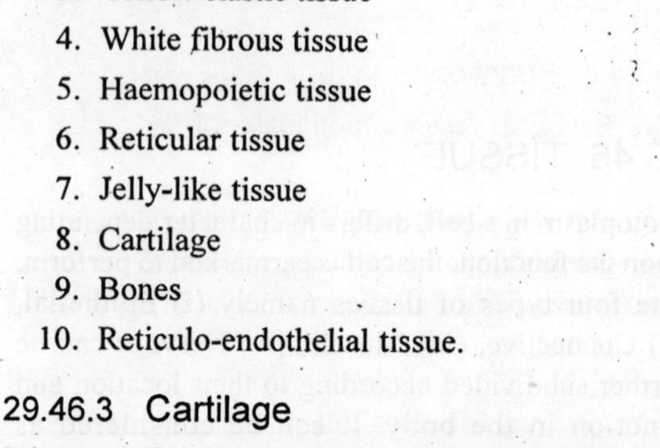

Fig. 29.20: Cartilage tissue

"ground substance" with yellow or white fibres. (Fig. 29.20). The tissue does not have blood supply.

29.46.3.1 Function

1. Mechanical function, e.g., support.
2. Acts as mechanical buffer.
3. It facilitates movement of joints due to its presence in between the two bones.

4. Acts as shock absorber in the joints.
5. Acts as strong, flexible substance between bones.
6. Acts as media for bone formation.

29.46.4 Adipose tissue

When plenty of fat cells are embedded in the loose fibrous connective tissue it is called adipose tissue. The fat cells contain fat which pushes the nucleus to the periphery and the cells look like "signet ring" (Fig. 29.21).

29.46.4.1 Function

1. Serves as reserve food material.
2. Conservation of heat.
3. Act as packing materials of the organs.
4. Supports the viscera.
5. Mechanical function as cushion in the soles of feet, palms of the hand.

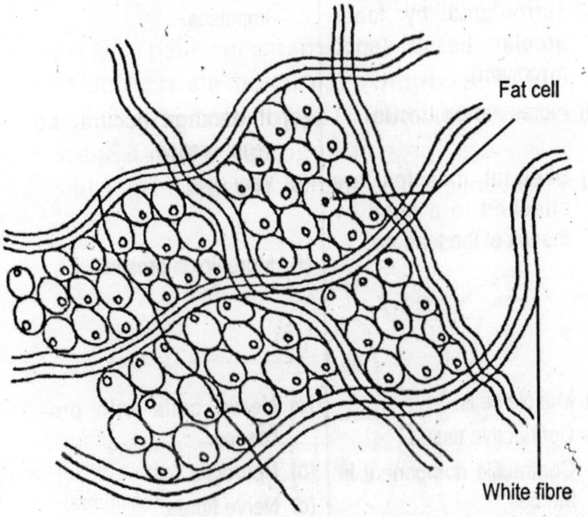

Fig. 29.21: Adipose tissue

29.46.5 Fibrous connective tissue

Loose fibrous connective tissue (Areolar tissue): It contains cells like fibroblasts, macrophages (wandering cells), fat cells, reticular cells and in the matrix there are two types of fibres: (1) white fibres—thin, non-branched from bundles, wavy bundles. (2) yellow elastic fibres—thin, branched, do not form bundles, elastic, non-wavy (Fig. 29.22).

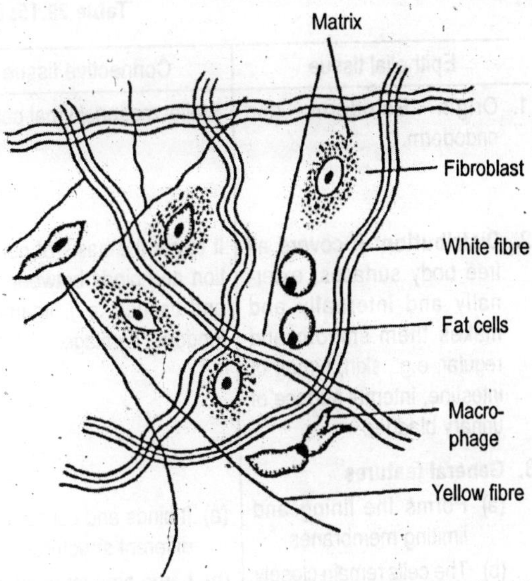

Fig. 29.22: Loose fibrous connective tissue (Areolar tissue)

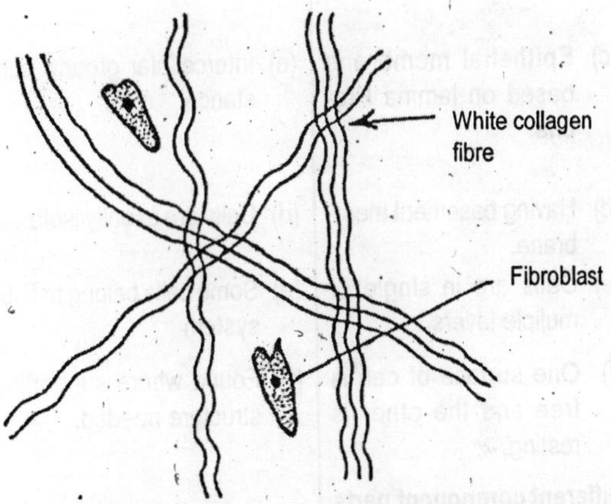

Fig. 29.23: Dense fibrous connective tissue

B. Dense fibrous connective tissue

It contains fibroblast cell and white collagen fibre (Fig. 29.23).

Table 29.14: Comparison of epithelial and connective tissue

Epithelial tissue	Connective tissue
1. It covers the outer or inner surface of the structure.	1. It supports or binds the different parts.
2. Intercellular substance minimum.	2. Maximum called matrix.
3. Cellular material maximum.	3. Minimum.
4. Fibres (white + elastic fibres) are absent.	4. Present.
5. One type of cell is present in the tissue.	5. One or more type of cells are present.

Table 29.15: Differences between the tissues

Epithelial tissue	Connective tissue	Muscular tissue	Nervous tissue
1. Origin Ectoderm and endoderm.	Stellet mesenchymal cells.	Mesoderm (except arrectors pilorum, iris muscle, myoepithelial cells of salivary, sweat and lacrymal gland).	Ectoderm (except microglia).
2. **Distribution:** It covers all free body surfaces, externally and internally and makes them smooth and regular, e.g., skin, lumen of intestine, internal surface of urinary bladder.	It forms the basis of connection and binds between cells and tissues. e.g. ligaments, tendons, cartilage.	It is for locomotion and movements of different parts of the body, e.g., skeletal muscle, cardiac muscle, muscle of intestine, uterus.	Brain, spinal cord, ganglia.
3. **General features**			
(a) Forms the lining and limiting membranes.	(a) It binds and connects the different structures.	(a) Muscle fibres in bundle and run parallel.	(a) Grey matter having nerve cells.
(b) The cells remain closely packed.	(b) Large amount of intercellular fibrous materials.	(b) Coarse connective tissue surrounds the whole muscle.	(b) White matter having nerve fibres.
(c) Epithelial membrane based on lamina Propria.	(c) Intercellular ground substance.	(c) Individual fibres is surrounded by loose areolar tissue (endo mycmium).	(c) Initiates and conduct impulses.
(d) Having basement membrane.	(d) Cells are usually isolated.	(d) Fibres are contractile.	(d) If damage occurs, no regeneration.
(e) Cells are in single or muliple layers.	(e) Some cells belong to R.E. system.	(e) Skeletal muscles are attached to a bone by means of the tendons.	(e) very soft.
(f) One surface of cell is free and the other is resting.	(f) Found where supporting structure needed.		(f) More lipids present.
4. **Different component parts**			
(a) Cells. (b) Intercellular substance. (c) Basement membranes.	(a) Connective tissue cells (b) Intercellular substance, ground substance, fibrous element, matrix.	(a) Muscle cells. (b) Connective tissue. (c) Contractile component in the cell.	(a) Nerve cells with processes. (b) Neuroglia (c) Nerve fibres.
5. **Functions**			
(a) Protection (b) Absorption (c) Diffusion (d) Secretion (e) Lubrication	(a) Mechanical (b) Defensive (c) Reparative (d) Regenerative (e) Nutritive (f) Metabolic (Adipose)	(a) Contraction. (b) Locomotion (c) Skill movement (d) Regulation of posture and equilibrium (e) Body heat.	(a) Controls and regulates and adjusts the body with external environment through sensory perception of vision, audition, taste, smell, skin, joints muscles, etc.

29.47 Mechanism of tissue fluid formation

Tissue fluid is formed from the plasma by the process of diffusion and filtration. It is produced from two sources:

1. Blood capillaries, 2. Tissue activities

1. **Blood capillaries**

Formation and amount of tissue fluid depends upon

i. Capillary permeability.

ii. Difference of pressure between the capillaries and the tissue fluid.

iii. Difference of colloidal osmotic pressure of blood and tissue fluid. But the arterial end of capillaries the average blood pressure is about 32 mm of Hg and the venous end, 10 mm of Hg. The colloidal osmotic pressure at both ends is same, i.e., 25 mm Hg. At the arterial end, the net filtration pressure is 7 mm of Hg towards the tissue fluid (Interstitial fluid). At the venous end, due to fall in blood or hydrostatic pressure, the filtration pressure is 15 mm of Hg to the opposite side, i.e., from tissue fluid to the capillary.

2. **Tissue activities**: The amount of tissue fluid formation depends upon the metabolic activity of the cells and breakdown metabolic products, water, etc. to the tissue spaces. Due to osmotic activity of the metabolites it draws fluid from blood.

Table 29.16: Comparison of Different Structures of Tissue

Epithelial tissue	Connective tissue
1. It covers the outer or inner surface of the structure.	1. It supports or binds the different parts.
2. Intercellular substance minimum.	2. Maximum intercellular materials, called matrix.
3. Cellular material maximum.	3. Minimum.
4. Fibres (white and elastic fibres) are absent	4. Present.
5. One type of cell is present.	5. One or more type of cells are present.

Areolar tissue	Adipose tissue
1. Abundant ground substance	1. Less.
2. Fibre: White fibres, yellow elastic fibres, a few reticular fibres.	2. White fibres.
3. Cells—Fibroblast Histiocytes, Fat cells, Plasma cells, Mast cells, Pigment cells.	3. Cells: Fat cells containing neutral fat. There is large fat globules which is surrounded by a thin layer of cytoplasm and flattened nucleus displaced to the periphery.

White fibrous tissue	Yellow elastic tissue
1. Shining white fibres	1. Yellowish colour
2. Fibres—Thin	2. Thick
3. Non-branching	3. Branching freely
4. Run in bundles and wavy	4. Runs singly and not wavy

Cartilage (Hyaline)	Bone
1. Matrix—Transparent, free from any fibre	1. Not transparent
2. Haversian system—absent	2. Present—Having lamellaes, lacunae, canal
3. The round cells "chondrocytes" are embedded in the matrix in groups within a space called "lacunae".	3. There is lacunae in the lamella having cells osteoblast and osteoclast.
4. Avascular	4. Vascular

Table 29.17: Comparison of Bone and ordinary Connective Tissue

Bone (specialised connective tissue)	Connective tissue proper
1. Specialized connective tissue	1. General connective tissue
2. Matrix—Hard	2. Matrix—Soft
3. Ground substance impregnated with calcium salt	3. No
4. Constantly changing its composition	4. No
5. Highly vascular	5. No
6. Fibres—White collagenous	6. Fibres—White and yellow fibres

Contd.

Bone (specialised connective tissue)	Connective tissue proper
7. Cells—Three types: (a) osteoblast (b) osteocyte (c) obsteoclast	7. Many types: Fibroblast, plasma cells, Blood cells, Histiocytes, Reticular cells, Pigment cells, Fat cells, Mast cells
8. Haversian sytem—Present	8. Absent

- **Cells present in the connective tissues**
  1. Mast cell
  2. Mesenchymal cell (undifferentiated).
  3. Macrophage
  4. Plasma cell
  5. Adipose cells
  6. Chondroblast
  7. Fibrocyte
  8. Osteoblast
  9. Osteocyte
  10. Osteoclast.

Fibres present in the connective tissue
1. Collagen fibre
2. Elastic fibre.

29.48 Extracellular Matrix

29.48.1 Components of extracellular matrix

ECM is made of three main types of extracellular macromolecules. They are:

I. Polysaccharide *glycosaminoglycans* (or mucopolysaccharide or GAG) is linked to proteins in the form of proteoglycans.

There are four main types of glycosaminoglycans:
1. Hyaluronic acid
2. Chondroitin sulfate
3. Heparan sulfate
4. Keratan sulfate.

They are distinguished by
(a) Sugar residue
(b) Type of linkage
(c) Number of sulfate group
(d) Location of sulfate group.

1. They are distributed in the matrix of different tissues. Hyaluronic acid facilitates cell migration during tissue morphogenesis and wound repair. Important component of joint fluid helps lubrication. It is broken down by the enzyme Hyaluronidase.
2. They exist in combination with protein called proteoglycans (Table 29.19).

29.48.2 Fibrous proteins

Structural Fibrous Proteins

(1) Elastin, (2) Collagen.

- **Elastin**

Characteristic
1. Elastin is a protein
2. Fibrillar
3. Cross-linked
4. Having random coil
5. Hydrophobic
6. Nonglycosylated
7. Gives elasticity of the tissue, e.g., skin, blood vessels, lungs.

29.48.3 Composition

It is rich in amino acids like proline and serine, and little amount of hydroxyproline and hydroxyserine.

29.48.4 Formation

It is secreted into the extracellular space and forms an extensive crosslinked network of fibres that can stretch and recoil like rubber band and gives elasticity to the ECM. This elastin fibre also contains a glycoprotein which is distributed on the microfibrils on the elastic fibre surface.

II. **Adhesive Fibrous Proteins**: The adhesive fibrous proteins called CAM's or Cell adhesion molecules. They are:

1. Fibronectin, 2. Laminin.

These are cell surface proteins mediating cell adhesion to matrix or adjacent cells and mediatory of transmembrane signals.

Fibronectin: It is a glycoprotein made of two polypeptide chains which are similar but not identical. It occurs in three forms:

1. A soluble form called *plasma fibronectin* which circulates in the blood and other body fluids. Its function is to
   (a) healing of wound
   (b) phagocytosis

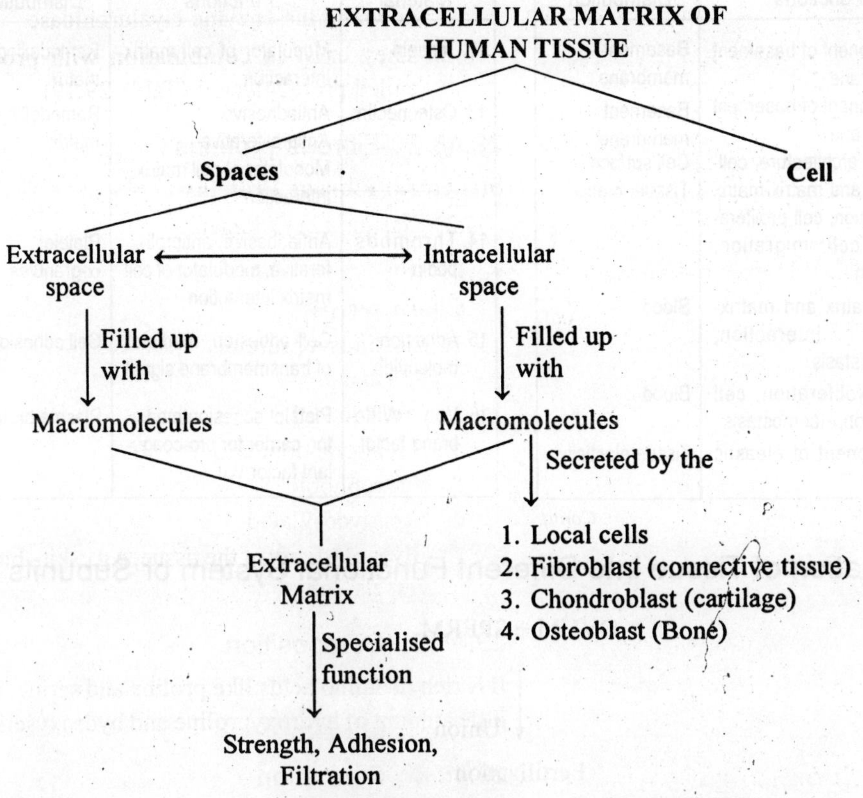

Fig. 29.24: Extracellular matrix and functions

i. In some tissue space between the cell membrane of two cells having a width between 200-300Å.
ii. In some tissue the extracellular space and matrix is a part of organised structure called *basement membrane* or *connective tissue stroma.*

2. This form is called *cell surface fibronectin* attached to the cell surface and functions for cell to cell attachment.
3. The third form is highly insoluble fibrillar fibronectin called *matrix fibronectin* which helps to adhere cell to the matrix.

Laminin: Laminin is an adhesive glycoprotein. It is a major part of the basal lamina. It binds the epethelia with the collagen. It contains more than 1500 amino acids. It is secreted from the epithelia cell. It has high affinity for binding sites for other components of the basal lamina. It is the first extracellular matrix protein to appear in the embryo. It acts as a major barrier of filtration in the kidney. When the protein laminin deposits in the kidney glomerular basement membrane, antibodies are produced against laminin and affect the function severely. Laminin is increased in basement membrane of diabetic patient. Antibodies are produced in Chagas disease.

Table 29.18: Components of extracellular matrix

Material	Functions	Distribution
1. Collagen	Tensile strength tissue architecture, interaction between cell and matrix	Tissue
2. Elastin	Elasticity of tissue	Elastic tissue, blood vessel, lung, skin
3. Proteoglycans	Imbraction of cell-matrix, matrix-matrix, cell proliferation, binding and storage of growth factor.	Tissue matrix
4. Hyaluronan	Cell-matrix interaction, matrix-matrix interaction, cell proliferation, cell migration	Tissue
5. Laminin	Component of basement membrane, cell migration	Basement membrane

Contd.

Material	Functions	Distribution
6. Epitigrin	Component of basement membrane	Basement membrane
7. Entractin	Component of basement membrane	Basement membrane
8. Fibronectin	Tissue architecture, cell-matrix and matrix-matrix interaction, cell prolifera-tion, cell migration, opsonin	Cell surface Tissue, matrix
9. Vitronectin	Cell-matrix and matrix-matrix interaction, Haemostasis	Blood
10. Fibrinogen	Cell proliferation, cell migration, Haemostasis	Blood
11. Fibrillin	Component of eleastic fibre	Tissue elasticity
12. Tensein	Modulator of cell-matrix interaction	Remodelling matrix
13. Osteonectin	Antiadhesive Antiproliferative Modulator of cell matrix interaction	Remodelling matrix
14. Thrombos-podin	Antiadhesive, antiproli-ferative, modulator of cell matrix interaction	Platelet α-granules
15. Adhesion molecules	Cell adhesion, mediator of transmembrane signal	Cell adhesion
16. Von Wille-brand factor	Platelet adhesion media-tor, carrier for pro-coagu-lant factor VIII	Plasma protein

Contd.

29.49 Organisation of Tissue into Different Functional System or Subunits

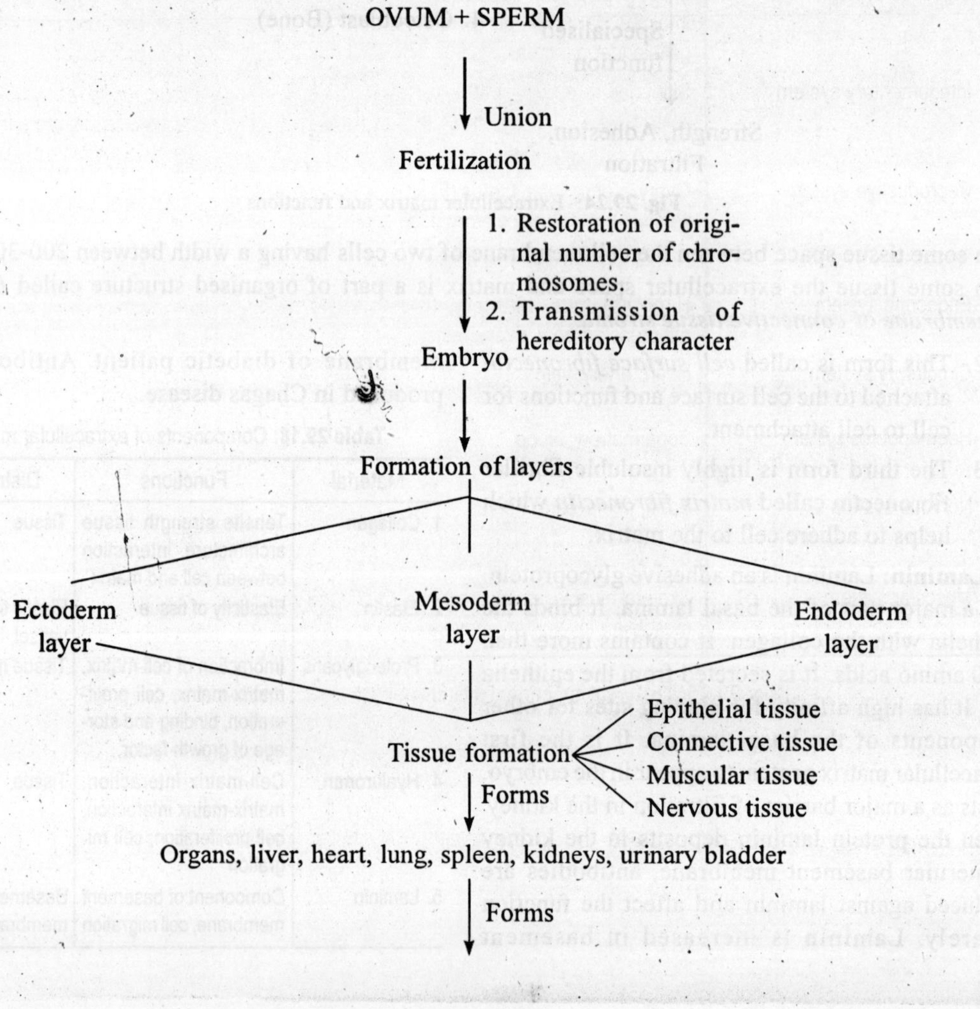

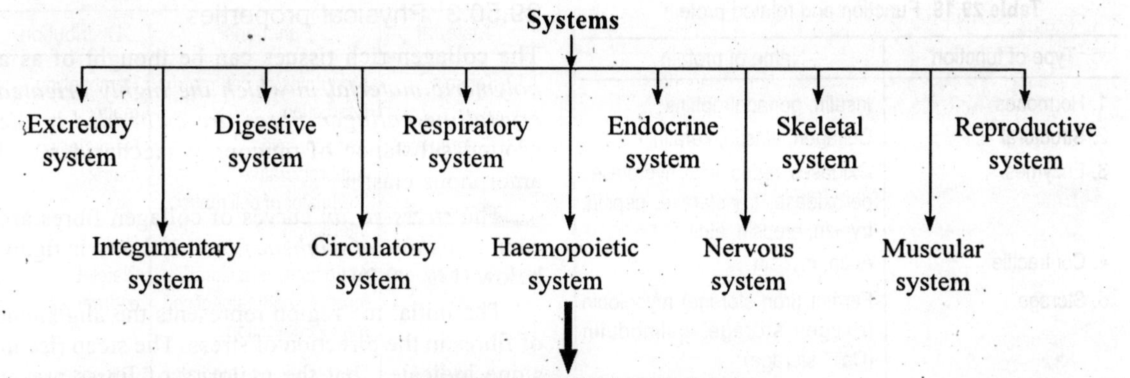

SYSTEM HAVING ORGANS

System	Organs forming	Function
1. Excretory system	Two kidneys, two ureters, one urinary bladder and urethra	Formation of urine, excretion of waste and toxic products.
2. Skeletal system	Bones of the body	Body framework, structure carry body weight, movement.
3. Digestive system	Mouth, pharynx oesophagus, stomach, small intestine, large intestine, rectum, anal canal, liver gallbladder pancreas	Digestion and absorption of digestive food and elimination of waste products.
4. Integumentary system	Skin, nail	Protection of the structure beneath the skin, temperature maintenance, sweat formation.
5. Reproductive system	Ovary, uterine tube, uterus (F) testes, vas deferense, seminal vesicle, ejaculatory duct, prostate (M)	Reproduction
6. Respiratory system	Nostril, pharynx, larynx, trachea, bronchus, lungs	Respiration.
7. Endocrine system	Hypothalamus, pituitary, thyroid, parathyroid, pancreas adrenal, ovary, testes	Secretion of hormones.
8. Circulatory system	Heart, artery, vein	Circulation of blood and supply of O_2 and nutrition.
9. Haemopoeitic system	Bone marrow, blood	O_2 carrying, body defense, production of immune substances.
10. Nervous system	Brain, spinal cord, nerves	Coordination of body function memory, postural equilibrium movement, etc.
11. Muscular system	Skeletal muscle for voluntary movement and visceral muscle for involuntary movement.	Movement, strength, posture, temperature regulation, storage.

Fig. 29.25: Different functional systems or subunits and having organs

Table 29.19: Function and related protein

Type of function	Name of protein
1. Hormones	Insulin, gonadotrophins
2. Structural	Collagen, elastin, keratin
3. Enzymes	Oxidases, reductorsis, hydrolases peroxidase, transferase, pepsin, trypsin, erepsin, etc.
4. Contractile	Actin, myosin
5. Storage	Ferritin (iron storage) myoglobin (oxygen storage) calmodulin (Ca^{++} storage)
6. Transport	Haemoglobin, Apoferritin, transferrin
7. Respiratory	Haemoglobin, cytochromes
8. Osmotic	Albumin
9. Clotting	Prothrombin, Fibrinogen
10. Storage food	Labile cellular protein
11. Cell membrane structure	Integral and peripheral proteins
12. Membrane receptor	Receptor protein
13. Ion channel of membrane	Channel protein
14. Cell adhesion	Adhesin, integrin, fibronectin, laminin, cadhesin
15. Messenger	G-protein
16. Toxic reaction	Toxin of bacteria of tetanus, diphtheria
17. Protection	Antibody protein (Immunoglobulins)

29.50 Collagen-Rich Tissue

29.50.1 Collagen-rich tissues

Collagen-rich tissues function mostly in load-bearing capacity. These tissues include skin, tendon and cartilage. Special functions such as transparency for the lens of the eye can also be born by collagenous tissues.

29.50.2 Composition and structure

The collagen-rich tissues are composed predominately of collagen. Collagen is made up of tropocollagen, a three chain coiled superhelical molecule. The collagen fibrils aggregate to form fibres. The fibrils and fibers are stabilized through intra-and intermolecular hydrogen bonding (C=O–HN).

29.50.3 Physical properties

The collagen-rich tissues can be thought of as a *polymeric material in which the highly oriented crystalline collagen fibres are embedded* in the ground substance of mucopoly saccharides and amorphous elastin.

The *stress-strain* curves of collagen fibres are *similar to those of synthetic fibres* as shown in figure below (Fig. 29.26).

The initial toe region represents the alignment of fibres in the direction of stress. The steep rise in slope indicates that the majority of fibres are in parallel directions. Individual fibres may be breaking well ahead of the final catastrophic failure. The highest slope of the stress-strain curve is about 1.0 GPa which is close to the modulus of collagen fibres but the tensile strength is much lower.

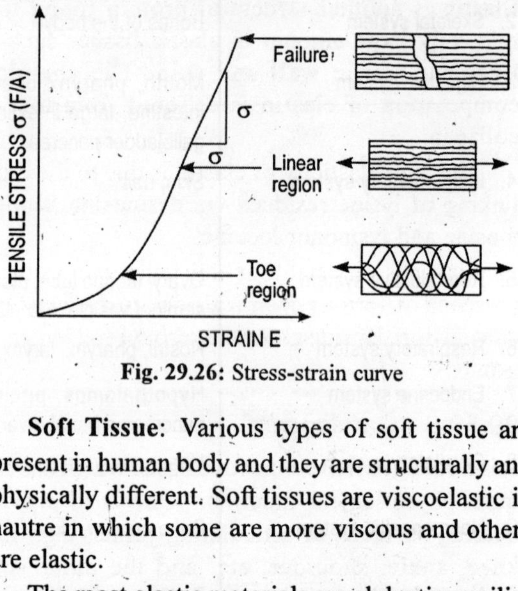

Fig. 29.26: Stress-strain curve

Soft Tissue: Various types of soft tissue are present in human body and they are structurally and physically different. Soft tissues are viscoelastic in nautre in which some are more viscous and others are elastic.

The most elastic materials are abductin, resilin, elastin and collagen. The skin, skelelal muscle, heart muscle, smooth muscle, ligaments, tendons, blood vessels are all different kinds of soft tissue.

29.50.4 Collagen

It is basic structural element for soft and hard tissues in animals. It is a fibrous protein and has the general amino acid sequence-Gly-Pro-Hyp-Gly-X-(X = any amino acid) arranged in a triple α-helix. This triple super helix is the molecular basis of tropocollagen, the precursor of collagen. The three chains are held together strongly by H bands between glycine residues and between hydroxyl (OH) groups of

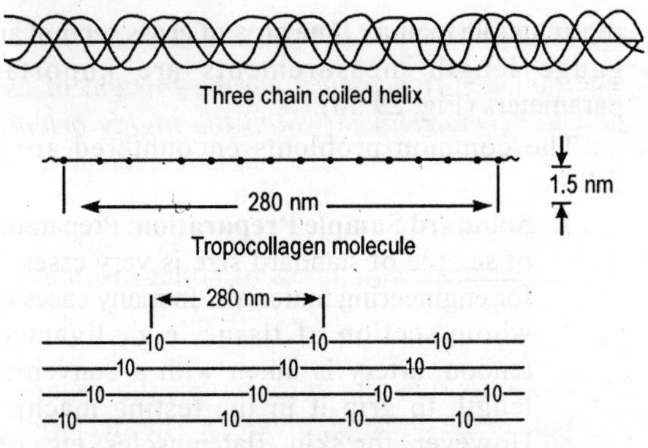

Fig. 29.27: Crosslink in collagen among the helices

hydroxyproline. In addition, there are cross-links via lysine among the helices (Fig. 29.27).

29.50.5 Elastin

Elastin is another structural protein found in a relatively large amount in elastic tissues such as ligament, aortic wall and skin. The chemical composition of elastin is differed from that of collagen.

The high elasticity of elastin is due to the cross linking of lysine residues via desmosine, isodesmosine and lysinonor leucine.

Elastin is very stable at high temperatures in the presence of various chemicals due to the very low content of polar side groups (hydroxyl and ionizable groups).

29.50.6 Cartilage

It is a collagen rich connective tissue which has two main physiological functions. One is to provide bearing surfaces of articulating joints, e.g., hip, knee, ankle, shoulder, etc. and the other is to maintain flexible shapes in ear, tip of nose, rings around trachea, in the walls of thorax and small masses in skull base.

These are three major types of cartilage in the body.

1. *Hyaline cartilage*—is most prominently found in diarthrodial joints covering long bones. In addition hyaline cartilage forms the growth plate by which long bones grow during childhood. Present in trachea.
2. *White fibrocartilage* (containing high collagen)—is permanently present in three major locations in the body.
   (a) the *intervertebral disks* of the spine, ligamentum Nuche.
   (b) as a covering of the mandibular condyle in the temporomandibular joint.
   (c) in the meniscus of the knee joint.
3. *Yellow elastic cartilage* (having high elastin network)—exists in the epiglottis and the eustachian tube.

29.50.7 Structure of cartilage

It consists primarily of intercellular matrix with cells chondrocytes interspersed in the matrix with small pockets called lacunae. Cartilage is a pliable material that allows for movement. The intercellular matrix is made up of protein and carbohydrate that forms a gel in which fibres and cells are found. The perichondrium is dense irregular connective tissue that is found on the surface of cartilage. Blood vessels run through the perichondrium and nutrients diffuse into the chondrocytes. So if cartilage becomes damaged, it heals slowly as there is no direct blood supply to each chondrocytes.

29.50.8 Function

Cartilage in the joints provides a smooth surface through the full range of movement during which it is continuously subjected to severe stresses such as the pull of gravity in weight bearing joints and muscle tension in all joints.

29.50.9 Tendons

Tendons are the cords of dense regular connective tissue that *attach skeletal muscles to bones and muscles to muscles.* (1) *It is composed of bundles of parallel white collagenous fibres.* (2) *Fibroblasts are present in continuous rows. Tendons are* (3) *tough* and *inelastic in nature.* Tendons have the same structural organization as ligaments. The major constituents of tendons are type-I collagen (~80% of dry weight), water (65-70% of the wet weight), elastin (< 3% of dry weight) and proteoglycans (~1%). Tendons vary considerably in size and shape. Some are wide and flat, others are cylindrical or fan shaped. The ultimate tensile strength of animal tendons has been reported to vary between 45 and 125 MPa. For humans, the ultimate tensile stress, ranges between 50 and 105 MPa, the elastic modulus ranges from 1.2-1.8 GPa and ultimate strain ranges from 9% to 35%.

29.50.10 Ligament

It *resemble* tendons, *but they connect bone to bone, Ligament are tough but elastic in nature because ligaments,* have significant amounts of elastic *as well as collagenous fibres.* It is composed of *yellow elastic fibres. Fibroblasts* are *scattered in ligament matrix.* Ligaments have similar properties of tendons. The structural organization of ligaments are similar to that of tendons.

29.50.11 Stress-strain relationship

If a piece of tendon is collected from a rabbit leg and tested under tensile load in an Instron machine fitted with a load-deformation recorder, it will produce a diagram as shown below when the rate of movement of the ram pulling tendon is kept constant which is in other words at constant deformation rate (Fig. 29.28).

The load deformation curve has three parts, the initial non-linear part OA, then the fibres get stiffened and present a linear part AB which makes an angle θ with the deformation axis. The slope of AB, tan θ give the elastic sliffness or Young's modulus. The part BC is again non-linear and ends up with rupture. The ultimate strength of the tendon varies between 50-100 MPa and strain at rupture is about 10-15%.

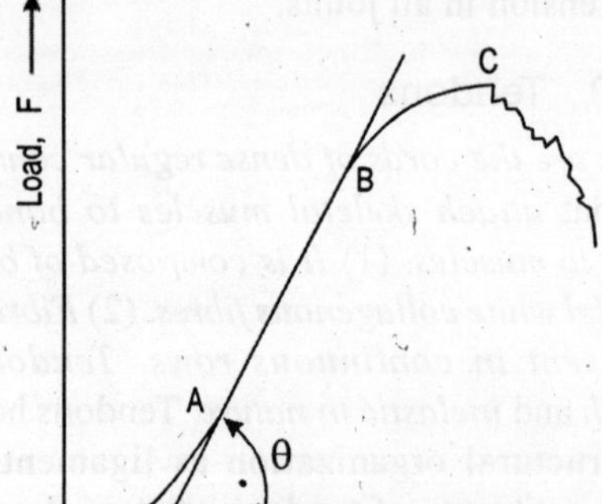

Fig. 29.28: Stress-strain relationship

29.50.12 Mechanical testing of soft tissues

Mechanical testing of soft tissue is not as simple as testing of other type of engineering materials because of its inherent characteristics and biological variation. The shape and sizes are variable and since they are viscous fluid filled solid or semisolid it is very difficult to measure their dimensions in a reproducible manner. The area of cross section and gauge length measurements are important parameters (Fig. 29.30).

The common problems encountered are as follows:

1. **Standard Sample Preparation**: Preparation of sample of standard size is very essential for engineering materials. In many cases the whole section of tissue, e.g., ligament, tendon, artery is taken with a convenient length to grip it in the testing machine. However, for skin, flat muscles, etc. one should use a convenient width of the sample and a length of at least 8-10 times its width to avoid the end-fixation effect.
2. **Cross section Measurement**: To quantify the strength we need to measure the area of cross section and divide the ultimate load to failure by the area to get the strength value. Therefore the area measurement is critical. The area has to be measured under a preset

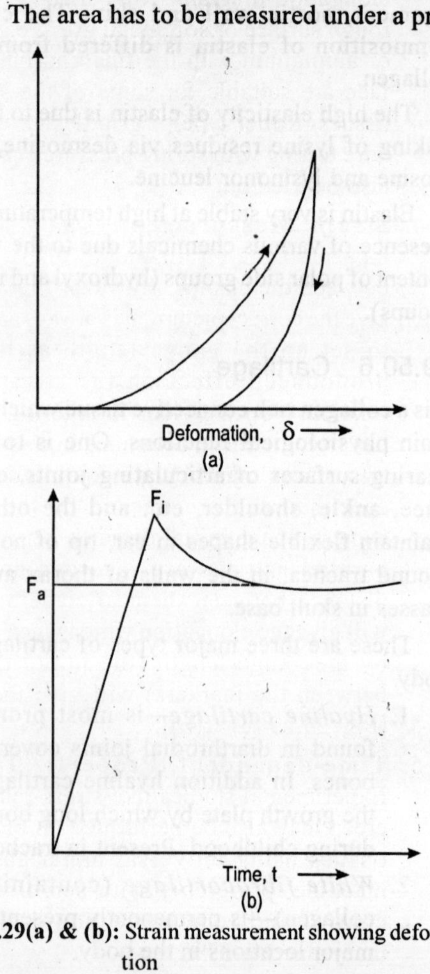

Fig. 29.29(a) & (b): Strain measurement showing deformation

pressure since the pressure between the anvils of a micrometer or jaws of vernier calipers increases the area which will continue to change with pressure.

Many authors have attempted to fix the section with fixing chemical and measured the section using microscope along with videographic technique. This also has some problems like shrinkage due to water loss and change of inherent character.

3. **Clamping of the Specimen**: Usual clamps at the grips of the material testing machines are serrated hardened steel jaws. These clamps are totally unsuitable for holding the soft tissue-samples. The tissue fibres gets crushed and tears under the pressure of the grip at the clamp and stress at this zone is unusually high. Also a very smooth metal grip causes slip and affects the load elongation diagram. Rubberised grip with rough surface or soft metal grips, e.g., brass or aluminium with the fine knurling on its face are suitable for some of the stronger tissues. Roller type of grip design used for wire testing with spring pressure clamp may also be tried out.
4. **Strain Measurement**: Many authors measured the strain or the displacement between the clamps. However this will not give the correct picture. It is always advisable to set up the gauge length as this has pronounced effect on the stress-strain relation. Now the video monitering method has become more common. In this method two sharp markers are installed on the specimen and are tracked by the video-camera. If the initial distance between two parallel markers are recorded and final distance between them after deformation will give the elongation which when divided by the gauge length (the initial distance between the markers) will give the strain [Fig. 29.29 (a) + (b)].
5. **Time-dependent Property Testing**: Experimental studies like stress-relaxation, creep may also be performed on biological tissues using universal material testing machine, preferebly with servohydraulic control and removable liquid chamber maintained at 37°C.

For testing of fatigue life of polymers, metals or composites to be used as load varying biomaterial in tension, campression or bending mode of loading. The workers developed a special machine which can perform such test very conveniently following German (DIN) standards. There are appropriate grips for various modes of testing, digital cycle counter, preload setter, digital temperature controller or the bath in which testing will be carried out.

All the soft tissues have these following features:
1. non-linear stress-strain relationship.
2. hysteresis loop in cycle loading
3. stress-relaxation at constant strain and preconditioning in repeated cycle of loading.

The above shown figures shows a ligament which was loaded at about 30% of its ultimate load then released at a constant speed, it never reaches the initial deformation. Fig. 29.29(b) shows the Force-time relation ligament is loaded to a level of F_i and then the elongated length is kept constant. Then the load will relax asymptatically to a limiting value F_a. This phenomenon was called as *stress relaxation*.

Soft Tissue Properties

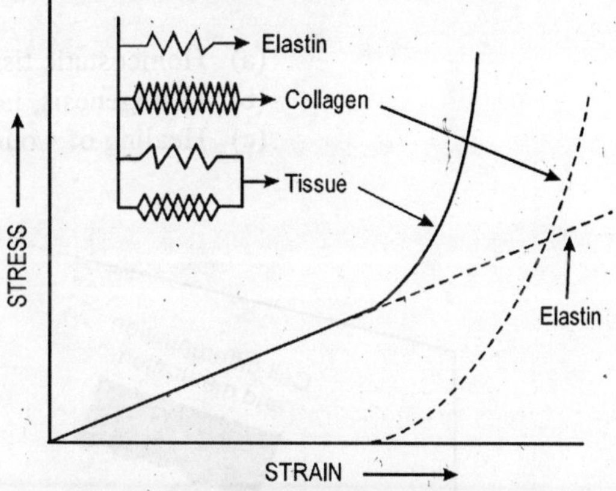

Fig. 29.30: Mechanical testing of soft tissue

PART IV

29.51 Tissue Dynamics

Tissue is made up of many cells having different developmental origin. Tissue formation, state and function depend on their dynamic behaviour and interaction between them. Individual cell activities are not so much significant but best cellular activities with tissue function is significant because of long period of time. The cellular processes which are related to dynamic formation of tissue are grouped into the followings:

1. **Hyperplasia**: For DNA synthesis and cytokinesis, i.e., cell prolification and replication cause and increase in number of cells due to complete cell division.
2. **Hypertrophy**: Increase in call volume due to DNA synthesis along with cytokinesis.
3. **Cell maturation and differentiation**: Changes in gene expression and acquisition of specific function.
4. **Cell adhesion**: Binding of cell physically with its immediate neighbour which may be cell; extracellular matrix or an artificial surface.
5. **Cell motility**: The motion of cell to a particular site.
6. **Programmed cell death or cell apoptosis**: Many normal cells including blood cell undergo apoptosis due to age. The survival of normal cells depends on the signals from the growth factors or from the extracellular matrix which inhibits apoptosis. It may undergo apoptosis following DNA damage.

These processes lead to three dynamic states at a tissue level. These are.

1. *Histogenesis*: The maturational lineages of the cells are derived from stem cell of tissue. Overall functions of tissue are combination or sum total function of all the cells within the maturational lineage of the tissue.
2. *Tissue formation*: There is variation in proportion in tissues regarding stem cells, diploid cells and polyploid cells depending on age.

   Tissue taken out from young donors (e.g., infant) has greater expansions in vitro and have greater potentiality for cell therapy.

   The explanation is that if the donor's tissue isolated from infant they show maturation lineages skewed toward the young cells in the lineage such as stem cells, diploid cells, somatic cells whereas isolated from adult or geriatric person will have tissue skewed toward the later stage of lineage, i.e., polyploid cells (Fig. 29.31).

(a) Homeostatic tissue function
(b) Histogenesis, i.e., tissue formation
(c) Healing of wound, i.e., tissue repair

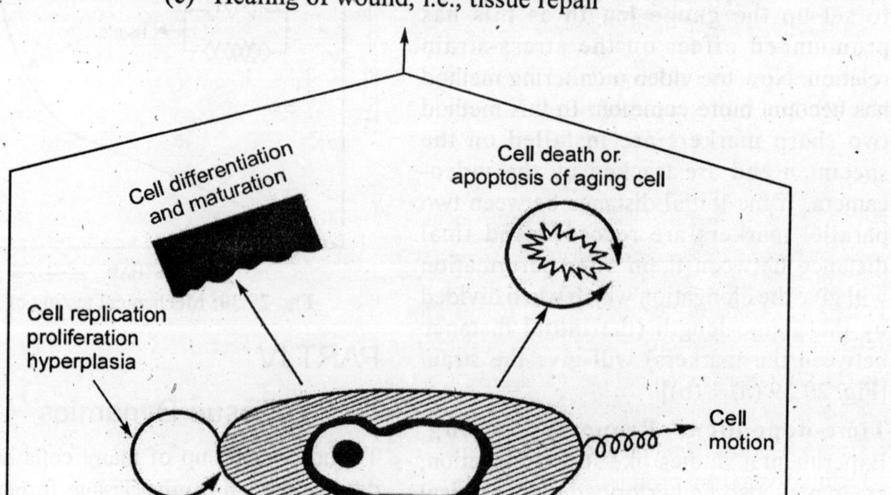

Fig. 29.31: Schematic representation of tissue dynamics

29.51.1 Oganization of Tissue at Cellular Level and Diversification

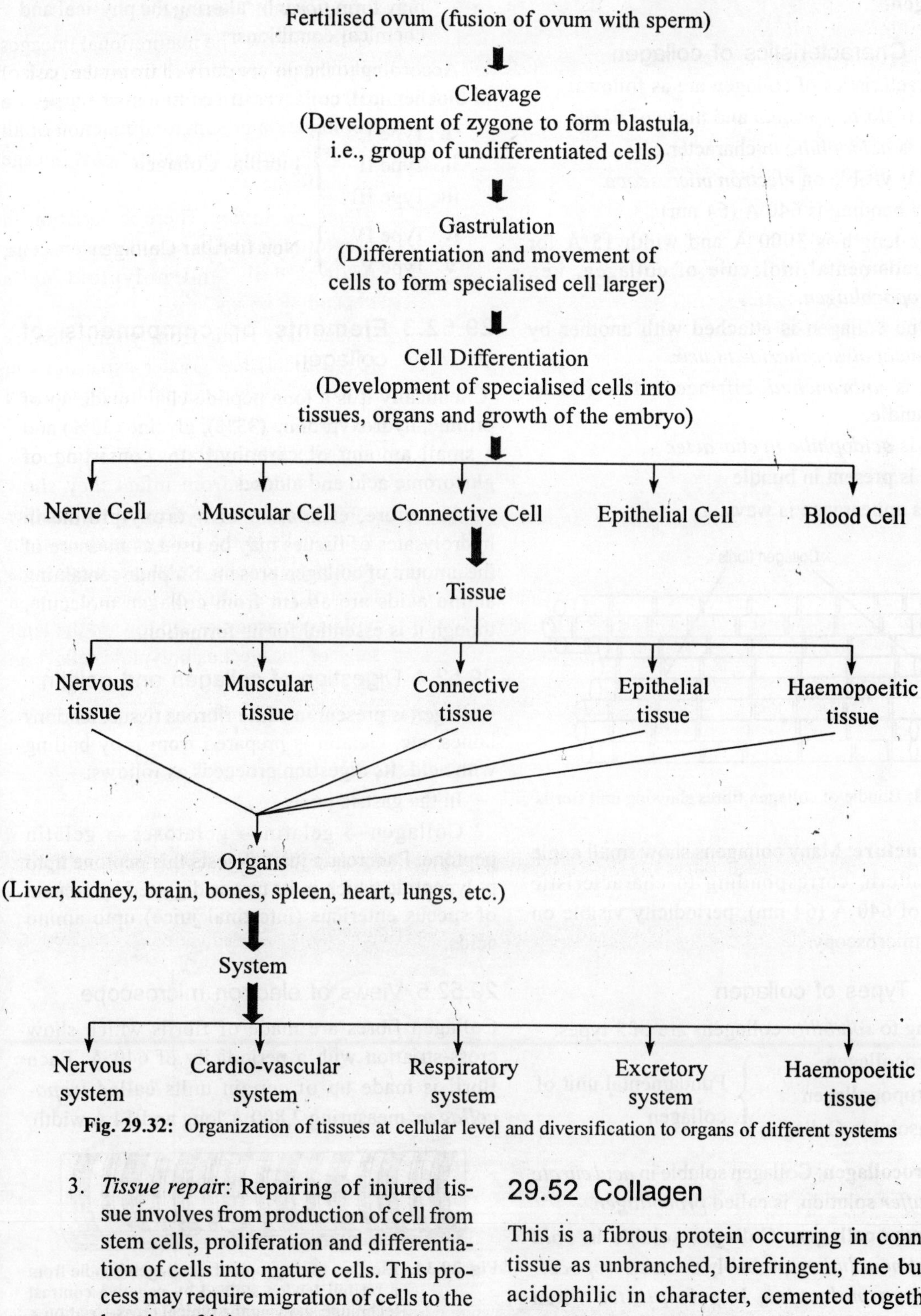

Fig. 29.32: Organization of tissues at cellular level and diversification to organs of different systems

3. *Tissue repair*: Repairing of injured tissue involves from production of cell from stem cells, proliferation and differentiation of cells into mature cells. This process may require migration of cells to the location of injury.

29.52 Collagen

This is a fibrous protein occurring in connective tissue as unbranched, birefringent, fine bundles, acidophilic in character, cemented together by mucopolysaccharide matrix, the ground substance.

White fibrous tissue is made of protein known as collagen.

29.52.1 Characteristics of collagen

The characteristics of collagen are as follows:

i. It is *fibrous protein* and made of *fibrils.*
ii. It is *acidophilic in* character.
iii. It is visible on *electron microscope.*
iv. Its bending is 640 Å (64 nm).
v. Its length is 3000 Å and width 15 Å for fundamental molecule of collagen, i.e., *tropocollagen.*
vi. One collagen is attached with another by *mucopolysaccharide matrix.*
vii. It is *unbranched,* bifringent, fine (thick) bundle.
viii. It is *acidophilic in character.*
ix. It is present in bundle
x. Its appearance is wavy.

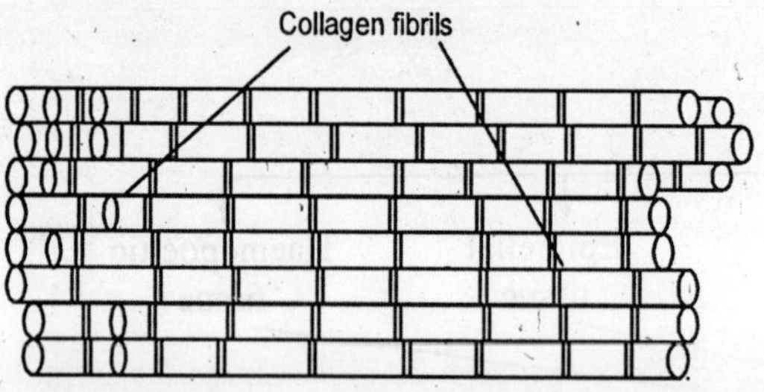

Fig. 29.33: Bundle of collagen fibres showing unit fibrils

- **Structure**: Many collagens show small angle X-ray pattern, corresponding to characteristic bending of 640 Å (64 nm), periodicity visible on electron microscopy.

29.52.2 Types of collagen

According to *solubility* collagens are of 3 types:

(a) Procollagen
(b) Tropocollagen
(c) Insoluble Collagen
} Fundamental unit of collagen

(a) **Procollagen**: Collagen soluble in *acid citrate buffer* solution, is called *procollagen.*

(b) **Tropocollagen**: Collagen soluble in *cold saline solution,* probably contains *tropocollagen molecules.* It *precipitates* when warmed.

(c) **Insoluble Collagen**: The soluble collagens may form *fibres* by altering the physical and chemical conditions.

According to the polypeptide chain on the basis of biochemical, collagens are of five major types:

i. Type I
ii. Type II
iii. Type III
} Fibrillar Collagen

iv. Type IV
v. Type V
} Non fibrillar Collagen

29.52.3 Elements or components of collagen

Chemically it is a long peptide chain made up of proline, hydroxyproline (33%), glycine (33%) and a small amount of carbohydrate, consisting of glucoronic acid and aldoses.

Therefore, estimation of hydroxy-proline in hydrolysates of tissues may be used as measure of the amount of collagen present. Sulphur containing amino acids are absent from collagen molecule, though it is essential for its formation.

29.52.4 Digestion of collagen and gelatin

Collagen is present in white fibrous tissue, tendons bones, etc, Gelatin is prepared from it by boiling with acid. Its digestion proceeds as follows:

In the gastric juice:

Collagen → gelatin → gelatoses → gelatin peptone. Pancreatic juice digests this peptone upto polypeptide which in its turn is digested by erepsin of succus entericus (intestinal juice) upto amino acids.

29.52.5 Views of electron microscope

Collagen fibres are made of fibrils which show cross-striation with a periodicity of 640 Å. Each fibril is made up of certain units called *tropocollagen,* measuring 2,800 Å long and 14 Å width,

Fig. 29.34: Electron micrograph of a collagen bundle from a rat tail tendon stained by negative contrast technique showing the typical cross-striation × 56,000

with a molecular weight of 340,000. These units join together end to end and by arrangement of their lateral neighbours from a periodicity of 640 Å (as described above). Thus the interchain bonds maintain the stability of primary polypeptide.

29.52.6 Formation of collagen fibrils

Fibroblast type of cells form *Soluble tropocollagen* secreted by these cells. It makes more dense (condense) into *mucopolysaccharide* in the extracellular space and forms fibrils of 640Å periodicity. The character of *silver* impregnation takes place according to the age of collagen when studied under electron micrography (Fig. 29.35, 29.36).

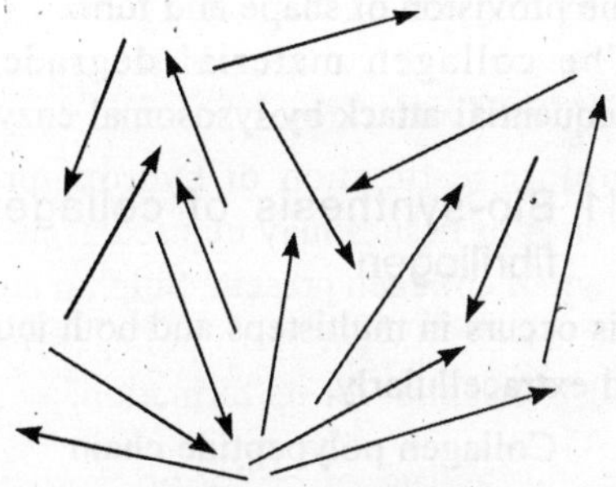

Fig. 29.35(a): In a solution of tropocollagen, the molecules are separated from each other

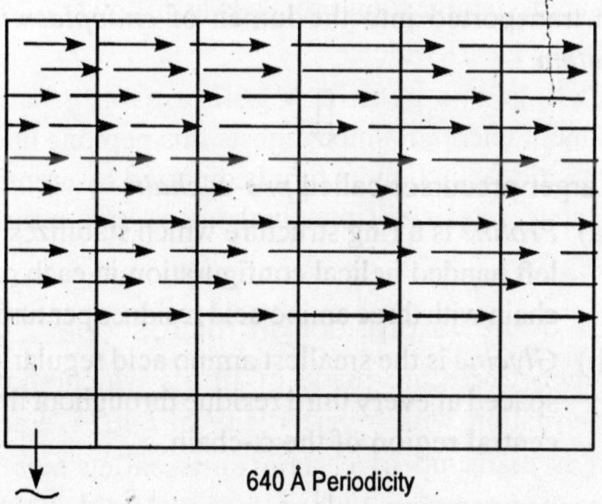

Fig. 29.35(b): Molecules are packed parallel to each other with a quarter overlap of fibril with 640 Å periodicity

The embryonic *fibrils* show irregular deposit of *silver* on the surface throughout. As maturation proceeds they are deposited mainly over the rigions of dark bands resembling reticulin fibres, and the fully matured fibrils show no silver deposit on the surface but internal silvering in the region of bands is observed as in the case of collagen fibres. In cornea, most fibrils are the *reticulin like fibrils* found in the *lymph nodes, lungs* and *spleen* and the matured ones in the *tendon sheaths* (Fig. 29.36).

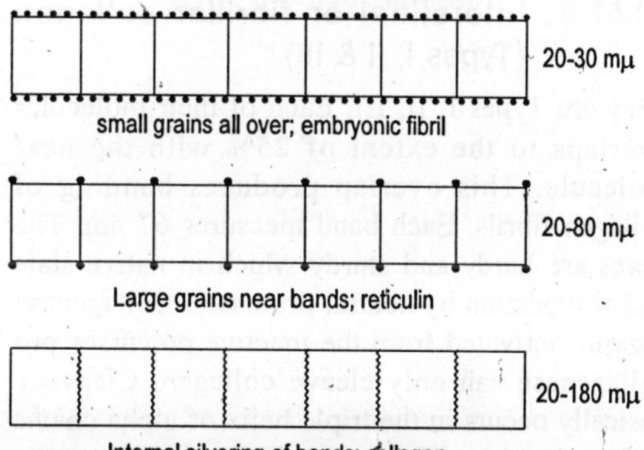

Fig. 29.36: Characters of fibril of different sizes under electron micrography *after silver impregnation*

29.52.7 Staining

(a) Collagen stain *Pink* with *eosin.*

(b) *Picric acid* stains it *red* as in *Van Gieson's stain.*

(c) *Aniline blue* combined with *Phosphotungstic* or *phosphomolybdic* acid *(Mallory's stain)* stains it *blue.*

(d) It is not blackened but assumes *brown* colouration when stained with *silver salts.*

29.52.8 Metabolic and biochemical aspects of collagen

The metabolism obviously include collagen biosynthesis (Anabolism) and its degradation (Catabolism). Three fundamental units of collagen are described above, they are—

(a) Procollagen
(b) Tropocollagen
(c) Insoluble collagen.

Tropocollagen seems to be of great importance.

Tropocollogen molecule is a long rod measuring 300 × 15 nm, mol. wt. is 285,000.

Biochemically, a molecule has 3 separate polypeplide (ALPHA) chains, tightly wrapped together to form a left-handed triple helix.

On the biochemical basis of the polypeptide chain, collagen has five major types, e.g.—

(a) Type I (b) Type II
(c) Type III (d) Type IV
(e) Type V

Each type has specific tissue distribution.

29.52.9 Interstitial or fibrillar collagen (Types I, II & III)

They are Types I, II, III. Each of their molecules overlaps to the extent of 25% with the next molecule. This overlap produces banding of collagen fibrils. Each band measures 67 nm. The fibres are hardy and sturdy which in native state resists digestion by neutral proteases. Collagenase enzyme activated from the inactive precursor pro collagenase can only cleave collagen. Cleavage basically occurs in the triple helix of alpha chains to form 2 unequal fragments. These fragments can now be digested by proteases. Procollagenase is found in macrophages, fibroblast, neutrophils, synovial cells and epithelial cells at some sites. So, in inflammation and healing stage, degradation of collagen occurs by over-accumulation of above mentioned cells.

Collagenase has harmful effects in collagen diseases, like R.A., S.L.E., etc. where collagen is broken down through various stages—

i. Stage of cloudy swelling
ii. Stage of collagenorrhexis
iii. Stage of collagenolysis
iv. Stage of fibrinoid necrosis.

Collagen Types IV and V have no fibril (non-fibrillar) and is degraded by certain selected leucocytic neutral proteases. But they are resistant to dermal collagenases (present in the skin).

Collagen Types IV and V are found in the basement membrane all through in association with some other proteins like fibronectin, heparan sulphate, proteoglycan and lamenin.

Table 29.20: Clinical Application of Collagens

Collagen form	Application
1. Solution	Plasma expander, drug delivery system
2. Gel	Vitreous body
3. Fibres	Suture material, vessel and valve prosthesis, and tissue support.
4. Membrane	Corneal replacement, haemodialyser, oxygenator, dressing.
5. Tubing	Wound dressing, cartilage substitute vessel prosthesis.

29.52.10 General features

1. Collagens are a family of structurally related proteins, which occur in all animals.
2. Collagens are extracellular proteins that are organized into insoluble fibres of great tensile strength.
3. Collagens provide the insoluble scaffold for the provision of shape and form.
4. The collagen material degrades a by sequential attack by lysosomal enzymes.

29.52.11 Bio-synthesis of collagen and fibrillogen

Synthesis occurs in multisteps and both intracellularly and extracellularly.

Collagen polypeptide chain

⇩ Synthesis on

Membrane bound *ribosomes*
Then transported into the lumen of *endoplasmic reticulum.*

⇩

Larger precursor called *pro-α-chain*

(a) *Proline* is a ring structure which stabilizes a left handed helical configuration in each α-chain with three amino acid residues per turn.
(b) *Glycine* is the smallest amino acid regularly spaced at every third residue throughout the central region of the α-chain.

⇩

In the lumen of endoplasmic *reticulum proline and serine* are *hydroxylated.*

⇩

Each pro-α-chain has an extra non-helical segment at their amino acid carboxyl terminal ends. The extra segment is called *telopeptides.*

⇩

Each pro-α-chain then forms triple stranded helical molecule called the *procollagen* or *Tropocollagen.*

⇩

The procollagen are then secreted to the extracellular space by the removal of telopeptides.

⇩

Several monomer collagen molecules or (procollagen or tropocollagen) combine with each other to form larger collagen fibril (10-300 nm in diameter)

⇩

Several collagen fibrils combine to form a *collagen fibre.*

Biosynthesis of collagen fibre and sequence of events occur intracellularly and extracellularly

Synthesis of pro-α-chain

↓ In Ribosome on Reticular endoplasmic reticulum

Prolines and lysines

↓ Hydroxylation

Selected Hydroxyproline

↓ Glycosylation

3 Pro-α-chain

Triple Helix formation | Cell membrane

↓ Secretion

Procollagen molecule

↓ Cleavage of propeptides

Assembly of fibrils

↓ +

Collagen fibrils

↓ Aggregation

Collagen fibre

Fig. 29.37: Flow chart showing biosynthesis of collagen fibre and events occur intracellularly and extracellularly

Collagen fibril arrangement in some tissues

Tendon	: Parallel fibrils
Bone	: Parallel fibrils
Skin	: Sheets of fibrils at different angles
Cornea	: Planar sheet arranged crosswise for reduction of light scattering.
Cartilage	: Not arranged specifically.

Presence of aminoacids in collagen: Glycine, proline, asparatic acid, glutamic acid, lysine, arginine, Histidine, etc.

29.53 Microenvironment

Regulation of tissue development is involved with many events in the surroundings. This undergoes interaction with cell. Apart from general growth factor, soluble specific growth factor and extra cellular matrix (ECM) etc. These interactions are necessary for:

1. Localization of cells within the microenvironment.
2. Directing cell migration.
3. Initiating growth factor mediated development programs.
4. *Developing tissue cells*: interactions are mediated by specific receptors-ligand interaction which is necessary for stimulation of proliferation and motility.
5. Both chemical and ECM which signal the cell move along a route to a definite tissue area.
6. Next due to *light concentration of contractants* or *other signals* localise the cell and stops at random movement.
7. These signals stop and direct proper region of microenvironments which is complex.
8. As it is seen in haematopoietic tissue where interaction of cell cytokines molecules ECM complexes serve the progenitor cells for localization and similar mechanism of interaction exists in the other developing tissue.
9. Thus, regulatory mechanism of cell development leading to tissue formation is the result of coordinated of stromal cells, paren-

chymal cells, ECM, growth factor, cytokine. Each component is the vital member of this coordinated microenvironment regulatory system.

Cellular interaction are grouped under 3 headings:

1. Cell-cell interaction
2. Cell-extracellular matrix interaction
3. Cell-growth factors.

Each of these is functional important for both mature and developing cells such as:

1. Blood cell: interact with each other as cells in other tissues.
2. Immunologic cell-cell interaction takes place when lymphocytes interact with antigen-presenting cells.
3. Aggression of neutrophil or lymphocytes migrate from blood vessels by the process of blood cell-endothelial cells of blood vessel recognition.
4. Interaction between cells and extracellular matrix.

29.53.1 Cells contributing to tissue microenvironment

→ **Stromal cells**—derives from a common precursor cell
- Fibroblast
- Myofibroblast
- Osteogenic/Chondrogenic cells
- Adipocytes, mesenchym

→ **Stromal associated cells**
- Endothelial cells
- Macrophages distinct from stromal cell

→ **Transient cell**
- B-lymphocytes, cytotoxic T-cells
- Natural killer (NK) cells
- Granulocytes Migrate to tissue during inflammation

→ **Parenchymal Cells**

29.53.2 Cells contributing hepatic environment

Kupffer cell
Vascular endothelial
Biliary endothelial cell
Fat storing cell
Fibroblast
Pit cell
Parenchymal Cell
Mononuclear (type I)
Binuclear (type II)
Acidophilic (type III).

29.54 Basal Lamina or Specialised Extracellular Matrix (ECM)

29.54.1 Basal lamina

Beneath the epithelial cell there is a continuous thin sheet on which the epithelial cells rest called basal lamina or specialised extracellular matrix (ECM).

29.54.2 Features

1. Structures wrapped by are individual muscle cells, fat cells, schwann cells.
2. Linked by the cell membrane of different types of cells by specific receptors.
3. It separates these cells from the connective tissue.
4. In the glomerulus of the kidney it lies between two sheets of the cells and forms a porous membrane having different sizes of pores which allow water, and molecules of different sizes and molecular weights are filtered in the glomerular filtrate as urine.
5. It influences cell metabolism.
6. It organizes the proteins in the neighbouring cell membrane.
7. It induces cell differentiation and helps in the cell migration.

29.54.3 Synthesis

The macromolecules which comprise the basal lamina are synthesized from the cells which rest on it.

29.54.4 Composition

Composition varies from tissue to tissue. Generally it is made of large quantity of type IV collagen.

29.55 Tissue Repair, Healing of Wounds

During injury or any wound due to inflammation imediate response of healing or tissue repair is observed. In prenatal life, i.e., fetal life healing occurs without any scar. But in post-natal life i.e., after birth tissue repair occurs making a scar mark. In fetal life healing occurs rapidly whereas end recovery is without any sign of healing. But in post-natal life tissue repair process is slow and restoration occurs but does not restore satisfactorily to normal tissue condition. Wound healing several processes involving cells e.g., stem cells compartment, their maturational lineage and their changes help in wound healing.

29.55.1 Events in tissue repair

1. *Haemostasis* (control of bleeding)
2. *Inflammatory* response
3. *Granulation*
4. *Remodelling.*

1. **Haemostasis**: Control or arrest of bleeding immediately following injury. It occurs by *adhesion* of circulatory platelets or thrombocytes to the injured site and vessel wall → Activation of platelets occurs → changes of shape, size and becomes motile → protrudes pseudopodia → release the contents, some disintegrates liberates a vasoconstrictor substance like *serotonin* or *5-hydroxy-tryptamine* which is an aid in haemostasis. Disintegration of platelets results in the release of substances which are essential for *thromboplastin* (from plalelets) and initiation of clotting. Release substances increase stickiness of platelets → adhesion of more platelets to vessel wall and to one another → plalelet plug formation → clot retraction leads to the formation of a firmer plug.

2. **Inflammatory process**: Neutrophil (white blood cell) is phagocytic in nature. They migrate from the circulation. The phagocyte means by throwing pseudopodia (false leg) it engulfs the foreign particles like tissue debris, bacteria, virus, protein etc. digests or kills them with the help of lysosomal enzymes. In the meantime more neutrophils rush to the site by the process of chemotaxis due to chemotactic agents produced in the site which are attractive to the phagocytic cells. Next macrophages start their activity. Two types of macrophages are involved—one is fixed macrophages present in the tissue which alone cannot combat for which wandering macrophages also come to help neutrophil and tissue macrophages. These factors initiate migration of endothelial cells and fibroblast to wound site to help for proliferation. During the process if the neutrophils fail to combat, the cell dies and the dead cell is called pus cell.

3. **Granulation of wound tissue**

During the process, granulation of tissue occurs. It comprises of:

(a) Fibroblasts
(b) Macrophages
(c) Developing vasculature.

These are embedded in the matrix which comprises of:

(a) Fibronection
(b) Collagen
(c) Hyaluronic acid

The fibroblats produce collagens type I, type II to increase the tensile strength. Myofibroblast contracts for shrinkage of the size of the wound and pulls the margin of the wound simultaneously.

4. **Remodelling of wound tissue**

The matrix undergoes remodelling involving coordinated process of biosynthesis and degradation of connective tissue proteins. Remodelling process causes some changes in the components of the matrix in due course. Such collagen III is in abundant in quantity at initial stage but later it gives way to type I. The balance of this mechanism determines the scar formation. During this period the wound is healed up but chemical and structural changes continue to occurs around the site of the wound. Finally, resolution of the scar occurs. The biosynthesis and degradation of the matrix compqnents and the injury site comes back to initial state.

PART V

TISSUE GRAFT, TRANSPLANTATION, CRYO-TECHNIQUE

1. Transplantation, Grafting, 2. Rejection of grafts, 3. Prosthetic heart value, 4. Cryo-technique or preservation technique.

29.56 Transplantation, Grafting

Definition: The replacement of a diseased organ or tissue by a transplant of healthy organ or tissue is known as transplantation or grafting.

Kidneys have been transplanted for a number of years with varying degrees of success. The first heart transplant was carried out in 1967 by Christian Barnard in Capte Town, and since then attempts have been made to graft lungs and livers. But it has created frustration due to uncooperative attemps of the body to reject graft from their individuals the phenomenon is known as *rejection phenomenon.*

29.56.1 Types of transplants between individuals and species

1. **Autograft**: Tissue grafted back to the original donor

   Autograph are of two types:

   (a) *Orthotropic*—If the part is placed in the anatomical region from which it is derived.

   (b) *Heterotropic*—If the region of grafting is different from the original location.

2. **Isograft**: Graft between sygenic individuals, i.e., of identical genetic constitution such as identical twins or mice of the same pure line strain.
3. **Allograft (or homograft)**: Graft between allogeneic (i.e., members of the same species but different genetic constitution) individuals such as *man to man* and one *mouse strain to another.*
4. **Xenograft (heterograft):** Graft between xenogeneic individuals, i.e., different species e.g., pig to man.

We have been most concerned with the *allograft* transplantation or grafting although it *should one day be possible* to use grafts from other species. The most common *allografting in liquid forms is probably blood transfusion* where mismatching may be fatal. Rejection of solid grafts such as *skin* or *organs* may be seen due to *immunological response* or because of the side effects of attempts to suppress the immune response, due to toxicity of the drugs or their depressive effect on resistance to infection.

Allografts may be several types clinically:

1. *Temporary free grafts*: Skin grafts and blood transfusion.
2. *Partially inert*: Provide framework for the growth of host tissue, e.g., bone, cartilage, nerve, fascia, tendon, etc.
3. *Permanent partially privileged structural free grafts*: Blood vessel, cornea, heart valves.
4. *Partially privileged functional free grafts*: such as ovary, testes, parathyroid.
5. *Whole organs grafts*: Pancreas, liver, kidney, heart, lung.

29.56.2 Clinical results in grafting

1. **Cornea**: Corneal grafts survive without the need for immunosuppression. Because they are avascular and do not sensitize the recipient although they become cloudy, if the individual has been sensitized.
2. **Cartilage**: Grafts of cartilage are succeeded in the same way but an additional factor is the protection afforded by the chondrocytes of the matrix.
3. **Bone and Artery**: It does not really matter if the grafts die because they can still provide a framework for host cells to colonize.
4. **Kidney**: Hundreds of kidneys have been transplanted and due to improved management, there is a high survival rate.

Matching	Percentage	Duration
1. Complete matching	90-100	4 yrs
2. Partial matching	60	6 yrs
3. Mismatching50	2 yrs	

Patients are partially immunosuppressed at the time of transplantation because uraemia causes a degree of immunological anergy.

29.56.3 Ways of rejection of kidney allografts

(a) *Hyperacute rejection* within minutes of trans-plantation.

(b) *Acute early rejection* accruing up to 10 days or so after transplantation is characterized by dense cellular infiltration, rupture of peritubular capillaries, cell-mediated hypersensitivity reaction involving T-lymphocytes.

(c) *Acute late refection* which occurs from 11 days onwards in patients suppressed with prednisone and azatheoprim.

(d) *Insidious and late rejection* associated with subendothelial deposits of immunoglobulin and C_3 on the glomerular basement membranes.

5. **Heart**: 40-50 per cent of transplant patients survive by one year.
6. **Liver**: Survival rates for orthotopic liner grafts are broadly in line with those receiving heart transplants.
7. **Haemotopoietic tissue**: The patients with certain immune deficiencies disorders and some forms of anaemia are treated *with lymphoid stem cells.* Successful result with *bone marrow transfers* needs highly compatible donors. If fatal grafts have reactions, they are to be avoided.

29.57 Transplantation Surgery

Transplantation surgery is an important field of medical endeavour in which great advances have been made in recent years. Kidneys have been transplanted for a number of years with varying degree of success. The first heart transplant was carried out in 1967 by Christian Barnard in Cape Town, and since then attempts have been made to graft lungs and livers. (For success, the tissues of the donor and recipient must be as *genetically similar as possible.* How close they are can be determined by *tissue-typing,* a procedure which is a necessary preliminary to surgery). Subsequent rejection of the transplanted organ or graft is prevented or at least slowed down, by *cobalt irradiation* and *immunosuppressive agents or drugs* which inhibit the body normal immune response e.g., cortisone and related steroids, drugs inhibiting cancer cells like endoxan fluorouracil, antilymphocyte serum (ALS). The trouble is that in so doing they inhibit the body's defences against micro-organisms and therefore, lay the patient wide open to disease. The high facilure rate in transplantation surgery is due to the patient succumbing to a disease, commonly pneumonia to rejection of the organ. In the future, success will depend on using drugs which abolish the graft-rejecting response but leave the disease-response unimpaired. If such drugs can be developed there is no reason why organs should not be transplanted freely between different individuals of the same species called *homografts* or between individuals of different species called *heterografts* with reasonable success.

However, it is as well to remember that time grafting the organ transplantation are highly unnatural events which never occur in nature. The fact that transplanted organs are rejected is merely the result of individual diversity, a reflection of the fact that every individual is genetically unique. Rejection is part and parcel of the body's normal defence against invading microbes one can hardly expect the body to distinguish between pathogenic micro-organisms and useful time introduced by a benevolent surgeon.

29.58 Rejection of Grafts

The rejection phenomenon is a by-product of immune response. The living tissue is transplanted from one person to another, the recepient recognizes the *foreign antigens* in the transplanted tissues (allograft) and destroy it. Destruction of a graft in this way is described as *rejection of graft.*

29.58.1 Tissues for allograft rejection

Avascular tissues, e.g., cornea and cartilage do not arouse the allograft reaction but grafts of tissues which possess a fibrous stroma and a blood supply are all liable to rejection. Some organs are better tolerated than others; a transplanted kidney arouses less reaction than an allograft skin. Successful organ transplantation can only be achieved by reducing the allograft reaction to a minimum.

29.58.2 Nature of allograft rejection mechanism

Graft rejection is an *immuno-response* and *cell-mediated.* Though similar in principle, the rejection of grafts involves somewhat different type of immune response. In this case the *antibodies are cell bound* and *not free in the blood plasma.*

Transplanted organs such as kidney and heart are rejected unless the two individuals *share exactly the same complement of antigens.* This will only be the case if they are *genetically identical,* i.e., *identical twins.*

29.58.3 Mechanism of graft rejection

There are two main types of grafts rejection process:

1. The so-called 'first set' response which occurs about 10 days after a first graft from an unrelated donor.
2. The 'second set' response occuring in about 7 days in an animal which had previously received a graft from the same unrelated donor.

It involves a sequence of separated phenomena developing from initial contact of the foreign antigens with the host immunological system through the rejection itself.

The sequence of events includes: Establishment of lymphatic drainage → antigen release → contact with cells of the immune system → proliferation of antibody-forming cells of the lymphoid tissues → inflammatory response → release of pharmacological mediators → cytotoxic effects of lymphoid cells → and/or antibody on the graft tissues.

29.58.4 Prevention of rejection (Recipients' acceptance of foreign tissue)

Under the following circumstances the recepient can accept the donor's tissue.

1. **Immunological tolerance**: (Introducing foreign tissue into the body). The principle is the formation of antibody against the foreign tissue. One way of achieving this is to introduce some of the foreign tissues into the recepient before or very soon after birth (i.e., at the time when according to Burnet's Clonal Selection Theory, antibody production against the individual's own antigens is being abolished. This was first discovered by Medaevar who infected cells obtained from an adult mouse into a foetus. He found that after birth the recepient would acept, parmanently, grafts of skin and other tissues from the donors. The recepient has been made immunologically tolerant to the donor's tissue.

   The maintenance of tolerance state requires the persistence of the antigen in the graft tissues but it does not occur. So tolerance persists indefinitely when the animal continues to be exposed in definitely to the tolerizing antigen.

   The immunological tolerance is the result of a change in the lymphoid cell population.

2. **Re-equipping the recepient with lymphocytes**: Such induced tolerance was found to be abolished by re-equipping the recepient with lymphocytes from another individual of the same strain.
3. **Anti-lymphocytic globulin**: An anti-lymphocytic globulin (ALG) or serum prepared by injection of lymphocytes of one species into an animal of another species, prevents lymphocytes from participating in immune reaction. ALG has been used to suppress reactions to human allografts, e.g., kidney transplant.
4. **Histocompatibility antigens and tests**: Attempts have been made to detect histocompatibility antigens (HLA) so that the most suitable donor can be selected. The cell surface antigens are known as histocompatibility antigens and the genetic region determines the expression of these antigens termed as the major histocompatibility complex. MHC exerts a predominant influence on the survival of grafts.
5. **Immunosuppressive agents**

   (a) X-ray irradiation prolongs the survival of allograft.

   (b) **Drugs**

   i. *Steroids*—Cortisone and related steroids.

   ii. Drugs inhibiting the growth of malignant cells.

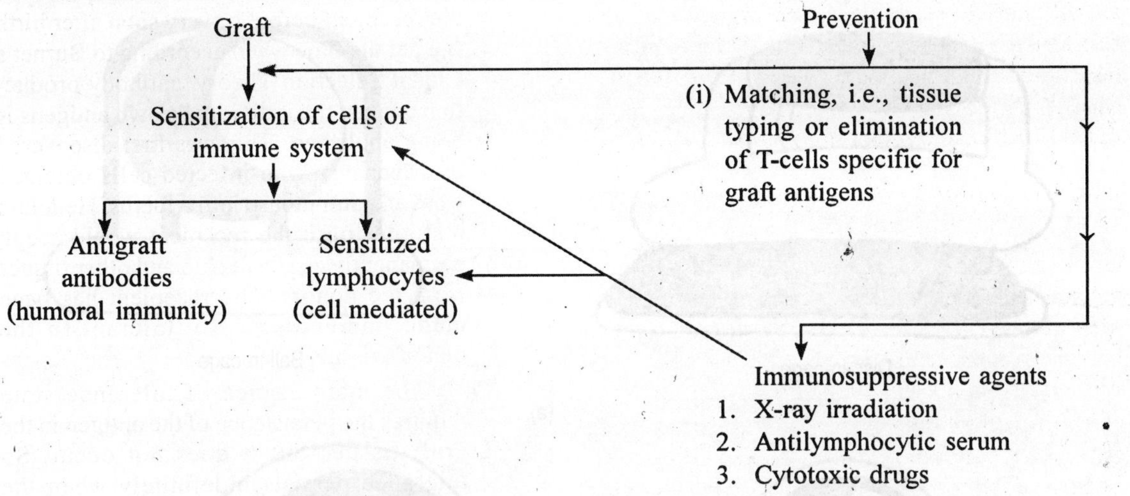

Fig. 29.38: Schematic diagram of rejection mechanism and their prevention

29.59 Prosthetic Heart Valves or Artificial Heart Valves

Majority of the population suffer from valve damages of the left ventricle. They are mitral valve at the junction of left atrium and left ventricle and the aortic valve at the opening of aorta in the left ventricle. Oxygenated blood passes through mitral valve from left atrium to the left ventricle and the blood from the left ventricle passes through aortic valve to the systemic circulation. These valves encounter high pressure of left ventricle due to forceful contraction of the heart. The pressure exerted by the left ventricle is higher than the right ventricle.

29.59.1 Deformities

Deformities of the cardiac valves generally occur from healed lesion of valvulitis or rarely from congenital malformation. The sequele of inflammatory changes in the heart valves are:

1. Thickening
2. Adhesion
3. Retraction
4. Shortening of valvular leaflets.

29.59.2 Structural components of heart valve

I. *Upper layer*: Fibrous, containing collagen fibrils running parallel to the edge of the leaflets, cross weaving of layers of collagen fibrils having 30-50 nm in diameter.

II. *Lower layer*: Spongiosa, contains collagen fibres arranged in radial direction, with fine elastic fibres.

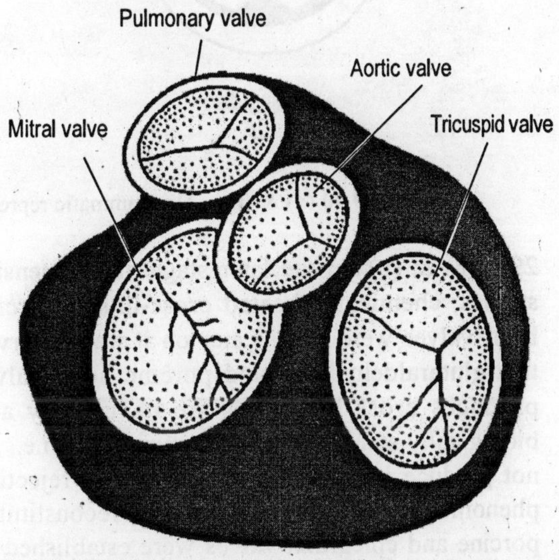

Fig. 29.39: Diagrammatic representation of transverse section of the heart showing view from above the relative position of the heart valves

The common defects are either due to:

1. stenosis or narrowing of the valvular orifice
2. or, regurgitate or insufficiency. Mitral stenoses is the most common of valvular disease and always due to rheumatic fever.

29.59.3 Replacement of the valve

In earlier days heart valve were replaced by mechanical device, e.g. a ball in metal cage (Fig.

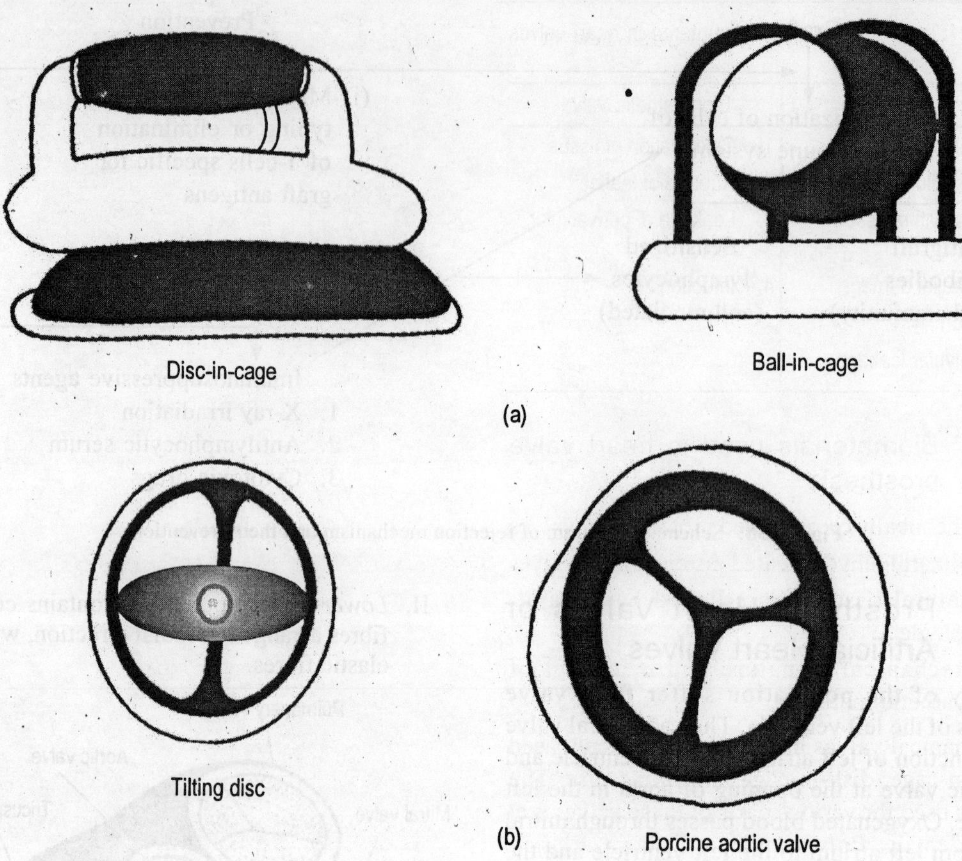

Fig. 29.40 (a) & (b): Diagrammatic representation of different design of prosthetic heart valves

29.29(a) & (b). In the later stage after extensive studies, chemically treated cross-linked porcine heart valves were used. Later on it was observed that glutaraldehyde treated porcine heart valves produces cross-links that are chemically and biologically stable and non-immunogenic, i.e., do not produce immunogenic reaction like rejection phenomena with body tissue. Thus reconstituted porcine and epicardial valves were established as useful prostheses and implanted into many patients. Major number of valves are made from natural tissue cross-linked by glutaraldehyde and the remaining portio are mechanical in nature and made of synthetic material.

Mechanical valves are of many types having different designs such as:

1. ball-in-cage
2. disc-in-cage
3. tilting disc
4. porcine aortic valve.

These are shown diagrammatically in (Fig. 29.40 (a) & (b). The material used for these valves are given in the section 29.59.4. Mechanical valves have disadvantage is that the patient to be maintained anticoagulant therapy regularly to prevent coagulation of circulatory blood.

Prosthetic valve made of biomaterial, i.e., bio-prosthetic valves are made of chemically treating porcine aortic valve or bovine pericardium with glutaraldehyde which cross-links collagen fibrils increasing resistance to enzymatic degradation by collagenase mechanical wear and rejection phenomena. Bio-prosthetic valves have great advantage that patient does not need anticoagulant therapy continuously.

In spite of various prosthetic valves which are used nowadays, the sum total result is not encouraging (Table 29.21).

Table 29.21: Troubles problems encountered in heart valves implantation

Mechanical Valve	Bioprosthetic Valve
1. Thromboembolism	1. Calcification of tissue
2. Suture failure	2. Rupture of leaflet
3. Overgrowth of tissue	3. Leakage of perivalvular region
4. Endothelial damage	4. Infection
5. Infection	
6. Perivalvular leakage	

29.59.4 Biomaterials used in heart valve prosthesis

1. Chemically cross-linked protein heart valves
2. Glutaraldehyde treated porcine heart valves
3. Natural tissues cross-linked by glutaraldehyde
4. Mechanical in nature and made out of synthetic materials.

Mechanical valves have different design based on ball-in-cage, disk-in-cage and tilting disc. The materials used for various parts of these valves are: Glutaraldehyde, Polypropylene, Dacron, Elgiloy wire, Nylon support band covered with polyester, Silicone rubber, Teflon, Prolytic carbon, Darcon, Titanium, Polyethylene (Ultra-high molecular, Bebrin, pirolytic carbon), Derline composite.

29.60 Cryo-Technique (Preservation technique for biological cell and tissue)

Nowadays biomaterials like protiens, cells, tissue, organs are used in medicine, surgery for *grafting, transplantation* and *blood transfusion, burn, wound repair* which depend on these materials but problem arises for availability. Natural materials are unstable and life span is short for deterioration. To overcome the difficulty, preservation technique has been developed considering their physical characteristics and chemical composition called cryo technique.

29.60.1 Techniques

I. **Non-freezing storage (Hypothermic)**: For organ transplantation preservation time is urgently needed. Hence hypothermic storage is used.

There are two techniques for hypothermic preservation for organs.

1. Static cold storage by ice immersion at –4°C to depress enzymatic activity thereby reducing metabolic activity without formation of ice crystals.
2. Continuous cold machine perfusion at –10°C to depress metabolism and removal of metabolic waste product.

   Discussion: Practically simple cold storage preservation is the method of choice since the storage time can be slightly increased by using continuous perfusion. Apart from that, the development of University of Wisconsin (UW) solution has extended the preservation time.

   As for example, in case of human liver the storage time can be increased from less than 10 hours to 30 hours, thus alleviating the need for continuous preservation storage.

   Extensive studies have shown the lactobionate is an essential component of UW solution because of an effective osmotic agent suppressing cell swelling when the cell or tissue is metabolically depressed.

3. The third non-freezing technique is used to depress chemical reaction of a biomaterial which is undercooled storage. The biomaterial is preserved in subzero temperature in the absence of ice crystal formation. This method is particularly useful for preservation of protiens for a medium.

II. **Freeze Thaw Technique**: This technique is most common for preservation of cell and tissue. It is very much useful in transplantation emergency, e.g., bioartificial liver. In the technique introduction of cryoprotective agents, i.e., glycerol has made improvement. The mechanism of action is not clearly understood but it has been explained that there are four main protective actions of *cryoprotective agents* (*CPAs*). These are:

1. Act to stabilize the proteins of biomaterials under temperature level for interaction of sugars with polar head of phospholipids.

2. Lowering the electrolyte concentration of the suspending medium of the cell by altering the phase relationship at a given temperatures during cooling.
3. CPAs reduce the temperature at which cells undergo lethal intracellular ice formation.
4. CPAs promote the formation of vitreous than crystalline phases inside the cell during cooling and help to prevent ice formation.

29.60.2 Cryopreservatives

1. Glycerol
2. Dimethyl sulfoxide
3. Polyvinyl pyrolidone
4. Hydroxyl starch.

29.60.3 Effect of Addition and Removal of CPAs

(a) In case of penetrating additive *dimethyl sulfoxide* and *glycerol*—less permeability to membrane than the permeability to the water.

Water entrapped in the cell biomaterial → supercooled → then at subzero temperature → thermodynamic equilibrium is established by intracellular ice formation → high electrolyte concentration and mechanical effect of ice formation → cell damage exosmosis in the cell → dehy-dration → shrinkage → high concentration of solutes → death may occur or may remain alive.

(b) During prefreeze addition to these components, the biomaterial recognizes CPAs as another extracellular solute. As a result the cell starts rapid transport of water through cell membrane to the extracellular fluid. At the same time CPA is permeable and gradually diffuses into the cell producing chemical unequilibrium. This process continues for a few minutes until equilibrium is established and by this time biomaterial regains its normal volume.

(c) Reverse process of cell structure and subsequent return to normal volume observed after removal of CPA.

(d) If the CPA concentration is high, the cell shows severe osmotic damage.

(e) Osmotic damage from addition or removal of CPA can be reduced by intermittent increasing or decreasing the concentration of CPA.

(f) Using stepwise CPA addition and removal in reducing osmotic damage is counter-

CHANGES IN CELL DURING CRYOPRESERVATION

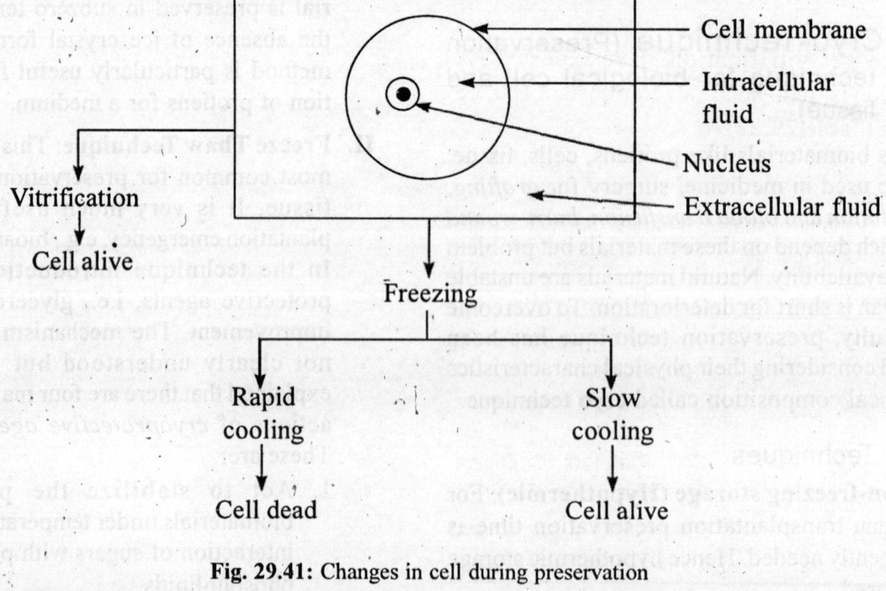

Fig. 29.41: Changes in cell during preservation

balanced by the increased risk of toxic damage in the biomaterial from longer exposure time.

(g) In case of single step CPA removal is shown to be effective.

(h) In case of impermeable CPAs—there is injurious effect due to exosmosis.

(i) If in extracellular medium—can be easily removed but unable to protect intracellular materials during freezing (Fig. 29.41).

29.60.4 Lyophilization (freeze-drying technique)

Freeze-drying technique or lyophilization is a dehydration storage technique. It is advantageous because it produces a final product which can be stored at temperature and reconstiuted with addition of solvents.

It is usually employed in case of proteins, e.g., collagen, sometimes RBC, platelets, and also applied for liposomes and drug delivery system.

29.60.5 Vitrification

The formation of ice crystals is disadvantageous which can be reduced by rapidly cooling the biomaterials at low temperatures at sufficient rates for production of amorphous solid. The viscosity plays a role in vitrification, such as the more viscous the liquid, slower it can be cooled to achieve the vitrification. Large amount of cryoproductans is required to get vitrification. But excess CPA concentration is toxic hence an equilibrium is to be maintained (Table 29.22).

Storage

Table 29.22: Various biomaterials under different preservation technique

Technique	Biomaterials
I. *Nonfreezing*	Liver, Heart, Heart valves, Protein solution
II. *Freeze-thaw*	RBC, Cartilage, Bone marrow, Skin
III. *Lyophilization* (freeze dry)	RBC, Platelets, Collagen, Liposomes
IV. *Vitrification*	Embryo, Islet of Langerhans, Corneal tissue

29.61 CRYOBATH (Cryoprecipitate bath)

Cryobath is a circulating thermostat used for thawing blood components. It has the following components (Fig. 29.42):

1. *Motor Pump*—For circulation of water for reduction of the thawing time.
2. *Refrigeration system*—For continuous cooling of the liquid in the bath using a compressor.
3. *Heater*—To maintain temperature of the liquid at 4°C ± 0.1°C.
4. *PID Controller*—For controlling the heater.
5. *An over-temperature cut off*—Disconnects the main supply, if the temperature goes beyond 6°C.
6. *A low level sensor*—To protect from heater in case of over-temperature or low water level.
7. *An audible Alarm*—is switched ON, in addition to the red LED being lit.
8. *A Timer*—Included to inform the operator the end of the thawing cycle.
9. *Yellow LED*—is along with audible beep indicating the end of thawing cycle.
10. *Green LED*—It is on the front panel. It starts blinking if the temperature of the water is less than +4 °C. This indicates the power is being supplied to the heater. The temperature displayed on the front panel display should begin to rise. When the temperature reaches +4°C and eventually drops down below +4°C when the green LED starts pulsing again. The water temperature once again crosses the set temperature of +4°C. However, the shoot up of this time is less than that during the previous heating cycle. This continues for a few cycles after which the temperature stabilizes. The bath is now can be used (Fig. 29.44).
11. *Cryoprecipitate bag and stainless steel basket*—These bags can be placed in the space provided in the stainless basket. Place the lid on the bath opening and switch ON the timer. An audible beep will be heard at the end of the thawing cycle, Switch OFF the timer and remove the bags.

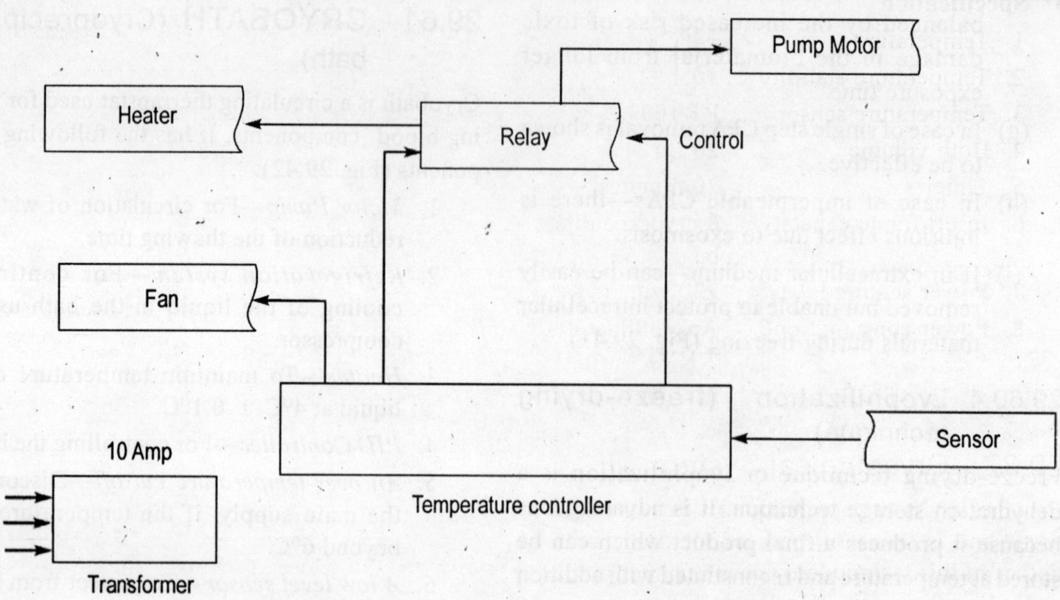

Fig. 29.42: Block diagram of control unit of cryobath

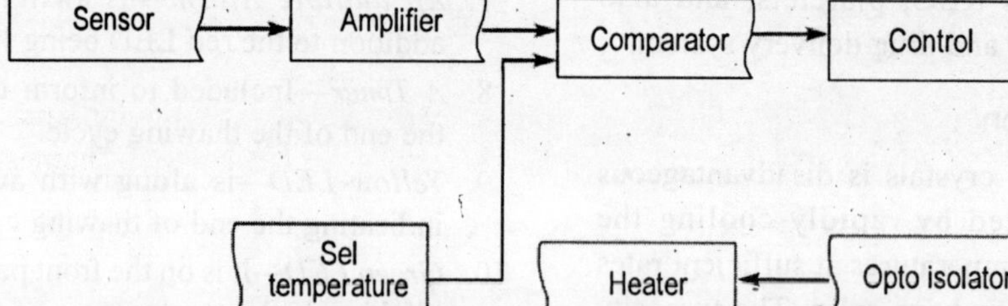

Fig. 29.43: Block diagram of circuit diagram

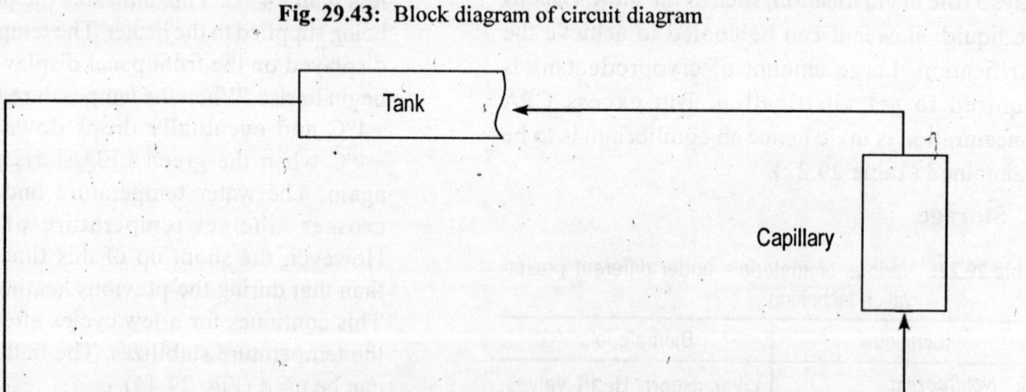

Fig. 29.44: Block diagram of refrigeration unit

- **Specification**

1. Temperature	+4°C
2. Temperature stability	+0.1°C
3. Temperature sensor	PT 100
4. Both volume	20 Ltr
5. Timer	20 minutes
6. Insert frame capacity	8 cryoprecipitate bags
7. Power acquired	230 volt
8. Power consumption	Approximately 2 KVA during heating to +4°C

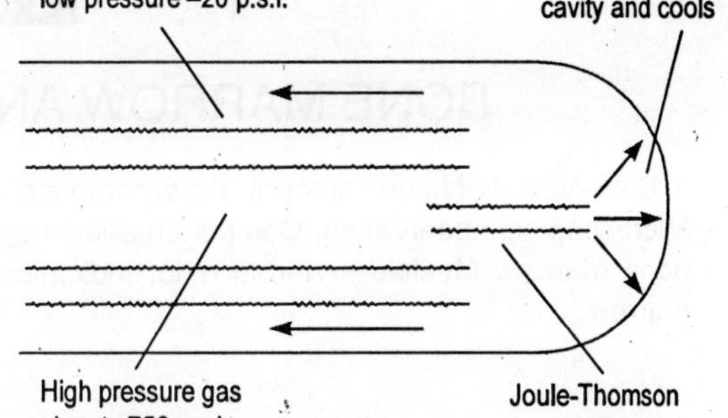

Fig. 29.45 Cut away section of cryoprobe

29.62 Cryosurgical System

Cryotherapy is the application of extreme cold to tissue and is an useful process for destruction of tissue. In this technique there is minimal bleeding and pain in the post-surgical period. This, technique has been successfully applied in various surgical methods such as:

(a) Dermatology
(b) Neurology
(c) General surgery
(d) Urology.

Modern cryosurgery was developed with a simple apparatus using nitrous oxide or carbon dioxide to get temperature –70°C. And it also uses high-pressure non-syphon cylinders of nitrous oxide.

The system consists of a flexible tube connecting the gas flow control unit to a cryoprobe (Fig. 29.45) having two concentric tubes. The inner tube supplies nitrous oxide or carbon dioxide at a pressure of 4000 Kpa and 6000 Kpa to a narrow aperture in the end called Joule-Thompson aperture.

PART VI

BONE MARROW AND TRANSPLANTATION

1. Bone Marrow-Nature, amount, Type, composition, Origin, Ratio of cells, Function, Distribution of marrow, Microscopical observation, Components, average percentage of different cells, Significance of study of bone marrow, Myeloid-erythroid ratio, Indications for bone marrow study. 2. Transplantation of bone marrow.

29.63 Bone Marrow (Myeloid Tissue)

Bone marrow is the *cellulovascular* (i.e., Cellular + Vascular tissue) *tissue* occupying the medullary cavity and the cancellous spaces of the bones. It is also called myeloid tissue (Myelos = Marrow).

29.63.1 Amount

Active marrow in the adult is estimated at 3.5-6.0 per cent of body weight. The volume of the marrow is 70 ml at birth and about 4000 ml in the adult.

29.63.2 Types, distribution, changes at different ages

There are two types of bone marrow:

A. *Yellow Bone Marrow* (half of the total marrow being *inactive*).

B. *Red Bone Marrow* (Half of the total marrow being *active*).

These two types are seen *in the adult only* because *before puberty all the bones contain red marrow only.*

A. **Yellow Marrow**: Yellow marrow consists of:

1. Fat cells (gives yellow colour of marrow)
2. Blood vessels
3. Reticulum cell
4. Reticular fibres.

In the adult life they appear and occupy the spaces where red marrow is absent. The *yellow colour* is due to *presence of enormous fat cells* and *not so richly supplied by the blood.* From birth upto before puberty all the bones contain highly cellular red marrow throughout the length of the bones. With increasing age, the marrow becomes more fatty, the process setting in:

1. First in the distal bones of the limbs (Tarsus and Carpus);
2. Then in the intermediate (Tibia, Fibula, Radius, Ulna);
3. Finally in the proximal bones (Femur, Humerus).

At the attainment of puberty, a patch of yellow marrow appears in the distal end of the shaft of the long bones and gradually extends on both sides. Ultimately, at the age of twenty, the entire red marrow of the long bones is replaced by yellow marrow except the two articular ends of long bones like femur, tibia, humerus, etc. Throughout the adult life this distribution persists but as age advances yellow marrow proportionally increases. By 70 years, half of the ribs and sternum contain yellow marrow which was full of red marrow during adult life.

Though the yellow marrow is inactive, yet it *has the capacity of resuming its activity to produce blood cells during urgent need.* Under such condition a certain portion of the yellow marrow is converted into the red marrow.

B. **Red Marrow**: It is red colour due to *high vascularity,* having a rich circulation from the blood entering via nutrient artery and *presence of RBC.* It consists of similar structure like yellow marrow having very few fat cells plus numerous blood cells of all kinds and their precursors—erythroid and myeloid cells are present. Hence, the structure or composition is as follows:

A. **Stroma**

1. Fat cells (very few)
2. Blood vessels
3. Reticular cells
4. Reticular fibres
5. Vasoconstrictor nerves
6. *No lymph vessels* in red and yellow marrow.

C. **Extracellular Matrix**: It contains: Collagen, laminin, fibronectin, vetronectin,

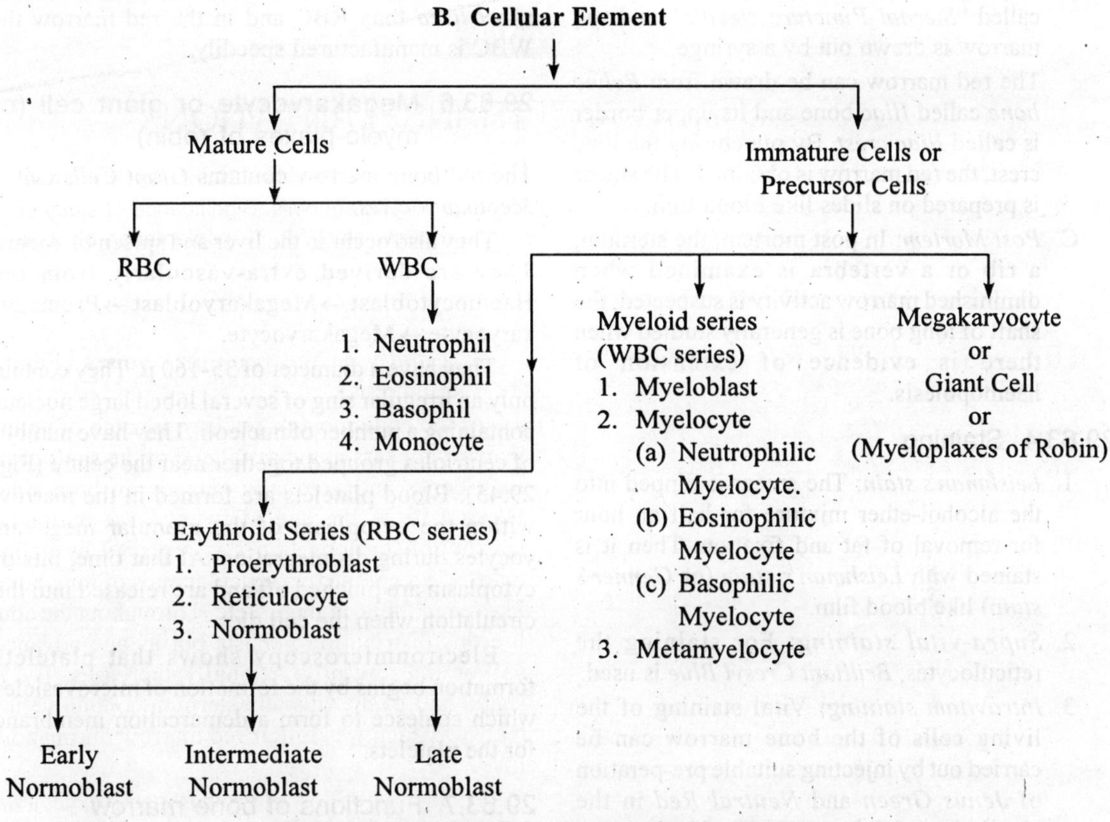

Fig. 29.46: Showing cellular elements of red bone marrow

hemonectin, thrombos-pondin, proteoglycans, glycosoamino-glycans—chondroitin sulphate, heparan sulphate, dermaten sulphate, Keratan sulphate, hyaluronic acid (1) ECM is secreted by stromal cells of the red bone marrow particularly by endothelial cells and fibroblasts.

From late foetal life to before puberty all the bones contain red marrow but after puberty it is only located in the following bones:

1. *Two ends* of the *long bones* (e.g., Femur, Tibia, Humerus, Ulna, etc.)
2. *Flat bones* (e.g., skull bones, Ribs, Sternum, Pelvic bones like Iliac bone, etc.
3. *Short bones*—Vertebrae, Carpal, Metacarpal, Phalanges, Tarsal, Metatarsal.

Marrow remains red throughout the life in the following bones: Ribs, Sternum, Skull, Clavicles, Vertebral bodies, Pelvic bones. Although only half the marrow is active in the adult, yet it has enormous functional capacity and considerable room for expansion. In old age, the red marrow of the skull bones undergoes degeneration and is called *"gelatinous marrow"*, *Vascular Arrangement.* The nutrient artery of the bones enters the medullary cavity through Volkman's Canal. It breaks up into smaller branches, which lead to a network of intercommunicating *"sinusoids"*.

These vessels (sinusoid) are lined by a thin *endothelium* (like capillaries elsewhere) but when dilated have the capacity of veins. The *endothelium of sinusoids* divide and give rise to RBC. These *sinuses,* where active erythrogenesis is going on, remain collapsed, thus creating the anoxic condition favourable for red cell formation. When the red cells are sufficiently mature, these collapsed sinuses open up, blood stream enters and the newly born cells are carried away into the circulation.

29.63.3 Method of Study

A. *In Animals*: The bone is taken out and the red marrow is collected after breaking the bone. It can be squeezed out of the bone by a bone crusher (such as ribs of goat). A smear is prepared like a blood smear.

B. *In Human* (*By Sternal Puncture*): The sternum is punctured by a special needle called "*Sternal Puncture Needle*" and red marrow is drawn out by a syringe.

The red marrow can be drawn from *Pelvic bone* called *Iliac* bone and its upper border is called *Iliac crest*. By puncturing the iliac crest, the red marrow is obtained. The smear is prepared on slides like blood film.

C. *Post Mortem*: In post mortem, the sternum, a rib or a vertebra is examined when diminished marrow activity is suspected; the shaft of long bone is generally studied when there is evidence of extension of haemopoiesis.

29.63.4 Staining

1. *Leishman's stain*: The smear is dipped into the alcohol-ether mixture for half an hour for removal of fat and fixation. Then it is stained with *Leishman's stain* (or *Genner's stain*) like blood film.
2. *Supra-vital staining*: For staining the reticulocytes, *Brilliant Cresyl Blue* is used.
3. *Intravitam staining*: Vital staining of the living cells of the bone marrow can be carried out by injecting suitable pre-peration of *Jenus Green* and *Neutral Red* in the circulation of living animals.

29.63.5 Ratio between myeloid cells and erythroid cells in the marrow

In the circulation the ratio between WBC and RBC is 1 : 700. But in red marrow it is reversed, i.e., ratio of myeloid cells and erythroid cells is 8 : 1. The reason is that in the circulation, the WBC is *short lived* than RBC and in the red marrow the WBC is manufactured speedily.

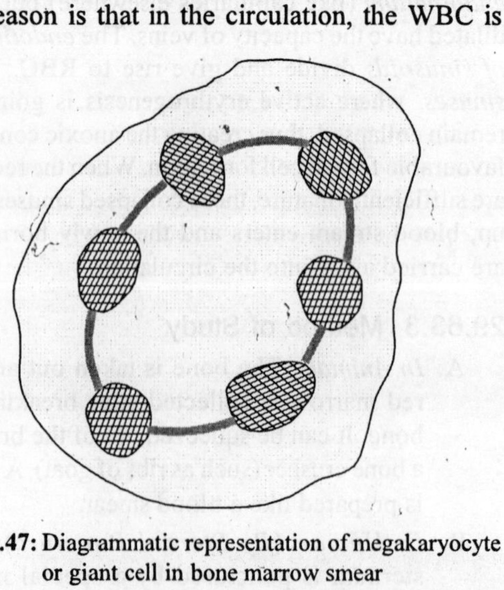

Fig. 29.47: Diagrammatic representation of megakaryocyte or giant cell in bone marrow smear

29.63.6 Megakaryocyte or giant cell (or myelo-plaxes of robin)

The red bone marrow contains *Giant Cells* called *Megakaryocytes.*

They also occur in the liver and spleen of foetus. They are derived extra-vascularly from the Haemocytoblast →Megakaryoblast →Promega-karyocyte →Megakaryocyte.

They have a diameter of 35-160 μ. They contain only an irregular ring of several lobed large nucleus containing a number of nucleoli. They have number of centrioles grouped together near the centre (Fig. 29.45). Blood platelets are formed in the marrow within the cytoplasm of the granular megakar-yocytes during disintegration. At that time, bits of cytoplasm are pinched off and are released into the circulation when the cell dies.

Electronmicroscopy shows that platelets formation begins by the formation of microvesicles which coalesce to form a demarcation membrane for the platelets.

29.63.7 Functions of bone marrow

1. **Production of Blood Cells**: All the formed elements of blood, i.e., blood cells like RBC, granulocytes, platelets (from megakaryocyte or giant cell), lymphocytes and monocytes to some extent are produced from red marrow. Erythropoietin acts in the red marrow for production of erythroid cells.
2. **Destruction of RBC**: Abnormal, imperfect, damaged and aged (senile) RBC are destroyed in the red marrow. These cells are sequestered or trapped and phagocytosed in the macrophages of reticulo-endothelial system of the bone marrow. Iron part is stored here (also in the liver, spleen, R.E. Cells) as *haemosiderin* and *ferritin* and the rest is ultimately converted into bile pigments.
3. **Storage Function**: Iron is stored in the bone marrow as haemosiderin and ferritin for future use of haemoglobin synthesis.
4. **Reticula-Endothelial Function**: The reticular cells and endothelial cells of the bone marrow are called macrophages. They

engulf the bacteria, foreign particles etc. and inactivate toxins or other toxic substances of the body. The free macrophages of the bone marrow are increased during the invasion of toxins or during rapid haemolysis.

29.63.8 Microscopical observation of red bone marrow

Observation

I. **Cells of Erythroid Series**
   (a) RBC
   (b) Normoblast: 7-10 μ diameter, cytoplasm reddish and piknotic nucleus.
   (c) Erythroblast: 10-15 μ diameter, cytoplasm shows mixture of acid and basic stains.
   (d) Megaloblast or pro-erythroblast: 15-20 μ diameter with large nucleus and deep violet cytoplasm.

II. **Myeloid Series**
   (a) Myeloblast: 12-18 μ diameter, large round pale blue nucleus with bluish cytoplasm.
   (b) Myelocyte: 12-18 μ diameter and classified as follows:
      i. Neutrophil myelocyte
      ii. Eosinophil myelocyte
      iii. Basophil myelocyte
   (c) Granulocytes: (i) Neutrophil, (ii) Eosinophil, (iii) Basophil.

III. **Megakaryocyte or Giant Cell**: 30-100 μ diameter containing irregular rings of multilobed large nucleus appears as annular or horse shape.

IV. Reticular cells with fibres and a few fat cells.

29.63.9 Functions of bone marrow

1. Production of blood cells
2. Destruction of RBC
3. *Storage function*: Iron as haemosiderin and ferritin.
4. Reticulo—endothelial function
   (a) Engulfing foreign particles
   (b) Engulfing bacteria
   (c) Inactivate toxins or other toxic substances of the body.

29.63.10 Components of red bone marrow

A. **Stroma**
   1. Fat cells (very few)
   2. Blood vessels
   3. Reticular cells
   4. Reticular fibres
   5. Vasoconstrictor nerves. *No lymph vessels in red and yellow marrow.*

B. **Cellular Element (Fig. 29.46).**

29.63.11 Distribution of red bone marrow

A. **Late foetal life to before puberty**: All the bones.

B. **After puberty**: It is only located in the following bones.

Long bones

1. *Two ends* (epiphysis) of the *long bones* (e.g., femur, tibia, humerus, ulna etc.)
2. *Flat bones*: Skull bones, ribs, sternum, pelvic bones like iliac bone (for diagnostic purposes, red bone marrow is obtained from iliac crest).
3. *Short bones*: Metacarpal, phalanges, tarsal, metatarsal, etc.

Bones red marrow throughout the life

Ribs, sternum, skull, clavieles, vertebral bodies, pelvic bones. Although only half the marrow is active in the adult, yet it has enormous functional capacity and considerable rooms for expansion.

Red marrow in the skull in old age

In old age, the red marrow in the skull bones undergoes degeneration and is called "gelatinous marrow".

Table 29.23: Average percentage of bone marrow cells in normal subject

Cell	Average
1. Myeloblast	2.5
2. Promyelocytes	6
3. Myelocytes Neutrophilic	15
Neutrophilic	12
Eosinophilic	1.5
Basophilic	0.3
4. Metamyelocytes	29
Neutrophilic	26.4
Eosinophilic	2
Basophilic	0.3
5. Segmented granulocytes	32

Contd.

Cell	Average
Neutrophils	27
Eosinophils	4
Basophils	1
6. Manocytes	2
7. Lymphocytes	13
8. Pronormoblasts	0.6
9. Early normoblast	2.4
10. Intermediate normoblast	10.6
11. Late normoblast	6.6
12. Megaloblasts	0.2
13. Megakaryocytes	0.4
14. Reticulum cells	1
15. Plasma cells	0.6
16. M.E. Ratio	3-4 : 1

29.63.12 Significance of studying bone marrow

RBC, WBC and platalets are produced in the bone marrow. When blood cell production is defective or abnormal cells are produced, the disease occurs. To assess the condition of bone marrow a smear of marrow is studied.

M : E ratio or myeloid-erythroid ratio

It is the ratio of white cells to nucleated red cells. As for example, if a bone marrow shows 80% white cells and 20% nucleated red cells, the M : E ratio is 4 : 1. The average M : E ratio is 4 : 1 with a normal range of 6 : 1 to 2 : 1.

Conditions in which M : E ratio is increased (7 : 1 or more)

Infections, granulocytic leukemia, leukemoid reactions and a decrease in the number of nucleated red cells.

Indications of bone marrow Study

I. **Absolutely indicated**
   1. Megaloblastic macrocytic anaemia
   2. Aleukaemic or subleukemic leukaemia

II. **Diagnostic indications**
   1. Multiple myeloma
   2. Aplastic anaemia
   3. Gauche's disease

III. **Confirmatory indications**
   1. Leukaemias of all types
   2. Haemolytic anaemia
   3. Idiopathic thrombocytopenic purpura
   4. Metastatic diseases
   5. Sideroblastic anaemia
   6. Disseminated lupus erythematosis.

29.64 Bone Marrow Transplan-tation (BMT)

Bone marrow highly vascular tissue present in the medullary cavity of bone. There are two types of bone marrow, i.e., yellow and red bone marrow. Red marrow is haematopoietic tissue i.e. it is responsible for genesis of blood cell. It is present in all bones in post-natal life before puberty. After puberty yellow marrow appears in the middle and their growth pushes red marrow to the ends of long bones and also present in flat bones and short bones. Red marrow produces 400 billion myeloid cells per day from small number of pluripotent cells. It is a composed of 100 to 1000 billion cells which are regenerated every 2-3 days. But sometimes there is variation in number such as in some systemic infection or sickle cell anaemia or when bone marrow is under stress such as anoxemia due to disease or during ascending high altitude there is increase in number of mature RBC called erythrocytosis but due to some disease or toxic substance if there is increase in number of both mature and immature RBC it is called polycythemia; two types vera or rubra In case of leukaemia there is increase in both mature and immature white blood cells.

Again red bone marrow depression due to diseases or drugs may cause anaemia, leucopenia, agranulocytosis, etc. But if there is damage of red bone marrow due to radiation (e.g., X-ray), chemotherapy, etc., bone marrow transplantation is performed to overcome the crisis,

29.65 Transplantation

Autologous Method: Before application of radiotherapy and chemotherapy, red bone marrow is collected froms the patient. It is preserved by cryotechnique (e.g., cryopreservation) in which it is preserved in a cryopreservative, e.g., glycerol maintaining subnormal temperature. After the treatment the bone marrow is thawed rapidly (freeze-thaw technique). Then it is infused into the patient which circulates in the blood and is lodged in the bone marrow and restores the function of bone marrow with the help of stem cells.

29.65.1 Advantages

1. It is one kind of cellular therapy.
2. Cells are cultured in vitro.

29.65.2 Disadvantages

1. It is a long procedure, takes several weaks for completion.
2. The patient is immune-compromised.

II. **Autologous BMT as cellular therapy**: It involves removal of cells from the patient and temporary storage in vitro (outside the body). The advantage is that there is growth of the stored cells as new therapy and the treatment is based on culture of haemotopoietic cells.
This method to culture in vitro of bone marrow stem cells have been developed. These methods use cytotoxic agents for mobilization of stem cells into circulation. The haematopoietic stem cells and progenitor cells and these cells are collected from circulation by the process of leukophoresis.

III. **Myoablative method**: This method is applied in allogenic cases. The donor's cells are placed to the marrow which help to increase population of the marrow cavity. The main disadvantage is high mortality (10-15%) due to reaction between graft and the recepients when transplanted cells recognize the recepients tissue as a foreign material. This procedure will be succeeded if the problem is removed.

29.65.3 Indication

1. Allogenic transplantation, i.e., leukaemia.
2. Autologous transplantation, e.g., lymphoma, carcinoma, of breast, testes.

I. **Analogous transplantation**: Analogous transplantation has been applied for the treatment of:
(a) Acute lymphoblastic leukaemia
(b) Acute myelogenous leukaemia
(c) Chronic myelogenous
(d) Lymphomas
(e) Breast Carcinoma
(f) Neuroblastoma
(g) Multiple Myeloma
(h) Gene therapy procedure (Insertion of therapeutic gene into haematopoietic stem cell)

The disadvantages are:

1. Long period of hospitalization till the effect is deserved.
2. Possibility of introduction of tumour cells along with cryo-preserved bone marrow cells.
3. Neutropenia, anaemia, thrombocytopenia require antibiotic therapy and repeated transfusion of blood components.

II. **Allogenic transplantation**: Allogenic transplantation has been used in patients having hematogenic malignancy, genetic defect in haematopoietic population.

Disadvantages are:

1. Difficult to get matched donors
2. Development of graft-versus-host diseases.

III. **Alternative haematopoeitic cells**: Stem cell as an alternative has been used. Pluripotent stem cells appear:

1. First in the yolk sac of the embryo
2. Then fetal liver
3. Umbilical cord blood during delivery
4. In peripheral blood in adult at very low concentration but high increase occurs in stem cell mobilization during chemotherapy and growth factors administration to haematopoietic cells, can be collected from umbilical cord, blood mobilized from peripheral blood, and bone marrow. The disadvantages of cord blood are:
    1. Very limited number of cells are available.
    2. Whether these small number of cells are capable to repopulate adult bone marrow.

But mobilized stem cells application, develops rapidly.

Production of large number mature blood cells: Limited production of stem and progenitor cells, led to productions of mature blood cells. Haematopoietic cultures provide large number of different blood cells, e.g., RBC, granulocytes platelets.

29.66 ECM of Bone Marrow

Extracellular matrix of red bone marrow contains: Collagen, laminin, fibronectin, vitronectin, homonectin and thrombosplastin, various proteoglycans with numerous glycosaminoglycan chains linked to a protein core. These glycosaminoglycans are chondroitin sulfate, heparan sulfate dermatan and keratan sulfate and hyaluronic acid.

1. ECM is secreted by stromal cells of the red bone marrow, particularly by endothelial cells and fibroblasts.
2. ECM functions are:
   - (a) Support and cohesion of marrow structure.
   - (b) Regulation of haematopoiesis
   - (c) Different glycosaminoglycans bind and present different growth factors to haematopoietic cells in an active form.
   - (d) Provides anchorage for immature haematopoietic cells. Fibronectin attaches to the receptors of erythroid precursor cells. When maturity progresses through BFU to the reticulocyte stage, adherence to fibronectin is gradually lost and the free cells enter the circulation.
   - (e) Binding of fibronectin makes the erythroid cell more responsive to erythropoietin.
   - (f) The other adhesion protein, hemonectin selectively bind immature granulocyte lineage acts like fibronectin.
   - (g) Binding of progenitor cells help for stem cell homing.

30

Biomedical Technique

1. Bioelectric Potential; Origin; Recording, 2. Concept of fibre optics, 3. Endoscopy-Fibre optic, 4. Heart-Lung Machine, 5. Centrifuge technique.

30.1 Bioelectrical Potential

30.1.1 Electric charge of the membrane

The membrane surface is having an electric charge to *selective adsorption of ions* and ionization of compounds in the membrane surface. Consequently, membrane *possesses Helmholtz* double ionic layer.

30.1.2 Potential difference of the membrane

There is a potential difference between the outside and inside of the cell membrane which is about 0.1 volt. The difference is *due to unequal distribution of K^+ ions in the intracellular and extracellular fluids.* Cell membrane is freely permeable to K^+ and its permeability is 50-100 times greater than Na^+ permeability. There is generally 30-40 times more K^+ ions in the intracellular fluid than the extracellular fluid.

According to the *Nernst equation* for a concentration cell, the potential is close to + 0.1 volt at 37°C:

$$E = 0.061 \log C_2/C_1$$
$$= 0.061 \log 40/1$$
$$= 0.061 \times 1.6 = 0.097 \text{ volt}$$

Where C_2 = the concentration of K^+ ion in the intracellular fluid

C_1 = the concentration in the extracellular fluid.

Therefore, it can be concluded that the potential across the membrane is due to the maintenance of unequal distribution of K^+ ions between the two sides of the membrane.

The living cell contains fluid, i.e., intracellular fluid which has different electrolyte concentration than the extracellular fluid such as Na^+ ion is very high outside the cell membrane and K^+ is very low. Inside the cell, the fluid contains more K^+, Cl^- and protein which is negatively charged. The cell membrane is semipermeable membrane, i.e., it has porses of different sizes for which some are allowed to pass and some are held back. Such as Na^+ cannot go inside the cell and K^+ which is positively charged can come out but is held back by negatively charged ion such as Cl^- and protein, and cannot go out. This causes significant difference in the ionic concentration and distribution due to establishment of Donnan *membrane equilibrium.*

Gibbs-Donnan effect can give to potential difference due to differential distribution of diffusible ions on both sides. The potential difference can be displayed by putting microelectrodes inside the cell and on the surface of the cell membrane and

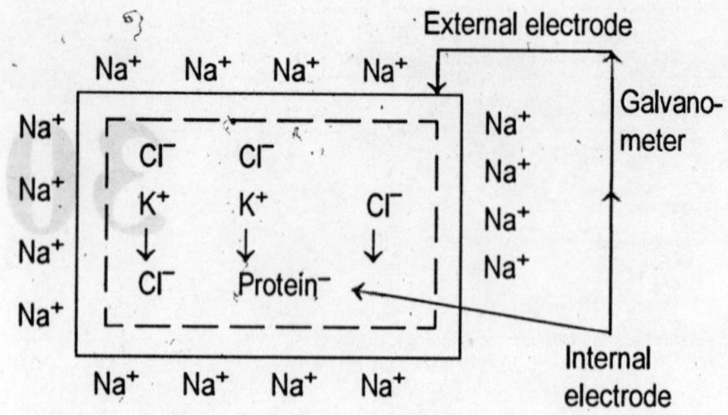

Fig. 30.1: Distribution and arrangement of electrolytes in the living cell

connecting to a galvanometer which is about –80 mV to –90 mV. according to the type of tissue. Due to semipermeable membrane Na^+ ions are arranged outside the cell membrane and negatively charged ions inside the membrane. As a result outside the cell is positive and inside the cell is negative. At this state the membrane is called *polarised membrane.*

When the surface of the membrane is stimulated, or injured, the site of the membrane becomes *depolarised* and is negative electrically in respect of adjacent site due to entry of Na^+ inside and K^+ comes out for *increase* in membrane permeability. Permeability of the cell membrane depends upon:

1. Size of the molecules
2. Size of the pores or channels
3. Lipid solubility
4. Electric charge on the cell surface
5. Thickness of cell membrane
6. Number of protein channels
7. Temperature
8. Molecular weight of the substance

The cell membranes are semipermeable, i.e., certain substances allow to pass through and some are prevented.

The events occur during depolarisation are:

1. Increase in permeability of the cell membrane
2. Entry of Na^+ inside the cell
3. K^+ comes out of the cell
4. Outside of the cell is negative
5. Inside of the cell is positive.

When the stimulus is withdrawn and the active part is given rest, the site becomes *repolarised* due to establishment of ionic equilibria. There is reversal of electric charge, i.e., outerside of the membrane becomes again positive and negative inside. These are called *membrane phenomena.* Any unequal distribution of the charged ion in a tissue on the two sides of the membrane causes difference of potential.

This reversal activity, i.e., repolarisation takes place due to sodium-potassium pump, which normally *transport* 3 *sodium ions inside for each* 2 *potassium ions outside* which gives a net transport of 30 per cent more positive charge outward than inward. Hence this pump is called an *electrogenic pump.*

The importance of sodium:

1. Maintenance of electrical potential of the cell membrane
2. Transmission of impulse in the nerve and muscle
3. Various glandular secretion
4. Prevention of swelling of cell.

This pump mechanism being *active transport requires energy* from ATP. The *carrier* for the mechanism is *sodium-potassium ATPase* which can split up ATP to provide energy. This ATPase is composed of two protein molecules, globulin and glycoprotein (Fig. 30.1).

At inner surface of the membrane Mg^+ binding with ATP releases high energy rich PO_4 (phosphate) bond to provide energy to Na^+-K^+ pump causes K^+ to split away from *carrier* and causes to bind Na^+ ion. At the outer surface of the membrane Na^+ splits away from the carrier and K^+ ions bind without any energy at the outer surface as both release of Na^+ and binding of K^+ is an exothermic process.

Some physiological activities related to potential difference:

1. Muscle contraction
2. Nerve impulse
3. Heart contraction
4. Brain activity
5. Vision
6. Glandular secretion.

30.1.3 Clinical importance of bioelectric potentials recording (Different methods for measurement)

Living cell in the human body has potential due to unequal distribution of electrolytes such Na^+, K^+,

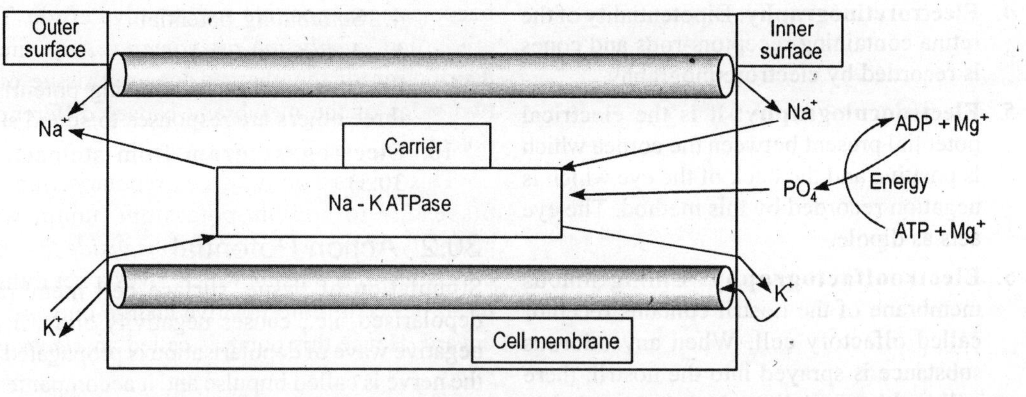

Fig. 30.2: Active transport of Na^+ and K^+ (Sodium-Potassium pump)

Cl^-, etc. between the two sides of cell membrane i.e., intracellular and extracellular concentration. In resting state, the inside of the cell is negative and the outside of the cell is positive. In active condition, there reversible reaction, i.e., outside the cell is negative and inside the cell is positive creating a potential difference. The electrical activity of the tissue is propagated along the tissue fluid and it can be picked up by the electrode from the surface and it can be displayed by the biomedical instrument graphically by the recorder. These are:

1. **Electrocardiograph (ECG)**: Bioelectric potential of heart muscle is picked up and displayed by electrocardiograph machine.
2. **Electroencephalograph (EEG)**: Recording of bioelectrical potential of brain called electroencephalography.
3. **Electromyography (EMG)**: Bioelectric potential recorded from the active muscle is recorded by the instrument called electromyograph.

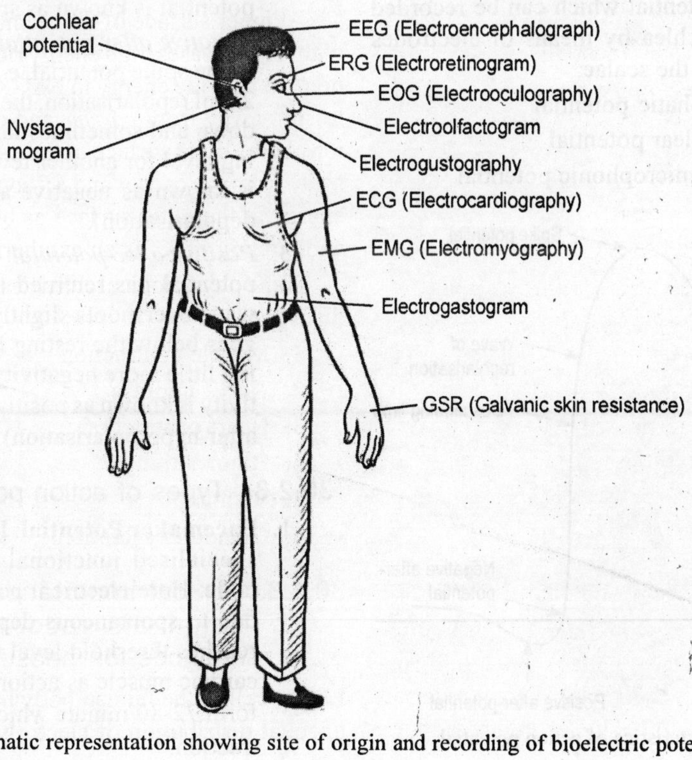

Fig. 30.3: Diagrammatic representation showing site of origin and recording of bioelectric potential in human body

4. **Electroretinography**: Bipotentiality of the retina containing receptors-rods and cones is recorded by electroretinography.
5. **Electrooculography**: It is the electrical potential present between the cornea which is positive and the back of the eye which is negation recorded by this method. The eye acts as dipole.
6. **Electroolfactography**: Lining mucus membrane of the nostril contains receptor called olfactory cell. When any odorous substance is sprayed into the nostril, there will be bipotentiality which is recorded in the method called electroolfactography. This potential remains as long as the odorous substance containing air continues to pass through nostril.
7. **Electronystagmography**: Recording of electrical potential during nystagmus e.g. rhythmic movement of eyeball (Rotatory, horizontal, vertical).
8. **Electrogustography** from taste buds of the tongue.
9. **Cochlear potential (hearing part of the ear) (cochlea)**: There are five types of electrical potential which can be recorded from the cochlea by means of electrodes inserted into the scalae.
   1. Cortilymphatic potential
   2. Endocochlear potential
   3. Cochlear microphonic potential

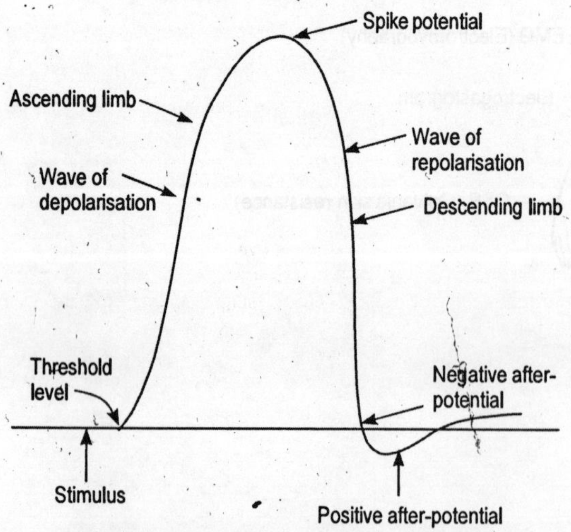

Fig. 30.4: Characteristics of action potential

   4. Summating potential
   5. Action potential

   First two of these are resting potential and three others are responses to sound stimuli.
10. **Electrogastogram** from stomach (Fig. 30.3).

30.2 Action Potential

Stimulation of tissue, surface cell membrane is depolarised, i.e., causes negativity and when this negative wave of depolarisation is propagated along the nerve is called impulse and it accompanies voltage change which is called action potential.

30.2.1 Recording

1. Galvanometer
2. Cathode ray oscilloscope.

30.2.2 Characterization

The graph of action potential shows the following characteristics (Fig. 30.4).

1. *Ascending limb* is the *depolarisation wave* having slow rise initially, then overshoots rapidly reaching the zero level or unipotential line, then to approximately +35 mV. This initial very large change in membrane potential is known as spike potential.
2. *Negative after-potential*: At the termination of the spike potential, e.g., at approximately 2/3 of repolarisation, the rate of fall is slowed down and sometimes fails to reach the resting level for another few milliseconds. This is known as negative after-potential (after depolarisation).
3. *Positive after-potential*: Once the membrane potential has returned to resting level, the wave overshoots slightly but slowly, e.g., if goes below the resting level due to becoming little more negativity. This excess negativity is known as positive after-potential (or after hyperpolarisation) (Fig. 30.4).

30.2.3 Types of action potential

1. **Pacemaker Potential**: In the heart there are specialised junctional tissue such as SA node. Here electrical potential is generated due to spontaneous depolarisation when it reaches threshold level transmitted through cardiac muscle as action potential in waveform 72-80/minute which causes heart contraction.

2. **Rhythmic action potential**: Repetitive discharges in the smooth muscle are responsible for rhythmic movements of the viscera like intestine and in the nerve cells of the respiratory centre in the medulla oblongata cause rhythmical respiration.
3. **Compound action potential**: When action potential is recorded from nerve such sciatic nerve having group of nerve fibres with different conduction velocities gives summated action potential called compound action potential.
4. **Receptor potential**: Any type of stimulus to the receptor (nerve endings) causes origin of electrical potential is called receptor potential. The characteristics are:
   1. Local and stationary
   2. Non-propagated
   3. Graded
   4. No refractory period
   5. Does not obey all or none law
   6. No effect by local anaesthetics.
5. **Nerve action Potential**: During stimulation of nerve depolarisation causes negativity. The negative site produces negativity to the next segment and this way negative wave of depolarizations is transmitted along the nerve known as impulse.
6. **Smooth or visceral muscle action potential**: They occur in smooth muscle but are absent in multiunit smooth muscle. The action potentials in smooth muscle are of two types: (i) spike potential, (ii) action potential with plateau.
7. **Synaptic Potential**
   (a) *Excitatory post-synaptic potential (EPSP)*: The initial depolarising response produced by a single stimulus to the proper input, begins about 0.5 ms after the afferent impulse enters the spinal cord and reaches the peak 1-1.5 ms later and then declines. During this potential the excitability of the neurone to other stimuli is increased and consequently the potential is called an excitatory post synaptic potential (EPSP).
   (b) *Inhibitory Post-synaptic Potential (IPSP)*: Stimulation of certain presynaptic fibres regulary initiates a hyperpolarization response in spinal motor neurons. This response begins 1-1.35 ms after the afferent stimulus enters the cord reaching the peak in 1.5-2 ms and declines exponentially with a time constant of 3 ms. During this period the excitability of the neurone to other stimuli is decreased consequently, it is called inhibitory post-synaptic potential (IPSP).
8. **End plate potential**: In resting condition there is leakage of small amount of acetylcholine which causes a low action potential of 0.5 mV which is not sufficient to excite the muscle fibre called *miniature end plate potential.* But due to motor nerve stimulation action potential develops which when reaches a critical level or threshold level called end plate (EPP) potential is propagated along the sarcolemma of the muscle fibre causing contraction. Therefore, the normal nerve-muscle function is said to have a very *high safety factor.*

Table 30.1: Bioelectric signals sensed by biopotential electrodes and their sources

Bioelectric potential	Biologic source	Abbreviation
1. Electrocardiogram	Heart muscle-picked up from body surface	ECG or EKG
2. Cardiac electrogram	Heart—as observed from within	Electrogram
3. Electromyogram	Muscle	EMG
4. Electroencephalogram	Brain	EEG
5. Electroretinogram	Retina of eye	ERG
6. Electrooptic-gram or Electrooculograph	Corneal—retinal potential associated with movement	EOG
7. Action potential	Nerve or muscle	ENG or EMG
8. Electrogastogram	Stomach	EGG
9. Galvanic skin reflex	Skin	GSR

Biopotential Body Surface Electrodes

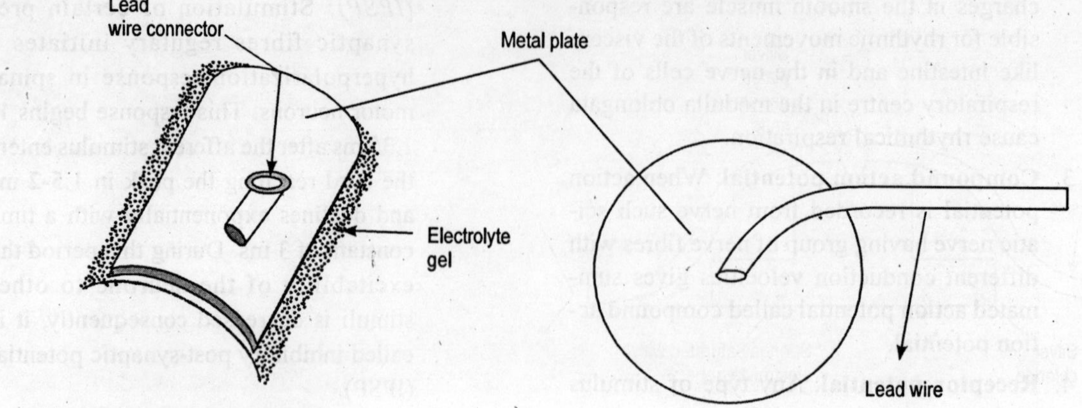

(a) **Metal plate electrodes**

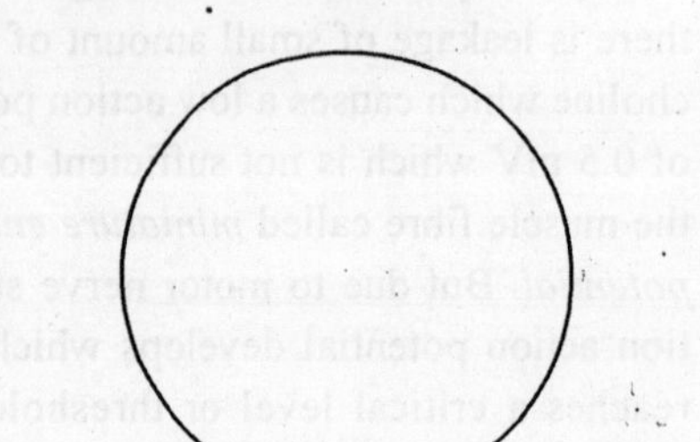

(b) **Suction electrodes for ECG**

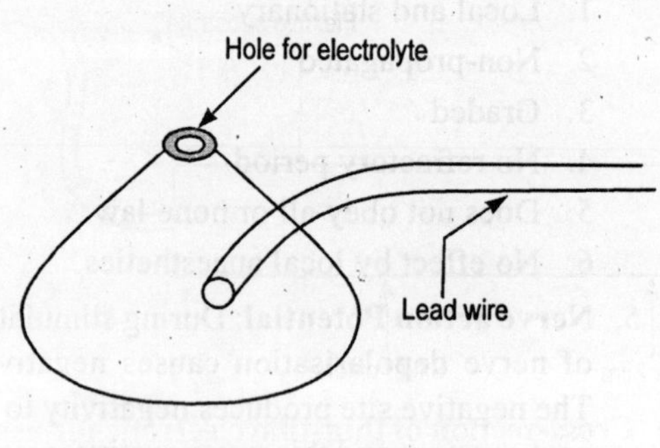

(c) **Metal conical cup electrode**

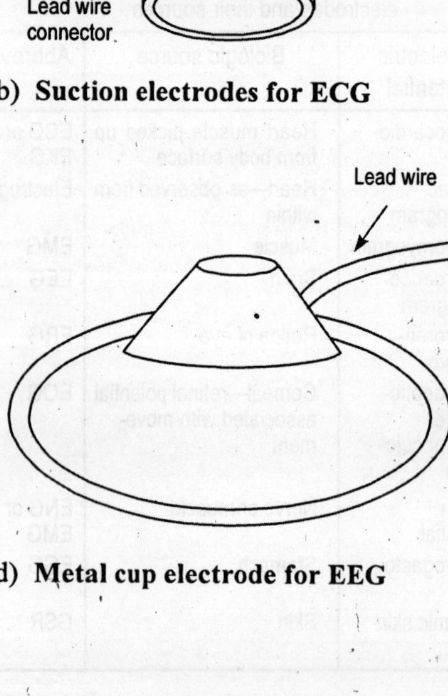

(d) **Metal cup electrode for EEG**

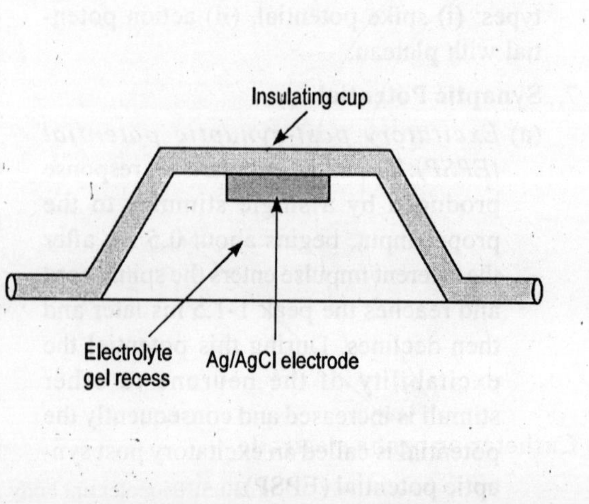

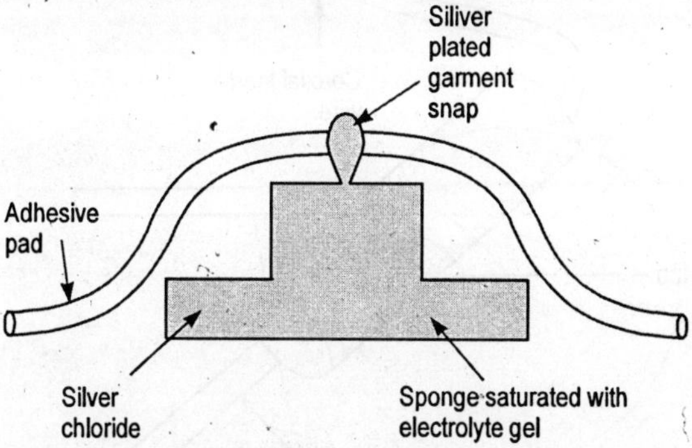

(e) Cross-section of disposable electrode with electrolyte-impregnated sponge

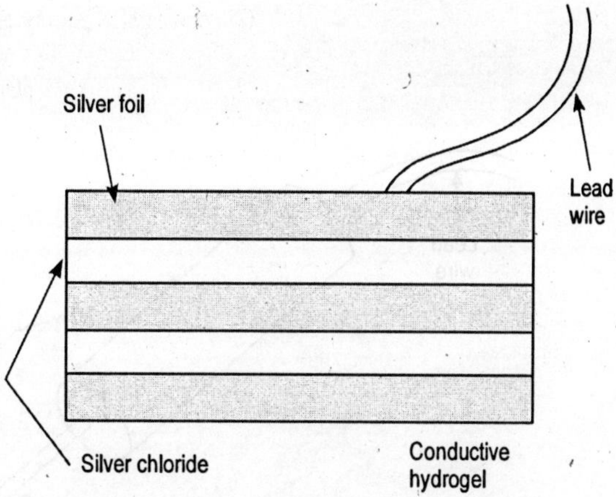

(f) Cross-section of disposable hydrogel electrode

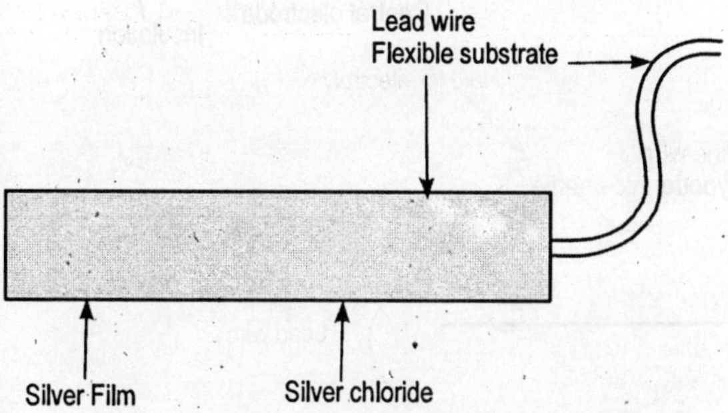

(g) Cross-section of thin-film electrode for use with neonates

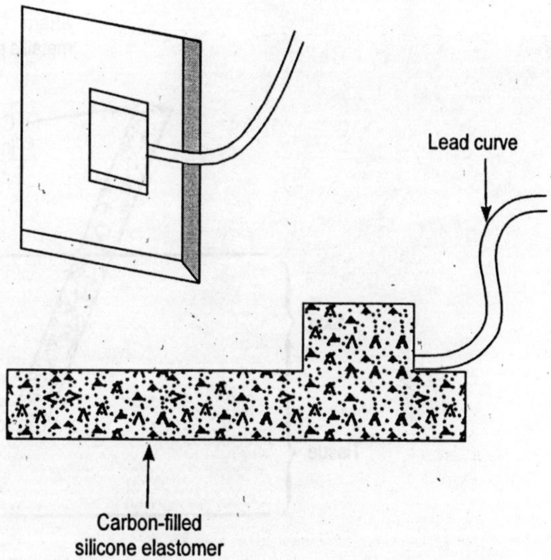

(h) Carbon-filled elastomer dry electrode

Intracavitary and intratissue electrodes for biopotential measurements

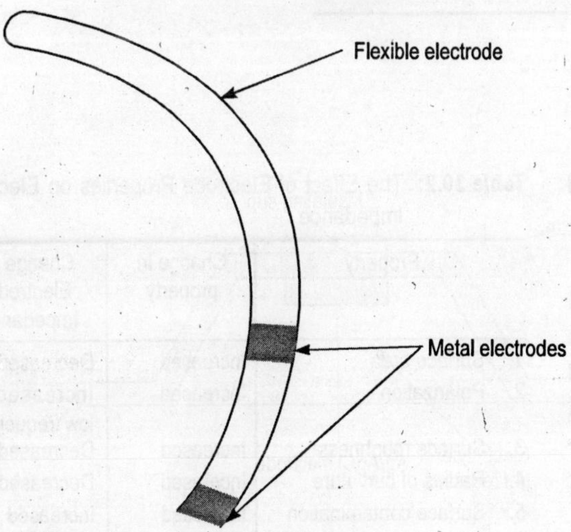

(a) Catheter or probe electrode

Fig. 30.5: Biopotential body surface electrodes

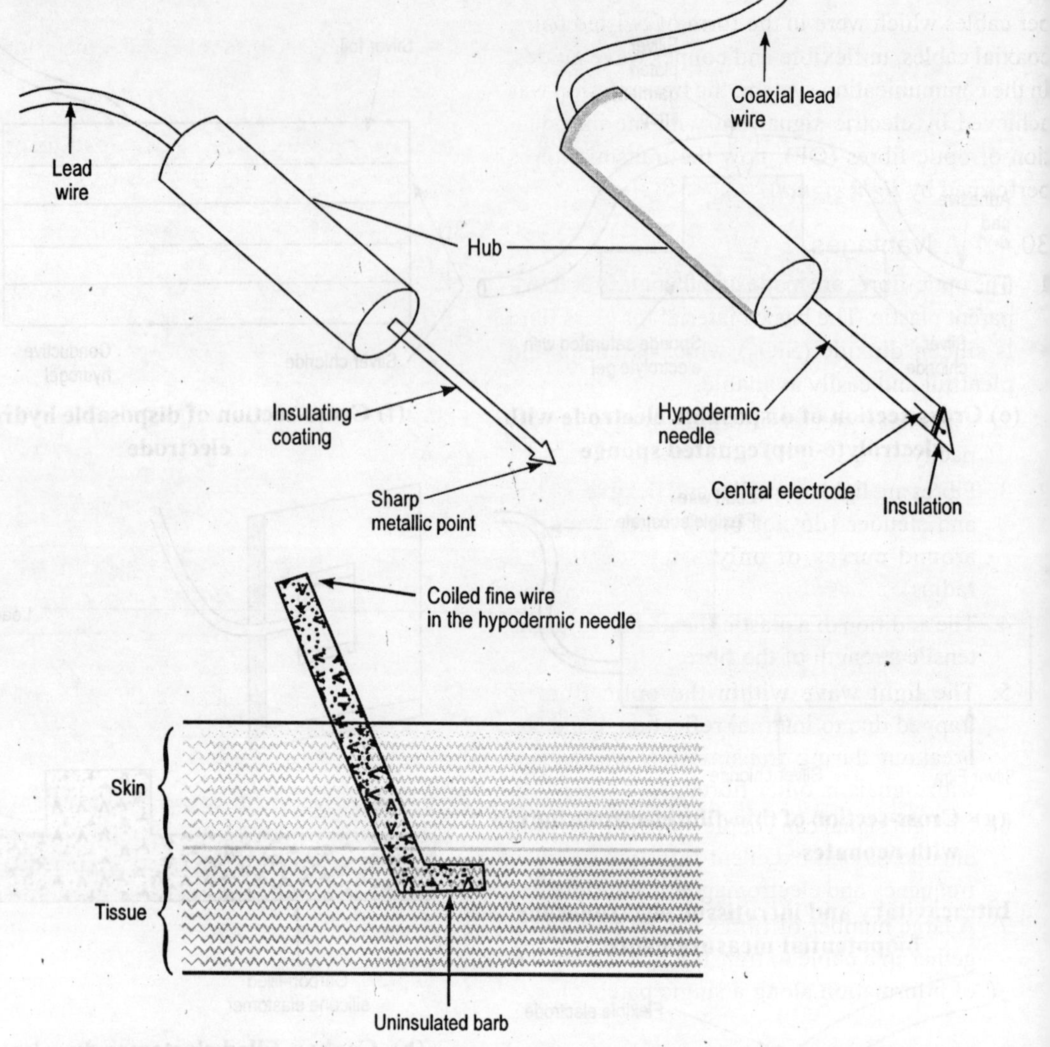

(b)Needle electrodes

30.3 Biopotential Electrode (Practical Electrodes for Biomedical Measurement)

This includes the electrodes which can be placed on the surface of the body for recording bioelectric potential, the integrity of the skin is not considered when the electrodes are applied and can be used for sharp term diagnosting recording, e.g., electrocardiogram and long term chronic recording, e.g., cardiac monitoring (Fig. 30.5).

Table 30.2: The Effect of Electrode Properties on Electrode Impedance

Property	Change in property	Change in Electrode Impedance
1. Surface area	Increased	Decreased
2. Polarization	Increased	Increased (At low frequencies)
3. Surface roughness	Increased	Decreased
4. Radius of curvature	Increased	Decreased
5. Surface contamination	Increased	Increased

30.4 Concept of Fibre Optic

Optical fibres are increasingly replacing the copper cables which were in the form of twisted pairs, coaxial cables, unflexible and copper wave guides. In the communication system, the transmission was achieved by electric signal but with the introduction of optic fibres (OF), now the transmission is performed by *light signal.*

30.4.1 Advantages

1. The optic-fibres are made of either glass or transparent plastic. The basic material for glass fibre is silicon dioxide (SiO_2) which is cheap and plentiful and easily available.
2. They have greater information carrying capacity.
3. Fibres are lighter, smaller and flexible, strong and slender (do not break when wrapped around curves of only a few centimetre radius).
4. The addition of a plastic sheath increases the tensile strength of the fibre.
5. The light wave within the optic fibre is trapped due to internal reflection, none can breakout during transmission to interfere with signals in other fibres.
6. No light signal can couple into the fibre from outside having excellent rejection of radio frequency and electromagnetic interference.
7. A large number of fibres can be packed together in a cable to transmit many channels of information along a single patch.

30.4.2 Disadvantages

1. High cost but gradually reducing as technology advances.
2. Difficult to splice to make them long or to repair any breaks.
3. Fibres do not radiate energy within them, hence it is difficult to switch or detect by intruder the transmitted signal.

30.4.3 Structure of optical fibres

The optical fibre is a thin strand of glass or plastic consisting of these parts: 1. Core, 2. Cladding, 3. Coating.

1. **Core:** The core is a transmission area of the fibre, similar to the hollow centre of a pipe. A larger core gathers and transmits more light. Typical core diameter ranges from 50 to 500 μm.
   The fibres are drawn from a furnace containing molten silica (SiO_2) with a small amounts of additives such as GeO_2 to permit the control of refractive index. Most of the cable bulk is made up of strengthening and buffering materials for mechanical, moisture and chemical protection
2. **Cladding:** A modern fibre consists of optical rod core surrounded with cladding which has a different refractive index than the core. The refractive index of the optical rod is higher than the refractive index of the cladding material in order to utilize the phenomenon of total internal reflection for the propagation of the light through the rod. The light

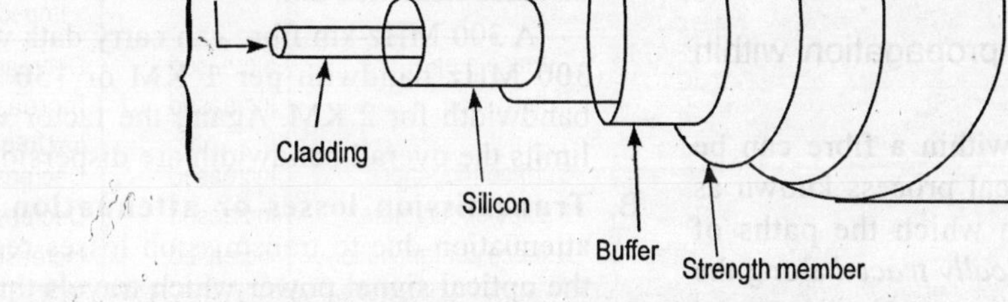

Fig. 30.6: Schematic diagram of optic fibre

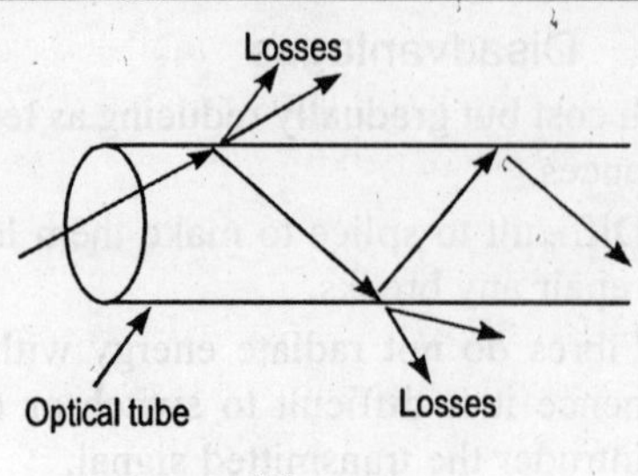

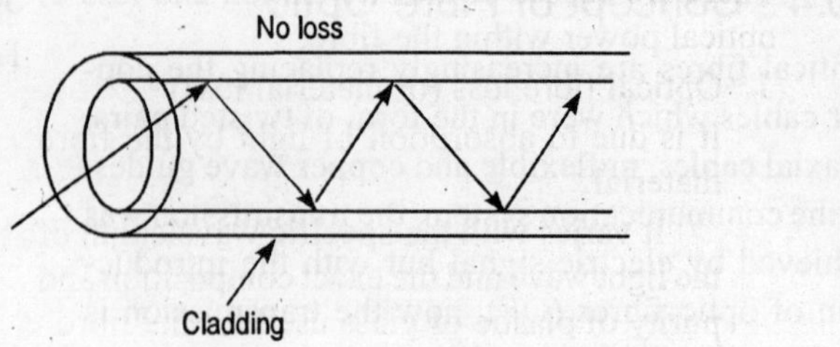

Fig. 30.7: (A) Shows an uncladded glass rod fibre where during travelling of light there is some loss of light energy (B) Shows cladded fibre without loss of light energy during travelling of light

used is coherent light and not the ordinary composite light (Fig. 30.6).

3. **Coating**: The core and cladding are surrounded by another layer called coating. This is specially formulated plastic coating that provides first level shock and abrasion resistance for the fibre. Coatings range in thickness is from 250-1000 μm. The coating is not involved in the actual transmission of light and its thickness is a function of the physical ruggedness needed.

Specification of optical fibre: Fibres are specified by the outer dimension of the core, cladding and coating. A 50/150/400 fibre means the core is 50 μm while the outside dimension of both core and cladding together is 150 μm and similarly 400 μm represents the outside dimension of core, cladding and coating.

Classification of optical fibre: Optical fibres are classified into three groups according to the way light propagates down the fibre core.

1. Multimode step—index fibre
2. Single-mode step—index fibre
3. Single-mode graded—index fibre.

The mode refers to the path the light will travel through the fibres. One of the basic differences among different types of three is their refractive index profile.

30.4.4 Principle of light propagation within a fibre

The propagation of light within a fibre can be understood from an analytical process known as *geometrical ray-tracing* in which the paths of undivided rays are *geometrically traced* along the guide path.

Light stays inside the fibre because it is totally reflected by the inside surface of the fibre. Generally light centers at the end of the fibre at a certain angle to the axis following a *zig zag path* through a series of reflection rays down the length of the fibre. Total internal reflection at the fibre wall can occur only if two conditions are satisfied such as:

(a) The fibre core must have a slightly higher refractive index (say, n_1) than the refractive index of the surrounding cladding (say, n_2) i.e., $n_1 > n_2$.

(b) The light must approach the wall with an angle of incidence (say, O i.e., with an angle between the ray path and the fibre wall) is greater than the critical angle (O_0) (Fig. 30.6).

30.4.5 Characteristics of fibre

The characteristics of the fibre depend on the type of material used and its physical shape and size. Things such as glass composition, diameter of the fibre, the variations of the refractive index within the fibre, imperfection and small bends (called microbends) directly affect its performance.

30.4.6 Performance of the optical fibre

The two most important parameters of optical fibres, performance are:

A. **Bandwidth**: Overall bandwidth of a fibre is usually expressed in the form of frequency times distance i.e. MHz-km.

A 300 MHz-km fibre can carry data with a 300 MHz bandwith per 1 KM or 150 MHz bandwidth for 2 KM. Again, the factor which limits the overall bandwidth are dispersion.

B. **Transmission losses or attenuation**: The attenuation due to transmission losses reduces the optical signal power which travels through the fibre and is measured in db/km.

There are causes of the attenuation and loss of optical power within the fibre:

1. Optical fibre loss (or material loss)
   It is due to absorption of light by the fibre material.

   It varies with the specific wavelength of the light wave and the exact composition and purity of plastic or glass used for the fibre.
2. *Microbending loss*: Microbending is due to minute fibre bends and deviation of the fibre. Even though the fibre may appear straight the effect of the pull of gravity and resultant stress within the fibre causes tiny and localised bending. Light may escape from total internal reflection at the bends and is absorbed by the cladding.
3. *Connector loss*: It is a function of the physical alignment of one fibre to the another fibre core. Three types of problems may arise:
   (a) Axial misalignment
   (b) Transverse misalignment
   (c) Excessive separation between pieces.
4. *Splice loss*: It occurs when two fibres are joined together to make a longer fibre.
5. *Coupling loss*: This occurs between the fibre and the optical signal source or optical signal detector and it is a function of both the source or receiver devices used and the type of fibre chosen.

30.4.7 Light sources for fibre-optics

There are two devices suitable for generating light as sources needed for the optical fibre:

1. Light emitting diode (LED)
2. Semiconductor Laser

Light sources for fibre optics act as *light transmitter* and just meet certain requirements, if they are to be acceptable for the purpose, i.e.,

(a) The light produced must be *monochromatic*, i.e., single frequency as far as possible. Light emitting diode (LED) and semiconductor laser emit light over a much narrower acceptable bandwidth though these sources are also not truly monochromatic.

(b) Light source should have such a light intensity output that sufficient energy is transmitted and the fibre to overcome the losses is encountered during transmission.

3. The source must be small and easily coupled to fibres so that excessive coupling loss do not occur.

30.4.8 Application of fibre optic

I. **Medical**: Fibre-optic endoscopy, e.g., esophagoscope, gastroscopy, colonoscopy, broncoscopy, laproscopy, arthroscopy, etc.

II. **Communication system**: Internation, intercity, inter-exchange, domestic, plant and traffic control, defence application (e.g., laser radar), laser instrumentation.

30.5 Fibre-Optic Endoscopy

Fibre Optics: Optical fibres are flexible, transparent glass rods. The surfaces of each fibre is coated with an intratransparent material so that light is trapped within the core of the fibre—the coating prevents leakage of light from one fibre to another (cross—communication) when large number of fibres are packed together in close proximity. Since the light rays are trapped within the fibres they can only emerge from the fibres at the end opposite to that by which they entered. This property of the optic fibres is employed in building instruments known as *endoscopes,* to examine internal cavities of hollow organs such as urinary bladder, stomach, intertisne, lung, bronchi, etc.

30.5.1 Use of endoscope

Endoscopes are used routinely these days in examination of the following parts of the body (Table 30.3).

ENDOSCOPE (Table 30.3)

Anatomical Part	Type
1. Oesophagus	Oesophagoscope
2. Stomach	Gastroscope
3. Duodenum	Gastro-duodenoscope
4. Lungs	Bronchoscope
5. Abdomen	Laproscope
6. Urinary bladder	Cystoscope
7. Kidney	Nephroscope
8. Large intestine	Colonoscope
9. Blood Vessel	Angioscope
10. Joints	Arthroscope
11. Thoracic cavity	Thorcoscope
12. Gall bladder and duct	Choledocoscope

In addition, a small instruments to the end of an endoscope allows for collection of biopsy material without operating on the patient.

Thus instruments have made the diagnosis of lesions of G. I tract and genito-urinary tract more easier and quicker since these lesions can be observed and confirm the diagnosis. High resolution images can be recorded on

(a) Colour film for documentation and for follow up.
(b) Cine film for arranging learning material and documentation.
(c) Television monitor for case of operator and learning by observers.

Fibre-optics lighting system has replaced conventional optic and light system.

30.5.2 Components of endoscope

An endoscope consists of a system of long optic fibres with a lens system of one end and a viewing head of the other for the observer and/or a camera of endoscope are used nowadays (Fig. 30.8).

1. The light source or generator unit (A)
2. The insertion tube with the optical image bundle (B).
3. The instrument or endoscope control section near the eye piece (C).

30.5.3 The light source

(a) The light source is a high intensity lamp in a box with an in-built parabolic reflector.
(b) This ensures that cold light is passed to the instrument or endoscope since the heat is left in the light source box.
(c) Special lamps such as *quartz, iodine, metal halide* and *xenon* have been developed to give a higher intensity of light of upto 500 watts. The fibre optic bundle is connected at one end to the light source and at the other end to the endoscope.
(d) It contains bundle of upto 20,000 flexible optical fibres which can conduct light but is not capable of carrying an image.

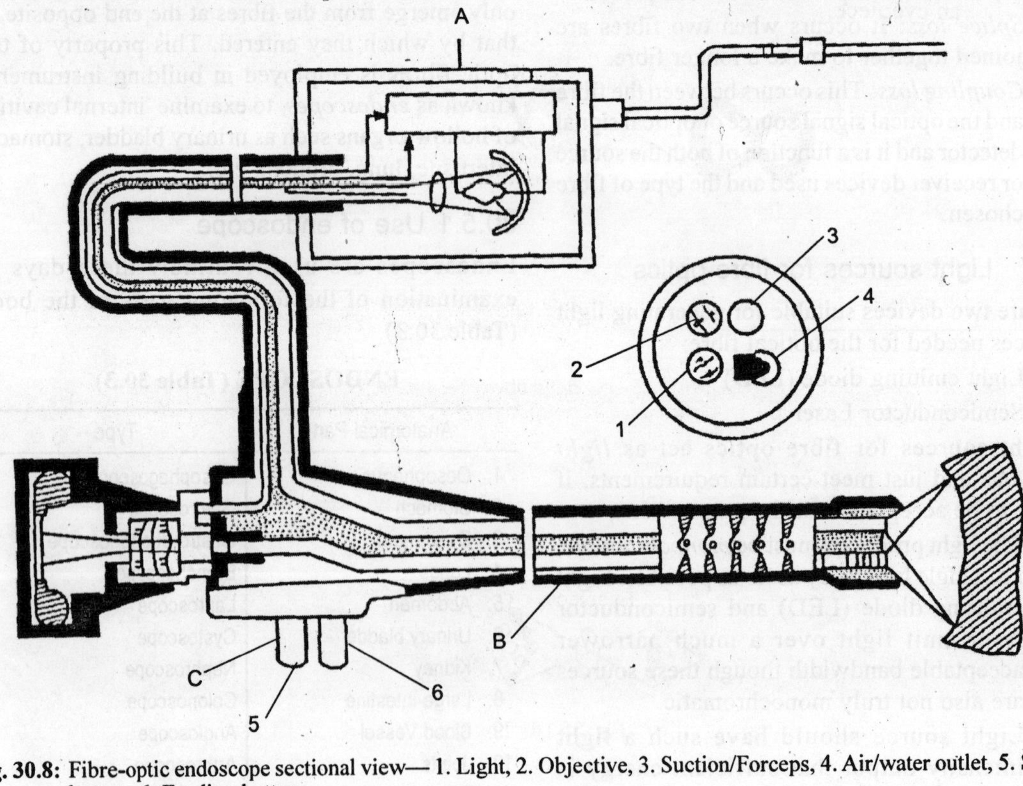

Fig. 30.8: Fibre-optic endoscope sectional view— 1. Light, 2. Objective, 3. Suction/Forceps, 4. Air/water outlet, 5. Suction button, 6. Feeding button

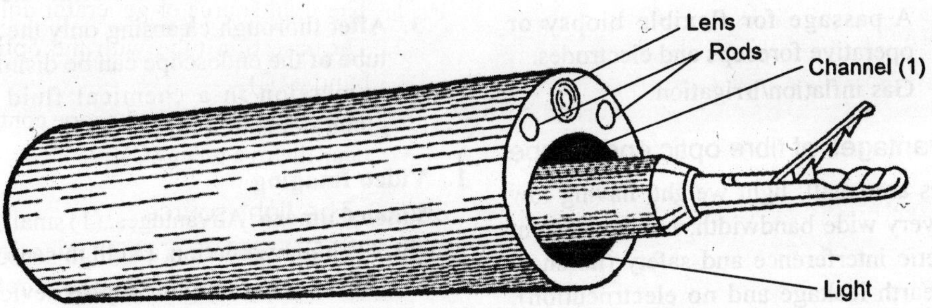

Fig. 30.9: The inserting end of the endoscope (sectional view)

(e) The bundles transmit a pre-focussed light intensity and uniform density to the endoscope.

(f) The bundles outlet covering is of metal and PVC, and the system is extremely flexible.

(g) In addition, there is a separate set of upto 16,000 optical fibres are capable of carrying an intagminating proximally at an eyepiece.

Fibre-optics refers to the system for transmitting of light rays and images from one point to another via separate bundles of optic fibres by total internal reflection (Fig. 30.9).

(h) Both these sets of optical fibres are contained in a protective sheath and distally a bus of the end of the endoscopes, generally. This sheath incorporates additional channels (Fig. 30.8) for:

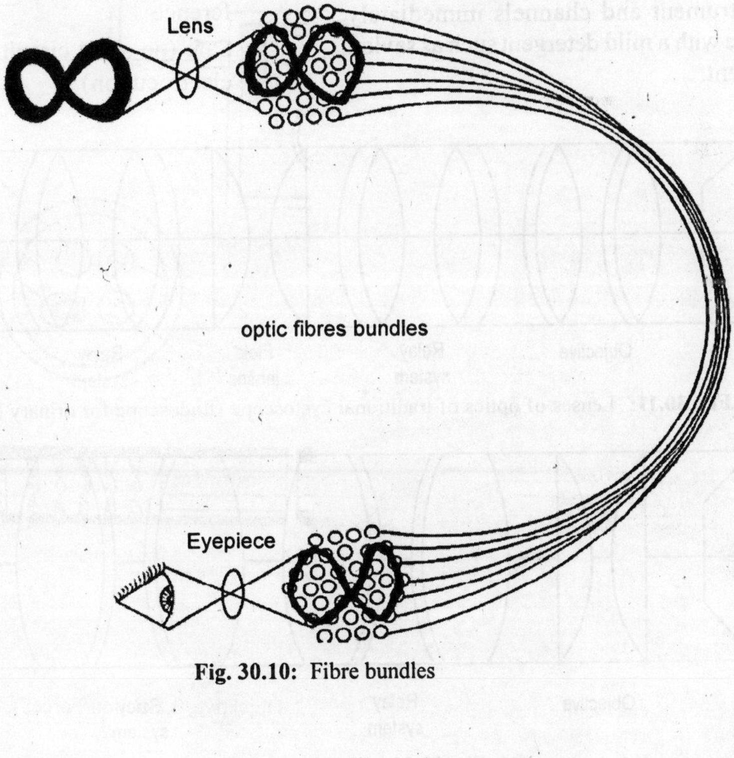

Fig. 30.10: Fibre bundles

1. Suction
2. A passage for flexible biopsy or operative foreceps and electrodes.
3. Gas inflation/irrigation.

30.5.4 Advantages of fibre optic endoscope

Optical fibres are small, light weight, having low attenuation, very wide bandwidth, immunity from electromagnetic interference and safety (no short circuits, no earth leakage and no electrocution). Fibres are non conductive, non-radiative and non-inductive.

The dimensions of fibre core 50 μm diameter 10 μm.

Attenuation	0.3 dB/μm
Optical windows	0.82 μm, 1.3 μm 1.55 μm

Fibre optic cables (Bundles)

10 μm diameter, flexible bundle

30.5.5 Special care of fibre-optic endoscope

Special care should be taken for maintenance such as:

1. Debris and mucus must be removed from the instrument and channels immediately after use with a mild detergent such as savlon 1 per cent.
2. Followed by rinsing in worm water.
3. After thorough cleansing only the insertion tube of the endoscope can be disinfected by immersion in a chemical fluid such as glutaraldehyde (cidex).

1. **Video Imaging**

Video Camera: Advantages: (1) small, (2) light weight, (3) high resolution, (4) high sensitivity.

CCD Camera (Charger coupled device) with a large number of photocell receptor. Each photocell produces an electronic signal that generates one *pixel* (17 μm × 13 μm). Present day CCD camera have 150,000 to 300,000 pixels. Picture resolution = 400 lines or more. Some of the cameras posess *built-in-zoom* facilitities and *freeze frame imaging*.

30.5.6 Salient feature about fibreoptic endoscope technology

1. Optical fibres are small
2. Light weight
3. Having low attenuation
4. Very wide bandwidth
5. No response from electromagnetic interference
6. Safe (no short circuit, no earth leakage, no electrocution)

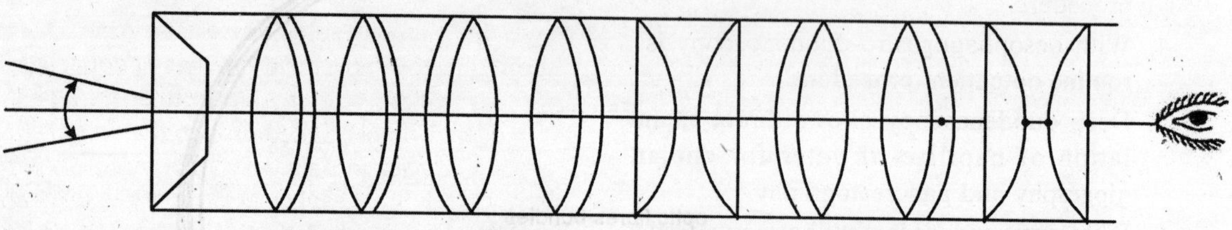

Fig. 30.11: Lenses of optics of traditional cystoscope (Endoscope for urinary bladder)

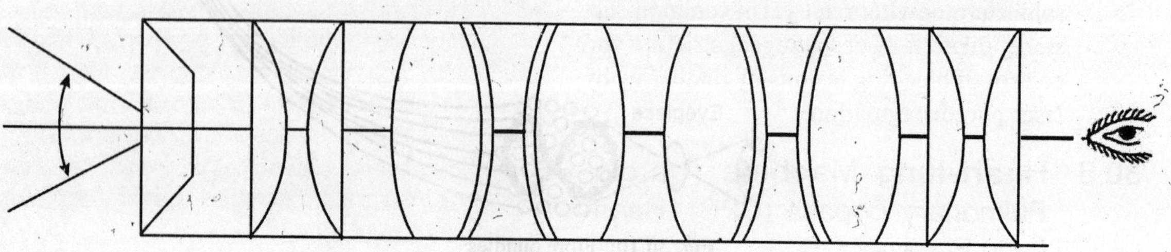

Fig. 30.12: Hopkins Rod Lenses

7. Fibres are non-conductive, non-radiative and non-inductive
8. The dimensions of fibres core 50 μm diameter
9. Attenuation. 3-dB/μm
10. Optical window 0.82 μm, 103 μm, 1.55 μm
11. Fibre optic cables/bundles
12. Diameter 10 μm
13. Flexible bundles.

30.5.7 The instrument

The basic instrument is a composite tool. The tip of the modern instrument can be angulated in both directions upto 180°. Finger tip controls are provided for suction, air insufflation and for water injection to clear the lens or mucosa (inner membranous layer of the intestine). A side channel permits passage of variety endoscopic tools, such as biopsy forceps, foreign body etc.

30.5.8 Advantages

1. The great advantage of fibre optic endoscope than the rigid endoscope is that a small biopsy specimen can be taken.
2. With G.I. endoscope it is now possible to examine the whole colon.
3. Oesophagoscope is nowadays outdoor procedure.
4. With oesophagogastro-duodenoscopy is a routine outpatient procedure.
5. Deep duodenascopy allows direct cannulation of papillae of vater for cholangiography and pancreatography.
6. Fibre-optic endoscopy has been extended to therapeutic area from diagnostic, e.g., removal of polyp from colon, foreign body, sphincterotomy for removal of common duct stone, dilation of oesophageal stricture and control of bleeding lesions by diathermy by laser photocoagulation.

30.6 Heart-lung Machine (Cardio-Pulmonary Bypass (CPB), Heart-Lung Bypass)

Cardiopulmonary Bypass (CPB), also called heart-lung bypass or heart-lung machine.

It allows the temporary

(a) replacement of the respiratory gas
(b) exchange function of the lungs
(c) blood pumping function of the heart.

In the procedure blood does not flow through the heart and lungs which makes the *operating field blood less* during cardiac surgery.

The terms pump-oxygenator and heart-lung machine graphically describe the equipment used.

30.6.1 Procedure

Open-heart surgery strictly speaking refers to interventions inside the heart cavities, which once provided the most frequent demand for cardiopulmonary bypass technology. By extension, the term is also applied to surgical procedures which take place primarily on the external aspects of the heart such as creating new routes for blood to reach the distal coronary arteries from the aorta, e.g., *coronary artery bypass grafting* (CABG) or popularly, *bypass surgery*.

In this operation the cardiac cavities are temporarily vented (i.e., open to atmospheric pressure) to avoid building up of pressure which could damage the cardiac muscle.

30.6.2 Extracorporeal circuit

As usually emplo-yed for cardiac surgery, the heart-lung machine is a part of a total vano-arterial cardiao-pulmonary bypass circuit or extra corporeale circuit meaning that all the venous blood returning to the right heart cavities is collected in the extracorporeal circuit. And circuited through the gas exchange device called *oxygenator* from which it is pumped back into the arterial branches thereby bypassing the heart cavities and the pulmonary circulation. On the blood side, this is the procedure usually involves modification, some degree of *hypothermia nonpulsatile* arterial perfusion at a flow rate near the resting cardiac output and continuous recirculation of blood in a extracorporeal circuit in series with the systemic circulation on the patient, on the gas side, oxygen- or oxygen enriched gas mixture (with or without a low concentration of CO_2) flows from a moderately pressurized continuous source in a recycling and is vented to the room temperature (Fig. 30.13).

30.6.3 Standard operating condition

Typical operating conditions for total procedure in an adult are summarized and represented in the Fig.

30.13. Most notables are the large differences in driving forces for O_2 and CO_2. An approximate comparison can be made using the inlet conditions. Thus, under normothermia (normal room temperature) and with pure oxygen fed to the gas phase the initial driving force is approximately 45-0-45 mm of Hg for CO_2 and 750-47-703 mm of Hg for O_2 yielding a ratio of CO_2 to O_2 driving force of about 0.06. The ratio of CO_2 to O_2 permeability for silicone rubber is roughly 5 and the corresponding flux ratio is 0.35 less than half that of the metabolically determined respiratory quotient. Under these conditions, CO_2 is the limiting gas, and the device should be sized on the basis of CO_2 transport rather than O_2 transport equipment. This is why silicone rubber membrane have been replaced by the micro processor polypropylene membranes where the solubility of CO_2 in the membrane material and consequently its permeability is no longer the limiting factor. Indeed, with some modern membrane oxygenator gas flows through the device may have to be controlled to avoid an excessive loss of CO_2 (Fig. 30.13).

30.6.4 Circulation

The blood is drained from the patient by *cannulation* of the *inferior vena cava* and *superior vena cava.* The heart and lungs are isolated by cross-clamping the aorta down stream of the aortic valve and vena cava at the entrance to the right atrium. The venous blood enters to the extra corporeal circuit and is transferred to the oxygenator where it is oxygenated and CO_2 is removed. It is direct contact type oxygenator used, then a defoamer and bubble trap are required to remove gaseous bubbles. A heart exchanger enables the blood to be cooled in a controlled manner. Before being returned to the patient via roller pump, the oxygenated blood is filtered. This arterial filter removes microaggregates and any residual gas microbubbles. Blood is then returned to the body by mean of a canula to the aorta (Fig. 30.13).

30.6.5 Bypass

Bypass means to indicate the fluid normally circulating through an organ is diverted around it, either to reduce the functional work load or to isolate the organ for the duration of a surgical procedure, blood does not flow through the organ like heart, organ etc. which makes the operating field blood less one at this time heart-lung machine artificially maintains the cardio-respiratory function.

30.6.6 Use

1. As a temporary substitutes for heart and lung function during surgery.

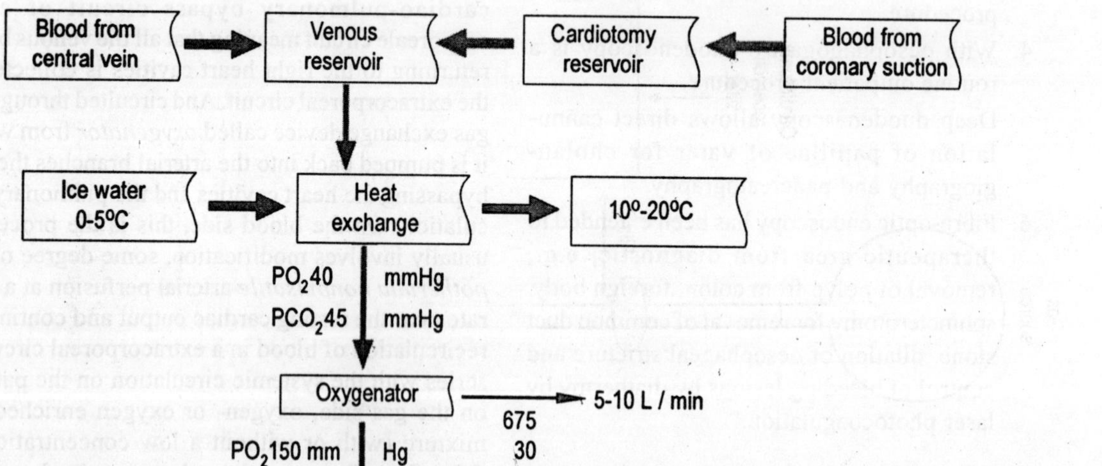

Fig. 30.13: Schematic representation of standard operating conditions during cardiopulmonary bypass procedure in an adult are summarised in this figure

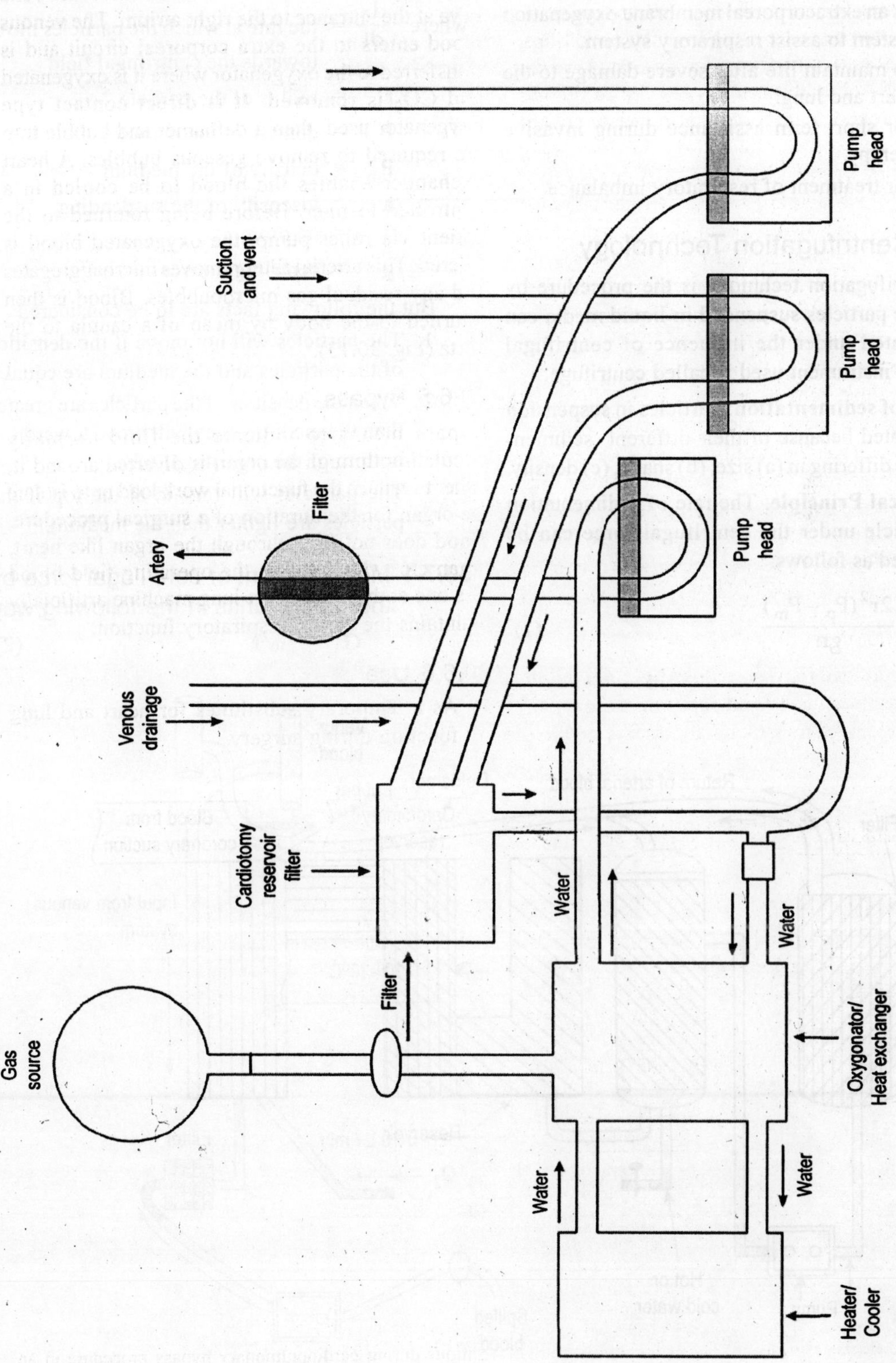

Fig. 30.14: Diagrammatic representation of heart-lung machine

2. As an extracorporeal membrane oxygenation system to assist respiratory system.
3. To maintain life after severe damage to the heart and lung.
4. For short term assistance during invasive therapy.
5. For treatment of respiratory imbalance.

30.7 Centrifugation Technology

The centrifugation technique is the procedure by which the particles suspended in liquid media can be separated under the influence of centrifugal force, the instrument used is called centrifuge.

Rate of sedimentation: Particles in suspension are separated because of their different sedimentation rate differing in (a) size, (b) shape, (c) density.

Physical Principle: The rate of sedimentation of a particle under the centrifugal force can be represented as follows:

$$\frac{dx}{dt} = \frac{2r^2(P_p - P_m)}{gn} \quad (1)$$

where $\frac{dx}{dt}$ = the rate at which the particles move towards the centrifugal field

r = radius of the particle

P_p = density of the particle

P_m = density of the medium

n = viscosity of the suspending medium.

g = centrifugal force

But the following facts are to be considered:

1. The particles will not move if the densities of the particles and the medium are equal.
2. When the densities of the particles are greater than the medium, they will move toward the bottom of the centrifugal tube.
3. The particles will remain at the top if the particles are lighter than the medium.

30.7.1 Measurement of centrifugal force by the gravity units in the following way

$$G = \omega^2 r \quad (2)$$

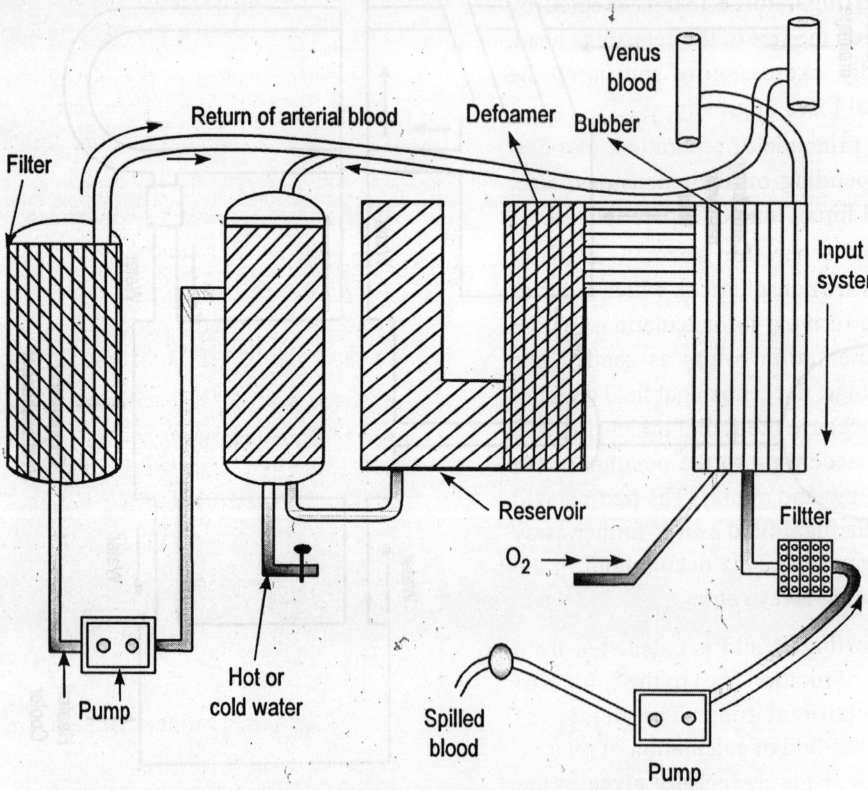

Fig. 30.15: Schematic diagram of heart-lung Machine

where G = centrifugal field
ω = angular velocity of the rotor
r = radial distance of the particle from the axis of the rotation.

The angular velocity of the rotor can be represented by the formula:

$$\omega = \frac{2\text{rev}/\text{min}}{60} \quad (3)$$

Introducing the value of ω in the equation (2), the contrifugal field (G) is derived as follows:

$$G = \frac{4(\text{revolution}/\text{min})^2 r}{60 \times 60}$$

$$G = \frac{4(\text{rev}/\text{min})^2 r}{3600},$$

which is shown as a multiple of, i.e., gravitational field of the earth. Sometimes it is expressed as RCF (Relative centrifugal field) which is the ratio of the weight of the particle in the centrifugal field of the same particle acted on by gravity. The speed of the centrifuge is ordinarily measured in revolution per minute (rpm) but this is not very standardized because the centrifugal force that is created by spinning depends on the size of the centrifuge head. Thus, scientifically, expression of rpm is by the relative centrifugal force in gravity 'g'.

Based on this principle of separation, particles are separated depending on their densities, size, shape, centrifugal force, time of separation, etc.

The method of separation becomes difficult when the particles are not spherical, which requires some complicated formula for calculation. In case of ordinary sedimentation rotors as used in the preparative centrifuge, the centrifugal field does not remain uniform, because the radial dimension of a particle will vary according to the position in the centrifuge tube (r min and r max). The particle will have a greater centrifugal field as it is further away from the axis of rotation. This occurs both in the fixed angle and swing away rotor.

Hence the centrifugal field is calculated from the average radius of rotation (r_{av}) of the column of liquid in the centrifugal tube. The details of maximum and the method of calculation of relative centrifugal field (RCF) are generally given in the manual of the centrifugal. The sedimentation rate of a particle can also be expressed as sedimentation coefficient (S) which is the sedimentation rate per unit of centrifugal field. The sedimentation values depend on the solvent-solute system.

As the sedimentation coefficient of many macromolecules is very small, the basic unit is taken as 10^{-13} seconds and is designated as Svedberg unit (S).

30.7.2 Component of the instrument

The centrifuge consists of the following components:

1. *Electronic motor*: To which are attached a head and accessories for carrying samples.
2. *Safety shield*: Around the rotating head to prevent any hazard.
3. *ON-OFF Switch*: For interruption of the current to the motor.
4. *Chamber*: In which the shaft rotates.
5. *Central shaft*: It rotates at high speed and is driven by the motor.
6. *Head*: It is fixed to the shaft and carries the centrifuge tubes in a bucket or cuff and hole.
7. *Timer*: Incorporated for either a simple spring driven clock-mechanism or an electric timer that performs as a switch for the present time cycle.
8. *Break*: Often supplied which are of electronic or mechanical break.
9. *Tachometer*: It is provided in many instruments to indicate speed in rpm by a cable or flexible shaft attached to the motor.
10. *Potentiometer*: Its function is to raise and lower the voltage supplied to the motor. For this purpose many centrifuges have a multipositioned switch that selects voltage steps by connecting taps on a transformer.
11. *Lid*: The centrifuge is provided with a lid which is closed during the rotation. The lid will remain closed all the time even when it is not working.
12. Centrifuge tube is either made of hard glass, plastics, poly-propylene, polycarbonate, stainless steel and nylon materials for withstanding the stress during centrifugation.

13. *Soft rubber cushion*: Present at the base of the socket. If these are not present, the centrifuge tubes will be broken.

30.7.3 Standardisation of centrifuge

Centrifugar force is the principal factor which determines the separation of the solid from the liquid phase in a suspension. The centrifugal force depends on both speed and on the size of the centrifuge head. The general expression of revolution per minute (rpm) indicates the centrifugal speed but does not actually express the centrifugal force. The centrifugal speed and the centrifugal force are, however, directly proportional. A better expression is the relative centrifugal force (RCF), which is as the acceleration due to gravity or G (dynes per cm). The relationship between the 'rpm' and 'G' is expressed in the following equation:

$$RCF\ (G) = r \times (rpm)^2 \times 118 \times 10^{-7},$$

where r = radius in cm

rpm = revolution per minute (shown on the tachometer provided with the centrifuge).

The use of nomogram is more convenient for determination of the RCF if the radius of the centrifuge head is known and the rate of spinning (rpm).

30.7.4 Time of centrifugation

The time of centrifugation is an important factor in applying centrifugal force for a period of time at a speed sufficient to cause the heavier particles to move to the bottom of the tube before the lighter particles. Hence, the tubes are spin for a specified period to obtain the desired effect. If the centrifuge is unable to accomplish the expected result, the time of centrifugation is changed. As for example, the ratio between RBC and plasma is 45 : 55 (haematocrit value) but if we get 48 : 52 by the centrifuge, then increasing the time of centrifuse until the standard value 45% reached.

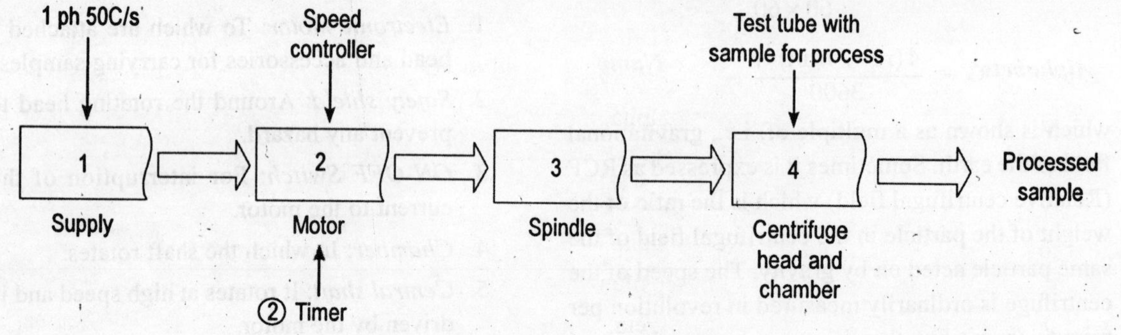

Fig. 30.16: Block diagram of centrifuge

1. Supply of power to the motor
2. Regulation of rotation of spindle at a desired speed and time
3. For holding the test tube holder bracket (rotor head)
4. For processing the sample kept in the test tube.

APPENDIX

GREEK LETTERS

Alphabets	Name	Alphabets	Name
α	alpha	ν	nu
β	beta	ξ	xi
γ	gamma	ο	omicron
δ	delta	π	pi
ε	epsilon	ρ	rho
ζ	zeta	σ	sigma
η	eta	τ	tau
θ	theta	υ	upsilon
ι	iota	ϕ	phi
κ	kappa	χ	chi
λ	lamda	φ	psi
μ	mu	ω	omega

BIBLIOGRAPHY

1. Adey, W.R., The Physiology of Weightlessness, Physiologist, 16, 178, 1974.
2. Ambrose, E.J. and Easty, D.M., Cell Biology, 2nd edition, Nelson, 1976.
3. Andreoli, T.E. *et al,* (eds.); Membrane Physiology, Plenum Press, New York, 1980.
4. Andrews, H.L., Radiation Biophysics, 2nd edition, Englewood Cliffs, N.J. Prentice-Hall, 1974.
5. Annine, J.S.—Clinical Chemistry, Little Brown Co. Boston, 1984.
6. Aston, R.—Principles of Biomedical Instrumentation and Mesurement, Merrill Publishing Co., ohio, U.S.A.
7. Bailey, G.I. (ed.), Haemodialysis, Principle and Practice, New York, Academic Press, 1973.
8. Bair, B.J.—Introduction to chemical Instrumentation, McGraw-Hill, 1962.
9. Baldwin, E., An Introduction to Comparative Biochemistry, 4th edition, Cambridge University Press, 1964.
10. Barna, A. and Porat, D. I.—Introduction to microcomputer and microprocessors, Jhon Wiley & Sons, Inc. 1976.
11. Bassett C.A.L. and Becker, R.O., Generation of electrical Potential in bone in response to mechanical stress, Science,137, 163, 1962.
12. Bates, R.G., Electrometric Determinations, John Willey & Sons, New York.
13. Bayliss, L.E., The Rheology of Blood. In Handbook of Physiology, Section 2, Circulation, Vol. I, Washington.
14. Bell, G.H., Davidson, J.N. and Emsliesmith. D., Textbook of Physiology and Biochemistry, Churchill Livingstone, 9th ed., 1976.
15. Benchimol, A.—Non-invasive Diagnosing Techniques in Cardiology Williams & Wilkins Co., Baltimore, 1977.
16. Berthiaume, F., Yarmush, M.L., Tissue Engineering in the Biomedical Engineering Hand Book, 2nd ed., C.R.C. Press, Boca Raton.
17. Bertholomeu D.—Electrical Measurement and Instruments, Allyn and Bacon, Inc. 1963.
18. Bernstein, E.F., (ed.), Non-invasive Diagnostic Techniques in Vascular Disease, St. Louis, C.V. Mosby, 1978.
19. Bier, M., Electrophoresis, 2nd edition, Academic, New York, 1968.
20. Billon, F.—Medical Equipment Service Manual, Englewood Cliffs, N-J—Prentice-Hall Inc., 1978.
21. Birshtein, T.M. and Ptitsyn, O.B. Conformations of Macromolecules, Interscience, New York, 1960.
22. Blandell, T.L. and Johnson, L.N., Protein Crystallography, Academic Press, London, 1976.
23. Bloom, W. and Fawcett, D.W., A Textbook of Histology, 9th edition, Saunders, 1968.
24. Bray, H.G. and White, K., Kinetics and Thermodynamics in Biochemistry, 2nd edition, Academic, New York, 1966.
25. Bradbury, S., Hewer's Textbook of Histology for Medical Students, William Heinemann Medical Books Ltd., London, 9th edition, 1969.
26. Brates, R.G.—Determination of pH—Theory and Practice, Jhon Wiley & Sons, Inc. 1984.
27. Brazier, M.A.B.—The Electrical activity of the nervous system, Wittiams & Wilkins, Baltimore, 1968.

28. Brecher, G.A., Venous Return, Grune and Stratton, New York, 1956.

29. Brown, J.H.U., (ed.), Engineering Principles in Physiology, Vols. I and II. Academic Press, New York, 1973.

30. Brown, J.H.U., Physiology of Man in space, New York, Academic Press, 1963.

31. Brown, J.H.U. et al, Biomedical Engineering, F.A. Davis Co., Philadalphia, 1971.

32. Bull, H.B., An Introduction to Physical Biochemistry, 2nd ed., F.H. Davis, Philadelphia, 1971.

33. Bullard, R.W., Physiological Problems of Space Travel, Annu. Rev. Physiol., 34, 205, 1972.

34. Burton, A.C., Physiology and Biophysics of the Circulation, Chicago, 1965.

35. Bushberg, J.U. et al—The Essential Physics of Medical Imaging, Williams & Wilkins, Baltimore, 1994.

36. Butter, J.A.V. and Katz, B., (eds.), Progress in Biophysics and Biophysical Chemistry, Vol. 1-9, Pregamon, London, 1959.

37. Carr. J.J., Brown, J.M., Introduction to Biomedical Equipment Technology, 4th ed., Pearson Education. Inc., 2007.

38. Catton, W.T., Mechanoreceptor Function, Physiol. Rev., 50, 297, 1970.

39. Chappel, J.B. and Rees, S.C., Mitochondria, Oxford University Press, 1972.

40. Christensen, H.N., pH and Dissociation, Saunders, Philadelphia, 1963.

41. Christensen, H.N., Body Fluids and the Acid base Balance, Saunders, Philadelphia, 1964.

42. Cobbold, R.S.C.—Transducers in Biomedical Instruments, Principles and applications, John Wiley & Sons, Inc, 1974.

43. Cold Spring Harbor Symposia on Quantitative Biology, Biological Clocks, Baltimore, The Waverly Press, Inc., 1960.

44. Cope, F.W. Nuclear Magnetic Resonance Evidence using D_2O for structured water in Muscle and Brain, Biophys, J., 9, 303, 1969.

45. Davis, H., Biophysics and Physiology of the Inner Ear, Physiol. Rev., 37, 1, 1957.

46. Davson, H. and Eggleton, M.G., Principles of Human Physiology, 14th edition, Churchill Livingstone Ltd., 1968.

47. Davson, H., A Textbook of General Physiology, 3rd edition, J & A Churchill Ltd., 1964.

48. Ducheyne, P., Vander Perre et al. Biomaterials and Biomechanic, Elsevier, Amsterdam, 1984.

49. Edsall, J.T. and Wyman, T., Biophysical Chemistry, Vol. I, Academic, New York, 1958.

50. Elliott, G.F., X-ray Diffraction Studies on Striated and Smooth Muscle, Proc. R. Soc. Lond. (Biol), 160, 469, 1964.

51. Enderlee, J., Blanchard, S., Broznino, J., Introduction to Biomedical Engineering, 2nd, ed, Academic Press.

52. Enderle, J. et al—Introduction to Biomedical Engineering 2nd Ed., 2006, Academic Press.

53. Erikson, C., Streaming potentials and other water dependent effects in mineralised tissues, Ann. N. Y. Acad. Sci., 238, 321, 1974.

54. Evans, C.L., Principles of Human Physiology, 9th edition J & A churchill Ltd., 1945.

55. Feigenbaum, H., Echocardiography, 3rd edition, Lea and Febiger, 1981.

56. Feigenbaum, H.—Echocardiography, Lee & Febiger, Philadelpia, 1976.

57. Florey, E., An Introduction to General and Comparative Physiology, Saunders, 1966.

58. Franklin, D.L.—Techniques for measure-ment of blood flow through intact vessel—Medical Electronics and Biological Engineering 3, 27-37, 1965.

59. Frey-Wyssling. A. (ed.), Deformation and Flow in Biological Systems, North-Holland, Amsterdam, 1952.

60. Fronek, A., Coel, A. M. and Bernstein. E.F., Quantitative Ultrasonographic Studies of Lower Extremity Flow Velocities in Health and Disease, 53, 957, 1976.

61. Fukuda, E., Yasuda, I., on the piezoelectric effect of bone, J. Phys. Res., Japan, 12, 1158, 1957.

62. Fuxe, K., Hokfelt, T., Jonsson, G., and Ungerstedt, U., Fluorescence Microscopy in Neuroanatomy. In Nauta, W.J. and Ebbesson. S., (eds.), Contemporary Research Methods in Neuroanatomy, New York, Springerverlag, New York Inc., 1970.

63. Ganong, W.F., Review of Medical Physiology, 12th edition, Lange Publications, 1983.

64. Garrahan, P.J. and Glynn, I.M., The Stoichiometry of the Sodium Pump, J. Physiol, 192, 217, 1967.

65. Geddes, L.A. and Baker, L.E.—Principles of Applied Biomedical Instrumentation, 2nd ed., John Wiley & Sons, Inc. 1978.

66. Gies, A.C., Cell Physiology, 4th edition, Saunders, 1973.

67. Glansdorf, P. and Prigogine, I., Thermody-namic theory of Structure, Stability and Fluctuation, Wiley—Interscience, New York, 1971.

68. Glasser, O. (ed.), Medical Physics, Vols. 1-3, Year Book Medical Pub., Chicago, 1944-60.

69. Gortner, R.A., Jr., and Gortner, W.A. (eds.), Outlines of Biochemistry, 3rd edition, John Wiley & Sons, New York, 1949.

70. Guyton, A.C., Textbook of Medical Physiology, W.B. Saunders Co., 6th edition, 1981.

71. Ham, A. W., Histology, 1965.

72. Harris, J.D., The Electrophysiology and Layout of the Auditory Nervous System, Indianapolis, The Boobs–Merrill Co., 1974.

73. Hastings, G.W. and Wheble, V.H., Evaluation of Biomaterials (ed.) G.D. Winter et al., John, Wiley & Sons, New York, 1980

74. Hench, L.L., Biomaterial Science, 208, 826-831, 1980.

75. Heftman, E., (ed.), Chromatography, 3rd edition, Reinhold, New York, 1975.

76. Hempleman, H.V. and Lockwood, A. P. M., The Physiology of Diving in Man and other Animals, Edward Arnold, London, 1978.

77. Henn. F.A., Decker, G.I., Greenawalt, J.W. and Thompson, T.E., Properties of Lipid Bilayer Membranes seperating two Aqueous Phases, Electron Microscope Studies, J. Mol. Biol., 24, 51, 1967.

78. Hope, A.B., Ion Transport and Membranes, Univ, Park Press, Baltimore, 1971.

79. Hubbell, W.I. and Bownds, M.D., Visual Transduction in Vertebrate Photoreceptors, Annu. Rev. Neurosci., 2, 17, 1979.

80. Huffaker, C.B., (ed.), Biological Control, Plenum Press, New York, 1974.

81. Hurry, S.W., The Microstructure of the Cells, 4th edition, John Murray, 1972.

82. Huxley, A.F. and Peachey, L.D., Interference Microscopy of Living Muscle Fibres, Nature, 173, 971, 1954.

83. Jaundrell-Thompson F and Ashworth, J.—X-ray physics and equipment, F.A. Davis Co., Philadelphia, 1970.

84. Karselis, T.C., The Pocket Guide to clinical Laboratory Instrumentation, F.A. Davis Company, Philadelphia, 1994

85. Ingram, D.J.E., Biological and Biochemical Applications of Electron Spin Resonance, Adam Hilger, London, 1969.

86. Keele, C.A., Neil, E. and Joels, N., Samson Wright's Applied Physiology, 13th edition, Oxford University Press, 1982.

87. Kotyk, A. and Janacek, K., Cell Membrane Transport, 2nd edition, Plenum, New York, 1974.

88. Kremkau, F.W., Diagnostic Ultrasound, Physical Principles and Exercises, New York, Grune & Stratton, 1980.

89. Kreel, I., Computed Tomography and the Cardiovascular System. In Dickson, C.J. and Marks, J., (eds.), Development in Cardiovascular Medicine, Lancaster, England. MTP Press, 1978.

90. Krieger. D.T. and Ascoff, I., Endocrine and other Biological Rhythms. In Degroot, I.H., et al (eds.). Endocrinology, Vol. I, Grune and Stratton, 1979.

91. Lehninger, A.L., Bioenergetics, 2nd edition, Benjamin 1972.

92. Lehninger, A.L., Biochemistry, Worth Publishers, 1970.

93. Leopold, A.C., Plant Growth and Development, 2nd edition, McGraw-Hill, 1975.

94. Linzer, M.(ed)—Ultrasonic Energy, Academic Press, New York, 1979

95. Lipscomb, D.M., (ed.), Noise and Audiology, Bultimore, University Park Press, 1978.

96. Lockwood, A.P.M., Animal Body Fluids and their Regulation, 1964.

97. Loewenstein, W.R., Biological Transducers, Sci. Am, 203, 98, 1960.

98. Lucy. J.A., Ultrastructure of Membranes, Micellar Organization, Br. Med. Bull. 24, 127, 1968.

99. Macdonald, D.A., Blood Flow in Arteries, Arnold, London, 1960.

100. Martin, D.W., Mayes, P.A. and Rodwell, V.W., Harper's Review of Biochemistry, 18th edition, Lange Medical Publications, 1981.

101. Miles. S., Underwater Medicine, 2nd edition, Philadelphia, J.B. Lippincott Co., 1966.

102. Murrary, R., Myocardial Imaging, New York, MSS Information corp., 1977.

103. O'Reilly, P.H., et al (eds.). Nuclear Medicine in Urology and Nephrology, Boston, Butterworths, 1979.

104. Parsons, D.S., Biological Membranes, Oxford University Press. London, 1975.

105. Pedley, T.J., The Fluid Mechanics of Large Blood Vessels, New York, Cambridge University Press, 1979.

106. Pepe, F.A., The Myosin Filament. II., Interaction between Myosin and Actin Filaments observed using Antibody Staining in Fluorescent and Electron Microscopy, J. Mol. Biol., 271, 227, 1967.

107. Pimentel, G.C. and McClellan, A.L., The Hydrogen Bond, W.H. Freeman, 1960.

108. Prigogine, I., Introduction to Nonequilibrium Thermodynamics, Wiley Interscience, New York, 1962.

109. Prosser, G.L. and Brown, F.A., Comparative Animal Physiology, 2nd edition, Saunders, 1961.

110. Richter, C.P., Inborn Nature of the Rat's 24 hour Clock, J. Comp. Physiol. Psychol., 75, 1, 1971.

111. Richter, C.P., Biological Clocks in Medicine and Psychiatry, Springfield, III, Charles C. Thomas, Publisher, 1965.

112. Ritman, E.I. et al, Three-Dimensional Imaging of Heart, Lungs and Circulation, Science, 210, 273, 1980.

113. Ruch, T.C. and Patton, H.D., Physiology and Biophysics, 19th edition, Saunders, Philadelphia, 1965.

114. Rusak, B. and Zucker, I., Neural Regulation of Circadian Rhythms, Physiological Rev., 19, 449, 1979.

115. Rushmer, R.F., Cardiovascular Dynamics, Saunders, Philadelphia, 1961.

116. Schmidt—Neilsen, K., Animal Physiology, 3rd edition, Prentice Hall, 1970.

117. Schneck, D.J. and Vawter, D.L., (eds.), Biofluid Mechanics, New York, Plenum Press, 1980.

118. Schoen. F.I., Cardiac Valve Prostheses, Review of clinical status and contemporary issues, T. Biomed. Material Research, 31, 91, 1987.

119. Secker—Walker, R.H. and Evens, R.G., The Clinical Application of Computers in Ventilation–Perfusion Studies, Prog. Nucl. Med. 3, 166, 1973.

120. Seigel, B.A. and Potchen, E.J., Radionuclide Studies of Pulmonary Function, Anatomic and Physiologic Considerations, Prog. Nucl. Med., 3, 49, 1973.

121. Setlow, R.B. and Pollard, E.C., Molecular Biophysics, Pregamon, New York, 1962.

122. Silver, F.H., Biomaterials, Medical Devices and Tissue Engineering, Chapman and Hall, London.

123. Silver, S, Radionuclide in Medicine and Biology, Lea and Febiger, Philadelphia, 1968.

124. Sloan, A.W., Man in Extreme Environments, Springfield, III, Charles, C. Thomas, 1979.

125. Spadaro, J.A., Electrically stimulated bone growth in animals and man, clin. Ortho Relat. Res., 133, 325, 1977.

126. Stacy, R.W., Williams, D.T., Worden, R.E. and McMorris, R.O., Essential of Biological and Medical Physics, McGrawHill, New York, 1961.

127. Starkebaum, W et al, stress generated potentials in bone on a microseale, Biomed. Res. Symp., Trans, 1, 82, 1977.

128. Stryer, L., Biochemistry, 2nd edition, K.H. Freeman and Company, New York, 1981.

129. Strauss, H.W., (ed.), Cardiovascular Nuclear Medicine, St. Louis, C.V. Mosby, 1979.

130. Strong, P.—Biophysical Measurement, Beaverton Ore : Tektronix, Inc., 1970.

131. Strumwasser, F., Neurophysiological Aspects of Rhythms. In Quarton, G.C., Melnechuk, T. and Schmitt, F.O., (eds.), The Neurosciences, A study Program, New York, The Rockefeller University Press, 1967.

132. Suttond., Radiology and Imaging for Medical Students, 6th ed., B.I. Churchil, Livingstone, 1994.

133. Swan, H.J.C. Ganz, W.et al,, Catheterization of the heart in Man with use of a flow directed Balloon Tipped Catheter, New England Journal of Medicine, 283, 437, 1970.

134. Tanford, C., Physical Chemistry of Macromolecules, Wiley, New York, 1961.

135. Taylor, M.G., Hemodynamics, Annu. Rev. Physiol,, 35, 38, 1973.

136. Timasheff, S.N. and Fasman, G.D., Structure and Stability of Macromolecules, Dekker, N.Y., 1960.

137. Tsvetkov, V.N., Eskin, V. Ye and Frenkel, S. Ya., Structure of Macromolecules in Solution, Butterworth, London, 1970.

138. Valentinuzzi, M., The Organ of Equilibrium and Orientation as a Control System, Harwood Academic Publishers, New York, 1980.

139. Vanof, H.M.—Biomedical Electronics, 2nd Ed., F.A.Davis Co., Philadelphia, 1972.

140. Volkenshtein, M.V., Molecular Biophysics, Academic Press, New York, 1977.

141. Wade, J.B., Kachadorian, W.A. and Discala, V.A., Freeze-fracture Electron Microscopy, Relation-ship of Membrane Structural Features to Transport Physiology Am. J. Physiol., 232, 1977.

142. Wells, P.N.R., Biomedical ultrasonics, Academic Press, New York, 1977.

143. West, J.B., (ed.), Best and Taylor's Physiological Basis of Medical Practice, 11th edition, Williams and Wilkins, Baltimore/London, 1984.

144. West, E.S. and Todd, W.R., Textbook of Biochemistry, 3rd edition, MacMillan Company, New York, 1961.

145. Weiss, M.D., Biomedical Instrumentation, Chilton Book Co., Philadelphia, 1973.

146. Whelpton, D., (ed.), Renal Dialysis, Philadelphia, J.B. Lippincott Co., 1974.

147. Whittingham, C.P., Photosynthesis, Oxford University Press, 1971.

148. Yapp, W.B., An Introduction to Animal Physiology, 3rd edition, 1970.

149. Zweig. G. and Sharma, J., Handbook of Chromatography, 2 Vols, CRC Press, 1972.

150. Carr. J.J. and Brown, J, M—Introduction to Biomedical Equipment Technology, Pearson Education, Inc., 2007.

INDEX

Acidification of Urine 113
Action potential 944
Active transport 324
Adhesion 882
Adiabatic process 9
Adsorption 377
Air embolism 768
Albot-Plateau Law 531
Algorithm 280
Alpha-particles 242
Aluminium filter 264
Amicrons 148
Ampholytes 100, 125
Analogous transplantation 939
Angiography 268
Anisotropic 555
Anisotropy 149
Antennal glands 350
Antiport 320
Aperture diaphragm 264
Apoptosis 869
Arrheniun type 21
Arrhenius plot 21
Artificial kidney 159
Asymmetric carbon atom 492
Audiometer 730
α-rays 201

Barometric pressure 762
Basal lamina 922
Beats 683
Beer's Law 922
Beta particles 242
Bingham body 788
Bioelectric electrodes 946
Bioelectric Potential 941
Biological clock 607, 612
Biological feedback 618
Biological light 550
Biological modifier 888
Biomagnetism 621
Biomaterials 894
Biomolecules 145
Bioreactors 892
Biorhythm 606
Birefringent 555
Blackman-Putter principle 21
Blood volume by isotope 448
Boltzman distribution 21
Bone scan 428
Bone marrow transplantation (BMT) 939
Bone-seeking isotopes 212
Brain scan 429
Brownian movements 122
Bypass (Heart lung) 955
β-rays 201

Calcification 600
Calvin Cycle 504, 510
Capillarity 396

Cardiac catheter 845
Cardiac Imaging 434, 439
Cardiopulmonary bypass (CPB) 955
Carnot cycle 23
Cell as the living unit 859
Cell to cell recognition 887
Cell receptors 648
Cell-matrix interaction 885
Cell refractometry 562
Cell signalling 885
Cellular differentiation 866
Cellular communication 835
Centrifugal acceleration 751
Centrifugal field 959
Centrifugal force 962
Chemical bonds 459
Chemiluminescence 550
Chloride shift 111
Chloride secretory cells 343, 351
Chlorophyll 503
Chromatography 186
Chromatophore 557
Clinical ultrasonography 694
Closed system 8
Cloud chamber 201
Coacervation 127
Coagulation 126
Coagulation rule 126
Cohesion 396
Cochlear microphonic potential 674
Collagen 917
Collagen-rich tissues 912
Collimator 266
Colorimeter 495
Common ion effect 601
Compton scattering 242
Computed tomography 282
Cone 264
Consonance or concord 685
Contractile vacuole 345
Contrast 260
Contrast media 270
Crenations 345
Critical fusion frequency 531
Critical potential 124
Cryo-technique 929
Cryobath 931
Cryoprotective 929
Cryosurgery 933
Cryotherapy 933
Curie 204
Cyclic photosynthetic phosphorylation 508
Cytolysis 344
Cytopempsis 367
Cytoplasm 142

Deafness 729
Decompression 767
Density gradient centrifugation 873
Dextrorotatory 492
Diagnostic nuclear medicine 458
Dialysis 171
Diapause 539
Dichroism 557
Dielectric constant 603
Donnan membrane equilibria 400
Doppler effect 690
Down regulation 649
DSA 271

Efficiency 18
Einstein equation 788
Elasmobranches 351
Electron-volt (eV) 206
Electro-osmosis 124
Electrophoresis 124
Electroretinogram 534
Emulsifying agent 133

Emulsion 133
Endocochlear potential 675
Endocytosis 361
Endoscopy 951
Energy 25
Engineering fluid 786
Enthalpy 13
Entropy 15
Epsilon potential 123
ERG 534
Excited molecule 501
Extracellular matrix (ECM) 908
Extracorporeal circuit 957

Ferry-Porter Law 531
Filtration coefficient 359
Filtration fraction 357
Filtration rate 359
Flocculation value 126
Fluidity 388
Fluorescence 256
Fluorescent 566
Fluorogram 255
Fluorography 255
Fluoroscopy 254
Free energy 18
Full turgor 336

G forces 750
Gamma-rays 242
Gap junction 902
Gel state 135
GFR 367
G^{-G} 751
Gibberellin 521
Glomerular filtration rate 357
Glomerular membrane 357
Goldman constant-field equation 410
Graded filters 412
Grafting 924
Grid 230, 265
G-suits 750
G-unit 751
Guttation 356
γ-rays 201

Haemolysis 342, 344, 525
Haemodialyser 159
Hagen 795
Hagen-Poiseuille's equation 796
Half-life 203
Hazards in the atmosphere 254
Hearing Aid 733
Henderson-Hasselbalch equation 95
Hess's law 22
Hofmeister series 126
Homeostasis 606, 618
Human centrifuge 752
Hydrotrophy 384
Hyperosmotic 338
Hypertonic 338
Hypotonic 338

Image intensifier 227
Imbibition 121
Impedance 604
Infrasonic 661
Interleukins 892
Invasive 219
Inversion 494
Iodine pool 443
Ion channel 638, 899
Isobars 205
Isoelectric pH 125
Isogravimetric 356
Isohydric Principle 109
Isohydric solutions 601
Isomers 205

Isosmotic 338
Isothermal process 9
Isotones 205
Isotonic 337
Isotonic solution 338
Isotopes 205
Isotopic effect 208
Isotropy and anisotropy 555
Isovolumetric 356

Junctional complex 870

Kelvin's postulate 23
Korotkow sound 838

Laplace's law 812
Laser beam 454
Leakage radiation 262
Levorotatory 492
Limiting factor 511
Liver scan 427
Loudness 663
Luciferin 551
Luminescence 256
Lyophilization 931

Macromolecules 139
Mega voltage 278
Magnetobiology 621
Magnetocardiography 625
Magnetoencephalography 622
Magnetography 624
Malignancy or cancer 866
Mammography 307
Marine birds 343
Marine reptiles 343
Marine teleosts 343, 351
Masking 664
Matrix 283
Mechanical valves 929
Medical imaging 219
Membrane fluidity 650
Membrane hydrolysis 410
Membrane phenomena 942
MeV 206
Micelles 117
Microcurie 204
Micro-dissection 147
Microenvironments 921
Microns 148
Migration 883
Millicurie 204
Mountain sickness 764
MRI 441, 294
Mutarotation 490, 494
Mutual Precipitation 126

Negative G 751
Negative G-forces 754
Neoplasm or tumour 866
Neuromagnetic field 622
Neutrons 243
Non-invasive 219
Newtonian and nonNewtonian fluid comparison 791

Open system 10
Optical fibres 951
Optically active 492
Organ scan 427
Organelles 861
Orthovoltage 278
Osmo-confirmers 345
Osmolality 335
Osmolarity 335
Osmole 335
Osmoreceptors 343, 348
Osmoregulation 345
Osmotic Work 343

Ostwald's dilution law 602
Oxygen toxicity 766

Pair production 243
Peritoneal dialysis 168
PET 437
Phagocytosis 361
Phosphor 256, 259
Phosphorescence 256
Photocathode 227
Photometer 498
Photo-chemical reactions 502
Photodynamic action 524
Photodynamic haemolysis 525
Photoelectric absorption 242
Photoelectric transducers 511
Photoperiodism 522, 538
Photoreactivation 546
Phototropism 520, 521
Photosynthesis 503
Physiological salt solution 339
Phytochrome 521
Pinocytosis 361
Pixel 283
Place principle 678
Plasmolysis 337
Poiseuille Law 794
Polarimeter 490, 491
Polarised light 491
Positive G 751
Power Law 681
Programmed cell deaths 870
Prosthetic heart valves 927
Proton pump 324
Protoplasm 141
Protoplasmic streaming 146
Pump mechanism 942
Piezometric circulation 828
Pz materials 713

Radioactive decay 202
Radiopharmaceuticals 440
Radiation barriers 262
for thyroid 442
Radioactive iodine 447
Radioactivity 200
Radioautography 218
Radiographic contrast 260
Radiography 227
Radionuclide 206
Radiopharmaceuticals 208
Radioscopy 227
Radiotherapy 275, 276
Reabsorption 375
Recycling techniques 761
Rejection phenomenon 924
Renal plasma flow (FPF) 357
Renal Imaging 424
Renography 430
Reverse osmosis 332
Reynolds number (Re) 806
Ringer's fluid 339

Salting-in 127, 598
Salting-out 127
Scattered radiation 262, 263
Semipermeable membrane 645
Shock wave 715
Signal receptor 884
Sodium-Potassium Pump 327
Soft tissue radiography 307
Sol state 135
Solarization 523
Solubility product 598
SPECT 440
Suspension 133
Spinning motion 294
Squid 624
Starling principle 404

Starling's Law of Ultrafiltration 356
Statoliths or graviceptors 743
Steady state 10, 17
Stem cell 863
Stern potential 123
Stomata 346
Streaming potential 124
Streamline flow 799
Stromal cells 864
Submicrons 148
Subsonic 688
Supersonic 688
Surface electrodes 946
Surface tension 394
Surfactant 482
Symport 320
Syneresis 121
System 1, 8

Tagging 207
Teleosts 351
Teletherapy 276
Terrestrial animal 352
Therapeutic nuclear medicine 458
Thermionic emission 220
Thoriated tungsten 230
Tight junctions 872
Timers 267
Tiselius apparatus 124
Tissue gel 139
Tissue repair (Healing) 923
Tomography 278
Tonicity 337, 339
Tracer 207
Tracer element 207
Tracer technique 207
Transplantation surgery 925
Transverse G 754
Troland 528
Tungsten filament 230
Turbulent flow 799, 806
Turgor pressure 336, 346
Tympanic membrane 671
Types of grafting 924
Tympanic reflex 672
Tyndal effect 122
Types of radiation 262

Ultracentrifuge 154
Ultrafiltration 173, 412
Ultramicroscope 148
Ultrasonic surgery 701
Ultrasonic therapy 700
Ultrasonic transducer 704
Ultrasound 688
Ultraviolet light 543

Valency 459
Van't Hoff's laws 332
Van't Hoff's rule 21
Vitrification 931
Viscosity 386

Weightlessness 758
Windkessel vessels 792

Xeroradiography 268
X-ray therapy tubes 278

Zeitgeber 611
Zeta potential 123
Zonograph 281
Zwitterion 126